Greek-English Lexicon

REVISED SUPPLEMENT

H. G. LIDDELL   ROBERT SCOTT
H. STUART JONES   RODERICK McKENZIE

Greek-English Lexicon

REVISED SUPPLEMENT

Edited by
P. G. W. GLARE

With the assistance of
A. A. THOMPSON

CLARENDON PRESS · OXFORD
1996

Oxford University Press, Walton Street, Oxford OX2 6DP
Oxford New York
Athens Auckland Bangkok Bombay
Calcutta Cape Town Dar es Salaam Delhi
Florence Hong Kong Istanbul Karachi
Kuala Lumpur Madras Madrid Melbourne
Mexico City Nairobi Paris Singapore
Taipei Tokyo Toronto
and associated companies in
Berlin Ibadan

Oxford is a trade mark of Oxford University Press

Published in the United States by
Oxford University Press Inc., New York

© *Oxford University Press 1996*

All rights reserved. No part of this publication may be reproduced,
stored in a retrieval system, or transmitted, in any form or by any means,
without the prior permission in writing of Oxford University Press.
Within the UK, exceptions are allowed in respect of any fair dealing for the
purpose of research or private study, or criticism or review, as permitted
under the Copyright, Designs and Patents Act, 1988, or in the case of
reprographic reproduction in accordance with the terms of the licences
issued by the Copyright Licensing Agency. Enquiries concerning
reproduction outside these terms and in other countries should be
sent to the Rights Department, Oxford University Press,
at the address above

This book is sold subject to the condition that it shall not, by way
of trade or otherwise, be lent, re-sold, hired out or otherwise circulated
without the publisher's prior consent in any form of binding or cover
other than that in which it is published and without a similar condition
including this condition being imposed on the subsequent purchaser

British Library Cataloguing in Publication Data
Data available

Library of Congress Cataloging in Publication Data
Data available
ISBN 0-19-864223-7

1 3 5 7 9 10 8 6 4 2

Typeset by Latimer Trend & Co. Ltd., Plymouth
Printed in Great Britain
on acid-free paper by
The Bath Press, Avon

PREFACE

The Greek–English Lexicon of Liddell and Scott was first published in 1843. In the 150 years which have passed since then it has gone through nine editions, the last being augmented by a Supplement published in 1968. This Supplement incorporated a considerable amount of new material, mainly taken from inscriptions and papyri, and also 'a good deal of information pertaining to literary sources, where the systematic reading of new editions, or ordinary consultation by scholars, has revealed the lexicon to be in need of revision'.

It had fallen to Dr M. L. West to see that Supplement through the press, and after its publication he kept custody of the slips from which it was printed and a collection of material that had been excluded from it. He did not undertake any further lexicographical work, but served as a receiving-point for any new material that happened to be sent to the Press. When he left Oxford in 1974 to take up a chair in London, he passed the archive to Mr (now Prof.) M. D. Reeve, then a Fellow of Exeter College.

In January 1979 Dr West presented to the British Academy a proposal for the preparation of a new Supplement. After some discussion a committee was set up to plan and cost the project in detail. It consisted of the following members: Prof. M. L. West (chairman), Prof. R. Browning, Dr J. Chadwick, Prof. A. M. Davies, the late Prof. D. M. Lewis, and Mr (now Prof.) P. J. Parsons. Early in 1981 the Council of the Academy, having come to an agreement with Oxford University Press, accepted the proposal and adopted the Supplement as one of the Academy's Major Research Projects. It was agreed that the project would run for ten years, and that a full-time editor would be needed. The post was advertised, and Mr P. G. W. Glare, who had amply proved his lexicographical competence and stamina by his labours on the *Oxford Latin Dictionary*, was appointed with effect from 1 August 1981. The Project Committee continued to monitor progress throughout with its original membership unchanged, except for the addition of Mr N. G. Wilson in 1989, and with the regular attendance at its meetings of representatives of the Press and of the British Academy.

In 1988, when it became clear that the undertaking was considerably greater than had at first been supposed, Dr Anne Thompson was appointed to the project as part-time Editorial Associate. Later in the same year Dr Carolinne White also joined the staff with the special function of checking the references, and proof-reading the print-out at all stages; she also became mainly responsible for the compilation of the lists at the beginning of the Supplement. Mrs Lorna Lyons acted as Secretary/Assistant throughout and, among multifarious duties, was entirely responsible for the keying-in of the text.

Much new material has become available since the publication of the first Supplement, and the incorporation of this is again the main aim of the new Supplement. The question of revision is altogether more complicated and re-examination of the entries in the first Supplement revealed the need to redraft many of them. How far revision of the main Lexicon should proceed is a more difficult question. It was evident from an examination of them that many articles require a more drastic rewriting than can be achieved by the addition, exclusion, or alteration of individual items. Minor changes may in fact only highlight shortcomings in the original articles.

For this reason and in view of the restricted staff and time available it has proved impossible to revise many articles which can now be seen to require changes. In a few cases new articles have been substituted for those appearing in the Lexicon, but these are mostly short ones. In longer articles it has been necessary to restrict changes to those which could be made within the existing framework of the article. It must also be explained that, although electronic retrieval facilities became available soon after the commencement of work on the Supplement, a systematic use of these was not made, since it would have involved a complete change in the scale and nature of the project.

At the time of the publication of the first Supplement it was felt that the Ventris decipherment of the Linear B tablets was still too uncertain to warrant the inclusion of these texts in a standard dictionary. Ventris's interpretation is now generally accepted and the tablets can no longer be ignored in a comprehensive Greek dictionary; however, for reasons explained in a note appended to this preface, Mycenaean forms are not included in the body of the articles, but attached at the end. Words written in the Cyprian syllabary are quoted in transcription as well as in a reconstructed alphabetical form, since the latter may be misleading. References to these have also been brought up to date.

The new inscriptions include many early ones in dialect, and advances in the classification and understanding of dialects since the last edition of Liddell and Scott mean that the information given in LSJ is not always reliable. While every effort has been made to explain new dialect forms and words, on many points recourse must be made to specialist works.

Words not already in LSJ but included in *A Patristic Greek Lexicon* (ed. Lampe, Oxford 1961) have not normally been added unless they help to explain related words which *are* in LSJ/Supplement. At the same time some words noted only from patristic sources, but used in a general, i.e. non-patristic/ecclesiastical sense, have been included if they have been overlooked by Lampe.

The occasional etymological notes in LSJ are frequently out of date. No attempt has been made to bring them up to date and readers are referred to the standard etymological dictionaries of Chantraine (*Dictionnaire Étymologique de la Langue Grecque*, Paris 1968) and Frisk (*Griechisches Etymologisches Wörterbuch*, Heidelberg 1960–70).

At the time of completion of this Supplement three fascicles of a new Spanish dictionary (*Diccionario griego-español* Madrid, 1980–) have been published and these have been consulted with profit for those parts of the alphabet which they cover. New abbreviations in the Supplement have, where possible, been chosen to match those of the Spanish dictionary.

New words in the first Supplement were distinguished by the prefix of a cross(*); this practice has been continued, and in addition an upright cross (✝) has been used to indicate those entries from the main dictionary which have been totally rewritten. A prefixed ° indicates a cross-reference to an article elsewhere in the Supplement, and ‡ means that readers should refer to the word so marked in both the Supplement and the main dictionary.

Vowel quantities are indicated on the same principles as in the main dictionary. To avoid misunderstanding all words in the new Supplement are so marked, not, as in

PREFACE

the old Supplement, only the new entries. An exception is made in the case of late borrowings from Latin. At the time the borrowings were made the distinction of length was breaking down in both Latin and Greek, and to mark them according to classical principles would be meaningless. A number of other small changes have been made in printing conventions in line with modern practice and in the interests of greater clarity. Some of these will be found in the list of abbreviations; the rest are self-explanatory and will cause no problems.

New editions of texts are noted in the lists of Authors and Works, etc. References to them in the articles are normally indicated by the initial of the editor. No attempt has been made to update the references in the main work, except where a substantial change of text is involved, or in order to make cross-references consistent, or very occasionally where the original source of quotation is particularly difficult of access.

Thanks must go to the many scholars who have supplied material for the new Supplement. In particular, the bulk of the epigraphical material was collected by Dr Thompson. Substantial contributions in the field of papyrology were received from Dr Helen Cockle, Dr Revel Coles, and Vincent McCarren. Ivars Avotins contributed legal material, and generously made available the typescript of his book *On the Greek of the Novels of Justinian* in advance of publication. Prof. J. O. Urmson sent the typescript of his book *The Greek Philosophical Vocabulary*. Dr Elizabeth Tucker undertook the task of searching periodicals published since the last Supplement for extra material. The text has been read all or in part by Dr L. Holford-Strevens, Prof. Robert Renehan, and Dr Rudolf Wachter. Technical help with computing was given in the early stages by Dr Timothy Ashplant and since 1986 by Dr Jonathan Moffett. To all of these much gratitude is due.

The British Academy, by making annual grants from 1981 to 1992, has borne by far the largest share of the research costs of the project. Various other bodies have given single or multiple subventions, namely: Oxford University Press; the Jowett Copyright Trust; the Cambridge Classics Faculty; the Craven Committee; the Gaisford Memorial Fund; All Souls and Christ Church Colleges, Oxford; Trinity, Gonville & Caius, and St John's Colleges, Cambridge. The generosity of all these is gratefully acknowledged.

NOTE ON MYCENAEAN AND CYPRIAN ENTRIES

The inclusion in an alphabetically arranged Lexicon of words spelled in a syllabic script raises obvious problems; this note is designed to explain the principles applied. The later editions of Liddell and Scott were able to record some of the new words and forms discovered by the decipherment of the Cyprian script by the simple expedient of devising a reconstruction which could be spelt in the Greek alphabet. This method is not satisfactory owing to the difficulty in some cases of selecting from a number of possible transcriptions. It was therefore decided to revise all such entries, and to insert as well as the alphabetic reconstruction the actual syllabic spelling, transliterated by the accepted system.

The decipherment of the Linear B script presented the same problem in a more acute form, so it was decided to include only a note of the corresponding Mycenaean form, usually at the end of an article, to warn the user that the word occurred in the Linear B texts. Here too the standard transliteration of the syllabic spelling is given, together with the probable reconstruction; but here the difficulty of choice is far greater, and such forms must be treated with suitable caution.

The two scripts are, despite considerable differences in both form and usage, clearly related, so it is convenient to describe them together. Both have signs for the five simple vowels, with no means of distinguishing vowel length, and for a range of consonants followed by each of the vowels. In both scripts the three orders of stops, unvoiced unaspirated, unvoiced aspirated, and voiced, are written with one series of signs, so that the choice is left to the reader. Thus a sign transcribed *ka* may be found in a reconstructed word as κα, χα or γα, *pa* as πα, φα or βα. However, although in Cyprian *ta*, *te*, etc. stand for τα, θα, δα, τε, θε, δε and so forth, Mycenaean has two series of signs, one restricted to the unvoiced, the other to the voiced dental stop. They are therefore transcribed *ta*, *te*, etc. in one case, but *da*, *de* in the other.

The other consonant series correspond to the appropriate Greek letter, with some exceptions. Mycenaean has only one series for liquids, transcribed *ra*, *re*, etc., but equivalent to λα, λε as well as ρα, ρε. In Cyprian these two sounds are allotted separate series of signs. The semi-vowel corresponding to the Greek digamma is regularly used in both dialects, and the signs are transcribed *wa*, *we*, etc. (in older publications also *va*, *ve* in Cyprian). This can be used, like Ϝ, to indicate a glide sound between υ and another vowel. Another semi-vowel, *ja*, *je*, etc. (also transcribed *ya*, *ye*) is used in both scripts to indicate the glide between ι and another vowel. But Mycenaean also uses this device to represent diphthongal ι before another vowel.

Another peculiarity of Mycenaean is the existence of a fourth series of signs for stops, which represents the inherited labio-velar sounds, which survived into Latin in the spellings *qu* and *gu*. These are conventionally transcribed *qa*, *qe*, etc. and must be understood to represent sounds such as $k^w a$, $k^w e$, $kh^w a$, $kh^w e$, $g^w a$, $g^w e$, etc. Since there are no letters in the Greek alphabet for such sounds, reconstructed forms have been given in the spelling which such words show in classical times. Thus Mycenaean *-qe* will be found under τε, *qa-si-re-u* under βασιλεύς.

Both scripts have some additional signs with special values, some of which are disputed. Cyprian has a sign previously transcribed *za*, but now known to have the value *ga*; see γῆ, ἀγαθός. The value of *zo* is still unclear, but *xa*, *xe* correspond to Greek ξα, ξε. Mycenaean has many more abnormal signs: a number of doublets, probably all with special values, such as a_2 = ἁ, a_3 = αι, ra_3 = ραι, λαι. There is also a sign with the value *au*. The exact value of the signs transcribed ra_2, ro_2 is still uncertain, but it seems historically to continue *rya*, *lya*, *ryo*, *lyo*. The sign pu_2 usually has the value of φυ, and there is a special sign with the value *pte* (πτε).

Clusters of consonants in both scripts are written by adding extra vowels: Cypr. *ka-si-ke-ne-to-se* = κασιγνέτος, Myc. *a-re-ku-tu-ru-wo* = Ἀλεκτρύων. Double consonants are not written. In Cyprian a nasal is regularly omitted in writing before another consonant; in Mycenaean this principle is extended to include liquids (λ, ρ), sibilants (σ) and diphthongal ι. Thus Myc. *e-ka-ra* = ἐσχάρα, *e-ko-si* = probably ἔχονσι as in Arcadian.

The peculiarities of grammar and the scripts may be seen in specialist handbooks, such as C. D. Buck, *The Greek Dialects*, Chicago, 1955 for Cyprian, and M. Ventris and J. Chadwick, *Documents in Mycenaean Greek*, 2nd edn. Cambridge, 1973 or E. Vilborg, *A Tentative Grammar of Mycenaean Greek*, Göteborg, 1960. In the vocabulary there are a number of new words, mostly compounds which have no direct representative in classical Greek. These have been listed under a reconstructed form in the Greek alphabet, prefixed by an asterisk. Thus *to-ko-do-mo* is entered under *τοιχοδόμος, which although hitherto unattested must have existed as the noun on which the verb τοιχοδομέω is built. A new verb related to κτίζω is recorded under that form, since its 1st person singular is unknown; it does, however, explain the formation of Homeric ἐυκτίμενος. Likewise a verb meaning 'to row' is related to ἐρέσσω, but lacks the suffix -σσ-. The more speculative additions to the vocabulary have not been entered.

A fuller account of the evidence for particular Greek words in Mycenaean will be found in J. Chadwick and L. Baumbach 'The Mycenaean Greek Vocabulary', *Glotta*, 41 (1963) 157–271, also L. Baumbach, *Glotta*, 49 (1971) 151–190. For words in Mycenaean spelling there is now the *Diccionario micénico* (*Diccionario griego-español, anejo*) by F. Aura Jorro, Madrid 1985–93.

I. AUTHORS AND WORKS

T(L)   Teubner (Leipzig)   T(S)   Teubner (Stuttgart)
T(B)   Teubner (Berlin)   OCT   Oxford Classical Texts
B   Budé, Paris

Achaeus tragicus, add 'Snell °*TrGF* 1 p. 115 [Achae. .. S.]'

Achilles Tatius scriptor eroticus, for 'iv AD' read 'ii AD'; add '*Leucippe and Clitophon*, E. Vilborg, Stockholm 1955 [Ach.Tat. .. V.]'

***Acta Alexandrinorum** [Act.Alexandr.] ii/iii AD, H. Musurillo, T(L) 1961.

***Acta Apostolorum Apocrypha**, R. A. Lipsius, M. Bonnet, Leipzig 1891–1903 [Hildesheim 1959] (3 vols.).

***Acta Petri et Pauli** v. °Acta Apostolorum Apocrypha.

***Acta Philippi** v. °Acta Apostolorum Apocrypha.

Acusilaus, add 'for Suppl. entries: Jacoby ‡*FGrH* no. 2: i (A) p. 47 [Acus. .. J.]'

Adamantius, add '*Vent.* = περὶ ἀνέμων, V. Rose, *Anecdota Graeca et Graeco-Latina* I, Berlin 1864 [Amsterdam 1963], p. 29 (cited by p.)'

Aelianus, end of line 2, insert '[Graz 1971]'; *NA*, add 'A. F. Scholfield, Loeb 1958–9 (3 vols.)'; *VH*, add 'M. R. Dilts, T(L) 1974'

Aelius Dionysius, add 'H. Erbse, *Untersuchungen zu den attizistischen Lexika*, Berlin 1950, p. 95'

Aeneas Tacticus, add 'A. Dain, A.-M. Bon, B 1967'

Aeschines orator, add 'V. Martin, G. de Budé, B 1927–8 [1962] (2 vols.); Sch. = Scholia, M. R. Dilts, T(L) 1992, [Sch.Aeschin. .. D.]'

Aeschylus, line 2, for 'A. Sidgwick' read 'D. L. Page' and for 'Oxford (OCT)' read 'OCT 1972'; Scholia, add 'O. L. Smith, T(L) i: 1976, ii (2): 1982'; line 6, for '*A.*' read '*Ag.*'; *Fr.*, add 'Radt °*TrGF* 3 [A.*fr.* .. R.]'

Aeschylus Alexandrinus, add 'Snell °*TrGF* 1 p. 312'

Aesopus, line 2, add 'A. Hausrath, H. Haas, H. Hunger, 1 (1) and (2), T(L) 1974⁴; B. E. Perry, *Aesopica* 1, Univ. Illinois 1952 [Aesop. .. P.]'

Aëtius, add 'Lib. i–viii, A. Olivieri, *CMG* viii 1, 2, Leipzig and Berlin 1935–50 [Aët. .. O.]'

Africanus, *Cest.*, add 'Les "*Cestes*" de Julius Africanus, J. R. Viellefond, Florence 1970 [Afric.*Cest.* .. V.]'

Agathias, line 2, add 'R. Keydell, Berlin 1967 [Agath. .. K.]'

***Agathodaemon** alchemista [Agath.Alch.] Berthelot °*CAlG*, pp. 115, 268.

***Agias Dercylus** historicus, Jacoby ‡*FGrH* no. 305: iii (B) p. 7 [Agias Dercyl. .. J.].

Aglaias, add '°*Suppl.Hell.* p. 7'

Albinus, add 'also known as ‡Alcinous'

Alcaeus lyricus, after 'iii p. 147' add 'E. Lobel, D. L. Page, *Poetarum Lesbiorum Fragmenta*, Oxford 1955 [1968 corr.], p. 112 [Alc. .. L.-P.]'

Alcidamas, *Soph.*, add '*Artium scriptores*, L. Radermacher, Vienna 1951 (= *SAWW* 227.3), p. 135 [Alcid.*Soph.* .. R.]'

Alcinous philosophus, add '(also known as Albinus q. v.)'; after '1892' insert '[1927]'; at end add '[Alcin.*Intr.* .. H.]'

Alcmaeon, after 'Diels' insert 'W. Kranz' and for 'i p. 131' read '⁷ i p. 210'

***Alcmaeonis** poema epicum, vi BC, Bernabé °*PEG* p. 32 [*Alcmaeonis fr.* .. p. .. B.].

Alcman, add 'Page °*PMG* p. 2 [Alcm. .. P.]'

Alexander Aetolus, for 'Elegiacus' read 'tragicus et lyricus'; for 'iii BC' read 'iv/iii BC'

Alexander Aphrodisiensis, *Pr.*, add 'Bks. iii, iv, H. Usener, Berlin 1859'

Alexander Ephesius epicus, add 'in *Suppl.Hell.* p. 9'

***Alexander Myndius** historicus [Alex.Mynd.] i AD Apud Athenaeum. Jacoby ‡*FGrH* no. 25: i (A) 189.

Alexander Trallianus, add 'repr. Puschmann [Amsterdam 1963], cited in Suppl. by vol., p. and line'

Amipsias, add 'Kassel, Austin °*PCG* 2 p. 197 [Amips. .. K.-A.]'

Ammonius grammaticus, add 'K. Nickau, T(L) 1966'

***Amphilochius Iconiensis** scriptor ecclesiasticus [Amph.] iv AD *hom.* = *homiliae*, Migne °*PG* 39.36, Corpus Christianorum, series graeca, vol. 3, Leiden 1978.

Anacreon, add 'West °*IEG* (2) p. 30 [Anacr. .. W.]; Page °*PMG* p. 172 [Anacr. .. P.]'

Anacreontea, line 1, add '?i/vi AD'; at end add 'M. L. West, T(L) 1984 [*Anacreont.* .. W.]'

Anaxagoras, after 'Diels' insert 'W. Kranz' and for 'i p. 375' read '⁷ ii p. 5'

Anaxandrides comicus, add 'Kassel, Austin °*PCG* 2 p. 236 [Anaxandr. .. K.-A.]'

Anaxarchus, after 'Diels' insert 'W. Kranz' and for 'ii p. 144' read '⁷ ii p. 235'

Anaximander philosophus, after 'Diels' insert 'W. Kranz' and for 'i p. 14' read '⁷ i p. 81'

Anaximenes, after 'Diels' insert 'W. Kranz' and for 'i p. 22' read '⁷ i p. 90'

***Anaximenes Lampsacenus** rhetor [Anaximen. Lampsac.] iv BC *Rh.* = *Ars rhetorica*, M. Fuhrmann, T(L) 1966 (= Arist.*Rh.Al.* in *LSJ*).

Anaxippus, add 'Kassel, Austin °*PCG* 2 p. 299'

Andocides, add 'G. Dalmeyda, B 1930 [1966]'

Andromachus, add 'Heitsch °*GDRK* 2 p. 8'

***Andromachus** minor medicus i AD Apud Galenum [Androm.min.ap.Gal.].

Anecdota Graeca e codd. MSS. Bibl.Oxon., add '[Amsterdam 1963]'

***Anecdota Graeca e codicibus regiis** [*An.Boiss.*] J. Fr. Boissonade, Paris 1829–33 [Hildesheim 1962] (5 vols.).

***Anonyma de musica scripta Bellermanniana** [Anon.Bellerm.] D. Najock, T(L) 1975.

***Anonymi Alchemistae** [Anon.Alch.] Berthelot °*CAlG*.

Anonymus, add '*Vit.Arist.* etc., v. Diogenes Laertius'

*Anonymus Argentinensis [Anon.Argent.] B. Keil, Strasburg 1902.

Anonymus Londnensis, read '**Londinensis**'; add 'W. H. S. Jones, Cambridge 1947'

Anthologia Graeca, line 6, after '1894–1906' add 'H. Beckby, Munich [1965–8]² (4 vols.; I–XV = *Anthologia Palatina*, XVI = *Appendix Planudea*)'; at end add 'A. S. F. Gow, D. L. Page, *The Greek Anthology* 1: *Hellenistic Epigrams*, Cambridge 1965 (2 vols.) [*HE* .. G.-P.]; 2: *The Garland of Philip and Some Contemporary Epigrams*, Cambridge 1968 [*Garl.* .. G.-P.]; *Epigrammata Graeca*, D. L. Page, OCT 1975 [*EG* .. P.]; *Further Greek Epigrams. Epigrams before* AD 50 *from the Greek Anthology and Other Sources*, D. L. Page (revised by R. D. Dawe, J. Diggle), Cambridge 1981 [*FGE* .. P.]; *Sch.AP* = *Scholia ad epigrammata arithmetica in Anthologia Graeca* (*scholia recentiora*), in *Diophanti Alexandri opera omnia*, P. Tannery, vol. 2 T(L) 1895 [(S) 1974]'

Antidotus comicus, add 'Kassel, Austin °*PCG* 2 p. 308'

Antimachus Colophonius, add 'West °*IEG* (2) p. 38 [Antim. .. W.]; B. Wyss, Berlin (Weidmann) 1936 [1974] [Antim. .. Wy.]; °*Suppl.Hell.* p. 20'

Antiphanes Macedo and **Megalopolitanus**, for these entries read '**Antiphanes Macedo** seu **Megalopolitanus** [Antiph.] i AD v. Anthologia Graeca'

Antipho orator, add 'L. Gernet, B 1954'

Antipho Sophista, after 'Diels' insert 'W. Kranz' and for 'ii p. 289' read '⁷ ii p. 334'

Antisthenes, after 'Rhetor' add 'et philosophus' and at end add '*fr.* = *Antisthenis Fragmenta*, F. D. Caizzi, Milan 1966'

Antoninus Liberalis, add 'M. Papathomopoulos, B 1968 [Ant.Lib. .. P.]'

Aphthonius, after 'p. 19' insert 'H. Rabe, T(L) 1926'

*Apollinaris Laodicenus scriptor ecclesiasticus [Apoll.] iv AD *Met.Ps.* = *Metaphrases in Psalmos* cited by volume and page of Migne °*PG*.

Apollodorus lyricus, line 1, add 'vi BC'; add 'Page °*PMG* p. 364 [Apollod.Lyr. .. P.]'

Apollonius Rhodius, add 'H. Fränkel, OCT 1961 [1986]; F. Vian, B 1974–81 (3 vols.); Sch. = *Scholia in A.R. vetera*, C. Wendel, Berlin (Weidmann) 1958²'

*Appendix Proverbiorum [App.Prov.] E. von Leutsch, F. G. Schneidewin *Paroemiographi* i p. 379.

Appianus, after line 3, insert 'P. Viereck, A. G. Roos, E. Gabba, T(L) 1939 [1968⁴]'

*Apuleius scriptor Latinus [Apul.] ii AD *Met.* = *Metamorphoses*, *Plat.* = *de Platone*, R. Helm, T. Thomas, T(L) 1931 [1963–70] (3 vols.).

Apuleius scriptor botanicus, after 'Basel 1560' insert '*Pseudo-Apulei Platonici Herbarius*, E. Howald, H. E. Sigerist, *CML* iv, Leipzig, Berlin 1927, p. 15'

Aquila, add 'L.-R.: v. °*Vetus Testamentum*'

Aratus, line 2, insert 'J. Martin, Florence 1956 [Arat. .. M.]'

*Archelaus philosophus [Archel.Phil.] v BC, H. Diels, W. Kranz, *Vorsokr.*⁷ ii p. 44.

Archilochus, line 2, after 'ii p. 383' insert 'West °*IEG* (1) p. 1. [Archil. .. W.]'

Archippus comicus, add 'Kassel, Austin °*PCG* 2 p. 538 [Archippus .. K.-A.]'

Archytas Tarentinus, after 'Diels' insert 'W. Kranz' and for 'i p. 322' read '⁷ i p. 421'

Aretaeus, line 2, add '1958²'

Ariphron, add 'Page °*PMG* p. 422 [Ariphron .. P.], cited by line number'

Aristaenetus, line 1, add 'v AD'; add 'O. Mazal, T(L) 1971'

Aristaeus Judaeus, add 'A. Pelletier, Paris 1962'

Aristarchus tragicus, add 'Snell °*TrGF* 1 p. 89 [Aristarch.Trag. .. S.]'

Aristides rhetor, *Or.*, add 'F. W. Lenz, C. A. Behr, Leiden 1976 [Aristid. .. L.-B.]'; line 7, (*Rh.*): after 'ii p. 457' insert '[Munich 1966], W. Schmid, T(L) 1926 [Aristid.*Rh.* .. S., referring to Spengel's numeration in margin]; Sch. .. D. = W. Dindorf, Leipzig 1829 [Hildesheim 1964]'

Aristides Milesius, add 'Jacoby ‡*FGrH* no. 286: iii (A) p. 163 [Aristid.Mil. .. J.]'

Aristides Quintilianus, after 'iii AD' delete '(?)'; add 'R. P. Winnington-Ingram T(L) 1963, cited by p. and line'

*Aristobulus Alexandrinus Judaeus ii AD A. M. Denis *Fr.Ps.Gr.* p. 217 [Aristobul.Alex. p. .. D.].

Aristodemus, for 'i BC' read 'iv AD'; add 'Jacoby ‡*FGrH* no. 104: ii (A) p. 493 [Aristodem. .. J.]'

*Aristodemus mythographus historicus i BC Jacoby ‡*FGrH* no. 22: i (A) p. 186 [Aristodem.Myth. .. J.].

Aristophanes comicus, line 3, add 'V. Coulon, M. van Daele, B 1923–30 [revised 1963–73]'; *Fr.*, add 'Kassel, Austin °*PCG* 3(2) [Ar.*fr.* .. K.-A.]'; Scholia, add 'Fr. Dübner, Paris 1877 [Hildesheim 1969] [Sch.]; *Scholia in Aristophanis Equites*, Groningen/Amsterdam 1969: *Scholia vetera* .., D. Mervyn Jones [Sch. .. J.], *Scholia Tricliniana* .., N. G. Wilson [Sch. .. W.]; *Scholia vetera in Nubes*, D. Holwerda, Groningen 1977 [Sch. .. H.]'

Aristophanes Byzantinus, for 'Philosophus' read 'grammaticus' and add '*fr.* = *Fragmenta*, W. J. Slater (post A. Nauck), Berlin 1986'

Aristopho comicus, add 'Kassel, Austin °*PCG* 4 p. 1 [Aristopho .. K.-A.]'

Aristoteles, *Ath.*, add 'M. Chambers, T(L) 1986'; *Cael.*, add 'P. Moraux, B 1965'; *de An.*, add 'W. D. Ross, OCT 1956 [1963]'; *EE*, add 'R. R. Walzer, J. Mingay, OCT 1990'; *EN*, add 'I. Bywater, OCT 1894 [1962]'; *Fr.*, add '[Stuttgart 1967]; *Fr.lyr.*, = *fragmentum lyricum*, Page °*PMG* p. 444'; *GC*, add 'C. Mugler, B 1966'; *HA*, add 'P. Louis, B 1964–9 (3 vols.)'; *Metaph.*, add 'W. Jaeger, OCT 1957 [1973]'; *Mete.*, add 'F. H. Fobes, Cambridge, Mass. 1919 [Hildesheim 1967]'; *Oec.*, add 'B. A. van Groningen, A. Wartelle, B 1968'; *PA*, add 'P. Louis, B 1956'; *Ph.*, add 'W. D. Ross, OCT 1950 [revised 1973]'; *Po.*, add 'R. Kassel, OCT 1965 [revised 1968]'; *Pr.*, add 'C. E. Ruelle, T(L) 1922'; *Rh.*, add 'W. D. Ross, OCT 1959 [1969]'; *Rh.Al.*, add 'v. °Anaximenes Lampsacenus'; *Top.*, add 'W. D. Ross, OCT 1958 [revised 1970]'

Aristoxenus, *Rhyth.*, add 'L. Pearson, Oxford 1990'; add '*fr.* = *Fragmenta*, *Die Schule des Aristoteles* 2², F. Wehrli, Basel 1967 p. 10 [Aristox.*fr.* .. W.]'

Arrianus, for '*Epict.* .. 1894' read '*Epict.* .. ed. 2, 1916 [T(S) 1965]'; *Fr.*, add 'Jacoby ‡*FGrH* no. 156: ii (B) p. 837 [Arr.*fr.* .. J.]; *Fr.Phys.* = *Fragmenta de rebus physicis*, A. G. Roos, G. Wirth, T(L) 1968 vol. 2, p. 186ff.'; other works, add 'A. G. Roos, G. Wirth, T(L) 1967–8 (2 vols.); also v.s.v. °Periplus Maris Euxini'

Artemidorus Daldianus, add 'R. A. Pack, T(L) 1963 [Artem. .. P.]'

Asclepiades Junior medicus, add 'sts. known as Asclepiades Pharmacion'

Asius, for 'Lyricus' read 'elegiacus et epicus' and at end for 'p. 406' read 'p. 23; West °*IEG* (2) p. 46; *Fr.Ep.* = *Fragmenta Epica*, Bernabé °*PEG* p. 127'

*Asterius sophista [Ast.Soph.] iv AD *Hom.* = *Homiliae in psalmos*, cited by volume and page of Migne °*PG*.

Astydamas, line 2, add 'Snell °*TrGF* 1 p. 198 [Astyd. .. S.]'

ˣ**Athanasius Alexandrinus** scriptor ecclesiasticus [Ath.] iv AD *Imag. Beryt.* = *Sermo in imaginem Berytensem*, two recensions, °*PG* 28.797, 805; *Sem.* = *Homilia de semente* °*PG* 28.144.

Athenaeus grammaticus, line 1, add ' = Ath. epigrammaticus (q.v.)'; line 2, at end insert '[Stuttgart 1965-6] (3 vols.)'

Athenaeus mechanicus, line 1, at end, insert 'ii/i BC'; add '*Griechische Poliorketiker*, R. Schneider, *Abhandlungen der Gesellschaft der Wissenschaften zu Göttingen* (Phil.-hist. Klasse), N.F. 12 no. 5, Berlin 1912: cited by Wescher's p., given in Schneider's margin'

ˣ**Athenagoras Atheniensis** scriptor ecclesiasticus [Athenag.] ii AD *Res.* = *De resurrectione mortuorum* (ascribed to Athenag. but may be iii AD), W. R. Schoedel, Oxford 1972.

Axionicus, add 'Kassel, Austin °*PCG* 4 p. 20 [Axionic. .. K.-A.]'

Babrius, add 'B. E. Perry, Cambridge, Mass., Loeb 1965; [Ps.-Babr. .. C.] refers to Crusius' T ed. of 1897 by no. and p.'

Bacchylides, after '1905' add 'B. Snell, H. Maehler, T(L) 1970¹⁰. [B. or B.*fr.* .. S.-M.]'

ˣ**Bardesanes** historicus Edessenus ii/iii AD Jacoby ‡*FGrH* no. 719: iii (C) p. 643 [Bardes. .. J.].

ˣ**Basilius Caesariensis** scriptor ecclesiasticus [Bas.Caes.] Migne °*PG* 29-32, 36.

ˣ**Basilius Seleucensis** scriptor ecclesiasticus [Bas.Sel.] v AD *HP* = *Homélies pascales*, M. Aubineau, Paris 1972.

Bianor, delete 'Idem .. q.v.'

Bion, for 'U. von Wilamowitz-Möllendorff², *Bucolici Graeci*, Oxford (OCT)' read 'A. S. F. Gow, *Bucolici graeci*, OCT 1952 [1969] p. 153'

Bito, add '*Greek and Roman Artillery: Technical Treatises*, E. W. Marsden, Oxford 1971 p. 65 [Bito .. M.]'

ˣ**Boeotica adespota** Page °*PMG* p. 345 [*Boeot.adesp.* .. P.].

ˣ**Bolus** philosophus et paradoxographus iii BC Giannini °*PGR* p. 377. See also °Pseudo-Democritus.

Callimachus, for 'Epicus' read 'poeta'; line 3, after '1921' insert 'R. Pfeiffer, Oxford 1949-53 (2 vols.)' to list of works add '*Dieg.* = Διηγήσεις (w. their own numeration and ref. to Call.'s text in parentheses), A. Vogliano in *PMil.Vogl.* i.18 (1937); Sch. = Scholia' at end for 'initial "P."' read 'letters "Pf." and those from his *Fragmenta nuper reperta* by "P."'; add '*VB* = *Victoria Berenices* °*Suppl.Hell.* p. 100'

Callinus, for 'Epicus' read 'Elegiacus' and add 'West °*IEG* (2) p. 47'

ˣ**Callisthenes Olynthius** historicus [Callisth.Olynth.] Jacoby ‡*FGrH* no. 124: ii(B) p. 631.

Callistratus Historicus, add 'Jacoby ‡*FGrH* no. 433: iii (B) p. 334 [Callistr.Hist. .. J.]'

Callixinus, read '**Callixenus**'; for 'iii BC (?)' read 'ii BC'; add 'Jacoby ‡*FGrH* no. 627: iii (C) p. 161'

Carcinus Tragicus, add 'Snell °*TrGF* 1 p. 210 [Carc. .. S.]'

Carmen Aureum, add 'D. Young, *Theognis* T(L) 1971² p. 86'

ˣ**Carmen Naupactium** Bernabé °*PEG* p. 123 [*Carm. Naup.* .. B.].

ˣ**Carmina Convivalia**, Page °*PMG* p. 472 [*Carm.Conv.* .. P.].

Carmina Popularia, add 'Page °*PMG* p. 450 [*Carm.Pop.* .. P.]'

Catalogus Codicum Astrologorum, for '1898' read '1898-1953 (12 vols.)'

ˣ**Cato** scriptor rei rusticae Latinus iii/ii BC *Agr.* = *De agricultura*, A. Mazzarino, T(L) 1962.

ˣ**Catullus, C. Valerius** poeta Latinus [Cat.] i BC R. A. B. Mynors, OCT 1958 [1972].

Cephisodorus, add 'Kassel, Austin °*PCG* 4 p. 63 [Cephisod. .. K.-A.]'

Cercidas, add 'Diehl °*Anth.Lyr.* 3³ p. 141 [Cerc. .. D.³]'

Chaeremon historicus, add 'P. W. van der Horst, Leiden 1987 [Chaerem.Hist. .. vdH.]'

Chaeremon tragicus, add 'Snell °*TrGF* 1 p. 215 [Chaerem. .. S.]'

Chares iambographus, add 'D. Young, *Theognis* T(L) 1906² p. 113; (app. the same as Chares Trag. q. v.)'

Charisius, add 'C. Barwick, T(L) 1925 [Charis. .. B.]'

Charito, add '*Charitonis Aphrodisiensis de Chaerea et Callirhoe amatoriarum narrationum libri octo*, G. Molinié, B 1979'

Choerilus Epicus, add 'Bernabé °*PEG* p. 187 [Choeril. .. B.]'

Choeroboscus in Heph., add 'C. refers to Consbruch's edition, [T(L) 1906]'; in Theod., add 'H. refers to Hilgard's edition, [Hildesheim 1967] (2 vols.)'

Choricius, add 'R. Förster, E. Richtsteig, T(L) 1929 [1972]'

Cicero, *Att.*, add 'W. S. Watt (part i), OCT 1965, D. R. Shackleton Bailey (part ii), OCT 1961'; before '*Fam.*' insert '*Div.* = *de Divinatione*, W. Ax, T(L) 1938 [1969]'; *Fam.*, add 'W. S. Watt, OCT 1982²'

Clearchus comicus, add 'Kassel, Austin °*PCG* 4 p. 79 [Clearch.Com. .. K.-A.]'

Clearchus historicus, line 1, add 'et philosophus'; at end add '*Die Schule des Aristoteles* 3², F. Wehrli, Basel 1969 [Clearch. .. W.]'

Clemens Alexandrinus, line 2, add 'O. Stählin, L. Früchtel, U. Treu, Berlin 1985, 1970 (2 vols., = °*GCS* 52, 17)'; add 'Sch. = Scholia: 1 = °*GCS* 12; *Ecl.* = *Eclogae propheticae* (°*GCS* 17)'

ˣ**Columella, L. Junius** scriptor rei rusticae Latinus [Colum.] i AD H. B. Ash, M. B. H. Förster, F. S. A. Heffner, Loeb 1941-55.

Coluthus, line 1, add '**Colluthus** in Suppl. [Colluth.]'; at end add 'P. Orsini, B 1972'

ˣ**Comarius** alchemista, Berthelot °*CAlG* p. 289.

Comica adespota, add 'Austin °*CGFP* p. 223 [*Com. adesp.* .. A.]; D. refers to Demiańczuk's *Supp.Com.*'

ˣ**Consentius** grammaticus v AD Keil *Gramm.Lat.* vol. v p. 338.

Corinna, for 'vi BC' read '?v/iii BC'; add 'Page °*PMG* p. 326. [Corinn. .. P.]'

Corpus Hermeticum, add 'A. D. Nock, A. J. Festugière, B 1945-54 (4 vols.)'

Cratinus comicus, add 'Kassel, Austin °*PCG* 4 p. 112. [Cratin. .. K.-A.]'

Critias, after 'Diels' insert 'W. Kranz' and for 'ii p. 308' read '⁷ ii p. 371. [Critias .. D.-K.]'; add 'Snell °*TrGF* 1 p. 172 [Critias .. S.]; West °*IEG* (2) p. 52 [Critias .. W.]'

ˣ**Critodemus astrologus**, v. *Cat.Cod.Astr.*

Ctesias, add 'Jacoby ‡*FGrH* no. 688: iii (C) p. 416 [Ctes. .. J.]'

ˣ**Cyranides** [Cyran.] i/ii AD de Mély °*Lapid.Gr.* (cited by p.); *Die Kyraniden*, D. Kaimakis, Meisenheim 1976 (cited by book, chapter and line).

Cyrilli glossarium, line 1, add 'v AD'

*Cyrillus Alexandrinus** scriptor ecclesiasticus [Cyr.Al.] v AD cited by volume and page of °*PG.*
*Cyrillus Scythopolitanus** scriptor ecclesiasticus [Cyr.S.] *Vit.Sab.* = *Vita Sabae,* E. Schwartz, *Texte und Untersuchungen zur Geschichte der altchristlichen Literatur,* Leipzig 1939.

Damascius, *Pr.,* add '[Brussels 1964]'; add '*in Phlb.* = *in Platonis Philebum scholia. Lectures on the Philebus Wrongly Attributed to Olympiodorus,* L. G. Westerink, Amsterdam 1959 (= Olymp. *in Phlb.* in *LSJ*)'
Damoxenus comicus, add 'Kassel, Austin °*PCG* 5 p. 1 [Damox. .. K.-A.]'
Demades, *Fr.,* add 'V. de Falco, Naples 1955²'
*Demetrius Laco** philosophus ii BC V. De Falco, Naples 1923 [Demetr.Lac. p. .. DF.].
Demetrius Phalereus rhetor, for 'iv BC' read 'i BC'; *Eloc.,* add '[Hildesheim 1965]'
*Demetrius Scepsius** grammaticus iii/ii BC R. Gaede, Diss.Greifswald 1880. Cited fr. Strabo or Athenaeus.
*Demioprata** [*Demioprat.*] cited from sources.
Democritus philosophus, after 'Diels' insert 'W. Kranz' and for 'ii p. 10' read '⁷ ii p. 81'; add '*fr.Orth* = *fragmentum Orth,* E. Orth, *Emerita* 26 (1958) p. 202'
Demophilus, *Sim.,* for '1900' read '1905'
Demosthenes orator, line 3, for 'Oxford (OCT)' read 'OCT 1903-31 [1967-74] (vols. 1, 2(1), 2(2), 3)'; add '*fr.* = *Fragmenta,* Baiter, Sauppe *Orat.Att.* ii p. 250'
Demosthenes Ophthalmicus, add 'cited also fr. Simon Ianuensis, *Clavis Sanationis,* Venice 1510'
Dialexeis, after 'Diels' insert 'W. Kranz' and for 'ii p. 334' read '⁷ ii p. 405'
Dicaearchus geographus, add '*Fragmenta, Die Schule des Aristoteles* 1², F. Wehrli, Basel 1967'
Didymus, *in D.,* add 'L. Pearson, S. Stephens, T(S) 1983 [Did.*in D.* .. P.-S.]'
Dinarchus, add 'N. C. Conomis, T(L) 1975'
Dio Cassius, add '*frr.* cited fr. Boissevain w. two figures, bks. w. three (i.e. book, chapter and paragraph).'
Dio Chrysostomus, add '*Orr.* 37 and 64 are now attrib. to ‡Favorinus philosophus'
Diocles medicus, add 'iv/iii BC'
*Diodorus Cronus** philosophus [Diod.Cron.] iv/iii BC *Die Megariker,* K. Döring, Amsterdam 1972 p. 28 (cited by no.) [Diod.Cron. .. D.].
Diogenes Apolloniates, after 'Diels' insert 'W. Kranz' and for 'i p. 416' read '⁷ ii p. 51'
Diogenes Atheniensis, add 'Snell °*TrGF* 1 p. 184 [Diog.Ath. .. S.]'
Diogenes Cynicus, after 'philosophus' add 'et tragicus [Diog.Sinop.]' and at end add 'Snell °*TrGF* 1 p. 253'
Diogenes Laertius, add 'H. S. Long, OCT 1964 [1966] (2 vols.)'
Diogenes Oenoandensis, add 'C.W. Chilton, T(L) 1967 [Diog.Oen.*fr.* .. C.]; new *frr.* ed. by M. F. Smith: nos. 1-4 in *AJA* 74 (1970) 56-62, nos. 5-16 in *AJA* 75 (1971) 358-89, nos. 39-51 in *Cahiers de philologie* (Lille) 1 (1976) p. 279 [Diog.Oen.*new fr.* .. S.]'
Diogenes Sinopensis, delete this entry (v. Diogenes Cynicus).
Diomedes grammaticus, line 1, add 'iv AD'
Dionysius Bassaricon auctor, v. Dionysius Periegeta.
Dionysius Chalcus, add 'West °*IEG* (2) p. 58'
Dionysius Comicus, add 'Kassel, Austin °*PCG* 5 p. 32 [Dionys.Com. .. K.-A.]'
Dionysius Halicarnassensis, *Opuscula,* add '[Stuttgart 1965] (2 vols.)'; *Vett.Cens.,* add '(= περὶ μιμήσεως B,

repr. Usener–Radermacher [Stuttgart 1965] vol. ii pp. 197, 200, 202-16), also known as *Imit.*'
Dionysius Periegeta, add 'i/ii AD Heitsch °*GDRK* p. 61; *Dionysii Bassaricon et Gigantiadis Fragmenta,* E. Livrea, Rome 1973 (attrib. dub.) [Dionys.Bassar.*fr.* .. L.]'
Dionysius Thrax, Scholia, add '[Hildesheim 1965] [Sch.]'
*Dionysius Tragicus** iv BC Snell °*TrGF 1 p. 240* [Dionys.Trag. .. S.].
*Dioscorus** Thebanus (Aegypti) epicus et lyricus vi AD Heitsch °*GDRK* 1² p. 127 [Diosc.epic.].
Diphilus comicus, add 'Kassel, Austin °*PCG* 5 p. 47ff. [Diph. .. K.-A.]'
Donatus, add '*Ter.* = *Commentum Terenti,* P. Wessner, T(L) 1902-8 [1966] (3 vols.)'
*Dorieus** poeta, before ii AD, in °*Suppl.Hell.* p. 182 (*fr.* 396).
Dorotheus, add '*Die Fragmente des Dorotheos von Sidon,* V. Stegemann, Heidelberg 1939 [Doroth.*fr.* .. S.]'
Dosiadas, *Ara,* add '*Bucolici Graeci* A. S. F. Gow, OCT 1952 [1969] p. 182'
Dromo comicus, add 'Kassel, Austin °*PCG* 5 pp. 124 [Dromo .. K.-A.]'
Duris historicus, add 'Jacoby ‡*FGrH* no. 76: ii (A) p. 136 [Duris .. J.]'

Empedocles, line 2, for '*PPF* p. 74' read 'W. Kranz, *Vorsokr.* i⁷ p. 276, cited by letter and no. [Emp. .. D.-K.]'; line 4, for 'pp. 154, 199' read 'Berlin 1898 [1958] p. 154'
*Ennius** poeta Latinus [Enn.] iii/ii BC *Sat.* = *Saturae* I. Vahlen, T(L) 1928 [Amsterdam 1976].
Ephorus, add 'Jacoby ‡*FGrH* no. 70: ii (A) p. 37. [Ephor. .. J.]'
Epicharmus, line 2, add 'Austin °*CGFP* p. 52'
Epictetus, line 2, for '1894' read '1916² [Stuttgart 1965]'; add '*Sent.* = *Gnom.D* (*Sententiae Codicis Vaticani* 1144) H. Schenkl 1916² [1965] p. 478'
Epicurus, line 4, add 'G. Arrighetti, Turin 1973²'; *Sent.Vat.,* add '= Arrighetti p. 139'; at end add '*Epicuri et Epicureorum Scripta in Herculanensibus papyris servata,* A. Vogliano, Berlin, 1969'
*Epiphanius Constantiensis** scriptor ecclesiasticus [Epiph.Const.] iv/v AD *Opera omnia,* W. Dindorf, Leipzig 1859; *De XII gemmis*: cited without title fr. de Mély °*Lapid.Gr.* p. 193 (= Migne °*PG* 43.293); *Vita Epiph.* = *Vita Epiphanii* (Migne °*PG* 41.23).
Erinna, line 1, for 'Lyrica' read 'poetria' and add 'iv or iii BC'; line 2, add 'Diehl °*Anth.Lyr.* 1²(4) p. 207; °*Suppl.Hell.* p. 186'
Etymologicum Genuinum, add 'also in *Mélanges de littérature grecque, contenant un grand nombre de textes inédits,* M. E. Miller, Paris 1868 [Amsterdam 1965] and modern editions of portions by Lasserre, Berger, Alpers'
*Euanthius** grammaticus Latinus iv AD (*De comoedia. Commentum Terenti* prefixed to that of °Donatus) p. 62 in Kaibel *CGF* [Euanthius .. K.].
Eubulus, add 'Kassel, Austin °*PCG* 5 p. 188 [Eub. .. K.-A.]'
Euclides, add '*Catoptr.* = *Catoptrica* (vii p. 286 Heiberg)'
Eunapius, *VS,* add 'G. Giangrande, Rome 1956 (cited w. Boissonade's numeration) [Eun. .. p. .. B.]'; *Hist,* add 'R. C. Blockley, *The Fragmentary Classicising Historians of the Late Roman Empire,* Liverpool 1983, vol. 2 p. 2 [Eun.*Hist.fr.* .. B.]'
Euphorio, line 2, add '°*Suppl.Hell.* p. 196'; line 3, add 'B. A. van Groningen, Amsterdam 1977 [Euph. .. vGr.]'

Eupolis, add 'Kassel, Austin °*PCG* 5 p. 294 [Eup. .. K.-A.]'

Euripides, line 2, for 'Oxford (OCT)' read 'OCT 1902–13 [1969–74] (3 vols.); J. Diggle, OCT 1981–94 (3 vols.); for *Hyps.*: G. W. Bond, Oxford 1963 [Eur.*Hyps.fr.* .. B.]; for *Ph.*: D. J. Mastronarde, T(L) 1988; for *Phaëth.*: J. Diggle, Cambridge 1970 [Eur.*Phaëth.fr.* .. D.]'; *fr.*, add 'B. Snell, suppl. to Nauck *TGF* p. 1027 [E.*fr.* .. N.-S.]'

Eusebius historicus, add 'Jacoby ‡*FGrH* no. 101: ii (A) p. 480 [Eus.Hist. .. J.]'

Eustathius episcopus, after 'Leipzig 1825–30' insert '[Hildesheim 1960] (3 vols.); *Commentarii ad Homeri Iliadem pertinentes*, M. van der Valk, Leiden 1971–87 (4 vols.) cited by page and line no. of ed. Rom. in margin'; add '*Eustathii Metropolitae Thessalonicensis Opuscula*, T. L. F. Tafel, Frankfurt am Main, 1832 [Amsterdam 1964]'

Eutychianus, for 'p. 365' read 'p. 369' and add 'Jacoby ‡*FGrH* no. 226: ii (B) p. 954 [Eutych. .. J.]'

*****Evagrius** scholasticus [Evagr.Schol.] vi AD *HE* = *Historia Ecclesiastica*, cited by volume and page of Migne °*PG*.

*****Excerpta Barocciana** [Exc.Barocc.] A. Nauck, *Lexicon Vindobonense*, Petersburg 1867 [Hildesheim 1967] p. 325.

Ezekiel, add 'Snell °*TrGF* 1 p. 288 [Ezek.*Exag.* .. S.]'

Favorinus philosophus, line 1, add 'ii AD '; add '*Exil.* = περὶ φυγῆς (= *PVat.* 11 p. 17, v. index III) (cited by column and line), A. Barigazzi, Florence 1966 [Favorin. .. B.]'

Firmicus Maternus, for line 2, read '*Math.* = *Mathesis*, W. Kroll, F. Skutsch, K. Ziegler, T(L) 1897–1913 [1968] (2 vols.)' and add '*De err.prof.relig.* = *De errore profanarum religionum*, K. Ziegler, T(L) 1907'

'Fragment eines griechischen Lexicons [*Fr.Lex.*] *Philologus* suppl. 15 (1922) p. 115: *Fr.Lex.* I = *ex Codic. Monac.graec.* 263; p. 142: *Fr.Lex.* II = *ex Codic. Berol.graec.quart.* 13(C); p. 145: *Fr.Lex.* III = *ex Eudemi Codic.Parisin.* n. 2634.

Fronto, M. Cornelius, add 'M. P. J. van den Hout, T(L) 1988 [Fronto *Ep.* p. .. vdH.]'

Galenus, line 2, after 'Leipzig 1821–33' insert '[Hildesheim 1964–5; 20 vols.]'; for '*CMG* .. (in progress)' substitute '*CMG* 5.4(1).1 = Kühn 5.1–148; 5. 4.(1).2 = Kühn 5.181–805; 5. 4.(2) = Kühn 6.1–831; 5. 9.(1) = Kühn 15.1–223, 418–919, 19.182–221; 5. 9.(2) = Kühn 16.489–840, 7.643–65, 18(2).1–317; 5.10.(1) = Kühn 17.(1).4–302; 5.10.(2).1 = Kühn 17.(1).480–792; 5.10.(2).2 = Kühn 17.(2).1–344; 5.10.(3) = Kühn 18(1).196–299; (Kühn's numeration in margin in *CMG*) R. Charterius (w. Hippocrates), 13 vols., Paris 1639 [1679]'; *Subf.Emp.*, add '*Die griechische Empirikerschule, Sammlung der Fragmente und Darstellung der Lehre*, K. Deichgräber, Berlin 1930, p. 42 (cited by Bonnet's p., given in Deichgräber's margin)'; after '*Vict.Att.*' add '*in Pl.Ti.* = *in Platonis Timaeum commentarii fragmenta*, H. O. Schroeder, *CMG* Suppl. i, (1934) p. 9 (cited by p. and line)'

*****Gargilius Martialis** scriptor de hortis [Garg.Mart.] iii AD, V. Rose, *Plinius* T(L) 1875 p. 133.

*****Gelasius Cyzicenus** scriptor ecclesiasticus [Gel.Cyz.] v AD *HE* = *Historia Ecclesiastica*, G. Loeschke, M. Heinemann, Berlin 1918, (= °*GCS* 28).

Gellius, Aulus, add 'P. K. Marshall, OCT 1968 (2 vols.)'

Geoponica, line 1, add 'x AD '

*****Gnomologium Vaticanum** [Gnomol.Vat.] L. Sternbach, *De Gnomologio Vaticano inedito* in *Wiener Studien* (*WS*), 9–11 (1887–9) [de Gruyter 1963].

Gorgias, after 'Diels' insert 'W. Kranz' and for 'ii p. 235' read '⁷ ii p. 271'

Gregorius Corinthius, for 'xii AD ' read '(?)x/xi AD '; *Trop.*, add '(an ancient work wrongly ascribed to Greg.), M. L. West in *CQ* N.S. 15.236; (cited by Spengel's p., given in West's margin)'

*****Gregorius Magnus** scriptor ecclesiasticus Latinus [Greg.Mag.] vi AD cited by volume and page of °*PL*.

*****Gregorius Nazianzenus** theologus [Gr.Naz.] iv AD *Carm.* = *Carmina* Migne °*PG* 37.397; *De vita sua* (= *Carmen* 11), C. Jungck, Heidelberg 1974.

*****Gregorius Nyssenus** theologus [Gr.Nyss.] iv AD *Ep.* = *Epistulae*, G. Pasquali, Berlin 1925 [1959].

*****Hegemon** epigrammaticus iii BC v. Anthologia Graeca.

Hegesippus epigrammaticus, line 1, add 'iii BC'

Heliodorus scriptor eroticus, add 'R. M. Rattenbury, T. W. Lumb, J. Maillon, B 1960² (3 vols.)'

Hellanicus, for line 2 read 'Jacoby ‡*FGrH* no. 4: i (A) p. 104 [Hellanic. .. J.]'

Hellenica Oxyrhynchia, add 'V. Bartoletti, T(L) 1959 [*Hell.Oxy.* .. B.], *Hellenica Oxyrhynchia*, P. R. McKechnie, S. J. Kern, Warminster 1988 [*Hell.Oxy.* .. M.-K.]'

Hemerologium Florentinum, add '*Die Kalenderbücher von Florenz, Rom und Leiden*, W. Kubitschek, in *Wiener Denkschriften* (*DAW*) 57 (1915) Abh. 3 [Hemerolog.Flor. .. p. .. K.]'

Hephaestio astrologus, add 'D. Pingree, T(L) 1973–4, cited by bk., chapter and numeration in the outside margin'

Heraclides Lembus historicus, add '*Pol.* = περὶ πολιτειῶν. *Excerpta politiarum*, M. R. Dilts, Durham 1971 (previously attributed to Heraclides historicus)'

Heraclitus, *All.*, add 'F. Buffière, B 1962'

Heraclitus philosophus, after 'Diels' insert 'W. Kranz' and for 'i p. 67' read '⁷ i p. 139 [Heraclit. .. D.-K.]'; *Ep.*, for 'p. 409' read 'p. 280 (v. *Epistolographi* in °v)'

*****Heras Cappadox** medicus i BC/i AD Apud Galenum. [Heras ap.Gal.]

*****Hermerius Claudius** [Claud.Herm.] iv AD *Mul.* = *Mulomedicina Chironis*, E. Oder, T(L) 1901.

Hermes Trismegistus, v. °*Corpus Hermeticum*.

Hero, for '?ii/i BC' read '?i AD '; after '1914' add '[1976] (5 vols.)'; *Bel.*, add '*Greek and Roman Artillery: Technical Treatises*, E. W. Marsden, Oxford 1971, p. 17'; add '*Mech.* = *Mechanica* in vol. 2'

Herodas, add 'I. C. Cunningham, T(L) 1987'

*****Herodes Atticus** sophista ii AD *Pol.* = περὶ πολιτείας, U. Albini, Florence 1968 [Herodes .. A.].

Herodianus grammaticus, line 3, add '[Hildesheim 1965]'; Hdn.*Epim.*, add '[Amsterdam 1963]'; add '*fr.* = *Fragmenta* H. Hunger, °*JÖByz* 16.1–33 [Hdn.*fr.* .. H.]'; *Philet.*, add 'A. Dain, Paris 1954'

*****Herodicus** epigrammaticus iii/ii BC °*Suppl. Hell.* p. 247.

Hesiodus, add 'F. Solmsen, OCT 1970 [1990³]; *Fragmenta Hesiodea*, R. Merkelbach, M. L. West, Oxford 1967 [1990³ w. Solmsen] [Hes.*fr.* .. M.-W.]; *Sch.* = *Scholia vetera in Hesiodi Opera et Dies*, A. Pertusi, Milan 1955 [Hes.Sch. .. P.]; *Scholia vetera in Hesiodi Theogoniam*, L. Di Gregorio, Milan 1975 [Hes.Sch. .. DiG.]'

Hesychius lexicographicus, add 'K. Latte, Copenhagen 1953–66 (*A–O*, 2 vols.). [Hsch. .. La.]'

Hierocles Platonicus, in *CA*, add 'F. G. Köhler, Stuttgart (T) 1974'

*****Hieronymus Eusebius Stridonensis** scriptor ecclesiasticus [Jerome] iv/v AD *Ep.* = *Epistulae*, J. Labourt, B 1949–63 (8 vols.); *in Ep.Eph* = *Commentaria in Ep.Eph.*,

Migne °*PL* 26.439; *in Ezek.*, Migne °*PL* 25.15; *adv. Rufin.* = *Apologia adversus libros Rufini*, Migne °*PL* 23.398.

Himerius, *Or.*, add '*Declamationes et Orationes*, A. Colonna, Rome 1951 [Him.*Or.* .. C.]'

Hippias Erythraeus, line 1, add '[?Hellen.]'; at end add 'Jacoby ‡*FGrH* no. 421: iii (B) p. 317 [Hippias Erythr. .. J.]'

Hippiatrica, add '[1971] [*Hippiatr.* .. O.-H.]; *Hippiatr.Paris.* = *Hippiatrica Parisina* in vol. 2 p. 1 O.-H.'

Hippocrates, *Jusj.*, *Lex*, *de Arte*, *Medic.*, *Decent.*, *Praec.*, *VM*, *Alim.*, *Liqu.*, *Flat.*, add '*CMG* 1.1, I. L. Heiberg, T(L & B) 1927'; *Oct.*, *Septim.*, add '*CMG* 1.2.(1), H. Grensemann, Berlin 1968'; *Aër*, add '*CMG* 1.1.(2), H. Diller, Berlin 1970'; *Superf.*, add '*CMG* 1.2.(2), C. Lienau, Berlin 1973'; *Nat.Hom.*, add '*CMG* 1.1.(3), J. Jouanna, Berlin 1975'; *Morb.*, (iii), add '*CMG* 1.2.(3), P. Potter, Berlin 1980'

Hippolytus, at end add '(= °*GCS* 26); M. Marcovich, *Patristische Texte und Studien* 25, Berlin/New York 1986'

Hippon, after 'Diels' insert 'W. Kranz' and for 'i p. 288' read '[7] i p. 385'

Hipponax, add 'West °*IEG* (1) p. 110 [Hippon. .. W.]; H. Degani T(L) 1983 [Hippon. .. D.]'

Homerus, for line 2, read 'D. B. Munro, T. W. Allen, OCT 1912 [1969-74] (5 vols.)'; *Od.*, add 'P. Von der Mühll, T(S) 1984'; add '*fr.* = *Fragmenta*, OCT op. cit. v, p. 147; Davies °*EGF* p. 104 [Hom.*fr.* .. D.]'; Scholia in Homeri Iliadem: add 'H. Erbse, Berlin 1969-88 (7 vols.) [Sch.Il.]; J. Barnes in editione Homeri, Cambridge 1711 [Sch.D]; *Les Scolies genevoises de l'Iliade*, J. Nicole, Paris 1891 [Sch.Gen.Il.]; *Scholia minora in Homeri Iliadem*, V. de Marco, Rome 1946 [Sch.min.Il.]'; Scholia in Homeri Odysseam: add '[Amsterdam 1962] [Sch.Od.]'

Horapollo, add 'F. Sbordone, Naples 1940 [Horap. .. S.]'

***Horatius Flaccus**, Q. poeta Latinus [Hor.] i BC E. C. Wickham, H. W. Garrod, OCT 1912 [1975] *C.* = *Carmina*; *Sat.* = *Saturae*.

Hymni Homerici, line 2, add 'T. W. Allen, W. R. Halliday, E. E. Sikes, Oxford 1936[2] [1963]'

Hymni Magici, add '*PMag.* 1, 2, 3, 4 (v. °iii)'

Hymnus ad Isim, add 'W. Peek *Der Isishymnus von Andros*, Berlin 1930 [*Hymn.Is.* .. P.]; Y. Grandjean *Une nouvelle arétalogie d'Isis*, Leiden 1975 [*Hymn.Is.* .. (Maroneia)], cf. *SEG* 8.548-51, ib. 26.821, °*IKyme* 41'

Hymnus Curetum, add 'M.L. West, *JHS* 85(1965) p. 149 [*Hymn.Curet.* .. W.]'

Hyperides, line 2, add 'Ch. Jensen, T(L) 1917 [Stuttgart 1963]'

***Iambica adespota** West °*IEG* (2) p. 16 [*Iamb.adesp.* .. W.].

***Iamblichus** alchemista [Iambl.Alch.] Berthelot °*CAlG* p. 285.

Iamblichus philosophus, *Comm.Math.*, add '[U. Klein, Stuttgart 1975]'; *in Nic.*, add '[U. Klein, Stuttgart 1975]'; *Protr.*, add '[Stuttgart 1967] (both works cited by page and line of Pistelli)'; *VP*, add 'L. Deubner, T(L) 1937 [U. Klein, Stuttgart 1975]'

Iamblichus scriptor eroticus, add 'E. Habrich, T(L) 1960 [*Iamb.Bab.* .. Ha.]'

Ibycus, line 2, add 'Page °*PMG* p. 144 [Ibyc. .. P.]'

Ilias Parva, add 'Bernabé °*PEG* p. 71 [Il.Parv. .. B.]'

Iliu Persis, add 'Bernabé °*PEG* p. 86 [Il.Pers. .. B.]'

Ion Chius, line 1, add 'v BC'; line 2, add 'West °*IEG* (2) p. 78 [Ion Eleg. .. W.]'; line 3, add 'Jacoby ‡*FGrH* no. 392: iii (B) p. 276 [Ion Hist. .. J.]'; line 4, add 'Page °*PMG* p. 383 [Ion Lyr. .. P.]'; line 5, add 'Snell °*TrGF* 1 p. 96 [Ion Trag. .. S.]'

Isaeus, add '[1963]; fragments cited by the numbering of Baiter, Sauppe *Orat.Att.* ii p. 228'

Isocrates, line 2, add 'G. Mathieu, É. Brémond, B 1928[1963]-1962'

Ister, for '409' read '418' and add 'Jacoby ‡*FGrH* no. 334: iii (B) p. 168. [Ister .. J.]'

***Jerome**, v. °Hieronymus Eusebius Stridonensis.

***Joannes** alchemista [Jo.Alch.] Berthelot °*CAlG* p. 263.

***Joannes Gazaeus** exegeta vi AD [Jo.Gaz.] P. Friedländer, *Johannes von Gaza und Paulus Silentiarius: Kunstbeschreibungen justinianischer Zeit*, Leipzig/Berlin 1912 [Hildesheim 1969].

***Joannes Malalas** chronographus [Jo.Mal.] vi AD *Chron.* = *Chronographia*, 9-12 A. Schenk, *Die römische Kaisergeschichte bei Malalas*, Stuttgart 1931; 1-8, 13-18 Migne °*PG* 97.65.

Juba, Rex Mauretaniae, add 'Jacoby ‡*FGrH* no. 275: iii (A) p. 127. [Juba .. J.]'

Julianus, add 'J. Bidez, G. Rochefort, Ch. Lacombrade, B 1932-64 (4 vols., 1 (1), (2[2]); 2 (1), (2))'

Juvenalis, D. Junius, for 'i AD' read 'i/ii AD'; for 'S. G. Owen' read 'W. V. Clausen' and add '1959'; add '*Sch.* = *Scholia*, P. Wessner, T(L) 1931 [1967]'

Lasus, add 'Page °*PMG* p. 364 [Lasus .. P.]'

Leonidas epigrammaticus, after '**Leonidas**' insert '**Tarentinus**' and add '[Leon.Tarent.] in Suppl.'

Leonidas Alexandrinus, add '[Leon.Alexandr.] in Suppl.'

***Leontius Byzantinus** scriptor ecclesiasticus [Leont. Byz.] vi AD *HP* = *Homélies pascales*, M. Aubineau, Paris 1972.

***Lesbiorum fragmenta adespota**, E. Lobel, D. L. Page, *Poetarum Lesbiorum fragmenta*, Oxford 1955 [1968], p. 292 [*Lesb.fr.adesp.* .. L.-P.].

Leucippus, after 'Diels' insert 'W. Kranz' and for 'ii p. 1' read '[7] ii p. 70'

***Lexicon Patmense** [Lex.Patm.] I. Sakkelion in *BCH* i (1877).

***Lollianus** scriptor eroticus ii AD A. Henrichs, Bonn 1972 [Lollian. p. .. H.].

Longinus, line 2, before 'cited without title' insert 'D. A. Russell, OCT 1968 [1974]'

Longus, add 'G. Dalmeyda, B 1960[2]; M. D. Reeve, T(L) 1982 [1971]'

Lucianus, after 'Sophista' insert '[Luc.]'; line 4, after '(T) 1906-' add 'M. D. Macleod, OCT 1972-87 (4 vols.)'; line 5, after 'Scholia .. 1906' add '[1971]; other scholia in the Bipontine Edition, Zweibrücken 1789-93'; after '*Electr.* = *Electrum*' insert '*Epigr.* = *Epigrammata*'

***Lucianus** interpres Veteris Testamenti [Lcn.] iii/iv AD *Origenis Hexapla*, F. Field, Oxford 1875.

***Lucilius Gaius** poeta satiricus Latinus ii BC F. Marx, T(L) 1904-5 [Amsterdam 1963].

***Lucretius** = T. Lucretius Carus, poeta Latinus [Lucr.] i BC C. Bailey, OCT 1947 [1978].

Lycophron, line 2, after '1908' add '[1958]; L. Mascialino, T(L) 1964 (cited by line no.)'; *fr.*, add 'Snell °*TrGF* 1 p. 275'

Lycurgus, add 'N. C. Conomis, Leipzig (T) 1970'

Lydus, before '[*Mag.*]' and '[*Mens.*]' insert '[Stuttgart 1967]; [Lyd. .. p. .. W.] refers to p. of this edition'

Lynceus, for 'Cetera .. Athenaeum' read 'Prose writings cited fr. Athenaeus'
Lyrica adespota, add 'Page °*PMG* p. 484 [*Lyr.adesp.* .. P.], Page °*SLG* p. 106 [*Lyr.adesp.S.* ..]'
Lysias, for 'v BC' read 'v/iv BC'; line 2, for 'Oxford (OCT)' read 'OCT 1912 [1978]'
Lysippus, add 'Kassel, Austin °*PCG* 5 p. 618 [Lysipp. .. K.-A.]'

***Macarius Aegyptius** scriptor ecclesiasticus [Mac.Aeg.] iv AD *Hom.* = *Homiliae* cited by volume and page of Migne °*PG*.
Macedonius Thessalonicensis, add '[Maced.Thess. II] in Suppl.'
Macho, add 'A. S. F. Gow, Cambridge 1965 [Macho .. G.]'
Maecius, for 'i AD (?)' read 'i BC'
***Marcellus Empiricus** medicus Latinus [Marcell.Emp.] iv AD M. Niedermann in *CML* v, Leipzig, Berlin 1916.
***Marcellus Sidetes** poeta medicus [Marc.Sid.] ii AD E. Heitsch °*GDRK* 2 p. 17.
Marcus Antoninus, add 'J. Dalfen, T(L) 1987'
Marcus Argentarius, for 'vi AD' read 'i AD'
Marinus biographus [Marin.] v/vi AD J. F. Boissonade, Leipzig 1814 [1966].
Martianus Capella, add 'A. Dick, T(L) 1925 [1969]'
Matro, add 'in °*Suppl.Hell.* p. 259, p. 266'
***Mela, Pomponius** geographus Latinus i AD C. Frick, T(L) 1880 [1968] (cited by bk. and short section).
Meleager, for 'i BC' read 'ii/i BC'
Melissus, after 'Diels' insert 'W. Kranz' and for 'i p. 176' read '⁷ i p. 258'
***Melito Sardianus** scriptor ecclesiasticus [Melit.] ii AD *fr.* = *Fragmenta*, in *The Apostolic Fathers*, E. J. Goodspeed, New York 1950.
Memnon, for 'i AD' read '?i BC/i AD'; add 'Jacoby ‡*FGrH* no. 434: iii (B) p. 336. [Memn. .. J.]'
Menaechmus, for 'iii BC' read 'iv/iii BC'; add 'Jacoby ‡*FGrH* no. 131: ii (B) p. 673 [Menaechm. .. J.]'
Menander comicus, lines 2/3, add 'F. H. Sandbach, OCT 1972, 1976 (revised w. appendix) [1990] [Men. .. S.]'; *Inc.*, add 'Sandbach [1990] p. 297'; *Mis.*, for 'Μισούμεναι' read 'Μισούμενος'; *Mon.*, add 'S. Jäkel, T(L) 1964 [Men. Mon. .. J.]'; to the list of plays add '*Asp.* = Ἀσπίς; *DE* = Δὶς Ἐξαπατῶν; *Dysc.* = Δύσκολος; *Sic.* = Σικυώνιος; *Th.* = Θεοφορουμένη'; at end add 'A. Körte, A. Thierfelder, T(L) 1938³ [1955, w. addenda 1957], 1959² (2 vols.); *Arg.* = *Argumenta* in vol. i p. 146; *fr.* = *Fragmenta* in vol. ii [Men.*fr.* .. K.-Th.] or quoted from Sandbach'
Menander rhetor, add 'D. A. Russell, N. G. Wilson, Oxford 1981 [Men.Rh. .. R.-W.]'
Menander Protector, see also Anthologia Graeca.
Mesomedes, line 3, add 'Heitsch °*GDRK* 1² p. 24 no. 2 [Mesom. .. H.]'
Metrodorus Chius, for 'v BC' read 'iv BC'; after 'Diels' insert 'W. Kranz' and for 'ii p. 140' read '⁷ ii p. 231'
***Mimi adespoti** [*Mim.adesp.*] in *Herodae mimiambi*, I. C. Cunningham, T(L) 1987 p. 36.
***Modestinus, Herennius** iurisconsultus Latinus [Modest.] iii AD *Dig.* = *Digesta* (q.v.).
Moero, for 'Epica iii BC' read 'poetria iv/iii BC'
Molpis, line 1, add 'ii/i BC'; add 'Jacoby ‡*FGrH* no. 590: iii (B) p. 704 [Molpis .. J.]'
Moschio paradoxographus, add '?iii BC Jacoby ‡*FGrH* no. 575: iii (B) p. 675 [Moschio .. J.]'

Moschio tragicus, for 'iv BC' read 'iv/iii BC'; add 'Snell °*TrGF* 1 p. 264 [Mosch.Trag. .. S.]'
Moschus, add 'A. S. F. Gow, *Bucolici Graeci* OCT 1952 [1969] p. 132 [Mosch. .. G.]'
Moses Alchemista, before 'ii AD' insert '[Moses Alch.]'; line 3, for 'p. 139' read 'p. 39'
Musaeus philosophus, after 'Diels' insert 'W. Kranz' and for 'ii p. 179' read '⁷ i p. 20'
Musonius philosophus, add '*fr. K.* = *Fragmentum*, in G. D. Kilpatrick, *CR* 63.94'

***Naevius** poeta Latinus [Naev.] iii BC *Trag.* = *Tragoediarum fragmenta*, E. V. Marmorale, Florence 1953.
Nausiphanes, after 'Diels' insert 'W. Kranz' and for 'ii p. 155' read '⁷ ii p. 246'
***Nautarum Cantiunculae** ii/iii AD Heitsch °*GDRK* 1² p. 32 nos. 3-4.
Neanthes, add 'Jacoby ‡*FGrH* no. 84: ii (A) p. 191 [Neanth. .. J.]'
Neophron, add 'Snell °*TrGF* 1 p. 92 [Neophr. .. S.]'
***Nepos, Cornelius** scriptor Latinus i BC *Eum.* = *Eumenes*; *Dat.* = *Datames*, E. O. Winstedt, OCT 1904.
Nessas, after 'Diels' insert 'W. Kranz' and for 'ii p. 140' read '⁷ ii p. 230'
Nicaenetus, add 'see also Anthologia Graeca'
Nicander, line 2, add 'A. S. F. Gow, A. F. Scholfield, Cambridge 1953'; for '*Alex.*' read '*Al.*' and for '*Ther.*' read '*Th.*'; after line 4 insert '*fr.* = *Fragmenta* [Nic. .. G.-S.]'; at end add '*Scholia in Nicandri Alexipharmaca*, M. Geymonat, Milan 1974 [Sch.Nic.*Al.* .. Ge.]; *Scholia in Nicandri Theriaka*, A. Cragnola, Milan 1971 [Sch. Nic.*Th.*]'
***Nicarchus I** epigrammaticus [Nicarch. I]. iv/iii BC v. Anthologia Graeca.
Nicarchus epigrammaticus, after 'Nicarchus' insert 'II'; add '[Nicarch. II in Suppl.]'
Nicolaus Damascenus, line 2, add 'Jacoby ‡*FGrH* no. 90: ii (A) p. 324. [Nic.Dam. .. J.]'
Nicolaus rhetor, line 1, add 'v AD'; *Prog.*, add 'sts. cited fr. *Rhetores Graeci* iii, L. Spengel, p. 447; sts. fr. *Rhetores Graeci* ii, Ch.Walz, p. 565, al.'
Nicomachus tragicus, line 1, add 'iii BC'; at end add 'Snell °*TrGF* 1 p. 285 (= Nicom. Alexandrinus, to be distinguished fr. Nicom.Atheniensis, v/iv BC, Snell *op.cit.* p. 154) [Nicom.Trag. .. S.]'
Nicostratus comicus, add 'Kassel, Austin °*PCG* 7 p. 74 [Nicostr.Com. .. K.-A.]'
Nonius Marcellus, add 'cited fr. Lindsay w. nos. fr. Mercerus' ed. [Non.Marc. .. M.]'
Nonnus, add 'R. Keydell, Berlin (Weidmann) 1959 (2 vols.); *Par.Ev.Jo.* = *Paraphrasis Evangelii Johannis*, A. Scheindler, T(L) 1881'
Nossis, for 'iii BC(?)' read 'iv/iii BC'
Novum Testamentum, line 2, add 'K. Aland et al., Stuttgart, 1968²'
Numenius Apamensis, add '*fr.* = *Fragmenta*, É. des Places, B 1973 [Numen.*fr.* .. P.]'

Olympiodorus philosophus, *in Phlb.*, now attributed to °Damascius; *in Grg.*, add 'L. G. Westerink, T(L) 1970 (cited by page number)'
Onosander, read '**Onasander**'; after '1923' add '1962 [Onas. in Suppl.]'
Oppianus Anazarbensis, for '**Anazarbensis**' read '**Cilix**'; add 'A. W. Mair, Loeb 1928 [1963] p. 200'
Oppianus Apamensis, C., add 'A. W. Mair, Loeb 1928 [1963] p. 1'

AUTHORS AND WORKS

Oracula Chaldaica, line 1, add 'ii AD '; at end add 'É. des Places, B 1971 [*Orac.Chald.* .. P.]'
***Oracula Sibyllina** [*Orac.Sib.*] ii/iv AD J. Geffcken, Leipzig 1902 [1967] (= °*GCS* 8).
***Oracula Tiburtina** [*Orac.Tib.*] P. J. Alexander, Dumbarton Oaks 1967.
Oribasius, line 1, add 'J. Raeder, *CMG* 6, Leipzig, Berlin 1926–33 [Amsterdam 1964]; *Collectionum medicarum reliquiae* 6.1(1–2), 2(1–2), cited without title'
Origenes, add '*Adnot. in Gen.* = *Adnotationes in Genesim* °*PG* 17.11'
Orphica, *A.*, add '*Argonautiques orphiques*, F. Vian, B 1987'; *H.*, add 'W. Quandt, Berlin (Weidmann) 1962³ [1973]; Εὐχή = Εὐχὴ πρὸς Μουσαῖον (v. *H.*)'; *L.*, add '[Hildesheim 1971], *Lithica Kerygmata*, G. Giannakis, Yannina 1987 [Orph.*Lith.Kerygm.* .. G.]'; *fr.*, add '[1963]'
***Ovidius Naso, P.**, poeta Latinus [Ov.] i BC/i AD *Am.* = *Amores*, E. J. Kenney, OCT 1961.

***Pachomius Tabennensis** scriptor ecclesiasticus [Pach.] *Poen.* = *Poenae monachorum*; *Reg.* = *Excerpta e regula*, L. Th. Lefort; in *Muséon* 37(1924) p. 9, 40(1927) p. 60.
Palladas, for 'vi AD ' read 'iv/v AD '
Palladius, add '*Febr.* = *de febribus*, Ideler °*Physici* i p. 107'
***Palladius** auctor Latinus [Pallad.] ?iv AD *Agric.* = *Opus Agriculturae*, R. H. Rodgers, T(L) 1975.
***Palladius** episcopus Helenopolitanus [Pall.] iv/v AD *V.Chrys.* = *Dialogus de vita Iohannis Chrysostomi* P. R. Coleman-Norton, Cambridge 1928 [1958]; *Gent.Ind.* = *De gentibus Indiae et Bragmanibus* περὶ τῶν τῆς Ἰνδίας ἐθνῶν καὶ Βραχμάνων apud Pseudo-Callisthenem, W. Berghoff, Meisenheim 1967.
Pancrates epicus, add 'Heitsch °*GDRK* 1 p. 52'
Pancrates epigrammaticus, line 1, add 'iii/ii BC'; at end add 'in °*Suppl.Hell.* p. 286'
Panyasis, add 'Bernabé °*PEG* p. 171 [Panyas. .. B.]'
Pappus, add '*in Alm.* = *Commentarium in Ptolemaei Magnam Constructionem sive Almagesto* in *Commentaires de Pappus et de Théon d'Alexandrie sur l'Almageste*, I.: *Pappus*, A. Rome, Rome 1931 [1967]'
Paulus Aegineta, line 2, after '1921–1924' insert 'H. refers to p. and line of Heiber's ed.'
Paulus Alexandrinus, for 'page' read 'leaf'; add 'E. Boer, T(L) 1958 (cited by p. and line number)'
Paulus Silentiarius, add '*Ambo* = *Descriptio Ambonis*, P. Friedländer, *Johannes von Gaza und Paulus Silentiarius*, Leipzig/Berlin 1912'
***Pausanius Grammaticus** [Paus.Gr.] ii AD H. Erbse, *Untersuchungen zu den attizistischen Lexika*, Berlin 1950, p. 152.
Pausanias periegeta, add 'M. H. Rocha-Pereira, T(L) 1973–81 [1989²] (3 vols.)'
Perictione, for 'Philosophus' read 'philosopha Pythagorica'; add 'ii BC'
***Periplus Maris Euxini** [*Peripl.M.Eux.*] (in LSJ s.v. Arr.) °Diller *Trad.min.Gr.geogr.* p. 102.
Periplus Maris Rubri, add '*The Periplus Maris Erythraei*, L. Casson, Princeton 1989 [*Peripl.M.Rubr.* .. C.]'
Perses, line 1, add '?iv BC'
Persius, for 'S. G. Owen²' read 'W. V. Clausen' and add '1959'
Phalaris, for 'p. 439' read 'p. 409'
Phanias epigrammaticus, add 'iii BC'
Pherecrates comicus, add 'Kassel, Austin °*PCG* 7 p. 102 [Pherecr. .. K.–A.]'

***Pherecydes Atheniensis** historicus [Pherecyd.Ath.] v BC Jacoby ‡*FGrH* no. 3: i (A) p. 58.
Pherecydes Lerius, for 'vi BC' read 'Hellen.'; for 'F. Jacoby .. 58' read 'Jacoby ‡*FGrH* no. 475: iii (B) p. 434'
Pherecydes Syrius, after 'Diels' insert 'W. Kranz' and for 'ii p. 198' read '⁷ i p. 43'
Philemo comicus, add 'Kassel, Austin °*PCG* 7 p. 221 [Philem. .. K.–A.]'
***Philetas Samius** epigrammaticus [Philet.Sam.] v. *Anthologia Graeca*.
***Philicus** lyricus [Philic.] iii BC °*Suppl.Hell.* p. 321.
Philippides, add 'Kassel, Austin °*PCG* 7 p. 333 [Philippid. .. K.–A.]'
***Philippus Pergamenus** historicus ii AD Jacoby ‡*FGrH* no. 95: ii (A) p. 446. [Philipp.Perg. .. J.].
Philistus, add 'Jacoby ‡*FGrH* no. 556: iii (B) p. 551 [Philist. .. J.]'
Philo mechanicus, before '1919, No. 12' insert '1918, no. 16' and for 'Wescher' read 'Thévenot'
Philo Judaeus, line 1, add 'i BC/i AD '; line 2, after '1915' add 'cited by vol. and p. of Mangey, given in margin of C.-W. ([de Gruyter 1962]) (Ph. 2.264–9 M. = v. 67–8 + i. 209–16 C.-W.; Ph. 2.437–44 M. = v. 324–35 C.-W.; Ph. 2.492.10–497.8 M. = vi. 89.11–97.3 C.-W.). References containing higher numbers than 2.600 will be found in T. Mangey's ed., London 1742, not in C.-W.'; line 3, after '1886' add 'cited by p.'
Philodemus philosophus, for '*D.* 1, 3 .. 1916, 1917' read '*D.* 1, 3 ..1915 no. 7, 1916 no. 4 [1916, 1917, Leipzig 1970]'; *Rh.*, add '[Phld.*Rh.* .. S.] refers to Sudhaus's edition, [1964] (2 vols. + Suppl.)'; at end add 'sts. cited by p. of A. Vogliano, *Epicuri et Epicureorum scripta in Herculanensibus papyris servata*, Berlin 1928. [p. .. V.]'
Philolaus, after 'Diels' insert 'W. Kranz' and for 'i p. 301' read '⁷ i p. 398'
Philoponus, Joannes, after line 2, add '*Aet.M.* = *de Aeternitate Mundi*, H. Rabe, T(L) 1899 [1963]; *Dif.Ton.* = *Collectio vocum quae pro diversa significatione accentum diversum accipiunt* P. Egenolff, Breslau 1880 [p. 361 in *Lexica graeca minora*, K. Latte, H. Erbse, Hildesheim 1965]'
***Philoxenus Leucadius** lyricus [Philox.Leuc.] v/iv BC Page °*PMG* p. 433.
Phlegon Trallianus, *Mir.*, add '*De Mirabilibus*, A. Giannini, *Paradoxographorum Graecorum reliquiae* Milan 1965, p. 170'; *Fragmenta Historica*, add 'Jacoby ‡*FGrH* no. 257: ii (B) p. 1159 [Phleg.*fr.* .. J.]'
Phoenix, add '*Phoinix von Kolophon*, G. A. Gerhard, Leipzig/Berlin 1909 [Phoen. .. G.]'
***Phoronis** carmen epicum vii/vi BC Bernabé °*PEG* p. 118.
Photius, *Lexicon*, add 'repr. Naber [Amsterdam 1965]; Chr.Theodoridis, Berlin 1982; vol. 1 (*Α–Δ*) [Phot. .. Th.]'; *Bibl.*, add 'R. Henry, B 1959–77 (8 vols., vol. 9: indices, J. Schamp 1991) [Phot.*Bibl.* p. (= paragraph) .. H.]'
Phrynichus Atticista, line 2, add '*Die Ekloge des Phrynichos* E. Fischer, Berlin 1974 (*Sammlung griechischer und lateinischer Grammatiker* I) [Phryn. .. F.]'
Phrynichus tragicus, add 'Snell °*TrGF* 1 p. 69 [Phryn. Trag. .. S.]'
***Physiognomici scriptores** [*Physiogn.*] R. Förster, T(L) 1893 (2 vols.).
***Physiologus** Graecus [*Phys.* A et B] ii AD F. Sbordone, Rome 1936.
Pindarus, line 2, add 'B. Snell, H. Maehler, T(L) 1971⁵'; *fr.*, add 'B. Snell, H. Maehler, T(L) 1975⁴ [Pi.*fr.* .. S.–M.]'
Pisander epicus, add 'Heitsch °*GDRK* 2 p. 44'

Pisander Camiraius epicus vii/vi BC Bernabé °*PEG* p. 164.

Plato comicus, add 'Kassel, Austin °*PCG* 7 p. 431 [Pl.Com. .. K.-A.]'

Plato philosophus, line 2, for 'Oxford (OCT)' read 'OCT 1900–7 [1967]'; *Epigr.*, add '°*Anth.Lyr.* 1³ p. 102 [Pl.*Epigr.* .. D.³]'; Scholia, add 'W. C. Greene, Haverford 1938 [Sch.Pl. .. G.]'

Plotinus, add 'P. Henry, H. R. Schwyzer, OCT 1964–82 (3 vols.); see also ‡Porph.*Plot.*'

Plutarchus, *Moralia*, line 2, after '1888–96' insert 'W. R. Paton et al., T(L) 1925 [1974]–1967 (7 vols.)'; line 3, for 'Wyttenbach' read 'Stephanus' edition 1599'; (*Vitae parallelae*) for '1881–65' read '1865–81' and add 'K. Ziegler, T⁴ 1964–73 (3 vols.)'; *Fr.*, for 'pp. 150–2' read 'pp. 150–82'; add 'F. H. Sandbach, T(L) 1967 (= *Moralia* 7); *Galb.* = *Galba*; *in Arat.* = *quaestiones de Arati signis* (vol. vii, p. 102 B.)'; *in Hes.*, for 'p. 5' read 'p. 51'; *Nob.*, now regarded as humanist forgery; Ps.-Plu.*Vit.Hom.*, for '*Vita* .. p. 192' read '*de Vita et Poesi Homeri* (vol. vii, p. 329 B.)'; add 'B. refers to Bernardakis' edition'

Poeta de herbis, add 'Heitsch °*GDRK* 2 p. 23'

***Poliorcetici** scriptores [*Poliorc.*] *Poliorcétique des grecs*, C. Wescher, Paris 1867, cited by p. and line.

Pollux, add 'E. Bethe, T(L) 1900–31 [Stuttgart 1967] (2 vols.)'

Polyclitus, after 'Diels' insert 'W. Kranz' and for 'i p. 294' read '⁷ i p. 391'

Polystratus Epicureus, add 'sts. cited by p. of Vogliano [p. .. V.], cf. °Philodemus'

Porphyrius, after '*in Harm.* .. Ptolemaeum)' add '*Porphyrios' Kommentar zur Harmonielehre des Ptolemaios*, I. Düring, Göteborg 1932'; *Plot.*, add '*Plotini opera 1*, P. Henry, H. R. Schwyzer, OCT 1964 pp. 1–38'; add '*Ep.Aneb.* = *Epistula ad Anebonem*, A. R. Sodano, Naples 1958; *Hist.Phil.* = *Historiae Philosophicae Fragmenta* in *Porphyrii Opuscula*, A. Nauck, T(L) 1886² [1963]; *in Prm.* = *in Platonis Parmenidem commentaria*, p. 64 in *Porphyre et Victorinus*, vol. 2, P. Hadot, Paris 1968'

***Posidonius** Apamensis et Rhodius philosophus ii/i BC [Posidon.] vol. 1² L. Edelstein, I. G. Kidd, Cambridge 1989; vols. II(i), II(ii) I. G. Kidd, Cambridge 1988.

Posidonius historicus, add 'Jacoby ‡*FGrH* no. 169: ii (B) p. 222 [in Suppl., Posidon.Hist. .. J.]'

***Possis** historicus iii/ii BC Jacoby ‡*FGrH* no. 480: iii (B) p. 444.

Pratinas, line 2, add 'Page °*PMG* p. 367 [Pratin. .. P.]'; line 3, add 'Snell °*TrGF* 1 p. 79. [Pratin. .. S.]'

***Probus Grammaticus**, Keil *Gramm.Lat.* vol. iv p. 3.

Proclus, *H.*, add 'E. Vogt, Wiesbaden 1957'; *in Alc.*, add 'L. G. Westerink, Amsterdam 1954 [Procl.*in Alc.* .. W.]'; *in R.*, add '[Amsterdam 1965], cited by vol. and p. [Procl.*in R.* .. K.]'; *in Ti.*, add '[Amsterdam 1965], cited by vol. and p. [Procl.*in Ti.* .. D.]'; *Inst.*, add '*Proclus. The elements of theology*, E. R. Dodds, Oxford 1963² [1977] [Procl.*Inst.* .. D.]'; for '*Par.Ptol.* .. Basel 1554' read '*Par.Ptol.* .. Leiden 1635'; add '*Phil.Chald.* = *Eclogae de Philosophia Chaldaica*, in *Oracles chaldaïques*, É. des Places, B 1971 p. 206; *Sacr.* = περὶ τῆς καθ' Ἕλληνας ἱερατικῆς τέχνης, J. Bidez in *Catalogue des manuscrits alchimiques grecs* vi (1928) p. 148'; *Theol.Plat.*, add 'H. D. Saffrey, L. G. Westerink, B 1968–87 (5 vols.)'

***Proclus Constantinopolitanus** scriptor ecclesiasticus [Procl.CP] v AD *Hom.* 1–2 = *Homiliae*, Migne °*PG* 65.833, *Hom.* 26–34 in *L'Homilétique de Proclos de Constantinople*, F. J. Leroy, Vatican 1967.

Procopius Caesariensis, line 2, add 'G. Wirth (post J. Haury), T(L) 1962–4 (4 vols.)'

Prodicus, for 'v BC' read 'v/iv BC'; after 'Diels' insert 'W. Kranz' and for 'ii p. 267' read '⁷ ii p. 308'

***Propertius, Sextus** poeta Latinus [Prop.] i BC E. A. Barber, OCT 1960².

Protagoras, after 'Diels' insert 'W. Kranz' and for 'ii p. 219' read '⁷ ii p. 253'

***Protagoras Nicaensis** astrologus [Protag.Nicae.] iii BC apud Hephaestionem astrologum.

***Pseudo-Acro** scholiasta [Ps.-Acro] vi/vii AD *Scholia in Horatium vetustiora*, O. Keller, T(S) 1902–4 [1967] (2 vols.) *C.* = *Carmina*.

***Pseudo-Archytas** philosophus [Ps.-Archyt.] cited fr. H. Thesleff, *The Pythagorean Texts of the Hellenistic Period*, Åbo 1965.

***Pseudo-Asconius** scholiasta [Ps.-Ascon.] i AD *in Verr.* = *Ciceronis Orationum Scholiastae*, T. Stangl, vol. ii (*commentarios continens*), Vienna/Leipzig 1912 [Hildesheim 1964].

Pseudo-Callisthenes, add 'W. Kroll, *Historia Alexandri Magni*, vol. i, Berlin 1926 [Weidmann 1959], cited by p. and line no. [Ps.-Callisth. .. K.]; *Recensio* λ, H. van Thiel, Bonn 1959 [.. λ.]'

***Pseudo-Chrysostomus** scriptor ecclesiasticus [Ps.-Chrys.] v AD *HP* = *Homélies Pascales*, M. Aubineau, Paris 1972.

Pseudo-Democritus, add 'Diels, Kranz °*Vorsokr.* ⁷ ii p. 207 frr. B 298 ff. (sts. attributed to °Bolus). *Symp.Ant.* = περὶ συμπαθειῶν καὶ ἀντιπαθειῶν W. Gemoll, Striegau 1884 (sts. attributed to Nephalius, or °Bolus)'

***Pseudo-Dioscorides** medicus [Ps.-Dsc.], Wellmann op. cit. [Weidmann 1958] (3 vols.), under text of Dsc.

Pseudo-Phocylidea, add 'D. Young, *Theognis* T(L) 1971² p. 95'

Ptolemaeus mathematicus, *Alm.*, cited by vol. and p. of Heiberg's ed. [Ptol.*Alm.* .. H.]; *Geog.*, for 'i–iii' read 'i–v'; add 'repr. Nobbe [Hildesheim 1966]'; *Harm.*, add 'I. Düring, Gothenburg 1930'; *Tetr.*, add '(= Ἀποτελεσματικά) F. Boll, E. Boer, T(L) 1940 [1957]'

Ptolemaeus Chennos, add 'sts. cited by p. of A. Westermann, Μυθογράφοι, Brunswick 1843'

Pythagoras, after 'Diels' insert 'W. Kranz' and for 'i p. 27, cf. p. 344' read '⁷ i p. 96, cf. p. 446'; add '*Pyth.Sim.* = *Pythagoreorum Similitudines*, Mullach *FPG* i p. 488'

Quintilianus, *Inst.*, add 'M. Winterbottom, OCT 1970 (2 vols.)'

Quintus Smyrnaeus, line 1, for 'iv AD?' read 'iii AD '; add 'F. Vian, B 1963–9 (3 vols.)'

***Res Gestae divi Saporis** [*Res Gestae Saporis*] iii AD *Recherches sur les Res Gestae divi Saporis*, E. Honigmann, A. Maricq, Brussels 1953, p. 11.

***Romanus Sophista** [Roman.] vi AD περὶ ἀνειμένου, W. Camphausen, T(L) 1922.

***Rufinianus, Julius** rhetor Latinus [Rufin.] *Fig.* = *de Figuris*, in *Rhetores Latini Minores*, C. Halm, Leipzig 1863, p. 38 (cited by p.).

Rufus Medicus, before '*Onom.*' insert '*fr.* = *Fragmenta*; *Interrog.* = *Interrogationes*, H. Gärtner, T(L) 1970' and before '*Ren.Ves.*' insert '*Oss.* = *de ossibus*; *Podagr.* = *de podagra*'

Sallustius, add 'G. Rochefort, B 1960'

Sappho, line 2, add '*Poetarum Lesbiorum Fragmenta*, E.

AUTHORS AND WORKS

Lobel, D. L. Page, Oxford 1955 [1968], p. 2 [Sapph. .. L.-P.]'

***Satyrius** epigrammaticus i BC *AP* 6.11 (author now dist. fr. Satyrus).

Satyrus historicus, *Vit.Eur.*, add 'G. Arrighetti, Pisa 1964'

*****Scaevola, Q. Cervidius** iurisconsultus Latinus [Scaev.] ii AD *Dig.* = *Digesta* (q.v.).

Scolia, add 'Page °*PMG* p. 472 [*Scol.* .. P.]'

*****Scribonius Largus** medicus, [Scrib.Larg.] i AD G. Helmreich, T(L) 1887.

Semonides iambographus, add 'West °*IEG* (2) p. 97 [Semon. .. W.]'

Semus, line 1, add '?iii/ii BC'; add 'Jacoby ‡*FGrH* no. 396: iii (B) p. 285 [Semus .. J.]'

Seneca, L. Annaeus, add '*Ep.* = *Epistulae*, L. D. Reynolds, OCT 1965 (2 vols.)'

*****Seneca, L. Annaeus** rhetor Latinus i BC/i AD *Contr.* = *Controversiae*, H. Bornecque, Paris 1932².

Servius, add '*A.* = *Aeneid*, Keil *Gramm.Lat.* vol. iv p. 403 (cited by vol., p. and line) [Serv. in *Gramm.Lat.* .. K.]'

Sextus Empiricus, for 'H. Mutschmann .. 1912' read 'H. Mutschmann, J. Mau, (indices by K. Janáček), T(L) 1912-62 (4 vols.)'

*****Simeon Metaphrastes** scriptor ecclesiasticus x AD *Vit. Luc.* = *Vita Luciani*, Migne °*PG* 114.397.

Simmias, add '*Securis*, A. S. F. Gow *Bucolici Graeci*, OCT 1952 p. 174; see also Anthologia Graeca'

Simon Atheniensis, before '*Eq.*' insert '[Simon.Ath.] J. Soukup, Innsbruck 1911 (*Commentationes Aenipontanae*, E. Kalinka, no. VI)'; *Eq.*, add 'K. Widdra, T 1964'

Simonides, add 'West °*IEG* (2) p. 113 [Simon. .. W.]; Page °*PMG* p. 238 [Simon. .. P.] see also Anthologia Graeca'

*****Socrates et Dionysius** de lapidibus scriptores *Lith.* = Περὶ Λίθων, *Les lapidaires grecs*, R. Halleux, J. Schamp, B 1985 p. 166 [Socr.Dion. .. H.-S.]; G. Giannakis, Yannina 1987 [Socr.Dion. .. G.].

*****Solinus, C. Julius** geographus Latinus [Solin.] ?iii AD T. Mommsen, Berlin 1895² [1958].

Solon, add 'West °*IEG* (2) p. 120 [Sol. .. W.] *Lg.* = *Leges*, E. Ruschenbusch, *Historia* Einzelschr. 9, Wiesbaden 1966'

Sopater, add '*Sopatros the Rhetor*, D. Innes, M. Winterbottom, *BICS* 48, London 1988 [Sopat.Rh. .. I.-W.]'

Sophocles, for 'A. C. Pearson ..' read 'H. Lloyd-Jones, N. G. Wilson, OCT 1990 [L.-J.-W.]'; add '*fr.* = *Fragmenta*, Radt °*TrGF* 4 [S.*fr.* .. R.]; *Ichn.* = Ἰχνευταί (= *fr.* 314 in °*TrGF* 4); *Pae.* = Scholia, W. Dindorf, Oxford 1852; Scholia Vetera, P. N. Papageorgius, T(L) 1888 [Sch. .. P.]; Sch.*Aj.*, G. A. Christodoulou, Athens 1977'

Sophron, add 'A. Olivieri, *Frammenti della commedia greca e del mimo nella Sicilia e nella Magna Graecia*, Naples 1930, p. 169 [Sophron .. O.]'

Soranus, after '1882' insert '*Gynaeciorum libri iv, de signis fracturarum, de fasciis, vita Hippocratis secundum Soranum*, J. Ilberg, *CMG* 4, Leipzig/Berlin 1927 [*Gynaecia* cited by Rose's numeration]'

Sosibius, line 1, for 'iii BC' read 'iii/ii BC'; add 'Jacoby ‡*FGrH* no. 595: iii (B) p. 713'

Sosiphanes, add 'Snell *TrGF* 1 p. 261 [Sosiph. .. S.]'

Sosylus, add 'Jacoby ‡*FGrH* no. 176: ii (B) p. 903 [Sosyl. .. J.]'

Statyllius Flaccus, for 'Idem .. q.v.' read 'v. Anthologia Graeca'

*****Pseudo-Stephanus** Alexandrinus [Ps.-Steph.] vi/vii AD °*Physici* 2 p. 199.

Stesichorus, before 'vii/vi BC' insert '(?)'; after 'iii p. 205' add 'Page °*PMG* p. 97 [Stesich. .. P.]; Page °*SLG* p. 5 [Stesich.*S.* 123 etc.]; Davies °*PMGF* p. 134 [Stesich. .. p. .. D.]'

Strabo, line 2, add 'G. Aujac et al., B 1966-89 (9 vols.), except for Bk. ix'

Strato Comicus, delete '(?)'; add 'Kassel, Austin °*PCG* 7 p. 618 [Strato Com. .. K.-A.]'

Suetonius, add '*Gramm.* = *de Grammaticis et Rhetoribus deperditorum librorum reliquiae*, I. Della Corte, Turin 1968; περὶ βλασφημιῶν (καὶ πόθεν ἑκάστη) and περὶ τῶν (παρ' Ἕλλησι) παιδιῶν, J. Taillardat, Paris 1967 [Suet.*Blasph.* .. T.; Suet.*Lud.* .. T.]'

Suidas, add 'A. Adler, T(L) 1928-38 [1967-71] (5 vols.) [properly *Suda*]'

*****Synesius Cyrenensis** scriptor ecclesiasticus [Synes.] iv/v AD Migne °*PG* 66. *Calv.* = *Calvitii Encomium* °*PG* 66.1167; *Hymn.* = *Hymni*, Chr. Lacombrade, B 1978; *Insomn.* = *De Insomniis* °*PG* 66.1281; *Ep.* = *Epistolae*, A. Garzya, Rome 1979.

Telestes, for 'iv BC' read 'v/iv BC'; add 'Page °*PMG* p. 419 [Telest. .. P.]'

Teucer Babylonius, before 'i AD' insert '[Teuc.Bab.]' and delete '(?)'; for '6' read '16'; add '[Hildesheim 1967]'

Thales, after 'Diels' insert 'W. Kranz' and for 'i p. 1' read '⁷ i p. 67'

Themistius, *Or.*, add 'H. Schenkel, G. Downey, A. F. Norman, T(L) 1965-74 (3 vols., cited by no. of speech and p. of Hardouin, exc. no. 34, cited by p. of Dindorf)'

Theocritus, line 3, add 'A. S. F. Gow, Cambridge 1952² [1965] (2 vols.)'; before '*Beren.*' insert '*Adon.* = εἰς νεκρὸν Ἄδωνιν A. S. F. Gow *Bucolici Graeci*, OCT 1952 [1969] p. 166'; for '*Coma Berenices*' read '*Berenice*'; Sch., add '[1967]'

Theodectes Tragicus, add 'Snell °*TrGF* 1 p. 227'

Theognis, add 'West °*IEG* (1) p. 172'

Theognostus, add '1-84 K. Alpers, Hamburg 1964 [Theognost. .. A.]'

*****Theon Alexandrinus** Mathematicus [Theon Al.] iv AD *in Ptol.* = *in Ptolemaeum*, in *Commentaires de Pappus et de Théon d'Alexandrie sur l'Almageste*. II. *Théon d'Alexandrie* A. Rome, Vatican 1936 (= Studi e Testi 72, 106); *Œuvres de Ptolémée*, M. Halma, vol. v, Paris 1821.

Theon Smyrnaeus, delete '*in Ptol.* .. Paris 1821.' N.B. in references, for 'Theo Sm. *in Ptol.*' read 'Theon Al. *in Ptol.*'

*****Theophilus Antecessor** iurisconsultus [Theophil.Antec.] vi AD *Institutionum graeca paraphrasis Theophilo antecessori vulgo tributa*, E. C. Ferrini, 1884-97 [Aalen 1967] (2 vols.).

*****Theophilus Antiochenus** scriptor ecclesiasticus [Thphl.Ant.] ii AD *Autol.* = *ad Autolycum*, R. M. Grant, Oxford 1970.

Theophrastus, *Char.*, add 'P. Steinmetz, Munich 1960-2 (2 vols.)'; *fr.*, add 'F. Wimmer, Paris 1855 [1931] (vol. 3)'; *Metaph.*, add 'W. D. Ross, F. H. Fobes, Oxford 1929 [Hildesheim 1967] [*Metaph.* p. .. R.-F.]'

Theopompus comicus, for 'v BC' read 'v/iv BC'; add 'Kassel, Austin °*PCG* 7 p. 708 [Theopomp.Com. .. K.-A.]'

Theopompus historicus, add 'Jacoby ‡*FGrH* no. 115: ii (B) p. 526, iii (B) p. 742 [Theopomp.Hist. .. J.]'

Thespis, add 'Snell °*TrGF* 1 p. 61 [Thespis .. S.]'

Thucydides, line 2, read 'H. S. Jones, J. E. Powell, OCT 1942¹ (rev.) [1974, 1976] (2 vols.)'; add 'Scholia, C. Hude, T(L) 1927'

Timaeus Locrus, add 'W. Marg, Leiden 1972 [Ti.Locr. .. M.]'

Timotheus Lyricus, line 2, add 'Page °*PMG* p. 399 [Tim. .. P.]'

Tragica adespota, add 'Kannicht, Snell °*TrGF* 2 [*Trag. adesp.* .. K.-S.]'

Tryphiodorus, line 1, at end add '[properly Triphiodorus; Triph. in Suppl.]'; add '*Oppian, Colluthus, Tryphiodorus*, A. W. Mair, Loeb 1928 [1963] p. 580'

Tyrtaeus, add 'West °*IEG* (2) p. 150 [Tyrt. .. W.]'

Tzetzes, Joannes, before '*Diff.Poet.*' insert '*Alleg.Il.* = *Allegoriae Iliadis* (*Prol.* = *Prolegomena*), J. F. Boissonade, Paris 1851 [Hildesheim 1967]'; *H.*, add 'A. M. Leone, Naples 1968 [Tz.*H.* .. L.]'

***Ulpianus, Domitius** iurisconsultus [Ulp.] ii/iii AD *Dig.* = Digesta q.v.

Varro, *LL*, add '[Amsterdam 1964]'; add '*fr.* = *Saturarum Menippearum fragmenta* R. Astbury, T(L) 1985 [Varro *fr.* .. A.]'

***Vegetius** veterinarius [Veg.] *Mul.* = *Mulomedicina*, E. Lommatzsch, T(L) 1903.

Vettius Valens, add 'D. Pingree, T(L) 1986'

Vetus Testamentum Graece redditum A, line 2, add 'A. Rahlfs, Stuttgart 1935 [1971] (2 vols.), L. Lütkemann, A. Rahlfs, *Hexaplarische Randnoten zu Isaias 1-16, aus einer Sinai-Handschrift* (*NGG* 1915 Beiheft); [.. (L.-R.)]'; after 'Heb. = Ἑβραῖος' insert 'Lcn. = Lucianus'; after 'Sm. = Symmachus (q.v.)' insert 'Syr. = ὁ Σύρος' **B**, after '*Ps.*' insert '*Ps.Sal.* = *Psalmi Salomonis*'

***Vita Aesopi** [*Vit.Aesop.*] *Aesopica*, B. E. Perry, Univ. Illinois 1952; (G) = *Vita Aesopi ex codice G Novaeboracensi Bibliothecae Pierponti Morgan* (Perry p. 35); (W) = *Vita Aesopi Westermanniana* (Perry p. 81).

Xenocrates, add '*Lap.* = Λιθογνώμων, M. Wellmann in *Quellen und Studien zur Geschichte der Naturwissenschaften und der Medizin*, P. Diepgen, J. Ruska, Band 4, Heft 4, p. 86 [426], Berlin 1935'

Xenophanes, for 'vi BC' read 'vi/v BC'; add 'West °*IEG* (2) p. 164; Diels, Kranz °*Vorsokr.*^{7} i p. 113'

Xenophon, line 2, for 'Oxford (OCT)' read 'OCT 1900–20 [1968–71]'

Xenophon Ephesius, for 'ii AD?' read 'ii/iii AD '; add 'A. D. Papanikolaou, T(L) 1973'

Zeno Eleaticus, after 'Diels' insert 'W. Kranz' and for 'i p. 165' read '^{7} i p. 247'

***Zenodotus Stoicus** epigrammaticus [Zenodotus] v. Anthologia Graeca.

Zonaras, line 1, add 'xii AD '; line 2, add '[Hildesheim 1967]'

II. EPIGRAPHICAL PUBLICATIONS

ABV = *Attic Black-Figure Vase Painters*, J. D. Beazley, Oxford 1956.

Amyzon = *Fouilles d'Amyzon en Carie.* I: *Exploration, histoire, monnaies et inscriptions*, J. and L. Robert, Paris 1983.

Antioch-on-the-Orontes = *Antioch-on-the-Orontes* I-III, G. W. Elderkin et al., Princeton 1934–41.

Ath.Agora = *The Athenian Agora* Princeton 1953–. vol. 10: *Weights, Measures and Tokens*, M. Lang, M. Crosby, 1964; vol. 15: *The Athenian Councillors*, B. D. Meritt, J. Traill, 1974; vol. 17: *The Funerary Monuments*, D. W. Bradeen, 1974; vol. 19: *Inscriptions: Horoi, Poletai Records, Leases of Public Lands*, G. Lalande, M. Langdon, M. Walbank, 1991; vol. 21: v. °Lang *Ath.Agora*.

Ath.Asklepieion = *The Athenian Asklepieion: The People, their Dedications and their Inventories*, S. Aleshire, Amsterdam 1989.

Baillet Inscr. des tombeaux des rois = J. Baillet, *Inscriptions grecques et latines des tombeaux des rois ou syringes*, 3 vols., Cairo 1920–6.

BCH suppl. 8, 9, v. index °IV.

Bernand Akôris = É. Bernand, *Inscriptions grecques et latines d'Akôris*, Cairo 1988.

Bernand IMEG = É. Bernand, *Inscriptions métriques de l'Égypte gréco-romaine: recherches sur la poésie épigrammatique des grecs en Égypte*, Paris 1969.

Bernand Inscr.Colosse Memnon = A. and É. Bernand, *Les Inscriptions grecques et latines du Colosse de Memnon*, Paris 1960.

Bernand Les portes = A. Bernand, *Les Portes du désert: Recueil des inscriptions grecques d'Antinooupolis, Tentyris, Koptos, Apollonopolis Parva et Apollonopolis Magna*, Paris 1984.

Bernand Pan = A. Bernand, *Pan du désert*, Leiden 1977.

Bile Dial.crét. = M. Bile, *Le Dialecte crétois ancien*, Paris 1988.

Bithynische St. = *Bithynische Studien. Bithynia İncelemeleri*, S. Şahin, Bonn 1978 (°*IGSK* 7).

BMC = *Catalogue of the Greek Coins in the British Museum*, F. G. Hill, London 1873–1927: the individual volumes are referred to as *BMC Ionia, BMC Caria*, etc.

Bonner Magical Amulets = Campbell Bonner, *Studies in Magical Amulets* (Univ. Michigan Studies, Humanistic Series, 49), Ann Arbor 1950.

Brixhe, Hodot AMNS = C. Brixhe, R. Hodot, *L'Asie Mineure du nord au sud. Inscriptions inédites*, Nancy 1988.

Buck Gr.Dial. = C. D. Buck, *The Greek Dialects: Grammar, Selected Inscriptions, Glossary*, Chicago 1955.

CCCA = *Corpus cultus Cybelae Attidisque*, M. J. Vermaseren, 7 vols., Leiden 1977–89.

CEG = 1: *Carmina epigraphica graeca saeculorum viii–v a. Chr. n.*, P. A. Hansen, Berlin 1983; 2: *Carmina epigraphica graeca saeculi iv a. Chr. n. Accedunt addenda et corrigenda ad CEG 1*, Berlin 1989.

CGIH = *Corpus der griechisch-christlichen Inschriften von Hellas:* I. *Die griechisch-christlichen Inschriften des Peloponnes*, N. A. Bees, vol. i, Athens 1941.

CID = *Corpus des inscriptions de Delphes.* I: *Lois sacrées et règlements religieux*, G. Rougemont, Paris 1977; II: *Les comptes du quatrième et du troisième siècle*, J. Bousquet, D. Mulliez, Paris 1989.

CIJud., add 'repr. [New York 1975]; vol. ii: *Asie–Afrique*, Rome 1952'

CIRB = *Corpus inscriptionum Regni Bosporani*, V. V. Struve, Leningrad 1965.

CISem. = *Corpus Inscriptionum Semiticarum*, E. Renan et al., Paris 1881–1951.

Comptes et inventaires = *Comptes et inventaires dans la cité grecque: Actes du colloque international d'épigraphie tenu à Neuchâtel du 23 au 26 septembre 1986 en l'honneur de Jacques Tréheux*, D. Knoepfler, N. Quellet, Neuchâtel/Geneva 1988.

Corinth, add 'vol. viii (3) *The Inscriptions 1926–1950*, J. H. Kent, Princeton, NJ 1966; vol. xv (3): Appendix I: *Inscriptions*, A. L. Boegehold, pp. 358 ff., Princeton, NJ 1984.'

de Franciscis Locr.Epiz. = A. de Franciscis, *Stato e società in Locri Epizefiri (L'archivio dell'Olympieion locrese)*, Naples 1972.

Delph., for '3(1), (2)' read '3(1)–(6)'; for '1909' read '1909–76'; add '(also known as *FD*)'

Demitsas Maced. = M. Demitsas, *Sylloge Inscriptionum Graecarum et Latinarum Macedoniae*, 2 vols., [Athens 1896] (Demitsas Μακεδ. in *LSJ*).

di Cesnola Salaminia = Alessandro Palma di Cesnola, *Salaminia, Cyprus: The History, Treasures and Antiquities of Salamis*, London 1882.

Didyma = *Didyma*, Th. Wiegand. Bd. II: *Die Inschriften*, A. Rehm, Berlin 1958.

Dubois Dial.arcad. = L. Dubois, *Recherches sur le dialecte arcadien*, 3 vols., Louvain-la-Neuve 1986.

Dubois IGDS = L. Dubois, *Inscriptions grecques dialectales de Sicile*, Paris 1989.

Dumont–Homolle Mél.Arch. = A. Dumont, Th. Homolle, *Mélanges d'archéologie et d'épigraphie*, Paris 1892.

*Dura*¹, *Dura*², add '*Dura*³, Third season, H. T. Rowell, A. R. Bellinger, 1932; *Dura*⁴, Fourth season, S. Gould et al., 1933; *Dura*⁹, Ninth season, I: *The Agora and Bazaar*, F. E. Brown, C. B. Welles, 1944; III: *The Palace of the Dux Ripae and the Dolichenum*, M. Rostovtzeff et al., 1952.'

EAM = Ἐπιγραφές ἄνω Μακεδονίας. τόμος Α': Κατάλογος Ἐπιγραφῶν, Th. Rizakis, G. Touratsoglou, Athens 1985.

Edict.Diocl., add '*Diokletians Preisedikt. Texte und Kommentare*, S. Lauffer, Berlin 1971'

Ephes., add '4 (3), 5 (1), Vienna 1951, 1944'

Expl.arch. de Délos = *Exploration archéologique de Délos*. XI: *Les sanctuaires et les cultes du Mont Cynthe*, A. Plassart, Paris 1928; XXX: *Les Monuments funéraires de Rhénée*, M.-Th. Couilloud, Paris 1974.

Fouilles de Byblos = *Fouilles de Byblos 1926–38*, M. Dunand, (Atlas 1, 2; Texte I-V), Paris 1937–73.

Friedländer Epigrammata = P. Friedländer, H. B. Hoffleit, *Epigrammata: Greek Inscriptions in Verse, from the Beginnings to the Persian Wars*, Berkeley 1948.

GDI, add '[Hildesheim 1971]'
Gonnoi = Gonnoi II: Les Inscriptions, B. Helly, Amsterdam 1973.
Guarducci EG = M. Guarducci, *L'epigrafia greca dalle origine al tardo impero*, Rome 1987.
GVAK = Griechische Versinschriften aus Kleinasien, W. Peek, Vienna 1980.
GVAThess. = Griechische Versinschriften aus Thessalien, W. Peek, Heidelberg 1974.
GVI = Griechische Versinschriften, W. Peek, Bd. I., Berlin 1955 [Chicago 1988].
Hallock Persepolis = R. T. Hallock, *Persepolis Fortification Tablets*, Chicago 1969.
Hatzopoulos Actes de vente = M. B. Hatzopoulos, *Actes de vente de la Chalcidique centrale*, Athens 1988.
Hierapolis = Inschriften in *Altertümer von Hierapolis*, W. Judeich, Berlin 1898.
IApam. = Die Inschriften von Apameia (Bithynien) und Pylai, T. Corsten, Bonn 1987 (°*IGSK* 32).
IAskl.Epid. = Inschriften aus dem Asklepieion von Epidauros, W. Peek, Abhandl. der Sächsischen Akad. der Wiss. zu Leipzig, 60(2) (1969).
IAssos = Die Inschriften von Assos, R. Merkelbach, Bonn 1976 (°*IGSK* 4).
IBith. = Inschriften und Denkmäler aus Bithynien, F. K. Dörner, Berlin 1941.
ICilicie = Inscriptions de Cilicie, G. Dagron, D. Feissel, Paris 1987.
IClaudiopolis = Die Inschriften von Klaudiu Polis, F. Becker-Bertau, Bonn 1986 (°*IGSK* 31).
ICS = Les Inscriptions chypriotes syllabiques, O. Masson, Paris 1964 [1985 with Addenda nova].
IEphes. = Die Inschriften von Ephesos, H. Wankel, R. Merkelbach et al., Bonn 1979–81 (°*IGSK* 11–17).
IEryth. = Die Inschriften von Erythrai und Klazomenai, H. Engelmann, R. Merkelbach, Bonn 1972–3 (°*IGSK* 1–2).
IG, *IG* 1³ = *Inscriptiones Atticae Euclidis anno anteriores* (1) *Decreta et tabulae magistratuum*, D. M. Lewis, 1981–94. *IG* 2², add 'fasc. ii (5220–13247) 1940'; *IG* 9², add 'fasc. ii = *Inscriptiones Acarnaniae*, fasc. iii = *Inscriptiones Locridis Occidentalis*, G. Klaffenbach, 1957, 1968'; *IG* 10.2(1) = *Inscriptiones graecae Epiri, Macedoniae, Thraciae, Scythiae*, pars 2: *Inscriptiones Macedoniae*, fasc. 1: *Inscriptiones Thessalonicae et viciniae*, C. Edson, 1972; *IG* 12 suppl. = *Supplementum*, F. Hiller von Gaertringen 1939.
IGBulg. = Inscriptiones Graecae in Bulgaria repertae, G. Mihailov, vols. 1–4, Bucharest 1956–66; vol. 1², Sofia 1970.
IGC = Inscriptions de Grèce centrale, F. Salviat, C. Vatin, Paris 1971.
IGChr. = Recueil des inscriptions grecques-chrétiennes de l'Asie Mineure, vol. I, H. Grégoire, Paris 1922 [Amsterdam 1968].
IGLBulg. = Spätgriechische und spätlateinische Inschriften aus Bulgarien, V. Beševliev, Berlin 1964.
IGLS = Inscriptions grecques et latines de la Syrie. Tomes 1–8(3), 13(1), 21(2), L. Jalabert, R. Mouterde, Cl. Mondésert, J.–P. Rey-Coquais, Paris 1929–86. (*Inscr.gr. et lat. de la Syrie* in *LSJ*).
IGSK = Inschriften griechischer Städte aus Kleinasien, Bonn (vols. listed separately acc. to provenance).
IHadr. = Die Inschriften von Hadrianoi und Hadrianeia, E. Schwertheim, Bonn 1987 (°*IGSK* 33).
IHistriae = Inscriptiones Daciae et Scythiae minoris antiquae. Series altera: Inscriptiones Scythiae minoris Graecae et Latinae. vol. I: *Inscriptiones Histriae et viciniae*, D. M. Pippidi, Bucharest 1983.
IIasos = Die Inschriften von Iasos I, II, W. Blümel, Bonn 1985 (°*IGSK* 28.1, 2).
IIlion = Die Inschriften von Ilion, P. Frisch, Bonn 1975 (°*IGSK* 3).
IKeramos = Die Inschriften von Keramos, E. Varinlioğlu, Bonn 1986 (°*IGSK* 30).
IKios = Die Inschriften von Kios, T. Corsten, Bonn 1985 (°*IGSK* 29).
IKnidos = Die Inschriften von Knidos I, W. Blümel, Bonn 1992 (°*IGSK* 41).
IKyme = Die Inschriften von Kyme, H. Engelmann, Bonn 1976 (°*IGSK* 5).
IKyzikos = Die Inschriften von Kyzikos und Umgebung. Teil I: *Grabtexte*, E. Schwertheim, Bonn 1980 (°*IGSK* 18).
ILabraunda = Labraunda. Swedish Excavations and Researches. vol. III(1) and (2): *The Greek Inscriptions*, J. Crampa, Lund 1969, Stockholm 1972.
ILampsakos = Die Inschriften von Lampsakos, P. Frisch, Bonn 1978 (°*IGSK* 6).
IMagnes.Sipylos = Die Inschriften von Magnesia am Sipylos, T. Ihnken, Bonn 1978 (°*IGSK* 8).
IMiletupolis = Die Inschriften von Kyzikos und Umgebung. Teil II: *Miletupolis. Inschriften und Denkmäler*, E. Schwertheim, Bonn 1983 (°*IGSK* 26).
IMylasa = Die Inschriften von Mylasa, W. Blümel, Bonn 1987–8 (°*IGSK* 34–5).
INikaia = Katalog der antiken Inschriften des Museums von Iznik (Nikaia). Teil I, II 1–2, S. Sahin, Bonn 1979–82 (°*IGSK* 9, 10 (1), (2)).
Inschr.Hierap. = Inschriften aus Hierapolis-Kastabala, M. Sayar, P. Siewert, H. Täuber, Vienna 1989.
Inscr.Cret., add 'II *Tituli Cretae occidentalis;* III *Tituli Cretae orientalis;* IV *Tituli Gortynii*, M. Guarducci, Rome 1939–50.'
Inscr.Délos, at beginning insert 'nos. 1–88, ed. A. Plassart, Paris 1950; nos. 89–104.33, J. Coupry, Paris 1972'; add '(6 vols.)'
Inscr.Dura = Inscriptions from Dura-Europos, R. N. Frye et al. in *YClS* 14.127–213.
Inscr.Gr. et Lat. de la Syrie, v. °*IGLS*.
Inscr.Perg. add '8(3) = *Altertümer von Pergamon* viii (3). *Die Inschriften des Asklepieions*, Ch. Habicht, Berlin 1969'
Inv.Inscr.Palm. = Inventaire des Inscriptions de Palmyre, J. Cantineau, Beirut 1930.
IPamph. = Le Dialecte grec de Pamphylie, C. Brixhe, Paris 1976.
IParion = Die Inschriften von Parion, P. Frisch, Bonn 1983 (°*IGSK* 25).
IPhilae = Les Inscriptions grecques de Philae, A. and É. Bernand. I: *Époque ptolémaïque*, II: *Haut et Bas Empire*, Paris 1969.
IPhrygie = Nouvelles Inscriptions de Phrygie, T. Drew-Bear, Zutphen 1978.
IPrusa ad Olympum = Die Inschriften von Prusa ad Olympum, T. Carsten, Bonn 1991 (°*IGSK* 39).
IPrusias = Die Inschriften von Prusias ad Hypium, W. Ameling, Bonn 1985 (°*IGSK* 27).
IRhod.Peraea = The Rhodian Peraea and Islands, P. M. Fraser and G. E. Bean, Oxford 1954.
ISalamis = The Greek and Latin Inscriptions from Salamis, T. B. Mitford, I. K. Nicolaou, Nicosia 1974.
ISelge = Die Inschriften von Selge, J. Nollé, F. Schindler, Bonn 1991 (°*IGSK* 37).
ISestos = Die Inschriften von Sestos und der thrakischen Chersones, J. Krauss, Bonn 1980 (°*IGSK* 19).

*ISmyrna = Die Inschriften von Smyrna, I, II (1), G. Petzl, Bonn 1982–90 (°IGSK 23, 24 (1)).

*IStraton. = Die Inschriften von Stratonikeia, M. Sahin, Bonn 1981–90 (°IGSK 21, 22).

*ITomis = Inscriptiones Daciae et Scythiae minoris antiquae. Series altera: Inscriptiones Scythiae minoris Graecae et Latinae. Vol. II: Tomis et territorium, I. Stoian, A. Suceveanu, Bucharest 1987.

*ITrall. = Die Inschriften von Tralleis und Nysa. I: Die Inschriften von Tralleis, F. B. Poljakov, Bonn 1989 (°IGSK 36(1)).

*ITyr = Inscriptions grecques et latines découvertes dans les fouilles de Tyr (1963–74), I: Inscriptions de la nécropole, J.-P. Rey-Coquais, Paris 1977.

*IUrb.Rom. = Inscriptiones Graecae urbis Romanae, L. Moretti, Rome 1968–79 (3 vols., the second in two parts).

*Jeffery LSAG = L. H. Jeffery, The Local Scripts of Archaic Greece, Oxford 1961. Revised edition with a supplement, A. W. Johnston 1990.

*JRCil. = Journeys in Rough Cilicia, G. E. Bean, T. B. Mitford, 1: 1962–1963, Vienna 1965; 2: 1964–1968, Vienna 1970.

*Kafizin = The Nymphaeum of Kafizin: The Inscribed Pottery, T. B. Mitford, New York 1980.

*Kalinka ADB = E. Kalinka, Antike Denkmäler in Bulgarien, Vienna 1906.

*Kerameikos = Kerameikos: Ergebnisse der Ausgrabungen Bd. 1–13 (Archäologisches Institut des Deutschen Reiches/Deutsches Archäologisches Institut), W. Kraiker, K. Kübler et al., Berlin 1939–88.

*Kontorini AER 1, 2 = V. Kontorini, 1: Inscriptions inédites relatives à l'histoire et aux cultes de Rhodes au 2ᵉ et au 1ᵉʳ s. av. J.-C., Louvain-la-Neuve/Providence 1983. 2: Ἀνέκδοτες Ἐπιγραφές Ῥόδου, Athens 1989.

*Kouklia-Paphos = Les Inscriptions syllabiques de Kouklia-Paphos, O. Masson, T. B. Mitford, Constance 1986.

*Kourion = The Inscriptions of Kourion, T. B. Mitford, Philadelphia 1971.

*Kretschmer GV = P. Kretschmer, Griechische Vaseninschriften, Gütersloh 1894 [Hildesheim 1969, Chicago 1980].

*La Carie = La Carie: Histoire et géographie historique avec le recueil des inscriptions antiques, L. and J. Robert, II, Paris 1954.

*Lane CMRDM = E. R. Lane, Corpus Monumentorum Religionis Dei Menis, Leiden 1971–8 (4 vols.).

*Lang Ath.Agora XXI = The Athenian Agora. vol. XXI: Graffiti and dipinti, M. L. Lang, Princeton 1976.

*Laodicée = Laodicée de Lycos; Le nymphée: Campagnes 1961–1963, J. Des Gagniers, etc., Paris 1969.

*Lefebvre RIGC = G. Lefebvre, Recueil des inscriptions grecques chrétiennes de l'Égypte, Cairo 1967.

*Leg.Sacr. = Leges Graecorum Sacrae I–II, J. Prott, L. Ziehen, Leipzig 1896–1906.

*Lindos = Lindos: Fouilles et recherches 1902–1914, Chr. Blinkenberg, K. F. Kinch, II: Inscriptions, Chr. Blinkenberg, Copenhagen 1941 (2 vols.).

MAMA, add 'vol. VII, W. M. Calder, Manchester 1956; vol. VIII, W. M. Calder, J. M. R. Cormack, Manchester 1962; vol. IX, B. Levick et al., London 1988'

*Meiggs–Lewis = R. Meiggs, D. M. Lewis, A Selection of Greek Historical Inscriptions to the End of the 5th Century B.C., Oxford 1969 [1988].

*Mitchell N. Galatia = S. Mitchell et al., Regional Epigraphic Catalogues of Asia Minor. II: The Ankara District: The Inscriptions of North Galatia, Oxford 1982.

Mon.Anc.Gr., add 'Res gestae divi Augusti: Das Monumentum Ancyranum, H. Volkmann³, Berlin 1969'

*Moretti IAG = L. Moretti, Iscrizioni agonistiche greche, Rome 1953.

*Moretti ISE = L. Moretti, Iscrizioni storiche ellenistiche. I: Attica, Peloponneso, Beozia (nos. 1–70). II: Grecia centrale et settentrionale (nos. 71–132), Florence 1967 and 1975.

OGI, add '[Hildesheim 1970]'

*Orac.Tiburt. = The Oracle of Baalbek: The Tiburtine Sibyl in Greek Dress, P. J. Alexander, Dumbarton Oaks 1967.

*Peek AV = W. Peek, Attische Versinschriften. Abhandlungen der Sächsischen Akademie der Wissenschaften zu Leipzig, Phil.-hist. Klasse, 69 (1980).

*Pouilloux Rhamnonte = La Forteresse de Rhamnonte: Étude de topographie et d'histoire, J. Pouilloux, Paris 1954.

*Ramsay The Social Basis = W. M. Ramsay, The Social Basis of Roman Power in Asia Minor, Aberdeen 1941.

*Rémondon Phoebammon = R. Rémondon et al., Le Monastère de Phoebammon dans la Thébaïde (ed. Ch. Bachatly), II: Graffiti, inscriptions et ostraca, Cairo 1965.

*Reynolds Aphrodisias = J. M. Reynolds, R. Tannenbaum, Jews and Godfearers at Aphrodisias: Greek Inscriptions with Commentary, Cambridge 1987.

*RIB = The Roman Inscriptions of Britain, I: Inscriptions on Stone, R. G. Collingwood, R. P. Wright, Oxford 1965; Epigraphic Indexes to RIB I, R. Goodburn, H. Waugh, Gloucester 1983; II: Instrumentum domesticum. 1: Military Diplomata, Metal Ingots, Tesserae, Dies, Labels and Lead Sealings, S. S. Frere et al., Gloucester 1990.

*Robert ATAM = L. Robert, À travers l'Asie Mineure, Paris 1980.

*Robert Castabala = L. Robert, La Déesse de Hiérapolis Castabala (Cilicie), Paris 1964, (= Bibl.Arch.Inst.Fr. Istanbul 16).

*Robert DAMM = L. Robert, Documents de l'Asie Mineure méridionale: Inscriptions, monnaies et géographie, Paris 1966.

*Robert Ét.épigr. = L. Robert, Études épigraphiques, 1ère série, 2ᵉ série (Repr. of BCH 52, 60), Paris 1928–36.

*Robert Hell. = L. Robert, Hellenica, 1–13, Limoges 1940–65.

*Robert Les Gladiateurs = L. Robert, Les Gladiateurs dans l'Orient grec, Paris 1940 [1971].

*Robert OMS = L. Robert, Opera minora selecta: Épigraphie et antiquités grecques, vols. 1–7, Amsterdam 1969–90.

*Robert Villes = L. Robert, Les Villes de l'Asie Mineure, Paris 1962.

*Rouèche Aphrodisias = C. Rouèche and J. M. Reynolds, Aphrodisias in Late Antiquity: The Late Roman and Byzantine Inscriptions, Including Texts from the Excavations at Aphrodisias, Conducted by K. T. Erim, London 1989.

*Salamine = Salamine de Chypre XIII. Testimonia Salaminia 2: Corpus épigraphique, J. Pouilloux et al., Paris 1987.

*Samothrace = Samothrace, Excavations Conducted by the Institute of Fine Arts, New York University, K. Lehmann, vol. 2 (1): The Inscriptions on Stone, P. M. Fraser, New York 1960.

Schwyzer, add '[Hildesheim 1960]'

*SEG = Supp.Epigr. in LSJ, q. v. (vols. 1–25, 1923–71); new series, H. W. Pleket, R. S. Stroud et al., vols. 26–36, Leiden, Amsterdam 1976–7 [1979]–1986 [1989].

*Side = Side Kitabeleri: The Inscriptions of Side, G. E. Bean, Ankara 1965.

SIG, add '[Hildesheim 1960]'

*Sokolowski 1, 2, 3, = F. Sokolowski, 1: Lois sacrées de l'Asie Mineure, Paris 1955; 2: Lois sacrées des cités grecques. Supplément, Paris 1962; 3: Lois sacrées des cités grecques, Paris 1969.

*Stoian *Tomitana* = I. Stoian, *Tomitana*, Bucharest 1962.
Suppl. epigr. ciren. = *Supplemento epigrafico cirenaico*, G. Pugliese Carratelli, D. Morelli, in *ASAA* N.S. 23-4 (1961/2), 218-375.
*Świderek *La Propriété foncière* = A. Świderek, *La Propriété foncière privée dans l'Égypte de Vespasien et sa technique agricole d'après PLond. 131ʳ*, Wrocław 1960.
TAM, add 'vol. iii: *Tituli Pisidiae*, R. Heberdey, Vienna 1941; vol. iv (i): *Tituli Bithyniae. Paeninsula Bithynica*, F. K. Dörner, M. B. von Stritzky, Vienna 1978; vol. v (i): *Tituli Lydiae: Regio septentrionalis ad orientem vergens*, vol. v (ii): *Regio septentrionalis ad occidentem vergens*, P. Herrmann, Vienna 1981, 1989'
Thasos I, II = *Recherches sur l'histoire et les cultes de Thasos* (École française d'Athènes, Études thasiennes III, V): I, J. Pouilloux, Paris 1954; II, C. Dunant, J. Pouilloux, Paris 1958.

Tit. Calymn. = *Tituli Calymnii*, M. Segre, *ASAA* N.S. 6-7 (1944/5) [Bergamo 1952].
Tit. Cam. = *Tituli Camirenses*, M. Segre, I. Pugliese Carratelli, *ASAA* N.S. 27-9 (1949/51), 141-318, 30-2 (1952/4), 211-46.
*Ugolini *L'Acropoli di Butrinto* = L. M. Ugolini, *Albania antica*; vol. iii: *L'Acropoli di Butrinto*, Rome 1942.
*Vidman *SIS* = L. Vidman, *Sylloge inscriptionum religionis Isiacae et Sarapiacae*, Berlin 1969.
*Wiegand *Palmyra* = T. Wiegand, *Palmyra: Ergebnisse der Expeditionen von 1902 und 1917*, Berlin 1932.
*Wolters *Kabirenheiligtum* = P. Wolters, G. Bruns, *Das Kabirenheiligtum bei Theben* I, Berlin 1940 (Kap. IV: *Inschriften*).
Xanthos = *Fouilles de Xanthos*, VII: *Inscriptions d'époque impériale du Létôon*, A. Balland, Paris 1981.

III. PUBLICATIONS OF PAPYRI & OSTRACA

BGU, for '1895–' read '1895–1983 (15 vols.), [1–9 Milan 1972])'

BKT, for '1904' read '1904–39 (8 vols.)'

**CPap.Jud.* = *Corpus Papyrorum Judaicarum*, V. A. Tcherikover, A. Fuks, M. Stern, Cambridge, Mass., 1957–64 (3 vols.).

CPR, for 'vol. i .. etc.' read '1895–1986 (10 vols.), [Milan 1974]'

Διηγήσεις, add 'v. index °I s.v. Call.*Dieg*.'

*Gallo *Fram.biogr.* = I. Gallo, *Frammenti biografici da papiri* I, II, Rome 1975, 1980.

*GLP = *Greek Literary Papyri*, D. L. Page, vol. i, Loeb 1942²; vol. iii, Loeb 1950.

**Gr.Roman-Papyri* = *Griechische Roman-Papyri und verwandte Texte*, F. Zimmermann, Heidelberg 1936.

*Milne *GSM* = H. J. M. Milne, *Greek Shorthand Manuals, Syllabary and Commentary; ed. from Papyri and Waxed Tablets in the British Museum, and from the Antinoë Papyri in the Possession of the Egypt Exploration Society*, London 1934.

*MPER = *Mitteilungen aus der Papyrussammlung der Nationalbibliothek in Wien (Papyrus Erzherzog Rainer)* N.S. Vienna 1932–.

*OAbu Mena = *Griechische Ostraka aus Abu Mena*, D. Wortmann (= *ZPE* 8.41).

**OAmst.* = *Ostraka in Amsterdam Collections*, R. S. Bagnall, P. J. Sijpesteijn, K. A. Worp, Zutphen 1976.

**OAshm.Shelton* = *Greek Ostraca in the Ashmolean Museum from Oxyrhynchus and Other Sites*, J. C. Shelton, Florence 1988.

**OBodl.* = *Ostr.Bodl.* in *LSJ*, add 'II: *Ostraca of the Roman and Byzantine Periods*, J. G. Tait, C. Préaux, London 1955; III: *Indexes*, J. Bingen, M. Wittek, London 1964 (sts. referred to as *OTait*).'

OCambridge, see *OBodl.* I pp. 153–73.

*ODouch = *Les Ostraca grecs de Douch*, H. Cuvigny, G. Wagner, Cairo 1986 (2 vols.).

*OFlorida = *The Florida Ostraka. Documents from the Roman Army in Upper Egypt*, R. S. Bagnall, Durham N.C. 1976.

**OHeid.* = *Griechische Papyrusurkunden und Ostraka der Heidelberger Papyrussammlung*, P. Sattler, Heidelberg 1963 (nos. 225–48 papyri, 249–88 ostraca).

**OLeid.* = *Greek Ostraka: A Catalogue of the Collection of Greek Ostraka in the National Museum of Antiquities at Leiden*, R. S. Bagnall, P. J. Sijpesteijn, K. A. Worp, Zutphen 1980.

*OLund = *Ostraka aus der Sammlung des Instituts für Altertumskunde an der Universität zu Lund*, Lund 1979.

**OMich.* = *Ostr.Mich.* in *LSJ*.

OPetrie, see *OBodl.* I pp. 82–152.

*OROM = *Ostraka in the Royal Ontario Museum*. I: *Death and Taxes*, A. E. Samuel et al., Toronto 1971.

Ostr., add 'now known as °*OWilck*.'

OStrassb. = *Ostr.Strassb.* in *LSJ*.

Ostr.Bodl., add ' = °*OBodl.* I'

*OTait = °*OBodl.*

**OWilck.* = *Ostr.* in *LSJ*; repr. w. addenda by P. J. Sijpesteijn [Amsterdam 1970].

**PAbinn.* = *The Abinnaeus Archive: Papers of a Roman Officer in the Reign of Constantius II*, H. I. Bell et al., Oxford 1962.

PAlex., add 'sts. referred to as *PAlex.Botti*'

**PAlex.* (1964) = *Papyrus grecs du Musée Gréco-Romain d'Alexandrie*, A. Świderek, M. Vandoni, Warsaw 1964 (nos. 1–40, pp. 47–79).

**PAmst.* = *Die Amsterdamer Papyri* I, R. P. Salomons et al., Zutphen 1980.

**PAnt.* = *The Antinoopolis Papyri*, C. H. Roberts et al., London 1950–67 (3 vols.).

**PApoll.* = *Papyrus grecs d'Apollônos Anô*, R. Rémondon, Cairo 1953.

**Pap.Brux.* (series) = *Papyrologica Bruxellensia*, Brussels 1962–. XVII: Pt. II, Textes inédits, 1979 (= *PCongr.* xv).

**Pap.Bub.* = *Die verkohlten Papyri aus Bubastos*, J. Frösén, D. Hagedorn, (*Pap.Colon* xv) 1990.

**Pap.Colon.* (series) = *Papyrologica Coloniensia*, Cologne/Opladen 1964–.

**Pap.Flor.* XIX = *Miscellanea Papyrologica*, M. Capasso et al., Florence 1990 (= *Papyrologica Florentina* XIX).

PBaden, add 'Heft 5, 1934, Heft 6, 1938'

**PBatav.* = *Textes grecques, démotiques et bilingues*, E. Boswinkel, P. W. Pestman, Leiden 1978. (= *Pap.Lugd.Bat* xix).

**PBeatty Panop.* = *Papyri from Panopolis in the Chester Beatty Library, Dublin*, T. C. Skeat, Dublin 1964, also known as *PPanop.Beatty*.

**PBerl.Möller* = *Griechische Papyri aus dem Berliner Museum*, S. Möller, Gothenburg 1929 (= *SB* 7338–50).

**PBerl.Zill.* = *Vierzehn Berliner griechische Papyri*, H. Zilliacus, Helsinki 1941.

**PBodm.* = *Papyrus Bodmer* Cologny-Geneva 1954– : 29 *Vision de Dorothéos*, A. Hurst, O. Reverdin, J. Rudhardt, 1984 (see also *ZPE* 75.82 ff.).

**PBon.* = *Papyri Bononienses*, O. Montevecchi, Milan 1953.

*PBouriant = *Les Papyrus Bouriant*, P. Collart, Paris 1926.

PBremen, add 'repr. in Berliner Akademieschriften zur Alten Geschichte und Papyruskunde II p. 193 ff. U. Wilcken, Leipzig 1970'

*PBrooklyn = *Greek and Latin Papyri, Ostraca and Wooden Tablets in the Collection of the Brooklyn Museum*, John C. Shelton, (= *Papyrologica Florentina* XXII) 1992.

**PCair.Isidor.* = *The Archive of Aurelius Isidorus in the Egyptian Museum, Cairo, and the University of Michigan*, A. E. R. Boak, H. C. Youtie, Ann Arbor 1960.

PCair.Zen., add 'vol v: O. Guéraud, P. Jouguet, Cairo 1940 nos. (59)801–(59)853 [Hildesheim 1971, all 5 vols.]'

*PCarlsberg cited by inv. no. and publication.

*PCharite = *Das Aurelia Charite Archiv*, K. A. Worp, Zutphen 1980.

**PCol.* = *Columbia papyri*. I: *Upon Slavery in Ptolemaic Egypt* (*PCol.* inv. 480), W. L. Westermann, New York 1929; VII: *Fourth-Century Documents from Karanis*, R. S. Bagnall, N. Lewis, Missoula 1979 (nos. 124–91).

*PColl.Youtie = Collectanea Papyrologica: Texts Published in Honor of H. C. Youtie, A. E. Hanson, Bonn 1975-6 (2 vols.).

*PCongr. xv = Pap.Brux. xvii.

PCornell, add '[1972]; sts. cited by inv. no. and publication'

*PDura = The Excavations at Dura-Europos Conducted by Yale University and the French Academy of Inscriptions and Letters. Final report V. i: The Parchments and Papyri, C. B. Welles, R. O. Fink, J. F. Gilliam, New Haven 1959.

*PEdfou = Tell Edfou 1937, 1938, 1939. Fouilles franco-polonaises. Les Papyrus et les ostraca grecs, J. Manteuffel et al. (Univ.Joseph Pilsudski de Varsovie. Institut français d'archéologie orientale du Caire. Fouilles franco-polonaises i–iii), Cairo 1937-50.

*PErasm. = Papyri in the Collection of the Erasmus University (Rotterdam), i P. J. Sijpesteijn, Ph. A. Verdult, Brussels 1986; ii : Ph. A. Verdult, Amsterdam 1991.

*PErlangen = Die Papyri der Universitätsbibliothek Erlangen, W. Schubart, Leipzig 1942 (sts. referred to as PErl.).

*PFam.Teb. = A Family Archive from Tebtunis, B. A. van Groningen, Leiden 1950 (= PLugd.Bat. vi).

*PFouad = Les Papyrus Fouad I (Publ. de la Soc. Fouad I de Papyrologie: Textes et documents iii), P. Jouguet, O. Guéraud et al., Cairo 1939 [Milan 1976].

*PFouad I Univ. = Fuad I University Papyri, D. S. Crawford (Publ. de la Soc. Fouad I de Papyrologie: Textes et documents viii), Alexandria 1949 [Milan 1976].

PFreib., add '1927, Abh. 7; repr. of the 3 vols. [Milan 1974]; vol. iv: Griechische und demotische Papyri der Universitätsbibliothek Freiburg, (nos. 45-75), R. W. Daniel, M. Gronewald, H. J. Thissen, Bonn 1986'

PGen., add 'ii Textes littéraires et documentaires nos. 82-117, C. Wehrli, Geneva 1986.'

PGiss., for '1910-12' read '1910-22 [Milan 1973]'

*PGiss.Univ. v. ‡PUniv.Giss.

PGoodsp.Cair., for 'Chicago 1904' read 'Chicago 1902 [Milan 1970] (sts. known as PCair.Goodsp.)'

PHamb., add '[Milan 1973]; ii: B. Snell et al., Hamburg 1954; iii: B. Kramer, D. Hagedorn, Bonn 1984'

*PHaun. = Papyri graecae Haunienses. i: Literarische Texte und ptolemäische Urkunden, T. Larsen, Copenhagen 1942; ii: Letters and Mummy Labels from Roman Egypt, A. Bülow-Jacobsen, Bonn 1981; iii: Subliterary Texts and Byzantine Documents from Egypt, T. Larsen, A. Bülow-Jacobsen, Bonn 1985.

PHeid., add 'N.F. iii, P. Sattler, Heidelberg 1963'

*PHels. = Papyri Helsingienses i Ptolemäische Urkunden, J. Frösén et al., Helsinki 1986.

*PHerm. = Papyri from Hermopolis and Other Documents of the Byzantine Period, B. R. Rees, London 1964.

PHib., add 'Part ii, E. G. Turner, M.-Th. Lenger 1955 (nos. 172-284)'

PIand., for '1912' read '1912-38 (8 vols.)'

*PIndiana Univ. = Indiana University Papyri, V. B. Schuman in CPh. 43 (1948) p. 110.

*PKöln = Kölner Papyri, B. Kramer et al., Cologne/Opladen 1976-87 (6 vols.).

*PKöln Ketouba = La Ketouba de Cologne: Un contrat de mariage juif à Antinoopolis, C. Sirat et al., Cologne/Opladen 1986 (= Pap.Colon. 12).

*PLaur. = Papyri from the Biblioteca Medicea Laurenziana, R. Pintaudi, Florence 1976-83 (4 vols.).

*PLeit. = Leitourgia Papyri, N. Lewis, Philadelphia 1963 (= TAPhA 53 pt. 9; repr. as SB 10192 ff.).

PLille, add 'sts. cited by inv. no. and publication'

PLond., add 'vi: Jews and Christians in Egypt, H. I. Bell, 1924 [Milan 1973] (= PLond. 1912-29 in LSJ); vii: The Zenon Archive, T. C. Skeat, 1974'

PLond., 1912-29, v. for eg.

*PLouvre cited by inv. no. and publication.

*PLugd.Bat. (series) = Papyrologica Lugduno-Batava, Leiden 1941–: i = PWarren; ii = °PVindob.Bosw.; xi = °PVindob.Sijp.; xiii = Papyri selectae, E. Boswinkel, P. W. Pestman, P. J. Sijpesteijn, Leiden 1965; xvii = Antidoron Martino David oblatum. Miscellanea papyrologica, E. Boswinkel, B. A. van Groningen, P. W. Pestman, Leiden 1968; xix = °PBatav.; xx = Greek and Demotic Texts from the Zenon Archive, P. W. Pestman, Leiden 1980; xxv = Papyri, Ostraca, Parchments and Waxed Tablets in the Leiden Papyrological Institute, F. A. J. Hoogendijk, P. Van Minnen et al., Leiden 1991.

*PLund = Aus der Papyrussammlung der Universitätsbibliothek in Lund, (6 vols.), Lund 1935-52.

*PMacquarie cited by inv. no. and publication.

PMag., add 'K. Preisendanz, A. Henrichs T(S) 1973-4²'

PMag.Lond., add '(= PMag. 5, 6, 7, etc.)'

PMasp., add '(also known as PCair.Masp.)'

*PMed. = ‡PMilan.

*PMed.Rez = MPER vol. xiii; H. Harrauer and P. J. Sijpesteijn, Medizinische Rezepte und Verwandtes, Vienna 1981.

PMerton, i, for '1939' read '1948'; add 'vol. ii, B. R. Rees, H. I. Bell, J. W. B. Barns, Dublin 1959; vol. iii, J. D. Thomas, London 1967 (= BICS Suppl. 18)'

PMich., add 'PMich. i: = PMich.Zen. q. v. (nos. 1-120); ii: Papyri from Tebtunis i, (= PMich.Teb.) A. E. R. Boak, Ann Arbor 1933 (nos. 121-8); A Papyrus Codex of the Shepherd of Hermas, C. Bonner, Ann Arbor 1934 (nos. 129-30); iii: Miscellaneous Papyri, J. G. Winter, Ann Arbor 1933 (nos. 131-221); v: Papyri from Tebtunis ii, (= PMich.Teb.) E. M. Husselman et al., Ann Arbor 1944 (nos. 226-356); viii: = Papyri and Ostraka from Karanis ii, H. C. Youtie, J. G. Winter, Ann Arbor 1951 (nos. 464-521); ix: = Papyri from Karanis iii, E. M. Husselman, Cleveland 1971 (nos. 522-76); xi: J. C. Shelton, Toronto 1971 (nos. 603-25); xii: G. M. Browne, Toronto 1975 (nos. 626-58); xiii: The Aphrodite Papyri in the Univ. of Michigan Papyrus Collection, P. J. Sijpesteijn, Zutphen 1977; xiv: V. P. McCarren, Chico, Calif. 1980; xv: P. J. Sijpesteijn, Zutphen 1982. Also sts. cited by inv. no. and publication'

*PMichael. = Papyri Michaelidae, D. S. Crawford, Aberdeen 1955.

PMich.Teb., v. °PMich. ii and v.

PMich.Zen., add 'now often known as PMich. i'

PMilan., i, add '2nd. ed. S. Daris, Milan 1967; ii: S. Daris, Milan 1966 (nos. 13-87); also known as PMil. or PMed.; sts. cited by inv. no. and publication'

*PMil.Vogl. = i: Papiri della R. Università di Milano i, A. Vogliano, Milan 1937 [Milan 1966]. (PUniv.Milan in LSJ). ii–vi: Papiri della Università degli Studi di Milano, various editors, Milan 1961-77. vii: La contabilità di un'azienda agricola nel ii sec. d. C., D. Foraboschi, Milan 1981.

*PMoen. cited by inv. no. and publication.

*PMon. = Veröffentlichungen aus der kaiserlichen Hof- und Staatsbibliothek zu München i: Byzantinische Papyri, A. Heisenberg, L. Wenger, Leipzig/Berlin 1914 [Stuttgart (T) 1986]; ii: Papiri letterari greci della Bayerische Staatsbibliothek di Monaco di Baviera, A. Carlini et al., Stuttgart (T) 1986; iii: Griechische Urkundenpapyri der Bayerischen Staatsbibliothek München, D. Hagedorn et al., Stuttgart (T) 1986.

*PNag.Ham. = Nag Hammadi Codices. Greek and Coptic Papyri

from the Cartonnage of the Covers, J. W. B. Barns, G. M. Browne, J. C. Shelton, Leiden 1981 (= *Nag Hammadi Studies* vol. xvi).

*PNess. = *Excavations at Nessana*, I: *Introductory Volume*, H. D. Colt 1962; II: *Literary Papyri*, L. Casson, E. L. Hettich, Princeton 1950; III: *Non-Literary Papyri*, C. J. Kraemer jr., Princeton 1958.

*PNew York = *Greek Papyri in the Collection of New York University*. I: *Fourth Century Documents from Karanis*, N. Lewis, Leiden 1967.

POsl., for '1925' read '1925–1936 (3 vols.)' and add 'I: *Magical Papyri*, S. Eitrem, Oslo 1925 (1–6), II: S. Eitrem, L. Amundsen, Oslo 1931 (7–64), III: S. Eitrem, L. Amundsen, Oslo 1936 (65–200).'

*POxford = *Some Oxford Papyri*, E. P. Wegener, Leiden 1942–8 (2 vols.).

*POxy.Hels. = *Fifty Oxyrhynchus Papyri*, H. Zilliacus et al., Helsinki 1979.

*PPalau Rib. cited by inv. no. and publication.

*PPanop. = *Urkunden aus Panopolis*, L. C. Youtie et al., Bonn 1980 (republ. fr. three articles in ZPE 7, 8 and 10; sts. referred to as *PPanop.Köln*).

*PPetaus = *Das Archiv des Petaus*, U. and D. Hagedorn, H. C. and L. C. Youtie, Cologne/Opladen 1969.

*PPhilad. = *Papyrus de Philadelphie*, J. Scherer (*Publ. de la Soc. Fouad I de Papyrologie: Textes et documents*, vii), Cairo 1947.

*PPrag. = *Papyri Wessely Pragenses*, L. Varcl, published in *Listy filologické* and *SB*.

*PPrag. I = *Papyri Graecae Wessely Pragenses*, R. Pintaudi et al., Florence 1988.

*PRainer Cent. = *Festschrift zum 100jährigen Bestehen der Papyrussammlung der Österreichischen Nationalbibliothek: Papyrus Erzherzog Rainer*, Vienna 1983.

PRein., add 'repr. vol. I [Milan 1972]; II: *Les Papyrus Théodore Reinach*, P. Collart, Cairo 1940'

PRev.Laws, add 're-edited by J. Bingen in *SB* Beiheft 1, Göttingen 1952'

PRyl. add 'vol. iv: *Documents of the Ptolemaic, Roman and Byzantine Periods*, C. H. Roberts, E. G. Turner, Manchester 1952 [1965]'

*PSAA = *Papyri Societatis Archaeologicae Atheniensis*, vol. i, G. A. Petropoulos, Athens 1939 [Milan 1972].

*PSakaon = *The Archive of Aurelius Sakaon: Papers of an Egyptian Farmer in the Last Century of Theadelphia*, G. M. Parássoglou, Bonn 1978. (v. ‡PThead.)

*PSarap. = *Les Archives de Sarapion et de ses fils*, J. Schwartz, Cairo 1961.

*PSelect. = *Papyri selectae*, E. Boswinkel et al., Leiden 1965 (= *PLugd.Bat.* XIII).

PSI, for '1912–' read '1912–79 (15 vols.)'

*PSorb. I = *Papyrus de la Sorbonne*, I, H. Cadell, Paris 1966.

*PSoterichos = *Das Archiv vov Soterichos*, Sayed Omar, Cologne/Opladen 1979 (= *Pap.Colon.* viii).

PStrassb., add '[Leipzig 1969]. II–VIII: *Papyrus grecs de la Bibliothèque Nationale et Universitaire de Strasbourg*, P. Collomp, J. Schwartz et al., Paris 1948 (vol. ii), Strasburg 1963–83'

PTeb., add 'iv: J. G. Keenan, J. C. Shelton, London 1976'

PThead., add '[Milan 1974]. II: *Aus den Archiven von Theadelphia*, L. Varcl *LFil.* (1961) p. 37 (nos. 1–61 re-edited as °*PSakaon*)'

*PThmouis = *Le Papyrus Thmouis* I, *colonnes 68–160*, S. Kambitsis, Paris 1985.

*PUG = *Papiri dell' Università di Genova*. I: M. Amelotti, L. Zingale Migliardi, Milan 1974; II: L. Zingale Migliardi, Florence 1980 also sts. cited by inv. no. and publication.

PUniv.Giss., for '1924' read '1924–39 (6 vols.); *Indices*, K. A. Worp 1975'

PUniv.Milan., add 'see °*PMil.Vogl.*'

*PUps.Frid = *Ten Uppsala Papyri*, B. Frid, Bonn 1981.

PVarsov., add '[Milan 1974]; also known as *PVars.*'

*PVatic.Aphrod. = *I papiri Vaticani greci di Aphrodito*, R. Pintaudi, Vatican 1980.

PVat., add 'also known as *PMarm.*'

*PVindob.Barbara cited by inv. no. and publication.

*PVindob.Bosw. = *Einige Wiener Papyri*, E. Boswinkel, Leiden 1942 (= *PLugd.Bat.* II).

*PVindob.Salomons = *Einige Wiener Papyri*, R. P. Salomons, Amsterdam 1976.

*PVindob.Sijp. = *Einige Wiener Papyri*, P. J. Sijpesteijn (= *PLugd.Bat.* XI), Leiden 1963.

*PVindob.Tandem = *Fünfunddreißig Wiener Papyri*, P. J. Sijpesteijn, K. A. Worp, Zutphen 1976.

*PVindob.Worp = *Einige Wiener Papyri*, K. A. Worp, Amsterdam 1972.

PWarren, add '*The Warren Papyri*, M. David et al., Leiden 1941 (= *PLugd.Bat.* I)'

*PWash.Univ. = *Washington University Papyri* I, V. B. Schuman, Missoula 1980 (= *ASP* 17).

*PWisc. = *The Wisconsin Papyri*, P. J. Sijpesteijn, vols. I and II; Leiden 1967 (= *PLugd.Bat.* XVI), Zutphen 1977 (= *Stud.Amst.* XI).

PWürzb., add 'repr. in *Berliner Akademieschriften zur Alten Geschichte und Papyruskunde* II, U. Wilcken, pp. 43–164, Leipzig 1970'

*P. XV Congr. = *Actes du XVe Congrès International de Papyrologie* II: *Papyrus inédits* (*Papyrologica Bruxellensia* 17), Brussels 1979.

*PYale = *Yale Papyri in the Beinecke Library*, I, J. F. Oates, A. E. Samuel, C. B. Welles, New Haven/Toronto 1967 (= *ASP* 2).

*PYale Inv. = Yale papyri, cited by inv. no.

PZen.Col., add 'vol. ii, W. L. Westermann et al., 1940 [Milan 1973] (= *PCol.* iv)'

*PZen.Pestm. = *Greek and Demotic Texts from the Zenon Archive*, P. W. Pestman, Leiden 1980 (= *PLugd.Bat.* xx).

*SB = *Sammelb.*, q. v., add 'VI–XVIII, E. Kiessling et al., Wiesbaden 1958–93'

*Schubart Gr.Lit.Pap. = W. Schubart, *Griechische literarische Papyri*, Berlin 1950.

Stud.Pal., for '1901' read '1901–24 (23 vols.) [Amsterdam 1965–83]'

*Suppl.Mag. = *Supplementum magicum*. I: R. W. Daniel, F. Maltomini, Cologne/Opladen 1990 (= *Pap.Colon.* xvi.1).

UPZ II, for '1935' read 'Berlin 1935–57'

*Youtie Scriptiunculae = H. C. Youtie, *Scriptiunculae posteriores*, Bonn 1981 (2 vols.).

IV. PERIODICALS

*AA = Archäologischer Anzeiger in *Jahrbuch des Deutschen Archäologischen Instituts* (*JDAI*) until 1982, then published separately, Berlin 1886– (*Arch.Anz.* in *LSJ*).

*AArch.Syr. = *Annales archéologiques de Syrie*, Damascus 1951–.

*AAWW = *Anzeiger der Österr.Akad. der Wissenschaften in Wien, Phil.-Hist.Klasse*, Vienna 1864–.

*ABSA = *Annual of the British School at Athens*, London 1895– (*BSA* in *LSJ*).

*AC = *L'Antiquité Classique*, Brussels 1932–.

*AD = Ἀρχαιολογικὸν Δελτίον, Athens 1915– (Ἀρχ.Δελτ. in *LSJ*).

*AE = Ἀρχαιολογικὴ Ἐφημερίς, Athens 1910– (Ἀρχ.Ἐφ. in *LSJ*).

*AEM = *Archäologisch-epigraphische Mitteilungen aus Österreich(-Ungarn)*, Vienna 1877–.

*AGWG = *Abhandlungen der Gesellschaft der Wissenschaften zu Göttingen*, Berlin 1843–.

*AHDO = *Archives d'histoire du droit oriental*, Brussels 1937–.

*AION(A) = *Annali dell'Istituto Universitario orientale di Napoli. Sezione di archeologia e storia antica*, Rome.

*AIPhO = *Annuaire de l'Institut de Philologie et d'Histoire Orientales et Slaves de l'Univ. Libre de Bruxelles*. Brussels 1932–.

*AJPh = *American Journal of Philology*, Baltimore 1880–.

*Anagennesis = Ἀναγέννησις, Athens 1981–.

*Anatolia = *Anatolia*, Inst. of Archaeol., Ankara Univ. 1956–.

*AncSoc = *Ancient Society*, Leuven 1970–.

*Ann.Mus.Gr.-R. d'Alex. = *Annuaire du Musée gréco-romain d'Alexandrie*, 1932–.

*APAW = *Abhandlungen der Preußischen Akademie der Wissenschaften*, Berlin (v. *Abh.Berl.Akad.* in *LSJ*).

*APF = *Archiv für Papyrusforschung und verwandte Gebiete*, T 1900– (*Arch.Pap.* in *LSJ*).

*AR = *Archaeological Reports*. Council of the Society for Hellenic Studies. The Management Committee of the British School of Archaeology at Athens, London 1959/1960–.

*Arch.Class. = *Archeologia classica*, Rome 1949–.

*ARW = *Archiv für Religionswissenschaft*, Leipzig 1943–.

*AS = *Anatolian Studies*, Journal of the British Institute of Archaeology at Ankara, London 1951– (*Anat.St.* in *LSJ*).

*ASAA = *Annuario della Scuola Archeologica di Atene e delle Missioni Italiane in Oriente*, Rome N.S. 1939–.

*ASAE = *Annales du Service des Antiquités de l'Égypte*, Cairo 1900– (*Annales du Service* in *LSJ*).

*ASMG = *Atti e Memorie della Società Magna Grecia*, Rome 1925–32, N.S. 1954–.

*ASNP = *Annali della Scuola Normale Superiore di Pisa, Classe di lettere e filosofia*, Pisa Serie III 1971–.

*Athenaeum = *Athenaeum. Studi periodici di letteratura e storia dell'antichità*, Pavia N.S. 1923–.

*BASO = *Bulletin of the American Schools of Oriental Research in Jerusalem and Baghdad*, Cambridge, Mass. 1919–.

*BASP = *Bulletin of the American Society of Papyrologists* New York, Columbia Univ. 1963/4–.

*BCAR = *Bullettino della Commissione Archeologica Comunale di Roma*, Rome 1872–. (*Bull.Comm.Arch.Com.* in *LSJ*).

BCH, add 'suppl. 8 = *Supplément* VIII: *Recueil des inscriptions chrétiennes de Macédoine du III*e *au VI*e *siècle*, D. Feissel, Paris 1983; suppl. 9 = *Supplément* IX: *L'antre corycien* II, Paris 1984 (ch. viii *Inscriptions*, J.-Y. Empereur)'

*BCO = *Bibliotheca Classica Orientalis*, Berlin 1956–.

*BCTH = *Bulletin archéologique du Comité des Travaux historiques et scientifiques*, Paris 1883–.

*BE = *Bulletin épigraphique* in *Revue des études grecques*, (*REG*; also sts. bound separately) 1938–.

*Belleten = *Türk tarih Kurumu, Belleten*, Ankara 1937–.

*Berytus = *Berytus, Archaeological Studies* published by the Museum of Archaeology of the American University of Beirut, Copenhagen 1934–.

*BICS = *Bulletin of the Institute of Classical Studies of the University of London*, London 1954–.

*BIFAO = *Bulletin de l'Institut Français d'Archéologie Orientale*, Cairo 1901– (*Bull.Inst.Franç.* in *LSJ*).

*BMB = *Bulletin du Musée de Beyrouth*, Paris 1937–.

*BSAA = *Bulletin de la Société Archéologique d'Alexandrie*, Alexandria 1898– (*Bull.Soc.Alex.* in *LSJ*).

*Byzantion = *Byzantion: Revue internationale des études byzantines*, Brussels 1924–.

*CE = *Chronique d'Égypte*, Brussels 1925–.

*Chiron = *Chiron: Mitteilungen der Kommission für alte Geschichte und Epigraphik des Deutschen Archäologischen Instituts*, Munich 1971–.

*CPh = *Classical Philology*, Chicago 1906– (*Class.Phil.* in *LSJ*).

*CPhil. = *Cahiers de philologie*, Lille 1976–.

*CRAI = *Comptes rendus de l'Académie des Inscriptions et Belles-Lettres* (cited by year; *CRAcad.Inscr.* in *LSJ*), Paris 1857–.

*CRIPEL = *Cahiers de recherches de l'Institut de Papyrologie et d'Égyptologie de Lille III*, Lille 1976–.

*CronArch = *Cronache di archeologia e di storia dell'arte*, Catania 1962–.

Dacia, add '48; N.S. Académie de la République populaire roumaine, Institut d'Archéologie, 1957–'

*DAW = *Denkschriften der Akademie der Wissenschaften in Wien*, Vienna 1850– (*Wiener Denkschr.* in *LSJ*).

*DOP = *Dumbarton Oaks Papers*, Cambridge, Mass. 1941–.

*EA = *Epigraphica Anatolica; Zeitschrift für Epigraphik und historische Geographie Anatoliens*, Bonn 1983–.

*EAC = *Études d'archéologie classique*, Paris 1955–.

*EHBS = Ἐπετηρὶς Ἑταιρείας Βυζαντινῶν Σπουδῶν. Athens.

*Emerita = *Emerita: Revista de lingüística y filología clásica*, Madrid 1933–.

*EPap = *Études de papyrologie*, Cairo 1932– (*Ét. de Pap.* in *LSJ*).

*Epigraphica = *Epigraphica*, Milan 1939–.

*GGA = *Göttingische gelehrte Anzeigen* (cited by year; *Gött. gel.Anz.* in *LSJ*).

*Gnomon = *Gnomon*, Berlin/Munich, 1925–.

PERIODICALS

*GRBS = Greek, Roman and Byzantine Studies, Durham, N.C. 1958-.
*Hellenica = Ἑλληνικά: φιλολ., ἱστορ. καὶ λαογρ. περιοδικὸν σύγγραμμα τῆς Ἑταιρείας Μακεδονικῶν Σπουδῶν, Thessalonica 1953-.
*HThR = Harvard Theological Review, Cambridge Mass. 1908- (Harv.Theol.Rev. in LSJ).
*IEJ = Israel Exploration Journal, Jerusalem 1950-.
*JA = Journal asiatique, Paris 1822-.
*JDAI = Jahrbuch des Deutschen Archäologischen Instituts, 1886- (Jahrb. in LSJ).
*JJP = Journal of Juristic Papyrology, Warsaw 1946-.
*JÖAI = Jahreshefte des Österreichischen Archäologischen Instituts in Wien, Vienna 1898- (Jahresh. in LSJ).
*JÖByz = Jahrbuch der Österreichischen Byzantinischen Gesellschaft, Verlag der Österr. Akad. der Wissenschaften, Vienna 1951-.
*JS = Journal des savants, Paris 1679-.
*JThS = Journal of Theological Studies, London 1900-57, Oxford 1958-.
*JWI = Journal of the Warburg and Courtauld Institutes, London 1937-.
*Kadmos = Kadmos. Zeitschrift für vor- und frühgriechische Epigraphik, Berlin 1962-.
*LFil. = Listy filologické, Prague 1874-1951, 1953-.
*Macedonica = Μακεδονικά: σύγγραμμα περιοδικὸν τῆς Ἑταιρείας Μακεδονικῶν Σπουδῶν, Thessalonica 1940-.
*MDAI(A) = Mitteilungen des Deutschen Archäologischen Instituts (Athenische Abteilung), Berlin 1876-. (Ath.Mitt. in LSJ).
*MDAI(I) = Mitteilungen des Deutschen Archäologischen Instituts (Istanbul Abteilung).
*MDAI(R) = Mitteilungen des Deutschen Archäologischen Instituts (Römische Abteilung), 1886- (Röm.Mitt. in LSJ).
*MEFR = Mélanges d'archéologie et d'histoire de l'école française de Rome, Rome/Paris 1881- (= Mél. de l'éc.fr. de Rome in LSJ).
*MH = Museum Helveticum, Basel 1944-.
*Mnemosyne: Bibliotheca Classica Batava, Leiden 4th Ser. 1948- (older series: Mnemos. in LSJ).
*MUB = Mélanges de l'Université Saint-Joseph, Beirut 1906-. (= Mélanges Beyrouth in LSJ).
*Muséon = Muséon: Revue d'études orientales, Louvain(-la-Neuve) 1882-.
*NC = Numismatic Chronicle, London 1838-.
*NGG = Nachrichten von der Gesellschaft der Wissenschaften zu Göttingen. (cited by year; Gött.Nachr. in LSJ).
*OA = Opuscula Archaeologica, Acta Instituti Romani Regni Sueciae, Lund 1935-.
*OMRL = Oudheidkundige Mededelingen uit het Rijksmuseum van Oudheden te Leiden, Leiden 1909-.
*Orientalia = Orientalia: commentarii trimestres editi a Facultate Studiorum Orientis Antiqui Pontificii Instituti Biblici, Rome 1920-.
*PAAH = Πρακτικὰ τῆς ἐν Ἀθήναις Ἀρχαιολογικῆς Ἑταιρείας, Athens 1872-.
*PalEQ = Palestine Exploration Quarterly, London 1869-.
*PASA = Papers of the American School of Classical Studies at Athens, Boston 1882-97. (= Papers of Amer. School at Athens in LSJ).
*PBA = Proceedings of the British Academy, London 1903-.
*PP = La parola del passato, Naples 1946-.
Philol., add 'suppl. = Supplementband'
*Phoenix = Phoenix: The Journal of the Classical Association of Canada, Toronto 1946-.
*QAL = Quaderni di archeologia della Libia, Rome 1950-.
*QIFG = Quaderni dell' Istituto di Filologia Greca (Università di Cagliari).
*RA = Revue archéologique, Paris N.S. 1966- (= Rev.Arch. in LSJ).
*RAL = Rendiconti della Classe di Scienze morali, storiche e filologiche dell'Accademia Nazionale dei Lincei, Rome 1892-.
*REA = Revue des études anciennes, Paris 1899-.
*Rec.Pap. = Recherches de papyrologie, Paris 1961-.
*REG = Revue des études grecques, Paris 1888- (Rev.Ét.Gr. in LSJ).
*RFIC = Rivista di filologia e d'istruzione classica, Turin 1873- (Riv.Fil. in LSJ).
*RIDA = Revue internationale des droits de l'antiquité, Brussels 1952-.
*RN = Revue numismatique, Paris 1838-.
*RPh. = Revue de philologie. 3rd series 1927- (= Rev.Phil. in LSJ).
*SAWW = Sitzungsberichte der Österreichischen Akademie der Wissenschaften in Wien, Vienna 1849- (Wien.Sitzb. in LSJ).
*SCO = Studi classici et orientali, Pisa 1951-.
*SDAW = Sitzungsberichte der Deutschen Akademie der Wissenschaften zu Berlin. Klasse für Philosophie, Geschichte, Staats-, Rechts- und Wirtschaftswissenschaften, Berlin 1948-68.
*SHAW = Sitzungsberichte der Heidelberger Akademie der Wissenschaften, Heidelberg 1910-.
*SIFC = Studi italiani di filologia classica, Florence N.S. 1920- (Stud.Ital. in LSJ).
*SO = Symbolae Osloenses, Oslo 1922-.
*SPAW = Sitzungsberichte der (Königlich) Preußischen Akademie der Wissenschaften, Berlin 1882-1938 (Berl.Sitzb. in LSJ).
*TAPhA = Transactions of the American Philological Association, Cleveland, Ohio 1869- (Trans.Am.Phil.Ass. in LSJ).
*T&MByz = Travaux et mémoires: Centre de recherches d'histoire et de civilisation byzantines, Paris 1965-.
*Tyche = Tyche: Beiträge zur alten Geschichte, Papyrologie und Epigraphik, Vienna 1986-.
*VDI = Vestnik drevnej istorii, Moscow 1938-.
*WJA = Würzburger Jahrbücher für die Altertumswissenschaft, Würzburg 1946-, N.F. 1975-.
*WS = Wiener Studien. Zeitschrift für klassische Philologie und Patristik, Vienna 1879- (Wien.Stud. in LSJ).
*YClS = Yale Classical Studies, New Haven 1928- (Yale Class. Studies in LSJ).
*ZNTW = Zeitschrift für die Neutestamentliche Wissenschaft und die Kunde der älteren Kirche, Berlin 1902-.
*ZPE = Zeitschrift für Papyrologie und Epigraphik, Bonn 1967-.
*ZVS = Zeitschrift für vergleichende Sprachforschung, Berlin 1851/-.

V. GENERAL LIST OF ABBREVIATIONS

*abbrev. = abbreviation, abbreviated.
*abst. = abstract.
*ACO = Acta Conciliorum Oecumenicorum, E. Schwartz, Berlin/Leipzig 1924–.
*add. = addenda.
*advl. = adverbial.
*Akkad. = Akkadian.
*An.Boiss., v. Anecdota Graeca in index °I.
*Anon.astr. = Anonymus astrologus.
*Anon.in Herc. = Scriptor Epicureus incertus, v. index °I s.v. Epicurus.
*Anon.Intr.Arat. = Anonymi Introductio in Aratum in Commentariorum in A. reliquiae, v. index °I s.v. Aratus.
*Anth.Lyr. = Anthologia Lyrica Graeca, E. Diehl, T(L) 1925; ed. 3, vol. i: fasc. 1–3, 1949–52; ed. 2, vol. ii: 1952.
Apoc., add 'v. Novum Testamentum'
*app. = apparently.
*Arg.Men., v. index °I s.v. Menander.
*arith. = in arithmetic.
*ASP = American Studies in Papyrology.
*assim. = by assimilation.
*astron. = in astronomy, astronomical.
*attrib. = attributed, attribution.
BCH, for 'BCH Suppl.' v. index °IV.
*BE = Bulletin epigraphique in REG (v. index °IV); cited by volume of REG and no. of Bulletin.
*Bile Dial.crét. v. index °II.
*Boll Sphaera, v. Teucer Babylonius in index I.
*c. = circa.
*CAlG = Collection des anciens alchimistes grecs, P. E. M. Berthelot, Paris 1888 [Holland Press 1963, Osnabrück 1967].
*CGFP = Comicorum Graecorum Fragmenta in Papyris reperta, C. Austin, Berlin 1973.
*CGL = Corpus Glossariorum Latinorum.
*chem. = in chemistry.
*Chr. = Christian.
*CID v. index °II.
*CIRB v. index °II.
*cl. = clause.
*CML = Corpus Medicorum Latinorum, Leipzig/Berlin 1915–.
Coll.Alex., add '[1970]'
*colloq. = colloquial.
*comm. = commentary, commentator.
*concr. = concrete.
*Copt. = Coptic.
*corr. = correction.
*CPap.Jud. v. index °III.
*cpd. = compound.
*def. = definition.
*Demioprata v. index °I.
*dial. = dialectal, in dialects.
*Dieg., v. Callimachus in index °I.
*Diller Trad.min.Gr.geogr. = A. Diller, The Tradition of the Minor Greek Geographers, American Philological Association 1952.

*dissim. = by dissimilation.
*dist. = distinguished.
*EAM v. °index II.
*EG = Epigrammata graeca, D. L. Page (OCT) 1975 [EG .. P.].
*EGF = Epicorum Graecorum Fragmenta, M. Davies, Göttingen 1988 [EGF p. .. D.]'
*eleg. = elegist, elegiac.
*Epiph.Const. v. Epiphanius Constantiensis in index °I.
*epist. = epistola (epistle, letter).
Epistolographi, add '[Amsterdam 1965]'
*expr. = expressing.
*FGE = Further Greek Epigrams, D. L. Page, Cambridge 1981.
FGrH, for 'Berlin 1923–' read 'Berlin 1923–9, Leiden 1926–58 [1954–60] (3 parts in 17 vols.); J. (= Jacoby) in Suppl. refers to this work'
*fig. = figurative(ly).
*foll. = following.
*Fr.Lex., v. Fragment in index °I.
*Fr.Ps.Gr. = Fragmenta Pseudepigraphorum quae supersunt Graeca, A. M. Denis, Leiden 1970.
*Fs.Theocharis = Διεθνές συνέδριο για την Αρχαία Θεσσαλία στη μνήμη του Δημήτρη Ρ.Θεοχάρη, Athens 1992.
*funer. = funerary.
*GDRK = Die griechischen Dichterfragmente der römischen Kaiserzeit, E. Heitsch, Göttingen 1963–4² (2 vols.).
*GLP, v. index °III.
*GCS = Die griechischen christlichen Schriftsteller der ersten drei Jahrhunderte, Leipzig 1897–.
*GVAK, v. index ‡II.
*HE v. Anthologia Graeca in index °I.
*Hellen. = Hellenistic.
*IEG = Iambi et elegi graeci ante Alexandrum cantati, M. L. West, Oxford 1971–2 (2 vols.), 1989 (vol. 1²), 1992 (vol. 2²).
*IGC, v. index °II.
*IGSK, v. index °II.
*incl. = including.
*Ind.Lect.Rost. = Index Lectionum in Academia Rostochiensi.
*inv. = inventory.
*iron. = ironically.
*JRCil. v. index °II.
*Kafizin, v. index °II.
*Lapid.Gr. = Les Lapidaires de l'antiquité et du moyen âge. Tome II: Les Lapidaires grecs, F. de Mély, C. É. Ruelle, Paris 1898.
*Lcn. = Lucianus interpres, v. index °I.
*lit.crit. = in literary criticism.
*Locr. = Locrian.
*LW v. LF in index II.
*Maced. = Macedonia(n).
*meton. = metonymical(ly).
*mil. = military.
*Miller Mélanges v. Etymologicum Genuinum in index °I.
*MPER v. index III.
*Myc. = Mycenaean.

GENERAL LIST OF ABBREVIATIONS

*NWGk. = North-West Greek.
*obj. = object.
*onomat. = onomatopoeic.
*O.Pers. = Old Persian.
 Orat.Att., for '1839' read '1845'; add '[Hildesheim 1967]'
*ostr. = ostracon. (For Ostr. v. index III).
*parenth. = parentheses.
 Paroemiographi, add '[Hildesheim 1958; w. Suppl., L. Cohn et al., Hildesheim 1961]'
*PCG 2, 3(2), 4, 5, 7 = Poetae comici Graeci, R. Kassel, C. Austin, Berlin/New York: II: Agathenor–Aristonymus, 1991; III 2: Aristophanes, 1984; IV: Aristophon–Crobylus, 1983; V: Damoxenus–Magnes, 1986; VII: Menecrates–Xenophon, 1989.
*PEG = Poetarum epicorum Graecorum testimonia et fragmenta 1, A. Bernabé, T(L) 1987.
*pers.n. = personal name.
*PG, *PL = Patrologiae cursus completus, series Graeca, J.-P. Migne, Paris 1857–68 (162 vols.); Patrologiae cursus completus, series Latina, J.-P. Migne, Paris 1844–55 (221 vols).
*PGR = Paradoxographorum Graecorum reliquiae, A. Giannini, Milan 1965.
*phr. = phrase.
*Physici = Physici et medici Graeci minores, J. L. Ideler, Berlin 1841-2 [Amsterdam 1963] (2 vols.).
*PMG = Poetae Melici Graeci, D. L. Page, Oxford 1962 [1967].
*PMGF = Poetarum Melicorum Graecorum Fragmenta, M. Davies, Oxford 1991.
*Poliorc. v. Poliorcetici in index °I.
*pr. = preface.
*pred. = predicate, predicative.
*presum. = presumably.
*prolog. = prologus, prologue.
*PSI, v. ‡index III.
*Pyth. or Pythag. = Pythagoras, Pythagorean.
*Pyth.Hell. = The Pythagorean Texts of the Hellenistic Period, H. Thesleff, Åbo 1965.
*quot(s). = quotation(s).
*redupl. = reduplicated.
*ref. = reference.
*repr. = reprint(ed).
*republ. = republished.
 rest., add 'restored'
*rev. = revised.
*Rhod. = Rhodian.
*RIB v. index °II.
*Rom.imp. = (of) Roman imperial (date).
*Salamine, v. index °II.

 Sch., add 'Sch.Gen.Il., Sch.min.Il., etc, v. Homerus in index °I.'
 Sext., v. Vetus Testamentum A in index I.
*sim. = similar.
*SLG = Supplementum Lyricis Graecis. Poetarum Lyricorum Graecorum fragmenta quae recens innotuerunt, D. L. Page, Oxford 1974.
 Sm., v. Vetus Testamentum A in index I.
*Socr.Rel. = Socraticorum reliquiae, G. Giannantoni, Rome 1983–5 (4 vols.).
*Sophocles Greek Lexicon = E. A. Sophocles, Greek Lexicon of the Roman and Byzantine Period (146 BC–1100 AD), New York 1887 [1951].
*sp. = spelling, spelt.
*spec. = specific(ally).
*subscr. = in subscriptione.
*Suppl.Hell. = Supplementum Hellenisticum, H. Lloyd-Jones, P. J. Parsons, Berlin 1983.
*Syr. = Syrus interpres, v. Vetus Testamentum in index °I.
*tab.defix. = tabella(-ae) defixionis. (For Tab.Defix. v. index II).
*Taillardat Images = J. Taillardat, Les Images d'Aristophane: Études de langue et de style, Paris 1965.
*transf. = (in) transferred (sense).
*transl. = translating, translation of.
*TrGF 1, 2, 3, 4 = Tragicorum Graecorum fragmenta. 1: Didascaliae tragicae, catalogi tragicorum et tragoediarum, testimonia et fragmenta tragicorum minorum, B. Snell, Göttingen 1971; Ed. correctior et addendis aucta, R. Kannicht, Göttingen 1986. 2: Fragmenta adespota. Testimonia volumini 1 addenda. Indices ad volumina 1 et 2, R. Kannicht, B. Snell, Göttingen 1981. [Trag.adesp. .. K.-S.] 3: Aeschylus, S. Radt, Göttingen 1985. 4: Sophocles, S. Radt (frr. 730a–g edidit R. Kannicht), Göttingen 1977.
*unexpld. = unexplained.
*unkn. = unknown.
*unpubl. = unpublished.
*van Brock Vocab.medic. = N. van Brock, Recherches sur le vocabulaire médical du grec ancien, Paris 1961.
*var. = variant (of).
*vbs. = verbs.
*Vit., for 'Anon.Vit.Arist.' and some other refs. containing 'Vit.' see Diogenes Laertius.
 Vorsokr., substitute 'Vorsokr.⁷ = Die Fragmente der Vorsokratiker⁷, H. Diels, W. Kranz, Berlin (Weidmann) 1954⁷ (3 vols.)'
*w. = with.
*wd. = word.
*WGk. = West Greek.
*WIon. = West Ionic.
*Youtie Scriptiunculae, v. index °III.

A

ἀ- **III**, delete 'v. ἀδάκρυτος'

†**ἄατος**, ον, (ἀαάω), in Il. ⏑--⏑, in Od., A.R. ⏑-⏑⏑, *admitting no error, infallible*, νῦν μοι ὄμοσσον ἀ. Στυγὸς ὕδωρ Il. 14.271; τόξα .. μνηστήρεσσιν ἄεθλον ἄαατον Od. 21.91, 22.5. **II** *invincible*, κάρτος ἄατος A.R. 2.77.

ἀαγής, for '*hard*' read '*unbreakable*'

ἀάζω, after 'Arist.' insert '*Mete.* 367ᵇ2'

ἄαπτος, omit 'cf. ἀπτοεπής'

ἄας, for 'ἀές' read 'ἄες'

ἀασιφρονία, after '*folly*' insert 'Apollon.*Lex.*'

*ἀάτη, ἡ, v. °ἄτη.

*ἄFατι, v. °ἄβατι.

ἀβάθματα, place after ἀβαθής.

†**ἀβακέω**, (‡ἀβακής), *take no heed*, οἱ δ' ἀβάκησαν πάντες· ἐγὼ δέ μιν οἴῃ ἀνέγνων τοῖον ἐόντα Od. 4.249.

ἀβακής, delete '(βάζω) *speechless* .. *gentle*' and substitute '*undisturbed*'; for 'Sapph. 72' read 'Sapph. 120 L.-P.'

ἀβάκητον, after 'Hsch.' insert '(ἀβάκητον La.)'

ἀβάκιον 1 a, add 'Ammon.*Diff.* 1'   **b**, for 'Plu.*Cat.Mi.* 90' read 'Plu.*Cat.Mi.* 70' add '**c** *kneading-board*, μάκτρα· ἀ. ἔνθα μάσσουσι τὸ ἄλευρον Hsch.; cf. °ἀβακίτης.'   add '**4** *dish*, *PCair.Zen.* 71.1 (-ειον, iii BC).'

ἀβακίσκος, for '*stone for inlaying*' read '*panel*' and add '(v. *REG* 80.325ff.)'

*ἀβακίτης, ὁ, *dough-kneader*, *SEG* 33.1165 (Side, iii AD); cf. °ἀβάκιον 1 c.

*ἄβακος, ὁ, *counting-board, abacus*, *OBodl.* i.262 (ii BC); cf. Lat. *abacus*.

*ἀβάκτης, ὁ, indecl., fr. Lat. *ab actis*, *records clerk*, *PFlor.* 71.509 (iv AD), ἀβάκτις *SB* 9106.3 (v AD).

ἄβακτον, delete the entry.

*ἄβακτος, ον, ἄβακα καὶ ἄβυκτον· τὸν μὴ μακαριστόν· Δωριεῖς δὲ τὸν ἀνεπίπληκτον καὶ ἀμεμφῆ *AB* 323, Phot. α 24 Th.

ἀβάκχευτος, delete '*generally* .. *Prose*'; for '*Lap.*' read '*Symp.*' add '**2** *without Bacchic celebration*, οὐδὲ ἀ. ἡ πηγή Philostr.*Im.* 1.23.2; in oxymoron, ἀ. θίασος E.*Or.* 319. **II** *not drinking wine*, Nonn. *D.*17.96, 40.295.'

ἄβαλε, read '**ἄβάλε**'; after '*O that* .. !' insert 'Alcm. 111 P.'; for 'Call.*fr.* 455' read 'Call.*fr.* 619 Pf.'

ἄβαξ 1 a, for this section read '*counting-board, abacus*, Arist.*Ath.* 69.1, *PTeb.* 793.6.4 (183 BC), *PFlor.* 1.50 (*RFIC* 1939.255)' delete '*sideboard* .. **4**' and after '(Delos, iii BC)' add 'Ammon.*Diff.* 1.' **II**, delete the section (v. °ἀβακος).

ἀβαρής II, add '**2** *free from burden* (of taxation), Just.*Nov.* 59.7.'

†**ἀβαρκνᾷ**· κομᾷ †τὲ Μακεδόνες Hsch. (La.).   **II** ἄβαρκνα· λιμός id.

*ἀβᾶς, v. °ἀββᾶς.

ἀβασάνιστος 1, for this section read '*not tortured*, ἀ. θνῄσκειν J.*BJ.* 1.32.3; (transf.) κημοῖς ὑπερῴαν ἀ. Ael.*NA* 13.9; (fig., of style) D.H.*Vett.Cens.* 3 (207.19) (prob. interpol.); adv. -τως *without torment*, ταῖς ἀκτῖσι τοῦ ἡλίου ῥᾳδίως καὶ ἀ. ἀντιβλέποντες Ael.*NA* 10.1 ; Aesop. 177.1 P.; *without vexation*, Ael.*VH* 9.1 (ἀβασκάνως cj. Dilts)'

*ἀβασίλευσία, *state of being without a king, anarchy*, *UPZ* 196.83 (ii BC).

ἀβάσκαντος, for '*secure* .. *harm*' read '*safe from the evil eye*'; line 4, after 'Dsc. 3.91' insert 'voc. ἀβάσκαντε *Vit.Aesop.*(W) 30 P., of a being w. magical powers'

†**ἀβάστακτος**, ον, (βαστάζω), *that cannot be carried*, fig., φορτίον ἀ. .. τὴν Καίσαρος διαδοχήν Plu.*Ant.* 16; transf., *insufferable*, χειρισμοῦ ἀ. *PTeb.* 758.15 (ii BC).   **II** *not removed*, *Inscr.Perg.* 8(3).37, (cf. σημεῖον ι 3 fin.); adv. -τως Hsch.

*ἄβατι, perh. for ἄFατι, dat. of *ἄFας, and related to ἄτη, *IPamph.* 3.6 (iv BC).

†**ἄβατος**, ον, also α, ον, Pi.*N.* 3.21, *affording no passage, impassable*, ὄρεα γὰρ ὑψηλὰ ἀποτάμνει ἄβατα Hdt. 4.25, 7.176, S.*OT* 719, ὕλη Str. 5.4.5; of water, ἅλς Pi.*N.* 3.21; ὡς δ' αὔτως καὶ ὁ Παρθένιος ἄβατος X.*An.* 5.6.9; fig., τὸ πόρσω σοφοῖς ἀ. κἀσόφοις Pi.*O.* 3.44; [τὸ ἀγαθὸν] ἐν ἀβάτοις ὑπεριδρυμένον Procl.*in Alc.* 319c W.   **2** of holy places, *not to be trodden*, S.*OC* 167, 675, ἕρπει πλοῦτος .. ἐς τἄβατα καὶ πρὸς †τὰ βατά† id.*fr.* 88.6 R.; ἀ. ἱερόν Pl.*La.* 183b, Porph.*Abst.* 4.11; ἀβατώτατος ὁ τόπος (i.e. cemetery) Arist.*Pr.* 924ᵃ5; transf., *undefiled*, ψυχή Pl.*Phdr.* 245a.   **b** as subst., ἄβατον, τό Theopomp.Hist. 343 J., *IG* 4².122.23 (Epid., iv BC); Διὸς καταιβάτου ἄ. *IG* 2².4965 (Athens, iv BC).   **3** of other places, *not affording access*, (οἰκίαι) ἄ. τοῖς ἔχουσι μηδὲ ἕν, Aristopho 4 K.-A.; ἀ. ποιεῖν .. τὰς τραπέζας Anaxipp. 3.   **4** fig., *inaccessible* to a quality, etc., φύσις .. ἄ. .. οἴκτῳ Ph. 2.53; κερδοσύνῃ ἄ. *ZPE* 6.187.2b (Nicopolis).   **II** *unavailable for sexual intercourse*, of fillies, ἀδμῆτες ἔτι καὶ ἄβατοι Luc.*Zeux.* 6; of a woman, id.*Lex.* 19.   **III** of land, cities, etc., *untrodden, desolate, waste*, Lxx *Je.* 31.9, 39.43, 51.6.   **IV** *marked by inability to walk*, w. ref. to gout, ἄ. ἄστατος πόνος Luc.*Ocyp.* 36.   **V** ἄβατον, τό, an unidentified plant, Gal. 6.623.

*ἀββᾶ or ἀββᾶς, ὁ, fr. Hebr. or Aramaic, *father*, of God, *Ev.Marc.* 14.36.   **2** title of respect given to monks, esp. *abbots*, *ITyr* 200 (AD 609), *SEG* 30.1689, written ἀβᾶ 31.1430 (both Palestine, vi/vii AD), *PKöln* 111.19 (v/vi AD), *POxy.* 2480.31 (vi AD); see also ἄππα, ἄππας.

ἀβέβαιος, delete 'of remedies, Hp.*Aph.* 2.27'

*ἀβεβαίωσις, εως, ἡ, = ἀβεβαιότης, ἀ. τῶν πραγμάτων Anon. *in Rh.* 117.34.

ἀβέβηλος, for 'of persons, *pure*' read 'transf., ἀβέβηλον καὶ τῆς τοιαύτης ἐπιθυμίας τόν ⟨τε⟩ καιρόν [ἐποίησεν] i.e. *made devoted to*'

*ἀβέλτης, v. °ἀβερτή.

*ἀβερτή, ἡ, = ἀορτή II, *knapsack*, Suid., ἀβέρτα *Edict.Diocl.* 10.1a. Also ἀβέλτης, ὁ, *PMich.*ix 576.4 (v AD).

ἀβήρ, for 'cf. αὐήρ' read 'cf. αὔηρ'

*ἀβιαστικός, όν, *irresistible*, ἀ. λόγος *Suppl.Mag.* 95.12→ (v AD).

ἀβίαστος 3, for 'τὸ ἀπαθεί' read 'τὸ ἀπαθές' and for 'Eus.*PE* 5.10' read 'Eus.*PE* 5,10.10'

*ἀβίᾶτος, ον, = ἀβίαστος, *not violent*, ἀνήρ Trag.adesp. in *NGG* 1922.25.

ἀβιδα, after lemma insert '(ἀβίλλιον La.)'

*ἀβιτώριον, ου, τό, fr. Lat. *abitorium*, cf. *abitus*, *latrine*, *IHistriae* 363.7.

ἀβλαβής, after 'ές' insert 'ἀβλάβη[ν] (Aeol. acc.), Sapph. 5.1 L.-P.'; add 'cf. °ἀβλοπές'

ἀβλέφαρος, for '*eyebrows*' read '*eyelashes*'

*ἀβλήχμων· ἀμβλύς Hsch.

*ἀβληχροποιός, όν, *taking away strength*, a suggested interpr. of ἀβληχρός in θάνατος ἀ. Eust. 1676.57.

*ἀβλίμαστον· ἄψαυστον Hsch.

†**ἀβλοπές**· ἀβλαβές. Κρῆτες Hsch.

ἀβλοπία, for 'cf. Hsch.' read 'v. °ἀβλοπές'

ἀβοηθησία, for '*helplessness*' read '*state of being without help*'

ἀβοήθητος, line 3, delete '*fatal*'   **II**, for '*helpless*' read '*lacking or cut off from help*'

ἀβολαία, delete the entry.

ἄβολλα, delete the entry (v. °ἀβόλλης).

*ἀβόλλης, ου, ὁ, Lat. *abolla*, *thick woollen cloak*, *POxy.* 1153.18 (i AD), *Stud.Pal.* 20.15.9 (ii AD); *Peripl.M.Rubr.* 6, *PMich.*inv. 3163 (*TAPhA* 92.258.5, iv AD); v. κάροινον II.

ἄβολος, for '(βολή)' read '(βολός)'   **1**, for 'also of an old horse .. *AB* 322' read 'cf. οὐδέπω ἔχων τι ἐπὶ τῶν ὀδόντων γνώρισμα, in Greek-Coptic glossary app. understood as "*packhorse*", *Aegyptus* 6.186.72 (v. *Glotta* 60.80)'

*ἄβολος (B), gender unkn., *woollen cloak*, Alciphr. 3.40.(1.23.)2 (cj.).

*ἀβόρᾶτος, ὁ, (perh. = ἀFόρᾱτος = ἀόρατος), kind of fish, *IGC* p. 98 A19 (Acraephia, iii/ii BC).

ἀβουλία I, after 'Id.*Ep.* 347a' insert 'D.*Ep.* 2.17'

ἀβούλητος I, transfer 'Pl.*Lg.* 733d' to section II after '*or will*'

ἄβουλος, line 4, for 'Th. 1.120.7' read 'Th. 1.120.5'

ἄβρα, for '*favourite slave*' read '*personal slave, maid*'; after 'Luc.*Tox.* 14' insert 'Ezek.*Exag.* 19 S. (pl.)'

ἀβραμυάς, add '(-ίας La.)'

ἄβραχος, delete the entry.

ἄβρεκτος, after 'Hp.*Aff.* 52' insert 'Call.*fr.* 384.34 Pf.'

ἀβροκόμης II, delete '(with play on both meanings)'

*ἀβρόπαις, παιδος, ὁ, ἡ, *of pretty children*, πάτρα *RFIC* 1937.54 (Ptolemais).

†**ἀβροπενθής**, ές, *mourning effeminately*, A.*Pers.* 135.

*ἀβρόπεπλος, ον, *wearing delicate robes*, Ναΐδες ἀβρόπεπλοι epigr. in *ZPE* 14.21 (Atrax).

*ἀβρόπους, ποδος, ὁ, ἡ, *delicate-footed*, gloss on σαυκρόπους, Hsch.

ἀβρός, line 5, for '*splendid*' read '*delicate, luxurious*'; line 11, for 'Stesich. 37 .. Sapph. 5' read 'Stesich. 37 P.; σύ .. ἄβρως ὀμ⟨μϵ⟩μϵίχμϵνον θαλίαισι νέκταρ οἰνοχόαισον Sapph. 2.14 L.-P.'; line 13, for 'Anacreont. 41.3' read 'Anacreont. 43.3 W.'

ἁβρότης, line 1, delete 'splendour'; line 5, for 'in the freshness of youth' read 'μεγάλας, in circumstances of great luxury'

ἁβρότονον, add 'also as fem. pers. n., Plut.Them. 1, Men.Epit. fr. 1'

ἁβροχικός, after ' = ἄβροχος' insert 'PVarsov. 26.30 (iv/v AD)'

ἁβροχίτων, after 'Inscr.Cos 5.11' insert 'Lindos 197f5 (written -κιτ-, cf. κιτών)'

ἁβρύνω, line 4, delete 'wax wanton'

ἄβρωμος, for 'Xenocr. 9' read 'Xenocr. 6'

ἀβρωσία, add 'prob. in E.Hipp. 136 (lyr.; codd. ἀμβροσίου)'

ἄβρωτος I 1, for 'ὀστᾶ Men. 129' read 'Men.Dysc. 452'

*ἀβστινατεύω, disinherit, ἀ. τῆς τῶν πατέρων κληρονομίας Just.Nov. 89.3.

*ἀβστινατίων, ονος, ἡ, the fact of disinheriting, PMasp. 97ᵛ.D82 (vi AD).

*ἄβυκτος, ον, v. ᵒἄβακτος.

*ἄβυρσος, ον, lacking hide, ἀ. .. ἡ τοῦ Ἀχιλλέως (sc. ἀσπίς) Eust. 911.62.

*ἀβύσσαιος, ον, lying at or coming from a great depth, ὕδωρ Anon.Alch. 403.1, (v.l. ἐναβύσσ-).

*Ἀγαγύλιος, ὁ, (sc. μήν), name of a month in Thessaly, IG 9(2).554 (Larissa), al.

ἀγάζηλοι, add 'cf. pers. n. Ἀγάζαλος SEG 24.383 (Delphi, iii BC)'

ἀγάζω, for 'exalt overmuch' read 'act without moderation' II, omit 'cf.'; add 'aor. part. ἀγασσαμένη Nic.fr. 74.15 G.-S.'

*ἀγαθημερία, ἡ, lucky day, PMich. in PBA 15.131 (cf. -μερος SEG 31.1223 (Pisidia, Rom.imp.) and -μερίς ib. 1037 (Lydia, AD 210/11) as pers. n.).

ἀγαθίδιον, delete 'Paul.Aeg. 2.57'

*ἀγάθιον, ου, τό, poultice, Paul.Aeg. 2.57 (124.20 H.).

Ἀγάθιος, ὁ, cult epith. of Zeus, INikaia 1061-7 (ii/iii AD).

*ἀγαθογνώμων, ον, of good opinions, of a person, Heph.Astr. 3.45.4.

*ἀγαθοδαιμόνημα, τό, astrol., position in propitious region, (v. ἀγαθοδαίμων III), Heph.Astr. 2.11.4, 11.

ἀγαθοδαίμων, delete 'less correct .. A.D.Adv. 60.15'

ἀγαθοειδής, at end add 'adv. -δῶς Procl.Inst. 25 D.'

ἀγαθοεργασία, delete the entry.

ἀγαθός, line 1, for 'ἀζαθός GDI 57' read 'a-za-ta-i prob. = ἀγαθᾷ, ICS 220.4, Kafizin 135d, 136, al.' I 6, at the beginning insert 'as epith. of Hermes, SEG 26.137 (Attica, iv BC), ib. 35.1118 (Ephesus)' and after 'δαίμων' insert 'ἥρωες ἀ. of deceased souls, SEG 36.854 (Selinous, iii BC), IUrb.Rom. 301, 373; at Thasos, of heroized fallen soldiers, οἱ Ἀγαθοί, Thasos 141.3, 8, 11' II 4, ἐπ' ἀγαθῷ, add 'in epitaphs, SEG 31.797, 1533, etc. (Rom.imp.)' III, sup., add 'ἀγαθέστατε (voc.) Hyp.fr. 219b J. (ascribed to Euripides in codd.Phot.)' IV, line 2, for 'Hp.Off. 9' read 'Hp.Off. 4'

ἀγαθύνω IV, for 'do well' read 'prosper'

ἀγαῖος (A), before 'Hsch.' insert 'ἀγαῖον· ἐπίφθονον' and at end add 'τὰ τῶι Λυκείωι δάρματα καὶ τὰν ἀγαίαν μόσχον CID I 9 (iv BC)'

ἀγαῖος (B), delete the entry.

ἀγακλεής, line 3, after 'Pi.I. 1.34' insert 'AP 9.26.5 (Antip.Thess.)'

*ἀγακλήεις, εσσα, εν, famous, ἀγακλήεντι MAMA 1.267 (Laodicea Combusta).

ἀγακλυτός I, for 'Il. 6.426' read 'Il. 6.436'

ἀγαλλίαμα, for 'transport of joy' read 'cause of joy' and add 'Lxx Ps. 118 (119).111; Is. 22.13, Si. 1.11'

ἀγάλλω, lines 7/9, for 'adorn .. Pass., glory, exult' read 'med., εὔιον ἀγαλλόμεναι θεόν E.Ba. 157. b honour an occasion, γαμηλίους εὐνάς E.Med. 1027; med., τὰ γενέσια ἀγάλλεσθαι D.C. 47.18.6. 2 med., glory, exult'; line 15, for 'Pl.Tht. 176b' read 'Pl.Tht. 176d '

ἄγαλμα after lemma insert 'Cypr. a-za-la-ma Kafizin 292' 3, before 'Hdt. 1.131' insert 'Jeffery LSAG p. 358 no. 49 (Egypt, Ion. letters, vi BC, written ἄκαλμα)' 5, for 'expressed by painting or words' read 'metaph.'

*ἀγαλματικός· ἀκουστής Hsch.

*ἀγαλματογλύπτης, ὁ, sculptor, IPhilae 182 (iii AD).

ἀγαλματογλύφης, ὁ, sculptor, Aesop. 90.2 P.

ἀγαλματογλύφος, for 'Rev.Ét. .. Aphrodisias)' read 'MAMA 8.574 (Aphrodisias), SEG 32.1311 (Lycaonia).'

*ἀγαλματοθήκη, ἡ, place for a statue, SEG 28.953.57 (Cyzicus, i AD).

*ἀγαλματομική, ἡ, (sc. τέχνη), the art of carving images, Call.Dieg. iv.29 (fr. 100 Pf.) (perh. shortened for *ἀγαλματοτομ-).

*ἀγαλματοπώλης, ου, ὁ, seller of images, Aesop. 99 tit. P.

ἀγαλματοφορέω, add 'of mankind, bear the image of God, τοῖς δὲ αὐτὸν ἐν ἑαυτοῖς ἀγαλματοφοροῦσι τὸν ποιητήν Athenag.Res. 12.6'

†ἀγαλματοφώρας, ὁ, thief of temple adornments, Schwyzer 424.13 (Elis, iv BC) (acc. sg. -φώραν, unless -φῶρᾶν, nom. -φώρ).

ἀγαλμοειδής, delete the entry.

ἄγαμαι I, add '6 w. imper., like Lat. amabo, in parenthesis, please, Cephisod. 3 K.-A.' II 2, after '(Od.) 23.64' insert 'w. dat. and gen., ἥιε, μή οἱ δῆμος ἐϋκλείης ἀγάσαιτιο A.R. 1.141; w. gen. only, Pi.Pae. 8.75 S.-M.'

Ἀγαμεμνόνεος, after 'ειον' insert 'also two termin., Ἀ. δαὶς ἢ τράπεζα, a fatal feast, Eust. 1507.60' and after 'ιον' for 'Pi.; A.' read 'Pi.P. 11.20, A.Ag. 1499, Ch. 861'

*Ἀγαμέμνων, delete 'Ἀγαμέμνονος .. 1507.60'

ἄγαμος I, after 'E.Or. 205' add 'in religious context, celibate, νικᾷ ἡ τύχη τῶν ἀγάμων Robert Hell. 11/12.494 (Didyma, early Byz.)' II, delete 'fatal marriage'

ἄγαν, add 'b οὐκ ἄ. not too, not particularly, oft. iron., εἰμὶ δ' οὐκ ἄ. σοφή E.Med. 305, 583, El. 1105; οὐκ ἄ. σφ' ἐπήνεσα id.Ph. 764, Ar.Eq. 598.'

ἀγανακτέω I, for this section read 'shudder, quiver, Hp.Liqu. 2, Heliod.ap.Orib. 46.7.8; of fermentation, Pl. 2.734e' II 2, after 'c. gen. rei' insert 'Lys. 14.39'

ἀγαννίφος, after 'Il. 1.420' insert 'Hes.fr. 229.6, 15 M.-W.; h.Merc. 325' and at end add 'PMed.inv. 71.82.18 (Aegyptus 52.90, i BC)'

*ἄγανον, v. ἤγανον.

ἀγανός 1, add 'b of things, of a crocus, h.Cer. 426; of water, IEryth. 224.12.'

ἄγανος, delete the entry.

ἀγανῶπις, for 'mild-eyed' read 'of gentle appearance'

ἀγάομαι, for 'only' read 'impf., ἀγᾶτο was angry, Hes.fr. 30.12 M.-W.' and delete 'and opt. .. Alc. 14'

ἀγαπάω II, add '2 w. inf., desire, MAMA 1.176 (Laodicea Combusta).' III 2, after 'Pl.R. 475b' delete 'cf.'; after 'Antiph. 169' insert 'Is. 8.43'

ἀγάπη I 1, after 'love' insert 'orig. in non-sexual sense but later w. some erotic connotation' and delete 'Lxx Ca. 2.7'; add 'IG 14.2317 (Padua). b loved one, Lxx Ca. 2.7; cf. as pers. n., SEG 19.422 (vi BC).' II, for 'love-feast' read 'the Christian agape' IV, delete the section (disputed text, v. JThS 18.142f., 19.209ff., 20.228ff.).

ἀγαπητός, lines 5/6, after 'αὕτη .. To. 3.10' insert 'τὸν υἱόν σου τὸν ἀ. Lxx Ge. 22.2, cf. Am. 8.10' II 2, after 'beloved' insert 'Ἄδων ἀγαπατέ Theoc. 15.149' III, adv., add 'with acquiescence, Luc.Am. 33'

ἀγάστονος, for 'much-groaning' read 'loud-roaring'; for 'loud-wailing, lamentable .. 123' read 'loud-wailing or -howling A.Th. 99, AP 14.123.3 (Metrod.), Nonn. D. 46.207; poet., of the pains of childbirth'

ἀγαστός, line 3, after 'An. 1.9.24' insert 'Pl.Lg. 808c'; at end, for '(Pure Att. θαυμαστός)' read '(rare in Att. prose; ἀ. θεοῖς Pl.Smp. 197d is in parody of Gorgianic style)'

ἀγατός, delete the entry.

ἀγαυρίαμα, for 'insolence' read 'pride, boastfulness'

ἀγαυριάομαι, for 'insolent' read 'proud, boastful'

ἀγγαρεία, after 'Arr.Epict. 4.1.79' insert 'οἴνου ἐγγαρία (cf. mod. Gk. ἐγγαρεία) BGU 21.3.16 (iv BC); of a soldier's duty, Cod.Just. 12.37.19 pr.; ἀγγαρεία· δουλεία Hsch. (ἀγγαρρία La.)'

*ἀγγαρεύω, = ἀγγαρεύω, PCair.Isidor. 72.32 (iv AD).

*ἀγγαρικός, ή, όν, concerning ἀγγαρεία, cj. for αιγα- in PCair.Preis. 33.6 (iv AD).

*ἀγγάριος, α, ον, impressed for public service, PCair.Isidor. 73.12 (iv AD); ἀπὸ τῆς ἀγγαρίου ib. 23, app. = ἀγγαρεία; masc. as subst. workman, gloss on Δάος Hsch.

ἄγγαρος I 1, delete 'Hdt. 3.126, X.Cyr. 8.6.17' at end for '(Assyr. agarru, "hired labourer")' read 'perh. cognate w. Aram. 'iggᵉrā "letter"; v. Glotta 49.97-100; earliest attestation in Gr. in pers. n. Ἀνγάρος, Corinth 15(3).1 (viii or vii BC)'

ἀγγεῖον I 1, line 10, after '1 Ki. 9.7' insert 'κρέως ἀγγία PRyl. 627.72'; at end add 'ἐπὶ τῶν ἀνγήων (= ἀγγείων), of a particular office, PPetaus 86.11 (ii AD)' 2, for this section read 'tank, reservoir, Pl.Lg. 845e; transf., of the hollow of the sea-bed, Pl.Criti. 111a' II, line 4, after 'id.PA 692ᵃ12' insert 'of the eyes S.E.P. 1.49'

*ἀγγειουργός, ὁ, maker of vessels, IG 2².1576.69 (iv BC).

ἀγγελία I 1, line 3, delete 'ἀ. λέγουσα .. 114'; line 4, delete 'Il. 18.17'; for 'πέμπειν' read 'πέμπει .. ἀγγελίην λέγουσαν τάδε Hdt. 2.114'; penultimate line, after 'Hes.Th. 781' insert '(dub., cod. ἀγγελίη)'; at end add 'applied to official delegations, πολλὰς δ' ἀνγελίας πρὸ πόλεως κατὰ φῦλα διῆλθεν ἀνθρώπων CEG 416 (Thasos, vi BC); epigr. in BCH 116.599 (Ambracia, vi/v BC)' I 2, delete the section adding exx. to section I 1 II, delete the section (see end of section I 1).

*ἀγγελίαρχος, archangel, AP 1.34, 35 (Agath.).

*ἀγγελίας, Ion. ἀγγελίης, ὁ, v. ‡ἀγγελία I 1 (at end).

ἀγγελιαφόρος, add ' = Lat. frumentarius, in sense of imperial agent, D.C. 78.15.1'

ἀγγελικός 1, add 'c a name for the dactylic hexameter catalectic, Diom. 1.512.23, Sacerd. 6.507.20; also, the trochaic dimeter, Mar. Vict. 6.85.27.'

2

ἀγγελιτείη, app. 3rd sg. aor. opt. of vb., perh. = ἀγγέλλω, *Inscr.Cret.* 4.146.4 (iv BC).

ἀγγέλλω, line 4, for 'ἠγγείλαμην' read 'ἠγγειλάμην'

***ἀγγελοδείκτης**, ου, ὁ, *revealer of* ἄγγελοι (v. ἄγγελος I 4), *PMag.* 4.1374.

***ἀγγελόεις**, εσσα, εν, = ἀγγελικός, τιμή *SIFC* 2.397 (Crete).

ἄγγελος I 2, line 5, for 'ἐμῶν' read 'ἐπῶν' **4**, add 'Ἀγαθὸς Ἄγγελος *IStraton.* 1118, Θεῖος Ἄγγελος Οὐράνιος ib. 1307 (prob. pagan, v. *SEG* 31.1689)' **II**, add 'and of a goddess at Didyma, perh. also Artemis, *Didyma* 406.9-10 (ii AD); title of Zeus in Arabia, *SEG* 32.1539 (ii/iii AD)' at end add 'Myc. *a-ke-ro*'

***ἀγγλάριον**, τό, unexpld. wd., τὸ ἐπὶ τὴν πομπὴν ἀνγλάριον *Inscr.Cret.* 3.iii 7.13 (ii AD).

ἄγγος, after '*vessel*' read '*receptacle*'; after '*IT* 953, 960' add 'for dry substances, Aen.Tact. 29.6, 8; *cinerary urn*, S.*El.* 1118, 1205; *sarcophagus*, *IAssos* 72 (sp. ἄνγος)' **II**, for this section read '*box, chest*, or sim., Hdt 1.113; S.*Tr.* 622; adv. ἄγγοσδε, *to the receptacle*, Emp. B 100.12 D.-K.'

***ἄγγουρα**· ῥάξ, σταφυλή Hsch. (s.v. ἄγγορ-).

***ἄγγρεσις**, ιος, ἁ, Thess. = αἵρεσις B I 3, *choice*, *IG* 9(2).504.4 (Larissa, ii BC), dat. ἀγγρεσι, *BCH* 59.56.40 (cf. ἐφάνγρενθειν s.v. ἐφαιρέομαι; presupposes *ἄγγρειμι = ἀγρέω).

ἀγγροφά, v. °ἀναγραφή.

ἀγγροφεύς, v. °ἀναγραφεύς.

***ἀγδαβάτης**, ου, ὁ, perh., *member of some section* or *class of Persians*, A.*Pers.* 924 (codd.).

Ἄγδιστις, ἡ, Phrygian title of the mother goddess Cybele, voc. Ἄγγδιστι Men.*Th.fr.dub.* 20 S., acc. Ἄγδιστιν Str. 10.3.12, dat. Ἀνγδίσση *SEG* 36.1201 (c.AD 200), dat. Ἀγδίσσιδι *IG* 12(2).524 (Methymna), Ἀνγδείσει Robert *ATAM* 239 (Pisidia, w. discussion of variants).

***ἄγεθλον**, τό, in pl., wd. of uncertain interpr., *IPamph.* 3.24 (Sillyon, iii BC).

ἀγείρω, line 4, after '*BC* 2.134' insert 'Aeol. (συν)αγάγρεται Alc. 119.10 L.-P.'; at end add 'Myc. 3 sg. *a-ke-re*'

Ἀγελάα, ἡ, = ἀγελείη, epith. of Athena, Sokolowski 2.19.89 (Athens, iv BC).

ἀγελάζομαι, after 'Act.' insert '*keep a flock*, Apoll.*Met.Ps.* 77.71 (*PG* 33.1429b)'

ἀγελαιοκομικός, add 'see also °ἀγελοκομική'

***ἀγελαιοκόμος**, ὁ, *keeper of a flock*, Pall.*V.Chrys.* 4 (v.l. ἀγελοκ-).

***ἀγέλαιος**, ἁ.· ὁ ἀμαθής Phlp.*Dif.Ton.* p. 361.2.

ἀγελαῖος II, add '**3** math., *cubic*, *POxy.* 3455.8 (iii/iv AD).'

ἀγελαρχέω, add '*hold the office of* ἀγελάρχης, *Xanthos* 21.5 (iii AD)'

ἀγελαρχέω, after '*Am.* 22' insert 'τῶν τράγων τὸν ἀγελάρχην Longus 2.31.2' at end add '**II** gloss on βουαγόρ, Hsch.'

ἀγελαρχικός, add 'cf. Procl. *in Ti.* 1.467.29'

***ἀγελάς**, άδος, ἡ, = φορβάς, Sch.A.R. 2.88b.

***ἀγελαστής**· ἔγχελυς Hsch.

***ἀγελαστοῦ**· ἀκακίας Hsch. († ἀγελαστοῦ La.).

ἀγελείη I, after ' = ἄγουσα .. *forager*' read 'or perh. *leader of the war-host* fr. *ἀγελήη' and add 'cf. °Ἀγελάα'

ἀγέλη I, at end omit 'but rare in early prose'

***ἀγελοκομική**, ἡ, (= ἀγελαιοκομική), *the art of keeping cattle*, Clem. Al.*Strom.* 1.7 (*GCS* 52.24.28).

ἀγελοκόμος, v. °ἀγελαιοκόμος.

ἀγένειος, for '*boyish* .. id.*Eun.* 9' read 'Alcm. 10(*b*).17 P.; Longus 1.16.2; τὸ ἀ. Luc.*Eun.* 9; (transf.) *puerile*, ἀγενειόν τι εἰρηκέναι id.*J.Tr.* 29'

***ἀγένειος** (B), ον, *not having children*, εἰ δὲ ἀγένειος εἴη Σαραπιὰς Ἀστοξένωι *GDI* 1891.29 (Delphi, ii BC); cf. ἀγενής III.

ἀγενής II, after 'S.*fr.* 84' insert '(anap.)'

ἀγερμός, after '(cf. ἀγείρω II 2)' insert 'Sokolowski 3.175.12 (Cos, iv BC); ib. 48.A8 (Piraeus, ii BC)'

***Ἀγερράνιος**, v. °Ἀγριάνιος.

ἄγερσις II, for ' = πανήγυρις' read ' = ‡ἀγερμός'; add 'cf. ἄγαρρις'

ἀγερωχία, after '*arrogance*' insert '*pride*, Sapph. 7.4 L.-P.'; delete 'in good sense .. *revelry*'; after 'Lxx *Wi.* 2.9' for 'pl., *feats of mastery*' read '**II** pl., *bold feats*'

ἀγέρωχος I, line 5, for 'Hes.*Fr.* 14 ' read 'Hes.*fr.* 30(*a*).12 M.-W., ἀ. ὄψις D.H.*Comp.* 16'; for 'of *noble* actions' read 'of *lofty* actions'

ἀγεσίφρων, for '(fort. -οφρύων)' read '(ἀγεσόφρυν La.)'

***ἄγεστα**, ἡ, *bank of earth* in siege warfare), cf. Lat. *aggestus, aggestum*, τὴν ἄγεσταν ἐργαζομένων Procop.*Pers.* 2.26.29. Also **ἀγέστον**, τό, Evagr.Schol.*HE* 4.27 (*PG* 86.2748c).

***ἀγέχορος** (or ἀγε-), όν, *leading the chorus*, Ar.*Lys.* 1281 (v.l. ἄγε χορόν).

ἀγεωργησία, for '*bad husbandry*' read '*lack of cultivation*'

ἄγη II, delete 'of the gods, jealousy'

ἀγή (A), after '*breakage*' insert '*PCair.Zen.* 15ʳ.27 (iii BC), etc., of pottery' and for '**1**' read '**b**' **2**, for '*place where the wave breaks, beach*' read '*breaking of the wave, surf*'

ἀγηθής, at end add 'S.*fr.* 583.10 R. (cj.)'

ἀγηλατέω, add 'Nicom.Trag. 14 S.'

ἀγηνόρειος, delete the entry.

ἀγηνορέω, for '*to be valiant*' read '*to be proud* or *boastful*' add '**II** trans., *treat arrogantly*, ἀ. τοκέας Euph. in *Suppl.Hell.* 415 ii 9.'

ἀγήνωρ 1, line 5, delete 'the Titans, Hes.*Th.* 641'

ἀγήραος, line 4, after 'Hes.*Th.* 277' insert 'ἀγείρω Corinn. 1 iii 25 P.' **1**, at end add 'θεῶν κῆρυξ ἀ. καὶ ἀθάνατος Hellanic.*fr.* 19(b) J.'

***ἀγηράσιος**, ον, = ἀγήρατος (A), βωμός *IUrb.Rom.* 128 (iv AD).

ἀγήρᾱτος (A), line 3, after '-ᾱτος' insert '*CEG* 548.4 (Attica), 721.2 (Aegeae, iv BC)'

***ἄγηρος**, ον, = ἀγήραος, *ageless*, Hes.*fr.* 25.28 M.-W.

***Ἁγησιτίμειοι**, ων, οἱ, name of a Rhodian guild, *Clara Rhodos* 2.203.

ἀγητός, line 5, after 'Sol. 5.3' insert 'abs., w. dat. of person affected, of a tomb, τὸ καὶ μακάρεσσιν ἀγητόν Theoc. 1.126; Opp.*C.* 1.364'

***ἁγιαστήρ**, ῆρος, ὁ, *consecrator*, *Tab.Defix.Aud.* 16 x 7 (Syria, iii AD).

***ἁγιαστί**, adv., *devoutly*, *Suppl.Mag.* 87.6 (iii/iv AD).

ἅγιος, after 'ἁ, ον' insert '(θώγιον = τὸ ἅγιον, Call.*fr.* 196.29 Pf.)' **I 2**, add 'as a title of Artemis at Ephesus, sup., *SEG* 31.960 (ii/iii AD), al.; of a god named Θεὸς Ἅ., *SEG* 32.1388 (Commagene), ib. 1551 (Arabia, iii AD); of (pagan or Jewish) ἄγγελοι, *SEG* 31.1080 (Galatia, ii/iii AD); of Christ, *SEG* 32.1617 (iv/v AD); of Mary, *Salamine* 234 (v/vi AD), *SEG* 30.1701 (Sinae, v AD); as a Chr. title, esp. of bishops, sup., *SEG* 30.1675 (vi AD), al.'

ἁγιστύς, for 'Call.*Aet.* 1.1.3' read 'Call.*fr.* 178.3 Pf.'

ἁγιωσύνη II, add '*Cod.Just.* 1.1.7 pr., al.'

ἀγκ-, for 'Poet.' read 'dial. or poet.'

ἀγκάζομαι, for 'λίθον .. 21.1' read '**II** *embrace* (cf. ἀγκαλίζομαι), καὶ δέ σ' .. ἀ[γκ]άσσαιτο Euph. in *Suppl.Hell.* 415 i 9.'

ἄγκαθεν, delete 'also .. 337'

***ἀγκαλιδοπώλης**, ου, ὁ, *seller* of produce *in armfuls* or *bundles*, in quot., of firewood, Phot. α 179 Th.

ἀγκαλίζομαι, after 'Semon. 7.77' insert 'Mosch.Trag. 9.3 S.'

ἀγκαλίς I 2, after '*armful*' insert '*bundle*' **II**, for '*word*' read '*use*'; transfer 'J.*AJ* 5.1.2.' to section I 2.

ἀγκάλισμα II, for 'metaph.' read 'Ach.Tat. 2.37.6; transf.'

ἄγκαλος, add '*POxy.* 3354.9 (AD 257)'

ἀγκής, add '(ἀγκήεις La.)'

ἀγκιστρεία, add 'Ael.*NA* 12.43; (fig.) οὐδὲ ἀπαγορεύσω τὴν ἐμὴν -αν, εἰ καὶ δυσθήρατος ἡ γυνή Aristaenet. 1.17'

ἄγκιστρον, add 'of an anchor, *Trag.adesp.* 379 K.-S.; of a grapnel, *Poliorc.* 214.1'

ἀγκιστρόομαι, delete the entry (v. °ἀγκιστρόω).

***ἀγκιστρόω**, *fasten with hooks*, ῥαφαῖς ὑπὸ τὰς ἐπιπτυχὰς τὴν συμπλοκὴν ἀγκιστρώσαντες Hld. 9.15.2. **2** fig., *grip*, ἠγκιστρωμένος πόθῳ Lyc. 67. **II** *fit with hooks* or *barbs*, Plu.*Crass.* 25.

ἀγκλάριον, delete the entry.

***ἄγκλιμα**, ατος, τό, v. °ἀνάκλιμα.

ἄγκοινα, for 'Hes.*fr.* 245' read 'Hes.*fr.* 43(a).81 M.-W., Matro in *Suppl.Hell.* 534.59' **II**, for 'Alc. 18.9' read 'Alc. 326.9 L.-P.' and add 'later ἀγκοίνη *PCair.Zen.* 756.1 (iii BC)'

***ἀγκοπάζω**, v. °ἀνακ-.

ἄγκος, after 'E.*Ba.* 1051' insert '*Trag.adesp.* 445a K.-S.' add '**2** ἄ. φρέατος *well-shaft*, Alex.Aet. 3.29.'

ἀγκτήρ 1, for '*instrument for*' read ' = Lat. *fibula*, *pin used in*' **4**, add 'ἀγκτῆρος δίκην shd. be restd. for ἀγκύρας δίκην in Lyd.*Ost.* 16 (cf. ἀκτῆρος (sic) δίκην id. in *Cat.Cod.Astr.* 11(1).146.10)'

ἀγκτηριάζω, add 'Archig.ap.Gal. 13.577, 651'

ἀγκύλη I, line 3, for 'B.*Fr.* 13.2' read 'B.*fr.* 17.2 S.-M., Anacr. 70 P.'; line 4, for 'Ath. 11.782d' read 'Ath. 15.667c; line 6, for 'cf.' read 'Asclep.Jun.ap.Gal. 13.968.7, 969.1' **II 1**, add 'on a fighting ship, *loop* to contain a pole used for fire-throwing, Plb. 21.7.1'

***ἀγκυλίδιον**, τό, dim. of ἀγκύλη, *loop, noose*, φιάλιον ἐπ' ἀγκ[υλ?]ιδίου *Inscr.Délos* 1442*B*59 (ii BC).

ἀγκυλιδωτός, delete '*for a handle*'

ἀγκυλοβλέφαρον, delete the entry.

***ἀγκυλοβλέφαρος**, ον, *suffering from contraction of the eyelids*, Cels. 7.7.6.

ἀγκυλόγλωσσον, after '*contraction of the tongue*' insert '*ankyloglossia*'

ἀγκυλοειδής, add 'adv. -δῶς Erot. 76.3 (cj. for ἀγκυρο-)'

ἀγκυλοκοπέω, delete the entry.

***ἀγκυλοκοπία**, ἡ, *hamstringing*, *PAbinn.* 15.15.

***ἀγκυλόπεζος**, ον, = στραβοπόδης, *having twisted feet*, *Philol.*suppl. 15.1.143 (= *Fr.Lex.* II).

***ἀγκυλόπρυμνος**, ον, *having a curved stern*, *Suppl.Hell.* 991.116 (poet. word-list, iii BC).

ἀγκύλος II 1, for '*intricate*' read '*circuitous*'; delete '*catchy*' and 'in good sense *terse*'; transfer Alciphr. quot. to section II 2.

***ἀγκυλοτόμον**, τό, *hooked instrument* used in tonsillectomy, Paul.Aeg. 6.30 (67.18 H.).

ἄγκῡρα **I**, add 'used as a grappling-iron, Plb. 21.27.5' **II**, add 'also unexpld. item in domestic inventory, PMichael. 18 iii 3 (iii AD)'

*ἀγκύρειος, α, ον, for an anchor, σχοινία IG 2².1609.101, Inscr.Délos 104-29.14 (both iv BC),; Ph.Bel. 100.34.

*ἀγκυρίς, ίδος, ἡ, perh. small anchor, IG 2².1550 (pl., iii BC). **II** gloss on κράδη III, some kind of theatrical machinery, Plu.Prov. 2.16.

ἀγκυροβολέω, for 'Hp.Dent. 18' read 'Hp.Oss. 18'

ἀγκυροβόλιον, add 'Peripl.M.Rubr. 7.24'

*ἀγκυροβόλος, -ῳ δείπνῳ· -α Φοίνικες τὰ δεῖπνα, ἃ παρεσκεύαζον τοῖς τελώναις ἐκ τῶν λιμένων Hsch. (La.).

ἀγκυροειδής, adv., add 'but v. °ἀγκυλοειδής'

*ἀγκύρωμα, τό, anchor, Sch.Ar.Eq. 762 J.

ἀγκών **I 1**, line 4, for 'Riv.' read 'Amat.' **2**, for 'arm' read 'the bent arm, forearm and elbow' **3**, for pres. def. read 'the corresponding part in animals' **II 1**, line 4, for 'ribs which support the horns' read 'shoulders'; line 5, for 'Semus 1, Hsch.' read 'Hsch., cf. Nic.Al. 562, Semus 1' **III**, line 3, delete 'Ath. 12.516a'

*ἀγκωνάριον, τό, an item of clothing, perh. w. elbow-length sleeves, or an armband, PVindob.G 16846.2 (Tyche 2.6; written ἀγκονάριον, c.AD 640).

ἀγκωνίζω, add '**III** perh. conduct canal water into side channels, PZen.Pestm. 20 Suppl. C (p. 268).'

ἀγκωνίσκιον, for 'dim. of ἀγκών' read 'piece projecting at right angles to a structure, key (in an organ)'

ἀγκωνίσκος, after 'Spir. 1.42' insert 'for fastening a structure'

ἀγκωνοφόρος, for 'IG 3.1280' read 'REA 32.5 (Athens, i BC), IG 2².2361.8 (iii AD)'

ἀγλάα, line 3, after 'Or. 4.148d' insert 'transf., of the Good, Numen.fr. 2.15 P.; of the intelligible world, Plot. 3.8.11'

*ἀγλάϊος, α, ος, = ἀγλαός, Σιμωνίδου ἀγλαΐῃ παῖς MAMA 3.793 (Cilicia); epith. of Zeus, Διὸς Ἀγλαΐῳ AA 1966.342.

*ἀγλαοδίνης, ου, ὁ, having bright eddies, Suppl.Hell. 991.106 (poet. word-list, iii BC).

*ἀγλαοειδής, ές, glorious in form, epic. in PLit.Lond. 38.26.

ἀγλαόθρονος, after 'bright-throned' insert '(or perh. to be derived fr. θρόνον flowers)'

ἀγλαόκαρπος **II**, delete the section.

*ἀγλαόκολπος, ον, fair-bosomed, Pi.N. 3.56.

ἀγλαομειδής, after 'PLG 3.639' insert 'ἀγαλμοειδ- codd.' and at end add '[ἀγ]λαομειδέσι Νύμφαις epigr. in IGBulg. 1579.5 (Augusta Traiana)'

*ἀγλαομήτης, ου, ὁ, with splendid wisdom, epigr. in SEG 36.1198.8 (Hierapolis, iv AD).

*ἀγλαόμολπος, ον, of beautiful song, PHarris 7 (ii/iii AD).

ἀγλαόπεπλος, add 'TAM 3(1).590.1 (Termessos, iii AD)'

*ἀγλαοφορέω, to be the bearer of brightness, Vett.Val. 344.8.

ἀγλαόφορτος, for 'proud of one's burden' read 'carrying a splendid burden'

†ἀγλαόφωνος, ον, producing a splendid sound, ἀ. ἀοιδός Orac.Sib. 12.173 (iii AD), of the Muses, Procl.H. 3.2.

*ἀγλαοχαίτας, α, bright-haired, Pi.fr. 52g(e) S.-M.

*Ἀγλᾱπιός, v. ‡Ἀσκληπιός.

Ἄγλαυρος, last line, read '(Hdt.) 8.53; μὰ τὴν Ἄγλαυρον Ar.Th. 533; Paus. 1.18.2, IG 2².3459, SEG 33.115'

ἀγλαφύρως, add 'Ath. 15.677f'

ἀγλευκής, add 'adv. -κῶς Philostr. VA 4.39'

ἀγλευκιτάς, add '(ἀγλευττας La.)'

*ἄγλευκος, ον, = ἀγλευκής (q.v.), Sch.Nic.Al. 171a, b.

ἀγλίς, at end for 'Call.fr. 140' read 'Call.fr. 495, 657 Pf.'

ἀγλύφος, add 'uncarved, κεφαλίς (column capital) PHib. 217.21, 29 (ii AD)'

ἄγμα, before 'Plu.Phil. 6' insert 'Inscr.Délos 104-10.6 (iv BC), al.' and after it insert '(pl.)'

*ἀγμίεις· παραθραύσεις Hsch. (La.).

*ἀγναβόν, ?τό, kind of plant or tree, Edict.Diocl. 36.111, 112; cf. ἀρναβόν.

*ἄγναθος, kind of fish, IGC p. 98 A10 (Acraephia, iii/ii BC).

*Ἀγναῖος, ὁ (sc. μήν), name of a month at Halos, IG 9(2).109a28, 71 (ii BC); cf. °Ἀγνεών.

ἄγναμπτος, for 'B.8.73' read 'B.9.73 S.-M.; ἀ. νόον A.Pr. 163 (lyr.)'; before 'Orph.L.' insert 'σθένος'

ἄγναπτος **I**, line 1, delete 'hence new'; line 2, delete 'cf.'; at end add 'as subst., ἀλλ' Ἰουδαῖοι σαββάτων ὄντων ἐν ἀγνάπτοις καθεζόμενοι Plu. 2.169c' **II**, delete the section.

*ἀγνᾰφάριος, maker of ἄγναφα, unfulled cloth or garments, MAMA 3.252b; Ἀλέξανδρος -ος ἐπίκλην Σακκᾶς Bithynische St. 7.II.2 (vi AD); (sp. ἀκν-) MAMA 3.622c, 767a.

ἄγναφος, for 'foreg.' read '= ἄγναπτος not fulled, PCair.Zen. 92.16 (iii BC), Peripl.M.Rubr. 6'

*ἀγνεάρχης, ου, ὁ, religious official at Ephesus, IEphes. 1044.15, 1045.5.

ἀγνεία **I**, add 'honourable public conduct, IEphes. 213.7, 790, SEG 7.825.15 (Gerasa, ii AD), IGLS 21.117; at Ephesus in formula in agoranomic inscrs., κόρος· ἀγνεία abundance and honesty, IEphes. 938.7, al. **b** the quality associated by musical theorists w. the number seven, Aristid.Quint. 3.6.' **II**, add 'sg., ritual purity, of a period of abstinence or fasting, Chaerem.Hist. 10.6 vdH. (= Porph.Abst. 4.6), Plu. 2.353b, IEphes. 3263 (ii/iii AD)'

*ἀγνεῖον· πηγαῖον Hsch. (perh. ἄγνειος 'pure', sc. ὕδωρ).

ἄγνευμα, add 'E.El. 256'

ἀγνεύω **I 1**, line 6, after 'Luc.Am. 5' insert 'transf., ἄχρι] σὸν γένειον ἀγνεύῃ τριχός Call.fr. 202.69 Pf.'

ἀγνεών, add 'Ἡ Ἀγνεών, ῶνος, ὁ (sc. μήν), name of a month at Magnesia, Inscr.Magn. 100a2, b20; cf. °Ἀγναῖος.'

†ἁγνίζω, fut. ἰῶ (ἀφ-) Lxx Nu. 8.6; pf. ἥγνικα 1Ep.Pet. 1.22: (ἁγνός):- make ritually clean, purify, πόντου σε πηγαῖς E.IT 1039; ἀ. πυρσῷ μέλαθρον ib. 1216; Diph. 125.1 K.-A.; Hp.Morb.Sacr. 1.45; Lxx Ex. 19.10; Act.Ap. 21.24; w. gen., χέρας σὰς ἁγνίσας μιάσματος E.HF 1324; med., Lxx Jo. 3.5; Plu. 2.1105b. **2** cleanse away impurities, λύμαθ' ἁγνίσας ἐμά S.Aj. 655; E.Or. 429. **3** purify by funeral rites, S.Ant. 545; σώμαθ' ἡγνίσθη πυρί E.Supp. 1211. **II** consecrate, Aristonous 1.17; sacrificial offerings, E.IT 705; A.R. 2.926; cf. E.Alc. 76.

ἄγνισμα, delete 'expiation'

ἁγνισμός, delete 'expiation'

ἀγνοέω **I**, add '**2** not to know a man (i.e. be a virgin), [ἡ ἀ]γνοήσασα ἄνδρα ἰς φθοράν Mitchell N.Galatia 468 (Byz.).'

ἀγνοητικός, for 'EE 1246ᵃ48' read 'EE 1246ᵃ38'

ἄγνοια, add '**III** loss of consciousness, Hipp.Epid. 7.85; Prorrh. 1.64.'

ἁγνός **I 2**, of Artemis, add 'unnamed, ἁγνῆς θεοῦ SEG 32.218.43, 84'; add 'of Aphrodite, SEG 31.731 (all Delos, ii BC); in Chr. usage, holy, 1Ep.Jo. 3.4, SEG 30.1479 (Hierapolis, iii AD)' **II 3**, add 'SEG 31.908 (Aphrodisias, iii AD); adv. -ῶς with integrity, IEphes. 712b (ii AD), al., MAMA 8.508 (Aphrodisias, c.AD 250)'

*ἁγνοτρᾰφής, ές, of pure nurture, pure, δέμας GVI 643.10 (Egypt, i BC/i AD).

*ἁγνόφῠτος, ον, of pure stock, GVI 1245.1 (Memphis, ii/iii AD).

ἄγνυμι, line 19, after 'Hdt. 1.185' insert 'cf. ἐαγώς Arat. 46'

ἀγνώμων **I 1**, after 'Pi.O. 8.60' insert 'Aristarch.Trag. 3 S.' **3**, for 'of judges' read 'of judging harshly'

ἄγνωστος **II**, delete 'Pi. .. cf.'

ἄγνωτος, add 'Call.fr. 620 Pf. (v.l. ἄγνωστον). **II** w. gen., ignorant of, φωνᾶν .. ψευδέων ἄ. Pi.O. 6.67.'

ἀγορά **I**, line 11, after 'AB 327' insert 'in Crete, SEG 32.908 (Phaestos, vi BC)' and after 'in late prose' insert '(as Lat. forum) ἀ. δικῶν, court of law of provincial governor, τὰς πόλεις τὰς ἐχούσας ἀγορὰς δικῶν Modest.Dig. 27.1.6.2' at end add 'Myc. a-ko-ra (sense dub.)'

ἀγοράζω **1**, delete 'occupy the market-place' **2**, add 'κοιμητήριον .. ἀγορασθέν .. παρὰ Φλαβίας Μαρίας SEG 29.641 (AD 507)' add '**4** meet in assembly, SEG 3.115.19 (Athens, iv BC).'

ἀγοραῖος **I**, line 3, after 'E.Heracl. 70' insert 'IG 1³.42.5, in Crete, SEG 23.566 (Oaxus, iv BC)'; line 5, before 'generally' insert 'of Heracles, SEG 32.672 (Thrace, ii AD)' **III 1**, line 3, after 'Str. 13.4.12' insert 'Πέργη ἡ πρώτη τῶν ἀγορέων (cf. conventus iuridici), SEG 34.1306 (iii AD)'; line 4, before 'IGRom. 4.790' insert 'also ἡ ἀγοραῖα, IG 12 suppl. 261 (Andros, i BC), JRS 30.148 (letter of Hadrian to Beroea); of a place used for market-day, τῷ τόπῳ τὴν καλουμένην ἀγορεῖον (sic) SEG 32.1149 (Magnesia ad Maeandrum, AD 209)' **2 b**, for this section read 'a weight, ἀγοραῖος λείτρα EA 16.141 (Nicomedeia, iii AD)' at end add 'Myc. a-ko-ra-jo'

*ἀγορᾱνομεῖον, τό, = ἀγορανόμιον, POxy. 75.14 (ii AD), etc.

*ἀγορᾱνομεύω, to be ἀγορανόμος, SEG 30.662 (Abdera, iii BC), unless error for -νομέω.

ἀγορᾱνομέω, add '**II** preside in the assembly, IG 9(2).517.10 (Larissa, iii BC).'

*ἀγορᾱσιαστικός, ή, όν, = ἀγοραστικός, ἀπὸ ἀ. δικαίου PLond. 1727.32 (vi AD), PMonac. 4.16 (vi AD).

ἀγόρᾱσις, for 'PRyl. 2.45.5' read 'PRyl. 245.5'

ἀγοραστής, add '**II** manager of a business, JEA 39.84, SEG 24.1179 (Egypt, iii BC), al.'

ἀγοραστός, add 'perh. also CIJud. 98*'

ἄγος (A) **1**, add 'Cypr. to-te-a-ko-se, prob. τόδε ἄγος, on a lead tablet to provide a curse against tomb robbers, ICS 311 (= Salamine 7)' **2**, add 'ἥαιι(α) (neut. pl.) IPamph. 3.13, SEG 34.989 (Sicily, i/ii AD)'

ἄγρα **III**, for 'Artemis' read 'Demeter' and for 'ib. 273' read 'IG 1³.369.91, cf. Pl.Phdr. l.c.'

*ἀγρακόμας, kind of bird (Pamphyl.), Hsch.

*ἀγραμματεύω, to be illiterate, SB 5174.17, 5175.21 (vi AD), al.

*ἄγραμμος, add '**II** = ἀγράμματος, illiterate, PMeyer 13.24 (ii AD).'

*ἀγραρεύω, lie in garrison, PGrenf. 2.95.2 (vi AD), SB 7656.8, etc.; cf. °ἀγραρία.

ἀγραρία | SUPPLEMENT | ἀγχιθάλασσος

*ἀγραρία, ἡ, Lat. *agraria*, *garrison*, PHerm. 75.2 (v AD), etc.   **2** as cult-title, Μητρὶ Θεῶν ἀγραρίᾳ IGRom. 1.92.

Ἀγραστυών, ῶνος, ὁ (sc. μήν), name of a month in Locris, IG 9²(1).721 C2 (ii BC); also Ἀγρεστυών, Ἀγροστυών, GDI 1757, 1880.

Ἀγρατέρα, v. °Ἀγροτέρα.

*ἀγράφιος, ον, perh. *not fit for writing on*, PMich.II 123ᵛ vii 25, 128.1(a)25 (both i AD).   **II** subst. *note*, *blank roll*, used in business transactions, PHarris 104.12; PMerton 24.17.   **2** v. ἀγραφίου γραφή.

ἀγρεῖος **1**, add 'Θεοί TAM 2(1).148.7 (Lydia)'   **2**, after '*boorish*' insert 'Alcm. 16.1 P.' and at end add 'neut. as adv. ἀγρεῖον .. ἐξεγέλασσε Call.fr. 24.13.Pf.'

ἀγρειοσύνη, delete '*clownishness* or'

*ἀγρελάτης, ου, ὁ, (ἄγρα, ἐλαύνω) perh. *game-beater*, POxy. 1917.41 (vi AD).

ἀγρέμιον, (*prey*), add 'SB 5301.2 (ἀκρ-), 8, 12 (pl., Byz.)'

*ἄγρενον, v. ‡ἄγρηνον.

*ἀγρεσιθήρα, app. epith. of Artemis, POxy. 3876 fr. 2.7 + fr. 6(b).2.

Ἀγρεστυών, v. °Ἀγραστυών.

Ἀγρετέρα, v. °Ἀγροτέρα.

ἀγρέτης, after 'Hsch.' insert '**b** title of magistrate or military leader in Crete, BCH 70.590 (Dreros, vii BC); cf. ἀγρεταί, ἀγρετεύω, °ἀγρήιον.'   **II**, for 'from .. *fields*' read '*hunter* (ἄγρα) or *assembler* (ἀγείρω)'

ἄγρευμα **II 2**, for '*net thrown over Agamemnon*' read '*net of fate*'; at end add 'τύχης ἀγρεύμασιν Gorg. B 11.19 D.-K.'

ἀγρεύς **I**, of Apollo, add 'SEG 35.491 (Atrax, c.200 BC)'

ἀγρευτικός, after 'Sch.E.Ba. 611' insert 'w. gen., Herm. *in Phdr.* p. 74'

ἀγρεύω, line 4, before '; of war' insert ', cf. ἀγρεύων αἷμα ταυροκτόνον ibid. 138' and after 'S.fr. 554' insert '*catch by stealth*, οὕτως .. ἂν .. ἥκιστα ὑπό τινων ἀγρευθεῖεν λάθρᾳ προσελθόντων Aen.Tact. 22.11; *capture booty*, id. 24.4'   **2**, delete 'thirst for .. 138'

ἀγρέω, at end after 'cf. ἄργειτε' add 'ἐξαγρέω, ἐφάνγρενθειν (s.v. ἐφαιρέομαι 2), °ἀνηάγληιμαι, ζωγρέω.   Myc. 3 sg. aor. *a-ke-re-se*'

*ἀγρήγορος, ον, *from which there is no waking*, τὸν ἀ. ὕπνον CIG 9449 (vi AD).

*ἀγρῆιον, τό, *meeting-place*, perh. in Inscr.Cret. 4.9 (Gortyn, vii/vi BC), cf. °ἀγρέτης I.

*Ἀγρήιος, ὁ (sc. μήν), name of a month at Locri Epizephyrii, de Franciscis Locr.Epiz. 13, 22, 31, 34.

ἀγρηνόν, add 'also perh. ἄγρενον, rest. in D.P. 9.20'

ἀγρία, add '**II** = ἄγρωστις, Sch.Theoc. 13.42b.'

*ἀγριαγγούριον, τό, *squirting cucumber*, expl. of ἀγριοσίκυον AB 1097 (ἀγραγγούρην cod.); cf. ἄγγουρον, ἀγγούριον *cucumber* in Sophocles *Greek Lexicon*.

*ἀγριάμπελος· ἡ βρυωνία βοτάνη Hsch.

*Ἀγριάνια· νεκύσια παρὰ Ἀργείοις. καὶ ἀγῶνες ἐν Θήβαις Hsch.; cf. ‡Ἀγριώνια.

*Ἀγριάνιος, ὁ (sc. μήν), Aeol. Ἀγερράνιος, IG 12(2).527.45 (Eresus), name of a month at Sparta, IG 5(1).18B8, at Rhodes, ib. 12(1).906, at Epidaurus, ib. 4².109.2.13, at Illyria, SEG 36.565.5, at Cyprus, ib. 23.685(g), at Caria, ib. 14.715, at Lesbos (Eresus) l.c.; cf. ‡Ἀγριώνιος.

*ἀγρίβροξ· ὀρίγανον Hsch.

*ἄγρινος, ὁ, = ἀγρονόμος, MAMA 3.663 (Corycus); cf. ἄγρινοι· ἀγρονόμοι. καὶ οἱ παιδεραστιαὶ οὕτως Hsch. (La.).

*ἀγρίοαιξ, αιγος, ἡ, *wild goat*, Rhetor.in Cat.Cod.Astr. 7.225.9.

ἀγριόεις **I**, after 'Nic.Al.' insert '30'   **2**, delete the section.

ἀγριολειχήν, add 'Aët. 8.16 (426.220 O.)'

ἀγριόμορφος, add 'ἀγριόμορφον ὄφιν PLille inv. 71 (CRIPEL 6.247, Hellen.)'

ἄγριος **I 2**, add '**b** subst. ἄγρια, τά, *weeds*, fig., ἄ. ψυχῆς Plu. 2.38c.'

ἀγριοσίκυον, add 'cf. ἀγριοσύκιον ἀγρανγούρην (read ἀγριοσίκυον ἀγριαγγούριον) καλεῖται AB 1097'

*ἀγριόσκορδον, τό, *wild garlic*, Paul.Aeg. 7.3 (260.25 H.).

*ἀγριοσύκιον, v. ἀγριοσίκυον.

*ἀγριοχηνοπρυμνίς, ίδος, ἡ (sc. ναῦς), *ship with wild goose as ἀκροστόλιον*, PMonac. 4.9 (vi AD; -πρήμνης pap.).

ἀγριόω **II 1**, add 'of way of life, τὸν ἀγριωμένον εἰς ἥμερον δίαιταν ἤγαγον βίον Mosch.Trag. 6.28 S.'

*Ἀγρίππαιος, ὁ (sc. μήν), name of a month in Cyprus beginning Nov. 2, Cat.Cod.Astr. 2.144.17, al.

*Ἀγρίππηνα, τά, name of an athletic festival, SIG 1065.3.

*Ἀγριππήσιοι, οἱ, name of a Jewish community at Rome, CIJud. 1.365, al.

*Ἀγριππίνειος, ὁ (sc. μήν), an Egyptian month in Caligula's calendar, APF 9.225 (i AD; -ηνίου), POxy. 3780.8 (written Ἀγριππῖνος, i AD), etc.

Ἀγριώνια, for 'etc.' read 'at Thebes, IG 7.2447 (prob. i BC)'

Ἀγριώνιος, add '**II** (sc. μήν) name of a month in Boeotia, IG 7.3348, SEG 24.343, al., see also °Ἀγριάνιος.'

*ἀγροβάτας, α, *ranging over the fields*, v.l. for -βότας in S.Ph. 214, E.Cyc. 54.

†ἀγροβότης, v. °ἀγροβάτας.

*ἀγρογειτνία, ἡ, *adjacent plot*, PMasp. 151.112, al. (vi AD).

ἀγροικικός, after '*rustic*' insert 'SB 7337.20 (inscr., 41 BC)'

*ἀγροϊωτικός, ή, όν, *rustic*, μοῖραν ἔχων ἀγροϊωτίκαν Alc. 130.17 L.-P.

*ἀγρόκηπον, τό, (or -πος, ὁ, masc.) *garden-plot*, IG 2².2776.145 (ii AD).

*ἀγροκωμήτης, ὁ, perh. *member of rural assembly*, TAM 4(1).45.14 (Bithynia, Rom.imp.).

*ἀγρομυρίκινος, ον, *of wild tamarisk*, PHamb. 12.19.

ἀγρονόμος **I 2**, add '**b** *farmer*, Opp.H. 4.602.'

ἀγρός **I**, after 'Od. 24.205' insert 'Call.fr. 489 Pf.'   **2**, line 4, after '(Od.) 1.185' insert 'φροῦδοι πάλαι εἰσὶν εἰς ἀγρόν *into the countryside*, Men.Dysc. 776'; line 6, for 'τὰ ἐξ ἀγρῶν' read 'τὰ ἐκ τῶν ἀγρῶν'   at end add 'Myc. *a-ko-ro*'

Ἀγροστυών, v. °Ἀγραστυών.

*ἀγρότας, α, ὁ, perh. = ἀγρέτης I, *leader*, Alcm. 1.8 P., A.Pers. 1002 (codd., ἀγρέτ- cj.).

*Ἀγροτέρα, ἡ, Ion. -τέρη, cult-title of Artemis, Il. 21.471, X.Cyn. 6.13, EAM 101, al.; worshipped at Agra in Attica, Arist.Ath. 58, IG 2².1028 (Athens, i BC), Paus. 1.19.6; ἁγρατέρα (app. for ἁ ἀγρατέρα) PP 42.47 (Tarentum, vi BC), Ἀγρετέρα IG 2².4573 (iv BC), ἡ Ἀγροτέρα alone, X.HG 4.2.20, Ar.Eq. 660.

†ἀγρότερος, α, ον, (ἀγρός; cf. ὀρέστερος) *living in the wild*, ἡμίονοι, σύες, αἶγες Il. 2.852, 12.146, Od. 17.295; ἀγροτέρης ἐλάφοιο Hes.Sc. 407; φὴρ ἀ. Pi.P. 3.4; abs., ἀγρότεροι Theoc. 8.58; ἀ. καὶ νέποδες AP 6.11 (Satrius); of the nymph Cyrene, Pi.P. 9.6.   **2** of rural inhabitants, AP 9.244 (Apollonid.), APl. 4.235 (i BC).   **3** of plants, *wild*, AP 9.384.8, cf. Nic.Th. 711, Coluth. 112, ὁδοῦ MDAI(A) 18.269.   **4** v. °Ἀγροτέρα.

ἀγρότης **II**, delete the section amalgamating quots. w. section I (omitting 'Alcm. 23.8')   **III**, delete the section.

ἀγρυπνέω **2**, after 'Lxx Da. 9.14' insert 'SEG 34.1243 (Abydos, v AD)'

*ἀγρυπνίστως, adv. = ἀγρύπνως, PLond. 1660.12 (vi AD, -πτίστως pap.).

ἄγρυπνος, adv., add 'Aët. 2.93 (183.8 O.), Paul.Aeg. 1.87 (62.25 H.)'

ἀγρυπνώδης, at end add '(ed. ἀγρύπνῳ)'

*ἀγρῴος, α, ον, line 3, after '*hunting*' insert 'Euph. 58.3 vGr.'

ἀγρώσσω, line 3, after '*hunting*' insert 'Euph. 58.3 vGr.'

ἀγρώστης **I**, for 'Theoc. 25.48' read 'Theoc. 25.51'   **II**, add 'adjectively ἀγρώ[σ]ταισιν ἐπιύζων σκυ[λάκεσσι Suppl.Hell. 939.21 (i/ii AD, hex.)'

*ἄγρωστος, kind of grass, PMil.Vogl. 214.1.3, 7 (ii AD).

*Ἀγυαία, ἡ, unkn. deity, Sokolowski 2.131 (Chios, v BC).

*Ἀγυεῖος, v. °Ἀγυίηος.

ἄγυια **1**, at end add 'Elean for Att. στενωπός acc. to Paus. 5.15.2'

ἀγυιάτης **2**, for '(Phars.)' read 'SEG 23.408 (both Pharsalus, iv BC; for doubts about the interpr. v. BE 1965.212)'

Ἀγυιεύς **1**, add 'Ἀπόλλωνι Ἀγυεῖ ITomis 116(1).i 1, ii 2 (Rom. imp.)'   **2**, add 'Cratin. 403 K.-A., Men.fr. 811 K.-Th. (both from Harp.)'

†Ἀγυίηος, ὁ, (sc. μήν), name of a month, Schwyzer 91.2 (iii BC), Ἀγυήιος ib. 92.2 (Argos, iii BC), Inscr.Cret. 4.197.11 (ii BC), Ἀγυεῖος IG 9²(1).638.6.2 (Naupactus, ii BC), GDI 1975.2 (Delphi, ii BC), SPAW 1936.371(b).4 (Potidania, ii BC).

ἀγυμνάσια, add 'Gal. 5.72.5'

ἀγυμναστία, add 'Gal. 5.91.3'

ἀγύμναστος **2**, at beginning insert 'ἀ. φρενὶ E.fr. 598, πόνοις δέ γ' οὐκ ἀ. φρένας id.fr. 344'

*ἀγυοτμήτος, *not divided into fields*, POxy. 3047.5, al. (iii AD).

ἄγυρις **2**, for '*gathering* of herbs' read '*mixture, collection*, ἐν μὲν δὴ βοτάναις ἀγύραν λυγρῶν τε καὶ ἐσθλῶν'

†ἀγύρτης, ου, ὁ, (ἀγείρω) *wandering beggar, mendicant*, δόλιος ἀ., of Tiresias, S.OT 388, E.Rh. 503, 715, Lysipp. 6 K.-A., Hp. Morb.Sacr. 1.10, Clearch. 5 (= 49 W.), ἀ. καὶ μάντεις Pl.R. 364b; of priests of Cybele, Μητρὸς ἀ. AP 6.218 (Alc.Mess.); Γάλλοις ἀ. Babr. 141.1.   **II** a throw of the dice, Eub. 57.5 K.-A. (On the accent cf. Hdn.Gr. 1.77)

ἀγυρτικός, for '*vagabond*' read '*mendicant*'; for '*juggling*' read '*of or belonging to mendicants*'; for '*jugglery*' read '*mendicancy*'

ἄγχαυρος, add '**II** subst., ὁ, *the time near dawn, morning twilight*, Call. fr. 260.64 Pf., Suid.; cf. Hsch. s.v. ἀγχουρος.'

†ἀγχεμωλία, ἡ, ἀνκεμō[λιαν] acc., prob. *law-suit undertaken for a close relative*, Inscr.Cret. 4.21.3 (Gortyn, vi BC); cf. °ἀντιμωλία, ἀμφιμωλέω.

*ἀγχιαλίτης, ου, ὁ, *one who lives near the sea*, Steph.in Rh. 269.15 (pl.).

ἀγχιβαθής **2**, add 'cj. for °ἀγχιβαφ- ib. 15.3'

*ἀγχιβαφής, ές, *half-submerged*, Nonn.D. 15.3 (cod., cf. °ἀγχιβαθής).

*ἀγχίδικος, Pythag. name for *six*, Theol.Ar. 38.

ἀγχιθάλασσος, add '-ος, ἡ, *coastland*, EA 14.22 l. 34 (Ephesus, i AD)'

ἀγχίθεος, after 'Od. 5.35' insert 'h.Ven. 200; of priests, *approaching god*, Luc.Syr.D. 31' deleting this ref. at end.
ἀγχίθυρος 1, add 'fig., w. gen., γήραος ἀ. AP 7.726.4 (Leon.Tarent.)'
ἀγχιμολέω, delete the entry.
ἀγχίμολος, add 'ἀ]γχίμολοι rest. in A.fr. 168.28 R.'
ἀγχινοία, line 3, for 'etc.' read 'PVindob.Tandem 2.5 (iii AD)'
*ἀγχιπύρα, ἡ, (ἄγχω, πῦρ) *instrument for extinguishing altar-fire*, SEG 17.146.10 (Argos iv BC; but see BE 1988 p. 385 no. 586).
Ἀγχίσαιος, ὁ (sc. μήν), name of a month in Cyprus beginning June 2, Cat.Cod.Astr. 2.146.12, 148.1.
*ἀγχιστήδᾱν, ἀνχιστέδαν ἐπινεμέσθō, *amongst near kin*, IG 9²(1).609.5 (Naupactus, v BC); cf. ἀγχιστίνδην.
ἀγχιστικός, add 'ἀ. νόμος Thasos 192.4 (i AD)'
ἀγχιστίνδην, after 'Hsch.' add '(ἀγχιστάδην La.)'
ἄγχιστος III, at beginning insert 'neut. pl. as adv.'
*ἀγχίτεξ, εκος, ἡ, *near the birth, soon to bring forth*, Theognost.Can. 40 A.; cf. ἀγχιτόκος, ἐπίτεξ.
*ἄγχνοος, v. °ἔγχνοος.
ἀγχόνη I, for this section read '*rope for hanging, noose*, οἱ δ' ἀγχόνην ἅψαντο δυστήνῳ μόρῳ Semon. 1.18 W., A.fr. 47a.14 R., Neophr. 3.2 S., Apollod. 3.13.3, Plu.Brut. 31; in pl., ἐν ἀγχόναις θάνατον λαβεῖν E.Hel. 200; HF 154; αἱ ἀ. μάλιστα τοῖς νέοις Arist.Pr. 954ᵇ35. 2 transf., *hanging*, ἀγχόνης .. τέρματα A.Eu. 746; ἔργα κρείσσον' ἀγχόνης deeds too bad for *hanging*, S.OT 1374; τάδ' ἀγχόνης πέλας E.Heracl. 246; ταῦτ' οὐχὶ .. ἀγχόνης ἔστ' ἄξια id.Ba. 246; ταῦτα .. οὐκ ἀ.; Ar.Ach. 125.'
*ἀγχονιμαῖα ξύλα, *gallows-trees*, Phot. s.v. ὀξυθύμια- (ἀγχονομ- cod.).
†ἀγχοῦρος· ὄρθρος· Κύπριοι. ἢ φωσφόρος καὶ οἱ σὺν αὐτῷ .. Hsch. (La.); see also °ἄγχαυρος.
ἄγχουσα, add 'Nic.Th. 638, 838; given as another name for ὀνοχειλές, Dsc. 4.24; for λυκαψός, id. 4.26'
ἄγω, line 4, for 'Sapph. 159' read 'Sapph. 169 L.-P.'; after 'Tim.Pers. 165' insert '(a barbarian speaks)'; after 'ἄξας' insert '(read ἄξας)'; line 5, after 'ἄξαι' insert '(f.l.)'; after 'Antipho 5.46' insert 'συνᾶξα GDI 1772.17, 1791.8 (both Delphi, ii BC), ἄξατε Nonn.D. 1.11, 34, etc.'; delete 'SIG 1 (Abu Simbel, vii/vi BC)'; line 6, after 'X.Mem. 4.2.8' insert 'ἄγωγα Dura⁴ 122'; line 13, delete 'also .. 8.20' I 3, line 11, after '(Locr., v BC)' insert 'Plb. 18.5.1' 6, after 'A.Ag. 406 (lyr.)' insert 'ἀπήμαντον ἄγων βίοτον Pi.O. 8.87' deleting this ref. in section IV 4. II, add '7 *handle* or *carry* on the books, POxy. 290.7 (AD 83/4), BGU 820.2 (AD 187/8), 84.4 (AD 242/3), PFay. 40.8.' V, line 4, delete 'θεοὺς .. 924' at end add 'Myc. 3 sg. a-ke'
ἀγωγαῖος, add 'II subst., ἀγωγαῖα, τά, perh. *entrance rites*, APF 16.175 (Chios, iv BC).'
ἀγωγεύς I, add '4 generally, *director, organizer* of a festival, Milet 1(7) no.263 (iii AD).'
ἀγωγή I 2, for 'ὑμῶν' read 'ἡμῶν' II 3, line 6, after 'youth' insert 'Sosib. 4 J.'; line 7, for 'Δακεδαίμονι' read 'Λακεδαίμονι'
ἀγωγός I, after 'aqueduct' insert 'SEG 29.1157 (Lydia, Rom.imp.), IEphes. 3217(b).24' and after 'without ὕδατος' insert 'SEG 31.953B (AD 113/20)'; at end add '*able to guide* or *influence*, ἀνθρώπων D.H.Dem. 22, cf. Th. 2'
ἀγών II, add 'καθεμένō τὠγῶνος SIG 38.32 (Teos, v BC), IG 5(2).113 (Tegea, v BC)' III 6 b, for '*power*' read '*urgency*, Cic.Att. 1.16.8'
*ἀγωνάριον, τό, dim. of ἀγών, *competition*, Inscr.Cos 43 (iii BC); εἰς ἀγωνάρια γυμνικά, SEG 38.1462.23 (Oenoanda, AD 124/6).
†ἀγωνάρχης, ου, ὁ, *organizer of contest* or *games*, S.Aj. 572, IHadr. 132.
*ἀγώναρχος, ὁ, an official in Boeotia, either in charge of games, or an equiv. of ἀγορανόμος in other cities (Sch.T Il. 24.1), IG 7.1817, SEG 23.271.3, cf. ib. 275, IGC p. 98 A1 (Acraephia, iii/ii BC, v. SEG 32.450).
ἀγωνία 1, add 'b *trial*, οὐ .. τὴν ἀγωνίαν ὑπεισελθεῖν βουλομένοις Just.Nov. 49.1 pr.' 3, for '*agony*' read '*anxiety*'
ἀγωνικός, add 'Ps.-Acro in Hor.C. 3.12.8; perh. also ἀγōγιʔὸν πόρο[ν] Lang Ath.Agora XXI K3 (vi BC)'
ἀγώνισμα I, add 'τὸ τελεώτατον τῶν ἀ. SEG 31.903.29 (Aphrodisias, iii AD, see also BE 1984.41)' IV, for '*plea*' read '*point at issue*'
*ἀγωνιστήρ, ῆρος, ὁ, *organizer of games*, perh. at IG 11(4).1053.30-1 (Delos; rest., v. ZPE 6.270).
ἀγωνιστήριον, for '*assembly*' read '*(judicial) contest*'
ἀγωνιστήριος, for ' = ἀγωνιστικός' read '*of* or *associated with contests*' and add 'σάλπιγξ Poll. l.c.'
ἀγωνιστικός I 1, at end delete 'τὸ ἀ. ib. 219c, e'
ἀγωνοδίκης, add 'perh. at SB 10493.2 (AD 228)'
ἀγωνοθεσία, add 'also ἀγωνοθετία, SEG 7.825.14 (Gerasa, ii AD)'
ἀγωνοθετέω I 2, for 'Plb. 9.343' read 'Plb. 9.34.3'
ἀγωνοθέτης 1, add 'POxy. 2611.14, al. (AD 192/3), Dor. ἀγωνοθέτας, SEG 30.1218 (Tarentum, iv BC); ib. 23.271 (Thespiae, iii BC)' 2, after '*judge*' insert 'in any contest'
*ἀγωνοθετία, v. ‡ἀγωνοθεσία.

ἀγωνοθετικός, after '(Sparta)' insert 'τιμαί SEG 13.244 (Argos, i BC/i AD), SEG 23.317 (Delphi, i AD)'
*ἀγωνοφῠλᾰκία, ἡ, perh. *office of warden of the games*, PRyl. 90.2, al. (iii AD, unless to be read as ἁλωνο-).
*ἀγώρις, v. ‡ἀώριος.
*ἄγωρος, v. ‡ἄωρος (A).
*αδα· ἡδονή Hsch.; cf. EM 16.45, Arc. 105.
*ϝᾱδᾱ́, ἁ, Cret., *decree, decision*, SEG 37.352.A6 (Lyttos, c.500 BC); cf. ‡ἅδος (A), ‡ἅδημα, °ἅδισμα, ἅδιξις.
Ἅδαδος, ὁ, the Syrian divinity Hadad, Inscr.Délos 2261, also Ἄδατος ib. 2258, (both ii/i BC); Adadus, Plin.HN 37.186; identified w. Zeus, SEG 31.731 (Delos, ii/i BC), ib. 28.1336 (Syria, ii/iii AD) (in some texts editors print Ἁ-).
ἀδαίδαλτος, for '*not carved*' read '*not ornamented*'
ἀδαίετος, add 'ZPE 1.185.11 (oracle of Apollo Clarios, ii AD)'
*ἀδαιτῄ, adv., perh. *without distinction*, Inscr.Cret. 4.51.13 (v BC).
ἀδαίτρευτος, for pres. def substitute '*with no meat carved*'
ἀδάκρῠτος I, line 4, after 'ib. 106 (lyr.)' insert 'cf. Isoc. 19.27, Herodes Pol. 16 A.' II 1, add 'AP 7.545 (Hegesipp.)'
*ἀδᾰμαντίς, ίδος, ἡ, a fabulous plant said to be found in Armenia and Cappadocia, Ps.-Democr.ap.Plin.HN 24.162.
ἀδᾰμαντόδετος, delete '(A.Pr.) 426' (v. °ἀκαμαντ-)
ἀδάμαστος, add 'adv. -τως Procl. in Ti. 2.314 D.'
ἀδάμᾱτος, last line, for 'restored by Elmsl.' read 'in all cases restored by edd.'
*ἀδᾰμής, ές, *invincible*, θεοί .. ἀδαμεῖς dub. in Swoboda Denkmäler 16 (Misthia in Pisidia).
*ἀδάμιος, v. ἀζήμιος.
*ἄδαμμα, *priest in cult of Agdistis*, Pouilloux Rhamnonte 140 no. 24.4 (prob. Phrygian, i BC).
*ἀδαξησμός, v. °ὀδαξησμός.
ἄδαστος, add 'SEG 39.1001.4 (Camarina, ii/i BC)'
ἄδδαυον, for 'i.e. ἄζαυον' read 'ἀδδανόν La.'; at end insert '(cf. ἀζαίνω)'
*ᾱ̓δε, adv., *in this manner, as follows*, Schwyzer 322 (Delphi, late v BC); ᾱ̓δ' ἔϝαδε Bile Dial.crét. no. 2 (Dreros, vii/vi BC); cf. ᾱ̓ δέ κα· ὅταν Κρῆτες Hsch. (La. p. 39).
†ἀδεαλτόω, app. *deface, erase*, αἰ δέ τιρ ἀδεαλτώhαιε τὰ(ν) στάλαν Schwyzer 424 (Olympia, iv BC); cf. ἀδηλόω, δεϝαλῶσαι s.v. °δηλόω.
ἀδεής (A), after 'Ep. ἀδειής, -ές' insert 'ἀδιής IG 12(3).552 (Thera); adv., ἀδειῶς SEG 9.3.36 (Cyrene, iv BC)'
ἀδετος II, after 'Ptol. Tetr. 159' insert '(cj.)'
Ἀδειγάνες, add 'v.l. πελιγᾶνες'
†ἄδεικτος, ον, *not indicated*, Ph. 1.197. 2 *invisible*, id. 1.618, al.
*ἀδελϊφήρ· ἀδελφός, Λάκωνες Hsch.
*ἀδελπιός, v. °ἀδελφός.
*ἀδελφεός, v. °ἀδελφός.
ἀδελφή, line 3, after 'Ep. -ειή' insert 'Call.fr. 43.57 Pf.'; line 4, after 'OC 535' insert 'ἀδευπιαί (nom. pl.) Inscr.Cret. 4.72 v 18; ἀδελφιά GVI 1275.9 (Thessalonica, ii AD)' add '4 as cult epith., Νύμφηι Ἀδελφῆι Kafizin 35, al.'
*ἀδελφιδεύς, έως, ὁ, *nephew*, Inscr.Délos 1993 (i BC), PGrenf. 1.47.6 (ii AD).
ἀδελφιδός, delete '*beloved one*'
ἀδέλφιον, add 'SEG 36.1182 (Galatia, v/vi AD), SEG 34.658 (Maced., iii AD, written -ιν)'
*ἀδελφιός, v. °ἀδελφός.
ἀδελφίς, add 'II = ἀδελφή, MAMA 3.598 (Corycus).'
*ἀδελφοθετία, ἁ, app. *festival of brotherhood* or *rite of fraternal adoption*, οἱ πολῖται ἑορταζόντω παρ' ἀλλάλοις κατὰ τὰς ἀδελφοθετίας SEG 30.1119 (Entella, Sicily, iii BC).
ἀδελφόπαις, add 'J.AJ 4.6.12'
ἀδελφός I, line 2, after '-εόο' insert '-ειωοῖο Call. in Suppl.Hell. 260A.5'; for 'Cret. ἀδελϝιός .. 18.319' read 'nom. pl., -ειοί A.R. 3.731, Locr. ἀδελφεός SEG 23.358 (vi/v BC), Boeot. ἀδελφιός SEG 22.416 (Thebes, iv/iii BC); ἀδελπιός Inscr.Cret. 4.72.2.21 (Gortyn, v BC); ἀδευφιός ib. 4.208.A2 (ii AD)' II 2, at end for 'cf. Pl.Smp. 210b' read 'Lys. 2.64, Pl.Smp. 210b, Aen.Tact. 24.14.'
*ἀδελφοσύνη, ἡ, *fraternal* or *friendly sentiment*, MAMA 5.91 (Dorylaeum); as term of address, *fraternal kindness*, CPR 5.23.5, 10 (?v AD), PMilan. 87.5 (vi AD).
ἀδελφότης III, add 'ἀσπάζομαι τὴν ὑμετέραν γνησίαν ἀδελφ(ότητα) καὶ τὰ παιδία αὐτῆς PVindob.Worp 14.1 (vi/vii AD)'
*ἀδεξίαστος, ον, (δεξιάζω) *not to be trusted in an engagement*, Ptol.Tetr. 166.
*ἀδεξιοβολία, ἡ, *unskilful casting*, Ps.-Callisth. 24.30.
ἀδέσποτος I, add 'prob. in Alc. 207.3 L.-P.'
*ἀδευπιά, v. °ἀδελφή.
*ἀδευφιός, v. °ἀδελφός.
*ἀδεφένδευτος, ον, = Lat. *indefensus, undefended*, Just.Nov. 155.1.
ἀδέψητος, for '(δεψέω)' read '(δέψω)'

*ἀδή, v. °Γαδά.
*ἀδηΐζω, app. *harbour, protect* (from the law), *SEG* 9.3.39 (Cyrene, iv BC); cf. ἀδεής (A).
*ἀδηληγάτευτος, ον, of taxes (app.) *not forming part of the delegatio* or *fixed rate*, *POxy.* 3424.5; cf. δηληγατεύω.
ἄδηλος, line 3, for 'hyaloid membrane' read 'capsule of the lens'
ἀδηλόω, add 'cf. perh. °ἀδεαλτόω'
ἄδημα, before 'Hsch.' insert '*decree*, *SEG* 28.1540 (Cyrene, i BC)'; at end add 'cf. °Γαδά'
ἀδημονέω, line 2, after 'D. 19.197' insert '*Ev.Matt.* 26.37, *Ev.Marc.* 14.33'; line 6, delete 'which is found only .. derivation'
†ἀδήμων, ον, gen. ονος, *distressed*, Gal. 17(1).177, Eust. 833.15; v.l. Hp.*Epid.* 1.18.
*ἀδής, v. ἀηδής.
Ἅιδης or ἅδης, line 1, after 'Dor. Ἀΐδας, α'' insert 'Ἁΐδα (gen.) *SEG* 34.994 (Apulia, late iv BC); Thess. Ἁΐδ[α]ο (gen.) *CEG* 120 (Demetrias), ΑΓίδαν (acc.) ib. 121 (Larissa, both *c.*450 BC)'; line 2, after 'anap.' insert 'voc. Ἄιδη, h.Cer. 347, Ἀΐδα epigr. in *Inscr.Cret.* 3.iv 39*B*11 (ii BC)'; line 3, after 'infr.' insert 'acc. Ἀΐδα (–∪∪) Arat. 299 M., Nic.*Th.* 181'; at end after 'Semon. 7.117' insert 'E.*fr.* 936', after 'prob. in' read 'E.*HF* 116 (lyr.)' and after 'S.*OC* 1689' insert '(lyr.); ἀίδεω *AP* 6.219.24 (Antip.Sid.), 7.711.8 (id.)'
ἀδηφαγέω, for 'of horses' read 'in part., of horses, = °ἀδηφάγος 1b'
ἀδηφάγος, lines 3/5, for 'ἀ. λύχνος' to end of article read '**b** t.t. for a category of horses at the games, Pherecr. 212 K.-A., Ar.*fr.* 758 K.-A. (both fr. Phot.), Theopomp.Hist. 250 J., *IG* 2².2311.55 (Athens, iv BC); cf. °ἀδηφαγέω; superl., Ach.Tat. 4.3.2. **2** fig., *eating up fuel, money*, etc., ἀ. λύχνος Alc.Com. 21; τριήρεις Lys.*fr.* 39, cf. Philist. 68 J.'
Ἀδιαβηνικός, ὁ, Lat. *Adiabenicus*, *of Adiabene* in Syria, an imperial title, *TAM* 4(1).27; *ICilicie* 78 (both iii AD), al.
*ἀδιάβροχος, ον, *waterproof*, D.P. 2.1.
*ἀδιάβρωτος, ον, *not eaten through*, Paul.Aeg. 6.85.1 (127.7 H.).
ἀδιάγνωστος, add 'hard to read, Ptol.*Tetr.* 47'
*ἀδιάγραφος, ον, perh. *without deduction*, δηνάρια *PDura* 29.8 (iii AD).
ἀδιάδοχος, add 'of Christ, *SEG* 32.1588 (Egypt, vi/vii AD)'
ἀδιάθετος, after 'Sch.Il. 22.487' insert '**b** perh. *not having one's schedule arranged*, Plu.*Cat.Ma.* 9.' **2**, delete 'Plu.*Cat.Ma.* 9' and after '(ii AD), al.' insert 'ἐξ ἀδιαθέτου (= Lat. *ab intestato*) Just.*Const. Δέδωκεν* 7, *Nov.* 1.1.1, *Cod.Just.* 1.5.15'
ἀδιαίρετος **I 1**, add 'ἐξ ἀδιαιρέτου equiv. of Lat. *pro indiviso*, *ISmyrna* 897 (ii/iii AD)'
*ἀδιάκλαστος, ον, *unbroken*, Poliorc. 257.9.
ἀδιακόνητος, for 'ἐπιστολήν' read 'ἐντολήν'
ἀδιάλειπτος, after 'adv.' insert '-τον Sch.S.*OT* 198 P.'; at end add 'λε[ι]τουργήσαντι ἀδιαλείπτως Mitchell *N.Galatia* 178 (Rom.imp.)'
ἀδιαλύτως **I**, after 'Hierocl. p. 17.23 A.' insert 'Procl.*Inst.* 48 D., al.; adv. -τως id. *in Ti.* 1.397.1'
*ἀδιαμέλητος, gloss on ἀφραδής, Hsch.
ἀδιαπάτητος, delete the entry (v. °ἀδιπάτητος).
*ἀδιάπλαστος, add 'Gal. 4.662.16, 10.987.5, Athenag.*Res.* 17.2'
ἀδιάπνευστος **I**, add 'οἶνος νέος ἀ. Sm.*Jb.* 32.19'
ἀδιάπταιστος, add 'adv. -τως Procl. *in Ti.* 1.193.19'
ἀδιάπτωτος, line 2, after '*PRyl.* 77.46 (ii AD)' insert 'Iamb.*Protr.* 21 κ' (118.24 P.); *unfailing, without deficiency*, χρήματα *SEG* 36.788 (Samothrace, iii BC)'
ἀδιαρίπιστον, delete the entry.
*ἀδιαρρίπιστον, gloss on ἄκροτον, Hsch.
*ἀδιασάφητος, ον, *not made at all clear, unexplained*, Sch.Pi. (*Eustathii prooemium*) ed. Drachmann vol. 3 p. 286.3.
*ἀδιάσηπτος, ον, *not penetrated by putrefaction*, Paul.Aeg. 6.85.1 (127.7 H).
ἀδιάστολος, add '**II** *undistributed*, *UPZ* 180a 15.9 (ii BC), al.'
ἀδιάστροφος **III**, adv., line 2, before 'ἀφίεσθαι' insert '*without evasiveness*' and add '(*Cod.Just.*) 10.27.2.13'
ἀδιάσωστος, after 'Ptol.*Tetr.* 47' insert '(v.l.)'
ἀδιάτμητος, for '*not cut in pieces*' read '*not to be cut*'
ἀδίαυλος, add 'of the journey there, τὴν ἀ. ὁδόν *IG* 9(2).648.1'
ἀδιάφορος, add '**VII** *free of interest*, ἀρτάβαι *POxy.* 2351.24 (ii AD).'
ἀδιαχώριστος, after '*undistinguished*' insert 'Arist.*EE* 1219ᵇ34' and after 'Suid.' insert 'w. gen., *inseparable from*, *SEG* 8.574 (Fayûm, prob. iii AD)'
*ἀδιαψήφητος, ον, *without a vote*, i.e. by acclamation, ἀ. νε[ικήσας] *Delph.* 3(1).550.5.
*ἀδιγόρ· τρωξαλλίς, ὑπὸ Σκυθῶν Hsch. (La.).
*ἀδιεγγύ[ητος], *not having security*, *IMylasa* 830.2 (or ?ἀδιεγγύ[ος]).
ἀδιεξίτητος, after 'Ph. 1.554' insert 'τὸ -ον Procl.*Inst.* 94 D.'
ἀδιήγητος, after 'Cic.*Att.* 13.9.1' insert 'ἀδιήγηθ' ὅσα *indescribably many*, Men.*Dysc.* 405'
*ἀδιής, v. ‡ἀδεής (A).

ἀδικέω **II 2**, after 'X.*Eq.* 6.3' insert '*Apoc.* 7.2, 9.4, etc.; ἀδικεῖν τὸ ὕδωρ *damage* the water supply, *IEphes.* 3217A.11 (ii AD); ὃς δὲ ταύτην [σ]τήλην ἀδικήσῃ· Mitchell *N.Galatia* 294'; at end add 'transf., τὸ μέτρον ἠδίκητο *violence was done to* the metre, D.H.*Comp.* 9'
*ἀδίκησις, εως, ἡ, *wrongdoing*, *PMasp.* 6.2 (vi AD).
ἀδικητικός, after 'Plu. 2.562d ' insert 'Ar.Byz.*Epit.* 2.144'
ἀδικία, add '**III** cessation of judicial and all other public business in the event of war, etc., Arist.*Oec.* 1348ᵇ10.'
*ἀδιοβάντης, ου, ὁ, perh. fr. Lat. *adiuvans*, an official, *PAnt.* 96.12.
*ἀδίοπτος, ον, *opaque*, Alex.Aphr.*de An.* 148.4.
*ἀδιούτωρ, ορος, ὁ, Lat. *adiutor*, *assistant*, *PLond.* 1711.89 (vi AD), al.
*ἀδιπάτητος, ον, perh. *not thoroughly trodden* or *crushed*, πυρός *POxy.* 1259.15, 2125.20 (both iii AD), *PSI* 1053.8 (ii/iii AD).
*ἄδισμα καὶ ἄδμα· ψήφισμα, καὶ δόγμα Hsch.
ἀδίστακτος, reverse the order of the first two refs.; before '*PTeb.*' insert 'ἀ. πάσης αἰτίας *not disputed* on any ground'
*ἀδιψάθεος, ὁ, (or -ον, τό), a shrub, app. = ἀσπάλαθος 1, Plin.*HN* 24.112 (v.l.).
*ἀδιψία, ἡ, *absence of thirst*, Poliorc. 203.4.
*ἄδμα, v. °ἄδισμα.
ἄδμητος, line 1, for 'poet. for ἀδάματος' read 'Dor. ἄδμᾱτος, B. 11.84 S.-M.' add '**3** *unconquered*, *AP* 7.723.'
*ἀδμινιστρατίων, ονος, ἡ, Lat. *administratio*, *official rank* or *function*, ἀπὸ κέρτας ἀ. *Cod.Just.* 6.48.1.10.
*ἀδμισσάριος, ὁ, Lat. *admissarius*, ταῦρος ἀ. bull *for breeding*, *Edict.Diocl.* 32.4.
ἀδμολίη, for 'Call.*Fr.* 338 .. Suid.' read 'Call.*fr.* 717 Pf. (v.l. -μωλίη, -μωλή)'
ἀδμωλή, after 'ἄγνοια' insert '*PAnt.* 60 (ii/iii AD), *CGFP* 204'
*ἀδμωλίη, v.l. for ‡ἀδμολίη, ‡ἀδμωλή.
*ἀδνοτατίων, ονος, ἡ, Lat. *adnotatio*, *imperial decision on a petition*, *SB* 9763.34 (v AD), *PMonac.* 14.85 (sp. ἀδνου-, vi AD).
ἀδοκίμαστος, add '**2** *unexamined, uninvestigated*, Just.*Nov.* 6.1.10.'
ἀδόκιμος **I**, after 'Pl.*Lg.* 742a' insert 'ἀ. ἀργύριον silver *that has failed the assayer's test*, *Lxx Pr.* 25.4, τὸ ἀργύριον ὑμῶν ἀ. *your money is not genuine*, *Is.* 1.22'
ἀδολεσχέω **II**, for '*Lxx Ps.*68(69).12' read '*Lxx Ps.*68(69).13'
ἀδολεσχία, for '*prating, garrulity*' read '*talk, chatter*' **II** and **III**, delete the sections transferring quots. to section **I**.
ἀδόνητος, add 'ἀστεμφεῖς καὶ ἀδόνητοι Agath. 1.21.8 K.'
ἄδοξος, line 2, after 'X.*Smp.* 4.56' insert '*Trag.adesp.* 423 K.-S.'
*ἀδόρωτος, ον, *not plastered*, prob. in *PMich.*v 253.5 (i AD, ἀδωρ- pap.).
ἄδος (B), after lemma insert '(or ἄδος)' and after '*decree*' insert '*sentence*'; for 'ἀδεῖν' read 'ἀδεῖν' and add 'perh. Γαδόν acc. *SEG* 37.743.12-13 (Axos, iv BC); cf. °Γαδά'
ἄδος, for 'ἡδος' read 'ἦδος'
ἀδούλωτος, after 'D.S. 1.53' insert '*Inscr.Perg.* 413 (i BC)'
*ἀδουσιάζομαι, *accept membership of*, φυλῆς καὶ δήμου καὶ φρατρίας ἀδουσιάσασθαι *IG* 2².553.15 (iv BC). **b** ἀδουσιασάμενοι· διελόμενοι, ὁμολογησάμενοι Hsch. (ἀδ- La.).
ἀδουσιασάμενοι, delete the entry.
*ἀδούσιον, *that is in agreement*, *BCH* 59.96 (Delphi, ii BC; unless read as pers. n. Ἀδούσιο[ς], v. *BE* 1953.93, ἀ.· ἐραστόν, σύμφωνον Hsch.
ἀδράνεια, before 'Hdn. 2.10.8' insert '*Demetr.Eloc.* 285 (-ία)'
ἀδρανής, line 5, after '(Simp. *in Ph.*) 815.24' insert 'Procl.*Inst.* 7 D., τὸ μὲν ἀδρανὲς ὂν καθ' αὑτό, τὸ δὲ ἀπαθές ib. 80 D.'
Ἀδράστεια **1**, at end add 'or a goddess linked with but distinguished from Nemesis, Men.*fr.* 266 K.-Th., Nicostr.Com. 35 K.-A., *SEG* 33.645 (Rhodes, i BC); identified w. Fate, ἀνάγκη, etc., Chrysipp.*Stoic.* iii 292.15, cf. Orph.*fr.* 54'
*ἀδραστοδάραν, title of Persian official, Procop.*Pers.* 1.6.18.
*ἀδρεσπόνσουμ, Lat. *ad responsum*, title of official on the staff of a praetor, Just.*Edict.* 13.2; also -σον, Athanasius Scholasticus *coll.* 4.4 (Heimbach p. 54).
Ἀδρεύς· δαίμων τις τῶν περὶ τὴν Δήμητραν, *EM* 18.36, cf. perh. cult-epith. in fragmentary dedication, *SEG* 31.745 (Naxos, Hellen.).
*Ἀδριάνειος, ον, *of Hadrian*, of a contest in his games, *SEG* 12.512 (Cilicia, ii/iii AD); Ἀδριάνεια, τά, festival at Athens named after Hadrian, *IG* 2².2050.18, 2067.17 (both ii AD), *TAM* 4(1).34; at Ephesus, *IEphes.* 618, 724, etc.
*Ἀδριανιών, ῶνος, ὁ, name of a month at Athens, *IG* 2².2050.14, 2067.19 (both ii AD).
*Ἀδριανονίκης, ου, ὁ, *victor in the Hadrianic Games*, *POxy.Hels.* 25.31 (AD 264).
*Ἀδριανός, ή, όν, *of or connected with Hadrian*, τῆς ἱερᾶς Ἀ. .. συνόδου *IG* 2².1350; of cities, Nicomedeia *TAM* 4(1).25, Tarsos *ICilicie* 30; Ἀ. γερουσία *IHadr.* 40; as a title of Zeus, *SEG* 33.1090 (Bithynia, i AD). **2** Ἀ. ὁ, sc. μήν, an Egyptian month = Choiak, *SEG* 31.1532 (Philae), *SB* 282, *BGU* 1616 (all ii AD); at Aphrodisias = Λῷος *ABSA* 59.29 no. 46.6. **3** neut. pl. subst. Ἀ., τά, *Hadrianic games*,

ἀδρομελής · SUPPLEMENT · ἀερόμικτος

in Cilicia, Moretti *IAG* no. 87.23, at Smyrna, *IG* 2².3169.26 (both iii AD).

*ἀδρομελής, ές, *strong-limbed*, cj. in Emp. B 67.2 D.-K. (comp.).

ἀδρός I, line 4, after 'id. 20.85' insert 'of a sieve, *coarse, large-meshed* (comp.), *Gp.* 3.7.1; of measure, *full*, μέτρον τετραχοινικὸν ἁ. *SB* 10532.27 (AD 87/8)' II 1, add '**b** *grand*, Ael.*NA* 10.50 (comp.), Hsch.' and transfer to this 'οἱ ἁ. *chiefs* .. *Ki.* 10.6' adding '*Je.* 5.5, *Jb.* 29.9, etc.' 2, for '*fine, fat*' read '*well-grown, sturdy*' 3 c, for '*ready to be laid*' read '*fully developed*'

ἀδρόσφαιρος, add 'Plin.*HN* 12.44'

*ἀδρότης, v. ‡ἀνδρότης.

ἀδροτής, after lemma insert '(or ?-ότης)'

ἀδρόχωρον, delete '*full, χῶρον*'

ἀδρυάς, add 'Nonn.*D.* 2.92, 22.14, al.'

ἀδρύνω, line 2, after 'Thphr.*HP* 3.1.3' insert 'of a young girl, εἰς γάμον *AP* 6.281 (cj. Leon.Tarent.)' and at end add 'of children, Lxx *Jd.* 11.2, 4*Ki.* 18, al.; *solidify*, of ice, Ael.*NA* 14.26'

ἀδρυφής, ές, *unscratched, undamaged*, *Iamb.adesp.* 38.14 W.

ἀδυναμία, line 2, after 'Hp.*VM* 10' insert 'Lys. 31.19, Pl.*Lg.* 646b'

ἀδύναμος, after '*weak*' insert *without strength*, of a woman, *SEG* 35.227.28, 30 (tab.defix., Athens, c.AD 250)'

ἀδυνασία, for '= ἀδυναμία' read '*physical weakness*'; after 'Hdt. 3.79' insert '*military weakness*'; after 'gen.' insert '*incapacity* for'

*ἀδύπotos, v. ‡ἠδύποτος.

Ἀδυτηνή, ἡ, name of a goddess, θεὰ Βρυζη ἀγνὴ Ἀδυτηνή *TAM* 5(1).533 (Maeonia, iii AD), al.

ἄδυτος, add '[ῠ *SEG* 8.551.3 (i BC)]'

Ἀδωνάρια, for 'kind of .. *gifts*' read '*lines of verse in a particular kind of metre*, ἁ. πέμψας ἄρρυθμα; cf. Ἀδώνιος II 2'

Ἀδώνια, after 'Pherecr. 170' insert 'Diph. 42.39 K.-A., *IG* 2².1261.9 (Peiraeus, 302/1 BC), etc.'

Ἀδωνιάζω, add 'Ἀδωνιάζοντες *SIG* 1113.2 (Rhodes, ii/i BC)'

*Ἀδωνιασταί, οἱ, *guild of worshippers of Adonis*, *SEG* 4.168 (Caria, prob. i AD).

Ἀδώνιος II, add '4 name of a month *SEG* 30.1119 (Sicily, iii BC).'

Ἀδωνις I, line 2, for 'ὤ' read 'ὦ' 2, for 'cuttings planted' read 'corn, etc., grown'

*ἀδωσιδικέω, *fail to pay one's due*, *PMich.*v 243.2, al. (i AD), *POxy.* 2351.58 (ii AD).

ἀδωσιδικία, add '*BGU* 1212.16 (i BC)'

ἀδώτης, line 2, for 'ἀ' read 'ἀ-'

†ἀέθλευμα, v. °ἄθλευμα.

†ἀεθλοθέτης, v. ‡ἀθλοθέτης.

†ἀεθλοσύνη, v. °ἀθλοσύνη.

ἀεί I 5, after 'ἀείδασμος' insert '(ἀΐδασμος)' and add 'Arc. αἴ Schwyzer 687.28 (Tegean decr. at Delphi, 324 BC)' 8, after 'Boeot. ἠί' insert 'v.s.v., also °ἠίν' add '10 ἀή *APAW* 1952(1).11 (Cos, in Lacon. dial., iii BC). 11 ἀείσε Steph. *in Hp.* 1.129.' at end for note in parenthesis read '(αἰϝεί, *CEG* 1, 344 (Phocis, vi BC); Cypr. *a-i-we-i ICS* 217.31; cf. καταιϝεί (°καταεί); see also °υϝαίς)'

ἀειβλύων, delete the entry.

*ἀείβουλος, ον, *that is a perpetual counsellor*, θεόκτιστον πρ(εσβύτερον) ὄντα, πάντων μύστην καὶ ἀείβουλον *Inscr.Cret.* 2.xxiv 7.5.

ἀειγενής 2, for '*everlasting*' read '*born again and again*'

ἀειγένητος II, for '= ἀειγενής' read '*eternal*'

ἀείγνητος, for 'ον, = ἀειγενέτης' read 'η, ον, = ἀειγένητος'

ἀείδελος I, for '*dark* .. etc.' read '*obscure*, Hes.*fr.* 67b M.-W., Opp.*H.* 1.86, *C.* 3.489, ἁ. ὠπήσασθαι ib. 3.160'

†ἀείδιος, v. °ἀΐδιος.'

ἀείδω, line 8, delete 'Ar.*Lys.* 1243 (Lacon.)' and after 'Poets, as' insert 'Theoc. 22.135, Call.*Ap.* 30, *Dian.* 186'; line 24, after 'Pl.*Tht.* 164c' insert 'Ar.*fr.* 101 K.-A., *Av.* 41; = μάτην λέγειν, Phot. α 551 Th.'; at end add 'opp. φράζειν Str. 1.2.6' II 1, add 'τὸ ἀδόμενον τοῦτο *as the saying is*, Ael.*NA* 5.11' 2, after 'B. 6.6' for 'etc.' read 'Arr.*An.* 1.11.2, 4.9.5, Philostr. *VA* 5.8'

ἀείζωος, line 3, after 'ib. 741' insert 'Pl.*Ep.* 8.356a'

ἀεικέλιος, after 'Il. 14.84' insert 'A.R. 2.1126, Nic.*Th.* 271'

ἀεικής 1, line 5, for '*meagre*' read '*unfitting*'; line 6, after '(Il.) 24.594' insert 'of persons, Od. 9.515 (v.l.), *h.Cer.* 83, 363, Call.*fr.* 344 Pf., A.R. 4.91' 2, add 'Critias 7.11 S.'

ἀεικίνητος, after 'adv. -τως' insert '(prob. f.l. for ἀκινήτως)'

*ἀείκλαυστος, ον, *to be for ever wept for, mourned*, *IG* 12(2).489.19 (Mytilene, i/ii AD). 2 *filled with perpetual lamentation*, μέλαθρα hymn ap.Hippol.*Haer.* 4.32 (= *GDRK* 53.4).

ἀείκλαυτος, delete 'μέλαθρα .. 4.36' and substitute 'ἀεικλαύτῳ παρὰ τύμβῳ *IKyme* p. 251 (*SEG* 29.1218, ii/i BC)'

*ἀεικοίματος, ον, *sleeping everlastingly*, *Lyr.adesp.* 119.8 P.

ἀείμνημα, τό, gloss on σκέλισμα, Hsch.

ἀείμνηστος, add '*TAM* 4(1).138 (Bithynia, Rom.imp.), *SEG* 33.1106 (Paphlagonia, iii AD)'

ἀειναῦται, for 'at Chalcis' read 'or perh. association of sailors at Eretria, *SEG* 34.898 (Eretria, c.510/500 BC)'

ἀειπάρθενος I, for 'Sapph. .. ἀϊπ-)' read '?Alc. 304.5 L.-P. (ἀϊ[∪∪]-παρθενος pap.)'; add 'of the Virgin Mary, *IEphes.* 4135.19; in epitaphs, of unmarried women, *BCH* suppl. 8 no. 23 (Edessa, v/vi AD), al.'

*ἀείπολος, ον, *ever-moving*, *Orac.Chald.* 61(f).2 P.

*ἀειπρέπεια, ἡ, *everlasting dignity*, *Corp.Herm.* 18.4.

ἀείρω, line 14, after 'E.*Hel.* 1597' insert '*Rh.* 54 (αἰρ- codd. in both places), Pl.*Lg.* 969a'; delete 'imper.' and after 'A.R. 4.746' insert '(v.l.)' I 1, penultimate line, after 'Hp.*Aër.* 6' insert 'but also of the *rising* of heavenly bodies, E.*Alc.* 450 (lyr.), Arat. 326, 405, 558 and so perh. in Alcm. 1.63 P.; intr., of heavenly bodies, *rise*, ἕως ἂν αὐτὸς ἥλιος ταύτῃ μὲν αἴρῃ S.*Ph.* 1331' 2, add '*pick up* lots, *SEG* 30.1119.22 (found at Entella, iii BC)' 7, add '*SEG* 30.1231 (med., Lugdunum, iii AD)' III, add '3 αἴ. ἀνά τι *divide* by, *PRyl.* 27 i 1, al.; αἴ. ἀπό τινος *subtract* from, ib. 8.' IV 1, after 'B. 2.5' insert 'Pl.*Lg.* 969a' 3, delete 'βοῦς .. 27.5' and add '**b** τοὺς βοῦς αἴ. *perform* the bull-*hoisting*, *IG* 1³.82.29 (v BC), *IG* 2².1028.10, 13, 28 (i BC), cf. ἄρσις I 1a; ταῦρον ἀράμενοι Paus. 8.19.2; cf. Thphr.*Char.* 27.5.' 4, add 'μὲ ἐναι πόλεμον ἄρασθαι *IG* 1³.105.35' V 2, add '**b** of rivers, X.*HG* 5.2.5, *AP* 9.568 (Diosc.).'

*ἀεισέβαστος, ὁ, = Lat. *semper Augustus*, *Ephes.* 4(1) no. 33, *SEG* 9.356 (Cyrenaica), *BCH* suppl. 8 no. 81 (Thessalonica), (all vi AD).

*ἀείσειστος, ον, *being always in a confused state of movement*, (opp. ἀεικίνητος) (cj. for ἄσειστος q.v.) D.L. 8.26 (cf. Thesleff *Pythagorean Texts* p. 235.1).

ἀεισιτία, add 'pl., ἀιαειτιῶν *SEG* 30.82 (Athens, c.AD 230)'

ἀείσιτος, for '(in form ἀϊσ-)' read '(in form ἀεισ- or ἀϊσ-)' and add '*SEG* 28.94'

ἀεκαζόμενος, after 'πολλ' ἀ.' insert 'Il. 6.458'

ἀεκητί, add 'Dor. -ατι B. 18.9 S.-M.'

ἀεκούσιος, line 3, after 'Democr. 240' insert 'ἡακόσι[α] *IG* 1³.6B.5 (v BC)'

*ἀελιδρόμος, ὁ, perh. *portico*, *BCH* 84.852 (Lissos, Crete).

ἄελλα, line 1, for 'Aeol. .. Hsch.)' read 'perh. Aeol. ἄυελλα Lesb. *fr. adesp.* 7 L.-P. (w. υ for ϝ; fr. Hsch., where cod. αυεουλλαι pl.), cf. °Ἀελλώ'; line 3, after '(Od. 5.) 304' insert '*Hes.Th.* 874'; line 4, omit 'cf.' and transfer '13.334' to follow 'Il. 13.795' (line 3).

ἀελλοπόδης, add 'of a wild ass, Opp.*C.* 3.184, θυγατέρες Λυκάβαντος ἀελλοπόδοιο τοκῆος Nonn.*D.* 11.486'

ἀελλόπος, add 'ἀελλοπόδεσσ' ἐλάφοισι Opp.*C.* 1.191'

Ἀελλώ, add 'perh. Αϝε[λλώ] on a vase, *SEG* 37.330 (Laconia, vi BC)'

ἀελπής, add 'neut. pl. as adv., Nic.*Al.* 125'

ἄελπτος, for '*unhoped*' read '*unlooked*'; line 4, delete '987 (prob. l.)' II, delete this section transferring 'h.*Ap.* 91' to section I (see also °ἄεπτος) III, for '*hope*' read '*expectation*'; after 'S.*El.* 1263' insert 'in bad sense, A.*Supp.* 987'

†ἄεμμα, τό, *bow*, Call.*Dian.* 10, *Ap.* 33, Philet. in *Suppl.Hell.* 673.4; cf. ἅμμα.

ἀέναος, for 'ἀείνως' read 'ἀείνως'

*ἀένιος, α, ον, *everlasting*, hymn in *Suppl.Hell.* 990.8 (perh. for ἀέναος).

*ἀεξίκακος, ον, *nourishing evil*, Nonn. *D.* 20.84.

ἀέξω, line 1, for 'twice in Trag.(lyr.)' read 'rare in trag.'; line 3, aor. add 'med., [ἢ]έξαντο Call.in *Suppl.Hell.* 287.9'; pass., add 'ἀεξήθησαν Nonn.*D.* 45.150' I 1, before '*increase*' insert '*cause to grow* or *flourish*, υἱὸν ἀ. Od. 13.360; of plants, Od. 9.111'; after 'Od. 17.489' delete 'υἱὸν .. 360'; after '(Od.) 15.372' insert 'δόμων ἀ. κάλλος A.*fr.* 451n.5 R.'

*ἄεπος, ον, *not uttering a word*, (dub. l.) Baillet *Inscr.des tombeaux des rois* 1402.

†ἄεπτος, ον, (ἔπω) *not to be dealt with, irresistible, fierce*, A.*Supp.* 908 (cj. ἄελπτ', *Ag.* 141 (v.l. ἀέλπτοις), *fr.* 213 R. (or ἄαπτ-).

ἀερικόν, for '*tax on lights*' read '*name of a tax*'

ἀέρινος 3, for this section read 'a precious stone, Ps.-Callisth. 1.4.13 (prob. f.l. for ἀερίτης)'

ἀέριος II, delete 'πῦρ Hp.*Vict.* 1.10' (v. ἠέριος); at end add '**b** epith. of Zeus, *TAM* 5(1).616 (Lydia, AD 223/4 or 277/8); perh. also θεός *Expl.arch. de Délos* xi p. 276.'

ἀέριπον, delete the entry.

ἀερτής, add 'Anon.Alch. 360.13'

ἄερκτος, add 'Hsch. (cod. ἀεριπ-)'

*ἀεροκέλαδοι· πιτυοκάμπται Hsch.

*ἀερόλευκος, ον, app. *dull white*, Anon.Alch. 387.14 (perh. f.l. for ἀκρό-).

*ἀερολόγος, ον, adj. without context in *PBrux.*inv. E7162.15 (*PP* 38.13; see also *Emerita* 56.236).

*ἀερομαντεῖον, τό, *air-oracle*, *PMag.* 3.278.

*ἀερόμικτος, ον, *mingled with the air*, φώναισί τε ἀερομίκτοις Orph.*fr.* 297; cf. ἀερομιγής.

8

*ἀερόμορφος, ον, *air-formed*, Orph.H. 14.11, 16.1, 91.6 codd.; cf. ἠερόμορφος.

*ἀεροπετέω, *cause to fly through the air*, ἀεροπέτησον τὴν ψυχὴν καὶ τὴν καρδίαν Λεοντίας Suppl.Mag. 43; pass., PMag. 36.111 (iv AD).

*ἀεροποτέομαι, *fly through the air*, Suppl.Mag. 38 (ii AD); cf. °ἀεροπετέω and ποτάομαι, ποτέομαι.

ἀερορίφης, delete the entry (read -φερής).

*ἀεροφερής, ές, *borne in the air*, PMag. 4.2508.

*ἄερσα, v. ἔρση.

ἀερσίλοφος, add '2 *lifting the crest of a mountain*, of a giant, Nonn.D. 48.44.'

ἀερτάζω, line 1, for '(Call.)*fr.* 19' read '*fr.* 261, 597 Pf.'

*ἄέρωμα, ατος, τό, *patch* of ether *converted into air*, πρόσγεια ἀ. τοῦ αἰθέρος Placit. 2.30.6 (pl.).

*ἀέτιον, τό, perh. dim. of °ἀετίς, Inscr.Délos 408.2 (ii BC); cf. ἀετός IV.

*ἀετίς, ίδος, ἡ, *gabled stele*, Sardis 7(1).167 (iv AD).

ἀετίτης, add 'also ἀετῖτις, ιδος, ἡ, Plin.HN 37.187.   2 *kind of convolvulus*, Plin.HN 24.139.'

*ἀετόμορφος, ον, *having the form of an eagle*, Procl.in R. 2.319 K.

ἀετός I 1, for 'S.*fr.* 885' read 'S.*fr.* 884 R.'   add 'b applied to vultures, Ev.Matt. 24.28, Ev.Luc. 17.37, Apoc. 12.14.'   3, after 'Arat. 591' insert '(see also ‡ἄητος (A))'   II, delete the section.

ἀετοφόρος, add '2 *bearing an eagle*, of a coin, Inscr.Délos 1442B51 (ii BC).'

ἀεχήνες, add 'cf. ‡ἀχήν'

†ἀϝλανεως, v. ἀλανής.

*ἀζαβαρίτης, v. °ἀζαραπατεῖς.

ἀζαθός, delete the entry.

ἀζαίνω, add 'cf. °ἄδδαυον'

ἀζάλη, add '(-ής La.)'

*ἄζαλμα, v. °ἄγαλμα.

*ἀζάμιος, v. °ἀζήμιος.

*ἀζαραπατεῖς· οἱ εἰσαγγελεῖς παρὰ Πέρσαις Hsch., corr. to ἀζαβαρίτης, Ctes. 15 J.

ἄζετον, add '(πιστόν La.)'

ἄζευκτος, add 'ἄ. πόθοι *not joined in a (natural) pair*, Opp.C. 2.425'

ἀζηλοπραγμόνως, add '(s.v.l.)'

ἀζήμιος, after lemma insert 'dial. ἀζάμιος, SEG 34.849.18 (Mytilene, v BC), 25.447.12 (Alipheia, iii BC), 36.502.13 (Delphi, ii BC), also ἀσάλ[μιος] Delph. 3(1).342; ἀδάμιος Schwyzer 503a note (Coronea, c.200 BC), Inscr.Cret. 4.146.2 (iv BC), ἀττάμιος Schwyzer 424.7 (Elis, iv BC), Inscr.Cret. 4.183.15 (ii BC)'   2, line 5, before 'Adv.' insert 'τὸ ἀζήμιον = Lat. *indemnitas*, Cod.Just. 1.3.55.4, 2.2.4.2'

*ἀζονίη, ἡ, an unidentified plant, Claud.Herm.Mul. 988.

†Ἀζόσιος, ὁ, a month name at Epidaurus, IG 4².103.127, al.; written Ἀζέσιος ib. 89; as epith. of divinities, θεῶν Ἀζοσίων ib. 434; Ἀζοσία, ἡ, ib. 410.13; cf. Αὐξησία s.v. °Αὐξησία.

*ἀζυγία, ἡ, app. *bereavement*, GVI 2093.15.

ἄζυγος, add '*unyoked*, βοῦν ἄζυγον Dorieus in Suppl.Hell. 396.7'

*ἀζύμτης, ον, *unleavened*, ἐπὶ κολλυρίων ἄρτων ἀζυμιτῶν Al.Le. 7.13.

ἄζυξ, for '*unyoked*' read '*that has not been yoked*'; after 'D.H. 1.40' insert 'B. 11.105, 16.20 S.-M.'

*Ἄζυρος, ὁ, cult epith. of Apollo, EA 10.102.10.

ἄζω (B), after 'S.*fr.* 980' insert 'cf. Gal.UP 7.13' deleting 'so perh. .. Th. 99'

ἄζωστος, after '(Dreros, iii BC)' insert 'cf. πανάζωστος'

ἀηδής, after 'ές' insert 'contr. ἀδής cj. in Thgn. 296 W. (Aeol. acc. ἀνάδην Sapph. 22.5 L.-P., nom. pl. ἀνάδεες Alc. 259.11 L.-P.)'

ἀηδία I 2, after 'opp. ἡδονή' insert 'Men.Dysc. 435'

ἀηδίζω, pass., add 'SB 4323.2 (Byz.).'

ἀηδόνιος, for 'Ra. 684 .. *light*' read 'Ra. 684; *like that of a nightingale*, Nicom.Trag. 13 S.; poet., *of brief or fitful sleep*'

ἀηδών, after 'ὁ, v. infr.' insert 'contr. ἀδών q.v.'

ἄημι, line 4, after '(Od.) 14.458' insert 'ἄεν A.R. 1.605, 2.1228'

ἀήρ I, line 3, after 'Thphr.Sens. 30' insert 'Triph. 669'   II, add '3 *space open to the air on a building, flat roof, terrace on a grave-monument*, etc., Hierapolis 158, Keil-Premerstein Erster Bericht no. 120, PMonac. 8.12, PLond. 1733.19 (both vi AD).'

ἀήσσητος 2, add 'as an imperial title (= Lat. *inuictus*), SB 4284.6 (AD 207), ἀηττήτων βασιλέων POxy. 3122 (AD 322)'   at end add 'see also ‡ἀνήσσητος'

†ἀήτη, ἡ, Dor. ἀήτα Simon. 90 P., B. 17.91 S.-M., later ἀήτης, εω, ὁ, A.R. 4.1537, ον, X.Eph. 3.2.13, Hld. 1.22: masc. forms regular in Alex. and later poetry, also in some edd. of Il. 15.626, Od. 4.567, *blast* of wind, ἀνέμοιο .. δεινὸς ἀήτη (v.l. ἀήτης) Il. 15.626; pl. ἀνέμων .. ἀήτας ib. 14.254, Hes.Op. 621, cf. 645; Ζεφύροιο Od. 4.567, Νότοιο Hes.Op. 675 (fem.).   2 *wind, breeze*, ἀήται id. 20.9 L.-P., μεγάλαις ἀήταις id. 20.9 L.-P., Tim.Pers. 107, Call.Del. 318, *fr.* 110.53 Pf., Theoc. 22.9, ἀήτης μείλιχος A.R. 1.423, Q.S. 1.537, Nonn.D. 32.158, etc.; poet. wd., v. Pl.Cra. 410b.

ἄητος (A), add '[ᾱη for ᾱε w. play on ἄηται in 323]'

†ἄητος (B), ον, dub. sens., θάρσος ἄητον Il. 21.395 (= θάρσος ἄᾱτον Q.S. 1.217); cf. ἀίητος.

†ἄητος (C), ον, = ἄᾱτος, ποηφάγος αἰὲν ἀ. Nic.Th. 783; ὁ ἀκατάπαυστος Hdn.Gr. 1.220; also ἄητοι· ἀκόρεστοι, ἄπληστοι παρὰ τὴν τροφήν, and ἀήτους· μεγάλας (A.*fr.* 3 R., ap. Hsch.).

*ἀηττησία, ἡ, *invincibility*, Afric.Cest. p. 19 V.

ἀθαλασσία, delete the entry.

*ἀθαλάσσιος, ον, of wine, *not mixed with sea-water*, SEG 39.959 (Eleutherna, ii/i AD); cf. ἀθάλασσος II.

ἀθαλής, after 'Ath. 12.524b' insert '(ἀειθαλές cod.)'

*ἀθανάτῐσις, εως, ἡ, *deification*, D.C. 60.35.3 (cj.).

*ἀθάνατος I, after 'Od. 24.47' insert 'in later cult Θεοὶ ἀθάνατοι SEG 33.1141 (Phrygia), 1187 (Lycaonia), of human beings in a continued existence after death, IGBulg. 796.4; οὐδεὶς ἀθάνατος formula in epitaphs, SB 3514, SEG 33.1276 (Jerusalem, i BC)'   II, add 'of a boundary-stone, SEG 37.1036.5 (Calchedon, ii/iii AD)'   III, add '3 of imperishable, i.e. inanimate, property, in the formula θνατὰ (τνατὰ) καὶ ἀθάνατα Inscr.Cret. 1.xvi 17.12 (ii BC), 4.76B 8 (v BC), cf. SEG 9.1.§1.11 (Cyrene, iv BC); v. °θνητός.'

ἀθάρη, after '*porridge*' insert '*made of wheat*'

†ἀθαρής· ἄφθορος ἐπὶ γυναικός, ἐπὶ δὲ σιδήρου στερεός Hsch.

*ἀθαροφόρος, τό, *kind of vessel*, presum. for holding ἀθάρη, Kafizin 133(a) (225/4 BC).

*ἀθεάφθειος, ον, *unsulphurated*, of wine-jars, BGU 2205.16 (AD 590); cf. θεάφιον.

ἀθεησίη, place after 'ἀθεεί'

ἀθελγής II, after '*having no power to soothe*' insert 'or *charm*' and before 'al.' insert '42.248'

ἀθέλητος, after 'Hsch.' insert 's.v. ἀβούλητον'

ἄθελκτος, add 'Trag.adesp. 665.29 K.-S.'

ἀθεμελίωτος, after 'Hsch.' insert 's.v. ἀτερμάτιστος'

*ἀθεμίζω, Cypr. aor. inf. *a-te-mi-sa-i* ἀθέμισαι perh. *treat improperly*, ICS 311.2.

ἀθέμιστος I, delete '(the former .. Prose)'

†ἄθεος (A), *that is without God*, ἄθεος ἄφιλος, S.OT 661, Ep.Eph. 2.12; of things, *lacking divine nature*, παράδειγμα ἀ. Pl.Tht. 176e, R. 589e, Procl. in Ti. 1.368.6 D.; *having no connection with the gods*, ἀ. ὀνόματα Clearch. 86 W.   2 *godless, impious*, of persons, A.Eu. 151, S.Tr. 1036, E.Or. 925, Ba. 995; of actions, things, Pi.P. 4.162, B. 11.109 S.-M., A.Pers. 808, Eu. 540, E.Andr. 491, Ar.Th. 671, Gorg. 11a(36) D.-K.   3 *not believing in the existence of gods, atheist*, Pl.Ap. 26c; as a sobriquet, Cic.ND 1.23.63, S.E.M. 9.51; adv. -ως *in an unholy manner, dreadfully*, ἀ. ἐφθαρμένη S.OT 254, El. 1181, sup. -ώτατα, El. 124.   II *impiously, godlessly*, Antipho 1.21, 23, Pl.Grg. 481a, 523b.

ἀθερίνη, add 'rest. in Hippon. 78.11 W.'

ἀθέρμαντος, at end add 'σώματα ἀ. Gal. 7.40.4'

ἄθεσμος, add 'Trag.adesp. 146a.5 K.-S.'

ἄθεστος, add '(= Suppl.Hell. 1066, cj. ἀπόθεστος)'

ἀθεόφατος, line 4, after '(Od.) 20.211' insert 'ἀ. φῦλ' ἀνθρώπων h.Ap. 298'

*ἀθετεία, ἡ, *cancellation*, PFam.Teb. 9.15 (ii AD).

ἀθετέω I 1, add 'perh. *let fall into disuse*, ἐπὶ ἠθετημένου χώματος SB 11654 (ap 146/7)'

*ἀθετησία, ἡ, *breach of faith*, Vett.Val. in Cat.Cod.Astr. 8(1).170 (acc. pl.).

*ἀθετητικός, ή, όν, *treacherous*, PMasp. 353ᵛ.A12 (vi AD)

ἄθετος II, adv., for '= ἀθέσμως' read '*so as to set at naught*'

*Ἀθήναδιον, τό, *small image of Athena*, POxy. 1802 *fr.* 3.55 (ii/iii AD).

Ἀθῆναι I, delete 'IG 1.373¹⁰⁷' and at end add 'Th. 4.5; Σούνιον .. ἄκρον Ἀθηνέων Od. 3.278'

Ἀθήναια, τά, for 'IG 3.1147' read 'IG 2².2119.22 (ii AD)' and add 'SEG 35.218 (Athens, iii AD), Dor. Ἀθάναια Schwyzer 12.10 (Sparta, v BC)'

*Ἀθηναϊσταί, οἱ, *guild of worshippers of Athena*, at Rhodes, IG 12(1).162.

Ἀθήνη, Ἀθηναίη, after 'ISmyrna 739 (Ἀθηναείη dat., c.600 BC), SEG 37.936 (Erythrae, vi BC)'; Ἀθηναία, add 'SEG 32.1086 (Spain, vi BC)'; Ἀθάνα, add 'SEG 35.419, 437 (Siphnian treasury at Delphi, vi BC)'; Ἀθαναία, for 'IG 1.373¹⁰⁸' read 'IG 1².499, SEG 30.380 nos. 1-4 (καθαναίαν, Tiryns, vii BC), CEG 375 (Laconia, vi BC)'; Ἀθανᾶ, for 'IG 1.351' read 'ib. 711'   add '6 *a type of aulos*, Poll. 4.77.'   at end add 'Myc. *a-ta-na*'

Ἀθήνησιν, (s.v. Ἀθῆναι), for this section read 'Ἀθήνησι(ν), Boeot. Ἀθάνασι Ar.Ach. 900, adv., *at Athens*, Th. 5.47, D. 17.28, 18.66'

ἀθήρ, for 'Lyr.adesp. 2 B.' read 'Call. in Suppl.Hell. 253.1'   II, after 'Cat.Mi. 70' add 'b *of a tongue of fire*, Eur.*fr.* 665a N.-S.'   III, for this section read '*supposed belly-fin of the female tunny*, Ath. 7.303d (quoting Arist.HA 543ᵃ14 where cod. has ἀφαρέα)'

*ἀθήριον, τό, *wheat porridge*, PPetaus 36.3 (ii AD), Stud.Pal. 22.56.29 (ii/iii AD).

†ἀθηρώδης (B), *gruel-like*, Paul.Aeg. 6.88.6 (133.16 H.).

9

ἄθλευμα, ματος, τό, in form ἀέθλευμα, = ἄθλημα, Eust. 1843.22 (pl.).

ἀθλέω I, at end add '2*Ep.Ti.* 2.5'

ἀθλιψία, ἡ, freedom from tribulation, Critodem. in *Cat.Cod.Astr.* 8(1).259.

ἀθλοθεσία, after 'ἀθλοθέτης' insert '*judging, awarding the prizes, Sokolowski* 1.49.B16 (Milet., ii BC)'

ἀθλοθέτης, add 'ἀεθλοθέτης *IG* 2².2193.2 (ii AD), ἀεθλοθέτας *SEG* 23.271.8 (Thespiae, iii BC)'

ἆθλον, ἄεθλον, add 'ἄϜεθλον *IG* 5(2).75 (Tegea), *SEG* 30.52 (Athens, 430/20 BC), [ἄ]Ϝεθλα *SEG* 26.472 (Tegea, vi BC), ἅϜεθλον *SEG* 30.366 (Argos, 460/50 BC)'; line 5, for '*CIG* 776' read '*CIG* 7761'

ἆθλος, line 2, insert 'ἄϜεθλος *SEG* 29.652 (Maced., 460/50 BC), *SEG* 30.1456 (Pontus, 470/50 BC)'; at end after '*IG* 5(2).75' insert '*SEG* 11.330 (Argos, v BC)'

ἀθλοσύνη, add '*BIFAO* 60.133; also ἀεθλοσύνη *AP* 5.293 18(Agath.)'

ἀθλοφορεύς, ῆος, ὁ, victor, μάρτυρος ἀθλοφορῆος Mitchell *N.Galatia* 211 (verse).

ἀθμονάζειν· τὸ εἰς δῆμον (i.e. Athmonon) *ἀφικνεῖσθαι τῆς Ἀττικῆς* Hsch.

ἀθραγένη, add 'Theophr.*Ign.* 64'

ἄθρακτος, add '*AB* 352'

ἄθραυστος 2, add '*invulnerable*, of a fortified place, *Pae.Delph.* 9; Plu.*Tim.* 18 (sup.)'

ἄθρεπτος, for 'Ar.Byz.*Epit.* 2.9.8' read 'Ar.Byz.*Epit.* 29.8'; for 'f.l. .. (Mel.)' read 'ἄ. καὶ ἀνάγωγοι γεγονότες ἄνδρες *Act.Alexandr.* 1.2.6'

ἀθρήματα, add 'cj. Snell (*Glotta* 37.283) in Sapph. 44.9 L.-P.'

ἀθροίζω, line 8, after 'E.*Ph.* 495' insert '*put together, compile*, (in quots. in pass.), τὴν νομοθεσίαν ἀθροισθῆναι ταύτην ἐγκελευόμενοι Just.*Const.* Δέδωκεν 17, 20, 21' **2**, add 'so act. w. reflex. pron., ἑαυτοὺς ἀθροίσαντες *having pulled themselves together*, Ael.*NA* 9.43, cf. 10.48'

ἄθροισις, after 'E.*Hec.* 314' insert 'τὰς ἀθροίσεις καὶ τὰς πέμψεις .. ποιεῖσθαι Aen.Tact. 4.12'

ἀθρόος II 2, for '*continuous*' read '*instantaneous*' and at end add 'v. *AJPh* 65.244/5' **V**, after 'Sup.' insert 'ἀθρωότατος Isoc. 15.107'

ἀθυμέω, lines 4/5, after 'id.*An.* 5.4.19' insert 'w. inf., ἀ. ἐπιχειρῆσαι Th. 7.21.3'

ἀθυμητής, ου, ὁ, one who is disheartened, Phot. s.v. ἄθυμος (a 488 Th.).

ἄθυρμα, line 2, for 'Sapph.*Supp.* 20a.9' read 'Sapph. 44.9, 63.8 L.-P., *Inscr.Délos* 1441*A*82 (ii BC)'; line 6, after 'Trag. and Com.' insert 'A.*fr.* 78c.50 R., Crates Com. 23 K.-A., Eup. 46 K.-A., E.*Hyps.fr.* 1 i 2 B.' and delete '*Com.adesp.* 839'

ἄθυρμα II, after '*unchecked*' insert 'ἄ. στόμα Simon. 36.2 P.'

ἀθυροστομία, after '= ἀθυρογλωττία' insert '*AP* 5.56 (Diosc.)' and for '*AP* 5.251' read '*AP* 5.252'

ἀθύρω, line 4, after 'A.R. 4.950' insert 'Pomp.Mac. 1.4 S.' **II 1**, add '**b** λαίφος ἀ. *play with* the bedding, *h.Merc.* 152.'

†**ἀθύρωτος**, ον, *not fitted with a door* or *doors*, οἰκία *POxy.* 1699.6 (iii AD); of a pinax, perh. *without apertures* or *not shut up*, *Inscr.Délos* 1417A.12 (ii BC). **2** metaph., ἀπύλωτον στόμα Ar.*Ra.* 838 is read as ἀθύρωτον σ., *unchecked* by Phot. a 495 Th., Suid.

ἀθυσίαστον, gloss on ἀφαρκίδευτον, Hsch. (ἀρυτίδωτον La.).

ἄθυτος I 1, after '*omitted*' insert 'πελανός E.*Hipp.* 147'

ἀθῷος, line 1, after 'ον' insert 'Ion. **ἀθώϊος** *IG* 12(8).265.6 (Thasos, iv BC), al., Call.*VB* in *Suppl.Hell.* 260.7'

ἀθῳόω, for 'Iamb.*Bab.* 223' read 'Iamb.*Bab.* 5 Ha.'

αἰ I, add 'αἰ δὲ μὲ ὑπερπαρσχ[ο]ιεν *SEG* 30.380 no. 7 (Tiryns, vii BC)'

αἴ, read **αἰ(ν)** or **αἴι(ν)**.

αἶα (A), line 2, for 'cf.' read 'Stesich. 222(b).205 (p. 214 D.), Anacr. 3.1 P., Simon. 7.1 P.'

αἶα (B), omit '· καὶ φυτόν .. ὁμώνυμος'

αἶα (C), add 'αἶα .. καὶ φυτόν τι· ἔτι δὲ ὁ καρπὸς αὐτῷ ὁμώνυμος, *EM* 27.24-26'

αἰακώς, ἐμβαρύθων· ἐμπίπτων αἰακώς Sch.Nic.*Al.* 541c.

αἰανής, at end delete 'is dub.' and add 'perh. *Trag.adesp.* 700.24 K.-S.'

Αἰάντειος, delete '[penult. short Pi.*O.* 9.112]'; for '*festivals held in his honour*' read '*a festival held at Salamis in his honour*, *IG* 2².1006.72, 1011.53, 1028.24'; after 'Zen. 1.43' insert 'Men. i p. 133 K.-Th. (cf. S.*Aj.* 303)'; add 'fem. Αἰαντεία name of a trireme, *SEG* 24.159 b.386-7 (Attica, iv BC); Ἀιάντειοι, οἱ, *descendants of* (*Locrian*) *Ajax*, Schwyzer 366'

Αἴας, at end, within the parentheses, add 'ΑἰϜας Schwyzer 122(3), Corinthian vase'

αἰβάλη, Theognost.*Can.* 6 A., Hsch. (La. αἰβάνη cod.), Suid.; αἰβαλέη, Cyr.

†**αἰβάνη**, v. °αἰβάλη.

αἰγάγριον, τό, *meat of the* αἴγαγρος (*wild goat*), *Edict.Diocl.* 4.45.

Αἰγαῖος, after '*Aegaean*' insert 'πόντος Ibyc. 1(a).28 P'; for '*mount Ida*' read 'a mountain near Lyctus'; after 'Hes.*Th.* 484' insert 'Αἰ. πεδίον, a plain adjoining Crisa, id.*fr.* 220 M.-W.' **II**, for this section read 'Αἰγαῖον (sc. πέλαγος), τό, *the Aegean*, Hdt. 7.55.1, Str. 2.5.21; also -ος (sc. πόντος), ὁ, Arist.*Mete.* 354ᵃ14 (v.l.), Plu.*Cim.* 8.5'

Αἰγαίων II, add '**2** = Poseidon, Call.*fr.* 59.6 Pf., Philostr.*VA* 4.6, Lyc. 135, Hsch.'

αἰγανέη, after 'Od. 4.626' insert '9.156, A.R. 2.829, Nic.*Th.* 170, Parth.in *Suppl.Hell.* 619, Opp.*H.* 1.712'

αἰγάς, άδος, ἡ, goat, Epich. 173 a Add. (p. vii K.), Hsch. (Δωριεῖς).

αἰγειοπράτης, ου, ὁ, seller of goatskins, Olymp.in Grg. p. 220.12 W.

αἴγειος, add 'cf. Myc. a₃-za (fem., <*aigyos); see also ‡αἴγεος'

αἴγειρος, at end for 'of a seat .. 339' read '*a view from a poplar*, ref. to spectators watching from a vantage-point outside a theatre, Cratin. 372 K.-A.'

αἰγεοθύτης, ὁ, app. *sacrificer of goats*, written ἐγεοθύτης *SB* 11003.2 (iv/v AD), written αἰγεωθύτης *POxy.* 1136.3 (v AD); cf. °αἰγοθύτης.

αἴγεος I, add 'κρέα *PCair.Zen.* 12.55 (iii BC); (sp. αἴγιος *Edict.Diocl.* 8.11); see also ‡αἴγεος'

αἰγιάλειος, add 'cf. Myc. a₃-ki-a₂-ri-jo'

αἰγιαλίτης, before 'ψῆφοι' insert '*of* or *belonging to the shore*'

αἰγιαλός, add 'also *lake-shore* or *river-bank*, *PTeb.* 826.21 (172 BC), 998.5 (ii BC), *PRyl.* 171.13 (AD 56/7), *PPrag.* I 22.7, etc.'

αἰγιαλοφόρητος, (sc. γῆ), ἡ, *shore-land washed away* (by flooding), *PTeb.* 701(a).12 (ii BC).

αἰγιαλοφυλακέω, hold office of αἰγιαλοφύλαξ, prob. in *BGU* 12.24 (ii AD).

αἰγιαλοφυλακία, ἡ, office of αἰγιαλοφύλαξ, *PPetaus* 49.4 (AD 185).

αἰγιαλοφύλαξ, add '*PMich.*III 174.6, XI 617.4 (ii AD), *PFay.* 222 (iii AD), *PSI* 460.5 (iii/iv AD), etc.'

αἰγιάριος, ὁ, goatherd or ?*seller of goat's meat*, *Corinth* 8(3).556, 587 (ἐγι-, vi/vii AD).

αἰγίκνημος, after 'ον' insert '(in quots., -κνᾱμος)'; at end add '*SLG* 387.4'

Αἰγικορεῖς, add 'Αἰκορεῖς (for ?Ἀϊκορεῖς) *ITomis* 164(49), al.'

Αἰγίκορος, ὁ, a son of Pan, named because he glutted himself on goat's milk, Nonn. 14.75.

αἰγίλουρος· κάραβος Hsch.; cf. °αἰγίπυρος.

Αἴγινα, section -ήτης, add 'non-Att.-Ion. -άτας *SEG* 32.412 (Elis, v BC)'

Αἰγιναιοπώλης, ου, ὁ, seller of Aeginetan wares, i.e. shoddy goods, *EM* 28.11; cf. °Αἰγιναῖος.

Αἰγιναῖος, line 2, after 'Th. 5.47, etc.' insert 'Αἰγιναῖα· τὰ ῥωπικὰ φορτία *EM* 28.10; cf. °Αἰγιναιοπώλης'

αἰγίνομεύς, add 'Nic.*Al.* 39'

αἰγιοπλάκινος, ον, perh., *of goathair felt*, χλανίδιν αἰγιοπλάκ(ινον) *Stud. Pal.* 20.245.4 (vi AD).

αἰγίοχος, for 'Alc. 85, .. etc.' read 'Alc. 343 L.-P., etc., *IG* 5(1).238 (vi BC); subst., *AP* 9.224 (Crin.)'

αἰγίπυρος, for '*rest-harrow, Ononis antiquorum*' read 'a thistle-like plant, perh. = σκόλυμος' and after 'Thphr. 2.8.3' insert '(codd.)' add '**II** glossed by κάραβος, Hsch., (dub., cf. °αἰγίλουρος).'

αἰγίς IV, add 'Gal. 19.127.14'

αἰγίσκος, add 'αἰγίσκον· αἶγα ἐκτομίαν Hsch.'

Αἰγλαπιός, v. °Ἀσκληπιός.

Αἰγλάτας, v. ‡αἰγλήτης.

αἴγλη, add '**III** καὶ ἡ θυσία δ᾽ ὑπὲρ τοῦ κατακλυσμοῦ {ἣ} εἰς Δελφοὺς ἀπαγομένη. ἐκαλεῖτο καὶ ποπάνου τι εἶδος, ἐν ᾧ διεπλάσσετο εἴδωλα. καὶ βόλος φαῦλος κυβευτικὸς αἴ. ἐκαλεῖτο. ἀλλὰ καὶ ἡ σελήνη. καὶ τοῦ ζυγοῦ τὸ περίμεσον καὶ παιδιά τις ἐκαλεῖτο αἴ. .. ἔνιοι δέ φασι σημαίνειν καὶ περιπόδιον κόσμον Phot. a 527 Th., Suid.'

αἰγλήεις, line 4, after **αἰγλᾶς** insert 'αἰγλάεντος ὠρανῶ Alcm. 3 *fr.* 3 ii 66 P.'

αἰγλήτης, add 'cf. pers. n. Αἰγλάτας *IG* 5(1).222 (Sparta, c.500 BC), *SEG* 31.348 (Mantinea, v/iv BC)'

αἰγοδόρας, ου, ὁ, goat-flayer, name of the month Ληναιῶν, Sch.Hes.*Op.* 504.

αἰγοθύτης, ου, ὁ, goat-sacrificer, prob. in *Stud.Pal.* 10.252.19 (vi AD).

αἰγόκερως, add 'sp. ἐνοκ- *POxy.* 3353.10 (iii AD)'

αἰγοκερωτή, ἡ, perh. = αἰγόκερας, *fenugreek*, *Hippiatr.Paris.* 958 O.-H.

αἰγοκράνιον, τό, goat's head, *Com.adesp.* 227 A.

αἰγόλεθρος, read 'αἰγόλεθρον, τό'

αἰγονομεύς, for 'Nic.*Al.* 39' read 'Opp.*H.* 4.313' and add 'cf. ‡αἰγινομεύς'

αἰγοτόμιον, τό, goat-sacrifice, Ramsay *Cities and Bishoprics* 1.150.

αἰγυλίς· λύγος Hsch.

Αἰγύπτιος, add 'Myc. a₃-ku-pi-ti-jo, pers. n.'

αἰγών I, after 'ὁ' insert '*goat-pen*, *PCair.Zen.* 771.14 (iii AD)'

αἰγώνυχον· πόας εἶδος Hsch.

†αἰγωπός, όν, *goat-eyed*, i.e. *yellow-eyed*, of men and animals, Arist.*GA* 779ᵇ1, 779ᵃ33, *HA* 492ᵃ3, Ptol.*Tetr.* 144 (αἰγοπ-).

*ἀΐδασμος, v. ἀείδασμος.

αἰδέομαι **I 1**, add '**b** αἰ. μή .. *shrink with shame* from the possibility that, αἰ. γάρ, μή τις ἐμὲ κτείνειεν ἀνάρσιος ἀπτόλεμος χείρ Nonn.*D.* 23.65.' **2**, add 'perh. w. inf., Γλαῦκος, ἔσπλιν αἴδεο *SEG* 30.869 (Leuce, v BC)'

αἰδέσιμος, line 3, after 'title' insert '*POxy.* 1969 (v AD), *PLond.* 1839.2'

*αἰδέστατος, ον, prob. sync. form of sup. of αἰδεστός, *Act.Alexandr.* 9A vi 22.

ἀΐδηλος **I**, add 'neut. pl. as adv., μωμεύειν ἀΐδηλα make *malevolent criticisms*, Hes.*Op.* 756' **II**, line 2, delete 'Hes.*Op.* 756'

*αἰδίλης, ὁ, Lat. *aedilis, aedile*, *Ann.Épigr.* 1927.173, *IGRom.* 3.238, *SEG* 6.555.

ἀΐδιος, line 1, add 'written ἀείδιος *TAM* 2.186a.13 (iv AD), Hsch.'; line 7, after 'Th. 4.63' insert '*APl.* 292'; line 8, after 'Plot. 2.1.3' insert 'κατ᾽ ἀΐδιον *SEG* 1.327.13, ib. 24.1026.(b).15 (both Callatis, i AD); neut. as adv. ἐπὶ φιλότατι πισταὶ καδόλοι ἀείδιον *Olympia Bericht* 7.207 (Sybaris at Olympia, v BC)'

ἀϊδνής (s.v. ἀϊδνός), delete the entry (see *Suppl.Hell.* 1097)

*αἰδοιόπληκτος, ον, gloss on σαννιόπληκτος, Hsch., cf. *POxy.* 3329 i 6 (iii/iv AD).

αἰδοῖος **II 3**, add 'A.*Supp.* 194'

Ἀϊδοκυνέα, ᾶς, ἡ, *Hades-cap* (v. κυνέη 2), S.*fr.* 269c.19 R.

Ἀϊδοναῖος, v. ‡Αὐδυναῖος.

ἀϊδρείη, add 'dat. ἀϊδρείηφι Nic.*Th.* 409'

ἀϊδρήεις, delete 'εσσα' and add 'hypothetical adv. -έντως Sch.Nic.*Al.* 415a'

Ἀϊδωνεύς, line 4, after 'later poets' insert 'Euph. 98.4'

*Ἀϊδωνίς, ίδος, fem. adj., *of Hades*, Λήθη *GVI* 1874.11 (Cnidos, ii/i BC).

αἰδώς, add '**IV** Chaldean name for the moon, Hsch.'

αἰδῶσσα, for 'αἴθουσα' read 'τῆς αὐλῆς τὰ τειχία'

αἰεναοιδός, delete the entry (read as two words).

*αἰέναος, v. ἀέναος.

αἰένυπνος, delete the entry (read as two wds.: αἰὲν ὕπνον, L.-J.-W.).

αἰζηός, for 'of a *stout* ..' to the end read 'more often as subst., *able-bodied man* (as engaged in fighting, hunting, ploughing, etc., usu. in pl.), Il. 2.660, Od. 12.440, Hes.*Th.* 863, *Op.* 441, Call.*Jov.* 70, A.R. 4.268, Nic.*Al.* 176, *Th.* 343, etc.; (in parody) κταμένοις ἐπ᾽ αἰζηοῖσι καυχάσθαι μέγα Cratin. 102 K.-A.'

αἰθάλη **I**, transfer 'Dsc. 5.75' to section II.

αἴθαλος, after '*smoke*' insert '*soot*'; at end add 'Myc. *a₃-ta-ro*, pers. n.'

αἰθαλόω, line 1, after '*El.* 1140' insert '*singe*, Arr.*Fr.Phys.* 3 (189.15 R.-W.)'

αἰθέριος **1**, add '*IG* 12(2).484.9 (Mytilene), Robert *Hell.* 10.20 n. 6' **2**, add 'ψυχὴ(ν) αἰθερείαις αἰ(ῶ)σι (?) θέτο, σῶμα δὲ γα(ί)η *BCH* suppl. 8 no. 5 (Edessa, v AD)'

αἰθερίτης, after 'precious stone' insert 'perh. *turquoise*'

*αἰθεροδινής, ές, *whirling in the ether*, orac. in *App.Anth.* 6.140.8.

αἰθερολαμπής, add '(cj.)'

*αἰθερολογία, ἡ, *study of the ether*, Pythag. in D.L. 8.50.

*αἰθερύσσω· τὸ ταχύνω Theognost.*Can.* 9 A.

αἰθήρ, line 1, after 'Hes.' insert '*Th.* 124'; line 2, after 'ἡ' insert 'Stesich. 32 i 4 P.' **1**, add '**b** personified, Ar.*Nu.* 578, Orph.*H.* 5.4, orac. in *ZPE* 1.185, II b line 11 (Phrygia, ii AD), Ἐθέρι ἀλεξιχαλάζῳ Robert *Hell.* 9.63 (Amasia).' **3 b**, delete 'cf. Arist.*Mu.* 392ᵃ5'

†Αἰθιοπία, s.v. Αἰθίοψ, Ion. -ίη, ἡ, *Ethiopia* (term loosely applied to the country southwards from Egypt), Hdt. 2.30.3, Th. 2.48.1.

αἰθιοπίς (s.v. Αἰθίοψ), **II**, *sage*, add 'Plin.*HN* 27.11; also app. another plant, ib. 24.163; αἰ.· φλομώδης πόα Hsch.'

Αἰθίοψ, line 2, for 'nom. Αἰθιοπεύς' read 'gen. sg. -οπῆος' and after '*Del.* 208' insert 'gen. pl. -οπήων Theoc. 17.87, A.R. 3.1192'; at end add 'Myc. *a₃-ti-jo-qo*, pers. n.'

*αἴθμα· delete Hsch.

*Αἰθοπία, ἡ, epith. of Artemis, Call.*fr.* 702 Pf., *IG* 12(2).92 (Lesbos), *AP* 6.269.3, 7.705.3 (Antip.Thess.).

αἰθός **I**, add 'perh. *black-* or *dark-complexioned*, S.*fr.* 269a.54 R.; cf. Αἰθίοψ' **II**, for '*shining*' read '*bronze-coloured*'

αἴθουσα, delete '(sc. στοά)'; for '*portico* .. *sun*' read 'some form of sheltered area on the outside of the μέγαρον'

αἴθοψ, delete '*flashing*'; for '*sparkling* (or *fiery*)' read '*fiery red*'

αἴθρη, add '**III** = °αἴθριον, *open court in a house*, τῆς πάσης οἰκίας ἀπό τε αἴθρας καὶ αὐλῆς .. *POxy.* 3355.6 (AD 535).'

αἰθρηγενής, add 'ἄνεμοι A.R. 4.765'

αἰθρία, after 'Ion. -ίη' insert 'poet. also -ιάα *Suppl.Hell.* 958.17 (prob.)'

αἰθριοκοιτέω, for 'Orib. 9.3.8' read 'Stob. 5.37.30'

αἴθριος **I**, after 'f.l. in S.*Ant.* 357' insert '(cj. ὑπαίθρεια)' **2**, epith. of Zeus, add '*SEG* 33.1052 (Cyzicus, ii BC), Robert *Hell.* 10.20/1 (Byzantium, i BC)' **III**, αἴθριον, for 'adaptation .. sense' read '*open court* in a house giving light to the surrounding rooms' and add 'cf. °αἴθρη III'

αἴθυγμα, line 1, delete '*glamour*'; line 2, delete 'cf.' and transfer 'Plu. 2.966b' to end of article adding 'Iamb.*Protr.* 5 p. 36.10 P.'

αἰθύσσω, line 4, after '(Nonn.*D.*) 48.689' insert 'φρένας B.*fr.* 20B.8 S.-M. (cod. Ath., cj. διαιθύσσω q.v.)'; after 'pass.' insert 'of winds, Euph.in *Suppl.Hell.* 415 i 23'

αἴθω, line 3, delete 'metaph.'; αἴθομαι, lines 6/7, after 'X.*Cyr.* 5.1.16, cf.' insert 'Theoc. 7.102' and after 'A.R. 3.296' insert 'ἐπί τινι Call.*fr.* 67.2 Pf.'

αἴθων **I**, for '*fiery, burning*' read 'of fire, etc., *dark red*' **II**, for 'of burnished .. Od. 1.184' read 'of bronze, bronze objects, *dark red* or *brown*'; at end add 'by poetic transf., of iron, Il. 4.485, Od. 1.184' **III**, after 'birds' insert '*reddish brown, tawny*' and for 'prob. of colour .. 559d)' read 'αἴ. θῆρες, i.e. drones, Pl.*R.* 559d; of bulls, Opp.*C.* 2.102; of a lion, ib. 3.54' **IV**, for 'metaph. .. *fiery*' read 'of complexion, *dark*, app. with the implication of manliness' and delete 'Hermipp. 46'

αἰκίζω **I**, add '**2** *inflict outrageously*, ἐμέτισαντ᾽ α[ἱ]κισθέντα φόνον epigr. in *BCH* 116.599 (Ambracia, vi/v BC).'

*Αἰκορεῖς, v. °Αἰγικορεῖς.

*αιλαι, wd. of unkn. meaning perh. for ἐλαί(ου), αιλαι οἴνου ξ(έσται) *PRyl.* 627.74 (iv AD).

*αἰλέω, v. ‡αἴρω.

αἴλινος **1**, after '*dirge*' insert 'ἁ μὲν εὐχαίταν Λίνον αἴλινον ὕμνει Pi.*fr.* 128c.6 S.-M.'; after '(Diod.)' insert '*weaving-song*, Epich. 14'; add 'on an altar of Linos [Λ]ίνου Αἰλίνου *SEG* 33.303 (Epid., iv BC)'

*αὔλιψ, unexpld. wd. perh. for αἰγίλιψ, Theognost.*Can.* 11 A.

*αἰλότριος, v. °ἀλλότριος.

*αἰλουρόμορφος, ον, *having the form of a cat*, (cf. ἱερακόμορφος), ξόανον Horap. 1.10 S.; εἰσὶ δὲ καὶ κανθάρων ἰδέαι τρεῖς. πρώτη μὲν αἰλουρόμορφος καὶ ἀκτινωτή ib.

*αἰλουροταφεῖον, τό, *burial-place of sacred cats*, *OWilck.* 1486 (ii AD).

αἷμα **III 1**, add 'of the Hyperboreans, πολυχρονιώτατον αἷ. Call.*fr.* 67.7 Pf., *Del.* 282' **2**, add 'Theoc. 24.73, αἷμα θεῶν Call.*VB* in *Suppl.Hell.* 254.2'

*αἱμάζομαι, *bleed*, *PWarren* 21.124 (iii AD).

*αἱμαλωπώδης, ες, *bloodshot*, Paul.Aeg. 2.57 (124.12 H.).

αἱμασιά, for 'to lay' read 'gather stones for' add '**II** *walled enclosure*, *IG* 12(3).248, (5).872, *IMylasa* 255, etc.'

*αἱματεύω, *furnish with blood*, ἓν τὸ ἄλλο αἱματευεὶ· καὶ ἓν τὸ ἄλλο γενεᾶ Anon.Alch. 22.4.

αἱμάτη, for 'Theognost.*Can.* 5.32' read 'Theognost.*Can.* 12 A.'

αἱμάρια, add 'cf. mod. Gk. αἱματιά, αἱματιά *blood-pudding*'

αἱματοειδής, add '*Orac.Tiburt.* 24, 25'

αἱματόεις, line 2, for '= αἱματηρός, Il. 5.82' read '*bloody*, Il. 5.82, Archil. 13.8 W., Tyrt. 10.25 W.' **4**, after 'etc.' insert 'Mimn. 14.7 W., φόνον Tyrt. 12.11 W.'

αἱματοσταγής, for 'cf. *Pers.* 816' read '*Pers.* 816 (v.l.)'

*αἱματοσφαγής, ές, *proceeding from bloody slaughter*, πελανὸς αἱ. A.*Pers.* 816; cf. °αἱματοσταγής.

*αἱματοχάρμης, ου, ὁ, *one who delights in blood*, *AP* 15.28.8.

αἱματώδης, add '*Trag.adesp.* 732 K.-S., *OBodl.* ii.2171; *blood-red*, χρῶμα Plu. 2.565c'

*αἱμαχάτης, ου, ὁ, *red agate*, Plin.*HN* 37.139.

*αἱμογενής, ές, *related by blood*, *SEG* 8.374 (Egypt, ii AD). **2** *producing blood*, *Suppl.Mag.* 90.A30, E↓6 (v/vi AD; sp. ημωγενη).

*αἱμοδρυφής, ές, (δρύπτω), *drawing blood by scratching*, ἄλγη .. αἱ., w. ref. to rending cheeks in mourning, *IEphes.* 2101A (epigr.); cf. ἀμφιδρυφής.

†αἱμοειδής, ές, *having the appearance of blood*, Ph. 2.244 (cj. ἠθμο-, v.s.v. II).

*αἱμοποιός, gloss on σκάραιβον, Hsch.

*αἱμοπότης, *drink blood*, gloss on δερκύλλειν (or δερμύλλειν), Hsch.

†αἱμοπότις, ἡ, *drinking blood*, epith. of Hecate, *PMag.* 4.2864.

αἱμοπτυϊκός, after '*blood*' insert 'Dsc. 2.85.3'

†αἱμόρρυτος, ον, (ῥέω) *flowing with blood*, αἱ. φλέβες *blood-vessels*, A.*fr.* 230 R.; *marked by efflux of blood*, νόσος *IG* 12(5).310.8 (Paros, ii AD, αἱμορυτ-).

*αἱμορυγχία, ἡ, *nosebleed*, Phryn.*PS fr.* 11.

*αἷμος, ὁ, αἵμους· ὀβελίσκους, Hsch.

*αἱμοστατικός, ή, όν, *styptic*, Alex.Trall. 2.199.18.

*αἱμοφθόρος, Thespis 5 S., gloss on βροτολοιγός, Hsch.

*αἱμοφυής, ές, *born of blood*, *Suppl.Mag.* 96.A25, E↓1, F*fr.* A4 (v/vi AD; sp. ημωφυη).

αἱμοφόρος, after 'Posidon. 8' add 'transf., of fighting, Lyc. 1411, Nicom.Trag. 15 S.'

*αἱμόχρωμος, ον, *blood-coloured, reddish*, of a cow, *PKöln* 55.5 (iii AD).

*αἱμοχυής, ές, *shedding blood*, *Suppl.Mag.* 96.A26, E↓2 (v/vi AD; sp. ημοχυη).

αἱμυλία, add '*plausibility* (in argument, etc.), Ael.*NA* 5.49' αἰμύλος, add '2 winning, seductive, Democr. B 104 D.-K., AP 7.643 (Crin.), GVI 840.2, 698.2.'

*αἰμωβόλιον, τό, sacrifice of blood, CIG 8558, cf. CIL 9.3015.

*ἄϊν, v. ἀεί I 5.

*Αἰνεάδαιος, ὁ, a month in a Cyprian calendar, Cat.Cod.Astr. 2.143 (see also YClS 2.213).

*Αἴνειος, v. °Αἰνικός.

αἰνελένη, add 'Nic.Th. 310'

*αἰνεσίθυμος, ον, prob. f.l. for ἀλγεσί- (q.v.), orac. in TAM 3(1).34D65 (Termessus).

αἰνετός, after 'praiseworthy' insert 'Cratin. 171.15 K.-A.'

αἰνέω I, line 9ff., delete 'ἐπαινέω .. usu.'; at end for '(both lyr.)' read '(anap., lyr.)'   III, delete the section transferring quots. to section II 2.

*Αἰνιάρχαι, οἱ, magistrates among the Aenianes, IG 9(2).5b15 (Hypata, ii BC).

*Αἰνιαρχέω, to be a magistrate among the Aenianes, IG 9(2).7 (Hypata, ii BC).

*αἰνιγμᾰτοειδής, ές, = αἰνιγματώδης, enigmatic, Zos.Alch. 241.26; adv. -ῶς, ἔμφατον· αἰ. εἰρημένον Hsch.

αἰνιγμᾰτώδης, at end for 'Arist.Rh. 1441b22' read 'Anaximen. Lampsac.Rh. 1441b22'

*Αἰνικός, ὁ, a month in a Cyprian calendar, Madrid Cod.Gr. 95 (v. YClS 2.213; Αἴνειος cj.); Hemerolog.Flor. p. 24 K. has Ἄννιος.

αἰνόγαμος, for 'fatally' read 'fatefully'; after 'E.Hel. 1120 (lyr.)' insert 'λέχος αἰνογάμου .. Ἑλένης Trag.adesp. 644.40 K.-S.'

†αἰνόγονος, ον, of terrible race (supposed etym. explanation of Chaldaean), Ph.Epic. in Suppl.Hell. 681.7.

αἰνόδακρυς, for '= foreg.' read 'terribly wept for'

αἰνοδρυφής, for 'Antim.[107]' read 'Suppl.Hell. 1002 (Antim. 156 Wy.)'

αἰνολέων, before 'Theoc. 25.168' insert 'Call.VB in Suppl.Hell. 257.21'

αἰνόμορος I, at end add 'Orph.H. 57.6'   II, after 'h.Merc. 257' insert 'of a mother who buries her sons, CEG 94.4 (?Athens, v BC)' and at end for '(Marcell.Sid.) 14' read 'Marc.Sid. 13'

*αἶνον· τὸ ὕψος, ἢ ὁ ὄχθος τοῦ ὄρους Theognost.Can. 13 A.

αἰνοπᾰθής, add 'SEG 26.405 (Corinth, Chr.).   2 causing terrible suffering, Orac.Sib. 5.185.'

αἰνός, line 3, after 'Od. 15.342' insert 'αἰνῷ φόβῳ Pi.P. 5.61, αἰ. ἄχος S.Aj. 706'; line 5, after '(Il.) 8.423' insert 'μᾱτρὸς ὑπ' αἰνᾶς Phryn.Trag. 6.3 S.'

αἰνοτᾰλᾶς, delete 'Antim. [106] =' and for '(Call.)fr. 506' read 'fr. 481 Pf.'

αἰνόφυτα, delete the entry.

*αἰνόφῡτος, ον, of terrible nature (cf. °αἰνογόνος), Ph.Epic. in Suppl.Hell. 681.5.

αἴξ, line 2, after 'IG 7.3171' insert 'acc. αἶγον dub. l. in AJPh 48.241 (Isauria, ii/iii AD)'   IV, add 'Longus 2.15; αἶγες· τὰ κύματα. Δωρεῖς Hsch.'

*Αἰόληος, α, ον, Aeolian, Alc. 129.6 L.-P.

αἰολοβρόντης, at end add '(cj. -βρέντ- Snell)'

†αἰολόνωτος, ον, having a variegated back or upper surface, Βρετανῶν αἰ. i.e. from staining or tattooing, Opp.C. 1.470, of fish (σάλπαι), id.H. 1.125; πόρδαλις αἰ. Nonn.D. 5.361; of patterning on a κρητῆρ, ib. 37.660.

αἰολόπρυμνος, for 'gleaming' read 'decorated' and for 'B. 1.4' read 'B. 1.114 S.-M.'

αἰολόπωλος, add '2 perh. having the nature of a lively filly, app. of a hetaera, Κυλιφάκ[η αἰ]ολοπώ[λ]ῳ rest. in SEG 27.672 (Etruria, v BC).'

αἰόλος, at end add 'Myc. a₃-wo-ro, name of an ox'

*αἰολωπός, όν, in dub. context, perh. of changeable appearance, S.fr. 269a.56 R.

ἆος, delete the entry (v. αἰώς).

*ἄϊπαις, παιδος, ἡ, ever a girl, GVI 1941.3 (Thisbe, ii/iii AD).

*ἤἄϊπερ, v. °ὅσπερ II 4.

†αἰπήεις, εσσα, εν, steep, precipitous, Il. 21.87, A.R. 2.721, Euph. in Suppl.Hell. 418.15, Q.S. 1.304; descending headlong, αἰ. καταιγίς AP 7.273 (Leon.Tarent.).

*αἴποθεν, from the heights, Thdt. in Suppl.Hell. 757.9.

αἰπόλιον I, add 'Schwyzer 721.14 (Thebes-on-Mycale, v/iv BC)'

αἰπόλος II, add '(αἰπόλος La., cf. ἐμπολή merchandise)'   add 'III ὁ Ἑρμαφρόδιτος ὑπὸ Σινωπέων οὕτω καλεῖται Phot. a 634 Th.'

*αἰπολοφύλαξ, ακος, ὁ, overseer of goatherds, SB 9415.7.2 (iii AD).

αἶπος, after 'A.Ag. 285, 309, etc.' insert 'E.Ph. 851, Theoc. 7.148'; delete 'hence αἰ.' to end.

αἰπός, delete 'αἰπόν, τό .. (Athens)'

αἰπύκερως, after '= ὑψίκερως' insert 'high-horned' and add 'Theognost.Can. 14 A.'

αἰπύς 1, line 4, delete 'on high .. 682'   2, add 'of sounds, high-pitched, αἰ. ἰωή Hes.Th. 682; θεὸς αἰ. of Diomede, dub. sens. in Lyc. 630'

αἷρα I, after 'hammer' insert 'Euph. 51.9' and for 'Call.fr. 129 .. Hsch.' read 'Call.fr. 115.12 Pf.'

*αἱράριον, τό, Lat. aerarium, public treasury, esp. of the Roman fiscus, IG 2².1119(a).10 (see also SEG 24.143), Mon.Anc.Gr. 9.12; of other provincial treasuries, Mitchell N.Galatia 414, στρατιωτικὸν αἰ. MAMA 4.139 (Phrygia, i AD); also ἐράριον, τό, BMus.Inscr. 4(2).1026 (Smyrna, c.ii AD); ἀράριον La Carie 177 (Sebastopolis).

*αἱράριος, ὁ, Lat. aerarius, citizen of the lowest class, D.C. 57.71.

αἵρεσις (Β) II 1, add 'good conduct in public life, SEG 33.1039 (Cyme, ii BC), 639 (Rhodes, c.100 BC)'   2, penultimate line, after '(Act.Ap.) 28.22' insert 'heretical sect, Cod.Just. 1.1.5.3, al.'   III 1, add 'condition attached to a bequest, Cod.Just. 1.3.52.13, al.'

αἱρετικός 2, add 'b heretical, of persons, Cod.Just. 1.5.18.8, al.'   4, add '-ῶς ἔχοντος πρὸς τὴν πόλιν being well-disposed to .., SEG 7.62 (Seleucia in Pieria, ii BC)'

*αἱρετισμός, ὁ, adherence to unorthodox practices, heresy, ἀργύριον αἱρετισμοῦ Lcn. 4Ki. 12.17.

αἱρετιστής 2, place 'τῶν λογῶν .. p. 12 C.' after 'Plb. 1.79.9, etc.'

αἱρέω, line 15, after 'εἰλάμην' insert 'Call.fr. 384.41 Pf.' and after '(Smyrna)' insert 'ἴλαντο, Ath.Agora xv 293.5 (i BC)'; line 18, Cret. forms, add 'inf. αἰλέν SEG 27.631.B7 (Lyttos, c.500 BC)   A II 4 b, after 'Isoc. 18.15' insert 'w. two accs., δίκας εἷλεν Εὔπολιν δύο Is. 7.10' and transfer 'Is. 7.13' to follow   5 b, at beginning insert 'of an amount or balance, to be due, αἱροῦντος (?sc. φόρου) γεωμετρίας BGU 2484.5 (ii AD)'   B 1, line 4, after 'Il. 22.119' insert 'receive, suffer, κῆδος ἐλέσθαι A.R. 2.858, 3.692'   C I 1, after '9.102' insert 'S.Ant. 406; ᾑρέθην .. ἐν μάχῃ E.Supp. 635'   II, at end for 'always' read 'mostly'

αἱρώδης, add 'ὄλυρα PTeb. 857.18, 35 (162 BC)'

αἶσα II 3, line 3, for 'Inscr.Cypr. 148' read '(a-i-sa) ICS 285.2; τὰν ψάφον τὰν τρίταν αἶσαν Schwyzer 84.17 (Argos, c.450 BC)'

αἰσθάνομαι, line 4, after 'Lxx Jb. 40.18' insert '(app. act. αἰσθάνω· οὐκ ἐσθάνετε ἀνθρώποις POxy. 3417.10, (iv AD), but could be misspelling of the middle)'; line 9, after 'E.Hipp. 603, etc.' insert 'w. cogn. acc., πᾶσαν αἴσθησιν αἰσθανομένῳ Pl.Phdr. 240d'

*αἰσθέσιον· τὸ νοσερόν (read νοερόν) Theognost.Can. 16 A.

ἀΐσθω, add 'w. internal acc., αἰνὸν ἀΐσθων Opp.H. 5.311'

αἰσιμία, after 'Eu. 996' insert 'fairness, ἐν αἰσιμίῃ .. ἄνυσσεν ἀρχήν Inscr.Olymp. 481 (iii/iv AD).   2 oracle, Call.fr. 18.9 Pf., cf. Theognost.Can. 16 A., EM 39.3, Suid.'

αἰσιμῶ, for 'Suid.' read 'Theognost.Can. 16 A., Zonar.'

αἴσιος I, add '2 epith. of Poseidon at Delos, Inscr.Délos 1562, 1581, 1582, 1902.'   II, for 'αἴ.' read 'αἰσία'

*ἄϊσιτος, v. ‡ἀείσιτος.

ἀΐσσω, line 13 (p. 43 line 1), for 'of one .. enemy' read 'of men'; line 14, after '(Il.) 18.506' insert 'D. 25.52, 47.53; fig., Plot. 5.5.4'; at end after 'ᾄ Lyr.' insert '(exc. μετ-ᾄξαις Pi.N. 5.43)'

αἰσυμνάω 1, for 'Call.Iamb. 1.162' read 'Call.fr. 192.6 Pf.'   2, add 'ᾐσύμνας Ἐφέσου Call.fr. 102 Pf.'

αἰσυμνήτης II 2, add 'SEG 31.984 (Teos, 470 BC)'

αἰσυμνῆτις, before 'Suid.' insert 'Call.fr. 238.10 Pf.; of a goddess, Syria 12.319 (Jerusalem)'

*αἰσύφιος· δεινός (δειλός Kuster), ψευδής, ἀπατεών Hsch.; cf. ἀσύφηλος.

*Αἰσχλαβιός, v. Ἀσκληπιός.

*αἰσχρία, ἡ, dirt, Eust. 1853.28.   2 dusk, id. 1402.29.

*αἰσχρίασμα, ατος, τό, dusk, Eust. 1401.29.

αἰσχροκερδής, add 'ἐσχροκερδοῦς ἕνεκεν .. PVindob.Tandem 4.18 (AD 312/314)'

αἰσχροπρᾰγέω, for '= αἰσχροποιέω' read 'behave basely'

αἰσχρός I and II, transpose these sections.   II, delete 'opp. καλός'

αἰσχροσεμνία, for 'avoidance of obscenity' read 'euphemistic treatment of something indelicate'

αἰσχρουργέω, for 'act .. masturbari' read 'perform sexual perversions'

αἰσχρουργία, combine examples under def. 'sexual perversion, obscenity'

αἰσχρουργός, for 'obscene' read 'sexually perverted'

αἰσχυντηρός, after 'αἰσχυντηλός' insert 'Lxx Si. 26.15'

αἰσχυντικός, at end add '(dub., v.l. ἀναίσχυντα)'

*Αἰσωπίτανος, α, ον, made in the manner of Aesop, i.e. having a sententious inscription, Αἰσωπίτανα κεραμίς SEG 39.1062 (Rhegium, i BC/i AD).

*αἰτάριον, prob., a small bird of prey, "little eagle", PAmst. 13 (v AD).

†ἀΐτᾱς, v. °ἀΐτης.

αἰτέω I, add '6 ask in marriage, Leg.Gort. 7.51.'   II, for 'ask for .. = the Act.' read 'in sense not clearly dist. fr. act.'; lines 7/8, delete 'αἰτησάμενος .. Men. 476'   add '2 borrow, Th. 6.46.3, And. 4.29, Lys. 19.27; οὐ πῦρ γὰρ αἰτῶν, οὐδὲ λοπάδ' αἰτούμεθ' Men.fr. 410 K.-Th.; ask for the services of, Λύσανδρον ἄρχοντα Lys. 12.59.'

12

*ἀΐτης, ου, ὁ, Dor. ἀίτᾱς, *beloved youth*, ὁ μὲν εἴσπνηλος .. τὸν δ' ἕτερον πάλιν, ὥς κεν ὁ Θεσσαλὸς εἴποι, ἀΐτην Theoc. 12.14; of the favourite of a goddess Χρύσας (sc. Ἀθανᾶς) δ' ἀίτας Dosiad.*Ara* 5; perh. also Lyc. 461 (ᾱ-, unless to be referred to ἀετός). **2** *of a fish*, (?*immature*), τὸν κεστρέα .. τὸν ἀΐτην καὶ ἄρσενα ἴσον *PTeb.* 701.44, 65 (iii BC).

αἰητής, after '*petitioner*' insert '*PTeb.* 894 *fr.* 6.13 (ii BC)'

αἰτία, add '**VI** *illness, disease*, ποίᾳ αἰτίᾳ οὗτος ἐτελεύτησεν *Acta Philippi* 1, Hsch. s.v. ἀναγωγή.'

αἰτιᾱτικός **I** 1, add 'adv. -κῶς, ib. (Sch.Il.) 22.329'

αἰτιᾱτός, after 'Plot. 6.2.33' insert 'D.L. 9.97, Procl.*Inst.* 11 D.'

*αἰτιολογισμός, ὁ, *explanation, justification*, Eus.*PE* 1.5.11 (cj., codd. αἴτιον λογισμόν, ἀπολογισμόν).

αἴτιος **II** 2, add 'prob. in Sapph. 67.6 L.-P.'

*ἆϊτις, ιος, ἁ, *a beloved girl*, Alcm. 34 P.; cf. °ἀΐτης.

αἰφνίδιος, line 1, after 'A.*Pr.* 680' insert '(s.v.l.)'; after 'Th. 8.14' insert 'of fate, *IG* 12(7).52.3 (Amorgos)' adv. -ίως, for 'ib.' read 'Th.' and add '*IG* 12(7).397.4, 401.5 (Amorgos)' adv. -ιον, add '*Peripl.M.Rubr.* 45, 46 C.'

*αἴφνως, = αἴφνης adv., *suddenly*, *GDRK* 49.5 (hymn to Sarapis), *SEG* 30.1480.3 (Phrygia, late Rom.imp.).

*αἰχμαῖος, α, ον, *of a spear-point*, Eratosth. in *Suppl.Hell.* 398.

*αἰχμᾰλόω, *take prisoner*, aor. pass. ἠχμαλώθη *SEG* 8.595 (Egypt) (back-formation fr. αἰχμάλωτος).

αἰχμή **I** 2, after '*C.* 2.451' add 'of the teeth of monsters, Opp.*H.* 5.141, Philostr.*VA* 3.7' **II** 1 b, add 'of a sickle, Hes.*Sc.* 289' at end before 'cf. Lith. *jiešmas*' insert 'form *αἰκσμά, Myc. *a₃-ka-sa-ma* (acc. pl.)'

αἰχμητής **I**, add 'αἰ. ἀλέκτωρ *a fighting cock*, *AP* 6.155 (Theodorid.)'

αἰχμοφόρος, at end add 'applied to praetorian guards, Hdn. 1.10.6'

*αἴχνινον, τό, prob. name of a cult-object, *Inscr.Cret.* 1.v 23.4 (Arcades, ii AD).

αἰψηροκέλευθος, for 'Poet.ap.Apollod. 3.4.4' read 'hence of a dog named Βορῆς Epic. in *Coll.Alex.* p. 72 (= Apollod. 3.4.4)'

ἀΐω (A), line 3, after 'Hdt. 9.93' insert 'Cypr. ἀϜίω, fut. med. *a-wi-je-so-ma-i* ἀϜιήσομαι, *Kouklia-Paphos* 237.5 (iv BC)' at end for 'ἄϊε .. 1.352' read 'ἄϊεις *MDAI(A)* 57.47 (i AD); ἀϊ- is v.l. for ἀκου- in Il. 2.486, Od. 1.352, 353, Hes.*Op.* 213'

αἰών, line 2, delete 'apocop. .. 350' **II** 1, lines 3/4, for 'οἱ ἀπὸ .. 63.20' read '*in all time*, πρῶτος πάντων τῶν ἀπὸ τοῦ αἰ. Ῥωμαίων D.C. 63.20; ἕνα ἀπ' αἰῶνος ἀπαγόμενον γυμνασίαρχον *Act.Alexandr.* 11.67 (ii AD)'; after '*life long*' insert 'Simon. 36.12 P.' **3**, for 'cf.' read 'Heraclit. 52 D.-K., Sotad. 15.1, Nonn.*D.* 7.10' **B**, delete the section.

*αἰωνακτῑνοκράτωρ, ορος, ὁ, *eternal lord of the sun's rays*, *PMag.* 1.201.

*αἰωνεργέτης, ου, ὁ, *eternal worker*, Αἰωνεργέτα, κύριε Σάραπι *SEG* 35.1051 (Rome, i/ii AD).

αἰώνιος 1, add 'app. w. dittography, εἰς τειμὴν αἰωνιαιαν *Inv.Inscr.Palm.* vii 6 (i AD)' **2**, add 'of an office, *held for life*, αἰώνιον στεφανηφορίαν *SEG* 30.1390 (Lydia, c.AD 150)'

*αἰωνογυμνασίαρχος, ὁ, *life gymnasiarch*, Κλαυδίου Συρίωνος ἐωνογυμνασιάρχου *POxy.* 2854.27 (iii AD).

*αἰωνοπῠρεῖον, τό, *place of everlasting fire*, *Epigr.Gr.* 1140 b 8.

αἰωνόφθαλμος, for '*seeing with eternal eyes*' read '*eye of Αἰών* (personified)'; at end add '(v. *CRAI* 1985.438)'

αἰώρα, add '**IV** as a measure of capacity, τειμ(ῆς) μεικρᾶς αἰώρ(ας) οἴνο(υ) (δραχμαὶ) δ' *PMil.Vogl.* 307.112 (ii AD).'

αἰωρέω, after '*raise*' insert 'ὁ δὲ (sc. eagle) κνώσσων'; delete 'of the eagle .. feathers'; for '*swing as in a hammock*' read '*hold aloft*' delete '**2** *hang*' **II** 3, before 'Epicur.' for 'ψυχῆς' read 'φύσεως' and for '*Nat.* 22 G.' read '[34, 17] 5 A.'

αἰώρημα, add '**3** *that which is poised ready to fall*, E.*Supp.* 1047.'

*αἰώριον, τό, dim. of αἰώρα, perh. *hoist*, *PUniv.Giss.* 10 ii 13 (ii/i BC).

*αἰώς· ὁ αἰών. παρὰ Στησιχόρῳ, prob. in *Cod.Bodl.Auct.T.* II(11) f. 90, cf. Stesich. 69 P.

ἀκᾶ, delete the entry.

ϝακάβα, v. °Ἑκάβη.

*ἀκαδήμαρχος, ὁ, *ephebic official at Cyrene*, *SEG* 20.742.11 (AD 161).

Ἀκαδήμεια, add 'cf. ‡Ἑκαδήμεια'

ἀκαθαρσία 1 b, after '(13 BC)' insert '*PSI* 787.22 (ii AD), *POxy.* 2109.45 (iii AD)' **3**, after '*impurity*' insert '*SEG* 19.427 (-ρτιαν Dodona, iv/iii BC)'

ἀκάθαρτος **I**, add '**4** *uncleared*, χώρα .. δύσορμος καὶ ἀκάθαρτος ῥαχίαις *Peripl.M.Rubr.* 20.'

ἀκάθεκτος, after 'Plu.*Nic.* 8' insert 'δρόμος ἀ. i.e. *stampede*, Hld. 4.21'

ἀκαθοσίωτος, add '*unholy*, *POxy.* 1865.8 (vi/vii AD); *dishonourable*, *PBerl.Zill.* 13.8 (vi AD)'

ἀκαθυπερτέρητος, after 'Ptol.*Tetr.* 157' add '190'

ἄκαινα **II**, add 'ἄκαιννα δεκάπος *SEG* 37.491 (Thessaly, v/iv BC)'

*ἀκαινιαῖος, ον, *measuring one ἄκαινα*, *IEphes.* 3217a.9 (ii AD).

ἄκαινον, for 'Olymp. *in Metaph.* 43.1' read 'Olymp. *in Mete.* 113.1'

ἀκαιρία **I** 1, for 'Pl.*Phd.* 272a' read 'Pl.*Phdr.* 272a'

ἀκαίριμος, for 'ὅ τι .. Lyr.adesp. 86 A.' read 'ὅττι κεν ἐπ' ἀκαιρίμαν γλῶσσαν ἴηι Lyr.adesp. 102 P.'

*ἀκαιροδᾰπᾰνία, ας, ἡ, *untimely expenditure*, ἐξ ἐρωτικῆς τινος ἀφορμῆς ἢ ἀκαιροδαπανίας κακοπραγμονοῦσιν Heph.Astr. 2.36.22.

ἄκαιρος **I**, after 'A.*Pr.* 1036' insert 'ἄκαιρα μωμένους id.*fr.* 451c.21 R.' **II** 1, for '*importunate, troublesome*' read '*acting in an inopportune or untimely manner*'; add 'ἀ. ἐν ταῖς μετωνυμίαις *using them inappositely*, D.H.*Dem.* 5, cf. *Lys.* 5'

ἀκακία (A) **I**, add 'ἀκακίῃ Hp.*Mul.* 2.189'

*ἀκακοποίητος, ον, *undamaged*, *SB* 8754.18; cf. ἀκακούργητος.

ἄκακος **I** 1, add 'ἀ. κόρην Men.*Dysc.* 222' **2**, add 'Pl.*Alc.* 2 140c'

ἀκαλλιέρητος, add 'Luc.*Bis.Acc.* 2, Philostr.*VA* 8.7.10'

*ἀκαλλίς, ἡ, *fruit of an Egyptian herb*, Paul.Aeg. 7.3 (189.21 H.).

Ἀκαμάντια, τά, *feast of the Akamantes*, *SEG* 34.1644 (Cyrene, iv BC).

Ἀκαμαντιάδες, αἱ, (sc. ἡμέραι), *days of the festival of Akamantia*, *SEG* 34.1645 (Cyrene).

ἀκᾰμαντόδετος, ον, *indissolubly bound*, A.*Pr.* 426 (v.l. ἀδαμ-).

*ἀκᾰμαντοπόδᾱς, 'α, ὁ, = ἀκαμαντόπους, χρόνος Synes.*Hymn.* 8.63.

Ἀκάμας, eponymous hero of the Acamantid tribe in Attica, Aeschin. 2.31, *SEG* 23.78(*b*)8 (iv BC); pl., guardian deities of the tribe, *IG* 2².1358 ii 32-3.

ἀκάματος, at end delete 'ἀκᾱμᾰτος .. but'

*ἀκαμπτέω, *behave uprightly*, ἐν τοῖς τόποις *PTeb.* 703.272 (iii BC).

ἀκανθέα, before '*PLond.*' insert '*PGoodsp.Cair.* 30.10.23 (ii AD)'

ἀκανθεών, after '*thorny brake*' insert '*POxy.* 1985.17 (vi AD)'

ἀκανθίας 2, for '*grasshopper*' read '*cicada*'

ἀκάνθιον, add 'see also °ἀχάντιον'

ἀκανθίς **I**, add 'Plin.*HN* 10.175, 205'

ἀκανθοβάτης, for '*walking* .. *nickname of grammarians*' read '*treading on thorns*, i.e. revelling in thorny passages, *of pedantic grammarians*'; after '-βάτις' insert 'ἀκρίδα .. ἀκανθοβάτιν'; for '(Leon.)' read '(Leon. Tarent.)'

ἀκανθοπλήξ, add 'also a play by Apollodorus, acc. to Suid.'

ἄκανθος, line 1, after 'ὁ' insert '(ἡ Dsc. 3.17)' add '**III** = ἀκανθίς I, Ael.*NA* 10.32.'

ἀκανθών, add '**2** *acacia*, *BGU* 2182.8 (AD 510).'

*ἀκάνθωτος, ή, όν, *decorated with ἄκανθαι*, ἧλοι *Inscr.Délos* 1439*Ab*á45, 1450.47 (ii BC).

ἀκάπνιστος, add 'ἔλαιον ἀ. Aët. 15.13, κυρύον (= κηρίον) ἀκάπνυστον pap. in *SCO* 27.98'

*ἀκαρδάμῠτος, ον, *unblinking*, v.l. in Opp.*C.* 1.208, 4.134; cf. ἀσκαρδάμυκτος.

ἀκάρδιος **I**, delete '*heartless* .. *Je.* 5.21' add '**b** *senseless, witless*, Lxx *Je.* 5.21, *Pr.* 10.13, 17.16, *Si.* 6.20(21).'

ἀκαρής **II**, for 'Men. 835' read 'Men.*Dysc.* 695'

ἀκαριαῖος, add 'of a drop of flavouring, ἀ. χυμός Arist.*Sens.* 446ᵃ10'

*ἄκαρον· τυφλόν Hsch.

ἀκαρφής, add '(dub., v. κατακαρφής)'

ἀκασκᾶ, after 'Cratin. 126' insert 'commentary on Pi. in *POxy.* 2451. fr. 14 i 10' and delete 'but .. fr. 28'

*ἀκαστόφρων· συνετός Hsch.

ἀκαταγνώστος, add '*TAM* 4(1).130 (Rom.imp.)'; for 'Keil-Premerstein .. (iii AD)' read '*TAM* 5(1).224 (AD 189/90)'

†ἀκαταιτίατος, ον, *not charged* or *accused*, J.*BJ* 1.24.8, 2.14.8, 4.3.10, al.

*ἀκατακλῑνής, ές, *unbending*, Procl.*in Ti.* 3.258.22 D.

*ἀκαταληκτικῶς, *indefinitely*, cj. in Arr.*Epict.* 2.23.46.

*ἀκατάληπτος, ον, *not gossiped about*, *POxy.* 3057.17 (i/ii AD).

ἀκαταληψία, for '*inability* .. term' read '*impossibility of direct apprehension*, term for Sceptic denial of Stoic doctrine of φαντασίαι καταληπτικαί, *impressions which carry a certainty of their truth*'; at end read 'S.E.*P.* 1.1, 236, 2.21'

Ἀκάταλλος, add '(iv BC)'

ἀκατάμακτος, add 'Hippiatr. 34; cf. κατάμακτος'

*ἀκατάμεμπτος, ον, *unexceptionable*, *SEG* 32.928 (Sicily, iv AD); adv. -ως, *IG* 12(7).231 (Amorgos).

ἀκαταμέμπτως, delete the entry.

ἀκαταπάτητος, delete the entry.

ἀκατάπαυστος, after 'D.S. 11.67, etc.' insert '*PMag.* 4.2364, *SEG* 27.1243.12'; after '-τως' insert 'Corn.*ND* 15' **II**, delete the section.

†ἀκατάποτος, ον, *not swallowed*, mistranslation of Hebr. *lōʾ yāʿlōs* in Lxx *Jb.* 20.18 (cf. Arabic *ʿalasa* "drank").

*ἀκαταργητί, adv., perh. *without being excused work*, διὰ τὸ ἀκατaργηθεὶ αὐ[τὸ]ν λαμβάνειν τὸν προκείμενον μισθόν *POxy.* 2875.26 (iii AD).

ἀκαταστασία **I**, line 3, after 'etc.' insert '*fickleness*, Lyr.Alex.adesp. 1.7, *SEG* 35.221.15 (tab.defix., Attica, c.AD 250)'

ἀκατάστατος **I**, add 'gramm., *irregular*, A.D.*Adv.* 134.19, *Synt.* 116.23' **II**, add 'Gal. 16.574.5, 17(1).557.10'

ἀκαταχρημάτιστος, for '*not encumbered with debt*' read '*not to be sold*

ἀκάτειος SUPPLEMENT ἀκονιτί

or *used as security for a loan*' and for '*Sammelb.* 364 (Alexandria)' read '*SEG* 24.1189 (Alexandria, i AD)'

ἀκάτειος II, for this section read 'ἀκάτειον, τό, *small sail* (= ‡ἀκάτιον), Aen.Tact. 23.4 (cj.); prob. in fig. phr., w. pun on sense of "cup" (cf. ἄκατος II), Epicr. 9.1 K.-A., Ar.*Lys.* 64'

*ἀκατεργᾰσία, ἡ, *lack of cultivation*, *SB* 5230.15 (i AD).

ἀκατηγόρητος, add 'ἀκατᾱγ- *SEG* 8.531.41 (Egypt, 57/6 BC)'

ἀκάτιον, for 'used by pirates' read '*CRAI* 1988.532-3.1, 7 (Ion. inscr., S. France, v BC)'; after 'Plb. 1.73.2, etc.' insert '**2** (unless ἀκάτειον should be read in these passages) kind of small sail, X.*HG* 6.2.27, Luc.*Lex.* 15, *JTr.* 46, *Hist.Conscr.* 45; in fig. phrs., Epicur.*fr.* 89 A.'

*ἀκαυμάτιστος, gloss on ἀπρόσειλος, Hsch.

ἄκαυστος 2, after 'Thphr.*Lap.* 4' insert 'subst., οἱ ἄ., name given to "carbuncles" (in practice, prob. rubies), Plin.*HN* 37.92'

*ἀκέζομαι, v. °ἀκέομαι.

ἀκενόσπουδος, for '*shunning vain pursuits*' read '*not courting useless trouble*'

ἀκέομαι, after '[ᾰ]' insert 'ἀκέζομαι Hld. 5.18.4 (v.l.)'; line 3, after 'Men. 863' insert 'ἀκεῦνται prob. in Hdt. 7.236'

ἀκέραιος, line 2, after 'Sch.Ar.*Pl.* 593' insert '*Dig.* 31.88.15'   **II 2**, after '*perfect*' insert 'Pl.*Plt.* 268b'   **II 3**, add '*naïve*, *SEG* 35.214.16 (tab.defix., Attica, c.AD 250)'

*ἄκερμος, ον, *without money*, prob. in Aesop.*Prov.* 39 P. (ἄκερος cod.).

ἄκερος, add 'Gal. 2.430.18, 10.21.9'

ἀκερσεκόμης, line 1, after 'Dor.' insert 'nom. -κόμας *SEG* 37.1175.11 (Pisidia, ii AD)'

*ἄκερσος, ον, *not shorn*, rest. in *PMasp.* 141 iii 11 (vi AD).

ἀκεσίας, delete the entry.

Ἀκεσίας, ου, comic name for a bad doctor, Ar.*fr.* 934 K.-A., Plu.*Prov.* 98, Phot. α 735 Th.'

*ἀκεστάλιος, α, ον, unexpld. adj., ἀ. ὄρνιθες Stesich. 70 P.

*ἀκέστριον, τό, *needle*, Eust. 1647.58.

Ἀκεσώ, ἡ, a goddess of healing, *IG* 4².135 i 3 (Epid.).

ἀκεύει, delete the entry ν. °ἀκεύω).

*ἀκεύθω, ἡ, = °ἀκεύω, Mitchell *N.Galatia* 234 (i AD).

*ἀκεύω, prob. *look after*, *Inscr.Cret.* 4.72.2.17; cf. ἀκεύει· τηρεῖ. Κύπριοι Hsch.

*ἀκεφαλαίως, adv. *without arrangement under headings*, *SEG* 8.694.9 (Luxor, iii/ii BC).

ἀκέφαλος 1, after 'J.*BJ.* 4.8.4' insert 'ζῷα Gal. 3.614, 631.7; of plants, τὸ δὲ γήτειον καλούμενον ἀκέφαλόν τι, i.e. *bulbless*, Thphr.*HP* 7.4.10'   **2**, for '*without beginning*' read 'fig., of badly constructed arguments or narrative'

ἀκέων, after 'Od. 11.142' insert '*h.Cer.* 194, A.R. 3.85'

ἀκή (B), for 'ἀκὴν ἔχεν Mosch. 2.18' read 'ἀ. ἔχειν Call.*fr.* 238.9 Pf., A.R. 3.521, etc.' and add 'cf. ἀκήν'

ἀκή (C), delete the entry.

*ἀκηδεί, adv., *carelessly* Theodos.Gr. p. 74.

ἀκηδέω 1, add 'w. acc., ἀ. τοὺς θεούς *FGrH* 153 F 7'

*ἀκηδιώδης, ες, *exhausted*, Vit.*Aesop.*(G) 76 (ἀκιδ- cod.).

ἀκήμων, add 'perh. *silent*, *still*, οὐρανὸς ἀ. cj. in Plot. 5.1.2 H.-S. (addenda)'

ἀκήριος (A) I, delete 'ψυχαὶ .. *Fates*' and for 'Ps.-Phoc. 99' read 'Ps.-Phoc. 105 V.'

ἀκήρωτος, after '*unwaxed*' insert '*Inscr.Délos* 507.13 (iii BC)'

ἀκιδνός, at end delete '(*insipid* .. Brandt)' and for 'ἰατρός' read 'λογισμός' adding 'Hp.*Mul.* 12, 52'

ἀκιδωτός I, after ' = foreg.' insert '*Inscr.Délos* 1421*Acd*18, 1450*A*21'

ἀκίναγμα, for pres. ref. read '*Suppl.Hell.* 1030'

†ἀκῑνάκης, ου, ὁ, acc. -άκεα Hdt. 3.118; gen -άκεος 4.62; acc. pl. -άκεας (v.l.) 3.128: *short straight sword* of Persian origin, ἀ. σιδήρεος, Pl.*R.* 553c, X.*An.* 1.2.27, D. 24.129, ἀ. ἐπίχρυσος; a Persian sword kept in the Parthenon, *IG* 1³.354.81, 2².1394.11. **b** as an object of worship, Σκύθαι .. ἀκινάκῃ θύοντες Luc.*JTr.* 42; in an oath, οὐ μὰ τὸν ἀ. id.*Tox.* 38 [*acinacēs* in Hor.*Carm.* 1.27.5] cf. κινάκης.

ἀκίνητος II 4, add 'gramm. *unmodified*, of a wd. in its basic form, i.e. nom. sg. or 1 sg., Choerob. *in Theod.* 363.27 H.'

*ϝακίνθια, v. Ὑακίνθια.

*ἄκιον, unidentified domestic object, *POxy.* 1290.6 (v AD).

ἀκιρός II, add '(perh. Lat. *aquilo*)'

ἀκίς I 2, add '*IG* 4²(1).122.58, 60 (Epid., iv BC)'   **4**, add 'of hostility, ἀκίσιν ὀφθαλμῶν ἐνήλατο Lxx *Jb.* 16.10'

ἀκίσκλη, after 'ἡ' insert '(or -ον, τό)'; add 'gen. pl. ἀκίσκλων or -ῶν, cf. Lat. *acisculus*'

*ἀκίτᾱλος, unexpld. wd., Hdn.*fr.* p. 21 H.

*ἀκκήσσος, ὁ, Lat. *accensus*, *supernumerary*, *IGRom.* 3.578, *Ann.Épigr.* 1929.89, *IEphes.* 1544.10 (ἀκκηνσ-).

ἀκκίζομαι 1, before 'Ael.*Ep.* 9' insert 'as v.l. in'

*ἀκκοθήκιον, τό, (?for ἀγγο-) *receptacle for vessels*, *PAnt.* 204.4 (vi/vii AD).

*ἀκκουβίκυλον, τό, Lat. *accubiculum*, *dining-room*, *POxy.* 2058.25 (vi AD).

*ἀκκουβιτάλια, τά, Lat. *accubitale*, app. *couch-coverings*, *PBerl.Zill.* 13 (vi AD).

*ἀκκουβιτᾶρις, Lat. *accubitalis*, τάπης ἀ. of or for a couch (°ἀκκούβιτος), *Edict.Diocl.* 19.34; neut. as subst., prob. *cover for a (dining) couch*, ἓν ἀκκουβιτάριν *POxy.* 3860.18 (iv AD).

*ἀκκουβίτιον, τό, *dining-room*, *MAMA* 6.84 (ἀκουβ-).

*ἀκκούβιτον, τό, *couch*, (sts. sp. ἀκουβ-), Ps.-Callisth. 1.21.9. **2** *bedroom*, *PLond.* 1724.30, 1733.19 (vi AD).   **3** *dining-room*, Ath. *Imag.Beryt.* 2 (*PG* 28.805).

*ἀκκούβιτος, ὁ, Lat. *accubitus*, *accubitum*, *couch*, *Edict.Diocl.* 19.34.   **2** *bedroom*, *BGU* 2202.19 (AD 565).

*ἀκκουμβίζω, = Lat. *accumbo*, *recline at table*, Vit.*Aesop.*(G) 40 (ἀκουμβήσωμεν cod.), Suid. s.v. πρόσκλιτον.

ἄκλαστος, for 'Thphr.*CP* 1.15.17' read 'Thphr.*CP* 1.15.1'

ἄκλαυ(σ)τος I, after 'A.*Eu.* 565' insert 'Luc.*Cat.* 5'

ἀκλεής 1, adv., after 'Antipho 1.21' insert 'Isoc. 4.84'

ἀκλήρημα, delete '*loss*' and 'Dicaearch. 1.25 (pl.)' and for 'etc.' read '(in pl.) *drawbacks* (including moral faults), Dicaearch. 1.25; *misfortunes*, *REG* 101.16.44, 61 (Xanthos, iii BC)'

ἀκληρονόμητος, after 'Eust. 533.32' insert '*not to be transmitted by will*, *SB* 9801.6 (ii/iii AD)'

ἀκλόνητος, after '*unmoved*' insert 'Palaeph. 52'

ἄκλονος, delete 'of a rider .. Palaeph. 52'

ἀκλοπεία, for '*administration*' read '*conduct*'

ἀκμάδιον, for 'Ps.-Mos.Alch.' read 'Moses Alch.'

ἀκμαῖος I, after '(A.)*Pers.* 441' insert 'Hippothoon 5 K.-A.'   add '**III** *pointed*, sup., Hld. 9.19.4; cf. ἀκμή I.'

*ἀκμαιότης, ητος, ἡ, *the prime of life*, Ptol.*Tetr.* 29.

*ἀκμάς, άδος, ἡ, fem. adj. *seasonable*, ἀ. βουλῇ *Orac.Chald.* 37.1 P.

ἀκμή III, add 'personified, *SEG* 34.1446 (Syria, Rom.imp.)'

ἀκμήν, lines 1/2, for 'A.*fr.* 451 G, Men. in *Cod.Vat.Gr.* 122' read 'A.*fr.* 339a R., Men.*fr.* 715a K.-Th.'; line 4, after 'Theoc. 4.60' insert 'Call.*fr.* 781 Pf.'; line 7, for '*Cod.Vat.Gr.* 122' read '*Cod.Vat.Gr.* 12 (Reitzenstein *Ind.Lect.Rost.* 1892/3 p. 4)'

ἄκμηνος, for 'Call.*Fr.anon.* 4' read 'Call.*fr.* 312 Pf.' and add 'A.R. 4.1295, cf. ἄκμα'

ἀκμής, after '(Alph.)' insert 'perh. *that is in new condition*, προτομὰν ἀκμῆτα Antip.Sid.in *GLP* 1.107(4).2 (= *HE* 48 G.-P.)'

ἄκμητος, delete '**II** *not causing pain*'

*ἀκμινάλιον, τό, cf. Lat. *agminalis*, *baggage-animal*, *POxy.* 3741.45 (AD 313).

*ἀκμονευτής, οῦ, ὁ, *one who hammers* in a forge, Sm.*Is.* 41.7.

*ἀκμονίσκος, ὁ, dim. of ἄκμων, *small anvil* on which coins were struck, *IG* 2².1408.12 (iv BC).

ἄκμων, lines 1/2, delete 'orig. ... **II**' transferring Hes. quots. to following sense; last line, delete 'sling-'

*ἀκναμπτεί, adv. *inflexibly*, Pi.*fr.* 70c.12 S.-M.

*ἀκναφάριος, v. °ἀγναφάριος.

*ἄκνη, ἡ, = ἴονθος II, *acne* or sim. skin condition, Aët. 8.14 (420.1 O.) (pl.).

ἄκνημος, for '*without calf*, of the leg' read 'of freaks of nature, *having no lower leg*'

ἀκοιλάντως, add 'ἀκυ- *Stud.Pal.* 142.21'

ἀκοινώνητος II, delete '*inhuman*, Cic.*Att.* 6.3.7' and 'cj. ib. 6.1.7 (v. ἀκοινονόητος)'

ἀκολάστασμα, delete 'Alciphr. 1.38'

*ἀκολαστότης, ητος, ἡ, *indiscipline*, Asp. *in EN* 3.15.

ἄκολος, after 'Od. 17.222' insert 'Call.*Cer.* 115'

ἀκολουθέω, line 8, for 'Men.*Adul.fr.* 1' read 'Men.*Kol.fr.* 1.1 K.-Th.'   **II 3**, add 'gramm., A.D.*Adv.* 163.16, *Pron.* 3.15'   **4**, line 2, for 'Hippon. 55' read 'Hippon.*fr.* 79.9 W.' and delete '(s.v.l.)'

*ἀκολουθικός, ή, όν, = ἀκόλουθος, *PCair.Zen.* 676.3 (iii BC).

ἀκόλουθος 3, adv., abs., add '*consequently*, *Cod.Just.* 1.4.32.3'

*ἀκομεντανήσιος, = °ἀκομενταρήσιος, *PHarris* 94.7 (v AD), *SB* 2253.12.

*ἀκομενταρήσιος, ὁ, fr. Lat. *a commentariis*, official in charge of public records, *IGRom.* 3.1264.

ἀκόμιστος, after 'S.*Ichn* 143' insert '*GLP* 1.122.17 (anon., ii AD); *PAnt.* 58.11 (iv/v AD)'

ἀκονάω, line 3, after 'Arist.*Pr.* 886ᵇ10' insert 'Sosith. 2.18 S.'

†ἄκονδος ἄχαρις. Κονδᾶς γὰρ χάρις ἐστίν Hsch. (cf. perh. °κόννος).

ἀκόνη, add '**4** *small mortar* used by doctors, *Lapid.Gr.* 194.19.'

ἀκονητής, after '*sharpens*' insert 'or *polishes*, = Lat. *samiator*'; after 'σπάθης' insert 'περικεφαλαίας, etc.'; after '*Edict.Diocl.* 7.33' insert '*MAMA* 3.414 ([ἀ]κωνιτοῦ gen.), perh. to be restored in *Corinth* 8(1).245 (ἀκον[-, v. Robert *Hell.* 11/12.37-38)'

ἀκονιτί, after 'D. 19.77' insert 'perh. Lacon. form ασσκονικτει *CEG* 372 (Laconian script at Olympia, vi BC), cf. perh. mod. Gk. σκόνη = *dust*'; add 'see also ἀκονητί'

ἀκόνῑτον **I**, delete 'dub. l. in Nic.*Al.* 42, cf.' and after '*AP* 11.123' insert 'perh. w. play on ἀκόνιτος (adj.)'
ἀκοντίζω, after 'Att. fut. -ιῶ' insert 'Boeot. inf. ἀκοντιδδέμεν *SEG* 32.496' **I 3**, for '*jettison* cargo' read 'πάντα .. ἠκοντίζομεν ἔξω τῆς νηός, i.e. jettisoned' **4**, for '*shoot forth rays*, of moon' read 'abs., of the moon, *dart its rays*'; after 'med.' insert 'of flames' add '**6** *direct* a stream of liquid, Ach.Tat. 4.18.6 V.; pass., *spout up* or *out*, τὸ αἷμα Alex.Trall. 1.13 p. 515.17 P.'
ἀκόντιον, add '**3** *goad*, ἀ. ἤτοι μάστιξ *Edict.Diocl.* 15.17. **II** = ἀκοντίας 1, kind of snake, Fronto *Ep.* p. 22.7 vdH.'
ἀκοντιστήρ **I**, add '**2** ἀκοντιστῆρες μόλυβοι perh. (*lead*) *jets* of a fountain, *IEphes.* 3214.16 (i AD).' **II**, lines 3/5, delete 'ἀκοντιστῆρες .. p. 89'
ἀκοντοδόκος, add 'Sch.Il. 16.361b, Didyma 496B9 (ii AD)'
ἀκοός, add '*Trag.adesp.* 279g.10 K.-S. **2** ἀκοό, οἱ, *witnesses* to a transaction, *BCH* suppl. 22 p. 236 (Argos, v BC).'
ἀκοπητί, delete the entry.
ἄκοπος **II 2**, for '*removing .. refreshing*' read '*relieving pain*' add '**4** ἀ. (sc. λίθος), ἡ, kind of stone, Plin.*HN* 37.143.' **III 2**, after '*whole*' insert 'κέγχρου ἀκόπου *Edict.Diocl.* 1.5'
ἄκοπρος **I**, add 'Gal. 15.699.18'
*ἀκοράστως, v. °ἀκόρεστος.
*ἀκορεσταίνω· ἀκόρεστα πράττω Phot. a 800 Th.
*ἀκόρεστος **I 1**, adv., add 'Cypr. *a-ka-ra-sa-to-se* perh. ἀκοράστως, *ICS* 264.2'
†ἀκόρετος, = ἀκόρεστος, edd. in A. *Ag.* 1117, 1143 (codd. -εστος).
ἀκορία **I**, add 'prob. *want*, *Tab.Defix.Aud.* 15.23 (Syria, iii AD)'
ἀκόρσωτον, add '(cod. ἀκόρωδον)'
*ἀκόρωδον, v. °ἀκόρσωτον.
ἀκοσκίνευτος, add '*PTeb.* 1029.5 (ii BC)'
ἀκοσμέω, add '*not to observe religious ritual* or *discipline*, *IG* 5(1).1390.39 (Andania, i BC), ἄν τίς τι ἀκοσμέ[ι] *IG* 1³.82.25'
ἀκόσμητος **3**, for '*unfurnished*' read '*undecorated*' and add '*SEG* 39.1055.9 (Neapolis, AD 194)'
ἀκοσμία **I 2**, add 'Plu. 2.926e' add '**III** *lack of adornment*, Gorg.*Hel.* B 11.1 D.-K.'
ἄκοσμος **I**, add '**b** *of loose morals*, χήρη *AP* 5.302.9 (Agath.); γυναῖκες Agath. 5.14.4 K.' **II**, after '(Jul.)' insert 'cf. καί νύ κε κόσμος ἀ. ἐγίνετο Nonn.*D.* 6.371'
ἀκοστή, add 'cf. κοσταί'
*ἄκοστος, v. Αὔγουστος.
*ἀκούβιτον, etc., v. °ἀκκουβ-.
*ἀκούρης, ?ου, ὁ, *one having long hair*, perh. in *IG* 12(5).225 (cj. in *Philol.* 65.633, Paros, v B.C.).
†ἀκούσιος, ηακόσιος, v. ‡ἀκούσιος.
ἄκουσμα **3**, after '*instruction*' insert 'Pl.*Ep.* 314a'
*ἀκοῦχος, ὁ, *some item of furnishing for a bedroom*, *PMich.*inv. 4001 (*ZPE* 24.83, i/ii AD), *PBremen* 21.8 (i AD).
ἀκούω **II 2**, add '**b** *receive an order*, w. inf., μετὰ δεῖπνον ἄγειν τὸν νεανίαν ἀκούσασα οὕτως ἔπραττε Hld. 7.26.1.' **III**, add '**6** pass., *to be given a judicial hearing*, *Cod.Just.* 10.11.8.1.'
ἄκρα **5**, add '*Inscr.Prien.* 4.51 (iv BC), Pouilloux *Rhamonte* no. 15.18, ἐν ταῖς τῶν ἄκρων πολιορ[κίαις] *MDAI(A)* 72.234 no. 64.28 (Samos)'
ἀκράαντος, add 'A.R. 1.469, 4.387; n. pl. subst., Hom.*Epigr.* 4.14'
ἀκράδαντος, after 'etc.' insert '*unaffected*, σῶμα οἴνῳ Clem.Al.*Paed.* 2.2.22'
ἀκραής, delete 'si ἀκραές .. 10.17.9'
ἀκραινές, after 'ἐγκρ-' insert 'ἀκρανές La.'
ἀκραῖος, for '= ἄκρος' read '*extreme, topmost*, etc.' **II**, before 'Aphrodite' insert 'Zeus, Sokolowski 1.56.13 (ii BC)'; after 'Paus. 1.1.3, 2.32.6' insert '*SEG* 32.1380 (Cyprus, AD 14)' add '**b** Ἀκρῆα, τά, a festival at Coronea, Schwyzer 503a.18 (*c.*200 BC).'
†ἀκραιφνής, ές, *pure, unadulterated*, ὕδωρ Ar.*fr.* 34 K.-A.; ἀήρ Hp.*Morb.Sacr.* 16, *Mul.* 1.11, etc., Anticl. 22 J., Lyc. 151. **b** of abstract things, ἀρετή *J.AJ Proem.* 4, ἀλήθεια Ph. 2.219, φύσις id. 2.374, πενία *AP* 6.191 (Corn.Long.); sup. Ph. 2.319; adv. -ῶς Ph. 1.100. **2** *unimpaired, intact*, ξυμμαχία Th. 1.19; of a fleet, id. 1.52.2, Pl.*Ax.* 2.366a, D.H. 6.14, Procop.*Aed.* 1.10; w. gen., *untouched by*, ἀ. τῶν κατηπειλημένων S.*OC* 1147, Lysipp. 9; adv. -ῶς Poll. 1.157. **3** *pure, innocent*, κόρης ἀκραιφνὲς αἷμα E.*Hec.* 537, *Alc.* 1052, Ph. 1.515.
†ἀκρασίων, ωνος, ὁ, *intemperate person*, Cerc. 4.3.
*ἀκράτευτής, οῦ, ὁ, = ἀκρατής II 2, *lacking self-restraint*, ἀ. εἰς μοιχείαν Anon. *in Rh.* 120.30 (pl.).
ἀκρατίζομαι, lines 5/6, after 'fut. -ιῶ' insert 'pf. part. ἠκρατικώς Didyma 286.8' and after '*breakfast*' insert 'τὴν πόλιν Didyma l.c.'
ἄκρατος **I 2**, add 'of myrrh, σμύρνης ἀκράτου *Suppl.Hell.* 1164, Emp. B 128.6 D.-K.' **6**, for 'ἀ. καῦμα *AP* 9.71 (Antiphil.)' read 'μανίην *AP* 12.115'
*ἀκρατοχολής, ές, *consisting of pure bile*, τὰ -ῆ διαχωρήματα Gal. 19.108; cf. ἀκρητόχολος.

*ἀκρέγιαιον, *an unknown bird*, *PAmst.* 13.14 (v AD).
ἀκρελεφάντινος, add '*Inscr.Délos* 1409Bai47 (ii BC)'
ἀκρεμών, add '**2** *tentacle* of a polyp, Opp.*C.* 3.181.'
*ἀκρεοπαώνι(ν), a bird, app. a kind of peacock, *PAmst.* 13.3 (v AD); cf. πάων.
*Ἄκρηα, v. °ἀκραῖος II b.
ἄκρηβος, add '*Trag.adesp.* 656.18 K.-S.'
*ἄκρηστος, v. °ἄχρηστος.
ἀκρηχολία, read 'ἀκρηχολίη'
ἀκρία, before 'of Athena' insert 'perh. of Hera, *IG* 9(1).698 (Corcyra, iv BC)'
ἀκριβάζω, for '*proud*' read '*proved*'
ἀκριβασμός, after '3*Ki.* 11.34' insert '(cod. A)'; delete 'pl., ἀ. καρδίας *searchings* of heart'; for '*portion, gift*' read '*fixed allowance*'
ἀκριβαστής, delete '*inquirer*'
ἀκρίβεια **I**, line 2, after 'Lys. 17.6' insert 'τὸν δὲ διαυγῆ χαλκόν, ἀκριβείης οὐκ ἀπολειπόμενον *AP* 6.210.4 (Philit.Sam.)'; line 6, after 'Arist.*Resp.* 478ᵇ1' insert 'ἐπ' ἀκριβείας Demetr.*Eloc.* 222'
ἀκριβής **I**, add 'adv. -βῶς, οὔτε ἀ. ὡπλισμένους of archers, Arr.*An.* 1.28.5'
ἀκριβολογέομαι, after '(Pl.)*Cra.* 415a' insert '*quibble*, Modest.*Dig.* 27.1.6.7'
ἀκριβόω **I**, delete 'ἀ. τάδε .. X.*Cyr.*' and transfer '2.3.13' to follow 'X.*Cyr.* 2.2.9' in section 2; after 'Arist.*Pol.* 1279ᵇ1' insert 'of a work of art, εὖ σμίληισιν ἠκριβωμένην Call.*fr.* 202.66 (Add. II p. 119) Pf.' **2**, for '*investigate thoroughly*' read '*make sure of*'; delete '*inquire carefully of*' and after '*Ev.Matt.* 2.7' insert 'Longus 3.12.4'
†ἀκρίδιον, τό, *a little tip, spike*, καρπὸν .. ἔχει (βρόμος) ἐπ' ἄκρῳ ὥσπερ ἀκρίδια δίκωλα etc., Dsc. 2.94.
*ἀκριδοθήκη, ἡ, *cage for locusts*, Longus 1.10.2 (v.l. -θήρα).
*ἀκριμακραγετα, unexpld. wd. on an amulet, ἀ. κύριε βοήθι *SEG* 14.883 (Egypt, Rom.imp.), a magical wd., or perh. for ἀκριδομακράγετα *one who drives locusts far away*, *AJPh* 76.308-9.
ἄκρις, line 1, after 'ιος' insert '-εως *IG* 5(1).1370.6 (i BC); -ιδος in place-name Ἄκρις *SEG* 28.103.4; τῶι Ἡρακλεῖ τῶι ἐν Ἄκριδι ib. 19 (Eleusis, iv BC)'
ἀκρίς, after 'sg., in collective sense' insert '*PTeb.* 772.2 (iii BC)'
ἀκριταγών, add '(ἀκριτάγωνος La., ἀκριτόγωνον Fix)'
ἀκριτοβάται, add '(ἀκροβάται La.) cf. °ἀκροβάτης'
*ἀκρῑτόφῡλος, ον, = ἄκριτος III 2, κλωστήρων γένη *SEG* 8.768.18 (Egypt); cf. φυλοκρινέω.
ἀκροάομαι **I**, after 'Pl.*Grg.* 499b' insert 'λέγοντος ἐμοῦ ἀκροάσονται id.*Ap.* 37d'; add 'perh. also act. form ἐλέλευσεν Πτολεμαίῳ .. ἀκροᾶσε (app. aor. inf.) τὸ πρᾶγμα *POxy.* 3398.12 (iv AD)' **2**, add '**b** *give a hearing, judge, Cod.Just.* 8.10.12.8.' **II**, for 'Th. 3.27' read 'Th. 2.37.3'
ἀκροαπίς should follow ἀκροάομαι
ἀκρόασις, for '**II** = ἀκροατήριον' read '**2** *lecture* (as a performance)'
ἀκροατήριον **II**, add 'prob. in D.H.*Dem.* 15'
†ἀκροβάτης, *acrobat*, *PRyl.* 641.22 (iv AD); as a functionary in the cult of Artemis at Ephesus, *IEphes.* 27.459, 537, 941.4, etc.
ἀκρόβατος, delete the entry.
ἀκροβελίς **I**, for '*dart*' read '*spit*'
ἀκροβηματίζω, before 'Hsch.' insert '*Iamb.adesp.* 35.2 W.'
ἀκροβολισμός, after 'Aen.Tact. 39.6, etc.' insert 'fig., Ach.Tat. 1.10.4; as a military exercise, Pl.*Lg.* 804c'
*ἀκρόγωνον, τό, *sharp corner, acute angle*, *PMasp.* 109.26 (vi AD).
ἀκροδίκαιος, for 'ἀκριβοδίκαιος' read '*eminently just*, Heph.*Astr.Epit.* 4.116.45; ἀ. τὸ ἔσχατον τῆς δίκης Hsch.; comp., ὃς πάσης ἀρετῆς ὢν ἀκροδικαιότερος *MAMA* 8.569'
*ἀκρόδικος, ον, = ‡ἀκροδίκαιος, *SEG* 35.1233.18 (Saittai, ii AD).
*ἀκρόθηκτος, ον, *sharpened at the point*, ἀ. ἐγχέων E.*fr.* in Snell *Alexandros* p. 80.24.
†ἀκροθίνιον, τό, *first-fruits of war, best of the spoils*, esp. as offered to the gods, E.*Ph.* 282, Th. 1.132.2, Pl.*Lg.* 946b; pl., Eumel. 12.1 B., Simon. 97 D., Pi.*N.* 7.41, Ἀθηναῖοι τ[ὀ]ι Ἀπόλλων[ι ἀπὸ Μέδ]ον ἀκ[ροθ]ίνια τες Μαραθ[ō]νι μάχες *Delph.* 3(2).1 (489 BC), Hdt. 1.86.2, 90.4; transf., *choicest offerings, firstfruits*, A.*Eu.* 834; fig., τὰ .. Ἑλλήνων ἀκροθίνια i.e. Orestes and Pylades; Philostr.*Ep.* 61, *VS* 512.
†ἄκροθις, ἡ, acc. -θῖνα (this has been taken by some editors to be a var. form of the neut. pl. ἀκροθίνια), = °ἀκροθίνιον, Pi.*O.* 2.4, 10.57, *GDI* 2561.D47.
ἀκροθώραξ, delete '= ἡμιμέθυσος, Hsch, cf.'; for '*πεπωκότ*' read '*πεπωκότα*'; for 'Diph. 46' read 'Diph. 45 K.-A., Hsch.'
*Ἀκροκαλλίστιος, ὁ, epith. of Zeus at Delphi, *Delph.* 3(1).362 iii 14.
ἀκρόκομος **I**, at end delete 'of goat's chin'
*ἀκροκωλοίφιον, τό, *tip of the hoof* (written *acrocolefio*, abl. sg.), Veg.*Mul.* 3.1.2.

*ἀκρόλευκος, ον, *white-tipped* or *-edged*, Vit.Aesop.(G) 92.

ἀκρόλιθος, for 'IG 4.558' read 'IG 4.558.14' and after '(Argos)' add 'cf. Vitr. 2.8.11'

*ἀκρόλλιν, v. °ἀκρούλιον.

ἀκρολογέω, for 'gather at top' read 'glean'

ἀκρολοφίτης, for 'mountaineer' read 'dweller on mountain tops'; for '(Leon.)' read '(Leon.Tarent.)' and add 'APl. 256'

*ἀκρόμεστος, ον, brim-full, cj. in POxy. 3722 fr. 1.8 (lemma in Anacr. commentary, ii AD;].κρομεστου pap.).

ἀκρόνυχος (A), line 4, after 'Nic.Th. 761' insert 'Posidipp. in Suppl. Hell. 698' and after 'Pr. 942ᵃ23' insert '2 = ἀκρόνυκτος, PMich.III 149 xi 9/11, cf. Cat.Cod.Astr. 8(2).84.5.'

*ἀκροξιφίδιον, τό, sword-point, PSI 756.14 (Oxyrhynchus, iv/v AD), Gloss.

ἀκροπενθής, delete the entry.

ἀκρόπολις, after lemma insert '(poet. acc. ἀκροπόληα, Procl.H. 7.21)' II, line 2, after 'Simon. 137' insert 'Ἑλλάδος ἀ., of Sparta, CEG 819.12 (c.340 BC), Amyntas in Suppl.Hell. 44.8'

ἀκροπόλος I, delete 'cf.' and add 'h.Ven. 54'

ἀκρόπρωρον, add 'Cyran. 1.21.56 K.'

ἄκρος, line 1, after 'ἀκή A' insert 'w. secondary aspiration in Corcyran καθ' ἄκρον IG 9(1).690 (ii BC); cf. °ἀκροσκιρίαι' I 2 a, line 11, after 'E.Ion 1166' insert 'ἐπ' ἄκρα βέβηκας APl. 275.3 (Posidipp.)' 3, delete the section adding example to section I 2 a, reading 'extreme i.e. inmost' for 'inmost' III 2, for 'Diph. 54' read 'Diph. 53 K.-A., μισθός Theoc.Ep. 8.5' V 1, add 'supremely, ἄκρα σοφαὶ χέρες APl. 262'

ἀκροσκιρία, delete the entry, v. °ἀκροσκιρίαι.

*ἀκροσκιρίαι, αἱ, heights covered with brushwood, Tab.Heracl. 1.65,71 (written ἡακροσκιρίαις dat., -ιᾶν gen.; cf. σκίρος).

*ἀκροσκόλιος, ον, curved at the end, Anon.Alch. 347.4.

ἀκροστιχίς, after 'D.H. 4.62' insert 'GVI 261.12 (Lycia, i/ii AD)' and after 'Cic.Div. 2.54.111' insert 'ἐπιθαλάμιον μετὰ ἀκροστοιχίδος τοῦ νυμφίου Dioscorus 23 tit.'

ἀκροστόλιον, add 'Inscr.Délos 1403Bai40 (re-edited SEG 37.692, ii BC)'

ἀκροσφαλής I, before 'Plu. 2.713b' insert 'ἀ. ἴχνος Nic.Al. 242; διάνοιαν ὑγρὰν ὑπὸ τῆς μέθης καὶ ἀ. γεγενημένην' II, add 'subst. τὸ -ές, hazardousness, Longin. 22.4'

ἀκρότης II, place 'D.H.Dem. 2' after 'Diog.Oen.fr. 38' and delete 'cf.' and 'etc.'

*ἀκρούλιον, τό, perh. woollen fringe or some kind of garment, Stud.Pal. 20.245.19 (vi AD, ἀκρόλλιν); PLugd.Bat. xxv 13.22 (vii/viii AD), PApoll. 104.10-12 (viii AD).

ἀκρουροβόρη, add 'also masc., -βόρος, PSI 28.30 (iii/iv AD)'

*ἀκροφυλάκιον, τό, guard-post on a citadel, OGI 254 (Babylon, ii BC) (pl.).

*ἀκρόχειρ, ὁ, murderer, Hymn.Id.Dact. 13, EM 53.37; cf. °ἀκρόχειρος.

ἀκρόχειρας, delete, and refer to °ἀκρόχειρ.

ἀκροχειρισμός, after 'wrestling with hands' insert 'at arms' length, opp. συμπλοκή'; add '(v. Robert Hell. 11/12.442 n. 4)'

ἀκρόχειρον, line 3, delete 'cf. Hymn.Id.Dact. 13'

*ἀκρόχειρος· ἀνδροφόνος Hsch.; cf. °ἀκρόχειρ.

*ἀκροχόρδονον, τό, kind of wart or top of a wart, Cyran. 35.5; cf. ἀκροχορδών.

*ἀκρόχρυσον, τό, gilding of the end(s), Anon.Alch. 378.17.

*ἄκρυτας, name of a bird, PAmst. 13.15 (v AD).

ἀκρωβέλια, after 'cod. -σβ-' insert 'edd. ἀκροβ-'

*ἀκρώμιος, ον, of the acromion (point of the shoulder), αὐτὴν τὴν σύνταξιν αὐτῶν ὀνομάζουσιν ἀκρώμιον ἁρμονίαν Gal. 2.766.

ἀκρωνυχία, for '= ἀκρώρεια' read 'rock, etc.'; add 'Philostr.VA 3.1'

ἀκρώνυχος, after 'Plu. 2.317e' insert 'in uncertain sense, ἔν τε τοῖς ἀκρωνύχοις καὶ τῇ ταυροδιδαξίᾳ Milet 1(7).205a (p. 302)'

ἀκρώρεια, for 'mountain ridge' read 'mountain heights' and for 'Theoc. 25.31' substitute 'Call.Dian. 224'

*ἀκρώρειος, ον, of mountain heights, Orph.H. 32.4.

ἀκρωρεῖται, for 'ridges' read 'tops'; add 'sg. Ἀκρωρείτης, cult-title of Dionysus at Sicyon, Apollod.ap.St.Byz. s.v. Ἀκρώρεια; epith. of Pan, HE 2277 G.-P. (Leon.Tarent.), ib. 491 (Antip.Sid.) (-ριται pap.)'

ἀκρωτερῆσαι, after 'Hsch.' insert '(ἀκρωτηριάσαι La.)'

ἀκρωτηριάζω I 2, add 'Philostr.VA 7.4'

*ἄκτα, τά, Lat. acta, official records, IGRom. 1.421.20, CIG 29.27, TAM 4(1).39, POxy. 2725.21 (AD 71), Inscr.Perg. 8(3).24; also sg., οὔτε δι' ἄκτου βουλῆς MAMA 8.584.

ἀκταινόω, for 'Pl.Com.' to end of article substitute 'Pl.Phd. acc. to gloss. in POxy. 2087.22 (where read ἐξᾶραι, ὑψῶσαι) and Phryn.PS p.39 B.'

ἀκταῖος, add 'Myc. a-ka-ta-jo (pers. n.)'

*ἀκτάριος, v. °ἀκτουάριος.

ἀκτέανος, add 'of Δίκη, i.e. incorruptible, BCH suppl. 8 no. 87 (= AP 9.686)'

*ἀκτεινοβολή, v. °ἀκτινοβολή.

ἀκτέον I 1, add 'Plot. 1.3.1'

ἀκτεροί, add '(ἀκτέϊνοι La.)'

ἀκτή (A), line 5, for 'usu. of sea, coast .. rivers' read 'of river-banks' and transfer 'χλωρὰ ἀ. (S.Ant.) 1132, ἀκταὶ ἔναλοι Tim. Pers. 98 P.' to follow 'Il. 12.284 (line 2)' 2, after 'into the sea' insert 'cape, peninsula' II, for 'generally, edge' read 'mound'

*ἀκτήνδε, adv., to the shore, A.R. 1.318 (v.l. ἀκτὴν δ').

ἀκτήρ, v. ἀγκτήρ.

Ἄκτια, τά, Actian games, TAM 4(1).34 (Delphi, Rom.imp.), SEG 34.1314 (Lycia, i AD), Ἄ. ἐν Νεικοπόλει IG 2².3169.17 (Athens, AD 253/7), Ἄ. ἐν Τύρῳ ib. 29, SEG 37.512 (Dodona, iii AD).

Ἀκτιάς, άδος, ἡ, festival of Actian Apollo, IG 9²(1).583.9, 45, 69 (Olympia, iii BC). 2 the four-year period between the celebrations of the games held in honour of Augustus' victory at Actium, J.BJ 1.20.4, SEG 37.512 (Dodona, iii AD).

*ἀκτινηβολία, ας, ἡ, emission of rays, Man. 1.322. 2 aspecting from the left, id. 4.166.

*ἀκτινοβολή, ἡ, perh. = °ἀκτινοβολία 2, Anon.astrol. in PMich.III 149 xiv 28 (ἀκτεινο-, ii AD).

ἀκτινοβόλος, add 'of the sun, Melit.fr. 8b.24; as name of a horse, SEG 8.213.26 (Berytus, ii/iii AD)'

ἀκτινοκράτωρ, add 'dub., v. °αἰωνακτινοκράτωρ'

*ἀκτινολαμπής, ές, shining forth with its rays, ἥλιος Orac.Tiburt. 22, 24, 29.

ἀκτινώδης, for 'like rays' read 'appearing to give out rays'; add 'Cat.Cod.Astr. 11(2).163.5'

*Ἀκτιονίκης, ου, ὁ, victor in the Actian games, BCH 9.68.8 (-νείκης, Aphrodisias).

ἄκτιος, before 'ον' insert 'α'; add 'of Apollo, IG 5(1).29 (Sparta), of Arsinoe, PEnteux. 26.6 (iii BC)'

ἀκτίς I 1, add 'ᵇ day, ἐν δὲ μονήρει .. ἀκτῖνι in a single day, Nic.Al. 401.' 3, for 'Intr.' read 'in Ptol.' II, for 'spoke' read 'cog'

ἀκτίτης, for 'dweller on .. II' substitute 'of or on a headland, ἀ. καλαμευτής AP 6.304 (Phan.)'

ἄκτιτος, add 'Myc. a-ki-ti-to (cf. °κτίζω)'

**ἄκτοινος, ον, Myc. a-ko-to-no, not possessing an estate.

*ἀκτουάριος, ὁ, Lat. actuarius, keeper of records, PLond. 237.20 (iv AD), PMasp. 229 (vi AD), IGChr. 211 (vi AD), PAnt. 190b31 (vi/vii AD), CodJust. 1.42.2, Just.Nov. 117.11; also ἀκτουάρις, Dura⁶ 292, ἀκτουριν PBeatty Panop. 1.21 (AD 298); also ἀκτάριος, SEG 26.1778 (Egypt), PHarris 96.14, 24 (i/ii AD), BGU 848.1 (iii AD), IGBulg. 1774, ἀ. σπείρης ib. 1835 (iii AD).

ἄκυθος, delete 'c. gen. .. s.v.l.'

ἀκυθῶν, add '(ἀκυθον· ἄγρυπνον La.)'

ἄκυλος, for 'ὁ' read 'ἡ' II, add 'IG 1³.386.62, 387.69'

ἀκύλωτός, add 'φιάλη IG 2².1421.48, 1425.93'

ἀκύμαντος I, line 2, for 'those of the stadium' read 'above the water-line'

ἀκύματος, add 'fig., calm, ἀ. βίους Lxx Es. 3.13b'

ἀκύμων (B), for 'of women' read 'of the womb'

ἀκυρολογέω, for 'Lex.Vind. 3.19' read 'Ps.-Hdn.Gr. in An.Boiss. 3.265, al.'

ἄκυρος II, after 'power' insert 'not sovereign'; after 'Pl.Tht. 169e' insert 'τούτων μὲν οὖν ἄκυρός ἐστιν ἡ βουλή Arist.Ath. 45.4'

*ἀκυρώσιμος, η, ον, subject to cancellation, μίσθωσις PMich.II 123ʳ viii 14 (i AD); πρᾶσις PSI 913ᵛ (add.) (i AD).

ἀκύρωσις, for 'BGU 1282.35' read 'POxy. 1282.35'

*ἀκώδωνος, ον, without sound of κώδων, μέλος ἀσάλπιγκτον καὶ ἀ. Steph.in Rh. 317.4.

ἀκώνητος, add 'PCair.Zen. 743.3 (iii BC); Dor. ἀκώνατος, Inscr.Cret. 1.xvii 2a8 (Lebena, ii BC); cf. °ἀχώνευτος, κωνάω'

*ἀκώριαι· ἄκανθαι Hsch.; cf. ἀκορραί.

*ἀλαβαντικός, ὁ, wd. occurring in inventory of personal military equipment (app. from Alabanda in Caria), PCol. 188.16 (AD 320).

†ἀλαβαστοθήκη, ἡ, case to contain alabastron vase, Ar.fr. 561 K.-A., IG 2².1425.265 (iv BC), D. 19.237, also ἀλαβαστροθήκη BCH 7.219 (Myrina); ἀλαβαστρουθήκη PLond. 402.28.

ἀλάβαστος, after 'ἡ' insert 'Poll. 7.177, 10.121'; for 'globular vase without handles' read 'vase'; line 4, for 'SIG 102' read 'IG 1³.421.207 (Athens, v BC)'

†ἀλαβάστρινος, η, ον, of alabaster, τράπεζα ἀ. POxy. 2058.25 (vi AD); neut. pl. as subst., alabaster quarry, PRyl. 92 i 1, ii 18 (ii/iii AD).

*ἀλαβαστροθήκη, ἡ, v. °ἀλαβαστοθήκη.

*ἀλαβαστροπώλης, ου, ὁ, seller of alabaster-ware or perh. perfumes, PAnt. 109.18 (vi AD).

*ἀλαβαστρωνίτης, ου, ὁ, worker in an alabaster quarry, PSI 822.5 (ii AD); ἀλαβαστ(ρωνιτῶν) SB 9904 (AD 154).

ἀλάβης, after 'a Nile fish' insert 'PTeb. 701.41 (iii BC)'

†ἀλαβώδης, ες, sooty, cindery, Antim. 151.5 Wy., cf. Hsch.

ἀλαζονεύομαι, line 3, for 'Isoc. 12.74' read 'Isoc. 13.10'

*ἀλάζω, v. °ἀλλάσσω.

ἀλαζών II 2, for 'by Men.' read 'used by Plautus (*Mil.Glor.* 86)'
*ἀλαία, v. ἀλαός II.
*ἀλαιθερές· χλιαρόν. ἡλιοθερές Hsch.
ἀλαιός, delete the entry.
ἀλάλαγμός I, add '*bleating*, τῶν προβάτων καὶ τῶν κριῶν Lxx *Je.* 32.36'
ἀλαλή, for 'μύρου' read 'μύρον'
ἄλαλος, after '*IG* 14.1627' insert 'in curse, *SEG* 35.214.16, 216.25 (Athens, iii AD)'
*Ἄλανδρος, ὁ, name of a god, θεὸς Ἄ. *SEG* 33.1172 (Lycia).
ἀλανές, ἀλανέως, delete the entries.
*ἀλανής, -ές, app. *pressed together, acting together*: ἀ.· ἀληθές Hsch.; adv. -έως· ὁλοσχερῶς. Ταραντῖνοι id., ἀϜλανέōs, Schwyzer 412 (Elis, vi BC); cf. Aeol. ἀολλής.
*ἀλᾱοσκόπος, ον, perh. *watching like a blind man*, cf. ἀλαοσκοπιά (-σκονος pap.) *PLond.* 1821.264 (*Aegyptus* 6.219).
ἀλαπάζω, line 2, for 'anap.' read 'dact., cf. λαπάσσω II'
ἄλαρα, add 'prob. the same as αὐαρά'
ἅλας, add 'prov., ὅτι οὐκ ἀνάνκη τῶν ἁλάτων ὦν ἐφάγαμεν ὁμοῦ κάδιον ἱστορῆσαι *PUniv.Giss.* 25.20 (iii AD), ἅλασιν ὕει, i.e. of great abundance, Suid.'
*Ἀλασιώτᾱς, ὁ, dat. a-la-si-o-ta-i, cult-title of Apollo, perh. *of Alasia*, an ancient name of Cyprus, *ICS* 216(b).4 (*c*.375 BC).
*ἀλαστοτόκος (dub.), ον, *giving birth in wretchedness*, ἐγὼν [μελέ]α καὶ ἀλασ[τοτόκος ..] Stesich. *S*13.2 (p. 158 D.).
*ἀλᾰτάριον, τό, *salt-cellar*, *BGU* 2360.10 (iii/iv AD).
ἀλάτιον, after 'ἅλας' insert '*IG* 4²(1).123.60 (iii BC)'
*ἄλβολον, Gallic name for βλήχων (q.v.), Ps.-Dsc. 3.31.
*ἀλβόμαυρον, τό, prob. *grey dye* or *grey medicine*, *PVindob.*G 39847.929 (*CPR* 5.110; iv AD).
*ἀλγενήσιος, ον, *dyed with seaweed*, *Edict.Diocl.* 24.9; cf. Lat. *algensis*.
ἀλγεσίθυμος, add 'orac. in *TAM* 3(1).34.D65 (Termessus) (v.l., v. °αἰνεσίθῡμος, the reading in *TAM*)'
ἀλγηδών I, transfer 'Hdt. 5.18' to section III.
ἀλγίων, comp., add 'E.*Med.* 234, Isoc. 10.34'; sup., add 'Ar.*V.* 1117, Lys. 6.1, Th. 7.68.2, Opp.*H.* 2.624; neut. pl. as adv., S.*OC* 1174'
ἀλδαίνω, add 'to an aor. 2 of this verb app. belong the foll. participial forms: act., ἀλδών A.*fr.* 47a ii 17 R. (lyr.); med. ἐναλδόμενον Nic.*Al.* 532 (v. ἐναλδαίνω), ἀλδομένη Q.S. 9.475 (s.v.l., v. ἀλθαίνω)'
ἄλδομαι, delete the entry.
*ἄλδω, τὸ αὔξω, Hdn.Gr. 1.440, perh. a coined form for °ἀλδαίνω.
ἀλέα (B), last line, after '*fomentations*' insert 'Gal. 11.60.6'
*Ἀλέα, ἡ, ḥαλέαι (dat.) *IG* 5(2).75 (Tegea, vi/v BC), Ion. Ἀλέη, title of Athena in Arcadia, Hdt. 1.66.4, X.*HG* 6.5.27, Men.*Her.* 84, Paus. 8.45.4.
ἀλεαίνω, for 'Archil.ap.Plu. 2.954f' read 'Archel.Phil. B 1a D.-K.'
*ἀλεαντρίς, ίδος, ἡ, name of a fish, *Gp.* 20.7.1.
*ἄλεαρ (B), ατος, τό, *coarse wheat meal*, Hdn.Gr. 2.472.12; pl., ἀλέατα *Milet* 3 p. 163 no. 31; w. metric lengthening, ἀλείατα Od. 20.108; late sp. ἀλήατος (gen. sg.) *SB* 5224.19, 40.
*ἀλέγη, ἡ, unexpld. term used in hydraulic operations, *PPetr.* 3 p. 42 Fb2 (iii BC).
ἀλεγύνω, for '*heed, care for*' read '*occupy oneself with, take care of*'; add 'ἀγλαὰ ἔργ' ἀλεγύνειν, *h.Ven.* 11, A.R. 4.1203'
ἀλέγω II 1, delete 'cf. Simon. 37.10' **2**, line 2, after '*Hes.Op.* 251' insert 'ἄχναν .. κύματος οὐκ ἀλέγεις, οὐδ' ἀνέμου φθόγγον Simon. 38.15 P.'
*Ἄλεια, v. °Ἁλιεία.
†ἀλέιατα, v. °ἄλεαρ (B).
*ἀλειμέντα, τά, v. °ἀλιμέντα.
ἄλειμμα, after lemma insert 'Aeol. ἄλιππα *EM* 64.40, perh. ἄλει[ππα] Alc. 45.7 P.' add '**II** *application of unguents, anointing*, Hp.*Morb.* 2.66, *IG* 5(1).1390.106 (Messenia, i BC), 12(3).330.128 (Thera, iii/ii BC); meton., *gymnastic training*, *SEG* 9.4.10, 19 (Cyrene, i BC).'
ἀλειμμάτιον, add '*BGU* 2357 ii 12 (sp. ἀλιμμ-, iii AD)'
ἀλεῖος, (ἀλεῖος La.), for 'Ἀλήϊος' read 'ἀλήϊος'
*Ϝαλεῖος, v. °Ἠλεῖος.
*ἀλειποτάκτως, v. °ἀλιποτάκτως.
*ἄλειπτα, v. °ἄλειμμα.
*ἀλειπτηρία, ἡ, perh. = ἀλειπτήριον I, *SEG* 36.1097 (Sardes, iv/v AD).
ἀλειπτήριον I, line 2, for 'or in Roman .. *sudatory*' read 'or *baths* (perh. also general term for baths, v. *GRBS* 16.217ff.)' **II**, for this section read 'ἀλειπτήριον· γραφεῖον. Κύριοι; cf. perh. διφθεραλοιφός'
ἀλείπτης 1, add 'ἀ. Καίσαρος Didyma 108, ἀ. παίδων Σεβαστοῦ *IG* 2².7155'
ἀλειπτικός, for '*trained under him*' read '*devoted to athletics*'; after 'Ti.Locr. 104a' add 'cf. Alex.Aphr.*in Top.* 152.25'
ἄλειπτος, before 'adv.' insert 'name of a horse, *ICilicie* 49.B3 (Ἄλιπτος, vi AD)'

ἄλειπτρον, delete the entry.
*Ἄλεισία, ἡ, epith. of Aphrodite, *Anecd.Stud.* p. 269.
ἄλεισον, for '*Call.Aet.* 1.1.13' read '*Call.fr.* 178.13 Pf.'
*ἀλειτηρός, ὁ, adj., *sinning, offending against*, ἀνὴρ .. ἀ. Alcm. 79 P. (ἀλιτηρός codd.), κὰξ ἀλειτηροῦ φρενὸς S.*OC* 371; masc. subst., *sinner*, *CEG* 439 (ostr., Athens, v BC); adv. ἀλιτἕρὸς Schwyzer 412 (Olympia, vi BC).
ἀλειτουργησία, for 'late word for Att. ἀτέλεια' read ' = Lat. *immunitas*'
ἀλειτούργητος, after 'D. 18.91' insert 'Is.*fr.* 143'
*ἀλείτουργος, ον, *exempt from λειτουργίαι*, *TAM* 2.224 (Sidyma; written ἀλιτούργου (gen.)).
ἄλειφα, for 'cf.' read 'Hippon. 58 W.' and for '*Call.fr.* 12' read '*Call.fr.* 7.12 Pf.'
ἀλειφαζόος, Myc. *a-re-pa-zo-o*, *a-re-po-zo-o*, *unguent-boiler*.
ἄλειφαρ, add 'Myc. *a-re-pa*'
ἀλείφω 2, after '(Ancyra)' insert '*train an athlete*, *PCair.Zen.* 60.2 (iii BC)' **3**, for '*polish*' read '*anoint objects*'
*ἀλειψανία, ἡ, *lack, state of not having any left*, *SB* 6319.56 (ii/i BC).
*ἀλείψανος, ον, *without relic*, κλῆσις Gr.Naz.*Carm.* 1.2.10.749 (*PG* 37.734).
†ἀλεκτόρειος, ον, (ἀλέκτωρ), *of a fowl*, ᾠὰ ἀ. Synes.*Ep.* 4.165; in the name of a laxative medicine, ἀ. καταπότιον Garg.Mart. 30; ἀλεκτορεία (sc. λίθος), ἡ, *stone found in the gizzards of cocks*, Plin.*HN* 37.144; neut. as adv., κοκκύζετ' ἀλεκτόρειον Gr.Naz.*Carm.* 2.1.11.1926 (*PG* 37.1164).
ἀλεκτόριον, add 'v. *BE* 1971.638'
ἀλεκτορίς, after '-ίδος' insert '(-ῖδος Herod. 6.100)' **I**, for 'used by Trag.' to end of section read 'Arist.*HA* 614ᵇ10, Diocl.*fr.* 141, Herod. 6.100; used generically, .. τὸ τῶν ἀλεκτορίδων γένος· ὀχεύουσι γὰρ οἱ ἄρρενες καὶ ὀχεύονται αἱ θήλειαι τῶν ἀλεκτορίδων καὶ τίκτουσιν ἀεί Arist.*HA* 544ᵃ31, 32'
ἀλεκτρυονώδης, for 'πρὸς ἡδονάς .. end' read 'Eun.*Hist.* p.2.18 B.'
*ἀλεκτρυώδης, ες, *like a cock*, πρὸς ἡδονάς Eun.*Hist. fr.*71.2 B.
ἀλεκτρύων, add 'Myc. *a-re-ku-tu-ru-wo* (pers. n.)'
*Ἀλεκτρώνα, ἁ, *divinity or heroine at Ialysus, Rhodes*, *IG* 12(1).677.4 (iii BC).
ἀλέματος, for 'Dor.' read 'Aeol. and Dor.'
*Ἀλεξάνδρεια, τά, *festival in honour of Alexander the Great*, *OGI* 222. 25/6 (Clazomenae, iii BC), Str. 14.1.31, *IG* 2².2226, (Athens, iii BC).
*Ἀλεξάνδρειος, ον, *of Alexander*, μναῖ, i.e. of his standard, *SEG* 9.1.8, 9 (Cyrene, iv BC); δραχμὰς Ἀ. *SEG* 30.1073 (Chios, ?189/188 BC), Didyma 38.7 (i BC), 446.9 (iii BC).
*Ἀλεξανδρῖνός, όν, *of Alexandrian workmanship*, *SEG* 33.946 (Ephesus, i AD).
ἀλέξανδρος I, delete the section **II**, add 'cf. Myc. *a-re-ka-sa-da-ra* (fem. pers. n.)'
*Ἀλεξεᾶτις, ιδος, ἁ, epith. of the goddess Enodia, *IG* 9(2).576 (Larissa).
ἀλέξημα, for '*defence* .. 479' read '*remedy, medicine*'; add 'transf. A.*Pr.* 479'
ἀλεξητικός, for 'Alex.Aphr.*de An.* 162.16' read 'Alex.Aphr.*de An.* 162.26'
†ἀλεξιάρη, fem. adj., (ἀρή) *protecting against ruin*, νειός Hes.*Op.* 464; ῥάμνος Nic.*Th.* 861; masc. ἀλεξιάρης Hsch., which shd. perh. be restd. in Hes. l.c.
ἀλεξίαμος, for 'Βάκχαι' read 'Νύμφαι'
ἀλεξίκακος, after 'Hes.*Op.* 123 (as v.l.)' insert 'οἶνος .. πυρὶ ἶσον ἐπιχθονίοισιν ὄνειαρ .. ἀλεξίκακον Panyas. 16.13 B.'; line 4, after 'epith. of Heracles' insert '*Hesperia* 5.400.108-9 (Athens, iv BC), *SEG* 17.451 (Rome, ii/iii AD)'; add 'pl., divinities at Byzantium, Robert *Hell.* 9.56; of a bath, *ICilicie* 22 (v/vi AD); see also *SEG* 36.1325'
ἀλεξίλογος, delete 'dub. in'
ἀλέξιον, for 'cf. Phot.' read 'cf. ἀλεξήματα· τὰ βοηθήματα, Phot.'
ἀλεξίπονος, for 'S.(?)*Eleg.* 7' read 'of Asclepius, S.*Pae.* 1(b).(i).1 P.'; add 'of a stone, *Lapid.Gr.* 195.21; λουτρόν *SEG* 33.773 (Ostia, iv AD)'
ἀλεξιφάρμακος I, add '**2** fig., παιώνειόν τινα καὶ ἀ. .. λόγον Longin. 16.2, 32.4.' **II 1**, add 'ὥσπερ ἀ. ἐστι τοῖς ἀδικεῖν βουλομένοις D. 24.85'
*ἀλεξίχαλαζος, ον, *warding off hail*, epith. of Αἰθήρ, Robert *Hell.* 9.63 (Amasia).
ἀλεόμαι, line 2, after '(v.l.)' insert 'A.R. 4.474' **2**, add '*Trag.adesp.* 705.10 K.-S.'
†ἀλεός· διάπυρος Hsch. **II** v. ἠλεός.
*Ἀλεπακίδαια, τά, name of a farm at Messene, *IG* 5(1).1434 (i BC/i AD), cf. mod. Gk. ἀλεπού *fox*.
*ἀλεπῐδος, ον, *not having scales*, ἰχθῦς Cyran. 1.1.8 K.
ἀλέπιστος, for '*not scaled*' read '*not having the scales removed*'
*ἀλεπύρος, ον, *free from husks*, κριθαί Hsch. s.v. ἀτυπηβές (-ίδες La.).
*ἀλεστός, *ground*, ἀλεστοῦ σίτου *POsl.* 155.16 (ii AD).
ἄλεστρον, after '*POxy.*' insert '736.8' and 'both' before 'i AD'; add 'also ἄλετρον *PPetr.* 3 p. 313 (iii BC), *SB* 7642.3 (iii BC)'

ἀλέτης, add 'of a man, *POxy*. 3169.91 (ii/iii AD), *PHib*. 268 int. (*c*.260 BC)'

*ἀλέτισσα, ἡ, = ἀλετρίς 1, *POxy*. 2421.31 (iv AD).

ἀλετός, after 'Plu.*Ant*. 45' insert '2.289f, Ath. 14.618d'

ἀλετρίς, add 'cf. Myc. *me-re-ti-ri-ja* (for *me-* v. °ἄλευρον)'

ἄλετρον, τό, v. °ἄλεστρον.

ἀλετών, after '*millstone*' insert '*IG* 1³.422.24, al. (Athens, v BC)'

*ἀλευραττίς· ἀγγεῖον εἰς ἄλφιτα Phot. a 932 Th. (ἀλεύραττις v.l.).

*ἀλευρητικός, ή, όν, perh. *made of flour*, dub. in *PTeb*. 894 *fr*. 10.9 (ii BC, perh. read ἀλευριτ-).

†ἀλευροδοῦντες, οἱ, (cj. ἀλευροῦντες), kind of wheatcakes, Anticl. 23 J.

*ἀλευροκαθάρτης, ου, ὁ, *one who sieves the flour*, *SEG* 33.1165 (Pamphylia, iii AD).

ἄλευρον (A), add 'see also ‡μάλευρον (Myc. *me-re-u-ro*)'

*ἀλευρόπωλις, ιδος, app. fem. adj., *used for the sale of flour*, ἀλε-[υρό]πωλις στωϊά *IG* 12(2).14.12 (Mytilene).

*Ἀλεύς, title of Apollo, *SEG* 30.1218 (Tarentum, iv BC).

ἀλέω, line 3, after 'Hdt. 7.23' insert '*BGU* 1858, also part. ἠλεσμένος *Edict.Diocl*. 1.13-14'

†ἀλεωρή, Att. -ρά, ἡ, *means* or *place of escape, refuge, protection*, Il. 24.216, ὡς καὶ αὑτοί τινα ἀλεωρὴν εὑρήσονται Hdt. 9.6, A.R. 1.694, J.*BJ* 3.163, Opp.*H*. 1.790; w. obj. gen. δηΐων ἀνδρῶν ἀ. Il. 12.57, 15.533, βελέων ἀ. Ar.*V*. 615; of escape from conditions, situations, etc., λιμοῦ ἀ. Hes.*Op*. 404, Hp.*Praec*. 7, D.S. 3.34, J.*AJ* 18.147; of bodily protection, τὴν περὶ τὸ σῶμα ἀ. Arist.*PA* 679ᵇ28, 687ᵃ29, *HA* 488ᵇ10.

*ἀλεωρία, gloss (with πολυωρία) on ἄλεαρ, Hsch. (prob. corruption of °ἀλεωρά).

ἄλη I 1, line 2, after '(pl.)' insert 'E.*Med*. 1285'   2, delete the section.

*ἄλη (B), ἡ, Lat. *ala*, *IPrusias* 104.2 (ii AD), etc.

ἀλή, line 1, for 'only' read 'normally'; add 'rarely sg., τῶι ἥρωι τῶι ἐπὶ τῆι ἀλῆι *SEG* 21.527.38, al., τὴν ἀλ[ή]ην καὶ τὴν ἀγορὰν τὴν ἐν Κοίληι ib. 17 (Attica, 363/2 BC); cf. St.Byz. s.v. Ἀλαί· ἔστι καὶ λίμνη ἐκ θαλάσσης'

*ἀλήατος, v. ‡ἀλείατα.

ἀληγός, add '2.912f, *BGU* 2353.14 (AD 115)'

ἀληθάργητος, for '*free from lethargy, energetic*' read '*unforgettable* (cf. ληθαργέω)'; for '*CIG* 2804 (Aphrod.)' read '*MAMA* 8.504.16, *IGChr*. 9'

*ἀληθέγγυος, ον, *that guarantees truth*, Phot. a 938 Th.

ἀλήθεια, end of line 1, for 'ἀλάθεα' read 'Aeol. ἀλάθεα' and before 'neut.' insert 'also interpr. as'   I 2, line 4, before 'Th. 4.120' insert '*in reality*'

ἀληθής II, add '3 perh. *legitimate*, *IEryth*. 2(c).9 (v BC).'   III 2, after 'Ar.*Ra*. 840' insert '(E.*fr*. 885)' and after '*Av*. 174' insert '*Pl*. 123, 429'

*ἀληθινοβάφος, ὁ, (sp. ἀληθεινο-), dyer or *maker of best quality purple*, *ITyr* 137 (Chr.).

ἀληθινόπινος, read ἀληθινόπῑνος; for '*patina*' read '*pearls* (cf. πίνη II)'

*ἀληθινοπράσινος, ον, *of real green*, *SB* 13597 (v AD).

ἀληθινός I 2, line 3, after 'ἰχθύς Amph. 26' insert '(cf. κάραβος ἀ. Macho 29 G.; the significance of the adj. is here uncertain)'

ἀληθοεπής, before 'Hsch.' insert '*Suppl.Hell*. 991.70 (poet. word-list, iii BC), ἀ. καὶ ἐτήτυμος *IStraton*. 1201.5'

ἀλήμων, for '(Leon.)' read '(Leon.Tarent.)' and after it insert 'D.P. 490, Opp.*H*. 3.455, Nonn.*D*. 1.528'; delete '- Ep. word'

ἀληνής, for 'prob. l. in Semon. 7.44' read 'Semon. 7.53 W. (codd., cj. ἀδηνής)'

*ἄληξ, v. ‡ἄλιξ.

ἀλήπεδον, for '= Ἀλήϊον πεδίον' read '*plain*, ἥξει δ' ἐρεμνὸν εἰς ἀλήπεδον φθιτῶν Lyc. 681'

ἀλησία, add 'perh. in *MAMA* 9.13 (i BC)'

ἀλήτης, for '*wanderer* .. *exiles*' read '*homeless wanderer, vagrant*, Od. 17.576, 18.18'   2, after 'ἀλῆτις, ιδος' insert '*GLP* 1.122.17' add 'II ἀλῆται *the planets*, incl. sun and moon, Nonn.*D*. 5.68, al., *AP* 9.822.3.'

*ἀλητικός (B), ή, όν, *of* or *for milling*, *PMil.Vogl*. 53.11, *PRyl*. 321.5 (sp. αλαι-).

†ἀλητύς, ύος, ἡ, *wandering* (cf. ἄλη), Call.*fr*. 10 Pf., Man. 3.379.

*ἀλητῶ· τὸ πλανῶ *EM* 822, *Et.Gen*. 459.

ἀλήτωρ, add 'as Cret. pers. n., Perdrizet *Memnonion* 60 (iii BC), *Inscr.Cret*. 1.xxii 47 (ii BC); cf. ‡ἀλείτωρ (λήτωρ)'

*ἀλήφατος, ον, *crushed*, ἀλήφατον ἄνθος ἐλαίης *SEG* 8.474.7 (Hermoupolis); ἀλήφατα (codd. ἀλί-)· ἄλφιτα ἢ ἄλευρα Hsch.

*ἀλθέσσω, *heal, cure*, Aret.*SD* 1.13.4, 2.9.9.

ἄλθεις, delete '645' (v. °ἀνθήεις).

*ἄλθησις, εως, ἡ, *healing*, ἐναλθῆ· τὸν χρῄζοντα ἀλθήσεως Sch.Nic.*Al*. 586i.

ἄλθος, add 'cj. in S.*fr*. 172 R.; pl. Nic.*Al*. 423 (v.l.)'

ἁλία (A), read ἁλία   I, in Sicily, add '*SEG* 30.1117.4 (Entella, iii BC)'; add '*assembly* of a phratry, *CID* I 9.A21, al. (iv BC)'

†ἁλία (B), ἡ, *salt-box*, Archipp. 13, Stratt. 15 K.-A., Poll. 10.169; ἁλίην τρυπᾶν ἐν Θέμιδος οἴκῳ dig into *the salt-box* in the house of Themis, a mark of honest poverty, Ap.Ty.*Ep*. 7, cf. Call.*Epigr*. 47.1 Pf.

*ἁλιάδιον, τό, kind of boat, *CPR* 10.2.4, 10.5 (early vii AD).

ἁλιαία, read ἁλιαία and after '= ἁλία (A) (ἀ-)' insert 'ἁλιαιίαν *SEG* 30.380 (Tiryns, vii BC)'; add 'in Arcadia, Schwyzer 666.6, *SEG* 25.443.10 (both Orchomenus, iii BC)'

*ἁλιάπους, ποδος, ὁ, a bird, perh. *stormy petrel*, Achae. 54 S.

ἁλιάς, add '(employed in postal service) *PBeatty Panop*. 1.252 (AD 298), *POxy*. 2675.9 (iv AD)'

ἁλίασμα, read ἁλίασμα and add '*SEG* 35.999.19, 30.1119.33 (both Entella, iv and iii BC), *IG* 14.952.8 (Agrigentum, iii/ii BC), ib. 612.5 (Rhegium, ii BC)'

†ἁλίασσις, ιος, ἁ, *act of the assembly* or ἁλιαία, *IG* 4.554.5 (Argos, vi/v BC).

*Ἁλιασταί, οἱ, religious association at Rhodes, *IG* 12(1).155 (ii BC); cf. Ἁλιάδαι.

ἁλιαστάς, read ἁλιαστάς

ἀλίαστος, for '*not to be turned aside*' read '*interminable*'; after 'Hes.*Th*. 611' insert 'A.R. 2.649; Musae. 318'; after 'Il. 24.549' insert 'Q.S. 4.17'   2, for this section read 'of things, perh. *restless*, κῦμ' ἀλίαστον A.R. 1.1326 (cj. κῦμα λιασθέν)'   II, substitute 'of persons, animals, *not to be stopped*, E.*Or*. 1479, Opp.*C*. 4.160, *H*. 2.590'; delete '-Ep. .. lyr.'

ἀλίβαπτος, for 'Nic.*Al*. 618' read 'Nic.*Al*. 605' and add '(v.l., ed. °ἁλίβλαπτος)'

ἀλιβδύω, delete 'expl. .. 269' (v. °ἁλιδ-).

*ἁλίβαπτος, ον, *hurt by the sea*, Nic.*Al*. 605 (v.l. ἀλίβαπτος).

ἁλίβρομος, for '*murmuring like the sea*' read '*sounding in the sea*'

ἁλιβρώς, add 'Diog.Oen.*fr*. 7 col. 2.6 (*AJA* 75.367)'

ἁλίβρωτος, for '*swallowed by the sea*' read '*gnawed by the sea*'

*ἁλιγενέτωρ, ορος, ὁ, compd. of ἅλς (B) and γενέτωρ, *Suppl.Hell*. 991.64 (poet. word-list, iii BC, ἁλιγνέτ- pap.); cf. °ἁλογενέτωρ.

ἁλίγκιος, line 1, after 'ον' insert '(also α, ον, B. 5.168 S.-M.)'

*ἁλίγνητος, η, ον, = ἁλιγενής, *Suppl.Hell*. 991.58 (poet. word-list, iii BC).

*ἁλίγονος, ον, = ἁλιγενής, *Suppl.Hell*. 991.66 (poet. word-list, iii BC).

*ἁλιδερκής, ές, compd. from ἅλς (B) and δέρκομαι, *Suppl.Hell*. 991.57 (poet. word-list, iii BC).

†ἁλιδίως, *sufficiently*, ἁ. πονηρός Epich. 84.35 A., cf. Hsch. (ἀλ-).

*ἁλιδόνητος, ον, *buffeted by the sea*, Gramm.ap.Ludwich *Aristarch* 2.665 (*Cod.Vindob*. 294), Trag.adesp. 720F K.-S. (= *Suppl.Hell*. 991.63, poet. word-list, iii BC).

†ἁλίδονος, ον, *driven by the sea*, codd. in A.*Pers*. 275.

ἁλίδροσος, after 'ον' insert 'app. *sprayed by the sea*' and for '*Lyr. Adesp.Oxy*. 219.11' read '*Lyr.Alex.adesp*. 4.11'

ἁλίεια, after 'Str. 11.2.4 (pl.)' insert 'Plu.*Tim*. 20, Ael.*NA* 14.20'

Ἁλίεια, delete the entry (v. °Ἁλίεια).

*Ἁλίεια, τά, *festival of the Sun* in Rhodes, Com.adesp. 336, *IG* 12(1).72a (ii BC), Ἁλίεια ib. 730.17 (i BC), contr. Ἁλεῖα ib. 12.4, 58.19, Ath. 13.561e.

ἁλιεινή, for '*Edict.Diocl*. 21.2' read '*Edict.Diocl*. 21.2A, (Ἀλτεινή Lauffer, cf. Lat. text *Altinata*)'

*Ἁλιεῖον, τό, app. *temple of the Sun*, ἐπὶ [τ]ὰν κορυφὰν τοῦ Ἁ. *IG* 4.926.12 (Megara, iii BC).

ἁλιευτής, read ἁλιευτάς, ᾶ' and for 'Cerc. 4.8' read 'Cerc. 7.9'

ἁλιευτικός, line 2, after '*An*. 7.1.20' insert '(ἁλιευτικόν, τό, *POxy*. 1846.1 (vi/vii AD))'

ἁλιεύω, after 'Luc.*Pisc*. 47' insert '*SEG* 24.1108'

ἁλίζω (A), line 3, for 'but ἡλ-' read 'ἡλ-'

ἁλίζωνος, after '*sea-girt*' insert '*Suppl.Hell*. 991.54, 67 (poet. word-list, iii BC)'; for 'Sos. 24' read '*fr*. 384.9 Pf.'; after 'Sid.)' insert 'Nonn.*D*. 37.152, al.'

ἁλικάκαβον, read ἁλικάκκαβον (v.l. -κακα-)

Ἁλικαρνασσίς, add 'also of a tribe, *SEG* 33.1103 (Paphlagonia, ii/iii AD)'

*ἁλίκιον, τό, perh. dim. of ἅλιξ 1, *groats*, *PHamb*. 192.24 (iii AD), *PRyl*. 629.293, 298 (iv AD, -κιν pap.).

*ἁλίκλα, ἡ, Lat. *alicula*, *light upper garment*, *SB* 9834b.10 (iv AD).

ἁλικνημίς, for 'ἀπήνη .. car' read 'perh. *having spokes* (i.e. *wheels*) *that go through the sea*, applied to Poseidon's chariot, ἀπήνη ἁ.'

†ἁλικός, ή, όν, *of salt*, τὴν ἁλικὴν ὠνήν *JHS* 74.97 no. 38.B8 (Caunus); ἁλική, ἡ, *salt-tax*, *PSI* 388.1 (iii BC), *PTeb*. 482 (iii, BC), *PHib*. 112.3, etc.   2 pl. = Lat. *salinae, salt-pans*, Charis. p. 33 K.

ἁλικρηπίς, for '*at the sea's edge*' read '*bordered by the sea*'

ἁλίκτυπος, for 'insert' ἁ. κῦμα E.*Hipp*. 754, Anacreont. 50.8 W.' and delete 'also .. (lyr.)'

ἁλίμενος 2, for '*shelterless, inhospitable*' read '*offering no refuge* or *protection*' and add 'γάμον A.*fr*. 154a.3 R.'

ἁλῐμενότης, for 'Peripl.M.Eux. 37' read 'Peripl.M.Eux. 25.2 (= 16ʳ.10 D.)'

*ἀλιμέντα, τά, Lat. alimenta, provisions, IEphes. 805.6 (Rom.imp.); also ἀλειμέντα, TAM 2.278.22 (Xanthus, ?iii BC).

ἀλιμενώτης, for 'cf. Hsch.' read 'v. sq.'

*ἀλιμένωτον· λιμένα μὴ ἔχοντα Hsch. (La.).

*ἀλιμμάτιον, v. ‡ἄλειμμ-.

*ἀλιμονικός, ή, όν, of or concerned with provisioning, fr. Lat. alimonium, ἀπὸ τοῦ ἀ. (λόγου) PMich.XI 613.3 (written ἐλιμ-, AD 415).

*ἀλίμοχθος, ον, toiling at sea, SLG 364.8.

ἁλιμῠρήεις, after 'εσσα' insert '(fem. only at Suppl.Hell. 991.50, poet. word-list, iii BC)'; for 'into the sea' read 'with salt water'

†ἁλιμῠρής, ές, flowing with salt water, ἀ. βένθη, πόντος GVI 1833.5 (Salamis, Cyprus, ii BC), Orph.A. 68, ἀφρός APl. 180 (Democr.); of sea-washed rocks, etc., A.R. 1.913, Phanocl. 1.17, Opp.H. 2.258; of rivers at their mouths, Orph.A. 344, 462.

†ἁλιναιέτας, α, masc. adj., dwelling in the sea, δελφῖνες B. 17.97 S.-M. (ἐναλι- pap.).

ἁλινδέω II 1, delete 'to be twirled .. 113' and substitute 'in erotic context, w. dat., Call.fr. 191.42 Pf.; μετά τινος, Herod. 5.30'   2, lines 2/3, delete 'having grovelled' and transfer 'frequent' before 'ἠλινδημένος' in line 2.   3, delete the section.

ἁλινήκτειρα, for 'the sea' read 'brine'

†ἅλινσις, ιος, ἡ, application of stucco or whitewash, τοῦ ἐργαστηρίου IG 4².102.39, SEG 24.277.A12 (both Epid., iv BC); cf. ἄλειψις (3).

ἁλίνω, for 'pound' read 'grind' and add 'cf. ἐναλίνω'

ἅλιξ I, add 'Dsc. 4.148, Edict.Diocl. 1.25; also written ἄληξ Hierocl.Facet. 222'   II, delete the section.

*ἅλιξ (B), ικος, ὁ, Lat. hallec, fish sauce, Gp. 20.46.2.

*Ἁλιονείκης, ὁ, = Ἁλειονίκης, victor in the Ἅλεια (°Ἁλίεια) at Rhodes, PLond. 1178.67 (ii AD), IGBulg. 1574.3 (Rom.imp.).

ἅλιος (B), line 3, after '(Od. 2.)318' insert 'h.Merc. 549'; add 'perh. also in Ion., fallow, εἰ νέον ἀροῖ το[ὺς] ἁλίους ἀρότους IG 12(7).62.8 (Amorgos, iv BC)'

ἅλιος (C), add '(Aeol. ἄλιος)'

Ἁλιοτρόπιος, add '(c.200 BC), perh. to be rest. at GDI 1338 (Dodona, (v. BCH 109.536))'

ἁλίπαστος, add '2 neut. subst., dub. in Erinn. in Suppl.Hell.401.24.'

ἁλίπλαγκτος, for 'IG 2.1660' read 'IG 2².4968'

ἁλιπλᾰνής, after 'Lucill.' add '6.223 (Antip.)'

ἁλίπλοος II 1, line 2, delete 'fisher'

†ἁλιπόρος, ον, travelling over the sea, ἀνά τε νηῦς ἁλιπόρους Lyr.Alex.adesp. 36.2; forming a sea-passage, ἁλιπόρου διασφάγος Luc.Trag. 24.

*ἁλιποτάκτως, adv., without leaving one's post, SB 9011.6 (ἀλειπο-, v/vi AD); cf. λιποτάκτης.

†ἄλιππα, v. °ἄλειμμα.

†ἁλίρραντος, ον, sprayed by the sea, ἀκτῇ ἐν ἁλιρράντῳ ἐπὶ πέτρῃ Epic.Alex.adesp. 8.1, Suppl.Hell. 991.55 (poet. word-list, iii BC), ἁλιρράντου .. πόντου AP 9.333 (Mnas.), 14.72.

*Ἅλις, v. °Ἦλις.

ἁλίσκομαι I 1, after 'Pl.R. 468a' insert 'Is. 6.1'   II 2, add 'of decrees, to be invalidated, τῶν ψηφισμάτων .. ἢ μενόντων κατὰ χώραν ἢ ἁλόντων D. 58.5'

ἁλισμός, add 'salt-incrustation, PZen.Col. 95.6 (iii BC)'

ἁλιστός, add 'also -όν, τό, bacon, Edict.Diocl. 4.7'

ἀλῐταίνω, lines 4/5, for 'ἀλιταίνεται Hes.Op. 330: aor.' read '1 aor. subj. ἀλιτήνεται Hes.Op. 330 (pap.); 2 aor.'; line 9, for 'ἀλίταινητ' read 'ἀλιτήνεται'

ἀλιτάνευτος, for 'PMag.Par. 1.1176' read 'PMag. 4.1776'

*ἀλιτεγγής, ες, perh. moistened by the sea, Suppl.Hell. 991.56 (cj. for ἀλιτεσγής, poet. word-list, iii BC).

*ἀλιτειχής, ές, walled by the sea, Euph. in Suppl.Hell. 442.4.

ἀλιτενής, for 'projecting into the sea' read 'extending along or into the sea, νῆσοι PHamb. 119 iii 15 (iii BC)'

ἀλίτημα, for '(Agath.) .. pl.)' read '9.154, 643 (all Agath.)'

ἀλῐτήμερος, add 'Archil. 196a.39 W.²'

*ἀλῑτήρ, ῆρος, ὁ, = ἀλείτης, EM 65.24; cf. ἀλίτης.

ἀλῐτήριος 1, add 'Lyc.fr. 2.8 S.'   II, add 'of Zeus, EM 65.41; of Demeter (fem. ἀλιτηρία), ib. 65.40'

†ἀλιτρός, v. °ἀλείτρος.

*ἀλιτουργησία, ἀλίτουργος, v. ‡ἀλειτ-.

*ἀλιτόφρων, ον, foolish, Hsch. s.v. ὠλιτόφρονας (for ὁ ἀ- or ὦ ἀ-).

ἀλιτραίνω, for 'Hes.Op. 243' read 'Hes.Op. 241'

*ἀλιτρεύω, sin, MAMA 1.235 (Laodicea Combusta).

ἀλιτρόβιος, for 'ἀλῑτρό-' read 'ἀλῐτρό-'

*ἀλίτροπος, ὁ, wd. of uncertain etym., app. meaning sinner or sim., Ps.-Phoc. 141 Y.

ἀλιτρός, for 'ἀλίτρός, όν' read 'ἀλῐτρός, ά, όν' and at end delete 'fem. .. Semon. 7.7'

ἀλῑτροσύνη, read ἀλῐτροσύνη, and add 'Orph.A. 1231, L. 62'

†ἀλίφατα, v. °ἀλήφατος.

*ἀλῐφή, ή, painting, τὴν ἀ. τῶν ξύλων IG 2².1682.29 (iii BC); cf. ἀλοιφή.

*ἀλίφησις, εως, ἡ, painting, Delph. 3(5).64.10 (sic ed., but perh. rest. ἀλιφή[s] gen., v. °ἀλιφή.

ἀλκαία I, for 'Call.fr. 317' read 'Call.fr. 177.23 Pf., Nic.Th. 123, 225'

ἄλκαρ, line 4, after 'h.Ap. 193' insert 'ἀναιδέος ὄθματος ἀ. Call.fr. 186.29 Pf.'; for 'Call.fr. 124' read 'id.fr. 304.2 P.'; after 'Aret.CA 1.1' insert 'Hp.ap.Gal. 19.75'

ἀλκή I 1, add 'b περὶ ἀλκῆς, athletic contest at Athens, IG 2².2113.57, 2130.90 (both ii AD).'   add '2 fig., strong point, ἔν τινι τὴν ἀ. ἔχειν, of an author, D.H.Th. 23.'

ἀλκήεις, delete 'of patients' and for 'al.' read '(Aret.) SD 4.5 (p. 71.29 H.)'

ἀλκηστής, add 'Trag.adesp. 585a K.-S.'

*ἀλκήτωρ, ορος, ὁ, protector, Ramsay Studies in Eastern Rom. Prov. p. 128.

ἄλκιμος, line 4, after '(Hdt. 1.)103' insert 'X.Oec. 4.15, 6.10, HG 7.3.1'

ἀλκτήριον, after 'against a thing' insert 'Call.fr. 346 Pf. (pl.)'

ἀλκυόνειον, after 'Dsc. 5.118' insert 'PColl.Youtie 87.10 (vi AD), ἀλκυόνυον PTeb. 273.34 (ii/iii AD)'

*ἀλκυονίτις, ιδος, fem. adj., halcyon, αἱ ἀ. (sc. ἡμέραι) halcyon days, Sch.Ar.Ra. 1309.

*ἀλκυονύτης, unexpld. wd. in Theognost.Can. 44.

ἀλλά, before 'I' insert 'usu. in first position, but occasionally postponed, e.g. Call.Jov. 18, AP 5.9 (Rufin.), 5.17 (Gaet.), etc.'   I 2 a, line 8, after 'Pl.Phdr. 262a)' insert 'ἀλλ' οὖν (without γε) E.Ph. 498, Dig. 50.6.6.2'   b, line 4, after 'id.Med. 912' insert 'And. 2.26'   II 1, line 11, after 'Od. 4.472' insert 'Hyp.Dem. 1, Eux. 1'   5, after 'it is not [so], but ..' insert 'D. 2.22, Pl.Grg. 453b, Arist.Pol. 1262ᵃ14'; after '(Pl.)Smp. 173b' insert 'Cra. 436d '; after '(Ar.)Ra. 58, 498' insert 'E.Ba. 785, Pl.Euthd. 286b, Call.fr. 191.1 Pf.'   III 3, ἀλλὰ γάρ, add 'after οὐ μόνον, Cod.Just. 1.40.13'

*ἄλλα, v. °ἄλλη.

ἄλλαγμα 3, delete the section transferring example to section 2.

ἀλλαγμός, for '= foreg.' read 'change, transition' and add 'Astrol.adesp. in PMich.III 149 v 33'

*ἀλλάζω, = ἀλλάσσω, ἀλαζέσθō δὲ ἀντὶ τô ἀρχô IG 9²(1).609.21 (Naupactus, iv/v BC).

ἀλλακτικός, line 2, for 'ἡ -κή .. exchange' read 'τὸ .. ἀλλακτικὸν (μέρος τῆς κτητικῆς τέχνης)' and after 'Pl.Sph. 223c' insert 'τῆς .. ἀλλακτικῆς (τέχνης) ib.'

*ἀλλαλο-, v. ‡ἀλληλο-.

ἀλλαμπᾶν, add 'ἀλλάμπαν Musurus' and for '-χωρ' read '-χωριον'

ἀλλάντιον, add 'POsl. 152.9 (i/ii AD), PMich.inv. 3731 ii 30 (Tyche 1.185, ii/iii AD)'

*ἀλλαξιμάριον, τό, change of clothes, Stud.Pal. 20.245.25 (vi AD, written ἀλαξ-); sp. ἀλαξαμάρνιον in PMasp. 6ʳ.66 (vi AD); cf. ἀλλάξιμα.

ἀλλάσσω III 1, line 9, after 'Pl.R. 371c' insert 'w. dat. pers.'; line 10, for 'Lg. 915d, e' read 'Lg. 915d; abs., ib. 915 d,e' and before 'τοῦ παντὸς ἀ.' insert 'fig.'   2, add 'med., change residence, move, εἰς .. PMich.III 203.9 (ii AD)'   add 'VI w. gen., differ from, ἀλλήλων D.H. Dem. 53; see also °ἀλλάζω.'

†ἀλλᾰχόθι, adv., elsewhere, AP 9.378 (Pall.), Jul.Or. 1.5c; in another passage, Demetr.Eloc. 156, A.D.Synt. 333.26, Plu. 2.20d.

ἀλλᾰχόσε, add 'elsewhere, Just.Nov. 53.6.1'

*ἄλλη, = ἄλλῃ, Schwyzer 148 (Megara, v BC), ἄλλη πη SEG 9.72.114 (Cyrene, iv BC); also Lesb. ἄλλα IG 12(2).645.49 (Nesos, iv BC).

*ἀλληλαναδοχή, ή, giving of mutual security, PMasp. 170.15 (vi AD); dub. in PLond. 1661.19 (vi AD).

ἀλληλεγγύη, after 'security' insert 'by which each debtor guarantees repayment of the whole debt, PMasp. 170.15'

*ἀλληλίσματα, τά, migration, Favorin.Exil. 10.15 B.

*ἀλληλόκακος, ον, Aeol. ἀλλᾱλο- wronging each other, τῶν ἀ. πολίτᾱν Alc. 130.22 L.-P.

*ἀλληλομῑσέω, ἡ, mutual agreement, POxy. 2343.21 (iii AD)'

*ἀλληλομολογία, ἡ, mutual agreement, PLond. 1727.49 (vii AD).

*ἀλληλοφῐλία, ἡ, Boeot. ἀλλᾱλο-, mutual love, SEG 37.389 (tab.defix., Hellen.).

ἀλληλοφόνοι, add 'adv. -ως Sch.A.Th. 734e S.'

*ἄλλιστος, after 'inexorable' insert 'Ἁϊδωνεύς Euph. 98 vGr.'

*ἀλλογράφον, τό, perh. copy of a document, SEG 24.530 (Maced.).

ἀλλοδᾰπός, line 4, ἐν ἀλλοδάπῃ, add 'SEG 30.1486 (Phrygia)'

ἄλλοθι II, add 'so perh. Od. 4.684' (ἄλλοθ' usu. referred to ἄλλοτε)

*ἄλλοι, to another place, οὐδέ τι μυνάμενος ἄλλοι τὸ νόημμα Alc. 392 L.-P.

ἀλλοῖος, after 'ἄλλοτε ἀλλοῖος' insert 'Hes.Op. 483, Semon. 7.11 W.'; line 4, for 'every .. billet' read 'i.e. if you throw often, the throw will sometimes be good ' and after 'Com.adesp. 448' insert 'cf. Arist.Div.Somn. 463ᵇ21'

ἀλλοιόω I 1, after 'Pl.R. 381a, etc.' insert 'Mosch.Trag. 6.19 S.'

*ἀλλοιωτέον, *a change must be made*, Gal. 6.242.6.
*ἀλλόκοιτος, ον, *sleeping elsewhere*, Theognost.*Can.* 95.
ἄλλομαι, line 7, after 'ἀλέσθαι' insert 'Call.*fr.* 177.33 Pf.'
ἀλλόμορφος, for 'Hp.*Morb.* 4.93' read 'Hp.*Vict.* 4.93'
*ἀλλοπάτριος, ὁ, prob. *member of an alien community*, in context, non-Christian, Ramsay *Studies in Eastern Rom.Prov.* p. 224 (written ἀλλω-, Philadelphia, iii/iv AD).
ἀλλοπίᾱς, add '*IGC* p. 98 A12 (Acraephia, iii BC)'
ἀλλοπολία, for '= ἀλλοδημία' read '*residence in a foreign city*'
ἀλλοπολιᾶται, add '*SEG* 37.752, (Lyttos, v BC)'
ἄλλος line 1, for 'Cypr. .. (Idalion)' read 'Cypr. αἶλος, ἄϝιλος (*a-i-la*, Kafizin 114a, *a-i-lo-ne ICS* 217.14, *a-wi-la Kafizin* 148, al.)' **II 3**, after '(X.)*Cyr.* 4.1.15' insert 'μὴ ἄλλα καὶ ἄλλα θορυβείτω Pl.*Ap.* 27b' **6**, line 2, after '*besides*' insert '*the rest of*, ἡ ἄλλη Ἑλλάς Th. 1.77.6, ἡ ἄλλη ψυχή Pl.*Men.* 88d, Procl. ad Hes.*Op.* 494' **III 2**, after 'c. gen.' insert 'ἄλλο τὸν γεγραμένον *IG* 9²(1).609 B 23 (Naupactus, vi/v BC)' **4**, after '*bad*' insert 'Hes.*Op.* 344'
ἄλλος, delete the entry.
*ἄλλος (B), v. °ἠλέος.
ἄλλοσε, at end for '- by .. ἀλλαχοῦ' read 'for locative, by attraction to following relative' and add '*for another purpose*, Arist.*Ath.* 29.5'
ἀλλοτριάζω, add '2 w. acc., app. *alienate*, *POxy.* 2267.8 (iv AD).'
ἀλλοτριοπραγέω, add 'Phlp. *in de An.* p. 528.23'
ἀλλότριος, after lemma insert 'αἰλότριος Schwyzer 411 (Elis); Cret. ἀλλόττριος, *Inscr.Cret.* 4.72 iii 12 (Gortyn, v BC)' **II 1 a**, add 'w. gen., φθόνου ἀ. *removed from it*, i.e. not the object of it, Luc.*Somn.* 7'
ἀλλοτρίωσις, add '**II** *alienation* of property, *BGU* 464.1 (ii AD).'
ἀλλοφανής, for 'Nonn.*D.* 14.156' read 'Nonn.*D.* 14.157'
†ἀλλοφυής, ές, *of a different kind* or *nature*, Nonn.*D.* 2.148, 4.419, *PMasp.* 19.7 (vi AD).
ἀλλόχρως, for '*looking strange* or *foreign*' read '*of a different colour* or *complexion*'; add 'in refs. to a theory (of Democritus and Anaxagoras) of perception by contrasts, Thphr.*Sens.* 27, 50, 54'
*ἀλύταρχος, = ἀλυτάρχης, *SEG* 17.199.3 (Olympia, i AD).
*ἀλύτης, v. ‡ἀλύτης.
*ἀλύω, v. ‡ἀναλύω (B).
ἄλλως **II 1**, after 'E.*Hec.* 302' insert '*Or.* 709' **II 3**, line 8, after 'Philem. 51, etc.' insert 'ὡς ἄλλως, D. 6.32, Is. 7.27'
ἅμα, for 'Epic. .. 32' read 'Euph. in *Suppl.Hell.* 453'
*ἀλμενιχιακά, τά, *almanac, astronomical list*, v.l. at Porph.*Ep.Aneb.* 2.12b (36), Eus.*PE* 3.4.1.
*ἁλμυροδῑνής, ές, *salt-eddying*, πόντος hymn in *Suppl.Hell.* 990.5 (iii BC).
ἁλμυρός **3 a**, for 'Cerc. 19.37' read 'Cerc. 17.37'
*ἁλμύρραξ, ᾱγος, ὁ, prob., *efflorescence of potassium nitrate*, Plin.*HN* 31.106.
†ἁλμυρώδης, ες, *salty* ῥεῦμα Hp.*Epid.* 1.26 ἐ'; χωρία Thphr.*CP* 3.17.2, J.*BJ* 4.8.2.
*ἁλμώδης, ες, *salty*, δάκρυον Hp.*Epid.* 4.35; πτύαλον id.*Coac.* 238; ἁλμωδεστέρα (οὖσα ἡ γῆ) πρὸς φυτείαν X.*Oec.* 20.12, Thphr.*HP* 8.7.6; of the bloom (χνοῦς) on certain fruit, id.*CP* 6.10.7.
ἀλοάω **I 3**, for '*destroy*' read '*batter down*'
*ἁλογενέτωρ, ορος, ὁ, = °ἁλιγενέτωρ, *Suppl.Hell.* 991.53 (poet. word-list, iii BC).
ἀλογέω **I 1**, line 4, for '*feel*' read '*be*'; add 'D.L. 1.32' **II 1**, delete '*to be .. indiscretion*' **2**, for '*out of one's senses*' read '*perplexed*'
ἀλόγητος, add '*UPZ* 110.205 (ii BC)'
*ἀλογί, adv. *irrationally*, Lib.*Decl.* 16.31 (s.v.l.).
ἀλόγιστος **I 1**, delete 'of persons, Phld.*Ir.* p.97 W.' (v. section 2) **II 2**, delete 'Men. 75'
*ἀλογόμορφος, ον, *having the form of an unreasoning animal* (cf. ἄλογος II), *Cat.Cod.Astr.* 8(1).173.
ἀλογόομαι, for '(prob. l.)' read '(dub. cj.)'
ἄλογος **I 2**, delete '*inexpressive* Pl.*Tht.* 203a' add '**3** *of which an account cannot be given, lacking explanation*, Pl.*Tht.* 203a.' **II**, after '*horse*' insert '*PKlein.Form.* 324.4 (vi AD)' **IV 1**, for '*incommensurable*' read '*irrational*' and for '(Euc.)*Def.* 10' read '*Def.* 3' add '**V** ἄλογος (sc. γραμμή), ἡ, *critical sign marking corrupt or doubtful passage*, Sch.A Il.16.613, Sch.Ar.*V.* 1282, *POxy.* 1788 p. 142, Isid.*Etym.* 1.21.27, Serv. ad Verg.*A.* 10.444.'
*ἀλογότης, ητος, ἡ, *lack of reason* (of animals), David *in Porph.* 209.24.
ἀλόητρα, after '*threshing*' insert '*SB* 7373.15 (AD 29); perh. ἀλώτρα (or ἀλώητρα) *PMich.*VIII 520.5 (iv AD)'
ἀλοίδορος, add '**II** = ἀλοιδόρητος 1 2, *IUrb.Rom.* 836.6 (prob. i AD).'
ἀλοιητήρ, after '*thresher*' delete '*grinder*'; for '*grinders*' read '*threshers*'; for 'λιμός' read 'λιμὸν ἀλοιητῆρα βρότειον'
ἄλοιμος, add 'also ἀλοιμμός, *IG* 2².1663 (iv BC)'
ἀλοιπηρός, delete the entry.
ἀλοίτης, delete '= ἀλείτης' and after 'Emp. 10' insert 'Call.*fr.* 271 Pf.'

ἀλοιφή **2**, after '(Od.) 18.179' add 'used by athletes, Hld. 10.31.5' at end add 'Myc. *a-ro-pa*'
*ἀλοιφίς, ἡ, *erasure*, *BGU* 2349.14 (ii AD) (ἀλυφις pap.).
ἀλοκίζω, after 'Ar.*V.* 850' insert 'of a wild boar attacking a hound, στήθη .. ἠλόκισ' *ploughed a furrow* in, epigr. in *Suppl.Hell.* 977.8 (iii BC'; for '*scratched, torn*' read '*furrowed*'
*ἀλοπέπερι, ρεως, τό, mixture of *salt and pepper*, *Vit.Aesop.*(G) 52 P.
*ἁλόποιος, ον, *manufacturing salt*, Sch.Nic.*Al.* 519a.
*ἁλοπόλιος, ὁ, = ἁλοπώλης, *Stud.Pal.* 3.141 (vi/vii AD) (unless for ὁλοπόλιος *grey-headed*, cf. Achmes *Oneirocriticon* (T) p. 14.22, 16.15).
*ἁλόπωλις, ιδος, ἡ, (*female*) *salt-seller*, *IG* 2².12073 (iv BC), 2².11244 (-πόλις, i/ii AD).
*ἄλορδος, ον, *free from inward curvature*, v.l. in Hp.*Fract.* 8 (sup.).
*ἅλος, ἡ, = ἅλως, *IG* 14.352 ii 28ff. (Halaesa).
*ἁλοτειχής, ές, perh. *walled in by the sea*, *Suppl.Hell.* 991.52 (poet. word-list, iii BC).
ἁλουργός, add 'as masc. subst., *dyer* (of purple cloth), *PPrag.* I 25.9 (vi/vii AD)'
*ἁλοφύλαξ, ακος, ὁ, *guard* or *watchman employed in saltworks*, *PHib.* 275.14 (i/ii AD).
*ἁλοχίς, ίδος, = ἄλοχος, *wife*, *Cat.Cod.Astr.* 3.45.
ἄλοχος **I 1**, add 'also masc., *husband*, *Epigr.Gr.* 720.5'
*Ἁλπαιωνία, v. °ἀλφ-.
*ἅπαρ, τό, perh. *pleasing thing, comfort* (cf. ἄλπνιστος), *Inscr.Cret.* 1.xvi 6 ivA5 (Lato, iii/ii BC).
ἄλπνιστος, read ἄλπιστος and for 'Sup. .. Pi.*I.* 5(4).12' read '*sweetest, loveliest*, rest. in Pi.*I.* 5(4.12), A.*Pers.* 981 (understood as pers. n. by some edd.)'
ἅλς (A), line 1, delete 'dat. pl. ἅλασιν (v. infr.)'; lines 14/15, delete 'ἅλασιν .. Suid.' **IV**, delete '*possible .. certain in*'; transfer 'ἄλες .. 685a' to section I 1.
Ἅλσειος, before 'ὁ' insert 'or Ἁλσεῖος'; add 'at Calymna, *Tit.Calymn.* 156, 172, al.'
ἀλσηίς, after 'A.R. 1.1066' add '4.1151'
ἄλσος **II**, at end, after '(A.)*Pers.* 111' insert 'δι' ἁλίρρυτον ἄλσος id.*Supp.* 868'
ἀτρός, add '(ἀτρόν La.)'
ἀλύζω, delete the entry (ghost-word).
*ἁλύκή, ἡ, *kind of salted food*, *BGU* 1069.2.9 (iii AD).
ἁλυκός, after 'Ar.*Lys.* 403' insert 'Pl.*Ti.* 65e'
†ἁλυκρός, ά, όν, *luke-warm*, Call.*fr.* 270 Pf., Nic.*Al.* 386 (comp.); cf. ἀλυκτρόν.
ἀλυκτέω, s.v. ἀλυκτάζω, after lemma insert '(A)'; after 'Hp.*Mul.* 1.5' insert 'Gal. 19.75'
ἀλυκτέω, (= ὑλακτέω), after lemma insert '(B)'
ἀλυκτοπέδαι, line 1, for '*bonds*' read '*shackles, fetters*'; after 'sg., *AP*' insert '12.160'
†ἄλυκτος, ον, *that cannot be escaped from*, φόνον .. ἄλυκτ[ον] *MAMA* 4.140.5 (*SEG* 30.1473) (Apollonia, Phrygia, AD 162/3), Suid., Zonar.
ἄλυπος **I**, add 'voc. freq. in epitaphs, *SEG* 23.262 (Zacynthus, ii BC), 31.1559 (Egypt, Augustan)'
ἀλυσθαίνω, after 'Nic.*Th.* 427' insert '*Al.* 141'
*ἀλυσθενέω, perh. *to be prostrated*, Call.*Del.* 212 (ἀλυσθενέουσα pap., ἀλυσθμαίνουσα codd.), cf. Hsch. ἀλυσθενεῖ· ἀσθενεῖ (ἀλυσθαίνει La.), cf. ἀλυσθένεια.
*ἀλυσθμαίνω, v. °ἀλυσθενέω.
*ἀλυσία, ἡ, *indissolubility*, ἀ. τοῦ κόσμου Procl.*in Ti.* 3.50.19 D.
ἀλυσίδιον, add '*Inscr.Délos* 1417Bii46 (ii BC), *BGU* 2328 (ἁλισείδιον, v AD)'
ἀλυσιδωτός, for 'ἀ. θώραξ' read 'θώραξ Lxx 1*Ki.* 17.5' and delete '(pl.)' after '6.23.15'
ἄλυσιον, add '*IG* 2².1533.20 (iv BC), Schwyzer 462.B51 (Tanagra, ii BC)'
ἄλυσις **1**, add 'fig., μιᾷ .. ἀλύσει σκότους πάντες ἐδέθησαν Lxx *Wi.* 17.16'
*ἀλυσιωτός, ή, όν, *chained*, Pi.*fr.* 169a.28 S.-M.
*ἀλυσκάνω, = ἀλύσκω, Od. 22.330.
ἀλύσκω, delete '- *Ep. verb* .. *Her.* 7):'; after 'Pi.*P.* 8.16' insert 'Phryn. Trag. 6 S.'; line 6, for 'etc.' read 'A.R. 4.1505'
ἀλυτάρχης, for '*BCH* .. (Tralles)' read '*ITrall.* 136' and add '*IEphes.* 502.6, 1114.6, al., *SEG* 31.1288.7 (Side, AD 249/52)'
ἀλυτάρχης, add 'written ἀλλυτάρχης, *AE* 1905.255.2 (Olympia, iii AD); cf. ἀλλύτης s.v. °ἀλύτης'
ἀλύταται, add '(ἀλυτάρχαι· παρὰ Ἡλείοις La.)'
ἀλύτης, add 'ἀλλύτης *AE* 1905.255.4 (Olympia, iii AD)'
ἄλυτος **1**, *not to be loosed*, add 'ἀ. .. δεσμοῖς Aristid.Quint. 1.3, *Cod.Just.* 8.10.12.7'
ἀλυχή, for 'Gal. 19.76' read 'Gal. 19.75'
ἀλύω **I 5**, delete the section. **III**, after 'trans.' insert '*to be distraught at*, τῶν σκελῶν .. τὴν ἧτταν ἀλυόντων Hld. 10.30.4'

*ἀλυώδης, ες, *marked by anxiety,* Hp.*Praec.* 14.
ἄλφα 1, add '*IG* 2².1425.95, al., *Schwyzer* 253.81 (Cos, iii/ii BC)'
ἀλφάβητος, add '*Orac.Tiburt.* 158'
*ἀλφαιωνία, ἡ, *peony*, Cyran. 1.21.42 K. (ed. ἀλπ-); written ἀλφαωνία *Cat.Cod.Astr.* 8(1).187.
ἄλφημα, ατος, τό, *produce*, Inscr.*Délos* 502*A*13 (iii BC).
ἀλφηστής I, for 'lit. .. men' read 'prob. *eaters of bread*'   **II**, add '*IGC* p. 98 A 8 (gen. ἀλφεισταο, Acraephia, iii/ii BC)'
ἀλφῐ, for 'cf. Str. 8.5.3' read 'Antim. 109 Wy.'
ἀλφῑτισμός, for the pres. ref. read '*Inscr.Délos* 442*A*220' and add 'Hsch. s.v. παλαία'
ἄλφῐτον I, for '*barley-groats*' read '*groats*, later sts. spec. *barley-groats*'; line 5, after 'Od. 14.429' insert '*Schwyzer* 725.4 (Miletus, vi BC)'   **II**, for 'generally *meal*, *groats*' read 'w. grain specified' and add 'Pl.*Lg.* 849c, Paus. 1.18.7; see also αὔφιτα'
ἀλφῐτοφάγος, delete '-*bread*'
ἀλωή III, for '(Arat.) 875' read '877'
ἀλωνία II, add 'written ἀλωνιεία PTeb. 727.21, 25 (ii BC)'
*Ἀλωνίτης, ὁ, cult-title of Zeus, *SEG* 33.1167 (Side), *TAM* 5(1).166a (Saittae, i AD).
*ἀλωνοθεσία, ἡ, *laying of a threshing-floor*, PKlein.*Form.* 810 (vi AD).
*ἀλωνοθέσιον, τό, *threshing-floor*, εἰς τὸ ἀλονοθέσιον PStrassb. 680.5 (early vii AD).
*ἀλωνοφῠλᾰκέω, *guard a threshing-floor*, PCair.Zen. 745.86 (iii BC).
ἀλωνοφύλαξ, add 'POxy. 2714.19, *SB* 4525 (both iii AD)'
ἀλοόφῠτος, after 'Nonn.*D.* 13.267' insert '(dub.; v.l., edd. °ἀλώφητος)'
ἀλωπός, after 'Hsch.' insert 'cf. ἀλώπα[(sic), Alc. 69.6 L.-P.'
ἀλωρῆται, for '*salt*' read '*threshing-floor*'
*ἀλωροί, οἱ, *guardians of the salt-pans*, *SEG* 32.869 (Hierapytna, Crete, ii AD).
ἅλως, after lemma insert '(perh. Cypr. *a-la-wo* (acc. and gen.), *ICS* 217.9, 18; see also ἄλων, °ἄλος)'; line 1, after 'ἄλωος' insert 'Nic.*Th.* 546, fr. 70.1 G.-S.'
ἁλώσιμος I 1, for '*easily beguiled*' read 'εὐεργεσίᾳ .. καὶ ἡδονῇ τὸ θηρίον τοῦτο (i.e. prospective friend) ἁλώσιμον .. ἐστιν'   **II**, after '*conquest*' insert 'Τροίας .. ἁλώσι[μο]ν [ἄμ]αρ Stesich. *S*89.11 (p. 186 D.), cf. Ibyc. 1(*a*).14 P.'
*ἀλωτήριον, τό, app. some component of a θυρίς, ἄλων (i.e. ἥλων) .. καὶ ἀλωτηρίων *IG* 4²(1).110.37.
ἀλώφητος, after 'Plu.*Fab.* 23' insert '2.1005e'; add 'Nonn.*D.* 13.267'
*ἁμᾰ, v. ‡ᾄμη.
ἁμᾶ A 2, delete definition 'partly .. partly .. ' (the examples belong to the preceding sense)   **II**, at end, for 'Arist.*Metaph.* 1028ᵇ 27' read 'Arist.*Metaph.* 1068ᵇ 27'   **C**, after 'D. 46.20' insert 'ἅμαν οἱ σταλογράφοι μόλωσι *SEG* 37.340.17-18 (Mantinea, early iv BC)' and at end for 'Α α II' read 'ἁ- II'
ἁμᾶ, for '*IG* 5(1)' read '*IG* 5(1).213.14, 86, ib. 1120.3, 10 (both v BC)'
*Ἀμαδρύες, αἱ, = Ἀμαδρυάδες, Nonn.*D.* 15.416, cj. in 5.440.
Ἀμαζών, delete 'II epith. .. Paus. 4.31.8'; after '-όνιος' insert '-η' and for 'Nonn.*D.* 37.17' read 'Nonn.*D.* 37.117'; add 'epith. of Smyrna, *SEG* 37.712.12 (epigr., Chios, ii AD)'
ἀμαθής, line 1, for '*stupid*' read '*lacking* or *incapable of understanding*'; line 5, for 'of .. unmanageable' read 'of class of animals (typified by wild boars)'; line 9, after 'E.*Ph.* 874' insert 'iron. ἐνδέχεται .. τὰς ξυμφορὰς τῶν πραγμάτων οὐχ ἧσσον -θῶς χωρῆσαι ἢ καὶ τὰς διανοίας τοῦ ἀνθρώπου Th. 1.140.1'   **1 b**, delete the section transferring quot. to section 1 a.   **2**, after 'of things' insert '*ignorant, uninformed*' and delete subsequent explanations in this section.   **II**, delete the section.
ἄμαθος, for '*links .. sea*' read '*stretch of sand, sands*'; after '*h.Ap.* 439' insert 'Euph. 44.2 vGr.'; add '(cf. Ἁμαθόῖ, name of Nereid, *Schwyzer* 122.7, Corinthian vase at Caere)'
*ἀμαθούσιος, ὁ, kind of jasper, Epiph.Const. in *Lapid.Gr.* 195.27.
*ἀμαιωτικός, όν, neut. pl. τὰ ἀ. *spirits that prevent successful childbirth*, *BCTH* N.S. 19B.111 (Tunisia, ii/iii AD); cf. ἀμαίευτος, μαιωτικός.
ἀμάλακτος I 1, add '*not softened*, Gal. 7.40.5'
Ἀμαλθεῖον, for 'country-house .. 18' read '*shrine of Amaltheia*, Cic.*Att.* 1.16.15, 18'
ἄμαλλα, after 'S.*fr.* 607' insert 'Call.*fr.* 186.27 Pf.'; add '[ἀμ- pap. in Call. l.c.; cf. °ἀμαλλεῖον]'
ἀμαλλεῖον, add 'written ἡαμ- in *IG* 1³.425.10 (Athens, v BC)'
ἄμαλλος (B), add '*cucumber*, PColl.Youtie 87.9 (vi AD)'
ἀμαλογεῖ, for '(prob. for ὁμο-)' read '(perh. for *ἀμαλολογεῖ)'
ἀμᾰλός, for 'παῖς Call.*fr.* 49 P.' read 'παῖς ἀμαλή Call.*fr.* 502 Pf.' and continue 'epith. of Zeus, *Lindos* 26.2 (c.400 BC); cf. Ἀμάλιος'
ἀμάμαξυς, read '**ἀμᾰμάξυς**' and before 'Matro *Conv.* 114' insert 'also ἀμαμ-'
ἀμαμιθάδες, read '**ἀμᾰμῐθάδες**'; for 'Phot. p. 86 R.' read 'Phot. *a* 1112 Th.' and add 'cf. Hsch. (ἀμμαμηθάδης cod.)'
*ἅμαν, v. °ἅμα.

ἀμᾱνῖται, add '*PVindob.* G40851 B 7 (*JÖByz.* 33.5-6); sg., *PL* I/3 in *Tyche* 1.163.22 (vi/vii AD)'
ἅμαξα I 1, for 'Hes.*Op.* 453' read 'Hes.*Op.* 426, 453, 456'   **II**, delete the section.   **III**, add 'pl., of the Great Bear and Little Bear, Arat. 27'   add '2 *wagon-load* as unit of capacity, *Suppl. epigr.Ciren.* no. 104 (iii BC), *ZPE* 25.155 (vi AD).'
*Ἁμαξᾶς, α, ὁ, *wagoner*, as a sobriquet, *IEphes.* 20 a 63.
*ἀμαξελάτης, v. °ἁμαξηλάτης.
ἁμαξεύς, add '*Salamine* 22 (i AD)'
ἁμαξηλάτης, after '*wagoner*' insert 'PCair.Żen. 176.281, 352 (iii BC); sp. ἁμαξελ- Agath. 2.6.4 K.'
ἁμαξήποδες, add 'see also ἁμαξόπ-'
ἁμαξιαῖος, for '*Com.adesp.* 836' read 'Polyzel. 7 K.-A.'
ἁμαξικός, add 'ὑποζύγια ἁμαξικά PTeb. 748.6 (iii BC)'
ἁμαξίς, after 'Hdt. 3.133' insert '*IG* 2².1673.11, 40 (iv BC)'
ἁμαξίτης, add 'II epith. of Hermes, *BCH* 85.846 (Paros, Rom.imp.).'
*ἁμαξολάτης, v. °ἁμαξηλάτης.
*ἁμάρα (B), v. °ἡμέρα.
ἁμαράκινος, for '*Edict.* .. 16' read '*Edict.Diocl.* 36.99'; add 'ἀ., τό, sc. μύρον, PCair.Zen. 536.18 (iii BC)'
ἀμαράντινος I, add 'of colour, PMasp. 6ⱽ.82 (vi AD)'
ἀμάραντος I, at end delete 'neut. pl. .. *Im.* 1.9'   **II**, after 'Poll. 1.229' insert 'Philostr.*Im.* 1.9'
ἀμαράσαι, for 'μαράσαι' read 'μαράσσαι, μαρίν'
*ἀμαρηγός, ὁ, *digger of trenches*, Teuc.Bab. p. 46 B.
ἀμάρηιος, for '*in a conduit*' read '*channelled* (for irrigation)'
Ἀμάριος, of Zeus, add '*SEG* 35.304.15 (Epid., i AD)'; **Ἀμάριον**, close parentheses after 'Plb. 5.93.10' and delete rest of entry.
ἁμαρτάνω II 1, at end for '*EN* 1126ᵇ1' read '*EN* 1126ᵃ1'   **2**, line 5, after 'ἁμαρτηθέντα' insert 'τὰ ἁμαρτανόμενα' and after 'X.*An.* 5.8.20' insert 'Isoc. 2.23, 8.39, D. 10.54, 17.21'
ἁμάρτημα, add '*delict*, Just.*Const.* Δέδωκεν 8'
ἁμαρτία, line 2, add 'of tragic heroes, Arist.*Po.* 1453ᵃ10, 16'   **2**, delete 'Arist.*EN* 1148ᵃ3, al.'
*ἁμαρτωλικός, ά, όν, app. = ἁμαρτωλός, Epich. 85.13 A.
ἁμαρτωλός I 2, add 'w. dat., ἔστω ἀ. θεοῖς χθονίοις *SEG* 33.1169 (Lycia, ii AD)'
ἀμάστικτος, for 'Sch.Pi.*O.* 1.32' read 'Sch.Pi.*O.* 1.33'
ἀμάτα, after '(Aetolia, iii BC)' insert 'adj. perh. in phr. ἀμάται τέχναι *Schwyzer* 309 (Dodona), nisi leg. ἄμα ταῖ τέχναι'
ἁματροχιά 2, delete 'by error for ἁρματροχιά'; for 'Call.*Fr.* 135' read 'Call.*fr.* 383.10 Pf.' and add 'cf. Porph. ad Il. 23.422'
†**ἀμαυρός**, ά, όν, (ός, όν Pl.Com. 20 K.-A.) *hardly seen, faint, dim*, εἴδωλον ἀ. Od. 4.824, 835, Sapph. 55.4 L.-P., ὄνειρος ἀ. Plu. 2.382f; of heavenly bodies, etc., Arist.*Mete.* 343ᵇ12, 344ᵇ29, 367ᵃ21, 375ᵃ30, Theoc. 22.21, Vett.Val. 6.15, Man. 2.117; of a lamp, Luc.*Tim.* 14, φλόξ Ph. 2.143; transf., of night, Luc.*Am.* 32.   **b** of sight, X.*Cyn.* 5.26, Ph. 1.33, Gal. 14.776.4, 16.609.6; -ὰ or -ῶς βλέπειν, *dimly*, Hp.*Acut.*(Sp.) 55, *AP* 12.254 (Strat.), *IG* 14.2111; fem. subst. = ἀμαύρωσις 3, *PMag.* 13.235.   **2** *inconspicuous, imperceptible*, ὅπου δ' ἂν ᾖ αὐταῖς ἀμαυρὸν τὸ ἴχνος X.*Cyn.* 6.21, Arist.*HA* 608ᵃ11, ἐντομὰς -οτέρας Thphr.*HP* 6.2.5, Lxx *Le.* 13.6; comp. adv. -οτέρως Gal. 9.249.8 (s.v.l.).   **3** of sound, *faint*, Arist.*Aud.* 802ᵃ19, Ph. 2.204; -ῶς Plu. 2.590c.   **II** *physically weak, feeble*, ἀμαυρῷ κώλῳ S.*OC* 182, φωτὶ 1018, χερσὶν 1639, E.*HF* 124, ἀ. σθένος 221, Pl.Com. 20 K.-A., Arr.*Cyn.* 26.4; transf., of disease, *AP* 7.78 (Dionys.).   **b** of faculties, ἀμαυρᾶς ἐκ φρενός A.*Ag.* 546, *Ch.* 157, ἀντίληψις Ph. 2.367.   **2** of wine, *weak*, Ath. 1.34c.   **3** of abst. things, *insubstantial, faint, vague*, κληδών A.*Ch.* 853, δόξα Plu.*Lyc.* 4, ἡδονή id. 2.125c.   **III** *obscure, unknown*, -οτέρη γενεή Hes.*Op.* 284, A.*Ag.* 466, E.*Andr.* 204.
ἀμαυρόω II, line 7, for '*IG* 12(9).1129.22' read '*IG* 12(9).1179.22'
ἀμαύρωσις 1, after 'Hp.*Coac.* 221' insert '*dulling*, of sight in old age, Arist.*de An.* 408ᵇ20, Diog.Oen.*fr.* 57 I 5-6 C.'   **2**, delete the section.
*ἀμαυρωτικός, ή, όν, *dulling*, τὰ -ὰ τῆς ὀρέξεως Philum.ap.Aët. 9.20.   **II** *invisible*, POxy. 3931.1 (iii/iv AD).
ἀμαχεί, delete '*without stroke of sword*'; line 2, after 'X.*An.* 1.7.9, etc.' insert '*Delph.* 3(1).506 (see also *SEG* 23.325, iv BC)'
ἀμαχητήρια, place after ἀμάχετος.
ἀμάχητος I, after '*unconquerable*' insert '*κέρδος* Simon. 36.9 P.'
ἄμαχος I, line 3, after '(Ar.*Lys.*) 1014 (lyr.)' insert '*unsurpassable*, ἀ. γεωργός Men.*Dysc.* 775'; line 8, after 'ἀ. κάλλος' insert 'Men.*Dysc.* 193'; line 9, after 'Ael.*NA* 16.23' insert '*intractable*, ἀ. τρόπος Men.*Dysc.* 869   **II 2**, add '*ISmyrna* 550.8, adv. τὴν γλυκυτάτην γυναῖκα συνβιώσασαν .. ἀ., ἡδέως, εὐσεβῶς ἐτίμησεν *TAM* 5(1).717.5 (ii AD)'
*ἀμβαθμός, v. °ἀναβαθμός.
ἄμβαρ, add 'also ἄμπαρ, Aët. 1.131 (66.17 O.)'
*ἀμβάριον, τό, = ἄμβαρ, PVindob. G10.734 (*Analect.Pap.* 3.121, vi AD).

ἄμβασε, for 'ἄμβᾰτος' read 'ἀμβᾰτός'
*ἀμβασμός, v. °ἀναβασμός.
*ἀμβῑκισμός, ὁ, operation of an alembic, Anon.Alch. 273.3.
ἄμβιξ 1, for 'spouted cup' read 'vessel narrowing towards the brim, cf. Hsch.'; for 'Posidon. 25' read 'Posidon. 67.25 E.-K.'; for 'Hsch., etc.' read 'EM 80.18'; add 'see also ἄμβυξ'
*ἀμβιτεύω, ?fr. Lat. ambitus, perh. to seek after public service, POxy. 2110.15 (iv AD), to solicit, intrigue for, Pall.V.Chrys. 61.9.
*ἀμβλατώριον, τό, Lat. ambulatorium, place for walking, walk or passage, IEphes. 437 (pl.), BE 1961.783 (ἀμβλατούριον, Phoenicia).
ἀμβλίσκω, line 5, after 'Tht. 150e' insert 'aor. 2 inf. ἀμβλῶναι Just.Nov. 22.16.1 (AD 536); cf. ‡ἐξαμβλόω'
*ἀμβλύχροος, ον, dull- or faint-coloured, Petos. p. 342.8.2.
ἀμβλυώσσω, after 'weak sight' insert 'A.fr. 25e.6 R.'
ἀμβόλιμος II, add 'v. Schwyzer 90.3, 91.2, 92.2 (perh. neut. subst., v. BCH 114.406); cf. ἀναβόλιμος, βόλιμοι'
*Ἀμβρακίδια, τά, dim. of Ἀμβρακίδες, Herod. 7.57.
*ἀμβρᾰκόομαι, quoted in conjunction w. ἀπαμβρακόομαι (q.v. for sense) Phot. α 1168 Th.
*ἀμβρότερος, α, ον, immortal, of Artemis, Orph.H. 36.9 (cj. ἀγροτέρα).
ἀμβρότιννον, add '(ἀμβροτίνον· ἄμοιρον La.)'
ἀμβροτόπωλος, add '(codd. ἀμβρότα πώλου)'
ἄμβροτος, line 1, after 'Tim.fr. 7' insert '(= 4 P.) Cret. ἄμορτος, Hymn.Curet. 17 (dub.); neut. pl. as subst., divine or undying things (?heavenly bodies), Emp. B 21.4 D.-K.'
ἄμβρυττοι, delete the entry.
*ἄμβρυττος, ἀμβρύττιος, = βρύσσος, kind of sea-urchin, Hdn.ap. Phot. α 1172 Th., Hsch.
ἄμβων, line 2, after 'Apollon.Cit. 1.7' insert 'acc. ἄββωνα IG 4.784 (Troezen)' 1, for 'Call.Aet. 3.1.34' read 'Call.fr. 75.34 Pf.' 5, for 'cf. JHS .. (Aspendus)' read 'platform in a synagogue, CIJud. 781.4 (Side)'
ἀμεθίστατος, read 'ἀμεθέστατος' and add 'PSarap. 22.9, 27.26, al. (ii AD)'
ἀμεθόδευτος 2, add 'without sound method, ἀμεθόδευτον .. ἐκκαλέσασθαι τὸν Περικλέα it is bad procedure to .., Sopat.Rh. 8.361.27 W.'
*ἀμεθυστίζω, resemble amethyst in colour, Plin.HN 37.93.
ἀμείβω A I 3, for 'in Att. .. hence' read 'pass over from one side to the other, Κρισαῖον λόφον ἄμειψεν Pi.P. 5.38, πορθμὸν ἀμείψας A.Pers. 70, E.IA 144' 5 b, for 'dislodge' read 'get rid of in exchange' and insert 'ἀμείψαι κερασφόρον ἄταν E.Hyps.fr. 1 iii 29 B.' II, before 'rafters' insert 'pitched'; for 'Nonn.D. 37.588' read 'Nonn.D.37.592' B II 1, after 'S.Tr. 737' add 'put on as a change (of clothing), E.Ph. 326'; add 'w. μετά, ἡ ψηφὶς μετὰ γῆρας ἀμιψαμένη νεότητα Salamine 202 (iv AD)'
ἀμειδής, for 'Orph.A. 1079' read 'Orph.A. 1075'
ἀμείλικτος I, add 'λοιμὸς ἀ. SEG 26.405 (Corinth, Chr.)' add 'III = ἄμικτος, Hsch. (3548 La.); combined w. ἄχραντος, ἀκήρατος, etc., Procl. in Ti. 3.258.22 D., al.; cf. μειλικτός.'
ἀμείλιχος I, after 'Sol. 32' insert 'πόντος h.Hom. 33.8, Anacr. 2fr. 1.16 P.'
ἀμείνασις, for 'ἡ δύοσμον' read 'ἡδύοσμον' and for 'EM 83.50' read 'EM 82.50'
ἀμείνων I, add 'used to denote curule (opp. plebeian) aediles, D.C. 53.33.3'
+ἀμειξία, ἡ, absence of communications, PLond. 401.20, PTeb. 72.45 (ii BC), Schwyzer 195.7 (Cret. inscr. fr. Delos, ii BC); cf. ἀμιξία.
*ἀμειπτός, ή, όν, (capable of being) switched about, app. of a siege-engine, Archil. 98.18 W.
ἄμειψις I, add '2 object received in an exchange, Just.Nov. 7.5.2 (AD 535).' II, add '3 commutation, i.e. money payment instead of payment in kind or payment in a different kind, PMich.XII 645.23, 28 (sp. ἀμιψ- and ἀμιμψ-, AD 304), POxy. 3795 (iv AD).'
*ἀμελγή, ἡ, app. milk-pail, πελλίσι· ἐν ταις ἀμελγαῖς Sch.Nic.Al. 77b Ge.
ἀμελητής, for 'Gal. 3.827' read 'Gal. 3.327.13'
ἀμελητί, add 'without delay, Just.Edict. 9 pr. (for ἀμελλητί)'
*ἀμελίτῑτις, acc. ιν, fem. adj., lacking honey, ἑορτή cj. in Herod. 5.85.
*ἀμέμαρον, gloss on βούθουτον, Hsch. (s.v.l., ἀνέκφορον La.).
ἄμεμπτος I 2, adv., add 'SEG 30.272 (Athens), TAM 4(1).208, 263 (Bithynia), PPetaus 24.29'
ἀμεμφής I, after '(A.)Supp. 581' insert 'Emp. B 35.13 D.-K.'; add 'adv. -φῶς Orph.H. 43.11, MAMA 8.413a17 (Aphrodisias); Ion. -φέως, Emp. B 35.9 D.-K.' II, delete 'adv. .. 11'
ἀμεμψιμοίρητος, adv. -τως, add 'SEG 39.606.13 (Maced., ii BC)'
*ἀμές, v. °ἐγώ.
ἀμενηνός, line 1, after 'ἡ, όν' insert 'Luc.Gall. 5'
*ἀμένης, -ητος, ὁ, one lacking bodily strength, child, Hdn.Gr. 2.684.
ἀμένητος, delete the entry (v. °ἀμενής).
*ἀμέρα, v. ἡμέρα.
ἀμέργω, for 'Com.adesp. 437' read 'Ar.fr. 406 K.-A.'

ἀμέρδω II, for this section read 'perh. = ἀμέργω, pluck, λειμώνιον ἄνθος ἀμέρσας (-ξας cj.) AP 7.657 (Leon.Tarent.), ἄμερσεν (-ξεν cj.) Nic.Th. 686' at end for 'cf. μέρδει' read 'cf. ‡μέρδει'
*ἀμεριμναῖος, ον, = ἀμέριμνος, dub. in Ibyc.ap.Hdn.fr. 6 H.
ἀμεριμνέω, add 'Bithynische St. III 2.10'
ἀμέριμνος I, line 2, after 'Eranos 13.87' insert 'εἰς Ἀίδαν ἀ. SEG 26.1808 (Egypt, Hellen.)' III, line 2, for 'either .. celebrated' read 'bringing more freedom from care'
ἀμέριστος I, add '2 rhet. ἀμέριστον (sc. σχῆμα), τό, repeated use of μέν without following δέ, Olymp. in Grg. 14.16 W. (p. 77 N.).'
*ἀμές, v. ἐγώ (IV, line 3).
*ἄμεστος, perh. not occupied, PRyl. 99.4 (iii AD).
*ἀμετάβατος II, for 'unextended' read 'having no extent'; at end for 'Epicur.Ep. .. U.' read 'τὰ ἀ. Epicur.Ep. [2].8 A. (cj., codd. -βολα)'
ἀμετάβλητος, after 'Ti.Locr. 98c' insert 'Sallust. I. 2, XVII.10'
ἀμετάβολος, for '= foreg.' read 'unchangeable' 2, add 'of pure vowels, unchanging, opp. diphthongs, S.E.M. 1.117, 118' add '3 -βολον, τό, gramm., liquid (λ μ ν ρ), D.T. 632.7, Hdn.Gr. 2.393, Choerob.in Theod. 1.298.35 H.'
ἀμετάκλητος, delete 'ὀργή Hld. 2.10 (v.l. -βλητος)'
ἀμετάληπτος, add 'of words, having no equivalent, Eust. 490.38, al.'
ἀμεταμίσθωτος, add 'PSoterichos 3.38, 42 (AD 89/90), PTeb. 378.29 (iii AD)'
*ἀμετανάτρεπτος, ον, unalterable, PLond. 1660.37 (vi AD).
*ἀμετάπρᾱτος, ον, not to be sold again, ἄσυλα καὶ ἀ. IGBulg. 995.3 (Philippopolis, iii AD).
*ἀμετάσχετον, gloss on ἀσχαδές, Hsch. (dub., ἀκατά- cj.).
ἀμετάτροπος, add 'IG 12(5).302.7 (Paros, i/ii AD)'
*ἀμετεπίγρᾰφος, ον, without change in its inscription, IG 9(2).32.6 (Aenis).
ἀμέτοχος, add 'adv. -χως without sharing (in an action), Eust. 1946.32'
*ἀμετρησίη, ἡ, = ἀμετρία, excessiveness, κερδέων ἀ., Philipp.Perg. 1 J.
ἀμετρία, add '3 absence of measure or rhythm, διὰ τὴν ἐν τῷ ποδὶ πρὸς τὴν λύραν ἀμετρίαν Pl.Clit. 407c.'
ἀμετροπαθής, add '(codd. ἀμετριο-)'
ἀμεύομαι II, for this section read 'acquire by exchange, aor. inf. ἀμεϝύσασθαι Inscr.Cret. 4.4.1 (Gortyn, vi BC), aor. subj. ἀμεύσονται Inscr.Cret. 1.xviii 1.3 (Lyttos, vi BC)'
ἄμη, after 'ἡ' insert 'also ἄμη Phryn.PS p. 107.10 (s.v. σκαφεῖον), ἅμα SEG 24.361.15 (Thespiae, iv BC)' 1, shovel, add 'PFreib. 53.31 (i BC)' 2, after 'pail' insert 'or scoop, SEG l.c.'
ἀμήνῑτος, after '(A.)Ag. 649' insert 'Plu. 2.90d, 413d, 464c'; after '-τως' read 'A.Ag. 1036, Achae. 15 S.'; add 'also -τεί Archil. 89.10 W.'
*ἀμητορίδαι, οἱ, = ἀμήτορες, (v. °ἀμήτωρ), Hsch. (cod.).
ἄμητος II, at end after 'Ammon.' insert 'Diff. 38 N.'
ἀμήτωρ I, add '2 ἀμήτορες, οἱ, designation of wandering minstrels in Crete, EM 83.15 (cf. Ath. 14.638b and °ἀμητορίδαι).'
ἀμηχανέω, line 5, delete 'relative'; line 7, after 'Epicur.fr. 203' insert 'χὤτι τὸ φάρμακόν ἐστιν ἀμηχανέοντος ἔρωτος Theoc. 14.52'
ἀμηχανία II, delete 'of things'; add 'Pi.N. 7.97'
ἀμία (A), add 'IGC p. 98 A9 (Acraephia, iii/ii BC)'
ἀμιγής I, after 'adv. -γῶς' insert 'Alcin.Intr. 10 (p. 164 H.)' III, for 'Proll.Ar.' read 'in Anon.Cram.Proll.Com. XI c. II 9 K.'
ἄμιθα, for 'kind of cake, perh. = ἄμης' read 'perh. provisions' and add 'PHaun. 22.11 (ii/iii AD)'
ἄμικτος, III 1 a, line 3, for 'Anaxil. 223' read 'Anaxil. 22.3 K.-A.'
*ἀμικτώριον, τό, Lat. amictorium, wrap, shawl, PFouad Inv. 45.17 (CE 27.196, ii/iii AD); perh. shroud, POxy. 1535ʳ.8 (iii AD).
ἄμιλλα 2, line 5, for 'Isoc. 10.15' read 'Isoc. 10.35'
ἁμιλλάομαι I 1, line 4, after 'contend' insert 'Schwyzer 726.15 (Miletus, v BC)'
ἁμιλλητήρ, add 'w. gen., competitor in, ἁμιλλητῆρας ἐρώτων Nonn.D. 6.12, 80'
*ἁμιλλητήριος, add 'contest, Schwyzer 726.14 (Miletus, neut. pl., v BC)'
ἄμιλλος, for the pres. ref. read 'Doroth.ap.Phot. α 1203 Th.' and add 'SEG 21.527.61 (Athens, iv BC)'
ἀμίμητος, after 'inimitable' insert 'τὸν Ἀννίβαν ἀμίμητον παρεισάγοντες στρατηγόν Plb. 3.47.7; of Antony as a god, OGI 195.2 (i BC)'
ἀμιναῖος, add 'also Ἀμιννέος, Ἀμίννιος, Edict.Diocl. 2.4, 4A., Ἀμινναῖος, Dsc. 5.19'
*ἄμιον, τό, perh. dim. of ἄμη, small shovel, περὶ τοῦ ἀμίου Aegyptus 15.273 (?iii AD).
ἄμιππος, after 'Xen.HG 7.5.23 (cj.)' insert 'Eq.Mag. 5.13'
ἀμίς, for 'be .. (in contempt)' read 'put something (vile or common) to its intended purpose'
ἀμίσαλος, after 'ap.Et.Gen.' insert 'v. Call.fr. 738 Pf.'
*ἀμίσαντος, ον, unexpld. wd. in magical formula in Inscr.Cret. 2.xix 7.18 (iv BC).

ἀμιχθᾰλόεις, for '= ἄμικτος III, *inhospitable*' read 'obscure wd. expl. by ancient commentators as *"inhospitable"*, *"misty"*, etc.' and after '*h.Ap.* 36' insert 'ἀμιχθαλόεσσαν .. ἠέρα Call.*fr.* 18.8 Pf., Colluth. 208'

ἀμμά, add 'also perh. ἀμμή *IGLS* 99 (v. *SEG* 7.50)'

ἄμμα I 5, add 'Opp.*H.* 3.317'   6, read 'as measure of length *POxy.* 669 (iii AD); = 40 πήχεις, Hero *Geom.* 23.14; = one-eighth of a σχοινίον, *VBP* IV 92 (*ZPE* 42.95), also as square measure, *SB* 6951; 11, al. (ii AD)'   add '7 *junction* of blood vessels, Pl.*Ti.* 70b.'

ἀμμαμηθάδης, for '= ἀμαμινθάδες (q.v.)' read 'corrupted fr. °ἀμαμιθάδες'

ἀμμάς, after 'ή' insert '*ISmyrna* 553.7 (i/ii AD), *PMich.*III 208.9 (ii AD)'; add 'cj. in Simm. 11'

ἀμματίζω, after '*bind*' insert 'Poet.*de herb.* 63, Hsch.'

ἀμμεδαπάν, add '(Aeol. for ἡμεδαπήν)'

ἀμμή, v. °ἀμμά.

*ἀμμηγία, ἡ, *conveyance of sebakh*, i.e. powdery earth used as manure, *PSoterichos* 1.24 (AD 69), κοπρηγία καὶ ἀ. *PFlor.* 143.6 (iii AD), *PBerl.Leihg.* 23.10 (-εία; iii AD).

ἀμμία, add 'cf. fem. pers. n. Ἀμμία'

ἄμμιγα, v. ‡ἀνάμιγα.

ἄμμινος, add '*SB* 7644.10 (iii BC).   2 app. *sandy-coloured*, κύτη (= κοίτη) ἀ. *POxy.* 3201.6 (iii AD); ἐρίων ἀ. *SB* 12314.16.'

†ἀμμίτης, ον, ὁ, name of precious stone, Isid.*Etym.* 16.4.29 ([*h*]*ammites*).

†ἀμμῖτις, ιδος, ἡ, name of precious stone, prob. = °ἀμμίτης, Plin.*HN* 37.167 (*hammitis*).

ἀμμόγειος, add 'as part of a village name, ἀ. Σέντρυφις *PVindob.Bosw.* 3.8 (iii AD)'

ἀμμοδύτης, delete 'διψάς'

*ἀμμόνιον, τό, (ἀναμένω), perh. *fine levied on defaulters*, *CID* I 9.A48 (Delphi, iv BC); for formation cf. καμμονίη.

ἀμμορία (A), lines 1/2, for 'Ζεύς .. ἀνθρώπων' read 'εὖ οἶδεν (Ζεύς) .. μοῖράν τ' ἀμμορίην τε καταθνητῶν ἀνθρώπων'; add 'perh. *what is not the portion* (of any party), epigr.ap.D. 7.40'

ἀμμορία (B), delete the entry.

*ἀμμοσκᾰφεῖον, τό, (ἄμμος, σκαφεῖον) prob. kind of scoop used by sand-diggers, *BGU* 1521 (iii BC).

*ἀμμόχρωμος, ον, *sand-coloured*, of a donkey, *CPR* 7.36.6 (iv AD).

ἀμμόχωστος, add 'Wilcken *Chr.* 1.227.1 (-χοστος; Fayûm, iii AD), *PCol.* 172.17 (iv AD)'

Ἀμμωνιακή, add '*PTeb.* 273.35 (iii AD)'

Ἀμμωνῖται, οἱ, *association of worshippers of Ammon*, *SEG* 33.1056 (Ἀμμωνειτ-, Cyzicus, ii AD).

ἄμμων, after 'etc.' insert 'Hsch., also ἄνᾱμμος, *grandson*, *SEG* 18.744.9 (Cyrene, ii AD), cf. Κυρηναῖοι τὰ ἔκγονα τῶν ἐκγόνων ἀνάμους καλοῦσι Ar.Byz.*fr.* 235 S.; αμ *CPap.Jud.* 428.4 (Egypt, ii AD) may be abbrev. for this wd.; also ἡ ἄμμαμμος *granddaughter*, *Inscr.Cret.* 1.98 (p. 212).B 1-2 (Lyttos, Rom.imp.)'

ἀμνάμων (A), after 'ὁ' insert '*grandson*, Call.*fr.* 338, rest. in 110.44 Pf.'

ἀμνάς, after '*lamb*' insert '*PCair.Zen.* 576.3, al. (iii BC), Theoc. 8.35 (pl.)'

ἀμνή, after '*ewe-lamb*' insert '*IG* I³.250.A16 (Attica, v BC)'

ἀμνημοσύνη, add '*loss of memory*, *Cat.Cod.Astr.* 8(1).189.24, 192.23'

ἀμνήμων I 1, add 'b *suffering from loss of memory*, *Cat.Cod.Astr.* 8(1).192.23.'

ἀμνησικᾰκέω, for '*bestow* .. τινός' read 'w. gen. of offence, *overlook an offence*, ἀμνησικακεῖν ὧν εἴργασται Ἀντώνιος'

ἀμνήστευτος, for '*E.fr.* 815' read '*E.fr.* 818'

ἄμνηστος, after lemma insert 'dial. ἀμνᾱστος'

ἀμνίς, after 'Theoc. 5.3' insert '5.139'

ἀμνοκόπος, add '(ἀμνοκόμος cj. La.)'

†ἀμνός, ὁ, (or ἡ, Theoc. 5.144, 149, *AP* 5.205), *lamb*, esp. *sacrificial lamb*, an older animal than ἀρήν; ἑτήρας ἀμνοὺς θεοῖς ἔρεξ' ἐπακτίοις S.*fr.* 551, Ar.*Au.* 1559, ἀμνὸν κριτόν *SEG* 26.136.19-20 (sacrificial calendar, Attica, iv BC), Arat. 1106, ἰσομάτορα ἀμνόν Theoc. 8.14, *AP* 6.282 (Theod.); ἐσόμεθ' ἀλλήλοισιν ἀμνοὶ τοὺς τρόπους, i.e. *meek*, Ar.*Pax* 935; metaph., ὁ ἀ. τοῦ Θεοῦ Ev.Jo. 1.36; fem. also ἀμνή (-ά), ἀμνίς, ἀμνάς qqv.

*ἀμόγως, adv. *without tiring*, *POxy.* 2331 iii 20 (pap. ἀλόγ-; verse, iii AD).

*ἄμοθος, *lacking a battle*, cj. in Nonn.*D.* 25.308; cf. μόθος.

*ἀμοιβᾰδίζω, gloss on Lat. *alternor*, w. ἀμφιβάλλω, Dosith. p. 430 K.

ἀμοιβάζω, add 'II *reward*, ἀμοιβάσασθαι αὐτὸ ταῖς πρεπούσαις τειμαῖς *IEphes.* 4337.12-13 (i AD).'

ἀμοιβή, after lemma insert 'dial. ἀμοιβά (Pi., ἀμοιβᾶ Sapph. 133.1 L.-P.); also ἀμοιϝά (v. infra); line 5, for 'χαρίεσσα .. (Corinth)' read 'χαρίϝεττα ἀμοιϝ[άν] (or ἀμοιϝ-) *CEG* 326 (Thebes, viii/vii BC), [χα]ρίεσ(σ)αν ἀμοιϝάν *CEG* 360 (Corinth, vi BC)'   add 'IV ἐπ' ἀμοιβῆς *on the other hand*, κείνων δ' οὔτε τι θῆλυ πέλει γένος, οὔτ' ἐπ' ἀ. ἄρσενες Opp.*H.* 1.765 (v.l. ἀπ' ἀ., ὑπ' ἀ.).'

ἀμοιβός I, add '*successor in office*, Just.*Nov.* 58'

*ἀμοίρᾱτος, ον, *exempt from fate*, ἀμοίρατον ἐς χρόνον ἔλθοις Dioscorus 4.10 H.

*ἀμοιρόγᾰμος, ον, *excluded from* or *having no part in marriage*, κοῦρος *IG* 12(5).1104 (Syros, ii AD).

ἀμόλγιον, after 'Theoc. 25.106' insert 'gloss on γαυλός, Sch.Od. 9.223'

ἀμολγός, add 'II *milking-pail* (cf. mod. Gk.), also wooden *container for wine*, *EM* 86.11-12.   III *milker*, *exploiter*, Paus.Gr. α 90, *EM* 86.13.'

*ἀμολυβδοχόητος, ον, *not fixed with lead*, λίθοι *IG* 7.3074.4 (Lebadea, ii BC).

ἀμόλυντος II, for '*not* .. *stain*' read '*solid enough not to make a mess*'; delete 'cf.' and insert 'Heras ap.Gal. 13.432.18'

ἀμόρα, add 'cf. Hsch.'

ἀμορβεύω, after '*attend*' insert 'ἀμορβεύεσκεν Call.*fr.* 271 Pf. (ἀμορμ- codd.)'; for '*let follow*, *make follow*' read '*give as burden*, *load on to*'

ἀμορβός I, for 'Call.*Hec.* 6' read 'Call.*fr.* 301 Pf.'

ἀμόργης, add 'acc. -ητα *PCair.Zen.* 839.3 (iii BC)'

ἀμορίτης, add 'see also ἀμορβίτης, ἀμοργῖτας, °ἀμυρίτης'

*ἀμορφέω, *to be shapeless*, Plot. 6.7.32.

ἀμορφία II, add 'of literary style, *inelegance*, D.H.*Comp.* 18, 19'

ἀμός (B), delete 'ἀμόθι'

*ἀμούμαντις, εως, ὁ, dub. sens. in Ps.-Callisth. 1.4.3 (cod. A; perh. ἀμμόμαντις, *diviner by sand*).

ἀμουσία I, for '*E.fr.* 1020' read '*E.fr.* 1033'

ἀμόχθητος, after '= sq.' insert 'ἀμόχθητον .. δίαιταν Alc. 61.12 L.-P.'

†ἄμπαιδες· οἱ τῶν παίδων ἐπιμελούμενοι παρὰ Λάκωσιν, Hsch. (perh. fr. *ἀμφίπαιδες or cf. ἀναπαιδεύω).

ἀμπαίνεθαι, add '(s.v. ‡ἀναφαίνω I 3)'

ἀμπαιστήρ, for '*IG* 4.1484' read '*IG* 4²(1).102' and add '(lapis ἀνπ-)'

ἀμπαλίνορρος, delete the entry.

*ἀμπᾰλίνυρος, ον, *returning*, Cratin. 78 K.-A., Philetaer. 11 K.-A.

ἄμπᾰλος, for '*auction*, *SIG* .. (Aetol., iii BC)' read '*contract*, *IG* 9²(1).188.15 (Melitea, iii BC)'

†ἄμπανσις, ἀ, acc. ἄνπανσιν, *act of adoption*, *Leg.Gort.* 4.10.33; cf. ‡ἀναφαίνω I 3.

†ἄμπαντος, ὁ, *adopted one*, *Leg.Gort.* 4.10.7 (ἀνπ-).

†ἀμπαντύς, ἀ, dat. -νῖ (or νι), *state of adoption*, *Leg.Gort.* 4.10.48-9.

*ἄμπαρ, v. ‡ἄμβαρ.

*ἀμπείθω, v. ‡ἀναπείθω.

ἀμπελεών, for '*poet.*' read 'non-Att.'; add '*AP* 6.226 (Leon.Tarent.)'

ἀμπελικός, line 2, after 'καρποί *PLond.* 163.16 (i AD), κτῆμα *BGU* 2127 (ii AD); also -κόν, τό, *OAbu Mena* 11'

ἀμπέλινος I, after 'Id. 2.37, 60' insert 'Antim. 85 Wy.'; add 'of colour, *PHamb.* 10.27 (ii AD)'

ἀμπέλιον, add '2 *vineyard*, *PStrassb.* 29.39 (iii AD).'

*Ἀμπελίτης, ὁ. cult-title of Zeus in Phrygia, *SEG* 33.1144, etc.

ἀμπελίων, after '*bird*' insert '*Edict.Diocl.* 4.34'

ἀμπελόεις, after 'Il. 2.561' insert 'and acc. fem. pl. ἀμπελόεις Nic.*Al.* 266'

*ἀμπελοῖνος, ὁ, app. *grape-wine*, *SB* 4486.6 (vi/vii AD, pl.).

ἄμπελος II, before '*vineyard*' insert '*land planted with vines*' and after it 'Hatzopoulos *Donation du roi* 17, line 19 (Cassandreia, iii BC), *PRev.Laws* 36.5 (iii BC), *PAvrom.* 1.A10, al., *Apoc.* 14.18, 19, *POxy.* 1631.21, 29 (iii AD)'

*ἀμπελοσκᾰφής, ὁ, *vine-digger*, A.*fr.* 46a.18 R.

†ἀμπελοτόμος, ον, *for pruning vines*, δρέπανον *TAM* 5(1).318.19-20 (ii AD), Hsch. s.v. βίσβης.

ἀμπελουργός, add 'ἀ. οἱ, association of worshippers of Dionysus, *SEG* 32.488 (Tanagra, *c.*100 BC).   2 ἀμπελουργός, όν, adj., δρέπανον ἀμπελοργόν *IG* I³.422.143 (Athens, v BC), cf. *IG* 2².1526.8.'

*ἀμπελοφόριμος, ον, *vine-bearing*, γῆ *PMasp.* 151.105 (AD 570).

ἀμπελόφυτος, add '*PMasp.* 151.125 (vi AD)'

*ἀμπελοχελώνη, ἡ, *pent-house* or *mantlet*, Poliorc. 214.5.

ἀμπελωνίδιον, add '*PCair.Zen.* 309.6 (iii BC)'

ἀμπεχόνη I, for '*fine shawl* .. *men*' read '*covering for the body*, *wrap*' and for 'Pherecr. 108.28' read 'Pherecr. 113.28 K.-A., Theoc. 27.59, 60, *APl* 306 (Leon.Tarent.)'   2, for 'etc.' read 'Pl.*Chrm.* 173b, *Lg.* 679a'

†ἀμπλᾰκιώτης, ιδος, fem. adj. applied to ἱερὰ νόσος, Poet.*de herb.* 175.

ἄμπνυτο, line 2, after 'al.' insert 'but see ἔμπνυτο'; add '[ἄμπνῦτο, Nonn.*D.*34.342]'

*ἀμπούλλιον, τό, *small flask*, dim. fr. Lat. *ampulla*, *PLond.* 191.16 (ii AD), *SB* 9238.19 (ii/iii AD).

†ἄμποχος, ὁ, (ἀμπέχομαι), *guarantor*, Sicilian wd., *SEG* 34.940.5, 39.996.6 (both Camarina, iv/iii BC and ii/i BC), ib. 4.62.4 (Morgantina, ii/i BC), Hsch.

ἀμπτᾶσα, ἀμπταίην, delete 'ἀμπτᾶσα'

ἀμπυκάζω, for 'bind front hair' read 'place a diadem or headband on, crown'

*ἀμπυκόω, tie a headband on, Phot. α 1256 Th. (ἀμπυκοῖς).

ἀμπύκωμα, add 'prob. in A.Supp. 235 (pl.)'

*ἄμπυλλα (or -η), ἡ, Lat. ampulla, BGU 40 (ἀνπ-, ii/iii AD); sp. ἄμβουλλα POxy. 3993.8 (ii/iii AD).

*ἀμπυλλάριον, τό, app. holder or container for flasks, CPR 5 Skar 455 (iv AD).

ἄμπυξ 2, for 'headband' read 'headstall' at end add 'Myc. a-pu-ke (pl.), sense II; cf. a-pu-ko-wo-ko = *ἀμπυκϝοργός, fem. occupational term'

ἄμπωτις, line 1, after 'ιος' insert 'dat. ιδι D.C. 39.40, pl. -ιδες Longin. 9.13' I 1, line 2, for 'πλημμυρίς' read 'πλημυρίς or πλήμη'; line 4, delete 'ebb and flow, tides' 3, add 'Longin. l.c.'

*ἀμτύριον, τό, (for ἀμιτύριον), cheese flavoured with caraway, PCair.Preis. 38.11 (iv AD).

ἀμυγδάλη I, after 'Theophr.HP 1.113' insert 'Men.fr. 133 K.-Th.' add 'forms ἀμυσγέλα, ἀμυσγύλα at Cyrene, SEG 9.32, 43 (iii/ii BC)'

ἀμύδις II, at end after 'late Ep.' insert 'Call.fr. 295 Pf.'

ἀμυδρός I, at end delete 'ἁ. ἔχειν .. 668ᵃ3' 2, for this section read 'of movement, indistinct, imperceptible, Aret.CA 2.3 (127.28 H.), SD 1.12 (53.18 H.); τυπαί Nic.Th. 358 (comp.); of creatures, etc., imperceptible in movement, ib. 195, 373; 158 (sup.)'

Ἀμυκλάδια, τά, dim. of Ἀμύκλαι, kind of shoe, IG 1³.422.244 (Athens, v BC).

Ἀμυκλαῖος, add 'cult-title of Apollo at Idalion, SEG 25.1071 (iii BC). 2 ὁ (sc. μήν), name of month at Argos, SEG 34.282 (iv BC), BCH 98.776 no. 2 (iii BC), at Gortyn, Inscr.Cret. 4.182.23 (ii BC).'

*ἀμυκτηρίστως, adv. without disparagement, ἀκούων ἁ. Vit.Aesop.(G) 87.

ἄμυλος II, delete 'cf.' and insert 'Pherecr. 113.17 K.-A., Metag. 6.11 K.-A.'

ἀμύμων, line 2, after 'Hsch.' insert 'hence traditional interpr. blameless'

ἀμυντήριον, line 3, for 'Plu. 2.714f' read 'Plu. 2.714e'.

ἀμύντης, for pres. refs. read 'Trag.adesp. 585b K.-S. (oxyt. acc. to Hdn. 1.78.5; paroxyt. in Phot. 1270 Th.); cf. pers. n. Ἀμύντας, -ης'

ἀμύντωρ I, for 'Il. 13.384 .. etc.' read 'Il. 15.610, 13.384 (v.l.), Od. 2.326 (pl.), Call.fr. 635 Pf.'

ἀμύνω (A) I 1 b, line 4, after 'Th. 3.67' insert 'w. ἐπί added, Ἀπόλλωνα .. αὐταί φασιν αἱ ποιήσεις ἀμύναι Κούρησιν ἐπὶ τοὺς Αἰτωλούς Paus. 10.31.3' 2, add 'c w. acc., aid, succour, Nic.Th. 868.'

ἄμυξις, for 'tearing, rending, mangling' read 'scratching, laceration' and delete 'irritation'

*ἀμυρίτης, ου, ὁ, prob. = ‡ἀμορίτης, Aq., Sm. 2Ki. 6.19.

+ἄμυρος, ον, interpr. by gramm. as watery but prob. malodorous, ἀμύρους τόπους S.fr. 512 R. 2 devoid of perfume, Orac.Sib. 5.128.

*ἀμυσγέλα, ἀμυσγύλα, v. ‡ἀμυγδάλη.

ἀμύσσω I, line 7, delete 'tear in pieces, mangle'

ἄμυστις, after 'ιδος' insert 'written ἄμυσσιν (acc.) SEG 30.938 (graffito, Olbia, vi/v BC) I 1, for 'Call.Aet. 1.1.11' read 'Call.fr. 178.11 Pf.'

ἀμυχή I 1, add 'γενύων τ᾽ ἀμυχάς E.fr. 925a N.-S.; μετὰ ἀμοιχῆς POxy. 3195.46 (AD 331)' add '3 scratched surface, Plu. 2.473e.'

+ἀμφαγνοέω, v. ἀμφιγνοέω.

*ἀμφᾳδά, adv. v. ‡ἀμφάδιος.

ἀμφάδιος I, add 'λέκτρα AP 5.219 (Paul.Sil.); φιλίη ib. 267.5 (Agath.)'

ἀμφαδόν, after 'A.R. 3.615' insert 'also neut. pl. ἀμφαδά, APl. 4.296 (Antip.Sid.), cj. in Posidipp. in Suppl.Hell. 705.17'

ἀμφαεικής, add '(ἀμφάϊκος La.)'

ἀμφακλής, add '(ἄμφακες La., v. ‡ἀμφήκης)'

ἀμφαλλάσσω, for 'change entirely' read 'transform' add 'II exchange, PMich.III 149.18.15 (ii AD).'

ἄμφανσις, ἀμφαντός, ἀμφαντύς, delete the entries (v. °ἄμπανσις, etc.)

ἀμφάνω, delete the entry (v. ‡ἀναφαίνω I 3)

*ἄμφαρμος, ον, archit., having joints on both sides, ἁ. λίθος MDAI(I) 19/ 20.238.12, 242 (Didyma).

ἀμφαϋτέω, add 'shout on all sides, στρατός Archil. 196a.51 W².'

ἀμφᾰφάω I, add 'part. ἀμφαφώμενος Archil.fr. 196A W.'

ἀμφεικάς, after '(Cos)' insert 'IStraton. 9.2 (ἀμφικ-, ii BC)'; at end delete 's.v. ἀμφ᾽ εἰκάς'

*ἀμφειλύω, wrap, enfold, Philet. 17.1. (tm.).

*ἀμφέλιξ, ικος, coiled, twisted, Erinn. in Suppl.Hell. 401.41 (rest.).

ἀμφελύτρωσις, delete the entry.

ἀμφήκης, after lemma insert 'Dor. ἀμφάκης (B., S. ll.cc.)' I, add 'b subst., τώμφακες axe, Sophr. in GLP 1.73.7.' II, after 'Luc.JTr. 43' add '(v.l. ἀμφήρης, °ἀμφήχης)'

ἀμφημερινός, after 'Hp.Epid. 1.6' insert 'Nat.Hom. 15, Morb.Sacr. 1.6'

*ἀμφήχης, v. °ἀμφήκης.

ἀμφί C II 2, add 'App.Syr. 20' E, after 'Od. 6.292' insert 'ἀμφὶ δ᾽ ἐκτύπουν πέτραι S.Tr. 787, E.Hipp. 770, Ion 224, Ph. 325' at end add 'Myc. a-pi'

ἀμφιάζω, line 9, for 'Them.Or. 13.235a' read 'Them.Or. 20.235a' and for 'Lxx Jb. 40.5' read 'Lxx Jb. 40.5(10)'; add '°ἐμφιέζω'

ἀμφίαλος, for 'S.Ph. 146' read 'S.Ph. 1464'

Ἀμφιαράϊα, τά, s.v. Ἀμφιάραος, add 'Ἀμφιάραα IG 2².2196.12 (c.AD 200), Ἀμφιαίρεια IG 2².2237.58 (iii AD); see also Ἀμφιεραία'

Ἀμφιάραος, line 1, -ηος, add 'Hes.fr. 25.34 M.-W.'; add 'also Ἀμφιέρ- SEG 31.438 (Boeotia, iv BC), IG 2².4452 (Rhamnous, iii BC)'

+Ἀμφιάρειον, τό, temple of Amphiaraus, Str. 9.2.10.

ἀμφίασμα, delete 'Luc.Cyn. 17' (read ἀμφίεσμα)

ἀμφιασμός, add 'SB 10228.2.16 (AD 125)'

*Ἀμφιασταί, οἱ, name of healing cult association, SEG 31.807 (Eretria, ii BC).

*ἀμφιάστωρ, ορος, ὁ, cloak, wrap, Sch.Gen.Il. 3.134.

*ἀμφιάτωρ, gloss on κόλωψ, Hsch.

Ἀμφίβαιος, for 'ἀμφίγαιος' read 'ἀμφίγειος'

ἀμφιβάλλω I 1 b, for 'built chamber over him' read 'built the chamber round it' c, line 6, for 'like a net' read 'like a mantle' IV 2, after 'Alciphr. 1.37' insert 'εἴ τις πρὸς μίαν τῶν εἰρημένων ἁγίων συνόδων ἀμφιβάλλει .. Cod.Just. 1.1.7.20, ἀμφιβάλλετε πρὸς ἐμέ PMasp. 295.19 (vi AD)'

+ἀμφιβαρής, ές, v. ‡ἀμφικέλεμνον.

ἀμφίβληστρον 1 b, add 'E.fr. 697'

ἀμφίβλητος, delete the entry (v. °ἀμφίβληστρον).

ἀμφιβόητος 2, for this section read 'celebrated all around, AP 9.241 (Antip.Thess.), APl. 278 (Paul.Sil.), Nonn.D. 26.141'

ἀμφιβολεύς, for 'fisherman' read 'one who fishes with a net' and add 'PCornell 46.6 (ii AD)'

*ἀμφιβολεύω, fish with a casting-net, PSI 901.13, 22 (i AD).

ἀμφίβολος I, after 'encompassing' insert 'A.Th. 298' deleting this ref. in section II 1.

*ἀμφιβῶτις, ἡ, fem. adj., celebrated all around, Ion Trag.fr. 35 K.-S.; cf. ἀμφιβόητος.

ἀμφίβωτος, delete the entry (v. °ἀμφιβῶτις).

Ἀμφιγυήεις, for pres. def. read 'with bent legs, bandy' and add 'epith. of the bird αἴγιθος, Call.fr. 469 Pf.'

+ἀμφίγυος, ον, curved on both sides (of the blade), leaf-shaped, ἔγχεσιν ἀμφιγύοισιν Il. 13.147, Od. 24.527, ἁ. δούρασιν, A.R. 3.1356. 2 fig., of antagonists, perh. keen on either side, S.Tr. 504.

+ἀμφιδαής, ές, two-edged, ἁ. ἡ ἀμφοτέρωθεν κόπτουσα μάχαιρα, Suid.

*ἀμφιδάνης, v. °ἀφιδάνης.

ἀμφιδεΐδιον, delete pres. ref. and read 'IG 2².1424a.183, 2².1428.75'

ἀμφιδέξιος, add '6 as epith. of Apollo, Cypr. ta-pi-te-ki-si-o-i τὰ(μ)φιδεξίοι ICS 335.'

+ἀμφιδεσφάγανον (sic), delete the entry.

ἀμφίδετος, for 'bound or set all round' read 'with a string attached to either end (of a bow-drill)'

+ἀμφιδημᾶ, ἁ, either personal adornment or footwear (cf. ἀμφιδήσασθαι ὑποδήσασθαι Hsch.) ἀνπιδήμας (gen.) Inscr.Cret. 4.43Ab11; ἀνπιδέμας ib. 72 v 40, 75B3 (all Gortyn, v BC).

ἀμφιδινεύω, after 'πυρίηα' insert 'codd. in'

ἀμφίδοξος II, line 3, delete 'causing doubt'

ἀμφιδρόμια, line 4, after 'Lys.l.c.' insert 'seventh, acc. to Hsch. s.v. δρομιάμφιον ἧμαρ (δρομιάφιον ἡ. S.) = Suppl.Hell. 1075'

ἀμφίδυμος, after 'λιμένες ἁ.' insert 'i.e. on both coasts' and after 'ἀκταί' insert 'i.e. accessible to ships from both sides'

ἀμφιέννυμι, line 1, after '-ύω' insert 'IG 1³.7.11 (v BC)'; line 6, for 'poet. part.' read 'part. ἠμφιεσμένη Hippon. 2 W., poet.'; line 7, delete 'ἀμφὶ .. Il. 19.393' (v. °ἴζω).

ἀμφιέπω I 1, add 'Hes.Th. 696' II 1, line 3, for 'dressed' read 'prepare'; for 'do honour .. to' read 'devote oneself to'; delete 'prob. in E.Med. 480, etc.'

ἀμφίεσμα, after '(Pl.)R. 381a' insert 'Luc.Cyn. 17 (also pl.)'

ἀμφίεστος, add 'SEG 33.1039.74-5 (Cyme, Aeolis), Didyma 38 (both ii BC), PDura 20.12 (ii AD), cf. 14.9'

ἀμφιετηρίς, after 'festival' insert 'OGI 51.27 (Egypt, iii BC)' and for 'SIG .. 69' read 'IG 2².1368.69'

ἀμφιετής, after 'Orph.' insert 'H. 53.1'

*ἀμφιθαλάσσιος, ον, surrounded by sea, Phot. s.v. ἀμφιπλήκτοις α 1363 Th.

ἀμφιθάλασσος, add 'as the name of a horse, ICilicie 49 (Rom.imp.)'

ἀμφιθαλής 1, for 'blooming' read 'flourishing'; for 'Call.Iamb. 3.1.3' to the end read 'esp. of young persons taking part in religious or agonistic rites, for auspicious reasons, Call.fr. 75.3 Pf., SEG 32.1243.4 (Cyme, Aeolis, i BC/i AD), Inscr.Magn. 98.19, MAMA 9.30 (ii AD)'

*ἀμφιθαλιτεύω, of a Laconian officer, app. perform the function of an ‡ἀμφιθαλής or be in charge of the ἀμφιθαλεῖς, SEG 11.677c (p. 250 Add., ἀπισαλιτ-, ii/i BC), 677 (ἀνφιθαλειτ-, time of Augustus)'

ἀμφιθάλλω, delete 'full'; for '(Antip.)' read '(Antip.Sid.)' and after it insert 'flourish about, ib. 9.221 (Marc.Arg.), 12.93 (Rhian.)'

ἀμφίθετος, for 'that will stand on both ends' read 'that can be used either way up'; add 'Antim. 24.1 Wy.'
†ἀμφιθηγής, ές, two-edged, ξίφος, S.Ant. 1309 (lyr.).
†ἀμφίθηκτος, ον, two-edged σάγαρις, AP 6.94 (Phil.).
*ἀμφικαπνίζω, make smoke round, Archil. 89.1 W.
ἀμφικάρηνος, for 'two-headed' read 'having a head at either end'
*ἀμφικατέργω, shut in all round, Orac.Sib. 2.295; cf. κατείργω.
†ἀμφικέλεμνον· ἀμφιβαρές, οἱ δὲ τὸν βασταζόμενον ὑπὸ δύο ἀνθρώπων δίφρον, οἱ δὲ ἀμφίκοιλον Hsch. (-κοιλον· ξύλον La.); cf. ἀμφικελέμνους· ἀμφιβαρεῖς Phot. a 1333 Th.
ἀμφικέφαλος, after 'ον' insert 'also η, ον (SEG 29.146.5, iv BC, in sense II)'   II, for 'IG 1.227d' read 'IG 1³.421.206'
ἀμφίκλαστος, after 'AP 6.223' add '(Antip.Sid.)'
ἀμφικλάω, after 'break all in pieces' insert '(in quots., fig.)'
ἀμφίκομος, add '3 subst., ὁ, precious stone, Plin.HN 37.160, also known as ἐρωτύλος (v.s.v. sense IV.)'
*ἀμφικουρία, Ion. -ρίη, ἡ, dub. sens., Archil. 146.9 W.
ἀμφικύπελλον, for '= ἀμφικάρηνος' read 'having heads on either side or all round'; add 'Trag.adesp. 586a K.-S.'
ἀμφίκροτος, for 'IG 3.82' read 'IG 2².3118'
Ἀμφικτύονες, line 3, after 'esp. at Delphi' insert 'Hdt. 7.200.2'; line 4, for 'IG 2.545.16' read 'CID I 10.16'
Ἀμφικτυονικός, add 'written Ἀμφικτιονικός, CID I 10.6, 42'
*ἀμφίκυρτος, ον, doubly convex, κεραμίδες Inscr.Délos 456A6 (ii BC).
ἀμφίλαλος, for 'Ar.Ra. 979' read 'χείλεσιν ἀ. Ar.Ra. 680'
ἀμφιλαφής 3, add 'ἔρις Nonn.Par.Ev.Jo. 7.167'   4, add 'ὄχλος ib. 12.81'   5, for 'great' read 'wide-ranging'
ἀμφιλειπής, after 'Mar.Vict. 2559' insert '(p. 110.31 K., amphilipes)'
ἀμφίλογος, after 'ον' insert 'dial. ἀμφίλλ- treaty in Th. 5.79.4, IG 5(2).343.A11 (Orchomenus, Arcadia, iv BC), SEG 25.447.5 (Aliphera, Arcadia, iii BC), 23.305 (Delphi, ii BC)'
†ἀμφιμήτορες, οἱ, αἱ, Dor. ἀμφιμάτ-, children with two rival mothers, i.e. a natural mother and a stepmother, (v. CQ 37.498ff.), A.fr. 73b.4 R., E.Andr. 466 (lyr.); sg. in Hsch.; cf. ‡ἀμφιπάτορες.
ἀμφιμυκάομαι, after 'Od. 10.227' add 'δῶμα ἀμφιμέμυκεν epigr. in SEG 30.172 (late Hellen.)'
ἀμφίον, before 'D.H. 4.76' insert 'Call.fr. 177.31 Pf. (pl.)'
ἀμφιπαίω, for 'spike, transfix' read 'impale' and for 'IG 4.951' read 'IG 4²(1).121.92'   add '2 = ἀμφισβητέω, Inscr.Cret. 4.80.12 (Gortyn, v BC).'
ἀμφιπάτορες, for 'cf. ἀμφιμήτορες' read 'Poll. 3.24 (but perh. wd. invented on the model of ‡ἀμφιμήτορες, v. CQ 37.498ff.)'
ἀμφιπιάζω, for 'squeeze all round, hug closely' read 'enfold in one's grasp'
ἀμφιπίπτω, line 4, for 'Parth. 8.4' read 'Parth. 7.3'
ἀμφιπλέκω, after 'Opp.H. 4.158' insert 'Anacreont. 42.6 W.'
ἀμφιπλήξ, after 'Id.Tr. 930' insert 'σφῦραι AP 6.205 (Leon.Tarent.)'
ἀμφιπολεύω 2, transfer '[τὰς κούρας] .. Od. 20.78' to section 3 and 'Hes.Op. 803' to section 1, line 2, after 'h.Merc. 568'   3, line 2, delete 'Q.S. 13.270'; add 'to be slave to, χείροσιν ἀ. Q.S. 13.270'
ἀμφιπολέω II, for 'roam with' read 'attend on'   add 'III abs. roam up and down, A.R. 4.1547.'
†ἀμφίπολος, ον, as subst., servant, attendant: fem., Od. 1.331, 6.199, etc.; λάβετ᾿ ἀμφίπολοι γραίας ἀμενοῦς E.Supp. 1115; w. other substs., ἀ. ταμίη, γραῦς Il. 24.302, Od. 1.191; masc., Pi.O. 6.32.   2 servant of god or cult: fem., θεᾶς ἀμφίπολον κόραν E.IT 1114, Διὸς IG 14.2111, D.H. 1.50; masc., SEG 11.314.12 (Argos, vi BC), Pi.Pae. 6.117, E.fr. 992, τῶν θεῶν Phld.D. 1.13, Plu.Comp.Demetr.Ant. 3, IG 9(1).683 (Corcyra).   3 ἀ. Διὸς Ὀλυμπίου title of magistrate at Syracuse, D.S. 16.70.   II frequented, τύμβος Pi.O. 1.93. Myc. a-pi-qo-ro.
ἀμφιπονέομαι II, for this section read 'to be troubled about, χὡς ὅρος ἀμφεπονεῖτο .. αὐτόν Theoc. 7.74; pass. τὰ ἀμφιπονεόμενα neighbouring parts affected by the disorder, Hp.Mul. 2.135'
*ἀμφιπυκάζω, wreathe, deck round, ἀμφιπυκάσσας Eleg.adesp. in POxy. 3723.3 (ii AD, unless to be read as two wds.).
ἀμφίπυλος, add '(or perh. -ον, τό, doorway)'
*ἀμφίροος, ον, that has streams flowing around or on both sides, attested in place-name Ἀμφίροος, SEG 36.336.16 (Argos, iv BC).
†ἀμφισαλεύομαι, rock up and down astride, AP 5.55 (Diosc.).
ἀμφισκέπαρνος, for 'λίθοι' to the end read 'archit., of a right-angle block in door-frame, λίθοι Didyma 25.A19, B26, 26.A29, 55, al.'
*ἀμφισταδόν, adv., on both sides, SEG 37.1537.9 (Arabia, c.AD 400).
ἀμφιτίθημι, line 2, for 'Thgn. 847' read 'Thgn. 848'
ἀμφιφαής, after 'Arist.Mu. 395ᵇ14' insert '(cj. -φανής)'   add 'II of the new moon, shining at either horn, Nonn.D. 4.281 (v.l. ἀρτιφαής), 22.349; cf. ἀμφιφανής 2.'
ἀμφιφανής 1, add 'Arist.Mu. 395ᵇ14 (v.l.)'
*ἀμφιφάων, οντος, ὁ, (cf. ἀμφιφῶν), visible on all sides, τόπον Orac. Chald.fr. 158.2 P.
ἀμφιφορεύς, add 'Myc. a-pi-po-re-we (pl.)'

ἀμφιχάσκω, for 'gape .. for' read 'open the mouth wide so as to swallow'; add 'w. cogn. acc., ἀμφιχανὼν ὀλίγον στόμα Opp.H. 1.223'
ἀμφίχωλος, add 'II in metre, halting at both ends, of an iambic line that has a long for the normal short in both the first and the last metron, Sacerd. 523.10, 531.6.'
ἀμφόδιον, after 'of sq.' insert 'BGU 1579.10, 1580.11 (both c.AD 119)'
*ἀμφοδογραμματεύς, έως, ὁ, secretary of an ἄμφοδον II, POxy. 2131.11 (iii AD), PLond. 935.1, 936.1 (both iii AD).
ἄμφοδον II, delete 'hence .. town' and after 'Lxx Je. 17.27' insert 'as administrative unit'
*ἀμφόδους, v. ‡ἀμφώδων.
†ἀμφόδων, v. ‡ἀμφώδων.
*ἀμφοκέραιος, ον, two-handled, POxy. 1343 (vi AD); -κέρυια pap.; for *ἀμφικ-).
*ἀμφοράριον, τό, dim. of ἀμφορεύς, PGot. 17ʳ.17 (vi/vii AD, sp. -φολ-); see also °ὀμφαλάριον.
ἀμφορεύς I 1, line 5, after 'wine in' insert 'Schwyzer 725.9 (Miletus, vi BC)' and for '[Ar.]fr. 299' read 'Ar.fr. 310 K.-Α.; for oil, given as prizes at the Panathenaea, IG 1³.422.41-60 (Athens, v BC), cf. Hesperia 27.178ff.'   at end add 'Myc. a-po-re-we (pl.)'
†ἀμφορίσκος, ὁ, dim. of ἀμφορεύς, Magn. 7 K.-Α., IG 2².1640.19 (354 BC), D. 22.76.
ἀμφορίτης I, for this section read 'a δίαυλος race at Aegina run by bearers of amphorae and called Ὑδροφόρια, Call.Dieg. viii.23 (fr. 198 Pf.), EM 95.3 (ἀμφιφορίτης)'
*ἀμφοροθύνω, provoke to strife, πεζὸν .. Ἄρη Polemon 2.57 (epigr., Thess., iii BC).
ἀμφότερος I 2, line 5, before 'Pi.P. 4.79' insert 'Il. 7.418'
ἀμφοτέρωθεν, add 'Myc. a-po-te-ro-te'
*ἀμφουριασμός, ὁ, deed of transfer of landed property, SEG 3.674.3, 46, al. (Rhodes, ii BC).
†ἀμφούριον, τό, (οὖρος = ὅρος) prob. copy of a declaration of transfer, PHal. 1.253 (iii BC), SEG 3.674.40 (Rhodes, ii BC).
*ἀμφύνω, v. °ἀναφύω.
ἄμφω, add '2 adv., like ἀμφότερον (s.v. ἀμφότερος I 2), Nonn.D. 11.187, AP 9.25.3 (Leon.Tarent.).'
ἀμφώδων, for 'οντος, ὁ, ἡ, (ὀδούς)' read 'ον, Arist.PA 663ᵇ36 (-ουν id.HA 507ᵇ34), gen. -οντος; after 'ruminants' insert 'Hp.Art. 8'; after 'al.' add 'neut. pl. subst., ἀμφώδοντα δὲ ἄνθρωπος ἵππος ὄνος Ael.NA 11.37'.   II, for 'ass' read 'applied to an ass'; for 'ἀμφόδων; cf. ἀμφόδους' read 'ἀμφοδ- as from ἀμφόδων or *ἀμφόδους'
ἀμφῶτις, after lemma insert '(ἄμφωτις La.)'
ἄμφωτις II, after 'A.fr. 102' insert 'Pl.Com. 256 K.-Α.'
*ἀμωλώπιστος, ον, not bruised, Plu. 2.1091e (cj.).
ἀμώμητος, after '-ον' insert '[also -η, -ον Hes.fr. 185.13 M.-W., SEG 37.908 (Ephesus)]'; line 3, after 'Ph.fr. 69 H.' insert 'ἐκ λειμῶνος ἀ. AP 4.1.31 (Mel.)'
ἄμωμον, add 'Apoc. 18.13'
ἄμωμος, add 'adv. -ως, IHadr. 120 (Chr.)'
*ἄμωροι (v.l. in Od. 12.89), = ἰχθυοφόροι, EM 117.26.
*ἄμωρος (B), dub. sens., Call.fr. 686 Pf.
ἄν (B), add 'also in Arc. inscr., ἂν δέ τις .. μὴ φάτοι SEG 37.340.21 (Mantinea, early iv BC)'
ἄν or ἀν, after 'Ep.' insert 'and dial.'
ἀνά, line 1, after 'ὀν' insert 'also Arc., Cypr. ὐν'   C III, after 'Ev.Luc. 9.14' insert 'PMich.III 145.3 iv 1, 11 (ii AD)' and after 'Ev.Matt. 20.10' insert 'Dioph. 5.12'
ἀνά (A), delete 'and always as address to gods' and add 'E.Rh. 828, Call.fr. 24.3 Pf., Epigr. 34.1 Pf.'
†ἀνᾱ (B), Dor. for °ἄνη.
ἀνάατος, after 'IG 5(2).357.177' insert '(v. SEG 31.351)'
ἀναβαθμός, add '2 Ep. ἀμβαθμός, ἀ. πετρώδεας rocky ascents, Nic.Th. 283 (cj.).'
ἀνάβαθρον, after 'Tralles' insert 'cj., but now read as τὸ πὰμ βάθρο[ν], ITrall. 240.4'; add 'ἀνά[β]αθρον PTeb. 793 viii 1 (ii BC), cf. anabathra (pl.), Juv. 7.46'
ἀναβαίνω A II 6, lines 4/5, delete 'ἀ. ἐπὶ τὸν ὀκρίβαντα .. Eq. 149' and transfer 'ἀνάβηθι [Ar.] V. 963' to follow 'of witnesses in court'   add 'b of the "entrance" of an actor, Ar.Ach. 732, Eq. 149, V. 1341; for ἀ. ἐπὶ τὸν ὀκρίβαντα Pl.Smp. 194b, v. ὀκρίβας.'
add '11 rise, be promoted (cf. Lat. ascendo), θεσπίζομεν μηδένα .. εἰς τὰς ῥηθείσας ἱερατείας ἀναβαίνειν Cod.Just. 1.3.52, οὐκ ἐπὶ πρὸ τὰ μείζω τῆς ἱερωσύνης ἀναβήσεται γέρα Just.Nov. 6.1.2.   12 astron., = ἀναβιβάζω II 10, Palch. in Cat.Cod.Astr. 6.63.8, al.'
III 1, add 'c of anger, rise or surge up, Lxx 2Ki. 11.20, al.'   IV, after 'of discourse' insert 'τοῦτο μέν νυν τοιοῦτόν ἐστι, ἀναβήσομαι δὲ ἐς τὸν κατ᾿ ἀρχὰς ἤϊα λέξιον λόγον Hdt. 4.82' and after 'Democr. 144ᵃ' insert 'Hp.Gland. 7, Morb. 4.45, X.Eq. 1.4'
ἀναβακχεύω, for '(E.) Or. 337' read '(E.) Or. 338'
ἀναβαλλαγόρας, add '(ἀναβαλλογῆρας La.)'
ἀναβάλλω A I 3, for 'Ctes.fr. 30' read 'Ctes. 1(p).γ J.'   4, after

ἀνάβαπτος | SUPPLEMENT | ἀναδέω

'*cause to spring up*' insert 'ὕδωρ Call.*fr.* 546 Pf.'   **7**, delete the section transferring quot. to section 1   **B 1**, line 3, after 'c. acc.' insert 'μέλος Theoc. 10.22'   **II 1**, add 'ἀ. προσελθεῖν Men.*Dysc.* 126'   **V**, delete the section.

*ἀνάβαπτος, ον, *tinged, dyed*, κάμβαπτον cj. in Philic. in *Suppl.Hell.* 680.60.

ἀνάβᾰσις, add '**V** astron., *rising*, Heph.*Astr.* 1.23.18; ἀ. ἡλίου *noon altitude* of sun, *Cod.Vat.Gr.* 1058.431ᵛ.'

*ἀνάβασμα, ατος, τό, *stair*, Hsch. s.v. σκάλα.

ἀναβασμός, line 2, after 'Paus. 10.5.2' insert '*ITrall.* 240.6'; at end for '*SIG*²587.308' read '*IG* 2².1672.308 (iv BC); also form ἀμβασμός *Didyma* 35.17 (ii BC); cf. ἀμβαθμός s.v. °ἀναβ-'

Ἀναβατηνός, ὁ, cult-title of Zeus, *IHadr.* 9.

*ἀναβέβροχε, v. ‡ἀναβέβρυχε.

ἀναβέβρῡχε, for '*gushed* or *bubbled up*' read '*gushes* or *bubbles up*'; after 'Zenod.' insert 'and edd.' and add 'cf. *βρόχω'

*ἀνάβηξις, εως, ἡ, *expectoration*, αἵματος ἀναβήξεως Gal. 8.287.6.

ἀναβήσσω, add '*Morb.* 2.49'

ἀναβῐβάζω **II 6**, for 'cf. *POxy.* 513.27' read 'pass., ἕνεκα τοῦ .. τὴν προκειμένην οἰκίαν .. ἀναβεβιβάσθαι εἰς δραχμὰς χειλίας because it *has had its price raised* (by overbidding), *POxy.* 513.27 (ii AD)'   add '**11** *reckon* time *back*, ἀπό γε τῆς ἐκβολῆς τῶν βασιλέων ἐπὶ τὸν πρῶτον ἄρξαντα .. ἀναβιβασθεὶς ὁ χρόνος D.H. 1.75.1.'

ἀναβῐβασμός, add '**4** *overbidding*, *PTeb.* 295.10 (ii AD).'

ἀναβῐβαστέον, add '**2** one must move a word *farther back* (in construing), Sch.Pi.*N.* 4.14a.'

ἀναβιώσκομαι **II**, after 'aor.' insert 'ἀνεβιωσάμην Crates Com. 52 K.-A.'

ἀναβλαστάνω, line 3, for 'Pl.*Lg.* 835d' read 'Pl.*Lg.* 845d' and after 'Plu. 2.366b' insert 'ἄρδω σ' ὅπως ἀμβλαστάνῃς Ar.*Lys.* 384'

ἀναβλέπω **I 3**, add 'ἀναβλέψατε εἰς ὕψος τοὺς ὀφθαλμοὺς ὑμῶν Lxx *Is.* 40.26'

ἀνάβλησις, for 'Call.*Ap.* 45' read 'Call.*Ap.* 46'

*ἀναβλητέον, *one must postpone*, ἀλλὰ ταῦτα μὲν ἐν τῷ παρόντι ἀναβλητέον, πρότερον δὲ .. Plot. 6.8.1.

ἀναβλύες, after 'Hsch.' add '(ἀνάβλυδες La.; cf. σύγκλυδες)'

ἀναβλύζω, delete 'fut. .. *Exag.* 137' (read ἀναβρυήσει)

ἀναβλυσθαίνω, (after ἀναβλυστάνω), add 'Procl. in *Ti.* 1.119, 120 D.'

ἀναβολεύς **II**, for pres. def. read '*lever used in surgical operations*'

*ἀναβολέω, *put on*, ἀνεβολήσατο ὡς ἔνδυμα ζῆλον Aq.*Is.* 59.17.

ἀναβολή **I 1**, add '**b** *laying* of bricks, rest. in *IG* 2².1661.7; cf. ἀναβάλλω I 5.'   **2**, line 2, for '*that which .. mantle*' read '*that part of a garment which hangs loose at the back*'; line 3, before 'of the *toga*' insert 'of the ἱμάτιον, ἐξ ὁ. τοῦ ἱματίου εἴλκυσεν (αὐτόν) Vit.*Aesop.*(G) 15, 28'; line 4, after 'Nic.*Dam.* p. 119 D.' insert 'ἀφεῖλε τῶν μανδυῶν αὐτῶν τὸ ἥμισυ ἕως τῆς ἀναβολῆς i.e. *up to just below the waist*, Lxx 1*Chr.* 19.4'

ἀναβολικός, ἀναβολικόν, τό, for this section read '*Egyptian tax*, used to cover costs of military clothing, *PAmh.* 2.131.15 (ii AD), *POxy.* 1135.2 (iii AD), *PThead.* 34.26, *Pap.Lugd.Bat.* xxv 62.8 (both iv AD)'

ἀναβόλιμος, add 'see also ‡ἀμβόλιμος'

*ἀνάβολον, τό, *wrap*, *POxy.* 936.24 (iii AD), *PTeb.* 413.10 (ii/iii AD).

*ἀναβομβέω, *blare*, λιγυρὴ δ' ἀνεβόμβεε σάλ[πιγξ] *GDRK* 32.80.

ἀναβράζω, add '**2** τήνδ' ἔωλον ἀναβεβρασμένην Ar.*fr.* 51 K.-A., interpr. in Phot. s.v. (a 1404 Th.) by ἀνακεκινημένη; prob. there is allusion to a *réchauffé*; cf. ἀναβρασμός.   **3** *toss up*, ῥάβδον Lxx *Ez* 21.26.'

*ἀναβροχέω, *flood*, τὸ πεδίον *PSI* 168.22 (ii BC).

*ἀναβρύχω, delete 'Eust. 1095.6'

ἀναβρύω, after lemma insert 'fut. ἀναβρυήσει (ἕλκη πικρά) Ezech.*Exag.* 137 S.'; add '*SEG* 31.1474 (Arabia, AD 575/6)'

ἀναβώνες, read 'ἀνάβωνες'

*ἀναγαργάλικτα, τά, *gargles* Hp.*Aff.* 4.

ἀναγγελία, for '*SIG* 598.11' read '*Inscr.Magn.* 91.11 (decr. at Delphi, ii BC)' and add '*PP* 42.121 (Cos, iii/ii BC), *IKyme* 13 i 9, *SEG* 37.1006 (Troas, both ii BC)'

ἀνάγειον, after 'ἀνάγαιον' insert 'Ph. 2.476'; for '*Reise*' read '*Reisen*'

ἀναγεννάω, after 'Pass.' insert 'Sallust. 4.10'

ἀναγεύω, add '*PMich.*VIII 473.28 (ii AD)'

*ἀναγίγνομαι, in tm., perh. *rise up*, ἀνὰ μὲν θυμὸς [ἐ]γεντο θε[ῆ]ς Call.*fr.* 63.6 Pf.

ἀναγιγνώσκω **I 2**, after 'Pi.*I.* 2.23' insert 'Philostr.*VA* 1.31'   **II**, line 2, delete 'dub. in *GDI* 5075'   add '**b** in textual criticism, *adopt as a reading*, Sch.Ar.*Pax* 594 H.'

ἀναγκάζω, after lemma insert 'ἀνακκάζω *IG* 9²(1).706A 16'   **I 1**, after 'Hdt. 5.101' insert 'in weakened sense *exert pressure on*, Th. 8.41.3, 76.1'

ἀναγκαῖος **I 1**, add 'ἐν καιροῖς ἀναγκαίοις *SEG* 37.957 (Claros, ii BC)'   **II 4**, for '*indispensable .. minimum*' read '*forced on one by necessity, indispensable*, D. 50.38, 54.17; w. ref. to a minimum standard, *absolutely necessary, essential*, οὐδὲ τἀναγκαῖα ἐξικέσθαι Th. 1.70.2, παρασκευή 6.37.2, Lys. 31.18, Is. 4.20'; delete '*less freq.* ..

6.37'   **5**, add '**b** prob. an honorary title, *kinsman* (cf. συγγενής III), in *OGI* 315.49, 763.31 (letters of Attalus II and Eumenes II).'   **7**, add 'Epiph.Const. in *Lapid.Gr.* 196.22 (comp.)'   add '**8** lit. crit., *sparing* of words, opp. περιττός, of Sophocles, D.H.*Vett.Cens.* 2.11; opp. ἀστεῖος and ἡδύς, of Lycurgus, ib. 5.3.'

*ἀναγκεπάκτης, ου, ὁ, *bringer of compulsion*, *PMag.* 4.1361.

*ἀναγκεπόπτης, ου, ὁ, *overseer of necessity*, *PMag.* 7.355.

ἀνάγκη, after lemma insert 'Dor. ἀνάγκα'   **I 1**, at end after 'c. inf.' insert 'D.Chr. 31.105, 114'   **2 a**, add 'of the rules of rhetoric, ῥητόρων ἀνάγκας Anacreont. 52.2 W.'   **II**, delete the section.

*ἀναγκίτης, ου, ὁ, name for ἀδάμας, *ob id quidam eum ananciten vocavere*, Plin.*HN* 37.61, τόν ῥα παλαιγενέες μὲν ἀναγκίτην (cj., codd. ἀνακτ.) ἀδάμαντα κλεῖον Orph.*L.* 192.

*ἀναγκῖτις, ιδος, ἡ, app. = °ἀναγκίτης, Plin.*HN* 37.192.

ἀναγκοφᾰγέω, after 'ἀναγκοτροφέω' insert '(?)Ephor. in *PLit.Lond.* 114.12'

*ἀνᾱλάγλημαι, Pamph. pres. imper. 3 sg. ἀηαγλέσθō *let him take upon himself, undertake*, *IPamph.* 3.15 (iv BC); cf. ‡ἀγρέω.

ἀναγλύφω, after '(Lydia, i AD)' insert '*SEG* 26.1835 (epigr., Cyrene, i AD)'

*ἀναγνέω, = ἀνάγω, ὕμνον Lasus 1 P.; cf. ἀγνέω, διεξαγνέω.

ἀνάγνιστος, delete '*unpurified*'

ἄναγνος, add 'J.*Ap.* 1.306'

ἀνάγνωμα, delete the entry.

ἀναγνώστης **I**, add '**2** *one who reads and expounds*, Lxx 1*Es.* 9.42.   **3** *reader*, office in the Christian church, *POxy.* 3787, *Cod.Just.* 1.3.44, *BCH* suppl. 8 no. 142 (?vi AD).'

ἀνᾱγόρευσις, after '(Cnidus)' insert '*SEG* 28.60 (Athens, iii BC), *IHistriae* 58.22, 59.23 (ii BC)'

*ἀναγορία, ἡ, *public announcement*, *PFay.* 66.4 (ii AD), *SB* 6951 (ii AD).   **2** *roll-call, muster*, *POxy.* 2902.24, 2903, etc. (all iii AD).

ἀναγραφεύς, after 'ἕως' insert 'Dor. ἀγγροφεύς *IAskl.Epid.* 42.65 (*c.*370 BC)'; line 2, for '*IG* 1.61' read '*IG* 1³.104.5 (409/8 BC)'   **II**, for '*IG* 2.192ϛ, cf. 191' read '*IG* 2².1700.215 (335/4 BC), al.'   **III**, for '*plan* .. 33' read '*template*, *IG* 2².1666 (iv BC), 2².244'   add '**IV** *secretary*, Phld.*Acad.Ind.* iii 37.'

ἀναγραφεύω, before '*IG* 14.757' insert 'rest. in'

ἀναγραφή, after lemma insert 'Dor. ἀγγροφά *IG* 4²(1).103.140 (Epid., iv BC)'

ἀναγράφω, after lemma insert 'Thess. ὀνγράφω *IG* 9(2).461, 517.21, 45'; ἀγγράφω, add '*SEG* 32.1586 (Naucratis, v BC)'   **I 1**, line 7, for 'ἀναγρψάσασθαι' read 'ἀναγράψασθαι'   **II 2**, add '**b** *enter in a list*, pass., of eclipses, εἰ [ἔκλειψις] ἐν ταῖς ἀναγραφείσαις εὑρίσκεται Hero *Dioptr.* 35.'   add '**V** *repaint* (ornamentally), *Inscr.Cret.* 3.ii 1.11 (ii BC).'

ἀναγρετόν, add 'cf. νήγρετος· ἄναγρετος, id.'

†ἀναγχίστευτος, ον, *having no kindred*, *GVI* 819.11 (Phrygia, iii AD).

ἀνάγχω, delete '*hang up*' and add 'transf., *deprive of life*, τῷ λιμῷ *PMasp.* 20.12 (vi AD)'

ἀνάγω **I 5**, for '*conduct* the choir' read '*lead* the dance'   **9**, line 3, after 'Id.*Aph.* 3.25' insert 'πνεῦμα .. ἀνηγάγετο Call.*Ep.* 43.2 Pf.'; after '*draw* a line' insert '*up to*'   **12**, add 'τῆς ἀναχθεί[ση]ς ἀμπέλου *PPetaus* 127.8 (ii AD)'   add '**13** *transfer* land to *a higher category, upgrade*, *BGU* 2060.13 (AD 180).'   **II 5**, for 'Pl.*Lg.* 915c' read 'Pl.*Lg.* 915d'   **10**, add '*Carm.Pop.* 5.1 P.'   at end add 'Myc. inf. *a-na-ke-e*, dub. sens.'

ἀναγωγή **I 1**, after 'X.*HG* 1.6.28' insert 'D. 33.25, 49.6'   add '**c** *bringing up* from the shore, quayside, or sim., *Didyma* 41.45.'   **II 7**, for 'Ath. 9.395a' read 'Ath. 9.394f'

ἀναγωνιατος, add '*PBaden* 48.12 (ii BC)'

*ἀναγώνιος, ον, *free from care*, ὥστε περί γε τούτου ἀναγώνιος ἴσθι *PAmst.* 89.1 (AD 3).

ἀναγώνιστος, add '*uncontested*, τῶν .. ἀ. δευτερείων τῆς αὐλῳδίας, *SEG* 19.335.50 (Tanagra, ii BC)'

ἀναδαίομαι, for 'v. ἀναδατέομαι' read 'pass., *to be distributed*, γᾶς ἀναδαιομένας orac. in Hdt. 4.159.3'

ἀναδᾰτέομαι, after '*redistribute*' insert 'ἀνδάξαθαι = ἀναδάσασθαι, *Inscr.Cret.* 4.5.2 (vii/vi BC); delete 'pass. .. 4.159'

ἀνάδειγμα, add 'see also ‡ἀνδεργμα'

ἀναδείκνῡμι **I**, after '*exhibit, display*' insert 'τὸν μάργον ὄνδειξαι θέλω, perh. *show up*, prob. in (?)Sapph. 99 i 24 L.-P.'   **II 2**, for '*dedicate*' read '*consecrate*'

ἀναδενδραδικός, add '*PSoterichos* 1.8 (AD 69), 2.8 (AD 71)'

ἀναδενδράς, add '(ἀναδενδρᾷς (pl. -ᾷδες), acc. to Seleuc.ap.Phot. a 1454 Th.)'

ἀναδέχομαι, at beginning insert 'poet. ἀνδ- Pi.*P.* 2.41, Orph. *A.* 1133'   **II 6**, add '**b** *recover* (the use of limbs), *MAMA* 4.266 (Dionysopolis).'

ἀναδέω **III**, for 'Plu. 2.222e' and '(Plu.) 343a' read 'Plu. 2.322e' and '(Plu.) 243a' *ἀναδημιουργέω, *reverse, annul, in one's capacity as* δημιουργός, *IG* 12(8).264.5 (Thasos, iv BC).

ἀναδίδωμι, line 1, add 'Cypr. aor. perh. ὐνέδωσε *Kafizin* 217 (221/200 BC)' **I 2**, add 'ἀναδοθεισᾶν ψάφων *SEG* 23.208 (Messene, i AD)' **II 5**, for '*present* by name' read '*nominate*' and add '*appoint*, *PMich*.XI 604.4 (iii AD, v. *BASP* 12.111-2)' **III**, add '**3** *assign* a liturgy, *BGU* 2475.5 (AD 138).' **V**, add '**2** *give back, return*, *BGU* 1149.23, *Cod.Just.* 1.2.17.3.'

ἀναδινεύω, for '*whirl*' read '*revolve, rotate*'

ἀναδιφάω, for '*grope after*' read '*bring to light by probing, unearth*'

*ἀναδιχάζω, v. ‡ἀνδιχάζω.

*ἀναδοσία, ἡ, *repayment, Stud.Pal.* 20.114.15 (v AD).

ἀνάδοσις **II 1**, add 'ἀ. σπερμάτων *POxy.* 1031, *SB* 9358 (both iii AD)' add **IV** *giving back, return*, *Just.Nov.* 7 pr., 46.6.'

*ἀναδούλωσις, εως, ἡ, *re-enslavement, Just.Nov.* 78.2 pr.

ἀνάδρομος, add '**II** as masc. subst., *path which turns back on itself*, of a labyrinth, σφαιρικοὺς ἀναδρόμους *Anon.Alch.* 39.19.'

*ἀναδυτήριον, τό, *niche*, app. *for statuette*, *SEG* 6.718 (Pisidia, Rom. imp.).

ἀναείρω, lines 3/4, for '*lift up in one's arms, carry off*' read '*raise from a position on the ground* or *other surface*, E.*Phaëth.* 81 D. (tm.)'

*Ἀναείτεια, τά, festival of Artemis Anaïtis, *IGRom.* 4.1634.7 (Philadelphia, Asia).

*Ἀναεῖτις, ιδος, ἡ, a Persian goddess, (cf. OPers. *Anāhitā-*, *the unblemished one*), *TAM* 5(1).172 (AD 93/4), *SEG* 29.1152 (ii/iii AD); identified w. Artemis, ib. 33.1007 (dat. Ἀναείτι, Lydia, ii AD), *IGRom.* 4.1611.9, 17.

ἀνάεξω, after 'Coluth. 247' insert '*raise in status, advance*, Nonn.*D.* 8.183, 9.100'

ἀναερτάζω, for '= sq.' read '*raise up, lift up*'

*Ἀναζαρβαῖος, ον, *from Anazarba* in Cilicia, a description of clothing, *PKöln Ketouba* 1. 14.

ἀναζεμα, read 'ἀνάζεμα'

ἀναζυγή, add '*PHamb.* 91.8 (ii BC)'

ἀναζωγρέω **2**, for 'Nonn.*D.* 19.102' read 'Nonn.*D.* 19.104'

ἀναζώννυμι, for 'Nonn.*D.* 19.73' read 'Nonn.*D.* 19.75'

ἀναζωπυρέω **II**, after 'act.' insert '*recover one's spirits*'

*ἀναθάλλωσις, εως, ἡ, *flourishing condition*, *PMasp.* 2 iii 22 (vi AD).

ἀνάθεμα **I 2**, add 'ἔσται ἡ ψυχὴ ἀν[αθε]μας (sic) *TAM* 5(1).21 (Chr.)' **II**, add '*SEG* 37.195 (Attica, Chr.)'

*ἀναθέσιμος, ον, *votive*, πινάκια *Inscr.Délos* 1442*B*35 (ii BC).

ἀναθεωρέω, after 'Thphr.*HP* 8.6.2' read '(pass.) D.S. 12.15, Longin. 7.3'

†ἀναθεώρησις, εως, ἡ, *close examination*, Plu. 2.19c; *further reflection*, D.S. 13.35, Cic.*Att.* 9.19.1; κατὰ τὴν ἀ. Longin. 23.2; ἐδόκει .. μεγάλην ἔχειν ἀ., i.e. *food for thought*, D.S. 13.34, cf. 35, Cic.*Att.* 14.15.1, 14.16.2.

ἀναθήκη, add '*loading* (on to wagons), ἐπ᾽ ἀναθήκει *IG* 2².1666A.35 (iv BC)'

ἀναθηλέω, add 'fig., *AP* 5.264 (Paul.Sil.)'

*ἀναθηρεύω, *hunt out*, ἀ[να]θηρ[ε]ύσαντες *Did.in D.* 12.15.

ἀναθρεπτικός, add 'adv. -κῶς Gal. 10.487.13'

ἀνάθρεψις, add 'Gal. 13.74.4, id.*Thras.* 7'

ἀναθρῴσκω, add 'ἄνθρωσκε S.*fr.* 422 R.; E.*Or.* 1416 (tm.)'

*ἀναθυμιάζω, gloss on ἀτμίζω, Hsch.

*ἀναιβασίη, ἡ, Ion. ίη, = ἀνάβασις, *ascent*, prob. in epigr. in *IG* 2².4831 (rest. [ἀν]αιβασίη, iv BC).

ἀναίδεια, for '*shamelessness*' read '*lack of proper restraint* or *consideration, intemperateness*' **II**, delete 'cf. ὕβρις'

ἀναίδην, delete the entry (read ἀνέδην).

†ἀναιδής, ές, (αἰδώς) *lacking in restraint, intemperate*, of Agamemnon, (ὢ μέγ᾽ ἀναιδές Il. 1.158, of Penelope's suitors, Od. 1.254, al., (ὢ θρέμμ᾽ ἀναιδές S.*El.*622, D. 8.68. **2** of conduct, character, etc., κυδοιμὸς ἀ. Il. 5.593, ἐλπὶς ἀ. Pi.*N.* 11.45, λόγοι τῶν ἀναιδῶν ἀναιδέστεροι Ar.*Eq.* 385, ἀ. γνώμη D. 21.91; neut. subst., ἐπὶ τὸ ἀναιδέστερον τραπέσθαι Hdt. 7.39, εἰς ἀναιδὲς .. δός μοι σεαυτόν S.*Ph.* 83, βλέφαρα πρὸς ταιανδες ἀγαγών E.*IA* 379, εἰνὰ τἀναιδὲς κρατεῖ *Trag.adesp.* 528 K.-S. **3** -δως adv. S.*OT* 354, E.*Alc.* 694, Ar.*Th.* 525, Pl.*R.* 556b; sup. -έστατα Heraclit. 15 D.-K. **II** of things, *unable to be stopped*, ὀστέα λᾶας ἀναιδὴς ἄχρις ἀπηλοίησεν Il. 4.521, ἀναιδέος .. πέτρης 13.139; of Sisyphus' stone, Od. 11.598; of abst. things, θάνατος Thgn. 207, Pi.*O.* 10.105.

*ἀναιδομᾰχέω, *quarrel insolently*, *POxy.* 2997.41 (ii/iii AD).

*ἀναιδόφθαλμος, ον, gloss on κυνάμυια, Sch.Gen.*Il.* 21.394.

ἀναίμακτος, for 'Pyth.' read 'Pythag.' and add '*AP* 6.324 (Leon. Alexandr.), etc.'

ἄναιμος **I**, add 'sup., Plu. 2.913f.'

ἀναιμωτί, after 'Ph. 1.323, al.' insert 'J.*BJ* 4.1.6' and after 'Gal. 2.604' insert 'Opp.*C.* 4.453'

ἀναίνομαι, line 2, after 'Alciphr. 3.37' insert 'ἀνήνατο Call.*fr.* 178.11 Pf. (v.l.)'

*ἀναιρετός (B), όν, *taken up*, of foundlings, *BGU* 1058.11 (i BC), *POxy.* 73.26 (i AD), etc.

ἀναιρέω, line 1, after '*Com.adesp.* 18.6 D.)' insert '*Pamph.*aor. ἀνῆελε *IPamph.* 3.1 (iv BC), see also °ἀνηάγλημαι, Aeol. aor. inf. ὀννέλην Alc. 130.27 L.-P.' **II 1**, add '**b** *worst, bring down* enemies, D. 6.15, 18.18.' **2**, add '**b** *get rid of*, στάσιν Alc. l.c., Pi.*fr.* 109; νεῖκος Theoc. 22.180.'

ἀναισθησία **2**, after '*obtuseness*' read 'Isoc. 7.9, D. 18.128, 22.64, al.'

ἀναίσθητος **I 1**, after '*feeling*' insert 'Hp.*VM* 15'; for 'Thrasymach. 1' read 'Thrasym. 1 D.-K.' and transfer to sense 2, adding there 'Isoc. 15.218, D. 5.15, 17.22, 18.120'

ἀναισιμόω, line 4, for '*he used up*' read '*she used up*'

ἀναισίμωμα, add 'Call.*fr.* 196.45 Pf.'

ἀναΐσσω, line 7, after 'Hp.' insert '*Morb.Sacr.* 1'

ἀναιωρέω, after 'Nonn.*D.*' insert '2.457'

ἀνακαθαίρω, after lemma insert 'Dor. ἀγκοθ-, v. infra' **II 3**, add 'εἴ τι πρότερον ἀτελὲς ἢ συγκεχυμένον ἐδόκει, τοῦτο καὶ ἀνακαθᾶραι καὶ τέλειον ἐξ ἀτελοῦς ἀποφῆναι *Just.Nov.* 7 pr. **b** *expound* or *declare*, [δόγματα] Porph.*Plot.* 3.'

ἀνακάθαρσις, after 'εως' insert 'Dor. ἀνκαθάρσιος (gen.) *IG* 4²(1).106.25; ἀγκ-, *IAskl.Epid.* 52.A25, rest. in 32 (both Epid., iv BC)'

ἀνακαινίζω, line 2, after 'Hsch.Mil. 4.33' insert 'ἀνεκενίσθη εἰ (= ἡ) γέφυρα *BE* 1972.293, *MAMA* 7.190 (both Hadrianoupolis), *SEG* 37.1442 (Dura, ii AD)'

ἀνακαλέω, line 1, after 'ἀγκ-' insert 'ὀγκ-' **II 1**, after '*appeal to*' insert 'Πάον ὀνκαλέοντες ἑκάβολον Sapph. 44.33 L.-P.' **2**, after 'Lys. 15.2' insert '*proscribe*, App.*BC* 4.25; *be invited*, ἀνακαλεῖσθαι δὲ αὐ[τὸ]ν εἰς προεδρίαν ὑπὸ τοῦ ἱεροκήρυκος *Amyzon* 3.13 (iii BC)' **III 1**, add '**b** *revoke* legal transaction, *Cod.Just.* 1.3.55.2.'

ἀνακαλυπτήρια **II**, add 'Pherecyd.Syr. 2 D.-K., Sch.E.*Ph.* 682'

*ἀνακάλυψις, εως, ἡ, *disclosure*, Plu. 2.70f, 518d.

*ἀνακάρδιον, τό, *upturned twig* of the mulberry tree, Cyran. 1.12.27 K.

ἀνακάς, delete the entry.

*ἀνακάταξις, εως, ἡ, *refracture*, Paul.Aeg. 6.109 (162.27 H.).

Ἀνάκεια, τά, add 'Ἀνάκια *IG* 1³.258.6 (v BC)'

ἀνάκειμαι **I 1 b**, transfer 'χρύσεοί κ᾽ .. 10.33' to follow '(Pl.)*R.* 592b' in previous sense **II**, line 7, for 'εἰς θάνατον .. 18.1.1' read '**b** w. εἰς, *to be assigned* or *given over to*, εἰς θάνατον ἦν ἀνακείμενα τοῖς ἀλογήσασι J.*AJ* 17.6.5, λιμὸς εἰς ὑστάτην ἀνακείμενος ἀναισχυντίαν 18.1.1, εἰς τὸ σωφρονεῖν ἀνέκειτο ἡ ἐπιτήδευσις τοῦ βίου 18.3.4; so w. ἐπί, τὴν ἐπὶ τῷ θανεῖν ἀνακειμένην ἐπιστολήν 18.8.9, D.C. 37.18; also w. dat., *to be assigned to*, Plu.*Arist.* 15, Lyc. 1.' **III**, add '**b** *to be laid up* by sickness, *PMich.*XI 624.27 (vi AD).'

Ἀνακεῖον, τό, add 'Ἀνάκιον *IG* 2².1400.44'

†ἀνάκεστος, ον, = ἀνήκεστος, perh. erron., on analogy of εὐάκεστος, Hp.ap.Erot. (ἀνήκ- in *Acut.* 39); for 'ἀνακ-' see °ἀνήκεστος.

ἀνακεφαλαιόομαι, add '**2** *recapitulate*, *Ep.Eph.* 1.10.'

ἀνακήρυξις, after '*proclamation*' insert '*SEG* 33.639 (Rhodes, ii/i BC)'

ἀνακηρύσσω, after 'Att. -ττω' insert 'dial. ἀνακάρ-, *IG* 5(2).16.6 (Tegea, iii BC), *SEG* 23.208.24 (Messene, i AD); ὀγκάρ-, *IG* 12 suppl. (2).2.6 (Mytilene, iii BC)'

†Ἀνάκια, v. ‡Ἀνάκεια.

†Ἀνάκιον, v. ‡Ἀνακεῖον.

ἀνακιρνάμαι **I**, for 'φιλίας .. friendship' read 'χρὴν .. μετρίας εἰς ἀλλήλους φιλίας θνητοὺς ἀνακίρνασθαι'

ἀνακλάζω, add '**b** of a cry, *arouse*, τίς ἠχ[ο]ς ἡμᾶς ἐκ δόμων ἀνέκλαγε; *Trag.adesp.* 649.11 K.-S.'

ἀνακλέπτω, after lemma insert 'perh. 3 sg. imper. in ἐμὲ μέδες ἀνκλετέτō *SEG* 34.1019 (Kylix, Salernum, vi BC)'

*ἀνακληρόω, *reallot*, Sch.Pi.*O.* 110a.

ἀνάκλητος, after lemma insert '**I** *recalled*, S.*fr.* 1008 R.; ἐν ἀνακλήτῳ (λόγῳ) formula in financial documents, app. of the amount carried forward at the end of each month, *CASA* 3.43ff. (Tauromenium, v. Dubois *IGDS* no. 186). **II**'

ἀνάκλιμα, add '**II** part of a ship where the κυβερνήτης reclines (κατακλίνεται), Poll. 1.90 (written ἄγκλιμα).'

ἀνακλιντήριον, delete the entry; v. °ἀνακλιτήριον.

ἀνάκλιτρον, delete 'also' and after 'τό' insert 'cf. °ἐπίκλιντρον'

ἀνάκλισις **II**, for 'at end for '*bench* .. 277*d*' read 'perh. *chair with back*, *IG* 1³.423.209, *JHS* 12.232 (Cilicia).'

*ἀνακλιτήριον, τό, *back* or *armrest of a chair* or *couch*, Erot. s.v. ἀνακλισμοῦ, *Hist.Aug.* 2.5.7; *chair with back* or *armrests*, *SEG* 17.545.5 (Pisidia, Rom.imp.).

ἀνακλύζω **2**, before 'Plu. 2.590f' insert 'Cleobul.ap.D.L. 1.90, ἀνακλύζῃ δὲ θάλασσα, vv.ll. ἀναβύζῃ, περικλύζῃ)'

*ἀνακολᾰφή, ἡ, = Lat. *subsumen*, Gloss.

*ἀνακολᾰφίς, ίδος, ἡ, = Lat. *replica*, Gloss.

ἀνακολλητικός, after '*glueing*' insert 'or *sticking*'

†ἀνακολυμβάω, *dive and fetch up*, Thphr.*HP* 4.6.4, *Inscr.Délos* 440*A*52 (ii BC).

ἀνακομβόομαι, delete the entry.

*ἀνακομβόω, *tuck up*, χιτῶνας *Vit.Aesop.*(W) 77a: med., *Gp.* 10.83.1.

ἀνακομιδή **4**, after '*bringing up*' insert '*conveyance*, ἀ. τοῦ .. σίτου *SEG* 24.344 (Oropus, ii BC)'

*ἀνακόμισις, εως, ἡ, *restoration*, *Stud.Pal.* 20.114.11 (v AD).

*ἀνακομιστής, ου, ὁ, *conveyer* (of goods), *POxy.* 3124.9 (AD 322).

*ἀνακομιστικός, ή, όν, *bringing back*, Hsch. s.v. νόστιμον ἦμαρ.

*ἀνακοπάζω, *check*, *restrain*, *Inscr.Cret.* 3.iv 37.14 (Itanos, i BC; ἀγκ-).

ἀνακόπτω **I 3**, after 'Thphr.*Char.* 25.2' add '*propel* a ship *back to land*, Arat. 346'

ἀνακοσμέω, add '**2** *decorate*, στεφάνυσιν .. ἀ. Corinn. 1(a) i 27 P. (dub.).'

ἀνακουφίζω, for 'of the ship of state' read 'fig., of a drowning man' add '**II** *subtract*, *Cat.Cod.Astr.* 8(1).146.15.'

ἀνακούφισμα, for '*a relief*' read '*raising the body* from a prone position by using the arms'

*ἀνακούω, *listen further*, Sch.S.*El.* 81 P. (S. codd. κἀνακούσωμεν, edd. κἀπ-).

*ἀνακραδαίνω, aor. part. -δάνας gloss on ἀμπεπαλών, *PBerol.* in E. Ziebarth *Aus der antiken Schule*² (Bonn 1913) p. 32.

ἀνακραδεύω, for '(ἀνακραδάω La.)'

ἀνακράζω **1**, line 5, after 'etc.' insert 'τηλικοῦτ' ἀνεκράζετε .. ὥστε D. 21.215'; line 7, for 'a relat.' read 'ὡς'; line 8, delete 'τηλικαῦτ' .. 215'   **2**, after 'animals' insert 'Stesich. *S*88 ii 21 (p. 185 D.)'

ἀνακρεμαστήρ, for 'Orib. 54.31.20' read 'Orib. 24.31.20'

*Ἀνακρεοντικός, όν, *Anacreontic*, of a metre, Isid.*Etym.* 1.39.7.

*ἀνακρίμνημι, = ἀνακρεμάννυμι, Pi.*Pae.* 8.79; cf. °κρίμνημι.

ἀνακρίνω **I 2**, after 'Antipho 2.3.2' insert 'γενήν Call.*fr.* 203.54 Pf.'

ἀνακροτέω, at end for 'in Hexam.ap.Diogenian. 3.67' read 'in hex.in Diogenian. 3.97'

ἀνακρούω **I 1**, after 'Aen.Tact. 18.6' insert 'Arat. 193'; line 3, after 'Plu.*Alc.* 2' insert 'cf. δοχμὸς ἀνακρούων θηρὸς πάτον Nic.*Th.* 479; *put* a ship *astern*, νῆα A.R. 4.1650; fig., ἀ. τοὺς μνηστῆρας τῆς ἐς τὴν Πηνελόπην ὕβρεως Aristid.Quint. 2.10 (p. 74.19 W.-I.)'   **2**, delete 'ἀπό .. 4.1650'   **II 1**, add '**b** *pull back*, κάλωας A.R. 1.1277; ἡνίας Sch.Ar.*Av.* 648b.'

ἀνακρύπτω, add '(cj. ἐνέκρυφε)'

*ἀνακρωτόφονος, gloss on γυρτεύς, Hsch.

*ἀνακτένιον, τό, kind of comb, *PRyl.* 627.168 (iv AD).

†ἀνάκτερος, α, ον, Myc. *wa-na-ka-te-ro*, *royal.*

ἀνάκτησις, add 'written ἄγχτησις *PRein.* 7.14 (ii BC)'

ἀνακτητικός, for '*recuperative*' read '*restorative*, w. gen., *from*, σίκυς .. ἀ. λειποθυμιῶν ὀσφραινόμενος Dsc. 2.135'

†ἀνακτίτης, v. °ἀναγκίτης.

ἀνακτορία, after lemma insert 'Ion. -ίη'; after 'A.R. 1.839' insert 'Parth. in *Suppl.Hell.* 640.1, *Orac.Sib.* 4.66, epigr.ap.Paus. 10.12.6'

*ἀνακτορίζω, *to be ruler*, of an emerald, ὁ δὲ .. ὑποχλωρίζων λέγεται σμάραγδος ἀνακτορίζων Socr.Dion.*Lith.* 1.13 G. (†ὑακτορίζων† codd., v. H.-S.).

†ἀνάκτωρ, ορος, ὁ, *ruler*, *lord*, usu. of gods, A.*Ch.* 357, E.*IT* 1414, ἀ. πόλεως (of Astyanax), *Tr.* 1217, ἀνάκτορι θήκατο τέχνας (i.e. Poseidon) *AP* 6.4 (Leon.Tarent.); pl., Cerc. 4.38, Ptol.*Tetr.* 122.

ἀνακυλίνδω, v. ἀνακυλινδέω

ἀνακύπτω, line 3, after 'Thphr.*Char.* 11.3' insert 'Men.*Dysc.* 537'

*ἀνακώσιος, ον, adj. fr. ἄναξ in dialect of Rhegium, Sch. D.T. p. 542 H., cf. Ibyc. 60(a) P., and χαριτώσιος.

ἀναλαλάζω, after 'E.*Ph.* 1395' insert 'Astyd. 2a.16 S.'

ἀναλαμβάνω **I 8**, add '*Cod.Just.* 1.3.45.14'   **II 1**, add '**b** w. ref. to confiscation by the Ptolemaic kings, *SB* 8008.31 (260 BC), τὰ ἴδια αὐτῶν ἀναληφθήσεται εἰς τὸ βασιλικόν *UPZ* 112 viii 18 (203/2 BC), *PTeb.* 1001.15 (ii BC).'   **3**, add '**b** med., *recover*, *be convalescent*, Gal. 10.679.7.'

*ἀναλδαίνω, *make to spring up*, Nonn.*D.* 40.390.

ἀναλδής, after 'Hp.*Aër.* 15' insert 'of stars, Arat. 394'; for 'ἄρουραι .. *fruiting*' read 'neut. pl. as adv., *feebly*, φυταλιαί .. ἀναλδέα φυλλιόωσαι'

ἀναλεαίνω, add '**II** ἀναλεαίνει· σχολάζει. Ταρεντῖνοι Hsch. (La.; cf. ἀναλεῖ).'

ἀναλέγω **II**, after 'Plu.*Lyc.* 1' insert 'μῆκος δὲ ἀμήχανον χρόνου ἐν βραχεῖ ἀναλεγόμενον Max.Tyr. 28(22).5e'   **III 1**, after '*read through*' insert '*AP* 9.63 (Asclep.)'   add '**b** act., *read out*, *recite*, *SEG* 31.985.D14 (Teos, v BC), D.C. 37.43.2, 53.11.1.'

*ἀναλείωσις, εως, ἡ, *covering with powder*, ὅταν μετὰ τοῦ κουφολίθου ἀναλείωσιν ξηραίνεται Pelag.Alch. 254.

*ἀναλεκτήριον, *bag* or other container, Aq. 1*Ki.* 17.40.

ἀναλέκτης, add 'Sen.*Ep.* 27.7'

ἀνάλεκτος, add 'παιδία ἀ. *SB* 4425.3.21 (ii AD)'

ἀνάλημμα **III**, for '*sun-dial* .. 9.7.7' read '*projection* on a plane of circles and points on the celestial sphere, Hero *Dioptr.* 35, Ptol.*Anal.* p. 202.26 H., al., Papp. 4.246 H.; used for the construction of a sun-dial, Vitr. 9.6.1, 7.6'; for '*CIG* 2681' read '*Ἴασος* 249' and transfer to section II.

*ἀναλημπτός, ή, όν, *confiscated*, *PSI* 104.14 (ii AD), *PMich.*inv. 148ᵛ 1.10 (ii AD, *ZPE* 27.127).

*ἀναλημπτρίς, v. ‡ἀναληπτρίς.

*ἀναλημψιακός, ή, όν, of or *belonging to a reception* (ἀνάληψις I 6), *corona analempsiaca CIL* 14.2215.11 (Nemi).

ἀναληπτρίς, before 'Gal.' insert 'Sor.*Fasc.* 41, 42 (-λημπ-)' and after '323' add '*Hippiatr.* 50 O.-H.'

*ἀνάληψιμος (written -λήμψ-), ον, of goods, *liable to be reclaimed*, *resumable*, *SB* 9016.1.20 (ii AD).

ἀνάληψις **I 1**, after '(Hp.) *Off.* 9' insert 'Gal. 18(2).414.1, 416.7, 417.1 al.'   **3**, before '*acquirement*' insert '*taking up* or *on*, φορτίων *Peripl.M.Rubr.* 55'   **6**, add 'δόντα καὶ ἀ. καὶ ταυροκαθάψια *SEG* 32.660 (Thrace, ii/iii AD)'   **II 1**, add '*recovery* of things previously paid out, Just.*Nov.* 1.3'

ἀναλίσκω, line 10, pf. pass., add 'ἀνέλοται *SEG* 35.13 (Attica, early v BC)'   **I 1**, line 17, after '(D.) 18.66' insert 'περὶ αὐτοὺς *Cod.Just.* 1.3.41.14'

*ἀνάλιφος, ον, τὸ -ον *lack of the means of anointing oneself*, *Ἴασος* 248.44 (ii AD).

ἄναλκις, add 'w. gen., ὅπλων *lacking strength of weapons*, Opp.*H*.1.578'

ἀναλοίωτος, at end for 'Thphr.*CP* 6.10.1' read 'Thphr.*CP* 6.10.3'

ἄναλλος, after '*topsy-turvy*' insert 'v.l. for ἔναλλος in Theoc. 1.134, *AP* 5.299.9 (Agath.)'

†ἀνάλμυρος, ον, ἀνάλμυροι· ἄναλοι ἢ οὐχ ἁλμυροί Diosc.Gloss.ap.Gal. 19.79.

ἀναλογία **I**, add '**3** *proportionate share*, *PMichael.* 42 A, *POxy.* 1892, *SB* 9153 (all vi AD).'

ἀναλογιστικός, after '*analogical*' insert 'Epicur.*fr.* [26]28.6, [31]16.23 A.'

*ἀνάλοητος, ον, *unthreshed*, *PMich.*inv. 3207.2 (*ZPE* 100.74, ii BC).

ἀναλύζω, delete 'lit. .. hence'; for 'vulg. ἀνωλύζεσκε' read 'ἀνωλύζεσκε codd.'

ἀνάλυσις **I 1**, add 'perh., from mortality, *POxy.* 3010.29 (ii AD)'

*ἀναλυτέος, *that must be dissolved*, Plot. 4.7.2.

ἀναλυτικός, add '**3** in magic, *releasing* from a spell, *PMich.*III 154 (iii/iv AD).'

*ἀναλυτρόομαι, *redeem* a pledge, *PMasp.* 23.21 (vi AD).

*ἀναλύτρωσις, εως, ἡ, *redemption*, *PMasp.* 167.13 (vi AD), *Stud.Pal.* 3.339.4 (vi AD).

ἀναλύω (A), delete the entry.

ἀναλύω (B) **I 2**, add 'med., ἐπὶ τοι ἀλλυσαμένοι ἔμεν *Leg.Gort.* 6.49, al., cf. Hsch. s.v. ἀναλυσάμενος'   **II 1**, transfer 'in med.' to follow '*unloose*' and after '*Del.* 237' insert 'μίτρας for sexual intercourse, Hes.*fr.* 1.4 M.-W.; μίτρην in child-bearing, Call.*Del.* 222'   **5**, before '*reduce*' insert 'math.' and after '*Geom.* 5.8' insert '*PMich.*III 145.3 iii 2, 3 vi 6'   **7**, add '[ἀλλ'] ὅτε δὴ τριτάτας] τεκέων ἀνέλυον ἀνάγκας *SEG* 25.752 (epigr., Callatis, ii BC)'   **II**, delete the section (v.II 2), and substitute '*unsettle*, Philostr.*VA* 5.35.'   **III 2**, add 'ἐξ εὐωχίας, Call.*Dieg.* iv.38 (*fr.* 102 Pf.)'

ἀναλφάβητος, add 'Ath. 176e, *App.Anth.* 5.12'

ἀνάλωμα, add 'see also ἄλωμα'

ἀνάλωσις **I**, add '*Bull.Soc.Alex.* 10.28.28 (ἀνήλ-, i BC)'

ἀνάλωτος, line 2, after '(Hdt.) 8.51' insert 'of the Nile, ἀ. .. καὶ δύσμαχος τοῖς ἐπιβουλεύουσιν Isoc. 11.13'

ἀναμαρμαίρω, for '*move quickly*' read '*cause to blaze*, ὁτὲ .. ἀναμαρμαίρουσιν πῦρ ὀλοὸν πιμπρᾶσαι' and after 'A.R. 3.1300' add '(pap. ἀναμορμύρουσιν)'

*ἀναμαρτέω, *to be sinless*, Herm.*fr.* 7.2 N.-F.

ἀναμαστεύω, after '(for fugitives)' insert '*IRhod.Peraia* 357.9 (ii BC)'

ἀναμασχαλιστήρ, for pres. def. read 'article of female dress, perh. the same as, or similar to, ἀναληπτρίς' and add '*IG* 2².1408.5 (prob.), Hsch.'

†ἀναμάχομαι, *renew the fight*, *fight over again*, Hdt. 5.121, 8.109.2, Th. 7.61.3.   **2** *fight* an argument *over again*, ἀ. τὸν λόγον Pl.*Hp.Ma.* 286d, *Phd.* 89c.   **II** *fight back and retrieve* a defeat, περιπέτειαν Plb. 1.55.5, νίκην Memn. 58 (38.7 J.), Jul.*Or.* 1.24c; transf., *make good* loss, damage, etc., ἡ φύσις τὴν φθορὰν ἀ. Arist.*GA* 755ᵃ31, ἀ. τὰ ἁμαρτανόμενα Thphr.*CP* 3.2.5, Aret.*CD* 2.6 (165.9 H.).

ἀναμείγνυμι, line 1, before 'B.*fr.* 16' read 'B.*fr.* 20B.9' and add 'ὀμ-μ(ὀν-) Sapph. 2.15, 44.24, 30'   **II 1**, at beginning insert 'μύρρα καὶ κασία λίβανός τ' ὀνεμείχνυντο Sapph. 44.30 L.-P.'

ἀναμερισμός, add '**II** *distribution* of burdens, *PSI* 684.12 (iv/v AD).'

ἀναμετρητής, add '*SB* 4295.3, Ptol.*Judic.* 4.5'

ἀνάμιγα, for '*promiscuously*, *confusedly*' read '*so as to be mingled together*'; after '*IG* 5(1).726' insert 'Theoc.*Ep.* 5.3, Κυκλάδες ἄμμιγα νῆσοι Heph.*Astr.* 1.1.102' and delete 'also τινός ib. 22'

ἀναμίγδην, add 'w. dat., Nic.*Al.* 557'

ἀναμιμνήσκω, line 2, for 'Sapph.*Supp.* 23.10' read 'Sapph. 94.10

ἀναμίξ　　　　　　　　　　　　　　　　SUPPLEMENT　　　　　　　　　　　　　　　ἀναπυρριάζω

L.-P.' **II 1**, add '*recollect*, τετάρτῃ .. ἡμέρᾳ ἀναμιμνήσκομαι ἐμαυτοῦ καὶ αἰσθάνομαι οὗ γῆς εἰμι Pl.*Mx.* 235c'

ἀναμίξ, after 'Th. 3.107' insert 'Arat. 104, *AP* 4.1.8 (Mel.)'

ἀναμίσγω, lines 3/4, for 'γέλως .. 7.3 P.' read '[πλοῦτος] ἀναμίσγεται ἄτῃ Sol. 13.13 W., Call.*fr.* 24.3 Pf., *Del.* 217'

ἀναμισθέομαι, delete the entry.

***ἀναμισθόω**, Dor. ἀμμ-, *let anew*, *IG* 11(2).142.5 (Delos, iv BC); pass. *Tab.Heracl.* 1.111.

ἀνάμνησις 2, for this section read '*commemoration* of a dead person, *TAM* 4(1).272 (Bithynia, Rom.imp.); cf. τοῦτο ποιεῖτε εἰς τὴν ἐμὴν ἀνάμνησιν Ev.Luc. 22.20, 1*Ep.Cor.* 11.24' transferring 'Lxx *Nu.* 10.10' to section 1.

ἀναμνηστικός I, add 'adv. -κῶς, *by way of recollection*, Alcin.*Intr.* p. 178.7 H.' **II**, for this section read '*suggestive, reminiscent*, Demetr.*Eloc.* 287. **2** *indicative of the past*, σημεῖα Gal. 1.313.'

†**ἀναμοχλεύω**, *unbar* gates, E.*Med.* 1317. **II** *raise by leverage*, ἀρθρεμβόλοις ὀργάνοις .. τὰς πόδας .. ἐξ ἁρμῶν ἀναμοχλεύοντες ἐξεμέλιζον Lxx 4*Ma.* 10.5, τὴν Ὄσσαν Luc.*Cont.* 4; in reducing dislocations, Gal. 18(1).403.7.

ἀνάμπυξ, add 'Myc. *a-na-pu-ke* (pl.)'

***ἀναμφιβόητος**, ον, perh. *not renowned*, οὐδ' ἀναφιβό[η]τον *GDRK* 27².22.

ἀναμφίβολος, adv. -ως, add '*without question*, *Stud.Pal.* 3.124.4 (*ZPE* 75.256), ἀ. καὶ ἀναντιρρήτως *PMasp.* 116.6 (vi AD)'

ἀναμφίλογος, adv., after 'X.*Cyr.* 8.1.44' insert '*Oec.* 6.3' and after 'id.*Ages.* 2.12' insert '*Oec.* 4.7'

ἀναμφισβήτητος, adv., add '*PGen.* 103 i 19 (ii AD)'

ἀναμφόδαρχος, for 'ἀμφόδαρχος' read 'ἀμφοδάρχης'

ἀνανδριεῖς, for 'cf. ἀναριεῖς' read 'v. Ἐνάρεες'

ἀνανεάζω, for 'Phyrn.' read 'Phryn.' and for 'Lxx 4*Ma.* 7.14' read 'Lxx 4*Ma.* 7.13'

ἀνανέμω 2, after 'over' insert '*GVI* 1210.2 (Eretria, v BC), perh. in graffito on vase, ho δὲ γράψας τὸν ἀννέμο(ν)τα πυγιξεῖ *SEG* 35.1009 (Sicily, early v BC)'

ἀνανεόομαι I, add '*restore* (buildings, etc.), *SEG* 24.880 (Thrace, iii AD), *TAM* 4(1).355, Salamine 212 (vi AD), *Cod.Just.* 8.10.12.2. **2** *reaffirm*, an obligation, *Zeitschr.d.Savigny-Stiftung* 56.101 (Dura, i AD).'

ἀνάνευσις, delete '(νέομαι .. Hsch.'

ἀνανεύω II, after '*throw the head up*' insert 'to God, in prayer, *PKöln* 111.6 (v/vi AD)'

ἀνανέωσις I, add '**2** *reaffirmation*, of an obligation, *POxy.* 1105.21 (i AD); so perh. ib. 274.20 (v. supr.). **3** *restoration* of building, *CPHerm.* 95.15 (iii AD), *POxy.* 1752.2 (iv AD), *SEG* 30.1787 A, Salamine 212 (both vi AD), *Cod.Just.* 1.12.17.1.'

ἀνανεωτής, add 'as imperial title, ἐπὶ τοῦ θεοφιλεστάτου καὶ ἀνανεωτοῦ τῶν ἱερῶν δεσπότου *SEG* 31.641 (iv AD)'

ἀνανήφω, transfer 'D.Chr. 4.77' to follow '2*Ep.Ti.* 2.16' and add to it 'M.Ant. 6.31'

***ἀνάνοικτος**, gloss on ἀχανής (ἀχανεῖ cod.), Hsch.

***ἀναντώδης**, ες, *uphill, steep*, Hsch. s.v. κνῆμαι.

ἄναξ line 2, after 'etc.' insert 'Cypr. *wa-na-xe ICS* 211, Myc. *wa-na-ka*' **I**, at end for 'The irreg. .. gods' read 'For the irreg. voc. ἄνα, v.h.v.' **II**, line 5, for '(' substitute ',' and lines 6/7, for 'cf. Isoc. .. 911)' read 'esp. in Cyprus, Clearch. 25, cf. Isoc. 9.72, *ICS* 211, 220(b)2; of Creon, S.*OT* 85, cf. 911' **III**, add 'Mitchell *N.Galatia* 25'

***ἀνάξιππος**, ὁ, ἡ, *master or mistress of horses*, Λάρισα B. 14B.10 S.-M.

***ἀναξίχορος**, ον, *ruling the dance*, κοῦραι prob. in B.*fr.* 65(a).11 S.-M.

***ἀναξοή**, Dor. -ξοά, ἡ, *polishing, smoothing*, rest. in *IG* 4²(1).102.66 (Epid., iv BC).

ἀναξυρίδες, line 1, for 'eastern nations' read 'Persians'; line 2, after 'X.*An.* 1.5.8' insert 'ἀ. Περσικαί Max.Tyr. 20(14).8c' and delete 'Hdt. 1.71, cf.'; line 3, for 'Hdt. 3.87, etc.' read 'Hdt. 7.64.2'

***ἀναπαιστήρ**, ῆρος, ὁ, *door-knocker*, *IG* 4²(1).102.79 (Epid., iv BC) (ἀπή-).

ἀνάπαλιν, add '**V** *upside down*, ἀ. πέτεσθαι Ael.*NA* 10.14.'

ἀναπαλλοτρίωτος, delete the entry (read ἀ[νεξαλλο]τριώτους, see *AJPh* 62.197).

ἀνάπαλλω, line 3, after 'Ra. 1358' insert 'so intr., ἀμπάλλοντι id.*Lys.* 1310'

ἀνάπαλος, delete 'κατ' ἄμπαλον .. (Thess.)' **II**, for 'Ath. 14.631d' read 'Ath. 14.631b'

ἀναπαριάζω, for 'Ephor. 107' read 'Ephor. 63 J., prob. in Lib.*Ep.* 555.3 (ἀνεπυρρίσσε codd.)'

ἀνάπαυλις, add 'perh. also ἀμπ-, rest. in *SEG* 30.1444 (?ii AD)'

ἀνάπαυμα I 2, add 'ἄμπαυμα μόχθων καὶ φύλαγμα σωμ[άτων] ἔθεντο *TAM* 4(1).303' **II 2**, add 'ἀπὸ ἀναπαύματος φακοῦ *PWürzb.* 14.18 (ii AD), ἀπὸ ἀ. λαχανοσπέρμου *PBouriant* 17.15 (iii AD), *PCair.Isidor.* 99.24, 100.18 (iii AD)'

ἀνάπαυσις 1, add 'of the repose of death, Mitchell *N.Galatia* 467, *SEG* 26.1672 (Jerusalem, v AD), *ICilicie* 21 (vi AD)'

ἀναπαυστήριος II 2, add 'τὸ τῶν νοσούντων ἀ. Procop.*Aed.* 5.3.20; of the tomb, *TAM* 4(2).269'

***ἀναπαύστρια**, ἡ, *giver of rest*, of Easter Day, Leont.Byz.*HP* 1.2.6.

ἀναπαύω II 2 c, ἀ. alone, add '*SEG* 29.640, 643, 644; ἀνεπάη ὁ μακάριος Τιμόθεος ib. 34.1466; ἀ. τῇ ψυχῇ ib. 31.1419 (both Palestine, iii/iv and vi AD)' add '**e** μετά τινος *sleep* with (sexually), E.*Cyc.* 582, Macho 286 G., Plu.*Alex.* 2.'

ἀναπείθω, line 1, for 'Arc. ἀμπ-' read 'Arc. 3 sg. aor. subj. ἀμπείσῃ (sense 3 infra)' **2**, add 'app. *exact, demand*, *Cod.Just.* 10.19.9'

ἀναπελάσας, for 'ὀλιγ-ηπελίη' read 'εὐηπελής'

ἀναπέμπω II 3, add 'τὸ δὲ μὴ νῦν κατὰ Ἀττικοὺς ἀναπέμπεται (i.e. μή νυν) Sch. E. *Med.* 584'

***ἀναπεσσεύομαι**, med. or pass. part., perh. of counter or sim., *move along*, PMichael. 4.4 (ii AD).

ἀναπετάννυμι, last line, after 'Stoic. 1.58' insert 'ἀναπεπταμένῃ .. βλασφημίᾳ Plu.*Them.* 21'

ἀναπηδάω, line 5, after 'ib. 1.3.9' insert 'in mounting a horse, ἐπὶ τοὺς ἵππους ἀναπηδᾶν id.*Eq.Mag.* 1.5, 17'

***ἀναπηδύω**, v. ‡ἀναπιδύω.

ἀναπηρτισμένως, add '(cj.); v.l. ἀναπηρτημένως (fr. 298a Marrone)'

ἀναπιδύω 1, add 'Lxx *Pr.* 18.4 (sp. -πηδ-)'

ἀναπιεσμός 1, for '= foreg.' read '*forcing up*'

ἀναπίμπλημι II 2, after 'Pl.*Ap.* 32c' insert 'τὴν πόλιν ἡμῶν πονηρᾶς δόξης ἀναπλήσει D. 20.50, 24.205, Aeschin. 2.88, Din. 1.25' and transfer here 'D. 20.28' fr. section II 1; after 'so in pass.' insert 'Aeschin. 2.72, Procl.*Inst.* 13 D.'

ἀναπίπτω 5 b, after 'sickness' insert 'Arr.*Epict.* 2.18.3'

ἀνάπλασις, add 'Gal. 13.824.5'

ἀναπλέκω 1, add 'ἀ. βοστρύχους D.H.*Th.* 19'; for 'ἀμπλ-' read 'ἀνπεπλεγμένας' **2**, after 'metaph.' insert 'διαλόγους D.H.*Comp.* 25'

ἀνάπλευσις I, add 'Crito ap.Gal. 13.794.10'

ἀναπληρόω I 6, add '*complete, finish*, τὰ αὐτὰ ἔργα ἀναπληροῦν *Cod.Just.* 8.10.12.9a'

***ἀναπληρωματικός**, όν, *expletive*, of particles, Charis. p. 226 K.

***ἀναπλωτικός**, ή, όν, *forming into a simple whole*, Procl.in R. 1.90.5 K.

ἀναπνέω IV, for '*breathe* the horse' read '*allow* the horse *to breathe*' add '**2** *rouse to consciousness*, Nic.*Th.* 547 (s.v.l.).'

ἀναπνοή IV, after '*breathing organ*' insert 'αὗται (sc. αἱ φλέβες) .. ἡμῖν εἰσιν ἀναπνοαὶ τοῦ σώματος τὸν ἀέρα ἐς σφᾶς ἕλκουσαι Hp.*Morb.Sacr.* 4'

ἀναποϊκός, delete the entry.

***ἀναπόγραπτος**, ον, *unregistered*, *IG* 2².1100.33 (Athens, ii AD).

***ἀναποδέω**, = ἀναποδίζω II, ἀναποδοῦσιν ἐπὶ τὴν μονάδα Plu. 2.876f; ἀπὸ τῆς μονάδος ἀναποδῶν ibid., cf. Pythag.ap.Stob. 1.10.12.

ἀναποδίζω II, delete 'ἐπὶ τὴν μονάδα .. (corr. Heeren)'

ἀναποδιστικός, before '*in retardation*' insert '*retrograde*, of planets, Ptol.*Tetr.* 113'

***ἀναποδόομαι**, *grow new feet*, in quot., of scorpions, Lyd.*Mag.* 1.42.

ἀναποδόω, delete the entry.

***ἀναπομπαζόμενον**· ἐν ἀναπολήσει γινόμενον Hsch.

ἀναπομπή, add '**III** *divorce* of wife by husband, *CPR* 24.30 (ii AD).'

ἀναπόμπιμος 1, add '*IG* 1³.21 (Athens, 450/449 BC)'

ἀναπομπός II, for '*distributor* of bread to soldiers' read '*one who delivers* supplies, etc.'

ἀναπορεύομαι, add '**b** πολλὴ ἀκαθαρσία ἀναπορεύεται καὶ λεύκη comm. in Alc. 306(14) ii 10 L.-P., where ἀ. glosses ὀνστείχει.'

ἀναπόρριφος, after '(ἀπορρίπτω)' insert '(sp. ἀναπόριφος *BGU* 446.5, 2049.12, al.)'; after '*free from blemish*' add '*δοῦλος* *SB* 7533.50 (ii AD). **2** *irrefundable*, δραχμαί *PSarap.* 10.8 (ii AD), ἀρραβῶνα *BGU* 446.5.'

***ἀναπότρεπτος**, ον, *not to be departed from*, *PMasp.* 98.4 (vi AD).

ἀναπράσσω, after '*OGI* 669.20' insert '(Egypt, i AD)'

***ἀναποζύμιον**, τό, *product of fermentation*, Iambl.Alch. 286.10.

***ἀναπτηνίζω**, *knock down, lay flat*, cj. in Nonn.*D.* 18.271.

ἀναπτύσσω, line 4, before 'δέλτων' insert 'ἀ. πίνακα A.*fr.* 281a.22 R.' and for 'E.*fr.* 370' read 'E.*fr.* 369'; line 5, after '*Ion* 39' insert 'φαρέτρας πῶμα B. 5.75 S.-M.'; line 7, after '*Im.* 2.17' insert 'fig., εἰ καὶ αὐτὸ τὸ εἶδος ἕκαστον πρὸς αὐτὸ ἀναπτύττοις Plot. 6.7.2'; line 12, after '*Tr.* 622' insert 'Antim. 22 Wy.'

ἀνάπτω I 3, at end after 'med.' insert 'like act., χάριν ἀ. τινί A.R. 2.214'

***ἀναπωτήριον**, τό, item of equipment app. associated with a *candelabrum*, *CIL* 14.100.

***ἀναπυγίζω**, intensive of πυγίζω, perh. to be understood fr. ἀπυγίζει (= ?ἀπυ-), *SEG* 37.817 (vase, Salerno, 480/470 BC), unless ‹ἐ›πύγιζε[ι] is to be read.

***ἀναπυριάω**, *re-subject to a vapour-bath*, Hp.*Mul.* 2.133.

***ἀναπυρριάζω**, v. °ἀναπαριάζω.

29

ἀναπωλέω, for 'CIG 2266.11 (Delos)' read 'Inscr.Délos 502A11 (ii BC)'
ἀναπωμάζω, add 'DAW 1896(6).64.125 (Cilicia) (v. ZPE 15.46)'
ἀνάρβηλα, after 'δέρματα' insert '‹ξέουσιν'
ἀναργὔρία, before 'non numeratae pecuniae' insert 'exceptio'; for 'Cod.Just. 4.21.16' read 'Cod.Just. 4.21.16.1; ἀναργυρία alone, Cod. Just. 4.21.16.4'
ἀνάργῠρος, add 'of physician, who practises without asking money, ἔγραψεν .. θαύματα τῶν ἁγίων ἀναργύρων Κοσμᾶ καὶ Δαμιανοῦ Suid., s.v. Χριστόδωρος Θηβαῖος'
*ἀναρδής, ές, moistureless, γούνατα Euph. in Suppl.Hell. 443.6; λοιμός hex. orac. in ZPE 1.185.20 (Hierapolis, iii AD).
ἄναρθρος II, add 'τῇ ἀνάρθρῳ φωνῇ Cels. in Dig. 33.10.7.2'
ἀναριστέω, for 'v.l. in Hp.Acut. 28' read 'Hp.Acut. 30, Gal.Consuet. p. 13'
ἀναρίτης, after lemma insert '(Hsch.), Dor. -τᾶς'; at end for '(ἀνηρ-)' read '(ἀνηρίτης)'
ἀναροτρίαστος, add 'gloss on ἄσκαλα, Sch.Theoc. 10.14'
ἀναρπάζω I, add 'pass., of garment, δίπλακες .. ἀναρπαζόμεναι τοῖς ὤμοις caught up at the shoulders, Lyd.Mag. 2.4' II 2, add 'b ἀ. εἰς ἐλευθερίαν calque of Lat. eripere in libertatem, take from the owner and set free, Just.Nov. 5.2.1; sim., ἀ. εἰς εὐγένειαν give status of being ingenuus to, ib. 22.11.'
ἀναρραψῳδέω, for 'begin singing' read 'recite again'
*ἀνάρρηγμα, τό, breach, EM (s.v. ἀνὰ ῥῶγας μεγάροιο); app. bursting out, ὠκυάλοις ποδῶν ἀναρρήγμασιν Trag.adesp. 450a K.-S.
ἀναρρίπτω II, after 'Ar.fr. 673' insert 'Critias Trag. 7.27 S.'
ἀναρρῐχάομαι, line 3, for 'ἀρριχάομαι' read '°ἀρριχάομαι'
ἀναρροιβδέω, line 3, after 'cf. (Od.) 236' insert 'A.fr. 217 R.'; line 4, after 'Paul.Aeg. 3.10' insert 'snort, AP 9.769.4 (Agath.)'
ἀναρροιζέω, for 'Plu. 2.979e' read 'Plu. 2.979d'
ἀνάρσιος I, add 'perh. in pass. sense, abhorred, hated, A.R. 2.343'
Ἀναρσιτική, ἡ, cult-title of Demeter, SEG 33.1166 (-ειτηκή, Side, Pamphylia).
*ἀνάρτᾶς, app. Dor. form of ἀναρίτης, Ath. 3.86b.
ἀναρτάω I 1, for 'χέρρας ὑμ'' read 'χέρρ᾽ ἀπύ μ᾽'' and for 'Alc.Supp. 4.21' read 'Alc. 58.21 L.-P.' 2, add 'w. gen., τῆς ἐκείνου πίστεως τὸ πᾶν ἀνηρτήσατο Just.Nov. 73.4'
*ἀναρτικός, ή, όν, used for suspension, gloss on ἀρτάνη, Sch.S.OT 1266 (perh. ἀναρτ‹ητικήν).
ἀναρχᾱΐζω, for 'πόλιν' read 'πατρίδ''
ἀναρχία III, add 'IG 2².1713.12, also at Thasos, Thasos 28.37 (-ίη, iv BC)'
ἀνασαβρῶσαι, add '(ἀνασαρῶσαι La.)'
ἀνασάξιμον, for pres. ref. read 'IG 2².1587.13, al. (iv BC)'
ἀνασειράζω, for 'hawser' read 'rope, trace, etc.' and after 'metaph.' insert 'E.Hipp. 237' 2, delete 'draw .. 237'
ἀνασείω I, add '4 shake up' and transfer here 'ὑδρίαν IG 2².204.36' fr. section 3 add 'III abs., make trouble, Men.Epit. 241; = ἐπηρεάζω, Phot. s.v. ἀνασείειν.'
*ἀνασελγαίνομαι, behave wantonly, codd. in Ar.V. 61 (ἐνασ- Herm.).
ἀνασεύομαι, for 'pass.' read 'med.'
*ἀνασήπω, leave to decompose, Anon.Alch. 21.13.
ἀνάσιλλος, add 'v. °ἀνάσιμος'
ἀνάσιμος 1, add 'Herod. 4.67 (v.l. ἀνάσιλλος)'
ἀνασῑμόω, add 'II pass., to be curved at the end, Hero fr. 2.1.294.'
ἀνασκάπτω 1, add 'ἀναμέτρησις χωμάτων· ἕκαστος τῶν ἐπιχωρίων ἀνασκάπτι πέντε ναύβια POxy. 2847.22 (iii AD)'
ἀνασκαφή, after 'digging up' insert 'PLille 1ʳ.8 (iii BC)'
*ἀνάσκαφος, ον, accursed, POxy. 1854 (vi/vii AD).
ἀνασκευάζω, line 3, for 'Philostr.VS 1.17.3' read 'Philostr.VS 1.17.2' I 2, add 'destroy a funerary monument, TAM 4(1).376 (aor. inf. ἀνασκεβάσ[ε]), v. Hellenica 11/12.389-9' 6, add 'reverse a judgement, PColl.Youtie 66.37 (AD 253/60)'
ἀνασκιρτάω, before 'Philostr.' insert 'perh. alert'
*ἀνασκολοπισμός, ὁ, impalement, Nigidius Figulus in Lyd.Ost. p. 74 W. (pl.).
*ἀνασκυβαλίζω, defile, IG 2².13221 (iii AD).
ἀνασοβέω, after 'Pl.Ly. 206a' insert 'Ep. 7.348a'; add 'transf., γυνή ἀνασεσοβημένη τοὺς τρόπους SB 9421.18 (iii AD)'
ἀνάσπᾰσις, add '2 raising, lifting up, κίονος, μελάθρων SB 10299.140, 144 (iii AD).'
ἀνασπαστός 1, for 'but mostly .. Asia' read 'uprooted from one's home, country, etc.' add '3 that can be tightened (of noose), AP 6.109 (Antip.Sid.).'
ἀνασπάω I 6, for 'pucker' read 'raise' and for 'οἱ τὰς ὀφρῦς .. Arist.' read 'οἱ τὰς ὀφρῦς ἀνεσπασμένοι πρὸς τὸν κρόταφον those whose eyebrows rise towards the temples, Arist.'
ἀνασπογγίζω, after 'Ulc. 4' insert 'Antipho 5.45 (v.l. ἀπο-)'
ἄνασσα, line 1, for parenthesis read '(Cypr. wa-na-sa-se Ϝανάσσας ICS 6; 7.4; [Ϝ]άναᶜʰα IPamph. 3.29, v/iv BC)'

ἀνάσσω, line 4, after 'Ϝανάσσω' insert 'SEG 11.336 (the Larissa at Argos, vii BC); ἐϜαϝασσαντο Jeffery LSAG p. 168 no. 7 (c.575/550 BC)'
†ἀνάστᾱμα, v. ‡ἀνάστημα.
*ἀναστᾰσία, ἡ, erection, construction, BGU 2375.20 (i BC). II removal, displacement, Anon.astr. in APF 1.495.16.
ἀνάστασις II d, add 'ἕως ἀναστάσεως BCH suppl. 8 no. 119 (Thessalonica, c.AD 300)'
ἀνάστᾰτος I 1, add 'Trag.adesp. 394 K.-S.' 2, line 2, for 'δόμους τιθέναι' read 'οἴκους τιθέναι' 4, delete the section transferring quot. to section I 1.
ἀναστείχω, add 'Aeol. ὀνστείχει, Alc. 306(14) ii 3 L.-P.'
ἀνάστημα, add 'also ἀνάσταμα SB 10771 (223/2 BC) add '6 garrison, Lxx 1Ki. 10.5. 7 hub of wheel, Sm.Ez. 1.18.'
ἀναστομόω, line 4, for 'med., φάρυγος ἀναστόμου' read 'φάρυγος ἀναστόμου (med. ἀναστομοῦ, cj.)'
ἀναστοφάγος, after 'Paus. 8.42.6' add '(v.l. °νασто-)'
ἀναστρέφω A I 1, line 7, for 'τῆς' read 'τὰς'; add 'τοῦτο ἔμπαλιν ἀνέστραπται is reversed or inverted, (X.) Hier. 4.5, cf. Cyr. 8.8.13, Arist.Mech. 854ᵃ10' II 2, add 'PCair.Zen. 815.4 (257 BC), PMich. Zen. 55.7 (240 BC), Lxx Ge. 8.11, 18.14' B II, for 'dwell in a place' read 'med., roam over a place, go up and down in it, sojourn' deleting 'go to a place and dwell there'; line 5, after 'Th. 8.94' insert 'στέρνοις AP 5.237.6 (Agath.)' III 2, delete the section.
ἄναστρος, add 'also ἀ. σφαῖρα, of the sphere beyond the fixed stars, Aristid.Quint. 3.12'
ἀναστροφή II 3, add 'Thess. ὀστροφά BCH 59.55 (Larissa)' 7, for 'Plu. 2.112c' read 'Plu. 2.112d'
*ἀνασυρτόλις, εως, ἡ, one who lifts her clothes (applied to a hetaera), Hippon. 135a W.
ἀνασύρω I 1, line 1, after 'up' insert 'ὕδωρ Aq.Is. 30.14'; line 5, for 'obscene' read 'exceeding the bounds of decency' and after 'Anacr.ap.Phot. p. 123 R.' delete 'lacking in decency' 2, delete the section, placing quot. after 'Clearch. 14' (line 2).
ἀνασώζω, lines 3/4, for 'rescue .. 1165ᵇ22' read 'ἐάν τι ἀπολωλὸς πυνθάνωνται, ἀνασῴζειν IG 1³.32.22 (Eleusis, v BC); ἀ. φίλον ἀλλοιωθέντα Arist.EN 1165ᵇ22; preserve from, ἀπὸ φόνου ἔρρυτο κἀνέσωσέ μ' S.OT 1351'
ἀνασωσμός, for 'Aq.Ge. 45.1' read 'Aq.Ge. 45.7'
*ἀνασωστέον, one must restore, ταῦθ᾽ ἡμῖν φυλακτέον ἐστι καὶ ἀ. διαφθειρόμενα Gal. 10.838.3.
ἀνασωφρονίζω, after 'sobriety' insert 'Sch.Gen.Il. 14.436'
*ἀνάτᾰλος, unexpld. wd. in Hdn.fr. 8ᵛ p. 21 H.
ἀνατᾰράσσω 1, for 'stir up the mud' read 'poet. ἀντᾰρ- Sol. 37.8 W.; stir up, a settled liquid (in quots., abs.), Sol. l.c.'; for 'thick' read 'turbid, cloudy'
ἀνάτᾰσις 2, after 'threats of violence' insert 'Clearch. 2'
ἀνατείνω I 4, add 'generally offer, πᾶσαν .. τὴν ἑαυτῶν ζωὴν εἰς τὸν φιλάνθρωπον ἀ. θεόν Cod.Just. 1.3.41.1; εὐχὴν καὶ πρεσβείαν ἀ. ὑπὲρ διαμονῆς ὑμῶν καὶ σωτηρίας PMasp. 2 iii 24 (vi AD)' 6, delete 'pucker' and add 'cf. ‡ἀνασπάω I 6' III, after 'food' insert '(medic.), Arr.Epict. 3.22.73: trans., impose abstinence, starve' and after 'Arr.Epict. 2.17.9' add 'generally, starve, ὄνους Sch.Call.fr. 1.43 Pf.'
ἀνατέμνω I 2, for 'Arist.Spir. 478ᵃ2' read 'Arist.Spir. 478ᵃ27' II, for 'Aeschin. 3.166' read 'D.ap.Aeschin. 3.166 (dub.)'
ἀνατεταμένον, for 'ἐλξίνη' read 'ἐλξίνη'
ἀνατίθημι, line 1, for 'pf. .. etc.' read 'poet. and dial. ἀντ-, Pi.O. 3.30, IG 7.3055 (Lebadea, iv BC), Inscr.Cret. 4.72 xi 14; Cypr., Thess. ὀντίθημι: 3 sg. aor. o-ne-te-ke ICS 265, al.; ὀνέθεικε IG 9(2).1027 (Larissa, v BC), ὀνέθεικε SEG 34.483 (Atrax, c.200 BC); Cypr., Arc. ὐτίθημι: 3 sg. aor. u-ne-te-ke ὐνέθεκε ICS 181, 3 pl. imper. aor. ὐνθεάντω SEG 25.447 (Aliphera, iii BC); for other inflexional forms see also s.v. τίθημι I 2, delete 'in Prose' and after 'person' insert 'πάντα θεοῖς ἀνέθηκαν Ὅμηρός θ' Ἡσίοδός τε Xenoph. 11.1 D.-K.' II 1, line 7, after '(Jaffa)' insert 'med., dedicate, Rev.Bibl. 34.579 (Jerusalem)' add 'b install in a priesthood, SIG 1011.12 (Chalcedon, iii/ii BC).' B I 1, line 3, freq. like act., at end add 'also dedicate, ναὸν .. ἄνθετ᾽ SEG 37.1175.9 (eleg., Pisidia, ii AD)' 3, add 'defer, put off, τὴν ζήτησιν Ath. 2.47a' II 2, after 'metaph.' insert 'ἀνατίθεσθαι ὥσπερ πεττὸν τὸν βίον οὐκ ἔστιν Antipho Soph. 52 D.-K., cf. Socr.ap.Stob. 4.56.39'; after '(Pl.)Men. 89d' insert 'Men.fr. 80 K.-Th.'
*ἀνατίθηνητέον, one must nurse, foster, θυμόν, πικρίαν, Muson.fr. K.
*ἀνατμητέος, α, ον, that is to be dissected, Gal. 2.481.5.
ἀνατοκισμός, add 'SEG 38.1462.21 (Oenoanda, ii AD)'
ἀνατολή I 4, add 'ἀντολαί .. Καύστρου, Nic.Th. 635' II, add '2 in pl., growth, of grass, PTeb. 703.51 (iii BC), Aegyptus 5.130.25 (ii BC).'
ἀνατομή I, add '2 breaking up of coins (prior to re-minting), τὰ πρὸς ἀνατομήν Inscr.Délos 461Bb49 (ii BC).'

ἀνατομία SUPPLEMENT ἀνδρεών

*ἀνατομία, transl. Lat. *apertio*, *dissection*, Cael.Aur. *CP* 1.8.57.

*ἀνατράπελος, ον, *overturned*, πάντα δ' ἀνατράπελα Call.*fr.* 7.30 Pf.

*ἀνατρεκιδδέτω, unexpld. wd. in *Inscr.Cret.* 4.95 (v BC).

ἀνατρέπω **II 4**, add '*CodJust.* 6.4.4.13; act., *make null and void, Cod.Just.* 7.62.35'

ἀνατρίζω, for '*chirp* aloud' read '*screech, squawk*' and for 'al.' read 'codd.'

ἀνατύπωμα, for '*Stoic.* 1.214' read 'Zeno *Stoic.* 1.19 (D.L. 7.61)'

ἄναυδος **I 1**, lines 4/5, after '*without speaking*' insert 'τέλεος καιρὸς ἄναυδος τάδ' ἐπαινεῖ Α.*fr.* 47a ii 25 R.'

†ἀναυλεί or ἀναυλί, *free of freight charge*, *POxy.* 3250.8 (*c.*AD 63); cf. ἀναυλεί χωρὶς ναύλου Suid.

ἄναυλος (A) for '*flute(s)*' in this article read '*aulos (auloi)*' **I 1**, after 'E.*Ph.* 791' insert 'ἀ. βρέγμα (dub., cj. ἄναυδον) A.*fr.* 451h.8 R.'

ἄναυλος (B), for 'αὐλίον' read 'αὔλιον'

ἀναυμαχίου, add 'Poll. 8.42-43'

ἀναφαίνω **I 3**, line 5, for 'ἀμφ-' read 'ἀμπ-'; for '*Leg.Gort.* 10.34,al.' read '*Inscr.Cret.* 4.72.10.34, 11.18'; add 'ἀνπ- ib. 10.43, 48' **II 2**, for '*to be declared* king' read '*to be seen manifestly to be* king'; line 4, for 'romancer' read 'speech-writer'

ἀναφαίρετος, line 2, after '(i AD)' insert 'Iamb.*Protr.* 5 (p. 36.14 P.)'; line 4, after '(iii AD)' insert 'Procl.*Inst.* 105 D.'

ἀναφάλακρος, for '*forehead-bald*' read '*having a receding hairline*'

ἀναφαλαντιαῖος, after 'foreg.' insert 'Ptol.*Tetr.* 143'; add 'Heph.Astr. 2.2.31, 32, 34'

ἀναφάλαντος, for '*forehead-bald*' read '*having a receding hairline*'

ἀνάφανσις, delete the entry.

*ἀνάφαυσις, εως, ἡ, *flash*, Anon. in Ptol.*Tetr.* 5 (v. *MH* 41.70).

ἀναφέρω **I 1 b**, add 'of archives, περὶ τῶν γραμμάτων τῶν μνημονικῶν τῶν τε ἀνενειγμένων εἰς τὸ ἱερόν .. *SEG* 33.679.4 (Paros, ii BC)' add '**c** astron., in pass., *rise above the horizon*, Hypsicl. 9.8, al.' **2**, line 3, after '(Plu.)*Alex.* 52' insert 'abs., *sob in breathing*, Hp.ap.Gal. 19.80' add '**7** pass., *to be seconded, detached* of troops, *Salamine* 75.3 (ii BC), *BGU* 2024.8 (AD 204).' **II 7**, line 3, for '(v. supr. I 2) Hdt. 1.116' read 'Hdt. 1.116.2 (cf. supr. I 2)'; line 4, after 'Theopomp.Com. 66' insert 'Men.*Sic.* 368, fr. 369 K.-Th.' **10**, add '**b** *recall, remember*, Isoc. 5.32, Plu. 2.607e, App.*BC* 1.121, al.'

ἀναφθείρομαι **I**, for 'cf. φθείρω' read 'cf. ‡φθείρω II 1'

ἀναφλάω, after '-φλάω' insert 'τὸ αἰδοῖον'

ἀναφορά **I 1**, add '**b** *lifting* or *carrying up*, λίθου Demetr.*Eloc.* 72.' **2 d**, for 'of a sign' read '*above the horizon*, Hypsicl. 9.8, al.' add '**3** perh. *duct carrying must from the wine-press to the fermentation vat*, δεῖ .. τὸν ληνεῶνα ὅλον καταλείφθαι πάντοθεν λειοτάτοις κονιάμασι, καὶ οὐχ ἥττον τὰς ἀναφοράς *Gp.* 6.1.3.' **II 4**, add '*SEG* 32.1611 (ii/iii AD)' **9**, delete the section. **III**, delete the section.

ἀναφορεύς, for '*bearer, bearing-pole*' read '*carrying-pole*'

ἀναφορέω, add 'Cypr. *u-na-po-re-i* perh. ὑναφορεῖ *Kafizin* 266b'

ἀναφορικός **II**, add 'νόσος ἀ. Ptol.*Tetr.* 87'

ἀναφρίζω, add 'app. describing a medical symptom, Men.*Asp.* 453'

*ἀναφυή, ἡ, *bud, shoot*, Aq.*Za.* 6.12.

*ἀναφυλάσσω, *keep, guard*, Epic.*Alex.adesp.* 9 ii 19, *PDura* 29.8 (iii AD).

ἀνάφυξις, add '(perh. read ἀνάψυξις, cf. Jul.*ad Them.* 258c)'

ἀναφυσάω **III**, for '*blow the flute*' read '*tune up* on the pipe' and before 'κύκνοι' insert 'cf., of swans'

ἀναφύσησις **I**, delete 'Plb. 34.11.17' **II**, for '*prelude in flute-playing*' read '*blowing* in playing the aulos'

ἀναφυσητός, after '*blown up*' insert 'Moses Alch. 312.7'

*ἀναφυτέω, *foster the growth of*, ἀναφυτοῦντος ὅλον τὸ προκείμενον πωμάριον *SB* 9907.24 (AD 388).

ἀναφύω **I 2**, add 'Sm.*Jb.* 14.9' **II**, for this section read 'ἀναφύομαι *come into being, grow up*, Pherecyd. 22(a) J., Hdt. 4.58, ἀναφυομένης ἐκ γῆς πόας Pl.*Plt.* 272a; also act. (intr.), ἢν γὰρ ἀποδαιμῇ εἶς τις πονηρός, δύ' ἀνέφυσαν ῥήτορες Pl.Com. 186 K.-A. **b** transf., διαβολαί Plu.*Thes.* 17, δίκαι id.*Per.* 37; act., πάλιν ἀπορία τις ἀναφύει Phlp. in *de An.* 195.12. **2** *grow again*, act. (intr.), ἀναφῦναι τὰς τρίχας Hdt. 5.35.3, τὸ ἀνδρεῖον ἀνέφυ Luc.*DMort.* 28.2. **b** transf., ἀναφύντος τοῦ δήμου Aeschin. 2.177; in form ἀμφύνει, οὐχ ἥκιστα γὰρ τῷ στομάχῳ ἐς ἔκλυσιν ἡ νοῦσος ἀμφύνει (s.v.l., cj. ἐμφ-) Aret.*CA* 2 (bk. 6).3.8.'

ἀναφωνέω **I 3**, for this section read '*proclaim*, intr., of slaves making declaration of liberty, πλὴν ὅτι τοὺς δούλους οὐ χάριτι τῶν δεσποτῶν ἀλλὰ λαμπρῶς ἀναφωνήσαντας ἐλευθερωθῆναι λέγει Artem. 1.56 p. 63.7 P.; cf. Plu.*Cic.* 27.6, *CodJust.* 6.4.4.6'

*ἀναφωνητικός, ή, όν, *exclamatory*, Eust. 1964.47; adv. -ῶς, *as an exclamation*, Eust. 1044.53, Sopat.Rh. 8.335.10 W. (cj. I.-W.; -φωνικῶς W.).

Ἀναφωνητικῶς, delete the entry.

ἀναχάζω **II**, for 'pass.' read 'med.' twice.

ἀναχαιτίζω **I 2**, after 'D. 2.9' insert 'Men.*Sam.* 209'

ἀναχάσκω, add 'of the cervix, Alcmaeon 3 D.-K.; Hp.*Superf.* 32, *Vict.* 1.30'

ἀναχέω, line 4, after 'Arr.*An.* 6.18.5' insert 'of water reaching inland, εἰς τοὺς ἐσωτάτους τόπους ὁ Περσικὸς κόλπος ἀναχεῖται *Peripl. M.Rubr.* 35'

ἀναχωρέω, line 1, for 'Locr., Cret. ἀνχ-' read 'Locr. inf. ἀνχορεῖν *IG* 9²(1).718.7, 27 (v BC), Cret. ἀνκορέν *Inscr.Cret.* 4.72 xi 10 (Gortyn, v BC)' **I 2**, line 7, for '*IG* 9 .. 334' read '*IG* 9²(1).718.7, 27'; add '*withdraw* to place of refuge, *go into hiding*, of strikers, *PTeb.* 26.18, 41.14 (both ii BC); of slaves, *PHib.* 71.6 (iii BC); of offenders, *PTeb.* 5.6 (ii BC)' **IV 1**, delete '= συγχωρέω .. *Arc.* 10' (read ἐνεχώρησαν) **2**, delete the section.

ἀναχώρησις **I**, add '**b** *turning back*, Ph. 2.539.' add '**V** ἀ. τῶν ὄντων, calque of "*cessio bonorum*" Just.*Const.Δέδωκεν* 7d.'

ἀναχωρητής, add '*SEG* 33.1366 (-ιτής, Nubia, vi AD), *PAnt.* 202b9'

ἀνάχωσις (s.v. ἀναχώννυμι), place the article after 'ἀναχωρίζω'; add 'rest. in *JÖAI* 3.56'

ἀναψησμός, add '*SB* 9699.622 (AD 78/9), *PWürzb.* 22.13 (ii AD), *BGU* 2264 I 5, II 7 (AD 198)'

ἀναψηφίζω, before 'Th. 6.14' insert '*IG* 1³.250*A*12 (Attica, v AD)'

ἀνάψυξις **1**, after 'Posidon. 72' insert '*refreshment*, *SEG* 36.1277 (Mesopotamia, iii AD)'

ἀναψύχω **I 3**, line 1, after 'metaph.' insert 'ὂν δ' ἔψυξας ἔμαν φρένα καιομέναν πόθῳ Sapph. 48 L.-P., cf. Thgn. 1273 W.'; line 2, for 'Call.*Hec.* 1.1.7' read 'Call.*fr.* 260.7 Pf.'

*ἀγγλάριον, v. °ἀγγλάριον.

*ἄγγρεσις, v. °ἄγγρεσις.

ἀνδαβάτης, for '*gladiator*' read '*a type of gladiator who fought blindfold*' and add 'cf. Cic.*Fam.* 7.10.2'

*ἀνδαιθμός, ὁ, *redivision, redistribution* of land, *IG* 9²(1).609.2 (Naupactus, vi/v BC); cf. δαιθμός, ἀνδασμός, ἄνδαιτος.

ἀνδάνω, lines 4/5, aor., add '3 sg. ἔFαδε *SEG* 23.530 (Dreros, vii BC), ἔβαδε *Inscr.Cret.* 1.xvi 1.2 (Lato, iii BC), ἇδε *SEG* 18.772 (Berenice, Cyrenaica, iv BC; cf. °ἅδημα)' **I**, line 9, after 'Od. 2.114' insert 'Alcm. 1.88, 45, 56.2 P.' add '**b** *rejoice in*, w. dat., *AP* 6.299 (Phan.).' **II**, line 4, for 'so' read 'cf.' and delete 'cf. Od. 2.114'

ἄνδεμα κτλ., delete 'ἀνδεσμός'

ἄνδεργμα, after lemma insert '(ἄνδειγμα La.; cf. ἀνάδειγμα)'

*ἀνδηρεύων, ὁ, app. = ἀνδηρευτής, *PUniv.Milan.*inv. 100 (*Acme* 9(2).59.17).

*ἀνδηρικός, όν, *determinative*, ὁ Δημόκριτος ἔλεγε ταῦτα δὴ τὰ ἀνδηρικὰ ὀνόματα, ἔν, μηδέν, ῥυσμόν, τροπήν, διαθιγήν Democr.*fr.Orth.*

ἀνδίκα, for 'ἀναδίκη' read '*ἀναδίκη'

ἀνδίκτης, for 'Call.*Fr.* 233' read 'Call.*fr.* 177.33 Pf.' and add 'cf. perh. ἀνδικτήρ, ῆρος, δούνακας ἀνδικτῆρας *AP* 6.296 (Leon.Tarent.) (cj. Lobeck, ἀντυκτῆρας cod.)'

†ἀνδῐχάζω (s.v. ἄνδιχα), *cut in two, cleave*, in a riddle about an oyster, *Epigr.adesp.* in *Suppl.Hell.* 983.7 (unless [ἂ]ν διχάσῃ is to be read). **2** app. *to be of divided opinion*, αἴ κ' ἀνδιχάζοντι τοὶ ξενοδίκαι *IG* 9²(1).717.10 (Khaleion, W. Locris, *c.*450 BC).

ἀνδραγαθία, line 3, after '*the character of an upright man*' insert '*integrity*' and after 'Phryn.Com. 1' insert 'Isoc. 3.44, 6.105, D. 22.72, 24.180'; add '*IG* 1³.97.13 (412/11 BC)'

ἀνδράδελφη, after '*husband's sister*' insert '*MAMA* 1.324 (Phrygia)'

ἀνδράδελφος, after '*brother-in-law*' insert '*MAMA* 1.363, 369, 8.167' and after 'Suid.' insert 's.v. γαλόω, also fem. *sister-in-law*, ib.'

ἀνδρακάς (B), delete '*man's*'

*ἀνδραποδική, ἡ, *tax on slaves*, MacDowell *Stamped Objects from Seleucia* pp. 41, 42, 64, *YCIS* 3.30-3, 37, 38 (Uruk, Babylonia).

*ἀνδραποδιστί, adv., *like a slave*, prob. in *PTeb.* 765.13 (ii BC).

*ἀνδράποδον **I**, add 'not dist. fr. δοῦλος Th. 1.139.2, 7.27.5'

*ἀνδρασώτειρα, ἡ, *saviour of men*, epith. of Isis, *POxy.* 1380.55 (ii AD).

*Ἀνδρέας, ὁ, cult-title of Zeus in Phrygia, *SEG* 26.1367, 33.1153, al.

ἀνδρεία **I**, add '**2** as honorific title, *PVarsov.* 17.9, 12 (ii AD), τὴν σὴν ἀνδρείαν *PAmh.* 2.82.5 (iii/iv AD). **3** personified, the deity *Virtus*, *SEG* 33.945 (Ephesus), 35.1377 (Hierapolis, iii AD), *BGU* 2150.13 (v AD).'

*ἀνδρεΐζομαι **II**, for '*stubborn*' read '*bold*'; after 'Luc.*Ind.* 3' insert 'sup. as honorific title, *SEG* 23.296 (Lebadeia, iii AD), *PLips.* 119 ii 3 (iii AD), *Aegyptus* 9.123 (v AD)' **IV**, delete the section.

*ἀνδρεϊστέον, *one must play the man*, Men.*Kith.* 76.

ἀνδρειφόντης, for 'ἀνδρότης' read 'ἀνδροτής'

*ἀνδρεοκαταμάκτης, ου, ὁ, temple-official, perh. one who rubs down men after bathing, or perh. who polishes statues, *SB* 7336.25 (iii AD); cf. καταμάκτης, καταμάσσω.

*ἀνδρεοπλάστης, ὁ, app. = ἀνδριαντοπλάστης, *modeller of statues*, *CPR* 14.47.1 (vi/vii AD).

ἀνδρεών, for '*IG* 14 .. (Segesta)' read '*IPamph.* 3.8 (ἀδριιόνα acc. sg., iv BC); Dor., *IG* 14.291 (Segesta), perh. here public meeting place; cf. °ἀνδρήιον'

31

*ἀνδρήιον, τό, building where public feasts (for men) were held, Inscr.Cret. 4.4.4 (Gortyn, vii/vi BC), 5.1.8 (Axos, vi/v BC), al.

Ἀνδριάντεια, τά, name of festival involving the dedication of statues, also called the Nea Adrasteia, MAMA 6.76 (Phrygia).

ἀνδριαντίδιον, for pres. ref. read 'Inscr.Délos 442B167 (ii BC)' and add 'SEG 34.1087.4 (Ephesus, i AD)'

ἀνδριαντίσκος, delete 'puppet'

ἀνδριαντουργός, delete '(ἔργον)' and add 'SEG 33.1139 (Hierapolis, Phrygia, ii/iii AD)'

ἀνδριάς, add 'Myc. a-di-ri-ja-te (dat. sg.)'

†ἀνδρίζω, med., come to manhood, Ar.fr. 772 K.-A., Hyp.fr. 228, τῷ σώματι ἀνδρίζεσθαι, Luc.Anach. 15. II make physically strong or manly, τοὺς γεωργοῦντας X.Oec. 5.4. 2 endow with moral strength, Pl.Tht. 151d; med., take courage, be resolute, X.An. 4.3.34, Arist.EN 1115ᵇ4, Lxx Jo. 1.6, 1Ep.Cor. 16.13, D.C. 50.24.7. III med., perform the man's part in sexual intercourse, D.C. 79.5, ἐπὶ τὸ παιδάριον Lcn. 4Ki. 4.35, Philostr.Ep. 54, Ach.Tat. 4.1.2 V.; cf. of a eunuch, ἐπὶ γυναῖκα Philostr.VA 1.37. 2 of a woman, behave (dress) like a man, Philostr.Im. 1.2.

ἀνδρικός I 2, for this sense read 'of things, suitable for a man, i.e. large, Eub. 56.1 K.-A.; also, having great force, σεισμοί Ael.VH 6.9'

ἀνδρισμός, add 'cf. SB 5948. II = Lat. capitatio, poll-tax, PRyl. 658.8 (iv AD).'

ἀνδροβᾶμων, for pres. ref. read 'Inscr.Cret. 3.3.25.2, 26.5, etc. (Hierapytna, i AD)'

*ἀνδροβλής, ῆτος, ὁ, struck by (or striking) a man, Trag.adesp. 645.12 K.-S.

*ἀνδρόβουλος, ον, having the will or purpose of a man, A.Ag. 11, Phryn.PS p. 31 B.; cf. γυναικόβουλος.

*ἀνδρογύναιος, ον, of a man, having a womanish nature, effeminate, Lxx Pr. 19.12(15).

*ἀνδρόγυνον, τό, married couple, Cyran. 1.2.26 K. (pl.), 3.22.3 K. (sg.).

ἀνδροδάμας II, add 'Anon.Alch. 5.12; variety of haematite acc. to Plin.HN 36.146, 37.144'

*ἀνδροκίδαλος, ὁ, or -ον, τό, ἐπώνυμον ἀνδροκιδάλου, gloss on κρίθων, Hsch., but La. reads ἀνδρὸς μοιχαλίου.

Ἀνδρόκλειος, ον, of Androklos, the founder of Ephesus, Ἀνδρόκλειος πόλις IEphes. 1064.

Ἀνδροκλωνεῖον, τό, app. club building or tomb of Androklos, rest. οἰνοδ[οτή]σαντα ἐν τῷ Ἀνδρόκλω. [..] SEG 34.1107 (Ephesus), v. °Ἀνδρόκλειος.

*ἀνδροκόμος, ον, caring for one's husband, Sch.Pl. p. 407 G.

ἀνδροκτόνος, for 'Hdt. .. Cyc. 22' read 'σὺν ἀ. B. 18.23 S.-M.; of Amazons, Hdt. 4.110.1; of Cyclops, E.Cyc. 22. 2 slayer of her husband, S.fr. 187 R.'

ἀνδρολέτειρα, as adj., add 'Suppl.Hell. 1168.3 (epic.)'

*ἀνδρολέτης, ου, ὁ, slayer of men, of Ares, GVI 1552.6 (Panderma, ii/i BC).

†ἀνδροληπτέω, capture men, IG 4².68.64 (Epid., iv BC).

*ἀνδρομᾰνία, ἡ, (male) homosexual lust, Orac.Tiburt. 123.

*ἀνδρόμαξ, ὁ, perh. kind of bird, OFlorida 15.4 (ii AD).

ἀνδρόμεος, after 'Il. 11.538' insert 'αἵματος ἀ. Hes.Sc. 256' and after '[A.R.] 4.581' insert 'μήδεά τ' ἀνδρομέοισι πανείκελα Opp.H. 1.581'

ἀνδρομητόν, for 'ἀνέδραμον' read 'ἀναδραμεῖν'

ἀνδροτής, after 'Il. 16.857, 24.6' read '(vv.ll. ἀδρ-, ἀδρ-)'

ἀνδροφόνος, lines 3/4, for 'rarely exc.' read 'in Hom. usu.' I 1, after 'murderous' insert 'ἀνδροφόνους .. Ἰλιάδας E.Hec. 1061'

*ἀνδροφῠλάκιον, τό, guardhouse, Sardis 7(1).17.10.

*ἀνδροφῠλαξ, guard, PCol. 2.4 ii 2 (sp. ἀνδρω-).

ἀνδρόω II, after 'E.HF 42' insert 'Pl.Smp. 192a, Arist.EN 1180ᵃ2' III, after 'sum' insert 'Cratin. 318 K.-A.'

ἀνδρών, after 'X.Smp. 1.4, etc.' insert 'in inscrs. freq. of the dining-hall of religious associations, SEG 33.639 (Rhodes, c.100 BC), Syria 26.55 (Palmyra)' and at end for 'Ion. -εών (q.v.)' read 'dial. -εών (‡q.v.)'

ἀνδρωνύμιον, delete the entry (v. JHS 106.184).

*ἀνδρωσις, εως, ἡ, arrival at manhood, PMich.III 149.5.27 (ii AD).

ἀνέβρᾰχε, for '*βράχω' read 'βραχεῖν'

ἀνεγείρω III, after 'of buildings' insert 'and other structures' and add 'SEG 32.1389 (Commagene, i/ii AD), ITomis 384.15 (iii/iv AD), SEG 32.1540 (Arabia)'

†ἀνέγερσις, εως, ἡ, waking up, rousing, Iamb.Protr. 20 (p. 103.16 P.); pl. ἀνεγέρσεις βακχεύοντες (i.e. performing rituals to rouse), Plu. 2.378f. II erection of buildings or structures, SEG 26.1672 (Jerusalem, v AD), PMich.XI 624.12 (vi AD).

ἀνεγκέφᾰλος, add 'b transf., brainless, PLond. 1075.19 (vii AD).'

ἀνέθιστος, add '2 of persons, unused, μαλακίαις (?)Ephor. in PLit.Lond. 114.16.'

ἀνείδεος, after 'Ael.NA 2.56' insert 'Sallust. 17.6'

*ἀνείκητος, v. ‡ἀνίκητος.

ἀνείλησις 3, for 'twisting of the body' read 'folding or doubling up of the body'

ἀνείλλω, after 'ἀνειλέω' read 'med., roll oneself up Pl.Smp. 206d. II unroll, unfold Pl.Criti. 109a.'

ἄνειμι I 1, lines 2/3, after 'Hdt. 3.85' insert 'πρὸς ἡλίο ἀνιόντ[ος] towards the East, SEG 32.161 (Athens, 402/1 BC)' 3, for 'v. supr. 1' read 'cf. ἀναβαίνω II 3'

ἀνείμων, add 'Call.fr. 7.9 Pf., Ph. 2.225, Nonn.D. 35.107, 47.281'

ἀνεῖπον, line 3, after 'proclaim' insert 'ὃν Ὠπόλλων ἀνεῖπεν ἀνδρῶν σωφρονέστατον πάντων Hippon. 63.2 W.'

*ἀνειρήνευτος, ον, irreconcilable, Hsch. s.v. ἀσύμβατον.

ἀνείσακτος, add 'Iamb.Protr. 21 (p. 106.17 P.); of those not properly taught medicine, Gal. 13.563.3'

*ἀνεισφόρητος, ον, exempt from taxation, SIG 612B5 (Delphi, ii BC).

ἀνεκάθεν I, add 'Plu.Thes. 33.3' II 1, add 'Plu.Num. 1.2'

*ἀνεκδυσώπητος, ον, quoted as equiv. of Lat. inexorabilis, Dosith. p. 392 K.

ἀνεκθέρμαντος, adv. -ως, add 'Paul.Aeg. 7.19.3 (375.24 H.)'

ἀνέκλῠτος, after 'indissoluble' insert 'IG 12(7).393 (Amorgos, i/ii AD)'

*ἀνεκπολέμητος, ον, app. invincible in war, Dosith. p. 392 K.

*ἀνέκτακτον, τό, in accounts, total not itemized, PMasp. 57 i 4, iii 8 (vi AD).

ἀνεκτέος, after 'Ar.Lys. 477' insert 'Trag.adesp. 382 K.-S.'

ἀνεκτός II, after 'without neg.' insert 'Isoc. 5.11, 8.126, 10.1, 12.110'

ἀνεκφοίτητος, add 'τῶν ἐκεῖ Syrian. in Metaph. 109.25'

ἀνέκφραστος 2, delete the section.

*ἀνέλεγχος, ον, irrefutable, SEG 26.788 (funerary epigr., Byzantium).

*ἀνέλκυσις, εως, ἡ, hauling up, Sch.Th. 7.25.

*ἀνελληνόστολος, ον, wearing non-Greek dress, cj. in A.Supp. 234 (v. ἀνέλλην).

ἀνεμομᾰχία, add 'Cat.Cod.Astr. 11(2).168.16, 17'

ἄνεμος, add 'Myc. a-ne-mo'

*ἀνεμόσυριν (acc.), name of whirlwind said to be derived from a fan similarly called used by ladies in Alexandria, Olymp. in Mete. 200.19 (Alexandrian wd.; perh. corrupt for *ἀνεμόσυριν (σύρω) or °ἀνεμούριον.).

ἀνεμόσυρις, delete the entry.

ἀνεμοτρεφής, last line, for 'cf.' insert 'ἀ. πυλάων Simon. 107 P.'

ἀνεμούριον, add '2 fan, ἀνεμούρ(ιον?) βιβραδ(ικόν?) dub. in PRyl. 627.165 (iv AD).'

ἀνεμόφοιτος, for 'v. sub ἠνεμ-' read 'v. ‡ἠνεμ-'

ἀνεμοφόρητος, after 'carried by the wind' insert 'prob. in comm. in Sapph. 90 iii 23 L.-P. (ἀν[εμ]οφορητο[pap.)'; for 'dub. sens.' to end read 'of spirits who have no rest, PMag. 15.8, 14 (iii AD)'

ἀνεμφᾰτος, lines 4/5, for 'p. 434' read 'p. 530' and for 'pp.434, 450' read 'p. 543'

*ἀνενάριθμιος, ον, not reckoned in, PMasp. 97ᵛ.D61 (v AD).

ἀνενδεής 1, after 'AP 10.115' insert 'αὐτὸ .. τὸ θεῖον ἀνενδεές Sallust. 15.1'; adv. -εῶς, for 'faultlessly, unexceptionably' read 'in a way that lacks nothing, generously' and add 'IEphes. 728.25'

ἀνενδοίαστος, adv., add 'Iamb.Protr. 20 (p. 96.8 P.)'

ἀνενεχύραστος, after 'distraint' insert 'PTeb. 817.22 (ii BC)'

*ἀνενήλιξ, ικος, ὁ, ἡ, of children, not of age, MAMA 6.225 (Apamea).

*ἀνενοχλησία, ἡ, freedom from disturbance, SEG 36.1051.9 (Miletus, ii AD).

ἀνέντατος, for 'Theopomp.Com. 71' read 'Theopomp.Com. 72 K.-A., κλίνη Inscr.Délos 1416A38 (ii BC)'

ἀνέντροπος, after 'Hsch.' insert 'irreverent, Sm.Ez. 7.24'

ἀνεξαλλοτρίωτος, for 'unalienated' read 'not assigned to anyone else, inalienable, of property'; for 'cf. Ath.Mitt. 3.58 (Lydia)' read 'IEphes. 3245.19; esp. of tombs, SEG 24.1206 (Egypt, i BC), AJA 36.453 (S.Galatia, iii AD)'

*ἀνεξάρπαστος, ον, perh. not to be robbed, POxy. 2712.14 (AD 292/3).

ἀνεξέλεγκτος, after lemma insert 'Cret. neut. pl. ἀνεξέλεντα SEG 35.989.19 (Cnossus, ii/i BC)'

ἀνεξέταστος, after 'examined' insert 'Isoc. 9.42' and after 'Aeschin. 3.22' insert 'Scaev.Dig. 33.8.22.2'

ἀνεξία, for 'endurance .. word' read 'dub. wd., app. requiring sense abstinence, forbearance, or sim.'

*ἀνεξιδίαστος, όν, not to be appropriated, dub. reading in Ramsay Cities and Bishoprics 2.475.

ἀνεξίλαστος, after 'implacable' insert 'SEG 29.1179 (AD 108/9)'

ἀνεξοδίαστος, for pres. refs. read 'ISmyrna 228.6, INikaia 283.4, 1231.8, IStraton. 258, 316, IGBulg. 1007 (Philippopolis, iii AD)'

ἀνέξοδος I 2, place after section II II, for this section read 'not going out of doors or in public, Vit.Aesop.(G) 59, BGU 2350.11 (?ii AD); because of illness, weather, Vit.Aesop.(G) 59, BGU 2350.11 (?ii AD); fig., of attitudes, etc., διάνοια Plu. 2.610a, λόγοι 1034b, βίος 1098d'

*ἀνέξοχα, gloss on ἔμπηρα, Hsch.

ἀνεπάγγελτος 1, add 'IGBulg. 388 bis 3 (Istria, ii BC)'

*ἀνεπάνοδος, ον, not returning, Heph.Astr. 3.47.

ἀνέπαφος, after 'ον' insert '(fem. -η IGRom. 1.892.10 (Panticapaeum, c.ii AD)'; line 2, after 'seizure' insert 'D. 56.38, 40'; line 4, after 'unencumbered' insert 'PMich.x 583.15 (i AD)'

*ἀνεπεύθυνος, ον, app. = ἀνυπ-, BGU 1262.19 (iii BC), PFrankf. 1.87 (iii BC).

ἀνεπηρέαστος, after 'ον' insert '(fem. -η CIJud. 1.65 (Panticapaeum, iii AD))'

ἀνεπίβλητος, for 'inattentive, heedless' read 'unheeded'   add 'II not required to make payment, PFlor. 323.12 (vi AD).'

*ἀνεπιγιγνώσκω, not to recognize as correct, BGU 874.7.

*ἀνεπιγραφέω, set down, record, Zos.Alch. 219.

ἀνεπίγραφος 1, line 3, for 'IG 2 .. al.' read 'IG 2².1514.28 (iv BC), [τρί]πους ἀ. SEG 34.95 i 86 (Attica, ii BC)' and add 'w. gen., ἀ. ὁλκῆς καὶ νομίσματος Didyma 468.13 (ii BC), cf. 469.7.   b of books, without title, λόγοι D.H.Dem. 13.'

ἀνεπιδάνειστος, after 'not mortgaged' insert 'PTeb. 817.22 (ii BC)'

ἀνεπίδετος, after 'Hp.Fract. 20' insert 'Gal. 16(2).543.6'

*ἀνεπικωλυτί, adv. without let or hindrance, MAMA 6.83 (Attouda).

ἀνεπικώλυτος, after 'unhindered' insert 'ἀ. τὰν κράναν SEG 23.398b.10 (Thestia, Aetolia, ii BC)'; adv., without let or hindrance, add 'PKöln 232.11 (iv AD)'

ἀνεπίληστος, add 'Sch.Call.Lav.Pall. 87'

ἀνεπίλυτος, for 'unbandaged' read 'not having one's dressing changed'

*ἀνεπιμέμπτως, adv., blamelessly, ζῶν ἀλύπως καὶ ἀ. IPrusa ad Olympum 108.

*ἀνεπιμηνίευτος, ον, free from the obligation of serving as ‡ἐπιμήνιος, SEG 8.529.40 (Egypt, ii BC).

ἀνεπίξεστος, for 'Them.Or. 26.388b' read 'Them.Or. 26.322b'

ἀνεπισήμαντος II, for 'without an attack of ἐπισημασία (q.v.)' read 'not showing the symptoms (v. ἐπισημασία)'

*ἀνεπίσπαστος, ον, not subject to seizure, PMasp. 151.143 (vi AD).

ἀνεπιστάθμευτος, after 'Plb. 15.24.2' insert 'IGBulg. 315.17 (Mesambria)'

ἀνεπιστρεψία, for 'want of regard' read 'inattention' and add 'ostr. in CGFP 318 (Philadelphia, iii BC), PSI 152 ii 13 (ii AD)'

ἀνεπίστροφος, add '3 whence there is no return, οἰκία (neut. pl.) .. νεκύων ἀ. Bernand IMEG 33.23 (ii BC).'

ἀνεπίτακτος, for 'control' read 'dictation'

ἀνεπίτατος, for 'not to be .. 2' read 'that does not admit of heightening' and after 'μᾶλλον' add 'S.E.M. 10.272'

*ἀνεπιτυχία, ἡ, failure, Cyran. 1.22.15 K.

ἀνεπίφαντος, before 'insignificant' insert 'inconspicuous, Ptol.Tetr. 170'

*ἀνεπίφορος, ον, not admissible in evidence, POxy. 1716.17 (iv AD).

ἀνεπιχείρητος 1, after 'Plu.Caes. 25' insert 'intact, IGRom. 4.661.17 (Acmonia, i AD)'

*ἀνεπίψογος, ον, blameless, Aristid.Quint. 3.9 (p. 108.5 W.-I.).

*ἀνεραυνάω, v. ‡ἀνερευνάω.

ἀνέργαστος, add 'raw, untreated, of hides, Edict.Diocl. 8.6a, 9, al.'

ἀνέρεικτος, add 'Gal. 19.81'

†ἀνερεύγω, throw up, disgorge, Men.Asp. 451, ἀνήρυγεν ἀτμόν (aor. 2) Nonn.D. 1.239, ἰωήν ib. 485; of rivers, εἰς θάλασσαν Arist.Mu. 392ᵇ16, A.R. 2.744.

ἀνερευνάω, before 'POxy. 1468.18' read 'ἀνεραυνῶσα .. τὰ ἔγγραφα'

ἀνερεύνητος 1, add 'unexplored, Peripl.M.Rubr. 18, 66'

ἀνερίθευτος, add 'adv. -τως OGI 7.4'

*ἄνερμος, ον, unexpld. wd., neut. pl. subst., ἄνερμα τοῦ ἱ[ερ]οῦ ἀργύρου IG 2².1544.24 (Eleusis, iv BC).

ἀνέρπω, for 'Call.Ap. 110' read 'Call.Ap. 111'; add 'w. acc., climb, τεῖχε' ἀ. Arat. 958'

ἄνερσις, εως, ἡ, (ἀνείρω) stringing up, cj. in Plu. 2.156b.

ἀνέρχομαι I 2, add 'b = ἀνέχω B 2, jut out, D.P. 400.'   add '4 rise in rank or status, τί γὰρ ὅτι πρὸς ἱερωσύνην ἅτερος αὐτῶν ἀνῆλθεν; Just.Nov. 89.9 pr.'   II 1, add 'come up from a village to a capital, CPR 8.29.4 (iv AD)'   3, for 'brought home to you' read 'brought before your bar' and for 'ἀνερχομένω is corrupt' read 'ἀναερχ- is v.l. for ἐπανερχ- and unmetrical ἀνερχ-'

ἄνεσις 2, line 3, after 'Plu. 2.102b, etc.' insert 'rest, of the dead, BCH suppl. 8 no. 180, ICilicie 35'; line 5, after '(Thisbe)' insert 'exemption from liturgy, BGU 2474.10 (ii/iii AD)'

ἀνέσουτο, for 'pass.' read 'med.'

ἄνεστιος, for 'metaph.' to end read 'of crayfish, Opp.H. 2.417; transf., of the soul, Max.Tyr. 14.8 (8.8 H.).   2 unoccupied, tenantless, ὄστρακον, οἰκητῆρος ἀνέστιον οἰχομένοιο Opp.H. 1.325.'

ἄνετος, add 'Myc. a-ne-ta (neut. pl.), of dues, remitted'

ἄνευ, last line but two, for 'as always in Arist.' read 'e.g. Arist.'

ἀνευρετής, add within parenthesis 'iv BC'

*ἀνευρησιλογήτως, adv., without subterfuge, SB 7663.17 (i AD).

*ἀνευσυνθετέω, fail to observe a contract, prob. in BGU 1738.32 (i BC).

†ἀνεφάλλομαι, leap up at, ἀνέπαλτο Il. 20.424, ἀνεπάλμενος A.R. 2.825 (see also ἀναπάλλω).

*ἀνεφάμιλλος, ον, unrivalled, καλοκἀγαθία prob. in PAmh. 2.145.6 (iv/v AD).

†ἀνέφαπτος, ον, that cannot be reclaimed, of manumitted slaves, SEG 26.720.13 (Epirus, iii BC), 32.623 (Bouthrotos, Illyria, ii BC), IG 9²(1).705.10, GDI 1684.6 (both Delphi, ii BC).

ἀνέφικτος, after 'unattainable' insert 'Aristeas 223'

*ἀνεφόρητος, ον, app. not watched over (by evil), unenvied, ἀνεφόρ[η]τος εὐδαίμων SEG 19.372(b).7 (Thespiae, Rom.imp.); cf. ἐφοράω 2.

ἀνεχέγγυος, for 'unwarranted' read 'not relied upon, not regarded as reliable' and delete 'because .. themselves'

ἀνέχραζεν, add '(ἀνέχραυεν· ἀνέχριμπτεν, ἀνήρειδεν, ἐκούφιζεν La.; cf. ἐγχραύω)'

ἀνέχω, line 12, after '(v. infr. C II)' insert '3 pl. pf. ἠνέσχονται PKöln 240.11 (vi AD)'   A I 5 b, at end for 'his anger' read 'a show of anger'   II, after 'Th. 6.86' insert 'Ar.fr. 632 K.-A., D.Prooem. 41.3'   C II 3, after 'Od. 22.423' insert 'E.Andr. 340, Tr. 101' and after 'D. 19.16' insert 'M.Ant. 5.33.6'

ἀνεψιαδοῦς, after 'ὁ' insert '(-ιδοῦς Sch. A.R. 3.359)'; for 'first cousin's son' read 'first cousin once removed (not always dist. fr. second cousin, cf. Poll. 3.28)'; delete all after 'Is. 11.12' inserting '45.54' after 'D. 44.26'

ἀνεψιός, add 'first cousin once removed, D. 43.41, 49; ἀνεψιῶν παῖδες second cousins, Is. 7.22, 11.2, D. 43.51; also sg. ἀνεψιοῦ παῖς Is. 11.10, al. (v. Glotta 48.76)'

ἀνεψιότης, after 'cousins' insert 'IG 1³.104.15, 21 (409/8 BC)'

ἄνεω, add 'but found w. sg. verb, Od. 23.93'

*ἄνεως· ἄφωνος καὶ τὸν νοῦν ἐκπεπληγμένος Gal. 19.81.

†ἄνη, Dor. ἄνᾱ Alcm. 1.83 P. (ἄν-, unless ἄνα = ἀ ἄνα is read), ἡ, fulfilment, Alcm.l.c., A.Th. 713, CallJov. 90.

ἄνηβος, line 2, after 'cf.' insert 'Sol. 27.1 W.'

ἀνθοποίητος 1, add 'n. sg. subst. Marcellin.Vit.Thuc. 57'   2, for 'unprincipled' read 'morally unformed' and for 'Cic.Att. 10.10.5' read 'Cic.Att. 10.10.6'

ἀνήκεστος, before 'incurable' insert 'Dor. ἀνᾱκ- B.fr. 20 D.9 S.-M.'

ἀνηλάκατος, for 'unable to spin' read 'lacking a distaff'

ἀνηλεγής, after 'Q.S. 2.414' add 'and read by Hdn. in A.R. 1.785'

ἀνήλειπτος II, add 'Nicom.Trag. 16 S.'

ἀνήλειπτος, ἀνήλιφος, at end after 'Hp.Ep. 17' add '(ἀνήλιπος cod.)'

†ἀνηλιοδείκτης, ου, one that points the direction without the aid of the sun, PMag. 4.1374.

ἀνήλιος, add 'Plu. 2.330d, Luc.Nec. 9, Luct. 2, ἀ. οἶκος Gal. 13.778.2'

ἀνηλιποκαιβλεπέλαιος, for 'unanointed (?)' read 'having a lamp-oil look (i.e. bleary-eyed from reading late)' and add '(v. FGE p. 476 P.)'

ἀνήλιπος, add 'EM 107.14, Hp.Ep. 17.6 (cod., cj. ἀνήλιφος)'

*ἀνηλίπους, ποδος, ὁ, ἡ, barefooted, IKios 19.3 (ἀνιλίποδες, i AD).

*ἀνηλόω, = ἀναλόω, v. ‡ἀναλίσκω.

*ἀνήλωσις, v. ‡ἀνάλωσις.

ἀνήλωτος, after 'not nailed' insert 'καλίκων στρατιωτικῶν ἀ. (Lat. caligae militares sine clabo) Edict.Diocl. 9.6'

ἀνημέρωτος, for 'untilled' read 'uncleared'

*ἀνηνεχυῖαν· ἀναφέρουσαν Hsch.

ἀνήνυτος 1, after 'E.Hel. 1285' insert 'Critias Trag. 1.14 S.'   2, neut. pl. as adv., add 'πονοῦσιν Pl.R. 531a, Theoc. 15.87'

ἀνηπελίη, for 'εὐηπελίη' read 'εὐηπελής'

ἀνήρ, line 1, after 'ἀνδρί' insert '(Pamph. ἀδρί Hsch.)'; line 11, before 'beast' insert 'god or'   III, add 'ἀνήρ, of statue personifying this athletic category, RA 1987.99 (Hierapolis, Phrygia, iii AD)'   VI 1, add 'transl. of Lat. vir in honorary titles and designations, ἀνδρὸς διασημοτάτου IEphes. 3217 (ii AD); τοὺς τρεῖς ἄνδρας (triumviros) Plb. 3.67.6'

*ἀηροσίη, ἡ, lack of cultivation, Orac.Sib. 3.542.

ἀνήσσητος, add 'ASAA 30/32(1952/4).290 no. 66.4 (Rhodes)'

ἀνθαιρέομαι I, add 'D.C. 43.46.2, 48.32.3'

ἀνθαιρετιστής, add 'PMich.III 149 xviii 16 (pl., ii AD)'

*ἀνθαιρετιστικόν, τό, astrol., exchange of qualities between the "conditions", PMich.III 149 viii 28; cf. αἵρεσις B II 4.

*ἀνθαιρετός, όν, favourite, as name of a horse, SEG 7.213.28 (tab. defix.; Palmyra, ii/iii AD).

ἀνθάπτομαι, line 4, after 'Th. 8.97' insert 'cf. περὶ τῆς μισθοφορᾶς .. μαλακωτέρως ἀνθήπτετο (sc. τοῦ πράγματος) 8.50.3 (or delete περί leaving μισθοφορᾶς as gen. constr. w. vb.)'   II 2, delete all after 'E.Med. 55, 1360'

*ἀνθαρπάζω, seize by way of reprisal, Hell.Oxy. 8.3 M.-K.

ἀνθέμιον 2, for 'IG 1.322' read 'IG 1³.474.47'   3, for 'of gold .. quality' read 'dub., Hebr. in this passage app. means bowl of a lamp'

ἀνθεμίς, add '3 in temple, perh. either flower-shaped support for dedication or floral ornament (cf. ἀνθέμιον 2), Ath.Asklepieion IV 66.'

ἄνθεμον, add '5 dub. item in temple account, cf. ἀνθεμίς, IG 1³.472.142, 146, 149.'

ἀνθεμωτός, add 'Inscr.Délos 1439Abci47 (ii BC)'

*ἀνθερίκη, ἡ, = ἀνθέρικος (asphodel-stalk), cj. in AP 12.121 (Rhian.) (codd. ἀθερ-).

ἀνθέρικος I 1, after 'Thphr.HP 7.13.2' insert 'Theoc. 1.52' **II**, for this section read 'applied to sim. flowering stalks, ἀνθέρικον νῦν εἴρηκε τὸ ἄνθος τῆς σκίλλης· κυρίως γὰρ τὸ ἄνθος τῶν ἀσταχύων Sch.Arat. 1060'

†ἀνθέριξ, ἴκος, ὁ, flowering stem of asphodel, Il. 20.227, Hes.fr. 62 M.-W.

Ἀνθεστήρια, line 3, for '(Teos)' read '(Teos, v BC), SEG 31.985.D6 (Abdera, v BC)'

ἀνθεστηριάδας, add '(ἀνθεστρίδες La.)'

*ἀνθεστρίδες, αἱ, perh. marriageable girls, Tit.Cam. 148; cf. ‡ἀνθεστηριάδας.

ἀνθέω I, add '3 have an efflorescence or secretion, Hp.Morb.Sacr. 5.2.'

ἄνθη, add '3 efflorescence, χαλκοῖο πάλαι μεμογηότος ἄνθην (i.e. verdigris or perh. cuprous or cupric oxide) Nic.Al. 529.'

ἀνθήεις, after 'εσσα, εν,' insert 'abounding in flowers, ἄκανθος Nic.Th. 645'; for 'bright-coloured' read 'coloured as though with flowers'

†ἀνθηλᾶς, ᾶ, ὁ, perh. flower- or papyrus-merchant, PLond. 387.21, PPrag. I 25.8 (both vi/vii AD) (v. °ἀνθήλη).

ἀνθήλη, after 'Dsc. 1.85' insert 'Str. 5.26, PUniv.Giss. 13.4 (refers in both of these to commodity, perh. head of the papyrus or similar plant, v. ZPE 75.155-6)'

*ἀνθηρεύομαι, to be in bloom, Hsch. s.v. χλοάζει.

ἀνθηρός **II 1**, line 3, after '(Plu.Pomp.) 2.50b' insert 'blooming, τά τε εἴδεα τῶν ἀνθρώπων εὔχροά τε καὶ ἀνθηρά ἐστι Hp.Aër. 5'

*ἀνθιεράομαι, to be ἀνθιερεύς, IG 2².1368.5 (Athens, ii AD).

*ἀνθιερεύς, έως, ὁ, deputy-priest, IG 2².1368.9, 27, al. (Athens, ii AD).

Ἀνθιηνός, ὁ, cult-title of Apollo, IGBulg. 1723.

*ἀνθίνης, ου, of wine, flavoured with flowers, Gal. 19.81 (s.v.l.), Hsch.

ἀνθίστημι **II 1**, line 7, after 'X.Smp. 5.1' insert 'ἐὰν δέ τις περὶ τούτου πράγματος ἀνθίστηται IEphes. 4101.4 (i/ii AD)'

Ἀνθιστήρ, ῆρος, ὁ, a divinity, perh. Dionysus, IG 12(3).329 (gen. Ἀνθισ[τῆ]ρος, Thera, iii/ii BC); cf. Ἀνθεστήρια.

*ἀνθοβατέω, walk on flowers, μαλακὴν ἀνθοβατεῦσα πόη[ν Honestus in SEG 13.344(i) (epigr., Thespiae, i BC/i AD).

ἀνθοβαφής, delete 'γῆ IG 7.1802'

ἀνθοκομέω, add 'Cyr.Al. in PG 73.620A.'

ἀνθοκόμος 1, add 'βελέμνῳ (of Eros), Nonn.D. 7.194, μάστιγι (of Dionysus) ib. 17.20'

*ἀνθοράω, = ἀντιβλέπω, ἀντοψόμεθα τὰ θεῖα ἐναργῶς Alb.Intr. 5 (p. 150.12 H.), dub. in PFreib. 2.5 in NGG 1922.33.

*ἀνθοριστικός, ή, όν, belonging to a counter-definition, Fortunat.Rh. 1.13, An.Ox. 4.15.25.

ἄνθος (A) **I 2**, after 'surface' insert 'κύματος ἄ. Alcm. 26.3 P.' **II 1**, line 3, after 'Thgn. 994' insert 'χροιῆς ἄνθος ἀμειβομένης Sol. 27.6 W.'; line 5, after 'id.R.601b' insert 'in epitaphs, metaph., of children, ἔνθα κατάκιτε νέον ἄνθος SEG 36.1183 (Galatia, v/vi AD), Ποπλικία .. ῥόδον ἐρχόμενον ἄνθος TAM 5(1).481.6' add '3 finest quality, flower, καὶ σὺ λαβὲ ἡδύσματα, τὸ ἄνθος Lxx Ex. 30.23.'

ἀνθοσμίας, for 'Pherecr. 108.30' read 'Pherecr. 113 K.-A., Longus 4.10.3'

*ἀνθοτόκος, ον, bringing forth flowers, Bernand IMEG 20.6 (ii BC).

ἀνθρακεύς, after 'charcoal-maker' insert 'Men.Epit. 81'

*ἀνθρακηγός, όν, charcoal-carrying, BGU 2353.12 (ii AD).

ἀνθρακηρός, for these refs. read 'Inscr.Délos 509.40 (iii BC), Alex. 211 K.-A. (m. pl. subst.)'

ἀνθρακίδες, add 'cf. ἐπανθρακίδες'

ἀνθράκινος 2, add 'cf. anthracina, charcoal-coloured garments, Varro in Non.Marc. 550 M.'

*ἄνθρακις, εως, ἡ, carbuncle, PLond. 77.28 (vi AD).

ἀνθρακίτης, for 'gem' read 'mineral'

ἀνθρακῖτις, for 'coal' read 'gem-stone'

*ἀνθρακοβάτης, ου, ὁ, one who walks on coals, Ps.-Steph. in Physici 2.201.15.

ἄνθραξ **I**, add 'ἄνθρακες ὁ θησαυρὸς πέφηνεν, of illusions destroyed, Zen. 2.1, cf. Luc.Philops. 32, Tim. 41, al.'

ἄνθρυσκον, for 'chervil, Scandix australis' read 'plant, perh. one or more species of Anthriscus'

ἀνθρωπεύομαι, for 'Herm.ap.Stob. 1.41.68' read 'Corp.Herm. 25.8 N.-F.'

*ἀνθρωπιμαῖος, α, ον, humane, Sch.Gen.Il. 23.98.

ἀνθρώπινος **I**, after 'Lys. 6.20' insert 'κατὰ τὸ ἀνθρώπινον Th. 1.22.4, οὐκ ἀνθρωπίνης δυνάμεως βούλησιν ἐλπίζει 6.78.2' **II**, adv., line 4, for 'And. 2.6' read 'And. 1.57, 2.6'; delete 'Of the three' to end.

ἀνθρωπόδηκτος, add 'n. pl. subst., wounds caused by the bite of a man, Gal. 13.432.3, 558.3, al.'

ἀνθρωπόθεν, before 'humanitus' insert '(parox.)'

ἀνθρωπόλεθρος, after 'Suid.' insert 'Sch.Gen.Il. 21.421'

*ἀνθρωπομορφίτης, ου, ὁ, anthropomorphic god, Isid.Etym. 8.5.32.

*ἀνθρωποπρεπής, ές, befitting a man, PMonac. 8.5 (vi AD).

ἄνθρωπος **3 a**, after 'Paus. 4.26.5' insert 'Sopat.Rh. 8.160.15 W.' **6**, at end for 'simply, brother' read 'without derogatory sense, ἄνθρωπε ἡὸ(ς) στείχε[ι]ς καθ' ὁδὸν CEG 28 (Attica, vi BC)' **7**, line 2, after '1Ep.Tim. 6.11' insert 'so prob. τῆς θεοῦ τὸν ἄ. Call.fr. 193.37 Pf.' **II**, last line, delete 'Aeschin. 3.137' at end add 'Myc. a-to-ro-qo'

ἀνθῠλακτέω, for 'Ael.HA 4.19' read 'Ael.NA 4.19'

*ἀνθυλλοπράτης, ου, ὁ, florist, PMasp. 156.8 (ἀνθυλο-, vi AD).

*ἀνθυλλοπρατική, ἡ, trade of florist, PMasp. 156.12 (ἀνθυλο-, vi AD).

*ἀνθυλλοπράτισσα, ἡ, fem. of -πράτης, PMasp. 156.5 (ἀνθυλο-, vi AD).

ἀνθυπαλλάσσω, add '**II** pledge as security for a debt, PIand. 142 i 13 (pass., ii AD).'

ἀνθυπηρετέω, add '**II** act as deputy ὑπηρέτης, IG 2².1945.5.'

*ἀνθυπήχησις, εως, ἡ, answering echo, [τῆς λύρα]ς PBerol.inv. 13426.18ᵛ in Gercke-Norden Einl. in die Altertumswiss. 1(9) p. 42 (T(L) 1924).

ἀνθυποφαίνω, delete the entry.

ἀνιάζω, at end delete 'metri gr.'

ἀνιαρίζω, delete 'Dor. for ἀνιερίζω' and add 'cf. ἱερίζω'

*ἀνιαροποιός, ον, causing painful things, ἀνιαρὲ ἀντὶ τοῦ ἀνιαροποιέ Sch.Theoc. 2.55.

*ἀνίασις, εως, ἡ, sadness, Sm.Ez. 23.33.

ἀνίατος **I 1**, add '**b** uncured, ἐξάρθρημα Gal. 10.220.15.'

ἀνιάχω, after 'cry aloud' insert 'E.Or. 1465 (lyr.)'; add '[ῑ except where augmented]'

ἀνιγρός, delete 'Call.Iamb. 1.164 (prob.)' and for 'Call.Aet. 3.1.14' read 'Call.fr. 75.14 Pf.'

*ἀνιεράω, dedicate, perh. in Myres Cesnola Coll. p. 548 no. 1903 (Cyprus, i AD).

ἀνιερόω, after lemma insert '(Arc. 3 sg. aor. opt. ὑνιερόσει, SEG 11.1112, vi/v BC)'

*ἀνίημα, ατος, τό, grief, Baillet Inscr. des tombeaux des rois 1087.

ἀνίημι **II 7 a**, line 1, after 'relax' insert 'τὰς ὀφρῦς ἄνες ποτ' Men.Dysc. 423' **8**, after 'Hdt. 2.113, 4.152' insert 'of snow, E.Ba. 662, χειμῶνος ἀνέντος Theoc. 18.27' **b**, after '(Hdt.) 2.121.β'' insert 'Th. 1.129.3'

*ἀνικάνέω, not to meet the requirements, SB 9066.23 (ii AD).

ἀνίκητος **I**, add 'as epith. of var. divinities, SEG 24.903 (Thrace, ii/iii AD), 36.616 (Edessa, iii AD), of Roman emperors (Lat. invictus), SEG 33.18 (Athens), 23.199 (Taenarum, both iii AD)'

*ἀνίκλιον, τό, unidentified garment, SEG 7.417 (Dura, iii AD).

*ἀνικμαίνω, = ἀνικμάζω, cj. in Nic.Al. 524.

ἀνίκμαντος, after 'Lyc. 988' insert '(cj.)'

*ἀνίλαστος, add 'PMag. 4.1776'

*ἄνίλατος (sp. ἀνείλατος), ον, implacable, IGLS 1.119, 212, 47 iii 7 (unless to be taken as ἀνήλατος) (i BC).

ἀνιμάω, read 'ἀνῑμάω (cf. καθῑμάω)'; for 'draw up, raise water .. (ἱμάντες)' read 'draw up from well '; after 'Sor. 1.93' insert 'Iamb.Protr. 21 (p. 105.24 P.)'

ἀνίμησις, add 'for the public baths, POxy. 2569.5-6 (iii AD).'

*ἀνιμητής, οῦ, ὁ, one who draws water, for the public baths, SB 10555.5, 8, al. (ἀνειμητ-, iii AD).

*ἀνίουλος, ον, not having the first growth of hair on the cheeks, Nonn.D. 11.373, 38.167.

ἄνιππος **1**, for 'without a horse to ride on' read 'not having a horse' and add 'Pi.Pae. 4.27'

ἄνις, add 'rest. in Call.fr. 3.1 Pf.'

*ἀνισόκυκλα, τά, app. system of gears, as used in a hoisting machine, Vitr. 10.1.3.

*ἀνισομεγέθης, ες, of unequal size, Hipparch. 2.1.12.

ἄνισος, add '**2** unequalled, ἄνισος ἀρεταῖς παναρχόντων γόνος Diosc. in GDRK 17.6.'

ἀνισοσθενής, for 'Gal. 5.415' read 'Gal. 4.415.11'

*ἀνισότονος, for 'not in unison, Ptol.Harm. 2.2' read 'music., of unequal pitch, φθόγγοι Ptol.Harm. 2.2, 3.2; varying in pitch, ψόφοι ib. 1.4'

ἀνίστημι **A**, line 2, aor. 1, add 'ἀνέστασε (3 sg., dial.) SEG 27.1116 (Egypt, v BC), ἠνέστησα SEG 6.377 (Lycaonia, Chr.)' **3**, raise from the dead, add 'Mitchell N.Galatia 87, cf. sense B 3' **5**, line 5, after 'Brut. 1' insert 'in funerary inscrs. sts. w. dat. of pers., acc. of monument, etc., not expressed, Mitchell N.Galatia 22' **B**, line 3, imper., add 'ἄστα SEG 34.994 (vase, Apulia, late iv BC), ἄνστα Theoc. 24.36' **I 5**, delete 'Τυφῶνα .. codd.'

*ἀνιστόρητος **II**, add 'unheard of, i.e. nothing like it before, [τῆς] σκληρᾶς σιτοδείας ἐκ τῆς γενομένης ἀ. ἀπορίας SEG 24.1217.10 (Thebes, Egypt, i BC = OGI 194)'

ἀνίσχυς, add 'as subst., ἐπὶ ἀφ[ανίσει] καὶ ψύξει καὶ ἀνίσχυϊ tab.defix. in SEG 35.216.18 (Attica, iii AD)'

†ἀνίσωμα (A), ατος, τό, equalizing portion or part, of wine, Ath. 10.447a; of masonry, Inscr.Délos 507.4, 5 (iii BC).

*ἀνίσωμα (B), ατος, τό, unevenness in the tooth of young horses, Hippiatr. 95.15 O.-H.

*ἀνκεμολία, v. °ἀγχεμωλία.

*ἀννάλιος, ον, Lat. *annalis, annual, CodJust.* 1.3.45.9, neut. pl. as subst., ib. 1.3.55.1.

ἄννησον, after 'Dsc. 3.56)' insert 'ἄννησσον *Inscr.Délos* 440*A*64 (ii BC)'

*Ἄννιος, v. °Αἰνικός.

ἀννίς, add '*IG* 7.3380 (Boeotia; parox. acc. to ed.)'

*ἄννω, ἡ, v. °νανννω.

*ἀννῶνα, (or ἀννώνα), ἡ, Lat. *annona, corn-supply, PTeb.* 404.12 (iii AD), *IGRom.* 3.409; also ἀννώνη *IGLS* 262.12, *OGI* 200.16, *POxy.* 1192.4 (iii AD); gen. pl. ἀννόνων *CodJust.* 1.4.18, 1.44.1; ἄννονα or ἀννόνα Suid.

*ἀννωνάριος, ὁ, *official administering the annona, ZPE* 22.101.26 (iv AD).

ἀννωναρχέω, add '(= *TAM* 4(1).189.8)'

*ἀννωναταμίας, ου, ὁ, = *praefectus annonae*, rest. in *IEphes.* 4330.10 (iii AD).

ἀννωνέπαρχος, add '*CodJust.* 1.44.2 (ἀννον-)'

†ἀννώνη, v. °ἀννῶνα.

ἀννωνιακός, add '-ιακή, ἡ, *corn-supply, PWisc.* 32.13 (i AD), *PMichael.* 40.20 (vi AD)'

ἀννωνικός, add '-ικόν, τό, *bread, BGU* 2358.7 (iv AD), *CPR* 7.42 i 2, al. (v AD), *Stud.Pal.* 8.978.2 (vii AD)'

*ἄννως, -ως, ἡ, kinship term, perh. *grandmother, IG* 9(2).877 (Larissa); cf. ἄννις, Lat. *anus*.

ἄνοδος (B), add 'V *advancement, promotion*, Just.*Nov.* 6.5.'

ἀνόδυρτος, add '2 *unmourned, Salamine* 199 (?i AD).'

ἀνόζεστα, after 'fort. ἀπόζεστα' insert 'ἀνόζωτα La.'

ἀνοηταίνω, after 'Pl.*Phlb.* 12d' insert '*Ep.* 359c'

*ἀνοητίζω, *to be stupid*, Aq.*Je.* 10.8.

ἀνόητος I 2, after '*province of thought*' insert 'Parm. 8.17 D.-K.'

*ἄνοθεν, adv. = ἄνωθεν, *PHib.* 110.66, al. (iii BC); cf. ἔξοθεν, κάτοθεν.

ἄνοια, after 'Thgn. 453' insert 'Ion. and poet. ἀνοίη, Hippon. 75.3 W.; Aeol. ἀνοΐα, Alc. 112.1, 119 L.-P.'

ἄνοιγμα I, add '*opening* from water-pipe, *tap, SEG* 31.953.A12, B5, 10 (Ephesus, ii AD), perh. *sluice-gate, Syria* 62.83.17'

*Ἀνοιγμοί, οἱ, festival at Miletus, *Didyma* 382.8, 385.1.3, al.

ἀνοίγνυμι, line 10, before 'v. infr.' insert 'also intr. **I 2**, for 'ἔργ᾽ ἀναιδῆ' read 'μὴ .. ἀνοίξῃς .. ἃ πέπονθ᾽ ἀναιδῶς'

ἀνοικοδομέω II, line 2, after 'etc.' insert '*SEG* 31.1548 (Egypt, ii AD)'

ἄνοιξις, add '2 *opening* of a business, πρὸς ἄνοιξιν καπηλείου *POxy.* 2109.10 (AD 261). **II** *aperture* of a garment, ἀνύξεως καὶ ὑπορραφῆς ὁλοσειρικοῦ ἱματίου *Edict.Diocl.* 7.49 (ἀνυξ-).'

ἀνολβέω, after lemma insert 'Ep. -είω'

ἀνόλβιος, after '= sq.' insert '*BCH* 85.848 (funerary epigr., Paros)'

ἄνολβος 1, line 4, before 'Thgn.' insert 'Archil. 88 W.'

ἀνολισθάνω, for 'Call.*fr.* 96' read 'Call.*fr.* 191.75 Pf.'

ἀνομία 1, add 'ἐξάλιψο(ν) τὰς ἀ. μου καὶ τὰ παραπτώματά μου *SEG* 31.1562.5 (?Egypt, Chr., iv/v AD); cf. °ἀνόμιον'

ἀνόμιμος, add '*PMag.* 36.140 (iv AD)'

*ἀνόμιον, τό, *transgression, wrong-doing*, ἐξάλιψον τὸ ἀνόμιον αὐτῆς Guarducci *EG* 4 p. 459 no. 2.12 (Nubia, Chr., v AD); cf. °ἀνομία.

ἀνόμοιος, after 'Arist.*Pol.* 1277a5' insert 'τῆς ἀνομοίου μίξεως σύμβολον Hld. 3.14.2' add '**3** *not equal to, not suited for,* ἄπορος .. καὶ ἀνόμοιος πρὸς τὴν λ[ει]τουργείαν καὶ τὴν ὑπηρεσίαν *PPetaus* 12.9 (ii AD).'

ἀνομοιόω I, after 'Pl.*Tht.* 166b, al.' insert 'Procl.*Inst.* 36, 110 D.; act. in this sense, Hsch. s.v. ἀπεοίκασιν'

ἀνομολογέομαι **I 1**, line 3, after '[Pl.]*Tht.* 164c' insert 'ἀνομολογησάμενος παρ᾽ αὐτοῦ *getting his agreement,* Pl.*Smp.* 199b' **2**, add '[Pl.] *Amat.* 136e'

*ἀνομολογέω, *disagree*, Porph.*Gaur.* 13.7.

ἀνομολόγητος **II**, after 'Ptol.*Tetr.* 47' insert '(v.l.)'

ἀνομόλογος, after 'S.E.*M.* 8.331' insert 'Ptol.*Tetr.* 47'

ἀνομολογούμενος, delete 'for a verb ἀνομολογέομαι .. does not occur'

ἄνομος **I 1**, line 2, after 'E.*Ba.* 995' insert '*IHadr.* 100 (epigr.)'

*ἀνόνειρος, ον, *dreamless*, ὕπνος Porph.*Gaur.* 12.7.

ἀνόνητος **I**, for 'Cerc. 4.4' read 'Cerc. 4.6'

ἄνοος, line 1, after 'ουν' insert 'also nom. pl. ἄνοες in *SEG* 35.220.17 (Attica, iii AD; cf. nom. pl. εὖνους)'

ἄνοπλος, line 2, after 'Pl.*Euthd.* 299b' insert 'Ion Trag. 53e S., Ezek.*Exag.* 210 S., *AP* 9.320 (Leon.Tarent.)'; at end for 'on the form v.' read 'cf.'

*ἀνοράσσω, var. of ἀνορύσσω, *Inscr.Cret.* 1.xviii 64.4 (Lyttos, iii AD).

ἀνοργίαστος **1** and **2**, for 'orgies' read 'mystery rites' **II**, add 'Iamb.*Protr.* 21 κη᾽ (p. 122.18 P.)'

†ἀνορίνω, Aeol. impf. ὀννώρινε, *disturb*, παίσαις .. ὁ. νύκτας Alc. 72.9 L.-P.

*ἀνορυγή, ἡ, *digging up, excavation, PRyl.* 95.8 (ἀναρ-, i AD); θεμελίου *SAWW* 149(5).14 (iv AD).

*ἀνόρυξις, εως, ἡ, *digging up, SB* 4774.9.

ἀνορύσσω **1**, add 'σάρκας ἀ. *EA* 13.21 no. 6 (Phrygia or Lydia)' at end add 'see also °ἀνοράσσω'

*ἀνορυχή, ἡ, *digging*, prob. in *PMasp.* 283.1.16 (vi AD).

ἀνοσία, after lemma insert '(A), Ion. -ίη' **I**, add 'Hp.*Praec.* 6' **II**, delete the section.

*ἀνοσία (B), (ἀνόσιος), ἀ, Cypr. *a-no-si-ja, impiety*, ἀνοσία Ϝοι γένοιτυ *ICS* 217.29 (Idalium, v BC).

ἀνοσιότης, add '*CodJust.* 1.11.10'

ἀνοσιουργία, add 'Iamb.*Protr.* 13 (p. 71.4 P.)'

ἄνοσμος, add 'of wine, *PZen.Col.* 108.5 (iii BC)'; for 'but ἄοσμος (q.v.) was preferred' read 'cf. ἄοσμος'

ἀνόστεος, add '**2** *shell-less*, of eggs, Nic.*Al.* 296 (codd., ἀνόστρακα G.-S.).'

ἀνόστητος **II**, add 'ὅροι *Trag.adesp.* 658.17 K.-S.'

ἄνοστος **II**, before 'Thphr.' insert 'γένοιτο αὐτῷ τὰ νόστιμα ἄ. *SEG* 6.802.23 (Salamis, Cyprus)'

ἀνούατος, for '*ear*' read '*ears*' and delete '*without handle*'

*Ἄνουβις, ιδος, (-ιος *Inscr.Délos* 2098, acc. -ιν D.S. 1.18, voc. Ἄνουβι *IKios* 21) ὁ, the Egyptian dog-headed deity *Anubis*, Call.*fr.* 715 Pf., *Inscr.Délos* 2039 (ii BC), *POxy.* 1256.12 (iii AD).

ἀνούλεγοι, after lemma insert '(ἀνούλιοι La.)'

ἀνοχ[ικ]ός **I**, after lemma insert '(ἀνοχαῖος La.)'

ἀνόχυρος, delete the entry.

*ἀνπανάμενος, v. ‡ἀναφαίνω I 3.

*ἄνπανσις, ἀνπαντός, v. ‡ἀμφαν-.

† ἄνπερ, v. °ἐάνπερ.

*ἀνπιδήμα, v. °ἀμφιδήμα.

*ἄνπυλλα, v. °ἄμπυλλα.

ἄνσατον, after lemma insert '(ἀνσάττεν La.)'

ἀνταγωνία, delete the entry.

*ἀνταγώνισμα, ατος, τό, gloss on ἅμιλλα, Phot. α 1201 Th.

ἀνταγωνιστής, add 'τῶν εὐτυχούντων ἀνταγωνιστὴς φθόνος *Trag.adesp.* 167 K.-S.'

*ἀνταγωνοθετέω, *act as deputy to* ἀγωνοθέτης, *IG* 10(2).133.5 (ii AD).

ἀνταίρω **I**, add '**2** ἀνταίρουσιν· ἀντιλέγουσι· Σοφοκλῆς Θυέστῃ (fr. 265 R.) Hsch.'

ἀντακαῖος 2, for 'ἄν καῖον' read 'ἀντακαῖον'

ἄνταλλος, add 'also as pers. n. Ἄνταλλος *IG* 2².6514, etc. (v. Robert *Hell.* 11.208 n. 2)'

ἀντἀμείβομαι **III**, delete '*again*'

ἀντάμειψις, after '*requital*' insert 'Aristeas 259, *BGU* 1816.18 (i BC)'

†ἀντἀμοιβός, όν, (epic form ἀντημ-) *substitute, given in exchange*, Call.*Del.* 52. **2** *requiting*, τὰ ἀ., *reprisals, PMasp.* 151.257 (vi AD).

†ἀνταναιρετέος, ον, *to be set off against* (a sum), *PTeb.* 61b.220 (ii BC).

ἀνταναιρέω, add '**5** *counteract with equal strength*, ἰσχυρὸν στόμα ἔχει, ὡς καὶ σιδηρᾶ ἀνταναιρεῖν, of parrot's beak, Cyran. 3.52.4 K.'

*ἀντανάπαιστος, ὁ, metrical foot (∪--∪∪), Diom. 1.481.29 K.

*ἀντανατέλλω, astrol., *rise in opposition, PMich.*III 149 xi 5 (ii AD).

*ἀντανἀφορά, ἡ, *official counter-report, BGU* 1859.B.11 (i BC).

ἀντάξιος **2**, add 'Ach.Tat. 3.23.1 V., Hld. 10.18.3'

†ἀνταπἀμείβομαι, *answer in turn*, Tyrt. 4.6 W., Call. in *Suppl.Hell.* 238.8.

ἀνταποδίδωμι **I**, delete '2 *take vengeance*' and include exx. in section I 1.

ἀνταπόδοσις **II 1**, add 'ἀ. τῶν ἐκλείψεων *return* of eclipses, *Cat.Cod.Astr.* 8(2).126.6'

†ἀνταπόποινα, v. ἀντίποινα.

*ἀνταπόμνυμι, 3 pl. imper. -υόντων, *make a counter-declaration on oath, SEG* 9.1.11 (Cyrene, iv BC).

*ἀνταποπάλλω, of earthquakes, *thrust back in the opposite direction* (opp. single shock), Arist.*Mu.* 396a8.

ἀνταποστέλλω, line 3, for 'an echo' read 'a reflection'

ἀνταποφαίνω, add 'med., ἀ. γνώμην *put forward a counter*-proposal, J.*AJ* 19.2.2'

ἀνταρκέω **II**, after 'c. part.' insert 'ἀκμάζων τῇ ὠκύτητι ἀντήρκεσε Paus. 6.13.4'

ἀνταρσία, add 'ἀνταρσίας γενομένης ἐν τοῖς Ἀθηναίοις Arist.*fr.* 667 R. (*Vita Arist.* vulg.)'

ἄνταρσις, add 'Sm.*Is.* 8.12'

*ἀνταρχιδικαστής, οῦ, ὁ, *deputy* ἀρχιδικαστής, *PSI* 1105.5 (ii AD), 1255.5 (iii AD).

*ἀνταρχιερεύς, έως, ὁ, *deputy high-priest, SB* 9016.1.1, 2.1 (ii AD), *POxy.* 3026.19 (ii AD).

ἀνταυλέω, for '*flute*' read '*pipe(s)* or *aulos*'

*ἀντἀχάτης λίθος, = ἀντιαχάτης, Socr.Dion.*Lith.* 40 H.-S., Plin.*HN* 37.139.

ἀντέγκλημα, after 'or *-charge*' insert 'Quint.*Inst.* 7.4.8'

*ἀντειρήναρχος, ὁ, *deputy* εἰρήναρχος, *SEG* 6.700, 711 (Pamphylia).

*ἀντειρωνεύομαι, *use* εἰρωνεία *in reply*, Sch.Gen.Il. 1.132.

ἀντεισάγω **I**, add '2 *bring in as replacement* for shortfall, Just.*Nov.* 43.1 pr.'

ἀντεισᾰγωγή, add '2 replacement of shortfall in funds, Just.Nov. 43.1 pr.'

ἀντεισοδιάζω, add 'SEG 14.639.E20 (sp. ἀντισ-, Caunus, i AD; in uncertain sense, app. w. ref. to entering port)'

*ἀντεισπράττω, exact in return, SEG 29.127.E82 (ii AD).

ἀντέκτιστος, delete the entry.

*ἀντέκτιτα, τά, gloss on ἀντίτιμα, Hsch. (ἀνέκτιστα cod.).

*ἀντελία, ἁ, = ἀντολίη, Boeot.adesp. 41 fr. 2.3 P. (pl.).

ἀντεμβάλλω, after 'Dsc. 2.49' insert 'ἀντεμβαλλόμενα καλοῦμεν τὰ ἀντὶ τῶν ἄλλων ἐμβαλλόμενα φάρμακα Gal. 19.721.4'

ἀντεμβριάζειν, add '(ἀντεμβιάζειν La.)'

*ἀντεμόνιον, τό, antimony, Anon.Alch. 334.2.

ἀντεμφᾰνίζω, add 'make out a case before, τῷ τε δήμ[ῳ] ἀντεμφανιοῦμεν PZen.Col. 11.10 (iii BC)'

*ἀντενᾰγωγή, ἡ, countersuit, Cod.Just. 7.51.5.1.

*ἀντένδειξις, after 'counter-indication' insert 'Gal. 10.630.14'

*ἀντέντευξις, εως, ἡ, counter-petition, summons, PHib. 203.7, 16 (iii BC).

ἀντεξανίσταμαι, add 'transf., of the principle of cold in relation to heat, Plu. 2.946d (v.l. ἀντεξίσταμαι, q.v.)'

ἀντεξισάζω, for 'Sch.Od. 11.308' read 'Sch.Od. 11.309'

ἀντεπιστᾰτης, for the ref. read 'Bernand Les Portes 32 (SB 7027.4)' and within parenth. add 'ii AD'

*ἀντεπιτροπεύω, astrol., to be associate-ruler, PMich.III 149 xvii 7, 11 (ii AD).

ἀντεραστής, after 'Ar.Eq. 733' insert 'X.Cyn. 1.7, Pl.Amat. 132c'

ἀντερείδω I, for 'clasping hand in hand' read 'thrusting out hand towards hand'

ἀντερῐδαίνω, = ἀντερίζω, Nonn.D. 36.28.

ἀντερῶ, after 'S.Tr. 1184' insert 'πρὸς τὰ ἀντειρημένα Stoic. 2.8'; add 'see also ἀντιλέγω'

ἀντέρως II, add 'pl., IG 12(5).917.6 (Tenos, i BC)' III, for 'name of a gem' read 'kind of amethyst'

*ἀντευγνωμονέω, perform a service duly in place of another, prob. in PSI 1037.42 (iv AD).

ἀντευεργέτημα, add 'PLit.Lond. 138 viii 40 (i AD, -γέμητα pap.)'

ἀντευποιέω, after '(Arist.)Rh. 1374ᵃ24' insert 'Arr.Epict. 2.14.18'

*ἀντεφάπτομαι, seize by way of reprisal, PMeyer 8.8 (ii AD).

*ἀντεφηβαρχέω, to be deputy overseer of ephebes, IGBulg. 14.7 (Dionysopolis, iii AD).

ἀντεφόρμησις, after 'attack' insert 'Th. 2.91.4 (v.l. ἀντεξ-)'; for 'Hld. l.c.' read 'Hld. 8.16'

ἀντήδην, delete the entry.

*ἀντηδίς· ἱκετευτικῶς Hsch. (La. cod. -δης), cf. Theognost.Can. 163.

†ἀντημοιβός, v. °ἀνταμ-.

ἀντηρίς I, add 'in building, ἀντειρίδας .. τοῦ βαλανείου Pap.Lugd.Bat. xxv 46.8 (late ii AD), unless in sense II'

ἀντί I, line 7, after 'witnesses' insert 'ἀντὶ τὸ ἀρχὸ BCH 87.3 (vi/v BC)' III 2, add 'Just.Const. Δέδωκεν 22; ἀνθ'οὗ, w. pers. n., alias, BGU 406.2.16, PBasel 10.5 (both ii AD)' B, add 'Hes.Th. 893' C 7, for 'counter, as ἀντίφορτος' read 'as ἀντίγραφος' add 'Myc. a-ti- in compds.'

ἀντιάζω I 1, after '(Hdt.) 4.80, etc.' insert 'φλέγει με περιβόητος ἀντιάζων (Ares) S.OT 192', for 'abs.' read 'transf.' and after '(Pi.)O. 10(11).84' insert 'πρὸς μέλος ἠντίασεν accompanied the song, sc. by dance and gesture, APl. 287 (Leont.)' 2, delete 'Ἄρεα ἀντιάζω S.OT 192'

ἀντίαος, ὁ, epith. of Zeus, god of suppliants, Alc. 129.5 L.-P., rest. in Sapph. 17.9 L.-P.; cf. ἀνταῖος II.

ἀντιᾰχάτης, for 'dub.' read 'prob.'

ἀντιβάλλω I, lines 1/2, delete '(the acc. pers. being understood)' II, of Mss., add '[ἐ]κγεγραμμένον καὶ ἀντιβεβλημένον ἐκ τεύχους χαρτίνου διαταγμάτων ἐκ τοῦ ἐν Μαγνησίᾳ ἀρχείου SEG 32.1149 (Magnesia ad Maeandrum, AD 209), cf. ISmyrna 598.1 (ii AD)'

ἀντιβίην, after '(Π.) 5.220' insert 'A.R. 1.1002, 2.758'

ἀντιβίος 1 b, after 'enemy' insert 'ἀντιβίοισι τύραννε h.Hom. 8.5'

*ἀντιβλεπόντως, adv. from pres. part. of ἀντιβλέπω, facing, καθεσθέντα ἀ. τῷ κάμνοντι Paul.Aeg. 6.114 (168.5 H.).

ἀντιβλέπω, add 'Pherecr. 224 K.-A.'

*ἀντίβλησις, εως, ἡ, expl. of ἀντιβολία, Hsch. s.v. κατ' ἀντιβολίαν.

*ἀντιβολάδιον, τό, a surgical instrument, gloss in Hermes 38.281.

*ἀντιβουκολέω, practise trickery upon in return, Aesop. 28 P.

*ἀντίβους, ὁ, ἡ, sacrificed instead of an ox, οἷς ἄν[τ]ίβους Sokolowski 3.18 i 64 (Erchiae, Attica, iv BC); cf. ἀντίβοιος, ἐπίβοιον.

Ἀντιγόνειος (s.v. Ἀντίγονος), add 'of a tetrachm, Ath.Asklepieion v 106 (244/3 BC)'

Ἀντίγονος, add 'b position in board game app. resembling backgammon, AP 9.482.17 (Agath.).'

*ἀντιγραμματεύς, έως, ὁ, deputy γραμματεύς, IG 2².2067.225 (ii BC).

ἀντιγρᾰφεῖον, for pres. refs. read 'IEphes. 14 (i BC), 1024.12 (ii AD, both sp. -ῖον)' and add 'SEG 33.1039.79 (Aeolian Cyme, ii BC, sp. -ῖον)'

ἀντιγρᾰφεύς, add 'legal and administrative official of the emperor, Just.Const. Δέδωκεν 9; senate official, Lyd.Mag. 3.27'

ἀντιγρᾰφή II, add 'legal term at Paros, SEG 33.679.27 (ii BC), significance uncertain' IV, add 'Modest.Dig. 27.1.12.1'

*ἀντιγράφιον, τό, medical prescription, PMerton 12.13 (i AD); cf. °ἀντίγραφος II.

ἀντίγραφος I, add 'ποιήσασθαι ἀ. ἐκ τῶν στηλῶν τὰ ἀναγεγραμμένα IG 2².120.22' II, after 'as subst.' insert 'Dor. ἀντίγροφον, IG 4².68.81 (Epid., iv BC), Inscr.Cret. 1.viii 10.5 (Cnossus, iii/ii BC), IG 12(3).248.22 (Anaphe, ii BC), SEG 25.640 (W.Locris, ii BC)' and add 'written reply, PSI 584.29, PCair.Zen. 375.14 (both iii BC), PKöln 166.14 (vi/vii AD); prescription, Orib.fr. 97'

ἀντιγράφω II 2, for 'keep a counter-reckoning .. (cf. ἀντιγραφεύς)' read 'of clerk, check' and delete 'simply check'

ἀντιγυμνᾰσιαρχέω, add 'PLund 4.9'

*ἀντιγυμνᾰσίαρχος, ὁ, deputy gymnasiarch, Str. 14.5.14 (v.l., al. ἀντὶ γυμνασιάρχου).

*ἀντιδείκνῡμι, counter-indicate, τὸ ἀντιδεικνύμενον Gal. 1.166.15, 18, 14.666.5.

ἀντιδέχομαι, add 'w. gen., accept, exchange one thing for another, Call.fr. 75.45 Pf. (tm.)'

*ἀντιδηλόω, inform in turn, reply, pap. in Tyche 1.163.19 (vi/vii AD).

ἀντιδιαγρᾰφή, after 'ἡ' insert 'PSI 571.4 (iii BC)'

ἀντιδιαίρεσις II, add 'Gal. 1.386.9, 11.128.11, 14'

*ἀντιδιαιρετικός, ή, όν, opposite, contrasted in a dichotomy, David in Porph. 214.13/14.

*ἀντιδιάλεξις, εως, ἡ, unexpld. wd., ASAA 3/5(1941/3).94 i 8 (Lemnos, ?ii BC).

*ἀντιδικαιολογέομαι, make a counter-claim, counter-plead, PMonac. 14.25, PMich.XIII 659.21 (both vi AD).

ἀντιδίσκωσις, add 'An.Ox. 3.405.8'

ἀντιδομή, delete 'opposed or'

ἀντίδοτος I, add 'b given in return, ἕλκος Nonn.D. 29.166.'

ἀντιδωρεά, add 'Just.Nov 2 pr. 1'

*ἀντιενείριος, ον, dub. sens., (cf. perh. ἀντιάνειρα) μητέρες ἀντιενίριοι, IKourion 127.2, 136.2, etc.

ἀντιεξάγω, expel in turn, BGU 1273.34 (iii BC).

ἀντίζυγος, after 'correspondent' insert 'Diog.Ath. 1.9'

*ἀντιζῶ, live instead, βίον κοινόν CIJud. 1.144 (Rome).

ἀντίθεμα, add 'Didyma 39.5, cf. 32.14, 15, 41.12, 17'

ἀντίθεος, add 'in place of a god, i.e. falsely similar to a god, Hld. 4.7.13, Iamb.Myst. 3.31, PMag. 7.635' II 2, delete the section.

ἀντίθεσις, line 6, for 'by negation' read 'involving use of the contradictory'

ἀντίθεω II, add 'Nonn.D. 1.498'

ἀντίθροος, before 'Coluth. 118' insert 'Nonn.D. 13.44'

ἀντίθυρος, add 'SEG 33.880 (Ephesus, perh. masc. subst.)'

*ἀντικαθιστάνω, substitute, IG 2².1099.24 (ii BC); also ἀντικαθιστάω Just.Nov. 16.1.

ἀντικαθίστημι II 1, add 'D.C. 48.53.1'

*ἀντικαταίνετοι, οἱ, financial officials at Epidaurus, IG 4².109 ii 150 (iii BC).

ἀντικατάλλαξις, for 'BGU 1210.177' read 'PGnom. 177'

ἀντικατάστᾰσις, add '2 preliminary hearing before a strategus, BGU 2241.3.'

*ἀντικατατίθημι, put back, replace, ἀντικαταθεῖναι ὁπόσον ἂν λάβοι Agath. 4.21.7 K.

*ἀντικατεργᾰσία, ἡ, tillage of plot in lieu of another, PMeyer 1.6 (ii BC).

ἀντικατηγορέω II, for this section read 'predicate by interchange, τινὰ ἀλλήλων D.H.Th. 24'; renumber old section II as III adding 'Sch. Iamb.Protr. 6'

*ἀντικάτοχος, ον, holding fast on the other side, Anon.Alch. 419.9.

ἀντίκειμαι III, add 'b ὁ ἀντικείμενος the Devil, Cod.Just. 1.4.34.9.'

*ἀντικήνσωρ, ὁ, Lat. antecessor, class of law-teacher under Justinian, Just.Const. Δέδωκεν 9, 11.

*ἀντικνήμη, ἡ, shin, PLond. 1821.179 (Aegyptus 6.179; vi AD).

*ἀντικνημία, ἡ, shin, SB 5274.9 (iii AD).

ἀντικονταίνω, v. ἀντικοντόω.

*ἀντικοσμητεύω (s.v. -ήτης in LSJ), for pres. ref. read 'IG 2².2079 (ii AD)' and add 'SEG 33.158.11 (Athens, iii AD)'

ἀντικρίνω, add 'Ael.NA 6.61'

ἀντίκριος, read 'ἀντικρῖος' and for 'Aen.Tact. 22.7' read 'Aen.Tact. 32.7'

*ἀντίκροκος, ὁ, kind of aquatic plant, Hippiatr.Paris. 712.

ἀντικρύς, line 3, for 'εἰς τὸ' read 'εἴσω (pap.)'

*ἀντικύριος, ὁ, app. one serving as his master's deputy, POxy. 3239.45 (gloss., iii AD).

†ἀντιλαβεύς, έως, ὁ, handle, included in inventory of a sailing-ship's furnishings, PCair.Zen. 756.4 (iii BC).

ἀντιλαμβάνω II 2, for this section begin 'give a helping hand to, οὐκ ἀντιλήψεσθ' E.Tr. 464; transf., come to the assistance of, help, w. gen., ἀντιλλαβέσθαι τᾶς πόλλιος, BCH 59.37.17 (Crannon, ii BC), ἀ. Ἑλλήνων, D.S. 11.13, etc.'

ἀντιλέγω, line 4, delete 'pf. ἀντείρηκα; fut. pass. ἀντειρήσομαι'; line 12, after 'Id. 1.28' insert 'urge in opposition' 2, delete 'πρὸς τὰ ἀντειρημένα .. Stoic. 2.8' (v. ἀντερῶ).

ἀντιληπτικός 1, line 4, after 'Stoic. 2.230' insert 'ἡ δὲ γευστικὴ αἴσθησίς τε καὶ δύναμίς ἐστι .. ἀντιληπτική τε καὶ κριτικὴ τῶν γευστῶν Alex.Aphr.de An. 54.1'

ἀντιλογία 3, delete the section.

ἀντιλογίζομαι, add 'X.Ath. 1.16'

***ἀντιλύω**, set free in turn, IG 2².1225.18 (iii BC).

***ἀντιμᾰχεία**, ἡ, = ἀντιμάχησις, An.Ox. 3.171.24.

***ἀντιμειδιάω**, smile in return, Vit.Aesop.(G) 32.

ἀντιμερίζομαι 2, after 'Hsch.' insert 's.v. ἀντιδιαιρεῖται'

ἀντιμεσουρᾰνέω, after 'Ptol.Tetr. 33' insert '(v.l., al. μεσουρ-)'

ἀντιμετάβᾰσις, delete the entry.

ἀντιμισθία, add 'ICilicie 35.8 (sp. ἀνταμισθεία, vi AD)'

***ἀντιμίσθωσις**, εως, ἡ, prob. contract of lease executed by the lessor as opposed to one executed by the lessee, PMichael. 43.23, 24 (vi AD), PMasp. 66.3, 107.18 (vi AD); cf. ἀντίπρασις.

ἀντιμνηστεύω, add 'Charito 8.7'

ἀντίμοιρος, after 'ον' insert 'having opposite fortunes'

ἀντίμολπος, for 'song, sleep's substitute' read 'i.e. song acting as a substitute for sleep'

†**ἀντίμορφος**, ον, corresponding in shape, τῶν ἀντιμόρφων χαρακτήρων ἀγράφους εἰκόνας Luc.Am. 44; adv. -φως, in a contrary form, Plu.Crass. 32.

***ἀντίμουσος**, ον, sounding in answer, μέλος Rh. 1.493.22.

***ἀντιμωλία**, ἡ, expld. as δίκη εἰς ἣν οἱ ἀντίδικοι παραγίνονται Hsch. s.v. μωλεῖ (ἀντιμολία cod.); ἀντ]ιμōλίαι rest. in Inscr.Cret. 4.13b (Gortyn, vi BC); cf. ‡ἀντίμωλος, °ἀγχεμωλία, ἀμφιμωλέω, ἀμφιμῶλος.

ἀντίμωλος, for '= ἀντίδικος' read 'ἀντίμολος adversary in law-suit' and add 'Inscr.Cret. 4.13a.1 (Gortyn, vii/vi BC), 2.v 10.4 (Axos, vi BC)'

***ἀντιναύτης**, ου, ὁ, substitute sailor, SEG.39.1180.64 (Ephesus, i AD).

ἀντινέμομαι, for 'bestow' read 'receive' and add 'act. prob. in SEG 11.343 (ἀ]ντιανέμων, Argos, iii BC)'

***ἀντινοέω** or **-νοέομαι**, hold opposing views, fut. part. ἀντινοησόμενος PMasp. 97.49 (vi AD).

***ἀντίνῠ**, τό, the letter N written in reverse, and representing a note in the musical scale, Alyp.Diat. 5, 13, Chrom. 5, 8.

ἀντίξοος, line 5, after 'A.R. 2.79' insert 'ἀ. .. διχο[φροσύνην] Call.fr. 43.73 Pf.'

†**ἀντίον**, τό, the upper cross-beam of a loom, Ar.Th. 822, POxy. 264.4 (i AD), Poll. 7.36, 10.125; as standard of comparison for a spearshaft Lxx 2Ki. 21.19, 1Ch. 11.23.

ἀντίος II, as adv., add '5 in opposite ways, ἀντία δεσμεύων Hes.Op. 481.'

***ἀντίουρος**, ὁ, = ἄνθορος, opposite boundary or boundary stone, IEphes. 3516.

Ἀντιόχειος, add 'II of the city of Antioch, of weights, IGLS 1071 (d), (f), etc.; neut. pl. subst., Ἀντιόχεια, τά, Inscr.Prien. 59.21 (decr. of Laodicea, c.200 BC).'

***Ἀντιοχεών**, ῶνος, ὁ, name of month at Smyrna, rest. [Ἀντι]οχεών, REA 38.26; (previously read as Ὀχεών); also rest. [Ἀντιοχ]εῶνι, in MAMA 6.5.16 (Laodicea, iii BC).

***Ἀντιοχήσιος**, α, ον, from Antioch, Lat. Antiochensis, στιχάριον Ἀ. PFouad 74.6 (v AD), POxy. 1978.4, al. (sp. -χύσιος, vi AD).

***Ἀντιοχισταί**, οἱ, partisans of Antiochus, Plb. 21.6.2.

***ἀντίπᾰνον**, τό, Lat. antepannus, embroidered cloth, Hsch. s.v. παρατούριον.

***ἀντιπᾰνουργέω**, practise trickery against someone, Sch.Gen.Il. 3.325.

ἀντιπαραβαίνω, for 'PLips. 298' read 'PLips. 29.8'

ἀντιπαράδοσις, add 'II return of hired property, SB 9305.5 (iv AD), Just.Nov. 64.1.'

ἀντιπαράθεσις, after 'contrast' insert 'D.H.Comp. 18'

ἀντιπαράκειμαι 1, add 'Str. 11.8.2, Peripl.M.Rubr. 61' 2, after 'τινί' insert 'D.H.Dem. 26'

***ἀντιπαραμένω**, remain in service as compensation, Stud.Pal. 22.36.14 (ii AD), PMich.inv. 5191a (BASP 22.255, iii AD).

ἀντιπαρίστημι, for 'Ptol.Geog. 8.1.14' read 'Ptol.Geog. 8.1.4'

***ἀντιπατρίδιον**, τό, dim. of ἀντιπατρίς, PCair.Zen. 38.11 (iii BC).

***Ἀντιπατρισταί**, οἱ, followers of the philosopher Antipater, Antip.Stoic. 3.246.

***ἀντίπατρος**, ὁ, perh. title of Mithraic initiate in one of the grades of initiation, Dura⁷·⁸ 119; cf. πατήρ v.

ἀντιπελαργέω, after 'cherish in turn' insert 'as a stork cherishes its parent, Cels.ap.Orig.Cels. 4.98'; delete 'cherish in place .. parent'

ἀντιπελάργωσις, before 'Com.adesp. 939' insert 'return of benefits'

ἀντιπεράν, after 'A.R. 2.177, al.' insert 'w. gen., ἀντιπέρην γὰρ ἀείρεται Ἀρκτούροιο Arat. 405'; add 'see also ἀντιπέρα'

***ἀντιπεράω**, cross to other side of a river, PFouad 87.25 (vi AD).

ἀντιπεριάγω, add 'transf., divert, ἀντιπεριάγοντα τὴν ψυχὴν τῆς ἐς τὸ πονεῖν σχολῆς Ach.Tat. 1.6.3'

***ἀντιπεριέχω**, enfold or encompass in turn, ἀντιπεριέχει γὰρ ἄλληλα τὰ ἴσα κατὰ τι ἀντίστροφον Dam.Pr. 377.

ἀντιπεριποιέομαι, add '2 part. as gloss on ἀρνύμενος, Hsch.'

ἀντιπεριωθέω, add 'Alcin.Intr. p. 175.24 H.'

ἀντίπηξ, for 'wheeled .. infants' read 'prob. a basket with a hinged lid'; after 'Eust. 1056.46' insert 'cf. Hsch. s.vv. ἀντίπηξ, ἀντίπηγα'

ἀντιπλάδη, add 'τειχῶν ἐπιστάται τῆς ἀντιπλάδης IEryth. 23.4 (iv BC)'

ἀντιποιέω I, after 'in return' insert 'τί A.fr. 78c.57 R.' II, lines 2/3, for 'τῶν σπουδαίων' read 'παιδείας'

†**ἀντίπους**, ποῠν, of position on the earth, situated diametrically opposite, Pl.Ti. 63a, Arist.Cael. 308ᵃ20, Cic.Acad.Pr. 2.39.123; εἰ γάρ εἰσιν ἀντίποδες ἡμῶν .. τῆς γῆς τὰ κάτω περιοικοῦντες Plu. 2.869c, Eratosth. 16.19; in the opposite hemisphere and under the opposite meridian, Cleom. 1.2; on the opposite side of the same hemisphere, Str. 1.1.13.

ἀντίπρᾱσις, for 'PLond.ined. 2227' read 'SB 6612.18'

ἀντιπροβολή, add 'τοῖς ἐξ ἀντιπροβολ[ῆ]ς καταστᾰθεῖσι [ταμίαις] ISmyrna 711.10 (late Rom.imp.)'

***ἀντίπροικον**, τό, = Lat. donatio propter nuptias, PLond. 1708.50 (vi AD).

ἀντιπροπίνω II, for 'προπίνω I 2' read 'προπίνω II 3'

ἀντιπρόσωπος, after 'X.Cyr. 7.1.25' insert 'τοῦ ἀ. θεάτρου id.Eq.Mag. 3.7'

ἀντίπτωμα, for 'stumble against .. 29' read 'cause of stumbling, Lxx Si. 35(32).20(25); conflict, ib. 34(31).29(39)'

***ἀντιπωλέω**, sell in opposition to (a state monopoly), PTeb. 709.14 (ii BC).

ἀντιρρητορεύω, add 'Lxx 4Ma. 6.1'

***ἀντίρριζον**, v. ἀντίρρινον.

***ἀντισέβεια**, ἡ, mutual reverence, PBremen 37.11 (ii AD).

***ἀντισέληνος**, ον, like the moon, A.fr. 204c.6 R.

ἀντισήκωμα, add 'ἀ. ψελλίων SB 1962 (ostr., iv/v AD)'

ἀντίσκηνος, for pres. ref. read 'IEphes. 2041' and add 'σινδ(όνιον) ἀντίσκ(ηνον) perh. cloth screen, PVindob. G25737.12 (ZPE 64.75, vi/vii AD)'

***ἀντίσκιον**, τό, azimuth, Ptol.Anal. pp. 193, 195 H.

***ἀντισκρίβας**, α, ὁ, deputy scribe (v. °σκρείβας), PSI 768.17 (v AD).

ἀντισπάω II, for '= ἀντέχομαι' read '(cf. ἀντέχομαι)'; after 'seize' insert 'and hold back'

***ἀντιστεφάνόω**, crown in return, IG 2².2969 (pass., iv BC).

ἀντιστήκω, add 'stand opposite or in front, Serv. in Verg.Georg. 2.417'

ἀντίστοιχος, add 'Anon.Lond. 20.5'

ἀντιστράτηγος, at beginning insert 'Aeol. ἀντιστρότᾱγος SEG 29.741'

***ἀντιστρᾰτιώτης**, add 'substitute soldier, SEG 39.1180.64 (Ephesus, i AD).'

ἀντιστρέφω IV, 3, for 'opposites' read 'contradictories'

ἀντιστροφή III 3, for 'opposite' read 'contradictory'

ἀντίστροφος VII, 1, for 'ἀπόστροφος' read 'ἀπόστροφος II 1'

***ἀντισύγγρᾰφος**, ον, executed by both parties, PMonac. 7.6 (vi AD). II **ἀντισύγγραφον**, τό, copy, counterpart of legal instrument, PDura 19.18 (i AD), 32.19 (iii AD).

ἀντισύγκλητος, for 'Marius' read 'P. Sulpicius Rufus'

***ἀντισχολαστής**, οῦ, ὁ, rival lecturer, Suid. s.v. Ἀδριανός.

***ἀντισυνίστημι**, set up in opposition, εἰ τὰ κατὰ σχέσιν ὑφεστῶτα ἀντισυνίστησιν ἄλληλα Dam.Pr. 82.

***ἀντιταγής**, ἐή, belonging to the opposite order, Dam.Pr. 56.

***ἀντιταμιεύω**, serve as deputy treasurer, inscr. in Studi off.a E.Ciaceri p. 257 (Rhodes, iii AD).

ἀντιτάσσω II 3, read 'set oneself against, match' and delete '(Pass.)'

ἀντίτεχνος, delete 'cf. Lg. 817b'; after 'w. gen.' insert 'τοῦ καλλίστου δράματος Pl.Lg. 817b'

ἀντιτιμάω II, add 'IG 12(2).526a.18-19, b.21-2'

ἀντιτιμωρέομαι, add 'pf. pass. κινδύνους ἐκφυγόντα ἀντιτετιμωρῆσθαι τοὺς πολεμίους SEG 13.206.12 (Messenia, ii/iii AD)'

***ἀντιτοξότης**, ου, ὁ, counter-archer, Afric.Cest. 50.69 V.

ἀντιτορέω, for 'τετορεῖν' read 'τορέω'

ἄντῑτος, line 1, for 'which occurs in Hsch.' read 'cf. Sch.Il. 24.213 τὸ τέλειον ἀντίτιτα, ἵν' ᾖ ἀντιτιμώρητα'; line 2, for 'requited, revenged' read 'done in requital or revenge'; at end delete 'cf. Call.Iamb. 1.160'; add 'A.Ag. 1429 (cj.)'

ἀντιτρέφω, add 'Phld.Epicur. p. 72 V.'

***ἀντιτῠπητικός**, ή, όν, resistant, solid, Taurus ap.Phlp.Aet.M. 520.10 R.

ἀντιτυπία, add 'II of sounds, clash, dissonance, D.H.Comp. 20, al.'

ἀντίτυπος I 1, line 4, for '(S.Ph.) 1460 (lyr.)' read '(S.Ph.) 1460 (anap.)'   2 c, line 2, after '(Berytus)' insert '[Εἰκ]όνος ἀντίτυπον καθορ[ᾶτε TAM 5(1).33'; line 3, after '(iv AD)' insert 'LW 1639.13 (Caria)'

*ἀντίφαντος, ον, perh. shining in one's face, A.fr. 451i.10 R.

†ἀντιφάρα, ή, dub. (φάρω), N.W. Greek for φέρω, perh. contraction fr. ἀντιφάρεα for *ἀντιφέρεια), ά.· ἀντιλογία. μάχη. ζάλη. οἱ δὲ μητρυιά Hsch., EM 114.19.

*ἀντιφαρής, ές, dub., perh. N.W. Greek for *ἀντιφερής, opposed to, the rival of, ἀντιφαρές· ἐναντίον Hsch.; perh. also ἀντιφαρίν (for ἀντιφαρήν = -φερῆ) Λάκωνι .. Ἀλκμᾶ[νι Alcm. 13(a).8 P.

ἀντίφᾶσις, for 'ἐξ ἀντιφάσεως συλλογισμός .. not day' read 'ἐξ ἀ. διαιρετικὸς συλλογισμός disjunctive syllogism w. contradictory alternatives'

ἀντιφᾰτικός, after 'contradictory' insert 'Alex.Aphr. in Top. 580.16'; delete 'only in'

ἀντιφέρω, line 5, for 'cogn.' read 'of respect'

ἀντιφιλοφρονέομαι. for pres. def. read 'treat with kindness or affection in return' and delete 'also, rival'

*ἀντιφρακτικός, ή, όν, obstructive, Phlp. in de An. 365.34.

ἀντιφωνέω II, add 'Just.Nov. 4.1'

ἀντιφώνησις, add 'II = Lat. constitutum, promise to pay an existing debt, PFlor. 343.3 (v AD), Just.Nov. 136.1.'

†ἀντιφωνητής, οῦ, ὁ, one who promises repayment of a debt, POxy. 136.39 (vi AD), Just.Nov. 4.1.

ἀντιχάλεπαίνω, for 'to be embittered against' read 'to show rancour in return'

*ἀντιχέλυσμα, v. ἀντιχάλημα.

ἀντίχρησις, add 'Dig. 13.7.33'

*ἀντίχωμα, ατος, τό, embankment, AE 1923.39.34 (Oropus, iv BC).

ἀντίψαλμος, add 'prob. rest. in A.fr. 451e.13 R.'

ἀντίψῡχος 1, for 'given for life' read 'given in exchange for life'; add 'ἀντίψυχον, τό, ransom given in exchange for a life, ἀντίψυχον αὐτῶν λαβὲ τὴν ἐμὴν ψυχήν Lxx 4Ma. 6.29, 17.21; cf. ἀντίλυτρα, ἀντίψυχα Hsch. s.v. περίψημα'

ἀντλέω I, add 'for irrigation, ἀντλοῦντ(ες) εἰ[ς] τὸ νεόφυτ[ον] PMil.Vogl. 308.30, al. (ii AD)'

ἀντλησμός, add 'sp. ἀντιλισμός in PBerl.Leihg. 23.15 (iii AD)'

†ἀντλητός, ὁ, irrigation, PMich.II 123ʳ ii 30 (i AD). PFlor. 369.6 (ii AD).

ἀντλήτρια, for 'DDeor. 2.1' read 'DMeretr. 2.1'

ἀντλιαντλητήρ, after 'bucket' insert 'Ar.fr. 486 K.-A.' and for 'Men. 30' read 'Men.fr. 269 K.-Th.'

ἄντλος I, neut. form τὸ ἄντλον, add 'Sch.A.Th. 796h' 2, after 'bilge-water' insert 'πὲρ μὲν γὰρ ἄντλος ἰστοπέδαν ἔχει Alc. 326.6 L.-P.'

ἀντοικοδομή, for 'IG 12(1)' read 'IG 12(3)'

ἄντοικος, for pres. def. read 'pl., those who live between the same meridian lines, but on the opposite side of the equator'

*ἀντολεύς, έως, ὁ, riser, of Apollo, PMag. 2.108.

ἀντολίη 1, after 'personified' for pres. ref. read 'PMag. 2.93, Orph.H. 12.12, 78.5'

ἀντόμνῡμι II, for 'Antipho 1.18' read 'Antipho 1.8'

ἄντομος, for 'stake or pale' read 'divider, hence unploughed land between plots, baulk, Tab.Heracl. 1.15, 2.13, al.; cf. ἀντόμους· σκόλοπας Hsch. (perh. erron.)'

ἀντορύσσω, delete 'metaph.' and after 'ὀφθαλμούς' insert 'gouge out each other's eyes'

ἄντρον I, add 'referring to a model of a cave, Ath. 5.200c'   II, for 'inner chamber, closet' read 'transl. Hebr. 'armôn (fortified tower of palace)'

ἀντροφυής, for 'born in caves, ἀνθίαι' read 'hollow by nature, πέτραι'

*ἀντροφύλαξ, ἄκος, ὁ, guardian of the cave, in Bacchic worship, IUrb. Rom. 160.3B.28 (Latium, ii AD).

*ἀντυγάς, άδος, ή, = ἄντυξ I 1, rest. in Call.fr. 115.16 Pf.

Ἀντώνεια, τά, shortened form of °Ἀντωνείνεια, IG 2².3015, al. (iii AD), perh. in SEG 30.1524.2 (Lycia), TAM 5(2).1024 (sp. -ήα, Thyatira), Ἀντώνηα Γετεῖα Ὀλύμπια Robert Laodicée p. 285 (sp. -ήα).

*Ἀντωνεινεῖα, τά, name of various festivals instituted in honour of Antonine emperors, Robert Hell. 11-12.350 (Ancyra), AS 39.50 (Balboura), CRAI 1970.20 (from Delphi, naming five different states); also Ἀντωνῖνα Robert Laodicée p. 293.

ἀντωνῠμικός, after 'D.H.Amm. 2.12' insert '(v.l. ἀντωνυματικόν)'

*ἀντώνυμον, τό, = ἀντωνυμία, A.D.Pron. 4.5.

ἀνυβριστί, delete the entry.

ἀνύβριστος I, after 'not insulted' insert 'esp. of property, free from outrage, inviolate'; add 'adv. -α, τὴν ἐξουσίαν ἐχούσης τῆς θεοῦ ἀνύβριστα SEG 24.498a (Leucopetra, Maced., ii AD)'   II, adv., add 'cj. in Anacr. (11(a).5 P.)'

*ἀνυγιαίνω, return to health, Bataille Hatshepsout no. 36.1 (v. Van Brock Vocab.med. 157 no. 1).

*ἀνυγραντέον, one must moisten, Aët. 7.20 (269.19 O.).

*ἀνυγραντικός, ή, όν, moistening, δύναμις Plu. 2.659c.

ἀνυμφής, delete the entry.

*ἀνύμφιος, ον, unmarried, τὴν ἀνύμφ[ι]ον κόρην rest. in Didyma 567.1.

ἄνυμφος I, add 'of Helen, τὴν ἄνυμφον πόρτιν Lyc. 102; of Europa, Mosch. 2.41 G.'   II, add 'of a man who died unmarried, Salamine no. 192 (ii/i BC)'

†ἀνυόδρομος, ον, perh. finishing the race, Lesb. fr.adesp. 11 L.-P.

*ἀνυπάλλακτος, ον, not pledged, PMasp. 309.35 (vi AD).

*ἀνυπάναγκαῖος, ον, voluntary, BGU 741.37 (ii BC).

ἀνύπεικτος, before 'Suid.' insert 'Paul.Aeg. 2.43 (125.8 H.)'

ἀνυπεξαιρέτως, for 'exception' read 'reservation'

ἀνυπέρβλητος 1, adv. -τως, add 'IHistriae 57.11'

*ἀνυπέρβολος, ον, not increased, of rent, PMil.Vogl. 267.32 (ii AD).

ἀνυπέρθεσία, for 'immediateness, haste' read 'immediate rejection'

ἀνυπερθετέω, for pres. def. read 'reject immediately' and after 'ib.' insert '57.21, 59'

ἀνυπεύθυνος I, add '3 not liable to reproach, Just.Nov. 25.3.'

†ἀνύπηνος, ον, Dor. -ᾱνος, app. having no moustache, Alcm. 10 (b).18 P., Eust. 1353.47; cf. ἀνύπηνον· ἀγένειον Hsch.

ἀνυπόδῐκος, line 1, delete 'Plu.Cat.Mi. 11' and before 'action' insert 'legal'; add 'SEG 33.1039.84 (Aeolian Cyme, ii BC); freedom from liability, ἀλλὰ καὶ τῷ γραφείῳ τὸ ἀνυπεύθυνον καὶ τὸ ἀ. ἐπίστευσεν Plu.Cat.Mi. 11, BGU 1273.35 (iii BC)'

ἀνυπόθετος I, adv., for 'not hypothetically' read 'not giving the conditions or situation'   add 'III not mortgaged, TAM 2.261b16.'

*ἀνυπόθηκος, ον, perh. without external suggestion or pressure, τοῦ βίου προαίρεσιν πεποιημένος -ον καὶ ἀναντι[.. Maiuri Nuova Silloge 443 (Cos).

ἀνυπόκριτος III, for 'in a .. sentence' read 'preceding clause which is not the apodosis' and for 'p. 24' read 'pp. 24, 27'

†ἀνυπόληπτος, ον, disreputable, discreditable, Anon. in Rh. 82.38, PIand. 132.8 (vi/vii AD).

†ἀνυπόμνηστος, ον, perh. not the subject of any petition or challenge, PFlor. 323.13, PMasp. 151.143 (both vi AD); dub. sens. in Phld.Piet. 98.

ἀνύποπτος 3, delete 'unhesitatingly' and transfer quot. to section 1.

ἀνυπόστατος I, line 4, after 'D. 54.38' insert 'πολέμιοι Isoc. 3.71'   II, add '5 in legal terminology, without a basis, invalid, of the institution of an heir, Cod.Just. 6.48.1.27, of a will, Theophil.Antec. 2.13 pr.'

ἀνυπόστολος, add 'τὰν τᾶς πόλιος ἀνυπό[σ]τολον εὐχαριστίαν SEG 26.1817.58 (Arsinoe, Cyrenaica, ii/i BC)'

ἀνυπόστροφος 1, for 'Orph.H. 56' read 'Orph.H. 57'

*ἀνυποσφράγις, ῖδος, (sic) not under seal, PMasp. 151.11 (vi AD).

*ἀνυποτελής, ές, not subject to charges, PMasp. 98.14 (vi AD).

ἀνυποτίμητος, -τως, add 'without calculating the cost, Didyma 287.20'

ἀνυτικός 1, line 3, for 'of persons .. (Sup.)' read 'J.BJ 5.9.2 (comp.), of persons, 1.17.8 (sup.)'

*ἄνυτος, ή, όν, feasible, S.E.M. 1.81.

*ἀνυφόρατος, ον, unexceptionable, BGU 1730 (i BC).

ἀνυψόω, for 'PGen. 51.27 (iii AD)' read 'PGen. 51.27 (iv AD)'

ἀνύω I 2, add 'AP 7.474; defeat, ἀθλητὰς ἤνυσε GVI 263.4 (Phrygia, ii AD)'   9, add 'b collect, χρήματα δημόσια Cod.Just. 10.30.4.6, 10.19.9.4.'

*ἀνφάδιος, acc. to Hsch. variant of ἀφάδιος (s.v.).

ἄνω (B) A II d, add 'τᾶι ἄνω πόλι καὶ τᾶι κάτω (referring to Gortyn) SEG 23.563.2 (Axos, iii BC); also out to sea, Il. 24.544'

*ἀνωβλεπής, ές, turned upwards, καρδίον ἀ. μορέας Cyran. 1.12.39 K.

ἄνωγα, line 3, after 'E.Or. 119' insert 'Call.fr. 628 Pf.'; line 14, for 'ἄνωγον Inscr.Cypr. 135 H.' read 'Cypr. a-no-ko-ne ἄνωγον ICS 217.2 (Idalium, v BC)'

ἀνωγή, add 'ἀχράντοισιν ἀνωγαῖς Salamine no. 202 (iv AD)'

ἄνωθεν I 2 b, for 'τῆς .. 12' read 'ὃς ἂν βάλλῃ τὰ ἐκ(κ)[αθάρματα ἄνωθεν τῆς ὁδὸ IG 12(5).107 (Paros, v BC), Ar.Ach. 433, Av. 1526'   II 1, after 'from farther back' insert 'Hp.VM 3' and add 'σύμμαχος ἄνωθε (sc. Νεικομήδεια) long-time ally, TAM 4(1).25.9'

ἀνωθέω 2, after 'Hp.Art. 80' insert 'Sch.Arat. 346'

ἀνώϊστος (A), after 'Il. 21.39' insert 'μύθον ἀ. A.R. 3.670, πότμῳ ἀ. id. 3.800, AP 7.564' and after 'A.R. 1.680' insert 'al.'

ἀνωμαλία III, add 'περιόδων D.H.Comp. 22'

ἀνωμαλοκράς, add '(ἀνωμαλόκουρος· ἀνωμάλως κεκαρμένη, οἱ δὲ λελυμένως La.)'

ἀνωμολόγητος, delete the entry.

ἄνωνις, after 'v.l. 485a' add '(= Suppl.Hell. 1138)'

ἀνωνυμία, add 'Alex.Aphr. in Metaph. 493.31'

*ἀνώνυμος, ον, nameless or unspeakable, ὄρνις Carm.Pop. 13 P. (dub.).

ἀνώνυμος I 3, after 'unspeakable' insert '[Τρο]ίας .. ἀλώσι[μο]ν [ἄμ]αρ ἀ. Ibyc. 1(a).14 P.'   II, at end for 'Herod. 6.14' read 'Herod. 5.45'

†ἄνωξις· βούλευσις Hsch.; cf. ἄνωγα.

†ἄνωρος, ον, not of age, minor, Leg.Gort. 7.29; untimely, ἀ. ἀποθανών

Hdt. 2.79 (v.l. ἄωρος), ἐνθάδ' ἄνωρος ἐὼν ἔθανον CEG 171.1 (Egypt, v BC); *unripe, sharp*, ἄνωρον· ὀξύν Hsch.; cf. ἄωρος (A) 2; adv. -ως, [πολλὰ γο]ῶσα, ὅτ' ἀνδρὸς ὤλετο ὃν ἀγαθός CEG 117.2 (Pharsalus, v BC).

*ἀνώρροπος, ον, *tending upwards*, of heat, Olymp. *in Phd.* p. 244 N.; κατὰ τὸ ἀ. Phlp. *in GC* 229.20.

*ἀνωτέρειος, α, ον, *belonging to the upper portion*, APF 1.64 (ii BC).

ἀνώτερος, after 'Nic.Dam. p. 25 D.' insert 'ποιήσει .. ναυαγίων ἀνώτερον will make him *superior* to shipwrecks, Dionysius in WS 20.319'

ἀνωφάλακρος, delete the entry.

ἀνωφελής, add 'form *νωφελής, Myc. *no-pe-re-a₂* (neut. pl.), of wheels, *unserviceable*'

ἀνωφέλητος **I**, add 'ἀ. καὶ θεοῖς ἐχθρός Stratt. 68 K.-A.' **II**, delete 'ἀ. .. 9 D.'

ἀνώχυρος **I**, delete ' = ἀνόχυρος' and add 'so prob. PZilliac. 1.24 (ii BC); cf. ἀνοχυρόομαι'

ἀξία **I 2**, add 'like Lat. *dignitas*: *rank*, *office*, κόμητος ἀξίαν CodJust. 12.33.8.2, *class*, τῇ τῶν μεγαλοπρεπεστάτων ἰλλουστρίων ἀξίᾳ Just.Nov. 15.1 pr.; in the church, *office*, CodJust. 1.3.41.19; *class*, τῆς ἱερατικῆς ἀξίας ib. 1.2.24.15'

ἀξιάγαστος, add 'PHerm. 2 (iv AD)'

*ἀξιάζω, = ἀξιόω, *think it right, expect that*, SEG 38.1476.99 (Xanthos), ILampsakos 34.33, IEryth. 121.5 (all iii BC).

ἀξιάω, delete the entry (v. °ἀξιάζω).

ἀξίνη **2**, add '*Alcmaeonis* 1.3 p. 33 B.' **3**, add 'Call.Cer. 6.35, S.Ant. 1109 (perh. here used for digging; cf. AB 62.9, as expl. of σκαφεῖον, mod. Gk. ἀξίνα *pickaxe*)'

ἀξινίδιον, after 'foreg.' insert 'PTeb. 794.13 (iii BC), PCair.Zen. 783.12 (pl., iii BC)'

*ἀξινογλυφία, *cutting into with an axe*, Phys. B. 164.2.

*ἀξινογλύφω, *cut into with an axe*, Phys. B. 164.1, 169.1.

*ἀξινομαντεία, ἡ, *divination by axes*, Plin.HN. 36.142.

*ἄξινος ὁ, = ἀξίνη, (distd. fr. πέλεκυς) SEG 24.361.14 (Thespiae, iv BC).

*ἀξιοδότας, α, ὁ, *having the entitlement to give*, IG 9²(1).609.17 (Locris, v BC).

ἀξιόζηλος, add 'Ael.NA 6.15, 43'

ἀξιοζήλωτος **1**, for ' = foreg.' read '*enviable, admirable*' and add 'TAM 2.767.23 (Lycia), IG 12(7).231.4 (Amorgos, ii/i BC); adv. -ως, BMus.Inscr. 925.21 (Branchidae, i BC)'

ἀξιόλογος, for 'ἀλίθοι' read 'ἀ. λίθοι' **2**, add 'as honorific title, of Apollo, POxy. 1578.iii.2, 2153.1.27; of prominent citizens, sup., MAMA 8.508, 509 (Aphrodisias), Didyma 156, 168; of women, IHadr. 68, ITrall. 240 (all iii AD)'

ἀξιόνικος, after '*victory*' insert 'Cratin. 171.2 K.-A.'; after 'X.Cyr. 1.5.10' insert 'τὸν ἀ. δεκάπρωτον SEG 31.372 (Olympia, ii AD)'; add 'adv. -κως, Didyma 194.12'

*ἀξιοπαθεῖ, gloss on βρενθύεται, Hsch. (ἀναξιο- La.).

*ἀξιοπάμων, ον, gen. ονος, perh. *worthily endowed*, Lyr.ap. Favorinus Exil. 11.6 B.

ἀξιοπρεπής, add 'adv. -πῶς, SEG 37.501 (Phthiotic Thebes, vi AD)'

ἄξιος **I 3 b**, line 4, for 'Lys. 22.18' read 'Lys. 22.8' and after it insert 'ἵν' ὡς ἀξιώτατον ὑμῖν πωλοῖεν id. 22.11, ἀξιώτερον τὸν σῖτον ὠνήσεσθε, εἰ δὲ μή, τιμιώτερον 22.22'

*Ἀξιοττηνός, ὁ, epith. of the divinity Μήν, TAM 5(1).251, SEG 37.1000, 1001 (Lydia, ii/iii AD).

ἀξιοχρεία, for 'CPH 97.13' read 'CPHerm. 97.13'

*ἀξιοχρέων, ονος, = ἀξιόχρεως, SPAW 1937.156 (Miletus, iii BC), Delph. 3(3).239.17 (ii BC).

ἀξιόχρεως **II**, adv., add 'Modest.Dig. 50.12.10'

*ἀξιοχρήματος, ον, *worth money*, οἱ ἀγαθοὶ οὐκ ἀ. λέγονται ἀλλ' ἀξιόλογοι Steph.in Rh. 304.17.

ἀξιόω **II 2**, line 6, after 'PEleph. 19.18' insert '*request* (someone), ἠξίωσα τοὺς δεσπότας INikaia 767 (iv AD)' **III 2**, line 6, after 'D. 21.101' insert '*consent, allow*, ὁ θεὸς ἀξιώσι ἡμᾶς προσκ(υνεῖν) ἐν οἰγία (= ὑγιείᾳ) POxy. 1837.16 (vi AD), 1857.4 (vi/vii AD)'; line 10, after 'X.Oec. 21.4' insert 'οὐκ ἠξιώθην καταφιλῆσαι τὰ παιδία μου I was not allowed .., Lxx Ge. 31.28' add '**4** ἀ. ὥστε w. inf., Macho 328 G.' **IV 2**, penultimate line, after 'II 2)' insert 'Pl.Phd. 86d' add '**V** ἀξιομένων (τῶν δεῖνα), official designation in BCH 50.401 (Thespiae).'

ἀξίωμα **I 2**, add 'Moschio Trag. 9.9' **3**, add '*office*, *post*, CodJust. 12.33.8.3, *rank, class*, Just.Nov. 38 pr. 3 (pl.)' **II 1**, add '*decree displayed in public* (as Lat. *libellus*), Αὔγουστος Σαμίοις ὑπὸ τὸ ἀξίωμα ὑπέγραψεν SEG 32.833 (Aphrodisias, i BC)' **3**, after '*request, petition*' insert 'SIG 656.6 (Abdera, i BC)'

ἀξιωματικός **I 1**, add '**b** neut. pl. subst., *meretricious payments, subsidies*, CodJust. 1.5.20.7.' **3**, delete the section transferring exx. to section I 2.

ἀξίωσις **I 2**, add 'ἀ. τῶν ὀνομάτων Th. 3.82.4' **II b**, add '*public petition* (for the holding of market-days), SEG 32.1149 (Magnesia ad Maeandrum, iii AD)' **IV**, delete the section.

ἀξουγγία, at end for 'ὀξούγγιον' read 'ὀξύγγιον'

ἀξυλία, after 'Hes.fr. 314 M.-W.' insert 'Call.fr. 176.4 Pf.'

ἄξυλος **I**, for '*with no timber* .. *wood* ' read '*having no timber*, i.e. devoid of large trees, ἄξυλος ὕλη'

ἄξων **I 1**, add 'sp. αὔξων POxy. 1986 (vi AD)' **II**, after '*axis*' read 'Plu.Sol. 23, 25, Plu. 2.779b, Ammon.Diff. 57, Sch.Ar.Av. 1354; sg., πρότος ἄχσον IG 1³.104.10, D. 23.31, Luc.Eun. 10' at end add 'Myc. *a-ko-so-ne* (pl., in sense I 1)'

ἀοζία, add '(of husband and wife)'

ἄοζος, for 'Call.Del. 249' read 'Call.fr. 563 Pf., Del. 249 (cj.)'; add 'fem., τὰν .. ἄ., SEG 15.385.2 (Dodona, v BC; unless erron, for τὸν); v. also ἆζος'

ἀοιδή **3**, add '**b** of a book of poems, AP 4.1.1 (Mel.).'

ἀοίδιμος, at beginning insert 'Boeot. ἀFύδιμος Corinn. 40 fr. 5(a).8 P.' **I**, line 2, before 'προφάταν (Pi.Pae. 6.6)' insert '*full of song, skilled in song*'; at end for '*notorious, infamous*' read '(cf. Od. 24.200 s.v. ἀοιδή 4)'

*ἀοιδολαβράκτας, α, ὁ, *boaster in song*, cj. in Pratin.Trag. 6.6 S.

ἀοιδοπόλος **2**, for 'Aus.Ep. 14' read 'Aus.Ep. 10'

ἀοιδός **I 1**, line 4, for 'Arist.Metaph. 983ᵃ4' read 'Sol. 29 W.' **II 1**, after '*musical*' insert 'τὴν ἀοιδὸν ἀκρίδα AP 7.198 (Leon.Tarent.); comp., ἀοιδοτέρα Alcm. 1.97 P.; sup.'; after 'cf.' insert 'Phanocl. 1.22'; add 'comp. adv. φθέγξετ' ἀοιδότερον AP 11.195 (Diosc.)' **2**, add 'AP 9.424 (Duris), sup. SEG 11.773 (epigr., Sparta, iv AD)' **III**, delete 'cf. δοῖδος'

ἀοίκητος **I**, add 'Isoc. 4.148, 15.22'

*ἀοῖος, v. ἠοῖος.

*ἀολλίζω **I**, add 'Τρώων ἀόλλιζον φάλαγγας B. 15(14).42 S.-M.'

*ἀόξος· ἀδιάγλυφος Hsch. (La.).

ἀόρατος, add 'see also °ἀβόρατος'

*ἀοργησία, ἡ, *deficiency in the passion of anger*, Arist.EN 1108ᵃ3, 1126ᵇ3. **2** *restraint in respect of anger, self-control*, Plu. 2.452 (title), Nic.Dam. 103w J., Andronic.Rhod. p. 575 M., Gal. 5.30.18.

ἀορισταίνω, for ' = sq.' read '*to be in an indeterminable state*, Alcin.Intr. 26 (179.28 H.), *to be indeterminate*'

ἀοριστία **2**, add 'Procl.Inst. 117 D.'

ἀορτήρ **1**, add 'Pherecr. 41 K.-A.'

ἀόρτης **1**, before '(Maced.)' insert 'PSI 858.37 (iii BC)'

ἀοσμία, for 'Thphr.CP 6.16.3' read 'Thphr.HP 6.6.5'

ἀοσσητήρ, after 'Od. 4.165' insert 'Call.Ap. 104, fr. 18.4 Pf.'

*ἀουνδίκιος, unexpld. wd., referring to land belonging to the ταμιεῖον rather than κτήτορες, POxy. 3205.14, 23, 35, etc. (iii/iv AD).

*ἄπ, v. °ἀπό.

*ἀπᾶ, v. °ἀββᾶς.

ἀπαγγελία **I 1**, delete 'in Psychology .. Plot. 4.6.3' and add '**3** *recitation* or *repetition from memory*, Plot. 4.6.3.'

ἀπαγγέλλω, add '**III** *recite* or *repeat from memory*, Synes.Ep. 111, Sch.Ar.V. 1109 (in dramatic rehearsals); Suid. s.v. Καλλιφάνης.'

ἀπάγγελσις, add 'rest. in Hesperia 26.52 no. 9 line 1 (iv BC)'

ἀπαγγελτικός **I**, after '*reporting*' insert 'Alcin.Intr. 4 p. 154.30 H.'

*ἀπαγγισμός, ὁ, (ἄγγος), *decanting*, PCol. 239 (iv AD).

*ἀπαγκάζομαι, *lift away*, ἀπὸ μὲν .. λίθον ἀγκάσσαθαι Call.fr. 236.1 Pf.

*ἀπαγκίστρωσις, εως, ἡ, *barbed form*, Sch.Gen.Il. 21.474 (pl.).

ἀπαγκωνίζομαι, for '*bare the elbows*' read '*thrust the elbows away from the body*'

ἀπαγλαΐζω, after 'Pers. 20' insert 'Agath. 2.15, 5.14 K.'

ἀπαγνῦμαι, add 'pf. part. act. ἀπαγνώς *broken*, IG 2².1447.15 (iv BC); [στέφανος] Inscr.Délos 385a23 (ii BC)'

ἀπαγορεύω **I**, line 11, after 'Arist.Pol. 1298ᵃ38' insert 'w. acc., ὁ νόμος .. αὐτῇ ἀπαγορεύει τὴν τοῦ παιδὸς διαδοχήν Just.Nov. 22.40'; at end add 'abs., *make a prohibition*, SEG 33.1468 (Ptolemais, Cyrenaica, ii/i AD)' **III**, for this section read '*speak from*, ἀπὸ τὸ λάο̄ (ὁ ἀπαγορευόντι, lit. *from the stone from which they speak*, Leg.Gort. 10.36, rest. in 11.13' add '**IV** (perh. transl. Lat. *renuntio*) *announce, proclaim*, Just.Nov. 117 epilogos.'

*ἀπάγριος, ον, perh. *domesticated, tame*, πρόβατο(ν) ἀπάγριν on mosaic depicting a sheep and a bird, SEG 34.1439 (Apamea, Syria, vi AD).

*Ἀπαγχομένη (ἀπάγχω), title of Artemis in Arcadia, Call.fr. 187 Pf., Paus. 8.23.7.

ἀπάγω **II**, add 'med., perh. *get oneself back, return*, OFlorida 1.4' **IV 3**, for 'And. 4.181' read 'And. 4.18' add '**4** *carry off to execution*, D.S. 13.102.3, J.BJ 6.2.7, Eu.Luc. 23.26, POxy. 33 iii 15.' **V b**, for 'Arist.APr. 29ᵇ9' read 'Arist.APr. 29ᵇ8' add '**VII** *weigh short, weigh less than the supposed weight*, Inscr.Délos 1417Aii68 (ii BC), al.; cf. ἄγω VI.'

*ἀπαγώγιμος, ον, *which is to be delivered*, σῖτον .. ἀ[πα]γώγιμον SIG 360.47 (Cherson.Taur., iii BC, v. BE 1950.151).

*ἀπαέτωμα, ατος, τό, = ἀέτωμα, IG 2².1685B4.4.

*ἀπαθῷος, ον, unexpld. wd., Hdn.fr. p. 19 H.

ἀπαίδευτος **I**, add '**3** *uncontrolled, unrestrained*, Trag.adesp. 523 K.-S., J.AJ 19.2.9; τὸ ἀ., *lack of control*, ib. 17.9.3.'

*ἄπαιδος, ον, *childless*, ἄπαιδος καὶ ἀδιάθετος PSI xx Congr. 16.8 (iv AD).

†ἀπαίνυμαι, v. °ἀποαίνυμαι.

ἀπαίρω **I**, line 4, delete 'in (E.)*IT* 967 .. πειρατήρια' and transfer quot. to section II 2 (*depart*).

ἀπαιτέω **I 1 a**, line 10, after 'Arist.*EN* 1164ᵃ17' insert 'ποινὰς ἀ. Jul.Or. 2.59a'; at end, w. inf., add 'ἀ. αὑτοὺς πάντα πληροῦν Cod.Just. 1.3.45 pr.; w. ἵνα, ib. 6.4.4.16a'   **b**, delete the section.   **II 2**, line 3, after 'Lxx *Wi*. 15.8' insert 'ποινὰς ἀ. Jul.Or. 2.58a'

*ἀπαιτήσιμος, ον, *subject to dues*, PMichael. 34.4 (vi AD); cf. ἀπαιτήσιμον, τό.

ἀπαίτησις, add 'Lxx *Ne*. 5.10, *Sir*. 34(31).31, 2*Ma*. 4.27, πληρῶσαι τὴν τῆς εἰκόνος ἀ. *satisfy* the *requirements* of the picture, Ael.*VH* 2.44'

ἀπαιτητής, add 'Cod.Just. 10.19.9.4'

ἀπαιωρέομαι **I 1**, delete 'hover about' and add 'Arr.*Tact*. 34.4'   **I 2**, delete the section.   **II**, for 'lift up a garment' read 'hold up, lift from the ground' and add 'pass. Luc.*Astr*. 19'

ἀπακρῑβόομαι, line 4, after 'of persons' insert 'ἐκ τοίης ὤνθρωποι ἀπηκριβωμένοι ὀστῶν ἁρμονίης AP 7.472.7 (Leon.Tarent.)'

ἀπάλαιστρος **I 1**, after '*unskilled in wrestling*' insert 'or perh. *incapable of physical exercises*'; add '*SEG* 27.261.B28 (Beroea, ii BC), IMagnes.Sipylos 34'

*ἀπαλάκιστος, ον, *unsweetened*, Anon.Alch. 11.2.

ἀπαλγέω, for '*put away sorrow for*' read '*work out* one's *grief for*'

*ἀπᾰλεξιάρη, prob. fem. subst. or adj., *she who wards off ruin*, POxy. 2819.11 (hex. fragment); cf. ἀλεξιάρη, ἀπαλεξίκακος.

ἀπᾰλεξίκᾰκος, delete 'f.l. in'

ἀπᾰλέξω **I 1**, add 'poet.ap.Pl.*Alc*. 2.143a'

†ἀπᾰλίας, ον, ὁ, *immature animal*, D.L. 8.20 (text dub.).

*ἀπάλιος, ον, of animals, *young, immature* (opp. τέλεος, *full-grown*), Sokolowski 2.22.8 (Epid., iv BC), Hsch.

ἀπάλλαξις, add '**2** *relief* from a disease, Hp.*Nat.Hom*. 8.'

ἀπαλλάσσω **A I 1**, line 3, after 'And. 1.59' insert 'τοῦ ζῆν Pl.*Ax*. 367c'   **II**, line 3, for 'E.*Med*. 786' read 'Ar.*Pax* 568'; line 8, delete 'τοῦ ζῆν .. 367c'   **B II 2**, delete 'Pl.*Phd*. 81c'

ἀπαλλοτριόω **3 b**, for 'Pl.*Ti*. 65d' read 'Pl.*Ti*. 65a'

*ἀπᾰλόθροος, ον, *softly-sounding*, ἦχος Nonn.*D*. 48.606.

ἀπαλοιφή **I**, before '*Gloss*.' insert '*SEG* 33.1177.10 (Lycia, i AD)'

ἀπᾰλοκροκώδες, add '*CIL* 13.10021.188'

ἀπᾰλός **I**, line 5, for 'Sapph.*Supp*. 23.16' read 'Sapph. 94.16 L.-P.'; line 6, for 'Alc.*Supp*. 14.5' read 'prob. in Alc. 39(a).5 L.-P.' and for 'Sapph. 76' read 'Sapph. 82(a) L.-P.'; line 7, for 'Ead.*Supp*. 25.13' read 'id. 96.13 L.-P.'; line 11, for 'of raw fruit .. X.*Oec*. 19.18' read 'of *fresh* fruit, κολοκύνταις .. ἀπαλωτέραις Alc. 117(b).9 L.-P., Hdt. 2.92.4, X.*Oec*. 19.18'

ἀπᾰλοχρώς, add 'Heras ap.Gal. 13.511.15'

*ἀπᾰλόψυχος, gloss on γλυκύθυμος, Hsch.

ἀπαλύνω **1**, delete '*make plump*, opp. ἰσχναίνω'; add 'Gal. 18(2).840.5, 841.7'   add '**3** *make gentle, calm* κῦμα θαλάσσης ἀπαλύνεται γαλήνῃ Anacreont. 46.4 W.'

*ἀπαμαύρωσις, εως, ἡ, *removing of dimness*, Zos.Alch. 211.1.

ἀπαμάω (A), add 'λαιμὸν σιδήρῳ Il. 18.34'

ἀπαμβρᾰκόομαι, for parenth. to the end read '(cf. °ἀμβρακόομαι) Pl.Com. 66 K.-A.'

ἀπαμέρδω, after '= ἀπαμείρω' insert '*deprive of*' and add 'w. acc. of pers. and gen. of thing, *GVI* 1547.7 (Rome, ii AD)'

*ἀπᾱμία, v. °ἀφαμία.

ἀπαναίνομαι, for 'Pi.*N*. 5.60' read 'Pi.*N*. 5.33' and add 'A.R. 1.611, Lxx *Jb*. 5.17'

*ἀπαναφέρω, *bring back*, aor. inf. ἀπονιγ[κ]ῆν *SEG* 37.340 (Mantinea, iv BC).

*ἄπανδρος, ον, *having no husband*, *BGU* 2462.14 (ii AD).

*ἀπανηγύριστος, ον, *unfit for a festival speech*, Eust. 1569.56.

ἀπανθίζω, after 'A.*Ag*. 1662' insert '(dub.)'

ἀπανθρᾱκισμός, ὁ, gloss on ἀποκραιπαλισμός, Hsch. (dub.).

ἀπάνθρωπος **II 1**, after 'S.*fr*. 1020' insert 'ἀ. ἄνθρωπος Men.*Dysc*. 6'

*ἀπάνονα, gloss on φόρβον, Hsch.

ἀπανουργευτος, read ἀπανούργευτος

*ἀπανούργητος, = ἀπανούργευτος, Sch.E.*Ph*. 469.

*ἀπανταίνω, *meet*, ἀνερχομένου ἔρχομε ἀπαντάνω σου PAmst. 56.6 (vi AD).

*ἀπαντᾰχόσε, *in every direction*, Gal.*UP* 12.5, 15.4.

ἀπαντάω **I 1**, add '**d** *approach with a request*, POxy. 1242 (iii AD).'   **2 a**, add 'αἰ[δέ] ἀπαντᾶι *if someone attacks*, *SEG* 23.563 (Axos, iii BC); *counter* an argument, ἴσως ἄν .. τοῖς εἰρημένοις ἀπαντήσειας; of a jury, *face, deal with* a line of argument, D. 21.24, 31.36'   **II 1**, add 'abs., Μενουθ[εσ]ιὰς ἀπαντᾷ νῆσος *comes upon one, comes next*, Peripl.M.Rubr. 15'

ἀπαντή, add '*SEG* 37.1186.41 (Pisidia, iii AD)'

ἀπάντησις **I**, for '= foreg.' read '*the action of going out to meet* an arrival, esp. as a mark of honour'; delete '*escort*' and add 'Cic.*Att*. 8.16.2, 9.7.2, 16.11.6'

ἀπαντητήριον, before 'PIand. 17' insert '*IGLS* 1.1316 (Apamea), *SEG* 32.1496 (Palestine), 36.1222 (Isauria, all v or vi AD)'

*ἀπαντοτρόφος, ον, *nurturing everything*, γαῖα Cyran. 1.7.40 K.

ἅπαξ **I 1**, line 1, for '*once only, once for all*' read '*on one single occasion*'; line 8, after 'Lxx 2*Ki*. 17.7' insert '**b** math., of a term taken *once*, Papp. 100.24, PMich.III.145.3 iv 6 (ii AD)'; after 'Pl.*Men*. 82c' insert 'ἅ. μετρεῖ Papp. 12.7'; add 'πρὸς ἅπαξ, of payment made *once*, a lump sum, Cod.Just. 6.48.1.14'   add '**III** *sometimes*, A.D.*Adv*. 170.18, Conj. 252.22, Thom.Mag. p. 200 R., al.'

ἁπαξαπλῶς, for '*in general*' read '*at one go, at one and the same time*'

ἀπαξία, add 'Iamb.*Protr*. 20 (94.24 P.)'

ἅπαξις, after '*arrest*' insert '*and taking before the magistrates*'

ἀπαράβατος line 7, for '3 act.' read '4 act.' and after 'adv. -τως' add 'Herod.Med.ap.Aët. 9.2'

*ἀπαραβίαστος, ον, *inviolable, indefeasible*, Polystr. p. 88 W.

ἀπαράβλητος, add '*IStraton*. 303.8; perh. *not to be set aside*, *CPR* 5.11.6 (?iv AD)'

*ἀπαράβροχος, ον, '*without excess dampness*, of a grain cargo, PRoss.-Georg. 2.18.129, al. (ii AD), PMich.inv. 3781.6 in *ZPE* 62.140 (v AD); cf. ἄβροχος, mod. Gk. παραβρέχω'

ἀπαραδειγμάτιστος, for '*not liable to censure*' read '*inconspicuous* (cf. °εὐπαραδ-)'

*ἀπαράδοτος, ον, *not transferred*, PFam.Teb. 15.62 (ii AD).

ἀπαραίτητος **I**, add '*Trag.adesp*. 495 K.-S.'

ἀπαρακάλυπτος, adv. -τως, add 'Hld. 8.5.6'

ἀπαράκλητος **I**, add 'adv. -τως, *SEG* 7.62 (Seleucia in Pieria, ii BC)'

ἀπαραλήκτως, delete the entry.

*ἀπαράπ‹ε›ιστος, gloss on ἄπιστος (app. in sense *disobedient*), Hsch.

*ἀπαράπτωτος, ον, *firm*, Poliorc. 200.4.

ἀπαράσημος **II**, for this section read '*without a title*, of a speech, Antipho 2.1 tit., Lys. 21 tit.'

ἀπαρατήρητος, before 'adv.' insert '*not requiring special precautions*, i.e. *favourable*, ἔστι δὲ εἰς τὸ πλεῦσαι ἀπαρατήρητος (ὥρα τοῦ Ἡλίου) Cat.Cod.Astr. 11(2).131.18'; add '*without supervision* or *hindrance*, Didyma 314.10 (ii AD, v. Robert Hell. 11/12.465)'

ἀπαργμα **I**, after '= ἀπαρχή (q.v.)' insert 'perh. in IHistriae 101 (vi BC, v. *SEG* 33.582), *IG* 9(2).1135.6 (epigr., Demetrias, i AD; see also *SEG* 23.451)'

ἀπαρενθύμητος, delete the comma after '*considering*'

ἀπαρήγορος, for 'Epigr.Gr. 344.2 (Mysia)' read 'epigr. in IHadr. 80 (ii AD)'

ἀπάρθενος, delete 'virgin wife and widowed maid'

ἀπαριθμέω **I**, add '**2** of an instrumentalist, *measure out* a rhythm, ἐπὶ τῶν ὀργάνων D.H.*Comp*. 25.'

ἀπαρκέω **II**, for this section read 'pass., *to be contented*, Cerc. 18 ii 13, Lyc. 1302'

ἀπαρκτίας, after '*Gloss*.' insert '*IEryth*. 304.4 (i BC), πρὸς ἀπαρκίαν Peripl.M.Rubr. 30'

ἀπαρνέομαι, lines 1/2, after 'Pl.*Grg*. 461c' insert 'also -ηθήσομαι S.*Ph*. 527' and after 'Att.' insert '(also Herod. 4.74)'   for '*deny utterly*' to end of article read '*say no* to an assertion, *deny* a fact, Hdt. 6.69.2, Antipho 2.3.4, S.*Tr*. 480, καὶ νῦν γέ φημι κοὐκ ἀπαρνοῦμαι E.*El*. 1057, Pl.*Tht*. 165a; w. inf., ἀπαρνήσονται δ' εἰκότως .. Ἀλέξανδρον αὑτοὺς ἠρεθικέναι Phld.*Rh*. 1.359.12; w. μή, τὸν τἄμ' ἀπαρνηθέντα μὴ χρᾶναι λέχη E.*Hipp*. 1266; w. double neg., τίνα οἴει ἀπαρνήσεσθαι μὴ οὐχὶ .. ἐπίστασθαι Pl.*Grg*. 461c; ellipt., φημὶ δρᾶσαι κοὐκ ἀπαρνοῦμαι τὸ μή S.*Ant*. 443.   **2** in logic, *deny* a proposition, opp. κατηγορεῖν, Arist.*APr*. 41ᵃ9; in pass. sense, ib. 63ᵇ37.   **II** *refuse to accept, reject*, ὅτ' Ἀχαιιάδες μιν ἀπηρνήσαντο πόληες ἐρχομένην Call.*Del*. 100; θεοῦς Luc.*Peregr*. 13; w. things as obj., Arist.*EN* 1127ᵇ25, Lxx *Is*. 31.7, Nonn.*D*. 47.478.   **b** *refuse to recognise, deny*, ὅτι .. πρὶν ἀλέκτορα φωνῆσαι, τρὶς ἀπαρνήσῃ με *Ev.Matt*. 26.34, *Ev.Marc*. 14.30, etc.; refl., *Ev.Matt*. 16.24, etc.   **2** *refuse* or *reject* a course of action, Hp.*Praec*. 12, τὸν .. Ἁρμόδιον ἀπαρνηθέντα τὴν πείρασιν Th. 6.56.1, Pl.*Sph*. 217c, Herod. 4.74, D.C. 51.7.1; w. inf., Hp.*Ep*. 9.328.12; abs., S.*Ph*. 527.'

ἄπαρσις, after '*departure*' insert 'Dicaearch.fr. 34 W.'

ἀπάρτησις **I 2**, delete 'metaph. .. 5.1.2'   **II**, add 'Plot. 5.1.2'

†ἀπαρτί, adv. *exactly, just*, ἡμέραι ἀ. ἐνενήκοντα, Hdt. 5.53, ἀ. ἐναρμόζειν πρός τι Hp.*Art*. 73, ἀ. ἐν καιροῖσι id.*Acut*. 41.   **II** (w. intensifying force) *altogether, absolutely*, φρόνιμος ἰν ἀ. ταύτης τῆς τέχνης Telecl. 39 K.-A., Pherecr. 77 K.-A., τί .. ἀποτίνειν τῳδ' ἀξιοῖς; – ἀ. δή που προολαβεῖν παρὰ τοῦδ' ἐγὼ μᾶλλον, id. 93, Ar.*Pl*. 388.   **III** often written as two words ἀπ' ἄρτι *from now on*, οὐ μή με ἴδητε ἀπ' ἄρτι, ἕως ἂν εἴπητε .. *Ev.Matt*. 23.59, 26.64, *Ev.Jo*. 14.7, *Apoc*. 14.13; w.

pres. tense, ἀπ' ἄρτι λέγω ὑμῖν πρὸ τοῦ γενέσθαι, ἵνα πιστεύητε .. Ev.Jo. 13.19.

ἀπαρτίζω II 1, line 1, after 'Plb. 31.12.10' insert 'ἡ σύμβιος .. ἀπήρτισαν ἑαυτοῖς .. of a funerary monument, SEG 30.1548 (Cilicia, iv/vi AD); Νέστωρ ἀπή[ρτι]σεν app. an artist's signature, Robert Villes p.169 (Lycia, Rom.imp.)'; line 6, after 'J.AJ 3.6.7' insert 'χαλεινοῦ ἱππικοῦ μετὰ τοῦ μασήματος ἀπηρτισμ(ένου) horse's bridle ending in a bit, i.e. with bit attached, Edict.Diocl. in SEG 37.335 iii 5; (cf. mod. Gk. ἀπαρτίζω form, constitute)'

ἀπαρύω I 2, for 'Epic.ap.Arist.Po. 1457ᵇ14' read 'Emp. 138'

ἀπαρχή 2, line 7, after 'Hdt. 4.88' insert 'τὰ ἐξ ἀπαρχᾶς δοθέντ[α SEG 23.566 (Axos, iv BC)'   **6**, for this section read 'app. *a board of officials* sent by a city to order the affairs of a colony, IG 12(8).273, 280, 283, 285 (Thasos, vi BC, v. esp. BCH 107.185); cf. ἀπάρχης'

⁺ἄπαρχος, ὁ, *commander*, A.Pers. 327 (v.l. ἐπ-), Ag. 1227.

*ἀπάσχαρος, v. °ἀπέσχαρος.

ἀπατάω, lines 1/2, impf., after 'Ion.' insert 'ἀπάτασκον Cercop. 3'; at end after 'Pl.Prt. 323a' insert 'w. inf., συνέσεσθαι ἀπατώμενον Hld. 1.30.6'

*ἀπατενίζω, gloss on *degenero*, Dosith. p. 436 K. (nisi leg. ἀπαγενίζω).

ἀπάτερθε I, adv., add 'h.Merc. 403'   **II**, add '*far short of, unlike*, οὐκ ἀ. Opp.H. 1.725'

ἀπάτη II, at the beginning insert '*the act of deceiving*, ἐν ἐκείνοις .. ἡγεῖται τὸ πιθανόν, κἂν ᾖ ψεῦδος, διὰ τὴν ἀ. τῶν θεωμένων Plb. 2.56.12, hence ..'

ἀπατήμων, for 'Zos. 1.52' read 'Zos. 1.57'

ἀπατήνωρ I, after 'Tryph. 137' insert 'Nonn.D. 26.118'   **II**, for pres. ref. read 'Euph. in Suppl.Hell. 418.25'

ἄπατος, before the pres. ref. insert 'Inscr.Cret. 2.xii 3.5, 11.15 (Eleutherna, vi BC); impers. w. dat. τὸι .. δικασσται .. ἄπατον ἤμην Inscr.Cret. 4.42 B.11-14' and after it add 'impers. w. acc., τοὺς πρειγ[ίσ]τους .. πράδδοντας ἄπατον ἔμεν Inscr.Cret. 4.80.11-12; cf. ἄνατος II 2'

Ἀπατουρεών, written **Ἀπατουριών**, add 'SEG 30.977 (Olbia, v BC), Thasos 18.5 (v BC), Inscr.Prien. 51 (ii BC), SEG 30.1353 (Miletus, ii/iii AD)'; written **Ἀπατοριών**, add 'Thasos 171.29 (i BC), SEG 33.679 (Paros, ii BC)'

Ἀπατουρία, for 'Aphrodite' read 'Athena'; after 'Ἀπατούρη' insert 'title of Aphrodite' and after '(Panticapaeum)' add '(iii AD), Ἀπατόρης on vase, prob. of Aphrodite, SEG 30.879 (Berezan, v BC)'

ἀπατρία, delete the entry.

ἀπαυαίνω, delete '*make to wither* .. 3.10.7'

*ἀπαυγή, ἡ, *radiance*, Call.fr. 273 Pf. (ἀγαυ- codd.).

ἀπαυδάω II, add 'app. in ICS 318 A II 3 (vi BC)'   **III**, at end for 'πόνοις' read 'πόθοις'

ἀπαυθαδιάζοντες, med., add '*refuse to admit one is in the wrong*, Arr.An. 4.9.6'

ἀπαύλια (ἀπαυλία), for 'EM 119.14 is confused' read 'EM 119.14 and Hsch. are confused'

⁺ἀπαυλόσυνος, ον, *away from one's shelter*, AP 6.221 (Leon.Tarent., s.v.l.).

ἀπαύξησις, delete 'hence, *disesteem*'

ἄπαυστι, for 'sq.' read 'ἄπαυστος'

ἀπαφρίζω, to act. exx. add 'Gal. 14.13.7'

*ἀπεγγόνη, ἀπέγγονος, v. °ἀπέκγονος.

*ἀπεγγυάω, *pledge, give security*, ἀπενγυάτω ὁ [κατειπὼν τὴν ἀπεγγύην] Thasos 7.7 (v BC, see also SEG 36.790).

*ἀπεγγύη, ἡ, *security*, rest. in Thasos 7.8 (v BC), v. °ἀπεγγυάω.

ἀπέδιλος, after 'Nonn.D. 5.407, al.' insert 'fig., [ἀπ]εδίλος ἀλκά i.e. perh. coming without delay, Alcm. 1.15 P.'

⁺ἀπειδοποιέω, *give final form to*, κατεξέσθη τὸ ὑπέρθυρον καὶ ἀπειδοπο[ι]ήθη Didyma 32.19.

ἀπειθής I 2, line 3, for '*impracticable*' read '*inexorable*'

*ἀπειθόω, *to be disobedient*, Inscr.Délos 1417Bii150 (ii BC).

ἀπειλέω (B) I, line 3, after 'Il. 23.863, cf. 872' insert 'Call.fr. 18.6 Pf.'   **II 3**, at end delete 'Theocr. 24.16' and add 'b *order with threats*, w. inf., Theoc. 24.16, A.R. 3.607.'

Ἄπειλων, ωνος, ὁ, Cypr. *a-pe-i-lo-ni* (dat.), = Ἀπόλλων, ICS 215.4 (Tamassos, elsewh. Ἀπόλλων); cf. ‡Ἀπέλλων.

ἄπειμι (A), add 'Myc. 3 pl. pres. *a-pe-e-si*, part. *a-pe-o*'

ἄπειμι (B) 1, at end, μηνὸς ἀπιόντος, add 'IG 4².83 (Epid., i AD), 7.506, Illion 51.30, 63.2, 4; μη(νὸς) Λῴον θ' ἀπιούσῃ TAM 5(1).129, μη(νὸς) Δύστρου ς'.. ἀ(πιούσῃ) ib. 93 (Saittae, iii AD)'

ἀπεῖπον, line 4, after 'med.' insert 'fut. ἀπεροῦμαι AP 12.120 (Posidipp.)'   **IV 1**, add 'Thgn. 89'

ἀπειράκις, add 'ἐξ ἀ. ἀπείρων, Procl.Inst. 1 (p.2.10, 11) D.'

*ἀπείραντος, ον, = ἀπείρων, Pi.P. 9.35.

ἀπείρατος I, add '(v.l., edd. ἀπειρίτῳ)'

ἀπειρέσιος, line 2, after 'Od. 11.621' insert 'Thgn. 8'; line 5, delete '(Q.S.) 3.386'

ἀπείριτος, after 'al.' insert 'cj. in Pi.O. 6.54 (v. ἀπείρατος)'; add 'also sg., A.R. 3.971, Q.S. 3.386; Lacon. ἀπήριτος Alcm. 7.14 P.'

*ἀπειρόμογος, ον, *unused to toil*, ἀ]πε[ιρ]ομόγῳ Διονύσῳ PMich.inv. 4926a (ZPE 93.157, iii AD).

ἄπειρος (B), line 3, delete 'πλῆθος Hdt. 1.204' and add 'πεδίον .. πλῆθος ἄπειρον ἐς ἄποψιν Hdt. 1.204' to previous group.

ἄπειρος, Ἄπειρος (C), v. ἤπειρος III.

*ἀπειρότεχνος, η, ον, *having unlimited skills*, Φοίβῃ ἀ. orac. ap.Lyd.Mens. 3.10.

ἀπειρότοκος, for 'Antip.Sid.' read 'Antip.Thess.'

ἀπείρων (B), add 'δι' ἀπείρονα γαῖαν E.Phaëth. 243 D., Chaerem.Hist. 10 vdH.'

Ἀπειρώτας [ᾱπ-], v. ‡ἠπειρώτης III 2.

ἀπέκ, add 'Q.S. 4.540, 14.230, ἀπὲκ ἀέρος PMich.XIII 665.21, 30, 60 (vi AD)'

ἀπέκγονος, for pres. ref. read 'AP 13.26 (?Simon., = Page FGE no. 36)'; add 'also ἀπέγγονος Cod.Just. 6.4.4.10, Just.Nov. 91.1; also fem., ἀπεκγόνη Theophil.Antec. 3.6.4'

ἀπεκδέχομαι I, add 'μαστῷ πόρτιν ἀπεκδέχεται AP 9.722 (Antip. Sid., = Page EG 3595, v.l. ὑπεκ-)'

*ἀπεκλείπω, *go away and leave*, τὸν βίον BCH 10.302.34 (Alabanda, ii BC).

*ἀπεκτίθημι, *remove and reject*, [μύρμηκες] τῶν καρπῶν τὰς ἐκφύσεις ἀπεκτιθέασιν Cels.ap.Orig.Cels. 4.83.

*ἀπελαστικός, ή, όν, *capable of driving away*, μιασμάτων Sch.Theoc. 2.35, Eus.PE 4.1.9.

ἀπελαύνω, line 5, after 'pf.' insert 'ἀπελήλαμαι Alc. 130.23 L.-P.'   **II**, after '*driven away*' insert 'ἀπὸ τούτων Alc.l.c.'

*ἀπελέκιστος, ον, = ἀπελέκητος, IG 2².1678ᵇA7 (iv BC).

ἀπελευθεριάζω, add '**II** prob. *emancipate, manumit*, τὰν .. ἀπηλευθερια(σ)μέναν ὑπὸ Σωτηρίχου Delph. 3(3).388.4 (i BC).'

*ἀπελευθεριόω, *emancipate, manumit*, Delph. 3(3).311.10 (i BC).

⁺ἀπελλάζω, *hold an assembly*, Plu.Lyc. 6, ἀπελλάζειν· ἐκκλησιάζειν. Λάκωνες Hsch.

ἀπέλλαι, after 'ἐκκλησίαι' insert 'τὰ δὲ ἀπελλαῖα ἄγεν Ἀπέλλαις καὶ μὴ ἄλλαις ἀμέραις CID I 9.A31, cf. D3 (Delphi, v BC)'

ἀπελλαῖα, after 'Delph.' insert 'v BC'

ἀπελλαῖος, for 'etc.' read 'in Lydia SEG 29.1184 (AD 152/3); also at Ephesus (beginning Oct. 24) and elsewhere, Hemerolog.Flor. p. 79 (p. 22 K.), Cat.Cod.Astr. 2.149.1'

*ἀπελλογαρίζω, *render an account*, PLond. 1708.104 (vi AD).

Ἀπέλλων, for 'cf. Ἀπείλων Inscr.Cypr. 140 H.' read 'cf. °Ἀπείλων'

*ἀπεμβλέπω, *look away and within*, i.e. *concentrate on looking within*, v.l. in Pl.Chrm. 160d.

ἀπεμπολάω, after 'pass.' read 'ἀπεμπολώμενοι, *disposed of by sale*, Ar.Ach. 374,'Ἰθακησίαν τινὰ ὑπὸ Φοινίκων ἀπεμποληθεῖσαν Certamen 26'

⁺ἀπεμπολή, ἡ, app. *sale*, Call.fr. 203.27 Pf.; ἀπεμπολήν· ἀπαλλαγήν. πράσιν. ἐμπορίαν Hsch., ἀ.· ἀπόστασις, ἡ μετὰ ἀπάτης πρᾶσις, καὶ ἐμπορία Suid.

*ἀπέμφασις, εως, ἡ, *false sense-impression*, opp. ἔμφασις, Carneades ap.S.E.M. 7.169.

ἀπεναρίζω, add 'ἀπηναρίσθη Hippon. 68.7 W.'

ἀπενεκτική, (sc. πτῶσις), ἡ, *ablative case*, Dosith. p. 392 K.

*ἀπένθεια, ἡ, *absence of mourning*, dub. cj. in A.Ag. 430 (codd. πένθεια).

ἀπενθής, after 'grief' insert 'or *lamentation*'

*ἀπενθησία, ἡ, app. *relief of misfortune* or *grief*, SEG 36.970.A6 (Aphrodisias, iii AD).

*ἀπέννοια, ἡ, prob. calque on Heb. *mᵉzimmāh* (evil devices) *craft, cunning*, Aq.Ps. 138(139).20.

ἀπεξαιρέω, add 'ἀπὸ δ' ἐξείλετο θεσμὸν μέγαν Anacr. 61 P.'

*ἀπεξαρθρέω, *dislocate*, PRyl. 529.37 (iii AD).

ἀπεξηγέομαι, for 'cj. in X.Eph. 5.9' read 'X.Eph. 5.9. (cod.)'

ἄπεπλος, delete 'i.e. *in her tunic only*, of a girl' and after 'Pi.N. 1.50' insert 'Pae. 20.14'

*ἀπεπνός, ὁ, or -όν, τό, name of a plant, Paul.Aeg. 4.22.2 (344.17 H.).

ἀπέραντος II, line 2, for 'A.Pr. 1087' read 'A.Pr. 1078'

ἀπεράω I, after '*disgorge*' insert 'ἀπὸ σφαγῆν ἐρῶν A.Ag. 1599'

ἀπεργάζομαι III 2, for 'Pl.Riv. 135e' read 'Pl.Amat. 135e'

ἄπεργος, add '**III** ἄπεργον, τό, *unworked part* of ashlar, *spall* or *offcut*, λίθοι ἀ. ἔχοντες IG 2².1666.A98, B48, 70 (Eleusis, iv BC), cf. ἀργός (B) II 1.'

ἀπερείδω I 1, add 'of a scorpion *planting* its sting, Ael.NA 16.27'

ἀπέρεισμα, after 'Hsch.' insert 's.v. ἀπόσκημμα'

ἀπερίγραπτος, add '*not marked for exclusion* or "*bracketed*" ἀπερί[ι]γρα(πτος) ἂν (αὕτη, sc. ᾠδή) Sch.Alcm. 3 fr. 1 P.'

ἀπερικάλυπτος, delete the entry.

*ἀπερίμετρος, ον, *boundless, infinite*, Apul.Plat. 190.

ἀπερίσπαστος, line 4, for 'Plu. 2.521c' read 'Plu. 2.521d'

*ἀπερίστροφος, ον, *unable to turn, immovable*, Afric.Cest. p. 8 V.

ἀπερίτμητος **I**, add '(non-Jewish instance) *PCair.Zen.* 76 (iii BC)'
ἀπερίτρεπτος, after 'adv. -τως' insert '*irrefutably*'
ἀπερίττωτος, for 'φύσις' read 'φυτόν'
ἀπερύκω **1**, med., last line, read 'later in act. sense στυγερὰς δ' ἀπερύκεο νούσους Maced.*Pae.* 31; w. gen., *rescue* from, ἀπερύκεο νούσου Nic.*Al.* 608'
ἀπέρχομαι **I 3**, delete 'κᾇτ' .. 689' **4**, read 'w. part. indicating outcome of action κᾇτ' ὀφλὼν ἀπέρχεται Ar.*Ach.* 689, πλέον ἔχοντες ἀπέρχεσθε Isoc. 17.57, Men.*Dysc.* 52, Plu.*Ages.* 7, Aristid.Mil. 2.2 J.'
*ἀπεστενωμένως, adv., *in a restricted manner*, Simp. *in de An.* 44.37.
*ἀπέσχαρος, ον, *scab-removing*, ζίριον (i.e. ξηρίον) ἀπάσχαρον (sic) *PMed.Rez.* 12.17.
*ἀπέταιρος, ὁ, one who is not a member of a ἑταιρεία, *Leg.Gort.* II 5, 25; cf. ἀφέταιρος.
ἀπέτηλος, add 'comp., *AP* 9.231 (Antip.Thess.)'
*ἀπευδοκέω, *satisfy*, τινὰ τῆς τιμῆς in respect of the price, *PLeid.*P. 3 (ii BC).
ἀπευλυτέω, add 'so prob. *PMich.*v. 243.9 (i AD)'
ἀπευτακτέω, to act. exx. add '*PSI* 360.19 (iii BC), *PTeb.* 40.22 (ii BC), *PRyl.* 578.9 (i BC)'
ἀπευχαριστέω, after '*show gratitude*' insert 'τοὺς ἀσπασαμένους τε αὐτὸν .. καὶ ἀπευχαριστήσοντας ἐπὶ τοῖς γεγονό[τ]οις εὐεργετημάτοις *Delph.* 3(3).239.9 (ii BC)' and after 'τινί' insert '*BGU* 2418.9 (ii BC), *GVI* 1232.7 (ii/i BC)'
*ἀπέφατο (A)· ἀπέθανεν Hsch. (v. θείνω).
*ἀπέφατο (B)· ἀπείπατο. ἀπεφήνατο Hsch.
†ἀπέφθιθον, v. ἀποφθίνω.
ἄπεφθος, line 2, after '*refined* gold' insert 'Ibyc. 1(*a*).43 P.'
ἀπέχω **III 1 c**, for this section read '*to be distant from* in time, ἀπόσχων τεσσαράκοντα μάλιστα σταδίους μὴ φθάσαι ἐλθών Th. 5.3.3; προειπὼν ὅτι οὐδὲ ἐκεῖνος ἀπέχει μακρὰν τῆς τοῦ βίου τελευτῆς Aeschin. 1.14.6, ἀπεχούσης δὲ χρόνον ἱκανὸν τῆς .. ἡμέρας D.S. 20.110.1' **IV**, *have or receive in full*, add '*D.fr.* 23 S.'
ἀπεψία, add 'ἀπεψίαις χρώμενος *IG* 4².126.3 (Epid., ii AD)'
ἀπηθέω, add 'also ἀφηθέω (q.v.); cf. ἠθμός'
ἀπηλεγέως, for '*A.R.* 4.687' read '*A.R.* 4.689'
†ἀπηλλαγμένως, adv., *in a way which is free from*, ὀργῆς ἀ. ἐπιψηφίζειν J.*AJ* 17.124, Alcin.*Intr.* p. 164.15 H.
ἀπήμαντος **I**, read '*free from harm* Od. 19.282; *free from sorrow or misery*, Hes.*Th.* 955, Simon. 526.3 P., ἀ. βίοτος Pi.*O.* 8.87, A.*Ag.* 378 (lyr.); in later prose, IStraton. 10.18 (Panamara, i BC), Sopat.Rh. 8.264.27 W.'
†ἀπηναῖος, α, ον, *belonging to a wagon*, ὀρῆες Call.*fr.* 85.5 Pf.
ἀπήνη **2**, after 'Str. 4.5.2' add '*state chariot*, poet. symbol of the *sella curulis*, καθαρῆς μετὰ μόχθου ἀ[π]ήνης *IEphes.* 1310'
ἄπηρος, for 'Hsch.' read 'Suid. s.v. ἄπηρα'
ἀπηρτημένως, after 'Plu. 2.105e' insert '(v.l.)' **II**, at beginning, insert '*discontinuously*, of hymns, μὴ ἀπηρτημένως ἀλλὰ συνεχῶς πλάττειν Men.Rh. 341.20 R.-W.'; after 'M.Ant. 4.45' add '(codd.)'
ἀπηχέω **I 2**, for this section read '*cause to ring out* or *resound*, φωνάς Arr.*Epict.* 2.17.8, Ph. 1.693, 2.44' **II**, delete 'to be discordant .. 2.44'
ἀπήχημα 1 and 2, for these sections read '*echo* (in quots., fig.) Pl.*Ax.* 366c, ὕψος μεγαλοφροσύνης ἀπήχημα Longin. 9.2, ἔτι γὰρ ἀπήχημα φέρουσι τῆς ἐκεῖ ζωῆς Procl.*in Alc.* p. 99 C., p. 135 C.' **3**, add 'Paul.Aeg. 6.90.2 (137.10 H.)'
ἀπηχής, for 'Aristid.*Or.* 40(5).8' read 'Aristid.*Or.* 29(40).8'
*Ἀπιδανῆες, οἱ, *Peloponnesians*, Call.*Jov.* 14, A.R. 4.263, Rhian. 13.
ἀπίθανος **III**, line 5, after 'Id.*Pseudol.* 16' insert 'ἀπίθανος ἐν τῷ λαλεῖν Men.*Dysc.* 145'
*ἄπικες, οἱ, Lat. *apices*, *conical hats*, τὰς καλουμένας ἄπικας ἐπικείμενοι ταῖς κεφαλαῖς D.H. 2.70.
ἄπικρος, after 'Ptol.*Tetr.* 158' add '(v.l.); app. *without sting, ineffectual*, γένηται καὶ ἀσθενὴς καὶ ἄπικρος *SEG* 35.223.7 (tab.defix., Athens, iii AD)'
*ἀπιομήδης, ες, (= ἠπιο-), *having kindly purposes*, rest. in Pi.*Pae.* 7.7 (pap. ἀπιομ[ήδ]ει).
ἄπιον **2**, delete the section. add '**II** Lat. *apium*, Paul.Aeg. 7.4; written ἄποιον in Orib.*fr.* 52.'
ἄπιος (A) **I 1**, for '*CP* 1.15.2' read '*CP* 1.15.1'
Ἆπις **I 1**, after 'Hdt. 2.153, etc.' add '(dat. Ἄπι); in Syria, *SEG* 31.1383 (dat. -ιδι, ii AD)'
ἀπίσσωτος, for 'Str. 11.10.2' read 'Str. 11.10.1'
ἀπιστέω **II 1**, after '*disobey*' insert 'Antipho *fr.* 21'
ἀπιστία **2**, add '(σημεῖον) τεράστιόν τε καὶ βροτοῖς ἀπιστία *source of disbelief*, Ezek.*Exag.* 91 S.'
ἄπιστος **II**, line 3, transfer 'ὦτα .. Hdt. 1.8' to the end of section I 1 and for '*credulous*' read '*trustworthy*'
ἀπίσωσις, add 'εἰς τὴν ἀφίσωσιν τῶν μερῶν *PDura* 19.14 (i AD)'

*ἀπίτυρος, ον, *free from husks*, κριθαί Hsch. s.v. ἀτυπῆδες (ἀτυπίδες La.).
ἄπλαστος, after '**II**' read 'app. = ἄπλατος, (w. which it frequently coexists as a variant), Hes.*Th.* 151, *Op.* 148'
ἀπλατής, line 2, after 'id.*Top.* 143ᵇ14' insert 'Euc. in *PMich.*III 143.3'
ἄπλᾱτος, line 3, for '(whence it must be restored .. A.*Pr.* 373' read 'cj. for ἄπληστος in A.*Pr.* 371'; line 7, delete 'E.*Med.* 151 (lyr.)' **2**, after '= ἄπλετος' insert 'χρόνου μῆκος Archestr. 59.9; σταδίων *EG* 1613 P. (Posidipp.)'
ἀπλετομεγέθης, for '*unapproachably great*' read '*immensely great*'
ἄπλετος, line 2, after 'S.*Tr.* 982' insert 'ἰχθύες A.R. 1.574'
ἄπληγος, after '(πληγή)' insert '*not bearing marks of blows*, *PRein.* 92.11 (iv AD)'
*ἀπληκεύω, *bivouac*, Suid. (perh. ἀπληκ⟨τ⟩εύω).
*ἄπληκτον, τό, *bivouac*, *PLond.* 1416.23, Suid.
ἄπληκτος, for 'Ep. .. *great*' read '= ἄπλετος, *immense, monstrous*, ἰσχύς Hes.*Th.* 153 (s.v.l.); ὄτοβος ib. 709; χόλος *h.Cer.* 83'; add 'neut. sg. as adv., ἄπληκτον κοτέουσα Hes.*Th.* 315, cf. *Sc.* 268, Semon. 7.34'
*ἄπλιος, ον, adj., applied to a textile, often coupled w. μονόβαφος, *Edict.Diocl.* 24.5, 29.20, al.; neut. subst., in inventory of items taken on a journey, *PRyl.* 627.161 (iv AD); v. °ἁπλόος III e.
*ἁπλόγραμμος, ον, app. *drawn in a single line*, Poliorc. 237.8.
ἄπλοια, after '(E.)*IT* 15' insert '*Trag.adesp.* 637.13 K.-S.'
ἁπλοϊκός, add '**b** *simple-minded*, Hld. 1.11.'
ἁπλόος, line 1, after 'ον' insert '(Cret. as two-termin. adj. acc. fem. pl. ἁπλόονς, *Inscr.Cret.* 4.72 I 48 (Gortyn); voc. ἁπλόε *Inscr.Délos* 1533.3 (Antisthenes of Paphos, hex.))' **I a**, add 'τὰς στοὰς τάς τε ἁ. καὶ τὰς διπλᾶς τοῦ ἱεροῦ perh. *single-aisled*, *SEG* 30.1535 3 (ii AD); neut. subst., perh. *original* of a document, τῆς ἐπιγραφῆς ἁπλοῦν ἀπετέθη εἰς τὸ ἀρχεῖον *SEG* 30.1352 (Miletus, iii AD), cf. τὸ δὲ συναίρεμα τοῦτο δισσὸ(ν) γρα(φὲν) ἐπὶ τῷ ἁπλο(ῦν) συνηγηθῆναι *PTeb.* 340.16 (iii AD)' add '**c** *flat, plane*, ἁπλοῖ πήχεις *POxy.* 2145.6 (ii AD); opp. καμαρωτικοί, ib. 921ʳ (iii AD).' **II c**, *simple-minded*, add '*Cod.Just.* 1.3.29.1' add '**d** *in good condition, sound*, ἐὰν οὖν ᾖ ὁ ὀφθαλμός σου ἁπλοῦς .. ἐὰν δὲ ὁ ὀφθαλμός σου πονηρὸς κτλ. *Ev.Matt.* 6.22, *Ev.Luc.* 11.34.' **III 1 a**, add 'neut. as subst., τὸ ἁπλόον *simple penalty*, *Inscr.Cret.* 4.41 I 4 (Gortyn)' **c**, for this section read 'of coinage, *unmixed*, i.e. of pure alloy, δραχμαὶ *SEG* 12.226.9 (Delphi, iv AD), *ABSA* 23.95 no. 20 (Thessalonica), χρυσοῦ νομισμάτια *PMasp.* 41' add '**e** of a quality of ἱματισμός, perh. of *single* size, *plain* or *fine* (unadulterated), *Peripl.M.Rubr.* 24, 28; cf. °ἁπλοπάλλιον, °ἄπλιος, °ἁπλουργός.'
*ἁπλοπάλλιον, τό, *cloak* or *cloth* of a certain quality, perh. *single* or *plain*, *PVindob.* G41673 (*Tyche* 1.88; vi/vii AD); cf. °πάλλιον, °ἁπλόος III e.
*ἁπλοπότιον, τό, wd. in list of foodstuffs, *PRyl.* 627.88 (iv AD), precise meaning uncertain.
ἁπλότης **II 1**, after 'D.H.' insert '*Is.* 4' add '**4** *ignorance, backwardness*, Just.*Nov.* 73.9.'
Ἄπλουν, for '*IG* .. Larissa)' read '*IG* 9(2).1027, 9(2).512.19 (both Larissa, v and ii BC)'
*ἁπλουργός, ὁ, kind of textile worker, *JHS* 56.79 (Laodicea ad Lycum), v. °ἁπλόος III e.
*ἄπλυντος, = ἄπλυτος, Sch.Nic.*Al.* 469a Ge.
ἄπλυτος, line 2, after 'Semon. 7.5' insert 'σήσαμον *IG* 1³.422.151 (Athens, v BC)'
ἄπλωμα, add 'perh. *open space*, in front of rooms of a monastery, *SB* 5174.5, 9, 16 (vi AD)'
ἁπλῶς **II 1 b**, after 'D. 18.308, etc.' insert 'ἁπλόως καὶ ἀδόλως *Inscr.Cret.* 4.174.8 (ii AD)'
*ἁπλωστί, adv. = ἁπλῶς II, cj. in A.*Ch.* 121.
ἄπλωτος, add '**2** *not transportable by water*, σῖτος *PTeb.* 703.73 (iii BC, s.v.l.).'
*ἁπλωτός, όν, *spread out*, δίκτυον .. ἁ. *AP* 6.185.3 (Zos., cj.); *prostrate*, Sisyphus *fr.* 1.6 (cj. for πλωτός, v. *Byzantion* 6.468).
ἄπνους **II 2**, after '*lifeless*' insert 'Call.*Epigr.* 5.9 Pf.' add '**4** *without scent*, Call.*fr.* 43.14 Pf.'
ἀπό, line 3, after 'etc.' insert 'w. apocope, ἀπ πατέρω[ν] Alc. 6.17 L.-P., ἀτ τᾶς πρεισβείας *IG* 9(2).517.12 (Larissa, iii BC)' **A I 2**, add 'indicating distance already travelled, εἰσπλεόντων ἀπὸ χιλίων ὀκτακοσίων σταδίων *Peripl.M.Rubr.* 1, al.' **9**, add 'ἀσπό(ρου) ἀπὸ κρι(θῆς) of unsown land *formerly* under barley, *BGU* 2441.118 (i BC)' **II**, line 10, ἀφ' ἧς, add 'Lxx 1*Ma.* 1.11, *Ev.Luc.* 7.45, *Act.Ap.* 24.11, 2*Ep.Pet.* 3.4'; line 18, after '(iv AD)' insert 'ἀ. προέδρων *PFlor.* 71.521, al.'; line 19, after 'Hdn. 7.1.9, etc.' insert 'ἀπὸ πριμιπιλαρίων = *e primipilaribus* (indicating soldier's rank on discharge), Roueché *Aphrodisias* 10.2 (iv AD)'; add 'ἐνδήμων ἐν τῇ πόλει ἡμῶν ἀπὸ .. years *since*, i.e. for years, *SEG* 31.576.12 (Larissa, ii BC)' **III 6**, lines 17/18, for '*by word* .. *Ag.* 813' read '*by words* (sc. by swearing falsely), Hes.*Op.* 322; *from tongues* (which may lie), A.*Ag.* 813; *orally*, Hdt. 1.123'; add 'ἀπὸ τριῶν *three times*, *BGU*

ἀποαίνυμαι   SUPPLEMENT   ἀποθύμιος

2157.15 (v AD)' **B**, line 2, for 'ἀπὺ τᾷ ζᾷ *Inscr.Cypr.* 135.8 H.' read 'Cypr. *a-pu-ta-i za-i ta-i-pa-si-le-wo-se* ἀπὺ τᾶι γᾶι τᾶι βασιλῆος *ICS* 217.8' **D 3**, for 'ἀπαλγέω .. ἀπανθίζω' read 'ἀπανθέω, ἀποζέω'

†ἀποαίνυμαι, also ἀπαίνυμαι, *strip off, remove*, τί τινος Il. 13.262; ἀπαιν- Il. 11.582, Mosch. 2.66 G. **2** *take away, deprive one of*, θεὸς δ' ἀποαίνυτο νόστον, Od. 12.419, 17.322.

ἀποβάθρα, add 'S.*fr.* 415 (unless ἀπόβαθρα, q.v., is read)'

ἀποβαίνω, after 'fut. -βήσομαι' insert 'Cret. fut. part. ἀποβασιόμενον *Inscr.Cret.* 1.xvii 10 A 7' **I 1**, for 'Lys. 2.24' read 'Lys. 2.21' **2**, for 'E.*Hec.* 142' read 'E.*Hec.* 140' **II 2**, for 'Plb. 26.6.15' read 'Plb. 25.2.15'; for 'Th. 5.4' read 'Th. 5.14'; add 'ἵν' εἰς βέλτιον ἀποβῇ τὸ φοβερόν Men.*Dysc.* 418' **3**, delete 'cf. 5.14'

*ἀποβάλσαμον, τό, = ὀποβ-, *BGU* 34.5.13.

ἀπόβασις **I 1**, add '**b** *fall* in river level, opp. ἀνάβασις I 3, *Peripl.M.Rubr.* 63.'

ἀποβατήριος **I**, add 'of Asclepius, *Iasos* 227'

ἀποβιβρώσκω, after '*eat off*' insert 'Zen. 6.44'

ἀποβιόω, add '*SEG* 24.911 (Thrace, iv AD)'

ἀποβλέπω **I 3**, add 'ἀκρωτήριον .. ἀπόβλεπον εἰς ἀνατολήν *Peripl. M.Rubr.* 30, καθ' ὃ μέρος ἀποβλέπει τὴν ἤπειρον ib.'

ἀποβλώσκω, add 'pf. ἀπὸ .. μέμβλωκεν Call.*fr.* 384.5 Pf.'

ἀπόβρασμα, *chaff*, add 'Gal. 19.116'

ἀποβρασμός, for '*throwing off of scum*' read '*boiling over, ebullition*; in quot. applied to *ejaculation* of semen'

*ἀποβρόχω, *swallow up*, ἀπέβροξεν δ' ἄχρις ἐπ' ὀμφαλίου AP 7.506 (Leon.Tarent.).

*ἀποβυρσόω, = Lat. *decorio, skin*, Dosith. p. 436 K.

ἀποβώμιος **II 2**, delete the section.

ἀπογαλακτισμός, add 'Aët. 4.29 (370.27 O.)'

ἀπογάλακτος, add '(but perh. to be read as ἀπὸ γάλακτος)'

ἀπόγειος **I 1**, add 'of persons, ἀπόγειος ὄψομαι Luc.*Lex.* 15' **II 2**, delete the section.

ἀπογεισόω, after '*crown with a cornice*' insert 'or *replace the cornice*'

*ἀπογείσωσις, εως, ἡ, *provision* or *replacement of a cornice*, *Inscr.Délos* 366*A*7 (iii BC).

*ἀπόγευμα, ατος, τό, *tasting*, Cyran. 1.7.88 K., etc.

ἀπογιγνώσκω, for '*IG* 2².457.30' read '*IG* 2².457.18'

*ἀπογόμωσις, εως, ἡ, *unloading*, *PAnt.* 108.4/5 (iv AD), *PVindob.Worp* 8.28 (iv AD).

*ἀπογομωτής, οῦ, ὁ, *stevedore*, *POxy.* 3867.17, 18 (vi AD).

Ἀπογονικός, for '*Hemerolog.Flor.*' read 'beginning Oct. 24, *OGI* 583.15 (Cyprus, i AD)'

ἀπόγονος, line 2, after 'ἀπογόνη' insert '*great-grand-daughter, female descendant*, *SEG* 1.399.5, *IEphes.* 3072.17'

ἀπογράφω, after lemma insert '(Arc. ἀπυ-, *SEG* 37.340.18 (Mantinea, iv BC))' **I**, add '**2** *alter* or *cancel in copying*, *CID* i 10.10 (iv BC).'

ἀπογυμνόω **2**, add 'Ezek.*Exag.* 47 S.'

*ἀπογυναικόομαι, pass., *to be made effeminate*, Agatharch. 101, Cyran. 1.10.52 K.

ἀποδακρύω **I 1**, add 'abs., Aristox.*fr.hist.* 90, cf. *AB* 427' **II**, delete the section.

ἀποδεής, for '*empty*' read '*not completely full*'; after 'Plu. 2.967a' insert '*SEG* 30.1663 (Ai Khanoum, ii BC)' add '**II** *of poor quality, inferior*, ἀποδεῆ χαρτία *Pap.Lugd.Bat.* XIII 6.6 (i AD).'

ἀποδείκνυμι, after lemma insert '(ἀπυ- *SEG* 34.1238, Aeolis, c.200 BC)' **A I 1**, line 5, for '*SIG* 134.2' read '*SIG* 134b.22' **I 5 a**, pass., add 'of places *assigned* for anchorage, ὁρμίζεσθαι .. ἐν τοῖς ἀ. τ[ό]ποις *PHib.* 198.112 (iii BC), τῶν ἀ. ὅρμων τῆς Ἐρυθρᾶς θαλάσσης *Peripl.M.Rubr.* 1' **II 1**, at end, *consul designatus*, add 'of other magistrates, *Laodicée* no. 5 p. 281, *IPrusias* 6.2'

ἀποδειλιάω **2**, add 'πράσσειν μὴ ἀ. J.*AJ* 19.1.19'

ἀπόδειξις **I 1**, add '**b** *display, demonstration*, ἐν τοῖς] ὅπλοις ἀπόδειξιν ἐποιήσαντο .. *SEG* 26.98.' **3**, add 'w. subordinate cl., ἀ. ὡς τὰ τούτων ἀδικήματα .. γέγον' αἴτια D. 18.42; *document providing proof*, *BGU* 1141.12, *POxy.* 257.19 (both i AD), *CodJust.* 4.21.16.3, 4'

ἀποδεκατεύω, add '*dedicate a tithe*, τῷ θεῷ *SEG* 9.72.56 (Cyrene)'

ἀποδέκτης, line 3, after '*IG* 12(8).608' insert '*SEG* 33.679 (sg., Paros, ii BC)'

*ἀποδεκτικός, ή, όν, *receptive*, Hierocl.*Prouap.Phot.Bibl.* 465 b 15.

ἀπόδεξις **I**, add 'ἀγάπησις ἀπόδεξις παντελής Pl.*Def.* 413b'

ἀποδέρω, line 2, after '(Hdt.) 4.60' insert 'Theoc. 25.278' **2** and **3**, for these sections read '**2** *strip back the foreskin of*, by causing an erection, Ar.*Lys.* 739, 953. **3** *strip the fibres from* stalks of mallow, etc., ibid. 739 (w. pun on section 2).'

ἀποδέχομαι **I 1**, add 'φόρτους *Peripl.M.Rubr.* 26'

ἀποδημέω **2**, at end for 'ἄλλοσε ἀ. Pl.*Lg.* 579b' read 'οὐδαμόσε ἀ. Pl.*R.* 579b'

ἀποδημία, add '**b** applied to the absence of a prefect from Alexandria on visits to the Egyptian χώρα (q.v. II 3), *PGiss.* 44, Act.Alexandr. vii A 134, 137, ix A 11.7.'

ἀπόδημος, for 'Plu. 2.799f' read 'Plu. 2.799e'

ἀποδιαστέλλω, for '*divide* ..' (ii BC) read '*assign, apportion*, *PTeb.* 740.30 (ii BC), *PTaur.* 8.22, 48 (ii BC)'

*ἀποδιατίθημι, *alienate*, Sopat.Rh. 8.169.5 W.

ἀποδιδάσκω, add 'Herodes Atticus *Pol.* 18'

ἀποδιδράσκω **1**, line 9, transfer 'of runaway slaves' before 'σώματα ἀποδράντα' in line 11, and in its place read 'οὔτε ἀποδεδράκασιν .. οὔτε ἀποπεφεύγασιν'

ἀποδίδωμι, line 2, after 'Hsch.' insert 'form ἀπυ-, Myc. 3 sg. aor. *a-pu-do-ke*; (for other forms v.s.v. ‡δίδωμι)'

ἀποδικάζω, add '*SEG* 31.358 (Olympia, v BC), rest. in *Inscr.Cret.* 4.22*B* (vii/vi BC)'

ἀποδικέω, for 'X.*HG* 1.7.21' read 'X.*HG* 1.7.20'

ἀποδιοπόμπησις, add 'Epict.*Diss.* 2.18.20'

ἀποδοκιμάζω, for 'τὸ ποιεῖν τι' read 'w. articular inf.' and add 'Isoc. 5.75'

ἀποδόσιμος **2**, for this section read '-μον, τό, *an order to hand over money or goods*, *PSI* 237.6 (v/vi AD), *PMich.*xv 742.2, *CPR* 5.18.12, al. (both vi AD)'

ἀπόδοσις **I 3**, delete the section. **II 2**, add '**b** in metre, *responding section* in antistrophic compositions, Sch. metr. Pi.*O.* 2 p. 58 D., al.' at end add 'form ἀπυ- Myc. *a-pu-do-si*'

ἀποδοχή **I**, add '**4** *acknowledgement of receipt*, ἀποσταλείσης ἀποδοχῆς *POxy.* 3121.15 (iv AD).'

*ἀποδοχικός, όν, μέτρον ἀ., a measure of capacity (= 42.25 choenices), *PLond.* 1940.11, 15; cf. δοχικός, παραδοχικός.

*ἀποδρακωνάριος, ὁ, *ex-standard-bearer*, *PAmst.* 45.7 (AD 501).

†ἀποδρομή, ἡ, *run-in* or *lay-by* for boats, ὁ δὲ τόπος ἀλίμενος καὶ σκάφαις μόνον τὴν ἀ. ἔχων *Peripl.M.Rubr.* 3.

ἀποδύρομαι **1**, after '*lament bitterly*' insert 'w. acc., *CEG* 13 (funerary, Attica, vi BC)'; line 3, for 'τινί .. (dub.)' read '*complain bitterly*, w. inf., πολλῶν ἀποδυρομένων πολειτῶν ὑπὸ τῶν ἐπαρχικῶν ἐξελαύνεσθαι *SEG* 30.568 (= *EAM* 186, ii AD)'

ἀποδυτήριον, in the palaestra, add '*SEG* 32.147 (Attica, iv BC)'

ἀποδύω **II 2**, lines 4/6, delete 'οἱ ἀποδυόμενοι .. Lys.*fr.* 45.1'; after '*IG* 14.256 (Phintias)' insert 'ἀπεδύσατο εἰς τὴν αὐτὴν παλαίστραν Lys.*fr.* 75.1'

*ἀποδώτης, ου, ὁ, = ἀποδοτήρ, *PMasp.* 126.7 (vi AD).

*ἀποέπαρχος, ὁ, *ex-eparch (prefect)*, *PLips.* 42 (AD 391).

ἀποζάω **1**, after '*to live off*' insert '*h.Ap.* 539 (tm.)' **2**, for this section read '*make a living*, πονηρῶς Luc.*Tox.* 59, μόλις Ael.*NA* 16.12, μόνον Lib.*Orat.* 11.253'

ἀποζεύγνυμι, after '*part*' insert 'τὰς ἀντονομασίας ἀπὸ τῶν ὀνομάτων D.H.*Comp.* 2, id. 7.67 (pass.)'

ἀπόζευξις, add '**b** *separation*, εἰ δὲ οὕτως ἐχούσης τῆς Ἀφροδίτης ἀπόζευξις γένοιτο Heph.*Astr.* 3.9.34.'

ἀποζέω **2**, add '**b** *to be cleaned by boiling, be boiled out*, ἡ ληνὸς πεμπταία ἀπέζεσεν καὶ κατηλείφθη *BGU* 1549 (iii BC), cf. 1550.'

†ἀποθεμιόω, gloss on corrupt form ἀπίθετο in Hsch.

*ἀπόθεμα, ατος, τό, *thing stored away*, Sm.*Ez.* 27.27 (pl.).

ἀποθεόω **I**, after 'μετὰ τὸ ἀποθεωθῆναι' insert 'i.e. after the burial' add '**b** *consecrate as a burial-place*, ἀπεθέωσα τὴν λάρνακα *IGRom.* 3.1480 (Iconium).'

ἀποθεραπεύω **2**, *cure*, add 'ἀ. ὀφθαλμού[ς] *Suppl.Mag.* 32 (v/vi AD)' add '**3** *restore, make good* a loss, εἰ δὲ βλάψει τὸ πρᾶγμα .. τὴν ζημίαν ἀποθεραπεύει *CodJust.* 1.2.24.6.'

*ἀποθερμαίνω, *warm up*, οὐ πολὺ μὲν ἔπιον, ἀλλ' ὅσον ἀποθερ[μ]ανθῆναι Lollian. B 1 23 (p. 122 H.), perh. in *PMed.inv.* 70.16.2 in *CE* 52.97.

ἀπόθετος **I 1**, add 'of a person, *hidden away*, Philostr.*Her.* 19.3'

ἀποθέωσις, after '*deification*' insert 'also *burial* (*CIG*, v. infr.)'

*ἀποθηκάριος, ὁ, *storekeeper*, Teuc.Bab. p. 51 B., *IGChr.* 10.4, *MAMA* 3.534, etc. (Corycus, var. sp., including ἠπο- 431, ὠπο-), *SEG* 37.620 (Thrace, Chr.).

ἀποθήκη **I 1**, after '*storehouse*' insert 'or *storeroom*' and add 'τὴν ἐξέδραν σὺν τῇ ἐν αὐτῇ ἀποθήκῃ διστέγῳ *Sardis* 7(1).12 (ii AD), *POxy.* 2729.31, 33'

ἀποθησαυρισμός, add 'Antioch.Astr. in *Cat.Cod.Astr.* 11(2).109.3'

*ἀπόθι, Arc. adv., *far away, apart*, *SEG* 36.376 (Lykosoura, ii BC); cf. °ἄπυθεν, ἄποθεν.

ἀποθλίβω **I 1**, after 'D.S. 3.62' read 'Nic.*fr.* 86, ἄρδην με τῆς οἰκείας ἀποθλίψεις χώρας Luc.*Jud.Voc.* 2' **3** and **4**, for these sections read '**3** *squeeze dry, wring*, τὰ κράσπεδα Diph. 43.30; cf. ἐνουρέω. **4** *press forcibly against, crush*, Nic.*Th.* 314, Lxx *Nu.* 22.25, *Eu.Luc.* 8.45.'

*ἀπόθραυμα, ατος, τό, = ἀπόθραυσμα, *Inscr.Délos* 1442*B*9 (ii BC).

ἀποθριγκόω, after '*furnish with coping*' insert 'or *replace the coping*'

*ἀποθρίγκωσις, εως, ἡ, *furnishing with coping* or *replacing the coping*, *IG* 4.823.39 (Troezen, iv BC)

*ἀποθρυόομαι, pass., (θρύον) pf. part. ἀποτεθρυωμένοι *bent like rushes*, v.l. for ἀποτεθρυμμένοι in Pl.*R.* 495e ap.Tim.*Lex.*, cf. Sch. ad loc., Suid.

ἀποθύμιος, after 'Hdt. 7.168' insert 'οὐδ' ἔστιν ὅπως ἀποθύμια ῥέξω

43

ἀποθυσία   SUPPLEMENT   ἀπόλαυσις

Call.*Del.* 245'; add 'used app. mistakenly in opposite sense, νεκρῶν δ᾿ ἀποθύμια ῥέζει(ν) epigr. in *SEG* 39.449.11 (Tanagra, v AD)'

*ἀποθῡσία, ἁ, pl., some kind of ritual, perh. *distribution after a sacrifice*, ἐποίησεν δὲ καὶ τοῖς κατοιχομένοις .. καὶ ταῖς ἀποθυσίας καὶ χονδρογάλα *SEG* 32.1243.35 (Aeolian Cyme, i BC/i AD).

*ἀποϊέρωσις, εως, ἡ, = ἀφιέρωσις, Albania 5.43 (Apollonia in Illyria).

ἀποίητος III, add 'of persons, *unsuitable* for office, *PPetaus* 93.126, [ε]ἰς χρίας ib. 26.9; w. dat. ἀποίητον τῇ πολιτικῇ ἐργασίᾳ *Vit.Aesop.*(G) 2'

ἀποικέω II, line 3, delete 'c. acc., .. (s.v.l.)'; for 'Corinth *was inhabited by me at a distance*' read 'Corinth *was dwelt far from* by me'

ἀποικία I 1, at end for 'Aeschin. 2.176' read 'Aeschin. 2.175' and add 'of Roman colony, Plb. 2.19.12, etc., *Ἀ. Πατρέων* as translation of *Colonia Patrensis, SEG* 18.64 (Athens, i AD)'

ἀποικιστής, delete ref. to Men.Rh.

ἄποικος II, for this section read '1 of cities, *settled as a colony*, πόλεις ὁπόσαι τῆς γῆς τῆσδ᾿ εἰσὶν ἄποικοι Ar.*Lys.* 582; πόλιν Σινωπέων ἄποικον ἐν τῇ Κολχίδι χώρᾳ X.*An.* 5.3.2, 6.2.1. **2** subst. ἄποικος, ὁ, *colonist*, Hdt. 5.97, Th. 1.25, 38, 7.57, *IG* 1³.46-7; poet., of iron, Χάλυβος Σκυθῶν ἄ. A.*Th.* 728.'

*ἀποίνιμος, ον, *carrying no penalty*, cj. in Hes.*fr.* 124.1 M.-W.

ἀποιωνίζομαι, add '*Dosith.* p. 430 K.'

ἀποκᾰθᾰρίζω, after '*purify*' insert 'wine, *CPR* 8.82.6 (vii AD), transf., ἑαυτόν'

ἀποκάθαρμα, add '*that which is sifted out* of grain, *PMasp.* 2.iii.11 (pl., vi AD)'

ἀποκαθαρτικός, after '*cleansing*' insert 'Thphr.*Sens.* 84'

ἀποκαθεύδω I, for 'of a woman .. 399' read 'used for ἀποκοιτέω by Eup.*fr.* 431 K.-A.; τουτέστι γυναῖκα χωρίζεσθαι ἀνδρὸς καὶ ἀφίστασθαι Suid. a 3332'

ἀποκαθιστάω, before 'v.l.' insert 'Duris 7 J., D.S. 1.78.2'

ἀποκαθίστημι I 1, add '*rebuild, restore* buildings, etc., *SEG* 34.1122 (Ephesus, i AD), *IHadr.* 44, *TAM* 5(1).517, *SEG* 31.1295 (Oenoanda, all ii/iii AD)' **2**, add '*deliver goods which have been ordered*, οἱ δὲ ἐργολαβήσαντες υἱκὸν ἢ οἰνικὸν μ[ὴ] ἀποκαταστήσαντες *SEG* 31.122 (Attica, ii AD)' **3**, add 'τὴν προγαμιαίαν δωρεὰν ἀποκατέστησε καὶ δέδωκε τῷ παιδί Just.*Nov.* 2 pr. 1' add '**5** app. in calque on Lat. *rationes reddere*, ἄχρι ἂν τὰς ψήφους ἀποκαταστήσωσιν Scaev.*Dig.* 40.5.41.4.'

ἀποκαίω 1, add '**b** *consume by fire, burn up*, Arist.*Pr.* 928ᵃ20, Luc.*Tox.* 61, *IG* 10(2).260.B9 (iii AD); cf. ‡ἀπόκαυσις III.' at end add 'form ἀπυ- Myc. pf. part. pass. *a-pu ke-ka-u-me-no*.'

*ἀποκαλά, ἡ, tentatively given as a gloss on φορμός in *Lex.Rhet.* in *AB* 315.

ἀποκάλυμμα, for '*a revelation*' read '*uncovering* (of the head)'

ἀποκαλύπτω 1, add 'ἀ. τὸ ὠτίον τινος *uncover* a person's ears, i.e. make a thing known to him, Lxx 1*Ki.* 9.15, 2*Ki.* 7.27, etc.'

ἀποκάμπτω 1, after 'X.*Eq.* 7.14' insert 'ἵνα ἀπ[ο]κάμψῃ καὶ ἀσχη[μονήσῃ] *SEG* 35.218.15 (tab.defix., Attica, iii AD)'

ἀποκαρπίζω II, in med., *harvest* (in quot., dates), *BGU* 2127.12 (AD 156).

*ἀποκάρπωσις, ἡ, ἀ. νεκρῶν perh. *exploitation of* or *profiting from* corpses, *BGU* 2462.8 (ii AD).

ἀποκαρτέρησις, add 'Charito 6.2'

ἀποκατάγνῡμι, add 'pf. part. intr. -εαγώς *broken off, Inscr.Délos* 1439A*b*ä47 (ii BC)'

*ἀποκαταράομαι, = *deprecor*, Dosith. p. 431 K.

ἀποκατάστᾰσις, add '= Lat. *restitutio*, restoration of property, *Cod.Just.* 1.3.52.13; ἡ εἰς ἀκέραιον ἀ. = Lat. *in integrum restitutio*, Just.*Nov.* 119.6'

*ἀποκατωρυξ, υγος, ἡ, reading of codd. in Thphr.*CP* 5.9.11, perh. due to confusion of ἀπώρυξ and κατῶρυξ.

ἀπόκαυσις, add '**II** *loss involved in melting down* gold votive offerings, prob. rest. in *IG* 2².1495.4, 13 (iv BC), cf. *NC* 1951.109-110, and °ἀφέψησις. **III** *burnt offering*, name of a type of sacrifice, prob. where the whole offering is consumed, *SEG* 32.1423.9, 11 (Syria), Mitchell *N.Galatia* 257 (both ii AD); cf. °ἀποκαίω 1, °ἀποκαυσμός.'

ἀποκαυσμός, for 'Judeich .. p. 142' read '*BCH* 38.52; cf. °ἀπόκαυσις III'

ἀποκείρω, line 3, for 'ἀπεικείρατο' read 'ἀπεκείρατο'

ἀπόκενος, after 'Hero *Spir.* 2.24' add '*PCair.Zen.* 680.3 (iii BC)'

*ἀποκενωτέον, one must evacuate, Archig.ap.Aët. 9.35.

†ἀποκηδεύω, *complete the funeral rites of*, Hdt. 9.31, *SEG* 6.220 (Phrygia).

ἀποκηρύξιμος, for the pres. ref. read '*IG* 2².1013.5 (ii BC)'

ἀποκηρύσσω, after 'Att. -ττω' insert 'dial. ἀποκᾱρ-' **II 1**, add 'also act., ἀποκηρύξας τὰ τέκνα *POxy.* 2342 (ii AD)' **IV**, before pres. ref. insert '*SEG* 32.623 (ἀπεκάρυξε, Buthrotum, ii BC)' add '**V** *proclaim, declare*, πατρογέροντας *IEphes.* 26.24 (ii AD).'

ἀποκλαίω, after '-έκλαυσα' read '*mourn for, lament*, θανόντα Thgn. 931, τύχας A.*Pr.* 637, ἐμαυτόν Pl.*Phd.* 117c; abs. Hdt. 2.121. γ'; w. internal acc., στόνον S.*Ph.* 695 (lyr.); med. ἀποκλαύσασθαι κακά S.*OT* 1467, τὴν πενίαν Ar.*V.* 564, E.*fr.* 563; abs. Luc.*Syr.D.* 6'

ἀποκλείω I 2, add 'as Lat. *excludere*, *exclude from inheritance*, w. acc., (τῆς κληρονομίας understood) *Cod.Just.* 6.4.4.14b-14c, 19a'

*ἀπόκλεμμα, τό, (*amount of*) *theft*, ἀποτεισάτω τὸ ἀπόκλεμμα πενταπλοῦν, τὸ δὲ νοσφισμὸν ἡμιόλιον *PMich.*x 587.29.

ἀποκλέπτω, add 'τὰ ἀποκλαπέντα Heph.Astr. 3.37 in *Cat.Cod.Astr.* 8(1).154'

*ἀποκλήζω, *discard*, τοῦτ᾿ (sc. ὄνομα) ἀποκληζομένη *IG* 2².3575 (Eleusis, ii AD).

ἀποκληΐω, delete 'later -κλήζω *IG* 3.900'

ἀπόκληρος II, add '*PCornell* 12.14 (iii AD), *Cod.Just.* 6.4.4.14'

ἀποκληρόω I 2, after '*assign by lot*' insert 'δικαστήρια Luc.*Bis.Acc.* 4, 12'

*ἀποκλησία, ἁ, *committee*, *IG* 9²(1).609.A11 (Naupactus, vi/v BC).

ἀποκλῑμάκωσις, for '*ladder*' read '*flight of steps*' and add '*SEG* 26.1449 (also Cilicia)'

ἀποκλίνω I, line 3, after '*h.Ven.* 168' insert 'τυτθὸν ἀποκλίνασα κάρηατα Call.*Del.* 236'; line 4, after '*slope away*' insert 'ἀπὸ δ᾿ ἐκλίθη ἔμπαλιν ὤμοις φοίνικος ποτὶ πρέμνον Call.*Del.* 209' **III 1**, add '**b** of ships, *tilt over*, *Peripl.M.Rubr.* 46.' **2**, add 'of ships, *JHS* 74.98 E19 (Caunus, i AD)'

ἀποκλύζω I, add 'Anacr.*epigr.* 3.4 P. (tm.)'

*ἀποκλώθω, perh. *bring about* (death) *by spinning*, μοῖραν .. οὐκ ἀπέκλωσε θεός *Inscr.Cret.* 1.v 42 (i/ii AD).

*ἀποκοιλαίνω, *hollow out*, *AE* 1923.45 (Oropus).

*ἀποκοιτία, ἡ, *absence for a night*, *PSI* 1120 (i BC/i AD).

ἀποκολυμβάω, add 'Clearch. 73 (ἀποκυμβ- codd. plerique)'

ἀποκομιστικός, add 'ἀποκομιστική (sc. πτῶσις), ἡ, Dosith. p. 401 K.'

*ἀποκομματικός, ή, όν, *abbreviated*, ἀ. λεξείδιον, Phot. s.v. ψό.

†ἀποκοντόω, perh. *bring near*. pass. ὥστε τέμνεσθαι τὰς παρακειμένας ἀγκύρας ἀντέχειν ἀποκοντουμένας ?*shortened*, *Peripl.M.Rubr.* 40 (acc. to others: *thrust out*); cf. in description of Theodora's lewd behaviour, τὰ ὀπίσω ἀποκοντωσα τοῖς τε διάπειραν αὐτῆς ἔχουσι καὶ τοῖς οὔπω πεπλησιακόσι Procop.*Arc.* 9.23 (v.l., codd., edd. ἀποκεντ-); cf. mod. Gk. ἀποκοντά *near*, ἀπόκοντος *very short*.

ἀποκοπή IV, delete the section.

ἀπόκοπος II, for '*precipitous*, ὅρη' read '*abruptly ending, sheer*, ἀκρωτήριον *Peripl.M.Rubr.* 12, ὅρη ib. 32, βυθός ib. 40'

ἀποκόπτω III, for 'E.*Tr.* 628' read 'E.*Tr.* 627' add '**IV** pass., ἀποκοπῆναι *lose the scent*, *AB* 428, cf. Hsch.'

*ἀποκορακόω, *unfasten*, *BE* 1971.647 (Hierapolis); cf. ‡κορακόω, κατακορακόω.

*ἀποκοσκίνημα, ατος, τό, *that which is sifted out of grain*, *PMasp.* 2.iii.11 (pl., vi AD).

*ἀπόκοσμος, ὁ, Cret. *one who has served in the office of κόσμος* (v.s.v. III), or *one who is not a κ.* (v. *SEG* 37.752), *SEG* 35.991.A5 (Lyttos, c.500 BC), 23.566.14 (Axos, iv BC).

ἀποκραιπᾰλάω I, after '*sleep off a debauch*' insert 'Men.*Dysc.* 457'

ἀποκρέμαμαι, delete the entry.

*ἀποκρημνίζω, add '**2** pass., of pulse, *to be precipitate*, Gal. 8.662.12, 942.18.'

ἀπόκριμα 2 b, for this section read '*judgement* pronounced by Emperor after hearing oral presentations, *PTeb.* 286.1 (ii AD), *PMich*ix 529 (iii AD)'

ἀποκρίνω I, add '**4** pass., w. εἰς, *to be classed as*, Luc.*Syr.D.* 10.'

†ἀποκρισιάριος, ὁ, *agent* or *envoy*, *POxy.* 144.14, *IEphes.* 1296, *Cat.Cod.Astr.* 5(3).93.22, 7.155.22 (all Byz.); οἱ ἐν τοῖς μοναστηρίοις διατρίβοντες μὴ ἐχέτωσαν ἐξουσίαν ἐξιέναι .. ὑπεξαιρουμένων τῶν καλουμένων ἀποκρισιαρίων *Cod.Just.* 1.3.29 pr., Just.*Nov.* 133.5.

ἀπόκρισις II 4, for the pres. ref. read 'Chor. 3.59' and add '*Cod.Just.* 1.3.29.1, Just.*Nov.* 123.25'

ἀπόκριτος, add 'neut. subst., φυλάττεσθαι .. τὸ ἀ. τοῖς ἀπολιμπανομένοις app. *opportunity for defence*, Lat. *servata .. absentibus defensione*, *Cod.Just.* 10.11.8.7a'

ἀποκρούω II, for '*knock off* .. 12' read '*break up* or *off, damage*, μηδὲ λωβήσασθαι μηδὲ ἢ ἀποκροῦσαι *IG* 2².13200.12 (ii BC), *SEG* 35.209.13 (Attica, ii AD)'

ἀποκρύπτω II, lines 4/5, delete 'cf. Th. .. αὐτούς'; line 6, delete 'ἀποκρύπτουσι .. 179' add '**b** abs., *disappear*, Hes.*fr.* 290 M.-W., Th. 5.65, Schwyzer 708.3, 6, 9.'

*ἀποκτανσις, εως, ἡ, *killing*, Anon. in Rh. 146.1.

*ἀποκτένισμα, ατος, τό, *combings*, τὰ ἀ. τοῦ στιππύου prob. in *PCair.Zen.* 176.42 (iii BC).

ἀποκυβιστάω, delete the entry.

ἀποκῠδαίνω, for '*IG* 3.1367' read '*IG* 2².11636'

ἀποκυλίω, after '*roll away*' insert 'λίθους *AE* 1923.39 (iv BC)'

ἀποκύπτω I, before 'X.*Hier.* 8.1' for 'ποιεῖν' read 'φιλεῖσθαι'

*ἀπόλαος, ον, *formally excluded* from an association, *banned*, ἀ. ἔστου [ἀ]πὶ τᾶς συγγενε[ίας] *SEG* 36.548 (Metropolis, Thessaly, iii BC).

ἀπόλαυσις II 1, add 'ἐπίτροπος ἀπὸ τῶν ἀπολαύσεων transl. of Lat. *a uoluptatibus*, *procurator, overseer of imperial entertainments*, *SEG*

44

36.556 (Epirus, ii AD); also ἐπὶ τῶν ἀ. IEphes. 852.18 (i/ii AD); Ἀπόλαυσις personified: SEG 26.438.4 (mosaic, Argos, v AD)'

ἀπολέγω I 2, add '*challenge* a juror, SEG 9.8.28 (edict of Augustus)' **II**, add '**2** *fail to appear*, of goods to be delivered, BGU 1564.12 (ii AD).'

ἀπολείβω, after 'Hes.Th. 793' insert 'A.Ag. 69 (anap., cj., codd. ὑπο-)' and delete 'metaph. .. Com.adesp. 39'

ἀπολείπω, line 1, aor. ἀπέλειψα, add 'Carm.Aur. 70' **I 2**, after 'S.Ph.' insert '1158' **C II** 2, after 'Id. 27.2' insert 'ἐμοῦ πολὺ ἀπολελειμμένου τῶν ἐμαυτοῦ κακῶν Lys. 1.15'

ἀπολειτουργέω, after '*service*' insert 'PCair.Zen. 35.3 (iii BC)'; add 'ἀ. τὸν βίον *fulfil one's obligations* to this life, PLond. 1708.29 (vi AD)'

⁺ἀπολείχω, *lick*, ἕλκη Ev.Luc. 16.21 (v.l.); *lick off*, αἷμα Dionys.Bassar. 20ʳ.6 L., Ath. 6.250a; w. partit. gen., φόνου A.R. 4.478.

ἀπόλειψις, add '**IV** *legacy, bequest*, SEG 32.1537.5 (Gerasa, ii AD), ἐξ ἀπολείψεως MAMA 8.451.12, (Aphrodisias).'

ἀπόλεμος I 1, delete 'cf.' and insert 'Hp.Aër. 16 (comp.)'

ἀπολεπίζω, add 'Hsch. s.v. ἀποσκόλυπτε'

*ἀπόλεσις, εως, ἡ, *loss*, Hippod.ap.Stob. 4.34.71.

ἀπολήγω I 2, for 'opp. γίνεται' read 'opp. ἐπιγίνεται' **3 b**, after '(ii BC)' insert 'Peripl.M.Rubr. 33'

*ἀποληκῠθίζω, app. *make a loud* or *hollow sound*, πλαταγωνίσας· ἀποληκυθίσας, ψοφήσας Hsch.

ἀπολιμπάνω, for 'ἀέκων .. 23.5' read 'ἀέκοισ' ἀ. Sapph. 94.5 L.-P.' and add 'in pass., *to be absent*, Just.Nov. 6.2'

ἄπολις I, add '**3** w. ref. to lack of Roman citizenship, Ulp.Dig. 32.1.3, Marcian.Dig. 48.19.17.1.'

ἀπολισθάνω 2, after 'Ar.Lys. 678' insert 'βίοιο AP 7.273 (Leon.Tarent.)'

ἀπολίτευτος II 3, delete the section.

ἀπόλλῡμι, line 5, 'freq. in tmesis in Ep.', add 'also occasionally in prose, e.g. Meliss. 7.5'

Ἀπόλλων I, line 5, delete 'Ἀπόλλων A.Ch. 559'; add 'pl., of Apollo as honoured in various cults, IG 2².1945.1; first syllable sts. lengthened in epic in oblique cases, e.g. Od. 9.198, h.Ap. 15' add '**III** in month name, μηνὸς Ἀπόλλωνος (app. gen. of god's name) at Chaleum (Locris), Delph. 3(3).38 (i BC); GDI 1931.1 (Delphi, end ii BC); cf. Ἀπολλώνιος II.'

Ἀπολλώνεια, after 'τά' insert 'festival at Myndos, Inscr.Cos 104.14' add '**2** Ἀ. Πύθια *games in honour of Apollo* at Hierapolis, JÖAI 30 Beibl. 203 (Ephesus, iii AD).'

Ἀπολλωνιασταί, add 'Inscr.Délos 1757 (i BC), etc.'

Ἀπολλώνιος III, -ώνιον, add 'also Ἀπολλωνιεῖον, SEG 9.73.4 (Cyrene, ii/iii BC)'

ἀπολογέομαι I 2, for 'Aeschin. 1.92' read 'codd. in Aeschin. 1.70'

*ἀπολογικόν, τό, v.l. for ἀπολογητικόν, Arist.Rh.Al. 1421ᵇ10 (PHib. 26.300), Syrian. in Hermog. 2 p. 11 R., Fortunat.Rh. 2.15.

ἀπολούω 2, med., add '*perform ritual ablutions*, Peripl.M.Rubr. 58'

ἀπολοφύρομαι, for '*bewail loudly*' read '*bewail to the full*' and for 'abs. *indulge one's sorrows to the full*' read 'of lamentation for the dead'

ἀπολύσιμος, after '**II**' insert 'ἀ. ἀρχαί *discharged, no longer subject to* εὔθυνα, D. in Lex.Min. 160' add '**b** of priests, temples, *tax-exempt*, PTeb. 292.6, 293.6.'

ἀπόλυσις I, add '**5** κατ' ἀπόλυσιν, *in an absolute construction or form*, A.D.Pron. 46.17, 81.25; id.Adv. 172.12.' **II 1**, line 2, delete '*separation*, .. Arist.GA 718ᵃ14' add '**2** *separation* on the completion of sexual intercourse, Arist.GA 718ᵃ14, 718ᵃ32, 756ᵇ3.'

ἀπολυτικός, add '**II** *absolute, simple*, prob. in Dosith. p. 406 K.'

ἀπόλυτος 3, add '**b** *absolute, independent*, A.D.Pron. 81.28.'

ἀπολύω A III 4, add '*defray the cost of*, τὴν .. διάβασιν (bridge) ἀ. AS 12.199' **II**, add '**b** of laying down an office, ἐκ τῆς ἀγωνοθεσίας SEG 3.367.31 (Lebadea, ii BC).' **B IV**, add '**2** *pass clear of*, Arist.Cael. 272ᵃ24, 26, 272ᵇ3, 10, 26.' **C I**, after 'X.Cyr. 6.2.37' insert 'ἀπολελυμένος λεγιῶνος IPrusias 31 (i AD)'; add '*to be exempted, have remission* from liturgy, PFlor. 312.4-5 (i AD), BGU 2474.9 (ii/iii AD)' **II**, add '**3** *to be separated after* sexual intercourse, Arist.GA 718ᵃ1, 731ᵃ21.'

ἀπολωτίζω, for '≈ ἀπανθίζω .. κόμας' read '*pluck off* a person or thing regarded as a flower, τίς ἄρα μ' .. ἀπολωτιεῖ;'

ἀπόμαγμα I 2, add '(pl.); transf., of a person, Vit.Aesop.(G) 14'

ἀπομαίνομαι, for '*go mad*' read '*recover from madness*' and add 'Men. Sam. 491, Aret.SD 1.6'

*ἀπομακκόω, *strike dumb*, [Ζεὺς Τρωσου] ἀπεμάκκωσεν αὐτὸν ἐπὶ μῆνας τρεῖς SAWW 265(1).58.4 (Lydia, ?i/ii AD); see also BE 1971.511; cf. μακκοάω.

*ἀπομάλακος, ον, *soft, effeminate*, ἐναντίως δὲ λάγονους, θηλυψύχους, φιλοκόσμους, ἀπομαλάκους Heph.Astr. 2.15.15.

ἀπομάσσω, penultimate line, delete 'c. gen.'

ἀπομαστίδιον, for '*suckling*' read '*weaned child*'

ἀπομειουρισμός, for '*curtailment*' read '*tapering off*'

⁺ἀπομειόω, *make less*, οὐδὲν παντελῶς τῶν ἄλλων παραγραφῶν ἀπομειούντες Just.Nov. 111.1, cf. 108.1; pass., [τὰ τέλη ..] ἀπομειωθῆναι IEphes. 38.7 (?v AD), ἡνίκα ἡ τῆς προσόδου μηδαμῶς ἀπομειοῖτο ποσότης CodJust. 1.2.17.1.

ἀπομείρομαι 1, for '*distribute*' read '*take as one's portion*'

*ἀπομελᾰνισμός, ὁ, = ᵒἀπομελανσις, Comarius Alch. 291.17.

ἀπομέλανσις, for '≈ λεύκωσις' read '*blackening*'

*ἀπομελάνωσις, εως, ἡ, = ᵒἀπομέλανσις, Zos.Alch. 210.15.

ἀπομέμφομαι, after 'E.Rh. 900 (lyr.' insert 'dub.'

ἀπομένω, after 'Polyaen. 4.6.13' add 'Just.Nov. 123.36'

ἀπομεριμνάω, add '**2** *to be free from anxiety*, pap. in JEA 21.53 (vi AD).'

*ἀπομεριστός, όν, *set aside*, PVindob.Boswinkel 6.4 (iii AD).

ἀπομέτρησις, after '*measuring out*' insert 'of land, SEG 34.477.5 (Atrax, ii BC)' and after '*distribution*' insert 'κλιμάτων καὶ ἐθνῶν'

*ἀπόμῑμος, ον, *imitator*, IMylasa 468.2.

*ἀπομίσθωσις, εως, ἡ, *letting out for hire*, ἀπομίσθωσιν ποιήσασθαί τινος Arch.f. Religionswiss. 10.211 (Cos, ii BC).

*ἀπόμματος, ον, *blind*, PLond. 1821.268 (Aegyptus 6.193).

ἀπόμοιρα, add '**3** *tax on produce*, OBodl. 43, 60, etc., OCambridge 8, OPetrie 52, BGU 1336-1346 (iii/i BC).'

*ἀπόμοσις, εως, ἡ, *denial on oath*, Hsch. s.v. μά.

ἀπομύζουρις, for '*obscene name of a courtesan*' read '*fellatrix* (cf. μύζουρις)'

ἀπομύσσω II 1, add '**b** *reject with contempt*, Hsch. s.v. ἀπέπτυσεν λόγους.' **III**, add '[κ]αὶ χαλκοῦν ἀπομυσσέτω [λυχνίσκον] SEG 31.851 (Italy, i AD)'

ἀπομύω, delete the entry (v. ἐπιμύω).

*ἀπομώρωσις, εως, ἡ, *the act* or *process of enfeebling*, τῆς διανοίας Archig.ap.Aët. 6.27 (170.16 O).

ἀποναίω II, add 'Call.fr. 43.51 Pf.'

ἀπονέμω I, pass., add 'ἀπονενεμῆσθαί τε αὐτῷ προεδρίαν ἐν τοῖς αὑτοῖς ἀγῶσιν IG 2².1064 (Athens, c.AD 230)'

ἀπόνοια 2, add '**b** *lack of constraint, impropriety*, Thphr.Char. 6.1.'

*ἀπονομάζω, *name for, dedicate*, w. dat. [Διὶ] ἀμπέλων δύο ὄρχους ἐκ τῶν πεκουλαρίων EAM 22.4 (ii/iii AD).

ἄπονος I 2, add 'Nicoch. 5 K.-A.' **b**, after 'Aret.SA 2.1' insert 'w. gen., τῆς πλε(υ)ρᾶς ἀ. SEG 6.213.10 (Eumenea)'

ἀπονοστέω, add 'Pi.N. 6.50'

ἀπονοσφίζω, add 'med., χῇ πικρῇ Μοῖρ' ἀπενοσφίσατο IG 10(2).368.6 (epigr., ii AD)'

*ἀπονουμεράριος, ὁ, = Lat. *exnumerarius*, POxy. 2004.2 (v AD) (-νομιρ-pap.).

ἀπονυχιστικός, for '*polishing to the nail*' read '*trimming the nails*' and after 'τέχνῃ', ἡ' insert 'An.Ox. 4.248.11'

ἀπονωτίζω, add '**II** med., *put down* a load *from one's back*, Hsch. s.v. νωτίσασθαι.'

ἀποξενόω I 1, line 3, add 'fig., τὸ γὰρ ὀρεγόμενον τοῦ ἐνδεές ἐστιν οὗ ὀρέγεται καὶ τοῦ ὀρεκτοῦ .. ἀποξενωμένον Procl.Inst. 8 D.' **4**, for '*disguise oneself*' read '*act as a stranger*'

ἀποξυλόομαι, after '*become hard like wood*' insert 'Men.Dysc. 534'

ἀποξύω 1, after 'φάρμακον' insert '(sc. ἔλαιον)' **2**, at end delete 'Med., .. D.Chr. 32.44'

ἀποπαιδαριόω, for 'dub. sens.' read 'prob. *treat as a small boy* or *slave boy*'

*ἀποπᾶν, adv., *totally*, Procl.in Ti. 2.111.29.

ἀπόπαξ, add 'IG 1³.435.50, 77 (v BC, cf. Hesperia 8.76)'

*ἀποπαραφέρω, *deliver, hand over*, ἀποπαρε[νε]χθῦναι (sic), SB 9906.16 (iii AD).

ἀποπειράομαι II, add '*make a sexual attempt* on, Sch.Ar.Nub. 1063a'

⁺ἀποπέκω, *comb out*, med. ὡς ἀπὸ χαίταν πέξηται Call.Lav.Pall. 32. **II** *shear off*, ἀποπέπεκται· ἀποκέκαρται Hsch.; *cut off* AP 6.155 (Theorid.).

*ἀποπεμπτήρια, τά, *means of getting rid of*, Hsch. s.v. καταστατήρια.

ἀποπεμπτικός, for pres. ref. read 'Men.Rh. p. 333.4, 10, p. 336.5, 6, 12 R.-W.'

ἀπόπεμπτος, delete 'cf. Hsch.'

*ἀποπετρόομαι, *turn into stone, petrify*, of coral, ἐγχρονίζουσαν δὲ πλέον ἀποπετροῦσθαι Orph.Lith.Kerygm. 20.8 H.-S.

*ἀπόπηγμα, ατος, τό, unexpld. wd. in broken context, PMich.III 149.13.32.

ἀποπίεσμα, for '*pressure* .. *bent*' read '*outward pressure*' and add '**II** = ἐκπίεσμα I, ἐλαιῶν Hsch. s.v. ψεαδερτῶν (written -πίασμα).'

ἀποπίμπλημι I, add 'perh. *fill up* a cup, [ἀ]ποπίπλη on shard, SEG 30.940 (Olbia, v BC)'

*ἀποπιτύρισμα, ατος, τό, = πιτύρισμα (which is v.l.), Arc. 20.

⁺ἀπόπλανος, ον, *wandering away*, ὄρπηξ Paul.Sil.Ambo 197. **II** as subst., ἀ., ὁ, = ἀποπλάνησις, κυκᾶν τοῖς ἀντιθέτοις, τοῖς πέρασι .. τοῖς ἀποπλάνοις .. νουβυστικῶς Cratin.Jun. 7 K.-A. **b** perh. *conjuror* or sim., ἀ. κακοί· γόητες Hsch.

ἀποπληρόω III 2, add 'POxy. 1255.16 (iii AD)' **b** pass., *receive in full*, PHamb.inv. 410.11 (JEA 34.100; ?vi AD).'

ἀποπλήρωσις, add '**3** *bleeding, blood-letting*, (cf. Lat. *depletura*), *Edict. Diocl.* 7.21.'

ἀποπλοκή, add '*separation* of married couple, prob. in *PRyl.* 154.31 (i AD); cf. ἀποπλέκω'

ἀπόπλοος (A), delete '**2** *voyage home* or *back*' (include quots. under '*sailing away*')

ἀπόπλυσις, add 'Pelag.Alch. 254.18'

*ἀποπλύτης, ου, ὁ, *washerman*, Teuc.Bab. p. 46 B.

ἀποπνέω **I 2**, for '*exhale, evaporate*' read '*evaporate from*, w. gen.' **3**, delete 'in Com.'

ἀποπνίγω **1 a**, at end after 'D. 32.6' insert 'X.*Eq.Mag.* 8.4, Men.*Dysc.* 668'

*ἀποποιμαίνω, *cherish, guide through*, βίον *POxy.* 3722 fr. 1.5, cf. fr. 28.3 (lemmata in Anacr. commentary, ii AD).

*ἀπόπομπον, gloss on ἀποτάξιον, Hsch.

*ἀποπραιπόσιτος, ὁ, formerly *praepositus*, *PMasp.* 127.23, *PLond.* 1687.23; ἀ. κάστρου *PMasp.* 296.3, *IPhilae* 224 (all vi AD).

*ἀπόπρᾱσις, εως, ἡ, *sale*, *Inscr.Délos* 353*A*38 (iii BC).

ἀποπρηνίζω, add '(cj. °ἀνα-)'

ἀπόπρισις, add 'Gal. 10.442.5'

ἀπόπρισμα, add '[θυ]ίνων ἀποπρισμάτων *Inscr.Délos* 1409*Ba* ii35 (ii BC)'

ἀποπροάγω, for '*in the second rank*, of things neither good nor bad' read 'of things neither good nor bad, but *negatively advanced*, i.e. *degraded* below the zero point of absolute indifference'

ἀπόπροθι, add '**b** prep. w. gen., A.R. 3.313, 372, 1065.'

ἀποπροθορεῖν, for '*spring far from*' read '*leap forward from*'

ἀποπροΐημι **1**, for '[κύνα]' read 'τὸν δὲ τέταρτον'

ἀποπτερνίζω, for pres. def. read '*trip up* (an opponent) *by twisting his heel*'

ἀποπτίσσω, add 'ἄρτοι ἀπεπτισμένοι loaves *made from husked grain*, Philostr.*Gym.* 44'

ἀποπτοέω, for 'Call.*Fr.anon.* 93' read '*Suppl.Hell.* 1046'

ἀπόπτυγμα, for '*IG* 2.652.*A*20' read '*IG* 1³.469.23 (v BC), 2².1388.20, etc.'

*ἀποπωλέω, = ἀπομισθόω, Πολέμων 1.32 (Demetrias, iii BC).

*ἀποπωνίω, v. ἀποφωνέω.

ἀπορέω (B) **I 1**, w. inf., add 'Pl.*Ti.* 44e'

ἀπόρθωμα, add '*Delph.* 3(5).74.15'

ἀπορία **III**, add '**4** *low yield, infertility*, of land, *Cod.Just.* 1.2.17.1.'

ἀπορνύμαι, add '*h.Ap.* 29, Mimn. 9.5 W.'

ἄπορος **III**, add '**4** *low-yielding, infertile*, of land, *PGen.* 67.7 (iv AD), *Cod.Just.* 1.2.17.1; neut. as subst., *unproductive land*, *PUniv.Giss.* 13.7 (i AD).'

ἀπορρέω, line 3, after 'Ath. 9.381b' insert 'ἀπέρευσα Opp.*C.* 2.193; 3 sg. pf. ἀπερρύηκε Archil. 196a.27 W.' **I**, add 'w. internal acc., γεννητικαὶ .. δυνάμεις τὰς ἀεννάους τῶν θείων προόδους .. ἀπορρέουσα Procl.*Inst.* 152 D.'

ἀπορρήγνυμι **I 2**, after 'App.*BC* 2.81' insert 'of *breaking* bad news, Demetr.*Eloc.* 216' **II 2**, add 'pf. part. fem., ἄλλην (sc. φιάλην) .. ἀπερρωγεῖαν (sic) μέρος τι *Inscr.Délos* 1432*Bb* ii19 (ii BC)'

ἀπορρήσσω, add 'ἀπορρήττω Ph. 2.304'

ἀπόρρητος **II 1**, penultimate line, after 'D.S. 15.20' insert 'φίλοι ἀ. *secret* friends, Aen.Tact. 11.7' **III**, adv., add '*surreptitiously*, Ael.*NA* 7.42'

ἀπόρροια **I 3**, after 'συναφή' insert 'Ptol.*Tetr.* 52' and after 'Gem. 2.14' insert 'Ptol.*Tetr.* 3'

*ἀπόρρυμα, ατος, τό, *drainings* from grapes, *PAvrom.* 1*B*34 (i BC). **II** liquid measure = ½ σαίτης of 22 ξέσται, used for consignment of coins, *PColl.Youtie* 84 (iv AD).

*ἀπορρυγή, ἡ, = ἀπόρρηξις, Hsch. s.v. ῥωγαί.

ἀπορρώξ **II**, lines 8/9, delete 'ἀπόρρωξ τῆς πόλεως .. 99d'

*ἀπορυτιάζων, οντος, ὁ, unexpld. title of ephebic official at Cyrene, *SEG* 9.51 (iii BC), 19.741.10 (i AD).

ἀποσαφηνίζω, delete this entry (transfer quot. to ἀποσαφέω).

*ἀποσείρωμα, τό, *something filtered off*, anon.medic. in *PColl.Youtie* 4.10 (iii AD).

*ἀποσειρωτόν (v.ll. -σιρ-, -σηρ-), τό, (σειρά) *linear measure* (opp. μέτρον, *liquid measure*), Lcn. 1*Ch.* 23.29.

ἀποσείω **1**, med., add '*unleash* a storm, Βορέου χείμ' ἀποσεισαμένου *SEG* 33.634 (epigr., Rheneia, ii BC)'

ἀποσεμνύνω, for '*extol, glorify*' read '*reverence*' and add 'Hld. 1.22, Charito 2.4' **II**, for '*give oneself solemn airs*' read '*adopt an aloof stance, move to a more exalted plane*'

ἀποσεύω, for '*chase away*' read '*rid oneself of*' and before '*AP* 9.642' insert 'w. ref. to excretion'

ἀποσημείωσις, add '*record* of inscription, ἀ. ἐτέθη ἐπὶ τὸ βασίλειον *MDAI(I)* 15.124.6 (Miletus, ii AD); v. *SEG* 30.1353)'

ἀποσιώπησις **3**, delete the section.

*ἀποσκάζω, *limp away*, Hsch. s.v. ‡κανάζοντα.

*ἀποσκαφή, ἡ, *side-trench*, *PRyl.* 583.62, al. (ii BC), Choerob. *in Theod.* 2.103.12.

*ἀποσκαφία, ἡ, *excavation*, *IG* 9(2).522.18 (Larissa, iii/ii BC).

ἀποσκευή **I**, add 'Anon. *in Rh.* 146.1' **II**, delete 'Lxx *Ge.* 34.29' add '**2** soldier's *encumbrances*, i.e. *family*, *PBaden* 48.9 (ii BC), *UPZ* 110.199 (ii BC); *dependants*, Lxx *Ge.* 46.5, al.'

ἀποσκλῆναι, after 'pf.' insert 'ἀπεσκληκός Hld. 8.8' and after 'fut.' read 'med. ἀποσκλήσῃ *you will perish with cold*, *AP* 11.37 (Antip.)'

ἀποσκληρύνω, after '*CP* 3.16.2' insert '*AP* 6.298 (Leon.Tarent.)'

†ἀποσκολύπτω, *strip the skin off*, S.fr. 423; (spec.) *circumcise*, Ael.Dion. α 162 E., Hsch., glossed by κεκακουχημένος *POxy.* 2328 ii 5/6 (i/ii AD); *pull back the foreskin*, Archil. 39 W.

ἀποσκορακισμός, for '*casting off utterly*' read '*execration*'

ἀποσμύχομαι, delete the entry.

ἀποσοβέω **I**, after '(Ar.) *V.* 460' insert 'μάρτυρας *PEnteux.* 86.6 (iii BC); *send away*, ἀ. με εἰς τὴν διόρυγα καὶ εἰς τὸ χῶμα *PKöln* 104.14 (vi AD)'

ἀποσόβησις, after '*scaring away*' insert '*PLond.* 1724.49 (vi AD)'

*ἀποσπάθιον, τό, perh. *knife*, *PHamb.* 223.11.

†ἀποσπαλακόω, *reduce to the condition of a* σπάλαξ (*blind-rat*), ὁ τᾶς Δίκας ὀφθαλμὸς ἀπεσπαλάκωται Cerc. 4.19.

ἀποσπάς **II**, for this section read '*as subst., something torn off*, σταφυλῆς ἀποσπάδα πεντάρρωγον, i.e. five grapes pulled off from a bunch, *AP* 6.300 (Leon.Tarent.); spec., *slip for propagation*, Gp. 10.23, 11.9, 16; transf., *branch* of a river, Eust. 1712.6'

ἀπόσπαστος, after '*separated*' insert '*PBerol.* in Gercke-Norden *Einl. in die Altertumsw.* 1(9) p. 42'

ἀπόσπληνος, for 'Apul.*Herb.* 79' read 'Apul.*Herb.* 80'

ἀποσπογγίζω, after 'Antipho 5.45' insert '(v.l. ἀνα-)'

*ἀποσπορά, ἡ, *sowing*, *PCair.Isidor.* 34.5, 38.8 (both iii AD), *PCol.* 136.40, 51.

*ἀπόσταθμον, τό, *weight remaining* after deduction, esp. of the part of a victim reserved for the god, *Thasos* I p. 451.

ἀπόσταξις, add '**2** *distillation*, Moses Alch. 303.10.'

*ἀποστάριος, ὁ, = °ἀποστασάριος, *PKlein.Form.* 1161 (v/vi AD).

*ἀποστασάριος, ὁ, perh. *butler*, *SB* 4640-2, 10990.16-26 (v/vi AD).

ἀποστασίου **I**, add 'D.H.*Din.* 12'

ἀποστατέω, line 1, after '*aloof from*' insert 'ἀπό τινος *Thasos* 174c5'

ἀποστέλλω, add '**V** *give off, convey* a mental or sense impression, ἡ καρὶς .. κινουμένη .. δόξαν τινὰ ἀποστέλλει μελλούσης .. παριέναι Ael.*NA* 1.15; τοὺς .. ὀφθαλμοὺς ἀποστέλλειν κυανοῦ χρόαν id. 4.52; ὁ ταὼς .. πρὸς τοὺς ἔξωθεν φόβον ἀποστέλλει id. 5.21.'

ἀποστεφανόω, delete '*discrown*' and insert '*Inscr.Olymp.* 225.21 (AD 49)'

*ἀποστέφω, *honour with libations*, ψυχῇ θ' ἡρώων πένθος ἀποστέφετε (= -ται) prob. to be read in *IGBulg.* 200, v. *ZPE* 9.186-7; cf. perh. ἀποστεφήσῃ, gloss on ἀποέρσει Hsch. (v. εἴρω (A).)

ἀποστηθίζω, after '*by heart*' insert 'Dam.ap.Suid. s.v. Σαλούστιος'

ἀπόστημα **3**, add '*PMich.*XIII 660.8 (vi AD)'

ἀποστηματώδης, add 'Paul.Aeg. 4.23 (345.9 H.)'

ἀποστήριξις, add '**2** *taking of a firm stance*, Gal.*Parv.Pil.* 2 (pl.).'

ἀποστολεύς **I 2**, for '*IG* 2.809ᵇ20' read '*IG* 2².1629ᵇ252' add '**3** *harbour official in charge of freight consignments*, *PLond.* 1940.3 (iii BC), Aen.Tact. 29.12.'

ἀποστολή **I 1**, for 'cf. *IG* 2.238.15' read '*sending of ambassadors*, *IG* 2².477.15, D.H. 8.64, = συνθεώροι, *SEG* 33.671.11 (= *OGI* 42, Cos, iii BC, in Dor. form -ά)' add '**b** *consignment* of goods, ὁ πρὸς ταῖς ἀ. *PLond.* 1963.6 (iii BC), *PCair.Zen.* 299 (pl., iii BC).' add '**5** prob. calque on Hebr. *shelah*, *shoot* (where context requires prob. "cheek"), Lxx *Ca.* 4.13.' **II**, after '**2**' insert '*exile*, Lxx *Je.* 39(32).36, *Ba.* 2.25' (renumbering pres. section 2 as 3)

ἀποστολικός, add '**2** of an apostle, Clem.Al.*Prot.* 1.44, with the quality of an apostle, id.*Strom.* 2.20.116; of the Christian church, *apostolic*, *BCH* suppl. 8.235 (iv AD), *PRyl.* 471.5 (v AD), *Cod.Just.* 1.1.5.1; neut. subst., *gospel quotation*, Clem.Al.*Ecl.* 25 (p. 143.23), esp. of St. Paul's letters, *Pap.Lugd.Bat.* xxv 13.52, 53 (vii AD).'

ἀπόστολος **II 1**, for '*IG* 2.809ᵇ190' read '*IG* 2².1629.243' add '**8** = Lat. *litterae dimissoriae*, *PCol.*inv. 1696 in *ZPE* 10.135, cf. *Dig.* 50.16.106.'

ἀποστόμωσις, after 'πόρων' insert 'perh. f.l. for ἀναστομ- in'; after 'Arist.*Pr.* 888ᵃ28' add 'but cf. *IG* 4.823.44 (Troezen, iv BC)'

ἀποστρατεύομαι, add '*Dig.* 27.1.8 pr., 27.1.8.1'

*ἀποστράτωρ, ορος, ὁ, *ex-στράτωρ* (q.v.), *Syria* 6.232 (Der'a, iii AD); but prob. ἀπὸ στρατόρων shd. be read; cf. ἀπό II.

ἀποστρέφω **A II**, add 'Din. 2.23' **B II 1**, w. acc., add 'Even. 2.5' add '**4** w. acc. and inf., *shrink from saying*, Plu. 2.387c.'

*ἀποστρεψίκακος, ον, *averting evil*, epith. of Zeus, Bonner *Studies in Magical Amulets* p. 172.

ἀποστροφή **III**, rhet., add 'of an adjuration' and transfer 'Longin. 16.2' to follow.

ἀποστυγέω, line 1, after '-έστυξα' insert '*AP* 6.48'

ἀποστυλόω | SUPPLEMENT | **ἀποφλισκάνω**

*ἀποστῡλόω, dub. sens., *Delph.* 3(5).85.8 (iv BC).

†ἀποσῡκάζω, *gather the fruit from fig-trees*, Amips. 33 K.-A. (pass.); fig., of cheats or the like, Ar.*Eq.* 259.   **2** ἀποσυκάζειν· σῦκα ἐσθίειν Hsch.

ἀποσυμβιβάζω, add '*PUps.Frid* 10.17 (iii AD), *PMilan.* 86.5 (v AD)'

*ἀποσυμμαλάσσω, *make into a paste*, ἐν κηρῷ ἀποσυμμαλαχθεὶς καὶ ἐπιπλασθεὶς μετώπῳ Cyran. 62 (v.l. συμμαλ-, v. ib. 2.16.5 K.).

ἀποσυνίστημι, add '**3** *reserve* money for a purpose, *Didyma* 488.22.   **4** ἀποσυνιστῶ gloss on Lat. *amando*, Dosith. p. 435 K.'

†ἀποσυντάσσω, *revoke an order, prohibit*, w. μή and inf., *PSI* 418, *PLond.* 1979.17 (both iii BC).

*ἀποσῡοκεφᾰλόω, *change so as to have a pig's head*, Cels.ap.Orig.*Cels.* 5.64.

ἀποσῡρίζω, for 'pass. .. *VH* 2.5' read '*give out whistling sound*, in pass., ἀπὸ τῶν κλάδων κινουμένων τερπνὰ .. μέλη ἀπεσυρίζετο Luc.*VH* 2.5' add '(ἀποσυριεῖς Lxx *Is.* 30.14 is prob. f.l. for ἀποσυρεῖς)'

ἀπόσυρμα **I 1**, add 'Erot. 93.8'

ἀποσύρω, add 'also perh. Lxx *Is.* 30.14, v. °ἀποσυρίζω'

ἀποσφάζω, pass., add 'Hp.*Nat.Hom.* 6'

ἀποσφήλωσις, delete the entry.

ἀποσφίγγω, add '*AP* 10.210 (Lucill.)'

ἀπόσχᾰσις **I**, add 'ἀ. τῶν σκελῶν *scarification*, Orib.*Syn.* 8.3.22'

ἀπόσχημα, delete the entry.

*ἀποσχῆναι· ἀπενεχθῆναι Suid.

†ἀποσχολάζω, *take one's recreation*, ἐν τούτοις (sc. amusements) ἀποσχολάζειν Arist.*EN* 1176ᵇ17, ἐν ταῖς ἀγοραῖς Ael.*VH* 9.25, τῷ οἴνῳ ib. 12.1, παρά τινι Ps.-Hdt.*Vit.Hom.* 5, 34.

*ἀποσχολέομαι, *to be busy* or *preoccupied*, *PKöln* 317.39 (vi AD).

ἀποσῴζω **II**, *get off safe*, add 'νὴ Δί' ἀπεσώθητέ γε Men.*Dysc.* 434'

*ἀποτακτάριος, ὁ, *assessor*, *SB* 9608.4.

ἀποτᾰμιεύομαι, add 'act., *An.Ox.* 3.195.11'

ἀπότᾰσις **2**, for 'Plu. 2.670d' read 'Plu. 2.670c'

ἀποτάσσω **I**, add '**b** *register*, ὁ γυμνασίαρχος .. ἀπέταξεν τοὺς ὑφ' ἑαυτῶι γενομένους ἐφήβους *CRAI* 1939.222.A13, 224.B11, 230 (Dalmatia, i AD).'

†ἀποταυρόομαι, pass., *to be changed into a bull* (or *cow*), w. internal acc. in mixed metaphor, τοκάδος δέρμα λεαίνης ἀποταυροῦται δμωσίν E.*Med.* 188; cf. ταυρόομαι; of Io, Erot. s.v. κερχνώδεα.

†ἀποτεκμαίρομαι, *find by following signs, trace*, πόρους A.R. 4.1538.

ἀποτέλεσμα **3**, after '*finished product*' insert 'Trypho *Trop.* p. 196 S.'

*ἀποτελεσμός, ὁ, *completion, result*, *SEG* 8.464.38 (Egypt).

ἀποτελέω **II**, delete the section.

ἀποτελωνέομαι, for '*PLond.ined.* 2092' read '*PLond.* 1979.10' and add '*PStrassb.* 238.13, 14'

ἀποτέμνω **I 1**, line 8, for '*PMag.Par.* 1.38' read '*PMag.* 4.38; abs. *execute* (by decapitation), *CodJust.* 10.11.8.2 (cf. ib. 9.1.20)'

†ἀποτηγᾰνίζω, also ἀποτᾰγᾱνίζω (Sotad.Com. l.c., Alex. 178.11 K.-A.), *eat off the frying-pan, eat broiled*, Pherecr. 128 K.-A., Phryn. Com. 60 K.-A.   **2** *broil*, Macho 421 G., Sotad.Com. 1.1 K.-A.

*ἀποτηγάνισμα, ατος, τό, *rendered-down fat, lard*, τοῦ κροκοδείλου Cyran. 2.22.6, 22.9 K.

ἀποτήκω, line 3, for 'τρία τάλαντα' read 'τέταρτον ἡμιτάλαντον'

ἀποτίθημι **I 1**, line 3, delete '*pigeon-hole*'   **II 3**, after 'a child' insert '*Leg.Gort.* 3.49, al.'   **II 3 b**, after '*bury*' insert 'Call.*Epigr.* 19.1 Pf., D.C. 73.5, 76.15 (Pass.)'   add '**c** *lay* charges, ὅταν παρ' αὑτοῖς ἀποτεθείη τις αἰτίασις κατά τινος *Cod.Just.* 1.4.29.8, τὸν κατ' αὐτοῦ τὴν αἰτίασιν ἀποθέμενον Just.*Nov.* 96.2.1.'   **6**, after 'Str. 10.5.2' insert 'ἀ. τὸ βάρος *PBremen* 63.4 (ii AD), Artem. 5.30'   add '**10** *put down, record*, τά τε γὰρ καλούμενα παρὰ πᾶσι πρῶτα ἐν τέσσαρσιν ἀπεθέμεθα βιβλίοις Just.*Const. Δέδωκεν* 3, 7c, 7e.'

ἀποτῑμάω **II**, add '**3** *reckon, assess* numbers, Phleg. 12.6 J.'

ἀποτίμησις **II 2**, after '*valuation*' insert 'of a ἡρῷον, *SEG* 30.1354 (Miletus, iii AD)'

ἀποτῑνῠμι, add '= ἀποτίνω'

ἀποτίνω **I 3**, add 'ἀ. πέντε δραχμάς *SEG* 23.76 (Attica, iv BC), ἀ. προστείμου .. δραχμὰς δέκα .. ib. 31.122 (Attica, ii AD)'

*ἄποτις, ιδος, app. fem. adj., *abstaining from drink*, Suid.

*ἀπότιστος, ον, *unwatered*, *PTeb.* 1126.1 (ii BC).

ἀποτμήγω **2**, after 'c. gen.' insert 'Parm.*fr.* 4.2 D.-K.'

ἀποτομᾱς **3**, delete the section transferring quot. to the beginning of section 2.

*ἀπότομον, τό, dub. sens., *BGU* 1546 (iii BC), perh. *offcut*.

ἀπότομος **I 2 b**, add 'Lxx *Wi.* 18.15'   **c**, add '(v. Robert *Les gladiateurs* 258ff.)'   add '**6** neut. subst. pl., of land *cut off* and enclosed for private use, τὸν ἀποτομὸν καὶ τὸν δαμοσίον *IG* 9²(1).609.2 (Naupactus, v BC); cf. ἀποτέμνω **II** 2.'

*ἀποτόσιτος, ον, *without food or drink*, dub. in Sophr. in *PSI* 1214d7.

ἀποτρεπτικός **2**, add 'τῶν δεινῶν Ps.-Luc.*Philopatr.* 8' deleting this ref. in section 1

ἀποτρέπω **II 2**, add 'τὸ ξύμπαν περὶ Πλαταιῶν οἱ Λακεδαιμόνιοι οὕτως ἀποτετραμμένοι ἐγένοντο Th. 3.68'   **3**, after '*E.IA* 336' delete '(lyr.), cf. Th. 3.68'

ἀποτρέχω, after lemma insert 'Dor. ἀποτραχ- *SEG* 26.700.7 (Dodona, iii BC), *Inscr.Cret.* 1.xxii 4 c 5 (Olous, iii/ii BC), see also τράχω s.v. τρέχω'   **I**, add 'w. inf., *avoid*, χρωτίζεσθαι ἀ. Lyr.*Alex.adesp.* 36'   **II**, delete the section.

ἀποτρίβω **II**, for this section read '*rub down* a horse, X.*Eq.* 6.2.   **b** ἀ. τὸ αἰδοῖον, *masturbate*, Macho 181 G., med., Plu. 2.1044b.'

ἀπότριπτος, add 'σφυρίς perh. *used, worn out*, *PSoterichos* 4.17 (i AD); κτήνη (beasts of burden) *SB* 7621.22, al. (iv AD)'

ἀποτρόπαιος, before 'ον' insert 'α'   **I**, line 3, for '*CIG* 464' read '*SEG* 21.541 i 33 (Attica, iv BC), *IG* 2².4852, Sokolowski 2.116.A3-4 (Cyrene, ii BC); also of Zeus and Athena, Sokolowski 1.25.59, 82, 146 (Erythrae, iii BC), Sokolowski 2.88*b*3 (Lindos, ii BC), cf. ib. a1 (iv BC); of Athena alone, *IG* 14.957 (Rome); of Zeus, *Inscr.Perg.* 8(3).161 p. 169 line 19 (both ii AD)' add 'neut. subst., the sanctuary of Apollo Apotropos, *SEG* 9.72.6 (Cyrene, iv BC)'

ἀποτροπάομαι, add 'act., = ἀποτρέπω, v.l. ἀποτροποῶμεν in Il. 20.119'

ἀποτροπία, -ίη, add 'perh. *distance, inaccessibility* of a divinity, IStraton. 543.16 (iii/ii BC)'

ἀπότροπος **I**, add 'εἴ τις .. ἀπότροπος οἴκαδ' ἀπέλθοι Panyas. 13.5 (nisi leg. ὑπο-)'   **II**, add 'of Apollo, *SEG* 9.72.6 (Cyrene, iv BC)'

ἀποτροφή, line 1, after '*PSI* .. pl.)' insert '*food for the poor*, *CodJust.* 1.2.25 pr., *maintenance* for soldiers, ib. 10.27.2.12'

ἀποτρόφιμος, add 'ἀ., τά, *subsistence*, rest. in *PMasp.* 151.268 (vi AD)'

*ἀποτρυπάω, *bore through* or *gouge out*, ὀφθαλμὸν ἀπετρύπησα τὸν τοῦ βασκάνου i.e. to avert evil, prob. w. sens. obsc., on phallic representation, *SB* 6295.   **2** ἀποτρυπῶν· λάθρα ἐξιών Hsch.

ἀποτρῠπῶν, delete the entry.

ἀποτρώγω **2**, w. gen. *nibble at*, for 'metaph. .. *swathe*' read 'transf., of a reaper, ὃς νῦν ἀρχόμενος τὰς αὔλακας οὐκ ἀποτρώγεις'

*ἀπότρωγμα, ατος, τό, *siftings, riddlings*, *IG* 2².1672.218 (pl.); cf. διαττάω.

ἀποτύμβιος, delete the entry.

†ἀποτυμπᾰνίζω, (later -τυπ- *PEnteux.* 86.6, 8 (iii BC), *UPZ* 119 (ii BC), *POxy.* 1798.1.7), *put to death, execute* (whether by judicial sentence or less formally; perh. originally *cudgel to death*, but this is not specified in the quots.), Lys. 13.56, D. 8.61, 9.61, 19.137, Arist.*Rh.* 1383ᵃ5, 1385ᵃ10, *PEnteux.* l.c., *UPZ* 119.37, Plu.*Dio* 28, *Sull.* 6, id. 2.523a, 778e, 1049d, Euph.*Fr.Hist.* 173 vGr., Lxx 3*Ma.* 3.27.

ἀποτυμπᾰνισμός, for '*crucifixion*' read 'prob. *destruction*'

ἀπότῠπος **1**, add '**b** *decorated with repoussé figures*, φιάλη *Inscr.Délos* 442*B*183, σκύφοι ib. 30 (ii BC).'

ἀποτύπτω **2**, for '*cease to .. mourning*' read '*finish beating the breast*'

*ἀπότυψις, εως, ἡ, perh. *repoussé work*, *Inscr.Délos* 1441*A*ii62, 1450*A*161 (both ii BC); perh. error for ὑπότυψις (q.v.).

*ἀπούᾰτος, ον, *ill-sounding, ill-omened*, ἄγγελος Call.*fr.* 315 Pf. (fr. misinterpretation of ἀπ' οὔατος Il. 18.272).

ἀπουλόω, add 'also ἀφουλ- Paul.Aeg. 6.47 (87.17 H.); pass., id. 6.40.3 (79.13 H.)'

ἀπούλωσις, add 'also ἀφούλ-, Paul.Aeg. 6.9 (54.5 H.), 6.58 (97.24 H.)'

ἀπουλωτικός, add 'also ἀφουλ- Paul.Aeg. 6.5 (48 H.)'

ἀπούλωτος, delete the entry (see ἀμωλώπιστος).

ἀπουρᾱς **I**, add '**4**' insert 'w. acc. and gen., νόστου τόνδε στόλον .. ἀπούρας A.R. 3.175.   **5**'

ἀπουσία **II**, add '*deficit* incurred in reminting old coins, *NC* 1950.1-22'

ἀπουσιάζω, after 'Sor. 1.87' add 'cf. ἀπουσία **III**'

ἀποφαίνω **A I**, for 'ἀ. παῖδας .. by her' read 'ὡς ἐκ ταύτης παῖδας ἀποφανῶν i.e. claim that children by her are his legitimate offspring'; after '*Is.*' insert '3.30, 73, 79'   **II 4**, transfer 'πρίν .. Ra. 845' to section 3, and after '*Lys.* 31.2' insert '**b** *pronounce*, of the Areopagus'   **B I 1**, add '**b** *appear as, be shown to be*, τῆς θαλάσσης .. κατὰ μικρὸν εἰς πέλαγος ἀποφαινομένης *Peripl.M.Rubr.* 26.'

ἀπόφᾰσις (A), line 1, delete 'opp. κατάφασις'

ἀπόφᾰσις (B), for 'διαίτης' read 'δίκης' and after 'cf. 33.21' insert 'of a *report* made by the Areopagus, Din. 1.1'

ἀποφέρω **III**, after '*returns*, etc.' insert 'ἀ. γραφὴν κατά τινος κακώσεως D. 58.32; ἀπηνέχθη ἡ κατὰ τοῦδε τοῦ ψηφίσματος γραφή (sc. παρανόμων) Aeschin. 3.219'

ἀποφεύγω, add '**4** *fail to meet an obligation, default*, Just.*Nov.* 120.8.'

ἀπόφημι, line 1, add 'pf. ἀποπέφηκεν Arist.*Met.* 1012ᵃ16 (s.v.l.)'   **II**, add '**3** w. acc. of predicate, κλεινὴν οὐκ ἀ. I do not *deny that you are* famous, *AP* 9.550 (Antip.).'

ἀποφθῐνύθω **I**, add 'νούσοι .. ἀποφθινύθουσι βροτοῖσι Orph.*H.* 68.3, καρποί .. ἀποφθινύθουσι orac. in Paus. 9.17.5'   **II**, delete '**2** *diminish*'

ἀποφθίνω **II 2**, for 'most freq. .. plpf. form ἀπέφθιτο[ι]' read 'most freq. in med. or pass., *perish, die*, esp. in aor. 2 form ἀπέφθῐτο'

*ἀποφλισκάνω, *pay off*, ἔχεις τὸ χρέος .. ἀποφληθέν Tz.*H.* 13.607.

47

ἀπόφλω, delete the entry.
ἀποφοιβάζω, add 'Certamen 35'
ἀποφορά **II**, for 'effluvia' read 'effluvium, bad smell' and add 'also scent ῥόδων Dsc. 4.45, οἴνου Gp. 7.7.5'
ἀποφόρησις, add '2 perh. removal, GDI 3362.45.'
ἀποφόρητος, after 'carried away' insert 'τὰ παρατιθέμενα ἔστω ἀ. IG 12(7).515.65 (Amorgos, ii BC)'; add 'δεῖπνον ἀ. IStraton. 270.13'
ἀπόφορος **1**, for 'not .. suffered' read 'that is given off (cf. ἀποφορά II 1)'
ἀποφράς, after 'Luc.Pseudol. 12' insert 'w. ellipsis of ἡμέρα Sallust. 18, p. 34.5'
ἀποφράσσω, add 'fig., ἁπλῶς εἰπεῖν ἀποφράττειν αὐτοῖς τὴν δεῦρο ἄνοδον Paul.Dig. 49.1.25'
*ἀποφροντιστής, ου, ὁ, ex-φροντιστής (local government official), PMich.xv 736.7 (vi AD).
ἀποφώλιος, line 4, after '(Od.) 11.249' insert 'uneducated, wild, ἐξ ὀρέων ἀ. ἀγροιώτης Philet. 10.1'
ἀποφωνέω, add 'cf. ‡φωνέω I 3'
ἀποχάραξις **I**, add 'of a footprint, Sch.S.Aj. 2' **II**, for this section read '?gap, Didyma 32.4, 34.17 (pl.)'
ἀποχειροτονέω **II 3**, for 'Ar.Pax 668' read 'Ar.Pax 667'
ἀποχή **I**, add '2 astron., elongation, Ptol.Alm. 6.2.462.6, al.' **III**, for 'PTeb. 11.14' read 'PHib. 162 (iii BC), PTeb. 11.18'
*ἀποχλωριαίνω, become pale, Vit.Aesop.(G) 43: also ἀποχλωριάω, ib. 54.
*ἀποχονδρόομαι, become cartilaginous, Paul.Aeg. 6.96.1 (149.21 H.).
ἀποχράω **A I 2 a**, add 'ἀποχρεῖ suffices, SEG 9.72.40 (Cyrene)' **3 a**, at end for 'D. 17.31' read 'D. 17.13' add '**C** (cf. χράω B) lend, 3 pl. fut. -χρήσουσι PCair.Zen. 107.5 (iii BC); med., borrow, PMerton 4.6 (iii BC).'
*ἀποχρηστεύομαι, to be unaccommodating, PTeb. 777.8 (iii BC).
ἀπόχυμα **1**, add 'drainage water, PMich.xi 617.9 (ii AD)'
ἀπόχυσις, add 'drainage ditch, BGU 2354.2 (ii AD)'
ἀποχωρέω **I 1**, add 'of slaves, run away, ἡμῶν PCair.Zen. 15ᵛ.41 (iii BC)'
ἀποχώρησις **II**, for '= ἀπόπατος 2' read 'latrine, PMich.i 38.31 (iii BC)'
ἀποψέ, add 'PLond. 1081.1 (pap. -α, vii AD); cf. mod. Gk. ἀπόψε'
ἀπόψημα, after 'refuse' insert 'PCair.Zen. 9 (iii BC)'
*ἀπόψηστος, ον, scraped off, made level, ἡμιχοινίκια IG 2².1013.21.
ἀποψηφίζομαι, add '**V** act., count off, i.e. (?)reject, (?)ἄχρ(ε)ι(α) Dura⁴ 127.'
*ἀποψηφοποιέω, gloss on praefragor (?disfranchise), Dosith. p. 432 K.
ἄποψις, add '**III** surface, Syria 18.372 (Palmyra).'
ἀποψοφέω, delete '**II** sound loudly' and for 'φωνή' read 'φώκη'
ἀπόψυγμα, add 'cf. ‡ἀποψύχω III'
ἀπόψυξις, delete 'evaporation'
ἀπόψυχος, for 'frigid' read 'prob. lifeless'
ἀποψύχω **I 1**, add '**b** evaporate, lose flavour, Sm.Ez 17.9.' **II 1**, delete 'they got the sweat dried off their tunics' **III**, add 'cf. ἐναποψύχω'
ἀποψωλέω, delete 'a lewd fellow' and add 'id.Pax 904, Lys. 1136, Th. 1187b, Pl. 295'
*ἄππαμα, τό, Boeot. for ἀναπ- (cf. ἀνακτάομαι, ἀππασάμενος), pl., τἀππάματα the debts to be recovered, IG 7.3172.163, al. (Orchomenus, iii BC; but some interpret as τὰ ππάματα = τὰ κτήματα).
ἄππας **II**, add 'διὰ τοῦ ἄππα καὶ ἐπιτρόπου μου Inscr.Cos 352.6'
*ἀππεισάτου, v.s.v. ἀποτίνω.
ἀπρᾱγία, add 'futility, Sm.Pr. 12.11, 28.19'
*ἀπρᾱγμᾰτικός, όν, perh. inalienable or not to be mortgaged, SEG 34.1339 (Iconium, iii AD).
ἀπραγμοσύνη **I 1**, for 'freedom from politics, love of a quiet life' read 'freedom from involvement in affairs, tranquillity' and add 'Just.Nov. 100.1, 147.1'
ἄπρᾱγος, for '= ἀπράγμων' read 'futile, ineffective'
*ἀπραίδευτος, ον, (cf. Lat. praedor) not sacked, Phot. α 344 Th.
ἄπρακτος **I 2**, line 4, after 'Th. 6.48' insert 'codd., edd. -τους'
ἄπρᾱτος, after 'Aeschin. 2.23' insert 'not for sale, ἄπρατον εἰς αἰεί SEG 34.1136, ISmyrna 228.6 (ii AD), v.l. in Plu.Galb. 17; neut. pl. subst., things left unsold, PPetaus 13.19 (AD 184/5)'
ἀπρεπής, add '**III** improper, illegal (use of water), IEphes. 3217.A12 (ii AD).'
*Ἀπρίλιος, α, ον, Lat. Aprilis, PLond. 130.44 (ii AD, -λει-), POxy. 899 i 7 (iii AD, -λλι-), al., SEG 34.1469 (Palestine, vi AD).
*ἀπροαισθήτως, adv. unforeseeably, Eurysus ap.Stob. 1.6.19.
†ἀπρόβουλος, ον, unforeseeing, ἀπροβούλῳ (cj., cod. -ως) .. ὕπνῳ A.Ch. 620 (lyr.).
*ἀπρόγραπτος, ον, which has not yet been published, PPetr.ined. in APF 33.24.8 (iii BC)
ἀπροϊδής, add 'δαίμων ἀπροιδής IG 14.1892'

ἄπροικος, after 'Lys. 19.15' insert 'Men.Dysc. 308'
*ἀπροϊσία, ἡ, seclusion, Sch.E.Hipp. 132.
ἀπροκοπία, add 'Heph.Astr. 2.32.10, 3.6.6'
*ἀπρόκρῐτος, ον, not pre-judged, free from praeiudicium, Cod.Just. 7.62.36.
ἀπροόρᾱτος **II**, for 'Ph. 2.268' read 'Ph. 2.269'
ἀπρόσβᾰτος, for 'νοῦσος' read 'νόσος'
ἀπρόσδεκτος **II**, transfer 'unacceptable .. Porph.Marc. 24' to section I, after 'S.E.P. 2.229'
ἀπροσδῐόνῡσος **I**, add '**2** of a person, not suited to Bacchic rites, Hld. 3.10.2.' **II**, transfer 'Plu. 2.612e, Luc.Bacch. 6' to section I 1.
*ἀπροσήκων, ον, app. no one's property, εἶναι τὸ .. σῶμα ἐλεύθερον, ἀπρόσηκον, μηδενὶ μηδὲν προσῆκον Delph. 3(6).116.8 (i AD).
ἀπρόσθετος, after 'added to' insert 'δηνάρια PDura 29.8 (iii AD)'
*ἀπροσκοπία, ἡ, (προσκόπτω) safety, freedom from accident, SB 7352 (ii/iii AD).
ἀπρόσκοπος (B) **I**, after 'unseeing' insert 'or perh. not seen beforehand'
ἀπρόσλογος, after 'not to the point' insert 'Artem. 1.11'
*ἀπροσμάχητος, ον, irresistible, Sch.Hes.Th. 295 (p. 58.16 G.), ib. 319b (p. 62.5 G.).
*ἀπροσοδίαστος, ον, not yielding income, IGRom. 3.422.19 (Ariassus); cf. ἀπρόσοδος II.
ἀπρόσοδος **II**, add '**b** unsalaried, BCH 83.363.25 (Thasos, i BC).'
ἀπροσποίητος, before 'in adv.' read 'SEG 3.226.14 (Athens, ii AD)'
*ἀπροσπολίη, ης, ἡ, loss of one's attendant, rest. in IG 12(8).92.9 (Imbros, ii/i BC).
*ἀπροσπόριστος, ον, of property, not acquired from the property of the paterfamilias, Just.Nov. 84.1.2.
ἀπρόσφορος, add 'not applicable, δικαστής i.e. not competent, Cod.Just. 3.1.12.2; w. gen., κἂν τοῦ ἐνάγοντος ἀπρόσφορος ib. 7.51.5.1'
ἀπροσωπόληπτος, add 'prob. voc. -λημπτε, to a divinity, IGLS 343.6 (lapis -λημιε)'
ἀπροτίελπτος, add 'χάρμα orac. in ZPE 1.184.22 (Guarducci EG 4 p. 103)'
ἀπροφάσιστος **I 1**, line 3, after 'Timocl. 8' insert '(cf. perh. ἀ. in καλοῖς inscr., SEG 35.252, v.l. codd.)'; adv., add 'neut. as adv., pl., AP 7.721 (Chaeremo), sg., ib. 5.250 (Paul.Sil.)'
*ἀπροφήτευτος, ον, not having a προφήτης, ἐνιαυτός Didyma 237 II 10 (?i AD).
ἄπταιστος, line 3, after 'Am. 46' insert 'unerring, D.H.Dem. 52'
ἀπτήν **I**, delete 'metaph. of men' and after 'Com.adesp. 1291' insert '= A.fr. 337 R.'
*ἀπτής, = ἀπτώς, Inscr.Olymp. 164 (iv BC).
†ἄπτιστος, ον, not winnowed or husked, Hp.VM 14; of loaves, made from unhusked grain, Philostr.Gym. 43.
ἄπτρα, for 'ἄπτριον' read 'ἀπτρίον' and add 'cf. ἀπτρήν (= -ίον) Gregorius Magnus in PL 77.178C, (mod. Gk. dial. ἀφτρί)'
ἅπτω **A I 1**, after '(to hang herself)' 11.278' insert 'οἱ δ' ἀγχόνην ἅψαντο Semon. 1.18' **B**, set on fire, after 'A.Ag. 295' insert 'w. gen., Th. 4.100.4' **III**, delete the section.
ἀπτώς, delete 'ἀ. δόλος .. 9.92'; add 'fig., of wrestlers and their art, σὸν κατ' ἀγῶνα ἀπτῶτ' ἀγγέλλω παῖδα κρατεῖν Lindos 699.b2 (epigr., ii BC), νεικήσας ἀγῶνα ἀπτώς SEG 31.1287 (Side, Rom.imp.)'
*ἀπτωτί, adv. without falling, in wrestling, φῶτας δ' ὀξυρεπεῖ δόλῳ ἀ. δαμάσσαις Pi.O. 9.92.
ἀπύ, add 'also Myc., Cypr. a-pu'
*ἀπυγίζω, v. °ἀναπυγίζω.
ἄπῡγος, add 'Comp., AP 11.327'
*ἀπύγων, gloss on διχόνδις, Hsch.
ἄπῡθεν, after 'Aeol.' insert 'or Arc., Cypr.'
ἀπύρετος, add at end '(interpol.)'
ἄπυρνος, after 'without' insert 'or with only a residual'; delete 'φοῖνιξ' and add 'Arist.Met. 1023ᵃ1'
ἄπῠρος **I 1 a**, add 'Alcm. 17.3 P.' **3**, for 'unfermented' read 'not heated' and for 'Alcm. 117' read 'Alcm. 92(a) P.' **6**, insert def. 'made without the use of fire' and add 'Pl.Plt. 287e'
ἄπυστος **I 1**, for 'Sapph.Supp. 25.19' read 'Parm. 8.21, Call.Del. 215, Cer. 9' **II**, add 'Call.fr. 611, 680 Pf.'
ἀπφᾶ, add 'Suid., AB 441'
ἄπωθεν, line 1, before 'Q.S. 6.647' insert 'Call.fr. 194.97, 197.25 Pf.'
ἀπωθέω, line 1, after 'ἀπῶσα' insert 'AP 9.326.5 (Leon.Tarent.)'
ἀπώμοτος, add 'one who denies on oath that he has done a thing, Inscr.Cret. 4.72 xi 28'
ἀπωνέομαι, for 'buy, purchase' read 'sell' and after 'ἀπωνηθήσεται' insert 'will be sold'
ἀπώρωτος, add 'Paul.Aeg. 6.110 (163.9 H.)'
†ἄπωτος, ον, perh. deaf, on votive offering in the shape of an ear, Cypr. to-po-to-e-mi, perh. τῷ ⸢πότῷ ἐμί ICS 289.
ἆρα **B 4**, for 'v. sub τοι II 2' read 'v. sub τοι III'
ἄρα **A**, at end for 'Id.fr. 931' read 'S.fr. 931' **B**, for 'almost always' read 'generally'

ἀρά **I 2**, for this section read '*ex-voto dedication, ISmyrna* 739 (Ion. acc. ἀρήν, *c*.600 BC), Cypr. *a-ra, Rantidi* 44 (vi BC), perh. in *ICS* 107, *Inscr.Cret.* 1.xxv 3, xxi 7 (both near Gortyn), *SEG* 23.593 (Gortyn, i BC)'

*Ἀραβάρχης, ου, ὁ, *ruler of Arabs*, applied by Cicero to Pompey, *Att.* 2.17.2. **II** title of official, *controller of customs*, esp. in Egypt, orig. E. of Nile, *OGI* 202.9, Juv. 1.130; dissim. **Ἀλαβάρχης**, J.*AJ* 18.6.3, al.; also in Lycia and Euboea, *OGI* 570, *IG* 12 suppl. 673 (Chr.).

Ἀραβαρχία, before '*office*' insert '*rule over Arabs*, J.*AJ* 15.6.2. **II**' and add '*OGI* 674 (i AD)'

ἄραβδος, delete the entry.

Ἀραβικός, s.v. Ἀραβία, add 'as Lat. *Arabicus*, title conferred on Septimius Severus, *TAM* 4(1).27, *ICilicie* 78, etc. **2** name of a god in Arabia, Θεὸς Ἀ. *SEG* 32.1540 (Gerasa, Rom.imp.).

Ἀράβισσα, s.v. Ἀραβία, before 'St.Byz.' insert '*Arabian woman*, *RDAC* 1975.142-3 (Cyprus, late Hellen.), *AD* 25.74 no. 22 (in Att. epitaph, i BC), *SB* 11169.13 (ii AD)'

Ἄραβος (B), ὁ, *Arab*, for phrase κατ' Ἀράβους v.s.v. °Ἄραψ.

ἀραγμός, add 'Plu. 2.594e'

ἄραδος, for '*palpitation* of the heart' read '*agitation*'

ἀραιόστυλος, for 'Vitr. 3.31' read 'Vitr. 3.3.1'

*ἀραιόφρυς, -υος, ὁ, *having scanty eyebrows*, in description of a slave, *POxy.* 3054.16 (sp. ἀρεο-, iii AD).

ἀραίωμα, after 'Plu. 2.980c' insert 'Longin. 10.7' and at end delete '*a little bit*, Longin. 10.17'

*ἀράκινος, η, ον, *of or from aracus*, χόρτος *PVindob. Worp* 3.19, perhaps = χορτάρακος· ἄχυρον *BGU* 2151.18.

*ἀράκομαι, med., *sow with aracus, PSI* 1021.19 (ii BC).

*ἀρακοφόρος, ον, *producing aracus, PMich.*I 31.26 (iii BC).

ἀράομαι **I**, line 1, after 'Aeol.' insert 'pres. ind. ἄρᾱμαι Sapph. 22.17 L.-P.' and for '*Supp.* 5.22' read '16.22 L.-P.'; line 2, for 'Sapph. 51' read 'Sapph. 141.7-8 L.-P., Cypr. *a-ra-wa-sa-tu*, ἀρϜάσατυ *ICS* 343a (p. 404); later ἠρασάμην *AP* 5.47 (Rufin.)' **2**, add 'h.*Hom.* 6.16' **3**, line 1, for 'Sapph. 51' read 'Sapph. 141.7-8 L.-P.' **4**, add 'abs. Cypr. ἀρϜάσατυ, *ICS* 343a; cf. °ἀρά I 2'

*ἀράριον, τό, v. °αἰράριον.

ἀραρίσκω **B I**, line 5, after 'etc.' insert '2 sg. ἠρήρεισθα Archil. 172.3 W.' **2**, line 2, for 'Il. 10.553' read 'Od. 10.553' **V**, last line, for 'Pl.*Epigr.* 6' read '*AP* 7.35 (Leon.Tarent.)' at end add 'Myc. pf. part. *a-ra-ru-ja* (fem. pl. = ἀραρυῖαι), *a-ra-ru-wo-a* (neut. pl. = *ἀραρϜόϜα), *fitted*'

*ἄρασιν· ἀράχνην Hsch. (after ἀράραι).

ἀράσσω **I**, for '*smite, dash in pieces*' read '*strike violently*' and for 'of any violent .. Pi. l.c.' read 'often w. implication of noise, χαλκέαις δ' ὁπλαῖς ἀράσσεσκον χθόνα (βόες) Pi. l.c.'; line 5, after 'συναράσσω' insert 'exc. in Od. 5.248 (v.l. ἄπρεν)'; line 7, for 'horses' read 'oxen' **2**, add 'Sch.Il. 2.801'

✝Ἀράτειος, ή, name of kind of fig-tree, Thphr.ap.Ath. 3.77a.

ἀρατός **II**, add 'neut. ἀρητόν as adv., *gladly*, prob. for ἄρητον in Call.*Del.* 205' add '**III** Ἄρητος, title of Heracles in Macedonia, *MDAI(A)* 27.311 no. 18 (Edessa), Hsch.'

*ἀραχνέω (or -όω), perh. *remove cobwebs, ZPE* 8.57.

ἀράχνη **V**, for '*sundial*' read '*hemispherical dial* devised by the astronomer Eudoxus of Cnidos'

*ἀραχνιάω, *to be covered with cobwebs*, Nonn.*D.* 38.14.

ἀραχνιόω **I**, delete 'act. in same sense, Nonn.*D.* 38.14'

*ἀραχνοποιέω, *make a (spider's) web*, Cyran. 2.16.12 K.

ἀραχνοϋφής, add '*Suppl.Hell.* 1071'

ἄραχος, add '*Schwyzer* 603 (Corope, vi/v BC)'

Ἄραψ, after 'Arab' insert '*IG* 2².8361-2 (iii and ii BC)'; after 'J.*BJ* 1.19.4' insert 'in dates, κατ' Ἄραβας referring to the Macedonian calendar adopted by the Arabs, *SEG* 30.1687, 34.1468 (both Palestine, vii and vi AD), also κατ' Ἀράβους ib. 34.1467 (Palestine, vi AD)'; after '*Pae.Delph.* 11' insert 'γένος Ἄραβα *IG* 9²(1).624(d).4'

Ἄρβακτις, εως, ὁ, name of an Egyptian divinity, *OGI* 52 (iii BC), Bernand *Les Portes* 23 (ii BC).

*ἀρβελλάριον, τό, kind of knife, used for pig-slaughter, *POxy.* 3866.3, 7 (vi AD); cf. ‡ἄρβηλος.

ἄρβηλος, for '*semicircular knife* .. leather-workers' read '*leather-worker's knife*, having a semicircular blade with two semicircles cut away from the straight edge'

ἀρβυλίς, add '**II** ἀρβυλίδα· λήκυθον. Λάκωνες Hsch.; cf. ἀρυβαλλίς II.'

ἀργαπέτης, add 'see also °ἀρκαπάτης'

*ἄργεθμον, τό, = ἄργεμον I, Cyran. 1.16.15 K. (v.l. ἄργεμος).

Ἀργεῖος, after 'α, ον' insert 'quadrisyll. (Ἀργείων) in E.*Hec.*'; add 'cult-epith. of Hera, Il. 4.8, Hes.*Th.* 12 (-είη) *SEG* 30.366, (Argos), 30.648 (Vergina), 30.1456 (Sinope, all v BC)'

*Ἀργείων, ή, *the Argive* (i.e. *Greek*) *woman*, sc. *Helen*, Hes.*fr.* 23(a).20, 217.6, 136.10 M.-W. (rest.), Theognost.*Can.* 700.

ἀργεννός, add 'prob. on chalky soil, Rhian. 54, Nic.*Th.* 67; form ἀεργενν- by false association w. ἀργός (B), *POxy.* 3536.3 (hex., ε erased but required metrically, iii AD)'

*ἀργεντάριος, ὁ, Lat. *argentarius, BGU* 781 vi 8 (i AD), *SEG* 2.421 (Maced.), etc.

*ἀργενταρίτης, ου, ὁ, *cashier, CPR* 14.41.7 (vi/vii AD).

ἀργέντινος, after '*silvery*' insert '*Stud.Pal.* 20.46.32 (ii/iii AD)'

ἀργέω, after lemma insert 'form ἀεργέω, *SEG* 30.1175.5 (tab.defix., Metapontum, iii BC; but ἀργ- in line 9)' add 'as legal term, abs., *to be inoperative, Cod.Just.* 6.4.4.11a, *to be nullified, cancelled*, ib. 1.3.55.2-3, Just.*Nov.* 110.1'

ἀργήεις, after 'εσσα, εν' insert 'nom. fem. pl. ἀργήεις Nic.*fr.* 74.26'; line 3, for '(v.l. ἀργινόεντι)' read '(cj.; ἀργινόεντι codd.)'

ἀργής, line 5, delete 'φύσις Orph.*H.* 10.10'

ἀργηστής **1**, for '*glancing, flashing*' read '*shining, bright*'; add 'of the wind, cf. ἀργεστής B. 5.67 S.-M.' **2**, for 'κύκνοι' read 'ταῦροι'

*ἀργητός, ή, όν, *lying idle, unused*, κέλλαι λ ὧν ἀργηταὶ οὖσαι δ *PMich.*XI 620.114 (iii AD).

ἀργιβρέντας, for pres. ref. read 'Pi.*fr.* 52m.9 S.-M.'

*ἀργίζω, *to be unemployed, idle, SB* 9699.197, 201, etc., rest. fr. abbrev. ἀργι, but could be ἀργῖ = ἀργεῖ.

*ἀργιζώστη, ή, (or -της, του, ὁ), *white bryony*, Cyran. 9 (1.1.108 K.); cf. ἀργεζώστις.

ἀργιλλοφόρητος, delete the entry.

ἄργμα, add '*CEG* 246 (v BC)'

Ἀργολικός, after '*Rom.* 21' insert 'Call.*fr.* 114.19, 384.22 (Add. II) Pf.'

*ἀργολογία, ή, *empty talk*, Hsch. s.v. βατ⟨τ⟩ολογία.

ἀργομέτωπος, add '*IG* 2².463.40 (rest.)'

Ἄργος, accentuate thus.

ἀργός (B) **I 2 a**, line 5, before 'adv.' insert '*Iamb.adesp.* 39 W.; ἀργοί (sc. ἡμέραι), *holidays*, Porph.*Plot.* 5' **II 1**, add 'ἀργύρου ἀργ(οῦ) *POxy.* 3628.10 (v AD)'

*ἀργυραμοιβήϊον, τό, *bureau de change*, Duchêne *La stèle du port, Fouilles du port* I p.20 I.42 (Thasos, v BC).

ἀργυραμοιβικός, add 'ἀργυραμο[ι]βικὴ τράπεζα *PRev.Laws* col. 73, 3-4'

ἀργυραμοιβός, delete '*banker*' and before 'Theoc. 12.37' insert '*assayer*'

*ἀργυρᾶς, ᾶ, ὁ, *silversmith, BGU* 1034.15 (ii AD).

ἀργυράφιον, add '*SB* 12084.6 (i AD)'

ἀργύρεος, line 1, after 'οῦν' insert '(fem. -εος, τὴν ἀργύρεον τράπεζαν Lanckoroński *Städte Pamphyliens und Pisidiens* 1.58.20, τὰς ἀργυρέους Μούσας *SEG* 32.1269 (Phrygia, Rom.imp.)); Lacon. ἀργύριος Alcm. 1.55 P.' **I**, line 4, for 'Pl.*Lg.* 801d' read 'Pl.*Lg.* 801b'; add 'fig., of a lover, *SEG* 31.847 (Thasos, rock inscr., iv BC)' **II**, add 'also neut. pl. ἀργυρά *POxy.* 2729.6 (iv AD), *PStrassb.* 330 (v AD)'

ἀργυρίζομαι, add '*earn money* (in quot., of prostitutes), Ath. 13.569d'

*ἀργυρικέλαιον, τό, lit. *silver oil*, perh. extract of *mercurialis annua* (λινόζωστις), *ODouch* 34.7.

*ἀργυριοθήκη, ή, *money-box, AB* 443.6, Suid.

ἀργύριον **I 1**, delete '(v. Poll. 9.89)' and transfer to sense I 2, *money*; add 'ἀ. Ἀττικόν *Attic coinage, SEG* 26.72.3 (iv BC; see also ib. 28.49); ἀ. μεγάλον *large silver coinage, SB* 5174.8 (vi AD), opp. μικρόν *PVindob.Sijp.* 10.10-12 (vi AD)' **II**, add 'app. this sense in Lang *Ath.Agora* XXI He 15, p. 78 (amphora graffito, ii AD)'

ἀργύριος **II**, delete the section (v. °ἀργύρεος).

ἀργυρίς, line 1, after 'ίδος' insert '(acc. ἄργυριν Alcm. 3 *fr.* 3 ii 77 P.)'

*ἀργυροζώμιον, τό, *silver wash*, Zos.Alch. 214.4.

ἀργυροκοπεῖον, -κόπιον, before pres. ref. insert 'rest. in *SEG* 26.6 (coinage decree, Athens, v BC)'

ἀργυροκόπος, add '*SEG* 26.72.54 (Athens, 375/4 BC)'

*ἀργυροκόραλλος, ή, *silver-coral*, name of a metal, Anon.Alch. 361.3.

*ἀργυρόκρανος, ον, *with silver head*, ἀνήρ (referring to the emperor Hadrian), *Orac.Sib.* 5.47.

ἀργυρολογέω, add '**2** *take money* (for consultations), Aesop. 161 P.'

ἀργυρολόγος, add '**II** ἀ., οἱ, financial officials, *Samothrace* II (1)5.14 (ii BC).'

ἀργυρόπαστος, for '*silver-broidered*' read '*silver-plated*' and add 'of a coin, prob. in *Inscr.Délos* 1442*B*50 (ii BC)'

*ἀργυροπλάστης, ου, ὁ, *silversmith, SB* 6259 (v/vi AD).

✝ἀργυροπράτης, ου, ὁ, *silver-merchant, IGChr.* 98 (v AD), *PSI* 76 (vi AD).

ἀργυροπρατικός, for pres. ref. read '*Cod.Just.* 4.21.22.5, Just.*Nov.* 4.3 (p. 27.43), 136'

ἀργυρόρρυτος, for '*beside a silver stream*' read '*silver-streaming*'

ἀργυρορύχή, for '*Mon.Ant.* 23.8' read '*Mon.Ant.* 23.78'

ἄργυρος, add 'Myc. *a-ku-ro*'

*ἀργυροστατήρ, ῆρος, ὁ, *silver stater*, dub. in Hsch. s.v. γλαῦκες Λαυριωτικαί.

ἀργυροταμίας, add 'τῶν φυλάρχων *TAM* 4(1).42'

ἀργυρότοξος, add 'CEG 326 (?Thebes, Boeotia, vii BC), 337 (the Ptoion, c.500 BC)'

*ἀργυροφύλαξ, ακος, ὁ, keeper of silver, IEphes. 4233 (iii AD).

*ἀργυροχοϊκός, ή, όν, of the ἀργυροχόος, ἡ ἁ. (sc. τέχνη) Phlp. in GC 70.14.

ἀργυροχόος, add 'IEphes. 585, SEG 34.1094 (Ephesus), 31.1592 (Asia Minor)'

ἀργυρόω I, add '2 make in silver, Cypr. a-ra-ku-ro-se ἀργύρωσε ICS 307.'

†ἀργυρωματικός, ή, ον, γῆ kind of earth used for polishing silver, IEphes. 27.542, 549.

ἀργυρωμάτιον, after 'dim. of ἀργύρωμα' insert 'PCair.Zen. 44.9, al. (iii BC), Inscr.Délos 1441Aii104'

*ἀργυρωματοφυλάκιον, τό, storehouse for keeping silver plate, PCornell 1.130 (iii BC).

*ἀργύρωσις, εως, ἡ, silvering, Zos.Alch. 214.6.

*ἀργυρωτός, ή, όν, silvered, rest. in IG 2².1473.11 (iv BC).

*ἀρδάλιον, τό, ἀρδάλια· τοὺς πυθμένας τῶν κεραμίδων, οὓς ἔνιοι γοργύρας καλοῦσιν Hsch.

ἄρδαλος, add 'cf. Lat. ardalio, busybody'

ἀρδάνιον, after '= ἀρδάλιον' insert 'used esp. for ritual purification at funerals'

ἄρδην II, after 'wholly' insert 'esp. w. ref. to destruction' and add 'D. 19.61, 27.26, Isoc. 14.19'

ἄρδις I, after 'Hdt. 4.81' insert 'Call.fr. 70 Pf.'

Ἄρειος, after lemma insert 'or ἄρειος'; line 7, after '(Hdt.) 4.23' insert 'Ἀραβίας τ' ἄρειον ἄνθος A.Pr. 420, στέφανος E.Ph. 832, ὅπλα D.C. 44.17.2'

Ἄρειος πάγος, line 2, delete '1.38a (prob.)'; line 3, before 'D. 18.133' insert 'SEG 12.87'; line 4, for 'βουλή Ἀρεία' read 'βουλή Ἀρεία'; line 6, for 'Isoc. 7.37' read 'Isoc. 7.38'; line 7, after 'id.Ath. 59.6' insert 'w. ref. to the Council of the Areopagus as the type of political integrity'

ἀρείων, add 'cf. Myc. a-ro₂-a (neut. pl.) perh. *ἄρροha or *ἄργοha'

*ἀρεόφρυς, v. °ἀραιόφρυς.

ἀρεσκεύομαι, add 'b please, oblige, BCH 83.499.51, 53 (i AD).'

ἀρέσκω I 3, add 'ἀρεσκομένου Χαρίτεσσιν AP 7.440 (Leon.Tarent.)' III, line 4, after 'Tht. 172d' insert 'Just.Const. Δέδωκεν 20a'

ἀρεστήριον, add 'AE 1923.39 (Oropus, iv BC)'

ἀρεταλόγος, at end after 'SIG 1133' insert '(Delos, i BC)' and after 'cf.' insert 'IG 11(4).1263'

*Ἀρέταρχος, ὁ, cult-title of Zeus, INikaia 1076 (Rom.imp.).

ἀρετηφόρος, add 'poet., ἀρεταφόρος, ή, of a temple-road (cf. λαοφόρος I 2), Lindos 487.23 (iii AD)'

ἀρηβῶ, delete 'Peripl.M.Rubr. 12'

ἀρήγω I 1, add 'Gal. 13.707.3'

ἀρηγών, after 'masc. in' insert 'Suppl.Hell. 1163'

*Ἀρηϊσταί, οἱ, members of an association of worshippers of Ares Enyalios, SEG 33.945 (Ephesus).

ἀρήν, line 1, delete 'only in Inscrr.' and insert 'Phryn.PS p. 9 B., Poll. 7.184, Aesop. 155 tit. P.'; line 4, after 'dat. ἀρνάσι' insert 'Arat. 1104' I, add 'ἀρέν κριτός IG 1³.234.22 (v BC), ἄρνα παμμέλαιναν SEG 21.541 col. i 9 (Attica, iv BC)'

Ἄρης, line 2, delete '(never contr.)' and in line 3, after 'fr. 16' insert 'contr. Ἄρευς AP 9.322.9 (Leon.Tarent)' add 'IV the Arabian Ares, identified w. a local war god, SEG 33.1301, 36.1374 (Hauranitis, Rom.imp.).' at end add 'Myc. a-re'

ἀρθμός, add 'Myc. a-to-mo, referring to group of men, exact sense uncertain'

ἄρθρον, τὰ ἄρθρα, genitals, add 'Ael.NA 3.47, μοιχὸς ἑάλω ποτέ, ὡς ὁ ἄξων φησί, ἄρθρα ἐν ἄρθροις ἔχων Luc.Eun. 10'

ἀρθροπέδη, after 'Phan.' insert 'prob. corrupt'

ἀρία, add 'Eup. 491 K.-A.'

ἀρίγνως, read 'ἀριγνώς'

ἀριδάκρυος, add 'Call.fr. 700 Pf.'

ἀρίζηλος, line 5, after 'ib. 519' insert 'Call.Epigr. 51.3 Pf.' II, delete the section.

ἀριθμέω 1, line 9, after 'AP 11.349 (Pall.)' insert 'the years of one's life, i.e. live so long, γήραι ἀριθμ[ή]σασ' ἐννέα ἐτῶν δεκάδας CEG 592.4 (Athens, c.300 BC)'; line 10, for 'πλίνθους' read 'τὰς ἐπιβολάς'

ἀρίθμησις I 1, after 'IPE 1².32B35 (Olbia)' insert 'Just.Nov. 18.10' add '3 report of collection made by tax collectors, PMich.x 577, 582 (i AD), PLond. 1157 (ii AD).'

ἀριθμητής, add 'PUG inv, DR61.c.3 (Atti Napoli III 900, 165 BC), Just.Nov. 73.7.1'

ἀριθμητικός III, for 'Sammelb. 4415.14' read 'SB 4415.4'

ἀρίθμιος II, after 'D.P. 263' insert 'Call.fr. 110.61 Pf.'

ἀριθμός I 2, add 'expenditure in cash, opp. γράμματα (estimate on paper), Pech-Maho lead in ZPE 82.161 (v BC), SEG 2.582 (Teos, iii/ii BC)' II, after 'Hdt. 8.7' insert 'ναῶν δ' εἰς ἀριθμὸν ἤλυθον

E.IA 231' VIII, for 'Dem. 52' read 'Dem. 54' IX, before 'line' insert 'number of lines in a book, Plb. 39.8.8, Luc.Hist.Conscr. 16. b'

ἀρικύμων, add 'Steril. 219'

†ἀρῑν, v. ἄρρις.

†ἀρῑς, v. ἄρρις.

ἀρίς II, for this section read 'bow-shaped sluice, also called φράκτης, Procop.Aed. 2.3.18, 21'

ἀριστεία, at end before 'Cic.Att.' insert 'Hdt. 2.116'

ἀριστεῖα 1, line 4ff., delete 'ἀ. τῆς θεοῦ .. IG 2.652A30, al.' and for 'ἀριστεῖον' to end of section read 'in sg., Ion. -ήιον, SEG 37.994 (-ήϊον, Priene, vi BC), MDAI(A) 87.153-4 no. 24 (Samos, v BC), Hdt. 8.11.2, -εῖον, SEG 31.1590 (ii BC, unkn. provenance, on bracelet); used esp. of golden crown, IG 2².1388.30, 1635aA32, SEG 18.200-202 (Samian inscrr. at Delphi, iv BC), D. 22.79, Amyzon 24, IStraton. 1321, Luc.DDeor. 2(22).3'

ἀριστεῖος, add 'SEG 17.584, ILS 8863 (both ii AD); of a person, MAMA 1.234 (Phrygia, -ῆον)'

ἀριστερός 1, add 'τούτου τοῦ ἥρωου ἡ ἐν ἀριστ<ε>ροῖς <κ>λείνη IEphes. 3456; without a prep., neut. pl. adv., ἀριστερὰ εἰσιόντων Inscr.Délos 1416Ai34 (ii BC). b ἐπ' ἀριστερὰ γράφειν write from right to left, Artem. 3.25; ἐπ' ἀριστερὰ περιβεβλῆσθαι, have dressed leftwards, id. 3.2.4.' 2, add 'simply ἀριστερᾶς, Inscr.Délos 1441Aii95 (ii BC)' 4, at end for 'τῷ ἀριστερῷ' read 'τῇ ἀριστερᾷ'

ἀριστεύς, at end delete 'CIG 2881 (Milet.)' and for '(Cibyra)' read '(Cibyra, v. SEG 32.1306), IGBulg. 150.3 (Odessus), 2².3733.15 (Athens, ii AD); an official at Miletus, Didyma 84.13'

*Ἀρίστη, ή, epith. of Artemis, Paus. 1.29.2.

*ἀριστήριον, τό, = ἀριστητήριον, BCH 28.262.13.

ἀριστητήριον, for pres. ref. read 'IStraton. 270.6, 17.7 (both ii AD), PZilliac. 6.26 (vi AD)'

ἀριστητικός, delete 'Eup. 130'

*Ἀριστ[ι]ασταί, οἱ, in Boeot. form -[ι]ασταί, devotees of Ariste, Schwyzer 463.3 (SEG 26.614, Tanagra, iii/ii BC) or Ἀριστ[η]ασταί, devotees of Aristaeus, cf. ZPE 25.135.

ἀριστίζω, after '(Ar.)Av. 659' insert 'Sosith. 2.21' and after 'Acraephia' insert 'SEG 26.1826.21 (Cyrene, ii AD)'

*ἀριστίνδα, adv. = ἀριστίνδην, IG 7.188.9 (Pagae, iii BC).

ἀριστίνδας, add '-δης SEG 11.501.1 (Sparta, ii AD)'

ἀριστίνδην, line 2, for '(IG) 9(1) .. -δαν' read '(IG) 9².717.12 (Locr., -αν, v BC)'

†ἀριστογαλατίας, ὁ, app. leading citizen of Galatia, Mitchell N.Galatia 287 (Chr.).

ἀριστόμαντις, after 'εως' insert '(-ιδος IG 9(1).645)'

ἄριστον, read 'ἄρ-'

ἀριστοπολιτεία, after '(Sparta)' insert 'SEG 31.372 (Olympia, ii AD)'

ἀριστοπόνος II, delete the section and add 'Nonn.D. 44.79' to section I.

ἄριστος, line 1, at end insert 'Thess. Ἄσ(σ)το-, for Ἀριστο- in pers. n., e.g. Ἀσ(σ)τόμαχος, Schwyzer 567.13, 569' III, adv., add 'SEG 33.1105 (Paphlagonia, ii/iii AD)'

ἀρίφρων, after 'prudent' insert 'ἡγητῆρες IEphes. 452.5 (iii AD)'

*ἄρκα, ἡ, Lat. arca, coffin, IG 14.2327.

Ἀρκαδία, Ἀρκαδικός, transfer to before and after Ἀρκαδίζω.

*Ἀρκάδισσα, ἡ, female Arcadian, Ann.Mus.Gr.-R. d'Alex. 1935-9.121 no. 7 (Alexandria), Iamb.VP 267.

*ἄρκανος, ὁ, an unidentified fish, perh. = ἄκαρναν, IGC p. 98 A11 (Acraephia, iii/ii BC).

*ἀρκαπάτης, ου, ὁ, = ἀργαπέτης, prob. as hereditary title, Dura 20.4 (ii AD).

ἀρκαρικός, add 'Just.Edict. 13.20'

ἀρκάριος, after 'Lat. arcarius' insert 'treasurer' and add 'Inscr.Perg. 8(3).99 (-ις), ib. 125, SEG 36.970.B56 (ἀρκά abbrev. Aphrodisias), τῆς ἐκκλησίας Cod.Just. 1.2.24.16'

*ἀρκεθεωρέω, = ἀρχιθεωρέω, rest. in IG 2².365b7.

ἀρκεθέωρος, for 'IG 2.181 a' read 'IG 2².365a7, 10, etc. (Athens, 323/2 BC)' and add 'rest. in Hesperia 37.375 (iv BC)'

*ἀρκέτιον, τό, name of a metal, Anon.Alch. 326.26.

ἀρκευθίς I, delete 'Thphr. .. -θος)'

ἄρκευθος, add 'V prob. = ἀρκευθίς 1, Thphr.Od. 5, Inscr.Cret. 4.184.16 (ii BC).'

ἀρκέω III 3, add 'Just.Const. Δέδωκεν 17'

†ἀρκήλα, (in LSJ proparox.), ⟨τὸ⟩ ζῷον. Κρῆτες τὴν ὕστριχα Hsch. (La.).

ἄρκηλος, for 'young .. panther, ibid.)' read 'an animal exhibited by Ptolemy II, Callix. 2; acc. to Ael.NA 7.47 young leopard (but some say a different species, ibid.)'

ἄρκιος (A) II, line 3, for 'he' read 'it'

ἄρκιος (A), add 'IV = ἄρκτος II, servant of Artemis, SEG 9.72.98 (iv BC).'

ἄρκος (B), add 'prob. in Nic.Al. 43'

*ἄρκος (C), ὁ, *sarcophagus, coffin, IG* 14.2326, al.; cf. °ἄρκα.
*ἀρκτεία, ἡ, *service* as ἄρκτος II, Hsch.
*ἀρκτόμῦς, υος, ὁ, *marmot,* Jerome *Ep.* 106.65.
ἄρκτος, add 'see also ‡ἄρκος (A)'
Ἀρκτοῦρος I, add 'ἁρκτοῦρος *IG* 1³.2.9'   add 'III ἀρκτοῦρος = ἄρκτιον, Dsc. 4.105, cf. Hsch.'
ἄρκῦς, metaph., add 'ὅταν δ᾽ ἔρωτος ἐνδεθῶμεν ἄρκυσιν Dicaeog. 1b.1 S.'
*ἀρκυστάσιον, τό, *line of hunting nets,* X.*Cyn.* 6.6.
ἅρμα I 1, add 'w. ref. to curule triumph, δὶς ἐ[πὶ κέλητος ἐθριάμβευσα καὶ], τρὶς [ἐ]φ᾽ ἅρματος *Mon.Anc.Gr.* 2.9'   add 'Myc. *a-mo-ta* (pl.), *wheels*'
ἁρμακιάς, add 'cf. perh. ἔρμαξ and mod. Gk. ἁρμακᾶς *dry stone wall*'
*ἁρμακίς, ίδος, ἡ, *section* or *parcel of land,* PNess. 31.15 (vi AD), al.
*ἁρμαμέντον, τό, *arsenal,* Just.*Nov.* 85.1, ζῷα διαφέροντα τοῦ θίου ἁρμαμέντου *BCH* 116.397 (?Constantinople, vi AD).
*ἅρμαρα, τά, kind of incense, *PMag.* 4.1.1294, 1990.
*ἁρμαραύσιον, τό, type of sleeveless military garment, cf. late Lat. (Goth.) *armilausa,* PMich.XIV 684.11 (-ιν), *PMonac.* 142.3 (ἐρμ[ε]λαῦσον, both vi AD).
*ἁρμάριον, τό, Lat. *armarium,* EM 146.56.
ἁρμαρίτης, read ἁρμαρίτης.
ἁρμάτἁρακτα, before 'τά' insert 'or ἁρμᾰτοτάρακτα' and for '(for ἁρμᾰτο-ταρ-)' read '(ἁρμᾰτορ- pap.)'
ἁρμάτειος, line 2, after 'X.*Cyr.* 6.4.9' insert 'D.H. 5.47 (-τίου)'
*Ἁρμᾰτεύς, έως, ὁ, epith. of Hermes, *IEryth.* 201d.31 (iii BC).
ἁρμάτιον I, after 'ἅρμα' insert '*PLond.* 1973.3 (iii BC); before '*Gloss.*' insert '*SB* 7263.3 (iii BC), *Inscr.Délos* 1441*A*i43 (ii BC)' and add 'title of poem by Theopompus of Colophon, Ath. 4.183b'
ἁρμᾰτίτης, add 'II of horses, *drawing chariots,* PCair.Zen. 673.5 (iii BC), *PLond.* 1930 (259 BC).'
*ἁρμᾰτοκολλιστής, οῦ, ὁ, *chariot-maker,* PHarris 97.3 (iv AD).
ἁρμᾰτοπηγός, add '*OAshm.Shelton* 119, 130, al.'
†ἁρμᾰτοτροχιά, ἡ, = °ἁρμᾰτροχιά, Luc.*Dem.Enc.* 23, Ael.*VH* 2.27.
*ἁρμᾰτοφορέω, pass., *to be carried in a chariot,* Hymn.Is. 37 (Maroneia).
*ἁρμᾰτροχιά, Ep. -ιή, ἡ, *wheel-track of a chariot,* Il. 23.505, Ph. 1.312, Q.S. 4.516.
*ἁρμένια, τά, small tools, Hero *Aut.* 24.2, Sch.Opp.*H.* 1.222.
Ἁρμενία, after lemma insert 'poet. -ίη, *AP* 16.61 (Crin.), *SEG* 34.1409 (Cappadocia, Rom.imp.)'
†Ἁρμενιακός, ή, όν, *Armenian,* Str. 11.14.2; adopted as title by Roman emperors, *victor in Armenia, IG* 2².2090a2, *ICilicie* 12, *ISelge* 13.10, *Salamine* 163 (all ii AD); μῆλα Ἀ. *apricots,* Dsc. 1.115; χρυσοκόλλα Ἀ., form of malachite, id. 5.89; also λίθος Ἀ., Alex.Trall. 1.427.16, Cyran. 6.6.1 K., *PHolm.* 88.
ἁρμενίζω, add '*Cyran.* 31, 86, al.'
†Ἁρμενικός, ή, όν, *Armenian;* μηλέα Ἀ. *apricot,* Gal. 12.76.16, also τὸ Ἀ. *PRyl.* 629.227 (iv AD); λίθος Ἀ. Dsc. 5.105.
†Ἁρμένιος, ον, *Armenian,* ἐξ Ἁρμενίου ὄρεος Hdt. 1.72; -ον, τό, a mineral, Dsc. 5.90; cf. λίθος Ἁρμενικός.
*ἁρμενοπετής, ές, gloss on Lat. *ueliuolus, P.XV Congr.* 3.50.
*ἁρμενορᾰφος, ὁ, *sail-maker, MAMA* 3.293, al. (Corycus), *SEG* 37.715 (mosaic, Chios).
ἁρμενοφόρος, before 'gloss' insert '*carrying sails*'
*ἁρμίγεροι, οἱ, Lat. *armigeri,* POxy. 1888.2 (v AD, ἐρμ- pap.), Lyd.*Mag.* 1.46.20.
*ἁρμιλλίγεροι, οἱ, Lat. wd. indicating military rank, equiv. of βραχιᾶτοι, ψελιοφόροι Lyd.*Mag.* 1.46.19.
*ἁρμίως, *at once,* Gal. 19.86.2; cf. ἁρμοῖ.
ἁρμόδιος, after 'α, ον' insert 'also ος, ον, Longin. 12.5, Just.*Nov.* 123.21.1'   II, after 'Pi.*N.* 1.21' insert 'παρεχόντων .. τἆλλα ἁρμόδια *CID* I 7.A18 (Andros, v BC)'
ἁρμόζω, line 10, after 'ἁρμόχθην' insert '*Olympia Bericht* 7.207 (3 pl. ἁρμόχθεν, Sybarite inscr. at Olympia, vi BC)'   I 1 a, for 'πόδα ἐπὶ' read 'πόδας ἐπὶ' and for 'foot' read 'feet'; line 15, for 'Simon. 182' read '*AP* 7.431.4 ([Simon.])'   4 b, after 'c. acc.' insert 'Ael.*NA* 13.21'   5, lines 4/5, for 'Simon. 184' read '*AP* 7.25.4 ([Simon.])'   II 1 a, line 1, after 'armour' insert 'or masonry'; add '*IG* 2².244.88, 100, 463.72'   add '6 as Lat. *competo,* to be *legally in force, to be due to someone, Cod.Just.* 1.5.13 pr., 6.4.4.23.'   at end add 'Myc. pf. part. pass. *a-ra-ro-mo-te-me-na* (fem. pl.)'
*ἁρμοκούστωρ, ὁ, v. °ἁρμοροκούστωρ.
ἁρμονία I 4, delete 'νεύρων καὶ'   IV 4, add 'πανηγυρικὴ τῆς λέξεως ἁ. D.H.*Dem.* 45, cf. *Isoc.* 3, al.'   add 'VIII ancient name for a *plane geometrical proportion,* Aristid.Quint. 3.6.'
*ἁρμοροκούστωρ, Lat. *armorum custos, SB* 1592 (Nubia), ἁρμοκούστωρ *OBodl.* 2022 (ii AD), ἐρμοκούστωρ PHamb. 88, ἁρμικούστωρ PWisc. 14.5 (all i AD).
ἁρμός 3, for this section read '*joint in the body,* τὰς χεῖρας καὶ τοὺς πόδας ἀπήρθρουν καὶ ἐξ ἁρμῶν ἀναμοχλεύοντες ἐξεμέλιζον Lxx 4*Ma.* 10.5, *Hippiatr.* 34'

ἁρμοστής I, add '**b** *harmonizer,* τῶν ὅλων (of Alexander), Plu. 2.329b.'   2, add '(dub., prob. reading δικασταί)'   3, delete '= *triumvir,* App.*BC* 4.7'; for '= *praefectus*' read '*governor* (in Roman contexts)' and add 'Luc.*Peregr.* 9, App.*Hisp.* 38'
ἁρμοστικός, add 'ἐνέργεια Procl.*in Ti.* 1.358.15, 2.216.22'
ἄρμυλα, add 'cf. ἀρβύλη'
ἀρναβόν, add 'cf. °ἀγναβόν'
*ἀρναβωράτιον, τό, product mentioned in price list, perh. dim. of ‡ἀρναβόν, *POxy.* 3766.104 (gen. pl. -βωρατιων), rest. in 3733.20 (both iv AD).
ἀρνακίς, for '*sheepskin coat*' read '*sheepskin*'
ἄρνειος I 2, add 'Call.*fr.* 26.1 Pf.'
ἀρνέομαι, line 4, aor. med., add '*Cod.Just.* 1.1.7.11, 7.62.36, al.'   3, add 'w. dat., εὐχωλῆσι Orph.*L.* 176; act. (in sense 2), Fronto *Ep.Gr.* 5.6 H.'
†ἀρνηΐς, ιδος, fem. adj., *connected with lambs,* ἀρνηΐδας (mutilated papyrus) Call.*fr.* 26.2 Pf.; ἐν ταῖς ἡμέραις ἃς καλοῦσιν ἀρνηΐδας Clearch. 79; θυσίαν ἄγουσι καὶ ἑορτὴν ἀρνηΐδα (ἀρνίδα codd.) Conon 19, all referring to an Argive festival, prob. in the lambing season, in which stray dogs were killed.
†ἀρνίς, v. °ἀρνηΐς.
*ἀρνοκτᾰσία, ἡ, *killing of lambs,* Rh. 3.607.9.
ἄρνῠμαι, line 2, delete 'Pl.*Lg.* 969a' (v. °ἀείρω); lines 3/4, for statement in parentheses read 'augm. 3 sg. ἤρετο *SEG* 33.716 (Histiaea, v BC); ἤρετο occurs as v.l. for ἤρατο; cf. ἀείρω'   I, four lines from end, delete 'δίκαν ἁρέσται .. (Locr.)'   add '2 w. gen., φήμης ἄρνυται ἀθανάτου Clara Rhodos 6/7.529 (Nisyrus).'
ἀροτήρ 1, add '*PTeb.* 886.69, 99 (ii BC), POxy. 2241.12, 41 (iii AD)'   2, add 'fig., *cultivator,* of poet, [Πιερίδων ἀ]ροτῆρι orac. in *SEG* 27.678.11 (Ostia, ii/iii AD).'
ἀροτήσιος, add '**b** epith. of Zeus, *Syria* 36.77-8 (Hippos, iii AD).'
ἀροτρεύς, after '= sq.' insert 'Arat. 1075, al.'
ἀροτριάζω, delete the entry.
ἀροτρίᾱμα, add 'pl., = γεννήματα, app. *fruits of the earth,* Hsch. (after ἀρώματα)'
*ἀροτρίασμα, ατος, τό, gloss on ἄρομα, Suid.
ἀροτριάω, for '= ἀρόω' read '*plough*'; after 'Babr. 55.2' insert '*IEphes.* 3217A.8 (ii AD)'; add 'fig. ψεῦδος Lxx *Si.* 7.12, τὰ ἄτοπα *Jb.* 4.8'
ἄροτρον II, add 'μήτε φυτοῖς μήτε ἀρότροις ἀνοίγεσθαι referring to water conduit, *IEphes.* 3217B.28 (ii AD)'
ἀροτρόπους, for 'Ju.' read 'Jd.'
*ἀροτρόω, *plough,* prob. in Alc. 120.8 L.-P. (ἀροτρώμμε[, athematic; cf. ἀροτροῦντος *Inv.OL* 1988.25.4 in *ZPE* 78.145 (sp. ἀρω-, v AD).
*ἀρουᾱλος, Lat. *arualis,* ἀδελφός ἀ. *Mon.Anc.Gr.* 4.7.
*ἄρουλλα, Lat. *arula,* gloss on ἐσχάρα, Sch.Ar.*Ach.* 888.
ἄρουρα, lines 2/3, for '*Inscr.Cypr.* 135.20 H.' read '*ICS* 217.20'   add '**IV** as goddess = Γῆ, Nonn.*D.* 1.154, al.'   at end add 'Myc. *a-ro-u-ra*'
*ἀρουρᾰτίων, ωνος, ἡ, *tax assessed in proportion to area of land,* PLips. 62 ii 21, POxy. 3397.22 (both iv AD), 3634.1 (v AD), PMasp. 329.11.8 (vi AD).
*ἀρουρίδιον, τό, dim. of ἄρουρα, *PSI* 476.1 (iii AD).
ἀρούριον, after 'ἄρουρα' insert '*PSI* 974.6 (i/ii AD)'
ἀρόω II 2, line 2, after '(Tanagra)' insert 'med., [Νέστορα ..] Πυλία .. ἀρόσατο χθών *GVAK* 47.1 (Cilicia, i BC)'
ἁρπᾱγή, for '*rape*' read '*forcible abduction*'
ἁρπάγη, for 'Men. 829' read 'Men. 657'; before 'Poll.' insert '*flesh-hook*'; after 'Poll. 6.88' insert '10.98'
*ἁρπαγηδόν, = ἁρπάγδην, gloss on *raptim,* Dosith. p. 412 K.
ἁρπάγιον, add 'II name of an eye-salve, *CIL* 13.10021.93.'
ἁρπάζω I 1 (p. 246a, line 7), for 'χἁρπάσαι' read 'χἁρπάσαι'; add 'fig., of death, λοιμὸς ἀμείλικτος κ(αὶ) ἀνάρσιος ᾽ἥρπασε᾽ Πλουτεύς *SEG* 26.405'   add 'III *take away,* ἁ. εἰς ἐλευθερίαν, of releasing a slave; cf. Lat. *eripere* or *ad libertatem, Cod.Just.* 6.4.4.7, Just.*Nov.* 144.2.4; cf. °ἀφαρπάζω.'
†ἁρπακτός, ή, όν, *taken without being earned* (i.e. by work), χρήματα δ᾽ οὐχ ἁρπακτά· θεόσδοτα πολλὸν ἀμείνω Hes.*Op.* 320.   2 *taken hurriedly, snatched* (i.e. by seizing an opportunity), πλόος Hes.*Op.* 684.   3 *taken illicitly, stolen,* ὑμέναιοι Nic.*fr.* 108.
ἁρπαλίζω 1, after 'med.' insert 'τόδ᾽ ἁρπαλ[ί]ζομ[αι] Archil. 24.4 W.'
ἅρπαξ II 1, add '(but perh. humorous personification, "Grabber")'
ἅρπασος, for 'a .. *prey*' read 'name of a bird, Call.*fr.* 43.61 Pf.'
ἁρπαστός 1, add 'by death, *SEG* 34.325.6 (Megalopolis, *c.*100 BC), cf. *Salamine* 193.10 (ii AD)'   2, add '**b** an eyesalve, *CIL* 13.10021.153.'
ἁρπεδόεις, for 'ἐρπεδίζω' read 'ἐρπεδόεσσα'
*ἁρπεδόνιον, τό, name of a gemstone, Socr.Dion.*Lith.* 23.2 G.
ἅρπεζα, at beginning add '(perh. ἅρπεζα; cf. ὑπάρπεζος)'
ἅρπεζος, for '*BCH* 46.405 (Mylasa)' read '*IMylasa* 254.2 in *EA* 19.16 (i BC)' and add '*LW* 327.6 (Olympus)'
ἅρπη, add 'Sosith. 2.19'

Ἁρποκράτης, ους, acc. ην, ὁ, Egyptian divinity, *AP* 11.115 (Lucill.), Plu. 2.358e; sp. -χράτας, *IG* 9(2).591 (Larissa, i BC), -χράτης, *IPhilae* 3, 4, *SEG* 24.413 (Ambracia, all iii BC); also Καρποκράτης Vidman *SIS* 88 (Chalcis, iii/iv AD).

***Ἁρποκρᾰτιακός**, ὁ, kind of physically deformed person (fr. °Ἁρποκράτης, who was premature and weak in lower limbs, acc. to Plu. 2.358d), Ptol.*Tetr.* 124.

***Ἁρποκρᾰτικός**, ή, όν, *with the nature of* °Ἁρποκράτης, prob. *weak or deformed*, Heph.Astr. 1.1.99, 2.9.5.

ἀρρᾰβών 1, add '*pledge*, of an object, perh. in *SEG* 38.1036.7 (Gaul, Pech-Maho, v BC), Lxx *Ge.* 38.17, 18' **2**, for 'Men. 697' read 'Men.*fr.* 688 K.-Th., ὀλέθρου κοὐχ ἑταίρας ἀρραβών Python 1.18', deleting 'Lxx *Ge.* 38.17, 18'

***ἀρρᾰβωνιακός**, ή, όν, *marking betrothal*, περιθέματα (*necklaces*) ἀ. Hsch. s.v. κάθορμα.

***ἀρρᾰβωνίζομαι**, add '*PCair.Zen.* 250.3 (iii BC); med. *betroth to oneself*, fut. ἀρραβωνίσομαι (-ησ- cod.) αὐτόν *Vit.Aesop.*(W) 30; ἀρραβωνίζεται· ἀρραβῶνι δίδοται Hsch.'

ἀρρενικός 1, form ἀρσ-, add 'Lxx *Ex.* 13.15, Ezek.*Exag.* 13'

ἀρρενογονία, add 'Heph.Astr. 2.7.3, 2.8.3. **2** *descent* or *relationship reckoned through males*, τοὺς ἐξ ἀρρενογονίας αὐτοῦ *Cod.Just.* 6.4.4.23, term expld. at Theophil.Antec. 1.10.1 (p. 40.28ff.).'

ἀρρενοκοίτης, after '*AP* 9.686' insert '(Maced., iv/vi AD, v. *BCH* suppl. 8 no. 87)'

ἀρρενόομαι, add 'act., **ἀρρενόω**, *make male* or *masculine, represent as male*, Heraclit.*All.* 71'

ἀρρενώδης, after '*brave*' insert 'Sch.BT Il. 8.39'

†**ἀρρενώπας**, (s.v. ἀρρενωπός in *LSJ*), ὁ, comic term for an androgynous person, Cratin. 417 K.-A., cf. Eust. 827.29.

ἀρρενωπός, after 'Luc.*Fug.* 27' insert 'ἀρσενωπέ (addressed to Athena), *PKöln* 245.10 (iii AD, verse)'

ἀρρεπής, delete 'of a balance' and after '*inclining to neither side*' insert 'Gal. 2.266.12, 760.6, *UP* 6.16, 12.9, *CMG* 5.10(2).1.21.2'

ἀρρευμάτιστος II, add 'Gal. 11.301.8, 18(2).844.14'

ἀρρεψία, for 'etc.' read 'rest. in *SB* 7183.3 (iii BC, ἀρεψ- pap.)'

ἄρρηκτος, after 'ον' insert 'also η, ον, *PASA* 2.352 (Cappadocia), Aeol. αὔρηκτος acc. to Hdn.Gr. 2.271, Eust. 548.31'

†**ἀρρητοποιέω**, *do unmentionable things to*, i.e. *fellate*, Artem. 1.79 passim.

***ἀρρητοποιΐα**, ἡ, βρίμη .. γυναικεία ἀ. Hsch.

ἀρρητοποιός I, for '*practising such vice*' read '*practising fellatio*' and add 'Sch.Ar.*Eq.* 1287' **II**, delete 'pedantically'

ἄρρητος I, line 2, delete 'ἄνδρες .. *Op.* 4'; at end after 'Id.*El.* 1012' add '*not spoken of*, ἄνδρες .. ῥητοί τ' ἄ. τε Hes.*Op.* 4; ἄρρητος τελετή *CEG* 317 (Athens, v BC); Arat. 2, 180'

***ἀρρητούργος**, ον, *indulging in fellatio*, γίνονται δὲ εὐνοῦχοι ἢ ἑρμαφρόδιτοι .. ἢ ἀρρητούργοι γυναῖκες Heph.Astr. 2.13.12.

ἀρρίγητος, delete '*daring*' and for '*AP* 6.219' read '*AP* 6.219.7 (Antip.Sid.)'

†**ἀρριχάομαι**, (ἀριχάομαι Arist.*HA* 624ᵃ34), *clamber, climb*, Hippon. 137 W., Arist. l.c.; cf. ἀναρριχάομαι.

ἄρριχος, line 2, after '(Diosc.)' insert 'cf. *EM* 149.30, *AB* 446.30, which give masc. in Ion., fem. in Att.'; line 3, for '162' read '62' and after 'Amorgos' insert 'iv BC'

***ἄρροπος**, ον, *not inclining the scale*, of a weight, *IGLS* 1272a (Laodicea ad mare, ii AD); *inflexible*, δικαστὴς ἄ. Gr.Naz. in *PG* 37.662A, cf. Att. pers. n. Ἄροπος, app. *unwavering, steady, SEG* 26.207 vi 2 (iii/ii BC), *IG* 2².2452.17 (ii BC), etc.; cf. ἀρρεπής, °ἔρροπος, °σύρροπος.

ἄρρυπος, add 'cf. ἀρύπαρος.'

†**ἀρρύσιος**, ον, (sp. ἀρυ-), *not liable to seizure*, *IG* 9²(1).706A.3 (Oianthea, iii BC); cf. ῥύσιον.

ἀρρωστία 1, after 'Hp.*VM* 6, etc.' insert 'τὴν τοῦ σώματος ἀ. D. 24.160, Isoc. 19.21'

ἄρρωστος 1, after 'Plu. 2.465c' insert 'comp. ἀρρωστότερος τῷ σώματι Isoc. 16.33'

***ἀρσατικός**, ή, όν, adj., otherwise unexplained, describing a quality of coinage, νομίσματα *Stud.Pal.* III 59.3, *CPR* 1.30 fr.ii 44, *BGU* 314.15 (all vi/vii AD).

ἄρσεα, v. °ἄρσια.

ἀρσενίκιον, add 'written ἀρσενίκην, i.e. ἀρσενίκιν, Anon.Alch. 318.7'

***ἀρσενοβάτης**, ου, ὁ, *sodomite*, Hsch. s.v. παιδοπίπας.

ἀρσενόθηλυς, add 'Serv.*Aen.* 10.89'

***ἄρσενος**, ον, = ἄρσην, *POxy.* 744.9 (i BC), *PMich.*III 203.6 (i/ii AD).

***ἀρσενόω**, trans., *turn to arsenic*, Anon.Alch. 269.3.

***ἀρσενωπός**, v. °ἀρρενωπός.

ἄρσην 1, at end after '*the male sex*' insert 'A.*Supp.* 951' add '**b** *directed towards males*, i.e. *homosexual*, ἔρως Cerc. 9.15; πῦρ *AP* 9.77 (Antip.Sid.); πυρσοί ib. 12.17. **c** Pythagoreans regarded odd numbers as male, Plu. 2.288c, cf. 264a.' **2**, delete 'Id.*Supp.* 951' **4**, delete 'but also .. 3.9.3'

***ἄρσια**, τά, designation of a place-name, ἀπὸ τᾶς λεγομ[έν]ας παρὰ τὰ ἄ. *Delph.* 3(4).42.13 (ii BC); cf. perh. ἄρσεα· λειμῶνες Hsch.

Ἀρσινόεια, add 'sg. τὸ Ἀρσινοεῖον, *temple of Arsinoe*, in Cyprus, *SEG* 25.1072 (Idalium, iii BC), at Philadelphia, *Aegyptus* 22.197, at Alexandria, Plin.*HN* 36.38, 37.108'

ἄρσις I 1 a, for 'as an athletic feat .. (pl.)' read '*bull-hoisting*, ritual act performed by ephebi at Eleusis, *IG* 2².1006.78 (pl., ii BC); cf. °ἀείρω IV 3 b, °Βοάρσαι, °Βοάρσιον' and add 'σημείου ἄρσις v. σημεῖον I 3'

ἄρσος, delete the entry (v. °ἄρσια).

ἀρτάβη II, for 'varying from 24 to 42 χοίνικες' read 'normally containing 40 χ.'

ἀρτᾰμέω, add 'prob. in A.*fr.* 281a.35 R.'

Ἄρταμις, after '-μίτιον' insert '-μάτιος'

ἀρτάμος 1, add '*Trag.adesp.* 148'

ἀρτάω II, add '**3** δικαὶ ἠρτημέναι, calque on Latin *lites pendentes*, Just.*Const.* Δέδωκεν 23.'

ἀρτεμής, for 'Call.*Iamb.* 1.227. -Ep. word' read 'Call.*fr.* 194.28 Pf., whence rest. in Hippon. 105.6 W.'

ἀρτεμία, after '*health*' insert 'Pi.*N.* 11.12 (cj.), id.*fr.* 52m.3 S.-M. (prob.)'

Ἄρτεμις, gen. -ιτος, add 'Myc. *a-te-mi-to*'; **Ἄρταμις**, gen. -ιτος, add '*SEG* 31.356A (Elis, vi BC);' dat. Ἀρτάμι, add 'cf. Ἀρτέμι app. in *SEG* 31.356B (Elis, iv BC)' at end add 'see also Ἄρτιμις'

***Ἀρτεμῐσιακόν**, τό, name of silver mine at Laurium, *IG* 2².1582.38, 114, al. (iii BC).

***Ἀρτεμῐσιάς**, άδος, ἡ, *period of the Artemisian games*, *IGRom.* 4.1609, 1610 (Hypaepa).

Ἀρτεμίσιος, Dor. Ἀρταμίτιος, add '*IG* 4².108.110 (Epid., iv/iii BC), also Ἀρταμάτιος, de Franciscis *Locr.Epiz.* 31.6 (iv/iii BC)'; after 'Plu.*Alex.* 16' insert '*SEG* 34.1221 (Saettae, i AD), 24.614 (Maced.), 32.1394 (Commagene, both ii AD), 36.1288b (Syria, ii/iii AD), etc.; answering to May, *PLond.* 229.30 in *Hermes* 32.274 (Seleucia, ii AD), cf. Hemerolog.*Flor.* p. 73 (p. 10 K.)'

Ἀρτεμισιών, add '*SEG* 24.574 (Torone, end iv BC), 33.1056.A3 (Cyzicus, ii AD)'

ἄρτημα II 1, for '*IG* 2.834 *c* 13' read '*IG* 1³.387.40'; add '(in *IG* l.c., al., ἄ. may mean *counter-poise weights*, v. *Hesperia* 13.186 and °ῥυμός)'

ἄρτι 1, line 3, after 'Pl.*Cri.* 43a' insert 'Theoc. 23.26', deleting this ref. fr. line 5; line 7, after '(iii AD)' insert 'for ἀπ' ἄρτι read °ἀπαρτί' **2**, add 'ᾤχετο φεύγων ἄρτι μὲν εἰς Μυτιλήνην, ἔπειτα δ' εἰς Χίον Charon.Lamps.*fr.* 9 J.'

ἀρτίγαμος, after '*just married*' insert 'A.*fr.* 168.20 R. (lyr.)'

ἀρτιγένειος, after '(Diod.)' insert '*GVI* 854 (Egypt), epigr. in *SEG* 26.456 (Laconia, both ii/iii AD); after 'Nonn.*D.* 18.135' insert 'Sch.Call.*fr.* 2 Pf.'

ἀρτιδαής, add '*SEG* 3.543 (Thrace, ?iii BC)'

***ἀρτίδμητος**, ον, *closely fitting*, ἀρτιδόμῳ δ' ἐκάθητο λιθοστρώτῳ παρὰ χώρῳ codd. Nonn.*Par.Ev.Jo.* 19.13.

ἀρτιλίθια, after 'in masonry' insert '*IG* 2².1671.36 (Attica, iv BC)'

***ἀρτίουλος**, ον, *just having the first growth of beard*, *Limes de Chalcis* p. 214 no. 50 (i AD).

***ἀρτιόφρων**, ονος, adj., *in full possession of one's faculties*, perh. in A.*fr.* 451r.3 R. (ἀρ[τιόφ]ρων, nisi leg. ἀρτίφρων).

ἀρτιπᾰγής II, add 'app. also in καὶ λοετροῦ πολὺς ὄλβος, ὃν ἀρτιπαγοὺς ἀπὸ γαίδης ὤπασεν *Milet* 1(9).343 (Chr. epigr.), but precise signif. uncertain'

ἀρτίπους I 1, add '**b** (ἀρτίπος) *in good health*, *AP* 5.287.4, 9.644.5 (both Agath.).' **2**, add 'S.*Tr.* 58' **II**, delete the section.

ἄρτισις, add '**2** *preparation*, *PCair.Zen.* 771.27 (iii BC).'

***ἀρτίτυπος**, ον, *newly made*, cj. for ἀντι- in Nonn.*D.* 39.11.

ἀρτιφανής, add 'E.*Phaeth.* 67 D.'

ἀρτίφρων, add 'see also °ἀρτιόφρων'

ἀρτιφυής, add '**III** app. = ἀρτίφρων, *GVI* 1917.9 (Cyme, ii BC).'

***ἀρτοδοτέω**, *give bread*, *Vit.Aesop.*(G) 19 (-δωτ- cod.).

***ἀρτοθέσιον**, τό, *bread-store*, *PBrooklyn* 15.2 (vi AD).

***ἀρτοκόλλυτος**, ὁ, *baker*, *PHamb.* 56.v.4, vi.8 (vi/vii AD).

***ἀρτοκοπέω**, *keep a bakery*, Cumont *Fouilles de Doura-Europos* 385 no. 22; cf. ἀρτοποπέω.

†**ἀρτοκοπία**, ἡ, *bakery*, *PMich.*x 586.7 (i AD), *OMich.* 257.2 (sp. ἀρτε-, iv AD); also *PThead.* 31.35, 36.21 (s.v.ll., iv AD).

ἀρτοκόπος, add 'cf. Myc. *a-to-po-qo* (no doubt the original form)'

ἀρτόκρεας, for '(Lydia)' read '(Sardes, v. Robert *Hell.* 11/12.480-1)'

†**ἀρτολάγυνος**, ον, *consisting of bread and bottle*, πτωχῶν πανοπλίη ἀρτολάγυνος *AP* 11.38 (Polem.).

ἀρτοποιέω, before 'Longus' insert 'Ctes.*fr.* 11 J.'

***ἀρτοπράτισσα**, ἡ, fem. of ἀρτοπράτης, *BCH* 4.205 no.27 (Isauria, v/vi AD).

ἀρτόπτης 2, after '*pan for baking bread*' insert 'Plaut.*Aul.* 400' and after 'Plin.*HN* 18.107' add '(perh. misunderstood by Pliny as meaning *baker*), Poll. 10.112'

***ἀρτοπτρίς**, ίδος, fem. adj., *used for baking bread*, ἐσχάρα ἀ. *PCair.Zen.* 692.12 (iii BC).

ἀρτοπώλης, for 'AJA 18.33' read 'AJA 18.68 (= Sardis 7(1).166)' and before it insert 'Arist.Ath. 51.3'

ἀρτοπωλικόν, for pres. ref. read 'IG 2².1707.4 (iii BC)'

*ἀρτοφαγία, ἡ, eating of bread, Sokolowski 3.177.37 (Cos, iv BC).

*ἀρτοψύγεῖον, τό, place for cooling bread, POxy. 3355.6 (ii AD), PSI xx Congr. 16.7 (iv AD; both written -ψυγιον; cf. ὑδροψυγεῖον).

ἄρτῦμα, line 3, after 'Anaxipp. 1.5' insert 'PPetaus 28.19 (ii AD)'   **II**, delete '(cf. ἄρτημα)'

*ἀρτῡματηρά, ἡ, tax on spices, PHels. 37.4, 16 (ii BC).

*ἀρτῡματοθήκη, ἡ, box for spices, SB 9509.3 (iii AD).

ἀρτῡματοπώλης, add 'BGU 1898.209 (ii AD), POxy. 3739.7 (iv AD)'

ἀρτύς, add 'rest. in Call.fr. 80.19 (Add. ii) Pf.'

ἀρτύω **I**, add 'ἐὰν δὲ τὸ ἅλας ἄναλον γένηται, ἐν τίνι αὐτὸ ἀρτύσετε; Ev.Marc. 9.50, Ev.Luc. 14.34'

†ἄρυα· τὰ Ἡρακλεωτικὰ κάρυα Hsch.

ἀρυβάσσαλον, after 'κοτύλη' insert 'ἡ φλάσκων'

*ἄρῠσις, εως, ἡ, drawing up of liquids, ποτοῦ Afric.Cest. p. 39 V.

ἀρυστήρ, line 1, for '= ἀρυτήρ' read 'kind of cup used as ladle, esp. for wine' and for 'Supp. 4.9' read '58.9 L.-P.'; line 3, for 'Call.Aet. 1.1.17' read 'Call.fr. 178.17 Pf.'; last line, after 'liquid measure' insert 'SEG 33.63 (amphora, Athens, c.500 BC)'

†ἄρυστις, εως, ἡ, perh. = °ἀρυστήρ, cup, τὰς ἀρύστεις ὧδ' ἔχουσ' ἐκώμασας S.fr. 764 R., (but ἀρύστεις· τὰς ἀπνευστὶ πόσεις Hsch., Phot. a 2918 Th. app. error for ἀμύστεις).

ἀρυστῖχος, for 'dim. of ἀρυτήρ' read 'kind of cup'; after 'Aegina' insert 'v BC'; add 'on an Ionian container, perh. for perfume, SEG 32.724 (Berezan, vi BC)'

ἀρυστρίς, for '= ἀρύταινα' read '= °ἄρυστις'

ἀρύταινα, for 'fem. of ἀρυτήρ' read 'kind of cup used as scoop in the baths' and add 'Ar.fr. 450 K.-A., PHels. 2.9 (ii BC)'

*ἀρῠταίνιον, τό, dim. of ἀρύταινα, Inscr.Cret. 1.xvii 2a9 (Lebena, ii BC).

ἀρῠτήρ 1, before 'ladle' insert '= °ἀρυστήρ'

ἀρύω (A), line 1, after '[ᾰ]' insert '[ῠ, in late poets also ῡ, as AP 9.37 (Stat.Flacc.), Nonn.D. 14.46]'

*Ἀρχάγαθος, ὁ, epith. of Zeus, app. source of all good things, INikaia 1071.2, 10 (i/ii AD).

ἀρχάγγελος, after 'archangel' insert 'ICilicie 116, POxy. 1151.42 (both v/vi AD), esp. of Gabriel or Michael'

ἀρχᾱϊκός, before 'interpol.' insert 'Inscr.Délos 1426Bi42, 1428ii50 (ἀρχαιϊκ-, ii BC)'

*ἀρχαιόθεν, adv. from early times, ἐξ ἀρχεόθεν MAMA 7.559.5.

ἀρχαιολογέω **I**, for 'discuss .. Th. 7.69' read 'say the same old things, Th. 7.69' and before 'ἀ. τὰ Ἰουδαίων' insert '**2** discuss antiquities'

ἀρχαιολογία, lines 3/4, title of works, add 'Dionysius of Halicarnassus'

*ἀρχαιολόγος, ὁ, dramatic performer, mime, IG 2².2153.7 (prob., cf. Robert, REG 49.235ff.), Gloss. s.v. Atellani.   **2** = Lat. antiquarius, perh. expert copyist of old texts, Edict.Diocl. 7.69.

ἀρχαιο-μελῐ-σῐδωνο-φρῡνίχ-ήρᾱτος, for 'ἀρχαῖα μελι- codd.' read 'v.l. ἀρχαῖα μελι-'

ἀρχαιόπλουτος, add 'Cratin. 171.70 K.-A.'

ἀρχαῖος **II**, after 'of persons' insert 'ancient, venerable'; add 'τῆς ἀρχέας (sic) ἁγίας Μαρίας ITyr 187'

ἀρχαιρεσία **I**, add 'in sg. = Lat. comitia, Plb. 1.52.5, but cf. section II'

ἀρχαιρεσιακός, add 'SEG 27.938.14 (Tlos, v AD)'

*ἀρχαιρέσιοι, οἱ, perh. = αἱρεσιάρχαι, SEG 16.696.8 (Caria, Rom. imp.).

*ἀρχαίρχων, οντος, ὁ, perh. chief magistrate, SEG 9.869 (Numidia).

ἀρχέβακχος, for '16' read '6'

ἀρχεδέατρος, for 'chief seneschal' read 'chief °ἐδέατρος'; add 'SEG 18.730 (Cyrene, ii BC), PTeb. 728.4 (ii BC); also ἀρχελέατρος UPZ 202.1.5 (ii BC)'

*ἀρχεδέκᾱνος, ὁ, v. °ἀρχιδέκανος.

ἀρχεζώστης, add 'also -ζώστρις Orib. 14.62.1; cf. °ἀργιζώστης'

*ἀρχεθεωρία, v. °ἀρχιθ-.

ἀρχεθέωρος, after '(Delos, iii BC)' insert 'SEG 28.60 (Athens, iii BC)'; add 'cf. ‡ἀρχιθέωρος'

*ἀρχεία, v. °ἀρχήια.

ἀρχεῖον **I 2**, after '(Dyme, ii BC)' insert 'SEG 23.305 iii 15 (treaty at Delphi, ii BC)'; add 'ὁ ἐπὶ τῶν ἀ. SEG 34.1107.7 (Ephesus), δι' ἀρχήου perh. = per tabularium, IEphes. 13 i 15, 16, etc. (i AD)'   **II**, pl. exx., add 'generally, public offices, Isoc. 7.24'

ἀρχεῖτις, add 'SEG 31.791 (Thasos, ii/iii AD), perh. late sp. for -ῖτις; cf. Lat. architidis (gen.), as Assyrian epith. of Venus, Macr.Sat. 1.21.1'

*ἀρχελέατρος, v. °ἀρχεδέατρος.

ἀρχέμπορος, add 'IG 10(2).564.7 (iii AD), SB 10529b.31 (iii/iv AD); voc., as title of respect, Vit.Aesop.(G) 12'

*ἀρχενδρομίτης, ου, ὁ, perh. chief runner or footman, BGU 1834.6 (i BC, -δρωμ- pap.).

ἀρχέπλουτος, for '= ἀρχαιόπλουτος' read 'having control of wealth'

*ἀρχέποδες, v. °ἀρχέφοδος.

*ἀρχεπολία, ἁ, perh. board of city magistrates, dub. rest. in SEG 9.72.132 (Cyrene, iv BC).

*ἀρχερᾱνεύς, έως, ὁ, = ἀρχερανιστής, rest. in ASAA N.s. 1/2.195, n. 2 (Rhodes, iii AD).

ἀρχερᾱνιστής, for pres. ref. read 'in Attica, IG 2².1339.4, SEG 37.103.4 (both i BC), in Rhodes (gen. sg. ἀρχερανιστᾶ), IG 12(1).9.1 (i BC)'

*Ἀρχερᾱνίστρια, τά, festival of the ἀρχερανισταί in Attica, BCH 84.658 (Acharnae, ?i BC).

*ἀρχέσκοπος, ὁ, title of religious official, IG 9(2).1322.1 (Halmyros, iv BC) (pl.).

*ἀρχέταιρος, ὁ, president of a religious association, Inscr.Dura 2.8 (i AD).

ἀρχέτας, add 'E.Phaëth. 100 D.'

ἀρχέτῠπος **II**, delete 'opp. ἀπόγραφον .. Is. 11, cf.'; add 'also masc. ἀρχέτυπος, ὁ, opp. ἀπόγραφος, D.H.Is. 11.   **b** ledger, Cic.Att. 12.5c (pl.).'

ἀρχεύω, w. dat., add 'Antim. 27 Wy.'

ἀρχέφηβος, add 'also ἀρχιέφηβος, inscr. in SB 9997.5 (?Memphis, iii AD)'

*ἀρχέφοδος, add 'app. athematic inflexion in ἀρχέφοδα (acc.), PAberd. 60.2 (i/ii AD); -δι (dat.), PGen. 107.11 (ii AD); cf. ἀρχέποδες app. same sense, as if fr. *ἀρχέπους, Ps.-Callisth. 1.31 β, γ, v.l. ἀρχέφοδοι'

ἀρχή, after lemma insert 'dial. ἀρχά Pi.O. 2.58, etc., Lesb. ἄρχᾱ IG 12(2).1.8 (Mytilene, iv BC)'   **I 1 a**, at end for '[ὁ ἄνθρωπος] .. 3.3.4' read 'ἡ ἀρχὴ ἐν τῷ πράττοντι Arist.EN 1110ᵇ4, cf. 1140ᵃ13, al., Plot. 3.3.4'; add 'μη(νὸς) Γωρπιέου (sic) ἀρχῇ at the beginning of the month, SEG 32.1442.4 (Syria, vi AD)'   **b**, add 'ἐν ἀρχῇ in the beginning, at the outset, inscr. in Phoenix 43.319.9 (Arsinoe, Hellen.); of all time, Lxx Ge. 1.1, Ev.Jo. 1.1'   **3**, for 'corner' read 'extremity'; add 'of road or street, Lxx Ez 16.25, 21.20(25)'   **5**, for 'branch' read 'head'   **6**, add 'al., Ps. 138(139).17 (pl.)'   **II 3**, line 8, after 'Antipho 6.42' insert 'ἐπὶ [τῆ]ς ἀρχῆς during his term of office, ICilicie 70'   **5**, for 'command, i.e. body of troops' read 'company, band'   add '**8** Dor. -ά, part of the νόμος κιθαρῳδικός, Poll. 4.66.'

*Ἀρχηγέσιον, τό, precinct of Apollo Archegetes, Inscr.Délos 316.115 (iii BC), 461Ab49 (ii BC).

*ἀρχηγετεῖον, τό, building for commemoration or worship of a founder or tutelary hero, SEG 12.373.15 (Cos, iii BC).

ἀρχηγέτης **1**, lines 3/4, title of Apollo, at Cyrene, add 'SEG 9.7.26 (ii BC)'; add 'at Apamea (Syria, in Dor. form), BE 1976.721'; line 7, for 'IG 2.1191' read 'CEG 314 (Rhamnous, v BC), IG 2².2849 (ii BC)'; line 11 (fem. ἀρχηγέτις), for 'cf. BMus. .. ii AD' read 'of Artemis, Inscr.Magn. 16.21 (iii BC), IEphes. 27.20 (ii AD), of Leto, SEG 38.1476.17 (Xanthos, iii BC)'; add 'pl. -γέται, applied to female deities (the Erinyes), Suppl.Mag. 42A.8 (iii/iv BC)'   **2**, for 'generally leader, chief' read 'chief ruler'

*ἀρχήγισσα, ἡ, Jewish title, fem. of °ἀρχηγός, μνῆμα Περιστερίας ἀρχηγίοις (i.e. -ίσσης) Robert Hell. 1.26 (see also SEG 33.1602, Thessaly, v/vi AD).

ἀρχηγός **II 2**, add 'ἀγγέλων ἀρχηγέ of archangel Michael, SEG 30.1266 (Caria, Byz.); as Jewish title, CIJud. 731g (see also SEG 33.1602, vi AD)'

†ἀρχηΐα, ἁ, Cret., magistracy, period of office, Inscr.Cret. 4.233.3 (Gortyn, iii BC), al.; ἀρχεία SEG 33.729 (Amnisos, i BC); sense uncertain in SEG 23.366.18 (Axos, iv BC).

ἀρχηΐς, add 'Plu. 2.364e, Delph. 3(1).466.5 (ii AD)'

ἀρχῐᾱριστάς, add 'also ἀρχιεριστάς Tit.Cam. 9.11, 11.14, etc. (iii BC), ἀρχιαριστής Tit.Cam. 53.3 (ii BC)'

*ἀρχιατρίνη, ἡ, app. fem. of ἀρχιατρός, MAMA 7.566 (-ιειατρηνα lapis); Robert Epitaphes p. 177.

ἀρχιᾱτρός, after 'of communities' insert 'τῆς [Ἐ]φεσίων πόλεως IEphes. 3055.16, SEG 27.1262 (Aphrodisias), ἀ. καὶ ἱεροφάντης TAM 5(1).268, τοῦ σύμπαντο[ς] ξυστοῦ Robert Hell. 9.25 (Thyatira, ii/iii AD)'

*ἀρχιβαλιστάριος, ὁ, chief ballistarius, SEG 7.989 (Philippopolis, Syria, ii AD).

*ἀρχιβάπτης, ου, ὁ, chief βάπτης (dyer), IEJ 7.76-7 (= Beth She'arim 188, iii/iv AD) (-βαφθ-).

*ἀρχιβασιλιστής, ὁ, chief of the °βασιλισταί, corr. to SB 1106.2 in SEG 34.1605 (?iii BC), previously read as ἀρχιβουλευτής.

ἀρχιβασσάρα, for the pres. ref. read 'IGBulg. 401.16 (Apollonia ad Pontum), IUrb.Rom. 160 i B24 (ii AD)'

*ἀρχιβάσσαρος, ὁ, chief of the βάκχοι, IUrb.Rom. 160 i B5 (ii AD).

*ἀρχιβούκολος **II**, add 'IGBulg. 1517.17 (iii AD), SEG 17.320 (Abdera, ?iii AD), Epigr.Gr. 1036a (Perinthus)'

†ἀρχιβουλευτής, v. °ἀρχιβασιλιστής.

*ἀρχιβωμιστής, οῦ, ὁ, chief priest, SEG 7.893 (Gerasa, i AD).

ἀρχιγάλλαρος   SUPPLEMENT   ἀρχογλυπτάδης

*ἀρχιγάλλαρος, ὁ, chief °γάλλαρος, prob. in IGBulg. 1517.19 (Philippopolis, iii AD).
*ἀρχιγερουσιάρχης, ου, ὁ, chief of Jewish board of elders, SEG 26.1178 (Rome, iii/iv AD).
*ἀρχιγερουσιαστής, οῦ, ὁ, chief member of a board of elders, IPrusias 25.3 (iii AD).
ἀρχιγέρων, after 'Sammelb. 2100.5 (i BC)' insert 'IFayoum 38.2 (i AD)'
ἀρχιγεωργός, add 'voc., as title of respect, Vit.Aesop.(G) 12'
ἀρχιγραμματεύς, add 'ἀ. ξυστοῦ SEG 23.395.4 (Corcyra, ?ii AD), PLond. 1178.82 (ii AD), IUrb.Rom. 246.B1, 2 (iv AD)'
*ἀρχιδέκᾱνος, ὁ, chief δεκανός, SEG 16.813.5 (Arabia, iii AD), IGChr. 269 (Aphrodisias); also ἀρχε- MAMA 8.46 (Lystra).
ἀρχιδιάκονος, before pres. ref. insert 'ICilicie 39, BCH suppl. 8 no. 281 (both v/vi AD), SEG 31.1446 (Palestine, vii AD), JRS 16.68 (Eumeneia)'
*ἀρχιδιάκων, = ‡ἀρχιδιάκονος, Inscr.Phryg. 77.
*ἀρχιδικαστικός, ή, όν, of the ἀρχιδικαστής, ἀρχή, PVindob.Salomons 5.15 (ii AD), ὑπηρέτης PSI 1328.56 (iii AD).
*ἀρχιδραγάτης, ου, ὁ, chief field-warden, JÖAI 30 Beibl. 24 (Ancyra, iii AD); cf. δραγατεύω.
ἀρχιεπισκοπή, ή, archbishopric, Epiph.Const. in PG 42.185A.
ἀρχιεπίσκοπος, before pres. ref. insert 'IEphes. 495 (iv AD), PBaden 65.20, SEG 30.1711 (Palestine, vi AD), BCH suppl. 8 no. 91, Salamine 206, 219 (vi/vii AD)'
ἀρχιεράομαι, after lemma insert 'ἀρχιαρ-, SEG 35.1416 (Pamphylia, Rom.imp.)'
ἀρχιερατεύω, line 1, within parentheses add 'also ἀρχιερετεύω, BCH 51.89 (Panamara)'
ἀρχιέρεια, line 1, after '(Olympia)' insert 'SEG 31.900 (Aphrodisias, i AD), IGBulg. 66.2, IEphes. 430.20 (ii AD)'
ἀρχιερεύς, line 1, before 'Ion.' insert 'Aeol. ἀρχείρευς, IG 12(2).239.6, ἀρχίρευς, ib. 249.5, al.'; line 5, after '(iii AD), etc.' insert 'ἀ. δι' ὅπλων of gladiatorial games, IGBulg. 1572.3, al.'; add 'in Chr. Gk., MAMA 1.208, SEG 37.501 (Thessaly, vi AD).   2 name of tenth month in a Cyprian calendar, YCLS 2.213f.'
*ἀρχιεριστάς, v. °ἀρχιαριστάς.
*ἀρχιεριστέω, to be ἀρχιεριστάς, Tit.Cam. 23.15, 27.15 (iii BC).
ἀρχιεροθύτης, after lemma insert 'Dor. -θύτᾱς'; add 'SEG 33.643 (Rhodes, ii/i BC), ISelge 17.7 (Rom.imp.), ἀρχιεροθύτας Lindos 70.5 (iii BC)'
ἀρχιερωσύνη, after '(Daphne, ii BC)' insert 'IHistriae 57.20 (pl., referring to a woman, ii AD)'; add 'also ἀρχιερ- SB 8267.12 (i BC)'
*ἀρχιέφηβος, ὁ, v. °ἀρχέφηβος.
ἀρχιζάκορος, delete the entry (read ἀρχινᾱ-).
ἀρχιζάπφης, for the pres. ref. read 'Inscr.Délos 2628a (ii BC)'
*ἀρχιζωγράφος, ὁ, master-painter, master-decorator, BE 1958.516 (Georgia, iv AD), Eust.Op. 307.23.
*ἀρχιζώστη, ή, v. °ἀργιζώστη.
ἀρχιθεωρία, add 'also ἀρχε- Din. 1.81, SEG 28.60.60 (Athens, iii BC)'
ἀρχιθέωρος, after '(Delos, ii BC)' insert 'SEG 37.709.A2 (Epirus, iii/ii BC)'; ἀρχιθέαρος, add 'SEG 26.487.(1)6 (Megara, ii AD)'; add 'also ἀρχεθέαρος CID I 7.A5, 8, 23 (Andros, v BC)'
ἀρχιθιᾱσεύω, delete the entry.
†ἀρχιθιᾱσῑτεύω, lead a thiasos, Inscr.Délos 1778.5, 1779.6, 1782.7 (ii/i BC).
ἀρχιθιᾱσῑτης, for 'IG 11(4).1228.4' read 'IG 11(4).1228.2' and add 'SEG 6.718.5 (sp. -ειτ-, Pisidia, Rom.imp.)'
ἀρχιθῠρωρός, add 'PTeb. 790.1 (ii BC)'
*ἀρχιθύτης, ου, ὁ, he who leads the sacrifice, perh. for archbishop, IEphes. 1356 (Chr.).
*ἀρχιατρίνη, v. °ἀρχιατρίνη.
*ἀρχιερωσύνη, ή, v. °ἀρχιερωσύνη.
ἀρχικός I 2, after 'Isoc. 4.67' insert 'emanating from or pertaining to a magistracy, ἀ. πρόσταξις Cod.Just. 1.2.20, ἀ. ὄχλησις ib. 1.3.45.70' add '4 in Roman contexts, curule, δίφρος ἀ. D.C. 43.14.5, 43.48.2, 53.30.6.'
ἀρχικῠβερνήτης, for 'pilot' read 'navigator' and add 'ἀ. ὢν τοῦ σύμπαντος στόλου D.S. 20.50.4'
ἀρχικῠνηγός, add 'SEG 23.271.26 (ἀρχικ[ου]ναγύ nom. pl., Boeotia, iii BC); app. chief, of gladiatorial rank, Illion 126.4 (Rom.imp.)'
*ἀρχίλαρχος, ὁ, written ἀρχιλί-, chief of the ἴλαρχοι, ASAA N.S. 61.342 (proto-Corinthian alabastron, c.700 BC; ?pers. n.).
*ἀρχιλᾱτόμος, ὁ, chief quarryman, Bernand Pan 28.3, (iv AD).
*ἀρχιμᾰγᾰρεύς, έως, ὁ, chief of the rite of the μάγαρον, IG 10(2).1.65.3 (Thessalonica, iii AD).
ἀρχιμαγειρεύς, delete the entry.
ἀρχιμανδρίτης, add 'IGLS 2143c, PKöln 112.12 (v/vi AD), CPR 10.122.12, PMasp. 242.4 (both vi AD)'
*ἀρχιμᾰχαιροφόρος, chief of the armed police, PMich.XI 656.6 (sp. -μ[α]χερ-, i AD).

*ἀρχιμεταλλάρχης, ου, ὁ, chief controller of mines, Bernand Pan 51.6 (AD 11).
ἀρχίμῑμος, for 'chief comedian' read 'chief actor in a mime'
*ἀρχιμονάζων, οντος, ὁ, chief monk, wax tablet in ZPE 56.90 (v/vi AD).
*ἀρχινᾱκόρος, v. °ἀρχινεωκόρος.
*ἀρχιναυπηγός, ὁ, head shipbuilder, PRyl. 640.19 (iv AD).
*ἀρχιναυφύλαξ, ακος, ὁ, chief ναυφύλαξ II, ASAA 2(1916).136 (Rhodes, i BC).
ἀρχινεᾱνίσκος, add 'IUrb.Rom. 160 i B22 (ii AD)'
*ἀρχινεωκορέω, hold the office of ἀρχινεωκόρος, IG 10(2).1.114.9, etc. (Thessalonica, i/ii AD); see also ἀρχινακορέω.
ἀρχινεωκόρος, add 'also ἀρχινᾱκόρος IG 10(2).1.244.7, prob. in IGLS 1263 (Laodicea ad mare, both ii AD)'
*ἀρχινεωποιΐα, ἡ, office of ἀρχινεωποιός, Wien.Anz. 1893.103 no. 11 (sp. -νεοπ-, Aphrodisias).
*ἀρχιοικιστᾱς, ᾶ, ὁ, commissioner for foundation of a colony, IG 9²(1).2.10 (Thermon, iii BC, pl.).
*ἀρχιοικοδόμος, ὁ, master-builder, Fouilles de Byblos II 76 no. 7186, PKöln 197.2 (v/vi AD), SEG 8.781.10 (Syene, vi AD).
*ἀρχιπαστοφορία, ἡ, the office of ἀρχοπαστοφόρος, PMich.inv. 5598.22 in ZPE 63.297 (ii AD).
*ἀρχιπερίπολος, ὁ, chief of patrols, Rev.Bibl. 1.246 (Caesarea, vi AD).
ἀρχιποίμην, add 'SEG 19.782 (ἀρχ[ιπ]οίμην, Pisidia, Rom.imp.)'
*ἀρχιποσία, ἡ, position of president of a drinking-party, Porphyrio ad Hor.Carm. 2.7.25, al.
*ἀρχιποτᾰμίτης, ου, ὁ, chief ποταμίτης, PHerm. 69.6, 21 (v AD).
*ἀρχιπρεσβύτερος, ὁ, chief presbyter, IGLS 21.57 (vi AD).
ἀρχιπροστᾰτέω, for 'συναγωγῆς' to end read 'in a military koinon, SB 626 (= IFayoum 16.4, ii BC)'
†ἀρχιπροστάτης, ου, ὁ, chief of a military koinon, SB 5959 (iii/iv AD) in CE 27.290.
*ἀρχιπροφήτης, ου, ὁ, chief προφήτης, PGen. 7.5 (i AD), PHerm.Rees 3.26 (iv AD).
*ἀρχιπρῠτᾰνεία, -ία, ἡ, office of chief president, Didyma 570, 157 1 a.
ἀρχιπρύτᾰνις, after 'έως' insert '(but -ιδος Didyma 252.7, etc., -ίδων ib. 272.5'
ἀρχιραβδοῦχος, before 'chief lictor' insert 'chief of the ῥαβδοῦχοι (q.v. sense 2), PWisc. 50.9 (ii AD)'
*ἀρχισῑτολόγος, ὁ, chief σιτολόγος, SB 6800.3 (iii BC), PTeb. 792.10 (ii BC).
*ἀρχισῑτωνέω, to hold the office of °ἀρχισιτώνης, JRCil. 2.19.6 (sp. -σειτ-, iii AD).
*ἀρχισῑτώνης, ὁ, chief officer in charge of grain supply, Side 114.4 (sp. -σειτ-); cf. σιτώνης.
*ἀρχισταβλίτης, ου, ὁ, chief stableman, perh. honorific title, PPrincet. 145.11 (vi AD), POxy. 1908.5 (vi/vii AD).
ἀρχιστολιστής, before 'keeper' insert 'chief' and add 'Bernand Les portes no. 1 (Antinooupolis, i BC)'
*ἀρχίστολος, ὁ, = °ἀρχιστολιστής, rest. in IEphes. 1244 ([ἀρχ]ί-, ii AD).
*ἀρχιστράτωρ, ορος, ὁ, title of an equestrian official, TAM 3.52.8 (Termessus, ii AD).
*ἀρχισύμμαχος, ὁ, chief messenger or courier, PFlor. 5.93.4, PMich.inv. 3706.16 in JÖByz 36.25, CPR 14.6.7 (both v/vi AD), POxy. 1866.4 (vi/vii AD).
†ἀρχισυνᾰγωγέω, to be leader of a religious association, IG 10(2).1.288 (Thessalonica, ii AD).
*ἀρχισυνᾰγώγης, ου, ὁ, = ἀρχισυνάγωγος, CIJud. 336 (Rome, iii/iv AD).
*ἀρχισυνᾰγώγισσα, ἡ, fem. of °ἀρχισυνάγωγος 2, CIJud. 731c (v. SEG 38.913, Crete, iv/v AD).
ἀρχισυνᾰγωγος I, for this section read 'ὁ, ἡ, leader of a religious association, SB 623 (i BC), IGRom. 1.782 (Perinthus).   2 leader or elder of a synagogue, Jewish honorary title, Ev.Marc. 5.22, al., IG 14.2304, Ramsay Cities and Bishoprics 2 no. 559; of women, ISmyrna 295 (Rom.imp.), CIJud. 756 (Caria, iv/v AD).'
†ἀρχιτεκτοσύνη, ἡ, art or skill of an architect, BCH 10.500 (Pisidia), IMylasa 468.7.
ἀρχιτέκτων I 1 a, add 'τῆς θεοῦ IEphes. 536, 1061, 1600.6 (Rom.imp.)'   add 'c ἔπαρχος -τεκτόνων, = Lat. praefectus fabrum, IG 2².3546 (Eleusis, i AD), cf. ABSA 56.23 (Paphos, ii BC).'
*ἀρχίτοκος, ον, perh. ruling childbirth, ὠδῖνες MDAI(A) 56.128 (Thespiae, Hellen.).
ἀρχυπηρέτης, after 'ἀρχυπερέτης' insert 'RA 6.31.8, 17 (Amphipolis, ii BC)' and transfer existing entries to sp. ἀρχυπ-
*ἀρχιφρᾱτωρ, ορος, ὁ, president of a phratry, IGLS 232.
ἀρχιφρουρέω, after '(Thess.)' insert 'al., SEG 36.635 (Maced., iii AD)'
*ἀρχιφρουρός, add 'PSI 938 (vi AD)'
ἀρχιφῠλᾰκίτης, add 'IFayoum III 209 (= SEG 33.1359, ii BC)'
*ἀρχιφύλετης, ου, ὁ, = ἀρχίφυλος, TAM 3(1).121 (Termessus).
ἀρχογλυπτάδης, for 'son of' read 'humorous patronymic for'

ἀρχόμαος, ὁ, perh. a religious official, *SEG* 38.945 (Megara Hyblaea, vi BC).

ἀρχοντεία, ἡ, *archonship*, *TAM* 2.612 (Teos).

ἀρχοντεύω, add '*IHadr*. 40.7 (ii AD)'

ἀρχοντικός 1, add '*of the authorities*, τὴν ἀρχοντικὴν τιμουρίαν ὑποστῆν(αι) *SEG* 37.200 (Athens, Chr.), εἰς ἀρχοντικόν *to the authorities*, *PMich*.XIII 660.5 (vi AD); ἀρχοντικῶν καὶ δημοτῶν app. *people of rank and commoners*, *POxy*. 2346.23 (iii AD); applied to angels, Cels.ap.Orig.*Cels*. 6.27'   **2**, for this section read 'in particular, referring to Roman offices: of *praefectus*, *SB* 9152.9; of *praeses*, *CPR* 5.17.8 (v AD); of *ex-consul*, *IG* 14.756a (Naples, i AD); of *duumviralis*, ib. 1789 (Rome)'

ἀρχός I 2, for this section read 'as ἄρχων, *chief magistrate*, ἀ. Τειχιδ(ο)ης *Didyma* 6 (Miletus, vi BC), *IG* 9².609.21 (Naupactus, c.500 BC), ἀ. καὶ Ϝοικιάται *SEG* 26.449.6 (Epid., v BC), *ICS* 2.1 (Paphos, iv BC), *IG* 7.3301 (Chaeronea, ii BC), *Inscr.Cret*. 4.250.3 (i BC)'

ἀρχωνία, v. °ἀρχωνεία.

ἀρχυλωρός, ὁ, *chief* ὑλωρός, Thessalian office, rest. in *SEG* 34.565 ([ἀρχυ]λουρός, Pherae, ii BC).

ἄρχω I 1, line 4, after 'and 2' insert 'ὁ ἄρχων τᾶς δίκας *Leg.Gort*. 11.51 (v BC)'   **4**, add 'so med., ἀρχόμεναι θυμέλας Posidipp. in *Suppl.Hell*. 705.4'   **6**, line 2, after '*begin*' insert '*mark the beginning of*' and line 3, before 'Lex ap.' insert '*Thasos* 18.5, 13 (v BC)'   **II 3**, add 'aor. part. designating past holders of the archonship, Εὐκλείδην τὸν ἄρξαντα Archipp. 27.3 K.-A., *SEG* 23.647 (Cyprus, i AD), *TAM* 4(1).238; cf. ἄρχων'

ἄρχων II 2, add 'πρῶτος ἄρχων *IPrusias* 7.7, 38.7 (Rom.imp.)'   **3**, *ruler* of a synagogue, add '*CIJud*. 347 (pl.)'   at end add 'see also ‡ἄρχω II 3'

ἀρχωνεία, ἡ, office of ἀρχώνης, form ἀρχοωνήαν (acc.), *IEphes*. 4101.17 (ii AD).

ἀρχωνέω, for '*BCH* 1.410 (Callipolis)' read '*IParion* 5.3, cf. Robert *Hell*. 9.81'

ἀρωγός I, add 'neut. pl. as subst. (sc. φάρμακα), Gal. 11.95.8'

ἄρωμα (A), line 2, delete 'prob. in *Suppl.Epigr*. 1.414 .. pl.)'

†ἀρωμάτης, ου, ὁ, *aromatic*, ἀ. οἶνος Dsc. 5.54, κάλαμος Gal. 11.405.

†ἀρωμάτιτις, ἡ, fem. of ἀρωματίτης, σχοῖνος ἀ. Str. 16.2.16; as subst., precious stone, Plin.*HN* 37.145.

ἀρωματοποιόν, gloss on ζειρίς, Hsch. (La., -ποιεῖν, ζειρεῖν cod.).

ἀρωματοφόρος 2, add 'rest. in *IEphes*. 1076, title of a religious official'

ἀσάμινθος, add 'S.*fr*. 204 R.; Myc. *a-sa-mi-to*'

ἀσάμιος, v. °ἀζήμιος.

ἀσάρίτη, = ὀροβάγχη, Sch.Cyr.

ἀσαρκέω, for 'causal, *make lean*' read 'intr. *to be lean*'

ἄσαρος, for 'Sapph. 77 (Comp.)' read 'Sapph. 103.11 L.-P., 91 L.-P. (comp.)'

ἀσάω, after 'Thgn. 593' insert 'Alc. 39.11 L.-P. *MAMA* 8.361.7'

ἀσβέστιον, τό, prob. *limestone*, *PNess*. 54.9 (vi/vii AD, pl.).

ἄσβεστος, after 'Il. l.c.' insert 'φάεος .. ἀσβέστου Call.*Dian*. 118'

ἀσβολάω, delete the entry.

ἀσβολόω, for 'Macho .. 3.16.3' read 'Macho 372 G., Arr.*Epict*. 3.16.3; as sobriquet of the descendants of Damon, Plu.*Cim*. 1'

ἀσεβέω, line 5, after '[Xen.]*Cyn*. 13.16' insert 'ἀσεβήτω πὸτ τὰρ Ἀθάναρ *SEG* 35.389.7 (Elis, iv BC)'

ἀσεβής, after 'A.*Supp*. 9 (anap.)' insert 'μήτ' ἔρδειν μήτε λέγειν ἀσεβῆ Thgn. 1180'

ἄσειρος, add '= ὁ ἄδετος χιτών, Cyr.'

ἀσείρωτος, add '**2** *not having a border* (or *worn without a belt*), τὸν φαιλόνην (for φαινόλην) μου τὸν ἀσίρωτον *PYale* 82.8 (ii/iii AD).'

ἄσειστος, after 'D.L. 8.26' insert '(but see also °ἀείσειστος)'

ἀσελγής, line 1, for '*brutal*' read '*unconstrained*'; line 3, for 'generally, outrageous' read 'of natural forces' and add 'οἷον αὖ τὸ πνῖγος, ὡς ἀσελγές Pherecr. 191 K.-A.'

ἄσεπτος, add 'E.*Ba*. 890, *IA* 1092'

ἄση I 2, for 'Sapph. 1.3 .. 14.11' read 'Sapph. 1.3 L.-P., Alc. 39(*a*).11 L.-P., Anacr. 2 *fr*.1.8 P.'   **3**, delete the section.

ἀσηκρῆτις, ὁ, Lat. *a secretis*, Procop.*Arc*. 14.4, cf. Lyd.*Mag*. 3.20.

ἀσημείωτος I, for 'Ph. 1.121' read 'Ph. 2.121'

ἀσήμινος, η, ον, *made of silver*, *PIand*. 103.15 (vi AD, ἀσίμ-), *PMich*.XIV 684.14 (vi AD, -ενος).

ἀσήμιον, τό, *silver plate*, *PHamb*. 227.15 (iii AD), *PSI* 825.13 (iv/v AD, both sp. ἀσήμιν), ἀργύρεον βάρος ἤγουν τοῦ ἀσημίου Sch.Nic.*Al*. 54a; cf. ἀσήμιος I 2 and mod. Gk. ἀσήμι.

ἄσημος II, add 'χρη]σμοὺς ἀσάμους Stesich. 222(b).247 (p. 215 D.)'   **III**, add '**e** of days on which critical signs are absent, Gal. 9.751.9, 776.10.'   **IV**, add 'Κόρινθος ἄστρον οὐκ ἄσημον Ἑλλάδος *Trag.adesp*. 128'

ἀσήμως, add 'adv. ἀσημόνως Phld.*Mort*. 37'

ἄσηπτος, for '*Acacia tortilis*' read '*Acacia* (*shittim*) *wood*'

ἀσήτωρ, ορος, ὁ, *sick at heart*, Antim. in *Suppl.Hell*. 65.1 (prob. nomen agentis fr. ἀσάω, but expld. by Sch. as τοῦ ἀνιωμένου τὸ (ἦ[το]ρ).

ἀσθενέω 2, add 'εἰ μὲν οὖν μὴ συνεβεβήκει τὰ κοινὰ τῆς πόλεως ἀσθενεῖν *SEG* 38.1476.A50 (Xanthos, iii BC)'   add '**II** trans., *weaken*, *cause to fall*, Lxx *Ma*. 2.8.'

ἀσθέταιροι, οἱ, unit of Macedonian infantry, Arr.*An*. 2.23.2, 4.23.1, 5.22.6, 6.21.3, 7.11.3 (codd., vv. ll. ἀσθέτεροι, cj. πεζέταιροι).

ἄσθιπποι, οἱ, *unit of Macedonian cavalry*, τοὺς ἀσθίππους ὀνομαζομένους D.S. 19.29.2 (v.l. ἀνθίππους); cf. °ἀσθέταιροι.

ἄσθμα III, line 3, after '(Agath.)' insert 'ἀ. πυρός ib. 7.210 (Antip.Sid.)'

ἀσθμάζω, add 'med. aor. part., *PMag*. 13.522'

ἀσθμάομαι, delete the entry.

ἀσθματώδης, add 'Gal. 7.949.8, 959.3, 16.662.14 (*CMG* 5.9.(2).85.17)'

Ἀσία, after '*Asia*' insert '(perh. orig. a name for Lydia and then extended to all the hinterland of Ionia and eventually over the continent) Hes.*fr*. 180.3 M.-W., Archil. 227 W., Mimn. 9.2, Sapph. 44.4 L.-P.'

Ἀσιᾱγενής, Ion. Ἀσιηγενής, add '*IEphes*. 600A'

Ἀσιανός, before 'ἡ, όν' insert 'Ion. -ηνός, Hp.*Aër*. 16, al.'; add 'Ἀσιανῶν θίασος a fraternity for worship of Dionysus, *SEG* 31.633B (Maced., ii AD)'

Ἀσιαρχία, add '*Milet*. 1(9).339 a 5, b 4 (iii AD)'

Ἀσιάς (s.v. Ἀσία in *LSJ*), line 4, after 'Euph. 34' insert 'Ἀσίδι .. αἴῃ Hes.*fr*. 165.11 M.-W.'

ἀσίδηρος I, add '**b** δόρατα ἀσίδηρα = Lat. *hastae purae*, D.C.ap.Zonar. 7.21.'   **II**, for '*sword*' read '*weapon*'; before 'βίος' insert 'w. ref. to the Golden Age'; after 'Max.Tyr. 36.1' insert '*without a tool*, χείρ *AP* 9.52 (Carph.)'

ἀσιδήρωτος, ον, ῥυμοί, *not strengthened with iron*, *IG* 1³.386.22; v. °ῥυμός IV.

ἄσιλος, delete the entry.

ἀσῑνής II, after 'Hdt. 1.105' insert 'X.*Eq*. 5.1'

Ἀσιονίκης, ου, ὁ, *victor in the Asian games*, *IG* 4.206 (i BC), *MAMA* 8.418.31 (Aphrodisias), *Inscr.Perg*. 8(3).71.

ἄσιος, add 'form ἄσϜιος, cf. Myc. *a-si-wi-ja*, pers. n., *a-si-wi-ja*, epith. of Potnia'

ἀσίρακος, delete '= τρωξαλλίς'

ἀσιτητέον, *one must fast*, Gal. 10.807.10.

Ἀσκαηνός, ὁ, cult-title of the deity Men, in Antioch, *SEG* 30.1503, 31.1138, etc., in Aphrodisias, Reynolds *Aphrodisias* 29.15, 32.7 (Ἀσκαιν-, both i BC).

ἀσκάλαφος, add 'cf. mythical name associated with an owl, Ἀσκάλαφον .. Δημήτηρ ἐποίησεν ὦτον Apollod. 2.5.12'

Ἀσκαληπιακόν, etc., v. °Ἀσκληπ-.

ἀσκαλία, add 'prob. in *POsl*. 48.6 (i AD, pl., -λει- pap.)'

ἄσκαλος, add 'orac. in *IEphes*. 1252.7 (ii AD)'

ἀσκαλώνιον, add '**II** ἀ., τό, vessel of unknown size used as measure of capacity, οἴνου ἀσκαλώνια δ' τραγημάτων ἀσκαλώνιον α'. *PHerm. Rees* 23 (iv AD), *POxy*. 1924.3 (v/vi AD), κοῦφον ἀσκαλόνιν ibid. 10, *PKlein.Form*. 1204.4 (vi AD).'

Ἀσκαλωνῖτις, ιδος, fem. adj., *of Ascalon*, γάστρα Zos.Alch. 210.15, Anon.Alch. 418.24.

ἀσκάνδης, add 'see also ‡ἀστάνδης'

ἀσκάντης I, after 'Ar.*Nu*. 633' insert 'Call.*fr*. 240 Pf.'

ἀσκαρίζω, for 'Att. form .. euph.' read '= σκαρίζω, *palpitate*' and before 'Hp.*Nat.Puer*. 30' insert 'Hippon. 33.2 W.'

ἀσκαύλης, add '*PSAthen*. 43ᵛ i 3, al. (ii AD)'

ἄσκαφος, add 'rest. in Pratin. 3 P.'

ἀσκέδαστος, after '*not scattered*' insert 'Alc.*Intr*. 25 (p. 177.21, 27 H.)'

ἄσκεπος, *defenceless*, for the pres. ref. read 'ἤριπε .. ἄ. of the city of Sparta, Amyntas in *Suppl.Hell*. 44.5'

ἄσκεπτος II 2, add 'ἀσκέπτους νεκύων εἰς θαλάμους epigr. in *Salamine* 192 (ii/i BC)'

ἀσκέρα, for 'Herod. 2.32' read 'Herod. 2.23' and add 'dub. in *IG* 1³.422.163 (Athens, v BC; perh. ἀσκηρά)'

ἀσκέω I 3, delete the section.   **II 2**, line 5, after 'metaph.' insert 'δαίμον' ἀσκήσω .. θεραπεύων Pi.*P*. 3.109; ἀσκεῖται Θέμις id.*O*. 8.22'

ἄσκημα, for 'in warfare .. chariots)' read '*form of warfare, military practice*'; add '*object of practice*, *art*, D.H.*Comp*. 3.5'

ἀσκητέος II, ἀσκητέον, add 'Muson.*fr*. K.'

ἀσκητήρ, add 'Myc. *a-ke-te-re* (pl.) *decorator, finisher*'

ἀσκητήριον, τό, *monastery*, *Cod.Just*. 1.3.53.3, al.; also *nunnery*, ib. 1.3.46.5.

†ἀσκητής, οῦ, ὁ, *one who is in training for athletics*, Ar.*Pl*. 585, Pl.*R*. 403e, Isoc. 2.11, Διόνυσος ἀ., title of comedy by Aristomenes, cf. οὗτός ἐστιν ὁ τῆς ἀληθείας ἀσκητὴς ὁ πρὸς τὰς τοιαύτας φαντασίας γυμνάζων ἑαυτόν Arr.*Epict*. 2.18.27.   **2** *one who exercises himself in any art* or *discipline*, ἀ. τῶν καλῶν κἀγαθῶν ἔργων X.*Cyr*. 1.5.11, λόγων D.H.*Is*. 2.1, σοφίης *IG* 2².11140.2 (ii AD), φρονήσεως Ph. 1.59, 1.643, al.; of things, καρτερίας ἀσκητὴν λογισμόν id. 1.89, ὁ νοῦς ὁ ἀσκητής 1.91.   **3** *monk, hermit*, Ath.*Apol.Const*. in *PG* 25.632a, al.

†**ἀσκητικός**, ή, όν, *concerned with* or *relating to physical training*, νόσημα Ar.*Lys.* 1085, βίος Pl.*Lg.* 806a, μελέται Ph. 1.646; of persons, id. 1.552; as subst., τοιαῦτα ὑφηγεῖται τῷ ἀ. ἡ ὑπομονή id. 1.551; neut., τὸ ἀσκητικόν *training* Epict. 3.12.6; adv. -κῶς Poll. 3.145.

†**ἀσκήτρια**, ἡ, *ascetic woman, nun, Cat.Cod.Astr.* 7.225.29, *MAMA* 1.174.1 (Laodicea Combusta), *SEG* 28.1576 (v/vi AD), *Cod.Just.* 1.3.45.9, Just.*Nov.* 59.3.25; Myc. *a-ke-ti-ri-ja, decorator, finisher.*

***ἀσκιᾱτρόφητος**, ον, *used to unsheltered life, hardy,* ?Com.ap.Phot. 264 Th.

ἄσκιος I, before 'αὐγή' insert '*Λακεδαίμων AP* 7.723'

ἀσκίπων, line 2, after '(Theodorid.)' insert 'Posidipp. in *Suppl.Hell.* 705.24'

***Ἀσκλᾱπ-**, v. ‡*Ἀσκληπ*-.

***ἀσκλατάριος**, ὁ, dub. sens., *SB* 6951ᵛ.34 (ii AD) (cf. perh. medieval Lat. *sclata, exclate* = *scindula, stick* or *rafter*).

Ἀσκληπιακός (s.v. *Ἀσκληπιός*), add 'σκάφιον Inscr.*Délos* 320*B*56 (iii BC); -κά, τά, *fund for the expenses* of Asclepios' festival, *IG* 12(5).544.B 2.9 (Ceos, iv/iii AD); neut. subst. sg., in form *Ἀσκαληπιακόν*, name of mine in Attica, *SEG* 32.233 (iv BC)'

ἀσκληπιάς, add '**4** pl., *haemorrhoids,* Cyran. 41 (1.21.64 K.).'

Ἀσκληπιασταί (s.v. *Ἀσκληπιός*), add '*Ἀσσκηπιαστῶν* (sic, gen. pl.) *ZPE* 70.152 (Chios, iii BC)'

Ἀσκληπίεια (s.v. *Ἀσκληπιός*), add 'also *Ἀσκληπεῖα SEG* 31.594 (Ephesus, ii AD), 32.1089 (Africa, c.300 AD), *Ἀσκλᾱπίεια SEG* 23.212 (Messene, i BC/i AD), *Ἀσκλᾱπεῖα IGRom.* 4.1064 (Cos)'

Ἀσκληπιεῖον (s.v. *Ἀσκληπιός*), add '*Ἀσκλᾱπιεῖον IG* 7.1780 (Thespiae), *Ἀσκλᾱπιῆον IGBulg.* 1².315.20 (Mesambria, i BC)'

Ἀσκληπιός, line 1, after '-πιός' insert '*Αἰσκλαπιός IG* 4².136 (Epid., vi/v BC); *Αἰσχλαβιός IG* 4.356 (Corinth); *Ἀσκλαπιός IG* 4.1172.3 (Epid.); *Ἀσκαλ[α]πιός IG* 9(2).397 (Scotoussa); *Ἀσκαλπιός Inscr.Cret.* 4.182.6 (Gortyn); Lacon. *Ἀγλαπιός IG* 5(1).1313 (v BC), *SEG* 12.371.5 (Cos, iii BC), *Αἰγλαπιός* ib. line 3, (cf. *Ἀγλαόπης* (Lacon.) and *Αἰγλάηρ* Hsch.); Cypr. *Ἀσκλαπιός SEG* 30.1678 (but *a-sa-ka-la-pi-o-i* in syllabic version, Syria, iv/iii BC); n. *Ἀσ(σ)κηπιάδης*, etc., v. *ZPE* 70.152-6); dat. sg. ηαισκλαπιεῖ (? from ηαισκλαπιεύς) *IG* 4².151 (Epid., vi/v BC)' and after 'etc.' add 'pl., *statues of A., Inscr.Délos* 1417*B*i147 (ii BC)'

***ἀσκόμισθοι**, οἱ, *those who let wineskins for hire,* συνεργασία ἀσκομίσθων *IEphes.* 444.9.

***ἀσκοναυτοποιός**, ὁ, *maker of rafts supported on inflated skins, BE* 1964.495 (Palmyra, iii AD).

***ἀσκοποιός**, ὁ, *maker of wineskins, BE* 1964.495 (Syria), *IGLS* 9158, 9159, 9160 (Bostra); = Lat. *utrarius, Gloss.* 3.307.

ἀσκός, add '**6** dub. sens. app. denoting some metal object, *IG* 2².1544 (iv BC), v. °ἐξάγιστος, ἀκκόρ.'

***ἀσκότεινος**, ον, *free from darkness,* gloss on ἀνέσπερον Hsch.

***ἀσκότιστος**, ον, gloss on ἀνέσπερον, Cyr.; cf. °ἀσκότεινος.

***ἀσκοφύσιον**, τό, *bellows made from skin,* Anon.Alch. 349.3.

***ἀσκόω**, *fit with leather pads, IG* 2².1604.13.29, 38 (iv BC); cf. ἄσκωμα 1.

ἀσκύλευτος, after '*stripped*' insert '*GVI* 1603.11 (Acraephia, iii BC)' and add 'Sopat.Rh. 8.51.11 W.'

ἄσκωμα 1, add '*IG* 2².1604.32, al. (iv BC)' **2**, for pres. def. read '*the fully-developed female breast*'

ἆσμα, for 'ἄττω' read 'ἄττομαι'

***ᾀσμᾰτογράφος**, ὁ, *song-writer*, Tz. ad Lyc. p. 1 S.

***ἀσμένεια**, ἡ, *pleasure, satisfaction, PFlor.* 294.13 (vi AD).

ἀσμενιστός, after '*welcome*' insert 'Cic.*Att.* 9.2a.2, 9.10.9'

ἀσοφία, for '*folly, stupidity*' read '*lack of skill* or *judgement*'

ἄσοφος, for '*unwise, foolish*' read 'lacking *skill* or *understanding*'; after 'Pi. *O.* 3.45' insert '*Φοίβου τ' ἄσοφοι γλώσσης ἐνοπαί* E.*El.* 1302, ἄσοφοι καὶ ἀκρατεῖς X.*Mem.* 3.9.4'

ἀσπαίρω, line 6, after 'of an infant' insert '*h. Cer.* 289'

ἀσπάλαθος 1, at end for 'Thphr. 9.7.3' read 'Thphr.*HP* 9.7.3'

***ἀσπαλακτής**, ὁ, name of a stone, Socr.Dion.*Lith.* 50 H.-S., al.

***ἀσπᾰραγυλιοκογχυλεύς**, έως, ὁ, *purple-fisher who uses a weel* (γυλιός) *made of* ἀσφάραγος, w. haplology, *MAMA* 3.681 (ἀσπαραγυλιωκονχ-, Corycus).

***ἀσπαράκιον**, v. °ἀσφ-.

***ἀσπερμολόγητος**, ὁ, epith. of Jesus Christ, app. *not humanly conceived, PMasp.* 188.1 (vi AD).

ἀσπιδεῖον III, for '= ἀσπίδιον (?)' read 'perh. *votive panel* (portrait)'

***ἀσπίδηος**, α, ον, Boeot. for *ἀσπίδεος *of shields,* ἄθλα ἀ. *IG* 7.2712.23 (Acraephia).

ἀσπιδίσκη (s.v. ἀσπιδίσκος), line 2, for '*SIG* .. ii BC)' read '*Inscr.Délos* 442B.32 (ii BC); as part of a brooch, πορπίον ἀ. ἔχον ib. 1417A.3'

ἀσπιδίσκος, after 'ἀσπίς' insert '*small shield,* votive in character, *IG* 2².47.7 (iv BC)'; after 'Sch.Il. 5.743' insert 'ornamental *boss* or *button,* ἀσπιδίσκοι χρυσαῖ ἑτέραις Aristeas 75; on high priest's ephod, Epiph.Const. in *Lapid.Gr.* 198.17; *disc* on each end of the crossbar of a cithara, Hsch.'

ἀσπίς 1, line 1, after 'al.' insert 'κοίλη Tyrt. 19.7 W. (cf. Mimn. 13a.2 W.), Alc. 357.6 L.-P.' add '**b** ἡ ἐξ Ἄργους ἀσπίς, name of games at the festival of Hera at Argos in Roman times, *IG* 2².3145 (i AD), v. *SEG* 33.296), etc., *IG* 14.746.18.(i/ii AD), *SEG* 34.1316 (Xanthos, i AD).'

ἄσπονδος, penultimate line, delete '(lyr.)'

ἄσπορος I, add 'γαῖα Moschio Trag. 6.25.5; as fem. subst., *SEG* 23.305 ii b8 (Delphi, ii BC)' **IV**, after 'Nonn.*D.* 2.221, al.' insert 'of Attis, ib. 25.311'

***ἀσπροειδής**, ές, *tending to whiteness,* Socr.Dion.*Lith.* 26.10 H.-S.

***ἀσπροπώλισσα**, ἡ, *seller of incense,* cf. ‡ἄσπρος III, *ITyr* 17B (Chr.).

ἄσπρος II, for 'ἀ. γράμματα .. very late' read '*colourless, white,* ἀ. γράμματα "invisible" writing, *Cat.Cod.Astr.* 1.108.1; *white, Stud.Pal.* 20.245.4 (v. *ZPE* 76.113, vi AD), dist. fr. λευκός, καρακάλλιον λευκὸν λινεγαῖον ἀ. *PSI* 14.1427.19 (vi AD)' **III**, add 'cf. perh. mod. Gk. ἀσπρούχι'

***ἀσπρόχρους**, -ουν, *white-coloured, Cat.Cod.Astr.* 5(4).170.

ἄσσα, line 5, after 'Cratin. 6' insert 'ποιεῖν ἄττα Pl.*R.* 339d'

***ἀσσκονικτεί**, v. °ἀκονιτί.

ἆσσον I, line 7, for 'A. *fr.*6' read 'A. *fr.*66 R.'

Ἀσσύριος, after 'adj.' insert '*Ἀ.* γράμματα, name given to various oriental scripts, id. 4.87 (prob. cuneiform), Th. 4.50 (prob. Aramaic); cf. °*Σύρος*; ξεῖνος' and delete 'al.'

ἀσταγής II, at end after '*in a stream*' insert 'Call.*fr.* 317 Pf.'

ἀστακός I, after '*lobster*' insert 'Epich. 30'

ἀστάνδης, after 'Plu.*Alex.* 18, 2.326f' add '340c, Ath. 3.122a (cj.) Hsch.'; add 'see also ἀσκάνδης'

***Ἀστάρτη**, ἡ, Dor. -ᾱ, Phoenician goddess, *SEG* 36.316 (Corinth, v BC), J.*AJ* 1.118, Plu. 2.357b.

ἀστασία, add '**II** *unsettled conditions,* in political sense, Sch.Th. in *DAW* 67 no. 2 p. 11.'

***ἀσταταρία**, ἡ, *candlestick, Stud.Pal.* 8.941, *PVindob.*G 25737.6 in *ZPE* 64.75; cf. στατάριον, medieval Lat. *cereostata statarii.*

***ἀστάτι**, adv. of ἄστατος, perh. in Sophr. in *Stud.Ital.* 10.249.

ἄστᾰτοι (or ἀστᾰτοί), read this for ἄστατοι and add 'Lyd.*Mag.* 1.46 (p. 48.8 W.), sg., *SEG* 33.1195 (Cappadocia, i/ii AD), ib. 7.86 (Syria, ii AD), *IGRom.* 3.1206 (Palestine)'

ἄστατος, after 'of persons .. Onos. 3.3' insert 'τύχῃ δ' ὡς ἀστάτῳ πιστευτέον ἑταίρᾳ Iamb.*Protr.* 2'

ἀσταφίς I, line 2, for 'Tegea' read 'Laconia'; line 5, delete 'σταφίς .. Theoc. 27.9' and substitute 'see also ‡σταφίς'

ἀστάφῡλος, for 'Aus.*Ep.* 12.24' read 'Aus.*Ep.* 8.24 (p. 243.24 P.)'

ἀστέγαστος, delete 'Gal. 17(2).153' (v. °ἀστέγνωτος)

***ἀστέγνωτος**, ον, *uncovered,* Gal.*CMG* 5.10.(2).2.208.10.

ἀστεῖος, adv. -ως, add *nicely, well,* Erasistr.ap.Gal. 9.206.17'

ἀστειότης, add '**2** in Rhetor Anon. in *PLit.Lond.* 138 ii 12 app. *citizenship.*'

ἀστεϊσμός, for '*wit*' read '*witticism*'; after 'D.H.*Dem.* 54' insert '(pl.)'

†**ἀστεμφής**, ές, *motionless, immobile,* σκῆπτρον .. ἀστεμφὲς ἔχεσκεν Il. 3.219, ἀ. οἵη νέκυς Opp. *H.* 2.70; neut. adv. ἀστεμφὲς ἔκειτο Mosch. 4.113. **II** *firm, immovable,* οὐδὸς Hes.*Th.* 812, θεμέλιος Amph.*Hom.* 1.1; of men, animals, Anacr. 22 P., ἀ. κύων i.e. Cerberus Trag.*adesp.* 658.2 K.-S., ἀστεμφής .. τίγρυν ἐλαύνων Nonn.*D.* 11.70, 28.14, *not giving way, unflinching,* Theoc. 13.37, Opp.*H.* 2.446, C. 4.174, Agath. 1.21, pr. n. of a Titan, Emp.B 123 D.-K.; adv. -έως Od. 4.419, 459. **2** of bonds, etc., *firm, fast,* ποδάγρα *AP* 6.296 (Leon.Tarent.), ζυγόν Opp.*H.* 1.417, δεσμός 2.84, Nonn.*D.* 40.324. **III** of abst. things, *unmoved,* βουλή Il. 2.344, βίη A.R. 4.1375, δύναμις Plot. 6.8.21; adv. ἀστεμφῶς τὸν βίον διενήξατο Marin.*Procl.* 15. **2** of atmospheric conditions, *unchanging,* νύξ *AP* 9.424 (Duris); neut. adv. ἀλωαὶ .. ἀστεμφὲς μελανεῦσαι Arat. 878.

***ἀστεοπρόσωπος**, ον, *with refined face,* epith. of a boy in paederastic rock inscr., *SEG* 32.847.A16 (Thasos, iv BC).

***ἀστέραρχος**, ὁ, *ruler of the stars,* Ἥλιος *Cat.Cod.Astr.* 9(2).162, 163.

ἀστεργάνωρ, for '*without love of man*' read '*refusing marriage*'

ἀστερίζω, add '*turn into stars, Placit.* 2.13.3'

ἀστερίας, add '**VI** ἀστερία, sc. λίθος, ἡ, a precious stone, Plin.*HN* 37.131.'

***ἀστερομαρμᾰροφεγγής**, ές, *gleaming with stars white as marble,* Lyr. *adesp.* 9 P. (iii BC).

ἀστεροπός, add 'as the name of a Cyclops, = *Στερόπης,* Euph. 51.11'

ἀστερόφοιτος II, add 'κόσμου φύσις *PMag.* 4.2552 (= *Hymn.Mag.* 20.26)'

ἀστή, add '*PMerton* 5.2'

***ἄστη**, ἡ, Lat. *hasta, spear,* δώροις δεδωρημένῳ .. [στε]φάνῳ τειχικῷ ἄστῃ καθαρᾷ οὐηξίλλῳ *IEphes.* 680.16 (i/ii AD). **2** *auction, PBeatty Panop.* 2.138, 139 (iv AD).

ἄστηνος, after '*miserable*' insert 'Call.*fr.* 275 Pf.' and add 'cf. δύστηνος'

ἀστήρ I 1, lines 3/4, for 'ἀ. Ἀρκτοῦρος .. etc.' read 'ἀ. Ἀρκτοῦρος

[Hes.Op.] 566, etc.; πλόον ἠελίῳ τε καὶ ἀστέρι τεκμήρασθαι A.R. 1.108'   **II**, add 'τὸν ἀγαπητὸν ἀστέρα τῆς οἰκουμένης Them.Or. 16.213a'   **V**, for this section read 'one or more plants of the Aster genus, Nic.fr. 74.66, ἀ. Ἀττικός Dsc. 4.119'   add '**X** app. part of a bath-house, SB 9921.16 (iii AD).'

*ἀστιάριος, ὁ, Lat. hastiarius (hastili-), grade of cavalry officer, SB 11591.17, 11592.17 (iv AD), BGU 1024.5, 8 (iv/v AD, cf. Berichtigungsl. 7.17).

*Ἀστιάς, άδος, fem. adj., epith. of Artemis at Iasos, Plb. 16.12, IIasos 88.3, 92.8, 248.6 (all Rom.imp.).

ἀστϊβής **1**, add 'X.Mem. 3.8.10 (sup.)'   **3**, for this section read 'of holy places, not to be trodden, ἄλσος S. OC 126, Ps.-Hdt.Vit.Hom. 21'

*ἄστϊγος, ον, unmarked, of a corpse, dub. in PRein. 92.12 (iv AD).

ἀστικός **I 1**, line 2, for 'epith. of Hecate' read 'epith. of Enodia Γαστικά'   **II 2**, delete 'ἀστικά .. ἀγροίκως'

*ἀστίλιον, τό, cf. Lat. hastile, shaft, spear, Edict.Diocl. 14.4 (ἀ- ed.), SB 9017.9 (i/ii AD).

*ἀστιοπόλων, gen. pl., dub. sens., IEphes. 454 (ii/iii AD).

ἄστολος, at end read '(sch. ad loc., cod. ἄστονος)'

ἄστοργος **1**, after 'without natural affection' insert 'Achae.fr. 2.4 S.'

ἀστός, lines 1 ff., for 'dist. from πολίτης, ἀστός' read 'dist., prob. wrongly, fr. πολίτης as'   add '**II** w. gen., dweller in, ἐρημάδος ἀ. ἐρίπνης Nonn.D. 25.272; cf. 17.40.'

ἀστοχέω, after 'rare in poetry .. Lyr.Alex.adesp. 4.21' insert 'Trag.adesp. 658.2 K.-S.'   add '**II** disregard, w. gen., τοῦ καλῶς ἔχοντος PTeb. 798.14 (ii BC); τοῦ προσώπου Callisth.Olynth. 44.'

†ἀστραβιστήρ' ὄργανόν τι, ὡς δίοπτρον Hsch.; cf. ἀστραφιστήρ.

ἀστράγάλη, after 'for ἀστράγαλος' insert 'Il. 23.88 (v.l.)' and after 'Herod. 3.7' insert 'AP 6.309 (Leon.Tarent.)'

ἀστραγαλίζω, add '**II** ἀστραγαλίσαι ἐπὶ τῶν ὑποποδίων perh. describes some method of fastening statues, Inscr.Cret. 3.ii 1 8 (ii BC).'

*ἀστραγάλιον, τό, dim. of ἀστράγαλος, IG 2².1533.32, 4².103.91 (Epid., both iv BC).

ἀστράγαλος **III**, for 'wrist' read 'knuckle' and for 'Lxx Da. 5.5, 24' read 'Thd.Da. 5.5, 24'   **VI**, add 'κατεγλύψαμεν ἀστράγαλον Didyma 39.20'

ἀστραγάλωτός, line 2, for 'Posidon. 9' read 'Posidon. 8'

*ἀστραγεύτως, adv., (στραγγενόμαι) without delay, prob. in BGU 1760.7 (i BC, cf. Berichtigungsl. 3.23).

*ἀστραπαία, ἡ, a precious stone, Plin.HN 37.189, Eust. 827.26.

ἀστραπαῖος, before 'α, ον' insert '[-πᾶος (< -παιος), Orph.H. 15.9, 20.5]'; for 'IGRom. .. (Bithyn.)' read 'INikaia 701, 702 (Rom.imp.); cf. °ἀστραποποιός'

ἀστραπή, lines 4/5, personified, add 'Pap.Lugd.Bat. xxv 8 i 8 (iii AD)'

ἀστραπηφορέω, for 'Ar.Pax 722' read 'E.fr. 312 N. (Ar.Pax 722)'

ἀστράπιος, delete the entry.

*ἀστραποποιός, ὁ, maker of lightning (title of Zeus), INikaia 1505 (Rom.imp.); cf. °ἀστραπαῖος.

ἀστράπτω **II**, after 'flash or glance like lightning' insert 'ὄμματα ἀστράπτοντα Il.Pers. 5.8 B.'

†ἀστράτευτος, ον, not having done military service (usu. w. implication of avoidance), Ar.V. 1117, Lys. 9.15, D. 24.102, 107, 119, Aeschin. 3.176, ἀ. καὶ λιποτάκτης Ph. 1.144; adv. -τως Poll. 1.159.

ἀστρατήγητος **II**, add 'Sopat.Rh. 8.193.18 W. (v.l. ἀστράτηγος)'

*ἀστρατηγικῶς, not in the manner of a good general, Sch.Gen.Il. 2.74 (v.l. ἀστρατηγητικῶς).

ἀστραφιστήρ, for 'dub. sens. .. ἀστραβιστήρ' read '= °ἀστραβιστήρ (cf. κισσύβιον, κισσύφιον), IG 2².1628.522, 1629.998 (both pl., iv BC)'

ἄστρεπτος, for 'without turning the back' read 'without turning round'

ἀστροθέτης, add 'PHarris 55.18 (ii AD)'

ἀστροΐτης, after 'astriotes' insert '(nisi leg. astrites, cf. Mart.Cap. 1.75 D.)'

*ἀστρολογογεωμέτρης, ου, ὁ, astrological geometrician, Phld.Epicur. 2.6.3 p. 60 V.

ἄστρον, line 7, for 'Arist.Cael. 290ᵃ20' read 'Arist.Cael. 290ᵃ23'; line 13, after 'S.OT 795' insert 'ὑπ' ἄστρα beneath the stars i.e. in the open air, Salamine 199 (i AD)'

ἀστρονόμημα, after 'observation of the stars' insert 'humorous personification applied to Thales' and add '(= Suppl.Hell. 797)'

*ἀστρόπληκτος, ον, star-struck, Gal. 14.402, Jo.Alch. 266.14.

ἀστροχίτων, add 'Suppl.Hell. 1051'

ἄστυ, line 5, before 'pl.' insert 'dat. ἄστυι SEG 30.622 (Maced., i AD)'   at end add 'Myc. wa-tu; also in compd. pers. n.'

ἀστύαρχος, delete the entry.

ἀστυνομικός, add 'as subst., Roman municipal official, Paul.Dig. 43.10.1 (cj. ἀστυνόμος).'

ἀστυόχος, at end for '(Γασστ-; cf. ἄστυ fin.)' read 'Γαστυνόχος IG 5(2).77 (Tegea, cf. ἄστυ fin.)'

ἄστυρον, for 'Call.fr. 19 .. Hec. 1.1.6' read 'fr. 11.5, 75.74, 260.6, 261.2 Pf.'

ἀσύγγνωστος **II**, for 'Gal. 1.13' read 'Gal.Protr. 7 (comp.)'

†ἀσυγκόλλητος, ον, not soldered on or together, PVindob.Salomons 2.2 (ii/iii AD), Sch.Il. 14.200.

ἀσύγκρϊτος **I 2**, add 'of a person, τῷ ἀσυγκρίτῳ ἀδελφῷ SEG 33.1196 (Cappadocia, i/ii AD), IHadr. 158'

ἀσῡλεί or -ί, for 'inviolably' read 'with immunity from σῦλαι (seizure on account of reprisals)'

ἀσῡλία, line 3, for 'in inscrr.' read '**b** immunity from σῦλαι'

ἀσυλλόγιστος **I 2**, add 'see also °ἀσυλλόχιστος'

*ἀσυλλόχιστος, ον, perh. f.l. for °ἀσυλλογιστ-, not calculated, of a debt, PRyl. 585.10 (ii BC).

ἄσυλος **I 1**, add 'epith. of Artemis at Perga, IGRom. 3.797, ABSA 17.231'

ἀσύλωτος, delete the entry.

ἀσυμφᾰνής, adv. -νῶς, add 'Phlp.in GC 55.1'

*ἀσυνᾰφής, ές, disconnected, Corn.Rh. 141 (p. 377 H.).

ἀσυνδεξίαστος, add '(v.l.)'

*ἀσυνειδησία, ἡ, lack of conscience, POxy. 3770.12 J. (iv AD).

*ἀσυνείσφορος, ον, not contributing, Rh. 3.573.5.

ἀσυνέξωστος, after 'of an athlete' insert 'perh. not thrown out of the arena of competition i.e. unbeaten'

ἀσύνετος, line 1, after 'Hp.Fract. 31' insert 'prob. in Alc. 67.2 L.-P. (ἀσύνν-)'

ἀσυνήμων, add 'Trag.adesp. 664 ii 23 (both ἀξυν-)'

*ἀσύνθετος **II**, line 3, after 'Ep.Rom. 1.31' insert 'Ptol.Tetr. 166'

*ἀσυνόδευτος, ον, expl. of incomitatus in Virgil glossaries, PSI 756.29, PNess. 1.965.

ἀσύντροφος, for 'βάκτος' read 'βάτος (A) 1'

ἀσῠρής, for 'lewd, filthy' read 'dirty, filthy, βρέγμα Herod. 4.51; τἀσυρὲ dirt, refuse, IG 1³.2.11 (Marathon, v BC)' and before 'ἄνθρωπος Plb. 4.4.5' insert '**2** morally debased, foul'

ἀσύστᾰτος, add '**9** ἀ. ὀνόματα names of persons not appointed, PLond. 1249.5 (iv AD).'

*ἀσυσχημάτιστος, ον, not in (astrological) relation, Heph.Astr. 3.7.

ἀσύφηλος, add 'ἡ νεότης ἀσύφηλος ἀεὶ θνητοῖσι τέτυκται Eleg.adesp. 25.1 W.'

*ἀσφαλάνθιον, τό, app. = ἀσπάλαθος, camel's thorn, POxy. 3733.20, 3766.103 (iv AD).

ἀσφάλαξ, add 'SEG 38.1237.11 (Lydia, iii AD)'

ἀσφάλεια, line 1, after 'ἡ' insert 'sp. ἀσπ- SEG 23.437 (Thessaly, iii BC)'   **I 1**, add '**b** safeguarding, security of structures, property, etc., ἐποίησαν τὴν τῶν θυρίδων ἀσφάλειαν CIJud. 766 (Phrygia, i AD), IEphes. 3217.26 (ii AD).'   **3**, after 'Lit. Crit.' for 'circumspection' read 'caution in use of words, D.H.Dem. 2'

†ἀσφάλειος, ον, also -ιος and -εος, that gives a guarantee, κατὰ τήνδε τὴν ἀσφάλειον PCair.Isidor. 105.9 (iii AD), BGU 96.6 (iii AD), PPanop. 21.18 (iv AD); neut. subst., ὑπὲρ ὑμετέρου ἀσφαλίου PRainer Cent. 84.6 (iv AD).   **II** epith. of Poseidon, Ar.Ach. 682, Paus. 3.11.9, 7.21.7, Plu.Thess. 36; form -εος, Didyma 132.2, 14 (ii BC); form -ιος, Opp.H. 5.680, IG 5(1).559.14 (Amydae), Aristid.Or. 46(3).1.

*Ἀσφάλεος, v. °ἀσφάλειος.

ἀσφᾰλής **I 5**, after 'rhythm' insert 'D.H.Dem. 24, al.'; adv., add 'D.H.Dem. 26'   add '**IV** epith. of Poseidon, = ‡Ἀσφάλειος, IG 4².555 (Epid.).'

ἀσφαλίζω **I**, for '**b** secure .. in Med.' read '**2** ensure the safety or security of, BGU 829.9 (i AD); esp. in med.'   **II 1**, add 'make provision for in will, POxy. 2348.41 (iii AD)'   add '**3** give security, Modest.Dig. 27.1.15.17, 50.12.10.'

†Ἀσφάλιος, v. °ἀσφάλειος.

ἀσφάλισμα, add '**II** in pl., gloss on λέπαδνα, Sch.Gen.Il. 5.730.'

ἄσφαλτος, after 'bitumen' insert 'Alc. 124.7 L.-P. (rest.)'

*ἀσφάργιον, τό, written ἀσπαράκιον, dim. of ἀσφάραγος (B), PMich.inv. 3731 ii 36 (Tyche 1.185; ii/iii AD).

ἀσφάραγος (B), at end (form ἀσπ-) add 'Lang Ath.Agora XXI Hd 11 (ii/iii AD)'

ἀσχημάτιστος **II**, after 'D.H.Pomp. 5' insert 'Rh. 9.13'

ἄσχημος, after 'late form for ἀσχήμων' insert '(but Ἀσχειμος as pers. n., SEG 25.662, Thessaly, iv BC)'

ἀσώδης **II**, delete 'slimy'; for '(lyr.)' read '(anap.)' and add 'cf. ἀσώδης· ἀμμώδης Hsch.'

ἀσώματος **I**, add '**2** lacking a body, νέκυς ἀ., i.e. a skull, AP 9.52 (Carph.).'

ἀσωτεῖον, for 'Longus 4.17' substitute 'Poll. 9.48'

*ἄτ, v. °ἀπό.

*Ἀταβῡριασταί, οἱ, Διὸς Ἀ. worshippers of Zeus Atabyrios, Lindos 391.31, 392a.12, IG 12(1).937.4 (near Lindos, i/ii AD).

*Ἀταβύριον, τό, or -ιος, ὁ (ὁ Ἀτάβυρις Str. 14.2.12), mountain in Rhodes, Ζεῦ πάτερ, νώτοισιν Ἀταβυρίου μεδέων Pi.O. 7.87, Str. l.c.

*Ἀταβύριος, epith. of Zeus in Rhodes and Agrigentum, ISelge T 52 (Rhodes, Mt. Ataviro, iv BC), Lindos 339 (i AD), IG 12(1).891.7 (Lindos, ii AD), Plb. 9.27.7, App.Mith. 26.

ἀτακτέω, line 5, for 'POxy. 275.24' read 'POxy. 275.25'
ἄτακτος II, line 5, after '(Pl.Lg.) 840e' insert 'of persons, immoral, ἄ. τὸν τρόπον Vit.Sapph. in POxy. 1800 fr.1 i 17' add '4 unassessed, πόλις IG 1³.277.31.' B adv., add 'ἄτακτα παίζει Anacreont. 59.26 W.'
ἀτάλαντος, for 'equal in weight, equivalent to, like' read 'equal in some respect to' and add 'ἀ. ἀπάντῃ, Emp.B 17.19 D.-K.'
ἀταλός, delete 'ἀ. χερσί .. A.Pers. 537 (anap.)'; at end for 'Sup. .. IG 1.402a' read 'sup. ἀταλότατα παίζει CEG 432 (viii BC)'
*ἀτανύω, = τανύω, aor. imper. ἀτάνυσσον, Diosc.epic. 4.16, aor. subj. ἀτανύσσῃς ib. 6.28, w. χεῖραν and dat., help, ib. ll.c., 3.25, 13.14, etc.
ἀτάομαι II, line 4, after 'Gythium' insert 'cf. ἀγατᾶσθαι (i.e. ἀϝατ-)· βλάπτεσθαι Hsch.'
ἀτάρ 1, line 11, delete 'sts. after ἐπειδή .. Il. 12.144' 2, for 'Pl. and Trag.' read 'Attic, except in the orators'; after 'S.OT 1052' insert 'Cratin. 200 K.-A., Ar.Pax 177, Av. 144, al.' 3, after '(Il.) 270' insert '12.144' and delete 'also in Com., Cratin. 188'
ἀταρπῖτός, for 'Ion.' read 'poet.'
ἄτε II, line 5, delete 'ἄ. .. A.Th. 140 (lyr.)'
*ἀτεγγής, ές, = ἄτεγκτος II, [οὔ]θε[ἱ]ς οὕτως ἐστὶν ἐν ἀν[θ]ρώποισιν ἀτενγής IG 2².12236.7 (iii BC).
ἄτεγκτος II, delete 'Ar.Th. 1047'
*ἀτειρήεις, εσσα, εν, = ἀτειρής II, Nonn.D. 35.226.
ἀτειρής II, add '2 patient in suffering, Anacreont. 56.1 W.'
*ἄτεις, ὁ, acc. sg. ἄτειν, perh. name of degree of kinship, MAMA 3.53 (Cilicia); cf. ἄττα (B).
ἀτείχιστος, after 'Lys. 33.7' insert 'τᾶς πόλιος ὑπαρχοίσας ἀτιχίστω SEG 28.1540.13 (Cyrenaica, i BC)'
†ἀτέκμαρτος, ον, lacking any demarcation, ἐρημία Plu.Luc. 14, θάλασσα Nonn.D. 13.537; neut. pl. as adv., fig., ἐπὶ μὰν βαίνει τι καὶ λάθας ἀτέκμαρτα νέφος Pi.O. 7.45. 2 having no sign or indication for the purpose of interpretation, prediction, etc., χρηστήριον Hdt. 5.92γ, μοῖρα (sup.) A.Pers. 910, δέος Th. 4.63.1, Pl.Lg. 638a, Orph.A. 1150; of persons, Ar.Av. 170; w. inf., ἀτέκμαρτον προνοῆσαι Pi.P. 10.63; adv. -τως without any indication, X.Mem. 1.4.4. II (dub.) boundless, unlimited, γαστήρ (i.e. appetite) Opp.H. 2.206 (s.v.l.).
ἀτεκνόω, after 'make childless' insert 'πάτρα γάρ μ' ἀτέκνωσε my country has robbed me of children (by depriving me of life in her service), Bernand IMEG 13.11 (Egypt)'
*ἀτεκτόνευτος, ον, expl. of infabricatus in Virgil gloss., PNess. 1.854 (vi AD).
ἀτέλεια, line 1, before 'Cret.' insert 'Ion. -είη SEG 36.982.C8 (Iasos, v BC)' I, add '2 ineffectualness, Schwyzer 167a.A4, B2 (Selinus, v BC).'
ἀτέλειος, add 'BMC Caria no. 20 (Alabanda, Rom.imp.)'
ἀτέλεστος I, add '2 of children cut off before reaching maturity, MUB 13.26.1, IHadr. 151 (Rom.imp.).' IV, add 'neut. pl. as adv., Arat. 678' add 'V unfinished, of buildings, POxy. 3691.6 (AD 138).'
ἀτελής I 2, line 3, after 'ib. 40' insert 'of a discussion, Pl.Prt. 314c' II 1, line 2, after 'ineffectual' insert 'καπνός Simon. 36.3 P.'; line 3, after 'Pl.Smp. 179d' insert 'of things, ἔρημον καὶ ἀ. φιλοσοφίαν λείποντες id.R. 495c' III 1 a, add 'τοῦ σώματος Inscr.Prien. 174.6 (ii BC)' b, add 'Cypr. a-te-li-ja (n. pl.), ICS 217.23'
ἀτενής I, add '5 of close texture, compact, γῇ ἀ. καὶ σκληρά Plu. 2.640e.' III, for 'Hp.Prorrh. 1.24' read 'Hp.Prorrh. 1.124'
ἄτερ II, line 4, for 'also in late prose' read 'in prose, ἄ. ἐμέο SEG 37.665 (N. shore Black Sea, c.400 BC), ἄ. τῆς Πτολεμαίου γνώμης ib. 9.1.42 (Cyrene, iii BC)'
ἀτεράμων, after 'Ar.Ach. 181' insert 'V. 730' and for 'Eub. 1 D.' read 'Eub. 22 K.-A.'
ἄτερθε II, add 'Nic.Th. 242'
ἄτερος, add 'IG 4².40.11, 15, 41.12, al.; Myc. a₂-te-ro (= ἅτερον)'
ἀτέρπης, line 4, for 'Simon. 37.6' read 'Simon. 37.6 P.'
*ἄτερυι, Aeol. adv., in a different place, Theognost.Can. 160; cf. ἀτέριγε.
*ἀτέρωτα, Aeol. for °ἑτέρωτε, at another time, κάτέρωτα Sapph. 1.5 L.-P., cf. A.D.Adv. 1.194.5.
ἀτεχνής, add 'adv. ἀτεχνέως in sense of Att. ἀτεχνῶς TAPhA 65.105 (Olynthus, iv BC)'
ἀτεχνία, after 'Hp.' insert 'VM 9'
†ἀτέω, only part. ἀτέων reckless, heedless, Il. 20.332, Hdt. 7.223.3; but indic., w. gen., Μουσέων Call.fr. 633 Pf.
ἄτη, line 1, for 'αὐάτα' read 'ἀυάτα' and after '(ἀϝ-)' insert 'Alc. 70.12 L.-P., Pi.P. 2.28, 3.24, Lyr.adesp. 55 P., poet. ἀάτη Call.fr. 557 Pf.'and delete 'v. infr.'
ἄτηκτος I, add 'Pl.Sph. 265c, Ti. 60e, al.'
*ἀτηρεύεσθαι, act mischievously, Hsch. s.v. σικελίζειν.
ἀτηρής, delete the entry.
Ἀθίς 1, line 2, after 'be read)' insert 'AP 12.55 (?Artemo)' add 'b Ἀτθίδες, αἱ, maidens of Attica, Call.fr. 178.4 Pf.'

ἀτίζω, last line, after 'A.R. l.c.' insert 'app. w. inf., ἐρέσθαι .. ἄτισσε id. 2.9'
ἀτιμάζω, line 6, after 'ἄκοιτιν' insert '(also ἠτίμασεν v.l. in Il. 1.11)'
*ἀτιμᾱσία, ἡ, = ἀτιμία, Favorin.Exil. 17.6, 22.45 B.
ἀτιμάω, line 4, for 'ἠτίμασεν' read 'ἠτίμησ' (v.l. ἠτίμασεν)'
ἀτιμία I 2, add 'as transl. of Lat. infamia, Just.Nov. 22.22 pr.'
*ἄτιμος, ον, worthless, PMasp. 1.25 (vi AD).
ἀτϊτάλτας, for the pres. ref. read 'Inscr.Cret. 4.15 (Gortyn, vii/vi BC)'
†ἄτλᾱς, αντος, ὁ, ἄτλας· ἄτολμος, ἀπαθής Hsch. 2 perh. insensible, unresponsive, ἀλλ' οὐκ ἄ. γὰρ βάσανος ἡ Λυδῇ λίθος S.fr. 91a R.'
ἄτλητος 1, add 'ἄτλητα πεπονθώς Thgn. 1029'
ἀτμένιος, for 'toilsome, prepared with trouble' read 'used by slaves, i.e. common'
ἀτμή, after 'Hes.Th. 862' add '(v.l.), Dieuch. 15.5 (Orib. 4.7.1)'
ἀτμήν, for 'Call.Aet. 1.1.19' read 'Call.fr. 178.19 Pf.'
ἀτοκία, add 'Gp. 12.38.1'
ἀτόκιος, after 'medicine for causing it, Hp.Mul. 1.76' read 'a contraceptive or abortifacient, Dsc. 1.77.2, 3.130, 3.134.2, Sor. 1.60 (CMG IV p. 45.1, 2, 4, 17)'
ἀτονέω, add 'b suffer loss of strength, tab.defix. in SEG 35.227.29 (Athens, iii AD). II transl. of Lat. deficio, fail to make a claim, Cod.Just. 6.4.4.14d. 2 of a condition, fail to occur, Theophil.Antec. 3.1.7.'
ἄτονος I, line 1, for 'of the limbs' read 'of the physique' and after 'Hp.Aër. 3 (comp.), 19' insert 'Arr.Epict. 3.16.7; deprived of strength, tab.defix. in SEG 35.215.3 (Athens, iii AD)' add '3 gramm., unaccented, Eust. 907.15.'
*ἀτοπέω, misconduct oneself, PTeb. 711.5 (ii BC).
ἀτόπημα 3, for 'offence' read 'outrage' and add 'PFlor. 5.60.11'
ἀτράκτιον, after 'pl.' delete '(written ἀτράκτεια)' and insert 'Inscr.Délos 1442 B 56 (ii BC)'
ἄτρακτος II, after '(S.)Tr. 714' insert 'E.Rh. 312, AP 5.188 (Leon. Tarent.)' and for 'specially Lacon.' read 'in Lacon. apophthegm'
ἀτρακτυλίς, for 'used for making spindles' read 'resembling a spindle covered w. wool' and add 'δοκίδας ἀτρακτυλίδος X.Cyn. 9.15'
*ἀτράκτυλον, τό, = ἀτρακτυλίς, prob. in Epich. 161.
ἀτραπός, line 2, after 'Il. 17.743' insert 'also Alcm. 102 P., ISmyrna 521.8 (ii/i BC), Nonn.D. 41.39'
†ἀτραυμάτιστος, ον, unwounded, free from wounds, διάθεσις Aët. 7.9 (261.20 O.); title given to successful boxers, Entretiens Hardt 14.236; poet. πόνοι, i.e. not arising fr. wounds, Luc.Ocyp. 36.
ἀτράχηλος I, add 'χιτῶνα .. ἀτράχηλον, i.e. without opening for the neck, Apollod.Epit. 6.23'
ἀτρεκής II 1, line 3, after 'ad loc.)' insert 'cf. AP 5.267 (Agath.)'
ἀτρέμα 3, add 'Lyc.fr. 2.8 S.'
*ἀτριᾱκοστολόγητος, ον, not subject to a tax of one-thirtieth, Illion 52.20 (ii BC); see Sokolowski 2.9.20.
†ἄτριον (A), v. ‡ἤτριον.
*ἄτριον (B), τό, Lat. atrium, Ἴσιδος ἐν ἀτρίῳ APF 2.439.42 (ii AD), IGRom. 1.1048, 1175 (both Egypt, ii AD) the Atrium Magnum at Alexandria, PFouad 21.4, SB 8247; written ἄτρειον IStraton. 15.7, 664.4 (both Caria, i AD); in a private house, POxy. 2406 (ii AD).
*ἀτρίχια, ἡ, hairlessness, Cyran. 35 (1.16.12 K.).
ἄτρομος, add 'of personified Φύσις, Orph.H. 10.26'
ἀτροπάμπαις, for 'dub. sens.' in' read '= °παῖς II'; add 'cf. πράτοπάμπαις'; for parenth. note at end read '(ἀτρο- perh. w. syncope fr. ἄτερο(ς))'
ἄτροπος 3, delete the section transferring quot. to section 2.
ἀτρύγετος, line 1, for 'later η, .. 900' read 'also η, ον Stesich. 32 i 4 P., IG 2².3575.9 (ii AD)'; for 'unharvested, barren' read 'sens. dub., acc. to sch. unharvested, barren; ἀτρυγέτοιο· ἀκάρπου ἀβύσσου καὶ ἀπείρου Hsch.'; line 4, after 'h.Cer. 67, 457' insert 'αἰθέρος ἀτρυγέτας Stesich. l.c.'; ἀ. χθών Nonn.D. 6.101'
*ἀτρυγόνιστος, ον, dub. sens., πρόβατα SB 8003.20 (iv AD).
ἀτρύπητος, add 'ἀ. ψῆφος Poll. 8.123, Phot. s.v. τετρυπημένη ψῆφος (v. τρυπάω 1); rest. in Lindos 410 iii 6 (i AD)'
ἄτρυτος 1, add 'ἄτρυτος ἐν πόνοις Trag.adesp. 163 K.-S.'
*ἄτρωγλος, ον, lacking an aperture, not perforated, ἀ. καὶ ἄτρητοι Ptol.Tetr. 150.
ἄτρωτος II, add 'ὅπλον AP 12.115, cf. ὅπλα .. ἄρρηκτα καὶ ἄτρωτα δι' ἅπερ σέ φασιν ἄτρωτον εἶναι Antisth.Od. 7'; ho metaph. exx. add 'μοῖρ' ὦ λιταῖς ἄτρωτε δυσθνῄων βροτῶν Moschio Trag. 2.2' III, add 'of institutions, Cod.Just. 1.2.17 pr., Just.Nov. 120.11.epilogos'
ἀτταγεινός, for 'Dorio ap.Ath. 7.322c' read 'Dorio ap.Ath. 7.322e'
*ἀτταγήνη, ἡ, fem. of ἀτταγήν (cf. Lat. attagena), Edict.Diocl. 4.30.
*ἀττακίτης (sp. -ήτης), ον, ὁ, kind of cake, PGoodsp.Cair. 30.7.21, al. (ii AD); cf. perh. ἀττανίτης.
*Ἀτταλικός, ή, όν, on the Attalic standard, δραχμή Inscr.Perg. 260.13. 2 name of medicinal compound, ἄλλη ἐκ τῶν Μαντίου δυνάμεων Ἀτταλική Asclep.ap.Gal. 13.162.16.
ἀττάμιος, v. ‡ἀζήμιος.

ἀττᾶνίτης, add 'PStrasb. 339.7, 12 (c. AD 200)'
ἀττάραγος, after 'Ath. 14.646c' insert 'Sch.Hippon.118 D12 W.'
Ἀττῐκίζω I, add '[D.] 58.37, Isoc. 8.108'
Ἀττῐκός I, add '2 of weight-standard, ταλάντων Ἀ. SEG 33.861 (Caria, ii BC); fem. subst., Attic drachma, Ἀττικὰς φ' IMylasa 455, Ἀτ(τ)ικὰς ιβ' TAM 4(1).250, etc.'
ἀτύλωτος I, delete the section.
ἀτύμβευτος, add 'EA 19.58 (Sinope, Rom.imp.)'
ἄτυπος I, add 'balbus et blaesus et atypus isque qui tardius loquitur, Ulp.Dig. 21.1.10; humorous transl. for pr. n. Balbus, Cic.Att. 12.3.2 (cj.)'
ἀτῠφία, add 'M.Ant. 11.6'
ἄτῡφος, after 'Timo 9.1' insert '(Suppl.Hell. 783), Cic.Att. 6.9.2, Ael. VH 2.20, 4.9'
ἀτυχέω 1, add 'w. internal acc., τῶν ἀτυχησάντων τὴν τῶν Μανιχαίων ἀσεβῆ πλάνην Cod.Just. 1.5.15'   3, line 3, after 'Eup. 114' insert 'οὐθενὸς ἀ. τοῦ δήμου τῶν δικαίων IG 2².275.5, 360.41'
*ἄτω, ἰάπτεται· ἄτει, βλάπτει Sch.Nic.Al. 251c Ge.
ἀτῶμαι, for 'v. ἀτάω' read 'v. ἀτάομαι'
αὖ IV, delete the section.
*ἀυάδεες, ἀυάδην, v. °ἀηδής.
*αὑαντήρ, ῆρος, ὁ, he that parches, epith. of Zeus, IG 2².2606 (iv BC) (prob.).
αὑαρά, add 'cf. ἄλαρα'
αὐγάζω I, line 3, before 'AP 9.221' insert 'Call.fr. 85.15 Pf.'   II 1, for 'sun' read 'heavenly bodies' and add '(ἀστέρες) αὐγάζοντες ἀεὶ νυκτὸς ζοφεειδέα πέπλον Orph.H. 7.10, AP 5.123 (Phld.)'
αὐγή 1, add 'also light (rays) of the moon, h.Hom. 32.12, Plu. 2.658b, f.'   2, add 'b ἐννέα αὐγὰς ἠελίοιο nine days, Nic.Th. 275.'   5, after 'eyes' insert 'h.Merc. 361'
*Αὐγουστάλης, ὁ, Lat. Augustalis (subst.), SEG 29.614 (pl., Maced., Rom.imp.).
*αὐγουσταλιανός, ή, όν, fr. Lat. augustalis, τάξις POxy. 1882.4, 8 (vi AD), PWash.Univ. 6.4 (vi/vii AD), Just.Edict. 13.2.
†Αὐγουστάλιος (s.v. Αὔγουστος), ὁ, priest in the cult of Augustus, MAMA 1.169, 216, 283 (Laodicea Combusta), Just.Edict. 13.1 pr.   2 title given to the prefect of Egypt, Lyd.Mag. 2.3 (p. 57.24 W.), PStrassb. 255.9; of Thebais, SB 7439.6 (vi AD).
Αὔγουστος I, add 'ἀεισέβαστος Ἄγουστος (sic) SEG 9.356.2 (Cyrenaica, AD 501); fem., as title of Empress, Εὐδοκίας Αὐγούστης SEG 32.1502 (Palestine, c.AD 455); as title of a legion, SEG 31.626 (Αὔγοστ-, Maced., c.AD 200)'   II, for this section read 'the month August, Plu.Num. 19; καλανδῶν Αὐ. Cat.Cod.Astr. 2.145, al., μη(νὶ) Ἀγούστου TAM 4(1).356, written Ἄκοστος SB 9529.12 (vi/vii AD)'
αὐδάζομαι 2, after 'name' insert 'w. double acc., Nic.Th. 464'; after 'Lyc. 892' and after 'Id. 360' insert '(v.l.)'
αὐδάω II 2, line 3, for 'A.Th. 1048' read 'A.Th. 1042, 1043'
αὐδή 1, after 'Il. 1.249' insert 'θεῷ ἐναλίγκιος αὐδήν 19.250, Od. 1.371, 9.4, Sapph. 1.6 L.-P. (αὔδα)'
αὐδήεις, add '(contr. αὐδῆς Hdn.Gr. 2.618)'
*αὐδῐτώριον, τό, Lat. auditorium, hall of justice, IEphes. 3009 (sp. αὐδειτ-), Just.Nov. 50 pr.
Αὐδυναῖος, after 'IG 12(3).254' insert '(Anaphe, iii BC) and add 'Αὐδον- SB 7341.2 (iii AD) and note on p. 53; Αὐδν- SEG 29.1161, Αὐδνίος SEG 31.989 (Lydia, ii AD), Αὐδαναῖος SEG 7.1.15 (Susa, i AD), Αἰδοναῖος SEG 25.712 (Thessalonica, ii AD)'
†αὐεούλλαι or ἄυελλα, v. °ἄελλα.
αὐερύω II, for this section read 'suck up, absorb, αὐ. τὸ φίλημα AP 5.285.5 (Agath.); of leeches, Opp.H. 2.603'
*Αὔξησία, v. °Αὐξησία.
αὐθάδεια, for 'wilfulness, stubbornness' read 'concern for one's own interests, self-interest, self-centredness' and add 'PCair.Isidor. 74.11 (IV AD)'
αὐθάδης 1, for 'self-willed, stubborn' read 'acting to please oneself, self-regarding' and add 'E.El. 1117'   2, add 'not concerned to please others, αὐ. κάλλος, of the style of Thucydides, D.H. Comp. 22'   3, after '-έστερον' insert 'Th. 8.84.2'
αὐθᾱδιάζομαι, delete 'J.BJ 5.3.4'
αὐθαίμων, add 'AP 6.14 (Antip.Sid.)'
αὐθαίρετος, at end delete 'independently, Luc.Anach. 34'
αὖθε, for 'αὐθέ περ' read 'ἀλ(λ) αὖθε πέρ γᾶς τᾶσδε ..'; after 'Cierium' insert 'v BC'
αὐθέκαστος 1, line 4, for 'Ph. 2.51' read 'Ph. 2.519'   3, at end before 'Phld.Vit. p. 30 J.' insert 'Aristo ap.'
*αὐθεντεύω, = αὐθεντέω, Cat.Cod.Astr. 8(3).196.12.
αὐθέντης 1 and 2, invert order of these sections; for 'murderer' read 'perpetrator of a murder or death' and delete 'suicide'
αὐθεντία 1, add 'b as honorary appellation (of praetorian prefect), Just.Nov. 111 epilogus.'
αὐθεντικός 2, line 4, after 'Ptol.Tetr. 182' insert 'σπουδῇ IMylasa 134.2, 6 (ii BC)'   add '3 subst., αὐ. τό, original copy, PFam.Teb. 31.13 (ii AD), v. °ἔκβασος.'

*αὐθεντόπωλος, ὁ, son (slave) of the master, Sch.Aristid. p. 54.10 D.
*αὐθῆμαρ, v. αὐτῆμαρ.
αὐθημερῐνός, add '5 τὸ αὐ. τοῦ ἡλίου solar longitude for given date, Cod.Vat.Gr. 1058.265ʳ.10.'
αὐθήμερος II, add 'also αὐταμερόν Herzog Heilige Gesetze von Kos 8B21, Tit.Cam. 218, Schwyzer 633.9 (Aeol., Eresos ii/i BC)'
αὖθι 3, after 'Call.Dian. 241' insert 'fr. 197.49 Pf.'; add 'see also αὖθε'
αὖθις, lines 2/3, for 'after '(id.OC) 1438' insert 'E.fr. 35, codd. in Supp. 679, al.; Ar.Av. 1326'; before 'Adv.' insert 'Cret. αὖτιν Leg.Gort. 4.3'   III, after 'Pl.Ap. 24b' insert 'sim. ἐπ' αὖθις ἀγωνοθετοῦντος καὶ προκαθεζομένου τοῦ ἀνδρός μου, ἐπ' αὖθις δὲ τῶν ἐξ ἐμοῦ γεννηθησομένων τέκνων Modest.Dig. 50.12.10'
*αὐθοπτικός, ή, όν, v. αὐτοπτικός II.
*αὐθόριστος, ον, self-defining, David Proll. 14.23.
αὐθυπόστᾰτος, adv., add 'Procl.Inst. 41, 86 D.'
αὖλαξ, line 2, for '(q.v.)' read 'A.Th. 593, S.OT 1211, E.Ph. 18, Emp.B 100.3 D.-K.'
*Αὐλᾰρίοκος, ὁ, epith. of Apollo, IGBulg. 1859, 1860 (both ii AD).
*Αὐλαρκηνός, ὁ, epith. of Apollo, IGBulg. 802, 841, SEG 37.652 (Bosphorus, c.AD 100), also Αὐλαρχ- IGBulg. 801.
αὐλέω I 1, for 'flute' read 'aulos'; line 4, after 'Plu.Alc. 32' insert 'w. ellipsis of μέλος (or sim.), αὔλει, Παρθενί, Πανός Men.Dysc. 432'   II, for 'generally, play' read 'play any wind instrument'
αὐλή I - III, for these sections read 'enclosed courtyard (in a farm, palace, private house, etc.), Il. 4.433, 6.247, 11.774, Od. 14.5, Hdt. 3.77.2, Ar.V. 131, Pl.Prt. 311a, SIG 1044.17 (Halic., iv/iii BC), Theoc. 25.99, 27.36; in a temple, ἱεροῦ IG 2².1299.28 (Eleusis, iii BC), Lxx Ps. 83(84).3.   2 w. ref. to the whole complex of buildings enclosing the courtyard, court, hall or sim., Ζηνὸς αὐ. Od. 4.74; τὴν Διὸς αὐλήν A.Pr. 122 (lyr.); ἀγρονόμοις αὐλαῖς S.Ant. 786 (lyr.); χαλκοδέτοις αὐλαῖς ib. 946; id.Ph. 153; used as equiv. of Lat. villa, D.H. 6.50.'
*αὐλήεις, εσσα, εν, living in an αὐλή, Arist.fr. 171, cj. for αὐδήεσσα in Od. 5.334.
αὔλημα, for 'piece of music for the flute' read 'performance on the aulos'
αὔλησις, for 'flute-playing' read 'aulos-playing'
†αὐλητήριον· τόπος παρὰ Ταρεντίνοις Hsch.; transf., of a noisy place, σύμμικτον ὥστε γλεύκος αὐλητήριον Trag.adesp. 420 K.-S. (s.v.l.).
αὐλητής, for 'flute-player' read 'piper'
αὐλητικός, passim, for 'flute' read 'aulos'
αὐλήτρια, after 'αὐλητρίς' insert 'Hesperia 37.368 (Athens, iv BC)' and add 'Arc. 95.15'
αὐλητρίς, for 'flute-girl' read '(female) piper' and for 'Simon. 178' read 'AP 5.159.1 ([Simon.])'
αὐλιάδες, for 'nymphs protecting cattle-folds' read 'prob. cave-dwelling nymphs'
*Αὐλιδεία, ἡ, title of Artemis, SEG 25.542 (Aulis, Rom.imp.).
αὐλίζομαι, line 8, for 'Eup. 322 (= 347 K.-A.)' read 'Eup. 347 K.-A., spend the night in the open, SEG 23.305 (Delphi, ii BC), act. form αὐλιζόντων'
αὔλιον, after 'τό' insert '(parox. in pap. at Call.fr. 181.6 Pf.)'; delete 'country house, cottage' and transfer 'h.Merc.' to follow 'stable, etc.'; for 'prov. .. Cratin. 32' read 'βοῦς ἐν αὐλίῳ, prov., of what is discarded as useless, Cratin. 32 K.-A., Longus 4.18.3'   add 'III any dwelling, AP 9.424 (Duris).'
αὔλιος, line 2, for 'Call.fr. 539' read 'Call.fr. 177.6 Pf.'
†αὖλις, ιος, ἡ, place for passing the night in, bivouac, ἐγγὺς .. νηῶν .. αὖλιν ἔθεντο Τρῶες Il. 9.232; meton., εἰ .. ἄγχι .. νὺξ αὖλιν ἄγει Nic.Th. 58.   2 accommodation for cattle, etc., h.Merc. 71, h.Ven. 168, E.Cyc. 363, Call.Dian. 87, Theoc. 25.18, AP 6.221 (Leon.Tarent.).   b resting-place for birds, wild animals, Od. 22.470, Theoc. 25.169, Arat. 1027.
αὐλο-, in compds. of αὐλός for 'flute' read 'aulos'
αὐλοβόας, 'IG 3.82' read 'IG 2².3118.6 (Attica, ii AD)'
αὐλοποιός, add 'SEG 18.36.A433 (Attica, iv BC)'
αὐλός 1, for 'pipe, flute, clarionet' read 'oboe-type musical instrument, reed-pipe, aulos'; lines 8/9, for 'pl., αὐλοί .. (Aegina)' read 'b pipe of a compound instrument: of organ, Simp.in Ph. 681.7; of syrinx, IG 4.53 (Aegina, Rom.imp.).   c Τυρρηνὸς αὐλός, = ὕδραυλις, Poll. 4.70.'   3, add 'pl., = Lat. tibiae, cannon-bones of a horse, shanks, Opp.C. 1.189'   at end add 'prob. Myc. au-ro, part of chariot (cf. section I 2)'
ἀϋότης, add '(cj. ταὐτότης)'
*αὐλοφύλαξ, ακος, ὁ, app. house-watchman, PCair.Zen 292.58 (iii BC).
*αὐλύδριον, τό, dim. of αὐλή, small house with courtyard, PSI 915.4 (i AD, = ὑπόμν- pap.), PMich.II 123ʳ.19.33, 21.36, al. (i AD), PSoterichos 25.20 (ii AD), PAbinn. 63.5 (iv AD).
αὐλών 3, add 'Peripl.M.Rubr. 25'
†αὐλωνίζω, -ίζουσα· ἐν αὐλῶσιν (ἐναύλως cod.) διάγουσα Hsch.
αὐξάνω, line 13, after 'Pl.R. 497a' insert 'later αὐξέω, ηὔξουν D.C.fr. 89.3 (ηὔξουν cod.), αὐξοῦνται Plu. 2.724f, αὐξούμενον dub. in

Αὐξησία   SUPPLEMENT   αὐτομολία

GVI 1903.3 (Megara, iii AD), αὐξῶν Ps.-Aristid.*Rh.* 1.505 S., cf. Procop.Gaz.*Ep.* 86' **I 1**, for '*increase* .. Pi.*fr.* 153, etc.' read '*cause to grow*, δενδρέων δὲ νομὸν Διώνυσος .. αὐξάνοι Pi.*fr.* 153 S.-M.; *increase*' **5**, after 'name of a *fallacy*' insert '(= σωρίτης)' **II 2**, line 2, delete '*grow* up'

Αὐξησία, add 'also Αὐζησία, *IG* 4.1588.28 (Aegina, v BC)'

αὔξησις **1**, add 'in logic, ὁ περὶ αὐξήσεως λόγος (= σωρίτης) Plu. 2.1083a'

*αὐξήτειρα, ή, *she who gives growth* or *increase*, of Artemis, oracle in *ZPE* 92.269 (Ephesus, ii AD).

αὐξητέον, before 'Men.' insert 'Arist.*Rh.* 1376ᵇ7'

αὔξι, add '*BCH* suppl. 8.291 (iv/v AD); also αὔξε *SEG* 34.1306 (Pamphylia, iii AD); αὔξησι *SEG* 34.1418 (Cyprus, ii/iii AD)'

αὔξις, add '*PMich*.XI 617.15 (AD 145/6)'

αὔξις, before 'Nic.*Al.* 469' insert 'but dub. sens. in'

*αὐξιφᾰής, ές, of the moon, *increasing its light, waxing*, Heph.Astr. 2.36.1, Man. 5.174, al., *Cat.Cod.Astr.* 8(4).217.19.

αὐξίφωνος, delete the entry.

*αὐξίφως, φωτος, of the moon, *increasing in light, waxing*, Heph.Astr. 2.33.9, 2.36.6, al.

αὐξίφωτος, add 'Σελήνη Heph.Astr. 2.35.4, 2.36.15'

αὐξομείωσις, add '**III** *gradation*, ἀ. τῶν ἀξιωμάτων Ptol.*Tetr.* 176.'

*αὐξομειωτικός, ή, όν, *varying in period*, *Cat.Cod.Astr.* 7.194.18.

*αὔξων, v. °ἄξων.

αὖος, line 2, for 'Philostr.*VS* 1.21.1' read 'Philostr.*VS* 1.20.2'

ἀϋπνία, after 'Pl.*Lg.* 807e' insert '*Trag.adesp.* 664.26 K.-S.'

ἄϋπνος, line 3, after 'S.*Aj.* 880' insert '[φρου]ρεῖν ἀϋπνοις φυλακαῖσιν orac. in *SEG* 30.175.14 (Athens, iv BC)'

αὔρα **1**, delete 'esp. a cool breeze .. morning' **2**, for 'metaph. .. Ar.*Av.* 1717' read '*exhalation, effluvium*, θυμιαμάτων αὖραι Ar.*Av.* 1717'; after 'Dionys.Com. 2.40' read 'transf. *influence*, αὔρη φιλοτησίη Opp.*H.* 4.114, δαίμονος αὔ. *AP* 6.220.9 (Diosc.)' **3**, add 'μικρά τις ἀπελείπετο αὔρα βοηθείας Ph. 2.559' **5**, for 'Gal. 8.94' read 'Gal. 8.194.17'

*αὐράριος, ὁ, Lat. *aurarius, goldsmith*, *MAMA* 1.214, 281, al. (Laodicea Combusta), 3.254, 348b (Corycus); pl., guild in Miletus w. reserved area in the theatre, τόπος αὐραρίων Βενέτω(ν) *BE* 1977.82 (see also *SEG* 36.1053 note); but some connect w. *aurarii sunt laudatores vel fauctores*, i.e. *supporters*, *Gloss.* 5.616.1, Serv.ad *Aen.* 6.816, 204; *qui favoribus splendidos, hoc est claros, faciunt*, Priscian.*Inst.* 3.509.33.

αὔρηκτος, for 'Hdn.Gr. 2.171' read 'Hdn.Gr. 2.271'

αὔριον **III**, line 3, after 'S.*OC* 567' insert 'also ἡ ἐς αὔριον Ps.-Hdt.*Vit.Hom.* 433'

*αὐροχάλκειος, ον, = ὀρειχάλκινος, θύραι *CISem.* 2.3914 (Palmyra, ii AD), also -κεος, λάμναι Hippiatr.*Paris.* 346 (II 56 O.-H.).

*αὐρόχαλκος, ον, *of gilded bronze*, Ἔρωτες *IEphes.* 3015.3 (iii AD), *Edict.Diocl.* 15.63a.

αὐσαυτοῦ, add '[Βακ]χυλὶς Δαματρίου αὐσαυτὰν καὶ τὸ[ν υἱὸν (?)] Εὔφορβον Νίκωνος θεοῖς *SEG* 23.224 (Messene, Rom.imp.); also cf. αὐτοσαυτόν'

αὐσόν, add 'prob. for αὐον; but cf. αὐσὸς δέ ἐστιν ἡ κάμινος παρὰ τὸ ἀτμόν τινα ἀφιέναι David *Proll.* 41.15'

Αὐσονία, line 3, after 'ib. 363, al.' insert 'Αὐσώνιοι, Hsch.' and after 'also' insert 'Αὔσων, ό, *AP* 11.24'; line 4, after 'aborigines):' insert 'Αὐσονιῆες, D.P. 333, al.'

*Αὐσονίδης, ου, ὁ, *Italian*, Αὐσονίδην .. ἀγακλυτὸν Ἀντωνῖνον *IG* 2².3411.7 (ii AD).

*Αὐσονικός, ή, όν, *Italian*, Mosch. 3.94.

*Αὐσονίτης, ου, ὁ, masc. ethnic, *Italian*, Lyc. 593; fem. adj. -ῖτις, -ιδος, id. 44, 702, 1355.

αὔτανδρος, line 3, after 'Sosyl. p. 31 B.' insert 'Call.*fr.* 7.33 Pf.'

αὐτάρ, after '*IG* 1².1012' insert 'vi BC'; for '*Inscr.Cypr.* 57 H.' read '*a-u-ta-ra*, *ICS* 235, 242, al.'

αὐταρχία, ή, *absolute rule*, D.C. 45.1.3, 53.4.3, 54.12.2.

αὐταυτοῦ, penultimate line, after 'Sophr. 19 (-τᾶς Pors.)' insert 'cf. ‡αὐσαυτοῦ, αὐτοσαυτόν'; delete 'also αὐτοῦτα' to end.

αὖτε **II 2**, after '(A.)*Ag.* 553' insert 'Democr.*fr.* 172 D.-K.'; antepenultimate line, delete 'not in prose'

αὐτεῖ, line 1, after 'αὐτοῦ' insert 'Alcm. 1.79 P.'; line 2, after 'αὐτί' insert 'Corinn. 39 *fr.* 7.5 P.'

*αὐτενεργητικός or αὐτοεν-, ή, όν, = αὐτενέργητος, Choerob. in *Theod.* 2.19.25.

αὐτενίαυτος, add 'w. advl. sense, *in the same year*, ἐπ[ὶ δ' ἔσ]πετο αὐτενίαυτο[ς] καὶ πόσις *SEG* 30.578 (Maced.)'

αὐτεξουσιότης, add '**2** *freedom from patria potestas*, Just.*Nov.* 81.1 pr.'

*αὐτεφόδιος, ον, *paying one's own travelling expenses*, *SEG* 33.861.20 (Euromos, ii BC), *IKeramos* 14.20 (i AD).

ἀϋτή, at end after 'Corc.' insert 'early vi BC'

αὐθῆμαρ, add 'Call.*Del.* 46, Coluth. 199, also αὐθῆμαρ *CEG* 815.2 (Epid., iv BC), 894.13 (Delphi, iv BC)'

†αὐτημερόν, v. αὐθημερόν.

αὐτίκα, line 1, after 'Adv.' insert 'Lesb. αὔτικα, Sapph. 31.10, 44.13, 30.2 L.-P.'; line 4, for 'Sapph.*Supp.* 20a.13' read 'Sapph. 44.13 L.-P.' **I 2**, add 'in phr. w. prep., ἐπωλήθη πρὸς αὐτίκα *SEG* 37.917.B3 (Erythrae, v/iv BC)' **3**, add '**b** perh., *at length*, A.R. 2.946, 3.23, 521, 4.1547.' **II**, line 3, for 'Pl.*Prt.* 395e' read 'Pl.*Prt.* 359e'

†αὖτις, αὖτιν, v. ‡αὖθις.

αὐτίτης **II**, after '*home-made*' insert '(or perh. for αὐτοέτης, cf. αὐτοετίτης)'

ἀϋτμή, add '[disyllabic in Hes.*Th.* 862, s.v.l.]'

*αὐτοανεξοδίαστος, ον, *absolutely unsaleable*, *INikaia* 127 (Rom.imp.).

αὐτογένεθλος, for 'sq.' read 'αὐτογενής' and for '*Orac.Chald.* 32' read 'πατρικὸς νόος *Orac.Chald.* 39 P.'

*αὐτογεννήτωρ, ορος, ὁ, *self-originator*, δεῦρό μοι, ὁ αὐτογεννήτωρ θεέ *Suppl.Mag.* 65.31 (iii AD).

αὐτογράφος, for 'τὸ αὐ. *one's own writing*' read 'ἔλεγχος'; add 'τὰ .. τῶν παλαιῶν αὐτόγραφα ψηφίσματα *original drafts*, Posidon. 253.152 E.-K.'

αὐτοδάκης, after 'Hsch.' insert '(= *Suppl.Hell.* 1072)'

*αὐτοδιαφορά, ή, *absolute differentia*, Simp. in *Cat.* 276.25, 30.

αὐτοδίδακτος, line 2, after '(lyr.)' insert 'ὅν Πιερὶς αὐτοδίδακτον θῆκ' *SEG* 37.1175.10 (Cremna, Pisidia, ii AD)'

αὐτοδόξαστον, after '*abstract*' insert 'Alex.Aphr. in *Top.* 572.17'

αὐτοέν, after 'τό' insert 'gen. αὐτοενός Procl.*Inst.* 4 D.'

*αὐτοενεργητικός, v. °αὐτοενεργητικός.

*αὐτοένωσις, εως, ή, *absolute union*, Plot. 6.1.26.

αὐτοεξούσιος, add 'see also αὐτεξούσιος'

*αὐτοεξουσιότης, v. ‡αὐτεξουσιότης.

αὐτοετής, after 'J.*AJ* 3.9.3' insert 'i.e. *holding an office in the same year* (as another office), προφήτης Didyma 270, 278.3, 283, etc.; cf. αὐτοετεις ἱερούς .. στεφάνους ib. 229 II 8 (metr., all Rom.imp.)'

*αὐτοζώη, ή, *absolute life*, Plot. 3.8.8.

αὐτόζυγος, add '*GVAK* 25.6'

αὐτοκάβδαλος **II**, for '*buffoons, improvisers*' read '*reciters of improvised verses*' and transfer 'Eup. 200 (for which read 192.195 K.-A.)' to section I

αὐτοκασιγνήτη, add 'λελίητο νέεσθαι αὐτοκασιγνήτηνδε A.R. 3.647, *Suppl.Hell.* 1168.4'

αὐτοκάσιγνητος, add '*IHadr.* 168.3 (ii/iii AD)'

αὐτοκέλευθος, at end for '[Nonn.*D.*] 21.167' read '[Nonn.*D.*] 21.169'

*αὐτοκῐβώτιον, *the ideal box*, Alex.Aphr. in *Metaph.* 553.23.

αὐτοκινησία, after '= sq.' insert 'Alex.Aphr. in *Top.* 297.23'

αὐτοκίνητος, after '*PMasp.* 122.3 (vi AD)' insert '*Cod.Just.* 1.3.43.4, al.'

αὐτοκράτωρ **I 3**, add '**b** the seventh month in Cyprus (in honour of Augustus), Hemerolog.*Flor.* p. 72 (p. 8 K.).'

*αὐτόκυβος, ὁ, *the ideal cube*, Alex.Aphr. in *Metaph.* 816.30.

αὐτόκυκλος, after '*form of circle*' insert 'Alex.Aphr. in *Metaph.* 816.29'

*αὐτολείπω, *leave behind*, τέσσαρας αὐτολιπὼν υἱούς *BCH* 25.21 (Bithynia, Rom.imp.).

*αὐτόλευκος, ον, τὸ αὐ, *ideal whiteness*, Alex.Aphr. in *Metaph.* 771.1.

†αὐτοληκύθος, ὁ, *one who carries his own oil-flask, a parasite* or sim., Antiph. 17 K.-A., Men.*fr.* 91, 182 K.-Th., Luc.*Lex.* 10, Plu. 2.50c; adopted as a sobriquet by a group of young men about town, D. 54.14.

*αὐτολόγος, ὁ, *the very word*, Cels.ap.Orig.*Cels.* 2.31.

αὐτολόχευτος, add 'Nonn.*D.* 37.68, *PMag.* 4.458'

*αὐτόλυρος, ον, ποιητὴς αὐ. *poet who accompanies himself upon the lyre*, *PBremen* 59.14 (ii AD).

αὐτομάθεια, delete 'also -μαθία'

*αὐτόμαργος, ον, dub. sens., prob. in A.*fr.* 451i.5 R.

*αὐτοματάρειον, τό, *vessel for spontaneous digestion* of minerals, dub. in Olymp.*Alch.* 91 (v.l. αὐτῷ τῷ βοταρίῳ, i.e. βωταρίῳ).

*αὐτοματάριος, ὁ, *maker of automata*, *POxy.* 2873.21 (-ις, iii AD).

*Αὐτομάτειος, ον, of °Αὐτομάτη, -είου sc. ὕδατος *PMilan.* 17.18.

*Αὐτομάτη, ή, name of a spring at Argos, Call.*fr.* 65.1, 66.8 Pf., *PMilan.* 17.14.

αὐτοματίζω, add '**5** *prophesy spontaneously*, ὁ Ἀπόλλων αὐτομάτιξεν Βάττῳ *SEG* 9.3.24 (Cyrene, iv BC), Aristid. *Or.* 28(49).103, Ath. 1.31b; pass., Sch.E.*Andr.* 445.'

αὐτοματισμός, for 'Phleg.*Mir.* 1' read 'Phleg. 36.1.4 J., D.H.*Comp.* 25'

αὐτόματος **II**, line 7, after 'X.*An.* 1.3.13' insert 'παρὰ τοῦ αὐ. Aen.Tact. 6.2' **III**, adv. -τως, add 'αὐ. ἤνθησα καὶ ἤκμασα *SEG* 33.1406.11 (Termessus, ii/iii AD)'

*αὐτομεγα, τό, *absolute size*, πρὸς αὐ. (gen. indecl.) Plot. 3.6.17.

*αὐτομενις, unexpld. item in temple inventory, αὐτομενις ξυλ(ιν-) περικεχ(ρυσωμεν-) α' *POxy.* 3473.13 (ii AD), *BGU* 387 ii 4 (ii AD).

αὐτομόλησις, add 'Arr.*fr.* 10 J.'

αὐτομολία, add '**II** pl., *suckers* or *shoots* of trees, Poll. 7.146; cf. μολεύω.'

αὐτόμολος | SUPPLEMENT | ἀφαμαρτέω

αὐτόμολος 1, add 'volunteer, in unpublished inscr. listing αὐτόμολοι, app. a Maced. garrison, SEG 37.280 (Argos, iv BC)'

αὐτονοέω, shd. precede αὐτονομέομαι.

*αὐτονομαστί, adv., by its very name, Gal. 17(2).26.

αὐτόξυλος, for 'of one piece of wood' read 'of wood in its natural state, i.e. rough, unpolished, etc.'

*αὐτοοικία, ἡ, ideal house, Alex.Aphr. in Metaph. 553.23.

*αὐτοουρανός, ὁ, the very heaven, Alex.Aphr. in Metaph. 198.15.

*αὐτοπατήρ, ὁ, the very father, Alex.Aphr. in Metaph. 126.13.

*αὐτόπερἀς, ἀτος, τό, abstract limit, Procl. in Prm. p. 875, Anon. in Cat. 67.32, Simp. in Cat. 337.31.

*αὐτόπλατος, τό, ideal plane, Alex.Aphr. in Metaph. 127.10.

*αὐτοποιέω, create by oneself, Numen.fr. 16.11 P.

αὐτοποιός, after 'S.OC 698 (lyr.)' insert '(αὐτόποιος codd., perh. fr. ποία = πόα)'; delete 'made by one's own hand' to end.   add 'II αὐτοποιόν, τό, abstract quality, Alex.Aphr. in Metaph. 563.1.'

*αὐτοποσότης, ητος, ἡ, quantity in itself, Simp. in Cat. 130.12.

αὐτοπραγία, add 'Iamb.Protr. 21 ιδ' (115.3 P.)'

*αὐτοπραξία, ἡ, privilege of collecting one's own taxes, IG 9²(1).137.20 (Calydon, ii BC).

αὐτόπρεμνος, add 'Eup. 260.25 K.-A. (s.v.l., pap. αὐτόπρυμνος)'

αὐτοπροαίρετος, add 'SEG 36.1051.6 (Miletus, ii AD)'

*αὐτοπυραμίς, ίδος, ἡ, the ideal pyramid (v. πυραμίς I 2), Alex.Aphr. in Metaph. 816.29.

αὐτοπυρτης, for 'Luc.Pisc. 44' read 'Luc.Pisc. 45'

*αὐτόρρευστος, ον, = αὐτόρρυτος, Ps.-Democr.ap.Moses Alch. 313.9; Anon.Alch. 20.2.

αὐτόρυτος, add 'μέλι Lyr.Alex.adesp. 37.10'

αὐτός, line 2, after 'Leg.Gort. 3.4, al.' insert 'ἀτός PHal. 1.130 (iii BC), PTeb. 812.9 (ii BC), SEG 29.771 (Thasos, ii BC)'   I 1, for 'one's true self .. Il. 1.4' read 'one's real (bodily) self, opp. an εἴδωλον or sim., Il. 1.4, Od. 11.602'   3, add 'οὐκ αὐτὸς ὁ Πλοῦτος not only .., Theoc. 10.19'   4, add 'absol., αὐτὸ .. ὅ ἐστι καλόν Pl.Smp. 211c, Cra. 432d, Arist.Met. 991ᵃ5, al., Iambl.Protr. 8'   8, add 'inclusive, of a date, ἕως Μεχεὶρ καὶ αὐτοῦ Wilcken Chr. 157.16 (Hermopolis, iii AD), cf. POxy. 270.42 (i AD), etc.'   10, add 'f combined w. ἐκεῖνος for greater precision or emphasis (v. R. Janko CQ 35.20ff.), Od. 24.321, Hdt. 2.115.6, Ar.Ec. 328, Pl.Smp. 192b, etc.'   11, add 'A.R. 1.199'   IV 1, add 'Ath. 6.270b'   2, delete the section.   3, add 'X.An. 1.9.21, Pl.Smp. 204a, Aen.Tact. 11.10; also pl. αὐτὰ ταῦτα Pl.Prt. 310e'   V 2, after 'Αὐτοθαῖς' insert 'cf. Herod. 6.59'   at end add 'Myc. au-to-jo (gen.); cf. au-to-te-qa-jo pers. n. = *Αὐτοθηβαῖος'

αὐτοσαυτόν, add 'cf. αὐσαυτοῦ, αὐταυτοῦ'

αὐτόσε, add '2 without spatial sense, αὐ. προστίθημι Pl.Men. 73d, Metag.fr. 6.4.'

*αὐτόσοφος, ον, endowed with innate wisdom, Rh. 3.530, Tz.H. 8.437; neut. subst., innate cleverness, ἥν αὐτόσοφον ὁ τέκτων PTurner 8 (ii AD).

*αὐτοστερεόν, τό, ideal solidity, Alex.Aphr. in Metaph. 127.10, 128.3.

*αὐτοσυμφυής, ές, naturally united by itself, Porph.Antr. 5 (cod. αὐτοφυής).

αὐτοσύστατος, add 'opp. ἑτεροσύστατος, Choerob. in Theod. 2.411.22 H.'

*αὐτοσφαῖρα, ἡ, ideal globe, Alex.Aphr. in Metaph. 636.16.

αὐτοσχεδιαστής, add 'αὐ. (pl.) πολέμων perh. engaging at will in wars, cj. in Vett.Val. 78.4 (75.1 P.)'

αὐτοσχέδιος II, add 'neut. subst. τὰ αὐ., extemporaneous speeches, Plu. 2.842c'

αὐτοσχεδόν 1, add 'Hes.Sc. 190'

*αὐτόσχημα, ατος, τό, ideal shape, Alex.Aphr.in Metaph. 742.25, Procl. in Ti. 2.136.14.

αὐτοτελής I 1, add 'c of a body of citizens, sovereign, independent, J.AJ 14.7.2.'   5, delete 'sufficing for oneself: also'

*αὐτοτετράγωνον, τό, ideal square, Alex.Aphr. in Metaph. 823.20.

*αὐτότευκτος, ον, self-made, natural, rest. in A.fr. 73b.2 R.

αὐτότης, delete the entry (v. Mnemosyne (4th ser.) 37.89-93, read ταὐτότης).

αὐτουργός I 1, for 'self-working .. 18.2; αὐ.' read 'self-reliant, unassisted, αὐτὸς αὐτουργῷ χερί S.Ant. 52'   2, for 'b metaph.' read 'αὐ. γίγνομαι do a thing oneself (not delegating it), Aen.Tact. 15.'

*αὐτοῦτα, Sicilian reflexive in -τα, gen. sg. αὐτοῦτα IG 14.287, 288b (both Segesta, iii/ii BC); pl. αὐτῶντα SEG 30.1119.19, 26 (Nakone, c.300 BC), Dubois IGDS no. 189.13 (Centuripa, ii BC), IG 14.316 (Thermae Himeraeae, ii/i BC); acc. pl. αὐτούστα Dubois l.c. lines 3, 5; dat. pl. αὐτοῖστα SEG l.c. line 27.

*αὐτοφαείνομαι, shine with one's own light, δαίμων ἀπηνὴς αὐτοφαεινομένην ἔσβεσα δᾷδα γάμων GVI 228.4 (Bithynia, ii/iii AD).

*αὐτοφάνεια, ἡ, appearance in person, Procl.Sacr. p. 152.6.

*αὐτοφθόνος, ον, radically or essentially malicious, κακία ἀ. An.Bachm. 2.352.14.

*αὐτοφἴλοτίμημα, ατος, τό, act of voluntary generosity, καὶ περ[ὶ π]ολλῶν αὐτοφιλοτειμημάτων εἰς ἡμᾶ[ς] IEphes. 22.15 (ii AD).

*αὐτόφλεψ, φλεβος, ἡ, an actual vein, opp. ἐοικός τι φλεβί, Ruf.Onom. 206.

αὐτοφόνος 1, add 'perh. in this sense, unexpld. wd. in list of sacred laws, SEG 9.72.132 (Cyrene, iv BC)' and transfer 'παλάμῃ .. Leont.)' to section 2

†αὐτόφορτος, ον, transporting one's own baggage or cargo, A.Ch. 675, S.fr. 251, Cratin. 266 K.-A.   2 of ships, together with their cargo, αὐτοφόρτους ὁλκάδας Plu.Aem. 9, 2.467d.

αὐτοφυής I 3, line 10, before 'Comp.' insert 'κόρη κάλλος αὐτοφυὲς καὶ ὅμοιον αὐτομάτῳ φυτῷ φέρουσα Aristaenet. 1.7.4'   II, adv., add 'Procl.Inst. 205 D.'

αὐτόφυτος 1, after 'self-engendered' insert 'grown spontaneously, of Prometheus' liver, which was renewed automatically, Nonn.D. 2.300'

αὐτόφωρος II, for 'mostly in the phrase .. in the act' read '(being) in the very act of crime, red-handed, κολάζων .. αὐτοφώρους Th. 6.38.4; esp. in phr. ἐπ' αὐτοφώρῳ λαμβάνειν or sim.'

αὐτοχειλής, for 'S.fr. 138' read 'S.fr. 130 R.'

αὐτοχειρία II, line 2, after '= αὐτοχειρί' insert 'τὸν νεὼν ἐξεποίησεν αὐτοχειρίῃ Robert Hell. 9.78 (Mysia, vi BC)'

αὐτόχθων I, add 'title of Μήτηρ θεῶν SEG 24.498, 26.729 (both Maced., ii AD; see also SEG 33.532)'   II, line 2, after '(ii AD)' insert '(πόλις) IStraton. 15.2 (Panamara, i/ii AD)'; line 4, for 'urbanitas, racy of the soil' read 'ἀ. urbanitas, authentic national, i.e. Roman, manners'

αὐτόχροος, -χρους 1, after 'colour' insert 'PCair.Zen. 92.6 (iii BC)'

αὐτόχυτος, line 1, for 'poured out of itself, self-flowing' read 'flowing spontaneously' and for 'θάλαμος .. 96.102' read 'Hes.fr. 204.140 M.-W.'

αὐχένιος I, add '3 neck-like, ὁλοσχοι Nic.Th. 871; αὐχένιαι κεφαλαί (of columns), MDAI(I) 19/20.238.14, 24 (Didyma, ii BC).'

*αὐχενοπλήξ, ῆγος, ὁ, ἡ, struck in the neck, Hippon. 102.6 W.

αὐχέω II 2, line 3, for 'A.Pr. 340' read 'A.Pr. 338'

αὐχή, after lemma insert 'dial.' and add 'Ibyc. 220.13, 221.5 S.'

αὐχήεις, add '[θυμοῦ ἐξ] αὐχήεντος SEG 24.1243 (Egypt, Chr.)'

αὔχημα I, for this section read 'cause for pride or boasting, S.OC 710, 713, Th. 7.75.6, Aq.Is. 3.18 (L.-R.)'

αὐχήν II 6, add 'ἀετο[ῦ α]ὐχένα SEG 31.1349.8 (Cyprus, i AD)'   add '7 part of spindle, prob. end to which thread is attached, ἀντία γ', ὦ Μοῖραι, γαμψοὺς ἐπεθήκατε ἀτράκτοις αὐχένας GVI 1681.8 (Rhenea, ii/i BC).'

αὔχησις, for 'Th. 6.16' read 'Th. 6.16.5, Aq.Pr. 4.9, al.'

αὐχμηρός 2, after 'squalid' insert 'χεῖρες Anacr. 2 fr. 1.4 P.'   add '4 of literary style, arid, D.H.Dem. 45, al.; ῥήτορες id.Din. 8.'

αὐχμός, add '5 disregard of niceties of toilet, unkempt state, Arr.Epict. 3.22.89.'

αὖχος, add 'SEG 30.1486 (Phrygia)'

αὔω (B) 2, line 2, for '(Il.) 13.475' read '(Il.) 13.477'; after 'Od. 9.65' insert 'Tim. 11 P.'; last line, after 'diphthong' insert 'except in Hymn.Is. 59 [ῡ]'

*αὐωρο-, v. °ἀωρο-.

*ἀφάγγρειμι, take away, subtract, τὸ χόυρον ἐμε[τρεί]θει ἀφανγρειμενᾶν τᾶν ὁδουὶν καὶ τᾶν ἐνόδουν IGC p. 11.14 (Larissa, iii BC); cf. ἀφαιρέω, ἐφαιρενθείν s.v. ἐφαιρέω.

*ἀφαγιστεύω, app. perform apotropaic rites, κἀφαγιστεύσας ἅ χρή S.Ant. 247 (but see °ἐφαγ-).

*ἄφαγος, ον, fasting, Sch.A.R. 4.1295.

†ἀφαδία, ἡ, enmity, Eup. 376 K.-A.

*ἀφάεσται, ἀφάεσοι, (?ε), med. inf. and 3 sg. subj. of Arc. vb. meaning pay, [ἀφά̣ε]σται δαρχμὰς τριάκοντα· εἰ δὲ μὲ ἀφάεσοι Jeffery LSAG p. 214 no. 2 (see also Dubois Dial. arcadien pp. 195ff., Pheneos, c.500 BC); cf. °ἐξάεσοι, perh. also ‡ἀφαιάσαι.

Ἀφαία, for 'IG 4.1580' read 'SEG 32.356, 37.260 (vi and v BC), Paus. 2.30.3, Ant.Lib. 40.4'

ἀφαιάσαι, add '(ἀφαιαμάσαι· δαπανῆσαι, ἀπολειτουργῆσαι, etc., La.)'

ἀφαίρεσις I, add '4 wrongful deprival, POxy. 3611.8 (iii AD).'   II 2, after 'Gramm., removal' insert 'of letters, words, etc., μετασκευῆς .. ἀφαιρέσεως λέγω καὶ προσθήκης καὶ ἀλλοιώσεως D.H.Comp. 6, 9'

ἀφαιρετικός I, add '2 ἀ. πτῶσις ablative case, Dosith. p. 392 K.'

ἀφαιρετός 1, after 'Pl.Plt. 303e' insert 'Arist.EE 1241ᵇ23, al.'

ἀφαιρέω, line 1, after 'Ion. ἀπαιρέω' insert 'cf. Thess. °ἀφάγγρειμι'   II 4, for 'cf. Lys. 23.10' read 'Lys. 23.9'; after 'Aeschin. 1.62' insert 'cf. ὥστε ἀφῃρεῖτ' αὐτὸν ὡς ἐλεύθερον ὄντα Isoc. 17.14, 49'

*ἀφἄκέομαι, aor. imper. ἀφ[α]κεσάσθω, repair damage, Sokolowski 2.27.11 (Argos, vi BC).

ἀφάλλομαι, line 1, for 'ἀφάλασθια' read 'ἀφάλασθαι'

ἀφαμαρτάνω, line 1, delete 'Orph.A. 643'

*ἀφᾰμαρτέω, wander off, Orph.A. 643 (v.l. ἀφομ-).

ἀφαμία, ἁ, Cret. term, perh. type of land-holding, ἐν ἀπαμίαις *Inscr.Cret.* 2.12.16 Ab 2 (Eleutherna, vi BC), app. same wd. as in τὰν Ἐξάκωντ[ος] ἀφαμίαν, landscape feature used as boundary mark, *SEG* 26.1049.72 (Lato, ii BC); cf. Ἀφαμιῶται.

ἀφάνεια II, ἀφανία, add 'καὶ ἐν ἀργίαι καὶ ἐν ἀφανίαιν *SEG* 37.215 (Attica, tab.defix., ?iv BC)'

ἀφανής 1, add 'w. gen., ‹φθίψμενοι κείμεθα γῆς ἀφανεῖς *CEG* 520.9 (Athens, c.360 BC)' **4**, add '**b** = Lat. *incertus* or *incerta persona*, one who is not precisely designated or whose existence is uncertain, ἐπίτροπος ἀφανέσιν οὐ δύναται δοθῆναι *Cod.Just.* 6.48.1.28, 6.48.1.2.'

ἀφανίζω I 6 b, add 'μῆλα .. ἠφάνισται *BGU* 38.11'

ἄφαντος, transfer 'θεοῖς .. Epimenid. 11' fr. section 3 to section 1; at end of that section, for '*invisible*' read '*blotting out, obscuring*' and add 'θύελλαι Alc. 298.12. L.-P.'

ἀφαρής, before 'Euph.' insert 'cj. in' and add 'cf. *Tav.Lign.Cer.* 66.1.20 (vi AD)'

ἀφαρκίδευτον, add '(ἄγρυπτον, ἀρυτίδωτον La.)'

ἀφαρπαγή, add 'Sol. 4.13 W. (s.v.l.)'

ἀφαρπάζω, aor. 1 pass., add 'ἀφαρπαχθεῖσα epigr. in *SEG* 37.489 (Larissa, iv/v AD), cf. aor. pass. forms given s.v. ἁρπάζω' add '**2** *transfer* to a changed status (cf. Lat. *eripere*, εἰς εὐγένειαν ἀφαρπάζονται *Cod.Just.* 6.4.4.3 (v.l. ἀφ' οὗ ἁρπάζονται).'

ἄφαρπαξ, αγος, ὁ, unclean bird, one of the *raptores*, Al.*Le.* 11.19.

ἀφαυαίνω 1, after 'Thphr.*HP* 3.18.9' insert '*CP* 3.10.8'

ἄφαυστος, ον, *inexplicable*, Plot. 6.6.7 (s.v.l.).

ἀφάω, add 'prob. in Ar.*Pax* 1144 (codd. ἀφευε)'

ἀφεδριατεύοντες, add 'also sg., *SEG* 23.271 (Thespiae, iii BC)'

ἀφειδέω I, for 'abs. .. E.*IT* 1354' read 'abs., ἀφειδήσαντες *putting aside restraint* or *inhibition*, Hp.*Art.* 37, E.*IT* 1354, τὸν ἄριστον ἀφειδήσαντες ἕλεσθε ὅρχαμον ἡμείων A.R. 1.338'

ἀφειδής I 1, line 3, for 'ἀ. πρὸς τὸν ἔρωτα Call.*Epigr.* 47.7' read 'ἀφειδέα ποττὸν Ἔρωτα Call.*Epigr.* 46.7 Pf.' add '**b** ae., *uncontrollable*, ἀφειδῇ ταύρον, ὃν οὐχ αἱροῦσ' ἀνέρες οὐδὲ δέκα Aristocl. in *Suppl.Hell.* 206.'

ἀφέλεια, line 2, after 'style' insert 'D.H.*Is.* 16'

ἀφελής, delete '(φελλεύς)' and section I. **II 1**, add '*AP* 5.42 (Rufin.)'; at end, adv., add 'comp. Ael.*VH* 7.1' **b**, for this section read 'w. pejorative force, Μένων ἀφελές ostr. in *MDAI(A)* 80.118 nos. 30-2 (Athens, v BC)' **2**, add 'cf., in apparent allusion to Cratinus' style, Κρατίνου .. ὃς πολλῷ ῥεύσας ποτ' ἐπαίνῳ διὰ τῶν ἀφελῶν πεδίων ἔρρει Ar.*Eq.* 527'; adv., add 'ὅτι φυσικῶς πως εἴρηται καὶ ἀ. D.H.*Is.* 7'

ἀφέλκω I, add '*take off* a lid, πῶματ' ἄφελκε κάδων Archil. 4.7 W.' **II**, delete 'κάδων .. Archil. 4'

ἀφελληνίζω, line 2, for '-ηλλήνισθη' read '-ηλληνίσθη'

ἄφεσις 2 a, add 'ἔντιμος ἄφεσις = Lat. *honesta missio*, *honourable discharge*, *CPR* 7.21.12 (iv AD)'

ἀφεσοφυλακία, read 'ἀφεσιοφυλακία' and add 'v. *PBremen* p. 44'

ἀφέταιρος, for 'cf. ἀπέταιρος' read '**II** subst., *one who is not a member of a ἑταιρεία or society of free citizens*, *Leg.Gort.* 2.5, al. (v BC); *one who is not a member of the association of Epilykoi*, *SEG* 35.989.11 (Knossos, ii/i BC).'

ἀφετήριος 5, for 'gate of a sluice' read 'ἀφετήριαι, αἱ, *sluice-gates*' and add '*outlet* of conduit, *POxy.* 2146.6 (iii AD), v. °ἐξομβριστήρ'

ἀφέτης I 1 c, add '*OAshm.Shelton* 174, 176'

ἀφεύω, line 1, for 'Semon. (v. infr.)' read 'Semon. 24.1 W.' **2**, delete the section.

ἀφέψαλος, add 'θ[οίν]ῃ δ' [εἴ]μὶ βροτοῖσιν ἀφέψαλος i.e. *uncooked*, *Suppl.Hell.* 983.6 (iii BC)'

ἄφεψησις, for 'Sch.Lyc. 156' read '**2** *boiling*, sc. of Pelops by Tantalus, Sch.Lyc. 157. **3** *refining*, of gold, *IG* 2².1496.201.'

ἀφέψω, add '**3** simply *boil*, ἀφεψήθη, of Pelops, Sch.Lyc. 157.'

ἀφή I, add 'Men.*fr.* 197 K.-Th.' **II 1**, delete the section. **2**, add '**b** pl., *the senses, perceptions*, Pl.*Ax.* 365a.'

ἀφήλικος, for '= sq.' read '= ἀφῆλιξ II, *POxy.* 2134.8, al. (ii AD)'

ἀφῆλιξ II, add '*BGU* 907.2, *POxy.* 2474.21, Modest.*Dig.* 26.5.21, *Just.Nov.* 100.2 pr.'

*****ἀφημερεία**, ἡ, *absence for a day*, *PSI* 1120.3 (i BC/i AD).

*****ἀφηρωϊσμός**, ὁ, *canonization as a hero*, *IG* 12(7).515.6 (Amorgos, ii BC).

ἄφθα (A), delete '*infantile*' and add 'Marc.Sid. 101'

ἄφθαρτος I, add '**2** *unaltered*, Ἑλληνικά 7.179 (Chalcis, iii BC).'

*****ἄφθαστος**, ον, perh. *not (to be) overtaken*, τῇ κακίᾳ ἄφθαστοι *Cat.Cod.Astr.* 11(2).136.16; cf. adv. ἀφθάστως.

ἄφθιτος, line 1, add 'fem. ἀφθίτα Mesom. 3.16' **2**, line 3, for 'Simon. 184' read '*AP* 7.25.1 ([Simon.])'; add 'adv. -ως, ἀφθίτως βιοτεύειν *Orac.Sib.* 5.303' **3**, add 'neut. as adv., *CEG* 862.6 (Olympia, iv BC)'

ἀφθώδης, add 'Crito ap.Gal. 12.933.15'

*****ἀφιδάνης**, ον, ὁ, name of gem, Xenocr.*Lap.* 108, cf. Plin.*HN* 37.147 (*amphidanes*).

ἀφίδρυμα I, for '*thing set up*, esp. *image* of the gods' read '*image set up in honour of a god*'; delete 'Str. 12.5.3' and add 'D.S. 15.49' **2**, delete the section. **II**, for this section read '*temple* or *shrine* copied fr. an original, Cic.*Att.* 13.29.1, Str. 6.2.5, ἱερὸν Ἀσκληπιοῦ ἀ. τοῦ ἐν Τρίκκῃ 8.4.4'

ἀφιδρύω, add 'med., ὃν .. ἀνέστησέν τε καὶ ἀφειδρύσατο *GVAK* 14'

ἀφίημι (A), line 9, after 'Il. 23.841, etc.' insert 'ἀφεώκαμεν (1 pl.) *SEG* 20.325.3 (Hyrcania, iii BC), ἀφέωκε (3 sg.) *PCair.Zen.* 502.4/5 (iii BC), *IG* 9(2).1042.11 (Gonnoi, i AD)'; line 15, transfer 'Arc.inf. .. (Tegea, iv BC)' to line 18 to follow '*IG* 5(2).6.14'; line 17, before 'plpf.' insert 'part. ἠφειμένους *ASAA* N.S. 3/4(1941-2).97.6 (Lemnos, i BC)' **II 1 b**, add 'of manumission, ἀφέω τοῦτον τὸν ἑπτάδουλον Herod. 5.75, cf. Hippon.*dub.fr.* 190 De., *SEG* 32.622, 623 (Illyria, ii BC)'

*****ἀφικετεία**, ἡ, *intercession*, *IKnidos* 220.6 (iii/ii BC).

*****ἀφικετεύω**, *intercede for a suppliant*, *SEG* 9.72.132, 138 (Cyrene, iv BC), *SEG* 39.729 (Lindos, iii BC); cf. ἀφίκτωρ.

*****ἀφιλανθρώπητος**, ον, = ἀφιλάνθρωπος, *BGU* 1785.10 (i BC).

*****ἀφιλόκερδος**, ον, *not devoted to gain*, *SAWW* 265 i 55 no. 12.4.

*****ἀφιλοπόνητος**, ον, *not given devoted attention*, *Vit.Aesop.*(G) 51.

*****ἄφιμος**, *unbridled*, ἵππος *Trag.adesp.* 328g K.-S.

ἀφιππεύω, after 'Hld. 4.18' insert '(v.l.), Charito 3.7.2'

ἀφιπποτοξότης, for 'v. ἀμφιππστ-' read '*bowman on horse-back*, D.S. 19.29 (nisi leg. ἀμφ-), cf. Plu. 2.197c (v.l. ἀμφ-)'

*****ἄφισμα**, ατος, τό, app. *image* set up in a temple, τοῦ ἀπολωλότος ἀφίσματος τοῦ Ἀπόλλωνος *BCH* 116.335 (Argos).

ἀφίστημι B, line 2, for 'Men. 375' read 'Men.*fr.* 158, 317 K.-Th.'; line 24, after 'Th. 7.28' insert 'w. gen., μὴ ἀποστήσεσθαι Ὀξυρύγχων πόλεως *depart from*, *PKöln* 148.6 (ii AD)'

*****ἀφίστησις**, εως, ἡ, *relinquishing*, τῶν ἑτέρων of his other claims, *Dura*⁷, ⁸428 (parchment 40; i AD).

ἀφίσωσις, v. °ἀπίσωσις.

ἄφλαστον, delete 'Asclep.Tragil. 31 J., Sch.'

ἀφλέγμαντος 1, adv., add 'Heras ap.Gal. 13.557.12'

ἀφνειός, line 4, after 'c. dat.' insert 'Hes.*fr.* 240.2 M.-W.'

ἀφνύει, delete the entry.

*****ἀφνύω**, v. °ἀφνύω.

*****ἀφνύω**, *enrich*, ῥυδὸν ἀφνύονται Call. in *Suppl.Hell.* 287.3 (ἀφνύνται Suid.); ἀφνύει· ἀφνύνει. ὀλβίζει Hsch.

ἄφοβος 1, at end after 'Pl.*Lg.* 682c' insert 'Arist.*EE* 1228ᵇ26'

*****ἀφοδευτήριος**, α, ον, ἀ. δίφρους *latrine-stools*, Hsch., s.v. λάσανα; neut. subst. *privy*, ἀπόπατος λέγεται τὸ ἀ. Sch.Ar.*Pl.* 1184.

ἀφολίδωτος, add 'rest. in *Inscr.Délos* 104-5.31'

*****ἀφομαρτέω**, v. °ἀφαμαρτέω.

ἀφομοίωσις, add '**2** perh. *levelling*, ἀ. τοῦ χωρίου *IG* 4.823.66 (Troezen, iv BC).'

ἀφοπλισμός, add '*Cod.Just.* 1.4.26.16'

ἄφοραο, line 1, after 'aor. ἀπεῖδον' insert 'ἀφίδω *Ep.Phil.* 2.23 (codd.)' **I 1 b**, add '*Peripl.M.Rubr.* 35'

*****ἀφορί**, adv. *without paying rent*, ἀφορὶ ἕξουσιν (sc. τὴν γῆν) *PTeb.* 737.27 (ii BC); ἀφορεί *PFlor.* 384.54 (*Berichtigungsl.* 2.60, v AD).

ἀφορίζω II 2 b, add '*Just.Nov.* 123.11 pr.'

ἀφόρισμα, add 'τὰ ὄρη καὶ τὰ ἀφορίσματα perh. *boundaries marked off* or *borders*, *IG* 2².30.18 (pl.; Attica, iv BC)'

ἀφορισμός II 1, add '**b** *excommunication*, *Cod.Just.* 1.3.38.2.'

*****ἀφορμάριος**, ὁ, *one who makes excuses*, *SB* 7168.4 (v/vi AD, pl., written ἀφορμαροι).

*****ἀφορμία**, ἡ, *deterrent*, *Vit.Aesop.*(G) 16.

ἀφορμίζομαι, add 'ἀφορμισάμενοι cj. for ὑφ-, Th. 2.83.3; act., perh. *push off* a boat from the shore, (τὸ πορθμεῖον) pap. in *SHAW* 1923(2).23 (pap. ἐφ-)'

*****ἀφορολόγιστος**, ον, = ἀφορολόγητος, Ps.-Callisth. 72.10 K.

ἀφοσιόω II 1, add 'D. 47.70' **2 c**, add 'Isoc. 12.269'

ἀφοσίωσις 2, after 'Plu.*Eum.* 12' insert 'ἀφοσιώσεως χάριν Modest.*Dig.* 27.1.13.6'

*****ἀφουλκόν**, v. ‡ἀπουλ-.

*****ἀφουλωτικός**, v. ‡ἀπουλ-.

ἀφρατίας, add '(ἀφραττίας La., i.e. ‹ ἀφρακτ-, perh. cf. Ἄφραττος)'

Ἄφριος, for 'Ἀρχ.Ἐφ. 1913.219' read '*AE* 1913.219, Ἄφροι (gen. sg.) *BCH* 59.56 (Larissa, ii AD)'

ἀφρογενής, add 'Hes.*Th.* 196 (prob. interp.; -γενειαν codd.)'

*****Ἀφροδῑσιακόν**, τό, name of a mine in Sunium, *IG* 2².1587.5 (iv BC).

*****Ἀφροδῑσιακός** (B), *of Aphrodisias* (in Caria), of a type of marble workmanship, ζῴδια Ἀ. *SEG* 33.946 (Ephesus, i AD), *IEphes.* 3803B.8 (iv AD).

*****ἀφροδισιαρχέω**, *preside over the festival of Aphrodite*, Ἀφροδεισία ἀ[φ]ροδεισιαρχήσασα Ἀφροδίτᾳ χαριστήριον *SEG* 25.595, 596, both of women (Phocis, i AD; ii BC acc. to *BE* 1970.305).

ἀφροδισιάς II, for 'ἄκορος' read 'ἄκορον' and add 'cf. Ps.-Dsc. 1.2'

Ἀφροδῑσιαστής 3, add '*SEG* 26.614 (Tanagra, iii/ii BC)'

*****Ἀφροδῑσιδεῖον**, τό, *sanctuary of Aphrodite and Isis*, *IG* 4².742.5 (Epid.,

Ἀφροδίσιος   SUPPLEMENT   ἄχυρον

ii/iii AD), but Ἀφροδισι{δ}είωι *Sokolowski* 2.25.A5 (= Ἀφροδίσιον).
Ἀφροδίσιος I, line 3, after 'Semon. 7.48' insert 'λόγοι ib. 91 W., cf. Ael.*NA* 6.1'   **II 1**, line 3, after 'X.*Mem*. 2.6.22' insert 'cf. 1.3.8' and delete 'also as concrete .. 1.3.8'   **III 1**, after 'PPetr. 3 p. 113' insert 'Inscr.*Délos* 1442*B*31, 33 (ii BC)'   add '**V** epith. of Zeus, *IG* 12(5).220 (Paros, iii BC).'
Ἀφροδισιών, add '*SEG* 30.533 (Phthiotic Thebes, ii BC), *Iasos* 47.1 (iv BC)'
Ἀφροδίτη, after lemma insert 'Dor. -ίτᾱ Alcm. 1.17 P., etc., Aeol. voc. -ῑτᾱ (proparox.) Sapph. 1.1. L.-P., etc; Ἀφορδίτᾱ Inscr.*Cret.* 1.9.1.27 (Dreros, iii/ii BC); cf. pers. n. Ἀφορδίσιυς *IPamph*. 21, etc.'   **I**, add 'pl., Call.*fr.* 200a Pf., cf. Ἀφροδείταις Καστνηΐτισιν (sic) *SEG* 17.641 (Aspendos, early Rom.); as statue of A., *SB* 9834a.7 (iv AD)'
***ἀφροδῖτολειτορεύω**, serve as priestess of Aphrodite, *SEG* 27.206 (sp. -λιτ-, Larissa, i BC/i AD).
***ἀφροειδής**, ές, *foam-like*, Sch.Hes.*Th.* 191a.
ἀφρόνιτρον, add 'Plin.*HN* 20.66, al.; also ἀφόνιτρον *PRyl.* 629.101, al. (iv AD)'
ἀφρός I 1, add 'personified as a Nereid, *SEG* 31.1387 (Syria, iv AD)'
***Ἄφρος**, α, ον, *African*, i.e. from the Roman province, τάπης *Edict.Diocl.* 19.35, al.; Ἄφροι, of the Carthaginians, Suid. s.v.
ἀφροσέληνος, add '**II** = λιβανωτίς, Ps.-Apul.*Herb*. 80.'
ἀφροτόκος, add '*Eleg.adesp.* in *POxy.* 3723.2 (ii AD)'
ἀφρούρητος, after lemma insert 'Dor. ἀφρούρᾱτος *IAskl.Epid.* 25.4 (iii BC)'; after '*ungarrisoned*' insert 'as a privilege, i.e. *free from military occupation*, *IAskl.Epid.* l.c., *SEG* 37.1003.19 (Sardis, iii/ii BC), *Ilion* 45.14 (ii BC)'
***ἀφρωραῖος**, α, ον, *beautiful in the foam*, Ἀφροδίτη *PMag.* 4.3232.
ἀφύη, after 'Ar.*Ach.* 640' insert 'id.*fr.* 520 K.-A., Hermipp. 14 K.-A., Call.Com. 10 K.-A., Aristonym. 2 K.-A.'; at end delete 'not used in sg. .. τιμή'
✝ **ἀφῡλισμός**, ὁ, *removal of solid matter*, of clearing of ditches, *PMich.*VI 380.6 (ii AD), *OMich.* 802.7 (AD 296), *PNew York* 2.5 (iv AD); cf. °ἐφυλισμός, παρυλισμός; of straining of wine, *OMich.* 12 (iii AD).
ἀφυπνόω II, after '*Ev.Luc.* 8.23' insert '*Vit.Aesop.*(G) 127, Aesop. 252 P.' and for 'v.l. ὑφυπν-' read 'cj. for ὑφυπν-'
ἀφύσσω I 2, for '*sound, probe*' read 'medic., *draw off fluid from a wound, drain*'   **II**, penultimate line, after 'σπειρήματ' insert '(codd. πειρήματ)' and at end read 'Trag. only in E.ll.cc., Ion Trag. 10 S., Ezek.*Exag.* 250 S.'
ἀφυστερέω II, add 'v.l. in *Ep.Jac.* 5.4 (pass.)'
ἀφωνία I, add 'στήλη δὲ φωνῶ ἀντ' ἀφ⟨ω⟩νίας βίου *IUrb.Rom.* 1322.6 (= *IG* 14.1977)'
***ἀφῶτε**, *from the time that*, *SEG* 37.340.12 (Mantinea, iv BC).
Ἀχαία I, add 'of Athena, Arist.*Mir.* 840ᵇ2'   **II**, transfer '(acc. to Hsch. .. id.)' to the end of section I.   add '**III** οἱ δὲ ἔρια μαλακά, Hsch. s.v. Ἀχαία; cf. perh. ἀπὸ ἀχάης *POxy*. 1978.4, al. (vi AD).'
Ἀχαῖα (s.v. Ἀχαιός), for this entry read '**Ἀχαΐα** later **Ἀχαΐα**, ἡ, *Achaea*, a region of the northern Peloponnese, τὴν νῦν καλεομένην Ἀχαιίην Hdt. 7.94, Th. 1.115.1, Paus. 7.1.1, etc.   **2** region of Thessaly (Achaea Phthiotis), Hdt. 7.173.1, Str. 9.5.6, etc.   **3** the Roman province of Achaea, J.*BJ* 1.26.4, D.C. 58.25.5, etc.'
✝**Ἀχαιϊάς**, άδος, fem. adj., *Achaean*, Ἀχαιϊάδες .. πόληες Call.*Del.* 100, γυναῖκες Nonn.*D.* 47.636.   **II** fem. subst., *Achaean (Greek) woman*, Il. 5.422, Od. 2.101.   **2** *Achaea*, *SEG* 13.226.6 (Corinth, ii AD), *BCH* 50.444.85 (Thespiae, iv AD).
Ἀχαιικός, add 'comp. -ωτέρα ὑπόθεσις Plb. 24.9.2'
✝**Ἀχαιΐς**, later **Ἀχαΐς**, ίδος, ἡ, *Achaean*, Ἀχαιΐδα γαῖαν Il. 1.254, προσβολὴν Ἀχαιΐδα A.*Th.* 28, νεῶς .. Ἀχαιΐδος E.*Hel.* 1544, Ἀχαιΐδας πόλεις X.*HG* 7.1.43, Ἀχαιΐδα .. κούρην A.R. 3.639.   **II** fem. subst., *Achaean woman*, Il. 2.235.   **2** *Achaea*, Il. 3.75, A.R. 3.1081.
***ἀχαιοπόρφῠρος**, ον, *of Achaean purple*, δελματικοφόριον ὀνύχινον ἀχαοπόρφυρον (sic) *SB* 11075.7 (v AD).
Ἀχαιός 1, add '**b** Θεὸς Ἀ. name of a god in Asia Minor, *SEG* 33.1542, 1543 (Rom.imp.).'
***ἀχαιόσημος**, ον, of clothing, *with an Achaean stripe*, *POxford* 15.10 (iii AD); also ἀχαο- *SB* 11075.10 (v AD).
ἀχάντιον, add 'Eust. 468.33'
ἄχᾰρις I 1, add 'of speech, D.H.*Lys.* 12, *Is.* 20'   **2**, line 3, after '(Hdt.) 7.52' insert 'ἄ. τιμή id. 7.36.1'   delete '**II** *ungracious* .. Hdt. 7.36.1' for 'χάρις ἄχαρις' to end read '**3** in phr. χάρις ἄχαρις *lacking the essential quality of* χάρις, A.*Ag.* 1545, κακῆς γυναικὸς χάριν ἄχαριν ἀπώλετο E.*IT* 566, *AP* 9.322.'
✝**ἀχάριτος**, ον, *lacking charm, unpleasant*, τὰ δέ μοι παθήματα ἐόντα ἀ. μαθήματα γέγονε Hdt. 1.207.1; δῆμον εἶναι συνοίκημα ἀχαριτώτατον id. 7.156.3; ἀχαριτωτάτου προσώπου (sc. of the Cyclops) Demetr.*Eloc.* 130; Plu.*Sol.* 20; of style, Demetr.*Eloc.* 139 (comp.); adv. οὐκ -τως ἔφη Ath. 7.281c; Hermog.*Id.* 2.11; D.C. 66.9.5.   **2** in phr., χάρις ἀχάριτος *lacking the essential quality of* χάρις, A.*Ch.* 44, E.*Ph.* 1757.

***ἀχασμώδητος**, ον, *not having hiatus*, Rh. 3.544.11.
ἄχειρ, add '**b** *having no holes for the arms*, χιτῶνα ἄχειρα καὶ ἀτράχηλον Apollod.*Epit.* 6.23.'
✝**ἀχειρής**, ές, *lacking hands*, καρκίνοι Batr. 298.
✝**Ἀχελωΐδες** (sc. πόλεις), dub. sens., app. of towns situated on or near rivers or lakes, A.*Pers.* 869.
Ἀχελῷος II, add 'Ar.*Lys.* 381, E.*Andr.* 167; Ἀχελῷον πᾶν ὕδωρ Εὐριπίδης φησὶν ἐν Ὑψιπύλῃ Macr.*Sat.* 5.18.12; as a divinity, object of cult worship, Sokolowski 3.18.B24 (Attica, iv BC), 96.35, 38 (Myconus, c.200 BC); at a fountain, *BCH* 113.610 (Messene)'
***ἀχέρνιπτος**, ον, *not to be used for ritual washing*, ὕδωρ A.*fr.* 273a.12 R.
***Ἀχεροντίς**, ίδος, fem. adj., *of Acheron*, Ἀ. λίμνην *GVI* 731.9 (ii/iii AD).
***Ἀχερουσίς**, ίδος, fem. adj., *of Acheron*, A.R. 2.728, *AP* 5.204 (Mel.), Λήθης Ἀχερουσίδος *IG* 10(2).1.368.1 (epigr., Thessalonica, ii AD).
ἀχεύω, line 3, for 'cf. Sapph.*Supp.* 1.11' read 'ἀχεύων Sapph. 5.11 L.-P.'
ἀχήν, line 1, delete '= ἠχήν (q.v.)'; for 'dat. pl.' to end read 'ἀχήν· ἄπορος Hsch., dat. pl. ἀχήνεσσιν *IStraton.* 543.7 (epigr., Lagina, iii/ii BC), see also ἠχῆνες, ἀεχῆνες'
***ἀχθαίνω**, *carry a burden of sorrow, ache*, πολλὸν δὲ περὶ φρεσὶν ἀχθήνασα Call.*fr.* 63.7 Pf.
ἀχθέω, add '(ἀχθίσας La., v. ἀχθίζω)'
ἀχθηρός, add 'v.l. in Phalar.*Ep.* 122'
ἄχθος I, add '**b** as a measure, φρ[υγά]νων ἄχθος, καὶ ξυλέων ἄχθος Inscr.*Cos* 39.14 (iv/ii BC).'
***ἀχθοφορικός**, ή, όν, *of or from bearing burdens*, Eust. 1577.44.
***ἄχι**, v. °ἤχι.
Ἀχίλλειος, line 1, before 'Ἀχιλλεΐος' for 'poet.' read 'Aeol.'; add 'τὸ Ἀχίλλειον *temple with tomb of Achilles at Sigeum*, Str. 13.1.39, 46, Ion. -ήιον Hdt. 5.94.2'
Ἀχιλλεύς, add '**III** *image of the sun thrown on a ceiling by a moving mirror*, Hero *Deff.* 135.12.'   at end add 'Myc. *a-ki-re-u* pers. n.'
ἀχλυόεις, line 1, delete 'δεσμός .. Hdt. 5.77' (v. °ἀχνυόεις)
ἀχλύς, line 1, after 'mist, Od. 20.357' insert 'Arist.*Mete*. 367ᵇ17, 373ᵇ12, Plb. 34.11.15, Str. 6.2.8'   **2**, add 'Iamb.*Protr.* 21 ιδ' (115.18 P.).'   at end delete 'Mostly poet. .. supra 2.'
***ἄχμα**, ματος, τό, app. *cargo*, τὰ δ' ἄχματ' ἐκπεπ[.].άχμενα Alc. 208 ii 7 L.-P., cf. perh. 167.7.
✝**ἀχνάζω**, ἀχνάζει· ἄχθεται, μισεῖ, ψέγει Hsch.; cf. °ἀχνάσδημι.
***ἀχνάσδημι**, Aeol., *to be miserable*, Alc. 349A L.-P.
ἄχνη, line 2, for '*foam, froth*' read '*spray*'
***ἀχνυόεις**, εσσα, εν, *grief-laden*, δεσμῷ ἐν ἀχνυόεντι σιδήρεῳ cj. in *CEG* 179 (c.506 BC), *GVI* 238.1 (Rom.imp.).
✝**ἀχνύς**, ύος, ἡ, *grief*, *Suppl.Hell.* 1031; personified, cj. in Hes.*Sc.* 264, cj. in Hld. 2.14.5.
***ἀχόνωδες**, dub. sens., φιδάκναι ἀχόν⟨ω⟩δες *IG* 1³.422.302 (v BC); cf. °χονδήν.
ἀχορτασία, for '*ravenous hunger*' read '*the state of being unsatisfied* (by food)'
ἀχόρταστος, for '*unfed, starving*' read '*unfilled, unsatisfied*' and add 'Cyran. 1.6.14 K.'
ἀχραής, after 'ψυχρόν ἀ.' insert '(cj., cod. ἀκραές)'
ἄχραντος, add 'Iamb.*Protr.* 21 α' (108.24 P.), Procl.*Inst.* 154, 156 D.'
ἀχρεῖος I 1, add 'of objects, *unserviceable*, φιάλη *IG* 7.303.10 (Oropus, iii BC); of garments, *SEG* 38.1210.8, al. (Miletus, ii BC), τὴν ἐσθῆτά μου διέρρηξεν καὶ ἀχρίαν ἀπέδειξεν *SB* 7449.12 (v AD)'   **II**, after 'as adv.' insert '*in a manner out of keeping with one's character or situation*'
ἀχρεοκόπητος, after lemma insert 'of debts, dues, *that cannot be remitted*, ἐπ[ικε]φάλια .. ἀχρε[ω]κόπητα *SB* 11379.8 (ii AD)
ἀχρημάτιστος III, for this section read 'perh. *having no business*, Bernand *Akôris* 28'
***ἀχρής**, ές, = ἄχρους, *pallid with fear*, acc. to *EM* 182.47, v. (?)Call.*fr.* 742 Pf.
ἀχρήσιμος, add '*PColl.Youtie* 77.6 (iv AD)'
ἄχρηστος, add '**V** perh. *deprived of civil rights*, Meiggs-Lewis 2 (Dreros, vi BC).'
ἄχρι II 1, after 'Lxx *Ge.* 44.28' insert '**b** *for the duration of, until the completion of*' and add 'ἄχρι τῆς ζωῆς Just.*Nov.* 7 pr.; also *by the end of*, ἤλθομεν πρὸς αὐτοὺς .. ἄχρις ἡμερῶν πέντε *Act.Ap.* 20.6'   **III 1**, delete '*so long as*' and add '**b** *so long as, whilst*, Call.*fr.* 195.23 Pf., *AP* 7.472.15 (Leon.Tarent.).'
***ἀχρύσωτος**, ον, *not gilded*, Inscr.*Délos* 1417*A*i150 (ii BC).
✝**ἀχρωμία**, ἡ, *effrontery, insolence*, *PPetaus* 26.14 (ii AD), *Gloss.* 2.254.
ἀχυρηγέω, before '*BGU* 698.22' insert '*PCair.Zen.* 176.145, al. (iii BC).'
ἀχύρικος τέλος, delete this entry (v. °ἀχυρός).
ἀχύριος, for '= ἀχυρός' read '*building to hold chaff* or *straw* (cf. °ἀχυρός)'; add 'prob. *chaff*, *PLand.* 146.6.8 (ii BC)'
ἄχυρον I, add '*OBodl.* i 237; app. used for grain and chaff together as separated from the straw, Theoc. 10.49, see Gow ad loc. and

ἀχυροπόρος   SUPPLEMENT   ἄωτος

CQ 49.227. **2** pl., *straw*, Hp.*Prog.* 4 (transferred from section I 1), *SEG* 9.11, 13, 41, 18.743, etc. (all Cyrene, iv/ii BC).'

*ἀχῠροπόρος, ον, ὁ, *chaff-seller*, cj. in *MAMA* 3.487 (Corycus, perh. erron. for -πώλης).

ἀχῠρός, for 'Ar.*fr.* 10 D.' read 'Ar.*fr.* 234 K.-A. (anap.)' and for 'should be read' read 'seems required by the metre in Ar.l.c.; elsewh. ‡ἀχύριος is possible; *storehouse for chaff* or *straw*, ἀ.· ὁ ἀχυρών. ἀχυροδόκη. ἀποθήκη τῶν ἀ. Hsch.'

ἀχῠρόω, after 'Polioch. 2' insert 'in building walls'

ἄχῠτος, add 'dub. sens., app. *not liquid, dry*, πλακοῦντες ἀ. *PCair.Zen.* 707.18 (iii BC)'

ἀχώνευτος, add '*not coated with pitch* (v. χωνεύω and cf. ἀκώνητος), *PCair.Zen.* 741.31; also -ητος, ib. 742.9 (both iii BC)'

ἀχώριστος **I**, add '**3** *irremovable*, πᾶσαν πόλιν .. ἔχειν ἐκ παντὸς τρόπου ἀχώριστον .. ἐπίσκοπον *Cod.Just.* 1.3.35 pr.'

ἀψευδέω **I**, add 'as a legal requirement, *not to commit fraud*, ὁ μὲν τοίνυν εἷς νόμος κελεύει ἀψευδεῖν ἐν τῇ ἀ[γορᾷ] Hyp.*Ath.* 14; D. 20.9' **II**, after '*observe faithfully*' insert 'ἀψευδήων (1 sg. subj.) ἂν τὰν συF(F)οικίαν'

ἀψεφέω, delete the entry.

ἀψεφής, add 'cf. pers. n. Ἀψεφίων And. 1.43, etc.'

ἀψίκορος, line 4, for 'Posidon. 41' read 'Posidon. 36 J.'

*ἀψίμοθος, ον, *kindling conflict, provocative*, Nonn.*D.* 28.92 (ὀψι- cod.).

ἀψινθᾶτον, add '*wine flavoured with wormwood*, *Edict.Diocl.* 2.18'

*ἀψινθοκραής, ές, *mixed with wormwood*, οἶνος *An.Boiss.* 3.410.

ἄψινθος, add '*BGU* 2358.3 (iv AD)'

†ἀψίς, Ep. and Ion. ἀψίς, ῖδος, ἡ, acc. ἄψιν Hes. l.c., *outer part of a wheel, felloe*, Hes.*Op.* 426, Hdt. 4.72.3, E.*Hipp.* 1233; fig., Ar.*Th.* 53; of water-wheel, *PLond.* 1177.200 (ii AD), *PFlor.* 218.4 (iii AD), *PThead.* 20.10 (iv AD); of potter's wheel, Nicaenet. 5. **b** *rim, band*, A.R. 3.138; perh. ξύλον εἰς τὰς ἀψῖδας τοῖς σφονδύλοις τοῦ κίονος *IG* 11(2).161.A70 (Delos, ii BC). **2** *segment* of circle, Arist.*Mete.* 371ᵇ28, 29; spec. *semicircle*, Hero *Geom.* 18.1. **3** *semicircle* of seats, D.C. 61.17.2, *GVI* 656.9. **II** *concavity, hollow* of a net, ἀψῖσι λίνοιο Il. 5.487, D.P.*Au.* 3.9, Opp.*H.* 4.146. **III** *arch*, Hld. 10.6.2, Lib.*Or.* 11.202; *triumphal arch*, D.C. 53.22.2, 55.2.3, 68.1.1, *CPHerm.* 127ᵛ.2.22; *arch* of bridge, *IGRom.* 3.887.7 (Cilicia). **2** *the vault of heaven*, Pl.*Phdr.* 247b, Archestr. in *Suppl.Hell.* 164.2, Luc.*Bis Acc.* 33. **3** the *curved* (visible) *course of the sun or moon*, τὴν ἡμερίαν ἀψῖδα E.*Ion* 88, σελάνας ἐς δεκάταν ἀψῖδα i.e. the tenth month, *Hymn.Is.* 38; also, the *course* of a planet in relation to the earth, Plin.*HN* 2.63, 64.

ἄψογος, after 'Poll. 3.139' insert '*ISmyrna* 550.9'

ἄψος, for '*juncture, joint*' read 'in pl., *limbs*' (deleting those wds. fr. line 3); add 'Od. 18.189, A.R. 3.676, Nic.*Th.* 332, τὰ δ' ἡμίβρωτα κέχυνται ἄψεα Opp.*H.* 2.294, Q.S. 1.252'

ἀψοφητί, before 'Pl.*Tht.* 144b' insert 'οἷον ἐλαίου ῥεῦμα ἀ. ῥέοντος'; add 'Plot. 3.8.5'

*ἄψυκτος, ον, *uncooled*, codd. in Pl.*Phd.* 106a. **2** fem., name of a precious stone, Plin.*HN* 37.148.

ἀψῡχέω, add '**2** *to be faint-hearted, lack courage*, *SEG* 35.989.20 (Cnossus, ii/i BC), Phot. α 3491 Th.'

ἄψῡχος **II**, add 'of dogs, X.*Cyn.* 3.2.3'

ἄω (A) **II**, delete the section (v. ἀέσκω).

ἄω (C) **I**, line 4, after '*satiate*' insert 'w. acc. and gen.' and at end after '(Il.) 5.289' insert 'cf. 9.489, 18.281, al.; w. acc. and dat., Il. 11.818; w. acc. only, ib. 24.211' add '**2** *administer liberally*, Nic.*Th.* 676, Al. 305, 331.' **II**, delete 'mostly'

*ἀωϊλιασταί, οἱ, *excavators*, *PCair.Zen.* 745.58 (iii BC); cf. °ἀωΐλιον 2.

ἀωΐλιον, for '**2** *cubic* πήχεις' read '*the cube of a royal double* πῆχυς' and after 'p. 118' insert '**2** the ἀ. being used to measure earth or sand removed, οἱ τὰ ἀ. ἐργαζόμενοι are *excavators of earth*, *PCair.Zen.* 745.61 (iii BC), cf. *PHib.* 1.100ʳ.3 (iii BC).'

ἀωρί, add 'ἔστιν ἀ. *AP* 12.116'

*ἄωρος, α, ον, *out of season, untimely*, of unripe fruits, Thphr.*CP* 2.2.2, Nonn.*D.* 2.78; *born at the wrong time*, of Dionysus, i.e. as a man, not as other immortals, Nonn.*D.* 20.206; of persons who die prematurely, *CEG* 75 (ἀhόριος, Athens, v BC), fem. pl. -ιαι ib. 696 (Peraea, iv BC); τὸν ἴδιον ἀγώριν (= ἀώριον) Side 114.9 (Rom.imp.); cf. mod. Gk. ἀγόρι *young boy*, ἄγωρος s.v. °ἄωρος; transf., ἀ. τύμβος *AP* 7.600 (Jul.Aegypt.).

*ἀωρόβιος, ον, cj. for δορόβιος, *SEG* 23.659 (Cyprus, iii AD).

*ἀωροθανής, ές, *of untimely death*, rest. in *ISmyrna* 533.6 (ii BC), *SEG* 28.1101, 1206.13, ἀνωρο- *AA* 48.124 (all Phrygia, iii AD).

ἄωρος (A), after 'ον' insert '[α, ον *MAMA* 7.345, *SEG* 26.1717 (Egypt, iii/iv AD)], ἀhōρος *CEG* 45 (Athens, vi BC), αὔωρος *MAMA* 7.313.3, ἄγωρος *SEG* 2.202 (Thespiae, iii/iv AD), cf. ἄγουρος, alleged by Eust. 1788.56 to be Thracian and Att. for *youth* (cf. ἄγωριν, s.v. °ἀώριος)'; line 2, after 'θάνατος' insert '*Carm.Conv.* 1.4 P.'; line 4, before 'ἀ. θανεῖν' insert 'of persons, οὐκ ἂν ἄωρος ἐὼν μοῖραν ἔχοι θανάτου Sol. 27.18 W.'; line 10, after 'Plu.*Sull.* 2' add 'neut. pl. as adv., *untimely*, Anacreont. 59.20 W.'; at end add 'see also ἄνωρος'

ἄωρος (C), delete this entry (v. °ὦρος C).

ἀωροσύνη, for 'dub. in *Epigr.Gr.* 414' read 'Bernand *IMEG* 73.4 (Abydus, ii AD)'

*ἀωρότης, ητος, ἡ, *immaturity*, τὴν σὴ[ν] ἀωρότηταν κὲ ἀθαλάμευ[τον] ἡλικίαν *SEG* 6.140.10 (Phrygia, iv AD), but perh. f.l. for ἀωροτάτην.

†ἄως, v. ‡ἕως.

ἄωτον, for '*the choicest .. wool*' read '*nap* or *pile* of wool'; for 'once of the *finest linen*' read 'of linen'; line 7, for '(Call.)*Hec.* 1.4.3' read '(Call.)*fr.* 260.57 Pf.'

ἄωτος, ον, for 'dub. in .. 1.138' read 'prob. in Call.*fr.* 399 Pf.'

B

*β-, for transcriptions of Lat. wds. beginning with 'v' see also οὐ-.
βαβάκινος, add '(βαβάκινον ‹καὶ βάκινον› La.)'
*βαβαλιστήριον, τό, cradle, Gloss. 2.361.
*βαβαλίστρια, ἡ, cradle, Leont.Byz.HP 1.2.4.
*βάβαλος, v. ‡βέβηλος.
*βαβούλιον, gloss on κύμβαλον, Hsch.; cf. °βακύλιον.
*βαβουτζικάριος, ὁ, name for person suffering from λυκανθρωπία, Cyran. 66 (2.23.18 K.), Suid. s.v. Ἐφιάλτης.
†βάβρηκες· τὰ οὖλα τῶν ὀδόντων. οἱ δὲ σιαγόνας. οἱ δὲ τὰ ἐν τοῖς ὀδοῦσιν ἀπὸ τῆς τροφῆς κατεχόμενα Hsch.; cf. βάρηκες.
*βαβύη· χείμαρρος. οἱ δὲ πόλις (?πηλός) Hsch.
*Βᾰβῠλωνάριος, ὁ, maker of Babylonian shoes or garments, SEG 8.138a (Palestine); cf. °καλιγάριος.
*Βᾰβῠλωνικός, ή, όν, Babylonian, Edict.Diocl. 8.1, al.
Βᾰβῠλώνιος, add 'λίθος σάρδιος ὁ Β. οὕτω καλούμενος Epiph.Const. in Lapid.Gr. 194.5, Cyran. 6.8.1 K.'
βαβύρτᾱς, after 'Hsch.' insert 'as pers. n., IG 1³.1157.18 (Athens, v BC), Plb. 4.4.5 (a Messenian), Maiuri Nuova Silloge p. 250 no. 3 (Hellen., ?Rhodes), Delph. 3(6).87.16 (ii BC), etc.'
*βαγεύει· πλανητεύει Cyr. (cf. Lat. vagor).
*βαγινάριος, ὁ, Lat. vaginarius, scabbard-maker, epitaph in ZPE 82.225 (Stobi, v/vi AD), Lyd.Mag. 1.46.
βάγος, add '= ἄγος (B)'
†βαγός, app. error by Hsch. for ἀγός.
βαδιστηλάτης, add 'carriage-driver, PWash.Univ. 49.11 (i BC)'
βᾰδιστικός II, add '-ὰ πορεία carriage-animals, PLond. 1973.3 (iii BC)'
βάδιστοι, add '(but βάϲρ›διστοι· βϲρ›αδύτατοι La.)'
*βάζιον, τό, a mineral (cf. πάζιον), SEG 20.670.7 (= Bernand Pan 51.6, i AD); sp. βασιον in OGI 660.3 (= Bernand Koptos 41, i AD); cf. Copt. basion.
βαθάρα, add '(dub., placed after βαταίνει β 320; cf. perh. °βαιθάρα)'
βαθμίς I, add 'ἔστρωσεν σὺν βαθμεῖσι IGLS 4034 (ii AD)'
βαθμός I 1, for 'Lxx .. 1127' read 'βαθμοῖς νυμφικοῖς ἐπεστάθη ὁ μοιχός Trag.adesp. 519 K.-S., ἐπὶ βαθμὸν οἴκου Lxx 1Ki. 5.5, πύργον [κα]ὶ βασμόν GDI 5524.10 (Cyzicus); rung of ladder, Lxx Trag. 221 (βασμ-), Edict.Diocl. 14.6' 2 and 3, delete the sections. II, of a genealogy, add 'Modest.Dig. 27.1.14.1, Just.Nov. 118.4'
βαθόημι, add 'β]αθόην[rest. in Alc. 288.2 L.-P'
βάθος 1, line 4, after 'Pl.R. 528b' insert 'μετρητικὴ δὲ μήκους καὶ ἐπιπέδου καὶ βάθους Lg. 817e' 2, after 'Plu.Pomp. 53' insert 'β. σεμνότητος Callistr.Stat. 10' 3, for 'bathos' read 'profundity'
βαθρικόν, for 'base' read 'pedestal with steps' and delete 'stairway'
*βαθρικός, ή, όν, of a βάθρον, τὸν δὲ ἀνδ[ριάν]τα ἀνέστησεν .. ἐπισκευάσας τὸ ὑπὸ πατρὸς γε[γον]ὸς β. ἔργον BCH 10.501 (Pisidia).
*[β]αθρόθυμα, ατος, τό, dub. sens., IG 1³.425.36 (v BC).
βάθρον 1, add 'cf. ἀλλὰ μὴν καὶ ἡ γῆ τούτου (sc. τοῦ ἠέρος) βάθρον Hp.Flat. 3' 3, add 'b horizontal timber (?sill) at the bottom of a door, ξύλα εἰς βάθρα ταῖς› θύραις τῶν πυλίδων IG 2².1672.149.' 4, add 'SEG 37.713 (Chios, Rom.imp.)' 6, add 'also β. Ἱπποκράτειον Gal. 18(1).303.12, 351.6, 747.34'
*βάθρος, ὁ, = βάθρον, Hierapolis 269.
†βᾰθύγλωσσος, ον, glossed by ἐλλόγιμος, Hsch., Suid. (also v.l. in Lxx Ez 3.5; cf. °βαρυγλ-).
βᾰθυδῑνήεις, add 'Il. 21.603'
βᾰθυδίνης, ου, add 'Orph.H. 38.17'
*βᾰθύκερως, ωτος, having deeply-curving horns, epith. of Isis, Lyr. Alex.adesp. 36.5 (= Mesom. 5.5 H.).
βᾰθυκνημῑς, after 'greaves' insert 'Ἄρεος θύγατρα' add '2 = βαθύκνημος, ἐρίπνη Nonn.D. 9.273.'
βᾰθυκύμων, for 'deep in waves' read 'having deep waves' and after 'Nonn.D. 23.320' insert 'βαθυκύμονος Ὠκεανοῖο D.P. 56'
*βᾰθυλιμενήτης, ου, ὁ, presiding over a deep harbour, epith., prob. local, of Apollo, SEG 15.766 (Artake nr. Cyzicus, ii/i BC), 33.1054 (Cyzicus, i BC).
βᾰθύνω II, intr., for 'sink deep .. 2.402' read 'go down or penetrate deeply, ὅταν αἱ .. ῥίζαι βαθύνωσι Ph. 2.402; of the mind's eye, id. 1.248; w. intent to hide, βαθύνατε εἰς κάθισιν Lxx Je. 30.2, 25'
βᾰθύπλουτος, add 'A.fr. 451g.3 R. (anap.); metaph., β. κραδίη APl. 40 (Crin.)'

†βᾰθύπορος, ον, wd. in broken context, perh. having deep paths, Lyr. adesp. 7(c).9 P.
βᾰθύρροος, add 'Ezek.Exag. 13'
βᾰθύς, line 2, add 'acc. fem. βαθέην Il. 17.466' I 2 b, comp., add 'Edict.Diocl. 25.2' 3 b, line 5, for 'AP 7.170' read 'AP 7.197 (Phaënn.)'; line 6, after 'Luc.DMar. 2.3' insert 'of death, v.l. in AP 7.170 (Posidipp. or Call.)' 4, add 'b not easily offended, of persons, Chrysipp.Stoic. 2.243.'
βᾰθύτης, for 'of mental profundity' read 'slowness to take offence'
†βᾰθύφωνος, ον, calque on Hebr. 'imkê-sāphâ, i.e. of unintelligible speech, Lxx Is. 33.19.
*βᾰθύχειλος, ον, = °βαθύφωνος, Lxx Ez 3.5.
βαῖα, add 'IG 14.839 (ed.: pers. n.), epitaph in Glotta 16.277 (Constantinople, vi AD), Suid. s.v. τεθή'
*βαίθ, or βέθ, representing Hebr. bath, Graecized as βάτος (C) q.v., β. ἐλαίου Lxx 3Ki. 5.25.
*βαιθάρα, ἡ, (perh. βαίθαρα, τά), dub. sens., ἐλαιῶν β. PAberd. 192.5 (ii/iii AD), prob. a measure; cf. °βαίθ, βάτος (C), ‡βαθάρα.
βαίνω, line 4, delete 'Eu. 76'; line 16, after 'Il. 1.428' insert '2 sg. aor. ἐβήσαο h.Ap. 141 (intr.), 3 sg. ἐπ-εβάσατο Call.Lav.Pall. 65 (trans.)'; line 19, after 'Th. 1.123, 8.98' insert 'περ-βέβαται Alc. 119.9 L.-P.' A II 1, delete 'in act. .. Them.Or. 21.248b)' and then read 'of the male, mount, cover, Pl.Phdr. 250e, Arist.HA 575ᵃ13; (fut.) med., Thgn. 185, Achae. 28'
*βαιοτελυπίου, Egyptian word for a kind of boat, PTeb. 701.260, PPetr. 3.129a.11, al., PLille 25.43 (all iii BC).
βαιός, line 14, after 'a little' insert 'Hes.Op. 418'; line 15, after '(S.)Tr. 335' insert 'also πρὸς βαιόν APl. 212.4 (Alph.)'
*βαίουλος, ὁ, dub. sens., GVI 1112.11 (Bithynia, ii AD); cj. in Hsch. s.v. οἴσυλος (= Lat. baiulus).
†βαίτη, ἡ, cloak made of skins, ἐκ τῶν ἀποδαρμάτων καὶ χλαίνας ἐπιέννυσθαι ποιεῦσι συρράπτοντες κατά περ βαίτας Hdt. 4.64.3, Sophr. 38, Theoc. 3.25, Herod. 7.128. II tent of skins, S.fr. 1031 R. 2 covered building in a market-place, IG 5(2).268.48 (Mantinea c.10 BC/c.AD 10), Inscr.Magn. 179.12, 15 (i AD), SEG 26.1652.5 (Syria, ii AD).
βαιτοφόρος, delete 'prob. for βαττ- in'
*βαιτύλιον, τό, dim. of βαίτυλος, in pl., Ph.Bybl.ap.Eus.PE 1.10, Dam.Isid. 94, 203.
βαίτῠλος, add 'prov. βαίτυλον ἂν κατέπιες, of greedy persons, Apostol. 9.24; a precious stone, Plin.HN 37.135; as title of Zeus, cf. Semitic bethel, Διὶ βετύλῳ SEG 7.341 (Dura, iii BC)'
βαιών I, for this section read 'kind of fish (= βλέννος acc. to EM 192.52), Epich. 64 (Ath. 7.288a)' II, add 'cf. βάϊς, βάϊον II'
*βακάνιον, τό, dim. of βάκανον, POsl. 48.10 (i AD).
βάκανον, for 'cabbage' read 'perh. Althaea cannabina' and delete 'also cabbage-seed'
*βᾰκέλας, α, ὁ, = βάκηλος, cj. in Alex.Aet. 9.2 (Coll.Alex. p. 127; codd. ‡μακ-).
βάκηλος I, add 'Suet.Blasph. 52 T.' at end add 'cf. κάβηλος, app. same wd. w. metath.'
βάκκαρις, dat., for 'Hippon. 41' read 'Hippon. 104.21 W., βακκάρει Achae. 10'
*βάκκερα, ἡ, plant sacred to Dionysus, Cyran. 22 (1.8.8 K.).
*βάκρον, v. °βάκτρον.
βακτηρία, see also βατηρία.
βακτήριον, for 'dim. of βακτηρία' read 'var. of βακτηρία'
*βακτηριοφόρος, gloss on καλαυρόφις, Hsch.
*βακτρᾶς, ᾶ, ὁ, dealer in βάκτρα, SEG 34.221 (?Athens, iv AD).
βάκτρον, add 'app. Cypr. var. pa-ka-ra (pl.) βάκρα, shafts of bronze lances, ICS 218, cf. 368a (Add. p. 415, v BC)'
*βακύλιον, gloss on κύμβαλον, Hsch.; cf. °βαβούλιον.
*Βακχάζω, in expl. of Διαγόρας, .. διὰ τὸ τοὺς μύστας βακχάζειν, τουτέστιν ᾄδειν τὸν Ἴακχον δι' ἀγορᾶς βαδίζοντας Hsch. s.v. Διαγόρας.
*Βακχεασταί, οἱ, worshippers of Bacchus, IGBulg. 20.3 (Dionysopolis, iii BC); cf. Βακχιαστής.
Βακχεία, line 3, after 'E.Ph. 21' insert 'Bacchic rite, Ael.VH 13.2'
Βακχεῖον 2 b, for this section read 'sanctuary or shrine of Bacchus, SEG 37.601 (Thrace, iii AD); cf. β.· τελεστήριον, νάρθηξ Hsch.'

65

Βακχειώτης, ου, ό, epith. of Dionysus, *Lyr.adesp.* 318 in *SLG*; cf. ‡*Βακχιώτης*.

Βακχιαστής, add 'cf. °*Βακχεασταί*'

Βακχικός, after 'D.S. 1.11' insert '*βαχχικὸς ὀρχηστής* epigr. in *IClaudiopolis* 83'

Βάκχιος (s.v. *Βάκχειος*) **II**, as subst., add 'Ar.*Ach*. 263, E.*Ba*. 67, 225, etc., *Ion* 550; *ὦ Βάχχιε* Philod.Scarph. 10'

Βακχιών, add 'at Ceos (Poessa), *IG* 12(5).1100.4 (v BC)'

βακχιώτης, after 'lyr.' insert 'Dor. -*τᾶς*' and add 'cf. °*Βακχειώτης*'

Βάκχος, after lemma insert '(sp. *Βάχχ-* *SEG* 26.683, Thessaly, iii BC), etc.'

βᾱλάβαθρον, v. °*μαλάβαθρον*.

βᾰλάναγρα, line 1, for '*key* .. *βάλανος* II 4' read '*key or hook for pulling out the bolt-pin of a door*'; line 2, for 'in pl. = *βάλανος* II 4' read 'also, *bolt-pin*' and add '*App.Anth*. 5.17 (Hedyl.)'

βᾰλᾰνάριον, add '*SEG* 7.417.12 (Dura, iii AD)'

βᾰλᾰνεῖον 1, add '*ἐπὶ βαλανείων* = Lat. *a balneis*, *IG* 5(1).669, *IG* 14.1052, *Ann.Épigr*. 1938.84, *PLond*. 1178.59 (ii AD)'

βᾰλᾰνειοφῠ́λαξ, *ακος, ὁ, caretaker of the baths*, dub. l. in *OMich*. 102 (iv AD).

βᾰλᾰνευτικός, add '*τὸ -όν* *bath tax*, *Pap.Lugd.Bat*. xxv 24.3 (AD 94)'

βᾰλᾰνικός, add '*-οῦ* sc. *ἐλαίου* *PSI* 481.6'

βᾰλᾰνίσσα, add '*Suppl.Mag*. 42 (Hermoupolis, iii/iv AD)'

βᾰλᾰνος II 7, for '*ballot-ball*' read '*lot*'

Βαλβίλλεια (-ηα), *τά*, *games in honour of Balbillus*, held in Ephesus, *IG* 14.746 (sp. *Βαρβ-*, Naples, i AD), etc., *SEG* 34.1314 (Lycia, c.AD 90), *IEphes*. 642, 686, al.; also held at Smyrna, ib. 1123.

βάλερος, add 'also *βάλλερ*[*ος*] *IGC* p. 99 B29 (Acraephia, iii/ii BC)'

βαλιδικός, delete the entry (v. *ZPE* 7.54).

βαλλαντᾶς, *ᾶ, ὁ, maker of βαλλάντια*, *ITyr* 95 (ed.: pers. n.).

βαλλάντιον I, for '[Simon.] 178' read '*AP* 5.159.3 (Simon.)'; add 'perh. *bag containing fixed sum of money*, *PBeatty Panop*. 2.93, 94, 97'

βάλλερος, v. °*βάλερος*.

βαλλίζω, delete 'in Sicily and Magna Graecia' and add '*GVI* 1112.11 (Bithynia, ii AD), *IHadr*. 101 (ii AD)'

βαλλιστάριος, *ὁ*, Lat. *ballistarius*, Lyd.*Mag*. 1.46, *SEG* 7.154 (E. of Palmyra); also *βαλλιστράριος* *IGChr.Russie* 7, Just.*Nov*. 85.2.

βάλλω A I, line 4, for '*ἐλβών*' read '*ἐλθών*' **II 1**, line 5, delete 'in prose abs.'; line 7, after 'Th. 4.33' insert 'cf. Call.*fr*. 191.79 Pf.' **2 b**, line 3, after '(S.)*Ph*. 1028' insert 'esp. of *throwing* into prison, *εἰς φυλακήν β*. *Ev.Matt*. 18.30, *Apoc*. 2.10, Arr.*Epict*. 1.1.24, al., *PTeb*. 567 (i AD)' **6**, line 6, before '*pour*' insert '*lay the foundations* of, *ἄστυ βαλεῖν* A.R. 2.849; cf. infr. B. I 4' **7**, line 3, after 'metaph.' insert 'abs. Call.*Epigr*. 8.4.Pf.'; for '*εὖ* or *καλῶς*' read '*καλῶς* or *εὖ*' **II**, add '**8** *bury illicitly* in a tomb, *SEG* 37.1086 (Pontus, ii/iii AD), *MAMA* 6.325 (iii AD), 1.167 (c.AD 400), *TAM* 2.1148, 4(1).283, Ramsay *Cities and Bishoprics* 1.233.80.' **B I 2**, add '**b** *cast lots*, *πάλους ἐβάλοντο* Call.*fr*. 119.2 Pf.'

βαλνικάριος, *ὁ*, *bath attendant*, Corinth 8(3).534 (v/vi AD).

βάλτιον, *τό*, *belt*, *PMich*.III 217.19 (-*ιν*, iii AD, cf. Lat. *balteus*).

βάλτιος, *ὁ*, *belt*, *Edict.Diocl*. in *SEG* 37.335 iii 17, al. (written -*ις*).

βαμβαίνω, for '*chatter with the teeth*' read '*make inarticulate sounds, babble, mumble*, etc.'

βαμβακύζω (s.v. *βαμβαίνω*), for pres. ref. read 'codd. in Hippon. 32.3 (v. °*βαμβᾰλ-*)'

βαμβᾰλϊαστύς, *ύος, ἡ, babbling*, v.l. for *κρεμβ-* in *h.Ap*. 162; cf. *βαμβαίνω, βαμβαλύζω*.

βαμβᾰλύζω (s.v. *βαμβαίνω*), before 'Phryn.' insert 'cj. in Hippon. 32.3 W., *Iamb.adesp*. 38.4 W.'

βάμμα II, for this section read 'something in which food is dipped, e.g. vinegar (glossed by *ὄξος*, Sch.Nic.*Al*. 49, al.), Nic.*Th*. 87, 622, *Al*. 369, al.; *β. σίμβλων* = *ὀξύμελι*, id.*Al*. 49'

βάνατα, *ἡ*, app. some kind of garment, *Edict.Diocl*. 19.55, 57, perh. cf. Lat. *pannus*.

βάναυσος II 1, add 'Iamb.*Protr*. 5' **3**, for 'later, *fastidious*' read '*rude, coarse-mannered*'

βάνδον, *τό*, also *βάνδα, τό*, Sch.Procl. *in Ti*. 1.462.11, *military standard* (Goth. *bandwa*), Procop.*Vand*. 2.2.1. **2** *company of infantry*, *IGLBulg*. 89.3 (vi AD), perh. also *IApam*. 136. **3** *military area*, Sch.Procl. l.c.

βανδοφόρος, *ὁ*, *standard-bearer*, Procop.*Vand*. 2.10.4.

βανιάτωρ, *ορος, ὁ*, Lat. *balneator*, *PKlein.Form*. 980.2 (vi AD).

βανωτός, for '*vase*' read '*jar*'

βάπτης, add '**II** name of gem, Plin.*HN* 37.149.'

βαπτίζω 1, line 7, after 'J.*BJ* 4.3.3' insert 'transf., *overwhelm, flood*, *ἐλευθέραν ἀφῆκε βαπτίσας ἐρρωμένως* Aristopho 13 K.-A.' **2**, delete 'Aristopho 14.5'

βαπτιστήριον, add '**II** *baptistery*, *SEG* 32.1065 (Rome, Chr.).'

βαπτιστός, *ή, όν*, dub. sens., *βαπτιστοῖο .. μελάθρου GVI* 134.3 (Tyana, iii AD).

βάπτρον, *τό, charge for dyeing*, *SB* 12314.16 (ii AD).

βάπτω I 1 - 3, for these sections read '*immerse* in a liquid, *dip*, *ὡς δ' ὅτ' ἀνὴρ χαλκεὺς πέλεκυν .. εἰν ὕδατι ψυχρῷ βάπτῃ* (so as to temper the red-hot steel) *Od*. 9.392; *ἔβαψεν ἰούς* S.*Tr*. 574; *χιτῶνα τόνδ' ἔβαψα* ib. 580; *β. εἰς ὕδωρ* Pl.*Ti*. 73e, cf. Emp. 100.11; *τάρια θερμῷ* Ar.*Ec*. 216; *εἰς μέλι, εἰς κηρόν* Arist.*HA* 605ᵃ29, *de An*. 435ᵃ2, *β. τὸν δάκτυλον ἀπὸ τοῦ αἵματος* Lxx *Le*. 4.17; pass., *βαπτόμενος σίδηρος* iron *in process of being tempered*, Plu. 2.136a. **b** of slaughter in trag., *ἐν σφαγαῖσι βάψασα ξίφος* A.*Pr*. 863; *ἔβαψας ἔγχος εὖ πρὸς Ἀργείων στρατῷ*; S.*Aj*. 95; *φάσγανον εἴσω σαρκὸς ἔβαψεν* E.*Ph*. 1578 (lyr.); in later prose, *εἰς τὰ πλευρὰ β. τὴν αἰχμήν* D.H. 5.15. **2** *colour by immersion, dye*, *β. τὰ κάλλη dye* the beautiful cloths, Eup. 333; *β. ἔρια ὥστ' εἶναι ἀλουργά* Pl.*R*. 429d; *εἵματα βεβαμμένα* Hdt. 7.67.1; *τρίχας βάπτειν AP* 11.68 (Lucill.); poet., *φάρος τόδ' ὡς ἔβαψεν Αἰγίσθου ξίφος* A.*Ch*. 1011; humorously, *βάπτειν τινὰ βάμμα Σαρδιανικόν dye* one in the [red] dye of Sardes, i.e. give him a bloody head, Ar.*Ach*. 112; fig., *βέβαπται β. κυζικηνικόν* he has been dyed in the dye of Cyzicus, i.e. is a thorough coward, id.*Pax* 1176 (v. Sch.); transf., of gilding and silvering, Ps.-Democr. p. 46 B. **3** *dip* a vessel in order to draw water, *ἀρύταιναν .. ἐκ μέσου βάψασα τοῦ λέβητος ζέοντος ὕδατος draw* water *by dipping* the bucket, Antiph. 25, cf. Thphr.*Char*. 9.8; *βάψασα ποντίας ἁλός* (sc. *τὸ τεῦχος*) *having dipped* it *so as to draw* water from the sea, E.*Hec*. 610; cf. *ἀνθ' ὕδατος τᾷ κάλπιδι κηρία βάψαι* Theoc. 5.127.'

βάραγχος, delete the entry.

βάραθρον, add '**IV** pl., gloss on *λάσανα*, Hsch. (*τὰ βάθρα* La.).'

βάρακος, after 'Hsch.' insert 'a freshwater fish in *IGC* p. 99 B 21 (Acraephia, iii/ii BC)'

βάρβαξ, add 'comic gloss in *CGFP* 343.7; cf. pers. n. *Βάρβακς* *IG* 12(3).543 (Thera, vii BC); *Βάρβακος* *Inscr.Cret*. 2.xi 3.6, 19 (Dictynna, i BC)'

βαρβαρικάριος, *ὁ, brocade-maker*, *Edict.Diocl*. 20.5, 7, *ITyr* 122.

βαρβᾰρικός I, add '*πρεσβείας .. βαρβαρικάς* i.e. *embassies to non-Greeks*, *IHistriae* 12.9 (iii BC)'

βαρβᾰρισμός, add '**II** *siding with non-Greeks*, in quot. spec. Persians, *SEG* 22.506.9 (Chios, letter of Alexander).'

βαρβᾰρόλεξις, *εως, ἡ, use of foreign speech*, Consentius p. 386 K.

βαρβᾰρόμῡθος, *uttered in a foreign tongue*, cj. in Ar.*Pax* 753 (v. *βορβορόθυμος*).

βαρβᾰρόστομος, *ον, speaking in a barbarous manner*, *Trag.adesp*. 696 K.-S.

Βαρβίλληα, *τά*, v. °*Βαλβίλλεια*.

βάρβιτον (s.v. *βάρβιτος*), after 'as' insert 'sts.'; delete 'Neanth. 5' and for 'etc.' read '*AP* 7.23b, ib. 588 (Paul.Sil.), Nonn.*D*. 42.253'

βάρβιτος, for '*musical instrument of many strings* .. Theoc. 16.45)' read '*type of bowl lyre with long arms*'; delete 'freq. used for the *lyre*'; line 3, for 'etc.' read 'Theoc. 16.45; in 4th century distd. fr. *κιθάρα* and *λύρα* Arist.*Pol*. 1341ᵃ40, Anaxil. 15 K.-A.'; line 4, for '(*Anacreont*.) 14.34' read '15.34 W. and *AP* 7.25 (Simon.)'; at end add '*βάρμιτος* Aeol. acc. to *EM* 188.21; cf. *βάρμος*'

βάρδοι II, delete the section.

βάρδος, *ὁ*, dub. sens., (refers to commodity which is purchased), *BGU* 276.11 (iii AD); cf. perh. °*βαρδόσημος*.

βαρδόσημος, *ον*, dub. sens., adj. applied to garment, *POxy*. 3860.20 (iv AD); cf. °*βάρδος*, or perh. = *παρδο-*, *with marks like a leopard*.

βᾰρέω, line 1, after 'Aeol. *βορ-*, v. infr.' insert 'line 7, but see also °*βορέομαι*' **I 1**, line 3, after 'Id.*Sol*. 7' insert 'Hdn.*Philet*. 212'; line 7, transfer '*κῆρ .. 25.17*' to section II line 8, after 'pass., pres.' add '**3** *charge with a legal obligation*, *Cod.Just*. 6.4.4.18, Just.*Nov*. 43 pr.'

Βαρζοχαρα, *ἡ*, epith. of the Iranian goddess Anaitis, *BE* 1968.538 (i BC) (see also *ZVS* 84.207 for etym.).

βαριοτομέω, *cut* (in quot., papyrus) *from a boat*, *BGU* 1121.20.

βᾶρις 2, at beginning for 'later' read 'wd. of uncertain origin, distinct. fr. *βᾶρις* 1'; for 'Kalinka .. 142' read '*IGBulg*. 400.5' and before it insert '*Didyma* 492.18, etc. (iii BC), *Inscr.Magn*. 122 d 4-8 (Rom.imp.)'; last line, transfer '(Egyptian word)' to the end of section 1.

βαρκαῖος, *ὁ*, kind of fish, Theognost.*Can*. 52.

βάρμῑτος, v. °*βάρβιτος*.

βάρμος, *ὁ*, = *βάρβιτος*, Alc. 70.4 L.-P., Phillis ap.Ath. 636c; also *βάρωμος* Sapph. 176 L.-P. (acc. to Euph. 180 vGr.) but perh. read *βάρμος*.

βάρνᾰμαι, for '*IG* .. (Corc.)' read '*CEG* 145.2 (Corc., c.600 BC), 155.2 (Paros, 476/5 BC), 6.2 (Athens, ?447 BC), al.'

βάρος II, add 'the *burden* of the womb, *ἐκύησε ἐγ γαστρὶ Κλεὼ βάρος* *IG* 4².121.9 (Epid., iv BC), *ἀποθέσθαι τὸ β*. *PBremen* 63.4 (ii AD), Artem. 5.30, *θηκαμένη τὸ β*. *SB* 5718' **VII**, add 'of language, D.H.*Dem*. 34, *Th*. 23 (in both places coupled w. *τόνος*)' add '**X** *legal obligation*, Just.*Nov*. 1.1.2.'

βαρουλκός, *ὁ*, "*weight-lifter*", an arrangement of interacting cogs

βαροφίτης described by Hero, and the title of a work by him, Papp. p. 1060.6 H.; also **βαρυολκός**, Olymp. in *Alc.* 191.15.

***βαροφίτης**, ου, ὁ, app. *crusher of snakes*, magical inscr., *SEG* 33.1551 (Egypt, iii AD), perh. also in *IGLS* 1098.

βαρυάης I, for '*breathing hard*' read '*marked by deep breathing*'

βαρύβρεμέτης, add 'epigr. in *SEG* 34.1308 (Side, i BC/i AD)'

***βαρυγαύτης**, ου, ὁ, garment, perh. the same as παραγαύδης, *PMich.*xv 752.42 (ii AD).

†**βαρύγλωσσος**, ον, *heavy of tongue*, i.e. *hesitant in speech*, λαός Lxx *Ez.* 3.5; *vitriolic*, of Hipponax, Suet.*Blasph.* 39 T.

βαρόδουπος, for last ref. read '*IHadr.* 80.13 (ii AD) and add 'Coluth. 55'

βαρυηκοΐα, add 'Gal. 12.533.6, Sever.*Clyst.* p. 18 D.'

βαρυκάρδιος, for '*heavy, slow of heart*' read '*having base inclinations*'

***βαρυκηδής**, ές, *causing great sorrow*, πότμος Bernand *IMEG* 67.7, ἄχθος *GVI* 1397.3 (Pisidia, ii AD).

βαρύλλιον, add '*levelling instrument*, Elias *in Porph.* 21.31, *in Cat.* 117.100'

βαρύμηνις, add 'of Pan, Orph.*H.* 11.12, of Zeus, ib. 20.4, of Dionysus, id. 45.5'

βαρύμοχθος I, add '**2** *greatly distressed*, *AP* 12.132 (Mel.).'

βαρύνω I 2, line 4, for 'Simon. 184.5' read '*AP* 7.25.5 ([Simon.])'; add 'ὄμματά τ' ἀστράπτοντα βαρυνόμενόν τε νόημα Il.Pers. 4.8 B.' add '**3** *charge with a legal obligation*, Just.*Nov.* 39.1 pr.'

†**βαρυολκός**, ὁ, v. °βαρουλκός.

βαρυπένθητος, for '*mourning heavily*' read '*deeply mourned*'

***βαρυπεψία**, ἡ, *indigestion*, Corp.Herm.*ad Amm.* in *Physici* 1.395.22.

βαρύς I 1, line 3, after 'Arist.*Cael.* 310^b25' insert 'Ath. 3.115e' **2**, lines 10/11, delete '*indigestible*, Ath. 3.115e' **4**, add 'of oaths, ὅρκος γὰρ οὐδεὶς ἀνδρὶ φηλήτῃ βαρύς S.*fr.* 933 R.; Call.*fr.* 75.22 Pf.' **II 4**, delete the section. **III 1**, penultimate line, delete '(οὐ opp. οὗ)'

***βαρυσμός**, ὁ, = βαρύτης I, β. τοῦ σώματος ὅλου Gal. 7.466.1 (s.v.l.).

***βαρυσταθμία**, ἡ, *full weight*, of coinage, *PLond.* 1405.3 (see also *ZPE* 85.298, AD 709).

βαρύσταθμος, after 'Ar.*Ra.* 1397' insert 'τὰ πράγματα .. βαρύσταθμα fr. 415 K.-A.'

βαρύστομος 3, for '*cutting deeply*' read '*heavy-bladed*'

βαρυτιμέω, for 'τιμουλκέω' read 'τιμιουλκέω'

βαρύφθονος, add '*EA* 13.19 no. 4, *GVI* 1375 (both Phrygia, Rom.imp.)'

***βαρύχειρ**, χειρος, ὁ, ἡ, *heavy for the hands*, παλά epigr. in *Lindos* 699b.1 (ii BC).

***βάρωμος**, v. °βάρμος.

***βαρών**, ῶντος, ὁ, Lat. *baro*, *blockhead*, *BGU* 836.1 (vi AD).

βασανίζω I, line 5, after '*prove*' insert 'τὸ πρᾶγμα Pl.*Euthd.* 307b' **II**, for 'αἰωνίοις' to end of section read 'fig., of the earth as being tortured to yield its produce, Philostr. *VA* 6.10. **b** *to be subjected to cruel treatment, be racked*, Ev.Matt. 8.6, θηρίοις βεβασανισμένοις Philostr.*VA* 1.38; of things, πλοῖον .. βασανιζόμενον ὑπὸ τῶν κυμάτων Ev.Matt. 14.24; *to be racked by emotions*, *SEG* 31.895 (Africa).'

βασανιστήριον I, delete 'of the stocks' and transfer 'Sm.*Je.* 20.2' to section II **II**, add 'Lxx 4*Ma.* 8.12, al.'

***βασανιστός**, ή, όν, *tested*, σφραγίδιον β. *IG* 2².1542.13 (iv BC).

βάσανος II, lines 4/5, for 'βάσανον .. *Lg.* 648b' read 'w. λαμβάνειν, *make a test*, τῶν πολιτῶν .. β. .. λαμβάνειν ἀνδρείας τε πέρι καὶ δειλίας Pl.*Lg.* 648b; δυνάμεις β. λαβοῦσαι *being tested*, D.H.*Dem.* 16'

***βασάρα**, v. °βασσάρα.

†**βασείδιον**, v. °βασίδιον, °βασίδιος.

***βασίδιον**, τό, dim. of βάσις, *base*, σὺν τοῖς ὑπὸ τοὺς πόδας βασιδίοις χαλκοῖς Inscr.*Délos* 1417B.i136 (ii BC), *SB* 9238.20, 9321.11 (both ii/iii AD).

***βασίδιος**, ον, *fitted with a base*, *BGU* 781 iii 6 (i AD), *CPR* viii 62.23 (vi AD).

βασίλεια, line 1, for 'βασιλέα' read 'βασίλεα'

βασιλεία I 2, add '**b** *position of queen*, Lxx *Es.* 1.19.' **4**, delete the section. at end add 'Myc. *qa-si-re-wi-ja*, precise sense obscure'

βασιλείδης, for '*prince*' read '*descendant of kings*' and add 'S.*Ant.* 941 (cj.)'

βασίλειον I 1, add '**c** *dominion, reign*, τὴν Τύχην .. τοῦ ἀνεικήτου βασιλείου *BCH* suppl. 8 no. 59 (Maced., iv AD).' **IV**, for 'at Olbia, *IPE* 1.105' read 'later of festivals founded by Hellenistic kings, *IG* 2².3779.19 (iii BC)'

βασίλειος, lines 4/5, delete 'used by Trag. in lyr.' **3**, after 'Crates Com. 2' insert 'of a type of fig, β. σῦκα Philem.Lex.ap.Ath. 3.76f, Poll. 6.81'

βασιλεύς, line 1, for 'Cypr. .. 135 H.' read 'Cypr. *pa-si-le-wo-se βασιλέϝος ICS* 15, 217.6, al.'; lines 2ff., now nom. pl., add 'Elean -άες (v. infra, II 2)' **I 1 b**, add 'Θεῷ Βασιλῖ *TAM* 5(1).167 (Saittae, iii AD)' **IV 2**, delete 'β. σῦκα .. Poll. 6.81' (v. °βασίλειος) and substitute 'of rivers, [ποταμῶν ἡμε]τέρων β. Call.*fr.* 7.34 Pf.: sup., π. βασιλεύτατος ἄλλων D.P. 353' at end delete words in brackets and add 'Myc. *qa-si-re-u*, *chief* (not *king*)'

βασιλεύω I 1 a, line 9, after 'Lxx 1*Ma.* 1.16' insert 'τῶν πρὸ ἡμῶν βεβασιλευκότων Just.Const.Δέδωκεν pr.'; at end transfer 'hence .. *Sull.* 12' to line 10 to follow 'Arist.*Pol.* 1284^b39, etc.' and add '(s.v.l.).' **c**, after 'cf.' insert '*IG* 2².4067 (βασίλει- lapis)'

βασιληΐς 1, add 'of Rome, 'Ρώμης βασιληίδος *IHadr.* 61 (ii AD), *IEphes.* 802'

***βασιλίδιον**, τό, name of an eye-salve, Asclep.Pharm.ap.Gal. 12.788.16.

βασιλικός I 2, line 4, after 'γεωργοί' insert 'i.e. cultivators of royal land leased to them'; add 'of coinage of Diocletian, νομίσματα β. *TAM* 4(1).352' add '**5** *of the master of a feast* (συμποσίαρχος), β. νόμους description of works by Xenocr., Speus. and Arist., Ath. 1.3f.' **II 2**, add '**b** *nave* of a church, *AD* 12.69 (Lesbos)' **3 d**, add 'perh. fig., of an Epicurean doctrine, Diog.Oen.*fr.* 51'

†**βασίλιννα**, ἡ, (person having some of the characteristics or functions of a) *queen*, D. 59.74, Men.*fr.* 652a K.-Th.

***Βασίλιος**, ὁ, name of a Cretan month beginning 23 Aug., Hemerolog.*Flor.* p.77 (p. 18 K.).

βασιλίς I 1 a, for 'Imperial *princess*' read '*empress*' add '**c** = Lat. *regina sacrorum*, *IGRom* 4.1687.' add '**III** ὑπόδημα γυναικεῖον καὶ αὐλητικόν Eratosth.ap.Hsch. s.v. βασιλίδες; cf. βασιλίσκος v 1.'

βασίλισσα 2, at end delete 'also βασίλιννα .. 907' add '**5** as title of goddesses, of Demeter, Mitchell *N.Galatia* 129, of Isis, *SEG* 24.1244 (both Rom.imp.).'

***βάσιον**, v. °βάζιον.

βασκαίνω, after 'evil eye, etc.' insert '*CEG* 455 (Amorgos, vi BC), Euph.*fr.* 175'

βασκανία 1, add 'personified, Call.*fr.* 1.17 Pf., *PMag.* 4.1451'

βάσκανος II 1, add '**b** *mean, niggardly*, Lxx *Prov.* 23.6, 28.22, *Si.* 14.3, 18.18, 37.11.'

***βασκαντήρ**, ῆρος, ὁ, = βάσκανος, Euph. in *Suppl.Hell.* 429.23.

βασκαύλης, for 'perh. = Lat. *vasculum*' read 'vessel, perh. = Lat. *bascauda*; cf. also μασκαύλης'

***βασκέλειον**, τό, prob. for βασκαύλειον, dim. of βασκαύλης, ἐλέου β. *PColl.Youtie* 84.10 (iv AD).

βάσκον, delete the entry (v. ‡βάσκω).

***βάσκυλα**, τά, Lat. *vascula*, *PRyl.* 27.82 (iv AD).

βάσκω, for 'only imper.' read 'usu. imper.', and at end for '(βάσκου .. Hsch.)' read 'ἔβασκε Alc. in *SLG* 262(a) ii 10, βάσκε· πορεύου Hsch., βάσκον· ἐχώρουν id.; see also ‡ἐπιβάσκω'

βασσάρα II 1, add 'Λυδῶν δὲ χιτών τις βασάρα (sic) Διονυσιακός, ποδήρης Poll. 7.60'

***βάσσος** (B), (βᾶσσος La.) οὐδετέρως· ἡ βῆσσα Hsch.

***βαστᾰγεύς**, έως, ὁ, perh. *carrier of sacred objects*, Inscr.*Délos* 2628.

βαστᾰγή, add '*Cod.Just.* 12.57.3'

βαστάζω I 2, add 'Sm.*Pr.* 4.8' **IV**, delete the section transferring exx. to section II 2 (act.) at end for 'Plb.' read 'Arist.'

βαστέρνιον, line 1, delete '*Cod.Just.* 8.10.12' add '**2** perh. some kind of covered passage in a building, *Cod.Just.* 8.10.12.'

***βαστέρνος**, ὁ, prob. = βαστέρνιον, Hdn.*fr.* p. 26 H.

βαστραχαλίσαι, add 'also βαστραχηλίζει Hsch. (La.)'

***βασυμνιάτης**, ου, ὁ, perh. *maker of βασυνίαι*, *MAMA* 3.645 (Corycus).

***βάταλον**, τό, *clapper for marking time*, cf. °κρούπεζα 2, Sch.Aeschin. 1.126, Phot. s.v. κρούπεζαι.

βατάνη, add '*PAlex.* (1964)31.4 (iii/iv AD)'

βατάνιον, add 'see also πατάνιον'

βατεία, delete the entry (v. °βατία).

***βατελλίκιον**, v. °πατελλίκιον.

***βατέρνος**, dub. sens., Hdn.*fr.* p. 26 H.

βατεύω, for 'perh. *trample, damage*' read '*cover* (of animals)'

βατήριον, delete the entry (v. °βατήριος).

***βάτηριος**, α, ον, *of or connected with* (animal) *copulation*, μηδ' ἀλόγοις ζῴοισι βατήριον ἐς λέχος ἐλθεῖν Ps.-Phoc. 188.

βάτης, delete 'treads or' and for 'expld. .. ἀναβάτης' read 'β. πίθηκος· ἀναβάτης Hsch.'

***βατιά**, ἡ, *thicket*, Pi.*O.* 6.54 (-εία, -ία codd.).

βατιάκιον, add '*PCair.Zen.* 120.2 (iii BC)'

βάτιον, add '*AB* 224'

βάτον, add '*SB* 11064 (i AD)'

βάτος (C), after 'ὁ' insert '(v. °βαΐθ'); delete '= Egypt. ἀρτάβη or Att. μετρητής'

βατός, for '*permissible*' read '*possible*' and add 'Marcellin. *Vit.Thuc.* 35, Just.Const.Δέδωκεν 12'

βατράχειος, after 'Ar.*Eq.* 523' insert '(edd. nonnulli βατραχειοῖς, v. βατραχειούς s.v. βατραχιοῦν)'

βατραχίτης, add 'Cyran. 39.27 (1.21.3 K.)'

***βατραχῖτις**, ιδος, ἡ, (sc. λίθος), = βατραχίτης, *PSI* 1180.52 (ii AD).

***Βατρόμιος**, v. °Βοηδρόμιος II.

*Βᾱτρομιών, v. ‡Βοηδρομιών.
†βαττολογέω, talk aimlessly, waffle, Ev.Matt. 6.7, Simp. in Epict. p. 91 D.
*βάττος· τραυλόφωνος, ἰσχνόφωνος Hsch. (false inference from Hdt. 4.155).
*βαυβύζω, bark, of a dog, PMag. 36.157.
βαυβώ, add '(acc. to Hsch., s.h.v.)'
βαυκάλιον, add 'II as a capacity measure, containing 3,000 bricks, POxy. 2197.3, al. (vi AD).'
*βαυκίδια, τά, dim. of βαυκίδες, Poll. 7.94.
*βαυκός, ή, όν, app. soft, effeminate or sim., Arar. 9 K.-A.; cf. βαυκά· ἡδέα Hsch.
*βαυκύων, ὁ, app. barking dog, PMag. 4.1912 (ref. to Cerberus).
βαῦνος, add 'the properisp. accent is Attic acc. to Trypho and Philemon ap. Hdn.fr. 53 H.'
βἀφή I, line 5, delete 'prob. poet. for σιδήρου β.' in' and 'v. Sch. ad loc.'   II, line 3, for 'the saffron-dyed robe' read 'sprinklings of saffron'
*βαφωρι-, article of women's dress (perh. cf. μαφόριον), PNess. 18.37 (vi AD).
*βαχχ-, v. ‡βακχ-.
βάψιμος, add 'PMich.inv. 3731 i 13 (Tyche 1.185, ii/iii AD)'
βδελύσσομαι I 1, delete 'to be sick'   2, after 'id.Ach. 586' insert 'Pl.R. 605e'   add '3 w. part., to be sick of doing, καρυκκοποιοὺς προσβλέπων βδελύσσομαι Achae.fr. 12 S.'   II, delete 'cause to stink'
βέβαιος I 1 b, add 'in Locr. as κύριος, valid, in force, τεθμὸς ὅδε περὶ τᾶς γᾶς βέβαιος ἔστō IG 9².609.1 (Naupactus, c.500 BC)'   II adv., add 'βεβάως Alc. 344 L.-P.'
βεβαιόω I 1, add 'b w. inf., affirm, β. ἀναπλεῖν ἐθέλειν D. 32.19.'
βεβαιωτήρ, add 'IG 9².612.9, etc. (Naupactus, iii BC), 394.9, etc. (Stratus, ii BC)'
βεβαιωτής 1, add 'ὧν ἁπάντων ἐναργέστατος ἦν βεβαιωτάς IAskl.Epid. 36.9 (iv BC)'
βέβηλος, line 1, after 'ov' insert 'also η, ον, Thasos 18.4 (v BC)'; after 'Dor.' insert 'βάβᾱλος SEG 9.72.9, 21 (Cyrene, iv BC)'; after 'βέβαλος' insert 'Theoc. 3.51, 26.14'
βέβρῡχε, for 'v. βρύχω' read 'v. βρυχάομαι'
*βέδοξ, ὁ, app. some kind of garment or covering, β. Νωρικὸς κάλλιστος ἤτοι βῆλον Edict.Diocl. 19.56, 58.
βέδυ, after 'Phryg.' insert 'Clem.Al.Strom. 5.8.46, 5.48.5'
*βεΐκουλον, τό, Lat. vehiculum, ἔπαρχος βεϊκούλων IGRom 4.1057.
*βεκάς, v. °ἑκάς.
βέκος, add 'β.· ἀνόητος Hsch.'
*βεκτοῦρα, ἡ, Lat. vectura, conveyance, transport, Edict.Diocl. 17.3-5.
*βελέαγρον, τό, or -ος, ὁ, instrument for extracting weapons from wounds, Aët. in J.G.Schneider ad Nic. Al. 511 p. 243 (Halle 1792).
*βελεβέκη· βελόνη Hsch.
βέλεμνον, add 'E.Andr. 1136'
*βέλλερα, ἄλλως. βέλλερα τὰ κακὰ λέγει Sch.Hes.Th. 325 (expl. of name Βελλεροφόντης).
βελλούνης, after 'τριόρχης' insert 'Λάκωνες'
βελοθήκη, for 'Lib.Decl. 30.9' read 'Lib.Descr. 30.9'
*βελομαντεία, ἡ, casting lots by shaking marked arrows from a quiver, Jerome in Ez. 21.24 (PL 25.206).
*βελονίστρια, ἡ, needlewoman, title of play by C.Decius Laberius, Nonius Marcellus p. 104.25.
βελοστασία, for 'range or battery of warlike engines' read 'artillery emplacement'; add 'Lxx 1Ma. 6.20'
βελόστᾰσις, after 'foreg.' insert 'Lxx Je. 28(51).27 (pl.)'
*βελότρωτος, ον, wounded by a missile, Cyran. 112 (4.28.7 K.).
βελουλκέω, add 'Paul.Aeg. 6.88.3 (131.22 H.)'
βελτίων, at end for '[ἴ Att., but βέλτῑον Mimn. 2.10]' read '[Att. usu. ῐ, but ῑ A.fr. 309.3 R., Eup. 336 K.-A.; also in Mimn. 2.10 W.]'
βελτίωσις, add 'ἀπὸ βελτιώσεως χέρσου SB 10891 i 18, al.'
*Βελφαῖον, τό, the Thetonian treasury at Delphi (or the sanctuary of Delphian Apollo at Thetonium), IG 9(2).257.10 (Thetonium, v BC); cf. Βελφοί, etc.
*Βελχάνια, v. Γελχάνια.
βέμβιξ I, add '2 spinning movement of dancers, Ar.V. 1531.'
*βενεφικιάλιος (A), ὁ, = °βενεφικιάριος, (in quot., one who provides medical aid for veterans), Lyd.Mag. 1.46.
*βενεφικιάλιος (B), ὁ, wd. incl. in list of criminal types, Just.Nov. 13.4 pr.
*βενεφικιάριος, ὁ, Lat. beneficiarius, IGRom. 3.110, etc; abbreviated βφ, SEG 32.1551 (Arabia, ii AD).
*βενεφίκιον, τό, Lat. beneficium, βενεφίκιν, PFlor. 296.49 (vi AD), Just. Edict. 4.1.
*Βενεφρανός, ον, of Venafrum, name of kind of olive oil, ἐλαίου Βενεφράνου Heras ap.Gal. 3.1042.12.

*βενέω, v. °βινέω.
*Βεννάρχης, ου, ὁ, official in the cult of Zeus Bennios, INikaia 1206 (-α⟨ρ⟩χης).
*Βεννεύω, dub. sens., MAMA 1.390 (Phrygia), perh. worship Zeus Bennios.
*Βέννιος, epith. of Zeus, IGRom. 4.535 (Phrygia), SEG 28.980 (Bithynia, AD 210).
*βένος, τό, dub. sens., SEG 6.550 (Pisidia).
*βεραιδαρικός, of or typical of a veredarius, Lyd.Mens. 1.32.
βερβέριον, for 'shabby garment' read 'kind of headdress'
*Βερενίκειος, α, ον, of Berenice, Theoc. 15.110, Call.fr. 110.62 Pf.
*βεριδάριος, = βερεδάριος, Hsch. s.v. οὐεριδάριος.
βερίκοκκον, add 'see also °πραικόκκιον'
*βέρνα(ς), v. °οὐέρνας.
*βεστιάριον, τό, Lat. vestiarium, Modest.Dig. 34.1.4 pr., Suid., etc.; also βιστ-, Hsch.
*βεστιάριος, ὁ, Lat. vestiarius, clothes-dealer, POxy. 3867.22, Stud.Pal. 3.50.1 (both vi AD).
*βεστιαρίτης, ου, ὁ, dub. sens., (perh. = °βεστιάριος, or person connected w. bestiae), CPR 8.56.15 (v/vi AD); sp. βηστ- Stud.Pal. 20.157.2 (vi AD).
βεστίον, add 'CPR 14.41.1 (vi/vii AD), cf. Lat. vestis'
*βετ(ε)ρανός, v. ‡οὐέτ-.
*βεττάριον, τό, dub. sens., perh. dim. of βέττον, PHib. 211.5 (iii BC).
βεττονίκη 3, for 'κέστρος' read 'κέστρον ι 1'
βεῦδος I, for 'Sapph. 155 .. 11.4 (pl.)' read 'Sapph. 177 L.-P., Call.fr. 7.11 Pf. (pl.), Parth. in Suppl.Hell. 646.6 (pl.)'
βήκη, add 'Theognost. Can. 109'
βηλόθυρον, add 'also οὐηλόθυρον PGrenf. 2.111.14 (v/vi AD)'
*βῆλον, τό, Lat. velum, covering or cloth, CIG 2758 ii B8, 4283.16; also οὐῆλον POxy. 2128.8 (ii AD), Edict.Diocl. 19.56, cf. Gloss. 3.92.58 β. pallium, 270.39 καταπέτασμα, κρήδεμνον velum.
βῆμα, line 3, for 'footfall' read 'walk, gait' and for 'Sapph.Supp. 5.17' read 'Sapph. 16.17 L.-P'   I 2, add 'b β. ποδός a step, as a very short distance, Lxx De. 2.5.'   add '4 imprint of a foot offered as a dedication, IG 11(4).1263, Inscr.Délos 2080.'   II 2 b, add 'D.C. 57.7.2, Modest.Dig. 27.1.13.10'   add '5 grade, rank, Cod.Just. 12.33.8.2.'
βημᾰτιστής I, for 'Ath. 10.442c' read 'Ath. 10.442b'
*βηρυλλίτης, ου, ὁ, a precious stone, Cat.Cod.Astr. 8(2).169.8.
βήσαλον, after 'brick' insert 'Moses Alch. 300.13'
*βησάρτης, ὁ, app. official in Bacchic cult, IG 10(2).1.244 i 16 (Thessalonica, ii AD), perh. also ib. 259.3 (i AD).
*βηστιαρίτης, v. °βεστιαρίτης.
*βήχω, to cough, χελύσσεται· ταράσσεται, ⟨βήχει⟩ Sch.Nic.Al. 81d Ge.
βηχώδης 2, add 'Gal. 13.56.12'
βία II 2, line 7, after 'etc.' insert 'πρὸς βίαν ἐπίνομεν Ar.Ach. 73'; line 8, after 'D.S. 20.51' insert 'εἰς βίαν Men.Dysc. 396'; add 'πὲρ βίαν πώνην Alc. 332 L.-P. (πρὸς βίαν cod.)'
*βιᾰθάνᾰτος, ον, = βιαιοθάνατος, SEG 6.803 (Cyprus, iii AD).
*βιαιοθᾰνᾰτέω, add 'cf. °βιοθανατέω'
*βιαιολεχής, ές, wedded by force, Dain Inscr.du Louvre 60.11 (Heraclea ad Latmum).
*βίαι]όπρᾱτος, or βι]όπρατος, ον, subjected to compulsory sale of one's property, prob. in PRyl. 617.10 (iv AD).
*βιάρπᾰγος, ον, life-robbing, βιαρπάγου Λήθης Suppl.Mag. 42.21.
*βίαρχης, ου, ὁ, = βίαρχος, PMich.xi 612.5 (vi AD).
*βίαρχία, ἡ, office of βίαρχος, Cod.Just. 1.31.1.
βίαρχος, add 'SEG 34.1292 (Phrygia), Cod.Just. 12.20.3; written βίορχος in LW 2037 (Arabia, iv AD)'
*βῑᾰσάνδρα, ἡ, coercing men, epith. of Ἄρκτος, PMag. 7.696, RPh 1930.249 (Egypt, tab.defix.).
βιᾱτᾱς, after 'mighty' insert 'Alcm. 1.4 P.'
*βιᾱτικόν, τό, Lat. viaticum, BGU 423.9, PGoodsp.Cair. 30 xli 18 (both ii AD).
*βιᾱφορέω, violate, τὸ παρὰ τὴν συγγραφὴν βεβιαφορημένον μέρος BGU 1844.25 (i BC).
βιβλιαφόρος, after 'D.S. 2.26' insert 'βυβλιαφόρος PRyl. 555.2 (iii BC), POxy. 710.2 (ii BC), SEG 24.1221, 1222 (Egypt, iv AD)'
†βίβλινος (A), v. βύβλινος.
*βίβλινος (B), η, ον, Bibline, kind of vine and wine, B. οἶνος Hes.Op. 589, E.Ion 1195, Theoc. 14.15; ὁ τῆς Βιβλίνης ἀμπέλου Ach.Tat. 2.2.2; from Thracian place-name acc. to Epich.fr. 174, but Βίμβλινος from a Naxian river acc. to Semus fr. 13 J.; cf. Βίμβλινος Lang Ath.Agora xxi L45 (iv AD), Hsch. (Βίβλ- La.).
βιβλιογράφος, add 'see also °βιβλιογράφος'
*βιβλιοκατᾰγωγεύς, έως, ὁ, βυβλ-, perh. forwarder of documents, PSI 1410.15 (ii AD).
βιβλιοπώλης, add 'βυβλιο-, POxy. 2192.37 (ii AD), see also βιβλοπ-'
βιβλιοφυλᾰκέω, for 'to be a librarian' read 'to be a keeper of archives'

βιβλιοφὔλᾰκικός, ή, όν, **βυβλ-**, of or for a keeper of archives, MacDowell Stamped Objects from Seleucia 39, YCLS 3.47 (Uruk).

βιβλιοφὔλάκιον, add '**βυβλ-** ABSA 42.202 (Cyprus, Rom.imp.)'

βιβλιοφύλαξ, add 'PGiss. 58 (ii AD), POxy. 1256 (ii AD)'

†βιβλίς, ίδος, ή, βιβλίδες· τὰ βιβλία ἢ σχοινία τὰ ἐκ βίβλου πεπλεγμένα EM 197.30; Aeol. βίμβλις (in broken context), ἐν βιμβλίδεσσι Alc. 208(a).ii 6 L.-P.

*βιβλογράφος, ὁ, Att. for βιβλιογράφος, acc. to Gramm. in Reitzenstein Ind.Lect.Rost. 1892/3 p. 4, Phryn.PS p. 52 B. cod.

*βιβραδ[ικός], ή, όν, fr. Lat. vibrare, vibratory, ἀνεμούρ[ιον] β., dub. in PRyl. 627.165.

*βίδυν, = βίδην, ψαλεῖ βίδυν Trag.adesp. 656.27 K.-S.

*βίετος, v. °βίοτος.

*βιζάκιον, τό, small stone, pebble, Suid., Zonar.

†βιζάριον, τό, prob. camel suckling her young, βιζάριν Aegyptus 6.188 (PLond. 1821); cf. °βυζαστρία.

*βίζια· mamillae Gloss. (perh. for βυζία, cf. mod. Gk. βυζί).

Βιθὔνιάρχης, after 'Bithynia' insert 'and of its festivals' and for 'OGI 528.10 (Prusias), al.' read 'IPrusias 3.1, 5.3, al. (Βει-, ii/iii AD), Inscr.Perg. 8(3).551, INikaia 726'

*βικαριανός, ή, όν, of a °βικάριος, Just.Nov. 26.2.2.

*βικαρία, ή, Lat. vicaria, office of vicar in an imperial diocese, [τοῦ ἁ]γνοῦ κόμητος καὶ ἀπὸ βικαρίας Salamine 207 (v/vi AD).

*βικάριος, ὁ, Lat. vicarius, governor of a diocese, Epigr.Gr. 929.2 (iv AD), Cod.Just. 12.37.19 pr., SEG 36.663 (Thrace, v/vi AD); see also °οὐικάριος.

βικεννάλιον, v. °οὐικεννάλια.

βῐκία, for 'Edict.Diocl. 17.6' read 'Edict.Diocl. 1.30, 17.6a'

βῐκίον, after 'βῖκος' insert 'PCair.Zen. 7 (iii BC)'

βῐκίον, add 'Edict.Diocl. 1.30A'

*βίκλα, ἡ, prob. watch-post (for *βίγλα, cf. Lat. vigilia), POxy. 1862.29; cf. °οὐίγιλ.

βῖκος 1, after 'cask' insert 'Hippon. 142 W.'   **3**, after 'measure' insert 'of land' and add 'ψιλοὺς τόπους ἀδ[ε]σπότους βίκων τεσσάρων POxy. 3334.8 (i AD)'

βιλλαρικός, for 'perh. = Lat. villaticus' read 'perh. for *βηλαρικός used for screening or veiling'

βιλλᾶς, add 'perh. pers. n.'

†Βίμβλινος, v. °Βίβλινος.

*βίμβλις, v. °βιβλίς.

*βίνδιξ, ικος, ὁ, Lat. vindex, in quots., app. official concerned with tax collection, Just.Nov. 128.5, 134.2.

†βῐνέω, (βενέω v. infra) coarse wd. for to have sexual intercourse, Hippon. 84.16 W., Ar.Ra. 740; w. acc. pers., γυναῖ]κα βινέων Archil. 152 W., Ar.Au 560, etc.; pass., Ion. impf. iterative, βινεσκόμην of male prostitute, id.Eq. 1242; of the woman, Eup. 385.2 K.-A., Philetaer. 9.4 K.-A.; app. understood as illicit, opp. marital, intercourse, Sol.Lg. 52b R. [βενέω Lang Ath.Agora xxi C2 (vi BC), C 14 (v BC), perh. in Inscr.Olymp. 7 (v BC; Elean), prob. represents βειν-; many codd. and papp. have βειν-].

*βιξιλατίων, ωνος, ἡ, Lat. vexillatio, military detachment, IGRom. 3.418.5 (pl.); see also °οὐιξ-.

βιοθάλμιος, for 'strong, hale' read 'productive of life'

*βιοθᾰνᾰτέω, = βιαιο-, Sch.A.Il. 13.393.

βίοκουρος, add 'also οὐιόκουρος Keil-Premerstein Dritter Bericht 129.2 (iii AD)'

*βιοκωλῡσία, ἡ, suppression of violence, POxy. 2046.56 (vi AD).

*βιολόγος, ὁ, kind of mimic actor or mime, IG 14.2342, POxy. 1025.7 (iii AD), IGRom. 1.552 (Salona), IGLS 9407 (Bostra, Chr.), GVI 515n (Citium, ii BC).

*βιοποριστέω, make a living, οὐ μικρὰ β. Aesop. 56 P.

βιοπρᾱγος, add 'POxy. 1477.14 (iii/iv AD)'

†βιόπρᾱτος, v. °βιαιόπρατος (cf. ‡βίοπραγος).

βίος II, line 8, after 'Iamb.VP 30.170' insert 'στοὰν ἐπισκευάσειν ἐκ τοῦ ἰδίου βίου i.e. at (his) own expense, SEG 23.207.19 (Messene, i BC/i AD)'   **III**, line 4, after 'Luc.DDeor. 13.1' insert 'οὐδέπω (τὰ γεγραμμένα) τῷ β. παραδέδωκας id.Hes. 2; χάριν τούτου ἐκαλούμην μέγας ἐν τῷ βίῳ Lyr.Alex.adesp. 4.19'   **V**, add 'βίους συνταξάμενοι D.H.Amm. 1.3; βίον ἀναγράψαντες ib. 6'

*βιοστερέτις, ιδος, ἡ, depriving of life, Μοῖρα GVI 845.2 (Panticapaeum, ii/i BC).

βιοτεία, add '**2** livelihood, maintenance, εἰς τὴν κοινὴν βιοτείαν POxy. 3491.12 (ii AD).'

*βιοτερπής, ές, delighting in life, orac. in IEphes. 1252.6 (ii AD), epigr. in Geel Catal.MSS.Lugd.Bat. p. 18 no. 54.

βίοτος, line 1, after 'Ep.' insert 'βίετος Inscr.Cret. 1.xvi 7.10 (Lato, ii/i BC)'

βιοφθόρος, add 'epigr. in BCH 85.848 (Paros)'

*βιόχρηστος, ον, useful for life, Menandro qui βιόχρηστα scripsit, Plin.HN 1.19, 20. al.

βιόω, line 8, delete 'v.l. for βιοίη in'; line 14, delete '(opp. ζάω live, exist)'   add '**3** aor. part., of one who has lived and is now dead, SEG 29.663 (Thrace, v/iv BC), δίδωσιν δὲ καὶ κηδευτικὸν τοῖς βιώσασιν Xanthos 67.30 (ii AD).

βιπίννιον, after 'Lat. bipennis' insert 'small axe'; for pres. ref. read 'Edict.Diocl. 7.36'

*βίρριον, τό, dim. of βίρρος, cloak, PUniv.Giss. 32.17 (iii/iv AD); βεί[ρι]ον).

βίρρος, for 'Edict.Diocl. 19.26, al.' read 'Edict.Diocl. 7.43, al., βίρος ib. 7.42'; for 'Suid. .. βύρρος.)' read 'also βίρρον, Suid.; **βύρρος**, BGU 814.8 (iii AD), SEG 7.431'

*βιστιάριον, v. °βεστιάριον.

βίττακος, for 'ψίττακος' read 'ψιττακός parrot'

βῐωφελής, line 1, after 'useful for life' insert 'Varro Sat.Men. 340'

*βῐωφέλιμος, ον, helping life, Heph.Astr. 2.21.15.

βλάβη 2, line 2, for 'cattle' read 'animals'

*βλᾰβοποιέω, damage, harm, ἐβλαβοποίησαν .. φοίνικας δύο PGen. 107.7 (Arsinoite, i AD).

*Βλαγανῖτις, ιδος, ἡ, epith. of Artemis in Macedonia, SEG 37.539 (Metochi); also Ἀ. ἐν Βλαγάνοις ib. 590, 592 (Vergina, both ii AD), fr. place-name, perh. meaning "place of frogs"; cf. βλαχάν, and θεᾶ τῶν βατράχων ib. 540 (Aegeae, iii AD).

*βλᾰδᾰρός, ά, όν, = πλαδαρός, flaccid, cj. in Gal. 19.88; β.· ἐκλελυμένον. χαύνον Hsch.

*βλᾰδεῖς· ἀδύνατοι, ἐξ ἀδυνάτων Hsch.

*βλᾰδόν· ἀδύνατον Hsch.

*βλαισοπόδης· βάτραχος Suid.; cf. βλαισόπους (s.v. βλαιτόνους).

βλαστέω, (s.v. βλαστάνω), lines 1/2, for 'interpol.' and 'corrupt' read 'dub.' twice.

*βλασφημολόγος, ὁ, blasphemer, Leont.Byz. HP 1.8.30.

*βλάττιος, α, ον, Lat. blatteus, purple, καμίσια β. Stud.Pal. 20.245.10 (ZPE 76.113, vi AD); -ον, τό, purple garment, Lyd.Mens. 1.21.

*βλαττόσημος, ον, having purple stripes, Edict.Diocl. 29.11, al.

*βλᾰχά, v. βληχή.

βλαχάν, after 'Hsch.' add 'cf. βλίκανος'

βλαψίτᾰφος, line 1, for 'IG 14.934.4' read 'IG 14.943.4'

*βλέθρα, v. πλέθρον.

βλεμεαίνω, for 'exult' read 'prob. glower'

βλέμμα, for 'Antiph. 235' read 'Antiph. 232 K.-A., Men.Dysc. 258'

βλέννα, add 'cf. πλένναι, πλεννεραί'

βλέπω, line 1, after 'etc.' insert 'Dor. **γλέπω** in ποτιγλέπω (v. °προσβλέπω), cf. γλέφαρον'   **I 2**, have regard to, after 'Arist.Pol. 1293ᵇ14' insert 'εἰς τὴν ταμειακὴν ἐξουσίαν βλέπον Just.Nov. 30.5 pr.'   **III 2**, after '(S.)Aj. 962' insert 'ζώντων φρονούντων βλεπόντων Aeschin. 3.94; X.Cyr. 8.1.22'

βλεφᾰρίζω, after 'wink' insert 'Clem.Al.Paed. 3.70'

βλέφᾰρον, line 1, after 'τό' insert 'dat. pl. -εσσι Nonn.D. 5.480, 15.408'

βλεφᾰροσπάξ, for 'arching the eyebrows' read 'prob. pulling out the eyebrows or eyelashes'

*βλησκούνιον, gloss on γλήχων, Sch.Nic.Al. 128b Ge.; cf. βληχώνιον.

*βλήσκω, iterative form of the pres. stem of βάλλω, τὴν .. βλησκομένην βοτρυῖτιν (βλισκ-, -την codd.) Zos.Alch. 207.6.

*βλήχησις, εως, ἡ, bleating, Alex.Aphr.Pr. 3.168.

βλήχων, line 4, after 'gen.' insert '[γ]ληχῶνος (-ῶ- pap.) Hippon. 84.4 W.'; line 6, after 'Ar.Ach. 874' insert 'γληχώ Nic.Al. 128'   add '**II** pubic hair, Ar.Lys. 89.'

βλιτομάμμας, add 'Sch.Pl.Alc. 1.118e'

βλίτον, add 'Antiph. 275 K.-A.'

*βλιτυρίστριαι, αἱ, female singers of a cult lament, Cyr., v. Univ. di Cagliari: Quaderni dell' Ist. di Filol. gr. 2.104.

βλῐχᾰνώδης, read βλῐχ- and delete 'of fish'

βλοσῠρότης, after 'grimness' insert 'Antioch.Astr. in Cat.Cod.Astr. 11(2).109.7'

βλοσῠρωπός, add 'AP 5.299.7 (Agath.)'

*βλόψ, onomat. wd. expr. sound of drops falling into water, cf. Eng. "plop", καὶ τῆς δικαστικῆς ψήφου ἦχος ὡς ὁ τῆς κλεψύδρας παρὰ Ἀττικοῖς βλόψ Hsch. s.v. κόγξ; Eust. 768.12.

*βλυάζω, produce, give birth to, Ar.fr. 609 K.-A.; cf. βρυάζω.

βλύζω, delete 'Orph.A. 599' and insert 'AP 11.24 (Antip.Sid.)'

βλυστάνω, for 'sq.' read 'βλύω'

βλύω, line 1, after '= βλύζω' insert 'ἔτι βλύοντι φόνῳ Nic.Th. 497'; after 'φόνῳ' insert '(v.l. φόνου)'

βλωθρός, after 'ά, όν' insert 'Ep. also **γλωθρός** Hes.fr. 204.124 M.-W.'

βλώσκω, line 2, aor. 2, after 'etc.' insert '(ἔμολα epigr. in INikaia 223.3)'; add 'used in prose in Arc., arrive, ἔβλωσκον ἰμ Μαντινέαν .. ἅμαν οἱ σταλογράφοι μόλωνσι SEG 37.340.16, 18 (Mantinea, iv BC)'

*βοᾰγεία, (for -γία), ἡ, conveyance of oxen, Αὐρηλί(ου) Ἀνεικήτου .. ἄρξαντος βοαγείαν POxy. 3565.4 (AD 245).

βοᾰθόος, add 'Βοαθοῖος Delph. 3(6).19.1, al.; Βαθοῖος ib. 134.3'

Βοάρσαι, οἱ, guild of *bull-hoisters*, *IG* 12(1).102.8, Maiuri *Nuova Silloge* 18.27 (both Rhodes); cf. °ἀείρω IV 3 b.

Βοάρσιον, τό, *contest connected with bull-hoisting*, ἀγωνοθέται Βοαρσίου *IG* 12 suppl. 646.21 (Chalcis, iii AD).

βοαύλιον, after 'dim. of sq.' insert '*AP* 7.717'

βόγγλωττος, v. ‡βούγλωσσος.

βόειος II 1, add 'Cret. βοῖα *Inscr.Cret.* 4.65.6 (Gortyn, v BC)'

βοεύς, add '*h.Ap.* 407'

βοηγία, ἡ, add '**II** *public service connected with such contests*, *GDI* 5633.3 (Teos).'

βοηγοί, delete the entry.

βοηγός, ὁ, *ox-driver*, title of certain religious officials at Miletus, *Didyma* 199.5, 16 (pl.), 19, 262.7, 263.8.

Βοηδρόμια, add 'also Βοαδρόμια *SEG* 25.445.32 (Arcadia, ii BC)'

Βοηδρόμιος II, add 'also **Βατρόμιος** at Calymna, *Tit.Calymn.* 79A.49, 88.38, 41, 94A.9; at Cos, *APAW* 1928(6).10.50'

Βοηδρομιών, add 'also Βατρομιών *SEG* 33.681 (Paros, ii BC)'

βοήθεια I, line 2, after '(Thermon)' insert '**βοάθεια** *SEG* 23.547 (Crete, c.200 BC)' **III**, add '**2** *staff* or *aides* of an official, *Cod.Just.* 10.30.4.12, Just.*Nov.* 7 epilogos.'

βοηθέω, lines 2/4, for 'Ion. **βωθέω** .. iv BC)' read '**βωθέω** Hdt. acc. to Eust. 812.59, *AP* 12.84 (Mel.), βωθέοντες· βοηθοῦντες Hsch.'; after '**βοᾱθοέω**' insert 'Pi.*N*. 7.33, *SEG* 30.1117, 1118, 1121 (Entella)'; line 4, after '(Thermon)' insert 'βοᾱθέω *SEG* 25.847 (Telos, c.300 BC), 23.547 (Crete, c.200 BC)' **1**, add 'w. gen., βοήθησον τῆς μεικρᾶς *Suppl.Mag.* 13 (iv AD)'

βοηθητικός, line 4, after '(Arist.)*HA* 515ᵇ9' insert 'of medic. treatment, *promoting a cure*, D.L. 3.85' add '**2** *of a* βοηθός, τὰ -ά, *fees for assistance*, *PMich*.XI 624.5 (vi AD).'

βοηθός, line 3, after 'Pl.*R.* 566b, al.' insert '= Lat. *adiutor*'; add 'as honorific title, β. καθόλου τῆς Ἀρσινοϊτῶν πόλεως *CPR* 10.25.5 (AD 526/7); written βοιθός, *SEG* 8.483.4 (Egypt, i BC), cf. *OWilck.* 1084.11 (ii BC)'

βοηλᾰτέω, after 'Opp.*C.* 4.64' insert '(v.l. °βροχηλατέω)'

βόησις, after '*shout for assistance*' insert 'Plu. 2.171d'

*βόθρευμα, ατος, τό, *pit*, *Poliorc.* 212.4.

βόθῡνος I, add 'Lxx 2*Ki.* 18.17, 4*Ki.* 3.16, *Is.* 24.17, etc.; *Ev.Matt.* 12.11, etc.'

βοίδιον, add 'also βουείδιον *SB* 9920 ii 19.9 (vi AD)'

*βοιθός, v. °βοηθός.

*βοῖπις· γῇ βουνώδης Theognost.*Can.* 19.

βοϊστί, after '*ox-language*' insert 'Iamb.*VP* 61'

*Βοιώταος, η, ον, (?for *Βοιωταῖος), = Βοιώτιος, Δήμητρος Βοιωτάης *SEG* 17.396 (Chios, iv BC).

Βοιωταρχέω, add '*SEG* 25.553 (Onchestus, iv BC): also **Βοιωταρχίω**, *IG* 7.2407.12, 2408.12 (iv BC)'

Βοιώτιος (s.v. Βοιωτός), for 'Hes.*fr.* 132, etc.' read 'Hes.*fr.* 181 M.-W., etc.; also as subst., ἀγῶνες Βοιωτίων Pi.*O.* 7.85'

*βοκάλιοι, οἱ, Lat. *vocales*, *singing rope-dancers*, *POxy.* 2707.7 [βοκ[άλιοι], vi AD).

βολαῖος, for '(βολή) *violent*' read '(βόλος) *caught in the net*'

*βόλβαξ, ακος, ἡ, app. edible bulbous plant, *PMich*.VIII 496.16 (ii AD).

†**βόλβῑτον**, τό or **βόλβῑτος**, ὁ (βόλβιθος *PMag.* 4.1440) = βόλιτον, *dung, excrement* (less acceptable form acc. to Phryn. 335), Hippon. 92.9; 144 W., Thphr.*HP* 5.5.3, Dsc. 2.167, Archig.ap.Gal. 12.173.

*βολένη· τὸ κρέας τὸ περὶ τὸ στόμα τῆς γαστρός Theognost.*Can.* 48 A.

*βολεός, όν, *heaped*; βολεοὶ λίθοι, series of cairns acting as boundary marks at a place in Epidaurus, *SEG* 11.377.16 (Hermione, ii BC), βολεοὶ λίθων κύκλοι *IG* 4².75.33 (ii BC); **Βολεοί** alone, treated as place-name, Paus. 2.26.3.

*βολευταί, gloss on δανδαρίκαι, Hsch.

βολή 4, line 4, for '*radiance*' read 'snow-*falls*'

βολίζη, delete 'Cretan word in'

†**βόλιμον**, τό, (s.v. βόλιμοι) *sacrificial victim* (slaughtered with an axe), *SEG* 22.508 (Chios).

βόλῐμος, add '*lead tablet*, τοὺς ἐν τῷ βολίμωι γεγραμμένους πάντας *SEG* 30.1175 (Italy, iii BC); cf. περιβολιβόω'

βολίς 3, delete 'ἀστραπὴ βολίς (sic)' and transfer '(Lxx) *Za.* 9.14' to section 1. **4 b**, delete the section transferring ref. to section 4a.

βολίτινος, delete 'σκέλη Cratin.*inc.* 17 Mein.'

*βολευτήριον, v. ‡βουλευτήριον.

βόλομαι, line 4, delete 'cf. A.R. 1.262'

*βόλτιον, gloss on °βότις, Hsch.

†**βομβαύλιος**, ὁ, comic conflation of βομβύλιος and αὐλός, "*drone-piper*", Ar.*Ach.* 866.

βόμβος, add 'of the sonorous utterance of tragedy, Μελπομένης βόμβον ἀπεπλάσατο (sc. an actress) *AP* 5.222.4 (Agath.)'

βομβῡκίας, for '*flutes*' read '*pipes*'

*Βομβῠλεία· ἡ Ἀθηνᾶ ἐν Βοιωτίᾳ Hsch.

†**βομβῠλία**· κρήνη ἐν Βοιωτίᾳ Hsch.

βομβῠλιάζω, for 'βορβορύζω' read '°βορβορύζω'

βομβῠλιός II, add '*Antisth.Protr.fr.*ap.Poll. 6.98, 10.68, Ion Trag. 64 S.'

βόμβυξ II 1, for '*flute*' read '*pipe*' and add 'cf. Dor. **Βομβύκα** Theoc. 10.27, name of a piper' **2**, for '*cap of a flute*' read 'included among parts of an αὐλός'

*βοο-, for βοῦς compds. see also βου-.

*βοοκλοπίη, ἡ, *theft of oxen*, Firm.*Err.prof.relig.* 5.2; cf. βουκλόπος.

*βοόκρημνος, ον, *having banks occupied by cows*, βοοκρήμνοιο .. ἴχνος Ἀράτθου Call.*fr.* 646 Pf. (s.v.l., see also °βούκρανος).

βοοκτᾰσία, after '*of oxen*' insert '*AP* 6.115 (Antip.), 263 (Leon.Tarent.)'

*βοόκτιστος, ον, = βούκτιτος, of Thebes, *founded through the action of a cow* (i.e. where a cow lay down), Ps.-Callisth. 55.10.

βοοσσόος, line 2, for 'βουσσόον' read 'βουσόον (v.l. -σσ-)'

*βοοτροφέω, *to rear oxen*, Heph.Astr. 3.5.24; cf. °βουτρόφος.

†**βόρασσος**, ὁ, name for the fruit enclosed in the spathe of the date-palm, Dsc. 1.109.5 (perh. Semitic).

*βοραι[.., kind of fish, cf. perh. °ἀβόρατος, *IGC* p. 98 A20.

*βορατίνη, ἡ, kind of juniper, cf. βόρατον, Aq.*Ca.* 1.17.

†**βόρατον**, τό, one or other species of *juniper*, D.S. 2.49, Dsc. 1.76, Sm.*Ps.* 103(104).17, *Ca.* 1.17, *Is.* 60.13 (cf. Hebr. *bᵉrôt*).

βορβορόθῡμος, add '(cj. °βαρβαρόμυθος)'

βορβορυγμός, add 'Gal. 14.122.5'

*βορβορύζω, *rumble*, seethe, resound esp. internally, ἐβορβόρυζε δ' ὥστε κύθρος ἔτνεος Hippon. 29a W.; Hp.*Int.* 6.

*βορδόνιον, τό, *mule*, Cyr.S.*Vit.Sab.* 44.

*βορεαῖος, ὁ, (sc. ἄνεμος) *north wind*, *IG* 12(3).357; cf. βορειαῖος.

Βορέας, after '**Βορρᾶς**, ᾶ' insert '(also Βορροῦ Aristonym. 7 K.-A.; cf. °Βορρόθεν)'; add 'see also βορεύς'

Βορέας II, add 'βορεᾶς .. ἀῆτα B. 17.91 S.-M.'

*Βορεασταί· Ἀθήνησιν οἱ ἄγοντες τῷ Βορέᾳ ἑορτὰς καὶ θοίνας, ἵνα ‹οὔριοι› ἄνεμοι πνέωσιν. ἐκαλοῦντο δὲ Βορεασμοί Hsch.

βόρειος I, add '**3** name of variety of *iaspis*, Plin.*HN* 37.116.'

*βορέομαι, *feed oneself*, v.l. (*POxy.* 2221) in Nic.*Th.* 394 (v. βοτέω); perh. to be understood in Sapph. 96.17 L.-P. (v. βαρέω).

*βόρεος, α, ον, = βόρειος, *Didyma* 27.A24, 67 (iii BC).

*βορίζω· ἐβόρισεν· ἐσίτισεν Hsch.; cf. perh. πορίζω.

*βορολίβας, α, ὁ, *the north-west*, ἐν τῷ βορολίβᾳ τοῦ οὐρανοῦ *PMag.* 4.1647 (s.v.l.).

*βορρόθεν, adv. *from the north*, Teuc.Bab. p. 67 B.

βορρόλιψ, delete '*PMag.Par.* 1.1646'

†**βόρυες**, οἱ, kind of animal found in desert or semi-desert country, Hdt. 4.192.2.

†**βοσκάδιος**, α, ον, *free-grazing*, χήν Nic.*Al.* 228.

βοσκάς I, for this section read '*greedy*, νηδύς Nic.*Al.* 782. **2** of birds which feed themselves, *not artificially fed*, ὀρταλίς Nic.*Al.* 293; Aët. 9.30; cf. βοσκός II.'

βόσκημα, sg., add 'coll. στερέσθω τοῦ βοσκήματος *IG* 12(9).90.12 (Tamynae, iv BC)'

βοσκός II, after '(*Edict.Diocl.* 4.) 22' insert 'φασιανὴ βοσσκή (sic), βοσσκὴ τρυγών ib. 20, 26'

Βοσπορᾱνός (s.v. Βόσπορος), add '*TAM* 4(1).239'

Βοσπόριος, after '(S.)*Aj.* l.c.' insert 'name of month at Rhegion near Byzantium, *Belleten* 23.552, al. (i AD)'

Βοσπορίτης, add '*SEG* 25.272 (Attica, iv BC)'

βοστρύχτης, add '**II** by-product of firing of copper ore, formed like bunches of grapes, Anon.Alch. 6.8.'

*βοστρυχῖτις, ιδος, ἡ, precious stone, perh. *moss-agate*, Plin.*HN* 37.150, 191.

βόστρυχος, line 1, pl. -α, add 'Nonn.*D.* 1.133, *AP* 5.218.4 (Agath.)'

*βοτανιάτης, ου, ὁ, *herbalist*, *MAMA* 4.93 (Robert *Noms indigènes* p. 141).

*βοτανοφᾰγία, gloss on ποηφαγία, Hsch.

βότειος, add 'δέρματα βότια *POxy.* 3505.4 (ii AD)'

*βότεος, α, ον, = βότειος, *of a sheep*, βόεα μηδὲ βότεα (sc. ἱερεῖα) μὴ ποτάγειν Maiuri *Nuova Silloge* 17 (Rhodes).

βοτέω, delete 'Ep. .. cf.'; for 'pass. .. 394' read 'med., *feed on*, Nic.*fr.* 74.46 G.-S., *Th.* 394; see also °βορέομαι)'

βοτήρ, add 'Theoc. 25.139, A.R. 3.592, 4.1248'

*βοτήραρχος (or ?-άρχης), ὁ, *chief herdsman*, Trag.adesp. 721.4 K.-S.

*βοτηρίδιον, v. ‡ποτηρίδιον.

†**βότις**, ιος, ἡ, Sophr. 64 (acc. to Ath. 286d kind of fish, or possibly plant); βότις· βόλτιον Hsch.

βοτόν, for '*beast*' read '*farm animal*'; add 'ἐν βοτοῖς Alcm. 1.47 P., βοτὸν τέλειον *SEG* 9.72.31, al. (Cyrene)'; last line, delete 'but also of birds'

βοτρυηφόρος, add 'Orph.*H.* 30.5 (-οφόρον codd.)'

βοτρυΐτης, delete the entry.

βοτρῦῖτις, ιδος, ἡ, form of *calamine* (sc. καδμεία), Dsc. 5.74, Plin.*HN* 34.101, Gal. 12.220.9.   **2** = ἀρτεμισία I 2, Poet.*de herb.* 26.

βοτρυοκαρποτόκος, ου, *bearing bunches of grapes*, Lyr.adesp. 9 P.

βοτρυοστέφᾰνος, for 'dub. in *IG* 3.3688' read '*IG* 2².11387 (*CEG* 550.3, iv BC)'

βοτρυοφόρος, ον, *bearing bunches of grapes*, epith. of the vine, Cyran. 11 (1.1.162 K.); of a coin, *bearing the representation of a bunch of grapes*, δραχμή Inscr.*Délos* 1450*A*108 (ii BC).

βότρυς, line 1, after 'ὁ' insert '(ἡ, Nic.*Al.* 185)' and for 'heterocl. pl. βότρυα, τά Euph. 149' read 'βότρυα (Euph. 149) cited as exceptional acc. sg. by Theodos.*Can.* p. 234 H.'

βοτρυώδης, for '= βοτρυοειδής, E.*Ba.* 12' read '*full of grapes*, ἀμπέλου .. βοτρυώδει χλόῃ E.*Ba.* 12; βοτρυώδη .. χάριν οἴνας ib. 534'

βοτρυωτός, ή, όν, *ornamented with bunches of grapes*, Inscr.*Délos* 1408*A*ii4, 1444*Aa*16 (ii BC).

βούβαλις, after 'A.*Fr.* 330' insert 'S.*fr.* 792 R.'   add '**II** v. βούταλις.'

βουβάρας, after '*EM* 206.18' insert '(= Eup. 436 K.-A.)'

Βουβαστεῖον, add '*PColl.Youtie* 19.15 (AD 44)'

βουβίλιξ, transfer the article before **βουβόσιον**.

βουβότης, add 'cf. Myc. *qo-u-qo-ta*, pers. n.'

βούβρωστις, after '*Epigr.Gr.* 793' insert '(= *SEG* 30.1473, ii AD)'

βουβωνίσκος, add 'Gal. 18(1).776.17, 827.1'

βούγλωσσος, add 'Boeot. βογχλώτ[τω] (gen.) *IGC* p. 98 A17 (Acraephia, iii/ii BC)'

Βουδιών, ῶνος, ὁ, month name, *SEG* 36.1116.A7 (Cyzicus, iv BC).

βουδόρος II 1, add 'prob. in *IG* 1³.405.9 (v BC)'

βουεργέτης, coined as expl. of βουγάϊε, Sch.Gen.Il. 13.824.

βουθοίνης, after 'ὁ' insert 'only in Dor. form **βουθοίνᾱς**'

βούθουτον, after 'ἀμέμαρον' insert '(ἀνέκφορον La.)' and add 'cf. βούθυτος'

βουθῠτέω, line 2, after 'E.*El.* 785' insert 'Aeschin. 3.77' and delete 'also in later prose'

† **βουκαῖος**, ὁ, *man in charge of oxen*, Nic.*Th.* 5; βουκαῖοι ζεύγεσσιν ἀμορβεύοντες ὁρήων id.*fr.* 90; prob. pers. n. in Theoc. 10.1, 57, but expld. as βουκόλος by Sch.; = ἄγροικος, acc. to Eust. 962.12; cf. ‡βοῦκος.

βουκελλάριον, τό, ingredient in the preparation of nard, Aët. 1.131 (66.26 O.).

βουκελλάριος, ὁ, Lat. *bucellarius*, *member of armed escort of military or civil functionaries in Egypt*, *POxy.* 150.1, 156.2, *BGU* 836.8, 12 (all vi AD).

βουκελλάτης (?-τᾶς), ὁ, prob. *baker of βούκελλαι*, *PErlangen* 81.49, cf. *CPR* 5.118.

βουκέλλιον, τό, *small loaf*, Paul.Aeg. 3.14.3 (158.21 H.), sp. βουκέλιον *PVindob*.G 39847.457 (*CPR* 5.97).

βούκερας, for '-αος' read 'gen. -αος Nic.*Al.* 424, -ατος *Edict.Diocl.* 1.18'; for '= τῆλις .. *Al.* 424' read '*fenugreek*, ll.cc.'

βουκέφᾰλος 1, add 'as pr. n. = Βουκεφάλας, *Gp.* 16.2 1, Hsch.'

βουκία, ἡ, or **βούκιον**, τό, *kind of cake or biscuit*, *POxy.* 397 (i AD), 155.4 (vi AD); (perh. = mod. Gk. βούκα 'mouthful" fr. Lat. *bucca*).

βούκινον, τό, Lat. *bucinum*, *trumpet*, Anon.Alch. 330.4.

βουκκᾶς, ᾶ, ὁ, *biscuit-maker*, *Stud.Pal.* 20.148 (vi AD), *PBaden* 31.21 (βουκᾶς), v. *ZPE* 54.93 (previously read as pers. n.); cf. βούκελλα, °βουκέλλιον.

βοῦκλεψ, v. **βόοκλεψ**.

βοῦκλος, ὁ, kind of jar, Zos.Alch. 140.15, Anon.Alch. 267.14.

βουκολᾱ, ᾱ, *female member of an association of βουκόλοι*, cf. -κόλος II, *IG* 4.207.3 (Cenchriae, Rom.imp.).

βουκολέω I 1, add '**c** fig., *feed* in one's mind, φροντίσιν πάθος A.*Ag.* 669; med., τόνδε πόνον id.*Eu.* 78.' **2**, delete ':– med. .. *Eu.* 78' **II**, delete 'πάθος .. 669'

βουκολιασμός, add 'see also °-ισμός'

βουκολικός II 1, add '**b** β. τό, prob. garment or ceremonial cloth used by a βουκόλος II 2, *BGU* 2427.6, 11 (?iii BC).'

βουκόλιον, after '*cattle*' insert '*h.Merc.* 288'

βουκολισμός, ὁ, = βουκολιασμός, Trypho ap.Ath. 14.618c; v.l. for -ιασμός, ib. 619a.

βουκόλισσα, ἡ, *cowherd*, *SB* 10447ʳ.ii.53 (iii BC).

βουκολιστής, οῦ, ὁ, *herdsman*, *PMil. Vogl.* 212ʳ.viii.9, xi.21.

βουκόλος, add 'Myc. *qo-u-ko-ro*'

βουκονιστήριον, for '*bullring*' read 'app. some kind of arena (the meaning of the prefix is disputed)'

† **βουκόπος**, ὁ, *butcher*, *IEphes.* 2.31, 32 (iv BC); gloss on βουπλήξ Hsch.

βοῦκος, add 'glossed by ἄγροικος, Eust. 962.12; by βουκόλος, Nic.ap.Sch.Theoc. 10.38'

βουκράνιον, for '*ox-head*' read '*ox-skull* (with the horns)'

βούκρᾱνος, for 'Call.*fr.* 203' read 'dub. cj. (βοοκράνοιο) in Call.*fr.* 646 Pf.; cf. °βοόκρημνος'

*** βούκρᾱς**, Myc. *qo-u-ka-ra*, *adorned with an ox-head*.

***βουκτέᾱνος**, ον, *possessing oxen*, Call.*VB* in *Suppl.Hell.* 260A.8.

***βουκτόνος**, ὁ, *ox-killer* (as a religious office), *IG* 2².4629.

βουλαῖος I, add 'as title of emperor Hadrian, *IG* 5(1).1352'

***βοῦλβα**, ἡ, Lat. *volva, vulva*, *sow's womb*, in list of meats, *Edict.Diocl.* 4.4, bulba μήτρα *Gloss.* 2.534.46.

βουλεῖον, add '*SEG* 23.207.19 (Messene, Augustus); Arc. βωλήιον *SEG* 37.340.21 (Mantinea, iv BC)'

βουλεύς, add '*SEG* 26.402 (Corinth), 723 (Illyria, both Hellen.)'

βούλευσις II 2, add '*IG* 2².1631.394 (Athens, iv BC)'

βουλευτήριον, after 'τό' insert 'Aeol. βολλ- *IKyme* 13.71 (ii BC)'   **II**, line 1, of local βουλαί, add '*Cod.Just.* 1.4.34.10, Just.*Nov.* 38 pr. 1, etc.'; line 3, after 'E.*Andr.* 446' insert '(so perh. in A.*Th.* 575, v. βουλευτήριος)'

βουλευτήριος, add '(but this may be °βουλευτήριον II)'

βουλευτής 1, add 'in a local council or *curia*, *Cod.Just.* 1.4.34.10, Just.*Nov.* 101.2, etc.'

βουλευτικός I 1, add 'in Roman provincial government, τάγμα Mitchell *N.Galatia* 195 (AD 126), *SEG* 33.1123 (Phrygia, iii AD)'

βουλευτός I, add 'Call.*Lav.Pall.*38'

βουλεύω B, for '5 rarely folld. by relat.' read '**b** *deliberate, consider*, folld. by clause' and transfer all to section 1.

βουλή, line 1, for 'Dor. βωλά .. 18.90' read 'Dor. βωλά (q.v.)' and add 'Cret. Fωλᾶς (gen., dub.) *SEG* 37.752.6 (Lyttos, v BC)' and after 'Aeol. βόλλα' insert 'Alc. 130.20 L.-P. (rest.)'   **I 2 and 3**, for these sections read '**2** *counsel, advice*, Il. 1.258, 2.202; κακὴ β. Hes.*Op.* 266; πρᾶτος .. καὶ βουλᾷ καὶ χερσὶν ἐς Ἄρεα *IG* 9(1).658 (Ithaca); νυκτὶ δὲ βουλὴν διδούς Hdt. 7.12.1; νυκτὶ βουλήν .. δίδου Plu.*Them.* 26; τούτοις οὐκ ἔστι κοινὴ β. Pl.*Cri.* 49d, X.*Cyr.* 7.2.6, βουλῆς ὀρθότης ἡ εὐβουλία Arist.*EN* 1142ᵇ16; ἐν νυκτὶ β. .. διδοὺς ἐμαυτῷ Men.*Epit.* 252; μαχέσασθαι οὐκ ἐποιεῦντο βουλήν Hdt. 6.101.2; in pl., οἵ τέ μοι αἰεὶ βουλὰς βουλεύουσι Il. 24.652, A.*Pr.* 221, *Th.* 842; ἐν βουλαῖς μὲν ἄριστον *Epigr.Gr.* 854, ἐν βουλαῖσι κράτιστος *IG* 2².3669.8 (iii AD). **3** *deliberation, consideration*, τὰ .. γενόμενα ἐν β. ἔχοντες Hdt. 3.78.1, 8.40.1, Arist.*EN* 1112ᵃ19, D. 9.46; οὐδεὶς περὶ τούτου προτίθησιν οὐδαμοῦ βουλήν id. 18.192; ὅπως δ᾽ ἂν βουλὴν ἀγάγοιεν Polyaen. 7.39, ἐν β. ἐγίνοντο πότερον .. D.H. 2.44.' **II**, line 7, after 'X.*HG* 5.29' insert 'in Lesbos, Alc. 130.20 L.-P.'   add '**2** *rank of membership of the βουλή*, Δελφοὶ .. Δελφὸν ἐποίησαν καὶ βουλῇ ἐτείμησαν *Delph.* 3(1).219, *IEphes.* 1615.19.'

βούλησις III, add 'ἐν τελευταίᾳ βουλήσει *Cod.Just.* 6.4.4.1, 27'

† **βουλογράφέω**, *carry out the enrolment of senators*, *OGI* 549.2 (Ancyra, iii AD).

† **βουλογραφία**, ἡ, *function of enrolling senators*, *IGRom.* 3.206 (Ancyra).

***βουλογράφος**, ὁ, *clerk to the council*, *GDI* 1172.37 (Elis, Hellen.; form βωλογράφορ).

βούλομαι I 1, line 26, w. inf. fut., add 'βουλόμενοι ἐξ αὐτέων παῖδας ἐκγενήσεσθαι Hdt. 4.111.2, Th. 6.57.3, Iamb.*Protr.* 16 (p. 83.27 P.)'; add 'w. legal instruments, etc., as subject, *Cod.Just.* 1.3.45.13, 1.3.52.5'   **2**, line 3, delete 'later' and after 'c. acc.' insert 'ἄνδρες τὰ Συρακοσίων βουλόμενοι Th. 6.50.3, cf. ib. 82.4'   **II**, for 'Att. usages' read 'in var. spec. usages'   **4**, add 'Democr. 173 D.-K.'

βουλῡτός, line 1, delete '(sc. καιρός)'

βούμαστος, after 'so in' insert '*PCair.Zen.* 33.15 (iii BC)'

***βουναῖος**, α, ον, *of or belonging to the hills*, app. of a kind of wine, Lang *Ath.Agora* xxi I45 (vi AD); cf. βουναία.

***βούνευρον**, (dub.), gloss on κίσσηρις, Hsch.

βουνιάς, add 'cf. μουνιαδικόν'

βούνιον I 1, add '*PMag.* 3.333'   **II**, delete the section (v. °βουνίον).

***βουνίον**, τό, dim. of βουνός, *hill*, Inscr.*Prien.* 42.41 (ii BC), as placename, Inscr.*Magn.* 122; cf. mod. Gk. βουνί.

βουνός I 2, before 'σίτου' insert '**3** as measure of quantity' and add '*JJP* 18.190 (ii/iii AD)'

βουπόρος, add '**2** subst., perh. = *ox-spit*, βουπόρος Ἀρσινόης μητρὸς σέο (app. ref. to Mt. Athos) Call.*fr.* 110.45 Pf.'

βούπρωρος I, after 'σημαίνει δὲ καὶ τὴν βουπρόσωπον Hsch. (*Trag. adesp.* 587b K.-S.)'

***βοῦργος**, ὁ, Lat. *burgus*, *tower, fort*, *BE* 1962.315 (Palestine, Byz.).

βουρδών, after '*Edict.Diocl.* 14.10' insert '(βουδρῶν ib. 11.4a)'

βουρδωνάριος, add '*BE* 1958.301 (Egypt, Chr.); also βουρδουννάριος, Teuc.Bab. p. 43.4 B.'

βουριχάλλιον, for pres. def. read 'some kind of carriage, prob. *ponycart*' and add 'perh. -χάλιον, i.e. dim. of late Lat. *buricale, cf. *buricus*'

***βουριχᾶς**, ᾶ, ὁ, *pony-dealer* or *drover*, *IEphes.* 551 (late Rom.imp.; cf. *manni equi dicuntur pusilli, quos vulgo burichos vocant* Porphyrio ad Hor.*C.* 3.27.7).

βοῦς I a, add 'ἐν δὲ Συρίᾳ .. οἱ βόες, ὥσπερ αἱ κάμηλοι, κάλας ἔχουσιν ἐπὶ τῶν ἀκρωμίων Arist.*HA* 606ᵃ15'   **b**, for this section read 'w. ἄγριος, applied to var. non-domesticated species, Arist.*HA* 499ᵃ4, *Mir.* 842ᵇ33'   **c**, delete the section.   **d**, for 'βούφθαλμος' read

βουστασία SUPPLEMENT βρυάθων

'βούφθαλμον' **IV**, add '**b** kind of cake, *Sokolowski* 1.42.*B*6, 43.3 (Miletus, v BC).' at end add 'Myc. *qo-o* (prob. acc. pl.)'
βουστασία, add '*Cat.Cod.Astr.* 11(2).181.14'
βοῦτις, add '*Edict.Diocl.* 13.17 ([βο]ῦττις)'
***βούτμημα**, ατος, τό, *furrow*, gloss on τμήγας, Hsch.
***βουτοί**· τόποι παρ' Αἰγυπτίοις, εἰς οὓς οἱ τελευτῶντες τίθενται Hsch.
†**βουτρόφος**, ὁ, *rearer of oxen*, *JHS* 54.143.69 (decr., pl., Delos, ii BC), Poll. 1.249, *EM* 209.54; cf. βοοτρόφος.
***βουτύπιον**, gloss on δάροσος (°δάρος), Hsch.
βούτυρον, add '*Peripl.M.Rubr.* 14, 41'
***βουφονία**, Ep. -ίη, ἡ, *sacrifice of oxen*, Call.*fr.* 67.6 Pf.
***βουφόντης**, ου, ὁ, *ox-slaying*, λίς Euph. in *Suppl.Hell.* 418.17.
***βοῦφος**, ὁ, *nocturnal bird*, Cyran. 86 (3.8.1 K.); cf. mod. Gk. μποῦφος.
βοῶν, add '*IG* 1³.425.42 (Athens, v BC; βο‹ῶ›νι lapis)'
βοώνης, add 'at Delos, *Inscr.Délos* 399*A*17 (ii BC)'
βοωτέω, add 'βωτέω Theognost.*Can.* 48 A., Suid. (Lacon.)'
***βοωφόρος**, ὁ, official in Bacchic cult, *IG* 10(2).1.244 i 4 (Thessalonica, ii AD).
†**βρά**· ἀδελφοί (Illyrian) Hsch.
βράβευμα, add 'Leont.Byz.*HP* 1.5.6'
βραβεύς I 2, delete the section transferring quots. to section 1, *judge, arbitrator*
βραβευτής II, add '*TAM* 5(1).234, 515 (Rom.imp.)'
βραγχώδης 1, add 'Gal. 13.4.14'
βράδινος, insert '(= ϝράδ-)' and for 'Sapph. 90, 104' read 'Sapph. 102, 115 L.-P., Alc. 304 ii 9 L.-P.'
***βρᾰδυγαμέω**, *be late in marrying*, Heph.Astr. 1.1.190.
βραδυδινής, add '(cj.)'
***βραδόθω**, *move slowly*, Nic.*Th.* 372 (v.l. μινύθ-).
βραδυλόγος, read βραδύλογος.
βραδυπειθής, for '*AP* 5.286 (Agath.)' read '*AP* 5.287.7, 299.7 (both Agath.)'
***βραδυπλοΐα**, ἡ, *slow voyage*, *POxy.* 2191.8 (ii AD).
βραδύς, line 3, βάρδιστος, add 'Theoc. 15.104'; line 9, comp. adv. βράδιον, add 'Plu. 2.459f, 460a'
***βραδυστομέω**, *be slow of speech*, Trag.adesp. 668.13 K.-S.
†**βράθυ**, τό, name of various trees or shrubs of the genus *Juniperus*, Dsc. 1.76, Cyran. 12.1 (1.2.1 K., v.l. βράθυος).
βρακαριος (s.v. βράκαι), for '*breeches-maker*, ib. 18' read '*maker of trousers, tailor*, *Edict.Diocl.* 7.42'; add '*IGChr.* 262 (Aphrodisias), βρικάριος (sic) *MAMA* 3.597 (Corycus) is perh. the same wd., see also βρεκάριος'
βράκια (s.v. βράκαι), dim. βράκια, for '*IG* 5 .. Asine' read '*Edict.Diocl.* 7.46'
†**βράπτω**, βράπτειν· ἐσθίειν, κρύπτειν, ἀφανίζειν. τῷ στόματι ἕλκειν. ἢ στενάζειν Hsch.; see also s.vv. ἔβραπτεν, ἔβραψεν in Hsch.; cf. μάρπτω.
βράσις, add 'fig. of anger, cj. in Plot. 4.4.28'
βράστης, add '**2** *winnower*, *PMich.*i 53.6 (pl. (s.v.l.), iii BC).'
***Βραυρωνία**, ἡ, title of Artemis, *IG* 1³.369.89, 2².1401 (iv BC), *SEG* 25.445 (Arcadia, ii BC).
***βραχιᾶτοι**, οἱ, *bracelet-wearers* (legionary rank in the late empire), Lyd.*Mag.* 1.46.
βραχίων, at end for 'as a .. 478' read 'of strength of arm, νέοι βραχίοσιν E.*Supp.* 738'
βραχύνω, for 'Pl.*Per.*' read 'Plu.*Per.*'
βραχύποτος, delete the entry (refer quot. to -πότης).
βραχύς, line 1, after 'ύ' insert 'Aeol. **βρόχυς** (q.v.); cf. βρόσσων'; line 8, for 'Men. 726' read 'Men.*fr.* 544 K.-Th., βραχύ *soon*, Anacreont. 52A.4 W.' **5**, after 'D.T. 631' insert 'S.E.*M.* 1.100ff.' and delete 'etc. .. S.E.*M.* 1.113'
βραχύτης 3, after 'of a syllable' insert 'or vowel' **6**, after 'Rhet.' insert '*conciseness of expression*, ἡ τοιαύτη β. κόμμα ὀνομάζεται Demetr.*Eloc.* 9'
βρέβιον, for 'Lat. *brevis*' read 'Lat. *brevium*'; **βρέουιον**, add '*POxy.* 3628.1 (v AD)'
βρέγμα I 1, after '*front part of the head*' insert '(anat.) *sinciput*' and add 'A.*fr.* 451h.8 R. (lyr.), Call.*fr.* 37.3, 177.28 Pf.'; delete '(prob. from βρέχω .. 24)'
βρεκτός, after '*soaked*' insert 'βρεκτῶν τε κομάων Euph. in *Suppl.Hell.* 442 *fr.* 2.7'
***βρεμέθω**, = βρέμω, Jo.Gaz. 2.145.
Βρετᾰνικός I, add 'as imperial title, D.C. 60.12, *SEG* 34.1268 (Bithynia, AD 189), *ICilicie* 78 (AD 209/211); βρυτανικῆς (sic) τέχνης prob. *tin-working*, *PKöln* 101.9 (iii AD)'
***Βρεταννίς**, ίδος, fem. adj., *British*, νήσοιο Β[ρ]εταννίδος *GDRK* 1².79 no 22.4.
βρέτας 1, add 'of a statue of Artemis, epigr. in *SEG* 23.220(b) (Messene, i AD)'

βρέφος II, for '*new-born babe*' read '*baby, child*'; after '[not in S.]' insert 'Theoc. 15.14, Call.*Cer.* 100, *fr.* 487 Pf.'; after 'in later prose' insert 'β. διετές *Delph.* 3(6).39.11, 57.10'
***βρεφοτροφεῖον**, τό, *institution for the maintenance of infants*, *Cod.Just.* 1.2.22 (pl.).
***βρεφοτρόφος**, ὁ, official in charge of βρεφοτροφεῖον, *Cod.Just.* 1.3 tit.
βρεφώδης, for 'Diog.Oen. 9' read 'Diog.Oen.*fr.* 9.12 C. (rest.)'
βρέχω, line 5, (aor. 2), after 'but' insert 'βρεχεῖσα Men.*Dysc.* 950'; line 6, after 'Wilcken *Chr.* 341.6 (ii AD)' insert 'and aor. pass. subj., βρεχῶσιν PHib. 90.8 (iii BC)'; lines 19/20, delete 'but also intr.' **II**, line 2, after '(ii AD)' insert 'ἔβρεξεν ὁ Ζε[ὺ]ς θεῷ οὐρανός orac. in *SEG* 31.1575 (Cyrene, ii AD)'
***βρεχώδης**, ες, = βροχμώδης, *damp*, rest. in Diog.Oen.*fr.* 9.8 C. (rest.).
†**βρία**, ἡ, expld. as πόλις (in Thracian), Str. 7.6.1, βρίαν· τὴν ἐπ' ἀγροῖς κώμην Hsch.; in false etym. expl. of Μεσημβρία, ἀπὸ [Μέ]λσα καὶ βρία epigr. in *IGBulg.* 345.4; so in Str. l.c. and St.Byz. s.v. Μεσημβρία.
***βριάριον**, τό, app. dim. of °βρία, *little settlement*, *Peripl.M.Rubr.* 58 (cj. φρούριον).
βρίζα I, after 'Gal. 6.514' insert '*Edict.Diocl.* 1.3'
***βριζόμαντις**· ἐνυπνιόμαντις Hsch.
βρίζω II, delete the section.
***βρίζω (B)**, perh. *pick, gather*, καρποὺς ξύλων D.Chr. 35.18; cf. βρίζει· ἐσθίει, πιέζει, κύει Hsch.
***βρῐθυεργός**, ὁ, *powerful worker*, of an architect, epigr. in *IHadr.* 132 (ii AD).
βριθύς, delete 'once in Trag.' and for 'id.*Eleg.* 5' read '(A.)*fr.* 353a R.'
βρίθω I 3, add '(= *SEG* 30.1473)' **III 2**, line 6, transpose 'συμποσίων .. 3.12' to section i 2 (after 'Od. 15.334')
***βρικάριος**, v. °βρακάριος.
***βριλών**, 'ωνος, ὁ, β.· ὁ βαλανεύς Theognost.*Can.* 35.
βριμός, before 'μέγας' insert '*mighty*, *Carm.Pop.* 16 P.'
***Βρίσαιος**, ὁ, title of Dionysus, *EM* 214.5, St.Byz. s.v. Βρίσα.
βρισόμαχος, add 'perh. in *POxy.* 3876 *fr.* 3.4 (Stesich.)'
***βρίσχος**, v. ὑρίχος.
***Βρῑτομάρπεια**, τά, *festival of Britomarpis*, *Inscr.Cret.* 1.xvi 5.43 (Latos, ii BC); cf. Βριτομάρτια (s.v. Βριτόμαρτις).
***Βρῐτόμαρπις**, ὁ, = Βριτόμαρτις, acc. -ιν, *Inscr.Cret.* 1.xviii 9 c 7 (Lyttos, ii BC), 1.vii 4 2 (Cret., Chersonesus, ii BC), al.
Βριτόμαρτις, etc., add 'cf. °Βριτομαρπ-'
***βριττάνδρα**, perh. fem. adj. *prevailing over men*, *Suppl.Mag.* 42.30 (iii/iv AD).
βρογχωτήρ, delete the entry (v. °βροχωτήρ).
βρόμος (A), line 4, after 'A.*Th.* 476' insert 'Aq.*Ez.* 23.20'
βρόμος (B) 1, add '*Edict.Diocl.* 1.17'
βρονταῖος, add '**2** βρονταία (sc. λίθος), ἡ, name of a precious stone, Plin.*HN* 37.150.'
βροντάω, after '*JHS* 5.258, etc.' insert 'Mitchell, *N.Galatia* 13, 77, *SEG* 32.1275 (Dorylaeum), in Bithynia, *INikaia* 1504, al.; *SEG* 35.1419 (Pamphylia)'
βρότεος (A), add 'Emp. 6.3 D.-K.'; form βρότεος, before 'Pi.*O.* 9.34' insert 'Simon. 76.6 P. (cj.)'
***βρότειος (B)**, α, ον, *gory*, λαιμοί E.*Heracl.* 822 (cj. βοείων), id.*IA* 1083; cf. βροτόρις.
βροτήσιος, after 'Hes.*Op.* 773' insert 'φῦλα Alcm. 106 P.'
βροτοβάμων, add '(cj. βοτοβάμων *mounting flocks*)'
***βροτοείκελος**, ον, *like a human being*, εἶδος *Inscr.Cret.* 2.xxiv 13.12 (epigr., Rhethymna, iv AD).
βροτόεις I, add 'Hom.*fr.* 7, Stesich. *S*15 ii 13 (p. 160 D.)'
βροτολοιγός, last line, 'ἔρως .. end' read 'Ἀΐδης *AP* 1.56, Ἄρης ib. 6.91 (Thallus), 9.323 (Antip.Sid.), ἔρως ib. 5.180 (Mel.), 9.221 (Marc.Arg.)'
βροῦκος, after 'Ion. acc. to Hsch.' insert 'but as pers. n. elsewh., *SEG* 9.46.8, 35 (Cyrene, iv BC), 32.281 (Athens, iii BC)'; βρεῦκος, after 'Cret. acc. to Hsch.' insert 'cf. pers. n. *SIG* 737.3, 11 (Cret. at Delphi, i BC)'; add 'cf. βραύκη, βρόκος, βρύκος'
βροῦνος, add 'Hdn.*fr.* p. 27 H.'
***Βροῦχος**, v. βροῦκος.
***βροχεύς**, έως, ὁ, *net-maker*, *PAmst.*inv. 21 (*ZPE* 9.49, iv/v AD).
***βροχηλᾰτέω**, *drive into snares*, v.l. in Opp.*C.* 4.64 (v. βοηλατέω).
βροχίς I, after 'Opp.*H.* 3.595' insert '(pl., *meshes of a fishing net*), *snare for birds*, *AP* 9.76 (pl., Antip.Sid.)'
βρόχυς, for 'Sapph. 2.7' read 'Sapph. 31.7 L.-P. (βρόχεα, neut. pl. as adv., *a little while*)'
***βροχωτήρ**, ῆρος, ὁ, *neck-hole* in a garment, J.*AJ* 3.7.2.
βροχωτός 2, for '*twisted, corded*, of chain work' read 'as transl. of Hebr. wd. app. meaning *finely-woven*'
βρυάζω, line 1, delete 'aor. ἀν-εβρύαξα Ar.*Eq.* 602'
βρυάθμον, delete 'cf. *Hymn.Is.* 89 (dub. sens.)'
***βρυάθων**, dub. wd. in *Hymn.Is.* 89 (perh. part. of βρυάθω = βρυάζω).

βρυάκτης, for 'jolly' read 'wanton'
βρῦγμα, delete 'gnawing'
*βρυδακίζειν· ἐκτείνειν Hsch. (La.); cf. ‡βρυλλιχίζειν.
†βρυδαλίχα· πρόσωπον γυναικεῖον Hsch. (remainder of gloss corrupt).
*βρύζα, ἡ (= °ὄβρυζα), χρυσοῦ β. refined gold, Edict.Diocl. 30.1a.
βρύζω, delete 'dub. sens. in Archil. 32.2' (v. °μύζω (B))
βρυκεδανός, after 'μακρός' insert '(cj. μάργος)'
βρύκω II, for 'grind' read 'snap'
βρυλλιχίζειν, after 'Hsch.' insert '(cj., cod. °βρυδακίζειν)'
†βρύλλω, dub. sens., βρύλλων Ar.Eq. 1126: expld. by Sch. as ἐξαπατώμενος, also ὑποπίνων, μεθύων, etc.
†βρυόεις, εσσα, εν, growing luxuriantly, νέον βρυόεντα θύμου στάχυν Nic.Al. 371, 478. **2** full of luxuriant growth, βρυόεντος .. ποταμοῖο id.Th. 208, ἐνὶ βρυόεντι .. κόλπῳ Nonn.D. 1.206, 21.180, al.
βρύον I, transfer 'Arist.HA 591ᵇ12' to follow 'β. alone' and after it insert 'Thphr.HP 4.6.6' and 'Theoc. 21.7' (from section II)
βρύσσος, add 'E.fr. 955b (cf. ἄμβρυττος). **2** pudenda muliebria (cf. βύττος), ὃς κατευδούσης τῆς μητρὸς ἐσκύλευε τὸν βρύσσον Hippon. 70.8 W.'
*βρυτανικός, v. °Βρετανικός.
†βρυχή, ἡ, roaring, bellowing, Opp.H. 2.530, Q.S. 5.392. **2** crashing, A.R. 2.83.
βρυχηδόν, for 'gnashing' read 'snapping'
βρύω 2, after 'c. gen., to be full of' insert 'βρύουσι φιλοξενίας ἀγυιαί B. 3.16'
βρῶμα II 2, for 'Ep.Je. 12' read 'Ep.Je. 11'
βρωμάομαι I, add 'βρωμωμένον· ὀγκωμένου Hsch.' for 'III etc.' read 'II ἐβρώμησε (ἐβρωμήσατο La.)· βρώσεως ἐδεήθη Hsch.'
βρωμᾰτίζω, add 'Vit.Aesop.(G) 45'
βρωμάτιον, add 'Pap.Lugd.Bat. xxv 73.3 (vi AD)'
βρωμήεις, add 'also βρωμέεις, Hdn.Gr. 2.921.1, al.'
βρωτός II, add 'Aen.Tact. 8.4'
*Βύβλιος, η, ον, from Phoenician Byblos, of a kind of wine, Archestr. ap.Ath. 1.29b.
*βυβλι-, see also βιβλι-.
βύβλος I 2 a, delete the section transferring 'Thphr.HP 4.8.4' and 'Hdt. 2.96.2, Plot. 2.7.2' to section 1. **b**, delete 'sg., strip of β. .. Ph. 2.522' and transfer 'Ph. 2.522' and 'Pl.Plt. 288e' (from 2 a) to section 3
*βυζάστρια, ἡ, gloss on τίτθη, Hdn.Philet. p. 81 Dain; cf. mod. Gk. βυζάστρα, wet-nurse.
βύθιος I, for 'Hymn.Is. 71' read 'Hymn.Is. 161'
*βυθοκλόνος, ον, convulsing the deep, PMag. 4.1363.
βυθός **a**, add 'Βυθός, personified, in mosaic, SEG 31.1387 (Apamea, Syria, iv AD)' **b**, add 'deep track, Nic.Th. 570'
βῡκᾰνητής, after 'Plb. 2.29.6' insert '30.22.11'
βῡκᾰνιστής, delete 'Plb. 30.22.11'
βύκτης I, add 'A.R. 3.1328, Orph.A. 125'
βυλλόω, add 'also ἐβύλλων· ἔβρυον, ἐπλήθυον Hsch.'
Βύνη, after 'Leucothea' insert 'Call.fr. 91 Pf.'
†βῦνις, εως, prob. = βύνη, malt, PMag. 12.426, 433, PMil.Vogl. 278.18, al.
*βυρίτιον, app. kind of garment, cf. perh. ‡βίρρος, στιχαρομαφόρην β. PWash.Univ. 58.3 (v AD).
βυρρός II, delete the section (v. °βύρρος).
*βύρρος, v. °βίρρος.
βυρσεῖον, after 'tan-pit' insert 'tannery, PTeb. 801.2 (ii BC)'
βυρσοδεψέω, add 'IG 1³.257.9 (Athens, v BC)'

*βυρσοφώνης, ου, ὁ, sounding with hides, of Salmoneus, imitating thunder with drums, S.fr. 10c.6 R.
*βύσαλον, τό, brick, Anon.Alch. 334.10.
*βυσσικός, ή, όν, made of βύσσος, Marcian Dig. 39.4.16.7.
βύσσινος 1, add 'fig., τὸν βασιλεῖ μέλλοντα μετὰ παρρησίας διαλέγεσθαι βυσσίνοις χρῆσθαι ῥήμασι i.e. with soft words, Plu. 2.174a'
βυσσοδομέω (s.v. -δομεύω), add 'SEG 29.1444 (Oenoanda, c.AD 200)'
βυσσόθεν, after 'Call.Del. 127' insert 'from the depths of the earth, id.fr. 202.59 Pf. (Add. II)'
βύσσος, add 'II a dye, Suid. s.vv. βυσσόν and βύσσινον.'
*βυσσοφαντεῖον (for *βυσσυφ-), τό, workshop for weaving linen, PMagdola 36 (BASP 22.131).
† βωθέω, v. °βοηθέω.
βωθύζειν, add 'βωθύσσειν Theognost.Can. 48 A., Zonar.'
βωλάκιον, add 'lump of earth, Afric.Cest. p. 82 V.'
βῶλαξ, add 'Lxx Jb. 7.5'
†βωλοκοπέω, break up soil, harrow, Ar.fr. 800 K.-A., Hp.Ep. 17, Ael.Ep. 19, PLond. 131ʳ iii 50 (i AD), PMil.Vogl. 305.96, al. (ii AD); w. acc., τὸν χοῦν IG 2².1672.45 (Eleusis iv BC), τὴν γῆν ib. 60; transf., app. beat up, of violent sexual assault, epigr. in IG 9²(2).253 (Thyrreum, iii BC); cf. ἀρόω II 2.
*βωλοκόπημα, ατος, τό, harrow, occa βωλοκόπημα Gloss.Philox.
βωλοκόπος, add 'occiliator (= occillator, harrower), Gloss.'
βωλόκρῑθον, add 'PCair.Zen. 292.437 (iii BC)'
βῶλος 3, add '**b** a precious stone, Plin.HN 37.150.'   at end after 'masc. in Arist. l.c.' insert '(elsewhere in Arist., fem.)'
βωλοστροφέω, for 'turn up clods in ploughing' read 'break up clods after ploughing' and add 'BGU 2126 ii 10 (iii AD)'
βωλοστροφία, after 'clods' insert 'PHib. 282.22 (i/ii AD), Gloss.'
*βωλοστροφικός, ὁ (or -όν, τό) perh. harrow, POxy. 3805.111 (vi AD).
*βωλοτρόφος, ον, feeding the tilth, or βωλότροφος, fed by silt, Hymn.Is. 176 (s.v.l.).
βωμικός, for '= βώμιος' read 'having a base or pedestal'
βωμίς, add '**2** prob. base (of sarcophagus), SEG 17.632, BE 1950.204 (both Perge, Rom.imp.).'
†βωμίσκος, ὁ, dim. of βωμός, altar (in quot., a model in a mechanical contrivance by Hero), Hero Spir. 1.38, al. **b** name for the constellation Ara, Ptol.Alm. 8.1. **2** app. kind of bandage shaped to the form of an altar, Gal. 18(1).823.11. **3** base of molar teeth, Poll. 2.93.   **II** geom., solid figure consisting of a truncated pyramid, Hero Deff. 114, Theo Sm. p. 41 H., Nicom.Ar. 2.16, Syrian. in Metaph. 143.7, al. **b** plane figure formed by the elevation of such a figure, quadrilateral, Papp. 878.
βωμολοχεύομαι, for 'play the buffoon, indulge in ribaldry' and 'play low tricks, in Music' read 'play antics, clown' and add 'Isoc. 15.284'
βωμός 1, delete 'but' and add 'platform or floor of a tomb, on which sarcophagi were placed, TAM 4(1).231, MAMA 6.191a (Apamea), 8.545 (Aphrodisias), etc.' **2**, delete 'mostly' **3**, for this section read 'funerary altar, MAMA 6.19 (Laodicea), etc.'
*βωμοφόρος, ὁ, altar-carrier, SEG 16.741 (Pergamum, Rom.imp.).
*βώνυσοι· ἄποικοι, βουκόλοι Suid., cf. Theognost.Can. 48 A.
*Βωρεῖς, οἱ, one of the Ionic tribes, SIG 57.3 (Miletus, v BC), etc.; sg. Βωρεύς, έος, ὁ, member of this tribe, IEphes. 945a, 1431, al.; also Βορεύς, ib. 1578a.7 (ii AD, also at Tomi).
βωρίδιον (s.v. βωρεύς), add 'POxy. 2728.33 (iii/iv AD), PRyl. 629.88 (βορ-, iv AD)'
*βώσεσθε, v. °βόσκω.
βωσίον, after 'dub. sens.' insert 'perh. = βωτίον'
*βωτέω, v. °βοωτέω.
βωτιάνειρα, for 'Hes.Cat. .. 1.16' read 'rest. in Hes.fr. 165.16 M.-W., Alcm. 77 P.'
βώτωρ, transfer 'βώτορες ἄνδρες' to precede 'Il. 12.302'

Γ

γᾰ, after 'Ar.*Ach.* 775' insert '(in the mouth of a Boeotian); *SEG* 11.244 (Sicyon, v BC)'
γαβαθόν, for 'cf. ζάβατος' read 'cf. ζαβατός 2, καβαθα'
γάβενα, add 'cf. mod. Gk. γαβάνη'
γαβεργόρ, add '(γαβεργός La.)'
γαγγᾰλίζω, for 'but the contrary is stated' read 'γαργ- said to be correct form'
*γαγγᾰλισμός, ὁ, γαργαλισμός· γαγγαλισμός, ἡδυπάθειά τις Hsch.
†γάγγαλος· ὁ εὐμετάθετος καὶ εὐρίπιστος τῇ γνώμῃ καὶ εὐμετάβολος Hsch.
†γαγγᾰμεύς· ἁλιεύς, ὁ τῇ γαγγάμῃ ἐργαζόμενος Hsch.
γαγγάμη, add 'σαγήνη ἢ δίκτυον ἁλιευτικόν. καὶ σκεῦος γεωργικὸν ὅμοιον τῇ κρεάγρᾳ Hsch.'
Γαγγητικός, delete 'a fragrant .. *Iwarancusa*'; after '*Peripl.M.Rubr.* 63' insert '(Γαγγιτικός Casson); cf. Γαγγῖτις νάρδος'; for a description of the plant see also Plin.*HN* 12.42.
γάγγραινα, add '2 = μέροψ (bird), Cyran. 92 (3.30.3 K.).'
γαγγραινικός, add '2 *used in cases of gangrene*, Gal. 13.739.13.'
*γᾰδαισία, v. °γεωδαισία.
Γάδειρα, etc., delete marks of quantity and add '[Γᾱ- Pi. l.c., cf. Lat. *Gādes*; Γᾰ- Theodorid. in *Suppl.Hell.* 744, *AP* l.c., D.P. 11, al., Anacreont. 14.25 W.]'
*γᾱϝεργέω, v. °γεωργέω.
*γᾱϝεργός, v. °γεωργός.
*Γαζαῖοι, οἱ, *inhabitants of Gaza*; in a dating formula, κατὰ .. Γαζαίους τοῦ ἐχ' ἔτους ἰνδικτιῶνος η' *SEG* 34.1467 (Palestine, vi AD).
*γαζαρηνοί, οἱ, (Aramaic *gazerin*) *astrologers*, Lxx *Da.* 2.27, 4.7, 5.7.
γαῖα I 2, delete 'the forms .. etc.'
†γαιᾶται· κερτομεῖ. καταμωκᾶται Hsch.; cf. γαιώ· τὸ κερτομῶ καὶ διαβάλλω Theognost.*Can.* 22 A.
*γαιάτης, ου, ὁ, *dweller on earth*, St.Byz. s.v. γῆ.
*γαιεῖος, v. °γαιήιος.
*γαιηϊᾰς, fem. adj. = γαιήιος, St.Byz. s.v. γῆ.
γαιήϊος II, add 'also γαιεῖος, βρότεον γένος ἠδὲ γαιεῖον Didyma 496B4 (ii AD)'
*Γαῖηος, ὁ, name of a month in Caligula's calendar in Egypt, *PTeb.* 492, *OStrassb.* 68 (AD 41); also Γάϊος, in full Γ. Σεβαστός *OBodl.* ii 469, 470, al.
*γάϊνος, v. °γήϊνος.
†γαιώδης, v. ‡γεώδης.
γᾰλᾰ I 3, transfer 'οὐδ' εἰ γ. λαγοῦ .. Alex. 123 (= 128 K.-A.)' to section I 1, after 'Theoc. 24.3'
γαλαθηνός, add 'Simon. 38.8 P.'
*γάλαιθος, unexpld. wd., anon.ap.*An.Ox.* 2.318.
γαλακτίτης I, add 'Plin.*HN* 37.162'
*γαλακτῖτις, ιδος, ἡ, = γαλαξίας II, Plin.*HN* 37.162.
*γαλακτοδότρια, ἡ, *giving of milk*, ἡ τῶν ἀναγεννηθέντων γ. Ps.-Chrys.*HP* 5.2, Leont.Byz.*HP* 1.6.2.
*γαλακτοποιός, όν, = γαλακτοποιητικός, βοτάνη Hdn.Gr. 1.395.
γαλακτοποτέω, for '(Written .. p. 111 V.)' read '(shd. be written -πωτέω acc. to Ammon. 399 N., s.v.l.)'
*γαλακτοπότος, ον, *sucking*, χοίρου γαλακτοπότο[υ] *Edict.Diocl.* 4.46 (v.l.).
*γαλακτορρύω, *flow with milk*, Procl.CP *Hom.* 29.18.
γαλακτοφόρος, add '2 γαλακτοφόροι, οἱ, *milk-carriers*, *OBodl.* i 304 (ii BC).'
γαλαξίας II, add '= γαλακτίτης I, Plin.*HN* 37.162'
†γαλαρίας· ἰχθύς, ὁ ὀνίσκος Hsch. (see also γαλλερίας, καλλαρίας, °γελαβρίας).
Γαλάται, add 'sg. -τας *IG* 9²(1).624c5 (Naupactus, ii BC)'
γαλεάγρα, after '*weasel-cage*' insert '*IG* 1³.422.139 (Athens, v BC)' add '**II** *winepress*, Hero *Mech.fr.* (Arabic text) 3.17 (2.238 N.-S.).'
*γαλεάριος, Lat. *galearius*, *soldier's servant*, *SEG* 19.787 (Pisidia).
γαλεός I, after '*small shark*' insert 'Philox.Leuc.*fr.* (b).11 P.'; add 'γαλιῶ μέδδονος *IGC* p. 99 A40 (Acraephia, iii/ii BC)'
*γαλεοψοποιέω, *prepare a dish of dogfish*, Archestr. in *Suppl.Hell.* 188.9 (γαλεοψοποιούντων cj., γαλῇ ὀψοποιούντων cod.).
γάλερος, after '*AB* 229' insert 'neut. as adv., *GLP* 123.1'
γαλεώνυμος, add 'Orib.*Syn.* 4.17.7'

γαληναίη, add 'Call.*Epigr.* 5.5 Pf., *AP* 7.640 (Antip.Thess.), Opp.*H.* 1.460'
*γαληνοποιός, ὁ, *one who makes calm*, gloss on στορεύς, Hsch.
γαληνότης, add '**II** as title, *Serenity*, *MAMA* 3.197 (Corycus), *ICilicie* 118.8, *PFlor.* 75.5 (all vi AD), Just.*Nov.* 124.4.'
*γαλίβδολον, τό, = °γάλιψις, Paul.Aeg. 7.3 (203.3 H.).
*γάλιψις, ἡ, plant said to resemble ἀκαλήφη (prob. in sense I), Paul.Aeg. 7.3 (203.3 H.).
*γάλλαρος, ὁ, member of a Dionysiac cult-society, prob. in *IGBulg.* 401.15-16 (Apollonia in Thrace); pl.); γ.· Φρυγιακὸν ὄνομα Hsch.; cf. °ἄρχιγ-.
γαλλερίας, add 'cf. γαλαρίας, καλλαρίας'
*γαλλεωρ, ὁ, accent unkn.; app. designation of man's trade or occupation, *BGU* 1614.Ci5 (i AD).
*Γαλλικός, ή, όν, *Gallic*, θάλασσα Ptol.*Geog.* 2.10.2, of garments, βάνατα γ. *Edict.Diocl.* 19.57, etc.; σαβάνων Γ. ib. 28.57; designating military units, εἴλης (sic) β' Γαλλικῆς ala II *Gallorum*, *SEG* 38.683 (Maced., ii AD), λεγεώνων Γαλλικῆς καὶ α' Ἰλλυρικῆς ib. 34.1598.5 (Egypt, iv AD).
γάλοως, γάλως, add 'prob. in *SEG* 31.1004.8 (Saittai, ii AD)'
γαμβρά, after '*sister-in-law*' insert '*PMich.*II 123ᵛ viii.10 (i AD), γαμρ-pap.)'
γαμβρός, line 1, after 'ὁ' insert 'also γαρβρός (sic) *SEG* 31.1031.5 (Saittai, ii AD); cf. γαβρός ib. 1004.7' and after 'A.*Ag.* 708 (pl.)' insert 'Plu. 2.620a'
γάμελα, for '*Michel* 995B36' read '*CID* 1.9A 24-25, B 36, equiv. of Att. γαμήλια'
γαμετή, add 'as name of a chariot horse, *SEG* 26.1837 (Leptis Magna, iv AD), sp. γαμητ- *BCH* 4.199, 7.237 (both Isauria), 503 (Philadelphia), *IG* 3.3479'
γαμέτης, before 'Euph. 107.3' insert 'Call.*fr.* 228.12 Pf.' and after 'Dor.' insert 'nom. γαμέτᾱς Call. l.c.'
γαμέω I 2, for this section read '*take as a lover* or *concubine*, Od. 1.36; οὐχὶ χατέρας πλείστας ἀνὴρ εἷς Ἡρακλῆς ἔγημε δή S.*Tr.* 460; βιαίως E.*Tr.* 44. **b** *have sexual intercourse with*, Luc.*Asin.* 32, *AP* 5.94 (Rufin.).' **3**, of the woman, add 'εἴθ' ὕστερον ἔγημε τὸν φθείραντα Arg.Men.*Her.* 3; abs., 1*Ep.Cor.* 7.34, Charito 3.2.17, Just.*Nov.* 97.6 pr.'
*γᾰμητή, v. ‡γαμετή.
*γαμητιάω, *be eager for sexual intercourse*, Vit.Aesop.(W) 103.
†γᾰμίζω, *give in marriage*, *Ev.Matt.* 24.38, A.D.*Synt.* 280.11; pass. *Ev.Matt.* 22.30, *Ev.Marc.* 12.25, *Ev.Luc.* 17.27. **2** *make* a woman *one's wife*, 1*Ep.Cor.* 7.38.
Γαμίλιος, after 'Epirus' insert 'and elsewhere' and add '*SEG* 30.990 (Corinth, iv/iii BC)'
*γαμμάτιον, τό, of an ornament, *a little* (representation of) *gamma*, *EM* 766.7.
†γαμμᾰτίσκιον, τό, = °γαμμάτιον, Lyd.*Mag.* 2.4.
γαμοκλόπος, add 'of Κύπρις, *Inscr.Cret.* 3.iv 37.5 (Hellen. epigr.)'
γάμος VI, Γάμος, add 'also at Argos, *SEG* 30.356 (300 BC)'
*γαμοστολίη, ἡ, perh. *wedding-dress*, *MAMA* 7.229.
*γαμότης, ητος, ἡ, *state of marriage*, σώφρων ἐν γαμότητι, περίφρων ἐν βιότητι *GVI* 1737.5 (Syria, iii AD).
γαμφηλαί, for 'once in sg.' read 'also in sg.' and insert 'v.l. in Moero 1.6'
+**γᾰνάεις**, εσσα, εν, *glorious*, prob. in Pi.*Dith.* 4(h).7 S.-M.
*γανᾰλός, unexpld. wd. in Hdn.*fr.* p.21 H.
γανάω II 2, for this section read 'γανάοντες *glorifying*, A.*Supp.* 1019 (cj., cod. γανάεντες)'
*γανίς, ίδος, ἡ, Egyptian liquid measure (in quots., of oil), *OMich.* 253, 254 (iii/iv AD).
γάννος, for '= γλάνος' read '*hyena*'; add '(Phrygian and Bithynian acc. to Hsch. (γάνος cod.))'
*γανοπετεῖν, v. γανυτελεῖν.
γάνος (A) 2, add 'στάζει ἐνὶ κραδίῃ γλυκερὸν γ. Opp.*H.* 1.275' **3**, add 'γάνος .. οἶνας *IG* 2².3783.5' **4**, delete the section (read γᾶν ὅς).
γάνος (B), for 'dub. in .. (Mytilene)' read 'τῶν γανων dub. interpr. in *IG* 12(2).58.(a)17 (Mytilene, Rom.imp.), cf. perh. *ka-no-se*, *ICS* 309.12, but this may be second element of a longer wd.'
γᾰνόω I, line 4, for 'ἑοῖς .. (Philae)' read 'νέοις ἐγάνωσεν ἰάκχοις

IPhilae 159.3 (i AD)'; line 6, for 'Ph. l.c., al.' read 'Ph. l.c. [1.121], al.' and after it insert 'enjoy, τί δεῖ (δη pap.) γανοῦσθαι τοῦτο; A.fr. 78c i 19 R.'   delete '**II** tin, lacquer' incorporating quots. in section I.

γάνυμαι, add 'γάνυται δέ τ' ἀκούων Hom.fr. 4'

γάνωσις 1, add 'fig., of style, lustre, cj. in Longin. 30.1'

γάνωτός, add 'Anon.Alch. 316.11'

γᾱοδίκαι, add 'BCH 116.200 (Phocis, iii BC)'

***Γᾱος**, ὁ, name of a divinity associated with Persephone, SEG 34.940.4 (Camarina, iv/iii BC).

†**γάπεδον**, v. °γήπεδον.

γᾱράριον, add 'prob. l. in Sch.Nic.Th. 526b C.'

γαργαλισμός, add 'see also °γαγγαλισμός'

γάργαρα, add 'χρημάτων .. γ. Trag.adesp. 442 K.-S.'

γαργαρίζω, add 'freq. in Lat. borrowing, Varro LL 6.96, Cels. 4.2.8, Plin.HN 28.129, etc.'

***γαργαρισμάτιον**, τό, gargle, Marcell.Emp. 15.19; written gargal- ib. 14.28.

γᾱροπώλης, add 'POxy. 3749.6 (iv AD)'

γᾱρος, neut., add 'PRyl. 629.88 (iv AD)'

γαστήρ I 2, line 9, for 'com. of one who has nothing to eat' read 'i.e. eat sparingly'   **II**, add 'by synecdoche, for the mother, Just.Nov. 54 pr., 156.1'   **III**, for this section read 'archit., precise significance uncertain but app. some form of bulge, τὸ κα[τά]ζευγ]μα τὸ ὑπὸ γαστέρα ἐπὶ τὸ ἐπ[ιστύλιον ..] IG 1³.474.253, τὰ τρήματα τῶν γαστρῶν Inscr.Délos 504 A 7 (iii BC), [τὴν κα]ταγλυφὴν τῶν ἐν αὐτοῖς γαστρῶν Didyma 39.4 (ii BC), but for these last two some assume nom. γάστραι'

γάστρα IV, add '2 womb, Sor. 1.3.9.'

γαστράφέτης, add 'Bito 64.4 M., al.'

γαστρίαν, after 'διάνοιαν' insert '(read διάρροιαν La.)' and after 'Hsch.' insert 'so γαστρίη Hippon. 118.9 W., expld. by Sch. ad loc. as γαστρὸς ἀλγηδών'

γαστρίμαργος, add '**II** -γα, ά, a kind of fish, IGC p. 100 B33 (Acraephia, iii/ii BC).'

γαστροκνήμιον, add 'BGU 975.11 (i AD, written καστρο-)'

γαστρόπτης, for 'vessel for cooking sausages' read 'perh. belly-shaped cooking vessel'; add 'IG 2².1638.67 (Delos, iv BC), etc.'

***γασυνδάνη**, ἡ, name of gem, Xenocr.Lap. 109.

***γᾱτομέω**, Dor. form of γεωτομέω, cleave the ground, A.R. 2.1005, Lyc. 268, 1396.

γᾱτόμος, add 'see also γεωτόμος'

γαυλός I 1, for 'water-bucket .. (Delos)' read 'water-bucket, A.R. 3.758, used for raising water from a well, Hdt. 6.119.3, γ. μοι χρύσεος φρείατος ἐκ μυχάτου Alex.Aet. 3.20, IG 11.146.A29, etc. (Delos, iii BC, v. BE 1960.168)'; add 'wine-cask, οἰνηρὸν ἀγγεῖον Suid.'

γαυριάω, line 3, for 'φυσῶντα καὶ γαυριώμενον' read 'γαυριώμενος'; line 5, after 'of persons' insert 'Semon. 10a W.'; at end for 'Anon.Oxy. 220iii3' read 'anon. metrical treatise in POxy. 220 v 3'

***γαυροειδής**· ὁ ὑπερήφανος Suid.

γαῦρος, line 4, epith. of ἔφηβοι, add 'Robert Hell. 1.127-31 (Eretria)'

γαυρόω, pass., add 'ταύρος ὡς γαυρούμενος Ezek.Exag. 268 S.'

***γαυρύνομαι**, pride oneself, δῶρα ἀμισῶς ἀποδίδων γαυρύνομαι IHadr. 24.7 (ii/iii AD), app. w. word-play on pers. n. Γαῦρος.

***γᾱωρύχιον**, v. °γεωρύχιον.

†**γέγειος**, α, ον, ancient, Hecat. 362 J.; βόες Call.fr. 277 Pf.; λόγος ib. 510; comp. -ότερος, γ. .. νίκης σύμβολον ib. 59.5.

γέγωνα, line 3, for 'ἐγεγώνειν' read 'ἐγεγώνει'; line 10, before '-είτω' for 'impf.' read 'imper.'   **1 b**, for 'speak .. sound' read 'make an effectual sound' and after 'γεγωνεῖ' insert 'ἂν μὴ λεῖον ᾖ τὸ πληγέν'

†**γέη**, v. °γῆ.

γείνομαι I, for this section read '(for the forms γεινόμενος, γεινόμεθα v. °γίγνομαι)'   **III**, line 2, delete 'in pass. sense' and after 'Call.Cer. 58' insert 'Del. 260, etc.'

***γέϊνος**, v. °γήϊνος.

γειόθεν, for 'Call.fr. 35c' read 'Call.fr. 110.49 Pf. (v.l γηόθεν)'

***γεισεπίστῡλον**, τό, cornice-architrave, MAMA 6.370 (Synnada, iii AD).

***γεισηποδίζω**, furnish a corbel for the cornice, ὅσ[αι δὲ] τῶν παρ[ό]δω[ν στ]ενότεραί εἰσι[ν] καὶ γεγεισηποδισ[σ]μέ[ναι λιθ]ίναι γεισηποδίσματι Hesperia 9.68.113 (= new fr. of IG 2².463.113, Athens, iv BC); cf. γεισηποδίειν· τὸ προσβάλλειν τὰ γεῖσα ἐν τοῖς τοίχοις Hsch.; also γεισιποδίζω Is.fr. 113.

†**γεισηπόδισμα**, τό, corbel supporting a cornice, IG 2².463.114 (Athens, iv BC); γεισιπόδισμα Is.fr. 113.

†**γεισήπους**, ποδος, ὁ, cornice-support, IG 2².463.51 (Athens, iv BC), app. referring to dentils beneath a cornice, Inscr.Délos 104.24.17.

†**γεισιποδίζω, -πόδισμα**, v. °γεισηπ-.

***γειτνεύω**, be a neighbour, border on, PMich.x 581.13 (ii AD), PCornell 11.8, see also °γειτνιεύω, γειτονεύω.

γειτνία, add 'BGU 94, PKöln inv. 1695 (ZPE 10.104, iv AD; sp. γιτν-); Γιτνεία Ἁμαξικὴ quarter near the chariot-gate, Robert Hell. 11/12.410 n. 1 (Apamea, Syria)'

γειτνίᾱσις I 2, add '**b** guild of neighbours, τῇ ἱερᾷ Λητοῦς γειτνιάσει TAM 3(1).765.11 (Termessus); cf. °γειτονίασις.'

***γειτνιεύω**, be a neighbour, border on, PFlor. 60.7 (iii AD); w. dat., POxy. 2190.57 (ii AD), BGU 2061.7, 30 (iii AD) (see also °γειτνένω).

γείτνιος, delete the entry (v. °γειτνία).

γειτονία 2, add 'Lang Ath.Agora xxi L8 181 (iv BC; sp. γιτ- ib. 31.1035 (Lydia, ii AD)), MAMA 7.301'

***γειτονιάρχαι**, οἱ, magistrates of a γειτονία or ward, Robert Hell. 11/12.410 n. 1 (Constantinople).

***γειτονίᾱσις**, εως, ἡ, group or guild of neighbours, IGRom. 4.548 (Orcistus, prob. iii AD); in connection w. a shrine, γ. Ἀχιλλέως Ἰητῆ<ρ>ος TAM 3(1).348 (Termessus); cf. °γειτνίασις 1 2b.

***γειτόνισσα**, ἡ, (female) neighbour, PSI 876.5 (v/vi AD).

γειτοσύνη, add 'INikaia 1202 (ii AD), REG 2.24 (Acmonia, iii AD)'

γειτόσυνος, for 'ον' read 'η, ον'; add 'τὰ γειτόσυνα IG 9(2).301.13 (Tricca, ii BC)'

γείτων I, lines 7/8, for 'ἐν γειτόνων .. Men.Pk. 27' read 'ellipt., ἐν γειτόνων δ' οἰκοῦσα living in the neighbourhood, Men.Pk. 27, cf. Dysc. 25'   **II**, at end after 'Jul.Or. 2.72c' insert 'fig., γείτονα πότμου ἡβητήν Nonn.D. 11.97; νέος .. Ἅιδι γείτων ib. 11.214' and after 'neut. γεῖτον' insert 'IG 1³.426.67 (Athens, v BC)'

***Γέλα**, ἁ, hoar-frost, cited by St.Byz. (γέλαν· πάχνην) as Sicilian or Oscan wd. in explanation of place-name Γέλα (s.v.); cf. Lat. gelu.

***γελαβρίας**, ὁ, name of a fish, IGC p. 99 A39 (Acraephia, iii/ii BC); cf. γαλλερίας, γαλαρίας.

γελάω, line 1, delete 'Ep.'   **I 1**, after 'laugh' insert 'smile'

†**γελγοπωλέω**, app. sell fancy goods, etc., Hermipp. 11 K.-A., Hsch.

γελγοπώλης, for 'garlic' read 'fancy goods'

Γελέοντες, add 'also Γλέοντες Sokolowski 2.10.A35, 47 (Athens, v BC)'

γέλλαι, add '(cf. ἑλλίζων)'

γελοιάζω, delete 'only pres.' and insert 'aor. inf. γελοιάσαι Hsch. s.v. γελυνμάξαι'

***γελοίασμα**, ατος, τό, gloss on ψιά, Hsch.

γελοῖος, line 3, after 'Thgn. 311' insert 'παρέξαι τὰ γελοῖα A.fr. 47a ii 15 R.'

***Γελώϊος**, ὁ, month name, Γελωΐου ἕκται ἐπὶ δέκα SEG 39.996 (Camarina, ii/i BC).

γέλως II, line 2, after 'food for laughter' insert 'πολὺς ἀστοῖσι φαίνεαι γ. Archil. 172.4 W., Semon. 7.74 W.'

γέμω 1 a, add 'of a tomb, SEG 34.1469 (Palestine, AD 588)'

†**γενάρχης**, ου, ὁ, founder or head of family or race, Call.fr. 229.1 Pf., Lyc. 1307; of Cronos, Orph.H. 13.8; epith. of Heracles, IG 5(1).497 (Sparta, al.); of Abraham, Ph. 1.513, of Abraham, Isaac and Jacob, id. 1.646.   **b** transf., of the 70 elders, Ph. 2.111; of the ἐθνάρχης at Alexandria, id. 2.527.

γενεά I, lines 10/11, for 'πατριὰ καὶ γ. .. (Elis)' read 'πατριὰν .. καὶ γενεάν gens and (immediate) family, Schwyzer 409 (Elis, vi BC), pl. children, Schwyzer 424 (Elis, iv BC)'   **3**, add 'SEG 33.765 (form γενιά, Italy, iii/ii BC)'   **II 3**, add 'birth, γ. ἑτέρη, of Dionysus, APL 257'

γενεᾱλογία, add 'γενεολογία dub. l. in Max.Tyr. 23.1'

γενεαρχικός, add 'Just.Nov. 21.1'

γενέθλη I 3, after 'birth' insert 'parturition, AP 6.272 (Pers.); metonym. that which gives birth to, [κ]υδαλίμων θεόν πάντων γενέθλαν Alc. 129.7 L.-P.'

***γενεθλιάς**, άδος, ἡ, (sc. ἡμέρα) birthday, GVI 2039.2 (Mytilene, ?i/ii AD).

γενέθλιος, after 'also α, ον Lyc. 1194' insert 'η, ον Call.fr. 202.21 Pf.'

†**γενειόλης**, ου, voc. γενειόλα, bearded, epith. of Hermes, Call.fr. 199.1 Pf.

γένειον, for 'part covered by the beard, chin' read 'beard' and combine w. section 2.   **5**, for 'dub. sens.' read 'perh. some form of clamp or fastening'   add '**6** barbel of a fish, Ael.NA 15.11.'

γενεολογία, v. °γενεαλογία.

***γενέρωσος**, η, ον, Lat. generosus, of noble birth, IEphes. 1540, JÖAI 23 Beibl. 171.118.

γενεσιακός, add 'κατὰ τὸ γ. BGU 1843.12 (i BC)'

γένεσις III, add '**2** of Γενέσια name of month, Hesperia 27.75 (c.200 BC).'   **III**, line 2, delete 'the birthday of'

γενεσιουργέω, add 'Herm.in Phdr. p. 169 A.'

γένεσις II 2, add 'SEG 7.904 (Gerasa)'

γενέτειρα I, add 'of Aphrodite, as mother of the imperial family, Venus Genetrix, SEG 30.1253 (Caria, i AD)'

γενέτης, add on °γονέτης.

γένημα, add '**b** perh. year, in respect of its produce, BGU 2269.15 (ii AD).'

***γενηματογράφος**, ὁ, sequestrator, PLond. 454.b4 (iv AD)

γενηματοφύλαξ, add 'PGen. 86 (?ii BC)'

***Γενιακός**, ὁ, epith. of Apollo, IGRom. 1.740.

γενικός I 1, line 3, after 'Phld.*Sign.* 18, 19, etc.' insert 'opp. to ἰδικός (*particular*), *Cod.Just.* 3.2.2.1, 6.48.1.8'; line 4, after '-κῶς' insert 'Cic.*Att.* 1.14.2'; add 'comp., Cic.*Att.* 9.10.6 (cj.)'

γέννᾰ I 1, add 'pl. πάντες δὲ βροτοὶ χ[άρεν σαῖς, ὦ Β]άχχιε, γένναις Philod.Scarph. 10'

γεννάδας, add 'w. non-Dor. inflexion, συμβίῳ γεννάδῃ καὶ ἑαυτῇ Mitchell *N.Galatia* 109'

γενναῖος I, add '4 sup., as honorific epith., *PLond.* 1645.25, *SB* 4299.28, *PRyl.* 177.16 (all iii AD), ὁ γενναιότατος καὶ ἐπιφανέστατος Καῖσαρ *PAmh.* 148.4 (v AD), *BCH* suppl. 8.63 (Maced., v/vi AD).' **II**, line 3, after 'Pl.*Lg.* 844e' insert 'μάζαι id.*R.* 372b; καρποί D. 18.309' **III 1**, at end after 'Ps.-Callisth. 1.38' insert '-ότερον D.H.*Dem.* 26'

γέννημα I 2, delete the section transferring ref. to section 1.

γεννῆται (s.v. γεννητής II), add 'D. 59.55, al.; transf. in sg., πῶς οὖν οὐ σοὶ πρώτῳ μέτεστι τῆς τιμῆς, ὄντι γεννήτῃ τῶν θεῶν Pl.*Ax.* 371d'

***γεννήτριος**, ον, *of childbirth*, ὠδῖνες γ. Ps.-Callisth. 13.6 (cj., cod. γεννητηρ-).

γεννήτωρ, add 'transf., ὧν δή εἰσι καὶ οἱ ποιηταὶ πάντες γεννήτορες Pl.*Smp.* 209a'

γένος V 3 c, *produce*, add '*Peripl.M.Rubr.* 10, 14' add '5 ἐν γένει *as concerning all, generally*, Just.*Nov.* 7.9.1.'

γέντα, for 'Call.*fr.* 309' read 'Call.*fr.* 322, 530 Pf.'

***γέντη**, ἡ, app. *flesh, meat*, θεὸς γέντης μὴ γενέσθ[αι] *IGBulg.* 2083.7 (Rom.imp.); cf. ‡γέντα.

γεοθαλπής, for '*CIG* 3769' read 'orac. in *TAM* 4(1).92.1'

***γεουχία**, ἡ, *landowning, SEG* 8.448*a*, 24.1191(*a*) (Egypt); *estate*, *PMich.*VIII 503.3 (ii AD), *PPrincet.* 69.1.

***Γεραιστιασταί**, οἱ, participants in the festival of τὰ Γεράστια (s.v. °Γεράστιος), *IG* 4.757B12 (Troezen, ii BC).

γέρᾰνος III, for the pres. def. read 'name of a dance instituted by Theseus on Delos' add '**b** cult-offering at Delos of unkn. nature, ἡ καλουμένη γέρανος *IG* 11(2).161.B61-2, γέρανος ἀργυρᾶ *Inscr.Délos* 296.B48-49, etc.'

γεραρός 1, line 2, for 'a table *of honour*' read '*grand, splendid table*' **3**, line 4, after '*IG* 2.2116' insert '12(3).420 (Thera)'

γέρᾰς, line 2, after 'γέρα' insert 'S.*El.* 443' at end add 'Myc. *ke-ra*'

γεράσμιος, after 'ον' insert 'or α, ον' **II**, after '*honoured*' insert 'μαῖαν ὡς γερασμίαν A.*fr.* 47a.6 R.'

Γεράστιος, add '**II** τὰ Γ. festival in Euboea, Sch.Pi.*O.* 13.159; masc. οἱ Γ. participants in it, ibid. (see also °Γεραιστιασταί).'

γερδιαινα (s.v. γερδιός), add 'Ταπᾶϊς γερδία(ινα) *PColl.Youtie* 36.5 (*PVindob.Gr.*inv. 39867; AD 184)'

γερδιακός, after '*PGrenf.* 2.59.10' insert 'γερδιακή, ἡ, *weaving*, *BGU* 2041.9 (AD 201)'

***γερδικός**, ή, όν, *of weaving*, *POsl.* 140.2, 13 (ii BC).

γερεαφόρος, add 'see also ‡γερηφόρος'

γερητηρία· ἀπώλεια Hsch. (γ 429 La.); cf. γέρρω].

γερηφόρος, add '(= *Tit.Calymn.* 250.3)'

γερῖνος, ὁ, kind of fish, Marc.Sid. 37; cf. γέρυνος.

***Γερμανίκεια**, τά, festival in honour of a Germanicus, *IG* 2².2067.115, 2068.207, etc. (ii AD).

***Γερμανίκειος**, ὁ, Egyptian month (= Pachon) in Caligula's calendar, *SB* 6705, *PLond.* 1171ᵛ.c13, *POxy.* 3780.9 (-νίκιος) (all AD 42).

***Γερμανικός**, ὁ, title of Roman emperors, D.C. 55.2.3, al., *Amyzon* 68 (Caligula), *SEG* 34.1122 (Nero), etc. **II** Egyptian month (= Thoth) under Domitian, *BGU* 1.260, *PLond.* 259.138, *PFay.* 110, *POxy.* 266.

†**γεροῖα**, τά, v. °Γεροῖα.

γερόντειος, add '**II** γεροντεῖον, τό, *body of elders, senate*, cf. γερόντιον II, [τῶν ἐν τῷ γερ]οντείῳ φερομένων rest. in *Berytus* 12.127/8 (Cyrene, ii/i BC).'

γεροντία, add 'Nic.Dam.*fr.* 103(3) J.'

γεροντικός, line 2, after 'Pl.*Lg.* 761c' insert 'ὅπλον Call.*Epigr.* 1.7 Pf.'

γεροντοδιδάσκαλος, add 'Varro *Sat.Men.*tit. p. 32 A.'

***γερός**, ά, όν, adj. used in description of buildings, γερὰ καὶ στεγνὰ καὶ τεθυρωμένα *Inscr.Délos* 1417C89, 58 (ii BC); cf. mod. Gk. γερός *strong, sturdy*.

γερουσία I 1, line 3, after 'BC)' add 'local γ. in Erythrae, *SEG* 37.938, in Phrygia, Ramsay *Cities and Bishoprics* no. 549 (Acmonia, sim. in other places)', 'γ. of Jews in Alexandria, Ph. 2.527' at end add 'Myc. *ke-ro-si-ja*'

γερουσιάζω, *to be a member of a γερουσία*, *IG* 12(8).389 (Thasos).

γερουσιακός, add '*Sardis* 7(1).17.2'

***γερουσιαρχέω**, *to be president of elders*, *IGBulg.* 1906, *IEJ* 4.252.

γερουσιάρχης, add '*SEG* 36.946 (Italy, iv AD), *CIJud.* 803 (Syria, iv AD)' add '**2** Dor. -ᾱς, *president of γερουσία*, *SEG* 1.327.16 (Callatis, i AD).'

***γερουσιάρχισσα**, ἡ, fem. of γερουσιάρχης, *IG* 10(2).1.177.10 (Thessalonica, iii AD).

γερουσιαστής, line 2, for 'iii BC' read 'iii AD'; add '*INikaia* 1242 (i/ii AD), *IHistriae* 57.26, 193.B1, 13 (ii AD), *SEG* 31.635 (Maced., ii/iii AD)'

***γερροφῠλᾰκία**, ἡ, form of guard, *SEG* 28.1484 (Egypt, 116 BC).

γέρων I 2, add 'in religious cult, οἱ τᾶς Οὐπησίας ἱεροὶ γέροντες *SEG* 23.215, 216 (Messene, ii/iii AD)' **II**, line 4, for 'πόνος' read 'πίνος', delete following parenth. and for '(S.)*OC* 1258' read '(S.)*OC* 1259'; line 7, after 'Plu.*Pel.* 2' insert 'transf., γέροντος ἤδη χρόνου πολιά Luc.*Am.* 12'; line 8, for '(for ἀρχαῖα)' read '*things characteristic or worthy of an old man*' and add '*EM* 227.9 (s.v.l.)' **III**, add 'Herod. 12.3' add 'Myc. *ke-ro-te* (pl., sense I), *ke-ro-ta* (neut. pl., sense II, of cloth)'

***Γετικός**, ή, όν, *of the Γέται*, a tribe in the region of the Istros, ἵπποι Ael.*NA* 15.24, ἀλαλαγμοί Arr.*Tact.* 44.1, Γετική χθ[ών] epigr. in *SEG* 25.823 (Dacia, ii AD).

γεῦμα I, add 'οἴνου παλαιοῦ πρώτου γεύματος app. *of first quality* or *first of the season, Edict.Diocl.* 2.8, 9, ἐλαίου ὀμφακίνου δευτέρου γ. ib. 3.2' **II**, add 'w. ref. to function not precisely explained, οἱ ἐπὶ τὸ γ. πραγματευόμενοι *IEphes.* 728.34 (iii AD)'

γεύω I, at end after 'Herod. 6.11' insert 'ναὸν ἱερουργημάτων J.*AJ* 8.4.5' **II 3**, add 'ὁ θανάτο(υ) γευσόμενος *ITyr* 126'

γέφυρα, line 1, after 'Stratt. 47.5' insert '(= 49.5 K.-A.) codd. βλ-' and after 'Hsch.' insert 'cf. also βουφάρας'; line 7, for '*limits of the battlefield*' read '*earthworks*'

***Γεφυραία**, epith. of Demeter at Athens, *EM* 229.4, St.Byz. s.v. Γέφυρα.

***Γεφῠραῖος**, epith. of Apollo, *IG* 2².4813.

***γεφυρίς**· πόρνη τις ἐπὶ γεφύρας, ὡς Ἡρακλέων. ἄλλοι δὲ οὐ γυναῖκα, ἀλλὰ ἄνδρα ἐκεῖ καθεζόμενον 'ἐπὶ τῶν ἐν Ἐλευσῖνι μυστηρίων συγκαλυπτόμενον ἐξ ὀνόματος σκώμματα λέγειν εἰς τοὺς ἐνδόξους πολίτας Hsch.

***γεωβᾰφής**, ές, adj. describing some method of dyeing, perh. with a kind of earth used as colouring agent or mordant, χλαμύς γ. *PCair.Zen.* 92.3 (iii BC) ποδείων γ. ib. 23; cf. γῆ IV.

γεωγράφος, for 'ον, *earth-describing*. Subst.' read 'ὁ' and add 'cf. γαιογράφος'

γεωδαισία, add 'also γαδαισία *distribution of land*, *IG* 9²(1).609.11 (Naupactus, *c*.500 BC)'

γεωδαίτης, for 'Call.*Oxy.ined.*' read 'pl., Call.*fr.* 43.64 Pf. (codd. γαιοδόται, cf. °γεωδότης)'

γεώδης I, line 1, after 'Pl.*Phd.* 81c' insert 'Plb. 2.15.8 (codd. γαι-)' and at end add 'fig., of faculties, *dull, torpid*, Plu. 2.625c'

***γεωδότης**, ου, ὁ, perh. *assigner of land*, 'Ἡγησίου γεωδότο[υ εὐεργ]εσίας ἕνεκεν *EA* 16.65.5 (Saittai, ii AD).

γεωμέτρης, for 'ib. 28' read '*BGU* 12.28' and after it insert '*SEG* 34.477 (Thessaly, ii BC); sp. γεωμ- *SEG* 32.1287 (Phrygia, iii AD), *ISmyrna* 893 (i AD), *TAM* 4(1).173 (Bithynia); cf. γαιομέτρης'

γεωμορία II, add '*Lyr.adesp.* in *SLG* 414(d).10'

***γεωμόριον**· τὸ τῆς γῆς μόριον Hsch., see also °γημόριον.

γεώπεδον, for pres. def. read '*holding of land, estate*'; add 'cf. γάπεδον'

γεωργέω, after lemma insert 'Thess. °γαοργέω, Boeot. 3 sg. fut. γαϜεργείσι *BCH* 60.182.11 (Thespiae, iii BC)]'

γεωργία I, add '3 *cultivation period*, *PRainer Cent.* 123.16 (AD 478).'

γεώργιον I, add 'Lxx *Ge* 26.14, *Pr.* 6.7, *Je.* 28(51).23' **III**, delete the section.

γεώργισσα, add '*PSI* 1021.31 (ii BC)'

γεωργός, after lemma insert 'Lacon. γαϜεργός (i.e. γαϜεργός)· ὁ ἀγροῦ μισθωτής Hsch. (La.), Boeot. γαεργός *BCH* 60.178.6 (Thespiae, iii BC)'

***γεωρῠχικός**, ή, όν, *of or concerning mining*, γεωρυχικὸν νόμον *SEG* 39.1180.78 (Ephesus, i AD).

***γεωρύχιον**, τό, *tunnel, mine*, *SEG* 39.1180.81 (Ephesus, i AD); in Dor. form γαωρύχιον, *Inscr.Cret.* 1.v 19 B 24 (Arcades, ii/i BC).

γεωτόμος, add 'see also γατόμος'

***γεωφάγος**, ὁ, *earth-eater*, περὶ γεωφάγων, Plu. title in Lamprias' catalogue (vii p. 477.191 B.); cf. γαιη-, γηφάγος.

γῆ, line 3, after 'γαῖα' insert 'gen. γέης Rouechè *Aphrodisias* 37.13 (v AD); acc. γέην Hatzopoulos *Actes de vente* II p. 23.5 (Chalcidice); γέαν *ICilicie* 49'; line 6, for 'Cypr. ζᾶς *Inscr.Cypr.* 135.50 H.' read 'Cypr. *za-i* γᾶι *ICS* 217.8, 17, al.' **I 1**, line 11, after '(A.)*Pers.* 629, etc.' insert 'ποιησάτω τὰ ποτὶ Γᾶν (καὶ) τὰς ἀλλαθεάδας *Delph.* 3(3).22.12, 3(6).40.3; as a goddess, Hdt. 4.59.1 (Scythia), Th. 2.15.4, *SEG* 31.748 (Tenos, iv BC), Πλούτων καὶ Γᾶ καὶ Ἑρμᾶ ib. 35.1011 (Sicily, i BC), 31.917b (Aphrodisias, i AD); line 13, after 'S.*OT* 416' insert 'ἐν γᾶι *SEG* 32.914 (Entella, iv/iii BC)' **2**, add 'as that to which dead bodies are reduced, Thgn. 878, Anacreont. 32.12 W.'

γηγενής I 3, for 'ib.' read 'Lxx'

γῆθεν, add '*Com.adesp.* 133.3, see also °γειόθεν'

γηθέω, 3 lines fr. end, after 'Sotad. 15.4' insert 'γηθομένη τειμαῖς

γήϊνος | SUPPLEMENT | γλῶσσα

MAMA 9.48.7 (Aezani, iii AD), abs., λωφᾷ τε τῆς ὀδύνης καὶ γέγηθεν Pl.*Phdr.* 251c, Plu. 2.372f'; add 'Pl.*Phd.* 85a'

γήϊνος, after 'η, ον' insert '(also γέϊνος Hsch. s.v. γηγενῶν, Dor. γάϊνος, ον, *SEG* 9.72.118 (Cyrene, iv BC))'

γηίτης, add 'II = αὐτόχθων, St.Byz. s.v. γῆ.'

γημόριον, add 'perh. mistake for μημόριον, v. *BE* 1987.400, but cf. °γεωμόριον, °γημόρος'

γημόρος, line 4, Att. γεωμόρος, add 'app. iron., καὶ νέκυς γεωμόρος εἴης epigr. in *SEG* 34.1247.3 (Miletoupolis, ii AD); cf. ‡γημόριον' and after 'μείρομαι)' insert 'see also γειομόρος'; line 5, οἱ γαμόροι in Sicily, add '*Marm.Par.* 52, Dubois *IGDS* no. 219.3 (v BC)'

γήπεδον, for '*plot of ground*' read '*estate, land*' II, for 'γάπεδον' read '°γάπεδον'

γήρας II, line 3, for 'Arist.*HA* 600ᵇ20' read 'Arist.*HA* 601ᵃ17'

γηράσκω, line 11, γηρείς, add '*Iamb.adesp.* 4 W.'

γηροβοσκός, add '*POxy.* 3555.8 (i/ii AD)'

γηροκομεῖον, add '*SB* 4845 (-ῖον Byz.)'

γηροκόμος, add 'II (in pass. sense, proparox.) *nursed in old age*, γενέτις *GVI* 1823.6 (Naucratis, ii BC).'

γηροτρόφος, add 'εἰ .. παῖδες ἔσονται [γη]ροτρόφοι Ἰσοδήμωι *JHS* 87.133 (lead tablet, Dodona)'

γήρυμα, after 'pl.' insert 'γάρυματα μαλσακά Alcm. 4 *fr.* 1.5 P.'

†**γητομέω**, -τόμος, delete the entry (v. °γεωτ-).

*†**γιγαντοπανορήκτης**, ου, ὁ, (in Gnostic mythology) *he who shatters all giants*, amulet in *MB* 18.32.

*†**γιγαντοπνικτορήκτης**, ου, ὁ, *he who throttles and shatters giants*, amulet in *MB* 18.32.

*†**γιγαντοπτορήκτης**, ου, ὁ, (perh. for γιγαντοπαντ-) *he who shatters all giants*, amulet in *MB* 18.32.

*†**γιγαντορήκτης**, ου, ὁ, voc. -ρηκτα, *he who shatters giants*, amulets in *MB* 18.32, *MUB* 15.76, *ARW* 28.269, *Hesperia* 20.326, *SEG* 33.1551 (Egypt, iii AD).

*†**γιγαντοφόντης**, ου, ὁ, voc. -φόντα, *giant-slayer*, amulet in *Hesperia* 20.326.

γίγγλαρος, for '*kind of flute or fife*' read '*kind of small aulos*'

*†**γιγγλίζειν** τὸ ἀπειλεῖν οἱ ἰδιῶται λέγουσι *AB* 1.88.

†**γιγγρί**· ἐπιφώνημά τι ἐπὶ καταμωκήσει λεγόμενον. καὶ εἶδος αὐλοῦ Hsch., Hdn.Gr. 1.506.

γίγνομαι, line 3, after 'ib. 7.3303' insert 'Cret. γίννομαι *Inscr.Cret.* 4.184.9, 232.2 (both Gortyn, ii BC)'; line 4, after 'ἐγενάμην' insert '*SEG* 34.1107 (Ephesus), *IEphes.* 3239 (ii/iii AD)'; line 8, after '(γη-το)' insert 'Ep. also has γειν- for γεν- in the forms γεινόμεθ' Il. 22.477, Hes.*Sc.* 88, γεινόμενος Il. 20.128, 24.210, Od. 4.208, Hes.*Th.* 82, cf. Alc. 39(a).8 L.-P.' I 1, add 'εὖ γεγονότας *free-born*, Just. *Const.* Δέδωκεν 7' II and II 1, 3, for 'folld. by' read 'with' 3 c, line 3, γ. εἴς τι, add 'εἰς ἄλλο τι γιγνόμενον Pl.*Ti.* 57a, ὅταν δὲ εἰς πύον γίνεται Gal. 16.71.4'; lines 6/7, delete 'ἐς Λακεδαίμονα .. 4.634)'; lines 15/16, after 'ἐπὶ ποταμῷ' insert 'ἐν τῷ προθύρῳ' and for 'Hdt. 1.189, etc.' read 'Hdt. 1.189.1, Pl.*Prt.* 313c'; three lines from end for 'Hdt. 7.22' read 'Hdt. 8.22.2'

γιγνώσκω, line 7, after 'Pi.*P.* 4.120' insert 'perh. 1 sg. *PAnt.* 58.27'; line 14, delete 'and in past tenses'

†**γίννος**, ὁ, *alleged offspring of mare by mule*, Arist.*HA* 557ᵇ25, *GA* 748ᵇ34; *of other mixed or defective parentage*, Hsch. s.v. (also s.v. ἴννος); prob. refers to horse or mule of less than normal size, Str. 4.6.2, *BCH* 66/67.181 (Abdera), γίνος *IG* 12(1).677.23 (Ialysus, both iv BC); also ἴννος (ἰννός cod.) Hsch.; cf. ἴννος, Lat. *ginnus, hinnus*.

γίνυμαι, v. γίγνομαι.

*†**γισγίνη**, γισγινόσημος, v. °υσγιν-.

γλαγερός, delete '2 *soft, plump*' transferring exx. to section 1.

†**γλάζω**, perh. *distil*, μέλι Pi.*fr.* 97 S.-M. (s.v.l.); cf. γλάγος.

γλᾶθις, for 'dub. sens. .. pl.)' read 'acc. pl. -ιας, epigr. in *Inscr.Cret.* 3.iv 38.5 (pl., i BC)'

*†**γλακτοπαγής**, ές, *rigid with milk*, γλακτοπαγεῖ μαστῷ epigr. in *ISmyrna* 541.7 (i AD)

γλαρίς II, delete the section.

*†**γλαυκάνεα**, τά, perh. *some product of mining operations*, *IG* 12(8).51.22 (Imbros, ii BC).

γλαυκινίδιον, read '-ῑνίδιον'

γλαύκινος, read '**γλαύκῐνος**' II, for this section read 'ἐλαίου γλαυκίνου *oil of* or *flavoured with* γλαύκιον I, *Edict.Diocl.* 36.100; neut. subst., a perfumed unguent, Mart. 9.26.2, Pompon.*Dig.* 34.2.21.1 (Lat. *glaucina* pl.).'

γλαυκόμματος, add '*FGE* 160 P., Pisander Lav. 18 H.'

γλαυκός II, after 'Hegesianax 1' insert 'αἱ δύο (sc. ζῶναι) μὲν γλαυκοῖο κελαινότεροι κυάνοιο Erastosth. 16.4' at end add 'cf. Myc. *ka-ra-u-ko* pers. n.'

γλαυκοφθαλμία, ἡ, *cataract*, Cyran. 106 (4.9.5 K.).

*†**γλαυκοφόρος**, ον, *bearing* (i.e. stamped with) *an owl*, τετράχμα Ἀττικὰ *Inscr.Délos* 1429*B*ii22, rest. in 1428ii76 (ii BC).

γλαυκοχαίτης, for 'Choerob. in *Cod.* .. f. 200' read '*An.Ox.* 2.317.24'

γλαυκῶπις, epith. of Athena, add '*CEG* 392 (Himera, vi BC), S.*OC* 706'

γλαύξ, line 3, before 'Epich. 166' insert 'γλαύχς *IG* 5(1).832 (Laconia, vi BC)'

γλαφῠρός II 2, of dishes, add 'Astyd. 4 S.' III 4, of music, add 'μέλη D.H.*Dem.* 26'

*†**Γλέοντες**, v. °Γελέοντες.

*†**γλεύδιον**, τό, perh. *mallet*, παύγλα ἤτοι γλεύδια (v.l. λεύδια) *Edict.Diocl.* 15.43.

γλευκάγωγός, for '*new wine*' read '*must*'

†**γλευκάω**, in gloss γλευκήσας· ⟨ὑπὸ⟩ γλεύκους γενόμενος ἔκλυτος καὶ παρειμένος, ἀπὸ τῶν οἴνῳ νέῳ μεθυσθέντων Hsch. (La.).

*†**γλευκίνης**, ου, ὁ, = γλευκίτης, Paul.Aeg. 7.3 (246.20 H.).

γλεύκινος 1, delete 'Androm.ap.Gal. 13.1039' and add 'γλεύκινον, τό, *a preparation of* γλεῦκος, Androm.ap.Gal. 13.1039'

γλευκοπότης, for '*new wine*' read 'γλεῦκος'

γλεῦκος I 1, for '*sweet new wine*' read '*unfermented grape-juice, must*' and add '*IG* 1³.237.4, 12 suppl. 347 (both v BC)' 2, delete the section adding 'Gal. 6.575.17' to section 1. at end add 'form *δλεῦκος, Myc. *de-re-u-ko* (cf. etym. note on γλυκύς)'

γλέφαρον, for 'Aeol.' read 'dial.'

γλίσχρασμα, for 'Aret.*CA* 1.9' read 'Aret.*CA* 5.1.7, 5.10.5-6'

γλίσχρος II 1, adv., sup. -ότατα, read 'ἕλκουσιν .. γλιχρότατα σαρκάζοντες ὥσπερ κυνίδια Ar.*Pax* 482' 2, adv., add 'οὐ γλίσχρως ἀλλ' ἀληθινῶς Isoc. 5.142'

γλίχομαι, after '*niggard*' read '*glutton*'

γλίχομαι, after lemma insert '(γλῐ- Ar.*fr.* 104 K.-A., *AP* 9.334 (Pers.))' and at end delete '(γλῐ- .. Hdn.Gr. 1.37)'

*†**γλοιαφίον**, τό, perh. *glue*, *SB* 9408(2).53, 9409(1).83, (6).38 (sp. γλυ-); cf. γλία.

γλοιός I, line 4, delete 'generally, *oily sediment* in baths'; add 'as a term of abuse, Ar.*Nu.* 449' II 1, delete the section. add '4 *sordid, mean*, Suid., adv. -ῶς prob. in Timocr. 1.10.'

*†**γλουθίον**, τό, dim. of γλουτός, prob. reading (unless γλουθρός as dial. form of γλουτός is accepted) in *ABSA* 21.172 (Lydia).

*†**γλουθρούς**, ὁ, = γλουτός, *buttock*, inscr. in *Talanta* 10/11.88 (Lydia, Rom.imp.).

γλουρός, after '*AP* 15.25.7' for '(Besant.)' read 'cj., cod. ταγχούρου, see τάγχαρας'

γλούτια II, delete the section (v. °γλουτός).

γλουτός II, for this section read '*the great trochanter*, Gal. 2.773.17; pl., γλουτοί· τὰ τῆς κοτύλης σφαιρώματα *the trochanters*, Hsch.'

*†**γλυάφιον**, v. °γλοι-.

γλυκάδιον, after 'Hsch.' add 'in form γλυκάδιν *Dura*⁴ 151 (iii AD)'

γλυκαίνω, pass., after 'Hp.*Aër.* 8' insert 'X.*Oec.* 19.19'

*†**γλυκασία**, ἡ, *kind feeling, affection*, *SB* 6263.29.

*†**γλυκελαία**, ἡ, *sweet olive*, prob. in *SB* 5747 (κεράμιον) γλυκυελεῶν; acc. pl. written κλοκελέας H.I. Bell, *Jews and Christians in Egypt* no. 1918.15.

γλυκέλαιον, delete '*Sammelb.* 5747.8 (γλυκυελ-)'

γλυκερός, add 'sup. -ώτατος *SEG* 31.846 (Italy, iii AD)'

γλυκίζω I, add '*SEG* 25.790 (Histria, ii BC), 32.1243 (Cyme, i BC/i AD)'

*†**γλυκυμείλιχος**, ον, *sweetly gentle*, *h.Hom.* 6.19.

γλυκύμηλον, for '= μελίμηλον, *sweet-apple*' read '*apple grafted on to a quince*' and add '*PMich.*XIV 680.13 (iii/iv AD)'

γλυκύνους, add 'epigr. in *SEG* 33.1110 (Paphlagonia, iii AD)'

γλυκύπικρος, before '*AP*' insert 'rest. in Pi.*fr.* 128b.7, Sophr. in *PSI* 1214*d*3'; line 3, delete ' "a gilded pill"'

γλυκύς I 1 a, add 'sup. adv. γλυκύτατα *most delightfully*, D.L. 4.59' 2, add 'w. ellipsis of ὀφθαλμός, οὐ τὸν ἐμὸν τὸν ἕνα γλυκύν, ᾧ ποθορῶμί ἐς τέλος Theoc. 6.22' II 1 b, delete the section.

*†**γλυκυφεγγής**, ές, *giving sweet light*, ἐηλιος *SEG* 8.548.10 (Hymn to Isis, i BC).

γλύπτης, add '*EA* 19.51 (Sinope, Rom.imp.), *Stud.Pal.* 20.260.9 (vi/vii AD)'

γλυπτός 2, after 'Lxx *De.* 4.25' insert 'εἰς θεὸν γλυπτόν *Is.* 44.17'; γλυπτόν, τό, for '*Is.* 44.10' to end read 'τὰ γλυπτὰ τῶν θεῶν *Ex.* 34.13, ἐπικατάρατος ἄνθρωπος, ὅστις ποιήσει γλυπτόν *De.* 27.15, *Jd.* 2.2, 4*Ki.* 17.41, al.'

γλυφικός, add '*Philostr.Jun.Im.proem.* 15'

*†**γλυφοποιός**, ὁ, *sculptor*, Porphyrio ad Hor.*C.* 4.8.6.

γλύφω II, for '*note down .. τόκους*' read 'τόκους γ. *mark down interest*, cf. τοκογλύφος'

*†**γλωθρός**, dial. for ‡βλωθρός, Hes.*fr.* 204.124 M.-W.; cf. γλήχων = βλήχων.

γλώξ, for '*beard of corn*' read '*ear of corn* (in quot., millet)'

γλῶσσα I 2, after 'E.*Hipp.* 612' insert 'τά τ' ἐκ ποδῶν σιγηλὰ καὶ γλώσσης ἀπὸ σῴζοντες, i.e. *not speaking*, id.*Ba.* 1049'; lines 6ff., for 'ἀπὸ γλώσσης .. A.*Ag.* 813' read 'ἀπὸ γλώσσης *by word of mouth*,

γλώσσαλγος   SUPPLEMENT   γόργιλον

orally, αἶνος .. ὃν ἐν δίκᾳ ἀπὸ γλώσσας Ἄδραστος .. φθέγξατ' Pi.*O.* 6.13; Hdt. 1.123.4, Th. 7.10, Arr.*An.* 2.14.1; in contexts contrasting wds. w. thoughts, actions, etc., Hes.*Op.* 322, Thgn. 63, τοῦ νοῦ θ' ὁμοίως κἀπὸ τῆς γλώσσης λέγω S.*OC* 936; contrasted with writing, Cratin. 122; οὐκ ἀπὸ γλώσσης not from *mere word of mouth*, but after full argument, A.*Ag.* 813'   **3**, for this section read 'meton., *speaker*, μεγίστη γ. τῶν Ἑλληνίδων i.e. Pericles, Cratin. 324 K.-A., S.*Ichn.* 151 R., Herod. 6.16, *AP* 7.345'   **III 1**, add 'Eup. 442 K.-A.; of trumpet, Poll. 4.85; cf. γλῶσσαί τινες σαλπίγγων ἢ αὐλῶν ταῖς ὀπαῖς (sc. of organ-pipes) προστιθέμεναι Simp.*in Ph.* 681.7'

γλώσσαλγος, delete '*itching* .. 510a'; add 'γ. ἦθος Trag.adesp. 562 K.-S.'

†**γλωσσάομαι**, pf. part. γεγλωσσαμένος, *tongued*, γεγλωσσαμέναν κακκαβίδων ὄπα prob. in Alcm. 39 P.

γλώσσημα II, for '*dart*' read '*spear*'

γλωσσίδιον, add '**III** *tenon*, Sch.Hes.*Op.* 426.'

(**γλωσσ**)**ίδος** (post γλωσσίδιον), delete the entry.

*γλωσσίς, v. ‡γλωττίς.

*γλωσσόζωμος, ὁ, *broth made from tongue*, Vit.Aesop.(G) 52.

γλωσσοκομεῖον, line 8, for '*cage*' read '*chest*' and delete 'rejected by Phryn. 79'; at end after 'also masc. -κομος .. (Pamphylia)' insert 'dub. gender, Aq.*Ge.* 50.26'

γλωσσόκομον, add '*box of lock into which key is inserted*, Gal. 14.638.12'

*γλωσσοπέτρα, ἡ, a (perh. imaginary) precious stone, Plin.*HN* 37.164.

γλωσσοποιός, add 'also γλωττοποιός· ὁ τὰς αὐλητικὰς γλωσσίδας ποιῶν Hsch.'

†**γλωσσοστροφεῖν**, v. °γλωττοστροφέω.

γλωττίς I, add 'form γλωσσίς, *Hippiatr* 130.138'

*γλωττοποιός, v. °γλωσσοποιός.

γλωττοστροφέω, after 'Ar.*Nu.* 792' insert '(prob. nonce-word), γλωσσοστροφεῖν· περιλαλεῖν καὶ στωμύλλεσθαι Hsch.'

*γλωττοτόμιον, τό, dim. of °γλωττοτόμον, Inscr.*Délos* 1450.180 (ii BC).

*γλωττοτόμον, τό, or **-τόμος**, ὁ, app. = γλωσσόκομον, *chest*, Inscr.*Délos* 1432*A*bi73, 1439*A*ba72, 1441*A*ii104 (ii BC); γλοττοτόμῳ ib. *A*i75, 86.

γλωχίν 2, add '*corner* of a chair-frame, Call.*Del.* 235'   **4**, for this section read 'topogr. *corner*, πυμάτην .. ἐπὶ γλωχῖνα νέμονται D.P. 184, θαλάττης γ. Agath. 5.22'

*γλωχῑνόομαι, *to be made barbed*, Eust.*Opusc.* 292.33.

*γνάθιος, ον, γ. λίθος, app. = °γνάθος III, Cyran. 14 (1.3.2 K.).

γνάθος, add '**III** *hard variety of stone*, Cyran. 13 (1.3.9 K.).'

γνάπτω, at end delete 'Nic.*Th.* 423' (read γναπτ-)

*γνάπτρα, τά, *cost of fulling*, *BGU* 1558.7 (iii BC), *PCair.Zen.* 398.7 (iii BC).

γνάπτω, for 'γνάμπτω, id.' read 'κνάπτω, q.v.'

γνάφαλλον, add 'see also s.v. κνέφαλλον'

γναφαλλολόγος, after '*flock-picker*' insert 'prob. incl. other stages in wool preparation' and add '*PFreib.* 60.5 (AD 181)'

*γνάφαλλος· πτίλον Hsch.

*[γ]ναφαλλοϋφάντης or [κ]ναφ-, ου, ὁ, *flock-weaver*, *IG* 2².7967 (v BC).

*γνάφησιος, ἰχθῦς, kind of fish, Cyran. 106 (4.12.1 K.).

*γνάψιμος, ον, *cleaned by fulling*, *POxy.* 4004.13 (v AD).

γνάψις, add '*SB* 9834b.3 (iv AD)'

γνήσιος I 2, line 7, before 'Gal. 15.748' insert 'D.H.*Lys.* 12'   add '**3** *dear*, τῇ γλυκυτάτῃ τεκούσῃ Μελ[τ]ίνῃ καὶ γνησίᾳ γυναικὶ Ἀμμίᾳ *MAMA* 4.305, 8.595; sup. *IG* 10(2).1.403 (Thessalonica, vi AD).'

γνησιότης I, add '*legitimate status* in marriage, Just.*Nov.* 89.9 pr.'

*γνόφαλλον, τό, v. κνέφαλλον.

*γνοφεντινάκτης, ου, ὁ, (ἐντινάσσω) *shaker* or *hurler in darkness*, of a thunder- or lightning-deity, *PMag.* 4.181.

γνοφεόν, for 'Id.' read 'Hsch.'

*γνοφερός, v. °δνοφερός.

*γνοφοειδής, = γνοφώδης, Gr.Roman-Papyri 8.43.

γνόφος, after '*darkness*' insert 'Lxx *Ex.* 10.22, al.'

γνοφώδης, add 'see also °γνοφοειδής, δνοφώδης'

γνύξ, for 'Arat. 921' read 'Arat. 591; αἰεὶ γνύξ id. 615'

γνύπονται, delete the entry.

*γνυπόομαι, (γνυπτ- La.) *to be depressed*, in pf. part. ἐγνυπωμένον· ταλαίπωρον. κατηφές, Hsch.; also ἐγνυπώθη· τρυφᾷ. καὶ τὸ ἐναντίον, id.

†**γνυπτέω**, *to be weak* or *feeble*, Hsch. s.v. γνυπτεῖν, γνυπτοῦντι (cod. γνυποντι).

γνώμα, delete '*test*'; for 'of an ass's teeth, Arist.*HA* 577ᵇ3' read 'applied to certain teeth as indication of age, τούτους (ὀδόντας) δὲ γνώμα καλοῦσι, τοὺς τετάρτους Arist.*HA* 577ᵃ22, 577ᵇ3, καὶ γνώμ' ἔχει. τὸ γνῶμα γοῦν βέβληκεν ὡς οὖσ' ἑπτέτις Com.adesp. 572, 573'

γνωμεισηγητής, delete the entry (v. *JEA* 21.238).

γνώμη II 1, line 13, for 'ἕτερον' read 'ἑτέρῳ'   **III 1**, add '**c** *consent, approval*, *PSI* 967.18 (i/ii AD), *PFam.Teb.* 38.10 (AD 168).'

*γνωμῐδιώκτης, ου, ὁ, (for γνωμιδιοδι-) *one who hunts after sententious maxims*, Cratin. 342 K.-A.

*γνωμοδίκᾱ, ἁ, *verdict, judicial decision*, *BCH* 93.76 (Phocis, ii BC).

γνωμολογία 2, for this section read 'usu. in pl., *aphoristic saying, maxim*, Plb. 12.28.10, D.H.*Dem.* 46, Plu.*Cat.Ma.* 2, *Fab.* 1; sg., title of collection of aphorisms, Suid. s.v. Θέογνις'

γνωμονικός II, after 'Procl.*Hyp.* 5.54' insert '*SEG* 36.1153 (sp. -ηκός, Bithynia)'

γνώμων I 1, add 'title of magistrate, *Inscr.Cret.* 4.14 (Gortyn, vii/vi BC, pl.)'   **II 1**, after '*square*' insert 'Thgn. 805'   add '**b** κατὰ γ. *perpendicularly*, Oenopides ap.Procl. *in Euc.* p. 283 F.'   for '**V**' read '**III**' and after '*AB* 233' add 'κατὰ γνώμονα *PMich.*III 145 iii 5.5 (ii AD)'

γνωρίζω I 3, add '*give legal recognition to* children, Just.*Nov.* 6.1.4'

γνώρῐμος, line 3, for '(Pl.)*Lg.* 798e' read '(Pl.)*Lg.* 798a'   **I 3 b**, for '*pupil*' read '*adherent*'   **c**, delete the section transferring 'Lxx *Ru.* 3.2' to section 3 a.

γνωριστής II, for '*diviner*' read '*wizard*, prob. as calque of Hebr.; cf. °γνώστης'

*γνωσία, ἁ, Arc. = γνῶσις I 1, to be read in *IG* 5(2).262.15 (Mantinea, v BC; v. *SEG* 39.393).

γνωσῐδίκα, delete the entry (v. °γνωσία).

γνῶσις 1, add 'recorded in writing, Arist.*Ath.* 53.2'   add '**VI** pl., *accounts*, ἐπίτροπον γνώσεων τῶν ἐξοχωτάτων καθολικῶν, = Lat. *procurator rationum summarum*, *APAW* 1932(5).46.'

γνωστεία, add '*Stud.Pal.* 22.50.3 (iii AD, -τια pap.)'

γνωστεύω, before '*to be witness*' insert '*to be personally acquainted with*, τινα *PTeb.* 816 i 8 (ii BC)'

γνώστης, for 'esp. *one who knows the future, diviner*' read '*wizard*'

γνωτοφόνος, for '*murderer of another's brother*' read '*murderer of a brother*'

*γοατήριον, v. °γοη-.

γοάω, see also °γόημι.

γόγγρος I, add '*IGC* p. 99 A36 (Acraephia, iii/ii BC)'

†**γογγυλίδιον**, τό, dim. of γογγυλίς, *turnip*, *CPR* 9.28 (written κογκυλίδ-, v/vi AD).   **II** = καταπότιον, Hp.ap.Erot. (γογγυλίδα codd.), Gal. 19.91.

γογγυλίς, add 'Crates Com. 30 K.-A.'

γογγύλλω, for '*round*' read '*turn round*' and add 'cf. συγγογγύλλω'

γογγύλος, line 2, after '*IG* 1².372.22' insert 'cf. Call.*fr.* 606 Pf.'   **II**, for '(proparox. .. *Th.* 855' read 'unexpld. use in Hdn.Gr. 1.164'

*Γόγγυλος, ὁ, name of a deity, Διονύσου Γογγύλου *IG* 10(2).1.259 (Thessalonica, i AD).

*γόημι, = γοάω, Erinn. in *Suppl.Hell.* 401.18.

*γοητήριον, τό, γοᾱ-, perh. *magic practice*, (ε)ἰσάγομεν εὖ γοατήρια terracotta in H. Kenner *Das Phänomen der verkehrten Welt* (Bonn 1970) p. 30.

γοι II, delete the section.

*γοί (B), v. °οὗ.

γοῖτα, delete '(leg. ὗς)' and add 'cf. γοτάν'

γοιταί, add 'cf. γοσταί s.v. °κοσταί'

*γομόρ, Hebr. measure, *homer*, Lxx *Ex.* 16.16, 1*Ki.* 16.20, *Ez* 45.11, *Ho.* 3.2.

γόμος I 2, add '*PCol.Zen.* 2.8 (iii BC)'

γομφιάζω, for '*have pain in the back teeth*, or *gnash* them' read '*grind the back teeth*'

γομφιασμός, after lemma insert 'prob. the same as γομφίασις'

γόμφος I 1, at end for 'χαλκοί' read 'χαλκοῖ'   **2**, add '**b** *eschar caused by cautery*, *Hippiatr.* 96.   **c** *projection on the shoulder of a horse*, ib. 26.'

γομφόω 1, add '*cause teeth to be embedded in the manner of nails*, Gal. 2.754.9'

*γονάγρα, ἡ, *gout in the knee*, Cyran. 95 (3.36.46), 109 (4.18.14 K.).

*γονέτης, ου, ὁ, = γενέτης, *father*, *IHadr.* 120.3 (Chr.epigr.).

γονή I 1, after '(S.)*Ant.* 641' insert 'Just.*Nov.* 22.20.1'

†**γονίζω**, *stock with offspring*, pass. pf. part. περιστερῶνα γεγονισμένον *PRyl.* 581 ii 8 (ii BC); so prob. *SB* 7814.32 (iii AD).

γόνιμος 3 b, for 'Antiph.' read 'Antiphil.'

γόνυ, line 8, delete 'but Γόννα .. Eust. 335.39' and for 'Alc.*Supp.* 10' read 'Alc. 44.7 L.-P., hyper-Aeol. acc. pl. γόννα acc. to St.Byz. s.v. Γόννοι, Eust. 335.39; dat. pl. γόννοις Theoc. 30.18'; before 'E. has γουνάτων' insert 'S. has γούνατα *OC* 1607'   **I 2**, line 5, after 'v. sub κάμπτω' insert 'ἐς γόνυ ἕζεσθαι *kneel*, Luc.*Syr.D.* 55'

γονυαλγής, add 'Gal. 17(2).605.4, 7'

*γόνυγρον· τὰ ἄκαρπα καὶ ξηρὰ πεδία Hsch. (cj. τὴν ἄνυγρον La.).

*γονύπεσος, ὁ, *Kneeler*, as a nickname, Δημήτριος ὁ γ., a grammarian, Hdn.Gr. 2.61, Sch.Gen.Il. 13.137, Sch.T Il. 15.683.

γοργεύω, add 'μέχρι τῆς ἀριθμήσαιως γόργευσ (= γόργευου) ὅσον δύνῃ ποεῖν καὶ ἀπετῆσαι *PMich.*x 577.9 (i AD), *BGU* 1097'

γοργίειος, add 'γοργίεια, τά, festival at Delos, *Inscr.Délos* 366*A*133 (iii BC)'

*γόργιλον, τό, = σέσελι, Paul.Aeg. 7.3 (205.25 H.).

*γοργοκτόνος, ον, *Gorgon-killing*, of Athena, poet. in *PKöln* 245.8 (iii AD).

*γοργόμματος, ον, *fierce-eyed*, Heph.Astr. 3.45.3.

*γοργόπλοος, ον, *swift-sailing*, *An.Par.* 4.200.22.

γοργός 1, after 'A.*Th.* 537' insert 'νῦν δ' ἐς γυναῖκα γοργὸς ὁπλίτης φανεὶς κτείνεις μ' E.*Andr.* 458, 1123, τὸ .. ἀντιμετώπους προσελαύνειν ἀλλήλοις γοργόν X.*Eq.Mag.* 3.11'; after '*IG* 3.1079' add '*Milet* I 7.308 no. 222, 9.176 no. 356' **2**, after 'of persons' insert 'Ar.*Pax* 565'

Γοργοτομία, for 'Str. 8.6.2' read 'Str. 8.6.21'

γόργυρα, add 'pl., ἀρδάλια· τοὺς πυθμένας τῶν κεραμίδων, οὓς ἔνιοι γοργύρας καλοῦσιν Hsch. **II** (in Ion. form γοργύρη) prob. as pr. n. = Γοργώ, τῆι Ἥρηι ἀνέθεσαν .. γοργύρην χρυσῆν *SEG* 12.391 (Samos, vi BC).'

*γοργύριον, τό, dim. of γόργυρα, *subterranean channel*, prob. in *ABSA* 26.220 (Sparta).

*γοργωπιάσκω, ἐγοργωπίασκεν· ἀτενὲς ἔβλεπεν Hsch.

γοργῶπις (s.v. γοργώψ), add 'also as designation of a lake, γ. λίμνη A.*Ag.* 302'

γοργωπός, after '*grim-eyed*' insert 'in some case w. allusion to the Gorgon'

*Γορδιάνεια, τά, festival in honour of Gordian, *IG* 2².2239.189, 2242.29 (iii AD).

*γόρνα or γόρνη, ἡ, = Syriac *gūrnā*, *amphora*, *burial urn*, γόρνης ἕνα ἥμυσυ *IGLS* 269.

*γοσταί· αἱ κριθαί Theognost.*Can.* 44 A.; cf. κοσταί, °κόσταια.

*γουνάριος, ὁ, Lat. *gunnarius*, *furrier*, Corinth 8(1).148, *IGLBulg* 99.4, 100.4, 102.8, al. (sts. γουνν-, all vi AD).

*γούντη, ἡ, (wd. of unkn. origin) *tomb*, *CIJud.* 767 (Phrygia).

*γουτάριον, τό, perh. *tomb*, *MAMA* 6.277 (Phrygia); cf. °γούντη.

*γουπνάριος, ὁ, = Lat. *gubernator*, *CPR* 10.57.

*γουργαθός, v. °γυργ-.

*γρᾶα, ἡ, αἱ λεγόμεναι γράαι, in context after mention of water-snakes, some kind of *dangerous aquatic creature*, fr. Skt. *grāhá- Peripl.M.Rubr.* 38.

*γράβακτον, v. °κράββατος.

γράβατος, delete the entry.

*γράδος, ὁ, Lat. *gradus*, *stepped pedestal*, *IGBulg.* 992, 993, *MAMA* 4.343.

γραῖα, after 'γραίη' insert 'Babr. 104.5' and after '**γραία**' insert 'Theoc. 7.126 (also sp. γρέα *POxy.* 2860.11, *OAmst.* 85.11)'

Γραικίτης (s.v. Γραικός), after 'ὁ' insert '*Greek style*, ἐν δὲ Γραικίταις πέπλοις'

*Γραικολατῖνος, ον, *Graeco-Latin*, of the name Zeno which is used in both languages, *Orac.Tiburt.* 159.

Γραικός, add 'adv. -ῶς, *in the Greek manner*, *Orac.Tiburt.* 158'

*Γραικόστἄσις, εως, ἡ, Lat. *Graecostasis*, platform or tribunal in the Roman forum, reserved for foreign envoys, Varro *LL* 5.155, Cic.*QF* 2.1.3, Plin.*HN* 7.212, etc.

γραιούδια· οἱ παρ' ἡμῖν ἐφθοὶ ὄμφακες Suid.; cf. γραῖα II, ‡γραῦς II.

*γραίοψις, εως, ἡ, *having the face of an old woman*, Dura⁹ 1.213.19 (iii AD, γρε-).

γράμμα, after 'Doric γράθμα' insert '*SEG* 30.380 (Tiryns, vii BC)' **II 5**, add 'as unit of currency, *TAM* 4(1).269.7 (Rom.imp.), *BCH* suppl. 8.159 (Thessalonica, v AD).'

γραμμάριον, add 'abbrev. ΓΡ on bronze weight (for pl.), *SEG* 32.1626 (v/vi AD)'

γραμματεῖον I 2, add 'also perh. of a receipt, *CPR* 5.12.5 (iv AD)' add '**V** app. some premises occupied by an association of artists, τὸ κοινὸν τῶν ἐν τῶι κατὰ Κύπρον γραμματείωι περὶ τὸν Διόνυσον τεχνιτῶν *Salamine* 83.5 (ii BC).'

γραμματεύς 1, line 3, after 'Th. 7.10' insert 'τὸν γραμματέα τὸν κατὰ πρυτανείαν καὶ τοὺς ἄλλους γραμματέας τοὺς ἐπὶ τοῖς δημοσίοις γράμμασιν *IG* 2².120.15 (Athens, 353/2 BC), γραμματῖ πολιτικῶν *POxy.* 3185.1 (iii AD)'; line 5, after '*UPZ* 110.145 (ii BC)' insert 'τοῦ γραμματέως τῶν δυνάμεων *Salamine* 89 (ii BC)' add '**b** as transl. of Hebr. *šōtēr*, Lxx *Ex.* 5.6, *Nu.* 11.16, *Jo.* 1.10, al.; *sop̄rim* or "*scribes*", 2*Ki.* 8.17, *Ne.* 8.1, *Ev.Matt.* 2.4, *Ev.Marc.* 1.22, al.'

γραμματικός II 2, line 3, for 'Cyrene' read 'Cumae'

γραμματιστής, after lemma insert 'dial. -ἁς'; after '(Thespiae)' insert '*SEG* 30.990 (?Corinth, iv/iii BC), 32.497 (Thespiae, iii BC)'

γραμματοεισἁγωγεύς, for '*schoolmaster: governor*' read '*minor official* (military or civil): transl. Hebr. *šōtēr*'

*γραμματοτρώξ, ὦγος, ὁ, "*nibbler of letters*", wd. in Hdn.Gr. 2.643.

*γραμματοφυλακεῖον, add '*SEG* 19.854.6 (Pisidia, ii AD)'

*γραμματοφυλακέω, app. *serve as γραμματοφύλαξ*, rest. in Xanthos no. 16 (Lycia).

γραμμή II, for '= βαλβίς .. *winning-point*' read '*finishing-line*'

γραμμοειδής, add '**2** perh. *having the form of a letter*, Afric.*Cest.* 1.2 p. 117 V.'

*γρανᾶτα, τά, Lat. *granata*, *pomegranates*, Anon. in *Rh.* 74.10, 176.5.

γραπτός II, add 'also sg., συνθέμενοι γραπτὸν πρὸς αὐτοὺς *IG* 7.4130.4 (Acraephia, ii BC), ἐν τοῖ γραπτοῖ *SEG* 35.665.B19, 27, al. (Ambracia, ii BC)'

γράπτρα, after '*copying*' insert '*PMich.*II 123ᵛ vi 9, ix 31 (i AD)'

γραστίζω, for pres. refs. read '*Hippiatr.* 98; pass., *Gp.* 16.1.11, *CPh* 19.234, *PCair.Zen.* 158 (both iii BC)'

*γραστολόγος, ὁ, *fodder-collector*, *Stud.Pal.* 20.213 (*Berichtigungsl.* II 2.165, vi AD) or *γραστολογία, (*Berichtigungsl.* VIII 471).

γραῦς I, line 6, after 'D. 19.283' insert 'γ. κορώνη Call.*fr.* 260.50 Pf.' **II**, for '*scum of boiled milk*' read '*scum forming on the top of heated liquids*' add '**b** perh. *froth* on fresh milk, Nic.*Al.* 91. **c** ἡ ἐν τοῖς χείλεσι τῶν ποταμῶν (ποτηρίων La.) γραμμή i.e. *the line of scum along river-banks*, Hsch.' add '**V** *a throw at dice*, Hsch.' at end add 'Myc. *ka-ra-we* (pl.); see also γρεῦς'

*γραφέας, ὁ, app. equiv. of γραφεύς, Πύρρο(υ) γρ[α]φέας· καὶ Φαρίξενος καὶ τοὶ μαστροὶ *SEG* 31.358 (v. *Olympia Bericht* 10.234 no. 30, v BC); cf. γραφής.

*γράφεία, ή, (?)*painting*, *CID* II 101 i 11 (iv BC; unless termination of longer wd.).

*γραφιάριον, τό, app. some item *connected with writing* (?*desk*), *PAmh.* 2.181 (iii AD).

γραφικός I 2, add '**b** *forming the perfect pattern of*, Plaut.*Ps.* 519, *Trin.* 936, 1024.'

*γράφιον, τό, *prescription*, τῆς διαλυτικῆς *PMerton* 12.23 (i AD); cf. °ἀντιγράφιον. **2** a kind of *poll-tax*, *IEphes.* 13 i 3, al. (i AD); see also γραφεῖον.

γραφίς II, for 'in pl. = *paintings*' read '*painting*, *APl.* 36 (Agath.), 80.4 (id.)' and after 'Nonn.*D.* 25.433' insert '(pl.)'

*γρᾰφύνα, ή, dub. sens., *IAskl.Epid.* 52 A 47 (iv BC).

γράφω, line 2, γεγράφηκα, add '*IG* 11(4).1026 (Delos, ii BC)'; line 10, after '*Leg.Gort.* 1.45, al.' insert 'ἠιγραμμένο *SEG* 38.13.B5 (Rhamnous, early v BC)'; line 11, after 'Dor.' insert 'γεγράβαται (pl.) Dubois *IGDS* 121 (Camarina, v BC, tab.defix.)'; line 18, before 'Pl.*R.* 420c' insert '*paint* a statue' **II 3**, κληρονόμον γράφειν, add '*Cod.Just.* 1.2.25 pr., 1.3.45.4' **6**, line 3, for 'etc.: abs. (sc. νόμον) D. 18.179' read 'D. 18.179, etc.' **8**, add 'ὁ δὲ γράφων ἡμεῖν ἀναγορευσάτω *IG* 12(5).655.12 (Syros, ii/iii AD)' **B 1**, lines 3/4, for 'ἐγραψάμην .. D. 56.6, etc.' read '*cause to be written*, *dictate*, Hdt. 3.128.2, ἐγραψάμην τότ' εὐθὺς ὑπομνήματα, ὕστερον δὲ κατὰ σχολὴν ἀναμιμνησκόμενος ἔγραφον Pl.*Tht.* 143a; συγγραφήν D. 56.6, etc.'; add '*cause to be painted*, Hdt. 4.88.1, *AP* 6.355 (Leon.Tarent.)' **2**, for '*IG* 1².374.16, ib. 2.115ᵇ21' read '*IG* 2².19.b8, al.'; add '*IG* 2².856.6, al.'

γράω, line 1, for 'Call.*fr.* 200' read 'Call.*fr.* 551 Pf. (v.l. κράω)'; line 2, for 'γράσθι .. (Golgoi)' read 'Cypr. *ka-ra-si-ti* perh. γράσθι (imper.) *ICS* 264.1'

γραώδης, for '= γραϊκός' read '*typical of* or *proper to an old woman*'

*γρεάγρα, γρέγρα, v. °κρεάγρα.

*γρηγορόφθαλμος, ον, *with watchful eyes*, Tz.*Alleg.Il.Prol.* 671, 675.

⁺γρήϊος, ον, Ion. for *γράιος, *of an old woman*, εἶδος Call.*fr.* 490 Pf.; μορφή Nic.*fr.* 62 G.-S.; ἴχνος prob. in Euph. in *Suppl.Hell.* 415 ii 6.

γρῖπος I, for '= γρῖφος' read '*net*'

*γρῐφάνη, ή, *rake*, Poliorc. 212.3.

*γρῑφεύς, έως, ὁ, = γριπεύς 1, Opp.*C.* 4.259.

γρῖφος 1, for '*basket*, *creel*' read '*net*'

*γρῖφος, ον, *obscure*, comp. and sup., γριφότερος, -τατος, *An.Boiss.* 4.61; cf. γρῖφον· τὶ ἀσαφές Hdn.*Epim.* 16.

γρονθονεύεται, after lemma insert '(γρονθων- La.)' and delete 'βρενθύεται'

⁺γρονθων· ἀναφύσησις, ἣν πρώτην μανθάνουσιν αὐληταὶ καὶ κιθαρισταί Hsch.

⁺γρουνός, ὁ, v. ‡γρυνός.

γρύζω, line 3, after '*mutter*' insert 'Hippon. 70.6 W.'; line 5, after '(Ar.)*Ra.* 913' insert 'Pl.*Euthd.* 301a'

⁺γρύλλιον· ῥωχμήν (cod. ῥωσμ-) Hsch.

γρύλλος 1, for '*PSorb.inv.* 2381 (*ZPE* 78.153, ii AD)' **2**, read '*painting by Antiphilus of such a person, subsequently used generically*, Plin.*HN* 35.114, *POxy.* 2331 ii 9 (iii AD)'

γρῡνός, add '(γρουνός is v.l. in all three places)'

*γρυπάλιον, v. γρυπάνιος.

γρυπάνιος, for 'also -άνιον .. Hsch.' read 'see also °γρυπάλιον'

*γρυτάρης, ου, ὁ, prob. = γρυτοπώλης, Sch.Ar.*Pl.* 17.

γρύτη 2, after '*frippery*' insert '*Peripl.M.Rubr.* 30'

γρυτοδόκη, add '(γρυτ- codd.)'

γρυτοπώλης, after '*small wares*' insert 'Maiuri *Nuova Silloge* 466 (Cos, i BC/i AD), *ISmyrna* 719.6 (i AD)'

γρύψ I, after '*griffin*' insert 'Hes.*fr.* 152 M.-W.'

γρῶνος II, γρώνη, after '*hole*' insert '*cavity*'; delete '**2** *hollow vessel, kneading-trough*' and for '(Leon.)' read '(Leon.Tarent.)'

γυάλας, after 'ὁ' insert 'or **γυάλα**, ή' and for '*a Megarian cup*' read '*a kind of cup*, attributed in quots. to Megarian and Macedonian sources'; add 'cf. γυλλάς' *γύαρχος or γυάρχης, ου, ὁ, official of unkn. function, PHib. 260.3 (iii BC).

γύης, after lemma insert 'dial. γύᾶς'   II 1, after 'field' insert 'SEG 39.996.2 (Camarina, ii BC)'   add '5 type of dike or embankment, SB 9699.82, 90 (i AD); cf. Swiderek La propriété foncière p. 60.'

γυήτης, after 'Hsch.' insert 'Theognost.Can. 19'

γυιοπέδη, add 'transf., in sg., of the paralysis inflicted by the electric ray, Opp.H. 2.85'

*γυιοφθόρος, ον, destroying the limbs, (v.l. θυμοφθ-) Nic.Th. 140.

γυιόω, line 2, for 'wound' read 'disable' and add 'Hp.VM 9'

γυλιός, γύλιος, add 'III γύλιον· χοῖρον ἢ λέοντα. σημαίνει δὲ καὶ τὸν Ἡρακλέα EM 244.24, Theognost.Can. 19.'

γύλλινα, add 'cf. Theognost.Can. 19'

γυλλός, after 'block of stone' insert 'sacred to Apollo'; after '(Milet., v BC)' insert 'γ.· κύβος ἢ τετράγωνος λίθος Hsch.; v. Pfeiffer ad Call.fr. 114.2 sq.'

γυμνάζω I 2, delete the section.   II 2, for 'investigate' read 'discuss' and add 'PFlor. 338.4 (iii AD), PTrinity College inv. D 28.53 (CE 44.313, iii AD), Cod.Just. 7.62.36'

γυμνᾱσιαρχέω, line 4, after '(Lys.) 6.60' insert 'of a female, IEphes. 3239A.10 (i/ii AD)'

γυμνᾱσιαρχία, after 'office of gymnasiarch' insert 'as a λειτουργία' and after 'Pl.Ax. 367a' insert 'Isoc. 16.35'

γυμνᾱσιαρχικός, add 'γυμνασιαρχικόν, τό, fund administered by the gymnasiarch, BCH 37.91.19 (Beroea)'

γυμνᾱσιαρχίς, add 'PMilan inv. 69.01 ii.6 (Aegyptus 60.126)'

γυμνᾱσίαρχος, fem., add 'TAM 3.58 (ii/iii AD), al.'

γυμναστικός I, add 'νεανίσκοι γ. members of an organization of young men participating in athletic exercises, SEG 29.1201 (Lydia)'

γυμνητής II, add 'Str. 16.4.17; transf., lacking the necessities of life, destitute, γ. βίοτος, of Diogenes the Cynic, AP 7.65 (Antip.Sid.)'

*γυμνιεύω, to be destitute, PRoss.-Georg. 3.28 (iv AD).

γυμνικός, add '2 -κός, ὁ, acrobat, of a child, SEG 30.1231 (Lugdunum, iii AD; sp. -κως).'

γυμνόλοπον, τό, variety of chestnut, τῶν δὲ καστάνων τὸ μὲν Σαρδιανόν, τὸ δὲ λόπιμον .. τὸ δὲ γυμνόλοπον Sch.Nic.Al. 271b Ge.

*γυμνομάχης, ὁ, = γυμνής II 1, Tyrt.23a.14 W.².

*γυμνοπᾱγής, ές, freezing because of (their) nakedness, Tim.Pers. 99 P.

γυμνοπερίβολος, add 'app. without covering garment, καὶ μαστειγοφόρους κ' .. ἐν ἐσθῆσι λευκαῖς γυμνοπεριβόλους καὶ σὺ [ν ἀσπ]ίσι καὶ μάστιξι SEG 38.1462.64 (Oenoanda, ii AD)'

γυμνός 5, for 'lightly-clad, i.e. in the undergarment only' read 'stripped, i.e. with one's main garment removed'

γυμνόω 1, add 'ἀτιμωθεὶς τῶν ὄντων γυμνούσθω Cod.Just. 1.3.35.1, 1.3.44.1'   3, for 'lay aside' read 'strip off a garment' and add 'so perh. in Call.fr. 191.30 Pf.'

*γυναικάδελφη, ἡ, wife's sister, SEG 33.1103 (Paphlagonia, ii/iii AD).

γυναικάδελφος, after 'wife's brother' insert 'Mitchell N.Galatia 385 (?iv AD)'

*γυναικ(ε)ιάριος, ὁ, manager of γυναικεῖον (°γυναικεῖος II 4), Cod.Just. 11.8.13.

γυναικεῖος I 1, delete 'Thphr.Char. 2.9'; add 'b ἀγορά γ. woman's market, i.e. where female clothing, ornaments, etc., are sold, τὰ ἐκ γυναικείας ἀγορᾶς Thphr.Char. 2.9, Men.fr. 390 K.-Th.'   II, add '4 γυναικεῖον, τό, (textile) factory employing women, Cod.Just. 11.8.2, al., JÖAI 23 Beibl. 205 (Heraclea-Perinthus).'

*γυναικεράστρια, ἡ, (female) lover of women, of Sappho, POxy. 1800 fr. 1 i 18.

*γυναικογένεια, ἡ, relationship through females, πρὸς -γένειαν on the female side, PSI 1016.26 (ii BC), κατὰ γ. PGen. 104.15 (ii AD).

γυναικοκρᾰτέομαι, add 'AP 10.55 (Pall.)'

γυναικονόμος, add 'PHib. 196.11, 18 (iii BC)'

γυναικοπίπης, for '(ὀπιπτεύω)' read '(ὀπιπεύω)'; add 'prob. in Sch.Hippon. 118.D9 W.'

*γυναικοτρᾰφής, ές, brought up by a woman, Hsch. s.v. τηθαλλαδοῦς.

*γυναικοϋφής, ές, woven by women, γυναικυυφῆ interpr. as acc. sg. in PSI 341 (ZPE 66.59).

γυναικοφίλης, for 'Theoc. 8.[60]' read 'Theoc. 8.60'

+γυναικόω, make a woman of by sexual experience, παρθένος οὐδέποτε γυναικουμένη Ph. 1.683.   2 make one with abnormal male characteristics into a complete woman, Hp.Epid. 6.8.32.   II make womanish or effeminate, τὰ σώματα μαλακότητι καὶ θρύψει γυναικοῦντες Ph. 2.21.

γύναιος I, add 'Myc. ku-na-ja (fem.), decorated with the figure of a woman'

γύνανδρος, for 'of doubtful sex, womanish' read 'of a man, womanish, effeminate'

γύος, for '= γύης II' read 'field (esp. as enclosed by dikes, opp. γῆ χέρσος)'; before 'etc.' insert 'BGU 1132.10, CPap.Jud. 142'

*γυπαιετούς, bird name, perh. error for γρυπ- or °ὑπαιετούς, Suid.

+γύπη, ἡ, hollow cavity, Call.fr. 43.71 Pf., γύπη· κοίλωμα γῆς. θαλάμη. γωνία Hsch. and id. s.v. γύπας; in Hsch. also, perh. by folk-etymology, vulture's nest.

γυπιάς, add '(or ?hollow, cf. °γύπη)'

γύπινος, add 'πτερά Edict.Diocl. 18.10'

*γυπίς, ίδος, ἡ, = γυπιάς, in place-name, ἐκ τᾶν γυπίδων πετρᾶν SEG 34.384 (Delphi, ii BC).

*γυπόμορφος, ον, having the form (head) of a vulture, epith. of Isis, hymn in POxy. 1380.66 (ii AD).

γύπωνες, after 'Sparta' insert 'οἱ δὲ γ. ξυλίνων κώλων ἐπιβαίνοντες ὠρχοῦντο, διαφανῆ ταραντινίδια ἀμπεχόμενοι' and add 'cf. ὑπογύπωνες'

γυργάθιον, add 'Anon.Alch. 360.15'

γυργαθός, after 'creel' insert '(or perh. net)' and after 'Luc.DMeretr. 14.2' insert 'esp. a bread-basket, cf. Hsch. s.v. γυργαθόν'; at end after 'Aristaen. 2.20' add 'also γουργαθός Vit.Aesop.(G) 18, al.'

*γυργαθώδης, ες, resembling a γυργαθός, Hsch. s.vv. σαργάναι and σεγάνιον.

γυρητόμος, for 'tracing a circle' read 'perh. trenched (cf. γῦρος 2)'

γύρῖνος, after 'tadpole' insert 'or frog'

γυροειδής, after 'round' insert 'PMag. 3.139'

γύρος, after 'for 'γ. πάλη .. Gym. 11' read 'of wrestling, contorted, Philostr.Gym. 11, 35'

γυρόω III, for 'coil oneself up' read 'curve or arch oneself'

γυρτόν, after 'σκύφον' insert '(κυφόν Vossius)'

γύρωσις, for 'γῦρος' read '(γῦρος 2)'   add 'II giddiness, reeling, Aq.Is. 19.17.'

γύψ, after 'Eus.PE 3.12)' insert 'gen. pl. γυπάων Opp.C. 4.392'

γυψίζω, add 'in sealing a jar, PMich.XII 657.17 (ii/iii AD)'

γύψινος I, after 'gypsum' insert 'στεφάνιον Inscr.Délos 1452B11 (ii BC)'   II, add 'Ἰσιδώρου τὸ γύψινον (on a plaster-cast of a horse's nose-piece), BE 1960.95 (iii/ii BC)'

*γυψοκόπος, one who powders gypsum, ITyr 31.

γύψος, for 'II cement' read '(II) 2 plaster made from gypsum'

*γωνά, ἡ, perh. = γωνία, SEG 7.1047 (Syria).

γωνία I, add '5 spike of corn, Sch.Hes.Sc. 398.   6 edge, point of a sword, Eust. 563.18.'

γωνιαῖος I, add 'and γωνιεῖος PRyl. 567.3 (iii BC)'   add 'III γωνιαία (sc. λίθος), ἡ, a precious stone, Plin.HN 37.164.'

γωνιόομαι, after 'Dsc. 3.7' insert 'Eust.Op. 292.33'

γωνίωσις, add 'II sharp point, Sch.Gen.Il. 8.297.'

γῶνος, add 'acc. to Theognost.Can. 50 A., oxyt. in the sense of ἕδος (cf. γουνός), properisp. as παιδιά παλαιστρική'

γωρυτός, for 'quiver' read 'bow-case'; line 2, for 'Od. 21.54, cf. Lyc. 458' read 'ὅς οἱ περίκειτο φαεινός Od. 21.54; so prob. in Lyc. 458.   2 quiver'   at end delete 'wrongly .. 1898.21'

Δ

δαγνόν, delete '(leg. ἀδινόν)'
δᾱγύς, delete 'used in magic rites'; add 'gen. pl. δαγ[ύ]δων Erinn. in *Suppl.Hell*. 401.21'
Δαδαφόρια, for pres. ref. read '*CID* I 9.D4 (Δαιδᾱφ[ό]ρια, iv BC)'
δᾰ́δινος, add 'σπάθη δ. Heras ap.Gal. 13.781.13'
δᾰ́διον 1, after '*little torch*' insert 'δ. χρυσοῦν ἐπὶ βάσεως *Inscr.Délos* 1417*A*i80 (ii BC), al.'
*δᾳδοσχίστης, ου, ὁ, *one who splits pine-brands*, *IG* 2².1557.29.
δᾳδουχία, after '*torch-bearing*' insert '*SEG* 30.93 (Eleusis, i BC)'
*δᾳδουχικός, ή, όν, *of a* δᾳδοῦχος 1 1, ἱεροφαντικῶν καὶ δ. οἴκων *IG* 4².84.30 (Epid., i AD).
δᾳδούχιον, delete the entry.
δᾳδοῦχος I 1, add 'also fem., *AJA* 37.239 (Latium, ii AD); of Persephone, *IG* 12(5).229.23 (Paros)'
δᾳδοφόρος, for 'B.*fr*. 23.1' read 'δαϊδοφόρε B.*fr*. 31.1 S.-M., *IG* 2².5146; δᾳδοφόροι μελανείμονες of the Furies, *Suppl.Hell*. 1154'
Δάειρα, Δαῖρα, add 'also, of Hecate, Δαῖραν μουνογένειαν A.R. 3.847'
*δαελόν· διάδηλον Hsch.; cf. δίαλον ib.
†δαελός, Syracusan form of δᾱλός, τὸν δαελὸν σβῆτε ?Sophr. in *PSI* 1214.13; cf. δᾱλός .. καὶ δαελὸς παρὰ Σώφρονι *EM* 246.35.
δαημοσύνη, add 'Max. 454'
δαήμων, line 1, after 'gen. ονος' insert 'contr. pl. **δῦμονες** (codd. δαήμονες) perh. in Archil. 3.4 W. (for v.l. v. °δαίμων B) **4**, after 'Democr. 197' insert 'w. κατὰ πόλεμον Arr.*Tact*. 12.1'
δᾱήρ, line 1, before 'dat.' delete 'Men. 135' and insert 'gen. written δῆρος *IKios* 53.3, cf. gen. pl. infra'
*δάθιος, v. °ζάθεος.
*Δαιδάλειος, α, ον, *of or named after Daedalus*, ἀγάλματα ἃ ἐκαλεῖτο Δαιδάλεια *Lindos* 2C.61, cf. Ar.*fr*. 202 K.-A., D.S. 4.30.1; neut. subst. *sanctuary of Daedalus*, *Ath.Agora* XIX P5, 11, 12, 22 (written -ειον and -εον, Athens, iv BC); cf. Myc. place-name *da-da-re-jo*(-*de*) (neut. acc.).
*δαιδαλικός, ή, όν, *skilfully wrought*, δαιδαλικοὺς λιβάνους Σμύρνης *Suppl.Hell*. 1062.
δαίδᾰλος II, last line, for 'Argos' read 'Plataea'
δαιθμός I, add '*IG* 9²(1).609.10 (Naupactus, v BC), cf. °ἀνδαιθμός, δασμός'
*δαϊθρᾰσής, ές, (δάϊς) gen. δαϊθρασέος *bold in battle*, Euph. in *Suppl.Hell*. 415 i 7.
δᾱϊκτήρ 1, for 'Alc. 28' read 'cj. for διακτήρ *Lesb.fr.adesp*. 6 L.-P.'
*δαιμονά, ἁ, *distribution*, prob. in Alcm. 65 P., cj. in A.*Eu*. 727.
δαιμονιόπληκτος, for '= δαιμονιόληπτος' read '*afflicted by evil spirits*' and after 'Ptol.*Tetr*. 169' insert 'Rhetor. in *Cat.Cod.Astr*. 8(4).165'
*δαιμονιόπλοκος, ον, *exposed to assaults of evil spirits, in their grip*, v.l. for δαιμονιόπληκτος, Rhetor. in *Cat.Cod.Astr*. 8(4).165; cf. θεόπλοκος.
*δαιμονοκλησία, ή, *invocation of spirits*, Anon.Alch. 397.15 (sp. δημ-).
*δαιμονώδης, = δαιμονιώδης, φαντασία *IGLS* 220 (iv/v AD).
δαίμων I, lines 4/5, for '*Divine* .. *Deity*' read 'unspecified agency affecting human fortunes' **2**, for '*the power*.. *lot or fortune*' read 'merging into the sense *fortune, lot*, but still often conceived as an active agency'; transfer 'στυγερὸς .. (Od.) 11.61' to section 1, line 6, after '11.792'; after 'Antipho 3.3.4' insert 'ἔδωκε τῇ τύχῃ τὸν δ. surrendered his *fortunes* to events, E.*Ph*. 1653' **B**, for this section read '= °δαήμων, *knowing*, δαίμονες μάχης Archil. 3.4 W., but δάμονες shd. prob. be written'
δαίξανδρος, after '*man-destroying*' insert 'δ. πολέμō *CEG* 798.1 (Delphi, iv BC)'; add '(= *GVI* 1153)'
δάϊος I 1, line 6, before 'φόβημα δαΐων' insert 'ἐγχέων'; add 'of armour, *warlike*, A.R. 1.635, *AP* 6.128 (Mnasalc.)'
*Δαῖρα, v. Δάειρα.
δαΐς, after 'ή' insert 'nom. sg. written **δαές** in *Inscr.Délos* 1442*B*23 (ii BC)' **1**, add 'of wedding torches, [οὐ]δὲ γὰρ οὐ δαΐδων καὶ παστάδος ἔλλαχεν ἦμ[αρ] epigr. in *IEphes*. 2101A.3'
δάϊς, before '(δαίω A)' insert 'ή'; add 'cf. Myc, *da-i-qo-ta*, compd. pers. n. (app. = Δηϊφόντης)'
*δαίστωρ, ο, perh. *consumer* or *?slaughterer*, δαίστορας ἀλλοτρίων epigr. in *SEG* 26.1215.
δαίτης, after 'E.*fr*. 472.12' insert '(dub.), δαΐτας· μεριστάς Hsch.; pers. n. Δαίτης a Trojan hero of feasting, Mimn.*fr*. 18 W.'

*δαιτῐκλῡτός, 'α, ον, *famed for feasts*, Κορίνθου δειράδ' ἐποψόμενος δαιτικλυτάν (codd. δαῖτα κλυτάν) Pi.*O*. 8.52.
δαῖτις, delete the entry.
*Δαιτίς, ή, epith. of Aphrodite, *IEphes*. 1202.5 (iii AD).
*δαιτόποινος, ον, unexpld. wd. in poet. word-list, *Suppl.Hell*. 991.27.
*Δαίττης, ου, ὁ, epith. of Apollo and Artemis, Ἀπόλλωνι καὶ Ἀρτέμιδι Δαίτταις *SEG* 7.17 (Susa, ii BC); τοῦ Ἀπόλλωνος καὶ τῆς Ἀρτέμιδος τῶν Δαιττῶν *OGI* 244.22 (Daphne, ii BC).
δαιτῠμών, after 'Od. 7.102, 148, al.' insert 'Alcm. 98 P.'
δαίω (A) I, lines 14ff., mostly metaph. sense, after '(II.) 2.93' insert 'οἰμωγὴ .. δέδηε Od. 20.353, Ὅμαδός τε Φόνος τ' Ἀνδροκτασίη τε δεδήει Hes.*Sc*. 155'
*Δᾰ́κεια, τά, games celebrating Trajan's conquest of Dacia, *Corinth* 8(1).77.
*Δᾱκικός, ή, όν, *Dacian*, τὸ ἀγώνισμα τοῦ Δακικοῦ ἅρματος *PColl.Youtie* 69.9 (AD 272); as title of Roman emperors, *SEG* 31.1124 (AD 104), 24.486 (AD 114).
*δακνᾶς, ᾶ, ὁ, *biter*, Phryn.*PS* p. 64 B., gramm. in Gaisford *Choerob*. 1 p. 43.
δάκος I, at end for 'β. δάκος' read 'βλοσυρὸν δ.'
δακρύοεις, add 'see also ‡ζακρυόεις'
δακρυρροέω 1, line 3, for 'δακρυροοῦν' read 'δακρυρροοῦν'; add 'ἀπ' ὀμμάτων κλάον πρόσωπον καὶ δακρυρροοῦν *Trag.adesp*. 447 K.-S.'
*δακρυστᾰγής, ές, *marked by flowing tears*, δακρυσταγεῖ [γ]όῳ Tim.*Pers*. 100 P.
δακρῡτός, add 'δακρύτ' ἄγαν E.*El*. 1182, πᾶσιν δακρυτός *Amyzon* 65 (ii BC)'
*δακρῠχοέω, *shed tears*, *GVI* 969.8 (Lydia, i AD).
δακρύω I 3, add 'τὸ κόμμι *Peripl.M.Rubr*. 29' **II**, before 'A.*Ag*. 1490' insert 'Simon. 48 P.'
*δακτῠλιδάριος, ὁ, *maker* or *seller of rings*, rest. in *ITyr* 166 (δακτυλιδ-, Rom.imp.).
δακτῠλίδιον, delete '[λῑ]'
δακτυλικός, for 'αὐλός .. Ath. 4.176f' read 'name of kind of aulos, Ath. 4.176f; also of a cithara (also called °Πυθικός), Poll. 4.66'
δακτύλιος II 2, add 'Plu. 2.518d'
δακτυλόδεικτος, delete 'cf. *PLond.ined*. 1821' and add '**b** in a Greek-Coptic gloss. given as name for the first finger, *PLond*. 1821.306 (vi AD).
*δακτῠλοκλείδιον, τό, *finger-ring-key*, *PFouad I Univ*. 8.12 (ii AD, δακτυλοκλίδιν pap.), cf. ib. 8.
δάκτῠλος I 1, add '**b** δ. θεοῦ, *the finger of God*, as symbol of divine agency, Lxx *Ex*. 8.19, *Ev.Luc*. 11.20.' **2**, add '**c** pl., *joints of a beetle's tarsi*, Horap. 1.10.' **VI**, for this section read 'kind of shell-fish, *dactyli, ab humanorum unguium similitudine appellati*, Plin.*HN* 9.184'
*Δακύτιος, ὁ, epith. of Hermes, *Inscr.Cret*. 4.174.60 (Gortyn, ii BC).
Δαλμᾰτεῖς, lines 3/4, Δαλματική, Δελμ-, add 'also Δελματικόν, τό, *PSI* 900.7 (iii/iv AD), *POxy*. 1741.15 (iv AD), var. Δελμάτιον *POxy*. 1026.16 (v AD)'; line 6, δερματίκιν, add '*PMich*.III 218.14 (-ιον, iii AD)'
*δαλμᾰτικομαφόριον, τό, attested only in form δελματικο-, *Dalmatian cloak*, *SB* 10988 (iv AD), 11075 (v AD), cf. °δαλματικομαφόρτης.
†δαλμᾰτικομᾰφόρτης, ου, ὁ, *Dalmatian cloak with hood*, *Edict.Diocl*. 22.5, al., δελμ- *POxy*. 1273.12 (iii AD); δελματικομαφόλτου *Edict.Diocl*. 22.13, δελματικομαφέρτ[η]ς ib. 19.43.
*δαλμᾰτικομᾰφόρτιον, τό, dim. of δαλματικομαφόρτης, δελματικομαφέρτιον *Edict.Diocl*. 19.8,, δερματικομαφόρτιν *POxy*. 114.5 (ii/iii AD), *PMichael*. 18.2, 4 (iii AD), cf. ‡μαφόρτης.
*Δαλμάτιον (Δελμ-), τό, app. = Δελματίκιον (for which it may be an error), *POxy*. 1026.16 (v AD).
δᾱλός II, for this section read '*smouldering ember*, fig., of a youth who no longer inflames a man with homosexual love, *AP* 12.41 (Mel.)'
*Δαλφοί, v. Δελφοί.
δᾰμάζω, line 4, after 'Od. 14.367' insert 'inf. δμῆσαι Hsch.'
δᾰμᾰλίζω, for 'codd.' read '*Lyr.adesp*. 415.7 S.'
δάμαρ, after 'A.*Pr*. 834' insert 'S.*OT* 930, al, E.*Alc*. 930, *Andr*. 4, al., Eup. 171 K.-A., Lex ap.Lys. 1.30, D. 23.55'
*δᾰμᾱσίβιος, ον, app. *subduing life*, δα]μαισιβίου (sic) epigr. in *Inscr.Cret*. 1.xvi 52 (i AD).

Δαμασκηνόν, add 'see also °Δαμασκηνός'

*Δαμασκηνός, ή, όν, of or from Damascus, οἱ Δ. Str. 16.2.20, ἡ Δ. (χώρα) 16.2.16; of cloth, damask, ἐνδρομίς .. Δαμασκηνή Edict.Diocl. 19.6, καμάσιον Δαμασκινόν SB 7033.41 (v AD). **II** Δαμασκηνοί, οἱ, damsons, οἱ Δ. PRyl. 630.80, al. (iv AD), PFreib. 4.67 (ii/iii AD, gen. pl. of masc. or neut.); cf. Δαμασκηνόν.

δαμαστικός, add 'δμήτειρα· δαμαστική Hsch.'

*δᾰμάστρια, gloss on δμήτειρα, Hsch., Sch.Gen.Il. 14.259.

δαμασώνιον **I**, for '= ἄλιμος' read '= ἄλιμον, s.v. ἄλιμος II'

δᾰμάτειρα, after 'δαμαντήρ' insert 'Call.fr. 267 Pf.'

Δᾱμάτριος, add 'see also Δημήτριος I'

†δᾱμεύω, v. °δημεύω.

δᾰμία, delete the entry.

Δᾱμία, add 'Λοκαία (for Λοχ-) Δαμία IG 12(3).361 (Thera)'

*δᾱμιεργέω, v. °δημιουργέω.

*δᾱμιοργία, ά, meeting of the δαμιοργοί, IG 4.493 (Mycenae), cf. δημιούργιον II.

*δᾱμιοργίζω, perh. serve as δαμιοργός, app. aor. part. δαμιοργίσᾱσα περτέδοκε IPamph. 17 (iii BC; unless to be read as δαμιοργὶς ὅσα).

*δᾱμιοργίς, ιδος, ά, perh. fem. of δαμιεργός, v. °δαμιοργίζω, δημιουργίς.

δάμνια, add '(δάμνιον v.l. for δ' ἀμνίον in Od. 3.444)'

*δᾱμονομέω, hold the office of *δαμονόμος (app. a local magistrate), IG 9²(1).138.14 (Calydon, iv/iii BC).

*δᾱμοσιεργός, ὁ, = δημιουργός, SEG 23.305 ii 20 (Delphi, 190 BC), cf. °δαμοσιοργία.

†δᾱμοσιοργία, ά, eligibility for public office, GDI 3052.10 (Chalcedon).

*δᾱμοσιόω, v. °δημοσιόω.

*δᾱμοτεύομαι, v. °δημοτεύομαι.

δᾱμώματα, add 'also δημώματα· παίγνια id.'

*δάμων, v. °δαήμων.

Δᾰνᾰΐδαι, after 'descendants of Danaus' insert 'used for the Greeks generally' and add 'E.Hec. 503, Tr. 447, al.'

δᾰνάκη, after 'ἡ' insert 'and δανάκης, ου, ὁ'; add '(see also note to Call.fr. 278 Pf.)'

*δάνας (B), Semitic word for wine-jar, Dura⁴ p. 122; cf. Akkadian dannu(m).

*δάνδηξ, ηκος, ὁ, kind of large dog, Ps.-Callisth. 2.33 cod. B.

δᾰνείζω **I**, before 'fut.' insert '(cf. δανίζω)'; line 6, after 'lend' insert 'Sol.Lg.fr. 68 R.' **2**, line 2, for 'ἀπό τινος' read 'παρά τινος'

δάνειον, for 'Men.Mon. 97' read 'Men.Mon. 759 J.' and add 'SEG 38.1476.A52 (Xanthos, iii BC)'

δάνεισμα, line 2, after 'Th. 1.121' insert 'δ. πράττειν Just.Nov. 4.1'

δᾰνειστής **I**, after 'creditor' insert 'D. 34.7, Plu.Sol. 13' deleting the latter ref. in section II; add 'also δανιστής TAM 5(1).231.11'

δᾰνειστικός, after 'etc.' insert 'ἐργασία money-lending business, Thphr.Char. 23.2'; ὁ δ., for '= δανειστής' read 'moneylender'

δᾰνίζω, for 'Lxx Pr. 19.14' read 'Lxx Pr. 19.17'

*δᾰνιοκαρπία, ή, usufruct, BGU 2338.9 (ii AD).

*δᾰνιστής, v. °δανειστής.

*Δάνκλε, Δανκλαῖοι, v. °ζάγκλη II (Ζάγκλη).

δᾱνός, add 'Call.fr. 243 Pf.'

δᾶνος (A) **II**, add 'Ammon.Diff. p. 82 N.'

δαόν, add 'prob. error for δοάν, cf. δήν'

δᾰπᾰνάω **I 1**, add 'w. περί, τῶν .. περὶ τοὺς κληρικοὺς .. δαπανωμένων Cod.Just. 1.3.42 pr., Just.Nov. 59.2'

δᾰπάνη **I**, for 'cost' read 'outlay' **II** delete 'money spent' and transfer exx. to section I

δᾰπάνημα, after lemma insert 'dial. δαπάναμα SEG 31.572.A8 (Thessaly, c.200 BC)' and for 'cost' read 'outlay'

δᾰπάνησις, add 'κράτησον τὴν ἐπιστολὴν εἰς δαπάνησίν σου Stud.Pal. 20.125.3 (v. ZPE 76.112)'

*δᾰπᾰνητέον, one must use up, consume, Aët. 9.108(98).

†δᾰπᾰνητής, ου, ὁ, defrayer of expenses, PMich.XII 658.7 (iii AD), EM 40.44.

δᾰπᾰνος, after 'Th. 5.103' insert 'κόλλοψ AP 12.42 (Diosc.)'

δᾰπεδον, line 4, ground, add 'Aen.Tact. 37.6'

δᾰπτης, for 'eater, bloodsucker' read 'devourer'

*δᾰράτα, ά, app. kind of bread, offered in religious ceremonies, μή δέκεσθαι μήτε δαρατᾶν γάμελα μήτε παιδήια CID I 9.A24 (iv BC), cf. °δάρατος.

†δάρατος, ὁ, or -ον, τό, kind of bread, τὸν ἄζυμον ἄρτον .. δάρατον Nic.fr. 134 G.-S.; said to be Thess., equiv. Maced. δράμις, Seleuc.ap.Ath. 3.114b; τὸ δά[ρατον] rest. in Schwyzer 603 (Coropa, Magnesia, vi/v BC), cf. °δαράτα, also δαρόν s.v. δαρός.

*Δαρδανικός, ή, όν, poet. Trojan, Δ. .. σκήπτροις AP 9.155 (Agath.); ἡ Δ. Dardania, a river in the Troad, Str. 13.1.43, in Illyria, ib. 7.5.1; χλαμὺς Δαρδανικὴ Edict.Diocl. 19.69, 70.

Δᾱρεικός, for 'a Persian gold coin' read 'gold coin, originally Persian, later Maced.' and add 'IG 7.3055.13, al. (Lebadeia), Delph. 3(5).61 ii A1, al. (both iv BC; v. SEG 29.445, 458)'; after 'Δᾰρικός' insert '[ῐ Herod. 7.102]'

*Δαρζάλεια, τά, games at Odessus in honour of the μέγας θεός Derzelas or Darzalas, BCH 52.395 (on coins); (for the god v. IGBulg. 47, 768, al.).

†δαρκνά, v. ‡δραχμή.

*δαρμός, a flogging, flaying, Hsch. s.v. μάστιγας.

*Δάρρων· Μακεδονικὸς δαίμων, ᾧ ὑπὲρ τῶν νοσούντων εὔχονται Hsch.

δάροσος, add '(δάρος La.)'

δαρτός, before 'Paul.Aeg. 6.61' insert 'Ruf.Anat. 61'

δασμός, add 'see also °δαθμός'

δασμοφορέω, add '**II** exact tribute from, w. acc., Call.Dieg. iv.7 (fr. 98 Pf.).'

†δασπλής, ῆτος, masc., fem. adj., horrid, frightful, δασπλῆτα Χάρυβδιν Simon. 37 P., Call.fr. 30 Pf., δασπλῆτες Εὐμενίδες Euph. 98 vGr., δασπλῆτε δράκοντε Nic.Th. 609, μάχαιρα Nonn.D. 22.219, γυναῖκες 46.210, al., Hsch.; also app. δασπλήτης An.Ox. 1.149.

†δασπλῆτις, fem. adj. = δασπλής, θεὰ δ. Ἐρινύς Od. 15.234, Ἑκάτα δασπλῆτι Theoc. 2.14.

δάσσω, for the pres. ref. read 'Suppl.Hell. 1068'

*δαστήρ, ῆρος, ὁ, land-commissioner, IG 9²(1).116 (pl., Aetolia).

δᾰσῠγρᾰφέω, for 'Hdn.Epim. 25' read 'Hdn.Epim.ap.Bast Epistula Critica, Appendix p. 25'

δᾰσύθριξ, for 'hairy' read 'having thick hair' and after it insert 'δασύτριχος .. τράγοιο Theoc. 7.15'

δασυκνήμις, before 'Nonn.D. 14.81' insert 'δασυκνήμιδι Φιλάμνῳ'

*Δᾰσυλλιεῖον, τό, sanctuary of Διόννυσος Δασύλλιος, Sokolowski 3.90.10 (Moesia, ii BC).

δᾰσύνω **IV**, for this section read 'in prohibition, μηδένα .. δασύνειν τὴν οἰνοποσίαν, μηδὲ ἀκος[μ]ίαν παρέχειν perh. gulp, BGU 2371.4 (i BC)' and renumber pres. section IV to V. **2**, of breathing, for 'rapid' read 'thick'

δᾰσῠπόδειος, delete 'τὸ δ. the species hare'

δᾰσύς **I 1**, add 'of snakes, rough, scaly, Arist.HA 607ᵃ32' **II 2**, after 'D.H.Comp. 14' insert 'comp., τὸ β τοῦ μὲν (sc. φ) ψιλότερόν ἐστι, τοῦ δὲ (sc. π) -ύτερον nearer to an aspirate, ib.' and after 'adv. -έως' insert 'λέγεται ib.'

δᾰτέομαι **I**, line 4, for 'Diog. .. infr. II' read 'in pass. sense Il. 1.125, etc., in act. sense Q.S. 2.57' **II**, line 6, for 'Diog.Apoll. l.c.' read 'Diog.Apoll. 3' add '**2** break up into lots for sale, sell off, κὰ(F) Ϝοικίας δάσασσθαι τὰς ἄνοδ' ἐάσ(σ)ας IG 5(2).262.17 (Mantinea, v BC = Dubois Dial.arcad. p. 95); prob. also in ἀλλὰ τὰ μὲν πολίων ἐξεπράθομεν, τὰ δέδασται Il. 1.125.' at end add 'Myc. aor. (ο-) da-sa-to, pf. (e-pi-)de-da-to (-δέδασται), cf. adj. e-pi-da-to (= *ἐπίδαστος)'

δᾰτήριος, for 'α, ον' read 'ον'

Δᾱτισμός, for 'the Median commander at Marathon' read 'the ref. is obscure'

*Δάτυιος, ὁ, name of a month at Dodona, SEG 15.384.19 (iv BC).

δαῦκος (A), at end delete 'also δαυχμός .. Al. 199' (v. °δαυχμός)

δαῦκος, read δαῦκος (B)

*δαυνός, perh. fiery, A.fr. 41a R. (*δαϜ-νός, v. δαίω (A)).

δαυχμός **I**, for 'v. δαῦκος' read '= δάφνη πικρά, Nic.Th. 94, Al. 199 (acc. to Sch.Nic.Th. 94)'

δαύχνα, after 'ἀρχιδαυχναφορέω' insert '‡συνδαυχναφόρος'

†Δαυχναφόριος, v. °Δαφνηφόριος.

δαφνήεις, add 'contr. δαφνῆς, ῆντος, Choerob. in Theod. 1.360, al.'

*δαφνηρεφής, ές, δαφνηρεφέων μυχάτων covered with laurel, Porph. ap.Eus.PE 6.3.239a.

*δαφνηφάγος, delete 'hence inspired'

δαφνηφορέω, line 2, delete 'Paus. 9.10.4'; line 3, after '(Hdn.) 7.6.2' insert 'serve as °δαφνηφόρος I 2, Paus. 9.10.4 (twice)'

Δαφνηφόριος, add 'Cypr. Δαυχναφόριος, ta-u-ka-na-po-ri-o (gen.) ICS 309.3'

δαφνηφόρος **I 2**, add 'δαφνᾱφόρος, wearing (crown of) bay, title of boy-priest of Apollo Ismenius, Paus. 9.10.4' **II**, delete 'at Thebes, Paus. 9.10.4'

δαφνίς **2**, delete the section transferring ref. to section 1; add 'Alex. Trall. 5.4'

δαφνόκοκκον, add 'Aët. 7.114 (385.2 O.)'

δαφνών, add 'as pr. n., POxy. 3917.10 (ii AD)'

*δάχμα, τό, = δῆγμα, Nic.Th. 119 (codd. also δηχ-, δαγ-), 128, al.

δαψιλής **I 1**, -εστέρως, after 'Ptol.Tetr. 58' insert '(v.l.)'

*δάω **II**, δάε, ἔδαε, add 'A.R. 3.529; perh. also δαείης h.Merc. 565'

δέ **II 3**, implying causal connection, add 'A.Supp. 190, S.Ph. 741, E.Ph. 689, Ar.Av. 935; also in prose Th. 1.86.2, Pl.Cra. 426a, Lys. 12.68' line 34 of article, for '**II**' read '**III**' add 'but cf. οὐ δὲ βίηφιν Od. 9.408, οὐ δ' εἰδὼς Pl.Smp. 199a' at end add 'Myc. -de, e.g. da-mo-de-mi pa-si, δᾶμος δέ μίν φασι'

-δε, at beginning for 'an enclitic post-position' read 'a post-position';

add '(It is uncertain whether -δε is enclitic, cf. A.D.*Adv.* 179.5, 181.13, Hdn.Gr. 1.498.) Myc. *-de*'

***δεγαλῶσαι**, v. °δηλόω.

δεγμόν, add '(ὅρμον La.)'

δέδηε, δεδήει, after 'v. δαίω (A)' insert 'and °δέω (A)'

δεδίσκομαι II, delete the section.

***δεδίσσω**, v. °δειδίσσομαι.

δέησις I 2, written *petition*, delete 'written'

***δεητός, όν**, *needed, necessary*, Plu. 2.687e.

δεῖ I 1, line 9, for '(X.)*Oec.* 7.20' read '(X.)*Oec.* 8.9, Pl.*Phlb.* 33b' add '**b** (in the protasis of a condition in which one apologizes for pressing a point) *if I had to go so far as to* .., or sim., πόσους εἴποιμ' ἂν ἄλλους, εἴ με μηκύνειν δέοι Ar.*Lys.* 1132, εἰ δεῖ τι καὶ σκῶψαι Pl.*Men.* 80a.'

Δεῖα, add '*Δ. Σεβαστὰ οἰκουμενικὰ ἐν Λαοδικείᾳ* I*Ephes.* 1605.5, 11 (ii AD); *Δ. Κομόδεια τὰ πρώτως ἀχθέντα ἐν Λαοδικείᾳ* I*Trall.* 135.15 (ii AD, v. *Laodicée* p. 283)'

***Δειγαία, ἡ**, Maced. cult-epith. of Artemis, *SEG* 37.590 (Vergina, ii AD), 591 (-έα, iii AD); see also ib. 539 (unpublished inscrs. from Metochi; also sp. *Διγαία*).

δεῖγμα 1, line 1, for '*sample, pattern*' read '*material example, sample*'; line 2, after '*POxy.* 113.5 (ii AD)' insert 'of a sealed sample of goods sent for approval, *PPetaus* 57.4 (ii AD)' and before 'τοῦ βίου' insert '*immaterial example, model*, etc.'

***δειγματοκαταγωγεύς, έως, ὁ**, *official charged with the delivery of sealed samples*, *PBerol.*inv. 7441 (*Eos* 58.63ff., iii AD).

***δειδία**· ἡ σκοτία νύξ Suid.

δειδίσκομαι, at beginning insert 'also δεδίσκομαι'; line 6, after 'Od. 7.72' insert 'Call.*fr.* 87, 186.12 Pf.' and after '*welcoming*' insert 'Od.' add '**III** = δειδίσσομαι I, Ar.*Lys.* 564.'

δειδίσσομαι, at beginning delete 'impf. .. *Lys.* 564'; line 9, delete 'D. 19.291, *Prooem.* 43'; line 10, for 'cf. δεδίσκομαι II' read 'act. in this sense, Sch.B Il. 24.569, Suid.'

δείδω, line 16, before '3 sg.' insert '2 sg. δείδιας *AP* 12.138 (Mnasalc.)'; line 23, after 'etc.' insert 'pf. part. fem. δειδυῖα A.R. 3.753'; line 27, after '(Nonn.D.) 35.30' insert 'late Ep. pres. part. fem. δειδιόωσα Jo.Gaz. 2.248'

δεικανάω II, at end after 'Od. 18.111' insert 'χερσὶ δὲ καὶ μύθοισιν ἐδεικανόωντο A.R. 1.884'

δείκελον 2, add 'Parth. in *Suppl.Hell.* 653, *AP* 5.260 (Paul.Sil.), 9.505'

†**δεικηλίκτᾱς, α, ὁ**, *actor in burlesque* or sim., Plu.*Ages.* 21 (codd. δικ-), 2.212f, Hsch. s.v. δίκηλον.

***δεικηλιστής, οῦ, ὁ**, Atticized form of δεικηλίκτας, Ath. 14.621e, f.

***Δείκνιος, ὁ**, name of a month, *SEG* 17.829.11 (provenance uncertain, *c.*200 BC).

δείκνυμι 2, add '**b** pf. part. pass. δεδειγμένος = *designatus*, ὕπατος [τὸ] γ', δεδειγμέ[νος] τὸ δ' *SEG* 9.165, cf. 167 (Cyrene, i AD).' **6**, line 2, for 'προθυμίαν' read 'δύναμιν'

δεικτήριον I, for '*place for showing*' read '*stage or room for a spectacle* or *exhibition*, *PPetr.* 3.142 (app. for the mysteries of Adonis), τὸ λογεῖον τοῦ δ. *SEG* 23.207 (Messenia, Augustus); applied to Theophrastus' σχολή D.L. 2.37; defined as ἄμβων or ἀκροατήριον by Bas.Sel. in *PG* 85.612D' **II**, delete the section.

δεικτικός, add '**III** subst. δεικτικός (sc. δάκτυλος), ὁ, *index finger*, Cael.Aur.*TP* 5.21.'

δεικτός, add '**3** *capable of being shown* (opp. described), *demonstrable*, Gal. 8.678.13.'

δειλαίνω, after 'aor. pass.' insert 'ἐὰν δὲ δειλανθῇ *if he shows himself a coward*, Lxx 1*Ma.* 5.41'

δείλαιος, line 1, within parentheses add 'Ion. fem. -η, epigr. in *SEG* 33.849.5 (Mauretania, i AD)'; line 3, after 'S.*OT* 1347' delete '(lyr.)'

***δειλανδρία, ἡ**, *cowardice*, Suid.

δείλαρ, for 'Call.*fr.* 458' read 'Call.*fr.* 177.17 Pf.'

δειλιάω, add 'Ezek.*Exag.* 267 S.'

δειμαίνω, line 1, for 'only .. 2.439)' read 'Ep. impf. δειμαίνεσκε Q.S. 2.439: aor. part. δειμή[ναντες Euph. in *Suppl.Hell.* 428.4'; line 2, for 'Pl.*R.* 330c' read 'Pl.*R.* 330e'

δειμός I, add 'Chrysipp.*Stoic.* 3.123 (pl.)'

***δείμυλος**, unexpld. wd. in Theognost.*Can.* 61.

δεῖνα I, line 2, gen. τοῦ δείνατος, add '*PMich.*II 122 passim (i AD)'; at end after '(D.) 20.106' add 'quasi-adj. κατὰ τὴν δεῖνα πόλιν *Cod.Just.* 1.12.3.5' **II**, at end after '(Ar.)*Pax* 268' insert 'Men.*Dysc.* 897'

***δεινόθυμος, ον**, gloss, w. σχέτλιος, on οὐλόθυμος, Hsch.

***δεινολέων, οντος, ὁ**, gloss on αἰνολέων, Call. in *Suppl.Hell.* 258.21, (see also 257.21).

δεινοπροσωπέω, for 'περὶ' read 'ὑπὲρ (codd. ὑπὸ)'

***δεινόφρων**, gloss on λυκόφρων, Hsch.

δεῖξις II, for '*display*' read '*public performance*'

***Δεῖος**, add '‡*Δῖος*'; also *Δεῖα*, *Δῖα*.

δειπνοκλήτωρ II, add 'Παβακ τοῦ δειπνοκλήτορος *Res Gestae Saporis* p. 17 l. 58 (Persepolis, iii AD)'

δειπνολόχος, for '*laying .. parasite*' read '*lying in wait for a dinner*, i.e. *greedy*' and add 'cf. βωμολόχος'

δεῖπνον 1, add '**b** *sacrificial meal*, *IG* 12(7).515.53 (Amorgos, ii BC).'

δειπνοποιέω, med., add 'Aen.Tact. 7.3'

δειπνοποιός, add '*IG* 12(1).579 (Rhodes)'

δειπνοσοφιστής, for '*one .. kitchen*' read '*learned man at a dinner party*'

†**δειπνοφορία, ἡ**, *solemn procession with food-offerings*, in honour of Herse, Pandrosos, and Aglauros, Is.*fr.* 151; in honour of Artemis, Men.*Kith.* 95.

***δειπνοφοριακός, ή, όν**, πομπή *marked by the carrying of food-offerings*, I*Ephes.* 1577.(a)9.

***δειραγχής, ές**, cj. for °δειραχθής, cf. δεραγχής.

†**δειραχθής, ές**, of a hunter's noose, *oppressing the neck*, ἄμμα *AP* 6.179 (Arch., s.v.l., v. °δειραγχής).

δειρή, line 2, after '(v. infr.)' insert 'ep. gen. δειρῆφι Opp.*C.* 1.176' **I 2**, add '**b** *mane* of a lion, Ael.*NA* 17.26.' **II**, for this section read '*col* of a mountain, Nic.*Th.* 502, Hermesian. 7.54 (prob.); pl., Pi.*O.* 3.27, 9.59' at end, line 2 of parenthesis, for 'Arc. δερϜά .. (Orchomenus)' read 'Arc. δερϜά Schwyzer 664 (Orchomenus, iv BC)'

δειρόπαις, for pres. def. read '*giving birth to young via the throat*'

†**δείς, δενός**, in quots. only in neut. δέν and in juxtaposition w. οὐδέν or μηδέν, *something, anything*, καί κ' οὐδὲν ἐκ δενός γένοιτο Alc. 320 L.-P., μὴ μᾶλλον τὸ δὲν ἢ τὸ μηδὲν εἶναι Democr. 156 D.-K. (back-formation fr. οὐδείς).

***δεισαής, ές**, *smelling of filth* (abusive term for a homosexual), Suet. *Blasph.* 64 T. (formed fr. δεῖσα on model of δυσαής; cf. δείσοζος).

***δεισακεία**, v. °δισάκκιον.

***δεισαλέος, α, ον**, *polluted*, Clem.Al.*Protr.* 4.55; δεισαλέον, gloss on μόλυχνον, Hsch.

δεισιδαιμονία 1, add '*PBonn.*inv. 2 (*JJP* 18.43, ii AD)'

***Δειφίλεια**, v. °*Διφίλεια*.

δέκα, line 1, after 'indecl.' insert 'but gen. δέκων Schwyzer 688.D13 (Chios, v BC); Arc. δέκο Dubois *Dial.arcad.* p. 196.6 (Pheneus, vi/v BC), cf. δυόδεκο'

***Δεκάβριος**, v. °*Δεκέμβριος*.

δεκαδαρχέω, add '*SEG* 19.742.9 (Cyrene, prob. ii AD); *SEG* 33.684 (Paros); pass. δεκαδαρχούμενοι *being ruled by a board of ten*, Sopat.Rh. 8.27.14 W.'

***δεκᾰδάρχης I**, = Lat. *decurio*, add 'Arr.*Tact.* 42.1'

δεκάδαρχος I, delete 'Plb. .. 42.1'; add '= Lat. *decurio*, Plb. 6.25.2 (gen. pl.; nom. sg. could be -άρχης), *EA* 17.26 (Cilicia), *SEG* 32.1473 (Syria, both iii AD)'

δεκαδιστής, add 'see also δεκατισταί'

δεκαδύο (s.v. δέκα), add 'also -δύω Milet 1(7).204*b*12 (i AD); neut. δεκαδύα *SEG* 26.672 (Larissa, *c.*200 BC)'

δεκαέξ (s.v. δέκα), add 'also δεκαέξε *SEG* 26.672 (Larissa, *c.*200 BC)'

δεκαεπτά (s.v. δέκα), add 'see also ‡δεχεπτά'

δεκαετηρία (s.v. -έτηρος), after 'ἡ' insert '*CIG* 8610 (Egypt, iv AD)'

δεκαέτηρος, add 'see also δεκέτηρος'

δεκάζω I, pass., add 'Ael.*VH* 2.8'

***δεκακορίνθιος, ὁ**, *coin with value of 10 (Corinthian) staters*, de Franciscis *Locr.Epiz.* 8 (iv/iii BC), cf. °πεντακορίνθιος.

†**δεκαμηναῖος, α, ον**, *amounting to ten months*, χρόνος Plu.*Num.* 12; of *children born in the tenth month of pregnancy*, βρέφη Alex.Aphr.*Pr.* 1.40, I*Kyme* 41.19 (i BC).

δεκάμηνος 3, delete 'subst. δεκάμηνον, τό' and leave quots. to follow '*Placit.* 5.18.1'

δεκάμνους, add '(prob. as adj.) δεκάμνων (νόμισμα) X.*Lac.* 7.5'

δεκαναία, add 'Plb. 24.6.1'

***δεκανεύω**, *hold the office of* δεκανός, *decurio*, *IGBulg.* 917, 1401 *bis* (both Philippopolis, iii AD).

***δεκάξυλος, ον**, *ten ξύλα* (v. ξύλον V) *in length*, δέσμαι *OTheb.* 144 (i AD).

***δεκαπάλαστος, ον**, *ten palms long*, *Inscr.Délos* 1442*B*67 (ii BC).

δεκάπεδον, delete the entry.

***δεκάπεδος, ον**, *measuring ten feet*, *IG* 4².106.25, 116.12, 13, 21 (Epid., iv and iv/iii BC), etc.; τὸ δ. *per ten foot length*, ib. 109 i 127, etc.

***δεκαπενταετής, ές**, *fifteen years old*, *GVI* 988.5 (Larissa, ii/iii BC).

***δεκαπεντάρουρος, ὁ**, *holder of fifteen* ἄρουραι, *PMich.*inv. 6167 (*ZPE* 33.200; AD 201), *Stud.Pal.* 17.23.369 (iii AD), *PVindob.Tandem* 18.16 (v/vi AD).

δεκαπέντε (s.v. δέκα), add 'δεκαπένδε *SEG* 6.728.66 (Perge, i BC)'

δεκάπλεθρος, add '*measuring ten* πλέθρα, *Inscr.Cret.* 1.v 21 (Arcades)'

***δεκαποδία, ἡ**, *space of ten feet*, *Delph.* 3(5).74 (iv BC).

δεκάπους, add 'also δεκάπος, ἄκαινα δ. *SEG* 37.491 (Magnesia, Thessaly, v/iv BC)'

***δεκάπρωσις** (sic), εως, ἡ, app. = δεκαπρωτεία, δ. τῆς κολω[νίας] *IGRom.* 4.222.

δεκαπρωτεία, for 'δεκάπρωτοι' read 'δεκάπρωτος' and for '(Syllaeum)' read '(Syllium, written -πρωτία)' δεκαπρωτεύω, add 'Salamine 128 (ii AD), SEG 34.1093 (Ephesus, ii/iii AD)'

*δεκαπρωτικός, ή, όν, having held the δεκαπρωτεία, γένος IGRom. 3.406 (Pisidia).

δεκάπρωτοι I, add '(unless dat. of *δεκαπρώτης)'

*δεκάργυρος, ὁ, or -ον, τό, name of coin, POxy. 3874.46 (iv AD).

*δεκαρουρία, ἡ, estate of ten ἄρουραι, PMich.v 238.79 (i AD).

δεκάρταβος, after 'ἀρτάβαι' insert 'BGU 1773.15 (i BC)'

δεκάς I 1, for this section read 'group or company of ten, Il. 2.126, Hdt. 3.25.6; ἡ Ἀττικὴ δ. the ten Attic orators, Luc.Scyth. 10; of ships, A.Pers. 340; of slaves, BCH 59.453 (Chios, v/iv BC).   b ἐτέων δ. decade, Call.fr. 1.6 Pf., CEG 477 (Athens, 400/390 BC), al., epigr. in IKyzikos 521 (i BC), etc.'   add '3 select group, ἧς καὶ σὺ φαίνῃ δεκάδος E.Supp. 219.'

†δεκάσημος, ον, consisting of ten units, in prosody, of short syllables, Mar.Vict. p. 49 K.; in music, of time-units, Aristid.Quint. 1.14.

δεκαστάδιον, for 'race-course .. IG 4.951.79' read 'sign marking the distance of ten stadia, IG 4²(1).121.79'

*δεκαστάσιος, ον, worth ten times its weight, of gold as compared with silver, v.l. in Poll. 9.76, prob. in IG 1³.376.110, 122 (NC 10(1930).24).

*δεκάστομος, ον, ten-mouthed, Trag.adesp. 653 K.-S. (or [δω]δεκάστομος twelve-mouthed).

δεκαταῖος I, for 'for ten days' read 'of or appropriate to the tenth day, i.e. having been in that condition for nine days' and add 'Gal. 9.935.14'   add 'III of or belonging to a tithe (δεκάτη), rest. in Call.fr. 186.3 Pf.'

δεκατάλαντος, line 2, after 'Ar.fr. 276' insert 'neut. as subst., Poll. 9.54'

*δεκαταρχέω, be a δεκάταρχος, SEG 33.683 (Paros, ii/i BC, new reading of IG 12 suppl. 210).

*δεκαταρχίς, ίδος, fem. adj., of a decurio, δεκαταρχίδι τειμῇ GVI 730 (Palestine, iii AD).

δεκάταρχος, add 'SEG 4.594.18 (Colophon, Rom.imp.)'

δεκατέσσαρες, s.v. δέκα, add 'δεκατέταρες Lang Ath.Agora XXI He 2'

δεκάτευμα, for 'Call.Ep. 40' read 'Call.Epigr. 39 Pf.' and add 'AP 6.290 (Diosc.)'

δεκάτευσις, for 'decimation' read 'offering of every tenth man to a god'

†δεκατευτής, οῦ, ὁ, tithe-farmer, Antipho fr. 10, IG 2².1609.97 (Athens), 7.2227.4 (Boeotia).

δεκατεύω I 1, add 'med., give as tithe, ἄσσ' ἀπὸ λικμητοῦ δεκατεύεται AP 6.225 (Nicaenet.)'   II 2, after 'divide into ten sections' insert 'or (as Lat. decuriare) organize into units (v. Athenaeum 46.273ff.)'   III, for 'D.ap.Harp.' read 'Did.ap.Harp. s.v. δεκατεύειν'

δεκατηφόρος, Dor. δεκαταφόρος, add 'Cypr. te-ka-ta-po-ro-se Kafizin 117b, al.'

δεκατισμός, add 'Cypr. te-ka-[ti]-si-mo-i δεκατισμοῖ prob. collection of tithe, Kafizin 266b.1'

δέκατος I, add 'τόκος δ. 10% interest, SEG 33.1041.27 (Aeolian Cyme, ii BC)'   II 1, (δεκάτη) after 'tithe' insert 'Eumel.fr. 12.1 B., τῆι Ἥρηι ἀνέθεσαν δεκάτην ἔρδοντες SEG 12.391 (Samos, vi BC)'   5, add 'SEG 24.486 (Maced., AD 114)'

*δεκατός, ή, όν, tithed, i.e. subject to payment of one-tenth of one's substance, SEG 9.72.34 (Cyrene, iv BC).

†δεκατόω, take tithes of a person, Lxx 2Es. 20.37(38), Ep.Hebr. 7.6; pass., ibid. 7.9.

*δεκατωνέω, act as tithe-farmer, TAM 5(1).195 (i/ii AD).

δεκατώνης, add 'TAM 2(1).1.19 (ii BC), Poll. 6.128'

δεκάχους, ουν, holding ten χόες, Arist.Ath. 67.2.

*Δεκέμβριος, ὁ, December, ἐν τῷ Δεκεμβρίῳ Plu. 2.272d; as adj., Δεκεμβρίαις εἰδοῖς id. 2.287a, Δ. καλανδῶν Jul.Ep. 376a, μεὶς Δ. SEG 25.346 (Corinth), 33.1270 (Palestine, vi AD); Δεκάβριος PLand. 654.3 (APF 1958.14).

δεκέτης, line 1, for 'ου, ὁ' read 'ες'; add 'δεχέτης GVI 678.8 (Crete, iii/ii BC)'

δεκήρης, for 'with ten banks of oars' read 'prob. having ten oarsmen to each "room"'

*δέκο, v. °δέκα.

*δεκόβολον, v. °δεκώβολον.

δεκουρία, ἡ, Lat. decuria, ἐπίλεκτον κριτὴν ἐκ τῶν ἐν Ῥώμῃ δεκουριῶν IGRom. 3.778.9 (Attalea).

*δεκουρίων, ωνος, ὁ, member of municipal council, IGRom. 1.499 (Lilybaeum).   2 cavalry officer, IGRom. 3.1231 (Arabia), 1.1336 (Egypt), 4.1156 (Lydia), al.   3 foreman of a slave household, IGRom. 4.102.

*δέκρετον, τό, Lat. decretum, decree, Just.Nov. 67.4, al.

δεκτήρ, after 'receiver' insert 'of offerings, etc.'; add 'θυμάτων δεκτῆρα βωμόν inscr. in SHAW 1971.2, no. 3 (Sparta, Rom.imp.)'

δέκτης, add 'epigr. in SEG 39.443.36 (Tanagra, v AD).   II heir, IG 9(2).522.27 (Larissa, iii/ii BC).'

δεκτικός 1, add 'b welcoming, hospitable, πηγὴ ζ[ω]τικὴ, δεκτικὴ BCH suppl. 8 no. 103 (v AD).'

δέκτρια, after 'fem. of δεκτήρ' insert 'receiver, welcomer'; for 'Archil. 19' read 'FGE p. 149 P. (ps.-Archil., iii BC or later)'

δεκώβολον, for pres. ref. read 'IG 2².1537.23 (δεκώβολ[ον], iii BC)' and add 'PCair.Zen. 111.12 (δεκόβολον, iii BC)'

δέλεαρ, add 'see also βλῆρ'

δέλλει, after 'leg. βάλλει' insert 'La.'; add 'cf. ἐσδέλλω s.v. ἐκβάλλω, ζέλλω ἔζελεν'

δέλλις (B), for 'Annuario 3.144 (Pisidia)' read 'SEG 2.710.13 (Pednelissus, i BC)'

*δελματικομάφόριον, etc., v. °δαλμ-.

δέλτα I, after 'δέλτα' insert 'Achae. 33'

*δελτικός, ή, όν, app. recorded on a tablet, [δημοσί]ᾳ δελ[τ]ικῇ διαθ[ήκῃ IEphes. 2061 ii 22 (Rom.imp.).

δελτογράφημα, for 'official receipt' read 'official document provisionally recorded on a wooden or wax tablet' and add 'Chiron 19.133.30 (Sardes, i BC), ABSA 59.22 no. 16 (Aphrodisias, Rom.imp.)'

δέλτος (B), line 1, for entry in parentheses read '(δάλτος Cypr. = δέλτος, ta-la-to-ne δάλτον ICS 217.26)'   I, add 'δ. Διός, of tablet on which men's offences are conceived of as recorded, A.fr. 281a.21 R., E.fr. 506, Luc.Merc.Cond. 12 (pl.)'

*Δελφαῖος, ὁ, epith. of Apollo, IGC p. 11.13 (Larissa, ii BC); see also °Βελφαῖον.

*δελφάκτιν, unexpld. wd. in Choerob. in Theod. 1.267 H.

†δέλφαξ, ακος, ὁ, ἡ, (cf. Ath. 9.375a) pig, acc. to Ar.Byz.ap.Ath. l.c. full grown, opp. χοῖρος; masc., S.fr. 671, Cratin. 155 K.-A., Epich. 100.4, Pl.Com. 118 K.-A., Sopat. 5, Anaxil. 12 K.-A.; fem., Hippon. 145 W., Hdt. 2.70, Ar.fr. 520.6 K.-A., Eup. 301 K.-A., Theopomp.Com. 49 K.-A., Arist.HA 573ᵇ13, IG 2².1367.7.

*Δελφικός, v.s.v. Δελφοί.

*δελφνειος, α, ον, of the dolphin, στέαρ Cyran. 108 (sp. -νιος 4.17.6 K.).

*δελφινέλαιον, τό, oil of dolphin, Cyran. 4.15.3 K.

*δελφινηρός, όν, abounding in dolphins, A.fr. 150 R. (s.v.l.).

δελφινοφόρος I, delete the section (v. °δελφινηρός).   II, add 'Poll. 1.85'

δελφίς II 2, for 'Opp.H. 3.290' read 'sinker placed in mouth of dead fish used as bait, Opp.H. 3.290, 4.81'

Δελφοί, at end delete entry for Δέλφιος.

*δέλφον, word alleged to be old Greek term for 'one', in etym. of ἀδελφός, Numen.fr.54 P.

*Δέλφυνα, ἡ, name for the serpent killed by Apollo, Call.fr. 88 Pf., also (?)Δελφύνης, ὁ, ibid.; Δελφύνη, ἡ, A.R. 2.706, Nonn.D. 13.28.

δέμα I, add 'b bundle, BGU 2208.15 (AD 614).'

δέμας I 3, for ' = πόσθη, Pl.Com. 173.10' read 'flesh, as euphem. for penis, Pl.Com. 189.10 K.-A.'

δεμάτιον, after 'δέμα' insert 'bundle, Edict.Diocl. 6.37'

δεμελέας, after 'δεμβλεῖς' insert '(δεμελεῖς La.)'

δέμνιον, add 'Myc. de-mi-ni-ja (pl.)'

*δεμνιοπετής, ές, taking to one's bed, Nic.Dam.fr. 44 J.

δέμω, after 'pres.' insert '(IEphes. 3.8 (iii BC))'; at end add 'Myc. fut. part. de-me-o-te δεμέοντες; cf. compds. na-u-do-mo, to-ko-do-mo (v. °ναυδόμος, °τοιχοδόμος)'

*δενδραχάτης, ου, ὁ, dendritic agate, Plin.HN 37.139, Orph.Lith.Kerygm. 3.1 G., Socr.Dion.Lith. 41; paraphrased as ἀχάτης δενδρήεις Orph.L. 236.

δενδρήεις I 2, after 'ἀχάτης' insert 'dendritic agate'

δενδρίτης, after 'Mart.Cap. 1.75' insert '(sc. λίθος) an Indian stone resembling coral, Cyran. 1.4.2 K., al.   II, delete the section

δενδρῖτις, add 'fem. subst., name of precious stone, Plin.HN 37.192'

*δενδροβάτης, ου, ὁ, acrobat who had to climb trees in the amphitheatre to escape wild beasts, Gloss. (Lat. arborarius).

*δενδρόκλων, ωνος, ὁ, twig, Socr.Dion.Lith. 19 G.

δενδρολίβανον, for 'Gp. 11.15 tit.' read 'Gp. 16.1'

δενδροτομέω, delete 'but usu.' to end of article and substitute 'as a hostile action, Th. 1.108.2; humorously, of a flogging, δ. τὸ νῶτον Ar.Pax 747'

δενδροτόμος, add 'ZPE 96.53 (Egypt, v/vi AD)'

δενδροφόρος, for 'Theodor.ap.Ath. 14.621b' read 'Sotad. 2'

δενδρύω II 2, for 'Plot. 3.4.1' read 'Plot. 3.4.2'

†δενδρυάζω, dive and remain under water (associated by popular etymology w. δρῦς), Eust. 396.29, Hsch., EM 256.4; fig., app. of constricted type of voice production, Ael.Dion. δ 7 E.

δενδρύφιον, after 'Dsc. 1.108' insert 'in statuary, Inscr.Délos 1416Ai91 (ii BC)'; for 'toy tree' read 'part of a working model made by Hero'; at end for 'Thphr.HP 4.7.2' read 'Thphr.HP 4.7.3'

†δενδρύω, dive under water, IG 4²(1).122.20 (Epid.); cf. °δενδρυάζω, δρυάσαι s.v. δρυάζειν, δρύεται.

*δενδρώτης, app. masc. adj. wooded, Theognost.Can. 44.

Δενδυρίτης· κροκόδειλος Hsch. (La.); (app. = Egyptian ethnic Τεντυρίτης).

δεξάριον, τό, perh. some kind of female ornament, δεξάρια ζυγὴν [μίαν] *PSI* 183.6 (v AD).

δεξίαμα, ατος, τό, v.l. for δεξίωμα, E.*fr.* 324.1 (*PRoss.-Georg.* 1.9), S.*OC* 619.

δεξιοκοιτέω, *sleep on the right side*, *Physici* ii p. 195.

δεξιόομαι I, at end after 'E.*Rh.* 419' insert 'Men.*Dysc.* 948'

δεξιός I, at end after 'δεξιά as adv.' for 'Plb. 3.82.9 (s.v.l.)' read 'E.*Hipp.* 1360'   **IV**, after 'Pl.*Hipparch.* 225c' insert 'τὴν καλλίπαιδα, τὴν τρόπον [τε] δεξιάν epigr. in *SEG* 36.293 (Athens, iii AD)'   at end add 'Myc. *de-ki-si-wo*, pers. n.'

δεξιόχωλος, ον, *lame in the right leg*, (*dexiocholus* in Lat.), Mart. 12.59.9 (s.v.l.).

δέξις I, add 'Pl.*Lg.* 761d'

δεξίωμα II, after 'S.*OC* 619' insert '(v.l. °δεξίαμα)'

Δεξίων, ὁ, title said to have been given to Sophocles after his death, ἀπὸ τῆς τοῦ Ἀσκληπιοῦ δεξιώσεως *EM* 256.7.

δέον, add 'τὰ δέοντα *necessaries, maintenance*, *PMich.*x 587.8 (AD 24/5), *UPZ* 42.45 (AD 162)'

δέος, line 12, after 'Th. 3.33' insert 'folld. by τοῦ μή w. inf. Hld. 1.33.4'

δέπας, line 3, after 'Stes. 7' insert 'A.*fr.* 74 R.'; add 'form *δίπας, Myc. *di-pa* (*di-pa-e*, dual), *vessel with up to four handles*'

δέπυρα, v. °γέφυρα.

δεράγχη, for '*collar*' read '*noose*'; for 'Antip.' read 'Antip.Sid.'

δεραγχής, add 'see also °δειραγχής'

δέραιον, collar, add 'Arr.*Cyn.* 5.8'

δεραιοπέδη, for '*collar*' read '*kind of snare for birds*'

δεραιοῦχος, ον, *of a noose, neck-holding*, *AP* 7.473 (Aristodic., cj.).

δερβιστήρ, add 'cf. δεριστήρ'

δέριον· *torquis*, Charis. p. 40 K. (var. of δέραιον).

δεριστήρ, add 'cf. δερβιστήρ'

δέρκηθρον, τό, unexpld. wd., Ph.Epic. in *Suppl.Hell.* 683.2.

δέρκομαι, line 1, after 'Hsch.' insert 'epigr. in Didyma 567.1 (ii/i BC) and perh. in Pi.*fr.* 52w(i) S.-M.'   **I 2**, add '**c** *of mental perception*, τὴν σὺ νόῳ δέρκευ Emp. 17.21 D.-K.'

δέρμα, after lemma insert 'Delph. δάρμα q.v.'

δερματηρά, add 'ὠνὴν δερματηρᾶς *PColl.Youtie* 32.7 (AD 199)'

†δερμάτίκιον, v. ‡δαλμ-.

δερμάτιον, add '*PCair.Zen.* 353.15 (iii BC)'

δερμάτόκολλα, ἡ, *glue extracted from skins*, Anon.Alch. 380.10.

δερμορράφιον, τό, *large kind of needle*, Sch.Ar.*Pl.* 301 (*Mnemos.* (ser. 4) 10.58).

†δέρρις, εως, ἡ, also δέρσις Th. 2.75.5, *curtain, screen of skin or hide*, later of any material, Eup. 357 K.-A., Pl.Com. 267 K.-A., Myrtil. 1 K.-A., Lxx *Ex.* 26.7, al., *IG* 5(1).1390.35 (Andania, i BC).   **2** *hung before fortifications to check missiles*, Th. l.c., Ph.*Bel.* 95.34, *AP* 12.33 (Mel.), D.S. 20.9, Apollod.*Poliorc.* 142.2, Polyaen. 3.11.13.   **b** *on shipboard as a protection against waves*, Cic.*Att.* 4.19.1.   **II** *cloak* or sim., perh. fr. similarity to Hebr. *aderet*, Lxx *Za.* 13.4.

δέρω, line 7, after 'Il. 23.167' insert 'of Marsyas, δεδαρμένον Timocl. 19.1 K.-A.'

δέσις I, add 'also perh. *soldering*, *MAMA* 8.430'   **III**, delete the section.

†δέσκαλος, v. °διδάσκαλος.

δέσμευσις, εως, ἡ, *binding*, τοῦ χόρτου *PCair.Zen.* 180 (iii BC).

δεσμεύω II, delete the section.

δέσμη 2, add '*PCair.Zen.* 782 b.5, *PZen.Col.* 113.12 (both iii BC), *PYale* 38.8'

δεσμίδιον, for '(= δέσμη 2b)' read '*small bundle*' and add 'Erasistr.ap.Gal. 11.215.5'

δεσμός I **2**, sg. collect., add 'A.*Pr.* 97, *fr.* 190 R., E.*Hipp.* 1237, (*the state of*) *being bound*, Pl.*R.* 378d, 390c'   **4**, add 'perh. also Call.*fr.* 191.41 Pf. (v. addenda)'   at end add 'Myc. *de-so-mo*, prob. *belt for sword*'

δεσμοφυλακέω, *to be a gaoler*, *PCair.Zen.* 354.11 (iii BC).

δεσμοφύλαξ, after '*gaoler*' insert '*PTeb.* 777.5, 791.17 (pl.) (both ii BC), *Dura* [7,8] p. 171 no. 875 (ii AD)'

δέσποινα 3, add 'pl. Δέσποιναι, of Demeter and Kore, Paus. 5.15.4, 10; *IG* 5(1).230 (Sparta)'

δεσπόσυνος II **1**, add '*AP* 12.169 (Diosc.), *Laodicée* p. 362 no. 19 (i AD); of a praetor, Bernand *IMEG* 24.8 (Egypt, ii AD)'

δεσποτεύω, add '**b** *rule as bishop*, *Ann.Épigr.* 1971.454 (Lesbos, v AD).'

δεσπότης I **1**, line 2, after '*lord*' insert 'Tyrt. 7.1 W., Sapph. 95.8 L.-P., etc.'; line 4, after '(A.)*Pers.* 169' insert 'δέσποτα πάτερ *POxy.* 3356.13 (i AD)'   **3**, add 'θεῷ δεσπότῃ Πλούτωνι *EAM* 15 (ii AD)'

δεσποτικός I **2**, *imperial*, add 'ἐπίτροπος χωρίων δεσποτικῶν *IG* 10(2).1.351.3 (Thessalonica, iv AD).   **b** neut. subst., *some form of tax or payment*, *JÖAI* 1 *Beibl.* 115; pl., *CPR* 7.26.20 (vi AD).'

δεσπότις, add 'epith. of the nymph at Kafizin, *Kafizin* 9 (iii BC), al.'

δεῦμα, delete the entry.

δεῦρο, after 'Aeol. δεῦρυ' insert 'prob. in Sapph. 2.1 L.-P., cf.' line 3, delete 'late δευρεί .. iv/v BC' (v.*Tyche* 7.225)   **I 2 d**, delete the section transferring quot. to section 2 c.   at end add 'in Myc. compd. *de-we-ro* (*-a₃-ko-ra-i-ja*), name of district'

δευσοποιός, after 'δ. φάρμακα' insert '*Trag.adesp.* 441 K.-S.'

δεύτατος, delete 'prob. f.l. in' and after 'Pi.*O.* 1.50' insert '*AP* 5.108 (Crin.)'

δεῦτε, line 4, delete 'rarely'; for 'Sapph. 60, 65' read 'Sapph. 53, 128 L.-P., *Lyr.adesp.* 286 ii 8 S.'; line 5, after 'E.*Med.* 894' insert 'and Com., δ. δή Men.*Dysc.* 866'

δευτεραγωνιστέω, for '*play second-class parts*' read '*act in the second role in a play*'

†δευτεραγωνιστής, ὁ, *actor who takes the second role*; transf., *one who speaks second in support of a case*, D. 19.10, Luc.*Peregr.* 36, Hsch.

δευτεραρχέω, *to be the second magistrate*, prob. in *JÖAI* 23 *Beibl.* 121 ([δευ]τερ-, Thrace, Rom.imp.).

δευτερεύω, add '**b** *hold the position of second-in-command* or *deputy*, *UPZ* 2.159.13 (iii BC).'

δευτεροβόλος, for '*shedding the teeth a second time*' read '*having shed the second group of teeth, i.e. the medial incisors*'

δευτερογαμέω, *marry for the second time*, Just.*Nov.* 2.2.1, al.

δευτερόγαμος, ον, *married for the second time*, Just.*Nov.* 137.1.

δεύτερος I **2**, line 6, for 'X.*Cyr.* 2.2.1' read 'X.*Cyr.* 2.2.2'   **II 1**, line 3, for 'very much *behind*' read '*easily second to*'; line 5, delete 'δ. παιδὸς σῆς E.*Tr.* 618'   **2**, neut. as adv., *twice*, add 'Just.*Nov.* 137.4'

†δευτερόω, *do a second time, repeat*, περὶ δὲ τοῦ δευτερῶσαι τὸ ἐνύπνιον Lxx *Ge.* 41.32, λόγον *Si.* 7.14, ὁδόν, i.e. *go back*, *Je.* 2.36; ellipt., 1*Ki.* 26.8, ἐὰν δευτερώσητε, ἐκτενῶ τὴν χεῖρά μου ἐν ὑμῖν 2*Es.* 23.21.   **II** *to be second*, dub. in Lxx *Je.* 52.24 (v.l. δευτερευ-).

δευτήρ, for 'kettle, cauldron, Demiopr.' read '*vessel for mixing a dry mass with liquid, Demiopr.*' and for 'cf. δεῦμα' read 'cf. δεύω (A) I 2'

δεύω (A), add '*Inscr.Perg.* 8(3).161 (ii AD)'

δεύω (B) **I**, at end after '(Eresus)' insert 'Alc. 117(*b*).30 L.-P.' and for 'Alc.*Oxy.* 1788.15 ii 3' read 'id. 119.5 L.-P.'   **II**, line 2, after 'Od. 6.192' insert 'part. (athematic inflexion) δεύμενος Sophr. 36, *BCH* 59.38 (Crannon)'; add 'οὐδέ τι θυμὸς ἐδεύετο δαιτὸς ἐΐσης Il. 1.468, al., Od. 16.479, al.'

δεχεπτά, add '*SEG* 33.1475 (Cyrenaica, i/ii AD); cf. δεκαεπτά s.v. δέκα'

δεχέτης, v. °δεκέτης.

δέχομαι, line 1, δέκομαι, add 'also Dor. *SEG* 30.1123 (Sicily, iii BC); Arc., *SEG* 37.340.10 (Mantinea, iv BC)'; line 2, for 'impf.' read 'aor.'; line 10, imper. δέξο, add '*h.Merc.* 312, Orph.*H.* 18.3'; line 11, inf. δέχθαι, add 'Il. 1.23'   **I**, line 6, *accept as legal tender*, add '*SEG* 26.72 (Athens, 375/4 BC)'   **II 1**, add '*receive sexually*, πρὸς ἔργον ἀφροδίσιον ἐλθόντ' ἑταῖρον ὀντινῶν ἐδέξατο Semon. 7.49 W.'   at end add 'Myc. aor. med. (*o-*)*de-ka-sa-to*, also *de-ko-to*, cf. δέκτο'

δέω (A), line 3, after 'And. 4.17 (prob.)' insert 'intr. pf. δέδηα, *to be bound*, Nic.*Al.* 436'; line 10, after 'Din.*fr.* 89.15' insert 'v. °καταδέω'   at end add 'Myc. pf. part. pass. *de-de-me-no*'

δέω (B), line 6, after 'ἰδεϝ' insert 'cf. δεύω (B) I'   **I 2**, add 'w. fut. inf. Pl.*Ap.* 37b'   **II**, delete 'part. δεύμενος id. 36' (v. °δεύω (B))   **1 b**, add 'w. acc. and inf., εἰ μὲν γὰρ μηδέν δέονται χρηστοὺς αὑτοὺς εἶναι Isoc. 11.43'

δή, line 6, after 'A.*Th.* 214 (lyr.)' insert 'also δὴ τότε δή, Opp.*C.* 2.271, Q.S. 10.224, Orph.*A.* 1270, Nonn.*D.* 22.299'

δῆγμα I, add '**2** *tooth* of a fish-hook, Opp.*H.* 4.444 (pl.).'

δηγμός 2, after 'Thphr.*HP* 4.4.5' insert 'Nic.*Al.* 119 (sp. δηχμός)'

δηθά, after '(Il.) 5.587' insert 'Od. 1.49, 120, 2.255, al.'

δῆθεν II, add 'Luc.*Alex.* 39, Ach.Tat. 2.1.1, Aen.Tact. 11.4, *Cod.Just.* 1.3.44.5, etc.'

δηιόω, line 12, delete 'said by Sch. .. Eumel.(*fr.* 9)' and insert 'δήϊον *AP* 6.122.4 (Nic.)'

δηκόκτα, ἡ, Lat. *decocta*, *decoction*, Gal. 10.467.17.

δηλάβρα, ἡ, fr. Lat., *winnowing-fan*, δ. ἤτοι πτύον *Edict.Diocl.* 15.44; cf. πτύον· *ventilabrum, delabrum*, *Gloss.* 2.425.47, *delabra: ption* ib. 522.25.

δηλαδή, add 'D.Chr. 9.13, *Cod.Just.* 1.3.52.15, etc.'

δηλάτωρ, ὁ, Lat. *delator, informer*, *BCH* 59.152 (Philippi).

δηληγατεύω, add '*SEG* 32.1554.A49, 56 (Arabia, vi AD), Just.*Nov.* 130.5 (cf. Lat. *delegatio, delego*)'

δηλητήριος, δ. τό, *poison*, add 'M.Ant. 6.36.2'

δηλητηριώδης, after '*noxious*' insert 'Zos.Alch. 201.15'

Δήλιος, add '**V** Δάλιος (sc. μήν) ὁ, name of a month, *Inscr.Cos* 30.1, 367.55, *Lindos* 465 i (ii AD), *Tit.Calymn.* 88.46, etc., *SEG* 25.1110 (Cyprus, Hell.).'

*Δηλίτης, καὶ Δηλίτης ὁ εἰς Δῆλον ἐρχόμενος χορός, Καλλίμαχος St.Byz. (s.v. Δῆλος).

δῆλος **III**, for 'Urim' read 'symbols of revelation of the Urim' and after 'Lxx' insert 'Nu. 27.21'

δηλόω, after lemma insert 'Arc. aor. inf. δεΓαλῶσαι SEG 37.340.21 (Mantinea, iv BC)' **II 2**, line 2, for 'δηλώσει .. R. 497c' read 'so perh. δηλώσει Pl.R. 497c'

δήλωσις **3**, for 'Urim' read 'symbol of revelation of the Urim' and add '(Lxx) Ex. 28.30, 1Es. 5.40'

*Δημαρχεξούσιος, ὁ, name of month (Apr./May) in an Augustan calendar from Cyprus, Madrid MS. in JÖAI 8.112 (-εξάσιος), IGRom. 3.930 (Soloi, prob. i AD); named Δήμαρχος acc. to Hemerolog.Flor., v. YCLS 2.213.

δημαρχία, add '**III** [θ]εοί· δημαρχία ἡ μέζων, heading of an Attic sacrificial calendar, precise significance disputed, Sokolowski 3.18 (Erchiae, iv BC).'

δήμαρχος **1 b**, add 'in Egypt, Hdt. 3.6' **2**, after 'plebis' insert 'SIG 601 (Teos, 193 BC), SEG 31.1300 (Lycia, ii AD)' add '**II** v. °Δημαρχεξούσιος.'

δηματρεύεσθαι, add '(δημαστ- La.; app. corruption of δηθὰ στρεύγεσθαι Il. 15.512)'

*δημεκδικέω, to be a °δημεκδικος, ACO 3.102.19 (vi AD).

*δημέκδικος, ὁ, = Lat. defensor civitatis, public advocate, PMasp. 353ʷ.A26 (vi AD).

δημεύω **I**, pass., later of persons, delete 'later' and insert 'αἴ τις .. ἐ δικάσ[ζοι] ἐ δικάσζοιτο .. τρέτō καὶ δαμευέσσθō ἐν Ἀθαναίαν .. have his property confiscated for the benefit of Athena, IG 4.554.5 (Argos, vi/v BC)' **III**, add 'τὰ κοινὰ καὶ δεδημευμένα Plu. 2.243d'

δημεχθής, for 'Call.fr. 472' read 'Call.fr. 486 Pf.' and add 'Trag.adesp. 337a K.-S.'

δημηγορικός, for 'suited to public speaking .. X.Mem. 1.2.48' read 'of, concerned with or typical of public (political) speaking, οὐχ ἵνα δημηγορικοὶ ἢ δικανικοὶ γένοιντο X.Mem. 1.2.48'; line 3, delete 'popular'

Δημήτηρ, line 1, after 'Arc.' insert 'Cypr.' **1**, add 'used in pl. for Demeter and Kore, Δαματέρων καὶ Διὸς Δαματρίου Lindos 183 (ii BC), Δαμάτερσιν οἶν κνεῦσαν ib. 671 (Camirus, i BC)'

*Δημητρίειος δραχμή, coin minted by Demetrius Poliorcetes, Inscr.Magn. 33.20.

Δημητριών, after 'SIG 380' insert 'at Chalcidice, SEG 37.576.5 (?Polichne, iv BC)'

*δημήτρουλος, ὁ, hymn in honour of Demeter, Semus 23 J.; cf. οὖλος (D), καλλίουλος.

δημιοπρᾶτα, before 'Ar.V. 659' insert 'δε̄μιόπρατα IG 1³.85.3'; add 'Ion. -πρητ- Inscr.Délos 72 (v BC)'

*δημιοπράτης, v. ‡δημοπράτης.

δήμιος, line 1, delete 'δημίην .. (Cypr.)' **I**, add 'advs. -ίως, -ιωστί, Theognost.Can. 160, Eust. 1899.57. **2** οἱ δάμιοι, magistrates at Dreros, Meiggs-Lewis 2 (Dreros, vii/vi BC).' **II 1**, add 'OAshm. Shelton 51' add '**3** δημίην Κύπριν· πόρνην Hsch.La. (cod. δημίην· πόρνην. Κύπριοι); cf. Δημίασι πύλαις· κοιναῖς, ἐπεὶ προεστήκεσαν ἐν ταῖς πύλαις αἱ πόρναι Hsch. (for other expls. v. Hsch. s.v.).' **III**, for this section read 'vulgar, coarse, δήμια λαβράζουσι Nic.Al. 160' at end add 'form δάμιος, Myc. da-mi-jo'

δημιουργέω, after lemma insert 'dial. δαμιοΓοργέω, δαμιοργέω, Cyren. δαμιεργέω' **II a**, hold office of δημιουργός, add 'ἐπὶ τōνδεōνὲν δαμιοργōντōν SEG 11.314 (Argos, vi BC); κὰ ὄζις τότε δαμιοΓοργε͂ Dubois Dial.arcad. p. 196 (Pheneus, c.500 BC); ἐν τῶι ὑστέρον Fέτ[ε]ι ἢ Νικῆς ἐδαμιόργη SEG 37.340.23 (Mantinea, iv BC); athematic part. δαμιεργέντων SEG 9.11.2, 12.2 (Cyrene, iv BC)'

δημιούργημα, line 3, for 'Jul.Or. 2.54b' read 'Jul.Or. 2.54a'; line 5, for 'Ph. 1.208' read 'Ph. 1.207'

δημιουργία **II**, add 'SEG 30.1518 (Side, iii AD)'

*δημιουργίς, ίδος, ἡ, (sc. ἀρχή) office of δημιουργός, ὑπὲρ τῆ[ς] δ[η]μιου[ρ]γίδος MAMA 3.103; see also °δαμιοργίς.

δημιουργός **I 1**, line 6, after 'Pl.R. 529e' insert 'Call.Dieg. vii.30 (fr. 196 Pf.)'; line 7, delete 'Men. 518.12 (fem.)' and after 'Alexandr.Com. 3' insert 'esp. fem. of the bridal attendant who made wedding cakes, Men.fr. 451.12 K.-Th., Poll. 3.41, Sch.Ar.Eq. 647' **II**, δαμιοργός, add 'Locr., IG 9²(1).609.22 (Naupactus, c.500 BC), 717.15 (Chalium, v BC), Arc., SEG 37.340.20 (Mantinea, iv BC); also of women, IGRom 3.794, 802 (Pamphylia)' **III**, delete the section.

*δημιωστί, v. °δήμιος.

*δημογραμμᾶτεύς, έως, ὁ, title of local government official, Cod.Just. 10.71.4.

δημοεγερτής, delete 'Suid.' and substitute 'Suet.Blasph. 83 T.'

δημόθεν **I 2**, after 'people' insert 'Call.fr. 93.15 Pf.' and add '**III** from the town, AP 9.316.2 (Leon.Tarent.).'

δημοθοινέω, add 'SEG 30.340 (Aegina, ii AD)'

δημοθοινία, add 'sp. δημοθυνία IStraton. 202.19, 266.23, al.'

*δημοκατάρατος, ον, cursed by the people, Thd.Pr. 11.26.

*δημοκήρυξ, 'υκος, ὁ, public herald, MAMA 4.351 (Eumeneia).

δημόκοινος **1**, add 'transf., ἀνδροφόνων καὶ δημοκοίνων θίασος Ph. 2.559'

*δημόκομπος· στωμύλος Suid. (perh. -κόπος).

*δημοκόπιον, τό, = δημοκοπία, POxy. 2400.5 (iii AD).

*δημοκοπίς, ίδος, ἡ, ἣν δὲ καί τι ὑπόδημα, δημοκοπίδες Poll. 7.89 (s.v.l.).

δημοκόπος, for 'Ph. 2.47' read 'Ph. 2.520'

δημοκρᾰτία **II**, for 'Paus. 1.3.2' read 'Paus. 1.3.3'

†Δημοκρίτειος, ον, of Democritus of Abdera, ἡ Δημοκρίτειος φιλοσοφία S.E.P. 1.213; masc. subst., follower of Democritus, Ael.VH 12.25; pl., Plu. 2.1108e.

δημοποίητος, add 'Aeschin.Ep. 12.13'

δημοπράτης, add 'vv.ll. δημιοπράτης, δημοπράτα'

*δημόπρᾱτος, ον, sold as state property, Ph. 2.539; cf. δημιόπρατα.

δῆμος **III 1**, personified, add 'IHistriae 19.14 (ii BC), Ἡλίωι καὶ τῶι Δήμωι ICilicie 83 (ii/i BC), εἰς τὸ ἐράριν δήμου Ῥωμαίων Laodicée p. 325, Δήμῳ Ῥωμαίων SEG 30.1253 (Aphrodisias, i AD), IEryth. 32, 33, al., etc.' **IV**, add 'transl. Lat. vicus, IEphes. 4101 (ii AD)' at end add 'form δᾶμος, Myc. da-mo, prob. a body of people, community'

*δημοσάριος, ον, app. fr. *δημοσιάριος, belonging to the state, τὸ δημοσαράριων POxy. 3423.12.

δημοσιεύω **I**, add '**6** make publicly known, expose, Cod.Just. 1.1.3.3, Just.Edict. 7.2 pr.' **II**, before 'Ar.Ach. 1030' insert 'IG 1³.164.13 (440/25 BC)' add '**2** appear before the people, Just.Nov. 161.1.1.'

δημόσιος **I 1 a**, line 5, after '(Halic.)' insert 'οἱ δ. θεοὶ Sokolowski 1.79.3 (?Pednelissus, i BC)'; add 'stamped on vessels, weights, etc,. to denote public ownership, SEG 31.374 (v/iv BC), Ἀργείων δαμόσιον 36.332 (Nemea, Hellen.), etc.' **2**, add '= δημοτικός I 1, γράμματα IG 12(5).14.6 (Ios, ii/iii AD)' **III 1**, neut., add 'e pl., public lands, IG 9²(1).609.3 (Naupactus, c.500 BC).' **2 a**, add 'sg., SEG 33.679 (Paros, ii BC)' **b**, add 'Just.Nov. 128.10, al.' add '**c** public sacrifices or victims, IG 1³.35.12 (Athens, v BC), SEG 31.416.29 (w. rhotacism, δημορίōν, Oropus, iv BC), etc. **d** ?public affairs or business, SEG 30.380 (Tiryns, vii BC). **e** public baths, sg., R. Stillwell Antioch-on-the-Orontes (1933-6) no. 112; (-ιν), pl., CRAI 1945.379.9 (Berytus, Byz.).'

†δημοσιουργία, v. ‡δαμοσιοργία.

δημοσιόω, after lemma insert 'Elean 3 sg. opt. δᾱμοσιοίᾱ, pres. inf. δᾱμοσιώμεν Schwyzer 424 (Olympia, iv BC)'

*δημοσιωνικός, ή, όν, νόμος law relating to revenue-leases, SEG 14.639.A11, al. (Caunus, i AD), 35.1439.16 (Myra, Lycia, ii AD).

δημοσιώνιον, after 'office' insert '(building)' and add 'SEG 35.1439.5 (Lycia, ii AD)

δημοσίωσις, add '**2** engagement in public affairs, Vett.Val. 1.7.'

*δημόσῡνος, η, ον, = δημόσιος; -η prob. at public expense, IG 2².4658 (Athens, iv/iii BC).

δημοσώστης, add 'SEG 16.428.7 (Callatis, i AD), ITomis 19(b).5 (ii/iii AD)'

δημοτελής **I 2**, add 'SEG 33.679 (Paros, ii BC)'

δημότερος, after lemma insert 'perh. Cypr. ta-mo-te-ro-ne δαμοτέρōν (gen. pl.) Kourion 218 (v BC)' **I**, for 'poet. for .. 3.606' read 'of the land, γυναῖκες A.R. 1.783; abs., id. 3.606 (v.l. δημογέρουσιν); Dor. δᾱμο- Call.fr. 228.71 Pf.'

*δημότευκτος, ον, furnished by the people, λίθοι Sardis 7(1).181 (i AD).

δημοτεύομαι, for 'pass.' read 'med.' and add 'Dor. δᾱμ- rest. in BCH 50.17 vii 4 (Law of Cadys, Delphi, iv BC)'

δημότης **II**, add 'opp. ξένος, SEG 31.416.9 (= IG 7.235, Boeotia, iv BC)'

δημοτικός **II 1**, add '**b** τὸ -όν, the populace, Just.Nov. 102.2.'

*δημοφίλητος, ον, as honorific title, beloved of the people, ἀρχίατρος καὶ δ. IGBulg. 150 (Odessus).

δημοχᾰρής, add 'epigr. in ICilicie 49 (iii/iv AD)'

†δήμωμα, v. ‡δαμώματα.

δημωφελής, adv. -λῶς, for '(Iotapata)' read '(Iotape)'

*Δήν, Δῆνα, v. Ζεύς, lines 15/6.

δήν, line 1, within the parentheses add 'see also δαόν' **1**, (p. 388), delete 'cf. 16.736, δ. δὴ .. φίλοι ὦμεν Thgn. 1243' add '**3** perh. in local sense far away, Thgn. 494, 597; w. gen., far away from, Il. 16.736.'

δηναιός **I 1**, neut. as adv., add 'A.R. 1.334, 3.590' **II**, for this section read 'occurring, coming, etc., after a long time, late, δηναιοὶ .. εἰσαφίκοντο A.R. 4.645, (τὰ ἰρὰ) δηναιὰ πατήρ ἐκόμισσεν Maiist. 8, δηναιὸν ἀδελφεὸν .. νοστήσαντα Opp.H. 4.154'

δηνάριον, add 'IG 9(2).1092 (Thessaly); also δῑνάριον Inscr.Délos 1439Abc 83, 1441Ai89, 1449Aα ii23, 36 (ii BC); δειν- IG 9(2).1104.14 (Thessaly), etc. **2** collect., Roman money, δ. χρυσοῦν καὶ ἀργυροῦν Peripl.M.Rubr. 8, 49.'

*δηπόσιτον, τό, Lat. depositum, PSI 1063.5 (ii AD).

*δηποτᾶτος, masc. adj. Lat. deputatus, sent from the provinces to the

δήποτε **SUPPLEMENT** διαϊπνίτης

Emperor, ἑκατόνταρχον δε[σ]ποτᾶτον *centurio d.*, *IApam.* 8.5 (iii AD).
δήποτε 1, delete αἰεί δ. Th. 8.73; *at length* add '**b** *sometimes*, Nic. *Th.* 683, *Al.* 133, 383.' **2**, add '**b** αἰεί δ., *always, whenever it might be*, Th. 8.73.5.'
δηριάομαι, penultimate line, for 'οὐκ ἂν τοί' read 'οὐκ ἂν οἱ θηρῶν'
δῆρις, after 'A.*Supp.* 412' delete '(lyr.)'; after '(only in acc.)' insert 'Hes.*Sc.* 241, 251, Ibyc. 30(*b*) P., B. 5.111 S.-M.; of athletic contests, Hes.*Sc.* 306 (all acc.)'
*δησέρτωρ, ορος, ὁ, Lat. *desertor*, PFlor. 362.3 (iv AD).
*δησιστάχυς, υος, ὁ, *binder of corn*, orac. in Paus. 8.42.6 (cj.).
δῆτα, line 4, after 'answers' insert 'or in corroboration of what has been said'; line 8, after '(lyr.)' insert 'E.*El.* 673, 676'; lines 11/12, delete 'also .. cf. 676'
*δητύω, ἐδήτυεν· δεῖλεν, διήρει Hsch.
*δηφήνσωρ, ορος, ὁ, Lat. *defensor*, PLips. 35.12 (iv AD), PHerm. 19.4 (iv AD), 69.3 (v AD).
*δηχμός, v. ‡δηγμός.
*Δηωΐς, ίδος, ἡ, *daughter of Deo or Demeter*, i.e. *Persephone*, Didyma 496*B*11 (ii AD); cf. Δηωΐνη.
διά, line 2, after 'anastroph.' insert '(but after its case in Hes.*Op.* 3, Tyrt. 4.2)'; lines 3/4, delete 'also .. which' **A I 1**, add '**b** *for a distance of*, φόρετρον δ. σχοίνων γ' *PMich.*III 145 III col.v.9 (ii AD); cf. 3, 5.' **III 2**, line 2, after 'thing is made' insert '*AP* 6.282 (Theod.), Plb. 14.1.15'; add 'in Marcell.Emp. 8.199, 210, al., collyria are named διὰ μίσυος, δ. χολῆς (diamisyos, diacholes), etc.' **IV a**, line 5, after 'δι' ἀπεχθείας γίγνεσθαι' insert '(i.e. *become an object of hatred*)'; add 'δι' εὐχῆς εἶναι *to be the object of one's prayer*, PBremen 20.6 (ii AD), Aegyptus 15.267 (ii/iii AD); διὰ σπουδῆς εἶναι Ael.*NA* 7.45, al.' **C**, for 'δ. πρό (v. supr. A I 1)' read 'δ. πρό (v. °διαπρό)'
διαβάλλω II, for 'αὐτῷ' read 'αὐτῇ' and for 'him' read 'it (the πόλις)' **III**, line 7, delete 'Hdt. 8.22'; line 9, after 'Isoc. 15.175' insert 'πρὸς τὴν ὠμότητα τοῦ υἱοῦ διαβληθείς Luc.*Macr.* 14, Arr. *Epict.* 2.26.3'; transfer '*to be brought into discredit* .. Lys. 7.27, 8.7' to section v 1 line 8 (after 'Pass.') **V 1**, lines 7/8, delete 'or πρός' and '*Macr.* 14' **VII**, delete the section.
διάβασις 2, delete '*bridge*'
διαβήτης 2, for 'cf. ib. 2.1054.10' substitute 'ὀρθοὺς πρὸς τὸν διαβήτην πανταχῆι Inscr.*Délos* 104-4.7 (iv BC)'
*διαβητικός, ή, όν, *made straight by rule*, Stud.Pal. 20.211.9 (v/vi AD, v. ZPE 71.117).
διαβήτινος, delete the entry.
*διαβήτρια, ἡ, a kind of priestess, Μητερείνη διαβητρίᾳ Περασίας *EA* 19.20 (Mopsuestia, ii AD).
†**διαβιάζομαι**, *force a way through*, τῶν φλεβῶν διαβιαζόμενον καὶ ξυνεπίστρεφον sc. πνεῦμα Pl.*Ti.* 84d; of plants penetrating the soil in germination, Thphr.*CP* 2.17.7; abs., *advance with force*, Lxx *Nu.* 14.44. **2** w. inf., *compel* to go, etc., E.*IT* 1365. **II** *overcome, master*, δ. τὴν ἀσθένειαν τῇ συνηθείᾳ τῇ πρὸ τοῦ Plb. 23.12.2.
διαβιβρώσκω, line 1, after 'Gal. 13.553' insert 'aor. 3 pl. διέβρον (?read -ων) Call.*fr.* 177.31 Pf.'
διαβοάω, line 5, for '(Plu.)*Per.* 9' read '*Per.* 19'
διαβουλία, delete the entry transferring refs. (both gen. pl.) to διαβούλιον I.
διαβρεχής, add '(cod., v.l. -βραχής, cj. -βραχείς, aor. part. of διαβρέχω).
*διαβρύκω or **-βρύχω**, *gnaw through* or *utterly devour*, πρῶτον γάρ μιν ἐλοῦσα γαλῇ μέσσον διέβρυξεν *PMich.*inv. 6946 (*ZPE* 53.13.6, Hellen. mock epic).
διάγγελμα, for '*message, notice*' read '*command*'
διάγγελος 2, delete the section transferring refs. to section 1.
διαγειτονία, delete the entry (v. °διαγονία).
διαγελάω 1, abs., add 'S.*fr.* 171'
διαγίγνομαι, line 9/10, delete 'he *was never anything* but a theorist' and preface quot. with '*spend one's life* doing'
*διαγκυλίζομαι, v. ‡διαγκυλόομαι.
διαγκυλόομαι, after '(v.l. -ισμένος)' insert 'cf. διηγκυλίσθαι Hsch.'
διαγλύφω 1, lines 3/4, for 'διαγλυφέντες .. athletes' read 'οἱονεὶ διαγλυφέντες καὶ διατορευθέντες, of men in good physical condition'
διαγνώμων, add 'subst. *arbiter*, *PMich.*XIII 659.56, al. (vi AD), κατὰ κρίσιν καὶ μεσιτίαν τοῦ κοινοῦ διαγνώμονος ib. 81, 99, *Cod.Just.* 8.10.12.7a, Just.*Nov.* 125.1'
διάγνωσις II, add 'διαγνώσεως = Lat. *in cognitione*, *Edict.Diocl.* 7.73'
διαγογγύζω, for pres. refs. read 'ἐπί τινα Lxx *Ex.* 15.24, *Nu.* 14.2; ἐπί τινος *Jo.* 9.18, *Si.* 31.24; κατά τινος *Ex.* 16.7, *Nu.* 16.11; πρός τινι Hld. 7.27; absol. Lxx *De.* 1.27, *Ev.Luc.* 15.2, 19.7'
*διαγονία, ἁ, some kind of family group, *ASAA* 2(1916).140 (iii BC), Maiuri *Nuova Silloge* 18, to be read in *IG* 12(1).922, *Clara Rhodos* 2.175 (ii BC), 203, *Lindos* 219.7 (ii BC), 454.9 (i AD).
διαγορεύω, line 5, after 'App.*BC* 1.54' insert 'w. ὥστε, Just.*Nov.* 58'
διάγραμμα I, add '**c** *enclosure*, ξένον μηδένα ἐντὸς τοῦ δ. πα-ρ[απ]ορεύεσθαι *BCH* 33.23 (Pontus, ii BC).' **III**, add '*BGU* 1053, 1054, etc.'

διαγραφάριος, add '*PKlein.Form.* 31 (vi AD)'
διαγραφή II, line 3, after '*description*' insert 'Longin. 32.5'
διαγράφω I 1 a, for 'Philostr.*VS* 2.2.7, *Her.* 2.1' read 'Philostr.*VS* 2.1.7, *Her.* 2.2' **IV**, after 'Ar.*Nu.* 774 (pass.)' insert 'διαγράφω τοὺς ἱππέας Ar.*Lys.* 676, ἡ .. λῆξις τοῦ κλήρου διεγράφη Is. 5.17' **V**, add 'pass., *be paid, receive money*, *SEG* 33.1039.54 (Aeolian Cyme, ii BC)'
διάγω I 1 a, add 'perh. also *take to court*, *Cod.Just.* 1.33.4, Just.*Nov.* 123.8, al.'
διαγωγή, line 2, after '*carrying across*' insert '*SIG* 135 (Olynthus, 393 BC)'
διαγωνίζομαι, line 4, after 'περί τινος' insert 'Aeschin. 1.132' and delete this ref., w. def. in line 6.
διάδεσμος, add 'Gal. 8.90.14, 11.181.4, 11.230.7'
διάδηλος, after 'Arist.*HA* 613ᵇ1' insert '*clearly visible, conspicuous*, *SEG* 23.207.39 (Messene, Augustus)' add '**2** *distinguishable as such, recognizable*, of the foetus in a miscarriage, *SEG* 9.72.107 (Cyrene, iv BC).'
διάδημα I, add 'φέρε τὸ διάδημα τὸ χάλκινον *OBodl.* i 262 (ii BC)'
*διαδιδασκαλία, ἡ, *teaching, education*, *PMasp.* 295 i 29 (v AD).
διαδιδράσκω 3, delete the section transferring ref. to section 1.
διαδικέω (A) 1, after 'Plu. 2.196c' insert 'D.C. 40.55' **2**, delete the section.
†**διάδικος**, ὁ, *opponent in a lawsuit*, Just.*Edict.* 7.1; cf. διάδικος (διὰ δίκης La.)· τὸ εἰς δίκην καλεῖν. Ἀττικοί Hsch.
διάδομα, for '*distribution of money*' read '*distribution of goods* or *money*' and add '*SEG* 32.1306.7 (Lycia, i AD), Didyma 360 (ii AD), *Kourion* 111.11 (AD 114)'
διαδοξάζω, after 'Pl.*Phlb.* 38b' insert 'Antisth.*Aj.* 9'
διαδοχή I 2, add 'dist. fr. κληρονομία, *Cod.Just.* 1.5.18.3, Just.*Nov.* 118.4. **b** *the assets of the deceased*, Just.*Nov.* 22.48 pr.'
διάδοχος 3, add '**b** dist. fr. κληρονόμος, *Cod.Just.* 1.3.55.4, 1.4.26.1.' **4**, w. dat., add 'neut. pl. as adv., διάδοχά σοι γόνυ τίθημι γαίᾳ *in turn*, E.*Tr.* 1307, *Andr.* 1200' **5**, for this section read 'abs. *taking over as a replacement*, πεντήκοντα τριήρεις διάδοχοι πλέουσαι Th. 1.110.4; also, *in succession to each other*, i.e. in relays, Hdt. 7.22.1; neut. pl. as adv., *in turn*, E.*Andr.* 200' **6**, add 'διάδοχος τῆς στρατηγίας *POxy.* 3975.3 (ii AD)'
*διαδρανής, ές, *very ineffective*, of planets, δ. εἰσὶν τοῖς ἀποτελέσμασιν *PMich.*III 149 xi.17 (ii AD); cf. ἀδρανής.
διαδρομή I 1, after 'Arist.*Mete.* 341ᵃ33' insert 'Ptol.*Tetr.* 102' **6**, add '*Cod.Just.* 1.3.45.1b'
*διαζευγίζω, *separate*, w. gen., ἣν .. μεσηλικίης διεζεύγισε .. Ἅδης *MAMA* 7.263.
διάζευγμα, for 'dub. sens. .. canal' read 'some sort of a connecting structure (cf. διάζωμα 4), *bridge, mole* or sim.' and add 'perh. spec. the *Mole* at the Piraeus, Thphr.*Char.* 23.2'
διαζεύγνυμι 1, line 8, after 'Pl.*Lg.* 784b' insert 'ἐὰν δὲ διαζυγῶσιν ἀλλήλ(ων) οἱ γαμοῦντ(ες) *POxy.* 3491.18 (ii AD)'
διάζευξις 1, after 'Pl.*Phd.* 88b' read '*separation, divorce*' and after '(Pl.)*Lg.* 930b' insert '*Cod.Just.* 1.3.52.15, Just.*Nov.* 22.14; *segregation*, ἡ δ. τῶν γυναικῶν in Crete, Arist.*Pol.* 1272ᵃ23'
†**διάζομαι**, *set the warp in the loom*, ἐδιάσατο τοὺς ἑπτὰ βοστρύχους τῆς κεφαλῆς αὐτοῦ (sc. of Samson) Lxx *Jd.* 16.14, *Is.* 19.10, Sch.Ar.*Av.* 4; fig., Aq.*Ps.* 138.13; app. in pass. sense, ὁ δ' ἐξυφαίνεθ' ἱστός, ὁ δὲ διάζεται Nicopho 13 K.-A. (s.v.l.).
διάζωμα, line 2, for '*girdle, drawers*' read '*loincloth*'
*διαζωμάτικός, ή, όν, of or *for a διάζωμα 3*, πίνακες *Inscr.Délos* 1442*B*42 (ii BC).
διαζώννυμι II, after '*encompass*' insert 'of the rainbow, (in tm.) ἔζωσε διά .. οὐρανὸν Ἶρις Arat. 940'
διάημι, add 'pass., *to be blown through*, Nic.*fr.* 74.41, see also ζάημι'
*διαθαλάμευσις, εως, ἡ, *seclusion in a chamber*, Eust. 782.48.
διάθεσις II 1 b, delete '*propensity*'; for 'πρός τινα Sch.E.*Hec.* 8' read '*partiality*, Favorin.*Exil.* 16.43 B., Sch.E.*Hec.* 886; in complimentary address, perh. *graciousness, condescension*, τὰ γράμματα τῆς σῆς ἱερᾶς διαθέσεως *CPR* 8.28.4, *PAmh.* 2.145.23, 26 (iv/v AD)'
διαθετικός, add '**b** unexpld. wd., perh. *relating to sale*, *PTeb.* 847 intr. (ii BC).'
διαθρυλέω I, add 'τοὺς πάλαι παρ' ὑμῖν διατεθρυλημένους (λόγους) Isoc. 15.55'
διαθρύπτω II 2, for this section read 'med., perh. *behave in an affectedly coy manner*, ἀ δὲ καὶ αὐτόθε τοι διαθρύπτεται Theoc. 6.15; of doctor's manner towards his patient, Gal. 17(2).148.12'
διαιθύσσω II, for 'B.*fr.* 16.4' read 'B.*fr.* 20B.8 (see also ‡αἰθύσσω)'
διαινέω, add 'δ. περί w. gen. perh. to be read in *IG* 5(1).1379.5 (Messenia, ii/i BC)'
διαίνω, add 'διάναι· σμῆξαι, πλῦναι Hsch.'
*διαιπετής, ές, *falling through*, ἀστὴρ ὠρανῶ δ. Alcm. 3 *fr.* 3 ii 67 P.
*διαϊπνίτης, ου, ὁ, app. *thoroughly oven-baked* (cf. ‡ἰπνίτης), διαϊπν[ίτης] τυρός *Sokolowski* 3.90.5 (Callatis, ii BC).

*διαίρᾱσις, v. °διέρασις.

*διαίρεμα, v. °διέραμα.

διαίρεσις **II**, *dividing, distribution*, add 'of land, Plb. 3.40.9'   **VI**, add '*division by punctuation*, Arist.*Po.* 1461ᵃ23'

διαιρετός **I 1 b**, for '*having divisions*' read '*set apart from each other*'

διαιρέω **II 1**, lines 6/7, for 'τοῖς δικάζουσι .. Lib.*Or.* 52.4' read 'οἱ παρακαθήμενοι .. τοῖς δικάζουσι διελόμενοι τὰ ὦτα, σφῶν δὲ μᾶλλον ἢ τῶν συνδίκων αὐτοὺς ἀναγκάζοντες ἀκροᾶσθαι, i.e. *dividing* their attention, Lib.*Or.* 52.4'; (pass.), add '(ποταμοὶ) διῃρημένοι κατὰ σταθμούς Peripl.M.Rubr. 15'

διαίρω **III**, delete 'intr. .. *over*' and add '*Peripl.M.Rubr.* 25, al.'

διαΐσσω, of sound, add 'Pl.*Ax.* 364a'; line 8, delete 'Anaxag.ap.'

δίαιτα **I**, line 1, after '*mode of life*' insert 'ἀμόχθητον .. δίαιταν Alc. 61.12 L.-P.; ἀνδρῶν δ. Pi.*P.* 1.93'   **III 2**, delete the section.

διαιτάω **I 2**, last lines before 'δίαιταν' insert 'in respect of diet, regimen, etc.'; add 'J.*AJ* 14.246'   **II**, for this section read '*act as arbitrator in, settle, arbitrate on* a dispute, etc., Pi.*P.* 9.68, Theoc. 12.34; νείκη D.H. 7.52, Hld. 10.10.2, w. cogn. acc., δ. δίαιταν Arist.*Ath.* 53.5, κρίσεις J.*AJ* 14.117; abs., Is. 2.29, Phld.*D.* 1.22, 24; w. dat., οὗτος διαιτῶν ἡμῖν D. 21.84; *act as mediator*, Ὀκταουία οὖν ἐχώρει πρὸς Καίσαρα διαιτήσουσα αὐτοῖς App.*BC* 5.93.   **2** *exercise office over* as arbitrator, οἱ .. τὴν Οἰνῄδα καὶ τὴν Ἐρεχθῇδα διαιτῶντες D. 47.12; *govern,* πόλιν Pi.*O.* 9.66.   **3** *administer, organize*, τὰ ἐκ τῶν διαθηκῶν δ. Luc.*Tox.* 23, τἆλλα οἷς τὸ πολυτελὲς τῶν ἐδεσμάτων ὁ Θεαγένης διήγησεν Hld. 3.10.3.   **b** *deal with, settle*, δ. τῆς ψυχῆς τὸ αἰδούμενον Hld. 4.6.1, 7.28.1.   **4** *pass judgement on* persons or problems, Str. 1.2.1, 2.2.1, 2.3.8.'

δίαίτησις, after '*way of life*' insert 'Plu. 2.500b'   add '**2** perh. *arbitration*, *SEG* 23.180.9 (Cleonaea, 145 BC).'

διαιτητέον **I**, add 'Gal. 11.35.12'

διαιτητικός **I**, for '*of* or *for diet .. dietetics*' read '*of* or *concerned with regimen*: ἡ δ. (sc. τέχνη) that part of medicine so concerned'; after 'Hp.*Acut.*(*Sp.*) 54' insert 'D.L. 3.85'

*δϊαιτία, ἡ, *way of life, SEG* 30.86.B36 (Athens, letter of Hadrian).

✝δίαιτός, ὁ, *arbitrator, BCH* 59.96 (Delphi), *APF* 15.75 (Dodona), Hsch.

διακαής, πυρετοί, add 'Gal. 10.759.11, 11.65.2'; metaph., add 'ὅλως ἐξηρέθιστο ἤδη καὶ δ. ἦν ὁ σατράπης Hld. 8.2.3'

διακάθαρσις **II**, add '*PSoterichos* 4.26 (AD 87)'

διακαθίζω **II**, for '= foreg.' read '*maintain a siege*'

✝διακαλίνδω, *roll along*, μισθωτοῖς το[ῖς δ]ιακαλίσασιν τὰ ξύλα *IG* 2².1672.158; cf. κυλίνδω.

*διᾰκάσιοι, v. °διακόσιοι.

διᾰκάτιοι, add '*SEG* 32.908 (Phaestus, Crete, vi BC), *IG* 9²(1).609.8 (Naupactus, c.500 BC)'

διακατοχή, = *bonorum possessio*, add 'Just.*Const.* Δέδωκεν 7, *Nov.* 18.5'

διακεάζω, for 'cf. A.R. 4.392' read '(understood by A.R. 4.392 as *burn*)'

διάκειμαι **II 2**, at end delete 'of a gift .. 7.3.17'   **3**, add '*CodJust.* 10.11.8.5, Just.*Nov.* 159 pr., al.'

*διακέλευσις, εως, ἡ, *exhortation*, λόγος δ. ἔχων τῶν εἰς τοὺς πολέμους Did.*in D.* 13.60.

✝διακέντησις, εως, ἡ, *the action* or *process of piercing*, τί ἐστιν δ.; [ἔστι]ν ἡ διὰ βελόνης τῶν σωμάτων τομή *APF* 2.2; of the teeth in dentition, Hp.*Dent.* 11.

*διακέντητον, τό, sc. κολλύριον, name of an eyesalve, Aët. 7.79 (326.11 O., v.l. δικέντ-), 7.114 (389.16 O.).

*διακιθαρισμός, ὁ, *competition in* κιθαρισμός, *Inscr.Cos* 59.

διακινέω **I 1**, for this section read '*move about, agitate*, Ar.*V.* 688, Hp.*Art.* 9, 30, *Vict.* 4.9; med., Hdt. 3.108.4'   add '**III** intr., *walk round*, διακινῶ μηθὲν ποιῶν *PMich.*VIII 465.16 (ii AD), *Vit.Aesop.*(G) 76.'

*διακλᾱρόω, v. ‡διακληρόω.

*διάκλειμα, ατος, τό, wd. of unkn. meaning (app. deriv. of διακλείω), τόμια ἀργυρᾶ καὶ διακλείματα *Inscr.Délos* 104.17 (iv BC).

διακλέπτω, before 'med., *steal away*' insert '*remove* a person *by stealth*, μία τοίνυν σωτηρίας ἐλπὶς διακλέψαι τὴν γυναῖκα Charito 5.2.9'

διακληρόω, after '(Dyme, iii BC)' insert 'w. εἰς, τὰν δὲ βουλὰν διακλαρῶσαι εἰς ἡμιόγδοα καὶ τριακάδα καὶ φυλὰν καὶ φάτραν *SEG* 30.990.21 (decr. at Delos, iv/iii BC)'; pass., add 'καθὼς ἁ χώρα διεκλαρώθη *BCH* suppl. 22 p. 237 (Argos, Hellen.)'

διακλήρωσις **1**, after '*apportionment*' insert '*PDura* 19.6 (i AD)'

✝διακλονέω, *shake violently*, ὅλος διακλονούμενος *Vit.Aesop.*(G) 18; διακλώνων· διασείων. Hsch.

*διακλυσμάτιον, τό, *little clyster*, Gal. 17a.610.4.

✝διακναίω, pf. διακέκναικα Pherecr. 155.20 K.-A., *grate to bits*, ὄψιν E.*Cyc.* 487; humorously, w. ref. to making of a salad, πόλις διακναισθήσεται Ar.*Pax* 251.   **2** *ruin utterly, destroy*, διακναιομένης κάμακος A.*Ag.* 65, E.*Med.* 164, *Alc.* 109; δ. Ὀρέστην (i.e. the play, of a bad actor) Stratt. 1.   **3** *make sore, ulcerate*, ἢν δὲ περυπωμένη ᾖ καὶ νέμηται, καὶ τὸν πελαστάτω χώρον διακναίῃ Hp.*Mul.* 1.66, 2.120.   **4** *wear* or *waste away*, ἡ ἀσιτίη δ. Hp.*Morb.* 1.13, οἵαις αἰκείαισιν διακναιόμενος A.*Pr.* 94, μόχθοις δ. ib. 541, πόθος μ' ἔχει διακναίσας Ar.*Ec.* 957; cf. τὸ χρῶμα διακεκναισμένος id.*Nu.* 120.

✝διακνημόομαι, διεκνημώσατο· διέφθειρε Hsch.

διακοιρᾰνέοντα, add '(prob. fr. Il. 4.230 πολέας διὰ κοιρανέοντα)'

διακολάπτω, for '*dress stone with a chisel*' read '*cut through with a chisel*'

✝διακολλάω, *stick throughout*, Luc.*Ind.* 16.   **2** *cover with a layer of decorative materials*, id.*Hipp.* 6, *POxy.* 3473 (AD 161/9).

*διακολπῐτεύω, *smuggle*, *PTeb.* 709.9, 14 (ii BC); cf. °κολπιτεύω.

διακομῐδή **I**, after '*carrying over*' insert '*transport*'   **II**, delete the section transferring quot. to section I, adding '*SEG* 34.558.39 (Larissa, ii BC)'

διακομπέω, for 'Posidon. 41' read 'Posidon. 36 J.'

διᾰκονέω **I 1**, antepenultimate line, after 'Luc.*Asin.* 53' insert 'τὰ μειράκια διακονούμενα Merc.Cond. 16'   add '**III** διακονῆσαι· κατεργάσασθαι. ἀπολέσαι. βλάψαι. Hsch. (s.v.l.).'

*διᾰκονητής, οῦ, ὁ, official of unkn. function, *PMasp.* 2 iii 19, 60.1, 126.73, *PAnt.* 95.2 (all vi AD).

διᾰκόνισσα, add '*MAMA* 7.69 (iv AD)', *CodJust.* 1.2.13, al.'

διᾱκονος **I 1**, line 3, for '(S.)*fr.* 133' read '(S.)*fr.* 137 R.' and add 'epith. of Hermes, Ἑρμεῖ διακόνῳ *SEG* 30.326.8 (Athens, i AD)'   **2**, line 5, fem., add '*TAM* 4(1).355, *SEG* 37.367 (Patrae, vi AD)'

*διᾰκοντισία, ἡ, = διακοντισμός, Robert *Ét.Anat.* 401 n.2 (Cos)

διακοπή **I**, line 2, after '(Gal.) 18(1).27' insert '*also the act of cutting, incision*'; after 'Plu.*Mar.* 19' insert 'Archig.ap.Gal. 8.90.15'

*διακοπτικός, gloss on δήϊος, δάϊος, Sch.Gen.Il. 14.422, Sch.E.*Andr.* 826.

διακόπτω **1**, line 3, after '*gash*' insert 'διακέκομμαι τὸ στόμα Men.*Sam.* 679' deleting this passage (*Sam.* 334) fr. section 3; at end delete 'so διακέκοπται .. Suid.' and insert '**b** *cut through* a base coin, so as to invalidate it, *SEG* 26.72 (Athens, 375/4 BC), Suid.'

διακόρευσις, add '*Lindos* 487.12 (iii AD)'

διακορεύω, delete 'Ar.*Th.* 480'

διακορέω, add 'Ar.*Th.* 480'

*Διᾰκός, ή, όν, prob. *of Zeus*, νόμος *IG* 12(2).58.8 (Mytilene, i BC).

*διᾱκοσιόδραχμος, ον, *of* 200 *drachmae*; neut. pl. subst., *loans of* 200 *drachmae* (sc. prob. δάνεια), *IG* 1³.248.36.

διᾱκόσιοι, after 'Ion. διηκ-' insert 'Cypr. διακάσιοι *ti-wi-ja-ka-si-a-se*, perh. δϜιjακασίας (acc. fem. pl.) = διακοσίας, *ICS* 318 A III 1'

διᾱκοσιοκαιτεσσᾰρᾰκοντάχους, for '*two-hundred-and-forty-fold*' read '*having a yield of two-hundred-and-forty-fold*'

διᾱκοσιοντάκις, for '*Suid.*' read 'Alex.Aphr. *in Top.* 586.8 W.'

διᾱκοσιοντάχους, for '*two-hundred-fold*' read '*having a yield of two-hundred-fold*'

διᾱκοσιοστός, after '*Written*' insert 'διακοσαστή *PCair.Zen.* 15ʳ.39 (iii BC)'

διακοσμέω **II**, pass., add 'of a building, *SEG* 37.1541.5 (Gerasa, vi AD)'

*διακόσμιος, ον, *pervasive throughout the universe*, Simp. *in de An.* 28.3.

διακρᾰτέω **I 2**, add '*CodJust.* 10.11.8.5'

διακρέκω, add 'prob. in Sapph. 99 i 4 L.-P.'

*διακρηνάω, Dor. -κρᾱνάω, perh. *mingle with spring water*, οἷον δὴ τόκα πῶμα διεκρανάσατε (v.l. διεκρανώσατε: see foll.), Νύμφαι Theoc. 7.154.

διακρηνόω, delete '*make to flow*' and after 'πῶμα' insert 'διεκρανώσατε (glossed by *EM* w. ἀνεῴξατε, but see °διακρηνάω)'

*διακρῐβεύω, *work out* or *determine exactly*, ἐπιλέγει .. δεήσει διακριβεύειν τὰς οὕτω τῶν ἐνιαυτῶν εὑρισκομένας διαθέσεις Heph.Astr. 2.11.120.

διακρῐδόν, for '*eminently*' read '*by a clear margin, decidedly*'

διάκρῐσις **I**, add '**3** *dividing up* (of land), αἱροῦνται δὲ τρεῖς ἄνδρες ἐπὶ τὴν διάκρισιν καὶ διανομὴν Plu.*TG* 13.'   **IV**, add '**2** *parting of hair*, Poll. 4.140.'   **VI**, for '*a bandage*' read '*a method of bandaging* (ἐπίδεσις)'

διακρούω **III**, for '*hinder, entangle*' read '*interrupt*' and add 'τὸ συνεχὲς τῆς ἁρμονίας D.H.*Comp.* 22'

διακρύπτω, med., add 'D. 41.17 (s.v.l.)'

*διακτήρ, v. °δαϊκτήρ.

διάκτορος, line 2, for 'δ. alone' read 'Ἑρμείαο διακτόρου'

διάκτος, delete the entry (v. ‡δρακτόν).

διακῠβεύω, after '(Plu.*Rom.*) 2.128a' insert '*Vit.Aesop.*(G) 81' and delete 'περί .. 70d'

*διακῡβιστάω, in broken context, app. *waggle the bottom*, Stratt. in *CGFP* 220.8(b) ii 96.

διάκων, add '*SB* 5124.207, al. (ii AD); as Jewish office, *CIJud.* 805 (Apamea, vi AD); fem., *IEphes.* 3415 (ii/iii AD)'

διαλαμβάνω **III 4**, add '*Peripl.M.Rubr.* 20'

✝διάλαυρος, ον, *bordered on all sides by streets*, fem. subst., διάλαυρος· οἰκία μεγάλη πανταχόθεν λαύραις διειλημμένη, ἡ λεγομένη περιάμφοδος Hsch., Eust. 1921.58; neut. subst., sc. χωρίον, *SEG* 23.398 (Aetolia, ii BC).

διαλεαίνω I, for this section read 'smooth, rub, massage, διαλεαινομένων τῶν ἄκρων Arch.ap.Gal. 13.169.13'

διαλέγω A II, after 'Ar.Lys. 720' insert 'διαλέξαντι τὸ θύρετρον BCH 32.69 (Delos)' B 1, three lines from end after 'Plu.Per. 7' delete 'reason .. Marc. 18'

*διαλειπτικός, ή, όν, intermittent, Gal. 9.284.17, 328.9.

διάλευκος, for 'quite white' read 'white, marked with white' and insert 'Pi. fr. 169.24 S.-M.'

διάληξις, for 'of an inheritance' read 'by lot'

διάληψις III 2, for 'Annales .. 42' read 'SB 6152.26, 6153.28 (i BC)'

διάλιθος, add 'as subst., app. a piece of jewellery set with stones, διόπας, διάλιθον, πλάστρα Ar.fr. 332.10 K.-A.'

διαλλάσσω IV 1, for this section read 'intr., w. dat., differ in some respect, δ. ταῖς ἡλικίαις Arist.EN 1161ᵃ5, κλήσει οὐ φύσει D.H. 1.29, τῇ ἐγκλίσει διηλλαχός A.D.Synt. 70.11; abs., τὸ διαλλάσσον τῆς γνώμης Th. 3.10.1, οὐ ταὐτὸ δ' ἐστὶ τοῦτο, πολὺ διήλλαχεν Dionys.Com. 2.10. **b** differ from, w. dat. and internal acc., διαλλάσσοντες εἶδος .. οὐδὲν τοῖσι ἑτέροισι Hdt. 7.70.1; w. gen., Plb. 2.37.11; w. πρός, Aristid.Or. 35(48).16; w. ἀπό, Modest.Dig. 27.1.13 pr.'

διαλογή 4, add 'Salamine 90 (ii BC)'

διαλογίζομαι III, delete the section.

*διαλόγϊσις, εως, ἡ, = διαλογισμός (in quot., II), Polystr. p. 81 W.

*διαλογογράφος, ὁ, writer of dialogues, Syrian.in Hermog. 1.95 R.

δίἄλον, διἄλας, add 'cf. δεϜαλῶσαι s.v. °δηλόω'

διαλυμαίνομαι II, pass., add 'D.Ep. 1.12'

διάλῠσις 6, add 'Charis. p. 283 K.'

διαλῡσίφῐλος, add 'epigr. in SEG 24.1238 (Egypt, Hellen./Rom.imp.)'

διαλύτης, add '2 mediator, Ath.Agora XIX 4b.6 (iii BC, pl.).'

διαλῠτικός II, add 'διαλυτική, ἡ (sc. ἔμπλαστρος), resolvent plaster, PMerton 12.22 (i AD)'

*διάλυτον, τό, asyndeton (= διάλυσις 6), Charis. p. 283 K.

διαλυτός II, add '2 capable of being taken to pieces, κλίμακες Plu.Arat. 6.'

διαλύω I 7, add 'b transf., pay for, τὴν παρρησίαν Macho 162 G.'

διαμανθάνω, for 'learn by inquiry' read 'learn in detail' and add 'E.Hyps.fr. 60.6 B.'

διαμαρτάνω 3, adv. διημαρτημένως, add 'Call.Dieg. ix.12 (fr. 201 Pf.)'

διαμαρτία II, after 'Luc.Sacr. 1' insert '(v.l.)'

διαμαρτῠρία 1, for pres. def. read 'sworn testimony to a fact that affects admissibility of an action at law'

διαμᾱσάομαι I, after 'chew up' insert 'Ar.Th. 494 (cj.)'

διαμαστῑγόω, for 'Pl.Grg. 524c' read 'Pl.Grg. 524e'

⁺διαμαστίγωσις, εως, ἡ, prolonged beating, name given to Spartan custom used to test the endurance of boys, Plu. 2.239d.

διαμάχησις, for 'Gal. 9.921 (ap.Aët. 5.24)' read 'Aët. 5.24 (16.14 O.)'

*διαμαχίζομαι, strive earnestly, διαμεμάχισται ἡ ψυχή μου ἐν αὐτῇ (sc. σοφίᾳ) Lxx Si. 51.19(25).

διᾰμείβω 2, add 'c transf., complete a period of time, δεκάδας δέκ' ἐτῶν διαμείψας CEG 477 (Attica, c.400/390 BC).'

*διαμελεϊστί, v. μελεϊστί.

*διαμελίστρια, ἡ, one that dismembers, (ἡμέρα), ἡ τοῦ λογικοῦ ἀμνοῦ δ. (i.e. the Passover) Leont.Byz.HP 1.2.5; cf. °διαμερίστρια.

διαμένω, add 'impers. διαμένει, -μεμένηκε, etc., the practice continues, has continued, etc., w. acc. and inf., Paus. 9.36.6, al.; w. dat. and inf., Arist.Ath. 8.1, Paus. 7.17.14, 27.8, al.'

*διαμερές, v. °διαμπερές.

*διαμερίστρια, ἡ, one that divides up, (ἡμέρα) ἡ .. τοῦ λογικοῦ ἀμνου δ. Ps.-Chrys. HP 2.6; cf. °διαμελίστρια.

διάμετρον, add 'perh. also payment in lieu, PHib. 110.14 (pl., iii BC), PCornell 3.25 (iii BC)'

διαμηρίζω, for 'femora diducere, inire' read 'part the thighs of'

*διαμήριον, τό, perh. payment for sexual intercourse, ἀπόδος τὸ δ. Kretschmer GV 89.

⁺διαμηρισμός, ὁ, sexual intercourse, Zeno Stoic. 1 (pl.).

διαμίγνῡμι, for 'Plu. 2.1131e' read 'Plu. 2.1132e'

*διαμίμνω, continue unchanged, persist, Democr.ap.Thphr.Sens. 55.

*διαμισθωτής, οῦ, ὁ, tenant of state land, BGU 2490.14 (ii/iii AD, s.v.l.).

διαμισθωτικόν, add 'PGiss. 64.7 (ii AD)'

διαμνημονεύω 2, for 'X.Cyr. 1.1.2' read 'X.Cyr. 1.2.2'

διαμονή, for pres. def. read 'continuance in good or unchanged condition'; add 'in prayers for the well-being of the imperial family, SEG 31.1124 (Phrygia, AD 104), 32.672 (Thrace, AD 144)'

διάμονος, after 'permanent' insert 'δ. καὶ βέβαια SEG 4.598.50 (i BC)'

διαμπερές, after lemma insert 'διαμερές CEG 108.9 (Eretria, vi BC)' II, for 'Supp.Epigr. 1.409 (Eretria)' read 'CEG l.c.'

διαμπερέως, add 'Stesich. 222(b).281 (p. 216 D.), Theoc. 25.120'

*διάμφοδος, gloss on διέλθυρις, Hsch.

διαμώκησις, for 'Ath. 5.200b' read 'Ath. 5.220b'

*διαμωμάομαι, strengthd. for μωμάομαι, Sch.Pi.O. 6.124.

διαναβάλλω, after 'med.' read 'Cod.Just. 1.3.45.7, Hsch. s.vv. διακρούεσθαι and διαπονδαρίζει'

*διαναίω, aor. διένασσα, establish in different places, Hes.fr. 33(a).3 M.-W.

*διαναρρίπτομαι, postpone, put off, Hsch. s.v. διαπονδαρίζει; cf. πυδαρίζω.

διαναυμαχέω, for 'maintain a sea-fight' read 'fight a sea-battle through to a decision'

διαναῦσαι (s.v. διανάω), after lemma insert '(διαναυσθλοῦσθαι cj. La.)'

⁺διανάω, flow through, percolate, Thphr.fr. 171.11; transf., of a container, Plu.Aem. 14.

διανέμησις, add '2 transfer of ownership, Just.Nov. 17.18.1.'

*διάνευμα, ατος, τό, app. gesture of the body, Φρυγίων διανεύματα Χαρίτων Ar.Th. 122 (dub., cj. δινεύματα).

διανεύω I, add '2 gesture with the body, Luc.Salt. 64.'

*Διάνη, ἡ, the goddess Diana, SEG 34.1306 (Pamphylia, iii AD).

*διανήφω, perh. lose potency, διένηφε δ' οὖν καὶ κατὰ μεικρὸν εἰς ὕδωρ Diog.Oen.fr. 7 col. 2.8 (AJA 75.367).

διανθής I, add 'ἀσφόδελος Nic.Th. 534' II, delete the section.

διανίστημι I, add 'fig., κἂν γὰρ εἰ .. πρὸς τὴν ὑπὲρ αὐτῶν διαναστῶμεν ἐκδίκησιν Just.Nov. 129 pr.'

διανοέω II, add 'w. ὡς, εἰ γὰρ καὶ διανοεῖταί τις ὡς δεῖ Pl.Alc. 1.109c'

διανόημα, line 4, delete 'esp. whim, sick fancy' and add 'Hp.Vict. 1.2'

διανόησις, add 'III ἀπελευθερωθεὶς κατὰ διανόησιν prob. by (declared) intention, IG 9(2).109b19 (Thessaly, ii BC), cf. ib. 1301.9, where [κατὰ διανό]ησιν δ[έ is distd. fr. κατὰ διαθήκην].'

*διάντομαι, διήντετο· διάζετο ἱστόν Hsch.

διανύω, line 9, before 'pass.' insert 'med., πέντε διαινυσάμην ἐνιαυτούς IGRom. 4.608 (Phrygia, iv AD).'

διαξέω, add 'διεξεσμένα, fig., of style, Poll. 6.141'

*διάξυστος, ον, perh. fluted, ἐπὶ στυλίδος διαξύστω ILesb. 6 B 10.

διαξύω, add '2 ruffle in passing, Γαλάτεια .. διαξύουσα γαλήνην Nonn.D. 39.258. 3 app. scratch through, διαγράφειν· διαξύειν, ἀπαλείφειν, ἀκυροῦν Hsch., AB 1.238.'

*διαπᾰλαιόω, strengthd. for παλαιόω, προπυλῶνος διαπαλαιωθέντος POxy. 2272.6 (ii AD).

διαπᾱσῶν, after 'divisim' insert 'Pherecr.fr. 31 K.-A.'

*διαπειλέω, add 'threaten a penalty, θάνατος τῷ παραβαίνοντι διηπειλεῖτο Hld. 5.25.2'

διάπειρα, add 'D. 44.58, 56.18, τὸ πρᾶγμα εἰς διάπειραν καὶ λόγον κατέστησαν Aeschin. 1.184, Just.Nov. 5.2 pr.'

διαπειραίνω, add 'perh. in reversed tm. in h.Merc. 48 (πειρήνας διὰ νῶτα); but cf. περαίνω IV'

διαπειράομαι, after 'proof of' insert 'ἔπεσιν δ. (sc. τοῦ πατρός) Od. 24.240 (v.l. in PRyl. 53, iii/iv AD)'

διαπέρᾱμα, for 'strait of the sea' read 'crossing'

διαπεράω III, delete 'ὕδωρ .. Eub. 151' and before 'Luc.DMort. 20.1' insert 'τὸν πορθμέα τοῦτον, ὅς σε διεπέρασε'

διαπέρθω, add 'διεπέρσατε Δύμιον ἄστυ Antim. 28.2 Wy.'

διαπεττεύω, for 'gamble' read 'subject to a throw of the dice' and delete 'try one's luck at play'

διάπηγα, for 'panels' read 'cross-pieces' and for '3Ki. 7.31, 32' read 'v.l. in 3Ki. 7.17 (31, 32)'

διάπηγμα, delete 'partition' and add 'Aq. 4Ki. 16.17'

διαπηδάω 2, delete 'medic.' and for 'ooze through' read 'let moisture through' add 'II contend in jumping, D.Chr. 8.12.'

διαπίνω, for 'drink one against another' read 'carry on drinking'

διαπίπτω III, perish, add 'Just.Nov. 15.5.1'

*διαπιστόομαι, confirm, prove, Aristodem. 16.2 J.

⁺διάπλασμα, ατος, τό, moulding, shaping, Sch.Ar.V. 614, τὸ δ. τοῦ τύπου Vit.Aesop.(G) 88.

διαπλαστικός, for 'Alex.Aphr.Pr. 1.47' read 'Alex.Aphr.Pr. 2.47'; add 'capable of being formed, ταῖς τῶν ἐμβρύων διαπλαστικῶν φύσεσι Alcin.Intr. p. 178.28 H.'

διαπλᾰτύνω, for 'X.Lac. 2.5' read 'X.Lac. 2.6'

διαπληκτίζομαι, for 'spar' read 'fight with the hands' and delete 'τινί' before 'Luc.Anach. 11'

διαπλοκή, add 'III intertwining, ἐπὶ ταῖς διαπλοκαῖς τῶν παλαισμάτων Philostr.Gym. 11.'

διάπλοκος, after 'plaited' insert 'Aristeas Judaeus 75'

διάπλοος II, add '4 strait, AP 7.666 (Antip.Thess.).'

διαπνέω I 2, delete the section transferring quot. to section 1, prefixed by 'abs.' add '2 break wind, Suid. s.v. ἀποψοφεῖν.'

διαπονέω, for 'PTheb.Bank' read 'Theb.Ostr.'

διαπολιτεία, for 'party-strife' read 'political conflict'

*διαπολύω, gloss on διολλύω, Hsch.

διαπονέω I 3, add 'Vit.Aesop.(G) 3'

διαπορέω, line 2, after 'aor. pass.' insert 'διαπορηθείς Pl.Sph. 217a' II, lines 5/6, delete 'Med. .. 217a' διαπορθμεύω **I 2**, line 3, after 'Iamb.*Myst*. 1.5' insert 'also Max.Tyr. 14.8 (of interpreters)'

διάπρᾶσις **I**, add '*SEG* 38.1462.9 (Oenoanda, ii AD)'

διαπράσσω **II 2**, add 'Plu. 2.404a'

διαπραΰνω, for 'Philostr.*VA* 6.14' read 'Philostr.*VA* 6.13'

διαπρέπω **I 2**, for 'ἐπί τινι .. D.C. 68.6' read 'πάνυ .. διαπρέψαντα ἐν τῇ ὀρχηστικῇ Luc.*Salt*. 9; w. ἐπί, πλεῖστον .. ἐπ' ἀνδρείᾳ .. διέπρεπε D.C. 68.6'

διάπριστος, for 'Demioprat.' read '*Demioprat*.'; add '*IG* 1³.422.13-14 (Athens, v BC), 2².2500.56'

διαπρίω **III**, after 'διαγοράζει' insert '(dub.)'

✝διαπρό, prep., w. gen. *straight through*, Il. 4.138, δόρυ δ' ὀφθαλμοῖο διαπρὸ καὶ διὰ ἰνίου ἦλθεν 14.494. **II** as adv., Il. 5.66, 7.260, etc., Od. 22.295, 24.524, *h.Ven.* 114.

*διαπροαιρέομαι, strengthd. for προαιρ-, Sch.Pi.*O*. 6.140.

✝διαπτερόω, *brush* or *clean out*, Hp.*Acut*. 58, Aret.*CD* 1.8; διαπτερῶσαι· ἀνευρῦναι, διαστῆσαι Hsch.

*διαπτυχής, ές, *split apart*, Philox.Leuc. (b).29 P.

διάπῠρος **2**, add 'Gal. 12.1003.8, 9, 12' **3**, after '(Pl.)*Lg*. 783a (sup.)' insert 'Men.*Dysc*. 183'

διαπύρωσις **II**, add '*scorching* by the sun, Ptol.*Tetr*. 56'

*διαπωτάομαι, *direct one's flight*, πᾷ μοι φθογγὰ διαπωτᾶται φοράδην; S.*OT* 1310 (pap., codd. διαπέταται).

*διάρᾱμα, v. °διέραμα.

*διᾰρᾰμᾰτία, ἡ, *lading* or *conveying of corn*, as a public service, *PSI* 1103.9 (ii AD); cf. °διέραμα.

*διᾰρᾱπίζω, *beat up*, *PTeb*. 798.15 (ii BC).

*διᾰράσις, εως, ἡ, v. °διέρασις.

*διαργυρόω, *decorate with silver*, ᾠὸν στρούθειον διηργυρωμένον *PMich*.II 9ʳ.3 (iii BC).

✝διάρημα, v. °διέραμα.

διάρθρωσις **3**, line 3, for '*distinctness*' read '*analysis*'

διαρίθμησις, add 'ψήφων *SEG* 9.8.32 (Cyrene, Aug.)'

διάριον, add '*Cod.Just*. 12.17 pr.'

διαρκέω **I 1**, line 4, for 'δ. πρός τινα .. etc.' read '*have resources to meet*, ἵνα καὶ διαρκέσῃς πρὸς τοῦ πένθους τὸ μέγεθος Luc.*Luct*. 24'

διαρκής **2**, for 'sup. .. Them.*Or*. 11.146a' read '*having staying power*, ἵπποι Them.*Or*. 11.146a; transf., of a race, Paus. 6.13.3'

δίαρμα **II**, add 'Cleonid.*Harm*. p. 206 J.'

διαρμόζω **2**, pass., add 'of musical instrument, *to be regulated* in pitch, διηρμοσμένον ἦν συρίγγιον, ᾧ τὴν φωνὴν οἱ ἁρμονικοὶ σχέδην ἐπ' ἀμφότερα διὰ τῶν τόνων ἄγουσι Plu. 2.456a'

*διάροσις, εως, ἡ, gloss on σφύρωσις, Hsch.

διαρπαγή, add '*SEG* 32.1128.13 (Ephesus, 39/8 BC)'

διαρπάζω **II 3**, for 'Nonn.*D*. 48.290' read 'Nonn.*D*. 48.920'

διαρραίνω, see at 'for 'Philostr.*Im*. 7.27' read 'Philostr.*Im*. 1.27'

διαρραίω, fut. med., add 'διᾰραίσομαι Tim. 791.133 P.' **II**, for this section read '**2** *divide* or *break up*, ῥωχμαὶ (wrinkles) σάρκα διαρραίουσι, i.e. break up the surface of, Marc.Sid. 80.'

*διάρ(ρ)αντος, ον, *speckled* or *brindled*, Lxx *Ge*. 30.32.

διαρραφή, add '*APF* 2.2.13'

διαρρέπω, for '*oscillate*: halt in one's gait' read '*tilt sideways* as one walks, of a person with one leg shorter than the other'

διαρρέω **II**, after '*waste away*' insert 'διέρρυε[ν] Sapph. 98(*b*).9 L.-P.'

διαρρήγνῡμι, add '**2** *rescind* will or other document, cf. Lat. *testamentum rumpere* and sim., Just.*Nov*. 107.2.'

διαρρήδην, line 3, for 'Lys. 1.20' read 'Lys. 1.30'

διαρρήσεις, after 'Pl.*Lg*. 932e' insert 'Poll. 2.128'

διαρρίπτω **I 3**, add '**b** *cause to go off in a wrong direction*, Arr.*Cyn*. 16.3, ὁ μὲν (sc. λαγωὸς) ἐξελίξας τὸν δρόμον καὶ διαρρίψας τὴν κύνα ib. 17.3.'

διαρριφή, for 'Dor.' read 'dial.' and add 'a (variously explained) rite at a feast, *SEG* 32.1243.38 (Aeolian Cyme, 2 BC/AD 2)'

διάρροος, before 'D.S. 13.47' insert '*IG* 7.4255.21 (Oropus, iv BC)'

διάρταβος, add '**b** *of two* ἀρτάβαι, σφυρίδες *SB* 6801ʳ.20, 24 (iii BC).'

*διάρτης, ου, ὁ, ποταμῶν δ. *one who raises* water from rivers, Teuc.Bab. p. 46.9 B.

*διαρτία, ἡ, *mould, form*, σωματικὴν διαρτίαν Sch.Nic.*Al*. 227c Ge.

*διασᾰλᾰκωνεύω, = διασαλακωνίζω, Hsch. s.v. διασαλακώνισον.

διασᾰλᾰκωνίζω, after 'Ar.*V*. 1169' insert 'Hsch.'

διασαλεύω, line 5, for '(Luc.)*Merc.Cond*. 35' read '(Luc.)*Merc.Cond*. 33'

διασᾰφέω, add '**b** *state plainly* something previously expressed figuratively, *Ev.Matt*. 13.36, J.*AJ* 5.293. **c** *make known officially, declare*, τὸ διασεσαφημένον ἐκφόριον *PTeb*. 105.39 (103 BC).'

*διασᾰφής, ές, *very clear*, Kafizin 146(a) (iii BC).

✝διασᾰφητικός, ή, όν, *declaratory*, διαφορὰ τοῦ ἢ συνδέσμου A.D.*Conj*. 221.16, 23; cf. *EM* 415.27, Sch.B Il. 1.117; adv. -κῶς, Sch.T ibid. **II** *explanatory*, Sch.Ar.*An*. 825, *An.Ox*. 1.118.

διάσεισις **II**, add '*UPZ* 1.9'

διάσημος **II**, sup., add 'adv., μεταπαρέλαβεν διασήμοτατα τὴν δαιδουχίαν *SEG* 30.93.5 (Eleusis, 20/19 BC)'

Διάσια, add 'also festival at Thasos, *SEG* 21.541.A38 (iv BC)'

*διασιαίνω, for *διασικχαίνω, *loathe utterly*, Sch.Luc.*Tim*. 7; cf. δικχαίνω.

διασκάπτω, line 2, for 'ἰσθμόν' read '*Μίμαντα*'

✝διασκᾰρῑφάομαι, *scratch up*, in fig. phr., τὰς εὐτυχίας .. διεσκαριφησάμεθα καὶ διελύσαμεν Isoc. 7.12; act., διασκαριφῆσαι· ἐπὶ ὀρνέων τῶν τοῖς ὄνυξι σκαλευόντων τὴν γῆν κυρίως λέγεται Hsch.

διασκευάζω **II**, line 2, after 'Plu.*Ant*. 24' insert 'τὴν θεὸν διεσκευασμένην καταπληκτικῶς D.S. 4.51'

διασκευαστής, add '**II** dub. sens., w. ref. to some activity of a person under the influence of Aphrodite, Ptol.*Tetr*. 164.'

*διασκεῦος, τό, pl. διασκεύη *equipment*, Roueché *Performers and partisans* no.1 (Aphrodisias, v/vi AD).

διασκηνέω **II**, delete the section transferring 'X.*Cyr*. 3.1.38' to section I.

διασκοπέω **I**, add 'w. inf., διεσκεψάμεθα μέσην τινὰ τῷ πράγματι τάξιν ἐπινοῆσαι Just.*Nov*. 22.26 pr.'

διασκορπίζω, line 1, after '*scatter abroad*' insert '*disperse*' and after 'al.' insert 'Ael.*VH* 13.46'

*διασκορπιστής, οῦ, ὁ, *winnower*, λικμηταί· διασκορπισταί Hsch.

διασπάθάω **I**, add '*SB* 10567.9 (iii AD)'

διασπᾰράσσω **I 1**, after 'A.*Pers*. 195' insert '*Suppl.Hell*. 996.7'; pass., add 'Memn. 8.8 J.'

διάσπᾱσις **II**, delete the section (v. °διάστασις).

διάσπαστος, delete the entry.

*διασπᾰτᾰλάω, *squander*, *PHarris* 67 ii 7 (ii AD).

διασπάω, add '**4** *squander, dissipate*, Just.*Nov*. 38 pr. 1.'

διασπείρω, line 4, delete '*squander*'

*διάσταθμον, τό, perh. *ration, allowance*, *PCair.Zen*. 320.8 (iii BC).

διασταλμός, add '*PWash.Univ*. 8.2 (vi AD)'

διασταλτικός **II**, for this section read 'v. °διαστατικός' add '**III** medic., *diastolic*, Gal. 9.298.3.'

διάστᾱσις **I 1**, add 'Plu. 2.721a' **III**, line 2, for 'Arist.*Top*. 142ᵇ5' read 'Arist.*Top*. 142ᵇ25'

διαστᾰτικός, after '**3**' insert 'of music, *exciting, exalting*, Ptol.*Harm*. 1.12, 3.11, Cleonid.*Harm*. 13, Aristid.Quint. 1.12, 19 (διασταλτικός occurs as ms. reading or variant in some of these passages)' **3**, renumber this section as **4** inserting before adv.' '*separating, distinguishing*, ἐν τῇ λαλιᾷ διαστατικός τῶν ὀνομάτων D.L. 4.33'

διαστέλλω **I 4**, for '*pronounce*' read '*state explicitly*'

διάστεμα, before '*PRyl*.' insert '*interval*'; add '*distance*, δύο ἡμερῶν ἐστι τὸ δ. *PBremen* 15.30'

διάστημα **I 1**, add 'ἐν τῷ τοῦ πλοὸς δ. in the *course* of the voyage, X.*Eph*. 1.14.7; see also ‡*διάστεμα*'

*Διαστής, οῦ, ὁ, *worshipper of Zeus*, *BCH* 46.342 (pl., Teos).

διάστησις, delete the entry.

διαστίζω **3**, for '*brand*' read '*distinguish, differentiate*'

*διαστικός, ή, όν, (διάζομαι) *of weaving*, ἡ δ. (sc. τέχνη) τῶν ἀραχνῶν Theodos.Gr. p. 53 G.

διάστιξις, for '*branding*' read '*distinction, differentiation*' and add 'Just.*Nov*. 112.1'

διαστολεύς **II**, for '*cashier*, title of official' read '*some local financial official*'

διαστολή **I 4**, for this section read '*itemized list, schedule* of payment, *BGU* 2493.13 (i/ii AD), *PTeb*. 363.1 (ii AD), *BGU* 652, al. (iii AD)' **II 2**, add 'Anon.Bellerm. 11'

διαστρέφω **I 1**, line 4, after '*to be warped*' insert 'of wood, Pl.*Prt*. 325d'

διαστροφή **2**, add 'ἐκ διαστροφῆς, opp. κατὰ φύσιν, Jul.*Or*. 6.202c' **4**, add 'δίχα ζημίας καὶ διαστροφῆς *Cod.Just*. 11.1.1.1'

διάστυλος **II**, for this section read '*intercolumniation*, *SEG* 35.1109, 1110 (Ephesus, iii AD), 37.851 (Aphrodisias, iv AD), *MAMA* 8.498'

*διασυλλαμβάνω, perh. *keep, preserve* in specified condition, *PGiss.inv.* 245.6 (*Pap.Brux.* 7.78, ii AD).

διασυρίζω, line 1, delete 'f.l. in' and substitute 'διασυριζόντων (ἰκτίνων)'

διασύρω, lines 2/3, delete '*tear in pieces* .. *pull to pieces*, i.e.' **II**, for '*break up, disperse*' read '*keep apart*' add '**IV** abs. *prolong one's stay, linger*, *POxy*. 3867.4 (vi AD).'

διασφαλίζομαι, after 'Orib. 8.7.3' insert '*SEG* 38.1462.B16 (Oenoanda)'

διασφάλλω, after 'Luc.*Abd*. 17' insert '*pervert*, νόον Neophr.(?) in *PLit.Lond*. 77 fr. 2.12'

✝διασφετερίζομαι, *appropriate, usurp*, v.l. in Ph. 2.130, *PLond.inv*. 2222 (*Mnemos.*(ser. 4) 40.413, AD 319).

διασωστής, for '*policeman*' read '*guide*'

διασωστικός, add 'Alcin.*Intr*. p. 182.32, 183.4 H.'

διαταγή, add '**b** perh. *order, requisition*, *BGU* 2347.5 (iii AD).'

διαταγματάριος, ὁ, an official in the service of a *dux* (δούξ), *PAnt.* 96.11 (vi AD).

διατάκτης, after '*assigner of posts*' insert 'Ptol. *Tetr.* 82'   add '**2** gloss on κοσμήτωρ 1, Sch.Hom. in *POsl.* 12 iii 16.   **3** title of Roman official in Egypt, *SB* 6026 (iii AD).'

διατακτικός, delete 'Ptol. *Tetr.* 82'

διαταξίαρχος, add '*MDAI*(*A*) 37.302 (Pergamum, ii AD)'

διάταξις II 2, add 'of other decrees and regulations, κατὰ τὴν Γεντιανοῦ διάτα<ξι>ν SEG 30.568.17 (Maced., ii AD)'

διατάσσω I 2, line 2, for 'Th. 4.103' read 'Th. 4.130.3'

διατείνω, line 1, for '*keep stretched out*' read 'of a surgeon, in setting a bone'; line 3, for 'τινὰ ὑπὲρ λεχέων' read '*Δωρίδα τὴν ῥοδόπυγον ὑπὲρ λεχέων διατείνας*' **B I 1,** line 5, after 'X.*Mem.* 3.7.9' insert '*strain*, of a woman in childbirth, Al.*Mi.* 4.10'; line 6, for '*prevented*' read '*exerted themselves to prevent*'

διατειχισμός, for '*fortifying*' read '*walling off*'

*διατεκταίνομαι, διετεκτήνατο· ἐμηχανήσατο Hsch.

διατελέω I, for '*bring quite to an end*' read '*carry through to the end*'   **II 1,** line 3, for 'And. 1.38' read 'And. 1.138'   **2,** add '**c** *remain* in a place, κατ᾿ οἶκον Aen.Tact. 10.13, cf. 3.6, 7.1.'

διατέμνω 1, line 3, after 'Hdt.' insert '2.41.3'; lines 5ff., delete 'τι ἀπό τινος' to end of article and substitute 'fig., διατετμηκότα τὴν πολιτείαν *split apart*, Aeschin. 3.207.   **2** *divide into two*, τὸ μέτρον D.H.*Comp.* 26; w. gen., *separate* from, τὴν μὲν διετέμομεν ἀπ᾽ αὐτῆς Pl.*Plt.* 280b.'

διατενής, add 'of the pulse, *strained*, Gal. 8.943.2 (unless διάτονος is read)'

*διατετυπωμένως, v. διεσκευασμένως.

διατήρησις, add '*SEG* 25.687 (Thessaly, Hellen.)'

διατίθημι A I, add '*h.Ap.* 254'   **II 1,** add '**b** w. double acc., οὐδὲν δ. αὐτὸν δεινόν *gave* him no ill-*treatment*, Aristodem. 5.3 J., cf. 8.1 J.'   add '**3** *subject to sexual practices*, Ptol.*Tetr.* 164, 166, 187, al.'   add '**IV** *compose, settle* (cf. B 4), Just.*Nov.* 159.3, ib. epilogos.'

†**διατίλλω,** *pull out the hair from,* κουραῖς .. διατετιλμένης φόβην S.*fr.* 659.7 R., λαβὼν με τῆς κόμης διέτιλεν Lxx *Jb.* 16.12.

διατίμησις, after '*valuation*' insert '*SEG* 34.558.52 (Larissa, ii BC)' and after '*assessed value*' insert 'διατιμήσεις ἀποτινέτων *RA* 6.31.16 (decr., Amphipolis, ii BC), *Cod.Just.* 6.4.4.4, al.'

*διάτιμος, ον, *honoured*, *OGI* 615.4 (iii AD) (sup.).

διατινάσσω II, after '(E.)*IT* 282' insert 'Aristo 13 iii W.'

διατμήγω, line 6, pass. διέτμαγεν, add 'διέτμαγον (act. intr.) v.l. in Il. ll. cc., also in A.R. 2.298, 3.1147 (cj. διέτμαγεν in both passages)'

*διατοίχιον, τό, dim. of διάτοιχος II, *Milet* 7 p. 56.

*διατόμιον, τό, dim. of διατομή in uncertain sense, *Inscr.Délos* 1442*B*41 (ii BC).

†**διάτομος, ον,** v.l. for διχότομος (in same sense) in Mart.Cap. 7.738, 8.864.

διατόναιον, after '(iii BC)' insert '*Inscr.Délos* 1417*A*i73'; for '*curtain-rod*' read '*cross-support*'; after 'so -τόνιον' delete '*curtain-hook* or *ring*'

διάτονος I 2, add '**b** διάτονα, τά, *cross-supports*, *Inscr.Délos* 290.216 (iii BC).'

*διατόξευσις, εως, ἡ, = διατοξεία, Robert *Ét.Anat.* 401 n.2.

διατοξεύω, line 2, for 'Hld. 5.32' read 'Hld. 5.33 (32.5; v.l. τοξ-)'

διάτορος I, add 'adv. *-ως by piercing*, S.*fr.* 314.316 R.'

διατραχηλίζομαι, delete the entry.

*διατραχηλίζω, *keep holding in a neck-lock*, in quot. pass., Teles p. 10 H.; transf., *hold in one's grasp*, ἐν ταραχῇ καὶ πλάνῃ δρόμοις ὀλεθρίοις καὶ παραφόροις διατραχηλιζόμενος Plu. 2.501d.

*διατραχηλισμός, ὁ, *state of being forced down in a neck-lock*, in quot. fig., Plu. 2.317b (v.l. v. διαταραχή).

διατρέπω I, line 5, for 'Epicur. p. xxviii U.' read 'Epicur. [1] 119.4 A., *turn to new business* (of one's own accord), Onas. 42.2'   **II 1,** delete the section.

διατρέφω, pass., add 'X.*Vect.* 1.1, D. 19.249, Isoc. 4.153, 6.78'

διατρέχω II 2, add '*Cod.Just.* 10.19.9.1, Just.*Nov.* 93 pr.'   add '**III** abs. *contend in running*, D.Chr. 8.12.'

διάτρησις, add '**3** *tunnel*, Ael.*NA* 16.15.'

διατρῐβή I 2 d, after '*school*' insert 'of rhetoric, Aeschin. 1.175'   **III,** add 'Men.Rh. p. 335, 336, al.'

διατρῑβω III, *put off by delay*, add 'Men.*fr.* 265 K.-Th.'

διατροπή 1, for '*fiasco, débâcle*' read '*upset*'

διατροφή, line 3, for '*means of subsistence*' read '*food, nourishment*'

*διάτροχος, ον, *running* or *extending through*, *Inscr.Délos* 500*A*19 (iii BC).

*διατρῠφής, ές, *pulverized*, Nic.*Th.* 709 (s.v.l.).

διατυγχάνω, delete the entry.

διατῠπόω 3, add '[δι]ετυπώθησαν ἐπιμελεῖσθαι i.e. were appointed as ἐπιμεληταί *SEG* 11.464 no. 24 (Sparta, AD 359), *SIG* 3.908 (Megara, AD 401/2)'

διατύπωσις I 3, for 'Alex.*Fig.* 3.25' read 'Alex.*Fig.* 1.24'   **II,** add '**2** term for tax assessed in money or kind in accordance with the schedule, = Lat. *delegatio*, *PCol.* 137.3 (AD 301/2), *PCair.Isidor.* 42 (AD 303).'

διαυγάζω II, after 'J.*AJ* 5.10.4' insert '(v.l. διυπνισθείς)'

διαυγής, line 4, after 'of stars' insert '*shining through* clouds'; line 5, transfer 'of gems .. 204' to follow '*AP* 9.227 (Bianor)';add 'of style, *transparent*, D.H.*Dem.* 5'

*διαύλη, ἡ, = δίαυλος I 1, *SEG* 29.147.8 (pl., Athens, ii/iii AD).

†διαυλία· ὅταν δύο ᾄδωσι Hsch., cf. *EM* 269.30.

δίαύλιον, for '*flute*' read '*aulos*'

δίαυλος I 1, for '*double pipe .. course*' read '*track running up one side of a stadium and back down the other* or the *race* along it'; add 'Hp.*Vict.* 2.63'   **b,** delete the section.   **2,** add 'of an outward and return voyage through a strait, E.*Tr.* 435'   **II 1,** delete the section.

διαφᾰνής I 2, for '*red-hot*' read '*glowing with heat, incandescent*' and for 'Hdt. 2.9' read 'Hdt. 2.92.5'

*διάφαργμα, v. ‡διάφραγμα.

διαφᾰσις, line 2, for 'ἐκφάσεις' read 'ἐμφάσεις' and for 'Plu. 2.354b' read 'Plu. 2.354c'; transfer 'Cic.*Att.* 2.3.2' to precede 'metaph.'   add '**b** *gap* or *space in a construction through which light can pass*, *Iasos* 22.9.'

διάφαυμα, add '*PSI* 939.4 (vi AD)'

διαφερόντως II 2, add '*Cod.Just.* 1.2.24 pr.'

διαφέρω II 3, after '*distract*' insert 'τινὰς λόγοισι E.*HF* 76'   **4,** for '*give each man* his vote' read '*vote on this or that side*'; add 'also ψῆφον δ. *put* a question *to the vote*, *BCH* 87.3 (Locr., v BC); pass., περὶ τούτου ψήφου διενεχθείσης *BCH* 86.58.25 (Maced., ii AD)'   **III 3,** line 6, after 'matters' insert 'E.*Hyps.fr.* 60.46 B.' line 3, after 'Th. 3.39' insert 'μηδὲν τῶν τυχόντων Isoc. 1.48, X.*Mem.* 4.2.2'; after 'Alex. 36.6' insert 'X.*Mem.* 4.2.1; ἐπί τινι ibid. (v.l.)'; line 9, after 'abs., *excel*' insert 'εἴς τι X.*Cyr.* 1.1.6'   **8,** line 2, before 'of persons' insert 'τινος *MAMA* 3.421 (Corycus), al., *IG* 10(2).1.790 (Thessalonica, iv AD)'   **IV,** line 8, after '(D.) 9.8' insert 'cf. Plu. 2.80d (w. nom. part.)'

διαφεύγω 2, delete 'δ. τὰ πολλά .. Jul.*Or.* 7.228a'   add '**b** *to be beyond the competence of*, αἴτιον .. τοῦ διαφεύγειν τοὺς παρὰ τοῖς Ἕλλησιν ἰατρούς τὰ πολλὰ νοσήματα Pl.*Chrm.* 156e, Jul.*Or.* 7.228a.'

διαφημίζω, after 'J.*BJ* 1.33.3' insert 'impers., *JÖAI* 23 Beibl. 16 (ii AD)'

διαφθείρω I 4, *lose by miscarriage*, after 'Is. 8.36' insert 'of dogs, X.*Cyn.* 7.2'

*διάφθερμα, ατος, τό, *abortion*, *SEG* 28.421 (Arcadia, before ii AD).

διαφθορά II, concr., add '**2** *means of ruin* or *destruction*, οὗτοί (sc. οἱ σοφισταί) γε φανερά ἐστι λώβη τε καὶ διαφθορὰ τῶν συγγιγνομένων Pl.*Men.* 91c, *Grg.* 464c.'

*διαφίστημι, *provoke*, βασιλέα εἰς ὀργήν *Sardis* 7(1).20.16 (vi AD).

διαφοιτάω I, for '*wander, roam*' read '*wander throughout* a place'; delete '*go backwards and forwards*'; after 'of a report' insert 'δ. τῆς πόλεως D.Chr. 34.20'

διαφορά VII, for this section read 'perh. *deficit*, *PEnteux.* 27.10 (iii BC)'

διαφορέω I 4, for '*tear in pieces*' read '*scatter in pieces*' and add 'fig., ὕψος τὰ πράγματα πάντα δ. Longin. 1.4'

*διαφορητέον, *one must disperse*, Archig.ap.Gal. 12.676.13.

διάφορος I 2, w. gen., add 'Is. 12.10'   **II 4 b,** at end transfer 'D.L. 6.9' to section 4 a.

διαφορότης, after 'Pl.*R.* 587e' insert '*Phlb.* 13a, *Prm.* 141c, etc.'

*διαφορόομαι, *transport*, *Vit.Aesop.*(G) 18.

διάφραγμα I, after lemma insert 'διάφαργμα *IAskl.Epid.* 52 A 10 (iv BC)' and after '*Inscr.Prien.* 99.19' insert '*SEG* 32.643.13 (Koroneia, ii AD)'

διαφράζω, for 'only in pf. διαπέφραδε' read 'usu. in redupl. aor. 2 διεπέφραδε; also unredupl. διέφραδον *Inscr.Cret.* 1.xvi 7 (Lato, ii/i BC)'

διαφυή III, add '*BGU* 2333.15 (AD 143/4)'

*διαφύλλω, διάφυλλε· διέτιλλε. διέτεμεν Hsch. (perh. corruption of *διεφύλλιζε).

διαφύομαι II, delete the section transferring 'Emp. 17.10' to section I.

διάφυσις II, add '= κτηδών, Sch.A *Il.* 21.169'; for '*division*' read '*divergent form* or *growth, bifurcation*'

διάφυσον, add '(διάφυσος La.)'

διαφωνέω 2, delete 'Plb. 21.43.23'   **3,** for '*desert*' read '*to be missing*'   **b,** of persons, add 'τῶν ἐπιλέκτων τοῦ Ἰσραὴλ οὐ διεφώνησεν οὐδὲ εἷς Lxx *Ex.* 24.11, *Nu.* 31.49, 1*Ki.* 30.9, *Ju.* 10.13'; of things, add 'Plb. 21.43.23'   add '**4** of wine, *be of poor quality*, *BGU* 1994.5 (240 BC).'

*διάφωσις, εως, ἡ, *the centre of a house open to the light* (cf. αἴθριον), *UPZ* 180a.1.10, al. (ii BC).

*διαχάλιγμός, ὁ, (χάλιξ) *intermediate layer of rubble*, *Inscr.Délos* 507 bis 6 (iii BC).

†διαχειριστικόν, τό, *charge for measuring grain*, *PLond.* 1940 (iii BC).

διαχέω I 4, after 'Philostr.*VS* 2.10.1' insert 'διαχεῖ .. τὸν Ἀπόλλω καὶ ποιεῖ χαίροντα id.*Im.* 1.26'   add '**b** of the effect of smooth sounds upon the ear, D.H.*Comp.* 15, cf. II 4.'   **II 2**, add 'fig., of facts, *leak out*, J.*AJ* 19.1.7'

διάχλωρος, add '*BGU* 2328.5 (v AD)'

διαχρέμπτομαι, add 'Theoc. 15.99 (pap., codd. διαθρύπτεται)'

διάχρυσος, add '**2** *gilded*, φιάλη ἔκτυπος δ. *IG* 11(2).161*B*69 (Delos, 279 BC).'

*διαχωννύω, = διαχόω, Hsch. s.v. διαχοῦ.

διαχωρισμός, add 'δίεσιν· διαχωρισμόν Hsch. (perh. in sense "divorce")'

*διαχωριστικός, ή, όν, *separating*, καιρώματα γὰρ τὰ δ. τῶν στημόνων πλέγματα Hsch. (s.v. καιροσέων).

*διάχωρον, τό, *section, division* of a building, *AD* 12.39 (Eresus).

*διαψαλμός, ὁ, *competition in* τὸ ψάλλειν, inscr. in Ziebarth *Gr.Schulwesen* p. 145 (Cos, ii BC).

διαψάω I, add 'med., διεψήσατο· διεκάθαρε Hsch.'

*διαψηλάφημα, ατος, τό, *unrhythmical passage of instrumental melody*, Anon.Bellerm. 3, 85.

διαψήφισις, add 'used to render Lat. *rogatio*, ἔφη διαψήφισιν προθήσειν περὶ τοῦ νόμου App.*BC* 1.12(51)'

διαψιθυρίζω II, for '*whisper among themselves*' read '*spread gossip by whispering*'

διάψιλος, before '*uncultivated*' insert '*bare, bald*, ψηνός· ψεδνός, διάψιλος Hsch.'

*διαψωμίζω, aor. διεψώμισε (-ησε cod..), compd. of ψωμίζω in corrupt entry in Hsch. s.v. διεψήσατο (perh. separate lemma).

*διβολίς, ίδος, ἡ, = διβολία II, Ps.-Callisth. 24.15.

δίβος, for '*square on the draught-board*' read '*position in board-game* app. resembling backgammon'

*Διγαία, v. °Δειγαία.

δίγαμμα, add 'the symbol used with uncertain significance, Cic.*Att.* 9.9.4 (s.v.l.)'

*Διγενής, epith. of Apollo, in uncertain significance (perh. variant of Διογενής II), *Tit.Cam.* 47.20 (ii BC).

δίγλωσσος II, for this section read '*double-tongued*, στόμα *AP* 9.273 (Bianor); fig., Lxx *Si.* 5.9, 14, 6.1.   **2** app. *loose-tongued*, Lxx *Pr.* 11.13.'

*δίγλωχις, ῑνος, *having two barbs*, neut. pl. -ῑνα, βέλη Paul.Aeg. 6.88.2 (130.1 H.).

διγονία, for '*double parturition*' read '*birth in two stages* (i.e. production internally of an egg and hatching)'

*δίγυον, τό, *measure of two γύαι*, Hsch. s.v. γύης.

διδακτικός, add '**II** [δάκτυλος] δ. *fore*-finger, *PLond.* 1821 in *Aegyptus* 6.194.304.'

διδασκαλία I 2, add 'ἕνεκεν διδασκαλίας λογικῆς Sch.Hes.*Th.* 720 p. 97.17'

διδασκαλικός 2, after 'sc. ὁμολογία' insert 'or συγγραφή'   at end add '*PMich*.III 121'ii 12, xi 13, xii 6 (AD 42); sp. διδεσκ- ib. 123'ii 34 (AD 45/7)'

διδασκάλιον II, for 'Plu.*Lyc.* 14' read 'Plu.*Lyc.* 13'

διδάσκαλος, lines 1/2, within parentheses, fem., add 'also ἡ δέσκαλος (sic) *BGU* 332.9 (ii/iii AD); cf. Mod.Gk. δάσκαλος'   **I**, add 'as ephebic office, Arist.*Ath.* 42.3'   at end add 'Myc. *di-da-ka-re* (?loc.)'

διδάσκω, add '**IV** med., *instruct an advocate*, ἐδιδαξάμην .. Χρυσάμμωνα ῥήτορα *POxy.* 2343.8 (iii AD).'

*διδεσκαλική, v. °διδασκαλικός.

*διδιπλοῦν, τό, a measure of capacity, = 2 διπλᾶ, *BGU* 2175.8 (v/vi AD), *SB* 9295.11 (vi AD), *CPR* 14.4.9 (vi AD): v. *Aegyptus* 55.54ff.

διδράσκω, line 2, for '*Tab.Defix.Aud.* 26' read 'tab.defix. in *RhM* 56.85'

διδραχμία, for '*BGU* 741 iii 3' read '*BGU* 748 iii 5' and before this quot. insert '*PTeb.* 281 (ii BC)'

διδραχμιαῖος, add '**2** δ. τόκος *interest at 2 drachmas per mina per month*, *PHamb.* 28.5 (ii BC).'

δίδραχμος IV, δίδραχμον, after 'τό' insert 'τὸ τέταρτον τῆς οὐγκίας Hsch.' and for '*half-shekel*' read '*shekel*'

Διδύμαῖος, line 3, after 'of Apollo' insert 'Call.*fr.* 191.57 Pf.'

Διδυμεύς (s.v. Διδυμαῖος), add '*SEG* 30.1352 (Miletus, iii AD)'

διδυμητόκος, add 'θεὰ διδυματόκε i.e. Leto, Orph.*H.* 35.1'

διδυμόζυγος, for '*with a pair of horses*' read '*twin-yoked*, i.e. consisting of two moving, acting, etc., together'

δίδυμος I, add '**b** Διδύμη as title of legion (transl. Lat. *Gemina*), ληγίωνος τρεῖς καὶ δεκάτης Διδύμης Mitchell *N.Galatia* 289.'   **II**, add 'cf. Myc. *di-du-mo*, pers. n.'

*διδυμωτός, ή, όν, *forked*, γλῶσσα Cyran. 14 (1.4.16 K.).

*δίδυφον, v. ‡δίζυφον.

δίδωμι, line 1, after '**δίδω**' insert '*POsI*.inv. 1460.7 (*Aegyptus* 31.179, i AD), δείδει (for δίδει) *PMich*.inv. 337 (*ZPE* 22.67, ii AD), line 8, inf., add 'δίδειν epigr. in Keil-Premerstein *Erster Bericht* No. 140 (i AD)'; line 16, for 'Lacon. ἔδον' read 'ἔδον Hes.*Th.* 30' and after '*B*1' insert '(Lacon.); Cypr. *e-tu-wa-n*(*o*) ἔδυϝαν *ICS* 217.6'; line 23, delete 'Cypr. .. ib. 6' and insert 'Cypr. inf. *to-we-na-i* δοϝέναι, *ICS* 217.5, also *to-e-na-i* δοέναι, *ICS* 306.6; Myc. 3 pl. *di-do-si*; pass. 3 sg. *di-do-to* δίδοτοι; fut. 3 sg. *do-se*, pl. *do-so-si*; aor. 3 sg. *do-ke*; pf. pass. part. *de-do-me-na*'   **I 4**, add '**b** *consign, deliver*, ἔδωκεν εἰς κναφεῖον Macho 413 G.'   **5**, line 4, after 'D. 21.87, 24.13' insert 'δ. ψήφισμα *propose*, Aeschin. 2.13'   **II 1**, lines 2/3, after 'Il. 5.397' insert 'cf. Hippon. 39.1 W., Pl.*Phdr.* 254e' and after 'Od. 24.65' insert 'cf. Lib.*Or.* 1.245'   **5**, add '*cause to be, make*, δώσω πάντας τοὺς ὑπεναντίους σου φυγάδας Lxx *Ex.* 23.27, δώσω τὴν Ἱερουσαλὴμ εἰς μετοικίαν *Je.* 9.10'

διέ (s.v. δίε III), add '*SEG* 31.572.B15 (Crannon, c.200 BC), 33.460.12 (Larissa, ii BC)'

διεγγυάω, after lemma insert '(aor. διενεγυήσαμεν· ἐνεχυρισάμεθα Hsch.)'

διεγείρω I, add 'ὄρωρεν· διήγειρε Sch.Nic.*Al.* 339d Ge.'

δίεδρος II 1, add 'also δίεδρος, ὁ or ἡ, *Didyma* 467.12'   **2**, for '*chaise-longue*' read '*double seat*, στρωματίου ὥστε τῷ μήκει ἐπὶ δίεδρον ἢ μικρῷ μεῖζον, διπρόσωπον *PCair.Zen.* 241 (iii BC)' and for 'Antyll.' read 'Herod.Med.'

*διεικάτιοι, v. °διακόσιοι.

διειλύομαι, for '*slip out of*' read '*wriggle through*'; add '*wind across*, χαίτη διειλυσθεῖσα καρήνου Nonn.*D.* 4.364'

διεῖπον II, med., for '*fix upon, agree*' read '*state fully*' and add 'Clearch. 2'

διείργω, line 1, after '**διέεργω**' insert 'aor. 2 opt. διεργάθοι Hsch.; cf. ἐργαθεῖν'

διέκ, after 'etc.' insert 'also w. acc., διὲξ τὸ μύρτον Archil. 32 W., διὲκ πέτρας ἐλάσειαν A.R. 2.558, 3.73'

διεκβολή II, for '*estuary*' read '*outlet of river*' and add 'Lxx *Ez.* 47.11'

διεκδρομή, after 'Ptol.*Tetr.* 102' insert '(v.l., v. °διαδρομή)'

διεκπεραιόομαι, add 'Sopat.Rh. 8.362.16 W.'

διεκπίπτω, line 2, after 'Ph.*Bel.* 57.3' insert 'J.*AJ* 14.15.3'

*διεκφυγγάνω, *come safely through* an illness, etc., ἡ δὲ νοῦσος σπερχνή τε καὶ θανατώδης, καὶ ὀλίγαι διεκφυγγάνουσιν αὐτὴν καὶ μελεδαινόμεναι Hp.*Nat.Mul.* 38.

διελίσσω, add '**2** *cause to revolve*, οὐρανόν *SEG* 7.14.8 (Susa, Hymn to Apollo, i AD).'

δίεμαι I, add 'A.R. 2.330, Nic.*Th.* 755'   **II**, for 'A.*Pers.* 701 (lyr., cod. Med. δείομαι)' read 'cj. in A.*Pers.* 701 (v. ‡δίω)'

*διενεγυήσαμεν, v. °διεγγυάω.

διένεξις, εως, ἡ, app. = διαφορά, Sch.Th. 2.37 (cj., cod. διενίξας).

διενοχλέω, after 'Luc.*Symp.* 14' insert 'πρὶν ἂν .. διενοχλήσειε τὸν ἄνδρα στέλλουσα γράμματα Just.*Nov.* 22.14 pr.'

*διεξάπτω, app. = ἐξάπτω B II, *set fire to, burn*, Gal. 8.415.8.

διεξαρκέω, after '*suffice*' insert 'ἐς τὸ ὀφθῆναι Hp.*de Arte* 11 (v.l.)'

διέξειμι II, line 2, after '(Hdt.) 7.77, etc.' insert 'ἐπαινῶν καὶ διεξιὼν τινας ὡς φιλανθρώπως ἔχουσι D. 23.13'

διεξελαύνω, add 'A.R. 3.879'

*διεξευκρίνοομαι, *make a thorough investigation of*, *PHeid.Gr.*inv. 1281 (*Scritti in onore di O.Montevecchi* ed. E. Bresciani, Bologna 1981, p. 406).

διέξοδος I, add '**6** *outcome, result*, D.S. 4.35.'

*διεξοχετεύω, *channel through and out*, τάφρον .. διεξοχετεύουσαν ἐκ τῶν .. ἀμπέλων τὸ καταφερόμενον ἐξ αὐτῶν ὕδωρ ἐπὶ τὸ χείμαρρον ὑποκείμενον ῥεῖθρον *SAWW* 265(1).8.9 (Lydia, i BC/i AD).

*διεπιβουλεύω, *compete in forming designs* on, τῷ γάμῳ Max.Tyr. 20.4 (v.l. δὴ ἐ.).

*διεπιτροπή, ῆς, ἡ, app. = ἐπιτροπή, the region administered by an ἐπίτροπος, *PNag.Hamm.* 22(h).1.

διέπω, line 2, after '(Il.) 24.247' insert 'χορούς *h.Pan.* 23'

†**διέραμα**, ατος, τό, (ἐράω (B), but sense I perh. influenced by διαίρω), kind of small boat, *POxy.* 3250.24 (AD 63), *BGU* 2027 (AD 296); also διάραμα *PBremen* 48.28 (ii AD), *PMil.Vogl.* 189.6 (AD 208); διάρημα Procop.*Aed.* 6.1; διαίρεμα *PThead.* 26.13, διέρεμα ib. 27.19 (both late iii AD).   **II** kind of vessel used in straining or decanting, Plu. 2.1088e.

†**διεραμάτιτης**, ου, ὁ, *owner* or *operator of διεράματα*, *POxy.* 1197.14 (iii AD); see also °διαραματία.

διέρασις, after lemma insert '(ἐράω (B), but perh. confused w. διαίρεσις)'; for '*lading*' read '*transport*' and add 'also διάρασις *PMil.Vogl.* 189.8, 11 (AD 208), διαίρεσις *POxy.* 2568.16 (iii AD)'

διερείδω II, after 'E.*Hec.* 66' insert 'D.H. 3.20'

διερέσσω 2, for 'E.*Tr.* 1258 (lyr.)' read 'E.*Tr.* 1258 (anap.)'

διερευνάω, after 'Plb. 14.2.1' insert 'ποία φροντίς, ποῖος δὲ λόγος διερευνᾶται παρὰ τοῖσιν; Epicr. 10.5 K.-A.'

διερμηνευτής, add 'θνατος δ᾿ οὐ[κ] ἐν ἀνὴρ διερο[ς] τὰ ἕκαστα εἴποι Ibyc. 1(*a*).26 P.'   **II**, line 1, delete 'after Homer' and after '*liquid*' insert 'αὐήν καὶ διερήν (sc. ἄροσιν) Hes.*Op.* 460; coupled w. τραφερός, Dionys.Bassar.*fr.* 34(b) + (c).3 L.'; line 3, for 'of birds .. *Nu.*

δίερσις   SUPPLEMENT   δικτυαρχέω

337' read 'of clouds, Ar.*Nu.* 337 (s.v.l.)'; line 11, after '(*soaked through*)' insert 'so Arist.*de An.* 423ᵃ24'

δίερσις, line 2, after 'Aen.Tact. 31.18' insert '(διαιρέσει cod.)'

διέρχομαι I 1, after 'Ar.*Au.* 181' insert 'διὰ παντὸς τοῦ μήπω μεμαχημένου τῶν ἐναντίων .. διελθεῖν Th. 7.43.7' **2**, for '*pass through, complete*' read '*complete the course of*' and add 'Sol. 36.17 W.'

δίεσις I 1, add '**b** *sifting, careful investigation*, Lxx *Wi.* 12.20 cod *H*.'

διεστραμμένως, after 'Hld. 2.19' add 'δ. τε καὶ ἀτάκτως Gal. 15.583.17'

διευημερέω, transfer the entry to follow διενεργετέω.

*****διευθυντής**, οῦ, ὁ, = διευθυντήρ II, Vett.Val. 42.24.

διευθύνω I, after '*keep straight*' insert 'εὐθείαν ὁδὸν εὐθύνων ἢ διευθύνειν γε ὀφείλων Modest.*Dig.* 27.1.10.3' and before 'Ph. 1.327' insert 'in fig. phr.' add '**IV** intr., euphem. = *relieve oneself*, Aq. 1*Ki.* 24.4.'

διευκρινέω II, for '*examine thoroughly, elucidate*' read '*subject to a critical judgement, appraise*' and add 'Cic.*Att.* 7.8.3, hoc διευκρινήσεις πρόβλημα sane πολιτικόν ib. 7.9.2'

διευλαβέομαι, for '*take good heed to*' read '*treat with caution*'; for 'δ. μὴ .. ib. 789e' read 'δ. μὴ .. στρέφηται ib. 789e' and delete following 'but'

διευλύτησις, add '*POsl.* 130.13 (i AD)'

διευλύτωσις, add '*CodJust.* 1.2.17.1'

διευτελίζω, for 'Ael.*VH* 14.49' read 'Ael.*VH* 14.48'

διέχεια, after 'Sch.Ar.*Pax* 938' insert 'κατὰ διέχειαν *after an interval*, Heph.*Poëm.* 10'

διεχθραίνω, add 'Aesop. 172 P.'

δίζημαι II, add 'Simon. 9, 37.22 P., Thgn. 83, al., Pi.*fr.* 51a.3 S.-M.'

δίζυφον, add '*PGen.* 117.6 (iii BC), *AP* 9.503.2 (Pall.), διζῡ- ibid. 1 (s.v.l.); also **δίδυφον** *SB* 9907.19 (AD 388)'

διηγηματικός, line 2, for 'ποίησις' read 'μιμητική'

διήγησις, add '*written* or *formal statement*, *POxy.* 3955.26 (AD 611); also to be read ib. 1892.42 (AD 581), 2420.21 (AD 610)'

διηθέω I 1, add '*isolate by filtering*, (ἔστιν) τόπος .. χρυσίῳ, ὅθεν διηθεῖται Lxx *Jb.* 28.1'

διήκω II, line 3, for '(but in an inverted constr.' read 'abs.' and delete closing bracket.

*****διημίεκτον**, τό, app. *a double ἡμίεκτον*, *SEG* 16.495 (Chios, iii/ii BC).

*****διημῖσυς**, εια, υ, *dealing in half-measures*, τέλεον καὶ οὐ διήμισυν δεῖν τὸν νομοθέτην εἶναι Pl.*Lg.* 806c.

διηνεκής, line 5, at end insert 'τῷ διηνεκεῖ βίῳ *in the whole course of his life*, *SEG* 32.1243.6 (Aeolian Cyme, i BC/i AD)'; adv. **διηνεκέως**, add '*SEG* 32.1243.29 (Aeolian Cyme, i BC/i AD)'; lines 14/15, εἰς τὸ διηνεκές, for '*JHS* 33.338' read '*SEG* 30.568'

διήνεμος, add '**2** *swift as the wind*, διηνέμοις τε ταρσοῖς Anacreont. 58.3 W.'

διήρης II, for '*with .. oars*' read '*having two files of rowers on either side*'

*****δίθηκος**, ον, *containing two coffins*, of a tomb (λατόμιον), *JÖAI* 23 Beibl. 204 (Thrace, Rom.imp.).

διθύραμβος, line 1, after 'ὁ' insert 'διθύραμφος Att. vase in *SEG* 16.40 (v BC)'

δίθυρος, line 2, after 'Porph.*Antr.* 3' insert 'κιβωτὸς δ. *IG* 1³.421.200 (Athens, v BC)'

*****διθύσανος**, ον, *having two tassels* or *fringes*, Hsch. s.v. κέρκυ.

*****διϊκανοδοτέω**, *give surety*, *SEG* 39.1180.102, 124 (Ephesus, i AD).

διϊκνέομαι, for '*go through, penetrate*' read '*penetrate* or *carry as far as*' **2**, add '*h.Cer.* 416'

διϊπετής, add 'cf. °διαιπετής'

διϊππεύω, add 'Arg. ii in Ar.*Pax*, τοῦ πότου διϊππεύοντος *the drinking-bout proceeding*, *Vit.Aesop.*(G) 68'

διΐπταμαι, after 'Arist.*Mir.* 839ᵃ23' insert '*SEG* 37.529 (Epirus, iii BC)'; add 'act. part. ὡς ὀρνέου διπτάντος ἀέρα Lxx *Wi.* 5.11'

διϊστάω, delete 'D.T. 642.31'

διΐστημι I 4, for 'Ath. 7.305d' read 'Ath. 7.303d' **III**, at end for '*spread*' read '*stretch from one point to another*'

διϊσχῡρίζομαι II, after '(Pl.)*Ep.* 317c' insert '*PMich*.XIII 659.14 (vi AD)'

*****διϋχύρισις**, εως, ἡ, *affirmation*, Epicur. 31[21].8 A.

Διϋσωτήρια, add 'cf. °Διοσωτήρια'

*****διϋτητικός**, ή, όν, *penetrable*, Phlp. *in GC* 214.27.

⁺δικᾰδία, ἡ, *vessel with the capacity of two kadoi*, *IG* 2².1695.3, *Ath.Asklepieion* III 17 (both iii BC).

δικάζω I 2 a, lines 10/11, after 'X.*Cyr.* 1.2.7' insert 'transf., δ. ἀνδρείᾳ τινὶ τὸ γύναιον *adjudges*, Procop.Gaz.*Ecphr.* p. 169 B.' **4**, add 'also w. dat. of thing, ὁ δικάζων τῇ ὑποθέσει *CodJust.* 4.21.22.6, *Just.Nov.* 15.3.2'

δικαιοδοτέω, line 2, after 'c. acc.' insert 'rei'; add 'of the person, *decide in favour of*, πρὸς τὴν κολωνείαν ἐδικαιοδότησα ὑμᾶς *Thasos* 186.3-4 (i AD)'

δικαιοδότης, after 'generally' insert '*OGI* 646'

δικαιοκρίτης, add '*Orac.Tiburt.* 101'

δικαιολογικός, after '*judicial*' insert '*PGrenf.* 2.7a.10 (iii BC)'

δικαιοπραγέω, add '**2** pass., *receive satisfaction*, *POsl.* 40.18 (ii AD).'

δίκαιος A 1, lines 3/4, for '[Γαλακτοφάγοι] δικαιότατοι' read 'Ἀβίων .. δικαιοτάτων ἀνθρώπων' **3**, delete 'euphem.' add '**4** as title of god, Ὅσιος καὶ Δίκαιος Mitchell *N.Galatia* 45, 242, etc., Ὅσιος Δίκαιος *SEG* 34.1294 (Phrygia), etc.' **B I 1**, add '**c** of measures, *exact*, Lang *Ath.Agora* XXI Ha 25, 27 (iii AD).'

δικαιοσύνη I 1, before 'Pl.*R.* 433a' insert 'Lys. 2.14, 12.5' **2**, delete the section transferring exx. to section 1. **III**, after 'personified' insert 'E.*fr.* 486'

δικαιότης, after lemma insert 'Dor. δικαιότας *SEG* 30.119 (Entella, Sicily, iii BC)' and after 'Pl.*Prt.* 331b' insert '*Gorg.* 508a'

*****δικαιοῦχος**, ον, *upholding justice*, *AA* 1934.168 (Olynthus, iv BC).

δικαιοφᾰνής, add 'Modest.*Dig.* 27.1.15.11'

*****δικαιόφρην**, ενος, = *δικαιόφρων, *righteous-minded*, epigr. in *AEM* 19.109 (Callatis).

*****δικαιοφροσύνη**, ἡ, *sense of justice*, *SEG* 38.1310 (δικεο- Phrygia, Rom. imp.).

δικαιόω I, delete the section. **III 1**, after '*punish*' insert 'νόμος .. δικαιῶν τὸ βιαιότατον Pi.*fr.* 169.2 S.-M.' and after 'pass.' insert 'δικαιωθείς *brought to justice*, A.*Ag.* 393 (lyr.)'

δικαίωμα 1 c, line 1, after '*documents* in a suit' insert '*SEG* 17.415.4 (Thasos, iv BC)'

δικᾰνικός II 1, add 'comp., D.H.*Vett.Cens.* 5.2'

δίκᾰσιμος, add 'D.*fr.* 11'

*****δικαστᾰγωγία**, ἡ, *function of δικασταγωγός*, *Delph.* 3(1).362 i 28.

δικαστᾰγωγός, for '*to their homes*' read '*to* and/or *from their homes*' and after '(ii BC)' insert '*SEG* 11.491 (Sparta, ii AD)'; add 'cf. Πολέμων 1.119ff. no. 422.6'

δικαστήριον, add '**3** pl., *hearings*, *POxy.* 3126.10 (AD 328).'

*****δικαστορεύω**, *serve as a magistrate*, (prob. δικάστωρ, unattested equiv. of δικαστής) ῥόλουρος δικαστορεύϜων ἔτευξε ὁ Παισιάδας τὸ τέγος *SEG* 29.548 (Thessaly, c.550 BC).

δικεῖν, line 2, for 'δίεπει' read 'διέπει' and add 'δικεῖν γὰρ τὸ βάλλειν Corn.*ND* 34 (cj., cod. δίκειν)' **1**, after '(E.)*HF* 498' insert 'w. advl. acc., μάκος δὲ Νικεὺς ἔδικε πέτρῳ ὑπὲρ ἁπάντων Pi.*O.* 10.72' deleting the ref. fr. section 2.

*****δικέλᾰδος**, ον, perh. *with two sounds*, Hsch. s.v. κέρκυ (q.v., διστέλεχος La.).

δίκελλα, for '*fork*' read '*hoe* or *mattock*' and add '*IG* 2².1673.51, Men.*Dysc.* 375, 390, al.'

*****δικελλευτής**, οῦ, ὁ, = δικελλίτης, *PCair.Zen.* 788.20 (iii BC).

*****δίκελλιον**, τό, dim. of δίκελλα, *two-pronged hoe* or *mattock*, *BGU* 2361a ii 13 (iv AD), *PVindob.* G14271 (*Aegyptus* 61.87, vi AD).

δίκελλον, add '[perh. read δίκελλον〈δίκελλα〉. δικραδές· τὸ ἐξ ἑνὸς πυθμένος δύο κλάδους ἔχον Hsch., cf. La. ad loc.]'

⁺δίκεντητον, v. °διακέντητον.

δικέραιος, after 'Antip.' delete '(?)'

δίκερας, for 'Callix. 2' read 'Callix. 2(34) J.' and add '*drinking-vessel*, Theocl. 3'

*****δικεράτιον**, τό, name of a tax, cf. °τετρακεράτιον *REG* 70.120 (Palestine, vi/vii AD).

δίκη I 2, after 'c. gen.' insert 'ἰκτίνου δίκην Semon. 12 W.' **II 1**, add 'referring to the constellation *Virgo*, Max. 208, Nonn.*D.* 6.249' **IV 2 b**, add 'δ. γράφειν *write a speech for the courts*, Demetr.*Eloc.* 229' **3**, line 11, for '(E.)*Heracl.* 852' read '*SIG* 167.37'

*****δικηλιστής**, v. °δεικηλιστής.

*****δικιτωνία**, ἡ, v. °διχιτωνία.

*****δίκίων**, ον, *having two columns*, [πρόθυρον] *IG* 1³.426.67 (Athens, v BC).

*****δικληρία**, ἡ, perh. *allotment in two sections*, *Ath.Agora* XIX L7.9 (iv BC).

δικόλλυβος, read 'δικόλλυβον, τό'

δικολογέω, add '*CodJust.* 2.7.19'

δικόλογος, after '*advocate*' insert 'Phld.*Rh.* 1.38.7'

*****δικορία**, ἡ, *double pupil* (eye disorder), Cyran. 34 (1.16.20 K.).

δικορίασις, add 'pl., Cyran. 75 (v.l. δικορία, 2.40.32 K.)'

δικόρυμβος, add 'Philostr. *VA* 2.3'

*****δικόρωνον**, τό, *pair of crows*, *Vit.Aesop.*(G) 77.

⁺δικράδεστος, prob. ghost-word: v. ‡δίκελλον.

δικρᾰδής, ές, *having two branches*, v. °δίκελλον.

δικραδισμός, *digging in* of manure, *PSoterichos* 4.27 (AD 87).

δίκροος, add 'also πρὸς τοῖς [δ]είκροις *on the zygomatic bones* (which are forked), *APF* 4.271 (iii AD)'

δίκροτος I 2, subst., δ., τό, add '*SEG* 33.684 (Paros)'

δίκρουνος, delete '*with two springs*' and for '*a vase from .. poured*' read '*vase with two spouts*, Antisth.Paph. in *Philol.* 101.105.16'

⁺Δικταῖος, α, ον, *of* or *belonging to Dicte*, mountain in Crete, A.R. 1.509, Call.*Jov.* 47, *Dian.* 199, *Epigr.* 22.3 Pf., Str. 10.4.12; Myc. *di-ka-ta-jo*.

δικτυαρχέω, delete '*in the cult .. (less prob.)*' and for '(Callipolis)' read '(Parium)'

δικτῠβόλος, ὁ, *fisherman* (one who casts nets), *AP* 6.4 (Leon.Tarent.), 6.105 (Apollonid.), 9.370 (Tib.Ill.), Opp.*H.* 4.578.

*****δικτυΐσκος**, ὁ, *railing in lattice-work*, *An.Par.* 4.21.14.

Δίκτυννα, add '*Inscr.Cret.* 2.xvii 1.15 (Lisos, iii BC), *Amyzon* 14.3, 15.3 (iii BC)'

*****Δικτυνναϊσταί**, οἱ, association of worshippers of Dictynna, *IMylasa* 179.4 (Rom.imp.).

δίκτυον 4, add '*IAskl.Epid.* 52A.42 (iv BC)'

δικτυοπλόκος, for 'Poll. 7.139' read 'Poll. 7.179' and add '*SPAW* 1934.1032 (tab.defix., Attica, iii BC)'

δικτῠουλκέω, *haul up with nets*, Call.*Dieg.* viii.14 (fr. 197 Pf.).

*****δίκτυς** (B), = δίκτυον (in quot., sense 4), Alex.Polyh.ap.Eus.*PE* 9.34.11.

δικτυωτός, after 'θυρὶς δ. *lattice-window*' insert '*PMich.*I 38.18 (iii BC)'; add 'also δικτυωτή, ἡ, Lxx *Jd.* 5.28, Hsch.'

*****δικωμία**, ἡ, *group of two villages*, *INikaia* 1551 (iii AD), *IPhrygie* 26.

δίλασσον, add '*PVindob.Worp* 24.4 (iii/iv AD)'

δίλημμα, after '*ambiguous proposition*' insert 'Serv.*Aen.* 2.675, 10.449, Jerome *adv.Rufin.* 3.3'

δίλιτρον, add 'χρηστοῦ δίλειτρον *IG* 14.2417.2'

δίλωρος, for 'dub.sens.' read '*having two* λῶροι (?*epaulettes or sim.*)' and add 'στιχαρίων δ' ἐρεῶν διλώρων *SB* 9305.7 (iv AD), δίλωρον ἐν ib. 11075.9'

⁺δίμαλλος, ον, perh. *made with two fleeces*, *Gloss.*; app. neut. subst., διμάλλων ζυγ(οῦ) *PMerton* 41.10 (v AD).

*****δῑμάτιον**, τό, *double* μάτιον, *PLond.* 1718 passim (vi AD).

διμηνία, delete the entry (v. *Anc.Soc.* 21.33).

*****δίμήνιος**, ον, *two months old*, *PMich.*XII 658.7.

*****διμισσωρία**, ἡ, Lat. *dimissoria*, *BGU* 27.13 (ii/iii AD).

δίμῑτος, delete 'καυσία'

⁺δίμιτρος, ον, *having a double headband*, καυσίαις διμίτροις Plu.*Demetr.* 41.

δίμνους, δίμνουν, τό, add '*SEG* 39.1752 (ii/i BC)'

διμοιραῖος I, add '**2** σιτηρέσια τὰ καλούμενα διμοιραῖα *double annona*, *SEG* 8.355 (Egypt, vi AD).' II, delete '*PMasp.* .. (vi AD)'

διμοιρία I 1, add '*IEphes.* 4337.15 (i AD)'

διμοιριαῖος, add 'τόκος *PMasp.* 126.38, al. (vi AD)'

διμοιρίτης I 1, add 'διμοιρίτην ἢ τριμοιρίτην Luc.*JTr.* 48' **2**, for 'Id.' read 'Arr.' II, delete '*mate* .. 48'

δίμορφος, add '*Com.adesp.* 386'

δινάζω, add 'δίνασεν ὄμμα B. 17.18 S.-M.'

*****δῑνάριον**, v. °δηνάριον.

δίνευμα, for 'prob. in Ar.*Th.* 122' read 'Ar.*Th.* 122 (cj.)' and after 'X.*Eq.* 3.11' insert 'cj., codd. and most edd. δὴ νεύματα)'

δῑνεύω, δῑνέω, line 8, for '*whirl*' read '*swing round in a circle*' I 2, add 'Λευκίππην ἔπι δίνεαι Anacr. 23 P. (ἐπιδίνεαι codd.)'

δίνη 4, add 'Opp.*H.* 4.420'

δῑνω, line 2, for '(v.l. περὶ)' read '(v.l. πέρι)'

Διο-, delete the entry.

διό, after '*on which account*' insert 'Hdt. 7.6.4'; διὸ καί, add 'Arist.*Ath.* 3.6, al., *Peripl.M.Rubr.* 10, al.'

Διοβλής, for '*hurled*' read '*struck*'

Διόγνητος, delete 'contr. for Διογένητος' and add 'Hes.fr. 60 M.-W.'

διοδεύω, line 3, for 'Arr.*Epict.* 2.23, 26' read 'Arr.*Epict.* 2.23.36'; add '*PCair.Zen.* 367.33 (240 BC); med., *BGU* 1273.56 (222/1 BC), *IEJ* 16.59.28 (c.200 BC)'

*****διόδιος**, ά, app. = διόδος, *BCH* 116.200.6, 7 (Phocis, iii BC).

δίοδος I, add '*right of way*, *SEG* 30.568 (Maced., ii AD); "*street*" in a military camp, δίοδον .. ἣν καλοῦσι πέμπτην Plb. 6.30.6'

διόδους, before '*Gloss.*' insert '*Edict.Diocl.* 15.47'

*****διόζια**, τά, perh. *forked branches*, Hippon. 92.5 W.

Διόθεν, add 'Il. 24.561, Hes.*Sc.* 22, Thgn. 197, Pi.*N.* 4.61, 6.13, A.*Ag.* 470, *Ch.* 306, *Pr.* 1089, *Supp.* 437, E.fr. 916.6'

διοικέω I 1, line 18, before 'pass.' insert 'in ref. to the imperial secretary *ab epistulis*, ἐπιστολὰς διοικεῖν D.C. 72.7.4, 78.13.4'; lines 21/3, delete 'pf. pass .. cf. 40' add '**c** *manage, control* a person, Alciphr. 2.2.' **2**, for this section read '*contrive, bring about*, τὸ εὐπρεπὲς δ. D.H.*Rh.* 9.3; med., D. 18.178; δ. οὕτως ἀδίκους πλεονεξίας id. 44.38, cf. 40; ὁ ἕτερα λέγων καὶ ἕτερα διοικούμενος λόγος id. 24.27, τόκους *SIG* 672.73 (Delph., ii BC), *SPAW* 1936.380 (Thestia, ii BC), Wilcken *Chr.* 167.26 (ii BC). **c** *provide for* a person, τὴν ἀδελφήν D. 24.202, διοίκει κἀμὲ καὶ τὴν μητέρα Men.*Dysc.* 739, Str. 14.2.24.' **5**, delete the section.

διοίκημα, add '*administrative act*, *IG* 12 suppl. 365 (pl., Thasos, ii BC)'

διοίκησις I, line 6, before 'ἔσπως' insert '*provision, funds*' add '**b** *ordering, government*, τῶν ἐνθάδε (i.e. in the world) Aristid.Quint. 2.2; τῶν ὅλων id. 3.26.' II, add '*group of provinces as an administrative unit of the church*, *Cod.Just.* 1.3.45.6'

διοικοδόμησις, for '*fortification*' read '*walling off*'

διολκή, add '**III** = παρολκή I, *delay*, *PMich.*VIII 486.8, *PFam.Teb.* 24.93 (both ii AD).'

⁺δίολκος, ὁ, *passageway for the transport of cargo-boats (and occasionally warships)*, spec. across the isthmus of Corinth, Str. 8.2.1.

⁺Διομανής, διομανεῖς· ὑπὸ Διὸς μαινόμεναι· ἢ τῷ Διὶ βουλόμεναι μάχεσθαι διὰ μανίαν Hsch.

*****Διόμβρια**, τά, name of festival, *SEG* 1.327.14 (Callatis, i AD).

διομολογέω II, add 'abs., Aen.Tact. 24.5'

Διονυσιάς I, add '**5** *a precious stone*, Plin.*HN* 37.157.'

Διονύσιος II, add '*SEG* 25.744 (Callatis, ii BC), 39.1001.1 (Camarina, ii/i BC)'

*****Διονῡσίς**, ίδος, ἡ, *female votary of Dionysus*, *SEG* 11.610.3 (Sparta, ii AD).

Διόνυσος, line 3, for 'and Δεύνυσος' read 'Ion. Δεύνυσος', before 'also Δίνυσος' insert 'Aeol.' and after '(Amorgos)' insert 'Δινύσō (gen., Berezan, v BC), Aeol. Δίννυσος *SEG* 32.1243.12 (Cyme, ii BC/i AD); Ζόννυσσος Alc. 129.9 L.-P., Ζόννυσος *IG* 12(2).69 (Mytilene, ii AD)'; at end add '[Δ]ιFονύσου *SEG* 29.360 (Argos, iv BC) cf. pers. n., ΔιFονύσεις *IPamph.* 77 (Aspendos, ii BC), etc.   Myc. *di-wo-nu-so*'

*****Διονῡσοτροφικός**. ή, όν, *concerned with the sustenance of Dionysus*, βωμός Robert *Ét.Anat.* 289 (Paphlagonia, Rom.imp.).

Διοπετής, line 1, after '*from Zeus*' insert 'i.e. *from heaven*' and after 'E.*IT* 977' insert 'ἀστήρ id.fr. 971'

διοπεύω, for '*captain*' read 'δίοπος' and add 'διοπεύειν· ἐπιμελεῖσθαι νεώς Hsch.'

διόπη, for '*IG* 2.652*B*26' read '*IG* 2².1388.76 (Athens, iv BC)' and add '*Inscr.Délos* 104.51 (iv BC)'

*****διοπλήξ**, ῆγος, ὁ, ἡ, *struck by Zeus*, Hippon. 19.1 W.

*****διόπομπος**, ον, *sent by Zeus*, αἰετός Sch.Pi.*I.* 6.53.

δίοπος (A) I, add 'cf. as pers. n. *SEG* 34.939 (Camarina, vi BC)' II, for '*captain of a ship*' read 'a ship's officer'

διοπτήρ, add 'also διοπτήρ· κατάσκοπος Hsch.'

*****διοργανισμός**, ὁ, *apparatus*, Zos.Alch. 252.17.

διοργάνωσις, add 'Iamb.*Protr.* p. 117.12 P.'

διορθόω II, line 7, for '(Plb.) 12.28.5' read '(Plb.) 11.28.5'

διόρθωμα II, add '*SEG* 23.305 (Delphi), 34.558 (Larissa, both ii BC)'

διορθωτής 1, as transl. of Latin *corrector*, add 'perh. δι]ορθωτὴν Γαλ[ατίας ἐπαρχείας] Mitchell *N.Galatia* 414'

⁺διορία, ἡ, *interval*, *Vit.Aesop.*(G) 82, cf. ib.(W).

διορίζω, line 1, fut., add '-ίσω E.*Melanipp.Capt.fr.* 6.26'

*****διορκόω**, *swear solemnly*, *SEG* 9.3.15 (Cyrene, iv BC).

διορμίζω, after 'D.S. 20.88' insert '*Peripl.M.Rubr.* 39, 55'

*****δίορον**, τό, *dividing-line* between day and night in an ἀνάλημμα, Hero *Dioptr.* p. 304 S.

*****Διορύγτης**, ου, ὁ, title of Asclepius, *IApam.* 5, 6 (-είτης).

δῖος I 2, add '*TAM* 4(1).48.8 (Rom.imp.); in word-play, cf. II, οἱ μὲν δὴ οὖν Διὸς δῖόν τινα εἶναι ζητοῦσι τὴν ψυχὴν τὸν ὑφ' αὑτῶν ἐρώμενον Pl.*Phdr.* 252e'   at end add 'Myc. *di-wi-jo* (variant *di-u-jo*), neut. subst., *shrine of Zeus*; *di-wi-ja* (var. *di-u-ja*), app. name of a goddess'

Δῖος, before 'ὁ' insert '(in inscr. freq. Δεῖος, sts. Δῆος'   at end add 'Myc. *di-wi-jo-jo* (gen.)'

Διόσδοτος, add 'διόσδοτον ἀρχάν Pi.fr. 137.3 S.-M.'

διοσκέω, for 'Anacr. 3' read 'Anacr. 14.3 P.' and delete 'διαπολέσαι, διαφθεῖραι'

Διόσκοροι, line 3, after 'h.Hom. 33.1, etc.' insert '*IG* 12.3.359 (Thera, viii/vii BC)'; dual, add 'Διοσκόρō *BCH* 102.51 (Attic vase, vi BC), oblique -οιν app. in *SEG* 30.1456 (Pontus, v BC)'   add '**IV** σημεῖον ἐν θυτικῇ, Hsch. s.v. Διοσκουροι.'

Διοσκουριασταί, add '*BCH* 10.425 (Cedreae, ii BC)'

*****Διοσσωτηριασταί**, οἱ, an association (κοινόν) connected w. the Διοσωτήρια, Lindos 683 (i BC).

*****Διοσ(σ)ωτήρια**, τά, *festival of Zeus Soter at Rhodes*, Maiuri *Nuova Silloge* 19 (iii/ii BC); cf. Διισωτήρια.

διοπτήρ, v. °διοπτήρ.

διότι I 2, before 'Henioch.' insert 'Pl.*Ion* 536d'   at end add 'see also διεκί'

⁺διού, v. °δύο.

διούγκιον, add '**2** *sixth of an estate*, *Cod.Just.* 4.6.6.16c, Just.*Nov.* 18.5.'

διουργεύω, delete the entry (cf. *Delph.* 3(1).457.9).

διοχλίζω, add 'Sch.Nic.*Al.* 452a Ge.'

δίπαις, for 'Cypr. .. 93 H.' read 'Cypr. *ti-pa-se*, δίπας *ICS* 84.3'

*****δῑπάλαστος**, ον, = διπάλαιστος, *Inscr.Délos* 1442*A*47 (ii BC).

⁺δίπαλτος, ον, "*doubly brandished*", δίπαλτα πολεμίων ξίφη, app. *swords brandished by two men*, E.*IT* 323; δίπαλτον .. κεραυνοφαές πῦρ (?forked), id.*Tr.* 1103.   **2** perh. *brandishing with two hands*, πᾶς .. στρατὸς δίπαλτος ἄν με χειρὶ φονεύοι S.*Aj.* 408.

*****Διπᾰνάμια**, τά, *festival at Rhodes*, *ASAA* 8/9(1925/6).317 (i BC), *Clara Rhodos* 6/7.437 (Camirus, i AD), Lindos 490.11 (ii AD); cf. Μάναμος.

*****δίπεμπτον**, τό, *two-fifths*, *BGU* 2331.12 (AD 91).

δίφωνος, add 'Peripl.M.Rubr. 20'
*δῐχαλκαῖος, α, ον, costing two chalci, PCair.Zen. 19.5 (iii BC).
*δῐχαλκηρός, όν, app. adj. fr. unkn. flower name, στέφανοι διχαλκηροί PAlex. 22.2 (ii AD).
διχαστός, add '2 divided in two equal parts, τὴν δὲ γῆν .. καὶ τὴν ἀγορὰν .. νείμασθαι διχαστὴν ἑκάτερος Ath.Agora XIX L4a.18 (iv BC).'
*δίχθα, ἡ, a perfume, Edict.Diocl. 36.85, 86.
διχθάδιος, line 1, after 'Il. 14.21' insert 'φονέας Call.fr. 177.32 Pf., A.R. 3.397'; line 2, delete 'simply, two'
*δῐχῐτωνία, ἡ, double (perh. lined) χιτών, PTeb. 514 (sp. δικιτ-, pl., ii AD).
*δῐχοινῑκον, τό, measure of two χοίνικες, Ar.Nu. 640.
διχοινικός, delete the entry.
*δῐχόνιον, τό, two-χοῦς jar, POxy. 3942.20, al. (AD 606).
†δῐχονοέω, to be in two minds, PMich.inv. 203 (ZPE 29.290).
*δῐχότης, ητος, ἡ, state of being cut in half, of a phase of the moon, PMich.III 149 xi 34.
διχοτομέω 2, add 'fig., τῷ υἱέῳ μου .. τῷ διχοτομήσαντί με τοῦ πολοέ(τι)ον ζῆν i.e. who has condemned me to live alone, etc., MAMA 8.252b'
*δίχρῡσον, τό, coin of value of two χρυσοί, Inscr.Délos 338Ba14.
*δίχῠτος, ον, formed by fusion of two (metals), πέταλον Maria Alch. ap.Zos.Alch. 146.16.
*δίχωρος, ον, app. having two compartments, [οὐ]εστάριον γυναίκιον δ. PMasp. 340ᵛ.41.
*διψάκιον, τό, = δίψακος I, Anon.Alch. 20.3.
διψάς II 1, add 'cf. δ.· ἔχις. ὕδρα. καὶ σημεῖον ἐν θυτικῇ ἐπὶ τοῦ ἥπατος Hsch.'
διψάω 2, line 5, w. inf., add 'AP 6.335 (Antip.Thess.)'
*διψέλλιον, τό, double armlet, PRainer Cent. 161.29 (?v AD).
δίψησις, for 'cf. δίψα' read '(v.l. δίψης)'
δίψιος, add 'III Δίψιος (sc. μήν), ὁ, name of month at Larissa, BCH 59.515 (iii BC).' at end add 'Myc. di-pi-si-jo-i (dat. pl.), prob. divinities'
δίω, last line, for 'f.l. for δίεμαι, id.Pers. 700' read 'w. inf., δίομαι μὲν χαρίσασθαι, δίομαι δ' ἀντία φάσθαι id.Pers. 700, 701'
*διωβελιαῖος, v. ‡διωβολιαῖος.
διωβολιαῖος, add 'also διωβελιεῖος (sic) JHS 74.87.30 (Caunus, ii BC)'
διώβολον, add 'IG 1³.236.10'
διωγμείτης, delete the entry.
*διωγμίτης, ου, ὁ, policeman in some Asian cities (v. SEG 33.1591), BCH 52.409 (Pisidia), IGRom. 4.580 (Aezani, sp. -είτης).
διωθέω I 1, at end for 'Pl.Ti. 67e' read 'Pl.Ti. 68a'
διώκω, line 3, after 'διώξομαι' insert 'A.fr. 204b.5 R. (lyr.)' 3, line 3, after 'X.Mem. 2.1.34' insert '(v.l. διώκει)' III 3, for 'Simon. 29' read 'Pi.fr. 107ab.(a)3 S.-M.' and add 'Ἰάονι τόνδε λαῷ παιᾶνα διώξω Pi.Pae. 2.4 S.-M.'
διωλένιος, line 1, for 'Antip.' read 'Antip.Sid.'
διωμοσία, add 'also app. during the proceedings, Cod.Just. 12.37.19.4, Just.Nov. 22.44.2'
Διώνη I, add 'with a cult at Dodona, SEG 23.474 (Διώνα, iv/iii BC); app. identified w. Hecate, ib. 34.1436 (Apamea, iii AD)'
διωνυμία, add 'Hsch. s.v. Βάκχου Διώνης'
†διορία, ἡ, interval, J.BJ 5.9.1, cf. διωρίαν· διαζυγίαν, διαχωρισμόν Hsch.; δ. ἐσχάτως ἀδόκιμον· ἀντ' αὐτοῦ δὲ προθεσμίαν ἐρεῖς Phryn. 16 F.; see also °διορία.
διώροφος, for 'with two roofs or stories' read 'having roofs at two levels'
δίωτος, add 'cf. fem. subst. diōta Hor.Carm. 1.9.8'
δμωίς, add 'Plu.Cam. 33.4'
δμῶος, after lemma insert 'or δμωός' and delete 'Call.Hec. 1.4.15 (pl.)'
δνοφερός, read 'δνόφεος'; for 'B. 15.32' read 'B. 16.32 S.-M.; v.l. for δνοφερός in Hes.Th. 736 (POxy. 2649); cf. γνόφεον, ζόφεος'
δνοφερός, at end for 'Hp.Morb.Sacr. 16' read 'Hp.Morb.Sacr. 13' and add 'see also °γνοφερός'
*δνοφοείμων, ον, gen. ονος, clothed in dark robes, rest. in IG 2².3606 (verse, Attica, ii AD).
*δοάκιστα· ἀσφαλέστατα Theognost.Can. 52 A., Cyr. (ineditus).
δόγμα 1, for 'that which seems to one' read 'that which seems right or reasonable'; add 'as a course of action, Pl.Ti. 90b, Lg. 854b' 2, add 'of other ordinances, etc., Ῥωμύλου .. τύμβος καὶ δόγματα ταῦτα SEG 32.1256 (Bithynia). b decree granting a privilege, μηδὲ δόγμα τινὶ διδόναι πολιτείας ἢ χρήσεως τόπων δημοσίων SEG 30.568 (i AD).'
†δογμᾰτογρᾰφέω, draft a decree, IEphes. 27.427, al. (AD 104), 4101a; serve as δογματογράφος, MAMA 9.15 (Rom.imp.).
δογματοποιΐα, add 'Alcin.Intr. 36'
*δοιάκι(ς), = δυάκις, epigr. in SEG 33.891.11 (Lydia, ii BC).
δοῖδυξ, add 'Lys.fr. 62a'

δοιοί, add 'form *δϜοιός, cf. Myc. du-wo-jo, dwo-jo, pers. n.'
*δοιώδεκα, poet. for δυώδεκα, twelve, Q.S. 2.595 (v.l.).
δοκεύω 1, add 'Lyr.adesp. 7 P.' 2, for 'Arat. 987, al.' read 'Arat. 1128' and after 'c. gen. .. id. 813' insert '1136'
δοκέω 1, add 'part. τὰ δεδοχμένα Delph. 4(4).357.14 (iii BC)' I, line 1, after 'δέχομαι II 3)' insert 'w. acc., ἀνέμοιο, ὑετοῖο κελεύθους Arat. 803/4' 1 a, lines 3/5, delete 'rarely .. X.An. 5.7.26' b, add 'A.Ag. 1649' 2, before 'mostly' insert 'w. two accs., τούτους τί δοκεῖτε; what do you think about them? X.An. 5.7.26'; line 3, delete 'Ar.Pax 47'; line 4, before 'πῶς' insert 'δ. μέν S.El. 61, 547, OC 995, Ar.Pax 47'; for 'to .. remarked' read 'you cannot think how much, how well, etc.; cf. οἴομαι VI 1' 3 b, lines 3/5, delete 'rarely .. 650' II 4, line 4, after 'Pl.R. 487d' insert 'without dat., X.An. 4.5.1, etc.'; at end delete 'without μοι, X.An. 4.5.1' and insert 'not parenth., opp. τὸ ἀληθές, in seeming, J.AJ 19.2.1; w. subj. ἐπὶ κώμον δοκεῖ ἰωμεν Antiph. 197 K.-A., ἢ δοκεῖ τοὺς θεοὺς ὑμνοῦντες σφόδρα τιμῶμεν; Pl.Epin. 980b' 5, line 6, after 'E.Hec. 295' insert 'Ep.Gal. 2.2; δοκέων, opp. ἀδόκητος, Pi.N. 7.31, Trag.adesp. 482 K.-S. (lyr.)'
†δοκή, glossed by ὑπόνοια, Hdn.Gr. 1.313; δόκαι· ἐνέδραι, παρατηρήσεις Hsch., δοκάν· θήκην id., δοκήν· δόκησιν, δοχήν id.
δοκιμάζω II 1, at end delete 'ἐκπονεῖν .. Mem. 1.2.4' 3, after 'think fit to do' insert 'Macho 371 G.'
δοκιμαστής, for 'money-changer' read 'scrutineer of currency, assayer' and add 'SEG 26.72.5, al. (Athens, 375/4 BC)'
*Δοκιμᾱτογλύφος, ὁ, carver of Dokimeion marble, Λιμναῖος καὶ Διομήδης ἀδελφοὶ ἀγαλματογλύφοι Δοκιματογλύφοι Δοκιμεῖς ἀνέθηκαν SEG 32.1311 (Iconium).
δοκιμεῖον, add 'II perh. test-sample of precious metal from which offerings have been made, CID II 102 iia 5ff., IG 2².1424a.313ff., 7.303 (Oropus, iii BC), Inscr.Délos 1449Aabii32 (ii BC).'
δοκιμή, add '3 characteristic feature, Epiph.Const. in Lapid.Gr. 194.18.'
*Δοκιμηνός, όν, made of marble from Dokimeion, BE 1971.642 (Hierapolis).
†δόκιμος, ον, (α, ον, Tab.Heracl. 1.103) reliable, dependable, Alc. 6.12 L.-P., Heraclit. 28 D.-K. (sup.), Democr. 67 D.-K.; w. inf., δόκιμος δ' οὔτις .. εἴργειν A.Pers. 87. 2 acceptable, approved, ὕμνος Pi.N. 3.11, κριθὰ καθαρά δ. Tab.Heracl. 1.103. b approved, examined, of naval stores, δόκιμα καὶ ἐντελῆ IG 1³.498-500 (v BC); of metal, coin certified, approved, ἀργύριον D. 35.24, PLond. 938.6 (iii AD), χρυσοῦ δοκίμου κεφαλαίου 991 (vi AD). 3 esteemed, distinguished, ἀνδρὸς δοκίμου ἐν Μήδοισι Hdt. 1.114.3, 124.3, 152.3, al., οἰκίης .. ἐὼν δοκίμου 5.66.1, ποιηταὶ Phld.Po. 1.11, 2.12, J.AJ 1.35, AJ 6.191, al. (all sup.); of rivers, Hdt. 7.129.2; w. dat. δοκιμώτατος Ἑλλάδι E.Supp. 277; w. παρά, Ἀρταχαίην, δόκιμον ἐόντα παρὰ Ξέρξῃ Hdt. 7.117.1. 4 adv. -μως, surely, for certain, Parm. 1.32 D.-K., A.Pers. 547, X.Cyr. 1.6.7.
†δοκιμόω, = δοκιμάζω, Hsch.; ἢν δοκιμώσῃς (sc. τὴν γραφήν) if you approve, Ps.-Pherecyd.ap.D.L. 1.122.
δοκίμωμι, add 'cf. δοκίμοι Theoc. 30.25, 26 (3 sg., cj.)'
δοκίς I 1, add 'Opp.H. 4.536'
δοκός, line 5, for 'one who has swallowed a poker' read 'of a stiff or inflexible speaker' and for 'Ar.' read 'Arist.'
δόκος I, add 'b perh. expectation, LW 1170.14.' II, add '(perh. III ἐς δόκον· ἐς ἐνέδραν Hsch.'
*δόλαντρον, τό, horn for carrying grease, δ.· κέρας γλοιὸν ἔχον, ᾧ ἁμαξεῖς χρῶνται εἰς τὸν ἄξονα Hsch.; also δόλατρον, Theognost.Can. 52 A.
δόλιος, line 6, after 'Hermes' insert 'SEG 37.1673 (Cyrenaica, vi BC)'
*δολιχαδρόμος, v. δολιχοδρόμος, °δολιχοδρομεύς.
*δολιχαῖος, ὁ, = δόλιχος, SEG 29.147.6 (Athens, c.AD 200).
δολίχαυλος, add 'Certamen 126'
*δολιχεύς, έως, ὁ, competitor in the δόλιχος (race), ABSA 26.213 (Sparta, ii AD), Pi.O. 12 tit. (v.l.).
δολιχεύω, add 'fig., of a runner in the race of life, SEG 35.1427 (Side, iii AD)'
*δολιχοδρομεύς, έως, ὁ, = δολιχοδρόμος; δολιχ[ο]δρομέ[α] rest. in PASA 3.413 (Kara Baulo, Asia Minor), but δολιχαδρόμον is possible.
δολιχόπους, delete the entry.
δολιχός I, add 'A.Pr. 284' at end add 'cf. Myc. pers. n. do-ri-ka-o, compd. do-ri-ka-no'
δόλιχος, after 'opp. στάδιον' insert 'IG 5(1).222 (Laconia, vi BC)' II, for pres. def. read 'a leguminous plant'
δολομήδης, for 'f.l. in Simon. 43' read 'Simon. 70 P. (cj.)'
δολοπλόκος, line 2, after 'Sapph. 1.2' insert 'L.-P., Simon. 36.9 P.'
*δολοποιέω, practise fraud, MAMA 3.225 (Corycus).
δόλος (A) 2, add 'Hdt. 9.90.3, Pl.R. 548a, X.An. 5.6.29, HG 7.1.46, Plb. 4.87.4, al.'
δολοφροσύνη, add 'A.R. 4.687'
δόλοψ, for 'lurker in ambush' read 'scout, spy, Theognost.Can. 52 A.' and add 'also glossed by μάστροπος Hsch.'

διπλάδιος I, for 'Antip.' read 'Antip.Thess.'
διπλάζω I, add 'δίπλαζ᾽ ἕκαστον κῶλον Trag.adesp. 166.3 K.-S.'
δίπλαξ I, line 1, after 'Il. 23.243' insert 'λώπη Theoc. 25.254'
διπλάσιος 1, add 'διπλάσιοι γίγνονται ταῖς προθυμίαις they become twice as eager, Onas. 23.2'   2, add 'ἐκ διπλασίων doubly, Cod.Just. 10.11.8.4b'   4, adv., add 'Gal. 13.872.10'
διπλεθρία, for 'IG 9(1).693.20 (Corc.)' read 'Inscr.Cret. 2.x 1.20, 22 (Cydonia)'
δίπλειον, for 'PPetr. 2 p. 42' read 'διπλείδι θōιέστō IG 9²(1).717.8 (Chalium, v BC)'
διπλῆ, line 1, after 'Gort. 2.7' insert '(v BC), διπλῆ Inscr.Cret. 4.13i (Lap. διπληι, Gortyn, vii/vi BC)'
*δίπλησον, aor. imper. *multiply by two*, perh. erron. for δίπλωσον (διπλόω) PCair. 10758 (vi AD).
*διπλόγραμμος, ον, *having double lines*, Poliorc. 233.5.
*διπλοείλητος, ον, *double-turned* (the precise sense is unclear) ἀστερίσκος, Hero Stereom. 1.77.
διπλόη I, delete 'but usu.' and add '2 *dual nature*, Plu. 2.441d, Aristid.Quint. 2.8.'   II 2, delete 'cf. 441d' and insert 'ἦν μὲν γὰρ ἐξ ἀρχῆς διπλόη τις ὕπουλος ὥσπερ ἐν σιδήρῳ Plu.Per. 11.3'
*διπλοία, ἡ, = διπλεία, Inscr.Cret. 2.xii 5.6, 13.3 (Eleutherna, vi BC).
διπλοῖς I, for 'Antip.' read 'Antip.Sid.'
διπλοκάριος, add 'cf. °δουπλικάριος'
διπλόος 1, delete 'properly of cloaks and articles of dress'   add '7 *having a double* or *composite nature*, ταύρου μεμίχθαι καὶ βροτοῦ διπλῇ φύσει E.fr. 997.'   II, add 'ἐν διπλῷ *doubly*, Cod.Just. 1.1.7.22, Just.Nov. 123.2.1'   III, delete 'in trag.' and add 'A.R. 1.588, AP 7.198 (Leon.Tarent.)'   add 'VI διπλοῦς, ὁ (sc. (?)πίναξ), *contract* or *copy of a contract*, MAMA 8.413.(c), al. (Aphrodisias, Rom.).'
διπλός, line 1, after '(cf. ἁπλός)' insert 'Emp. 17.1, 16 D.-K.'
διπλόω, line 2, after 'Vett.Val. 159.27' insert '*duplicate*, διπλόσσαι (sic) τὴν κλεῖδαν ostr. in ZPE 62.69 (Mons Claudianus, ii AD)'
δίπλωμα II, for this section read '*double document* (a certificate written out twice, on papyrus or on a diptych, and closed or rolled with one copy forming the interior which was sealed), BGU 1113.9 (i BC), ISmyrna 236b.18 (i AD), δ. γάμων BGU 388.31 (ii AD), SB 10530 (ii AD), δ. Ἑλληνικόν PVindob.Bosw. (iii AD); prob. of a licence, δ. ὄνων BGU 213 (ii AD), δ. ἵππων PAmh. 2.92.21 (ii AD); of a travel permit, "passport", Cic.Att. 10.17.4, Fam. 6.12.3; of an order enabling a traveller to use the public post, Plu.Galb. 8, OGI 665.25 (Egypt, i AD)'
*διπλωμάτιον, τό, dim. of °δίπλωμα II, POxy. 2730.3 (iv AD).
διποδιάζω, for 'fut.' read 'Dor. (or Lacon.) aor. 1 subj.'
*Διπολιασταί, οἱ, association connected w. the Διπολίεια, Hesperia 9.331 (Athens, iv BC).
διπρόσωπος 1, add 'στρωμάτιον .. δ. PCair.Zen. 241 (iii BC)'
δίπτυχος II, add 'δίπτυχον ἄτην, of the death of a brother and sister, epigr. in SEG 34.1290 (Phrygia, Rom.imp.); δίπτυχα ὀρχεῖσθαι (sc. σχήματα) Aristid.Or. 28(49).129'   III, add 'Cod.Just. 1.1.7.22'
*δίρ, v. °δίς.
δίρκος, for '= φθείρ III' read '*louse*'
*δί(ρ)ρυτος, ον, perh. *flowing in two ways*, Poliorc. 214.9.
δίς, add 'δϝίς SEG 26.407 (Isthmia, vi BC); also Lacon. δίρ IG 5(1).302.5'
*δῐσακκία, ἡ, = δισάκκιον, POxy. 2424.34 (ii/iii AD), δεισακεία μία PColl.Youtie 84.12 (iv AD), POxy. 1923.6 (v/vi AD), Stud.Pal. 20.67.38.
*δῐσάρχων, οντος, ὁ, *twice ἄρχων*, CIJud. 1 p. 505 (v. BE 1959.259).
*δῐσεπίτρῐτος, ον, διπλάσιος καὶ δ. [λόγος] *two and two-thirds times as great*, the ratio 8:3, Theo Sm. p. 56 H.
*δῐσευποσιάρχης, ου, ὁ, *one who holds the office of εὐποσιάρχης twice*, Dacia N.S. 1.182.7 (Callatis, i AD).
*δίσεφθος, ον, *twice boiled*, Gal.CMG 5.4.(2).206.4, 312.19, 326.12.
δίσημος I-III, for these sections read 'I *in prosody or music, containing two time-units*, χρόνος Aristox.Rhyth. 2.10, μέγεθος ib. 2.31, πούς Aristid.Quint. 1.14; ἔστι γὰρ μακρὰ παρὰ τοῖς μουσικοῖς τεσσάρων χρόνων, ἣν καὶ δίσημον καλοῦσιν ὡς διπλασίαν τῆς παρὰ τοῖς μετρικοῖς μακρᾶς Elias in Cat. 189.9.   II (of vowels), *having two quantities*, i.e. long or short, Sch.D.T. p. 38 H.   III *having two meanings*, Sch.Od. 9.106.'
δισθανής, for '*twice dead*' read '*dying twice*'
δισκεύω, pass., add '*to be hit by a quoit*, GVI 815.4 (Pamphylia)'
δισκέω, add 'Anacr. 19 P.'
δίσκος II 4, after 'gong' read 'Lxx 2Ma. 4.14, Plu.Per. 6, S.E.M. 5.28, 69'
*δισκούσσωρ, ὁ, Lat. *discussor, auditor*, (= λογοθέτης), SEG 8.310.6 (Palestine, vi AD).
*δισκοφᾰνής, ές, *disc-shaped*, ἄρτους δισκοφανεῖς δύο SEG 39.449.9 (Tanagra, v AD).
δισκοφόρος, for '*bringing .. discus*' read 'τὴν δ. (sc. χεῖρα) *holding the quoit*'

†δισμῡριεπτᾰκισχῑλιοστός, ή, όν, *twenty-seven thousandth*, Gal. in QSGNM Bd. 3 [Heft 4].336[128].
*δίσορος, ον, on stele, perh. *tomb having two σοροί*, SEG 6.101 (see also 26.1835, Cotiaeum).
δισπερίοδος, delete the entry.
*δισποντάρχης, ου, ὁ, *one who holds the office of ποντάρχης twice*, ITomis 116 ii 6 (ii AD).
δισσός I 1, add 'διττόν, neut. as adv., *twice*, ITrall. 89'
διστάζω, line 5, after 'Plu. 2.62a' insert 'ὡς οὐ .. Longin. 28.1'
*δίσταθμος, ὁ, *double camp*, given in Hebr. letters as equiv. of Mahanaim, Midrash to Ps. 33 (Tarbiz ii.507).
*διστάξιμος, ον, *doubtful*, Papp. in Alm. p. 93.
*δίστεγος 1, add 'SEG 6.672 (Pamphylia), MAMA 8.498 (Aphrodisias), Sardis 7(1).12; neut. subst. *two-storey building*, BE 1971.644 (Hierapolis)'
*διστέλεχος, cj. for °δικέλαδος, Hsch.La. (s.v. κέρκυ).
δίστοιχος, add 'prob. of a χορός, A.fr. 78c.ii2 R.'
δίστομος II, add '-ος, ἡ, *double axe*, Nonn.D. 30.141'
*δίστῡλον, τό, *colonnade with two rows of pillars*, Princeton Exp.Inscr. 419a.
*δισφυλαρχία, ἡ, *tenure of φυλαρχία twice*, ITomis 123.2 (ii AD).
*δισφύλαρχος, ὁ, *holder of the office of φύλαρχος twice*, ITomis 123.1 (ii AD).
δίσχοινος, add 'τὸν γύαν .. δ. τὸ εὖρος SEG 39.996.3 (Camarina, ii BC)'
δίσωμος, add '2 *designed to contain two bodies*, καμάρα Sardis 7(1).163, κοιμητήριον δίσωμον T&MByz 7.316 n. 10 (Thessalonica, AD 535).'
διτάλαντος, add 'Poll. 9.54'
†δῐτοκέω, *bear two at a birth*, Arist.GA 772ᵃ35.   2 *bear twice*, id.HA 558ᵇ23, Nic.fr. 73.1 G.-S.'
*δῐτομή, ἡ, *splitting* of reeds, POxy. 3354 (iii AD).
δῐτομία, for '*second cutting*' read '*splitting*'
†διττάμενον· ἀρνούμενον. Κρῆτες Hsch.; perh. read ἀνύμενον (v. PP 23.365).
διτταχῶς, add 'PHamb. 128.60 (iii BC)'
*διττότης, ητος, ἡ, *ambiguity*, Anon. in SE 57.25.
διϋδᾰτίζω, for 'Sch.Il. 6.302' read 'Sch.Il. 2.307'
διύλισμα, add 'Ps.-Democr. p. 41 B.'
*διυπερβάλλομαι, *outdo, surpass*, SEG 38.1462.9 (Oenoanda, ii AD).
*διυπέχω, *submit oneself*, διυπ[έχ]ω ἐμαυτόν SB 10724.18 (iii AD).
*διυπόκειμαι, *to be detailed below*, PSorb. 20 (AD 253).
*διυποκολλάω, *attach a document at the end*, POxy. 3365 (AD 241).
διυφή, add 'PCair.Zen. 423 (iii BC)'
*δίφᾰρετρος, ον, epith. of τόξον, *equipped with two quivers*, epigr. in Inscr.Cret. 4.243 (ii BC, -αλε- lap.).
διφάω, line 1, for 'only pres., *search after*' read '*probe, poke into, seek for by delving*'
διφθέρα I, at end before 'used for' insert '*page of a codex*, Afric.Cest. p. 16 V. (pl.)' and for 'bindings' read 'casings'   II, for '*anything made of leather, leather jerkin*' read '*upper-garment made from animal skins*, typically worn by country people, slaves, etc.'   at end add 'Myc. *di-pte-ra*; also in compd. *di-pte-ra-po-ro*'
†διφθερᾰλοιφος, ὁ, Cypr. *ti-pe-te-ra-lo-i-po-ne*, διφθεραλοιφόν (gen. sg.) ICS 143, *schoolmaster*, γραμματοδιδάσκαλος παρὰ Κυπρίοις Hsch.; cf. perh. °ἀλειπτήριον II.
†διφθερίας, ου, ὁ, *person wearing a °διφθέρα II*, Luc.Tim. 8, Poll. 4.137; on the stage, Varro RR 2.11.11.
διφθέριον, add '*piece of leathery tissue*, Hippiatr. 58'
*διφθεροποιός, ὁ, *leather-worker* or *parchment-maker*, MAMA 6.44 (Colossae; sp. -πύς for -πύος).
*Δῐφίλεια, τά, Τραιάνεια, a festival at Pergamum, IEphes. 4114, 1605.9 (Δει-).
διφορέω II, (pass.), add '2 *to be repeated*, Sch.Il. 9.26.'
δίφορος I 1, add 'humorously applied to one Ephorus who paid his fee to his teacher twice, Hsch.'   II, delete the section.
δίφραξ, add 'in the original version of A.R. 1.789 acc. to Sch. ad loc.'
διφρηλᾰτέω, after 'E.Rh. 781' insert 'Theodect. 17'
διφρίον, before 'Tim.' insert '*chair* or *stool*, διφρία ξύλινα δύο Inscr.Délos 1417B ii77 (ii BC)'
δίφρος I, line 8, delete '*travelling-car*' and transfer '*litter* .. D.C. 60.2' to section II.   II, line 4, after '*seat of office*' insert 'Inscr.Cret. 4.160B3 (iv/iii BC, pl.)'; after '*throne*' insert 'Call.Jov. 67'; add 'δ. γυναικεῖος *obstetric stool*, POxy. 3491.8 (ii AD)'
διφρυγής, subst., add 'Heras ap.Gal. 13.779.17'
δίφυής, line 6, after 'δ. Ἔρως' for '*sexual intercourse*' read '*of double nature*, i.e. *androgynous*'   add '3 *bearing two crops* (in the year), [γῆς] χόρτῳ PWarren 2.16 (vi AD), cf. PLond.ined. 1769aᵛ.2.   4 δ. (sc. λίθος), ἡ, *a precious stone*, Plin.HN 37.157.'
δίφυιος III, after 'ζίφ-' insert 'ib. 409.6'   add 'IV δίφυιον νυκτός· τὰ δύο μέρη, τὸ δίμοιρον Hsch.'

δολόω, line 1, after '*take by craft*' insert 'Hes.*fr.* 33(a).18 M.-W.' **II**, line 2, for 'Dsc. 1.81' read 'Dsc. 1.68'; for '*alloy*' read '*give a disguised harmful effect to*'

δόλωσις **2**, for '*alloying*' read '*making noxious with a disguised admixture*'

δόμα (A) **1**, add 'δ. ἀναπόδοτον *PCair.Zen.* 825.3 (252 BC), *UPZ* 2.8 (ii BC), *SEG* 37.851 (Aphrodisias, iv AD)'   **2**, delete the section.

*δομεστικός, ὁ, Lat. *domesticus*, title of palace official, *SEG* 32.1554 (Arabia), *ICilicie* 89, Just.*Nov.* 30.70.2 (all vi AD), etc.

*Δομιτιανός, ὁ, name of month in Egypt (= Phaophi), μὴν Δ. *POxy.* 237 viii 43, *PLond.* 259.99 (both under Domitian).

δομοτέκτων, add '*JÖAI* 23 *Beibl.* 183 (Thrace), *IGBulg.* 690'

*Δονάκεια, τά, festival founded by Δόναξ at Delos, *Inscr.Délos* 366A54, al. (iii BC).

δονακοφοίτης, delete the entry.

δόναξ, line 2, delete '(δονέω "a reed *shaken with the wind*")' **II 3**, delete 'or *limed reed*' and after '(Apollonid.)' insert '6.27 (Theaet.), 28 (Jul.), 29 (id.), Opp.*H.* 3.74; *limed reed*, *AP* 6.109 (Antip.Sid.)' **4**, for this section read 'used in the construction of a lyre, ἕνεκα δόνακος, ὃν ὑπολύριον ἔνυδρον ἐν λίμναις τρέφω Ar.*Ra.* 232'

δονέω **I 1**, line 5, before 'pass.' insert 'A.*fr.* 311.3 R.; med., Βορέης .. δεδονημένος Nonn.*D.* 1.69, 2.80'

δόξα **III 1**, line 2, for 'first in Sol. .. 34' read 'πρὸς ἁπάντων ἀνθρώπων αἰεὶ δόξαν ἔχειν ἀγαθήν Sol. 13.4, 34'   add '**5** *an honour conferred on one, distinction*, τὰς .. ᾿Ελληνικὰς δόξας καλλίστας ἡγούμενοι Lxx *2Ma.* 4.15. **b** accorded to things, δόξαν .. ἀπομέρισον τῇ προτετιμημένῃ .. ἡμέρᾳ ib. 15.2.'

δοξάζω **II**, for '*magnify*' read '*hold in honour*' and add '*OFlorida* 17.10'

δοξαστής **I**, add 'Antisth.*Aj.* 8' **II**, add '*AB* 1.242'

*δόραντον, prob. corruption for °δόλαντρον, Hsch.

*δορατοδέξιος, ον, *holding a spear in the right hand*, poet. in *PKöln* 245.9 (iii AD).

δοριάλωτος, after 'Plb. 23.10.6' insert '*SEG* 32.833 (Samos, 38 BC)'

*δορίδαπτος, ον, "*consumed by the spear*", *Suppl.Hell.* 991.98 (poet. word-list, iii BC).

δορίκρανος, for '*spear-headed*' read '*forming a spearhead*' and add '(cj. δουρικρανοῦς, as fr. δουρικρανής)'

δορίμαχος, add 'cf. as pers. n., *SEG* 34.281e (Corinth, vi BC)'

†δορίπληκτος, ον, Ion. δουρί- A.*Th.* 278, *struck by the spear*, λάφυρα δαΐων .. δουρίπληχθ' A.l.c., Sch.E.*Andr.* 653.

δορίς, for 'Call.*Aet.* 3.1.11' read 'Call.*fr.* 75.11 Pf.'

*δορίσκηπτρος, Dor. -σκαπτρος, ον, *ruling by the spear*, Antip.Sid. in *EG* 3616 P.

*δορίστεπτος, ον, = δοριστέφανος, ῾Ρωμαίων ἀρχάν prob. in Limen. 46.

δορκάδιον, after 'τό' insert 'ζορκάδιον, *Cat.Cod.Astr.* 8(2).164.12'

*δορκᾱδοθήρας, ου, ὁ, *gazelle-hunter*, *PCair.Zen.* 744.1 (iii BC).

δορκάς, after 'ἴορκος Opp.*C.* 2.296, 3.3' insert 'ἴορκες Hsch.'   add '**II** *dice made of deerhorn*, Herod. 3.63; cf. δορκαλίδες.'

δορκών I (s.v. δορκάς), add '*PColl.Youtie* 55.20'

*δοροδόκιον, τό, *beam, transom*, *PMerton* 39.5, 12 (iv/v AD, pl.); cf. δουροδόκος.

δόρπον, add 'S.*fr.* 734'

δορποφόρος, add 'Δορποφόρων ἱερόν *SEG* 28.708 (Paros, iv BC)'

δόρυ, lines 6 ff., for 'dat. δορί or δόρει .. Choerob.*in Theod.* 1.346' read 'dat. δορί; δόρει A.*Supp.* 846, S.*OC* 620, 1314, 1386, Achae.29 (all edd. for metrical requirements, codd. δορί)' **I 2**, *ship*, add 'δ. νήϊον A.R. 3.582; δόρυ alone, *AP* 7.665 (Leon.Tarent.), Opp.*H.* 3.213, *C.* 4.265' **II 1 a**, add 'as symbol of judicial court, [δέκ]α ἀνδρ[α ἐπὶ τῶν] ὑπὸ δόρυ κρι[τηρ]ίων, cf. Lat. *decemvir ad hastam*, *Xanthos* 48' add '**II** Boeot. pl. δόρα, as measure of length, *REG* 10.29 (Thespiae, iii BC).'

δορυθαρσής, for 'also -θρασής' read 'also δοριθρᾱσής' and add 'epigr. in *SEG* 37.751.9 (Lato, *c.*100 BC)'

δορύκνιον, after 'τό' insert '(also sp. δορυχν-)'

*δορύπαλος, gloss on ἐγκεφαλος, Hsch.

δορυσσόος, add 'Nonn.*D.* 13.49, 14.305, etc.'

*δόρωμα, ατος τό, prob. = δόρωσις, *PMich.*1 37.9 (iii BC).

δόρωσις, add '*IG* 2².1682.30 (iii BC)'

δόσιμος, after 'η, ον' insert 'Boeot. δότιμος, *liable to be handed* or *paid over*, τὸ ἄλωμα ἅπαξ δότιμον *SEG* 25.556.25 (Haliartus, *c.*210/200 BC)'

δόσις **I 1 b**, add '*Peripl.M.Rubr.* 32' **II 1**, add 'ἐν δόσει *as a gift*, ἔδωκεν ἐν δόσει τοῖς τε νῦν αὐτοῦ καὶ ἐσομένοις μύσταις *IG* 10.2(1).259 (Thessalonica, i AD)'

**δοσμός, ὁ, Myc. *do-so-mo*, *contribution*; deriv. adj. *do-si-mi-jo* (cf. ἀπυδοσμός, -δόσμιος, s.v. ἀπο-).

δοσοληψία, add '*PLond.* 1727.45 (vi AD, -ληψι- pap.)'

δότειρα, add 'Aeol. πολέμου δότε[ρ]ραν (or -τε[ι]-) Alc. 298.9 L.-P.'

*δότιμος, v. °δόσιμος.

*δουακα, unkn. aromatic, *Peripl.M.Rubr.* 8.

δουκηνάριος **I**, add '*BCH* suppl. 8.267 (Thasos, v/vi AD), δωκενάριος *SEG* 7.1097 (Arabia)'

*δουκιανός, ή, όν, *of* or *belonging to the dux*, ἐξουσία *PMasp.* 283.1.3 (vi AD).

δουκικός, add '*Epigr.Gr.* 446.6 (Arabia, iv AD), δουκικῆς τάξεως *PMasp.* 23.5 (AD 569); *SEG* 32.1554a.20 (Arabia, vi AD)'

*δουκτάριον, τό, cf. Lat. *ductarius*, *a rope used for hauling*, *Edict.Diocl.* in *SEG* 37.335 iii 4.

δουλᾱγωγέω **2**, add '*Suppl.Mag.* 38.9'

δουλᾱγωγός, add '*Vit.Aesop.*(G) 91 s.v.l., cj. δουλαγωγία'

δουλεία, line 1, after 'also' insert 'δουλίη Sol. 36.13 W.'   add '**IV** *legal servitude*, *Cod.Just.* 8.10.12.3.'

δούλειος, line 1, delete 'Pi.*fr.* 223'

δουλελεύθερος, add '*POxy.* 2238.11 (vi AD)'

δουλεύω **3**, for '*render a service*, τινά' read '*behave as a servant*, ἀλλήλοις' add '**4** pass., *to be served*, *Vit.Aesop.*(G) 31, al.'

*δουλίη, v. ‡δοῦλος II.

δουλικός **1**, add 'app. written δουλιδός, *BGU* 2548.6 (ii AD)' **2**, add 'neut. as subst., *slave*, *Peripl.M.Rubr.* 13 (pl.); *slave* or *place of burial of a slave*, [Α]ὐρηλίου Ποπλίου .. ἡ μάκρα σὺν τῷ δουλικῷ *MAMA* 3.795 (Rough Cilicia)'

δουλοκοίτης, add 'μὴ πρόσιθι μοι, δουλοκοῖτα *Vit.Aesop.*(G) 49 (-κόπα cod.)'

*δουλοπαράσῑτος, ὁ, *parasite of slave origin*, *Dura*⁹ 1.217 (iii AD).

δοῦλος (A) **I**, line 4, after '(q.v.)' insert 'for the distinction between δοῦλος and οἰκέτης cf. Chrysipp.*Stoic.* III.86, Ammon.*Diff.* 45'; line 7, after 'A.*Pers.* 242' insert 'of the subject of Roman emperor, *IG* 4.204 (Corinth, vi AD)'   **II**, line 6, after 'etc.' insert 'δοῦλα τοῦ θεοῦ τρία ἀδέλφια *SEG* 36.1182; fem. as subst., ἡ δούλη τοῦ θεοῦ ib. 1186 (both Galatia, v/vi AD)'   at end add 'Myc. *do-e-ro*, *do-e-ra* (fem.)'

δοῦλος (B), delete the entry (v. °δοῦμος).

*δουλοχείρων, ονος, ὁ, *one worse than a slave*, *PMasp.* 353.18 (vi AD).

*δουμοπύραιθοι, οἱ, *members of an association connected w. worship of Cybele*, in Lat. transcription *dumopiretis* (dat. pl.), *JDAI* 44.132-6 (Moesia, ii AD); cf. πύραιθοι.

*δοῦμος, ὁ, *religious association*, perh. a Maeonian wd., Hippon. 30 W., *AP* 7.222.3 (Phld.); *TAM* 5(1).179 (Lydia, ii AD); ἱερὸς δ. ib. 536 (iii AD), etc.; app. an association of women in *IGBulg.* 1925.20 (Hadrian); perh. δούμος (cj., δοῦλος cod.)· ἡ οἰκία ἢ τὴν ἐπὶ τὸ αὐτὸ συνέλευσιν τῶν γυναικῶν Hsch.

δούξ, after '= Lat. *dux*' insert '*commander of a military district*'; add 'ἀπὸ δουκῶν *of ducal rank*, *PMich.*XI 611.4 (AD 412)'

*δουπλικάριος, ὁ, Lat. *duplicarius*, *military title indicating the receipt of double pay*, *PAmh.* 108.3 (ii AD), *SEG* 8.346 (sp. δοφλικάρις, Palestine, v AD); also δουπλικιάριος (var. as in Lat.) *PGrenf.* 2.51.5 (-ιάρις, ii AD), *SEG* 8.608.9 (Egypt, ii/iii AD), *PBerol.*inv. 7347 (*JJP* 19.93, iii AD)'

δοῦπος, at end, rare in prose, add 'Hld. 1.30, 9.8'

δούρειος, add 'form *δόρϝειος, Myc. *do-we-jo*'

δούριος, add 'Paus. 1.23.8, Q.S. 12.110, Hsch.'

δουρίπηκτος, delete the entry (v. °δουρίπληκτος).

*δουρίπληκτος, v. °δορί-.

δοχαῖος, for 'κράδην .. δοχεῖον' read 'subst. δοχαίη, ἡ, *receptacle*, Nic.*Al.* 21 (s.v.l.)'

δοχεῖον, add '**b** *reservoir*, *Lindos* 289, 290 (i BC), *SEG* 37.282 (Argos, ii AD).'

δοχεύς, add '**II** = ὑποδοχεύς, *receiver, host*, *IG* 12 suppl. 365.11 (Thasos)'

*δοχμᾱς, άδος, fem. adj. *slanting*, ὁδός *Delph.* 3(4).42.8.

δόχμιος **I**, for 'ἐμβάνειν' read 'ἐμβαίνων'

δράγμα, after '(δράσσομαι)' insert 'also δράχμα Nic.*Th.* 667; Cypr. *ta-ra-ka-ma-ta* δράγματα *Salamine* 2 (*c.*600 BC)'; for '*handful*; esp. .. *truss*' read '*the amount that can be grasped in the hand*, δράχμα χερὸς πλήσας Nic. l.c.; cf. δράγμα· τὸν τῆς σταφυλῆς βότρυν, καὶ τὰς φοινικίνας βαλάνους Hsch., esp. in reaping'   for '**II** later, *uncut corn*' read 'applied to crops still standing'

δραγματηγέω, add '*BGU* 1511.8 (iii BC)'

δραγματηγία, add '*BGU* 1513 (iii BC)'

*?δραγματολογία, ἡ, app. *gleaning*, rest. fr. ἀπὸ δραγματολ(..) *POxy.* 3473.32 (iii AD)

†δραγματολόγος· ὁ τὰ δράγματα συνάγων Hsch.

†δραφεός, ὁ, *cauldron*, *CEG* 344 (pl., Phocis, vi BC); cf. δραιόν.

δράκαινα, for '*dragon*' read '*serpent*, applied esp. to mythical beings'

*δράκιον, τό, word of dub. sense in inventory of clothing, *PRyl.* 627.14 (iv AD)

δρακονθόμιλος, for '*of dragon brood*' read '*swarming with serpents*'

*δρακονῖτις, ιδος, ἡ, = δρακοντίας 4, Plin.*HN* 37.158.

δράκόντειος, for 'Luc.*Philops.* 4' read 'Luc.*Philops.* 24'

*δρακοντέλιξος, ον, *covered with coiling snakes*, *Suppl.Mag.* 42.2, 63 (iii/iv AD).

*δρᾰκοντιοῦς, perh. *snake-coloured*, -οῦντα Ps.-Callisth. Γ 1.7.1 (unless to be taken as part. of *δρακοντιόω *"turn into a snake"*).

*δρᾰκοντόστηθος, ον, *having the breast adorned with snakes*, of Athena, φένγασπι, δρακοντόστηθε poet. in *PKöln* 245.9 (iii AD).

δράκος II, add 'Poll. 2.147'

δρακτόν, for '*small vase*' read '*measure of capacity*, perh. a quarter of a ξέστης' and add '*MAMA* 8.492, *SEG* 6.185 (Sebaste), *Laodicée* p. 265.15'

*δρακτός, ή, όν, adj. applied to unguents used by athletes, app. *grasped by the hand*, i.e. *solidified* (opp. *liquid*), δρακτοῖς ἐλαίοις *MAMA* 8.484, 485, δ. ἄλειμμα *BCH* 10.520 no. 19 (Nysa).

δράκων I, line 2, delete '*dragon*' and 'interchangeable with ὄφις' IV 4, for '*dragon*' read '*snake*'

*δρᾰκωνάριος, ὁ, *bearer of the serpent-standard*, *MAMA* 1.218 (Laodicea Combusta), *PLond.* 113.1.86 (vi AD), *SEG* 32.1554.A36 (Arabia, vi AD).

*δρᾰμᾰτοθέτης, ου, ὁ, *one who arranges a dance*, rest. in *SB* 7336.

δράξ I, *handful*, add 'Lxx *Le.* 2.2, 5.12, 3*Ki.* 17.12, *Ez.* 13.19' II, *hand*, add 'Poll. 2.144, Lxx *Ez.* 10.2'

†δράπανον, v. ‡δρέπανον.

*δρᾰπετᾰγωγεῖον, τό, *prison for runaway slaves*, *BGU* 1881.7 (sp. -ώγιον, i BC).

δραπετεύω, add '*PStrassb.* 612.23 (Arsinoe, ii AD), *PKöln* 281.10 (vi AD)'

δρασείω, add 'Ar.V. 168'

†δράσιμος, ον, *that can be done*, A.*Th.* 554.

δράσσομαι I 1, after 'Il.l.c.' insert 'Archil. 223 W.' II, line 2, after '*catch*' insert '*APl.* 275.10 (Posidipp.)'

δραστέος II, δραστέον, add 'Pl.*Phlb.* 20a, *Criti.* 108d, *Lg.* 626a, al.'

δραχμή, for 'δράσσομαι .. Plu.*Lys.* 17)' read '(δράσσομαι), δαρκνά *GDI* 4985, *Leg.Gort.* 1.32, al.; δαρκμά *GDI* 5071 (Knossos); δαρχμά ib. 1154 (Elis), *IG* 5(2).3 (Tegea), *SEG* 24.361 (Khorsiai, iv BC), Hsch.; θαρχμά *SEG* 23.392 (Corcyra, vi BC): *handful*, ὀβελίσκων δαρχμαί τριάκοντα πέντε *SEG* 24.361 (Khorsiai, 386/380 BC))' for 'I' read 'II' 2, after 'Pl.Com. 174.17' insert 'Men. *Epitr.* 159'; at end delete 'δαρχμή .. δαρκνά'

δραχμιαῖος 2, add 'Heras ap.Gal. 14.203.14'

δραχμίον, delete the entry.

*δράψ, ἀπός, ὁ, perh. = δραπέτης, Ar.*fr.* 809 K.-A.

δράω (A), line 10, after 'freq. in trag.' insert 'and in prose, e.g. Antipho 2 δ 5, al., Hp.*Epid.* 4.43, al., D.S. 5.55.6, al., Ph. 1.44, al., etc.' 2, add '**b** abs., *work*, of poison, Plu. 2.258c.' II, delete '*IG* 1².4'

δρεπανηΐς, for 'poet. for foreg.' read '*of the sickle* or *sickle-shaped*, Ζάγκλης .. δρεπανηΐδος ἄστυ'

δρέπανον, line 1, for '(q.v.)' read '*SB* 9834b.37 (iii AD)' 2, after '*pruning-knife*' insert '*IG* 1³.422.143 (Athens, v BC), Ar.*Ra.* 576' 3 & 4, for these sections read '3 *weapon with a curved blade*, Hdt. 5.112.2, 7.93; cf. of blades attached to a war-chariot, X.*Cyr.* 6.1.30.'

δρεπᾰνουργός, for '*sword-maker, armourer*' read '*sickle-maker*'

δρηστήρ I, after 'Od. 16.248' insert '*server*, δ. κυπέλλων, of Ganymede, Nonn.*D.* 10.259, al.' and after '(Od.) 19.345' add 'A.R. 3.700'

*δριλοπότης, ου, ὁ, *cup in the form of* δρῖλος, Sch.Juv. 2.95 (codd. drillopotae).

†δρῖλος, ὁ, perh. *worm*, in pers. n. Βροῦκος Δρίλου *SEG* 32.281 (Attica, iii BC); cf. κροκόδιλος, δρίλαξ. 2 *limp penis*, *Suppl.Hell.* 975.1 (ostr., iii BC), *AP* 11.197 (Lucill.), *IG* 9²(1).733 (δρειλ-, mosaic, Amphissa, iii AD); cf. δ., δρίλλος· *verpus*, Gloss.

δριμύλος, for 'ὄμμα .. eye' read 'ὄμματα δ. *piercing* eyes'

δριμύς II, line 5, after '*Cond.* 18' insert '*PCair.Zen.* 33.15 (iii BC; gen. masc. δριμίος)'

δρίφος, add 'Theoc. 15.2 (*PAntin.*, codd. δίφρ-); δ. Συρακούσιοι *EM* 287.50, Hsch.'

*δρίωτος, v. °τρίωτος.

*δρόγγος, ?*troop*, cf. late Lat. drungus, Theognost.*Can.* 46 A.

δρομάς, line 2, for 'ἄντυξ δ.' read 'ἄμπυξ δ.'

*δρομάω, add 'pres. part. δρομέων, *CEG* 815.2 (Epid., iv BC)'

δρομεύς, line 3, for '*Call.fr.* 555' read '*Call.fr.* 441 Pf., cf. Simon. 1 P.' 3, delete the section (v. *ZPE* 93.149).

δρομιάφιον, add '(δρομιάμφιον Salmasius, cf. *Suppl.Hell.* 1075)'

δρομικός, comp., add 'Φήμη ὕδατος ὑγροτέρα, πνεύματος δρομικωτέρα, Ach.Tat. 6.10'; sup., add 'Ach.Tat. 7.16'

δρομικότης, for '*fleetness* .. 18' read '*hypothetical derivative of* δρομικός, Simp. *in Cat.* 214.18'

*δρομίσσω, unexpld. wd. in Theognost.*Can.* 14.

*δρόμος I 1, line 4, after 'Luc.*Dom.* 10' insert 'δρόμους ἀπελαύνω τὸν ἵππον *ride off at a gallop*, Aristid.*Or.* 49(28).5'; line 7, after 'D. 19.273' insert '*a day's run* (for ship), *Peripl.M.Rubr.* 9, 15, al.' 2, add 'in fig. phrs. referring to the course, Pl.*R.* 613d, *Cod.Just.* 1.3.55.3, *SEG* 31.1440 (Palestine, AD 614)' add '5 perh. *runner, courier*, *PBodm.* 28.19.' II 3, line 4, after '(Itanos, iii BC)' insert 'cf. Call.*fr.* 261.3 Pf.'; line 7, after 'Str. 17.1.28' insert 'cf. Call.*fr.* 715 Pf.'

δρόμων I, add 'prob. in *SB* 9855.4 (ii AD)'

δροσερός 1, add 'δ. Γαλατία *SEG* 32.1502.11 (Palestine, v AD)' 2, add '*AP* 5.292.4 (Agath.)'

δροσίζω I, line 3, after 'D.L. 7.152' insert 'δροσιζέσθω .. πρόποσις *AP* 5.134 (Posidipp.)' II, after 'Arist.*Pr.* 939ᵇ38' insert 'ἐὰν μὴ δροσίσῃ *PMil.Vogl.* 60.4 (ii AD)'

δρόσος I 4, for this section read '*surface moistness*, τὸ καθαρόν τε καὶ Ἀττικὸν ὥσπερ δρόσον καὶ χνοῦν ἀποδρεπομένους Plu. 2.79d; on the genitals, Ar.*Nu.* 978' II, delete 'in sg. .. Call.*Hec.* 1.2.3' and add 'cf. δρόσους· ἀχρείους. Κύπριοι Hsch.'

*Δρούσαιος, ὁ, *month in Cyprus beginning 2 Apr.*, prob. in *Cat. Cod.Astr.* 2.144.14.

*Δρουσιεύς, έως, ὁ, *Egyptian month* (= Epeiph) in Caligula's calendar, *PLond.* in *PRyl.* ii p. 381 (AD 40), *CPR* 242.16, 36 (AD 40); also Δρουσεύς, *Atti* XVIII *congresso intern. di papirologia* (1984) pp. 1107-13, *PSI* 908.19 (AD 43; dat. Δρυσί).

*Δρουσιλλῆος, ὁ, *Egyptian month* (= Pauni) in Caligula's calendar, μὴν Δ. *BGU* 1660.12, *PMich.inv.* 622 (*JEA* 13.185, both AD 41).

Δρυάς I, before 'Plu.' insert 'Pl.*Epigr.* 26 D.³'

δρύϊνος, line 3, for 'Antip.' read 'Antip.Thess.'

*δρύμιος, α, ον, Cypr. *tu-ru-mi-o-ne*, *of* or *belonging to a copse*, perh. proper name of stream, *ICS* 217.19 (Idalium).

δρυμός I, add 'spec. of wet or irrigable land, *PWisc.* 31.5, 34.6, al. (ii AD)' II, add 'parox. δρύμα, acc. to gramm. in Reitzenstein *Ind.Lect.Rost.* 1890/91 p. 9'

*δρῡμόχορος, ον, *having wooded dancing-places*, Κιθαιρὼν δ. eleg. in *POxy.* 3723.12.

δρύος, delete the entry.

*δρυοσάνδραξ, gloss on σαννίς, Hsch.

*δρυοστέφανος, ον, *wearing crown of oak-leaves*, δρυοστεφάνοις .. Λυκαίοις *CEG* 814.8 (Argos, 350/325 BC).

*δρυοφόρος, ον, *bearing oak-branches*, of a thiasus in a Bacchic cult, *IG* 10(2).1.260c.11 (sp. δροιο-, Thessalonica).

δρύοχοι I, add 'defined as *ribs* of a ship, = νομεῖς, Procop.*Goth.* 4.22'

δρῦς, line 2, after '(Thyrrheum)' insert 'masc. also in *POxy.* 2113.18 (iv AD)' and after '*Schwyzer* 664.23' insert '(Arcadian Orchomenus, iv BC)'; line 6, for 'gen. δρυός .. verse, Hes.*Op.* 436' read 'δρυός scanned as monosyllable in Hes.*Op.* 436'; line 7, for 'originally *tree* .. trees*' read 'etym. *tree*, cf. δρῦν ἐκάλουν οἱ παλαιοὶ πᾶν δένδρον Sch.Il. 11.86, Hsch., but usu. denoting any of various kinds of oak' and add 'S.*Tr.* 766, Paus. 8.1.6'; line 16, before 'prov.' insert 'applied to unspecified kinds of tree, E.*Cyc.* 615' II, delete the section. IV, delete the section.

δρύσσομαι, delete the entry.

δρυτόμος, add 'Myc. *du-ru-to-mo*'

*δρύφακτος 1, add '*IG* 12(3).326.25 (Thera, ii AD)'

δρύφειν, delete the entry (v. °δρύφω).

*δρύφω, app. = δρύπτω, κῆπὶ ταῖ μύλαι δρυφήται κῆπὶ ταῖς συναικλίαις Alcm. 95(a) P.; δρύφειν· περαίνειν Hsch., δρυφόμενοι· φθειρόμενοι id.

†δρύψελον, τό, app. *flake, paring* of bark, peel, etc., ῥίζης δρύψελα Ποντιάδος Parth. in *Suppl.Hell.* 642, σελίνου δ. ib. 643; cf. δρύψελ[λ]ον· τὸ λέμμα, ὁ φλοιός *Et.Gen.*, δρύψελα (cod. -αλα)· πέταλα δρυώδη Hsch.

†δρύψια, τά, app. *scrapings*, δ. τυρῶν *AP* 6.299 (Phan.).

δρωπᾰκιστής I, add '*SEG* 37.1434 (Apamea, ii AD)' add 'II applied to one who uses wax for peeling the gold from gilded statues, Sch.Juv. 13.151.'

*δρώπακος, ὁ, *depilator*, *SB* 10447ʳ.ii.51.

δρώψ, add 'cf. Clem.Al.*Strom.* 5.48.5, δ. δὲ ὁ λόγος ὁ δραστικός ib. 9'

*δῠανδρία, ἡ, = Lat. duumviratus, *SEG* 19.830, 37.1176, 1177, 1179, etc. (Pisidia, ii/iii AD).

*δύασμα, ατος, τό, *duplication*, Schubart *Gr.Lit.Pap.* 36.31.

*δύβρις, = θάλασσα, Asclep.Myrl.ap.Sch.Theoc. 1.118.

*δύγαστρον, v. °ζύγαστρον.

δυερός, after '*miserable*' insert 'Call.*VB* in *Suppl.Hell.* 257.24, δυερὴν μοῖραν *EA* 13.3-4 no. 496 (Mylasa, Rom.imp.)'; for '*IG* 3.1337' read '*GVI* 1029 (Athens, Rom.imp.)'

*Δύμαινα, fem. adj., *of the* Δυμᾶνες, Δ. φυλή Call.*fr.* 703 Pf.; pl. subst., Alcm. 4 *fr.* 5.4 (rest.), 10(b).8 P., Euph. 47.

*Δῠμᾶνᾱται, οἱ, *the people of the tribe of* Δυμᾶνες (at Sicyon), Hdt. 5.68.2.

*Δῠμᾶνες, οἱ, *name of tribe in Dorian cities*, rest. in Tyrt. 19.8 W., *SEG* 25.394 (Epid., iv BC), St.Byz.

*Δυμᾱνίς, ίδος, = °Δύμαινα, Δυμανίδος ἠπείροιο *Suppl.Hell.* 1173.

δύνᾰμαι I 2 b, add '*Cod.Just.* 1.3.44.1, Just.*Nov.* 47 pr.' II 3 a, line 3, after 'Th. 7.58' insert '(prob. interp.)' and add 'λοιδορίαν οὐκ ἔχουσαν ἔλεγχον ὁ λόγος αὐτῷ δύναται Gorg.Pal. 29, σκοπεῖν ἕκαστον τί δύναται τῶν ῥημάτων Strato Com. 1.44 K.-A.'

δύναμις II 5, add 'D.S. 4.51' and for 'PMag.Leid. v 8.12' read 'PMag. 12.257' **V 2**, for this section read 'math., *square root*, Pl.Tht. 147d, e; limited to surds, ibid. 148a'

δυναστεία II, line 4, for 'Roman Senate' read 'triumvirates'

δυναστευτικός, after '(Arist.Pol.) 1272^b3' insert '*holding a position of power*'

δυναστεύω I, p. 453, line 5, before 'Pass.' insert 'of χαρακτῆρες τῆς λέξεως, *stand out, be most important*, D.H.Dem. 8'

δυνατέω 1, add 'Ep.Rom. 14.4, 2Ep.Cor. 9.8; w. impers. constr., δυνατί .. τῷ κυρίῳ θεῷ .. παρασχεῖν POxy. 3819.9 (iv AD)'

δυνατός, after lemma insert '(Boeot. sp. διουνατόν SEG 28.449.13)' **I 3**, add 'of a goddess, TAM 5(1).250' **II**, line 6, after 'ἐς τὸ δ.' insert 'Simon. 36.14 P.'

***δύνω**, v. δύω.

δύο, line 3, after 'Inscrr.' insert '(Boeot. sp. δούο BCH 60.179.31; διούο, διού IG 7.3193.1, 5; SEG 3.356.4)'; line 4, after 'Thess. fem. δύας ib. 9(2).517' insert 'neut. δύα SEG 26.672.10 (Larissa, 200/190 BC);'; line 6, dual δυεῖν, add 'Men.Dysc. 327'; line 13, dat. δυοῖσι, add 'Hippon. 92.6 W.'; line 14, δυοῖς, add 'Hsch.' at end add 'Myc. *dwo*, instr. *du-wo-u-pi*'

***δυόβολος**, ὁ, *double obol*, PPetr. ii.44.25, SB 5729.9 (both iii BC); also **δυόβολον**, τό, OBodl. 643, 984 (both AD 36); see also διώβολον.

δυοκαίδεκα, add 'gen. -δέκων Alc. 349(c) L.-P.'

δυοκαιδεκάδελτος, add 'Cod.Just. 6.4.4 pr., etc.'

***δυοτριακοστόν**, τό, *thirty-second part*, PGen. 116.13 (iii AD), PCornell 20(a).54, 75 (iv AD); written δυτριακοστόν PMich.v 322.(a)23, al. (i AD).

***δυοτρίαντον**, τό, *thirty-second part*, PCornell 20.13, al. (iv AD).

†**δύπτης**, ὁ, *diver*, Lyc. 73, Opp.H. 2.436; of birds, καύηκες Call.fr. 522 Pf., κηρύλος Lyc. 387.

δυσάγγελος, before 'Nonn.D. 20.184' insert 'Call.fr. 125.3 Pf.' and after it add 'al.'

***δυσάθλιος**, add 'epigr. in SEG 38.605.4 (Maced.)'

δυσάλυκτος, add 'GVAK 16 (Hadrianopolis)'

δυσανακόμιστος, add '**2** *involving a difficult return*, ἐπάνοδοι Ptol.Tetr. 196. **3** *having difficulty in convalescing*, Gal. 18(1).718.13.'

δυσανάληπτος II 1, add '*finding difficulty in convalescing*, Gal. 13.213.7, 1048.6'

***δυσανάφορος**, ον, *slow-rising*, of constellations, Serv. ad Virg.G.1.32 (Daniel).

†**δυσάνεμος**, v. °δυσήνεμος.

***δυσανεύρετος**, ον, *hard to find* or *get at*, ἐπειδὴ δυσανεύρετός ἐστιν ἡ ἐσχάτη ῥίζα τοῦ ἠρυγγίου Cyran. 21 (1.7.90 K.).

δυσανιῶν, delete the entry.

***δυσανίως**, = δυσάνιος, Critias 42 D.-K.: neut. δυσανίων, τό, Plu. 2.106d (-ιῶν codd.).

***δυσάνυστος**, ον, *difficult in execution*, πρᾶξις Cat.Cod.Astr. 8(1).250.

δυσαπάλλακτος, add 'Hp.Nat.Hom. 15, etc., Pl.Ti. 85a'

***δυσαπόπλυτος**, ον, *hard to wash out*, Sch.Pl.R. 429e.

***δυσαπόρρυτος**, ον, *flowing away with difficulty*, Gal. 11.529.9.

δυσαπότροπος, for 'IG 2.1660' read 'orac. in IG 2².4968 (Athens, iv BC)'

δυσαπούλωτος, add 'also **δυσαφούλωτος** Paul.Aeg. 6.40.3 (79.8 H.)'

δυσαρεστέω, line 3, for 'δ. ὅτι' read 'δ. ὅτε τις ..'

δυσαυλία, add 'prob. rest. in A.fr. 78c i 7 R. (pl.)'

δυσαυχής, for '*idly boasting, vain-glorious*' read '*haughty, unapproachable*'

***δυσαφούλωτος**, v. ‡δυσαπούλωτος.

***δυσβάϋκτος**, ον, *marked by doleful howling*, τείνε δὲ δυσβάϋκτον βοᾶτιν τάλαιναν αὐδάν A.Pers. 575.

***δύσβροχος**, ον, *poorly-watered*, BGU 1185.20 (i BC).

δυσγενής, add '**III** *cowardly*, adv., δυσγενῶς ἔχειν πρὸς ψυχρολουσίαν Agathin.ap.Orib. 10.7.5.'

δυσγρίπιστος, add '(the sophist Stagirius in Gr.Nyss.Ep. 26)'

δυσδιαφορησία, add 'Gal. 1.220.2'

δυσδιαχώρητος II, add 'Gal. 15.760.18'

δυσεγκαρτέρητος, add 'prob. in Plu. 2.36b, sp. δυσεκκ- in Phld.D. 1.12; cf. ἀνεκκαρτέρητος'

δυσειδής II, delete the section.

δύσεικτος, add 'adv. -τως, Paul.Aeg. 6.112.3 (166.5 H.)'

†**δυσεκκαρτέρητος**, v. °δυσεγκ-.

δυσέκλειπτος, add 'Ach.Tat. 6.11.1'

δυσεκλήπτως, delete '(leg. δυσεξάλειπτον)'

δυσέκφευκτος, line 1, delete 'Tim.Pers. 130'; line 2, for 'Tim.Pers. 140' read 'Tim. 15.129 P., *hard to reach in escaping*, ib. 119 P.'

Δυσελένα, add 'E.IA 1316'

δυσέμβολος II, add 'γῆν .. πυρὸς γέμουσαν ῥεύμασιν δυσεμβόλοις Carc. 5.7 S.'

***δυσεμπέλαστος**, ον, *making approach difficult*, θάλασσα An.Par. 1.179.

δυσεντερία, line 1, after 'ἡ' insert 'Ion. -ίη' and add 'in horses, BCH 66-7.181 (Abdera, iv BC)'

δυσεξάλυκτος, after 'avoid' insert 'λοιμός Keil-Premerstein Erster Bericht p. 9 (Troketta), cf.'

***δυσεξέταστος**, ον, *hard to search out*, Ptol.Harm. 1.8.

δυσεξύβωτός, for '*not easily displaced outwards*' read '*not easily rid of its bumps*'

δυσεύρετος 2, add 'τὸ στόμα τοῦ .. ποταμοῦ δυσεύρετόν ἐστιν Peripl.M.Rubr. 43'

***δύσεφθος**, ον, *hard to boil*, Aët. 3.8 (267.14 O.).

δύσζηλος I 1, for '*exceeding jealous*' read '*harmfully jealous*' **2**, for '*eager*' read '*eager to no purpose*' **II**, for '*rivalling in hardship*' read '*unenviable*' and delete 'αἰθυίῃσι'

***δυσηβόλιος**, ον, unexpld. wd. in Call.VB in Suppl.Hell. 257.29.

δυσηκοΐα, after 'Vett.Val. 109.31' insert 'Heras ap.Gal. 12.610.10'

δυσηλεγής, for 'Thgn. 795' read 'Mimn. 7.1 W.'

δυσήλιος, add 'δυσηλίους φάραγγας Mosch.Trag. 6.5 S.'

δυσημερία, add 'Plu. 2.168c, 467e, 741a'

δυσήνεμον, delete the entry.

***δυσήνεμος**, ον, Dor. δυσάνεμος, *afflicted by adverse winds*, δυσάνεμοι στόνῳ βρέμουσιν ἀντιπλῆγες ἀκταί S.Ant. 591, Plu. in Hes. 82 S., δυσήνεμος· δυστάραχον, τὸ κακοὺς ἀνέμους ἔχον Hsch.

δυσήνυτος, delete 'also δυσήνυκτος'

***δυσθεραπεία**, ἡ, *difficulty of treatment*, Hp.Aff. 20, 22, Aët. 7.45 (297.17 O.).

δυσθετέω I, add '*to be in difficulty*, Call.Dieg. iv.40 (fr. 102 Pf.)'

***Δύσιος**, ὁ, month-name, SEG 34.1209 (Maeonia).

δυσκάθαρτος I, *hard to purge*, add 'of patients, Gal. 11.353.13'

***δυσκατάσχετος**, ον, *difficult to stop*, of haemorrhage, Afric.Cest. p. 61 v¹.

δυσκέλαδος, add 'perh. w. ref. to arrows, Suppl.Hell. 939.4'

***δύσκερκος**, ον, *having a harmful tail*, ἔστι δὲ αὕτη (sc. σαύρα) δ. Sch.Theoc. 2.58.

***δυσκληρία**, ἡ, *misfortune*, Just.Nov. 22.22.1.

***δύσκλωστος**, ον, *spun with adverse effect*, μίτος Μοιρῶν MAMA 5.30.

δυσκρασία, line 2, for 'Plu.Stoic. 3.216' read 'Plu.Stoic. 2.216'

***δυσκράταιος**, ον, *harmfully mighty*, δυσκράτεε Τυφῶν SEG 35.221.11 (tab.defix., Athens, iii AD).

δυσκράτητος, add 'Hymn.Is. 21 (Maroneia)'

***δύσμα**, ατος, τό, *sunset, west*, Inscr.Délos 1417 C73 (ii BC).

δυσμάχος, add '**II** *fighting in vain*, Nonn.D. 48.452.'

δύσμεικτος II, after 'unsocial' insert 'subst. δύσμικτον, τό, *unsociability*, Gal. 8.689.11'

δυσμεναίνω, add 'Trag.adesp. 535 K.-S.'

δυσμενής, add 'δυμενέες SEG 11.1112 (Arcadia, vi/v BC)'

***δυσμετάγωγος**, ον, *hard to tame* or *break in*, θρέμμα Afric.Cest. p. 22 v¹.

δυσμή II, add 'πρὸς δυσμήν Amyzon 28.11 (ii BC)' at end, δυθμή, add 'Pi.I. 4.65 S.-M.'

***δυσμηνήτης**, ου, ὁ, *wrathful man* (cf. μηνιτής), Ptol.Tetr. 159.

δυσμήνιτος, delete 'ψυχαί Ptol.Tetr. 159 (-ίτας)'

***δυσμιαῖος**, ον, *western*, τὸ δυσμιαῖον μέρος PNess. 16.12.

***δυσμιγής**, ές, *harmfully mixed*, Sch.Call.VB in Suppl.Hell. 251.25.

***δυσμογέω**, app. *suffer distress*, Lyr.adesp. in SLG 458 i 3.

***δυσμοιρία**, v. ‡δυσμορία.

δύσμοιρος, add 'SEG 34.1247 (Miletoupolis, ii AD)'

δυσμορία, add 'also δυσμοιρία Salamine 193 (ii AD)'

***δυσμορφής**, ές, = δύσμορφος, rest. in Hymn.Is. 115 P. (Andros).

***δύσορμος I**, add 'Peripl.M.Rubr. 10, 20, 43'

***δυσουριακός**, ή, όν, *suffering from suppression of urine*, Firm. 4.15.2.

δυσπαίπαλος, after 'rough' insert 'of animals, their skins, etc.' and add 'of snakes, Opp.C. 2.270'

δυσπαρηγόρητος, add 'IG 12(7).239 (i/ii AD)'

δυσπέμφελος, line 4, after '(Hes.)Op. 618' insert 'Herodicus in Suppl.Hell. 494.5' and after 'Nonn.D. 2.550' insert '13.75, 22.171, 24.64'

δυσπιστία, after 'disbelief' insert 'Zos.Alch. 209.8'

δύσποτμος, add 'Anacr. 1 fr. 7 P. (s.v.l.)'

δυσπραγία, add 'SEG 38.340 (tab.defix., Camarina, v BC)'

δύσριγος, add 'prov. δυσριγότερος χελώνης· ἐπὶ τῶν εἰς ὁτιοῦν γλισχρευομένων Macar. 3.41'

δύσσοος, for 'Riv. .. 266' read 'ἧοι δὲ πάντες δύσσοοι SEG 38.340 (tab.defix., Camarina, v BC)'

***δύσστατος**, ον, (στέλλω) *hard to check*, Hippiatr. 9.1.

***δυσστρατοπέδευτος**, ον, *ill-suited for encamping*, Aen.Tact. 8.1.

***δύσσφαλτον**· δύσμαχον Hsch.

δύσταλτος, delete the entry (v. °δυσσ-).

δυστερπής, add 'κεῖμαι .. δυστερπεῖ τῇδ' ἐνὶ πέτρῃ GVI 959.3 (Thespiae, ii/iii AD)' δύστηνος **I 2**, before 'μόχθος δ. Pi.P. 4.268' insert 'δ. .. νοῦσοι Semon. 1.12 W., δ. μόρῳ ibid. 18'

δυστλήμων, add 'S.fr. 555.8 R.'

δυστοκία **II**, for ' = δυστεκνία' read 'unsuccessful or unlucky childbirth' and add 'epigr. in SEG 25.752 (Moesia, ii BC)'

*δύστοπος, ον, difficult, Sch.Clem.Al. 1.296.31; adv. sup., Suid.

δυστρᾰτοπέδευτος, delete the entry (v. δυσσ-).

Δύστρος, add 'cj. in Anaxipp. 1.41 K.-A., SEG 32.1395 (Commagene, ii AD), 31.990 (Lydia, iii AD), Kafizin 267a (iii AD)'

δυστῠχία, after 'ill fortune' insert 'Thgn. 1188'

δυσυπνήτως, delete the entry (v. ‡δυσύποιστος).

δυσύποιστος, add 'adv. -ως, ἔχειν πρός τι Agathin.ap.Orib. 10.7.27 (-υπνοιστ-)'

δύσφαλτος, delete the entry (v. °δύσσφαλτον).

δυσφημέω, add 'Hippon. 78.8 W. (broken context)'

δύσφθαρτος, for 'not easily spoilt' read '**2** of (ingested) food, not easily broken down' and add 'Gal. 8.42.5'

δύσφορος, line 5, for 'δύσφορόν [ἐστι]' read 'πειρῶμαι μηδέποτε ὑπερπίμπλασθαι· δύσφορον γάρ' **I 3**, delete the section.

†δυσφρόνα, ἁ, = δυσφροσύνη, cj. in Pi.O. 2.52.

δυσφροσύνη, delete 'E.Tr. 597' add '**II** ill-will, malice, E.Tr. 597.'

*δυσχερασμός, ὁ, irritation, anger, Phld.Lib. p. 80 O.

δυσχερής, line 1, delete '(χείρ)'

δύσχρηστος, after 'adv. -τως' insert 'ἔχειν to be intractable, Men.Dysc. 249'

δυσώνης, for pres. def. read 'one who is unskilled at shopping'

δυσωπέω, line 3, after 'X.Eph. 4.5' insert 'so perh. δυσωπήσας λίνα Μοιρῶν, of Apollo, Didyma 217.13 (hex. poem or oracle)' **III**, add 'through being dazzled, Ael.NA 8.10, 10.14'

δύτη, for 'shrine(?)' read 'well or reservoir'; after 'Cabireum' insert 'Thebes, ii BC'; add 'AE 1948/9.136, A17, 18, 20 (Epid., iv BC)'

*δῠτορύκτᾱς, α, ὁ, ?excavator, IAskl.Epid. 52A.24, 28-32, al. (iv BC).

*δυτριᾱκοστόν, v. °δυοτριακοστόν.

δυώδεκᾰ, add 'gen. pl. δυωδέκων SEG 12.391 (Samos, vi BC); δοώδεκα SEG 24.361 (Thespiae, iv BC)'

δῠωδεκάβοιος, add 'Suppl.Hell. 991.80 (poet. word-list); see also δωδεκάβοιος'

δῠωδεκαῖς, delete 'Att.' and insert 'also δωδεκαῖς; for 'SIG² .. ii BC)' read 'CID I 9.D34 (v BC)'; after 'etc.' insert 'δωδεκαῖδα βούπρῳρον SIG³ 604.9 (Delphi, ii BC)'

*δῠωδεκάπεδος, ον, twelve foot long, CID II 113.26 (iv BC).

*δυωδεκάσημος, v. °δωδ-.

δῶ, add 'Myc. do-(de), acc. w. -δε, to the house (of)'

δωδεκάβοιος, add 'see also δυωδεκάβοιος'

*δωδεκᾰδραχμία, ἡ, twelve-drachma tax on boats, PMich.Zen. 60.2 (iii BC).

δωδεκάθεος **II 1**, add 'perh. τὸν Ἀξιοττην[ὸν ἐν] τῷ ἐκεῖ δωδεκάθην καθήμ[ενον SEG 29.1179 (Lydia, AD 108/9)'

*δωδεκαῖος, ὁ, or -αῖον, τό, coin worth twelve (?obols), RA 3.40 (Amphipolis, iii/ii BC), 6.32 (ib., ii BC).

*δωδεκαῖς, ίδος, ἡ, v. °δυωδεκαῖς.

*δωδεκᾰκέφᾰλος, ον, twelve-headed, δ. δράκοντα IGChr. 210 ter 6 (Arcesine).

*δωδεκάκιστος, η, ον, perh. one of a group of twelve, Suppl.Mag. 47 (ii/iii AD), 42.30, 49 (iii/iv AD), BMus.Inscr. 949.

δωδεκάμηνος, for '-μηνον, τό' read '-μηνος, ἡ'

*δωδεκάπλησον, aor. imper. multiply by twelve, perh. erron. for δωδεκάπλωσον (°δωδεκαπλόω) PCair. 10758 (vi AD).

*δωδεκαπλόω, multiply by twelve; v. °δωδεκάπλησον.

*δωδεκᾰπόδια, ἡ, length of twelve feet, SEG 25.394.C15.

δωδεκάπους, after 'long' insert '(sc. σκιᾶς or στοιχείου)' and for 'Men. 364' read 'Men. 304.3 K.-Th.'

δωδεκάσημος, add 'in prosody, of the length of twelve short syllables, Mar.Vict. p. 49 (duod-)'

*δωδεκάστολος, ον, forming a squadron of twelve, Αἰνιάνων δὲ δωδεκάστολοι νᾶες ἦσαν E.IA 277 (s.v.l.).

*δωδεκάστομος, v. °δεκάστομος.

δωδέκατος **I**, add '-ον, neut. as adv., for the twelfth time, SEG 36.1207.3 (Milyas, Pisidia, 5/4 BC)' **II**, after 'ἡ' insert 'the twelfth day of the month, SEG 32.117 (Athens, iii BC), 33.1039 (Aeolian Cyme, ii BC), etc.; δυωδεκάτη Hes.Op. 774, δυωδεκᾰτᾶ SEG 25.744 (Callatis, ii BC)'

*δωδεκαχαλκία, ἡ, tax of twelve χαλκοί, PHib. 112.8 (iii BC, abbrev.).

*δωδεκετηρος, ον, twelve years old, GVI 665.5 (Maced., i AD).

†δωδεκέτης, ες or -ετής, ές, twelve years old, Call.Ep. 21, Plu.Aem. 35, παῖδα δυωδεχέτη epigr. in IG 10(2).1.368 (Thessalonica, ii AD); δωδεχετῆ CEG 709.5 (Halic., iv BC); fem. -έτις, ιδος, CEG 591.2 (c.350/325 BC), AP 11.70 (Leon.Tarent.); see also δωδεκαέτης.

Δωδωναῖος, after 'α, ον' insert 'of Dodona, esp. as epith. of Zeus'; after 'Cratin. 5' insert 'SEG 34.592 (Epirus, i BC); of Zeus identified w. Hadrian, SEG 37.521 (Epirus, ii AD)'; prov., add 'τὸ Δωδωναῖον ἄν τις χαλκίον ὅ λέγουσιν ἠχεῖν .. Men.fr. 60.3 K.-Th.'

*Δωδωνεύς, έως, ὁ, of Dodona, epith. of Zeus, Hsch.

Δωδώνη, line 3, heterocl. forms, add 'dat. -ῶνι S.fr. 455 R., Call.fr.483 Pf.; acc. -ῶνα Euph. 4 vGr.'

*δωκενάριος, v. ‡δουκηνάριος.

δωλένετος, delete 'cf. sq.'

δωλοδομεῖς, delete '(cf. δοῦλος .. to be born)'

δωμάτιον, delete 'IG 12(8).442.8' (v. Glotta 50.68)

*δωμᾰτουργία, ἡ, structure, ἐνηλλάγη ἡ δ. πᾶσα τῆς βασιλικῆς IEJ 11.184 (Sepphoris, vi AD).

δώμημα, add 'IKnidos 303.1 (i BC)'

†δώμησις, εως, ἡ, building, AA 19.8 (Milet., pl.), δώμησις· οἰκοδομή Hsch.

†δωμητύς· κατασκευή Hsch.

δωμήτωρ, add 'epigr. in Fs. Theocharis p. 399 l. 7 (Nemea, ii AD)'

*δωνατίουον, τό, Lat. donativum, donative, gratuity, POxy. 1047.4 (iv AD); also δωνάτιον PBeatty Panop. 2.162, 164, al. (AD 300).

δωράκινον, add 'PMich.XIV 680.9 (-εινον, iii/iv AD); cf. ῥοδάκινον'

*δωράκιον, τό, = δωράκινον, PRyl. 630.419, al. (iv AD).

δωρεά, line 1, for 'IG 1².77' read 'IG 1³.131.20' **I 1**, add 'κατὰ mortis causa δωρεάν Cod.Just. 1.3.45 pr., τὴν πρὸ γάμου δωρεάν ib. 1.3.52.15' **II**, line 4, after '(iv BC)' insert 'also δωρεᾶς ISmyrna 600.17 (ii AD), Robert Ét.Anat. 388 (Limyra in Lycia)'

*δωρεακός, ὁ, official or employee of a δωρεά 1 2, BGU 1540.3 (iii BC); cf. χρειακός.

δωρεαστικός, add 'CPR 10.122.8 (vi AD)'

δωρέω **I**, line 3, after 'pass.' insert 'ἐκ θεῶν .. μηχαναὶ δωροῦνται Sch.Pi.P. 1.74'

Δωριάζω **I**, add 'Sch.E.Hec. 933, Hsch.'

Δωριεύς **II**, add 'as the name of phratry at Argos, SEG 30.355 (iv BC)'

Δωρικός, add 'w. ref. to architecture, γεῖσα Δ. IG 2².1666.A55 (iv BC)'

†Δώριος, ον, (fem. α, Pi.O. 1.17, 3.5, N. 5.37, I. 2.5) Dorian, Δ. μέλος Pi.fr. 67 S.-M., Δώριον χορείαν Pratin.Lyr. 1.17, AP 7.436 (Hegemon), Δώριος (ἁρμονία) Arist.Pol. 1276ᵇ9, 1290ᵃ22, D. γλυφάς Ph. 1.666, Δ. τόνου Plu. 2.1135a, D.Chr. 33.42, Iamb.VP 24.1, Nonn.D. 25.21. **b** of the Isthmus of Corinth, Ἰσθμὸς Δ. Pi.N. 5.37; w. ref. to the Isthmian games, id.I. 2.15, 8.64.

*Δωρίσκος, ὁ, compound foot scanned -⏑⏑-⏑, Diom. 482.2.

δωρίτης, add 'Sch.Pi.O. 8.101'

δωρογραφία, delete the entry.

δῶρον **I 1**, w. gen., add 'σχολῆς τὸ δῶρον E.Tr. 911' add '**b** δῶρα στρατιωτικά (= Lat. dona militaria), TAM 2.1201A (ii AD).' at end add 'Myc. do-ra (pl.)'

δωροτελέω, add 'w. acc., make a benefaction of, δωροτελεῖ μναμεῖο[ν] SEG 33.735 (= Bile Dial.Cret. no. 35, Cydonia, c.400/350 BC)'

δωροφόρος, add 'Callix. 2.32 J.'

*δωρύφιον, τό, wedding-gift, PHamb. 87.11 (ii AD), Mitteis Chr. 290ii7 (vi AD).

δώς, add '(perh. to be read as a pr. n.)'

δωσίβιος, for 'Mus.Belg. 16.70' read 'eleg. in BCH 50.529 (Marathon, ii AD)'

†δωσίπῠγος (or δοσί-), ον, adj. applied to sexually compliant woman, Suid., s.v. ἀφελές.

δοτήρ, line 1, for 'ἐάων' read 'ἑάων' add '**2** donor, of the dedicator of a building, IGLS 9027; see also δοτήρ.

δωτινάζω, for pres. def. read 'collect donations'

*δωτίνᾱσις, ιος, ἁ, letting of land, BCH suppl. 22.237 (Argos, Hellen.).

*δωτῐνᾱτήρ, ῆρος, ὁ, officer in charge of °δωτίνασις, BCH suppl. 22.237 (Argos, Hellen.).

*δωτινάω, let land, in pass., ἁ χώρα .. κατεμερίσθη κατὰ γύας δωτιναμένας BCH suppl. 22.237 (Argos, Hellen.).

δωτίνη, after 'gift, present' insert 'w. notion of obligation or reciprocity, as to a ruler or host'; line 2, delete 'give as a free gift' **II**, for this section read 'Dor. δωτίνα rent for land, Sokolowski 3.59.8, 11 (Calauria, iii BC), BCH suppl. 22.237 (Argos, Hellen.)'

δώτωρ, line 1, for 'ἐάων' read 'ἑάων'

*δώω, v. ζῶ.

E

ἐάν, line 2, after '(iv BC))' insert 'ἰάν *Hesperia* 33.385 (Eretria, vi BC), Dubois *IGDS* 15.5 (Sicily, vi BC), ἐάμ (before labial), ἐά (before μ), *IG* 1³.1 (*c.*500 BC)'

ἐᾱνός **I**, at end for 'Sapph. (?)122' read 'Sapph. 156 L.-P.' **II 1**, add 'cf. ἰανόν· ‹λεπτὸν› ἱμάτιον Hsch. La.'; at end add 'Myc. *we-a₂-no-i* (dat. pl.)'

⁺ἐάνπερ, ἤνπερ, ἄνπερ, strengthd. for ἐάν, ἤν, ἄν (cf. εἴπερ), *that is to say if*, ἐάνπ. A.*Pers.* 529, E.*Med.* 727, Pl.*Tht.* 166b, etc.; ἐάνπερ γε id.*Phd.* 89b; ἤνπ. Ar.*Lys.* 551, X.*Cyr.* 4.6.8; ἤ. γε id.*Eq.* 10.11; ἄνπ. Pl.*Ti.* 90c, *Lg.* 631c, D. 20.22.

ἔαρ (A), add 'see also ‡ἦρι'

ἐαρῑνός, after 'εἰαρινός' insert 'Ϝειαρινός *IG* 7.1919 (Thespiae, pers. n.)'; after 'Plb. 3,34.6' insert 'ἠρινοῦ X.*HG* 3.2.10'; after 'E.*Supp.* 448' insert 'φθέγμασιν ἠρινοῖς Ar.*Au.* 683'; after '*PCair.Zen.* 33.13 (iii BC)' insert 'ἠρινῶν· τῶν θυννίδων τῶν τῷ ἔαρι ταριχευμένων *AB* 1.263.10'; after 'neut. as adv., *in springtime*' insert 'Alc. 115(a).10 L.-P.' at end, adv., read 'ἦρις ὡς (ἠρινῶς La.)· ἐαρινῶς Hsch.' add '**2** metaph., of beauty, ἀμεθύσου .. ἐ. τις ὥρα Hld. 5.13.'

*ἐαρίτης, ου, ὁ, name given to a precious stone, ἀγλαίας δ' ἐαρίτην τὸν αἱματίτην λίθον φησί Sch.Nic.*Al.* 314e Ge.

ἑαυτοῦ, line 5, after '*SIG* 774.2 (Delph., i BC)' insert 'ἀτοῖς, ἀτῶν *SEG* 29.771.18, 23 (Thasos, ii BC)'; lines 8/9, for 'ϝιαυτοῦ .. p. 34' read 'ϝιαυτῶ *Inscr.Cret.* 4.47.5 (Gortyn, v BC)'; lines 9/10, after '*IG* 9(2).517.16' insert 'acc. pl. fem. εὐτάς, *BCH* 59.55.18 (Larissa, ii BC)'; line 11, for 'Alc. 78' read 'Hes.*Th.* 126'

ἐάω, add '**III** *leave behind one*, τέκνα *MAMA* 6.215 (Apamea, iii AD), cf. ib. 124 (Heraclea Salbace; aor. med.), θυγατέρας *SEG* 24.911.18 (Thrace, iv AD); *bequeath*, ἀνδριάντα πλατ(ε)ίᾳ *MAMA* 6.176 (Apamea).'

ἐάων, read 'ἑάων'

ἑβδέματος (s.v. ἑβδεμήκοντα), for 'ον' read 'α, ον'

ἑβδεμήκοντα, add 'ἐτδεμεικοντα *SEG* 26.672.34 (Larissa, 200/190 BC)'

*ἔβδεμος, v. °ἕβδομος.

⁺ἑβδομᾱγέτης, ου, ὁ, *ruler of the seventh*, τὰς δ' ἑβδόμας (sc. πύλας) ὁ σεμνὸς ἑβδομαγέτας ἄναξ Ἀπόλλων εἷλετ' A.*Th.* 800; Procl. *in Ti.* 3.200d.

ἑβδομαῖος **I**, add '**3** *belonging to the seventh generation*, Sm.*Ge.* 4.24. **4** epith. of Apollo, Ἑτδομαῖος *BCH* 113.638 (Atrax).'

*ἑβδομαϊστής, οῦ, ὁ, *one who celebrates the ἑβδομαῖον*, *SEG* 32.244 (Attica, iv BC).

*Ἑβδομαιών, ῶνος, epith. of Apollo, *IEryth.* 207.87 (ii BC).

*ἑβδομάριος, ὁ, app. title of functionary in the imperial palace, *ITyr* 88.

ἑβδόματος, for 'ον' read 'η, ον' and add 'Hes.*Op.* 805, *fr.* 362 M.-W., etc.'

ἑβδομήκοντα, add 'see also ‡ἑβδεμήκοντα'

*ἑβδομηκοντάδραχμος, ον, *having a salary of seventy drachmas*, *PMich.Zen.* 66.11 (iii BC).

*ἑβδομηκονταετής, v. °ἑβδομηκοντούτης.

*ἑβδομηκοντάρουρικός, η, ον, *consisting of seventy* ἄρουραι, *PVindob.* G39919 (*ZPE* 50.135, 24 BC).

*ἑβδομηκοντοστός, ή, όν, *seventieth*, Mitchell *N.Galatia* 325 (AD 247); cf. ἑβδομηκοστός.

ἑβδομηκοντούτης, add 'also ἑβδομηκοντᾱετης *SEG* 6.138.3 (Phrygia, -μαικ-)'

*ἑβδομοκούρης, ὁ, *temple official*, *IEphes.* 10.24, 1042.11 (iii AD).

ἕβδομος, after lemma insert 'ἔβδεμος, α, ον *SEG* 9.72.101 (Cyrene, late iv BC)' **3**, for this section read 'ἕβδομα, τά, *festival celebrated on the seventh day after birth*, Call.*Dieg.* ix.26, cf. ib.*fr.* 202.22 Pf.; *seven-day marriage feast*, Lxx *Ge.* 29.27'

*ἑβδομοστάτης, ου, ὁ, *religious functionary of the seventh rank*, *SEG* 32.1484 (Sidon, i/ii AD).

*ἐβένωσις, εως, ἡ, *the process of turning ebony-black*, Anon.Alch. 419.22.

*ἐβί, v. °ἐπί.

*ἐγγαΐδιον, τό, *small plot of land*, cj. rest. in *PAmh.* 2.36, *SB* 4638 (v. *ZPE* 41.256, both ii BC).

ἐγγαιέω = ἐγγαέω, *SEG* 17.197 (Olympia, i BC), 37.370 (Augustus; = *Inscr.Olymp.* 335).

*ἐγγᾰμιστός, ή, όν, *betrothed*, ἐδνωτήν· ἐγγαμιστὴν νυμφίῳ Hsch. (ἢ νύμφιον Cyr.).

ἐγγαρέω, delete the entry (v. ἐγγαιέω).

ἐγγαστρίμῡθος, for '*ventriloquist* .. means' read 'one who speaks from the belly'

ἐγγάστριος, add 'ἐγγάστριον, τό, *foetus*, *Cat.Cod.Astr.* 8(1).175, al.'

*ἔγγαστρος, ον, *situated in the womb*, Hsch. s.v. συλλήψεται.

ἐγγέαβλος, after lemma insert '(-κλος La.)'

*ἐγγένημα, v. °ἐκγένημα.

ἐγγεύομαι, for 'pass.' read 'med.'

ἐγγηράμα, delete 'Cic.*Att.* 12.25.2' and add '**b** *place to grow old in*, Cic.*Att.* 12.25.2, 29.2, 44.2.'

ἐγγίζω **II 1**, after 'Plb. 18.4.1' insert 'τοῖς τελείοις D.H.*Comp.* 14'

ἐγγλοφούμενα, after lemma insert '(ἐγχλοιούμενα La.)'

ἐγγομφόω, delete 'pass. .. 25.6.4)'

*ἐγγονία, ἡ, *grandchildren*, *SEG* 33.765 (Heraclea, Lucania, iii/ii BC) (for ἐκγ-, cf. ἔγγονος).

*ἐγγόνιον, τό, *grandchild*, *IGLS* 1335 (ἐγον-, iv AD), *CIJud.* 590 (sp. ινγόνιν, v/vi AD) (for ἐγκ-, cf. ἔγγονος).

ἔγγονος **1**, add 'ἔγγονοι *grandchildren*, *Cod.Just.* 1.3.41.2, παῖδες .. καὶ ἔγγονοι καὶ ἀπέγγονοι ib. 6.4.4.15'; 3 lines fr. end, after 'ca.300 BC' insert 'and are synonymous in *IG* 1³.426.62 (Athens, v BC)'; last line, after '(Samos, iv BC)' insert '*IG* 12(3).1296.22 (Thera, ii BC)'

⁺ἔγγραυλις, εως, ἡ, *anchovy*, ἐγγραύλεις, οἱ δὲ ἐγκρασιχόλους καλοῦσιν αὐτάς Ael.*NA* 8.18, Opp.*H.* 4.470.

ἔγγραφος **I**, line 2, subst., add '*SEG* 33.1039.90 (Aeolian, Cyme, ii BC); also sg., *SEG* 30.1349 (Miletus, ii BC)'

ἐγγράφω, at end for '(perh. written ἐκυρ- *SIG* 742.29)' read '(app. written ἐκυρ- *SIG* 742.29, *GDI* 5496.10, 15)'

ἐγγυάω **2**, med., add '*PHarris* 65.7 (AD 342)' **II 1**, add 'spec., *act as fideiussor*, *Just.Nov.* 4.1'

ἐγγυεύω, add 'Arc. pf. ἰνγεγύευκε *SEG* 25.447.18 (Aliphera, iii BC)'

ἐγγύη, line 1, for 'Delph.' read 'Delos'

*ἐγγυητήριον, τό, *security, guarantee*, *SEG* 38.1036 (lead tablet, Gaul, v BC).

ἐγγυητής, add '*POxy.* 3576.16 (AD 341); spec., = Lat. *fideiussor*, *Just.Nov.* 4.1, 99 pr.'

ἐγγυητικός, add '*PLond.* 1494.11 (Byz.)'

ἐγγυητός, add 'ἀστὸν ἐξ ἀστῆς ἐγγυητῆς αὐτῷ γεγενημένον D. 57.54; fem. subst. ἐξ ἐγγυητῆς ἢ ἐξ ἑταίρας Is. 3.6, 24'

ἐγγυήτρια, add '*POxy.* 3938.15 (AD 601)'

ἐγγύθεν **4**, add '**b** οἱ ἐ. *those within a given legal relationship*, *Just.Nov.* 91.2.'

ἐγγυθήκη, after 'Hegesand. 45' insert '*Inscr.Délos* 372B30 (iii/ii BC)'

ἐγγυμνάζω, lines 2/3, delete 'ἐν σοὶ .. *Phdr.* 228e' add '**II** ἐκκλήτους ἐγγυμνάζειν calque on Lat. *appellationes exercere*, *launch appeals*, *Just.Nov.* 41 pr., 112.3 pr.

*ἐγγυοβεβαιωτής, οῦ, ὁ, *guarantor*, *PKöln* 232.2, 3, al. (iv AD).

ἔγγυος **II**, lines 4/5, for '*IG* 9(2).4' read '*IG* 9(2).5b14'

ἐγγύς **I**, line 5, after '(A.)*Eu.* 65' insert 'τὸν ἐνγιστά σοι μένοντα *POxy.* 3314.23 (iv AD)'

*ἔγγων, ονος, ὁ, ἡ, *grandchild*, *CIJud.* 1.140 (Rome); (prob. for ἐκγ-, cf. ἔγγονος, etc.).

ἐγγώνιος **I**, adv., add 'Gal. 18(1).815.1'

ἐγδάκτῠλος, for '*IG* 2.809ᵇ195' read '*IG* 2².1629.428, 429, *SEG* 3.137 iii 12 (Athens, iv BC)'

ἐγείλησαν, add '(ἐγείλησαν· συνήλασαν La.)'

ἐγείρω, lines 5/6, aor. pass., add 'redup. γεγέρθη κρῖνε (sic) ζῶντας καὶ νεκρούς, of Christ, *Suppl.Mag.* 23 (v AD) **I 1**, delete 'ἐ. τινά .. (lyr.)' and add 'med. τόν γ' ἔγρεο Nic.*Al.* 456' **2**, line 3, after 'ib. 510' insert 'cf. λόγος .. ἐ. ὀργάς Lxx *Pr.* 15.1'; add 'w. gen., *cause to rise from*, μὴ .. εὐνᾶς ἐγείραιεν E.*HF* 1050 (lyr.); τὸν μὲν λέχους ἤγειρ[α *Trag.adesp.* 664.29 K.-S. **4**, after 'Luc.*Alex.* 10' insert 'τίτλον πε[ρὶ] τύμβῳ ἔγειρεν Mitchell *N.Galatia* 338' **III**, after 'oneself' insert 'Ar.*Ra.* 340, E.*IA* 624 (both imper. ἔγειρε)' and add 'ἔγειρε ἐς τὸ μέσον Ev.Marc. 3.3'

ἐγερτί, add '(ἐγερτί· γρηγόρως La.)'

ἐγέρσιμος, add '*PAnt.* 57 fr.(a)ʳ.35'

*ἐγέρτης, ου, ὁ, *one who rouses up*, τὸν τῶν νεκρῶν ἐγέρτην Leont.Byz. *HP* 1.4.5.

⁺ἐγϝηληθίωντι, v. ἐξείλλω 2.

ἐγκαθαρμόζω, for 'Ar.*Lys.* 682' read 'Ar.*Lys.* 681'
ἐγκαθίζω I 2, line 2, for 'pass.' read 'med.'
*ἐγκαθιστάω, appoint, *Cod.Just.* 1.3.45.3, *Just.Nov.* 57.2.
ἐγκαθίστημι I, add '2 put a cargo on board, *PWarren* 5.6 (*Aegyptus* 13.242, ii AD).'
ἐγκαίριος, for '*PGrenf.* 1.64' read 'in *PGrenf.* 1.64 εὐκαιρείαν (= εὐκαιρίαν) is prob.'
ἐγκαίω IV, add 'w. acc., Διὶ .. ἀμνὸν καὶ ὄρνιν *Mél.Glotz* 872 (Maced., ii AD)'
ἐγκαλέω II 1, c. gen. rei, add 'Demad. 61 B.'   2, line 3, for 'ἐ. τινὶ περί' read 'ἐ. περί'
*ἔγκαλυμμα, ατος, τό, wrapping, cover, *SEG* 38.1210.4 (Didyma, ii BC).
ἐγκαλύπτω I, pass., transfer fr. section II 'And. 1.17' and add 'Thgn. 1045'
*ἐγκαμψικήδαλος, ον, onion-eating, Luc.*Lex.* 10.
ἐγκανάσσω, after 'E.*Cyc.* 152' insert '(cj.)'
ἐγκαναχάομαι, read 'ἐγκαναχέομαι'
ἐγκάπτω, for 'gulp down greedily' read 'take a mouthful of'   II, delete the section (v. °ἐγκαπύει).
*ἐγκαπύει· ἐμπνεῖ Hsch. (cj. Toll for ἐγκάπτει· ἐκπνεῖ.
ἐγκάρδιος, after 'ον' insert '(ος, α, ον Horap. 1.7)' and after 'in the heart' insert 'ψυχή Horap. l.c.'   II 2, for 'Roussel .. 236' read 'χαλκοῦν *Inscr.Délos* 1442*B*58; ξύλινον ib. 59' add 'III ἐγκαρδία (sc. λίθος), ἡ, gem with the shape of a heart on it, Plin.*HN* 37.159.'
*ἐγκαρπίζω, w. acc. and gen., make fruitful with; fig. τοὺς δὲ καὶ ὑπνώοντας ἑῆς ἐνεκάρπισεν ἀλκῆς *Orac.Chald.* 118.2 P.
ἔγκαρπος, add 'ἔγκαρπον· ἔγκυον Suid.'
*ἐγκαταβλέπω, ἐγκατάβλεψον, gloss on ἰνκαπάταόν, Hsch.
ἐγκαταγράφω, add 'write a curse against *SEG* 16.573.14 (Selinous, v BC)'
ἐγκαταδαρθάνω II, for '(Plu. 2.)688f' read '(Plu. 2.)688e'
ἐγκαταλαμβάνω I, add 'transf., ἔθανον λοιμοῦ ν[έ]φει ἐνκαταληφθείς *MAMA* 9.79 (ii AD).   2 come and see a person, ἐὰν ῥᾷον ἔχῃς ἐνκατάλαβε ἡμᾶς *PMich.*XI 624.7 (vi AD).'
ἐγκαταλέγω I 1, add 'Call.*fr.* 64.7 Pf.'
ἐγκατάλειμμα 2, add 'Alex.Aphr. 63.3'   3, delete the section transferring exx. to section 1.
ἐγκατάλειψις, add 'b excess, residue, *SB* 9066 ii 13 (ii AD).'
*ἐγκατάλοιπον, τό, deficit, suggested reading in *IGBulg.* 2265.18 (?i AD).
*ἐγκαταλοχίζω, enter in a register, v.l. in Lxx 2*Ch.* 31.18 (v. °καταλοχία).
ἐγκαταναίω, add 'perh. in hymn in *SEG* 32.1020.3 (Italy, ii/iii AD).'
ἐγκαταπήγνῡμι 1, line 3, after 'Il. 9.350' insert 'ἐγκαταπήγνυσι (τῷ Χρυσίππῳ) τὸ ξίφος Plu. 2.313e'   2, delete the section.
ἐγκαταπλέκω, add 'fig., w. acc. and dat., D.H.*Comp.* 12'
*ἐγκαταρδεύω, water or irrigate in, τινι, fig., Polystr. p. 87 V.
*ἐγκαταρρήγνῡμι, come violently down upon, of a wind, *Gr.Roman-Papyri* 8.24.
ἐγκατασκήπτω I, add 'ὅταν τι ναοῖς ἐγκατασκήψῃ μύσος *Trag.adesp.* 466.2 K.-S.'
ἐγκατασπείρω, after 'Ph. 2.673' insert '(cj. for κατέσπειρεν 2.420)'
ἐγκαταταράσσω, add '(dub., edd. ἕλικα τεταραγμένην)'
ἐγκατατίθημι II, add 'of burial, ζῶντες ἐνκατεθόμεθα τὴν θήκην *TAM* 4(1).374'
*ἐγκαταφυτεύω, plant in, implant, Hsch. s.v. ἰνκαφότενε.
*ἐγκατέρχομαι, come down into a place, εὐθέως οὖν μνησθήσῃ αὐτῷ ἵνα ἐνκατέλθῃ *PMil.Vogl.* 279.10 (i AD).
ἐγκατέχω II, pass., add '2 to be caught up in, τέσσαρσι χρείαις ἐγκατασχεθήσεται παρὰ τοὺς νόμους Modest.*Dig.* 27.1.4 pr.'
ἐγκατοικοδομέω I, for this section read 'to build on or into, στρωτῆρας *IG* 2².463.59, 60 (iv BC); impers. pass., φρούρια δ' ἔστιν ᾗ ἐπὶ τῶν καρτερῶν ἐγκατῳκοδόμηται Th. 3.18.4'
*ἐγκαυλέω, v. °ἐκκαυλέω.
ἔγκαυμα, add 'IV fuel, *PHels.* 12.12, 20 (AD 163), *POxy.* 2206.9 (vi AD), (in this sense app. the same as ἔκκαυμα).'
*ἐγκαύσιμος, ον, painted in encaustic, *IG* 4².109 iii 63 (Epid., iii BC).
ἔγκαυσις, add 'III heating, τῇ τῶν δημοσίων βαλανείων ἐγκαύσει *Just.Edict.* 13.16 (in this sense app. the same as ἔκκαυσις).'
ἐγκαυστήρια, add 'sg., *Ath.Asklepieion* iv 84'
ἔγκαυστος, add 'fem. -ᾱ, *Lindos* 420*a*24, *b*34 (i AD)'
ἐγκαυχάομαι, for 'pride oneself on' read 'glory, exult in' and add 'Lxx *Ps.* 51(52).1, 96(97).7, 105(106).47 (perh. the same as ἐκκαυχάομαι)'
ἐγκαψικίδαλος, delete the entry (v. °ἐγκαμψι-).
*ἔγκειμαι, to lie on or in, ἐπεὶ οὐκ ἐγκείσεαι αὐτοῖς (sc. εἵμασι) Il. 22.513; Hdt. 2.73.   b to be situated on or in, Pl.*Cra.* 402e, *R.* 616d, Thphr.*HP* 5.3.6, Arat. 138, δοιὰ μετώπῳ ἔγκειται .. φάη Nic.*Th.* 292, *IG* 14.1389 i 27, J.*AJ* 8.83(3.6), βελῶν .. πολλῶν ἐπὶ τοῖς τραύμασιν ἐγκειμένων Plu.*Fab.* 16, Hld. 1.14.   c to be in a plight, situation, etc., πόθῳ Archil. 193.1 W., βλάβαις S.*Ph.* 1318, E.*And.* 91, *Hel.* 269, μόχθοις id.*Ion* 181, Plb. 14.9.5.   2 to exist or be present in, ἐγκείμενον ἔχοντας ἔρωτα Plot. 4.4.40, 5.7.2, 6.1.13, al.   b to be included or involved in, ἐπειδὴ ἐν σώματος λόγῳ ἔγκειται μέγεθος Plot. 3.6.16, 6.7.3.   II of hostile forces, to press hard on, Th. 1.49.7, 144.3, X.*HG* 3.5.20, 5.2.1; τοὺς κατὰ πλευρὸν ἐγκειμένους J.*BJ* 2.543(19.7), *AJ* 14.417(15.4), Hld. 8.16.5; in argument, policy, etc., Ar.*Ach.* 309, ἐνέκειντο τῷ Περικλεῖ Th. 2.59.2, 5.43.1; of afflictions, περὶ τῆς ἐγκειμένης σοι ἀλγηδόνος Sch.S.*OC* 513, ὑπέκκαυμα γίνεσθαι τοῦ ἐγκειμένου πόθου Hld. 1.24.2.   2 to press one's point, be insistent or urgent, X.*Eph.* 2.13, ἰδὼν αὐτὴν σφόδρα ἐγκειμένην Ach.Tat. 5.16.7, Hld. 10.30.8; w. adj., πολλὸς ἐνέκειτο λέγων Hdt. 7.158.1, Th. 4.22.2, D. 18.199.   3 to apply oneself to task, policy, etc., ἰσχυρῶς ἐγκείσονται Th. 1.69.3, ἐ. ἐπὶ τὰ πονηρά Lxx *Ge.* 8.21, ἐγκεῖσθαι πρὸς τὴν ζήτησιν Ach.Tat. 6.10.2.   b to be devoted to a person, οὔνεκ' ἐγὼ μὲν τὶν ὅλος ἔγκειμαι Theoc. 3.33, Herod. 5.3.
ἐγκεκαροῦται, add '(ἐγκεκλάρωται· ἐγκαταλέγει La.)'
ἐγκέλευμα, add 'word of command, ἀφ' ἑνὸς ἐ. J.*AJ* 19.1.14'
*ἐγκενίδες, αἱ, side-planks, *EM* 310.38; cf. ἐπηγκενίδες.
ἐγκεντέω II, delete the section transferring quot. to section I and add 'cf. ἐκκεντέω'
ἐγκεντρίζω I, for 'goad, spur on' read 'sting, bite'; add 'cf. ἐκκεντρίζω'
ἐγκεραυλέω, for 'Phrygian flute' read 'Phrygian aulos'
*ἐγκεύθω, poet. ἐνικεύθω, conceal, contain, σορὸς ἥδ' ἐνικεύθει Ἑρμαῖον *SEG* 6.635 (Termessus, ii AD)
ἐγκεφάλιον, for 'ἐγκέφαλος I ' read 'ἐγκέφαλος II ' and add '*PCornell* 50.11 (i AD)'
ἐγκηδεύω, add '*SEG* 31.1304 (Lycia)'
*ἐγκήρυκτος, ον, verb. adj. fr. ἐγκηρύσσω (q.v.), Theognost.*Can.* 83.
ἐγκιλικίζω, add 'cj. in Ar.*fr.* 107 K.-A.'
ἐγκινδῡνεύω, add '*Just.Nov.* 134.2'
⁺ἐγκισσάω, to come into heat, Lxx *Ge.* 30.38, 39, 41, 31.10.
ἐγκλαστρίδια, add 'cf. ἐκκλαστρίδιον'
ἐγκλάω, form ἐνικλάω, add 'Call.*fr.* 75.22 Pf.'; line 2, for 'σωομένους ' read 'σωομένοις'
*ἔγκλειστος, ὁ, hermit, *SEG* 8.39 (Scythopolis, vi AD), *PHaun.* 26.3 (vi/vii AD).
ἐγκλέφωνος, add '(ἐγκλησίφωνος La.)'
ἔγκλημα I, lines 6/7, for 'γίγνεται .. *Lys.* 10.23' read 'τίνος ὄντος ἐμοὶ πρός ὑμᾶς ἐγκλήματος; with what complaint against me of an offence respecting you? Lys. 10.23, 16.10; γίγνεται .. παισὶ πρὸς ἀλλήλους .. ἐγκλήματα X.*Cyr.* 1.2.6'
*ἐγκληρονομέω, inherit, dub. in *PDura* 24.31 (ii AD).
ἐγκλιδόν, line 1, for 'ἐ. ὄσσε βαλοῦσα aslant or askance, A.R. 3.1008' read 'ἐ. ὄσσε βαλοῦσα casting her eyes down, A.R. 1.790, 3.1008'
ἔγκλιζε, add '(ἐγέλλιζε La.; cf. γελλίζειν)'
ἐγκλίνω I, add '8 pass., lie down, pass the night, perh. of incubation in a temple, *PLit.Lond.* 53ᵛ.10 (Ionic iambics).'
ἔγκλισις I 4, after 'a singer's' insert 'or speaker's' and add 'D.H.*Dem.* 54'
ἐγκλυστέον, add 'Aët. 16.64, 89 Z.'
*ἐγκοιλόομαι, to be hollowed out, Paul.Aeg. 6.118.2 (176.10 H.).
ἐγκοιμάομαι 1, sleep in a temple, add '*Inscr.Perg.* 8(3).161.14 (ii AD)'
ἐγκοιμηστός, add '2 gloss on ἔναυλος, Sch.Hes.*Th.* 129 (29.8 Di G.).'
ἐγκοιμητήριος, after '-τήριον, τό' for 'grave' read 'dormitory for performing incubation'; add '*Inscr.Perg.* 8(3).161, 11, al. (ii AD), *IG* 4².127.7 (iii AD)'
ἐγκοιμήτριον, add 'also adj. -ιος, -ιον, ἐγκοιμήτριν ὀθόνιον *PPar.* 53.8'
ἐγκοίμητρον, line 2, after 'Ammon. p. 140 V.' insert 'written ἐκκ- *PBaden* 48.4 (ii BC)'
ἐγκόλαψις, delete '4.1484.265 (Epid.)'
ἐγκολήβαζω, for 'κόλοις' read 'κόλαις (sic)'
ἐγκολλάω, add '2 cement on, τῶν λίθων τὰ κροιά *AE* 1923.39.68 (Oropus, iv BC).'
*ἐγκόλοβος, ον, rather short in stature, *POxy.* 3477.9 (AD 270).
ἐγκολπίας, add 'Seneca *QN* 5.8.1'
ἐγκολπόω, med., add 'appropriate, ἅπαντα τὰ κατέλιψεν ἐνκολφωσάμενος (sic) *PCair.Isidor.* 64.6 (C.AD 298)'
ἐγκόμβωμα, for 'a sort of frock .. clean' read 'kind of over-garment' and for 'Thd.*Is.* 3.20' read 'Sm.*Is.* 3.20'
ἐγκονέω, line 5, after 'Ar.*Av.* 1324' insert 'ep. impf. ἐγκονέεσκον Euph. in *Suppl.Hell.* 415 i 16'
ἐγκονίω, delete the entry.
ἐγκότημα, for ' = sq.' read 'object of anger' and before 'Hsch.' insert 'ἐ.· ὀργή'
*ἐγκότιος, ον, = ἔγκοτος I, ὑμεῖς δαίμονες ἐνκότιοι αὐτῷ γένοισθε *SEG* 6.802 (Cyprian Salamis, ii/iii AD).
*ἔγκοψις, εως, ἡ, app. some form of enamelling process, *Anon.Alch.* 323.7, 22.
ἐγκρατεύομαι, delete 'force oneself to do a thing'
ἐγκρατής III, add '3 member of the Encratite sect, *SEG* 6.348, 349

(Lycaonia, iv AD).' **IV** adv. **2**, line 2, after 'φέρειν τι' insert 'Men.Dysc. 770'

*ἐγκρᾰτία, poet. -ίη, ή, *sect of Encratites, SEG* 6.488.6 (Isauria, iv AD).

ἐγκρίνω 2, line 3, after '*HG* 4.1.40' insert 'w. dat., ἐνικρινθῆναι ὁμίλῳ A.R. 1.227'

ἐγκρίς, for '*a cake made with oil and honey*' read 'kind of cake' and before 'Ph. 1.214' insert 'δύο ἐγκρίδες, ἡ μὲν ἐκ μέλιτος, ἡ δ' ἐξ ἐλαίου'; for 'also expld. .. Hsch.' read 'ἐγκρίδες· πέμμα ἐλαίῳ ἑψόμενον καὶ μελιτούμενον. ἔνιοι δὲ ἀμανίτας (ταγηνίας La.) Hsch., ἐγκρίς· γλύκασμα (?-υσμα) ἐξ ἐλαίου ὑδαρές id.'

*ἐγκρῐτής, οῦ, ὁ, *judge who admits* entrants to athletic contests, *DAW* 44.30 (Hieropolis Kastabala, pl.).

ἐγκροστόω, for '*Suppl.Epigr.* 2.698' read '*SEG* 17.596.7 (Pamphylia, ?iii AD), perh. in *Corinth* 8(1).318 (v. Robert *Hell.* 11/12.52)'

ἐγκρύπτω 2, add 'Arist.*Juv.* 470ᵃ16 (pass.)' **3**, add 'pass., ἑρπετὰ δ' ἰλυοῖσιν ἐνέκρυφεν (-φθεν v.l.) Call.*fr.* 336 Pf.'

*ἐγκρυψις, εως, ἡ, *covering over*, of fire with ashes, Arist.*Juv.* 470ᵃ12; of planets, *PMich.*III 149 x 40 (ii AD).

ἐγκτάομαι, add '*IG* 2².43.37 (Athens, 378/7 BC)'

ἔγκτησις, lines 1/2, for 'Dor. .. εως, ή' read 'εως, ἡ, **ἔκτησις** *IG* 9²(1).19.12 (Thermus, iii BC), Dor. **ἔγκτᾱσις**, Thess. **ἐντᾱσις** (q.v.), see also ἔμπᾱσις'; line 5, after '(Sparta)' insert '*IG* 1³.227.21 (Athens, v BC)'

ἐγκῦδον, add '(ἐγκυδές La.)'

ἐγκυέομαι, delete the entry.

*ἐγκυέω, w. acc., *to become pregnant with*, χόλον *AP* 7.385 (Phil.); pass., *to be carried in the womb*, Theon *Prog.* 2.

*ἔγκυθρον, τό, in pl. app. contents of χύτραι, perh. grape-juice, *SEG* 34.1213 (ἐνκ-, Lydia, ii AD); cf. ἐγκύτριαι, κύθρος, etc.

ἐγκυκλέομαι III, add 'Philod.Scarph. x 124'

*ἐγκυκλιακός, ή, όν, *of* or *for* the ἐγκύκλιον (ἐγκύκλιος IV), λόγος *BGU* 914.5 (ii AD), cf. *PMich.*II 123ʳ xxii 31, 123ᵛ vii 16, 19 (i AD); ὁ ἐ. the official collecting the ἐγκύκλιον, *POxy.* 2281.3 (ii AD).

ἐγκύκλιος I, add '2 ἐγκύκλιον, τό, *semicircular exedra*, Petersen-Luschan *Reisen in Lykien* 257.' **III 3**, add 'ἐπιστολὴ ἐ. Lxx *Da.* 4.34ᵇ'

ἔγκυκλος II, before 'ἔγκυκλον' insert '*having a border all round*, ταινία *PCair.Zen.* 696.5 (iii BC); ἱμάτιον Phot. s.v. παράπηχυ'

ἐγκῡμονέω, for '*conceive*' read '*carry in the womb*'; add 'pass., *Cod.Just.* 6.48.1.3'

ἐγκύπτω, add 'ἐγκύψας κάτω Ar.*Ra.* 804, ἐγκύψας καὶ ἐκτείνας τὴν κέρκον Pl.*Phdr.* 254d'

*ἐγκυριεύω, *to be in control*, Callistr.Hist. 5 J.

ἐγκῡσίκωλος, read '**ἐγκῡσίχωλος**'

*ἔγκωλον, τό, v. δαῖμα.

ἐγκωμιαστής, add '*SEG* 32.502 (Thespiae, ii AD)'

ἐγκωμικός, delete the entry.

ἐγκώμιος I, delete the section.

*ἐγώ, v. ἐγώ.

ἐγρηγορέω, delete the entry.

ἔγρηνται, add 'cf. ‡ἀγρέω'

ἐγρήσσω, after 'A.R. 2.308' insert '*Trag.adesp.* 664.21 K.-S.'

*ἐγροάς· στόμα ἢ πόρου .. Hsch.; cf. ἐκροή, ἔκροια.

ἔγρω, delete the entry.

ἐγχᾰδές, after 'ἐγχαλές' insert 'or ἔγχαλις (La.)'

ἐγχᾰλῑνόω 3, for 'metaph. of reins' read 'of veins'

ἔγχαλκος I, delete '*in* or *with brass*' and add 'Men.*Mon.* 492 J.'

ἐγχάραξις, add 'Gal. 11.322.6'

ἐγχᾰράσσω I, line 5, after 'Jul.*Or.* 7.217a' insert 'fig., ἐνεχάραξε .. τὸ θεῖον αὐτῷ τὸ τῷ παντὸς λόγῳ σύσταμα Archyt.in Iamb.*Protr.* 4 (p. 18.26 P.)' **II**, *scarify*, add 'Gal. 11.321.18'

ἐγχᾰρίζομαι, add 'ἐνκεχαρισμένος dub. sens., in tomb curses, Mitchell *N.Galatia* 129, 362'

ἐγχάσκω II, add 'Ar.*Nu.* 1436, *Lys.* 271'

ἐγχείη, add 'Myc. e-ke-i-ja, cf. e-ke-i-ja-ta, pers. n.'

ἐγχειρέω 2, add 'of tomb robbers, *SEG* 30.1546 (Cilicia, iv/vi AD)'

*ἐγχειρής, ές, *that is to hand*, cj. in B. 10.11 S.-M.

ἐγχειρίδιον 1, for this section read '*hand-weapon, sword*, Hdt. 1.12, 214, Th. 3.70.6, Lys. 4.6, Lxx *Ez* 21.3, 4, 5. **b** rendering Hebr. *hereb* as used in stone-cutting, Lxx *Ex.* 20.25. **c** rendering Hebr. *kidōn* (spear), Lxx *Je.* 27(50).42.' **4**, delete the section.

ἐγχειρογάστωρ, for 'Cleanth. .. 16' read 'Anon.ap.Ath. 1.4d'

*ἔγχειρον, τό, pl. ἔγχερα *wages, IG* 2².1126.4; cf. °ὔχηρος.

**ἐγχειρόποινος, Myc. e-ke-ro-qo-no, *wage-earner*.

ἔγχελυς, line 3, for 'ἐγγέλεων' read 'ἐγχέλεων'; line 5, after 'Arist.*HA*' insert 'gen. sg. ἐγχέλιους *IGC* p. 100 B31 (Acraephia, iii/ii BC)' add '**b** τυφλὰς ἐ. *penises*, Archil. l.c.'

*ἔγχερα, v. °ἔγχειρον.

ἐγχηρωτύλει, after lemma insert '(ἐγχήροντι δύη La.)'

ἐγχλίω, for '[ι]' read '[ῑ]'

ἔγχνοος, add 'Nonn.*D.* 22.25 (cj., ἄγχν- cod.)'

ἔγχος, add 'Myc. e-ke-si (dat. pl.)'

ἐγχοῦν, add '(ἔγχουτον Meineke)'

ἔγχουσα, add 'Amips. 3 K.-A.'

ἐγχρίμπτω I, add 'ἐ. αὐτοῖς τὰ κέντρα of bees *planting* their stings, Ael.*NA* 17.35'

ἔγχρισις I, add 'Asclep.Jun.ap.Gal. 12.746.9, 753.2'

ἐγχρονίζω I, line 4, for 'ἐν τόπῳ' read 'ἐν αὐτῇ, sc. τῇ Ἀλεξανδρείᾳ'

*ἐγχρωτάζω, *to be smeared on*, ἡ βούπρηστις ἐγχρωτάζουσα εἰς τὰ χαλινὰ λίτρῳ ὁμοίαν ἔχει τὴν γεῦσιν Sch.Nic.*Al.* 337a Ge.

**ἐγχῠσεύς, ὁ, Myc. e-ku-se-we (pl.), *funnel*.

ἐγχυτρίστρια I, for '*EM* 313.41' read 'Suid. s.v.; καὶ ὅσαι τοὺς ἐναγεῖς καθαίρουσιν αἷμα ἐπιχέουσαι ἱερείου, καὶ θρηνητρίας, .. ib.'

*ἐγχωνεύω, *fill up*, of a method of sealing tombs, ἐγχωνεύσει τοὺς πελεκείνους *SEG* 17.633.9 (Perge).

ἐγχωρέω I **1**, add '**c** pres. part., *fitting*, τὸν ἐγχωροῦντα τρόπον Just.*Nov.* 4.1, *Cod.Just.* 12.60.7.2.'

ἐγχώριος II, for '*of .. rustic*' read '*of* or *belonging to one's estate, domestic*'

ἐγώ I, line 3, after '(lyr.)' insert 'Cypr. e-ko-ne ἐγών before consonant, *ICS* 213a' and after 'A.D.*Pron.* 51.4' insert 'ἐγό, verse in *POxy.* 2331.17, 19 (iii AD; for text see *CR* 7.191)'; line 6, after 'ἐμίνγα' insert 'Sophr. 86, A.D.*Pron.* 81.20, ἐμίγγα' and transfer to section II. **II**, line 3, for 'Sapph. .. 23.7' read 'Sapph. 94.7 L.-P.' and after 'ἐμείο' insert '*Il.* 1.174, al., read by Zenodotus at *Il.* 14.118, al., A.R. 1.829, al.'; line 5, after 'dat. ἐμοι' insert '(strengthened w. γε, ἔμοιγε)' and after 'enclit. μοι' insert 'Cypr. o-mo-i-po-si-se ὅ μοι πόσις *ICS* 84.2'; line 7, for 'ἐμίν' read 'ἐμίν'; line 9, at end for 'Cypr. .. H.' read 'Cypr. mi = με *ICS* 234, 235' **III**, line 4, after 'νῶν' insert 'A.*Ch.* 234' **IV**, line 3, Dor. ἁμές, add 'also ἀμέν ib. viii 10.7 (Cnossus, iii/ii BC)'; line 5, delete 'Herod. 1.46'; line 9, after '*Pron.* 95.3' insert 'cf. A.R. 2.616'

ἐδαλάχθη, for 'Id.' read 'Hsch.'

*Ἐδαλιεύς, ῆϝος, ὁ, *inhabitant of Idalion*, Cypr. e-ta-li-e-we-se Ἐδαλιέϝες, *ICS* 217.2.

†ἐδᾰνός, όν, unexpld. epith. (cf. ἐδανή), ἐλαίῳ ἀμβροσίῳ ἐδανῷ *Il.* 14.172, *h.Ven.* 63 (cj.), ἐδανοῖο .. οἴνης Nic.*Al.* 162, 181.

ἐδάφιον, add '**2** *small plot of ground*, rest. in *IG* 7.2808a.29.'

ἔδαφος 1, add 'fig., *foundation*, Longin. 8.1' **2**, add 'fig., [πάθη ψυχῆς] καταλύειν εἰς ἔ. Plu. 2.515c'

ἐδέατρος, for '*among the Persians .. seneschal*' read '= ἐλέατρος, *steward* or sim. domestic officer'; add 'cf. ‡ἀρχεδέατρος'

ἔδεος, add '(ἔδος La.)'

*ϝhεδιεστᾱς, ᾶ, ὁ, *private citizen*, = ἰδιώτης, Buck.*Gr.Dial.* 83 (Argos, vi BC).

ἔδικτον, v. °ἤδικτον.

*ϝέδιμνος, v. ‡μέδιμνος.

*ἐδνῆστις, ἡ, acc. sg. -ιν, *paid for by a bride-price*, Call.*fr.* 67.10 Pf.

ἔδνον IV, add 'χαρίσιον ἔ. Call.*fr.* 383.1 Pf.'

ἔδνον II, med., for pres. defs. read '*seek in marriage*'; for 'Hes.*fr.* 94.47' read 'Hes.*fr.* 200.7 M.-W.'

ἕδος I **3**, delete '*seated*' and after 'Paus. 8.46.2' insert 'D.C. 51.1, 59.28'; for 'ἕ. ὑπαίθριον D.C. 51.1' read 'D.C. 48.14'

ἕδρα I **3**, line 7, after 'D.C. 57.21' insert 'archit., *Inscr.Délos* 104-4.5 (iv BC)'

ἑδραῖος II, line 6, after 'Ath. 11.496a' insert 'Ἑδραία epith. of γῆ, *SEG* 12.513 (Cilicia, i AD)'

ἕδρασμα, add 'meton., of Cerberus, στυγοῦ σκότους ἕδρασμα *Suppl.Mag.* 42 (iii/iv AD)'

*ἕδρευμα, ατος, τό, some unidentified fitting in a building, *Pap.Lugd.Bat.* XIII 21.6, 12 (i AD).

ἑδρικός, add '**2** masc. subst., perh. *stone for foundations, BE* 1966.512 (Tauromenium, Sicily).'

*ἑδρύσκη, v. ὑδρίσκη.

ἑδωδός, after 'Hp.*Aër.*' insert '4'

*ἑδωλή, ἡ, *seat, Naukratis* ii.68.

*ἐέ, exclam., v. ἔ.

ἐέλμεθα, ἐελμένος, add 'but ἐελμένοι is for ἐελδόμενοι in Keil-Premerstein *Erster Bericht* 9 (Lydia, ii AD)'

†ἐητύς, v. °ἐυτής.

ἔθειρα II, a bird's *feathers*, add 'B. 5.29 S.-M.'; add 'of pubic hair, ἐπὶ κτενὸς ἔσκον ἔθειραι Call.*fr.* 343 Pf.'

*ἐθειρᾰς, ᾰδος, ἡ, = ἔθειρα (in quot., sense II), Sch.Theoc. 1.34.

ἐθέλω, line 10, after '46.47' insert 'med. (or pass.) θέλοιτο *Cod.Just.* 1.4.19.11'; line 12, delete 'nor in Aeolic'; line 14, after '(Milet., iv BC)' insert '*IG* 2².851.15 (224/3 BC)'; line 15, after 'both forms in' insert 'Sapph. 1.17, 24 L.-P.'; line 16, delete 'trag. never use ἐθέλω exc. in augmented forms ἤθελον, -ησα' **I 6**, add 'S.*El.* 585, Macho 383 G., Herod. 8.6' **7**, folld. by subj., add '*E.Ba.* 719, Anacreont. 10.1 W., *Ev.Matt.* 13.28, *Ev.Luc.* 9.54'

*ἐθετῶ· τὸ σκοπῶ Theognost.*Can.* 9 A.

ἐθημολογέω, delete the entry.

ἐθήμων, delete 'c. dat., ἐλπίδι'

ἐθίζω I, pass., add 'Diog.Sinop.fr.dub. 9 S.'

ἔθιμος, line 4, after 'etc.' insert 'ἔθιμοι ἡμέραι days fixed by custom for slave's service to a divinity, SEG 26.729, 37.590 (both Maced., ii AD)'

ἐθμοί, after 'δεσμοί' insert '(ἐσμοί Solmsen)'

*ἐθνιάρχης, ου, ὁ, = ἐθνάρχης I 2, CIJud. 1.719 (Argos).

ἔθνος I 3, after 'PPetr. 3 p. 67 (iii BC)' insert 'PKöln 260.3 (iii BC)'

*ἐθνυμών, όνος, ὁ, app. = ἔθνος, Hdn.Gr. 1.33, 2.735.

ἔθος, line 7, ἐξ ἔθους, after '(Arist.EN) 1103ᵃ17' insert 'SEG 31.122 (Attica, ii AD), according to custom, τὸν ἐξ ἔθους τριῶν μυριάδων τόκον Modest.Dig. 50.12.10'

ἔθω, line 12, for 'εἰώθοσιν' read 'εἰωθόσιν'

εἰ, before 'indecl.' insert 'τό'

εἰ, line 2, for 'ἤ Inscr.Cypr. 135.10 H.' read 'e-ke ĕ κε (= ἐάν) ICS 217.10, 23' A 2 b, εἴθε, w. inf., add 'Ps.-Phoc. 45' add 'e εἴθε γάρ, = εἰ γάρ, Diog.ap.D.L. 6.52, Hsch., Suid. s.v. εἰ γάρ, Phot. α 504 Th., Sch.Vet.A.Th. 550; condemned by Moer. p. 161 P.' B II, line 5, after 'εἰκ alone ib. 3.21' insert 'also perh. orac. in Hdt. 1.174, B. 13.228 S.-M.'; line 7, after '(Pi.)P. 4.266, al.' insert 'εἴ τι ἐάν (= ἐάν τι) BGU 2494.16 (iii AD)' VII 3 a, line 5, εἰ μή τι, ἀλλά, add 'Pl.R. 502a, 509c' e, delete 'siquis alius'

εἶα, at end after 'S.Ichn. 87' insert 'fr. 221.4 R. (v. POxy. ix Pl.iv)'

*εἰαριήτης, ου, ὁ, (ἔαρ B) = αἱματίτης, λίθος Aglaïas in Suppl.Hell. 18.19; cf. Ἀγλαΐας δ' ἐαρίτην τὸν αἱματίτην λίθον φησί Sch.Nic.Al. 314e.

εἰβάτας, read 'εἰβᾱτάς'

εἴβω, add 'cf. εἰφθῇ· εἴβηται Hsch.'

*εἰδάλιμος (B), η, ον, v. °ἰδάλιμος.

*εἶδε, v. °ἤδη.

*εἴδεος, ον, (εἶδος) having form, Zos.Alch. 205.8. (Formed after ἀνείδεος, πανείδεος.)

εἰδήμων, after 'AP 9.505.4' insert '9.496 (Ath.)'; before 'Hermog.' insert 'Rhet.Anon. in POxy. 410.27'

εἰδογράφος, for 'classifier of literary forms' read 'classifier of forms (in quots., musical categories)'

εἰδομαλίδας, add 'also -ίδης Suet.Blasph. 63 T.'

εἰδοποιία 2, for 'Rhet. descriptive quality .. 18.1' read 'transf. ταῖς τῶν σχημάτων εἰ. by means of the specific characters of the figures, Longin. 18.1'

εἶδος I 2 b, after 'Apollon.Perg.Con. 1.14, 21, al.' insert 'perh. also applied to a section of a city laid out on a rectangular grid, POxy. 2975.5 (AD 198)' III, add '4 special proposition, i.e. one belonging to particular subject, opp. τόπος, Arist.Rh. 1358ᵃ31.' add 'V unexpld. use, χιλίαρχοι λογχοφόροι εἴδους Βιθυνῶν δευτέρων BSAA 7.64.'

*εἰδοσύνη, ἡ, knowledge, IEphes. 452 (iii AD).

εἰ δ' οὖν, for 'v. εἰ VII 4c' read 'v. εἰ VII 3c'

†εἰδοφόρος, ὁ, frieze, Inscr.Délos 442B232 (ii BC), MAMA 8.560.

*εἴδω A, line 3, after '(Balbilla)' insert 'Cypr. e-wi-te ἔϜιδε ICS 379 (addenda p. 423)'; line 4, after 'ἰδεῖν' insert 'Ϝιδέν CEG 375 (Laconia, vi BC)' II, add '4 think that something is so, w. acc. and inf., A.R. 1.718, 1024.' B, line 1, delete 'I see with the mind's eye'; line 11, for 'ἴσᾱμι' read '‡ἴσᾱμι'; line 19, transfer 'also ἰδέμεν Pi.N. 7.25' to section A, line 5, after 'Ep. ἰδέειν'; line 21, for 'Pl.Smp. 119a' read 'Pl.Smp. 199a' 1, w. gen. (12 lines fr. end), add 'perh. also Men.fr. 434 K.-Th., Dysc. 385' 3, at end after 'with part. omitted' insert 'D. 4.18' 5, add 'b w. ὅτε, Il. 14.71, Od. 16.424, E.Hec. 110, Tr. 70.' at end add 'Myc. 3 sg. aor. (o-)wi-de'

εἰδωλολατρία, add 'Cod.Just. 1.11.10.4'

εἴδωλον I 3, add 'b image imprinted in sand, Ar.Nub. 976.' add 'VI Εἴδωλον the constellation Hercules, Arat. 64, Nonn.D. 1.256.'

εἰδωλόπλαστος, for 'ideal' read 'imagined'

εἰθάρ, for 'Antim. 16.5' read 'Antim. 20.5 Wy., Call.fr. 31b Pf. (Add. II)'

εἰθισμένως, add 'POxy.Hels. 37.6 (s.v.l., AD 176)'

*εἴθυνα, v. °ἴθυνα.

εἰκάζω II, line 4, after 'cf. (Hdt.) 4.31' insert 'Pl.R. 488a'

εἰκαθεῖν, line 2, after '(S.)Ph. 1352' insert 'opt. εἰκάθοι A.R. 3.849'

εἰκαῖος 1, add 'casual, offhand, σφίγξις Heliod.ap.Orib. 50.9.10' 3, for this section read 'ordinary, taken at random, Call.fr. 334 Pf., Nic.Th. 394, ξύλα Iamb.Comm.Math. 4'

εἰκάς I, line 1, for 'B.Scol.Oxy. 1361 fr. 1.5' read 'B.fr. 20B.5 S.-M.'; line 2, delete '(Hes.Op.) 820' (v. °μετεικάς); lines 3/7, for 'Epicur. .. Lg. 849b' read 'Ar.Nu. 17, And. 1.121, Epicur. [1]18, 8 A.; μετ' εἰκάδα, μετ' εἰκάδας are used in referring to days in the month later than the 20th, Men.fr. 265.3 K.-Th.: in Att. inscrr. (cf. Arat. 1149-52) they are usu. counted backwards fr. the end of the month, thus δευτέρα μετ' εἰ. = the 29th (or in a 29-day month the 28th), τρίτη μετ' εἰ. = the 28th (or 27th).' at end add 'see also ἰκάς'

εἴκασμα I, after 'A.Th. 523 (lyr.)' insert 'Γύ[γην γὰρ ὤ]ς ἐσεῖδον, [ο]ὐκ εἴκασμά τι Trag.adesp. 664.18 K.-S.'

εἰκασμός, add 'b estimate of yield on land, PCair.Zen. 147.3 (iii BC); of price, Aq.Ge. 26.12.'

*εἰκαστήριον, τό, perh. place for an image, Nonn.D. 13.517 (cod. ἰκ-).

εἰκῇ I 1, delete 'D. 28.5' add '2 without good cause, D. 28.5, 30.20, UPZ 108.24 (i BC).' II, add 'ITyr 17' add 'IV perh. = εἰκότως, probably, Antiph. 216.7 K.-A.'

εἰκοβολέω, for 'talk at random' read 'guess'

εἰκοβολία, for 'talking at random' read 'guessing'

*εἰκόνη, ἡ, image, PMil.Vogl. 150.11 (ii AD).

εἰκονίζω 1, for 'from a pattern' read 'a document' 2, for 'draw up an official description' read 'verify the identity of' and add 'BGU 2475.6 (ii AD)'

εἰκόνιον, add 'εἰς ἰκόνια τῶν Σεβαστῶν POxy. 3792.19 (iv AD)'

εἰκόνισμα, line 1, for 'S.fr. 573' read 'Trag.adesp. 700a.3 K.-S.'

εἰκονισμός II, after 'PLond.ined. 2196' insert '(JEA 52.135)'; at end for 'Sen.Ep. 95' read 'Sen.Ep. 95.66'

εἰκονιστής, for 'registrar' read 'scribe, copyist'

*εἰκονοφόρος, ὁ, = Lat. imaginifer, carrier of the emperor's effigy, Lyd.Mag. 48.4, MAMA 9.131.

εἰκός II, add 'ἐὰ[ν μὴ ἐπι]θῶσιν κατὰ τὸ εἰκός SEG 30.61.A39 (Athens, iv BC)'

εἰκοσάβοιος, add 'εἰκοσόβοιον Suppl.Hell. 991.79 (poet. word-list, iii BC)'

εἰκοσάγωνος, after 'τὸ εἰ.' insert 'i.e. the dodecahedron'

*εἰκοσαδράχμιος, ον, Cret. ἰκᾰτιδάρκμιος SEG 27.631.A13 (-ίος acc. pl., Lyttos, vi BC).

*εἰκοσάδραχμος, ον, of twenty drachmae, τιμή PLond. 1157ᵛ.8 (iii AD); also εἰκοσίδραχμος PTeb. 373.12 (ii AD).

εἰκοσαετής, add 'see also εἰκοσέτης'

εἰκοσαετία, add 'PCol. 175.44 (AD 339), IKeramos 31'

*εἰκοσάκοτυλον, τό, measure of twenty κοτύλαι, SEG 16.496 (Chios).

εἰκοσάμηνος, for '(Leon.)' read '(= Theoc.Ep. 16)'

εἰκοσάπρωτοι, for 'OGI 629.10 (Palmyra)' read 'Petersen-Luschan Reisen in Lykien 38' and add 'SEG 38.1462.15, al. (Oenoanda, ii AD); also in sg., ib. 28'

*εἰκοσέτηρος, ον, = εἰκοσέτης, Corinth 8(3).305 (iii AD).

εἰκοσέτης, for '(Cypr.)' read '(Cyprus)' and add 'GVI 1081 (Ephesus, i AD); fem., SEG 3.543 (Thrace, iii BC)'

*εἰκοσήμερος, ἡ, period of twenty days, SEG 17.829, Pap.Lugd.Bat. XX 3.14 (iii BC).

εἴκοσι, line 4, Ϝίκατι, add 'ἴκατι BCH 61.334 (Drerus, vi BC), SEG 37.422 (Phocis, v BC), ἴκατιν (before consonant) Call.fr. 196.32 Pf., ϝίκατι IPamph. 17, 18 (iii/iv BC)'

*εἰκοσίδραχμος, v. °εἰκοσάδραχμος.

*εἰκοσιεννάετης, ες, twenty-nine years old, GVI 816.4 (Egypt, iii AD).

*εἰκοσίεπτάς, άδος, ἡ, the number twenty-seven, Procl. in Ti. 2.213.32, 215.22 D.

*εἰκοσίπεντάετηρίς, ίδος, ἡ, period of twenty-five years, Cat.Cod.Astr. 11(2).112.1, 113.12 (both pl.).

εἰκοσιπεντάρουρος, add 'OMich. 90.2 (ii BC)'

*εἰκοσιποδία, ἡ, space of twenty feet, prob. in IG 2².1654.38, see J.M. Paton The Erechtheum (Harvard 1927) 420.

*εἰκοσόβοιος, v. °εἰκοσάβοιος.

†εἰκόσορος, ep. ἐεικ-, ον (εἴκοσι, ἐρ-/ὀρ- as in ἐρέτης), having twenty rowers, Od. 9.322, Teles p. 27 H.; applied derogatorily to courtesan, ὁλκάδος AP 5.161 (Asclep.), ib. 204 (Mel.); as subst., εἰ. (sc. ναῦς), ἡ, D. 35.18, AP 6.222 (Theodorid.).

εἰκοστός I, add 'neut. subst., twentieth year (sc. ἔτος), SEG 26.482 (Dyme, ii/iii AD)' II, add 'SEG 30.979 (Olbia, c.500 BC)'

εἰκοτολογικός, ή, όν, based on εἰκοτολογία, λόγοι Procl. in R. 1.284.5.

†εἰκότως, adv. fr. εἰκώς (v. ἔοικα), Ion. οἰκότως Hdt. 2.25.2, 7.50.1. I suitably, in keeping with, ἀπουσία .. εἶπας εἰ. ἐμῇ A.Ag. 915. II reasonably, fairly, A.Supp. 403, S.OC 432, 977, αὐτοὶ οὐκ εἰ. πολεμοῦνται Th. 1.37.1, 2.93.1, Isoc. 12.101, And. 1.142, Pl.Epin. 979d, X.Cyr. 8.8.20, Plu. 2.409f; at the end of sentences, Th. 1.77.5, D. 1.10, Pl.La. 183b; in constr. w. acc. and inf., σθένειν τὸ θεῖον μᾶλλον εἰ. ἔχει E.IT 911, cf. Or. 737, τοῦτ' εἰ. δὴ δοκεῖ ἀνδρῶν ἀγαθῶν .. ἔργον εἶναι And. 1.140.

εἴκω I 1, line 6, for 'Gal. 18(1).97' read 'Gal. 18(2).97' 2, add 'Alcm. 83 P., Ps.-Phoc. 220' III, delete the section (v. D.Page Sappho and Alcaeus (Oxford 1955) 23).

εἰκών, line 4, after 'Maiist. 15' insert '(these forms may come fr. *εἰκώς, declined like αἰδώς)' and after 'acc. pl. εἰκούς' insert 'A.fr. 78a.1 R. (rest.)'; line 5, for 'Ϝεικ- Inscr.Cypr. 151 H.' read 'Cypr. we-i-ko-na Ϝεικόνα ICS 276'

*Εἰλαῖος, v. Ἰλαῖος.

Εἰλείθυια I, pl., add 'SEG 24.226 (Attica)' add 'b transf., ὃ δὲ καλοῦσιν οἱ μάντεις Εἰλειθυίας ἄφεδρον Theoph.HP 5.9.8; cf. τὰ ἐκ τῶν ξύλων ἐκφυόμενα καὶ μάλιστα ἐκ τῶν ἐλατίνων ἃ καλοῦσιν οἱ

Εἰλειθυιαῖα   SUPPLEMENT   εἰς

μάντεις -ας id.CP 5.4.4.' at end add 'also Ἐλίθιουια IG 7.3385, Εἰλειθούα 7.3391 (both Boeotia, ii BC); Εἰλειόθυια SEG 24.1163 (Gortyn, ii/iii AD), ἡιλέθυια SEG 35.37 (Attic vase, vi BC), Ἰλυθει[Lang Ath.Agora XXI G8 (iv BC), Ἐλυθία SEG 31.1586, Lacon. Ἐλευσία ABSA 12.348 (ii/i BC), Myc. e-re-u-ti-ja'

*Εἰλειθυιαῖα, τά, festival at Delos, Inscr.Délos 401.22, 440A69, 461Bb53 (all ii BC).

Εἰλείθυιον, add 'Ἰλειθυεῖον rest. in Ath.Agora XIX L6.98, Ἰλύθειον Inscr.Délos 1403Bbi97, ii37; also Ἰλύθυιον ib. 1421Bbii2 (both ii BC)'

εἰληδόν, after '= ἰληδόν' insert 'εἰληδὸν ἐσμεύουσιν rest. in Call.fr. 191.28 Pf.'

εἰληθερής, for 'cf. ἐλαθερής' read 'also ἐλαιθερής, ἐλαιθερὲς ὕδωρ Suppl.Hell. 1019; ἐλάθερής Hsch.'

εἴλημα I, add 'ἐν ἰλήματι καινῷ λεντίῳ POxy. 3060.6 (ii AD)' II 1, delete the section (read ἀνείλημα). 2, add 'perh. also book-roll, Aq.Je. 43(36).2' III, add 'JRCil. 1.24 (late Rom.imp.)' add 'V Lacon. Γήλημα, v. βήλημα.'

εἰλῆς εἶ, after 'εἰλῆς' insert '(εἴληος La.)'

*εἴληφα, εἰληφώς, v. λαμβάνω.

εἰλίονες, add 'cf. °ἰλίων'

εἴλον, v. αἱρέω.

†εἰλόπεδον, τό, θ' εἰλόπεδον read by some edd. for ‡θειλόπεδον in Od. 7.123; cf. EM 449.29, Eust. 43.38.

εἴλυμα, add 'form Γέλυμα, Myc. we-ru-ma-ta (pl.)'

εἰλυτά, delete the entry.

*εἰλύτας, v. °ἐλλύτης.

εἴλω, line 6, after 'Il. 13.408' insert '3 pl. ἄλεν 22.12'; line 8, after '(Il.) 13.524' insert 'κατα-Γελμένος Leg.Gort. 10.35 (v BC); also Γευμέναν Schwyzer 177.14 (Gortyn, v BC)'; lines 13/14, for 'J.AJ 12.1.9' read 'J.AJ 12.2.9' B, line 7, after 'ib. 8' insert 'cf. ἀμφὶ κῆρες εἰλεῦνται Call. in Suppl.Hell. 253.1' C, line 4, for 'Call. Iamb. 1.144' read 'Call.fr. 191.83 Pf.' and add 'perh. unroll, μὴ .. ἐπ' ὀμφαλὸν εἴλεε (imper.) βίβλον AP 9.540.1'

*εἴλως, v. °ἴλαος.

εἷμα, line 2, for '(but gen. fem. Γήμας 5.40)' read 'see also °Γήμᾱ'; after '(Od.) 10.542' insert 'ἀπὸ κρόκεον ῥίψαις Ἰάσων εἷμα Pi.P. 4.232'

*εἱματιστής, οῦ, ὁ, servant in charge of clothes, La Carie 18 (ii AD).

εἰμί (sum), lines 1/2, for 'Cret. .. 4959a' read 'Dor. ἠμί Schwyzer 273 (Rhodes, vi BC), Inscr.Cret. 2.x 7.2 (ἐμί, Cydonia, v BC), ib. xii 31a.3 (ἠμί, Eleutherna, ii BC), etc.'; line 10, after 'Sapph. 1.28' insert 'L.-P., AP 9.318 (Leon.Tarent.)'; line 12, for 'CIG 2664' read 'BMus.Inscr. 918 (Halicarnassus, ii/iii AD) and after 'al.' insert 'pl. ἤτωσαν Mitchell N.Galatia 129; cf. ἐξήτω AS 12.209, 211 (Cilicia)'; line 13, after 'ἔστωσαν' insert '(rare, and often altered to ἔστων; but cf. E.Ion 1131, Pl.Sph. 231a, Lg. 737e, 762d)'; line 15, for 'ii BC, ib. 2².1328' read 'iv BC, SEG 9.1.13, cf. IG 2².1328; so codd. in Pl.R. 352a, codd. BT in id.Sph. 231a' and after 'ib. 1126' insert 'Boeot. ἔνθω IG 7.3172.88'; line 17, delete 'IG 7.3172.165'; line 18, after 'εἴη 9.245, etc.' insert 'ἤν POxy. 1061.13 (i BC), etc.'; line 25, inf., εἶναι, add 'Stesich.fr.suppl. 15 i 7 P.'; line 27, ἔμεν, add 'Theoc. 25.116, Call.fr. 67.20 Pf.', ἤμεν, add 'SEG 30.356 (Argos, c.300 BC)', εἶμεν (in IG 12(1).155.100 written ειμμ), add 'Stesich.fr.suppl. 102.5 P.'; lines 25/27, inf. εἴμεναι, Aeol., add 'IKyme 13 i 18 (iii BC)', ἔμμεν, add 'also Thess., Schwyzer 590.20 (Larissa, iii BC)', εἶμεν, add 'also Boeot., SEG 32.496 (Thespiae, iii BC)'; line 30, εἴμειν, add 'Stesich.SLG 102.5 P.', εἶν, add 'Stesich.SLG 15 i 7 P.'; lines 31/4, for 'SIG 135.4' read 'SIG 135.3'; part., after 'Ep. ἐών, ἐοῦσα, ἐόν' insert 'gen. ἐόντς SEG 32.356 (Aegina, vi BC)', for 'ἰών .. 23 H.' read 'i-o-ta ἰόντα ICS 217.23', for 'IG 7.3172.15' read 'IG 7.3172.116'; after 'Theoc. 28.16' insert 'ἤσσα SEG 9.11.17 (Cyrene, iv BC)' and after 'ἔασσα' insert 'PAE 1931.89 (Dodona, vi/v BC), ἔασα SEG 11.112, 37.340.6 (vi/v and iv BC, both Arcadia); gen. ἐόνσας SEG 29.529 (Larissa, iii/i BC); acc. sg. ἔντα Lyr.adesp. 62 P.'; line 36, before 'ἔασα .. (Messene)'; line 38, after 'dat. pl. ἔντεσσι ib. 104' insert 'Thess. εἴντεσσι SEG 36.548.2 (Matropolis, iii BC)'; line 47, delete '(wh. is v.l. in Pi.I. 1.26)'; line 48, delete 'Erinn. 4.4'; line 50, after 'Sophr. 59' insert 'Pi.I. 1.26 (v.l.), Erinn. 4.4'; line 51, after '(Delph.)' insert 'also Arc. IG 5(2).6.37 (iii BC), Cypr. e-se ICS 398'; line 56, after 'Epidamnus, iii BC)' insert 'cf. ἐξῆι· ἐξεγένοντο Hsch.', 'ἔξειμι (B)'; p. 488a line 1, after 'GDI 1696' insert 'SEG 16.255.17 (Argos, ii BC)'; line 2, after '(lyr.)' insert 'ἦσκε Alcm. 74 P.' at end of para. add 'Myc. pres. 3 pl. e-e-si (form ἔhενσι), part. e-o ἐhών (masc. sg.), e-o-te ἐhόντες (pl.); a-pe-a-sa ἀπ-ἑhασσαι (fem. pl.); fut. 3 pl. e-so-to' A IV, line 2, for 'X.Cyr. 2.3.16' read 'X.Cyr. 2.3.18'; lines 10/11, w. relat. particles, add 'w. whose, E.Alc. 113, w. ἦ, Pl.Lg. 721b'; line 13, delete 'in questions expecting a neg. answer' VI, lines 6/7, delete 'imper.' and 'ἔστω .. 10.7' C I, add 'καλῶς ἔσται Men.Dysc. 571' E I, for 'phrases .. do a thing' read 'limiting phrases' and after 'Pl.Plt. 300c' insert 'ἀριθμὸν εἶναι id.Ep. 337c'

εἶμι (ibo), line 2, after '3 sg. εἶσι' insert 'εἶτι SEG 9.72.57, 88 (Cyrene, iv BC)'; line 3, after 'compd.' insert 'πρόσει Epict.Ench. 32.2'; line 5, after 'Sophr. 48' insert 'εἴη (lapid. ΕΙΕΙ) Inscr.Cret. 4.81.7 (Gortyn, v BC), (ἐπ-) SEG 9.72.3 (Cyrene, iv/iii BC)'; line 9, for 'Il. 24 .. (Crete)' read 'dub. in Il. 24.139, Od. 14.496, Hes.Op. 617'; line 11, after 'Str. 9.2.23' insert 'SB 6152.22, 6153.25 (Egypt, i BC)'; line 12, delete '(προσ-εῖναι .. 353)' I, add 'Myc. fut. part. i-jo-te ἰόντες (masc. pl.)' II 1, add 'SEG 9.72.88 (Cyrene, iv BC)' VI 2, add 'E.Ba. 365, Heracl. 455, etc.'

*εἰμιναῖος, v. °ἡμιμναιαῖος.

εἴν, v. οὔ.

εἰναετής, add 'also εἰνέτης Call.Dian. 14; cf. ἐνναετής, °ἐννεέτης'

εἰνάτερες, line 4, after '(acc. sg.), add 'cf. τὴν ἰανάτερα TAM 5(1).682, 754 (prob. loan-word)'

*Εἰνατίη, from the city Εἴνατος in Crete, epith. of Ilithyia, Call.fr. 524 Pf., St.Byz. s.v. Εἴνατος; Ἐλεύ]θυιαν Βινατίαν Inscr.Cret. 4.174.76 (Gortyn, ii BC), cf. ib. 61.

*εἰνέτης, v. °εἰναετής.

εἴπερ II, line 2, after 'etc.' insert 'εἰ μή πέρ γ' ἅμα αὐτὴ γένοιτ' ἂν γραῦς τε καὶ νέα γυνή Ar.Nu. 1183' III, for this section read 'w. an ellipse, if what precedes is inevitable, is a valid assumption, etc. Ar.Nu. 227, Pl.Lg. 667a, 900e; ἀλλ' εἴπερ, Pl.Prm. 150b, R. 497e, Arist.EN 1101ª12, 1155ᵇ30, 1174ª28'

εἶπον, line 3, after 'Pi.O. 4.25' insert '3 sg. ἧπε Clara Rhodos 9.221 (Lindian decree, v BC); line 5, for 'Dor.' read 'Aeol.'; line 6, ἔειπα, add 'Pi.N. 9.33'; lines 8/9, for 'persons' read 'person' and delete 'and imper.'; line 11, εἰπάτω, add 'X.Ath. 3.6'; line 13, before 'part.' insert 'opt. εἴπαιμεν Pl.Sph. 240d; εἴπειεν Arist.Ath. 75ª24, Top. 159ᵇ35' and after 'εἴπας' insert 'Hdt. 8.102.1, 9.42.4, al., Longin. 1.2'; line 17, Γείπαι, add 'SEG 32.908 (Phaestus, vi BC)' I 1 b, add 'μέλος Call.Del. 257, Nonn.D. 43.392; ὑμηναίους Call.fr. 75.43 Pf. (ειδον pap., εἶπον Pf.)' III, at end after 'Herod. 6.26' insert 'folld. by ὅπως and fut. ind., Men.Dysc. 237'

εἶπος, delete the entry (v. °ἴπος).

Εἰραφιώτης, line 2, for 'Call.fr.anon. 89' read 'Suppl.Hell. 1045'

εἰρεσία I 1, line 5, for 'close to her throbbing breast' read 'beside the rhythmic beating of her breast (cf. ἐρέσσω II 1)' II 1, line 2, for 'Antip. (?)' read 'Antip.Thess.'; insert 'Opp.H. 5.301'; transfer 'Th. 7.14' to section I 1. 2, delete the section transferring exx. to section I 1.

εἴρην, for 'who .. year' read 'aged between thirteen and nineteen, teenager'; add 'Call.fr. 487 Pf.'; ἰρήν in Hdt. 9.85.2; ἴρανες· οἱ εἴρενες. οἱ ἄρχοντες ἡλικιώταις. Λάκωνες Hsch.; cf. μελλείρην, ‡τριτίρενες'

*εἰρναρχία, ἡ, office of εἰρηνάρχης, εἰρηναρχίαι ε', i.e. five tenures of office, Didyma 157 II (b).

εἰρήνη I, line 3, after 'E.Med. 1004' insert 'καὶ πολέμῳ καὶ εἰρήνης in peace and war SIG 110 (Rhodes, 410/8 BC), SEG 31.969 (Erythrae, iv BC), 30.990.17 (decree from Delos, iii BC), etc.. 2 assurance, making safe, ἐξ ἀλληλενγύης τῇ εἰρήνῃ τοῦ αὐτοῦ Ἀμουλοῦ pap. in Mnemosyne (series 3) 3.236 (iv AD). 3 as the condition of the dead, cf. section IV, TAM 4(1).357, SEG 33.764 (Italy, iv AD).' fourth line from end, ἰράνα, add 'SEG 30.360 (Argos)'; after '(IG) 7.2407' insert 'Rhod. ἰρήνα Schwyzer 278 (v BC)'; for 'but Cret. .. 508.5' read 'w. aspirate χἰρῆνας (= καὶ ἰ-) Inscr.Cret. 186.B5, 184.a7, Thess. ἡιρένα SEG 23.415 (Pherae, v BC), ἰιράνα SEG 26.461 (Sparta, v BC)'

εἰρηνικός 3, add 'οἱ δὲ κατοικοῦντες εἰρηνικώτεροι Peripl.M.Rubr. 8'

*εἰρηνόφρων, ονος, ὁ, ἡ, peaceably-minded, MAMA 8.321.

*εἰριπόνος, ον, = εἰροπόνος, Simon. 113 P.

*εἰρκτήριον, τό, = Lat. carcer, Charis. p. 32 K.

*εἰροπλόκος, ὁ, wool-weaver, IG 2².13178 (iv BC).

*εἴρους, εἰρούισσα, v. ‡ἧρως, ‡ἡρώισσα.

*εἰροφόρος, ον, wearing wool, Νίοβος εἰ., title of play attributed to Ar., Hdn. 29 H. (cf. ἐριοφόρος).

εἴρω (A), line 2, after '(v. infr.)' insert 'εἰρμένος Call.fr. 657 Pf. (v.l. εἰργμ-), prob. in Poll. 6.75 (codd. εἰργμ-)'; line 6, delete 'τὸ .. Plot. 2.3.7'

εἴρω (B), line 3, after '(Od.) 11.137' insert '3 pers. εἴρει Arat. 739'; impf. εἴρεν B.17.20, 74 S.-M.'

εἰρωνευτικός, add 'οἱ λόγοι τοῦ Γοργίου εἰ. εἰσιν ἅπαντες Anon. in Rh. 190.5'

*εἰρωνων, gen. pl., word occurring in price-lists of spices, POxy. 3733.25, 3766.107 (both iv AD).

εἰς, line 8, after 'vowels' insert '(exc. Sapph. 44.23, 26 L.-P.)'; line 11, after '(Crete)' insert 'also ἰος (for *ἰνς) Inscr.Cret. 2.v 1.12 (Oaxos, vi/v BC), but ἰς ib. 9.13, IPamph. 3.4' I 1 b, delete 'and Ion.' and for 'Hdt. 4 .. Ar.Av. 619' read 'in Pindar e.g. ἐς ἄνδρας O. 2.38, ἐς θεόν ib. 7.31, but not freq.; in Hdt. and Att. where persons stand for country or region, ἐς τοὺς συγγενέας Hdt. 4.147.3; ἐς ἀνθρώπους ἀπόρους Th. 1.9.2; ἐς Πισίδας X.An. 1.1.11; also of coming before an assemblage, εἰς ὑμᾶς before your court, Pl.Ap. 17c, D. 18.103,

105

etc. (cf. ἐς τὸν δῆμον Th. 5.45.1, etc.); also χωρεῖν ἐς and the like, of attacking, Th. 4.95.3, X.*An.* 3.2.16, etc. ἕξει ἐς τὸν ἄνδρα Ar.*Eq.* 760. (in εἰς Ἄμμωνα id.*Av.* 619, Ἆ. is the place, cf. Str. 17.1.42.)' **2**, line 12, after 'D.S. 14.117' insert 'τὰ παιδία εἰς τὴν κοίτην εἰσίν Ev.*Luc.* 11.7; εὑρέθη εἰς Ἄζωτον Act.*Ap.* 8.40' **II**, of time, add 'w. ellipsis of acc., ἐς ἔτους φανέντος ἄλλου Anacreont. 56.14 W.' **2**, at end for 'Th. 4.63 (v. εἰσαῦθις)' read 'v. °εἰσαῦθις'; after 'Ar.*Pax* 367' insert 'cf. Arat. 770, 1103 (prob.)'; add 'ἐς ἄχρι, v. ἐσάχρι' **IV 3**, line 4, after 'Ar.*Ach.* 686' insert 'ἐὼς βίαν Men.*Dysc.* 396' **V 2**, add '**b** as, by way of, εἰς μεγάλην .. χάριν AP 5.287.10 (Agath.), cf. 7.614.14 (id.); εἰς μισθόν Ps.-Hdt.*Vit.Hom.* 4.'

εἷς, line 1, after 'μιᾶς, ἑνός' insert '(acc. εἷνα only Hdn.Gr. 1.546), Myc. *e-me* (masc. dat. = ἑνί)' **1 g**, line 6, after 'Luc.*Salt.* 12' insert 'so ἕνα παρ' ἕνα SEG 31.825 (Sicily, ii BC), μίαν παρὰ μίαν on alternate days, Suppl.Mag. 10 (iii/iv AD); so παρὰ μίαν Plb. 3.110.4'; line 9, after 'Ael.*NA* 5.9' insert 'ἐν ἑνί taken together, total (of prices), BCH 60.119.6 (Delph.)'; add 'διὰ μιᾶς after a day's break, Suppl.Mag. 34 (vi AD)' **h**, line 2, after 'Th. 8.109' insert '(s.v.l.)' **i**, for this section read 'μίαν, as adv., once, POxy. 1593, BGU 984 (both iv AD); rptd. μίαν μίαν, app. from time to time, S.*fr.* 201' add '**j** εἷς εἷς τῷ Ἐλεαζάρ καὶ εἷς εἷς τῷ Ἰθαμάρ one each .. (app. in imitation of Hebr. constr.), Lxx 1*Ch.* 24.6.'

εἰσαγγελία I 1, add '(s.v.l.; cj. προσαγγελία)'

εἰσαγγέλλω I 1 b, add 'for honorific purposes, τῇ βουλῇ Amyzon 35, IMylasa 126.2 (iii/ii BC)'

εἰσάγω I 3, add '**b** bring in water supply, SEG 31.1363 (Cyprus, i AD), 26.784.9 (Thrace, ii AD), IEphes. 3217.(b)21 (ii AD), etc.' **II 3 a**, transfer 'A.*Eu.* 580, 582' to follow 'of the εἰσαγωγεύς'; for 'D. 24.10' read 'οὐδ' αἰσχύνει φθόνου δίκην εἰσάγων D. 18.121; cf. εἰ γραψάμενοι τὸν νόμον καὶ εἰσαγαγόντες εἰς ὑμᾶς λῦσαι δυναίμεθα id. 24.10'

*εἰσαγωγεῖον, v. ‡εἰσαγώγιον.

εἰσαγωγεύς I, add 'official under the ἀγωνοθέτης who admitted the contestants, SEG 16.258 (Argos, ii AD)'

εἰσαγωγή I 3, add 'ἐσαγωγὰν καὶ ἐξαγωγάν SEG 32.1586 (Egypt, v BC)'

εἰσαγώγιμος, line 2, for 'Arist.*Oec.* 1345ᵃ21' read 'Arist.*Oec.* 1345ᵇ21'

εἰσαγώγιον, add 'SEG 14.639.(E)9 (Caunus, i AD): also -εῖον SEG 33.115.26 (Athens, iii BC), Hesperia 11.295.16 (Athens, ii BC). **b** import duty, SEG 37.859.B16 (Caria, ii BC), PHels. 36.5 (ii BC), IEphes. 13 i 24 (i AD). **2** office of the εἰσαγωγεῖς at Samos, BCH 59.478 (Samos, ii BC).'

εἰσαεί, add 'also ἐσαεί IG 2².1064 (c.AD 230), Hesperia suppl. 6, no. 31.14 (Athens, iii AD)'

εἰσαΐω II, add 'ἐσαΐειν Hp.*Mul.* 1.4; ἐσαΐει ib. 1.9'

εἰσακοντίζω, line 1, delete 'at, τινά'

εἰσακούω I 1, at end after 'Ev.*Matt.* 6.7' add 'SEG 34.1220 (Lydia)'

εἰσάκτης, add 'gloss on εἰσηγητής Hsch. **2** tax-collector, PMich.VIII 989.4 (ostr., iii AD).'

εἰσαλείφω, delete 'anoint, Aristid. 2.292 J.'

εἰσαμείβω, for ' go .. τεῖχος' read 'allow to enter'

*εἰσάμην, v. ἴζω I 2.

εἰσάμην III, delete the section.

εἰσαναβαίνω, add 'of the dead, ψυχὴ δ' αἰθέρα εἰσανέβη epigr. in SEG 37.198 (Attica, ib/ii AD), Corinth 8(3).658 (v/vi AD)'

⁺**εἰσανδρόω**, Ep. ἐσ-, fill with men, A.R. 1.874 (v.l. ἐπ-).

εἰσανέχω, jut out, A.R. 1.1360, 4.291; w. acc., jut out into, id. 4.1578.

*εἰσανύω, prob. f.l. for ἀνύω in Cat.Cod.Astr. 8(1).168 (Vett.Val.), cj. in Suppl.Hell. 962.17.

*εἰσαπαντάω, meet, encounter, PMerton 65.8 (ii AD).

*εἰσαπόλλυμι, lose into (in quot., a well), Men.*Dysc.* 681 (pap.; perh. read ἐξαπ-).

⁺**εἰσαράσσω**, Att. -ττω, drive in disorder into, Hdt. 4.128.3, ἐσαράξαντές σφεας ἐς τὰς νέας 5.116, D.C. 43.40.3, τὴν .. ἵππον .. ἐς τοὺς πεζοὺς ἐσήραξε id. 51.26.1.

⁺**εἰσαῦθις, ἐσαῦθις** (or εἰς, ἐς αὖθις), for another (later) occasion, ἐ. ἀναβάλλεσθαι Ar.*Ec.* 983, Th. 4.63.1, Pl.*Euthphr.* 15e, *Smp.* 174e; on another occasion, later, εὐτυχοῦσι δέ οἱ μὲν τάχ', οἱ δ' ἐσαῦθις, οἱ δ' ἤδη βροτῶν E.*Supp.* 551.

εἰσβαίνω I 2, add '**b** embark on, begin an action χορεῖον εἰσέβαινε ῥυθμόν began, Men.*Dysc.* 951.' **II**, line 2, after 'E.*Alc.* 1055' delete '(lyr.)'

εἰσβάλλω I 1, line 6, for 'country' read 'region, place' **4**, after 'begin' insert 'D.H.*Lys.* 17'

εἰσβατικόν, after 'τό' insert 'entrance fee, POxy. 2239.21 (vi AD)'; for 'tax in Egypt' read 'unexpld. sense'

⁺**εἰσβιάζομαι**, force one's way in, ὁ μὲν γὰρ ἂν οὐκ ἀστὸς εἰσβιάζεται Ar.*Av.* 32, OGI 736.6 (Fayûm), PPetr. 3 p. 39 (iii BC); w. πρός, D.S. 14.9; w. ἐς τὸν Βόσπορον D.C. 42.47; transf., D. 39.33 εἰς τὰ πρῶτα γένη Plu. *Num.* 1.

*εἰσβοηθέω, bring help to, SEG 38.1476.96 (Xanthos, 206/5 BC).

εἰσβολή 2 b, add 'entrance to a bay, Peripl.M.Rubr. 5' **3**, proem, preface, add 'Arg.E.*Med.*'

εἰσγραφή, add '**II** inscription, τιμῶν (on statues), IGRom. 3.739 ix 60 (Rhodiapolis, ii AD, pl.).'

*εἰσδημέω, pay a visit to another country, Ἀθήναζε Eun.*VS* p. 491 B.

εἴσειμι, add 'see also ‡ἔνσειμι'

*εἰσείρω, app. fasten on, Hsch., s.v. ἵρεται.

εἰσελαστικός, for 'celebrated by a triumphal entry' read 'carrying the privilege of a triumphal return to one's native land or city'

εἰσελαύνω, line 1, after '-ελάω' insert 'Eretrian εἰρελάω q.v.'

*εἰσένεκτον, τό, perh. entrance-passage, τὸ περίβολον καὶ τὸ ἐν αὐτῷ εἰσένεκτον MAMA 4.85 (Phrygia), 7.323 (Lycaonia), Ramsay Cities and Bishoprics I 163 (Pisidia).

*εἰσεργάζομαι, perform rites to, Κρονίω(νι) παιδίο(ις) εἰσεργα(ζομένοις) PMil.Vogl. 212ʳ 12.

εἰσέρχομαι, line 11, after 'visit' insert 'Men.*Asp.* 428'

⁺**εἰσέτι**, adv., still, yet, Call.*Del.* 189, Mosch. 2.19.45, AP 6.271 (Phaedim.), Theoc. 27.19; εἰσέτι νῦν Phanocl. 1.28, A.R. 1.1354; εἰ. νῦν γε id. 2.717; εἰ. καὶ νῦν Call.*Dian.* 77.

εἰσευπορέω, after lemma insert 'Thess. ἐνευπορέω IG 9(2).66a.5, 13 (Lamia, ii BC)'

εἰσέχω I, add 'Arr.*Ind.* 32.11, 43.2'

εἰσηγέομαι 2, add 'pass., διάταξιν, σωφρόνως μὲν εἰσηγηθεῖσαν κοσμίως δὲ τεθειμένην Just.*Nov.* 87.7'

*εἰσθεάομαι, gaze upon, τὸ .. εἰσθεᾶσθαι γῆν ὅλην τ' οἰκουμένην Ezek. *Exag.* 87 S.

εἰσθλίβω, for 'Them.*Or.* 14.197a' read 'Them.*Or.* 15.197a'

εἰσιτήριος, εἰσιτήρια (sc. ἱερά), add 'or before a battle'; at end, sg., add 'entrance charge, Didyma 314.10 (ἰσ-, ii AD); cf. mod. Gk. εἰσιτήριο ticket; see also °ἐνετήρια'

⁺**Εἰσιτυχή, v. °Ἰσιτύχη.

*εἰσκήρυξις, εως, ἡ, proclamation by sending a herald, SEG 33.1039.37 (Aeolian Cyme, ii BC)

*εἰσκλεῖστρον, τό, perh. rod, BGU 2361a ii 11 (iv AD; written εἰσκλίστρον).

εἴσκλησις, for 'summons' read 'invitation' and add 'εἴσκλησις εἰς τοὺς Διονυσιακοὺς ἀγῶνας SEG 21.506.20 (Athens, iii AD), 30.82.11 (Athens, c.AD 230)'

*εἰσκόλαψις, ιος, ἁ, carving, SEG 25.383 (Epid., iv BC, ἰσκ-).

εἰσκομίζω I, add 'οὐδέ[να δὲ ἄλλον τινὰ] ἐξέσται εἰσκομισθῆναι σορ[ῷ SEG 34.1401 (Lycaonia)'

εἴσκρισις II, for 'enrolment, admission' read 'examination for admission to priesthood, guild, etc.' and add 'PLond. 329.7, PTeb. 598 (both ii AD)'

εἰσκριτικόν, add 'by priests on admission, Stud.Pal. 22.184.25 (ii AD), PTeb. 294.20 (AD 146), SB 9320.33 (AD 171)'

*εἰσκρύπτω, app. intr. hide, go into concealment, SEG 24.1224 (form εἰσκρύβω; Egypt, v AD).

εἰσκυκλέω I, for 'wheel in .. spectators' read 'wheel in (metaphor prob. derived fr. the use of stage machinery)'

εἰσνέομαι, delete '[Thess.]'

εἰσόδιος I, after 'PPetr. 2 p. 54 (iii BC)' insert 'entrance fee of society, Inscr.Délos 1521.17 (ii BC)'

εἴσοδος II 2, add 'entrance on a magistracy, OGI 458.15 (i BC)' **II 4**, after 'Lys. 1.20' insert 'SEG 35.219.4 (ἰσ-, tab. defix., Athens, iii AD)' and after 'of a doctor' insert 'Hp.*Decent.* 12, 13, Men.*Mon.* 659 J.'

εἰσοιχνέω, add 'epigr. in IG 10(2).1.368.7 (Thessalonica, ii AD)'

εἰσόκε I, w. subj., add 'epigr. in BCH suppl. 8.5 (Edessa, iii AD)' add '**III** when, w. aor. ind., A.R. 2.857 (v.l. εἰσότε).'

εἴσομαι, after lemma insert '(A)' and for 'II' read 'εἴσομαι (B) adding 'Ἀμφίνομος δ' Ὀδυσῆος ἐείσατο κυδαλίμοιο Od. 22.89'

*εἰσόμνυμι, aor. ἐσώμο[σ]αν, app. swear to a thing, SEG 23.271.61 (Thespiae, 220/208 BC).

*εἰσονομάζω, name, designate, SEG 32.1423 (Syria, ii AD).

εἰσοπτρίζω, line 2, for '(Plu. 2.)141c' read '(Plu. 2.)141d'

*εἰσοπρομαντεία, ἡ, divination by mirrors, rest. in PMag. 13.752 (iv AD).

εἰσορμάω, line 1, read 'impel into, E.*IA* 151 (cj.); transf., πάλιν εἰσώρμησα τὸν ἄρσενα Δωρίδι Μούσῃ ῥυθμόν AP 7.707 (Diosc.)'

εἰσορμίζω, line 1, after 'bring into port' insert 'epigr. in SEG 23.395.A10 (Corcyra, ii BC)'

εἴσοψις, for 'spectacle' read 'observation, attention'

εἰσπλέω 2, after 'Pl.Com. 183' insert 'sail in (with merchandise), import, Lys. 22.17, 21'

εἴσπλοος 2, add 'ἔσπλον καὶ ἔκπλον SEG 36.982.C9 (Iasos, v BC)'

εἰσποιέω I 2, add 'pass., Arr.*An.* 4.11.5'

εἰσπορεύω, for 'pass.' read 'med.' and add 'Aen.Tact. 18.1, 15, al.'

εἴσπραξις I, add 'τῆς ἰσπράξεως τοῦ πυροῦ PPetaus 53.12 (ii AD)'

*εἰσπρίαμαι, *buy up*, aor. εἰσεπρίατο *IG* 2².1629.698, ἐσπριάσθω *SEG* 26.72.40 (iv BC).

*εἰστελής, ές, dub. sens., ἡμέραι εἰστελεῖς, followed in context by ἡμέραι ἀργυρικαί, *SB* 7551.24, 34 (ἰστ-, ii AD).

*εἰστήλην, for εἰς στήλην (v. °στήλη).

εἰστίθημι **3**, delete the section.

*εἰσφέννω, perh. for εἰσφαίνω or εἰσφέγγω, *shine in*, *PAnt.* 64.18 (vi AD).'

εἰσφέρω, line 2, aor., add 'εἰσήνικα *SEG* 7.381, 382 (Dura, ii AD)'
 I 1, add 'καὶ μηδένα ἰσφέριν ἀλλότριν νεκρόν into the tomb, *SEG* 24.1189 (Egypt, i AD)'

εἰσφορά, line 1, after '(εἰσφέρω)' insert 'dialect form ἐμφορά *SEG* 23.398 (Aetolia, ii BC)' **III**, after '*proposal*' insert 'Arist.*Pol.* 1322ᵇ14'

εἰσχέω, line 1, after '(s.v.l.)' insert 'ἐσχέαι *Thasos* 10.8 (iv BC)'

*εἰσχύνω, = εἰσχέω, πετρῶν ἀφ' ἧς οὐκέτ' αὐτὸν εἰσχύσεν Diog.Oen. new fr. 7 ii 1 S. (*AJA* 75.367).

εἰσωπός **2**, add 'Simon. 13.12 W.²'

εἰσώστη, for '*tomb*' read '*niche*'; delete '(Caria)' and add 'oft. written ἰσ-, e.g. *MAMA* 8.537.3, 582.1'

*εἰσωφόριος, ον, *inner*, of garment, στιχάριον *POxy.* 1684.4, 8 (iv AD; εἰσο-, ἰσο- pap.).

εἶτα **I 1**, line 8, after '*SIG* 1171' insert 'εἶτ' αὖθις D.H.*Comp.* 18' **II**, add 'κᾆτα after part., Ar.*Eq.* 392, cj. in Pl.*Lg.* 861c'

εἰτακεῖν, add 'v. °ἰτάω'

ἐκ, line 1, for 'also .. 135.5 H.' read 'also in Cypr. before consonant, if *xe* is accurate transcription, *e-xe-to-i wo-i-ko-i* ἐξ τôι Foίκōι *ICS* 217.5, al.; Cypr. also *e-se* ἐς *Kafizin* 218b, 266b, al.'; line 2, form ἐξ, add 'also before δ *IPamph.* 3.19'; line 3, ἐς, add '[*H*]ερακλέος ἐς Θεσπίας *SEG* 30.541 (vase, Epirus, v BC); ἐς Λαρίσας *SEG* 24.571 (Maced., iv BC); ἐος Corinn. 1 col. 3.34 P.'; line 8, w. dat., add 'ἐξ ἐ[πι]τερίια *IPamph.* 3.4' **I 4**, line 5, after 'cf. (Il. 18.) 432' insert 'ἐξ ἀλλᾶν Pi.*O.* 6.25' **5**, line 4, for 'sate .. Hdt. 3.83' read 'remained *in* the middle, i.e. neutral, Hdt. 3.83.3' and transfer to the end of section I 6 (cf. μέσος III 1 c). **III 2**, add 'w. pers. n. to express metronymic, τῶ ἐκ Πεισῶς *SEG* 33.724 (Amnisos, Crete, i BC) **9 d**, after 'Space' insert 'ἐκ δέκα ποδῶν *by* ten feet, Eup. 102.3 K.-A.' add '**10** indicating proportionate share of inheritance, εἰ μέντοι τοὺς μὲν παῖδας ἔκ τινος μοίρας γράψαιεν κληρονόμους, ἐξωτικοὺς δὲ ἐξ ἑτέρας Just.*Nov.* 22.20.2, al. (cf. Lat. *ex*).'

*Ἑκάβη, ἡ, v. °Ἑκάβη.

*Ἑκάβη, ἡ, Dor. Ϝακάβα Schwyzer 122(4), *Hecabe*, the wife of Priam, Il. 6.293, al.; app. meton. for a pig because of her fertility, ἐν τοῖς Ὀρφικοῖς οἱ χοῖροι ἑκάβαι προσαγορεύονται Orph.*fr.* 46 K.

Ἑκαδήμεια, add '[*h*]όρος τε̑ς Ηεκαδημείας *SEG* 24.54 (vi BC) cf. pers. n. Ϝhεκάδαμος Schwyzer 452.5 (Tanagra, v BC)'

Ἑκαέργη (s.v. Ἑκάεργος I), add 'epith. of Artemis, *SEG* 31.934 (Caria)'

ἑκάλιθμος, add '(ἑκάλιμος La.)'

ἑκάς, add '**III** *severally* or *apart*, Nic.*Th.* 345 (perh. by misinterpr. of Hes.*fr.* 233.2 M.-W.).' at end add 'ϝεκάς· μακράν Hsch.'

ἕκαστος **II 1**, add 'τὰ ἕκαστα Il. 11.706, Od. 12.16, τὰ ἕκαστα διαρρήδην ἐρέεινον h.*Merc.* 313, Ibyc. 1(*a*).26 P., A.R. 1.339' **III 2**, add '**b** w. ellipsis of ἡμέρα, ἐφ' ἑκάστης *each day*, Just.*Nov.* 40 epilogus; sim. καθ' ἑκάστην ib. 43.1.1.'

ἑκάστοτε, add 'ἐπιμέλειαν ποιήσασθαι .. τοὺς πρυτάνεις τοὺς ἑκάστοτε γινομένους *IPriene* 59.24 (v. *Laodicée* p. 250)'

ἑκαταβόλος, add 'of Apollo, Simon. 6 fr. 1(*a*).6 P., Pi.*P.* 8.61, fr. 2.2 S.-M.; ἑκαταβόλων Μοισᾶν ἀπὸ τόξων id.*O.* 9.5'

Ἑκαταῖος **II**, for 'or Ἑκάτειον .. 804' read '(-ειον in Ar.*V.* 804, Lys. 64 is v.l. in Sch.' and for 'Ar. l.c.' read 'Ar.*V.* 804'

†ἑκατερέω, ἑκατερεῖν· τὸ πρὸς τὰ ἰσχία πηδᾶν ἑκατέραις ταῖς πτέρναις Hsch. (but cf. ἑκατερίς).

Ἑκάτη, line 1, delete 'lit. *she who works her will*'

*Ἑκατηφόρια, τά, festival at Eretria, *IG* 12 suppl. 646.12 (Chalcis, iii AD).

ἑκατόγγυιος, line 1, delete 'or *bodies*'; line 2, for '100 maidens' read '50 maidens'

ἑκατόμβαιος **II**, delete the section (v. *BCH* 98.562 no. 24).

Ἑκατόμβιος **I**, add '*SEG* 36.548.20 (Thessaly, iii BC), *SEG* 38.665 (W.Maced., *c.*100 BC)'

Ἑκατόμβοια, add "Ἑκατόμβοια δόλιχον ἐν Ἄργει *SEG* 30.489 (at Delphi, vi BC)'

*ἑκατονέξε (= ἑκατονέξ), *one hundred and six*, *SEG* 26.672.18 (Larissa, ii BC).

*ἑκατονεττά (= ἑκατονεπτά), *one hundred and seven*, *SEG* 26.672.44 (Larissa, ii BC).

*ἑκατοντᾱδικός, ή, όν, of the ἑκατοντάς, Sch.Procl. *in Ti.* 2.332.20, 26.

ἑκατοντάμᾰχος, add 'also ἑκατοντομάχος, Bonner *Studies in Magical Amulets* 163'

*ἑκατοντάρουρικός, όν, consisting of 100 ἄρουραι, -ὸς κλῆρος *PMoen.* inv. 17.9.

ἑκατοντάρχης, add '*SEG* 33.1306 (Arabia, i AD), *TAM* 4(1).285 (Rom.imp.); written ἑκατονθάρχης *IGRom.* 3.1367 (Gerasa)'

ἑκᾰτόνταρχος, add '*INikaia* 1551.14, *SEG* 31.905 (Aphrodisias), *POxy.* 3029.4 (iii AD)'

ἕκᾰτος, line 2, after '(Il.) 20.295' insert 'Alcm. 46 P., ἱερῆος Ἀπόλλωνος ἑκάτοιο epigr. in Mitchell *N. Galatia* 74a (i AD)' and after 'ὁ' insert 'Il.'; line 3, for 'Simon. 26 A' read 'Simon. 68 P.'

*ἑκᾰτοστάριος, ὁ, *collector of the* ἑκατοστή, *BMB* 7.78 (v. ἑκατοστός II) (Berytus, v AD).

ἑκᾰτοστηρία, delete '*PCair.Zen.* 12.76, al.'

ἑκᾰτοστός **I**, add '*IG* 1³.182.8 (*c.*420/405 BC) **II**, after 'Plu.*Luc.* 20' add 'cf. τὸν ἀπὸ τρίτου μέρους τῆς ἑκατοστῆς τόκον Just.*Nov.* 2.4'

ἑκᾰτοστύς **II**, add 'as military unit, Arr.*An.* 6.27.6'

*ἑκβακχεύω, line 7, intr. in act., add 'Ε.*Tr.* 169'

ἐκβάλλω **I 1**, after '*carry out* to sea' insert 'of a river, ὕδωρ ἐς θάλασσαν ἐκβάλλων *Peripl.M.Rubr.* 38' **2**, add '*cause to be thrown* from chariot, *SEG* 34.1437 (Dyria, v/vi AD). **b** *send out* to a task, *PMich.*xi 618.15 (ii AD), *OMich.* 655.4.5, *PRyl.* 80.1.' **III**, line 8, after 'Thphr.*HP*4.8.4' insert '*vomit*, τὰ σπλάγχνα Plu. 2.831c' **IX 2**, delete the section.

ἐκβαρβᾰρόω, pass., add 'Isoc. 9.47'

ἐκβᾰσᾰνίζω, line 2, for 'Philostr.*VA* 2.31' read 'Philostr.*VA* 2.30'

ἔκβᾰσις, add '**VI** *withdrawal*, in legal sense, αὐθεντικὸν ἐκβάσεως τόπων *PFam.Teb.* 31.13 (ii AD).'

†ἐκβᾰσμίδωσις, εως, ἡ, *stepped pedestal*, *IEphes.* 1627.5.

†ἐκβάσμωσις, εως, ἡ, = °ἐκβασμίδωσις, *IGRom.* 4.514, *BCH* 4.381 (both Pergamum).

ἐκβᾰτήριος **I**, add '**2** Ἐκβατηρία, epith. of Artemis in Siphnos, *IG* 12(5). p. xxvii no. 1454, Hsch. s.v.' **II**, add 'Them.*Or.* 4.61a (βακτηρίας cod.)'

*ἐκβδάλλω, *drain off*, ᾗὲ καὶ ἐκβδήλαιο καταχθέος ἔρματα γαστρός Nic.*Al.* 322.

ἐκβιάομαι, delete '(Act., .. Jd. 14.15'

ἐκβιαστικός, for '(s.v.l.)' cf. ἐκβιβ-' read '(v.l., ed. ἐκβιβ-)'

ἐκβιβάζω **II**, add 'ἐζήτει ἐκβιβασθῆναι τὴν συνθήκην *Vit.Aesop.*(W) 72, cf. τὴν συνθήκην ἐκβίβασον (ἐμ- cod.) ibid.(G)'

ἐκβῐβασμός, delete 'Aq. .. 23' add '**II** *expulsion*, μορφωμάτων Aq. 1*Ki.* 15.23.'

ἐκβιβαστικός, after '*oppressive*' insert 'Ptol.*Tetr.* 155'

ἔκβιος, add 'Sch.Od. 11.134'

ἐκβλύω, after 'ἐκβλύζω' insert 'I'

†ἐκβλώσκω, v. ἐκμολεῖν.

ἐκβόησις, add 'πολλάκις ἐκβοήσεσι κεχρῆσται κατὰ τὴν ἁγίαν ἐκκλησίαν κατὰ Ἰωάννου *PMich.*xiii 659.42 (vi AD)'

*ἐκβοητικός, ή, όν, of musical συστήματα, *suited to shrill cries*, Aristid. Quint. 2.14.

ἐκβολεύς, for '*inspector of dikes*' read 'ἐ. χωμάτων *one who assigns work on the dikes*' and add '*PPetaus* 86.6 (ii AD), *PMich.*xi 618.14 (i AD), *POxy.* 1301 (iii/iv AD)'

ἐκβολή **VII 1**, add '(perh.) Sch.E.*Med.* 1 (v.l. ὑπερβ-)'

*ἐκβουτυπόομαι, *to be changed into a cow*, S.*fr.* 269a.37 R.

ἐκβράζω **IV**, for '*boil over*, of water' read '*surge out*, dub. in'

ἔκβρᾱσις **1**, after 'Suid.' add 's.v. Καλλισθένης'

*ἐκγαίζεσθαι, wd. of uncertain sense, *MDAI(A)* 68.4 (Samos, i AD).

ἐκγένημα, add 'also ἐγγέννημα *DAW* 85.31 (Cilicia, pl.)'

*ἐκγόνιον, τό, dim. of ἔκγονος, τέκνοις καὶ -ίοις *TAM* 3(1).378 (Termessus).

ἔκγονος, ἔσγονος, add '*SEG* 29.502 (Thessaly, iii BC) **2**, at end after 'id.*Phdr.* 275d' insert 'of animals, *IG* 1³.426.62 (but ἔγγονα ib. 63) (Athens, v BC)'

ἐκγρᾰφω **II**, delete '*IG* 1².84.28, Decr.ap.And. 1.77 (Pass.)'

†ἐκγρυτεύω, *rummage out*, ἐκγρυτεύσῃ· ἐξερευνήσῃ Hsch.

*ἐκδαίνυμαι, *eat out of* the pot, imper. -δαίνυσο Nic.*fr.* 68.8 G.-S. (cod. -δαίνεο, v.l. ἐξαίρεο).

*ἐκδᾰμάζω, *make tender*, Aesop.fab.Syntip. 5 p. 530 P.

ἐκδᾱνείζω, add 'ἐγδανείζεσθαι *IG* 12(7).515.10 (Amorgos, ii BC)'

ἐκδανεισμός, *IG* 12(7).515, add 'sp. ἐγδ-' and after '*BGU* 362 xiv 21 (iii AD)' insert 'οἱ ἐπὶ τοῦ ἐκδανεισμοῦ *POxy.* 2848.5 (iii AD)'

*ἐκδᾱνειστεία, ἡ, = ἐκδάνεισις, ἐκδανεισμός, rest. in *ASAA* 30/32(1952/4).295, no. 67.26, 31 (Rhodes).

*ἐκδάπτω, *devour out of*, in tm., ἐκ μέλαν εἶαρ ἔδαπτεν Call.*fr.* 523 Pf. (cj. ἔλαπτεν).

ἐκδάω, *learn*, pres. not in use, aor. pass. ἐξεδάην A.R. 4.1565, Man. 6.469.

ἐκδέκτωρ, for '*one who* .. toil A.*fr*. 194' read 'neut. pl. -ορα, ἐκδέκτορα πόνων (animals) *relieving* men from toils, A.*fr*. 189a R.'

ἐκδέχομαι 7, add 'ὁ στέφανος τῷ βασιλεῖ, ὃν ἐξεδέξατο, Ἀπολλώνιος Ἐπικύδει PCair.Zen. 36.26, 636.4 (iii BC), PPetr. 3.64.(b)6 (iii BC). **b** *stand surety on behalf of*, PSI 349.1 (254/3 BC), PZen.Col. 121.3 (181 BC).'

ἐκδημητής, οῦ, ὁ, *one who goes abroad*, Rhetor. in Cat.Cod.Astr. 7.205.11.

ἐκδιδύσκω, add 'AP 5.309 (Diophan.)'

ἐκδίδωμι I 5, add 'ἐκδότω στεφάνους τις ἡμῖν, δᾷδα Men.Dysc. 963'

ἐκδικαιωτήρ, ῆρος, ὁ, Cret. ἐσζικαιωτήρ, *official responsible for exacting penalties*, SEG 37.752.8 (Lyttos, c.500 BC).

ἐκδικέω I 1, for '2Ki. 4.8' read '4Ki. 9.7, Ho. 1.4'

ἐκδίκησις, add 'Lxx Ex. 7.4, Jd. 11.36, Sir. 5.7, Je. 11.20, al.'

ἐκδικία 1, add 'SEG 30.1384.6 (Lydia, AD 301)' **2**, delete the section.

ἔκδίκιον, τό, *penalty*, δώσει .. ἐγδεικίου δηνάρεια πεντεκειχείλεια (sic) INikaia 1331.14.

ἔκδικος II 1, before 'cf.' insert 'ἔ. ἀνάγκη prob. in B.*fr*. 20a.13 S.-M.' **3**, add '= *defensor civitatis*, CPR 5.9.4 (iv AD), Cod.Just. 1.4.22.1'

ἐκδιοικέω, for '*collect* dues, etc.' read '*alienate* (property)' and before 'al.' insert '700.38 (ii BC), Iasos 4.20 (ἐγδ-, ii BC)'

ἐκδιοίκησις, for '*collection* of dues' read '*alienation* (of property)'

ἔκδίομαι, = ἐκδιώκω, subj. ἐδδίεται Inscr.Cret. 4.88 (Gortyn, v BC), cf. ἐπιδίομαι.

ἐκδίφάω, after 'aor. 1' insert 'Hippon. 85.8 W.'

ἐκδιώκω, add 'fig., πάντα τὰ αἰσχρὰ ἐξεδίωκεν X.Ages. 3.1. **II** *prosecute, perform*, τὸ ἔργον UPZ 81 iv 18 (ii BC).'

ἔκδοσις I 3, add 'τὴν δὲ ἔγδοσιν τῶν στηλῶν i.e. perh. of the contract for, SEG 29.1089.19 (Caria, i BC)' **4**, add 'Hyp.Dem. 5' **5 a**, after 'Ael.Tact.Praef. 4' insert 'of decree, τῆς ἐγδόσεως τῶν προειρημένων SEG 23.447.31 (Thessaly, ii AD)'

ἐκδοχεία, ἡ, = ἔκδοχεύς, PMich.I 26.3 (ἐγδ-).

ἐκδοχεῖον, before 'Peripl.M.Rubr. 27' insert '*entrepôt*'

ἐκδοχεύς, after '*agent*' insert 'or *receiver*'

ἐκδοχικός, ή, όν, *explicative*: adv. -κῶς, Demetr.Lac. pp. 3-4 DF.

ἔκδοχος, ὁ, = ἀνάδοχος II, *surety*, PSI 584.14 (iii BC).

ἔκδραχμον, τό, = ἐξάδραχμον, *the sum of six drachmas*, Hsch.

ἔκδραχμος, delete the entry.

ἐκδρομάς, for '*one who has outrun the age of youth*' read '*one who has gone to excess*'

ἐκδύγηρας, adj., *shedding its slough* (cf. γῆρας II), of a serpent, cj. in Dosiad.Ara 14.

ἔκδῦμα, ατος, τό, *that which is taken off the body, spoils*, expl. of *exuviae* in Virgil gloss., PNess. 1.1020, PSI 756.47 (both v AD), AP 5.199 (Hedyl., v.l. ἐνδ-).

ἐκδύσια, add '(cf. the ephebic ceremony cited s.v. ἐκδύω I 3)'

ἐκεῖ, line 1, for 'Sapph. 51' read 'Sapph. 141.1 L.-P.' and after it insert 'Ion. κεῖ, q.v.' and delete 'Dor. τηνεῖ' add '**IV** perh. *in that matter*, SEG 30.568.6 (Maced., i AD).'

ἐκεῖθεν, for 'Dor. τηνώθεν, τηνώθε' read '(for τηνώθεν, τηνώθε v.hh.vv.)'

ἐκεῖθι, for 'Dor. τηνόθι Theoc. 8.44' read '(for τηνόθι v.h.v.)' **II**, delete the section.

ἐκεῖνος, line 5, Aeol. κῆνος, for 'Sapph. 2.1' read 'Sapph. 31.1 L.-P., Thess. κένō (gen.) SEG 23.415 (Pherae, v BC) (also Dor., Inscr.Cret. 4.41.4 (Gortyn, v BC), SIG 1025.25 (Cos, c.300 BC))'; for 'Dor. τῆνος' read '(for τῆνος v.h.v.)'

ἐκεῖσε II, after 'Hp.Vict. 2.38' insert 'A.R. 4.1217, al.'

ἐκεχείριος, α, ον, of games, etc., *marked by a cessation of public business*, SEG 31.1288.5 (Side, iii AD), Inscr.Magn. 100.25.

ἐκζητέω, add '**III** *search out, weigh, observe*, Lxx Ps. 60(61).8, 118(119).94.'

ἐκζήτησις, for '*research*' read '*speculation*'

ἐκζωόομαι, for '*worms*' read '*living creatures*'

ἐκζωπυρέω, for '*rekindle*' read '*cause to blaze up*'

ἐκζωπύρησις, for '*rekindling*' read '*causing to blaze up*'

ἐκηβόλος, line 1, for 'Dor.' read 'dial.' and add 'Ϝεκαβόλōι ἀργυροτόξσōι CEG 326 (Boeotia, early vii BC), Ϝεκαβόλōι Ἀπέλ(λ)ονι CEG 370 (at Delphi, vi BC, perh. Lacon.), Πάον' .. ἐκάβολον Sapph. 44.33 L.-P.'; line 3, after 'of Artemis' insert 'CEG 425 (ηεκηβώ[λωι], Chios, vi BC)'; line 7, after 'Agath. 3.17' insert 'neut. pl. as subst., prob. *slings*, J.BJ 2.17.5'

ἔκθαμβος, for 'Tab.Defix. 5.20' read 'Tab.Defix.Aud. 271.20 (Hadrumetum, iii AD)'

ἐκθέβεν, v. ‡ἐκθέω.

ἐκθεῖος, ὁ, *uncle's son, cousin*, Rev.Bibl. 41.577 (Syria).

ἔκθεμα, add '**II** (as transl. of Hebr. *rāmāh*, perh.) *platform for displaying goods*, Lxx Ez. 16.24.'

ἐκθέμεναι, delete the entry.

ἐκθεόω I, add 'ὑπὸ τῆς εὐτυχίας τῆς περιλαβούσης αὐτὸν τότε ἐκθεούμενος *assuming the attributes of divinity*, Ael.VH 2.19'

ἐκθεράπευτέον, *one must cure completely*, Gal. 6.440.9.

ἐκθερίζω, for 'PEdgar 27' read 'PCair.Zen. 155'

ἔκθεσις I, add '**3** *giving out, allocation*, ὁ αἱρεθεὶς ἐπὶ τὰν φυλα(κὰν) καὶ ἔχθεσιν τοῦ ἐλαίου SEG 39.774.7 (Rhodes, i BC).' **X**, for this section read '*arrears of payment*, PMich.XII 656.11 (i AD), POxy. 583.136 (vi AD), Hsch.'

ἐκθέτης, add '**2** app. *a kind of basket*, PMich.XIV 680.22 (iii/iv AD).'

ἐκθέτωσις, εως, ἡ, perh. *projection* on a building, La Carie 185.3 (Kidrama); cf. ἐκθέτης 1, ἔκθεσις VII 1.

ἐκθέω, add 'ἐκθέβεν· ἐκτρέχειν Hsch.'

ἐκθηλύνω, for '*soften, weaken*' read '*make soft* or *weak like* (*that of*) *a woman*' and add 'Gal. 13.392.15, 949.18'

ἐκθῑνόω, *silt up completely*, SEG 14.615.11 (ἐχθεινο[ί]η[τ]ε. tab.defix., Rome, ii/iii AD); cf. ἀποθινόομαι.

ἐκθλίβω, add 'γένος ἀνθρώπων ἐκτεθλιμμένων τὴν ῥῖνα (i.e. with flattened noses) Peripl.M.Rubr. 62'

ἐκθνῄσκω II, for 'Luc.Hist.Conscr. 27' read 'Luc.Hist.Conscr. 20'

ἐκθοράξει (ἐκθοράψει La.)· ἐκδιώξει. ἀπὸ τοῦ ἐκθορεῖν Hsch.

ἐκθρῴσκω, for '*start up* .. DMar. 2.3' read '*start up*, Luc.DMar. 2.3; ἀπὸ τοῦ ὕπνου id.Herm. 71'

ἐκκᾰθᾰρίζω, add 'cf. Lat. *excatarisso*, colloq. *clean out* of money, Petron. 67.10'

ἔκκάθαρμα, ατος, τό, *offscouring*, τὰ ἐκ[α]θάρματ[α] (sic) IG 12(5).107.2 (Paros, early v BC).

ἐκκαθοράω, *look down from*, w. gen., Q.S. 8.430.

ἑκκαίδεκά, add 'Ϝεκαίδεκα PP 42.42.8 (Tarentum, vi BC)'

ἑκκαιδεκαέτης, add 'cf. ἑκκαιδεκέτις, °ἑκκαιδεχέτης'

ἑκκαιδεκάπεδος, ον, *sixteen feet long*, IG 4².109 ii 139 (Epid., iii BC).

ἑκκαιδεκάταιος, add 'Afric.Cest. p. 22 V.'

ἑκκαιδέκατος, add 'fem. subst. (sc. ἡμέρα), *the sixteenth day*, SEG 34.396.2 (-δεκέτα, Delphi, i AD); neut. subst., *name of weight*, BMB 8.62 no. 28, p. 63 no. 35, al.'

ἑκκαιδεχέτης, ου, ὁ, = ἑκκαιδεκαέτης I, GVI 1352.1 (Britain, iii AD).

ἑκκακαβίζω, perh. *pour from a κακκάβη*, Anon.Alch. 441.10.

ἐκκαλέω IV, after 'Med.' insert '*appeal to*, ἐκκαλέσθω ἐς βολήν SEG 16.485.13 (Chios, vi BC)' and after '(Plu. 2.)178f' insert 'SEG 29.127 ii 15, 20 (Athens, c.AD 100); w. ἀπό, SB 11222.12 (AD 332); ἐκκαλούμενος = Lat. *provocator*, Hdn. 1.15.6'

ἐκκαλύπτω, line 5, after 'E.IA 872' insert 'med., οὐδεὶς ὄκνος πάντ' ἐκκαλύψασθαι λόγον Critias 1.5 S.'

ἐκκαυλέω, after lemma insert '(ἐγκ- codd. in Arist.Pr. 926ᵃ26, Thphr.HP 1.2.2)'

ἐκκηρύσσω II 2, add 'Lys. 3.45'

ἐκκλαστρίδιον, add 'cf. ἐγκλαστρίδια'

ἐκκλείω, add '**5** *shut off, enclose* a place, SEG 36.935 (Rome).'

ἔκκλημα, for 'Jahresh. 14.168' read 'Schwyzer 366A21' and for 'Foed. .. B 20' read '328ᵃii B 20'

ἐκκλησία, after lemma insert 'Thess. ἐκκλεισσία BCH 59.38 (Crannon)' **II 2**, add 'SEG 33.1272 (Palestine, vi AD)'

ἐκκλησιάζω I 1, add 'abs., ἐξοπλασίαν .. ποιησάμενος ἐκκλησιάζειν ἐπεχείρει sc. the president of an assembly, Arist.Ath. 15.4'

ἔκκλησις 1, after '*appeal*' insert 'BGU 1756 (i BC)'

ἔκκλητος 3, after '*appeal*' insert 'δίκας ὁκόσαι ἂν ἔκκλητοι γένωνται Schwyzer 687B12 (Chios, c.600 BC)'; after 'IG 2².111.74' insert '(Athens, iv BC)'

ἐκκλυστέον, for 'Aët. 16.89' read 'Paul.Aeg. 6.73.2 (114.31 H.)'

ἐκκοίτιον, τό, = ἐκκοιτία, RA 3(1934).40 (Amphipolis, iii/ii BC).

ἐκκολυμβάω, at end for 'App.Syr. 6' read 'App.Syr. 56'

ἐκκομίζω I, add '**5** *export*, SEG 35.1439.11 (Lycia, ii AD).'

ἔκκοπος, add 'ταῖς πράξεσιν Cat.Cod.Astr. 8(1).184 (s.v.l., cj. ἐγκοπτικός)'

ἐκκόπτω I 2, after 'X.HG 6.5.37' insert 'ἐκκοπτομένης ἀμπέλου POxy. 2847.5 (iii AD)'. **b**, add 'Lys. 28.6' **7**, add '**c** abs., *make meaning clear, make sense*, οὐκ ἐκκόπτει ἡ γραφή Sch.Pi.P. 4.195a.' **II**, delete the section (v. ἐγκόπτω IV).

ἐκκορέω, for 'prov. .. Horap. 1.8' read 'ἐκκόρει, κόρει, κορώνας, marriage cry of dub. form and significance, Carm.Pop. 35 P., given as ἐκκορὶ κορὶ κορώνη Horap. 1.8; see also °κορικορώνη'

ἐκκρέμαμαι II, add '**2** *pay attention to, hang on the words of*, ὁ λαὸς .. ἐξεκρέματο αὐτοῦ ἀκούων Ev.Luc. 19.48.'

ἔκκρουσις I, for '*beating out, driving away*' read '*the act* or *process of forcing* a thing *from its position*'

ἐκκρούω I 1 b, add 'pass., ἐὰν δὲ δυοῖν φερομένοιν ἀπὸ τῆς αὐτῆς ἰσχύος τὸ μὲν ἐκκρούοιτο πλεῖον τὸ δὲ ἔλαττον Arist.Mech. 849ᵃ7'

ἐκκῠέω, add '(cj. °ἐγκυέω); τοὺς δ' ἑτέρους τρεῖς πέδας (= παῖδας) ὅτ' ἐξεκύησα epigr. in SEG 33.1082.12 (Nikaia)'

ἐκκῠκάω, *stir up*, (tm.), ἐκ δ' ἀφάντοις .. ἐκύκα θυέλλαις Alc. 298.12 L.-P.

ἐκκυλίω   SUPPLEMENT   ἐκπλέκω

*ἐκκῡλίω, = ἐκκυλίνδω, roll out, τὰς ἐμπροσθίους ὁπλὰς μετεωρίζοντες ἐξεκύλιον τοὺς ἐλατῆρας Agath. 3.27.4, 4.18.5; (see also ἐκκυλίομαι).

ἐκκῡμαίνω II, for 'Pass., to be cast up by the waves' read 'of the sea, to cast ashore, Heraclit.All. 79.7; pass.'

ἐκκῠνηγέσσω, add 'fut. -έσω, cj. in A.Eu. 231'

ἐκκῠνηγετέω, for 'prob. in A.Eu. 231' read 'cj. in A.Eu. 231' and add 'Lyc. 1025'

ἐκκύπτω II, delete the section transferring ref. to section I.

*ἐκκῠρόω, confirm completely, SB 9121.8 (pap. -κοιρ-, i AD).

*ἐκλακένω, gloss on ἐκκολλαβήσαντα, Hsch.

ἐκλαμβάνω IV, for 'ἔργα ἐ. = ἐργολαβέω, contract to do work' read 'ἔργα, etc., ἐ., contract to perform' and add 'τοῖς ἐξειληφόσι τὴν ἐξεδραχμίαν τῶν ὄνων POxy. 1457 (4/3 BC), PRyl. 95.5 (AD 71/2), PTeb. 40.4 (ii AD); ἐξέλαβον τὴν ζυτηρὰν PHib. 133 (iii AD)' V 2, after 'PGen. 74.8 (iii AD)' add 'Just.Nov. 90.4 pr.'

ἔκλαμψις I, add '2 inflammation, Sm.Le. 13.26.'

ἐκλανθάνω II, line 2, after 'ἐκλεθάνω' insert '(Aeol. ἐκλᾱθάνω rest. in Sapph. 25.5 L.-P.)' 1, for 'make one .. c. gen. rei' read 'make one forgetful of, w. gen.' and after 'Od. 7.220' insert 'Ἥρης ἐκλελαθοῦσα κασιγνήτης ἀλόχου τε h.Ven. 40'

*ἐκλαχμός, ὁ, apportionment by lot, PNess. 21.19 (vi AD).

ἔκλειγμα, for pres. def. read 'medicine to be licked from a spoon, electuary'

ἐκλειγματώδης, for 'lozenge' read 'linctus'; add 'v.ll. ἐλιγμ-'

ἐκλεικτικός, for 'made into a lozenge' read 'made into a linctus'

ἐκλειόω, before 'Alex.Trall.' insert 'Asclep.Jun.ap.Gal. 13.744.14'

ἔκλειψις II 2, add 'ἔκλειψις ὑγροῦ Arist.Pr. 937ᵃ13'

*ἐκλείωσις, εως, ἡ, complete pulverisation, Anon.Alch. 7.7.

ἐκλεκτικός I 1, delete 'moral' and add 'w. gen. D.H.Lys. 15' II, delete 'Gal. 14.684 (v. ZPE 72.241)'

*ἐκλέπτυνσις, εως, ἡ, reduction to a fine state, Zos.Alch. 251.16.

ἐκλέπω, add 'ὅς (sc. eagle) τρία μὲν τίκτει, δύο δ' ἐκλέπει, ἓν δ' ἀλεγίζει Musae.fr. 3 D.-K., Trag.adesp. 328.1 K.-S.'

ἔκληψις 2, after 'PTeb. 38.11 (ii BC)' add 'CodJust. 1.2.24.2'

*ἐκλικμόω, = ἐκλικμάω, dub. in PTeb. 727.27 (ii BC).

ἐκλιστρόω, for 'slap' read 'lightly touch, graze, given as colloq. equiv. of λίζω'

ἐκλογεύς, add 'Antipho fr. 52; ἐ. φόρων applied to Roman procurator, Ph. 2.575'

ἐκλογή I 5, delete the section. add 'III ὑπὲρ ἐκλογῆς in PRyl. 157.6 (ii AD) of payment for the superior value (perh. orig. the right of choosing), made by the recipient of the better portion in division of property, cf. PFlor. 47.14 (iii AD; exchange of property). 2 perh. balance in accounts, σὺν καὶ τῇ ἐγλ(ογῇ) BGU 362 vi 10 (iii AD; also read as τῇ ἐκλόγῳ and as τῇ ἐκ λόγου w. fem. noun understood); cf. ib. 64.10 (iii AD).'

*ἐκλόγισμα, ατος, τό, salary, SEG 20.180.8 (Paphos, ii BC, ἐγλ- lapis), BGU 1749.12 (i BC).

ἐκλογιστής 1, add 'SEG 31.122.30 (Attica, ii AD)' 2, add 'φόρου διοικήσεως ἐξ ἀναπομπῆς ἐκλογιστοῦ POxy. 3170.257 (iii AD)'

ἐκλογιστία, add 'PLond. 1708.159 (vi AD). 2 office of ἐκλογιστής, POxy. 1436.23 (ii AD), PGiss. 48.1 (iii AD).'

ἔκλογος (A) II, delete the section.

*ἔκλογος (C), ὁ or ἔκλογον, τό, perh. payment in arrears, balance, Ostr. 47 in PFay., but the compd. is doubtful. Cf. °λόγος I 1 c.

*ἐκλοιδορέω, abuse verbally, καὶ ἐξελυδόρησεν καὶ ἀνέσ{ο}υρεν αὐτήν· POxy. 2758.11 (AD 110/2).

ἐκλοχεύω, line 3, after 'E.Hel. [258]' insert 'Lyc. 88'

ἐκλοχίζω, add 'II μαιούμενος· ἐκλοχίζων Hsch.'

ἔκλυτος III, weak, add 'of an athlete, SEG 35.213.16 (ἐγλ-, tab.defix., Athens, iii AD)'

ἐκλύω I, at end delete 'abs. .. 531' II 3, at end for 'cease' read 'lose force'

ἐκμαγεῖον I, for this section read 'cloth for wiping, Meyer Ostr. 62.5 (ii BC), Archig.ap.Gal. 12.621.7, Paul.Aeg. 1.57 (38.9 H.); used in simile w. ref. to the spleen, οἷον κατόπτρῳ παρεσκευασμένον καὶ ἕτοιμον ἀεὶ παρακείμενον ἐκμαγεῖον Pl.Ti. 72c, cf. ὅτιπερ ἐκμαγεῖόν ἐστι αἵματος μέλανος (sc. ὁ σπλήν) Aret.SD 1.15'

ἐκμάθησις, add '2 learning by heart, εἴ τις .. ἀμελοίη .. τῆς ἐκμαθήσεως τῶν ψαλμῶν Bas.Caes. in PG 31.1305c; cf. Ἑλληνικά 8.66 (Lydia).'

ἐκμαίνω, line 8, after 'passion' insert 'Τροίω ὑπ' ἄνδρος (or -ῳ ἐπ' ἀνδρί) ἐκμάνεισα, of Helen, Alc. 283.5 L.-P.'

ἐκμάκτος, for 'express' read 'moulded, modelled' and add 'τύπος IG 2².1534.64 (iii BC)'

ἐκμαλάσσω, add 'pass., ἡ αἴσθησις of the effect of pleasing sounds, D.H.Comp. 12 (v.l. μαλ-)'

ἐκμαρτυρέω I, line 2, after '(A.)Ag. 1196' insert 'ἐπὶ τοῖς δικάζοσιν before those judging the case, IEphes. 1678B (vi BC)'; line 3, after 'Aeschin. 1.107' insert 'ἐγμαρτυρῶν ὑπὲρ τῆς εὐνοίας τῶν πολιτῶν Amyzon 15.12 (215 BC)' add '2 acknowledge publicly, med., ἐκμαρτυρεῖσθαι τὴν αὐτὴν ὁμολογοῦσαν τὸν γεγονότα αὐτῇ ἐκ τοῦ Χαιρήμονος υἱόν POxy.Hels. 35.25 (AD 151).'

ἐκμάσσω I, add '3 polish, scour, ἀπεικονίσματα γῇ ἀργυρωματικῇ (q.v. in Suppl.), IEphes. 27.542, 545 (AD 104).'

*ἐκμήδομαι, devise, Αἰγύπτῳ κακὰ σημεῖα καὶ τεράστι' ἐξεμήσατο Ezek.Exag. 226 S.

ἔκμηνος 1, subst., add 'fem., ἁ ὑστέρα ἔγμεινος BCH 60.183.31 (Boeotia, iii BC)'; add 'cf. ἑξαμηνός'

*ἐκμηρίζω, remove thigh-bones, IEphes. 10.8, 1201a.7.

*ἔκμισθος (B), ὁ, perh. leasing, Θεοφίλῳ βοηθ(ῷ) ὑπὲρ ἐκμίσθου καὶ μετρήσεως PVindob.Tandem 17.29 (vi/vii AD).

*ἐκμογέω, labour at, ἱμερτὰν ἐξεμόγησα τέχναν epigr. in SEG 34.325 (Arcadia, c.100 BC); cf. ἐκμοχθέω.

ἐκμοχθέω 3, add 'φιλοσοφοῦντες ἐκμοχθοῦσί τι Diog.Sinop.fr.dub. 6.1 S.'

*ἐκμῡθέομαι, speak out, Theoc. 25.3 (tm.).

*ἐκνεόω, renew, τὸν σῖτον καὶ τὰ ξύλα, Ἑλληνικά 7.179 (Chalcis, iii BC).

*ἔκνευμα, ατος, τό, gloss on κλειτύς· κλιτύν· τὸ ἀπόκλιμα, ἔκνευμα, ἐξοχήν Sch.Nic.Al. 34c Ge.; cf. ἀπόνευμα.

†ἐκνεύω, turn aside by a movement of the head, πάντα γὰρ ὁπόσα ἂν δύσκολ' ἢ περὶ τὸ πρόσωπον ὁ ἵππος ἐκνενείκεν πέφυκεν ἄνω X.Eq. 5.4, μὴ ἐκκρούσῃ (τὸ προβόλιον) ἐκ τῶν χειρῶν τῇ κεφαλῇ ἐκνεύσας, sc. the boar, id.Cyn. 10.12. 2 cause to move aside by a movement of the head, nod aside, ἡμᾶς .. ἐξένευσ' ἀποστῆναι πρόσω E.IT 1330. II turn the head, look round, Lxx 4Ki. 23.16. 2 transf., turn aside, bend one's course to, E.IT 1186, εἰς θάνατον id.Ph. 1268; πρός τι τῶν ἡδονῆς φίλτρων ἐκνενευκότα Ph. 1.297; w. gen., τῶν παρόντων Plot. 6.7.34. III w. acc., avoid by turning, dodge, Phld.Sign. 27, Ph. 1.146, Orph.A. 458, ξίφος Hegesias ap.D.H.Comp. 18, πληγήν D.S. 17.100. 2 shun, avoid, Themist.Or. 34.6.

ἐκνῑκάω I 1, add 'prevail over so as to compel, w. inf., αὐτοὺς ἀναστῆναι ἐξενίκησαν Ael.NA 17.41; so pass., τὰ θηρία ἐκνικᾶται ἐμπεσεῖν ib. 8.10' add 'III perh. recover possession of at law (= Lat. evincere), IGRom. 4.914.6 (v. ZPE 48.269); cf. °ἐκνίκησις.'

†ἐκνίκησις, εως, ἡ, recovery of possession at law, eviction (= Lat. evictio), CodJust. 1.3.38(39), Just.Const. Δέδωκεν 5.

ἐκοντήν, for '= foreg.' read '= ἑκοντί, SEG 18.343.45 (Thasos, i BC/i AD)'

*ἐκουβίτωρ, v. °ἐξκ-.

ἐκπαγλέομαι II, after 'A.Ch. 217' insert 'κάλλος ἐκπαγλούμενον fr. 451n.5 R.'

ἐκπαθής I, adv., add 'Ath. 10.443d'

ἐκπαλής, add '2 (carried) out of orbit, Plu.Lys. 12 (cj., Coraes for ἐκ παλμῶν).'

ἐκπαλιγκοτεῖν, for '(prob.)' read '(cod. ἔκπαλιν κοτεῖν)'

*ἐκπάρθενος, ον, deprived of virginity, Sch.Theoc. 2.40.

*ἔκπεισμα, ατος, τό, (written ἔκπισμ-), consideration, inducement, PCair.Isidor. 80.6, 13 (AD 296), 81.13 (AD 297).

ἐκπέμπω I 1, at end for 'etc.' read 'Trag.adesp. 664.29 K.-S.'

ἐκπεράω I 1, add 'reach, A.R. 4.329 (tm., s.v.l.)'

ἐκπέσσω, after '-ττω' insert 'later -πτω Plu. 2.683d, Ath. 3.83f'

ἐκπέτομαι, line 3, after '(Chios)' insert 'ἐξεπταμένη (sic) IEryth. 302.2 (iii BC); καὶ ἐπεὶ ἐκ τᾶς ἰδίας ἐξεπέτομες καὶ ἐπλανώμεθα SEG 30.1123.10 (Entella, Sicily, iii BC)'

*ἐκπέφαται, pf. pass. of *ἐκθείνω cf. θείνω II, remove by killing, ἐκ δ' αἰὼν πέφαται Il. 19.27.

ἐκπηδάω 1, add 'in a frenzy, Suppl.Mag. 42.17 (iii/iv AD)'

ἐκπίνω 2, line 3, after 'Antiph. 3, etc.' insert 'ἐκπῖθι τὸ φρέαρ εἰσπεσὼν Men.Dysc. 641'; add 'w. dat., ἔχων σκύπφον Ἐρξίωνι τῷ λευκολόφῳ ἐξέπινον Anacr. 88 P.'

ἐκπιπράσκω, delete 'cf. Poll. 7.9'

ἐκπίπτω, line 6, delete 'After Hom. .. ἐκβάλλω' 2, add 'δίκης ἐκπίπτειν lose one's case (cf. Lat. causa cadere), CodJust. 2.2.4.2; abs., lose one's position, εἰ .. ὑπὲρ τὸν εἰρημένον ἀριθμὸν γένηται χαρτουλάριος, ἐκπίπτει ib. 1.2.24.13' 3, add 'b of a wrestler, be thrown, SEG 35.213.8 (tab.defix., Athens, iii AD); fig., of an unsuccessful lover, ib. 219.10.' 8, add 'ὡς ἤδη ἥ τε ποίησις ἐξεπεπτώκεε καὶ ἐθαυμάζετο ὑπὸ πάντων Ps.-Hdt.Vit.Hom. 36' add '18 of time, run out, Just.Nov. 158.1.'

ἐκπλαγής, add 'possessed, maddened, AP 9.603 (Antip.Sid.)'

†ἐκπλεθρίζω, to carry out form of exercise in which one runs up and down within a πλέθρον, reducing the distance each time until a single step is reached, Gal. 6.133.18.

ἐκπλέθρος, for 'ἐν ἐ. ἀγών .. narrowing' read 'ἀγὼν E.El. 883; κῶλον ἐκπλέθρου δρόμου prob. in id.Med. 1181 (v.l. ἐκπλέθρον)'

†ἐκπλέκω, unfold a document, ἐκπέπλεκται PDura 31.59 (AD 204); fig. διάνοιαν Alex.Fig. 2.1. 2 PTeb. 768.17 (ii BC, dub.), arrange, settle, ἔχω ἕως οὗ ἐκπλέξω ὃ ἐν Ἀλεξανδρείᾳ μετέωρον PBremen 17.10,

109

ἐκπλεονεξία   SUPPLEMENT   ἕκτος

cf. ib. 11.35 (pass.; both ii AD). **b** without object, *to be ready*, PStrassb. 73.18 (iii AD).

*ἐκπλεονεξία, ἡ, *wrong* done to someone, Pap.Lugd.Bat. XIII 18.7 (iv AD).

ἔκπλεος **1**, add 'εἰκόνων Fronto Ep. 1.4'

ἐκπλέω **I**, line 3, delete 'ἔξω τοῦ Ἑλλησπόντου id. 5.103' and add 'ἐκπλώσαντες .. ἔξω τὸν Ἑλλήσποντον Hdt. 5.103.2' to section II 1.

ἐκπλήγδην, for '*terribly*' read 'glossed by ἐκπληκτικῶς' and after 'Suid.' insert '*in terror* or *amazement*, Theoc. 24.56 (PAntin., συμπλήγδην codd.)'

ἐκπληρόω **I**, add '**6** *complete* task or project, Cod.Just. 8.10.12.9a, 9b.'

ἐκπλήσσω **I**, add '*throw off course*, in quot. fig., S.Aj. 33'

ἔκπλοος **I**, add 'καὶ ἔσπλον καὶ ἔκπλον ἀσυλ[εί] SEG 36.982.C9 (Iasos, 500/450 BC)'

ἐκποιέω **II**, add '**2** dub. sens., Hippon. 7 W.' **IV**, for '*cause* .. 2 *permit*' read '*permit*, Thphr. CP 1.14.1, 2'

ἐκποίησις **IV**, add 'PMich.XIII 659.102 (vi AD)'

*ἐκπολυωρέω, *take care of*, κατὰ πάντας τρόπους -ηθείς Anon. in Herc. 176 p. 48 V.

ἐκπομπή **I**, add 'Cod.Just. 12.37.18'

ἐκπονέω **6**, for this section read '*exert oneself to obtain*, E.Ion 1355, Hel. 1514' **7**, after '*to digest*' insert '*by taking exercise*'; after '(X.)Cyr. 1.2.16' insert 'Arist.Pr. 877ᵃ18' add '**11** *exert oneself over* or *in the matter of*, E.Andr. 1052.'

ἐκπορεύω, line 3, after 'etc.' insert 'w. acc., ἐξόδους SIG 1219.15 (Gambrea, iii BC); τὰ λόχια Milet 1(7).204b9 (v. °λόχιος III)'

*ἐκποριστικός, ή, όν, *providing*, w. gen., τῶν ἀναγκαίων Procl.in R. 1.216.14.

†ἐκπορνεύω, *commit fornication*, Lxx Ge. 38.24, Ez 16.16, Ep.Jud. 7; med., Poll. 6.126; w. εἰς, *resort to for fornication*, Lxx Num. 25.1. **b** *resort immorally* to false gods, ἐκπορνεύσωσιν ὀπίσω τῶν θεῶν αὐτῶν ib.Ex. 34.15, Jd. 2.17, Ez. 20.30. **II** *prostitute* or *cause to fornicate*, τὴν θυγατέρα ib.Le. 19.29, Ez 16.33. **b** *seduce into immoral practices*, ib. 2Ch. 21.11, 13.

ἐκπράκτης, add 'Cret. ἐσπράττας, Inscr.Cret. 4.87.1 (v BC), al.'

ἐκπρεμνίζω, add 'see also ἐσπρεμμίττεν'

ἐκπροφέρω, after 'Man. 6.733' insert 'cj. in Orph.H. 71.7'

ἐκπτύω **I**, line 2, before 'AP l.c.' insert 'Ar. V. 792' **II 1**, delete the section.

ἔκπτωσις, line 2, for '*projection* of rays from the sun' read '*projection* of vision from the eyes'

ἐκπυέω, add 'Heras ap.Gal. 13.775.3'

ἐκπυητικός, add 'Gal. 12.328.1, 771.8'

*ἐκπωλή, ἡ, Dor. -πωλά, *selling, sale*, χαλκωμάτων Lindos 419.143 (i AD).

ἐκπωματοποιός, after '*cup-maker*' insert 'LW 2741 (Citium)'

*ἐκπώνω, Aeol. for -πίνω, ποτήριον prob. in Alc. 376 L.-P. (tm.).

ἔκρηγμα **II 1**, add '*breach* in a dike or sim., Wilcken Chr. 386.6 (iii BC), Lxx Ez 30.16 (v.l. ἔκρημα), SB 7174.18 (i AD) (ἔκχρηγμα)' **3**, add 'cf. perh. φλεγματόεν ἔ. Suppl.Hell. 1116'

ἐκρήσσω, before 'Theano' insert 'Gal. 11.786.4'

ἐκρίπτω, after 'A.Pr. 932' insert '*cast off* a garment, App.BC 2.126'

ἐκρομβέω, add 'ἐξέσται τῷ [βου]λομένῳ τὸ εἰσενεχθὲν ἀλλότριον πτ[ῶμα] ἢ ὀστᾶ ἐκρομβῆσαι TAM 5(2).1143.16'

ἔκροος **II**, add '**2** *outlet channel*, τὸν ἔγρουν SEG 39.442.2 (Oropus, iv BC).'

ἐκσάω, add 'epigr. in Salamine 204 (v AD)'

*ἐκσειασμός, ὁ, app. *purgation*, OLeid. 1.12 (ii BC).

*ἐκσπάστης, ου, ὁ, *one who draws out*, cj. for σκεπαστής in Lxx Ps. 70(71).6; cf. ὁ ἐκσπάσας με in Ps. 21(22).10.

ἐκσπεύδω, after 'Ar.Th. 277' insert '(s.v.l.)'

†ἐκστάδιος, ον, prob. *going outside the bounds of the stadium*, Luc.Nav. 39 (v.l. ἐκστάδιος, *six stades long*).

*ἐκστάσιον, τό, *surrender* of property to a creditor (cf. Lat. *cessio bonorum*), Just.Nov. 135.1.

*ἐκστράνιος, ὁ, *stranger*, opp. συγγενής, TAM 3(1).481 (Termessus); also ἐκστράνιος ib. 608 (ἐξτρ- 541); ἐκτράνιος SEG 3.208 (Athens, iii AD); cf. Lat. extraneus.

ἐκστρατεύω **II 2 a**, delete the section.

ἐκστρέφω, after lemma insert 'perh. also ἐκστράφω SEG 30.380 no. 6 (ἐξστρ- Tiryns, vii BC)'

ἐκστροφή, add '**V** *diversion from the proper purpose*: ἐπ' ἐκστροφῇ to the *prejudice* of the rightful owner, Studi in onore di P.Bonfante (Milan 1930) 3.64 (ii AD), prob. in PGnom. 10 (ii AD), teste Schubart; PBerl.Möller 2.17 (i AD).'

*ἔκστρωσις, εως, ἡ, perh. *preparation of bedding* (or *laying of roads*), PBeatty Panop. 1.260 (AD 298).

*ἐκστρώστης, ου, ὁ, perh. *organizer of bedding* for imperial visitations (or *of road-laying*), PBeatty Panop. 1.256, 259, 262, 263 (AD 298).

ἐκσύρω, add 'pf. pass., PVat. 11ʳ viii 22, al. (ii AD).'

*ἐκσφηκόω, τοὺς θύρσους ἐξεσφηκωμένους φοροῦντα, perh. *wasp-waisted*, Hsch. s.v. κάθαπτος (expl. of E.fr. 752).

*ἐκσφούγγευσις, εως, ἡ, *discharge* of soldiers, POxy. 1204.6 (iii AD).

*ἐκσφουγγεύω, = Lat. *expungo*, *discharge* soldiers, POxy. 1204.19 (iii AD).

ἐκσῴζω, line 6, after 'Pass.' insert 'Eup. 260.24 K.-A. (parody of S.Ant. 1128)'

ἐκταδόν, add 'epigr. in SEG 37.1537.8 (Arabia, c.AD 400)'

ἔκτακτος **II**, add 'ἐν ἐκτάκτῳ PBeatty Panop. 1.361 (AD 296); adv. -τως, *separately, individually*, ib. 231, 233, 264, 266'

ἐκταμιεύομαι **I**, add '*entrust* to the keeping of a subordinate, ἐρίφους PCair.Zen. 429 (iii BC)'

ἐκτανθαρύζω, add '(ἐκτονθορύζω La.)'

ἐκτανύω **3**, add 'Εὐψύχι Ταῇσι μητρῷον μόρον ἐκτανύσασα σωφροσύνῃ καὶ φιλανδρίᾳ SB 5037'

ἐκταρσόομαι, for '= ταρσόομαι' read 'pass., *to be stretched*'

*ἐκτασμός, ὁ, unexpld. agricultural operation (cf. °ἐκτάσσω II), PMil. Vogl. 52.50, 87 (ii AD).

ἐκτάσσω **I**, add '**2** *muster*, λαόν Lxx 4Ki. 25.19.' **II**, for this section read '*carry out* (an unexpld.) agricultural operation, ἐργάτης ἐκτάσσων PMil.Vogl. 69.A94, B55, 62 (ii AD), SB 9379.2.12 (ii AD), 9410(5).3 (iii AD); cf. ἐκτάσσοντα· χαράσσοντα, γράφοντα Hsch.' add '**III** ἐξετάγη .. εἰς, *was appointed* or *assigned* to a liturgy, BGU 2251.2-3 (ii AD).'

ἐκτείνω **I 2**, after 'X.HG 6.5.19' insert 'abs. ἐξέτειναν ἐπὶ τὴν Γαβαα Lxx Jd. 20.37' add '**3** *extend to*, ἡ προσηγορία .. καὶ ἐπὶ τοὺς ἐγγόνους ἐκτείνεται Modest.Dig. 27.1.2.8.' **IV**, for this section read '*pronounce* a vowel or syllable *long*, A.D.Adv. 159.21, interpol. in D.H. 2.58; pass., id.Comp. 14, A.D.Pron. 27.2, al.' **V**, delete the section.

ἐκτεκνόω, med., add 'Cret. ἐστετέκνοται Inscr.Cret. 4.72 viii 24'

ἐκτένεια, line 2, delete '*gush, empressement*'

ἐκτέος **II**, ἐκτέον, one must have, add 'Pl.R 535b, Amat. 138e, D. 58.60' **2**, add 'Pl.R. 468a'

ἕκτη, after lemma insert 'Fέκτα SEG 23.392 (Corcyra), 31.397 (Haliartus, both vi BC)' **III**, add 'SEG 36.790.5 (Thasos, c.480 BC)' add '**IV** *sixth day* (in a month, etc.), ἡ πρώτη ἕκτη Hes.Op. 785, opp. ἕκτη ἡ μέσση 782; ἕκτη ἐπὶ δέκα SEG 21.541 i 45 (Attica, iv BC), ἕκται ἐφ' ἱκάδα SEG 30.1117.3 (Entella, iii BC).'

ἐκτήκω **I 2**, add 'Mosch.Trag. 9.9 S.' **II 2**, add 'Cratin. 196 K.-A.'

ἐκτημόριοι, delete all after 'Plu.Sol. 13' (v. °ἐκτήμορος)

*ἐκτημόριον, τό, a *sixth part*, S.E.M. 10.140, Protag.Nicae. ap.Heph.Astr. 3.30.37.

*ἐκτήμορος, ον, οἱ ἐκτήμοροι = ἐκτημόριοι, Arist.Ath. 2.2. **II** ὁ ἐκτήμορος (sc. κύαθος), *a liquid measure*, Herod. 1.80, cf. μέτρῳ ἕ. PSI 30.5, 10.

ἐκτίθημι **II 1**, add '**c** *grant* a benefit, Cod.Just. 10.30.4.8.'

ἐκτίλλω **II 1**, after 'Hippon. 84' insert '(dub.; = 114A W.)'

ἐκτιμάω **II**, add 'χρήματα SEG 9.72.43 (Cyrene, iv BC)'

†ἐκτιμήτρα, τά, Dor. -ατρα, some form of honorific offerings, IKnidos 138.3 (iii BC).

ἐκτιναγμός, for 'perh. *winnowing* or *threshing*' read '*harvesting* of olives *by shaking from the tree*' and add 'PPrag. I 110.7 (iii AD)' add '**2** *disposal, sale* of stock, PFlor. 209.13 (iii AD).'

ἐκτίνακτρον, for '*winnowing-shovel*' read 'pl., *payment for* the work of *harvesting olives* (cf. ἐκτιναγμός 1)'

ἐκτινάσσω **I 1**, add '**b** *scatter, shower* missiles, ἐξετίναξαν τὰς σχίζας εἰς τὸν λαόν Lxx 1Ma. 10.80.' **II**, for '*make a disturbance*' read 'of the bowels, *be in violent motion*'; delete '*make a thorough* .. (ii BC)'

ἐκτίνω **1**, add '**b** *render accounts for*, ἐκτίσαι τὰ τῆς κληρονομίας Modest.Dig. 19.2.49 pr.' **II 1**, add 'abs., Inscr.Cret. 4.14, al. (vii/vi BC)'

*ἐκτίον, τό, perh. dim. of ἕκτη *a sixth*, PUniv. Giss. 25.6 (iii AD).

†ἐκτιστής, οῦ, ὁ, *payer in full*, PSI 1435.6 (i AD), Hsch.

ἐκτοκίζω, for '*exact interest*' read 'w. cogn. acc., οὐκ ἐκτοκιεῖς τῷ ἀδελφῷ σου τόκον ἀργυρίου'

*ἐκτοκισμός, ὁ, *interest*, SEG 33.1039.72 (Aeolian Cyme, ii BC).

†ἐκτομάς, άδος, ἡ, πυλὶς ἐ. *wicket-gate*, Aen.Tact. 24.5, 28.2; also ἐ. alone Stud.Pal. 20.211.9 (v/vi AD).

ἐκτομή **II**, add '**4** *piece* of weaving, PTeb. 703.95 (pl.), 113 (iii BC).'

*ἐκτονέω, perh. *unstring*. fig., ἐκτετόνημαι POxy. 3724 fr. 1 iv 30 (list of epigrams, i AD).

ἐκτοπίζω **I**, add 'perh. ἐκτοπισμοὺς στρατοπέδων Scymn. 26 (GGM 1.196, s.v.l.)'

ἔκτοπος **II 2**, line 3, after 'Thphr. CP 6.18.12' insert 'οἰκία Men.Dysc. 624; θέα ib. 690' line 5, after '(Arist.)Mir. 833ᵃ14' insert 'Men.Dysc. 824'

Ἑκτόρειος (s.v. ἕκτωρ), add 'irreg. fem. Ἑκτόρεια χείρ E.Rh. 762 (cj.)'

ἕκτος, after 'Il. 2.407, etc.' insert 'λεγιῶνος ἕκτης IGRom. 4.266 (iv

ἐκτός
SUPPLEMENT
ἐλάσσων

AD), μηνὸς ἕκτου SEG 34.389 (Delphi, ii BC)'; after 'Plu. 2.268a' insert 'ἕκτον, τό, *sixth part*, SEG 24.486 (Maced., AD 114)'

ἐκτός II, for 'abs. .. Plb. 2.4.8' read 'w. definite art., τὰ ἐκτός *external things*, Plb. 2.4.8, etc., ἐκ (or ἐγ) τῆς ἐκτός = ἐκτοσθε, Didyma 25A9, B21, al. (ii BC)'

ἔκτοσθεν, line 5, for '*delirious*' read '*wandering in mind*'

ἔκτοτε, after 'Luc.Sol. 7' add 'ἔ. κατὰ μηδένα τρόπον εἴπῃς, ἀλλ' ἐξ ἐκείνου Phryn. 31'

*ἐκτράνιος, v. °ἐκστρανήιος.

*ἐκτραορδινάριοι, οἱ, Lat. *extraordinarii*, name of specially selected troops, Plb. 6.26.6.

ἐκτραπελόγαστρος, after 'ον' insert '(or -γάστωρ, ορος)'

ἐκτράπελος II, delete the section, incorporating quots. in section I.

ἐκτρέπω I 1, line 6, after '*turn off* or *aside*' insert 'w. acc. of route taken' and after 'Hdt. 1.104' insert '*turn aside from*, τῶν ἐκτρεπομένων τὰς βασιλικὰς .. ὁδοὺς στρατιωτῶν SEG 13.492 (Caria, iii AD)'

ἐκτρέχω 1, add '**b** *leave the* ἀγέλη *on completion of ephebic training* (*become a* δρομεύς 2), ἐπεί κ' ἐγδράμωντι Inscr.Cret. 1.xvi 5.21.'

ἐκτρίβω III, add '**2** *ease by rubbing, massage*, συνέκαμψέ τε τὸ σκέλος καὶ ἐξέτριψε τῇ χειρὶ Pl.Phd. 60b.' **V 1**, for 'Class.Phil. 19.234' read 'PCornell 1.194' **2**, for '*wipe out*' read '*wipe clean*'

ἐκτρυχόω, add 'Arr.An. 4.28.7'

ἔκτυπος I 1, add 'sg., Inscr.Délos 104.96 (iv BC)'

ἐκτύπωσις I, add 'Lxx 3Ki. 6.35, Alcin.Intr. p. 162.35 H.'

*ἔκτυφος, ον, *free from delusion, veracious*, Μοῦσα Oenom.ap.Eus.PE 5.21.

ἐκτύφω, line 1, for '(sic, post ἐξήια) read '(ἐξέθυμεν La.)'; line 4, for '*swelled up* with weeping' read '*was inflamed* with weeping' and add 'Men.fr. 439 K.-Th.'

*ἐκτυχίζω, *dress with the* τύχος (*mason's hammer*), in Att. fut. ἐκτυχιεῖ, ὁμαλῶς IG 2².1670.20.

ἕκτωρ II, add 'cf. Myc. *e-ko-to, e-ko-to-ri-jo*, pers. nn.'

ἑκυρά, after '*mother-in-law*' insert '(at first only *husband's mother*, acc. to Ar.Byz.ap.Eust. 648.55)'; add 'MAMA 7.321, 576'

*ἐκυρεύς, έως, ὁ, = ἑκυρός, GVI 1422 (Antioch, i AD).

ἑκυρός, after '*father-in-law*' insert '(at first only *husband's father*, acc. to Ar.Byz.ap.Eust. 648.53)'; after 'Jul.Or. 3.127c' insert 'ὑκερός (w. metathesis of vowels or misspelling), SEG 31.1007 (Lydia, ii AD)'

ἐκφαίνω, line 2, after 'Il. 13.278' insert 'aor. 2 -εφάνη Il. 19.46, etc.; 3 pl. -έφανεν Pi.O. 13.18' add '**III** intr., *appear*, Nic.Th. 855.'

ἔκφανσις, add 'Aristobul.Alex. p. 216 D. (cf. °ἔκφαυσις)'

*ἐκφαντεύω, *reveal, make known*, SEG 37.1001.18 (Katakekaumene, Lydia, ii/iii AD).

ἐκφαντικός, add 'perh. adv. -ικῶς *clearly*, Aristobul.Alex. p. 217 D. (v.l. ἐμφ-)'

*ἐκφαύλισμα, ατος, τό, expl. of σκυβαλισμός Hsch.

*ἔκφαυσις, εως, ἡ, *radiation*, v.l. in Aristobul.Alex. p. 216 D. (v. °ἔκφανσις).

ἐκφέρω II 3, at end for 'Isoc. 5.36' read 'Hdt. 5.36.2' **11**, add 'perh. μόνης τῆς ὑπεροχῆς, ἐάν τις ᾖ, ἐκφερομένης POxy. 2411 (AD 173)' add '**VI** math., *divide*, παρὰ τῶν ϛ' (by six) Cat.Cod.Astr. 8(1).173.'

ἐκφεύγω 3 a, add 'δίκην *be acquitted*, A.Eu. 752'

*ἐκφευκτέον, *one must avoid*, Archig.ap.Gal. 13.168.4.

ἐκφλαυρίζω, for 'f.l.' read 'cj.'

*ἐκφλυαρίζω, *treat contemptuously, slight*, Eust. 3.3, 1416.37, 1675.56; also in codd. of Plu. 2.680c, Pomp. 57 (v. ἐκφλαυρ-).

ἐκφοιτάω, add '**II** *divulge*, Ael.fr. 45 (quoted by Suid. l.c.).'

ἐκφορά I 1, add '**b** *bier*, SEG 9.4.16 (Cyrene, i BC).'

*ἐκφόρησις, εως, ἡ, *carrying out*, Arg.E. Cyc., prob. in BGU 1774.6.

ἐκφόριον II, add 'SEG 38.1462.29 (Oenoanda, ii AD)' add '**III** *export*, Just.Edict. 13.5 (unless neut. of *ἐκφόριος, *exportable*).'

+ἐκφορτίζομαι, *unload a cargo*, POxy. 36 (ii/iii AD); fig., ἐξημπόλημαι κἀκπεφόρτισμαι πάλαι S.Ant. 1036.

*ἐκφορτισμός, ὁ, *unloading*, POxy. 36 (ii/iii AD).

ἐκφυής, after '*abnormally developed*' insert 'v.l. in AP 5.56(55).6 (Diosc.)'

ἐκφυλάσαι, add '(perh. for ἐκφυλάξαι, cf. °ἐκφυλλάζω)'

*ἐκφυλλάζω, *strip of leaves*, fig., ἐξεφύλλασεν γένος A.fr. 154a.13 R.; cf. perh. ἐκφυλάσαι.

ἐκφυλλοφορέω, add 'Δείναρχος ἐν τῇ κατὰ Πολυεύκτου ἐκφυλλοφορηθέντος ἐνδείξει Harp. s.v. παλιναίρετος'

ἔκφυσις II 2, add 'of horse, *spring of the neck*, Simon. Ath.Eq. 6, Hippiatr. 115'

*ἐκφωτίζομαι, *to be illuminated*, Plu. 2.922e.

ἐκχέζω, for '= Lat. *ecacare*' read '*excrete*'; add 'Inscr.Cret. 1.xvii 9.12 (Lebena, ii BC, sp. ἐσχ-)'

ἐκχέω, line 2, for 'fut. .. χέω') read '(-χέω in E.Supp. 773 is pres. subj.)' **I 1 b**, line 2, after 'Men. l.c.' add '*empty*, κελέβην Call.fr. 246 Pf.' **3**, add '[εὐπραξίαν] *see it slip away*, A.fr. 154a.20 R.'

ἐκχλοιόομαι, add 'Gal. in CMG 5.9.(2).151.14, 13 (vv.ll. ἐκλύεσθαι, ἐνοχλεῖσθαι)'

ἐκχοΐζω I, after '*dig out*' insert 'POxy. 2272.66 (ii AD)'

ἐκχράω I, add 'φωνήν Posidipp. in Suppl.Hell. 705.12'

*ἐκχυμεντας, v. παλινεκχυμενίτας.

ἐκχύμωμα, add 'Gal. 12.804.5, 13.385.12, 17'

*ἐκχυσία, ἡ, app. = ἔκχυσις 1, PKöln 234.9 (AD 431).

ἐκχωρέω II, after '*cede*' insert 'cf. °ἐκχώρησις II 2' and add 'PMich.II 123 (AD 45/7), SB 10880, 10881 (AD 302)'

ἐκχώρησις II 2, for '*deed of surrender*' read '*withdrawal from a property, cession*' and add 'PSI 1144 (AD 100), PMich.VI 238 (AD 145/6), SB 10880,10881 (AD 302)' add '**3** *transfer of right to sue* (Lat. *cessio actionum*), Cod.Just. 4.35.24, Just.Nov. 72 pr.'

ἐκχωρίζω I, add '**2** *confiscate, take away*, Lxx 1Es. 4.44, 57.'

ἑκών, after 'ἑκόν' insert 'Dor. fem. ἕκασσα SEG 9.72.87 (Cyrene, iv BC), ἑκοῖσα ib. 89'

ἔλα I, after 'γέλαν' insert 'εἴλη' and add 'also ἔλα: dat. ἔλᾳ cj. in Pi.fr. 123.10 S.-M.'

*Ἐλαγάβαλος, epith. of the god Helios, Θεῷ Ἡλίῳ Ἐλαγαβάλῳ SEG 33.1254 (Emesa, iii AD); also Ἐλεγάβαλος D.C. 78.31, etc.; cf. Ἐλαγάβαριος.

*ἐλαγμάτιον, v. ‡ἐλασμ-.

ἐλάδιον II, for '*a little oil*' read '*olive oil*'

+ἐλᾱθερής, v. ‡εἰληθερής.

ἐλαία, line 1, after 'Att. ἐλάα' insert 'also Aeol., Alc. 296(b).2 L.-P. [-ā-]' **II**, for 'but ἐλάα is simply .. etc.' read 'while ἐλάα is generally the Attic form, cf. IG 1³.84.33, 2².1013.21, 2492.36, etc., both forms appear in IG 1³.422.84, 89, 118 (Athens, v BC)' at end add 'form *ἐλαίϝα, Myc. *e-ra-wa*'

ἐλαιήεις I, insert 'θαλλὸν ἐλαιήεντα Nonn.D. 11.510' and transf. 'φλοιός Nic.Th. 676' to section II.

ἐλαιηρός 3, delete the section.

*ἐλαιθερής, v. ‡εἰληθερής.

ἐλαϊκός, fem. subst., add '*olive-oil tax*, PHamb. 182.10, 16 (iii BC)'

ἐλάϊνος 3, add 'ἐλαίου ἐλάϊνο[υ] SEG 30.1663 (Bactria, ii BC)'

*ἐλαιοβαστάκτης, ου, ὁ, *olive-carrier*, ITyr 35 (sp. ἐλεο-).

ἐλαιόγαρον, add 'PWash.Univ. 59.14 (v AD)'

*ἐλαιόκαρον, τό (written ἐλεο-), perh. *caraway oil*, PVindob.G 39847.456 (CPR 5.97, iv AD)

ἐλαιοκονία, after '*oil*' insert 'Zos.Alch. 141.11'

ἔλαιον, after 'τό' insert 'Cypr. *e-la-i-wo* ἐλαίϝō (gen.) Salamine 1 (c.600 BC)' **I**, add 'pl., αἱ δὲ κόμαι θυόεντα πέδῳ λείβουσιν ἔλαια (i.e. drops) Call.Ap. 38' at end add 'form ἔλαιϝον, Myc. *e-ra₃-wo, e-ra-wo*'

*ἐλαιονοπαράδεισος, v. ‡ἐλαιωνο-.

*ἐλαιοποιέω, *manufacture oil*, PSI 1030.12 (ii AD).

*ἐλαιοπράτισσα, ἡ, *female oil-dealer*, PPanop. 24.2 (ZPE 10.117; AD 323/6).

*ἐλαιοπωλέω, *carry on trade of oil-merchant*, Θατρὴν ἐλαιοπωλοῦσαν PSI 1239.10, ἐλεοπωλοῦ(ντας) PLaur. 187.14 (ii AD).

ἐλαιοπώλης, add 'IEphes. 2.44 (iv BC), EA 16.47 (Sinope, iv BC)'

*ἐλαιόπωλις, ιδος, ἡ, *female oil-seller*, rest. in PMasp. 287 iv 23, OBodl. 2355, (?iii AD; sp. ἐλε-).

*ἐλαιοπώλισσα, ἡ, *female oil-seller*, SB 10978 (AD 320).

ἐλαιοτρίβιον, after '*oil-press*' insert 'IGLS 1509 (ἐλεο-)'

*ἐλαιουργικός, ή, όν, *belonging to oil-manufacture*, μέτρον POxy. 2350 i 9 (iii AD), etc.

*ἐλαιούργισσα, ἡ, *female oil-worker*, PGot. 22 (vi AD).

*ἐλαιοφοινικών, ῶνος, ὁ, *grove of olive-trees and palms*, CPR 7.46.9 (Hermopolis Magna, early ii BC).

ἐλαιοχρήστας, α, ὁ, = ἐλαιοχρίστης, as a member of a ship's crew, Clara Rhodos 8.228.9 (Rhodes, i BC).

ἐλαιοχρίστης, for '*municipal official responsible for supply of oil*' read '*one who rubs a person down with oil after exercise*'

*ἐλαιοχρίστιον, add 'SEG 37.859.10 (Heracleia on the Latmos)'

ἐλαιρός, after 'Edict.Diocl.' insert 'SEG 37.335 iii 15'

ἐλαιών for 'PCair.Zen.57.2' read 'PCair.Zen.157.2'

ἐλαιωνοπαράδεισος, add 'also ἐλαιονο- Pap.Lugd.Bat. xxv 21.13 (78 BC)'

*ἐλαιωνοφοινικοπαράδεισος, ὁ, *olive and date orchard*, PVindob.Bosw. 8+9.9 (BASP 14.96, AD 331).

+ἐλάσιος, ὁ, *one that drives away*, τοὺς δὲ τὰς ἐπιληψίας ἀποτρέπειν δοκοῦντας ἐλασίους μέν ὀνομάζουσι (sc. at Argos) Plu. 2.296f.

ἔλασις, add '**5** *rowing*, Ael.NA 13.2.'

ἐλασμάτιον, before 'Dsc.' insert 'Inscr.Délos 1443Ai148, ii61, 1449Aabii21 (ii BC)'; add 'also ἐλαγμ- Inscr.Délos 1441Aiii85 (ii BC)'

*ἐλασοδαφία, ἡ, prob. sim. to λειψεδαφία, *loss of soil*, PRyl. 677.9 (i AD).

ἐλασσονέω, delete the article.

ἐλασσόω I 1, add 'Cod.Just. 10.27.2.10'

ἐλάσσων I 3, add 'ἐπὶ τὰ ἐλάσσονα *in the least detail*, SEG 33.1034.B13 (Aeolis, iii BC)' **III**, add '**2** ἐλάσσων τὴν ἡλικίαν *under age, minor*, Just.Nov. 1.4.1, 22.19.'

111

Ἐλάστερος, ὁ, epith. of Zeus, *AE* 1948-49.1 (Paros, v BC).

ἐλάτη III, delete '(but .. 5)' and add '**2** the fruit enclosed by the spathe, Dsc. 1.109.5.'

ἐλατήρ II, transfer 'Call.*Jov.* 3' to section I.

ἐλαττονέω 2, for this section read '*grow less*, Lxx 3*Ki.* 17.14; *be missing*, *PMagd.* 26.9, 12 (iii BC), *BGU* 1195.19 (i BC)' **3**, for '(*PMagd.*) 11.22' read 'Lxx 3*Ki.* 11.22'

ἐλαύνω, line 3, after '(Hp.)*Nat.Mul.* 32' insert 'inf. ἐλάσσειν *AP* 7.427 (Antip. Sid.)' and for '(ἐλάσσω (παρ-)' read '(παρελάσσεις'; line 21, pres. ἐλάω, add 'Tim.*Pers.* 210 P.'; line 25, after 'A.R. 3.872' insert '*GVI* 1844.3 (Rome, ?ii AD)' **I 1 b**, add 'ἐλαύνων ἐργ(άτης) *drover*, *SB* 9379 ii 11, al.' **III 2**, at end for 'generally .. 3.74' substitute '**b** metaph., *drive as a team*, Pi.*N.* 3.74; *set in motion*, A.*Ag.* 701 (lyr.).'

ἐλάφειος, after 'ον' insert '(perh. also 3 termin., πήραν ἐλαφείαν v.l. Longus 3.15.3)' and after '(iii AD)' insert '*of deerskin*, Longus l.c. Myc. *e-ra-pe-ja* (fem.)'

ἐλάφιος, add '**II** = ἐλάφειος, κέρας *Inscr.Délos* 104.21 (iv BC), *Edict. Diocl.* in *SEG* 37.335 ii 17. Myc. *e-ra-pi-ja* (fem.) *of a deer*'

ἐλάφοκεράτίτης, ου, ὁ, *a precious stone*, Orph.*Lith.Kerygm.* 4.

ἔλαφος, add 'cf. Myc. *e-ra-po ri-me-ne* = ἐλάφων λιμένει (dat.), placename = *Deer-harbour*'

ἐλαφρία II, add '*ICilicie* 36 (vi AD)'

ἐλαφρός III 1 b, for '*gentle, mild*' read '*adaptable*' and delete 'cf.'

ἐλαφρόω, after '= sq.' insert 'Plu. in Hes. 41'

*****ἐλαφώδης**, ες, *deer-like*, comp. -έστερον Eun. in Phot.*Bibl.Cod.* 77.

ἐλάχιστος I 1, add 'adv. -τως, *to the least degree*, Damocr.ap.Gal. 13.1057.15' add '**5** *least in rank* or *estimation*, *Ev.Matt.* 25.40, etc., *BCH* suppl. 8.225 (Philippi, v/vi AD), *SEG* 33.1316, al.'

ἐλαχύς, line 4, for 'Archyt.Amphiss. 2, Euph. 11' read 'Archyt.Amph. 2 = Euph. 188.2 vGr.'; line 5, after 'Nonn.*D.* 37.314' insert 'masc. ἐλαχὺν δόμον Call.*fr.* 525 Pf.; ἐγὼ δ᾽ εἴην οὐλαχύς id.*fr.* 1.32'

ἐλαών, after '= ἐλαιών' insert '*PCair.Zen.* 788.18, 27 (iii BC)'

*****ἐλβούνιον** or **ἐλβύνιον**, τό, *some kind of comestible*, *PSI* 862.9 (iii BC), *PHerm.* 23.8 (iv AD); cf. perh. ἔλπος, ἔλφος.

ἔλδομαι and **ἐέλδομαι**, w. gen., add 'σοὶ δ᾽ εἰ πλούτου θυμὸς ἐέλδεται Hes.*Op.* 381'

ἐλέατρος, for '= ἐδέατρος' read '= ‡ἐδέατρος'

ἐλεάω, add '*IKyme* 41.37 (*Hymn.Is.*)'

†**ἐλεγαίνω**, glossed variously by τὸ παραφρονεῖν, τὸ ὁπωσδήποτε ἀκολαστᾶινειν, ἀσελγαίνειν *EM* 152.50, 327.6 (perh. invented, as etym. of ἀσελγαίνειν).

†**ἐλεγεία**, ἡ, *literary poem in elegiacs*, Arist.*Ath.* 5.2, 3, Thphr.*HP* 9.15, etc. **2** *elegy* as genre, Μίμνερμος .. ποιητὴς ἐλεγείας Str. 14.1.28, cf. 13.1.48, 4.8; δι᾽ ἐλεγείας (gen. sg.) *in elegiacs*, Plu.*Cim.* 4.9, Suid. s.v. Θέογνις, al. **3** in pl., = ἐλεγεῖα, *elegiacs*, Parth.*praef.*, Plu.*Cim.* 4.10.

ἐλεγεινή, add 'cf. ἀλεγεινός'

ἐλεγείνω, delete the entry.

†**ἐλεγεῖον**, τό, *distich consisting of hexameter and pentameter, elegiac couplet*, Critias 4.3 W., Th. 1.132.2, Arist.*Po.* 1447ᵇ12, D.S. 19.1; in pl., *elegiacs*, whether forming complete poem or not, Pherecr. 162.10 K.-A., Pl.*R.* 368a, D. 19.252, Arist.*Rh.* 1375ᵇ32; of poet's whole output, ib. 1405ᵃ33, etc.; of inscriptions, Lycurg. 142; of single couplet, D. 59.98. **2** *poem in elegiacs*, *CEG* 819.13 (Lacon. epigr. at Delphi, iv BC), D.S. 10.24, 11.14, Plu.*Flam.* 9, *Them.* 8, Sch.Pi.*O.* 13.32 (all of four-line dedications); of literary poems, Str. 14.6.3, D.H. 1.49, Paus. 7.18.1. **II** *pentameter*, Heph. 1.5, 15.14, 15, Sch.D.T. 20.13, al.; cf. Pl.*Hipparch.* 228d. **III** *without metrical connotation*, *epitaph*, Ps.-Hdt.*Vit.Hom.* 36, D.Chr. 4.135. **2** in pl., *lament, song of woe*, Luc.*Tim.* 46.'

ἐλεγειοποιός, before 'ὁ' insert '(Dor. -γηο- *AP* 13.21 (Theodorid.))'

†**ἔλεγος**, ὁ, *sung lament*, inscr.ap.Paus. 10.7.6 (pl.), Ar.*Av.* 217 (lyr., pl.), E.*IT* 146 (lyr., pl.), *Hel.* 185 (lyr.), A.R. 2.782 (pl.). **2** ἔλεγοι name of an aulodic nome, Ps.-Plu. 2.1132d. **II** in pl., = ἐλεγεῖα, *elegiacs*, Call.*fr.* 7.13 Pf., *AP* 4.1.36 (Mel.), 11.130.3 (Poll.); ἱλαροὶ ἔ. id. 10.19.5 (Apollonid.).

ἐλεγχής I, add 'comp. ἐλεγχότερος *Eleg.adesp.* in *Suppl.Hell.* 964.20 (s.v.l.)'

ἔλεγχος (B) **III**, add '**2** app. = ὁ ἐλέγξας, *informer*, *TAM* 2(3).991.6 (Lycia).'

ἐλέγχω II 5, add 'Call.*fr.* 84 Pf., Nonn.*D.* 1.42'

*****ἐλεεινολόγημα**, ατος, τό, *appeal to pity*, Sch.T Il. 21.70.

ἐλεθαινομένη, add 'cf. °ἐλεγαίνω'

Ἐλειθυιαιών, ῶνος, ὁ, *last month of the year at Tenos*, *IG* 12(5).872.75 (iii BC).

ἐλειοδίακτος, delete the entry.

*****Ἐλείτας**, ὁ, Cypr. *e-le-i-ta-i* (dat.), *title of Apollo*, *ICS* 215(b).

ἐλείτης I, add '*PMil.Vogl.* 69.50, 122, ib. B107 (ii AD)' **II**, delete the section.

ἐλελίζω (A) **III**, line 4, for 'Simon. 29' read 'Pi.*fr.* 107a.3 S.-M.'

†**ἐλελίζω** (B), *utter a wavering* or *warbling sound* (ἐλελεῦ): of the noise made by troops going into battle, ἐφθέγξαντο. πάντες οἷον τῷ Ἐνυαλίῳ ἐλελίζουσι X.*An.* 1.8.18, τῷ Ἐνυαλίῳ ἠλέλιξαν ib. 5.2.14; of bird's trill, E.*Ph.* 1514; med., Ar.*Av.* 213, E.*Hel.* 1112; of inanim. things, ἡ (sc. ἀσπίς) δ᾽ ἐλέλιξεν ἐνόπλιον Call.*Del.* 137.

ἐλελίσφακος, before '*salvia*' insert 'plant, uncertainly identified w.' and add 'Nic.*Th.* 84. **II** app. = ἐλελίσφακον II, πόα τις ὁμοία δικτάμῳ Hsch.'

†**ἐλελύζω**, = ὀλολ-, γύναικες δ᾽ ἐλέλυσδον ὅσαι προγενέστεραι Sapph. 44.31 L.-P. (v.l. ὀλόλ-).

ἔλεμος, add '(ἐλεμόσπερμα La.); cf. °ἐλίμαρ'

*****ἐλενίδιον**, τό, dim. of ἐλένιον 2, *elecampane*, *POxy.* 2570 iii b 11, 3733.26, 3766.108 (all iv AD); cf. *helenidi* (gen.) *Edict.Diocl.* 34.88 in *ZPE* 34.183.

*****ἐλεόπολις**, v. °ἑλέπολις.

*****ἐλεορέω**, *to be overseer of marshland pasture*, *IEryth.* 17 (iv BC).

ἑλέπολις, add 'also ἐλεόπολις Lxx 1*Ma.* 13.43, 44' **II**, delete 'invented by Demetrius Poliorcetes' and add 'Lxx 1*Ma.* l.c.'

*****Ἐλευθέρα**, ἡ, *Liberty* (as goddess), *SEG* 17.680, *EA* 16.116, also perh. *IGRom.* 3.700 (all Lycia).

Ἐλευθεραί, add 'Ἐλευθεράθεν *IG* 1².943.96'

ἐλευθερία 1 a, add 'implying freedom from taxation, *SEG* 32.833 (Samos, 38 BC)' **b**, add '*IG* 9²(1).624.a8, al. (Naupactus, ii BC), Just.*Const.* Δέδωκεν 7b, Theophil.Antec. 1.12.6; *manumission-document*, *MAMA* 4.279 (Dionysopolis)' add '**5** personified as goddess, *Epigr.Gr.* 903 (Sardis, iv AD).'

ἐλευθερικός, add 'ἔδωκε τὸ ἐλευθερικὸν τέλες (sic) Πολυξένῳ *SEG* 37.450 (Thessaly, ii/iii AD), *ITyr* 45'

ἐλευθέριος I 2, line 6, after '(Arist.)*Pol.* 1340ᵇ10' insert 'τύμβος *AP* 7.178 (Diosc.), 185 (Antip.Thess.)' **III**, add '*IMylasa* 207.5, etc., *SEG* 34.558.15 (Larissa, ii BC)'

ἐλευθερόπαις, add 'eleg. in *AD* 10(1926) παράρτημα 49 (Phalanna, iii BC)'

ἐλεύθερος I 1 b, delete 'married woman, Ath. 13.571d'; after '*POxy.* 1872.8 (v/vi AD)' insert '*PVindop.Worp* 14.2 (vi/vii AD), *PGron.* 10.17'; delete 'but, *freedwoman* .. (Vienne)' **2**, after 'D.S. 4.46' insert 'also ἐλευθέρα τήρησις *POxy.* 3346.5 (early iii AD)' **II**, add '**2** in cultural or moral sense, opp. δημιουργός Pl.*Prt.* 312b, *Lg.* 848a, *R.* 405a, 431c.' at end add 'Myc. *e-re-u-te-ro*, *free*, of contributions, *not payable*'

ἐλευθερόω, add '**3** *manumit*, *CodJust.* 6.4.4.10, also perh. *emancipate*, ib. 6.4.4.25.' at end add 'Myc. aor. *e-re-u-te-ro-se*, *allowed free*'

ἐλευθερωτής, add '**2** *manumissor*, *CodJust.* 6.4.4.24, Just.*Nov.* 78.4.1, etc.'

Ἐλευθία, delete 'Ἐλευθώ'

*****Ἐλευθώ**, = Εἰλείθυια, Call.*Del.* 276 (cj.), *AP* 7.604 (Paul.Sil.), 9.268 (Antip.Thess.), cf. Hsch.

ἐλεύθω, line 2, for 'Ibyc.*Oxy.* 1790.18' read 'Ibyc. 1(*a*).18 P.; 3 sg. ἔλευσεν Pi.*fr.* 70d.39 S.-M.'

*****Ἐλευσία**, v. °Εἰλείθυια.

Ἐλευσίνιος II, for '*Eleusis*' read '*Athens*'

Ἐλευσίς I, add 'Ἐλευσεῖν *GVI* 1058 (Eleusis, iii/iv AD)'

†**ἐλεφαντάρχης**, ου, ὁ, *master of elephants*, Phylarch. 31 J., Plu.*Demetr.* 25, Lxx 2*Ma.* 14.12; spec., *commander of squadron of sixteen elephants*, Ascl.*Tact.* 9, Ael.*Tact.* 23.

ἐλεφάντειος, add 'Myc. *e-re-pa-te-jo*, *made of ivory*'

*****ἐλεφαντιᾱκός**, ή, όν, *suffering from elephantiasis*, Firm. 3.5.30; also -τικός, ib. 8.26.13.

ἐλεφάντινος 1, after 'Alc. 33.1' insert '(= 350.1 L.-P.), λύραι ἐλεφάντιναι *IG* 1³.345.47' **2**, add 'perh. so used by Alcm. 10(*b*).6 P.'

*****ἐλεφαντιώδης**, ες, *suffering from elephantiasis*, Antyll.ap.Aët. 3.9 (p. 269.10 O.).

ἐλεφαντουργός, subst., add '*SB* 10258 ii 14 (iv AD)'

ἐλέφας II, add '*IG* 1³.457.22 (442/1 BC), *CID* II 62 ii A 5, Lxx *Ez.* 27.6; *ivory statue*, χὸ βῶμὸς χἐλέφας *SEG* 32.356 (Aegina, vi BC)' **III**, add 'Al.*Jb.* 2.7' at end add 'Myc. *e-re-pa*'

*****Ἑλήγηρις**, εως, ἡ, epith. of Demeter (διὰ τὸ ὑπὸ τῆς τοῦ ἡλίου ἕλης γηρᾶν), Eust. 1197.53.

ἕλιγμα I, add 'perh. *hairband*, πορφυρεῦν .. κόμης ἕλιγμα *AP* 6.211 (Leon.Tarent.)' **III**, delete the section.

ἑλίκη III, add 'Myc. *e-ri-ka*, *willow*'

ἑλικοβλέφαρος, for '*with ever-moving eyes, quick-glancing*' read '*having curling eyelashes*'

ἑλικός, for '*eddying* .. *fr.* 290' read 'perh. *meandering* (but *black* acc. to Sch.), of river, Call.*fr.* 299 Pf.' and before 'χορεία' insert '**II** *circling*'

ἑλικτός I, lines 4/5, for 'a *wheeled* ark' read 'prob. *plaited basket*'; line 6, before 'Theoc. 1.129' insert 'i.e. wrapped round with bands'; line 7, before 'comp.' insert 'ἑλ(ε)ικτὸν κεφαλίδα Jos.*Jer.* 43(36).2' add '**b** ἑλικτά, τά, app. kind of vessel, *Inscr.Délos* 442B210 (ii BC).'

Ἑλικών, add 'III experimental instrument used by musical theorists, Ptol.*Harm.* 2.2.'

Ἑλικωνίδες, add 'Ibyc. 1(*a*).24 P.'

Ἑλικώνιος II, add '*SEG* 15.784 (Sinope, iii BC)'

ἕλικωψ, for '*with rolling eyes* .. spirits' read 'epith. of uncertain sense, perh. *with rolling eyes*'; line 4, for 'Sapph.*Supp.* 20a.5' read 'Sapph. 44.5 L.-P.' and after it insert 'Alc. 283.16 L.-P.'; after '*P.* 6.1' insert 'ἑ. .. λαγωοί Nonn.*D.* 48.900'; add 'expld. as *black-eyed* by Sch.Il. 1.98, *Et.Gud.*, Eust. 57.1, etc., cf. °ἑλικός'

*ἕλιμαρ· κέγχρῳ ὅμοιον ἢ μελίνη ὑπὸ Λακώνων Hsch. (La.); cf. ‡ἔλεμος, ἔλυμος III.

*ἑλιμονικός, v. °ἁλιμονικός.

ἔλινος, for 'later ἐλινος' read 'later ἔλινος' and after 'Nonn.*D.* 12.299' insert '(cod. σελίνοις); cf. ‡ἐλινοφόρος'

ἐλινοφόρος, delete 'Ep. εἰλ-' and after 'ib. 17.333' insert '(v.l.)'

ἐλινύες, add 'cf. ἡ παρὰ τὸ ἐλινύειν τὸ ἡσυχάζειν γέγονεν ἐλινὺς καὶ Ἐρινὺς Sch.Hes.*Th.* 472 (p. 75 Di G.)'

ἐλινύω, line 5, delete 'μὴ ἐλινύειν Hdt. 1.67'   **3**, line 2, after 'cf.' insert '1.67'   add 'II trans., *bring to rest, halt*, πότμον ἐλινύσειεν Call.*fr.* 330 Pf.'

ἕλιξ (A), for 'epith. of oxen .. *rolling*' read 'epith. of oxen: the original significance is unknown; it was interpreted by Alexandrian scholars as "*black*" and perh. used by Theoc., etc. in this sense'

ἕλιξ (B) III 1, line 2, before 'Thphr.' insert 'Hes.*Sc.* 295'   **2**, add '**b** of palm, *AP* 4.1.50 (Mel.).'   **4**, at end for '*feelers*' read '*tentacles*', for '(Antiphil. Byz.)' read '(Antiphil.)' and after it insert '*feelers of the prawn*, Ael.*NA* 16.13'

ἑλίσσω I 3, add 'ἑ. τὸ μέλος trill its song, of the nightingale, Ael.*NA* 5.38'   **4**, add '**b** = ἀνελίσσω I 1, i.e. *read*, γράμματα (= συγγράμματα) Call.*fr.* 468 Pf.; βίβλον Posidipp. in *Suppl.Hell.* 705.16, cf. *AP* 9.161.1 (Marc.Arg.).'

ἐλίχρῡσος, after 'ὁ' insert '(-ον, τό Dsc. 4.57)'

ἑλκητήρ, for 'harrow' read 'rake'

ἑλκόω I 2, add '**b** *produce ulceration by cautery*, πέλματα *PMerton* 12.20 (i AD).'

ἑλκυστός II, for '*refined, fine-drawn* oil .. Stratonicea' read 'oil *which may be drawn* by the users *IStraton.* 15.10, 197.12, al.' and add '*SEG* 11.492.11 (Sparta, ii AD); ἀλείμματα ἑ. *IGRom.* 3.804 (Aspendus); γυμνασιαρχία ἑ. *IStraton.* 210.8, 219.3'; transfer 'cf. ἑλκυστῷ· λείῳ Hsch.' to section I 1.

ἕλκυστρον 2, for 'φορβει(ά' read 'φορβεωά'

ἕλκω, line 5, delete '*IG* 11(2).287.B61 (Delos, ii BC)'   **I 2**, add 'med., *APl.* 306, 307 (Leon.Tarent.)'   **II 1**, for 'barge-pole' read 'paddle'   **4**, add '**b** καθ' ὃν τόπον .. καὶ ὁ ῥοῦς ἕλκει *towards which place the current sets*, *Peripl.M.Rubr.* 13.'   **6**, at end for 'εἰκλυσμένων' read 'εἰλκυσμένων'

ἑλκώδης I, add 'adv. -δῶς, *so as to produce ulceration*, Gal. 17(2).620.16, 621.3'

ἑλκωτικός, add 'II ἑλκωτική, ἡ, (sc. ἔμπλαστος), kind of plaster producing ulceration by its caustic properties, *PMerton* 12.15 (i AD).'

ἐλλᾰθῑ, after 'ἴληθι' insert '(v. *ἴλημι)'; after 'B. 10.8' insert 'Simon. 54 P. (s.v.l.)'; for 'Call.*fr.* 121' read 'Call.*fr.* 7.13 Pf.'

ἐλλαμβάνω I, for '*Supp.Epigr.* .. ii BC)' read 'ψάφος ἐλλάβοιεν *BCH* 66/7.144 (Thess. inscr. at Delphi, iii BC)'

†ἐλλαμπρύνομαι, *distinguish oneself, shine*, μηδὲ τούτῳ ἐμπαράσχητε τῷ τῆς πόλεως κινδύνῳ ἰδίᾳ ἐλλαμπρύνεσθαι Th. 6.12.2, App.*BC* 3.66; without dat., Luc.*Dom.* 1.   **2** *glory in*, τῷ ἔργῳ D.C. 73.10, πρὸς τὰς φίλας ἑ. λόγοις J.*AJ* 18.3.4.

Ἑλλᾰνοδίκαι I, after 'Paus. 5.9.5 sq.' insert 'at the Olympic games of Alexandria, *SEG* 8.658 (sg., Coptos, iii AD)'

*ἕλᾰτε, v. ‡ἔλλαθι.

ἔλλειμμα, line 3, after '(D.) 22.44' insert 'ostr. cited *OBodl.* 49 note'

ἐλλειπόντως, after 'Plot. 1.3.6' insert '*deficiently, with insufficient force*, Simp. *in Epict.* p. 38 D.'

ἐλλείπω I 1, after 'E.*El.* 609' insert 'Th. 5.103.1'

ἐλλείχω, add '*take in by licking*, Hp. *Mul.* 1.8'

ἔλλετε, for 'Call.*fr.* 292' read 'Call.*fr.* 1.17 Pf.' and delete 'cf. ἔλλατε (v. ἔλλαθι)'

Ἕλλην III, lines 4/5, delete 'Πυλῶν Ἑλλήνων D. 18.304'

Ἑλληνάρχης, add '*IPrusias* 5.1, 46.2'

Ἑλλήνιος II, after 'Hdt. 2.178' insert 'ἐν τῶι Ἑλλανίωι *SEG* 32.1586.17 (Naucratis, v BC)'

Ἑλληνίς II 1, add 'pl., Hyp.*Epit.* 36'   **2**, add '*Ev.Marc.* 7.26'

Ἑλληνιστί, for '*PTaur.* 1*4' read '*Mitteis Chr.* ii.31 v 4'

Ἑλλήσποντος, line 5, after 'Aegean' insert '*AP* 7.705 (Antip.Thess.)'

ἐλλιμενίζω, for 'Ar.*fr.* 455' read 'ἐλλιμενίζεις ἢ δεκατεύεις Ar. 472 K.-A.' and add 'also ἐνλ- Hsch.'

ἐλλιμένιος I, add 'as epith. of Hera, *IG* 2².5148, cf. λιμένιος'

ἐλλιμενιστής, for '*farmer*' read '*collector*'

*ἐλλίμην, ενος, ὁ, perh. an area of the agora, τὰ ποτ' ἐνλίμενα (cf. ‡λιμήν III) *SEG* 37.494.11 (Matropolis, Thessaly, iii BC).

ἐλλῐπής II 1, add 'ἑ. ἐπὶ τῆς τραπέζης αὐτοῦ Lxx *Si.* 14.10'   **2**, add '-εστέρως J.*AJ* 19.1.17'

ἐλλῐτές or ἐλλιστές, wd. of unkn. meaning in Epich. 183.

ἐλλόγιμος I, add 'sup., as honorific title, ὁ ἐλλογιμώτατος συνήγορος τοῦ Θηβαίων φόρου *SB* 7033.21 (v AD), *POxy.* 1886.1 (v/vi AD)'

ἐλλογιμότης, add 'II as complimentary address, παρακαλῶ τὴν σὴν ἑ. *your notability, POxy.* 1885.11 (vi AD).'

ἐλλοπίης, delete the entry.

ἐλλότης, add 'also εἰλότᾱς (acc. pl.), Sokolowski 3.74.4, 6 (Lebadea, iv BC)'

*ἐλλυχνιδᾶς, ᾶ, ὁ, *lamp-wick maker*, Παύλου ἐνλυχνιδᾶ *SEG* 32.375 (Argos, Chr.).

ἐλλύχνιον, add '3 τὸ τοῦ ποδὸς ἑ. = θέναρ, Cyr.'

ἔλμινς, add 'Λίμινθες· ἔλμινθες Hsch.'

ἕλος, line 3, for '*Inscr.Cypr.* 135.9 H.' read 'Cypr. *e-le-i*, *ICS* 217.91'   **2**, add 'glossed by σύμφυτος τόπος, ἢ χεῖλος ποταμοῦ, Hsch.; cf. ἕλη· σύνδενδροι τόποι id.'   at end add 'cf. Myc. *e-re-e* (dat.), place-name'

*ἑλοτρεφής, ές, gloss on ἑλεόθρεπτος, Hsch.

*Ἐλουσία, title of Demeter, Ἐλουσία· Δημήτηρ παρὰ † Ἀλφουσίοις Hsch.; also Ἐλουία, Δαματ[ρὸ]ς ἱεροὶ Ἐλουίας *SEG* 30.1174 (Corinthian jug, iii/ii BC).

*ἐλπῐδηφόρος, ον, *bringing hope*, Theognost. *Can.* 95; also as pers. n., *IG* 12(2).76.e7 (Mytilene), etc.

ἐλπῐδοκοπέω, add '*PKöln* 166.13, *SB* 10525.3 (both vi/vii AD)'

ἐλπίς, line 1, before 'v. ἔλπω' insert 'ἑελπίδα *CEG* 10.9 (Athens, v BC)' and for '*hope, expectation*' read '*expectation* (whether of good or bad)'   **I 1**, at end before 'S.*Ant.* 330' for 'beyond *hope*' read 'i.e. otherwise than I expected'   **2**, for '*object of hope*' read '*basis of one's hopes* or *expectations*'   **II**, delete the section, transferring quots. to section I 1.

ἐλπωρή, before 'ἡ' insert '(-ᾰ Alc. 119.11 L.-P.)'

*Ἐλυγεύς, ὁ, epith. of Dionysus in Samos, Hsch. (post ἐλίγαινον; cj. Ἐλεγ-, Ἐλελ-, al.).

ἔλυμα, read 'ἔλῡμα, in obl. cases by metr. lengthening ἐλῡματ-'; for '*stock*' read '*share-beam*' and after 'Hsch.' insert '(cf. εἴλυμα)'

ἐλύμνιαι, add 'Myc. *e-ru-mi-ni-ja*'

ἔλυμος, for 'ὁ' read 'ον'   **I** and **II**, for these sections read '*curved*: αὐλοὶ ἔλυμοι pipes, also acc. to Ath. 4.176f called Phrygian, of which the left-hand one of the pair was crooked at the end, S.*fr.* 450, 644 R., Call.Com. 23 K.-A., Cratin.Jun. 3 K.-A., Poll. 4.74; cf. ἔλυμοι· τὰ πρῶτα τῶν αὐλῶν, ἀφ' ὧν ἡ γλωσσίς Hsch.   **2** perh. a stringed instrument, Ath. 14.636f.   **II** ἔλυμοι· .. καὶ ἡ τῆς κιθάρας καὶ τοῦ τόξου θήκη Hsch.'

ἐλυτροειδής, for 'Antyll.ap.Orib. 44.23.75' read 'Antyll.ap.Orib. 44.20.75'

ἔλυτρον 2, *shell* of crab, add 'ἀρτιφύτοισιν .. ἐλύτροις Opp.*H.* 1.303'

(F)ελχάνιος (s.v. Fέλχανος), add 'cf. Cypr. pers. n. *wa-la-ka-ni-o*, perh. gen. Fαλχανιῶ *ICS* 299.4'

Fέλχανος, after 'Crete' insert '*Inscr.Cret.* 1.xxiii 5 (Fενχ-, Phaestus)'

*ἔλψ, ἐλπός, ἡ, = ἐλπίς, φίλον ὄλεσεν ἔλπ' ἀγαθέν *CEG* 51 (Athens, *c.*510 BC).

ἐλώριον, pl., add 'Opp.*H.* 4.429'

ἐμαυτοῦ, line 7, for 'in pl. .. etc.' read 'pl. υἱῷ καὶ ἐμαυτοῖς *PASA* 2.278 (Armenia); (ἡμῶν αὐτῶν, etc. are normally used)'

*ἐμβαθμός, ὁ, perh. *entrance*, ἑ. τοῦ .. κάστρου *PNess.* 24.3 (vi AD, pl.).

ἔμβαθρα, delete the entry.

*ἔμβαθρον, τό, *right of entry*, *PAvrom.* 2B9 (i BC).   **II** pl., ἔμβαθρα, τά, kind of shoes, Poll. 7.93.

ἐμβαίνω I 5, add '**c** *enter into possession*, εἰς κτήματα *IEphes.* 4.75, al. (iii BC); of the lessee of sacred land, w. acc., τὰν γᾶν *IG* 7.1739.5 (Thespiae): abs., ib. 9, al.   **d** *enter* or *join* a society, *SEG* 31.122.39 (Attica, ii AD).   **e** perh. *reach the age of*, ἐμβαίνοντες εἰς τὰ ιζ' ἔτη *IGLS* 607, cf. °ἐπαναβαίνω.   **II**, add 'in fut., of the ἀρχά that assigns sacred land to a lessee, *IG* 7.1739.10, *Mél.Navarre* p. 353 (both Thespiae)'

ἐμβάλλω I 1, add '**b** *put into* an already occupied tomb, Ramsay *Cities and Bishoprics* 1.233.80, *SEG* 6.219 (Phrygia), *TAM* 4(1).231.'   **3**, line 3, after 'Pl.*R.* 344d' insert 'Men.*Dysc.* 352'   **III 1**, after 'one's own' insert 'ἐν δὲ κλήρους ἐβάλοντο Il. 23.352'   **3**, for '*fall upon*' read '*throw in*, i.e. stuff oneself with'   **4**, add '*SEG* 34.558.37 (Larissa, ii BC), *Peripl.M.Rubr.* 32'

*ἔμβᾰμα, ατος, τό, *statue*, *IRhod.Peraia* 13.3 (iv/iii BC).

ἔμβαμμα, add '*Inscr.Dura* 198'

ἐμβάπτω, add 'X.*Cyr.* 2.2.5'

*ἐμβᾰρέω, *to be reluctant, hesitate*, *BGU* 1816.9 (i BC).

ἔμβαρος I, delete the section.

*Ἔμβαρος, ὁ, a proverbially cunning hero, οὐκ Ἔμβαρος εἶ, Men.*fr.* 368 K.-Th., cf. id.*Phasm.* 80 S., Paus.Gr.*fr.* 163, Suid. (Hsch. ἔμβαρος·

ἠλίθιος, μῶρος is due to abridgement of οὐκ Ἐ. εἶ· ἠλίθιος εἶ, cf. id. s.v. οὐκ ἔ. εἶ).

†ἐμβᾰρύθω, to be heavy, Nic.Th. 324; to be oppressive, ib. 468, 512, id.Al. 541 (v.l.).

*ἐμβᾰσίλευμα, ατος, τό, realm, Μίλατος .. Φοίβου κλυτὸν ἐ. epigr. in Didyma 229 ii 9 (66 BC).

ἔμβᾰσις I 5, add 'lease of sacred land, IG 7.1739.18; rent under such a lease, ib. 12, 13, Mél.Navarre p. 353' II, for this section read 'means of walking, ὑπαί τις ἀρβύλας λύοι τάχος, πρόδουλον ἔμβασιν ποδός A.Ag. 945; in periph. for cows' legs, δίχηλον ἔμβασιν E.Ba. 740' III, add '2 tomb, BE 1964.631 (cf. °ἐμβατός II b).' add 'IV pedestal, Javolenus Dig. 32.100.3.'

ἐμβᾰτεία, add 'BCH 22.402'

*ἐμβᾰτευτικός, όν, ius ἐμβατευτικόν "the right of setting foot on", perh. the right of a superficiarius to use what is on the surface of another's land, Ulp.Dig. 27.9.3.4.

ἐμβᾰτεύω I, line 3, for 'E.El. 595' read 'E.fr. 696.3'; line 6, after 'S.OT 825' insert 'enter, πόλιν E.El. 595' add 'V glossed by ζητέω, Hsch. s.vv. ἐμβατεῦσαι, ἐμβατεύσας.'

†ἐμβᾰτέω, = ἐμβατεύω I, Nic.Th. 147 (v.l. -βροτ-), 804 (v.l. -βοτ-); med., w. acc., Lyc. 642. II prob. f.l. for °ἐμβοτέω.

*ἐμβᾰτή, v. ‡ἐμβατός.

†ἐμβᾰτήρ, ῆρος, ὁ, perh. footway, terrace, prob. in IG 4.481.2 (Nemea), Ath.Agora XIX L4b.15 (iii BC), cf. Hsch.'

ἐμβᾰτήριος I, add 'ἐ. αὐλούς Poll. 4.82'

ἐμβάτης, add 'III perh. ditch or fosse, (cf. ἐμβατή) τοῦ τύχου (= τοίχου) τούτου σὺν ὅλον τοῦ πέλματος καὶ τοῦ ἐμβάτου ἐπὶ .. SEG 37.727 (Samos, Byz.).'

ἐμβᾰτός I, after 'accessible' insert 'APl. 95 (Damag.)' II, fem. subst., add 'b tomb, AD 21(1966).335.'

*ἐμβᾰφεία, ἡ, dipping in, Moses Alch. 313.24.

*ἐμβᾰφής, ές, dipped in, Moses Alch. 309.9.

ἐμβάφιον, after 'Hdt. 2.62' insert 'SB 9158.6 (v AD), CPR 8.66.9 (vi AD)'

†ἐμβῐβαστέον, one must cause to enter, Herod.Med.ap.Orib. 10.38.3, Gp. 14.7.18.

ἐμβιόω, line 2, after 'Philostr.Her. 2.3' insert 'ἐνδεδιωκότα (= ἐμβεβιωκότα) Schwyzer 62.120 (in sense II, Heraclea, iv BC)'

ἐμβλέπω, line 3, after 'cf. D. 19.69' insert 'ἐ. ἀγάλματι Men.Dysc. 677, παρθένῳ ib. 682'

ἔμβλημα 2, add 'written ἔμβληθμα (s.v.l.) in IG 11(2).287B134 (Delos, iii BC). b fig., ἔμβλημα σωφροσύνης epigr. in SEG 36.399 (Dyme, c.100 BC).'

*ἔμβλητος, ον, used as inlay, inlaid, πίνακας ἐμβλήτους γραφὰς ἔχοντας Inscr.Délos 1403Bbii18, 1417Aiii36 (ii BC).

†ἐμβολάδιον, τό, little portico, JHS 18.308 (Mopsuestia, v. ICilicie p. 129).

*ἐμβολάρχης, ου, ὁ, official in charge of loading of ships, (ἐμβολή I 3), POxy. 3612.4 (iii AD), CPR 7.26.11 (vi AD).

*ἐμβολαρχία, ἡ, control of the loading of ships (ἐμβολή I 3), PMerton 90.11, 22 (iii/iv AD).

*ἐμβολάτωρ, ορος, ὁ, collector of dues, POxy. 126.15, PMasp. 54 i 7 (both vi AD), etc.

*ἐμβολευτικόν, τό, wd. of uncertain reading and sense, PTeb. 847.17 (ii BC).

ἐμβολεύω, add '2 perh. colloq., appropriate, τῆς πράσεως τὴν τιμὴν ἐνεβόλευσε POxy. 2342.9 (ii AD).'

ἐμβολή I 3, after 'POxy. 62.11 (iii AD)' insert 'Peripl.M.Rubr. 32' II 2, after 'E.HF 869' insert 'ταῦρος λέοντος ὡς βλέπων πρὸς ἐμβολήν id.fr. 689.4' 4, line 3, after 'ἐσβάλλει' insert 'ἡ τῶν ὑδάτων εἰς τὰ πεδία ἐμβολή PHels. 6.4 (ii BC)'

ἐμβόλιμος 1, add 'Ξανδικοῦ ἐμβολίμου ιγ' SEG 31.1046 (Sardis, before 9 BC)'

ἐμβόλιον IV, add 'Inscr.Délos 372B30 (200 BC)'

Ἐμβολῖται, οἱ, a guild at Ephesus centred on the Ἔμβολος (v. °ἔμβολος 8), IEphes. 3059.11.

ἔμβολος 3 a, before 'Pi.P. 4.191' insert 'Hippon. 28.3 W.' 4, for 'wedge-shaped order of battle' read 'military or naval formation in the form of ram' 8, line 1, after 'portico' insert 'or colonnaded street'; line 2, delete 'Ephes. 3 No. 8'; add 'spec., a street in Ephesus, IEphes. 3008.12, al.'

ἐμβόσκομαι, add 'also act., MDAI(A) 68.5 (Samos, i AD), IStraton. 513.55 (iii AD)'

*ἐμβοτέω, pasture animals in a place, AP 7.657 (Leon.Tarent., cj.).

ἔμβρεφος, for 'boy-like' read 'having a child in the womb, in quots., of women who died in pregnancy' and add 'epigr. in LW 116 (Teos)'

*ἐμβρῑμέω, to be indignant, w. dat., Stilpon in Lex.Patm. in BCH 1.151 (Rh.Mus. 32.477).

ἐμβρύκω, add 'ὅτ' ἐμβρύξῃσιν (v.l. ὅταν βρύξῃσιν) Nic.Th. 271'

*ἔμβρύμιον, τό, papyrus mat or cushion, PFouad I Univ. 26.5 (pl., i AD), PPetaus 33.7 (AD 184/7), PCol. 240 (iv/v AD).

†ἐμβύθιος, ον (fem. -η AP 9.227 (Bianor), cj. in 423 (id.)), existing or situated at the bottom of the sea, ἐ. θαλάμας Isid.Char. 20, πέτρα AP 7.504 (Leon.Tarent.). 2 of water, deep, κρηνῖδες D.H. 1.32, 6.13.

*ἔμβυσμα, ατος, τό, perh. stopper, plug, Pland. 144ᵃ2 (pl., ἐνβ-; iii AD); cj. in CGFP 289.4.

*ἐμέτερος, α, ον, = ἡμέτερος, POxy.Hels. 49.12 (iii AD).

ἐμετήριος, add 'Gal. 11.173.1'

ἐμετοποιέομαι, add 'act., dub. in IG 14.2577.13 (Xanten)'

ἐμμᾰνής, line 5, for '-έστερος' read '-έστερον' and transfer 'Comp. .. Luc.Am. 14' to last line after 'Eun.VS p. 455 B.'

*ἐμμέλεια (B), ἡ, authority of a local official, CPR 5.12.5 (AD 351).

*ἐμμελής II 3 b, after 'suitable' insert 'A.fr. 78c ii 22 R. (sup.)' III 3, add 'σὲ γάρ φη Ταργήλιος ἐμμελέως δισκεῖν Anacr. 19 P.'

ἐμμενής, adv. -νέως, add 'epigr. in IStraton. 543.8'

ἐμμενύτρωτος, add '(ἐμμεσότροπος La.)'

*ἐμμένω, line 1, fut., insert '-μενίω SEG 33.638 (Crete, late ii BC)' and after 'Th. 1.5' insert '3 sg. aor. opt. ἐμ[μ]ένᾱι SEG 36.548.8 (Matropolis, iii BC)'

*ἐμμεσότροπος, v. °ἐμμενύτρωτος.

*ἐμμετροποιός, όν, composing in regular metres, Phld.Po. in Eos 29.19 (ἐνμ-).

*ἔμμηνος II, add '4 of an official, serving for a month, SEG 33.1039.77 (Aeolian Cyme, ii BC).'

*ἔμμομφος, Arc. (?)ἴνμομφος, v. ‡μόμφος.

ἔμμονος I, add 'of dyes, fast, Hsch. s.v. δευσοποιόν'

*ἐμόρμησεν· ἐπενόησεν Hsch.; cf. μερμαίρω.

*ἐμμόχλιον, τό, socket for a bar, K. Kourouniotes Ἐλευσινιακά i.190 (Eleusis), Poll. 10.23.

ἐμός II 2, add 'of lovers, Μυΐσκος ἐμὸς ἡδύς graffito in BCH 106.12 no. 57 (Thasos, iv BC)' II 3, delete 'PEdgar 4.6 (iii BC)' and insert 'BGU 37.3 (i AD)'; add 'ἐξ ἐμοῦ from my funds, Mitchell N.Galatia 240'

ἐμπᾰγή, for 'suretyship' read 'trap'

ἐμπάθεια II, after 'Ptol.Tetr. 92' insert '(v.l.)'

ἔμπαιγμα, for 'jest, mocking, delusion' read 'raillery, mocking'

ἐμπαίζω I 1, add 'w. εἰς, ἐμπαῖξαι εἰς οἰανδήποτε ἐκκλησιαστικὴν κατάστασιν Just.Nov. 123.44'

†ἔμπαις, παιδος, ἡ, with child, ἡ παῖς γὰρ ἔμπαις ἐστίν Cratin. 318 K.-A., ἔμπαις· ἐγκύμων Hsch. 2 having children, Muson.fr. 15A (p. 78 H.).

ἔμπαισμα, before 'Eust.' insert 'Inscr.Délos 1412a7 (ii BC)'

ἐμπαίω II, add 'εἴ τε ῥᾳθυμός τις ἐμπέπαικεν Plu. 2.52c' add '2 fall into, εἰς ἄρκυν Lyc. 105 (v.l.).'

ἔμπᾰλιν I a, add 'ἡ ἐ. θύρα door leading out again, Luc.Merc.Cond. 42'

ἔμπαμα, delete the entry (v. °ἅπαμα).

ἐμπαρέχω, line 4, for 'give oneself up as his tool ' read 'hand oneself over into his power'

ἐμπαρίσταμαι, delete the entry (v. °ἐμπαρίστημι).

*ἐμπαρίστημι, pf. part. act. intr., stand by, Hld. 7.19.1 (v.l.).

ἔμπᾱς (A), line 4, for 'Call.Ep. 14' read 'Call.Epigr. 12, fr. 726 Pf.'

ἔμπᾱσις, Boeot., add 'and also ἔμπασσις SEG 32.476 (Onchestus, iv BC)'

*ἐμπασμός, ὁ, strewing, SEG 33.1056 (Cyzicus, ii AD).

ἐμπαστήρας, after lemma insert '(ἔμπιστ- La.)'

ἔμπατον, add '(ἔμπαστον· κατανθιζόμενον La.)'

ἐμπέδιος, add '2 = ἔμπεδος 2, constant, unfailing, LW 1522 bis (Cyme) (s.v.l., cf. IKyme 4).'

*ἐμπεδομητις, ιδος, ὁ, ἡ, of steadfast counsel, IG 2².12318.7 (verse, iii BC).

†ἐμπελάδην, ἔμπλην· ἐμπελάδην, συνεγγύς Apollon.Lex.; app. in temporal sense, just, Nic.Al. 215.

ἐμπελάνα, add 'cf. perh. εὐχὰν ἐμπελάνον, rest. in Dubois IGDS 54 (Selinus, vi/v BC)'

ἐμπελάω, delete 'IG 14.271 (Selinus)'

*ἐμπεπείρακται· ἐμπεπόδισται Hsch. (cj. ἐμπεπέδηται La.).

ἐμπέρᾱμος, add 'Spartan pers. n. (vii BC), Paus. 4.20.5'

*ἐμπεριπλέκω, enfold, Alcm. 13(b).10 P. (dub.).

ἐμπερονάω I, med., add 'οἵμαι σε τὸν ἐπ' ἀριστέρ' ἐμπερονώμενον Men.fr. 691 K.-Th.'

ἐμπήγνῡμι I, line 7, after '(Ar.)V. 437' insert 'στήλαις .. ἐνπηγνυμέναις ἐκεῖ SEG 34.1243 (Mysia, v AD)'

*ἐμπήκτης, for pres. def. read 'one who inserts the πινάκια of potential jurors, etc., into the slots of the κληρωτήριον'

ἐμπηνός, add '(ἐμπηγός La.)'

ἐμπήσσομαι, add 'Aët. 7.21 (269.22 O.)'

ἔμπιλα, for 'bandage .. νακτά' read 'νακτά· τοὺς πίλους. καὶ τὰ ἐμπίλια Hsch.'

†ἐμπιπίσκω, aor. ἐνέπῑσα, Pi.fr. 111.1 S.-M.: give to drink, Nic.Al. 519; transf., Pi.l.c.; med., soak in ὕδατι, ὄξει, Nic.Th. 573, Al. 320; pass., to be administered in, νύμφαις (sc. water) ἐμπισθέν id.Th. 624.

ἐμπιπράσκω, delete 'Hsch. (Pass.)'

ἐμπίπτω 1 b, add 'of adjoining houses, *ICilicie* 124 (i/ii AD)'   3, add 'c of things, *come one's way*, ἀντιφορτίζονται τὰ ἐμπεσόντα *Peripl.M.Rubr.* 14.'   10, for '*desert*' read '*go over to*'

ἐμπίσιον, add '(ἐμπίσειον La.)'

*ἐμπίστευτος, ὁ, *trustee*, Vett.Val. 379.17 P.

ἐμπιστεύω, add 'III pass., w. inf., *to be assured*, ἐνπιστευθεὶς ὑπὸ τούτου ἔχειν *POxy*. 2347.4/5 (iv AD).'

ἐμπιστῆρας, v. ‡ἐμπαστῆρας.

ἐμπλανάομαι, line 2, after '-πλανωμένη' insert 'ἑαυτῇ'

ἐμπλάσσω I 5 a, after '(Gal.) 15.204' insert 'Nic.*Al.* 79'

ἔμπλαστρον, add 'in Latin transcription *emplastrum*, Cato *Agr.* 39.2, etc.'

*ἐμπλείω, *fill up*, Nic.*Al.* 613.

*ἔμπληγος, ον, = ἐμπληγής, *An.Bachm.* 1.43.30.

⁺ἐμπλήδην, adv., *together with*, w. dat., Nic.*Al.* 129.

ἐμπλήθομαι, add 'act. intr., Q.S. 13.22 (cj.)'

ἐμπλόκιον 1, delete the section.   2, add 'Macho 257 G.'

ἔμπλουμος, add '*PVindob*. G25737 (*ZPE* 64.77, vi/vii AD)'

*ἐμπνευματώδης, ες, *flatulent*, Androm.Min.ap.Gal. 13.982.8.

ἐμπνευμάτωσις 2, add 'Heras ap.Gal. 14.203.4'

ἔμπνευσις, for '*on-breathing*' read '*the action of breathing on*'

ἐμπνέω II 2, add 'ἐπιθυμίαν Jos.*Jer.* 28(51).11c'

ἔμπνοος 1, add '2 of blood, *aerated*, opp. θολερός, Philostr.*Gym*. 30.'

ἐμποδίζω, line 2, after '*Gp*. 2.49.1' insert 'pf. inf. ἐνπεποδικέναι *IEphes.* 4101.19 (ii AD)'

ἐμπόδισμα, for 'D. 3.4' read 'D. 3.7'

ἐμποδοστατέω, delete 'also ἐμποδιοστατέω'

ἐμποιέω III, at end after 'etc.' insert '(in pass. aor. *PAvrom*. 2.*B*13)'

ἐμποίησις II, for this section read '*installation*, κλαι[κός] *SEG* 25.383.193 (Epid., iv BC)' and renumber present section II as III; after '*claim to*' insert '*PMich*.III 121ʳ ii 9.5 (i AD)'

ἐμπολάω I 2, line 2, after '(S.)*Ant.* 1037' insert 'νυνὶ δὲ πεντήκοντα δραχμῶν ἐμπολῶ i.e. *sell* for .. , Ar.*Pax* 1201' deleting this ref. in section II 1.

ἐμπολέμιος 1, add 'τὰ ἐ. *war-material*, prob. for τὰ ἐν πολεμίᾳ in Afric.*Cest*. 32 V.'

ἐμπόλημα I, add 'ἡ Μιλησία σμάραγδος, ἐ. τιμηέστατον *Trag.adesp.* 109 K.-S.'

ἐμπόλιον, after lemma insert '= ἐμβόλιον'; for '*casing for dowel*' read '*dowel*' and add '*Inscr.Délos* 104-4.6 (iv BC)'

ἐμπομπεύω, after '*walk in procession*' insert 'Sopat.Rh. 8.178.19 W.'

ἐμπονέω I, for this section read '*work on*, ἀργὸς δὲ ἡ γῆ χηρεύουσα τῶν ἐμπονούντων Alciphr. 3.25.1; w. acc., ἢ ἐνπονεῖν ἢ ἀγοράζειν ἢ κατέχειν δημοσίαν γῆν *SEG* 30.568 (Maced., ii AD)'

ἔμπονος I, delete 'Ezek.*Exag*. 208'

*ἐμπορητικός, ή, όν, = ἐμπορικός, Isid.*Etym*. 6.10.5.

ἐμπορία I 1, at end delete 'cf. D. 56.8'   II, line 2, after 'X.*Vect.* 3.2' insert 'D. 56.8 (pl.)'

ἐμποριακός, ή, όν, prob. f.l. for -ίαρχος (= ‡-ιάρχης), in *SIG* 880.28 (Maced., ii/iii AD).

*ἐμποριαρχέω, *act as* ἐμποριάρχης, *SEG* 34.1107 (Ephesus).

ἐμποριάρχης, for '*supervisor of trade*' read '*headman of an* ἐμπόριον 1 b' and add '*INikaia* 1071.11 (ii AD)'

*ἐμποριδονήτας (ἐμποριοδ- La.)· ἐνοικίου πρακτῆρας Hsch.

ἐμπόριον I 1 a, add '*ISmyrna* 713'

ἔμπρακτος, line 6, after '(D.S. 13.)70' insert 'w. gen., ἔ. παιδείας *educated*, *Vit.Aesop.*(G) 81'

*ἐμπράσσω, perh. *bring about, effect*, ἀπὸ .. αὐτῆς (sc. Ἄτροπος) ἐνπραχθέσεαι [ἐν τῶι βίωι καλὰ *Kafizin* 219 (iii BC).

ἐμπρέπω, line 2 for 'cod. Med.' read 'codd.'

*ἐμπροθεσμί, adv., perh. *punctually*, *PCol*. 176.8 (-εί, AD 325).

*ἐμπρολείπω, *leave behind, predecease*, *Bithynische St.* III 16.7 (iii AD).

ἔμπροσθεν, line 4, after 'E.*Hipp*. 1228' insert 'Arcad. ἴνπροσθε *SEG* 25.447.17 (Aliphera, iii BC)'   II, as prep., 1, add 'Lxx *Ge*. 24.7, 16, etc.'

ἐμπύλαι, delete 'cf. τιτύπαι' and add '(ἐμπύλαιαι (sc. κρῆναι)· νυμφαῖα La.)'

ἐμπυρίζω, after 'ἐμπυρεύω' insert '*PCair.Zen*. 387.3 (iii BC)'

ἐμπυρισμός, add 'used to render Hebr. wd. meaning some sort of blight of cereal crops, Lxx 3*Ki*. 8.37'

ἐμπυριστής, add 'ἐ. ἄγγελον Lxx 4*Ma*. 7.11'

ἐμπυρίφοιτος, for '*dwelling*' read '*going about*' and for 'Orph.*H*. 1.33' read 'Orph.*Εὐχή* 33'

ἔμπυρος I, add '*made with the aid of fire*, Pl.*Plt*. 287e'   III 2, line 4, delete 'hence .. S.*El*. 405'; last line, delete 'dub. sens. in *PCair.Zen.* 14.17'   add 'b *vessel for burning incense*, or sim., S.*El*. 405, *PCair.Zen*. 14.17, *IG* 2².1534.94, *SEG* 29.1205 (Sardis, iv BC).'

ἐμπυρόω, add '*SEG* 37.1003.8 (Sardis, iii/ii BC)'

*ἐμπυρώδης, ες, *marked by inflammation*, τῶν ἐμπυρωδῶν ἑλκώσεων ἢ νομῶν Heph.Astr. 2.13.19.

⁺ἐμύς or ἐμύς, ύδος, ἡ, (also ὁ Arist.*HA* 600ᵇ22, s.v.l.), *marine* or *fresh water chelonian*, *turtle*, Arist.*HA* 558ᵃ8, al., Thphr.*fr*. 171.1 (codd. μῦς).

ἐμφαίνω I 2, after '*display*' insert 'λόγον Emp. 3.4 D.-K., ἐρατὸν δέμας id. 62.7'   4, add '*IG* 5(2).6.24 (ἰνφ-, Tegea); (pass.), *SEG* 37.340.22 (Mantinea, iv BC)'

*ἐμφάλσωμα, v. °ἐμφάρσωμα.

ἐμφανής II 3, add 'sup. θεῶν ἐνφανέστατον, of Antoninus Pius, *AS* 7.147 (Derbe, AD 157), *IGRom*. 3.704 III B.16'   b, line 3, after 'cf.' insert 'ἐμπανία (masc. acc. sg.) δειξάτω *Inscr.Cret.* 4.47.32 (v BC)'   add 'd of documents, (*made*) *public* or *official*, τῶν .. δημοσίων ἀποδείξεων ἐμφανῶν ἐν ὑπομνήμασι γινομένων Just.*Nov*. 46.1.'

ἐμφανίζω, line 2, after '*PSI* 4.400.2 (iii BC)' insert 'Thess. ἐνεφανίσσοεν (= ἐνεφάνιζον, 3 pl.) Schwyzer 590.12 (Larissa, iii BC)'

ἐμφάνισις 3, for this section read '*entering* of documents *into official records*, Just.*Nov*. 15.3 pr.'

ἐμφανισμός, add 'dub. sens., *AAWW* 1962.5.33 (Lycia, i AD)'

*ἐμφανόω, = ἐμφανίζω, *enter private documents in public records*, ἐν ὑπομνήμασιν ἐμφανωθῆναι *Cod.Just.* 4.21.22.7, Just.*Nov*. 134.3.

ἐμφαντάζομαι I 1, before 'M.Ant.' insert 'Alcin.*Intr*. p. 164.13 H.'

*ἐμφάρσωμα, ατος, τό, perh. *inset woodwork*, Didyma 254.4 (ii AD; pl., ἐνφ-); also ἐμφάλσωμα *ABSA* 51.154.

ἔμφασις I 1, line 3, (κατ' ἔμφασιν), after 'id.*Mu*. 395ᵃ29' insert '*obliquely*, Sopat.Rh. 8.229.13 W., al.'   2, after 'Arist.*Div.Somn*. 464ᵇ12' insert 'Men.*fr*. 722.2 K.-Th.' and add '*true sense-impression*, opp. ἀπέμφασις, Carn.ap.S.E.*M*. 7.169'

ἐμφερής III 1, for 'τὰ ἐμφερόμενα .. *matters*' read 'τῶν ἐμφερομένων τοῖς πράγμασι μορίων καὶ τόπων'

*ἐμφθείρω, *cause to perish in*, ἐμφθορέων· ἐμφθειρομένων Sch.Nic.*Al*. 176d Ge.

*ἐμφιέζω, or -άζω, *clothe*, perh. coined (like ἀμφιάζω) as back-formation from ἠμφίεσμαι, etc. (ἀμφιέννυμι); pf. part. ἀγάλματα .. ἐμπεφιεσμένα *ISmyrna* 753.22 (i AD); κλεῖν .. ἐμπεφιασμένην ib. 25.

ἐμφιλόσοφος, add 'ἡ ἐμφιλόσοφος ἐπιστήμη Sch.Iamb.*Protr*. 5; adv. -φως Sch.Luc.*Pisc*. 26, Sch.Aristid. p. 482 D.'

ἐμφιλοχώρως, for 'Ptol.*Harm*. 3.11' read 'Ptol.*Harm*. 2.11'

ἐμφοβέω, add 'οὐκ ἐμφοβήσῃ *you will not be terrified*, Favorin.*Exil*. 25.29'

ἔμφοβος II, adv. -βως, add '*SEG* 26.1717 (Antinoupolis, iii/iv AD)'

*ἐμφορά, ἡ, Cypr. i-po-ra-se, ἰ(μ)φοράς *ICS* 318 (Salamis, c.600 BC), *special contribution*, *SEG* 32.456.26 (Boeotia, iii BC), ib. 23.398 (Thestia, ii BC).

ἐμφόρβιος II, for this section read 'ἐμφόρβιον, τό, *pasture-tax*, *AJP* 56.375 (pl., Colophon, iv BC), Hsch.'

⁺ἐμφορβισμός, ὁ, Arc. ἰνφ-, *imposition of pasture-tax*, *IG* 5(2).3.2 (Tegea, iv BC).

⁺ἐμφορβίω, Arc. ἰνφ-, *impose pasture-tax*, *IG* 5(2).3.3, al. (Tegea, iv BC).

ἔμφορος I, add '2 *paying rent*, *PCair.Zen*. 310.3, 328.133 (iii BC).'   add 'III *full of*, *EM* 677.30.'

*ἐμφράγνυμι, = ἐμφράσσω, Ael.*NA* 4.15 (pass.).

*ἐμφρόνιμος, ον, = ἔμφρων, prob. in *OGI* 383.106 (Commagene, i BC).

ἔμφρων II 2, delete Thgn. 1126; line 4, after '*Hipparch*. 226c' insert 'θεὸς ἔμφρων (app. = Σωφροσύνη) epigr. in *INikaia* 752'   add '3 *mindful*, w. gen., v.l. in Thgn. 1126.'

ἔμφυλος I 1, line 4, after '*kinsman*' insert 'Hes.*fr*. 190.2 M.-W. (prob.)'; line 7, after '(Hierapytna)' insert 'transf., πέτρῃσι βαθείαις ἔ. *at home in*, Opp.*H*. 1.249'

*ἐμφύνω, *arise*, or *grow in*, Aret. p. 22.7, 28.17, 29 H., al; cf. ἐμφύω II 1 (intr.).

ἐμφυσιόω, after lemma insert 'ἐνιφ- *GLP* 123.42'

ἐμφυτεία, for 'Thphr.*HP* 1.6.1, 2.1.4, al.' read 'Thphr.*HP* 2.1.4, *CP* 1.6.1: sg., ib. 1.6.5, 5.6.10'

ἐμφύτευσις, for '*tenure of such a holding*' read '*tenure with right to plant*'   add '2 *planting in* a place, *SEG* 24.614.11 (Maced., ii AD).'

ἐμφυτευτής, add 'ἀναγνώστης καὶ ἐμφυτευτὴς τῆς κτήσεος *Inscr. Olymp*. 656 (v/vi AD)'

ἐμφυτευτικός, add '*si ius* ἐμφυτευτικὸν *vel* ἐμβατευτικὸν *habeat pupillus*, Ulp.*Dig*. 27.9.3.4 (perh. because of early date not connected w. ἐμφύτευσις 1, q.v., but some provincial institution)'

ἐμφυτεύω, line 1, for '*implant*' read '*plant* in a place, ἀμπέλους D.Chr. 7.27; *SEG* 24.614.17 (Maced., ii AD)   II, for this section read '*grant* ἐμφύτευσις *on* land, *PMasp*. 298.39 (vi AD), Just.*Nov*. 7.3.3, *Cod.Just.* 1.2.24.5, Theophil.Antec. 3.24; med., *hold land on such tenure*, *Cod.Just.*, Theophil.Antec. ll.cc.'

ἔμψογος, for 'Call.*Fr*. 121' read 'Call.*fr*. 7.13 Pf.'

*ἔμψογος, ον, = ἐπίψογος, *PTeb*. 276.1 (ii/iii AD).

*ἐμψυκτέον, *one must cool*, Gal. 2.127.14.

*ἐμψυχοποιέω, *endow with life*, Heph.Astr. 3.7.13.

ἔμψυχος 2, add 'Alcid.*Soph.fr*. 15(28) R.'

ἐμψῡχόω, after 'AP 9.774 (Glauc.)' insert 'enliven, Sopat.Rh. 8.58.28 W.'

ἔμψῡχρος, before 'Thphr.' insert 'cj. in'

ἐν, line 3, for 'Inscr.Cypr. 135.9 H., al.' read 'ICS 217.9, 27' and add 'also sts. in Crete, Inscr.Cret. 2.xii 16 Ab 2, 3 (Eleutherna, vi BC), 2.v 1.5 (Axos, vi/v BC)' **A I 1 a**, delete 'ἐν αὑτῷ εἶναι' to end of section and add 'w. place-name indicating place of residence, Νικοβούλη ἐν Ἐφέσοι καλή SEG 31.763(a) (Thasos, iii BC), al.; also w. gen., ἐν κώμης Φεβίχεως μεγάλης CPR 5.15.3, PVindob.Sijp. 7.2, Stud.Pal. 20.127.6 (all v AD)' .. Ar.V. 642' line 9, for 'ἐν αὑτοῦ .. Ar.V. 642' read 'in fig. phr., ἐν αὑτοῦ εἶναι to be in control of oneself, ἐν σαυτοῦ (v.l. σαυτῷ, cf. sense I 6) γενοῦ S.Ph. 950' **5 a**, at end for 'v. ὁ' read 'v. ὁ A VIII 6' **6**, line 8, after 'E.Alc. 278' insert 'ἐν τῷδε κἀχόμεσθα σωθῆναι λόγῳ; id.Heracl. 498 (v. °ἔχω C v)'; add 'ἐν ἑαυτῷ in control of oneself, Hp.Morb. 1.30, X.An. 1.5.17' **11**, line 2, after 'Th. 1.9' insert 'ἐν Διομήδεος ἀριστείῃ Hdt. 2.116' **II 3**, w. subst., (ll. 6 ff.) add 'ἰ(ν) τύχαι i-tu-ka-i Kafizin 169(a).1' **IV 1 b**, add 'IG 9²(1).748 iii 49 (Myanian treaty at Delphi, c.190 BC)' **B**, w. acc., add 'Pi.P. 2.11, 5.38, N. 7.31, etc., Corinna fr. 1 col. i.21 P.; in Boeot., Thess., NWGk. inscr., IG 9²(1).748 iii (Delphi), 31.572 (Krannon, ii BC), IG 9(2).1226.21 (Larissa), etc.; Cypr. i-ta-ti-o-ne ἰ(ν) τά(ν) θιόν ICS 217.27' **C 3**, before 'S.Aj.' insert 'Pi.fr. 70b.10, 11, 15 S.-M.' at end add 'Myc. only in compds., e.g. e-ne-e-si (v. ‡ἔνειμι)'

*ἐναβύσσαιος, v.l. for °ἀβύσσαιος.

ἐναγής **II**, add 'involving such a curse, θεοῖς δ' ἐναγέα τέλεα πελομένων καλῶς ἐπίδρομ' A.Supp. 123'

ἐναγισμός **II**, add 'also sg., SEG 23.207.13 (Messene, Augustus)'

*ἐναγκῡλίζω, med. = ἐναγκυλάω, Poll. 1.136; pass., to be fitted into a loop, Plb. 27.11.5.

ἐναγλαΐζομαι, add 'act., set gloriously in, τὸν νέον εὐσεβέων χώρῳ ἐναγλάϊσον GVI 1154 (Samos, ii/i BC)'

ἔναγχος, add 'Pl.Chrm. 155b, Tht. 147c, D. 11.5, 13.32, etc.'

ἐναγώνιος, add '**IV** alarmed, distressed, Arg.Ar.Pax 1.7.'

ἐναείρομαι, add '(v.l. ἀν-)

ἐναέριος, before 'in the air' insert 'inhabiting or situated' and add 'M.Ant. 12.24'

ἐναιθέριος, after 'in upper air' insert 'Arat. 532'

ἐναίθομαι, add 'BSAA 8.60 (Gizeh)'

*ἐναιμής, ές, covered with blood, Ἄρηος .. ἐναιμέος orac. in Robert DAMM 92 (Syedra, i BC, hex.).

*ἐναίμιος, ον, = ἔναιμος, Trag.adesp. 627 K.-S.

ἔναιμος **I 1**, line 2, after 'Hdt. 3.29' insert 'A.fr. 451m.33 R.'

ἐναίρω, line 1, after 'ἤναρον' insert 'Ibyc. 1(a).2 P.'

ἐναίσιος **I**, before 'D.C. 38.13' insert 'Pl.Lg. 747d' **II**, for ' = foreg. II 1' read 'favourable'

ἐνάκις, for 'ἐννεάκις is' read 'ἐννεάκις Lindos 421a5 (i AD)'

*ἐνᾰκισμύριοι, αι, α, ninety thousand, App.Hann. 6.

*ἐνᾰκόλουθος, ον, accordant, in conformity with, Simp. in de An. 250.16.

ἐνᾰκόσιοι, add 'ἠνακάτιοι SEG 9.2.59 (Cyrene, iv BC)'

ἐνᾰκοσιοστός, delete the entry.

ἐνάλειπτος, for 'anointed with' read 'applied as ointment'; add 'form *ἐνάλιπτος, Myc. e-na-ri-po-to, of a chariot frame, prob. painted'

ἐναλθής, ές, undergoing medical treatment, Nic.Al. 586 (v.l. ἀν-).

*ἐνᾰλῐναιέτης, v. ‡ἁλιναιέτας.

+ἐνᾰλίνω, Cypr. pf. part. pass. i-na-la-li-si-me-na ἰναλαλισμένα write on, inscribe, ICS 217.26.

ἐνάλιος, line 7, after '(lyr.)' insert 'Ζεῦ ἐ. prob. in A.fr. 46a.10 R.'; at end, prose exx., add 'Gal. 10.128.18, 131.3, 15.754.3'

ἐναλλαγή **II**, after 'Lyd.Mag. 2.16' add 'Cod.Just. 6.4.4.23'

ἐναλλοίωσις, after 'alteration' insert 'PSI 483.3 (iii BC)'

ἐνάλλομαι **2**, add 'fig., ὑμῖν ῥήμασιν Lxx Jb. 16.5'

+ἔναλλος, ον, contrary, other, v.l. in Theoc. 1.134 (v. ἄναλλος), AP 5.299 (Agath.); adv. -λως Plu. 2.1045e.

ἐνάμαρτος, after 'faulty' insert 'Cyran. 5 (prolog. 81 K.), Gloss.'

ἐνᾰμείβω, add '**2** intr., alternate, prob. in Pi.N. 11.42.'

ἐνᾰμοιβᾱδίς, add '(codd., edd. ἐπ-)'

ἐνᾰναστρέφομαι, for 'Stob. 3.1.49' read 'Stob. 4.1.49'

*ἐνανατέλλω, rise in, ἐναντέλλουσα Nonn.D. 28.231 (s.v.l.).

ἐνανθρωπέω, add 'Aegyptus 55.60 (version of Creed, cf. homo factus est), cf. Cod.Just. 1.1.5.1'

*ἐνανθρώπησις, εως, ἡ, incarnation, TAM 4(1).372, Cod.Just. 1.1.5.3.

ἐναντίβιος, adv., add 'Od. 14.270, 17.439'

ἐναντιοβουλία, for 'contrary purpose' read 'the fact of intending the opposite of what one says'

ἐναντιόβουλος, for 'of contrary purpose' read 'intending the opposite of what one says'

ἐναντιολογέω, before 'abs.' insert 'Epicur. [31.9]7 A.'

*ἐναντιολόγος, ὁ, person who maintains the contrary, Simp. in Ph. 131.31.

ἐναπερείδω **II 2**, delete 'Pass., to be so fixed .. ib. 23'

ἐναπόγρᾰφος, add 'γεωργὸς ἐ. POxy. 2479.7 (vi AD)'

*ἐναποθέρομαι· ἐνδέχομαι Hsch.

*ἐναπόκλειστος, ον, recluse, Cat.Cod.Astr. 8(1).264.

ἐναπολαμβάνω **I**, after 'Pass.' insert 'Hp.Prog. 11'

ἐναπολαύω, after 'enjoy' insert 'Sopat.Rh. 8.237.24 W.' and after '(vi AD)' insert 'Favorin.Exil. 20.26 B.'

ἐναπομάσσω, line 2, after 'Plu. 2.99b' insert 'rub off an image upon, κατόπτρῳ Ach.Tat. 5.13'; line 4, delete 'to be .. 5.13'

ἐναπονίζω **I**, med., add 'ἐκπώματα Poll. 6.100'

ἐναποσπάω, add 'transf., ἐναποσπωμένων τῶν ἀγροίκων Syria 34.281.28 (Hama, i AD)'

ἐναποτελέω, after 'produced' insert 'in a place'

ἐναποτίθεμαι, line 4, include (in a written work), add 'Just.Const. Δέδωκεν 5'

ἐνάπτω **II**, add 'fig., perh. λυγρὸν θρῆνον ἐναψαμένη epigr. in IG 10²(1).368 (Thessalonica, ii AD).'

ἐνάργεια **II**, add '**b** of the manifestation of a god, ἐποιήσαντο (sc. Zeus and Hecate) προφανεῖς ἐναργείας IStraton. 1104.4 (ii AD).'

ἐναργής **I 1**, line 7, ἐ. τινὰ στῆσαι, for 'to set him bodily before one' read 'to set him in clear view' **3**, add '**b** of style, vivid, D.H.Isoc. 2, Is. 3.'

ἐνάρετος, after lemma insert 'also ἰνάρετος Hsch.' **I**, add '**b** in honorific addresses, τῷ δεσπότῃ μου τῷ τὰ πάντα ἐναρέτῳ Ἀπφοῦτι νο[τ(αρίῳ)] POxy. 1834, 1872, 1873 (all v/vi AD).'

ἐναρηφόρος, add 'Epigr.Gr. 856a7 (Hypate)'

*ἐνᾰρίστερος, α, ον, situated on the left, Inscr.Délos 1441Ai71 (ii BC); adv. -τερα, PCol.Zen. 81.15, al. (iii BC), Inscr.Délos 1439Abá55 (ii BC), Roussel Cultes Égyptiens p. 213 (Delos, ii BC).

ἔναρος, for 'Rev.Ét.Gr. 24.415' read 'Inscr.Cret. 3.4.6.6, 7 (Itanos, iii BC); before 'Hsch.' insert 'Clara Rhodos 2.171.12'

*ἔναρσις, εως, ἡ, uprooting, συκαμίνου Inscr.Délos 356bisA27 (iii BC).

+ἐναροφόρος, ον, = ἐναρηφόρος, Hes.Sc. 192, Alcm. 1.3 L.-P.

*ἐνάρτησις, εως, ἡ, fitting up, installation, μηχανῆς Didyma 39.44 (ii/i BC).

*ἐναρχή, ἡ, period or term of office, SEG 34.1304 (Pamphylia, Rom. imp.).

ἐνάρχομαι **I 2**, add 'pass., περίβολον ἐναρχθέντα SEG 6.424 (Iconium)' **II**, (act.), add 'POxy. 3350.11 (AD 330)'

*ἐνάτειρα, ἡ, sister-in-law, cf. ‡εἰνάτερες, JRS 18.176 no. 51 (Gerasa).

ἐνᾰτεύω, add 'Thasos 10a.1 (iv BC)'

ἐνάτηρ, v. ‡εἰνάτερες.

ἔνᾰτος, line 2, after 'εἴνατος' add 'Ion., Thasos 18.5'; line 3, after 'iii BC)' insert 'Cret. ἤνατος, SEG 9.72.102 (Cyrene, iv BC), Inscr.Cret. 4.181.5 (ii BC); εἴνατον ἡμιτάλαντον 8½ talents, Hdt. 1.51.2; fem. subst. (sc. ἡμέρα), Hes.Op. 772, SEG 21.541 ii 27 (Attica, iv BC); as market-day (cf. Lat. nundinae), ib. 1149 (Magnesia on the Maeander, AD 209)'

ἐναυγής, delete the entry.

ἐναύλιος, add '**II** epith. of Zeus, perh. as god of the farm, Robert Hell. 10.34 (Thrace).'

ἐναυλιστήριος, for 'habitable' read 'providing lodging or shelter' and for 'Antip.' read 'Antip.Sid.'

ἔναυλος (A) **II**, line 3, after 'HF 371' insert 'of mountain caves, A.R. 1.1226 (pl.)' **III**, delete the section.

ἔναυλος (B) **I 1**, for 'flute' read 'aulos' **2**, line 6, after 'Plu. 2.17d ' insert 'cf. Call.fr. 384.6 Pf.'

+ἔναυσις, εως, ἡ, kindling, ὑδάτων τε πηγαίων ‹ὀχετείαν› καὶ πυρὸς ἔναυσιν Plu.Cim. 10. **2** incitement, stimulus, Did.ap.Porph. in Harm. p. 26 D.

*ἐναυτόθι, adv. at the said place, PTeb. 798.5 (ii BC).

ἐναύω (A), line 9, delete 'of poets, draw inspiration' and for following quotation read 'Ἔφεσον ὅθεν περ οἱ τὰ μέτρα μέλλοντες τὰ χωλὰ τίκτειν .. ἐναύονται Call.fr. 203.14 Pf.'; penultimate line, delete 'ἐπαύω'

*ἐναφῆλιξ, ικος, ὁ, ἡ, = ἀφῆλιξ II, minor, PFam.Teb. 7.22 (ii AD)

ἐναφίημι **II**, add 'ἀπελθοῦσα τὴν μορφὴν ἐναφῆκέ μου τοῖς ὀφθαλμοῖς left the imprint of her beauty in my eyes, Ach.Tat. 1.19 (v.l. ἐπ-)'

*ἐναφωρισμένος, η, ον, separated, Plot. 6.4.12; cf. ὁρίζω, ἀφωρισμένως.

ἔνβενος, add '(ἔνβεννος La.)'

ἐνδᾱεῖ, add '(ἐνδαμεῖ· ἐμμένει La., but perh. dat. of *ἔνδα(F)ής, cf. δάος)'

+ἐνδαίνῠμαι, feast on, ἐνδαινυη[.. Thasos 128 (iv BC); Ath. 7.277a (s.v.l.).

ἐνδαίω (B), delete the entry.

ἔνδακρυς, add 'Ezek.Exag. 211 S.'

*ἐνδαναία· ἐρημία Hsch. (La.).

ἐνδᾰπιος, line 1, before 'Mosch. 2.11' insert 'Nicaenet. 6.4'

+ἐνδατέομαι, assign, apportion, βέλεα θέλοιμ' ἄν .. ἐ. S.OT 205, ἐ. λόγους ὀνειδιστῆρας E.HF 218, λόγου ὑπόστασίν τινι Aristid.Quint. 2.2; pass. ἐνδεδασμέναι ἡλικίαι Iamb.VP 31.201. **2** mention specially, τὸ δυσπάρευνον λέκτρον ἐ. S.Tr. 791; in dub. context, δίς τ' ἐν ἐνδαψιλεύομαι SUPPLEMENT ἐνθεμολογέω

τελευτῇ τοὔνομ᾽ (sc. of Polynices) ἐνδατούμενος A.*Th.* 578; pass. Nic.*Th.* 509.   **II** *divide into pieces*, Lyc. 155.
ἐνδαψῐλεύομαι I, add 'Suid. s.v. Φιλίσκος Κερκυραῖος'
ἐνδεής 6, line 3, after 'E.*fr.* 898.8' insert 'ἐ. πράττειν Men.*Dysc.* 280'
ἕνδεκα II 2, add 'also at Thespiae, *SEG* 32.504 (iv BC)'
*ἐνδεκάβριος, app. = *November*, *SB* 9529.4 (vi/vii AD).
*ἐνδέκαρχος, ὁ, *member of a board of eleven magistrates*, *SEG* 23.271.54 (Thespiae, iii BC).
ἑνδέκατος, add 'in legionary title, ἐν λεγιῶνι ἑνδεκάτῃ Κλαυδίᾳ *SEG* 24.911 (Thrace, iv AD)'
ἔνδενδρος, for 'prob.' read 'rest., nisi leg. Ἐ[λαστέ]ρο, *c.*500 BC'; add 'γῆς ἐνδένδρου *wooded*, *SEG* 38.619.7, 13, 18 (Cassandrea, iii BC)'
ἐνδέξιος I 2, line 4, for 'ἐνδεξία' read 'ἐνδέξια' and after '*Inscr.Prien.* 19.46 (iii BC)' insert 'τῆς ἐνδέξια φλιᾶς *Inscr.Délos* 1413*b*27, 1439*Abc*i54 (ii BC)'
*ἐνδέσμιος, ον, *tied in a bundle*, Sor.ap.Gal. 13.42.15.
ἔνδεσμος I, for this section read '*anything tied up, bundle, package*, Lang *Ath.Agora* XXI B9 (v BC), ἐ. ἀργυρίου Lxx *Pr.* 7.20, Dsc. 3.83' and transfer 'Lxx 3*Ki.* 6.10, al.' to section II.
*ἔνδεσμος (B), ον, *tied* or *bundled up*, Luc.*Lex.* 10.
ἐνδεύω (A), add '*soak* with, πατρὸς κόλπους ἐνιδεύσας αἵματος .. νοτίσιν *GVI* 874.7 (Smyrna, ?ii AD)'
ἐνδέχομαι III 3, acc. abs., add 'Arist.*de An.* 407ᵇ21, etc.'
ἐνδέω (A), after lemma insert 'pf. part. pass. ἐνδεμένον *SEG* 37.1001.9 (Katakekaumene, Lydia, iii AD)'   **I**, line 8, for 'also οὐρανὸς [ἀστράσιν]' read 'κύκλοις οὐρανὸν'   **III**, for this section read 'med., *bind* by spells, ἐνδούμενοι τὰ δαιμόνια J.*AJ* 8.2.5'
ἐνδέω (B), line 4, for 'indentical' read 'identical'; line 5, after '(Pl.)*Phd.* 74d' insert 'abs., Aeschin.Socr. 8 D.'
ἐνδημία, form ἐνδαμία, add 'also other non-Att.-Ion. dialects, *SEG* 31.574.7 (Larissa, ii BC), etc.'; add 'of life on earth, opp. death, *ICilicie* 59 (v/vi AD)'
ἐνδιαβάλλω 2, for '*stand in the way as an adversary*' read '*cause to turn aside*'
ἐνδιαίτημα, add '*deck-house* on a boat, Arr.*fr.* 19 (pl.)'
*ἐνδιαιτητήριον, τό, *dwelling-place*, Sch.Hes.*Th.* 129 (p. 29.8 G.).
ἐνδιαλλάσσω, after 'pass. -αγμένος, ὁ' read '*male prostitute*, Aq. 3*Ki.*22.47; also by analogy -αγμένη, ἡ, (*female*) *prostitute*, Aq.*Ge.* 38.21, *De.* 23.18'
ἐνδίδωμι V 2, add 'Aeschin.*Ep.* 9.1'   **3**, add 'Al.*Le.* 14.56'
*ἐνδικτίων, v. °ἰνδ-.
ἐνδῑνεύω, for '= sq.' read '*go round and round* in a place'
ἔνδιος II 2, delete the section.
ἐνδίφριος, after 'αὐτῷ' insert 'ἱκέτης'
ἔνδοθεν 2, add 'Simon. 74.6 P.'
ἔνδοθι 1, add 'κεῖται .. τῷδ᾽ ἔνδοθι τύμβῳ *SEG* 34.1271 (Paphlagonia, iii AD)'   **2**, add '**b** w. acc., *to within*, A.R. 4.1235.'
†**ἐνδοθίδιος**, v. °ἐνδοσθ-.
*ἐνδομῡχία, ἡ, (?) *secretiveness*, Antioch.Astr. in *Cat.Cod.Astr.* 11(2).109.6; pl., expl. of *latebrae* in Virgil gloss., *PNess.* 1.278, 360 (vi AD).
ἐνδόμυχος I, add '**3** *domestic*, Cic.*Att.* 5.14.3, 5.21.14.'
ἔνδον I, line 2, before 'D. 27.10' insert 'Pl.*Smp.* 213c' deleting this ref. in line 6   add '**6** like ἐντός I 3, *on this side of*, Plu.*Cim.* 13.4.'
ἐνδοξάζομαι, add '(first two exx. perh. med., *display one's glory*)'
ἔνδοξος I 1, add 'as honorific form of address, app. = Lat. *illustris*, *Cod.Just.* 1.4.22 pr.; sim. sup. ἐνδοξότατος ib., *Just.Nov.* 22.14, *POxy.* 1974.1, 1982.3 (v AD), *PLond.* 1780.80 (vi AD); = μεγαλοπρεπέστατος (Lat. *magnificentissimus*), *Cod.Just.* 4.59.1.1; also = Lat. *gloriosissimus*, *Just.Nov.* 7 epilogus'
†**ἐνδόρωμα**, ατος, τό, app. *plastering of lime* on the inside of a sarcophagus, σὺν τῷ ἐνδορώματι (unless in error for ἔν‹δον› δορώματι) *ISmyrna* 243 (see also ib. II(2) p. 369).
ἔνδος, after '(Delph.)' insert '*SEG* 9.11ff. (Cyrene, iv BC)'
ἐνδοσθίδια, delete the entry.
*ἐνδοσθίδιος, α, ον, *belonging inside* the building, *Leg.Gort.* 2.11 (sp. ἐνδοθ- for ἐνδοσθ-); neut. pl. subst. = ἐνδόσθια, *entrails*, *IG* 4².40.17 (Epid.); see also ἐντοσθίδια.
ἐνδόσιμος I, add '**2** neut. subst., *period of grace, respite* (= Lat. *laxamentum*), *Just.Nov.* 72.6.'   **II**, add '**2** ἔστω δὲ ἐνδοσιμώτερα τὰ γυμνάσια *less strenuous*, Philostr.*Gym.* 52.'
ἐνδοτέρω I 1, add '*Peripl.M.Rubr.* 40'
ἐνδρομίς II 2, after '(Herod.Med.ap.Orib.) 38.1' insert 'κρεβατταρία *Edict.Diocl.* 19.5'
ἔνδῠμα, add '[ἔνδυμα *AP* 6.280; cf. ὑπένδυμα q.v.]'
*ἐνδῠνᾰμόω, = ἐνδυναμόω, Heph.Astr. 3.4.4.
ἐνδυναστεύω I, add 'Procl.*Inst.* 159 D.'
ἐνδυτή, ἡ, *dress*, *PHarris* 88.22 (v AD)
ἐνδυτός 2, line 2, for 'Simon. 179.10' read '*AP* 6.217.10 ([Simon.])'
ἐνδύω I 1, line 10, after '*Ep.Eph.* 4.24' insert 'ἐνέδυ θεόν *APL* 290 (Antip.Thess.)'   **2**, line 7, after 'Pl.*Phd.* 89d' insert '*make an* *indentation* in coastline, κόλπος .. ἐπὶ βάθος ἐνδύνων εἰς τὴν ἤπειρον *Peripl.M.Rubr.* 32, 42'
*ἐνδωρότερον· ὠμότερον. ἀκρατέστερον Hsch.; cf. ζωρός.
ἐνέδρα, add '**IV** *anus*, Hsch. s.v. ῥινοβόλους ἀνέμους; cf. ἔνεδρος II.'
ἐνεδρεία I, add '*Vit.Aesop.*(G) 55 P.'
ἔνεδρος I, add '**2** *holding office*, *SEG* 33.1177.44 (Myra, i AD).'
ἐνείλω, for '*wrap up*' read '*roll up tight*'
ἔνειμι, line 1, after 'ἔνι' insert '(written ἔνει *Inscr.Délos* 4442*A*8, 68 (ii BC))'; line 2, delete '3 sg. ἔνι freq. for'; line 4, after 'Od. 21.288' insert '*repeated*, S.*fr.* 314.191, Ar.*Lys.* 545'   **II 2 a**, at end for 'Plb. 21.4.14' read 'Plb. 21.2.14'   at end add 'Myc. 3 pl. pres. *e-ne-e-si*, part. *e-ne-o*'
ἕνεκᾰ, line 5, after 'Sch.Pi.*O.* 7.10)' insert 'also ἕνεχεν *MAMA* 1.197 (Laodicea Combusta)'; line 14, after 'Hdt. 3.122, etc.' insert 'Mitchell *N.Galatia* 322'   before '**1**' insert 'also app. w. acc. εὐχὴν ἕνεκεν Mitchell *N.Galatia* 198, μνήμην ἕνεκεν ib. 277'   **II 1**, for 'Call. .. 287' read 'Call.*fr.* 75.6 Pf. (ἕνεκ᾽), ib. 1.3 (εἵνεκ᾽), cf. Hes.*fr.* 180.10 M.-W.'   **2**, add 'Call.*fr.* 6 Pf. (ἕνεκ᾽)'   at end add 'Myc. *e-ne-ka*, prep. w. gen., preceding case'
ἐνελίσσω 1, line 3, delete 'ὀλίγῳ .. Nic.*Al.* 287'   add '**3** med., *whirl round* in, ὀλίγῳ ὄγμῳ Nic.*Al.* 287.'
ἐνεματίζω, add '*use as an enema*, Herod.Med.ap.Aët. 9.2'
*ἐνενηκονθεκταῖος, ον, (occurring) *on the ninety-sixth* (*day*), Gal. 7.501.10.
*ἐνενηκονθήμερος, *of* or *lasting ninety days*, *PMich.*III 149 xi 19 (ii AD); ἐνενηκοντάμερος Petos. 359.2.
*ἐνενηκονταέξάπηχυς, υ, *ninety-six cubits long*, *PSI* 905.6 (i AD), represented by symbols in *PTeb.* 382.9 (i BC) and *PMich.*II 121ʳ.3 xi 1 (i AD).
*ἐνενηκοντάμερος, v. °ἐνενηκονθή-.
*ἐνενηκοστόεκτος, η, ον, *ninety-sixth*, *PWisc.* 9.15, 22 (ἐνενηκόσεκτος pap., AD 183), *PKöln* 94.27, 40 (prob. ἐνενηκόσθεκτος pap., AD 213).
ἐνεξουσιάζω II, for '*usurp authority*' read '*stand on one's rights*'
ἐνεός 1, add '*PMich.*XV 723.8 (iv AD)'
ἐνεοστᾰσία, for '*standing dumb*' read '*inability to utter*, Κύπριν δ᾽ ἐνεοστασίη λάβε μύθων'
*ἐνεπίσκημμα, ατος, τό, *claiming as owed to oneself some part of property confiscated from another by the state*, Harp., Poll. 8.61; *such a claim on behalf of a tribe*, *IG* 1³.429.8 (v BC), *Ath.Agora* XIX P26.512, 528 (iv BC).
ἐνεπισκήπτομαι 1, for '*claim .. state*' read '*claim as owed to oneself some part of property confiscated from another by the state*'; after 'Poll. 8.61' insert '*make such a claim for a tribe* or *religious group*, κοινὸν ὀργεώνων *Ath.Agora* XIX P5.31; ib. P26.515 (both iv BC)'
ἐνέπω, line 3, delete '(anap.)'
ἐνεστιάομαι, for '*give an entertainment*' read '*feast*'
ἐνεύχομαι, add 'w. ὥστε and inf., *SEG* 30.1382.C20, *IG* 12(3).330.14 (Thera); w. inf. alone, *IG* 5(1).1208.50'
ἐνεχυρασμός, after '= ἐνεχυρασία' insert '*PEnteux.* 87ᵛ.3 (iii BC)'
*ἐνεχῠρέω, = ἐνεχυράζω, prob. in *SB* 9834 b 48 (iv AD).
ἐνεχῠριάζω, add 'aor. inf. -ιάξαι'
*ἐνεχυρίαξις, ιος, ἁ, *taking of a pledge*, *BCH* 50.16 (Delphi, iv BC).
*ἐνεχυρίζομαι, = ἐνεχυράζομαι (v. -άζω), Hsch. s.v. διενεγ(γ)υήσαμεν.
*ἐνεχυρῐμαῖος, α, ον, *pawned, held in pawn*, βοῦς ἐνεχυριμαῖος τὰ πολλὰ ἔξω βλέπει· ἐπὶ τῶν ἀμελουμένων καὶ πρὸς τοὺς ἐπιμελουμένους ἀφορώντων *Paroemiographi* suppl. p. 75 no. 30.
ἐνέχῠρον, after 'τό' insert 'Boeot. ἐννέχυρον *Mél.Navarre* 353 (Thespiae), Cret. ἰνέκυρον inscr. in *JHS* 69.34 (Oaxos, vi/v BC)'; add 'in Rom. law = Lat. *pignus*, *Cod.Just.* 4.24.1, *Just.Nov.* 8.5 pr., al.'
*ἐνήκοος, ον, *obedient*, w. dat. and gen., *JA* 1958.3.10 (Kandahar, Inscr. of Aśoka).
ἐνήλᾰτον I, for 'S.*Fr.* 315' read 'S.*fr.* 314.316 R.'
*ἐνήλᾰτος, gloss on γλίσχρος, Hsch.
ἔνηλιξ, for '*in the prime of manhood*' read '*adult, grown-up*' and add 'μήπω ἐ. οὖσα *SEG* 37.1000 (Katakekaumene, Lydia, ii AD)'
ἔνημαι, add **ἔνηαι**.
*ἐνήμῑσυ, v. °ἰνίμῑνα.
†**ἐνήνοθε**, v. ἀνήνοθε; also in compds. ἐπ-, κατ-, παρ-ενήνοθε.
*ἔνηφι, v. ἔνος (B).
ἔνθᾰ I 1, add 'in epitaphs, τίθησί με .. ἔνθα *ITrall.* 219 (Tralles, i AD), *SEG* 30.1479 (Phrygia, before AD 216), ἔνθα κίτη Θεώδωρος *ITyr* 9 (Chr.)'
ἐνθάδε, after lemma insert 'Cypr. *i-ta-de* ἰνθάδε *ICS* 11, 213a'
*ἐνθᾰλᾰμεύω, *embed*, (λίθος) ᾧ ἡ βήρυλλος ἐνθαλαμεύεται Eust.ad D.P. 1010.
ἐνθαστικός I, line 3, after 'Syrian. *in Metaph.* 42.14' insert 'ἔχειν ἐ. Men.*Dysc.* 44'
ἐνθεμέλιοι, add '*IEphes.* 1073, *SEG* 34.1127 (Ephesus)'
*ἐνθεμολογέω, *collect (financial) deposits*, or perh. *produce*, *SEG* 14.547 (Poeëssa, Ceos, v BC, see also *ABSA* 84.295 n. 39); cf. θημολογέω.

117

ἐνθένδε, add 'late ἐνθένδεν Chor. 28.2 cod. (p. 312 F.-R.)'
ἐνθερμαίνω, add 'ἐξ ὧν ἂν διψῶεν οἱ ἐνθερμανθέντες Gal. 17(2).103.8'
†ἐνθηκάριος, ὁ, perh. *storeman, SEG* 19.553 (-ις, Cos, Chr.).
ἐνθήκη **I**, add '*SEG* 16.754 (Phrygia)' **II**, after '*capital*' insert '(financial)' and add 'Hsch. s.v. ἀφορμή. **2** perh. *security or deposit*, δοὺς ἐνθήκην ὅσα ἔχω ἐν χερσίν *PHaun.* 21.9.'
ἔνθηρος **I**, add 'ἐσθημάτων τιθέντες ἔνθηρον (i.e. verminous) τρίχα A.*Ag.* 562, ἐνθήρου (i.e. maggoty) ποδός S.*Ph.* 698, πᾶν ὅσον ἔνθηρον ἦν τοῦ τόπου Ael.*NA* 6.63' **II**, delete the section.
*ἐνθρήνιον, τό, *wild bee's nest*, Sch.Nic.*Al.* 547b.
ἐνθρίζειν, for 'ἐνέθριξε' read 'ἐνέθριξεν (ἐνέθρεξεν La.)'
ἔνθρυπτος, τὰ ἔ., delete 'sops or' and add 'Aristid.*Or.* 46'
ἐνθρώσκω, line 4, delete '*kicked* him on the hip'; after 'D.C. 74.14' insert 'comm.in Alc. 306.9.4 L.-P.'
ἐνθύμημα **II**, for '(X.*HG*) 5.4.52' read '(X.*HG*) 5.4.51'
ἐνθυμία, add 'v.l. in Longus 4.17.1'
ἐνθυμιστός, delete '(nisi leg. -ητόν)' and add 'ἑ. αὐτῷ ἔστω Thasos 141.5 (iv bc); ἑ. εἶναι *BCH* 64-5.176 (Thasos, i bc)'
*ἐνθώϊος, ον, = ἐπιθώϊος, Thasos 150.14 (iv bc).
ἐνιαύσιος **II**, after '*annual*' insert 'ὀλολύγα Alc. 130.35 L.-P.' at end add 'cf. Myc. *e-ni-ja-u-si-jo*, pers. n.'
ἐνιαυτοκράτωρ, add 'Heph.Astr. 2.27.6'
ἐνιαυτός **I 2**, line 10, for 'five' read 'four' **II**, delete the section.
ἐνίημι **I 8**, delete the section.
*ἐνικεύθω, v. °ἐγκεύθω.
*ἐνικλάω, v. ‡ἐγκλάω.
ἐνικνέομαι, add 'cod. in Thphr.*CP* 5.13.1'
*ἐνικρίνω, v. °ἐγκρίνω.
*ἐνίλλω (B), v. ἐνείλλω.
†ἐνιλλώπτω, *eye lasciviously*, Ael.Dion.ι 8 E., Clem.Al.*Paed.* 3.70 (who equates with βλεφαρίζω).
ἐνίστημι **A 2**, add '*Cod.Just.* 1.2.25 pr., med., Just.*Nov.* 101 pr.' **3**, line 7, after 'Plu.*Arat.* 16' insert 'ἐνστήσας βασκανίην *AP* 5.218.10 (Agath.)'; add 'also pres. οὐκ ἀγνοῶ .. ὅσον ἔργον ἐνίσταμαι Isoc. 12.36'
*ἐνίψογος, v. ἔμψογος.
*ἐνλιμενίζω, v. °ἐλλιμ-.
ἐνναέτης (A), add 'also ἐννηέτης *IHadr.* 69 (Rom.imp.); ἐννεέτης *GVI* 1118.3 (Piraeus, iv bc); see also ἐννεαέτης'
ἐνναέτης (B), after '= ἐνναετήρ' insert 'Epigr.ap.D. 7.40, Call. in *PAnt.* 113.1(a)13'
*ἐνναία, ἡ, perh. *spring, fountain*, *SEG* 19.181.1, 182.1, also perh. in *IG* 2².2491.7 (Attica, iv bc), Phot.; see also °ἐνναί.
*ἐνναούγκιον, τό, = Lat. *dodrans, three-quarters* of an inheritance, Just.*Nov.* 18.2, al.
*ἐννάπηχυς, v. ἐννεάπηχυς.
*ἐνναύων· πρὸς τῷ ναῷ διάγων, ἱκετεύων Hsch. ε 2715 (La.); cf. ἐναύω C, ναεύω, ναύω II.
ἐννέα, line 1, for 'Dor... (q.v.)' read '*hεννέα Tab.Heracl.* 2.17, al., *IG* 4.1588.16 (Aegina, v bc), WGk. ἐννή *CID* II 5 ii 43, 52 (iv bc), *SEG* 9.1.32, ἔννηα ib. 9.3.16 (both Cyrene, iv bc); cf. ἐννῆ· θ' Κυρηναῖοι Hsch.' at end add 'Myc. cpd. form *e-ne-wo-*'
*ἐννεάδραχμος, ον, *costing nine drachmae, IG* 2².408.13 (iv bc).
ἐννεακαιδέκατος, add 'fem. subst. (sc. ἡμέρα) *the nineteenth day*, *SEG* 32.1149 (Magnesia on the Maeander, ad 209); neut. as adv. *for the nineteenth time*, *AS* 36.137ff., no. 1 (Pisidia, 5/4 bc)'
ἐννεακαιδεκέτης, add 'form -δεχέτης *GVI* 1214.3 (Pholegandros, ? iii/ii bc)'
*ἐννεάκεντρος, ον, *having nine* (i.e. *many*) *stings*, Sch.Nic.*Th.* 781 = Nic.*fr.* 37; cf. ἐννεάδεσμος.
ἐννεάκις, after '= ἐνάκις' insert '*Lindos* 421*a*5 (i ad)'
ἐννεάπηχυς, add 'also ἐννάπηχυς *PFlor.* 167.14, 215.3 (iii ad)'
*ἐννεάπυλον, τό, *precinct with nine gates*, of the Pelasgikon at Athens, *AB* 419.
*ἐννεάχίλιοι, αι, α, *nine thousand*, (see also ἐννεάχιλοι), epigr. in *SEG* 30.483 (Thespiae).
*ἐννεέτης, v. °ἐνναέτης (A).
ἐννεόβολον, delete the entry.
*ἐννεόβολος, ον, *accruing at the rate of nine obols*, τόκοι *BGU* 1161.10 (i bc).
*ἐννεόμφαλος, ον, "*having nine knobs*", name of kind of sacrificial cake, πόπανον ῥαβδωτὸν ἐννεόμφαλον *Inscr.Perg.* 8(3).161.
*ἐννεός, v. ἐνεός.
†ἐννῆ, v.l. for ‡ἐννέα.
*ἐννηετής, v. °ἐνναέτης (A).
*ἔννιος, wd. of unkn. meaning, Hdn.*fr.* p. 27 H.
*ἐνναί· πηγαί Hsch. (La.) (cod. ἐννοιαί, cf. °ἐνναία, νόα).
*ἔννοσος, ον, *morbid*, σαπρία Aët. 7.11.
ἐννύμι, line 7, after 'Od. 14.529' insert 'Arc. 3 sg. aor. subj. Ϝέσετοι

(unless = Ϝέσετοι, subj. of pres. *Ϝέστοι, cf. ἐπίεσται s.v. ἐπιέννυμι) *SEG* 11.1112 (Pheneos, *c.*500 bc)'
ἐνό, ἔνο, add 'graffito in *ZPE* 12.268 (Sicily, *c.*420 bc)'
*ἐνοδιάζω, app. = ἐφοδιάζω, τὸ ὀφίλον τούτοις ἐνοδιάζζεσθαι ἀργύριον παραδόντω (sic) *Lindos* 419.70 (i ad).
ἐνόδιος **I 1**, add 'οἳ πρῶτοι κακοεργὸν ἐχαλκεύσαντο μάχαιραν εἰνοδίην i.e. the highwayman's knife, Arat. 132' **II**, Ἐνοδία alone, add 'Ἐπάγαθος δῶρον Ἐννοδίᾳ *SEG* 31.1584; w. epiths., Ἐννοδίας Μυκαικᾶς *IGC* p. 11.28 (Larissa), Ἐ. Ἀστικά *IG* 9(2).575, etc.'
ἔνοδμος, for '*sweet-smelling, fresh*' read '*strong-smelling, pungent*'
†ἔνοδος, ἡ, prob. Thess. for εἴσοδος, *way in, entrance*, *IGC* p. 11.14 (Larissa).
ἐνόζυγος, delete the entry.
*ἐνοικήτειρα, ἡ, *inhabitant* (fem.), [πό]ντου .. ἐνοικήτιραι .. [Ν]ηρῇδες epigr. in *Inscr.Cret.* 3.iv 37.1.
*ἐνοικιάζω, *let*, *PLond.* 1735.11 (vi ad), v. *Kapsomenakis* 97.
ἐνοίκισμα, add '*PMich.*III 188.16 (ii ad)'
*ἐνοικοδόμημα, ατος, τό, *building in a place*, cj. in Pl.*Lg.* 760e (pl., codd. ἐν οἰκοδ-).
ἔνοικος **1**, add '**b** *worker in an establishment*, Just.*Nov.* 43.1.2.'
*ἐνόλμιον, τό, part of the ὅλμος (in a reed-pipe, v. ὅλμος II 5), Hsch. s.v. καταστομίς (ἐνόλβιον cod.).
*ἐνομματίζω, = ἐνομματόω, *Hymn.Is.* 18 (Maroneia).
ἐνοπή **2**, add 'Emp. 62.8 D.-K.'
*ἐνόπλιος **I**, after 'Anon.Vat. 64' insert 'δρόμος Didyma 201.14'
ἔνοπλος **III**, for '*portrait-statue in armour*' read '*representation on a shield*'
ἐνοπτρικοί **I**, after '*geometers*' insert 'Alex.Aphr. *in Mete.* 141.17'
ἐνοργείας, add '(ἐνορταλίας La.)'
*ἐνόρκιος **I 1**, for this section read '*bound by an oath*, Pi.*O.* 2.101, *Inscr.Olymp.* 22(f).6 (decree of Megara Hyblaea and Selinous, vi bc); *faithful to an oath*, *SEG* 30.1117.8, 1118.10 (both Entella, iii bc)'
ἔνορκος **II a**, add 'used to express Latin *foederatus*, (πόλεις) ὅσαι ἑαυτὰς ἐγκεχειρίκεσαν ἐπὶ συνθήκαις ἔνορκοι App.*BC* 1.102'
ἐνορμάω, 'rush upon, w. dat., ἐνθορεῖν καὶ ἐνορμ[ῆσαι τ]οῖς τοῦ Φιττακοῦ νώτ[οις comm.in Alc. 306.9.5 L.-P.'
ἐνόρμιον, after '*harbour-dues*' insert '*BGU* 1834.11 (i bc), *PFay.* 104.8 (iii ad)'
*ἐνορταλίας, v. °ἐνοργείας.
ἔνορχις, add '**II** an unidentified precious stone, *enorchis* (sc. λίθος) *candida est divisisque fragmentis testium effigiem repraesentat*, Plin.*HN* 37.159.'
ἔνος (C) **2**, line 3, after 'acc. to' insert 'Plu.*Sol.* 25.4'; lines 4/5, for 'Hes.*Op.* 770' read '*IG* 2².1241.26, 28, Sch.Ar.*Nu.* 1131; but in Hes.*Op.* 770 *the first of the month*
*ἐνουλίζομαι, add 'perh. also of a woollen textile, [ζώμ]ατα δύο ἐνουλισ[μένων] Τηνίων *SEG* 37.692.43 (Delos, ii bc)'
ἐνούσιος **3**, delete the section.
ἐνοχή, add 'Just.*Const.* Δέδωκεν 7e; also in uncertain sense, Διονύσις τὴν ἐ. ἀνέθηκεν Robert *Hell.* 10.14 (Phrygia)'
ἐνοχία, add '*PMich.*xv 740.25 (vi ad)'
ἐνοχλέω **I 1**, line 9, *be unwell*, add '*PPetr.* 2.25(a).12 (iii bc), *PCair.Zen.* 812.5 (257 bc)' add '**b** *subject to pressure*, ἐνοχληθέντες παρὰ τῶν .. ἐπισκόπων *Cod.Just.* 1.3.45.7; w. dat., *put pressure on*, Just.*Nov.* 80.3.'
*ἐνοχλής, ές, *troublesome*, *PLond.*inv. 2226.17 (*Pap.Flor.* xix 517, ad 308).
ἔνοχος, after lemma insert 'Arc. ἴνοχος *SEG* 25.447 (Alipheira, iii bc)' **I 1**, add 'abs., *held fast*, of anchor, *AP* 7.506 (Leon.Tarent.)'
ἐνόω, add '*cause* a wound *to close*, Hld. 1.8.5'
ἐνραβῶς, after lemma insert '(ἐνραβδώσας La.)'
*ἐνρευματίζομαι, *to be full of or affected by rheum*, *APF* 4.270 (iii ad).
ἐνρυθμος, *possessing rhythm*, add 'τραγῳδίας epigr. in *SEG* 31.1072 (Herakleia, Pontus, Rom.imp.)'
ἔνσειμι, after 'Cret.' insert 'and Arg.'; after '*Leg.Gort.* 5.36' insert '*Schwyzer* 84.21 (cj., ἐνς[lapis, v bc)'
ἐνσεισμός, delete 'of engines of war'; add 'cf. ἐνσείω II'
ἐνσείω, line 2, after 'ἐνσέσεικα' insert 'Men.*Dysc.* 581' **4**, add 'Gal. 18(2).922.9, 924.3'
ἔνσιμος **I**, delete 'f.l.' add '**III** of garments, *striped* or perh. *embroidered*, *POxy.* 1273.13, 14 (iii ad).'
*Ἐνσιτάρχιος, epith. of Zeus, perh. *ruler of the guests*, Διϝὸς Ἐνσιταρ[χ]ίου *SEG* 30.351 (Kleonaia, iv bc).
*ἔνσκαμβος, ον, *crooked*, gloss on ἔγγαυσος, Hsch.
*ἐνσκηνόω, v.l. for σκηνόω, Lxx *Ge.* 13.12.
†ἐνσκολιέύομαι, perh. *twist oneself, twist and turn*, Lxx *Jb.* 40.19(24) (passage corrupt in Hebr.).
ἐνσκοπέομαι, substitute 'in Hld. 8.10.1 is f.l. for ἀνασκ-'
†ἐνσόριον, τό, *niche* in a sepulchre, *ISmyrna* 190, 192, al., ib. 211 (sp. ἐσσόριν), *AJA* 18.68 (Sardes, iii ad).

ἐνσπόνδια   SUPPLEMENT   ἐξάδραχμον

*ἐνσπόνδια, τά, = σπονδαί, ἐ. ποίμεν Schwyzer 491.14 (Thespiae, ?ii BC).
ἔνστᾰσις **I 3**, *inheritance*, add '*Cod.Just.* 1.2.25 pr.' **II 2**, at end delete 'ὀνύχων'
*ἐνστρᾰτεύομαι, *serve in a (military) unit*, οὐδὲ χρήματα εἰσῆγον ἐφ' ᾧ ἐνστρατεύσασθαι Agath. 5.15.3 K.
ἐνστρέφω **2**, delete the section.
ἐνστρηνές, after lemma insert '(ἐνστηνές La.)'
ἐνσχίζω, add '*PMich*.IX 576.3 (iii AD)'
ἐνσωμᾰτόω, pass., add 'Alcin.*Intr.* p. 178.10, 36 H.'
*ἐνταλματικός, ὁ, unexpld. wd., app. expr. some rank or occupation, *BCH* 33.86 (Cappadocia, Byz.); cf. perh. ἐντολικάριος.
ἔντᾰσις **II**, after 'Hp.*Epid.* 3.1.β'' insert 'Gal. 11.207.11'
ἐντᾰτός, after '*stretched*' insert '*IG* 2².1541.23-25 (Athens, iv BC)'
ἐντᾰφή, add '**2** *tomb*, τεθῆναι αὐτὸν ἐν τῇ ἐ. *IEphes.* 614c.26 (i AD).'
ἐντᾰφιαστής, for '*undertaker, embalmer*' read '*one who prepares for burial*' and for 'of the Bactrian dogs' read 'of Bactrian dogs, reputed to be kept for the disposal of corpses'
ἔντεα, sg., read 'Archil. 5.2, 139.5 W., Hes.*fr.dub.* 343.18 M.-W.'
ἐντείνω **III 2**, after 'pass.' insert 'ἐντέταμαι *APl.* 236 (Leon.Tarent.)'
ἐντέλεια **I**, for '*GDI* 1339.11 (Dodona)' read '*SEG* 23.471 (iv BC), 26.701 (c.205 BC), 37.511 (iv BC, all Dodona)'
⁺ἐντελέω, *complete* term, πεντεάδας δισσὰς ἐντελέσας ἐτέων epigr. in *SEG* 33.848 (Mauretania, i AD); in uncertain context, Phld.*Ir.* 12 W.
ἐντέλλω **II**, delete the section transferring exx. to section I.
ἐντέμνω **II 2**, for this section read '*tap* (the root of) plant to extract the sap, Thphr.*CP* 6.11.14, 15, al.: so fig., ὕπνον τόδ' ἀντίμολπον ἐντέμνων ἄκος (also expld. as "shredding") A.*Ag.* 17'
ἐντενής, delete 'only' and add 'adv. ἐντενίως *Inscr.Cret.* 4.168.10 (iii BC)'
⁺ἐντέριον, τό, dim. of ἔντερον, in quot. used in a slighting description of sexual intercourse, ἐντερίου παράτριψις M.Ant. 6.13.
*ἐντερίς, ίδος, ἡ, = αἱμορροΐς I, Cyran. 29.27 (pl.).
ἐντερειδής, add '(s.v.l.)'
ἔντερον **II**, delete 'but *worm-casts*'
ἐντερόνεια, delete '= ἐντεριώνη'
ἐντεσίεργός, add 'also –ουργός Hsch.'
*ἐντετραίνω, *pierce*, aor. 1 part. ἐντετράνας *IG* 2².1665.18, 1672.176 (both Athens, iv BC).
ἐντεῦθεν, after 'ἐνθεῦτεν' insert 'ἐντεῦθε dub. in Chor. 29.50 (p. 327 F.-R.)'   **I**, at end delete 'ἐντεῦθεν .. Men.*Pk.* 184'
ἐντευκτικός, after '*affable*' insert 'Phld.*Rh.* 1.222.10, 13'
ἔντευξις **2 b**, for '*manners, behaviour*' read '*manner of encounter* or *converse*'
ἐντευτλανόομαι, delete 'Aret.*CA* 1.2'
ἐντεύχω, after 'ἐρυθήματα' insert 'Aret.*CA* 1.2 (p. 101.18 H.)'
ἐντηρέω, delete 'Procop.*Arc.* 4'
ἐντίθημι, line 1, for 'aor. 1 inf.' read 'aor. 2 inf.'   **2**, add '**c** *place in the tomb*, *TAM* 4(1).264 (Rom.imp.), *SEG* 30.1557 (Cilicia, iv/vi AD).'
*ἔντῐμος, ον, = ἔντιμος, sup. -ώτατος *IG* 12(7).410.8 (Aegiale).
ἔντιμος **I 1**, add 'ἔνθα κῖτε ὁ 'Ρεββὶ Ἀββᾶ Μάρις ὁ ἔντιμος *SEG* 29.968 (Naples, iv/v AD)'   **III**, add '*highly-priced*, ἔντιμον κατέλιπον τὸν σῖτον D. 56.9'
*ἐντίναγμα, ατος, τό, *a hurling, shower*, χαλάζης Aq.*Is.* 28.2, cf. 32.2; v. ἐντιναγμός.
ἐντολή, add '*instruction* in will, *TAM* 5(1).18; also representing Lat. *mandatum*, Just.*Nov.* 4.1'
ἐντολικάριος, add '*PCol.* 175.50, 63, 70 (AD 339)'
*ἐντόλιον, τό, dim. (in form) of ἐντολή, *PBremen* 20.8 (ii AD).
⁺ἐντόλιος, ον, prob. = φιλέντολος, *SB* 2654 (Jewish).
ἐντομίς **I**, add 'Lxx *Je.* 16.6; perh. also in this sense, ὁ πρῶτος ῥυμὸς σὺλ λίνοις καὶ πίττει καὶ κηρῷ ὥστε μέσον εἶναι τῶν ἐντομίδων τῶν ἐν τεῖ θεραπείαι *Inscr.Délos* 1444*A*a19 (ii BC)'   **II**, for this section read '*niche* or *cavity* in a tomb, *IG* 10²(1).308, 478, al. (Thessalonica)'
ἔντομος, lines 4/5, for 'Call. .. x 38' read 'Call.*fr.* 43.80, 694 Pf.'
ἔντονος **I 2**, line 5, after 'Pl.*Tht.* 173a' add 'ἔ. τι φθεγξαμένη Nic.Dam. 68 J.'
*ἐντορνεία, ἡ, app. *defensive breastwork* of warship, *SB* 9215.11 (250 BC), v. ἐντορνία.
ἐντός **2**, add 'Isoc. 4.144, 15.110'   **II**, lines 3/4, after 'id. 2.76' insert 'cf. ἐκ (or ἐγ) τῆς ἐντός *Didyma* 25*A*22, *B*20, al. (ii BC)' and for 'ib. 49' read 'Th. 2.49.2'
⁺ἐντοῦθα, = ἐνταῦθα, Schwyzer 792 (Cumae, v BC), 811.17 (Oropus, iv BC, ἐντόθα); Schwyzer 792a (Cumae, iv BC, v. *SEG* 4.93).
ἐντράπελος, for '*shameful*' read '*false, deceitful*'
ἐντρέφω, lines 2/3, delete 'Med. .. :–'; line 4, after '*in*' insert 'Hp.*Aër.* 12' add '**2** med., φυτὰ ἐνθρέψασθαι *bed* them *in*, Hes.*Op.* 781.'
ἐντρέχεια, line 2, after 'Gal. 14.213, cf. 306' insert 'as honorific title,

POxy. 3350.6 (AD 330), διὰ τοῦτο ἐπιδίδωμι τῇ ὑμῶν ἐντρεχείᾳ *PMasp.* 91, 92, al. (vi AD)'
ἐντρῐβής **2**, add 'Max.Tyr. 12(6).2'
ἐντροχάζω **II**, delete the section.
⁺ἐντρυγάω, *gather in* grapes or other fruit, Moeris s.v. ἄρριχος, *PPrincet.* 39.7 (iii AD, ἐντρυκ-).
ἐντρύχομαι, delete the entry.
*ἐντρῡχόω, *harass, wear down*, [πόλις] πολεμίοις -ωθεῖσα Memn. 20.3 J., D.C. 38.46 (codd.).
ἐντύλη, delete the entry (read 'ἐν τύλῃ')
*ἐντυλιγμός, ὁ, *swaddling cloth*, gloss on σπείρημα, Sch.Nic.*Al.* 417b.
ἐντύνω, line 4, after 'A.R. 1.235' insert '4.1191'   **I**, at end transfer 'A.R. 4.1191' to follow 'A.R. 1.235'
ἐντῠπάζω, for '*enwrap* .. (Pisidia)' read '*carve* or *mould on*, *TAM* 3(1).922.2 (Termessus, ii/iii AD)'
ἐνυαίνειν, after lemma insert '(ἐνυᾱνεῖν La.; cf. ὑηνέω)'
Ἐνυᾱλιος, line 2, for 'written' read 'τὸνυFαλιδ (gen.) ἱαρά *SEG* 11.327 (Argos, vii BC), ἘνυFαλίο ib. 23.187 (Mycenae, vi BC)'; line 4, after 'cf.' insert '*SEG* l.c.'   **II**, line 3, for '*Lyr.adesp.* 108' read '*Lyr.adesp.* 109(b) P.'; at end for 'ῠ .. metri gr.' read 'ῠ, exc. *Lyr.adesp.* l.c.; cf. Ἐνυώ' at end add 'Myc. *e-nu-wa-ri-jo*'
*ἐνυβριστής, οῦ, ὁ, gloss on λωβητῶν, Hsch.
ἔνυγρος, add '**V** ἔ. (sc. λίθος), ἡ, a precious stone, perh. *chalcedony*, Plin.*HN* 37.190 (v.l. *enydros*).'
ἔνυδρις **I**, read 'cj. in Ar.*Ach.* 880'
ἔνυδρος, add '**5** v. °ἔνυγρος.'
ἐνυπνιόμαντις, add 'cj. in Semus 20 J. (ἡ ἐν ὕπνῳ μάντις in Ath. 335a)'
Ἐνυώ, after 'Il. 5.333' insert 'Ἔνhυο[ῖ] *SEG* 35.1014 (Naxos, Sicily, c.600 BC)' and for 'ib. 592' read 'Il. 5.592'
*ἐνφ-, v. ἐμφ-.
ἐνωδάς, after lemma insert '(ἐνώλας La.; cf. οὖλος B)'
ἐνώδιον, read 'cf. ἐνώδιον (cf. *IG* 2².1544.20 (iv BC))' and add 'written ἐνόδιον *SB* 7260.2*b*8, 3*a*2 (pl.) (i AD)'
ἐνωμοτάρχης, lines 2/3, transfer 'Arr.*Tact.* 6.2' to follow 'Ascl.*Tact.* 2.2'
ἐνωνά, after 'Chaeronea' insert 'ii BC' and add '*SEG* 34.355.8 (Leuctra, iv BC), *SEG* 39.400 (Orchomenos, 225/200 BC)'
ἐνώπιος **I**, add '*SB* 7817.56 (AD 201)'   **II**, as prep. w. gen., add 'Lxx *Ge.* 11.28, etc.; w. ref. to personal service, ib.*Jd.* 20.28, 1*Ki.* 16.21; *in the eyes* or *opinion of*, translating Hebr. b° ʾênê, ὁδοὶ ἀφρόνων ὀρθαὶ ἐνώπιον αὐτῶν ib.*Pr.* 12.15, εἰ εὗρον χάριν ἐνώπιον τοῦ βασιλέως *Es.* 5.8'
ἕξ, of Thess. ἔξε, *SEG* 13.394 (iii BC); Myc. in cpd. *we-pe-za*, prob. *six-foot*'
ἐξάγγελος **II**, add 'Hsch.'
ἐξάγιον, after 'τό' insert '(Lat. *exagium*)'   add '**II** *part payment*, *POxy.* 3955 (AD 611).'
⁺ἐξάγιον, τό, = Lat. *sextula*, Orib. in *CMG* vi 2.2 p. 231 R., Suid. s.v. στατήρ.
*ἐξάγισις, εως, ἡ, perh. *purification*, *IG* 1³.8.23 (Sunium, v BC).
ἐξάγιστος **II**, for this section read '*taboo*, ἃ δ' ἐξάγιστα μηδὲ κινεῖται λόγῳ S.*OC* 1526, cf. Hsch.'   add '**III** perh. *deconsecrated*, *SEG* 34.116.26 (Eleusis, 394/3 BC), σίγλοι καὶ ἀσκοὶ ἐξάγιστοι *IG* 2².1544.22 (329/8 BC), cf. ib. 1453.10; v. *Hesperia* 25.100/1.'
*ἑξάγκαλος, ον, (ἀγκάλη III) *containing six sheaves*, *PBerol.* 13062ᵛ in *PMilan.* pp. 27-29.
ἐξᾰγορεία, for '*excantation* .. *confession*' read '*confession of sin* as means of obtaining cure of disease, coupled with θεοφορία and with ἐνθουσιασμός' add 'Lxx *Ps.Sal.* 9.6'
ἐξᾰγόρευσις **II**, after 'Ptol.*Tetr.* 154' add '(v.l., ed. ἀγορία)'
ἐξᾰγορευτής, for 'one .. sins' read '*one who practises* ‡ἐξαγορεία, δεισιδαίμονας ἱεροφοιτοῦντας ἐ.'
ἐξᾰγορευτικός, add '**2** perh. *of confession-cures*, δρόμου ἐξαγορευτικοῦ Ἀφροδείτης *POxy.Hels.* 23.30 (AD 212).'
ἐξάγω **I 2**, line 5, after 'Chrysipp.*Stoic.* 3.188' insert 'Plu. 2.242d'   **II**, add '*IG* 1³.61.37, al. (v BC)'   **III**, add '**2** *utter a judgement*, *Cod.Just.* 7.45.15.'
ἐξᾰγωγεύς **II**, add 'perh. in *Inscr.Délos* 399*A*98 (ii BC)'
ἐξᾰγωγή **I 3**, *exportation*, add 'ἐσαγωγὰν καὶ ἐξαγωγάν *SEG* 32.1586 (Lindian decree, Naucratis, iv BC)' add '*IG* 1³.236.9 (v BC)'
ἐξᾰγωγίς, after '*drain*' insert '*IG* 4².116.2 (Epid., iv/iii BC)'
ἑξάγωνος **I**, add '**2** ἑξάγωνον, τό, *hexagon*, *PKöln* 52.15, 17, 19, etc. (AD 263).'
ἑξαδάκτυλος, add 'see also ‡ἑγδάκτυλος'
*ἑξαδελφιδῆ, ἡ, *great-niece*, rest. in Dain *Inscr.du Louvre* 56.4 (Miletus).
*ἑξαδελφιδοῦς, ὁ, *great-nephew*, *BCH* 87.203 (Delphi, iii AD).
*ἑξάδελφος **I**, fem., add ''Ηραῒς ἡ ἑξάδερφος τοῦ ἱερέος *IEphes.* 3415 (ii/iii AD); cf. mod. Gk. ξάδερφος.'
*ἑξαδραχμιαία, ἡ, = ἑξαδραχμία, *OBodl.* 1078 (ii AD).
ἑξάδραχμον, add 'see also °ἑκδραχμον'

119

ἑξάδραχμος, ον, *costing six drachmae*, συρίας ἑ. rest. in *PHib.* 51.6 (iii BC).

ἐξᾰδῠνᾰτέω, line 2, after 'Plu.*Alc.* 23' insert 'w. gen., ἑ. τοῦ προτιθέμεν *IG* 9²(1).583.9 (Acarnanian decree at Olympia, iii BC); περί τινος ib. 69'

ἐξάδω **I 2**, for '*sing the* ἔξοδος .. *chorus*' read '*sing aloud*' and add '(s.v.l.)' **II 1**, add 'περίπτωμα σχοῦσα καὶ ἐξασθεῖσα *TAM* 5(1).331'

*ἐξαέτοι, 3 sg. subj., perh. *pay a fine, IG* 5(2).261.10 (Mantinea, v BC); cf. °ἀφάεσται, ἀφαέτοι.

ἐξαιθρᾰπεύω, add '*Amyzon* 2 (321/320 BC)'

ἐξαιθριάζω, after 'pass.' insert 'ὕδωρ' and for '*Com.adesp.* in *PLond.ined.* 2294' read '*Com.adesp.* 274.20 A.'

ἐξαιμάτωσις, add '*PMich.*III 149 iv 21 (ii AD)'

ἐξαίρεσις **II**, before '*place where cargoes are landed*' insert '*unloading*, *Didyma* 39.37 (ii BC)'

ἐξαιρετός **I**, line 2, for 'ἐξαίρετα' read 'ἐξαιρετά' **II 2**, after 'And. 3.7' insert 'ἑ. [νῆες] *reserve squadron, IG* 2².1612.39, al. (Athens, iv BC)' **3**, add 'ἐξαιρετόν, perh. as adv. in *PWarren* 19.1 (ii AD)'

ἐξαιρέω **II**, add '**3** *make an exception of, exempt, CodJust.* 1.3.41.8, 11.1.1 pr., al.' **IV 1**, add 'Isoc. 17.14'

ἐξαίρω **II 1**, line 2, delete 'ἐκ'

ἐξαΐω, delete the entry.

ἐξᾰκανθίζω, for '*pick out .. holes in*' read '*deck with thorns;* fig., *denigrate*'

ἐξᾰκανθόομαι, add 'Gr.Nyss.*Ep.* 28 (p. 83.10 P.)'

*ἐξᾰκεράτιον, τό, *measurement of weight* = 6 κεράτια, *PFlor.* v 112.6, 113.8, 116.10, 122.9.

*ἐξᾰκεστήρ, ῆρος, ὁ, *the one who remedies evil*, τρεῖς θεοὺς ὀμνύναι κελεύει Σόλων, ἱκέσιον καθάρσιον ἑ. Sol.*Lg.* 44b R.

ἐξᾰκεστήριος, after '*remedying evil*' read 'as a divine epith., v.l. in Sol.*Lg.* 44b R. (v. °ἐξακεστηρ), θεοί D.H. 10.2.6, Ἐξακεστήριος· ὁ Ζεύς. καὶ ἡ Ἥρα Hsch.'

*ἐξᾰκισχῑλιοστός, ή, όν, *six-thousandth*, Hsch. s.v. κοδράντης.

ἐξᾰκολουθέω **1**, add '**b** *take up* an action, e.g. a fight, instituted by another, *SEG* 31.122 (Attica, ii AD).'

ἐξᾰκόσιοι, form Ϝεξακάτιοι, add '*SEG* 23.393 (Corcyra, late vi BC)'

ἐξάκουστος, add '**3** *heard out, listened to*, πᾶσιν ἀνθρώποις Cyran. 25.'

*ἐξακτορεύω, *hold office of* ἐξάκτωρ, *POxy.* 2110.18 (iv AD).

*ἐξακτορικός, ή, όν, *of an* ἐξάκτωρ, τάξις, rest. in *POxy.* 126.4 (vi AD).

ἐξᾰλᾰπάζω, at end after 'Theoc. 2.85' insert '(v.l.)'

*ἐξάλειπτρον, for '*unguent-box*' read '*unguent-flask* or *-basin*' add '**2** perh. *trinket-box*, ἐξάλειπτ[ρον τ]ορευτό[ν] χρυσοῦ[ν] ἐν ὧι βασ[ίλει]ον *PLond.* 1960.12.'

ἐξαλλᾰγή **2**, add '**b** *departure from common idiom*, D.H.*Dem.* 13.'

ἐξαλλάκτης, add 'Hsch. s.v. κλοτοπευτής'

*ἐξάλοβος, ον, *six-lobed*, Ar.Byz.*Epit.* 2.168.

ἐξᾰλος, add '**II** *away from the sea*, Eust. for ἐξ ἁλός in Od. 11.134, 23.281.'

ἐξᾰμαρτάνω **II**, for 'ἐξημαρτήθη' read 'ἐξαμαρτηθῇ' and add to the ref. '(s.v.l.)'

ἐξᾰμάω (B), for '= ἐξαφύσσω' read '*scrape out*'

ἐξαμβλόω, line 2, delete 'prob. f.l. for -ῶσαι in'

ἐξαμβλύνω, for 'Dsc. 1.88' read 'Dsc. 1.69.4; τοὺς δικαστάς *PLit.Lond.* 138 viii 7'

*ἐξᾰμεινόω, *amend, improve* an expression, Cratin. 171.72 K.-A.

ἐξάμετρος, add 'ἐξάμετρον, τό, *measure of six metra, CPR* 6.74.7 (AD 301)'

*ἐξᾰμηνιεῖος, α, ον, = ἐξάμηνος **II**, *PCair.Zen.* 340.5, 27 (iii BC).

ἐξάμηνος **2**, for 'ἑ. (sc. χρόνος), ὁ' read 'ἑ. (sc. περίοδος), ἡ' and after '*half-year*' insert 'Hdt. 4.25'; delete this ref. at end of section and substitute '*SEG* 33.464 (Larissa, 27 BC)'

✝ἐξαμναῖος, α, ον, = ἐξάμνους: πολῖται ἑ., citizens who have acquired their status for six minae, or perh. who have a potential annual income of six minae, *IEphes.* 2001.9 (iii BC; v. *SEG* 37.882).

*ἐξάμυξος, ον, *having six wicks*, λύχνοι *IUrb.Rom.* 106.9; cf. μύξα II 2.

ἐξαναβαίνω, add 'Il. 24.97 (v.l. εἰσανα-)'

*ἐξάναγκος, ον, *necessary*, ἐξάνανκα πραττέσθω *SEG* 31.122 (Attica, ii AD).

*ἐξαναπάλλω, med. sync. aor. κορυφᾶς .. ἐξανέπαλτο *sprang up out of*, Ibyc. 17.4 P.

ἐξαναριθμέω, for '*IGRom.* 4.661.34' read '*IGRom.* 4.661.3'

*ἐξαναστησείω, *want to get up again*, ἱλυσπώμενον δὲ αὐτὸν ἔτι καὶ ἐξαναστησείοντα Agath. 3.4.6 K.

*ἐξαναστροφή, ἡ, *turning upside-down, total change*, Sch.Hes.*Th.* 253b (p. 52.7 G.).

ἐξανατέλλω **1**, add '**b** *cause to arise, bring into view*, νῦν δὲ Τιμόθεος μέτροις ῥυθμοῖς τ(ε) ἑνδεκακρουμάτοις κίθαριν ἐξανατέλλει Tim. *Pers.*231 P.'

*ἐξανατρέφω, *bring up, rear*, πεδὰ τὸν κοινὸν ἐξαναθρέψας epigr. in *SEG* 26.645 (Thessaly, *c.*AD 400).

ἐξαναφαίνω, add '**2** *give birth again to*, of a mother, *CEG* 605.1 (Piraeus, iv BC).'

*ἔξανδρος, *bereaved of a husband*, Hsch. s.v. χήρωσε.

ἐξανέχω **II**, add 'Ar.*Nu.* 1373'

ἐξανθέω **II 1**, after 'of wine' insert 'perh. *lose its bouquet*'

ἐξάνθημα, add '**II** ἀκάνθης ἐξανθήματα *thistledown*, Hsch. s.v. γήρεια.'

ἐξανίημι **I 1 a**, after 'Call.*Del.* 207' insert 'ἐξανέηκεν ἄναξ .. πηγήν epigr. in *Didyma* 159 iii 7 (iii AD)'

ἐξάντλησις, add '*SB* 10283.10 (ii BC)'

*ἐξαξεστιαῖος, ον, = ἐξάξεστος, οἴνου ἐξαξεστιαῖα ἐβδομηκονταεννέα *PMich.*XIII 674.5 (vi AD).

ἐξάξεστος, add '*PMich.*XIII 667.17 (vi AD)'

*ἐξαούγκιον, τό, *six-twelfths*, i.e. *one-half* (of an inheritance), Just.*Nov.* 18.1; also -ούγγιον ib. 2 pr. 1.

ἐξαπάτη, add 'D. 20.98, 45.46'

*ἐξαπελεύθερος, ὁ, *freedman*, Syria 27.238 (iii AD); cf. ἐξαπελευθερόω.

*ἐξαπηθέω, *strain off* from something, Nic.*Th.* 707-8 (tm.).

*ἐξαπλησον, aor. imper. *multiply by six*, perh. irreg. for ἐξάπλωσον (ἐξαπλόω) *PCair.* 10758 (vi AD).

Ἐξάπολις, add 'of Greek cities on the Euxine, *IGRom.* 1.634.3, al.; in Cyrenaica, *SEG* 20.727.3 (ii AD)'

ἐξαπόλλυμι **II**, add 'Emp. 12.2'

ἐξάπρυμνος, for '*stems*' read '*sterns*'

*ἐξάπτερος, *six-winged*, Theognost.*Can.* 89.

ἐξαπτέρυγος, add '*Orac.Tiburt.* 44'

ἐξᾰράομαι **II**, delete the section (v. °ἐξαρέσκομαι).

*ἐξάρεον, τό, *central area in a temple enclosure*, Jo.Mal. 9.287 (v. *ZPE* 43.140).

ἐξᾰρέσκομαι **2**, add 'Aeschin. 3.116'

*ἐξαριθμοζῠγοκαμπανοτρῠτᾰνίζω, *count and weigh out with a steelyard*, (?)Jul.*Ep.* 205 (s.v.l., ed. ἐξ ἀριθμοῦ ζυγῷ Καμπανῷ τρυτανίσας).

ἔξαρνος, line 4, before 'ἑ. ἦν τοῦ φόνου' insert 'τοῦ δὲ τρίτου (sc. ταλάντου) ἔξαρνος γίγνεται Isoc. 21.3'

ἐξαρπάζω, line 7, after 'S.*OC* 1016 (s.v.l.)' insert 'D. 18.133'

ἐξαρτίζω **I**, after '(Mytilene)' insert 'a statue, *SEG* 24.1038 (Callatis, iii AD)' **II**, for '*equip and dispatch*' read '*furnish for a voyage*'

*ἐξαρτιστήρ, ῆρος, ὁ, *fitter*, Ὀλυμπίῳ ἐξαρτιστ(ῆρι) *PRyl.* 641.11 (iv AD).

ἐξαρτιστήριον, after '*equipment*' insert '*PRyl.* 641.8 (iv AD)'

ἔξαρχος **2**, line 2, after '*chorus*' insert 'E.*Ba.* 140'

ἐξάρχω **4**, *rule*, add 'Just.*Nov.* 6 pr.'

ἐξάς, add '*SEG* 32.1601.24 (Nubia, iv/v AD)'

ἐξασελλάνωμεν, add '(ἐξασελγαίνωμεν La.)'

*ἔξασθμα, ατος, τό, *exhalation*, *PMag.* 13.10.

*ἐξατίλιον, τό or -ιος, ὁ, perh. a fish (?cf. Lat. *saxatilis*), *PRyl.* 630*. 332, al. (iv AD).

*ἐξάτμῐσις, εως, ἡ, = ἐξατμισμός, Zos.Alch. 138.5 (lemma).

*ἑξάτροχος, ον, *having six wheels*, Poliorc. 239.9.

✝ἐξαυλέω, *wear out by playing the pipe*, ἐξηυλημέναι γλῶτται αἱ παλαιαί Poll. 4.73.

ἐξαυστήρ, for '[ἐξ]αυστήρ .. ib. 689' read '*IG* 2².1640.27, 1641.39'

*ἐξαύστης, ου, ὁ, = ἐξαυστήρ, *Inscr.Délos* 372*B*26, 379.15 (ii BC).

ἐξαυτῆς, delete 'Thgn. 231, Aen.Tact. 22.29'

ἐξαῦτις **III**, add 'A.R. 4.455'

ἐξάχαλκος, delete the entry.

ἐξάχειρ, after '(Luc.)*Tox.* 62' insert 'epith. of Hecate, *PMag.* 4.2119'

ἐξᾰχοίνικος, add '*PTeb.* 210, *PTurku* 1.34 (*Tyche* 6.101; both ii BC); -ον, τό, measure of six choenices, *IEphes.* 3437'

ἐξᾰχῠρόω, read 'ἐξᾰχῠριόω'

ἑξᾰχῶς, add 'ἅρματα ἑ. ἁμιλλώμενα *six at a time*, D.C. 75.4'

*ἑξδέκατος, η, ον, *sixteenth*, *GVI* 440.4 (Cappadocia, ?ii/iii AD).

*ἔξε, v. °ἕξ.

*ἐξέβενος, ἡ, precious stone, *exhebenum .. candidam, qua aurifices aurum poliant*, Plin.*HN* 37.159.

*ἐξεγγόνη, ἡ, *great-granddaughter, BCH* 87.202 (Delphi, iii AD).

✝ἐξεγγύη, ἡ, *surety, security*, Is. 5.3.

ἐξεγείρω, add '**II** *erect* a building, *Salamine* 204 (v AD).'

ἐξέδρα, ἡ, delete separate definitions and substitute '*an open recess* or *alcove with seating* (whether accommodating small group or large gathering)'

ἔξεδρος **I 1**, for 'id.*Rh.* 1406ᵃ31' read 'Alcid.ap.eund.*Rh.* 1406ᵃ31' **2**, metaph., add 'Ath. 5.187f'

ἐξεικάζω, line 2, after 'X.*Hier.* 1.38' insert 'med., *model*, τὰ πόπανα πρῶτος ἐξηκάσατο *IG* 2².4962.17 (iv BC)'

ἐξεικάττοι, for '*Supp.Epigr.* .. ii BC)' read '*BCH* 66/7.144 (iii BC)' and add 'also ἔξεικατι (sic) *SEG* 26.672.27'

ἐξείλησις, for '*release, escape* from' read '*twisting free* of'

ἐξείλλω, add 'cf. ἐγείλασαν'

*ἐξειλώτισεν· ὡς ἐπὶ †τοῦ λωτοῦ† ἐξεπόρθησεν Hsch. (but prob. fr. ἐξειλωτίζω, make into a helot).

ἔξειμι (B), add 'ἐξῆν· ἐξεγένετο· ἢ δύνατον ἦν Hsch. (La.)'

ἐξείρω II, delete the section transferring quots. to section I.

*ἐξελαίωσις, εως, anointing with oil, Anon.Alch. 345.4.

ἐξελεύθερος, line 1, after 'freedman' insert 'SEG 22.509 (Chios, vi BC), IEryth. 2.B21 (v BC)'

ἐξέλικτρον, add 'PCornell 29 (iii/iv AD)'

ἐξελίσσω I 1, line 5, after '(Plot.) 2.4.9' insert 'Procl.Inst. 93 D.'

*ἐξενεχυρασία, ἡ, = ἐνεχυρασία, APF 15.93.5 (ii AD); cf. ἐξενεχυριάζω.

*ἐξενιαυτέω, serve to the end of one's year of office, -ῆσαι τὴν πρυτανείαν Sch.Pi.N. 11.10.

*ἔξεο, = ἐκ σέο, IG 12(5).472.13 (Oliarus), 14.2012Cb2 (Rome), SEG 19.456.7 (Thrace, ii AD).

ἐξεπομβρέω, for 'rain on' read 'send rain'

ἐξέραμα, add 'Philum.Ven. 4.13'

*ἐξέρανος, ὁ, expelled from the ἔρανος, SEG 31.122.44 (Attica, ii AD).

ἐξεράω II 2, add 'med., pour out for oneself, ἔλαιον ἐξηρασάμην Com. adesp. 289.17 A.'

ἐξεργάζομαι I 5, add 'POxy. 2812.19'

ἐξεργάτης, delete the entry.

ἐξέρεισμα, add 'in pl., buttresses, SEG 4.270 (Panamara)'

ἐξερεύγομαι I, add 'λόγους APl. 328 (Barb.)'

ἐξερέω (B) II 2, add 'A.R. 2.695, 4.1546'

*ἐξέρκετον, τό, = Lat. exercitus, army, POxy. 3872 (vi/vii AD).

ἐξέρπω, line 1, aor., add 'ἐξήρψα Lxx Ps. 104(105).30' III, for this section read 'w. internal acc., crawl, swarm with, ἐξῆρψεν ἡ γῆ αὐτῶν βατράχους Lxx Ps. 104(105).30'

ἐξέρχομαι I 1 c, add 'in chariot-racing, leave the starting traps, Hesperia 54.221 no. 6, Tab.Defix.Aud. 234-238 (all iii/iv AD)' III 1, add 'of conditions, be fulfilled, CodJust. 6.4.4.17a'

ἐξεσία, add 'nom. pl. prob. in Call.fr. 80+82.22 Pf. (Add. II)'

ἔξεστι, line 1, imper., add '(ἐξήτω SEG 20.94.13 Cilicia, ?ii AD)'; inf., add '(WIon. ἐξεῖν TAPhA 65.105.10 Olynthus, iv BC)'; add 'ἐξόν in periphrastic constr. w. εἶναι (expr. or understood), ἡ .. δούλη .. ἐτόλμησεν .. ἰδεῖν ἃ οὐκ ἐξὸν (sc. ἦν) αὐτῇ Is. 6.50, καὶ οὐκ ἦν ἐξὸν καταπεσεῖν Luc.As. 16, Modest.Dig. 27.1.13.5'

ἐξετάζω, after 'ἐξετάσω' insert 'Ar.Ec. 729, etc.'

*ἐξεταιρέω, prostitute, ἐπὶ τῶν ἐκδεδιητημένων καὶ ἐξηταιρημένων ἀρρένων Suet.Blasph. III (tit.) T.

*ἐξετάσιμος, ον, subject to scrutiny; n. pl. subst., of documents, SB 7173.29 (ii AD).

ἐξέτασις, line 4, after 'D. 18.246' insert 'Call.Epigr. 59.3 Pf.'

ἐξέτεροι, delete the entry (v. °ἐξέτερος).

*ἐξέτερος, η, ον, some other, ἐξετέρην θανάτου φύξιν Nic.Th. 588; pl., νοῦσοι ἐξέτεραι ib. 744; masc. pl. subst., ib. 412.

ἐξέτι, add '2 app. in local sense, from, A.R. 1.976 (s.v.l.).'

ἐξευλαβέομαι, add 'Plu. 2.31b, 85e, al.'

*ἐξευχαριστέω, give token of gratitude, prob. in MAMA 4.288 (Dionysopolis).

ἐξέφηβος, add 'Milet 1(7).203b26 (ii BC)'

ἐξέχω I 2 b, for '(Ar.)fr. 389' read '(Ar.)fr. 404 K.-A., Stratt. 48 K.-A.'

ἐξηγητεία, add 'POxy. 2127.6 (iii AD)'

ἐξηγητής II, add '4 local official in Egypt, Ἑρμοπολίτου (sc. νομοῦ) PAmh. 85.2, cf. 86.1 (both i AD); of Alexandria, Str. 17.1.12, BGU 1073.3 (iii AD).'

ἐξηγητικός, add 'III of or belonging to an ἐξηγητής, ὑπηρέτης PTeb. 397.28 (ii AD); ἐξηγητικόν, τό, the board or body of ἐξηγηταί, POxy. 1413.9 (iii AD).'

ἑξηκονθημερίσια, delete the entry (v. °ἑξηκονθημερήσια).

*ἑξηκονθημερήσια, τά, provision of sixty days' quarters and forage, PCair. Zen. 341(b).5 (iii BC; sp. -ίσια)

ἑξήκοντα, line 1, for 'Ϝεξ-' read 'Ϝεξέκοντα' and add 'Ϝεξέϟοντα SEG 23.393 (Corcyra, 525/500 BC)'

*ἑξηκοντάδραχμος, ον, of sixty drachmae, Stud.Pal. 5.101.7.

*ἑξηκοντᾰείς, μία, έν, sixty-one, SEG 26.672.35 (Larissa, early ii BC).

ἑξηκοντάς II, before 'Str.' insert 'Eratosth.ap.'

ἐξήμαρε, add '(ἐξημάρευσε La.; cf. ἀμαρεύων ib.)'

ἑξῆς, line 1, after 'ἑξείης' insert '(ἑξέης perh. in SEG 37.575.6 (Chalcidice, iv BC), ἑξείας Isyll. 79 = IG 4².128.74)' and after 'Dor.' insert 'ἑξᾶς ASAA 14-16(1952/4)290.663 (Rhodes)'; line 5, for 'πάντας ἑ. .. 657.2' read 'κατὰ τὸ ἑξῆς Peripl.M.Rubr. 7, 15, al.' 3, delete 'καὶ τὰ ἑ. .. etc.' 4, add 'c καὶ τὰ ἑ. and so on in sequence, PTeb. 319.34 (iii AD), Longin. 23.4, etc.'

ἐξηχευη, delete the entry (v. °ἐξηχεύομαι).

*ἐξηχεύομαι I = βακχεύομαι, Hsch. s.v. βακχευθεῖσα. II = stupeo, Gloss.

ἐξιδιάζομαι 1, add 'Phryn. 172 F. (cited as non-Att.)'

ἐξιδιόομαι, add 'J.AJ 1.6.2, al. 2 = ἐξιδιάζομαι 2, ib. 14.1.3.'

*ἐξιεριστέω, Dor. ἐξιαρ-, = ἐξιερισετύω, Tit.Cam. 40.15 (iii BC).'

ἐξίημι I 2, delete 'ἐς θάλασσαν'

ἐξικνέομαι II 3, for 'abs.' read 'w. adv. or advl. phr.' and transfer 'E.Ba. 1060' to exx. w. gen. in section II 2.

*ἐξιόντως, adv., perh. in course of time, MAMA 6.83 (Attouda).

ἐξισόω I 1, line 7, after 'rival' insert 'Sapph. 96.22 L.-P.'

ἐξίστημι B II 1, add 'abs., = Lat. bonis cedere, Cod.Just. 9.4.6.8' 6, for 'of language .. usage' read 'of a subject, etc., to be removed from everyday concerns'

ἐξισχύω II, delete the section.

†ἐξίταλα, v. ‡ἐξίτηλος.

ἐξίτηλος, add 'II neut. pl. subst., ἐξίταλα· ἀναλώματα Hsch.'

ἑξίτης, add 'also ξείτης, ξεῖτος, ξεῖθος, BCH 8.501ff.'

ἐξιτητήρια, add 'SEG 26.98.25 (Athens, iii BC)'

*ἐξκέπτωρ, ὁ, Lat. exceptor, a minute-clerk or sim. attached to a magistrate, PKöln inv. 1699.4 (ZPE 10.143, AD 332), PMich.XI 624.31 (vi AD), POxy. 943.6 (vi/vii AD); also ἐξκήπτωρ PMich.XIV 683.1 (v AD).

*ἐξκουβίτωρ, ορος, ὁ, Lat. excubitor, soldier of the imperial guard, Corinth 8(3).541, 558a (Chr.), SEG 34.927 ([ἐ]κουβίτωρ, Crete, all v/vi AD).

*ἐξκουσατεύω, excuse (from a duty), τοῦ πρώτου .. καλουμένου ἐξκουσατεύοντος ἑαυτὸν ὁ μετὰ ταῦτα ἐκαλεῖτο Cod.Just. 6.4.4.20a.

*ἐξκουσατίων, ονος, ἡ, Lat. excusatio, grounds for being excused, Modest.Dig. 27.1.13 pr.

*ἐξκουσᾶτος, ὁ, Lat. excusatus, excused, exempt, Just.Nov. 59.2; also ἐξκ- PAnt. 33.37 (iv AD).

*ἐξκουσεύω, = °ἐξκουσατεύω, Just.Nov. 43.

ἐξοδεία, add 'III passing away, Lxx (A) Ez. 26.18; cf. ἐξοδεύω II.'

ἐξοδεύω, add 'III expend (money), SEG 30.1383.A10 (AD 301).'

ἐξοδιασμός I, delete the section (v. ἐξιδιασμός). II, before 'payment' insert 'expenditure, Lycurg.fr. 22 B.'

*ἐξοδιαστικός, ή, όν, used for payment, μέτρον prob. in POxy. 494.17 (ii AD).

ἐξόδιος II 3, for 'a feast .. Exodus' read 'outgoing, terminating day of a feast'

ἔξοδος (A) III 1, after 'death' insert 'Lxx Wi. 3.2'

ἐξόθεν, delete the entry (read ἐξ ὅθεν).

ἔξοθεν, add '2 apart from, besides, SEG 26.437 (Argos, Chr.).'

ἐξοιδίσκομαι, delete the entry.

*ἐξοιδίσκω, = ἐξοιδέω, Gal. 9.521.7,; med., Hp.Morb. 2.57, Gal. 6.790.1.

ἐξοκέλλω I 2, for 'metaph. drift into' read 'fig., run aground (on the reef of)' II, metaph., add 'A.fr. 154a.3 R.'

ἐξολεθρεύω, add 'SEG 32.1601.32 (Nubia, iv/v AD)'

ἐξολισθάνω, line 4, for 'of leaves' read 'of stags' horns'

*ἐξομβριστήρ, ῆρος, ὁ, conduit for carrying off rain-water, PMich.V 252.4 (i AD), POxy. 2146.6 (pl., iii AD).

ἐξόμνυμι II 1, add 'b (Boeot. 3 pl. aor. ἐσσώμοσσαν) of magistrates swearing themselves in, SEG 23.271.61 (Thespiae, iii BC).'

ἐξομόργνυμι, line 6, delete 'parodied by'

ἐξονομάζω II, for 'call by name' read 'give a name to persons or things' and add 'Carm.Naup. 1.1 B.'

*ἐξονυχιστής, οῦ, ὁ, scrutinizer, λέξεων ἐ. Poliorc. 200.14; cf. ὀνυχίζω III.

*ἐξονυχιστικός, ή, όν, connected with paring the nails, ἡ ἐ. (sc. τέχνη) Sch.D.T. 298.22.

*ἐξοξέω, give accurate edges to, ἐ. ἁρμούς Inscr.Délos 500A44, 46 (iii BC).

ἔξος, after '-ος, ον' insert '(ος, η, ον freq. in pap.)'; after 'Hp.VM 14' insert 'PAmh. 99a.9, PRyl. 164.7'

*ἐξορέγομαι, app. stretch out, orac. in IEphes. 1252.12 (ii AD).

†ἐξορθρίζω, dub. ἀπὸ μήτρας ἐξωρθρισμένης Aq.Ps. 109(110).3; cf. ἐκ γαστρὸς πρὸ ἑωσφόρου Lxx, ἐκ μήτρας ἀπὸ πρωΐ Th., etc.

ἐξορία (s.v. ἐξόριος II), add 'Cod.Just. 9.47.26 pr., etc.'

ἐξορίζω (A), line 1, after 'aor. subj.' read 'Cypr. e-xe o-ru-xe, perh. ἐξορύξε (fr. -ορϜιξ-) ICS 217.12, al.'; line 2, after 'banish' insert 'expel'

ἐξορκίζω 1, add 'Aeschin. 2.85, 87' 2, add 'sp. ἐζορκ- SB 11247'

ἐξορμέω, for pres. def read 'lie offshore'

ἔξορμος II, delete the section.

ἐξορνῡμι, add 'ἐξώρετο θεῖος ἀοιδός Suppl.Hell. 1185'

*ἐξόροφος, ον, that belongs outside the house, outdoor, [τὸν δ' ἐ]ξόροφον οὐ μιανεῖ SEG 9.72.17 (Cyrene, iv BC).

*ἔξορρος, ον, (ὀρός) sapless, dry, cj. in Thphr.HP 1.11.3 (codd. ἔξ-ορθος).

*ἐξόρυξις, εως, ἡ, excavation, τοῦ ὄρους BCH 44.252 (Ptoion, i BC).

ἐξορύσσω 2, add 'so perh. in Hippon. 104.35 W.'

†ἐξοῦσα, Dor. for ἔξω, CID II 34 ii 58 (iv BC), SEG 9.11ff. (Cyrene, iv BC), cf. An.Ox. 2.164.

ἐξοστείζω, line 1, for 'prob. l.' read 'cj.'

ἐξότε, add 'orac. in *Didyma* 496*A*3'

*ἐξουδένημα, ατος, τό, *object of contempt*, ἐ. ἀνθρώπων Lxx *Ps.* 21(22).7, *Da.*(Thd.) 4.14(17).

ἐξουθένημα, delete the entry (v. °ἐξουδένημα).

ἐξουσία I 1, add '*licence* conferred on teachers of Jewish law, *Ev.Marc.* 1.22'; add '*πατρικὴ* ἐ. *patria potestas*, Modest.*Dig.* 26.3.1.1, *CPR* 6.12.3 (AD 300/1). **b** = Lat. *imperium*, μηθενός ἐν ἐκείναις (ὑπαρχείαις) ἐξουσίαν μείζω εἶναι τῆς σῆς *PKöln* 10.10 (Augustus).'

ἐξουσιάζω 1, transfer 'D.H. 9.44' after 'Lxx *Ec.* 8.4'

*ἐξοφθαλμέω, perh. = ἐξοφθαλμιάζω, *PAlex.* p. 36, no. 318.

ἔξοχος II 1 a, add 'Hes.*Op.* 773' add '**d** ἐπίτροπος .. τῶν ἐξοχ[ωτά]των καθολικῶν (= Lat. *procurator summarum rationum*), *Inscr.Perg.* 8(3).44.'

*ἐξπεδῖτος, ον, *ready for action*, ἐξπεδῖτοι, εὔζωνοι, γυμνοί, ἑτοῖμοι πρὸς μάχην Lyd.*Mag.* 1.46; ἐν ἐξπεδίτῳ *in a state of readiness*, Just.*Nov.* 117.11.

ἐξπελευστής, delete the entry (v. °ἐξπελλευτής).

*ἐξπελλευτής (ἐξπελευστής *Cod.Just.* 10.19.9.1), οῦ, ὁ, *collector* of taxes, κώμης Ἀφροδίτης *PFlor.* 291.6, cf. *PLond.* 1038 (both vi AD); of arrears, *Cod.Just.* l.c., Just.*Nov.* 128.6; (Lat. *expello*; cf. *compulsor*).

*ἐξσκούσατος, v. °ἐξκουσ-.

ἐξυβρίζω II, add 'of water in flood, Lxx *Ez.* 47.5'

*ἐξυδατάω, = ἐξυδατόω, Heph.Astr. 3.6.10.

*ἐξυδατισμός, ὁ, *changing into water*, Zos.Alch. 197.10.

ἐξυδρίας, before 'Ach.Tat.' insert 'cj. in'

*ἐξυπτιόω, = ἐξυπτιάζω, pass., *live luxuriously*, Phot. s.v. πεταχνοῦνται; cf. Hsch. s.v. πεταλοῦνται.

ἐξυφαίνω I 1, line 3, delete 'Nicopho 5' and after 'pass.' insert 'ὁ δ᾿ ἐξυφαίνεθ᾿ ἱστός, ὁ δὲ διάζεται Nicopho 13 K.-A.'

ἐξυψόω, for '*elevate*' read '*transform into an elevated style*'

ἔξω I 1 a, add 'w. ref. to the movement of dislocated bones away from the body, Hp.*Mochl.* 16, *Art.* 64' **2 a**, add 'ἔξω ἐγένετο *became distracted*, Hp.*Epid.* 5.80' **b**, line 8, ἐ. τοῦ πράγματος, add 'Is.*fr.* 22'

*ἐξώβολος, ον, *consisting of six obols*, Hsch. s.v. λεπτὰς καὶ παχείας; neut., *sum of six obols*, *OStrassb.* 67 (i AD, ἐξόβ-).

ἔξωθεν II a after '1*Ep.Ti.* 3.7' insert '*those outside a (particular) relationship*, Just.*Nov.* 1.1.3' **b**, for 'X.*An.* 5.7.24' read 'X.*An.* 5.7.21' **c**, add '*Cod.Just.* 1.2.17.15; abs., besides, *MDAI(A)* 51.13 (lapis ἐσ-, Cos), *JHS* 15.112 (Lycia), *TAM* 2.247.11, etc.'

ἐξωθέω I 1, penultimate line, after 'S.*OC* 428' insert '**b** *debar, exclude*, αὐτὸν .. ἐξωθείτω καταλόγων *Cod.Just.* 1.4.34.9.'

ἐξώλεια I, add 'ἐξξόλειαν *IG* 9²(1).609.15 (Naupactus, vi/v BC)' add '**II** *abandoned* or *abominable conduct*, Horap. 2.65.'

*ἐξώλεος, ον, = ἐξώλης, *IMylasa* 476 (Rom.imp.).

ἐξώλης II, add 'ἐ. κίναιδος Hsch.'

ἐξωμίς, line 4, after 'X.*Mem.* 2.7.5' insert '*SEG* 34.122.47 (= *IG* 2².1673, Eleusis, iv BC)'

ἐξωμοσία I, add '2 = Lat. *cautio iuratoria*, *Cod.Just.* 10.11.8.7, Just.*Nov.* 134.9.1.'

ἐξωπυλῖται, after 'organized body' read 'app. connected w. burial work, *TAPhA* 71.650 (ostr., iii AD), *PGrenf.* 2.72.4 (iii/iv AD), *BGU* 34 ii 21 (iv AD)'

ἔξωρος 2, w. gen., add 'ἔξωρον .. τῶν ἐρωτικῶν Philostr.*Her.* 11'

*ἐξώρροπος, ον, *inclining outwards*, Php. in *APo.* 439.7.

ἐξώστης 4, add 'Theophil.Antec. 4.6.2'

ἐξώτατος, (s.v. ἐξωτάτω), add '*Peripl.M.Rubr.* 4 C. (cod. ἔσω-, v. *CQ* N.S. 30.495)'

ἐξωτικός 1, before 'ή, όν' insert '(ἔσω- *RPh* 10.121)'; after '*foreign*' insert '*unguenta exotica* Plaut.*Most.* 42'; line 4, after '(Iasos)' insert '*RPh* l.c. (Perinthus)' add '**b** *Graecia exotica, overseas* i.e. *colonial Greece*, Plaut.*Men.* 236. **c** *being outside a (particular) legal relationship*, Just.*Nov.* 48.1 pr.'

*ἐξωφάκαι (or -άκες), αἱ, *kind of haemorrhoids*, Cyran. 29 (v.l. ἐξωχάδας, v. °ἐξωχάδες), 103 (v.l. ἄνθρακας).

*ἐξωχάδες, αἱ, *external haemorrhoids*, Cyran. 1.12.31, 2.30.13 K., al.

ἐοικότως, delete the entry (v. °εἰκότως).

*ἐόργη, ἡ, τορύνην, ἣν καὶ εὐέργην ὠνόμαζον καὶ ἐόργην Poll. 6.88, 10.97.

ἑορταστικός, add '*PSI* 791 (vi AD)'

ἐός, Boeot., add 'ἑϜοῖ (dat.) *CEG* 444(i) (at Delphi, ?*c.*550 BC)'

ἐπαγγελία, after lemma insert 'Thess. ἐπαγγελλία *BCH* 59.38 (Crannon, in quot., sense 3)' **3**, ἐξ ἐπαγγελίας, add '*SEG* 30.1274 (Caria, *c.*AD 200)'

ἐπαγγέλλω 6, for this section read 'med., *propose, ask* as a concession or favour, D. 19.41, 193'

ἐπάγγελμα 4, add '*Cod.Just.* 1.3.41.26'

ἐπαγλαΐζω I, add 'Ps.-Hdt.*Vit.Hom.* 192'

ἔπαγρος, line 2, for 'Call.*Hec.* 1.4.10' read 'Call.*fr.* 260.64 Pf.'

ἐπάγρυπνος, delete 'Aristaenet. 1.27' and add 'Ἄμμων ἐπάγρυπνος ὀπτ<ήρ> *IUrb.Rom.* 141; adv. -ως *SEG* 29.250 (Attica, Chr.)'

ἐπάγω I 8 a, line 4, after 'D. 47.28' insert 'abs., ἐφόρων καὶ γερόντων ἐπαγόντων *BSAA* 39.133 (Euesperides, iv BC)'

ἐπαγωγή 1, add '*bringing in, introduction* of a water-supply, *PAvrom.* 1*A*27' **3**, add '**b** *visitation*, Ἑκάτης φάσκων ἐπαγωγὴν γεγονέναι Thphr.*Char.* 16.7.' **4 b**, delete 'Ἑκάτης .. 16.7' **7**, for this section read '*trouble, distress* Lxx *Si.* 2.2, 3.28, al., 23.11 (pl.), *Is.* 14.17'

ἐπαγωγός, add '**III** οἱ ἐ., *officials in tax-farming company*, *Michel* 1225.16 (Cyzicus, i BC). **IV** *supply-pipe*, *IG* 4².116.19 (Epid., iv/iii BC, pl.).'

†ἐπάγων, οντος, ὁ, *kind of pulley*, = ἀρτέμων II, *the guiding pulley of a system*, Vitr. 10.2.9.

ἐπαείδω 2, add '*πείθειν* ἐπάδουσ᾿ ὥσθ᾿ ὁμαρτεῖν μοι πέτρας E.*IA* 1212' add '**3** *relate* a story *about*, w. dat., μύθων τῷδε τῷ ζῴῳ ἐπᾷσαι Ael.*NA* 6.51, 16.5, λόγον ib. 12.30.'

*ἐπαέτιον, τό, *top of gable*, ἐ. ξύλινον μεμολυβδωμένον *Inscr.Délos* 421.17, 442*B*168 (ii BC).

ἔπαθλον, line 1, for 'etc.' read '1262 (cj.)' and at end add 'cf. παρ᾿ οὐδενὶ κεῖται τὸ "ἔπαθλα" ἢ μόνῳ τῷ Εὐριπίδῃ Sch.E.*Ph.* 52'

ἐπαΐκλα, delete the entry.

*ἐπαϊκλον or ἐπάακλον, τό, (αἶκλον) *additional meal, dessert*, Dor. for ἐπιδειπνίς, ἐπιδόρπισμα, Ath. 14.664f; pl., Pers.*Stoic.* 1.101.454, Sphaer.*Stoic.* 1.142.630, Molpis 2b J.; also ἐπαίκλεια or ἐπαΐκλεια, Apion ap.Ath. 14.642e.

ἐπαινέτης II, delete the section.

ἐπαινέω, line 8, after '(*Schwyzer*) 623.34)' insert 'also Dor., *REA* 33.210.16 (Theangela, decree of Troezen, iii/ii BC)'; line 13, delete '= αἰνέω .. Att.)' **I 4**, delete the section transferring quots. to section I 2. **II**, for '= παραινέω .. *advise*' read '*express one's approval of* a course of action, *recommend*' **IV**, delete the section.

ἐπαινός, add '**2** *praiseworthy*, Triph. 52, Mitchell *N.Galatia* 103.'

ἐπαινουμένως, add 'Just.*Nov.* 82.1.1'

ἐπαινύων, line 1, delete '(intr.)'

ἐπαίρω I 2, add 'in fig. phr., ἐπᾶραι κύριος τὸ πρόσωπον αὐτοῦ ἐπὶ σέ (transl. Hebr. idiom, i.e. *show favour*) Lxx *Nu.* 6.26' **4**, add '*raise oneself* in bed, prob. in Call.*fr.* 191.43 Pf.' **5**, line 1, for 'Gal. 6.264' read 'Gal. 6.265.1'

ἐπαισθάνομαι 1, add 'Hes.*fr.* 204.120 M.-W.'

ἐπαιτιάομαι, penultimate line, delete 'to it'

ἔπαιτον, delete 'τό' and for 'dub. sens. in' read 'perh. adv., *approximately*' and add '*PGen.* 113.8 (sp. ἐπαίετον), 114.8 (sp. ἐπάαιτον) (both iii AD)'

*ἐπαιχμάζω, *attack with a spear*: transf., of mating wild boars, Opp.*C.* 1.389.

†ἐπαιωρέω, *hold suspended* over, ξίφος .. τοῖς αὐχέσιν ἐπαιωρούμενον Hdn. 5.2.1, πέτρον ἐπηώρησε καρήνων Nonn.*D.* 4.456. **b** med., *poise threateningly* above or over, Plu.*Fab.* 5; Σκύθαι τοῖς μέσοις ἐπαιωροῦντο Them.*Or.* 8.119c; of conditions, τοῖόν σφιν ἐπὶ δέος ἠώρειτο A.R. 1.639, τὰ ἐκτὸς ἐπηωρημένα Ph. 1.650, κίνδυνον οὐκ ἐπαιωρούμενον ἀλλ᾿ ἤδη παρόντα Hdn. 1.9.10. **2** *hold poised for action*, ἄκουε .. τὰ ὦτα ἐπαιωρήσας Ph. 2.125; med., w. dat., *hold oneself poised* for, ἐπαιωρεῖσθαι πολέμῳ Plu.*Pel.* 29, τοῖς πράγμασι id.*Tim.* 2. **3** *cause to rest on the top* or *surface* of, τρίχας .. καὶ τὰ ὅμοια ἐπαιωρηθήσεσθαι διὰ τὴν κουφότητα Dsc. 5.75.11, 5.92, ἐ. πτερῶν ἠέρι πολλῷ *GVI* 1765.5, fig., αἰσιν (εὐτυχίαις) ἐπαίρεται ἀεὶ βίον *AP* 7.645 (Crin.), ἐλπίσιν ἐπαιωρούμενοι Luc.*Alex.* 16, Hdn. 2.9.1. **II** med., *rise, swell*, ὄγκος ἐπαιωρεύμενος ἔξω Aret.*CA* 1.7, *SD* 1.14.

ἐπακμάζω, line 2, after '*come to its height*' insert 'Longin. 13.4'

*ἐπακολουθητής, οῦ, ὁ, *concurring party*, *PColl.Youtie* 67.8 (AD 260–1).

*ἐπακολουθία, ἡ, = ἐπακολούθησις, Sch.Hes.*Th.* 245b.

ἐπάκριος I, add '*SEG* 21.541 v 60 (Attica, iv BC)' **II**, add '**2** ἐπάκριοι, οἱ, *a party at Athens in the time of Solon*, Plu. 2.763d; cf. πεδιεῖς, πάραλος.'

ἐπακροάομαι, line 2, for 'Hld. 2.17' read 'Hld. 2.16'

ἐπακτήρ, after 'Il. 17.135' insert 'Call.*Jov.* 77, Opp.*C.* 1.481' and for 'later' read 'also'

ἐπακτός I 2, line 2, after 'Pi.*O.* 10(11).89' insert 'E.*Ion* 290, 592'; lines 6/7, delete 'ἐ. πατήρ .. *Ion* 592'

ἐπαλειπτέον, add 'Gal. 10.498.11'

ἐπαλείφω, after lemma insert '(app. ἐπᾱλ- Alcm. 80 P.)'

ἐπάλης, for '*open to the sun, sunny*' read '*exposed to the fire*'; for '(nisi .. ἀλής)' read '(or perh. *crowded*, cf. ἀλής)'

ἐπάλλαξις 3, delete 'διαιτημάτων .. 385 (pl.)'

ἐπαλληλία, line 4, for 'Gal. 19.679' read 'Gal. 19.680.1'

ἐπάλληλος II 1, for 'by *one another's* hands' read 'by *hands used against each other*'

†ἐπάμερος, v. ἐφήμερος.

ἐπαμύντωρ, after 'Od. 16.263' insert 'as title of Apollo, *SEG* 31.559 (Delphi, iv BC)'

*ἐπαμφιάζοντες· ἐπενδύοντες Hsch.

ἐπαμφόδιος, delete the entry.

ἐπάμων, for 'cf. Hsch. (pl.)' read 'ἐ. (cod. ἐπάλλων)· δοῦλος λάτρις Hsch. (La.), ἐπάμονες· ἀκόλουθοι id. (ἐπ- Schmidt)'

ἐπαναβαίνω **I 3**, add '**c** perh. *reach the age of*, w. εἰς, *IGLS* 607; cf. °ἐμβαίνω.'

ἐπαναβάλλω **III**, add 'pass., of *deferred* payment, *IG* 11(2).142.1, 3, 4 (Delos, iv BC)'

ἐπαναγκάζω, omit '*by force*' and add '*IG* 12(7).515.85 (Amorgos, ii BC)'

ἐπἀνάγκης, line 2, for 'And. 1.12' read 'And. 3.12'   **2**, add 'ἡ δὲ δημοθοινία [γε]νέσθω ἐν τῷ γυμνασίῳ ἐπάναγκες *as is required*, *IG* 12(7).515.106 (Amorgos, ii BC)'

ἐπάναγκος **I**, add 'neut. as adv., *PMich*.inv. 1410.11 (*JJP* 18.158)'

ἐπανάγω **III 2**, for this section read 'εὖ ἐ. *enjoy health* or *prosperity, get on well*, τῷ σώματι Apollon.Perg.*Con*. 1 pr.; sim., w. other advs. μετρίως, ἱκανῶς ἐ., *PTeb*. 755.6 (ii BC), *UPZ* 110.6 (ii BC); στενῶς ἐ. ib. 60.15 (ii BC)'

*ἐπαναδύω, dub. sens., cf. ἀναδύομαι, Dor. aor. 2 part. ἐπανδύς, prob. = *ἐπαναδύς, Sophr. in *PSI* 121d11.

ἐπανακαλέω **I**, delete the section.

ἐπανάληψις, add '**III** *recapture*, Just.*Const*. Δέδωκεν pr.'

*ἐπαναμισθόω, *let afresh*, *BCH* 60.182 (ἐπαμμ-, Thespiae, iii BC).

ἐπανανεόομαι, add '*restore*, ἐπανανεωσάμην τὴν (sic) ἐκτησάμην ποίαλον (= πύαλον) *TAM* 4(1).352'

ἐπαναποδίζω, add 'ἐπαναποδιστέον Arist.*GC* 317ᵇ19'

*ἐπανάπωλος, ον, *resold to a further party* (in quot., of contracts), *IG* 7.3074.3 (Lebadea, ii BC).

ἐπανάστασις **I 3**, after '(Th.) 8.21' insert '*SEG* 31.985A (Teos, v BC), ib. 28.60.12 (Attica, iii BC)'

ἐπαναστρέφω **II**, for 'Pass., *return to the surface*' read 'med., *turn oneself over*'

ἐπανατέλλω **II**, lines 6/7, delete '*show oneself* .. A.*Ch*. 282' and add 'w. dat., *rise close behind*, Arat. 341; *grow up in consequence of*, A.*Ch*. 282'

ἐπανατίθημι, for '*lay upon*' read '*shift* a load, weight *on to*' and after 'metaph.' insert '*PPetr*. II 4.1+4.9 (256 BC; v. *ZPE* 59.62)'

ἐπαναφέρω **I 3**, add '*report*, ἐπαναφέρω σοι περὶ τῆς αὐτῆς γεωργίας *POxy.Hels*. 13.4, 10 (i AD)'

ἐπαναχέω, add 'also ἐπανχ-, v.l. in A.*Ag*. 1137 (cf. ἐπεγχέω)'

ἔπανδρος, add 'sup. *La Carie* p. 98 no. 5'

ἐπανδρόω, add '**2** *fill with men*, Λῆμνον A.R. 1.874 (v.l. ἐσ-).'

ἐπανερωτάω, after '*question again*' insert 'or *further*' and after 'Pl.*Clit*. 409d' insert 'Demetr.*Eloc*. 288'

ἐπάνεσις **I**, add '**2** *relaxation, remission* of dues, καθόλου ἐπάνεσις ἔστω τοῦ τελωνοῦ *SEG* 39.1180.47 (Ephesus, i AD)'

ἐπανήκω, add 'of an estranged lover returning, Heph.Astr. 3.9.34'

ἐπανθέω **II**, add 'w. internal acc., [ὃς] μὲν γὰρ παριαῖσιν ἐπήνθει παῦρον ἴουλον epigr. in Mitchell *N. Galatia* 392.17'   **III**, delete the section.

ἐπάνθισμα, add '**II** dub. sens., ἐ. [ἱα]ρεῖ Διονυσίωι *IAskl.Epid*. 51.7 (p. 44) (iii BC).'

ἐπανοίγω, for '*open*' read '*open up again*' and add 'εἰ δέ τις τολμήσει ἐπανῦξε τὴν σορόν *TAM* 4(1).267, *IEphes*. 3327 (both ii/iii AD)'

ἐπανορθόω, line 2, for 'Lys. 1.70' read 'Lys. 2.70'   add '**4** *repair, restore* artefacts, τὰ δημόσια ὅπλα .. ἐπανορθοῦν τε καὶ ἀνανεοῦν Just.*Nov*. 85.2.'

ἐπανόρθωσις, add '**b** *restoration, reconstruction* of buildings, etc., *IG* 9²(1).583.59 (Acarnanian decr. at Olympia, iii BC), *IG* 12(5).1030 (Paros, ?i BC).'

ἐπανορθωτής, add 'ἐπανορθωτὴν Ἀσίας *SEG* 31.910 (Aphrodisias, iii AD)'

ἐπαντλησμός, for '= foreg.' read '*artificial irrigation*, *PBremen* 30.4 (ii AD)'

*ἐπάντλιον, τό, kind of machine for irrigation, *SB* 12524.21 (i BC).

ἐπάνω **I 1**, add 'οἱ ἐπάνω *those above the earth, the living*, *INikaia* 1282, 1450, al.'   **2**, line 3, after '(Tenos, iii BC)' insert 'οἱ ἐπάνω γῆς = οἱ ἐπάνω (v. supra) *INikaia* 1395'; at end after 'Val. 48.5' insert 'ἐ. εἴλης, = Lat. *praepositus alae*, *SEG* 6.167 (Temenothyrae, Phrygia)'   add '**4** *up-country, inland*: ἐπάνω τοῦ λιβός *inland in a westerly direction*, *Peripl.M.Rubr*. 15; τούτων ἐπάνω ib. 47.'

ἐπαποθνήσκω, for '*die after* another' read '*add oneself to in death, die in addition* to, οὐ μόνον ὑπεραποθανεῖν ἀλλὰ καὶ ἐπαποθανεῖν τετελευτηκότι'

*ἐπαπολισθάνω, of perspiration, *stream off* a person, *POxy*. 1381.130 (ii AD).

ἐπαπορέω, line 3, ἐπαπορεῖταί τι, omit 'τι' and substitute 'impers.'

ἐπαποστέλλω **I**, add '**2** *send in addition* on a mission, expedition, or sim., pass. ἐ. ὑπὸ Ἀγησιδάμου *Inscr.Cret*. 1.xvi 35 (Lato, ii BC).'

ἐπαράομαι, add '**7** w. acc. and inf. with μή, *conjure* (deities) to prevent an occurrence, *SEG* 6.784 (Cilicia, ii/iii AD).'

ἐπαράρίσκω **I**, add '[λίθον] γυίοις ἐπήραρε Euph. 9.14'   **II**, lines 5/6, for '*well-fitted* .. Nonn.*D*.' read '*made fast*, Hes.*Op*. 601, 627: ἐφάρμενος, *suited*, w. dat., Nonn.*D*. 7.78'

*ἐπάρδεια, ἡ, *irrigation*, *POxy*. 3354.27 (AD 257).

ἐπάρδια, delete the entry (v. *ZPE* 41.256).

*ἐπᾰρετέω, *requisition* or *take for use*, κτήνη, πλοῖα *PTeb*. 5.182, 252 (ii BC); cf. ἡ ἐν ἀρετῇ κειμένη γῆ (v. ἀρετή I 2 b).

ἐπάρκεια, add '*SEG* 25.653 (Thessaly, ii BC)'

ἐπαρκής **I**, delete the section transferring quots. to section II.

ἔπαρμα **I**, add '**2** *raised bank* or *platform*, ἔθηκεν ἔπαρμα ὕψος πήχεις ἑξήκοντα, πλάτος αὐτοῦ πήχεων ἑξήκοντα Lxx 2*Es* 6.3; ὄχθοι δὲ γῆς ἐπάρματα Ammon.*Diff*. p. 108 V. (369 N.).   **3** app. *a weight lifted* by a strong man, ἐ. Φαβατίωνος *SEG* 23.666 (Cyprus, iii AD).'   **II 2**, delete the section.   add '**III** wd. occurring in inventory of utensils, *SEG* 24.361.10 (Thespiae, 386/380 BC).'

†ἐπαρτής, ές, *ready for action, use*, etc., ἐπαρτέες εἰσὶν ἑταῖροι Od. 8.151, 14.332, ἐπεὶ δμώεσσιν ἐπαρτέα πάντ᾽ ἐτέτυκτο A.R. 1.234, 3.299.   **II** app. *hanging*, Opp.*H*. 5.359, πολλαὶ δ᾽ οὐρανόθεν καὶ ἐπαρτέες ἐκ νεφελάων .. ἐπόρνυνται φηγοῖς .. πηγυλίδες Orph.fr. 270.

ἐπαρχεία **I**, add 'also app. of a bishop's reign, *SEG* 31.1474 (Arabia, vi AD)'

ἐπαρχεῖον, delete the entry.

*ἐπάρχειος, ον, *of* or *pertaining to an ἔπαρχος*: -ειος, ἡ, *province*, *IPE* 1².54 (Olbia), *IGBulg*. 1690.e30 (Pizus); sp. -ιος, *IG* 14.1078a.7, *IGRom*. 1.580 (Nicopolis ad Istrum), *POxy*. 2106.4 (iv AD).   **2** cult-title of Zeus, *of the province*, *ICilicie* 109 (AD 99).

ἐπαρχή, after 'Oropus' insert 'v. *SEG* 31.416 (iv BC)'

ἐπαρχία, line 14, for 'of Carthage, *empire*' read 'in general, *subject territory*' and add 'Plb. 1.15.10, 1.75.5, al.'

*ἐπαρχιακός, ή, όν, *provincial*, Modest.*Dig*. 26.5.21.4; neut. pl. subst., *dowry*, Cod.Sinait. in *BCH* 4.452.

ἐπαρχικός **I**, add '**2** ἐπαρχικός, ὁ, *ex-prefect* (cf. ἔπαρχος I 2), *OGI* 578.14 (Tarsus, iii AD), *SEG* 8.647 (Egypt, iv AD).   **b** employee in the office of an ἔπαρχος, *ITyr* 12, *MAMA* 3.374, 416, 691.'

*ἐπάρχιος, v. °ἐπάρχειος.

*ἐπαρχῑτικός, ή, όν, *of a province* (Lat. *provincialis*), ἐπαρχειτικῶν ἀγρῶν *Mon.Anc.Gr*. 8.23.

ἔπαρχος **I 1**, delete '*commander* .. ἄπαρχος' (v. °ἄπαρχος).   **2**, line 3, after '(Plu.)*Brut*. 51' insert 'ἐ. ἀρχιτεκτόνων *IG* 2².3546 (Eleusis, i AD)'; line 7, after '(Sidyma)' insert 'ἐπάρχῳ λεγ(ιῶνος) ια' Κλαυδίας *SEG* 33.1194 (Cappadocia, i/ii AD), ἔπαρχος ἱππέων *ITomis* 127 (ii/iii AD)'   **II**, add 'ἐ. ἐξουσία *BCH* 11.351 (Laodicea), cf. ἐπαρχικός I'

ἐπασσύτερος **I**, add 'Stesich. S139.7'

*ἐπασχαλάω, *to be indignant at*, w. dat., ἐπασχαλόων .. κυδοιμῷ Nonn.*D*. 36.413; Aeol. pres. part. ἐπα[σχάλαντες] συμφόραισι rest. in Alc. 69.1 L.-P., cf. 306 i 18/19 L.-P.

*ἐπάτερθεν· ἐπέκεινα Hsch.

ἐπαυλέω, ἐπαύλημα, for '*flute*' read '*aulos*'

ἐπαύλιον **III**, delete the section.

*ἐπαύλιος· ἡ τῆς αὐλῆς ὁδός Suid., Zonar.

ἔπαυλος, after 'Od. 23.358' insert '(v.l. ἐν-)'; delete 'A.R. 1.800' and add 'Hes.*fr*. 66.1 M.-W.'   **2**, after '*home*' insert 'Hes. l.c.'

*ἐπαυξητικός, ή, όν, *increasing, intensifying*, Sch.E.*Hipp*. 518.

ἐπαυρέω **II 1 a**, lines 2/3, delete 'μόχθων .. Pi.*N*. 5.49'; line 9, transfer 'Hdt. 7.180' to section b; add 'w. acc., μόχθων ἀμοιβὰν Pi.*N*. 5.49; κτῆσιν *AP* 9.332 (Noss.)'

ἐπαυχμέω, for '*send drought* .. ἐπαυχμήσας' read '*cause a drought*'

ἐπαφάω, line 3, after 'c. gen.' insert 'Ἄρκτου Arat. 93'

ἐπᾰφή **I 1**, add '**b** ἐπαφῇ τῆς σελήνης as cause of the birth of Apis, Plu. 2.718b.'   **III**, for 'prob. *external claim*' read 'perh. *skin-disease*' and add '*BGU* 2111.14 (ii AD)'

*ἔπαφησις, εως, ἡ, *touching*, cj. in Nicostr.ap.Stob. 4.22.102 (pl.).

ἐπαφίημι **1**, line 5, after 'Alciphr. 1.22' insert 'transf., τὴν ὀργὴν εἴς τινας J.*AJ* 19.2.2'   **2**, after 'Thphr. *CP* 2.5.5' insert '(Schneider)'

Ἐπάφιος, delete the entry (v. °Ἐπάφριος).

*ἐπαφορμίζομαι, *look for a pretext*, *Vit.Aesop*.(W) 58.

*ἔπαφος, ον, app. *attached*, i.e. *staked* or *trained*, ἐὰν .. ὀλιγωρήσῃ τὴν ἄμπελον καὶ μὴ ποιήσῃ αὐτὴν ἔπαφον *PAvrom*. 1*A*26, 1*B*27 (i BC; v. *JHS* 35.55).

*Ἐπάφριος, epith. of Dionysus, Orph.*H*. 50.7 (ἐπάφιε codd.), 52.9; cf. Ἄφριος.

ἐπάχθεια, add '*PKöln* 110.4 (v/vi AD)'

ἐπβάσκω, v. °ἐπιβάσκω.

*ἐπεγγράφω, *register in addition*, Plu. 2.278d.

ἐπεγκλίνω, add 'ἐ. νοῦν *incline* the mind, *direct* it to something, *Orac.Chald*. 1.2 P.'

*ἐπεγρήγορος, ον, *alert, watchful*, Plu.*Brut*. 36.2, Horap. 1.60 S.

*ἐπεγρία, ἡ, *wakefulness*, Iamb.*VP* 3.13.

*ἐπέγχωσις, εως, ἡ, *the action of heaping on top* (in quot., in building up a dike), *PBremen* 14.7 (ii AD).

ἐπεί, line 1, after 'ἐπειδή' insert '(Thess. ὁπειδεί *SEG* 27.202 (Larissa, iii BC), and so perh. ὁπεί κε = ἐπειδάν, BCH 59.55, SEG 31.575.31 (Larissa, both ii BC)' **A II**, line 5, add 'w. κα in Doric ἐπεί κα πέντε Ϝέτεα hεβōντι IG 5(2).159.4 (at Tegea)' **B 1**, for otherwise, add 'so also ἐπειδή Mitchell N.Galatia 257 (AD 140)'

ἐπείγω IV 3, add 'Gal. 9.309.3, SEG 30.1390B (Lydia, ii AD)' add '**V** impose a penalty, θέματα, ἐφ' ὧν ἐξ ἀρνήσεως διπλάσια τὰ τῆς καταδίκης ἠπείγετο CodJust. 1.3.45.7a.'

ἐπεῖδον, after 'ἐπιδεῖν' insert 'late aor. inf. ἐφιδῆσαι PTeb. 751.10 (ii BC)'

ἐπεὶ ἦ, add 'and ἦ adv. I 2'

ἐπείκτης, add '**II** collector, χρυσοῦ στεφάνου POxy. 1413.25 (iv AD); written ἐπίκτης POxy. 1428.3 (iv AD).'

*****ἐπείλῡμα**, ατος, τό, wrapping, POxy. 1765.16 (iii AD; pl., sp. -ειλύμματα).

ἔπειμι (A) **II**, add 'Call.Epigr. 50.3 Pf.'

ἔπειμι (B) **III 1**, delete 'χώρους Hdt. 5.74'

ἐπείσακτος, for 'SIG 1231' read 'TAM 4(1).276' and add 'Procl.Inst. 201 D.'

†ἐπεισβαίνω, to go into so as to make an approach or attack, ἐπεσβαίνοντες ξὺν τοῖς ὅπλοις ἐς τὴν θάλασσαν Th. 2.90.6, 4.14.2, ἵππῳ εἰς θάλασσαν X.HG 1.1.4; fig., ἐχόμενοι .. ὥς τινος ἀσφαλοῦς πείσματος ἐπεισβαίνωμεν εἰς τὸν νῦν λόγον Pl.Lg. 893b.

ἐπεισκρίνομαι, delete the entry.

*****ἐπεισκρίνω**, bring in on top of or in succession, pass., Hp.Alim. 5; med., οὐσία ἀεὶ ῥεῖ τε καὶ ἑτέρα ἀνθ' ἑτέρας ἐπεισκρίνεται S.E.P. 3.82.

†ἐπεισκυκλέω, roll or trundle in one after another, θεὸν ἀπὸ μηχανῆς ἐπεισκυκληθῆναί μοι τοῦτον ᾤμην Luc.Philops. 29; fig. ὁ δ' Ἄττης .. καὶ ὁ Κορύβας .. ἡμῖν ἐπεισεκυκλήθησαν; id.Deor.Conc. 9, Hist.Conscr. 13, ἀσάφειαν ἡμῖν τοσαύτην S.E.P. 2.210, Gal. 8.575.5, ἕτερα ἑτέροις ἐπεισκυκλούμενα Longin. 11.1, 22.4.

ἐπεισπαίω, after 'Com.adesp. 439' insert 'ἀγῶσι thrust oneself into, prob. in PLond. 1912.92 (i AD)'

ἐπεισφέρω, line 5, add 'εἰ δέ τις ἕτερον νέκυν ἐπισενέγκ[ῃ] TAM 4(1).249 (ii/iii AD)'

*****ἐπείτοιγε**, introducing a final alternative or resort, or else, otherwise, CodJust. 1.3.45.10, al., Just.Nov. 1.1.2, 17.3, al.

ἐπεκδικέω, add 'cf. TAM 3(1).418 (ii AD), al.'

*****ἐπεκκεντέω**, stab, pierce, Sm.Za. 12.10.

*****ἐπεκονίθη·** κατωρύχθη Hsch.

ἐπεκτείνω I 1, after 'Arist.EN 1097ᵇ12' insert 'τὴν αὐτὴν .. νομοθεσίαν καὶ ἐπὶ γυναιξὶ παρθενευούσαις ἐπεκτείνομεν CodJust. 1.3.52.14' add '**6** expand in value, τὰς προσόδους Str. 17.1.15.' **II 2**, add 'Macho 208 G.' **III**, delete the section.

ἐπελαύνω I 1 b, add 'ἐπελ[ά]στō SEG 30.380 (Tiryns, vii BC)'

ἐπελέγχω, add 'disgrace, put to shame, Thgn. 1011 (tm.)'

†ἐπελεύθω, aor. 1 ἐπήλευσα, bring to, Inscr.Cret. 4.41 i 9, al. (Gortyn, v BC), Leg.Gort. 3.45, al., GVAK 32.15.5.

ἐπέλευσις, add '**4** assault, PMich.VI 423-424.4 (ii AD), POxy. 69.15 (ii AD).'

ἐπεμβάλλω 3, line 5, for 'thou .. intrudest thyself' read 'you impose yourself'

ἐπεμπηδάω II, for this section read 'leap upon, Gal. 8.556.15'

*****ἐπεμπόδων**, adv., = ἐμποδών, MAMA 4.279 (Dionysopolis).

ἐπεμφέρω, add 'bring in, serve, IG 12(7).515.49 (Amorgos, ii BC)'

*****ἐπεναντίον**, contrary to, ἐπεναντίον τούτῳ PFlor. 294.43 (vi AD).

ἐπενδύτης, line 3, after 'Lxx 1Ki. 18.4' insert 'PMich.inv. 1648.12 (AJP 65.257; ii AD)'

*****ἐπενδύτιον**, τό, dim. of ἐπενδύτης, SEG 39.1278 (Katakekaumene, Lydia, ii AD).

*****ἐπενεκτέον**, one must give a name to, Gal. 9.813.5; cf. ἐπιφέρω I 6.

ἐπενήνοθε 2, for 'had passed' read 'has passed'

ἐπενθάπτω, add 'TAM 4(1).264 (Rom.imp.), INikaia 766'

ἐπενίημι, for 'compress the pulse' read 'exert on, πόσον ἀρκεῖ τῆς θλίψεως ἐπενεῖναι τοῖς οὕτως ἔχουσιν'

ἐπεντίθημι, after 'pass., to be put in besides' insert 'of additional (unauthorized) burial in a tomb' and add 'TAM 4(1).239 (Rom.imp.)'

ἐπέξειμι, add 'παλαιᾶς ἀπεχθείας Arr.An. 1.9.6'

*****ἐπεξελέγχω**, denounce, TAM 3(1).823 (Termessus).

ἐπεξεργάζομαι 4, add 'ὑπέρ τινος Ptol.Tetr. 117'

ἐπεξεργασία, after 'Ptol.Tetr. 117' insert '(v.l.)'

ἐπεξορκίζω, add '**b** conjure, w. acc. of deity invoked, Swoboda Denkmäler 18 (Pisidia).'

ἐπέραστος, line 2, after '(Luc.)Im. 10' insert 'AP 5.299.1 (Agath.)'

ἐπεργάζομαι 3, delete the section. add '**III** contrive in addition, εὔνους γὰρ γεγονὼς τοὐμῷ πατρὶ .. κατὰ πόδας θανάτου μοῖραν ἐπειργάσατο epigr. in SEG 32.611 (Thessaly, i BC).'

*****ἔπεργος**, ον, aiding the business, useful, PSI 619.8 (iii BC). **2** -ος, ὁ, assistant. SB 5680.3 (iii BC). **II** -ον, τό, work, effort, ὃν (νηόν) ἀκαμάτοισιν ἐπέργοις .. ἐξετέλεσσεν epigr. in SEG 37.1537 (Arabia, c.AD 400). **2** work done in addition to payment of rent, τοῦ μισθώματος καὶ τῶν ἐπέργων ἁπάντων ἀπότεισμα IG 12(7).62.15 (Amorgos).

ἐπέρεισις, line 2, for 'Gal. 2.386' read 'Gal. 2.387.17'

*****ἔπερθα**, adv., = ὕπερθεν, Alc. 208(a).ii.8 L.-P.; cf. °κατέπερθεν.

ἐπέρχομαι, line 2, after 'ἐπῆλθον' insert '(Cret. part. ἐπευθών, Inscr.Cret. 4.168.17 (iii BC))' **I 3**, add 'neut. part. used abs. w. ἄλλως, haphazardly, as it came to mind, ἄλλως ἐπελθόν, οὐκ ἐξεπίτηδες ἤρετο Luc.DDeor. 20.4, cf. Apollod. 1.9.16'

ἐπερωτάω 5, add 'CodJust. 2.12.27.2, Just.Nov. 162.1 pr.'

ἐπερώτησις 1, add 'τὰν μαντείαν καὶ ‹ἐ›περώτασ[ιν SEG 30.85.18, 25 (Delphian letter, i BC)'

ἐπές, add 'w. gen., app. w. assim., ἐπὲ Ϝέργō Dubois Dial.Arcad. p. 196 (Pheneus, c.500 BC)'

ἐπεσθίω I 2, for 'Thphr.CP 6.4.7' read 'Thphr.CP 6.4.6'

ἐπέτης, masc., add 'PRyl. 627.146, al. (iv AD); Myc. e-qe-ta, title of high official; deriv. adj. e-qe-si-jo'

ἐπετήσιος, for 'Epigr.Gr. 815' read ''νχρονίσας ἐφετήσιον (annual sacrifice) οὐκ ἀπέδωκε epigr. in SEG 33.736 (Crete, ii AD)'

ἐπευάζω, add 'cf. ἐφευάζω'

*****ἐπευθών**, v. °ἐπέρχομαι.

ἐπευνάζω, read 'ἐπευνάζομαι'

ἐπευωνίζω, after 'D. 23.201' insert 'τὴν ἀγορὰν ἐπευωνίζων SEG 38.1462.9 (Oenoanda, ii AD)' add '**2** intr. of prices, get lower, PCair.Zen. 363.14 (iii BC; ἐπεων-).'

*****ἐπηβολέω**, dial. ἐπᾱβ-, achieve mastery, knowledge, etc., of, w. acc., Iamb.adesp. 38.11 W.; w. gen., Pi.Pae. 6.182.

ἐπηβολή, line 1, after 'Leg.Gort. 5.50' insert 'also Boeot. SEG 3.342.16 (Thebes, iii BC)'

ἐπήβολος I 1, line 3, after 'Hdt. 9.94' insert 'ἀμφοτέρων ἐ. Call.fr. 384.44 Pf.'; last line, after 'Hld. 10.20' insert 'hitting the mark, effective, ὁρμή A.R. 2.1280; μῆτις id. 4.1380' **2**, last line, delete 'πάντεσσιν .. 4.1380' **II**, for this section read 'having been attained, ἐπήβολός ἐστ' ἀλεωρή A.R. 1.694, ἐπήβολος ἅρματι νύσσα id. 3.1272'

ἐπήκοος II, lines 5ff., abs., add 'SEG 30.519-526 (Corcyra, c.500 BC); of Apollo, Mitchell N.Galatia 154,; of Hermes, Salamine 44 (ii/i BC)'

ἐπηλύς II, after 'neut. sg.' insert 'ἔπηλυ πλῆθος Heraclit.Ep. 9.6'

ἐπηλύσις, for 'assault' read 'onset' and after it insert 'Archil. 196A.50 W., Call.fr. 331 Pf.'; add 'ποταμοῦ (sc. the Nile) GDRK 60 ii 1'

ἐπήλυτος, add 'Lxx Jb. 20.26'

*****ἐπήμενοι**, v. ἔφημαι III.

ἐπηρεάζω I and **II**, combine these sections under definition 'treat vexatiously, obstruct' **III**, add 'Gal. 9.283.9, 14'

ἐπηρεασμός, line 1, for 'despiteful treatment' read 'vexatious or obstructive conduct'; line 3, after 'cf. (Arist.Rh.) 1382ᵃ2' insert 'Men.Dysc. 178'

ἐπηρεαστής, for 'insolent person' read 'ill-disposed or hostile person'

ἐπήρεια, line 1, delete '(ἐπί, ἄρος)' and for 'insulting treatment, abuse' read 'hostile or obstructive attitude or conduct (so in subsequent glosses)'; add 'μείζονα ἐπήριαν προσάγουσιν οἱ πρακτῆραις PCol. 174.7 (iv AD)'

†ἐπήρης, ες, of ships, equipped with oars, νῆες A.R. 1.235 (pap.), πλοῖα Agatharch. 83, κελήτιον Arr.An. 5.7.3; transf., ἐ. πτερύγεσσιν Max. 415.

ἐπητύς, for '(This .. ἐδ-η-τύς)' read '(etym. obscure)'

ἐπί, line 1, after '(iii BC)' insert 'sp. ἐβί GVI 1990.9 (Egypt, ii/i BC), 817.5 (Arcesine, Amorgos, ii/iii AD)' **B I c**, line 2, after '(lyr.), etc.' insert 'δῶρα δεχόμενον ἐπὶ τοῖς τῆς πατρίδος συμφέρουσιν Din. 2.26' **II 1**, add 'ἐπὶ τοῖς ἄστροις .. καὶ μάλιστα ἐπὶ κυνί Arist.HA 600ᵃ3-4' **III 3**, line 4, for 'on condition that .. ' read 'on condition that, w. fut. indic.' and delete 'in orat. obliq.'; line 5, after '(Hdt.) 7.154' insert 'Pl.Ap. 29c'; line 6, after 'Th. 1.126' insert 'ὁρκίσας ἐφ' ᾧ .. ἔσται SIG 684.26 (Dyme, ii BC)' **6**, at end delete 'ἐ. θυγατρί .. 4.154' add '**8** in the presence of, ἐπὶ τοῖς δικάζοσσιν IEphes. 1678 (vi BC).' **C I 1**, add 'c arith., multiplied by, PMich.III 145.3.5.6, al.; ἐπὶ ι' τέσσαρες μονάδες ἐπὶ η' γίνονται λβ' Papp. 26.10; ἐπόησα τὰς ε' ἐπὶ τὰς β' I multiplied five by a thousand, PMich.III 145.3 vi 4.' **5**, line 8, for 'Th. 1.50, cf. 62' read 'Th. 1.62.6, cf. 50.2' **II 1**, add '**b** ἐφ' ἔτος, v. ‡ἔφετος II.' **2**, add 'up to now, τῶν γ' ἐπὶ τάδε γεγενημένων Isoc. 9.37' **III 2**, at end delete 'κρείσσων ἐπ' ἀρετήν Democr. 181' **E I**, add 'also item, introducing an entry in accounts, ἐπί· ἔδωκα κτλ. SEG 7.387, cf. 381, al. (Dura, iii AD)' at end add 'Myc. e-pi (w. dat.)'

*****ἐπιάλης**· ὁ ἐφιάλτης Hsch.

†ἐπιάλτης, v. ‡ἐφιάλτης.

ἐπιάομαι, delete the entry (cf. Schwyzer 369.22).

*****Ἐπίασσα**· Δήμητρος ἐπώνυμον Hsch.

*****Ἐπιβάθρα 2**, for this section read 'landing-place, Plb. 3.24.14 (pl.), Ael.NA 12.15. **b** transf., place offering means of approach, "stepping-stone", "gangway" or sim., ἵνα, ἐὰν πρόθηται διαβαίνειν αὖθις εἰς τὴν Ἀσίαν, ἐπιβάθραν ἔχοι τὴν Ἄβυδον Plb. 16.29.2; Plu.Demetr. 8; fig.,

ἐπίβαθρον SUPPLEMENT ἐπιδημέω

γάμον ἐ. τισὶ γενέσθαι J.AJ 11.8.2, τῷ ἑξῆς λόγῳ Arr.Epict. 1.7.22, Plot. 1.6.1, Gal. 9.149.18.'

✝**ἐπίβαθρον**, τό, *embarkation fee, passage-money*, καὶ δέ κεν ἄλλ' ἐπίβαθρον .. δοίην Od. 15.449, Call.Del. 22 (pl.), PCair.Zen. 753.34, 36 (pl., iii BC), D.S. 1.96, Et.Gen., Suid.; transf., of a sacrifice before embarkation, A.R. 1.421 (pl.), ὅ γε πελαργὸς ἐπίβαθρόν τι τῆς ‹διαίτης› δίδωσιν Plu. 2.727f.   **II** *something on which to sit or rest*, of a stool, AP 9.140 (Claudian.), of a perch, ib. 9.661 (Jul.Aegypt.).   **III** *ladder or steps*, PSI 171.27 (pl., dub., ii BC); perh. τὠπίβαθρον τῶ θρόν[ω] τὸ χρύ[σι]ον (part of Zeus' statue at Olympia) Call.fr. 196.23 Pf.

ἐπιβαίνω II 1, add 'c *take possession*, w. gen., ἐπιβαίνειν τῶν προσηκόντων αὐτοῖς πραγμάτων Cod.Just. 1.3.52.7; w. dat., ib. 10.11.8.7a.'

ἐπιβάλλω I 1, add 'med., *cast lots for*, καὶ ἐπὶ κλήρους ἐβάλοντο Od. 14.209'   **4**, delete 'in Med. .. 23.27'   **6**, delete 'βλαστούς ib. 3.5.1'   **II 5**, add '**b** ὁ ἐπιβάλλων in law, *the next in succession*, Leg.Gort. 7.36, 11.42, al.'   **6 b**, delete the section.   **III 2 a**, line 4, delete '*take possession* .. 14.209'   **IV 1**, for 'Pass. .. *put upon*' read 'med., *place* upon (sc. ὀϊστούς)'   **2**, before 'to be set over' insert 'pass.'

ἐπιβαρύνω, add 'Hsch. s.v. ἐπεζάρηκεν'

ἐπιβάσκω, add '**2** perh. *encroach on*, Thess. 3 sg. pres. ἐπβάσκει IGC p. 10.8, al. (Larissa, ii BC).'

ἐπιβατήριος III 2, after '*disembarkation*' insert 'IGRom. 4.1542 (Smyrna, ii AD)'

ἐπιβάτης 4, for '*male quadruped*' read '*stallion*, w. ref. to its function in breeding'

ἐπιβατός II, for 'παίων' read 'παιών (but cf. Παιάν ad fin.)'

*****ἐπιβημάτίς**, ίδος, ἡ, *step* (at the foot of a structure), ISmyrna II(2). p. 355 ix (ii/iii AD).

✝**ἐπιβήτωρ**, ορος, ὁ, *one who sets foot on or dwells in*, ὕλης οὐρανίας Orph.fr. 353.   **b** *one that moves upon*, θοῶν ἐπιβήτορα κύκλων ἵππον (i.e. the Trojan horse) Triph. 307.   **2** *one who goes on board*, νεώς ἐ. λαόν (i.e. crew) AP 7.498 (Antip.Sid.); νηῶν ἐ. ἄνδρας (= ἐπιβάτας) Opp.H. 5.298.   **b** *one who mounts* a horse, etc., ἐ. ἵππων Od. 18.263, Simm. 1.3; of tamer of wild animals, θηροδιδασκαλίης ἐ. Man. 4.245.   **b** transf., adj. ἐ. παλμῷ Nonn.D. 20.113.   **2** of a male animal used for breeding, συῶν ἐπιβήτορα κάπρον Od. 11.131; Theoc. 25.128.

ἐπιβιάζομαι, for '*constrain besides*' read '*use force against* (in tomb robbery)' and add 'SEG 17.632.10 (Perge, Rom.imp.)'

✝**ἐπιβιβάζω**, (fut. -βιβῶ Lxx Ho. 10.11), *cause to go on board*, ἐπ' ὀλίγας ναῦς τοὺς ὁπλίτας .. ἐπιβιβάσαντες Th. 4.31.1.   **2** *cause to mount on*, ἐπιβιβάσατε τὸν υἱόν μου .. ἐπὶ τὴν ἡμίονον Lxx 3Ki. 1.33, ἐπιβιβάσας .. αὐτὸν ἐπὶ τὸ ἴδιον κτῆνος Ev.Luc. 10.34, pass., Apollod. 3.1.1; fig., Aristid.Or. 30(10).23.   **b** app. *put to the plough* (calque Heb.), Lxx Ho. 10.11.   **c** *set or place* on, ἐπιβιβάσαν τὴν χεῖρά σου ἐπὶ τὸ τόξον Lxx 4 Ki. 13.16.   **d** *cause to ride* or *tread* over, ἐπιβίβασας ἀνθρώπους ἐπὶ τὰς κεφαλὰς ἡμῶν Lxx Ps. 65(66).12, ἐπεβίβασας εἰς θάλασσαν τοὺς ἵππους σου Lxx Hb. 3.15, 19.

ἐπιβίοω, for 'only in' read 'impf. ἐπεβίουν Just.Nov. 66.1.4, aor. 1 part. ἐπιβιώσασα SEG 31.342 (Sparta, Rom.imp.)'

ἐπιβλέπω, after 'Lxx Le. 26.9' insert '(ἐπιγλέποι· ἐπόψεται Hsch. La., cod. ἐπιπλέγοι; cf. ποτιγλέπω)'   **I 1**, add 'abs. SEG 36.552.14 (Thessaly, ii BC)'   **2**, add '**b** w. acc., of the gods, *watch over*, Call.fr. 602.2 Pf.; cf. ἐφοράω 1.'   **3**, for this section read '*face in a particular direction*, ἐπιβλεπέτω δὲ τὸ μὲν κοῖλον αὐτῶν ἄνω, κάτω δὲ ἡ ἀποτομή Dsc. 5.120'

✝**ἐπίβλημα**, ατος, τό, *cloth thrown over as a covering*, Nicostr.Com. 13.3 K.-A., Gal.UP 11.12; as coverlet, bedspread, IG 12(5).593.4 (Iulis, v BC), Plu.Cat.Ma. 4, Arr.An. 6.29.5, Gal. 14.638.18, Sor. 1.85.   **b** included among items of clothing, IG 1³.403.61, 2².1514.31, Lxx Is. 3.22.   **c** used to patch clothing, Ev.Matt. 9.16, Ev.Marc. 2.21, etc.   **d** as an outer bandage, Paul.Aeg. 6.92 (146.4 H.).

*****ἐπίβλησις**, εως, ἡ, chemical operation expld. as συλλείωσις κατασπωμένη, Anon.Alch. 7.5.

ἐπιβοάω I 4, add 'med., TAM 2(3).838e19 (Lycia, ii AD)'

ἐπιβοήθεια, add 'SEG 32.1128.11 (Ephesus, i BC)'

ἐπιβόητος I, for 'Aeschrio 8' read 'AP 7.345'

*****ἐπιβολᾰδοποιός**, ὁ, *maker of* *ἐπιβολάδες (perh. mantles or wrappers) or of *ἐπιβολαίων (dim. of ἐπιβόλαιον), IG 2².11175 (iv BC).

ἐπιβόλαιον, add 'Hsch. s.v. καννάκαι'

ἐπιβολεύς, add '**2** gen. pl., dub. sens., IG 4².110A40 (Epid., building accounts, iv/iii BC).'

ἐπιβολή I, add '**6** *landfall*, Peripl.M.Rubr. 55.'   **II 3 b**, add '*extra payment*, APAW 1952(1).12'   add '**4** *compulsory allocation* of land to proprietors on which they were obliged to pay taxes, BGU 2023.11-12 (AD 198/201).'

*****ἐπίβολον**, τό, *linch-pin*, Gorg.ap.Poll. 1.145.

ἐπιβουλεύω I 4, add 'μὴ ἐπιβουλεύσῃς› τῷ τάφῳ SEG 33.311 (Laconia, Chr.)'

*****ἐπιβρίζω**, = ἐπιβρίθω, Nonn.D. 20.347, al.

ἐπιβρίθω, after 'Porph.Abst. 1.43' insert 'Plot. 5.3.15'

ἐπιβρύκω, at end after 'Herod. 6.13' insert 'cf. ἐπιβρύκων (pap. -βρυχον ante corr.) Hippon. 104.15 W.'

ἐπιβωμίς, add 'TAM 4(1).45.9'

ἐπίγαιος, add 'of evil spirits, PMasp. 188ᵛ (vi AD)'

ἐπιγαμβρεία, add 'also ἐπιγαμβρία, PHarris 202.8'

ἐπιγάμβρευσις, add 'gloss on κηδεία Exc.Barocc. 324'

ἐπιγαμέω, add '**II** med., of the woman, *marry as a second husband*, w. dat., Sch.Pl.Mx. 235e.'

ἐπιγαμία II b, add 'οἱ κατ' ἐπιγαμίαν οἰκεῖοι Modest.Dig. 26.6.2'

ἐπιδουπέω, add 'w. dat., κύμβαχος .. ἐπεγδούπησε κονίῃ *fell with a thud* on, Nonn.D. 36.218'

ἐπίγειος, line 2, before 'opp.' insert 'ὕδατα ἐ. *on the surface of the ground*, Thphr.CP 2.5.1'

ἐπιγεννάω, add 'σὺν τοῖς ἐπιγεννωμένοις *with* (*her*) *descendants*, SEG 24.498(a) (Maced., ii AD)'

ἐπιγεννητός, add '**b** gramm., *agnomen ex aliqua virtute forinsecus quaesitum, quod ἐπιγεννητόν Graeci dicunt*, Diom. p. 321 K.'

ἐπιγίγνομαι II 3, add 'Cod.Just. 1.2.17.2a'   add '**7** *come additionally into being*, Emp. 17.30 D.-K.'

ἐπιγιγνώσκω IV 2, add 'Cod.Just. 3.10.1.1'

*****ἐπιγλέποι**, v. °ἐπιβλέπω.

ἐπιγλισχραίνω, add 'Gal.CMG 5.9(1).245.17'

✝**ἐπιγλωσσάομαι**, Att. -ττάομαι, *utter forebodings*, μηδ' ἐπιγλωσσῶ κακά A.Ch. 1045, cf. Hsch. s.v. ἐπιγλωσσῶ· περὶ τῶν Ἀθηνῶν δ' οὐκ ἐπιγλωττήσομαι τοιοῦτον οὐδέν Ar.Lys. 37; w. gen., *against*, ταῦτ' ἐπιγλωσσᾷ Διός A.Pr. 928.

ἐπιγνώμων, after '*judge*' insert 'SEG 30.380.6, 7 (Tiryns, vii BC)'

ἐπίγνωσις I 1, add '**b** *recognition* (as mark of honour), *approval*, SEG 34.553.8 , 23.447.13 (both Demetrias, ii BC).'

ἐπιγογγύζω, add 'gloss on ἐπιμύζω Sch.Gen.Il. 8.457'

*****ἐπίγομος**, ὁ, *additional cargo*, PUG inv. DR 48 (ZPE 65.173).   **II** measure of capacity, Ἰωάνω ἐλαίο(υ) ἐπιγό(μους) γ' μετρητὰς ς' SB 7365 (AD 104).

ἐπιγόνειον, for 'Egyptian .. μάγαδις' read '*musical instrument with forty strings*, prob. a board zither' and add 'perh. also ἐ. (codd. ἐπιτόνιον) ψαλτήριον Ath. 10.456d'

ἐπιγονή I 1, line 4, after 'θρεμμάτων' insert 'Androt. 55 J.'   **I 2**, add '**b** πυρὸς συντέχνου μυρίας ἐπιγονάς *benefits resulting from* fire, Anon.ap.Suid. s.v. Ἀρίσταρχος Τεγεάτης.'   **II**, for 'later apptly. .. Πέρσης τῆς ἐ.' read 'Πέρσης τῆς ἐπιγονῆς *member of the lowest class of Greek settlers in Egypt*; later, applied to Hellenizing Egyptians'

ἐπίγονος II 2 c, at end delete 'τῷ Ἐπιγόνου .. (Cnidus)'

ἐπιγράβδην II, for '*like lines*' read '*in the form of letters*'

ἐπιγραμματογράφος, add 'SEG 28.1493 (Egypt, ii BC)'

ἐπιγραμματοποιός, add 'of Posidippus, IG 9²(1).17A24 (Thermus, iii BC); Ps.-Plu.Vit.Hom. 84 (of Antipater of Sidon).'

ἐπιγραφή, after lemma insert 'ἐπιγροφά IAskl.Epid. 52 A 16 (iv BC), cf. γροφά'   **1 a**, add 'εἰκόνα χαλκέαν ἐπὶ τῶ αὐτῶ βάματος ἐπιγραφὰν ἔχοισαν IKyme 13'   **3**, add 'ἐν τῶι δ[ι]κρότωι ὧι ἐπιγραφὴ Δη[μή]τηρ SEG 33.684 (Paros)'

ἐπιγυμνάζω, for 'Pass., .. abs.' read 'med., *take exercise*, Hp.Insomn. 88'

*****ἐπίγυος**, ον, *imminent*, PEnteux. 15.5 (iii BC), cf. ὑπόγυιος I.

ἐπιγώνιος I, for '*at the angle*' read '*situated at a corner*' and after 'Nicom.Ar. 1.19' add 'of buildings, etc., οἰκία SB 9902A ii 30 (iii AD), ψιλός ib. K 10, πύργος ib. Q ii 6.'

✝**ἐπιδαίομαι** (A) (δαίω A), *to burn with passion* for, w. dat. and acc. of respect, κίχλης δ' ἐπιδαίεται ἦτορ (κόσσυφος) Opp.H. 4.173.

ἐπιδαίομαι (B), for 'dub. sens.' read 'prob. corrupt'; delete 'Pass. .. Hes.Th. 789' (v. °ἐπιδατέομαι)

ἐπιδάκνω, delete 'med.' and transfer 'Nic.Al. 19, 121' to follow 'pass.'

ἐπιδαμιοργός, after 'al.' insert '(-ουργός BCH 52.174 (Delphi, i BC))'

*****ἐπιδατέομαι**, *allot*, Ὠκεανοῖο κέρας, δεκάτη δ' ἐπὶ μοίρα δέδασται Hes.Th. 789.

ἐπιδέρκομαι, add 'w. dat., AP 12.87.5 (Mel., tm.)'

ἐπιδερμίς I, add '**2** *conjunctiva*, Ruf.Onom. 28.'

ἐπίδεσμος, add 'cf. Myc. *o-pi-de-so-mo* (pl.), *bindings*'

ἐπιδετόν, v. Hp.Hum. 5.'

ἐπιδεής I, *in need of*, w. gen., add 'Theoc. 25.50'

ἐπιδεύομαι II, for 'later c. acc. rei' to end read 'perh. w. acc., ἀλκήν (codd., ἀλκῆς cj., ed.) A.R. 2.1220; act. τεθνάκην δ' ὀλίγω 'πιδεύσην φαίνομαι cod. in Sapph. 31.15 L.-P. (v. ἐπιδεύης II); ἀρχόμενος φείδου στρώμνας, μὴ τέρμ' ἐπιδεύης Ps.-Phoc. 138'

ἐπιδέω (Β) **II 2**, add 'Ἐννοδία Θεᾷ Νίκανδρος Παρμενίωνος εὐχὴν ἐπιδεδωμένος EAM 98 (ii AD)'

ἐπιδημέω III, line 8, after 'Philostr.VS 1.22.4' insert 'BCH 52.172

(Delphi, Hellen.), τὸς ἐγ Κῶι ἐπιδαμεῦντας SEG 33.675.7 (Cos, ii BC)'

ἐπιδημητικός II, add 'Just.Nov. 134.1'

ἐπιδημία 1, add 'as the title of a work by Ion Chius, Ath. 13.603e; visit of inspection, PLond. 1259 (iv AD); transf., βραχυτάτου δὲ τοῦ τῆς ἐπιδημίας ὄντος ἐν τῷ βίῳ χρόνῳ Plu. 2.117f' **4**, add 'SEG 39.733 (Rhodes, i BC)'

ἐπιδήμιος, add '5 epith. of Zeus, MDAI(A) 19.372 no. 4 (Bithynia, ii AD).' at end add 'cf. Myc. o-pi-da-mi-jo (pl.)'

ἐπίδημος 2, line 2, before 'οἱ ἐπίδαμοι' insert 'τὸν ἐπιδέμὸμ (sic) IG 1³.3.5'

*ἐπιδίδημι, bind, Hsch. s.v. λαμπάδιον.

ἐπιδίδωμι I, add '**8** nominate to an office, BGU 1022.16 (AD 196).'

*ἐπιδιετὲς ἡβῆσαι, written by some editors for ἐπὶ διετὲς ἡβ., e.g. Hyp.fr. 192; cf. διετής 1.

*ἐπιδικεύομαι, = ἐπιδικάζομαι, SEG 37.340.23 (Mantinea, iv BC).

ἐπίδικος II 1, add 'w. παρθένος Men.Asp. 349'

ἐπιδίμερής, for 'containing 1⅔' read 'standing in the proportion of 5:3'

ἐπιδίομαι, add 'A.Eu. 357 (tm., s.v.l.)'

ἐπιδίφριος II 1, add 'cf. τοὺς ἄγαν ἀχρείους καὶ μόνον τοῦ λαιμοῦ ὄντας ἀνθρώπους ἐπιδιφρίους καλοῦσι Arethas ad D.Chr. 7.110, p. 105 Sonny (Kiev, 1896)'

*ἐπιδοκεύω, watch, tm. ἀλλ' ἐπὶ καὶ τὰ δόκευε περισκοπέων ὑετοῖο Arat. 987. **2** look out for, expect, ἀλλ' ἐπὶ χεῖμα δόκευε id. 1018.

ἐπίδομα, add 'SEG 1.276.14 (Maced., ii AD) (pl.)'

†**ἐπιδόρπιος**, ον (α, ον Ath. 4.130c), used for the purposes of dinner or as an accompaniment to it, ὕδωρ Theoc. 13.36, ἐ. τρύφος μάζης Lyc. 607, 661, θήρη Opp.C. 2.7; cf. of the stomach, τεύχεος ἐπιδορπίου Nic.Al. 21; app., postprandial τράπεζαι Ath. l.c.

*ἐπιδοτέον, one must administer, Gal. 12.516.16.

ἐπιδρομή II, ἐξ ἐ. cursorily, add 'Ptol.Tetr. 55'

ἐπιδυσ‹ω›χεῖν, add '(ἐπιδυοχεῖν· ἐπιπωμάζειν La.; cf. δυοχοί Hsch.)'

ἐπιείκεια, after lemma insert 'ἐπείκεια ISelge 17 (Rom.imp.)'

ἐπιεικής III 1, line 2, after 'Hdt. 2.92' insert 'ἐ. πάλαι Pl. Tht. 142a'; line 4, after 'Hell.Oxy. 13.5' insert 'ἐ. ἔντιμον D. 56.9'

*ἐπιενεχυρέω, med. pf. ἐπιενεχύρειμαι receive as additional pledge, SB 9834.b56 (iii AD); cf. ἐπενέχυρον.

ἐπιέννυμι, line 9, after 'Il. 14.350' insert 'γὰν ἐπιέμμενοι Alc. 129.17 L.-P.'

*ἐπιζαής, ές, blowing strongly against, prob. in Gr.Roman-Papyri 8.28.

*ἐπιζαμενής, ές, violently angry; perh. neut. as adv., ἐπιζαμενὲς κοτέουσα Nic.Th. 181 (read by edd. as ἐπὶ ζαμενὲς κοτέουσα); cf. ἐπιζάφελος.

ἐπιζάρέω, after '= ἐπιβαρέω' insert 'fall heavily upon'; add 'Sch.Od. 22.9: pf. ἐπεζάρηκεν Hsch.'

ἐπιζάω, add 'PMil.Vogl. 207.33 (iii/ii BC)'

ἐπιζέω, line 2, for 'Stob.App. p. 9 G.' read 'Stob. 1.31.8'

ἐπιζητέω 1, add '**c** miss, regret the dead, IUrb.Rom. 452.9.' add '**5** examine medically, τινά Ptol.Tetr. 9.'

ἐπιζήτησις 3, for 'rendering .. examination' read 'claim, demand'; add 'for taxes, μετὰ μετόχ[ων τὴν] ἐπιζήτησιν PHels. 38.2, ἐ. παντοπωλίου OBodl. 81.2 (ii BC)'

*ἐπιζήτητος, ον, missed, regretted, IUrb.Rom. 1012.

ἐπίηρα I, line 4, delete 'Antim. 87'; at end of article after 'ἦρα' insert '(B)'

ἐπιθαλάμιος II, add 'also neut., Sch.Sapph. 103.17 L.-P. (s.v.l.), D.H.Comp. 25, Serv.A. 1.31'

†**ἐπιθάλλω**, bloom upon, Plot. 5.3.11 (v.l. ἐπιβάλλειν).

ἐπιθάνατος I, for 'sick to death' (twice) read 'terminally ill'; -τως ἔχειν, add 'Alex.Aphr.Pr. 3.175'

ἐπιθέατρον, after 'theatre' insert 'or perh. seats above the διάζωμα'

ἐπιθέτης II, for 'IG 3.1280a' read 'Sokolowski 2.26.10 (Tiryns, vi BC), IG 2².2361A17'

ἐπίθετος I, add '**4** rhet., adventitious, artificial, κόσμοι D.H.Dem. 1; φράσις ib. 4.'

ἐπίθημα 2, add 'SEG 35.209.17 (Marathon, ii AD)'

ἐπιθλίβω, for 'press upon the surface' read 'press down on'; after 'tread' insert 'Σκορπίον Arat. 84'; after 'Nonn.D. 7.91' insert 'of a gravestone pressing on the dead, AP 7.655 (Leon.Tarent.)'

ἐπιθολόω, before 'Max.Tyr.' insert 'cover with muddy water'

ἐπιθραύω, for 'cripple, Sopat.Rh. 8.332.3 W.'

ἐπίθυμα, after 'victim' insert 'SEG 19.335.55 (Tanagra, i BC)'

ἐπιθυμέω, line 3, delete '[Men.] ap.Clem.Al.Strom. 5.119' and after 'Lys. 20.3' insert 'w. acc. of thing, μηδὲ βελόνης ἔναμμ' ἐπιθυμήσῃς Men.fr. 683.11 K.-Th., Teles p. 42.12 H.'

ἐπιθύμιος, add '**2** νῦ]ν δέ μοι οὔτε .. [ἐστ'] ἐπιθύμιον οὔτε .. w. inf., it is not in my desires or on my mind, Ibyc. 1(a).11 P.; cf. καταθύμιος II.'

ἐπιθύριον, ἐπίθυρον, for 'lintel' read 'ornamental fitting on a door'

*ἐπιθυσία, ἡ, = ἐπίθυσις, rest. in IG 12(1).762A23.

ἐπιθύω II, add 'PMag. 4.1497'

ἐπιθώϊος, for 'under penalty of a fine' read 'involving the penalty of a fine'

*ἐπικαθάπτω, aor. pass. ἐπικατήφθη, gloss on ἐπὶ ἑάφθη, Sch.Gen.Il. 13.543; v. ἑάφθη.

ἐπικαθίστημι 3, after 'Plb. 2.2.11' insert 'ὁ μετὰ τὴν Φλαμινίου τελευτὴν ἐπικατασταθείς (sc. ὕπατος) id. 3.106.2'

*ἐπικαίνισμα, ατος, τό, novel or strange event, Cat.Cod.Astr. 8(3).195.6.

ἐπικαίριος I 2, after 'X.Oec. 15.11' insert 'AP 7.477 (Tymn.) (s.v.l.)'

ἐπικαίω I, add '**2** metaph. in pass., be inflamed with passion for, τινί Ath. 1.23d; ἐπί τινι Sch.Ar.Lys. 221.' **II 1**, line 2, after 'Pl.Ep. 340d' insert 'Men.Dysc. 754' **3**, add 'make a mark by branding on, ἐπικεκαῦσθαι βουσὶ καὶ ἡμιόνοις ῥόπαλον Str. 15.688'

*ἐπικάλυψις, εως, ἡ, covering over, Plu. 2.266e.

ἐπικαμπής, add 'of a promontory, Peripl.M.Rubr. 40'

ἐπικαρπία 1, add '**b** perh. amount of output imposed as a condition on a cultivator, Inscr.Cret. 4.43Ba9 (v BC).' **2**, usufruct, transfer 'D.H. 3.58' to section 1, and substitute 'Modest.Dig. 31.34.7, Just.Nov. 18.3'

ἐπικάρσιος I 1, after 'at an angle' insert 'ἀπέκλινε δ' ἄρ' αὐχένα .. ἐπικάρσιον Stesich. S15 ii 15' and add 'fig. σάφ' οἶδ' ὅτι πάντα βροτοῖς Ζεὺς ἐπικάρσια τέμνει Trag.adesp. 482 K.-S.' **2**, delete the section transferring ref. to section III. **III**, for 'striped garment' read 'transversely woven garment'

ἐπικαταβολή, add 'PEnteux. 14, 15 (iii BC), PTeb. 817 (182 BC)'

ἐπικαταλαμβάνω 1 a, line 3, after '(Epid.)' insert 'fig., of misfortunes, Sch.E.Hipp. 732' add '**c** reach a place, ἐπικαταλαμβάνομεν τὴν Ὀξυρυγχιτῶν POxy. 3932.4, ἐπικαταλήμψομαι τὴν πόλιν PMasp. 82.3 (both vi AD).'

ἐπικαταλείπω, for pres. ref. read 'SEG 6.550 (Pisidia)'

ἐπικατάραομαι 1, add 'SEG 34.1212 (Lydia)'

*ἐπικατασκευή, ἡ, building extension, τοῦ θεάτρου TAM 2.420 (Patara).

*ἐπικατατομή, ἡ, carrying of mine-workings beyond one's boundaries, Ath.Agora XIX P26.304, P27.97, 106 (iv BC).

ἐπικατέχω, add 'Heras ap.Gal. 12.594.11'

ἐπικατέχω, add 'pass., of land, to be subject to a further claim, PMich.II 121²ii 1.9 (i AD)'

*ἐπικατήφθη, v. °ἐπικαθάπτω.

ἐπικατορύσσομαι, delete the entry.

*ἐπικατορύσσω, bury, SPAW 1934.1030 (tab.defix.).

*ἐπικαυχάομαι, vaunt oneself, Hsch. s.v. ἐναροκτάντας.

ἐπίκειμαι II 2, add 'insist or harp on, τοῖσδε Emp. 113 D.-K.' **4**, for 'to be imposed' read 'to be applied to'

ἐπικείρω I, add 'ἀνδρῶν οὓς νῦν δαίμων ἐπέκειρεν A.Pers. 921 (v.l. ἀπ-)'

ἐπικερδαίνω, add 'Plu.Ant. 93'

*ἐπικερδία, ἡ, interest on money, BGU 2140.9 (AD 432).

*ἐπικεφαλίς, ίδος, ἡ, perh. bearing or axle-box, Poliorc. 220.22.

*ἐπικέφαλος, ον, allocated per head, ἐ. ὀβολός CID II 10A i 14.

ἐπίκηρος 1 a, line 4, delete 'βίος Call.Ep. 59' **2**, after 'hazardous' insert 'βίος Call.Epigr. 58.3 Pf.'

ἐπικηρυκεύομαι I 2, add 'πρὸς Λακεδαιμονίους ἐπικηρυκεύεσθαι D. 20.52'

ἐπικηρύσσω I, add '**4** propose, ψήφισμα rest. in SEG 19.124.21 (Athens, ii BC).'

ἐπικίχρημι, delete the entry (v. ‡ἐπιχράω (C)).

*ἐπικίων, ονος, ὁ, architrave (or a section of it), SEG 20.142 (Cyprus, ii BC).

ἐπικλάζω, for 'sound to' read 'add one's sound to'; add 'shout to, Arr.Cyn. 16.8'

†**ἐπικλαίω**, Att. -κλάω, add tears to an action just completed, Ar.Th. 1063, App.Pun. 53; w. dat., weep in response to, Nonn.D. 30.114.

ἐπικλασμός, for 'dub. .. (ii AD)' read '**2** additional taxes or dues, extra levy, ὁ ἐσόμενος ἐ. τοῦ ἐνεστῶτος γ' (ἔτους) PTeb. 391.27 (i AD), POxy. 899.9 (ii/iii AD), BGU 920.22 (ii AD), etc.'

ἐπικλάω II 2, add '(μουσικὴν) ἐπικεκλασμένην τοῖς μέλεσιν Plu. 2.397b' add '**III** perh. transfer a charge to someone, PPhilad. 1.47 (ii AD), POxy. 3792.31 (iv AD), τὸ .. λάχανον ἀποτιμῶνται καὶ τὴν τούτου διατίμησιν ἐπικλῶσι τῷ λαμβάνοντι κηπουρῷ Just.Nov. 64.1.'

ἐπικλεής, line 1, after 'famous' insert 'Simon. in POxy. 3965 fr. 2.14'

ἐπικλείω (B) 2, after 'A.R. 1.18' insert '(cj.)'

ἐπικλεσαιδόνα, after lemma insert '(ἐπικλεηδόνα La.)'

ἐπίκλην 1, add 'w. τό, Ἀνδρέας οὗ τὸ ἐπίκλην Κομιτᾶ BCH suppl. 8.247 (Philippi, ?v AD)' **2**, at end after 'IG 14.1018.6' insert 'Poll. 9.104'

ἐπίκληρος, lines 1/2, read 'ἐπίκληρος, ον, Dor. -κλᾶρος: ἡ -ος, heiress' etc. **3**, for this section read '**II** astrol., occupying a κλῆρος II 4, χρηματίζοντες ἀστέρες λέγονται οἱ ἐπὶ δυνάμεως ἐπίκεντροι ἢ ἐπίκληροι Cat.Cod.Astr. 8(4).215.'

ἐπικληρόω 1, line 5, after '(Samos, iv BC)' insert 'BCH 57.493 (Temnos, iii BC), cf. 496)'

ἐπίκλητος II 1, for this section read 'subst., invited guest, Ar.Pax 1266,

Men.*Dysc.* 608. **b** *supernumerary guest* (i.e. one brought by an invited guest), Plu. 2.707a.'

†ἐπίκλιντρον, τό, part of couch, perh. *elbow- or head-rest*, Ar.*Ec.* 907 (lyr.), *fr.* 41 K.-A., *IG* 1³.422.286 (v BC), 2².1541.26, 11(2).144.66 (both iv BC), Gal. 18(1).344, *Gp.* 13.14.9, Poll. 6.9, 10.34; cf. ‡*ἀνάκλιντρον.*

*ἐπικλοπάδᾶν, *slily, craftily*, σιγᾷ δ' ὅ γ' ἐπικλοπάδαν ἐνέρεισε μετώπῳ Stesich. *S*15 ii 6.

ἐπίκλοπος, add 'transf. epith., ὠκεῖαν ἑλὼν καὶ ἐπίκλοπον ἄγρην Opp.*H.* 3.270'

ἐπικλύζω I 2, line 2, for 'E.*Tr.* 1327' read 'E.*Tr.* 1326'

*ἐπίκλωσμα, ατος, τό, variant for °μετάκλωσμα (q.v.), Cyran.prol. l. 82 K.

*Ἐπικνίσιος, ὁ, (κνῖσα) epith. of Apollo, *Tit.Cam.* 120.

ἐπικοινάομαι, before '*consult*' insert 'w. dat.' and add '*SEG* 23.474 (Dodona, iv/iii BC)'

ἐπικοινόω II, after 'Pl.*Lg.* 631d' insert '(but perh. med. governing γάμους)'

*ἐπικονδύλιον, τό, *knuckle-ring*, *POsl.* 46.17 (iii AD).

ἐπικόπτω 5, delete the section.

*ἐπικορύσιος, ον, (*located*) *on the helmet*, Myc. *e-pi-ko-ru-si-jo* (dual); cf. *o-pi-ko-ru-si-ja* (neut. pl.).

ἐπικοτέω, add 'unless to be read as ἐπιζαμενὲς κοτέουσα, v. °ἐπιζαμενής)'

ἐπικουρία, add '**IV** form of land-tax paid in barley, *PCair.Isidor.* 11 ii 27 (iv AD).'

*ἐπίκουρος (B), ον, prob. *ready for shearing*, *PCair.Zen.* 771.6 (iii BC); cf. κουρά.

ἐπικουφίζω II 2 a, add 'χρήμασι Arr.*An.* 2.18.4'

ἐπικουφισμός, add '*PAlex.* p. 40 no. 271 (v AD)'

ἐπικραδαίνω, for '*wave on high*' read '*brandish at*'

ἐπικραίνω, add '**2** *confirm, guarantee*, w. inf., A.*Supp.* 13 (anap.); w. acc., id.*Eu.* 949 (anap.); med., w. acc., ib. 969 (anap.); pass., ib. 347 (lyr.).'

*ἐπικράνιον, τό, = ἐπίκρανον I, *headdress*, Lindos 487.7 (iii AD, pl.).

*ἐπικράνιος, ον, *placed on the head*, στεφάνωμα Hsch. s.v. κράδεμνον.

*ἐπικραστίζω, *pasture* horses, *PTeb.* 724.2 (ii BC).

ἐπικρατής, for '*master* of a thing: only Comp.' read '*having the mastery*, Sch.E.*Ph.* 1058; esp. comp.'; after 'Hes.*Sc.* 321' insert 'Stesich. *S*40.24'

ἐπικρατητικός, delete the entry.

†ἐπικρατύνω, *increase the force, potency*, etc., *of, strengthen*, νοῦσον Hp.*Morb.* 4.49, ἐπικρατύνειν· ἐπισχύειν, ὀχυροῦν Hsch.

ἐπίκριμα, add '*IMylasa* 132.2'

ἐπικρίνω I 1, line 8, after '2*Ma.* 4.47' insert 'ib. 3*Ma.* 4.2' **II 2**, after '*POxy.* 39 (i AD)' insert 'ἐπικριθείς one included in a category of recipients of grain, ib. 2892 i 7 (AD 269), 2894 ii 8 (AD 270), etc.'

*ἐπίκριος, ον, *roosting*, prob. in Nic.*Th.* 198.

ἐπίκρισις II, add '*PMich.*XIV 676.5, 19 (AD 272)'

†ἐπίκροκον ἐπανθητόν (σπαθητόν La.) Hsch.; the name of a woman's garment, Paul.ex Fest. p. 82 M., etc.

ἐπικρούω I, for '*jeer at* .. Ath. 13.579b' read '*impugn*, Macho 240 G.; cf. Ἀριστοφάνης δ' ἐν Πλούτῳ καὶ τῷ ἐπικρούσασθαι ἐπὶ τοῦ νουθετῆσαι κέχρηται Arist.*fr.* 432 K.-A.'

ἐπικτηνίτης, for '*drover*' read '*head ostler*' and add '*IGBulg.* 1519 (sp. -είτης)'

ἐπικυδής, add 'sup. οἱ ἐπικυδέστατοι Ἀθηναίων Philostr.*VA* 8.15'

*ἐπικύρημα, ατος, τό, gloss on κύρμα, Sch.Gen.*Il.* 17.151.

*ἐπικώκυτος, ον, *lamented over*, *GVI* 1279 6.6 (Callatis, ii/iii AD).

ἐπικωκύω, add 'w. tm., ἐπ' οὐλοὰ κωκύσαντες *GVI* 1990.3 (Egypt, ii BC)'

*Ἐπικωμαῖος, ὁ, epith. of Apollo at Thurii, Thphr.*fr.* 97.3.

ἐπίκωμος, add '*BGU* 2430.43 (i BC)'

†ἐπίκωμος (B), *staying* or *residing in a κώμη*, Call.*fr.* 384.49 Pf.

ἐπιλαλέω, for '*interrupt in speaking*' read '*speak against*'; for '*charm*' read '*utter spells against*'

ἐπιλαμβάνω II 1 a, after 'Th. 2.51' insert 'of sleep, Hp.*Ep.* 5.28' **b**, line 3, after 'Pl.*Epin.* 974a' insert 'abs., ὡς νὺξ ἐπέλαβεν Memn. 40.2' **III 1**, line 3, before 'c. gen.' insert 'usu.'; add 'w. acc., Ev.*Luc.* 9.47, 23.26' add '**b** of taking hold of a person to help him, Lxx *Si.* 4.11, *Ep.Heb.* 2.16, Sch.A.*Pers.* 742 codd.; cf. ἀντιλαμβάνω II 2.'

*ἐπιλάσκω, pf. -λέληκα, *fill with squawking*, Opp.*H.* 3.247 (tm.).

ἐπιλέγω I 3, pass., add 'ἀπὸ χωρίου ἐπιλεγωμένου Πίβρου (i.e. *known as*), *SEG* 26.790 (Byzantium, vi AD)' **III 1**, line 2, after 'ταῦτα' insert 'B.5.136 S.-M.'; delete 'rare in trag. .. A.*Ag.* 1498 (anap.)' (passage corrupt)

ἐπιλεκτάρχης, for '*commander* .. *band*' read 'title of certain Aetolian officers'

ἐπίλεκτος 2, add '**d** ἐ. κριτής = *selectus iudex*, *IGRom.* 3.778.9 (Attalea).'

ἐπιλήθω II 2, add 'w. acc., ἵνα τοὺς μεταξὺ κινδύνους ἑκὼν ἐπιλάθωμαι Demad. 11'

ἐπιλήκητος, after lemma insert '(ἐπιλήμητος La.)'

ἐπιληκυθίστρια, for '*comic nickname* .. *bombastical*' read '*one who utters with a booming voice*, applied derogatorily to the muse of Mnasalcas'

ἐπιλήνιος I, delete 'ἐπιλήνια .. *C.* 1.127' **II**, add '**3** ἐπιλήνιον, τό, perh. *wine-vat*, Opp.*C.* 1.127; cf. ὑπολήνιον and Suid. s.v. τριπτῆρα.'

ἐπιληψία II, add '*SEG* 30.1794 (iii AD)'

*Ἐπιλιμένιος, α, ον, *dwelling by the harbour*, title of Hera, *IG* 12 suppl. 409 (Thasos); of Aphrodite, *SEG* 28.1596 (Aegina, v BC).

Ἐπιλίμνιος, add 'cf. Myc. *o-pi-ri-mi-ni-jo*, pers. n.'

*ἐπιλινεύω, = ἐπιλινάω, Hsch. s.v. λινοπτάζει.

*ἐπιλίτραις (ἐπίλιτρίς La.)· τὸ μέσον τοῦ ζυγοῦ Hsch.

ἐπιλίζω 2, for '*blink*' read 'perh. *fix the gaze*'

*ἐπιλλύζω, v. ἐπιλύζω.

ἐπιλογιστέον, for '*reckon*' read '*consider*'

ἐπίλογος I, add 'prob. also Aeol., cf. κα]τ' ἐπίλλογ[ον Alc. 204.2 L.-P. (κατ' ἐπιλογισμόν Sch.)' **II**, add '**4** last part of the νόμος κιθαρῳδικός Poll. 4.66.'

ἐπίλογχος (B), add '*POxy.* 2894.13, 2896.2 (both iii AD)'

*ἐπιλοιπογράφέω, perh. *carry over additionally*, (cf. λοιπογραφέω), *PTeb.* 718.9 (ii BC) (dub.).

ἐπίλοιπος, add 'cf. Myc. *o-pi-ro-qo*'

*Ἐπίλυκοι, οἱ, *members of a religious group*, *SEG* 35.989 (Knossos, ii/i BC).

ἐπίλυσις, add '**6** perh. *loss of skin* by peeling, or sim., ἄλφος καλεῖται ἡ ἐ. ἡ καλουμένη μελανία ἡ περὶ τὰς παρειὰς γινομένη ἐκ τοῦ ἡλιακοῦ καύσωνος Lxx Sch.*Le.* 13.39 (perh. read ἐπίχυσις).'

*ἐπιλυτέον, *one must solve the question*, ἐ. οὕτως Sch.Pi.*O.* 6.23.

*ἐπιλώγεον, v. ἐπιλόγεον.

ἐπιμάζιος, add 'Nonn.*D.* 3.380, Triph. 345, Poll. 2.8'

ἐπιμαίομαι I, at end after 'Timo 5.7' insert 'w. acc., *search for*, Arat. 89' **II 1**, add 'transf., of darkness, abs. Orph.*A.* 121' **III**, delete the section.

ἐπιμαρτυρία II, add 'Heph.*Astr.* 2.32.11'

ἐπίμαρτυς, line 2, for 'Call.*Aet.* 3.1.48' read 'Call.*fr.* 75.48 Pf., *AP* 12.129 (Arat.)'

ἐπιμείγνυμι II, line 3, for '(Hld.) 5.33' read '5.34'

ἐπίμεικτος 2, add 'εἰσὶν δὲ ἐπίξενοι καὶ ἐπίμικτοι Ἀράβων τε καὶ Ἰνδῶν Peripl.M.Rubr. 30'

ἐπιμελαίνομαι, delete the entry.

*ἐπιμελαίνω, *blacken on the surface*, Arist.ap.Stob. 1.29.1; med., as symptom of mortification, Hp.*Fract.* 35; of the tongue, id.*Morb.* 3.6; of ripening fruit, Thphr.*HP* 3.15.6.

†ἐπιμέλας, αινα, αν, *having a black surface*, Thphr.*HP* 3.8.7, 6.5.3; subst., kind of gem, *epimelas est, cum candida gemma superne nigricat* Plin.*HN* 37.161.

ἐπιμέλεια, line 13, for 'Isoc. 6.154' read 'Isoc. 5.154'; line 15, after 'Is. 7.14' insert 'ἵνα ἐν ἐπιμελείᾳ τῆς κόρης γενώμεθα Men.*Dysc.* 228'

ἐπιμελέομαι, form -μέλομαι, add 'ἐπιμέλεσθον *IG* 12(2).6.23 (Mytilene, iv BC)'

*ἐπιμελετής, οῦ, ὁ, = ἐπιμελητής, εἴλης *IGRom.* 3.642 (Temenothyrae); pl., Schwyzer 491.15 (-τάς, Thespiae, ?ii BC).

ἐπιμελής I, line 1, for '*careful or anxious about*' read '*careful about* or *attentive to*' add '**3** adv. -λῶς χλωροί *remarkably* sallow, Str. 14.2.3 codd.' **II**, line 5, for 'made him *anxious*' read '*became a matter of interest or concern*'

ἐπιμελητεύω, add 'τῆς πόλεως *IG* 2².1103.14 (ii AD), 3546.17 (Eleusis, i/ii AD), etc.'

ἐπιμελήτρια, for 'fem. of ἐπιμελητής' read '*woman in charge of domestic arrangements, welfare*, or sim.' and add '*PMasp.* 97.(D)35 (vi AD)'

†ἐπιμερής, ές, *superpartient*, of ratios which are not multiple (n:1), superparticular (n+1:n), multiple superparticular (nm+1:n), nor reciprocal to these, Theo Sm. p. 76 H., Nicom.*Ar.* 1.17, al.; cf. ‡ἐπιμόριος.

ἐπιμερισμός, line 4, delete '*parsing*'

†ἐπίμερος, ον, = ἐφίμερος, Alcm. 1.101 P., Semon. 7.51.

ἐπιμετρέω I, delete the section. **II**, for pres. def. read '*measure out in addition*' and add 'Hes.*Op.* 397, Hdt. 3.91.3'

†ἐπιμηθής, ές, *thinking after one acts*, i.e. *hasty*, Theoc. 25.79; adv. -θέως *on second thoughts*, Herod. 3.94.

*ἐπιμηκάζω, *bleat at*, Eust. 1761.26.

*Ἐπιμηλίδιος, ή, *protectress of sheep*, epith. of Apollo at Camirus, *Tit.Cam.* 135 (iii BC).

†ἐπιμήνιος, Ion., *magisterial college of the* ἐπιμήνιοι, (cf. *JHS* 82.4), *SIG* 58.11 (Miletus, v BC).

ἐπιμήνιος II 2 b, for 'Plb. 31.12.13' read 'Plb. 31.20.13'

ἐπιμηχανάομαι II, add '*devise to meet the need*, Opp.*H.* 1.322' ἐπιμιμνήσκομαι (not -μνήσκ-) **1**, add 'σταφυλαὶ δρεπάνης ἐπιμιμνήσκονται AP 11.37 (Antip.Thess.)'

*ἐπίμισθος, ον, *in receipt of payment, paid,* ἡ πρώτη σύνοδος οὐκ ἐ[πί]μισθος συνήχθη IEphes. 17.58 (i AD).

*ἐπίμιτρον, τό, *part* of a loom, perh. *rod-heddle,* POxy.Hels. 34.5 (AD 101), POxy. 264.4, 2773.14 (i AD; v. Berichtigungsl. VIII).

*ἐπιμνήμων, ονος, ὁ, official at Alabanda, BCH 10.312, 313.

ἐπιμοιράομαι, for '*receive* .. *grave*' read '*grant as one's due share,* ἐ. κόνιν sc. for burial'

*ἐπιμοιρασία, ἡ, *allocation of one's due share,* Sch.Hes.Th. 565 (p. 86.12 DiG.).

†ἐπιμορίασμός, ὁ, *formation of a ratio of the form n+ 1:n,* Iamb.*in Nic.* p. 108 P.

ἐπιμόριος, for pres. def. read '*having a ratio of the form n+ 1:n, superparticular*'

ἐπιμορμύρω, for '*murmur*' read '*froth, bubble*' and add 'med., ἐπιμορμύρεται An.Ox. 3.220'

ἐπιμύρομαι, delete 'An.Ox. 3.220' (v. ἐπιμορμύρω).

ἐπιμύω **II**, add '*also of ranks in battle,* PBerol. 6926.B iii. 22'

ἐπιμωκεύω, delete the entry.

ἐπίμωμος, after '*blameworthy*' insert 'Ptol.Tetr. 163'

*ἐπίναιον (A), τό, = ἐπίνειον, Inscr.Cret. 4.146.

*ἐπίναιον (B), τό, *addition to a temple,* ICS 1(b), sp. Cypr. e-pi-na-e[-a] ICS 1(a).

*ἐπίνακτον (-νάκτιον La.)· τὸν ἐπιδιδόμενον ἔξωθεν ναύτην Hsch. (cod. ἐπίνακτιν).

ἐπινάστιος, delete '*taken as a stranger into a country*'

*ἐπινεβεύω, aor. part. fem. ἐπινε[β]εύσασσα, *perform some (particular) service to Artemis;* cf. °νεβεύω, SEG 34.493 (Atrax, c.200 BC).

*ἐπινειόθι or ἐπὶ νειόθι, *towards the bottom,* A.R. 4.1615.

ἐπίνειος (s.v. ἐπίνειον), add '*τὰς ἐπινείους κώμας* PBeatty Panop. 2.47, 102 (iv AD); -οις .. τόποις ib. 110'

ἐπινέμω **I 1**, add '**c** *give extra* or *further grazing to* a flock, Longus 1.8.'

ἐπινίκιος **I**, add '**2** epith. of Zeus, ICilicie 16; of Hermes, ib. 17.' **II 2 a**, add 'perh. particular festival at Athens, SEG 26.184 (iii AD); games celebrating a victory, IEphes. 671, 721, al.'

ἐπινίκος, line 2, after '(dub. 1.)' insert 'τὸν ἐπινίκον ἀεθλ[ον] SEG 30.499 (Delphi, vii BC); as epith. of Heracles, Ἡρακλεῖ ἐπινίκ[ῳ] SEG 29.569 (Thessaly; unless -[ίῳ] is read)'; add 'rest. in Call.Dieg. viii.21 (fr. 198 Pf.)'

*ἐπινοστέω, *return,* τῇ πατρίδι Sch.Pi.O. 7.36 (v.l.).

ἐπινυκτίδιος, add 'cj. in (?)Call.fr. 775 Pf.'

ἐπινυμφεύομαι, after '*contract a second marriage*' insert 'or perh. *become betrothed* (v. SEG 33.669)'

ἐπινωμάω **I**, for '*bring* .. τινί' read '*visit,* w. dat.' and before 'σώματα' insert 'w. acc.'

†ἐπινῶς· τὸ λίαν Suid.; also v.l. (for ἐπιμανῶς) and Sch. in Luc.VH 2.25.

†ἐπινωτίζω, *put on one's back,* E.HF 362; med., Paus.Gr.fr. ε 6 E. **II** acc. to Hsch. = ἐφορμάω, Archipp. 5 K.-A.

*ἐπιξενεύω, *lodge with* or *at,* στρατιωτῶν ἐπιξενευσάντων τῷ οἴκῳ μου POxy. 3581.13 (iv/v AD).

ἐπίξενος **2**, for '*stranger*' read '*person residing away from his normal or registered domicile*' and add 'SB 4251.3 (i AD), PFay. 24 (AD 158), Peripl.M.Rubr. 30'

ἐπίξηνον, delete '*executioner's block*'

ἐπιξοά, for 'IG 4.1484.84 (Epid.)' read 'IG 4².103.17 (Epid., iv BC)'

ἐπιξύω **1**, add 'Crito ap.Gal. 13.863.6'

*ἐπιφοικίζω, v. ἐποικίζω.

*ἐπιφοικοδομέω, v. ἐποικοδομέω.

*ἐπίολπος, ον, *expected,* Q.S. 14.291, 295 (codd., cj. ἐπίελπτος).

*ἐπιονειδίζω, v. °ἐπονειδίζω.

†ἐπιοπτεύω, v. ‡ἐποπτεύω.

ἐπιορκέω, line 2, forms ἐφι-, after '(Smyrna, iii BC)' for 'etc.' read 'PTeb. 78/7 (ii BC), IG 5(1).1390.6 (Andania, i BC); sp. -ίω SEG 33.638 (Crete, ii BC)'

ἐπιορκία, add 'ἐφιορκίας γυναικίας SEG 33.1119 (Phrygia, iii AD)'

ἐπιόσσομαι, add '**2** abs., *keep watch,* GVI 1178 (Rhod.Peraea, ?ii BC).'

ἐπιόψομαι, line 2, after 'aor. 1' insert 'inf. ἐπιόφσασθ[αι IG 1³.3.4 (Marathon, v BC)' add '**II** *supervise,* IG 1³ l.c.'

*ἐπιπαιδειάζειν· τὸ μὴ ⟨ἐν καιρῷ⟩ θύειν φρατρίαν. Λάκωνες Hsch. (La.; cod. ἐπιπαίζειν).

ἐπίπαιμα, after lemma insert '(ἐπίπταιμα La.)'

*ἐπιπαλλακεύομαι, *take concubines,* Sch.E.Andr. 216.

*ἐπιπαρωθέω, for '*thrust aside, deflect,* PMich.III 149 xii 26, 37 (pass., ii AD).

ἐπίπας, after 'AP' insert '7.490 (Anyt., v.l.)'; delete '(Strat.)'; add 'οὐ μείους ὥσπερ χείλιοι ὡς ἐπίπαν *a thousand in all,* Xenoph. 3.4 W.'

*ἐπιπασσαλεύω, *peg upon, nail upon,* A.fr. 78a.19 R.

ἐπιπαστέον, add 'Aspasia ap.Aët. 16.94'

*ἐπιπατρίδιον, τό, *patronymic,* τὰ ὀνύματα κὴ τὰ ἐ. AD 1931-2(14) Pl. iii 4 (Thespiae, iii BC).

ἐπίπεδος **I**, add '**2** τὰ ἐ. (sc. γῆς) *the surface of the earth,* opp. τᾶς γᾶς ὑπένερθε (= Pi.fr. 292), Pl.Tht. 173e.' **III 1**, add '**b** τὸ ἐπίπεδον *face* of a solid figure, Simp.*in de An.* 68.7.'

ἐπιπέλομαι, for '*so of a storm*' read '*of blindness*'

*ἐπίπεμπτος **II**, for this section read 'ἐπίπεμπτον, τό, *one-fifth* (whether or not considered as an additional sum): of the votes in a trial, Ar.fr. 212 K.-A., Eup. 75 K.-A.; as a fine, penalty, or sim., Lxx Le. 5.16, IG 7.3073.1 (Lebadea, ii BC)'

ἐπιπέμπω **II 1**, add 'as an °ἐπίπλοος (A) III a; pass., ἐ. ὑπὸ το[ῦ] τριηράρχου GDI 4335 (Rhodes, ii BC)'

ἐπιπεντεκαιδέκατος, add '**b** ἐ. τό, *one-fifteenth in addition,* PLille 29.1.8 (iii BC).'

ἐπιπήγνυμι **II**, insert at beginning '*plant* or *fix on top,* σῆμα τύμβ[ῳ] MAMA 1.370 (Phrygia)'

ἐπιπλέκω **II**, add '**2** prob. *swindle,* τινά PEnteux. 48.7, 10 (iii BC).'

*ἐπιπλευστής, οῦ, ὁ, perh. = °ἐπίπλοος (A) III b, PIand. 150 ii 16.

ἐπιπλέω **III**, add '**b** *to be an* °ἐπίπλοος (A) III a, Plb. 16.5.1 (Rhodian ship).'

ἐπίπληξις **2**, transfer 'Lxx 2Ma. 7.33' to section 1.

ἐπίπλοος (A) **I 3**, delete the section. add '**III** ἐπίπλοος, ὁ, gloss on δίοπος Harp. **a** *officer in charge* of a ship, appointed by a trierarch to command in his stead, Clara Rhodos 8.228 (i BC), Arr.ap.Suid. s.v.; ἐπίπλοι τριήραρχοι SEG 29.799 (Rhodes, Hellen.). **b** *agent of the state in charge* of a cargo of corn, in Egypt, POxy. 276.3 (i AD), PLond. 301.10 (ii AD), PGrenf. 2.46.7 (ii AD); cf. °ἐπιπλέω III.'

*ἐπιπλωΐα, ἡ, the office of an °ἐπίπλοος III b, POxy.Hels. 20 i 9 (AD 138).

ἐπιπνέω **I 1**, line 4, after 'Od. 4.357' insert 'w. acc., Call.Del. 318' **II 3**, delete 'c. acc., .. 3.121' **III**, before 'Pass.' insert 'w. acc. *inspire,* Call.fr. 260.50 Pf., A.R. 3.937, Nonn.D. 3.121'

*ἐπιπολέω, *go up to* or *upon,* Sch.Hes.Th. 2 (p. 2.14 DiG.).

ἐπιπολή **I 1**, add '**b** sg., *some defined area* in Nemea, ὥρος ἐπιπολᾶς SEG 34.285 (Nemea, iv BC).'

*ἐπιπολιορκέω, *besiege in addition,* Arr.fr. 10 J.

*ἐπιπολυπραγμονέω, *inquire further into,* w. acc., Ptol.Tetr. 120.

ἐπιπομπή **2**, for '*enchantment*' read '*charm*' and add 'tab.defix. in SEG 35.214.15, etc. (Athens, iii AD)'

*ἐπίπομπος, ον, perh. *outlawed* or *banished,* ἀποτεισάτω .. τριάκοντα μνᾶς .. καὶ ἐ. ἔστω IG 9²(1).138.11 (Calydon, iv/iii BC).

ἐπιπορπίς, delete 'νυμφᾶν' and after 'AP 6.274' insert '(cj.)'

ἐπιπρεπής, after '*becoming*' insert 'τὰν ἐπιπρεπέα χάριν SEG 23.220 (Messene, i AD)'; add 'adv. -έως, epigr. in Lindos 177 (ii BC)'

†ἐπιπρίω, *gnaw at,* τὸ γένειον AP 7.531 (Antip. Thess.). **2** *grind* the teeth, ἐπιπρίσησιν (-βρι- codd.) ὀδόντας Call.fr. 332 Pf., Hsch.

ἐπιπρό, for '*right through, onwards*' read '*further, beyond that*' and add 'Call.fr. 238.22 Pf.'

*ἐπιπροσδέομαι, *ask in addition,* PColl.Youtie 26.2 (AD 156).

ἐπιπροσθετέω **2**, delete the section.

ἐπιπροσθέω, line 6, for '*stands* .. *view of*' read '*the centre is in line with the extremes*'; line 10, for 'Longin. 32.1' read 'Longin. 32.2'

ἐπιπροστίθημι, add 'pass. aor., Hp.Alim. 4'

†ἐπίπταισμα, ατος, τό, *stubbing* of the toes, τὰ δὲ ὑπὲρ τοὺς δακτύλους κρούματα πταίσματα. Ἀριστοφάνης δὲ καὶ ἐπιπταίσματα αὐτὰ καλεῖ Poll. 2.199 (Ar.fr. 818 K.-A.; see also ἐπίπαιμα).

*ἐπιπτερύσσομαι, aor. ἐπεπτερύχθην, *fly in pursuit of,* Cyran. 86 (3.6.3 K.).

ἐπίπτησις, add 'ὕπνου Afric.Cest. p. 38 V.'

ἐπιπτύσσω, line 3, for 'abs. .. *folds*' read '*corrugate*'

Ἐπιπυργιδία, add 'epith. of Artemis at Eleusis SEG 30.93 (20/19 BC)' and 'also masc., ἥρωι Ἐπιπυργιδίοι Sokolowski 2.19.86 (Athens, iv BC)'

ἐπιπώλησις, add 'also applied (in one ms.) to the middle section (85-152) of Theoc. 25'

ἐπίρραμμα, after '*Gloss.*' insert 'ἐπίραμμα Inscr.Délos 1409Baii118 (ii BC)'

ἐπιρράπτω **2**, add 'Hsch. s.v. κάθαπτος'

*ἐπιρραχῖτις, ιδος, fem. adj. *spinal,* ἀρτηρίαι Hippiatr. 33.5.

*ἐπίρρεγμα, ἐπίρεγμα· ἐπίθυμα Hsch.

ἐπιρρέζω **1**, add 'ἐπιρέξαι· ἐπιθῦσαι. ἐπαγαγεῖν Hsch.'

ἐπιρρέπω **I 2**, add '**b** *be impending,* δί[κ]ην ἐπιρρέ[π]ουσαν ἐδε[δοίκ]ειν Antipho fr. 1a col. 1.'

ἐπιρρέπω **2**, line 7, for '(Pl.) Tht. 177e' read '(Pl.) Tht. 177c'

ἐπιρρήγνυμι, after 'A.Pers. 1030 (lyr.)' insert 'ἵνα καὶ σοὶ ἐπιρρήξαιμι χ[ιτῶνα Call. in Suppl.Hell. 287.13'; add 'pass., of a storm, *burst,* ἐπιρραγέντος ὑετοῦ Ael.NA 7.8'

ἐπίρρημα **II**, add '**2** ἐ. σχετλιαστικόν *interjection* expressing distress,

D.T. 642.2, A.D.*Pron.* 34.30, *Adv.* 127.19, Sch.Ar.*Nu.* 1; [ἐ.] θαυμαστικά D.T. 642.7; ἐ. θρηνητικόν Tz. ad Lyc. 31.' **III**, delete the section.

ἐπίρρητος I, add 'Philostr.*VA* 1.12, 5.7, *Ep.* 38; sup., Origenes *Cels.* 3.50'

ἐπίρριν, ινος, long-nosed, POxy. 3617.9 (ἐπίριν, iii AD); cf. ἐπίρρινος.

ἐπιρρῑπίζω, for 'dub. sens. in' read 'w. dat., *fan into flame*, in quot. fig.'

ἐπιρρίπτω I, add '**6** *put* oneself *forward*, τοὺς βουλομένους ἐπιρρίπτειν ἑαυτούς (sc. to become tutors) Modest.*Dig.* 26.5.21.6.' **II**, delete the section.

ἐπίρριψις, add '**2** *imposition*, τελῶν *IMylasa* 601.7 (sp. ἐπιρειψιν).'

ἐπίρροια, after 'D.S. 5.25' insert '*POxy.* 2341.5, *PLond.* 934 (*Tyche* 1.8; both iii AD); ὑδάτων .. ἐπιρροίαις *conduits*, Luc.*Phal.* 1.3'

ἐπιρροιβδέω 1, for '*croak so as to forebode rain*' read '*make a whirring or rushing sound*'

ἐπιρροφέω, line 4, delete 'Archig.ap.'

ἐπιρρυθμίζω, delete '*dress* oneself *simply*' and insert 'Chaerem. 1.3 S.'

ἐπίρρῠσις I, add '**2** in irrigation, *water-intake*, *PTeb.* 703.31, 37 (iii BC).'

ἐπιρρύσμιος, for '*adventitious*' read '*remoulding*'

*ἐπισᾰλᾰγέω, move violently towards one, tm. σαλαγεῦντος ἔπι δροσεροῖο Νότοιο Opp. C. 4.74 (v.l. σελαγ-).

*ἐπισεβάζω, consecrate, εὐχαριστήριον ἐπισεβάσας τὰ τῶν προγόνων TAM 4(1).76 (Bithynia; cf. κατασεβάζω).

ἐπίσειστος I, add '**b** σειστὰ ἐξ ἀμφοτέρων τῶμ μερῶν ἐπίσειστα *earrings to be worn pendant* on both sides, *Inscr.Délos* 461*B*a5 (ii BC).'

ἐπισείω 1, add 'φόβον ἐ. Lib.*Or.* 56.11'

ἐπίσημα (form ἐπίσαμα), for '*Schwyzer* 607' read '*grave-marker*, *SEG* 24.405'

ἐπισημαίνω II, add '**2** *give a signal for action*, Men.*Pk.* 476, Phot. p. 153.17.'

ἐπισημασία I, line 4, after 'cf. (Cic.*Att.*)14.3.2 (sg.)' insert '*TAM* 5(1).48 (ii/i BC)'

ἐπισημειόομαι, add '**II** *indicate*, w. acc. and inf., Memn. 60.3.'

ἐπίσημον I, line 4, after 'Plu.*Thes.* 6' insert '*Peripl.M.Rubr.* 47'

ἐπίσημος I, add '**2** ἐπίσημα, τά, *badges* or *tokens of office* (app. calque of Lat. *insignia*), Just.*Ed.* 8.3.4.' **II 3**, add 'ἐπίσημοι ἡμέραι *feast-days*, *IGRom.* 4.860 (Laodicea ad Lycum)' add '**5** *conspicuous*, ἐν ἐπισήμοις τόποις *SEG* 24.1217 (Egypt, 39 BC).'

*ἐπισιλλαίνω, ridicule, Sch.Pi.*N.* 4.60.

ἐπισκαλμίς, add 'Agath. 5.22.2 K.'

ἐπισκάπτω II, for this section read '*cover by digging* or *hoeing*, *PSoterichos* 1.25 (AD 69), 2.21 (AD 221), τὰ σπαρέντα τὸ μὲν κάλλιστον δι' ἀνθρώπων ἐπισκάπτεσθαι Gp. 2.24.1, Hsch. (s.v. ἐπισκαφεύς)'

⁺**ἐπισκᾰφεύς**, έως, ὁ, *one who hoes in seed*, Hsch.

ἐπισκέπτης, add 'of official dealing with the determining of land under cultivation within a nome, *PMich*.inv. 341 (*ZPE* 36.80; ii/iii AD)'

*ἐπισκεπτίτης, ου, ὁ, app. = ἐπισκέπτης, *inspector*, *MAMA* 7.190, *IApam.* 130.

ἐπισκέπτομαι, after lemma insert '(also act., Hsch.)'

ἐπισκέπω, for 'med.' read 'pass.'

ἐπισκευάζω I, add '**4** *prepare, construct*, Mitchell *N.Galatia* 117, ἐπισκευάσαντες τὸ μνῆμα τῷ Δομετιανῷ ib. 179.'

ἐπισκευή I, add 'πὲρ τᾶς ἐπισκευᾶς τοῖ γυμνάσσοι *SEG* 33.460 (Larissa, ii BC)' **II**, add '**2** sg., *furniture*, Memn. 4.5 J.'

ἐπισκευόω, delete 'Ἐφ.Ἀρχ. .. (Crete)'

ἐπίσκεψις 1, add 'w. implication of calling to account or punishing, Lxx *Nu.* 16.29, *Je.* 9.23, 23.12, etc.' add '**4** *oversight, charge*, Lxx *Nu.* 3.36, 1*Ch.* 24.3.'

ἐπισκηνόω, after '*to be quartered in*' insert '*SEG* 33.870 (Caria, *c*.203 BC)'

ἐπισκήπτω II 3, add '**b** *rely, base oneself* on, τοῖς τῆς διαθήκης ἐπισκήψαντες ῥήμασιν Just.*Nov.* 159.1.' **III**, add 'also of accusations for homicide, Lys. 3.39, Is.*fr.* 137'

ἐπισκιάζω, add '**5** *follow closely in pursuit*, Arr.*Cyn.* 16.3.'

ἐπισκιρρόομαι, add '(dub., ed. σκιρ-)'

*ἐπισκοπάζω, *inspect*, dub. rest. in *POxy.* 3410.9 (iv AD).

ἐπισκοπεία, add '**2** (sp. -σκοπία) *office of* ἐπίσκοπος (in quot., *inspector of weights*), *BE* 1971.61 (Palestine).'

ἐπισκοπεύω, for '= sq.' read '*keep watch over*' add '**b** *serve as ἐπίσκοπος* in the supervision of building operations, *Syria* 29.317 (Arabia, iii AD). **c** *to be a bishop*, *BCH* suppl. 8 no. 2 (Edessa), *SEG* 37.479 (Thessaly, both v/vi AD).'

ἐπισκοπέω 5, add '*to be a bishop*, *SEG* 37.479 (Thessaly, v/vi AD), *Cod.Just.* 1.3.41.6'

ἐπισκοπή I, add '**2** *visitation, punishment*, Lxx *Is.* 24.22, 29.6.'

*ἐπισκοπία, v. °ἐπισκοπεία.

*ἐπισκόπιον, τό, *office of the* ἐπίσκοπος, *PSI* 1310.26 (ii BC).

ἐπίσκοπος (A) **3**, line 2, delete '*municipal*'; add '*inspector of weights and measures*, *BE* 1971.61 (Palestine)' **4**, add 'in the more fully developed ministry, *bishop*, Ἐπιφάνης ἐπι[σ]κόπου Εὐγενίου Mitchell *N.Galatia* 135, *SEG* 31.1396 (Syria, v AD), etc.'

ἐπισκοτέω, add 'ὑπὸ .. τῆς αἰσθήσεως ἐπισκοτεῖσθαι Iamb.*Protr.* 3 (p. 13.6 P.)'

⁺**ἐπίσκυρος**, ὁ, *ball game played between two teams*, Hsch., Poll. 9.103, Sch.Pl.*Tht.* 146a; also ἡ ἐπίσκυρος (sc. παιδιά) Poll. 9.104; see also °σκύρος.

*ἐπίσκυρος, ὁ, sens. dub., Κεκροπίης τευμήσατ' ἐπίσκυρος Εὐρύκλεια *Suppl.Hell.* 1044.

⁺**ἐπισμᾰρᾰγέω**, *crash upon*, Opp.*H.* 2.159; *resound*, id.*C.* 2.78; w. cogn. acc., ἐ. ὕμνον τινί Nonn.*D.* 48.965.

⁺**ἐπισμήχω**, *smear* on, κωκνέει ῥοδαλῆσιν ἐπισμήχουσα (sc. dust) παρειαῖς Opp.*C.* 1.501.

*ἐπισμικρύνω, *belittle*, Corn.*Rh.* p. 378 H.

ἐπίσπαστρον I 2, add '*Inscr.Délos* 1417*A*ii10'

ἐπισπάω, line 2, after '(E.)*Andr.* 710' insert 'Tim.*Pers.* 15.144 P.' **3**, add '*draw in* a net, Sol. 23.3'

ἐπισπέρχω II 1, intr., for '*rage furiously*' read '*hasten on*'

⁺**ἐπισπορά**, ἡ, *sowing of one* (*parasitic*) *plant on another*, Thphr.*CP* 2.17.10. **II** *second sowing, sowing of an after-crop*, *PTeb.* 27.37, al., 375.14 (both ii AD).

ἐπισπουρία, for '= foreg.' read '*oversowing*'

ἐπισπουδάζω I, delete '*further*' and before '*Pr.*' insert '*bring about in haste*, ὕπαρξις ἐπισπουδαζομένη'

ἐπισπουδαστής II 2, add '*PColl.Youtie* 16.25 (109 BC)'

ἔπισσαι, **ἔπισσον**, delete the entries.

*ἔπισσος, α, ον, *later, subsequent*, Hecat. 363 J., Μνημοσύνης ἧδ' ὧδε γόνου χαρίεντος ἔπισσα (?)Call.*fr.* 735 Pf., ἔπισσον· τὸ ὕστερον γενόμενον Hsch.; cf. μέτασσαι.

*ἐπίσσοχον· ἀκόλουθον Hsch.

ἐπίσταθμος I, line 3, delete 'neut. pl. .. 4.173' and insert 'perh. humorously, of an image, Call.*Epigr.* 24 Pf.' **II**, for '*quartermaster .. AB* 253' read '*governor appointed over a city or state*, Isoc. 4.120; ἐ. Καρίας ib. 162, cf. *AB* 253, *EM* 364.36; in uncertain application, Plu. 2.612c: fig., ψυχὴν ἐ. σωμάτων Aristid.Quint. 2.2' **1 b** and **2**, delete these sections.

ἐπίσταλμα II, after '*Cod.Just.* 7.37.3.1c' insert '*requisitioning order*, *CPR* 8.37.12 (iv AD), *PMerton* 100.2, 6 (vii AD)'

ἐπίσταμαι IV 1, after 'Od. 21.406' insert 'cf., by back-formation, οὔπω .. νείκεος ἠπίσταντο Arat. 108'

ἐπιστάσιον, add '*EAM* 87.8 (ii BC)'

ἐπιστάσιος, after 'Plu.*Rom.* 18' insert 'fem. ἐπιστασίη, epith. of Aphrodite, perh. as patron of the ἐπιστάται, Thasos 1 p. 234 no. 24 (iv BC); delete etym. note and substitute '(ἐφίστημι, cf. perh. °ἐπίστασις VI)'

ἐπίστασις II 1, add '**c** *remission of fever*, Erasistr.ap.Gal. 11.208.13, 16.' **3**, delete 'κατὰ τὴν ἐ. .. (Samos. vi BC)' **V**, after 'Cypr.' insert '*e-pi-si-ta-i-se*' and for '*Inscr.Cypr.* 144 H.' read '*ICS* 264.3' add '**VI** *apparition, presence* (divine or heroic), *SEG* 30.1080 (Samos, *c*.500 BC), ib. 1517 (Perge, v/iv BC), Robert *Hell.* 11/12.544 (Miletus, ii AD).'

ἐπιστατεία, add '**IV** app., *fee for the superintendence of cargo*, *BGU* 2274 i 4 (AD 155).'

ἐπιστεγάζω, add '*PMerton* 76.32 (ii AD)'

*ἐπίστεγος, ον, *roofed*, οἰκήματα *IRhod.Peraea* 352 A27, 353 A13, B10 (all iii/ii BC).

ἐπιστείχω, add 'χείματι δ' οὔποτε φασὶν ἐπιστείχειν ἁλὸς ὕδωρ πουλύποδας Opp.*H.* 2.241'

ἐπιστένω 1, add 'ἐπέστενε δ' αἶα νέκυσσι Q.S. 8.88'

*ἐπιστέργω, *give one's love to*, *GVI* 728.8 (Cappadocia, ii/iii AD).

*ἐπιστήκω, *superintend*, ἐπίστηκε (imp.) *PMich*.inv. 1610 (*ZPE* 35.104); ὁ -ων, *superintendent*, *PMich*.VIII 515.2 (iii AD).

ἐπιστήμη I, add '**3** perh. *discipline, sense of order*, Zos. 1.7.2, 2.32.2, 33.5, cf. δημοσία ἐ. ?*public order*, *POxy.* 3123.8 (AD 322).' **II**, line 3, for 'πλέως' read 'πλέων'

ἐπίστιος II, delete the section (v. ‡ἐφέστιος)

*ἐπιστολᾱφορία, ἡ, *office of letter-carrier*, *PPetaus* 84.3 (AD 185), *POxy.* 3095.10 (AD 217/8).

ἐπιστολεύς, after '*secretary*' insert '(= Lat. *ab epistolis*)' and add '*OGI* 679, Phryn. 203, 356'

ἐπιστολή 2, add '**b** *letter of appointment*, *Cod.Just.* 12.33.8.2.'

ἐπιστολῑοφόρος, add 'also ἐπιστολοφόρος, ἐ. πρὸς Κλυταιμήστραν Weitzmann *Illustrations in Roll and Codex* (Princeton, 1947) p. 20 (Megarian bowl, iii BC), *PMich*.III 217.21 (iii AD), *PSI* 887.4 (vi AD)'

ἐπιστολῐαῖος, line 2, after 'Ph. 2.533' insert '*POxy.* 3296.14 (iii AD)'

ἐπιστολογραφεῖον, form ἐπιστολᾱ-, add '*BCH* 58.291 (ἐπιστωλαγραφίον, Caria, iii BC)'

*ἐπιστολογρᾰφέω, *to be a secretary*, *SB* 7638.15 (iii BC).

129

*ἐπιστολοφόρος, ὁ, v. ‡ἐπιστοληφόρος.
ἐπιστομίζω I, at end after 'ἐπεστομίσθη' insert '(sc. Πῶλος, with a play on his name)'
ἐπιστρατεία, add 'D. 13.24, Paus. 8.25.4'
ἐπιστρεπτικός, add 'II capable of turning one from mental aberration, Horap. 2.117.'
*ἐπιστρέφειον, τό, app. = ἐπιστρέφεια, POxy. 3304.19 (AD 301).
ἐπιστρέφω I 1 b, after 'Plb. 1.47.8, 50.5' insert 'in a διέκπλους manoeuvre, Sosyl. 1 J.' II 4, for 'E.Andr. 101' read 'E.Andr. 1031' 5, at beginning for 'pf. part. pass.' read 'pf. pass. ἐπέστραμμαι to be vehement, Longin. 12.3'
ἐπιστρωφάω, add 'II spin, rotate, ἀχθινὸν ἐπιστρωφήσατ' ἄτρακτον epigr. in SEG 37.990 (Miletus, ii BC).'
ἐπισῡκοφαντέω, add 'PHels. 1.23 (ii BC)'
*ἐπισύμβᾱμα, ατος, τό, supervenient accident, Anon. in Cat. 48.3.
*ἐπισυμβῐόω, live in second marriage with, SB 7333 (s.v.l.; AD 186/7).
ἐπισυνάγω I, pass., add 'of contributions of money, SEG 31.122 (Attica, ii AD)'
ἐπισυνάπτω I 1, add 'b intr., follow on without a break, ἐπισυναπτούσης τῆς ἡμέρας Onas. 42.10.'
*ἐπισυναρμόζομαι, med., of a woman, join (herself) in a second marriage, ἀνδρί PFam.Teb. 13.56 (ii AD).
*ἐπισυνέρχομαι, come together towards, ἐκ τῶ ἀδήλω ἐς τὸ ὁρατὸν ἐπισυνερχόμενα Hippod.ap.Stob. 4.34.71.
*ἐπισυντῑμάομαι, make an additional valuation of, τοὺς τόπους PBremen 24.11 (ii AD).
+ἐπισῡρίζω, make a whistling sound at, Arr.Tact. 35.4, Nonn.D. 1.71, 170.
ἐπισύρω I 1, add 'b w. dat., cause to go along with, attach to, ὁρίζω τὸ ταύτης τέλος ἀεί ποτε βασιλικὸν ἐνέλκεσθαι καὶ ἐπισύρεσθαι καὶ ἐπαναστρέφεσθαι τῇ ἐμῇ περιουσίᾳ PMasp. 151.135, PFlor. 294.41 (both vi AD).'
*ἐπισφᾰλερός, ά, όν, stumbling, Nic.Al. 33.
ἐπισφίγγω, line 3, for 'ἐ. τοὺς ἀναγωγέας .. tight' read 'χρυσοῖς .. ἀνασπαστοῖς ἐπέσφιγγε τῶν βλαυτῶν τοὺς ἀναγωγέας'
ἐπισφρᾱγίζω I 2, after 'Vett.Val. 354.19' insert 'ἐ. τῇ ἰδίᾳ δυνάμει Vit.Aesop.(G) 91' II 2, line 3, before '(Pl.)Phd. 75d' delete 'prob. cj. in'
*ἐπισφράγισμα, ατος, τό, appendix, postscript, Afric.Cest. p. 48 V. (title of section).
ἐπισφρᾱγιστής, add 'sealing inspector who placed official seal on state granaries, POxy. 2841.12 (AD 85), Stud.Pal. 20.32.10, PKöln 94.9, 29, SB 10270(42).7 (all iii AD)'
*ἐπισχολάζω, study with, Πυθαγόρας .. τοῖς Μάγοις ἐπισχολάσας SEG 33.802B ii 27 (i AD).
*ἐπισχῡρίζω, enforce, Suppl.Mag. 45.53 (v AD).
ἐπίσχω I, line 2, for '[σελάννα] .. Sapph.Supp. 25.9' read 'intr., reach or extend over, φάος ἐπίσχει θάλασσαν ἐπ' ἀλμύραν Sapph. 96.9 L.-P.' III 1, add 'μικρὸν δ' ἐπίσχες Men.Dysc. 255 (dub.)'
*ἐπισᾠζω, continue to save, Εἰλειθυίῃ σωζούσῃ ἐπισωζούσῃ εὐχήν IG 2².4793 (ii AD).
ἐπίσωτρον, line 3, ὀπίσσωτρον, add 'PMasp. 279.20 (vi AD)'
ἐπιταγή 2, add 'κατ' ὀνίρου ἐπιταγήν TAM 4(1).60'
ἐπιτάδε, for 'in Mss.' read 'written'
ἐπιτάκτης, add 'as the name or title of a god, Pap.Lugd.Bat. xxv 8 iii 1 (after AD 231)'
ἐπιτακτικός, at end after '-κῶς' insert 'Arist.EE 1249ᵇ14'
ἐπίταλον, after lemma insert '(ἐπιταδόν La.)'
ἐπιτᾱξ III, for 'Call.Aet. 1.1.9' read 'Call.fr. 178.9 Pf.' and delete 'dub. in Iamb. 1.239'
ἐπιταξίδια, add 'cf. perh. ‡ἐπιτοξίς'
*ἐπιτᾰρᾰχώδης, ες, troubled, disordered, Ἄρης Κρόνῳ ἐπιμερίζων μῆνας καὶ ἡμέρας ἐπιταραχώδεις τοὺς χρόνους σημαίνει Heph.Astr. 2.32.10.
ἐπίτᾱσις, add '7 development of the plot of a play, between πρότασις and καταστροφή, Donat. in CGF p. 69.'
ἐπιτάφιος I, line 4, after 'Luc.Eun. 4' insert 'Certamen 63' add '2 neniae ἐπιτάφια Charis. p. 33 K.' II, for 'ἐπιτάφια .. ἀγών' read 'Ἐπιτάφια, τά, festival in honour of the dead'
ἐπιτεγξις I, add 'Gal. 18(2).570.4'
ἐπιτείνω I 1, line 3, for 'Id. 4.201' read '(Hdt.) 4.201.1; βαρὺν ζυγὸν αὐχένι Call.fr. 4 Pf.' 2 c, line 2, after 'Arist.Pol. 1308ᵇ4' insert 'ἐπετάθη πάντων .. τιμή J.AJ 9.14.2'
ἐπιτέλεια II, add 'SEG 23.448 (Thessaly, ii/iii AD); personified, καὶ Δίκαι καὶ Ἐπιτελείαι τῶν ἀγάθων SEG 36.750 (Lesbos, iv BC)'
ἐπιτέλειος, add '2 brought to fulfilment, ἐπιτελειᾶν δὲ τᾶν εὐχᾶν γενομενᾶν SEG 23.547.17 (Crete, 201/200 BC).'
*ἐπιτελευτάω, die in addition, SB 8979 (AD 180).
ἐπιτέλλω (B), for 'Pass.' read 'med.'; line 3, after 'intr. in Act.' insert 'Il.Parv. 9 B.'
*ἐπιτελωνέω, app. pay (tax), IMylasa 601.8 (i BC); cf. τελωνέω II.

ἐπιτέμνω I, add '3 = τέμνω II 2, ὅρκια ἐπιταμνέτω SEG 16.485.30 (see also ib. 35.921) (Chios, vi BC).'
ἐπίτερα, after lemma insert '(ἐπιτερῇ La.)'
+ἐπιτέρμιος, ον, marking the end, final, ..] δ' ὁρᾶτε τοὐπι[τ]έρμιον γάμου A.fr. 154a.5 R.; ἐπιτέρμιον· ἐπὶ τοῦ τέρματος, οἷον ἐπὶ τοῦ τέλους Hsch.; Ἐπιτέρμιος title of Hermes, id.
ἐπίτεταρτος, for 'ratio of 4:3' read 'ratio of 5:4'
ἐπιτετραέβδομος, for 'one plus four-sevenths' read 'standing in the proportion of 11:7'
*ἐπιτετραίνω, perforate, bore, τὰς παραετίδας (sc. δοκοὺς) ἆραι καὶ τὰς δοκοθήκας ἐπιτρῆσαι IG 11.161A55 (Delos, 279 BC).
ἐπιτετρᾰμερής, for 'one plus four-fifths' read 'standing in the proportion of 9:5'
ἐπιτετράπεμπτος, for '= foreg.' read 'standing in the proportion of 9:5'
ἐπίτευγμα, line 2, delete 'coup'
*ἐπιτευτάζειν· πραγματεύεσθαι ἢ σκαιωρεῖν Did.Plat. p. 245.
ἐπιτήδειος, after 'α, ον' insert '(ος, ον Th. 5.112.3)' I, add 'εἰ μὴ .. ἡλικίας καὶ τέχνης ἐστὶν ἐπιτήδειος Cod.Just. 11.8.16 pr.'
ἐπιτηδειότης I 1, add 'potentiality, capacity, Alex.Aphr. in APr. 184.7, in Top. 400.3'
+ἐπιτηλίς, ίδος, fem. adj. μήκων .. ἐ., perh. the same as μήκων κερατῖτις (Dsc. 4.65), Nic.Th. 852.
*ἐπίτηνα, adv., Dor. = ἐπέκεινα, δωρεὰν τὰν ἐπίτηνα τῶ ἅληκος de Franciscis Locr.Epiz. 23.11, 30.14, 31.9.
ἐπιτηρέω I, line 2, after 'καιρόν' insert 'Men.Dysc. 291' II 1, after 'App.BC 4.39' insert 'of guardian deities, Peripl.M.Rubr. 32' 2, add 'attend to business, ὅταν τις τῶν τριῶν ἀνδρῶν ἐπὶ τῆς καταστάσεως τῶν δημοσίων πραγμάτων ἐπιτηρῷ SEG 31.952.10 (Ephesus, Trajan)'
ἐπιτηρητής 2, add 'ἐπιτηρητοῦ οὐσιακῶν PPetaus 75.7 (AD 184)'
+ἐπιτηρία, ἡ, = ἐπιτήρησις (in quot. sense 2), ἐξ ἐ[πι]τερίᾱ IPamph. 3.4 (iv BC).
ἐπιτίθημι A III 1, add 'b add statements, etc., D.H.Isoc. 14, Is. 7 (pass.); ἐ. ὅτι ib. 4.' IV, line 6, after '(D.) 49.42' insert 'Men.Dysc. 308'
ἐπιτῑμάω II 1, add 'abs., inflict punishment, Decr.ap.D. 18.74' 2 b, add 'D. 18.294'
ἐπιτίμησις II 2, for 'heightening .. term' read 'rejection of one term in favour of another (usually a stronger term)' add 'III estimated value, PWisc. 15 (AD 236).'
ἐπίτιμος, delete the entry (v. BCH 58.497).
ἐπίτῑμος I, add 'adv. -ως, w. ref. to completion of, or discharge from, duty, with honour, La Carie 172, IGRom. 1.648, Mitchell N.Galatia 178, Modest.Dig. 27.1.8 pr., 6' II 1, add 'PCol.Zen. 100 (iii BC)' 2, add 'PTeb. 38 (113 BC)'
*ἐπιτιμωρέω, avenge, πατρί Sch.E.Or. 775; med., avenge oneself, Sch.E.Med. 465.
*ἐπιτίτλωσις, εως, ἡ, title of a criminal charge, Steph.in Rh. 286.1, v. ἐπίγραμμα 4.
*ἐπιτοιχογράφος, ὁ, writer of graffiti, SEG 29.974 (s.v.l., Ostia, Rom.imp.).
ἐπιτόνιον, line 2, delete 'prob. in' and after 'Ath. 10.456d' insert '(cj. °ἐπιγόνειον)'
ἐπιτοξίς 1, for 'dub. sens. in' read 'name of some ritual object, cf. prob. ἐπιτοξίδες· ἀγκυρίδες σιδηραῖ δίβολοι Phot. ε 1768'
*ἐπίτοπος, ον, positioned in a place, Plb. 3.40.4.
*ἐπιτρᾰπεζίδιον, τό, perh. small tray or table-ornament, ἐ. ἀργυροῦν ἔχον στεφάνην Inscr.Délos 1439Cb20 (ii BC).
ἐπιτρᾰπέζιος I, add 'neut. pl. subst., tableware, ἐπιτραπέδι[α (sic) Lang Ath.Agora xxi B13 (iv BC)'
*ἐπιτρᾰπεζόω, med., perh. provide extra food, SEG 18.21.9 (Athens, ii BC); cf. °ἐπιτραπέζωμα.
+ἐπιτρᾰπέζωμα, ατος, τό, dish served as an extra, prob. dessert, Pl.Com. 76 K.-A.
ἐπιτρέπω, lines 2/3, for 'Cret. fut. inf. .. 5024.12' read 'Cret. 1 sg. -τραψίω Inscr.Cret. 1.xviii 9 c 13 (Lyttos, ii BC), inf. -τραψῆν ib. 4.174 A12, 14 (Gortyn, ii BC)'
ἐπιτρέφω II, pass., add 'Hp.Nat.Hom. 12'
ἐπιτρέχω II 2, add 'b pervade, of literary qualities, D.H.Dem. 13, 41.'
ἐπιτρῐμερής, for '1 + 3/4' read 'standing in the proportion of 7:4'
ἐπιτροπεύσιμος, for 'subject to wardship' read 'capable of acting as guardian in Roman law'
+ἐπιτροπεύω, govern, w. acc., βυθίην Κύπρον ἐπιτροπέων IG 2².3662.10 (Eleusis, ii/iii AD); w. gen., ἐπιτροπ[έω]ν Θηβηίδος SEG 8.724 (Egyptian Thebes, ii AD), Bernand Inscr.Colosse Memnon 36.
ἐπιτροπή II 1, add 'used to render Lat. tutela, Cod.Just. 6.4.4.20a'
ἐπιτροπία, add 'τοῦδ' ἔλαχ' ἀθανάτων ὅστις ἐπιτροπίην epigr. in IMylasa 496.11 (EA 13.4). 2 guardianship, POxy. 2133.13 (iii AD); cf. ἐπιτροπή II.'

ἐπιτροπικός **II**, add 'IGBulg. 514'   add '**III** -ική, ή, power of attorney, PPhilad. 16.7, 14 (ii AD) (s.v.l.).'

ἐπίτροπος **I 1**, add 'PMich.xv 733.1, POxy. 1973.5, 2033.1 (all vi AD)'   **II**, add 'Modest.Dig. 19.2.49 pr., 27.1.13 pr.'

ἐπιτρώγω **II**, for 'generally, eat' read 'eat as an accompaniment, ἄλας'

ἐπιτυγχάνω, line 1, aor., add 'ἐπέτυχα SEG 34.1212 (Lydia)' **III 1**, line 2, for 'εὐχωλὰς Inscr.Cypr. 134 H.' read 'ta-se e-u-ko-la-se [e]-pe-tu-ke τὰς εὐχολὰς ἐπέτυχε ICS 220.4'; at end after 'abs.' insert 'Com.adesp. 357'   **4**, add 'ποιήματα D.S. 16.92.3'

*ἐπιτυρόομαι, turn cheesy, curdle, Nic.Al. 364 (v.l. ἐπιθρομβ-).

ἐπιτυχία **2**, add 'Alcid. 15.4 R.'

*ἐπιΰζω, yell at, w. dat., Suppl.Hell. 939.21.

ἐπιφαιδρύνω, after 'A.R. 4.663' add '(v.l. περι-)'

ἐπιφαίνω **II**, delete 'dawn' and insert 'μήτε δὲ ἡλίου μήτε ἄστρων ἐπιφαινόντων Act.Ap. 27.20; transf., θεὸς κύριος καὶ ἐπέφανεν ἡμῖν Lxx Ps. 117(118).27'

ἐπιφανής **II 1**, add 'ἡμέραι Mél.Glotz 290.28 (sup.; Delphi, ii BC)'   **3**, add 'title of Aphrodite, SEG 13.458.28 (Thasos, ii/i BC); of Artemis Iakinthotrophos, SEG 38.812 (Cos, iii/ii BC); sup., of Roman Emperors, SEG 31.932, 940 (Caria, AD 293/305)'   **III 1**, add 'comp. adv. ἐπιφανεστέρως Gal. 9.249.7, 9'

ἐπιφαύσκω, line 2, after '(Lxx Jb.) 31.26' insert 'Orph.H. 50.9 (s.v.l.)'

*ἐπιφερία, ή, dub. sens. in context relating to land allocation, perh. attached or marginal land, PAlex. 14.4, 13 (ii/iii AD).

ἐπιφημίζω **I 1**, add 'Arr.An. 5.3.1'   **III 1**, add 'med. aor. ἐπεφημίξαντο Arat. 442'

ἐπιφθίνω, add '**2** die in addition, ἁ δὲ πολυθρήνητος ἐπέφθιτο μήτερι κούρα epigr. in ICilicie 41 (i BC).'

ἐπιφιλοτιμία, delete the entry (v. SEG 29.139).

*ἐπιφλυκτίς, ίδος, ή, pimple or blister, Hsch. s.v. ὀλοφυκτίς (perh. f.l. for ἐπινυκτίς, cf. Phot.).

ἐπιφορά **I 1**, add 'of land granted to soldiers, PHamb. 168.7, 9 (iii BC)'   add '**b** as pay category (indicating rank), BGU 2367.8 (iii BC).'

ἐπιφόρημα, add '**II** additional fee, prov. Ἀβυδηνὸν ἐ., of a petty nuisance, Ath. 14.641a.'

*ἐπιφόριμα· ἐπὶ δέρμα La.) Hsch.

ἐπιφροσύνη, line 2, after 'Od. 5.437' insert 'Thgn. 1100'

*ἐπιφυτευτικός, ή, όν, i.e. held by the tenure of ‡ἐμφύτευσις, Dura⁶ 429 (parchment).

ἐπιφύω **II**, penultimate line, after '(Plu.)Pomp. 51' insert 'PMonac. 1.46 (vi AD)'

ἐπιφώνησις, add '**V** interjection, A.D.Adv. 121.15.'

ἐπίχαρις, adv., add 'Isoc. 15.132, id.Ep. 6.6'; delete 'Boeotian ἐπιχαρίτως dub. l. in Ar.Ach. 867'

ἐπίχαρμα, add '**III** cause for joy, E.Phaëth. 93 D., cf. Hsch.'

ἐπίχειρον **I**, add 'Lxx 2Ma. 15.33'   **II 1**, line 2, after 'Theoc.Ep. 18.8' insert 'εὐδοξίας ἐ. Pi.Pae. 14.31 S.-M.'; line 3, after 'ironically in' insert 'Trag.adesp. 664.20 K.-S., Call.fr. 260.59 Pf.'

ἐπιχειροτονέω **2**, delete 'τοὺς προέδρους .. hence   add '**3** put to the vote, lex ap.D. 24.39.'

*ἐπιχιτώνιος, ον, Myc. e-pi-ki-to-ni-ja (?neut. pl.), prob. ornaments on a tunic.

ἐπίχνοος, add 'Gal. 16.553.2'

ἐπιχράω (B), line 1, after 'ἐπέχραον' insert '(3 sg. ἐπίχρα Nic.Th. 14)'

ἐπιχράω (C) **I**, for this section read 'lend, τάγματα ὧν ἐπέχρησε δύο Καίσαρι Plu.Pomp. 52; ἐπιχρήσας ἑαυτὸν εἰς ἀπαλλοτρίωσιν ISmyrna 212.11'   **II 2**, add '**b** w. dat., occupy oneself with, Iamb.Protr. 20 (96.4 P.).'

*ἐπιχρεμετίζω, neigh at, Q.S. 8.57.

ἐπίχρισις, add '**2** coating with paint, IG 4².109 i 130 (Epid., iii BC).'

ἐπίχυμα **II**, add 'BGU 2333.13 (ii AD)'

ἐπιχύνω, for 'JHS 19.73 (Galatia)' read 'Mitchell N.Galatia 78'

*ἐπιχυσίδιον, τό, dim. of ἐπίχυσις III, Inscr.Délos 1408Ai32, 1443Aii45 (ii BC).

ἐπίχυσις **III**, for 'beaker or wine-jug' read 'pouring-vessel'

ἐπιχυτήρ, for '= ἐπίχυσις III' read 'vessel for pouring oil into a lamp'

*ἐπιχώρημα, ατος, τό, perh. leave, permission, PRyl. 222 fr. (ii AD), PWash.Univ. 1.22 (rest.), 23.

*ἐπιψέλλως, adv. incoherently, γράφειν PTeb. 763.14 (ii BC).

ἐπιψηφίζω **I 1**, line 2, before 'D. 22.9' insert 'Th. 6.14.1'   **2**, delete 'Th. 6.14, etc.'   **III**, add 'ἐπιδεὶ ἐπεψαφίτταιο ὁ δᾶμος ἀποδόμεν Νικαρέτηω IG 7.3172 viii 112'

*ἐπιψήχω, dub. sens., βῶλον ἐπιψήχων πυροφόροις βοτάναις GVI 1165.2 (near Sardes, ii AD), v. RPh. 31.19.

*ἐπιώβολος τόκος, interest at 16⅔%, Inscr.Délos 442C61 (ii BC).

*ἐπιωπής, ές, watchful, vigilant, κλυτὴν ἐπιωπέα κούρην orac. in ZPE 92.269 (Ephesus, ii AD).

ἐποδύρομαι, add 'cj. in Apollod. 1.9.8 (codd. ἀποδ-)'

ἐποίζω, delete the entry.

ἐποικία **I**, add 'v.l. for ἀπ- in Pi.O. 1.24'

ἐποικίζω **III**, for 'bring into cultivation' read 'settle on'; add 'ἐπιFοικίξ[ε]ιτη BCH 60.182.25 (Thespiae, iii BC)'

ἐποίκιον **II**, add 'w. place-names, ἀπὸ ἐποικίου Γεννέσυ ὅρων Ἀπαμέων IG 14.2327, 2329, etc.'

*ἐποικιοφύλαξ, ακος, ὁ, settlement-guard, POxy. 3518 (iii AD).

*ἐποικιώτης, ου, ὁ, inhabitant of an ἐποίκιον, PFlor. 180.7 (iii AD).

ἐποικοδομέω, after lemma insert 'ἐπιFοικοδομέω BCH 60.182.25 (Thespiae, iii BC)'; line 2, after 'D. 55.25' insert 'Men.Dysc. 376 (cj.)'

ἔποικος **I 3**, last line, for 'Call.Aet.Oxy. 2080.69' read 'Call.fr. 43.67 Pf.'

ἐποικτίζω, med., add 'ὃν [ὁ] πᾶς δῆμος ἐπῳκτίσατο epigr. in SEG 39.972.9 (Lato, ii BC)'

ἐποιμώζω, add 'τέκνοις τοῖς τεθνηκόσιν A.fr. 154a.7 R. (s.v.l.)'

ἐποίχομαι **I 2**, add 'A.Ch. 956'

ἐποκέλλω **2**, add 'fig. ἐποκέλλοντες δὲ εἰς τὸ ῥωπικὸν καὶ κακόζηλον Longin. 3.4'

*ἐπονειδίζω, insult, στίγματα μὴ γράψῃς ἐπονειδίζων θεράποντα Ps.-Phoc. 225; also ἐπιον- Pland. 97.4 (iii AD).

†ἐπονήμενοι, ἐπονάμενοι, dub. lect. and sens., Alc. 5.9, 33(b).4, 119.17 (L.-P.).

ἐπονομάζω **3**, line 2, after 'so and so' insert 'κἀπωνύμασσαν ἀντίαον Δία they entitled Zeus god of suppliants, Alc. 129.5 L.-P.'

ἐποξίζω, add 'Vit.Aesop.(G) 63'

ἐποπτεύω **I 2**, add 'IEryth. 2C.1 (v BC), ἐπιοπτευέτω καὶ ὑποζυγὴν ἔναι ib. 4'

*ἐπορθοβοάω, lift up a cry, ἵνα πατρὶ γόους νυχίους ἐπορθοβοάσω E.El. 142 (s.v.l.).

ἔπος, line 1, after '(v. infr.)' insert 'Cypr. we-po, perh. Fέπο(s), ICS 264.1'

*ἐπούλων, ωνος, ὁ, ἑπτὰ ἀνδρῶν ἐπουλώνων (transl. Lat. septemvirum epulonum, board of seven in charge of public feasts), IEphes. 3033, 3034.

ἐπούλωσις, add 'Heras ap.Gal. 13.765.15'

ἐπουλωτικός, add 'Heras ap.Gal. 13.765.13'

ἐπουράνιος **1**, add 'τῷ μεγάλῳ Θεῷ Ὑψίστῳ καὶ Ἐπουρανίῳ SEG 31.1080 (Galatia, ii/ii AD), cult-title of Zeus, INikaia 1114, 1115 (ii/iv AD)'

ἐπουρίζω **I**, add 'S.fr. 442.7 R.'

ἔπουρος **I**, after 'S.Tr. 954 (lyr.)' add 'carried by a favourable breeze, id.OT 194'

ἐποφλισκάνω, add '**2** simply owe, Just.Nov. 121.1, Edict. 9 pr.'

ἐποχέομαι, line 1, after 'med.' insert 'athematic pres. part. ἐποχήμενος Nonn.D. 8.229, al.'   **3 a**, line 2, after 'transcend the lower' insert 'τὸ ἀγαθὸν ἐποχούμενον ἐπὶ τῇ οὐσίᾳ Numen. fr. 2 P.'

ἐποχή **II 2 a**, add 'position or orbit of planets, Nicom.Harm. 3 (pl.)'   **3**, delete the section.

*ἐπόχησις, εως, ἡ, = ἐποχή III 2, Iamb.VP 15.65 (s.v.l.).

*ἐπόχθων, ονος, ὁ, app. error for ἐπίχθων, = ἐπιχθόνιος 1, opp. ὑπόχθων, of a δαίμων, SEG 7.213.6 (tab.defix.; Beirut, ii/iii AD).

*ἐπόχιμος, ον, suspended, of payments, PTeb. 337.3 (ii/iii AD); cf. ἐποχή II 2.

ἐπόψιος, line 1, for 'Arat. 258' read 'Arat. 81, 258'   **II**, add 'Ἐ.· Ζεύς. καὶ Ἀπόλλων Hsch.'

ἔποψις **I**, add '**2** perh. façade, οἰκοδόμησαν τὴν ἔποψιν Θεῷ Κρόνῳ M.Dunand Le Musée de Soueïda (Paris 1934) no. 198 (ii AD).'

ἔπρεσε, for 'πρήθω' read 'πίμπρημι'

ἑπτά, indecl., insert 'Cret. ἑττά Inscr.Cret. 1.viii 5A.9 (iii BC)'

ἑπταγράμματος, add 'PMag. 12.8.6'

*ἑπταέτηρος, ον, seven years long, φύλοπιν ἑ. Nonn.D. 25.3.

ἑπταετής **I**, after 'seven years old' insert 'IG 9²(1).431.1 (Acarnania, ii/i BC)'

ἑπταέτις (s.v. ἑπταετής), add 'GVI 1508 (Gaza, Ptolemaic)'

ἑπτακαιδέκατος, add 'Thess. ἑττακαιδεκότα SEG 29.529 (BCH 104.643)'

ἑπτακάτιοι, add 'SEG 25.387.A22 (Epid., iv BC)'

*ἑπτακελλάριον, τό, perh. chest, or sim., with seven compartments, PAnt. 93.31 (iv AD, -ρον pap.).

*ἑπτάκλαδος, ον, having seven branches or shoots, Ps.-Callisth. 131.2.

*ἑπτάμετρον, τό, of capacity, seven (standard) measures, PFlor. 356.11 (i/ii AD).

ἑπτάμηκης, delete the entry.

ἑπτάμηνος **II**, add 'PTeb. 342.30 (ii AD; v. AncSoc 21.33)'

*ἑπτάξυλος, ον, containing seven sticks, or seven ξύλα (v. ξύλον v) in length, δέσμαι OTheb. 144 (i AD).

†ἑπτάπάλαστος, ον, seven palms long, S.E.M. 9.321; also -πάλαστος IG 1².373.237, PPetr. 3.41ⁿ.5 (iii BC), Inscr.Délos 1442B 66 (ii BC).

*ἑπτάπλησον, aor. imper. multiply by seven, perh. erron. for ἑπτάπλωσον (°ἑπταπλόω) PCair. 10758 (vi AD)

*ἑπταπλόω, perh. multiply by seven, v. °ἑπτάπλησον.

ἑπτάρουρος, add 'PColl.Youtie 16.3 (109 BC)'

*ἑπτάστεγος, ον, having seven stories, POxy. 2719.9 (iii AD).

*ἑπτάστολος, ον, *wearing seven στολαί*, of a devotee of Isis, ἱεροφόρος ἑ., *AE* 1931.174 (Samos).

*ἑπτάστροφος, ον, *consisting of seven strophes*, Heph.*Sign.* 4.

*ἑπτασφόνδυλος, ον, *having seven segments*, of the scorpion's tail, Cyran. 46 (1.24.7 K.); transf., ἑ. μοι ῥήματα εἶπας *Vit.Aesop.*(G) 31.

ἑπτάτονος, for 'B.*Scol.Oxy.* 1361 *fr*. 1.2' read 'B.*fr.* 20*B*.2 S.-M. (of a βάρβιτος)'

*ἑπταύχενος, ον, app. *seven-necked* (in broken context), Dain *Inscr.du Louvre* 60.28 (Heraclea ad Latmum).

ἕπω (A), line 2, after 'Il. 6.321' insert 'ἀμφὶς ἔπουσιν = ἀμφέπουσιν honour, Nic.*Th.* 627 (v.l.)'

ἕπω (B), **I 10**, add 'part., w. month name, Ἀμυ⟨κ⟩λαίου ἑπομένου *SEG* 34.282 (Nemea, iv BC)'

ἐπωβελία **1**, add 'Aeschin. 1.163'

ἐπῳδή, line 5, after 'A.*Pr.* 174' insert 'id.*fr.* 281a.20 R., Hp.*Morb.Sacr.* 1' **II**, add 'Zen. 5.68 (Archil. 201 Q.)'

*ἐπωδύνιος, ον, = ἐπώδυνος, *GVI* 1675.2 (Tomi, ii/iii AD).

*ἐπῴζω, *wail* over, τέκν]οις ἐπώζει ζῶσα τοῖς τεθνηκόσιν perh. to be read in A.*fr.* 154a.7 R. (pap. ἐποιμώζουσα, Αἰσχύλος Νιόβη μεταφορικῶς· ἐφημερίῳ τάφον τέκνοις ἔπωζε .. τοῖς τεθνηκόσιν Hsch. s.v. ἐπῳζειν); cf. also Ar.*Av.* 266.

ἐπῴζω, for 'Cratin. 108' read 'Cratin. 115 K.-A. (ἐπωάζ- cod. Ath.)' and delete 'cluck .. Ar.*Av.* 266'

ἐπωλένιος, after '(*h.Merc.*) 510' insert '(ὑπ- codd.)'

ἐπωμάδιος **I**, for 'on the shoulders' read 'situated on the shoulders' and add 'Arat. 249'

ἐπώμιος, add 'Myc. *e-po-mi-jo* (dual), *shoulder-pieces* of armour'

ἐπώμοτος **I**, after '*Tr.* 427' insert 'in broken context -- δε]κα λεβήτōν ἐπόμοτον ἤμ[ὲν --, *Inscr.Cret.* 4.8 (vii/vi BC)'

ἐπώνια **I**, sg., for pres. ref. read '*SEG* 37.917.A5, al. (Erythrae, v/iv BC), *IEryth.* 201 (iii BC)'

ἐπωνυμία, add '**II** *office of* °ἐπώνυμος II 2 c, τὴν ἑ. τῶν Σαραπιαστῶν *BCH* 51.220.3 (Thasos, prob. ii BC).'

ἐπωνύμιον, after 'τό' insert 'title, ?Alc. 304 i 10 L.-P.'

ἐπώνυμος **II 2 a**, add 'later, of a patron or benefactor, *Hesperia* suppl. 6, no. 24.1 (Athens, ii AD)' add '**c** official of Sarapiastae whose name appeared in their decrees, *BCH* 51.220, 221 (Thasos, prob. ii BC).'

Ἐπωπετής, before 'Hsch.' insert '*SEG* 21.541 iii 20 (Erchiae, Attica, iv BC)'

*ἐπώπια, τά, perh. *eyebrows*, Call.in *Suppl.Hell.* 238.7.

*ἐπωστός, όν, *capable of being pushed forward*, Eratosth.ap.Eutoc.*in Archim.* p. 94 H.

ἐπωτίδες, add 'also applied to similar fittings on a trolley used for transporting marble drums, ἐπωτίδες εἰς τὴν λιθαγωγίαν *IG* 2².1673.34. **II** sg. ἐπωτίς, ίδος, ἡ, *bandage for the ear*, Gal. 12.488 Charterius (σπωτίς cod.).'

ἔραμαι, lines 5/6, delete 'poet. .. 19'

ἔρανος **I 2**, add 'Epich. 87.1'

ἐραπίδα, add '(ἐράπεδα· ἃ ἡμεῖς δαπέδα La.)'

*ἐράριον, v. αἰράριον.

ἐρασίχρηματος, add 'Philostr.*VA* 1.35'

*ἐράσκομαι, = ἐράω, *GVI* 280.2 (Thrace, iii AD).

*ἐρασμίβωμος, ον, *having lovely altars*, ἐρασμιβώμοις θύμασιν δωρουμένα *PKöln* 125 i 3 (s.v.l.).

ἐράσμιος, after 'X.*Mem.* 3.10.3' insert 'Pl.*R.* 402d, al.'

ἐραστής, add '**3** name given to white spots on the finger nail (cf. ψεῦδος II 2), Alex.Aphr.*Pr.Anecd.* 2.58.'

ἐραστός, line 3, for '[Simon.] 178.1' read '*AP* 5.159.1 ([Simon.])'

ἐρατεινός, after '(Pi.)*fr.* 122.7' insert 'σωφροσύνην ἐρατεινήν epigr. in *SEG* 25.299 (Attica, iv BC)'

ἐρατός, line 6, delete 'παίδων' **2**, delete 'ἀνδράσι .. Tyrt. 10.29' and substitute 'epigr. in *ZPE* 44.102 no. 12 (Pisidia)'

*ἐρᾱτόφρων, ονος, ὁ, *friendly*, rest. in epigr. in *SEG* 15.620.6 (Ostia, iii AD).

ἐραυνητικόν, add '*PCornell* 3.5 (iii BC), *PMerton* 15.28, 33 (AD 114), also ἐρευνητικόν *PCair.Zen.* 753.35, 40, *PTeb.* 867 (iii BC)'

ἐράω (A), line 2, form ἐρέω, add 'Anacr. 14.1, 83.1 P.'; line 4, after 'ἐρᾶσθαι' insert 'Men.*Epit.* 432'; lines 5/6, for 'also .. ἔραμαι' read 'med. ἐρᾶται Plu. 2.753b is suspect' and add 'pf. pass. Cypr. *e-re-ra-me-na*, ἐρεράμενα, *ICS* 264.2' **I 2**, delete 'without sexual reference'

*ἔρβουλον, τό, ?*vetch*, *Edict.Diocl.* 1.8a; cf. Lat. *ervum*. **2** an Italian wine, Ath. 1.27c.

ἐργάζομαι, lines 4/5, aor., add 'Thess. 3 sg. ἐργάξατο *IG* 9(2).1027b (?Atrax, v BC); ἐϝεργά(σ)σατο *SEG* 11.379 (Hermion, v BC); line 6, after 'iv BC)' insert 'also ἠργάσετο Robert *Hell.* IX.40, εἰργάσετο Ramsay *Cities and Bishoprics* 1.338 no. 186, both in sculptors' signatures' **I**, add '**2** w. dat., *work for*, *serve*, Lxx *Je.* 34(27).6, 35(28).14, *Ba.* 1.22.' **III 1**, add 'of skins, *treated*, *tanned*, *Edict.Diocl.* 8.10, *BCH* suppl. 8 no. 85 (Thessalonica, v/vi AD)'

ἐργάνη, line 3, for '*APr.*' read 'A.*Pr.*'

*ἐργανοφύλαξ, ακος, ὁ, name of an occupation, app. some kind of watchman, *BGU* 1988 b 3 (s.v.l. ?ὄργανο-, iii BC).

ἐργασία **II 3**, at end transfer 'of sexual *intercourse*, Arist.*Pr.* 876ᵃ39' to section **I 1 b**, add 'Demad. 8, *Act.Ap.* 16.16, 19' **II 5**, line 2, delete '(non legit Sch.)'

ἐργάσιμος **II 2**, add '*energetic*, of a person, Nic.Dam.*fr.* 61 J.'

ἐργαστηριακός, for '*practising a handicraft*' read '*working in an ἐργαστήριον*'

ἐργαστήριον **1**, add 'of a local θησαυρός with its branches, *PTeb.* 722.7 (ii BC), al.; app. of a medical clinic, *SEG* 30.1175 (Italy, iii BC)'

ἐργαστής, after '-τής' insert '(Cret. Ϝεργαστάς Bile *Dial.crét.* no. 27, Axos (vi/v BC)'; after '(Thyatira)' insert '*Pland.* 8.150 ii 16 (iii AD)'

ἔργαστρα **II**, after '*IG* 2².839.85' insert '(but this may belong under section I)'

*ἐργάστρια, ἡ, fem. of ἐργαστής, prob. in Sch.E.*Med.* 408.

*ἐργατηγός, ὁ, app. *foreman, overseer*, Μουσ.Σμυρν. 1884/5 p. 79.

*ἐργάτης, form Ϝεργάτας, Myc. *we-ka-ta*, (pl., *-ta-e* dual), of oxen.

ἐργατικός, add '**2** -κόν, τό, *payment for labour*, *PCornell* 3.15 (iii BC), *Inscr.Délos* 440A72, 79 (ii BC).'

ἐργεπιστάτης, add '*SEG* 30.1254.10 (Aphrodisias, AD 102/116)'

*ἐργεόπτης, ου, ὁ, *overseer*, *JHS* 12.263 (Cilicia).

ἔργμα, add 'Hes.*Op.* 801, *Th.* 823'

*ἐργμός, = εἱργμός 2, δεσμῶν ἑ. *SPAW* 1937.156 (Aetolian decree, Miletus, iii BC).

ἐργοδιωκτέω, for '*to be a taskmaster*' read '*to be an overseer*'

ἐργοδιώκτης, for '*taskmaster*' read '*overseer*' and add '*PHarris* 100.11, *POxy.* 2195.128, 2197.176, al. (all v/vi AD); perh. also τῷ ἐργοδιώτῳ (sic) Mitchell *N. Galatia* 161.8'

ἐργοδότης, add '*IMylasa* 895 in *EA* 13.7.17 (ii BC)'

*ἐργολᾰβέω, add '**III** *victimize*, ἠργολάβησέν με *PMich.*VI 425.13 (ii AD).'

*ἐργολᾰβικόν, τό, *contractor's fee*, *PMich.Zen.* 62.13 (iii BC).

ἐργολάβος, line 3, after 'Them.*Or.* 21.260b' insert '*PRyl.* 577.11, 16 (ii BC)'

ἔργον, line 1, Ϝέργον, add '*SEG* 28.37.3(b) (amphora from S. Italy, vi BC); ib. 32.496.19 (Thespiae, iii BC)'; line 2, after 'τό' insert 'see also °ἔργος' **III 1**, add 'of a monument, *SEG* 30.1383.A11 (Lydia, AD 301), Mitchell *N.Galatia* 142, al.' **III 2**, add 'D. 28.13'

†ἐργοπόνος, ὁ, *one who toils* or *labours*, *AP* 11.9 (Leon.Tarent.), Nic.*Th.* 831; w. gen. indicating occupation, ἐργοπόνοι κρατεροὶ θήρης ἐρικυδέος Opp.*C.* 1.148; ἐ. ἐλέφαντος Man. 1.298; epigr. in *SEG* 31.1284 (Pisidia); fem., of Athena, as patroness of crafts, Coluth. 195. **2** one destined for toil, *drudge*, *Cat.Cod.Astr.* 7.198.

*ἔργος, τό, = ἔργον, gen. sg. [ἐ]ργεος *Epigr.Gr.* 321.3 (Lydia, ii/iii AD), dat. pl. ἔργεσι ib. 343 (Germae).

*ἐργοστασιάρχης, ου, ὁ, *workshop foreman*, *BpW* 1910.310 (Rhodes).

*ἐργοτόχιος, ου, perh. = ἐργοδόχος, ?*contractor*, *MAMA* 3.487 (Corycus).

ἔργω **I**, add '**2** *hold together*, *keep shut*, ἄλλικα .. ἐεργομένην ἐνετῇσιν Call.*fr.* 253.11 Pf.' **II 3**, add '**c** *shun*, *Inscr.Cret.* 4.176.25 (ii BC).'

*ἐργωνικός, ὁ, prob. = ἐργώνης, *MDAI(A)* 24.204 (near Pergamum).

ἔρδω, line 9, for 'aor. 1 ἔϜερξα *Inscr.Cypr.* 146 H.' read 'aor. 1 Cypr. *e-we-re-xa*, perh. ἔϜερξε, *ICS* 261' **2**, line 4, after 'Hdt. 1.131' insert 'δεκάτην *SEG* 12.391 (Samos, vi BC)'

ἐρέα, add '**b** applied to other materials resembling wool, ἐρέας λαγείας νωτιαίας *Edict.Diocl.* in *SEG* 31.911.10.'

ἐρεβίνθιον, add 'sg. in collective sense, *PVindob.* G40805 (*ZPE* 50.129 (iv AD))'

ἐρέβινθος **II**, delete '*Ach.* 801'

ἐρεβῶπις, for '*gloomy-looking*' read '*having the face of a creature from Erebus*, of a gorgon'

ἔρεγμα, for '*bruised corn*' read '*crushed corn, pulse*, etc.' and add 'ἐμφύσσα· ἐρέγματα διδοῦσα Hsch.; also ἔριγμα, φακῶν ἢ ἐρεβίνθων ἐρίγμασι Hp.*Coac.* 621, ξὺν κριθέων ἐρίγματι id.*Mul.* 2.195, Sch. Gen.Il. 17.295, Al.*Le.* 2.16'

ἐρέγμινος, for '*made of bruised beans*' read '*made into ἔρεγμα*'

ἐρεγμός **I**, add 'ἔριγμός· ὁ λεγόμενος διεσχισμένος κύαμος Apollon.*Lex.* s.v. ἐρικόμενος'

ἐρεείνω **I 3**, delete the section transferring '*h.Merc.* 533' to line 3 after 'Il. 6.145'

*Ἐρεθειβιάζω, = Ἐρεθεμιάζω, *Tit.Cam.* 87.7.

*ἐρεθύδανον, v. °ἐρυθρέδανον.

ἐρείδω **I 1**, line 20, for 'τι' read 'τινά' **III 2**, add 'transf. of words in a sentence, D.H.*Comp.* 22'

ἐρεικτός, add 'also ἐρεικτή, ἡ, Phot. s.v. πολφοί'

ἐρείπω, line 6, aor. 1 pass., add 'ἠρείφθην, epigr. in *IEryth.* 308 (i AD)'; line 14, after 'metaph.' insert 'ἄνεμός τέ μιν .. δείματι ἐρείπεν Simon. 38.5 P.' **II**, lines 5/6, delete 'metaph., .. 37.3'

ἐρέκτης, add '*PPrag.* I 25.12 (vi/vii AD)'

*ἐρεοπλυτικος, v. °ἐριο-.

*ἐρεουργός, ὁ, = ἐριουργός, MAMA 3.275, al. (Corycus): written -ωργός, ib. 435.

ἐρεοῦς, add 'app. fem. subst. *woollen garment*, παραθήκην ἔδωκί τινι ἐρεᾶν π[ρά]σινον Mitchell *N.Galatia* 242.10; form *ϜερϜέεος, Myc. *we-we-e-a* (neut. pl.)'

ἐρέπτομαι, line 3, after '(*AP*) 7.20' insert '([Simon.])'

ἐρέπτω II, delete the section.

ἐρεσιμέτρην, after lemma insert '(ἐρεσιμετρίην La.)'

*ἐρεσκῴος, ον, unexpld. wd., Hdn.*fr*. p. 19 H.

ἐρέσσω, line 2, delete '(earlier ἐλαύνω)' I, add 'ἔρεσσε *go on your way* (by land), Leon.ap.Stob. 4.52.28' II 1, add 'pass., in fig. phrase, of a woman in sexual intercourse, *AP* 5.54 (Diosc.)' add 'cf. Myc. inf. *e-re-e* fr. stem ἐρε-, *row*'

ἐρέτης, add 'Myc. *e-re-ta*'

ἐρετμόν II, add 'Pl.Com. 3.4 K.-A.'

ἐρετμός, add 'II *oar*, Orph.*A*. 278.'

ἔρετο, add 'v. ὄρνυμι, line 23'

Ἐρέτρια, line 2, after 'etc.' insert 'Ἐρετριᾶθεν, adv., *from Eretria*, *IG* 12(9).272 (Eretria, v BC)'

Ἐρετριάς, after 'ἡ' insert '*the land of Eretria*, *SEG* 31.804 (Eretria, iv BC)'; penultimate line, before '*a kind of clay*' insert 'γῆ Ἐρετριάς'

Ἐρετρικός, add 'τὸ Ἐ., an unguent, *PPetr*. 2.34.8 (iii AD)'

ἐρευθέδανον, add 'see also °ἐρυτρέδανον'

ἔρευθος, add 'pl., Arat. 837'

*ἐρεύνησις, εως, ἡ, *investigation*, *EM* s.v. Ἐριούνιος.

*ἐρευνητικόν, v. °ἔραυν-.

ἐρεφύλλινον, delete the entry.

*ἐρεφύλλινος, ον, adj. referring to an unidentified plant, στέφανοι ἐ. *PAlex*. 22.5 (ii AD), ἄνθος ἐ. *PMag*. 13.25.

ἐρέχθω, add '(ἐριχθ-, as read by Apion in the Homeric passages)'

ἐρημαῖος, line 2, for '*silent*' read '*empty*'; add 'neut. sg. as adv., ἐρημαῖον βοῶντες Arat. 1003 (cj.), Q.S. 12.513'

*ἐρημεῖος, α, ον, = ἔρημος, χώρη *EAD* 469.3 (Myconus).

ἔρημος II, add '4 of a vessel, *empty*, Lang *Ath.Agora* XXI Hb6 (ii AD).' add 'Myc. *e-re-mo* (subst. describing land)'

ἐρημοτελωνία, for pres. def. read '*farming of taxes for passage through the desert*'

ἐρημοφυλακία, for '*maintenance of this force*' read '*tax paid for protection of desert travellers*'

ἐρημόω II, add '3 *clear out, get rid of*, ὕρακας Nic.*Al*. 37.'

"Ἐρησίειον, τό, *temple of the Egyptian god* Ἐρῆσις, Bernand *Les Portes* 1 (i BC).

ἐρητύω 3, add 'also w. part. ὅς μιν θαρσύνεσκεν, ἐρητύων ἀχέουσαν A.R. 4.1054'

*ἐρθυρίς, v. ἐριθυρίς.

*ἐρίβρυχμος, ον, = ἐρίβρυχος, Q.S. 3.171 (codd.); cf. βρυχμή, βρυχμός.

ἐρίβρυχος, after 'Q.S. 3.171' insert '(cj.)'

*ἔριγμα, v. ‡ἔρεγμα.

+ἐρίγμη, ἡ, = ‡ἔρεγμα, Sch.Ar.*Ra*. 508.

*ἐριγμός, v. ‡ἔρεγμός.

ἐρίδιον, add 'prob. in Cerc. in *POxy*. 1082 fr. 32 (ἐρίδια τριβ[)'

*ἐριδισμός, ὁ, *strife, contention*, Cat.Cod.Astr. 8(3).196.3.

*ἐρίδρομος, ον, *fast-running*, Nonn.*D*. 23.28 (cod.).

ἐρίζω, line 8, delete 'Pass.' and '(in act. sense)'; line 16, after 'c. inf.' insert '*urge*, Plu.*Comp.Per.Fab*. 3; w. acc. & inf.' I 2, line 8, after 'Hdt. 5.49' insert 'w. gen. pers., οὐδὶς γὰρ ἐδύνατο τούτου ἐρίζειν epigr. in *WS* 53.152 (near Iconium, prob. iii AD)' II 1, line 3, for 'also in pf. pass.' read 'abs., Pi.*O*. 1.95' 2, delete the section.

ἐρίθακος, after 'ἐρίθυλος' insert '°ἐρύθακος'

ἐριθεύομαι II 1, after 'Arist.*Pol*. 1303ᵃ16' insert 'μηδὲ ἠρειθεῦσθαι ἐπὶ κακοσχολίᾳ μηθέν *Delph*. 3(1).362 i 31'

ἐριθευτός, add 'εἴνεκεν τοῦ λαβεῖν ἐριθευτοὺς (τοὺς) δικαστάς *Delph*. 3(1).362 i 33'

ἐρίθεχνα, delete the entry.

ἔριθος, add 'III epith. of Artemis, *IG* 2².5005 (Ἐρει-; Athens, ii AD).'

ἐρίκτυπος, add 'of Zeus, Archil. 91.42 W.'

ἐρινεός, line 1, after 'ὁ' insert 'ἡ, Apollod.*Epit*. 6.3)'

*ἔρινος (B), η, ον, *woollen*, Dura⁴ 30 (iii AD); cf. ἐρινοῦς.

Ἐρινύς, line 5, before 'E.*Med*.' insert 'A.*Th*. 700 (lyr.)'; after '(anap.)' insert 'Choerob. *in Theod*. 1.331' II, add 'Thebais fr. 2.8' add 'Myc. *e-ri-nu*'

+ἐριοκάρτης, ου, ὁ, (κείρω) *shearer* or worker who shaved roughness from woollen cloth to provide a smooth finish, *PMich*.II 123ʳ iii 9, xvii 35 (i AD), *PFreib*. 60 (ii AD), *PFlor*. 71.438, al. (iv AD).

ἐριοκαρτία, ας, ἡ, *wool shearings*, *BGU* 2295.5 (AD 157/8).

ἔριον, add 'pl., meton., of the part of the market where wools are sold, Teles p. 13 H.; ἔ. Σηρικόν perh. *raw silk*, Peripl.*M.Rubr*. 64'

ἐριόξυλον, before 'cf. ἐρεόξ-' insert 'τὰ ἐ., *SB* 9026.11, 13 (ii AD)

*ἐριόξυλος, ον, *made of cotton*, τὸν χιτῶνα .. τὸν ἐ. *POxy*. 3991.14 (ii/iii AD).

*ἐριοπλυτικός, όν, *of* or *for fulling*, κόπανον· ξύλον. ὄργανον ἐρεοπλυτικόν Hsch.

ἐριοπώλης, add '*IEphes*. 454 (ii/iii AD), *SEG* 30.1382 (Lydia, AD 301), *TAM* 4(1).174 (Rom.imp.)'

*ἐριοπωλικός, όν, *of wool-selling*, *POxy*. 3455 fr. (iii/iv AD).

*ἐριοραβδιστικός, όν, ἐργαστήριον ἐριοραβδιστικόν *fulling workshop*, *PBon*. 24a 8-9, b 16-17, c 11 (AD 135).

ἐριουνιος, line 2, after 'of uncertain meaning' insert 'perh. *speedy* (cf. οὖνει, οὖνιος, οὖνον) or *thieving* (cf. οὔνης, οὔνιος)'; add '*Trag.adesp*. 588 K.-S.'

ἐριουργός, add 'ἡ συνεργασία τῶν ἐριοργῶν *SEG* 29.1198 (Saittai, AD 223/4); see also °ἐρεουργός.'

ἐρίπνη, add 'ἐκ .. τῶν λόφων τῶν ὑπεράκρων, οὓς ἐρίπνας οἵ τε νομευτικοὶ φιλοῦσιν ὀνομάζειν καὶ ποιητῶν παῖδες Ael.*NA* 14.16'

+ἐριπτοίητος, ον, *wildly excited*, Nonn.*D*. 17.198, 28.13.

*ἐρισκός or ἐρισκός, ὁ, perh. = ὑριχός (or possibly ῥίσκος), *PMich*.II 121ʳII.ii 8 (i AD), Suid. s.v. κώθωνες; v. °πλυτάριος.

*ἐρισφαλής, ές, *very unsteady*, ἴχνος Nonn.*D*. 47.63.

ἐρίτιμος, line 3, after 'Ar.*Eq*. 1016' insert '(in mock oracle)'

ἐρίφειος, add 'δέρμα ἐρίφειον ἀνέργαστον Edict.Diocl. 8.17 (δέρμα ἐρίφιον ἄνεργον *SEG* 37.335 ii 11)'

*ἐρίφλοισβος, ον, *loud-roaring*, Nonn.*D*. 39.295.

ἔριφος I, line 1, for '(ἡ .. Crete)' read '(ἡ, *Inscr.Cret*. 4.260 (ii BC), *POxy*. 2887 i 3 (?i/ii AD))'; line 2, for 'Alc. l.c.' read 'Alc. 71.1 L.-P.' add '2 *cinaedus pilosus*, *AP* 11.216.6 (Lucill.).'

ἑρκίτης, after 'Ath. 6.267c' insert '*An.Ox*. 2.45.7' and for 'written ἑρκίταται in Hsch.' read 'Hsch. cod. ἑρκίταται'

ἕρμα II, for 'Ael.*NA* 17.35' read 'Ael.*NA* 17.25' at end add 'Myc. *e-ma-ta* (pl.), *shoe-laces*'

ἑρμάγένη, for '*Hermae*' read '*Hermeses*' and add '(humorously, for the offspring of a promiscuous woman)'

Ἑρμάδιον, for 'Keil-Premerstein .. 117' read '*IEphes*. 3334.11 (i AD)'

ἑρμάζω I, add 'pass., id.*Fract*. 26 (v.l. ἡρμόσθαι)'

Ἑρμαϊκός I, add '-όν, τό, name of a mine at Laurium, *IG* 2².1588.4, 5 (iii BC)'

ἕρμαιον I 1, add 'Men.*Dysc*. 226' 3, for this section read '*tomb*, *MAMA* 4.178 (Apollonia), 4.250 (Tymandos)' II, add '3 app. = Ἑρμῆς I 2, *herm*, Str. 17.1.50 (pl., s.v.l.).'

Ἑρμαιών, for 'at Halicarnassus' read 'in Caria'; after '*SIG* 45.4 (v BC)' insert '*SEG* 29.1089 (i BC)'

*Ἑρμάνιος, ὁ, name of a month at Scarpheia, *Delph*. 3(4).159.2.

ἑρματίζω II 1, after 'Plu. 2.967b' insert '*load up with ballast*, τοῖς τὴν ἅμαξαν ἑρματίσασιν καὶ ἀπαγαγοῦσιν *Inscr.Délos* 372A101 (200 BC)'

*ἑρμελαῦσον, v. °ἁρμαραύσιον.

ἑρμή, after 'Hsch.' insert '*Didyma* 486.26'

ἑρμηνεία, add 'II *office of* ἑρμηνεύς in an Egyptian sanctuary, *SB* 9355.1, 2 (ii AD, -νια pap.).'

ἑρμηνεύς I, add 'transf., οὗτος (sc. ὁ ἐγκέφαλος) γὰρ ἡμῖν ἐστι τῶν ἀπὸ τοῦ ἠέρος γιγνομένων ἑ. Hp.*Morb.Sacr*. 16'

ἑρμηνεύω I, add 'transf., διό φημι τὸν ἐγκέφαλον εἶναι τὸν ἑρμηνεύοντα τὴν σύνεσιν Hp.*Morb.Sacr*. 17'

Ἑρμῆς, line 6, for 'later' read 'also' and before 'Call.' insert 'Hes.*Op*. 68, h.*Pan*. 28' I 1, add 'identified with the Emperor Tiberius, Τιβερίωι Καίσαρι Ἑρμεῖ *IEphes*. 3420' 3, at end delete 'hence, .. 37.19' II 1, add 'Ἑ. ἡ τελευταία πόσις Poll. 6.100; Phot.' at end add 'form Ἑρμάhας, Myc. *e-ma-a₂*'

ἑρμίν, before 'Herod.' insert 'Hippon. 79.8 W. (perh. w. pun on Ἑρμῆς; cf. *EM* 376.40)'

*Ἑρμογένειοι, οἱ, a Rhodian guild, Ἀφροδισιαστᾶν Ἑρμογενείων κοινόν *SEG* 3.674.34, al. (ii BC).

Ἑρμοκοπίδης, add 'Epicur. 104.8 A.'

*ἑρμόλυχνον, τό, *lighting of lamps before herms* in honour of the dead, cj. in *IG* 2².1368.151 (τό θ' ἑρμ. for τὸ θερμόλυχνον): see *JÖAI* 24.168.

ἔρνος I 1, add 'Alcm. 3.68 P.' II 1, line 3, delete 'Ἡρακλέος .. 2.121'

*ἐρογλέφαρος, ον, *showing love in one's eyes*, Χάριτες Alcm. 1 i 21 P.

ἐρόεις, line 2, after '(lyr.)' insert 'μορφά Ibyc. 1(*a*).44 P., πηκτίς Anacr. 28.2 P.'

*Ϝεροῖα, τά, *tales*, καλὰ Ϝεροῖ' ἀϊσομ[έναν Corinn. 2 fr. 1(b).2 P.; written Ϝεροῖα as title of her poems, prob. in Ant.Lib. 25 P.

ἔρος (A), line 4, after 'Thgn. 1064' insert 'εἰς ἔρον ἦλθε Sapph. 15(*b*).12 L.-P.'

*ἐρουθρός, v. °ἐρυθρός.

ἐρπετώδης, add '*Corp.Herm*. 10.7'

+ἑρπηστήρ, v. °ἑρπυστήρ.

+ἑρπηστής, v. °ἑρπυστής.

+ἑρπυστήρ, ῆρος, ὁ, *creeping thing*, whether insect, small mammal, reptile or mollusc, Androm.ap.Gal. 14.37, Opp.*H*. 1.305, al., id.*C*.

3.110; as adj., ὄφεις ἑ. id.C. 3.411, Orph.L. 49. (Freq. ἑρπηστ- in codd., but wrongly.)

†**ἑρπυστής**, οῦ, ὁ, = °ἑρπυστήρ, Nic.Th. 9, Androm.ap.Gal. 14.38, Opp.H. 3.345; of a mouse, AP 9.86 (Antiphil.).   **b** *guinea-worm*, Hippiatr. 58.   **2** adj. *crawling*, of a baby, AP 9.302 (Antip.Thess.); *creeping*, of ivy, ib. 11.33 (Phil.). (Freq. ἑρπηστ- in codd., but wrongly.)

ἕρπω 1, for '*move slowly*' read '*move on the ground*'; of infants, add 'Arist.HA 501ᵃ3'; add 'of flocks, ib. 610ᵇ24, of snakes, ib. 696ᵃ9'; at end for 'an animal that *walks* on its teeth' read 'humorously, of the belly'   **2**, line 4, after '(Cos)' insert 'ἕρπεν ἐπὶ τὰ[ν] προκειμέναν χρῆιαν SEG 35.989.22 (Cnossus, ii/i BC)' and for 'cf. καθέρπω' read 'cf. καθέρπω II'

*****ἔρρηγμα**, ατος, τό, unexpld. wd. in PCair.Zen. 499.38 (iii BC).

*****ἔρροπος**, ον, quoted with °σύρροπος in unspecified sense by Sch.D.T. p. 465.5 H.

ἔρρω (A), line 17, after 'opt.' insert 'ἔρροι νὺξ αὗτα καὶ δαίμων E.Tr. 204'

ἐρρωμένος, lines 3/4, after 'Lys. 24.7' insert '*strong, muscular*, ἄνδρες Hp.Fract. 15'; after 'Pl.Phdr. 268a' insert 'exclamatory, ὦ βίας ἐρρωμένης Men.Inc. 23 S., ὦ πολυτίμητοι θεοί, ἐρρωμένου πράγματος ib. 57 S.'

ἔρση, line 5, after 'Hes.Sc. 395' insert 'ἔρσῃ δέ θαλερὸς .. ἀμαράκος Chaerem. 14.16 S.'

ἑρσήεις, for 'metaph., of a corpse' read 'transf., *moist, wet*'; add 'ὄστρεά θ' ἑρσήεντα Opp.H. 1.317'

ἔρσην, comp., add 'ἐρσέντερος BCH 81.584 (Dodona)'

ἐρυθαίνω, after '*blush scarlet*' insert 'AP 9.322 (Leon.Tarent.)'

*****ἐρύθακος**, ὁ, prob. = ἐρίθακος, Hdn.fr. p. 20 H.

*****ἐρυθρέδανον**, τό, = ἐρευθέδανον, PCair.Zen. 326 bis 24 (app. misspelt ἐρεθρύδ-, iii BC).

ἐρυθρῖνος 2, add 'SEG 23.326.14 (Delphi, iii BC)'

*****ἐρυθρονεφής**, ές, *surrounded by red clouds*, ἥλιος Cat.Cod.Astr. 8(1).138.

ἐρυθρός, line 3, for 'but the metre .. Choerob. in Theod. 2.76' read '-ότερος Anaxandr. 23 K.-A., Dromo 1 K.-A., Choerob.in Theod. 2.76 H.'   **I 1**, delete 'a ship painted with vermilion' and after 'Orac.ap.Hdt. 3.57' insert '(interpreted as a red-painted ship)'   **2**, add 'ἐρυθρά, ἡ, *redness* (as a medical complaint), Suppl.Mag. 88' add '**4** ἐρυθρός, ὁ, kind of fish, IGC p. 99 B1 (ἐρουθρ-, Acraephia, iii/ii BC).'   at end add 'Myc. e-ru-ta-ra (fem.)'

*****ἐρύθρω**, = ἐρυθαίνω, Sm.Is. 63.1 (pass.).

ἐρύθω, after '= ἐρεύθω' insert 'part. -ουσα (-οισα) Call.fr. 80.10 Pf.'

ἔρυμα 2, add 'E.Ba. 55'

*****ἔρυμος**· ὁ ῥυμός Theognost.Can. 64.

ἐρυσίπτολις, add 'Call.fr. 626 Pf.'

†**ἐρυσίχθων**, ονος, ὁ, ἡ, *grubbing up the earth*, v.l. in Strato Com. 1.19; cf. °ῥηξίχθων.

ἐρύω (A), line 5, delete '(in Hdt. .. εἴλκυσα)'   **2**, add '**b** *draw up*, ἐρύσσαι ὀφρύας AP 5.216.3 (Agath.).'   **B I 3 a**, add 'αὐτμάν *sniff, inhale*, AP 6.219.10 (Antip.Sid.)'

ἐρύω (B), line 12, after 'A.R. 2.1208' insert 'part. ῥυμένη hymn in SEG 8.548.27 (Egypt, i BC)'; line 13, after '(Il.) 20.195' insert 'Aeol. pres. or fut. inf. ῥύεσθαι Alc. 129.20 L.-P.'   **5**, line 13, after 'c. gen.' insert 'ἀνθρώποι[ς] θα[ν]άτω ῥύεσθε Alc. 34.7 L.-P.'   at end add 'Myc. athematic 3 pl. (o)-u-ru-to, *Ϝρυντοι.'

ἔρχομαι, lines 7/8, after 'Lys. 22.11' insert 'fr. 47'; lines 14/15, delete 'imper. ἐνθέ Aristonous 1.9'; line 16, after '(iv BC)' insert 'ἐνθών (ἐπ-) Inscr.Cret. 4.168*.17 (III BC), cf. ib. 1.xxiv 2.5, ἤνθον (πορτ-) Schwyzer 186.11 (Gortyn, ii BC)'; line 20, after 'AP 14.44' insert 'ἐρτ(ε)ῖν = ἐρθεῖν, POxy. 1069.31, PMich.VIII 516.10 (both iii AD)'; line 24, after '(Cyrene)' insert '1 sg. plpf. ἠληλούθειν Call.fr. 265 Pf.'   **IV**, add 'w. inf. ἤλυθεν .. μαθεῖν Neophr. 1.1 S.'

ἐρῶ, line 2, delete 'Nic.Th. 484' and for 'Ath. 9.400a' read 'Ath. 9.400b'

ἐρωή, before 'ἡ' insert '(Aeol. ἐρωΐα Theoc. 30.6)'   **II**, before '*escape*' insert '*respite*, Theoc. 30.6'

*****ἐρωμεναγοράστης**, ου, ὁ, gloss on *amicarius, procurer*, Diom. p. 326 K.

ἐρωμένιον, add 'Lucr. 4.1166'

*****ἐρωμενοπάροχος**, as °ἐρωμεναγοράστης.

*****ἐρωμενοπώλης**, as °ἐρωμεναγοράστης.

ἔρως, add '**V** οἱ Ἔρωτες, members of a religious association, JÖAI 14 Beibl. 46 (Lydia, ii AD).'

ἐρωτάω III, after '*entreat*' insert '(use disapproved of in Hermog.Meth. 3)'

Ἐρωτίδιον, τό, = Ἐρωτάριον, IG 11(2).287B7 (Delos, iii BC, pl.).

ἐρωτικός II, add '**b** -ός, ὁ, *lover*, AP 5.216.7 (Agath.).'

ἐρωτίς II, add 'Nonn.D. 32.28'

ἐς II, = ἐκ, add 'also Thess.GDI 1329.1a.15, Boeot. BCH suppl. 6(1980).211-212 (v BC), SEG 30.567 (iv BC); ἔσ GDI 713b.8, etc.'

*****ἔσαν**, v. °ἵζω.

*****ἐσζικαιωτήρ**, v. ἐκδικ-.

ἐσθίω, line 2, after 'Hes.Op. 147' insert 'εἴσθιον Antiph. 166 K.-A., Timocl. 35 K.-A.'

ἐσθλός, line 2, at end insert '[ἐσλ- scanned short Pi.O. 13.100, al.]'

ἔσθος, add 'h.Hom. 31.13'

*****ἐσίταμον**· δηλοῖ δὲ τὴν πρόσοδον Theognost.Can. 16 A.; prob. = ‡ἐσσίταλα.

†**ἔσκε**, = εἰς ὅ κε (i.e. ἔστ' ἄν), *until*, ἔσκε μάχηται Archil. 15 W.

*****ἐσκλητόρ**· ὁ δοκιμαζόμενος Hsch. (La.).

*****ἐσμεύω**, (ἑσμός) *swarm*, rest. in Call.fr. 191.28 Pf. (pap. [..] μένουσιν).

*****ἐσοχάδες**, αἱ, *internal piles*, Ps.-Gal. 14.495 (lemma, ἐσωχ-), Cyran. 2.30.13, 4.28.26 K.

*****ἐσπασμένως**, adv., *with convulsions*, Gal. 7.810.17.

Ἑσπερία, for '*Western* .. 1.49' read '*West*, Epigr.Gr. 823.3'

Ἑσπερινός, line 3, for "Ἑσπέρινος" read "Ἑσπερινός'

ἑσπέριος, line 2, after 'E.HF 395 (lyr.)' insert 'Peripl.M.Rubr. 18'   **I**, line 6, after 'Pi.P. 3.19' insert 'Ἔριφοι *setting in the evening*, Theoc. 7.53'   **II**, line 2, delete 'ἔριφοι Theoc. 7.53'

ἑσπερίτης, for 'D.L.' read 'D.S.'

ἕσπομαι, for 'ἕπομαι q.v.' read 'ἕπομαι (v. ἕπω)'

ἑσπόμην, add '(v. ἕπω)'

*****ἑσπράττας**, ὁ, v. °ἐκπράκτης.

ἐσσήν (A), lines 3 and 5, for 'Call.Aet. 1.1.23' read 'Call.fr. 178.23 Pf.'

ἐσσηνεύω, for pres. ref. read 'IEphes. 969.1, 1578b.7, al'

ἐσσηνία, for pres. ref. read 'IEphes. 956.2, 957.13, al.'

*****ἔσσηπρος**· ὁ μάντις Theognost.Can. 16 A.; cf. ἐσσήτιοι.

ἐσσήτιοι, add 'cf. °ἔσσηπρος.

ἐσσίταλα, for '(cf. ἐξίταλα)' read '(cf. ἐξίτηλος)'

*****ἐσσόριον**, v. ‡ἐνσόριον.

ἔσσων, delete the entry.

ἔστασαν, for '(but the v.l. ἵστασαν is to be preferred)' read '(v.ll. for ἵστασαν)' and add 'E.Heracl. 937 (cod., edd. ἵστασαν)'

*****ἔστασις**, ιος, ἡ, *placing*, θυρᾶν IAskl.Epid. 52A.14, 49.

ἔστε, lines 5/6, delete 'ἔσκε .. f.l.' (v. °ἔσκε).   **III**, add '2 w. gen., of time, ἔ. τᾶς τριακάδος Clara Rhodos 2.171.'

*****ἐστεκνοόμαι**, v. ‡ἐκτεκνόω.

ἑστία I 2, add 'Modest.Dig. 27.1.12.1, PLips. 41.10, PGrenf. 78.10'   **II**, line 2, add 'Ἱστία SEG 30.1117 (Entella, Sicily, iii BC)'

*****Ἑστιαιικός**, ή, όν, v. °Ἱστιαϊκός.

*****ἑστιασμός**, ὁ, = ἑστίασις, TAM 2.201 (Sidyma, pl.).

ἑστιᾶτικός II, for '*fund* .. Delos' read '*a fund for temple-expenses deposited in the sanctuary of Hestia*'; add 'Inscr.Délos 365.5'

ἑστιᾱτορία 2, before 'ii AD' insert 'ἱστια-'; add 'also εἱστια- PTeb. 598, ἰοτα- PMich.II 123' xvii 25'

ἑστιᾱτόριον, before 'IG 11(2)' insert 'SEG 11.244 (Sicyon, vi/v BC)'

ἑστιοῦχος I 3, for 'ψόλος .. (prob.)' read 'σέλας A.fr. 204b.4 R. (lyr.), cf. fr. 492b.2'   add '**II** functionary in the cult of Hestia, IEphes. 1060.12, 1070.3, al. (iii AD); cf. °ὑπεστιοῦχος.'

ἕστωρ, after lemma insert '(A)'; for '*peg* .. *reins through*' read '*peg near the end of the chariot-pole, over which was passed a ring* (κρίκος), prob. *for holding the yoke in place*'

*****ἕστωρ** (B), ορος, ὁ, *founder*, IUrb.Rom. 1155.88.

ἐσχάδις, adv., unexpld. wd. in Theognost.Can. 163.

ἐσχάρα, line 2 to end, read '**I** *place for a fire*, Τρώων πυρὸς ἐσχάραι Il. 10.418; esp. *in domestic use*, πῦρ μὲν ἐπ' ἐσχαρόφιν μέγα καίετο Od. 5.59, ἡ δ' ἧσται ἐπ' ἐσχάρῃ ἐν πυρὸς αὐγῇ 6.305, 7.153, 20.123, E.Cyc. 382, Plu.Marc. 17.11.   **2** transf., *hollow scab formed over wound caused by cautery*, etc., Hp.Morb. 2, Arist.Prob. 863ᵃ12, Pl.Com. 200.4 K.-A., Diosc. 1.56, Gal. 10.315.3.   **b** *hollowed out wood in which fire-drill is rotated*, Thphr.HP 5.9.7, Ign. 64.   **c** *external female genitals*, Ar.Eq. 1286.   **II** *container for fire, brazier, fire-basket*, etc., *used for domestic heating*, Plu. 2.180e; *for cooking*, Ar.Ach. 888, V. 938, PCair.Zen. 692 (iii BC); *in religious ritual*, πρὸς ἐσχάρῃ Φοίβου A.Pers. 205, νυκτίσεμνα δεῖπν' ἐπ' ἐσχάρᾳ πυρὸς ἔθυον id.Eum. 108, S.Ant. 1016, E.Andr. 1240, Ph. 274; *carried in procession*, X.Cyr. 8.3.12, Callix. 2; *not clearly distinguished from sense "altar"*, D. 59.116, Inscr.Délos 104.142, al.   **2** transf., *grid or lattice-work forming the base of var. structures*: καὶ ποιήσεις αὐτῷ (sc. θυμιαστηρίῳ) ἐσχάραν ἔργῳ δικτυωτῷ χαλκῆν Lxx Ex. 27.4; *as the base of a ballista or sim.*, Vitr. 10.11.9, Ph.Bel. 92.13.'   add 'Myc. e-ka-ra, *brazier*'

ἐσχαρίς, after '*brazier*' insert 'Ar.fr. 946 K.-A.'   add '**2** app. *platform on a trolley used for transporting marble*, ἐσχαρὶς ἐπὶ τὸν λίθον παγεῖσα IG 2².1673.63.'

ἐσχαρίτης, for '*over the fire*, Antidot. 3' read '*in the ashes*, Hp.Vict. 2.42, Antid. 3 K.-A.'

*****ἐσχαρωθεν**, unexpld. wd. in Theognost.Can. 156.

*****ἐσχάρως**, unexpld. wd. in Theognost.Can. 156.

ἔσχατος, line 1, after 'Arat. 625' insert '628'

ἐσχημένως / SUPPLEMENT / εὔαρχος

*ἐσχημένως, *at once, immediately*, PMag. 4.1876.

ἐσώτατος, add 'ἐν .. τοῖς ἐσωτάτοις τόποις Peripl.M.Rubr. 42'

†ἐσωτικός, v. °ἐξωτικός.

ἐτάζω 1, line 5, after '(prob. 1.)' insert 'ἔτασον ἐκεῖσε Cat.Cod.Astr. 8(1).190'; at end for '*reveal, unmask*' read '*prove by test*' and after 'τινα' insert 'cj. in' 2, after 'Ge. 12.17' insert 'pass., ἐν ἀσθενείαις ἢ κινδύνοις Cat.Cod.Astr. 8(1).256'

ἑταιρεύομαι, add 'act. pres. part., Hsch. s.v. σκαφίον, pf. part. ἡται[ρ]ευκώς SEG 27.261 (Maced., ii BC)'

ἑταίρα I, for '*keep company with*' read '*act as a courtesan* (or the male equivalent)' and add 'Ar.Pax 11'

ἑταιρίζω, after lemma insert 'ἐτάρ- Il. 13.456, Call.Dian. 206'

ἑταιρικός I 3, add 'cf. *equitum alae quae Hetaerice appellabatur* Nepos Eum. 1.6'

ἑταιρίστρια, for '= τριβάς' read '*lesbian*'

ἑταιροποιέομαι, add 'also act., as expl. of Lat. *sociare* (Verg.A. 1.600), pap. in Aevum 1(1927).65'

*ἑταρίζω, v. ἑταιρίζω.

ἐτασμός, add 'PSorb. 1.9 (iii AD)'

ἐτέα, for 'Theognost.Can. 7' read 'Theognost.Can. 17 A.'

Ἐτεόκρητες, add 'Str. 10.4.6, D.S. 5.64.1'

*ἐτεόλβος, ον, perh. *having honourable wealth*, epigr. in JRS 57.43 no. 8 (Lycia).

ἐτεός, add 'form ἐτεϜός, cf. Myc. *e-te-wo-ke-re-we-i-jo*, patronymic of ἘτεϜοκλέϜης (Ἐτεοκλῆς)'

*ἑτεροβάμων, ον, gen. ονος, perh. *one-legged* or *limping*, κόρη orac. in DAW 42(2).3 (iv AD).

ἑτερόγλαυκος, add 'PPetr. 2 p. 115 (iii BC); of the eyes, Ps.-Callisth. 1.13 codd. B,C'

*ἑτερογνωμονέω, *to be of a different way of thinking*, τῇ ἐκείνου ἐννοίᾳ PRyl. 463.9 (iii AD; fragment of a Gnostic gospel).

ἑτεροειδής, line 2, after 'Arist.HA 508ᵇ11' insert 'v.l. for ἐντερο-)'

ἑτεροκρανικός, add 'adv. -κῶς, Paul.Aeg. 7.4.4 (276.11 H.)'

ἑτερομήκης 2, add 'b of ὁ ἄρτιος ἀριθμός, *made up of one species of length only*, being the sum of two even or of two odd numbers, Iamb. *in Nic.* p. 12 l. 17 P.; cf. ἀμφιμήκης.'

*ἑτερόμματος, ον, app. = ἑτερόφθαλμος (in one or other sense), in quot., of horses, Afric.Cest. p. 24 V.

ἑτεροπλατής, add 'Poliorc. 263.6'

*ἑτερόριστος, ον, *defining other things*, opp. °αὐθόριστος, David Proll. 14.23.

ἕτερος, lines 1/2, for 'Dor. .. v. infr.)' read 'non.-Att.-Ion. ἄτερα IG 4².40.11 (Epid., iv BC), etc.; Myc. *a₂-te-ro*' I 1, add 'ἅτεροι πότεροι, v. ‡πότερος III' II a, add '(acc. to Sch.Aeschin. 2.116 the use of ἕτερος for ἄλλος is Attic)' III, line 6, after 'Ep.Gal. 1.6' insert 'ἑτέραν καὶ ἑτέραν ὁδόν *two different ways*, D.Chr. 42.3' 2, add 'οὔτερος δαίμων Call.fr. 191.63 Pf., εὔχεται μὲν ὁ νοῦν ἔχων τὰ βελτίονα, προσδοκᾷ δὲ καὶ θάτερα Plu. 2.474c' IV 1 b, c, for 'θατέρᾳ' read 'θητέρᾳ' V 1, τοῦ σκέλους .. Philostr.VA 3.39, delete '= ἑτεροσκελὴς εἶναι' and transfer ex. to section 3.

*ἑτεροσεβέω, *depart from established forms of worship*, Vett.Val. 4.15.1 (p. 174 P.).

ἑτερόσκιος, read '*throwing shadows only one way* (only north or only south), of those who live between the polar circles and the tropics (cf. ἀμφίσκιος I, περίσκιος)'

*ἑτεροσύστατος, ον, *subsisting by means of something else, dependent*, of the subjunctive mood, opp. αὐτοσύστατος, Choerob. *in Theod.* 2.411.22 H.

ἑτερότης, opp. ταυτότης, add 'Plot. 5.3.10, 6.2.15'

*ἑτεροϋπόστατος, ον, *subsistent in something else* (cf. ὑπόστατος II), Elias *in Cat.* 162.2.

ἑτερόφθαλμος I, for 'metaph., of the proposed destruction of Athens' read 'fig., of Greece, envisaged as deprived of one of its chief cities'

ἐτέρσετο, for 'τερσαίνω' read 'τέρσομαι'

ἑτέρωθι III, add '(s.v.l., v. °ἑτέρωτε)'

ἑτερώνιος, for '*another's property*' read '*foreign, introduced from outside*, Theognost.Can. 17 A.'

*ἑτέρωτε, *at another time*, A.D.Adv. 1.193.14, 194.4, cj. in Hdt. 3.35.5 (Aeol. ἑτέρωτα acc. to A.D. ib., cf. °ἀτέρωτα).

*ἐτησιακός, ή, όν, *etesian, kind of vine* at Arretium, Plin.HN 14.36.

ἐτησιάς, for 'poet. fem. of sq.' read 'fem. adj., *etesian*'

ἐτήσιος, line 1, for 'and in Hp. η, ον' read 'also Ion. η, ον' 2, after 'Arr.Ind. 2.21' insert 'predicatively σὺ μέν, φίλη χελιδών, ἐτησίη μολοῦσα Anacr. 25.2 W.'; after 'adv. -ίως' insert 'PBerl.Leihg. 23.9 (iii AD)'

*ἐτητύμέω, *to be true*, ἐ]τητυμέοντα[ς] .. ὀνε[ίρους Suppl.Hell. 922.1 (iii BC).

ἐτητυμία, for 'Call.Aet. 3.1.76' read 'Call.fr. 75.76 Pf.'

ἐτήτυμος II, line 2, for 'Archil. 62' read 'h.Ap. 64, Archil. 110 W., A.R. 1.142, al.'

*Ἐτηφίλα, ά, title of Persephone in Mytilene, IG 12(2).222.3, 263.4, cf. Ἐταιφίλα· Περσεφόνη Hsch.; in pl. = Demeter and Persephone, IG 12(2).255.3.

*Ἐτηφίλιος, epith. of Hadrian as devotee of Persephone, Hesperia 32.78/9 no. 164 (Mytilene, ii AD), v. °Ἐτηφίλα; dub. reading and accent, Ἐτηφίλια IG 12(2).239.9 (Mytilene), cf. ib. suppl. p. 10.9.

ἔτι I 2, after 'Pl.Prt. 310c' insert 'X.HG 2.4.11'

ἐτοιμόδακρυς, add 'Sch.E.Med. 903'

*ἐτοιμολογία, ἡ, *talkativeness*, Vit.Aesop.(G) 88a, Hsch. s.v.προφορά.

ἕτοιμος III, adv. -οτέρως, add 'Gal. 2.312.13, 11.622.14'

ἔτος 1, add 'πολλὰ τὰ ἔτη (expression of wish for a long life), IEphes. 1192(3)' at end, for 'Ϝέτος, for 'Inscr.Cypr. 135.1 H.' read 'Cypr. *we-te-i* (dat.) ICS 217.1, al., Myc. *we-te-i-we-te-i* (redupl. dat.), *annually*' and add 'Ϝετεόν SEG 30.380 (Tiryns, vii BC), Ϝετ[ι]ια IPamph. 3.5 (Pamphylia, iv BC); ἔτος is found in inscriptions, papyri and codd., e.g. IG 9²(1).2.11, 31, 32 (Thermus, iii BC, written ΗΤΕΩΝ), πένθ' ἐτέων GVI 1576.7 (Capri, i/ii AD), καθ' ἔτος BGU 538.31 (AD 100); cf. °δεχέτης, ἐφέτειος, etc.)'

ἐτός (B), for 'ἐτά .. anon. 283' read 'ὡς ἐτά *how truly*, (?)Call.fr. 780 Pf.; adv. ἐτῶς id.fr. 75.39 Pf. and perh. fr. 203.16 Pf.'

ἔττακαν, after lemma insert '(ἔττασαν cj.)'

*ἑττία· ἑστία Hsch.

*ἐτύμόγλωσσος, ον, *true of tongue*, Suppl.Hell. 991.71 (poet. word-list, iii BC).

*ἐτύμόμαντις, εως, ὁ, *prophet of truth*, Suppl.Hell. 991.69 (poet. word-list, iii BC).

*ἐτύμόφανος, ον, unexpld. wd., Suppl.Hell. 991.72 (poet. word-list, iii BC) (perh. -φαμος = -φημος; cf. ψευδόφημος).

*ἐτύμοφάς, dub. reading and accent, Suppl.Hell. 991.73 (poet. word-list, iii BC).

εὖ I 1, at end after 'D. 5.2, etc.' add 'εὖ .. ὁ ἀποφηνάμενος .. εἴπας .. Longin. 1.2, cf. Cerc. 5.12. b εὖ δίδωμι, of gods, etc., bestow *good*, S.OT 1081, E.Alc. 1004; w. dat., S.OC 642.' III, add 'in pl., πολλῶν γὰρ τῶν εὖ δύναται τυχεῖν Arist.Cael. 292ᵇ3' at end add '*e(h)u-* in Myc. compd. pers. names, as *e-u-me-de* (= Εὐμήδης)'

εὐαγγελίζομαι, line 8, after 'J.AJ 18.6.10' and after 'Alciphr. 3.12' insert '(codd.)'; after 'Hld. 2.10' insert '(v.l. σοι)'

εὐαγγέλιον I, line 6, after 'Aeschin. 3.160' insert 'Laodicée p. 265.12 (see also p. 273)' II 1, add 'b esp. *announcement of an emperor's accession*, J.BJ 4.10.6 (pl.), cf. PBerol. in POxy. 7 p. 150 (sg.).'

Εὐαγγέλιος I, add 'fem., epith. of goddess Μήτηρ, IMiletoupolis 17 (sp. -εία, ii/i BC)' II, after 'month' insert 'beginning in late Apr.' and add 'p. 72 (p. 8 K.)'

*Εὐαγγελίς, ίδος, ἡ, (prob.) title of priestess of Hera, MDAI(A)68.47.22 (= GDI 5702, Samos, iv BC).

*εὐάγγελμα, ατος, τό, = εὐαγγέλιον, ἐπὶ τοῖς τῶν ἡμετέρων ἀγαθῶν [εὐ]αγγέλμασι MDAI(A) 48.100 (Nicopolis ad Istrum, late ii AD).

εὐάγγελος, at end, title of Hermes, add 'IG 12(5).235 (Paros, i AD). II -ος, ὁ, *gospeller*, POxy. 3958.13, 36, 42 (AD 612).'

εὐαγής (A) 3, line 3, delete 'ὕμνοι AP 7.34 (Antip.Sid.)'

*εὐαγία, ἡ, *holy offering*, εὐαγίας δ' ἐπὶ τοῦδε (sc. τοῦ βωμοῦ) τελεῖετε μηνὸς ἑκάστου orac. in Ramsay *Studies in Eastern Rom.Prov.* p. 128.

εὐαγκής, for '*with sweet glades*' read '*having lovely glens*'

*εὐάγκριτος, ον, = εὐανάκριτος *easy to judge*, of an indisputable victor, IG 7.2470 (Thebes, iv/iii BC).

εὐαγρέω, for 'Ath. 7.297f' read 'Ath. 7.297e' and add 'AP 9.337 (Leon.Tarent.)'

εὐάγωγος II, add 'of parts of the body, *easily controlled*, Philostr. Gym. 35' III, add '(cj. for εὐαγῶς; εὐλαβῶς Lambinus e cod.Tornes.)'

εὔαερος, add 'εὐϜάερος app. in Lacon. pers. nn. Εὐβάβερος IG 5(1).154, cf. Εὐβαβερίσκος SEG 11.552.6, 528 (all ii AD)'

εὐαής I, add 'Ezek.Exag. 244 S.'

*εὔαθλος I, add 'εὔϜαθλος, on a terracotta ball, SEG 39.940 (Eretria, vii BC)'

εὐαί, after 'εὐοῖ' insert 'SEG 32.779 (Berezan, graffito, vi BC)'

*εὐακία, ἡ, *easy healing*, PAnt. 66.i.7.

εὐάμπελος, after 'Str. 3.3.1, al.' insert 'ὕλη Nonn.D. 12.300, Φρυγίη ib. 34.214'; add 'Οἰνεύς Nonn.D. 43.54'

εὐανακόμιστος, for 'of health' read 'to health'

*εὐανδρησία, ἡ, *manliness*, Trag.adesp. 193 K.-S. (s.v.l.).

εὐανθής, after 'ἐς' insert '(fem. εἴα, Orph.H. 10.11)' II 1, add 'εὐ. ῥόδα Chaerem. 13 S.'

*εὔανθος, ὁ, *name of a stone*, Cyran. 16 (1.5.2 K.) (but τὸν εὐάνθη λίθον ib. 17 (1.5.27 K.)).

εὐάνιος, add 'as pers. n., IG 12(3).783, SEG 26.476 (vi BC), ib. 11.244 (c.AD 500)'

εὐάντητος I, add 'Orph.H. 2.5, 3.13, etc., Nonn.D. 27.178, 35.316, 39.207'

εὐαρχία, after '*government*' insert 'IEphes. 44.8 (c.AD 440)'

εὔαρχος I 1, delete the section. II, after '*beginning well*' insert 'μύλος Lyc. 233'

135

*εὐάσκητος, ον, well-crafted, [εὐ]άσκητον μνῆ[μα] rest. in MAMA 1.171 (Laodicea Combusta).

εὐαστήρ, for 'Arch.Pap. 7.4' read 'Dionys.Bassar.fr. 19.17 L.' and add 'AP 6.154 (Leon.Tarent. or Gaet.)'

εὐβαφής I, add 'of mosaic, εὐβαφεέσσι λίθοισιν Gerasa p. 484, no. 327 (vi AD)'

*εὐβίαστος, ον, easily constrained, Simp. in Cael. 267.23.

*εὐβοή, ἡ, name for the nightingale, Cyran. 16 (1.5.1, 4 K.).

εὐβολέω, add '2 to be lucky (in hunting), Call.Dieg. ix.4 (fr. 200b Pf.).'

*εὐβότᾱνος, ον, having plentiful grass, Hsch. s.v. ποιήεντι.

εὐβουλεύς, add 'GVI 2030 (Syros, ii/iii AD)'

εὐβρῐθής, for 'laden with fine yarn' read 'well-weighted'

εὔβωλος, delete the entry.

*εὔβους, ὁ, ἡ, acc. εὔβων (v.l. εὔβουν), rich in cattle, h.Ap. 54.

*εὔγᾰμος, ον, having a happy or successful marriage, AP 9.59 (Antip. Thess.), Heph.Astr. 1.1, Nonn.D. 1.27; transf., εὐνή ib. 13.352, ὕδωρ (sc. of the river Adonis) 20.144.

εὐγένεια, add '6 free-born status, Cod.Just. 6.4.4.3, Just.Nov. 89.9 pr.; also legitimate birth, ib. 74.2 pr.'

εὐγένειος, add 'b having a sturdy jaw, δελφὶς δ᾽ ἠυγένειος Opp.H. 2.565.'

εὐγενής I 1, add 'ἄνδρα εὐ. prob. = Lat. patricium, Inscr.Perg. 8(3).21.21 (ii AD). b freeborn, Modest.Dig. 27.1.1.4, Just.Nov. 22.11; also, of legitimate birth, ib. 89.15.1. c sup. in honorific address, masc. POxy. 1664.15 (iii AD), PLond. 1023.1 (v/vi AD), 1319.4 (AD 544/5); fem., ITyr 37.'

εὐγενία, after 'AP 7.337.6' insert 'Didyma 496B10 (oracle, ii AD); εὐγενίη epigr. in SEG 32.1608 (Cyrene, iii AD)'

*εὐγηρασία, ἡ, = εὐγηρία, PMag. 13.783.

†εὐγλᾰγής, ές, also εὔγλᾰγος, ον, Lyc. 307, metaplast. dat. εὐγλαγι AP 9.744 (Leon.Tarent.); abounding in milk, Q.S. 13.260, v.l. for περιγλ- in Il. 16.642 ap.Ath. 11.495c; epith. of Hermes, AP l.c.; milky, fresh, θάλος Lyc. l.c.; with milky juice, Nic.Th. 627.

*εὔγλῠφος, ον, easily engraved, of a gem, Xenocr.Lap. 90.

εὐγνωμονέω II, for 'reward' read 'pay what is owing to' and add 'PSI 303.14, PMasp. 243 (Byz.), POxy. 3584.7 (v AD)'

εὐγνωμοσύνη, add '3 repayment, SB 9770.6 (vi AD).'

*εὔγνωτος, ον, = εὐγνωστος (in quot., sense 1), epigr. in SEG 31.335 (Laconia, ii/iii AD, s.v.l.).

*εὔγραφος, ον, = εὐγραφής I, στήλη AEM 19.99 (Moesia Inferior).

εὔγυρος, for 'tortuous (= γυρός, q.v.)' read 'well-curved, (in quot., w. ref. to wrestler's stance)'

εὐδαιμονισμός, add 'b perh. funeral celebration, PHib. 202.2 (iii BC).'

εὐδαιμοσύνη, add 'b app. personified (cf. εὐδαιμονία 2 b), ἡμέρα Εὐδαιμοσύνης IGRom. 4.661.5 (Acmonia, i AD).'

Εὐδαίμων 3 add 'in epitaphs, SEG 25.1143 (Cyprus, ii/iii AD), v. BE 1967.659'

εὐδάμνας, add '(εὐδάμας La.)'

εὐδείελος I, add 'in prose, ἐπὶ τὰν κολουάδα τὰν εὐδείελον decr. in BCH 116.200.9 (Phocis, iii BC), cf. Str. 9.2.41'

*εὔδειλος, ον, = εὐδείελος, τέμενος Alc. 129.2 L.-P.

εὔδενδρος, add 'in lyr. also ἠΰδενδρος, B.17.80 S.-M. (εὐδ. pap.)'

εὐδερκής, add '2 app. easily seen, A.fr. 281a.30 R.'

*Εὐδιήμεια, τά, festival at Delos, Inscr.Délos 371B 51 (iii BC).

*εὐδιάδοχος, ον, gloss on εὔκηλος, Sch.Gen.Il. 17.371.

εὐδιάκρῐτος, add '3 perh. having its parts well separated, i.e. open, ὄμμα Simp.in Cael. 75.1.'

*εὐδιάλεκτος, ον, easy to talk to, PMag. 8.28.

εὐδιάφθορος I, add '2 of food, easily broken down, Xenocr.ap.Orib. 2.58.145, opp. δυσφθαρτότερα, Dsc. 1.105.' II, delete 'easily going bad .. 105'

εὐδιάω, for 'of persons' read 'of living creatures' and add 'εὐδιόωντι .. ὄρνιθι Arat. 278, δελφῖνες .. εὐδιόωντες A.R. 4.933'

εὐδίδακτος, for 'docile' read 'receptive to teaching' and add 'CIJud. 1.190 (Rome)'

εὔδιος I, delete 'εὐδίᾳ O Berl.Sitzb. 1911.639, cf.'; after 'neut. .. as adv.' insert 'τὸν εὐδι' ἰανονθ' E. HF 1049 (cj.)' at end for '(For εὐδιϝος' to end of article read '[ῐ in verse, except Orph. l.c., Arat. l.c.; cf. ἔνδῖος]'

*εὐδῐφής, ές, carefully exploring, χείρ prob. in Androm. in GDRK 62.102.

†εὐδοκῐμάζω, consider good, Origenes.Adnot.in Gen. 7.4 (PG 17.13a).

εὐδοκίμησις, add 'SEG 31.903.8 (Aphrodisias); in an honorific address, καταξειώσῃ ἡ σὴ εὐδοκίμ(ησις) PNess. 24.4, PBerolinv. 25022 (Aegyptus 70.43, both vi AD)'

εὐδόκιμος, add 'sup. as honorific epith., Mitchell N.Galatia 186, POxy. 1898.38, PLond. 1708.167, PMasp. 94.5, etc. (all c.vi AD)'

*Εὐδοσία, ἡ, Benefaction, name of a ἡρωίνη, Πολέμων 1.246 (iv BC).

εὐρώνεια, for 'bodily strength and health' read 'vigour' and add 'GGA 1897.407 (Phrygia, ii/iii AD; εὐδρανίην)'

εὔδρομος, add 'III (δρόμος II 3) of a city, having fine public walks, App.Anth. 3.281.4 (εὔ-)).'

εὐδυκήμερος, delete the entry.

*εὐδώμητος, ον, well-built, ἄγυια epigr. in IParion 52.9 (ii AD).

*εὐδωσιδῐκέω, pay one's dues satisfactorily, POxy. 2351.33 (ii AD), opposite of °ἀδωσιδικέω.

*εὐέγερτος, easily aroused, Hierocl.in CA 8 p. 32 K.

εὐέγρετος, delete the entry.

εὔθειρα, for 'Anacr. 76' read 'Anacr. 73 P.' and add 'Simon. 14 fr. 84.6 P.'

εὐεκτος, after 'soft, yielding' insert 'Alcin.Intr. p. 175.2 H.'

εὐέκβᾰτος, add 'Gal. 15.709.7'

*εὐεντέλιος, α, ον, fully developed, ἄμπελος PNess. 34.2 (vi AD).

εὐέντευκτος, adv. -τως, add 'IMylasa 603.7'

εὐεπίβᾰτος II, after 'Ph.Bel. 94.40' insert 'of a single person, J.AJ 19.1.14' and for 'id.' read 'Ph.'

εὐεπίθετος, add '2 easy to apply oneself to, Alcid.Soph. 3.'

*εὐέπιος, = εὐεπής, Et.Gud.

εὐεπιχείρητος II, for 'readily attempting' read 'easily persuaded or tempted'

εὐεργεσία, after '-εσίη' insert 'Cypr. e-u-we-re-ke-si-a-se, εὐϜεργεσίας ICS 261'

*Εὐεργεσιασταί, οἱ, an association celebrating the Εὐεργέσια, CRAI 1951.256 (Syria, i AD).

εὐεργέτεια, for 'iv BC' read 'ii BC'

εὐεργέτης, add 'Dor. -ας SEG 31.306 (Argos, Hellen.)'

εὐεργέτις, after 'E.Alc. 1058' insert 'IG 9²(1).582.19, 23 (Magnesia, 207 BC), Salamine 81.5 (ii BC)' add '2 = τορύνη (ladle), Miller Mélanges p. 405, EM 762.34.'

εὐεργέω, for 'BGU 1118.27, al. (i BC)' read 'PKöln 144.31 (ii BC); perh. also in BGU 1119.30 (i BC)'

*εὐέργη, ἡ, app. = °εὐεργέτις 2, Poll. 10.97; cf. °εόργη.

εὐέργημα, for 'JHS 22.366' read 'MAMA 8.317'

*Εὐέρνειος, name of a month in Sicily, SEG 30.1120.17 (Entella, iii BC or later).

εὐετηρία I, after 'X.HG 5.2.4, etc.' insert 'D. 10.49'; add 'SEG 32.1243.33 (Aeolian Cyme, 2 BC/AD 2)'

εὔζηλος II, for 'enviable' read 'glorious'

εὔζυξ, read 'ἐύζυξ' and for 'AP 5.55' read 'AP 5.56 (Diosc.)'

εὐζωΐα, add 'Iamb.Protr. 5 p. 36.24 P.'

εὐηγορία, after lemma insert '(εὐᾱγ-)' and add '-αγορία Pi.Pae. 2.67 S.-M.'

εὐήθης I 1 b, delete the section.

εὐήκοος 3, add 'epith. of Aphrodite, Ἀφροδίτῃ Εὐακόδι SEG 31.515 (Boeotia, iv BC)'

εὐηλᾰκᾰτος, add 'Pi.Pae. 7(a).4 S.-M.'

*εὐήλικος, ον, = εὐῆλιξ, Vit.Aesop.(G) 32.

†εὐῆλιξ, ικος, ὁ, ἡ, of or characteristic of the prime of life, i.e. early adulthood, εὐ. προσώπου Men.Dysc. 950. 2 of good stature, Polem.Phgn. 5, Lyd.Mag. 1.23.

εὐημερία 3, add 'Certamen 94'

εὐήμερος 1, add 'ἄγεν .. εὐάμερον keep holiday, Inscr.Cret. 1.xix 3.39 (Malla, ii BC), cf. ib. 3.iii 3.2 (Hierapytna, iii BC)' 2, after 'happy' insert 'ξὺν τύχαις εὐ. A.fr. 451k(a).3 R.'

εὐήνωρ I, add '2 involving a man of honourable status, γάμων εὐ. θεσμός Orph.A. 885.'

εὐθάλεια, add 'ὑπὲρ εὐετηρίας καὶ εὐθαλείας πόλεως SEG 36.1095 (Lydia, iii AD)'

*εὐθελγής, ές, bewitching, dub. in Hymn.Is. 104 P., v. πολυθελγής.

*εὐθέμεθλος, ον, set on good foundations, Γαῖαν h.Hom. 30.1 (ἠΰ-).

εὐθενέω I, line 4, after 'D. 18.286' insert 'Men.Dysc 275'

εὐθενιακός, add 'POsl. 83.5, 6 (iii/iv AD)'

*εὔθερα· μέρος τι τῆς νεώς Hsch.; cf. ἐνθύριον.

εὔθετος 2, at end before 'adv.' insert 'w. gen., εὐ. τῆς ἀρχιερωσύνης inscr. in SB 8267.12 (i BC)'

εὐθημονέομαι, delete the entry.

*εὐθημονέω, keep in good order, θῶπλα λάζευ καὶ γνάθους εὐθημόνει Trag.adesp. 381 K.-S.; med., Pl.Lg. 758b. 2 intr., be in good order Simp.in Ph. 1067.24.

εὐθηνία I 1, add 'POxy. 2479.26 (vi AD)'

*εὐθηνιάρχης, ὁ, = εὐθηνιάρχος, POxy. 1417.28 (iv AD).

εὔθρονος, after 'with beautiful seat or throne' insert '(or adorned with beautiful flowers, cf. θρόνον)'

εὐθῠμᾰχος, for 'Simon. 137' read 'AP 7.442.1 ([Simon.])'

εὐθῡμέω II, pass., add 'εἰρήνης .. γενομένης εὐθυμέονται make merry, Arr.Ind. 12.4'

εὐθυμία, add 'Hp.Lex 4, Hymn.Is. in SEG 8.549.33'

*εὐθυνέω, act as εὔθυνος II, INikaia 1083, 1153 (see also SEG 31.1069).

εὐθύνω I, add '4 w. internal obj., drive a straight path (in quots. fig.), Modest.Dig. 27.1.10.3, εὐθύνει τὴν ἀτραπὸν τοῦ παντὸς βίου Ph. 1.271, 297.' III 3, censure, add 'A.D.Pron. 81.6, Adv. 171.1'

εὐθυόνειρος, after 'Arist.Div.Somn. 463ᵇ16' insert 'EE 1248ᵃ40'

εὐθῠΓορία SUPPLEMENT εὐόνυξ

†εὐθῠΓορία, εὐθυορία, v. °εὐθυωρία.
εὐθύπορος I, after 'metaph.' insert 'with a straight course, εὐ. λά[χος] A.fr. 168.11 R. (lyr.)'
*εὐθύρ(ρ)ῑνος, ον, straight-nosed, PLips. 2.6 (i AD), 5.2.7 (iii AD).
εὐθύς A 2, line 10, ἐξ εὐθείας, add 'Cod.Just. 1.3.42.1' **B II**, line 14, after 'Pl.Tht. 186b' insert 'Modest.Dig. 27.1.6.8'
εὐθύτης II, add 'rightness, τὴν ἄκραν εὐ. τοῦ θείου νόμου D.Chr. 36.23'
*εὐθύτοκος, ον, new-laid, of an egg, Cyran. 101 (3.55.3 K.).
*εὐθυφορικός, ή, όν, directed in a straight line, κίνησις Phlp. in GC 134.18.
εὐθυωρία, before 'ἡ' insert '(εὐθυορία in SEG 11.405.36 (Epid., iii/ii BC) and SEG 11.377.18 (Hermione, ii BC), copies of the same text)'
εὐθύωρος, neut. as adv., add 'Antipho fr. 39 T.; Ion. neut. pl. as adv., ἰθύωρα SEG 26.845.13 (Berezan, c.500 BC)'
εὐθώρηξ, well-mailed, read 'equipped with a good corslet' and for 'μύες Marcell.' read 'of mussels, Marc.'
εὐιάς (s.v. εὐιακός), add 'τὰς ἱερὰς προφυγὼν εὐιάδας εἰς Ἅιδην SEG 31.633B (AD 171/2; εὐιάδας scanned as trisyll., v. ZPE 52.288)'
εὐίερος, line 1, for 'Pae.Oxy. 675.14' read 'POxy. 675.14'; line 3, after 'holy' insert 'σάκος Theoc.Ep. 4.5'; add 'Orph.H. 7.2, 12, al.'
†εὐϊλᾰσία, ή, app. propitiation, SEG 32.1269 (sp. εὐειλ-; Phrygia, Rom.imp.); cf. εὐιλασία· εὐπειστία Hsch.'
εὐΐλατος, add 'SEG 30.1180 (Pompeii, i BC)'
Εὔιος I, add '2 transf., wine, Εὔ. γέροντα πολιὸν ἤδη Men.Dysc. 946.' II, add '2 masc. or neut. as subst., Bacchic cry, Εὔιον ἀείσειε .. Διονύσῳ Nonn.D. 15.131, Βάκχον .. Εὖια παππάζοντα ib. 48.954, ταῖς βάκχαις ἐκδιδόντα τὸν Εὔιον Him.Or. 46.4.7 C.' add '**III** fem. Εὐία (sp. -εία), name or title of a priestess, IG 10(2).1.260B2 (Thessalonica, iii AD).'
εὐϊώτης, after 'Lyr.Alex.Adesp. 22' insert 'cf. An.Ox. 1.86.29'
*εὐκᾰλέω, v. °εὐκηλέω.
†εὐκᾰμία· ἡσυχία, ἤτοι εὐφημία (Dor.), EM 392.5; εὐκαμίαν νυν παρέχεσθε Sophr. in GLP iii 73.14; cf. κημός.
εὐκαμπής, line 5, after 'τὸ εὐ. τῶν μελῶν' insert 'i.e. the fine modulations of the melodic line (v. καμπή III 1)'
εὔκαρπος, line 1, delete 'of women, h.Hom. 30.5'; line 4, after 'S.Aj. 671' insert 'of men, blessed with good crops, h.Hom. 30.5'
*εὐκατάακτος, ον, = εὐκάτακτος, easily shattered, δόξα Sch.Pi.N. 8.58.
†εὐκατάγνωστος, ον, easily recognizable, εὐκατάγνωστον εἶναι ἑαυτῷ συνιστοροῦντα .. it is easily perceived that he is conscious .. (cf. καταγιγνώσκω I) Mitteis Chr. II.31 viii 11 (ii BC), cf. EM 400.6.
εὐκατασκεύαστος 2, add 'b adv. -τως, in a well-contrived way, Sch.E.Hec. 1288.'
εὐκατάσπαστος, ον, easily pulled down, SEG 7.265 (Sidon, ii BC).
εὔκεραστος, add 'γλῶτταν prob. in Cratin. 171.63 K.-A.'
*εὐκηλέω, Dor. -κᾰλέω, εὐκαλεῖ· ἀτρεμίζει Hsch.
εὐκηλήτειρα, add '(s.v.l.)'
εὐκλεής, line 3, before 'dat.' insert 'gen. shortened to εὐκλέος Inscr.Délos 1658.6 (i BC); line 4, for 'Id.' read 'Pi.'; line 14, before 'adv.' insert 'sup. in honorific address, POxy. 1983.3 (vi AD)'
εὔκλεια I, line 5, delete 'ἄγαλμα'
*Εὔκλειος, ὁ, name of a month, SEG 24.1021 (Callatis, iii BC).
εὐκνήμῑς, add 'III = εὐκνήμος I, Nonn.D. 18.60 (s.v.l.).'
*εὐκολοδιάβλητος, ον, easy to slander, Anon. in Rh. 72.28.
εὔκολπος, add '3 having a fair bosom i.e. a welcoming embrace, GVI 2020.6 (Corinth, ii/i AD).'
εὐκόμιστος 2, add 'Simp. in Cael. 267.23'
εὐκοσμία, add 'adornment, decoration, SEG 31.1472 (Arabia, AD 603)'
*εὐκόσμιος, ον, decent in appearance Hsch. s.v. κίδαρις (adv. -ίως, q.v.)
εὔκουρος, add 'in uncertain sense, Rev.Bibl. 42.250 (Syria)'
εὐκταῖος 2, add 'Ἀρτέμιδι εὐκτέᾳ SEG 32.1260 (Paphlagonia, ii AD)' 3, adv., add 'GVI 788.9 (Pontus, ii/iii AD)'
εὐκτέανος (A), after '(A)' insert 'Ep. εὔκτ-' add '2 costly, Q.S. 14.271.'
*εὐκτερής, ές, who has been given rich funeral honours, epigr. in Lindos 698.7 (c.200 BC).
εὐκτήριος I, add 'SEG 34.1292 (Phrygia, late Rom.imp.)' II, add 'SEG 37.1541 (Gerasa, vi AD)'
εὔκτιμενος, for '= εὖ .. dwell in' read 'epith. of land, or sim., of unknown meaning' and at end after 'Od. 24.226' insert 'in Homeric imitation, 'Ρώμης .. εὐκτιμένης SEG 11.773 (Sparta, iv AD), cf. Myc. ki-ti-me-na'
εὔκτιτος, before '= εὐκτίμενος' insert '(εὔκτ- in Anacr. 13.5 P.)'; after 'B. 3.46' insert 'δόμος εὔκτιτος Lyr.adesp.S 414.13'
εὐλαβέομαι II 3, delete 'quietly' III, add 'imper. εὐλάβει Com. adesp. 239.7 A.'
*εὐλαλία, ή, eloquence, Hsch. s.v. προφορά.
εὐλή, for pres. def. read 'larva of the fly or other insect, maggot, grub, worm, etc.'; line 4, delete 'of common worms' and insert 'οἷον εἰ εὐλαὶ ἐν σαπέντι μέρει τοῦ φυτοῦ γίγνοιντο Plot. 4.3.4'
εὐλίβανος, after 'Aristonous 1.23' insert 'SB 6699.2 (inscr., iii BC)'

εὐλιπής 1, add 'ὄρνεον PLond. 2741 iv 71 in SHAW 14(1923).16 (iii AD)'
*εὔλιστος, ον, favourably received, θυηλαί orac. in Didyma 496B3 (oracle, ii AD).
*εὐλόβρωτος, ον, worm-eaten, An.Par. 4.182.21 (εὐλοβρώτειος cod.).
εὐλογέω II 1, add 'pres. part. as a form of address in a letter, τῷ εὐλογο[υ]μένῳ .. υἱ[ῷ]'
εὐλογία II, add 'Lycurg. 46, Isoc. 7.76' III 1, add 'TAM 4(1).374, 375'
†εὐλοιδόρητος, ον, open to abuse, Men.Sic.fr. 2 S., Plu. 2.757a.
*Εὐλοχία, = εὔλοχος, epith. of Artemis, Gonnoi 173.3 (iii BC).
εὔλοχος, add 'II of women, fertile, fruitful, Nonn.D. 31.111, 38.134, 44.309; transf., of the earth, ib. 1.294.'
εὐλῠτέω, add 'SEG 39.1180.48, 87 (Ephesus, i AD).'
εὔλῠτος I, add '6 fig., mentally supple, ἀγχίνοια Alcid.Soph. 16, cf. ib. 20, 34.'
εὐλῠτόω, delete the entry.
*εὐλύτωσον (εὐλύτρωσον La.)· ἀπάλλαξον Hsch.
εὐμάθεια, line 3, after 'Ion. -ίη' insert 'Call.Epigr. 48.1 Pf.'
εὐμᾰρέω, add 'δυνάμεις τὰς εὐμαρεούσας, of the plant θύρσιον, Cyran. 23 (1.8.10 K.)'
εὐμᾰρής II 1 b, delete the section transferring quot. to follow 'Trag.adesp. 383' in section II.
Εὐμένειος II, add '-ειον, τό, sanctuary of Eumenes, Inscr.Perg. 8(1).240 (ii BC)'
εὐμενέτειρα, add 'SIFC 2.389 (Crete, Chr.)'
*εὐμενητικός (εὐμενιστικός La.): gloss on μειλικτήριος Hsch.
*Εὐμενίδειος, ὁ, name of a month in Sicily, SEG 30.1117 (Entella, iii BC), ib. 1118, 1120.
εὐμετάβολος, add 'adv. -λως, Hsch. s.v. ἐμπλήγδην'
εὐμήκης, add '4 εὐ. (sc. λίθος), ἡ, a precious stone, Plin.HN 37.160.'
*εὐμήνυτος, ον, perh. easily detected, Theognost.Can. 83.
*εὐμίτρης, ου, ἡ, a precious stone, Plin.HN 37.160.
εὔμνηστος, line 2, before 'χρηστήριον' insert 'II of which it is good to make remembrance, or celebrated'
εὐμοιρία, add 'personified, SB 642, 1625 (ii/iii AD)'
εὐμοιρίτης, add 'perh. also fem. εὐμοιρῖτις, ιδος, rest. in SEG 36.573 (Illyria, vi AD)'
εὔμοιρος, add 'euphem., of the dead, PHaun. 17.9 (ii AD), PUniv.Giss. 20.10'
εὔμολπος, for 'sweetly singing' read 'melodious' and after it insert 'εὐμόλποις ὑμ[εναίοις] A.fr. 168.19 R. (lyr.), εὐμόλπῳ .. ἄνακτι λύρης'
εὔμορφος, add 'of style, elegant, D.H.Dem. 18: adv. -φως, Luc.Salt. 71 (comp.), Sch.Luc.JTr. 12'
εὔμωλος, after '-ότατον' insert '(-ύτατον La. app. fr. *εὔμωλυς, cf. μῶλυς)'
εὐνάζω, line 2, for 'Simon. 184.10' read 'AP 7.25.10 ([Simon.])' 2, line 5, for 'Simon. l.c.' read 'AP l.c.'; last line, for 'Call.Aet. 3.1.1' read 'Call.fr. 75.1 Pf.'
εὐνᾱής, delete the entry.
*εὐνᾱής, ές, (νάω) fair-flowing, B. 1.75, 9.42 S.-M., Call.fr. 65.1 Pf.
εὐναῖος II 2, add 'Trag.adesp. 589 K.-S., Call.fr. 727 Pf.'
*εὐναστήριον, τό, grave, E.Or. 590 (codd.), Lyc. 583; cf. εὐνατήριον.
εὐνατήριον, delete '(εὐαστήριον .. E. ll.cc.)'
εὐνέτης, for '= εὐναστήρ' read 'bedfellow, husband' and after '(lyr.)' insert 'E.El. 803'; fem. -ις, add 'GVI 1262.2 (Telus, ii/i BC)'
*Εὐνίκειον, τό, app. sanctuary of a hero Eunicus, IG 9²(1).757b (Amphissa, ii BC).
εὖνις (B), after '(E.IA) 807' insert 'Call.fr. 55.1 Pf.'
εὐνομία I 1, at end for 'οἱ ἐπὶ τῆς εὐνομίας .. (Lato)' read 'b name of a Cretan magistracy, Inscr.Cret. 1.xiv 2.2, al. c w. gen., peace from, τόξων εὐ. APl. 212.4 (Alph.).' 4, add 'pl., epigr. in SEG 29.1139 (Miletus, iv/v AD)'
εὔνομος I 1, add 'b law-abiding, SEG 6.796 (Cappadocia, iii AD).' transferring 'ἄνδρες Pl.Lg. 815b' to this section.
εὔνοος 1, line 4, after 'friendly' insert 'θυμός Alc. 129.9 L.-P.'; line 9, comp. Ion. εὐνοέστερος, Amyzon 186.8 (ii BC)'; at end before 'EM 394.5' insert 'IG 12 suppl. 693.10 (Eresus, iii AD)' 2, adv. εὐνόως, after 'διακείμενος' insert 'SEG 30.533.A3 (Thessaly, ii BC)'
*εὐνουχικός, ή, όν, of or relating to eunuchs, μοῖραι Vett.Val. 14.26.
εὐνοῦχος I 1, before 'Hdt. 3.130' insert 'Hippon. 26 W.' II, add 'app. in complimentary sense, of a faithful wife, SEG 28.1536 (Egypt, Rom.imp.)'
εὐξεΐς I, add 'app. a title of the descendants of Antenor, founder of Padua, Εὐξείνοι[s] Ἀντηνορίδα[ις] SEG 30.1132 (Aquileia, ii AD)'
εὐοδία, after 'journey' insert 'E.fr. 308 cod. Sch.Ar.Vesp. 757'; after 'A.fr. 36' insert 'PVindob. G39995 (CE 56.305, iii AD)'
*εὐόκερως, v. †αἰγόκερως.
εὐόνυξ, for 'claws' read 'operculum'

εὐοπλία, add 'II pl., *feats of arms*, Aristid.Quint. 2.10.'

εὔοπλος, add 'of Athena, εὔοπλε καὶ ἀρσενωπέ, poet. in *PKöln* 245.10 (iii AD)'

εὐοργία, for 'sic' read 'εὐπειστία La.'

✝**εὔοργος**, ον, *good-tempered*, S.*fr.* 33a R.; adv. -ως E. in *POxy.* 3317.14 (= E.*fr.* 165 where εὐλόγως or cj. εὐλόφως is read).

*εὐοχθεία, ἡ, *abundance, plenty*, ὄρσει δὲ πολύλλιτον εὐοχθείαν Robert *DAMM* p. 92 (Syedra, i BC; hex.).

εὐπᾰγής, for '*well-woven*' read '*close-woven*' and after '*BGU* 1564.10 (ii AD)' insert '*PBeatty Panop.* 2.22 (AD 300); of words, *well-assembled*, i.e. with a euphonious arrangement of letters, Demetr.*Eloc.* 176'

εὐπαιδευσία, add '*SEG* 24.491 (Maced., AD 202)'

εὔπαις, add '(so perh. in Ar. l.c., *noble son* of Apollo, cf. καλλίπαις II)'

εὐπᾰλής I, after 'A.R. 2.618' insert 'οὔ]τι γὰρ εὐπαλές ἐστι *Lyr.adesp.* 14.9 P.' II, add '*BCH* 107.875 (unless pers. n.)'

*εὐπαραδειγμάτιστος, ον, *well-exhibited*, i.e. *clearly showing*, Ptol.*Tetr.* 170.

✝**εὐπάρθενος**, ον, *having a beautiful maiden* or *maidens*, εὐ. ἄστυ Triph. 51, Nonn.*D.* 39.188, 42.462, 43.430; εὐ. εὐνήν ib. 16.311. II (*that is a*) *beautiful, blessed*, or sim., *maiden*, εὐ. Δίρκα E.*Ba.* 520, *AP* 6.287 (Antip.Sid.).

εὐπαρόξυντος, add 'Paul.Aeg. 7.4.9 (278.16 H.)'

*εὐπάταγος, ον, *fine-sounding*, ἠχῆς εὐπατάγου perh. of drums and cymbals, *POxy.* 3723.14 (elegy, ?ii AD); cf. εὐκέλαδος.

εὐπάτειρα, for 'Men. 616 (with v.l. εὐπατέρεια)' read 'Men.*Dysc.* 968'

*Εὐπᾰτοριστᾰί, οἱ, guild of worshippers of Mithridates Eupator at Delos, *Inscr.Délos* 1567 (i BC).

εὐπατρίδης I, add '*Kafizin* 50'

εὐπάτωρ, add 'III epith. of kings, e.g. Antiochus V, App.*Syr.* 46 §236.'

εὐπειθής, add '2 *satisfied*, of a contracting party, *PMich.*XI 604.22, *POxy.* 2769.26 (both iii AD).'

εὔπεπτος 1, add 'adv. -τως, εὐ. ἔχειν πρός τι *have an appetite* for, *Vit.Aesop.*(G) 3'

εὐπερίπᾰτος, add 'II of persons, *able to walk easily*, *Cat.Cod.Astr.* 11(2).191.31.'

εὔπῐδαξ, read 'εὔπῑδαξ'

εὐπῐθής, act., add 'E.*Andr.* 819 (codd. εὐπειθ-)'

✝**εὔπλητος**, ον, expl. of εὔστιπτος Sch.A.R. 2.30; dub. l. in Arist.*Sens.* 438ᵃ15 (comp.).

εὔπλεκτος, delete 'also η, ον .. (cj. for ἀπλ-)'

Εὔπλοια (εὔπλοια II), add 'of Isis, *Inscr.Délos* 2153 (ii BC)'

*εὔπλουμος, ον, *well-embroidered*, *PAnt.* 44.13 (iv/v AD).

εὐποδία, add '*SEG* 23.109 (Attica, ii AD)'

εὐποιΐα, line 3, before 'εἰς πλῆθος' insert 'Thphr.*fr.* 584A.100 F.'

εὐπόλεμος, after '*h.Mart.* 4' insert '*CEG* 10.4 (Athens, v BC), Ἀρετή ib. 102.2 (Athens, *c.*400 BC)'

εὐπορέω II, after '*furnish*' insert 'Hp.*Cord.* 11'; line 5, after '(iii AD)' insert 'also abs., *make provision*, rest. in *JRS* 27.19 (Galatia)' and omit 'hence in'

εὐπορία I, add '3 *power, might*, Aq.*Ps.* 109(110).3; concr., of an armed force, Lxx 4*Ki.* 25.10(cod. A), Aq.*Is.* 36.2.'

εὔποτος II, after '*good to drink from*' insert 'ποτέριον *CEG* 454 (Ischia, viii BC)'

εὔπους I, line 3, after '*fleet of foot*' insert 'εὐ. νύμφαν Sapph. 103.5 L.-P.'; for '*Call.fr.* 48' read '*Call.fr.* 302 Pf.' and add 'Ὦραι Nonn.*D.* 38.131, 331'

εὖπραξις, add '(cf. pers. n., Εὔπραξις, Lang *Ath.Agora* XXI D7 (vi BC))'

*εὐπρᾱτικός, ή, όν, *of a cheap kind*, ζῦτος *BGU* 1069ᵛ9 (iii AD).

*εὐπρόσεκτος, ον, *cherished*, τέκνον *BCH* 58.343 (Caria).

*εὐπροσώνυμος, ον, *of good name, honoured*, τὴν εὐ. ἡμῶν πόλιν Act. Alexandr. ix col.vi.17.

εὐπροσωπία, add '*BGU* 1787.12'

εὐπρόσωπος, line 1, after 'Anaxandr. 9.5' insert 'of Pan, hymn in *IG* 4².130.19' 3, delete the section.

*εὐπτέρυξ, ῠγος, ὁ, ἡ, = εὐπτέρυγος, ἐϋπτερύγεσσι πελείαις *GVI* 655.9 (Trachonitis, ii/iii AD).

εὐρέκτης, add 'also ἐϋρρέκτης epigr. in *Inscr.Olymp.* 481 (iii/iv AD); (cf. pers. n. *IG* I³.319.14, Athens, 432/1 BC)'

✝**εὕρεμα**, v. ‡εὕρημα.

εὑρετός, add 'w. θεός in uncertain significance, Mitchell *N.Galatia* 138'

εὑρέτρια, add '*Hymn.Is.* in *SEG* 8.548.3 (Fayum, i BC)'

εὕρημα, line 1, delete '(q.v.)' I, add '(form εὕρεμα) Hp.*Vict.* 1.2, Str. 16.2.24, *AP* 7.411 (Diosc.), Babr.*Prooem.* ii 2, *PMag.* 13.299' III, add '*sum specified in a tender*, *IG* 7.3074.3 (Lebadea, ii BC)'

εὑρησιλογία, add '*casuistry, quibbling*, *PAlex.* 10.11 (i AD), *BGU* 2042.15 (AD 105), *PNess.* 26.21 (AD 570)'

εὔρῑπος, add 'Myc. *e-wi-ri-po*, place-name; cf. deriv. adj. *e-wi-ri-pi-ja*'

εὑρίσκω I, add '7 pass., *to be found, be present*, *Cod.Just.* 1.2.17.2.'

εὐροέω II, add 'τοῖς εὐρεοῦσι καὶ θέουσι prob. in Plu. 2.375d'

*εὔροιζος, ον, *ringing true*, of gold, Ps.-Callisth. 2.41.1 cod.C.

εὔροπος, line 1, εὐ. ἅμμα, for 'an *easy-sliding* noose' read 'clasp *that easily overturns* (a bull)'

*εὐρράθᾰμιγξ, ιγγος, adj. *copiously dripping*, Nonn.*D.* 5.258, 33.101.

εὐρυάναξ, read 'εὐρῠάναξ'

εὐρύβατος II, add 'often coupled w. Φρυνώνδειος (q.v.)'

*εὐρύβοτος, ον, *having broad pastures*, πατρίδος εὐρυβότου epigr. in *SEG* 38.734 (Thrace, iii/ii BC).

εὐρυθμοκάρηνος, -κερως, delete the entries.

εὔρυθμος 3, add 'κείσιν εὐρύθμοις epigr. in *SEG* 37.1537 (Arabia, *c.*AD 400)'

*εὐρῠκάρηνος, ον, *having a broad head*, σιγύνης Opp.*C.* 1.152, Πίθος Nonn.*D.* 20.127.

*εὐρύκερως, ωτος, *having spreading horns*, Opp.*C.* 2.293, v.l. in Mosch. 2.153.

εὐρῠκόων, ωντος, fem. κόωσα, unexpld. wd. (cf. Hsch. s.v. -κόωσα) used as epith. of οὐρανός, poet. in *PAnt.* 56B(a)ʳ.6; of νύξ, Hsch.; of the sea-goddess Ceto, Euph. 116 vGr. (for suffix, cf. perh. Λαοκόων).

εὐρυκόωσα, delete the entry.

εὐρύνωτος, add '*Suppl.Hell.* 991.46 (poet. word-list, iii BC)'

εὐρύπρωκτος, for '*wide-breeched*' read '*having a wide anus*' and add '*SEG* 26.1708 (Egypt, early v BC)'

*εὐρύρροος, ον, = εὐρυρέεθρος, cj. in A.R. 4.269.

εὐρύς, lines 3/4, for 'Asius 13' read 'Asius *Fr.Ep.* 13.3 B.' at end add 'in Myc. compd. names, as *e-u-ru-da-mo* (= Εὐρύδημος)'

*εὐρύσῐμωλος (v.l. εὐρύσωλος), unexpld. wd., Arc. 57.14 B.

εὐρύσορος, for '*with wide bier* or *tomb*' read 'of a tomb, perh. *holding a wide sarcophagus*, or sim.'

εὐρύχορος, add '2 *widespread, far-flung*, οὐκ ἔθανεν γάρ, ζώσης εὐρυχόροιο τέχνης ἀρεταῖσι μαθητῶν *GVI* 742.7 (Rome, ii/iii AD).'

✝**εὐρώδης**, ες, prob. *mouldy* (ἐϋρώεις) S.*Aj.* 1190 (lyr.).

Εὐρωπαῖος, after '*European*' insert 'Hp.*Aër.* 16, 23'

Εὐρώπη, after 'Pi.*N.* 4.70' insert '(Εὐρώπαν ποτὶ χέρσον perh. as adj.)'

*εὐρωτίας, ου, ὁ, a precious stone, Plin.*HN* 37.161.

ἐΰς, add 'used for Lat. *pius*, Ἀντωνῖνος ἐΰς *SEG* 12.1502 (Palestine, v AD)'

εὐσέβεια 1, add 'εὐσεβίας χάριν *SEG* 31.1533 (Egypt, *c.*AD 200)' 3, add 'b as an honorific title, *SEG* 34.1243.6 (Abydus, v AD), etc., τὸ τῆς ἡμετέρας εὐσεβείας ἐπώνυμον συντεθείκαμεν βιβλίον Just.*Const.Δέδωκεν* 1.'

✝**Εὐσέβεια**, τά, games founded at Puteoli by Antoninus Pius in honour of Hadrian, *IG* 2².3169/70.16, 14.737.8, 7.49.23, Artem. 1.26.

εὐσεβέω, line 4, for 'εὐ. τὰ πρὸς θεούς' to end of article read 'θύουσα καὶ εὐσεβοῦσα τοῖς θεοῖς *PRyl.* 112(a).4 (iii AD); w. internal obj., εὐσεβεῖν τὰ πρὸς θεούς S.*Ph.* 1441; εὐ. τὰ περὶ τοὺς θεούς Isoc. 3.3; pass., ἵνα τοῦτο εὐσεβηθῇ Pl.*Ax.* 364c. 2 trans., *reverence*, εὐ. θεούς A.*Ag.* 338, E.*Tr.* 85, *Ph.* 1321; pass., Ph. 2.201.'

εὐσεβής I 2, add 'also of other monarchs, *SEG* 30.1697 (prob. sup.; Palestine, AD 585). c as the title of a legion (= Lat. *Pius*), λεγ(ιῶνος) ια′ Κλαυδίας εὐσεβοῦς πιστῆς *SEG* 33.1194 (Cappadocia, i/ii AD), *Xanthos* 50.7.' II, add '2 *venerable*, *AP* 9.360 (Metrod.).' add 'IV εὐ. (sc. λίθος), ὁ or ἡ, a precious stone, Plin.*HN* 37.161.'

*εὐσεβουργός, όν, *doing pious service*, *SEG* 6.66 (Ancyra).

εὔσειστος, add '2 w. ref. to the eyes, εὔσειστοι τὰ ὄμματα gloss on κλαδαρόμματοι, Hsch.'

εὔσελμος, add 'σωτῆρες εὐσέλμων νεῶν *Trag.adesp.* 463 K.-S.'

*εὔσεμνος, ον, *august*, σπείρης *IG* 14.925 (Portus Trajani, ii/iii AD).

εὔσημος II, after 'Hp.*Mochl.* 16' insert 'Gal. 9.776.16'

εὐσῐτέω, add 'Gal. 17(2).526.11, 13'

*εὔσκηνος, ον, *well-staged, beautiful on the stage*, χορεία *AE* 1931.117 (Miletus, ii BC).

εὔσοια, add 'dub. in Alc. 286(a).6 L.-P.'

*εὐσόλοικος, ον, *irregular but permitted* (in regard to speech), Eust. 1287.37.

*εὔσοφος, ον, prob. *very learned, well-skilled*, τὸν εὐ. καὶ πανάριστον *PASA* 3.175 (Isauria).

εὐσπάρτεος, add '(*Suppl.Hell.* 1078)'

εὔσπειρος, for '*well-wreathed*' read '*well-coiled*'; for 'Antip.' read 'Antip.Sid.'

εὐστάθεια, line 2, after '(Strat.)' insert '*Didyma* 496*B*13'

εὔσταθμος, line 3, after 'νομίσματα' insert '*POxy.* 1932.6 (v AD)'

*εὐστᾰσία, ἡ, = εὐστάθεια 1 personified, poet. in *Inscr.Perg.* 8(2).324.15.

ἐϋστείρη, delete the entry.

*εὔστειρος, α, ον, poet. εὔστ-, *having a good keel*, Call.*fr.* 18.4 Pf. (uncertain gender in broken passage), ἐϋστείρης .. νηός A.R. 1.401.

εὔστεκτος, delete the entry.

*εὐστέφιος, α, ον, = εὐστέφανος II, γαῖα BCH 59.148 (Philippi).
εὐστόν, transfer the entry before εὔστοργος.
εὔστοργος, add 'II loving, μήτηρ JRS 18.30 (Upper Tembris valley).'
εὐστόρθυγξ, for 'consisting .. branch' read 'perh. well-spiked'
εὔστοχος II 3, after 'ἄγρη' insert 'AP 6.13 (Leon.Tarent.)'
εὔστρεπτος I, add 'close in texture, expl. of εὔστιπτος, Sch.A.R. 2.30'
εὐστροφάλιγξ, for 'curly' read 'whirling'; for 'Antip.' read 'Antip.Sid.'
εὔστρωτος, after 'h.Cer. 285' insert 'Alc. 283.8 L.-P.'
†εὐσύγκριτος, ον, well-constituted or -compounded, ὄμμα Simp.in Cael. 75.2; in Epicurean philosophy, Diog.Oen. 1 ii 14, 2 iii 4.'
*εὐσυμβούλευτος, ον, easy to advise, Ps.-Callisth. 30.7.
εὐσύνδετος, add 'Heph.Astr. 3.20.5'
εὐσυνείδητος I, add 'adv. -τως, εὐσεβῶς .. καὶ -τως ἔχων MAMA 8.413e'
εὐσύνθετος I 2, add 'b well put together, κέκλεικε τοῦτο μαρμάροις εὐσυνθέτοις CIJud. 1.653 (Syracuse), SEG 34.327 (Tegea, Chr.).'
*εὐσυννόητος, ον, easy to comprehend, Simp.in Cael. 264.16.
εὐσύνοπτος II, for 'easily taken in by the mind' read 'easily comprehended' and add 'Is.fr. 161'
*εὔσυριγξ, ιγγος, perh. melodious with the sound of pipes, μέλαθρον Nonn.D. 3.320.
*εὐσχεθής, ές, good to handle, εὐσχεθὲς .. τόξον Hes.fr. 33(a).32 M.-W.
εὐσχημονίζω, for 'train, educate' read 'treat properly, maintain adequately'
*εὐσχημονισμός, ὁ, proper maintenance, GDI 1708.16 (Delphi, ii BC).
εὔσχημος, after 'adv. -μως' insert 'Chor. 29.77 p. 335 F.-R.'
εὐσχήμων II 2, add 'sup. in honorific address, ἀγωνοθετοῦντος τοῦ εὐσχημονεστάτου Αὐρ. Διοτίμου SEG 30.1524 (Lycia, iii AD)'
*εὐ]ταξία, ἡ, app. good order, εὐ]τακτίη rest. in SEG 26.551 (Coronea, iii BC).
εὔτακτος I 2, add 'in epitaphs, esp. of slaves, SEG 34.1483 (Palestine, iii/ii BC)'   II, after 'of payments' insert 'PStrassb. 228 (iii BC)'
εὔταρσος 1, for 'delicately winged' read 'well-operculated' and for 'grasshopper' read 'cicada'
*εὐταφία, ἡ, good burial, ζῶον ἐλευθερίη [καὶ] νέκυν εὐταφίη MAMA 9.76.
εὖτε I 1, line 5, delete 'δὴ τότε γε 22.182'; line 7, after '(Od.) 20.56' insert '22.182'
εὐτεκνέω, add 'Trag.adesp. 681.9 K.-S.'
εὐτεκνία I, add '2 easy birth, Epiph.Const. in Lapid.Gr. 197.4.'
εὐτέλεια I 2, add 'b as term of self-depreciation, ἡ εὐτέλειά μου POxy. 1165.1, 8 (vi AD).'
εὐτελής I, add '3 of words, style, etc., undistinguished, commonplace, Arist.Rh. 1408ᵇ13, Demetr.Eloc. 70 (comp.); of the speaker, ib. 100; adv. -λῶς, ib. 167 (sup.), D.H.Dem. 18 (comp.).   4 belonging to the lower orders, CodJust. 11.41.7, Just.Nov. 12.1.' II, add 'οἱ εὐτελεῖς κληρικοί SEG 37.195 (Attica, Chr.); freq. as pr. n. IG 12(9).126 (Styra, v BC), SEG 26.368, etc. (Attica, iv BC)'
*εὐτερπίη, ἡ, delight, Hymn.Is. in SEG 8.549.18 (Egypt, i BC).
*εὑτής· ἀγαθότης Hsch. (La.; v. ἐητύς).
εὐτλήμων, after 'δόξῃ' insert '(v.l. ἐν τλήμονι)'
εὐτονέω, add 'Iamb.Protr. 21 λβ' p. 124.9 P.'
εὐτονία a, add 'εὐ. τοῦ ἔργου w. ref. to a circus athlete, Delph. 3(1).216.5'
εὐτράπεζος 2, add 'good to eat, Xenocr. 9'
εὐτραπελεύομαι, add 'Sch.Pl.Euthphr. 24c G.'
εὐτραπίζω II, delete 'conciliate' and 'pass. in med. sense'; add 'humorously, w. ref. to appropriation of another's possessions, AP 9.316 (Leon.Tarent.)'
εὐτρίαινα, add '(εὐρυτρ. codd.)'
εὔτριχος, add 'ἄνδρα εὔτριχον Cat.Cod.Astr. 11(2).190.10'
εὔτροπος II, after 'Sch.Od. 1.1' insert 'Sopat.Rh. 8.56.33 W.'; of diseases, add 'Gal.in CMG 5.9(2) 114.25, 129.17, 142.18'
εὐτρόχαλος I, after 'quick-moving' insert 'ἄμαξαι A.R. 1.845'   II, for 'ἐτροχάλῳ' to end of section read 'III well-rolled, ἀλωή Hes.Op. 599, 806.'
εὔτροχος I, line 4, delete 'εὐ. κύκλος .. 19'   III, after 'round' insert 'E.Ion 19'
*εὐτρύγιον (or ?εὐτρύτιον), τό, item in list of comestibles, etc., otherwise unexpld., BGU 2358.15 (iv AD).
εὐτυκάζομαι, add 'act., Call.fr. 177.32 Pf.'
εὐτυχέω II, for this section read 'act., have the good fortune to obtain, attain to, παρὰ τῶν Σεβαστῶν στέφανον IEphes. 3070.13, Men.Rh. 439.10 R.-W., Sopat.Rh. 8.247.18, 328.19 W.'
εὐτύχημα, add 'SEG 31.903.9 (pl., Aphrodisias, iii AD).'
εὐτυχής I, lines 2/3, comp., after 'S.Aj. 550' insert 'masc. ending for fem., SEG 30.1485 (Phrygia, iv AD)'   add '2 sup. as honorific title (= Lat. felicissimus), SB 4678.11, POxy. 1042, 1896, PLond. 1723 (all vi AD).'   II, after 'at close of letter, D.H.Amm. 2 fin.' insert 'SEG 30.812'; add 'at the end of an inscr., SEG 31.635 (Maced., ii/iii AD), etc.'

εὔυμνος II, for this section read 'making beautiful song, ῥήματα Suppl.Hell. 980.3'
†εὐυπάντητος, ον, easily approached, IGBulg. 390.6; adv. -ως, VDI 1960(3).154 (Chersonese).
*εὐυπόστατος, ον, staunch, steadfast, Afric.Cest. p. 18 V.
*εὐφανής, ές, illustrious, IGRom. 3.739ᵛ.5 (Rhodiopolis, ii AD; sup.).
*εὐφαρέτρειος, α, ον, having a beautiful quiver, Ἄρτεμιν εὐφαρέτρειαν orac. in ZPE 92.267 (Ephesus, ii AD).
εὐφαρέτρης, add 'SEG 23.126.3 (Attica, i BC)'
εὔφημος I 1, lines 4/5, for 'moving the lips of reverent thought' read 'uttering the words of auspicious thought'
εὔφθαρτος I 2, II, for these sections read 'II of food, easily broken down, Diph.Siph.ap.Ath. 2.68f, Gal. 8.34.5.'
εὔφθογγος, line 2, after 'λύρη' insert 'Margites 1.3 W.'
εὐφραδής 1, adv., for 'eloquently' read 'prudently, wisely' and add 'IG 2².5201, CRAI 1968.423 no. 6'
εὐφραίνω, line 7, after 'Ar.Ach. 5' insert 'pf. part. εὐφραμμένος Hsch. s.v. κεκραιπαληκώς'   II, add 'w. ref. to sexual fulfilment, Ar.Lys. 165, 591'
εὐφημός 1, for 'Sch.E.Hec. 100' read 'Sch.E.Hec. 98'
*Εὐφράνωρ, ορος, ὁ, epith. of Zeus, ABSA 49.12 (Dorylaeum).
*εὐφρονέω, be in good heart, CIJud. 303.
εὐφρονέων, delete final parenth.
εὐφρόνη II, add 'cj. in Orph.H. 9.8'
*Εὐφρονίσιοι, οἱ, name of group of Corybantes, IEryth. 201a.62 (iii BC).
εὐφροσύνη II, add '2 title of Isis, Ἴσιδι Εὐφροσύνῃ Inscr.Délos 2107 (i BC).'
εὐφωνία I 1, after 'X.Mem. 3.3.13' insert 'D. 19.339'
*εὐφώτιστος, ον, well-lighted, Simp.in Cael. 457.12.
εὐχαίτης, add 'of Dionysus, AP 9.524'
εὐχάρακτος, add 'Procl.CP Hom. 28.29'
εὔχαρις I, add '2 graceful, of diction, Demetr.Eloc. 173; of style, D.H.Din. 81.'
*εὐχαριστεύς, έως, ὁ, bestower of favours, benefactor, nom. pl. -ῆς SEG 38.1476.108 (Xanthos, iii BC).
εὐχαριστέω 2, line 5, after '1Ep.Cor. 1.4, etc.' insert 'med., show gratitude, μεταξὺ ἀχαρίστων τε .. καὶ εὐχαριστουμένων παίδων Just.Nov. 22.48 pr.'
εὐχαριστητικός, for '= -ιστικός' read '= εὐχαριστικός' and after '(Ph.) 177, 371' insert 'IGRom. 3.704iiiB 6 (Cyanae)'
εὐχαριστία 1, add 'Θεῷ Ἡλίῳ Ἐλαγαβάλῳ Μαιδονας Γολασου εὐχαριστίας ἀνέθηκεν Philol. 127.257 no. 2 (Emesa, iii AD)'
εὐχάριστος I, for 'εὐχάριστα .. (Cyprus)' read 'Cypr. e-u-ka-ri-ta εὐχάριστα, acceptable gifts, Kafizin 117b, 303' and add 'πλείονα εὐχάριστα πεποίηκεν τῇ συνόδωι SEG 32.453 (Boeotia, ii BC)'   III, add 'cult-title of Zeus, INikaia 1085 (iii AD).'
εὐχαριτος, add 'fem. εὐχαρίτη, ἀρχήν epigr. in Inscr.Délos 36 (iv/iii BC); adv., εὐχαρίτως ἄρξας rest. in epigr.ib. 37'
εὐχέρεια IV, add 'εὐ. τῆς βολῆς poet.ap.Ath. 15.667e (εὐχειρία codd.)'
εὐχερής I 1, line 8, for '(Pl.) Tht. 184c' read '(Pl.) Tht. 184b'   III, before 'Batr. 62' insert 'πρᾶγμα Damox. 2.10 K.-A.' and after it 'μελέτη Plu.Dem. 2'; after 'adv. -ρῶς' insert 'ἐργάζεσθαι Hp.Fract. 30' and add 'διὰ τὴν ἔνδειαν τῆς τροφῆς εὐχερῶς ἀπολλύμενοι Peripl.M.Rubr. 29'
εὐχή 1, line 5, delete 'εὐχὴν ἀνέστησεν .. i/ii AD'   add 'b votive offering, καλλίγραπτον εὐ. A.fr. 78a.12 R. (lyr.); εὐχὴν ἀνέθηκεν IG 12(3)458, etc., cf. Samothrace p. 49; without verb, in a dedication, Δάματρι εὐχάν SEG 33.765 (Heraclea, Italy, iii/ii BC), Γεννάδης Ἀπόλλωνι εὐχήν SEG 30.1474 (Phrygia), 1604 (Cyprus).'
†εὔχλοος, ον, Ep. εὔχλ-, characterized by green vegetation, πόλις Βερόη Nonn.D. 41.15 (cj.); epith. of deities assoc. w. vegetation, εὐχλόου Δήμητρος S.OC 1600.
εὔχομαι II 1, line 6, delete 'cf. .. codd.'   add '4 aor. part. in dedications, Κυρίῳ Ἀσκληπιῷ Ἀμφείων .. εὐξάμενος ἀνέθηκεν SEG 30.719 (Thrace, ii/iii AD).'   III 3, after 'declare' insert '(acc. to Sch.Pi.O. 6.88a so used in Laconian)' add 'Myc. 3 sg. pres. ind. e-u-ke-to (= *εὔχετοι)'
*εὐχόρευτος, ον, excelling in the dance, of Pan, poet. in IG 4².130 19 (Epid.).
εὔχορος, delete the entry (v. ‡εὔρυχ-).
*εὔχους· χώνη, Σαλαμίνιοι Hsch.
*εὐχρήσιμος, ον, performing beneficial service, GDI 3011.5 (Megara, iv BC).
εὐχρηστία I 1, after 'Chrysipp.Stoic. 3.168' insert 'PMil.Vogl. 11.7 (ii AD)'   2, add 'Inscr.Magn. 58.12 (iii BC); εὐχρηστίας ποιεῖσθαί τινι advance him money or money's worth, BGU 1731.8, 1732.8 (both i BC)'
εὔχρηστος, add 'τὸ τρ[ύβ]λιον τό[δ]ε εὔχρεστον Kafizin 49 (221/0 BC)'
εὐχρωτέω, delete the entry.
εὐχυμία II 2, delete the section.

εὐχωλή **I 1**, for 'Inscr.Cypr. 94 H.' read 'Cypr. *e-u-ko-la* εὐχōλά *ICS* 85, 220(b).3'   **2**, after 'offering' insert 'εὐχōλέν *IG* 1³.618 (Athens, vi BC), *Naukratis* ii p. 65.776, 777; εὐχōλᾱν ἀνέθε̄κε *SEG* 30.1150 (Heracleia, Italy, *c*.iv BC), *AP* 6.137 (Anacr.)'

εὔψοφος, add 'adv. -ως Sch.Theoc. 11.57'

εὔψυχος, adv. -χως, for this section read '*courageously*, X.*Eq.Mag.* 8.21, Plb. 5.23.9, al., Lxx 2*Ma.* 7.20; *magnanimously*, *IGRom.* 4.860.12'

εὐώδης, add '**2** εὐώδες, τό, *sweet-smelling medicament*, Hp.*Epid.* 4.30, *Superf.* 24.'

εὐωνίζω, add '**2** *make cheaper*, τὴν ἀπο[ρίαν] *LW* 3.1661 (Lydia).'

εὐώνυμος (A) **III**, line 2, after '*from the left*' insert 'εὐ. πλευρωμάτων A.*Th.* 888'   add '**IV** εὐώνυμοι, οἱ, a body of troops in Cyprus, *ABSA* 56.21 (ii BC).'

εὐωρία **I**, add '*Lib.Ep.* 434.4'

*εὐωχήτρια, fem. adj., *festal*, (ἡμέρα) ἡ τῶν ποτῶν εὐ. Leont.Byz.*HP* 1.2.5.

εὐωχία, add 'Astyd. 4.1 S.'

†ἐφαγιστεύω, *perform rites over* the dead, S.*Ant.* 247 (κἀφα-, also interpreted as καὶ °ἁφα-).

*ἐφαγνίζω, *perform rites over* the dead, τὰ πάντα S.*Ant.* 196 (v.l. ἀφ-).

ἐφάλλομαι, at end delete 'εἰς τοὐπίσω .. Gal. 6.145'

ἔφαλος, add 'form *ὀπίhαλος, Myc. *o-pi-a₂-ra* (neut. pl.) *coastal regions*'

*ἐφάνγρενθειν, v. ἐφαιρέομαι.

ἐφαπτίς, for '*soldier's upper garment*' read '*kind of cloak, worn esp. by soldiers*' and add 'Str. 11.14.12'

*ἔφαπτον, τό, kind of cloak, cf. ἐφαπτίς, Callistr.ap.Sch.Ar.*Av.* 933.

ἐφάπτω, line 5, for 'she *had made fast* (i.e. perpetrated)' read 'he (sc. Hyllus) by his anger *had fastened on her* (i.e. doomed her to)'   **II 5, III**, delete the sections.

ἐφαρίξαντο, at end of parenth. add 'ἐφατίξαντο· ἐψηφίσαντο La.'

ἐφαρμογή, add 'ἐφαρμογάν Ps.-Archyt. p. 30.19'

ἐφαρμόζω **II 1**, add '**c** σῆμα ἐ. *build* or *set in its place*, *AP* 7.295.7 (Leon.Tarent.)'

ἐφέδρα **II 1**, delete the section.   **2**, add 'orac.ap.Phleg.*Mir.* 3'

ἐφέδρανον **3**, add 'Gal. 18(1).747.12'

*ἐφεδρευτής, οῦ, ὁ, *one who lies in wait*, Sch.Pi.*N.* 4.155.

ἐφεδρεύω **II 2**, for '*draw a bye*' read '*wait one's turn* (having drawn a bye in the first round)'   add '**5** *attend* at a meeting, [ταῖς ἐκκλησί]αις ἐφεδρεύοντες *SEG* 29.116 (Athens, iii BC).'

ἐφεδρίς, ίδος, ἡ, *throne*, Call.*fr.* 196.37 Pf.

†ἐφεκτός, όν, *on which judgement is to be suspended*, S.E.*P.* 3.55.

ἔφεκτος, line 2, after 'D. 34.23' insert '*BCH* 80.53 (Sigeum, ii BC)'

*ἐφελιώμενος, app. epith. of oxen, of unkn. meaning, cf. perh. ἔφηλις II 1, *OGI* 456.22 (Mytilene, i BC).

ἐφέλκω **I 1**, add '**b** of circumstances, *bring* to a place, εἰς ἃς δήποτέ σε ὑπαρχείας τὰ κοινὰ τῶν Ῥωμαίων ἐφέλκοιτο *PKöln* 10.7-9 (i BC).'   **3**, delete the section transferring 'E.*Cyc.* 151' to section I.   add '**IV** w. inf., dub. in *JRS* 40.78.6 (Cyrene; letter of Hadrian), v. °ἐφολκέω.'

*ἐφενέπω, *proclaim*, *SEG* 12.371.3 (Spartan decree engraved at Cos, iii BC) (ἐπι- + ἐνέπω, w. aspirate transferred fr. ἐνhέπω > ἐνέπω).

ἐφεξῆς, after lemma insert 'non-Att.-Ion. -ᾶς *BCH* suppl. 22 p. 287 (Argos, Hellen.)'   **II 1**, for 'esp. with πᾶς' read 'combined w. πᾶς, all *without exception*'   **3**, add '**b** *subsequently*, κρατήσει .. οὗτος ὁ νόμος .. ἐπὶ τοῖς ἐφεξῆς Just.*Nov.* 18.5, *Cod.Just.* 1.3.55.1.'

ἐφέπω **II 1**, line 2, for 'Simon. 142.2' read '*AP* 7.296.2 ([Simon.])'   **B II 2**, add 'Thgn. 217, 1073'

ἐφερμηνευτικός, after '*explanatory*' insert 'Alex.Aphr.*in Metaph.* 745.10'

*Ἐφεσηΐς, ίδος, ἡ, *a celebration of the* (μεγάλα) Ἐφέσεια, used in dating, *BMus.Inscr.* 605.10 (Ephesus, ii AD).

Ἐφέσια, τά, add 'also Ἐφέσηα *IEphes.* 859a'

Ἐφέσιος, add 'epith. of Artemis esp. at Ephesus, *SEG* 33.939, etc.; at Alea, Paus. 8.23.1'

ἔφεσις **I 2**, after '*appeal*' insert 'or *referral* (v. *SEG* 28.4)'   **II**, after 'Pl.*Lg.* 864b' insert '(s.v.l., cj. ἄφεσις)'

ἐφέσπερος, add 'ἐφέσπερον δαίουσα λαμπτῆρος σέλας Trag.adesp. 407 K.-S.'

ἐφέστιος **I**, add '**2** of sacrifices *offered at the* (*public*) *hearth*, οἷς *SEG* 12.371.17 (Spartan decree engraved at Cos, iii BC); cf. ἐνέστιος.'   **IV**, for this section read 'fem. subst., a drink, perh. particular mark of hospitality, πίνουσα τὴν ἐπίστιον Anacr. 82.4 P. (expld. by Ath. as ἀνίσωμα)'

ἐφέτειος, after '(Apollonis)' insert '*TAM* 5.1206.9 (-τήϊος; Apollonis); *of the present year*, *PFam.Teb.* 3.23 (AD 92)'

ἐφέτης **III**, add 'as adj., παρὰ τῷ ἐφέτῃ δικαστῇ Just.*Nov.* 93.1'

†ἐφετινός, ή, όν, *of the present year*, χόρτος *POxy.* 1482.12 (ii AD), αἶγες, ἐρίφια *PMasp.* 141.6ᵛ.9, 12 (vi AD); as name of a category of ephebes, *IG* 12 suppl. 690.10.

ἐφετός **II**, delete the section.

*ἐφέτος (unless to be taken as two words ἐφ' ἔτος), adv. *this year*, *IG* 5(2).433.7 (Megalopolis, ii BC), so ἐφέτους (or ἐφ' ἔτους) *Stud.Pal.* 22.23.9 (i AD; pap. εφετους), *PMich.*VIII 473.10 (ii AD); mod. Gk. φέτος.

*ἐφεύδω, = ἐγκοιμάομαι 1, part. fut. ἐφευδησίονσαν *Inscr.Cret.* 1.xvii 3, 9 (Lebena).

ἐφεύρεμα, for '*IG* 2².1119.6 (iii AD)' read '*SEG* 24.143 (Attica, Augustus)'

ἐφευρίσκω **I 1**, line 3, for 'Sapph.*Supp.* 4.9' read 'Sapph. 15(b).9 L.-P.'; line 9, after 'Sapph.l.c.' insert 'Πενία, τί σ' ἡμεῖς τηλικοῦτ' ἐφεύρομεν; Men.*Dysc.* 209'

*ἐφευροκλέψ, unexpld. wd., Theognost.*Can.* 97 (cf. νακοκλέψ; perh. better parox.).

ἐφηβεία **1**, add '*AP* 7.467 (Antip.Sid.)'

*ἐφήβειον, τό, = ἐφήβαιον, *SPAW* 1934.1049 (tab.defix.).

ἐφηβεύω, after lemma insert 'aor. ἠφήβευσαν *Inscr.Délos* 2594.4 (ii BC), pf. part. ἠφηβευκότων *OGI* 178.5 (Egypt, i BC)'

ἐφηβικός **I**, add '**2** ἐφηβική (sc. παιδιά), ἡ, alternative name for the game °ἐπίσκυρος, Poll. 9.104.'

ἐφηβος, after lemma insert 'ἔπηβ- *SEG* 9.72.40 (Cyrene, iv BC)'   **I**, add '**3** as adj., ἐκ νηπίας ἡλικίας καὶ οὔπω τὴν ἔφηβον ἐκβάσης *Cod.Just.* 1.3.52.1, Just.*Nov.* 72 pr.'

ἐφηβότης, add 'cf. perh. Pamph. ἐφ[ι]ιεhιόται[*IPamph.* 3.9'

ἐφηβοφύλαξ, add '*AJA* 58.236 (Nicopolis)'

ἐφηγέομαι, add '**II** *be in charge of*, ἐφηγείσθω γε αὐταῖς (sc. ταῖς ἐπαρχείαις) ἀνὴρ εἷς Just.*Nov.* 28.2.'

ἐφήγησις, add '**II** *conducting* of arbitrators along disputed boundary, *Delph.* 3(1).362 i 14, 15, 25.'

ἔφηλις **II**, for '(Hp.)*Mul.* 2.215' read '(Hp.)*Mul.* 3.215'

ἔφημαι, line 5, for 'Id.*fr.* 157' read 'id.*fr.* 154a R.'; delete '(ἐφιμένη cod. Hsch.)'

*ἐφημερησία, ἡ, perh. *day's work*, *PHeid.* 328.9 (iii AD).

ἐφημερία, add '**II** *division of guards* on duty, οἱ ἐκ τῆς μέσης ἐ. *SEG* 7.29 (Susa, i BC), cf. ib. 30.'

ἐφημέριος, add '**II** subst. -ιον, τό, *space of a day*, *MDAI(A)* 56.125 (nr. Sardis).'

ἐφήμερος, Dor. ἐπάμ-, add 'Theoc. 30.31, *CEG* 139 (Troezen, *c*.500 BC)'   **III 2**, add 'neut. sg. adv., *within the same day*, *CEG* l.c.'

ἐφθέος, delete 'to be'

*ἔφθιον, τό, some kind of boiled food (?soup), *OAshm.Shelton* 74, 76, al. (sp. ἔπτιν).

*ἐφθοπώλης, ου, ὁ, *seller of cooked food*, *PUG* II 71.3 (vi AD).

ἐφθός **I 1**, add 'of perfumed oil, *BCH* 59.440 (Acraephia, i AD)'

ἐφιάλτης, line 2, add 'Dor. Ἐφιάλτᾱς (as pr. n.), *BCH* 109.94 no. N7'; line 3, after 'Alc. 129' insert '(= 406 L.-P.), Macr.*Somn.* 1.3.7'; line 4, after 'Od. 11.308' insert 'Salamine 355 (amphora, vi BC)'

ἐφιδύη, for '(Theognost.*Can.*) 7.30' read '18 A.'

ἐφίερος, for '*IG* 3.74' read '*IG* 2².1366.24'

ἐφίημι **B II 1**, after 'Isoc. 2.25' insert 'τυραννίδος Pl.*Ep.* 8.354c'

ἐφικνέομαι **I 1**, line 3, delete 'with a stick' and insert 'οὐκ ἐδύναντ' ἐπίκεσθαι (i.e. reach the apple) Sapph. 105(a).3 L.-P.'   add '**5** ἐ. ἐς *approach* in similarity, *resemble*, Luc.*Syr.D.* 15.'   **III**, delete the section.

ἐφίμερος, line 1, after 'ον' insert 'also °ἐπίμερος'

*ἐφινίους· τὰς ἐπὶ τοῦ ἰνίου σάρκας Hsch. (La.; ἐφίνους cod.).

ἐφιορκέω, after '= ἐπιορκέω' insert '*SEG* 23.320(a) (Delphi, iv BC)'

ἐφιππάζομαι **3**, delete 'Palaeph. 52'   add '**4** *drive*, Palaeph. 52.'

ἐφιστάνω **II 2**, line 3, for 'Ammon.*in APr.* 68.10' read 'Ammon.*in APr.* 60.18'

ἐφίστημι **A I**, add 'esp. *place* gravestone *over*, Ἀγαθόκλεια .. τῷ ἀνδρὶ .. ἐπέστησεν *BCH* 37.202 (Chios, ii BC), *SEG* 23.640 (Paphos, iv BC), cf. *GVI* 97'

†ἐφιστορέω, *inquire of* an oracle, ἐ. τὸν Δία .. περὶ .., Ἠπειρωτικὰ Χρονικά 1.254 (Dodona, iv BC, ἐπιστ-); cf. ἐφιστορεῖν· ἐπερωτᾶν Hsch.

†ἐφοδευτής, οῦ, ὁ, *one who goes the rounds*, of one who patrols roads as a guard, *POxy.* 1033.10, 15 (iv AD); as a spy, transl. Hebr. *rāgal*, Aq.*Ge.* 42.9.

ἐφόδιος **I**, add '**2** fem. Ἐφοδία, title of Artemis (or Hecate), *SEG* 36.329 (Nemea, archaic).'

ἔφοδος (C) **I 2**, add '**c** *approach* to a piece of land, *access-road*, *Mél.Navarre* p. 354.6, cf. *IG* 7.1740.6 (both Thespiae, iii BC).'   **II**, line 1, for 'A.*Eu.* 375' read 'A.*Eu.* 370'

*ἐφολκέω, *drag one's feet*, *delay*, *IG* 12 suppl. p. 213 (= *FGrH* 502) iv.19-20; w. inf., *delay to*, dub. in *JRS* 40.78.6 (Cyrene; letter of Hadrian).

ἐφόλκιον **2**, for '*generally appendage*' read '*baggage*' and transfer 'Plu.*Pomp.* 40' to section 1.   **3**, for this section read 'ἐ. διὰ τοῦ ι τὸ πηδάλιον Hsch.'

ἐφολκίς, add '**3** *part of a ship*, = ῥινωτηρία, Poll. 1.86.'

ἐφομαρτέω, add '(Arr.)*Cyn.* 16.8, *Fr.Phys.* 6 (p. 194.11 R.-W.)'

ἐφορατικός, add '**II** prob. *obvious, apparent*, BGU 2380.8 (265 BC).'

ἐφορμαίνω, add '**II** *meditate, ponder*, Opp.H. 3.503.'

ἔφορος II, line 4, after 'Test.Epict. 4.1' insert 'at Cyrene, Heraclid.Lemb.Pol. 10, 18 D.'

ἐφυβριστής, after 'Ptol.Tetr. 165' add '(v.l., v. sq.)'

ἐφύβριστος I, after 'Vett.Val. 71.18' insert 'Ptol.Tetr. 165'; add 'adv., -τως, *intemperately*, Posidon. 7 J. **2** *scornful, contemptuous*, ἔλεγχος ἐφύβριστος Lxx Wi. 17.7' **II**, delete the section.

ἐφυδριάς, add 'sg., Call.fr. 66.2 Pf.'

*****ἔφυλος**, ον, *wooded*, τόπος SEG 37.100.76, 96, al. (Athens, iv BC), prob. also in SEG 3.117.15 (Oropus, iv BC).

ἐφύμνιον, for '(Call.)Sos. 8.4' read '(Call.)fr. 384.39 Pf.'

ἐφύπερα, add 'SEG 23.678 (Cyprus, ii BC)'

ἐφύπερθε, line 3, for 'Simon. 183.7' read 'AP 7.24.7 ([Simon.])'

*****ἐφυπερῷος**, α, ον, *upper*, οἶκος Inscr.Délos 1416B ii1 (ii BC).

ἐχεμυθέω, add 'PCair.inv. 3733(17A)(19).5 (*Atti Napoli* III 840, vi AD)'

ἐχενηΐς II, for 'a small .. 505ᵇ19' read 'name of various fish. **1** *blenny* or *goby*, Arist.HA 505ᵇ19. **2** *sucking-fish* and *lamprey* (authors confuse the two).'

ἐχετογνώμονες, add '(codd.; cj. ὀχετο-)'

*****Ἔχετος**, epith. of Apollo, SPAW 1927.8 (Locris, v BC).

ἐχεφροσύνη, add 'SEG 36.629 (Maced., Chr.)'

⁺ἐχέφρων, ον, gen. ονος, *sensible, prudent*, ἀνὴρ ἀγαθὸς καὶ ἐ. Il. 9.341, Od. 13.332; freq. as epith. of Penelope, 4.111, etc.; adv. -όνως D.S. 15.33. **II** of (normally) inanimate things, animals, etc., *endowed with a mind*, ἐχέφρονα νῆα Nonn.D. 1.91, νεβρὸς ἐχέφρων 5.538, 7.227, 16.226, πέτρον ἐχέφρονα Orph.L. 369; cf., of the gestures of a mime, ἐχέφρονι .. σιγῇ Nonn.D. 19.218.

*****ἔχθεμα, ἔχθεσις**, v. ‡ἐκθ-.

ἐχθές, line 5, for 'only form used' read 'best-attested form'

ἔχθιστος 2, line 3, after 'X.An. 3.2.5' insert 'cf. οἱ ἔχθιστοι οἱ ἐμοί Antipho 5.85'

ἐχθρᾱ, line 1, before 'Ion.' insert '(ἐχθρᾶν AP 11.340 (Pall.))'

*****ἐχθαλέος**, α, ον, app. *hateful*, Nic.Al. 594 (v.l. ἐχθομένη).

ἐχθρία, add 'PMich.VIII 516.10 (iii AD; ἐκθ-)'

*****ἔχθυμα**, v. °ἔκθυμα.

ἔχθύσῃ, after lemma insert '(-σσῃ La.)'

ἔχθω (A), line 5, after '(Od.) 14.366' insert 'σφετέρῳ δ' ἤχθοντο τοκῆι Hes.Th. 155'

*****ἐχίδηκτος**, ον, *bitten by a viper*, Theognost.Can. 96.

ἔχιδνα, line 2, delete 'prob. of a *constrictor snake*' and after 'Act.Ap. 28.3' insert 'spec., of the female, Arist.Mir. 846ᵇ18'

ἐχινέες, for pres. def. read '*spiny mice, Acomys sp.*' and after '(v.l. ἐχῖνες)' insert 'Ael.NA 15.26'

ἐχινόπους, add 'Plin.HN 11.18'

ἐχῖνος II 2, for pres. def. read '*jar in which were sealed various documents relating to impending court cases* (v. *Hesperia* suppl. 19.3f.)'

ἐχιόδηκτος, for '= ἐχιδνόδηκτος' read '*bitten by a viper*'

ἔχϊς II, delete '636'

ἔχμα, add 'Myc. *e-ka-ma-te, -ma-pi* (instr. sg. and pl.), prob. *strut* (of table)'

*****ἐχόμενα**, adv. pres. part. of ἔχομαι, *in succession* or *next*, SB 4325 vi 9, viii 5 (iii AD), PPetr. 2 p. 118 (iii BC), cf. PLond. 267.3 (ii AD). **2** w. gen., *near*, Lxx Jd. 19.14, Ps. 139(140).6, Am. 2.8; *next to, adjoining*, PPetr. 3 p. 2.22 (iii BC), PStrassb. 29.36 (iii AD). **b** w. gen. of person, *with, beside*, δόξον ἐμὲ εἶναι ἐ. σου PCol. in CPR 30.145.24 (i/ii AD), cf. PGiss. 77.11 (ii AD). **II** *immediately*, PSI 514.8 (iii BC).

ἐχομένιον, before '*coriander*' insert 'plant, sts. identified as'; add 'also ὀχομ- POxy. 2284B6 (iii AD)'

*****ἐχόνομα**, = °ἐχόμενα, as prep. w. gen., OFlorida 14.7 (ii AD), PPetaus 29 (ii AD), PMich.VIII 510.15 (ii/iii AD), etc.

⁺ἐχόντως, adv. fr. part. of ἔχω, *in the manner of one having*, εἰ μὲν οὖν ἀφρόνως ἢ καὶ νοῦν ἐ. ταῦτ' ἐδόξαζου Isoc. 5.7, Pl.Lg. 686e; cf. ἆρ' οὐκ .. ἐχόντως ἑαυτὸν τὸν νοῦν φήσομεν .. ἀποκρίνασθαι; δικαίως καὶ λόγον ἐ. Isoc. 7.60.

ἐχυρός I 1, after '*secure*' insert 'ἐς δ' ἐ. λίμενα Alc. 6.8 L.-P.'

*****ἐχύρωσις**, εως, ἡ, *strengthening*, πρὸς ἐχύρ[ωσιν πυ]λῶν PBeatty Panop. 1.386, 390, 406 (AD 298); cf. ὀχύρωσις.

*****ἐχυσία**, ας, ἡ, app. for *ἐκχυσία (= ἔκχυσις), perh. *scoop*, PKöln 234.9 (AD 431).

ἔχω (A), line 4, after 'inf.' insert 'ἐξέμεν Il. 5.473' and for 'Call.Aet. 3.1.27' read 'Call.fr. 75.27 Pf.'; seven lines before 'A', after '(Il.) 21.345' insert 'cf. B.26.15 S.-M.' and after 'Od. 11.279' insert 'Pi.fr. 52u.17'; after 'Isoc. 19.11' insert 'Pl.Prt. 321c (s.v.l.)' **A I 1**, add '**b** *have* an ἀγών, i.e. a victory in it, *to one's credit*, Delph. 3(1).554, IG 14.1102.14 (Rome); Ἀσκλήπεια ib. 737.9 (Neapolis; all ii AD).' **II 1**, add '**b** of hunters, fishermen, *to have caught*, Certamen 326, cf. Ar.Nu. 733.' **3**, line 5, after 'etc.' insert 'but ὀϊστούς, ἄρδιν ἐ. *have in one's flesh, in one's heart*, Call.Epigr. 37.5 Pf., fr. 70 Pf., cf. AP 6.9.3 (Mnasalc.)' **4 a**, after 'Arist.Pol. 1335ᵇ18' insert 'Sm.Mi. 6.14' **5**, delete the section transferring exx. to section 9 (*check, stop*) **8**, line 21, delete 'ἕξει .. (Laodicea)'; add '**b** *support, sustain* an action, πόλεμον Il. 14.100.' **B I 4**, add '*be concerned with*, [ἕ]ξι πρὸ[ς] τὸν Θεὸν [ὅσ]τις κρεινῖ δ[ικαίους καὶ ἀδίκους] Mitchell N.Galatia 246; JRS 14.88 (Laodicea)' **II 2**, add 'ὡς ἔχω w. vb. of movement, *immediately, quickly*, Hdt. 2.121. δ, Th. 3.30.1, Ar.Eq. 488, Lys. 376, Men.Dysc. 559; w. vb. of speaking, *in a straightforward manner, directly*, Isoc. 9.30, 15.311' **III 3**, add 'Paus. 9.38.10, 10.28.2.' **IV 1**, add '**b** πιὼν .. τρίτην ταύτην ἡμέραν ἔχω, i.e. it is two days since I drank, Alciphr. 3.32.' **2**, add 'ληρεῖς ἔχων Cratin. 208 K.-A.' **C I 3**, line 6, of time, add 'τῇ ἐχομένῃ ἡμέρᾳ SEG 31.122 (Attica, cAD 121/2); τῆι ἐχομένηι εἰς Κροκοδίλων πόλιν καταπλέωμεν PKöln 262.5 (iii BC)'; at end for 'τὰ ἐχόμενα .. Isoc. 6.29' read 'τὰ ἐχόμενα, *what follows*, Pl.Grg. 494e (s.v.l.), Isoc. 6.29' **V**, for 'pass. .. B I' read '*stand, be*, cf. B I 1, II' and add 'see also °ἐν A I 6' at end add 'Myc. 3 sg. and pl. *e-ke, e-ko-si*; inf. *e-ke-e*, part. *e-ko, e-ko-te* (masc. sg. and pl.); cf. compd. pers. n. *e-ke-da-mo, e-ke-me-de*'

ἔχω (B), for '3 sg. aor. 1 .. Inscr.Cypr. 66 H.' read 'Cypr. *e-we-xe*, perh. ἔϜεξε, *brought as an offering*, ICS 245'

ἑψητός II, add 'Eup. 5, 16 K.-A., IGC p. 99 B2 (Acraephia, iii/ii BC)'

*****ἔωθεν**, = ἔωθεν, Didyma 384.4.

ἔωθεν 1, add 'ἔωθεν ἑκάστης ἡμέρας SEG 37.1019 (Pergamon, ii AD)'

ἑωθῑνός 1, line 2, after 'Ar.Ach. 20' insert 'IG 1³.68.30'; after 'Bato 5.3' insert 'Alciphr. 1.34, cf. perh. Macho 314 G. (s.v.l.)'; line 8, delete 'cf. .. (dub. l.)'

ἕωλος 2, at end delete 'of payments .. AD)' (v. OBodl. i 49).

ἑώρα, after 'αἰώρα' insert 'ι 2'; for 'Ael.Dion.fr. 23' read 'Ael.Dion. α59 Erbse, cf. Phot., Suid. **II** after 'Arist.fr. 515' insert 'codd. Ath.' and after 'αἰ- codd.' add 'Poll.'

ἕως (B) **A I 2**, at end delete 'ἐ. οὗ .. 8.32' **6**, line 3, after 'X.Cyr. 5.1.25' insert 'ἕως ὅτε περίεισι *for as long as they live*, Cod.Just. 6.48.1.15'; line 4, after 'later Gr.' insert 'Lxx Jd. 3.30, al.', for 'Gem.l.c.' read 'Gem. 8.32' and after 'etc.' insert 'ἕ. ἂν οὗ Macho 454 G.' **III b**, add 'exceptionally ἕως ἂν ὅσον ζῷ *as long as she may live*, Sokolowski 1.79.7 (?Pednelissos, i BC)'

Ἑωσφόρος, line 2, after 'Pi.I. 4(3).24' insert 'Trag.adesp. 664.27 K.-S.' add '**II** transf., of illustrious persons (cf. ἀστήρ II), Ἀονίης Πολύδωρον 'Ε. ἀστέρα πάτρης Nonn.D. 5.208.'

Z

ζᾶ, delete the entry (v. γῆ).   for 'Cypr. za-ne' v. °ὐFαις.
*ζαβέρνα, ἡ, bag, Edict.Diocl. 11.2, 7, 7a.
*Ζαγκλαῖοι, οἱ, inhabitants of Zancle (v. sq.), Hdt. 6.23.2, 3, al.; Δανκλαῖοι Dubois IGDS 2 (Elis, vi BC).
ζάγκλη II, for this section read 'old name of Messene in Sicily, Hdt. 6.23.2, Th. 6.4.5, etc.; also Δάνκλεν Dubois IGDS 3 (Elis, vi BC), Δάγκλᾱ, perh. in SEG 26.1122 (Syracuse, vii BC)'
ζάημι, after 'Hsch.' insert 'fem. ζάεισαι dub. in (?)Alc. 261(b).7 L.-P.'
ζαής, add 'Q.S. 3.619'
ζάθεος, after '(E. Tr.) 1075 (lyr.)' insert 'Boeot. δάθιος Corinn. 1(a).i.13 P.'; line 6, delete 'ἄνεμοι Hes. Th. 253'; at end delete 'later' and after 'of persons' insert 'Corinn.l.c., Philod.Scarph. 139'
*ζάκαρπος, ον, fruitful, Cyr., v. QIFG 2.104.
ζακορεύω, add 'w. gen., τοῦ .. Ἀσσκληπιοῦ IG 2².4521a (ii AD)'
*ζακορία (ζακόρια La.)· θυσία Ἀφροδίτης Hsch.
ζακρυόεις, after 'freezing' insert 'or (= δακρυόεις) full of lamentation'
†ζαλάω, rage, transf. of a skin eruption, ἀμφὶ δὲ γυίοις χειμερίη ζαλόωσα πέριξ βέβριθε χάλαζα Nic. Th. 252.
†ζαλλεύω, v. ζηλεύω.
*ζαμβύκη· μουσικὸν ὄργανον Hsch., Phot.
ζαμενής, delete 'neut. as adv. .. Nic.Th. 181' (v. °ἐπιζαμενής).
ζαμίλαμπις, transfer to follow ζαμία and add 'Plin.HN 37.185'
ζάπεδον, after 'Paros' insert 'c.500 BC'; add 'also cj. in h.Cer. 283, Stesich.S 15 i 17 P.'
ζάπλουτος, add 'Lat. saplutus, Petron.Sat. 37.6'
*ζάρωμα, ατος, τό, wrinkle, ῥυτίδας ἤτοι ζαρώματα Cyr.
*ζάτραφος, ον, app. well-fed, Alcm. 134 P.
ζαχρεῖος, after 'needy' insert 'ἔπη of sore need, prob. in A.Supp. 194.
†ζέα, v. ‡ζειά.
ζειά, lines 1/2, after 'ἡ' insert 'also ζεά D.H. 2.25, Dsc. 2.89, 3.74, Hippiatr. 1 p. 8; ζεή PPetr. 2 p. 69 (iii BC) and for 'one-seeded .. monococcum' read 'emmer, Triticum dicoccum'; lines 7ff., for 'ζειὰ .. Gal. l.c.' to end of article read 'Thphr.HP 8.9.2, al. (where ὄλυρα is a cultural variety); including also one-seeded wheat, Triticum monococcum, ζέα δισσή· ἡ μὲν γὰρ ἁπλῆ ἡ δὲ δίκοκκος καλεῖται Dsc. 2.89 (v.l. ζειά) = Gal. 6.517.17.   2 = λιβανωτὶς κάρπιμος, Dsc. 3.74.   II the roof of a horse's mouth, Hippiatr. l.c.'
*Ζειρήνη (-νίς La.)· Ἀφροδίτη ἐν Μακεδονίᾳ Hsch.
ζεστάκρατα, add 'also sg., Paul.Aeg. 7.5.13 (283.14 H.)'
ζετραία, add '(v.l. ζεταία)'
ζευγίον, for '= ζύγον III 2' read 'yoke'
ζεύγλη I 2, delete '–Not .. Prose.'
ζεῦγμα I 2, for this section read 'applied to a bridge of boats, Plb. 3.46.2, 4, τὰ ζ. τῶν ποταμῶν D.H. 9.31, Plu. 2.174e; also, to boats lashed together to form a floating platform, id.Marc. 14, 15. b applied to a normal bridge, AP 9.147 (Antag.).'   add 'III a constellation, Vett.Val. 10.2.'
ζευγματικόν, delete the entry (v. °ζευγματικός).
*ζευγματικός, ή, όν, πλοῖον ζ. some kind of cargo transport, POxy. 2415.44 (iii AD); w. ellipsis of πλοῖον, ib. 56.   2 -όν, τό, app. a type of customs-duty, PLond. 3.1157.6, al., POxy. 2129.4, 3180.5 (all iii AD).
ζεῦγος II 2, for this section read 'double-reed mouthpiece, Arist.Aud. 804ᵃ13, Thphr.HP 4.11.4, 6'   IV, at the beginning insert 'a land measure, SEG 37.859.C6 (Heracleia on the Latmos, late 2 BC); cf. ζυγόν x'   at end add 'Myc. ze-u-ke-si (dat. pl.), pair'
*ζεύκτρα, ἡ, app. some form of bond or fastening, ζεύ[κ]τρας σχοίνιναι perh. plaited-reed ropes, BGU 2361a i 2 (iv AD).
*Ζευξάνθιος, epith. of Poseidon, AE 1933 Chron. p. 1 (Crannon).
ζεῦξις I, for 'oxen' read 'animals'   add '2 yoked beasts, Sch.E.Ph. 847.'
Ζεύς, dat., add 'Ζί SEG 31.364 (Olympia, c.500 BC)'; line 24, for 'Aeschrio 8.5' read 'AP7.345'   IV, Διὸς ἡμέρα, add 'SEG 30.1212 (Rome)'   add 'Myc. di-we (dat.), di-wo (gen.); (see also δῖος)'
ζεφύριος III, add 'cape in Egypt, St.Byz.'
ζεφυρῖτις, for '= foreg. III, .. ζεφυρηῒς 1' read 'epith. of Arsinoe-Aphrodite, from her temple at Zephyrion in Egypt, St.Byz. s.v. Ζεφύριον, Call.fr. 110.57 Pf., Epigr. 5.1 Pf., Posidipp. 12.7, 13.3 (EG 1624, 1630 P.) (prob. corr. of Ζεφυρηΐδος, v. ζεφυρηῒς 1)'

Ζέφυρος, add 'cf. Myc. ze-pu₂-ro, pers. n.; ze-pu₂-ra₃, description of women, ?from place-name'
ζέω I 3, add 'ὅθεν καὶ Ὅμηρος .. "ἔζεσεν αἷμα" Arist.EN 1116ᵇ29, Theoc. 20.15'   at end add 'Myc. fut. ?pass. part. ze-so-me-no (dat.)'
ζηλεύω, after '= ζηλόω' insert 'Aeol. imper. 3 pl. ζαλλευόντων Alc. 5.10 L.-P.'
ζηλοδοτήρ, for 'giver of bliss' read 'as epith. of Dionysus, giver of success, good fortune'
ζῆλος II, add 'ζῆλον ἔχōσ' .. τὸμ μακαριστότατον CEG 538.2 (Attica, iv BC)'   III 2, for 'esp. .. style' read 'of oratorical style, showiness'
ζηλόω, after '(ζῆλος)' insert 'Thess. pf. inf. ἐζᾱλουκέμεν BCH 59.55 (Larissa, ii BC)'   I, line 5, after 'Lxx Si. 9.1' insert 'ἐπί τινι ib.Ps. 72(73).3'   II, line 3, after '(D.) 20.141' insert 'ἔργον .. φαῦλον ζ. Men.Dysc. 289'
ζημία, line 1, after 'Dor. ζᾱμία' insert '[τᾱ]ν ζᾱμίαν SEG 30.380 no. 7 (Tiryns, vii BC)'
ζημιόω, after lemma insert 'Cret. pres. inf. ζᾱμιόμεν Inscr.Cret. 2.v 1.4 (Axos, vi BC); δᾱμιόμεν ib. 4.80.6 (Gortyn, v BC), Boeot. pres. part. δᾱμιώοντες Schwyzer 528.14 (Orchomenos, iii/ii BC)'   II, line 3, after 'Pl.Lg. 936a' insert 'w. acc. δαρκνᾶν Inscr.Cret. l.c.'
*ζήνη, ἡ, goldfinch, Cyran. 89 (3.14.1 K.).
*ζητάριος, ὁ, Lat. cetarius, fishmonger, SB 9152.9 (AD 492).
ζητητής II, add 'also a sim. officer at Mylasa, IMylasa 132.1'
ζιγγίβερις, for 'perh. .. singaber)' read '(cf. Skt. śṛñgavera-, prob. orig. Dravidian.'
*ζινίχια, ἡ, unkn. substance in list of metals, Anon.Alch. 25.3.
ζμάραγδος and derivatives, v. ‡σμαρ-.
*ζμηνών, v. °σμηνών.
*ζόμβρος, ὁ, kind of bull or bison, AP 9.300 tit.
*Ζόννυσος, v. °Διόνυσος.
*ζορκάδιον, τό, = δορκάδιον, Cat.Cod.Astr. 8(2).164.12.
ζοφερός 1, add 'E.fr. 868'
ζοφοειδής, add 'μολίβου ζοφοειδέος Nic.Th. 256, νυκτὸς πέπλον Orph.H. 7.10, cf. 18.8, 71.9'
ζόφος 1, add 'E.Hipp. 1416'
*ζτεραῖος, α, ον, Arc. adj. of uncertain sense (cf. ζειρά), used in description of a garment, ζ. λόπος SEG 11.1112 (Pheneos, vi/v BC; v. Dubois Dial.arcad. pp. 197/8).
ζύγαινα, after 'shark' insert 'A.fr. 46a.9 R.'
*ζυγάς, άδος, ἡ, v.l. for ξυστάς, Poll. 7.147 (v. °ξυστάς).
ζυγάστριον, after 'of sq.' insert 'IAskl.Epid. 52.B63-66'; add 'IG 4².118.61, 63'
ζύγαστρον, after 'τό' insert 'Boeot. δύγ- SEG 24.361.23 (Thespiae, 386/0 BC)'
*ζυγία (B), ἡ, = ζεῦξις, νυμφίδιαι ζ., i.e. wedlock, GVI 653 (Syros, ii/iii AD).
ζύγιος II, before 'epith. of Hera' insert 'of marriage, nuptial, ζ. .. θαλάμων GVI 1431 (Athens, ii AD)'   III, for 'κώπη' to end of section read 'b of a ζυγίτης or ζυγῖται, κώπη IG 2².1604.71; κῶπαι ib. 1607.59, 1609.51 (all iv BC), Polyaen. 5.22.4.'
ζυγομαχέω 1, add 'Plu. 2.445c'   2, after 'Com.adesp. 207' insert 'Men.Dysc. 17'
ζυγόν, line 1, after 'in various senses' insert 'in Lxx masc. where determinable, exc. neut. pl. Le. 19.36'   I 2, add 'w. ref. to a bridge, A.Pers. 72'   II 1, after 'φόρμιγξ' insert 'and sim. instruments' and add 'Thphr.HP 5.7.6'   III 2, for 'panels of a door' read 'cross-pieces of a door'; delete 'cf. ζευγίον'   IV, add 'PColl.Youtie 92.28 (AD 569)'   VIII, of the chorus, add 'Sch. Ar.Pax 733 D.'   X, add 'cf. ‡ζεῦγος 4'
ζυγοποιός, add 'AD 24.300 (Thessalonica; see also SEG 33.495)'
ζυγοστασία, add 'b = ζυγοστάσιον, weigh-house, PLond. 301.11 (ii AD).'
ζυγοστάσιον 2, for 'Cod.Just. 11.28.1' read 'Cod. Theod. 14.26.1'
ζυγοστατέω I a, add 'weigh out, allot, τοῖς θεοῖς τὰς τιμάς Max.Tyr. 39.5'
ζυγοστάτης, add 'PAlex. 40.2 (iv/v AD), Corinth 8(1).158; of Hermes (as patron of ἀγορανόμοι), Illion 4.9'
*ζύγωσις, εως, ἡ, weighing of money, prob. in CPR 8.40.3 (iv AD).
†ζῡθοπώλης, v. °ζυτο-.

ζυμάριον, τό, *ferment*, Anon.Alch. 20.3 (mod. Gk. ζυμάρι *dough*).

ζυμουργός, after '*leaven*' insert 'POxy. 754 (i AD)'

ζυμόω 2, delete 'γῆν Gal. 10.964'

ζῦτικός, ή, όν, *of beer*, τιμή PMich.II 121ʳ4.vi 2 (i AD); -κόν, τό, *beer-tax*, PTeb. 337 intr. (ii/iii AD), PLond. 254ᵛ.70 (ii AD).

ζῦτοπώλης, ου, ὁ, *beer-seller*, PMil.Vogl. 278.22 (ii AD), BGU 2280b.1 (AD 276); ζυθο- POxy. 85 iv 4 (iv AD).

ζῶ, line 17, before 'Cret. **δώω**' insert 'Boeot. **δώω** Schwyzer 509.14 (Lebadea, iii BC), etc.'; line 20, after 'ἔζωσα' insert 'Call.fr. 191.39 Pf.' **I 1**, add 'part., on a tablet set up to a person in his/her lifetime, ζώσῃ φρονο[ύ]σῃ Πρωτογένης ἀνέστησεν συμβίῳ Ἀργυρίδι Mitchell N.Galatia 93, cf. φρονέω IV lines 13ff.; also of a survivor in a joint epitaph, ib. 108, SEG 30.595 (Maced., ii/iii AD)'

ζώγιος, v. °ζῶκος.

ζωγράφημα, add 'Plu.Tim. 36, id. 2.64a, etc., SEG 36.1287 (Syria, Chr.)'

ζωγράφησις, εως, ἡ, *painting*, ISmyrna 685.3 (iv AD).

ζωγράφητός, add 'prob. in Duraᵗ p. 93'

ζωγραφία II 1, add 'Inscr.Magn. 107.13 (ii BC), Phld.Rh. 2.166 S.'

ζωγράφος, delete '(ζωγρ- without iota' and 'so ζωγραφία' to end of article.

ζωγρέω I, pass., add 'Isoc. 12.194' **II**, line 2, after '(quoted by Aret.CA 2.3)' insert 'perh. *activate*, ἐπὶ φλογὶ ζωγρηθεῖσα χαλβάνη Nic.Th. 51 (v.l. μοιρηθεῖσα)'

ζῳδιακός, after 'Adv. -κῶς' insert '*according to*, or *by the zodiac*' and after 'Vett.Val. 22.12' insert '137.27, al.'

ζῴδιον II, add 'Vett.Val. 22.12'

ζῳδιωτός, after '= ζῳωτός' insert 'λάρναξ Inscr.Délos 1409Ba ii36 (ii BC)'

ζωή I 2, add 'ἐν ζωῇ *during (his) lifetime*, Cod.Just. 1.3.45.8' **II**, add 'Myc. *zo-a*'

ζωητόκος, ον, = ζωοτόκος, Theognost.Can. 87.

ζωθαλπής, after 'Nonn.D. 1.454' insert 'ζωθαλπέες Ὧραι ib. 16.397' and delete 'fem. ζώθαλπις' to end.

ζωϊτός or **-τόν**, app. some form of sculptured figure(s), Inscr.Délos 104-24.26 (iv BC); cf. ζωοφόρος I.

ζῶκος, ὁ, ζ. ἐστὶ πτηνόν· οἱ δὲ ζώγιον (vv.ll. ζύγγιον, etc.) φασι, οἱ δὲ ἴρηκα. ἔστι δὲ εἶδος λευκοῦ γυπὸς νεκροβόρου Cyran. 17 (1.6.9 K.).

ζωκρός, = ζωρός, comp. ζωκρότερον GVI 1815.7 (Naxos, i/ii AD).

ζωκτήρ, add 'PAbinn. 4.8 (iv AD; but perh. for ζευκτ-, cf. °ζωκτήριον)'

ζωκτήριον, τό, unexpld. item in an account, perh. = ζευκτήριον, BGU 2357 ii 15 (iii AD).

ζωμάρυστρον, add 'PAlex. 31.8 (iii/iv AD)'

ζωμός, after 'fish, etc.' insert 'Asius 14.3 W.'

ζωνάριον, add 'Inscr.Délos 1442A52 (146/5 BC)', PMich.xv 740.8 (vi AD)'

ζώνη, line 1, after '(ζώννυμι)' insert 'Cret. τώνᾱ, Hsch.' **I 1 c**, transfer this section to follow section II 1 b **3**, prov., add 'Philostr.VA 2.31'

ζώννυμι I, for pres. def. read '*surround with a belt* or *girdle, gird*' and add 'σεαυτόν EuJo. 21.18' **II**, for pres. def. read '*fasten* or *secure one's dress with a belt* or *girdle* as a prelude to action' and add 'Act.Ap. 12.8'

ζωνοβαλλάντι(ο)ν, τό, *belt-purse*, pap. in Eos 32.30 (v/vi AD).

ζωνομάχαιρα, ἡ, *a knife carried in the belt, dagger*, Sch.B Il. 19.252.

ζωνοπώλης, ου, ὁ, *belt-seller*, SEG 39.1176B 10 (BE 1993 p.489 no.196, Ephesus, i AD).

ζωογόνος, for '*animals*' read '*living things*'

ζωολογικός, ή, όν, *concerning animals*, τὰς πραγματείας Ἀριστοτέλους τὰς ζ. Sch.Luc. 131.9.

ζῷον I, add 'extended to include plants, Pl.Ti. 77a; cf. also Arist.PA 681ᵃ33' **II**, add 'sp. δῶια (pl.), SEG 32.1612 (iii BC)' **III**, for 'till after the middle of the fifth cent. BC' read 'before Semonides'

ζωοπλαστέω, delete '*mould to the life*' and '*analogous to ζωγραφέω*'

ζωοπλάστης II, add 'Heph.Astr. 2.19.15, Ptol.Tetr. 180'

ζωοποιός, όν, *bringing about (supernatural) life*, of a holy person, PLond. 1303.2 (Tyche 6.198, AD 498).

ζωοπώλης, ου, ὁ, ζωοπώλᾱς· ὁ τὰ ἱερὰ ζῷα πιπράσκων. καὶ ὁ τόπος ζωοπωλίς (-πώλιον L.Dindorf) Hsch.

ζωοπωλίς, ᾶ, (sc. ἀγορά), *animal-market*, v. ζῳοπώλης; gen. sg. ζωπωλίδος rest. in GDI 5224 I 12 (Tauromenium; v. Dubois IGDS 186).

ζωός, line 1, after '**δωός**' insert '(or δοός)'; line 2, for 'ΖωΓόθεμις Schwyzer 684' read '*zo-wo-te-mi-se*, perh. ΖωΓόθεμις ICS 354'; line 5, after 'A.Eleg. 3' insert 'of a surviving relative, on an epitaph, Διοσκουρίδης Μουκασου .. ζωὸς ἑαυτῷ καὶ Σουρᾳ Μουκασου συνβίῳ SEG 30.596 (Maced., AD 106); masc. pl. as subst., *the living*, Il. 23.47, etc.; ἐν ζωοῖσι καὶ ἐν Ἀΐδαο δόμοισι epigr. in ISmyrna 550.2'

ζῳοτόκος, before 'opp. φοτόκος' insert '*viviparous*'

ζωοφθόρος καὶ σαλαμάνδρα. καὶ ἀγγεῖον νεκροῦ Hsch. (La.; cod. -φόρος).

ζωπύρωσις, for '*kindling*' read '*fanning into flame*'

ζώπωλις, v. °ζωοπωλίς.

ζωροποτέω, for 'Call.fr. 109' read 'Call.fr. 178.12 Pf. (v.l. ap. Ath. 10.442f)'

ζωρός, line 5, after 'Arist.Po. 1461ᵃ14' insert 'Plu. 2.677c-678c'; line 6, after 'Ephipp. 10' insert 'Philum.Ven. 2.3, 4.2'; at end for note in parenth. substitute '**2** app. understood by Empedocles (from Homer) as "*mixed*", Emp. 35.15 D.-K.'

ζωστήρ, line 4, after 'Pi.fr. 172' insert 'E.HF415, Heracl. 217' **III 1**, for '*stripe marking certain height in the ship*' read '*one of the horizontal rows of planking forming the external shell of a boat*' **IV 2**, after 'Zoster' insert 'IG 2³.369.67' and delete '(sed leg. ζωστήριος)'

Ζωστήριος, add '**3** Ζωστήρια, τά, *festival of Apollo Zoster*, AD 11.40 (Attica).'

ζώστρα, for '*head-band, fillet*' read '*encircling band* or *ribbon*'

ζώφυτος I, add 'sp. σόφυτος PMich.IX 540.8 (AD 53)'

ζώωσις, add 'Aq.Ge. 45.5'

H

ἤ (A), line 1, after 'Ep. also ἠέ' insert '(q.v. in Suppl.)' **A I 3**, add '(ἤ ..) ἢ ἔπειτα either .. or (failing that), Il. 13.742-3, 20.120, 24.356; but in Alc. 129.19 L.-P. ἤπειτα (= ἢ ἔπειτα) introduces the more desirable alternative' **3**, add 'S.Ph. 983, E.Alc. 628, And. 1.33, X.Ath. 2.12' add '**4** ἢ καί and/or, PPetaus 25.8 (ii AD), CPR 5.11.12 (?early iv AD).' **B 1**, line 7, after 'Tab.Heracl. 1.121' insert 'ἰν τῶι ὕστερον Ϝέτ[ε]ι ἢ Νικῆς ἐδαμιόργη in the year after .. SEG 37.340.23 (Mantinea, iv BC)'

ἤ (C) **1**, for this section read 'Cypr. e-ke, v. εἴ' **2**, for 'Cret. for when' read 'Cret. ἔ, starting from the time when'

ἦ, adv., **I**, line 7, delete 'ἢ δῆτα S.OT 429'

ᾗ **I**, Dor. ᾇ, add 'ᾇ ὕδωρ ῥεῖ Inscr.Cret. 4.182.11 (Gortyn, ii BC); ᾇ μὲν .. ᾇ δὲ on this side .. and on that, Tab.Heracl. 1.81' add '**2** of time, ᾇ ὄκ(α) from that time when, Inscr.Cret. 4.72 v 4.' **II**, add '**4** ἵνα, in order that, Inscr.Cret. 4.41 ii 11 (v BC), 168*.24 (iii BC).'

*ἡβαδόν, v. °ἠβηδόν.

*ἡβάζω, = ἡβάω, cum vero ἡβάζοντα (significat), pubes puberis (declinatur), Charis. p. 542 K.

ἡβάω **1**, for 'attain or have attained puberty' read 'to have passed puberty, be a young adult' **2**, penultimate line, for 'Simon. 183.3' read 'AP 7.24.3 ([Simon.])'

ἥβη **I 1 c**, for pres. def. read 'legal puberty, at Athens attained at the age of sixteen' **II**, delete 'Dor. .. Theoc. 5.109'

ἡβηδόν, after lemma insert 'Dor. ἡβαδόν Tit.Calymn. xii 10'

*ἡβοκάτος, ὁ, v. °ἡουοκάτος.

*ἡβότᾱ, ᾱ, young men (collect.), perh. in ἡϵιόταισι (dat. pl.) IPamph. 3.7 (iv BC).

ἠγάθεος, Πυθώ, add 'Od. 8.80, h.Hom. 24.2, B. 3.62, 5.41 (ἀγ-); Ἑλλάς Carm.Pop. 21.1 P.'

ἤγανον, add 'also Boeot. τυροκνασστίδες τρῖς, Ϝαγάνω δύο SEG 24.361.19 (Thespiae, iv BC)'

ἡγέμαχος, add 'Dor. ἀγέμαχος, Simon.fr.eleg. 11.14 W².'

ἡγεμόνεια, add 'Dor. ἀγ-, ὄρος Ἀρτέμιδος Ἀγεμονείας Ὀρθωσίας IG 12(5).894 (Tenos, ii BC)'

ἡγεμονέω, add 'τὸ ἡγεμονοῦν ruling principle, Plot. 4.7.7'

ἡγεμόνη, add 'of Hecate, Orph.H. 1.8, of Physis, ib. 10.12'

ἡγεμονικός **II 2**, before '= consularis' insert 'of a ‡ἡγεμών II 1 c, οὔτε διὰ ψηφίσματος οὔτε δι' ἐντεύξεως ἡγεμονικῆς MAMA 8.554'

ἡγεμών **II 1 a**, line 5, after 'Th. 8.89' insert 'στρατηγοῖς ἱππάρχαις, πεζῶν ἡγεμόσι Amyzon 10 (iii/ii BC)' **c**, add 'title of officials at Istria, IHistriae 6.1, al. (iii/ii BC); leader of ephebes at Athens, SEG 26.176.61 (AD 170/1-175/6); magistrate at Chalcis, SEG 29.806 (120/100 BC)' add '**d** as divine title, τ[ὸν Ἑρ]μῆν τὸν Ἡγεμ[όνα] SEG 23.547.53 (Crete, 201/200 BC)'.

ἡγέομαι, line 1, for 'irreg.' read 'Cret.'; line 4, after 'cf. περιηγ-' insert 'καθηγ-' **I 1 f**, before 'of logical priority' insert 'come earlier, precede, CodJust. 1.2.24.9' **II 3 d**, add 'SEG 30.1688 (Palestine, AD 576/8), 30.1704 (Arabia, vi AD)'

ἠγερέθομαι, line 2, after 'impf.' insert '(exc. ἠγερέθεσθε A.R. 2.632)'

⁺ἡγεσία, ἡ, leading the way, guiding, Amph.hom. 4.7.231 (117 D.), Gr.Naz.Carm. 1.1.8.91 (PG 37.453a), ἡγεσίης· ὁδηγίας Hsch.

ἡγέτης, fem. -έτις, add 'πολύσκιος ἡγέτις ὀρφνης Jo.Gaz. 2.289'

ἡγηλάζω, at end for 'ὑφηγηλάζω' read '°ὑφηγηλάζω'

"Ἡγησιαστής, οῦ, ὁ, imitator of Hegesias of Magnesia, Didyma 181.6.

*ἡγητορεύω, hold office of ἡγήτωρ II, ABSA 56.37 no. 99 (Cyprus, Ptolemaic).

ἡγήτωρ **I**, add 'E.Med. 426 (ἀγ-)' **II**, for 'ἀγήτωρ' read 'ἀγήτωρ (ἀγ- cod.)'

ἡγός, after '= ἡγεμών' insert 'Hesperia 5.95 (Athens, iii AD)'

*ἡγουμενικός, ή, όν, of or for a leadership, PMich.II 123ʳ viii 5 (i AD).

ἤγουν, last line, after 'or' insert 'Vett.Val. 138.12' and after 'POxy.' insert '2085 fr.3.15, 16 (ii AD)'

ἠδέ **II**, lines 7/8, after 'AP 9.788.9' insert 'also ἠδὲ .. τε Orph.H. 10.27'; line 9, delete 'Ch. 1025'

*ἡδέοσμον, τό, or -οσμος, ὁ, v. °ἡδύοσμον.

ἤδη, before 'adv.' insert 'Thess. εἶδε BCH 59.37 (Crannon, ii BC)'

*ἤδικτον, τό, Lat. edictum, D.H. 5.73.1, BCH suppl. 8.85 (Thessalonica, v/vi AD), Just.Const. Δέδωκεν 5, 21; also ἔδικτον Plu.Marc. 24.

ἥδομαι **II**, add 'w. dat., Chrysipp.Stoic. 3.108.4, 115.39, 116.19'

ἡδονή **I 1**, fourth line from end, for 'take pleasure in them' read 'regard them with favour' **II**, taste, flavour, add 'Hp.Vict. 1.23; also, scent, Heraclit. 67 D.-K.'

*ἡδονίς· οἱ δὲ ἀφύδιον Cyran. 18 (1.7.2, 15 K.).

ἦδος **I**, add 'Lib.Or. 1.274'

*ἡδύβολος, ον, thrown, shot, precipitated, etc., with agreeable effect, ἡ. ὄϊστῷ Nonn.D. 48.472 (cj. for °ἡδυμόλος), ἡδυβόλου νιφετοῖο Jo.Gaz. 2.129.

Ἡδύλειος, add '(nisi leg. κύλιξ ἡδυ(πότις) λεία, v. RPh. 64.45)'

ἡδυλίζω, add '2 ἡδυλίσαι· συνουσιάσαι Hsch.'

ἡδυλισμός, add '2 ἡδυλισμός· συνουσία Hsch.'

ἡδυλογία, delete 'in pl.'

ἡδυλύρης, delete 'epith. of Apollo' and for 'Philol. 71.6' read 'CEG 816.2'

*ἡδυμόλος, ον, coming with sweetness, dub. in Nonn.D. 48.472 (v. °ἡδύβολος).

*ἡδύμοχθος, ον, to whom toil is sweet, γεωργὲ ἡ. epigr. in SEG 18.456.7 (Caria, Rom.imp.).

*ἥδυνσις, εως, ἡ, causing of pleasure, Olymp.in Grg. 242.11 W.

*ἡδυοινέω, produce sweet wine, PCair.Zen. 446.8 (iii BC).

ἡδύοσμον (s.v. ἡδύοσμος II), after 'Str. 8.3.14' insert 'Ev.Matt. 23.23'; after 'Dsc. 3.34' insert 'also -οσμος, Hsch.; ἡδέοσμος or -ον ISmyrna 728.14 (ii/iii AD), SB 4483.12, 4485.3'; after 'as trisyll.' insert 'ISmyrna l.c. (hex.)'

ἡδύπνοος **1**, after 'Pi.I. 2.25' insert 'of a poet, AP 4.1.11 (Mel.)'

*ἡδυποτέω, imbibe delicious drink, epigr. in SEG 26.1835.3 (Cyrene, ii AD).

ἡδυπότις, for 'something .. cup' read 'cup for delicious drink, a kind of κύλιξ' and after 'ib., ii BC)' insert 'called κύλιξ, IG 11(2).287B⁵'

ἡδύποτος, after 'of a cup' insert 'ἡαδύποτος SEG 34.370 (Boeotia, c.500 BC), 34.462 (graffito, Phocis, Hellen.)'

ἡδύς **II**, add 'attached to the name of a loved one, Πυθίων ἡδύς SEG 32.847 no. 30, Μυΐσκος ἡδύς ib. no. 31, al. (Thasos, erotic graffiti, all iv BC)' **III**, adv., add 'Dor. ἁδέως, SEG 26.426.10 (Argos, c.AD 200)'

*ἡδυϋεία (or -ϋΐα), ἡ, sweetness of disposition, rest. in Vit.Philonid.fr. 55b.1 (p. 949).

ἡδύφωνος, add 'adv. -ως, Poll. 2.113'

ἡδυχαρής, add '**II** (he) who delights in luxury, title of comedy by Theopompus, Theopomp.Com. 14 ff. K.-A. (p. 715).'

ἡδύχροος **I**, add 'name of a bath-house monument (pleasant to the skin, w. ref. to water, or of a pleasant colour w. ref. to marble), Laodicée 19 (pp. 362-3)'

ἠέ, after 'whether' insert 'but for ἤ (A) B 1 in μελεδαντὸς ἀνδράσιν ἠὲ πάρος (without μᾶλλον) epigr. in BCH 50.529.2 (Marathon, ii AD). [In some passages of Hellen. and later verse, where the second syll. is long by position, the first syll. is short: Numen.ap.Ath. 7.328a, Nic.fr. 50, 74.19 G.-S., Nonn.D. 34.47, v.l. in Max. 127.]'

ἠέριος, add 'see also ‡ἀέριος'

ἠερόθεν, for 'from air' read 'from the air' and add 'Nonn.D. 34.284'

⁺ἠερόμικτος, v. °ἀερο-.

⁺ἠερόμορφος, v. °ἀερο-.

ἠερόποιταν, add '(ἠεροποίναν La.)'

*ἠεροποῖτις, ιδος, ἡ, wd. of uncertain meaning, Suppl.Hell. 991.31 (poet. word-list, iii BC); cf. perh. ἠερόποιταν, εἰαροπώτις.

ἤην, add 'Men.Dysc. 465 (ην pap.)'

ἡθάς **I 2**, after 'E.Andr. 818' insert 'τῶν ἠθάδων καὶ συνεστηκότων ῥητόρων D. 22.37'

*ἠθητής, οῦ, ὁ, one who strains wine, rest. in PHib. 268.15, CPR 13.20.3 (both iii BC).

*ἠθητός, ή, όν, strained, PCair.Zen. 436.2 (iii BC).

ἠθικός **II 1**, add 'b ἠθικοί, οἱ, types of moral character, typical characters, τοῖς ἰδίως ὀνομασθεῖσιν ἠ., οἷον λίχνοις ἢ δειλοῖς Hermog.Id. 2.2.' **2**, at end after '(Plu.)Alex. 52' add '= ἐν ἤθει (v. ‡ἦθος), Sch.E.Hipp. 307, etc.'

ἠθμός, read 'ἡθμός' and for 'SIG 2 .. ἠθμός, ὁ' read 'written ἡεθμόν (acc.) Schwyzer 731 (Attic text, Sigeum, vi BC), ἰθμός Stud.Pal. 20.46.18 (ii/iii AD)' *ἠθμωτός, ή, όν, *equipped with a filter*, ὀξίδες ἠθμωταί graffito in *RAL* 1993 p.72 (Attic crater, iv BC).

ἠθοποιέω II, add 'D.L. 3.18'

ἦθος II 2 b, add 'ἡ δ' ὀφρὺς ἐπίκειται τῷ τοῦ ὀφθαλμοῦ ἤθει Philostr.*VA* 7.28' **4**, add '**b** ἐν ἤθει (λέγειν) in *an assumed character*, ἐν ᾗ. εὐνοίας Call.*Dieg.* vii.23 (*fr*. 195 Pf.); also *so as to convey a meaning indirectly*, ἐν ᾗ. καὶ εἰρωνείᾳ Sch.E.*Hec.* 26, cf. Sch.E.*Or.* 750; μετὰ ἤθους Sch.E.*Ph.* 388.'

ᾖ, add 'also ἠΐν *Boeot.adesp.* 39 i 3 P.'

ἠϊα (A), after '**II**' insert '(prob. a different word)'; line 4, after 'Pherecr. 161' insert 'ἄχυρα καὶ ἔια *IG* 1³.422.85 (Athens, v BC)'; add 'cf. εἰαί, εἴοι'

ἠίθεος I 1, add 'D. 59.22'; as adj., add 'τὸν ᾐ. (βίον) Antip.*Stoic.* 3.255'

ἤϊος, add 'Orph.*Εὐχή* 7'

ἠκής, add '(perh. only inferred fr. ἀμφήκης, εὐήκης)'

ἥκω, line 3, transfer 'Gal. 6.56' to line 13 (fut.); line 6, after 'ἥκατε' insert 'Call.*fr*. 177.13 Pf.'

ἠλᾰκάτη, add 'cf. Myc. *a-ra-ka-te-ja, spinning-woman*'

*ἠλᾰκάτιον, τό, dim. of ἠλακάτη, *Inscr.Délos* 1442*B*56 (ii BC; pl.); cf. ἠλεκάτιον.

ἠλάριον, after '*nail*' insert 'Zos.Alch. 236.9'

ἠλάσκω, add 'Lyc. 575'

*Ἠλεῖος, α, ον, *Elean*, Il. 11.671, etc.; Elean Ϝᾱλεῖος Schwyzer 409.1 (vi BC), *SEG* 31.364 (*c*.500 BC).

*ἠλέκτραι· τὰ ἐν τοῖς κλινόποσι τῶν σφιγγῶν ὄμματα Phot.

ἠλεκτρίς II, add '*sg*., of one such island, A.R. 4.505'

ἠλεός, for 'ἄλλος .. 110' read 'ἄλλος dub. in *Lesb.fr.adesp.* 5.3 L.-P., Pi.*N.* 4.39 (cj.)'

Ἡλιάδης, add 'see also Ἁλιάδαι (s.v. ἁλιάδης)'

ἡλιαία I 1, add 'at Delos, *Inscr.Délos* 442*B*113 (ii BC)' at end add '(perh. ἠλ-, v. Dover on Ar.*Nub.* 863, etc.)'

ἡλιακός, add 'ἡλιακόν, τό, *place exposed to the sun, solarium*, *PKöln* 230.13 (ii AD)'

ἡλιαστήριον, for '*place for drying fruit*' read '*room open to the sky used for drying fruit*' and add '*PVindob.Salomons* 12.9 (AD 334/5)'

*Ἡλίεια, τά, *festival of the Sun*, *IG* 2².3779.21 (iii BC); see also °Ἁλίεια.

ἤλιθα II, at end for 'cf. 140' read 'see also ἤλιθος'

ἠλιθιάζω, add '*to be foolish* or *senile*, Procop.*Arc.* 9.50'

ἠλίθιος, after lemma insert 'hελίθιον *CEG* 37 (Attica, vi BC)' **I**, add '-ον as adv., *in vain, CEG* l.c.' **II**, line 4, of persons, after 'X.*Smp.* 3.6' insert 'Simon. 37.37 P., Is. 9.11, D. 7.26, 31.11, 47.30'; for 'ἠλίθιόν [ἐστι] .. Antiph. 58' read 'as complement of inf., ὥστε ἐν ὁποιοῦν τέχνῃ τὸ κατὰ γράμματ' ἄρχειν ἠλίθιον Arist.*Pol.* 1286ᵃ12, *Rh.* 1395ᵃ6, τῷ γὰρ καθορῶντι τῶν ἀϊδίων τι ἠλίθιον περὶ ταῦτα σπουδάζειν id.*fr*. 59'

*ἤλιθος, ον, *useless*, Nic.*Al.* 140.

ἡλικία I 1, line 7, for 'X.*HG* 6.1.4' read 'X.*HG* 6.1.5' **II 1**, add 'κατ' ἁλικίαν *according to age group*, *SEG* 37.340.17 (Mantinea, iv BC), καθ' ἁλικίαν *Inscr.Magn.* 20.23'

ἧλιξ, line 3, after 'A.R. 2.479' insert 'ᾗ. χαίτην Call.*Del.* 297' add '**3** Cypr. *wa-li-ka* Ϝάλικα *of the same kind, similar*, Kafizin 117b, 159, al. (unless to be taken as Ϝαλίκα (ἡλίκος)'.

*ἡλιοβολή, ἡ, ἁλιοβολή· σύνοδος ἡλίου ἅμα καὶ σελήνης Λάκωνες Hsch.

*ἡλιόγονος, ον, *born of the sun*, *PMag.* 3.331.

ἡλιοδρόμος, add '**2** an Indian bird, Cyran. 89 (3.15.1, 2 K.).'

ἡλιοδυσία, ἡ, *sunset*, Hsch. s.v. γελοδυτία.

ἡλιοθερής, add 'Hsch. s.v. ἀλαιθερές'

ἡλιοκάμινος, add '*TAM* 5(1).517 (ii AD)'

*ἡλιοκογχύλιον, τό, wd. of uncertain meaning, Anon.Alch. 32.6.

*ἡλιοκόσμιον, τό, wd. of uncertain meaning, Anon.Alch. 32.6.

*ἡλιορόδιος, ον, ἀγών the festival Helieia at Rhodes, dub. in Sch.Pi.*O.* 7.146a Dr.

ἥλιος, line 4, Dor. ἅλιος, add 'Schwyzer 173.50 (Chersonesus Taurica, iv/iii BC), *Inscr.Cret.* 1.ix 1A28 (Dreros, iii/ii BC)'; lines 5/6, Aeol. ἀέλιος, add '*SEG* 34.492 (Atrax, *c*.200 BC)' **II**, add '**3** ἡμέρα ἡλίου a day of the week, "*Sunday*", *SEG* 31.830 (Catane, iv/v AD), *IG* 14.142 (Syracuse).'

ἡλιοσέληνος, at end for 'Procl.*de sacrificio* .. p. 8)' read 'Procl.*Sacr.* p. 149 B.'

Ἡλιοσέραπις, add 'also Ἡλιοσάραπις (dat. -σαράπει) *IGRom.* 3.93 (Sinope)'

*ἡλιοφέγγος, ον, = ἡλιοφεγγής, εἶδος *Lapid.Gr.* 175.14 (cf. Plin.*HN* 37.181).

*Ἦλις, ιδος, ἡ, *Elis*, Il. 2.615, etc.; Elean Ϝᾶλις, dat. Ϝάλει *SEG* 12.371.38 (iii BC).

ἡλίτης, for 'Procl.*de sacrificio* .. p. 8)' read 'Procl.*Sacr.* p. 149 B.'

ἡλοκόπος, add '*EA* 19.27 (Pessinus, Chr.)'

*ἧλος· τόπος οὕτω καλούμενος, ἐν ᾧ οὐδὲν φύεται Hsch.

ἧλος I 2, add '*IG* 2².1673.33, 41'

ἠλοσύνη, for '= ἠλιθιότης' read '*irrational behaviour, witlessness* (cf. ἠλεός)' and after 'Nic.*Al.* 420' insert '*madness* (of the Proitides), Hes.*fr*. 37.15 M.-W. (ἠλ- pap.)'

*ἡλουργός, ὁ, *nail-maker*, *Stud.Pal.* 8.955 (*ZPE* 76.110, v/vi AD).

ἡλόω, add '**III** *bristle*, app. by confusion of Heb. *sāmar* and *sᵉmōr* (cf. καθηλόω II), Aq.*Ps.* 118(119).120.'

*Ϝημᾶ, ἡ, = εἷμα, τά .. τρίτρα τᾶς Ϝήμας *Inscr.Cret.* 4.43*Ab*10, 4.72 v 40.

ἧμαι, line 2, for 'ἧστε' read 'ἧσθε'; line 5, after 'κάθημαι)' insert 'irreg. 3 sg. εἴατο (παρ-) Call.*fr*. 497 Pf.'

ἤμαιθον, add '*IG* 12(1).891.1 (Rhodes, ii BC)'

ἦμαρ, line 1, ἄμαρ, add 'Cypr. *a-ma-ta* (pl.), *ICS* 318 B vi' **I 1**, add 'fig., of life, τριέτης λίπεν ἦμαρ *SEG* 23.632 (hex., Cyprus, ii BC)' **II**, add 'Cypr. *a-ma-ti-a-ma-ti*, ἄματι ἄματι *day by day*, *ICS* 318 B v2 (*c*.600 BC)'

*ἡμάτιον, v. °ἱμάτιον.

ἡμεδᾰπός, line 3, after '(sc. γῇ)' insert 'Ath. 4.138f'; add 'see also °ἀμμεδαπάν'

ἡμέρα, line 2, after 'etc.' insert 'Arc. ἀμάρα *IG* 5(2).3.9, al. (*c*.400 BC)'; line 5, after '*IG* 1².49.6, al.' insert 'cf. κατ' ἠμέραν *IG* 2².1656.3 (394 BC)' **I 5**, add 'διὰ τὸ ἐνλιπέσθε ἡμέρας *days of observance*, *SEG* 34.1210.5 (Saittai, ii AD)' **II**, add '**4** ὅσαι ἡμέραι, v. ὁσημέραι.' **III**, line 10, after 'ἐν τρισὶν ἡ.' insert 'Hdt. 8.66.1' **III**, line 16, ἐπί w. acc., add 'ἐφ' ἡ(μέρας) ε´ *for five days*, *PGen.* II 92 (i AD)' **IV**, for 'as pr. n., *the goddess of day*' read 'personified as a goddess' and add '*SEG* 31.922 (Aphrodisias, i/ii AD)'

*ἡμερείσιος, v. ‡ἡμερήσιος.

ἡμερήσιος II, after lemma insert '(sp. -είσιος *IG* 4².742.1 (Epid., ii/iii AD)' and add 'δρόμους ἡ. *Peripl.M.Rubr.* 15' **III 1**, after 'D.L. 7.181' insert 'ἡμερείσια sc. ἱερά, *daily sacred rites*, *IG* l.c.' add '**4** ἡ. κύκλος the circle of the sun's course *on a given day*, Hero *Dioptr.* p. 304 S.'

*ἡμερία (B), Ep. -ίη, ἡ, = ἡμερότης, *culture* of plants, hex. in *POxy.* 1796.19 (= Heitsch *GDRK* 50, ii AD).

ἡμερινός, add 'φύλακες ἀμερινοί *SEG* 30.703A (Mesambria Pontica, ii/i BC)'

ἡμέριος I, line 4, after '*mortals*' insert 'E.*IA* 1331, Nic.*Th.* 346'

*ἡμεροκωμία, ἡ, *daytime revelry*, Epicur. [104] 1 A.

ἡμερολεγδόν, line 1, after 'A.*Pers.* 63 (anap.)' insert '*IG* 2².458.6 (iv BC)'

ἥμερος I 2, line 2, after 'δένδρεα (Hdt.) 4.21, 8.115' insert 'w. ellipsis of noun, τῶν ἡμέρων *SEG* 33.1034.A3 (Aegae, Aeolis, iii BC)'

ἡμεροσκόπος, for '*day-watcher*' read '(*daytime*) *look-out*'

ἡμερούσιος, add 'adv. -ιον, pap. in *Tyche* 1.163 (vi/vii AD)'

ἡμέτερος, ἡμετερόνδε, add '*h.Cer.* 163, A.R. 1.704'

ἡμιαρούριον, add '*PAlex.* 26.16 (ii/iii AD)'

*ἡμιαρούριος, ον, *measuring half an aroura*, prob. in *PLugd.Bat.* xxv 21.5 (i AD).

ἡμιγύναιξ, after lemma insert 'or -γύνη'; after '*half-woman*' insert 'i.e. eunuch' and for 'Simon. 179.9, Suid. s.v. ἄρρεν' read '*AP* 6.217.9 ([Simon.])'

*ἡμιγύνης, acc. -ην, = ἡμιγύναιξ, *OBodl.* 2171.6 (Elephantine, ii AD).

ἡμιδεής, for '*AP* 5.182 .. -δαής in' read '*PSI* 428.24, cj. for -δαής in *AP* 5.183 (Posidipp.)'

*ἡμίδιον, τό, *one half*, *PMerton* 39.3 (iv/v AD); v. *PFreib.* iv p. 34).

*ἡμιεκατοστιαῖος, α, ον, *half one-hundredth*; ἡ. τόκοι *interest of 0.5% monthly*, i.e. 6% p.a., *PNess.* 46.6 (AD 605).

*ἡμιεκτάνιον, τό, half *ἐκτάνιον, a monetary value (cf. perh. ἕκται Φωκαΐδες), τρίτον ἡ., = 2½ ἑκτάνια (cf. ἡμιτάλαντον), *SEG* 38.1036.4 (lead tablet, Gaul, ii AD); also ἡμιοκτάνιον ib. 3, 12, either var. or = half *ὀκτάνιον (an eighth fraction).

ἡμίεκτον, after '*half* ἑκτεύς' insert 'Hippon. 21 W. (scanned ⏑⏑)'

ἡμιθᾰλής, for '*half-green*' read '*half-fresh*, i.e. wilting'

ἡμιθᾰνής, add 'epigr. in *SEG* 30.1421.6 (Bithynia, i BC)'

ἡμιθέα, add 'a Carian goddess: see J.M. Cook and W.H. Plommer *The Sanctuary of Hemithea at Kastabos* (1966) p. 58 (*c*.300 BC)'

ἡμιθωράκιον, for '*front plate of the* θώραξ' read '*half-*θώραξ'; add '*RA* 6(1935).31.7 (Amphipolis, ii BC)'

*ἡμικάβινος, ον, app. *measuring half a* κάβος, οἴνου ἡμικαβίνου *PRyl.* 629.247 (iv AD).

*ἡμικάβιον, τό, dim. of °ἡμίκαβος *PRyl.* 629.91 (iv AD).

*ἡμίκαβος, ὁ, half a κάβος, *PRyl.* 629.186 (iv AD).

*ἡμικεράτιον, τό, *half a* κεράτιον (v. κεράτιον II), *CRAI* 1945.379.4 (Berytus, v AD), *BGU* 2142.3 (v AD).

ἡμικόλλιον, add '**II** *half-strip of papyrus*, *PMich.*II 123ʳ vii 39 (i AD; ἱμι-).'

ἡμικρανία, add 'cf. ἑτεροκρανία'

ἡμίκρανον, add '*PMasp.* 141.11ʳ.20 (vi AD)'

*ἡμίκρανον, add 'also *pain on one side of the head* (? *migraine*), *Suppl.Mag.* 32 (v/vi AD)' *ἡμικτεύς, ὁ, *half a medimnus*, ἡμικτέος (gen.) de Franciscis *Locr.Epiz.* no. 23 (formed after ἑκτεύς).

ἡμικύλινδρος, add 'of an apse in elevation, Procop.*Aed.* 1.1.32'

ἡμιλίτριον, add '*SEG* 36.1342A (Palestine)'

ἡμίλιτρον 2, add '*IEphes.* 558 (iii AD), ib. 3437a (both -λειτρ-), *Edict.Diocl.* 19.15'

ἡμίμετρον, add 'rest. in *IG* 4.523 (Argive Heraeum)'

ἡμιμναῖον, add 'Ion. ἡμιμνήϊον *IG* 12(5).123b (Paros), cf. *IEphes.* 1b.1 (vi BC), ἡμιμνοῦν *IEphes.* 3437a'

*ἡμιμναιαῖος, ον, Boeot. εἰμιμναῖηος, *weighing half a mina, IGC* p. 99 B18 (Acraephia, iii/ii BC); also ἱμιμναῖηος ib. B22.

*ἡμιμύριος, ον, *five thousand*, παρέχοντος ἐτησίως κάλαμον ἡμιμύριον *PBerl.Leihg.* 23.9 (iii AD).

ἡμίνα **I**, add '(ἰνιμίνα· ἐνήμισυ cod.; cj. ἐν ἱμίνᾳ (or ἡμίνᾳ)· ἐν ἡμίσει)' **II**, add '**b** ἱμίνα· χοῖνιξ Hsch.'

*ἡμίνομον, τό, *half a νόμος* III, rest. in *SEG* 4.48 (Tauromenium, i BC).

ἡμίξεστον, delete '(Alexandrian .. 121b)'

ἡμιοβόλιον, transfer the article before ἡμιόγδοον and add 'cf. ἡμιωβόλιον'

ἡμιόγδοον, add '**II** a division of the citizen body, τὰν δὲ βουλὰν διακλαρῶσαι εἰς ἡ. καὶ τριακάδα καὶ φυλὰν καὶ φάτραν *SEG* 30.990.22 (decree found at Delos, prob. of Phlius, iv/iii BC).'

ἡμιόδελος, delete the entry.

*ἡμίοδος, ἡ, *field-path*, *BGU* 2159.8 (AD 485).

*ἡμιοκτάνιον, v. °ἡμιεκτάνιον.

ἡμιόλιος **I**, after 'half as much or as large again' insert 'ἡμιόλια .. τὰ χρήματα Isoc. 17.19'; line 7, after 'Arist.*HA* 629ª13' insert 'as nickname of the general Theodotus, Plb. 5.42.5, 79.5'; to neut. exx. add '*IG* 12(7).515.32 (Aegiale, ii BC); fem. σὺν ἡμιολίᾳ *with*, i.e. *including, 50% interest, PLugd.Bat.* XVII 4'

ἡμιονικός, delete 'ὁδὸς .. Str. 6.3.7'

ἡμιονίτης, add '*CPR* 13.11.41 (iii BC)'

ἡμίοπος **I**, for this section read '*half-holed*: ἡ. αὐλός type of small aulos (app. w. half the normal number of holes), Anacr. 30 P., A.*fr.* 91 R.; equated with αὐλὸς παιδικός in Ath. 4.182c; cf. ἡμίοπος· αὐλὸς ὁ ὑποτεταγμένος τῷ τελείῳ. μεταφορικῶς δὲ ἡμίοπος θράσος Hsch.' **II**, add 'ἡμίοπ(α) (neut. pl.), in list of kitchen equipment) perh. *half-size*, Lang *Ath.Agora* XXI B14 (iv/iii BC)'

ἡμίοπτος, add '*Trag.adesp.* 327f.2 K.-S.'

*ἡμιόρυκτος, ον, *half-dug*, λάκκος *PVat.* 11ᵛ 19 v 39, v 9 (ii AD).

ἡμιούγκιον, after 'Gal. 13.558' add '-όγκιον in *BGU* 781.5.17, 18'

*ἡμιροδία, ἡ, = ἡμιρόδιον, *Inscr.Délos* 1442B52 (ii BC).

ἡμιρρήνιον, delete 'fem. -ρρηνιαία, ἡ, ib. 35'

*ἡμίρροπος, ον, *half turning the scale*, in quot. transf., οἷον ἡμίρροπός τις ἐκ χαλκίτεως εἰς μίσυ μεταβολή Gal. 12.228.2.

ἡμιρρόπως, add 'Gal. 17(2).507.8'

ἡμισάκιον, add '*IG* 1³.422.152-4 (Athens, v BC)'

*ἡμίσιος (B), = ἥμισυς, *MAMA* 1.301 (Phrygia); also ἡμύσιος *AJA* 36.455 (S. Galatia).

*ἡμισοαγκωνοειδής, ές, *shaped like an elbow*, Bito 58.9.

*ἡμῖσον, v. ἥμισος.

*ἡμιστάθμιον, τό, *half a στάθμιον, PMich.*II 127 ii 4 (AD 45/6).

ἥμισυς, line 11 at end add 'also acc. sg. fem. ἡμίση *PTeb.* 815 fr. 10 ii 8 (iii BC); Boeot. fem. ἡέμιττα *SEG* 24.361 (sp. ἡμ-, Thespiae, iv BC; cf. ἡμίτεια)' p. 774a, line 3, ἥμισσον, add '*SEG* 31.825 (Halaesae, Sicily, ii BC)'; line 5, after 'Aug.)' insert 'Aeol. αἴμισσος *IG* 12(2).1.9, 11 (Mytilene, iv BC), Theoc. 29.5' **II 1 a**, line 14, after 'Pi.*N.* 10.87' insert 'Call.*Dian.* 90'; lines 15/16, transfer 'regul... 601c' to the end of section I 1, and for '*half-done*' read 'ῥηθέν *half said*'

ἡμισφαγής, after '*half slain*' insert 'Ps.-Callisth. 21.13'

+ἡμισφήκιον, τό, archit., *half-σφήξ, half-chevron, Inscr.Délos* 403.17 (ii BC); cf. °σφήκιον, σφηκίσκος.

*ἡμίσωμα, ατος, τό, *half-solid* (i.e. *vaporous*) *substance*, Anon.Alch. 7.14.

*ἡμισωράκιον, τό, a *half-σώρακος*, τοξευμάτων *IG* 2².1424a.344 (Athens, iv BC).

ἡμίτεια, after 'ἡ' insert '(fem. of *ἥμιτυς)'; add 'also ἡμίτεα *IG* 7.2712 (Acraephia; cf. *BCH* 62.156); ἡμιτία *SEG* 16.848 (Caesarea, ημμ-lapis)'

*ἡμιτέλειος, ον, *half-complete*, ἀριθμητικόν *SB* 4415.12, *BGU* 330.6 (both ii AD).

*ἡμιτετάρτεον or -τετάρτεων, τό, *half a τεταρτεύς, *IG* 1³.250.A20, 25, B3 (Attica, v BC).

ἡμιτέταρτον 1, add '*SEG* 31.154(b) (ἡεμι-, Attica, v BC)'

*ἡμιτία, v. ‡ἡμίτεια.

*ἡμίτονον, τό, *semitone*, Alex.Eph. in *Suppl.Hell.* 21.15, 19.

*ἡμιτρῐβᾰκός, όν, *half-worn*, στρῶμα Pach.*Reg.* in *PG* 40.952a.

ἡμιτρῐβής, add 'written εἱμιτριβ-, *PMich.*inv. 3163.9, 13 (*TAPhA* 92.258, iii AD)'

ἡμιτριταῖος, for '*half .. fever*' read '*of a fever, semitertian*'

ἡμιτύβιον, after 'Samos, iv BC)' insert '[ἡ]μιτυβίων ἐπ[πορικῶ]ν ἱστούς *PHels.* 7.6 (ii BC)'

ἡμιφανής, add 'Gr.Naz. in *PG* 37.764a'

ἡμιχοαῖος, add 'w. dissim. of aspirates, ἡμικοαῖα (n. pl.) *SEG* 35.134.31 (Athens, iv BC)'

ἡμίχοον, line 2, after 'ἡμίχα' insert '*Inscr.Délos* 104.132 (364 BC), Lang *Ath.Agora* XXI B15 (iv BC)'

ἡμιχρῡσους, for '-χρῡσος' read '-χρῡσον, τό'; add '*SEG* 34.95.36, al. (Athens, ii BC)'

ἡμιωδέλιον, add '*SEG* 30.1222.B1 (Tarentum, iv/iii BC)'

*ἡμοιρικοί (-ρηκώς La.)· μὴ μετέχοντες (-χων La.), Hsch.; cf. ἄμοιρος.

*ἡμουλίτρινος, ον, = ἡμιλιτραῖος, κόβαθρον ἡ. *PCol.* 188.9 (AD 320).

*ἡμύσιος, v. °ἡμίσιος (B).

ἤν, for 'φημι' read 'ἠμί'

*Ἡναῖος, ὁ, name of month, *Hesperia* 27.75.14 (c.200 BC, unknown provenance).

*ἡνᾰκάτιοι, v. ‡ἐνακόσιοι.

*ἤνατος, v. °ἔνατος.

ἠνεμόεις 1, add 'οὐρανὸν ἠνεμόεντα h.*Ven.* 291'

ἠνεμόφοιτος, add 'also ἀνεμόφοιτος Sch.Lyc. 1119'

*ἠνεμόφωνος, ον, *having the sound of rushing wind*, βροντή Jo.Gaz. 2.164.

ἠνία (B), line 3, after 'sg.' insert 'A.*fr.* 132b.4 R.' at end add 'form ἀνία, Myc. *a-ni-ja* (fem. pl.)'

ἡνίκα 2, add 'without ἄν, *CodJust.* 1.3.52.1, 6.4.4.1, al.'

ἡνιορράφος, add '*MAMA* 3.741 (ινιοραφ-, Isauria, Chr.)'

+ἡνιοχᾰράτης (ἡνιοχάρτης La.)· διδάσκαλος ἱππικῆς τῶν νέων. Λάκωνες Hsch.

ἡνιοχεύς, add '**2** transf., *ruler, governor*, *SEG* 33.940 (Ephesus, ?v AD); cf. ἡνίοχος I 4.'

ἡνιοχεύω, line 3, for 'ἀνιόχευεν' read 'ἔπι .. ἀνιοχεύων'

ἡνιοχέω, delete 'prose form of ἡνιοχεύω'; line 4, *drive*, add 'abs. ὁμολογῶ ἐκουσίως .. ἡνιοχῖν [σ]οι ἐ[π]ὶ τοῖς σοῖς ἵπποις *POxy.* 3135 (?AD 273/4)'

ἠνορέη, add 'Hes.*Th.* 516'

ἤνπερ, add 'etc.; v. °ἐᾱνπερ'

**ἦος, v. ἕως.

*ἠουοκᾶτος, ὁ, Lat. *evocatus* (mil. rank), Βάλβιλλος ἠουοκᾶτος *IGRom.* 1.78, *MAMA* 6.376; also ἠβοκᾶτος *IGBulg.* 1570.4 (iii AD).

ἠπᾰνία, add '*AP* 9.368.4 (Jul.)'

*ἠπᾰτημένως, adv. (ἀπατάω), *mistakenly, erroneously*, Olymp. *in Grg.* 251.13 W.

ἤπειρος **I**, after 'Timocr. 8' insert 'pl., Theoc. 17.77, epigr. in *IPhilae* 142.1 (7 BC), Nic.*Th.* 827' **II**, add 'w. spec. ref. to its coastline, *Peripl.M.Rubr.* 38, 41, al.'

ἠπειρώτης **III** 2, add 'Isoc. 4.157, al., Φίλων Ἠπειρώτας καλός *SEG* 31.770(a) (Thasos, Hellen.); also Ἀπειρώτας *IG* 9²(1).17.63 (iii BC, etc.)'

ἤπερ, for 'v. ἤ (A)' read 'ἀρείοσιν ἠέ περ ὑμῖν Il. 1.260, al; πρότερον ἤπερ πρὸς Λακεδαιμονίους Hdt. 1.77.2, cf. ἤ (A) B 1; also Amphis 33.2 K.-A., but doubtful in Attic prose (codd. Th. 6.40)'

ἠπεροπεύς, after 'Od. 11.364' insert 'Hom.*Cercop.fr.* 1.1' and add 'as adj., δόλον ἠπεροπῆα Nonn.*D.* 2.7, ἠπεροπῆι .. μύθῳ ib. 11.116'

ἠπεροπεύω, delete 'used only in pres. and impf.'

*ἠπητικόν, τό, *repair shop*, *OBodl.* 1046 (ii AD).

ἠπητρικόν, after '*wages*' insert '*OMich.* 1.8 (iii BC)'

*ἠπιόβουλος, ον, *of gentle counsel*, *GVAK* 10 (Halicarnassus, Hellen.).

ἠπιόδωρος, add 'app. of a doctor, epigr. in *IEphes.* 3821'

ἤπιος **I** 1, add 'epigr. in *TAM* 5(1).19'

ἠπιόχειρ, delete 'prob... 84.8' and add 'epigr. in *Bithynische St.* III 3.5, Ἀσκληπιὸν ἠπιοχεῖρα *SEG* 37.840 (Chester, iii/iv AD)'

*ἠπιόχειρος, ον, = ἠπιόχειρ, Orph.*H.* 23.8, 84.8.

*ἠποθηκάριος, v. °ἀποθ-.

ἤρα (B) **I**, add 'μηδοσύνας ἤ. τίνων Simm.*Securis* 1'

Ἥρα 1, delete 'an oath of Athen. women' and add 'ἥεραι (dat.) *SEG* 36.341 (Argos, vi BC), *SEG* 30.1176.B2 (Metapontum, vi/v BC), 30.1456 (Pontus, 470/450 BC)' at end delete etym. note and add 'Myc. *e-ra*; Cypr. *e-ra-i* (dat.) *ICS* 90.5'

Ἡραῖος **I**, τὰ Ἡ., add '*CID* I 9 D4 (iv BC), *IStraton.* 316.2, etc.' **II**, Ἡραῖος, add '*SEG* 34.940 (Camarina, iv/iii BC), *SIG* 279.17 (Mysia, iv BC)'; Ἡραιών, add 'Amyzon 28 (ii BC)'; add 'also Ἡραών, ἥεραόνος (gen.) *IG* 12 suppl. 549 A6 (Eretria, v BC); Ἡρεώνος (gen.) *SEG* 26.691 (Phthiotic Thebes)' add '**III** Ἡραία, ἡ, = Ἥρα, *SEG* 13.236 (Mycenae, c.500 BC).'

Ἡρακλέης, line 3, after 'Prose' insert 'Cypr. [*e*]-*ra-ke-le-we-se*, Ἡρακλέϝης *ICS* 415'; line 14, after 'Ἡρακλῆν' insert '*Inscr.Délos* 1416A17, al. (ii BC), *SEG* 31.122.31 (Attica, c.AD 121/2)'; line 16, for 'later Ἥρακλες' read 'also Ἡρακλές *CEG* 396 (Metapontum, 525/500 BC)' add '**3** name of a plaster used to staunch blood, Gal. 13.858.5.'

*Ἡρακλειασταί, οἱ, guild of *worshippers of Heracles*, *SEG* 31.122.4

(Attica, c.AD 121/2), IHistriae 57.32 (ii/iii AD); also **Ἡρακλεϊσταί**, IG 12(1).162 (Rhodes), SEG 24.1037 (Moesia, ii/iii AD).

Ἡρακλεῖδαι, add 'SEG 34.487 (Atrax, c.200 BC)'

***Ἡρακλείδιον**, τό, dim. of Ἡρακλέης, ὦ κάλλιστον Ἡρακλείδιον Achae. 26 S.

Ἡράκλειος, line 2, add '-ηος INikaia 1202' and add ref. to section VI **II 1**, for 'temple .. Hdt. 2.44, al.' read 'temple or sanctuary of Heracles, Hdt. 2.44.5, al., Paus. 9.11.6' **2**, add 'SEG 31.985.D8 (Teos, v BC), Inscr.Perg. 8(3).3 (ii BC)'

***Ἡρακλεϊσταί**, v. °*Ἡρακλειασταί*.

Ἡρακλείτειος, for 'α, ον' read 'ον'; add 'Arist.Ph. 185ᵃ7, Metaph. 987ᵃ33'

Ἡρακλεῶτις, add 'SEG 25.274 (Attica, Rom.imp.)'

***Ἡραών**, v. °*Ἡραῖος*.

ἠρέμα 2, for 'dub. in Luc.Merc.Cond. 28 codd.' read 'Luc.Merc.Cond. 28, Icar. 9'

ἠρεμέω, line 1, for 'hyperdor.' read 'Dor. (cf. ἀράμεν, ἀραμέναι; Tsakonian αραμού remain)'   **1**, add 'app. euphem., of the dead, SEG 30.976 (graffito, Olbia, iv BC)'

ἠρέμησις, for 'hyperdor.' read 'Dor.'

ἠρεσίδες, before 'EM' insert 'παρθένοι αἳ καλοῦνται ἠ. (or 'Η.) Agias Dercyl. 4 J., cf. Call.fr. 65 Pf. adn.; Hsch.'

***ἠρεσιώνης**, ὁ, some kind of priestly official, (cf. perh. εἰρεσιώνη), Hesperia 40.316 line 5, ib. 32.48 no. 69.6; to be rest. in IG 2².1825.71, 3680.11 (all Attica, iii AD).

-ήρης 2, add 'when compounded w. numerals (e.g. τριήρης) it indicates the number of files of rowers along each side of the ship'

ἦρι, after '*in the morning*' insert 'Call.Jov. 87'; at end for '(Cumae)' read '(Cumae, vii/vi BC); (acc. to some edd. here ἦρι is dat. of ἔαρ (Α))'

ἠρίγένειον, for '= ἠρύγγιον, Hsch.' read 'ἠ. (-νιον La.)· τὴν ἠρύγγιον πόαν Hsch.'

†**ἠρινός**, v. ‡ἐαρινός.

***ἠρινοτόκος**, ἠρινοτόκου· ἀπὸ τῶν ποιμνίων τῶν κατ' ἔ⟨τ⟩ο⟨ς⟩ φόρους τελούντων ἔαρι Hsch.

***ἠριπότην** ἡμέραν ἐξ ἡμέρας .. Hsch.

***ἠρογάτωρ**, ὁ, Lat. *erogator*, official in charge of distribution of supplies, SEG 32.1554 (Arabia, vi AD).

***ἠρόδοτος**, ον, given in spring, An.Ox. 3.350.7.

***Ἡροξένια**, τά, *feast in honour of the Heroes*, Sokolowski 2.69.3 (Thasos, iv BC); also **Ἡρωιξένια**, Thasos 192.23 (i AD).

***Ἡροσούρια**, τά, an Attic festival, perh. connected with fair winds in spring, Sokolowski 3.18.B28 (Erchiae, iv BC); cf. Ἡροσάνθεια, Ἡροφάνεια.

***ἠρτυλημένος** (-λιμ- cod.)· ἠρτυμένος Hsch.

ἦρυς, add 'but perh. sp. of ἠρωίς'

***ἠρωάς**, άδος, ἡ, = ἠρωΐνη I, Corinn. 11(*b*).2 P.

***ἠρῴδιον**, τό, dim. of ἥρῳον, TAM 1.73 (Cyanae).

ἡρωϊκός I 2, add 'ἔδοξεν τῇ πόλει τειμάς τε αὐτῷ ἡρωικὰς ψηφίσασθαι SEG 23.319 (Delphi, ii AD)'

ἡρωΐνη I, add 'pl. *Ἡρωῖναι*, Attic deities, Sokolowski 3.18.A19 (Erchiae, iv BC)'

ἡρωΐς I 2, add 'Ἀφροδεισίᾳ ἡρωΐδι SEG 34.1028 (Italy, ii/iii AD)' **II 1**, after 'ἡρωϊκός' insert 'θεαρίας Pi.Pae. 14.36 S.-M.'

ἡρωϊσμός (s.v. ἡρωΐζω), for '*worship*' read '*cult*' and add 'SEG 33.946.8 (Ephesus, i AD).

†**ἡρώϊσσα**, ἡ, contr. ἡρῷσσα, (Thess. εἰρούισσα AE 1931.178) *female hero, demigoddess*, A.R. 4.1309, 1358, AP 6.225 (Nicaenet.). **2** *deceased woman*, IG 5(1).610 (Sparta), al., 12(5).325 (Paros), SEG 24.398 (Thessaly, iii/ii BC).

***ἥρων**, ωνος, ὁ, Syracusan for ἥρως, ἡρώνεσσι Sophr. 154.

ἡρῴνα, add '(also read as εἰρώνα = *ἱερωνή (i.e. ἱερωνία), v. ZPE 49.187ff.)'

ἡρῷον 1, before 'Hdt. 5.47' insert '*Ἑρόον τὸν ἐν Θέβαις* SEG 37.283 (Argos, c.550 BC)'

ἡρῷος, for 'πούς .. etc.' read 'ἡ. πούς AP 7.9 (Damag.)'

ἥρως, line 1, for 'signf. III' read 'signf. II', add '(Thess. εἶρους AE 1931.178)' and, gen. ἥρωος, add 'hέρōος SEG 37.286 (Argos, v BC)'; line 4, after 'Orac.ap.D. 43.66' insert 'ἥρωτι IGBulg. 362, 1727, 1750'; line 8, delete 'ἡρώνεσσι .. 154.1' **I 3**, add 'a Thracian horseman divinity, IGBulg. ll.cc., SEG 31.638 (Maced.), etc.' add '**4** of great writers of the past, Phld.Rh. 1.200, Longin. 4.4, 14.2, 36.2.'

ἡρωστής, delete the entry.

Ἡσιόδειος, add 'τῶν σ[υν]θυτάων τ[ἆμ] Μωσάων τῶ]ν Εἰσιοδείων IG 7.1785 (Thespiae)'

Ἡσίοδος, for 'Aeol. *Αἰσ*-' read 'hyperaeolic *Αἰσ*-'

ἡσσάομαι, line 1, delete 'Th. 3.57'; line 6, after 'etc.' insert 'Hellen. ἑττηθήσεσθε PSI 340.21, ἐσσηθείς PHib. 197.53 (both iii BC)'

ἡστός, add 'Dam. in Phlb. 124.1, al.'

ἡσῠχάζω I, add '**c** of rights, *lapse, be suspended*, αἱ δὲ λοιπαὶ πατρωνικαὶ διακατοχαὶ ἡσυχαζέτωσαν Cod.Just. 6.4.4.21a.' **II a**, add 'Sol. 4c.1 W.'

ἡσῠχῇ 1, line 3, after '*gently*' insert 'Hippon. 26.1 W.'

ἡσῠχία 4 a, at end, pl., for these entries read 'τὰς ἡ. ἄγειν Ephor. 236 J., Ath. 3.114a; ἔχειν id. 11.493f'

ἦτορ, line 2, after 'codd. Ath.' insert 'Pi.Pae. 6.12 S.-M.'

ἤτριον, line 2, for '(Leon., pl.)' read '(Leon.Tarent., pl.), (AP) 15.27 (Simm.)'

†**ἤτω**, v. ‡εἰμί.

***Ἡφαίστειος**, α, ον, *of or belonging to Hephaestus*; neut. pl. subst., *works of Hephaestus*, Call.fr. 202.57 Pf. (Add. II). **2** the name of a month, cf. Ἡφαίστιος, SEG 37.453.19 (sp. -ειος, Thessaly, ii/iii AD).

***Ἡφαιστιασταί**, οἱ, *worshippers of Hephaestus*, SEG 30.1004 (Rhodes, ii/i BC).

Ἡφαιστείον (s.v. *Ἡφαιστεῖον*), before '*temple of Ptah*' insert '*temple of Hephaestus* at Athens, SEG 26.98 (iii BC)'

***Ἡφαίστιον**, τό, = Ἡφαιστεῖον, IG 1³.472.2 (Athens, v BC).

***ἡφαιστίτης**, ου, ὁ, λίθος, a mineral, Cyran. 18, 19 (1.7.2, 17 K.).

***Ἡφαιστόπους**, *having a foot like Hephaestus*, i.e. *lame*, Olymp. in Alc. 160.9 sch. W.

Ἥφαιστος, line 1, at end insert 'Ηέφαστος Jeffery LSAG 77 no. 24 (vase, Athens, c.550 BC), Kretschmer GV p. 127' add 'cf. Myc. a-pa-i-ti-jo, pers. n. (cf. Ἀφάστιος IPamph. 149, ii BC)'

Ἡφαιστότευκτος, add 'Η. πανοπλίαν Procl.Chr. p. 106.2'

Ἡφαιστόχειρος, for 'Choerob.Orth. in AB 1380' read 'An.Ox. 2.317'

ἠχέω, line 5, for 'grasshopper' read 'cicada' **II**, at end after 'S.OC 1500' insert 'of long vowels, πολὺν ἠχεῖται χρόνον D.H.Comp. 14; (ἁ βάρβιτος ..) Ἔρωτα μοῦνον ἠχεῖ Anacreont. 23.4 W.'

ἦχι, add 'Dor. ἆχι Suppl.Hell. 1034-1035'

ἠῷος, add 'cf. Myc. a-wo-i-jo, pers. n.'

ἠώς I 4, at end for 'IG 7.235' read 'SEG 31.416'

Θ

θαάσσω, add 'prob. in Sapph. 73(a).7 L.-P.'
θᾶκος, after 'θῶκος' insert '(also Men.*Dysc.* 176)' **I 3**, add '*IEphes.* 455 (ii AD)'
θαλάμη I 1, line 3, after 'polypus' insert '*h.Ap.* 77'; line 6, after '(E.)*Ion* 394 (pl.)' insert 'of the *underground shrine* of Rhea, Nic.*Al.* 8 (pl.) and Sch., cf. Phot. and Hsch. s.v. θαλάμαι' at end add 'cf. Myc. *ta-ra-ma-ta*, pers. n.'
θαλαμίς, add '**2** prob. *funerary monument*, *SEG* 31.1126 (Phrygia, AD 150/200).'
°**θαλαμίσκος**· κοιτωνίσκος Cyr.
θαλαμίτης, for pres. def. read '*a member of the lowest file of rowers* in a trireme'
θάλασσα, line 1, after 'Att.' insert 'Boeot.' and after parentheses insert 'Lac. θάλαθ(θ)α *SEG* 26.461.7 (Sparta, v BC), Cret. θάλαθθα (ii BC) **4**, for this section read 'applied to a trench (perh. fr. similarity to Hebr. *t⁽ᵉ⁾alāh*), Lxx 3*Ki.* 18.32, 35, 38' **5**, for '*laver*' read '(*artificial*) *pool made of bronze*' and add '3*Ki.* 7.23'
θαλασσαῖος 1, add 'Boeot. θαλαττῆον, τό, *produce of the sea*, *IGC* p. 98 A4, 5 (Acraephia, iii/ii BC)'
θαλασσίγονος, delete the entry (read °θαλασσό-).
°**θαλάσσινος**, η, ον, *of the sea*, ἐπὶ θαλασσίνηι ὀδ[?ῶι] (i.e. *leading to the sea*) *Delph.* 3(1).362 iii 15. **2** *sea-like*, prob. *dyed purple, thalassina vestis*, Lucr. 4.1127; cf. θαλάσσιος III.
θαλάσσιος, line 2, after 'E.*IT* 236' insert 'Luc. *VH* 2.46'; line 7, before 'πεζοί' insert 'μόσχος θ. *seal*, *Edict.Diocl.* 8.37, θ. νεκρός of a shell used as a trumpet (cf. Ath. 10.457f), Thgn. 1229'; line 9, delete 'θ. .. 1229' **3**, for '= ἀλουργής' read '*dyed purple*'
θαλασσίτης, add '**2** kind of hyacinth stone, Epiph.Const. in *Lapid.Gr.* 196.24.'
θαλασσοβαφής, add '*sea-coloured*, of a stone, Epiph.Const. in *Lapid.Gr.* 197.21'
°**θαλασσόγονος**, ον, *born from the sea*, θαλασσογόνου Παφίης Nonn.*D.* 13.458.
°**θαλασσοδίαιτος**, ον, *dwelling in the sea*, *APAW* 1943(14).8 (Chalcis, iii AD).
θαλασσοειδής, add 'epith. of (Jewish) God, *sea-like*, prob. w. ref. to his immensity, *PMag.* 4.3068'
°**θαλασσόζωνος**, ον, *sea-girt*, prob. in *Suppl.Hell.* 991.68 (poet. word-list, iii BC).
θαλασσομαχέω, for '*fight by sea*' read '*combat the sea*, in quot., of a κυβερνήτης'
θαλασσομάχος, for '*fighting by sea*' read '*combating the sea*'
θαλασσοπόρος, add 'Nonn.*D.* 20.376, 21.187, 40.531, 43.425'
°**θαλασσοσέρις**, ἡ, name of a plant, Aët. 7.114 (388.7 O.).
θαλασσουργός, before 'ὁ' insert 'also **-οεργός** *GVI* 1859 (Teos, ii/i BC)'
θάλαττα, etc., after 'Att.' insert 'Boeot.'
θάλεα, line 3, for 'Alcm. 10' read 'Alcm. 15 P., cf. Pi.*Parth.* 2.36 S.-M.' and for 'θαλέεσσιν .. *Fr.anon.* 31' read 'τὼ μὲν ἐγὼ θαλέεσσιν ἀνέτρεφον Call.*fr.* 337 Pf.'
°**Θάλειοι**, οἱ, *devotees of* Θάλεια III 2, *IEryth.* 201 a 62 (iii BC).
θαλερός II, line 6, for 'the thick and frequent sob' read '*mighty lamentation*'; line 8, after 'A.*Th.* 707 (lyr.)' insert 'but v. °θελεμός'
θαλία I, pl., add 'X.*Hier.* 6.2 (cf. Lac. pers. n.)'
°**θαλικτάριον**, τό, = θάλικτρον, *PMed.Rez.* 12.13 (sp. θαλκιτάριον, vi/vii AD).
†**θαλιοποιοί**· οἱ τὰ σκυτούμενα κιβώτια, καὶ τοὺς δερματίνους ῥίσκους ἐργαζόμενοι Hsch. (cf. σαλία, θολία; but **θαλλικοποιοί** La., cf. °θάλλικα).
°**θαλκιτάριον**, v. °θαλικτάριον.
°**θαλλέομαι**, *put out shoots*, *BGU* 2157.15 (AD 485).
°**θάλλικα**· σάκκου εἶδος Hsch.
θαλλικοποιοί, v. °θαλιοποιοί.
θάλλινος, στέφανος, add '*SEG* 30.1892 (Maced., iii BC)'
°**θαλλίς**· μάρσιππος μακρός Hsch., Afric.*Cest.* p. 47 V.
°**θαλλισμός**, ὁ, *special monetary provision in will*, *PMil.Vogl.* 84.9 (ii AD); cf. θαλλός III.
°**θαλλοδοτέω**, *distribute branches* at a shrine, *POxy.* 3094.40, 43 (AD 217/8; cf. θαλλός III).
°**θαλλόομαι**, prob. *produce shoots*, of a date-palm, *BGU* 2158.3-4 (v AD).

θαλλός I, line 7, after 'etc.' insert 'also w. apposition, θαλλῷ στεφάνῳ *IG* 12(1).160, 161, *ASAA* 8/9(1925/6).322 (Rhodes)' **I**, add '**2** as cult-title of Zeus, Δεὶ Θαλλῷ εὐχήν *SEG* 32.1282; Διὶ Θαλῷ 37.1171 (both Phrygia, Rom.imp.), al.'
θάλλω 3, after '*to be fresh, active*' insert 'πατρίδ᾿ ἂν ἱμερτὰν πένθος ἔθαλλε τότε epigr. in *BCH* 116.599 (Ambracia, vi/v BC)'
θάλος, add '**II** *victor's wreath, crown*, Pi.*I.* 7.24, *fr.* 70a.14 S.-M.; fig. κλεινᾶν Συρακοσσᾶν θάλος Ὀρτυγία id.*N.* 1.2, *Lyr.adesp.* 1029.3 P. **III** *well-being*, cf. θάλεα, epigr. in *BCH* 85.849 (Paros, Rom.imp.).'
°**θαλπίζω**, *cherish, look after*, *PNag.Ham.* 70.17.
°**θάλπιον**· θερμότερον Cyr.
θάλπω II 3, for '*hatch*' read '*brood over*' **III**, add '**4** *prefer, favour*, Theophil.Antec. 2.20.3.' **IV**, line 2, after 'θάλψαι' insert '(s.v.l.)' and for '*to live*' read '*keep oneself warm*'; for 'Leon.' read 'Leon. Tarent.'
θαλυκρός, for first quot. read 'ἂ πάντη πάντα θαλυκρὸς ἐγώ Call.*fr.* 736 Pf.'
θαλύσια 1, for '*offerings .. Artemis*' read '*harvest-offerings to the gods*'; line 2, delete 'later'; add '*offerings of any first fruits*, Nonn.*D.* 47.493, 48.224'
°**θαλυσμοσύνη**, ἡ, dub. sens., epigr. in *ITomis* 241(77).10 (Rom.imp.).
θάλψις II, add 'Just.*Const.Δέδωκεν* 1'
θαμέες, line 7, after '(Il.) 12.287' insert 'θαμειαὶ σφενδόναι Archil. 3.1 W., Panyas. 4.2'
°**θαμνίον**, τό, dim. of θάμνος, Gal. 12.108.18.
θαμνομήκης, for '*a long .. bush*' read '*of the length equal to the height of the bush*'
θάμνος, add 'Emp. 9.2, 20.6, 117.2 D.-K.'
°**θαμνοῦχος**, ον, *full of bushes*, ἐν ὑψηλοῖσι θα[μν]ούχοι[ς cj. in A.*fr.* 73b.3 R.
†**Θαμυρίζω**, prob. *celebrate the cult of Thamyris*, Θαμυριδδόντων Πισάνδρῳ, Δαμοκλείος *SEG* 32.503 (Thespiae, 400/350 BC); cf. θαμυρίζει· ἀθροίζει, συνάγει Hsch.
θανάτησιος, for 'Afric.*Cest.* 14, 16 .. Thévenot' read 'Afric.*Cest.* p. 28 V. (v.l. -ήσιμος), p. 30 V.'
†**θανατιάω**, = θανατάω I, anon.ap.Suid. **II** = θανατάω II, Luc. *Peregr.* 32, S.E.*M.* 9.153.
°**θανατοσυνάρτης**, ου, ὁ, *joiner-together of deaths*, *PMag.* 4.1372.
°**Θάξιος**, ὁ, month-name (app. = Θάσιος II), *SEG* 33.1039.57 (Aeolian Cyme, ii BC).
†**θάομαι**, line 6, after 'Theoc. 15.23' insert '(v.l.)' **II**, add '1 pl. aor. subj. θασώμες Theoc. 15.23 (*PAntin.*)'
θάπτω, line 8, after 'Hes.*Sc.* 472' insert 'θάπτουσι κατακαύσαντες ἢ ἄλλως γῇ κρύψαντες Hdt. 5.8'; line 10, after 'cremation' insert 'E.*Antiop.* p. 21 A.'
Θαργήλια, add 'Θαργήλιος as title of Apollo, *SEG* 30.977(a) (Olbia, v BC)'; form Ταργ- as pers. n., add '*SEG* 33.659 (Rhodes)'
Θαργηλιών, after 'Athens' insert 'and elsewhere' and add '*SEG* 30.977(c) (Olbia, v BC), 33.679'
°**Θαρησιεῖον**, τό, prob. = °Ἐρησιεῖον, *SB* 9628 (*SO* 57.77).
θαρσέω I 4, add 'ἐξ ἀνθρώπων as far as men are concerned, Epicur.*Sent.* 6; ἀπὸ τῶν ἔξωθεν ib. 39'
°**Θαρσικάριος**, v. ταρσικάριος.
°**Θαρσικός**, v. Ταρσικός.
°**θᾶς**, Lesb. = Ion. τέως (q.v. I 2), Alc. 70.8, 206.6 L.-P.; cf. ᾆς.
Θάσιος II 1, add 'cf. °Θάξιος'
θάσσω, after '(E.)*Hec.* 36' insert 'τίς ἐπ᾿ αὐλείοισι θύραις θάσσει; Ar.*V.* 1482'
Θαύλιος, add 'epith. of Zeus at Atrax, *SEG* 34.490 (*c.*200 BC), Θαύλιος (cod. Θαῦμος) ἢ θαυλός· Ἄρης Μακεδόνιος Hsch.'
θαῦμα I 2, for '*mountebank-gambols*' read '*acrobatic feats*, *IG* 1³.757 (500/480 BC)' and after 'cf. (X.*Smp.*) 7.3 (sg.)' insert 'Max.Tyr. p. 344 H.'
θαυμάζω 2, add 'pass., w. deponent force, θαυμασθῆναι βασιλέα σάρκινον Lxx *Es.* 4.17p.'
θαυμάσιος I 1, add 'sup. as honorific epith., γραμματεύς *PLond.* 1842.4, ἰατρός *PMasp.* 6.ii.14, προνοητής *POxy.* 206.6 (all vi AD)'
θαυμασιότης II 2, add '*PGrenf.* 1 56.7 (vi AD)'
θαυμαστός, line 1, insert 'σαυμαστά (n. pl.) Alcm. 4 *fr.* 1.4 P. (v. *POxy.*

2388)'; line 7, after 'c. gen.' insert 'ῥᾳστώνης Pl.*Lg.* 648c'; lines 8/9, delete 'πλέοσι .. 9.122'

θαυσήκρι, add '(**θαυσίκριον** La.)'

θεά I, add '**2** as the title of empresses; of Julia Domna, *SEG* 24.953 (Moesia, AD 198).' at end, monosyll., add 'also Herod. 4.11'

θεᾱ́ I 1 a, line 5, for 'τινός' at the *sight of* read 'τῇ αὑτοῦ angry at his *gazing* (at the fallen enemy)'

***θεαγεία,** ἡ, the office of θεαγός, *POxy*. 3974.18 (AD 165/6).

θεᾱγός, for '*priest .. gods*' read '*"god-bearer"*, a priest concerned with the transporting of sacred images'; delete '*PTeb*. .. (ii BC)' and for 'etc.' read 'θ. Θοήριος *PTeb*. 61(*b*).59 (ii BC; pl.), θ. Σούχου ib. 121.76 (pl.), 133 (i BC), θ. Θοηρείου ἐξαγορείων καὶ ἑτέρου Σιντάνω λεγομένου pap. in *JEA* 20.21.9 (iii AD; pl.), cf. ib. 28'

***Θεᾰδέλφεια,** τά, a Ptolemaic festival, *SEG* 1218.9 (Xanthos, iii BC), *Inscr.Olymp.* 188 (-εα, ii BC; cf. *Hesperia* 4.90), *PSI* 431.

θέαινα, after 'θεά' insert '**θεαίνη** Nonn.*D*. 6.123' and after 'Od. 8.341, al.' insert '*h.Ap.* 311'

θέᾱμα, add '**2** in act. sense, *gazing at*, αἰθέρος θεάμασιν Chaerem. 14.4.'

θεάομαι, line 7, after '(Od.) 5.74' insert 'perh. Cypr. *e-ta-we-sa-to* ἐθαϜέσατο *ICS* 319' **I,** add '**5** *see,* i.e. consult, βουλόμενος ἰατρὸν ὀφθαλμῶν θεάσασθαί τινα Luc.*Nigr.* 2.' **II,** line 3, after 'Hsch.' insert 'dub. in Men.*Epit.* 564'

***θέαπνον,** τό, unkn. product in a medicinal recipe, *PAmst.inv.* 148 (*Mnemos.* series 4, 30.146); cf. θεάφιον.

***θεατράλιος,** α, ον, cf. Lat. *theatralis, of the theatre,* Just.*Nov.* 63.1.

θέᾱτρα, add '*SEG* 25.306 (Attica, i BC)'

θεᾱτρίζω I, for this section read '*perform in the theatre,* οἵ τε ἀγων-[ιζ]όμενοι πάντες καὶ οἱ κατὰ καιρὸν θεατρίζοντες Gerasa 192.18 (ii AD), Suid.'

θεᾱτρικός 2, add '**b** of style, *showy,* D.H.*Isoc.* 12.15.'

θέᾱτρον, line 4, for '*Act.Ap.* 19.20' read '*Act.Ap.* 19.29'

θεατροτορύνη, for '*stage-pounder*' read '*one who rouses the theatre-audience*' and delete '*who was a clumsy dancer*'

θεάφιον, add 'Charis. p. 32 K.; cf. θειάφιον'

θέεινος, add 'written θόειν- *PMich.inv.* 3163.7 (*TAPhA* 92.258; iii AD)'

θεηγόρος, add 'παντὸς ἔργου καὶ θεηγόρου λόγου *Trag.adesp.* 118b K.-S. (perh. Chr.)'

***θεηδόχος,** ον, *recipient of deity,* οὖδας ἀρούρης Nonn.*D*. 13.96.

***θειασμός,** ὁ, *divine possession* or *frenzy,* θειασμοῖς κάτοχοι γυναῖκες D.H. 7.68; θειασμοῦ (ἐπιρρήματα) D.T. 642.17. **2** *inspired utterance,* ἄγαν θειασμῷ προσκείμενος Th. 7.50.

***θειάφιον,** see also ‡θεάφιον.

***θεικᾱ,** ά, v. *θήκη.*

θειλόπεδον, for pres. def. read '*area exposed to the sun used for drying grapes*' and add '*IMylasa* 843.6'

***θειογρᾰφικός,** ή, όν, *theological,* ἱστορίαι Rh. 3.541.22.

θεῖον (A), add '**II** = ἐλελίσφακον, Dsc. 5.107, Poet.*de herb.* 93.'

θεῖος (A), line 2, after 'Bion.*fr.* 15.9' insert 'Cret. θιήιος *SEG* 27.631.4, 6'; delete 'late' and after '**θήιος**' insert 'Alc. 45.8 L.-P.' **II 1,** add 'in Lydian dedications, Keil-Premerstein *Dritter Bericht* no. 30 (ii AD), cf. *TAM* 5(1).186, *IStraton.* 519' at end add 'Myc. *te-i-ja* (fem.)'

θεῖος (B), after 'ὁ' insert '**θῖος** *IG* 12(1).72b (Rhodes, i BC), perh. δεῖος *SEG* 31.1004 (Lydia, AD 101/102)'; after '(Balbilla)' insert 'Dor. θῆος (q.v.)'

θειόστεπτος, delete the entry.

θειοτελής, delete the entry.

θειότης II, add '**2** attribute of Roman emperors personified as deity, *SEG* 30.1253 (Aphrodisias, i AD).' at end add 'see also θεότης'

***θείς (A),** θεῖσα, θέν, gen. θέντος, aor. part. of τίθημι (q.v.).

***θείς (B),** θέν, gen. θενός, = °δείς, ἡ δὲ γῆ ἦν θὲν καὶ οὐθέν Thd.*Ge.* 1.2.

θειώδης (B), for '*divine*' read '*supernatural, marvellous,* ἔτι δὲ τούτου θειωδέστερον συνεμαρτύρουν Eun.*VS* 459' and for '*PMasp.* 451.42, 56' read '*PMasp.* 317.27'

***θείωσις,** εως, ἡ, *consecration,* by initiation, Plu. 2.351f; see also θέωσις.

†**θελεμός,** ον, app. *quiet, placid,* ποταμοὺς δ᾽ οἳ διὰ χώρας θελεμὸν πῶμα χέουσιν A.*Supp.* 1027, θελεμωτέρῳ πνεύματι id.*Th.* 707 (cj.); θελεμόν· ἥσυχον Hsch.; glossed by θελημός Hdn.Gr. 1.171; adv. -μῶς· ἡσύχως Hsch.

θελήμων, add 'of persons, *willing,* A.R. 4.1657'

***θελησμός,** ὁ, *volition,* Hdn.*fr.* p. 24 H.

***θέλκαρ·** θέλγμα Hsch.

θέλκταρ, delete the entry.

***θελκτῶ·** κολακευτική Phot., Suid.

θελξιμελής, for '*IG* 3.400' read 'hex. in *IG* 2².5200 (iii AD)'

***Θελξίνη,** ἡ, epith. of Hera, *SEG* 26.1211 (Velia, iv BC).

θέλξις, add 'prob. in D.H.*Th.* 33'

θελοντής, add '(in S.*Aj.* 24 prob. reading κἀγὼ 'θελοντής)'

θέλυμνα, delete the entry.

***θέλυμνος,** η, ον, *close-packed, dense,* θ. τε καὶ στερεωπά Emp. 21.6 (codd. θέλημνα, θέλημα; cf. προθέλυμνος, τετραθέλυμνος).

***Θελχίνια,** v. *Τελχίνιος.*

†**θέλω,** v. *ἐθέλω.*

θέμα I 1, after 'Plu. 2.116a, b' insert '*bank account, deposit, PCharite* 38.4 (iv AD)'; after '*PTeb.* 120.125 (i BC)' insert '*receipt* acknowledging deposit of grain, *PMich.*xi 604.16 (iii AD), *POxy.* 2769.14 (iii AD)' **6,** for this section read '*platform serving as base for monument,* ἡ σορὸς καὶ τὸ βαθρικὸν καὶ τὸ ὑποκείμενον θέμα Hierapolis 208, cf. 124, al., *TAM* 4(1).140, rest. in *MAMA* 6.19'

θεματοποιέω, add '**II** *place in position,* θ]εματοποήσας τὰς γεφύρας καὶ κατασκευάξας τὰς ὁδούς *IRhod.Peraia* 601.4 (Hellen.).'

θέμεθλα, at end delete 'Call.*Dian.* 248' (v. θέμειλον, s.v. °θεμείλια) and insert 'θέμεθλα· ἕδραι, βάθρα, θεμέλια Hsch.'

θεμείλια, line 3, after 'Opp.*H*. 5.680' insert 'fig., θ. πήγνυτο χάρμης *the foundations of battle were being laid,* Nonn.*D.* 17.135, cf. 29.324, 43.3: sg. **θεμείλιον** Nic.*Th.* 608, *Inscr.Délos* 290.202 (iii BC)'; line 4, after '**θέμειλον**' insert 'Call.*Dian.* 248'

θεμελιόθεν, add 'also **θεμελιᾶθεν** *IGLS* 9121 (v AD); **θαιμηλιῶθε** *BE* 1959.459 (Syria, Chr.)'

θεμέλιος I, add 'οἰκοδομήθη ἐκ θεμελίων *SEG* 31.1473 (Arabia, AD 562)'

θεμελιοῦχος, of Poseidon, add '*SEG* 30.93.17 (decr., Eleusis, 20/19 BC)'

θεμελίωσις, add '*IG* 11(2).199*A*88 (Delos, iii BC)'

θεμερός, (not θέμ-), after 'Hsch.' insert 'θ. ὀπί Pi.*N*. 7.83 cod. D'

***θεμιονείκης** (i.e. **-νίκης**), ου, ὁ, *winner of a* θέμις II, *TAM* 2.688 (Cadyanda).

θέμις, line 4, after 'Pi.*O*. 13.8' insert 'Pl.*R*. 380a (cf. Θέμιτι *IG* 7.1816.2)' **II,** add 'gen. θέμεως *MAMA* 4.124, 132 (Metropolis), Schwyzer 686a (Pamphylia, ii BC)' **IV,** add '*SEG* 37.491 (dat. -ιστι, Magnesia, v/iv BC), 26.717.8 (dat. -ιτι, Epirus, iv BC)'

***θεμισσόνος,** ον, *preserving justice, SEG* 36.1198.12 (Phrygia, iv AD).

θεμιστεύω I, after 'Od. 11.569' insert 'λαοῖς θεμιστεύσοντα *Trag.adesp.* 664.30 K.-S.' **II,** add '**b** w. dat. and inf., *in* (oracular) *answer enjoin* on one to .., *PAAH* 1932.52.2 (Dodona, iv BC).'

θεμιστοπόλος I, add 'Hes.*fr.* 10(a).25 M.-W., θεμιστοπόλῳ δέ τε βουλῇ epigr. in *SEG* 13.277.11 (Patrae, iv/v AD)'

θεμιστός I, after '= θεμιτός' insert 'Archil. 177.3 W.'; add 'as pers. n., *SEG* 32.756 (Berezan, vi/v BC)'

θεμιτεύω, for 'orgies' read 'rites' and delete 'metri gr.'

θεμιτός, line 6, for 'τὰ μὴ θεμίτ᾽ ἦν .. dub. l. in' read 'εἶδε τὰ μὴ θεμιτά'

θεοβλᾰβής, add '**2** *heaven-inflicted,* ἄχη θεοβ[λαβῆ rest. in S.*fr.* 269c.30 R.'

θεογέναιος, add 'also **-γένιος** *POxy.* 3780.5 (AD 40/2)'

θεογενής, add 'as epith. of Caesar, *SEG* 30.1245 (Aphrodisias, i BC/i AD)'

***θεογηθής,** ές, *rejoicing in god,* Didyma 344.5, epigr. in *IGBulg.* 2086.2.

***θεόγνητος,** ον, *born of a god, Suppl.Hell.* 991.89 (poet. word-list, iii BC).

θεόγνωστος, add '**2** fem. subst., another name for βράθυ, Cyran. 12 (1.2.1 K.).'

***Θεοδαισία,** ἁ, = *Θεοδαίσια,* τά, *IG* 12(2).68.9 (Lesbos, ii AD).

Θεοδαίσιος (s.v.*Θεοδαίσια*), as the name of a month, add 'ἐπὶ Ἀλεξάνδρου, Θεοδαισίου *SEG* 30.1637 (Rhodian amphora, Paphos, Hellen.)'; form Θευδ-, add '*SEG* 39.1008 (Morgantina, iv/iii BC)'

***Θεοδέκτρια·** θεὸν δεδεγμένη Hsch. (La.; -δέκτορα cod.).

Θεοδέκτωρ, delete the entry.

θεόδμητος, after '(Pi.)*fr.* 87.1' insert 'Lacon. σιόδματος Alcm. 2(iv).5 P.'

Θεοδόσιος, add '(as pers. n., Θεōδοσιο *IG* 1³.1151, Attica, v BC, etc.)'

θεοειδής, add 'see also θεειδής, σιειδής, °σιοειδής'

***θεοθελής,** ές, *willed by the gods,* Sch.B Il. 14.120.

***θεοκοίμητος,** ον, *resting in God, IG* 14.88 (Syracuse, Chr.).

θεοκολέω, add 'see also θευκολέω'

θεοκόλος, add '**θευκόλος** de Franciscis *Locr.Epiz,* 21.6; see also °σιοκόλος'

***θεοκόρος,** ὁ, gloss on σιοκόρος, Hsch.

†**θεόκραντος,** ον, *ordained by the gods,* A.*Ag.* 1488.

θεόκριτος, add '**2** gloss on Δάν (Hebr. = *judge*), J.*AJ* 1.19.7.'

θεολογία, for '*science of things divine,* Pl.*R*. 379a' read '*talk about gods,* Pl.*R*. 379a; *science of things divine*'

θεολόγος, lines 6/7, for '*BMus.Inscr.* .. ii AD' read 'as an official title, θεολόγου ναῶν ἐν Περγάμῳ *IEphes.* 22.4, 645.6, 1023.4, etc.'

θεομᾰχέω, add 'X.*Oec.* 16.3, D.S. 14.69, Philostr.*VA* 4.44'

***θεομήδεα,** τά, *counsels of God,* Gerasa 327 (vi AD).

***θεομμᾰτος,** ον, hyperdor. = θεόμιμος, Hippod.ap.Stob. 4.1.95 (p. 102.12 Thesleff).

θεομῑσής I, after 'θεοφιλής' insert 'Ar.*Av.* 1548' **II,** delete 'Ar.*Av.* 1548 (ubi v. Sch.)' and '(θεομίσης v.l. in Ar. l.c.)'

*Θεοξενιακά, τά, *fund for the expenses of the Theoxenia*, IG 12(5).544B2.9 (Ceos, iv/iii BC).

θεοπάτωρ, add 'of a Roman Emperor, MDAI(A) 75.121.21 (Samos, i AD)'

θεοπειθής, after 'to God' insert 'IG 4².424.5 (iii AD), 551.2'

θεόπλοκος, before 'Cat.Cod.Astr.' insert 'Rhetor. in'

θεόπνοος, add 'perh. of a priest of Apollo at Claros, Laodicée 12c p. 337'

θεοποιός I, add 'maker of statues of gods, Ἐπαφρᾶς θεοποιὸς ἐποί[ησε] SEG 32.1381 (Paphos, ii BC)'

θεοπόλος, delete the entry.

θεοπροπέω, for 'prophesy' read 'explain the will of the gods' II, add 'Ephes. 4(3) p. 294 no. 62 (θαιο-)'

θεοπροπία, for 'prophecy' read 'statement of the will of the gods'

θεοπρόπος I 1, line 1, for 'prophetic' read 'divining'; delete 'ἔπος .. lyr.' add 'b τοὖπος τὸ θ. the oracle, S.Tr. 822 (lyr.); πολλὰ .. θεοπρόπα .. χρησεῖ Call.Lav.Pall. 125.' 2 b, delete the section.

*Θεοπρόσπλοκος, ον, very religious, Ptol.Tetr. 71 (where Proclus paraphrases προσπλεκόμενοι πρὸς θεούς), 155, 159.

*Θεοπρόσπολος, ον, dedicated to the service of God, v.l. for prec. in Ptol.Tetr. 71, 155.

θεός, line 1, Lacon. σιός, after 'v. infra' insert 'and s.v. ‡σιός' and θιός, add 'Arg. θιιδι (dat.) SEG 36.341 (c.550 BC)'; line 2, for 'Inscr.Cypr. 135.27 H.' read 'ICS 217.27, 219 (both fem.), 267.2, al.'; line 5, after 'Τιμόθεε' insert 'ep. gen. and dat. sg. and pl. θεόφιν Il. 17.101, 14.318, al.' I 1 a, add 'κατὰ πρόσταγμα σὺν θεοῖς πᾶσι SEG 35.989 (Cnossus), in dedications, etc., θεοῖς πᾶσιν ib. 25.867 (Telos, both ii/i BC), etc. (cf. Myc. pa-si-te-o-i)' add 'g as a statue, τὸν θεὸν .. ἐκ τῶν ἰδίων ἀνέσ(σ)τησε TAM 4(1).70.' III, line 5, after 'not in Com.' insert 'exc. Men.Pk. 397' at end add 'Myc. te-o'

Θεόδοτος, add 'Θεόζοτος (as pers. n.) SEG 32.610 (Pharsalus, c.350 BC); cf. Θεοζοτίδης SEG 28.46 (Attica, 403/2 BC)'

*Θεοστεφής, ές, crowned by god, Epigr.Gr. 1064.10 (Constantinople, vi AD).

θεοστυγής, for 'hated of the gods' read 'hated by the gods (God)' and add 'S.fr. 269a.22, E.Cyc. 396'

θεοσύλης, after 'sacrilegious' insert 'Alc. 298(a).4 L.-P., Iamb.adesp. 35.10 W.'

*Θεόσυλις, fem. adj. sacrilegious, ῥῖνα θεόσυλιν rest. in Hippon. 118.1 W.

*Θεοταρβής, ές, god-fearing, Suppl.Hell. 991.38 (poet. word-list, iii BC).

θεότης, add 'see also ‡θειότης'

θεοτόκος, add 'SEG 30.1701 (Sinai, v AD). 2 giving birth to gods, γῆ rest. in orac. in TAM 2.174B11 (Sidyma, ii AD).'

*Θεότροπος, ον, godlike in character, of a queen, BMC Bactria p. 43.

θεουδής, at end after '(Q.S.) 3.775' insert 'where the adj., applied to ὄμβρος and νῆσος respectively, perh. = θεσπέσιος'

θεοφάνια, add 'SEG 30.1073 (Chios, perh. c.189/8 BC)'

*Θεοφάνικά, τά, prob. fund for the expenses of the Theophania, IGRom. 4.950 (Chios, i BC/i AD), cf. Robert OMS 1.523 n. 3.

θεοφιλής I, line 4, as honorific epith., delete 'in Egypt' and add 'SEG 31.641 (Thessalonica, AD 361/3), 37.367 (fem. sup., Patrae, vi AD)'; lines 7/8, delete 'θεοφιλές .. Plu. 2.30f'

θεοφίλητος, add 'Socr.Ep. 35'

θεόφιλος, add 'of a person, SEG 32.1442 (Syria, AD 542)'

θεοφορέω, name of play of Menander, add '(OCT ed. Sandbach p. 143)'

θεοφορία, add 'Myc. te-o-po-ri-ja, perh. name of a festival'

θεοφόρος I, add '-φόρος, ὁ, as title of functionary in a Bacchic association, AJA 37.244 (pl.; Latium, ii AD)'

θεοφροσύνη, add 'Euph.in Suppl.Hell. 443.4 (s.v.l.)'

*Θεοφύλακτος, ον, protected by God, τῶν γαληνοτάτων καὶ θεοφυλάκτων ἡμῶν βασιλέων Μαυρικίου καὶ Θεοδοσίου ICilicie 118 (AD 596).

θεόχρηστος, add 'pers. epith., APAW 1925.5, p. 18 no. 2.5 (Cyrene, iii AD)'

†θέπτανος· ἁπτόμενος Hsch. (θεπτάνων La., cf. θεπταίνων· ἁπτόμενος Cyr.; perh. misreading of θειγγάνων = θιγγάνων).

θεράπαινα, add 'also θεράπεινα SEG 31.1387 (Apamea, iv AD)'

θεραπεία II, add '2 treating, ἐγχώριος θ., according to native (Egyptian) custom i.e. embalming, CPR 6.1.14 (AD 125). 3 remedying of a situation, Just.Nov. 4.1, οὐδὲ τοῦτο ἀρκεῖ πρὸς τελειοτάτην τοῦ πράγματος θεραπείαν ib. 69.3.1. b financial compensation, remuneration, id.Edict. 9.2 pr.'

θεράπευμα II 2, add 'b thing to be treated, case for treatment, δύσκολον θ. τὴν ἀδολεσχίαν Plu. 2.502b.'

*Θεραπεύσιμος, ον, repaired, POxy. 3595.36, 3596.33, 3597.33 (all iii AD).

θεραπευτέον II, add '4 one must remedy a legal difficulty, Just.Nov. 4.1.'

*Θεραπευτήριον, τό, wd. of uncertain sense, app. relating to the preparation or supply of wine, περὶ δὲ τῶν οἰναρίων ὑπερεθέμεθα τὰ θ. εἰς τὸ μέλλον POxy.Hels. 50.17 (iii AD).

θεραπευτής I, add 'worshipper of Asclepius, Inscr.Perg. 8(3).71' II, add '3 perh. caulker (cf. °θεραπεύω), θ. ναυπηγός OGIS 674.'

θεραπευτικός II 2, add 'δύναμις θ. Epiph.Const. in Lapid.Gr. 194.8'

θεραπεύω II 10, add 'of embalming, ζῷον Horap. 1.39 S., cf. UPZ 162 ii 22 (ii BC), APF 13.76; also, of some process connected with shipbuilding, perh. caulk, PBeatty Panop. 2.271 (AD 300)'

*Θεραπνεύω, to be a θεράπων, Ἀθηνᾶ 20.216, 217 (Chios).

θεράπνη II, add 'Orph.A. 950, 1208 (pl.); cf. θεράπναι· αὐλῶνες, σταθμοί Hsch.'

θεράπων I, line 9, for 'Cypr.' read 'Cyprus' II, after 'servant' insert '(whether slave or free)' and delete 'in Chios, slave' at end add 'perh. Myc. te-ra-po-ti (dat.) (?)pers. n., cf. te-ra-po-si-jo, pers. n.'

†θερειγενής, ές, produced or coming into being in summer, θερειγενέος .. κυμίνου Nic.Th. 601; of the Nile flood, ὑδάτων Nonn.D. 26.229, Νείλοιο θ. οἴδμα 26.238.

θέρειος II, after 'Plb. 5.13, al.' insert 'Delph. 3(3).237.10 (ii BC)' III, for 'Nic.Th. 460' read 'Nic.Th. 469'

θεριστής, add 'Sosith. 2.21 S.'

*Θέριστον, τό, = θέριστρον, Phot.

θερίστριον, add 'Eub. 101 K.-A., Vit.Aesop.(G) 32'

θερμαντήριος II, after '= θερμαντήρ' insert 'IG 1³.421.96 (Athens, v BC)'

θερμαντικός, add 'II hot, lustful, of a bull, Horap. 1.46 S. (sup.).'

θέρμαστις, for 'perh. .. garment' read 'name of a garment, παρυφὴν ἔχει θέρμαστιν'

*Θερμάστριον, τό, app. some vessel or implement, θ. σιδηροῦν παλαιόν Inscr.Délos 1417Aii58, cf. Hsch. s.v. σχίνδαν.

θερμαστρίς, line 5, after 'Hsch.' insert 'used by a painter, prob. for encaustic work, PCair.Zen. 782(a).50, 61 (iii BC)'

*Θέρμαυστις, εως, ἡ, = θερμαστρίς III, IG 1³.421.97-8 (rest.; Athens, v BC), IG 2².Add. et Corr. 1424a.287.

θερμαύστρα, for 'f.l. .. (q.v.)' read '= θερμάστρα, Call.Del. 144 (codd.), IG 11(2).144B19 (late iv BC)'

*Θέρμαυστρον, τό, or ος, ὁ, portable brazier, Heracleo and Ar. Byz.ap.Hdn.fr. p. 28 H.

θερμέλατος, ον, struck hot, Moses Alch. 304.7.

*Θερμέλη· ἡ θέρμη Suid.; cf. ἕλη.

θέρμη II 1, add 'also sg., Orph.H. 59.3'

θερμημερίαι, add '2 warm days in winter, Diocl.fr. 141 W. (ap.Orib.inc. 40.41).'

θερμηνός, ή, όν, of hot springs, Μητρὶ Θ. εὐχήν Robert Hell. 10.78 (S.E. of Dorylaeum).

θέρμιον, add 'OFlorida 14.12'

*Θέρμιος, ὁ, name of month, Hesperia 27.75 (c.200 BC, unkn. provenance). 2 as title of gods: of Apollo, IG 12(2).104 (Mytilene); also fem. Θερμία of Artemis, ib. 12(2).67.14 (Mytilene), 544 (Methymna).

θερμοδοσία, after 'ἡ' insert 'administering of hot drinks'

θερμοδότις, for 'bath-attendant' read 'one who serves hot food or drink'

*Θερμόλοια, τά, festival at Gortyn, Inscr.Cret. 4.143.3 (Gortyn, iv BC).

θερμόλυχνον, add 'also read as °ἑρμόλυχνον (i.e. τό θ' ἑρμ.)'

θερμομιγής, for 'half-hot' read 'mixed with heat'

θερμοπερίπατος, for pres. ref. read 'IGBulg. 615 (AD 184/5)'

*Θερμοπώλης, ου, ὁ (written -πολ-), keeper of a cook-shop, MAMA 3.165 (Corasium), 719 (Corycus).

θερμός I, add '2 warm, of living creatures, in quot. "still alive" Lxx Je 38(31).2.' II 1, add 'b of a situation, dangerous, τὰ γὰρ πράγματα θερμὰ γε[ίνεται SB 10556.28 (iii AD). c of speech, vehement, Philostr. VS 1.25.10.' III 3, delete the section. 4, add 'hot baths, Thermae, Ἁδριανῶν θ. POxy. 54.14 (iii AD); τῶν μειζόνων θ. ib. 473.5 (ii AD).'

θέρμος, add 'Lyc.fr. 2.10 S.'

θερμοῦθις (to precede θερμουργία), add 'as title of Isis, Bernand Les Portes 24 (i BC), 3 (ii AD)'

*Θερμοφυλάκιον, τό, app. a cooking utensil for keeping food hot, θερμοφυλάκιον ὀλκ(ῆς) λι(τρῶν) ζ΄ PWash.Univ. 59.12 (v AD).

*Θερμοψύχέω, to be hot (in temper) or agitated, POxy. 3860.7 (iv AD).

*Θερμοκαυσώδης, ες, causing summer heat, PMag. 4.1359.

*Θερσιτοκτόνος, ον, killer of Thersites, Ἀχιλλεὺς Θ. title of play by Chaeremon, Stob. 1.6.5 (1.85 W.) (cf. SEG 33.322).

*Θερσόλη, ἡ, unexpld. wd., Arc. 109 B., Theognost.Can. 111.

θέρσος, after 'θάρσος' insert 'Alc. 206.2 L.-P.'

*Θερσύς, v. °θρασύς.

*Θεσίδιον, τό, position (for coffin or sarcophagus), burial-place, ITyr 72, 130, al. (sp. -ιν).

θέσις, line 2, after 'Pi.O. 3.8' insert '= ἡ ποίησις παρὰ Ἀλκαίῳ, EM 319.31' III, line 4, after '(Amorgos)' insert 'οἱ εἰς θέσιν .. δεδομένοι (cf. Lat. in adoptionem dare) Cod.Just. 6.4.4.21' add 'IX *burial-place, grave*, SEG 30.1349 (Miletus, c.AD 180/200), ib. 1479 (Phrygia, before AD 216), etc.'

θέσκελος, delete 'perh. *set in motion by God* (κέλλω) and so'; line 3, after 'Call.*fr.anon*. 385' insert '*divine*, εἶδος, opp. βροτοειδέα μορφήν, Nonn.*D*. 47.718'

θεσμοθέτις, for ' = θεσμοφόρος' read 'fem. of θεσμοθέτης'

θεσμοπόλος, add '*GVAK* 38 (Pisidia, v AD)'

θεσμός, line 2, for 'Locr. τετθμός .. v BC) read 'W.Locr. τεθμός *IG* 9²(1).609.A1 (vi/v BC)' **I 2**, add '**b** *jurisdiction*, ἐν ἁπάσαις ταῖς πόλεσιν ὅσας ὁ ἡμέτερος κατέχει θεσμός Just.*Nov*. 2.5 epilogos, 7 epilogos.' **3**, add '**b** ἐρόεις καὶ ἄλκιμος εἰν ἑνὶ θεσμῷ *in combination, at once*, Nonn.*D*. 29.29 (s.v.l.).' **III**, add '**2** perh. = θήκη 2, *ARW* 10.403 (Cos, iii BC).'

*θεσμοτόαρος, ὁ, *guardian of the law* (functionary in Arcadia), *BCH* 111.169.19 (Mantinea, iv BC).

*θεσμοφοριακός, ή, όν, *belonging to the Thesmophoria*, St.Byz. s.v. Κάλατις.

*θεσμοφοριαστής, οῦ, ὁ, *one who celebrates the Thesmophoria, IG* 12(1).157 (i AD).

Θεσμοφόριος II, month-name, add '*SEG* 30.1637 (Rhodes)'

Θεσμοφοριών, add '*Amyzon* 14 p. 146 (202/1 BC)'

θεσμοφόρος, after 'of Demeter' insert 'Dubois *Dial.arcad*. p. 196 (Pheneos, c.500 BC)'

*θεσμοφυλακέω, act as θεσμοφύλαξ, Petersen-Luschan *Reisen in Lykien* no. 19 *o* 5.

θεσμοφύλαξ, at end, Boeot. τεθμοφούλαξ, add '*SEG* 32.456.21 (Haliartus, c.235/0 BC)'

θεσπέσιος III, after 'θεσπεσίηθεν .. Emp. 96.4' insert '(prob. f.l. for θεσπεσίησιν)'

θεσπίζω I, add '**2** *reveal*, δίκταμον .. ἀθάνατοι τὴν αὐτοὶ θέσπισαν ἡμῖν Poet.*de herb*. 75.' **II**, after 'Ph. 2.38' insert '(s.v.l.)' **III**, add 'of other rulers, in quot., of Lycurgus, w. inf., Sch.B Il. 1.534'

θέσπισμα 2, add 'generally, *ordinance*, τὰ τῶν ἀρχόντων θ. Sch.B Il. 8.12'

*θεσπιστήρ, ῆρος, ὁ, *prophet*, epigr. in *SEG* 15.620.5 (Ostia, iii AD).

*Θεσσαλιώτας· ἐναγι[α]σμός τις παρὰ Λάκωσι, Hsch. (Θεσσαλώπας· ἐναγισμός La.).

*θέστας, α, ὁ, *suppliant*, prob. to be understood in vase inscr. θεστᾶν μνάμων *SEG* 34.966 (nr. Gela, c.350 BC).

θέσφατος I 2, add 'Luc.*Philops*. 38, *Syr.D*. 36'

θετικός III, lines 5/6, for '*addressed* .. *feeling*' read '*more abstract*'

θετός II, add 'perh. also *TAM* 4(1).276'

*θευκόλος, v. ‡θεοκόλος.

θέω (B), after '*IG* 14.1389 ii 24' insert 'cf. χλωραθέω'

*θεώδης, ες, = θεοειδής, Heph.Astr. 3.7.13 (comp.).

θεώρημα II 1 c, add 'Plot. 3.8.4, τὸ θεωροῦν μου θεώρημα ποιεῖ ὥσπερ οἱ γεωμέτραι θεωροῦντες γράφουσιν Plot. 3.8.4, 3.8.6, Iamb.*Protr*. 2 p. 9.12'

θεωρητικός II, add '*Cod.Just*. 11.1.2'

θεωρία, line 2, after 'Dor. θεᾱρία (v. infr.)' insert 'also Arc. *BCH* 111.168.10 (Mantinea, iv BC)' **III 2**, add 'ὁκοίη .. τῶν ἐν γῇ φυομένων θεωρία Hp.*Lex* 3' **3**, add 'Didyma 279a11, 152.10 (θυορ-), 329.12 (θεορ-) (all pl.)'

θεωρικός II, add '**b** *fee paid to* θεωροί, *SEG* 12.372*B*10 (Cos, iii BC).'

θεωρίς I 1, add 'S.*fr*. 765 R.' add '**3** by confusion = θυωρίς (sc. τράπεζα), Poll. 4.123.'

θεωροδοκία, (θεαρο-), add '*SEG* 31.535 (Delphi, c.320 BC)'

θεωρός, lines 11/12, form θεαρός, add 'Ion. inscr. from Andros in *Hesperia* 18.59.A33 (Delphi, 425 BC)'; penultimate line, after 'θιαρός' insert '*Inscr.Cret*. 2.xii 11 (Eleutherna, vi/v BC)'

θεωροσύνη, add 'also **θεωρωσ-** rest. in *SEG* 23.180 (Cleonaea, 145 BC)'

Θῆβαι, line 1, after 'Il. 9.381' insert "Ἐρόōν τὸν ἐν Θέβαις *SEG* 37.283 (Argos, vi BC)'; line 2, after '(Il.) 4.406' insert '*CEG* 787, Call.*Del*. 87, 88 (Egyptian), *GVI* 943 (Demetrias, iii BC), 870 (Tanagra, ii BC); Myc. *te-qa*'

Θηβαϊκός, after 'ή, όν' insert '*of or from the Thebais*' and after 'Hdt. 2.4, etc.' add '*SEG* 33.946.12 (Ephesus, i AD), of wine, *POxy*. 3740.17, 3762.16, 3765.4 (iv AD)'

Θηβαῖος, add 'epith. of Zeus (*of Egyptian Thebes*), Jeffery *LSAG* p. 415 no. 49 (Naucratis, vi BC); cf. Myc. *te-qa-ja*, fem. pers. n., *au-to-te-qa-jo* masc. pers. n.'

θηητός, after 'Tyrt. 10.29' insert 'cj.'

*θηήτωρ, ορος, ὁ, = θηητήρ, Nonn.*D*. 22.57.

θηκαῖος, add 'of a bronze hydria used as an urn for ashes, θηκαία Αὐτονοεία *SEG* 33.472 (Thessaly, v BC)' **II**, add '*IMylasa* 470 (sp. θηκαῖν)'

θήκη 1, after '*chest*' insert 'for money or other things' and after 'X.*Oec*. 8.17, etc.' add 'θ. καλάμων *Edict.Diocl*. in *SEG* 37.335 iii 17 (θήκην καν[νῶν ib. 10.17 L.)' **2**, add '*GVI* 1613 (Attica, ii AD), *TAM* 4(1).354, 374, 376; perh. also masc. θῆκος *SEG* 32.1530' **3**, at beginning insert 'θ. σπάθης *Edict.Diocl*. 7.37'

+**θηκίον**, τό, *chest*, *PFay*. 104.5 (iii AD). **2** *tomb*, *IG* 12(3).1238 (Melos, iii/iv AD), *SEG* 32.317 (Athens, Chr.); cf. Hsch. (both senses).

θηκοποιέω, pass., add '*PHels*. 31.14 (160 BC)'

*θηλαστήριον, τό, *establishment of wet-nurses*, *BGU* 1854.6 (?i AD).

θηλή II, add '**2** part of the θυμιατήριον, Inscr.*Délos* 443*Bb*143 (ii BC).'

θηλητήρ, add 'cf. θηρατήρ'

*Θηλούθιος, ὁ, v. Θειλ-.

*θηλυκάρδιος, ή, sc. λίθος, a precious stone, Plin.*HN* 37.183.

*θηλυκόσωμος, ον, *having the body of a woman*, Ps.-Callisth. 1.5.

θηλυμᾰνέω, add 'sim., of animals, Sch.B Il. 23.295' add '**II** *to be over-luxuriant*, of trees, as etym. of προθέλυμνος, Andromachus in Sch.B Il. 13.130; cf. μάχλος 2a and ‡καθυλομανέω.'

θηλυπρεπής, after 'Chor.*Lyd*. 7' insert 'θηλυπρεποῦς φωτός, perh. Heracles, Didyma 501.9'

*θηλύρριζος, ή, sc. λίθος, a precious stone, Plin.*HN* 37.183.

θῆλυς, line 5, delete 'acc. fem. .. Nic.*Al*. 42' **I 1 c**, add 'n. pl., *females*, τοῖς θήλεσι (opp. τοῖς παιδικοῖσι) E.*Cyc*. 584' **2**, at end for 'murder *by women*' read 'murder *of women*' **II 3**, add 'of stones, Theophr.*Lap*.(*fr*. 5) 30' **III**, for 'Seleuc.ap.Ath. 650d' read 'Seleuc.ap.Ath. 658d'

*θήμνες, gloss on θιάλλαι, Hsch. (but perh. distortion of entry ib. s.v. θίλα).

θημολογέω, add '(prob. corrupt; ἐχμολόγει Meineke)'

+**θημωνία** (θιμ- v.ll. and edd. in Lxx) *heap*, *pile*, Lxx *Ex*. 8.10, *Jb*. 5.26, *Ca*. 7.2, *Si*. 20.28, *Ze*. 2.9, etc., Aq., Th.*Jb*. 21.32, Eust. 1539.16; also **θειμωνειαί** and **θημονιά** Hsch. (-μων- Schmidt).

*θημωνιάζομαι, *to be heaped up*, Al.*Ex*. 15.8.

θην, add 'Pi.*fr*. 203.1 S.-M., A.R. 2.915, Cerc. 4.35'

θήνιον, add 'prob. in *Inscr.Cret*. 1.xvii 18 (Lebena, i BC)'

θήρ, line 1, after 'also ἥ' insert 'Nic.*Th*. 814' **2**, line 3, after 'Ar.*Au* 1064 (lyr.)' insert 'of a centipede, Nic. l.c.'

θήρα, add '**V** *snare* (transl. Hebr. *rešet* "net"), Lxx *Ps*. 34(35).8, *Ep.Rom*. 11.9.'

θηραγρέτης, delete 'E.*Ba*. 1020 (lyr., s.v.l.)'

*θηραγρευτής, οῦ, ὁ, *hunter*, cj. in E.*Ba*. 1020 (cod. θηραγρότ-).

Θηραϊκόν or **Θήραιον**, add '*IG* 2².1415.25, 1421.126, al. (iv BC, Θήραιον or Θήραια)'

θήραμα, add '**2** *snare*, E.*IA* 963.'

θηρατήρ, add 'see also θηλητήρ'

θήρατρον, line 3, after 'Max.Tyr. 16.5' insert 'of the tentacles of an octopus, Ael.*VH* 1.1'

*θήραφος, ὁ, *spider*, Cyran. 62 (2.16.1 K.).

θηράω I 2, line 4, after 'A.*Ag*. 1194' insert '(cj.)'

θήρειος I, add '**2** = θηριακός, φάρμακα Aristid.*Or*. 48(24).64.'

*θηρέστατος, η, ον, perh. *most devoted to hunting*, ὁ γὰρ θηρέστατος ἦεν *Suppl.Hell*. 970 i 20.

θηρευτικός 2, add 'ἄλλαι πᾶσαι ἐπιστῆμαι θηρευτικαί τινές εἰσι .. τῶν ἀγαθῶν Iamb.*Protr*. 5 p. 27.3'

θηρεύω II 2, at end after 'Phld.*Rh*. 2.5 S., al.' insert 'θ. ἀκοήν, of birds, *Lyr.adesp*. 118.2 P.'

θηρίδιον, for '*animalculae*' read '*animalcules*'

*Θηρικλείδιον, τό, (*small*) *vessel made by Thericles* or *in his style*, Θ. οὐκ ἔχον οὔτε ὦτα οὔτε πυθμένα Inscr.*Délos* 1450*A*137 (ii BC).

Θηρίκλειος, add 'also sp. Θυρεικλείος *Kafizin* 40, 41, 46'

θηριοδεῖκται, for '*exhibitors of wild beasts*' read 'perh. *snake-charmers*'

θηριοδήκτης, delete the entry.

*θηριοδιώκτης, Marsus, Gloss.

*θηριομάχιον, τό, *contest with wild beasts*, *TAM* 2.508.12 (Pinara, pl.).

θηρίον I 2, add 'of snakes, *Act.Ap*. 28.4, of vermin, etc., Dsc. 1.75' **3 b**, delete the section. **IV**, add 'also, any of constellations symbolized by animals, v.l. in Lxx *Je*. 10.2' at end add 'see also °σηρίον'

θηριοτρόφος I, after '*keeping wild beasts*' insert 'Ptol.*Tetr*. 179'

θηριώδης II 2, add 'Hp.*VM* 7, Critias 19.2 S., D.S. 1.8.1; adv. Isoc. 3.6, 4.28'

*θηριωνυμία, ή, *name derived from a wild beast*, *Orac.Tiburt*. 143.

θηριώνυμος, add '*Orac.Tiburt*. 136'

θηρόβοτος, add 'θηροβότου .. κολώνης Nonn.*D*. 16.259; θηρόβοτος, ή, *beast-haunted wilderness*, Phalar.*Ep*. 147.4'

*θηροδίαιτος, ον, *living like wild beasts*, Didyma 496*B*6 (ii AD).

+**θηροκόμος**, ὁ, *keeper of a wild animal*, Hld. 10.27.3.

θηροκτόνος, add 'see also σηροκτόνος'

θηρομαχία, delete '*IGRom*. 3.631 (Xanthus)'

*θηρομάχια, τά, *wild beast fights*, *TAM* 2(2).287.14.

θηρόπεπλος, for 'Cerc. 10 .. Timae. 80' read 'Stratonic.ap.Timae. 16 J.'

θής I, add '**4** *temple servant*, prob. in Call.*fr*. 186.14 Pf.'

θησαυρός II 1, add 'of the imperial treasury, ὁ θησαυρὸς τοῦ κυρίου SEG 34.1306.12 (Perge, iii AD)'

θησαυροφύλαξ, transfer 'Vett.Val. 85.23' to section II.

Θησεύς, add 'Myc. *te-se-u*, pers. n.'

*θίας, α, ὁ, Rhodian = θεῖος (B), *Lindos* 198.7 (ii BC), al., *Clara Rhodos* 2.193 (ii BC), al.

θιᾰσῑτικός, add '*PEnteux.* 20.5, 9 (iii BC)'

θίᾰσος I, add 'Alcm. 98 P.'

*θῐβοῦχος, ὁ, *basket-carrier*, Swoboda *Denkmäler* 217 (Palaia Isaura, s.v.l.).

θῐβρός II, after 'Euph. 81' insert 'cf. θίρρον· τὸ τρυφερόν, Theognost.*Can.* 15.20'

θῑγάνα, for '*Schwyzer* .. (Delph.)' read '*CID* I 9.639 (iv BC)'

θιγγάνω I 1, w. acc., add 'Alcm. 58.2 P., S.*Ant.* 546, Pherecr. 10.4 K.-A.' add '**4** w. dat., *be adjacent to*, τὰ θιγγάνοντα τῷ ἱαρῷ τεμένια *REA* 44.35.45 (Olus, ii BC).'

θικέλιον, after lemma insert '(**θίκελιν** La.)'

*θιμωνιά, ἡ, v. °θημωνιά.

*θῖος, v. ‡θεῖος (B).

θίς I 2, add '**c** *hill*, Lxx *Ge.* 49.26, *De.* 12.2, *Jb.* 15.7.'

θλίβω II, lines 5/6, delete 'θλιβομένα .. Theoc. 21.18'

θλῖψις 3, add 'from famine, τὰν τῶν ἰδίων συνπολιτᾶν θλῖψιν SEG 26.1817.38 (Cyrenaica, ii/i BC)'

θνῆσις, add 'προβάτων καὶ βοῶν θνῆσις γενήσεται *Orac.Tiburt.* 125'

θνητογᾰμία, add 'Sch.Il. 1.5'

*θνητόγονος, ον, = θνητογενής, epigr. in *SPAW* 1932.862.

θνητός 1, after 'Hdt. 1.216, 2.68' insert 'παντοῖα .. εἴδεα θνητῶν Emp. 115.7' add '**b** τὰ τνᾱτά (= θνητά), *property in cattle, animals*, *Leg.Gort.* 5.39 (v BC), etc., v. °ἀθάνατος.'

θοάζω (C), delete the entry.

*θοάω, v. °θωάζω.

*θοιναρμοστρέω (?), *serve as θοιναρμόστρια*, aor. part. in Lacon. form σειναρμοστρῄάά perh. to be read in *IG* 5(1).229.2 (i BC/i AD), cf. *ABSA* 45.266, n. 13.

θοιναρμόστρια, after 'σειναρμόστρηα ib. 229' add '(dub., v. °θοιναρμοστρέω)'

*θοινᾱτᾱς, ᾶ, ὁ, = θοινάτωρ, τᾶς Δάματρος *Dacia* 3/4.451 (Callatis, i BC, pl.); also **θοινητής**, οῦ, SEG 24.975 (sp. θοινειτ-, Moesia, ii AD).

θοινητής, v. °θοινατᾶς.

*θολέρησις, εως, ἡ, *turbidity*, cj. in Plu. 2.383d.

θολερός II, after 'adv. -ρῶς' insert 'Man. 6.178'

*θολίδιον, τό, dim. of θόλος, *IG* 2².1534.280 (Athens, iii BC).

*θόλιος, α, ον, *of the Tholos* (at Athens), *Hesperia* suppl. 4.145 (ii BC).

*θολοποιέω, *make muddy*, τὸ ὕδωρ Aesop. p. 542 P.

θόλος, add '(θόλος as wd. of two genders, S.E.*M.* 1.148; for masc. v. sense II)' **II 2**, add 'Gal. 18(1).777.5; also fem. in this sense SEG 12.1503 (Palestine, c.AD 455)'

θοός (A), line 2, after 'etc.' insert 'Βιττίδα .. θοήν Hermesian. 7.77'; line 12, after 'motion' insert 'Ἄϊδος' and for 'Antim. 71' read 'Antim. 187.2 W.'; add 'comp. θανάτοιο θοώτερος ἵξεται αἶσα Nic.*Th.* 120' add '**II** interpreted as "*sharp*" by Hellen. writers (cf. Str. 8.3.26 w. ref. to Od. 15.299, where it may be a proper name) θοοῖς .. γόμφοις A.R. 2.79, θοῶν .. ὀδόντων 3.1281, θοοῖς πελέκεσσιν 4.1683, *AP* 9.157, Q.S. 4.417, 14.305.' at end add 'in Myc. cpd. *pe-ri-to-wo*, pers. n. = Πειρίθοος'

θοός (B), delete the entry (v. °θοός (A)).

†θοόω, perh. *urge on*, Od. 9.327. **II** *excite*, ἐν πυρὶ .. φωνὴν τεθοωμένως, of Cerberus, Hermesian. 7.11, Nic.*Th.* 228, λύσσῃ τεθοωμένος Opp.*H.* 1.557, 2.525.

θοράνας, after lemma insert '(**θόρανδε** La.)'

θορή, add 'Aret.*SD* 2.5.1 (71.11 H.), al.'

*θορηνεύς· ὁ ξιφίας ἰχθύς Hsch. (cod. θορνευῦσαι).

*θόρνη, ἡ, perh. *mating*, *PDerveni* xvii 1 (*ZPE* 47.*10; late iv BC).

θόρνυμαι, after 'Nic.*Th.* 130' insert 'thematic forms, θορνύεται Theognost.*Can.* 46 A.'; add 'act. part. θορνύς *mating*, Nic.*Th.* 99.'

θορῠβέω II, after '(Th.) 6.61' insert '*Cod.Just.* 1.3.52.9'

*θουννόκειτ(?ος), v. θυννόκητος.

θουράω, for '*leap upon*, c. acc.' read '*leap impetuously* upon, ἐπ' ἀρσένων .. λέκτρα θουρῶσαι βροτῶν'

θούριος, add 'epith. of Ares, SEG 31.1285.7 (Pisidia, Rom.imp.)'

*θράγανα, τά, perh. *mortar and pestle*, θ. διπλόα SEG 24.361.16 (Thespiae, 386/0 BC).

θραγμός, for '*crackling*' read 'perh. *ground flour*'; for 'cf. θραύω' read 'cf. °θρωγμός, perh. also °θραττεύομαι'

*Θρακαρχέω, *to hold the office of* °Θρακάρχης, *IGBulg.* 1183.

*Θρακάρχης, ου, ὁ, *president of the provincial council of Thrace*, *IGBulg.* 1170, 1559, al., SEG 31.677.B6 (Thrace, ii/iii AD).

Θράκιος, add 'Θρεῖκιος Hippon. 72.5, 127 W.'

θρᾱνίδιον, add '*IG* 1³.421.140 (Athens, v BC)'

†θρᾶνις, ὁ, *sword-fish*, = ξιφίας, τὸν θράνιν ἑλόντε *AE* 1937(3).833ff. (Eleusis, v BC); Xenocr. 8; cf. θρανίας.

θρᾱνίτης I, transfer 'Ar.*Ach.* 162' to section II inserting before it 'ὁ θρανίτης λεώς'

θρᾱνογράφος, for '= τοιχογράφος' read 'app. *writer of graffiti*'

θρᾶνος I 1, add 'defined as *tanning-bench*, Sch.rec.Ar.*Eq.* 369' **II 1**, add 'S.*fr.* 269a.41; *roof-beam*, ἀπὸ θράνω λέλακα γλαῦξ Alcm. 1.86 P.' **2**, before 'θ. ποικίλος' insert 'in the προστάς of a house'

Θρᾷξ, for 'Call.*Aet.Oxy.* 2079.13' read 'Call.*fr.* 1.13 Pf., al.' and for 'Ion. dat. pl. .. Archil.*Supp.* 4.48' read 'Θρέϊξ rest. in Archil. 42.1, 93a.6 W.'

*θρᾰσύαιγις, ιδος, perh. *headlong, impetuous* or sim., *Lyr.adesp.* 7(*e*).12 P.

*θρᾰσύδης, ες, *bold*, *PMich.*III 149 xi 23 (ii AD).

θρᾰσυμήχᾰνος, add 'Ἄρης cj. in Simon. 70.2 P.; cf. αἰγίς II, ἐπαιγίζω'

θρᾰσυπτόλεμος, add 'epigr. in *IUrb.Rom.* 69 (iii/iv AD)'

θρᾰσύς, line 1, after 'Philem. 20 (s.v.l.)' insert 'Aeol. **θερσύς** (2 termin.), gen. ύος *IGC* p. 11 l. 24 (*c.*iii BC; cf. °Θροσία)' **1**, line 2, after 'Il. 8.89, etc.' insert 'as epith. of Athena, Pi.*N.* 3.50, *IGC* l.c.'

*θρᾰσῠτολμία, ἡ, *effrontery*, Act.Alexandr. 10.34 (ii/iii AD).

θρᾰσύφωνος, add '*Suppl.Hell.* 986.4'

Θρᾷττα, add '*IGC* p. 99 B3 (Acraephia, iii/ii BC)'

*θραττεύομαι συντρίβομαι, συγκόπτομαι Hsch. (La., s.v.l.; cf. θράσσω).

θραυμάτιον, τό, dim. of θραῦμα (in quot., sense I): pl., ἀργυρᾶ *Inscr.Délos* 1450*A*115 (ii BC).

θραῦσμα II, for 'in leprosy, *scab*' read '*break in the skin, sore, ulcer*'

θραύστης, delete the article (read pers. n. [Τι]θραύστης).

*θραυστικός, ή, όν, *that breaks up*, θραυστικὰ .. τῶν λίθων Aët. 3.152 (324.13 O.).

θραύω, line 4, after 'Pl.*Lg.* 757e (v.l. -τεθραυσμένον)' insert 'cf. ἀποτεθραυμένην *Inscr.Délos* 1450*A*33, but περιτεθραυσμένην ib. 35 (ii BC)'

*θρειστίον, τό, = °θρισσίον, *PWisc.* 6 (AD 210/1).

θρέμμα, add '**II** *nourishment*, Pl.*Plt.* 289b.'

θρέομαι, line 1, for 'only in pres.' read 'usu. pres.' and add 'impf. θρεύετο poet. in *IG* 4².616.4 (Epid., iv BC)'; after 'women' for 'θρέομαι' read 'θρεύμαι'

θρεπτάριον, add '*INikaia* 1376.9 (Rom.imp.), *BCH* suppl. 8.62 (Beroea, iv/v AD)'

θρεπτήριος III 2, after 'Hes.*Op.* 188' insert 'A.R. 1.283'

θρεπτός I, for 'as Subst. .. Lxx *Es.* 2.7' read '*brought up in another household, fostered*, παῖς θρεπτή Lxx *Es.* 2.7. **2** masc., fem. as subst., *domestically raised slave* (= Lat. *verna*), Lys.*fr.* 215 S., Pherecr. 130 K.-A.'; after 'Plin.*Ep. ad Traj.* 65, etc.' add 'fem., SEG 37.453.11 (Thessaly, ii/iii AD); also neut., (?sc. παιδίον), SEG 31.1004.9 (Lydia, AD 101/2); fig., *fosterling* (of the Nile), describing Sosibius, a Greek born in Egypt, Call.*fr.* 384.28 Pf.' add '**3** cult-name of Triptolemus, Sokolowski 3.10*A*69 (Athens, *c.*400 BC).'

θρεσκός, add 'cf. φιλόθρεσκος'

*θρηναύλης, ου, ὁ, *player of aulos for dirges*, *PSAA* 43ᵛ i 13 (ii AD).

θρηνήτρια, add 'SEG 8.621.18 (Egypt, ii AD)'

*θρηνία, ἡ, ἐνταῦθα δὲ τὴν ἀληθῆ μέλισσαν λέγει, ἣν ἔνιοι θρηνίαν φασὶ καὶ πληθυντικῶς θρήνια Sch.Nic.*Al.* 547b G.; cf. perh. τενθρήνη.

*θρηνικόν, (sc. μέτρον), τό, *anapaestic monometer*, Serv. in *Gramm.Lat.* 4.461.31 K.

*θρηνοτόκος, ον, *engendering lament*, μολπὴ *GVI* 1244 (Athens, ii/iii AD; less prob. to be read as θρηνοτόκος *born of lament*).

θρῆνυς, add 'form *θρανυς, Myc. *ta-ra-nu*'

θρησκεία 1, add 'in honour of the dead, τὴν τοῦ ἥρωος θ. *SAWW* 265(1).12.51 (Nakrason, i BC/i AD), *TAM* 2.247 (AD 146), al.'

θρησκώδης, add 'sup. [θρη]σκωδέστατα rest. in *IG* 2².1074.15 (ii AD), cf. *AJP* 70.300'

θρῖαι II, add 'Call.*fr.* 260.50 Pf.'

θρίαμβος I 1, after 'Cratin. 36' add '*AP* 13.6 (Phal.)'

θριγκός, at end after 'θριγγός v.l. in Plu. l.c.' insert 'Longus 4.2'; for '*SIG* 1231.6' read 'τριχὸς *TAM* 4(1).276.6'

*θρίγκωμα, for 'cj. .. 15.11.3' read 'J.*AJ* 15.11.3, cj. for τριχώματα in E.*IT* 73'

θρῐδᾰκηΐς, for '*of the lettuce*' read '*lettuce-like*'

θρῖδαξ, line 4, after 'θίδραξ' insert 'Arr.*Epict.* 2.10.9, 3.24.44' and after 's.v. θιδρακίνη' insert '**θύδραξ**, *PRyl.* 627.151, 629.233'

†θρίζω, aor. ἔθρισα, v. θερίζω.

θρίνᾰκη, for 'Call.*fr.* 46 P.' read '(?)Call.*fr.* 799 Pf.' and for 'θρῖναξ ἦν' read 'θρίναχ' ἦν'

Θρῑνάκιος, add '*AP* 7.714'

θρίξ, add '**IV** (pl.) kind of seaweed, μνία καλεῖταί τινα .. καὶ τρίχες Ael.*NA* 13.3.'

*θρισσίον, τό, dim. of θρίσσα, *POxy.* 1923.9 (v/vi AD).

*θρίσσος (B), ὁ, Thessalian name for kind of snake, Afric.*Cest.* p. 16 V.

θρίψ II, add 'cf. perh. pers. n. Θριφόνδας SEG 2.192 (Tanagra, vi BC; v. MH 43.256)'

θροέω, line 7, after '(troch.)' insert 'of the wind, ἁδύ τοι ἐν χλωροῖς πνεῦμα θροεῖ πετάλοις APL 228 (Anyt.)' **II**, for 'causal' read 'trans., *boo* a speaker, ἐν ἐκκλησίᾳ -ούμενος Gnomol.Vat. in WS 10.224' and before 'scare' insert '2'

θρόμβωσις 2, add 'Gal. 18(2).446.14'

θρονισμός, after 'enthronement' insert '(in mystery rites)'

*θρονιστήριον, τό, app. *place for a throne*, Ps.-Callisth. 38.1 (v.l.).

*θρονοθήκη, ἡ, *box of herbs* or *drugs*, Babylonian Talmud, Kelim ch. 13 p. 42 (in Hebr. letters).

*θρονόμαντις, εως, ὁ, *diviner by θρόνα, magic herbs*, APAW 1943(14).8 (Chalcis, iii AD).

θρόνος I 1, add 'of an honorific seat in the theatre, SEG 30.82.7 (Athens, c.AD 230)' **2**, add 'θ. ἀνθυπάτων = Lat. *sella curulis*, SEG 33.940 (verse, Ephesus, ?v AD). **b** transf., for the occupant of a seat of authority, *Cod.Just.* 1.3.52.11, *Just.Nov.* 82.1.1.' at end add 'Myc. *to-no* (prob. in sense I 2 = *θόρνος; in compd. also θρονο-, *to-ro-no-wo-ko* = °θρονοϜοργοί)'

θρόνωσις, add 'cj. in Orac.Sib. 8.49'

*Θροσία, ἡ, epith. of Artemis, perh. = θρασεῖα, Πολέμων 1.249 (Larissa), SEG 34.481 (Atrax, c.150 BC).

θρυαρίς, add 'θρυγανίς· ψίαθος La.)'

θρύϊνος, add 'PLond. 122.103 (iv AD)'

θρύμμα, add 'θρύμματα· κλάσματα ἄρτου Hsch.'

θρυμματίς, add 'cf. θρυματίς· κρηπίς Theognost.Can. 20'

*θρυοκόπος, ὁ (or -κοπεύς, έως, ὁ), *cutter of rushes*, POxy. 2243a84 (vi AD).

*θρυοτιλτής, οῦ, ὁ, *rush-gatherer*, PHarris 97.3 (iv AD).

*θρύπτειρα, fem. adj., *that breaks up and disperses*, θ. κονίη, *lye*, Nic.Al. 370 (v.l. ῥύπτ-).

*θρυπτεύεται· ὑπερηφανεύεται Hsch. (La.; cod. θρημνεύεται).

θρύπτω II 2 b, delete the section. **c**, delete '*bridle up*' and add 'ὄμματι θρυπτομένῳ *with a look of feigned reluctance*, AP 5.287.8 (Agath.)'

θρύσκα, add '(due to wrong division of ἄν/θρυσκα Sapph. 96.13 L.-P.)'

†**θρύψιχος**, prob. *effeminate*, θ.· τρυφερός Hsch.; θ.· φοβερός Theognost.Can. 20.

*θρωγμός· τρίβος Theognost.Can. 20.

*θρωμός· ᾠᾶ δικτύου Theognost.Can. 20.

θρώσσει, add '= ἄλλεται, Theognost.Can. 20; cf. θρῴσκω'

θυαλόν, after lemma insert '(θυαλοῦν La.; cf. θυηλέομαι)'

θυάω, add 'θυᾶν· καπρᾶν. ἐπὶ ὑός Hsch.'

θυγάτηρ, line 1, insert 'Lacon. συγάτηρ SEG 11.677a, c'; at end add 'Myc. *tu-ka-te*'

θυγατριδοῦς, after 'OGI 529.23 (Sebastopolis)' insert 'τοῦ -οῦς ABSA 45.277 (Laconia, i/ii AD)'

θυγατροποιία, add 'SEG 31.937 (Caria, iii/ii BC)'

*θύδραξ, v. °θρίδαξ.

θυεῖον, add 'perh. also θῖον POxy. 3354.15 (AD 257)'

*θυή, ἡ, *burnt sacrifice*, Philoch. 194 J.

*θυηδόκος, ον, = θυοδόκος, Antip.Sid. in Inscr.Délos 2549.3, v. Hermes 76.411.

θυηκόος, add 'IGRom. 3.73 (Claudiopolis)'

θυηπολία, after 'Ion. -ίη' insert 'dial. θυᾱ- SEG 23.639 (Paphos, iv/iii BC), BCH 10.424 (Cedreae)'

θυηπολικός, add 'Θυηπολικόν, title of work attributed to Orpheus, Suid. s.v. Ὀρφεύς'

θυιάς, line 2, for 'θύω' read 'θύω (B)' **I**, add '**b** pl., *nymphs associated with Dionysus' revels*, Alcm. 63 P.'

θυίω II, after 'Hes.Th. 131 (pap.)' insert 'Anacr. 2.17 P.'

θυλάκιον I, add 'of money-bag, Lxx To. 9.5'

θύλακις, delete 'Nic.Th. 852' add '**II** adj., = θυλακῖτις, μήκων θ. Nic.Th. 852.'

θύλακος, line 1, before 'Hdt. 3.46' insert 'Epich. 113'

θῦμα, last line, before 'Supp.Epigr.' insert 'Babr. 97.12'

θυμαίνω, after 'Ar.Nu. 610' insert 'Call.fr. 24.2 Pf.'; add 'med., θυμαίνεται· (θυμαίνει La.) ὀργίζεται Hsch.'

θύμαλλος, for 'an unknown fish' read 'a fish, perh. *grayling*'

θυμάλωψ, add '**2** οἱ δὲ καλούμενοι νῦν τῶν ἀμπέλων ἐπίτραγοι θυμάλωπες ἐλέγοντο Poll. 7.152.'

θυμαρής, line 3, after '(Od.) 17.199' insert 'δαῖτας θ. Call.Cer. 55'; line 6, after 'A.R. 1.705' insert '(cj. θυμηδές)'

θυμέλη II d, add 'ἐὶν θυμέλαισι κλυταῖς epigr. in SEG 35.1427.8 (Side, iii AD)'

θυμηδής, add 'Q.S. 14.312, 340'

*θυμιαντήριον, v. °θυμιατήριον.

*θυμίασμα, ατος, τό, = θυμίαμα, *incense*, PMag. 4.2575, 2643.

*θυμιατηρίδιον, τό, dim. of θυμιατήριον (in quot., sense I 1), Inscr.Délos 1416A̓i34 (ii BC).

θυμιατήριον, after 'θυμιητ-' insert 'also θυμιαντήριον EAM 104' and for 'censer' read '*incense-burner*' and add 'Kafizin 302(b) (iii BC), EA 13 p. 7 no. 895 (ii BC)'

θυμίατρον, add 'IEphes. 1004, al.'

θυμιάω I 2, add 'βωμοὶ δὲ τεθυμιάμενοι λιβανώτῳ cj. in Sapph. 2.3/4 L.-P. (δεμιθυμ- ostr.)'

θυμοειδής 2 a, add 'Hp.Aër. 16'

θυμόεις, add 'Call.fr. 238.23 Pf.'

*θυμοιδής, ές, *irascible*, A.fr. 281a.32 R. (cj.).

*θυμοκόρυμβος, ὁ, *spike of thyme*, PVindob. G10734 (Analect.Pap. 3.121, vi AD).

θύμον, line 3, after 'θύμος, ὁ' insert 'Nic.fr. 92'; for 'Cretan thyme, *Thymbra capitata*' read 'some kind of thyme or sim. plant' **2**, delete the section transferring exx. to section 1.

θυμός I 1, add '**b** as exhaled upon something, Κυρίου Lxx Is. 30.33; as the vehicle of snakes' venom, ib. De. 32.33 (bis), Am. 6.12.' add '**4** *membrum virile*, Hippon. 10 W. (but cf. Hdn.Gr. 1.169 θυμὸς δὲ τὸ μόριον ἢ ἡ βοτάνη and θύμος (B)).'

θυμοφθόρος, after 'Nic.Th. 140' for '(v.l. γυιοφθ-)' read '(v.l., v. °γυιοφθ-)'

θύννα, for 'f.l. in Hippon. 35.2' read 'Hippon. 26.2 W.'

θυννάς, add 'θυννάδες· τεμάχη ταρίχου Hsch. (cf. °θυννίς 2)'

θυννίς, line 2, after 'ἡ' insert 'θυνίδω[ν] (gen. pl.) SEG 23.326.15 (Delphi, iii BC), θυννίδων IGC p. 99 B7 (Acraephia, iii/ii BC)' and delete 'prob. .. Hippon. 35.2' add '**2** θυννίδες· θύννων τεμάχη, ὑποκοριστικῶς Hsch.'

*θυννόκητος, ὁ, *name of a sea-fish*, IGC p. 98 B4 (sp. θουννόκειτ-, Acraephia, iii/ii BC).

θύννος, add 'cf. θύννον· τὸν ὄρκυνον λέγουσι· τὴν δὲ πηλαμίδα θυννίδα Hsch.'

†**θυννώδης**, ες, *like a tunny*, εἶδος θ. ἰχθύος Hsch. s.v. πρημάδες (v. ‡πρημνάς); *typical of a tunny*, ἄπαγε, θυννῶδες τὸ ἐνθύμημα (i.e. stupid) Luc.JTr. 25.

θύος I, for 'burnt sacrifice' read 'a substance producing a fragrant smell when burnt, incense, or sim.' and add 'Hp.ap.Gal. 19.104' **2**, for this section read '*fragrant oil*, Nic.Al. 203, 452' at end add 'Myc. *tu-wo*, aromatic substance'

*θύρ, θυρός, ὁ, θύρ· πτηνόν Cyran. 22 (perh. misspelling of θήρ).

θύρα I 1, add 'applied to symbolic door on tomb, Mitchell N.Galatia 242, MAMA 7.323, IPhrygie 4.24, 32' **3**, line 7, after 'S.E.M. 1.43' insert 'cf. παρὰ θύρας ἀπαντᾶν answer beside the point, Olymp. in Grg. 23.2, 26.11, 25 W.' add '**10** *leaf* of a writing tablet, Poll. 4.18.'

θυραῖος I 4, add 'ἀρετὴ .. οὐκ ἐκ -ων τἀπίχειρα λαμβάνει E.fr. 908a'

†**θύρεθρα**, τά, = θύρετρα, Maiist. 28, Hsch.

θυρεός I, for '*stone* .. shut' read '*stone used to block an entrance*' **II**, line 2, after 'Callix. 2' insert 'AP 6.129 (Leon.Tarent.)'

θυρίδιον, add '**2** *niche*, POxy. 2058.24 (vi AD).'

θυριδωτός, after 'κιβωτός' insert 'IG 1³.425.19 (Athens, v BC)'

θύριον, add 'ἀνέστησε θύριν μν[ήμης χάριν Mitchell N.Galatia 245'

θυρίς I, add '**4** *wall niche*, SB 7574 (ii AD), PRoss.-Georg. III 1 (iii AD).'

θυρξεύς, for 'Achaea' read 'Lycia'

*θυροκρουστέω, app. *knock on a door*, PMich.III 149.18.9.

*θύρος, ὁ, kind of fish, Marc.Sid. 10 (s.v.l.).

*θυρουρικός, v. °θυρωρ-.

*θύρσις, ιδος or εως, ἡ, = °βάκερα, Cyran. 22 (1.8.1 K.).

θυρσίτης, add '**2** θ. λίθος *stone resembling coral*, Cyran. 22 (1.8.1 K.).'

*θυρσοκλόνος, ὁ, = θυρσοτινάκτης, of Dionysus, GVAThess. 29; applied to Egyptian god, prob. identified w. Dionysus, APAW 1943(14).8 (Chalcis, iii AD).

θυρσοκόμος, after 'keeper' insert 'Διόνυσος θ. Ps.-Callisth. 7.9'

θυρών, add '**2** pl., perh. *tablets made to resemble doors* for stage purposes, PFouad I Univ. 14; cf. θυρώνας· τὰς σανίδας. καὶ τὰς εἰσόδους Hsch.'

*θυρωρικός, ή, όν, θυρουρ-, *of a door-keeper*, ἄρτος POxy. 1890.11 (vi AD). **II** subst., θυρουρικόν, τό, *porter's lodge*, SB 9898.9 (c.AD 220), PMich.XI 620.9 (AD 239/40).

θυρωρός, for 'Cypr. .. 215 H.' read 'Cypr. *tu-ra-wo-?* perh. θυρά-ϝο[ρος] ICS 417'; add 'as ephebic rank, SEG 29.152 (Athens, AD 175/6)'

θυσία I, line 2, before 'v.l. in Batr.' insert 'h.Cer. 312, 368'

θυσιαστήριον, add '**2** *sanctuary*, AD 12.27, 69; IHadr. 121.'

θυσσός, add 'BCH 8.49 (Maced.); pl., members of a cult-society, IGRom. 1.832 (Abdera)'

*θύσιν, τό, v. °θύσιον.

*θύσιον, τό, perh. *censer*, CPR 8.66.5 (vi AD).

*θύσκον, τό, object for religious use, CPR 8.66.8, 15 (vi AD); cf. perh. θυΐσκος.

θύσσομαι, add '(perh. due to wrong division of ἐσκίαστ' αἰθυσσομένων Sapph. 2.7 L.-P., but cf. Sch.Pi.P. 4.411)'

θύτης, after 'or *diviner*' insert 'Call.*fr.* 194.25 Pf.'

***θῦτόν**, τό, *sacrificial meat*, Hsch. s.v. θυαλόν.

θύω (A), line 14, after 'Pi.*O.* 13.69' insert 'θύεται [ῡ] id.*Pae.* 6.62 S.'; line 15, delete '*Cyc.* 334'

***Θυωνίδας**· ὁ Διόνυσος παρὰ Ῥοδίοις. τοὺς συκίνους φάλητας Hsch.; cf. Θυωναῖος.

***Θυωνοφόρος**, ὁ, religious official of uncertain function, ἐλθόντι ἐπὶ τὸ Καπετώλιον μόνῳ ἄνευ τοῦ θυωνοφόρου *SEG* 29.807 (Chalcis, Rom.imp.).

θυωρός, for '*BCH* 11.161' read 'epigr. in *SEG* 30.1272.7'

θωάζω, for '*pay the penalty*' read '*penalize, fine*', deleting this in line 2; after '*IG* 1².4.7, 12' insert 'also **θωιάω** *Thasos* 141.6 (v/iv BC)'; line 3, after '*Michel* 995 D 19' add '(v BC)'; add '(fut. θοάσει *IG* 2².1362.14 may be copied fr. an earlier version w. o = ω)'

θωή, line 3, for 'Ion. also' read 'Arg. **θωϊά** *IG* 4.555; Ion.'; after 'Archil. 109' insert '= Call.*fr.* 195.22 Pf. (θωίη pap.; cf. ἀθῷος)'

***θωπάζω**, app. = θωπεύω, scholiast in *JThS* 47(1946).70.

***θωπεῖον**, τό, a nocturnal bird, Cyran. 89 (3.16.1 K.).

θώπτω, add 'θώπτει· σκώπτει. θεραπεύει Hsch.'

***θωρᾱκαῖος**, a, ον, equipped with a θώραξ (sense I 1), [λιβανωτίδα ἔχουσαν νίκη]ν ἐπὶ τοῦ πώματος θωρακαίαν *Inscr.Délos* 1417*B*ii53 (ii BC).

θωρᾱκεῖον I 1, add 'supporting a sarcophagus, *MAMA* 8.556b'

θώραξ II, add '**c** pl., *breast* of a chicken, Nic.*Al.* 388.'   **III**, add 'Archimel. in *Suppl.Hell.* 202.10 (pl.)'   at end add 'Myc. *to-ra-ke* (pl.)'

θώρηξις, add 'cf. θώρηξις· οἰνοποσία. καθόπλισις Hsch.'

θῶσθαι, line 1, for 'A.*fr.* 49' read 'A.*fr.* 47a.2.20 R.'

I

ἴ, for 'Inscr.Cypr. 135.24 H.' read 'ICS 217.34'
ἰά, add 'E.Hipp. 585'
ἰαίνω, line 4, after 'later poets' insert 'Call.fr. 80.8 Pf.'
ἰακχάζω, add 'III ἰακχάζει· φυλλολογεῖ Hsch.; cf. perh. ἰάκχα.'
ἰακχαῖος, add 'of person, Salamine 43 (ii/iii AD)'
Ἴακχος I 1, after 'Ar.Ra. 398' insert 'SEG 30.914 (Olbia, iv BC)'
ἰάλεμος I, after 'dirge' insert 'Pi.fr. 128e.(b)6 S.-M.'
ἰάλλω, line 2, after 'Dor. ἴᾱλα' insert 'aor. pass. hιάλε Dubois IGDS 11 (Himera, c.475/50 BC), Lacon. decr. in Inscr.Délos 87.2 (v BC)'
 I 1, at end after 'so later' insert 'utter a cry, ἰάλλων φρικαλέον βρύχημα Nonn.D. 6.182, μυκηθμὸν ἰ. ib. 198'
Ἰᾱλυσός, at end, adj. Ἰηλύσιος, before 'D.P. 505' insert 'Anacr. 4 P.'
ἴαμαι, v. °ἰάομαι.
*ἰᾱμᾰτικός, ή, όν, curative, Cyran. 10 (1.1.118 K.), Epiph.Const. in Lapid.Gr. 197.20 (cf. mod. Gk. ἰαματικός).
ἰαμβεῖος II 1, delete 'in pl., iambic poem, Luc.Salt. 27' 2, delete the section.
ἰαμβίζω I, delete 'assail in iambics'; add 'ἰαμβίζειν· τὸ λοιδορεῖν, κακολογεῖν· ἀπὸ Ἰάμβης τῆς λοιδόρου Hsch.'
ἰαμβικός, line 3, after 'Ath. 15.629d' insert 'comp. -ώτερος SEG 15.517 (iii BC)'
ἰαμβιστής, for pres. def. read 'performer of ἴαμβοι (v. °ἴαμβος II)'
ἰαμβοποιέω, for 'parody' read 'satirize'
ἴαμβος I, add 'AP 14.15.3. 2 iambic trimeter, AP 14.15.1.'
 II–III, for these sections read 'II poem written in iambic, trochaic or epodic metre, esp. of a scurrilous or satiric nature, καί μ' οὔτ' ἰάμβων οὔτε τερπωλέων μέλει Archil. 215 W., ἐν ἰάμβῳ τριμέτρῳ Hdt. 1.12.2, ἴαμβον Ἱππώνακτος Ar.Ra. 661, Pl.Ion 534c, Lg. 935e, Arist.Rh. 1418ᵇ29, Po. 1448ᵇ33, Clearch.ap.Ath. 14.620c, Str. 8.3.30, Ath. 14.645f. b applied app. to prose pieces by Asopodorus, ἐν τοῖς καταλογάδην ἰάμβοις Ath. 10.445b. 2 transf., a person as the subject of such a poem, Luc.Pseudol. 2. 3 (see quot.) ὕστερον δὲ ἴαμβοι ὠνομάσθησαν (οἱ αὐτοκάβδαλοι) αὐτοί τε καὶ τὰ ποήματα αὐτῶν Semus 24 J.'
†ἰαμβύκη, ἡ, musical instrument, prob. = °σαμβύκη ι 1, Eup. 148.4 K.-A., prob. in Arist.Pol. 1341ᵃ41 (speculatively dist. fr. σαμβύκη by later antiquaries, Phillis ap.Ath. 14.636b, Hsch., Phot., Suid.).
*ἰάν, = °ἐάν.
*ἰανάτηρ, v. °εἰνάτερες.
†ἰανογλέφᾰρος, ον, dark-eyed, Alcm. 1.69 P.
ἰανόν, v. ‡ἐανός.
Ἰανουάριος, α, ον, Lat. Ianuarius, εἰδ]οῖς Ἰανουαρία[ις the Ides of January, SEG 18.495.8 (Smyrna, ii AD), μη(νὸς) Ἰανουαρίου Mitchell N.Galatia 441.
ἰάομαι, line 1, before 'Hp.Loc.Hom. 24' insert 'Pi.P. 3.46'; line 2, for 'ἰᾶσθαι Inscr.Cypr. 135.3 H.' read 'i-ja-sa-ta-i perh. ἰjᾶσθαι but more likely R. athematic ἴαμαι ICS 217.3, cf. pr. n. Ἰαμενός'; II, for 'act. .. 1236' read 'act. only in ff.ll.'; for 'Ev.Luc. 6.17' read 'Ev.Luc. 6.18'
Ἴαονες, after 'Ἰάων rare' insert 'Hes.fr. 10a.23 M.-W. (as eponym, rest.), A.Pers. 950, 951 (s.vv.ll., app. ∪∪—), Pi.Pae. 2.3 S.-M.'; add 'Ἰήονες Call.fr. 7.29 Pf.; Ἰηονίη AP 16.295.2; cf. Myc. i-ja-wo-ne (dat.), pers. n.'
ἰάπτω (A), at end after 'Theoc. 2.82' insert 'ἰαφθῆναι· ἀποθανεῖν. πεσεῖν. φθαρῆναι Hsch.'
ἰάπτω (B) I 1, add 'b w. acc. and dat., inflict on, βουβῶσι τυπὴν ἀλίαστον ἰάπτει Nic.Th. 784, Al. 187.'
†ἴαραξ, v. ‡ἱέραξ.
*ἰαρεῖον, v. ‡ἱερήϊον.
†ἰαριγμόν, χαράν. καὶ θροῦν. Κρῆτες Hsch.
Ἰάς, at end for '[ῑ, .. 2.21.]' read '[ῑ, cf. AP 7.83, but ῐ in arsi Nic.l.c., App.Anth. 2.21]'
*Ἰᾱσίς, ίδος, ἡ, daughter of Iasus, Io, Call.fr. 66.1 Pf.
*ἰασμός, ὁ, shouting, crying aloud, v.l. in Aq.Je. 32.16.
ἴασπις, line 2, after 'ἡ' insert '(also ὁ, AP 9.750 (Arch.))'
*ἰασσεῖν· θυμοῦσθαι. δάκνειν Hsch.
ἰᾱτήρ, for 'Cypr. .. 135.3 H.' read 'Cypr. to-ni-ja-te-ra-ne τὸν ἰjατε̄ραν ICS 217.3; at end add 'Myc. i-ja-te'
ἰᾱτής, add 'perh. also νούσων εἰη[τήν] rest. in epigr. in IG 2².5935 (ii AD, cf. SEG 26.284)'

ἰᾱτικός, after 'healing' insert 'Pl.Ti. 87b'
ἰᾱτρεῖον II 1, add 'Inscr.Perg. 8(3).161 (pl., ii AD)'
ἰᾱτρίνη, add 'as epith. of Μήτηρ θεῶν, IG 2².4714 (i BC/i AD), al.'
*ἰᾱτρίσκος, ὁ, contemptible physician, quack, Sch.D.T. 228.3.
ἰᾱτρόμαια, add 'MAMA 3.292 (-μεα, v. SEG 37.1854)'
ἰᾱτρός, after lemma insert '(written hιατρ-, SEG 31.834, Megara Hyblaea, vi BC)'; after 'ὁ' insert '(ἡ)' I, of Apollo, add 'SEG 30.880 (Berezan, vi BC), 30.977 (Olbia, v BC)' and add 'of Asclepius, Paus. 2.26.9, SEG 26.1818 (Cyrene, ii AD)'; at end for 'midwife ..' s.v. μαῖα' read 'μαῖα καὶ ἰατρὸς Φανοστράτη CEG 569 (Acharnae, iv BC), Hellad.ap.Phot.Bibl. 531a'
ἰᾱτροσοφιστής, add 'of one who practises magic arts and divination, Ps.-Callisth. 1.3'
ἰαύω II, after 'c. acc. and gen.' insert 'cause to rest from'
ἰάχω 3, line 6, after 'E.El. 707' insert '(dub., cj. ἰαχεῖ)'
*Ἰαώ, indecl., Yahweh, D.S. 1.94, orac.ap.Macr.Sat. 1.18.20, PMag. 3.149, 211, al., SEG 32.1082 (Spain, iii AD).
*Ἴβηρ, ηρος, ὁ, Iberian, Hdt. 7.165, Th. 6.2.2, GVI 1001.12 (Rhodes, c.100 BC); cf. Hsch. Ἴβηρ· χερσαῖόν τι θηρίον· ἀφ' οὗ καὶ Ἴβηρες.
ἰβιοτάφος, add 'PFouad 16 (ii BC)'
ἶβις, for pres. def. read 'one or other species of ibis'
*ἰβύκη· εὐφημία Hsch.; cf. ἴβυς.
*ἴγα· σιώπα. Κύπριοι Hsch. (= σίγα).
*ἰγμαμένος, v. ‡ἰκμάω (B).
*Ἰδαλιάνιος, ὁ, a month at Termessus, TAM 3(1).4.15 (ii AD).
ἰδάλιμος, add '(prop. εἰδ- or εἰδάλιμος)'
*Ἰδάλιον, τό, city in Cyprus, Theocr. 15.100 [initial ῑ perh. metri gratia, cf. Verg.Aen. 1.693], Cypr. e-ta-li-o-ne Ἐδάλιον, ICS 217.1, see also Ἐδαλιεύς.
ἰδαλίς, add '(ἰδάλιος La.)'
ἰδᾱνός, for 'χάριτες .. 535' read 'Χάριτες Call.fr. 114.9 Pf., Musae. 76 (prob.), Hsch.'
*Ἰδάτης, ὁ, title of Zeus in Crete, SEG 23.547.51 (Olous, c.200 BC).
ἰδέ I, add 'Cypr. i-te, introducing a new sentence, ICS 217.26' II, for this section read 'Cypr. introducing an apodosis, in that case, then, ICS 217.12, 24'
ἰδέατος, for 'ἰδήρατος' read 'καλὸς ἀνήρ'
ἴδη I, line 3, before 'in sg.' insert 'ἴδηφιν· ἴδαις (-ες cod.), Βοιωτοί Hsch.'
ἰδήρατος, add 'as pers. n., IG 2².10366 (Skione, iv BC)'
*ἰδιαστικός, ή, όν, app. having a peculiar nature, ἰδιαστικῶν ἤγουν ἰδιοποιῶν Eustr. in APo. 82.34.
*ἰδιόκοιτον· ἰδιόρρυθμον Hsch. (perh. read -κοπον).
ἰδιόκτητος, line 3, after 'Cod.Just. 10.3.7' insert 'χω(ρίον) ἰ. ICilicie 33 (iii/iv AD)'; after 'PTeb. 5.111 (ii BC)' insert 'land belonging to one's own people, Onas. 6.13'
*ἰδιόλογος, ὁ, = ὁ ἴδιος λόγος (v. ἴδιος II 1 b), cj. in Str. 17.1.12, cf. CIL 10.4862.
ἰδιοξενοδόκος, for pres. ref. read 'SEG 26.670.16 (Doliche, late ii BC)'
ἰδιοποιός, add 'II acting on one's own initiative, POxy. 2407.12 (iii AD).'
ἰδιοπρᾰγέω, add 'Phlp. in de An. p. 455.30, al.'
ἴδιος I 3, add 'b τὰ ἴδια, one's own funds, one's private resources, ἐκ τῶν ἰδίων, at one's own expense, AS 35(1985).50 (Pisidia, AD 150), SEG 31.167.7 (Eleusis, AD 162/9), Mitchell N.Galatia 193; also sg., ἐκ τοῦ ἰδίου SEG 30.1617 (Cyprus, 44/31 BC), IHadr. 24 (ii/iii AD), ἀπὸ τῶν ἰδίων ICilicie 124 (i/ii AD). 4, add 'τὸν ἴδιον συμβιωτήν fellow-townsman, SEG 29.1185 (Lydia); add 'Mitchell N.Galatia 1, TAM 4(1).77' 5, add 'SEG 25.539 (Aulis, iii BC)' 6 b, after 'no. 133' insert 'ἰδίῳ (sc. θανάτῳ) ἔθανον BCH 52.391 (Thasos)' and add 'cf. Phalar.Ep. 147.4' II 3, add 'b of words, = κύριος A II 5, opp. τροπικός, Aristid.Rh. 1 p. 468 S.' V, add 'ἰδιαίτατος· ἴδιος, ὑπερθετικῶς Hsch.' VI 1, add 'b = ἰδίᾳ, τῶν ἰ. (sc. ἐρδομένων) Schwyzer 728.13 (Miletus, c.400 BC).' at end add 'καθ' ἰτδίαν BCH 59.37 (Crannon)'
ἰδιοσπορέομαι, after 'labour' insert 'PBaden 90.39 (iii AD)'
*ἰδιοσυστᾰτως· καθ' ὑπόστασιν ἰδίαν Hsch.
ἰδιοφυής, before 'Archelaus' insert 'a writer called '
ἰδιόχειρος, add 'adv. -χείρως with his own hand, Sch.AP 7.432'
ἰδιώτης II 3, after 'D.H.Dem. 2' insert '(without λόγος, id.Lys. 3,

155

ἴδμων | SUPPLEMENT | ἱερουργέω

4)' **III 1**, of prose writers, add 'Pl.Lg. 890a' **IV**, delete the section. at end add 'see also °Fhεδιέστας (after °ἕδεος)'

ἴδμων, add 'w. inf., ἴδμων .. σημήνασθαι ἀϋτμήν Opp.C. 1.480'

ἰδνόομαι, act., add 'ἰδνῶν· κάμπτων Hsch. La. (cod. εἴδεος)'

ἴδος, for 'Call.fr. 124 (prob.)' read 'Call.fr. 304 Pf. (cod. εἴδεος)'

ἰδού II 3, for this section read 'in answer to a summons with following question, "yes", "here I am", (what do you want?, or sim.) ἰδού· τί ἐστιν; Ar.Nu. 825, Eq. 157.'

ἴδρις, line 9, after 'Vett.Val. 4.19' insert 'w. acc. of respect, οὐδὲν ἴδρις S. OC 525 (lyr.), ἴδριες οὐδέν id. Tr. 649 (lyr.); ταῦτ' οὐκέτ' ἴδρις id.fr. 269a.31 R.'; lines 9/10, delete 'οὐδέν .. (lyr.)'

†**ἱδροσύνη**, ἡ, the state of being fixed, GVI 1487.2 (pl., Phrygia, iii AD).

ἴδρυμα I 1, for 'Call.Aet. 3.1.73' read 'Call.fr. 75.73 Pf.' **2**, add 'of a funeral monument, ISmyrna II(2) p. 357, no. xii'

ἰδρύω I 2, add 'of words in a sentence, to be placed, D.H.Comp. 6'

ἰδρῶον, for 'cloth .. (ii BC)' read 'cloth for covering horses, donkeys, etc., when heated, PTeb. 796.11 (ii BC), PSI 527, etc.; ὀνικά PCair.Zen. 720.4 (iii BC)'

ἰδρώτιον, add 'pl., Aët. 3.3 (261.19 O.)'

*ἰδύαι· τρίχες Hsch.

ἱεραγέω, add 'PVat.Gr. 65 in Tyche 5.102.2 (iii BC)'

ἱεράζω, add '**II** trans., = ἱερόω, Princeton Exp.Inscr. 653 (Syria).'

*ἱερακάδιον, v. °ἱερακίδιον.

ἱερακάριος, add 'MAMA 3.17, 79 (Cilicia)'

ἱερακίδιον, for pres. ref. read 'Inscr.Délos 1416A19 (ii BC); also written -άδιον ib. 1452A9 (ii BC)'

*ἱερακῖτις, ιδος, a plant, = ἱεράκιον I, Cyran. 75, PMag. 4.902.

*ἱερακόμορφος, add 'ἀνδριάντες Ἀπόλλωνος ἱερακομό(ρφου) χαλ(κοῖ) γ' POxy. 3473.10 (ii AD)'

*ἱεραμφοδίτης, ου, ὁ, inhabitant of the sacred quarter, SEG 30.1449 (-είτης, Pontus, AD 257/8).

ἱερανθεσία, for 'only Dor.' read 'dial. ἰαρ-'; add 'SEG 36.516, 517 (Delphi, i AD); (second element < -αναθεσία), cf. °ὠνανθεσία'

ἱερανομέω, delete the entry (v. Glotta 50.77).

ἱέραξ II, add 'cf., ἱάραξ· ἰχθὺς ποιός, Δωρικώτερον· διὰ τὸ ἐοικέναι τῷ πτηνῷ. καὶ λύχνος ὁ πρὸς τὰ ἱερά Hsch.'

*ἱεραπολία, ἡ, office of ἱεραπόλος, epigr. in SEG 13.422 (Delos, iii BC).

ἱεραπόλος, add 'ἰαρᾱ- SEG 39.1008 (Morgantina, c.iii BC)'

ἱερατεύω, line 2, after '(perh. i BC)' insert 'IG 10(2).1.95 (ii/i BC), 114 (ii AD), etc.'; line 7, after 'Hdn. 5.6.3' insert 'ἱερατεύσαντα πρὸ πόλεως MAMA 7.406'; line 8, med., add 'SEG 31.635 (Maced., ii/iii AD)'

ἱερατικός, line 5, after 'Dam.Pr. 399' insert 'ἱ. γράμματα, βίβλος PTeb. 291 ii 41, 43 (ii AD)'

ἱεραύλης, for 'flute-player' read 'aulos-player'

*ἱεραφάντρια, v. °ἱεροφάντρια.

ἱεραφόρος, after 'SIG² 754 (Pergamum)' add 'also fem., IG 7.2681 (Thebes)'

ἱέρεια, line 1, after 'ἡ' insert 'ἤρεια epigr. in IG 2².3606.15 (ii AD); line 5, after 'al.' insert '(acc. pl. τὰς ἱερῆς Inscr.Cos 386.9)'; line 6, after '(Thebes)' insert 'ἰάρηα SEG 23.566.9, al. (Axos, Crete, iv BC), Aeol. ἴρεα JÖAI 5.141.19 (Eresus, ii/i BC)'; after 'BCH 6.24 (Delos, ii BC)' insert 'in Jewish context, CIJud. 1007 (iv AD)' at end add 'Myc. i-je-re-ja'

ἱερεία III, for this section read 'Cypr. ta-ni-e-re-wi-ja-ne τὰν ἱερεϜίαν τᾶς Ἀθάνας prob. sanctuary, ICS 217.20'

ἱερεῖον, Ion. ἱερήιον, add 'SEG 30.1283 (Didyma, vi BC)' **II**, of sucking pigs, add 'PMich.Zen. 84.3, PZen.Col. 46.5 (both iii BC)' at end add 'Myc. i-je-re-wi-jo'

ἱερειτεύω II, add 'IG 10(2).1.156'

ἱερεύς, line 1, for 'Cypr. .. 59 H.' read 'Myc. i-je-re-u; Cypr. i-je-re-u-se ἰρηεύς ICS 7.3, gen. i-e-re-wo-se ἱερεϜος ICS 234'; line 3, before 'Ion. nom.' insert 'Lesb. nom. ἴρευς, acc. εἴρεα, ἴρεα, IG 12(2).102, 242.4, 239.8, al.'; line 5, after 'GDI 4841 (Cyrene)' insert 'ἰαρεύς SEG 29.361 (Argos, c.400 BC)'; line 9, for 'Inscr.Cypr. 100 H.' read 'Cypr. i-je-re-se ἱρηές ICS 4'

ἱερεύω 1, add '**b** to be a priest, w. dat. ἱερεύοντος δὲ τῷ Ἀσκλαπιῶι SEG 32.622 (Illyria, ii BC); w. gen. ἱερε[ύ]οντος δὲ τοῦ Ἀσκλαπιοῦ SEG 32.623 (Illyria, ii/i BC).'

ἱερεωτική (sc. τέχνη), ἡ, perh. art of priesthood, Poll. 7.210.

ἱέρισσα, add 'in Jewish use, Μάριν ἱέρισα χρηστή CIJud. 15.14 (Egypt, 28 BC)'

*ἱερόγαλλος, ὁ, title of priest, TAM 3(1).740 (Termessus).

*ἱερογραμματεία, ἡ, work-place of a ἱερογραμματεύς, Pap. Lugd. Bat. XIII 21.16.

*ἱεροδούλη, ἡ, = (ἡ) ἱερόδουλος, AS 10.48 no. 96 (Pisidia), TAM 2(3).1023, 3(1).567.

*ἱεροεθνής, οῦς, ὁ, person of priestly stock, POxy. 3470.16, 3471.14 (both AD 131), CPR xv 32.8 (ii/iii AD).

*ἱεροζωμουργοί, οἱ, makers of mash for the Apis bull, PRoss.-Georg. 5.16.16 (ii/iii AD; -ζομ- pap.).

*ἱεροθαλής, ές, having sacred branches, Orph.H. 40.17.

ἱεροθαλλής, delete the entry.

*ἱεροθυσία, ἡ, in Dor. form ἰαρο-, sacrifice, SEG 9.13.20 (Cyrene, iv BC).

ἱεροθύτης, after '(Euboea, iii BC)' insert 'SEG 32.330 (Athens, Rom.imp.)'; add 'Arg. ἰᾱροθύται BCH suppl. 22 p. 235 (v BC)'

ἱερόθυτος, subst., add 'sg., SEG 18.596.15 (Babylonia, iii AD)'

*ἱεροίατροι, οἱ, priestly doctors who attended the Apis bull, PRoss.-Georg. 5.16.15 (ii/iii AD).

*ἱεροκαλλίνικος, ον, holy and victorious, Lyr.adesp. 19 P.

*ἱεροκηρυκεία (-κᾱρυκ-), ἡ, office of a ἱεροκῆρυξ, ASAA 30/32(1952-4).295, no. 67.23 (Rhodes).

ἱεροκηρυκεύω, add 'SEG 37.973.7 (Claros, AD 172/3)'

ἱεροκῆρυξ, line 2, for 'prob. in IG 1².6.89' read 'IEphes. 2.53, 10.22, al., IStraton. 503.6 (318 BC)'

*ἱεροκώμη, ἡ, sacred village, CIG 5069 (Nubia, iii AD).

ἱερόμαντις, add 'fem., τὴν -ιν, AE 1945-7.106 (Thessaly)'

ἱερομνημονεία, ἡ, office of ἱερομνήμων II 2, IEphes. 4324.5 (i BC, -ηα lapis).

ἱερομνημονέω, add 'athematic part. ἰαρομναμονέντες L.Gasperini Le laminette iscritte dal repostiglio dell'Agorà di Cirene (Giornata Lincea 3 Nov. 1987, Cyrene, iv BC)'

ἱερομνήμων, after 'ονος, ὁ' insert 'ἰαρομμνάμονα SEG 30.380 (Tiryns, vii BC), ἰαρομνᾱμ- SEG 34.282 (Nemea, iv BC), 23.271 (Thespiae, iii BC)' **II 1**, add 'Ar.fr. 335 K.-A.' add '**4** a precious stone, Plin.HN 37.160.'

ἱερομοσχοσφραγιστής, add 'PGrenf. II 64.1 (ii/iii AD, ἰαιρο- pap.)'

*ἱεροναύτης, ου, ὁ, sailor on a sacred vessel, Inscr.Délos 50 (iv BC), ITomis 98.4 (Rom.imp.).

ἱερονίκης, after 'Luc.Hist.Conscr. 30, etc.' insert 'w. ref. to exemption from taxes, τῶν ἱερονικῶν καὶ ἀτελῶν BGU 2122.1 (AD 108)'; at end add 'Aeol. εἰρονεικ- IG 12(2).68.11 (Mytilene, ii AD)'

*ἱερονικοτελοῦσα, ἡ, giving victory in sacred games, epith. of Isis, Hymn.Is. in POxy. 1380.78 (ii AD, written ἱερω-).

*ἱερονόμᾱς, ὁ, Aeol. ἰρο- = °ἱερονόμος, Θεοφάνης ὁ εἰρονόμας SEG 29.741 (Mytilene, after AD 138).

*ἱερονομία, ἡ, office of °ἱερονόμος, Robert Hell. 6.70 (Thyatira).

ἱερονόμοι, delete the entry.

*ἱερονόμος, ὁ, Aeol. ἰρονόμος SEG 34.1234 (Aeolis, c.200 BC), official in charge of sacred rites, SIG 982.23, IGRom. 4.461 (Pergamum), Illion 31.24, 32.20, al.; of the pontifices at Rome, D.H. 2.73; as pers. n., SEG 31.348 (Mantineia, iv/iii BC).

*ἱεροπλατεῖται, οἱ, occupants of a °ἱεροπλατίη, IHistriae 57.32 (early iii AD).

*ἱεροπλατίη, ἡ, = ἱερὰ πλατεῖα, sacred street or square, ἱ. τῶν φιλόπλων Milet 2(3).134 no. 403.

*ἱεροποικόν, τό, fund for the ἱεροποιός, Inscr.Délos 1521.23.

ἱεροπρεπής, adv., add 'neut. pl. sup. -έστατα IHistriae 57.36 (early iii AD)'

*ἱεροπρόσπλοκος, ον, devoted to religious matters, Ptol.Tetr. 159 (v.l.), 181.

ἱεροπρόσπολος, for 'Ptol.Tetr. 159' read 'Ptol.Tetr. 181 (v.l.)'

ἱερόπτης, add 'as pers. n., IG 2².12237 (c.400 BC), SEG 23.155 (Attica, mid iv BC)'

ἱερός, line 2, after 'Orac.ap.Hdt. 8.77' insert 'cf. ἱερὸς ἕδρη Arat. 692'; line 3, ἱαρός, add 'SEG 30.1176.F5 (Metapontum, vi BC), 31.368 (Olympia, v BC)' and Aeol. ἶρος, add 'εἶρ- SEG 32.1243.45 (Cyme, i BC/i AD)' **I**, add 'Stesich. 8.3 P.' **II 3 b**, after 'Od. 24.81' insert 'ἰαρὸς Χαροπ[ί]νος· ἰᾱρ[ὸς] Ἀρισσοτόδαμος Schwyzer 66.6 (Messenia, early v BC)' **c**, add 'τοῦ εἱεροτάτου βαφίου ΠYr 28 (Rom.imp.)' add '**d** sup. as honorific epith. of Caesars, PLond. 948, POxy. 1114.20, Stud.Pal. 20.51 (all iii AD).' **III 1**, add 'ὁ ἐπὶ τὰ ἱερά SEG 32.218.203 (Athens, 89/8 BC)' **2**, line 1, after 'Ion. ἱρόν' insert 'B. 3.15, E.Hel. 1002, IT 969, etc.' **5**, add 'IPamph. 3.1 (Sillyon, iv BC); fem., add 'ἱαρὰν μίστωμα de Franciscis Locr.Epiz. 23' **IV 4**, add 'also ἀφ' ἱεροῦ Pl.Lg. 739a' **10**, add 'cf. ἱ. ῥάχις AP 9.644.7 (Agath.)' line 6 fr. end, after 'Theoc. 5.22' insert in iambics, Lyc. 950, 1350 (καθιερώσει)' at end add 'Myc. i-je-ro (in sense II 2, owned by a deity)'

ἱεροσαλπικτής, for '(CIG) 2983 (Ephesus)' read 'IEphes. 1034, al.'

ἱεροσκοπέομαι, add 'act., Hsch. s.v. ἱερᾶται'

ἱεροσκόπος II, add 'IKeramos 31, IEphes. 1004, al.'

ἱεροσυλία, add 'Is. 8.39'

*ἱερόσυλις, fem. adj., expl. of °θεόσυλις, Sch.Hippon. 118.1 W.

ἱερόσυλος, for 'sacrilegious person' read 'pilferer'

ἱεροταμιεύω, add 'med. aor., BCH 1.291 (Ephesus, ii AD)'

*ἱεροτίθηνοι, οἱ, tenders of the Apis bull, PRoss.-Georg. 5.15.12, 16.1 (ii/iii AD).

ἱερουργέω I, add 'Dor. ἰᾱρωργ- Inscr.Cret. 1.23.4 (Phaestus, ii BC)' ἱερουργός, add 'PVindob.Worp 21.1 (v/vi AD); form ἱεροϝοργός Myc. *i-je-ro-wo-ko*'

*ἱερουσιάρχης, ου, ὁ, late sp. for γερουσιάρχης, CIJud. 1.405 (Rome): εἱεροσάρχης, ib. 408 (Rome).

*ἱεροφάντειος, ον, = ἱεροφαντικός, rest. in epigr. in TAM 2(2).418.5 (Lycia).

ἱεροφάντης, add 'IEphes. 10.11, 47.39, al., POxy. 2782.2 (ii/iii AD)'

ἱεροφαντικός, after 'Alex. 60' insert 'οἶκοι IG 4².84.30 (Epid., i AD)'

ἱεροφάντρια, add 'also ἱεραφάντρια ISelge 15.4 (Rom.imp.)'

ἱεροφύλαξ 1, after 'cj. Markl.' insert ', ἱεροῦ φύλακες Diggle'; add 'Sokolowski 3.155.5 (Cos, iii BC), IG 9²(1).1.95, etc. (Phystium, ii BC)'

ἱερόφωνος, for 'with sacred voice' read 'making a holy utterance'; for 'f.l. for ἱμερό- in Alcm. 26.1' read 'Alcm. 26.1 P. (cj. ἱμερο-)'; after 'IG 14.914' insert 'D.Knibbe Der Staatsmarkt (Vienna, 1981) p. 170/1 no. 10 (Ephesus, iii AD; v. SEG 31.950)'

ἱερόω, add 'hold to be sacred, οἱ πρῶτοι τοῦτο τὸ νόσημα ἱρώσαντες v.l. ἀφιερω- Hp.Morb.Sacr. 1'

ἱέρωμα, for 'consecrated object, offering' read 'sacred image'; line 2, for 'ἰαρώματα .. ἀρώματα)' read 'ἰαρώματα Inscr.Cret. 4.145.7 (Gortyn, iv BC)'; line 3, delete 'ἰαρ[ώ]ματα IG 4.917 (Epid., iv BC; read ἰαρ[ε]ῖα τά)'

*ἱερωνᾶς, α, ὁ, buyer of sacrificial victims, IG 12 suppl. 120.13 Rhodes, (iii BC); rest. in Inscr.Cret. 3.3.3A91 (iii/ii BC); cf. βοώνης.

*ἱερωνέω, act as °ἱερωνᾶς, Lindos 449.12 (i/ii AD).

ἱερωνία, for 'dub. sens. in' read 'purchase of sacrificial victims'

ἱερωσύνη, line 3, after 'etc.' insert 'ἱεροσύνη SEG 34.1095 (Ephesus, iii AD), ἱερειοσύνα IG 5(1).1114.21, 25 (Laconia)'; line 4, after 'D. 59.92' insert 'in Chr. use, Mitchell N.Galatia 493 (εἱερ-), CodJust. 1.3.43.10, etc.'; at end, pl., add 'SEG 32.825 (Paros, ii BC)'

ἵζω, line 3, after 'etc.' insert '3 pl. ἔσαν Il. 19.393'; line 4 (and I 1, line 7), for 'ἔσσαντα' read 'ἵσσαντα' (changing ref. to SEG 9.71.134 (Cyrene, iv BC)) **I 1**, add '**b** set or settle in place, λέπαδνα Il. 19.393.' **2**, line 6, for '[ἥ]σσαντο .. iii BC)' read 'ἵσσαντο BCH 81.477 (Argos, iv BC)' and add 'w. deity as obj., set up in form of cult-statue, etc.' and transfer here 'Thgn. 12' fr. line 4, adding 'Call.fr. 200b Pf., Del. 309' **III 1**, line 3, for 'Berl.Sitzb. 1927.169 (Cyrene)' read 'SEG 9.72.122 (Cyrene, iv BC)'

ἵημι, at end of the note on quantity (lines 29ff.) delete 'with variation .. Carm.Pop. 1' **I 2**, add 'w. partit. gen., οἱ δὲ ὄνοι .. οὕτω δὴ μᾶλλον πολλῷ ἵεσαν τῆς φωνῆς Hdt. 4.135.3' **5**, line 2, after '(Il.) 5.513' insert 'E.Rh. 291' **II 2**, line 3, after '(Il.) 2.589' insert '(so in act., εἷεσαν ἐκτελέσαι poet. in BCH 50.406 (Thespiae))'; line 4, before '11.168' insert 'Il.'

ἰήρια, delete the entry.

*ἰήρια, τά, unexpld. wd. in Inscr.Cret. 4.145.4 (Gortyn, iv BC).

*ἰήτε, 2 pl. imper. formed fr. ἰή, dub. l. in Pi.Pae. 6.122 S.-M.

*ἰθαινάθυμος, ον, cited as compd. of ἰθαίνω Theognost.Can. 81 (cf. sq.).

†ἰθαίνω, perh. gladden, ἴθαινε θυμόν anon.ap.An.Ox. 1.61, ἰθαίνειν· εὐφρονεῖν (? εὐφραίνειν) Hsch., med. ἰθαίνεσθαι· θερμαίνεσθαι id.; supposed etym. of ἰθαγενής A.D.Adv. 187.25 (prob. cogn. w. ἰθαρός).

ἰθαρός I, for 'Alc.Supp. 4.18' read 'Alc. 58.18 L.-P.; ἰθαραῖς .. λογάσιν with glad eyes, Call.fr. 85.15 Pf.; ἰθαρὸν (v.l. ἱκανὸν) γόνυ id.Cer. 132'

ἴθι, line 1, for 'Adv.' read 'exclam.'

*ἴθμη, ἡ, way, passage, Theognost.Can. 112, Sch.Opp.H. 1.738; cf. εἰσίθμη, ἴθμα. (For the accent see C.A. Lobeck Paralipomena grammaticae Graecae (Leipzig, 1837) p. 395.)

*ἰθμός, v. °ἠθμός.

*ἰθουλίς, ίδος, ἡ, Boeot. name of a fish, perh. = ἰουλίς, IGC p.99 B10 (Acraephia, ii BC).

*ἰθῠβάτης, ου, masc. adj. running in a straight line, κανών cj. in AP 6.62 (Phil.).

*ἰθῠβέλεια, ἡ, straight-shooter, of Artemis, oracle in ZPE 92.269 (Ephesus, ii AD).

ἰθύδικος, for 'righteous' read 'giving right judgement'

†ἴθυμβος, ὁ, the name of a type of performer or performance, ἴθυμβος· γελοιαστής. καὶ τὸ σκῶμμα. ἀπὸ τῶν ἰθύμβων ἅτινα ποιήματα ἦν ἐπὶ χλεύῃ καὶ γέλωτι συγκείμενα. καὶ ᾠδὴ μακρὰ καὶ ὑπόσκαιος Hsch.; καὶ ἴθυμβοι ἐπὶ Διονύσου καὶ καρυατίδες ἐπὶ Ἀρτέμιδι Poll. 4.104; cf. θρίαμβος, ἴαμβος, etc.

ἴθυνα, add 'εἴθῦνα SEG 17.377.14 (Chios, v BC)'

*ἰθύνοος, ον, perh. fair-minded, ἰθυνόων .. θεσμῶν Nonn.D. 41.353.

ἰθυντήρ, line 2, delete 'IG 9(1).390 (Naupactus)'

ἰθύνω 3, add 'CodJust. 1.4.33.2 (AD 534)'

ἰθῠπόρος, add '2 giving a straight voyage, of winds, Opp.H. 5.677.'

ἰθυπτίων, line 1, after 'Il. 21.169' insert '(also read by Zenod. at 20.273 s.v.l.)'; line 3, after 'Zenod.' insert 'and Callistr.'

ἰθύς (A), after 'cases)' insert 'perh. also εἰθεῖα, ἁ ὀδὸς ἁ εἰθεία (nisi leg. ἀὲ ἰθεῖα) SEG 35.991.B5 (Lyttos, c.500 BC)' **II 2**, at end after 'Hp.Off. 3' insert 'ἐς ἰθύ in length, Call.fr. 196.26 Pf.'

ἰθύς (B) **1**, for this section read 'straight line, πρὸς ῥόον ἀΐσσοντος ἂν' ἰθύν Il. 21.303, ἐπεὶ δὴ σφαίρῃ ἂν' ἰθὺν πειρήσαντο Od. 8.377, μῆκός τε καὶ ἰθύν in outstretched length, Nic.Th. 398'

ἰθυτενής, delete 'upright, perpendicular' and 'metaph.'

ἰθύφαλλος **III**, for 'metaph., lewd fellow' read 'adopted as a group-name by an association of young men'

*ἰθύωρα, v. ‡εὐθύωρος.

*Ἰθωμαῖα, τά, games at Messene in honour of Zeus Ithometas, SEG 23.208 (Messene, i AD), Paus. 4.33.2.

ἰθών, add 'perh. in POxy. 3729.19 (AD 307)'

ἱκανοδοσία, add 'POxy. 3807.36 (AD 28)'

ἱκανοποιέω, add 'SEG 4.648.11, 37.1001 (Lydia, ii AD)'

ἱκᾰνός **I 1**, at end delete 'ὁ 'I. the Almighty Lxx Ru. 1.21' **2**, delete the section. **III 1 a**, lines 3/4, for 'later' read 'also' and after 'amply' insert 'Hp.Epid. 5.49'

ἱκάς, after 'twentieth of the month' insert 'SEG 23.530 (Crete, vii BC)'

ἱκαστός (s.v. ϝίκατι), add 'ϝικαστῆ (sc. ἀμέρα) IG 7.3172.109 (late iii BC)'

*ἱκᾰτιδάρκμιος, v. °εἰκοσαδράχμιος.

*ἱκᾰτιείς, = εἰκοσιείς, SEG 26.672.47 (Larissa, c.200 BC).

*ἱκᾰτιεννέα, = εἰκοσιεννέα, SEG 26.674.4 (Larissa, ii BC).

*ἱκᾰτίπεμπε, = εἰκοσιπέντε, SEG 26.672.23 (Larissa, c.200 BC).

ἴκελος, add 'as pr. n. ϝίκελος BE 1990.863 (Selinous, c.500 BC)'

ἱκεσία, line 4, for 'AP 5.215' read 'AP 1.34.8, 5.216.2' **2**, for 'AP l.c.' read 'AP ll.cc.'

ἱκέσιος **II 1**, of Zeus, add 'SEG 33.244d, e (ἡικ-, Attica, archaic)' at end after 'A.R. 2.215' add 'AP 5.300.5 (Paul.Sil.)'

*ἱκετεύσιμος, ον, of a suppliant, Hsch. s.v. προστροπαίων.

ἱκετευτέος, add 'X.Mem. 1.5.5 (s.v.l.)'

*ἱκετέω, = ἱκετεύω (in quot., sense 4), PTeb. 2dᵛ.9.

*ἱκετηριάς, άδος, ἡ, = ἱκετηρίς, perh. eleg. in POxy. 3723.10 (ii AD).

ἱκετήριος **II 1**, line 3, after 'A.Supp. 192' insert '(ἱκετ- cod.)'

ἱκέτης, line 2, after 'ὁ' insert 'Arg. ἱκέτᾱς Schwyzer 97 (Argos, vi BC), cf. Lacon. Διοϝικέτα (= Διὸς ἱκεσίου) Schwyzer 1'; add 'cf. Myc. *i-ke-ta*, pers. n.'

ἱκετικός, before 'Adv.' insert 'BGU 1053 ii 6 (i AD); ἱερὰ ἄσυλα καὶ ἱ. IStraton. 1101.3 (ii AD)'

ἱκέτις, add 'E.Hel. 1238'

ἱκμαίνω, add 'II express (liquids), cause to be exuded, Nic.Al. 97.'

ἰκμᾰλέος **1**, add 'of the noise emitted by the parrot-wrasse, φθέγγεται ἰ. λαλαγήν Opp.H. 1.135'

ἴκμαρ, add 'Antim.in Suppl.Hell. 57.4'

ἰκμάω (A), after 'Id.' insert 'cf. ἀνικμάω, ἀπικμάω'

ἰκμάω (B), for 'pf. part. .. Inscr.Cypr. 135.3H.' read 'part. *i-ki-ma-me-no-se*, ἰκμαμένος wounded, ICS 217.3'

ἰκνέομαι **II 3 b**, add 'A.Supp. 333'

ἴκρια, before 'τά' insert '[ῐ by nature, see Ar.Th. 395, Cratin. 360 K.-A.]'; for 'sg. v. infr. III' read 'sg. Hsch., see also infr. III' **II 1**, add 'perh. balcony, PDura 19.9 (i AD)' **3**, delete the section adding quots. to section II **2**, line 18, after 'Nic.Th. 198' insert 'so perh. pl. in PRein. 2065.35 (JJP 11/12.66; ii AD)'

ἰκριοποιέω, for 'Rev.Phil. 50.69' read 'SEG 4.448.5, 449.21'

*ἴκταιον· τὸ τρόφιμον, Theognost.Can. 15 (perh. read ἰκμαῖον).

ἰκταῖος, delete 'with penult. short'

ἴκτερος **I**, add '**2** rust on plants, Lxx 2Ch. 6.28, al.' **II**, add 'cf. χαραδριός'

ἰκτίς, after 'marten' insert 'A.fr. 47a.2.10 R. (lyr.)'

ἴκω, line 2, after 'Trag.' insert 'exc. in A.fr. 6 R.'; line 7, insert 'asigmatic aor. ἥκαι IPamph. 3.9 (iv BC)'; at end delete '- ἵκοντ' .. 36'

*ἶλαξ· ἡ πρῖνος, ὡς Ῥωμαῖοι καὶ Μακεδόνες Hsch. (cf. Lat. ilex).

ἱλάομαι, add 'act. ἱλάοντες· ἐξευμενιζόμενοι, ἐξιλεούμενοι Hsch.'

ἵλαος, line 2, after 'Ar.Th. 1148)' insert 'ἥιλαος IG 9².609.A16 (Naupactus, c.500 BC)'; line 7, after 'Pl.Phd. 95a' insert 'εἵλως (= ἵλεως) Sokolowski 1.29.13 (Metropolis in Ionia, iv BC)' **III**, adv., add 'Kafizin 291 (iii BC)' line 3 fr. end, after 'Theoc. 5.18' insert 'AP 6.334 (Leon.Tarent.)' at end add 'see also ‡εἰλής'

ἱλαότι, v. ἴλημι.

ἱλαρός **I**, add 'sup. -ώτατα (adv.), epigr. in SEG 23.206.20 (Messenia, ii/iii AD)'

*ἱλαροφυΐα, ἡ, cheerful nature, Dioscorus in Byz.-neugr.Jahrb. 10.342.

ἰλάρχης, line 3, for 'praefectus turmae' read 'decurio'

ἴλαρχος, add 'ϝίλαρχος SEG 23.271.16 (Thespiae, 220/208 BC)'

ἱλάσκομαι **I 1**, line 10, before 'ὁ 7.9' insert '**b** reverence sacred things, ἱερόν A.R. 2.808, Posidipp.ap.Ath. 7.318d.' **2**, add 'νέκταρ χυτόν .. ἀεθλόφοροις ἀνδράσιν πέμπων .. ἱλάσκομαι Pi.O. 7.9'

ἱλαστήριος **II**, add '**4** app. some item of water-raising equipment, POxy. 1985.11 (vi AD); perh. a separate wd. ἱλαστ- for ἐλαστήριος = ἐλατήριος.'

157

*Ἰλαστηριών, ῶνος, ὁ, name of a month at Caunus, Robert *Hell.* 7.174.60 (Smyrna, ii BC).

Ἰλάων· ἥρως, Ποσειδῶνος υἱός, ἀφ' οὗ Ἀριστοφάνης ἐν Τριφάλητι (fr. 567 K.-A.) Ἰλάονας ἔφη τοὺς φάλητας μεταφέρων Hsch.

ἴλημι, for 'Dor. ἴλᾰθι' read 'Dor. and Ep. ἴλᾰθι, Call.*Cer.* 138, *fr.* 638 Pf., A.R. 4.1014' and after 'Luc.*Epigr.* 22' insert 'etc.'

ἴλια, add 'cf. ἴλιον'

Ἰλιακός I, after 'Trojan' insert 'Call.*fr.* 114.25 Pf.'   add '2 Ἰλιακά, τά, festival at Ilium, *Illion* 52.17 (ii BC).'

ἰλιγγιζόμενον· συστρεφόμενον Hsch.

ἰλιγγιώδης, transfer the article before ἴλιγγος.

ἴλιγγος 3, add '*Peripl.M.Rubr.* 40'

*Ἰλίεια, τά, festival of Athena at Ilium, *IG* 2².3138 (Athens, iv BC), Hsch.

ἴλιον, add 'cf. ἴλια and Lat. *ilia*'

ἰλίων, ονος, ὁ, = σύγγαμβρος, Choerob. in *An.Ox.* 2.221; cf. εἰλίονες.

ἴλκα· γλοιός, ῥύπος Theognost.*Can.* 15, Hsch.

ἰλλᾰτίζω, app. = ἰλλάζω, *PAlex.* 26.19 (ii/iii AD).

ἰλλός, add 'of the feet, ? *twisted*, ἰλλοί τε πόδες *Suppl.Hell.* 1026'

*ἰλλούστριος, ὁ, cf. Lat. *illustris*, as rank, *SEG* 36.1326 (Palestine, vi AD), *Cod.Just.* 40.20.16 pr., Just.*Nov.* 15.1 pr.; sp. ἴλλυστρος *MAMA* 3.504.

ἴλμη, add 'cf. εἴλεα, εἶλος'

Ἰλύθειον, Ἰλύθυιον, v. °Εἰλείθυιον.

ἰλυσπάομαι, line 1, for 'crawl, like a worm' read '*move like a snake or worm*'; line 3, before 'Ael.*NA* 8.14, 9.32' insert 'of other creatures, *move convulsively, wriggle from side to side, squirm*' and after it insert 'act. ἰλισπῶντες· συνειλοῦντες Hsch.'

ἰλυώδης, last line, for 'τὸ -ώδες' read 'εἰ μηδὲν ἔχουσα διεφθορὸς ἐν ἑαυτῇ μηδ' ἰλυῶδες'

*ἰμαγινιφέρ, ὁ, Lat. *imaginifer*, *PBeatty Panop.* 2.297 (AD 300), *SB* 8430.3.

†ἰμαῖος, ον, *of hauling* or *drawing*, ἰμαῖον (sc. ᾆσμα) kind of work-song, ἀείδει καί πού τις ἀνὴρ ὑδατηγὸς ἱμαῖον Call.*fr.* 260.66 Pf.; so ἰμαῖος (sc. ᾠδή) ἡ ἐπιμύλιος καλουμένη, ἣν παρὰ τοὺς ἀλετοὺς ᾖδον Trypho ap.Ath. 14.618d, cf. Sch.Ar.*Ra.* 1297, Hsch.

ἱμαντόδετος, add 'κολλήματα Ps.-Callisth. 120.2 K.'

*ἱμαντοπαικτική (sc. τέχνη), ἡ, *sparring with boxing-thongs* (opp. serious boxing), Eustr.*in EN* 10.17, 16.9.

ἱμαντόπους, for '(ἱμάς .. 1' read 'having a malformation or abnormality of foot or leg, the nature of which is uncertain, = Lat. *loripes*'

ἱμαντοσκελής, for 'Tz. l.c.' read 'Apollod. l.c.'

ἱμάς I 1 b, after 'etc.' insert '(in 23.363 has also been taken as *lash of a whip*)'   d, for this section read '(naut.) *lifts*, *PZen.Col.* 100, *PCair.Zen.* 754, 756 (all iii BC)'   2, add 'k title of book of problems by one Anaxagoras, Anaxag.A 40 D.-K.'   at end for 'always ῑ' read 'regularly ῐ, but v. ἱμάω'

ἱμάσθλη, line 2, after 'Eranos 13.88 (pl.)' insert 'fig., δαιμονίης κακότητος ἐβακχεύθησαν ἱμάσθλῃ Nonn.*D.* 9.39, al.'

*ἱμᾱσιοπώλης, v. °ἱματιο-.

ἱμάτιον, at end after 'εἱματισμός' add 'Dor. ἡμάτιον *SEG* 9.13.15 (Cyrene, iv BC)'

ἱμᾰτιοπρᾱτης, add '*MAMA* 3.619 (Corycus), *BCH* suppl. 8.157 (vi AD), *PKlein.Form.* 969 (vi/vii AD)'

ἱμᾰτιοπώλης, add 'ἱμασιοπώλης *UPZ* 7.8'

ἱμᾰτιοπωλικόν, add 'also -ική, ἡ *PErasm.* 5'

ἱμάω, after 'from a well' insert 'Men.*Dysc.* 191'

Ϝιμβάναι, add 'cf. ἴμμας'

Ἴμβρος, line 1, for 'ὁ' read 'ἡ'; before 'epith. of Pelasgian Hermes' insert 'masc.'

ἱμείρω II, add 'Democr. 223 D.-K.'

*ἵμερα· τὰ πρὸς τοὺς καθαρμοὺς φερόμενα ἄνθη καὶ στεφανώματα Hsch.; cf. °ἵσμερα.

ἱμερόεις, at end after 'of persons' insert 'Hes.*Th.* 359'

ἵμερος I 1, line 2, for 'raised' read 'roused'; line 5, after '(Od.) 4.113' insert 'τῶν ἀντερώντων ἱμέρῳ πεπληγμένοι A.*Ag.* 544' deleting this ref. in lines 12/13   II, line 2, delete 'only in' before 'neut. as adv.' and add 'sup. *APl* 16.182 (Leon.Tarent.)'

*ἱμερόφοιτος, ον, *wandering in a frenzy of desire*, Nonn.*D.* 15.227.

ἱμερτός, add 'fem. as pr.n. Ἱμερτή· τὸ πάλαι ἡ Λέσβος Hsch.'

*ἵμεστος (ἱμέσιτος La.)· δίκη (Sicel) Hsch.

*ἱμητήρ, ῆρος, ὁ, κάδον ἱμητῆρα bucket *for raising water*, dub. in *Inscr.Délos* 1417A 146 (ii BC).

ἱμιμναῖος, v. °ἡμι-.

*ἵμιτραον· ὑπόζωσον, Πάφιοι Hsch. (perh. = ἐμμίτρασον, fr. *ἐμμιτράω; ἱμίτραιον· ὑπόζωστρον La.).

ἱμπερᾰτωρ, ορος, ὁ, Lat. *imperator*, *IG* 5(1).1454.3 (i BC), *Thasos* 175 i 11.

*ἱμφειβλᾱτώριον, τό, *cloak fastened with a fibula*, *Rev.Phil.* 1937.106 (Pessinus, ii AD); cf. °φιβλατώριον, Lat. *infibulo*.

*ἱμφορά, ἁ, perh. Cypr. for εἰσφορά, *contribution*, i-po-ra-se ἰ(μ)φοράς *ICS* 318 A III.

*ἴμψας· ζεύξας. Θετταλοί Hsch.; cf. Ϝιμβάναι, °Ἴμψιος.

*Ἴμψιος, ὁ, epith. of Poseidon in Thessaly, Fs.*Theocharis* p. 381.

ἵνα B II 3 c, line 3, delete 'cf.' and for 'Pl.*Ap.* 26d' read 'Pl.*Ap.* 26c'   add 'IV w. τοῦ and inf., ἵ. τοῦ τὰ ὅλα συντελεσθῆναι *OGI* 5.15 (letter of Antigonus Monophthalmus, iv BC).'

ἰνάρει· μαστεύει Hsch.; cf. νάρειν.

ἰνάρετος, v. °ἐνάρετος.

ἰνάσσω, for 'Call.*Fr.anon.* 126' read '*Suppl.Hell.* 1036'

*Ἰνάχεια, τά, festival of Leucothea in Crete, Hsch.

*Ἰναχίδαι, οἱ, *descendants of Inachus*, Argives, E.*IA* 1117, *SEG* 35.267c (Argos, late iv BC), *Suppl.Hell.* 1088.

*Ἰναχίη, ἡ, poet. for Argos, epigr. in *SEG* 11.325 (Argos, iv/v AD).

*Ἰνάχιος, = Ἰνάχειος, Call.*Epigr.* 57 Pf.

ἰνδάλλομαι, add '3 trans., *deem like*, D.Chr. 11(12).53, S.E.*M.* 11.122.'

ἴνδαλμα, for '*IG* 3.1403' read 'epigr. in *IG* 2².12142 (iii AD)'

*Ἰνδία, ἡ, *India* (the actual area is not defined), Luc.*Alex.* 44, *SEG* 31.1116 (Phrygia, Rom.imp.).

*ἰνδικοβᾰφος, ὁ, *indigo-dyer*, Anon.Alch. 418.22.

ἰνδικοπλάστης, delete the entry.

*ἰνδικοπλύντης or -πλύτης (-πλεύστης cod.), ου, ὁ, *dyer*, *Gloss.*, cf. *BCH* 77.658.

*ἰνδικοπλύτιον, τό, *dyer's shop*, *PAmst.inv.* 62 (*Mnemosyne* ser. 4 30.146, ii AD).

Ἰνδικός I, after 'Hdt. 3.98' insert 'Ἰνδική alone, Hdt. 3.106.2, Lxx *Es.* 3.12, *SEG* 24.1225 (Philae, i BC); σίδηρος Ἰνδικὸς καὶ στόμωμα καὶ ὀθόνιον Ἰνδικόν *Peripl.M.Rubr.* 6'

*ἰνδικτίων, ωνος, ἡ, Lat. *indictio*, *SEG* 26.1697 (Palestine, c.AD 299/300), 31.1389(a) (vi AD), al., also ἐνδικτίων *ICilicie* 118 (AD 596).

*Ἰνδοκτόνος, ὁ, = Ἰνδοφόνος, applied to Egyptian god, *APAW* 1943(14).8 (Chalcis, iii AD).

*ἰνιμίνα· ἐνήμισυ Hsch.; v. ‡ἡμίνα I.

*ἰνιοράφος, v. ‡ἡνιορρ-.

ἶνις, masc., add 'A.*Supp.* 251, *AP* 15.26 (iii BC), *IG* 14.1374 (ii AD)'; fem., add '*Inscr.Délos* 1533.9'; line 3, delete 'Trag. only in lyr.'; at end for '*Inscr.Cypr.* 101, al.' read '*ICS* 15c (vi BC), 7.5 (iv BC), al. (masc.)'

*ἰνκαφύτευε, v. °ἐγκαταφυτεύω.

*ἴνκολας, ὁ, acc. -αν, Lat. *incola* in sense *resident alien*, Modest.*Dig.* 27.1.13.12, 50.1.35.

*ἴνσπεκτον, τό, *inspection*, *review*, *SEG* 27.1139.34 (Ptolemais in Cyrenaica, vi AD).

*ἰνστρουμεντᾰριος, ὁ, Lat. *instrumentarius*, *keeper of documents*, ἰ. ταβουλαρίων *MAMA* 1 p. xiv, Lyd.*Mag.* 3.19, 20.

*ἰντερκᾰλάριος, α, ον, Lat. *intercalarius*, *Inscr.Prien.* 71, 76 (late i BC).

*Ἰνύνια· ἑορτὴ ἐν Λήμνῳ Hsch.

ἰνώδης, line 3, delete 'sinewy, X.*Cyn.* 4.1'

ἰξαλῆ, at end after 'Theognost.*Can.* 14' insert 'ἰττέλη prob. in Poll. 7.211, 10.57'

ἰξευτής, after '*bird-catcher*' insert '*AP* 9.337 (Leon.Tarent.)'

ἰξία III, add 'Plu.*Mar.* 6'

ἰξύας· ἰχθύς τις Hsch.

ἰξώδης, add 'adv. -δῶς Archig.ap.Aët. 6.8 (137.11 O.), Gal.ap.Aët. 9.10'

ἰόβας (ἰόβλης La.)· κάλαμος παρὰ Κρησίν Hsch.

*ἰοβρυχέουσα· ἀνιωμένη. πικραινομένη Hsch.

*ἰόβρωτος, prob. *devoured by rust*, Hsch.

*ἰοδερκής, ές, *looking with dark eyes*, Κύπρις (?)B.*fr.* 61 S.-M.

†ἰόζωνος, ον, *having a purple girdle*, Call.*fr.* 110.54 Pf., Hsch.

*ἰόκολπος, ον, *wearing a purple robe*, Sapph. 21.13, 30.5, 103.6, 7 L.-P.

Ἰόνιος, for 'of, called after, Ἰώ' read 'Ionian'; before 'across which Io swam' insert 'explained as that'

ἴορκες, αἱ, v. ‡δορκάς.

ἰός (B), line 2, after 'Ruf.*fr.* 118' insert 'Gal. 16.621'

ἰός (C), line 3, after 'Gal. 12.218' insert 'applied loosely to gold and silver, *Ep.Jac.* 5.3'

*ἰός (D), ὁ, = υἱός, *BICS* suppl. 10.34, *SEG* 23.640 (both Cyprus, iv BC).

ἰοστέφανος, line 2, after 'h.Hom. 6.18' insert 'Ϝιοστεφάνοι Ἀφροδίται *SEG* 32.395 (Laconia, archaic)'

ἰού 2, line 2, delete 'Grg. 499b'

*ἰούγερα, τά, Lat. *iugera*, *IMylasa* 272, 273, al. (iii/iv AD).

*ἰούγον, τό, Lat. *iugum*, *unit of land* (for assessment purposes, the area depending on use and quality), Just.*Nov.* 17.8 pr., al.

Ἰουλαῖος, line 3, Ἰούλιος, add 'the month *of July*, Lang *Ath.Agora* xxi Hc9, Mitchell *N.Galatia* 424'

*Ἰουλιεύς, έως, ὁ, name of a month in Egypt, = Choiak, *Stud.Pal.* 22.173.16 (AD 40), *PMich.ined.* 1285 (v. *JEA* 13.185).

ἰούλιος, ὁ (or -ον, τό), app. fiscal unit (in contexts in conjunction w. ζυγοκέφαλον and °ἰούγον), Just.*Nov.* 17.8 pr., al.

ἰουλίς, add 'cf. °ἰθουλίς'

ἴουλος IV, for 'Arat. 959' read 'Arat. 957'

ἰουλοφόρος, for pres. ref. read '*GVI* 385'

ἰουλώδης, for '*scolopendra-like*' read '*millipede-like*'

*****Ἰούνιος**, μὴν Ἰ., the month *June*, *Corinth* 8(1).145 (AD 446), *SEG* 30.1711.4 (Arabia, AD 596); perh. also Ἰόνιος *Inscr.Cret.* 4.181.3 (ii AD).

*****Ἰουστινιανός**, ή, όν, *of Justinian*; Νουμίδαι Ἰουστινιανοί, body of soldiers stationed in Egypt, *BGU* 2197.7 (vi AD).

*****ἰοχάλκιον**, τό, *verdigris*, rest. in Anon.Alch. 20.2.

*****ἰόχαλκος**, ὁ, *verdigris*, Anon.Alch. 281.1.

ἰπνασία, add '(ἰπναστά La.)'

ἴπνασμα· κάπνη Hsch. (La.) (v. ἴπαμα).

ἰπνευτής, delete 'prob. for ἰπνίτης .. (Phan.)'

*****ἰπνεύτης**, ου, (-ᾱς, ᾱ) ὁ, = ἰπνίτης, ἰπνεύτα φθόϊος *AP* 6.299 (Phan.; ἰπνέστα cod.).

ἰπνίτης, at end delete 'ἰ. φθοῖς .. -ευτής)'

*****ἰπνιών**, ῶνος, ὁ, Cret., = ἰπνών, *Inscr.Cret.* 4.73A9 (Gortyn, v BC).

*****ἰπνοκοδόμαν** (cj. for ἰπνοδόμαν, q.v.)· τὴν φρύκτριαν. Κρῆτες Hsch. (La.).

†**ἰπνός**, ὁ, *domestic oven*, Semon. 7.61, Hdt. 5.92 η', Hp.*Mul.* 1.220.4, Antiph. 174.4 K.-A., Archestr. in *Suppl.Hell.* 177.4, Diph. Siph.ap.Ath. 2.54a; as a source of warmth, προσιόντες εἶδον αὐτὸν θερόμενον πρὸς τῷ ἰπνῷ Arist.*PA* 645ᵃ20; used for heating water, Dsc. 5.88. **b** *kiln*, ἀπὸ τοῦ κεραμέου ἰπνοῦ Hp.*Epid.* 4.20, perh. also id.*Morb.* 2.47. **2** *room containing an oven* (a large *oven* could be understood in some cases), Ar.*V.* 837, *Av.* 437, App.*BC* 4.2.22. **3** a kind of *brazier* filled with glowing charcoal to provide light out of doors, Ar.*Pax* 841, *Pl.* 815, *SIG* 1027.13 (Cos, iv/iii BC), Ael.*NA* 2.8. **4** Ἀριστοφάνης (369 K.-A.) δὲ ἐν Κωκάλῳ καὶ τὸν κοπρῶνα οὕτως εἶπεν Hsch. s.v. ἰπνός, perh. based on a misunderstanding; Myc. *i-po-no*.

ἶπος 1, delete '(815 ?)' and after 'id. 7.41' insert 'ἶπον Call.*fr.* 177.33 Pf.'; delete 'cf. εἶπος'

ἴππα, delete 'as pr. n. .. *H.* 48.4'

†**ἰππαπαῖ**, comic adaptation of the rowers' cry (ῥυππαπαί) for horses imagined as rowing warships, Ar.*Eq.* 602.

ἱππάριον, add '**4** name of an eye-disease, Cyran. 35 (1.16.20 K.). **5** bird resembling the χηναλώπηξ, Hsch.'

ἱππάρχης, lines 1/2, for 'Samothrace' read 'Cyzicus'

ἱππασία I 1, add '*SEG* 26.121.42 (Athens, i BC)'

ἱππαστήρ, for 'ὁ, .. μύωψ' read 'masc. adj., *used in riding*, μύωψ'

*****ἱππᾰφεσία**, ή, = ἱππάφεσις, prob. in Ps.-Callisth. 20.1.

ἵππειος 1, add '**b** ἱππεία, ή (sc. νευρά), *horsehair bowstring*, Hsch.' at end add 'Myc. *i-qe-ja* (fem.), epith. of goddess Potnia'

ἱππελάφος, for '(Arist.*HA*) 499ᵇ2' read '(Arist.*HA*) 499ᵃ2' and add 'Tim.Gaz.ap.Ar.Byz.*Epit.* 131.15'

ἱππεύς I 1, after '(Il.) 23.262' add 'Hes.*Sc.* 305'

ἱππεύω IV, add 'w. perlative acc., *drive over*, οὐρανόν ἰ. Nonn.*D.* 23.239'

ἱππιάναξ, for 'king of horsemen' read 'commander of horsemen'

ἱππικός I 2, add 'of a cavalry unit (σπεῖρα), *SEG* 33.1266 (Palestine, i AD)' add '**4** -κός, ὁ, *groom*, *POxy.* 922.6 (vi/vii AD).'

ἵππιος II, line 3, after 'ἀγών' insert '*IG* 11(2).203A67 (Delos, iii BC)' add '**IV** fem. as subst. in Myc., = *chariot*, *i-qi-ja*.'

ἱππιοχαίτης, for 'shaggy with horsehair' read 'of a helmet-plume, horsehair'; add 'also ἱπποχαίτης, Hsch.'

*****ἱππιόχαρμος**, ον, = ἱππιοχάρμης, *Suppl.Hell.* 991.20 (poet. word-list, iii BC).

*****Ἱπποδαμάντειος**· οἶνος ποιὸς ἐν Κυζίκῳ Hsch.

*****Ἱπποδάμεια**, ἡ, *tamer of horses*, χεῖρα ἱπποδάμειαν (w. ref. to charioteer) Euph. 127 vGr.; as pers. n., Il. 2.742, etc.

ἱππόδᾰμος, delete 'fem. Ἱπποδάμεια' to end.

ἱπποδίνητος, add '*Suppl.Hell.* 991.19 (poet. word-list, iii BC)'

†**ἱπποδιώκτης**, ου, ὁ, perh. = ἱππηλάτης, Theocr. 14.12, ἱπποδιώκτας· ἡνίοχος Hsch. **2** a kind of mounted gladiator, *ISmyrna* 404, *Sardis* 7(1).162. **3** functionary in the circus (sts. acting also as ἀφέτης "*starter*"), *OAshm.Shelton* 93, 97, al.

Ἱπποδρόμιος I, add 'in Thessaly, *BCH* 79.446.36, 447.19 (ii BC)'

ἱππόδρομος I, add 'in poet. ref. to the sun's chariot, E.*IT* 1138'

*****ἱπποζύγιος**, ὁ, gloss on ἐρυσάρματας (acc. pl.), Sch.Gen.Il. 15.354.

ἱππόθεν, add '*from horseback*, *IG* 7.1828.6 (Thespiae, c.AD 125, metr. dedication of Hadrian); of one falling *off his horse*, ἰ. ὠλίσθησε Nonn.*D.* 36.208'

*****ἱπποθοίνην**· τὴν μεγάλην εὐωχίαν Hsch. (La.).

ἱπποθόρος, after 'mules' insert 'Anacr. 32 P.'

*****Ἱπποκαθέσια**, τά, (καθίημι I 2) *Horse-Races*, name of a festival at Rhodes, *Tit.Cam.* 153.8 (iii BC), *ASAA* 30/32(1952/4).256, 258 (i BC).

*****ἱπποκενταυροδελφίς**, ῖνος, ἡ, imaginary creature (depicted on a seal), *PBatav.* 30.2 (ii AD).

ἱπποκοινάριον, for '*Raccolta* .. 374' read '*SB* 7182.45 (late Ptolemaic)'

*****ἱπποκομικός**, ή, όν, *of a groom*, *POxy.* 1858.4 (vi/vii AD).

ἱπποκύων, add 'title of Menippean satire, Varro *fr.* 220 A.'

ἱππόλοφος, add '*Suppl.Hell.* 991.15 (poet. word-list, iii BC)'

*****Ἱππολύτειον**, τό, *shrine of Hippolytus*, Ἀφροδίτης ἐν Ἱππολυ[τείῳ *SEG* 22.47.66 (Attica, v BC); app. incorporating a gymnasium, *IG* 4.754 (Troezen).

ἱππομανής I, add '**2** *mad on horses*, Nonn.*D.* 37.275, Sch.BT Il. 5.25.'

*****ἱππόμαυρος**, ὁ, app. *Moorish horseman*, ἱππόμαυροι σκουτάριοι *Stud.Pal.* 20.98 (AD 348).

ἱπποπάρηος, read 'ἱπποπάρηος'

*****ἱπποπείρης**, ου, ὁ, *experienced in horses*, Anacr. 72.6 P.

*****ἱππόπορνος**, ὁ, ή, *vulgar prostitute* (cf. ἵππος VII), Men. i p. 101 K.-Th., Alciphr. 1.38, al.; facetiously interpr. as *mounted prostitute*, Diog.ap.Ath. 13.565c.

ἱπποπόταμος, for '-ποτάμις' to end read '-ποτάμιος, ὁ, *PMag.* 13.319, *POxy.* 1220.21 (-ις, iii AD)'

*****ἱππόριζος**, ὁ, *a medicament for horses*, Hippiatr. 1.88.10.

ἵππος I 1, line 11, after 'Lys. 19.63' insert 'of the west wind imagined as a horse, Call.*fr.* 110.54 Pf.; applied as sobriquet to sexually intemperate women, Ael.*NA* 4.11. **b** name given to small boats at Gades having a horse as figure-head, Posidon.Hist. 28 J.' **2**, add '**b** ἵ. ἀργυροῦς, Corinthian coin bearing a figure of Pegasus, E.*fr.* 675.' **IV a**, delete the section. at end add 'Myc. *i-qo*'

ἱπποσείρης, delete the entry (v. °ἱπποπείρης).

ἱπποστασία, add 'Ps.-Callisth. 18.16 K.'

ἱπποστάσιον, after 'Lys.*fr.* 56 S.' insert 'Sm.*Da.* 11.45'

ἱππόστασις, after '*stable*' insert '*IG* 4².109 iii 91 (Epid., iii BC)'

ἱπποσύνη, line 3, for '(= *IG* 1².946)' read '(= *IG* 1³.1181)' and insert 'E.*Or.* 1392'

ἱπποτέκτων, add 'Call.*fr.* 197.3 Pf.'

ἱππότης (A) I, line 3, after 'Il. 2.336, etc.' insert 'ἱππότα as gen., Arat. 664'; after 'S.*OC* 59' insert 'X.*Eq.* 8.10'; line 5, after 'Ascl.*Tact.* 10.2' add 'dial. pers. n. Ἰκκότας *SEG* 36.626 (Maced., iv BC)'

ἱππότιγρις, for '*a large kind of tiger*' read '*zebra*' delete 'cf. ἵππος VII' and add 'Tim.Gaz. in Haupt *Opusc.* iii 283'

†**ἱππότις**, ιδος, ἡ, (*female*) *driver of horses*, Triph. 670, Nonn.*D.* 1.172. **2** adj. *used by drivers*, ζώνη *APl.* 336.

ἱπποτροφέω, add '**2** *perform the liturgy of* ἱπποτροφία, *SEG* 33.1053 (Mysia, ii BC).'

ἱπποτροφικός, add '-κά, τά, prob. *literary works on horse-rearing*, Hsch. s.v. σφυροδέται'

ἱπποτρόφος II 2, add '*JRS* 18.174 (Gerasa, prob. iii AD)'

ἵππουρος 1, add '*IGC* p. 99 B11 (Acraephia, iii/ii BC)' add '**4** gloss on σκίουρος, *squirrel*, Hsch.'

*****ἱπποφάτης**, ου, ὁ, *slayer of horses*, *Suppl.Hell.* 991.16 (poet. word-list, iii BC).

*****ἱπποφονία**, ή, *sacrifice of horses*, Ps.-Callisth. 125.3.

ἱπποφόρβιον I, for 'Arist.*HA* 576ᵃ20' read 'Arist.*HA* 576ᵇ20'

ἱπποφορβός 2, for 'flute' read 'aulos' at end add 'Myc. *i-po-po-qo-i* (in sense 1, dat. pl.)'

*****ἱπποχαίτης**, v. °ἱππιοχαίτης.

*****Ἵπτα**, Hipta, nurse of Dionysus, Orph.*H.* 48.4, 49; μητρὶ Ἵπτα *ABSA* 21.169 (Maeonia), al., cf. Ἵπτα· ὁ δρυοκόλαψ ἐθνικῶς. καὶ Ἥρα Hsch.

ἵπταμαι, before 'Mosch. 3.43' insert '*APl.* 275.4 (Posidipp.)'; add 'Phryn. 297 F.'

*****ἰρᾶν**, v. °εἴρην.

ἰρέα, delete the form.

*****ἴρην**, v. °εἴρην.

†**ἰρίτης**, ου, ὁ, name of a stone, Socr.Dion.*Lith.* 51.1 G.; cf. *iritis*, Plin.*HN* 37.138.

ἰρμοφόρος, delete the entry (v. °φορμοφόρος).

ἰρών, delete the entry (v. °οἰρών).

ἰρωστί, add 'Semon. 24.2'

ἴς (B), line 2, after 'Od. 9.538' insert 'cj. in Pi.*P.* 4.253'

ἰσάζω I, add '*to make level*, Nonn.*D.* 43.132' **II 1**, add 'A.R. 3.1045'

*****ἰσαίων**, ωνος, ὁ, ή, *having an equal span of life*, Hes.*fr.* 1.8 M.-W.

ἴσᾱμι, add 'part. ἴσαις *AP* 7.718 (Noss.)'

*****ἰσαντινόϊος**, ον, *equivalent to the Antinoan games*, ἀγών *JEA* 37.87 (Memphis, iii AD), *PGiss.Univ.*inv. 252 (*ASP* 1.20).

ἰσαριθμέω, add 'Aristid.Quint. 3.19'

*****ἰσατέον**· γνωστέον. ἰστέον Hsch.

*****Ἰσαυρικός**, ή, όν, *made in Isauria*, στιχάριον Ἰ. *PVindob.* G16859 (*Tyche* 2.10, vi AD).

Ἰσεῖον (s.v. Ἰσιεῖον), add 'sp. Ἰσῖον *SEG* 26.1777 (Egypt, Ptolemaic)'

ἰσημερινός, add '-ή, sc. τροπή, on a sundial, *SEG* 31.931 (Aphrodisias, v AD)' ἰσήρης, line 2, for 'ῥαιβοῖσιν ἰσήρεες' read 'ῥοικοῖσιν ἰσήρεες ἄντα παγούροις'

*ἰσθμή· φρόνησις Theognost.Can. 14; cf. ἰσμή.

ἰσθμιάζω I, after 'Isthmian games' insert 'A.fr. 78a.34, 75 R.'

Ἰσθμιακός, add 'πίτυν -ήν epigr. in SEG 37.712 (Chios, ii AD)'

Ἰσθμιάς, line 1, before 'άδος' insert '(also Ἰθμιάς, SIG 36A26 (Delphi, v BC))'

*Ἰσθμιάτης, ου, ὁ, = Ἰσθμιαστής, Inscr.Délos 1441A53 (pl., ii BC).

Ἰσθμικός, add 'II παῖδες 'I. boy competitors of the age fixed for the Isthmian games, SIG 1065.9, al. (Cos, i BC/i AD); without παῖδες, SEG 3.335.7, al. (Thespiae, ii AD).'

ἴσθμιον II 2, add 'ἴσ[θ]μιον <φ>ρεατ(ος) Lang Ath.Agora xxi K1 (vi BC)'

Ἰσθμιονίκης, after 'B. 9.26' insert add 'Lang Ath.Agora xxi L12'

ἰσθμός I 2, after 'fauces' insert 'Nic.Al. 80, 508' II 4, transfer 'Inscr.Délos ll. cc.' to section II 1.

*Ἰσιάστησις, εως, ἡ, the worshippers of Isis collectively, PColl.Youtie 14.24 (132/1 BC).

Ἰσιδεῖον, add 'sp. Εἰσιδεῖα IG 12(5).606 (Ceos)'

ἰσικιάριος, add 'MAMA 3.343 (Corycus), εἰσικι[άριος] SEG 35.1110 (Ephesus, iii AD), PRyl. 640.10, 641.30, al. (iv AD, εἰσ-, εἰσσ- papp.)'

†ἰσικιομάγειρος, ὁ, sausage-maker or -seller, PMich.inv. 3780 (ZPE 62.133, AD 458), SB 9456 (AD 594).

ἰσίκιον, line 2, for 'Ath. 9.376b' read 'Ath. 9.376d'

*ἰσικιοπώλης, ου, ὁ, sausage- or mince-seller, PLond. 1028.12 (vii AD).

Ἴσις, line 3, after 'etc.' insert 'Εἴσεως (gen.) ASAA 30/32(1952/4).264, no. 11 (Rhodes)'

*Ἰσιτύχη, ἡ, goddess combining the natures of Ἴσις and Τύχη (Fortuna), SEG 30.708, CIL 14.2867 (in Roman letters); sp. Εἰσιτύχη CIL 4.4138.

ἴσκω (B) III, add '2 call, name, w. double acc., A.R. 4.1718.'

ἴσμα, delete the entry.

*ἰσμαίνει, v. ἰσθμαίνω (s.v. ἴσθμα).

*ἴσμερα· τὰ εἰς τοὺς καθαρμούς Hsch.; ἰσμέρα· τὸ εἰς τοὺς καθαρμούς, Theognost.Can. 14; cf. °ἵμερα.

ἰσμή, add 'cf. °ἰσθμή'

*Ἰσμήνιος, epith. of Apollo, [Ἀπόλλον]ι ἰισμ[ενίοι ..] SEG 22.417 (Thebes, vi BC), Hdt. 1.52, Paus. 2.10.5, Hsch.

ἰσόγραφος, for 'Timo 30.2' read 'Timo in Suppl.Hell. 804.2' and add 'see also °ἰσόκραγος'

ἰσόδρομος, line 2, delete 'abs.' and insert 'νάεσσιν'

ἰσοδυνάμέω, line 3, after 'A.D.Pron. 41.15, al.' insert 'Aristid.Rh. 1 p. 485 S., al., Hermog.Id. 1.11 p. 284 R.'

ἰσοκέφαλος, for 'Ibyc. 16' read 'Ibyc. 4.3 P.'

*ἰσόκλητος, ον, prob. similarly named, epigr. in Lindos 698.17 (c.200 BC; v. Hermes 77.208).

*ἰσόκραγος, ον, equally noisy, cj. for ‡ἰσόγραφος.

*ἰσολεξία, ἡ, app. use of a succession of words of equal length, Rh. 6.328.1.

ἰσόμοιρος, line 1, after 'ον' insert 'or α, ον'; line 2, after '(Gort.' insert 'vii/vi BC'; line 6, after '(Melos)' insert 'λαμβάνωσιν διανομὴν .. ἀνὰ δραχμὰς ἰσομοίρας in equal portions, IEphes. 4123.11' 2, delete 'κίβισιν'

ἰσόνεκυς, for 'dying .. Sch.' read 'virtually dead, E.Or. 200 (lyr.)'

ἰσονέμειος, add 'ISmyrna 574.9 (iii BC)'

ἰσοπαλής 2, for 'equivalent, equal' read 'equal in force, value, effect, etc.' and add 'Pl.Ti. 62e'

*ἰσοπάρθενος, ον, resembling a maiden, in an address to the moon, κύων PMag. 4.2251.

ἰσοπᾶχής, at end after 'codd.' add 'cf. °ἰσόπηχυς'

ἰσόπηχυς, add 'in Sch.Pi.O. 6.154, Sch.Ar.Av. 1283, ἰσοπήχεις is perh. confused w. ἰσοπαχεῖς'

ἰσοπολῖτις, add 'τῆς ἰ. καμίνου Lxx 4Ma. 13.9 (s.v.l., sense unclear)'

ἰσοπύθιος, add 'SEG 31.1287 (Pamphylia, Rom.imp.)'

ἰσόρροπος I 2, add 'ἐπιχείρημα Democr. 26 D.-K.; comp., Sch.B.Il. 12.421' II, line 2, for 'πορεύεσθαι' read 'εὐήνια ὄντα'

ἰσόρυθμος, delete the entry.

ἴσος, line 2, after 'Hsch.' insert 'Fίσος SEG 37.340.4 (Mantinea, early iv BC)'; line 10, after 'Pl.R. 441c' insert 'θυγατέρας ἐξ καὶ ἴσους ἄρρενας (i.e. as many) Plu. 2.312c, ἡμέρας .. ἐπτὰ καὶ ἴσας νύκτας Luc.VH 2.1' IV 2, line 2, after 'D. 14.6' insert 'ἄχρι τῆς ἴσης id. 5.17'; line 8, ἐξ ἴσης, add 'SEG 30.1382 (Lydia, AD 301)'; line 18, after 'Plb. 1.18.10' insert 'ἐπ' ἴσῃ, etc.; ἐπὶ τᾶι FίσFαι [καὶ τ]ᾶι ὁμοίαι on fair and equal terms, Schwyzer 175.2 (Gortyn, v BC or earlier); ἐπὶ τοῖς FίσFοις καὶ τοῖς ὑμοίοις id. 665 A¹.4 (Orchomenus Arcadiae, iv BC); ἐφ' ἴσηι καὶ ὁμοίηι id. 708ª (Ephesus, iv BC)'

ἰσοστάσιος 1, line 5, after 'Dam.Pr. 91' insert 'ἰ. μύρον Hsch.'

*ἰσόστυλον· τὸ στοιχηδόν Hsch. (La.) (after ἰστνάζει, v. ἰστυλόν).

ἰσότης III, add '2 equability of weather, Cat.Cod.Astr. 8(3).196.4.'

ἰσότυπος II, add 'Nonn.D. 2.553'

ἰσουράνιος, add 'γένεθλα MDAI(A) 56.122 (Smyrna), epigr. in SEG 28.541.6 (Maced., Hellen.)'

ἰσοφόριος, delete the entry; v. °εἰσωφόριος.

ἰσοχρόνιος, line 3, after 'Ptol.Tetr. 36' insert '(v.l.)'

*ἰσόχωρος, ον, gloss on ἰσόπεδος, Hsch.

ἰσόψυχος, add '3 precious as life, Sch.E.Andr. 419.'

Ἰσπανός, for "Ἰσπανόν" (line 1) to end read "Ἰσπανόν, τό, kind of oil, Gal. 10.790.13, 822.8, 12.513.9; see also ‡Σπανός'

*Ἰσπνιᾶται, οἱ, ropes, Theognost.Can. 14.25.

*ἰστάκη· δρέπανον. Βοιωτοί Hsch., Theognost.Can. 14.

ἵστημι I, line 2, after 'E.Supp. 1230' insert 'καθίστη Ar.Ec. 743' and after 'Il. 9.202' insert 'προσίστα Macho 20 G.' II 1, line 3, for 'Dor. στᾶθι' read 'Dor. and Aeol. στᾶθι' and for 'Sapph. 29' read 'Sapph. 138.1 L.-P'; line 9, after 'Hdt. 7.152' insert 'Aeol. pf. part. fem. παρεστάκοισαν Alc. 298.7 L.-P.'; line 11, after 'ἔσταθι' insert 'Od. 22.489, Ar.Au. 206'; line 17, before 'POxy. 68.32' insert 'Inscr.Délos 1443O6 (ii BC)'; line 3 fr. end, delete 'hence' and after 'Hom.Epigr. 15.14' insert 'Ar.Lys. 634, Pl.Smp. 220d, D. 20.37, cf. Th. 3.37.3, 102.6 (καθ-), Pl.R. 587b (ἀφ-)' A III 5, add 'ἐστάθησαν ἕως ἐνταῦθα οἱ λόγοι Ἰερεμίου have been established, Aq.Je. 51(28).64' B II 1, line 6, after 'Arist.HA 588ª8' insert 'ἔστη τὸ αἷμα the blood was staunched, Lxx Ex. 4.25'

Ἰστιαϊκός, add 'also Ἰστιαιικός, Inscr.Délos 1429Βii35, al.; ib. Ἑστιαικός, ib. 1441Aii108 (ii BC)'

*ἰστιᾱτικός, ή, όν, of a fund for temple-expenses deposited in the sanctuary of Hestia, ἀργύριον Inscr.Délos 449A33 (ii BC).

*ἰστιορραφεῖον (-φῖον), τό, sail-mender's workshop, Inscr.Délos 1416Βi92 (ii BC).

†ἰστιορράφος, ὁ, sail-stitcher, Poll. 7.160; as a term of opprobrium, Ar.Th. 935; also ἰστιαρράφος PCair.Zen. 754.1 (iii BC; -αράφος), gramm. in Reitzenstein Ind.Lect.Rost. 1892/3 p. 4.

*ἱστιοφόρος· ἁρμενοφόρος. καὶ ἰστοφόρος Hsch.

ἰστοβοεύς, line 4, after 'Paus. 9.37.4' insert 'also, pole of a waggon, Opp.C. 1.532.' and delete 'Acc. ἰστοβόην' to end.

*ἰστοβόης, ὁ, acc. -βόην, plough-tree, AP 6.104 (Phil.).

**ἰστόεις, εσσα, εν, form ἰστόFεις, Myc. i-to-we-sa (fem.), fitted with an upright part.

*ἰστοπένδιον, v. °στιπένδιον.

*ἰστοποιός, ὁ, loom-maker, MAMA 3.693 (Corycus).

ἰστοπόνος, after 'working at the loom' insert 'ἰστοπόνοι μείρακες Lyr. adesp. 57(a) P.'

ἰστός II 1, add 'στίχης ἀπὸ ἰστοῦ Edict.Diocl. 7.56' III, for 'leg' read 'shank, κώλων .. ἰστοί'

*ἰστοτέλεια, fem. adj. mistress of the loom, of Athena, Nonn.D. 6.154, 37.312, 45.49.

ἰστοφόρος, after 'Hsch.' add 's.v. ἱστιοφόρος'

Ἰστριᾱνός, line 3, for 'Ar.fr. 88' read 'Ar.fr. 90 K.-A.'; add 'm. pl. subst., inhabitants of Istria, D.C. 38.10.3, etc.'

*Ἰστριεύς, έως, masc. adj., of Istria, Lyc. 74.

ἰστῶν, add 'epigr. in SEG 30.1429 (Nicaea, ?i AD)'

ἵστωρ I, add 'glossed by συνθηκοφύλαξ, Hsch. (s.v. ἵστορας)'

ἰσφαίνειν, add 'Theognost.Can. 14'

*ἴσφατον· βίαιον πεπραγμένον Hsch. (βιαίως πεπληγμένον La.); cf. °ἰσφαίνειν.

†ἰσχάδιον, τό, dim. of ἰσχάς, fig, Ar.Pl. 798, IG 4².742.45 (Epid.).

ἰσχαδοκάρυον, add 'written σχαδο- Mél.Glotz 872.14 (Maced., ii AD)'

*ἰσχαίνω, v. °ἰσχναίνω.

ἰσχαλέος, form ἰσχν-, add 'Hp.Mul. 1.17'

ἰσχάς I 1, after 'fig' insert 'Hippon. 8 W.'; add 'in epigrams based on pun on the senses of fig and, transf., anus, APl. 240.1, 8 (Phil.), 241.2, 5 (Marc.Arg.); pl., meton., stalls where dried figs are sold, Teles p. 13 H.'

ἰσχέπλινθα, after '(perh. door-jambs)' insert 'or perh. sockets to secure lintel and sill'

*ἰσχιαλγικός, όν, suffering from pains in the hips, Inscr.Cret. 1.xvii 9 (Lebena).

ἰσχναίνω 2, line 2, for 'σφυδῶντα' read 'σφριγῶντα' and for 'A.Pr. 382' read 'A.Pr. 380' at end delete 'In the metaph. sense'

*ἰσχνοπρεπεῖς, gloss on συναγέσκεο, Hsch.

ἰσχνός 3, adv. -νῶς, add 'slightly, τὸν ἰ. ὀξύν Archig.ap.Gal. 8.87.7, 106.15, 107.5'

ἰσχνόφωνος, line 2, delete 'Gal. 17(1).186' (v. °ἰσχο-).

*ἰσχόφωνος, ον, having a constriction in one's voice, Gal.in CMG 5.10(1).94.10; also v.l. for ἰσχνο- in Hdt. 4.155.1, cf. AB 100.

ἰσχυρίζομαι II 1, line 4, for 'persist .. Th. 7.49' read 'τοσαῦτα λέγων ἰσχυρίζετο kept insisting on these points in discussion, Th. 7.49.1; cf. abs. μή τι καὶ πλέον εἰδὼς ὁ Νικίας ἰσχυρίζηται i.e. lest N.'s insistence might not be based on superior knowledge, ib. 4' 2, add 'Din. 1.8'

ἰσχυροπαίκτης, for 'one who plays valiantly' read 'app., one who performs feats of strength for entertainment'

ἰσχυρός, line 3, after 'Hp.Art. 50' insert 'of drink, potent, ἰ. ποτόν Luc.Nigr. 5'

ἰσχῡρόστομος, ον, *having a strong mouth*, i.e. beak, Cyran. 87 (3.12.2 K.).

ἰσχύς Ι 2, add 'personified, *Pap.Lugd.Bat.* xxv 8 i 5 (after AD 231)' **3**, line 2, after 'A.*Pr.* 214' insert 'κατ' ἰσχύος τρόπον id.*fr.* 281a.20 R.' **5**, add 'also of works of art, D.H.*Is.* 4'

ἰσχύω 2, add 'c *to be able to, succeed in*, εἰ δὲ οὐκ ἰσχύσειε παρὰ τοῦ δανεισαμένου λαβεῖν ἢ εἰς μέρος ἢ εἰς ὁλόκληρον Just.*Nov.* 4.1, 22.15.1.'

ἴσχω ΙΙΙ 3, add 'εὖ κτερέων ἴσχοντα A.R. 4.1536'

ἰσωνία, after 'Ar.*Pax* 1227' insert 'Lys.*fr.* 48 S., *PSI* 670 (iii BC)'

ἴσωρος, ον, *equal in age*, θυγατέρες μόνωροι καὶ ἴσωροι *IEryth.* 525.6 (cj., Rom.imp.).

ἴσως IV, add 'Arist.*Ath.* 33.1'

ἰσώστη, v. °εἰσώστη.

Ἰτᾰλίδης, delete 'Call.*fr.* 448' and substitute '*Orac.Sib.* 4.104'

***Ἰτᾰλίης**, ητος, ὁ, Ion. for *Italian*, Antioch.Hist.ap.D.H. 1.12.

Ἰτᾰλίς, add '**2** *period of the games called* Ἰταλικά, *IG* 14.748 (Naples, ii AD).'

Ἰτᾰλός, after 'as Adj.' insert 'Call.*fr.* 669 Pf. (prob.; ἰτ-)'

***ἰτᾰμότης**, ητος, ἡ, *initiative, enterprise, boldness*, Pl.*Plt.* 311a, Plu. 2.715e, [Simon Ath.]*Eq.* 11; in pejorative sense, Jul.*Or.* 7.225, συγγραφέως Plb. 12.9.4.

***ἰτάω**, *go*, inf. pf. ἰτάκειν Hsch. (εἰτακεῖν cod.); cf. ἐπανιτάω, ἰτητέον, εἰσιτητήριον.

ἰτεῖνος, add '*PAlex.* 27.4 (ii/iii AD; sp. ειτοειν-)'

ἰτεόφυλλος, for '*Annuario* 4/5.463' read '*SEG* 4.187.16, 19'

ἰτεών, for '*willow-ground*' read '*willow-plantation*' and add '*PKöln* 163.10 (iii AD)'

ἴτη· συρισμός. ῥοῖζος Hsch.

ἴτηλος, η, ον, app. *effective, operative*, εἰ καὶ παραμείν[ηι τέ]ως ἁ ὠνὰ ἴτηλος ἔστω *IG* 9²(1).621 (Naupactus, ii BC); cf. ἴτηλον· τὸ ἔμμονον καὶ οὐκ ἐξίτηλον Αἰσχύλος Γλαύκῳ Ποτνιεῖ (*fr.* 42 R.) Hsch.

***ἰτράριος**, ὁ, *maker of ἴτρια*, *MAMA* 3.459, 598 (Corycus), *ITyr* 33B.

ἴτριον, transfer to follow ἰτρίνεος.

ἰτριοπώλης, add 'cj. in *CIG* 4434 (Pompeiopolis in Cilicia)'

***ἰτρόγᾰλα**, ακτος, τό, app. *kind of rich cake*, Olymp. *in Grg.* p. 164.10 W., Suid.

ἴττα, delete the entry (v. °Ἴπτα).

ἴττιον, after 'Hsch.' add '(dub., cod. ἴττεο)'

Ἰτῠλος, add 'app. also = Ἴτυς, *SEG* 30.1142 (Italy, Augustan)'

ἰτῦς, line 5, for '*arch of the eyebrows*' read '*rim of the eyes*' and add 'βλεφάρων .. πυρόεσσαν ἴτυν *APl.* 140'; last line, after '(Gal.) 10.448' insert '*vault of heaven*, or perh. *orbit of stars*, ἐς ὑψιπόρων ἴτυν ἄστρων Nonn.*D.* 2.575, al.; of the Milky Way, γαλαξαίην ἴτυν ib. 6.338; ἀν' οὐρανίαν ἴτυν rest. in Theoc. 24.172'

Ἰτωνία, add 'at Coronea, *SEG* 28.458 (vase, vi/v BC); at Larissa, ib. 34.558.64 (ii BC); at Haliartus, ib. 32.456.8 (iii BC)'

Ἰύγγιος, add 'Ὑνγιος *BCH* 79.449-51 (Skotoussa)'

ἴυγμα, ατος, τό, *shout*, A.*fr.* 46a.17 R. (pl.).

ἴφθῐμος, line 1, for '*stout, strong, of bodily strength*' read 'wd. of uncertain origin, app. implying power, virility and sim. qualities'

ἴφι (A), line 1, after 'ἴς' insert '(A)'; add 'Myc. *wi-pi-no-o*, pers. n.'

***ἰφίμωλος**· δυσχερής Hsch. (La.) (cod. ἴφικλος q.v.).

ἴφυον, add '(app. also an edible plant, Hsch.; cf. Ar. l.c.; v. *JS* 1988.165-6)'

ἰχαίνω, for 'Call.*Aet.* 1.1.22' read 'Call.*fr.* 178.22 Pf.'

ἰχάλη· ἧπαρ υός, ἐσκευασμένος ἰχθύς. ἢ κίχλη τὸ ὄρνεον Hsch., cf. °ἴχλα.

ἰχθύα ΙΙ, for pres. def. read '*vessel*, prob. type of *lekanis*'

ἰχθυᾰκός Ι, for '*Cat.Cod.Astr.* 1.160' read '*Cat.Cod.Astr.* 1.166'

ἰχθῠβολέω, for '*strike, harpoon fish*' read '*fish*'

ἰχθῠβόλος ΙΙ, for this section read '(pass., proparox.) *consisting of a catch of fish*, θήρα *AP* 6.24, δεῖπνα Opp.*H.* 3.18'

ἰχθύβοτος, for '*Epic.Oxy.* 213ᵛ.15' read '*Epic.Alex.adesp.* 3.36'

ἰχθύδιον, for 'Archestr.*fr.* 45.18' read 'Archestr. in *Suppl.Hell.* 176.15' add '**ΙΙ** *the constellation Pisces*, *SEG* 7.364 (Dura-Europos, AD 218).'

***ἰχθύειος**, α, ον, *made of fish*, γάρον τὸν ἰχθύειον S.*fr.* 799a R.

ἰχθυοβόλος, add 'in *PTeb.* 868.5 ἰχθυοβ[may represent a noun ἰχθυοβόλον, τό, *harpoon, trident*'

***ἰχθυογρῖπος**, ὁ, prob. *basket-trap for fish*, *PTeb.* 868.4 (ii BC); the wd. may be ἰχθυογρῑπεύς = γριπεύς.

ἰχθυοθήρας, after '*fisherman*' insert 'Plu. *in Hes.* 8'

ἰχθυοφάγος, line 3, for 'Arabian Gulf' read 'Red Sea'

ἰχθυοφόρος 2, for '-φόρος, ὁ, .. (Epid.)' read '-φόρος, ὁ, *fish-carrier*, i.e. *itinerant fishmonger*, *IG.*4².123.21 (Epid.)'

ἰχθῦς, line 4, before 'Alex. 261.9' insert 'Emp. 21.11 D.-K.'

ἰχθυσιληϊστήρ, for '*a stealer of fish*' read '*fish-pirate*, humorous term for a fisherman'

***ἴχλα**, ἡ, *kind of fish* = κίχλη ΙΙ, *IGC* p. 99 B8 (Acraephia, ii BC), Hsch.; cf. °ἰχάλη.

ἴχματα, after 'Id.' insert 'read for ἴχνια by Zenod. and Ar.Byz. in *Il.* 13.71'

***Ἰχναῖος**, α, ον, *of Ichnae* (in Macedonia), of Themis, *h.Ap.* 94, of Nemesis, *AP* 9.405 (Diod. Sard.), Lyc. 129; (in Thessaly) Ἴχναι, ὅπου ἡ Θέμις Ἰ. τιμᾶται Str. 9.5.14.

ἰχνάομαι, for '= ἰχνεύω' read 'glossed by ἰχνοσκοπέω'

ἰχνευτής Ι 1, add 'also of a work on plagiarism, Porph.ap.Eus.*PE* 10.3; adj., ἰ. κινωπέτον Nic.*Th.* 195' **ΙΙ**, after 'Hdt. 2.67' insert 'S.*fr.* 314.305 R.' and delete 'Nic.*Th.* 195'

ἰχνηλατέω, add 'ἰχνηλατῆσαι· ἐκ τῶν ἰχνῶν ζητῆσαί τινα ψηλαφῆσαι, ἢ τὰ ἴχνη ἐλάσαι Hsch.'

ἰχνηλάτης, for '*APl.* 4.289' read '*APl.* 289'

ἴχνιον 2, after '*remnant*' insert 'νομογραφίης Call.*fr.* 43.91 Pf.; σπινθῆρος *AP* 12.31 (Phan.)'

ἴχνος 1, add '**b** *of sound*, ἴχνος αὔτης μαιομένη Opp.*H.* 3.391, ἴχνος .. βληχῆς C. 4.96.' **2**, delete 'Herod. 7.20' and insert '*AP* 6.219.11 (Antip.Sid.)' **3**, *sole of a shoe*, add 'perh. also Herod. 7.20 (in damaged text)' **6**, for this section read '*track, route*, Lxx *Ge.* 42.9, 12; in the name of a tax, ἴχνους ἐρημιοφυλακίας *PFay.* 75, 76, al., *PLond.* 1266, *PLips.* 82 (all ii/iii AD)' **7**, add '*TAM* 5(1).524 (AD 184/5)'

ἰχώρ Ι, delete 'later .. A.*Ag.* 1480 (anap.)' transferring ref. to section ΙΙ 2. **ΙΙ**, line 2, delete 'of the blood'

ἴψον, after lemma insert '(ἰψών La.)'

ἰώ, add 'Sapph. 86.7 L.-P., X.*Cyn.* 6.17'

ἰωά (B), for 'Call.*fr.* 1.40 P.' read 'Call.*fr.* 228.40 Pf.'

ἰώδης Ι 2, delete the section transferring the refs. to section Ι 1.

ἰωή, line 6, after 'Hes.*Th.* 682' insert '(s.v.l.)'

ἰωκή, add 'ἰωκαί· διώξεις. ὁρμαί Hsch.'

Ἰωνικός 1, add '**b** Ἰ. ῥῆσις *long-winded* (opp. Spartan terseness), Ath. 13.573b.' add '**4** archit., γεῖσα *IG* 2².1666*B*9 (iv BC), κεφαλὴ Ἰ. *Ionic capital*, Didyma 39.53, cf. 39.24 (ii BC).'

***Ἰωνογενής**, ές, *Ionian*, epith. of Ephesus, *Inscr.Olymp.* 225.11 (AD 49).

ἴωψ, after lemma insert 'Boeot. ϝίωψ *IGC* p. 99 B12 (Acraephia, ii BC)'

K

†κᾰ, form of κάτ (= κατά, v. κατά F), used before τ, mainly in inscrs. which do not write double letters, κὰ(τ) τὸν θεθμόν Schwyzer 57.A8 (Tegea, v BC), κὰ(τ) τόνδε IG 9²(1).718A1 (Naupactus, v BC); also occasionally in later inscrs., κὰ(τ) τοὺς νόμους SIG² 860.9 (Delphi, ii BC), but usu. the result of haplography, κατὰ ‹τὰ› εἰω[θότα] IG 2².334.15 (Athens), κατὰ ‹τὰ› δόξαντα .. τῇ βουλῇ Inscr.Magn. 179.33 (ii AD); also in compds., v. καβαίνων.

*κᾰ (B), Cypr. = °κάς and, usu. before vowel, ICS 217.5, 220.1, but also before consonant, ka-to-pa-ti-ri κὰ(s) τό(ι) πατρί ICS 167, al.

καβαλλαρικός (s.v. καβάλλης), after '(Edict.Diocl.) 19.22' add 'PNess. 18.28 (vi AD)'

*καβαλλάριος, ὁ, horse-driver, Teuc.Bab. p. 42 B.

†κᾰβάλλειον, or κᾰβάλλιον, τό, working horse, RA 86(1925).259 (Callatis), Hsch.   2 ἡ πρώτη τοῦ τρικλίνου κλίνη, διὰ τὸ ἀνάκλιτον id.   3 = κόλλοψ I 1, Sch.Ar.Ra. 510 (colloq.; written καβάλιον).

κᾰβάλλης, after 'caballus' insert 'AP 9.241 (Antip.Thess.); cf. pers. n. Καβαλλᾶς IEphes. 1437 (iv BC)'

*καβάτωρ, ορος, ὁ, Lat. cavator, gem-cutter, PKlein.Form. 607, 813 (both vi AD).

*καβιαία, ἡ, the area which can be sown with a qab (v. κάβος) of grain, PNess. 24.5, 12 (vi AD).

καβιδάριος, add 'MAMA 3.118 (Corasium)'

*καβίδιον, τό, app. small jar or sim. vessel, BGU 2359.9 (iii AD), PRyl. 627.346 (iv/v AD), PStrassb. 35.7 (iv/v AD), etc.; cf. χαβίτια, χαβότια.

*καβιθακάνθιον, τό, an unknown musical instrument, Anon.Alch. 438.10.

*καβικλάριος, ὁ, = Lat. clavicularius, keeper of a prison, MAMA 3.648 (Corycus), IEphes. 1347 (v. SEG 37.915); cf. Lyd.Mag. 3.8 cod.

†καβλή· μάνδαλος τῶν θυρῶν. Πάφιοι Hsch. (= καταβλής).

*καβόνιον, τό, a measure, ἄρτων καβόνιον PMag. 13.1013; (pap. ἄρτων χαβωνίων); cf. κάβος.

*καβουρᾶς, ᾶ, ὁ, crab-fisher, IEphes. 4282 (cf. mod. Gk. κάβουρας crab).

†κάγ, sp. of shortened form of κατά before γ, κὰγ γόνυ Il. 20.458 (v.l. κάκ); perh. also Hes.Op. 533, Sapph. 101.2, 5 L.-P. (cf. κατά F).

καγκαίνω, delete '(Cf. κέγκω)'

†κάγκαμον, τό, kind of gum, Dsc. 1.24, Plin.HN 12.98, Peripl.M.Rubr. 8, Hsch.

κάγκανον, after 'κακκαλία' insert '(στρύχνον (q.v., sense 4) ὑπνωτικόν)'

†καγκελλάριος, ὁ, financial official in the late imperial service, Lyd. Mag. 3.36, 37, PMasp. 5.19 (vi AD).

*καγκέλλιον, τό, lattice, railing, fr. Lat. cancelli, Stud.Pal. 20.151.18 (vi AD, -κελλιν pap.)

†καγκελ(λ)οειδῶς, adv. in the form of a lattice, criss-cross, Hippiatr. 117.

κάγκελλον, delete the entry (v. sq.).

*κάγκελλος, ὁ, and -ον, τό, Lat. cancellus, latticed barrier or balustrade, masc. pl. POxy. 2146.12 (iii AD), Side 58, Hsch. (s.v. δρύφακτοι, sp. -ελοι); neut. sg., Sch.Theoc. 8.58, Sch.Ar.Eq. 641 (sp. -ελον); pl., ib. 675 (sp. -ελα); gender indeterminate, sg., PRyl. 233.4 (ii AD, sp. καγγ-); pl., IG 7.1681; cf. κάγκελλον αὐτὸ οἱ Ῥωμαῖοι καλοῦσιν ὑποκοριστικῶς ἀντὶ τοῦ δικτύδιον ὅτι πρωτοτύπως κάγκρους αὐτοὶ τὰ δίκτυα λέγουσιν, ὑποκοριστικῶς δὲ καγκέλλους Lyd.Mag. 3.37.   II a system of measure of capacity (gender indeterminate), μέτρῳ τῷ κ. ἀρτάβας ἕνδεκα τέταρτον POxy. 1447 (i AD), σί(του) καν(κέλλῳ 127.1 (vi AD), 3936.22, etc.

*καγκελλωτός, ή, όν, trellised or latticed, καγκελωτὴ θύρα Sch.Ar.V. 124, Poll. 8.124, καγγελωτὴ διαβάθρα PRyl. 233.3 (ii AD), [θυ]ρίδα κανκελλωτήν SEG 17.545.8 (Pisidia, Rom.imp.), Hsch. s.vv. κιγκλίδες, δικτυωτή.

κάγκελος, delete the entry (v. °κάγκελλος).

καγκελωτή, delete the entry (v. °καγκελλωτός).

*κάγχαλος· κρίκος ὁ ἐπὶ ταῖς θύραις. Σικελοί Hsch.

*κάγχασος, ὁ, name of throw at dice, Poll. 7.204.

*καδᾶς, ὁ, maker of κάδοι, ὁ υἱὸς τοῦ κατὰ Κολοτσε Aegyptus 10.73-5.

*καδδίζω, reject on a vote, τὸν δὲ οὕτως ἀποδοκιμασθέντα κεκαδδίσθαι (codd. κεκαδεῖσθαι, cj. (ἐκ)κεκαδδιχισθαι λέγουσι Plu.Lyc. 12.

κάδδιχος, delete 'hence, voting-urn .. ibid.'; for 'also' read '2'; after 'Tab.Heracl. 1.52' insert '(gen. κάδδιχος, fr. athematic κάδδιξ)' add '3 οἱ τοῖς θεοῖς θυόμενοι ἄρτοι κάδδιχοι Hsch. (s.v. κάδδιχον)'

κᾰδεστής, read κᾰδεστάς.

Καδμῖλος, line 3, after 'Κασμ-' insert 'Iamb.adesp. 58 W.'; line 4, for 'Call.Fr. 409' read 'Call.fr. 723 Pf.'

Καδμίς, ίδος, ἡ, descended from Cadmus, Ibyc. 21 P.

καδμῖτις, ιδος, ἡ, precious stone, perh. calamine, Plin.HN 37.151.

κάδος, after 'δ' insert 'Cypr. ka-to-se ICS 318 Aiv, Bv, al.'   III, delete the section (v. CEG 438).

*κάδουκος, η, ον, Lat. caducus, of a bequest, that becomes void (because of the legal incapacity of the legatee), caducary, Just.Const. Δέδωκεν 6b.

*καγοικία, v. °κατοικία.

κᾰθά, after 'καθ᾽ ἅ' for 'according as, just as' read 'in all its senses'; for 'Men.Mon. 551' read 'Men.Mon. 848 J., ἔδοξε τᾶι ἁλίαι καθὰ καὶ τᾶι βουλᾶι SEG 30.1117 (Sicily, iii BC)' add 'b where, Paus. 8.42.13, Gal. 2.82.7.'   II, line 3, after 'D. 37.16, etc.' insert 'Thess. καττάπερ IG 9(2).234 (Pharsalus, iii BC)'; line 5, after '(nisi leg. καίπερ)' insert 'καθάπερ ἐνμανεῖς ὄντες IStraton. 10.17'; at end add 'see also °καθάσσα'

καθαίρεσις I 4, for this section read 'drawing down, τὰς ἐκλείψεις ἡλίου καὶ σελήνης καθαιρέσεις τῶν θεῶν Sch.A.R. 3.533'

καθαιρετικός, add 'adv. -κῶς putting an end to, Sopat.Rh. 8.383.14 W.'

καθαιρέω II 1, add 'b lay flat, fell (in quots., in boxing), δὶς τοὺς ἴουλον ἀνθεῦντας, ἄνδρας δὲ Πίσῃ δὶς καθεῖλε πυκτεύσας Herod. 1.53, Theoc. 22.115.'   3, for 'raze to the ground' read 'knock down buildings, etc.'; add 'of natural forces, οἰκίας καθειρημένης PColl.Youtie 13.6 (170 BC)'

καθαίρω I 6, for 'metaph. = μαστιγόω' read 'transf., beat up' and add 'Men.Dysc. 901'

καθάπαξ, after 'Adv.' insert 'also κατάπαξ SEG 24.151 (Attica, iv BC)'   I, after 'once for all' add 'irrevocably'; line 6, before 'οὐδὲ κ.' insert 'II once'   at end delete 'singly, Plb. 3.90.2'

καθαπτής, delete the entry (v. sq.).

καθαπτός II, for this section read '2 plucked, of stringed instruments, ἔντατον .. καὶ καθαπτὸν sc. ὄργανον Aristocles ap.Ath. 4.174c.   II hanging, suspended, fem. subst. of type of vessel, γάστρας καὶ καθαπτάς PSI 420.26 (iii BC).'

κᾰθάρειος, line 8, after 'Plb. 11.9.5' insert 'ἱματίου καθαρίου Edict.Diocl. 7.48 in SEG 37.335 i 9; of wood, trimmed, ῥαβδίων καθαρίων id. 12.19a'; line 9, for 'Sammelb. .. 230 (pl.)' read 'PMag. 5.230; without ἄρτος SB 5730 (iv/v AD)'

καθαρειότης 1, line 2, after 'X.Mem. 2.1.22' insert 'PHarris 193.7 (ii AD)'

*κάθαρεοσύνη, ἡ, = κάθαρσις, SB 10278.21 (i/ii AD).

κάθαρεσις, delete the entry (read καθαίρεσις).

*κάθαρεύς, έως, ὁ, purifier, dub. in IG 1³.250.B37 (Athens, v BC).

κᾰθᾰρίζω I, line 6, after '(Andania, i BC)' insert 'clear a building site, Pap.Lugd.Bat. xxv 46 (ii AD)'

κᾰθᾰρισμός, add 'w. ref. to polishing of marble, PMil.Vogl. 304.16 (AD 166)'

κάθαρμα I 1, line 3, after 'Str. 3.2.8' insert 'σιδήρου (καὶ) καθάρμ(ατος) μνᾶς κδ´ SB 7365.92 (iii AD)'

καθαροποιέω II, after 'encumbrances' insert 'PDura 25.10, 32 (ii AD)'

καθαροποίησις, add 'PLond. 1724.50 (vi AD)'

*κᾰθᾰροπώλης, ου, ὁ, seller of pure bread, PTeb. 872.19, 22 (?iii BC); cf. ‡καθαρός I 2.

*κᾰθᾰροπώλισσα, ἡ, fem. of καθαροπώλης, τῇ καθαροπολίσσῃ PPrag. I 97.1 (iv AD).

κᾰθᾰρός I 2, line 6, before 'κ. ἄρτος' insert 'of flour, bread, free from bran, bolted (sts. combining sense of ritual purity)'; delete 'of white bread '; line 7, after '(Gal.) 19.137' insert 'also καθαροί alone, PTeb. 884.12, 16 (iii BC) and after 'PTeb. 93.36 (ii BC)' insert 'Tab.Heracl. 1.103'   3, line 14, before 'c. gen.' insert 'κ. οὔασιν, with clear (unblocked) ears (in quot., fig.), Posidipp. in Suppl.Hell. 705.2, cf. auribus puris Prop. 2.13.12'   b, line 4, after 'POxy. 633 (ii AD)' insert 'Cod.Just. 6.4.4.16, Just.Nov. 1.2.2'   II 5, after 'correctly' insert 'κ. γράφειν ἢ λέγειν D.H.Lys. 2'   add '7 perh. of weight, contents, net, καθαροῦ λ(ίτραι) Lang Ath.Agora xxi Hd 10 (ii AD).'

κᾰθᾰρότης 6, after 'style' insert 'Hermog.Id. 1.2, al.'

καθαρουργεῖον, add 'Stud.Pal. 10.233.3.7 (v AD), PAlex. 32.10 (v AD)'

καθαρουργία, add 'PAlex. 32.11 (v AD)'

καθαρουργός, add 'PAlex. 32.4 (v AD)'

κάθαρσιος III, add 'so prob. in Ael.VH 14.7'

κάθαρσις III, add 'pruned wood, prunings, Lxx Ez 15.4'   V, add

162

'τῇ τῶ[ν] ἐνόντων φυτῶν πάντων καὶ φοινίκων δικ[αίᾳ] καθάρσει PVindob.Salomons 8.15 (?AD 325)'

*κἀθάσσα, i.e. καθ' ἄσσα, = καθά, Milet 3.136 (iv BC).

καθέδρα II 3, delete the section transferring quot. to section II 1. add 'V καθέδρα· θυσία Ἀδώνιδος, and καθέδραι· πένθους ἡμέραι ἐπὶ τετελευτηκόσι Hsch., cf. AB 1.268.'

καθέζομαι II 1, add 'b remain inactive, ἐκαθέζετο ὁ Κῦρος ἀμφὶ τὴν περὶ τὸ φρούριον οἰκονομίαν X.Cyr. 5.3.25.'

καθείργνῡμι 2, after 'Ar.Nu. 751' insert 'transf.'; add 'καθεῖρξα βοήν Trag.adesp. 664.24 K.-S.'

καθεῖς, for 'εἷς καθεῖς Ev.Marc. 14.19, etc.' read 'cf. εἷς 1f' add '2 each individual, ὁ δὲ καθεὶς ἄνθρωπος δεήσεις περὶ τοιούτων ἐννοιῶν μὴ ἐπιδότω CodJust. 10.16.13 pr.'

καθεκτός, add 'III καθεκτόν· ἐφικτόν, καταληπτόν Hsch.'

καθελίσσω, line 1, after '(v. infr.)' insert 'Aeol. aor. (?med. part.) κατελιξαμε[ν- Sapph. 98(a).4 L.-P.'

*καθελκτικός, ή, όν, downward-drawing, virtus κ. peristalsis, Macr.Sat. 7.4.14.

καθέλκω I 2, line 2, for 'δρῦν' read '[..]'; for 'Call.Aet.Oxy. 2079.9' read 'Call.fr. 1.9 Pf.' and add 'Zeno Stoic. 1.23.18'

κάθεμα, after 'collar' insert 'PTeb. 761.12 (iii BC) and after 'Antiph. 319' add 'POsl. 46.11 (iii AD)'

κάθεσις, add 'III κάθεσιν· καταγωγήν, οἴκησιν Hsch.'

κάθετος, add '4 κάθετον, τό, perh. a chamber beneath the surface or the floor of a grave monument, MAMA 6.335.3 (Acmonia).'

καθεύδω I, line 9, after 'Timocl. 16.2' insert 'prov. ἐπ' ἀμφότερα τὰ ὦτα κ. Aeschin.Socr. 54, cf. Men.fr. 333.2 K.-Th.'

*καθεψητέον, one must boil down, Gal. 13.613.5.

καθηγεμών, add 'κ. ἐφήβων inscr. in JEA 37.87 (Memphis, iii AD)'

καθηγέομαι 2, add 'pass., καθηγηθείς (written -ηκηθ-) having been told, PMich.VIII 497.12 (ii AD)' add '7 = ἡγέομαι, think, Is. 5.14 (s.v.l.).'

*καθηγέτις, ιδος, ἡ, leader, guide, κ. θεά MAMA 8.419.

κάθημαι 4, add 'of a god, θεὸν τὸν καθήμενον ἐπάνω τοῦ ὄρους παλαμναίου, etc., SEG 31.1594'

*καθημερήσιος, α, ον, daily, μισθός Stud.Pal. 22.36.10 (ii AD).

*κάθηκε, v. κατατίθημι.

†καθηρατόριον, v. °κασσ-.

καθιγνύσαι, for '(Apptly. .. καθαγνίσαι)' read '(perh. corrupt for καθαγνίσαι, but cf. ἴγνυς s.v. ἴκνυς)'

καθιερόω, line 1, add 'Boeot. καθιᾱρόω SEG 23.271.33 (220/08 BC)'; line 6, after 'καθιερωμένος' delete '[ῑ]'; at end add 'also [with ῑ] by Lyc. 950, 1350; see also °καταϊερόω'

*καθιέρωμα, ατος, τό, dedicated offering, Phot.

καθίζω I 1, add 'prov. ἐπ' οὐδεὶ φῶτα καθίσσαι i.e. robbed him of all his possessions, h.Merc. 284' 5, after 'X.Cyr. 2.2.15' insert 'Smp. 3.11'

κάθημαι I 2, of plays, add 'Plu. 2.839d'

*καθίκω, go down, καθίκ[ειν] prob. rest. in Call.fr. 191.38 Pf.; cf. ἵκω, παρίκω.

καθιππεύω 1, after 'Opp.H. 2.515' insert 'w. gen., Nonn.D. 2.646, ποταμοῖο 23.156, 40.348, etc.' 2, add 'w. gen., ἐλεφάντων Nonn.D. 1.25'

κάθισις, add 'III app., place of refuge, βαθύνατε εἰς κάθισιν Lxx Je. 30.2, 25.'

καθιστάνω, add 'pass. inf. καταστάνεσθαι SEG 31.122 (Attica, c.AD 121/2)'

καθίστημι A I 1, at end, set up, erect, for 'Inscr.Cypr. 94, 95 H.' read 'Cypr. ka-te-se-ta-se κατέστασε ICS 85, 86' and add 'πύργους SEG 24.154.9 (Attica, iii BC), βωμόν TAM 4(1).56' add '4 pay fines, prices, etc., Inscr.Cret. 4.14 (vii/vi BC), al.' B, line 1, for 'aor. 2' read 'aor. 1 and 2' and after 'S.OC 23' add 'οὔτε καταστήσαντες (οἱ ἁλιεῖς) ἐπὶ θέαν X.Oec. 16.7, CodJust. 1.2.25 pr.' 1 b, add 'E.Hec. 531' 4, add 'of style, D.H.Lys. 9' 8, after 'πρός τινα' insert 'κατέστην πρὸς αὐτοὺς καὶ ἐδίδουν ἀντίγραφα τῶν δικαιωμάτων PPetr. in APF 33.24.6 (iii BC)'

καθό, add 'III where, Str. 2.5.31, J.BJ 3.7.7, Poll. 2.185.'

κάθοδος, line 1, for 'Demeter' read 'Persephone'

καθολικός I, line 10, after 'Dam.Pr. 310' add 'of the Chr. church, SEG 34.1341 (Lycaonia), BCH suppl. 8.62, 233, al. (Maced., iv/v AD), CodJust. 1.1.5 pr.'

*καθολικότης, τητος, ὁ, the office of the καθολικός (in quot. w. ref. to the sum owing to him), POxy. 3408.27, 3423.20 (iv AD).

*κάθολον, τό, the total sum, ὁ δὲ μὴ δοὺς τὸ κάθολον ἐξέρανος ἔστω SEG 31.122.44 (Attica, AD 121/2).

καθοπλίζω I, line 4, after 'Lxx 4Ma. 3.12' insert 'fig., τὸ μὴ καλὸν (cj. ἄκος καλὸν) καθοπλίσασα δύο φέρειν S.El. 1087' II, delete the section.

καθορμίζω 1, add 'fig. in med., lay to rest, σῶμα GVI 788.10 (Bithynia, ii/iii AD)'

καθόρμιον, at end delete 'κάθορμον Hsch.'

*καθόρμιον (B), καθόρμια· τὰ ἐνόρμια Hsch.

καθοσιόω 1, add 'b pf. pass. part. qualifying titles of imperial officials, δομεστικοί POxy. 1982.4 (v AD), ἀκτουάριος PMasp. 320.2 (vi AD), κεντηνάριος Stud.Pal. 20.139.5 (vi AD), Φλαβιάλις Mitchell N.Galatia 450.'

†καθότι, Ion. κατ-, for καθ' ὅ τι, as, κ. γέγραπται SIG 577.18 (Miletus, iii/ii BC), PKöln 219.15 (iii/ii BC), 193.8 (v/vi AD). II for the reason that, Plb. 4.25.3. III where, Paus. 6.20.10. See also ‡κατά B IV 1.

καθῡλομανέω, add 's.v.l.; cf. °θηλυμανέω'

*καθυπεμφαίνω, give a faint indication of, τέλος Eust. 1568.28.

*καθυπερδέξιος, ον, possessing superiority, epith. of Zeus, Robert Hell. 10.63; cf. °ὑπερδέξιος.

καθύπερθε, line 2, after 'Ion. κατύπερθε' insert 'Aeol. °κατέπερθεν'

*καθυπερῷος, α, ον, perh. upper, of a room, Inscr.Délos 1406B14 (ii BC; force of prefix unclear in broken context).

*καθύπο, adv. underneath, Ps.-Democr.Alch. p. 51 B.

καθύφεσις, before 'Poll. 8.143' insert 'Sopat.Rh. 8.280.25 W.' add 'II decline, recession, ἡ τιμὴ τοῦ οἴνου ἐνταῦθα πάνυ ἐν κατυφέσει ἐστίν POxy. 3507.18 (iii/iv AD).'

καθυφίημι II 2, add 'Men.Epit. 402'

καθυφίσταμαι, for 'Jul.Or. 4.163d' read 'Jul.Or. 5.163d'

καθώσπερ, add 'Ep.Hebr. 5.4, v.l. in 2Ep.Cor. 3.18'

καί A V, after 'correlative' insert 'both .. and, as .. so' add '2 τε .. καί and καί .. τε, v. τε A II.' B 7, for 'assent' read 'consent' 8, after '(Pl.)Lg. 663d' insert 'at the beginning of a law, καὶ ἐὰμ μὴ 'κ [π]ρονοία]ς [κ]τ[ένει τίς τινα IG 1³.104.11, D. 24.39' C 1, for 'ἔγνωκα .. etc.' read 'Pi.P. 10.58, Call.fr. 1.15 Pf., al., Euph. 51.7' D, add 'καὶ ὁ, written as χο Schwyzer 80², κο SEG 31.696 (N.shore, Black Sea, v BC)'

καιετάεσσαν (s.v. καιάδας), after 'Od. 4.1' insert '(so Eust., but καιτάεσσαν Sch. ad loc.)'; for 'Call.Fr. 224' read 'Call.fr. 639 Pf. (so Eust. ibid., but καιτα- Sch.Od. l.c. and POxy. 2377 Front 6)'

*καιλούριον, τό, app. = κολλούριον, occurring in context with ψωμία (ψωμία), SB 1975 (pl., v AD).

*καίμιον, v. °κέμιον.

καινέω, delete the entry.

*καινοκέραμος, ὁ, new wine-jar, PSI 1249.27, 1250.3 (both pl., iii AD); cf. °παλαιοκέραμος.

καινόκουφον, add 'PKlein.Form. 968 (iv AD; v. Tyche 7.230), CPR 14.2.15 (vi/vii AD)'

καινολόγος, add 'gloss on εὑρεσιεπής, Sch.Pi.O. 9.120'

καινοπαθέω, delete the entry (v. °κενο-).

καινός II, line 3, delete 'οὐκ .. Tim.fr. 21'

*καινούργημα, ατος, τό, innovation, Just.Nov. 84 pr.

καίνυμι I, line 10, after '(Od.) 2.158' insert 'τό[ξα ..] τάδε δώσω παλά[μα]ισιν ἐμαῖσι κεκασμένα .. [ἐ]πικρατέως βάλλειν Stesich. 40.23 P.' at end add 'w. acc. of thing, ὀδμή .. λειμῶνος ἐκαίνυτο λαρὸν ἀυτμήν Mosch. 2.92'

καίνω, line 2, after 'Theoc. 24.92' insert '(codd., κανεῖν cj.)'; add 'ἐκάνετ' ἐκάνετε E.fr. 588'

καιρικός 1, for this section read 'suitable for the occasion, IG 2².3800.8'

καίριος II 1, add 'b of the moment, extempore, GVI 1924 tit. (Rome, i AD).' II, add 'vital, essential, Longin. 1.1; sup., id. 10.1'

καιρός III 1 a, add 'καιρὸν εὑροῦσα τοῦ γραμματοφόρου finding an opportunity to avail oneself of, PMich.inv. 430 (Glotta 58.177); personified, Ὀλυμπίο SEG 26.1211 (Velia, v BC), Καλοὶ Καιροί SEG 34.279 (Corinth, iv AD), 34.1448 (Syria, Rom.imp.)' b, line 4, after 'Pl.Cri. 44a' insert 'ἐμ παντὶ καιρῷ at every opportunity, SEG 30.990 (Delos, c.325/275 BC)'; line 6, after 'BGU 15.10 (ii AD)' insert 'CodJust. 1.4.22.2; also κατὰ τὸν καιρόν SEG 31.575 (Thessaly, 171 BC)' IV, line 2, for 'to his advantage' read 'to your advantage'

*καιρωτός, ή, όν, app. well-woven, Call.fr. 383.13 Pf.; cf. καῖρος.

*καίσαπος, ὁ or ἡ, or -ον, τό, Greek name of kind of lettuce, acc. to Plin.HN 20.59 (caesapon acc.).

Καισάρειος, τὸ K., after 'temple' insert 'or shrine' and add 'at Xanthos, Lycia, SEG 30.1535 (after AD 152), POxy. 1683.19 (sp. κησ-, iv AD)'

*Καισαρήσιος, α, ον, of or from Caesarea, ὀθόνιον PMasp. 6ᵛ.85 (vi AD).

*Καισαρογερμάνικεια, τά, games in honour of Germanicus, SEG 23.638.7 (Cyprus, AD 18 or 19).

καιτάεσσαν, for 'f.l. .. Od. 4.1' read 'v. °καιετάεσσαν'

καί τοι II, line 6, after 'E.fr. 953.10' insert '(= Men. i p. 143 K.-Th.)'

καίω II, add '6 part. as epith. of a fire-god, Καίοντος Μάνδρου IKyme 37.5.'

κἀκαγγελία, add 'Hp.de Arte 1'

κἀκάγγελος, add 'Call.fr. 260.48 Pf.'

*κἀκαγωγία, ἡ, bad behaviour, BGU 1816.13 (i BC)

κἀκανδρία, after 'unmanliness' insert 'A.fr. 132a.4 i 2 R.'

*κἀκεπίτροπος, ὁ, felonious guardian, διὰ τὴν τῶν κ. πλεονεξίαν PMed. Bar. 15ʳ (Aegyptus 66.7.42, ii BC)

κάκη, add '3 pl., app. *troubles*, κἀγὼ ἔχο μου τὰς κάκας *POxy.* 3417.13 (iv AD).'

*****κἀκήμερος**, ον, *experiencing a bad day* (opp. καλήμερος), *AP* 9.508 (Pall.).

κακιθά, add '**κακιθή** Theognost.*Can.* 109'

κακιθής, delete the entry.

*****κἀκινκάκως**, adv., ⟨κακὴν κακῶς⟩ *with much trouble*, *Vit.Aesop.*(G) 19.

κακκάβιον, add '*PAlex.* 31.3 (iii/iv AD)'

κακόβουλος I, add 'adv. -ως, ἀφραδέως· κακοβούλως, ἀνοήτως Sch. Nic.*Al.* 502b Ge.'

κακοδοξία I, after 'Pl.*R.* 361c' add 'Sopat.Rh. 8.15.9 W.'

κακόδουλος I, add '*Vit.Aesop.*(G) 26, 28'

κακοδρομία, for '*bad passage* (by sea)' read '*unlucky journey*, alluding to Icarus' flight'

κακόδωρος, before 'Suid.' insert 'Sch.S.*Aj.* 665' and after it add 'Πανδώρη κ. Euph. in *Suppl.Hell.* 415 ii 1 (prob.)'

κακόζηλος, add '**2** *jealous, spiteful*, φθόνου κακοζήλου *IMEG* 114 iv 13 (Panopolis).'

κακόηχος (s.v. κακοηχής), add '*Cat.Cod.Astr.* 11(2).189.9'

κακομιλία, for '*bad intercourse* or *society*' read '*evil association* or *company*'

*****κἀκομνήμων**, ονος, ὁ, prob. = μνησίκακος, title of mime by D.Laberius, Gell. 16.7.8.

κακομουσία, for '*corruption of music*' read '*the quality of offending against the principles of art*'

κακονοέω, add 'w. acc., *bear malice towards*, *PHaun.* 10.21'

κακοπαθέω, line 3, after 'D. 18.146' insert 'Men.*Dysc.* 348, 371'

*****κακοπάθημα**, ατος, τό, gloss on ὄτλημα, Hsch.

†**κακοπάρθενος**, ή, *evil maiden*, Μοῖρα *AP* 7.468 (Mel.).

*****κακοπίαστος**, ον, *hard to hold, unmanageable*, gloss on ἀμήχανος, Sch.Gen.Il. 10.167.

*****κακοπόδινος**, ον, *whose coming brings bad luck*, *REA* 62.357 (cf. mod. Gk. κακοπόδαρος).

*****κακοποιεία**, ή, *evil-doing*, Jos.*Jer.* 9.3(2).

κακός, add 'Myc. comp. ka-zo-e (pl.) < *kakyos-es*'

κακόοιτος 2, for 'Ἀρχ.Δελτ. 2 App. 47' read '*IG* 9²(1).253 (iii BC)'

*****κακοστομᾰτίζω**, *speak ill of*, κακ]οστοματισθήσε[ται *PMerton* 11.56.9 (ii AD).

κακοσυνθεσία, after 'Hsch.' add 'Sch.bT Il. 15.16'

*****κακοσύστατος**, ον, in Lat. *cacosystatae* (sc. *controversiae* or *materiae*), *scarcely forming a coherent whole*, Fortunat.*Rh.* 1.3 (distd. fr. *asystatae*).

κακοτεχνέω I 1, add 'Men.*Dysc.* 310' **II 2**, add '*IG* 1³.21.48 (Athens, ?450/49 BC)'

*****κακοτέχνησις**, εως, ή, *fraud*, *SB* 9109 (AD 31), rest. in *PColl.Youtie* 19.26 (AD 44).

*****κακοήιος**, α, ον, *evil*, oracle in *ZPE* 92.269 (Ephesus, ii AD).

κακότροπος I, for '*malignant*' read '*having evil ways, evil-living*' and insert 'Sapph. 71.4 L.-P. (prob. rest.), Ar.*fr.* 717 K.-A., decr. in *SEG* 30.80.11 (Athens, i BC)'; before '*PMasp.*' insert '*POxy.* 2342.12 (AD 102)'

κακουργέω I 1, line 4, after 'Pass.' insert 'τάδε κακουργεῖται Aen.Tact. 18.2' **II 4**, after 'Pl.*R.* 416a' add '(s.v.l.)'

κακουχία, line 4, transfer 'Alex. 80' after 'Pl.*R.* 615b' in line 2 and before 'Vett.' insert 'Plu. 2.112c'

κακόψογος, delete 'cf. Ptol.*Tetr.* 166'

κακόω, line 5, for 'A.*fr.* 156' read 'A.*fr.* 154a.9 R.'; line 10, for 'ἐκάκωτο' read 'ἐκεκάκωτο'

*****κᾰλᾰθᾶς**, ᾶ, ὁ, *basket-maker*, καλαθᾶτες *PAmst.inv.* 21 (*ZPE* 9.49, iv/v AD).

κᾰλᾰθηφόρος, add 'καλαθηφόρος, ἡ, (*female*) *basket-bearer*, καλατηφόρῳ Νεσμείμεως *POxy.* 2781.2 (ii/iii AD), *IEphes.* 1060, 1070a, al. (iii AD)'

κᾰλᾰθίσκος II, for 'Men. 1018' read 'Men.*fr.* 855 K.-Th.' and transfer to section I 1 before 'Theoc. 21.9'

κάλᾰθος I 1, after 'esp. for wool' insert 'καλάθου μείμημα τρόπαιον as a symbol of wifely virtue, *SEG* 36.1260 (Paphos, late ii AD)'

*****κάλᾰθρον**, τό, *basket*, = κάλαθος, *SB* 13273 (Ptolemaic).

*****κᾰλᾰθωνία**, ή, *provisions* (in a basket), *CRAI* 1945.378 (Beirut, Byz.).

*****κᾰλᾰϊκός**, ὁ, *name of a stone*, Socr.Dion.*Lith.* 50.1 G.

κᾰλάϊνος I, line 4, after 'καλαεινον)' insert '*PCair.Isidor.* 58.14 (iv AD, καλλιείνων pap.)'

κάλᾰϊς II, for this section read 'app. *some kind of sacrificial animal*, *IG* 4².40.5, 41.6 (Epid., c.AD 400)'

*****κᾰλᾰκἀγᾰθιος**, = τῶν καλῶν κἀγαθῶν, epith. of Zeus, *SEG* 6.550 (Pisidia).

κᾰλᾰμαῖος, delete 'καλαμαία, ή, *a kind of grasshopper*'

κᾰλᾰμαύλης, add '*PSAAthen.* 43ᵛ i 9 (ii AD), *PSorb.inv.* 2381 (*ZPE* 78.153)'

†**κᾰλᾰμευτής**, οῦ, ὁ, *angler*, *AP* 6.167 (Agath.), 10.8 (Arch.). **2** perh. *catcher with a limed reed* (also understood as *gleaner*), Theoc. 5.111.

*****καλαμεύω**, κεκαλαμευμένοι (cod. κεκαλαμινθευμένοι)· καλάμη γεγονότες Hsch.

κᾰλάμη, line 1, after '*straw of corn*' insert '*whether cut or left standing*' **I 1**, add 'metaph., as a typically fragile material, μεμνημένος .. ἐξ οἵης ἡρμόνισαι καλάμης *AP* 7.472.16 (Leon.Tarent.)' **2**, delete the section, transferring material to section 1.

*****κᾰλᾰμίδιον**, τό, *reed-crop*, *PFreib.* 56 (i/ii AD), *PVindob.Worp* 5.24 (AD 169).

κᾰλᾰμίζω, add '**II** *grow reeds*, *PMich.*XIII 666.21 (vi AD).'

*****κᾰλᾰμίνθη**, ή, *name of var. kinds of mint, or sim. plant* (three varieties mentioned by Dsc. 3.35), Ar.*Ec.* 648, Thphr.*CP* 2.16.4, Gal. 11.882.18, 19.731.5, ἡ ἔνδροσός τε καὶ νοτερὰ καλαμίνθη Ael.*NA* 9.26, Hsch.; also -μίνθα Philum.*Ven.* 7.9, 14.6, Phot.

κᾰλάμινος II, add 'μέλι τὸ καλάμινον τὸ λεγόμενον σάκχαρι *Peripl.M.Rubr.* 14'

κᾰλάμπτης I, add '**2** *name of kind of green frog*, Plin.*HN* 32.122.'

κᾰλᾰμῖτις, for 'ἡ, = καλαμαία' read '*of* or *associated with the cornstalks* (ἀκρίδα) τὴν καλαμῖτιν; cf. ‡καλαμαῖος'

κάλᾰμος, line 1, after '*reed*' insert 'Alc. 115.9 L.-P.'; line 2, after 'Th. 2.76' insert 'in building, *IG* 2².463.68, 1663.1 (Athens, iv BC), cf. καλαμίς 5' **II 1**, add '**b** = δόναξ ὑπολύριος (v. ὑπολύριος), S.*fr.* 36 R., cf. h.*Merc.* 47. **8**, add '**b** used in hairdressing, κ. τινα ἔχουσιν ἀεὶ ἐν αὑτῇ τῇ κόμῃ ᾧ ξαίνουσιν αὐτὴν ὅταν σχολὴν ἄγωσι D.Chr.*Enc.Comae* p. 386 B.'

κᾰλᾰμοστεφής, for '*covered*' read '*crowned*'

κᾰλᾰμουργία, add '*POxy.* 729.4 (ii AD), *PVindob.Salomons* 8.30 (AD 325)'

κᾰλᾰμών, add '*IMylasa* 803.10, 814.9'

*****καλανδαρικά**, τά, = καλανδικά, *SEG* 9.356.69 (Ptolemais in Cyrenaica, vi AD).

κᾰλάσιρις, add '*SEG* 38.1210.5 (-σειρις, Didyma, ii BC)'

*****κᾰλαυδάκη**, ή, *headband*, gloss on ἀναδέσμη, Sch.AT Il. 22.469-70 (Sch.T -δεύκη).

*****καλαυδάκιον**, τό, *headband*, *SB* 9122.10 (i AD).

κᾰλαῦροψ, for '*shepherd's* .. *herd*' read '*herdsman's staff*, which was thrown to control cattle'; at end for '*BSA* .. Pamphylia' read '*SEG* 17.552 (Pisidia)'

*****Καλαφωνία**, ή, perh. = Κολοφωνία (s.v. Κολοφώνιος), *BMB* 7.78 (Berytus, V AD).

*****καλεα**, ?τά, perh. *some kind of surgical appliances or instruments*, καλεα μοτεα στερεα *SEG* 29.972 (Magna Graecia, iv BC).

*****καλενδάριον**, τό, Lat. *calendarium*, *Ann.Épigr.* 1910.169 (Laodicea Combusta).

†**καλέχες**, v. καταλέχομαι.

κᾰλέω I 2, line 4, after 'E.*Ion* 1140' insert 'ἐπ' ἔριφον καὶ χοῖρον κ. prov. for an invitation to a choice meal, Alc. 71.1 L.-P.' add '**7** *designate as heir* to, Just.*Nov.* 22.48 pr., 53.6 pr.; w. πρός, *Cod.Just.* 1.5.18.3; w. εἰς, ib. 1.5.18.9.'

*****καλεων**, ό, prob. *non-Hellenic name of some cult-object*, *Inscr.Cret.* 1.v 23.9 (Arcades, ii AD).

κᾰλήμερος, for '*bringing a fair day*' read '*enjoying a good day*'

κᾰλιά, line 3, transfer 'A.R. 1.170, 4.1095' to follow 'Hes.*Op.* 301, 307' in line 2; for '*Anacreont.* 25.7' read '*Anacreont.* 25.3, 7 W.'

*****καλιγαρικός**, ή, όν, *of* or *pertaining to boots*, περὶ φορμῶν καλικαρικῶν (sic) *Edict.Diocl.* 9.1.

*****καλιγάριον**, τό, = καλίγιον, Sch.Luc.*Cat.* 15.

*****καλιγάριος**, ὁ, Lat. *caligarius*, *bootmaker*, *MAMA* 3.235 (Corycus), *SEG* 8.45 (Palestine, iv/v AD); κ. Βαβυλωνάριος *maker of Babylonian shoes*, *MAMA* 3.616; sp. καλικ- ib. 3.131 (Corasium), *SB* 10258 ii 17 (iv AD); καλκ- *MAMA* 3.30 (v/vi AD).

καλίγιον, add 'καλλίγιον *SEG* 7.423 (Dura, iii AD); καλίκιον *Edict.Diocl.* 9.5A; also prob. καλλίκιν *PRyl.* 627.34 (iv AD)'

καλίζομαι, add '(καλια- La.)'

κάλικα, delete the entry (v. °κάλιξ).

*****κάλιξ**, ιγος, ή, = Lat. *caliga*, *Edict.Diocl.* 9.5, οἱ τὰς ἀπὸ κάλιγος στρατείας .. στρατευσάμενοι i.e. *common soldiers*, Modest.*Dig.* 27.1.10 pr.

κᾰλιός, add '**4** perh. *shrine*, *PVindob.Salomons* 2.20 (ii/iii AD).'

*****καλκουλάτωρ**, ὁ, Lat. *calculator*, *accountant*, Modest.*Dig.* 27.1.15.5. **2** *arithmetic teacher*, καυκουλάτορι (sic) ὑπὲρ ἑκάστου παιδὸς *Edict.Diocl.* 7.67.

†**καλλαϊνοποιοί**, οἱ, *makers of a green dye*, *OBodl.* 45 (ii BC); also καλλαϊνιο- *PBodl.ined.* c.88(P).

καλλᾰρίας, add 'cf. γαλαρίας, γαλλερίας'

*****καλλεανός**, v. καλανός.

*****καλληλακανία**, ή, app. *shrub*, perh. same as °καλωλακάνθη, *PCornell* 25ᵛ.10 (28/3 BC).

†**καλλίας**, ου, (Lacon. **καλλίαρ** Hsch.) ὁ, *humorous or euphemistic term for an ape*, Din.*fr.* 6.2, Gal. 18(2).236, 611; Ion. καλλίης Herod. 3.41. **II** *name for* ἀνθεμίς (v.s.v. 2), Dsc. 3.137.1.

καλλίγονος, masc., add 'of Zeus, epigr. in *SEG* 31.962 (Ephesus, Rom.imp.)'

***καλλίγραπτος**, *beautifully drawn* or *painted*, A.*fr.* 78a.12 R.

***καλλιγράφισσα**, ἡ, *female calligraphist*, *SEG* 7.196 (Beirut, v/vi AD).

***Καλλίδρομος**, ὁ, name of a month in Crete, *IG* 12(5).868.25 (decr. at Tenos).

καλλιέλαιος, add '[Arist.] *de Plantis* 820ᵇ40'

καλλιεπέω, before 'Them.' insert 'D.H.*Dem.* 5'

⁺**καλλιεργέω**, *work* or *construct beautifully*, Phlp.*in Ph.* 327.1; *beautify with mosaic, paved work* or sim., καλλιεργῶν καὶ σκάπτων Quint.*Ps.* 140.7; τὴν στρῶσιν *make the beautiful* paved-work, *Inscr.Olymp.* 656.8 (v AD); τὴν πᾶσαν ἐκαλιέργησεν (sic) τρίστῳον *DOP* 6.87 (Nicopolis, vi AD); πόλεις καὶ ναούς *An.Par.* 1.168. **2** *improve land by cultivation*, *SB* 5168.27 (ii AD).

***καλλιεργικός**, ή, όν, *characterized by good cultivation*, πρὸς ἐργασίαν καλλιεργικήν *PCornell* inv. II.38.15 (*Rec.Pap.* 3.33; AD 388).

καλλίεργος, add 'epith. of Athena, *IG* 4².408 (Epid., iii AD), 485 (Epid.), Procl. *in Ti.* 1.169.4; cf. ἐργάνη'

***καλλιέτης**, ες, *having a prosperous year*, *SEG* 9.173, 186 (both Cyrene, ii AD), 18.750 (Cyrene, iii AD).

***καλλιθέμειλος**, ον, *having fine foundations*, epigr. in *SEG* 37.1537 (Arabia, vi AD).

***καλλίθρονος**, gloss on χρυσόθρονος, Hsch.

***καλλιθύγατηρ**, *having a beautiful daughter*, Δηὼ καλλιθύγατρα *Didyma* 496 (*ZPE* 7.207).

***καλλιθύεσσα**, Ἰὼ καλλιθύεσσα· καλλιθύεσσα ἐκαλεῖτο ἡ πρώτη ἱέρεια τῆς Ἀθηνᾶς (Ἀνθείας ‹Ἥρας› La.) Hsch.

καλλίκαρπος II, for this section read 'epith. of Dionysus, *ICilicie* 78 (AD 209/11), Mitchell *N.Galatia* 155 (Rom.imp.); identified w. Domitian, *JÖAI* 18 *Beibl.* 55 (Anazarba)'

καλλικέρας, add 'Pi.*fr.* 169a.50 S.-M. (Sch. -κερως)'

***καλλικίθων**, v. °καλλιχίτων.

***καλλίκλωνος**, ον, *having beautiful twigs* or *sprays*, Ast.Soph.*Hom.* 1.4.

***Καλλικόραι**, αἱ, title of nymphs, *SEG* 34.639 (Maced., ii AD).

***Καλλικράτειοι**, οἱ, name of a Rhodian guild, *Clara Rhodos* 2.203.

κάλλιμος, add 'Certamen 222, *h.Hom.* 31.5'

καλλίνικος I, after '(Paros)' insert 'epith. of Heracles, *Salamine* 45 (ii BC)'

***καλλιπάρηος**, add 'Λατωΐδι καλλιπαράωι *SEG* 37.1175 (Pisidia, ii AD)'

καλλιπρόσωπος, after '*face*' insert 'Anacr. 1 *fr.* 1.3 P., graffito in *SEG* 31.847.28 (Thasos, iv BC)'

καλλίρροος, add 'of water-nymphs, καλλιρόοισι θεαῖς epigr. in *SEG* 37.1239 (Lycaonia)'

καλλιστεῖον, line 2, after 'Sch.Il. 9.129' insert 'τῷ κρίναντι τὰ κ. Πριάπῳ *AP* 6.292 (Hedyl.)'

καλλιστέφανος, line 1, after '*beautiful-crowned*' insert 'of Aphrodite, *CEG* 454 (Pithecusae, viii/vii BC; cf. Jeffery *LSAG*² p. 235)'

καλλίσφυρος, line 3, after 'Od. 5.333' insert 'Alcm. 1.78 P.'

καλλιτέχνης, delete 'pl. -τέχνεις Epigr.Gr. 796' (v. °καλλίτεχνος)

καλλίτεχνος, add 'of Athena, epigr. in *ZPE* 15.226 (Attica)'

***καλλιχίτων**, ωνος, *wearing a beautiful χιτών*, καλλικίθωνι [χο]ρίδι (prob. reading) *CEG* 785 (v BC).

καλλίχοιρος, before 'Arist.' insert 'interpol. in'

καλλιώνυμος, for 'sens. obsc., *Com.adesp.* 1023' read 'μεταφέροντες δέ τινες τὴν λέξιν καὶ ἐπὶ τοῦ αἰδοίου ἔτασσον ἀνδρός τε καὶ γυναικός Hsch.'

κάλλος 3, line 3, after 'Pl.*Phd.* 110a' insert 'Call.*fr.* 7.11 Pf.'

καλλυνθρον, for '*sweeper*, *duster* made of palm-leaves' read '(palm-)frond'; add 'cf. °κάλυτρα'

***Κάλλων**, ωνος, ὁ, epith. of Dionysus, *SEG* 18.279, 280, al. (Rhegion nr. Byzantium, i AD).

καλλωπίζω II 2, line 2, for 'also κ. ὅτι ..' read 'ἐνδείξασθαι καὶ καλλωπίσασθαι ὅτι ..'

καλλωπισμός II 2 b, after '*embellishment*' insert 'D.H.*Th.* 29, al.'

καλλωπίστρια, add 'transf., ἡ ἡμέρα .. ἡ τῆς ἀναστάσεως ἔθιμος καὶ τῆς χάριτος καλλωπίστρια Ps.-Chrys.*HP* 2.5, Leont.Byz.*HP* 1.2.4'

***καλόδουλος**, ον, *treating slaves well*, *Vit.Aesop.*(G) 26.

***καλόζηλος**, ον, *eager for beauty, having good taste*, Ptol.*Tetr.* 165.

***καλοίδιον**, v. °καλῴδιον.

καλοκἀγαθία, after '*goodness*' insert 'Ar.*fr.* 205.8 K.-A.'

καλόκαιρος, after ' = *bonum tempus*' insert '*fair season*' and add 'epigr. in Robert *Hell.* 9.51 (Attalea, i/ii AD); cf. mod. Gk. καλοκαίρι *summer*'

***καλοκοίμητος**, ον, (in quots. sp. -κυμ-) *resting well*, of the dead, *IG* 14.2290, 2293, al. (Italy, v AD), *BCH* suppl. 8.166 (Thessalonica, v/vi AD), etc.

***καλοοὐνυμος**, v. °καλώνυμος.

⁺**καλοπέδιλα**, τά, (καλον), *wooden shoes, clogs* (also interpr. as *hobble* for cows during milking), Theoc. 25.103 (codd. κωλ-).

***καλοπόδινος**, ον, *whose coming brings good luck*, *REA* 62.357 (Syria, v AD); cf. mod. Gk. καλοπόδαρος.

***καλοποίητος**, ον, *well-made*, Moses Alch. 314.27.

***καλοποιός**, v. °καλωποιός.

καλός A I 2, add '*SEG* 32.847 (erotic graffiti, iv BC)' **b**, for this section read 'as epith. of Artemis, ἁ καλά A.*Ag.* 140 (cj.); καλλίστη, Paus. 1.29.2, 8.35.8, *IGLS* 182 (Beroea); also of Hera, *SEG* 33.704 (Thasos, iv BC)' **II 1**, line 11, after 'Th. 5.59, 60' insert 'ὅπου ἂν δοκεῖ ἐν καλλίστῳ εἶναι *SEG* 25.486.19 (Boeotia, iii BC); [ὑπὲρ] τοῦ γενέσθαι τῶν διασαφουμένων τὴν διεξαγωγὴν κατὰ τὸ κάλ(λ)ιστον *SEG* 26.677 (Larissa, ii BC)' **B**, add 'cf. mod. Gk. καλύτερος' **C II 1**, add 'ἱερατεύσαντες καλῶς *properly*, *SEG* 30.1420 (Bithynia, Rom.imp.). **b** τὸ καλῶς ἔχον *what is right and fair*, *PPetr.*2 p. 19(1) (iii BC), *UPZ* 12.46 (ii BC).' **2**, add '**b** in expression of welcome, χέρετε παροδῖτε καὶ καλὸς ἤλθατε *SEG* 26.791 (Byzantium, iv AD); cf. mod. Gk. καλώς ἤρθατε *welcome*); as written at the end of an epitaph, *SEG* 31.1041 (Lydia, iii AD).' **5**, after 'Aeschin. 3.232' insert 'καλά γ' ἐπόησε he *deserved* it, Men.*Dysc.* 629' **10**, comp., add 'perh. also καλιτερōς *GDI* 1156 (Elis, vi BC)'

κάλπασος, add 'sp. καλπασσ- *POxy.* 3931.27 (iv/v AD)'

***καλπίδιον**, τό, dim. of κάλπις, graffito on oenochoe, *SEG* 35.33 (Athens, viii BC).

κάλπις, line 6, after '*cinerary urn*' insert '*AP* 7.444.6 (Theaet.)'

***καλτάριος**, ὁ, *shoemaker*, *BCH* 7.243 (Chr.).

***κάλτις**, ὁ, an Indian gold coin, *Peripl.M.Rubr.* 63.

***κᾰλῠβός**, after '*chamber*' insert 'decr. in *SEG* 14.656.14 (Caunus, ii BC)' and for '*Epig.Gr.* 260 (Cyrene)' read '*GVI* 1254 (Cyrene, iii/ii BC)'

κᾰλύδριον, add '*Inscr.Délos* 1429*B*84 (ii BC)'

***κᾰλυκοειδής**, ές, = °καλυκώδης, *of the chrysanthemum*, Cyran. 44 (1.22.8 K.).

⁺**κᾰλυκώδης**, ες, *having the form of a bud*, τὰ τοῦ καρύου καλυκώδη περικάρπια Thphr.*HP* 3.5.6, ὅταν ᾖ καλυκώδες (τὸ ἄνθος) ib. 3.10.4; transf., ἐνθάδε Κλειτόριος κεῖται δρίλον καλυκώδες .. (i.e. either "immature" or "not erect") epigr. in *Suppl.Hell.* 975.1 (iii BC).

***κᾰλυκωπός**· εὐόφθαλμος Hsch. (La.; cod. καλυκοντος).

***κάλυκωσις**, εως, ἡ, perh. *budding flower*, Aq.*Is.* 35.1, *Ca.* 2.1.

κάλυμμα 3, for '*skull*' read '*dura mater*' **10**, delete the section (v. °καλυμμάτιον) and substitute 'perh. *lid of a dish*, Lang *Ath.Agora* XXI L19 (Hellen.)'

⁺**κᾰλυμμάτιον**, τό, dim. of κάλυμμα (in quots. app. sense 9), Ar.*fr.* 70 K.-A., *Didyma* 39.55 (καλυμματα lapis; ii BC).

κᾰλυξ I 2, line 2, delete '.. κισσοῖο .. Theoc. 3.23'; line 5, after '*h.Cer.* 427' insert '(στέφανον) ἀμπλέξας καλύκεσσι Theoc. 3.23' **II**, add 'Call.*fr.* 80.5 Pf.' add '**IV** καλύκων· τῶν ὀμματοφύλλων Hsch. **2** κάλυξ .. ἔνιοι ἔμβρυα ἀποδιδόασι κάλυκας id.'

⁺**κᾱλυξις**, εως, ἡ, (s.v.l., κάλυξι (dat. pl.) La.) **I** = κάλυξ I 2, Hsch. **II** = κάλυξ II, Id.

***κάλυτρα**· σπάθαι φοινίκων. σκόλοπες, χάρακες, σταυροί Hsch.

κάλυψ, delete the entry.

κάλυψις, add 'perh. also *IAskl.Epid.* 52.A43 (iv BC)'

***κάλφομαι**, verb cited in explanation of ἀκαλήφη (*nettle*) and κνίδη, Sch.Nic.*Al.* 201a Ge.

κάλχη I 1, add 'Lyc. 864'

κᾰλῴδιον, after 'Th. 4.26' insert 'Men.*Dysc.* 580'; add 'καλοίδιον *PCol.Zen.* 43 (iii BC)'

***καλωλακάνθη**, ἡ, perh. shrub of the genus *Acacia*, *PColl.Youtie* 24.20 (AD 121/2); cf. °καλληλακανία.

***κάλων**, v. °κήλων.

κᾰλώνυμος, add 'Εὐφροσύνη καλοούνομε (sic) *GVI* 1856 (Aegiale, ii/iii AD)'

***κᾰλῶπις**, ιδος, fem. adj. *having a beautiful face* (or *eyes*), Περσεφόνην δὲ καλῶπιδ[α] *SPAW* 1934.1046 (tab.defix.).

***κᾰλωποιός**, ὁ, app. *rope-maker*, Φιλιστίδας .. ho [κ]αλοποιό(ς) Dubois *IGDS* 130 (Gela, c.500 BC); (also interpr. as κᾰλο-, i.e. *shipwright*)

κάλως, line 5, after 'κάλωας' insert 'A.R. 1.566'; line 6, for '*reef*' read '*halyard*'

***κᾰμᾰκίς**· κοσμάριον, ὃ τοὺς πλοκάμους περιέχει. ἔνιοι σύριγγα Hsch.

κάμαξ, line 1, for 'infr. 3' read 'infr. 2, 3'

κᾰμάρα, line 1, after 'Ion. -η' insert 'also *κάμερα IGRom.* 3.1057.6 (Syria, ii AD)' **I**, line 4, for '*vault* of a tomb .. (Teos)' read '*burial chamber*, ἡ θύρα τῆς καμάρας *IEphes.* 3704, *IHadr.* 75, *TAM* 4(1).188, etc.'; line 7, for '*tester-bed* Arr.*An.* 7.25.4' read 'perh. *meeting-room*, *LW* 2220, 2240 (Syria)'

***καμαράριος**, ὁ, Lat. *camerarius*, *personal servant*, *POxy.* 1300.7 (v AD, καμαλ- pap.).

***κᾰμάρια**· κοιτῶν καμάρας ἔχων Hsch.

***κάμαρος** (B), app. = καμάρα I, *LW* 2426*b* (Syria).

κᾰμάρωσις I, add '*SEG* 16.470.16 (Thera, i/ii AD)'

καμάσιον, add '*PIand.* 125.2 (iv AD), *PHeid.* 333.28 (v AD); cf. ‡καμίσιον'

καματηρός II, delete 'Pass.' **2**, for 'toiling' read 'patient of toil, ψυχαί Max.Tyr. 39.3'; add 'gloss on φάλαγγες, Sch.Ar.Ra. 1349'

*καμβαών, v. ‡καμπαγών.

*κάμβειν, τό, perh. = °κομβίον, TAM 5(1).706.

*κάμβειος, ό, a kinship term (cf. °κάμβειν), SEG 31.1031 (Lydia, ii AD).

*καμελαύκιον, τό, cap, SB 9754.3 (AD 647).

*καμέρα, v. ‡καμάρα.

*καμηλαῖος, ό, camel-driver, PBaden 31.22 (v. ZPE 54.93).

καμηλάριος, after 'camel-driver' insert 'Edict.Diocl. 7.17 (rest.), POxy. 1870.7 (v AD)'

*κάμηλία, ή, perh. camel-load, PVindob. G39847.803 (CPR 5 p. 107, iv AD).

*κάμηλινος, η, ον, of a camel, τριχῶν .. καμηλίνων Edict.Diocl. in SEG 37.335 iii 11.

*κάμηλις, εως, ή, female camel, BGU 2106.3-4 (AD 142).

*καμηλιών, ῶνος, ό, unkn. object, costing 3 obols, SB 10241ᵛ.6 (i AD).

κάμηλος, add '3 camel's load, PGrenf. 50 (ii/iii AD), PWisc. 47 (iv AD), etc.'

κάμῖλος, for 'Sch.Ar.V. 1030' read 'Sch.Ar.V. 1035'; line 2, after 'τρυπήματος' insert '(v.l. τρήματος)'; add 'perh. also ICilicie 108 (sp. καμηλ-, v/vi AD)'

*καμῑνάριος, ό, furnace-man, ITyr 111.

καμίνιον, add 'BGU 2361a i 4 (iv AD)'

κάμῑνοκαύστης II, add 'POxy. 2272.22 (ii AD)'

κάμῑνος, add '2 part of a ship, Hsch. II κάμινοι· εὔπλευροι βόες, ἰσχυροὶ καὶ εὐίσχιοι id.'

*κᾰμῐσᾰγοραστής, ό, seller of shirts, Corinth 8(3).522 (καμισογ-, iv AD).

κάμίσιον, at end for 'κάμασος .. different' read 'cf. ὑποκαμίσιον, late Lat. camīsia; see also κάμασος, ‡καμάσιον'

*καμμορέων· κακοπαθῶν Hsch.

κάμνω I 1, add 'οἴκους Philet. 8' (from section 3). **3**, for this section read 'aor. med., (w. pred.), render by toil, οἵ κέ σφιν καὶ νῆσον ἐϋκτιμένην ἐκάμοντο Od. 9.130'

*καμπάγια, τά, gloss on ξυρίδες, Suid. (-άκια), Phot.; cf. Lat. campagus, καμπαγών, ξυρίς II.

καμπαγών, for pres. ref. read 'Edict.Diocl. 9.11 (καμβαών ib. 11A0)'

καμπεσίγυιος, add 'Suppl.Hell. 1082'

κάμπη I **2**, after 'ornament of this shape' insert 'IG 2².1425.251 (Athens, iv BC)' II, delete 'Indian'

*καμπιδούκτωρ, ορος, ό, Lat. campiductor, -doctor, drill-master, MAMA 1.168 (iv AD).

*κάμπιστρον, τό, Lat. campestre, loin-cloth, Ann.Épigr. 1907.29 (Aphrodisias), PRyl. 627.341, al. (iv AD).

καμπτήρ, add 'III prob. = κάμπτρα, Inscr.Délos 104-28bB19.'

*καμπτίον, τό, some kind of case or container, SB 9834b.23 (iv AD).

κάμπτρα, add 'cf. καρδοπεῖον .. ἡ κάμπτρα ὅπου τὰ ἄλευρα μάσσουσιν Hsch.'

κάμπτω I, line 13, after 'bend the knee in worship' read 'Lxx 1Ch. 29.20, 1Es. 8.73, etc., Ep.Rom. 11.4; intr., of the knee, ἐμοὶ κάμψει πᾶν γόνυ Lxx Is. 45.23; of a person, ὃς ἂν κάμψῃ ἐπὶ τὰ γόνατα Lxx Jd. 7.5, 4Ki. 1.13, ἔκαμψεν ὁ βασιλεύς 2Ch. 29.29'

καμπυλόπρυμνος, after 'stern' insert 'Suppl.Hell. 991.115 (poet. wordlist, iii BC)'

†καμπύλοχος, ον, compd. of καμπύλος and -οχος (ἔχω), in uncertain sense, κερκίδες Orph.fr. 33.

κάμψα, add 'PVindob. G25737 (ZPE 64.77, vi/vii AD)' and after 'κάψα' insert 'PLaur. 188.7 (iii AD), PHeid. 333.8 (v AD)'

καμψάκιον, after 'καψάκιον' read 'POxy. 2273.6 (iii AD), κ.· γλωσσόκομον Hsch.'

καμψάριος, add 'see also καψάριος'

*καμψίγουνος, ον, bending the knee, Suppl.Hell. 991.29 (poet. wordlist, iii BC).

καμψίον (s.v. κάμψα), add 'καμψίν PCornell 29 (iii/iv AD)'

καμψίουρος, for 'v. σκίουρος' read 'as subst., = σκίουρος, Hsch.'

*καμψίχειρ, χειρος, adj., bending the hand, Suppl.Hell. 991.28 (poet. word-list, iii BC).

†καναβιουργός, v. °κανναβ-.

κάναβος, at end for 'cf. κίναβος' read 'cf. κινάβευμα, κίνναβος'

*καναθρέω, beat with rods, PMich.inv. 3690 (ZPE 1.97, ii/iii AD).

κάναθρον, for 'cane or wicker carriage' read 'carriage furnished with wicker-work' and after 'X.Ages. 8.7' insert 'Plu.Ages. 19'

*κανάλιον, τό, culvert, dim. of Lat. canalis, pap. in AHDO 1.267 (Dura).

*κανανικλάριος, ό, app. some minor official, perh. form of Lat. canalicularius, PColl.Youtie 66.28, 38 (AD 253/60), POxy. 2925 (c.AD 270) (also interpr. as corrupt form of ‡κανονικάριος).

κανιά, add 'Myc. ka-pi-ni-ja, chimney'

καναστραία, add 'sg., κανασ[ραί]ον δριωτόν Inscr.Cret. 4.145.6 (Gortyn, v/iv BC)'

κάναστρον, delete 'dub. .. Crete' (v. °καναστραία)

κανᾰχέω, add 'φωνήν Posidipp. in Suppl.Hell. 705.12'

*κανᾰχισμός, ό, = καναχή, Orac.Chald. 61c P. (pl.).

*κανδηλάπτης, ου, ό, candle-lighter, Teuc.Bab. p. 42 B.

κανδήλη, add 'Corinth 8(3).618'

*κανδιδάριος, ό, Lat. candidarius, baker of white bread, MAMA 5.254 (-άρις, Nacolea), SEG 39.649 (-άρις, ii/iii AD).

*κανδιδᾶτος, ό, Lat. candidatus, candidate for office, κανδιδᾶτον αὐτοκράτορος IG 4.588.9 (Argos, ii AD), etc.; κυαίστορα κ. IEphes. 677 (ii/iii BC), IGRom. 1.134, SEG 30.1556 (Cilicia, iv/vi AD).

*κανδύλη, v. °κανδυτάνης.

*κανδῠτᾰλις, delete the entry (v. °κανδυτάνης).

*κανδυτάνης, ό, clothes-press, Diph. 39 K.-A., Men.Sic. 388; κανδυτάναι καὶ κανδύλαι· ἱματιοθῆκαι Hsch.; pl. -ᾶνες Poll. 7.79, κανδύτανες· ἱματιοφορίδες· οἱ δὲ εἶδος ἰχθύος· ἔστι δ' ὅτε τὸ αἰδοῖον Phot. **2** the name of a kind of rat found in Babylonia, Ael.NA 17.17.

*κάνειος, α, ον, Myc. ka-ne-ja (neut. pl.), made of basketry.

κάνεον, line 2, pl. κανᾶ, add 'Hld. 3.2, X.Eph. 1.2.4'

†κάνης, ητος, ό, reed-mat, D.H. 2.23, Plu.Sol. 21, in gnomic remark, ὁ κάνης δὲ τῆς κοίτης ὑπερέχειν μοι δοκεῖ, app. of trivialities being given precedence over important things, Crates Com. 14 K.-A.; cf. Phot.s.v.; used for winnowing, Poll. 6.86.

*κανθαρίας, ου, ό, gem in scarab form, prob. = κάνθαρος VI, Plin.HN 37.187.

κανθᾰρίς I, line 3, delete 'pl.' and insert 'Gal. 12.363.14'; line 5, delete 'so .. 363' add 'III = καπνός II, Ps.-Dsc. 4.109.'

κάνθαρος IV, add 'IGC p. 98 A6 (Acraephia, iii BC)'

κανθήλια I, at end add 'Myc. ka-tu-ro₂ (gen. pl.), cf. ka-tu-re-wi-ja-i (dat. pl. of deriv.), perh. saddle-bags, may reflect a form κανθυλ-, cf. κανθύλη'

κανθός I **2**, for 'Call.fr. 150' read 'Call.fr. 177.28 Pf.' II, after 'wheel' insert 'Polyaen. 7.21.3'

κάνισκος (s.v. κᾰνίσκιον), after 'Gloss.' insert 'κανίσκον, τό, Inscr.Délos 372B25 (iii/ii BC)'

*καντης, ό, basket-maker, rest. in IEphes. 454.

†κανναβάριος, ό, worker in hemp, IEphes. 454; = stupparius, Gloss.

*κανναβᾶς, ᾶ, ό, tow-seller, in quot., as pers. n., TAM 5(2).1298.18, cf. Κανναβίων, pers. n. IMylasa 463.

*κανναβιουργός, ό, app. = κανναβάριος, Tab.Defix. 87ᵃ7 (sp. καναβ-; iv BC).

κανονικάριος, add 'see also °κανανικλάριος'

κανόνιον II, for 'compass' read 'ruler or measuring-rod'

κανονίς I, after '(Phil.)' insert 'cf. °ἰθυβάτης' II, read 'perh. upright of a door-frame, IG 2².1672.155' add 'IV column of slots in a °κληρωτήριον (sense I), Arist.Ath. 64.2.'

κανονισμός, add '2 ordering, regulation, PHamb. 234.2 (vi AD).'

κανονωτός 1, after '(iii BC)' insert 'cf. PCair.Zen. 847.5 (w. note)'

*κανοῦν, v. κάνεον.

*κάνψη, gen., name of a relation, MAMA 3.745; gen. pl. κανψίων LW 1784 (Tarsus); perh. also gen. sg. [κ]ανψίου Rott.Kleinas.Denkm. 374 no. 89 (SEG 34.1411); cf. Tsaconian kambzi 'child'.

Κάνωβος, add 'in title of Zeus, identified with Helios and Sarapis, SEG 24.1192 (Egypt, ii AD)'

κανών I 3, lines 8/10, for 'μολίβδινος κ. .. κῦμα' read 'μολίβδινος κ., flexible rule that can be adjusted to curved outlines, Arist.EN 1137ᵇ31' II 7, add 'b rent, Cod.Just. 1.4.32 pr.' add '8 in athletics, τὸ μέτρον τοῦ πηδήματος Poll. 3.157; prob. also in SEG 15.501 (Rhodes).'

*καπανεύς, v. °σκαπανεύς.

*καπβολαία, v. °καταβολαία.

Καπετώλια, add 'Καπετώλ[εια ἐν 'Ρώμῃ] SEG 37.712 (Chios)'

*Καπετωλιάς, άδος, ή, a celebration of the Ludi Capitolini; meton., victory in these games, δύ' ἔχω καὶ Καπετωλιάδας epigr. in SEG 37.712.4 (Chios, ii AD).

Καπετώλιον, add 'citadel in any town, SEG 29.807 (Chalcis, late Rom.imp.)'

κᾰπηλεία, add '2 shop, ἐπρίατο οἴκησιν καὶ τὰν καπη[λ]είαν τὰν Δίων[ος] SEG 34.940 (Camarina, c.400 BC).'

κᾰπηλικός I, line 2, for 'ἀργύρωμα .. 111' read 'τὸ -ικόν, ή -ική, kind of cup, IG 11(2).110.24, 124.39, al. (all Delos, iii BC)'

†κάπητον, v. °καπιτόν.

*κάπιστρον, τό, Lat. capistrum, halter, κ. ἱππικόν Edict.Diocl. in SEG 37.335 iii 10.4, rest. in Edict.Diocl. 10.4.

*κάπιτατίων, ωνος, ή, Lat. capitatio, allowance of food or fodder, Just.Nov. 8.2, al.

*κάπιτον, τό, (perh. formed as sg. of Lat. capita) daily ration of fodder, PHerm. 39.2 (v AD), Lyd.Mag. 1.46; in general, ration allowance, PHerm. 78.3 (v/vi AD); cf. κ.· παράβλημα ἀλόγων Hsch.

καπνία, add 'Myc. ka-pi-ni-ja, chimney'

καπνίζω, add 'III heat over steam, βαλανεῖον PBremen 56b.5 (ii AD).'

*καπνισμός, ό, smoking (in quots., process employed to give pottery a dark grey colour), POxy. 3596.15, 3597.20 (iii AD).

†καπνοβάτης, ου, ό, in pl., walkers through smoke, name given to the

Mysians, app. in respect of some religious observance, Posidon.ap.Str. 7.3.3.

καπνόομαι, delete the article (v. °καπνόω).

***καπνόω**, *smoke* (a beehive), *AP* 9.226 (Zon.). **II** pass., *to be turned to smoke*, i.e. burnt up, Pi.*P.* 5.84, E.*Supp.* 497, *Tr.* 8.

καπνώδης 1, add '**b** *producing smoke* when burned, φύλλον D.Chr. 66.5.' **2**, delete 'φύλλον .. 66.5'

κάπος, add '**II** καὶ ὁ τοῦ φοίνικος φλοιός, ἐν ᾧ κέκρυπται ὁ καρπός. καὶ ἡ πρώτη ἔκφυσις Hsch. s.v.'

κάπουπλος, add '(καπουστάς· φάρυγξ La.)'

***καπουστάς**· φάρυγξ Hsch. (La.) (v. κάπουπλος).

***Καππαδοκαρχία**, ἡ, *Presidency of Cappadocia* (of its council and festival) as part of Imperial cult, *Dig.* 27.1.6.14.

καππάριον, before '*Gloss.*' insert '*BGU* 227.19 (ii AD)'

κάπρος, add '**III** disease of bees, Hsch.'

κᾰπύρια, add '*Ath.* 3.113d'

κᾰπῠρόομαι, after '*become crackly*' insert 'Str. 11.13.2'

***καπύσσων**· ἐκπνέων Hsch.

κᾰπύω, after 'aor. 1' insert 'ψυχὴν οὔ τι'; for 'κεκαφηώς' read 'καπύσσων· ἐκπνέων Hsch.; cf. ἀποκαπύω (from which it may be a back-formation), perh. also κεκαφηώς'

κᾱρᾶ (A), line 1, after 'τό' insert '(v. infr.)'; line 19, after '*Anacreont.* 50.9' insert 'nom. κάρη is fem. in Q.S. 11.58, and so acc. κάρη (nisi leg. -ην) id. 13.241'   at end add 'Myc. ka-ra-a-pi (instr. pl.)'

κάρα (B), for 'Id.' read '*Inscr.Cret.* 1.xvii 12*A*1 (Lebena), Hsch.'

***κᾰρᾰβιάριος**, ὁ, *fisher of crayfish* (?or *boatman*, cf. κάραβος III), rest. in *ITyr* 24A.

Κᾰραιός, add 'also Κεραιός *IG* 2².2360 (ii BC), Καραός *IG* 9²(1).434.1 (Astacus, ii BC)'

κᾰρᾰκάλλιον, τό, add '*PL* ι/3 l. 26 (-ιν, *Tyche* 1.164, vi/vii AD)'

***κᾰρακέριον**, τό, perh. corruption of foreg., *PVindob.* G39847.846 (*CPR* 5.108, iv AD).

κᾰρᾱνιστήρ, delete '*touching the head*'

κάρβανος, after 'A.*Supp.* 914' insert 'S.*fr.* 269a.54'

***καρβᾶς**, ὁ, occupational term, perh. *charcoal-merchant*, *OGI* 697 (Egypt) v. *Amyzon* p. 136 no. 32.

καρβάτινος, add 'cf. *crepidas .. carpatinas* Cat. 98.4'

κάρδακες, add 'cf. Nepos *Dat.* 8.2'

***καρδᾰμέα**, ἡ, = κάρδαμη or κάρδαμον, *OBodl.* 2183.4 (iv AD).

***καρδᾰμογλύφος**, ὁ, *one who chops* κάρδαμον (humorous term for a miser), Hsch. s.v. κυμινοπρῖσται; cf. κυμινοπριστοκαρδαμογλύφος.

κάρδαμον, add 'cf. Myc. ka-da-mi-ja'

καρδία I 1, line 4, after '(E.)*Hipp.* 1274' insert 'κραδία epigr. in *SEG* 37.1175.13 (Pisidia, ii AD)' and for 'Sapph. 2.6' read 'Sapph. 31.6 L.-P., Alc. 207.9 L.-P.'   add 'καρδίαν μὴ ἐσθίειν (and sim. phrs.) D.L. 8.17, 18, Iamb.*Protr.* p. 108.5, 123.3 P., Pythagorean saying meaning (acc. to D.L.) "*not to waste one's soul in pain and grief*", (acc. to Iamb.) "*not to break up the unity of the universe*"'   **III**, add '**2** of a golden crown, perh. *core* or *framework*, *Inscr.Délos* 1449*Aa*bii 13 (ii BC).'   **V**, add 'cf. Ἡελίου κραδίην (= the planet Mercury) Nonn.*D.* 38.392'

καρδιᾰκός II, after '*heart disease*' insert 'Cic.*Div.* 1.81'

***καρδίδιον**, τό, *twig* of the mulberry tree, Cyran. 29 (1.12.21 K.).

καρδιοειδής, add 'adv. -ῶς, *PWarren* 21.82 (iii AD)'

***καρδιοστάλακτος**, ον, *dropping from the heart*, κ. δάκρυον epigr. in *IGChr.* 295 (Megiste).

***καρδιοτομέω**, *cut the heart from*, tab.defix. in *SEG* 30.326 (Athens, v/vi AD; v. *Glotta* 58.64).

καρδιουργέω, after 'καρδιουλκέω' insert '*draw out the heart*' and add '*IEphes.* 10.7 (iii AD)'

***καρδόπιον**, τό, dim. of κάρδοπος, κ. λίθινον ἐπὶ βάσεως τετραγώνου *Inscr.Délos* 1417*Ai*70 (ii BC); Hsch. (sp. καρδοπείον).

κάρδοπος, line 1, after 'Ar.*Ra.* 1159' insert 'κ. λιθίνη *IG* 1³.422.4, 11, κεραμεία ib. 9 (Athens, v BC)'

Κάρειος, after 'ὁ' insert 'epith. of Apollo at Hierapolis, *ASAA* 41/42(1963/4).353 (?ii/i BC), 360 (?ii AD)'; before '(sc. μήν)' insert '**II**'

***Κᾱρία**, ἡ, *Caria*, a region of Asia Minor, later Roman province, Hdt. 1.142.3, al., Paus. 1.29.7, Str. 2.5.31, *SEG* 31.1116.

Καρικός, add '**VI** name of a coital posture, ὄντα γ᾽ ἐν Ἀθήναις Καρικοῖς χρῆσθαι σταθμοῖς Macho 310 G.; δηλοῖ δὲ καὶ ἀφροδίσιον σχῆμα αἰσχρόν Hsch.'

κάρκαρον, after 'Sophr. 147' insert 'cf. Lat. *carcer* (*carcar*)'

⁺**καρκῐνάς**, άδος, ἡ, *crab*, Artem. 2.14, Opp.*C.* 2.286.   **b** spec. *hermit crab*, Gal. 6.717.12, Ael.*NA* 7.31, Opp.*H.* 1.320, 542.

⁺**καρκίνηθρον**, v. ‡καρκινώθρον.

καρκίνος IV 1, add '**b** *crane* = μηχανὴ λιθαγωγός, Poll. 10.148.'   **5**, add 'instrument similar to the διαβήτης, Papp. *in Alm.* p. 70 R.'

καρκινόω I, for 'Pass. .. ib. 3.23.5' read '*cause* roots *to spread out crabwise*, i.e. *intertwined on the surface*, ὁ χειμὼν πιλώσας καὶ καρκινώσας τὰς ῥίζας Thphr.*CP* 3.23.5; καρκινωμένος, of roots, so *spread out*, id.*HP* 1.6.3, *CP* 3.21.5, *PPetaus* 22.31 (AD 184/7)'

καρκῐνώδης II, add 'n. pl. as subst.' and transfer 'Dsc.*Eup.* 2.72' from beginning of section.

καρκίνωθρον, add 'Plin.*HN* 27.113; cf. καρκίνηθρον = *polygonos*, *Gloss.*'

⁺**κάρμα**· γλεῦκος. τὸ πρῶτον ἀποθλιβόμενον διὰ τῶν χειρῶν. καὶ κούρευμα Hsch.

***καρνάριος**, ὁ, Lat. *carnarius*, *butcher*, *PFlor.* 207.5, 214.3, al. (iii AD), *POxy.* 2331 ii 12 (-ις, iii AD).

Κάρνειος, add '*IG* 5(1).222 (Laconia, c.530/500 BC), *Inscr.Cos* 38.11, al.'

κάρνον II, delete 'hence .. al.'

κάρνυξ, add 'τὴν σάλπιγγα Γαλάται Hsch. (La.) (κάρνον, q.v., cod.)'

***καροῦσθαι**· ὠνεῖσθαι Hsch.; also καρούμενος· ὠνησάμενος ib.

καρουχάριος, add 'χαρουχα[.. *ITyr* 205 (unless χαρουχᾶ, gen. of χαρουχᾶς, is read in same sense)'

κᾰρόω, line 2, of wine, add 'τὴν ψυχήν μου κάρωσον Anacreont. 52A.3 W.'

καρπεύω, line 2, for 'Corc.' read 'Cret. decree at Corcyra, iii BC'

⁺**κάρπιον**, τό, tree found in India, Ctes.*fr.* 45.47 J.

Κάρπιος, add 'of Dionysus, Πολέμων 6.17/18, *Rev.Phil.* 35(1911).124 (both Larissa)'

καρπιστής, for '*emancipator*' read '= Lat. *adsertor*, *vindex* (cf. *RIDA*, ser. 3, 6, pp. 190-3)'; add '*SB* 9801.7 (ii/iii AD)'

καρπογόνος, add '*IG* 12(1).783.6 (Lindos) = *AP* 15.11.6'

καρποδαιστάς, after 'ᾶ' insert 'or **-δαίστᾱς**, ᾶ'

***καρποδότης**, ὁ, *giver of fruit*, Διὶ Βροντῶντι καὶ Διὶ Καρποδότῃ *INikaia* 1085 (iii AD), 1084.

***καρποθάλεια**, fem. adj., *rich in fruit*, cj. in epigr. in *Didyma* 496*B*3 (ii AD; v. *ZPE* 7.207).

***καρποθηκείτας**, ου, ὁ, perh. *granary superintendent*, *IRhod.Peraia* 603.2 (ii BC).

***Καρποκράτης**, v. °Ἁρποκράτης.

⁺**καρπολογέω**, *prune the fruiting boughs* of: pass., of trees, Thphr.*CP* 1.15.1.   **II** serve as καρπολόγος II, *SIG* 1000.29 (Cos).

καρπολόγος II, add 'also at Colophon and elsewhere, decr. in *AJPh* 56.362.37; cf. °καρπολογέω II'

***καρπόμετρον** (?), τό, (rest. fr. abbrev. καρ.) an undefined measure, *SEG* 31.374 (Olympia, vi/v BC).

καρπός (A), line 7, after 'of grapes, *Il.* 18.568' insert 'of fruit from trees, *Od.* 11.588, 19.112'   at end add 'Myc. ka-po'

καρπός (B), line 2, after '(E.*Ion*) 891' insert 'καρποὶ χειρῶν Hp.*Aër.* 20'

***καρπότεξ**, τεκος, adj. *bearing fruit*, epith. for a month, Dionys.Trag. 121 S.

καρποτρόφος, after '*Milet.* 7.64' insert '*ICilicie* 78 (iii AD)'

καρποφόρος, of Demeter, add 'rest. in *SEG* 30.1341.5 (Miletus, ii/i BC)'

καρπόω I 1, line 2, after 'A.*Pers.* 821' insert 'med., *Inscr.Cret.* 4.43*Aa*3 (v BC)'

κάρπωμα II, for this section read '*an offering* (properly of fruits, but used by Septuagint translators to cover offerings of all kinds, mainly animal victims), Lxx *Ex.* 29.25, *Le.* 1.4, *Jo.* 22.26, etc.'

καρπωνία, for '*fruit-buying*' read '*purchase of a crop*'; add 'sp. -εία, *PMich.*v 238.34 (AD 46), *PLond.* 168.7 (AD 162), etc.; (meton.) *the crop so purchased*, *PSoterichos* 4.5 (AD 87), *SB* 9132 (iii/iv AD)'

κάρπωσις I, add '*SEG* 37.77.10 (Athens, iv BC)'   **II**, after '*offering of fruits*' insert 'including animal offerings (cf. °κάρπωμα)'; after 'Lxx *Le.* 4.10' insert '22.22, *Jb.* 42.8, *Si.* 30.19'

***καρρᾱρικός**, ή, όν, *of a wagon*, τροχός *Edict.Diocl.* 15.30.

***καρσανάριος**, ὁ, unkn. occupational term, *MAMA* 3.421, 422 (Corycus).

καρτερέω I, transfer 'ἀκούων .. 241' fr. line 10 to follow 'Arist.*Pol.* 1287*b*27' in line 5 of section II, prefixing 'w. part.'

***καρτερόθροος**, ον, *loud-voiced*, κ]αρ[τ]ερόθρουν βριαρό[ν τ]ε *PRyl.* 15.10 (= Heitsch *GDRK* no. 11; ii AD).

καρτερός I 2, after 'Thgn. 480' insert 'ὅκου δὲ μὴ αὐτοὶ ἑωυτῶν εἰσι καρτεροὶ ⟨οὐ⟩ ἄνθρωποι Hp.*Aër.* 16'   **3**, sup., add 'Thrasym. 1 D-K.'   **6**, add '**b** w. part., *having discretion in*, οἱ πολέμαρχοι θωϊόντων κ. ἔστων *Thasos* 141.6 (iv BC).'

***καρτέω**, v. ‡κρατέω.

***κᾰρῠδᾶς**, ὁ, *nut-seller*, *SEG* 32.1611 (ii/iii AD).

κᾰρύδιον, add 'written καροιδ- *PRyl.* 629.185, al. (iv AD)'

***κᾰρύζω**, v. κηρύσσω.

κᾰρύκινος, for '*dark-red*' read '*brown*'

κᾰρῡκοποιέω, after 'καρύκη' insert 'in quot. fig., i.e. *stir things up*'

***κᾰρυοφύλαξ**, ακος, ὁ, *guard set over a nut-plantation*, *PSI* 297.19 (v AD).

κᾰρυόφυλλον, add '*PAlex.* 36.6 (iv/v AD), *PColl.Youtie* 87.4 (vi AD)'

***κάρυσσα**, Aeol. fem. of κῆρυξ, *BCH* 59.473 (Mytilene), prob. in *IG* 12(2).255 (ib.).

κᾰρυωτός 1, for '*date-*palm, *date*' read '*name of variety of date*' and add 'D.S. 2.53, *PCornell* inv. II 38.18 (*Rec.Pap.* 1964 p. 32, n. 4)'

†**κάρφη**, ἡ, *dry straw, hay*, or sim. material, X.*An.* 1.5.10, *SEG* 9.11, al. (Cyrene, iv BC), Arr.*An.* 1.3.6, κάρφην· φορυτόν Hsch.

καρφίον, add '**3** *fenugreek* (= κάρφος v), *ICilicie* 108 (v/vi AD).'

κάρφος I, line 5, after '*AP* 10.14 (Agath.)' insert 'κάρφη· ξύλα λεπτά, καὶ ξηρά Hsch.' add '**2** *hay* (cf. κάρφη), *SEG* 9.35 (Cyrene, iii BC).' **II**, add 'cf. καρπίζω (B)' **V**, add 'Plin.*HN* 24.184'

καρχᾰρίας, add '*IGC* p. 98 A5 (Acraephia, iii/ii BC)'

καρχᾰρόδους, line 1, after 'neut. -όδουν' insert 'Choerob.*in Theod.* 1.347' and after 'Plot. 6.7.9' insert '(v.l.)'

καρχᾰρόδων, add 'Nonn.*D.* 41.210'

*καρχᾰρόπεπλος, ον, *having a saw-toothed peplos*, of Hecate, χαρχαρόπεπλε *PMag.* 7.701.

*καρχᾰρόστομος, α, ον, *jagged-toothed*, χαρχαροστόμα σκύλαξ *Suppl.Mag.* 42.1, 63 (iii/iv AD).

καρχήσιον II, for '*mast-head of a ship*' read '*naut., truck*'

κάς II, delete the section.

*κάς (B), Arc. and Cypr. = καί, *IG* 5(2).261, 262 (Mantinea, v BC), Cypr. *ka-se*, *ICS* 92, 217.1, al., also *ka*, v. °κᾰ (B).

κασάνδρα, delete the entry.

*κασαπανα (pl.), transcription of Middle Iranian *karshapana*, type of punch-marked coins, *SEG* 33.1223(c) (Bactria, ii BC).

κᾰσία, lines 1/2, for '*Cinnamomum iners*' read '*Cinnamomum cassia* or sim. species' and after '(Thphr.) *Od.* 30' insert '*Peripl.M.Rubr.* 8, al.'

κᾰσιγνήτη, line 2, for 'Hippon. 34, cf. 70ᵃ' read 'Hippon. 48, 103.10, 144 W.'; line 3, for 'Cypr. ... (q.v.)' read 'Cypr. *ka-si-ke-ne-ta* κασιγνέτα *ICS* 164, var. *ka-si-ne-ta-i ICS* 153; cf. καινίτα'

κᾰσίγνητος I, line 5, after 'Ps.-Luc.*Philopatr.* 11' insert 'applied to a half-sister, E.*Ion* 467 (lyr.)' **II**, line 5, after '(Eresus)' to end of article read 'Cypr. *ka-si-ke-ne-to-ne* κασιγνέτον *ICS* 217.14, al.; Thess. κατίγνειτος rest. in *IG* 9(2).894, *SEG* 31.575.16 (Larissa); Lesb. κασίγνᾱτος *IKyme* 13 i 16 (ii BC)'

*κᾰσιεύς, ὁ, app. acc. κασιέα, *brother*, *SEG* 37.494.4 (Thessaly).

*κάσινος, η, ον, *perh. of* κασῆς, ἱδρώου κασίνου *APF* 5.392.31 (i AD).

κάσιοι, delete the entry.

*Κάσιος (A), ὁ, cult-title of Zeus, Ach.Tat. 3.6, *SEG* 24.1196, etc. (Egypt); in Syria, *AP* 6.332, *SEG* 36.1301; in Sicily, *SEG* 34.980 (Syracuse, ii BC); also Κάσσιος epigr. in *SEG* 23.477 (Epirus, i BC).

*κάσιος (B), ὁ, *brother*, app. in Mitchell *N.Galatia* 14 (Rom.imp.). **b** κάσιοι· οἱ ἐκ τῆς αὐτῆς ἀγέλης ἀδελφοί τε καὶ ἀνεψιοί Hsch.

κάσις, line 3, after '*sister*' insert 'Anacr. 25 P.' and for 'Call.*Aet.* 3.1.23' read 'Call.*fr.* 75.23 Pf.'; add 'also πάν[τες γὰρ πέλομε]ν κάσιες poet. in *Hesperia* 5.95.28 (iii AD) '

*Κασιωτικός, ή, όν, *of or made in Casiotis*, κάτοπτρον δίπτυχον Κασιωτικόν *POxy.* 3491.7 (ii AD).

*Κασμῖλος, ὁ, v. ‡Καδμῖλος.

*κάσσος, v. °κάσσος (A) and (B).

*Κασσανδρίζω, *side with Cassander*, Polyaen. 4.7.6.

*Κασσεῖα, τά, *games in honour of Cassius* (prob. Cassius Apronianus, father of Cassius Dio), *TAM* 2.428 (Patara).

*κασσηρατόριον, Lacon. = *καταθηρατόριον, *hunt*, an athletic contest at Sparta, *IG* 5(1).279 (i/ii AD), al.; also **καθηρα-** ib. 274 (-τόριν), 288 (-τόριον).

Κασσιέπεια, add 'Luc.*Salt.* 44, *SEG* 31.1394 (Palmyra, iii AD), 31.1387 (Apamea, iv AD), Nonn.*D.* 33.296'

*Κάσσιος, v. °Κάσιος.

κασσῐτέρινος, add 'of coins (prob. with core of tin, plated with silver), *Inscr.Délos* 1442*B*51 (ii BC)'

κασσῐτεροποιός, add 'κασιδεροποιός *PTeb.* 414.34 (ii AD), κασειδεροποιός *SB* 9375 (iii AD)'

κασσῐτερουργός, for '*tinker*' read '*tin worker, tinsmith*'; add 'Heph.*Astr.* 2.19.16'

κάσσος, add 'written κάσος, prob. in *BGU* 759.17 (ii AD)'

*κάσσος (B), ὁ, also κᾶσος, Lat. *casus*, *occasion, occurrence*, Just.*Nov.* 53.5.1, 97.4. **II** *part, portion* of an estate, ib. 2.4, 123.40.

κασσύω, add '**III** καττύεσθαι· Ὑπερείδης· τὸ ὑποδεδέσθαι, ἀπὸ τῶν καττυμάτων Phot.'

*Κασταλίς, ίδος, fem. adj. *of Castalia*, νύμφαι Theoc. 7.148.

*καστελλίτης, ου, ὁ, perh. = Lat. *castellarius*, *man in charge of a reservoir*, *PLond.* 1652.6 (iv AD).

*κάστελλος, ὁ, or κάστελλον, τό, Lat. *castellum*, *fort*, Res Gestae Saporis 12, Procop.*Aed.* 2.5.9. **2** *water-reservoir*, *PLond.* 1177.65, al. (ii AD), *Stud.Pal.* x 205 (vi AD), Hsch.; (masc. forms in Res Gestae Saporis, Procop. and Hsch., elsewhere indeterminate).

*καστελλοφύλαξ, ακος, ὁ, *guardian of the fort*, app. a Persian court-official, Res Gestae Saporis 63.

*Καστνιῆτις, ιδος, ἡ, title of Aphrodite, *of Mt. Kastnion* in Pamphylia, Call.*fr.* 200a Pf., Str. 9.5.17; in pl., *JHS* 78.65 (Aspendos).

Καστόρειος II, after 'Dsc. 2.24' insert 'δέρμα καστόριον *Edict.Diocl.* 8.31'

καστορίδες II, for '*sea-calves, seals*' read '*beavers*'

*κάστρα, τά, and later κάστρον, τό, Lat. *castra, castrum*, *SEG* 34.309 (Sparta, c.AD 200), *IGRom.* 3.237, *PFay.* 50, etc.

*καστρένσιος, v. °καστρήσιος.

*καστρησιανός, ὁ, Lat. *castrensianus*, *soldier of the frontier guard* stationed in a fort or permanent camp, *SEG* 9.356.46, 80 (Ptolemais in Cyrenaica, vi AD), *PMasp.* 126.62 (vi AD).

*καστρήσιος, α, ον, also καστρένσιος *IEphes.* 852 (Trajan), *of a military camp* or *station*, ἐπίτροπος κ. = Lat. *procurator castrensis*, *Inscr.Dessau* 8856; pl. subst. = Lat. *castrenses milites*, prob. in *SEG* 8.643 (Ptolemais, Egypt); cf. γαστρισι[.. *POxy* 1001 (AD 572). **II** *of the (Byzantine) imperial court*: μόδιος (ἡμιμόδιος) κ. Hero *Mens.* 1.204.4, *Hippiatr.* 1 p. 425.27.

*καστροκνήμιον, v. ‡γαστρο-.

*κάστυ, τό, scribe's *equipment*, Aq., Thd.*Ez* 9.2 (Hebr. qeset fr. Egyptian gstj); perh. to be read in *IG* 14.2413.17.9, 25 (Jewish amulet fr. Sicily), cf. *Aegyptus* 33.172.

Κάστωρ, add 'Myc. *ka-to, ka-to-ro* (gen.), pers. n.'

κάτ, add 'κατά (A) F'

κᾰτά, line 1, after 'A.D.*Synt.* 309.28' insert 'cf. κ. δονακώδεος ὕλης epigr. in *SEG* 16.702 (Caria, ii/iii AD), and ' **A II 3**, after 'Hdn. 6.7.8' insert 'cf. Luc.*Pisc.* 7' add '**8** Locr., *in accordance with*, κα τῶνδε *IG* 9(1)².718.1; κα τᾶς συμβολᾶς ib. 717.15; καθ' ὧν ib. 267.9.' **B II 3**, add 'κατὰ μηδὲν *in nothing, in no respect*, *SEG* 33.1041 (Aeolian Cyme, ii BC)' **III 2**, add 'κ. πόδα ὑπέλαβες of *immediately* seizing what has been said, Pl.*Sph.* 243d' **IV 1**, add 'καθ' ὅ τι *in what way*, *IG* 1³.35.7, Th. 1.82.6; κατ' ὅ τι *for what reason, for the reason that*, Hdt. 6.3, 7.2.3'

καταβαίνω, line 5, after '(Ar.)*Ra.* 35' insert 'Men.*Dysc.* 633 (-βαι pap.)' **II**, add '**6** *pass to a less lofty style*, D.H.*Dem.* 13, 25. **7** *descend* (genealogically), Just.*Nov.* 3.3 epilogus.'

καταβάλλω I 7, pass., after 'Isoc. 12.8' insert 'καταβεβλημένα ἔπη Philostr.*Her.* 2.19'

καταβατικός, add '**2** κ. κύκλος a vertical great circle passing through the zenith and a star, Ptol.*Anal.* pp. 191, 205 H. (Lat. *descensivus*).'

καταβατός I, after 'Porph.*Antr.* 23' insert 'subst., καταβατή, ἡ, perh. *path down, descent*, κατεσκεύασα τὴν καταβατὴν σὺν τῇ ἐπικειμένῃ σορῷ Dumont-Homolle *Mél.Arch.* 378 (Perinthus)' **II**, add 'esp. *column* of days in an ἐφημερίς, Cod.Vat.Gr. 1058.328ʳ.6, al.'

*καταβαφικός, ή, όν, *effected by* καταβαφή, καῦσις Zos.Alch. 208.5.

καταβιβρώσκω, add 'σίδηρος καταβεβρωμένος ὑπὸ τοῦ ἰοῦ *IG* 2².1672.310 (iv BC)'

*καταβίων, app. a kind of garment, βάνατα Νωρικὴ διπλῆ ἤτοι καταβίων *Edict.Diocl.* 19.55.

†**καταβίωσις**, εως, ἡ, the *living out of one's life, spending one's days*, Cic.*Att.* 13.1.2, D.S. 18.52, App.*BC* 4.16.

καταβλάπτω, pass., add 'inf. καταβλάπεθαι (= -βλάπτεσθαι) *Inscr.Cret.* 4.42*B*11 (Gortyn, v BC)'

*καταβλάστημα, ατος, τό, *shoot, plant*, Hsch. s.v. πρέμνα.

κατάβλημα III, for '*payment*, dub. in' read 'perh. *support* for a statue'

*καταβολαῖα, ἆ, in Thess. form καπβολαία, a land measure, subdivision of a πελεθρ(ι)αία, *SEG* 26.672, 675, 676 (Larissa, iii BC).

καταβολαῖον, add '*PMich.*ix 540 (AD 53)'

καταβολή I 1 a, add 'δρέπανον (or δρέπανον) εἰς καταβολὴν *sickle for reaping* or *pruning hook*, *SB* 9834b37 (iv AD)' **2**, before 'D. 59.27' insert '*SEG* 26.72.7 (Athens, 375/4 BC)'

κατάβολος, add '**III** *payment of an instalment*, de Franciscis *Locr.Epiz.* p. 16 n. 2, p. 24 n. 10, etc. (iv/iii BC), *Mél.Navarre* 357 (Thespiae, iii BC).'

καταβυθίζω, for 'πολεμίοις .. i BC) ' read 'πολέμοις καταβυθισθεῖσαν τ[ὴν πόλιν] *IPE* 1².34.7 (*SEG* 37.670, Olbia, c.200/175 BC)'

καταγγίζω, add 'καταγγίζω ε[ἰς σ]άκκους *PMil.Vogl.* 214ʳ.i.16 (AD 154)'

καταγγίσιμος, η, ον, *packed in bottles* or *jars*: n. pl. as subst., *BGU* 2355.10 (ii/iii AD).

καταγγισμός, add '*PMil.Vogl.* 250.2 (AD 167)'

*καταγγιστής, ὁ, *bottler, packer*, *SB* 10258 ii 11 (iv AD).

κατάγειος I, add 'of the gods of the underworld, Θεοῖς καταγαίοις *SEG* 33.844 (Africa Proconsularis)'

καταγῑνέω II, delete the section transferring quot. to section I.

*κατακτηρία, ἡ, app. some form of fastening, μοχλοὶ ταῖς κατανκτηρίαις παλαιοὶ κατεχρήσθησαν εἰς σφῆνας καὶ σφύρας καὶ πιεστῆ[ρας *IG* 2².1672.304 (Eleusis, 329/8 BC).

καταγλαΐζω, after '*glorify*' insert 'τινὰ μνημείοις *SEG* 4.633.10 (Sardis, iii/iv AD), cf.'

καταγλισχραίνω, pass., add 'Gal. 15.654.10, 11, 680.8'

*κατάγλυφος, v. °*Iasos* 8.2 (AD 135/6)'

*κατάγμάτιον, τό, dim. of κάταγμα (B) I, *Inscr.Délos* 1441*A*i67 (ii BC).

κατάγνυμι, line 7, after '(Hes.*Op.*) 693' insert 'also part. κανάξαντες Call.*fr.* 260.53 Pf.'

κατάγνωσις II, add '*the record of a judgement*, law in D. 24.63'

καταγοράζω, pass., add 'ἐάν τι τῶν καταγορασθησομένων σωμάτων πάθῃ SEG 33.1039.75 (Cyme, ii BC)'

καταγορασμός, add 'τὸν .. καταγορασμὸν τῶν σωμάτων SEG 33.1039.69 (Cyme, ii BC)'

*****καταγράφιον**, τό, perh. form of poll-tax, IEphes. 13 i 3, 23 (i AD, cf. SEG 37.884).

κατάγραφος II, for 'drawn .. 1.4.5' read '(seen as) drawn from the side, of stars, Hipparch. 1.4.5; of a portrait-bust, GVI 979 (Panticapaeum, i AD) add 'III (as Lat. loan-word) figured, catagraphosque Thynos Cat. 25.7.'

καταγράφω II 5, add 'SEG 16.573, 39.1020A.2 (both Selinous, v BC), ib. 31.837 (Phintias, Sicily, ii/i BC)'

*****καταγρέω**, overcome, σιγὰ δ' ἐν νεκύεσσι, τὸ δὲ σκότος ὄσσε καταγρεῖ (cj., cod. Stob. κατέρρει) Erinn. in Suppl.Hell. 402.2 cf. κατάγρημμι.

καταγρυπόω, for 'Schneid.' read 'Winckelmann'

καταγυιόω, add 'Gal. 15.665.11, 12'

κατάγχω I, add 'Plu.Dio 57.2 (cj.)'

κατάγω I 1, line 8, after 'PGrenf. l.c.' insert 'κατάξω δὲ ὑμεῖν καὶ ὕδωρ SEG 32.460.11 (Boeotia, AD 125)'

καταγωγεύς, after 'BGU 92 (ii AD)' add '(dub., cj. °χοιροκαταγωγεύς) OLund. 10.2 (iii AD). **2** transporter of goods, PHels. 7 (ii AD).'

καταγωγή I 2, add 'b forwarding to the coast, SEG 34.558 (Larissa, ii BC), πρὸς παράλημψιν καὶ κ. βιβλίων PRyl. 83.4 (ii AD).' **4**, shelter for cattle, add 'PVindob.Salomons 12.5 (AD 334/5)'

καταγώγιον III, after 'festival of the return' insert 'of Aphrodite' and after 'SIG 1109.114 (ii AD)' add 'IEphes. 661.20 (ii AD)'

καταγωγίς, add 'III epith. of Artemis, SEG 9.13.12 (Cyrene, iv BC); v. SEG 39.1714. **IV** σκεῦος πεντηρικόν. καὶ κράσπεδον. καὶ παράλωμα Hsch., i.e. = παράρρυμα 1.'

καταγωνίζομαι, line 1, delete 'τινας' and substitute 'κ. τὸν ἐργώδη γέροντα Men.Dysc. 965'; line 3, after 'Dam.Isid. 122' insert 'abs., J.AJ 4.6.12'

†**καταδάκνω**, seize with the teeth, bite, ἡ μὲν (sc. μύραινα) τοῦ ἀντιπάλου τὰ κέντρα (i.e. spines or prickles) .. οὐκ ἐννοοῦσα καταδάκνει Ael.NA 1.32.

καταδακτυλίζω, for 'feel with the finger, sens. obsc.' read 'make an obscene gesture at with the middle finger'

†**καταδακτυλικός**, ή, όν, w. gen., inclined to make an obscene gesture at, Ar.Eq. 1381 (prob. nonce-wd.).

καταδαρθάνω, line 4, after 'A.R. 2.1227' insert '(v.l. -ον)'

καταδεής (A) I 2, (compar.), add 'used to describe plebeian (opp. curule) aediles, D.C. 53.33.3'

καταδείκνυμι 1 and 2, combine these sections under def. 'make known something that one has discovered, devised, invented, etc., introduce'

*****καταδέκτρια**, ἡ, app., one who admits to a place, epith. of sea-goddesses, Βύνης καταδέκτριαι cj. in ?Call.fr. 745 Pf.

κατάδενδρος, after 'thickly wooded' insert 'Dicaearch. 1.1'

καταδέρκομαι, after 'aor. 2 κατεδράκον' insert 'Pi.N. 4.23'

κατάδεσις II, for this section read 'binding by spells, in pl., Pl.Lg. 933d'

κατάδεσμος II, for '-δέσμοις τοὺς θεοὺς πείθοντες' read 'βλάψει ἐπαγωγαῖς τισι καὶ καταδέσμοις'

καταδέχομαι, after lemma insert 'κ]αδδέκεται Lesb.fr.adesp. 27 L.-P.' **3**, for this section read 'accept into a class or category, Str. 16.1.6'

καταδέω (A), before 'fut.' insert 'contr. -δῶ Tab.Defix. 101, 100a (both iv BC), al.'

†**καταδημαγωγέω**, attain by the arts of a demagogue, w. acc. and inf., Arr.fr. 150 J.; pass., to be won by such arts, Plu.Cleom. 13. **2** disadvantage by demagogic means, ἀδελφούς id. 2.482d; pass., id.Per. 9.

*****καταδημεύω**, confiscate, rest. in Thasos 150.17, 21, see REA 61.289.

*****καταδιαβαίνω**, come down across, POxy. 2331 ii 11 (= Heitsch GDRK no. 11, iii AD).

καταδιαιρέω 3, add 'med., interpret, σημείον Vit.Aesop.(G) 81'

καταδίδημι, after 'al.' insert 'SEG 37.215 (Athens, ?iv BC)'

*****καταδιδράσκω**, overrun, μέλαθρα τῷ πυρὶ καταδέδρακεν Sch.V E.Tr. 1303.

καταδίδωμι I, add 'τοῖλ Λατοσίοις Inscr.Cret. 4.58 (v BC)' add '**2** distribute among, δέρμα τοῖς [ἀεὶ ἱερασο]μένοις SIG 624.37 (ii BC).'

καταδικάζω I, line 6, after 'Luc.DMort. 29.2' insert 'w. acc. of person convicted, Cod.Just. 4.20.13.2, Just.Nov. 69.2.1'; line 16, after '(Artem.) 5.21' insert 'w. πρός, condemn to, κατεδικάσθησαν .. πρὸς θηρ[ία PPetaus 9.10 (AD 185)'

καταδίκη, after lemma insert 'Arc. καδίκας acc. pl., SEG 25.447.8 (Aliphera, iii BC, in sense 2)' **1**, add 'καταδίκαν γράψαι SEG 31.351 (Arcadia, c.300 BC)'

καταδίψιον, add 'see also Suppl. Hell. 1083'

καταδρέπω, add 'Pl.Ti. 91d'

καταδρομή I 2, delete 'cf. D.H.Th. 3' and after 'Plb. 12.23.1' insert 'w. obj. gen., D.H.Th. 3, Pomp. 1'

*****καταεικοβολέω**, conjecture from probabilities, Sch.Pi.I. 2 inscr.

*****κατάερος**, ον, situated in the air, Hymn.Is. 28.

†**καταέρρω**, v. ‡καταίρω.

*****καταέσσας** (cod. -έσσας)· κατακοιμηθείς Hsch., cf. ἀέσκω.

κατάζευγμα, add 'rest. in IG 1³.474.252'

*****καταζώγραφος**, ον, painted over with pictures, ἰστήλη GVI 133.1 (Galatia, ii/iii AD).

*****κατάζωσις**, εως, ἡ, girding, name of Bacchic rite, IUrb.Rom. 160 ii A.1 (Latium, ii AD); cf. κατάζωσμα II.

καταζωστικός, for 'τὸ κ. .. robes' read 'Κ., τό, an Orphic poem app. concerning the girding of initiates'

*****καταζώστρη**, ἡ, lace for fastening the κόθορνος, Herod. 8.33 (s.v.l.).

†**κατάημι**, fut. καταήσεται· καταπνεύσει Hsch.; in undetermined sense, καταήσσατο Alc. 296.2 L.-P.

κατάθεσις 1, for 'κ. κλάδων .. Gp. 9.5.1' read 'Gp. 4.3.4, 9.5.1, 6.1' **2**, add 'SEG 33.169B i 3 (Athens, iv BC)' **4**, add 'θείων μηνιμάτων (v.l. καθέσεις) Aesop. 56 P.' **7**, add 'SEG 34.1262 (Calchedon, AD 452)' add '**8** laying down on the ground, περικομιδὴ καὶ κ. χελώνης, as a charm, Gp. 1.14.9. **9** arrangement, διπλῇ τῇ καταθέσει τῶν κλάδων D.S. 2.53 (perh. read καθέσει). **10** mortgage, Inscr.Cret. 4.43Ba7 (v BC), cf. κατατίθημι I 4b.'

καταθλέω, add 'III spend on athletic contests, Inscr.Délos 316.114 (iii BC), 372A117 (iii/ii BC), al.'

καταθνητός, add 'Thgn. 897'

καταθύμιος 1, line 1, after 'Eumel. 1.13' insert 'Call.Lav.Pall. 33, 69, Muson.fr. 14 p. 74 H.' **II**, for 'according to one's mind' read 'that is in accordance with one's wishes, close to one's heart' and add 'ἐμοὶ καταθύμιον ἄνδρα SEG 30.1565 (Cappadocia), etc.'

*****καταθυτός**, ά, όν, to be used in sacrifice, ζέκα μναῖς κα ἀποτίνοι .. κα(τ)θυταὶς τοῖ Ζὶ Ὀλυνπίοι Schwyzer 409.4, 410.5 (rest.), 418.6 (all Elis, vi/v AD).

κάται, add 'cf. κ. δονακώδεος ὕλης epigr. in SEG 16.702 (Caria, ii/iii AD)'

καταιβάτης 3, for pres. def. read 'downward-plunging'

καταιβάτις, add 'see also °καταβάτις'

καταιβατός, add 'see also ‡καταβατός'

*****καταιρεί**, for ever, αὐτὸν καὶ τὸ γένος κ. Schwyzer 362.4 (Locris, v BC).

*****καταϊερόω**, = καθιερόω, JÖAI 26 Beibl. 13 (Ephesus, ii AD), SEG 37.812 (Rome, sp. καταειερόω).

καταινέω 1, add 'abs., A.Ch. 706'

κατάϊξ, for 'Eumel. 9' read 'A.R. 3.1376'

καταίρω I, for this section read 'take down (in Aeolic form, in tm.), κὰδ δ' ἄερρε κυλίχναις μεγάλαις Alc. 346 L.-P.' **II** 1, before 'ἐς Δελφούς' insert 'cf. ὡς ἔθνος τι ἄπειρον κοράκων κατῆρε τότε' and transfer this quot. to follow 'Plu.Rom. 9'

*****καταίσχυντος**, shameful, Hsch. s.v. κατηφόνες.

*****καταιτιατικός**, ή, όν, noxious, rest. in Vett.Val. 208.22, cf. αἰτιατικός I 2.

κατακαίνω, line 2, after '(X.)An. 3.2.12' insert 'BCH 33.452 (Argos)'

κατακαλέω I, add '**2** invite to participate in games, SEG 30.1117, 1121, 1122 (all Entella, iii BC).'

*****κατακάρδιον**, τό, down-turned twig of the mulberry-tree, Cyran. 29 (1.12.29 K.).

κατακαρφής, after 'Nic.fr. 70.9' add '(cj.)'

κατάκειμαι, line 2, after 'Pl.Smp. 213b' insert 'Cypr. part. ka-ta-ki-me-na κατακίμενα Kafizin 270 (sense 3, 225/18 BC)' **1**, add '**b** lie in a grave, IG 10(2).1.352 (Thessalonica, iv AD), SEG 29.310 (Corinth, AD 529), ἐνθάδε κατάκιτε .. TAM 4(1).353.' add '**10** to be given in pledge, Inscr.Cret. 4.47.1, 10, al. (Gortyn, v BC); cf. κατατίθημι I 4 b.'

κατακένωσις, εως, ἡ, the process of emptying, CPR 7.20.13 (iv AD).

κατακέφαλα, add 'παραφέρειν κ. sweep away head over heels, POxy. 1853.5 (vi/vii AD); perh. also of blows on the head, PVindob.Salomons 15.6 (v/vi AD)'

*****κατακήδευσις**, εως, ἡ, burial, TAM 2.620 (Tlos).

*****κατακήλησις**, εως, ἡ, enchantment, Cels.ap.Orig.Cels. 1.6 (pl.).

κατακηλιδόω, after 'strengthd. for κηλιδόω' insert 'Phryn. 393 F.'

*****κατακληρουχία**, ἡ, apportionment, rest. in BGU 2444.93, 2445.24 (i BC).

*****κατακλιτικός**, ή, όν, ἡ κ. ὥρα, ἡμέρα, the hour, day of taking to one's bed, Vett.Val. 205.3, 339.20.

κατακλύζω I 1, add 'humorously, with wine, ὅς μοι δοὺς τὸ πῶμα κατέκλυσεν E.Cyc. 677 (cj.)'

*****κατάκλυστος**, ον, inundated, πρὸς τῇ οὔσῃ κατακλύστῳ κλήρου ἀρούρας τρῖς PColl.Youtie 27.10 (AD 165); perh. also PMil.Vogl. 105.20 (ii AD). **II** κατάκλυστον, τό, floor which can be washed down, Inscr.Délos 2420.

*****κατακνάπτω**, tear to pieces, in quot. fig., μελέα μήτηρ, ἣ τὰς μεγάλας ἐλπίδας ἐν σοὶ κατέκναψε βίου E.Tr. 1252 (cj.).

κατακνάω, add 'scratch all over, κατὰ μὲν χρόα πάντ' ὀνύχεσσι δακνόμενος κνάσαιο Theoc. 7.110'

*κατακόλλησις, εως, ἡ, glueing together, τραπέζης Delph. 3(5).68 i 3 (iv BC).

κατᾰκολουθέω, add 'act in conformity with, live up to, κατακολουθοῦσα τῇ ἑαυτῆς καλοκἀγαθίᾳ SEG 33.1036.5 (Aeolian Cyme, ii BC)'

κατάκομος, after 'with falling hair or beard' insert '(?)Pi.fr. 356 S.-M.'; add 'transf., of trees, Ps.-Callisth. 3.6 cod. B.'

κατάκοος, delete the entry.

κατακορᾰκόω, add 'SEG 17.632 (Perge)'

κατάκορος II, adv. -ως, add 'abundantly, freely, Androm.Jun.ap. Gal. 13.71.6'

κατακράζω, add 'κατέκραξαν κατ' αὐτῶν SEG 26.1813.23 (Nubia, iv/v AD)'

κατακρᾰτέω, line 1, delete 'pers.' and insert 'γαστρὸς οὐ κατακρα[τεῖς Hippon. 118.2 W.'

κατακρέμᾰμαι, delete the entry (v. sq.).

κατακρεμάννῡμι, line 5, pass., add 'Hdt. 4.72.4, Cratin. 175 K.-A.; w. gen., κώδωνες .. κατακρέμανται τῆς ἐσθῆτος Plu. 2.672a'

*κατακρεμάς, άδος, fem. adj., overhanging (precise sense uncertain in broken context), Trag.adesp. 653.7 K.-S.

κατακρημνίζω, line 1, for 'Carm.Pop. 46.33' read 'Hermocl. 33'

*κατακρημνῶν, gloss on κατακρημνίζων, Hsch.

κατάκρῐμα 1, delete 'judgement' and transfer 'Ep.Rom. 5.16, 8.1' to this section.   2, for this section read 'money paid under penalty, PAmh. 2.114.8, PCol.v 3.170 (both ii AD).

κατακρίνω I, add '3 place under obligation, bind, αὐτὴν τὴν βασιλείαν κατακρίνομεν .. ὥστε .. προνοεῖν Just.Nov. 59.7.'

κατάκρῐτος, after 'ον' insert '(also η, ον, Hsch.)'; add 'κεφαλῆς κατά-κριτον, = Lat. capitis damnatum, SEG 8.13 (Nazareth, rescript of Augustus or Tiberius)'

κατακρούω 4, for 'Perh. = διακρούω' read 'knock or hammer downwards'

*κατακρῠβῇ, = κατακρύβδην, PMich.VIII 520.9 (iv AD).

κατακτείνω, line 3, aor. 1, add '3 sg. aor. subj. κατασκένηι Inscr.Cret. 4.41 I 14 (Gortyn, v BC)'; line 4, after '(Il. 6.)164' insert 'Aeol.part. κακκτάνοντες Alc. 129.19 L.-P.'

κατακτενίζω, add 'ψαρὸν .. ἵππον Suppl.Hell. 996.9 (i BC)'

κατακύπτω, line 2, after 'Il. 16.661' insert 'Men.Dysc. 538'

κατάκυψις, transfer the article after κατακυρόω.

καταλαμβάνω I 1, line 7, after 'arrive at a place' insert 'Ἀθήνας Synes.Ep. 54'   add 'b lay claim to, Σῖμος κατέλαβε Ἀσκαληπιακόν (on the boundary stone of a mine), SEG 32.233 (Attica, c.350 BC).'   add 'VI = συλλαμβάνω IV, conceive, Aq.Mi. 6.14.'

καταλαμπρύνω, after 'splendid' insert 'or bright'; add 'Sch.Pi.O. 3.35'

καταλάμπω I, add '2 illuminate mentally or intellectually, καταλάμπεται .. πάντα ἀθρόως ὑπὸ τῶν θεῶν Procl.Inst. 143 D.'

*κατάλαμψις, v. °κατάληψις.

*καταλᾰπαξῐκοίλιον, τό, evacuation of the stomach, ὄνασις (ε)ἶ τοῦ κ. Glotta 34.45/7 and 297 n. 1 (graffiti on vases, Emporion and Apulia, iv BC).

καταλέγω (B), line 6, after 'Vit.Hom. 21' insert 'cf. καταλέγεσθαι· ὀδύρεσθαι τὸν τεθνεῶτα Hsch.; cf. κατάλεγμα.'

καταλείπω I 2, add 'c τὰ καταλελειμμένα instructions, Just.Nov. 1.1.1, al.' III 1, c. inf., add 'w. article, Arist.Cat. 7ᵃ37, fr. 58'   2 c, at the beginning insert 'κ. τι ἐν τοῖς ἀκούουσι leave to the readers' intelligence, Aristid.Rh. 2 p. 523 S.' and after 'c. inf.' insert 'ib. p. 524 S.'

καταλέκτρια, delete 'Βύνης .. 217.5'; add 'v.s.v. καταδέκτρια'

*κατάλευσμα, ατος, τό, stoning, Sch.Lyc. 1181.

καταλέχομαι, add 'act., put to sleep, Hsch. s.vv. καταλέξας and κατέλεξας'

καταλήγω, line 4, for 'Arr.Epict. 6.20.21' read 'Arr.Epict. 1.6.20, 21'   2, add 'in phonology, of words, εἰς τραχὺ γράμμα D.H.Dem. 40'

καταληκτικός, at end delete 'Arr.Epict. 2.23.46' (v. °ἀκατα-)

καταληπτικός 2, after 'Phld.Rh. 2.120 S.' insert 'Alex.Aphr. de An. 71.11'

κατάληψις, after lemma insert 'Dor. κατάλαμψις, Lindos 160.5 (ii BC)'   I 4, add 'Alex.Aphr. de An. 71.12'

καταλιμπάνω, add 'Sapph. 94.2 L.-P., IGBulg 2236.45 (Thrace, iv AD).'

καταλῐπᾰρέω, after 'entreat earnestly' insert 'Men.Sam. 721'

κατάλλαγμα, add 'II something given in exchange, Acta Joannis fr. in MH 31.102.15.'

καταλλάσσω, line 1, after 'change money' insert 'SEG 26.13 (Attica, c.440/20 BC)'

κατάλληλος II, add '2 in music, of tetrachords, conjunct, Aristid.Quint. 1.8; cf. °παράλληλος.'

καταλογεῖον, before 'POxy. 73.34' insert 'PTeb. 770.13 (iii BC)'

καταλογή II, line 2, delete 'codd. (-δοχή Reiske)'

*καταλογισμός, ὁ, reckoning, account, PRyl. 627.90 (iv AD).

*καταλογιστής, οῦ, ὁ, local official, perh. registrar or sim., PGrenf. 79.2.1 (iii AD).

κατάλογος 2 a, add 'pl., enlisted men, troops, Just.Nov. 102.2'

κατάλοιπος, add 'κατὰ τὸν κ]ατάλοιπον χρόνον τοῦ ἐνιαυτοῦ SEG 32.118.11 (Athens, 244/3 BC)'

καταλοκίζω, add 'κατηλόκιζε Suppl.Hell. 977.18'

καταλούομαι, for 'spend in bathing' read 'wash away, i.e. squander in bathing'   add '2 bathe, prob. of ceremonial washing, JÖAI 23 Beibl. 24 (Maeonia, ii AD; κατελούσετο).'

καταλουστικοί, for pres. ref. read 'TAM 5(1).490.2, 8 (Lydia, ii AD)'

+καταλοχία, ἡ, register, οἱ Λευῖται .. ἀπὸ εἰκοσαετοῦς καὶ ἐπάνω ἐν διατάξει ἐν καταλοχίαις .. Lxx 2Ch. 31.18.

καταλοχίζω 2, add 'b enrol into a group, class, etc., PMich.v 2.338.9 (i AD).   add '3 register as owner of, διὰ τωι καταλελοχίσθαι αὐτὸν ἐπὶ τῶν αὐτῶν ἀρουρῶν PMich.XI 621.11 (AD 37).'

καταλοχισμός 2, add 'in official title, Εὐδαίμονι τῶι πρὸς καταλοχισμοῖς POxy. 3482.6 (73 BC)'

καταλυμαίνομαι, add 'of cattle destroying crop-land, PColl.Youtie 77.4 (AD 324)'

κατάλυσις II, add '4 accommodation for animals, in quot., sheep, Lxx Je. 30.14(49.20).'

καταλυτήριον, add 'Sch.Pi.O. 10.57'

καταλύτης I, add 'καταλύτου μονοημέρου Lxx Wi. 5.14'

καταλύω II 2, add 'Macho 350 G.'

καταμακτος, add 'PVindob.Salomons 2.24 (ii/iii AD)'

καταμαντεύομαι 2, add 'D. 60.34'

+κατᾰμάω (A), mow down, κατ' αὖ νιν φοινία θεῶν τῶν νερτέρων ἀμᾷ κοπίς S.Ant. 601 (lyr.); lay low, kill, Euph. 84.3 (med.).

*κατᾰμάω (B), draw or scrape down on to something (in quots., med.), τήν ῥα (sc. τὴν κόπρον) κυλινδόμενος καταμήσατο χερσὶν ἑῇσι Il. 24.165, τὸν χοῦν καταμήσονται (cj.) Pherecr. 126 K.-A., J.BJ 2.15.4, 21.3.

*καταμείνας, ὁ, perh. resident, sitting tenant, POxy. 2244.65, 3640.2 (vi AD).

*καταμελητέος, α, ον, not demanding attention, An.Boiss. 5.381.

καταμέμφομαι, add 'Anacr. 13.7 P.'

*καταμέριμνος, ον, κ. [sollicitus], gloss. in PVindob. L150 (Tyche 3.141ff., v AD); cf. ἐμμέριμνος.

*κατάμερος, τό, partial payment, ἐκ τοῦ καταμέρους (prob. to be understood as κατὰ μέρους) PMich.XV 748.8 (vii AD).

καταμετρέω 2, line 5, after '(Arist.)Cat. 4ᵇ33' insert 'ἑξακοσιάκις καὶ πεντηκοντάκις -εῖται ὁ κύκλος οὗτος ὑπὸ τῆς διαμέτρου τῆς σελήνης Papp. 6.556.15'

καταμήνιος II 1, delete the section.

καταμιεῖ, add '(καταμνιεῖ .. μνιεῖν La.)'

καταμιμνήσκομαι, add 'ἐμνήσαντο γάμου κάτα (= γ. κατεμνήσαντο) prob. in Call.fr. 75.18 Pf.'

καταμίσγω, add 'Emp. 93 D.-K.'

καταμνημονεύω, add 'SEG 33.1183.19 (Lycia, 260/59 BC)'

καταμοσχεύω, add 'Anon.Alch. 364.20'

+καταμπῠκόω, bind as with a headband, στεφάνοισι κρᾶτα καταμπυκοῖς S.fr. 402 R.

*καταμφωτοί· αὐλοί τινες οὕτω καλοῦνται Hsch.

καταναλίσκω 1, line 8, after 'Ath. 8.345d' insert 'ταῦτα .. περὶ ἴδια τέκνα καὶ συγγενεῖς καταναλίσκων Cod.Just. 1.3.41.3'

καταναυμαχέω, pass., add 'Isoc. 9.56'

κατανομοθετέω, after 'Pl.Lg. 861c' add '(v.l. κᾰτα νομ-)'

κατάντημα, add '2 destination of a property, etc. PMich.inv. 4962.10 (Aegyptus 71.20, iii AD).'

καταντης I, line 2, after 'Ar.Ra. 127' insert 'ἄντρον AP 6.220.5 (Diosc.)'

καταντλέω 1, metaph., add 'Com.adesp. 28 D.'

κατανύσσω, last line, for 'keep silence' read 'be silenced'

καταξαίνω 2, pass., add 'A.fr. 132c R.'

*κατάξεσμα, τό, = κατάξυσμα, Hsch., Suid. s.v. μύγματα (for ἀμύγματα).

καταξέσματα, delete the entry.

*καταξιοπιστεύομαι, claim belief for, ἑτέραν τέχνην Ptol.Tetr. 6; (cf. καταξιοπιστέομαι).

κατάξιος, line 1, fem. -αξία, add 'SEG 32.453.17 (Boeotia, ii BC)'

καταξιόω II, delete the section, transferring quots. to section I.

καταξύω I 2, add 'b leg., scratch out, cancel, Sch.Ar.Nu. 774b K.'

καταπαλαίω, line 2, for 'τὰ ῥηθέντα' read 'ἐμέ'

καταπάλλομαι, line 2, after 'Il. 19.351' insert 'Stesich. 32.i.4 P.'

+κατάπαξ, v. ‡καθάπαξ.

*καταπαρίζω, perh. "live in the house of Paris", Sch.Od. 8.517 (s.v.l.).

καταπάσσω I, line 5, pass., add 'καππεπάδμ[Alc. 143.9 L.-P.'

καταπάτησις I, add 'λεκάνη τῆς κ., app. of a basin for washing the feet, Syr.Ps. 59(60).10'   2, add 'PErlangen 121 (iii AD), SB 11213.6 (AD 310)'

*καταπελτιστής, οῦ, ὁ, *catapult-man*, Lyd.*Mag.* 48.19.
*καταπέντε, wd. of unkn. meaning, perh. trade designation of Egyptian origin, Χαρίδημ(ος) καταπέντε PMich.IV 224.4059, 5533 (ii AD), OMich. 324.
†κατάπερ, v. καθάπερ s.v. ‡καθά.
καταπέτασμα, after 'Hld. 10.28' insert '(v.l. for παρα-)' add '3 as personal possession, perh. item of clothing, POxy. 3150.37 (vi AD).'
καταπίνω I, add 'hyperb., καὶ μὴν ὁ Παφλαγὼν οὑτοσὶ προσέρχεται .. ὡς δὴ καταπιόμενός με Ar.*Eq.* 693, V. 1502, cf. ὁ ἀντίδικος ὑμῶν διάβολος ὡς λέων ὠρυόμενος περιπατεῖ ζητῶν τινα καταπιεῖν 1*Ep.Pet.* 5.8'
καταπιπράσκω, add 'SEG 35.1439.9 (Lycia, ii AD)'
καταπίπτω, line 3, after '(Epid.)' insert 'Aeol. part. καππέτων Alc. 130.14 L.-P.'; line 4, after 'pf. -πέπτωκα' insert 'inf. -πεπτώκειν SEG 38.1476.95 (Xanthos)'; line 10, after 'οἰκίαι καταπεπτωκυῖαι' insert '*collapsed*' **2 a**, line 4, before '*base*' insert '*disheartened*, D.Chr. 18.15'
καταπλάσσω I **3**, add 'τὸ καταπεπλάσθαι τὸν βίον (καταπεπλῆχθαι codd.) D. 37.43'
καταπλαστός I, after 'Ar.*Pl.* 717' insert 'Hippiatr. 130.185.'
καταπλέκω I **1**, add '**d** *entwine oneself round* a person, prob. in Nic.*Th.* 475.'
καταπληκτικός, add '**2** in pass. sense, gloss on ἔμπληκτος, Hsch.'
καταπλήξ **2**, add 'Iamb.*Protr.* 21 λγ' (p. 124.13 P.)'
κατάπληξις **1**, add '**b** *object of admiration*, *wonder*, Lucr. 4.1163.' **2**, after 'Arist.' insert '*EE* 1221ᵃ1, 1233ᵇ27'
καταπλίσσομαι, add 'act., καταπλίξας dub. rest. in Hippon. 104.16 W.'
*καταπλόϊν, τό, perh. some sort of cloth (cf. καθαπλόω), POxy. 2729.37 (iv AD).
καταπλοκή I, add 'fig., εἰς κ. ἰέναι *arrive at a deadlock*, PRainer Cent. 161.17 (?v AD)'
καταπνέω **1**, add 'ῥοώδης, καταπνεόμενος ἀπὸ τῶν παρακειμένων ὀρῶν, ἐστὶν ὁ κατ' αὐτὴν διάπλους Peripl.*M.Rubr.* 25'
*καταποδίδωμι, perh. *trade away*, ἐὰν δέ τις .. καταποδῷ τὰ δημόσια JHS 33.338.28 (Maced., ii AD).
καταπονέω I **1**, line 5, delete 'πάντα .. Men. 744' add '**b** πρᾶγμα καταπονεῖν *get down to*, *put one's back into* a task or business, Men.*Dysc.* 392, fr. 526 K.-Th.'
*καταπόντισμα, ατος, τό, app. = καταποντισμός, var. for καταπάτημα Lxx La. 2.8.
†καταποντισμός, ὁ, *submersion under the sea*, καταποντισμοὺς (*drownings*) καὶ τυφλώσεις Isoc. 12.122, ὁ κ. τῶν χρημάτων App.*Mac.* 16, καταποντισμοὶ πόλεων καὶ χωρῶν Orac.*Tiburt.* 137; transf., *destruction*, Lxx Ps. 51.6.
†καταπράϋνσις· *mitigatio*, Gloss.
καταπρηνίζω, add "Ἡφαίστου πυρόεσσα κατεπρήνιζεν ἀϋτμή epigr. in IUrb.*Rom.* 1342.3'
*καταπρίσσομαι, *tear off*, A.fr. 451h R.
καταπρολείπω, for '*forsake utterly*' read '*leave behind* in a place'
καταπτύω, add 'Iamb.*Protr.* 21 λβ' (p. 108.7 P.), al.'
*καταπύγαινα, ἡ, fem. form of καταπύγων, Hesperia 22.215 (Athens, vi/v BC).
*καταπῡγόω, καταπυγῶν· κατασελγαίνων Phot., Suid.
καταπύγων I, for '*given .. lewd*' read '*male passive homosexual*, sts. loosely as a term of abuse'; add 'AJA 38.11 (Attic cup, Hymettus, ?vii BC), Lang *Ath.Agora* xxi C5 (vi BC), C18 (v BC), al.'
*κατάρᾱσις, εως, ἡ, *tamping*, CID II 139.16 (iii BC).
κατάρατος, add 'PHamb. 192.9 (iii AD)'
καταργέω I **2**, *to be rendered* or *lie idle*, add 'PMerton 79 (ii AD), PFlor. 218 (AD 257)'
*κάταργος, ον, *quite unwrought*, ξύλον, Call.*Dieg.* iv.28 (fr. 100 Pf.).
κατάρδω **2**, for '*besprinkle*' read '*soak*, *dowse*'
†κατάρης, v. ‡κατώρης.
καταρίγηλός, add 'perh. also Euph. in *Suppl.Hell.* 442.10'
*κατάρίθμητος, ον, *numbered* in a class, category, etc., ἐν τοῖς ζῶσι Ps.-Callisth. 132.12.
*κατάρίθμιος, *numbered* in a class, category, etc., ἐν φθιμένοις GVI 1984.3 (Ancyra, ?iv/v AD).
καταρνέομαι, add 'AB 1.85.31'
κάταρξις, add 'PGrenf. 2.87.21 (AD 602)'
†καταρρᾱκόω, *tear to shreds*, ἄναρθρος καὶ κατερρακωμένος sc. Heracles, S.*Tr.* 1103.
καταρράκτης **3**, delete the section transferring quots. to section 5 (*sluice*) add '**7** *some means of punitive restraint* (whether dungeon, stocks, or other means), Lxx Je. 20.2, 36(29).26.'
καταρρεπής, for '= ἑτερορρεπής' read '*weighing* or *pressing down*' and add 'A.R. 2.593, Plu. 2.952e'
καταρρέω I **1**, add 'fig. in part., of style, *flowing*, D.H.*Comp.* 20'
καταρροφέω, add 'Dor. -ρυφέω, Sophr. in PSI 1214*d* 2'

καταρρυθμίζω, line 2, for '*passages over-rhythmical*' read '*parts given rhythmical form*'
*καταρρύθμῐσις, εως, ἡ, the process of *bringing into rhythm*, Ps.-Archyt. p. 32.3 T.
κατάρρυθμος I, for '*very rhythmical*' read '*having rhythmical form*'
καταρτίζω II, med., add 'φωνὴ κυρίου καταρτιζομένου ἐλάφους Lxx Ps. 28(29).9'
†κατάρτιος, ἡ, app. *ship's mast with yard*·(or the *yard* alone), ἔοικε γὰρ τῷ ἱστίῳ καὶ τῇ καταρτίῳ τῆς νεὼς ὅλης διὰ τὰς βύρσας καὶ τὰ κέρατα (sc. ταύρος) Artem. 2.12, 53, 3.36, *EM* 478.23.
καταρτισμός III, add 'SEG 17.545 (Pisidia, Rom.imp.), 32.1269 (Phrygia). **2** *completion*, *perfection* (of dentition), ἑτέροις δὲ μησὶ δυοκαίδεκα τὸν καταρτισμὸν φύει Hippiatr. 95.'
κατάρτῡσις, add 'Hippod.ap.Stob. 4.1.94 (*Pyth.Hell.* p. 100.22)'
*κάταρχος, ὁ, app. title of an official, οἱ κάταρχοι καὶ ταβελλάριοι BCH 47.382.18 (Notium).
κατάρχω I **1**, w. acc., add 'Alcm. 98 P., Carm.Pop. 5(b).5 P.' **2**, delete 'with reference to the religious sense infr. II 2' **II 2**, line 4, after 'Od. 3.445' add 'τὸ πρόθυμα .. κ. Hesperia 7.4 (Athens, iv BC)' **III**, add 'Lxx Jl. 2.17, Na. 1.12, etc.' **IV**, delete the section.
*κατασεβάζω, *consecrate*, Sch.Pi.O. 3.35.
†κατᾰσελγαίνω, *to treat libidinously*, Phot., Suid.
κατασημαίνομαι II, delete the section transferring exx. to section I.
κατασήπω **2**, add 'Plu. 2.231a'
κατασκαφή, add 'CPR 6.1.14 (AD 125)'
κατασκελετεύω, line 2, after 'Sch.Ar.*Ra.* 153' insert 'transf., *cause resources to waste away* (in quot., abs., Onas. 1.5'; line 3, delete 'Onos. 1.5 (Act.)'
†κατασκελής, ές, of style, *over-loaded*, D.H.*Isoc.* 2. **2** *elaborate*, *complicated*, τὸ τῶν ἐπιτεχνημάτων κ. Ptol.*Alm.* 13.2; μέθοδος id.*Harm.* 2.13 (comp.); τοῦτο κ. ἔχει [ὁ] τρόπος ib. 2 (comp.).
*κατασκενῶ, v. °κατακτείνω.
κατασκευάζω, line 4, after '(Tanagra, iii BC)' insert 'Cypr. ka-te-se-ke-u-wa-se κατεσκεύασε ICS 2.3'; line 5, after 'D. 42.30' insert 'SEG 37.528 (Epirus, i BC)' **I 3**, line 4, after 'D. 18.71' insert 'τοῦτο τὸ ἡρῶιον κατεσκεύασε IEphes. 3327 (ii/iii AD); ἑαυτῷ τὸ μνημεῖον κατεσκεύασεν Mitchell *N.Galatia* 201 (AD 218)'; line 11, for 'Id. 1.93, 2.17' read 'τἆλλα ic. ib. 1.93.8; abs., id. 2.17.3' **11**, delete 'ὡς πολεμήσοντες Th. 2.7'
*κατασκεύασις, εως, ἡ, = κατασκευή (in quot., III), Cat.Cod.Astr. 11(2).131.14.
κατασκεύασμα I, add 'of a piece of pottery, rest. in Kafizin 309 (225/18 BC)' add '**b** *construction* of a literary work, τὸ τραγικὸν κ. Sch.E.*Tr.* 1129.'
κατασκευή I **1**, line 3, after 'Pl.*Grg.* 455b' insert 't]οῦ θεάτρου Illion 1.10 (iv BC), τειχέων SIG 569.15 (Cos, Halasarna, c.201 BC)' **IV**, add 'Din. 1.53'
κατασκευόω, add 'εἰ δέ τι κατασκεώσαιτο SEG 25.606.11 (Doris, ii BC)'
κατασκηνόω, add 'w. acc., *occupy*, κατασκήνου τὴν γῆν Lxx Ps. 36(37).3, Pr. 2.21; transf., ἐγὼ ἡ σοφία κατεσκήνωσα βουλήν ib. 8.12'
κατασκήνωμα I, for 'A.*Ch.* 985' read 'A.*Ch.* 999'
κατασκιάζω, line 3, after 'Archil. 29' insert 'Anacr. 2 fr. 1.1 P. (tm.); προσώπου ἄνθος κατεσκιασμένη Men.*Dysc.* 951'
κατασκοπέω, line 5, after 'X.*Mem.* 2.1.22' insert 'ὑπτιάζων καὶ κατασκοπούμενος ἑαυτόν Aeschin. 1.132'
κατασμικρύνω II, after 'Hierocl. p. 59 A.' insert 'Demetr.*Eloc.* 123'
κατασπασμός **2**, add 'PSoterichos 4.27 (AD 87)'
κατασπάω IV, *pull down*, add 'IG 9(2).1229.27 (Thessaly, ii BC)' **V**, add 'τὴν φράσιν D.H.*Comp.* 20 (pass.)' add '**VIII** gloss on ξύει, *plane* or *shave wood*, Hsch., cj. in Thphr.*HP* 5.1.7.'
κατασπείρω III, line 1, for 'πλούτῳ Ἑλλάδα κ.' read 'κ. πλούτῳ Ἑλλάδα γῆν'
κατάσπεισις I, add 'Plu. 2.437b (sg.)'
κατασπένδω I, line 4, after 'Hdt. 2.151' insert 'AP 7.260 (Carph.)' deleting this ref. from section II 2.
κατασπεύδω **2**, delete the section.
*κατασπορεία, ἡ, office of κατασπορεύς, SB 9050 vi 12 (i/ii AD).
κατασπορεύς, after '*sower*' insert 'Ath.*Sem.* 3 (PG 28.148a). **2** *inspector of crops*'
*Κατασπόρια, τά, festival at Caunus, JHS 74.87.22, al.
*καταστάγμα, ατος, τό, *drop of liquid*, Anon.Alch. 381.4.
καταστάσιάζω I, pass., after 'D. 44.3' read '52.7, 58.22, etc.'
κατάστᾰσις II **2**, add '**b** *personal status*, Modest.*Dig.* 27.1.6.18, Cod.Just. 6.4.4.23.'
καταστάτης **2**, add 'of a tax official, POsl. 88.16, 35 (iv AD)'
καταστεγάζω, after '*cover over*' insert 'rest. in Panyas. 15 B.'
καταστέλλω I, add 'fig., σὲ .. οἶκτον περιβαλὼν καταστελῶ E.*IA* 934' **II 2**, delete 'οἶκτον E.*IA* 934'; line 4, after 'Plu. 2.207e, cf. 547b.

etc.' insert 'καταστίλατε τὴν ὀργὴν καὶ τὸν θυμὸν τοῦ (δεῖνος) PPrag. I 4.4 (v AD)'

*κατάστεμμα, v. ‡κατάστημα.

καταστενάζω, after 'al.' insert 'w. acc. of person, Sch.E.Tr. 318'

*κατάστερος, ον, perh. sprinkled as with stars, κ. δὲ πόντος .. ἐγάργαιρε σώμασιν Tim.Pers. 791.94 P. (s.v.l.).

κατάστημα, line 1, after 'Lxx 3Ma. 5.45' insert 'κατάστεμμα POxy. 3817.11 (iii/iv AD)' 1, add 'POxy. l.c.' 6, add 'καὶ τοῖς μεταξὺ καταστήμασιν τῆς σελήνης Peripl.M.Rubr. 45'

*καταστοχαζόντως, gloss on καταχρηστικῶς, Hsch.

*καταστράτεια, ἡ, offensive expedition, BMCParthia p. 40.

καταστρατοπεδεύω II, for pres. def. read 'go to a place and camp there' and add 'Lxx 2Ma. 4.22; cf. κατεστρατοπέδευσε· κατεπολέμησεν Hsch.' III, delete the section.

*καταστρωτέος, gloss on στορνυτέος, Hsch.

κατασφαγή, add 'Heph.Astr. 1.21.15'

κατασφρᾱγίζω, penultimate line, after '(ii BC)' insert 'τὰ ὦτα Cels. ap.Orig.Cels. 5.64' add '2 shut up, confine, Βασσαρίδων .. φάλαγγα εὐρώεντι κατεσφρήγισσε μελάθρῳ Nonn.D. 45.267.'

κατάσχεσις II, add '2 taking possession, ἑτέρων Memn. 36 J.'

*κατασχετικός, ή, όν, retentive (?), οἰκείωσις Hsch. s.v. σχετική.

κατασχίζω, line 4, after 'pass.' insert 'κὰδ δὲ λῶπος ἐσχίσθη Anacr. 96(b) P.'

κατασχολέομαι, add 'II act. release from the concerns of, (ὕπνος) ψυχὰς ἀπὸ τῶν σωμάτων κατασχολῶν Afric.Cest. 1.17 p. 38 V.'

+κατατᾰράσσω, gloss on καταπλήσσω, Hsch.

κατατάσσω I 3, for this section read 'allocate funds, εἰς τὸ βασιλικόν PSI 510.13 (iii BC); med., εἰς ἀπόδοσιν τῷ θεῷ IG 11(2).224A7 (Delos, iii BC); pass., SIG 459.6 (Beroea, iii BC)'

κατατείνω I 1, add 'b pass., of stringed instruments, Aristid.Quint. 2.16.' 4, add 'ταῦτα (sc. δόγματα) .. ἐν ἔπεσι κατατεῖναι Arist.fr. 7'

*κατατελίσκω, app. dedicate, ἱερεῖα κατατελισκόμενα IEphes. 10.8 (iii AD).

+κατάτεχνος, ον, full of art, ὦ κατατεχνοτάτου (v.l. κακοτ-) κινήματος AP 5.132 (Phld.); artificial, τὸ κ. Plu. 2.79b.

κατατίθημι, line 6, after 'Od. 19.17' insert 'pres. inf. act. καττιθέν SEG 9.4.38 (Cyrene, i BC), imper. κάθες Lang Ath.Agora XXI B1 (vi BC), Cypr. aor. act. 3 pl. ka-te-ti-sa-ne κατέθισαν ICS 94, ka-te-ti-ja-ne κατέθιjαν ICS 217.27' I 3 b, for this section read 'dedicate, Cypr. ka-te-ti-ja-ne i-ta-ti-o-ne κατέθιjαν ἴ(ν) τὰ(ν) θιόν ICS 217.27; ka-te-te-ke κατέθεκε ICS 205, al.' 4 b, after 'Leg.Gort. 6.19' insert '(ὁ καταθένς who gives a slave as surety, ὁ καταθέμενος who receives one, cf. Inscr.Cret. 4.47.8, 4 (v BC))' 5, add 'INikaia 766, TAM 3(1).479, SEG 31.1425 (Palestine, vi AD), etc.' II 4 b add 'so perh. λύπην κ. αἰφνιδίῳ θανάτῳ pay a tribute of grief to .., Inscr.Cret. 2.iii 44.8 (Aptera, iii or iv AD)' 7, add 'enter in one's accounts, μισθούς IG 11.110.17 (Delos, iii BC)'

κατατομή, add 'V fraction or section, POxy. 3465.14 (i AD).'

κατατρέχω II 4, pursue, add 'UPZ 68.6 (152 BC)' add 'III as term of accountancy, prob. run on, be carried forward, PAnt. 32.7, 8, al. (iv AD), cf. PHib. 255.5 (iii BC).'

κατατρῐδομέω, add '(reading uncertain, perh. κατ‹οικο›δομήσας, v. BE 1989.101)'

κατατροχάζω I, for 'cause to run .. promote' read 'perh. run to the attack in a form of sham fight (v. SEG 32.678)'

*κατατρῠχόω, = κατατρύχω, pass. aor. part. -ωθείς Memn. 40.4 J.

κατατυμβοχοέω, delete the entry (v. °τυμβοχοέω).

καταυγάζω, line 2, after 'Hld. 1.1' insert 'fig., Ὀποῦντα κ. ὕμνοις Sch.Pi.O. 9.33' II, for 'Hld. 5.31' read 'Hld. 5.32'

καταυλέω, passim, for 'flute' read 'aulos'

καταύλησις, for 'flute' read 'aulos'

*καταΰτμενος, ον, epith. of fine garments, perh. floating with the breeze, Sapph. 44.9 L.-P. (s.v.l.), cf. ib. 101.2.

καταφέρω, line 1, fut. -οίσομαι, read 'Il. 22.425 (med.), Arat. 871 (pass.)' I 1, add 'g hand down property, CodJust. 1.5.18.5, 7.' 2 c, add 'of wind, come down, Arat. 871' d, add 'of destruction by an earthquake, SEG 30.1254.B2 (Aphrodisias, AD 102/16)' V, add '2 ὁπεί κε ὁ καιρὸς κατενέκει when the occasion requires, BCH 59.55 (Larissa, ii BC).'

καταφθείρω 1, at end for 'PMagd. 11.9' read 'κατεφθαρμένην τὴν ὁδὸν ὑπὸ χρόνου ἀποκατέστησεν TAM 4(1).11' add 'b pass., remain idle, waste one's time in a place, PEnteux. 27.7, 9.'

*καταφλεγμαίνω, to be inflamed, Gal. 11.320.15.

κατάφορτος, add 'heavily laden, of ships, Memn. 36 J.'

καταφράζω, line 3, after 'Hes.Op. 248' insert 'Hippon. 79.13 W.'

*καταφρακτάριος, ον, Lat. cataphractarius, mail-clad, ἱππεῖς BGU 316.6 (iv AD).

καταφρόνητος, add 'Sch.E.Or. 1156'

καταφρυάττομαι, delete 'τινι'

καταφυγή I, add '3 personified, Θεῷ Ὑψίστῳ καὶ Ἁγείᾳ Καταφυγῇ SEG 19.852 (Pisidia).'

καταφύγιον, add 'of a person, X(ριστ)ὲ ἡμῶν ἡ ἐλπὶς καὶ καταφύγιον SEG 35.733 (Beroea, iv AD)'

*καταφύρω, foul, defile, aor. 1 κατέφυρσε, Hsch. s.v. κατέδευσε.

καταφυτεύω I, add 'PCair.Zen. 157.3 (256 BC)'

κατάχαλκος, for 'a serpent lapt in mail, i.e. scales' read 'of a snake covered with bronze (-coloured scales)'

καταχαρίζομαι, lines 1/2, for 'corruptly make .. thing corruptly' read 'give as a favour, make a present of material or abstract things'; line 5, after 'D. 26.20' insert 'πάντα ταῦτα κ. 41.12' (deleting from section 3); after 'Arist.Pol. 1271ᵃ3' insert 'D.H. 6.30, 7.63'

καταχέω I 2 c, add 'pour out song, ἠχέτα τέττιξ .. λιγυρὴν καταχεύετ' ἀοιδήν Hes.Op. 583, κακχέει λιγύραν ἀοίδαν Alc. 347(b).2 L.-P.'

καταχθονίζω, add '2 fig., ἐλπίδας .. Μοῖρα κατεχθόνισεν GVI 1989.14 (Panticapaeum, ii/i BC).'

καταχθόνιος, add 'of Hecate, SEG 30.326 (i AD); of Hermes, SEG 26.1717 (tab.defix., Egypt, iii/iv AD)'

*καταχνύω, oppress, Nic.fr. in Suppl.Hell. 562.8.

καταχορεύω, after 'Ael.NA 1.30' insert 'gloss on κατακωμάζω, Sch.E.Ph. 352'

καταχράομαι II 1, add 'Aeschin. 2.70' 2, add 'μοχλοὶ .. κατεχρήσθησαν εἰς σφῆνας IG 2².1672.304 (Eleusis, 329/8 BC)'

κατάχρεος, add 'III καταχρέως· ἐπαρκῶν εἰς μαρτυρίαν Hsch.'

κατάχριστος, after def. insert 'Heras ap.Gal. 13.779.4'; add 'also subst., ἡ κ. Cat.Cod.Astr. 8(3).148.11'

κατάχρῡσος 4, for 'Phld.Po. 5.15' read 'Phld.Po. 15.16'

*καταχρωμένως, adv., (καταχράομαι) = καταχρηστικῶς, opp. κυρίως, Gal. 17(2).790.

*καταχύσιος, ον, app. adj. fr. κατάχυσις in unkn. sense, IMylasa 308.4.

κατάχυσμα 2, for 'a bride .. Theopomp.Com. 14' read 'a bridegroom and bride, τὰ κ. κατάχει τοῦ νυμφίου καὶ τῆς κόρης Theopomp.Com. 15 K.-A.'

+καταχωρέω, come in, of revenue, SEG 2.481 (Scythia, iii AD).

καταχωρίζω, line 2, after 'Apollon.Cit. 2' insert 'spellings w. η for ι, e.g. κατεχώρηκα Plu. 2.312b, κατεχώρηκεν APF 3.134, by itacism; cf. -ρεῖν, prob. for -ριεῖν BGU 981, etc.'

καταψάλλω 2, for this section read 'play music over, in quot. pass., Procop.Pers. 2.23'

καταψάω 2, add 'καταψᾶν καὶ τιθασεύειν τὴν ὀργὴν τοῦ βασιλέως Demad.fr. 20 B.'

καταψηφίζομαι I 1, penultimate line, delete 'pf.'; last line, after '(D.H.) 5.8' add 'κατεψήφ[ισα]ν Call.Dieg. i.44 (dub., fr. 84 Pf.), καταψηφίζων Hsch. s.v. καταστίζων'

καταψήχω II, for 'stroke, caress' read 'smooth down'

καταψύχω I 1, after 'Epicur.Fr. 60' insert 'in curses, κραταιὲ Βετπυτ, παραδίδωμί σοι Εὐτυχιανόν, .. ἵνα καταψύξῃς αὐτόν tab.defix. in SEG 35.213.3 (Athens, iii AD)'

*κᾰτέ, = κατά, κ. ἐπιταγήν TAM 2(3).729, 730 (Lycia).

κατέδω, line 2, after 'Il. 24.415' insert 'applied to cannibalism, Emp.fr. 137.6 D.-K.'

κατείβω II, delete 'trans.' and for 'Ἔρος .. 36' read 'Ἔρως .. κατείβων καρδίαν ἰαίνει Alcm. 59(a) P.'

κατεικονίζω, add '2 register with an official description, ἐγράφη κατεικονισθ(ὲν) διὰ γραφ(είου) [Τεβ]τύνεως διά τε Κρονίωνος γραμματέως PMil.Vogl. 145.28 (AD 174).'

κάτειμι I, add '2 to be descended, οἱ κατιόντες descendants, CodJust. 6.4.4.14c.'

κατείργω, line 1, for 'Cypr. .. H.' read 'Cypr. ka-te-wo-ro-ko-ne κατέϜοργον (perh. -ē-) ICS 217.1.'; line 3, after 'shut in' insert 'τὰς μὲν ἐν ἡσυχίῃ κατέερξε h.Merc. 356 (v.l. κατέρεξεν)'; line 7, for 'πτόλιν Inscr.Cypr. l.c.' read 'po-to-li-ne πτόλιν ICS l.c.'

κατεισαγωγή, delete the article.

κατελέγχω I, line 1, for 'νόον' read 'νόος' II, delete the section, transferring quots. to follow 'Pi.O. 8.19' in section I. III, for 'betray' read 'convict'

*κατελευθερόω, liberate, τῆς τυραννίδος Sch.Pi.O. 12.1 (pass.).

*κατέλευσις, gloss on κατήλυσις, Hsch.

*κατέμπαλιν, adv. backwards, h.Merc. 78 (tm.).

*κατεμπᾰτέω, trample, κατεμπατοῦνται POxy. 3021 i 16 (i AD, s.v.l.).

*κατεναντίος, α, ον, face to face with, w. gen., ἀνθρώπων κατεναντίη Arat. 102 (v.l. -ίον).

*κατενέργεια, ἡ, activity, study, ὡς ἀποδείξω ἐκ πασῶν τῶν γραφῶν ἐν τῇ ἐμῇ κ. περὶ τοῦ σταθμοῦ Zos.Alch. 178.3.

*κατενωπός, όν, right opposite, Theognost.Can. 69.

*κατεξανιστάω, = κατεξανίσταμαι, gloss on καταπλήσσω, Hsch. s.v. καταπλήσσει.

κατεπᾴδω I 1, for 'soothe' read 'blandish, cajole' II, for 'to be always repeating' read 'to recite in detail or at length'

κατεπείγω I 2, add 'urge on an action or performance, κατεπείγειν τὴν ἐξάνυσιν τῶν κεχρεωστημένων τίτλων CodJust. 10.19.9 pr.'

*κατέπερθεν, Aeol. adv., = καθύπερθεν, Alc. 357.3 L.-P., cf. °ἔπερθα.
κατεπίθυμος, for 'inf.' read 'gen.'
*κατέρᾱσις, εως, ἡ, *pouring off*, Agath.Alch. 270.24.
*κατέργω, v. κατείργω.
κατερείπω, after 'Hdt. 7.140' insert 'Nic. *Th.* 724' and before 'Max.Tyr. 1.3' insert 'κατήριπεν δὲ ὁ χειμὼν τὰ θαυμαστὰ ἐκεῖνα πάντα'
κατερυθραίνομαι, add 'act., *dye red*, Hsch. s.v. καταβάπτει'
κατέρχομαι, line 8, after 'etc.' insert 'Ἀϊδονῆος Q.S. 3.15'
κατεσθίω 2, add 'τὸν κλῆρον Hippon. 26.4 W.; (cf., w. play on literal sense of θαλλός), ὅτι τὸν .. κατέφαγεν ἐραστήν ποτε Θαλλόν Macho 425 G.'
κατευφραίνω, add 'κατ[ευ]φραίνει μάλιστα τὴν [ψυ]χὴν Diog.Oen. new fr. 12.7 S.'
κατευχή, after '*vow*' insert 'Simon. 32 tit. P. (pl.)'
κατέχω **A I 1 a**, add '*embrace*, Anacreont. 41.7, 50.20 W.' add '**g** pass., *to be under a legal obligation*, οἶδας γὰρ ἀκριβῶς ὅτι καὶ ἡ ταβέρνα καὶ οἱ δοῦλοί μου οὐδενὶ κατέχονται ἢ σοί Scaev.*Dig.* 20.1.34.1, *Cod.Just.* 6.34.4.'
κατηγορέω **I 3**, add 'w. cogn. acc., ταῖς κατηγορίαις ἃς .. κατηγοροῦσι D. 8.8' **b**, add 'Gorg.*Pal.* 28, 32'
κατήλῠσις **I**, after '*AP* 10.3' insert '*way down into*, κύρτοιο Opp.*H.* 4.116' and for 'Simon. 179.1' read '*AP* 6.217.1 (Simon.)'
κατήορος, line 3, for '*AP* 5.259' read '*AP* 5.260'
κατηρεφής **1**, line 5, delete 'χθονός'
κατῆτος, before '= κατὰ ἔτος' insert 'adv., prob.'
*κατθῠτός, v. °καταθυτός.
*κάτια, τά, unexpld. wd. found in household inventory, *BGU* 34.iii 4, al., iv 24 (?error for ἀκάτια).
†κατϊαραίω, Elean form of καθιερεύω, *utter an imprecation against, accuse*, Schwyzer 424.5 (iv BC); aor. opt. -ιαραύσειε (τινος) ib. 409.2 (v BC).
κατιερόω, κατιέρωσις, delete the entry.
κατῑθύς, for '*opposite*' read '*straight ahead* or *down from*'
*κατιλλάριος, ὁ, Lat. *catillarius*, *dish-maker*, *SEG* 35.1024.A3 (Arretium, i BC/i AD).
κατίλλω, for 'dub. sens.' read 'app. describing a distorted or constricted condition of the voice'
†κατιλλώπτω, perh. *look sideways at* or *through the slits of one's eyes*, εὖ κατιλλώψας ἄθρει A.fr. 226 R.; in erotic contexts, τινι Philem. 115 K.-A., θῆλυ κ. *AP* 5.119; cf. κατιλλώπτειν, τὸ καταβλέπειν ἐπὶ χλευασμῷ Poll. 2.52.
*κατῑρόω, v. καθιερόω.
κάτισχνος, add 'fig. of style, *jejune*, Sch.Flor. ad Call.fr. 1 Pf., lines 8/9 (rest.)'
κατίσχω **I**, add 'κάτισχε λῆμα καὶ σθένος θεοστυγές Neophr. 2.4'
*κατοικάδιος, ον, = κατοικίδιος, *domestic*, ἥρως *IGBulg.* 1874 (Thrace).
κατοικέω, after lemma insert 'Aeol. pres. part. κατοίκηντας *IKyme* 13.76 (ii BC)' **I 2**, add 'fig., ὁ κατοικὸν ἐν βοη(θείᾳ τοῦ ὑψίστου) *SEG* 31.1598, 32.1573' **III**, for '*lie, be situated*' read '*exist as a settlement*' and add 'τὰς πόλεις τὰς ἐν Πελοποννήσῳ κατοικούσας Isoc. 12.166'
κατοικία 2, = *colonia*, add '*IEphes.* 3327, 3239A, etc.'
κατοικίδιος, after 'α, ον, only in Gp. 1.3.8' insert 'Just.*Nov.* 69.1.1'; line 2, for 'Theopomp.Hist. 258(a)' read 'Antig.*Mir.* 137' add '**2** *domiciled*, *BGU* 1816.8 (i BC).'
κατοικίζω **I**, line 7, for 'A.*Pr.* 252' read 'A.*Pr.* 250'
*κατοκλάζω, *squat down*, tm., κατά τ' ὀκλάζουσιν ὁμοῖοι Opp.*C.* 3.473.
*κατόμοσις, εως, ἡ, *affirmation on oath*, Hsch. s.v. μά.
κατομφάλιος, after '*navel*' insert '(or analogous part)'
*κατόνησις, Dor. -όνᾱσις, εως, ἡ, *enjoyment, right of profit from*, καρποῦ *SEG* 23.474 (Dodona, iv/iii BC).
κατόνομαι, add 'imper. κατόνοσσο Arat. 1142'
κατόπιν (s.v. κατόπιθεν) **II**, add 'κ. γίγνεσθαι *to fall behind, be remiss*, Just.*Nov.* 59.3'
κατόπισθεν **II**, add '*h.Merc.* 407'
*κατοπτάζομαι, *discern, see*, Aesop. 266 P.
*κατοπτευτήρ, ῆρος, ὁ, *overseer*, Colluth. 54 (s.v.l.).
κατοπτεύω **I**, pass., add 'ὑπὸ λουτροῦ ἀληθινῶς κατοπτεύθησαν Χ.*Oec.* 10.8' add **III** w. dat., *adopt an unfavourable attitude towards*, ἡμῖν *POxy.* 2342.29 (ii AD).'
κατορύσσω, line 7, after 'X.*Mem.* 1.2.55' insert 'καταδῶ, κατορύττω, ἀφανίζω ἐξ ἀνθρώπων *SPAW* 1934.1023 (tab.defix., Attica, iv BC)'
*κατορύχω, = κατορύσσω, *SPAW* 1934.1041 (tab.defix.).
κατουλέω and κατούλη, delete the entry (v. *JÖAI* 32.72).
*κατοχεία, ἡ, *fertilizing*, τοῦ .. φοινικῶνος τὰς διακαθάρσις καὶ κατοχείας καὶ κατασπασμοὺς *PSoterichos* 4.27 (AD 87).
κατοχή **II 1**, add '**b** pl., concr., *possessions*, *SEG* 30.568 (Maced., ii AD).'
*κατόχημα, ατος, τό, *residue*, Anon.Alch. 348.15.
†κατοχμάζω, *fasten down*, Opp.*H.* 5.226.

κάτοχος, line 2, after '*IG* 3.1425a' insert '*SEG* 8.13.12 (Nazareth, rescript of Augustus or Tiberius)'
*κάττάπερ, v. °καθά.
*κάττῠσις, εως, ἡ, *stitching*, ὑποδημάτων *IG* 2².1672.190, 230 (Eleusis, 327 BC), v. κασσύω.
κάτω **II c**, add 'ἁ κάτω πόλις *the lower quarter* of the city, *SEG* 23.563 (Crete, iii BC); cf. ἄνω (B) A II 1 d'
κατώβλεψ, add 'Theognost.*Can.* 97'
κατωκάρα, add '[κα]τωκάραι Alc. 58.22 L.-P.'
κατωμάδιος **I**, add '*lying with shoulders down*, of the Great and Little Bears, κατωμάδιαι φορέονται, ἔμπαλιν εἰς ὤμους τετραμμέναι Arat. 29'
κατώμοτος, add '**2** *bound by an oath* (made against oneself), κ. κατ' αὐτοῦ μὴ ἔστω Thasos 18.4 (v/iv BC)'
*κατωμόχᾰνος, ον, (= χαίνων κατ' ὤμου Sch.) Μιμνῇ κατωμόχανε Hippon. 28 W.
*κατωνυμία, ἡ, *designation*, Socr.Dion.*Lith.* 46.2 G.
κατώρης, add '*rushing downwards*, ἄνεμος Alc. 412 L.-P. (v.l. κατάρης), Sapph. 183 L.-P.'
κατώτερος **1**, after '*Ep.Eph.* 4.9' insert 'κώμης Μαγαράτων κατωτέρας *SEG* 32.1061 (AD 414)'
*κατωχεία, ἡ, v. °κατοχεία.
κατωχεύει, for 'πηδᾷ' read 'πηδᾷ'
καυάζοντα, add '(acc. to La. corrupt for ὀκλάζοντα)'
καῦδος, delete the entry.
†καύης, ὁ, Lydian wd. for *priest* or *priestess*: masc. Hippon. 4 W., Sardis 6(2).23.6 (iv BC), al.; fem. καῦις κάυειν (for καῦιν, acc.) Sardis 7(1).51.3 (ii AD), al.
*καυκεών, ῶνος, ὁ, perh. deriv. of καῦκος; cf. κυκεών, Theognost. *Can.* 28.
καυκίον, add 'Lxx Sch.*Ge.* 44.2'
*καυκουλάτωρ, ορος, ὁ, Lat. *calculator*, *teacher of arithmetic*, *Edict.Diocl.* 7.67.
*καυληδόν **II**, add 'Gal. 18(2).759.4, 788.17, 888.8'
*καυλοκόπος, ὁ, *stalk-cutter*, *PMil.Vogl.* 69.B23 (ii AD, κολο- pap.).
καῦμα **I 2**, add 'Pl.*Ti.* 86a (pl.)'
*καυματινός, ή, όν, *burning hot*, κ. ὥρης ὑπαρχούσης Vit.Aesop.(G) 28.
καυνάκης, for 'thick cloak' read '*covering* (blanket, rug, or sim.)'; at end for '(Assyr. .. mantle)' read '(cf. Pahlavi *gonaka*, rug, bed-cover, of Iranian origin, Avest. *gaona*, hair, colour, *APAW* 1936(3).5)'
*καύνη, ἡ, cf. Lat. *canabae*, *settlement of camp-followers*, Dura⁶ 434 (iii AD).
*καυνιάζω, τὸ κλῆρον μακρόθεν ἡγοῦμαι Hdn.fr. p. 27 H.; cf. διακαυνιάζω.
*καυνιάρι(ο)ν, τό, *gaming-table*, Hdn.fr. p. 27 H.
Καύνιος, add 'in a cult-title, βασιλεῖ Καυνίωι *SEG* 27.942.7 (Xanthus, Lycia, 337 BC)'
*καυνός (B)· κακός, σκληρός, κλῆρος Hsch., gloss contaminated with καυνός (A).
καυσία, delete 'forming part of the regalia of their kings' and before 'Plu.*Ant.* 54' insert 'κ. διαδηματοφόρῳ'
καύσιμος, line 2, κ. ξύλα, add '*IG* 1³.425.9'; at end add 'in uncertain sense, A.fr. 73b.5 R.'
καῦσος (A) **I**, add 'Hp.*Acut.* 5.66, *Aph.* 4.58, *Morb.* 1.3, 29, etc.'
καύστειρα, at end for 'in the form .. Nic. l.c.' read 'taken to be fr. καυστειρός, hence καυστείροιο Opp.*H.* 2.509 (v.l. καυστηρ- both here and in Nic. l.c.), and perh. so Call. in *POxy.* 2375 ii 2'
†καυστηρός, v. ‡καύστειρα.
καῦστις, add 'παρὰ δὲ τοῖς τραγικοῖς καῦστις εἴρηται μεταφορικῶς ἡ μάχη. παρά τισι δὲ ἡ ὀσφῦς καὶ τὰ ἰσχία, παρ' ἐνίοις δὲ ὁ ξηρὸς χόρτος. ἔνιοι δὲ τὸ αἰδοῖον Phot. s.v. ἀμφίκαυστις'
†καύστρα, ἡ, *place of cremation*, Str. 5.3.8, *Rev.Phil.* 13.212 (Faustinopolis), *GVI* 1751.1 (Ion. -η, Termessus, ii/iii AD), *IEphes.* 2123, al. **b** *funerary urn*, *ITrall.* 203.
καυσώδης, add 'w. ellipsis of πυρετοί, Hp.*Coac.* 128, 134'
καυτήριον, after 'τό' insert 'also καυστ-' **I**, add 'ἀκοῆς καυστήρια, of persons, Cels.ap.Orig.*Cels.* 5.64' **III**, add 'Plin.*HN* 1.35'
καύχημα 2, add 'of Christ, *SEG* 32.1588 (vi/vii AD)'
*καυχητικός, ή, όν, *boastful*, ᾠδή Sch.Pi.*N.* 9.16.
καυχός, delete 'καυχοῦς' and 'χαλκοῦς'
*καῦχος, εος, τό, *boast, subject of boasting*, Princeton *Exp.Inscr.* III A no. 160 (v AD).
καχεξία 2, add '*Amyzon* 20 (κακ-)'
κάχρυς, line 1, after 'ὕδος ib. 20' insert 'pl. acc. κάχρῡς Ar.*Nu.* 1358'
*καψάριον, τό, *cupboard for clothes*, *PHarris* 79.10 (iii AD).
καψάριος, after 'ὁ' insert '*slave in charge of clothes*, esp. at baths'; after '*capsarius*' insert '*Edict.Diocl.* 7.75 (v.l. καμψάριος), *SB* 10258 i 11 (iv AD), *ITyr* 151 (καμψ-)'
καψοί, add '(κάψαι· τεύχη cj. La.)'
κε, line 1, for '*Inscr.Cypr.* 135.10 H.' read '*ICS* 217.10'

κεβλήπυρις, delete 'nickname .. Hermipp. 72'
κεγχρεών, for pres. def. read 'some part of a foundry'
κεγχρίδιον, for 'κιχρηδῶν' read 'κιγχρηδῶν'
***κέγχρον**, τό, prob. for κέγχρος, *granulation* on gold work, *CPR* 5.22.8 (?v AD).
κέγχρος II, add '**4** *grain of sand*, Aq.Sm.Theod.*Is.* 48.19.'
κεδνός I 1, add 'pl. subst. ὦ μοῖρα κεδνῶν καὶ κακῶν κυνηγέτι *Trag.adesp.* 504' **II**, add 'E.*Tr.* 683, ἔν τι κεδνὸν αἱροῦ Thal. ap.D.L. 1.35'
κεδρία, line 1, after 'κεδρελάτη' insert 'or a preparation made from it' and add '*PSorb.* I 34'
***Κεδρίτας**, ὁ, title of Hermes, *SEG* 26.1046 (Crete, iii AD).
κέδρον, add '**III** = στρούθειον (s.v. -ειος III), Ps.-Dsc. 2.163.'
κέδρος 2, delete 'for a beehive'
κεῖ, for '= κεῖθι, ἐκεῖθι' read '= ἐκεῖ' and after 'Herod. 1.26' add 'Call.*Del.* 195 (v.l., in *POxy.* 2225)'
†**κεῖθεν**, **κεῖθι**, v. ‡ἐκεῖ-.
***κειλαιδίκηνος**, ὁ, title of Asclepius, *SEG* 30.717; also **κειλαιδέκηνος**, ib. 736, **κειλαίσκηνος**, ib. 727 (all Serdica, Thrace, ii/iii AD).
κεῖμαι, line 2, after '*IG* 1².94.25' insert 'Cypr. *ke-i-to-i* κεῖτοι, *ICS* 11'; line 5, for 'Il. 24.567' read 'Il. 24.527'; line 6, after 'Aret.*SD* 2.4' insert 'Cret. κίαται *Inscr.Cret.* 4.174*A*22 (Gortyn, ii BC)'; line 14, after 'Od. 6.19' insert 'κέοντο Q.S. 3.728, (v.l.); κατεκείαθεν· κατεκοιμήθη Hsch.' **I 6**, for 'have a fall' read '*lie flat out* after a fall (wrestling), fig.' and after 'Ar.*Nu.* 126' add 'Pl.*R.* 451a; Call.*fr.* 194.80 Pf.'
κειμηλιάρχης, add '*MAMA* 3.349 (Corycus)'
κειμήλιον I, add 'κε̄μέλιον *EAC* 7.128 (Acanthus, vi BC)'
κειμηλιφῠλάκιον, add 'Just.*Nov.* 74.4.2'
κειρία II, add 'ribbon, *Inscr.Délos* 1439*Aba*70 (ii BC)'
κείρω, before 'fut.' insert 'Aeol. κέρρω Choerob.*in Theod.* 1.126'; line 9, after 'Luc.*Lex.* 5' insert 'part. κεκραμένος Schwyzer 686 (Pamphylia, iv BC)' **II 2**, add 'Thgn. 892 W.' **4**, for this section read 'Cypr. *e-ke-re-se* ἐπίβασιν .. ἔκερσε *made by cutting, hewed out*, *ICS* 3'
κεκαδδίχθαι, delete the entry.
κεκᾰφηώς, lines 3/4, for 'fordone .. ib. 26.108' read '*exhausted*, of persons, Nic.*Al.* 444; δέμας κεκαφηότα λιμῷ Nonn.*D.* 26.108'; line 6, after 'Opp.*H.* 3.572' insert 'κεκαφηότα γυῖα id.*C.* 4.206, Nonn.*D.* 2.539'
κεκῆνας, add 'cf. Κηκήν as a pr. n. in *Inscr.Cret.* 1.xxii 52 (ii/i BC)'
***κεκραγῖναι** (**κεκραδῖναι** La.)· ἄγριαι θρίδακες Hsch.
***Κεκροπηΐς**, ίδος, = Κεκροπίς, *SEG* 21.748 (Athens, ii/iii AD).
Κεκροπία, add 'Ion. -ίη *Suppl.Hell.* 1044'
***κελαδαία**, unexpld. wd. in damaged inscr., *SEG* 19.115.17 (Athens, ii/iii AD).
κελαδέω, line 4, after 'E.*Hel.* 371 (lyr.)' insert 'inf. κελαδέσαι prob. in Pi.*I.* 5.48 S.-M.' **I 3**, line 2, for 'grasshopper' read 'cicada'
κελαινός, add 'Myc. *ke-ra-no*, name of an ox'
κελέβειον, delete 'Ion. -ήιον' and after 'Antim. 17' add '18, 19 Wy.'
κέλευθος IV, after 'of life' insert 'Simon. 36.13 P.'
κελευστής, add '**II** *house-steward*, Mitchell *N.Galatia* 88.'
κελεύω I 9, add '**b** w. ὥστε or inf., ἡ διάταξις κελεύει ὥστε πάντα δικαστὴν ἐν τῇ ἀποφάσει αὐτοῦ κελεύειν .. *Cod.Just.* 7.51.5 pr., 7.62.35; sim. w. ἵνα, ἡ διάταξις κελεύει ἵνα .. μὴ λαμβάνειν .. ib. 12.37.19 pr.'
***κελεφία**, ἡ, a form of leprosy or sim., Cyran. 15; cf. ‡κελεφός.
κελεφός, add 'Epiph.Const. in *PG* 42.43'
κέλης I, add 'ἐπὶ (τοῦ) κέλητος in celebration of an *ovatio*, D.C. 54.8.3, 55.2.4. **b** obsc., w. ref. to a coital posture (cf. °κελητίζω), ἐπὶ τῶν κελήτων διαβεβήκασ' ὄρθριαι Ar.*Lys.* 60; cf., of a facetiously invented deity, ἥρῳ Κέλητι Pl.Com. 188.18 K.-A.; interpreted as female pudenda, Eust. 1539.34.' **II**, for 'fast-sailing .. oars' read 'a light *fast-sailing* ship' **III**, delete the section.
κελητίζω I, delete 'of one who leaps from horse to horse'; add 'pres. part., pl., of statues, figures *riding on horseback*, Plin.*HN* 34.75, 78' **II**, for this section read 'of a woman, *to adopt a coital posture astride* a man, Ar.*V.* 501, *Th.* 153, w. acc., Macho 171 G.'
***κελητισμός**, ὁ, in obscene text, app. a coital position, cf. °κελητίζω II, *AE* 1957.48 (Attica, iii BC) (pl.).
***Κελκαία**, ἡ, title of Artemis (identified w. Empress Sabina), *SEG* 37.522-5 (Epirus, ii AD).
***κελλαρεύω**, *to be a* ‡κελλάριος, inscr. in *ZPE* 33.187 (Troad, v/vi AD).
***κελλαρικάριος**, ὁ, *cellarman*, *PKlein.Form.* 1000 (v AD), *CPap.Jud.* 513.20 (AD 586).
***κελλαρικός**, ή, όν, *of a cellar* or *store-room*, κελλαρικῶν εἰδῶν, *provisions*, *PBeatty Panop.* 1.219.
κελλάριος, after 'cellarman' insert 'Lat. *cellarius*, *PRyl.* 228.24 (i AD)'; at end add 'in pl., gloss on ταμίαι, Sch.Gen.Il. 19.44; fem. κελλαρία, ἡ, Hsch. s.v. ταμίη'
***κελλία**, ἡ, = κέλλα, κελλίον, *CPR* 4.44.3, 7, 8 (v/vi AD).
κελλίβας, before '*PRyl.* 136.10' insert '*PTeb.* 793 vi 4 (ii BC)'

κελλικάριος, delete the entry.
κέλλω, line 2, fut., delete 'E.*Hec.* 1057' and after 'ἔκελσα' insert 'ἐπέκειλα *Act.Ap.* 27.41'
***κελλώνιον**, unexpld. wd. cited as derived fr. κέλλω, Sch.E.*Or.* 800.
***κέλοξ**, οκος, Lat. *celox*, *small boat*, πλοίῳ κ., gloss on κέλητι, Sch.Th. 8.38.
κέλῡφος 2, for 'sheath' read 'περὶ τοῦτον δ' (sc. καυλὸν) οἷον κέλυφός ἐστι τὸ καλούμενον αἰδοῖον'; line 6, for 'crustaceous fish' read 'crustacea'
κεμάδειον, delete '(prob.)'
κεμαδοσσόος, after 'chasing deer' insert 'Call.*fr.* 186.31 Pf.'
κεμάς, delete 'so Ar.Byz. ap.'
***κέμβερος**, another name for λάλαξ, Hsch. s.v. λάλαγες (dub., v. ‡Κέρβερος).
***κεμήλιος**, ὁ, prob. cult-title of Dionysus, Alc. 129.8 L.-P.
***κέμιον**, τό, prob. a kind of vegetable, *PRyl.* 629.266, al. (iv AD); also καίμιον *POxy.* 1656.14 (iv/v AD); cf. °κεμοράφανος.
***κεμιοπώλης**, ὁ, *seller of* κέμιον, *SB* 9902 B iii 6, T 7 (iii AD).
***κεμοράφανος**, ὁ, a kind of vegetable, *PRyl.* 629.37, al. (iv AD), v. °κέμιον.
***Κεμπηνός**, ὁ, title of Asclepius, *IGBulg.* 1².354 (Rom.imp.).
***Κενδρείσεια** or **-είσια**, τά, name of a festival in Thrace, *IGBulg.* 889, al. (Philippopolis); written **Κεντρ-** *IG* 2².3169.20 (Athens, iii AD).
κενεάριον, v. ‡κενήριον.
κενεμβᾰτέω I 1, after 'Plu.*Flam.* 10' insert 'of the sun passing over an uninhabited region, ὥσπερ κ. Xenoph. in *Placit.* 2.24.9'
***κενεμεσία**, ἡ, *retching*, Gal.ap.Aët. 9.10 (pl.).
***κένευρον**, ?τό, Cypr. *e-pi-ke-ne-u-wo-ne* perh. ἐπὶ κενεύϝον, *cenotaph*, *ICS* 94.
κενεών I 1, add 'of boats, Opp.*H.* 3.555'
κενήριον, add 'Dor. κενεάριον *SEG* 35.295 (Kynouria, late v BC; see also ib. 39.369)'
κενοδοξέω 2, add '**b** *glory in empty distinctions*, D.Chr. 38.29.'
†**κενολόγος**· *vanilocuus, Gloss.*
κενοπᾰθέω, add 'Plu. 2.1106a'
κενός, line 9, for 'κενευϝός Schwyzer 683.4' read '°κένευϝον' **I 2**, line 16, delete 'ἐν κενοῖς S.*Aj.* 971' **II 2**, line 7, after 'Lxx *Ge.* 31.42' insert '*bereaved*, ἐν κενοῖς S.*Aj.* 971' **b**, add 'of style, *hollow, pretentious*, D.H.*Dem.* 44; of an author, id.*Is.* 20; comp., ib. 19'
κενόσπουδος, line 3, for 'mere curiosity' read '*useless anxiety*'
κενοτάφιον I, delete '*CIG* 4340d, e' and add '**2** *place for corpse* or *sarcophagus, tomb chamber*, *SEG* 2.702, 708, 6.667, Robert *Hell.* 10.172, *SEG* 17.601, 602, 630 (Pamphylia, Rom.imp.).'
***κενοφόβος**, ον, gloss on ψοφοδεής, Hsch.
κενόω I 1, at end add '*make barren* or *desolate*, ἐκενώθη ἡ τίκτουσα ἑπτά Lxx *Je.* 15.9'
***κενσίτωρ**, ωρος, ὁ, Lat. *censitor*, *registrar* or *taxation officer*, *PHerm.* 32.11 (vi AD).
***κένταρχος**, ὁ, *centurion*, *IG* 12(5).712.98 (Syros).
κεντέω, add '**II** *adorn with mosaics* (cf. κέντησις II), Βυζάντιον 4.715 (Eresus, iv/v AD), *REG* 79.229 (Aetolia, ib. 82.462.'
κεντηνάριος, add '*SEG* 24.911 (-άρις Thrace, iii/iv AD), *BGU* 2141.5 (AD 446)'
κέντησις II, add '*BCH* suppl. 8 no. 226 (Philippi, iv AD)'
κεντητικός, add '**2** *engraved* (κεντιτικός pap.), *BGU* 2359.2 (iii AD).'
***κέντινος**, unexpld. adj. describing a horse, *POxy.* 922.11 (vi/vii AD), cf. ib. 1289.9 (v AD).
***κέντουκλον**, τό, Lat. *centunculus*, *a piece of cloth*, *Edict.Diocl.* 7.52.
***κεντουρία**, ἡ, Lat. *centuria*, *CIG* 5046, written -τυρία *PLond.* 142.4 (i AD), *Aegyptus* 62.171, Just.*Nov.* 128.1.
***κεντριστής**, οῦ, ὁ, *goader, driver*, Hsch. s.v. κέντορες.
κεντρίτης I, for 'v. κεντρίνης III' read '*a venomous serpent*' add '**IV** perh. occupation term, *maker of goads*, *MAMA* 3.278.'
κέντρον 1, *goad*, as symbol of sovereignty, add 'Sol. 36.20 W.' **5 d**, for '= πόσθη' read '*penis*' add '**e** in pl., *antennae* of the crayfish, Opp.*H.* 2.324.'
***κεντρορρᾰγής**, v. κεντρομανής.
κέντρων, add '**III** perh. *goad*, *PMich.*xv 717.4 (iii AD).'
κεντρωνάριον, for 'case for κέντρωνες' read 'perh. *case for* κέντρωνες II (*pen-wipers*)'
κεντρωτός 1, add 'φιάλαι *Inscr.Délos* 104-12.8 (iv BC)'
***κενωτέος**, α, ον, *needing to be purged*, Gal. 11.211.8.
κέπφος 2, add 'Call.*fr.* 191.6 Pf. (cf. Sch.ad loc.)'
κεπφόω, for 'ensnare like a κέπφος' read '*make like a κ.*' and after 'feather-brained' insert '(or *changeable*)'
κεραία II 3, for 'apex of a letter' read '*mark placed over letter to indicate length*' and add '*stroke of a letter*, *SEG* 32.1256 (Nicomedia)'
***Κεραιάτης**, ὁ, v. ‡Κερεάτας.
κεραιός, add 'Myc. *ke-ra-ja-pi* (fem. instr. pl.)'

κεραιοῦχος, for 'upholding the right, Hsch.' read 'κεραιοῦχον· δικαιοδότην ἀπὸ τοῦ ἐν τοῖς πλοίοις κεραιούχου Hsch.'

***κεραμάρχης**, ου, ὁ, prob. some annual magistrate, *BCH* suppl. 5 pp. 274-5 (three inscr., Thasos, iv/iii BC).

κεράμβηλον, add '*Et.Gen.* in Miller *Mélanges* 183'

κεραμεύς, add 'Myc. *ke-ra-me-u*'

κεραμεύω 2, after 'c. acc.' insert 'ἐκεράμευσεν ἐμέ *ABV* p. 349'

κεραμεών, after 'Ar.*Lys.* 200' add '(cj. κεραμών)'

***κεραμηγός**, όν, *carrying pottery*, of a boat (κύδαρος), *BGU* 2353.5 (AD 115).

***κεράμιδᾶς**, ᾶ, ὁ, *potter*, *PSI* 899.35 (iii AD, -ειδας pap.).

κεραμικός, after 'of or for pottery' insert '*used in making pottery*' and add 'λίθου κεραμικοῦ *PMich.*v 238.143, 181'

***κεραμοδέτης**, ου, ὁ, kind of craftsman, perh. *tiler* or *pot-mender*, *Kerameikos* 3 p. 98.9 (iv BC).

κεραμοπώλης, add '*IG* 2².1673.21 (iv BC)'

κέραμος II 1 b, insert at beginning 'χαλκέῳ δ' ἐν κεράμῳ Il. 5.387 (said to be used as a prison by the Cyprians: Theon.*Prog.* 13), cf. epitome Monac. of Thphr.*Char.* 6.6; pl., Nonn.*D.* 16.162' **2**, add 'also κέραμος, ἡ, *BCH* 56.293 (Stobi, ii AD)' **III**, delete the section.

***κέραμος**, τό, *pot*, κεράμε πάντα *BE* 1989.305.

✝**κεραοξόος**, ον, *carving* or *working in horn*, κ. τέκτων Il. 4.110, *AP* 6.113 (Simm.); masc. subst., *worker in horn*, Opp.*C.* 2.509.

κεραός I, before 'τράγος' insert 'Call.*fr.* 25.1 Pf.'; after 'Theoc. 1.4' insert 'Διώνυσος Nic.*Al.* 31'; add 'Bernand *IMEG* 107.1'

κέρας, line 6, after 'Q.S. 6.225' insert 'dat. κεράασι [∪–∪∪] A.R. 4.978' **III 1**, delete 'τόξον .. cf.' **2**, for '*flute*' read '*aulos*' **3**, for '*OGI* 214.43 (Didyma, iii BC)' read 'Didyma 424.43 (iii BC)' **V 5**, add 'perh. in wider sense, cf. κεραία II, *Peripl.M.Rubr.* 36' **7 b**, for 'πόσθη' read '*penis*' add '**IX** = κεράτιον II, as a coin, Epic. in *BKT* 5(1).120.61 (vi AD).' at end add 'Myc. *ke-ra-a* (pl.), *ke-ra-e* (dual)'

κερασβόλος, add '**III** κερασβόλα· οἱ περὶ τῶν κεράτων βοῶν δεσμοί, καὶ οἱ ἐν ταῖς ἀρχαίαις λύραις κόλλαβοι Hsch.'

κεράστης I, add 'κοχλίας κεράστας Achae. 42. **b** λύκος κ. *lynx* (from its tufted ears), *Edict.Diocl.* 8.35 (v.l.). **c** = °κερασφόρος I b, Nonn.*D.* 45.43.'

κεραστίς I, add 'of Cyprus (fr. shape of coastline), Hdn. 1.104.15'

κεραστός, add 'alloyed, κ. νόμισμα Cels.ap.Orig.*Cels.* 6.22'

κερασφόρος I, add 'κ. κριόν Ph.Epic. in *Suppl.Hell.* 682' add '**b** ὁ κ. αὐλός the Phrygian pipe *having a horn-shaped addition* at the end, Aristid.Quint. 3.21; cf. κέρας III 2, °κεράστης, κεραύλης, ἐγκεραύλης.'

✝**κερατίζω**, *butt with horns*, *gore*, ἐὰν κερατίσῃ ταῦρος ἄνδρα Lxx *Ex.* 32.2, al.; fig., ἐν σοὶ τοὺς ἐχθροὺς ἡμῶν κερατιοῦμεν Ps. 43(44).5; abs., Ph. 1.57, Sch.Theoc. 3.5; w. ref. to a crocodile, perh. in mistranslation fr. Hebr., σὺ ὡς δράκων ὁ ἐν τῇ θαλάσσῃ καὶ ἐκεράτιζες τοῖς ποταμοῖς σου Lxx *Ez.* 32.2.

κεράτινος, add 'ον, τό, *vessel made of horn*, Gal. 13.139.5'

κεράτιον II, add 'abbrev. κρ *PVindob.Worp* 11.3, al. (vi AD)'

***κερατόμορφος**, ον, *horned*, or *horn-shaped*, epith. of Egyptian god, *APAW* 1943(14).8 (Chalcis, iii AD).

κερατουργός, for '= κερατοξόος' read '*worker in horn*' and add 'prob. *CPR* 7.50.16 (AD 636)'

κερατῶπις, add '*PMag.* 4.2548'

***Κεραυνιασταί**, οἱ, *association of worshippers of Zeus Keraunios*, *SEG* 35.1483.B27-8 (Antioch, i AD).

κεραύνιον II, add '2 some kind of design or pattern, πρόσωπον καὶ κύκλωι κεραύνια Inscr.*Délos* 1443Aii35, 1450A187.'

κεραύνιος I 1, at end for 'Call.*Aet.* 3.1.64' read 'Call.*fr.* 75.64 Pf.' and after '*VP* 17' insert 'without λίθος, Sm.*Ex.* 28.17' **II**, after '*Milet.* 1(7).278' insert '*SEG* 20.99 (Cilicia), 30.1617 (Cyprus, 44/31 BC); *POxy.* 885.44 (ii/iii AD), Paus.Ant. 10.2 J.; cf. Θεῷ μεγίστῳ κεραυνίῳ Βη‹το›υχιχι *SEG* 32.1445 (Syria, iii AD).'

***κεραυνίτης** λίθος, perh. = κεραυνία λίθος, Cyran. 26.

***κεραυνοβιᾶς**, ᾶ, ὁ, *mighty by thunder*, prob. in B.*fr.* 20E.7 S.-M., *Coll.Alex.* p. 84, vi 9 (cj.).

κεραυνοβολέω II, add 'pass., Sch.E.*Med.* 144'

κεραυνοβολία, add '*hurling of a thunderbolt*, *stroke of lightning*, *SEG* 7.980 (Arabia, iii AD)'

κεραυνός II, for 'as a name .. Arist. 6' read 'nickname of Ptolemy, (?) elder son of Ptolemy Soter and Eurydice, Plb. 2.41.2, Plu.*Arist.* 6; given διὰ τὴν σκαιότητα καὶ ἀπόνοιαν, Memn. 5.6 J.'

κεραυνοφόρος, after 'D.C. 55.23' insert '*SEG* 31.1300 (AD 145/6)'

Κερβεροκίνδυνος, delete the entry.

Κέρβερος, add '**III** kind of toad or frog, Sch.Nic.*Al.* 578 Ge., Hsch. s.v. λάλαγες (κεμβ- cod). **IV** = ὠχρός, Hsch.'

***κερβήσιος**, ὁ, *beer*, fr. Lat. *cervesia*, *Edict.Diocl.* 2.11.

***κερβικάριον**, τό, Lat. *cervicarium*, *pillow*, *POxy.* 1269.37 (ii AD), 921.8 (κερπ-; iii AD), *Stud.Pal.* 20.67.29 (ii/iii AD, κεβρ-), *BGU* 814.11 (iii AD); κ. θρόνου *PWisc.* 30 ii 14; as receptacle for dates, *PSI* 1331.12-14 (iii AD).

κερβολέω, add 'cf. σκερβόλλω'

κερδαίνω I 1, add '**b** *gain*, *win over* a person, *Ev.Matt.* 18.15, 1*Ep.Cor.* 9.19, 1*Ep.Pet.* 3.1.'

κερδαλέος 1 b, after 'Archil. 89.5' insert 'cf. ἄγρην κερδαλέην Opp.*H.* 2.119'; add 'of other creatures, κερδαλέαι .. σηπίαι Opp.*H.* 4.160'

***Κερδοῖον**, τό, *sanctuary of Apollo Κερδοῖος*, *SEG* 27.202.25 (Larissa, iii BC).

***Κερδοῖος**, ὁ, v. ‡κερδῷος.

κερδοσύνη, add 'sg. in epigr., *IGBulg.* 656 (Moesia, iv AD)'

***κέρδων**, ωνος, ὁ, attested by Lat. *cerdo*, *artisan*, Pers. 4.51, etc.

κερδῷος I, before '*IG* 9(2)' insert 'Thess. Κερδοῖος'; after '(Phalanna)' insert 'et al.'

***κερεάλιος**, α, ον, Lat. *cerealis*, οἱ κ. ἀγορανόμοι D.C. 43 (arg.), of an aedile, *SEG* 6.555 (Pisidia, ii AD).

Κερεάτᾶς, add 'also Κεραιάτης *SEG* 20.138 (Cyprus, iii BC)'

κερητίζει, delete the entry.

***κερητίζω**, prob. *play a game with a stick curved at the end and a ball*, Plu. 2.839c (cj. κελητ-, but v. *AD* 6 (1920/1) pp. 56-59). **2** κερητίζει· βασανίζει Hsch.

κέρθιος, for '*tree creeper, Certhia familiaris*' read '*short-toed tree-creeper, Certhia brachydactyla*'

κερκίς II 1, after 'Poll. 1.252' insert '*strut*, *IG* 2².1668.52' and delete '*hair-pin* or' **7**, delete the section. **IV 1**, add 'Call.*fr.* 284 Pf. (cf. *Suppl.Hell.* 287.7)'

κερκορώνος, delete 'perh. f.l. for κερκίων'

κερκούριον, delete 'only'; after 'Dim. of κέρκουρος' insert '*cargo vessel*, *BGU* 1933.2 (ii BC)'

κέρκουρος, line 2, delete 'esp. of the Cyprians; add 'invented by the Cyprians acc. to Plin.*HN* 7.208'

Κέρκυρα, add 'also *SEG* 23.474 (Dodona, iv/iii BC); Κέρκυρ, athematic form, Alcm.*fr.* 114 P.; adj. Κερκυραϊκός, ή, όν, Th. 1.118.1'

κέρκωψ 2, for '*knave*' read '*teller of false tales*'

κέρμα, line 1, for 'dub. l.' read 'prob.'

κερματίζω I, after 'Pl.*Men.* 79a' add 'οὐκ ἀναλίσκουσα τὸ ἐν οὐδὲ κερματίζουσα Plot. 5.5.4' **III**, add 'ἀργύριον κεκερματισμένον *small change*, Ar.*fr.* 215 K.-A.'

κερμάτιον, *cash*, add '*POxy.Hels.* 48.13, 19 (ii/iii AD)'

κερματιστής, add '*PUniv.Giss.* 30.11 (iii/iv AD)'

κέρνας, read 'κερνᾶς'

κερνί[ον], read '**κερνίον**' and add 'Theognost.*Can.* 123'

κερόεις, for 'Simon. 30' read 'Pi.*fr.* 107a.4'

***κέρρω**, v. ‡κείρω.

***Κέρσουλλος**, ὁ, *title of Zeus*, *IHadr.* 3.3, also Κέρσουσσος ib. 4.7 (ii AD).

***κερτόκορος**, ὁ, perh. f.l. for κερκόκορος, *brush* (cf. κόρος (C)), made from an animal's tail (κέρκος), *SB* 9834b.28 (iii AD).

κερτομέω, lines 4/8, delete 'σὲ δὲ .. Ph. 1235' and 'c. dupl. acc. .. 1.62' add '**2** *mock by false statement*, *make game of*, σὲ δὲ κερτομέουσαν δίῳ ταῦτ' ἀγορευέμεναι Od. 13.326, S.*Ph.* 1235, E.*Hel.* 619, *IA* 849; κοὔτι τυ κερτομέω Theoc. 1.62.'

***κέρχμων**, ονος, unexpld. wd. in Theognost.*Can.* 35.

κέρχνος (A), add 'also χέρχνος *IG* 1³.422.153 (Athens, v BC)'

κέρχνωμα, add 'also cj. in E.*Ph.* 1386'

***κεσσωνάριοι**, οἱ, app. Lat. *quaestionarii*, perh. for pap. κεσσωπαρ-, *SB* 2253.5 (*Berichtigungsl.* VII p. 180).

κεστρεύς, add 'κεστρεῖος (gen.) τῶ μέδονος *IGC* p. 99 B17'

κέστρος II, add '*SEG* 35.581 (Thessaly, ii BC)' add '**III** κέστρος· ὄχημα δίχα τροχῶν *Gloss.* II 200.5.'

***κευθμωνοδίτης**, ου, ὁ, *traveller in the underworld*, of Cerberus, *Suppl.Mag.* 42.3, 64 (iii/iv AD).

κεύθω, line 3, after '*Th.* 505' insert 'med., redupl. aor. κεκύθεσθε Call.*fr.* 238.6 Pf.'

κεφάλαιος II 1, add '**b** transf., *highest point*, τὸ κεφάλαιον τοῦ ἀντιπροσώπου θεάτρου X.*Eq.Mag.* 3.7.'

κεφαλή I 1 b, line 5, for 'in Archit., *upright*' read '*along the top*'; add 'also ἐπὶ κεφαλῆ Hyp.*Lyc.* 17' **d**, lines 5/6, delete 'οὐ βουλόμενος .. *Lyc.* 17' **2**, lines 7 ff., delete 'periphr. .. sense' and 'ἡ κ. .. (iv AD)' inserting 'periphr. .. Hdt. 9.99' after 'Vett.Val. 74.7' add '**b** esp. as the unit on which personal taxation and compulsory service are imposed, ὑπὲρ συντερίας (i.e. συντελείας) τῆς καιφαλῆς *POxy.* 1331 (v AD); τὰ δημόσια τῆς αὐτοῦ ib. *PLond.* 1793 (v AD); τὰ ναύβια τῆς κ. αὐτοῦ καὶ τῶν αὐτῶν *PRein.* 57.8 = Wilcken *Chr.* 390; iv AD). **3**, add 'κ. κινδυνεύσει *PCair.Isidor.* 1.19 (AD 297)' add '**5** in leg. sense, = Lat. *caput*, *IGLS* 718.61 (i BC), see also °κατάκριτος.' **II b**, last line, after '*origin*' insert 'Hp.*Coac.* 498' **c**, add '*crest* of wave, *Peripl.M.Rubr.* 46' **e**, for '*PPetr.* 3 p. 72' read '*PPetr.* II p. 121' add '**f** *summit*, of mountains, Orph.*H.* 36.16, Q.S. 7.558.' **V 3**, add 'Philostr.*VA*

4.32' **4**, for 'band of men' read 'column of troops' add '**6** capital city, Μετούλον, ἥ τῶν Ἰαπόδων ἐστὶ κεφαλή App.Ill. 54.'

*κεφᾰληγόνος, ον, producing heads, Nic.fr. 74.25 G.-S.

*Κεφαλήν, ῆνος, ὁ, epith. of Dionysus in Lesbos, Paus. 10.19.3.

κεφᾰλίζω, add 'POxy. 2339.6 (i AD)'

κεφᾰλικός **IV**, delete the section.

*κεφᾰλιόω, hit on the head, Ev.Marc. 12.4 (v.l., v. κεφαλαιόω II).

*κεφᾰλιτιόνη, ἡ, = °κεφαλιτίων, Cod.Just. 10.16.1.

*κεφᾰλιτίων, ωνος, ἡ, (cf. Lat. capitatio, also capitum, κᾰπητόν) food or fodder allowance (sts. in cash), Cod.Just. 12.37.19.3.

Κεφαλληνία, add 'Dor. Κεφαλλᾱνία, SEG 23.189 (Argos, iv BC)'

κεφᾰλοδέσμιον, after 'τό' insert 'Edict.Diocl. 28.7, SB 9746.31 (iv AD)'

*κεφᾰλόω, pass. pf. part. κεκεφαλωμένος, headed, having a head, Simp. in Cat. 187.36.

*κεφᾰλώτιον, τό, dim. of κεφαλωτόν, head of leek or other vegetable, PRyl. 627.81, al. (iv AD).

κεχρημένος, for 'χράω C vi' read 'χράω (B) C i'

*κεχρημένως, adv., (χράομαι) w. dat., in a manner requiring, ἐπιτεύγμασι Phld.Po. 5.27.

κεωρεῖν, add 'κνεωρεῖν La.; cf. κνέωρον ii'

κῆβος, add 'cf. cephos, Plin.HN 8.70'

*κήδαρ· πένθος Hsch.

κηδεμονεύω, after 'guardian' insert 'EAM 19.5-6 (Maced., ii/iii AD)' and before 'παίδων' insert 'spec., be a tutor'

κηδεμονία, after 'ἡ' insert 'Aeol. καδ- SEG 32.1243' add '**b** as legal t.t., the position of a tutor or curator, Modest.Dig. 19.2.49 pr., Cod.Just. 1.4.30 pr., Just.Nov. 72.2; of a curator alone, ib. 72.3.'

κηδεμών, line 1, after '(κήδω)' insert 'Aeol. καδ- SEG 32.1243' **I 2**, line 2, after 'X.Mem. 2.7.12' insert 'Men.Dysc. 737, al.'; lines 5/6, for 'of a legal guardian, POxy. 888.2 (iii/iv AD)' read 'legal t.t. for a guardian, encompassing tutor and curator, Modest.Dig. 26.6.2 pr., 4, POxy. 888.2 (iii/iv AD); tutor alone, Modest.Dig. 26.6.2.1, Just.Nov. 72.2'

κηδεστής, add '**4** καδεστάς in Crete, relative on the mother's side, Leg.Gort. 2.18, 2.29 (v BC).'

κηδέστρια **2**, for this section read 'mother (or daughter)-in-law, INikaia 557.7, Gloss.'

*κηδευτικόν, τό, funeral fund, SEG 30.1535 (Lycia, after AD 152).

κῆδος **I 2 a**, after '(v. infr. ii)' insert 'Pi.O. 1.107, N. 1.54 S.-M.'; add 'Archil. 128.1 W., Pi.I. 8.7 S.-M., B. 19.36 S.-M., Thgn. 656 W.' **b**, delete 'cf. (Pi.)N. 1.54'

κηδοσύνη, add 'sg., anxious care, solicitude, Sch.E.Or. 1017'

κήδω, line 5, pf. κέκηδα, for '(in pres. sense)' read '(in sense of pres. med.)' **II 2**, add 'SEG 32.1149 (AD 209). **b** care for as a tutor, Just.Nov. 94.1.'

*κηθάριον, τό, cup or small basket from which balls, pebbles, or sim. were poured into a voting-urn, Ar.V. 674, cf. Sch.ad loc.

κηθίς, add 'Myc. ka-ti, a type of jug'

κῆλας, read 'κηλᾶς'

*κήλεα, τά, = κᾶλα (v. κᾶλον), κήλεα νηῶν, ship's timbers, Hes.fr. 314 M.-W. (cj. κήελα).

κηλέω, at end after 'in good sense' for Pl. quot. read 'Trag.adesp. 566 K.-S.'

Κηληδόνες, delete 'but harmless'; for 'Pi.fr. 53' read 'Pi.Pae. 8.71'

*κηλησμός, ὁ, = κηληθμός, Hdn.fr. p. 24 H.

*κηλητής, ὁ, beguiler, Hippon. 79.15 W. (κ[η]λητ[ῇ]).

κηλικτᾶς, after lemma insert 'or κηλίκτας'; add 'unless δεικηλίκτας is read'

κῆλον, for 'shaft .. arrow' read 'means by which a god's miraculous power is manifested or deployed'; line 4, after 'Hes.Th. 708' insert 'cf. id.fr. 204.138 M.-W.'; delete 'metaph.' and line 6 fr. 'also' to the end; add 'perh. of Bacchic madness, Euph. in Suppl.Hell. 430 ii 18'

*κηλοτομέω, operate on for hernia, Gal. 1.197.13 (pass.).

κήλων, after '(κῆλον)' insert 'Dor. κάλων, AE 1948/9.136 (Epid., iv BC)' **I**, delete 'PLond. 1.131ʳ.303 (i AD)'

κηλώνειον, line 2, after 'Aen.Tact. 39.7' insert 'Men.Dysc. 536 (pl.)'

*κηλωνοστάσιον, τό, dim. of °κηλωνόστασις, PBerl.Leihg. 13.14, PLond. 131ʳ.303 (i AD).

*κηλωνόστᾰσις, εως, ἡ, support or base for a swing-beam (κήλων ι), SB 12524 (17 BC).

κημός **II**, add '**3** funnel-shaped device on a fire-thrower, Plb. 21.7.1, 4.'

κηνεῖ, Dor. adv. there, APAW 1928(6).6 (Cos, iv BC).

*κηνσιτορεύω, serve as censitor, SB 8246 (AD 340), PColl.Youtie 78.8 (AD 342).

κῆνσος, for '= Lat. census' read 'assessment (for tax purposes)'; after 'Ev.Matt.' insert '17.25' **II**, delete the section.

*κηνσουάλιος, ὁ, Lat. censualis, MAMA 3.29, 206 (Cilicia), Just.Nov. 17.8 pr.

κηνῶ, (s.v. κῆνος), for 'ἐκεῖθεν' read 'ἐκεῖ'

κήξ, delete 'and καύηξ .. q.v.)'; add 'transf., καύαξ· πανοῦργος Suid.; cf. λάρος'

κηπαῖος **II 1**, after 'back-door' insert 'IG 1³.425.42 (Athens, v BC)'

*κηπεῖον, τό, garden, SEG 33.168 B fr. a.1, 6 (Athens, iv BC), al.

*κηπεργός, ὁ, gardener, MAMA 3.348, 687 (Corycus).

*κηπικός, ὁ, gardener, BGU 1151.42, 51 (i BC).

*κηπίων, ωνος, ὁ, name of a musical nome, Plu. 2.1132d.

*κηποκόμος· κηπουρός Hsch.

κῆπος **I**, line 4, for 'also of heaven' read 'prob. the garden of the Hesperides'; line 5, transpose 'cf. Pl.Smp. 203b' to follow 'S.fr. 320 (lyr.)' and for 'eastern' read 'northern' **III**, add 'ὁ ἀφροδίσιος κ., Archipp. 2 D.'

κηποτάφιον, for 'tomb in a garden' read 'garden with a tomb'

κηπότᾰφος, after '(Ilias)' insert 'IUrbRom 836'; add 'SB 9801.2 (ii/iii AD)'

*κηπυριστής, wd. without context in SEG 9.824 (Carthage). (Prob. for κηπουρ-; cf. κηπουρός).

Κήρ, line 1, after 'acc. Κῆρα' insert 'Dor. Κάρ, epigr. in BCH 116.599 (Ambracia, vi/v BC), acc. κᾶρα, Alcm. 88 P.' **II 2**, add 'οὐκ ὀλίγας κῆρας ἐν τῷ βίῳ διώσεαι, φθόνον καὶ ζῆλον καὶ δυσμενίην Democr. 191, 285 D.-K.'

κῆρ, line 2, before 'Trag.' insert 'in Pi., B., and'

*κηρᾱφίς, ίδος, ἡ, some kind of sea-creature, prob. crustacean, Nic.Al. 394; κ.· κάραβος Hsch.

κηριαπτάριον, (s.v. κηριάπτης), add 'container for wax, PVindob.G 25737 (ZPE 64.75; vi/vii AD)'

*κηρίνθη, ἡ, Lat. cerintha, honeywort, Verg.G. 4.63, Plin.HN 21.70 (-e).

κήρινος **II 1**, after 'as wax' insert 'Hippon. 79.5 W.'

*κηρίολος, ὁ, Lat. cereolus, wax taper, IEphes. 2227, 3216.5 (ii AD); gloss on κηρίνη φρυαλλίς Hdn.Philet. 215 D.

κηρίον **I 2**, after 'AP 9.190' insert 'Babr.Prooem. 18' **II**, add 'Hp.Aff. 35'

*κηριτρόφος, ον, (κήρ) death-breeding, deadly, ὄφις Nic.Th. 192 (dub.; v.l. κηρο-).

*κηρογρᾰφεῖον, τό, implement for writing on wax, stylus, Babylonian Talmud, Kelim, ch. 16 p. 48 (in Hebr. letters).

*κηροδόχος, ὁ, honeycomb, Hsch. s.v. σμήναι.

κηρόκλυστος, add 'perh. Cypr. ke-ro-ku-lu-su-to-se ICS 208 (addenda)'

κηροπλάστης, add 'PCair.Zen. 782(a).64 (iii BC), Plot. 3.8.2'

κηροπλαστικός, add 'κηροπλαστικοῖς .. τύποις Procl. in Prm. 841.28'

κηροποιός, for 'Sch.Ar.V. 1075' read 'Sch.Ar.V. 1080'

κηρός, line 4, after '(IG) 14.1320' insert 'Anacreont. 16.34, 17.25 W.'; after 'writing-tablets' insert 'Hdt. 7.239, AP 4.1.10 (Mel.), 7.36 (Eryc.)'

κηροτρόφος (A), after 'Nic.Th. 192' add '(v.l.)'

*κηρόω (B), (?κήρ) pass. pf. part. κεκηρωμένη· κεκακωμένη Hsch.

κήρυγμα, line 1, after '(κηρύσσω)' insert 'Aeol. κάρυγμα SEG 32.1243'

κηρῡκικός, add 'ἑκατοστὴ κηρυκικῶν, = κηρύκειον II, OBodl. i 41 (ii BC): -κόν, τό, perh. = κηρύκειον ii, Inscr.Délos 1408Aii50 (ii BC)'

*κηρῡκίσκος, ὁ, perh. crier, ephebic official, IG 2².1723.7 (Athens, i AD, w. new fr., SEG 26.166.18).

κῆρυξ line 1, after 'Pi.N. 8.1' insert 'also Dor. IG 4².1.102 (Epid., iv BC), etc., Cypr. ka-ru-xe, alphabetic Κάρυξ pers. n., ICS 260' **I 4**, add 'τοὺς ἐρινάζοντας τοὺς ἐρινοὺς κήρυκας λέγουσι Hsch. s.v. κήρυκες' at end add 'Myc. ka-ru-ke (dat.)'

κηρύσσω **II 1**, lines 2/3, delete 'κ. τινά .. Ar.Ach. 748'

κήρωμα **2**, after 'mud or clay' insert 'mixed with oil'

κηρωμᾰτιστής, before 'Sch.Ar.' insert 'suggested as equiv. of παιδοτρίβης'

*κηρωμᾰτίτης, ου, ὁ, prob. the same as κηρωματιστής, Edict.Diocl. 7.64 (-είτης), MAMA 8.605 (κηρο- lapis).

κηρών, for 'bee-hive' read 'unexpld. deriv. of κηρός' and add 'POxy. 3412.6 (c.AD 360)'

κήρωσις, add '**II** waxing, a process applied to statues (cf. γάνωσις 1), εἰς τὴν κ. τοῦ ἀνδριάντος Inscr.Délos 290.130 (iii BC).'

*κηρωψός, ὁ, wax-maker, IGLBulg. 106 (vi AD, cf. °μυροψός).

κήτειος **II**, add 'Ἀλκαῖος δέ φησι τὸν Κήτειον ἀντὶ τοῦ Μυσόν Sch.Od. 11.521 (= Alc. 413 L.-P.)'

κητώεις, for 'perh. full of ravines' read 'full of sea-creatures'

κηφήν, line 2, for 'vagabond' read 'parasite'; line 3, after 'Ar.V. 1114' insert 'Timo in Suppl.Hell. 840.8' and after 'Pl.R. 552c' insert 'κηφήνων βοτάνη ib. 564e'; line 4, delete 'Plu. 2.42a'

*Κηφηνίδης, name for a rich idler, Phld.Rh. 1.236 S.

*κῆφος, v. °κῆβος.

*κιβαριάτωρ, ορος, ὁ, fr. Lat. *cibariator, army official issuing food and wine to the troops, SB 9457.2 (ii AD), OWilck. 1265 (ii AD; κιβαράτ-), 1142 (iii AD), SB 9230.3 (iii AD).

*κιβάριον, τό, Lat. cibarium, provisions, PLond. 1159.8 (ii AD); pl. IG 4²(1).92.10 (Epid., iii/iv AD).

*κιβάριος, ον, κ. ἄρτος, household bread, IEphes. 910, 3010, PPetaus 45.1 (ii AD); sim. ψωμίων κιβαρίων PRyl. 627.71.

κιβδηλιάω, for '*look like adulterated gold*' read '*be pale* or *covered with* °*κίβδηλις*'

***κίβδηλις**, εως, ἡ, *dross* of metals, Hsch. s.v. κιβδηλιῶντας (-ίς cod.), *EM* 512.53.

κίβδηλος II, add 'κ. φίλος *Trag.adesp.* 638.17 K.-S.'

κίβῑσις, add 'Alc. 255.3 L.-P.' and for 'Call.*fr.* 177' read 'Call.*fr.* 177.31, 531 Pf.'

***κίβον**· ἐνεόν. Πάφιοι Hsch.

κιβώριον, add '**III** *tomb*, *MAMA* 6.339 (Acmonia; κιβώρεν); prob. to be read in *IGLS* 684 for κ(αὶ) Ἰβο[υ]ίου.'

κιβωτός, at end after 'opp. κίστη q.v.' insert 'perh. *sarcophagus*, *Explor.arch.de Délos* 30.148 (i AD)'

κίγκλος, for '*dabchick*, *Podiceps ruficollis*' read '*wagtail*, *Motacilla*'

κιθάρα I, add '2 the constellation *Lyra*, Nonn.*D.* 8.388, 13.359.'

κιθαρίζω, line 6, after '*Nu.* 1357' insert 'distd. fr. ψάλλω as using a plectrum, opp. the fingers, *SIG* 578.18 (Teos, ii BC)'

κιθάρισμα, add '2 = κιθάρισις, *SPAW* 1934.1040 (tab.defix., Boeot.).'

κιθαρισμός, add 'distd. fr. ψαλμός, *SIG* 959.10 (Chios, ii/i BC)'

κιθαριστής I, after lemma insert 'non-Att.-Ion. -τάς *SEG* 35.989 (Knossos, ii BC); add 'distd. fr. ψάλτης, *SIG* 578.15 (Teos, ii BC)'

κίθαρος II, add '*IGC* p. 99 B14 (Acraephia, iii/ii BC)'

***κιθαρῳδίστρια**, ἡ, *woman who plays and sings to the cithara*, Heuzey-Daumet *Mission Arch. de Macédoine* no. 10.

***κιθαρῳδός I**, after 'as fem.' insert 'Plu. 2.397a, 972f (Glauce)'; after 'Alciphr. 3.33' add '*AP* 5.98'

***κιθωνάριον**, τό, dim. of κιθών, *small tunic*, *IEphes.* 4106.3 (iv AD).

***κίκερροι**· ὠχροί. Μακεδόνες Hsch.

***κικεών**, ῶνος, ὁ, *castor-bean*, *Ricinus communis*, Aq.Thd.*Jn.* 4.6.

***Κικήλια**, τά, festival at Alexandria (equiv. of the Saturnalia, acc. to Epiph.Const. p. 482 Dindorf), *OGI* 56.64 (iii BC).

κίκι, after 'κίκιος Hdn.Gr. 2.767' insert '*SB* 9667.3 (iii BC), al.'; add '(cf. Hebr. *qiqāyôn*)'

***κίκιννα**· τριχόπλαστοι Hsch. (κικινᾶς· τριχοπλάστης La.).

κικιοφόρος, add '*PCair.Zen.* 629.3, 5 (iii BC)'

***κικίς**, ίδος, ἡ, *castor-oil berry*, κ. φρυγείσας ἐμβάλλοντες τῷ οἴνῳ *Gp.* 7.12.9 (v.l. κηκίδας).

***κικκάμη**, ἡ, *screech-owl*, Gloss.; cf. κικυμίς.

***κικκάς**, perh. = κίκκαβος, Hsch. s.v. οὐ μάλα κικκάς (cf. Stratt. 10).

***κικουργικός**, ή, όν, *used in making castor-oil*, ὄργανον *CPap.Jud.* 452*b*.4 (ii AD).

κικυμίς, delete the entry.

***κικυμωΐς**, ίδος, ἡ, *screech-owl*, Call.*fr.* 608 Pf.; also -μηΐς (-μωνίς La.) Hsch.; cf. κίκυμος Lat. *cicuma*.

***Κιλαδηνός**, ὁ, cult-title of Asclepius in Thrace, *SEG* 30.720, 725, 742, etc.

***Κιλικάρχης**, ου, ὁ, *president of the provincial council of Cilicia*, *SEG* 26.1457.10 (Tarsus, ii/iii AD).

Κιλικαρχία, thus, not Κιλικιαρχία.

Κιλίκιος, add 'neut. pl. *Cilician cloths*, *PMich.*inv. 4001 (*ZPE* 24.84, i/ii AD)'

κιλλακτήρ, add 'Hsch.; cf. pr. n. Κιλλάκτωρ (-κτήρ edd.) *AP* 5.28, 44'

***κιμαῖος**, ὁ, or -ον, τό, *mulberry juice*, τῶι κιμαίωι Hippon. 78.14 W.; cf. κιμαί, κιμαός.

***Κιμιστηνός**, ὁ, title of Zeus, *SEG* 25.825 (Dacia, ii/iii AD), al.

***κιναίδιος**, name of a plant, a fish, a bird, a stone, Cyran. 24.

***κιναιδολόγος**, add 'Varro ap.Non. p. 79 L.'

κίναιδος I 2, for '*CIG* 4926 (Philae)' read '*IPhilae* 154'; add 'Plb. 5.37.10'

***κιναῖος**, ὁ, (or adj. ος, α, ον?), κ. κάλαμος kind of reed, *PRyl.* 583.15, 61 (ii BC).

***κινάρᾶς**, ὁ, *artichoke-seller*, *ZPE* 65.91 (*SB* 12497.39).

***κιναφεύειν**· πανουργεύεσθαι Hsch.

***κινδαύει** (κινδάνει cj.)· κινεῖται, κερατίζει Hsch.; cf. κίνδαξ.

κινδῡνεύω 2, w. preps., add 'w. εἰς, *Cod.Just.* 10.27.2.4, 12.60.7.7; w. περί and acc., ib. 1.40.13; w. ἐπί and dat., Just.*Nov.* 2.2.1' **4**, add 'w. fut. inf., ἡγούμην .. κινδυνεύειν .. πάντων ἐνδεὴς γενήσεσθαι Isoc. 17.6' add '**6** *to be on trial*, D. 30.2, 47.5, etc., Hyp.*Lyc.* 10, Is. 1.6, etc.'

κίνδῡνος 1, line 15, for 'A.*Th.* 1033' read 'A.*Th.* 1048' **2**, add 'metaph., κ. as a throw of the dice, ἀναρρῖψαι κ. παρὰ τὸ ἀναρρῖψαι κύβον .. Phryn.*PS* 29.1' add '**II** naut., ἡ ἐν πρώρᾳ σελίς Hsch.'

κῑνέω I 1, add '(ellipt.) *move* (troops, household, etc.), Plb. 2.54.2, 9.18.6; ἐκίνησεν ἐκεῖθεν Ἀβραάμ Lxx *Ge.* 20.1' add '**5** *rack off*, *decant* wine, *POxy.* 1631.17 (iii AD), al., *SB* 9778.9 (vi AD)' **II 5**, last line, for 'Alc. 82 (s.v.l.)' read 'Alc. 351 L.-P. (πύκινον codd.), cf. κίνεις πάντα λί[θον id. 306(14) ii 31, v. λίθος v, πέτρος I'

κίνηθρον, delete the entry.

κινησίγαιος, after 'ον' insert '*earth-shaking*, *PMag.* 4.1356'

κινητήρ, for '= κινητής' read '*shaker*, of Poseidon'

***κῑνητήριον**, τό, colloq. wd. for *brothel*, Eup. 99.27 K.-A.

***κῑνητιάω**, = βινητιάω, Men.*Dysc.* 462, *Vit.Aesop.*(G) 32 (but βιν- ibid. W).

κινητικός I 1, add '**b** *suited to movement*, of iambics and trochaic tetrameters, Arist.*Po.* 1460ª1.'

κιννάβαρι, for 'τεγγάβαρι' read 'τιγγάβαρι'

κινναμωμοφόρος, add 'Hld. 9.16, 19'

***κῑνόφθαλμος**, ον, *rolling the eyes suggestively*, cited in support of reading κινάμυια (for κυνάμυια), Sch.Gen.Il. 21.394.

κίνύρομαι 1, after '*lament*' insert 'A.*fr.* 47a.ii 6 R.'

***κιονοκέφαλον**, τό, *capital of a column*, *CIJud.* 781.5 (Side, iv AD).

***κίρκινος**, ὁ, Lat. *circinus*, *pair of compasses*, Gal. 1.47.4.

***κιρκίτωρ**, ορος, ὁ, Lat. *circitor*, *inspector* of frontier posts, *SEG* 9.356.35 (Ptolemais in Cyrenaica), 32.1554 (Arabia, both vi AD).

κιρροειδής, after 'Apollod.Hist. 214 J.' insert '(v.l. κηρο-)'

***κίρρος**· ὀρός (cod. ὄρος), καὶ αἷμα, καὶ πόμα γάλακτος· Λάκωνες Hsch.

κιρρός, add '**b** subst. κιρρόν, τό, app. as the colour of gold, Aq.*Pr.* 8.19, *Is.* 13.12 (Lxx τὸ χρυσίον τὸ ἄπυρον).'

***κιρρότης**, ητος, ἡ, *orange colour*, Physici 2.295.22.

***κιρσουλκός**, ὁ, *sharp-pointed hook* used in operation for varicocele, ἄγκιστρα τῶν σφόδρα μικροκαμπῶν, καλουμένων δὲ κιρσουλκῶν γαμμοειδῆ κατὰ τὴν καμπήν Orib. 45.18.5, Gal. 14.790.15.

***κίρυλος**· ἰχθύς ποιος. καὶ ὀρνέου εἶδος Hsch.; cf. κηρύλος.

***κίρσιον**, τό, *type of thistle with soft spikes* used for treating varicose veins, Dsc. 4.118, Plin.*HN* 27.61.

***κίρων**· ἀδύνατος πρὸς συνουσίαν. καὶ αἰδοίου βλάβη. καὶ ἀπεσκολυμμένος. καὶ κυρίως μὲν ὁ σάτυρος, καὶ ἐντεταμένος, ὁ γυναικίας, καὶ μὴ δυνάμενος χρῆσθαι Hsch.

***κίσευρις**, ἡ, some kind of herb, *PColl.Youtie* 87.8 (vi AD).

κίσηρις, after 'κίσηλις' insert '*AE* 1948/9.137 (Epid., iv BC)', *PSI* 1180.34 (ii AD)'; line 3, after 'Choerob.*in Theod.* 1.319 H.' insert '-ιος, *SEG* l.c.'

***κισηροειδής**, ές, *resembling pumice-stone*, Thphr.*HP* 3.7.5; of heavenly bodies, as in some sense porous, Diog.Apoll. in *Placit.* 2.13.5, 2.20.10, 2.25.10; adv. -δῶς (s.v.l.), Epicur. in *Placit.* 2.20.14.

κίσθος, line 2, after '*HP* 6.1.4, 6.2.1, 2' insert 'Theoc. 5.131'

κίσιρνις, add 'cf. κίσσιρις· εἶδος ὀρνέου, Suid.'

κίσσινος, after '*of ivy*' insert 'στέφανοι Pi.*Dith.* 3.7'

***κισσῖτις**, ιδος, ἡ, precious stone with markings resembling ivy-leaves, Plin.*HN* 37.188.

κισσοφόρος, add '**3** of a coin, *carrying a device of ivy*, *Inscr.Délos* 1449*Aa*bii23, 1450.102 (both ii BC).'

κισσόφυλλον I 1, after '*ivy-leaf*' insert 'τροχίσκ[οι δύο ἔχοντε]ς κιττόφυλλ[α of decoration, *SEG* 34.95.46 (Athens, ii BC)'

κισσοχαίτης, add '*SEG* 32.552.2 (Delphi, iv BC)'

***κισσύβιον**, τό, = κισσύβιον, *IG* 2².1424a.265 (Athens, iv BC).

***κισσώδης** (B), ες, *decorated with ivy-leaves*, Nonn.*D.* 1.17.

***κίσταρχος**, ὁ, *official in charge of the κίστη*, emendation fr. κτισαρχ-, *BCH* 14.538 (pl.).

κισταφόρος, add 'as fem., *AJA* 37.246 (Latium, ii AD)'

***κιστέρνα**· λάκκος φρέατος Hsch.; sp. κινστ- Just.*Nov.* 159 pr. (Lat. *cisterna*).

***κιστιόκοσμοι**, οἱ, app. functionaries who arranged baskets for religious ceremonies, *SEG* 39.381 (Messene); also to be read in *SEG* 23.209.2, 210.2 (both Messene, iii BC).

***κιστιπλινθουργός**, ὁ, app. *maker of some kind of brick*, *Stud.Pal.* 8.909.2 (*ZPE* 77.186).

***κιστοπλόκιον**, τό, *basket-maker's shop*, *PAmst.*inv. 62 (*Mnemosyne* 30.146).

κιστοφόρος I, after 'processions' insert '*Sardis* 7(1).195 (i BC)' **II**, add 'κ. τέτραχμον *Inscr.Délos* 1443*Ai*149 (ii BC), *SEG* 31.983 (ii/i BC)'

***κιτάτωρ**, ὁ, Lat. *citator*, *summoner*, *PHamb.* 39.59 (p. 173) (ii AD).

***κιτρειαβολή**, ἡ, (= κιτρια-), *waving of the ethrog* (citron), a ritual act in the Jewish Feast of Tabernacles, *SB* 9843.9 (ii AD).

***κίτρις**, εως, ἡ, *citrus-fruit*, Al.*Le.* 23.40.

***κίττανος**· ἡ κονιακὴ τίτανος Hsch.

κίχλη II, after 'Epich. 60' insert '(Dor. κίχλᾱ)'; add 'cf. ἴχλα'

κιχλιδιάω, add 'v. χλιδιάω'

***κιχυβεῖν**· δυσωπεῖν Hsch.; cf. κίκυμος, κίκκυβος.

κίων, line 3, before 'al.' insert 'Cypr. *ta-se-ki-jo-na-u-se*, app. τὰς κίjονᾱυς (acc. pl.) *ICS* 10a (addenda)' **I**, add '**3** of persons, Νάξου κίονας [Archil.] 325 W.; Τροίας κίονα, of Hector, Pi.*O.* 2.90.' at end add 'Myc. *ki-wo*'

κλᾰδᾰρός, for 'δοράτια' read 'δόρατα'

κλᾰδευτήριον II, add '*cut branches*, Gloss.'

κλάδιον, add 'as *sticks* for burning, *Edict.Diocl.* 14.12'

κλάζω 2, for '*bark*' read '*howl*' and add 'Nic.*Th.* 674'

κλακοφορέω, hold office of κλακοφόρος (of a woman), *SEG* 36.558 (Illyria, iii/ii BC).

κλᾳκτός, read **κλαϊκτός**; after κλειστός insert 'closed, locked'; add 'cf. κλῄζω (B) and forms fr. κλᾴζω s.v. κλείω'

κλάμμα, for pres. ref. read 'Alc. 119.11 L.-P.'

κλανίον, add 'also **κλάνον** or **κλανόν**, *PHamb.* 10.46 (ii AD), *POsl.* 46.9 (iii AD); cf. χλάνος, °χλανίαι'

κλαουικουλάριοι, οἱ, Lat. *clavicularii*, officials of some kind, *POxy.* 2050.3 (vi AD), *SB* 2254.3.

κλαρία· κλήματα ἀμπελόφυλλα Hsch.; cf. κλάριοι.

Κλάριος, before '*Rev.Phil.* 22.268 (iii AD)' insert '*SEG* 33.973 (Notion, iii/ii BC), ib. 26.1288.9 (Klaros, ii AD), al.'; after 'hence' insert '**Κλάρια**, τά, name of a festival, *SEG* 4.479 (Colophon, iii BC)'

κλᾱρογράφέω, v. °κληρο-.

κλᾱροπάληδόν, v. °κληρο-.

κλάσσα, ἡ, Lat. *classis, fleet*, *BGU* 455.8 (i AD), 2492.7 (ii AD), *PColl. Youtie* 53.7 (ii AD), *IGRom.* 1.623 (Tomi, iii AD), al.

κλασσικός, ή, όν, Lat. *classicus*, *naval*, *SEG* 34.1243.24, al. (Abydos, v AD).

κλαστήρ, ῆρος, ὁ, = κλάστης, *vine-dresser*, *CPR* 10.56.3 (v AD).

κλαστήριον, add '**II** ξύλινα· κλίνης πόδες δύο καὶ κλαστήριον *Inscr.Délos* 1452.*A*30 (ii BC), where κλασ- may be for κρασ- v.s.v. II.'

κλαυδιανός, ὁ, *alloy of copper and lead*, Anon.Alch. 14.6, 24.3.

Κλαύδιος, α, ον, *Claudian*, λεγ(ιῶνος) ιαˊ Κλαυδίας *SEG* 33.1194 (i/ii AD).

κλαύθομαι, *cry, weep*, epigr. in *PTeb.* 3.7.

κλαυμῡρίζομαι, for 'κλαυμαριόμενον' read 'κλαυμυριόμενον'

κλαυστικός, after '*mourning*' insert 'τὸ κ. opp. τὸ γελαστικόν, David in Porph. 203.19'

κλάω (A) **3**, for '*enfeebled* eyes' read '*drooping eyelids*'

κλειδᾶς, add '*MAMA* 3.689 (κλιδ-, Corycus)'

κλειδοποιός, after lemma insert '*PPetr.* II 39d 15 (iii BC); add '*SEG* 31.1517 (sp. κλιδ-; Egypt, Chr.)'

κλειδουχέω I, for '*to be her priestess*' read '*to be the guardian of her temple*' **II**, delete the section.

κλειδουχικός, ή, όν, *of* or *for the* κλειδοῦχος, κλείς *Inscr.Délos* 1442*B*56, 1443*B*i163, 1444*A*a47 (ii BC).

κλειδοφορία, ἡ, *office of key-bearer*, τῆς Ἑκάτης *BCH* 51.97 (Panamara).

κλείδωσις, add '*AS* 9.113 (Pisidia)'

κλεῖθρον II 4, for 'Κλάθροις' read 'Κλᾱίθροις' and for '*Mnemos.* 42.332' read '*SEG* 33.336'

κλεινοστατέω, perh. *close the lock-gates*, *PRein.* 117.8 (iii AD); cf. κλεινία (perh. = κλειδ-).

κλείς, line 2, after 'contr. κλεῖς, v. infr. III' insert 'also I 3 (Suppl.)'; line 8, for 'E.*Med.* 212 (anap.)' read 'E.*Med.* 212 (lyr.)' **I 3**, add 'acc. pl. κλεῖς *Inscr.Délos* 1450.199 (ii BC), *POxy.* 729.23 (ii AD), al.' **IV**, for '*rowing bench* in a ship' read '*hooked thole-pin*' at end add '(*κλᾱϜίς; Myc. compd. *ka-ra-wi-po-ro*)'

κλεισιάδες, at end for 'Usu. .. but' read 'written κλισ- *IG* 1³.425.42 (Athens, v BC) and often in codd.'

κλεισούρα, after 'ἡ' insert '(after Latin *clausura*)'

κλειτός, add 'Cypr. *ke-le-wi-to* κλεϜίτō, *ICS* 402'

κλείω (A) **III 2**, delete the section.

Κλεοπατρεῖον, τό, *shrine of Cleopatra*, *Bull.Soc.Alex.* 10.27 (i BC), Fraser *Ptol.Alex.* II 379 no. 319.

κλέος, line 2, for 'only nom. and acc. sg. and pl.' read 'usu. nom. or acc. (sg. and pl.); gen. sg. κλέους Antiph. 163.3, Corn.*ND* 14'; add 'cf. Myc. compd. n., as *e-te-wo-ke-re-we-i-jo* (possess. adj.)'

κλέπος, add '*Iamb.adesp.* 56 W.'

κλέπτω, line 6, for 'Ar.*V.* 57' read 'κέκλαμμαι Ar.*V.* 57 (sch.), Ath. 9.409c (cod. A), Choerob. *in Theod.* 2.188 H.'

κλέψημα, gloss on κοία, Hsch.

κλεψίαμβος, before '*musical instrument*' insert '*stringed*'

κλεψιγαμέω, gloss on κλοτοπεύειν, Hsch.

κλεψιγαμία, ἡ, *illicit love*, Sch.Luc.*DMeretr.* 7.4 p. 280.16 R., Eust. 152.3.

κλεψίγαμος, add 'epigr. in Robert *Ét.Anat.* p. 97 (Troas)'

κλεψικοίτης, delete 'Ismenias ap.'

κλεψιμαῖος, adv. -αίως, add '*PMich.*x 581.7 (*ASP* vol. 6, ii AD)'

κλεψιποτέω, for '*drink unfairly*' read '*drink secretly*' and add 'epigr. in *IEphes.* 1062.7 (i AD)'

κλεψίρρυτος, ον, *secretly flowing*, κλεψίρρυτον ὕδωρ (?Call. fr. 771 Pf.)· τὸ τῆς κλεψύδρας Hsch. (v. κλεψύδρα).

κλεψοπασχῑτης, ὁ, *one who celebrates Easter in secret*, Leont.Byz.*HP* 2.2.10.

κλεψύδρα III, after 'Ar.*Au.* 1695 (lyr.)' insert '*Lys.* 913'

κλέω (A), line 8, before 'E.*fr.* 369.7' insert 'prob. in'

†**κληδονισμός**, ὁ, *soothsaying*, Lxx *De.* 18.14 (v.l.), *Is.* 2.6 (both pl.).

κλήδρος, ὁ, = κλήθρα, *alder*, Gloss.

κλῄζω (A), line 6, before 'ἔκλεισα' insert 'ἐκλήισσα *Suppl.Hell.* 953.16, Maiist. 38' and after '(Etruria)' insert 'κλῆξα Orph.*A.* 1004'; line 8, after 'Man. 6.571' insert 'aor. 3 pl. ἐκλήϊχθεν Orac.ap.Porph.*Plot.* 22'

κλῄζω (B), delete 'Pass., *AP* 9.62 (Even.)'

κληΐς, Ion. for κλείς.

κλήθρα, add 'applied to a writing-tablet made of alder, Philet. 10.2'

κλημάτιον, τό, dim. of κλῆμα I, *PPrag.* 109.4 (iii AD).

κλημοφόρος, ὁ, = Lat. *vitifer*, i.e. *centurion*, *Gerasa* 219 (ii AD); cf. κλῆμα ι 2.

κληρικός, after '*cleric*' insert 'freq. of lower cleric as opposed to bishop'

κληρογράφέω, Dor. **κλᾱρ-**, *write* a name *on a lot*, οἱ δὲ ἄρχοντες τὰ ὀνόματα κλαρογραφήσαντες χωρὶς ἑκατέρων ἐμβαλόντες ἐς ὑδρίας *SEG* 30.1119 (Nakone, iii BC).

κληροδοτέω, line 2, after '*Ps.* 77(78).55' insert 'cf. κληροδοτήσει υἱοὺς υἱῶν Lxx *Pr.* 13.22 (v.l.)'

κληρονομέω III, add '**b** w. acc., *give an inheritance to*, Ἰσραήλ Lxx *Si.* 46.1.(2).'

κληρονόμος, after 'ὁ' insert '(also ἡ, *SEG* 30.615, ib. 33.939 (of Artemis, Ephesus, ?iii AD), cf. D. 31.11)'; for '*heir*, freq. *the heir in possession*' read 'of an heir, (*designated*) *holder* or *possessor of an estate*'; line 7, after 'Mosch. 3.96' insert 'w. ἐξ, Scaev.*Dig.* 34.4.30.1, *Cod.Just.* 1.2.25 pr.'; add 'ἀπὸ κληρονόμων ποιεῖσθαι *disinherit*, *Cod.Just.* 6.4.4.24; so ἀπὸ κ. ποιεῖν, γράφειν, λέγειν Just.*Nov.* 115.3 pr., 115.3.14, 115.3.15'

κληροπάληδόν, Dor. **κλᾱρο-**, *by the shaking of lots*, Stesich.*fr.* in *PMGF* 222b.223.

κλῆρος (A) **I 1**, add 'transf., of the dispensation of fate, πρωτότοκος Λοῦκις, δισσῷ κλήρῳ Θεόδοτος, παρθένος ἡ Δόμνα κλήρον τρίτον ἐξετέλεσσεν *EG* 4.363 (c.AD 300)' **II 3 a**, add 'spec. *legacy*, opp. inheritance or *fideicommissum*, *Cod.Just.* 1.2.25.1, 1.3.45.1a'

κληρόω, add 'imper. κληρούσθωσαν *SEG* 31.122.25 (Attica, ii AD)'

κλήρωσις, add 'Dor. **κλᾱρ-**, *SEG* 31.825 (Halaesae, Sicily, ii BC)'

κληρωτήριον III, for '*list of citizens* .. *lot*' read '*device in the form of a kind of hopper for drawing lots*, (see *Hesperia* suppl. 1(1937) pp. 198 ff., *IG* 2².972, and so in Arist.*Ath.* ll.cc.)'

κληρωτής, add '*SEG* 26.120 (Athens, ii BC)'

κλήσιος, α, ον, perh. either *fame-bringing* or one *who locks*, as epith. of Artemis, Orph.*H.* 36.7.

κλῆσις I, add '**7** designation as an heir, *Cod.Just.* 6.4.4.12, 19b.' **IV**, for this section read 'cited in an etym. discussion of Lat. *classis*, D.H. 4.18'

κλῇσις, for 'cf. 7.70' read 'pl., *fastenings*, ἐπειρῶντο λύειν τὰς κλήσεις 7.70'

κλητήρ II, for this section read 'generally, *one who calls* or *summons*, esp. as a function of a herald, A.*Supp.* 622, Ἐρινύος κ. id.*Th.* 574, Ion Trag. 49, *Trag.adesp.* 664.33 K.-S.'

κλίμα II 2, add '**b** *geographical position*, Hero *Dioptr.* pp. 302, 304 S.' add '**6** *precinct*, *zone of a city*, Just.*Nov.* 43.1.1.' **IV**, for this section read '= κλιμακτήρ II, *SEG* 35.1060 (= *IG* 14.2431, Gaul, i AD)'

κλιμάκιας, add 'but cf. Κλιμακίας Sicilian pers. n., Cic.*Verr.* 2.2.118'

κλιμάκις, add 'Amips. 12'

κλιμακτηρικός, after 'Vett.Val. 148.20' insert 'κ. νόσος id. 289.27'

κλῖμαξ I, add '**3** in pl., *terraces*, E.*IT* 1462: sg. as pr. n., of a district, D.S. 19.21.' **IV**, after '*climax*' insert 'figure of speech in which the principal wd. of each clause is caught up and added to the next'

†**κλῑνάρχης**, ου, ὁ, *ruler of a feast*, Ph. 2.537, *OBodl.* III 372.

κλῑνέα, ἡ, *small couch*, *PSoterichos* 3.28 (s.v.l., AD 89/90).

κλίνεον, v. °κλινίον.

κλίνη I 3, add '*IEphes.* 3456' **II**, add 'κ. ἀμφικεφάλη *SEG* 29.146 (c.325 BC)'

κλινίον, add 'also κλίνεον, pl. κλίνεα *PFlor.* 369.2, [κλίν]ει̣α̣ *PSoterichos* 1.27, κλείψεα 2.22 (i AD)'

κλινοπετής, add '*PHels.* 2.22 (ii BC)'

κλῖνος, τό, archit. term occurring in description of base of a temple, *Inscr.Délos* 500*A*8 (iii BC).

κλιντήρ, for '*couch*' read '*kind of couch*, sts. dist. fr. κλίνη' and add 'κλῖναι πέντε· κλιντῆρες πεντέκοντα *SEG* 24.361 (Boeotia, 386/380 BC)'

κλιντηρία, ἡ, *dining-couch*, Ps.-Callisth. 21.10.

κλιντηρίδιον, for 'foreg.' read 'sq.'

κλίνω, before 'fut.' insert 'Aeol. **κλίννω** (Choerob. *in An.Ox.* 2.227.19)' **IV 1**, add 'κλίνουσιν .. εὐθὺς ἀπὸ τοῦ στόματος τοῦ κόλπου *Peripl.M.Rubr.* 44'

κλισία, for '*place for lying down* or *reclining*' read '*lean-to* or *temporary shelter*' **I**, combine quots. fr. sections 1 and 2 under this def. **II 3**, delete '*nuptial*' and add '*AP* 5.127.2 (Marc.Arg.), 7.207.7 (Mel.) (pl.)'

κλισιάδες, for 'κλεισιάδες' read '‡κλεισιάδες'

*κλῖσιάρχης, ου, ὁ, = κλινάρχης, PMich.v 246.14, 19 (i AD).
κλίσιον, for 'outbuildings round a κλισία or herdsman's cot' read 'prob. lean-to shed or shelter'
κλίσις I, add 'perh. *slope* of a hill, IRhod.Peraia 401 (ii BC)'
κλισμία, delete the entry.
*κλισμίον, τό, *couch*, Call.fr. 75.16 Pf. (pl.), IG 11(2).287B20 (Delos, iii BC).
κλοιός, add '**4** κ.· μέρος τι τῆς νεώς Hsch.'
*κλοκελέα, v. °γλυκελαία.
κλόκιον, after 'ἀμίς' insert 'Alex.Trall. 11.2 (v.l.)'
κλοπῐμαῖος, add '**II** astron. κ. ἡμέραι *epagomenal* days, cod. Vat.Gr. 1058.275ᵛ22, al.'
κλουβίον, add 'also χλου-, PFay. 72.4, PTeb. 413.14 (ii/iii AD); χλι-, SB 7365.16, al. (AD 104)'
κλύμενος, add 'Myc. *ku-ru-me-no*, pers. n.'
κλυντήρ, for 'IGRom. 1.730' read 'IGBulg. 903' and after 'Philippopolis' insert 'ii AD'
*κλῠσιδρομάς, άδος, fem. adj. *drenching as it speeds along*, αὔρα Tim.Pers. 92 (81 P.).
*κλυστηρίζω, *apply a clyster*, in Lat., Cael.Aur.Acut. 3.4, Veg.Vet. 2.15.5 (both v AD).
*κλυστικός, ή, όν, *used for clyster pipes*, Paul.Aeg. 7.3 (219.19 H.).
κλῠτοεργός, add 'also -γής, γέος (gen.), epigr. in SEG 37.1537 (Arabia, c.AD 400)'
*κλῠτοηχής, ές, *of resounding fame*, Ph.Epic. in Suppl.Hell. 681.2.
κλῠτός 2, line 3, after '(Od.) 15.472' insert 'ὄρθρος Ibyc. 22(b) P.'; penultimate line, after 'Pi. O. 8.52' insert 'v. δαιτοκλυτός'
κλῠτότοξος, add 'hymn in SEG 39.355 (Epidaurus, Rom.imp.)'
κλῠτόφημος, after 'fame' insert 'epigr. in IHistriae 1.171 (iv BC; -φαμ-)'
*κλυτωνος, unkn. wd., perh. an error for *κλυτόνηος, SB 10769.99 (iii/iv AD).
κλωγμός II, add 'in uncertain sense, Cratin. 171.15 K.-A.'
κλώθω II, for this section read 'intr., in uncertain sense, ἀμφοῖιν κλώθοντος ἐν ἀρπέζῃσιν ἐρίνου Nic.Th. 647, cf. Al. 93, 528'
†κλῶνος, ὁ, *bundle, bunch*, or sim., in quot., of beans, Edict.Diocl. 6.33 (in Latin, *fascis*); cf. *ramus*, Gloss.
κλῶσμα 1 and 2, combine these sections under def. 'thread'
κλωστήρ I, for this section read 'distaff, Theoc. 24.70' II 1, add 'A.R. 4.1062'
κλωστήριον, add 'Suid. s.v. νήτρον'
κλωστός, after 'spun' insert 'E. Tr. 537' 2, add 'τὸ κ. *the thread of fate*, BCH 112.451 no. 4.8 (Maced.)'
*Κνακᾰλησία, ή, epith. of Artemis at Caphyae in Arcadia, Paus. 8.23.3; cf. κνῆκος.
*Κνακεᾱτις, ιδος, ή, epith. of Artemis at Tegea, Paus. 8.53.11; cf. κνήκος.
κναπ-, κναφ-, κναψ-, see also γναπ-, etc.
κνάπτω 1, add 'dress leather, ὅλον ὅτε .. γναπτόμενοι μυδόωσιν ὑπ' ἀρβήλοισι λάθαργοι Nic.Th. 423'
*[κ]ναφαλλοϋφάντης, v. °[γ]ναφ-.
κναφεύς, add 'Myc. *ka-na-pe-u*'
κναφικός, line 2, after 's.v. κνάφος' insert 'κναφικός, ὁ, *fuller*, PColl. Youtie 83.8 (AD 353)'; line 3, after 'sc. ἐργασία' insert 'or τέχνη' and after '(i AD)' add 'cf. GDI 1904.6 (Delphi, ii BC)'
κνάφος, delete '**II** *carding-comb*' and append following part of article to section 1.
κνάψις, form γνάψις, add 'SB 9834b.3 (iv AD)'
κνέφαλλον, add 'κνάφαλλον also in Hsch. s.v. γνάφαλλον'
κνέφας, after 'τό' insert 'ep. gen. κνέφαος Od. 18.370'; line 4, delete '(only in nom. and acc.)' **2**, add 'ὑπὸ κνέφας A.R. 2.1032'
κνήδιον, for 'κνίδια .. nettle-seeds' read 'Κνίδια, perh. *wine-jars*, v. Κνίδιος III'
κνηκίς, for 'Call.fr.anon. 36' read 'Suppl.Hell. 1084, Call.fr. 238.17 Pf.'
κνῆκος, add 'Myc. *ka-na-ko*'
*κνηκουργός, ὁ, *maker of oil from the plant κνῆκος*, PMed.inv. 68.38 R. ii 11 (Aegyptus 69.24, 13 BC/AD 32).
κνήμη II, after 'Eust. 598.4' insert 'cf. ὀκτάκνημος' add '**III** καὶ ὁδοὶ ἀνώμαλοι καὶ ἀναντώδεις Hsch. s.v. κνήμαι.'
κνημίδιον, add 'sg., IGLS 1292 (Laodicea ad mare)'
κνημίον, delete the entry.
κνημίς I, add 'X.An. 1.2.16, al., Arist.HA 548ᵇ2'
κνησμονή, add '**2** transf., *annoyance*, Sch.E.Hipp. 14.'
κνησμός, add 'cf. Hp.Aff. 35'
κνηστήρ I, for 'scraping-knife' read 'grater'
κνήστριον, delete 'ἰχθύων .. (Geronth.)'
κνῆστρον, for '(expl. .. 4.172' read '**b** = θυμελαία, prob. *Daphne Cnidium*, Dsc. 4.172. **2** *grater*, Edict.Diocl. 13.9, 10, Erot.'
*κνηστώδης, ες, *grated*, ζειαὶ κνηστώδεις, ὁ κατειργασμένος σῖτος Hsch. s.v. πύρνοι.

*κνήσων, dub. sens., ἐν τῷ κιβωτίῳ κνησῶνας τρεῖς Inscr.Délos 1444Aa37 (ii BC).
*κνίδειος, α, ον, *of the nettle*, Theognost.Can. 54.
Κνίδιος III, after 'of wine' insert 'PInd. 8.6 (ii AD)'; add 'hence perh. *wine-jars*, Stud.Pal. 22.75.7, 16 (iii AD), v. °κνήδιον; jar of perfumed oil, POxy. 3748.15 (AD 319)' **IV**, delete the section. add 'Myc. *ki-ni-di-ja* (fem.)'
κνῑδόσπερμα, add 'Paul.Aeg. 7.17.14 (350.24 H., 351.2 H.)'
κνίζω **I** 2, for this section read 'dub. reading in passage where context requires *chop up*, *grate* or sim., Thphr.HP 9.20.4 (perh. κνησθεῖσα)' **II**, for 'tickle' read '*scrape* or *scratch* with the fingernails (without necessarily breaking the surface)' **2**, for 'chafe, tease' read 'bother, irritate' **b**, after 'provoke' insert 'an emotion'
κνῑπολόγος, after 'ὁ' insert 'ή'; for pres. def. read 'common treecreeper'; add 'κ. πιπώ Ant.Lib. 14.3'
κνῑσάω, delete '(never τὰς ἀγυιάς)' and add 'Orac.ap.D. 43.66, Luc. .Prom. 19'; delete 'Orac.ap.' before 'D. 21.51'
*κνῑσευτήρ, ῆρος, ὁ, *priest who carried out burnt sacrifices*, ABSA 42.206 (Cyprus).
κνίσμα, for 'in pl. *scratches*' read '*scratch*' and add 'AP 5.157 (Mel.)'; for 'of lovers' quarrels' read 'of scratching, tickling, etc., used in love-play' and add 'AP 12.209 (Strat.)' add '**b** *a small fragment broken off, chip, splinter* or sim., κ. πυρὸς θραύσας AP 12.82 (Mel.).'
*κνισμώδης, ες, *marked with scratches*, Procl.CP Hom. 26.11.
κνίψ, line 2, delete 'and devour the fig-insect (ψήν)'
κνοή, η, v. °κνόος.
*κνόος, ὁ, κνοῦς· ὁ ἐκ τοῦ ἄξονος ἦχος. λέγεται δὲ καὶ κνοή. καὶ ὁ τῶν ποδῶν ψόφος, ὡς Αἰσχύλος Σφιγγί (fr. 237 R.) Hsch.; perh. also (sp. χνόος) ἔτι χνόον .. ἄξονος .. ἵππος ἔναυλος ἔχει Call.fr. 384.5 Pf.
*κνούφιον, τό, *substance named after the Egyptian god* Κνοῦφις, Anon.Alch. 9.14.
κνύζα (B), delete '(pl.)' and '(sg.)'
κνύζα (C), after 'Anacr.' insert '(432 P. κνυζή)'; delete 'cf. κνυζός'
κνυζός II, delete the section.
*κνωδάκιον, add 'PUG inv. 1164 (ZPE 58.92, iii AD)'
κνώσσω, for 'Simon. 37.6' read 'Simon. 38.9 P. (κνωοσσ-, v.l.)'
*κοάζω, of frogs, *croak*, Aesop. 307 H.
*κόαμος· ὁ βασιλεύς Theognost.Can. 56 A.
*κόβαθια, v. κωβάθια.
*κόβαθρον, τό, = κύβεθρον, PCol. 188.9 (AD 320).
*κοβάλευμα, ατος, τό, = κοβαλίκευμα, Et.Gen. in Miller Mélanges 191.
*κοβᾱλικεύομαι, *cheat by tricks*, prob. in Ar.Eq. 270 (codd. ἐκκοβ-).
*κοβᾱλικός, ή, όν, *knavish, scoundrelly*, ἀργυρίοισι κοβαλικοῖσι πεισθείς Timocr. 727.6 P.
κόβᾱλος I, add 'cf. κόβαλοι· δαίμονες .. περὶ τὸν Διόνυσον, Sch.Ar.Pl. 279, interpol. in Harp.'
*κογκυλίδιον, v. °γογγ-.
*κογνᾱτικός, ή, ον, *of* or *involving blood-relations* (*cognati*), Just.Nov. 84 pr.
*κογνᾱτος, ὁ, Lat. *cognatus*, *blood-relation*, Just.Nov. 115.3.14.
κόγχη I, add 'applied to the pearl-oyster, Arr.Ind. 8.11' II, add '**3** *pudendum muliebre*, Sophr. 25, 26, cf. κογχυλαγόνες, and Plaut.Rud. 704.'
*κόγχισμα, ατος, τό, *kind of vessel or container*, POxy. 2729.29, 34 (iv AD).
κόγχος, after 'ή' insert 'Epich. 42.8, 43'; line 2, after 'Crates Theb. 7' insert 'Call.Epigr. 5.1 Pf., Hedyle ap.Ath. 297b' II, add '**6** dub. sens., building inscr., IAskl.Epid. 52 B60 (iv BC, gen. pl. or fr. κόγχη).
*κογχεύς, έως, ὁ, *purple-worker* (or *purple-fisher*), MAMA 3.309, al. (Corycus), κογχυλέως λιμένο(ς) Ἀστρονόης ITyr 8, al.
κογχυλευτής, add 'ITyr 7, al.'
*κογχυλεύω, app. *extract purple dye*, Hsch. (κογκ- cod.).
†κογχύλη, ή, = κόγχη; spec. *murex*, v.l. in Ph. 1.536, AP 9.214 (Leo Phil.); cf. κογχύλαι· κηκίδες Hsch.
κογχύλιον 1, after 'Epich. 42.1' insert '(ῡ, cf. Lat. *conchylia*)'
*κογχυλοκόπος, ὁ, *producer of purple dye*, ITyr 72, 95 (abbreviated).
*κογχυλοπλυτής, οῦ, ὁ, *purple-dyer*, ITyr 28, 198.
κογχωτός, for 'having a boss' read 'having a boss *or ornamentation in the form of a shell*' and add 'Inscr.Délos 1444Aa6, 11 (ii BC)'
κοδομεύς, delete 'perh. to be read in OStrassb. etc.'
Κόδρος, cf. Myc. *ko-do-ro*, pers. n.'
κοδύμᾱλον, add 'cf. Alcm. 90' read 'Alcm. 100 P.' and for 'καρύαι Περσικαί' read 'κάρυα Περσικά' add '**II** οἱ δὲ ἄνθους εἶδος, οἱ δὲ κόσμος περιτραχήλιος Hsch.'
†κόθουρος, ον, dub. sens., epith. of drones (κηφῆνες), Hes.Op. 304.
κοία I, read 'expld. by Hsch., etc., as *balls* or *stones*, κοίας ἐκ χειρῶν σκόπελον μέτα ῥιπτάζουσιν Antim. 89 Wy.'
*κοιαίστωρ, ορος, ὁ, v. °κουαίστωρ.

*κοίας· στρογγύλος Theognost. Can. 21.
*κοϊκᾶς, ὁ, *maker of baskets from palm-leaves*, PLaur. 24 (ZPE 35.111-2, iii AD); cf. κοίξ.
*κοΐκιον, τό, *basket*, PWash.Univ. 30.6, POsl. 159.13 (both iii AD); cf. κοίξ.
κοιλαίνω **I**, after 'Opp.H. 4.19' insert 'ἕλκος Numen. in Suppl.Hell. 590' add '**b** *open the recesses of*, μή ποτε κοιλήνῃς Παφίῃ νόον AP 9.443 (Paul.Sil.).'
κοίλασμα, after '*hollow*' insert 'of a pit hollowed out as trap'
κοιλία **II 4**, add 'meton., *offspring*, Orac.Tiburt. 148' **III**, add 'of a gulf, Peripl.M.Rubr. 40'
*κοιλίας, gloss on κοστίας, Hsch.
κοιλιολυσία, for '*looseness* .. *medicine*' read '*emptying of the bowels*, περὶ κ. γίνεσθαι'
κοιλιολυτέω, for '*suffer from looseness of the bowels*' read '*empty one's bowels*'
*κοιλίον, τό, perh. = κοῖλον (v. κοῖλος IV 1), point in a boundary, κατὰ τὸ κ. Inscr.Cret. 1.v 19B23 (Arcades, i BC).
*κοιλιόστροφος, ον, *colicky*, Sch.Nic.Al. 597a Ge.
κοῖλος **1**, line 16, after 'Hero Bel. 75.15)' insert '*curved*, perh. *hollow*, κλῆθρα S. OT 1262; σταθμὰ θυραῖον Theoc. 24.15'; line 18, before 'Ael.' insert 'Inscr.Délos 1417Ai102 (ii BC)' **III 2**, delete the section.
κοιλόσταθμον, add 'PMich.Zen. 38.9, 20 (iii BC)'
κοιλόσταθμος, for '*with coffered ceilings, panelled*' read '*with curved or hollow supports*, precise signf. uncertain, but app. of wood'; after 'Hg. 1.4' insert '*with wooden frames*' and after 'p. 143 (iii BC)' insert 'PCair.Zen. 764.3 (iii BC)'; line 3, for '*coffered ceiling*' read '*curved or hollow door-jamb*'
*κοιλουργός, ὁ, *maker of hollow ware*, perh. of gold and silver, PSAAthen. 1.1 (iii BC); < κοFιλοFοργός, Myc. ko-wi-ro-wo-ko.
κοίλωμα **I 1**, add 'of the black cavity or cup in the infundibulum of the teeth of horses, Gp. 16.1.16'
κοιμάω, line 2, after 'Med.' insert 'imper. κοιμοῦ CIJud. 1.281, cf. 150'; line 6, for 'Aeschrio 8.2' read 'AP 7.345' **II 3**, add 'κοιμοῦ CIJud. 1.281 (κομοῦ ib. 150)'
κοίμησις **II**, add '**2** *laying to rest*, or *resting-place* i.e. tomb, κύμεσις Εὐφημίας δούλης Mitchell N.Galatia 468; κύμησις ib. 469.'
*κοιμητηρία, ἡ, = κοιμητήριον, EM 550.56.
*κοιμητήριον, τό, *dormitory* for performing incubation in a temple, SEG 31.416 (Oropus, in BC), Dosiad.ap.Ath. 4.143c. **II** *burial-place*, SEG 26.434, etc., 29.250, 31.286 (all Chr.); sp. κυμ- SEG 31.263, ib. 34.238 (-ιν), INikaia 284.
*κοίμητρον, τό, *bed*, Sm.Jd. 4.18.
*κοινάζω, *participate* in, w. gen. SEG 37.340.4 (Mantinea, iv BC).
κοινανέω, add 'SEG 12.371.24 (Thelphoussan inscr. at Cos, iii BC)'
κοινεῖον **II**, after '*common fund*' insert 'BCH 78.322.17 (Ceos, iii BC)'
κοινεών, for '= κοινωνός' read 'form of κοινῶν'
*κοινοβίωσις, εως, ἡ, *cohabitation*, PDura 32.10 (AD 254).
κοινόβουλος **II**, for 'IGRom. 3.7' read 'TAM 4(1).42' and add 'SEG 30.1534 (Lycia, after AD 152)'
*κοινογράφέω, *write a word in the ordinary way*, Eust. 1553.28 (pass.).
κοινοδίκιον **2**, for 'PMagd. 21.12, 23.9' read 'PEnteux. 65.19, al.' and delete 'abbrev.'
*κοινόθακος, ον, perh. epith. of an ancestral tomb, *offering a common burial-place*, S.fr. 212.6.
*κοινοπάτωρ, ορος, ὁ, ἡ, *having the same father*, Philicus in Suppl.Hell. 680.27.
κοινοπραγία, after lemma insert 'SEG 23.547.6 (Crete, iii BC)'
κοινός **I**, line 21, after 'Pl.Mx. 241c, etc.' insert 'ἐν Περγάμῳ κοινὸν Ἀσίας (sc. ἀγῶνα) SEG 34.1316.16 (Xanthos, i AD)'; line 23, before 'Iamb.' insert 'D.H. Th. 3' **IV 3**, *neutral*, add 'Isoc. 14.28'
κοινοτάφής, for 'Ath.Mitt. 10.405' read 'CEG 563'
κοινῶν, add '*accomplice*, sp. κονῶνα (acc.) PVindob.Salomons 15.12 (v/vi AD)'
κοινωνέω **I 2 a**, add 'rarely w. dat., ἐκοινώνησαν τῇ στρατείᾳ Sch.A Il. 2.339' **4**, add 'w. ἐν, ἐν τῇ ἁγιωτάτῃ ἐκκλησίᾳ κοινωνεῖν Just.Nov. 115.3.14'
*κοινωνησία, ἡ, = κοινώνησις 2, EA 8.29 (Ephesus, Rom.imp.).
κοινωνία **I 1**, add 'ἀπὸ τῆς Ζήν[ω]νος κοινωνί[ας Kafizin 119' **2**, add 'ἐπὶ προεδρίᾳ καὶ κοινωνίᾳ θυσιῶν SEG 30.82.14 (Athens, c.AD 230)'
κοινωνιμαῖος, add 'κοινωνιμέων κτημάτων POxy. 2954.15 (iii AD)'
κοινωνός **2**, add 'κοινωνοὶ οἱ περὶ Μέ[να]νδρον SEG 24.976 (Thrace, ii/iii AD)'
κοΐξ **2**, add 'μικρὸν .. κόϊκα ἄρτων PMich.III 212 (ii/iii AD), al.'
κοιόλης, add 'cf. κοίολις· ἱερεύς, Theognost. Can. 21'
κοιρανία, after '*sovereignty*' insert 'ἐπὶ τὰν κ[λειτὰ]ν κ. ὕπατον *dignity of the consulship*, (cf. ὕπατος III 1 at end), epigr. in IG 9(2).1135.8 (Demetrias, i AD)'
*κοιρανίδες, αἱ, v. °κυρανίδες.

*κοιρανόμοιρος, ὁ, *Lord of Fate*, PMag. 4.1360.
κοίρᾰνος **2**, add 'of Hades, νυκτὸς ἀϊδνᾶς ἀεργηλοῖό θ' ὕπνου κοίρανος Lyr.adesp. 78 P.; κώμου καὶ .. παννυχίδος AP 7.31 (Diosc.); ὕμνων (of Homer) ib. 213 (Arch.)'; after 'Orph.fr. 38' add 'IG 9(1).270'
*κοισυροῦται· κοσμεῖται Theognost. Can. 21; κεκοισυρωμένη .. περιεσταλμένη Hsch.; cf. ἐγκοισυρόομαι.
κοιτάριον, after 'dim. of κοίτη' insert '*small couch*, PUniv. Giss. 20.35 (ii AD)'
κοιτασμός, after '*folding*' insert 'προβάτων POsl. 33.10, PMich.II 121ʳ iii 14 (both i AD)'
κοίτη, line 1, for '= κοῖτος 1' read '*bed*'
κοῖτος **I** and **II**, combine these sections under def. '*sleep, going to rest*' add (new) '**II** *resting-place*, of a sarcophagus, Λοχιγοῦ κοῖτος SEG 32.1515.'
κοιτών, line 4, after 'IG 14.2143, al.' insert 'Θεοῦ Τ[ί]του ἐπὶ κοιτῶνος IEphes. 852 (c.AD 100)' **2**, add 'MAMA 7.323 (Lycaonia), etc.'
*κοιτωνάριον, τό, *resting-place*, PPrincet.inv. AM 8963.1 (Tyche 3.29).
κόκκᾰλος, add 'PLund 11.1.19 (ii AD)'
*κοκκίνόω, *dye scarlet*, Hsch. s.v. ἠρυθροδάνωται.
κοκκίζω, for 'A.fr. 363, Ar.fr. 610' read 'A.fr. 363 R. (= Ar.fr. 623 K.-A.)'
κόκκος, after 'ὁ' insert '(ἡ, Lxx Si. 45.10)' **I 1**, add 'of mustard, Ev.Matt. 13.31' **II 1**, for '*berry (gall) of kermes oak*' read '*kermes* (thought to be a berry produced by the kermes oak)'; add '*scarlet thread*, κεκλωσμένῃ κόκκῳ, ἔργῳ τεχνίτου Lxx Si. 45.10'
*κοκκοῦσα· συκῆ Hsch.
*κοκκοφάδιον, τό, perh. *hoopoe*, PMag. 7.411.
*κόκκυγος, ὁ, = κόκκυξ, Alc. 416 L.-P.
κόκκυξ **I**, line 5, after 'Hsch.' insert 'AB 1.27.24' **II**, add 'SEG 32.450 (Boeotia, iii/ii BC)' **III**, for this section read '*early fruit*, κόκκυγας ἐρινάδος Nic. Th. 854' **V**, for this section read 'part of the συνωμία of an ass, κατάγραφος κ. (s.v.l.) Hippiatr. 14.3; of a horse, ib. 26.9, 115.3'
*κοκκώ, v. °κωκώ.
*κόκκωρα, τά, perh. *pomegranate*, Semus 3.
κολάζω, line 2, after 'Pl.Lg. 714d, etc.' insert 'κολῶ Hsch.'
*Κολαινιασταί, οἱ, members of a cult-association of Artemis, MDAI(A) 67.165 (Attica, ii/iii AD), v. Κολαινίς.
*κολᾰκευμάτιον, τό, dim. of κολάκευμα, Hsch. s.v. κοσκυλματίοις.
*κολανδιοφωντα, τά, some kind of Indian ships, Peripl.M.Rubr. 60.
κόλαξ **I 1**, add 'ψόφῳ κ. ποιμένων S.Ichn. 154 (= 160 R.)'
κολάπτω **2**, add 'med., *have engraved*, τὸ δὲ ἁλίασμα τόδε κολαψάμενοι οἱ ἄρχοντες ἐς χάλκωμα SEG 30.1119 (Entella, iii BC)'
*κολάσιμος, η, ον, *deserving of punishment*, Suppl.Mag. 62.5 (v/vi AD), PMich.inv. 490 (ZPE 84.41, vi AD).
κόλᾰσις **2**, add '**b** *that which brings about punishment*, Lxx Ez 14.4, al.'
κολασμός, after '= κόλασις' insert 'Call.fr. 114.12 Pf.'
†κολεάζω, *sheathe* a sword; transf., of sexual intercourse, κολεάζοντες· ὠθοῦντες ‹εἰς› κολεόν[τες] περαίνοντες Hsch.
†κολεασμός· τὸ περαίνεσθαι Hsch., v. °κολεάζω.
*κολέντερον, τό, = Lat. *longao* (*longavo*), kind of sausage, Gloss. (perh. for *κωλέντερον).
κολεός, before 'ὁ' insert '(κουλεός Sch.E.Ph. 276, al.)' **I 2**, for this section read '*vagina*, Hsch. s.v. κολεάζοντες, v. °κολεάζω (also in mod. Gk.)'
κολεός (**B**), v. °κολιός.
κολετράω, add 'Hsch., s.v. κολετρῶσι'
κολίανδρον, for 'Sch.Ar.Eq. 679' read 'Sch.Ar.Eq. 682b'; add 'Al.Nu. 11.7, Ex. 16.31'
*κολιάσαι· ὀρχήσασθαι Hsch., perh. also impf. ἐκολίαζε (written ἐϙολιαδῃ) IG 12 suppl. 244 (Syros, vi BC; unless obsc., cf. °κολεάζω).
*κόλιξ, ικος, ἡ, v. °κύλιξ.
κολιός, add 'κολεοί Edict.Diocl. 4.41A, mistransl. Lat. *coturnices* (ὄρτυγες ib. 4.41), perh. for κολοιοί'
*κολίσκιον, τό, perh. dim. of καυλίσκος, some kind of vegetable, SEG 7.434 (Dura).
κόλλα **1**, after 'Hdt. 2.86' insert 'E.fr. 472.7 (anap.)' and add 'PKöln 52.12,59 (AD 263)' add '**b** fig., ἁρμονίης κόλλησιν Emp. 96 D.-K.; δεσμοῦ τινος ἢ κόλλης ἴδια D.H.Dem. 40; Δημάδης κόλλαν [ὠνόμαζε] τὰ θεωρικὰ τῆς δημοκρατίας Plu. 2.1011b.'
κολλάριον, add 'PSI 1116.8 (ii AD)'
κολλάω **I 1**, line 2, before 'ἐπιστύλια' insert '*attach firmly, join by mortising*, χσύλα ἐπιστύλια .. ἐγ δυοῖν κεκολλεμένον IG 1³.386 ii 107' **II**, lines 8/9, after 'Act.Ap. 5.13' insert 'Vit.Aesop.(G) 30' **III**, after 'Pi. O. 5.13' insert 'ἀντίθετα Plu. 2.350d'
*κολλεκτάριος, ὁ, Lat. *collectarius, banker*, PStrassb. 35.11 (?vi AD).
κόλλημα **I**, line 3, after 'Antiph. 162' insert '*fastening*, ἀνάκλιντρα ἱμαντοδέτοις κολλήμασιν ἐπηγμένα Ps.-Callisth. 120.2 K.'; add 'also *column of writing*, prob. in pap. in APF 4.97.3, al. (i AD), PIand. 7 ii 2, al. (ii/iii AD), see Youtie Scriptiunculae II 718'

κόλλησις, after lemma insert 'Dor. [κ]όλλᾱσις, SEG 24.277B.51 (Epid., iv BC)'   add '**III** name for *holy vervain, Verbina supina*, περιστερεὼν ὕπτιος Ps.-Dsc. 4.60.'

⁺**κολλητίων**, ωνος, ὁ, perh. *filing clerk* employed by military police, *PFlor.* 91.27 (rest., ii AD), *POxy.* 1100.19 (AD 206), *BGU* 23.5, 6 (AD 207), *OBodl.* ii 1934.7 (iii AD).

κολλόροβον I, after '*crook*' insert 'gloss on καλαῦροψ, Sch.A.R. 4.974 (v.l.)'   **I**, add '**2** *harpoon* or *gaff* for catching fish, *POxy.* 2234.15 (i AD).'

κολλούρα, omit '(?)' and add 'Sm. 1*Ki.* 10.3'; after 'cf. κολλύρα' add 'mod. Gk. κουλούρα, κουλούρι'

*κολλουρίδιον, τό, dim. of κολλούρα, *PPalau Rib.*inv. 66 (*APF* 23.272, i AD).

κόλλοψ, for '**II 2** metaph.' read '**III** (prob. a different word)'

*κολλῠβίζω, *cut into small pieces*, Suid. s.v. κολλάβους.

κολλῠβιστής, delete '*small*' and '*PPetr.* 3 p. 173 (prob. l.)' and substitute '*PTeb.* 1079.49 (iii/ii BC), *BGU* 1303.18 (i BC)'

*κολλῠριακός, ή, όν, *used in eye-salves*, Aët. 7.106 (369.6 O.).

*Κολλυριών, ὁ, month name, *SEG* 36.982c (Iasos, v BC).

κολλώδης, after 'Pl.*Cra.* 427b' insert 'Clearch.Com. 2.1 K.-A. (neut. pl. subst.), Hp.*Carn.* 3, 4, al.'

*κολοβᾰφής, v. ‡χολοβαφής.

*κολοβᾰφινος, v. ‡χολοβάφινος.

κολοβιομαφόριον, add '*PKöln Ketouba* line 17'

*κολοβίων, ωνος, ὁ, = κολόβιον 1, Sch.Aeschin. 1.131 S.

κολοβόρῑνος, delete the entry (read κολοβάφινος).

κολοβός 2, before 'ὄνος κ.' insert 'of animals, perh. spec. *with teeth worn away* (as an indication of age) (cf. *CPR* 6.20-1, 25), or more generally *docked* or *stunted*'

κολοβοῦρος, add 'as a pers. n., Κολοβούρῳ (dat.) *BGU* 2348.2 (iii AD)'

*κολοβοφυής, ές, *low-growing*, of grass for hay, *BGU* 2198.11 (vi AD).

*κολοδιδός, ή, όν, *yellow*, Anon.Alch. 330.13.

κολοιάρχης, read κολοίαρχος.

*κολοιδορόω, 3 sg. -οῖ, = ταράσσει, Theognost.*Can.* 56 A. (-δωροῖ codd.).

*κολοίσιππος, ὁ, prob. non-Hellenic wd. in a cult inscr., *Inscr.Cret.* 1.v 23.8 (Arcades, ii AD).

κολοιτία I, add 'cf. Lat. *colutea* (n. pl.), its fruit, Plaut.*Per.* 87 (cj.)'

κολοίφρυξ, after 'ἀλεκτρυών' insert 'καὶ ὄρος Βοιωτίας'

*κολοκυνθᾰρύταινα, ἡ, *scoop* or *dipper made of a gourd*, *Suppl.Hell.* 960.7.

κόλον II, add 'cf. °κωλικός'

*κολοός, perh. = κολοιός, Hdn.*fr.* p. 17 H.

κολοσσιαῖος, after '*colossal*' insert 'εἰκόνα κ. *SEG* 33.1035 (Cyme, ii BC)'

κολοσσός, add '(v. *REA* 62.5 ff.)'

*κολουᾶν· θορυβεῖν Hsch.; cf. κολοάω.

*Κολουάς, άδος, ἁ, an unexpl. topographical feature, ἐκ τοῦ Πετράχου ἐπὶ τὰν κολουάδα τὰν εὐδείελον· ἐκ τᾶς κολουάδος ἐπὶ τὰν σκοπι[άν] *BCH* 116.200.8, 9 (Phocis, iii BC).

*Κολούρα, ἡ, name of a hill, *SEG* 11.377.17-18 (Hermione, ii BC), app. referred to in Paus. 2.36.3 (v. Βολεοί, s.v. °βολεός), cf. κολουραῖος (Call.*fr.* 235 Pf.).

κόλουρος, add '᾿Ρόλουρος pers. n., *SEG* 17.287 (Magnesia, vii BC)'

κολοφών I, add 'for a different expl. cf. Sch.Pl.*Tht.* 153c G.'

*Κολοφωνιακός, ή, όν, = Κολοφώνιος, Μιμνέρμου τοῦ Κ. *Suppl.Hell.* 1060.

*κολπῑτεύω, *smuggle*, *PVindob.Barbara* 8.4 (*ZPE* 40.139, iii BC), *PPhilad.* 35.22 (ii AD); cf. °διακολπιτεύω.

*κολπῑτικός, ή, όν, *smuggled, contraband*, *PTeb.* 38.12, 125, 1094.3 (-εικ-, all 114/13 BC).

κολπόω, line 1, after '*belly*' insert 'ἱστίον B. 13.130 S.-M.'

κολυμβάς I, last line, for 'Call.*Iamb.* 1.273' read 'Call.*fr.* 194.77 Pf.'

κολυμβήθρα IV, add 'Procop.*Arc.* 17.9'

κόλυμβος II 1, for ' = κολύμβησις' read '*swimming*'; add 'humorously, of boiling food, Theodorid.ap.Ath. 6.229a'   add '**III** *vat*, Pelag.Alch. 255.'

κολχικόν, add 'Plin.*HN* 28.129'

κολῳάω, for '*brawl, scold*' read '*cry out, shout*, cf. κολοιός, °κολουᾶν'

κόλων, add '*SEG* 34.630 (Maced., iii AD), al., *POxy.* 2476.32, 48'

κολώνη, line 1, for '*Lyr.adesp.* 74' read '*Lyr.adesp.* 65 P.' and add 'Ἀκράγαντος Pi.*P.* 12.3'; line 2, after '*peak*' insert 'A.R. 1.601'

κολῳός, for '*brawling, wrangling*' read '*tumult, uproar*' and add 'A.R. 2.1064; cf. κολοιός, κολοιή'

*κωμακτορία, ἡ, *auction tax*, *PKöln* 83.12-13 (κωμακτορεία, AD 167), *POxy.* 1523.4 (iii AD).

κομάκτωρ, for 'dub. sens.' read 'prob. = Lat. *coactor* (cf. *comactores: argentarii* Gloss.)' and add '*PStrassb.* 79.3 (i BC, pl.)'

*κόμανος, ὁ, unexpld. wd., Hdn.*fr.* p. 28 H.

κομάω I 2, for 'of her lover' read 'of interpreters (of Erinna)'; for 'Opp.*C.* 3. 192' read 'Opp.*H.* 2.534'

*Κόμβα, ἡ, cult-name of Artemis, *TAM* 2(1).4 (Lycia).

κόμβαλα, for 'παίγματα' read 'πήγματα' and add 'perh. also Euph.in *Suppl.Hell.* 450 i 2'

*κομβάλιον, τό, part of the framework of a window, perh. *window-catch*, *PCair.Zen.* 847.18 (iii BC).

*Κομβική, ἡ, title of Artemis in Lycia, Xanthos 2, 3, *TAM* 2(2).407.

*κομβίον (B), τό, dim. of °κόμβος (B), *IIasos* 394 (cf. Robert *Ét.Anat.* 470); cf. °κάμβειν.

κόμβος (A), for '*roll, band, girth*' read '*knot, fastening*'

*κόμβος (B), ὁ, prob. *grandchild*, *IKeramos* 26.12, al., *Gnomon* 18.667 (*Didyma* 349).

κόμβωμα, add '(in pl.) *knots* or *buds* on a branch'

*κομεᾶτος, ὁ, Lat. *commeatus*, *supplies*, *PGiss.* 41.4, *POxy.* 1666.14 (iii AD); also κομι- *POxy.* 1477.7 (iii/iv AD), Hsch., Suid.

*κομενταρήσιος, v. °κομμεντ-.

*κομεντάριον, v. °κομμεντ-.

κομέω (A), at end after '(Karanis, iii BC)' insert 'A.R. 1.780, of the dead, *AP* 7.707 (Diosc.), of beehives, ib. 7.717'

κομήεις, add 'Myc. *ko-ma-we*, *-we-to* (gen.), pers. n. ("*long-haired*")'

κόμης, add 'πρὸς τὸν κόμιτα *POxy.* 3150.16 (vi AD), al.'

*κομιᾶτος, v. °κομεᾶτος.

κομίζω, line 1, before 'κομίσω' insert 'κομίσσω A.R. 1.419, al., Nonn.*D.* 1.446, al., *AP* 5.278.5 (Agath.)'; line 2, for 'only late, as' read '*IG* 11(4).1027.6 (Delos, iii BC)'; line 7, after '8.284' insert 'Lesb. [κο]μίσσασθαι (inf.) *IG* 12(2).29.9. Boeot. κομιττάμενοι (part.) *IG* 7.2406.8'   **II 4**, delete 'ἔξω ... E.*Tr.* 167 (lyr.)'

κόμιστρον I 1, add 'A.*fr.* 154a.11 R.'   **3**, for pres. def. read 'perh. *payment for porterage*'   **II**, add 'cf. Poll. 6.186'

*κομιτᾶτος, ὁ, Lat. *comitatus*, *staff*, *SB* 7181.48 (AD 220), *PLips.* 34.6 (iv AD).

*κομμέντα, τά, Lat. *commenta*, *records*, *PSI* 951.2 (?iv AD), *POxy.* 1877.5 (v AD).

*κομμενταρήσιος, ὁ, Lat. *commentariensis*, *secretary* or *accountant*, w. var. spellings, e.g. κομεντ-, *PBerol.*inv. 7347 (*JJP* 19.93, iii AD), κομμεντ- *BGU* 2162.2 (AD 491); κομμετ-, *SEG* 32.1554A.18 (Arabia, vi AD).

*κομμεντάριον, τό, Lat. *commentarium*, *system of shorthand*, *IEphes.* 3054 (s.v.l.), sp. κομεντ-, *POxy.* 724.8 (ii AD).

*κομμερκιάριος, ὁ, Lat. *commerciarius*, *merchant*, *SEG* 32.1554A.5 (before vi AD), Just.*Nov.* 154.

*κομμέρκιον, τό, Lat. *commercium*, *trade*, ἀπὸ κομερκίων *BGU* 972.1 (vi/vii AD).

κόμμι, add 'acc. pl. κόμμιδες (for κόμμιδας) *BGU* 2357 ii 5 (iii AD)'

*Κομμόδεια, τά, *festival in honour of Commodus*, *Delph.* 3(6).143, καὶ ἐν Σμύρνῃ Κομόδεια Ὀλύμπια *TAM* 4(1).34; at Athens, *IG* 2².2113.53, al.

*Κόμμοδος, ὁ, name given by Commodus to the eighth month, D.C. 72.15.

*κομμονιτώριον, τό, Lat. *commonitorium*, *letter of instructions*, *PMerton* 45.1 (v/vi AD), *PLond.* 1680.22, *SEG* 35.1360.5 (κομονητουρ-, Hadrianopolis, iv AD), Just.*Nov.* 128.17, Suid. (-μονητορ-); also κομμων- *Cod.Just.* 1.4.26.5, al.

*κομμωτίζω· ἐπιμελοῦμαι Suid.

*κομόδιον, τό, *gratuity* paid to officials, cf. Lat. *commodum*, *POxy.* 3358.4, 3424.3 (both iv AD).

*κόμορος, gloss on κοστίας, Hsch.

*κομοτροφία, ἡ, *letting the hair grow long*, Porph. in *Mél.Bidez* 149, prob. in Critodemus in *Cat.Cod.Astr.* 8(1).260 (codd. κωμο-).

*Κομπεταλιασταί, οἱ, guild which celebrated the Compitalia, *Inscr.Délos* 1761.18, al.

*κομπλεύσιμος, η, ον, *relating to the filling up of documents*, Lyd.*Mag.* 3.25.

*κομπλεύω, *fill up*, *complete* documents (as the function of a special clerk), *PUniv.Giss.* 33.2 (vi AD).

⁺**κομπολᾰκύθος**, ὁ, *boaster*, an imaginary bird, Ar.*Ach.* 589, 1182; also κομπολήκυθος, nickname for tragic poets, acc. to Choerob. *in Heph.* p. 230.21 C.

*κομπολύρας, α, ὁ, perh. *with noisy lyre*, rest. in A.*fr.* 451c.28 R. (lyr.).

κόμπος (A) II 2, add 'Pi.*fr.* 94b.13 S.-M.'

κομπός (B), after 'E.*Ph.* 600 (troch.)' insert 'Epicur.*Sent.Vat.* 45 A. (κόμπους cod., *WS* 49.33), Call.*fr.* 96 Pf.'

*κομπρομισσάριος, α, ον, *of* or *connected with a compromissum*, Just.*Nov.* 113.1.1.

*κομπρόμισσον, τό, Lat. *compromissum*, *an arbitrated compromise*, *SB* 10733.7 (iii or v AD), 9775.11 (vi AD); τὸν -ον (erron.) *CPR* 6.8.4 (AD 509).

κονᾰβίζω, line 3, for 'cf.' read 'trans., γαῖαν κ. ποσσίν'

*κοναύθιον, τό, type of river boat, *PHels.* 7.3 (ii BC).

*κόνβεντος, ὁ, Lat. *conventus*, *assembly*, cf. κονβενταρχέω, *IGRom.* 4.1169.

*κονδέα, ἡ, Sicil. word for "*fox*", *AB* 272.

*κονδικτίκιος, Lat. *condicticius*, ὁ ex lege κονδικτίκιος = ex lege conditio, *Cod.Just.* 1.3.45.6, Just.*Nov.* 162.1.2.

*κονδιτάριος, ὁ, *maker of spiced wine*, *PMed.* I 71.1, *PMich.*xv 740.14 (vi AD); also -ία, ἡ, Βικτωρίας κονδειταρίας τόπος *Not.Scav.* 1893.309 (Syracuse, v AD).

*κονδῖτον (or κονδεῖτον), τό, Lat. *conditum* (sc. *vinum*), *spiced wine*, *AP* 9.502 (Pall.), Alex.Trall. 8.2 (ii p. 341.5 P.), 11.1 (ii p. 469.12 P.), *PRyl.* 629.367 (iv AD, -διτ- pap.).

κονδοκέρατος, add 'cf. κοντός (B)'

*κονδόρῑς, ῑνος, *short-nosed*, κονδόρεινα (acc.) *POxy.* 3054.16 (AD 265); cf. °κοντόρρινος.

*κονδουκτορία, ἡ, cf. Lat. *conductor*, *office of contractors*, *POxy.* 900.6, 2110.4 (both iv AD).

*κονδουκτώριον, τό, *board of contractors*, *PCornell* 52.10 (iii AD), *POxy.* 2115.3 (iv AD).

*κονδούκτωρ, ορος, ὁ, Lat. *conductor*, *contractor*, *PBeatty Panop.* 1.60 (AD 298), *POxy.* 2115.6 (pl., iv AD), *PMich.*xi 624.24 (vi AD).

κόνδυ, line 3, delete 'as a measure'

κονδυλωτός, after '*knobby*' insert 'perh. *embossed*, ποτήριον *Inscr.Délos* 104.16 (364 BC)'; neut. subst., add '*Inscr.Délos* 104.23'

*κόνητρον, τό, unexpld. wd. in Chr. inscr., Roueché *Aphrodisias* 80.

*κονθηλαί· ἀνοιδήσεις Hsch.; cf. κανθύλη.

*κονθιναρχειμι, *to be in command of the* *κονθινοι (dub. sens. cf. perh. κοντός (A) 2), κονθιναρχέντ[ου]ν *AE* 1932 suppl. 17 (Crannon, iii BC).

κονία I 1, add '*earth* (of the grave), τὸν ὀθνείῃ κείμενον ἐν κονίῃ *IHadr.* 61.4 (ii AD)'; at end after 'κονίῃσι' insert 'cf. κονίαισι Alc. 283.15 L.-P.'

*κονιᾱκόπος, ὁ, *plaster-grinder*, *POxy.* 2272.29 (ii AD).

*κονιᾱκός, ή, όν, *powdered*, κίττανος· ἡ κονιακὴ τίτανος Hsch.

κονιᾱσις, after '*Gp.* 2.27.5' insert '*POxy.* 3793.10 (AD 340)'

*κονιᾱτης, for '= foreg.' read '*plasterer* or *whitewasher*' and add '*ITyr* 186.2 (κονε-, Chr.), *PMich.*xi 620.137 (AD 239/40)'

*κονιδισμός, ὁ, *disorder of the eyelids*, Cyran. 35 (1.16.11 K.).

κονίζω, for 'v. κονίω' read 'pass., = κυλίεσθαι, prob. *roll in the dust*, Hsch.; fut. κονιεῖσθαι Ph. 2.173 (v.l.)'

†κόνικλος, v. °κουνίκολος.

κόνιμα, for 'v. κόνισμα' read '= κονίστρα 2, *arena*, *SEG* 27.119.15 (Delphi, iii BC). 2 = κονίαμα, *plaster*, *Inscr.Délos* 365.48 (iii BC).'

κονιορτόω, add 'pass., Sch.Gen.Il. 23.764'

κόνιος II, add 'also of Demeter, *SEG* 31.368 (perh. unconnected with κόνις, *dust*)'

κόνισμα, for 'also κόνιμα .. Delph., iii BC' read 'v. °κόνιμα)'

κονίστρα 2, add 'b in pl., Call.*fr.* 328 Pf. (app. referring to παλαίστρα Κερκυόνος Paus. 1.39.3).'

κονίω, line 2, after 'fut.' insert 'κονίσομαι *APl.* 25 (Phil.)' and delete 'κονιοῦμαι .. κονίζεσθαι'; lines 5/6, for 'in Mss. .. 128' read 'mss. sts. incorrectly give forms fr. κονίζω' I 1, add 'οὖδας A.*Pers.* 163' II, last line, delete 'cf. *Pers.* 163 (troch.)'

κόνναρος, after 'κόνναρον, τό, *its fruit*' insert '*PHaun.* 20.10 (sp. κόναρ-; iv/v AD)'

κόννος 1, add '*Inscr.Délos* 1421*Ab*19 (ii BC)' 2, for this section read '*fringe of hair, beard*, Luc.*Lex.* 5; cf. ἱέρωμα· τὸν κόννον Λάκωνες, ὅν τινες μᾶλλον ἢ σκόλλυν Hsch. κ.· ὁ πώγων, ἢ ὑπήνη. ἢ χάρις id.' 3, delete the section.

*κονσιστώριον and κων-, τό, Lat. *consistorium*, *imperial assembly*, *POxy.* 140.5, *PMasp.* 32.15 (both vi AD).

*κονσουλάριος, a, ον, *consular*, Just.*Nov.* 8 not.inscr.

κοντάκιον, add 'Alex.Aphr.*in Metaph.* 548.18'

*κοντηλίτιον, τό, perh. dim. of κοντός (A), *PVindob.* G39847.916, 931 (*CPR* 5.110, iv AD).

*κοντίλος, ὁ, *pole*, perh. used in vulgar sense for *penis*, cf. κ.· εἶδος ὀρνέου, ἢ ὄρτυξ. ἔστι δὲ καὶ ὄφις Hsch., Eup. 364 K.-A.

*κοντοβερνάλιος, ὁ, Lat. *contubernalis*, *comrade, aide*, *PGen.* 79.2, *PLips.* 40 ii 22 (both iv AD); also written κοντοβ- *Dura*⁵ 39.

κοντοκυνηγέσιον, delete the entry.

*κοντόρρινος, ον, *short-nosed*, Rhetor. in *Cat.Cod.Astr.* 7.202.7; cf. °κονδόρις.

κοντός (A) 1, delete '*punting-pole*' and add 'Plb. 21.7.2' 2, before 'Luc.' insert 'Arr.*Tact.* 43.2'

*κοντροκυνηγέσιον, τό, *wild-beast hunt with pikes*, *IGRom.* 4.1632.7 (Philadelphia, v. *SEG* 6.608).

*κοντώδης, ες, app. *armed with a pike*, Ἄρης, ἄφωνον πῆμα, κοντῶδες τέρας Gr.Naz.*Carm.* 2.1.11.

κόνυζα, add 'see also κνύζα, σκόνυζα'

*κοξάλια, τά, late Lat. *coxalia*, *loin-cloths*, *Edict.Diocl.* 27.2, 5.

*κόον, acc. (gender unkn.) κόον γάρ φασι κατὰ γλῶσσαν τὸ πρόβατον λέγεσθαι (sc. in Carian expl. of etym. of ἡ Κῶς) Eust. 318.40.

*κοορταλῖνος, ὁ, Lat. *cohortalinus*, *member of the lowest level of the staff of provincial governors*, *Cod.Just.* 1.5.12.6, 15.

*κοόρτη, ἡ, = κοόρτις, *IEphes.* 737.5 (iii AD).

κοπάζω, add '2 trans., *curb, moderate*, Lxx *Si.* 39.28, 43.23, 46.7, 48.10.'

κόπανον I, after '*pestle*' insert '*Inscr.Cret.* 1.xvii 2*b*2 (ii BC)'

κοπή 3, for this section read '*beating of the web*, i.e. *fulling*, as part of the weaving process; in quots., only in the title of officials to whom tax on fullers and weavers was paid or due, *PFay.* 58.7 (ii AD), *PAmh.* 2.119.4 (AD 200), *BGU* 2547.2 (iii AD)' 7, add '*SEG* 32.794.10 (Olbia, *c.*325 BC)'

κοπιάτης, for 'also κοπιᾶς .. Philippi' read '*SEG* 34.1330 (Lycaonia), perh. also Τίμωνος κωπιά[τ]ας χρηστέ, χαῖρε ib. 30.1622 (Cyprus, ii/iii AD)'

κοπιάω, line 2, for '*Apoc.* 2.3' read '*Onos.* 22.1'

*κοπιδᾶς, ᾶ, ὁ, (κοπίς (B)) *maker of cleavers*, *MAMA* 3.573 (Corycus).

κοπίδερμος, add 'Aesop.*Prov.* 15 tit. P.'

κόπις (A), for '*prater, liar, wrangler*' read '*glib talker*'

κοπίς (B), line 3, for '*broad curved knife*' read 'applied to any chopping or cutting weapon'; line 4, after '(X.*Cyr.*) 6.2.10' insert 'by Lucanians, *AP* 6.129 (Leon.Tarent.)' II, add 'Molpis ap.Ath. 4.140b (= Molpis 1 J.), Hsch.'

κόπος II 3, add '*fruits of one's labour*, *Orac.Tiburt.* 116'

κοπόω, for '*weary*' read '*exhaust* with physical effort'; add '*SEG* 31.1474 (Arabia, AD 575/6)'

κοπρηγία, add '*PSoterichos* 1.24 (AD 69), *PFlor.* 143.5 (iii AD)'

κοπρηγός, after '*dung-cart*' insert '*PLond.* 131.30 (pl., AD 78/9)'

*κοπριαναίρετος, ον, *taken from the dunghill*, *PAmst.* 41.88, 114 (10/8 BC).

*κοπρίδιον, τό, ἀφόρδια· κοπρίδια Sch.Nic.*Al.* 140d Ge.

κοπρίζω, add 'πρασιαῖς κεκοπρισμέναις Dsc. 1.81, *BGU* 2354.12 (ii AD)'

*κοπριών, ῶνος, ὁ, Cret. for κοπρεών, κοπρών *Inscr.Cret.* 4.73*A*9-10 (Gortyn, v BC).

κόπρος I 1, add 'applied also to compost, X.*Oec.* 16.12, 20.11'

κοπτή II, for 'κοπτός II 2' read 'κοπτός II 1'

†κοπτικός, ή, όν, *of* or *for striking, chopping*, or sim., μέρη (sc. ξίφους) Sch.E.*Hec.* 543; adv. -κῶς Hdn.*Epim.* 134.

*κοπτοπώλης, ου, ὁ, *seller of pastries*, *AS* 20.43 no. 13.

κοπτός II 2, delete this section transferring exx. to section II 1.

κόπτω I 6, add 'b *cut letters* (on wood), γράμματα Call.*fr.* 73.1 Pf.' 9, add 'with the oars, in rowing, κ. ὕδωρ Call.*fr.* 18.11 Pf., A.R. 1.914'

*Κοραγωγός, ὁ, *religious official*, *IG* 2².1247.20 (iii BC).

κορακῖνος II, add '*IGC* p. 98 A8 (Acraephia, iii/ii BC)'

*κορακοβρωσία, ἡ, *food for ravens*, *PMasp.* 353.A19 (vi AD, -οσια pap.).

*κορακόπους, ποδος, ὁ, gloss on κορωνόπους, Hsch.

κορακόω, add '*SEG* 17.630, 635 (Perge)'

*κορακώδης, v. κορακοειδής.

κοράλλιον, add '*PSI* 1128.27 (iii AD)' and delete 'Alciphr. 1.39' (line 2), 'Luc.*Apol.* 1' (line 4) and at end 'sts. .. κωράλιον'

*κοράλλιον (B), τό, *figurine*, Alciphr. 1.39, Luc.*Apol.* 1 (κουρ-); cf. κωράλιον.

κοραλλιοπλάστης, for pres. def. read '*maker of figurines*'

*κοραλλίς, ίδος, ἡ, *a precious stone* (?*red jasper*), Plin.*HN* 37.153.

*κοραλλοαχάτης, ου, ὁ, *a kind of agate resembling coral*, Plin.*HN* 37.139, 153.

*κόραλλος, *coral*, *PMag.* 4.2304, Theognost.*Can.* 56 A.

κόραξ, line 1, for '(not in Hom.' read 'place-name in Hom., Κόρακος πέτρη Od. 13.408' I 4, add '*IUrbRom* 106, 107'

*κόραξαι· ἄγαν προσλιπαρῆσαι Hsch.; cf. κοράσσει.

κοραξός, add '*BGU* 1666.12 (i AD)'

κορασίδιον, add '*SEG* 36.616 (Edessa, iii AD)'

*Κορασιοδρόμος, ὁ, *courier serving the town of Corasium*, *MAMA* 3.415 (Corycus).

κοράσιον, line 2, after 'Philippid. 36' insert '(acc. to Phot. introduced as foreign word '; line 3, before '*IG* 7.3325' insert 'esp. of slave girls in manumission inscrs.' and before '*PStrassb.*' insert '*PCair.Zen.* 28.10 (iii BC), *SEG* 24.498, 530, 34.658 (Maced., ii/iii AD, sts. -ιν)'; add '(Maced. wd. acc. to Sch.bT Il. 20.404)'

*κορβάσει· τὸ καταπνεῖ, τὸ βλέπει Theognost.*Can.* 46 A.

*Κορδακά, ας, ἡ, title of Artemis in Elis, Paus. 6.22.1.

*κόρδαμον, = κάρδαμον, in pl. *Dura*⁴ 129.

κόρδαξ, add '2 name given to the trochaic metre, Cic.*Orat.* 193; cf. κορδακικός.'

*κορδούνια, τά, unexpld. wd., prob. a measure of (or receptacle for) grain, *OBodl.* III 295 (i AD).

*κορδυλίς [?ῡ], ίδος, ἡ, *name of a fish*, Ath. 7.306c; cf. κορδύλη III.

κορδύλος, add 'also *some kind of creeping marine animal*, mentioned with the octopus, Opp.*H.* 1.306'

κορεία (B), after 'D.Chr. 7.142' insert 'Nonn.*D.* 1.350 K., al.'

κόρειος II 2, for '*Ath.Mitt.* .. iii BC)' read '*SEG* 24.151.21-2, ib. 153.8 (Attica, iv BC)'

κόρη, line 5, after '*h.Cer.* 439)' insert 'voc. κοῦρα Call.*Dian.* 72, Naumach.ap.Stob. 4.23.7' and after '**κώρα**' insert '(also v.s.v.)'; line 6, after 'Theoc. 6.36' insert '(voc. κῶρα id. 27.52)'   **I 1**, add 'of Athene, γλαυϟόπιδι ϟόρει *CEG* 182, al. (Athens, vi BC); τὴ[ν ἡμ]ετέραν Κόρην (perh. identified w. Aphrodite) *SEG* 18.578.2 (Cyprus, AD 14)'   add '**5** the constellation *Virgo*, Doroth. 78, Nonn.*D.* 2.655 K.'   at end add 'Myc. *ko-wa*'   **B Κόρη**, after '**Κώρα** *GDI* 5047' insert 'Κούρα, Schwyzer 262 (Cnidus, iii BC), Κόρρα, *SEG* 35.826 (Thrace, iv/iii BC)'

*****Κορθιᾶτᾶς** (also **Κροθιᾶτᾶς**), ὁ, title of Pan, *Schwyzer* 67.3, 5 (Messenia, archaic).

+**κορθίλη**, ἡ, κορθίλας καὶ κόρθιν· τοὺς σωρούς. καὶ τὴν συστροφήν Hsch.; cf. -ας ποιεῖν a farming operation, *IG* 2².2493.16 (iv BC).

κόρθιλος, for 'βασιλίσκος' read 'βασιλίσκος III'

κόρθυς, delete 'lengthd. form of κόρυς'

*****Κορία**, ἡ, title of Artemis in Arcadia, Call.*Dian.* 234 (-ίη).   **2** of Athena, Paus. 8.21.4, Cic.*ND* 3.59.

*****κοριανδρόκοκκος**, ὁ, *coriander-seed*, Lxx Sch.*Ex.* 16.31.

κόριαννον, after 'κορίαμβλον Hsch.' insert 'also κολίανδρον (v.s.v.)'   **I**, add '**b** given as alternative name for κώνειον, Sch. Nic.*Al.* 186a Ge.'   at end add 'Myc. *ko-ri-ja-do-no* (*κορί(h)αδνον), *ko-ri-a₂-da-na* (pl.)'

κορίαξος, after 'a kind of *fish*' insert '*Stud.Pal.* 20.224'

κορίδιον I, add '*Schwyzer* 462 (Tanagra, iii BC)'

κορίζομαι, for '*fondle, caress*' read '*speak fondly to*, w. acc.'

*****κορικορώνη**, ἡ, ἐκκόρει κορικορώνην perh. in Horap. 1.8 (ἐκκορί κορί κορώνη(ν) codd.), Sch.Pi.*P.* 3.32 (ἐκκόρει κόρει κορώνας), cf. *Carm.Pop.* 35(b) P., a marriage cry; cf. χελιχελώνη.

Κορίνθιος, add 'Myc. *ko-ri-si-jo* (*Κορίνσιος)'

Κόρινθος I, add 'cf. Myc. *ko-ri-to*, place name'   **III**, Κορινθόθεν, for '*Michel* 1087' read '*SEG* 24.310'

κόριον (A), after '*little girl*' insert 'Eup. 30 K.-A.'

κόρις II, add '*SEG* 32.450 B.13 (Acraephia, iii/ii BC)'

*****κόρκομα**, ἡ, Lat. *curcuma*, *basket*, Anon.Alch. 330.13.

+**κόρκορα**· ὄρνις. Περγαῖοι Hsch. (κορκόρας cj.)

*****κορκορὔγέω**, *rumble*, of the bowels, Sch.Ar.*Nu.* 386.

κορκορυγμός, read '**κορκορΰγισμός**' (v.l. -υγμός)'

*****κορμηταί**· κοσμηταί Hsch.; cf. κόρμος (B).

κορμίον II 2, add 'cf. mod. Gk. κορμί *body*'

+**κορμολογία**, ἡ, perh. *gathering of logs* or *removal of stocks*, *SB* 5126.25 (AD 261), *PCol.* 179.18 (AD 300).

+**κορμός** (A), ὁ, *the trunk* or *stock* of a tree, Od. 23.196; κ. ἐλάϊνοι *PCair.Zen.* 431.4 (iii BC), *PLond.* 1972.2; in the measurement of a vineyard, ἀπὸ κορμοῦ εἰς κορμόν *PFlor.* 50 i 2, al. (iii AD).   **2** *sawn-off trunk, log*, κορμοὶ ξύλων Hdt. 7.36, E.*Hec.* 575, *HF* 242, κ. ἐλάας Ar.*Lys.* 255, *PCair.Zen.* 154.2 (iii BC), poet. κ. ναυτικοί, i.e. oars, E.*Hel.* 1601.

κορνικουλάριος, add '*SEG* 31.1583 (iii AD); κορνικολάριος *POxy.* 2004, κορνοκλάριος *IGRom.* 3.59, *TAM* 4(1).112, κολλικλάριος *BGU* 435.8'

κόροιβος, after '*fool*' insert 'Suet.*Blasph.* p. 59 T.'; after '**Κόροιβος**' insert '*IG* 1³.1147.44 ([K]όροιβος, v BC), Call.*fr.* 587 Pf.'

κοροκόσμιον I, for this section read '*small votive image of a nymph*, Clem.Al.*Protr.* 4.58 and local loc. (false etym. in Sch.Theoc. 2.110)'   **II**, add '(calque on Coptic expression)'

*****κορορᾶτωρ**, ορος, ὁ, Lat. *colorator*, perh. *dyer* or *polisher*, *Edict.Diocl.* 7.54, cf. *Gloss.*

κόρος (B), add 'Myc. *ko-wo* (κόρϝος)'

κόρος (C), before 'Hsch.' insert 'perh. Bion. 2.17'; add 'cf. κορέω (A)'

κόρος (D), add '*PSI* 554.14 (259/8 BC)'

*****κορόσπερμον**, τό, unexpld. wd., *PRyl.* 630*.8 (iv AD): perh. read κορ⟨ι⟩όσπερμον, *coriander seed*.

*****κορριγία**, ἡ, Lat. *corrigia*, *thong*, part of a harness, *Edict.Diocl.* 10.19 (sp. χορηγία in version in *SEG* 37.335).

κόρση I 4, add 'of the head of a plant, Nic.*Al.* 253'

*****κορότης**, ου, ὁ, precious stone, prob. = κορσοειδὴς λίθος, *Lapid.Gr.* 171.12, 189.23.

*****κορτήγ(ι)ο)ν**, τό, wd. of unkn. meaning, perh. in a medical context, *PVindob.*G 39847.927 (*CPR* 5.110, iv AD).

*****κορτιανός**, ὁ, fr. Lat., *member of a cohort*, *POxy.* 1253.4 (iv AD).

*****κορτίνα**, ἡ, also κορτίνη, *PMasp.* 6 ii 48 (vi AD), Lat. *cortina*: **I** (archit.) *arch, vault*, *IGLS* 13.1.9135-6 (Bostra, vi AD).   **II** as late Lat., *curtain*, *PMasp.* l.c., *PNess.* 180.7 (vi/vii AD).

*****κόρτον**, τό, wd. occurring in inventory of utensils, perh. *sieve*, *SEG* 24.361.15 (Thespiae, 386/380 BC); cf. κύρτη, κύρτος.

κορυδός, line 6, after '**κορυδαλλή**' read '**κορὔδαλλά** or **κορὔδαλλᾰ**' and add '*ASAA* 21/2 NS 275-8 (Sicily, iv/iii BC)'   at end add 'cf. Myc. *ko-ru-da-ro-jo* (gen.), pers. n.'

*****Κορύδων**, ονος, ὁ, name of a feast, *SEG* 32.1243.36 (Aeolian Cyme, 2 BC/AD 2).

Κορυθαλίστριαι, add '(-θαλίστ- La.)'

*****κορυθήκη**, ἡ, *helmet-case*, *Inscr.Délos* 1417Ai121 (ii BC).

κόρυθος II, add 'Paus. 4.34.7, also Κόρυνθος (q.v. II)'

*****κορύθων**· ἀλεκτρυών Hsch.; cf. κορυνθεύς II.

κορυμβίας, add '**II** kind of giant fennel, Plin.*HN* 19.175.'

κορύμβιον, add '**III** curled wig, Petron. 110.1, 5; cf. κόρυμβος II.'

*****κορυμβίς**, ίδος, ἡ, = κόρυμβος II, Antim. 175 Wy.

κορυμβῖτης, add '**II** kind of spurge, Plin.*HN* 26.70.'

κόρυμβος I 1, add 'of a hunting-net, ἐπ' ἀκροτάτοισι κορύμβοις *at the extreme ends*, Opp.*C.* 4.125'

κορυνήτης, add 'Parth. in *Suppl.Hell.* 634'

κόρυς, add 'Myc. *ko-ru*, *ko-ru-to* (gen.), *ko-ru-pi* (instr. pl.)'

κορύσσω I 1, line 3, after 'Pi.*I.* 8(7).58' insert 'ἐμφύλιον αἷμα Hes.*fr.* 190.2 M.-W., Semon. 7.105, Ibyc. 30(b) P.'

κορυστής, add 'Alcm. 1.5 P. (-τάς)'

κορύφαιον III, for '*ridge-beam*' read '*main rafter*'

κορύφαιος I, add '**2** = κορύφαινα, Cyran. 110.'   **II 2**, sup., add '*Suppl.Hell.* 1176, Favorin.*fr.* 134 B.'

κορυφή I 1, add 'in fig. phr., ἀπὸ κορυφῆς ὡς ὀνύχων ἔπεσεν τὸ Βένετον *SEG* 31.1492 (Egypt, AD 608/10)'   **II 4**, add '*the Chief*, i.e. the emperor, *CPR* 5.17.4 (v AD)'

κορυφιστήρ 2, add 'pl., gloss on ἀμπυκτῆρες, equated with προμετωπίδια, Sch.A.*Th.* 461'

+**κορυφιστής**, οῦ, ὁ, κ.· κόσμου γυναικείου τὸ περὶ τὴν κεφαλὴν χρυσίον. καὶ κεκρυφάλου τὸ μέσον ῥάμμα Hsch., prob. also κορυφισ[τής] *SEG* 21.556 (Athens, iv BC).

κόρυφος III, add '*POxy.* 3298.2 (iii AD)'   add '**IV** a hair-style, Hsch.'

κορώνη I 2, after '*crow*' insert '*Corvus corone*'; line 2, for '*C. corone*' read '*C. frugilegus*'   **II 6**, add 'κορώνας ἀναδούμενοι Sophr. 163, cf. κορωνίς II 1, χορωνός'   add '**8** τὸ ἄκρον τοῦ αἰδοίου, Suid., cf. perh. [Archil.] 331 W.'   add '**III** a kind of fish, Hsch.'

κορωνιάω II, for 'leaves' read 'sheaves (of corn, fr. the ripeness of the ears)'; add 'also v.s.v. κορυνιόεις (v.l.)'

κορώνιος I, for this section read 'μηνοειδῆ ἔχων κέρατα βοῦς Hsch.'

κορωνίς II 2 a, for 'Heph.*Poem.*' read 'Heph.*Sign.*'; at end for 'etc.' read '*D.H.Comp.* 4 (p. 21)'

κορωνισταί, add '**II** sobriquet of gangs of young men in Cumae, οὓς κ., ὡς ἔοικεν, ἀπὸ τῆς κόμης ὠνόμαζον Plu. 2.261e.'

*****κορωνοβόλον**, τό, app. *sling for shooting crows*, *AP* 7.546.

κορωνοβόλος, delete the entry.

κορωνός, line 1, after '*crooked*' insert 'Il. 2.746, al., as pers. n.'

κοσκινευτής, add 'included among employees at the circus, *OAshm.Shelton* 190'

*****κοσκινοειδής**, ές, *sieve-like*, adv. -ῶς, Zos.Alch. 139.3.

κόσκινον II, for this section read 'κ. Ἐρατοσθένους, method of finding prime numbers by elimination, Nicom.*Ar.* 1.13'

κοσμαρίδιον, add '*SB* 11075.15 (v AD)'

*****κοσμάριον**, add '*BGU* 2217.12 (pl., written -ιοι, ii AD)'

κοσμέω II 1, line 5, after 'Pl.*Phd.* 97c' insert 'κοσμήσας τὴν διοίκησιν *PThead.* 14 (iv AD)'   add '**b** pass., *conform to, be guided by*, w. dat., D.H.*Amm.* 1.2; κατά τι ib. 1.1.'   **2**, after 'κόσμος III' insert '*Inscr.Cret.* 4.14 (vii/vi BC); after 'Plb. 22.15.1' insert 'w. acc., *perform an act as* κ., Meiggs-Lewis 2.3.3 (Dreros, 650/600 BC)'   **III 2**, add 'πάσῃ ἀγαθῇ ἀρε[τῇ] κεκοσμημένη Mitchell *N.Galatia* 79, *MAMA* 8.449; also w. gen., πάσης ἀρετῆς κεκοσμημένος *MAMA* 1.228 (Laodicea Combusta); *embellish* a building, *SEG* 35.1471.(2)18 (Cyprus, vii AD)'

κόσμησις, line 3, after 'Arist.*Oec.* 1344ᵃ19' insert '*SEG* 30.1786, 1787A (Cyrenaica, vi AD)'

κοσμητεία, add '*IEphes.* 4337.8 (-ῄα, i AD)'

κοσμητήρ II, for this section read 'κοσμητῆρες, οἱ, magistrates at Itanos, *Inscr.Cret.* 3.iv 3.22, 4.14'

κοσμητής II, add '**3** = ἐνταφιαστής (Κουριεῖς), Hsch.'

+**κοσμίζω**, *adorn, embellish*, *Cod.Just.* 8.11.3; cf. σαρῶ· κοσμήσω (cod. κοσμίζω) Hsch.

κοσμικός I, add '**2** *world-wide*, ἀφορία Longin. 44.1; σεισμός *IGRom.* 3.739 xiii 48 (Rhodiapolis, ii AD).'   **II 1**, add 'neut. pl. as subst., Mitchell *N.Galatia* 465'   add '**III** *fashionable*, Mart. 7.41.1,2.'

κόσμιον, add '*adornment*, Just.*Nov.* 14 pr. 1.   **2** *votive image*, *AP* 9.326 (Leon.); cf. °κοροκόσμιον.'

κοσμογένεια, add '*creation*, applied to *Ge.* 1-2.7, Cels.ap.Orig.*Cels.* 6.29'

*****κοσμογενής**, ές, *native of the universe*, Gnomol.Vat. in *WS* 10.237.

*****κοσμογόνος**, ον, *producing the physical world*, κ. στοιχεῖον τὸ ὕδωρ Heraclit.*All.* 22.

κοσμοκράτωρ, delete 'epith. .. *H.* 4.3'

κοσμοποίησις, after '*ornamentation*' insert '*Inscr.Délos* 1443Bii 104 (ii BC)'

κόσμος, line 1, after 'ὁ' insert '(heterog. pl. κόσμα *POxy.* 494.10 (in sense II, ii AD); for form with rhotacism v.s.v. κόρμος)'   **II**, add 'τὸν ναὸν καὶ τὸν ἐν αὐτῷ κόσμον *IEphes.* 3757; of the equipping of a religious festival, τῶν μυστηρίων SEG 32.1243.14 (Aeolian Cyme, 2 BC/AD 2)' **IV 2**, add 'Arist.Ph. 252ᵇ 26'

*κοσμοτόκος, ον, giving birth to the world, Heraclit.All. 64.7, 85.18.

*Κοσμοτορύνη, ἡ, World-stirrer (of war), title of a satire by Varro, Nonius pp. 8, 231 L.

κοσμοτρόφος, after 'Man. 1.2' insert 'SB 4313.2 (Alexandria, i/ii AD)'

*κοσούλλιον, τό, unexpld. wd. in Mim.adesp. 15.32.

κόσσυφος, line 1, after 'Att.' insert 'Boeot., Thess.' **II**, add 'κοττούφω (gen.) SEG 32.450. B9 (Acraephia, iii/ii BC), cf. pers. n. Κόττυφος D. 18.151, Aeschin. 3.124, al.'

*κόσταια· ἄρτος κρίθινος Theognost.Can. 46 A.; cf. κοσταί, °γοσταί.

κοστάριον, for 'prob. = κόστος' read 'spice derived from κόστος' and add 'PHaun. 20.11 (iv/v AD)'

κοστόϊνος, η, ον, perh. having the colour of κόστος, pale yellow, SB 9307.3,6,9 (ii/iii AD).

κόστον, (s.v. κόστος), after 'Thphr.Od. 32' insert 'pl. κόστα Dura⁴ 129' and delete '(but κόστα .. 15.19)'

*κόστον (B), τό, in pl., wooden parts of a cart, Edict.Diocl. 15.19, cf. Lat. costae.

κόστος, add 'SB 9834 b 22 (iv AD), cf. Skt. kúṣṭha-'

⁺κοσύμβη, ἡ, app. some kind of over-garment, D.Chr. 72.1, Poll. 2.30 (v.l. κορσύμβην, κοσσάμην), Hsch. (κοσσύμβη, κόσσυμβος), EM 311.5, 349.15.

⁺κοτεινός, ή, όν, dub. wd. in PHib. 225 iii 2 (perh. = κοτήεις); (also cj. for σκοτεινός in Pi.N. 7.61).

⁺κότθυβος, an article of a soldier's equipment, RA 6.31 (Amphipolis, ii BC); perh. = °κοσύμβη.

κοτίκας, read 'κοτικᾶς'

κοτίλιον, v. °κοτύλιον.

κότινος, line 1, after 'Theoc. 5.32' insert 'al.' add '**b** olive-garland, SEG 31.903.33 (Caria, iii AD).'

*κοτόνιον, τό, app. kind of fig, PKöln 318.10 (vi/vii AD); cf. κόττανον.

*κότος, τό(?), prob. for κύτος, jar, in inventory of utensils, SEG 24.361 (Thespiae, 386/380 BC).

κόττα̣βος 3, add 'κόττα̣βον βαλεῖν Pi.fr. 128'

κοττάναθρον, for 'κοττόβαθρον' read 'κοττανάβαθρον'

κότταν̣ον, add 'fig., καὶ ἡ παρθένος παρὰ Κρησί Hsch.'

*κοττιδιανός, ή, όν, Lat. cotidianus, daily, POxy. 2408.10 (iv AD), Lyd.Mag. 107.18, 160.1.

⁺κοττιστής, οῦ, ὁ, dice-player, Dura⁹ 1 p. 217 (iii AD), Gloss.

κόττος **III**, add '2 κόττῳ in a lump sum, ἐὰν κόττῳ πραθῇ οἶνος Basilica 53.7.10 Sch.'

κοτύλη, after lemma insert 'non-Att.-Ion. -ᾱ, SEG 25.343 (ϙοτύλλᾱ Corinth, viii/vi BC)' **3 b**, delete 'prob. also a smaller measure .. ὀξύβαφον' and transfer 'Hp.Mul. 1.6' to section 3 a.

κοτυληδών 3, add 'Hp.Int. 18, Oss. 16'

*κοτύλιον, τό, dim. of κοτύλη, (κοτιλ- Inscr.Délos 1429Bii 25 (ii BC)) small vessel or receptacle, Inscr.Délos l.c., SB 1.22 (iii AD), PLond. 1657 (-ιν, iv/v AD).

κοτυλίς, delete the entry.

κότυλος, add 'Schwyzer 440.4 (Boeotia, vi BC)'

*κουαδράριος, ὁ, Lat. quadra(tar)ius, relating to stonemasonry, BGU 21.1.5, PGoodsp.Cair. 12.i.6 (both iv AD), al.

*κουαιστορεία, ἡ, quaestorship, IGLS 9112 (κυαι-; Bostra, iv AD)

*κουαίστωρ, ορος, ὁ, Lat. quaestor, IGRom. 3.238.4; also κυαίστωρ ib. 4.1307.15; κοιαίστωρ Procop.Arc. 14.3; κοιαίστωρ Cod.Just. 1.12.8 pr.; κυέστωρ Syria 29.317 (Arabia, iii AD).

*κουαιστώριος, ὁ, Lat. quaestorius, σκρεῖβα κ. λιβράριος ILS 8833, Ramsay The Social Basis 60.

*κουαττόρουιρ, ον, ὁ, Lat. quattuorvir, IG 12(2).235.4 (Mytilene, ii AD).

*κουβαρίζω (v.l. -άζω), gloss on μηρύω (v. μηρύομαι), Sch.Theoc. 1.29, as mod. Gk. roll into a ball.

κουβηζός, after 'στηβεύς' insert '(perh. read στιβεύς)'

*κουβικουλαρία, ἡ, chamberlain (female), SEG 34.1262 (Bithynia, v AD), 8.175 (Jerusalem, vi AD).

*κουβικουλάριος, ὁ, chamberlain (male), IMylasa 620.6.(sp. κουβουκλ-, v AD).

*κουβουκλεῖον, τό, (Lat. cubiculum), bedchamber, Just.Nov.Ed.Not. 1, al. **2** tomb, κουβούκλιν IG 14,1451 (Rome, ii AD).

*κουδισάμιος, ὁ, perh. tool-grinder, (cf. Lat. cudis and samiarius), MAMA 3.724 (Corycus).

*Κουέριος, title of Poseidon, Schwyzer 559 (Thessaly, iii/ii BC).

*κούκεον, τό, fruit of the doum-palm, Ostr.Wilbour 76 (prob. ii AD); also κούκιον PHeid. 333.29 (v AD, sp. κούκιν); cf. κόϊξ.

*κουκκούλλος, ὁ, Lat. cucullus, hood, κουκκ[.]λον POxy. 3060.5 (ii AD).

*κουκούβη, ἡ, species of owl, (cf. Lat. cucubare, to hoot), Eust. 1523.59.

*κουκουλάρι(ο)s, ὁ, maker of cuculli, SEG 39.649 (Augusta Traiana, ii/iii AD).

*κουκούλλιον, τό, dim. formed on Lat. cucullus (-lla), hood, PMich.VIII 482.4 (ii AD, written κοκκούλιον), POxy. 1300.9 (v AD, written κούκλιν).

*κούμουλον, τό, a measure, one thirtieth of an artaba (cf. Lat. cumulus), inscr. in ZPE 15.176 (v/vi AD), PLond. 1718.8, 9, al.; as tax or levy, PFlor. 75.21, PVindob. G13933ᵛ.10 (ZPE 32.255, κούμολ-), POxy. 3395.12 (κούμηλ-), 3481.10 (κούμελ-, all iv AD).

⁺κουνίκλος, v. °κουνίκολος.

*κουνίκολος, ὁ, Lat. cuniculus, rabbit, Edict.Diocl. 4.33A, also κουνίκλος Ath. 9.400f, κονίκλος Ael.NA 13.15, κυνίκλος Plb. 12.3.10 and prob. Gal. 6.666.

*κούνιον, τό, cradle, Sch.Call.Jov. 48, cf. Lat. cunae, mod. Gk. κούνια (fem).

*κουνόπρειστις, v. κυνόπρηστις.

*κοῦπα, ἡ, Lat. cupa, grave, niche, IUrbRom 300, Not.Scav. 1931.369 (Catania, not before iii AD).

κουρά **II** 1, add 'ἀπονυχίσμασι καὶ κουραῖς Pythag.ap.D.L. 8.17'

⁺κουράλιον, v. ‡κοράλλιον (A) and (B).

*κουρατορεύω, add 'trans., act as guardian of (a minor), in pass. οἱ ἐπιτροπευόμενοι ἢ κουρατορευόμενοι Modest.Dig. 27.1.15.3; also κουρατωρ-, Cod.Just. 3.10.1.1'

κουρατορία, add 'position of guardian, Modest.Dig. 27.1.2; also κουρατωρεία, Cod.Just. 3.10.1.2'

κουράτωρ, after 'ορος' insert '(ωρος Cod.Just. 3.10.1.1)' and add 'JJP 18.190 l. 3 (ii/iii AD)'

κουρεῖον **II**, delete the section.

*κούρειον, τό, victim offered for boys and feasted on by the φράτερες at the feast κουρεῶτις, S.fr. 126, Is. 6.22 (κούριον codd.), IG 2².1237.28. **b** kid or lamb offered in spring to Hermes fr. each herd or flock, Schwyzer 721.13 (Thebes on Mycale, iv BC).

Κούρεος, delete 'from foreg. II'

*κούρευμα, hair cut off, Hsch. s.v. κάρμα.

κουρεύομαι, add 'ἔγκαρτα· τοὺς κεκουρευμένους πυρούς Hsch.'

κουρεύς **I** 1, add '**b** Cypr. ko-ro-u-se perh. κόρους or κο(ρ)ρούς, app. as title of official, Kafizin 117b, 118b, al. (in alphabetic script written κουρεύς).' **2**, add 'Edict.Diocl. 7.22, 23'

κουρευτής, after 'barber' insert 'UPZ 96.10 (ii BC)'

κούρητες **II** 1, line 5, after 'worshipped in Crete' insert 'and elsewhere'; lines 7/9, for '(sg. only late)' read 'sg., IG 12(3).350, al. (ϙōρες, Thera, perh. vii BC), SEG 9.107, 108, 110 (κωρ-, Cyrene, iv/iii BC)'; at end delete ')' after 'Dam.Pr. 267' **3**, add 'sg. κούρης IEphes. 47.30, al.'

*κουρητεύω, serve as κουρῆτες II 3, IEphes. 47.7, 1060.9, 1061.8.

*κουρία, ἡ, Lat. curia, D.C.fr. 5.8.

*κουριάτιος, α, ον, Lat. curiatius, applied to an assembly, D.H. 9.41.2, BMus.Inscr. 3.544.

κουρίδιος **I** 2, after 'Archil. 18' insert 'κωριδίας τ᾽ εὐνᾶς Alcm. in POxy. 2443.17'

κουρικός **I**, after 'hair' insert 'σίδηρος PMich.Zen. 54.2 (iii BC)' add '**b** κουρική (sc. τέχνη), D.Chr. 7.117.'

*κουρῖτις, ιδος, ἡ, a plant, = περιστερεὼν ὕπτιος, Ps.-Dsc. 4.60, Ps.-Apul.Herb. 4, 72.

*κουριωσος, ὁ, late Lat. curiosus, inquiry agent, informer, PColl.Youtie 74.4 (iii AD), SEG 35.1523 (-ιόσοι pl., Seleucia Pieria, vi AD), BCH suppl. 8 no. 150 (κουρίος[-, Thessalonica, v/vi AD).

*κουρκούτη, ἡ, gruel, Sch.Ar.Pl. 673; cf. mod. Gk. κουρκούτι neut.

*κουροπερσονάριοι, οἱ, fr. Lat. cura and personalis, officials of some kind, POxy. 2050.5. (vi AD), cf. SB 2254.4.

κοῦρος (B), for 'loppings .. tree' read '= κορμός (A) 1' and add 'Inscr.Délos 442A157 (ii BC)'

κουρότερος, line 2, for 'Hes.Op.[447]' read 'Hes.Op. 447'

*Κουροτρόφιον, τό, shrine of Γῆ Κουροτρόφος, IG 2².4756 (i/ii AD).

κουροτρόφος, line 4, after 'Od. 9.27' insert 'Εἰρήνη .. κ. Hes.Op. 228'

κουρούλλιος or -ούλιος, add 'also κουρούλης IGRom. 3.238.7 (Galatia), κορούλης JRS 49.96.11 (Cyrene, ii AD)'

*κουρσόριος, α, ον, τροχάδια κ. fr [gal]licae cursuriae, Edict.Diocl. 9.14, (but 14 A: τροχάδων κουρσόρων).

⁺κούρσωρ, ορος, ὁ, Lat. cursor, as a rank, BCH suppl. 8 no. 152 (-ουρος gen., Thessalonica, v/vi AD).

κουσούλιον, before 'cloak' insert 'a garment, perh.' and add 'PHeid. 333.7'

κουστούμηνα, add 'sg. (-μίνου), Hsch. s.v. ὄχνη'

κουστωδία, add 'al., PRyl. 189.2 (ii AD); also κοστωδία, prob. in POxy. 294.20 (AD 22)'

*κουφάριον, τό, app. cask or sim., PAmst. 79.3 (iv/v AD); cf. κοῦφος 6.

*κουφήρης, ες, unexpld. wd., Lyr.adesp. 5.3 P.

κουφίζω **II 2 b**, line 4, delete 'τῆς .. 16 J.] **3 b**, for this section read 'cancel an entry in a register of tax-payers, κουφισθείσης αὐτῆς PBremen 24.7 (ii AD); κουφίσαι τὸ ὄνομα .. τοῦ πατρός POxy. 126.8 (vi AD); an item of property, τὰς γᾶς κουφισθῆναι ἀπὸ τοῦ ὀνόματος PWürzb. 18.10 (iv AD); med., make an abatement of claim, PMasp. 95.10 (vi AD); reduce a price, BGU 2332.13, 24 (AD 374)'

κουφοκεράμειον, τό, *workshop* of a κουφοκεραμεύς, *POxy.* 1917.102 (vi AD).

κουφοκεραμεύς, έως, ό, kind of potter, *POxy.* 1917.22 (vi AD), *PVindob.Tandem* 17.2 (vi/vii AD); cf. κοῦφος 6, neut. as subst., prob. referring to kind of lightweight ware.

κουφοκεραμούργιον, τό, *pottery*, rest. in *CPR* 14.2.20 (vi/vii AD).

κουφοκεραμουργός, ό, = κουφοκεραμεύς, *Stud.Pal.* 612.3 (vi AD), *BGU* 368.13, 29 (AD 615).

κοῦφος I 1, line 1, delete 'pl.' (v. section II 2) **6**, add '**b** transf. w. gen., συνέσεως κούφη Sch.S.*El.* 403.'

κοφαλο-, incomplete wd., name of trade or profession in *PIand.*inv. 488 (*Aegyptus* 27.50, iii AD).

κόφινος II, after 'Boeotian' insert 'and Thess.' and add '*SEG* 34.558.28 (Larissa, ii BC)'

κοφορέω, shortened form of κοπροφορέω, *convey manure*, *PHarris* 95.

κοφορία, ή, *conveyance of manure*, *POxy.* 1220.8 (iii AD), v. °κοφορέω.

κοχλαζοκύμων, ον, *with splashing wave*, *PMag.* 4.184.

κοχλιάζων, delete the entry.

κοχλιάξων, ονος, ό, app. *shaft with screw-thread*, Orib. 49.20.6.

κοχλίας II, add '**6** *the pinna of the external ear* (opp. σκάφος), Poll. 2.85.'

κοχλιοειδής, delete 'spiral .. πολύδονος'

κοχλίον, τό, dim. of κόχλος, = τελλίνη, a type of mollusc, *Com.adesp.* 292.20 A.; in pl., Hsch. s.v. ξιφύδρια.

κοχλιώδης, ες, *conchoid*, Ath. 3.86b.

κόχλος 1, line 3, after 'fem.' insert 'Theoc. l.c.' add '**b** κ. ναυτικός, type of mosaic design, *PCair.Zen.* 665 (iii BC).' **3**, for 'Eust. 728.47' read 'Eust. 728.48'

κοχῡδέω, add 'also **κοχυδεύω** dub. in Sophr. in *PSI* 1214d 6'

†**κόψα**· ὑδρία Hsch.; cf. κοψία.

κραβάτιον, add 'also **κραββ-** Hsch. s.v. λέκτρα, also v.s.v. κράββατος'

κραβατοπόδιον, τό, *leg of a bed*, Sch.Od. 8.278 (gloss on ἑρμῖνες).

†**κραβάτριος**, ό, (κραβατάριος), masc. as subst., perh. *chamberlain*, *IPE* 2.297; cf. κραβακτήριος s.v. κράββατος, °κρεβαττάριος.

κράββατος, line 6, after 'Virg.*Mor.* 5' insert 'γράβακτον *PMasp.* l.c.' and add 'also v.s.v. ‡κραβάτιον, °κραβάτριος, °κρεβαττάριος'

κραδάω I, for '= κραδαίνω, only in part.' read '*brandish*' and after 'Il. 20.423' insert 'κραδάω τοῦδ' ὕπερ αἰγανέην epigr. in *IStraton.* 41(a)' add '**2** intr., *wave, quiver*, ἐπεὶ κελάδοντος ἀήτεω ταινίαι .. ἐφύπερθε διηέριαι κραδάουσι Opp.*C.* 4.410.' delete etym. note at end of article.

κράδη I 1, delete 'quivering spray at the' **III**, for this section read 'device for suspending actors above the stage, Ps.-Plu.*Prou* 2.16, Poll. 4.128'

κράζω, p. 989, line 1, after 'Men.*Sam.* 204' insert 'κεκραγήσομαι Hsch. s.v. κεκραγήσει'; line 4, after 'ἔκραγον (ἀν-)' insert 'Od. 14.467'; line 6, for 'late κέκρᾱγα' read 'l. κέκρᾱγ' in'; line 7, after 'κεκράγετε' insert '(κεκράγατε codd. RV)'; line 8, delete 'post-Hom.' **1**, add 'transf., perh. *babble*, κραζόμενα φάσκεις καὶ ἐγὼ φάσκω (pass.) *POxy.* 2353.2 (i AD)'

κραίνω I 1, line 3 fr. end, delete 'λάχη .. A.*Eu.* 347 (lyr.)' **III 1**, line 2, after 'Od. 8.391' insert 'Ephebic oath in *REG* 84.370 (κριν- in Poll. 8.106, Stob. 4.1.48)' **2**, add 'pass., *AP* 6.114'

κραιπᾰλόβοσκος, for 'which draws on drunkenness' read '*provoked by a hangover*'

κραιπνός, add 'ἔβαινε κραιπνὸν βῆμα βαστάζων ποδός Ezek.*Exag.* 269 S.'

κράκτης, add 'v.l. in Ar.*Eq.* 304'

κράμβη, add '**4** prov., of stale repetition, Juv. 7.154, δὶς κ. θάνατος Sch.ad loc., Suid.'

κραμβήεις, for '*like a cabbage*' read '*of a cabbage*'

*κραμβιτάριος, ό, *cabbage-seller, greengrocer*, *ITyr* 31.

*κραμβιτᾶς, ᾶ, ό, perh. *vegetarian*, *AE* 1929.152 (κραμβ-, Thessalian Thebes), *Corinth* 8(3).563, *EHBS* 38.48 (pap., κραμπ-, all Byz.).

*κραμβοκέφαλος, ον, *cabbage-headed*, *PSI* 1259.7, *SB* 7997.7 (both ii/iii AD).

*κράμμη, ή, = κράμβη, *PCair.Zen.* 702.26 (iii BC).

κράνα II, delete the section.

*Κραναΐδαι, οἱ, *sons of Cranaos*, i.e. *people of Attica*, E.*Supp.* 713 (cod. δαν-).

*Κραναΐος, α, ον, *Attic*, St.Byz. s.v. Κραναή.

*Κρᾰναῖος, v. ‡κρηναῖος.

κρᾰνᾰός 2, for 'hard' read 'applied to things having a rough appearance' delete '**3** *stinging*' and add 'πίτυς *AP* 6.110'

κράνεια, line 3, after 'Od. 10.242' insert '(eaten by Diogenes, D.Chr. 6.62)'; line 4, for 'cherry-wood' read '*cornel-wood*'; line 6, for '= spear' read '*spear of cornel-wood*' and before '*AP* 6.123 (Anyte)' insert '*AP* 6.122 (Nicias)'

κράνειον, delete 'Amphis .. (prob.)'

*κράνειος (A), α, ον, *of cornel-wood*, ἀστίλιον κράνειον *Edict.Diocl.* 14.4.

*Κράνειος (B), ό, name of a month at Buthrotum, P.Cabanes *L'Épire* no. 43 (iii/ii BC); perh. = Κάρνειος.

κράντωρ, at end for '*AP* 6.116 (Samos)' read '*AP* 6.166 (Samius)'

κράς I, add '*top, upper rim* of a *crater*, κρατῆρες .. ὧν κρᾶτ' ἔρεψον S.*OC* 473'

κράσπεδον 1, for this section read '*edge, border, fringe*, esp. of a garment, κράσπεδα στεμμάτων Ar.*V.* 475, Diph. 43.30, Theoc. 2.53, χρυσᾶ κ. Chamael.ap.Ath. 9.374a, Chrysipp.*Stoic.* 3.36.37; of a sail, E.*Med.* 524; in Jewish context, Lxx *Nu.* 15.38, 39, Sm.*Ez.* 5.3, al., Ev.Matt. 9.20; transf., τοῦ κρασπέδου τῆς κορυφῆς μου (app. = forelock), Aq.*Ez.* 8.3'

*κρασπέδωσις, εως, ή, *seaming of edges*, *PZen.Col.* 15.3, 5 (iii BC).

κραστήριον II, add 'perh. also *Inscr.Délos* 1452*A*30 (written κλαστήριον v.s.v.)'

*κραστιφόρος, ον, *producing fodder*, Pallad.ap.Ps.-Callisth. 3.13.

*κραστός, ά, όν, prob. past pass. part. of °κράω, cf. γράω, κράστις, etc., *fed at grass* (horses), *SEG* 25.556 (Haliartus, iii BC).

*κραταιγίς, ή, a name of the plant σατύριον, Plin.*HN* 26.99.

κραταιόω I 2, add 'πολεμεῖν αὐτὸν ἐκραταιώθη Lxx 2*Ch.*35.22'

*Κρατεᾱνός, ό, cult-title of Apollo in Mysia, Robert *Hell.* 10.137ff.; -ή, ή, of Artemis, Ἀρτέμιδι Κρατιαν[ῇ] *SEG* 33.1101.3 (Paphlagonia).

*κράτεραυγής, ές, (in dub. context, app.) *having powerful rays*, *Lyr.adesp.* 7(d).8 P.

κρατερόφρων, for 'Διὸς .. *IG* 1².503' read 'Διὸς κρατερόφρ[ονι παιδί *CEG* 243 (cf. ..]φρονι παιδί ib. 206)'

κρατεύω, add 'in sense *hold in the hand*, *SEG* 23.220 (Messene, i AD)'

κρᾰτέω II 2, lines 1/2, delete 'prevail .. 104' **4**, after '*master*' insert 'Pherecyd. 22(a) J.'; add '*to be restrained, held back*, οἱ δὲ ὀφθαλμοὶ αὐτῶν ἐκρατοῦντο τοῦ μὴ ἐπιγνῶναι αὐτόν Ev.Luc. 24.16' **IV 3**, add 'μὴ κρατεῖσθαι τὸν εἰς τόπον αὐτοῦ χειροτονηθέντα Modest.*Dig.* 27.1.13.5'

κράτημα 3, add 'hilt of a sword, prob. in Sch.E.*Hec.* 543'

κρατήρ, add 'Myc. *ka-ra-te-ra*'

†**κρατηρίαρχος**, a religious official in celebration of the Mysteries, *IGBulg.* 1.401.6 (-α[ρχ]ος, Thrace).

κρατηρίζω II, add 'rest. in *IEryth.* 206.7-8 (iv BC), v.s.v. °κρατηρισμός'

κρατηρίσκος, after 'ό' insert '*Inscr.Délos* 104.27 (364 BC)'

*κρατηρισμός, ό, Ion. κρητηρ-, *serving of drink from a κρατήρ in a religious ceremony*, *IEryth.* 206.11 (iv BC).

κρατηροφόρος, add 'of a coin, as subst., *Inscr.Délos* 1432*B*bi52 (ii BC)'

κρατητής, add '**2** *pommel of a sword*, μύκης· τοῦ ξίφους ‹ὁ› κατὰ τὴν λαβὴν ὑπὸ κ. καλούμενος Hsch.'

κράτιστος 2 b, = *clarissimus*, add 'of a woman, *INikaia* 1062 (iii AD)'; add 'of a φυλή, *TAM* 4(1).238, 258; as cult-title of Zeus, Κράτιστος Μέγιστος Φροντιστής (= *Iupp.Opt.Max.Tutor*) *INikaia* 1141'

κράτος II 3, add 'as honorific appellation of the emperor, πρὸς τὸ ἡμέτερον κράτος Just.*Nov.* 113.1 pr.'

κραυγάζος, after 'Ptol.*Tetr.* 164' add '(v.l.)'

κραύγασος, after 'shouter' insert 'Ptol.*Tetr.* 164 (v.l.)'; add 'Hsch. s.v. βαβάκτης'

κραυγαστής, after '*AB* 223' insert 'Ptol.*Tetr.* 164'

*κρᾰφάγος, v. °κρεοφάγος.

*κράω, v. °γράω.

κρέαγρα, add 'written γρεάγρα *PLond.* 191.10; γρέγρα *PMich.*inv. 3163.44 (*TAPhA* 92.258, iii AD)'

κρέας, line 3, after 'etc.' insert 'κρεῖα Schwyzer 721.24 (Thebes-on-Mycale, iv BC), contr. κρῇ *ASAA* 33/4(1955/6).165, no. 14 (Ialysus)' and after 'κρεῶν .. 191' insert 'Cret., inscr. in *Kadmos* 9.126'; line 7, after 'κρέᾱ' insert 'Timocr. 1.11'

*κρεβαττάριος, α, ον, = κραβακτήριος (s.v. κράββατος), ἐνδρομίς *Edict.Diocl.* 19.5; cf. mod. Gk. τὸ κρεββάτι *bed*.

κρείων, line 4, after '(Il.) 11.751' insert 'w. gen. θεῶν κρείοντι Rhian. 1.21'; line 9, delete '3.10' and after '(lyr.)' insert 'οὐρανοῦ κ. Pi.*N.* 3.10'

κρεμάννυμι, line 3 (form κρεμνάω), add 'also κριμν- *SEG* 33.1039 (Aeolian Cyme, ii BC)' **II 4**, delete the section.

κρεμασμός, line 3, for 'generally' read 'as employed in the reduction of dislocations'

κρέμβᾰλα, add 'Ath. 14.636c'

κρεμβᾰλιαστύς, add in final bracket 'also °βαμβαλιαστύς v.s.v.'

κρέξ II, after '*hair*' insert 'Call.*fr.* 288 Pf.'

†**κρεοβόρος**, ον, *feeding on meat*, cj. in A.*Supp.* 287 (cod. -βροτ-; cf. °κρεόβοτος).

κρεόβοτος, ον, *fed on meat*, A.*Supp.* 287, rest. in *fr.* 451*l* R.

*κρεοδελός, ό, *spit for meat*, κρεοδ[ελός F]έξ inscr. in *PP* 42.42 (Tarentum, vi BC).

*κρεονομία, ή, *distribution of meat*, *Syria* 18.372 (Palmyra).

κρεοπώλιον, after '*butcher's shop*' insert '*meat market*' and add '*SEG* 23.207.34 (Messenia, ii AD)'

κρεουργέω, after '*mangle*' insert 'J.*AJ* 13.12.6, al.'

κρεοφάγος, add 'Dor. **κραφάγος**, Σώσι Νύμφων κραφάγε χαῖρε *IG* 14.351 (Kephaloidion)'

***κρεοφυλάκιον**, τό, perh. for χρεωφυλάκιον (rather than *meatstore*), *AE* 1936.32 (Cavalla, ii BC).

***κρεΰλλιον**, τό, dim. of κρέας, Theognost.*Can.* 126.

κρήγυος 1, for this section read '*good, honest*, οὔ πώ ποτέ μοι τὸ κ. εἶπας Il. 1.106, ποτὶ οὐδὲν κράγυον σχολάζοντες Lysis *Ep.* 3; of persons, παρ' οἴνῳ κ. AP 7.355 (Damag.), Herod. 4.46, 6.39; adv. κρηγύως ἐπαιδεύθην Call.*fr*. 193.30 Pf., νομίμως καὶ κ. Perict. ap.Stob. 3.28.21. **b** without moral connotation, *serviceable, desirable*, ἄλλο μὲν οὐδὲν AP 7.284 (Asclep.), οὐδὲ γουνάτων πόνος κρήγυον *a good symptom*, Hp.*Coac.* 31; of persons, *proficient*, οὐκ ἐπίστανται, οὐδέ κ. διδάσκαλοί εἰσι Pl.*Alc.* 1.111e, εἰ δ' εἰσὶ κ. τε καὶ παρὰ χρηστῶν Theoc.*Ep.* 19.' **3**, delete the section.

κρήδεμνον I, add '**b** poet., of the tomb, πέτρινα κ. E.*Tr.* 508.' **II**, for this section read 'fig., of the walls or other defensive works of a city, pl., Τροίης ἱερὰ κ. Il. 16.100, Od. 13.388, *h.Cer.* 151, *h.Hom.* 6.2, B.*fr.* 20B.11 S.-M., Euph.*fr.* 54; sg., Θήβης κρήδεμνον Hes.*Sc.* 105. **b** of the stopper of a wine-jar, Od. 3.392.'

†**κρήδεσμον**, τό, *headband*, κράδεσμα ἀργύρεα *SEG* 19.618 (Metapontum, vi BC), Attic form in Hsch.

***Κρήζιμος**, ὁ, title of Zeus, *IEphes.* 3415 (ii/iii AD).

κρηΐνον, for '*larder*' read '*store-room for meat*'

***κρηματίς**, ίδος, ἡ, name of an article (perh. a cup) in a temple-inventory, κ. ἱερά in *IG* 7.3498.15, 20 (Oropus); cf. κρημοφόρος.

κρήμνημι, after lemma insert '(**κρίμν**-)'; line 2, for 'κρημνάντων' read 'κριμνάντων'; line 6, for 'κρημναμεναν' read 'κριμναμεναν'

κρημνισμός, add 'Heph.Astr. 2.24.8'

***κρημνοβάμων**, ον, gen. ονος, *cliff-walking*, *PMag.* 4.1365.

κρημνοβάτης 2, add 'Gr.Naz.*Carm.* 2.1.17.8 (*PG* 37.1267a)'

κρημνός (A) **1**, add 'prov. ἐκ κρημνῶν γεννᾶσθαι Heraclit.*Ep.* 4.2'

κρηναῖος, add 'Κρᾶναῖος, title of Poseidon, *SEG* 35.590 (Larissa), fem., of Athena, *IG* 9(1).109 (Phocis)'

κρήνη, line 2, after '(Mytil.)' insert 'Alc. 150.5 L.-P.'; line 6, after '(Dreros, iii BC)' insert 'κράνα· κεφαλή *source*, Hsch.'; add 'in title of Athena, ἐν κράναις *IG* 9(1).97.20 (Phocis, iii BC)'

***Κρηνίς**, ἡ, Dor. **Κρᾶν**-, ίδος, ἡ, *water-nymph*, Mosch. 3.29 (pl.).'

κρηνῖτις, add 'as epith. of nymphs, [Νύμφαις ὑδρο]χόοις κρεινείτισιν inscr. in *Gnomon* 59.612 (Lycaonia)'

***κρηπιδάριος** (written κριπ-), ὁ, Lat. *crepidarius*, *maker of* κρηπῖδες, *AE* 1929.157 (Thess. Thebes, Chr.).

κρηπιδιαῖος, for '*Rev.Phil.* 50.67 (Didyma, ii BC)' read '*Didyma* 41.27 (ii BC)'

†**κρηπίδιον**, τό, archit., *floor-slab*, *SEG* 25.394 (Epid., iv BC), *Didyma* 39.33, 54, al. (all pl., ii BC).

***κρηπιδόσφυρος**, ὁ, ἡ, *one who wears boots covering the ankles*, of Athena, *PKöln* 245.10 (iii AD, verse).

κρηπίς I 1 a, for '*man's .. half-boot*' read '*shoe with a platform sole attached by straps to the feet, worn by men and women*'; after '*Parod.* 4' insert '(also interpr. as II 1 and as II 2)'; after 'Hist. 8' insert 'but εἶδος ὑποδήματος, *AB* 273, cf. Poll. 7.85; a man's shoe, *AB* l.c., cf. Plu. 2.760b; but there were women's κρηπῖδες, Luc.*Rh.Pr.* 15, cf. ‡ὀπισθοκρηπίς' **b**, for 'κρηπῖδες .. themselves' read 'worn by Macedonians, Plu.*Ant.* 54, cf. Plu. 2 l.c.; by soldiers' **II**, line 4, after 'E.*Hel.* 547' insert '*step of an altar*, Lxx *Jl.* 2.17, 2*Ma.* 10.26'; line 7, before 'Onos. 4.4' insert 'Pl.*Lg.* 736e, *Plt.* 301e' **2**, add '*river-bank*, Lxx *Jo.* 3.15, 4.18, 1*Ch.* 12.15'

Κρής, fem. Κρῆσσα, add '*SEG* 33.713 (Chalcis, iv BC)'

***Κρησίς**, ίδος, fem. adj. *Cretan*, σίδη Κ. Nic.*Al.* 490.

†**Κρῆσσα**, ἡ, v.s.v. **Κρής**. **2** as subst., = κρήτη, *chalk*, *Hippiatr.* 1.130.57, Hor.*Carm.* 1.36.10.

κρηστήριον, add '*IG* 2².1424*a*277 (add.)'

κρησφύγετον, add 'Poll. 1.10'

***Κρηταγενέτας**, = °Κρηταγενής, *Amyzon* 14 (202/1 BC).

***Κρηταγενής**, ές, *born in Crete*, epith. of Zeus, *Inscr.Cret.* 2.xvii 1.18 (Lisos, iii BC), al.

Κρητάρχης, for '*CIG* 2744' read '*MAMA* 8.426' and add '*Inscr.Cret.* 4.250 (i BC)'

Κρήτη, add '(personified) *SEG* 31.923 (Caria, i/ii AD)'

***Κρητηνεύς**, ὁ, title of a sacred ox, *Schwyzer* 768 (written Κρετενέος gen., ?v/iv BC).

***κρητηρισμός**, v. °κρατηρ-.

Κρητικός II 2, add 'neut. as subst., Κρητικόν .. καὶ γένος ὀρχήσεως Hsch.' add '**3** Κρητική (sc. γῆ) kind of earth used for fulling, *PHerm.* 12.8 (iv AD), cf. Lat. *creta*.'

Κρητογενής, delete the entry.

***κρήφιον**, τό, wd. of unkn. meaning (perh. a tool, cf. κράφα), included among articles given as security, *PMich.*III 173.11 (iii BC).

κριβανάριος, add 'perh. related to Persian *gribān* "*coat of mail*" rather than κρίβανος, cf. κρίβανον *breastplate* (Lampe)'

κριβάνη, for 'Alcm. 20 .. Ath.)' read 'Sosib. 6 (Ath. 3.115a, 14.646a)'

κρίβανος I 2, add '*part of a hot-bath system*, *SEG* 32.1502 (κλιβ-, Palestine, v AD)'

κριβανωτός, for 'Alcm. 20 (codd. Ath.)' read 'Alcm. 94 P.'

κρῐγή I, after 'also κριγμός' insert 'Ael.*NA* 5.51 (pl.)' **II**, delete the section transferring 'Hippon. 54' to section III and adding 'in this sense perh. parox.'

***κρίγκιον**, v. ‡κρίκιον.

κρίζω, for 'κεκριγότες' read 'κεκρῑγότες' and add 'κρίζει· ὀξὺ αὐλεῖ Hsch.'

κρῑθή, add 'also χριθή *IG* 1³.232.16, 42, 250.22 (Attica, v BC); Myc. *ki-ri-ta*'

κρίθινος, after '*beer*' insert 'X.*An.* 4.5.26'

***κριθόβωλον**, τό, perh. *barley cake for fodder*, *PCair.Zen.* 658.6 (iii BC); cf. βωλόπυρος.

***κριθολογέω**, *check (consignments of) grain for purity*, κρι[θ]ο[λ]ογηθῆναι τὸν πυρόν *PPetaus* 53.11 (ii AD), v. sqq.

***κριθολογία**, ἡ, *tax paid in compensation for adulteration of barley*, *POxy.* 2021.3 (vi/vii AD); also **κρίθηλογία**, *Theb.Ostr.* 113 (ii/iii AD), *PIand.* 150 (iii AD).

κριθολόγος, delete 'hence .. 26.1'

***κρίθον**, τό, v. °κριθός.

***κριθοποιός**, όν, *used in preparing barley*, κόσκινον Poll. 10.114.

κριθόπυρον, add '*EA* 13.40 (?Lydia, iii AD)'

κριθοπώλης, add '*SEG* 25.180 ii 55 (Athens, iv BC), *ITyr* 39C, 178 (Rom.imp.)'

***κριθός**, ὁ (or -όν, τό), error or variant for κριθή, ἐν κριθῷ *CPR* 7.45.7 (v AD), *PMich.*xi 608.6 (vi AD).

***κριθοσπορέω**, *sow with barley*, τὴν κριθοσπορουμένην γῆν *PThmouis* I 86.4 (ii AD).

κριθοφόρος, add '*PCair.Zen.* 728.4 (iii BC, fem. as subst. sc. γῆ)'

κρίθων, add '(ἐπώνυμον ἀνδρὸς μοιχαλίου La.)'

†**κρικέλλιον**, τό, *hoop, ring*, Sch.A.*Pr.* 74 (-έλιον), Alex.Trall. 8.2.

κρίκιον, after lemma insert '(**κρίγκιον** Sch.Od. 7.90, of a handle)'; add '*articles of jewellery*, cf. Pontic κρικί *ear-ring*'

***κρικολάγιον**, τό, wd. of uncertain meaning, perh. *vessel with ring-shaped handle*, *PVindob.* G16694 (-ιν, *ZPE* 35.130, vi AD).

κρίκος 1, for 'on a horse's breast-band' read 'on the middle of a yoke' **8**, add 'Str. 5.3.8' add '**10** κρίκος· κίρκος ἔνθα ἡ κώπη εἰσέρχεται Hsch.'

κρίμα I 2, add 'κρίματι βουλῆς *TAM* 4(1).32' at end of article for 'ἰ .. *poetry*' read 'ἰ in inscr. in *JWI* 2.369 (i AD)'

***κρίμνημι**, v. °κρήμνημι.

κρίμνον, after 'τό' insert '(in codd. also κρίμον)' **2**, delete the section. **3**, after '*crumbs*' insert '*AP* 6.302 (Leon.); *grains*'; after 'Herod. 6.6' insert 'Babr. 108.9, 32'

κριμνός, (in parenth.) for 'κριμνον' read 'κριμμον'

***κριμνοσπορέω**, *sow (?)barley*, *PStrassb.* 864.4 (ii AD).

κρίνον, add '**3** a perfumed unguent, Pompon.*Dig.* 34.2.21.1.' **IV**, add '*MAMA* 9.61'

***κριντήρ**, ῆρος, ὁ, *judge*, epigr. in *Inscr.Cret.* 4.325 (Gortyn, v AD, pl.); cf. κριτήρ, -ής.

***κρίνω**, line 9, after 'Aeol. **κρίννω**' insert 'Alc. 130.32 L.-P.' **II 1**, line 8, after 'Hdt. 3.31' insert 'οἱ κεκριμένοι *classical authors*, Poll. 9.15, 153'; line 9, after 'Pi.*N.* 7.7' insert 'w. acc. resp., Λεσβιάδας κριννόμεναι φύαν Alc. l.c.'; line 11, after '(Leon.)' insert '**b** *admit to a class, number* in it'; line 12, after 'E.*Supp.* 969 (lyr.)' insert '*GVAK* 3.6'; line 13, after 'Luc.*Am.* 2' add 'esp. of admitting as a competitor in games, κριθέντα Πύθεια *JRS* 3.295 (Pisidian Antioch), κ. ἐν Δελφοῖς *IG* 12(2).388 (Mytilene)' **2 a**, add 'so perh. also A.R. 2.148, τὸ κριθέν = (*senatus*) *consultum*, D.S. 14.113.7' **3 c**, add 'fig., *to be in a critical condition*, τῆς πόλεως κρινομένης *SEG* 24.1217 (i BC)' **7**, line 2, delete 'cf. *Supp.* 396' **8**, line 2, after '(ii BC)' insert 'Lxx *Is.* 41.6, al., Nic.Dam.*fr.* 47.8 J.'

κριός IV, add '**2** archit., *ram, buttress*, *PEnteux.* 8.11 (iii BC, pl.)'

κριοτάφος, add '*PTeb.* 61b.401 (118/7 BC), ostr. in *CE* 55.311'

†**κριοφάγος**, ον, *that eats ram's flesh*, Θρήϊκες ἀμορβοὶ κριοφάγοι Nic.*Th.* 50; cf. κ. θεός τις, ᾧ κριοὶ θύονται Hsch.

κρίσις II 1 b, after this section insert '**c** = (*senatus*) *consultum*, D.S. 14.113.6; = *decretum*, D.Chr. 37.9; = *iurisdictio*, Str. 14.5.6; ἐπὶ τῆς τῶν διαφορῶν κρίσεως = *stlitibus iudicandis*, *IEphes.* 701 (i AD); for pres. '**c**' read '**d**' and add 'cf. τὴν ὥραν τῆς κρίσεως *SEG* 31.1419 (iii/iv AD); *the judgement* of God, *TAM* 4(1).375 (Jewish)' **III 2**, add '**b** *critical stage in life, climacteric*, Sen.*Ep.* 83.4.' add '**V** *class or category* of competitors in games, ἀγωνισάμενον τὰς τρεῖς κ. παῖδα ἀγένειον ἄνδρα *CIG* 2810*b*9 (Aphrodisias), *IGLS* 1265 (Laodicea, iii AD); παῖς κρίσεως τῆς Ἀγησιλάου *IG* 5(1).19.10, al. (Sparta), ib. 14.754 (Naples).'

κρισσός I, add 'X.*Eq.* 1.5'

κριτεύω, serve as κριτής (iudex), SEG 35.714 (Maced., ii/iii AD).
κριτήριον 2 b, add 'of Fate, ISmyrna 541 (i AD)'
κριωπός, delete '= κριός VII, POxy. 1801.26'
κροαίνω, delete 'only pres. part.' (-ειν in Philostr.Im. 1.30 and VS 25)
*Κροθιᾶτᾱς, v. °Κορθιᾶτᾱς.
*Κροκᾱγόριος, ὁ, name of a month, Hesperia 27.75 (unkn. provenance, c.200 BC).
*κροκαλλίς, ίδος, ἡ, a precious stone, Plin.HN 37.154.
κροκᾱτόν, delete 'Asin.' and add 'cf. Lat. crocatus'
*κροκάω, to be yellow, κροκόωντες κόρυμβοι Nic.fr. 74.22 G.-S.
*κροκεανόν, τό, alternative name for κώνειον, Sch.Nic.Al. 186a Ge.
κρόκη, line 2, for 'only in Hsch.' read 'Hsch., Theognost.Can. 40' for 'II' read 'κρόκη (B), ἡ,' and add 'cf. κρόκκαι'
*κρόκη (C)· πέμματος εἶδος Hsch.
†κροκίας, ου, ὁ, saffron-coloured, of a cock, Plu. 2.375e. 2 name of a precious stone, Plin.HN 37.191.
κρόκκαι, add '(κρόκαι La.; cf. °κρόκη (B))'
κροκοβαφής, for 'metaph., .. dying men' read 'of blood, as the colour associated w. fear'
*κροκόδεσμος, ον, bound with saffron, Theognost.Can. 21.9.
*κροκοδίανον, τό, a species of teasel, An.Boiss. 2.396.
κροκοδιλέα, for 'dung' read 'a sweet-smelling substance from the intestines'
κροκόδιλος, line 5, for 'cf. κροκύδιλος Hippon. 119 .. Indogerm.Forsch. 15.7)' read 'κερκύδιλος Hippon. 155, 155a W. (var. κρεκυ- and κροκύδειλος)' and add 'cocodrillus, Sen.QN 4a.2.13'; add 'sp. κορκόδειλος Peripl.M.Rubr. 15'
κροκόεις, add 'Tyrt. 18.2 W.'
*κροκοπώλης, ου, ὁ, app. saffron-seller, SEG 37.214 (iv BC).
κρόκος, line 1, insert in bracket 'also τό, v. section 4'
κροκόττας, for 'an Indian wild beast .. hyena' read 'a wild beast found in Ethiopia and India, prob. some kind of hyena'
κροκόω I, for 'crown with yellow ivy' read 'make yellow'
κροκύδιον, place before 'κροκυδιαμός', deleting 'hence'
*κροκύδιτος, gloss on οὖλος, Sch.Gen.Il. 16.224, (cod. -δωτος).
κροκύς 1, add 'b pl., woollen cloth, AP 9.567 (Antip.).' add '3 a plant, Hsch.'
κροκύφαντος, after 'woven' insert 'BGU 1300.23 (iii/ii BC)' and after 'Subst.' insert 'Aq.Is. 3.19'
*κροκύφαντωτον, τό, network on a capital, Aq.Je. 52.22, 23.
*κροκών, ῶνος, ὁ, saffron-bed, Hdn.Gr. 1.29.
κροκώτινος, add 'IG 2².1529.17 (Attica, iv BC)'
κροκώτιον, add 'IG 2².1529.18 (Attica, iv BC)'
*κροκωτοβᾰφής, ές, saffron-dyed, Sch.E.Hec. 471.
κροκωτός 2, for 'worn by gay women' read 'worn by women on special occasions'; add 'also fem. in Lat., Naev.Trag. 43, Plaut.Aul. 832 (corc-)'
κρομμύδιον, add 'Rémondon, Le monastère de Phoibammon no. 21 (?iv AD)'
κρόμμυον I 1, after 'Bias ap.D.L. 1.83' add 'cf. Plu. 2.153e' II, line 5, for '(perh. metri gr.)' read 'Philem. 122'; add '[κρομμύων [ῠ] Nic.Al. 413]'
κρομμυόφακον, delete the entry.
Κρονίδης I, add 'of Chiron, Poet.de herb. 115, Orph.L. 11'
*Κρονίσκος, ὁ, dim. of Κρόνος; Κρονίσκοι ἑπτὰ ἐν ἑνί, title of a work by Galen, Gal.Libr.Propr. 12.
Κρονίων II, add 'BCH 116.279.1 (Colophon, ii BC)' add 'III Κρονίωνας· παλαιοὺς ἀνθρώπους Hsch.'
Κρόνος I 2, ἡ Κρόνου ἡμέρα, add 'T&MByz 7.323f. no. 15 (Thessalonica), SEG 31.830 (Sicily, both iv/v AD)'
κροσσοί, add 'sg., Hsch. s.v. θυσανόεσσα; Lxx Ex. 28.22'
κροτάλια, after 'each other' insert 'BGU 1300.25 (iii/ii BC)'
κρόταλον I, after 'in dances' insert 'Sapph. 44.25 L.-P.'; add 'Ath. Asklepieion v 75 (iii BC); see also °κρούπεζα 2'
*κροτεῦσαι· τὸ μὴ πλήρες assure Theognost.Can. 21.
κροτέω II 2 a, after 'Thphr.Char. 11.3' insert 'cf. κροτήσατε Suid., = plaudite' 4, line 2, after 'weld together' insert 'AP 6.117 (Pancrat.)'
κρουματικός, after 'Plu. 2.1138b' insert 'Phot. s.v. νιγλαρεύων'
κρουματογρᾰφία, ἡ, music., instrumental notation, Anon.Bellerm. 11.
κρουνός 1, add 'ὁ ἐν Κρουνοῖς, of Asclepius, IG 9²(1).631 (Naupactus, ii BC), al.' 4, after 'nozzle' insert 'App.Anth. 3.67 (Hedyl.), IG 11(2).287A.79'
*κρουνοφόρον· οὕτω καλεῖταί τι τῶν ἐν ταῖς ναυσίν Hsch.
*κρούπαλα, τά, perh. = °κρούπανα, ἀμφίλινα κρούπαλα S.fr. 44.
*κρούπανα· ξύλινα ὑποδήματα. καὶ κλεῖς Hsch.
*κρούπεζα, ἡ, usu. pl., clogs or wooden shoes worn by Boeotians and used in crushing olives, Paus.Gr. κ48 E., Phot. 2 clapper of wood or metal pressed by an aulos-player with his foot to mark the rhythm, Poll. 7.87 (sg.); οἱ δὲ κρόταλον, ὁ ἐπιψοφοῦσιν οἱ αὐληταί· τὸ βάταλον Phot.

†κρουπέζιον, τό, dim. of κρούπεζα, τὰ ξύλινα σανδάλια κρουπέζια λέγεται, καὶ ὑποδήματα ξύλινα, μεθ' ὧν τὰς ἐλαίας πατοῦσι Hsch. s.v. κρουπεζούμενος. 2 clapper used by αὐληταί, Poll. 7.87.
*κρούπετα· ὑψηλὰ ἢ ξύλινα ὑποδήματα, ἢ γυναικεῖα Hsch.
κρούστης, add 'for formation cf. Προκρούστης X.Mem. 2.1.14, etc.'
κρουστικός II 2, for this section read 'of a speaker, writer, or of style, forcible, emphatic, Ar.Eq. 1379, Luc.Dem.Enc. 32, Hermog.Id. 1.12'
κρούω 1, add 'of a scorpion, κρουστὶς (i.e. κρουσθεὶς) ὑπὸ σκορπίου Lefebvre RIGC no. 120' 5, line 2, for 'Simon. 183' read 'AP 7.24.6 ([Simon.])' 8, add 'perh. at SEG 35.915 (Naxos, vi BC)'
κρῠερός, line 4, after 'E.fr. 916.6 (anap.)' insert 'Phryn.Trag. 6 S.'; line 5, after 'icy-cold' insert 'Alc. 286(a).3 L.-P.'
κρῡμᾰλέος, after 'chilly' insert 'Eratosth. 16.10'
κρυπτάδιος, line 3, after 'Il. 1.542' insert '(neut. pl. as adv.)' and for 'Regul.Adv. .. 6.182' read 'also advs. -ίως Man. 2.195, 6.182, -ίῃ in hiding, Max. 339'
κρυπτή I, after 'Callix. 1' insert 'PSI 547.18 (iii BC), Str. 17.1.37, Ev.Luc. 11.33, Labeo Dig. 43.17.3.7' add 'b covered passage, arcade, Sen.Ep. 57.1, Juv. 5.106, CIL 1.1505.3.'
κρυπτήρ, add 'as subst., Inscr.Délos 1403Abii79 (ii AD)'
κρυπτός, add 'archit., τὸν κ. περίπατον Ath. 5.206a'
κρύπτω, line 3, after 'κρύψα 11.244' insert 'aor. 2 ἔκρυφε AP 7.423 (Antip.Sid.), 700 (Diod.), Q.S. 1.393, Nonn.D. 7.45, al.'
κρυσταίνομαι, delete 'with cold, freeze'
κρυστάλλινος I, add 'νίπτρα AP 9.330 (Nicarch.)' deleting this ref. in section II.
κρυσταλλοειδής I, at end delete 'Adv. .. 2.11.2' II, add 'Paul.Aeg. 3.22.30 (184.17, 20 H.); adv. -δῶς Placit. 2.11.2'
κρύσταλλος, after 'ὁ' insert 'also ἡ, v. sense II with Suppl.' II, after '(Claudian)' insert 'Q.S. 10.415'
*κρυσταλλοφόρος, ον, producing rock-crystal, Ps.-Callisth. 118.30.
*κρυφιμαῖος, α, ον, = κρύφιμος, IEphes. 4106.6 (iv AD).
κρύφω, delete the entry.
κρυψίδομος, delete the entry.
*κρυψίδρομος, ον, running secretly, Orph.H. 51.3.
κρωσσός 1, add 'also for unguents, Plu.Arist. 21' 2, after 'Mosch. 4.34' insert 'Lyc. 369 (pl.)'
†κρωτάνεροι (dub.)· βάναυσοι πολῖται, καὶ ἐξελευθεριωταί Hsch. (βάναυσοι. πολῖται· καὶ ἐξελεύθεροι. ἰδιῶται La.).
*κτάμιον, τό, special type of cultivated land, PMich.II 123ʳ iv 33 (AD 45), PMil.Vogl. 83.25, 105.19, 170.6, al. (all ii AD).
*κτάρα· ἰχθὺς βραχύτερος πάντων Hsch.
*Κτᾱρος, ὁ, epith. of Hermes, Lyc. 679.
κτέανον, lines 8/10, for 'Hom. .. "wealth"')' read 'dat.pl. κτεάνεσσι epigr. in IG 2².11120.8 (c.iii AD) (unless old dat. pl. of κτέαρ q.v.)'
†κτέαρ, ατος, τό, = κτέανον (which is prob. after old pl. of κτέαρ, *κτέανα), Hom. only in dat. pl. κτεάτεσσι, Il. 23.829, Od. 14.115, cf. Pi.O. 5.24, E.fr. 791.3: nom. and acc. sg. in later poetry, Maiist. 33, AP 9.52 (Carph.), 9.752 (Asclep. or Antip.Thess.), 11.27 (Maced.), Q.S. 4.543; dat. pl. κτεάτοις Hdn.Gr. 2.936.
*κτέατισμα, ατος, τό, possession, hymn in BE 1991.419 (Berezan, i AD).
κτείνω 2, add 'τοῦ κτείνειν .. ἐξουσίαν = ius gladii, J.BJ 2.117'
κτείς, line 1, delete 'Edict.Diocl. 13.3' and transfer to end of section 1 6, after 'Art. 51' insert 'Call.fr. 343 Pf.' and delete 'Call.fr. 308' before 'AP 5.131'
*κτενᾶς, ᾶ, ὁ, comb-maker or wool-carder, MAMA 3.327, 739 (Corycus); cf. κτενιστής II.
κτενίζω, add 'II card, pass. pf. part., dub. in PIand.inv. 314 in Aegyptus 27.49 (ii/iii AD), ἐκτενισμ[.. pap.).'
κτένιον 1, add 'of κτείς, hair-comb, Edict.Diocl. 13.7'
†κτενιστής, οῦ, ὁ, hairdresser, Gal. 13.1038, Gloss. II wool-carder, PTeb. 322.23 (ii AD), BGU 1021.5 (iii AD), v. Aegyptus 26.41.
*κτενιστρίδιον, τό, hairdresser's shop, PAmst.inv. 62 (Mnemosyne 30.146).
κτῆμα 2, at end, estate, farm, etc., add 'Men.Dysc. 328, 737 S.'; vineyard, add 'SB 11240 (vi/vii AD)'; pl., SB 11009 (iii/iv AD)'
*κτηνάρχης, ου, ὁ, one in charge of providing transport animals, SB 11223.9 (AD 332).
*κτηναρχία, ἡ, provision of transport animals, SB 10202 (iii/iv AD), 11223.5, 9, al. (AD 332).
κτηναφαίρεσις, add 'PLond. 1677.35 (vi AD)'
†κτηνίτης, ου, ὁ, one having charge of (transport) animals, cattle-driver, Θησσεὺς κτηνείτης ὁ λαλούμενος MAMA 8.569; as title of Hermes, Schwyzer 721.9 (Thebes-on-Mycale, iv BC).
κτῆνος, τό, line 1, before 'flocks and herds' insert 'possessions, esp.'; line 2, after 'h.Hom. 30.10' insert 'Hes.fr. 200.9 M.-W.'
*κτῆνος (B), ὁ, property, wealth, Hes.fr. 193.5, 198.6. M.-W.
κτηνοστάσιον, for 'cattle-stall' read 'place for cattle-breeding' and add 'Heph.Astr. 3.5.71'
κτηνοτροφέω, for 'keep cattle' read 'rear cattle' and add 'Heph.Astr. 3.5.35'
κτηνοτροφία, for 'cattle-keeping' read 'rearing of cattle'

κτηνοτρόφος, for 'keeping cattle, pastoral' read 'raising cattle' and for 'cattle-keeper' read 'cattle-rearer'

κτήσιος **I**, for 'belonging to property' read 'that belongs to one, one's own' and add 'transf., ἐπεὶ κτησίων φρενῶν ἐξέδυς S.fr. 210.36' **II**, add 'cf. Πάσιος'

κτῆσις **II 1**, for '(from pf.)' read 'thing possessed, possession' **2**, add 'PPrincet. 166.9 (ii/iii AD), κωμητική κ. ib. 134.i.1 (iv AD)' add '**3** title of possession, καθὼς περιέχ(ε)ι ἡ κ. Hierapolis 88.1, cf. 216.8, 262.2.'

*κτήτης, ου, ὁ, proprietor, Hsch. s.v. κτήτορες οἰκιῶν.

κτίδεος, after 'ἰκτίδεος' for '(which is not in use)' read '(Suid.)' and at end add 'prob. originally fr. misdivision of ἐπ' ἰκτιδέην in pre-Homeric version of 10.335'

κτίζω, add 'cf. Myc. athematic 3 pl. ki-ti-je-si (κτίενσι), part. ki-ti-me-na, cf. ἐϋκτίμενος'

κτίλος **I**, lines 3/4, delete 'ἱερέα .. Pi.P. 2.17' and for 'perh. their cherished eggs' read 'dub. sens.' **II**, for 'ram' read 'the leading animal (ram) of a flock' and add 'AP 9.72 (Antip.)'

κτίσμα **1**, building, add 'SEG 31.1472, 34.327 (Chr.)'

κτίστης **I 1**, of Apollo, add 'SEG 31.1287 (Side), 31.1575 (Cyrene)'; add 'of Heracles, SEG 35.842 (Moesia, ii/iii AD); of Hadrian, identified with Zeus Olympios, SEG 33.943, al.'; add 'granted as a title to outstanding citizens, SEG 31.910 (Caria, iii AD), etc.' add '**IV** name given to certain Thracian ascetics, οἳ χωρὶς γυναικὸς ζώωσιν, Posidon.ap.Str. 7.3.3.'

*κτιστικός, ή, όν, of a κτιστής, foundry, IHadr. 132 (ii AD).

κτιστύς, for 'Ion. for κτίσις' read '= κτίσις' and add 'AAWW 1948.305/6 (Egypt, Ptolemaic)'

κτίστωρ, add 'of Constantine, as "restorer", Salamine 202'

κτίτης, add 'Myc. ki-ti-ta'

κτοίνα, add 'Myc. ko-to-na, ko-to-i-na, estate (exact meaning uncertain)'

κτοινάτης, read 'κτοινάτᾱς' and s.v. κτοινέτης, add 'Myc. ko-to-ne-ta'

*κτουπων (?parox.), Cret. wd. of uncertain meaning and inflection, SEG 23.566.19 (Axos, iv BC).

κτώ, delete the entry.

κυάθειον, before 'Nic.Th. 591' insert 'IG 1³.405.10 (Athens, v BC)'

κυαθίζω **II**, add 'repeated at Plu.Marc. 17.1'

κυᾱθίς, add 'Inscr.Délos 104.24 (pl., 364 BC)'

κύαθος **I**, delete 'Anacr. 63.5' add '**2** vessel for cupping, Inscr.Perg. 8(3).72.9 (sp. κύεθος).' **II**, add 'Anacr. 11(a).5 P., Nic.Th. 582, Al. 58'

*κυαίστωρ, v. °κουαίστωρ.

*κυαιστώριος, v. °κουαιστώριος.

κυάμιαῖος, add 'IIasos 20.9 (iv BC)'

*κυᾱμικός, ή, όν, unexpl. deriv. of κύαμος, IEphes. 4123.19 (Rom.imp.).

κύᾱμινος, add 'ἄχυρα Gp. 9.10.1'

*Κυᾱμίτης, ου, ὁ, god or hero of the bean, prob. Iacchus, Paus. 1.37.4, Hsch.

*κυᾱμοπώλισσα, ἡ, seller of beans, PPanop. 15 (iv AD).

κύᾱμος, add '**V** name of a small monetary unit, Rh.Mus. 60.331 (Tauromenium, i BC).'

κυᾱμώντης, for 'bean-grower' read 'one who works in bean-fields'

*κυάνασπις, ιδος, ὁ, ἡ, having a dark shield, Anacr. 4.2 P.

κυαναυγής, add 'of Circe, Lyr.adesp. 7(c).5 P.; of Persephone, SEG 35.1683 (Egypt, Hellen.)'

*κυαναυγίς, ίδος, ἡ, app. = κυαναυγής (in fem.), Suppl.Hell. 991.12 (poet. word-list, iii BC).

*κυάναυλαξ, delete 'Orac.ap.'

κυάνεος **II 1**, add 'late dat. pl. as fr. *κυανής, τίτλον περὶ τύμβῳ ἔγειρεν γράμμασι κυάνεσσι Mitchell N.Galatia 338' **2**, add 'applied to night, κυανέας [πο]λυόμματον [ποί]κιλμα νυκτ[ός lyr. in POxy. 2879 i 1'

*κυανόθειρα, fem. adj., dark-haired, Suppl.Hell. 991 (poet. word-list, iii BC).

*κυανόζυγος, ον, having a dark yoke, Suppl.Hell. 991 (poet. word-list, iii BC).

*κυανόκολπος, ον, having a dark recess, dark-bosomed, Suppl.Hell. 991 (poet. word-list, iii BC).

κυανόπεπλος, after 'Hes.Th. 406' insert 'ὀμίχλη, of evening twilight, Musae. 113, 232'

κύανος **I 4**, after 'cornflower' insert 'AP 4.1.54 (Mel.)' at end add 'Myc. ku-wa-no; cf. compd. ku-wa-no-wo-ko-i (dat. pl.) cyanus-worker'

*κυανόσελμος, ον, having dark benches, δόρυ poet. in POxy. 2625 fr. 1.4.

*κυανοχίτων, ωνος, ὁ, ἡ, dark-robed, Pi.Dith. 3.5.

Κυανοψιών, after 'month' insert 'in Chios, SEG 17.379.6 (v BC)'; after 'Κυανεψιών' insert 'SEG 30.977 (Olbia, v BC)'

*κυβαλία ἡ, app. some form of sexual perversion, PMasp. 97ᵛ(D)45 (vi AD); cf. κυβαλής.

κύβδα, after 'Ar.Eq. 365' insert 'εἰς γόνατα κύβδ' ἑστάναι id.Pax 897'

*Κύβδασος, ὁ, one of several invented names for gods of coition, Pl.Com. 174.17; cf. κύβδα.

κύβεθρον, for '= κυψέλη II' read 'a kind of container' and add 'see also °κόβαθρον'

κύβελα, delete the entry (gloss refers to Κύβελα, v.s.v. Κυβέλη).

*Κυβελείη, ή, title of Mother Goddess, Μητρὶ Κυβελείη(ι) SEG 22.511 (Chios, iv BC).

Κυβέλη, add 'archaic form Ϙυβάλα SEG 35.1820 (Locr.Epiz., vii/vi BC)'

κυβερνάω, after lemma insert 'Cypr. κυμερέναι, s.v. κυμερνήτης, κυμερῆναι' **I 3**, after 'govern' insert 'Heraclit. 41, Parm. 12.3'; add 'w. gen. κυβερνᾶν θεῶν τε καὶ ἀνθρώπων Pl.Smp. 197b'

κυβερνήτης, line 1, add 'Dor. κυβερνάτας Alcm. 1.94 P., SEG 33.640 (Rhodes, ii/i BC)' **2**, add 'κ. Νείλου, title of a priest, SB 4100 (Philae), OGI 676 (Silsilis, ii AD)'

κυβερνητικός **1**, line 4, after 'Grg. 511d' insert 'Arist.EE 1220ᵇ24, 1247ᵃ6'

κύβερνος, add 'Hdn.fr. p. 25 H.'

κυβεύω **I 2**, add 'Hell.Oxy. 1.2 B.'

κῡδάλιμος, after 'also η, ον' insert 'Alc. 129.6 L.-P.'; add 'of a goddess, Ὄμπνια κυδαλίμη SEG 30.1272 (Caria, late Hellen.)'

*κυδάω, unexpld. wd. in POxy. 2741 fr. 1b col.2.

κύδιμος, after '= κυδάλιμος' insert 'Doroth.fr. 38.11 S.'; add 'w. dat., Βιτιανόν .. κύδιμον εὐνομίαις SEG 29.1139 (Miletus, iv/v AD)'

κύδιστος **1**, add 'comp. κυδίων, ἕτερον δ' ἑτέρου κυδίον' ἔθηκεν Il.Pers. 5.2' **2**, add 'v. κυδρός at end'

*κύδνος· κύκνος Hsch.

†κῡδνός, ή, όν, v.l. for κυδρός (in same sense), Hes.Th. 328, A.R. 4.1333, perh. also GVI 743.7.

κυδοιμός, lines 4/5, delete '5.593'; after 'Ar.Pax 255' insert 'as the weapon wielded by Eris, Il. 5.593'

κυδρός, add 'X.Eq. 10.15'

†Κυδωνία, Cret. town, Th. 2.85, Myc. ku-do-ni-ja, (mod. Khania). **2** quince-tree, cf. Κυδωνέα, Gp. 4.10.24.

*κυεστωνάρ(ιοι), οἱ, app. Lat. quaestionarii, POxy. 2050.2 (vi AD).

*κυέστωρ, ορος, ὁ, v. °κουαίστωρ.

Κυζικηνός, add 'also as ethnic of persons, Hdt. 4.76.3, etc.'

*κυητικός, ή, όν, relating to conception, Clem.Al.Paed. 2.10.

*κυήτωρ, ορος, ὁ, ἡ, parent, of a bird, Cyran. 96 (pl.), cf. Eust. 1546.21.

*κυηφόρος, ον, σωτήριος κ. δίφρος, app. kind of chair used for childbirth, Ps.-Callisth. 12.2.

Κυθέρεια, form Κυθήρη, add 'SEG 31.963 (Ephesus)'

κύθρα, etc., for 'κυθρόκαυλος' read 'κυθρόγαυλος'

*κυθροβρόχος, ὁ, perh. person who prepares clay for pots, (cf. χύτρα), PPrag. 25.10 (vi/vii AD).

κυΐσκομαι, line 1, for 'pass.' read 'med.' **II**, after 'Gal. 4.513' insert 'pass., to be conceived, Ptol.Tetr. 121'

κυκάω **II**, penultimate line, after 'Archil. 66' insert 'καρδίας κυκωμένης Trag.adesp. 664.23 K.-S.'

κυκεών, line 2, after 'acc. κυκεῶνα' insert 'Hippon. 39 W.'

κυκλάνεμον, after 'τό' insert 'fan' and delete '(dub. sens.)'

κυκλάς **I**, add '**3** κυκλάδες αὖραι whirling breezes, Nonn.D. 1.133.'

κύκλευμα, add 'applied to the κόσμος, τί ἐστι κόσμος; ἀκατάληπτος περιοχή .. ἀπλανὲς κύκλευμα Secund.Sent. 1'

†κυκλευτήριον, τό, app. equipment or plant operated by a water-wheel, (ἀρ.)ζ' ἐν αἷς τροχὸς καὶ κυκλευτήριον PColl.Youtie 65.49, 68 (c.AD 241), PGiss. 56.8 (vi AD).

κυκλεύω **I 2**, add 'SEG 31.1116'

κυκλέω **I 3**, add 'give currency to, λέξιν D.H.Dem. 56'

κυκλικός **I**, add '5 of style, rounded, εὐρυθμία τῶν περιόδων D.H.Pomp. 6.' **II-IV**, for these sections read '**II** of or belonging to the Epic Cycle, θέλγει· ἡ -κή (sc. ἔκδοσις), θέλγεις Sch.Od. 16.195, 17.25; ἡ κ. Θηβαΐς Ath. 11.465e; κυκλικοί, οἱ, poets of the Epic Cycle, Sch.Il. 3.242, al.; adv. -κῶς, in the manner of the Cyclic poets, κ. 6.325, οὐ κ. τὰ ἐπίθετα προσέρριπται Sch.Od. 7.115; comp. -κώτερον Sch.Il. 9.222. **2** transf., commonplace, conventional, ἐχθαίρω τὸ ποίημα τὸ κ. Call.Epigr. 28.1 Pf.'

*κύκλιον, τό, perh. small reel of thread or ball of wool, PKöln 124.3, al. (iv AD); SB 11289.6 (iv/v AD).

κύκλιος **I**, add '**2** metaph., τῆς ἀγωγῆς τῶν περιόδων τὸ κ. D.H.Isoc. 12.' **II 1**, line 8, before 'invented' insert 'κυκλίοις SEG 25.177.3 (Athens, 331 BC)'

κύκλος **II 1**, delete 'on the janker' **7**, add 'E.Ph. 1382' add '**12** an article of attire, PZilliac. 11.19, 22 (iii AD); cf. κυκλάς II.' **III**, add 'c roundabout phrasing, circumlocution, Plu. 2.408f.'

κύκλωμα, line 4, after '4' insert 'kerb running round base of altar, Lxx Ez 43.17. **5**' at end delete 'κόσμος ἀπλανὲς κ. ib. 1' (v. °κύκλευμα)

*Κύκνεια, fem. adj., of Cycnus, μάχα Pi.O. 10.15.

κύκνειος **II**, delete the section.

†κυληβίς· ἄκαρπος κράμβη Theognost.Can. 121; cf. κ.· κολοβή Hsch.

*κυλίβη (v.l. κυλίνη)· ἡ ἄκαρπος κράμβη Suid., v. °κυληβίς.

κύλινδρος, add '**6** cylindrical seal, Ath.Asklepieion IV 124.99 (Athens, κυλίνδω iii BC); app. also of a "sliced" cylinder, κ. τετράγωνος πανταχεῖ ὑά[λινος] ib. 101; cf. κύλινδρος· σφραγῖδος εἶδος Hsch.'

κυλίνδω II 1, line 3, for 'Alc. 18' read 'Alc. 326.2 L.-P.'

κύλιξ, after 'ἡ' insert 'κόλιξ AE 1953-4.205' and after 'wine-cup' insert 'Ἐπαμείνονος ha ϙύλιξς SEG 31.838 (Sicily, vii/vi BC), [ϙ]ύλυιξ Corinth 15(3) p. 360, no. 17'

*κυλίφακτος, ?ά, prob. kind of vase, SEG 34.852 (Thera, vi BC).

κυλίχνη I, for 'small cup' read 'cup or bowl' and add 'ABSA 59.42' II, for this section read 'κ.· φιάλη καὶ ἡ ἰατρικὴ πυξίς Hsch.'

κυλίω, line 6, delete 'Call.'

Κυλλήνιος, insert 'Dor. Κυλλάνιος Inscr.Cret. 1.16.7 (Lato, ii/i BC)'; after 'of Hermes' insert 'Od. 24.1'; delete 'BCH 27.295 (Crete)'

κυλλός I 2, for this section read 'of plants, πὸτ τὰς ῥάπα[ς] κυλλάς SEG 39.996.5 (Camarina, ii BC); in dub. sens. of some irrigation equipment, κυλλῆς κυκλάδος μιᾶς PLond. 3.776.10 (vi AD)' II, add '2 fig., of a month of 29 days, Gal. 9.907.18.'

κυλλόω, pass., add 'Gal. 13.562.6, 576.3, 6'

*κυλοιάζω, κυλοιάζειν· τὸ τοὺς ὀφθαλμοὺς ἐπικλίνειν χλευάζοντα Theognost.Can. 21.

*κυλτίς, ίδος, ἡ, app. kind of filter for a water-wheel, (cf. κυρτίς), POxy. 3354.16 (AD 257).

κῦμα I 2 a, for '(A.Th.) 1083' read '1077' b, line 2, delete 'κακῶν' and 'Th. 758 (lyr.)' II, add 'A.R. 4.1492'

κυμαίνω I 2, add 'ἰδών σε .. ἐκύμηνε τὰ σπλάγχνα Herod. 1.56'

*κυμάτιαῖος, α, ον, having a wave pattern, voluted, MDAI(I) 19-20.238.25, 245 (Didyma).

κυμάτιον 1, for 'of the volute' read 'a form of moulding, esp. of the echinus of an Ionic capital '

κυματωγή, add 'Ps.-Hdt.Vit.Hom. 268, 488'

*κυμβαλία, ἡ, prob. = κυμβάλιον, PColl.Youtie 87.12 (vi AD).

κυμβαλίζω, add 'transf., κυμβαλίζει γὰρ ἡ κοίλη (ὁπλὴ ἵππου) μᾶλλον ἢ ἡ πλήρης καὶ σαρκώδης [Simon.Ath.]Eq. 5'

κύμβαλον, add '2 stoup for lustral water placed at the door of a house in which a corpse lay, Sch.E.Alc. 98.'

*κύμβαλος, ὁ, = κύμβαλον 1, prob. rest. in Call.fr. 194.106 Pf.

κύμβαχος, line 2, after 'Il. 5.586' insert 'Call.fr. 195.29 Pf.'

*Κυμβελλείτης, ου, ὁ, name of a kind of marble, ISmyrna 697.28 (ii AD).

κύμβη (A) I, add '2 shell of a crab, Opp.H. 1.335.' add 'IV head, EM 545.27.'

+κύμβη (B), ἡ, generic term for a bird (or class of birds), πτεροβάμονες κύμβαι Emp. 20.7.

*κυμβίδιον, τό, dim. of κύμβη (A), in quot., cup, IG 2².1534A119 (Athens, iii BC).

κυμβίον, add 'Astyd. 3.2 S., [D.]47.58'

κύμβος, for '= κύμβη (A), cup' read 'bowl used for mixing' and add 'Sophr. 165'

*Κύμη, ἡ, kind of wild cabbage, Nic.fr. 85.5 G.-S., Plin.HN 20.90 (cj.), (prob. identified w. place-name Cyme).

*κυμήνασθαι· κείρεσθαι Theognost.Can. 21; cf. ‡κωμαίνω.

κυμητήριον, v. °κοιμητήριον.

*κυμῖνᾶς, ᾶ, ὁ, seller of cummin, SEG 8.143 (Jaffa, iii/iv AD), ZPE 11.15 (Cilicia, Chr.).

κύμινδις, last line, after 'cf.' insert 'Hippon. 61 W.'

κύμινον, add 'Myc. ku-mi-no'

κυμινοπρίστης, add 'also quibbler or logic-chopper, D.C. 70.3.3'

κυμοπλήξ, add 'v.l. for κυματο- in AP 10.7.1 (Arch.)'

κυμόρροον, add '(or κυμόρροον, cf. Theognost.Can. 121 (κύμορον))'

+Κυνᾱγίδᾱς, title of Heracles in Macedonia, EAM 6, 20, al. (i BC/iii AD).

κυναγχικός, add 'II κυναγχική, ἡ, another name for the plant ἀπόκυνον, Ps.-Dsc. 4.80.'

*κυναγχῖτις, ιδος, ἡ, a plant used as a remedy for κυνάγχη, Ps.-Dsc. 3.24.

*κυνάδακνος, ὁ, dog-bite, SEG 6.802.36 (Salamis, Cyprus, ii/iii AD; pl., tab.defix.).

*κυνανᾱιδής, ές, shameless, rest. for κυνάπαιδες in Sophr.ap.Sch.Gen.Il. 21.394, also rest. in Euph. in Suppl.Hell. 415 i 12.

*κυνανθρωπία, ἡ, disease in which a man imagines himself to be a dog, Aët. 6.11 (151.21 O.).

κυνάπαιδες, add 'v. °κυναναιδής'

κυνάριον, add 'in later Gk. without dim. sense, Rémondon Le monastère de Phoibammon p. 34 no. 2 (?iv AD)'

κύνειος, after 'Ar.V. 898' insert 'cf. κύνειον θάνατον· ἄγαν φοβερόν Hsch.'

κύνειρα, for 'dog-leash' read 'perh. = πόρνη'

κυνηγέσιον II 2, add 'Plb. 30.26.1'

*κυνηγέσιον, ὁ, title of Hadrian identified w. Zeus, BCH 102.437.

κυνηγέτης, add 'Myc. ku-na-ke-ta-i (dat. pl.)'

κυνηγέτις, of Artemis, add 'SEG 33.1174 (Lycia)'

*Κυνθήιος, ον, app. = Κύνθιος, δόμον Κυνθήιον SEG 30.1237 (Africa, ii/iii AD).

*Κύνθιον, τό, name of a sanctuary, Inscr.Délos 1417Aii47 (ii BC); cf. Κύνθιος.

Κύνθιος, add 'also epith. of Zeus at Delos, SEG 32.218.45, 85 (Athens, 103/1 BC), Κυνθίη of Athena, Schwyzer 769 (Paros, vi BC), cf. Lat. Cynthia, of Artemis'

*κυνιατοα, prob. non-Hellenic name of a cult-object, Inscr.Cret. 1.xvii 2b2 (ii BC).

+κύνικλος, v. °κουννίκολος.

κυνοβάτης, add '[Simon Ath.]Eq. 5'

*κυνοκεφάλειος, α, ον, = κυνοκέφαλος, PMag. 4.2651.

κυνοκέφαλος, for 'dog-faced baboon, Simia hamadryas' read 'name for baboons and, prob., other kinds of monkey' and add 'Peripl.M.Rubr. 50'

*κυνοκοίτης, ου, ὁ, having sexual intercourse with dogs, Vit.Aesop.(G) 49.

*κυνοπόταμος, ὁ, beaver, Cyran. 64.

κυνόπρηστις, add 'II kind of fish, IGC p. 98 A4 (κουνόπρειστις, Acraephia, iii/ii BC).'

κυνόρροδον, add 'III kind of lily, Plin.HN 21.24.'

κυνόσορχις, ἡ, kind of orchid, Plin.HN 27.65.

κυνοσουρίς I, add 'b κυνοσουρίδες· οἱ ἀπὸ ἀλωπέκων καὶ κυνῶν τικτόμενοι παῖδες AB 452, Sch.Call.Dian. 94.'

*κυνοτάφος, ὁ, dog-burier, priest in an Egyptian temple, PHib. 213.9 (iii BC).

κυνουλκός, add 'PSAAthen. 2.2 (iii BC)'

κυνούχιον, delete the entry.

+κυνοῦχος, ον, dog-restraining, τραχηλοδεσπότας κλοιοὺς κυνούχους AP 6.107 (Phil.). II masc. subst., bag, sack, X.Cyn. 2.9, Ael.Dion.fr. 206, Poll. 10.64, Phot., Hsch.; for carrying money, PCair.Zen. 22.22 (iii BC), Inscr.Délos 442A7, 461Aa7 (ii BC), AP 6.298 (Leon.Tarent.).

κόος, delete 'IG 12(5).646 (Ceos)'

*κυοφόρημα, ατος, τό, that which is carried in the womb, Sch.Gen.Il. 3.17.

κυοφόρος, add '2 subst. womb (or perh. applied to the penis as 'producer of pregnancy'), PLond. 1821.161.'

*κυπαιρίσκος, ὁ, dim. of κύπαιρος, Alcm. 58 P.

+κύπαιρος, ὁ, cyperus, galingale, Myc. ku-pa-ro, ku-pa-ro₂, Alcm. 60 P.; cf. κύπειρος, κύπερος.

*κυπάρίσσειος, α, ον, Myc. ku-pa-ri-se-ja neut. pl., made of cypress wood.

*Κυπάρισσία, ἡ, title of Athena at Messene, SEG 23.209 (iii BC), SEG 39.381; cf. Κυφάρισσία.

κυπαρισσόκομος, add 'Lyr.adesp. in POxy. 2736 fr. 2(b).12'

κυπάρισσος I, line 3, delete 'ἐλαφρά' and for 'Pi.fr. 154' read 'Pi.Pae. 4.50' add 'III ποίη κυπάρισσος, prob. = χαμαικυπάρισσος, Nic.Th. 910.' at end add 'cf. Myc. [ku-]pa-ri-so, place name, ku-pa-ri-si-jo, ethnic'

κύπασσις, line 1, for 'Alc. 15.6' read 'Alc. 357.7 L.-P.'; line 2, for pres. def. read 'kind of chiton, of varying length and worn by both sexes'; line 3, after 'Lys.fr. 58 S.' add 'Herod. 8.31'; at end (κυπασσίσκος), for 'Hippon. 18' read 'Hippon. 32 W.'

*κυπειρόεις, εσσα, εν, form κυπαιρόϝεν Myc. ku-pa-ro-we, scented with cyperus.

κύπειρος, add 'Myc. ku-pa-ro (masc. or neut.), variant ku-pa-ro₂'

*κυπελλοδόκος, ον, receiving cups (cj. for κυπελλοτόκ-), τράπεζα Nonn.D. 47.62.

κυπελλοτόκος, after 'Nonn.D. 47.62' add '(cod.; cj. -δόκος)'

*κύπερις, ιδος, ἡ, = κύπερος, PColl.Youtie 87.14 (vi AD).

Κυπρία, (v. Κύπριος 2) add 'SEG 30.1571, 32.1318 (Cyprus, iv BC)'

*Κυπριάρχης, ου, ὁ, governor of Cyprus, Lxx 2Ma. 12.2.

Κύπριος 1, add 'Κυπρίας τε πόλεις A.Pers. 891' at end add 'cf. Myc. ku-pi-ri-jo'

Κύπρος, add 'personified, SEG 31.924 (Caria, i/ii AD)'

κυπτόν, delete the entry.

*κυπτός, ή, όν, distorted, κυπτὸς ὀρθοῦται λόγος A.Ch. 773 (v.l. κρυπτός).

κύπτω 1, last line, after 'Id.Pax 33' insert 'εἰς ἑαυτὸν κύψαντα ἐνωχεῖσθαι Ath. 1.6a'

κύρα, delete the entry.

*κῦρά, v. °κύριος.'

*κυρανίδες or κοιρανίδες, (sc. βίβλοι), αἱ, title of treatise, Cyran. 3.4.

Κύρβας, gen. pl., add 'Κυρβάνθων inscr. in PP 4.73 (Rhodes, ii/i BC)'

κυρβᾱσία, add 'Ar.fr. 559 K.-A.'

κύρβος, delete the entry.

Κυρηναϊκός, add 'as the title of a Roman legion, TAM 2.1201 (Lycia, AD 145/6)'

Κυρήνη, after 'Ar.Th. 98' insert 'Call.Dian. 206, Epigr. 21 Pf.'

*Κυρηνιακός, ή, όν, = ‡Κυρηναϊκός, SB 8802 (AD 82/3).

Κυρήτειος   SUPPLEMENT   κωπίς

*Κυρήτειος (?ῠ), ὁ, epith. of Poseidon, Tit.Cam. 31 (Camirus, iii BC), al.

κῡριακός **I**, κ. φίσκος, add 'SEG 30.1349 (Miletus, c.AD 180/200)'; at end add 'ἰατροῦ τῆς κυριακῆς οἰκετείας IEphes. 3233.17 (iii AD)'   **II**, κ. ἡμέρα, add 'POxy. 3407.16 (iv AD)'

*κῡρίευσις, εως, ἡ, gaining possession of, ἀπὸ τῆς Πτολεμαίου Αἰγύπτου κυριεύσεως Marm.Par. 109.

κῡριεύω **I 1 b**, delete 'Pass. .. Ptol.Tetr. 112'

κύριος **A II 3**, add 'also of place, S.OT 1453'   **5**, add 'τὸ κύριον, ordinary language, ἐργῶδες ἐν τῷ κυρίῳ τραγικῶς ἅμα καὶ συμπαθῶς γράψαι'   **B 1 a**, line 4, after 'PTeb. 5.147 (ii BC) insert 'owner or secure possessor, εἰ τριακονταετὴς παρέλθοι χρόνος καὶ ἡ κατοχὴ κυρίους τοὺς λαβόντας καταστήσειε Just.Nov. 22.24'; add '(in later Gr. freq. written κῦρις or κύρις w. late vowel shortening, e.g. Vit.Aesop.(G) 30), SEG 31.830 (iv/v AD)'   **2**, line 4, for '**κύρα**' read 'κυρά, Vit.Aesop.(G) 32'   **3**, add 'of a comes, AS 35.96'   add '**5** trustee (of a mortgage), IG 12(7).515.34 (ii BC).'

κῡριότης, line 1, after 'Ep.Eph. 1.21' insert 'w. gen., Memn. 4.6 J.'

κυριόω, after '= κυρόω' insert 'ἐκυριώθη IGRom. 4.661.31 (Acmonia, i AD)'

*κῦρις (or κύρις), v. ‡κύριος.

*κῡριώνῠμος, ον, properly so-called, αὕτη ἡ ἡμέρα ἡ κυριώνυμος Leont.Byz.HP 1.2.3.

κῡρίως **I**, for 'SIG 1004' read 'SEG 31.416'   **IV**, last line, after 'EN 1098ᵇ6' insert 'in more straightforward language, D.H.Th. 29'

κῡρόω **1**, line 13, after 'PRev.Laws 48.17 (iii BC) insert 'cf. Hsch. s.v. προσθεῖναι, where κυρῶσαι (κῦρσαι cod.) is explained as τὸ παραδοῦναι τῷ ἐωνημένῳ ὑπὸ κήρυκι'

*Κυρτᾶς, ᾶ, ὁ, nickname of a fishmonger, SEG 11.169 (Corinth, vi AD); cf. κυρτοβόλος.

κυρτίς, add 'cage, κ.· ὀρνιθοτροφεῖον Hsch.; cf. κύρτος 2'

*κυρτοπλοκεῖον, τό, basket-weaving shop, POxy. 2719.11 (-ῖον, iii AD).

κυρτός **1**, add 'full, brimming, ὑπὸ κυρτοῖσι κυπέλλοις Anacreont. 50.22 W., (or ?curved); cf. κύρτωμα 2, κύρτωσις 2'

κυρτωτός, add 'rest. in Didyma 50.1A60'

κῡρωτής, add 'Hesperia 5.401.2 (Athens, iv BC)'

*κῡρωτικός, ή, όν, ratifying, affirming, κύριε· κυρωτικὲ καὶ τελεστικέ Sch.Pi.P. 2.106.

κύσθος **II**, add 'see also χύστος'

*κῡσοκνησιάω, suffer from anal pruritus (or subst. -ία, ἡ in analogous sense), POxy. 2811.5a.13.

κῡσοχήνη, add 'rest. [in undefined sense], Hippon. 82.2 W.'

κύστεροι, add 'cf. κυρσερίδες'

κύτισος, for '(ἡ, Theoc. 5.128, 10.30)' read '(ἡ, Gal. 14.23, prob. in Theoc. 10.30, indeterminate id. 5.128)'; add 'form κύτεσος: Myc. ku-te-so, kind of wood; adj. ku-te-se-jo'

κύτος **1**, add 'Alcm. 17.1 P.'

κύτταρος **1**, add 'cf. κυρσερίδες, κύστεροι'

⁺Κυφάρισσίτας, α, ὁ, epith. of Pan (or perh. Hermes), Inscr.Cret. 1.xvi 7 (Lato, ii/i BC).

κῡφός, line 4, for 'shrimps' read 'prawns'; line 5, before 'Eub.' insert 'Philox. 2.17 (cj.), Alexis 115.13 K.-A.'; line 6, delete 'shrimps .. squilla'

κύφων **I**, add 'perh. transf., of a heavy load, Men.Dysc. 102'   **III**, for 'part of a woman's dress' read 'women's clothes, or kind of tunic' and after 'Posidipp. 44' add '(pl.)'

κυψέλη **I**, add 'S.fr. 441a.5 R.'

κύψελος, add 'Myc. ku-pe-se-ro, pers. n.'

κύω **II**, for 'in aor. act. .. A.fr. 44.4' read 'act., of the male, impregnate, transf., ὄμβρος ἔκυσε γαῖαν A.fr. 44.4 R.; procreate, οἱ δὲ (Ἐρωτιδεῖς) τραφέντες εὐθὺς πάλιν κύουσιν ἄλλους Anacreont. 25.16 W.'; line 9, after 'Lxx Is. 59.13' insert 'ὕε κύε, a cry at the Eleusinia, Hippol.Haer. 5.7, Procl. in Ti. 3.176 D., cf. IG 2².4876 (Athens, ?i AD)'   at end delete 'The causal sense .. aor. ἔκυσα.'

κύων **II 1**, line 7, for 'also of offensive persons .. dogs' read 'as a derogatory term for non-Jews'; line 8, after 'Ep.Phil. 3.2' insert 'Apoc. 22.15; of male sacred prostitutes, Lxx De. 23.18(19)'   **IV**, line 3, after 'Ael.NA 1.55' insert 'κ. ποτάμιος, prob. otter, id. 14.21'   **V**, add '**2** κύων πρότερος = the star Προκύων, Nonn.D. 16.202.'   **IX**, add 'Nicostr.ap.Simp. in Cat. 26.24'   add '**XIII** colloq. term for large kind of fly, = στρατιῶτις 3, Luc.Musc.Enc. 12.   **XIV** νυκτερινοὶ κύνες acc. to Hsch. (s.v. νυκτ-) kind of women's shoes, Suppl.Hell. 1090.'

κῶας, add 'form *κῶϜος: Myc. ko-wo, sheepskin'

κωβάθια, add 'SB 1049ᵛ.26-7 (sp. κοβ-, ii AD)'

*κωδικίλος, ὁ, also κωδικέλλος Arr.Epict. 3.7.30, Cod.Just. 6.48.8, usu. pl., Lat. codicillus, official letter from the emperor, Arr. l.c., OGI 543, IGRom. 3.175.   **II** pl., codicil to a will, TAM 2.77, Inscr.Cret. 4.300, Cod.Just. l.c.

*κωδικιλλόω, app. add as a codicil, POxy. 2283.11 (κοδ-, AD 586).

κῴδιον **I**, add '**2** pl., strips of fleece on the upper end of the forearm of the "sharp-thonged" type of boxing-glove, Philostr.Her. 6.'

*κώδιστρον· ὀπὸς τῆς μήκωνος Theognost.Can. 56 A. (perh. ὁ ⟨καρ⟩πός, cf. EM s.v. κώδη).

κώδων, line 1 (fem.), add 'κόδων[α] νέαν inscr. in PP 42.42 (Tarentum, vi BC)'   **2**, for 'crier's bell' read 'bell attached to warhorse or (perh.) other animal '; delete ' "is his own trumpeter"'

κώδων **I**, add '(cf. Italian borrowing qutun, Testimonia Linguae Etruscae 28, 63, vii BC)'   **II**, line 3, for 'Plu.Ant. 4, etc.' read 'Hegesand.ap.Ath. 11.477e; feast, Lxx Es. 8.17'

*κωθωνικός, ή, όν, drunken, PVindob.Tandem 2.9 (iii AD).

κωθωνιστής, add '(v.l. φιλοκωθ-)'

*κωκῡμός, ὁ, = κώκυμα, Call.fr. 177.14 Pf. (s.v.l.).

*κωκώ, οῦς, ἡ, colloq. name for the penis at puberty, AP 12.3 (Strat.; cj. κοκκώ).

κωλακρέτης, add 'also κωλοκράτης SEG 39.148 (Attica, iv BC)'

*κωλάνεμος, ὁ, wind from the bowels, on magical gem in SO 19.76.

*κώλαργος· λευκόπους Theognost.Can. 56 A.

κωλεός, add 'Sokolowski 1.71.6 (Casossus), ib. 63.6 (-ειός, Mylasa)'

κωλῆ, uncontracted form κωλέα, add 'SEG 29.1088 (Caria, iii BC)'; κωλία, add 'SEG 32.456 (Boeotia, c.235/0 BC)'

κωλίζω, after 'Olymp.Hist. p. 463 D.' insert 'Procl. in R. 2.218 K.'

*κωλιοθήρας, ου, ὁ, perh. mackerel-fisher as pers. n., (for κολιο-), IEphes. 20.B27.

⁺κωλοειδής, ές, formed or arranged in members, Sopat.in Rh. 8.56.14 W., al.; adv. -δῶς ib. 8.9.24 W.

κῶλον, line 1, for 'A.Pr. 325' read 'A.Pr. 323'   **2**, add '**b** bone, Juba 82 J., Poll. 4.75.'

*κωλοπέδιλα, τά, codd. in Theoc. 25.103 (v. ‡καλ-).

*κῶλος, ὁ, Lat. culus, buttocks, backside, Vit.Aesop.(W) 77a.

κώλῡσις, add '**2** prohibition, Cod.Just. 1.4.25, 4.59.1.3.'

*κωλυσμός, ὁ, = κώλυσις, Heph.Astr. 2.30.16.

κωλῡτήριος, after 'preventive' insert 'A.fr. 47a.15 R.'

κωλύτωρ, ορος, ὁ, = κωλυτής, A.fr. 78a.20 R.

κωλύω, add '**II** prohibit, forbid, Cod.Just. 1.12.3.1a, 4.20.16 pr.'

κῶμα, line 1, after 'deep sleep' insert 'trance'

κωμαίνω, after 'κωμική' insert 'κωμήνασθαι Theognost.Can. 56 A.'

κωμαῖος, add '**2** Κωμαῖα, τά, a festival at Thasos, BCH 82.195 (iv BC).'

*Κωμαιών, ῶνος, ὁ, name of a month at Colophon, AJP 56.361 (iv BC), REG 47.29 (iii BC).

*κώμαρις, ἡ, app. var. κόμαρος, Anon.Alch. 9.19, 278.16.

κωμαρχέω, add '**II** to be leader of a κῶμος or κώμη, IG 2².3104 (Attica, iv BC).'

κωμάρχιος, ον, name of a musical nome, Plu. 2.1132d.

κώμαρχος **I**, for 'Πολέμων 1.45' read 'IG 2².3103.3' and add 'cf. ARV 26 no. 1'

κωμασία, add '**II** celebration of a κῶμος in honour of a victor at the games, Sch.Pi.N. 4.17.'

κωμαστήριον **I**, after '(Taposiris)' add 'PBremen 23.47 (ii AD)'

κωμᾶται, add 'κ.· λοιδορεῖται Theognost.Can. 56 A.'

*κωμᾰτικός, v. °κωμη-.

κωματώδης, add 'adv. -δῶς, Paul.Aeg. 3.6.1 (144.32 H.)'

κώμη, add 'Dor. κώμα, BCH 111.168 (Mantinea, iv BC)'   **II**, add 'κώμας ἐκάλουν τοὺς στενωπούς Aphth.Prog. 8'

κωμητικός, after 'of a κώμη' insert 'BGU 802.8.9 (i AD)'; after 'PTeb. 340 i 10 (iii AD)' insert 'κ. κτήσεως PPrincet. 134.1 (iv AD); Κωμᾰτικός, epith. of Zeus, Robert Hell. 10.38 (Thrace)'

*κωμογραμματικός, ή, όν, of or for a κωμογραμματεύς; κ. κλῆρος, a κλῆρος whose profits supported a κωμογραμματεύς, BGU 2437.41 (i BC).

κωμομισθωτής, add 'PSI 554.13 (iii BC)'

*κωμοπράκτωρ, ορος, ὁ, village tax-collector, PKöln 137.37 (AD 88), PAmst. 29.2 (i/ii AD).

κωμύδριον **I**, add 'Suid. s.v. Καλλίμαχος'

κῶνος **II 2**, for 'cone or peak of a helmet' read 'conical helmet'; add 'RA 6.31.3 (Amphipolis, ii BC)'

*κωνοψώμιον, τό, kind of bread, prob. in the shape of a cone, or made fr. pine-kernels, κωνοψωμίω[ν] ὀψάρει(ον) Περσικόν SEG 26.382 (Athens, Rom.imp.).

κώνωψ, after '(Arist.HA) 552ᵇ5' insert 'Plu. 2.663d, AP 12.108 (Dionys.), Gp. 6.12, al.'

Κῷος, after 'Κώϊος' read 'Call.fr. 532 Pf., TAM 4(1).1.2 (iii BC)'

*κωπαστής, οῦ, ὁ, rower or pilot, title of play by Aeschylus, A. T78.2d R. (cf. REG 100.33).

κωπεύς, add '**II** = κωπηλάτης, AB 273.32.'

⁺κωπεών, ῶνος, ὁ, handle, haft, Thphr.HP 4.1.4, 5.1.7, PMerton 73.18, κωπαιωνες PReinach inv. 2065.33, 38 (JJP 11/12.59; all ii AD).

κωπήεις, add 'hafted, αἰγανέη Opp.H. 2.497'

*κωπίς, ίδος, ἡ, handle, Theognost.Can. 56 A.

190

*κωπιών, = °κωπεών, Pap.Lugd.Bat. XXVII 1.6 (ii AD).
*κωποδέτης, ου, Dor. -τᾱs, ὁ, one who binds an oar to the thole, Clara Rhodos 8.228 (i BC).
*κωρᾰκίδιον, τό, a fish found in the Nile (perh. = κορακινίδιον), PAmst. 92.4 (ii/iii AD).
κωράλιον, for 'cf. κοράλλιον' read 'v. °κοράλλιον'
*Κωρής, Dor. for Κουρής, v. Κουρῆτες II 1.
*κώρτη, v. °χώρτη.
κώρῠκος I 2, line 2, before 'Sor. 1.49' insert 'Dionys.Eleg. 3.3'; line 5, delete 'metaph.'
*κωταρχέω, hold the office of κωτάρχης, Didyma 305.4, 451.1.

*κωτίαι, ταί, perh. name of a priesthood or a festival, de Franciscis Locr.Epiz. p. 22 no. 8, p. 49 no. 35.
κωτίλλω I, after 'Thgn. 852' insert 'Sol. 34.3 W.'
*κωφαίνω, 1 aor. ἐκώφηνα, deafen, TAPhA 68.54 (Beisan, tab.defix., iv AD).
*κωφεύς, έως, ὁ, deaf man, Call.fr. 195.34 Pf., Lexicon in An.Boiss. 4.386.
κωφίας, for 'τύφλωψ' read 'τυφλώψ'
κωφός II 1, add 'b of speech, lacking sonority, flat, σύνθεσις Demetr. Eloc. 68.' 2, lines 3/5, delete 'οὐ .. Cratin. 6' b, w. gen., add 'φύσει γάρ ἐστ' ἔρως τοῦ νουθετοῦντος κωφόν Men.fr. 53 K.-Th.'

Λ

†λᾷα, v. ‡λεία (B).
λᾱάρχης, add 'BGU 1763.11'
λᾶας, line 5, for 'also masc. .. 93 H.' read 'also Cypr. masc. λᾶος o-la-o o-te ὁ λᾶος ὅδε ICS 84.1'; add 'cf. Myc. adj. ra-e-ja'
†λαβδᾰκισμός, ὁ, excessive use of the letter lambda in pronunciation or composition, labdacism, Quint.Inst. 1.5.32, Diom. p. 453 K., Donat. p. 393 K., Mart.Cap. 5.514.
*λαβέλλιον, τό, prob. dim. of Lat. labellum, small basin, bowl, CPR 8.65.18 (vi AD).
λάβιον, read 'λᾰβεῖον (v.l. λαβίον)'
λᾰβίς II 3, add 'POxy. 3473.21 (ii AD)'
*λᾱβόλιον, τό, = λιθοβολία, Alc. in SLG 262.3 (PKöln 59.3); cf. °λήβολος.
*λάβολος, v. °λήβολος.
λαβράκτης, delete the entry (v. °ἀοιδολαβράκτης).
*λάβριχος, ὁ, a fresh-water fish, IGC p. 99 B24 (Acraephia, iii/ii BC).
λάβρος III, adv., after 'furiously' insert 'Alc. 72.3 L.-P.'
λαβροσύνη, line 1, delete 'greed'
λᾰβύρινθος I 1, lines 4/5, for 'name of a building at Rome' read 'transf., of the underworld' 2, line 5, after 'Icar. 29' insert 'λαβυρίνθους σοφίας ἀνελίττων Aristid. 2.79 J.' add 'Myc. da-pu₂-ri-to-jo (gen.)'
λᾰβῠρινθώδης, add 'τὸ λ. τῆς φράσεως Eust.Proem. 9 in Sch.Pi. iii p. 289 D.'
†λᾰγᾰνίζω, perh. be thin, blow weakly, of the South wind, Hp.Morb.Sacr. 13 (γαληνίζω cj.).
λᾰγᾰνον, line 2, after 'Fr. 116' insert 'PCair.Zen. 569.89, 707.6 (both pl., iii BC)'
λᾰγᾰρύζομαι, add 'also act. form cited without expl. as example of -ύζω termination, Theognost.Can. 142'
*λάγγουρος, ὁ, name of stone, Socr.Dion.Lith. 41.1 G.
λᾱγέτας, add '(form λαϜαγέτας) Myc. ra-wa-ke-ta, title of high official; adj. ra-wa-ke-si-jo'
*λαγκίον, τό, cf. Lat. lanx, Charis. p. 42 K.
λαγνεία II, pl., add 'X.Oec. 1.22'
λάγνης, add 'non-Att.-Ion. gen. λάγνα SEG 34.370 (Boeotia, c.500 BC)'
*λᾰγῡνάριος, ὁ, maker of flagons, MAMA 3.236 (Corycus).
λᾰγύνιον, add 'PVindob.Worp 11.7 (vi AD)'
*λᾰγῡνίσκος, ὁ, dim. of λάγῡνος, Vit.Aesop.(G) 87.
λαγχάνω, line 3, after 'al.' insert 'Aeol. λόχον Bernand Inscr.Colosse Memnon 29.17 (Balbilla)'; line 11, add 'med. aor. ἐλάχοντο Porph.Plot. 22'
λᾰγωβολία, for 'hare-shooting' read 'hare-hunting (using a throwing-stick)'
*λᾰγωβόλος, ον, hunting (with a throwing-stick), κούρη Nonn.D. 16.14.
λᾰγών I 2, after 'womb' insert 'AP 7.168 (Antip.Thess.), SEG 32.1615 (iv AD)'
*λάη, app. indecl. wine-measure, SB 1969 (vi AD), 1960 (vi/vii AD), PStrassb. 680.4 (early vii AD).
*λάθησις, εως, ἡ, given as expl. of Λητώ, Sch.Gen.Il. 1.36.
λᾰθῐπήμων, delete the entry.
†λᾰθίφρων, ον, gen. ονος, out of one's senses, λύσσα Nonn.D. 47.741, Hsch.
λαθρίδιος, after '= λάθριος' insert 'AP 7.457 (Aristo)' and after 'Luc.Bis.Acc. 33' insert 'epigr. in SEG 39.449.23 (Tanagra, v AD)'
λάθριος II, for 'Aphrodite' read 'Artemis' and after '(Leon.)' insert 'cf. Antim. 182 W., where expld. as = Προθυραία'
*λαθροδάκτης, gloss on σιγέρπης, Hsch.
λαθρόν, for 'Cyr.' read 'Hsch.' and add '(βάπτειν Cyr.)'
λάθῠρος, add '2 nickname of Ptolemy VIII, Plu.Cor. 11.'
*λαΐα, v. °λεία.
λαιβα, add 'cf. λαιός (B) end, °λαίδας'
λαίγματα, add 'cf. λαίγμα· τὸ ἱερόν, θῦμα Theognost.Can. 27 A.'
†λᾶιγξ, ιγγος, ἡ, rounded stone (of any size), pebble, etc., Od. 5.433, 6.95; A.R. 1.402, βαρείας .. λάιγγας id. 4.1678; πυρσοτόκους λάιγγας Nonn.D. 37.59.
*λαίδας· ἡ ἀσπὶς ἡ ἀπὸ βύρσης Theognost.Can. 27 A.; cf. ‡λαίβα.
†λαίθαργος, ον, (λάθ- Phryn.PS p. 87 B., ληθ- Hsch.) app. making sudden surprise attacks, of an apparently friendly dog, κοὐκ ὡς κύων λαίθαργος ὕστερον τρώγει Hippon. 66 W., S.fr. 885, Ar.Eq. 1068; λαιθάργῳ ποδί Trag.adesp. 227 K.-S.

*λαιθυράζω· τὸ διὰ τοῦ στόματος ψόφον τελεῖν ἐπὶ τῷ μαστῷ Theognost.Can. 27 A.; = χλευάζω Suid.
†λαικάζω, practise fellatio on, Ar.Eq. 167, Th. 57; Θειοδοσία λαικαδε[ι] εὖ Lang Ath.Agora xxi C33 (iv BC); καὶ πυγιζέσθω καὶ ληικαζέτω graffito in Glotta 62.167; in colloq. obscenities, med. λαικάσομ' ἄρα, i.e. I'll do anything rather, Cephisod. 3; οὐ λαικάσει φλυαρῶν; Men.Dysc. 892 S., Strato Com. 1.36; cf. frigori laecasin dico Petron. 42. II λαικάζω· ἀπατῶ Suid.
*λαικάς, -άδος, ἡ, = °λαικάστρια, Robert Collection Froehner p. 14 (Attic, iv BC), Aristaenet. 2.16.
λαικαστής, for 'wencher' read 'fellator'
λαικάστρια, for 'strumpet .. 235' read 'fellatrix, Pherecr. 152 K.-A.; loosely, whore, tart, Ar.Ach. 529, 537, Men.Pk. 485 S. II ἡ τίτθη παρὰ Λάκωσιν Orus ap.Et.Gen.' delete 'also .. (s.v.l.)'
λᾱϊκός I 1, add 'λαϊκὴ σύνταξις, poll-tax, PMich.v 241.35, 355.6, PStrassb. 522.6 (all i AD)' add '3 adv. -κῶς communally, λαϊκῶς πανδημεὶ δειπνίζων SEG 35.747 (Maced., i AD).'
λαῖλαψ, at end after 'occurs in' insert 'PMag. 4.182'
λαιμάω, after '= λαιμάσσω' insert 'λαιμᾷ δέ σοι τὸ χεῖλος ὡς ἐρῳδιοῦ Hippon. 118.3 W.'
λαιμώσσω, delete 'Hippon. .. causa)'
λάϊνεος, add 'TAM 4(1).48, Nonn.D. 37.68'
λαΐνθη, delete the entry.
*λᾱϊνοουργός, ὁ, stone-mason, JRS 57.43 no. 8.9 (Lycia, ?early i AD).
λάϊνός 1, after 'Trag.adesp. 44' add '(= Ar.Ach. 449)'
*λᾱϊνότευκτ(ος), ον, built of stone, perh. in epigr. in OA 6.43 (Cyprus, iii AD).
λάϊον, delete the entry.
*λάϊον, τό, ploughshare, A.R. 3.1335 (s.v.l.).
λάϊον, v. λήϊον (A).
λαιός (A), after 'Ant.Lib. 19.3' insert '(λάϊος)'
*λαίσθα, λαίσθη, Theognost.Can. 27 A. (prob. the same as λάσθη).
*λαισκίδης· ὁ βούπαις Theognost.Can. 27 A.; cf. λαόπαις.
λαιφάσσω II, add 'Hsch.'
λαῖφος I, for 'shabby, tattered garment' read 'blanket or sim., used as a cloak'
*λαιφύη· τὸ πρυτανεῖον Theognost.Can. 27 A.; cf. λαιμώρη II.
λαιψύς, after lemma insert '(‡λάφυξ, La.)'
†λαίω, λαίεται· καταλεύεται Hsch.; λαίειν· φθέγγεσθαι id., λαίω· τὸ βλέπω καὶ τὸ φονεύω Theognost.Can. 27 A.
Λάκαινα, line 2, for 'Helen' read 'Hermione'
λάκᾰνη, add 'POxy. 1269.23 (ii AD)'
*λᾰκάνιον, τό, = λαγάνιον, SB 7572.3 (ii AD).
*λᾰκᾱνιοργός, ὁ, for *λεκανιουργός, maker of λεκάναι, MAMA 3.367 (Corycus).
Λᾰκεδαιμόνιος, add 'ἁ -ία, sc. γᾶ or πόλις, AP 9.320.4 (Leon.)'
Λᾰκεδαίμων, line 3, for 'Hdt. 1.67' read 'Hdt. 6.58.2, 7.234.2'
λᾰκέρυζα, line 2, after 'Op. 747' insert 'Stesich. 32 i 9 P.'
*λάκες, pl., bottles, CPR 8.66.2 (vi AD).
*Λακευτής, ὁ, epith. of Apollo, perh. utterer or sim., cf. ληκητής, SEG 23.621 (Cyprus, iii BC).
λάκημα, for 'dub. sens.' read 'in accounts, in uncertain sense, perh. lot' and add 'PRyl. 706ᵛ.1'
*λάκησις, εως, ἡ, clucking with tongue round palate, Hsch. s.v. κλωγμός.
*λᾰκίνιον, τό, fragment, cf. λακίς and Lat. lacinia, Stud.Pal. 20.244.18 (vi/vii AD).
λᾰκίς, line 2, for 'ἐμπίτνω .. 131' read 'λ. λινοσινεῖ A.Supp. 120'; line 3, after 'cf. 903' insert 'λακὶς χθονός Trag.adesp. 228 K.-S.'
*Λακκίον, dim. of λάκκος, as name of the Little Harbour at Syracuse, D.S. 14.7.
λακκόπρωκτος, for 'loose-breeched' read 'compd. of °λάκκος (A) and πρωκτός as term of abuse' and add 'Lang Ath.Agora xxi C23 (v BC)'
†λάκκος (A), ὁ, (written λάκος, PCair.Zen. 176.276, iii BC) pit, tank, cistern, vat used for storing water, wine, or other things, (πίσσαν) ἐσχέουσι .. ἐς λάκκον ὀρωρυγμένον ἀγχοῦ τῆς λίμνης Hdt. 4.195, ἔτρεφον .. ὄρνιθας .. λιμναίους ἐν .. λάκκοισι ib. 7.119, Ar.Ec. 154,

192

λάκκος — SUPPLEMENT — λατίδιον

λάκκος, D. 29.3; οἶνος πολὺς ἦν ὥστε ἐν λάκκοις κονιατοῖς εἶχον X.An. 4.2.22; Lxx Ge. 37.20, Thd.Da. 6.7(8), Alex. 174.9; contemptuously, of the Sea of Galilee, Porph.Chr. 55; obsc., of the female pudenda, Macho 282 G.; of the pit of death, Sheol, Lxx Ps. 27(28).1, al.; fig., ἀνήγαγέ με ἐκ λάκκου ταλαιπωρίας ib. 39(40).2. **b** Κούρτιος λ. = Lat. *lacus Curtius*, D.H. 2.42.

***λάκκος** (B), ὁ, λ. χρωματικός kind of dye, *lac*, *Peripl.M.Rubr.* 6, cf. Prakrit *lakkha*.

λακκόω, for '*hollow out*' read '*dye with lac*', v. °λάκκος (B).

λάκτιμα, after 'Hsch.' insert '(-ημα La.)' and after '*PGen.* 56.27' insert 'dub.'

***λακχάϊνος**, η, ον, *dyed with lac*, λακχά *Edict.Diocl.* 8.4.

Λάκων II, add 't.t. for method of forming line, Arr.*Tact.* 31.4, 32.1'

Λακωνίζω II, add 'D. 59.36, Isoc. 4.110, etc.'

Λακωνικός II, after 'as subst.' insert '1 οἱ Λακωνικοί *Spartans*, Ar.*Pax* 212, *Nu.* 186.' and renumber pres. sections 1-4 as 2-5.

Λακωνισμός II, add 'Isoc. 14.30, 15.318'

λαλαγέω, line 2, for 'grasshoppers' read 'cicadas'; line 3, for 'dub. l.' read 'cj.'

λαλαγή, for '*prattle*' read '*babble*'

†**λαλάγημα**, ατος, τό, *chattering*, in quot. meton., of tambourine, *AP* 6.220.15 (Diosc.).

†**λαλάζω**, *babble*, μηδ᾽ ὥστε κῦμα πόντιον λάλαζε Anacr. 82 P.

***λαλαθάνατος**, wd. of uncertain meaning occurring in *defixiones*, *Kourion* 127.41, 131.29, etc.

λαλαχεύομαι, for ' = λαχνόομαι' read 'perh. *live dissolutely* (cf. °λαλαχός)'

***λαλαχός**, gloss (with μοιχός) on τογέρα, Hsch.

λάλησις, add '*POxy.* 1083.16'

λαλιά I 1, add '**b** *a talk*, as a comparatively short and informal ἐπίδειξις of a sophist, Men.Rh. p. 434 S., al., Aristid.*Rh.* 2 p. 538 S.: pl., ib. p. 539 S.'

***λαλοβαρύοψ**, οπος, *loud-chattering*, Pratin. 3.13 S.

λαλοβαρυπαραμελορυθμοβάτης, delete the entry.

λάλος, line 2, after 'Theoc. 5.75' insert 'Call.*fr.* 192.14 Pf., *Epigr.* 16.3 Pf., Φάμα λ. *ISmyrna* 513.3 (ii BC)'; line 7, after '*VS* 2.30' insert 'D.H.*Dem.* 5'

***λαλοῦ**, colloq. wd. for the *penis* of boy before puberty, *AP* 12.3 (Strat., s.v.l.).

λαμβάνω A I 1 **b**, add 'w. captor indicated by εἰς, τοῦ πάππου .. ληφθέντος εἰς τοὺς πολεμίους Is. 7.8' **4**, line 5, for 'with Adj.' read 'w. predicative adj.' and add 'E.*HF* 223' **9 c**, line 6, after 'Plu.*Alc.* 18' insert 'w. εἰς, λαμβάνει οὐκ εἰς ἀδικίαν ὅσα πέπονθε Men.*Dysc.* 297, Aristid.*Or.* 26(14).76, Lib.*Or.* 11.161' **II 1 f**, after '*receive* an oath' insert 'Is. 2.39' **g**, line 2, after 'Hp.*Prorrh.* 2.24' insert 'in *double entendre* w. sense of *eat*, Macho 50, 52 G.' **i**, add 'λ. ἐξετασίν τινος D. 18.246, 20.139, Call.*Epigr.* 59.3 Pf.'

***λαμιώδης**, ες, *like a Lamia*, Hsch. s.v. Λάμια.

†**λάμνα**, ἡ, Lat. *lamina, lamna*, *(metal) plate*, *Edict.Diocl.* 30.5, *PMag.* 4.2153, *PLond.* 124.26.

λαμνίον, τό, dim. of λάμνα, *PMag.* 4.3014, *PAnt.* 66.37, Iambl.Alch. 287.12.

***λαμπαδάριος**, ὁ, Lat. *lampadarius*, *torch-bearer*, Teuc.Bab. p. 42 B.

λαμπαδηφόρος, add 'of Hecate, *SEG* 26.819 (iii AD); as adj., λ. ἀνδριάντας Didyma 346.11, 16'

λαμπαδίας, add '**II** *torch-race*, *IG* 2².2119.230 (ii AD), *SEG* 29.147.9 (Athens, ii/iii AD).'

λαμπάς (A), add '**IV** Λαμπάδες, αἱ, nymphs attending Hecate, Alcm. 63 P.'

λάμπη II, delete the section.

***λάμπη** (B), v. °λάπη.

λαμπηδών, add '**3** app. *an inflamed swelling*, Hsch., s.v. φαύσιγγες.'

λαμπρός II 1, line 6, after '*IG* 14.911, 7.91, etc.' insert 'spec., the lowest of the three senatorial ranks, οὐδένα ἐμβάλλεσθαι ἐν φυλακῇ δίχα προστάξεως τῶν .. ἐνδόξων ἢ περιβλέπτων ἢ λαμπροτάτων ἀρχόντων *Cod.Just.* 1.4.22 pr.; applied to non-senatorial persons, *SEG* 31.1081'

***λαμπροτράπεζος**, ον, *keeping a fine table*, poet. in *SEG* 6.796.6 (Cappadocia, iii AD).

***λαμπροφοίτης**, ου, ὁ, *brightly-moving*, of the sun, *PMag.* 12.177.

***λαμπροφυής**, ές, *brilliant*, *PGron.inv.* 66.12 (*ZPE* 41.72).

***λαμπτηροῦχος**, ὁ, Boeot. λαμπτηρώχος, *torch-holder*, λ. σιδάριοι τρῖς *SEG* 24.361.20 (Thespiae, iv BC).

λαμπτηροφόρος, add '*PCair.Zen.* 782(a).69 (iii BC)'

λάμπω I 4, add 'βέλτιστος πᾶσι δ᾽ ἔλαμψα φίλος *SEG* 26.1808 (Egypt, Hellen.)' **II**, line 3, after '*Trag.adesp.* 33, etc.' insert 'fig., μή τις .. κῦδος ἔλαμψε νόθον *AP* 7.430 (Diosc.).'

λαμυρία, add 'of both men and women, Str. 17.1.16'

***λαμυρόω**, gloss on λαιθαρύζω, Hsch.; cf. λαμυρός.

***λαμψωδης**, ες, *resembling the plant* λαμψάνη, Hsch. s.v. ῥαφίς.

***λανάριος**, ὁ, Lat. *lanarius*, *wool-worker*, *TAM* 5(1).85 (AD 145/6), *PStrassb.* 309ʳ.5 (iv AD), Βασσιανὸς λανάρις *IGBulg.* 1922.5, *Edict.Diocl.* 21.1, 1a.

λανθανόντως, for '*secretly*' read '*surreptitiously*' and add 'Demetr.*Eloc.* 181'

λανθάνω, line 10, after 'Sol. 13.27' insert 'fem. part. λελαθυῖα Hes.*fr.* 343.13 M.-W.' **A 3**, add 'so perh. ὄμμα .. θέλγειν οὐ λάθε *AP* 5.282 (Agath.)' **4**, line 1, for 'relat. clause' read 'noun clause'; add 'without acc., οἱ πολλοὶ λανθάνουσιν ὅτι κολάζονται Plu. 2.554c' add '**6** w. acc. of respect, λ. τὴν ἀπόβασιν Th. 4.32.1; freq. w. neut. pron. or adj., id. 8.17, E.*IA* 516 (combined w. acc. pers., Th. 7.15.2).' **C I 1**, line 11, after '*Phdr.* 252a' insert 'plpf. w. acc. πάντ᾽ ἐλέλασο Erinn. in *Suppl.Hell.* 401.28; w. inf., λάθοιο .. ἀπενθεῖν Theoc. 11.63'

λάξ, add 'cf. λ.· λάκτισμα Hsch.'

λαξευτήριον, add 'Sch.Th. 4.4; Hsch.'

***λάξιον**, τό, deriv. of λάξις, *allotment*, written λακσιον, *Kadmos* 9.145 (Crete, v BC).

λάξις, line 3, for 'so prob.' read 'so also'

λαογράφος, after '*Sammelb.*' insert '5661.2 (pl., i AD)'

***Λαοδικεών**, ῶνος, ὁ, name of month at Smyrna, *ISmyrna* 578.34, 709.13, al.

λαοδόκος, delete 'dub. in' and '(δαμοδόκων Bgk.)'

***λαοκρίσιον**, τό, *court-house of the* λαοκρίται, *PTeb.* 795.9 (ii BC).

λαοξόος (λαξός), add '*SEG* 33.1321 (Egypt, i/ii AD)'

λαός, add 'Myc. form λαϝο- in compds., e.g. ra-wo-do-ko, pers. n. = Λαόδοκος'

λαοτύπος, for '*cutting stones*' read '*of* or *for stone-working*' **II**, add '*SEG* 29.1187 (AD 165/6); cf. λατύπος'

λαοφόρος I 1, add 'transf., of a prostitute, λεωφ]όρε λεωφόρ' Ἡροτίμη Anacr. 1 *fr.*1.13 P.' **II**, delete the section

λάπαθον, add '**III** in pl., *faeces*, Sch.Gen.*Il.* 5.166.'

***Λάπατος**, name of month at Orchomenos, Arcadia, Schwyzer 667; Myc. month name ra-pa-to.

***Λαπηθιασταί**, οἱ, name of a guild, *IG* 12(1).867 (Lindos), *Clara Rhodos* 2.203.

***λαπιδόρχας**· ὁ μεγάλους ὄρχεις ἔχων Hsch.

***λαπίθης**, ου, ὁ, λ.· ὁ αὐχηματίας, παρὰ τὸ τοὺς λαοὺς εἰς ὄπιν ἄγειν καὶ ἐπιστροφήν, ὅς› ‹ἐκπείθει τινὰς περιαυτολογούμενος Suet.*Blasph.* 132 T., from the character of the Λαπίθαι, *Il.* 12.128, etc.

***λαπρός**, α, ον, perh. = λαπώδης, Hdn.*fr.* p. 18 H.

***λαργιτιονάλια**, τά, perh. in *CPR* 7.26.19, 24, 33, 37, Lat. *largitionalia*, money tax on landed property.

***λαργιτίων**, ωνος (ονος), ἡ, Lat. *largitio*, *distribution of gifts*, *IEphes.* 38.11 (v AD); τῶν θείων λαργιτιόνων Just.*Const.* Δέδωκεν 9.

***λαργιτιωναλικός**, ή, όν, Lat. *largitionalis*, *concerned with the distribution of gifts*, πρόσοδοι *Stud.Pal.* 20.143.9 (v/vi AD), *PFlor.* 377.15 (vi AD).

λάρδος, for '*salted meat*' read '*bacon*, cf. Lat. *laridum, lardum*'; add '(cf. Λαρδᾶς as pers. n., *Amyzon* 53, Caria, ii BC)'

Λάρισα, add 'see also °Λᾶσα'

†**λάρναξ**, ακος, ἡ (ὁ, v. infr. 2) *rectangular box, chest*, Il. 18.413, λάρνακι ἐν δαιδαλέαι Simon. 38.1 P., Hdt. 3.123.2, B. 5.141 S.-M., A.R. 1.622, D.S. 5.62, Plu. 2.968f, Luc.*Syr.D.* 12, Apollod. 1.7.2; used to hold the bones of the dead, (ὀστέα) χρυσείην ἐς λάρνακα θήκαν Il. 24.795, λάρνακας κυπαρισσίνας ἄγουσιν ἄμαξαι .. · ἔνεστι δὲ τὰ ὀστὰ κτλ. Th. 2.34.3; of the Ark of the Covenant, *AP* 1.62. **2** *trough*, ὁ λ. οὗτος *IG* 12(1).961 (Chalce).

λάρος, add 'cf. λάρος· ὄρνις. καὶ ἰχθῦς ποιός Hsch.'

λαρός, lines 5/6. for 'Simon. 183.10' read '*AP* 7.24.10 ([Simon.])' **3**, add 'so λαροῖς ποσίν Hes.*fr.* 315 M.-W. (s.v.l.) acc. to Sch.A.R. 1.456 (glossed by ἁπαλός in *Et.Gen.*)'

λάσα, delete the entry.

***Λᾶσα**, = Λάρισα, Hsch.

†**Λασαῖος**, = Λαρισαῖος (v. °Λᾶσα), *IG* 9(2).517.19 (Larissa, iii BC), also Λασσαῖος *BCH* 59.37 (Crannon).

λάσανα I, for '*trivet .. pot*' read '*separate supports designed to hold any round-bottomed cooking-pot over a fire* (cf. *Hesperia* 54.393ff.)'

λάσειος, delete the entry.

λασθαίνειν, add 'ἐλασθαίνομεν· ἠκολασταίνομεν id.; μὴ δοκῇ με λασθαίνειν Hippon. 104.14 W.'

***λασικόμας**, masc. adj. *shaggy*, Κριός cj. in Mesom. 8.10 H. (λασιοκόμαι cod.)

λάσιος II, add 'of laden table, λ. τράπεζα· πληρεστάτη Hsch.'

λάσταυρος, after 'Phryn. 173' insert 'Men.*Sic.* 266 S.' and add 'Hsch. s.v.'

***λατερκουλίσιος**, ὁ, Lat. *laterculensis*, *employee of a laterculum*, Just.*Nov.* 23 epilogos.

***λατέρκουλον**, τό, Lat. *laterculum*, *registry of public offices*, Just.*Nov.* 25.6, al.

***λατερπής**, ές, *delighting the people*, Pi.*fr.* 346b.3.

***λατίδιον**, τό, kind of fish (cf. °λάτις), *PAmst.* 92.5 (ii/iii AD).

Λατινίς SUPPLEMENT λεντιάριον

*Λατῑνίς, ίδος, ἡ, fem. adj. *Latin*, δέλτος Nonn.*D*. 41.160.
Λατῖνος, add 'adv. Λατίνως, *Orac.Tiburt*. 162'
*λάτις, ιδος, (gender uncertain), kind of fish, prob. *Nile-perch, Tilapia nilotica*, *PMich*.II 123ᵛ vii 19 (i AD), *Et.Gud.*; cf. λάτος.
*Λᾱτοῖος, ὁ, (sc. μήν) name of month at Byzantium, *Milet* 1(3).153 (ii BC); cf. °*Λητῶιος*.
λᾱτομεῖον, line 1, before 'Str.' insert '*SEG* 26.134 (Attica, iv BC)' add 'II *tomb hewn out of rock*, *BCH* 36.618-20 (Thrace; -ιον, -ιν), etc.'
λᾱτομικός, add 'λατομικῇ .. τέχνῃ *SEG* 30.1481'
λᾱτομίς, after '*stone-chisel*' insert '*PCair.Zen.* 759.1, 782(*a*).54 (both iii BC)'
*λᾱτρευτής, οῦ, ὁ, *hired labourer*, *IEphes*. 1247b.8.
λάτρον, line 2, after '*EM* 557.35' insert 'cf. Varro *LL* 7.52' and before 'λάτρων' insert 'Call.*fr*. 276 Pf.'
λᾱτύπη, for '*Rev.Phil.* 50.67' read '*Didyma* 40.28, 41.54'
*Λαυδικηνόν, τό, prob. *Laodicean garment*, *Dura*⁴ 153 (cf. Λαδικηνός *Edict.Diocl.* 19.25, al., *Laudicea CIL* 3 p. 847).
*λαυκάνιον, τό, part of the throat, τὸ μεταξὺ τοῦ λ. καὶ τοῦ αὐχένος ἠχῶδες given as expl. of λήκυθος II, *Adam's apple*, Sch.Pl.*Hp.Mi.* 368c; cf. Hsch. s.v. λήκυθος, cf. λαυκανίη.
λαύκη, delete the entry.
*λαυνός, unexpld. wd. (noun in -αυνός), Hdn.*fr.* p. 27 H.
λαύρα I, add '*district* or *quarter* of a city, *SEG* 34.940.4 (Camarina, iv/iii BC)' II, for 'Hippon. .. 1089.10' read 'Hippon. 61, 92.10 W., Sotad. 2.1'
*λαύραρχος, ὁ, *chief of a district* or *quarter*, *BE* 1966 p. 445 no. 512.19, 25 (Tauromenium, i BC).
Λαφρία, after 'Ant.Lib. 40.2' insert 'τᾶι Ἀρτέμιτι τᾶι Λαφρίαι *SEG* 25.621.8 (Aetolia, ii AD), cf. τᾶι Βασιλείαι τᾶι ἐν Λάφρδι ib. 25.640 (W. Locris, ii BC)'
Λαφραῖος, add 'also in W.Locris *IG* 9²(3).624g.2, 638.3.3, 639.4.1; cf. °*Λοφριαῖος*'
†Λάφριος, ὁ, (after Λαφρία in *LSJ*), of Hermes, Lyc. 835, *SEG* 24.231 τὸν Λάφριν (Attica, ii AD). II month in Phocis, *GDI* 1719, al.; at Gytheion, *IG* 5(1).1145.28, etc.
Λάφριος, (after Λαφριάδαι in *LSJ*), delete the entry.
λαφύκτης, add 'Ath. 11.485a'
*λάφυξ· δάπανος, Drachmann *Cyrillglossar* (1936) p. 123; cf. λαιφύς.
*λᾰφῡρικός, ή, όν, *of booty*, de Franciscis *Locr.Epiz.* p. 41 no. 27.
†λᾰχᾱνάριον, τό, *vegetable pan*, *BGU* 2359.7, *PFam.Teb.* 49a ii 1, b ii 1 (iii AD), *Gloss*.
*λᾰχᾰνάρις, ὁ, *vegetable-grower*, *BIFAO* 70.42 no. 2.
λᾰχᾰνᾶς, add '*POxy.* 2421 (iv AD), *BGU* 2194.2 (vi AD)'
λᾰχᾰνευτής, after '= foreg.' insert '*PUniv.Giss.* 3.6 (ii BC)'
λᾰχᾰνεύω I 2, add 'Dsc. 1.91'
λᾰχᾰνιά, for '*garden-bed*' read '*vegetable-plot*' and add '*POxy.* 1913.17 (vi AD)'
†λᾰχᾰνικός, ή, όν, *of vegetables*. -κή, ἡ, *tax on vegetables*, *Inscr.Magn.* 116.42 (ii AD); so -όν, τό, *SB* 2085; perh. also *Ostr.* 787.1 (i AD).
λάχᾰνον 1, add 'perh. *vegetable-seed*, *SB* 10402.4 (AD 125)'
λᾰχᾰνόπωλις, add '*CPR* 13.16.10 (iii BC), *PMich.*IV 224.2120 (ii AD)'
†λᾰχειδής, ές, epith. of kind of toad, perh. *green-hued*, Nic.*Al.* 568.
λάχη, delete 'τάφων .. A.*Th.* 914 (lyr.)' add '2 collection of dues on *allotted rights of pasture*, *SEG* 3.357.2 (Acraephia, iii BC).'
*λᾰχή, ἡ, *digging*, A.*Th.* 914 (pl.).
λάχνη I, line 3, after '*O.* 1.68' insert 'of *hair on the chest*, Call.*fr.* 24.2 Pf.'; line 4, after '*Il.* 2.219' insert 'cf. A.R. 4.1531'; at end before 'Opp.*H.* 2.369' insert 'so in sg.' II, delete 'Nic.*Al.* 410'
λάχος, add 'also Cypr. la-ko-se λάχος *ICS* 318 B vii'
*λαψάνιον, τό, = λαψάνη, *PAnt.* 92.26 (iv/v AD); λαμψανεια pap.).
λεαίνω I 3, add '*smooth so as to erase*, τὰς δέλτους *POxy.* 2741.1a i 19'
λέβης I b, for '*coin* .. Crete' read 'as monetary unit, *Inscr.Cret.* 4.1, al. (vii/vi BC)' and add '*SEG* 37.752 (Lyttos, v BC)'
*λεβητίσκος, ὁ, dim. of λέβης, *IG* 2².1424*a*.147 (iv BC).
λεγεών, after 'λεγιών ib. 214.3, al.' insert '*TAM* 4(1).189; also written ληγιών Mitchell *N.Galatia* 289'; at end delete 'hence λεγιονάριος .. al.'
*λεγεωνάριος, ὁ, *legionary*, Modest.*Dig.* 27.1.8.6; also λεγιωνάριος *INikaia* 1551, *POxy.* 1419.7 (AD 265), *PLond.* 1254.5 (iv AD), λεγιονάριος *IGRom.* 3.214.3; ληγιωνάριος *BGU* 344 col. 2.4.
*λεγίτιμος, η, ον, Lat. *legitimus*, *based on law*, Just.*Nov.* 84.1 pr., al.
*λεγουμενᾶλε, Lat. neut. adj. *leguminale*, of a sieve, *Edict.Diocl.* 15.60.
λέγω (B) I after '2' insert 'act., *choose, pick*, Pl.*Lg.* 738a; pass., Il. 13.272, Pl.*Lg.* 737c'
λεία (B), line 1, after '(Pi.*O.* 10(11).44)' insert 'λαία *SEG* 23.324(*a*) (Delphi, 277 BC), sense 4, cf. Hsch., s.v. λαίαν' add 'form λᾱ-Fία in deriv. adj. Myc. ra-wi-ja-ja (fem. pl.); cf. ληϊάς'
λειαύστηρος, before 'Poll. 6.15' insert 'v.l. in'
*Λειβηθριάς, άδος, *of Libethrum*, τῶν Λειβηθριάδων νυμφῶν Str. 10.3.17,

καὶ πηγαί—τὴν μὲν Λειβηθριάδα ὀνομάζουσιν Paus. 9.34.3; epith. of the Muses, κούρη Λειβηθριὰς ἔννεπε Μοῦσα Orph.*fr.* 342, *Suppl.Hell.* 993.7.
*Λειβηθρίς, ίδος, = °Λειβηθριάς, τῶν Λειβηθρίδων νυμφῶν Str. 9.2.25, *Suppl.Hell.* 980.1.
Λείβηθρον II, at end delete 'the Λειβηθρίδες .. 342'
*Λειβῆνος· ὁ Διόνυσος Hsch.
λειεντεριώδης, add 'Gal. 16.800.5'
λεῖμαξ I 1, line 2, after '(both lyr.)' insert '*IA* 1544 (interp.)'
*λείμψᾰνον, τό, = λείψανον, *ABSA* 59.19 (Aphrodisias).
λειμών III, line 2, delete '1.19'; last line, after '*Praef.* 6' insert 'λόγοι τῶν καλῶν λ. ἀφθόνους προτείνοντες Procl. *in R.* 1.161 K.'
λειμωνιάς, add 'Orph.*H.* 29.12'
λειμωνιάτης, delete the entry.
*λειμωνῐᾱτις, ιδος, ἡ, a stone, = *smaragdos*, Plin.*HN* 37.172.
λειμώνιος, add 'as a divine epith., Κόρηι Λειμωνίαι *RA* 6.67 (Amphipolis, ii BC)'
*λειμωνῖτις, ἡ, a plant, = λειμωνία (s.v. λειμώνιος), Suid.
λεῖος, after '*a, ov*' insert '(Ion. λέος, η, ον, *Didyma* 434.23, 25, 437.4 (both iii BC))' I 3, line 7, delete '3.14' II, line 2, before 'Dsc.' insert 'Arist.*HA* 534ᵇ23, *PA* 674ᵇ13'
*λειουργέω· καλλωπίζω Phot.
*λειπογᾰλαξία, ἡ, *absence of milk, agalactia*, νόσων γὰρ καὶ λειπογαλαξίας αἴτιος ἔσται Heph.Astr. 3.5.49.
λειπογνώμων, add 'also λειπε- *SEG* 26.136 (Athens, iv BC)'
*λειποτελής, ές, *in arrears with taxes*, *PTeb.* 711.4 (ii BC).
λείπω I 3 a, line 2, after 'ib. 903' insert 'ὡς ἡ μὲν λίπε μῦθον *thus she ceased talking*, Call.*fr.* 43.84 Pf.' II 2 a, add 'of prosecutor, *drop the case*, Arist.*Ath.* 16.8' II 2, line 8, before 'inf.' insert 'substantival' b, add 'med., *wane*, Orph.*H.* 9.4, Nonn.*D.* 5.164' B II 3, add 'τῶν κτεάνων .. τὸ λειπόμενον *deficiency* in possessions, *AP* 5.267.8 (Agath.)' add 'Myc. med. part. *re-qo-me-no*'
λειριόεις 1, add 'ὀδμὴ .. λειριόεσσα Opp.*H.* 3.410'
λείριον II, before 'Dsc. 4.158' insert 'perh. *Narcissus poeticus*' add 'IV flower of the plant ἔχις, Nic.*Th.* 543.'
*λειριόπρυμνος, ον, *lily-(?white-)prowed*, *Suppl.Hell.* 991.114 (poet. word-list, iii BC).
*λείριος, ον, *like a lily*, of white coral, λ. ἄνθεμον ποντίας ἐέρσας Pi.*N.* 7.79; ὄμματα B. 17.95 S.-M.; κηρία Opp.*C.* 1.128. 2 applied to melodious voice, ἵεσαν ἐκ στομάτων ὄπα λείριον (Σειρῆνες) A.R. 4.903, Orph.*A.* 253; cf. ‡λειριόεις.
*λειριόχροος, ον, *having lily-white skin*, Hdn.*fr.* p. 15 H.
λείστριον, delete '= λίστριον'; add '*IG* 2².1678*a*A5 (λιστρ- lapis; Athens, iv BC)'
*λείτορας, ὁ, = λείτωρ, (in quot., priest of Apollo), *SEG* 36.548 (Thessaly, iii BC).
λειτορεύω, for '= ἱερεύω' read 'serve as λείτωρ'; line 2, at beginning insert '*SEG* 8.714 (Thebes, Egypt, ii BC)'
λειτουργέω I, add 'of a house or family, οἴκων λητουργούντων D. 42.23'
λειτουργός II 1, add 'b at Athens, *clerk* named at the end of list of magistrates and their assistants, *IG* 2².1728, 1729, 1731 (all i AD).'
λείτωρ, for '*priest*, ἀρωγός λ. *MDAI(A)* 12.283' read 'a kind of priest, Ἀρωγος λ. *IG* 2².4817.25, cf. λήτωρ, ἀλήτωρ (cf. Ἑλληνικά 8.229)'
λειχήν 3, delete 'of animals'
λείχω, add 'in sens. obsc., *SEG* 34.1015 (Rome, ii/iii AD); cf. λείκτης'
*λειψᾰνοθήκη, ἡ, *repository for the remains of the dead*, *SEG* 31.630 (Rom.imp.).
λείψανον 2, add 'also sg., *BCH* 102.413-4 fig. 3b (iii AD), etc.'
λέκκη, add 'also λέκτη, Theognost.*Can.* 27 A.'
λέκος, for '*dish, pot, pan*' read '*dish, bowl*' and insert '*SEG* 33.996 (Smyrna, vi BC), *IG* 1³.422.131 (Athens, v BC)'
*λεκτεῖκα, ἡ, Lat. *lectica*, *litter*, *SB* 7348.3 (i AD).
*λεκτικάριος, ὁ, Lat. *lecticarius*, *bearer at funerals*, Just.*Nov.* 43 pr., al., *PIand.* 8.154 (vi/vii AD), *ITyr* 29*B*, al.
λεκτίς, add 'II *tomb*, masc. w. acc. λεκτεῖκα *SEG* 26.1314, 1320 (Sardis, iv/vi AD); cf. °λεκτεῖκα.'
λεκτός I, add 'βολὴ .. λεκτὴ πεντήϙοντ' ἀπὸ φυλῆς *council chosen* fifty from each tribe, *SEG* 16.485C.7 (Chios, vi BC)'
*λελεγίζω· τὸ κιθαρίζω Theognost.*Can.* 27 A.
†λεμβαρχέω, *command a λέμβος*, *IParion* 5.
*λεμησία, ἡ, wd. of unkn. meaning, *PYale* 902 + 906 (ii AD) in *YClS* 10.217, line 53; cf. λέμυσος, λεμεῖσα.
λέμμα 1, add '*shell of an egg*, Ael.*VH* 10.3'
*λεμόνη, ἡ, *lemon*, Anon.Alch. 328.23.
*λεμψάνη, v. °λειψάνη.
*λεντιαρία, ἡ, *female linen-dealer*, *AAWW* 1956.230.10 (Lydia).
*λεντιάριον, τό, perh. *recess* or *cupboard for linen*, *ITomis* 389 (iii/iv AD)

194

†**λεντιάριος**, ὁ, Lat. *lintearius*, perh. *cloakroom attendant* at gymnasium, *IG* 2².2130.221, 14.2323. **2** *linen-dealer*, Σευῆρος λεντιάρις *IG-Bulg.* 1922.7.

λέντιον, delete 'hence λεντιάριος .. 14.2323'

***λέντιος**, ον, *made of linen*, ἰλήματι καινῷ λεντίῳ *POxy.* 3060.7 (ii AD).

***λενῶ**· τὸ λυποῦμαι Theognost. *Can.* 27 A.

***λεξιγράφος**, ὁ, = °λεξογράφος, *AB* 1094.

†**λεξίδιον**, τό, *verbal expression*, Arr.*Epict.* 2.1.30, 2.23.43, etc., Gal. 13.575.8, etc.

λεξιθηρέω, delete 'Plu.ap.'

***λεξογράφος**, *one who records* λέξεις, *lexicographer*, Sch.Hes.*Op.* 633-40, Lyd.*Mag.* 11.16, 21.25 W; (see also λεξικογράφος).

***λεοδράκων**, οντος, ὁ, *mythical creature*, (*lion-serpent*), *Inscr. Cret.* 2.xix 7.19 (Phalasarna, iv BC).

λεοντάγχης, add 'of type of engraved agate, *Hippiatr.* 2.148.5'

λεοντάριον I, for 'Dim. of λέων' read '*statue* or *image of a lion*'

***Λεοντεῖον**, τό, app. *meeting-place of the* λέοντες *in the cult of Mithras*, λεοντίῳ *PBerol.* 21196.8 (*Tyche* suppl. 1992 p. 18, iv AD).

λεόντειος, add 'Myc. *re-wo-te-jo*'

λεοντίς, for pres. ref. read '*SEG* 33.946'

λεοντοβότος II, delete the section transferring 'Str. 16.1.24' to section I and add 'Νεμέης .. λεοντοβότου *Inscr.Magn.* 181.9 (ii AD)'

λεοντοδέρης, for 'like .. tawny' read 'lionskin, name given to agate'

***λεοντόθῡμος**, ον, *lion-hearted*, Hsch. s.v. θυμολέοντα.

λεοντοπρόσωπος, add 'Heph.*Astr.* 2.2.32'

***λεοντορήκτης** or **λεοντορήκτα**, ὁ, *lion-breaker*, Bonner *Studies in Magical Amulets* 169.

λεοντόχασμα, for '= λεοντόκρουνον' read '*spout in the form of a lion's mouth*' and add 'Aristo in *Gnomol. Vat.* in *WS* 10.23'

***λεοντοχασμάτιον**, τό, dim. of λεοντόχασμα, rest. in Sch.Pi.*Pae.* 6.7.

***λεοντοχασματύπανον**, τό, *kind of wheel for raising water, IGLS* 645.

λέπαργος I, for 'A.*Fr.* 304.5' read 'S.*fr.* 581.5' and for 'sheep or goat' read 'calf'; add 'βοὸς λεπάργου *Trag.adesp.* 231 II, add '**2** λέπαργος· ἡ χιών, Theognost.*Can.* 27 A., Zonar.'

λεπίδιον I, add 'part used in the construction of a boat, *POxy.* 2195.141 (vi AD)'

λεπιδίσκη, delete the entry.

***λέπιδος**, η, ον, Lat. *lepidus*, *charming*, *agreeable*, Λεοσθένης λέπιδος καὶ ἐράσιμος *IG* 14.40.

***λεπίδωτις**, ιδος, ἡ, *gem*, prob. the same as λεπιδωτός II 2, Plin.*HN* 37.171.

λεπιδωτός I 2, add 'Sm., Thd. 1*Ki.* 17.5'

λεπίς 3, add 'Ptol.*Tetr.* 153'

λεπιστής, add '(or λαπιστής, see Hsch. s.h.v.)'

***λεπταμικτόριον**, τό, *a fine amictorium*, *PFouad* inv. 45.18 (*CE* 27.196, ii/iii AD).

***λεπτόδομος**, ον, perh. *slenderly fashioned*, πείσματα A.*Pers.* 112 (s.v.l., edd. °λεπτότονος, etc.).

***λεπτοκεραμικός**, ή, όν, *used for fine pottery*, λίθος *CPap.Jud.* 452b.5 (ii AD).

***λεπτόκλωνος**, ον, *having fine* or *thread-like branches*, βεττονική λ. Paul.Aeg. 7.3 (200.19 H).

λεπτοκοπέω, add 'Heras ap.Gal. 13.558.12'

***λεπτοκυμία**, ἡ, gloss on φρίξ, Sch.Gen.Il. 23.692, Hsch.

***λεπτοπηνος**, ον, *made of fine fabric*, λεπτοπήνοις ὑμέσιν Eub. 67.5, 82.4 K.-A.

***λεπτοποιός**, ὁ, perh. *joiner* or *cabinet-maker*, Ramsay *Cities and Bishoprics* I p. 142 no. 31.

***λεπτοπρόσωπος**, ον, *thin-faced*, Heph.*Astr.* 1.1.58.

***λεπτοπῠρέτιον**, τό, *slight fever*, Cyran. 59 (pl.).

λεπτός III 2, add 'in *Ev.Marc.* 12.42 equated with half a *quadrans*'   add 'Myc. *re-po-to* (of linen, cf. I 3)'

***λεπτόσφυρος**, ον, *having thin ankles*, Hsch. s.v. τανίσφυρος, Rhetor. in *Cat.Cod.Astr.* 7.209.14.

λεπτότης II, add '**2** *of literary or artistic style, refinement, delicacy*, Hermog.*Id.* 2.12, D.H.*Isoc.* 3.'

***λεπτότονος**, ον, *finely-stretched*, πείσματα cj. in A.*Pers.* 112 (but v. °λεπτόδομος).

***λεπτόχειλος**, ον, = λεπτοχειλής, Heph.*Astr.* 3.45.9.

λέπτυνσις, add 'Gal. 11.111.14'

λέπῠρον, add 'of an eggshell, Ach.Tat.*Intr.Arat.* p. 33.19'

λέπω II, add 'τῷ ῥοπάλῳ τὰν κεφαλὰν λέπομες *AP* 9.330 (Nicarch.)'

***λερνός**, ή, όν, unexpld. wd., Hdn.*fr.* p. 25 H.

***Λεσβιακός**, ή, όν, *Lesbian*, Λ. λόγοι, work by Dicaearchus, Cic.*Tusc.* 1.77; Λεσβιακά, τά, by Myrsilus (477 F 1-3 J.), also by Hellanicus (4 F 34-35), (or Λεσβικά (id. 33)).

Λεσβιάς, add '**2** Λ. (sc. λίθος), ἡ, *a precious stone*, Plin. *HN* 37.171.'

λεσχαῖος, add 'as title of Apollo, *IG* 9(2).1027 (Larissa, v BC)'

λέσχη I 1, line 2, after 'Camirus' insert 'vi BC, Dor. -ᾱ' **3**, lines 5/6, for 'at Cnidus, *council-chamber*' read '*council-chamber* of the Cnidians at Delphi'

***λεσχήνευμα**, ατος, τό, *place for conversation*, *Inscr. Cret.* 2 v 51 (i AD).

†**λεσχηνευτής**, οῦ, ὁ, = °λεσχηνώτης, ὁ Ποντικὸς λ. "Talker", i.e. author of Λέσχαι Ath. 14.649c.

Λεσχηνόριος, line 4, after 'al.' insert 'written Λεσχηνούριος, *AE* 1927/8.123 (ii BC)'

†**λεσχηνώτης**, ου, ὁ, *one who takes part in a discussion*, Thales ap.D.L. 1.43, Anaximen.ap.D.L. 2.4.

***λεσωνις**, ου, ὁ, = λεσῶνις, *SB* 6154.31 (i BC), *BGU* 37ᵛ (i AD), *PTeb.* 313.6 (AD 210/1).

λεσῶνις, lines 2/3, delete 'later gen. .. *BGU* 37ᵛ (i AD)'; line 3, after '(ii BC)' insert 'nom. pl. -ῶνες *BGU* 916.9, 25 (i AD)' (v. °λεσώνης)

***λεύδια**, v. °γλεύδιον.

Λευκᾱθέα I, add 'in Syria *SEG* 31.1392'

***λευκάνθεμος**, ή, = λευκάνθεμον, Sch.Nic.*Th.* 849 C.

λεύκανσις, add '**2** *whitening, whitewashing*, *ABSA* 61.306.122 (Epid., iii BC).'

λευκαντής, add '*SEG* 34.1124 (Ephesus, ii AD), *POxy.* 3743.7-8 (iv AD), *PWash.Univ.* 37.4 (v AD), *Corinth* 8(3).522 (v/vi AD)'

***λευκάριον**, τό, λ. κωβαθίων (κο- pap.), perh. *white powder*, *SB* 10492ᵛ.26 (ii AD).

λευκάς II 2, for this section read 'epith. of ποίη, Nic.*Th.* 849, λ. π. acc. to Sch. ad loc. being ἡ λευκάνθεμος'

λευκασία, add '**2** *name of a skin disease*, Cyran. 15.'

λεύκασπις, line 2, after 'X.*HG* 3.2.15' insert 'of a Tarentine corps, D.H. 20.1'

λευκόγεως, after 'λευκόγαιος' insert 'cf. *leucogaeam*, Plin.*HN* 37.162'

***λευκογράφῑτις**, ιδος, another name for °γαλακτῖτις, Plin.*HN* 37.162.

***λευκόζωνος**, ον, dub. l., *white-girdled*, of Ino (Leucothea), *SEG* 26.683.5 (Thessaly, iii BC).

Λευκοθέα I, add 'pl., Alcm. S 5(b).12, *SEG* 32.1538 (Gerasa, ii/iii AD), 33.1262 (Syria, AD 268/9)'

λευκόλινον, for '*white flax*' read 'app. a kind of flax or hemp' and add '*PLond.* 1965.2 (iii BC)'

***λευκομέννιον**, τό, *white sprat*, *BGU* 2358.13 (iv AD).

λευκομέτωπος II, after 'bird' insert 'prob. *coot*'

***λευκοπώων**, gen. ονος, *fat and white*, prob. in *POxy.* 1631.25 (iii AD); cf. Sch.Ar.*Ra.* 1124.

λευκόπῡρος, add '*BGU* 1067.16 (ii AD)'

***λευκοπωρινός**, ή, όν, *made of* °λευκόπωρος, *BE* 1971.642 (Phrygia).

***λευκόπωρος**, ὁ, *a kind of white marble*, *BE* 1971.642 (Phrygia).

λευκός I 1, add '*Peripl.M.Rubr.* 38' **2**, line 3, after '(Phil.)' insert 'λ. ῥήσει Babr. ii prooem. 13' **II 1 a**, add '*b of person, white-haired*, Call.*fr.* 194.52 Pf.' **2**, add '**b** *of tortoise-shell, pale*, *Peripl.M.Rubr.* 3, 30.' **3**, line 1, for 'Call.*Aet.* 1.1.2' read 'Call.*fr.* 178.2 Pf.'; line 4, for 'cf. Sch. .. 1094.39' read 'Call.*fr.* 191.37 Pf. (pl.)'; line 5, after 'Per. 27' for 'ἡ λ. .. Hsch.' read '**b** *white*, i.e. *favourable*, *of a voting pebble*, ἡ λ. καὶ σώζουσα ψῆφος Luc.*Harm.* 3; ψήφου ἐνεχθείσης ἐγένοντο πᾶσαι λευκαί *SEG* 24.614 (Maced., ii AD), cf. *SEG* 9.354.25, *REG* 62.284.28, 286.21 (all Cyrenaica, i BC/i AD). **c** *of place, fortunate* or *glorious* (cf. 1), ἡ τότε λ. Δῆλος *AP* 9.421.5.' add 'Myc. *re-u-ko*'

λευκότης, add '**2** *of style, clearness, limpidity*, Eun.*VS* p. 458 B.'

λευκουργός, for '*BCH* 32.500 (Aphrodisias)' read '*La Carie* 162 (Apollonia of Salbake, c.AD 200)'

λευκόφυλλος, after 'Dsc. 4.103' insert '*PSoterichos* 4.17 (AD 87)'

***λευκόχαλκος**, ὁ, τὸ δὲ λευκὸν τοῦ αὐτοῦ ᾠοῦ καλοῦσιν ὑδράργυρον, ὕδωρ ἀργυρικόν, λευκόχαλκον, etc. Anon.Alch. 19.19.

***λευκοχίτων**, ωνος, ὁ, ἡ, = λευκοχίτωνος, ζειά *Lyr.adesp.* 11(d).5 P.

λευκόχρωμος, add 'ὄνος *POxy.* 1708.10 (iv AD)'

λευκόχρως, for 'colourless' read 'pale-skinned'

λευκώλενος, of Hera, add 'λευρδλενδι ἐραι *JHS* 106.196 (vi BC)'

λεύκωσις, add 'III *whitewashing*, *IG* 4².102.305 (Epid., iv BC).'

λευκωτής, add '*SB* 10258 i 5 (iv AD)'

λευρός I, at end for 'Call.*Aet.Oxy.* 2080.67' read 'Call.*fr.* 43.65 Pf.'

λεύσσω, line 5, delete 'Poet.Verb .. *AB* 1096' **3**, for 'c. acc. cogn.' read 'w. cogn. or adverbial acc.'

λευτο-, Arcad. vb. see part. λεύτον *IG* 5(2).3, λεύτοντες ib. 5(2).16.

λεύω, line 1, after 'λεύσω' insert 'A.*fr.* 132c.1 R.'; line 3, after '*stone*' insert 'Hippon. 37 W.' and after 'Th. 5.60' insert 'τοὐμὸν σῶμα A. l.c.'

λεχεποίη, delete '*h.Merc.* 88'

λεχεποίης, add '*h.Merc.* 88'

****λεχεστρωτήριον**, τό, Myc. *re-ke-to-ro-te-ri-jo*, *name of a festival*.

λεχήρης, add '*IEphes.* 2109'

λέχομαι, line 3, before 'Hsch.' insert 'Antim. 178 Wy.'

λεχώ, add 'on a funerary monument perh. indicating death in childbirth, *Inscr. Cret.* 2 x 12'

***λεχώζω**, *hatch*, Suid. s.v. σελάχια, cf. Phot. s.v. σελάχιον.

λεψάνη SUPPLEMENT λιθάριον

*λεψάνη, ή, = λαιψάνη, PLond. 1771.10 (vi AD); also λεμψάνη, PHamb. 68.41 (vi AD).
*Λέψυνος, ὁ, epith. of Zeus, Ilasos 151.
*λεώκερας, τό, word in magical formula, Inscr.Cret. 2 xix 7.18 (Phalasarna, iv BC).
λέων III, after 'Th. 643' insert 'Afric.Cest. p. 16 V.' VI, add 'Rev.Hist.Rel. 109.63 (Rome)' add 'Myc. re-wo-pi (instr. pl.)'
*Λεωνίδαια, τά, festival at Larissa, BCH 59.515 (iii BC); also Λεωνίδεια, at Sparta, in honour of the hero of Thermopylae, IG 5(1).18.8, al.
*Λεωνίδαιον, τό, shrine of Leonidas, Paus. 5.15.2, 6.17.1.
*λεῶρες· τὸ ἄωρον Theognost.Can. 27 A.
+λήβολος, ον, Aeol. λάβ- Alc. 68.3 L.-P. (s.v.l.), condemned to stoning, λήβολε· λιθοβολε, ἄξιε λιθασθῆναι Hsch. (ληβόλε· λιθοβόλε cod., La.).
*ληγατάριος, ὁ, Lat. legatarius, legatee, Cod.Just. 6.4.4.18a; also λεγ- Modest.Dig. 26.6.2.3.
*ληγατεύω, bequeath in the form of a legacy, Cod.Just. 1.5.15.
*ληγᾶτον, τό, Lat. legatum, legacy, BGU 1.327.6, IGRom 3.828, Scaev.Dig. 33.8.23.2, Cod.Just. 1.2.25 pr., al.
*ληγᾶτος, ον, Lat. legatus, past pass. part. bequeathed, SEG 34.1213 (Lydia, ii AD).
*λεγιών, v. °λεγεών.
λήγω II 1, line 5, after 'Th. 7.6' insert 'abs., ληγούσης Πελοπηΐδος Call.fr. 384.11 Pf.'
ληδεῖν and ληδήσας, delete the entry.
+λῆδος, εος, τό, Dor. λᾶδος, a kind of dress (cf. ληδίον), λᾶδος Γημένα καλόν Alcm. 117 P., λάιδος· λῆδος, τριβώνιον Hsch.
λήθαιος III, add '= °ἀναγκίτης Orph.L. 197'
ληθαργέω, after 'forget' insert 'SEG 34.1065 (Aphrodisias, vi AD)'
ληϊβοτήρ, add 'Plu. 2.730b'
ληΐτιδιος, after 'captive' insert 'Euph. in Suppl.Hell. 415 ii 18'
*ληϊστής, v. λῃστής.
λήϊτον, line 4, before 'λαιετόν' insert 'λαίετον' and before 'Suid.' insert 'λαιτρόν, Theognost.Can. 27 A.'
λεκάω, for '= λαικάζω' read 'have sexual intercourse with'
*ληκήτρια, ἡ, λ. θεά Lyc. 1391, fem. of ληκητής.
ληκυθίζω, line 2, for 'Call.fr. 10.13 P.' read 'Call.fr. 215 Pf.'
*ληκυθίς, ίδος, ἡ, = ληκυθος, Hesperia 33.83.
λήκυθος I 1, after 'ή' insert 'Δηρίππō ἐμὶ λήρυθος SEG 33.995 (Smyrna, vi BC)' 2, add 'cf. Sch.Heph. p. 230.18 C.'
λῆμα II 1, at end after 'Ra. 463' insert 'Call.fr. 43.59, 345 Pf.'
*λημᾱτίζομαι, to be eager for, λελημάτισθαι γὰρ τὸ τῇ διανοίᾳ πρὸς πᾶν ὁρμητικῶς ἔχειν Hsch. s.v. λελημένοι.
λημάω, line 3, after 'Ar.Nu. 327' insert 'χύτραις λημῶντες Luc.Ind. 23'
λημματίζω, add 'also λημματ- PLond. 995, 996, etc., λυματ- PHerm. 24.3 (iv/v AD)'
λημματικός, add 'Zos.Alch. 213.2'
*λημματισμός, ὁ, entry, crediting, SB 9772.1 (pap. λιματ-; vi AD).
Λήμνιος, add 'Myc. ethnic adj. ra-mi-ni-ja'
Ληναϊκός, for 'ἀγῶνες .. 7.414e' read 'Ληναϊκήν .. Καλλιόπην Posidipp.ap.Ath. 10.414e (s.v.l.)'; add 'οἱ Ληναϊκοί authors only qualified to compete at Lenaea, Eratosth. in POxy. 2737; also perh. τὸ Ληναϊκ[όν], Lenaean competition, ib.'
*Λήναιος, ὁ, form of month Ληναιών given at Hemerolog.Flor. p. 70 (p. 4 K.).
Ληναιών, add 'SEG 34.1175 (Miletus, ii AD), also Ληνεών SEG 30.977 (Olbia, v BC), 30.968 (Olbia, ii/iii AD)'
*ληνίον, τό, perh. box, Cyran. 26, 27.
ληνίς II, add 'in some similar sense, λ. ἀργυρᾶ Inscr.Délos 104.96'
*ληνοβατέω, = ληνοπατέω, Cyran. 23; pass., ib. 22.
ληνοβάτης, add 'Anacreont. 4.16 W.'
*ληνοβατικῶς, adv., πατεῖν λ. i.e. in the wine-vat, Eust. 1574.8.
ληνός 1, line 2, after 'are pressed' insert 'IG 1³.422.189, 425.34, 426.148 (cf. Poll. 10.130)' add 'b storage vessel for wine, POxy. 1569, PFlor. 253 (iii AD).' add '9 bathing-tub, λ. δημοσίας MDAI(A) 32.274 (Pergam., ii BC), Inscr.Délos 1423Baii4. 10 part of door, Maier Griech.Mauerbauinschr. 19.32 (Eleusis, iv BC).'
+λῆνος, εος, τό, flock of wool, A.Eu. 44, A.R. 4.173, 177, Call.Suppl.Hell. 257.6 (cf. fr. 722 Pf.), Nic.Al. 452, Epic.in APF 7.4, ?D.P.fr. 19ᵛ.20 L., Nonn.D. 6.146.
*λῆνος (B), ὁ, prob. = βεβακχευμένος, Schwyzer 791 (Cumae, v BC); cf. λῆναι.
λῆξις (A) I 1 c, for 'pl., fortunes' read 'portion, lot' and add 'Just.Const. Δέδωκεν 18, Cod.Just. 1.3.52.4,5, PMasp. 19.6 (vi AD)'
λῆξις (B) I 2, delete the section.
Ληξοπύρετος, v. ληξιπ-.
ληπτός I, add 'Trag.adesp. 168'
λῆρος (B), after '(Hedyl.)' insert 'Inscr.Délos 1433.6 (ii BC)'
*ληρότης, ητος, ἡ, gloss on ὕθλος, Hsch.

λῃστεία, after 'piracy' insert 'raiding'
λῃστεύω, after 'practise .. piracy' insert 'make raids'
λῃστήρ, add 'of Hermes, h.Merc. 14'
*λῃστογνώστης, ου, ὁ, associate of thieves, Just.Nov. 13.4 pr., al.
+λῃστοδιώκτης, ου, ὁ, = Lat. latrunculator, official or agent appointed to catch thieves, Just.Nov. 8.13, al.
*λῃστοδίωκτος, ον, pursued by pirates, prob. cj. in X.Eph. 1.6.
*λῃστολογέω, app. recruit λῃσταί as soldiers, IGBulg. 1126 (iii/iv AD).
λῃστρικός 2, add 'comp. λῃστρικώτερον ἀφιγμένος i.e. more concerned with plunder, Plu.Pyrrh. 10'
Λητοΐδης, after 'Alc.Supp. 30.3' insert 'CEG 785.1 (sp. Λετ-, v BC), AP 12.55'; after '(trisyll.)' add 'so Λατοΐδα CEG 302.1 (vi BC); also Λητωΐδης SEG 23.126 (c.i BC)'
Λητώ, line 4, acc., add 'Λατών SEG 33.638, Λητοῦν REG 101.14 (Xanthos, iii BC)'
*Λητωΐδης, ὁ, = Λητοΐδης, son of Leto, i.e. Apollo, SEG 23.126 (c.i BC).
*Λητωΐνη, ἡ, daughter of Leto, i.e. Artemis, Suppl.Hell. 962.9.
*Λητώϊος, ὁ, son of Leto, i.e. Apollo, TAM 4(1).48.
λήτωρ, for 'priest, prob. in' read 'a kind of priest (perh. as pers. n.)' and add 'cf. ἀλήτωρ'
λιάζω (B), for '[παρά .. 34.27' read 'λίην λιάζεις Archil. 113.8 W., Anacr. 85 P.'
λιᾰρός, line 2, after 'Od. 24.45, etc.' insert 'λιαροῖο (v.ll. λαρ-, λιπαρ-) .. γάλακτος Theoc. 25.105'
*Λίβαιος, ὁ, month in Cyprus beginning 2 Dec., Cat.Cod.Astr. 2.147.15, 148.13.
λίβᾰνος I, add 'fem. in Pi.fr. 122.3 S.-M., Nic.Al. 107' II, line 2, delete 'Pi.fr. 122.3'
*λιβανοφλόγος, ον, incense-burning, expl. of turicremus in Virgil gloss., PNess. 1.935 (vi AD).
λιβανόχροος, add '2 λιβανόχρους (sc. λίθος), ἡ, a precious stone, Plin.HN 37.171.'
λιβανωτίς (B), add 'see also °λιβανωτρίς.'
*λιβανωτοπώλιον, τό, incense-shop, IEphes. 4102.2 (iii BC).
λιβάνωτός, line 2, before 'Xenoph.' insert 'Sapph. 2.4 L.-P.'
λιβανωτρίς, for 'censer' read 'box or casket for incense'
*λιβελλήσιος, ὁ, Lat. libellensis, official in the imperial bureau dealing w. petitions, Just.Nov. 20.7.
λιβέλλος, after 'ὁ' insert '(neut. pl. λιβέλλα Edict.Diocl. 7.41)' and add 'Salamine 29.4 (iii AD)'
*Λιβερνάριος, ὁ, = Λιβυρνάριος, POxy. 1902.4 (vi AD).
*Λιβέρνιος, α, ον, Liburnian, πλοῖον POxy. 2032.52, 54 (vi AD).
λιβέρτος, add 'fem. λιβέρτα SEG 35.1011 (Sicily, i BC)'
*Λιβιανός, ή, όν, Livian; κριθὴ Λ. barley from Livia's estate, PKöln 116.3 (iii AD).
*λιβράριος, ὁ, Lat. librarius, scribe, BCH 7.275 (ii AD), SB 6971, Edict.Diocl. 7.69, written λειβράρις SEG 26.1600 (Commagene).
*Λιβύανδε, adv., to Libya, SEG 9.3.19 (Cyrene, iv BC).
Λιβυάρχης, add 'SEG 30.1785.1(e) (Cyrene, iv AD).
λιγάνταρ, insert '(λιγάντωρ La.)'
*λίγγα, ἡ, Lat. ligula, spoon (μυστίλη), Poll. 6.87; also λίνγλα (lingula, as measure), BGU 781 vi 3, 16 (i AD).
*λιγμάτιον, τό, dim. of λικμός (= λίκνον), winnowing-fan, BGU 2359.8 (iii AD).
*λιγνύζω or -νυΐζω, have a smoky colour, of carbuncles, Plin.HN 37.94.
λιγνύζων, delete the entry.
λιγνύς 1, line 1, for 'Call.fr. 1.57 P.' read 'Call.fr. 228.57 Pf.'; add 'w. ref. to the colour of the Ethiopians, ἥλιος .. σκοτεινὸν ἄνθος ἐξέχρωσε λιγνύος εἰς σώματ' ἀνδρῶν Theodect. 17.2'
*λίγξε, add '(cf. λίγγω· ἠχῶ Theognost.Can. 16)'
+λιγουρά, v. °λιγυρός.
+λίγουροκώτιλος, v. °λιγυρο-.
Λιγυαστάδης, before 'ον' insert '(or -αστ-)'
λιγύριον, add 'Epiph.Const. in Lapid.Gr. 196.14, al.'
*λιγυροκώτ[ι]λυ[ς (∪∪∪-∪-) ἐνοπῆς Corinn. 2(b).5 P. (Boeot.).
λιγυρός, line 1, after 'λιγουρά' for '(q.v.)' read '[∪∪-] Corinn. 11(a) P.' I 1, line 7, for 'poets' read 'poetry' add '3 of style, clear, lucid, D.H.Dem. 5; adv. -ρῶς Plu. 2.874b; alternative expl. (= γλυκερῶς) offered by Sch.D.T. p. 173 H.' II, for 'pliant, flexible' read 'fine, slender'
*Λιγυστίς, ή, Ligurian, in quots., of Circe, E.Tr. 437, A.R. 4.553.
*λιγυφεγγέτις, ιδος, fem. adj., shining clearly, σεληναίη App.Anth. 6.140.6.
*λιθᾰγωγέω, transport stones, SEG 34.122 (Attica, iv BC).
λιθάζω II, after 'Pass.' insert 'Hsch. s.v.ληβόλε'
λιθᾰκός, line 2, for 'Stesich.Oxy. 1087.48' read 'Stesich. 37 P.'
λιθάριον, add '3 gravestone, SEG 28.1582 (c.vi AD).'

λιθάρτης, ου, ὁ, (αἴρω) *stone-lifter*: as adj., καρκίνος λ. *IG* 2².1424*a*.272 (Athens, iv BC).

λιθάς, line 2, after '(Od.) 23.193' insert 'perh. as used in mosaic construction, *SEG* 19.408'

Λιθήσιος, ὁ, epith. of Apollo, St.Byz.; hence Λιθέηια, τά, festival in Laconia, *IG* 5(1).213.54, 60 (Sparta, v BC).

λιθία, v. λιθεία.

λιθίζω, delete the entry.

λίθιος, add 'sp. λίτθ- *SEG* 31.572 (Crannon, *c*.200 BC)'

λιθογνώμων, add 'title of work by Xenocr. (*Lap*. 89)'

λιθόδμητος, after 'stone-built' insert '*PHib*. 172.90 (iii BC)'

λιθοκόλλητος II, delete the section.

λιθόκολλος, for '*CIG* 2852.47' read '*Didyma* 424.47' and add '*Lindos* II 2 c 86, 87'

λιθοκονία, ἡ, app. *pulverized stone*, Hsch. s.v. στῖα.

λιθοκόπος, add '*IHadr*. 103.6 (Rom.imp.)'

λιθολογέω, for 'build with unworked stones' read 'build a stone foundation of' **II**, for this section read '**2** perh. *reduce to the foundations*, Aq.*Mi*. 3.12.'

λιθολόγητος, ον, *built with piled-up stones*, *Inscr.Délos* 1416*Bi*11 (ii BC).

λιθόξεστος, ον, *made of polished stone*, *IG* 12(1).842.9 (Lindos).

λιθοξόος, line 1, after 'ὁ' insert '-ξοιος *IEryth*. 535' **1**, add '*SEG* 30.1397 (Lydia, ?iii/iv AD)'

λιθοποιέω, add 'have inscribed on stone, τὰ τοσαῦτα ὑμεῖν ἐλιθοποιήσαμεν γράμματα *SEG* 29.1476 (*c*.AD 200)'

λιθοπρίστης, add '*EA* 12.118 no. 70 (Caria)'

λίθος III, for '*Epigr*. 8.1' read '*fr*. 64.7 Pf., *AP* 9.67.1, *SEG* 30.1269' **V**, for 'Alc. 82' read 'Alc. 351 L.-P.'; add 'cf. Aristid. 2.55 J.'

λιθοσσόος, ον, *stone-moving*, of the sound of Amphion's lyre, Nonn.*D*. 25.428.

λιθόστρωτος 1, add 'λ., ἡ, *paved road*, *Kourion* 111.7, 9'

λιθουλκός II, for pres. def. read 'a kind of hook for extracting bladder-stone'

λιθουργός 1, add '*gem-engraver, seal-cutter*, App.*Anth*. 3.79 (Posidipp.)'

λιθουρικός, ἡ, όν, *suffering from λιθουρία*, Cyran. 116, al.

λιθοφόρος 1, for '*IG* 3.296' read 'of a priest at Athens, *IG* 2².3658, 5077 (w. ἱερεύς); λιθοφόρος τοῦ ἱεροῦ λίθου Clinton *Sacred Officials* pp. 50-1, line 15 (i BC)'

λίθωμα, ατος, τό, prob. *stone-work*, *AE* 1948/9.136 (Epid., iv BC).

λικμάς, add '*PStrassb*. 680.5 (early vii BC)'

λικμάω, line 4, delete '*make away with*' and '*crush, destroy*'

λίκμησις, εως, ἡ, *winnowing*, *PPetaus* 53.12 (ii AD), *PMich*.XI 609.22, *SB* 9409 iii 7.78 (both iii AD).

λίκμητρα, τά, *payment for winnowing*, *SB* 7373.15 (AD 29).

λικναφόρος, ον, = λικνοφόρος, *AJA* 37.250 (Latium, ii AD); in Bacchic θίασος, *IGBulg*. 1².401.2 (Apollonia in Thrace).

λιμάσσω, v. ‡λιμώσσω.

λιμεναρχέω, add '*SEG* 34.1093 (Ephesus, ii/iii AD)'

λιμέναρχος, ὁ, *harbour-master* (or, in quot., perh. = ἀγορανόμος, cf. λιμήν III), *SEG* 23.271 (Thespiae, iii BC).

λιμένιον, after 'λιμήν' insert '*IG* 4².76.27 (ii BC, λιμιν- lapis)'

Λιμένιος, add 'also of Hera, Ναύμαχός με ἀνέθεκε τᾶι Ἥραι τᾶι Λιμενίαι *SEG* 11.226 (Perachora, vi BC)'

λιμενίτης, add '**3** *dweller by the harbour*, *MAMA* 3.424, al. (Corycus).'

λιμήν, add 'prob. Myc. *ri-me-ne* (dat.), (as part of place-name)'

λιμηρός, add 'perh. in place-name Ἐπίδαυρος Λιμηρά Th. 4.56.2, 7.26.2, *barren*, expld. by Apollod.ap.Str. 8.6.1 as = εὐλίμενος'

λιμπρός, delete the entry.

λιμιταναῖος, ὁ, = °λιμιτάνεος, *SEG* 32.1554 (vi AD).

λιμιτάνεος, ὁ, Lat. *limitaneus*, m. pl. *frontier troops*, Just.*Nov*. 103.3.1

λίμιτον, τό, fr. Lat. *limes, frontier*, Just.*Nov*. 13.23, al., *SEG* 32.1554 (vi AD), 8.296.1 (Palestine), 782 (Syene, vi/vii AD).

λιμιτοτρόφος, ον, Lat. *limitotrophus*, *furnishing subsistence to troops stationed on the frontier*, *Cod.Theod*. 5.13.38, *Cod.Just*. 11.60.

λιμναγενής, for '*BMus.Inscr*. 1009' read '*IKyzikos* 507'

λιμνάζω, lines 1/2, for '*form stagnant pools*' read '*form a lagoon* or *pool*'

λιμναῖος I 1, add 'of fish, *fresh-water*, *IGC* p. 99 B20 (λιμνήων: Acraephia, iii/ii BC)' **II**, delete the section.

Λιμναῖος, α, ον, *of Limnae*, on the borders of Laconia and Messenia; as epith. of Dionysius, Call.*fr*. 305 Pf.; of Artemis, Paus. 2.7.6; Λιμναῖον, τό, *temple of Artemis at Limnae*.

λιμνήτης II, add '[Ἀρ]τέμιτι Λιμνάτιδι *SEG* 36.558 (Illyria, iii/ii BC)'

λιμνίον, after 'Arist.*Mir*. 840ᵇ33' insert '*PAAH* 1955.172 (Dodona)'

λιμός, line 6, before 'Th. 2.54' insert 'Hdt. 7.171.2'

λιμπάνω, add '*GVAK* 19'

λίμψανον, τό, = λείψανον (in quot., 2), *SEG* 31.1419 (iii/iv AD).

λιμώσσω, add 'also λιμάσσω Orac.*Tiburt*. 117, Suid.'

λιναϊκός, ἡ, όν, *made of linen*, *Inscr.Perg*. 8(3) p. 110 (Miletus, ii BC).

Λίνδος, after 'adv.' insert 'Λίνδοι (loc.) *at Lindos*, *Clara Rhodos* 9.212 (Lindos, v/iv BC)'

Λινδοῦχος, ἡ, epith. of Athena, epigr. in *Lindos* II 177.4 (ii BC).

λινέλαιον, τό, *oil of flax*, Moses Alch. 311.18.

λίνεος, line 5, delete 'A.*fr*. 206'; line 6, after '(v.l. λιναία)' insert 'contr. λινῆ, τείναντες λινᾶς ἑξαχῶς Didyma 43.24'

λινεργός, όν, *spinning* or *working flax*, S.*fr*. 269a.43 R. (unless λινεργής q.v. is read); cf. λινουργός.

λινεύς, add '*SEG* 23.326 (Delphi, iii BC)'

λινεψός, add '*BGU* 2471.5 (ii AD)'

λινοκατάγωγεύς, έως, ὁ, *linen-transporter*, *POxy*. 3111.3 (AD 257).

λινοκριθή, delete the entry.

λίνον II, add '**6** λ. Καρπάσιον *asbestos*, Paus. 1.26.7.' at end add 'Myc. *ri-no*, cf. *ri-ne-ja* (fem.) *flax-workers*'

λινοξός, ὁ, perh. *instrument for beating flax, scutcher*, *MAMA* 3.40 (Meryemlik), 457, al. (Corycus).

λινόπεπλος, add '*SEG* 28.1585 (epitaph of Isis mystes)'

λινόπηξος, ον, *of combed linen*, *SB* 9570.8 (vi AD); sp. -πιξ-, *PMich*.XIV 684.8 (vi AD).

λινοπλυτής, read 'λινοπλύτης'; add '*PVindob.Tandem* 19.15 (v/vi AD); also λινοπλύστης (s.v.l.) *PRyl*. 640.8 (iv AD)'

λινοπῦρος, delete the entry.

Λίνος I, add '*SEG* 33.303 (Epid., iv BC)'

λινοσινής, ές, *damaging linen*, λακίς, prob. in A.*Supp*. 120, 131.

λινόστημα, ατος, τό, *cloth woven of flax and wool*, Isid.*Etym*. 19.22.17.

λινούδιον, add '*SEG* 37.1001 (Lydia, ii/iii AD)'

λινόϋφος, add 'subst., *linen-weaver*, *BGU* 2471.4, etc. (ii AD)'

λινοφακός, delete the entry.

λινυφαντάριος, ὁ, *linen-weaver*, *MAMA* 3.450 (Corycus, written λεν-).

λινυφαρία, ἡ, *(female) linen-weaver*, *Vita Epiph*. 1 (*PG* 41.24).

λινύφαριος, for ' = λίνυφος' read '*linen-weaver* or *owner of linen-weaving business*'

λιπαίνω I 2, add 'pass., of man, *grow fat*, τοῦ κατὰ φύσιν πολὺ πλέον ἐλιπάνθη Memn. 4.7 J.'

λιπαραία, ἡ, *liparaea*, a precious stone, Plin.*HN* 37.172, presum. fr. Lipara.

λιπάραμπυξ, υκος, ὁ, ἡ, *having a shining* or *gleaming headband*, Μναμοσύνα Pi.*N*. 7.15; humorously, of an oily sauce, Ar.*Ach*. 671.

λιπαρόθρονος, add 'Theoc. 2.165 (*PAntin*. -χροε codd.)'

λιπαρός I 2, line 6, before 'adv.' insert 'also of flavours, Arist.*de An*. 422ᵇ12'

λιπαρόσκηπτρος, ον, *having a shining sceptre*, rest. in Simon. 14 *fr*. 60(*b*).4 P.

λιπαρόχροος, line 2, after 'Theoc. 2.165' insert '(codd., -θρονε *PAntin*.), Παφίη λ. Anacreont. 20.7 W.'

λιπαρών, ῶνος, ὁ, Alexandrian name for χνῆστρον, *Daphne oleoides*, Aët. 4.22 (368.28 O.), 6.68 (218.24 O.).

λιπερνέω, line 3, after 'λιφερνοῦντες' insert '(cj. φιλερν-)'

λιπερνής, for 'forlorn, outcast' read '*deprived, destitute*'; for 'context doubtful in *BCH* 11.161' read '*GVAK* 11' and add 'perh. in A.*Supp*. 362'

λιποδαμία, ἁ, *failure to serve the community*, *SEG* 25.447.9 (Arcadia, iii BC).

λιπόναυς, delete '(or .. *fleet*)'

λιποναύτης, add '*IG* 12(2).646*c*50 (Nesus)'

λιπόπνοος I, for this section read '*with failing breath, dying*, *AP* 12.132 (Mel.), *APl* 4.110.5 (Philostr.), 133.5 (Anon.)'

λιποτακτέω, add '**2** *abscond*, *PHerm*. 21.17, *PLond*. 1247.14 (iv AD).'

λιποτριχία, ἡ, *loss of hair*, γενείων Cyran. 52 (pl.).

λιπόχρως, acc. masc. -οα, *sleek*, ταῦρος Nonn.*D*. 19.67.

λιπυρία, add 'Gal. 11.586.7'

λίς, λινός, ἡ, app. *linen thread*, οὐ λίνας ἱστῶσιν, ἀλλὰ πίστει ζωγροῦσιν Leont.Byz.*HP* 1.7.9; cf. λῖτα.

λίσπος, add '(oxyt. acc. to A.D. in Suid. s.v. λίσπη); cf. λίσφος'

λίστριον, τό, *tool for smoothing stone*, *IG* 2².1678 (Delos, iv BC); in Ar.*fr*. 847 K.-A. = κοχλιάριον, acc. to Phryn.*Ecl*. 292.

λῖτα, add 'Myc. adj. *ri-ta* (neut. pl.), *of linen*'

λιτανευτικός, add 'Sch.Pi.*P*. 4.385'

λιτή I, at end after 'E.*Or*. 290' insert 'ὑπὲρ λειτᾶς perh. = *ex voto*, *IG* 4.584 (i AD)' add '**2** *religious procession* or *other ceremony*, Just.*Nov*. 123.32.'

λιτήσιος, ον, *entreating*, Nonn.*D*. 43.137.

Λιτοάμφοδον [ῐ], τό, *name of block of houses*, ἀπὸ τοῦ οἰκειδίου Λιτοανφόδου τοῦ καλουμένου *Hesperia* 6.390 (Athens, tab.defix.).

λῖτός I 1, line 11, after 'Call.*Lav.Pall*. 25' insert 'cf. id.*fr*. 110.78 Pf.' **III**, line 3, after 'Call.*Ap*. 10' insert 'comp., id.*fr*. 384.32 Pf.'

λῖτός (A) and **λιτός (B)**, delete the entries.

λιτός, ή, όν, *suppliant, supplicatory*, θυσίαι Pi.*O*. 6.78, ἐπαοιδαί id.*P*. 4.217. **II** *prayed for*, Ἀώς id.*fr*. 21 S.-M. **III** dub. sens., perh.

λιτότης   SUPPLEMENT   λοπάδη

invoked in prayer or *venerable*, γαῖα Alex.Aet. 1, Orph.A. 92; λιτῇ χθών· ἀπὸ τοῦ προσκυνεῖσθαι καὶ λιτανεύεσθαι Hsch.'

λιτότης I, add 'of style, D.H.*Vett.Cens.* 5.2'   **II**, delete '(cf. μείωσις)'

***λῑτούργιον**, v. λειτούργιον.

λίτρα I, after 'Posidipp. 8' insert 'in early use, metal bar used as currency, *Archeologia Classica* 38-40.13 (Sicily, vi BC)'   **II**, after 'J.AJ 14.7.1' insert 'inscribed on weights, *IEphes.* 3493, 4364, etc.' and after 'AP 10.97 (Pall.)' add '*Cod.Just.* 10.72.5'   **III**, add 'written on amphora, λῖ(τραι) *SEG* 31.813 (i BC/i AD)'

λιτρισμός, after '*delivery by weight*' insert '*Edict.Diocl.* 19.6'

λίτρον I, line 2, after 'Ar.fr. 320.1' insert '*IG* 1³.422.150 (Athens, v BC)'

***λίφερνέω**, v. ‡λιπερνέω, also φιλερνέω.

***λιχνικάριος**, ὁ, perh. for *λικνικάριος, *winnower*, or *λυχνικάριος, *lamp-maker*, *MPER* xv 141.3 (vi AD).

λίχνος I 1, delete 'metaph. .. Pl.R 579b'   **2**, for this section read '*greedy* (for knowledge), *inquisitive*, Pl.R. 579b, Crates Theb. 4; λ. ἔσσι [γὰρ] καὶ τό μεν πυθέσθαι Call.fr. 196.45 Pf., of lover's eyes, id.fr. 571 Pf.'

λίψ (C), delete the entry.

λοβός II 1, add 'used as binding material with dried mud in building, *IG* 2².463.68'

λογᾰριάζω, line 1, after 'Ar.Pl. 381' insert 'pass. λογαριασθῆναι, *have one's accounts audited*, expl. of εὐθύνας δοῦναι, Anon. *in Rh.* 204.20'

***λογᾰριαστής**, οῦ, ὁ, *calculator*, Alex.Aphr. *in SE* 12.36.

***λογᾰρίτης**, ου, ὁ, *cashier*, *MAMA* 3.280 (Corycus).

λογάς (A), add '**3** λογάδες, αἱ, *the eyes* (perh. as "chosen companions"), Sophr. 49, Call.fr. 85.15 Pf., Nic.Th. 292, *AP* 5.269 (Paul.Sil.); sg. expld. as *white of the eye*, Poll. 2.70.'

λογάς (B), delete the entry.

λογγάζω, add 'possible reading in *PMich.*VIII 486.18 (ii AD)'

***Λογγᾶτις**, ιδος, ἡ, epith. of Athena, Lyc. 520, 1032.

λογεῖον, add 'τὸ λογεῖον δεικτηρίου *SEG* 23.207 (Messene, i BC/i AD)'

λογευτήριον, add '*PHels.* 24.1 (ii BC)'

***λογέω**, v. λογάω II.

***λογή**, ἡ, *attention*, *heed*, λογήν μου μὴ ἔχουσαν *PMich.*III 217.6 (iii AD), *PAmst.* 95.18 (iii AD); cf. mod. Gk. τί λογῆς.

***Λογῖνα**, ἁ, female counterpart to Λόγος in title of play by Epicharmus, Epich.fr. 87-9.

***λογίσκος**, ὁ, *a little conversation* or *debate*, cj. in Antiph. 207.

λογισμός I, add '**3** pl., = Lat. *rationes*, *property*, *assets*, Just.Nov. 117.10.'

***λογιστήρ**, ὁ, = λογιστής, de Franciscis *Locr.Epiz.* no. 32.

λογιστήριον I 1, add 'used as place of detention, *PBeatty Panop.* 1.228, 346, 350 (AD 298)'

***λογιστορικόν**, τό, any one of collection of dialogues by Varro, Gell. 4.19.2, 20.11.4.

***λογιστός**, ή, όν, (λογίζομαι) *to be counted* or *calculated*, Call.fr. 196.47 Pf.

λογοθεσία I, *audit*, add '*Cod.Just.* 10.30.4.4'

***λογοθέσιον**, τό, *audit of accounts*, *POxy.* 3627.2, *CPR* 7.21.11 (both iv AD), Just.Nov. 128.18.   **2** pl., *account-books*, Palchos in *Cat.Cod.Astr.* 1.94.17.   **II** *documents containing* or *dealing with contracts*, Just.Nov. 136.5 pr.

⁺**λογοθέσιος**, ὁ, *accountant*, Palchos in *Cat.Cod.Astr.* 1.95.26.

***λογομάχος**, gloss on ὑπήρατος, Hsch.; cf. ἐπήρεια, etc.

λογοποιέω II, add '**2** *remonstrate*, *PMich.*v 229.18, 230.16, *POxy.* 2234.17 (all i AD).'

***λογοπρᾱγία**, ἡ, function of a λογοπράκτης, *ODouch* 57.

⁺**λογοπράκτωρ**, ορος, ὁ, app. kind of *accountant*, *PBaden* 26.40 (iii AD), perh. also *PHarris* 96.26 (i/ii AD), *POxy.* 3564 (iii AD).

λόγος I 1 a, at end, metaph., add 'εἰ δέ τις ἐγχειρήσει, ἕξει τὸν λόγον πρὸς ⟨τὸν Θεόν *SEG* 30.1546 (iv/v AD); also of other things than money, οἰνικὸς λ. *Stud.Pal.* 20.85 ii 1, λ. κριθῆς *PCair.Zen.* 464.1 (iii BC)'   add '**c** ἐκ λόγου, in accounts, *brought forward*, freq. in papyri, ἐγ λ. λήμματος τοῦ δεκάτου ἔτους *PLond.* 131 (i AD); ἐγ λ. τοῦ μηνὸς ἐλογοπραφήθησαν there were entered as balance from the month's *account*, *BGU* 362 vi 9 (iii AD); τὸ ἐγ λ. τοῦ Φαμενώθ the *balance* from Ph., *PCair.Zen.* 333.1 (iii BC); ὃ ἔχεις ἐγ λ. the *balance* you have in hand, ib. 593.12 (iii BC): see also °ἔκλογος (C).'   **2**, line 4, after 'Hdt. 8.100' insert 'ἵ τις δὲ τολμῇ ἀνύξεν, δώσι Κυρίῳ λόγον *BCH* suppl. 8 no. 58 (Beroea, iii/iv AD)'; line 13, after 'al.' insert 'εἰς μισθοῦ λ. Plu. 2.240d'; line 14, after 'Arist.*HA* 517ᵇ27' insert 'Plb. 9.20.3'   **4**, line 15, after 'v. infr. VI 2 e)' insert 'οὐδ᾽ ἐν λόγῳ ἄνδρα τιθείην Tyrt. 12.1'   **II 2**, add 'λόγον ἔχειν πρός τι *be in proportion to*, Plu. 2.147a'   **VI 1 a**, at end after 'also in sg.' insert 'τὰ μὲν ἂν Μοῖσαι .. ἐμβαίεν λόγ[ῳ] Ibyc. 1(a).24 P.'   **3 a**, add 'παρεῖναι ἐν λ., i.e. where there is *speaking*, οἵ π. ἐν λ. who hear my words, E.Rh. 149 (v.l.), Ar.Ach. 513'; in direct address to the audience, ὦνδρες οἱ παρόντες ἐν λ. id.Au 30; ἀπὸ λόγου app. *orally*, *PMich.*VIII 492.5, 19 (ii AD)'   **e**, add 'also *rhetoric*,

as *discipline*, *GVI* 1081.3'   **VII 2**, line 4, after 'Luc.*Alex.* 9, etc.' insert 'ἵνα τὸ τοῦ λόγου πάθω Men.*Dysc.* 633 S.'   add '**6** *declaration of legal immunity*, τοὺς καλουμένους λόγους Just.Nov. 17.6; cf. οἷς ἂν ἐθέλοιεν λόγον ἀσυλίας παρέχοντας id.*Edict.* 2 pr.   **7** ὁ διὰ λόγων, expression used to add a person's official name, *according to official documents called* .., *BGU* 2263.7 (ii AD).'

***λογχοβόλος**, ὁ, *spearman*, Ps.-Callisth. 53.22.

λοιβή I, add 'λυβά *SEG* 25.556 (Haliartus, iii BC, Boeot. decr.)'

***λοιγίζω**, λοιγισθῆναι· ἀπολέσθαι Theognost.*Can.* 22.

***λοιγωπός**, όν, (one) *whose eyes spread destruction*, of Athena, *PKöln* 245.8 (iii AD).

λοιδορέω II, line 3, delete 'τινος Ach.Tat. 1.6'; line 4, after 'πρός τινα' insert 'Macho 215 G.'

λοιμώδης, add 'adv. -δῶς, ὁπότε λ. ἐνόσησεν ἐν Συρίᾳ Gal. 12.285.6'

⁺**λοιπογρᾰφέω**, *enter as the amount of arrears*, *PGoodsp.Cair.* 7.7, *PHamb.* 3*A*3, *B*2 (i AD); in granting deferment of payment, δύο συνγραφὰς ἐλ. χωρὶς ἀργυρίου κομιδῆς *IG* 12(5).860.23 (Tenos, i BC).   **b** of the debtor who acknowledges the debt, *PGiss.* 46.5 (ii AD).   **2** *carry over as balance*, *BGU* 362 vi 9, xiv 17, al. (iii AD): metaph., χρόνους -ουμένους the *balance*, Nech.ap.Vett.Val. 279.12.   **II** w. acc. pers., *enter as in arrears* or *in debt*, *PFay.* 109.7 (i BC/i AD); so prob. in pass., μηδὲν αὑτὸν -εῖσθαι *PLond.* 940 (iii AD).   **b** *allow* person *to defer payment*, ἀφ᾽ ὧν ἐλ. αὐτόν *BGU* 362iii21, *PPetr.* 3 p. 154 (iii BC).

⁺**λοιπογρᾰφή**, ἡ, *balance carried over*, *Stud.Pal.* 20.85ᵛ *B*5 (iv AD); λοιπογρα[φὰς] ἑαυτοῖς προσῆψαν (of magistrates leaving office) i.e. acknowledged *balance owing* from them, *IEphes.* 15.5 (ii AD).

⁺**λοιπός**, ή, όν, (-ός, όν *IG* 1³.365.32) *that which remains* (after other parts have been accounted for), *the rest of*, οἱ λοιποὶ Λυδοί Hdt. 1.13.1, ἡ λοιπὴ πᾶσα Ἀσίη id. 1.192.1; Th. 3.95.2, τὴν λοιπὴν (sc. ὁδόν) πορευσόμεθα X.An. 3.4.46, J.BJ 1.4.6, δύο τὰ λοιπὰ (sc. τάγματα) id.AJ 17.10.9; w. partit. gen., ἡ λοιπή τῆς Λιβύης Hdt. 4.191.2, τὰς λοιπὰς τῶν νεῶν Th. 7.72.3.   **b** οἱ λοιποί, *the rest*, *the others*, Hdt. 1.207.7, Th. 3.18.2, J.AJ 12.7.1.   **c** τὰ λοιπά, *the rest*, τὰ λ. μου κλύουσα A.Pr. 476, τὰ δ᾽ ἄλλα, οἶδας γὰρ ὁποῖα τὰ λ. Aristaenet. 1.16.   **2** *that remains* (to make up the full total), τὸ λ. γένος Hp.Aër. 23, Apollod. 3.8.10, τὸν δὲ τρίτον (sc. πύργον) .. J.BJ 4.9.11, τὰ γένη τὰ τρία τὰ λοιπά Plot. 6.2.19, Heph.Astr. 1.2.11.   **b** *the other parts of*, *the rest of* (apart from the one specified), τοῦ δὲ λοιποῦ ἔργου *IG* 1³.474.40, μέλι .. οὐ πολλῷ τοῦ λοιποῦ χεῖρον J.BJ 4.8.3; subst. τὰ λοιπά, Ἰουδαίους .. πρὸς τὰ λ. καρτερῶς ἀντέχοντας ἑκάλκωσαν οἱ πύργοι ib. 5.7.2; ἐν τοῖς λοιποῖς *among other things*, *SEG* 26.426.18 (c.AD 200).   **c** τὸ λοιπόν, *for the rest*, i.e. to complete the account, argument, etc., Dsc. 2.83, λ., ἀδελφοί, χαίρετε 2*Ep.Cor.* 13.11, Arr.*Epict.* 1.24.1, *BGU* 969.19 (ii AD).   **3** *the remaining* (unspecified), τῇ .. λοιπῇ τῇ κατὰ τὴν δίαιταν ἐπιμελείᾳ Plb. 1.59.12, παρασκευή id. 4.63.2; neut. pl. subst., εἰ δὲ ταῦτα, καὶ βῆχα καὶ πταρμὸν .. καὶ τὰ λοιπά Plu. 2.1084c, Heph.Astr. 1.23.3.   **4** *left over*, ὄλωλα κοὐδέν λοιπόν .. κακῶν E.Hec. 784, τί λοιπὸν ἔσται τοῖς ἀντιλέγουσιν ..; Isoc. 5.57, cf. Pl.R. 466b.   **b** w. inf., λοιπὸν οὖν ἐστιν οὐδὲν ἄλλο πλὴν .. ἐπανελθεῖν nothing *remains* but to .., Isoc. 12.88, οὐκοῦν λοιπὸν ἂν εἴη .. ἀποδεικνύναι X.Smp. 4.1.   **c** λοιπόν, *in respect of the time remaining*, *by that time*, *by then*, ὅτε λ. ἐνόσει ἐπὶ θανάτῳ Ael.*VH* 8.14, ἦσαν δὲ λ. μέσαι νύκτες Ach.Tat. 2.26.1, ἑσπέρα δὲ ἦν λ. Jul.Or. 1.24c.   **d** expr. the consequence, *as a result*, Plb. 1.15.11, 1.30.8, al.   **5** *that is yet to come*, *future*, λ. βίοτος Pi.O. 1.97, γένος ib. 2.15, εὐχαί ib. 4.15, μηκέτι τῆς λοιπῆς φιλίας κοινωνεῖν D. 21.118; *second men of the future*, Pi.I. 4.39.   **b** of time, ὁ λ. χρόνος Pi.N. 7.67; w. partit. gen., πρὸς τὸν λοιπὸν τοῦ χρόνου D. 15.16; in advl. phrs., τὸν λ. χρόνον S.Ph. 84; τοῦ λ. χρόνου id.El. 817; ἐν τῷ λοιπῷ χρόνῳ *IG* 1³.101.36; εἰς τὸν λ. χρόνον Pl.Ep. 358b; ἐκ τοῦ λ. χρόνου D. 59.46; neut. subst. τὸ λοιπὸν τῆς ἡμέρας X.An. 3.4.16.   **c** τὸ λοιπόν, *henceforward*, *for the future*, Pi.P. 5.118, N. 7.45, A.Eu. 1031, S.OT 795, *IG* 1³.14.14, al.; εἰς ἅπαντα χρόνον A.Eu. 763, τὸ λ. ἤδη Plb. 5.4.5, καθεύδετε τὸ λ. *Ev.Matt.* 26.45; also τὰ λοιπά A.Th. 66, E.Hel. 698, Th. 8.21, X.HG 1.1.27; τοῦ λοιποῦ Hdt. 1.189, Ar.Pax 1084, D. 4.15.   **d** w. preps., in advl. phrs., ἐκ τοῦ λοιποῦ X.Smp. 4.56; ἐκ τῶν λοιπῶν Pl.Lg. 709e; ἐς τὸ λοιπόν A.Pers. 526, Eu. 708, Th. 3.44.3; εἰς τὰ λοιπά J.BJ 3.6.1; πρὸς τὸ λοιπόν J.AJ 8.12.3.

***λοκόπινος**, ὁ, (?for *λευκο-), λ. ἐστιν ὁ βάπτων εἰς βάθος καὶ μὴ ἀποπτύων Anon.Alch. 10.19.

***λομας**, app. adj. in agreement with ἀρούρας, *BGU* 2245.11 (i AD).

⁺**λοξόβᾱμος**, ον, *walking slantwise*, Gr.Naz.Carm. 1.2.1.714, Hsch.

λοξοκέλευθος, after 'Nonn.D. 5.233' insert '(cj.)'

λοξός, add '**3** *collateral relationship*, Theophil.Antec. 3.6 pr.'

λοξόφθαλμος, after '*oblique-eyed*' insert 'Ptol.*Tetr.* 144'

***λοξώδης**, ες, *oblique*, Gal. 2.297.3.

***λοπάδη**, ἡ, = λοπάς, *dish* in list of kitchen equipment, Lang *Ath.Agora* XXI B14 (iv/iii BC).

λοπάδιον, add 'in list of kitchen equipment, Lang *Ath.Agora* XXI B12'

λοπάς I and **II**, for these sections read 'kind of cooking-pot, *casserole* or sim., Ar.*Eq.* 1034, *V.* 511, Eub. 108 K.-A., Pl.Com. 173.12, Men.*Sam.* 365 S., Dsc. 2.142. **b** *food served in a casserole*, AP 12.44, Gal. 6.653.13. **2** (see quot.) λοπάς· παρὰ Συρακοσίοις τὸ τήγανον· παρὰ δὲ Θεοπόμπῳ (Com. 2 K.-A.) ἡ σορός, καὶ παρὰ τοῖς κωμικοῖς Suid.'

λοπάω I, add 'of the peeling of the skin, A.*fr.* 73b.6 R.; cf. λοπῶντα· λεπιζόμενον ἢ λοπιζόμενον Hsch.'

λοπίδιον, add '*Inscr.Délos* 1441Aii62 (ii BC)'

λοπίζω, after 'Hsch.' insert 'hence pf. part. pass. *having the gilding flaked off*, θυμιατήριον .. περικεχρυσωμένον, τινὰ δὲ λελοπισμένον *Inscr.Délos* 1429Aii12, 13 (ii BC)'

*****λόπιμον**, τό, *sweet chestnut*, Nic.*fr.* 76 G.-S., Ath. 2.53b, Dsc. 1.106, Hsch.

†λόπιμος, ὁ, = λόπιμον, Gal. 6.621.12, 12.420.7.

†λοπίς, ίδος, *scale* of fish or reptile, Ar.*V.* 790, Nic.*Al.* 467, *Th.* 154. **2** (*metal*) *plate*, Aen.Tact. 20.3.3. **3** *fragment* of ἀκρόβασις, *Inscr.Délos* 1432Abii14. **4** *dish*, Schwyzer 89.20 (Argos, iii BC).

λορδόω, line 2, after 'Hp.*Art.* 46' insert 'Men.*Dysc.* 533 S.'

λόρδων, for '*the demon of impure* λόρδωσις' read 'humorously invented erotic deity'

*****λουδάριοι**, οἱ, perh. the staff of a *ludus* other than the gladiators, φαμιλία μονομάχων καὶ λουδαρίων *IGRom.* 4.1453 (Smyrna), see Robert *Les Gladiateurs* pp. 209, 285 n. 2.

*****λοῦδος**, ὁ, Lat. *ludus*, *gladiatorial school*, *IGRom.* 4.1072, *PLips.* 57.11, *SEG* 30.1308 (Trajan), *Inscr.Perg.* 8(3).99.

*****λουέτιον**, τό, perh. = λουτήριον, λοέτιον χαλκὸν τετρεμένο[ν] *IG* 4.1588.17 (Aegina), unless λοετρόν (λουτρόν) is to be read; cf. χαλκίον ἐγλοτέριον *IG* 4.39.18.

*****λουκάνικον**, τό, Lat. *lucanicum*, *sausage*, *PLond.* 1259.30, *PRyl.* 627.208, al. (iv AD), Charis. 94.12 K.

*****λούπα**, ἡ· λέγεται δὲ οὕτως παρὰ Ἰταλιώταις ἡ λύκαινα, Lat. *lupa*, Suet.*Blasph.* 32 T.

λοῦσις, after 'Gloss.' insert 'see also λῶτις'; line 2, for '*cleaning*' to end, read '*free bathing* (granted as an amenity to a city), *BCH* 76.655.17 (Delphi, iv AD), *IGRom.* 3.584 (Lycia); for meaning see *BE* 1954.146'

*****λουσώριον**, τό, *place for games*, inscr. in *RAL* 1993 p.267.12 (Iasos, iii AD). **2** *pleasure-ship*, cf. Lat. *lusoria navis*, *IGRom.* 3.481 (Termessus, iii AD).

λουτήρ, delete '*Supp.Epigr.* .. i AD)' and '*IGRom.* 4.454 .. i AD)'; add 'as container for oil, *IGRom.* 4.454.10, *SEG* 4.263.10 (both i AD), *OGI* 479.10 (Dorylaeum)'

*****λούτρα**, ἡ, *sarcophagus, coffin*, *MAMA* 3.210, al. (Corycus); cf. μάκρα, πυρία.

†λουτρίς, ίδος, ἡ, *temple servant employed for washing*, Hsch., Phot. **II** ᾠα λ., app. *loin-cloth* or sim. worn during bathing, Theopomp.Com. 38 K.-A.

λουτρόν I 2, add '*BCH* suppl. 8 no. 5' **II**, line 2, delete 'E.*Ph.* 1667' add 'form λεϝοτρ- in Myc. adj. *re-wo-te-re-jo*'

λουτροφορέω, add '*hold office of* λουτροφόρος, λουτροφορήσαντα δίς *IIasos* 115.8'

λουτροφόρος 1, add '**b** *temple assistant who carried water for ritual purposes*, Didyma 330.4, *IIasos* 628.3.'

λουτροχόος, add 'Myc. *re-wo-to-ro-ko-wo* (= λεϝοτροχόϝος)'

λουτρών, for 'baptismal font' read 'baptistery'

*****λουτρωνίδιον**, τό, (*small*) *bathing establishment*, *PHels.* 12 (163 BC).

λουτρωνικός, add '*Cod.Just.* 10.30.4 pr.'

λούω I, add '**c** *provide free baths*, *IGRom.* 4.555, Demitsas Μακεδ. 51, τὸ βαλανεῖον λούειν μελέτω τοῖς ἐπιμελη[ταῖς *SEG* 30.1382 C (AD 301); v. °λοῦσις.'

λόφα, add '(λοφαδίσκος La.)'

λοφαδίσκος, v. ‡λόφα.

λοφίδιον, after 'λόφος II' insert 'Men.*Dysc.* 100, *Asp.* 59 S.'

λοφορρῶγα, add 'cf. Hippon. 104.39 W.'

λόφος, line 1, for 'of a horse, *withers*' read 'in a horse, including the mane'; after 'Il. 23.508' insert 'cf. Ael.*NA* 16.20'

λόφουρος, add 'λ.· ἐπίσημος Hsch.'

*****Λοφριαῖος**, = Λαφρ., *IG* 9²(1).100, 105, 9²(3).634.12, (Aetolia, ii BC); cf. ἐν Λοφρίῳ *SIG* 366.4.

λοχᾱγέτας, after 'A.*Th.* 42' insert '*fr.* 451k.4 R. (pl.)'

λοχᾱγός, after 'D.H. 2.7' insert '**3** an ephebic office at Athens, *IG* 2².2976 (iv BC), al. **4** *leader of an* ἄγειμα (ἄγημα), *SEG* 23.271 (Thespiae, iii BC).'

λοχαῖος I, line 1, for '= λόχιος' read '*set in ambush*' **II**, for this section read '*teeming*, σχῖνος Arat. 1057; σῖτος Phot. (expld. as βαθύς); so prob. in Thphr. *CP* 3.21.5, 23.5; as epith. of Δαμία, *IG* 12(3).361'

λοχεῖος 1, line 3, delete 'θυέτωσαν .. 204*b*9'; line 5, delete 'cf. λοχαῖος'

*****λοχευτικός**, ή, όν, *connected with childbirth*, ἡ λ., *midwife*, in quot. fig., λ. τῶν γενεσιουργῶν λόγων Procl.*in R.* 1.18.28.

λοχεύτρια II, add 'Λοχεύτριαι, in the cult of Hera at Argos, Hagias-Derkylos 4 J.'

λοχεύω, add '**IV** λοχεύοντες· ἐνεδρεύοντες Hsch.'

†λοχή, ἡ, *thicket*, = λόχμη, *IMylasa* 254.7.

*****λοχιάς**, άδος, ἡ, *midwife*, dub. in *POxy.* 3642.16 (ii AD); epith. of Hecate, *PMag.* 4.2285.

λοχίδιον, for 'dub. sense' read 'perh. *couch used in childbirth*'

λοχίζω, add '**IV** λοχισθέν· γεννηθὲν ἢ σφαγέν (σφαλέν La., as "caught in ambush").'

*****λόχιον**, τό, perh. *birth-house* (as shrine of Isis), *PColl.Youtie* 51 (ii/iii AD).

λόχιος II, add 'of Isis, *SEG* 34.622, al. (Maced.)' **III 1**, add 'offered in sacrifice, Antim. 182 Wy.; αἱ τὰ λόχια ἐκπορευόμεναι καὶ ζωννύμεναι Milet 1(7) 204*b*9219'

*****λόχις** λῆξις Theognost.*Can.* 58 A. (prob. Aeol. for *λαχίς).

λοχισμός, add '**II** in pl., *formations into* λόχοι, cj. in A.*Ag.* 404 (lyr.; λογχίμους codd.); cf. λοχίζω II.'

*****λόχον**, v. °λαγχάνω.

Λυαῖος, line 2, after '(cf. Lat. *Lyaeus*)' insert '*AP* 6.154 (Leon. or Gaet.)'

λυγγούριον, for 'a kind of *amber*' read 'perh. *yellow* or *brown tourmaline*'

*****λύγγουρος**, ὁ, = λυγγούριον, Cyran. 28.3.

λυγγώδης, add 'Gal. in *CMG* 5.9.(1).329.10, 12 al.'

λύγδινος 1, line 2, after '*marble*' insert 'στάλα *SEG* 9.3.17 (Cyrene, iv BC)'

λύγκιος, for '(Λυγκεύς) *of Lynceus*' read '(λύγξ) *of a lynx*'; add 'δέρμα λύνγιον *Edict.Diocl.* 8.35'

*****λυγκίδιον**, τό, dim. of λύγξ (A), Aesop. 37 H. (cj., λυκ- codd.).

λύγξ (A), after 'gen. λυνκός' insert 'also λυγγός Ael.*NA* 7.47, al; acc. λύγγα, E.*fr.* 863'; delete '(λύγγα .. λύγγιος)' add '**III** perh. species of ape, Gal. 2.430.9, 535.11.'

*****Λῡδηΐς**, ίδος, fem. adj., *Lydian*, Hermesian. 7.41.

*****Λῡδιεργής**, ές, *of Lydian workmanship*, Call.*fr.* 196.29 Pf.

λύζω I, add 'Archig.ap.Gal. 11.836.15, 13.176.8'

*****λυήεις**, εσσα, εν, *discordant*, τὸ δὲ λυῆς λυήντος ἀπὸ τοῦ λυήεις Hdn.Gr. 1.59.

*****λύημβρος**, ὁ or ἡ, kind of fish, = βεμβράς (v. μεμβράς), Cyran. 46.15.

*****λυῆς**, v. °λυήεις.

*****λυθίραμβος** (v.l. λυθίραμμος), v. διθύραμβος at end.

λύθρον, lines 3/4, for 'Ph.ap.' read 'Ph.Tars.ap.'; add '*Trag.adesp.* 235 K.-S.'

*****λυκαιχμίας**, ὁ, unexpld. wd., perh. *wolf-battle*, i.e. wolf-like or guerrilla fighting, οἷος ἐοίκησα λυκαιχμίαις Alc. 130.25 L.-P, commentary on Alc. in *POxy.* 3711 ii 32.

Λύκειον I, add 'also temple of Apollo at Argos, *SEG* 30.355. **2** app. an offering to Apollo in the form of a monument or sim., *IG* 12(3).389, 12 suppl. p. 86.'

*****Λυκιάδες** (cj. Λυκηιάδες)· κόραι τὸν ἀριθμὸν λ΄, αἱ τὸ ὕδωρ κομίζουσαι εἰς τὸ Λύκειον Ἀθηναίων Hsch.

λυκιδεύς, add 'Plu. 2.462e'

λύκιον I 2, line 2, before 'Dsc.' insert '*Peripl.M.Rubr.* 49' **II**, add 'also λύκιος, ὁ, *SEG* 32.1618 (iii/ii BC)'

Λύκιος II, add 'Pi.*P.* 1.39'

*****λυκοπρόσωπος**, ον, *wolf-faced*, Teuc.Bab. p. 47.4 B.

λύκος I, add '**b** λ. κεράστης *lynx*, *Edict.Diocl.* 8.35.'

†λυκοσπάς, άδος, ὁ, ἡ, app. *breed of horse found in southern Italy or round the Adriatic coast*, Call.*fr.* 488 Pf., Ael.*NA* 16.24; in conjectural explanations of its etym., Plu. 2.641f, Choerob.*in Theod.* 1.287, Hsch.; in unexpld. context, Nic.*Th.* 742.

†λυκόσπαστος, ον, *torn by wolves*, Hsch. s.v. λελυκωμένα.

*****Λυκούλλιος**, α, ον, name of kind of marble, *Edict.Diocl.* 33.4.

*****Λυκωρεῖος**, ὁ, = sq., A.R. 4.1490.

*****Λυκωρεύς**, έως (έος), ὁ, cult-title of Apollo at Delphi, Call.*Ap.* 19, Euph. 80.3.

λῦμα (A) III, after 'A.*fr.* 692' insert 'Orph.*H.* 14.14'

*****λῡμᾱγωνεία**, ἡ, app. *offence against the rules of a contest*, *BGU* 1823.24 (i BC).

*****λῡμᾱγωνέω**, app. *offend against the rules of a contest*, *SEG* 27.261.69 (Beroea, ii BC).

λῡμαίνομαι (B), line 3, after 'A.*Ch.* 290' insert 'fut. λυμανθήσομαι A.*fr.* 47a.12 R.'

λῡμάντωρ, add '*SEG* 9.1.70 (Cyrene, iv BC)'

*****λῡμᾰτίζω**, v. °λημματίζω.

λυμνός, add '*SEG* 29.308 (v AD)'

*****λῡπητροτόκος**, ον, perh. = λυποτόκος, *ZPE* 48.279 (Cilicia).

λύρα, for 'a stringed instrument .. shell of a tortoise' read 'in earlier period applied equally to a bowl lyre (with tortoise-shell soundbox) and to a box lyre, but later spec. to the former; cf. κιθάρα, φόρμιγξ' λῠρᾰοιδός, delete 'Plu.Sull. 33'

λῠριστής, add 'IG 14.2030 (Rome), SEG 17.438 (Malta, iii AD)'

*λῠροκτῠ́πος, ον, striking the lyre, GVI 1522a, Bernand IMEG 167.2, 170.2.

†λῠρόκτῠπος, ον, sounding like a lyre, of a bowstring, Lyc. 918.

Λῡσάνδρια, add 'AA 1965.440 (Samos, written -εια)'

*λύσειος, v. λύσιος.

λῡσίζωνος II, add 'IG 12(5) ii p. 332, no. 582'

λῡσιμέριμνος, add 'of eternity, αἰῶνα .. τὸν λ. SEG 30.1256'

*λῡσίνοσος, ον, curing disease, φάρμακα REG 69.127.

λῡσιπόλεμος, after 'ὁ' insert 'ender of wars, of a person, SEG 26.1835.3 (Cyrene, iv BC)'

λῡσίπονος, add 'ὕπνος AP 12.127 (Mel.), SEG 33.563 (iii AD), of baths, Robert Hell. 4.76, al.'

λῡσιῳδός, line 2, after 'Plu.Sull.' insert '33 (codd. λυρῳδ-)'

λύσσα II 1, add 'cases of frenzy in human beings, Hp.VM 19'

λῠτήρ, add 'III destroyer, Byzantion 5.10 (Stobi, iv AD).'

λῠτήριος I, line 6, delete 'τό .. 758'   II, after '= λύτρον' insert 'Stesich. 222b.226 (p. 214 D.)'; line 3, after 'Pi.P. 5.106' insert 'πημονῆς, of drunkenness, S.fr. 758 (where τὸ μεθύειν are words of Ath.)'

λύτρον I 1 a, add 'also sg. SEG 23.460'

λυχναπτέομαι, delete the entry.

*λυχναπτέω, light lamps (in quot. pass.), Sokolowski 1.28.13.

*λυχνάπτης, ου, ὁ, lamplighter (in temple), written λυχνάτ- MAMA 3.437; ?erron. pl., λυχνάπτοι POxy. 1453.4, 8 (i BC); cf. Hsch. s.v. δαδοῦχος.

λυχνάπτρια, add 'SEG 34.655 (Maced.)'

†λυχναψία, ἡ, lamplighting, app. the same as λυχνοκαΐα, TAM 4(1).16 (AD 122/3), PAmh. 2.70.11 (ii AD), IGRom. 4.1176 (Aegae), Ath. 15.701b.

λυχνεύς, delete 'cf. Ath. 15.699d'

*λυχνιάζω, burn a lamp, ἐλαίῳ λ. (sc. λύχνον) PWarren 21.1 (iii AD).

*λυχνικόν, τό, perh. lampholder, MAMA 6.361.

λύχνιον, add '3 kind of eyesalve, Asclep.Jun.ap.Gal. 12.744.6.'

λυχνίς I, add '3 app. for λυχνίτις candle wick, Edict.Diocl. 18.5.'

λυχνίτης II, for this section read 'λ. λίθος, kind of Parian marble, prob. "of a luminous quality or translucent" but supposed to be so called because quarried by lamplight, Varro ap.Plin.HN 36.14'

λυχνίτις, add 'II made from λυχνίτης II, λ. ζώνη Suppl.Hell. 978.6.'

*λυχνοδότης, ου, ὁ, lantern-giver, priest in Egyptian temple, PHib. 213.12 (iii BC).

λυχνοκαΐα, add 'written λυχνοκαιία, SEG 9.13.16 (Cyrene, iv BC)'

†λυχνοκαυτία, ἡ, lamplighting, app. the same as λυχνοκαΐα and °λυχναψία, Cephisod. 11 (Ath. 15.701b).

λύχνον, add 'CEG 463 (vi BC)'

*λύχνος, ους, τό, = λύχνος, ὁ, Lxx Da. 5.1.

λύχνος, ὁ, I 1, add 'pl., used as title of book, sunt etiam qui λύχνους inscripserint, Gell.Praef. 7'

*λυχνουρέοντες, οἱ, unknown objects (made of bronze, ?parts of a lamp), Inscr.Délos 1417Bi37 (ii BC), al.

λυχνοῦχος, for 'lampstand' read '(portable) lampholder'

λύω, line 8, for 'opt. plpf.' read 'opt. pf.'   I 2 c, at end delete 'buy from a pimp'; after 'Ar.V. 1353' insert '(under influence of Rabbinical phr.) ὃ ἐὰν λύσῃς ἐπὶ τῆς γῆς, ἔσται λελυμένον ἐν τοῖς οὐρανοῖς Ev.Matt. 16.19, 18.18'   II 1, line 3, after 'Od. 2.69, etc.' insert 'λ. τάξιν, break formation, Polyaen. 7.28.2'   IV, add 'med., pay in quittance from a vow, Πύθερμός με ὁ Νέλωνος ἐλύσατο τῆς Ἔσιος ἄγαλμα Schwyzer 749 (c.500 BC)'

λῶ, line 7, subj., add 'λείει SEG 36.855 (Sicily, vi/v BC)'; med., line 3, after 'Hsch.' insert 'pf. λέληνται POxy. 2256 fr. 8.7'

*λώβηξ, ηκος, ὁ, bird identified with γύψ, Cyran. 28 (1.11.1,7 K.).

†λώβησις, εως, ἡ, maiming, impairment, Ptol.Tetr. 151, Sch.E.Hec. 1098.

*λωγεῖ· μαίνεται Theognost.Can. 58 A.

λώγη, read λωγή.

*λωγήρυχος· ὁ ἀναλαμβάνων τὰ πίπτοντα τοῦ σίτου ἐν τῷ ἀμητῷ Theognost.Can. 58 A.

*λωδικάριος, ὁ, maker of coverlets or blankets, Teuc.Bab. p. 45 B.

λωΐων I, add 'ἐπὶ λωιτέρῳ SEG 34.1628 (Egypt, ii/iii AD)'

*λώλια· σῦκα κεκομμένα Theognost.Can. 58 A.

*λώλωμα· παιδικὸν βρῶμα Theognost.Can. 58 A.

λῶμα, add 'Sch.Call.Dian. 12; Myc. wo-ro-ma-ta (pl., = Ϝλώματα) (?)wrappings'

Λῷος, add 'in Asia Minor, Hemerolog.Flor. p. 75 (14 K., Ephesus), p. 77 (18 K., Tyre), SEG 31.1389 (Syria); in Egypt, SEG 34.1598 (AD 323)'

*λωπάς, άδος, ἡ, kind of bottle, Anon.Alch. 33.1.

λωποδυσία, for 'highway-robbery' read 'clothes-stealing' and add 'PMilan. 30.1 (ii BC)'

*λωποδυτία, ἡ, = λωποδυσία, Poll. 7.42.

λῶπος (s.v. λώπη) before 'Herod. 8.36' insert 'SEG 11.1112 (λδπος Arcad. decr., vi/v BC)'

*λωραμέντα, τά, Lat. loramenta, harness, Edict.Diocl. 8.8, 10.1.

*λωρίκα, ἡ, Lat. lorica, breastplate, POxy. 812 (i BC), PBeatty Panop. 1.343 (AD 298).

λωρίκιον, add 'Just.Nov. 85.4'

λῶρος I, add 'neut. pl. λῶρα, Hsch. s.vv. ἡνίαι, ἡνία σιγαλόεντα'

*λωροτόμος, ὁ, strap-cutter, MPER xv 111.30 (vi/vii AD).

λωτέω, for 'play the flute' read 'play the aulos' and add 'Theognost.Can. 58 A.'

*λωτίζω, ?wash (cf. °λῶτις), εἴ τις .. τὰ πρόβατα ποτάγοι πρὸ τᾶ[ς λ]ώτιος λωτίξας ἀπαγέτω Delph. 3(4).352 (190 BC).

λώτινος I, add 'λ. ποίαις Anacreont. 32.2 W.'   II 1, add 'αὐλίσκων ὑπὸ λωτίνων Pi.fr. 94b.14'   add '4 perh. lotus-coloured, τοῦ ἐριδίου τοῦ λωδίνου POxy. 3060.11 (ii AD).'

λῶτις, for 'dub. sens.' read 'perh. (ritual) bath, app. NWGk. form for λοῦσις (< *λόϜετις)' and add 'gen. λώτιος Delph. 3(4).352 (190 BC); cf. λωτίζω'

λωτός III 1 a, for 'flutes' read 'auloi'   b, line 2, delete '(lyr.)' and for 'flute' read 'aulos'; add 'E.HF 11'

*λωτοσπορεύς, έως, ὁ, sower of fodder plants, PTeb. 893 (ii BC).

M

μά (A) **I**, at end for 'μὰ ναί *Inscr.Cypr*. 109 H.' read 'Cypr. *ma-na-i* μὰ ναί, *ICS* 8.6'

✝**μᾰγᾰδίζω**, *produce an* (*octave*) *concord*, of singers, μαγαδίζουσι ταύτην (sc. τὴν διὰ πασῶν συμφωνίαν), ἄλλην δὲ οὐδεμίαν Arist.*Pr*. 918.40; μ. ἐν τῇ διὰ πασῶν συμφωνίᾳ ib. 921ᵃ12; metaph., Theophil. 7.2.

μάγαδις, for '*magadis*' to end read '*an octave concord*, ψάλλω δ' εἴκοσι χορδαῖσι μάγαδιν †ἔχων Anacr. 29 P.; πηκτίδων ἀντιζύγοις ὁλκοῖς κρεκούσας μάγαδιν Diog.Ath. 1.10 S.; κέρασι .. καὶ σάλπιγξιν .. οἷον μάγαδιν (v.l. μαγάδι) σαλπίζοντες X.*An*. 7.3.32; of an effect on the cithara, Philoch. 23 J., Hsch.; in related senses, perh. Alcm. 101 P., Telest. 4.2 P., Canthar. 12 K.-A. **II** of musical instrument, Λυδός τε μ. αὐλός Ion Trag. 23 S. (dub. sens.); Aristox.*fr*. 97-99 W. and later antiquarians (v. Ath. 14.634c-7a, Poll. 4.61, Hsch., Phot.) interpreted earlier refs. as referring to some instrument (harp or aulos); in S.*fr*. 238, πηκταὶ δὲ λύραι καὶ μαγάδιδες, the last two wds. do not scan normally and are prob. a gloss.'

✻**μᾰγᾰρεύς**, έως, ὁ, *initiate of a group who gathered in a megaron*, *SEG* 39.649 (Thrace, ii/iii AD); cf. °ἀρχιμαγαρεύς.

✝**μάγαρον**, τό, *underground pit* into which young pigs were thrown in worship of Demeter, Men.*fr*. 870 K.-Th., Phot.; also sp. μέγαρα (always pl.) Paus. 9.8.1, Porph.*Antr*. 6, Sch.Luc.*D.Meretr*. 1, cf. Hebr. *mᵉ ārāh, cave*, mod. Gk. μαγαρίζω *make foul, dirty*, see also °ἀρχιμαγαρεύς.

μαγγανάριος II, add 'in context of building, *IGRom*. 3.1165 (AD 485)'

✻**μαγγανικός**, ή, όν, *of* or *for a pulley-block*, μ. ξύλον *PMich*.XIII 660.10 (vi AD).

μάγγανον II, add 'φεσκάσιον· μάγγανον πλοϊκόν Phot., Suid.'

✻**μαγδάλλει**· τίλλει, ἐσθίει Hsch.

✻**μαγειρηΐα**, ἡ, prob. *tax on* or *licence for butchers* (cf. μαγειρικός 4), *IG* 12 suppl. 125.18 (Eresus).

μᾰγειρικός 1, for '*fit for a cook* or *cookery*' read '*of a butcher* or *cook*'; after 'Pl.*Grg*. 500b' insert '*SEG* 31.983.8-9 μαγε[ιρικοὺς] ἐργάτας (Priene, ii/i BC)' **4**, for this section read '*tax on butchers, SB* 7645.13 (iii BC), *PUniv.Giss*. 2.5 (ii BC), *Inscr.Magn*. 116.42 (ii AD)' add '**5** name of kind of plaster, μ. ἔμπλαστρος *PSI* 1180.44 (ii AD).'

✻**μᾰγείριος** or -ίριος, epith. of Apollo in Cyprus, *to-a-po-lo-ni to-ma-ki-ri-o ICS* 304 (iii BC).

μᾰγευτής, add 'cf. μαγευτὰν αὐλόν· τὸν μαγεύοντα τοὺς ἀκροωμένους Hsch.'

μᾰγεύω III 2, for '*call forth by magic arts*' read '*apply magic arts to*'; before 'Luc.*Asin*. 11' insert 'ἔρωτα'

μᾰγιᾰνός, for '*inscribed with charms*' read '*magic*'

✻**μάγιστερ**, ὁ, Lat. *magister, PSI* 481.10 (v/vi AD), Just.*Nov*. 30.2.

μαγιστράτη, ἡ, *magistracy, IGRom*. 1.599 (Istros, AD 201).

μαγίστρατος, ὁ, *magistrate, BE* 1970.398 (Tropaeum Traiani, iii AD).

✻**μαγιστριανός**, ὁ, Lat. *magistrianus*, official on the staff of the *magister officiorum, POxy*. 904.2 (v AD), *Cod.Just*. 12.60.7.2.

✻**μαγιστροκήνσος**, ὁ, Lat. *magister census, registrar of the senate and city of Constantinople, Cod.Just*. 1.2.17.2, 2a.

✻**μάγιστρος**, ὁ, Lat. *magister, BGU* 927.5 (iii AD), *MAMA* 1.216 (iv AD), *Cod.Just*. 10.11.8.12.

✻**μαγιστρότης**, ητος, ἡ, *office of the magister, PAmh*. 138.11 (iv AD).

✻**μαγίστωρ**, ορος, ὁ, cf. Lat. *magister, SEG* 34.1095 (Ephesus, iii AD), *PLond*. 1790.10 (v/vi AD, -σσ- pap.); written μαΐστωρ Suid.

✻**μαγκίπισσα**, ἡ, *female baker*, Lxx Sch. 1*Ki*. 8.13.

✻**μάγκιψ**, ιπος, ὁ, Lat. *manceps*, contractor, *MAMA* 3.409, al. (Corycus).

μάγμα, for '*thick unguent*' read '*mouldable sediment deposited in unguents*, etc.' and before 'Plin.*HN* 13.19' insert '*faecem unguenti magma appellant*'

Μαγνησίη, add 'also Μαγνησία Th. 1.138.5, 8.50.2, *SEG* 23.189 (Argos, iv BC)'

Μάγνησσα, add '*Μάγνησσαν κόραν Lyr.adesp*. 107 P., A.R. 1.584, etc.'

Μάγος I 2, add '**b** perh. *a priest in Mithraic worship*, Mitchell *N.Galatia* 404.' **II**, after '*magical*' insert 'ἐπῳδαί Sosiph. 1'

✻**μάγουλον**, τό, *cheek*, as mod. Gk., Melamp. p. 503 F.

✻**μαγουσαῖος**, ὁ, = μάγος I 2, Bardes. 3.16 J., Suid. s.v. γοητεία.

✻**μάγωζ**(?a), τά, *treasuries*, ἐν μαγώζοις Aq.Sm.*Ez* 27.24 (v.l. μαγούζοις).

μᾰδᾱγένειος, for 'Dor. for μαδηγένειος' read '*smooth-chinned*, (non-Att.-Ion., cf. μαδιγένειος)'

μᾰδᾰρός 2, add 'applied as a nickname, Cic.*Att*. 14.2.2, cf. perh. Μάδρος *IG* 5(2).387 (Lusi, v BC)'

μᾰδᾰρόω, after '*bald*' insert 'rest. in *Inscr.Cret*. 1.xvi 6.i-iii.24 (Lato, ii BC)'

μᾰδάω 1, for 'of a disease in fig-trees' read 'of the wood of diseased trees' **2**, add '**b** *become bare* by chafing, πᾶς ὦμος μαδῶν Lxx *Ez* 29.18.'

μαδηγένειος, delete the entry.

μάδησις, add 'Gal. in *CMG* 5.10.(2).2 62.15'

μᾰδῐγένειος, line 2, delete 'prob. μαδηγένειοι'

μαδωνᾱΐς, add '(μαδωνία cj.)'

μᾶζα, line 1, for '(μάσσω .. 258 P.)' read 'or μάζα (on the accentuation see Hdn.Gr. 2.937, Moeris p. 258 P.)'; add 'see also μάδδα'

μαζονομεῖον, add '*SEG* 29.146b II 5 (Attica, iv BC)'; form -νόμιον, add 'Poll. 6.87'

μαζονόμον, after '(Didyma, iii BC)' insert 'Varro *RR* 3.4.3, Hor.*Sat*. 2.8.86'

✻**μαθόω**, *agglomerate*, Anon.Alch. 333.17.

μᾰθητής, add 'perh. of a trainee athlete, *Hesperia* 54.217 no. 3 (iii AD); μ. ἱππέων, transl. Lat. *discens equitum, SEG* 31.1116; see also μαθετάς'

✻**μαθκων** or **μαθκωνον**, pl. μαθκωνα, name of a garment, *SEG* 7.417, 419 (Semitic wd.).

μαῖα I 3, add 'Robert *Les stèles funéraires de Byzance* 176'

✻**Μαιαδεύς**, έως, ὁ, a name of Hermes, Hippon. 32.1 W., see Μαῖα, Μαιάς.

Μαίανδρος I, add 'as a divinity, Schwyzer 721.11 (Thebes at Mycale, v/iv BC)' **II**, add 'expld. as κόσμος τις ὀροφικὸς παρὰ τοῖς ἀρχιτέκτοσι, Sch.*AP* 6.286; as εἶδος ἱππασίας παρὰ τοῖς ἱπποδαμασταῖς, ib.'

Μαιμακτήρ, delete 'prop. = Μαιμάκτης' and add 'at Ephesus, *IEphes*. 690, al.; at Cyme, *SEG* 33.1040. **II** pl. Μαιμακτῆρες, divinities at Mytilene, *IG* 12(2).70.'

✻**Μαιμακτήρια**, τά, autumn festival in Thasos, *BCH* 82.195 (iv BC).

Μαιμακτηριών, add 'also in Samothrace, *SEG* 31.803'

Μαιμάκτης, add 'at Naxos, Διὸς M. *IG* 12(5).47; M.· μειλίχιος, καθάρσιος Hsch.'

μαινάς II, for 'esp. of love' read 'applied to the ἴυγξ'; after '*P*. 4.216' add 'μαινάδα βότρυν *AP* 4.1.25 (Mel.)'

μαινόλης I, line 1, after lemma insert 'Dor.Aeol. -λᾱς' and after 'Sapph. 1.18' insert 'L.-P.; θίασος Phot., Suid. s.v. θίασος'; line 3, after 'gen.' insert 'Archil. 196a.30 W.' and delete 'ἀσέβεια .. *Or*. 823 (lyr.)'; add 'pl. μαινόλιδες (μεν- lapis) Bacchants, *SEG* 17.772 (iii AD)' **II**, delete '(From μαίνομαι .. φαίνομαι)'

μαινόλιος, add 'epith. of Zeus, *IG* 12(2).484.16 (Mytilene)'

μαίνομαι, line 4, after 'also' insert 'μεμάνηκα Arg.Men.*Oxy*. 1235.66 (μεμενηκέναι pap.), and '; before 'Theoc.' insert 'Men.*Epit*. 879'; line 5, after 'aor. med.' insert '(ἐπ)εμήνατο Il. 6.160'

✻**μαῖοι**, οἱ, *adoptive parents, IG* 12(5).199 (Paros); cf. μαῖα 2.

✻**Μαιουμάρχης**, ὁ, *one who presides over the Maioumas*, (v. °Μαιουμᾶς) Roueché *Aphrodisias* 40 (v AD).

✻**Μαιουμᾶς**, ᾶ, ὁ, festival in Syria and elsewh., Jul.*Mis*. 362D, *INikaia* 63, *Gerasa* 279 (vi AD).

✻**μαιουμίζω**, *celebrate the Maioumas*, (v. °Μαιουμᾶς) *ITyr* 151.

✝**μαιριάω**, Ταραντῖνοι δὲ μαιριῆν τὸ κακῶς ἔχειν Hsch. s.v. Μαῖρα.

✻**μάκαμον**, τό, *beer*, κερβησίας ἤτοι μακάμου Edict.Diocl. 2.11.

μάκαρ, line 7, after 'Sol. 14' insert 'Hippon. 43, 117 W.' **III**, add 'in epitaphs, *GVI* 795 (iii/iv AD)'

μᾰκάρι, add 'cf. mod. Gk.'

μᾰκᾰριστός, adv. -ῶς, add '*GVI* 788 (ii/iii AD)'

μακαρῖτις, add 'acc. -ῖτιν *MAMA* 9.75 (ii/iii AD)'

μᾰκεδνός, line 1, after 'Od. 7.106' insert 'prob. rest. in A.*fr*. 451*l*.13 R.'

✻**Μᾰκεδονιαρχέω**, *exercise the office of* Μακεδονιάρχης, *SEG* 24.497 (Maced., ii AD).

✻**Μᾰκεδονιάρχης**, ου, ὁ, app. some religious officer in Maced., *AA* 1942.176, al., *SEG* 24.479 (Maced., iii AD).

✻**Μᾰκεδονιαρχικός**, ή, όν, *of the office of* Μακεδονιάρχης, ἐν προβολαῖς M. *AA* 1942.176.

✻**Μᾰκεδονιάρχισσα**, ἡ, fem. of °Μακεδονιάρχης, *AA* 1942.176.

Μᾰκεδών, add 'ἐξελιγμὸς *M*., a form of military manœuvre, Arr.*Tact*. 23.1, 24.1; w. ellipsis of ἐξελιγμός, ib. 31.4'

***μάκειρ**, v. ‡μάκιρ.

μᾰκέλας, read 'μᾰκελᾶς' and add 'see also °βακέλας'

μᾰκελλάριος, (s.v. μάκελλον), add '*BASP* 12.153 (?iii AD; sp. μακελάρειως), *SEG* 29.327 (Corinth, v/vi AD)'

μᾰκελλον II, line 4, after 'Aesop. 134' insert '*SEG* 29.327 (Corinth)'

μᾰκελλωτός, add 'cf. *macellotae* fem. pl., for *ostia*, Varro *LL* 5.146'

***Μᾰκηδονίς**, ίδος, ἡ, *Macedonian*, Hom.*fr*. 24 D., *GVI* 1015.1 (Alexandria, i/ii AD), Nonn.*D*. 2.400.

μάκιρ, after '*malabarica*' insert 'or sim. tree'

μακραίων 1, add 'Emp. 115.5 D.-K.'

***μάκριον**, τό, dim. of μάκρα (*sarcophagus*), *ITyr* 81, 82 (v/vi AD).

***μακροβιοτεία**, ἡ, *longevity*, Phld.*Sign*. 17.

μακροημέρευσις, add '*SEG* 34.1515 (Arabia, vi AD)'

μακροκέντης, add '*JJP* 19.110 (uncertain sense in broken context)'

***μακρόκνημος**, ον, *long-shanked*, Heph.Astr. 3.45.9.

***μακρόουρος**, ον, *long-tailed*, Cyran. 65 (2.22 K.).

***μακροπαράληκτος**, ον, *having the penultimate syllable long*, Sch.Il. 6.268, Eust. 407.36.

μακρόπορος, before '*ravelling*' insert '*having a long course* or *orbit* (opp. βραχύπ-), Procl.*in R*. 2.20 K.'

μακροπώγων, add '*Cat.Cod.Astr*. 11(2).138.10'

μακρός I 1, add 'app. w. ellipsis of δρόμος *CEG* 374.2 (Laconia, vi BC)' **II**, add '**4** μακρόν, τό, part of the parabasis of a comedy = πνῖγος, Sch.Ar.*Ach*. 659, *Nu*. 518, Poll. 4.112.' **III 2**, add 'ἐς τὸ μακρόν *in height*, Call.*fr*. 196.31 Pf.'

μακρότης, add 'τοῦ ῥυθμοῦ, i.e. predominance of long syllables, Demetr.*Eloc*. 40'

***μακρόχειρον**, τό, = Lat. *tunica manicata, long-sleeved tunic*, *Dura*⁴ 98.

***μακροχρονία**, ἡ, *long duration*, βίου Moses Alch. 315.17.

***μακροψῡχία**, ἡ, *fact* or *power of taking a long view*, Cic.*Att*. 9.11.4.

μακτήριον I, delete '= μάκτρα Plu. 2.159d' and for 'Call.*fr*. 7.32 P.' read 'Call.*fr*. 23.11 Pf.'

***μακτήριος**, α, ον, *of a kneader*, Plu. 2.159d.

μάκτρον, add '*IEphes*. 456 (iv AD; written μακρον)'

μάλα II, line 2, delete 'only in' and after 'Tyrt. 12.6' insert 'Call.*fr*. 67.13 Pf.' **III 1**, line 3, after 'id.*Grg*. 510b' insert 'ὡς ἔνι μάλιστα Men.*Dysc*. 699 S.'

μᾰλάβαθρον, after 'Plin.*HN* 12.129' insert '*SB* 9834b.22 (iv AD, written -βατρα pl.); also μηλόβαθρον Androm.ap.Gal. 14.41, βαλάβαθρον *SB* 9804.5 (ii AD), *BGU* 953.2 (iii/iv AD)'

μάλαγμα II, add 'cf. μάλαγμα *moecharum* (applied by Augustus to Maecenas), Macr.*Sat*. 2.4.12'

***μάλαθρον**, τό, gloss on ἄνηθον, Sch.Theoc. 7.63; cf. μάραθρον.

μαλάκια, for 'i.e. .. *shells*' read 'ὅσα ἄναιμα ὄντα ἐκτὸς ἔχει τὸ σαρκῶδες, ἐντὸς δ' εἴ τι ἔχει στερεόν'

μαλᾰκίων, for 'term of endearment, *darling*' read 'term of contempt, *softy*'

***μᾰλᾰκόμματος**, ον, *soothing the eyes*, ὕπνος Lyr.adesp. 11(g).1 P.

μᾰλᾰκός I 1, add 'w. implication of rottenness, δρῦν *AP* 6.254 (Myrin.); of ἀμυσγέλαι almonds, *fresh* (opp. σκληραί), *SEG* 9.41.17, al. (Cyrene, ii BC)' **III 2 d**, for 'παθητικός' read '*pathic*' add '**3** of style, *gentle*, D.H.*Dem*. 20; adv. -κῶς id.*Pomp*. 6.'

μᾰλᾰκόσωμος, add 'Paul.Aeg. 7.3 (237.10 H.)'

μᾰλᾰκόφρων, for '(Orph.*H*.) 69.13' read '69.17'

μᾰλᾰκόφωνος, add 'Sch.Pi.*O*. 9.34c'

μαλακτήρ, delete 'χρυσοῦ μ. καί'

μᾰλάχη 1, for 'μολόχη .. (cod. F)' read 'see also μολάχη, ‡μολόχη'

***μαλάχιον**, τό, dim. of μαλάχη, *mallow*, *PIndiana Univ*. 4.1 (*CPh* 43.112, iii AD). **II** sp. μαλάκιον, *a woman's ornament worn round the neck*, Ar.*fr*. 332.10 K.-A. (ap. Poll. 7.95; μαλάχιον Phot. s.v.), Poll. 5.87, Hsch.; μολόχιον Clem.Al.*Paed*. 2.124.2.'

***Μᾰλεᾶτας**, ὁ, cult-name of Apollo, *IG* 4².128 (*c*.280 BC), *SEG* 33.306, Paus. 2.27.7.

***Μᾰλεάτεια**, τά, *festival of Apollo of Malea*, *IG* 5(1).213.57 (Sparta, v BC).

μᾰλερός II, for this section read 'μαλερὰς φρένας *Suppl.Hell*. 1087 (μαλερὰς φρένας· ἀσθενεῖς Hsch.)'

μάλευρον, delete 'Alc. 70, Achae. 51' and after 'Theoc. 15.116' insert 'Call.*fr*. 177.18 Pf.'; add 'cf. Myc. *me-re-u-ro* (form *μέλευρον)'

***μαλεων**, gen. pl., app. some implement or object used in construction work, ὑπὲρ τι(μῆς) κάμπτρας μαλεων *PWash.Univ*. 28.7 (vi/vii AD).

μάλη 1, add 'perh. Myc. *ma-ra-pi* (loc. pl.)' **2**, for '*underhand, secretly*' read 'ὑπὸ μάλης (done) *in an underhand manner, secretly*' and add 'αἱ ὑπὸ μάλης πράξεις Plu. 2.64e'

μάληκος, after 'pr. n. in inscrr.' insert 'e.g. Μαλε͂ρο̄ (gen.) *Corinth* 15(3).1 (*c*.700 BC), Μάλεκος *IG* 5(2).425 (Phigaleia, v BC)'; add 'cf. μαληκῳ παιδίῳ· χοιριδίῳ δεσμῷ Hsch. (μαλήκῳ πέδα· μοιριδίῳ δεσμῳ La.)'

***μάλημπτος**, Lat. *male emptus, bought as a bad bargain*, *SEG* 39.1062 (Rhegium, i BC/i AD).

***μαλθᾰκιάζω**, *make soft* or *effeminate*, Cyran. 25 (1.10 K.).

μαλθᾰκός, after lemma insert 'Lacon. μαλσακός Alcm. 4 *fr*. 1.5 P.'

***μᾱλία**, Aeol. for μηλέα, Sapph. 2.3 L.-P.

***μᾰλίς**, v. μηλίς (A).

μαλιώτερα, after 'Hsch.' insert 'fr. a comp. form of μάλα'

***μαλκιόεις**, εσσα, εν, = μάλκιος, *Suppl.Hell*. 1167.

μάλκιος, last line, delete 'the latter .. Nic.*Th*. 382'

μαλλός 1, add '(cf. Myc. *wool* ideogram app. *ma+ru*)' **2**, for this section read 'of men's hair, πλοκάμων μαλλοί E.*Ba*. 113, cf. Hsch.; of a bird's neck-feathers, Ezek.*Exag*. 260 S.'

***μαλλουργέω**, *work the flock of wool*, i.e. prob. *card*, *SB* 10209 (ii/i BC).

***μαλλοφρονέω**, *to be wiser*, τῆς γὰρ ἐπιστήμης μαλλοφρονεῖν ἔμαθες *GVI* 1934 (unless error for μᾶλλον φρονεῖν).

μαλλωτός, add 'of shoes, *Edict.Diocl*. 9.25'

***Μᾱλόεις**, εντος, ὁ, (cf. μῆλον (B), -ϝεντ- suffix) epith. of Apollo in Lesbos, Th. 3.3.3, etc., as place-name, *IG* 12(2).74.5 (Mytilene, iii BC). **II** poet. for *Lesbian*, Μαλόες (Dor. = Μαλόεις) ἦλθε χορός Call.*fr*. 485 Pf.

μᾱλοκόμος, μᾶλον, for 'Dor.' read 'hyperdor.'

μᾱλοπάραυος, for 'Aeol. .. Hsch.' read '*white-cheeked* (μᾱλός (A); cf. Hsch. μαλλοπάραυος· λευκοπάρειος) or perh. *apple-cheeked* (μῆλον (B)), Alc. 261(b)i 5 L.-P. (rest.)'

μᾱλός (A), after 'Hsch.' add 's.v. μαλλός'

***μάλουρις**, fem. of μάλουρος, Hsch.; as subst., of an animal, in quot. prob. a cat or sim., Call.*Cer*. 110.

***μάλουρος**· λεύκουρος Hsch.

***μαλφιον**, τό, unexpld. item in accounts of a business establishment, *PMich*.inv. 1933 (*BASP* 16.82, ii AD).

μάματα, after 'ποιήματα' insert '(πέμματα cj. Meineke)'

μαμμάκυθος, after lemma insert '*one who hides in his mother's skirts* (κεύθω)'

μάμμη I, add 'voc. μάμμᾰ Ar.Byz.' **III**, add 'written μάμη, *AAWW* 1961.124 (Lydia, iii AD)'

μαμμία, add '*IG* 2².10743 (iv BC)'

μαμμοπάτωρ, for '*Inscr.Cypr*. 159 H.' read 'Cypr. *ma-mo-pa-to-re* μα(μ)μοπάτορ *ICS* 277'

μανδάκης, after 'ὁ' insert '(or μανδάκη, ἡ)'

μανδάκιον, add '*PHamb*. 21.5 (iv AD, μανταk- pap.)'

μάνδαλος, add 'Hsch. s.v. καβλή⟨ς⟩'

***μάνδαξ**, ακος, ὁ, = μανδάκης, *PRein*. 110.8, 10 (iii AD).

***μανδᾶτον**, τό, Lat. *mandatum*, form of *consensual contract*, Just.*Edict*. 9.3, al.

***μανδάτωρ**, ωρος, ὁ, Lat. *mandator, guarantor*, Just.*Nov*. 4.1, al.

***μανδατωρεύω**, = Lat. *mandare, mandate*, Just.*Nov*. 4.1.

μανδήλη, after '(v AD)' insert 'also **μαντ**- Poll. 7.74'

***μάνδιξ**, ικος, ὁ, = Lat. *mantica, wallet*, *Vit.Aesop*.(G) 4, al.

μάνδρα 1, after 'Plu. 2.648a' insert 'for sheep, Nonn.*D*. 34.252' add '**b** *monastery*, Epiph.Const. 42.340a M., etc.' **2**, for this section read 'a place of human habitation, *Peripl.M.Rubr*. 2, 20, *POxy*. 984 (i AD)'

***μανδράγόριον**, τό, dim. of μανδραγόρας, Cyran. 48.

***μανδράρχης**, ὁ, person in charge of a μάνδρα, *PHib*. 211.6 (iii BC).

***Μάνδρος**, ὁ, a divinity in Asia Minor, with the title Καίων, *IKyme* 37.5-6, also in theophoric pers. n. Ἀναξίμανδρος etc.

***μανέανον**, τό, an unidentified tool or instrument, *BGU* 544.25 (ii AD).

μάνης I, line 3, read 'μάνᾱς' **II**, for '*small bronze figure*' read 'part of stand' **III**, at beginning insert '*Μάνης* (perh. diff. wd.), also acc. pl. Μάνας Ar.*Av*. 522'

***μᾰνιάκης**, cf. Iran. *mani-, necklace*)

***μανίζω**, prob. *damage*, fut. μανίσει Ramsay *Cities and Bishoprics* 1.157, unless fr. μηνίω.

μάννα, line 2, delete 'but .. λίβανος'

μαννάριον, delete 'perh. .. μαμμάριον' and add 'cf. mod. Gk. μάννα *mother*'

μανός I, add '**2** metaph., *mentally weak*, Suid. s.v. μανόν.' **II 2**, after '-νῶς' insert 'Hermog.*Id*. 2.12'

***μανόσπορος**, add 'A.*fr*. 113a.1 R.'

***μανούβριον**, τό, Lat. *manubrium, handle*, *BGU* 544.22 (ii AD, pl.).

μανόφυλλος, add 'read by Zenod. in Od. 13.346 (w. Att. scansion ⏑⏑−⏑)'

μαντεία, after lemma insert 'Arg. μαντήα Schwyzer 89 (iii BC)'

μαντευτικός, add 'Diog.Oen.*fr*. 122 i 5 S.'

***μαντηλαρία**, ἡ, = °μαντηλάριος (ἡ), *IEphes*. 1078.14 (iii AD).

***μαντηλάριος**, ὁ and ἡ, *slave who brought towels* or *napkins* at a banquet (cf. Verg.*G*. 4.377), *SEG* 34.1126 (Ephesus), *IEphes*. 1060.14; cf. ‡μανδήλη.

***μαντήλη**, v. °μανδήλη.

μαντιαρχέω SUPPLEMENT **μάτημι**

***μαντιαρχέω**, serve as μαντιάρχης, aor. part. Cypr. alphabetical μανζιαρχήσαντος, *Kafizin* 258(b) (iii BC).

μαντιάρχης, add 'Cypr. *ma-ti-a-*[, perh. μα(ν)τια[ρχō, *Kourion* 9'

✝**μαντίαρχος**, ὁ, = μαντιάρχης, *SEG* 20.162 (vii/vi BC), 23.621 (iii BC, both Cyprus).

μαντικός I 1, line 1, for '*prophetic*' read '*of prophetic utterances*'; add 'μαντικώτερα εἰρῆσθαι Plu.*Cat.Mi.* 52, *Pomp.* 60' **II**, line 3, after 'Luc.*Hes.* 7' insert 'μαντικώτατος of Romulus, *devoted to divinations*, Plu.*Cam.* 32.5'

μαντίον, add 'Lyd.*Mag.* 69.1'

μάντις I 1, add 'as a civic office, *SEG* 26.694 (Ambracia, *c.*150 BC)' **II**, delete 'a kind of *grasshopper*'

***μάντος**, ὁ, Lat. *mantus*, *short cloak*, *Edict.Diocl.* 19.71.

μαντοσύνη, add '**2** in pl., *oracular response*, *APl.* 296 (Antip.).'

μαντῷος, add 'Olymp. *in Alc.* p. 201'

***μᾱνῡ́τειρα**, ἡ, Dor. fem. of μηνυτήρ, στάλα *SEG* 8.482 (Egypt, i BC/i AD).

μάνωσις, add 'ἡ τῶν ἄστρων ἀνταύγεια καὶ μ. *the stars' power of rarefying the light they reflect*, Hp.*Hebd.* 1.2'

***Μαξιμιάνειος**, α, ον, *named in honour of Maximianus*, ἀγών *JRS* 3.289, al. (Antioch in Pisidia).

***μάππα**, ἡ, Lat. *mappa*, *napkin*, *Vit.Aesop.*(G) 44.

μαππάριος, add 'written μαμπ- *PGrenf.* 2.111.12 (v/vi AD)'

✝**μαππίον**, τό, dim. of °μάππα, *Gloss.*; sp. μαπιν *POxy.* 1051.17 (iii AD); neut. pl. μαπα (dub.) ibid. 19, μαμπία ib. 1741.17, μαπία *PRyl.* 627.20 (iv AD).

μάραγδος, after 'σμάραγδος' insert 'Men.*fr.* 315 K.-Th.'

***μαραθᾶς**, ᾶ, ὁ, *seller of fennel*, dub. in *IG* 12(9).522 (Euboea) (cf. *Rev.Phil.* 18.52).

μάραθον, add '(form μάραθƑον) Myc. *ma-ra-tu-wo*'

Μᾱραθῶνάδε, add 'Euc.ap.Arist.*Po.* 1458ᵇ9'

μαραίνω, line 5, delete '(leg. -αμμ-)' and after 'Plu.*Pomp.* 31' insert 'cf. μεμαραμένη *GVI* 1801.2 (Bithynia, late Rom.imp.)'

μαράσσαι, add 'cf. ἀμαράσαι, μαρίν'

***μαργάρεος**, ὁ, = μαργαρίτης, Gr.Naz.*Carm.* 2.1.38.34.

μαργαρίτης, add '**III** an Egyptian plant, Arist.*Plant.* 1.4.1 (819ᵃ11 ed. Apelt).'

μαργάω, after 'only in part.' insert '(except μαργᾷ· μαργαίνει Hsch.)'

μάργος 1, after 'of wine' insert '*maddening*'; add '(s.v.l.)' **3**, add 'Arist.*Phgn.* 808ᵇ6'

μάρδος, for '*flute*' read '*reed-pipe*'

μάρη, delete the entry (v. °μάρος).

Μᾱριανδῡνοί, (s.v. Μαριανδυνία), add 'as name of a helot class at Heracleia, Str. 12.3.4'

μαριεύς, for 'Hsch. (μαριζεύς cod.)' read 'cf. °μαριζεύς'

***Μαριεύς**, έως, ὁ, *of the town Marion* in Cyprus, S.*fr.* 69, D.S. 19.62, 79.2.

***μαριζεύς**· λίθος τις, ὃς ἐπισταζομένου ὕδατος καίεται Hsch.; cf. μαριεύς.

✝**μᾰρικᾶς**, μαρικᾶν· κίναιδον. οἱ δὲ ὑποκόρισμα παιδίου ἄρρενος βαρβαρικόν Hsch.; Μαρικᾶς represented Hyperbolus in a play of this title by Eup., cf. Ar.*Nu.* 553.

μαρίν, add 'cf. μαράσσαι, ἀμαράσαι'

***μάριον**, τό, dim. of μάρις, *POxy.* 1297.3 (iv AD).

μάρις, add 'Hallock *Persepolis Fortification Texts* p. 2 (*c.*500 BC). **2** μ. .. καλεῖται δὲ ὁμωνύμως καὶ τὸ μακρὸν πέπερι Hsch.'

μαρίω, delete the entry.

***μαρκαζῆτα**, ἡ, *marcassite*, *white iron pyrites*, Anon.Alch. 333.28.

μαρμαράριος, after 'ὁ' insert 'cf. Lat. *marmorarius*'; add '*Edict.Diocl.* 7.5, *MAMA* 3.21 (Seleucia), *Inscr.Olymp.* 657 (v/vi AD), *ITyr* 152 '

μαρμάρεος II, add 'see also °μαρμάριος (A)'

μαρμαρίζω, line 2, for '-ιζούσας Pi.*fr.* 123.2' read '-ιζοίσας Pi.*fr.* 123.3 S.-M. (v.l. -υζοίσας)'

μαρμαρικός, add 'Zos.Alch. 186.2, cf. (for sense) Μαρμαρικοῦ .. πολέμοιο *SEG* 26.1835 (Cyrene)'

μαρμάρινος, line 3, after 'λίθος' insert 'kind of stone wrongly identified with onyx, *Lapid.Gr.* 198.2'; add 'neut. pl. subst., (?)*marble tablets*, *SEG* 33.1040 (Cyme, ii BC)'

***μαρμάριος** (A), α, ον, Aeol. for μαρμάρεος II, *IKyme* 19. **II** Μαρμάριος, ὁ, epith. of Apollo at Delos, *Inscr.Délos* 2473.

***μαρμάριος** (B), prob. = μαρμαράριος, *MAMA* 3.683 (Corycus), *MAMA* 3.25 (-άρις, Seleucia).

μαρμαρῖτις, add '**III** the plant *fumitory*, Ps.-Dsc. 4.109.'

μαρμαρόεις, for '= μαρμάρεος' read '*gleaming*' and add '**II** *of marble*, μ. στήλη *IG* 9(2).650, cf. 14.1603.'

✝**μαρμαρόπαιστος**, ον, *struck out of marble*, epigr. in *JHS* 73.139 (*c.*ii AD).

***μαρμαρύζω**, v. μαρμαρίζω.

μαρμάρωσις, add 'Keil-Premerstein *Dritter Bericht* 64 (iv AD), *ICilicie* 34.4 (v/vi AD). **2** *marble paving*, *JHS* 28.195 (Side).'

***μάρος**, ους, τό, *hand*, Pi.*fr.* 310 (Sch.Il. 15.137); cf. εὐμαρής, εὐμάρεια.

Μαρσήλλιος, ὁ, name of a month, *Amyzon* 2 (321/0 BC).

μάρσιππος, add 'a kind of sackcloth, *PHels.* 7.7 (ii BC)'

***μαρτῠρητικός**, ή, όν, *providing evidence, testimonial*, ψηφίσματα μαρτυρητικὰ καὶ τε[ι]μητικά *La Carie* 78.18.

μαρτῠρία I, lines 4/5, delete 'μαρτυριῶν .. 1316'; line 8, after '*commendation*' insert 'D.H. *Th.* 35'

μαρτῠ́ριον III, add '*AS* 35.96 (iv AD), *SEG* 34.1212 (-ρειν for -ριν, Lydia)'

***μαρτῠρολόγος**, ὁ, *reader at the commemoration of a martyr*, *PLand.* 154.8 (v/vi AD).

μαρτῠ́ρομαι, add '**5** *give testimony*, Modest.*Dig.* 27.1.13.8.'

μαρτῠροποιΐα, after 'Ptol.*Tetr.* 183 (pl.)' insert '(v.l.)'

μάρτῠς I, add 'ᵇ as adj., μάρτυρι σιγῇ Nonn.*D.* 3.123; μ. πομπῇ ib. 4.207.'

Μάρων, line 2, after 'Od. 9.197' insert '(great-grandson of Dionysus, Hes.*fr.* 238 M.-W., a companion of D., pap. in *Aegyptus* 6.192)'

μᾰσάομαι I, line 5, after 'Cass.Fel. 32' insert 'as a gesture of contempt, Philostr.*VA* 7.21' and delete 'or Att. Prose' **II**, delete the section.

μάσημα, add '**2** *part of the head-harness of a horse*, = Lat. *salivarium*, *Edict.Diocl.* in *SEG* 37.335 iii 10.'

***μᾰσητήρ**, ῖδος, ἡ, fem. of μασητήρ, Hsch. s.v. νάρθη.

***μασθέλιον**, τό, in pl., some part of a pack-camel's equipment, perh. *straps* or *harness*, *PNess.* 74.7 (vii AD).

μάσθλης, line 3, for 'Sapph. 19' read 'Sapph. 39.2 L.-P., perh. in Alc. 143.12 L.-P.'

***μασθοδοσία**, v. °μαστο-.

***μασθός** (B), Dor. for μαδός, Heraclid.ap.Eust. 1562.4.

μασκαύλης, add '?cf. βασκαύλης'

***μάσλης**, v. ‡μάσθλης.

***μασονάφι(ο)ν**, τό, perh. = μασουάφιον, *PVindob.* G39847.928 (*CPR* 5.110, iv AD).

***μασσανδάνια**, τά, unexpld. wd., ostr. in *BASP* 23.25.

***μάσσινος**, ὁ, *rope*, *ICilicie* 108.5 (v/vi AD).

μάσσων, line 6, after 'I' insert '(μαεν Schwyzer 230 (Arcesilas vase, vi BC) has been interpreted as = μαγέν, sc. τὸ σίλφιον)' **II**, add 'see also *SEG* 30.1364' **III**, insert 'fig.' before '*take the impression of*'

μάσσων, add 'see also μασσότερον'

μάσταξ II, add 'unless = sense III *locust* (or its grub), also perh. in Theoc. l.c.' **III**, add 'Artem. 2.21'

μαστιγοφόρος I, add 'λύθρον μ. *Trag.adesp.* 235 K.-S.'

μαστίζω, line 2, after 'Il. 5.768' insert 'ἐμάστιξα Lxx 3*Ma.* 2.21'; line 3, after '(Leon.Alex.)' insert 'also μαστισθείς *SEG* 8.246.17 (Palestine, ii AD)'

***μαστικτήρ**, ῆρος, ὁ, *that stabs* as with a goad, ἤκουσα μαστικτῆρα (cj.) καρδίας λόγων A.*Supp.* 466.

μάστιξ II, add 'ᵇ of persons, transl. Hebr. wd. from root meaning "smite", but Hebr. perh. corrupt, Lxx *Ps.* 34(35).15 (πλήκται in Aq., Sm. ad loc.).'

μαστίχη, add 'also μαστύχη, *Edict.Diocl.* 36.63'

***μαστοδοσία**, ἡ, (μασθο-), *suckling*, ἐν ταῖς μασθοδοσίαις τῶν βρεφῶν Heph.Astr. 3.5.49.

***μαστοδοτέω**, *suckle*, Cyran. 74.

***μαστρεῖον**, τό, *assembly of the* μαστροί, *Tit.Cam.* 110.44, rest. in *Lindos* 419.25 (i AD).

***μαστρεύω**, serve as μαστρός, *Lindos* 420ᵃ6, al. (i AD).

μαστρικός, add 'masc. subst., perh. = μαστρός, *Tit.Cam.* 89.7'

✝**μαστροπεύω**, of a pimp, *procure*, X.*Smp.* 4.57, Luc.*Tim.* 16, Ach.Tat. 6.3; fig., οὐκοῦν σύ με .. μαστροπεύσεις πρὸς τὴν πόλιν ..; (i.e. public life), X.*Smp.* 8.42; μαστροπεύουσι δὲ αὐτῇ (sc. ἡδονῇ) .. τὸν ἔρωτα αἰσθήσεις μ. ἡδονῇ Ph. 1.40, 156.

μαστροπός I, after '*pimp* or *procuress*' insert 'Sophr. 69 (dub. gender)'; line 2, delete 'metaph.'

μαστρός, add 'at Olympia, *SEG* 31.358 (v BC)'

***μαστύς**, ύος, ἡ, *search* (cf. μάστευσις), Call.*fr.* 10 Pf.

***μαστύχη**, v. °μαστίχη.

μᾰσύντης, add 'cf. Μασυντίας Ar.*V.* 433'

***μασχάλην** (or -ᾱν)· τὸ τοῖς λευκίνοις σχοινίοις τὰς ἀγκύρας σχάσαντες περὶ τὸν ἀγκυρίτην λίθον περιθεῖναι Hsch.

μασχαλίζω, add 'perh. Boeot. form in Hsch.: μασχαλίττει· ὑπὸ κόλπον καὶ μάλην φέρει.'

μασχαλίσματα 2, add '*SEG* 36.206.16, 17 (Attica, *c.*300 BC)'

***μάσχιον**, τό, = μασχάλη II 2, *OBodl.* II 1756 (ii AD, pl.).

μάταιος I, add '**3** μάταια, τά, *empty beings*, *phantoms*, Lxx *Le.* 17.7, 2*Ch.* 11.15.'

✝**μᾰταϊσμός**, ὁ, *miscarriage*, Seleuc.ap.Ath. 2.76f, Plato Com. 61 (ibid.).

***μᾰτερία**, ἡ, Lat. *materia*, *dough*, Chrysipp.Tyan.ap.Ath. 3.113b,c.

***μάτημι** (A), line 2, before ':– Pass.' insert 'elsewhere ματέω, ματεῖ ζητεῖ Hsch., aor. part. ματίσας (for -ήσας) PUniv.Giss. 32.16 (iii/iv AD), cf. inf. ματίσαι Hsch.'; at end delete the entry in brackets.

μάτημι (B), Aeol. for πατέω, tread, μάτει Alc. 74.3 L.-P., pres. part. fem. pl. μάτεισαι Incert.auct. 16.3 L.-P., fut. inf. ματήσην Alc. 200 L.-P; ματεῖ· πατεῖ Hsch.

μᾱτίδιον, τό, a measure (app. dim. of μάτιον), PCol. 188.18 (iv AD).

ματίζω, delete the entry.

ματίς· μέγας. τινὲς ἐπὶ τοῦ βασιλέως Hsch.

μάτλα, ἡ, Lat. matula, chamber-pot, SB 1160.6 (ostr.).

ματρικάριος, ὁ, public servant employed app. in maintaining order, Just.Nov. 13.5.

μάτριξ, ικος, ἡ, Lat. matrix, roll, list, PBeatty Panop. 1.17, al. (AD 298), edict of Anastasius I in SEG 9.356.6, 13 (Cyrenaica, vi AD), PMonac. 2.8 (vi AD), LW 1906d, e (Bostra, prob. vi AD), Lyd.Mag. 3.2.

ματρώνα, ἡ, Lat. matrona, married woman, IGRom. 3.244, PFlor. 16.2 (iii AD).

ματρωνίκιον, τό, women's part of baths, PFlor. 384.7, 14 (v AD); pl., women's quarters, attached to monastery hospice, PNess. 79.29, al. (early vii AD, ματρων- pap.).

μαυλάκι(ο)ν, τό, prob. dim. of μαῦλις (B), PFouad 84 (ii AD).

μαῦλις (B), for 'Call.Aet. 3.1.9' read 'Call.fr. 75.9 Pf.'

μαυρός, add 'cf. mod. Gk. μαῦρος'

μαυρόω, add 'v.l. for ἀμαυρόω (q.v.), Hes.Op. 693'

Μαύσωλλος, for 'Μαυσωλεῖον' read 'Μαυσώλειον'

***μᾰφάρι(ο)ν**, τό, = μαφόριον, Dura⁴ 93, 129.

μαφόριον, add 'Lyd.Mens. 1.20; see also °βαφωρι-'

μαφόρτης, for 'veil .. priests (cf. Gloss.)' read 'short cloak with a hood'; add 'cf. Hebr. maʽăpōret'

μαφόρτιον, add 'Edict.Diocl. 29.29'

μάχαιρᾶς, add 'MAMA 3.628 (Corycus)'

μαχαιρίδιον, delete 'Luc.Pisc. 45'

***μάχαιρον**, τό, = μάχαιρα, POxy. 1289.4,7 (v AD, μαχερ- pap.).

μαχάτης, ὁ, or -άτη, ἡ, or -άτι(ο)ν, τό, a measure of cubic capacity, SB 9303.5 (iii AD).

μάχη I 1, add 'b brawl, affray, SEG 31.122 (c.AD 121/2); μ. παροίνους Anacreont. 42.13 W.'

μαχητής, add 'cf. Myc. ma-ka-ta, pers. n.'

***μᾰχῐκός**, ή, όν, = μάχιμος, An.Ox. 4.266.4.

μᾰχῐμικός, ή, όν, belonging to a μάχιμος, [κλῆρος] PVarsov. 3.35 (ii AD); so also PRyl. 202.5 (i AD); γῆ μ. perh. to be read in BGU 958b (ii/iii AD); cf. PLond. 193.34 (i AD): see APF 12.95.

μάχιμος, after 'PTeb. 61(a).109 (ii BC), etc.' insert 'sg. μ. ἑπτάρουρος PColl.Youtie 16.3 (109 BC)'

***μαχλίς·** ἑταίρα, πόρνη Hsch.

μάχλος 1, line 3, for 'Aeschrio 8.6' read 'Aeschrio in Suppl.Hell. 4.6'

***μαχλῶντες·** πορνεύοντες Hsch.

***μέ**, v. °μετά.

***μεγαβρόντης**, ου, m. adj., loudly-thundering, codd. in Ar.V. 323.

μεγαίνητος, add 'IG 2².3632.10'

***Μέγαιρα**, ἡ, name of one of the Erinyes, Corn.ND 10.

⁺μεγακήτης, ες, having the form or appearance of a monster (usu. marine), δελφίς Il. 21.22, νηῦς 8.222, 11.5, 600; ὅρκυνοι μεγακήτεες Opp.H. 3.132, 546; of the Trojan Horse, Q.S. 12.151. **2** of the sea, full of monsters, μεγακήτεα πόντον Od. 3.158, A.R. 4.318.

μεγακλεής I, add 'gen. -κλέος Euph. in Suppl.Hell. 416.1'

μεγαλάμπρως, delete the entry.

μεγαλάμφοδος, add 'Sch.Gen.Il. 16.635'

Μεγαλάρτια, add 'at Delphi, CID 1.9D10 (v BC)'

***μεγάλατος**, ον, involved in great ruin, A.Pers. 1016, Eu. 791, 821.

μεγαλαυχής, for ' = μεγάλαυχος' read 'lofty, proud' and before 'Vett.Val. 272.8' insert 'in pejorative sense, haughty, conceited'

μεγαλαυχία, add '3 in good sense, pride, SEG 18.293.14.'

***μεγάλευκτος**, ον, greatly prayed for, sup., πίστιν .. τὰν μεγαλευκτοτάταν cj. for μεταλευκοτάταν in paean ap.Plu.Flam. 16.

***μεγαλία**, ἡ, as honorific title, highness, [τ]ὴν [σὴν] μεγαλίαν PCol. 173.7 (iv AD).

***μεγαλοευπώγων**, ωνος, ὁ, having a large fine beard, Cat.Cod.Astr. 7.217 (perh. μεγαλοπ- or εὐπ-).

⁺μεγαλόηχος, ον, stirring up tumult, turbulent, Sch.Pi.P. 12.38, I. 8.45, Hsch. s.v. ἐριβρεμέτης, ἐρίβρομον.

***μεγαλόθρονος**, ον, mightily enthroned (unless to be derived from θρόνον), Ἥρη hymn in SEG 8.548.21 (Egypt, i BC).

μεγαλόθῡμος, add 'Plu. 2.614b'

***μεγαλόκλονος**, ον, loud-sounding, Trag.adesp. 109d K.-S.

μεγαλομέρεια III, add 'SEG 23.447 (ii AD)'

***μεγαλοναύτης**, gloss on βουβάρας, Hsch.

μεγαλοπάρηος, thus, not -πάρηος.

***μεγαλόπλευρος**, ον, with big flanks, gloss on ἐρίπλευρος, Sch.Pi.P. 4.419.

μεγαλοπολίτης, add 'also citizen of Megalopolis, SEG 23.226, 227'

***μεγαλοπόρως**, adv. bountifully, SB 8267.30.

μεγαλοπρεπής I 2, sup. as honorific title, add 'SEG 31.1401 (Palestine, v AD)' **3**, add 'of a writer, D.H.Isoc. 3'

μεγαλοσώματος, after 'large-bodied' insert 'Plu.fr. 149 (p. 177 B.)'

μεγαλότης, add 'Plu. 2.441b (pl.)'

μεγαλοτράχηλος, add 'Sch.D.Il. 10.305'

μεγαλόφθαλμος, after 'Ptol.Tetr. 143' insert '(v.l.)'

***μεγαλοφροσύνως**, adv. generously, IGRom. 3.739 xvi 49, xviii 13 (Rhodiapolis, ii AD).

μεγαλύνω I, add '2 make long, μεγαλύνουσι τὰ κράσπεδα i.e. wear long tassels, Ev.Matt. 23.5.'

μεγαλώνυμος I, delete 'giving glory' and insert '(?)Alc. 304.3 L.-P.'

μεγαλωστί II 1, before 'v.l.' insert 'greatly, Phld.Oec. p. 6 J.'

μεγάνωρ, add 'GVAK 11.3'

***μεγαρίζω** (B), perform the rite of the μέγαρα, (v. °μάγαρον), Clem.Al.Protr. 2.17.1.

μέγᾰρον, read 'μέγαρον (A) **IV**, delete the section.

***μέγαρον** (B), v. °μάγαρον.

μέγας, line 3, after 'μεγάλα, etc.' insert 'Pamph. fem. acc. πόλι μhε[ι]άλα IPamph. 3.5 (Sillyon, iv BC)'; line 4, for 'and only once .. (anap.)' read 'voc. masc. μεγάλε A.Th. 822, AP 14.100, Them.Or. 13.163c' **C 1**, line 6, after 'Plu.Lyc. 19' insert 'μέδδων IGC p. 98 A24, 27, al. (Acraephia)'; line 7, μειζότερος, after 'elder' insert 'SEG 33.1299 (Tiberias, iv AD)' **2**, add 'as title of gods, Διὶ τῶι Μεγίστωι SEG 33.675 (ii BC), etc.; cf. Demetrius Poliorcetes, Moretti ISE 7.2 (303/2 BC); of Emperors, Αὐτοκράτορα μέγιστον Νέρωνα Καίσαρα SEG 32.251, etc.; transl. Lat. pontifex maximus, ἀρχιερέα μέγιστον SEG 30.1635, etc.' add 'Myc. form of compar. *μέζως, me-zo, me-zo-e (pl.); sup. me-ki-ta (neut. pl.)'

μεγασθενής, line 2, for 'also' read 'of non-personal agents' and add 'οἴστρος Ἀφροδίτας Simon. 36.10 P.'

μεγαυχής, for ' = μεγάλαυχος' read 'worthy of pride, glorious'

***μεγαύχητος**, add 'epigr. in Inscr.Cret. 1.viii 33 (Cnossus, ii BC), Ph.Epic. in Suppl.Hell. 681.3'

μέγεθος II 1, add 'strength, force, ἀνέμου Thphr.Sign. 29' **III**, add '3 body having magnitude, mass, Arist.GC 321ᵇ16, al.'

μεγεθύνω I 1, line 2, after 'Iamb.Protr. 21 ιζ'' insert 'κόμην let .. grow, Aq.Nu. 6.5'

***μεγιστοῦχος**, holder of the highest office (app. high-priest), Ph.Epic. in Suppl.Hell. 683.2.

***μεδιανόν**, τό, Lat. medianum, central hall, ISmyrna 192.

μέδιμνος, before 'ὁ' insert 'Cret. ϝέδιμνος Inscr.Cret. 4.184.16 (Gortyn, ii BC)'; line 3, after 'corn-measure' insert 'SEG 30.380 (Tiryns, vii BC)'

***Μεδίμῳ·** ἥρωι Hsch.

μεθαιρέω, line 1, delete 'only' add '**II** change over to, ἥν περ μεθείλες τὴν τέχνην A.fr. 78c.56 R.' **III** remove, μεθελόντω τὰ ἀναθήματα ἐς ἄλ[λον τ]όπον ASAA 30/32(1952/4).249, no. 1, line 21 (Rhodes). **IV** med., change, τὰ οἰκία SIG 344.72 (Teos, iv BC); [τὰ ἄρμενα] Teles p. 10 H.'

μεθάλλομαι II, add 'fig., ἐπ' ἄλλα ἀσύνδετα Longin. 20.2'

***μεθαύριον**, v. °μεταύριον.

μεθελῑτης, for 'dub. sens.' read 'perh. mutton-butcher (cf. μέθλην)'; line 2, after 'ib. 674' insert 'SB 11003.7 (iv/v AD)'

μεθέπω, line 1, for 'Sapph.Supp. 23.8' read 'Sapph. 94.8 L.-P.' **I**, add '4 look after, Il. 10.516 (tm.), Sapph. l.c.' **II**, for this section read 'pursue, follow a course, occupation, γεηπονίην Ps.-Phoc. 161, αἶαν Pi.N. 6, μοῦσαν App.Anth. 3.157.3; go for, undertake, ἑκόντι .. νώτῳ ὑπέδυμον ἄχθος ἄγγελος ἔβαν Pi.N. 6.57. **2** possess, have, Nonn.D. 48.355, 362, al.'

***μεθερμηνεία**, ἡ, interpretation, Zos.Alch. 118.14.

μεθημοσύνη, add 'TAM 4(1).382'

μεθιδρύω, line 2, delete 'pass.'

μεθίημι, line 8, for 'Colluth. 127' read 'Coluth. 128' **II 2 b**, delete the section.

μεθιστάνω, form -άω, add 'Cod.Just. 1.3.55.2'

μεθίστημι A II 2, place 'so in med. .. 18.9.5' in a parenthesis and add 'ἑαυτὸν ἐκ τοῦ ζῆν D.S. 3.5, cf. 4.55; τοῦ ζῆν μ. [τινα] BGU 36.13 (ii AD)' **B I 2**, line 5, for 'σκότος' read 'σκότον' and after 'Pl.R. 518a' insert 'ἐκ τοῦ ζῆν PLond. 354.10 (i BC)'

μεθοδεία, add '**III** trade, employment, Just.Nov. 122 pr.'

μεθοδευτής, add '3 investigator, τῆς λεκιθώδους ὕλης Zos.Alch. 144.16.'

μεθοδεύω 3, after 'get round' insert 'POxy. 2342.27 (ii AD)' **5**, add 'SEG 32.1554 (Arabia, vi AD)'

μεθόδιον II, for 'cf.' read '**III** ingenious device'

μέθοδος II, add '**6** occupation, trade, Just.Nov. 122.1, τὴν μέθοδον κρανβιτᾶς AE 1929.151 (Thessaly, v/vi AD).'

μεθόριος, line 10, after 'Plu.Crass. 22' insert 'ἐν μεθορίᾳ γῆς καὶ οὐρανοῦ Max.Tyr. 14(8).11; ἡ μεθορία banishment, Philostorgius in PG 65.480A'

Μεθυμναῖος, add 'perh. jocular version of °Μηθυμναῖος'

μεθυποτίθημι, pawn, mortgage, PPanop. 21.20 (iv AD), POxy. 3355.13 (vi AD).

μεθύσκω, line 1, at end insert 'first at Hdt. 1.106 (κατ)εμέθυσα'

μεθύστερος II, line 2, for 'in a moment' read 'never thereafter'

μεθυτρόφος, for 'ἡμερίς Simon. 183.1' read 'AP 7.24.1 ([Simon.])'

μείγνυμι, line 19, after 'S.fr. 271 (anap.)' insert 'Critias Trag. 5.11 S.' **B 4**, add 'perh. also act. ellipt. in same sense, Ἐγέστρατος μοί μίσγη (?for μίσγει or ἐμίγη) Lang Ath.Agora xxi C8'

Μειλινόη, add 'cf. Jahrb.Ergänzungsheft vi(1905).13.26 (Pergamum, iii AD) (Μηλ-)'

μειλίσσω II, delete 'to be subdued .. A.R. 3.531' **III 2**, line 2, after 'subdue' insert 'πυρὸς μειλίσσετ' αὐτμήν A.R. 3.531'

μειλίχιος I, add 'μειλίχιόν τε καὶ αἰνετὸν ἔργον SEG 31.1288 (iii AD)'

μείλιχος, line 1, for 'cj. in Sapph. 100' read 'Sapph. 2.11, 112.4 (cj.) L.-P.'

μειξόκρουστος, ον, adj. of uncertain meaning describing clothing, PVindob. G16846 (Tyche 2.6; v. also ib. 7.63; vii AD).

μειονεκτέω, line 4, before 'c. gen. rei' insert 'Call.fr. 196.44 Pf., to be at a disadvantage, Arr.An. 3.8.7'

μειότης I, for 'minimizing' read 'reduction to a minimum (in context indicated by γέ in τοῦτό γέ μοι χάρισαι)'

μειόω II, line 2, after 'Pl.Cra. 409c' insert '(cited in supposed etym. of μείς)'

μειρακιώδης I, delete 'τὸ μ., of style .. Longin. 3.4' **II**, after 'characteristic of youth' insert '(as lacking in restraint, taste, etc.), juvenile' and add 'D.H.Isoc. 12, Pomp. 2, Longin. 3.4; adv. D.H.Dem. 29'

μείρομαι (A), line 5, after 'ἐμμόραντι' insert '(ἐμμόρατι La.)' **II 2**, for this section read '2 fall to one's lot, σε .. ἔμμορε κῦδος A.R. 4.1749. **3** w. predicative adj., become by fate such and such, πᾶν δὲ νόημα ἔμπληκτον μεμόρηκε Nic.Al. 213.' **III**, line 10, after 'destiny' insert 'Antipho 1.21'; at end after 'μεμόρηται' insert 'A.R. 1.646' and after 'Man. 6.13' insert 'μεμόρητο A.R. 1.973'; delete 'but μεμορημένον .. μορέω (q.v.)' **IV**, add 'pf. part. μεμορημένος Nic.Al. 229'

μείς 3, add 'cf. Myc. adj. me-no-e-ja' at end add 'Myc. me-no (gen.)'

μείχμα, for 'Alc.Supp. 13.7' read 'Alc. 34(b).7 L.-P.'

μείων, line 4, after '(Tegea)' insert 'Boeot. μίων BCH 60.28 (Acraephia, ii BC)'; line 6, after 'less' insert 'Il. 2.528, al., smaller, μείων κεφαλῇ Ἀγαμέμνονος Il. 3.193'; line 7, after 'A.Supp. 596 (lyr.)' insert 'Epich. 62'; line 8, delete '(not in other works of Hp.)'; line 9, delete 'or Com.'; line 10, before 'younger' insert 'pl., fewer, Xenoph. l.c.; χρόνῳ μ. γεγώς'; at end add 'form μεϜjως(?), Myc. me-wi-jo, me-u-jo'

μείωσις, add '**b** = °μειότης I, A.D.Synt. 267.25.'

*****μελαγκάλαμον**, τό, ink and pen, PFouad 74.9 (cj., cf. Glotta 35.299; v AD).

†**μελαγκρηπίς**, ῖδος, fem. adj., perh. having black shoes, Suppl.Hell. 991.3 (poet. word-list, iii BC), Eust. 174.9, 1437.53.

μελάγχιμος, line 3, after 'A.Pers. 301' insert 'μελάγχιμον ἰόν A.R. 4.1508'

*****μελαγχολαίνω**, = μελαγχολάω, Sch.Gen.Il. 6.202.

*****μελαγχρινός**, ή, όν, = μελάγχροος, Cat.Cod.Astr. 12.149.1; cf. mod. Gk. μελαχρινός.

*****μελαιναῖος**, α, ον, dark (= μελανός), Orac.Sib. 5.349 (s.v.l.).

μελαινάς, delete the entry.

μελαίνω I 1, line 4, after 'Nic.Al. 472' insert 'darken, Plu. 2.373d'; line 6, for 'D.H.Pomp. 2' read 'D.H.Dem. 5'; add 'of the moon, to be darkened, Plu.Aem. 17'

μελαμπέταλος, add 'rest. in Suppl.Hell. 991.1 (poet. word-list, iii BC).'

Μελαμποδεῖον, add 'Schwyzer 664.5 (Orchomenos, Arcadia, iv BC)'

μελάμπυγος, after 'Archil. 110' insert 'Philostr. VA 2.36'

μελαμφαρής, add 'Trag.adesp. 660.6 K.-S., Suppl.Hell. 991.4 (poet. word-list, iii BC)'

*****μελάμφαρος**, ον, = μελαμφαρής, Hymn.Is. 43, ISmyrna 728 (ii/iii AD).

μελάμφυλλος II, after 'Subst.' insert 'Μελάμφυλλος, ἡ, old name of Samos, Str. 10.2.17, 14.1.15, Iamb. VP 2.3'

μέλαν I 2, for this section read 'μ. Ἰνδικόν indigo or Indian ink, Peripl.M.Rubr. 39, cf. Plin.HN 35.43, 46' add '**3** lamp-black, Ael.NA 17.25.'

μελανάθηρ, for '(-αίθηρ Hsch.)' read 'also μελαναίθηρ PCair.Zen. 731.11 (iii BC), Hsch.'

μελανδίνης, after 'dark-eddying' insert 'μελανδῖναι .. ῥόες (-δεινη .. ροσσ lapis) GVI 1684.9 (Chersonesus, i/ii AD)'

μελάνζοφος, before 'EM' insert 'Simon. 125 P.'

μελανία I, line 2, before 'Str. 12.8.18' insert 'Thphr.HP 5.3.1' add '**2** darkening (of the sky), X.An. 1.8.8; (morbid) darkening of the skin or flesh, Lxx Sch.Le. 13.39; pl., (in quot. fig.) Plb. 1.81.7.' **II**, delete the section.

μελάνιον, after 'ink' insert 'Edict.Diocl. 18.11, 11a'

*****μελανόβαφον**, τό, perh. ink-bottle, AJA 63.275 (Sicily, iii BC).

*****μελᾰνοπτεροφαιολοσώμᾰτος**, ον, having dark wings with a wholly grey body, χελιδών GLP 1.95.9.

*****μελᾰνοχαίτης**, ου, masc. adj. dark-haired, Theognost.Can. 85.

*****μελαντικός**, ή, όν, blackening, ὕδωρ μ. τριχῶν Paul.Aeg. 3.2.3 (132.25 H.).

*****μελάνωσις**, εως, ἡ, = μέλανσις, Moses Alch. 309.10 (lemma).

μέλας I, line 7, after 'Od. 4.359' insert 'πόντος μ. E.IT 107'

μέλασμα I, add 'Gal. 12.266.13'

μελασμός I 1, add 'Gal. 17(2).803.2, 5 (pl.), 18(2).556.18'

μέλε (B), lines 5/6, delete '(who says .. women only)' and for 'Eq. 668' read 'Eq. 671'

*****Μελέαγρεια**, τά, festival in honour of Meleagros, AS 39.50 (Balbura), SEG 29.1439, 1441.

†**μελεδαίνω**, be concerned about, worry or care about, w. gen., πενίης Thgn. 1129, Theocr. 9.12; w. acc., Archil. 14 W., οὐ μελεδαίνει τὸν τό πιεῖν ἐγχεῦντα Theocr. 10.52. **II** devote care to, take care of, w. acc., SIG 2 (Sigeum, vi BC); τοὺς νοσέοντας Hdt. 8.115.3, τὰς ὑστέρας Hp.Mul. 1.17; id.Morb.Sacr. 8 (pass.), Aret.CA 6.10; see also μελεταίνω.

*****μελεδών** (B) (or -ών)· φροντιστής, μεριμνητής, ἐπίτροπος, οἰκονόμος, προεστώς, φύλαξ Hsch.

*****μελεκοπέω**, = μελοκοπέω, Heph.Astr. 2.18.66, 2.25.11.

μελεταίνω, add 'cf. °μελεδαίνω'

*****μελετητής**, οῦ, ὁ, trainer, coach, Aristid.Or. 50(26).28.

μελετητικός, add '**III** μελετητικὴ ποιότης ῥημάτων desiderative (e.g. lecturio), Dosith. p. 406 K.'

μέλι I, add 'Myc. me-ri, me-ri-to (gen.); adj. me-ri-ti-jo' **II**, for pres. def. read 'perh. honeydew'

μελία I, add 'μελίας καρπός· τὸ τῶν ἀνθρώπων γένος Hsch.'

μελίαμβοι, after 'Cercidas' insert '(Κερκιδᾶ κυνὸς μ. title in POxy. 1082 fr. 4)'

*****μελιανθής**, ές, honey-scented, μελιανθέος οἴνης Nic.Al. 58.

μελιβόας, for 'sweet-singing' read 'with honeyed tones, ὕμνος Lasus 1 P.'

μελίεφθον, for 'honey-jar' read 'app. pan for boiling honey'

*****μελιέψιον**, τό, boiled honey, something boiled with honey, BGU 2355.5 (ii/iii AD).

*****μελιεψός**, ὁ, honey-boiler, SB 11003.8.

μελίζω (B), line 3, after 'Med.' insert 'κάλλα μελισδομέναι Alcm. 35 P.'

μελίθροος, add 'SEG 33.563 (Thrace, iii AD)'

*****μελικηρίδιον**, τό, dim. of °μελικήριον, POxy. 3406.11 (iv AD).

μελικήριον, for 'honey-comb' read 'honey-cake'

μελικτής, for 'singer, player, esp. flute-player' read 'musician'

μελίνη I, after 'S.fr. 608' insert '-ην ἀντὶ πυρῶν ἀλλάττεσθαι Aristid.Or. 34(50).6, Edict.Diocl. 1.6' **II**, for 'Edict.Diocl. .. Aeg.)' read 'Edict.Diocl. 8.29'

μελιουργός, ὁ, honey-maker, Gp. 15.3.7.

μελίσκιον, delete 'Alcm. 65 .. A.D.)'

*****μελίσκον**, τό, dim. of μέλος, a little song, Alcm. 36 P.

μέλισσα III, line 2, after 'S.OC 481' insert 'Nic.Al. 374'

μελίσσειος II, add 'cf. mod. Gk. μελίσσι'

μελισσοβότανον, for 'balm, Melissa officinalis' read 'expl. of μελίτεια'

*****μελισσοβός**, ὁ, name of a bird, Cyran. 92 (3.27.1 K.).

*****μελισσόφυλλον**, τό, name given to one or more fragrant herbs, Thphr.HP 6.1.4, Dsc. 3.103, 104.

μελισσόφυτον, τό, read '-φῠτος, ἡ'

*****μελιτᾶ(ς)**, ὁ, honey-merchant, Παύλου τοῦ μελιτα[. SB 13036 (Tyche 6.232).

†**μελίτεια**, ἡ, a fragrant herb, perh. the same as °μελισσόφυλλον, Theoc. 4.25, 5.130.

μελιτερπής, for 'Simon. 184.9' read 'AP 7.25.9 ([Simon.])'

*****μελιτευχής**, ές, honey-producing, παγά B.fr. 28.14 S.-M.

μελίτινος, for pres. def. read 'made or flavoured with honey'; after 'Diog.ap.D.L. 6.51' insert 'οἴνου μελιτ(ίνου) Lang Ath.Agora xxi He30'; before 'στεφάνια' insert '**2** honey-coloured' and add 'POxy. 3201.3 (iii AD)'

*****μελίτριχος**, ον, having honey-coloured hair, Cat.Cod.Astr. 10.187, 190, 216.

μελίφρων I, add 'of persons, AP 13.12 (Hegesipp.)'

*****μελίφωνος**, ον, honey-voiced, Sapph. 185 L.-P., AP 9.66 (Antip.Sid.).

μελίχλωρος, add '**2** μ. (sc. λίθος), ἡ, a precious stone, Plin.HN 37.191.'

μελίχροος I 2, before 'in gen.' insert 'Lat. melichrus, Lucr. 4.1160; τὸν ὠχρὸν ὑποκοριζόμενος μελίχρουν Plu. 2.45a' add '**III** μελίχρους (sc. λίθος), ἡ, a precious stone, Plin.HN 37.191.'

μέλλαξ, add 'SEG 26.1717 (tab.defix., Egypt, iii/iv AD)'

μελλάρχων, add 'inscr. in Syria 29.326 (Arabia, iii AD)'

μελ(λ)είρην, after 'at Sparta' substitute 'boy about to become an εἴρην, in sixth year of public education, i.e. in thirteenth of his age, Plu.Lyc. 17, Ἡροδότου Λέξεις in Stein Hdt. ii p. 465 (Berlin 1871)'

*****μελ(λ)έπαρχος**, ὁ, eparch-designate, Ann.Épigr. 1984.839b (?v AD).

μελλέφηβος, before 'Censorin.' insert 'IG 2².2986 (ii BC), 2991 (i BC)'

μελλησμός I, add 'PVindob.Tandem 2.22 (iii AD)'

μελλητικός, add '**II** τὸ μ. app. = τὸ μέλλον, κατὰ τοῦ μ. μηδὶς ἀνοίξῃ CIJud. 1.652 (Syracuse; -τεικοῦ).'

*μελλίχόμειδος, ον, voc. -ε (or -ες fr. μελλιχομείδης), gently smiling, Alc. 384 L.-P.

†μελλόγαμβρος· μελλονύμφιος Hsch.

*μελλογραμμᾶτεύς, έως, ὁ, γραμματεύς-designate, CIJud. 1.121, 279 (Rome).

μελλογυμνᾰσίαρχος, add 'POxy.Hels. 15.6 (i AD)'

*μελλολέων, λέοντος, ὁ, one about to be a λέων VI, the final grade in Mithraic initiation, MDAI(R) 49.206 (Dura, pl.).

*μελλονύμφη, ἡ, = μελλόνυμφος, ἡ, CIJud. 1.106 (Rome), cf. Poll. 3.45.

*μελλονύμφιος, = μελλόνυμφος, of a man, Phryn.Com. 78, CIJud. 1.148 (Rome), Hsch. s.v. μελλόγαμβρος.

μελλόνυμφος, lines 3/4, delete 'Phryn. .. -νύμφιος)'

μελλοπρύτανις, after 'πρύτανις-designate' insert 'IEphes. 1051.4 (i/ii AD)'

μέλλω IV, add 'with day of the week, ἐν τῇ μελλούσῃ παρασκευῇ Hesperia 54.214 no. 1 (iii AD)'

μελοποιέω II 1, add 'SEG 11.52c (Isthmus, ii AD)'

μέλος A, add '**4** member of a group, (Lat. membrum), Just.Nov. 109 pr.'

μέλπω II, line 5, after 'Paus. 1.2.5' insert 'IG 2².5056, 5060'

*μελύγιον, intoxicating drink made from honey, gloss. in POxy. 1802.36 (ii/iii AD); μελύγιεον EM 578.8, μελίγυον Zonar.

μέλω A III 2, at end for 'aor. .. 200' read 'also καὶ μέλονται πρός τινος ἢ Διὸς ἢ .. Ἀθάνας Trag.adesp. 167c.4 K.-S., cf. μεληθὲν βάρβιτον AP 5.200 (s.v.l.)' **B II**, line 4, after 'c. dat.' insert 'τῇ παιδὶ .. μεμέλησο AP 5.220.7 (Agath.)'

μελῳδία II, add 'μ. ἡ τραγῳδία τὸ παλαιὸν ἐλέγετο Call.fr. 462 Pf.'

†**μελῳδικός**, ή, όν, of or involving singing, μ. κίνησις Cleonid.Harm. 2, Ptol.Harm. 3.5; πειθώ Aristid.Quint. 2.10; opp. λογικός, Aristox.Harm. 10.

μέμνεο, etc., for 'μιμνήσκω' read 'μιμνήσκω'

Μεμνόνειον I, add 'pl., Suppl.Hell. 984.11'

*μεμοριάλιος, ὁ, Lat. memorialis, class of clerk in the civil service, Cod.Just. 4.59.1.2.

†**μεμόριον**, τό, grave-monument, (fr. cross-influence between μνημεῖον and Lat. memoria), IGRom. 4.1650 (Philadelphia), Ramsay Cities and Bishoprics 2.736 (iii/iv AD), T&MByz 7.323 no. 15 (iv/v AD); also μημόριον SEG 2.393, 404 (Maced.), μιμόριον IG 10(2).353 (Thessalonica, iv AD); μημήμορίον SEG 31.855; μημηόριων BCH suppl. 8 no. 294 (v AD); applied to a martyrium, BCH suppl. 8 no. 1 (v/vi AD); see also ‡μνημόριον.

μεμπτός, after 'blameworthy' insert 'Alc. 1.8 L.-P.'

*Μεμφίτης, ου, ὁ, v. °τεφρίας.

μέμφομαι 3, add '**b** w. dat. of thing censured only, AP 5.299.6 (Agath.), 6.71.10 (Paul.Sil.).' **6**, for 'GDI 4998 (Gortyn)' read 'Inscr.Cret. 4.41 vii 13 (Gortyn, v BC), al., Just.Nov. 100.1.1, al.'

*μεμψίμοίρημα, ατος, τό, perh. document of public censure, μ]εμψιμοιρηματ[.. SEG 33.888 (Ephesus).

μέμψις, add '**3** legal complaint, charge, Cod.Just. 1.4.29.8, 1.4.34.7; κινῆσαι τὴν κατὰ τῆς διαθήκης μέμψιν ib. 6.4.4.16.'

*Μενδήσιον, τό, temple of Mendes (prob. identified w. Pan), POxy. 3332.5 (i AD).

μενεδήϊος, after 'Il. 12.247' insert 'τὸ πάρος μενεδήϊος ἦσθα'

μενέδουπος, add 'Hsch. s.v. βρυαλίκται'

Μενέλᾱος, line 2, after 'Dor.' insert 'nom. Μενέλᾱς Jeffery LSAG pl. 69 no. 47 (c.600 BC)'

μενετός, after 'patient' insert 'κριταί Cratin. 171.6 K.-A.'

μενέχαρμος, add 'GDRK 17ʳ.9'

μενοεικής, add 'A.R. 3.984, Opp.H. 5.374'

μένος, add 'as element in pers. n., already in Myc. e-u-me-ne Εὐμένης'

μένω, line 2, after '(Tegea, iv BC)' insert 'BCH 111.168.6 (Mantinea, iv BC)' **I 2 b**, add 'in pregnant sense, CR 46.250 (Asia Minor)' **II 1**, line 5, after 'E.Ph. 740' insert 'wait for the end of, χειμῶνα Hes.Op. 652; χεῖμα AP 6.221 (Leon.Tarent.), Q.S. 7.137; λαίλαπα id. 8.379'

*μεραρχέω, perh. to be governor of a district, Robert Hell. 10.24 (?Byzantium).

μέρδει, add 'act. aor. μέρσε, IParion 52.10; cf. ἀμέρδω'

μέρεια, after lemma insert '(or μερεία)'

μερίζω I 2, add 'distribute by testamentary disposition, PUps.Frid 1.6, 9,14 (AD 48)'

μεριμνάω, 4 lines fr. end, before 'pass.' insert 'med., μεμεριμνημένους περὶ τὸν πορισμόν Onos. 1.20'

μεριμνητής, add 'gloss on °μελεδῶν, Hsch.'

μερίς I 1, add 'βελτίων μερίς Men.Dysc. 283' **II**, for this section read 'category, division, E.Supp. 238, D. 18.64; faction, ἡ Σύλλα μερίς Plu. 2.203b'; transfer 'Pl.Lg. 692b' to section I 1 and 'Jul.ad Them. 253c' to section I 3 d.

*μέρισις, εως, ἡ, = μερισμός, PLond. 394.13, 15 (vi/vii AD).

μέρισμα, add '**II** prob. inlaid work, IGLS 3.733.'

μεριστής, add '**II** pl., financial officials at Istria, IHistriae 6, al. (iii BC).'

μερῑτεία I, add 'PMich.II 121ᵛ i 9 (i AD), al.; written μεριτία in APF 10.214 (ii AD)'

μέρμερος I, line 1, delete '(only in Il.)'; line 3, delete 'in Hom. always' **II**, add 'cf. pers. n. Μέρμερος Il. 14.513, -ίδης Od. 1.259'

μερμηρίζω II, line 4, after '(Od.) 16.256' insert 'Call.Epigr. 8.5 Pf.'

μερμηρικοί, for 'πειρᾶται' read 'πειραταί'

μερμίλλων, after 'μορμίλλων' insert 'SEG 32.1145 (Erythrae)'

μέρμῑς, for 'dat. pl. .. Zonar.' read 'also **μέρμῐθος**, ὁ, -α, ἡ, -ον, τό, μερμίθαις Agatharch. 47, μέρμιθα· μέρμιθον, σπαρτίον, λεπτὸν σχοινίον, ἢ ἀργυροῦν δεσμόν Hsch., μέρμιθος· ἡ σχοῖνος Zonar.'

*μερμνάδαι· οἱ τρίορχοι (Lydian), Andron ap.POxy. 1802.46; as name of Lydian dynasty, Hdt. 1.7.1.

μέρμνος, delete 'Call.Aet.Oxy. 2080.68'; add 'gen. μέρμνου (nom. incert.), Call.fr. 43.66 Pf.; cf. pers. n. Μέρμνων Theocr. 3.35'

*μεροῖς, ίδος, ἡ, plant with lettuce-like leaves found at Meroë, also called αἰθιοπίς, Plin.HN 24.163.

*μέροξος, ὁ, = λευκογραφίς, Paul.Aeg. 7.3 (237.7 H.).

μέρος II 2, add 'Ἀρχὸν ἐξ Ὑπνίας κὰτ τὸ μέρος ἐλέσθ[ων] SEG 23.305 (Delphi, 190 BC)' **IV 1**, add 'act of a play, M.Ant. 12.36, Platon.Diff.Com. p. 4' add '**6** the part facing in a particular direction, side, Hdt. 4.101.1, Th. 7.80.2; ἐξ ἀμφοτέρων μερῶν τοῦ δρόμου OGI 56.52 (Canopus, iii BC), cf. PPetr. 3 p. 43(2)ᵛ iv 11 (247/6 BC), LXX Ex. 32.15.'

μέροψ, add 'of devotees in relation to a god, σῶν μερόπων Luc.Trag. 193, IG 2².4533.5 (iii AD)' add '**2** spec., of the supposed original inhabitants of Cos descended from Merops (Il. 2.831, al.), Κόως .. πόλις Μερόπων ἀνθρώπων h.Ap. 42; substr., Pi.N. 4.26, I. 6.31, Suppl.Hell. 903A.1, 13. **3** μέροπες· οἱ ἄφρονες ὑπὸ Εὐβοέων, in POxy. 1802.47.'

*μές (B), Thess. until, w. gen. μὲς μὲν τᾶς τετράδος SEG 31.577, μὲς τὰς πέμπτας BCH 59.55 (both Larissa, ii BC); cf. μέστε, μέσφα, etc.

μέσᾰβον, line 3, for '(Call.)Fr. 513' read 'fr. 651 Pf.' and after 'γλυφαί)' insert '177.5 Pf.'

μεσᾱβόω, for 'yoke, put to' read 'fasten to the yoke'

μεσάγκῠλον, add '**II** the thong of such a javelin, Philostr.Gym. 31.'

μεσάζω, line 1, after 'D.S. 1.32' insert 'of a person, Aq. 1Ki. 17.4' add '**III** be in control, PMerton 46.7, 12 (vi AD).'

*μεσαιπόλος, ον, perh. going in the middle, Ἀδρία βαθύπλου .. μεσαιπόλε πόντου Mesom. 6.2 H.

μέσακλον, for 'weaver's beam' read 'heddle-rod' and after '1Ki. 17.7' insert '(vv.ll. μέσακνον, μεσάντιον)'

*μεσᾱλουργής, ές, middling purple, χιτωνίσκος IG 2².1524.189 (Brauron, iv BC).

*μεσᾱρίστερος, ον, of a horse, prob. the one in a team of four next to the left-hand side of the pole, SEG 34.1437 (Syrian, Apamea, v/vi AD).

*μεσάρκειος, v. °μεσερκ-.

*μεσαυλικός, ή, όν, of °μεσαύλιον (B), μεσαυλικὰ κρούματα Aristid. Quint. 1.11.

†μεσαύλιον (A), τό, inner courtyard, Vit.Aesop. in POxy. 2083.27.

*μεσαύλιον (B), τό, piece of aulos music played in the intervals of a choral ode, Eust. 862.19.

μέσαυλος I, add 'Ar.fr. 387 K.-A. **b** μέσαυλον καὶ νῦν κατοικία ἀγροτική, τουτέστιν ἔπαυλις Eust. 1664.26.' **II**, for 'Att. .. in full' read 'adj., of the door between the αὐλή and the inner part of the house' add '**III** μέσαυλον, τό, colonnaded court of a church, synagogue, etc., Rev.Phil. 32.36 (Side, v/vi AD).'

μεσεγγυάω, after 'party' insert 'PAnt. 35 ii 14 (iv AD)'

μεσέγγυος II, add 'in trust, μ. τὴν μείρακα καταθέσθαι Ar.fr. 746 K.-A.'

*μεσεγγύωσις, ἡ, depositing of security, prob. in Charis. 33 K.

*μεσεμβρίη, v. ‡μεσημβρία.

μεσέρκειος, add 'written μεσσάρκειος Tit.Cam. 126 (vi BC), μεσάρ- ib. 127'

μέση I, add '**2** perh. the name of a Delphic Muse, Νήτας. Μέσσας. Ὑπάτας (from the strings of the lyre) SEG 30.382 (Argos, c.300 BC), cf. Plu. 2.744c.' add '**IV** (sc. ἡμέρα) midday, SEG 32.1149 (Magnesia on the Maeander, AD 209), the south, Call.Del. 280.'

μεσηγύ, line 2, delete 'only in' and insert 'A.R. 4.602' **I 2**, add 'μεσσηγὺς τοῦ τε ὤμου καὶ τοῦ τραχήλου J.AJ 19.1.14' **II**, add 'A.R. 3.723, 930'

*μεσηλικία, ἡ, middle age, SB 6133.9, MAMA 7.263.

μεσημβρία, line 2, after 'Arr.Ind. 3.8, al.' insert 'also μεσεμβρίη SEG 37.576 (Maced., iv BC)' μεσημβρίζω, add '**2** of stars, *shine at midday*, Nonn.*D.* 7.297.'

μεσημέριος, add 'cf. mod. Gk. μεσημέρι'

μεσήρης, after '*midmost*' insert 'μεσέρες perh. on vase in *AION (A)* 6.281ff.'

μεσιτεία, after '*mortgaging*' insert '(used in connection with catoecic land)' and add '*BGU* 2473.11 (c.AD 100)'

μεσιτεύω **I 3**, for this section read '*pledge, mortgage* in connexion with catoecic land, *CPR* 1.1.19 (ii AD)'

μεσίτης **I 1**, add '**b** official entrusted with collection and disbursement of taxes, *CPR* 6.9.1, 2 (v/vi AD), *REG* 70.120 (Palestine, vi/vii AD).'

μέσκος, add '(in Hsch.), unless mistake for πέσκος'

*μεσμίρω, *become runny, flow*, Anon.Alch. 323.24.

μεσόγαιος **I**, add 'μεσόγειος, ον, *Peripl.M.Rubr.* 2' **II 1**, add 'μεσόγειος, ή, spec., *the interior region of Attica*, Arist.*Ath.* 21.4'

*μεσοκώμιον, τό, *centre of a village*, Mitchell *N.Galatia* 181 (AD 145).

μεσονύκτιον, at end, μεσανύκτιον, add '*SEG* 30.622 (Maced., i AD); cf. mod. Gk. μεσάνυχτα, neut. pl., μεσόνυχτι, τό (poet.)'

*μεσοπλᾰτείτης, ου, ὁ, *resident in "the middle street"*, cj. in *IMylasa* 403; cf. °ξυστοπλατείτης.

μεσοπόρος, read 'μεσόπορος' and for '*going .. middle*' read '*of the mid-passage*: μ. πελάγεσσιν *in the deep-sea ways*'

*Μεσοποτᾰμηνός, όν, *of Mesopotamia*, *POxy.* 3053.16 (AD 252).

*μεσόπυργος, ή, *a construction between towers*, *AArch.Syr.* 15.75 (Syria, vi AD), rest. at *SEG* 25.733 (Moesia, i BC).

*Μεσορή, Μεσορί, name of a month, *AP* 9.383.12, *PSI* 635.15 (iii BC), *SEG* 24.1237 (Egypt), 33.1445 (Cyrenaica), al.

μέσος **I 3 b**, add 'applied to a go-between, ὁ διδοὺς καὶ ὁ λαμβάνων καὶ ὁ μέσος γενόμενος Just.*Nov.* 123.2.1, al.' **III 1 a**, line 8, after 'id.*Hec.* 1150' insert 'ἐν μήεσσι Κεφαλές τε καὶ ἄστεος *CEG* 304.1 (Attica, vi BC)' **c**, line 2, after '(D.) 18.294' insert 'ἄπελθ' ἐκ τοῦ μ. Men.*Dysc.* 81 S.' **e**, add 'also ἀνὰ μ. = μεταξύ, *SEG* 9.8.64 (Cyrene, i AD)' **3**, add 'τὸ μ. *moderation*, E.*Hyps.fr.* 11 iii 33 B.' **IV**, add 'ἡ μέση τῶν ποταμῶν (sc. χώρα) *Mesopotamia*, Hdn.Gr. 1.331, Philostr.*VA* 1.20'

μεσοστροφώνιαι, add '(perh. read μεσσοτροφώνιαι, see *REA* 62.303/4)'

μεσόστῡλον, before 'Sch.Od.' insert '*MDAI(A)* 31.431 (ii BC)'; add 'also app. a structure erected in such a space, *Cod.Just.* 1.4.26.8'

μεσόσφαιρος, add '*Plin.HN* 12.44'

μεσότακτος, delete the entry.

*μεσοτέλεστος, ον, *half-finished*, Aetolian acc. to gloss. in *POxy.* 1802.51 (ii/iii AD).

*μεσοτοίχιον, τό, *party-wall*, *PDura* 19.13 (i AD).

μεσουρᾰνέω **II**, add 'Ptol.*Tetr.* 33'

μεσουράνησις, delete 'Plot. 3.1.5 (pl.)'

μεσόφθαλμος, after '*eyes*' insert 'Ptol.*Tetr.* 143'

*μεσόφρυς, υος, ἡ, = μεσόφρυον, *PSI* 1140.12 (ii AD); in pl., ib. 907.21 (i AD), *PSoterichos* 25.9 (AD 109), *POxford* 11.7 (ii AD).

*μεσοχορέω, *serve as* μεσόχορος ι, *IHistriae* 167, 207 (ii AD).

μεσόχορος **I**, add '*IHistriae* 100 (iii AD), v. °μουσόχορος'

μεσόω **1**, of time, add 'μεσο(ῦντος) Νο(εμβρίου) *SEG* 30.1648 (ii/iii AD), *Salamine* 27 (Cyprus, ii/iii AD)'

μέσπιλον, for 'ἴ Archil. .. Eub. 74.4' read 'ἴ *APl.* 255, ῑ Eub. 74.4 K.-A.'

+μέσσᾰβον, v. ‡μέσαβον.

μέσσατος, line 3, after 'D.P. 204' insert 'ἐν μεσάτοισι κλύδωνος ἀνωίστου τε κυδοιμοῦ orac.ap.Porph.*Plot.* 22'; add 'Myc. *me-sa-ta* (?nom. pl. fem.), *of middle size* or *quality*'

*μεσσίκιος, v. °μισσίκιος.

*μεσσοδόμα, v. μεσόδμη II.

μέστα, line 2, after '(Cyrene)' insert 'v.l. in Call.*Cer.* 92, 111'

μέσφα **2**, line 2, for '(Call.)*Hec.* 1.1.4' read '*fr.* 260.4 Pf.' add 'see also °μές, μεσπόδι, μέστα, μέστε, μέττα'

μετά, line 2, after 'πεδά (q.v.)' insert 'which is a different word; μέ *IG* 7.2712.95 (Acraephia, i AD) (cf. μεταυτα for μετὰ ταῦτα)'; line 3, after 'acc.' insert 'Myc. *me-ta* (w. dat.)' **C II 3**, lines 6/7, for 'A.*Th.* 1080' read 'A.*Th.* 1074' **IV**, add 'ἔχων μετὰ χεῖρα τὴν Ἀνθίαν *holding A. by the hand*, X.Eph. 1.12.1'

μεταβαίνω **II**, after 'μεταβῆσαι' insert 'and perh. in fut. μεταβήσειν' and after 'Pi.*O.* 1.42' insert 'μεταβάσοντας ἐλθεῖν Ποίαντος υἱόν id.*P.* 1.52 (cj.)'

μεταβάλλω **A III 1**, at end delete 'c. gen. rei' and after 'καιναὶ καινῶν' insert '(or καίν' ἐκ καινῶν)'

*μεταβαστάζω, expl. of Lat. *convecto* in Virgil gloss. (cf. Verg.*A.* 7.749), *PNess.* 1.865 (vi AD).

μεταβλέπω **II**, for '*look after*' read '*look towards*'

μεταβλητός, add 'Procl.*Inst.* 76 D., al.'

μεταβολή **II**, add '**8** *translation*, Modest.*Dig.* 27.1.1.1.'

+μεταβουλία, ή, *change of heart*, Simon. 38.23 P.

μεταγγίζω, line 2, after 'Gal. 11.215' insert '*SEG* 33.1221a (Bactria, ii BC)'

Μεταγειτνιών, add 'Μεταγειτονιῶ[νος] *Inscr.Délos* 338Aa42 (iii BC)'

μεταγενής **2**, after 'D.H.*Th.* 9' insert 'Modest.*Dig.* 1.4.4'

+μεταγίγνομαι, *fall as a share* to, ᾧ δ' αὐτε γάμου μετὰ μοῖρα γένηται Hes.*Th.* 607. **2** *migrate*, Lxx 2*Ma.* 2.1.

μετάγω **I 1**, add 'pass., *to be diverted*, of funds, *IEphes.* 17.52 (i AD)'

μεταδετέον, for '*one must untie*' read '*one must change the tethering* (from one place to another)'

μεταδιαταγή, after '*POxy.* 899.40' insert 'al.'

μεταδιατίθεμαι, add '*PMich.inv.* 4719 (*BASP* 22.327, i AD)'

μεταδίδωμι **2**, add 'πάντα τῇ ψυχῇ καλά *GVI* 1113a.5 (Apamea, ii AD), 1978.17'

μεταδιεράω, add '**2** as t.t. in wrestling (precise sense unclear), *POxy.* 466.11 (ii AD).'

*μετάδοτος, η, ον, *shared*, prob. in *Stud.Pal.* 22.184.32 (ii AD).

μετάδουπος, for '*falling at haphazard, indifferent*' read '*thundering* (?)*changeably*, i.e. *of uncertain portent*'

Μετάδως, for 'αἰδώς' read 'δώς'

μέταζε, for '= μεταξύ' read 'τὰ μ. *afterwards*' and delete 'but .. Δωριεῖς'

μεταθύω, for '*appease by sacrifice*' read '*repeat an offering* (only med. attested)'

μεταιβολία, delete the entry.

μεταίρω **II**, for '*depart*' read '*move (elsewhere)*'

μεταίτης, for 'Ph. 2.516' read 'Ph. 2.526'

μεταίχμιος, line 8, after 'Arist.*PA* 676ᵃ2' insert '*Iamb.adesp.* 2 W.'

μετακατασκευάζω, add '*SEG* 33.1040.4 (Aeolian Cyme, ii BC), *TAM* 5(1).242 (AD 209/210)'

*μετακηδεύω, *move to another tomb*, *TAM* 2(3).1166.11 (Lycia) (pass.).

μετακίαθω **I**, add '**2** *follow up, continue the example of*, πατρὸς λώβην Nic.*Th.* 132.'

μετακῑνέω **1**, after 'Hdt. 9.74' insert '*IG* 2².13200 (Attica, AD 160)' and transfer '*IG* 5(1).1390' to section 2. **2**, add '*PVindob.* G25945.12 (*Tyche* 2.48, iii AD)'

μετακίνησις **1**, add '*change of formation* of troops, τὰς μεταβολάς τε τῶν τάξεων καὶ τὰς μ. Arr.*Tact.* 9.4'

μετακινητέος, add '**2** *to be changed*, ἔθος Gal. 6.410.12.'

+μετακινητός, ή, όν, *able to be changed*, ὁμολογίᾳ Th. 5.21.2; νόμοι Solon ap.Plu. 2.152a.

*μετάκλωσμα, ατος, τό, *that which is interwoven*, fig., μοιρῶν ἀνάγκης τε μετάκλωσμα Cyran. 5 (ἐπίκλωσμα K.).

+μετακομιδή, ή, *moving from one place to another, transportation* of grain *SEG* 39.1180.74 (Ephesus, i AD); *moving from one bed to another*, pl., Gal. 18(2).503.14.

μετακομίζω, add '**2** *transpose* into a different dialect, D.H.*Dem.* 41.'

μετακύνιον, add 'Simon.Ath.*Eq.* 5'

μεταλαμβάνω **V**, add '**3** *write down from dictation*, Porph.*Plot.* 8.'

*μετάλημψις, εως, ή, = μετάληψις, *POxy.* 1200.36 (iii AD).

*μετάληξις, subst. corresponding to μεταλήγω, only as pers. n., *SEG* 23.87 (Attica, iv BC).

*μεταληπτός, gloss on πεδάγρετος, Hsch.

μετάληψις **II 4**, line 2, after 'Trypho *Trop.* 5, etc.' insert 'D.H.*Th.* 31'

μεταλλεύω **I 4**, add '*IKyzikos* 560' add '**6** fig., *exploit as if a mine*, prob. in Phld.*Oec.* p. 26 J.; cf. mod. Gk. ἐκμεταλλεύομαι *exploit.*'

μεταλλίζω, add '*Basilica* 35.1.9 Sch.'

μέταλλον **II**, add '**b** Lat. *metallum*, *marble paving-slab*, *JS* 1988.30 (Apamea, Chr.).'

μεταλλωρύχος, ὁ, *sapper*, *FGrH* no. 533 2.20 J. (pl., ii AD).

*μεταμειπτός, όν, *exchangeable*, rest. in Hes.*fr.* 43(a).43 M.-W.

*μεταμελλοδῡνά, ά, *remorse*, Cerc. 2(b).13 D.³.

μετάμελος **II**, add 'δύνασθαι τὸν μετάμελον .. ἀνακαλεῖσθαι (sc. τὴν δωρεάν) Just.*Nov.* 87 pr.'

*μεταμπελεύω, ?*transplant vines*, Heph.Astr. 3.5.69.

*μεταναγραφή, ή, *re-registration, transfer in the books*, prob. in *IEphes.* 14.34 (i AD).

μεταναστεύω **2**, for 'ib. 61(62).6' read 'ib. 61(62).7'

μετανάστης, for '(μεταναστῆναι, cf. ὑπερανάστης)' read '(sts. taken as μετανά-στης or for *μεταναστάτης, but prob. to be analysed as μετα-νάστης; cf. μεταναίω, °νάστης)'

+μετανέομαι, *visit, approach*, εὐνήν Musae. 205, Nonn.*D.* 14.89, ἄλλον ἐνηείῃ μετανεύμενος, ἄλλον ἀπειλῇ ib. 29.7, 36.161.

μετανίσομαι, after lemma insert 'or μετανίσσομαι'; line 1, for '*pass over*' read '*change position, cross over*'; line 2, after 'Od. 9.58' insert 'of the polar axis, οὐδ' ὀλίγον μετανίσεται Arat. 21'

μεταξάβλαττα, read 'μεταξαβλάττη'

μεταξάριος, add '*SEG* 32.1439 (Syria, AD 516), *ITyr* 22, 98'

*μετάξενια, τά, app. some form of offerings, *CID* 7 A 26 (v BC).

*μέταξος, ή, = μέταξον, Just.*Nov.* app. 5.

μεταξύ **I 2 b**, add 'Plu. 2.240a,b; μ. ἐβίω Ruf.*Ren.Ves.* 2.4' **II 1**, after '*POxy.* 914.8 (v AD)' insert '*Cod.Just.* 1.3.45.13, 1.4.33.2'

*μεταπαραβολή, ή, astron., *conjunction*, Simp. *in Cael.* 471.9.

*μεταπερισπάω, prob. *divert*, τοῦ Δημητρίου μεταπερισπασθέντος εἰς δημοσίαν χρείαν PFam.Teb. 24.54 (ii AD).

μεταπιπράσκω, add 'PVatic.Aphrod. 10.13 (vi AD)'

μεταπίπτω **I 1 a**, line 4, after 'A.D.Adv. 188.25' insert 'ὁ λόγος θαμινὰ μ., of changes of construction, D.H.Dem. 39'

*μεταπλανάομαι, *disperse* or *wander off into*, w. acc., ζῷα ἄλογα μεταπλανώμεναι λοιπὸν διατελέσετε, of metempsychosis, Herm. ap.Stob. 1.49.44.

μεταποιέω **I**, line 4, after 'Sol. 20.3' insert 'cf. Sch.A Il. 20.273, al.'

*μεταποικίλλομαι, *embroider*, in quot. fig., Agath. pr. 17.

*μεταπολέω, *go along* with, τῶν ἑτέρων τοῖς ἑτέροις ἐπακολουθούντων καὶ μεταπολουμένων Chrysipp.Stoic. 2.293 (unless this is fr. °μεταπόλλυμι *perish in succession*).

*μεταπόλλυμι, v. °μεταπολέω.

*μεταπολογίζομαι, *transfer to another account*, MAMA 8.413(b).4.

μεταπορεύομαι **I 2**, add '**b** *claim at law*, SEG 9.8.121 (Cyrene, i BC, senatus consultum).'

μετάπρᾱσις, add 'ἔρια εἰς μετάπρασιν ὠνεῖσθαι Heph.Astr. 3.5.39, 43'

μεταπρέπω, add 'pass. form in same sense, μεταπρεφθεὶς ἑτάροισιν GVI 1519.5 (Moesia, i BC); cf. ἐμπρέπω'

μεταρίθμιος, add '**2** *of account among*, ἰχθύσι δ' οὔτε δίκη μ. οὔτε τις αἰδώς Opp.H. 2.43.'

μεταρρυθμίζω **I 2**, add 'τοὺς ἀδικοῦντας μεταρυθμίζειν ἐπὶ βελτίονα βιοτήν, Βυζάντιον 6.366 (Sardis, vi AD)'

μέταρσις, add '**2** *arrogance, conceit*, Aq. 1Ki. 2.3.'

*μεταρυθμίζω, v. °μεταρρυθμίζω.

*μετασαλεύω, *remove, alter*, οὐδ[ενὶ ἐξέσται] μετασαλεῦσαί τι (exhortation against disturbing a grave) SEG 34.1144 (Ephesus).

*μετασκούτλωσις, εως, ἡ, prob. *replacement panelling* or *paving* in a theatre, JRA suppl.2 pp. 20, 28 (Aphrodisias, ii/iii AD).

μετάστασις **II 1 a**, add '*change of position*, τανυπτερύγου μυίας μ. Simon. 16.4 P.' **b**, delete 'τοῦ βίου μεταστάσεις id.fr. 554' and 'Simon. 32' **2**, after 'γνώμης' insert 'βίου' and after 'Andr. 1003' insert 'fr. 554'

✝μεταστείχω, *go in search of, go after*, E.Hec. 509, Supp. 90, A.R. 3.451, Ἰνδὸν Ἄρηα μετέστιχε θυιὰς Ἐννώ Nonn.D. 29.279.

μετατάσσω, line 2, for '*adjourn*' read '*rearrange, change* (the date of)'

μετατίθημι **II 1**, add '**b** *transfer* to another purpose, τὸ δὲ ἀρχαῖον τὸ ἐπιδιδόμενον .. εἰς ταῦτα μὴ ἐξέστω μεταθεῖναι εἰς ἄλλο μηθέν IG 12(7).515 (ii BC). **c** *transfer* to another *the right to* or *possession of*, Cod.Just. 1.3.41.11, Just.Nov. 22.20.2.'

μετάτροπος **2**, for this section read '*having a change of mind* or *attitude*, δαίμων γὰρ ὅδ' αὖ μετάτροπος ἐπ' ἐμοί A.Pers. 943, A.R. 3.818' add '**3** μετάτροπα ἔργα *reversal, restitution*, Hes.Th. 89.'

*μετατροπόω, *convert*, τούτους (sc. τοὺς ἐχθροὺς) εἰς φιλίαν Gnomol.Vat. in WS 11.225.

μεταυγάζω, for '*look keenly after .. τινα*' read '*discern from a distance, spot*'

μεταύριον, add 'διὰ τῆς μεθαύριον POxy. 1844.4, 5 (vi/vii AD); cf. mod. Gk. μεθαύριο'

μεταυτίκα, add 'Theoc. 25.222 (v.l. παρ-)'

μεταφέρω, add '**5** *transport*, PKöln 274.4, PPetr. 3 p. 46 i 16, PCair.Zen. 520.10, 620.11 (iii BC).'

*μεταφορεῖος, ον, *of* or *concerning transport*, Kafizin 227.

*μεταφορεύς, έως, ὁ, *courier, messenger*, Ἁρποκρατίωνι μεταφορ(ε)ῖ PPetaus 34.11 (AD 184).

✝μεταφορέω, *convey* from one place to another, *shift*, Hdt. 1.64.2, 2.125.4.

μεταφυτεύω, add 'Pythag.ap.Iamb.Protr. 21 λη' p. 125.12 P.'

μεταχάσκω, for 'Apostol. 7.20 (Journ.Philol. 4.320)' read 'Com.adesp. 41 D.'

μεταχειρίζω **5 a**, line 2, delete 'τὸν .. 1.20'

μεταχέω, add 'Myc. pf. part. pass. me-ta-ke-ku-me-na (fem.), app. *taken to pieces*'

μεταχθόνιος, delete the entry (read μεταχρόνιος).

μεταχρόνιος **II**, lines 3/4, delete 'μεταχθόνιος .. μεταχρόνιος'

*μετεγγυητής, οῦ, ὁ, *joint-security*, SB 4884.

*μετεικάς, άδος, ἡ, (sc. ἡμέρη) *twenty-first day of the month*, Hes.Op. 820 (s.v.l.).

μέτειμι (εἰμί *sum*) **II 1**, lines 5/6, delete 'οὐδὲν μᾶλλον .. cf.'

μέτειμι (εἶμι *ibo*), for 'Att. fut. of μετέρχομαι' read 'used as fut. of μετέρχομαι'

*μετεκβάλλω, *cause to change* from a practice, μετεκβάλωσι τοῦ νυνὶ τρόπου ?Cratin. in PCGF 76 fr. 1 col. i.3.

*μετέκθεσις, εως, ἡ, *supplementary list*, PMich.inv. 4607 (TAPhA 83.78.1, iv AD, -εχ- pap.), PMasp. 138 iiʳ.47 (-εχ- pap.).

*μετεκποιέω, *finish making*, μετεξεποίησεν τρίχωμα Pap.Lugd.Bat. xxv 16.5.

μετεξέτεροι, for '= ἔνιοι, *some among many, certain*' read '*some among others*' and add 'Nic.fr. 76, Th. 414 (interp.)'

μετέπειτα, add 'Heph.Astr. 1.1.9'

μετεπιγράφω **I**, add 'SEG 17.729 (Lycia, Rom.imp.)'

μετέρχομαι, line 1, after 'Theoc. 29.25' insert 'πεδελθέτω Alc. 129.13 L.-P.' **IV 1**, add 'esp. *go to fetch* a bride, AP 7.367 (Antip.Thess.), διὰ κηρύκων Plu. 2.297c' **3**, line 5, delete '*narrate* them'

μετεωρισμός **I**, add 'μετεωρισμὸν ὀφθαλμῶν (i.e. a haughty look) μὴ δῷς μοι Lxx Si. 23.4, 26.9'

μετεωρολέσχης, add 'perh. w. ref. to sophists, Ar.fr. 401 K.-A.'

μετέωρος, line 16, for '*more rapid*' read '*shallower*' **II 3**, add 'μέλη .. μετέωρα ἐντόνοις ὅμοια Zeno Stoic. 1.58' **III 3**, add 'ἐν μετεώρῳ *in suspense*, Just.Nov. 30.8 pr.' add 'cf. Dor. πεδάορος, πεδήορος'

μετῆλαι, for 'Poll. 1.243' read 'Poll. 1.143'

μέτηλυς **I**, delete 'PFlor. .. iii AD)'

μετοικεσία, for 'μετοικία ι' read '*removal to a foreign country*' and for 'also .. leal'' read '**2** *place of residence abroad*, πλεόνων μ. *the abode of the majority*, i.e. the dead'

μετοικέω **II**, add 'transf., of the status of citizens under an oligarchy, Isoc. 4.105'

μετοίκιον **I**, after 'X.Vect. 2.1' insert 'Hesperia 5.401.126 (Athens, iv BC)'; after 'Luc.Deor.Conc. 3' insert 'πρὸς τὸ μ. τινα ἀπάγειν in connection with, i.e. for not paying, *this tax*, Plu. 2.842b, cf. id.Flam. 12'

μετοικοδομέω, after '*build elsewhere*' insert '[με]τοικοδομήσειν Men. Dysc. 446 S.'

μέτοικος **1**, after 'esp. at Athens' insert 'SEG 34.47 (verse, Attica, 506 BC; form μετάοικος)' **2**, delete the section.

*μέτοπις, ιδος, ἡ, *vengeance*, Διὸς Hom.Epigr. 8.4 (nisi leg. μέτ' ὄπις).

*μέτοπον, τό, *metope* (archit.), IG 1³.474 i 30; cf. μετόπη, μεθόπιον.

μετόρχιον, add 'prob. rest. in PRyl. 583.20, 71 (ii BC)'

μετουσία **I**, add 'w. πρός and acc., Just.Const. Δέδωκεν 20a, Cod.Just. 1.3.43 pr.' add 'PHamb. 128.59 (iii BC)'

μετούσιος, add '**II** *participating* in, Alcin.Intr. p. 164.9 H.'

μετοχετεύω, line 1, for '*convey in a channel*' read '*channel*' and after '*divert*' insert 'D.Chr. 18(35).20'; line 4, after 'p. 165 A.' insert 'fig., D.H.Vett.Cens. 1'

μετοχή **III**, delete the section.

*μετόχλησις, εως, ἡ, *removal by means of a lever*, θύρας μετοχλήσει Heph.Astr. 3.46.

μετρέω **III 4**, add 'SEG 30.1121 (iii BC)' add '**5** *adjust, accommodate to*, ὁ νόμος πρὸς .. τὸ παρασυμβαῖνον μετρηθείς Just.Nov. 7.2 pr.

μετρηδόν, after 'adv.' insert '*by measure*, Nic.Al. 203'; after 'Nonn.D. 7.115' insert '22.271, al.'

*μετρηματιαῖος, ὁ, *hired labourer*, SB 9406.16, PFlor. 322.20 (both iii AD).

μετρητής **II**, add 'abbrev. με. SEG 31.983 (ii/i BC)'

μετριάζω **I 1**, add 'of (the use of) figures of speech, D.H.Dem. 4'

μέτριος **A II**, add 'μετρίων ἡμερῶν Thphr.Sign. 24' **B I 1**, line 6, after 'Pl.Euthd. 305d' insert 'w. adj., μετρίως .. ἐπίπονον Pl.R. 329d, D. 6.19, Theoc. 30.3'

μετριότης **II**, add 'freq. in a self-depreciatory sense, PSI 449.9, PBeatty Panop. 1.69 (AD 298)'

μετρονόμοι, add 'SEG 24.157 (Attica, iii BC)'

*μετροποιός, ὁ, *maker of measures*, PTeb. 277.2 (iii AD, pl.).

μέττα, add 'μετ(τ) as prep. or w. ἐς = *up to* (a place), BCH 109.163B.8 (Crete)'

μετωνυμία, line 2, before 'Cic.' insert 'D.H.Dem. 5 (pl.)'

μετωπηδόν, line 2, after 'Hdt. 7.100' insert 'Sosyl. 1 iii 13 J.'

μέτωπον **II 1**, lines 6/7, delete 'dub. sens. .. 30' (v. °μέτοπον); penultimate line, after '(X.Cyr.) 2.4.3' insert 'ἐπὶ μετώπου ἐλαύνειν id.Eq.Mag. 3.6'

μέχρι **III 1**, line 4, delete 'Call.Sos. 5.4 and 5.5' **2**, line 3, for 'μέχρις .. Men. 633' read 'μέχρι .. Men.fr. 525 K.-Th.'; line 4, after 'Ench. 11' insert 'μέχρις κε μένῃ Call.fr. 388.9 Pf.'

μή **A 8**, add 'μὴ προφάσεις AP 5.193(192) (Diosc.), cf. 53(52) (id.; πρόφασις cod.); also w. nom., v. βαιῶν' **C Π 1**, lines 1/2, for 'with .. *apprehension*' read '*where fear is implied*'; line 3, for 'περισκοπῶ .. El. 898' read 'εἰσόμεσθα μή τι .. καλύπτει S.Ant. 1253' **D 2**, after 'μή is sts. repeated' insert 'cf. οὐ C'

μηδάμα, line 2, delete 'and of Manner, *not at all*' and insert 'μηδάμα μηδ' ἕνα *never anyone*, Alc. 129.16 L.-P., and so'

μηδέ, line 9, for 'οὐδέ' read 'οὐ' and for 'A II 3' read 'A II 8'

*Μήδεια, ἡ, epith. of Artemis (prob. ethnic), SEG 32.1612 (iii BC).

μηδείς, line 2, after '(Mytil.)' insert 'also fem. μεδενία SEG 26.461 (Sparta, v BC); acc. μαδέμιναν (for μηδένα) SEG 36.548 (Metropolis, Thessaly, iii BC)'

μηδεπώποτε, add 'Plu.Ages. 19.6, D.L. 6.54'

μηδέτερος, line 3, after 'id. 7.5.9' insert '[Isoc.] 1.42'

Μηδικός **I**, add '**2** as an imperial title, ISelge 13 (after AD 175).'

*Μηδίσκιον, τό, *silken garment*, PDura 30.20 (iii AD); cf. Μηδικός I.

*Μήδισσα (sc. γυνή), ἡ, = Μηδίς, IG 2².9354 (iii/ii BC).

μήδομαι, at end after 'only in lyr. exc. A.Pr.l.c.' insert 'Ch. 991'
μῆδος (A), add 'in Myc. compd. n., as e-ke-me-de = Ἐχεμήδης, pe-ri-me-de = Περιμήδης'
Μῆδος, add 'Cypr. ma-to-i Μᾶδοι ICS 217.1'
μηθείς, line 4, delete 'but rarely .. NT' and add 'Dura⁵ (1)12.8'
μηθέτερος, add 'Lxx Pr. 24.21 (B)'
*Μηθυμναῖος· ὁ Διόνυσος Hsch., app. from Methymna, but see ‡Μεθυμναῖος.
*μηιονιστί, adv., in the Maeonian language, Hippon. 3a W.
*μὴ καί, like μὴ ὅπως I 1, not just not, folld. by ἀλλά, οὐδέ τις ἔτλη, μὴ καὶ λευκανίηνδε φορεύμενος, ἀλλ' ἀποτηλοῦ ἑστηώς A.R. 2.192; after a neg. sentence, much less, id. 3.589.
μηκόθεν, add 'διαγνῶναι ἀπὸ μ. Sch.E.Ph. 1118'
μήκων IV, add '2 socket of the eye, Hsch. s.v. κρατηρίσκοι.'
*μηκωνίας, Dor. μᾱκ-, α, ὁ, = μηκώνειος, μακωνιᾶν ἄρτων rest. in Alcm. 19.2 P.
μηκωνίς II, delete the section.
μηλἄφάω, for 'Sophr. in Cod. .. 82' read 'Sophr. 146b O.'
μηλέα, line 5, after 'Thphr.HP 4.13.2' insert 'μ. ὀξεῖα crab-apple, Pyrus acerba, ib.'
μήλινος II, for 'μύρον μ.' read 'μήλινον, τό, quince-perfume' 2, add 'Ezek.Exag. 262 S.'
Μηλίς, (s.v. Μηλιεύς), after '(Hdt.) 8.31' insert 'cf. Μαλίς· Ἀθηνᾶ Hsch., Hippon. 40 W.'
*μηλῖτις, ιδος, ἡ, (μῆλον B) a precious stone (named from its colour), Plin.HN 37.191.
*μηλόκερως, ων, having sheep's horns, Ps.-Callisth. 27.20, 33.12.
μηλολόνθη, add 'see also μηλάνθη I'
μηλονόμος, after '(lyr.)' insert 'PMasp. 2 iii 4, al. (vi AD)'
μηλοπάρειος, delete the entry.
*μηλοπεπόνιον, τό, dim. of μηλοπέπων, melon, OAshm.Shelton 217 (written μελοπεπόνιν).
μηλοσκόπος, add 'II watcher over sheep, of Pan, GDRK 17ʳ.11.'
μηλοσόη, read 'μηλοσόα'
*μηλόται, v. μηλάτης.
μηλοφόρος, line 4, after 'IG 14.268 (v BC)' insert 'ib. 271; also θεᾷ Μαλοφόρῳ IGBulg. 1².370 bis'
μήλοψ, read 'μῆλοψ'
μηλόω II, for this section read 'dye wool, Hsch. (s.v. μηλῶσαι), Poll. 7.169; med., Eust. 1394.32'
μήλωθρον II, add 'μήλωθρα· βάμματα. οἱ δὲ τὸ τῶν δερμάτων βάμμα. ἄλλοι τὸ παρύφασμα τῆς πορφύρας. οἱ δὲ καλλωπίσματα Hsch.'
μηλώσιος, add 'or perh. dressed in a sheepskin (μηλωτή)'
*μήμη, prob. great-grandmother, Didyma 345.12 (i BC/i AD), SEG 30.1286 (Didyma, i AD).
*μημόριον, v. °μεμόριον.
μήν, read 'μήν (A)' II 2, line 6, after 'S.Ant. 626 (anap.)' delete 'etc.'; line 10, for 'καὶ μὴν καί' read 'καὶ μὴν .. καί'
μήν, ὁ, read 'μήν (B), ὁ' II, delete the section.
*Μήν (C), an Anatolian divinity, IG 2².1365, 1366, etc.; nom. sg. Μείς, SEG 4.647.2, 648.3 (Lydia).
μηναγύρτης, for 'Rhea' read 'perh. of °Μήν C'
μηνιαῖος I 1, add 'τὰ μ., also monthly payment, salary, Edict.Diocl. 7.64, al.'
*μηνιακός, ή, όν, recurring monthly, ἡμέραι SB 5959.10 (iii AD).
μηνιάρχης, after 'prefect' insert 'OBodl. II 1986 (ii/iii AD)'
μηνιεῖος, for 'λοιπογραφομένων' read 'λοιπογραφουμένων'; add 'cf. Myc. me-ni-jo (*μήνιον), monthly ration'
μῆνις, line 1, for 'gen. μήνιος Pl.R. 390e' read 'Ep. gen. μήνιος Od. 3.135 (used by Pl.R. 390e in ref. to Iliad)'
*μήνισις, εως, ἡ, anger, Psalm.Salom. 2.25(23).
*μηνίσκος, τό, dim. of μηνίσκος, as an ornament, PMich.II 121ʳ II ii 2.8 (i AD), PHamb. 10.45 (ii AD).
*μήνισμα, ατος, τό, = μήνιμα, Πολέμων 1.213 (Iolcus, iii BC).
μηνιτής, after lemma insert '(or μηνίτης)'
*μηνσώριον, τό, Lat. mensorium, basket, Stud.Pal. 20.151.3, 14 (vi AD, μηνσωρρ-).
*μήνυον· εἶδος ἄνθους Theognost.Can. 130.
μηνυτής, ὁ, informer, SEG 29.980 (Italy, late Rom.imp.).
μηνυτής II, after 'Cratin. 428' add 'PKöln 266.20 (iii BC)'
μηνύω I 4, add 'Arist.de An. 403ᵃ19' II, line 1, after 'Athens' insert 'and elsewhere'; line 5, after 'D. 24.11' insert 'w. εἰς, lay information before, CodJust. 1.4.26.4'
*μηνωπός, όν, perh. moonlit, cj. in POxy. 3724 fr. 1 vii 9 (list of epigrams, i AD).
μήποτε I 2, add 'Cypr. ICS 264'
Μηριόνης II, delete 'pudenda muliebria'; after 'AP' insert '12.97 (Antip.)'
μηρός, after lemma insert 'also neut. pl. μῆρα q.v.' add '4 transf., flank, ἐν μηροῖς ὅρους Lxx Jd. 19.1. 5 archit., the segment between the grooves of a triglyph, Vitr. 4.3.5.'
μηρυκάζω, add 'med., τῶν μαρυκαζομένων ζῴων Hsch. s.v. ἤνυστρον'
μήρῡμα I, line 1, after 'strand' insert 'of rope, IG 2².1627.70, 150, 1128.335, al.'
μήρυξ, read 'μῆρυξ'
μηρύομαι I 2 a, add 'Carm.Pop. 30(c).2 P.' II, add 'also μηρύω δὲ τὸ κουβαρίζω Sch.Theoc. 1.29; πλατύνειν· μηρύειν Hsch.'
*μητᾶτον, τό, Lat. metatum, pl., quarters, billet, edict of Anastasius I in SEG 9.356.43, 44 (Cyrenaica, vi AD), ib. 40 (μιτᾶτα), Just.Nov. 130.9.
*μητατορικός, ή, όν, (for -ωρικός) concerning measuring, fr. Lat. metatorius, ἐνδοματικὰ μ. Lyd.Mag. 3.70.
μήτε 1, add '(μήτε .. ᾖ in codd. of Ph. 2.137, Porph.Abst. 4.8)'
*μητεύω, commander for military quarters, Just.Nov. app. 4 rubric, 4.1.
μήτηρ I 1, add 'Ἰουλίαν Δόμναν .. μητέρα κάστρων SEG 24.953, Amyzon 71, etc., μήτηρ συναγωγῆς CIJud. 496 (ii/iii AD)' 2, Μήτηρ θεῶν, add 'SEG 32.268 (ii AD), TAM 4(1).90, etc.; Μήτηρ ἀπὸ σπηλέου SEG 34.1293' add 'form μάτηρ, Myc. ma-te, ma-te-re (dat.)'
μητίετα, after '(Tegea)' insert 'gen. μητιέταο Max. 445'
*μητρίδιον, τό, app. dim. of μήτηρ on the model of πατρίδιον, Ar.Lys. 549.
μητρίδιος, delete the entry.
μητροκάσιγνήτη, add '2 mother's sister, Nonn.D. 3.425, 21.182.'
*μητροκάσίγνητος, ὁ, uterine brother, CEG 166 (Sicinus, v BC).
μητροκωμία, add 'PBeatty Panop. 2.228 (AD 298)'
μητροπάτωρ, add 'title of Apollo at Clarus, AC 35.417 n.1'
μητρόπολις, line 3, after 'ἑως' insert '(Μητροπόλιος, REG 101.14-16, 89, Xanthos, iii BC, place-name)' I 2, for 'Ph.ap.Plu. 2.718e' read 'Philol.ap.Plu. 2.718e'; add 'τὴν φιλαργυρίαν εἶπε μητρόπολιν πάντων τῶν κακῶν Diog.ap.D.L. 6.50'
μητροπολίτης II, add 'SEG 30.1713 (Arabia, AD 624)'
μητροπολιτικός I, add 'POxy. 1521.3 (ii AD)'
μητρυιά, line 1, after '(Lesbos)' insert 'μητρυά Pl.Lg. 672b, 930b, Com.adesp. 12.4 D.'
*μητρωϊκός, ή, όν, perh. = μητρικός, τύπος μ. Inscr.Délos 1409Baii 100 (ii BC).
μητρῷος I 1, at end after 'τὰ μ.' insert 'Leg.Gort. 4.44, Inscr.Cret. 4.20 (vii/vi BC)' add '3 fem. subst., maternal aunt, ἁ ματρόα Mitchell N.Galatia 28.' II 2 b, after 'Duris 16 J.' insert 'Plu. 2.763a'
μήτρως, line 2, after 'acc. ωα and ων' insert 'ως SEG 34.1226 (Lydia, AD 158/9)' 2, add 'Leg.Gort. 8.52 πὰρ τοῖς [μ]άτρōσι τράπε(θ)θαι (= τρέφεσθαι); ib. 9.4 τὸ]νς ματρόανς'
μηχανάριος, after 'engineer' insert 'mechanic'
μηχανεύομαι, after '= μηχανάομαι' insert 'Trag.adesp. 573'
μηχανεύς, add '2 engineer, PPetaus 103.38 (ii AD, written μιχ-).'
μηχανή I 1, add 'for grinding corn, μ. σιταλητική PMich.inv. 257.3 (BASP 13.3, iii AD), BGU 405.7 (iv AD)' II 1, at end after 'Pl.Phd. 72d' insert 'Alc. 1.130b' 2, add 'ἐκ πάσης μηχανῆς CodJust. 1.3.45.3a'
*μηχᾰνία, ἡ, trickery, Hsch. s.v. μαγγανεία.
μηχᾰνικός II, add 'w. ref. to architecture, σοφίᾳ τῇ καλουμένῃ μηχανικῇ Procop.Aed. 1.1.2 K.'
*μηχάνιον, τό, dim. of μηχανή, apparatus, equipment, PHarris 112.10 (v AD).
*μηχᾰνοδέτης, ου, ὁ, workman who assembles machines, MAMA 3.752 (Corycus).
⁺μηχᾰνοποιΐα, ἡ, construction, engineering, of engines of war, Ath.Mech. 10.9; of a clock, Inscr.Perg. 8(3).103.
μηχανοποιός 1, add 'b architect, builder, Procop.Aed. 1.1.24.'
μηχανουργός, ὁ, artificer, mechanic or sim., APl. 5.382, POxy. 1970.14, 34 (vi AD); transf., as gloss on ῥᾳδιουργός, Hsch.
μιαίνω 1, delete the section transferring exx. to section 2. 3, at end after '(Iulis' insert 'v BC' and add 'PMich.v 244.17 (i AD)'
*μιαιόω, defile, ἀνθ' ὧν αὐταὶ ἐμιαίωσαν αὑτὰς ἐν φυρμῷ ἀναμείξεως Psalm.Salom. 2.15(13).
μιαιφονία, line 2, after 'murder' insert 'AP 5.215 (Mel.), 9.157'
μιαιφόνος, line 2, after '(Il.) 844, al.' insert 'of human beings, A.Pr. 868, Arist.EN 1177ᵇ10'; line 3, after 'B.Scol.Oxy.fr. 5.1' insert '(= fr. 20A.16 S.-M.), μ. γάμων S.El. 492'; add 'see also μιηφόνος'
*μιακαιεικάς, άδος, ἡ, the twenty-first of the month, BGU 2042.2, 3 (AD 105).
*μῐᾱρολόγος, ον, foul-mouthed, Sch.Luc. 205.7.
μῐᾰρός 4, penultimate line, after 'Pl.R. 562d' insert 'D. 18.296, al.; Aeschin. 1.42, al.' add 'see also μιερός'
μίγα, add 'Orph.A. 791'
μίγδην, after 'Nic.fr. 68.4' insert 'cf. μιγός (perh. read μιγῇ)'
μίγμα 1, add 'transf. of style, D.H.Dem. 5' 2, line 2, for 'μ. σμύρνης' read 'sg., Lxx Si. 38.8; μίγμα σμύρνης' add 'cf. ‡μείχμα'
*μιγός, ή, όν, not sorted out, ἐρέα Edict.Diocl. 25.11, cf. ‡μιγής.

μιεῖν, add '(dub., v. μνίειν and °καταμιεῖ)'

*μἴεστήρ, prob. = *μιαστήρ (= μιάστωρ), gloss in POxy. 1802.61 (ii/iii AD); (cf. χιέζω = χιάζω, πιάζω = πιέζω).

μῐκιχιζόμενος, for 'under age' read 'in third year of public education, i.e. nine years old'; lines 3/4, for 'is expld. .. year' read '= foreg.'

μικκός, line 1, after 'μικρός' insert 'A.fr. 47a i 23, ii 15 R.'

μῐκός, add 'cf. stem of Myc. pers. n. mi-ka-ri-jo'

*μῑκροκέρᾰμον, τό, small pot, PBremen 22.7 (ii AD).

μικρολογία II 2, add 'Ptol.Tetr. 192'

*μῑκρολῡπία, ἡ, annoyance at trifles, pap. in Aegyptus 63.154-5.

*μῑκρόπρος, adv., about, more or less, ζήσας .. τὸν [πάντα βίον ἔτη π]εντήκοντα μικρόπρος SEG 32.1064 (late Rom.imp.); also μικροῦ πρός SEG 31.1431 (AD 582), μικρῷ πρός, ACO 2(1).186.31.

μικρός I 1, add 'in place-names, ἀπὸ Διοπολείτου μικροῦ SEG 30.1720, μηκρᾶς Ἀ[σ]ίας ib. 26.790 (Byzantium, vi AD)' II 1, line 2, after 'Pl.R. 498d' insert 'πρὸς μικρόν GVI 1842.3 (Egypt, i/ii AD)' III 5 a, for 'Antipho .. 4.129' read 'Hdt. 4.129.5; in detail, Antipho 6.18 (s.v.l.)' at end delete 'ἴ only .. fr. 36.17 J.'

μικρόσφαιρον, read '-σφαιρος, ον' and add 'Plin.HN 12.44'

μικρόφθαλμος, after 'Hp.Epid. 6.7.1' insert 'Ptol.Tetr. 143'

μικρόφωνος I, add 'comp., Plu. 2.963a'

μικροψυχία, line 3, delete 'Cic.Att. 9.11.4'

μικύθειος, for 'α, ον' read 'ον'; add 'ἀργυρίς, κύλιξ, ib. 442B142, 172 (ii BC)'

μῖλαξ II, delete the section.

*μῖλαξ (B), = μέλλαξ, Hermipp. 33, Hsch.

Μῑλήσιος, add 'form Μιλάτιος, Myc. mi-ra-ti-ja (fem. pl.)'

Μίλητος, line 1, after 'Theoc. 28.21' insert 'Μίλατος SEG 23.189 (Argos, c.330 BC); also a town in Crete, Μίλητος II. 2.647, ethnic Μιλάτιοι, Inscr.Cret. 1.8.6.36'

μιλιάριον II, add 'b mile, τέσσαρες (sic) μειλιάρια ἡμεῖν ἐπίκεινται JRS 46.46 line 6 (Phrygia, iii AD).'

μέλιον, add 'ἀπὸ Ἁλικαρνασσοῦ μί(λια) β΄ SEG 31.932 (iii/iv AD)'

μίλλος, add 'also in pers. n., e.g. Μίλων, Afric.Euseb. I 201 (vi BC)'

*μιλτόεις, εσσα, εν, form μιλτόϝενς, Myc. mi-to-we-sa (fem. pl.), painted red.

μιλτοκάρηνος, for 'red-headed' read 'red-topped, λόφων ἀπὸ μιλτοκαρήνων'

μιλτοπάρῃος, after 'ον' insert 'Dor. -πάρᾱος Suppl.Hell. 991.36 (poet. word-list, iii BC)'; line 3, for 'Macho ap.Ath. 3.135b' read 'Matro ap.Ath. 4.135b'

μιλτόπρεπτος, for 'A.Fr. 116' read 'A.fr. 47a.24, 116 R.'

μιλτόπρῳρος, add 'Suppl.Hell. 991.33 (poet. word-list, iii BC)'

μίλτος I 2, delete the section transferring exx. to section 1.

μιλτόω, add 'ψιάθους .. μεμιλτομένας (for -ωμενας) POxy. 3505.5 (ii AD)'

μῑμαίκῡλον, after 'fruit of κόμαρος' insert 'Ar.fr. 698 K.-A.'; add 'cf. pers. n. Μυμαικύλη SEG 30.238 (Attica)'

†μίμαρκυς, ἡ, dish composed of offal and blood, acc. to Hsch. usu. of hare, Ar.Ach. 1112, Pherecr. 255 K.-A., Diph. 1 K.-A.

μίμαυλος, for 'flute' read 'aulos'

μιμέομαι II, line 5, delete 'of μῖμοι .. X.Smp. 2.21'

*μίμερά (or -ηρά)· ἡ μιμητικὴ τέχνη, καὶ ἡ μίμησις Hsch.

μιμηλός II, delete this section transferring exx. to section I and adding 'Plu. 2.215a, Nonn.D. 42.217, al.'

μίμημα 1, for 'anything imitated, counterfeit, copy' read 'copy, imitation, whether intended to represent, counterfeit, or for any other purpose' 2, delete the section transferring exx. to section I and adding 'Ph. 2.146, al., καλάθου μίμημα τρόπαιον SEG 36.1260 (late ii AD), Plot. 1.2.2, al. Nonn.D. 1.373, al.'

μιμνήσκω B I 3, line 4, for 'etc.' read 'μέμνημαι .. σου λέγοντος ib. 8; μ. Κριτία συνόντα σε Pl.Chrm. 156a' and for 'relat.' read 'clause' III, line 4, at end insert 'κ(ύρι)ε, μνήσθητι τοῦ δούλου σου SEG 31.1474 (Arabia, vi AD)' add 'Μνᾶσι- Μνησι- in pers. n., Myc. ma-na-si-we-ko'

μιν, add 'Myc. enclitic -mi'

μίνθα, for 'also μίνθος, ἡ, .. 732b' read 'μίνθα· τὸ ἡδύοσμον. καὶ ἀνθρωπεία κόπρος Hsch.' add 'Myc. mi-ta' (see also °μίνθος)

μινθόβαψ, delete the entry.

†μίνθος, ὁ, ἡ, mint, (masc.), Mnesim. 4.63, Plu. 2.732b; (fem.), Thphr.CP 2.16.2; used in comedy for κόπρος acc. to Eust. 1524.12, cf. μίνθα, μινθόω.

*μίνθων, ωνος, ὁ, app. a kind of sexual pervert, Phld.Vit. p. 37 J., Luc.Lex. 12, Hsch. s.v. κικκίδαι.

μίνθωνος, delete the entry.

*μινίκιον, τό, bangle, necklet, armlet, SEG 2.776 (Dura, iii AD).

*μινοδολόεσσα· ἀριθμῶν σύνταξις παρὰ Χαλδαίοις gloss. in POxy. 1802.67 (ii/iii AD); cf. Hsch. s.v. μινδαλοεσσα(s).

μινύζηον, read 'μινύζωον'

*μῐνυθάνω, form of μινύθω, PMich.I 11.7.

μινύθω, after lemma insert '[ῠ, but ῡ in μινύθει B. 3.90 S.-M., cf. 5.151 (cj.)]' II, line 6, after 'Mochl. 19' insert 'πόθῳ μ. h.Cer. 201, 304'

μινυθώδης, add '2 app. weak-willed, Suet.Blasph. 111 T.'

μίνυνθα, line 4, after 'B. 5.151' insert '(μίνυθεν cj.)'

μινύρομαι, for 'of the nightingale, warble' read 'in quot., of the nightingale'

μινυώριος, after 'AP 9.362.26' insert 'ἀστέρα λέκτρων Musae. 305'

μίνυωρος, delete 'cf. Musae. 305'

*μῐνῴδες, v. °Μίνως.

Μῑνώϊος, add 'also Μῑνόϊος, SEG 36.731.22, 33 (Delos, iii BC)'

Μίνως, delete 'but also ἴ Pl.Com. 15 D.' (prob. pr. n. Μίνων); line 3, after 'Hdt. 1.173' insert 'Pl.Lg. 624a, AP 7.727 (Theaet.)'; line 6, after 'Hsch.' insert 'cf. μινῴδες· ἄμπελοί τινες οὕτω λέγονται παρὰ Ῥοδ(ίοις, gloss. in POxy. 1802.71 (ii/iii AD)'

μιξοβόας, add 'Simon. 14 i 5 P.'

*Μιρίϙυθος, graffito, both pers. n. and n. of an insect, REA 49.36 (Delphi, vii/vi BC).

μίρμα, add '(μίργμα La.)'

μῑσάνθρωπος, before 'Pl.Phd. 89d' insert 'οὔτε μισόδημος ὢν οὔτε μισάνθρωπος Isoc. 15.131'

μίσηθρον, add 'POxy. 433.27 (sp. μεῖσ-)'

μῑσητός II, (μίσητος), for this section read 'lecherous, promiscuous, μίσετος ho πα[ῖς Lang Ath.Agora XXI C1 (vii BC), γυνή (?)Archil. 206 W., Cratin. 354 K.-A., Suet.Blasph. 24 T., Ammon. 322 N.'

μισθαποδότης, add 'IGLBulg. 207 (iv AD)'

*μισθογέωργος, ὁ, perh. hired farm-manager, PLond. 1076 (vii AD).

μισθοδοτέω, line 4, pass., for 'receive pay' read 'be paid' and add 'Hell.Oxy. 14.2'

μισθοφορία II, add 'income, Luc.Nav. 13; rent, BGU 2139.12 (AD 432)'

*μισθωσίδιον, τό, dim. of μίσθωσις, SB 7530.4, 18 (i BC).

μισθώσιμος, add 'PMerton 24.13 (ii/iii AD)'

μισοβάρβαρος, before 'Luc.Dem.Enc. 6' insert '2 hating barbarism (in language)'

μισόδημος, after 'And. 4.16' insert 'Pl.R. 566c'

*Μῑσοκύων, κύνος, ὁ, title of a book on misfortunes by Hermagoras of Amphipolis, Stoic. 1.102.

*μῑσομήτωρ, τορος, hating one's mother, Orac.Tibult. 174.

μισοπονηρία 1, line 3, after 'etc.' insert 'b severity in dealing with wrongdoing, τυχεῖν τῆς προσηκούσης μ. UPZ 8.30 (ii BC).'

μισοπόνηρος, after 'adv. -ρως' insert 'with severity towards wrongdoing'; delete 'simply, with hostile sentiments' and add 'ICilicie 70 (iii BC)'

*μῑσόχρυσος, ον, hating gold, i.e. not mercenary, of a physician, Not.Scav. 1941.193 (Rome, ii AD).

*μισσιβίλια, τά, Lat. missibilia, throwing-spears, Just.Nov. 85.4.

*μισσίκιος, ὁ, Lat. missicius, a discharged soldier, PGnom. 53, 54 (ii AD), sp. μεσσίκιος, TAPhA 90.139.

*μίσσος, ὁ, Lat. missus, μίσσος ἡνιόχων chariot-race (as an event in a programme), POxy. 2707.3,6, al. (vi AD).

*μισσώριον, τό, Lat. missorium, dish, bowl, CPR 8.66.14 (μυσσ- pap., vi AD).

*μίσχη· πιλήματα, ταινίαι, μαλλοὶ οἱ τῶν ἐρίων Hsch.

μῐτόομαι, line 3, for 'let one's .. a string' read '(describing the stridulation of a grasshopper or sim.)'

μίτρα I 2, after 'girdle' insert 'Hes.fr. 1.4 M.-W.' and add 'AP 5.199, 6.272' 3, for 'wrestlers' read 'charioteers'

*μιτράνα, ἁ, headband, Sapph. 98(a).10, (b).3 L.-P.

μιτροφόρος, delete the entry.

μίτυλος, after 'μύτιλος' insert 'or μύταλος' II, before 'μίτυλον' insert 'αἷμα πιεῖν μύταλον Call.fr. 691 Pf.'

μιχθαλόεις, add 'μιχθαλόεσσα Suppl.Hell. 991.65 (poet. word-list, iii BC)'

μνᾶ I, add 'IEphes. 3437a'

*μνᾶ (B), Aeol. for μνεία, IEryth. 122.28.

μναῖος, line 4, after '(iv AD)' insert 'κιθάρῳ μέδδονος μναίῳ IGC p. 99 B15 (Acraephia, iii BC); line 5, after 'POxy. 265.18 (i AD)' insert '905.6 (ii AD, μναγαῖον)'

μνανόοι, for 'extra ordinem, fort. μνακόοι' read 'dub., cj. μνακόοι or μναμονόοι'

μνάομαι I, line 3, after 'Call.Ap. 95' insert 'w. acc., μνώεο .. οὔνομα Μηδείης A.R. 3.1069; Προμηθέα .. μνώμεναι, ὡς κείνοιο θεοπροπίησι Κρονίων δῶκε Θέτιν Πηλῆι Q.S. 5.338'

μνάσον, add 'fr. Egyptian mnw'

*μνᾱσιχολέω, v. °μνησιχολέω.

μνεία I, for 'ἐμὴν μ. dub. in Ael.VH 6.1' read 'μ. τὴν ἐμήν Ael.NA 12.32'; add 'μνείας χάριν IG 9(2).1311, etc.; μνίας χ. SEG 30.1044; μνείας ἕνεκεν SEG 34.1028 (ii/iii AD)' add 'III αἱ Μνεῖαι local name for the Muses, Plu. 2.743d.'

μνῆμα I 2, after 'D. 18.208' insert 'pl. as a place-name, SEG 23.236 (Arcadia, iii BC)'

μνημεῖον 2, add 'of a stele, ζῶν φρονῶν ἑαυτῷ τὸ μνημίον κατεσκεύασεν Mitchell N.Galatia 204 (iii AD)' μνήμη, after 'Dor. **μνάμα**' insert '(late sp. **μνήσμη** *SEG* 6.390 (Lycaonia, iv/v AD))'   **I 1**, of the dead, add 'εὐσεβοῦς μνήμης *Cod.Just.* 1.1.3.3'

*μνῆμις, ἡ, *memorial*, μνῆμις Λόκρου κελευστοῦ Mitchell *N.Galatia* 88 (Byz., s.v.l.).

*μνημόδουλος, ὁ, app. *slave looking after a monument*, *TAM* 2.794 (Arycanda).

μνημοδόχος, delete the entry.

μνημονευκτικός **II**, after 'Ptol.*Tetr.* 155' insert '(v.l., v. °μνημονικός)'

μνημονεύω **I 1**, add 'w. part., Pl.*Ep.* 319a' **III**, add 'μναμονεύϝεν (inf.) *SEG* 27.631 (Lyttos, c.500 BC)'

μνημονικός **I 2**, for this section read 'of or connected with the μνημονεῖον, περὶ τῶν γραμμάτων τῶν μνημονικῶν *SEG* 33.679 (Paros, ii BC), μ. συγγραφῆς *CPap.Jud.* 142 (14 BC)' **II**, add 'Ptol.*Tetr.* 155 (v.l. μνημονευτικός)'

μνημόριον, add '*IG* 3.3513, *SEG* 25.806 (Moesia, Rom.imp.); cf. °μεμόριον'

μνημοσύνη, after 'Pi.*O.* 8.74' insert 'E.*Hyps.fr.* 1 ii 27 B.' and delete 'in Att. only as pr. n.'

μνημόσυνον **1**, add '**b** *funerary monument*, *TAM* 4(1).179.'

*μνημόσυνος, η, ον, *of record* or *remembrance*, γράμματα records, Lxx *Es.* 6.1; ἡμέραι ib. 9.27 cod. A; ἀσπάσματα πικρᾶς μνημόσυνα τύχης *SEG* 34.1259 (i AD). **2** of monuments, *memorial*, βωμόν μνημόσυνον Mitchell *N.Galatia* 50.

μνήμων **II 2**, for this section read 'see quot.: οἱ γὰρ ἐν Σικελίᾳ Δωριεῖς ὡς ἔοικε τὸν ἐπίσταθμον μνάμονα προσηγόρευον Plu. 2.612c' **3**, after '*Leg.Gort.* 11.16' insert '*SEG* 33.679 (Paros, ii BC)'

μνησίκακος, add 'adv. -ως, μ. ἔχοι Onos. 42.22'

*μνῆσις, εως, ἡ, *memory*, orac. in *SEG* 33.1056B.10 (Cyzicus, ii AD).

*μνησιχολέω, μνᾶσ-, *harbour a grudge*, *SEG* 25.447.4-5 (Alipheira, iii BC).

*μνηστεῖα· *sponsalia*, Gloss.; cf. μνή<στεια· γάμου δῶρα Hsch.

μνήστειρα, for '*bride* .. (Agath.)' read '*betrothed*, *AP* 5.276.1 (Agath.)'

μνηστεύω, line 9, after 'censured by' insert 'Hdn.*Philet.* 161' **II**, delete '*promise in marriage* .. E.*El.* 313'

μνηστήρ, add '**III** *mindful*, Hsch.; cf. μνήστειρα II and μνήστωρ.'

*μνηστηροφονία, ἡ, *slaughter of suitors*, esp. w. ref. to book 22 of the Odyssey, Str. 1.2.11, Plu. 2.194c, Ath. 5.192d, Longin. 9.14.

μνηστός **I**, after 'abs.' insert '*h.Ap.* 208; *betrothed*'; add 'Procop.*Goth.* 3.1.44, *Arc.* 4.37, Just.*Nov.* 123.40' **II**, before '*memorable*' insert '(fem. -ός)'

*μνῆστρα, τά, *betrothal*, = Lat. *sponsalia*, Charis. p. 34.7 K.

μνῆστρον, delete the entry.

μνιαρός, for '*mossy*' read '*covered in seaweed*' **2**, for '*soft as moss*' read '*soft as seaweed*'

μνίον, last line, after '[ῖ', insert 'Nic.*Th.* 787'

μνιός, for 'Hsch. s.v. μνοῖον' read '°μνοῖος'

*μνόιος or μνοῖος, α, ον, (μνόος) *soft, downy*, Hsch.; ὦ μνοίων μαστῶν *AP* 5.132(131) (Phld.) (cj., μοιν cod.); cf. μνιός.

μνόος **II**, delete '(codd. Ath., s.v.l.)' and add '*IMylasa* 302.12'

*μνωιονόμοι· τῶν Εἱλώτων ἄρχοντες Hsch. (La.; μονονομοιτῶν cod.); cf. μνοία.

*μογγιλάλος, ον, *talking hoarsely*, Ptol.*Tetr.* 150 (v.l.), v. μογι-; cf. μογγός.

μογγός, add 'cf. mod. Gk. μουγγός *dumb*'

μογερός **I**, for 'Ar.*Ach.* 1207' read 'Ar.*Ach.* 1209' and add '*AP* 7.457 (Aristo)'

μογέω **I 2**, line 3, after 'E.*Alc.* 849' insert 'w. acc., τὰς κόρας μ. *suffer from eyestrain*, *HE* 3173 G.-P. (Posidipp.)' **II**, delete the section.

μογιλάλος, line 2, after '*Tetr.* 150' insert '(v.l.)'

μόγις, after 'μύγις' insert 'Sapph. 62.7 L.-P.'; last line, delete '[ῖ metri gr., Il. 22.412]'

*μογόεις, εσσα, εν, *painful*, κακόν Q.S. 4.402.

μόγος **2**, after '(lyr.)' insert 'Nic.*Th.* 428'

*μοδεράτωρ, τορος, ὁ, *governor of a province*, Just.*Nov.* 20 pr., al.

*μόδινος, ὁ, = μόδιος, *EA* 13.44 no. 3 (W.Phrygia, ii AD).

μόδιος **1**, add 'Plb. 21.43.19, Lang *Ath.Agora* xxi Ha 44 (v/vi AD)'

*μοικίω, v. °μοιχάω.

*μοικλός, v. °μυκλός I.

μοῖρα **A I 3**, after 'army' insert 'J.*BJ* 4.9.1.487'; add '*class, grade* in a religious society (the Essenes), J.*BJ* 2.8.10.150' **5**, after 'etc.' insert 'μ. τοπική, μ. χρονική, 1/360 of the ecliptic, or of time of the daily revolution, Hypsicl. p. 5 M.'; before 'Procl.*Hyp.* 3.52' insert '*SEG* 30.1795 (AD 327)' **III 1**, at end for 'Alc.*Supp.* 14.10' read 'Alc. 39.10 L.-P.'; ζῴω μοῖραν ἔχων ἀγροιωτίκαν a rustic's *lot*, id. 130.17 L.-P.' **B**, line 7, delete 'of the Furies, Id.*Eu.* 172'; line 8, after '(Halic., iv/iii BC)' insert '*SEG* 24.1128 (early ii BC)'

μοιραγέτης, after '*IG* 1².80.12' insert '*Schwyzer* 696 (Chios, ?iv BC)'

μοιραῖος, add 'adv. -ως, τὸν ἄνδρα μ. ἀντιτιμηθῆναι *VDI* 1960(3).154 (Chersonese)'

*μοιράφιον, τό, dim. of μοῖρα, Theognost.*Can.* 127.

μοιράω **I**, add 'τέτραχα μοιρηθέντα .. ἕδρανα κόσμου Nonn.*D.* 48.385' **II**, for '*divided* .. hair' read '*gave a share of* their locks, i.e. cut off their locks and laid them on the corpse'

*μοιρηφόρητος, = μοιρο-, Sch.Gen.Il. 8.527.

*μοιρίζω, *share*, (cf. μοιράω), ὅταν μοι[ρί]ζωμεν *SEG* 17.415.1 (Thasos, iv BC).

μοιχάω, after lemma insert 'properly Dor., also *μοιχέω: Cret. μοικίδν, pres. part. *Leg. Gort.* ii 21'

*μολάχη, v. °μολόχη.

μόλιβος, line 1, delete 'Ep.'; add 'cf. μόλυβος'

μολίβουργός, add '*POxy.* 915.1 (vi AD); written μολυβ- O*Wilck.* 1188'

μόλις, line 8, after 'Philem. 88.8' insert 'Plu.*Alc.* 2.5'

Μολίων, add 'cf. perh. Myc. pers. n. *mo-ri-wo*'

†μολοβρίτης, εω, ὁ, σῦς μ. *wild boar*, Hippon. 114b W.

μολοβρός, line 2, after 'Lyc. 775' insert 'as Spartan pers. n., Th. 4.8.9, cf. Myc. pers. n. *mo-ro-qo-ro*'

*μολοκός· ἀνίδρως Theognost.*Can.* 60 A.

+μολουρίς, ίδος, ἡ, *locust*, unless related to ‡μόλουρος, Nic.*Th.* 416; pl., βατραχίδες καὶ τῶν σταχύων τὰ γόνατα Hsch.; cf. μελουρίς, μολυρίς.

*μολοῦρις· αἰδοῖον· κολοβὴ λόγχη· ἢ μόλις οὐρῶν Hsch.

μόλουρος, after 'a kind of *serpent*' insert 'or some small reptile'; add 'cf. pers. n. Μόλουρος *SEG* 33.460 (Larissa, ii BC)'

μολόχη, (s.v. μαλάχη 1), add 'cf. Cret. place-name ἐμ Μολοχᾶντι *SIG* 940; also μολάχη *Epigr.Gr.* 1135'

μολόχινος, for '*made of mallow-fibre*' read '*of mallow*, trade term applied to garments made from a type of fine cloth (prob. cotton) which were imported from India'

*μολόχιον (B), τό, dim. of μολόχη, *mallow*, P*Indiana Univ.* 4.1 (iii AD.).

μολπάζω, add 'ὕμνον ἀμφί τινα A.*fr.* 204b.9 R. (lyr.)'

μολπαρχέω, for '*lead the song and dance*' read 'to be leader of the μολποί'

μολπαστής, add 'of Apollo, μολπαστής· συμπαίκτης Hsch.'

μολποδώρα, delete the entry.

μολποί, add 'at Olbia, *MH* 31.211 (v BC); at Hierapolis, *ZPE* 1.185.23 (ii AD)'

μολύβδιον, add '**III** *lead tablet*, *SEG* 38.13.B4 (Rhamnous, c. 500/480 BC).'

μολυβδόδετος, after '*lead*' insert '[μ]ύκε χο͂ μ. *IG* 1³.425.38 (Athens, v BC)'

μόλυβδος **II 1**, delete the section transferring exx. to section I. add 'Myc. *mo-ri-wo-do*; see also βόλυβδος'

μολυβδοχοέω **2**, add '*TAM* 2.437 (Patara; μολυβο-)'

μολυβδόω **2**, add 'ἐπαέτιον ξύλινον μεμολυβδωμένον *Inscr.Délos* 442*B*168 (ii BC)'

*μολυμός [?ῠ], ὁ, perh. = μολυσμός, *MAMA* 4.280 (Dionysopolis).

μόλυχνον, add '(cj. δεισαλέον)'

*μολυχυάζω, *thicken* (intr.), Anon.Alch. 334.28.

μόμφος, after '(Mantinea, v BC)' insert 'ἓν μόνφον interpr. by some as adj. ὕνμομφον, i.e. ἔμμομφον, *liable to blame*, but cf. ἐμμεμφής (ὐνμεμφές line 23)'

*μονάδιον, τό, *little monastery*, [.. τῷ εὐαγεῖ μο]ναδίῳ rest. at *BCH* 98.780.1 (Argos, v/vi AD).

μονάζω **I 1**, add '**b** *to be a monk*, P*Köln* 151.7 (AD 423).'

μόναξ, delete the entry.

*μονάρταβος, ον, *taxed at one ἀρτάβη per ἄρουρα*, *POxy.Hels.* 9.12 (AD 26), *PSI* 1328.47 (ii/iii AD), [ἐκ] μοναρτάβ[ο]υ *POxy.* 2143ʳ (iii AD).

μοναρχία, add 'fig., opp. ἰσονομία, τὴν δ' ἐν αὑτοῖς (sc. δυνάμεσιν) μοναρχίαν νόσου ποιητικήν Alcmaeon 4 D.-K.'

μονάς **II 3**, of weight, add 'ἀργυρίου μονάδαν μίαν *POxy.* 3402.4 (iv AD)'

μοναστήριος **II 1**, for '*hermit's cell*' read '*room set aside for solitary religious exercises*' **2**, after '*monastery*' insert '(for male or female communities)' and add '*Cod.Just.* 1.3.46.6'

μονάστρια, add 'also μονήστρια *MAMA* 3.45'

μοναυλέω, for '*play a solo on the flute*' read '*play the single aulos*'

μοναυλία (A), for '*solo on the flute*' read '*playing the single aulos*'

μοναύλιον, for '*solo instrument*' read '*single aulos*'

*μοναύλιος, α, ον, *living alone, celibate*, *EA* 12.151 no. 5.16 (Lydia or Mysia, iii AD), Suid.

μόναυλος (passim), for '(single) *flute*' read '*single aulos*'

μοναχός **1**, add '**b** of a garment, *single*, made with a single cloth, or perh. unlined, PHamb. 10.26 (ii AD), *POxy.* 1273.13 (iii AD); μ., ἡ, *Peripl.M.Rubr.* 6.' **II**, add '*SEG* 37.1058 (Bithynia)'

*μονείμων, ον, gen. ονος, *wearing one garment*, Sch.A.R. 3.646; see also μονοείμων.

μονή **II 2**, add '*SB* 9683.12 (iv AD), *BCH* suppl. 8 no. 255 (v/vi AD)'

μονήρης **I 1**, add 'of a (Christian) ascetic, *Cod.Just.* 1.3.52.15. **b** *single*' and transfer 'Nic.*Al.* 400' here **II**, for '*with one man to each oar*' read '*having a single bank of oars*' and transfer 'Plb. 21.43.13' fr. section I 1.

*μονήτης (gen.), *mint*, ἱερᾶς μ. *POxy*. 3618.15 (iv AD), cf. Lat. *moneta*.

μονία (A) 1, for '*changelessness*' read '*abiding, rest*' and add '(also interpr. as μονία (B))'; for 'Emp. 27.4' read 'Emp. 28.2 D.-K.'

μονία (B), add 'v. °μονία (A)'

μονόβολον, add 'also name of a game of chance, *Cod.Just.* 3.43.1.4'

μονόγαμος, add 'μουνό-, *AS* 5.31.5 (N.Phrygia, iv AD)'

⁺μονόγραμμος, ον, *existing only in outline*, in mocking ref. to Epicurus' gods, Cic.*ND* 2.59; of very thin men, Lucil. 59, 725 M. (Non.Marc. 37.9 M.). II *marked by a single streak*, of a variety of *iaspis*, Plin.*HN* 37.118.

*μονογραφεῖον, τό, *office of a notary*, *APF* 33.24,27 (?iii BC).

μονοδεσμία, for 'a tax of uncertain nature' read 'a kind of tax on land originally paid in kind' and add 'μ. χόρτου *BGU* 2285.4 (AD 194/5)'

μονοδραχμία, delete the entry.

μονόδραχμος, delete the entry.

*μονόειδος, ον, = μονοειδής, Zos.Alch. 113.6.

μονόζυξ II, add '2 of a limb, *containing one bone*, Paul.Aeg. 6.107 (v.l. ὁμοζύγων).'

μονοθρηνέω, for '*mourn in solitude*' read '*sing a solo dirge*'

*μονοικίδιον, τό, *detached house*, *PKlein.Form.* 239.3, al. (vi/vii AD).

*μονοκάδιον, τό, a liquid measure, *PPrincet.* in *Tyche* 3.29ff.

*μονόκερος, ὁ, *unicorn*, *PMag.* 3.504.

μονόκερως I, for 'poet. .. 181' read 'μουνο- Archil. 276 W. (Hsch.)' II, for '*wild ox*' read 'used mistakenly to translate word meaning *wild ox*'

*μονόκοπος, ὁ, *a single cutting*, χόρτου μ. *PMichael.* 43.8 (vi AD).

*μονοκόρωνον, τό, *a single crow*, *Vit.Aesop.*(G) 77.

μονόκροτος, for pres. def. read '*manned by a single bank of oars*'

*μονόλεκτρος, ον, *sleeping alone*, λέχος εἰ μονόλεκτρον ἴαυες *SEG* 30.1142.

μονόμαλλος, add '*Dura*⁴ 93, 153'

μονομάχης, add '*ITrall.* 100'

μονομάχος II, add 'sg. J.*AJ* 19.1.15'

*μονόμεσος, ον, *having only one intermediate*, Olymp.*in Cat.* 137.32, Elias *in Cat.* 243.31.

*μονόμυξος, ον, *having one wick*, *Inscr.Délos* 1417*A*i59, 72 (ii BC).

*μονόνυμφος, ἡ, *having one husband*, metrical epitaph, *IGLS* 1366 (Apamea).

μονόξυλος, for '*πλοῖα canoes*' read '*πλοῖον dug-out canoe*' and add 'Aeschin. 2.124'; line 3, delete 'cf. Pl.*Lg.* 956a'

μονοπάλη, after '-πάλη' insert 'Ion. μουνο-'

μονόπελμος, for '*Edict.Diocl.* 9.16' read '*Edict.Diocl.* 9.13,16' and add 'Callistr.ap.Harp. s.v. ἁπλᾶς'

*μονόπληγος, *involving a single stroke*, τῷ δ᾽ ἑτέρῳ (σκαφητῷ) θερινῷ μονοπλήγῳ *PSoterichos* 1.22 (AD 69), 2.18 (AD 71).

μονοπρόσωπος I 1, add 'as epith. of Hecate, Ἑκάτη τρίμορφε Ἑκάτη μονοπρόσωπε *SEG* 30.326 (i AD)' 2, add '*PCair.Zen.* 764.3 (iii BC)'

*μονοπωλάριος, ὁ, app. = μονοπώλης, *PHamb.* 228.4 (vi AD).

*μονοπώλης, ου, ὁ, app. *trader, retailer*, Mitchell *N.Galatia* 418.

μονοπωλία, add '*PMich.*i 60.6 (iii BC)'

μονοπώλιον, line 2, delete '*PSI* 619.10 (iii BC)'

⁺μονόπωλος, ον, *driving a single horse*, E.*Or.* 1004 (dub.).

μονορύχης, after lemma insert 'Ion. μουν-'

μόνος, line 4, delete 'by E. only in μούναρχος' I, add '3 of a magistrate (normally having colleagues) *sole*, *IEphes.* 3493.' add 'V *ace*, throw of one at dice, *BCH* 8.501-3 (μοῦνος).'

*μονοσάμβαλος, ον, = μονοσάνδαλος, Sch.Pi.*P.* 4.133.

μονόστεγος, for '*of one story*' read '*having a single continuous roof*' and add '*PMich.*xii 627.6 (AD 298)'

*μονοσύνθρονος, ον, *only sharer of the throne*, of Christ, μ. παῖς πατρὸς ἀθανάτου *PKöln* 172.12 (iv/v AD).

*μονόσωμος, ον, *for one body*, κοιμητήριον (lapis κυμ-) *BCH* suppl. 8 no. 153 (v AD), *SEG* 19.443 (Philippi).

*μονοτάφιον, τό, *single tomb*, Rott *Kleinas.Denkmäler* p. 369 no. 74; perh. also *SEG* 31.1419 (Palestine, iii/iv AD).

*μονόφακος, ὁ, *lentil*, *CPR* 10.52.9 (iv/v AD).

*μονοχίτων, ωνος, ὁ, kind of tunic, perh. unlined, -ων μελίτιν(ος) *POxy.* 3201.3 (iii AD).

*μονοχίτωνος, ον, adj. of uncertain sense, cf. °μονοχίτων, στολὴν πορφυρῆν μονοχίτονον (sic) *PFam.Teb.* 21.20 (ii AD); cf. διχίτωνος, λευκοχίτωνος.

μονόχορδος, add '(*Arabian*) *one-stringed lute*' and transfer 'Poll. 4.60' to follow.

μονόχωρον, add '2 *single room*, *POxy.* 3057 (i/ii AD), 1957 (v AD), 1964 (vi AD).'

μονόω II 2 b, for 'c. gen. rei' read 'without sense of desertion, *deprived of*' and add 'εἰ μετὰ μὲν μαρτύρων .. μονούμενος δὲ μαρτύρων Antipho Soph. 44 A col. 1.20 D.-K.'

*μόνωρος, ον, *of a single age*, τῶν δύο ἀδελφῶν αἱ θυγατέρες, μόνωροι καὶ ἴσωροι *IEryth.* 525.5.

μόργιον, add 'unless f.l. for μόρτιον; cf. μορτή'

μόργος, add 'III μ.· (corr. for μόριος) ἄπληστος Hsch.; cf. μοργίας.'

*μοργός, ὁ, = ἀμοργός, (cf. ἀμαυρός, μαυρός) Suid.

μορέω, delete the entry.

μόριον III, add '3 (gramm.) *particle*, Aristid.*Rh.* 2.532, 533 S.'

μόριος, for '*of burial*' read '*allotted by fate*'

*μορίτης, ου, ὁ, perh. a wine made from mulberries (μόρα), Zos.Alch. 184.16.

μορμολύκειον 1, line 1, after '*Phd.* 77e' insert '(pl.)'

*μορμολύκιον, τό, = μορμολυκεῖον, as term of opprobrium, *Vit. Aesop.*(W) 77b, (unless late sp. of μορμολυκεῖον).

μορμολύττομαι, delete 'Act. μορμολύττω is f.l. in'

*μορμόρυξις, εως, ἡ, *frightening, scaring*, rest. in Pi.*Pae.* 20.6 S.-M.

μόρμυλος, read μορμύλος

μορμύρω, line 3, after 'A.R. 1.543' insert '*APl.* 182 (Leon.Tarent.)'

μορμωτός, add 'also as pers. n. Μόρμωττος, L.Robert *Monnaies antiques en Troade* pp. 120-121'

⁺μορόεις, εσσα, εν, app. *mulberry-like*, perh. *having a shiny, bubbled pattern*, like a mulberry or blackberry, (μόρον, but referred by ancient lexicographers to μόρος "*fate*"), ἕρματα ..τρίγληνα μορόεντα Il. 14.183, Od. 18.298; in unexpld. sense, perh. *glistening*, τεύχη Q.S. 1.152; ποτόν Nic.*Al.* 130, 136; ἐλαίη ib. 455; φρυνός ib. 569; expld. by Hsch. as μετὰ πολλοῦ καμάτου πεπονημένος, cf. Eust. 976.40; by Apollon.*Lex.* as ἀθάνατα, μόρου μὴ μετέχοντα.

*μόρρα, perh. hyperaeolic form for μοῖρα, Sappho commentary, *SLG* 261A *fr.* 2 i 8.

μορτή, add 'cf. μόργιον (= ?μόρτιον)'

μορτός, add 'Hsch. (μόρτος La.), also glossed μέλας, φαιός; cf. °ἄμβροτος and non-Att.-Ion. pers. n. Ἀγέμορτος, Κλεόμορτος, Χαρίμορτος, etc.'

Μόρυχος I, add 'also name of a glutton in Att. comedy, Ar.*Ach.* 887, *Pax* 1008, etc.'

μορφάζω, add 'II *make specious* or *adorn*, τὰ ψευδῆ Eust. 1691.8.'

*μορφίζομαι, *simulate, make a show of*, μ. τὴν εὐσέβειαν Gel.Cyz. *HE* 3.16.6.

*μορφόλυκος, ον, *having the form of a wolf*, *PMag.* 4.2812.

μόρφωμα, add '2 pl., *idols*, Aq. 1*Ki.* 15.23.'

*μόσθιον, τό, dim. of μοῦστος, a measure of wine, *POxy.* 1589.17 (iv AD); cf. μουστάριον.

*μοσμένιον, τό, *little horse*, pap. in *Aegyptus* 6.187.

*μοσμονάριος, ὁ, (?) *keeper* of μοσμένια, pap. in *Aegyptus* 6.187.

μόσσυν, at end for 'A.R. 2.1017, Call.*Aet.Oxy.* 2080.70' read 'A.R. 2.381b, Call.*fr.* 43.68 Pf.'

*μοσχάλιον, τό, = μασχάλιον, *OStrassb.* 677 (ii AD; -λει-).

μοσχανοσῖτος, add 'read perh. two words μοσχανὸς σῖτος (‡μόσχος 1)'

μοσχέλαιον, add 'μουσχελ[.. *PLouvre* inv. 6745 app. 177 (*WS* 96 (NF 17).69)'

*μοσχευτικός, ή, όν, *used for (cutting) suckers*, δρέπανα *PCair.Zen.* 851*a*24 (iii BC).

*μόσχημα, = μόσχευμα, *PCair.Zen.* 839.2 (sp. βοσχ-), 4 (iii BC).

μόσχινος, add '-ον, τό, prob. *calfskin*, *CPR* 10.52.8 (iv/v AD); perh. also *calf* or *calfmeat*, *PNess.* 85.6 (vii AD).

μοσχίον, add 'II ἁπαλὰ φυτά .. ἢ κρομύουν τὸ σπέρμα Hsch.'

μόσχιος, after lemma insert '= μόσχειος' add 'II name of a month, *SEG* 17.829.5 (c.200 BC; unkn. provenance).'

μοσχοθύτης, add '*PMich.*iv 225.1814 (ii AD), *SB* 9902 L6, M6 (iii AD)'

*μοσχολόγος, ὁ, a kind of *actor* or *mime*, *Inscr.Cret.* 4.223 (Gortyn).

μόσχος (A) and (B), read as sections I and II under same lemma II 3, delete 'any young .. even'

μοσχοτρόφος, line 2, for 'τιθηνός' read 'τιθήνη'

*μοσχών, ῶνος, ὁ, *byre for calves*, *PCair.Zen.* 642.3 (iii BC).

μοτός, I, add 'cf. 'Call.*Fr.* 7.40 P.' read 'Sch.Call.*fr.* 23.21 Pf.' and after 'Hsch.' insert 's.v., *Et.Gen.* s.v. ἄμοτον'

μοτόω, add '*BGU* 1903.3 (iii BC)'

μουγκρίζω, for '*slobber*, or perh. *snarl*' read '*growl, groan*'; add 'cf. mod. Gk. μουγγρίζω'

*μουζίκιον, τό, *box inlaid with mosaic*, *PColorado* inv. 2 (*ZPE* 93.213, v AD).

*Μουκίεια, τά, *games in honour of Q.Mucius Scaevola*, *IGRom.* 4.188.4 (Poemanenum, i BC).

*μουλάγορας, ου, ὁ, perh. *mule-seller*, *MAMA* 3.86 (Diocaesarea).

μούλη, read μούλα

*μουλιατρός, ὁ, = Lat. *mulomedicus*, *mule-doctor*, prob. in *Edict.Diocl.* 7.20A.

*μουλικός, ή, όν, *of* or *for mules*, χαλεινοῦ μουλικοῦ *Edict.Diocl.* in *SEG* 37.335 iii 10.6, al.

μουλίων, after '*muleteer*' insert '*TAM* 4(1).39 (iii AD)'

*μοῦλος, ὁ, Lat. *mulus*, *mule*, *PMich*.xi 620 vᵛ 284, *SB* 9409 fr. 7.110, *PCornell* 39.3.

μουνᾰδόν, for '= μόνον' read '*individually*' and after 'Opp.' insert '*H*. 1.444'

μουνάξ, add 'Arat. 119, Euph. 98'

*μουνικίπιον, τό, Lat. *municipium*, *a self-governing community*, *IEphes*. 3048.13.

*μουνίφεξ, ικος, ὁ, Lat. *munifex*, *private soldier without exemption from military duties*, *PBeatty Panop*. 2.28 (AD 300).

Μουνυχία, read 'Μουνῐχία' (and so in all derivatives; sts. sp. Μουνυχ-, e.g. in Sch.E.*Hipp*. 759, St.Byz.)

*μούργος, ὁ, prob. the *brown* one, of a horse or mule, (unless ὁμουργός is read), *POxy*. 922.19 (vi/vii AD); cf. mod. Gk. = *watchdog*.

Μοῦσα II 1, add 'b ἀπὸ μούσης = ἀπόμουσον, ἄμουσον, οὐκ ἔστιν ἀπὸ μούσης it is not *out of place*, Ael.*NA* 12.34, al.'

*μουσαναγωγός, ἡ, *leader of the Muses*, title of Isis at Canopus, *POxy*. 1380.62 (ii AD).

Μούσαρχος, delete 'Dor. Μώσαρχος' and for 'Terp. .. Diehl)' read '*Lyr.adesp*. 23 P. (Μωσ- Bgk.); of a man, *IHistriae* 167 (ii AD), 100 (iii AD)'

Μουσεῖον I 1, add 'E.*Melanipp.Sap*.prol. 19' 3, add 'at Ephesus, *IEphes*. 3239' II, before 'Paus. 1.25.8' insert '*SEG* 28.60 (Athens, 270/69 BC)' IV, add '*SEG* 23.271 (Thespiae, iii BC)'

μουσϊκεύομαι 1, add 'Sch.Pi.*P*. 4.526'

μουσικός I, line 3, after '*Lg*. 828c' insert 'εὐφωνία D.H.*Isoc*. 3' II, lines 3/4, delete '*lyric poet* .. 243a (but .. p. 96 K.)'

+μούσμων, ωνος, ὁ, Lat. *musmo*, *wild sheep found in Sardinia, perh. mouflon*, Str. 5.2.7.

*μουσόγραφος, ον, app. *written in verse*, μουσογράφους φοιτῶν ἀρτιμαθεῖς σελίδας *SEG* 28.541.

*μουσοεπής, ές, *speaking in verse*, φάμα καρύσσω μουσοεπεῖ στόματι *GVI* 1179.2 (Smyrna, ii BC).

*μουσόθετος, ον, *set up by music*, Θήβης τείχεα μ. *SEG* 8.528 (Egypt, ii/iii AD).

μουσοποιέω II, add 'abs., *Trag.adesp*. 496.1 K.-S.'

μουσοπόλος I, add 'ὡς δ' ὅτε μουσοπόλων ἔργων ἄπο παῖδες ἴωσιν, i.e. from school tasks, Opp.*H*. 1.680'

*Μουσότροφος, *nurtured by the Muses*, in quot., name of a horse, *SEG* 7.213.22 (Beirut).

μουσουργία, for '*singing, making poetry*' read '*the work of the Muses*, i.e. *music, poetry*, etc.' and add 'Hld. 2.24'

μουσόχορος, for '-χαρής' read 'μεσόχορος (q.v.) rather than for μουσοχαρής'

*μουστάριος, α, ον, *of must*, perh. in *CPR* 8.63.2 (vi AD).

*μουσχάτιον, τό, *muscatel*, *PMed.Rez*. 17 (vii AD); cf.μόσχος (C).

*μουσχοροσᾱτον, τό, ?*wine flavoured with rose-water*, *PMed.Rez*. 17 (vii AD).

*μούσωσις, εως, ἡ, *mosaic*, *SEG* 37.367 (Patrae, ?vi AD).

*μουσωτής, οῦ, ὁ, *worker in mosaic*, *Syria* 1.302 (vi AD); cf. °παξαμᾶς II.

μόχθος, line 4, delete 'of the *labours* of Heracles'; add 'ἐκ τῶν ἰδείων μόχθων, by one's own *efforts*, i.e. at one's own expense, *BCH* suppl. 8 no. 62'

μοχλεία, line 1, for '= sq. 1' read '*leverage*', for '*Supp.Epigr*. 2.569.19 (Didyma, ii BC)' read '*Didyma* 40.25 (ii BC)' and add '*Plot*. 5.9.6'

μῦ, add 'cf. Hebr. *mēm*, see also μῶ' 2, delete the section.

*μῦ (B), used to represent *a muttering sound* made with the lips, μῦ λαλεῖν to *mutter*, Hippon.ap.S.E.*M*. 1.275 (read as μοιμύλλειν fr. 124 W.); to imitate the sound of an aulos lament, μῦ μῦ μῦ μῦ Ar.*Eq*. 10; cf. μύζω (A).

μυάγρα, after '*mouse-trap*' insert 'Ar.fr. 576 K.-A.'

*μυάκανθα, ἡ, = μνάκανθος, *Gp*. 10.21.6.

μυάκιον, after 'dim. of μύαξ' insert 'II *spoon*' and add '*BGU* 2360.6 (iii/iv AD)'

*μυγερός, ὁ, = νυκτικόραξ, Cyran. 29.

μύγματα, delete the entry (v. ἀμύγματα).

μυδαλέος, read 'μῠδᾰλέος' and delete note on quantity.

μυδάω, after lemma insert 'aor. μυδῆσαι Hsch., pf. μεμύδηκα Dsc. (v. infra)'

*μυδροστᾰσία, ἡ, *place of an anvil, forge*, Roueché *Aphrodisias* 208 (Chr.).

μυελός 1, add 'perh. as the vital fluid, A.*Ag*. 76' deleting this ref. fr. section 4.

μυέω I, add 'in this sense Myc. *mu-jo-me-no*, but for form see °μύω' II, add 'w. gen., οὔτε γὰρ εὐεπίης Μου[σ]ῶν φιλίων ἐμνήθης Mitchell *N.Galatia* 146'

μύζω (B), add 'Archil. 42.2 W. (cj.), Hsch.'

*μύθαρ· μύθος Hsch.

μυθίζω, med., after 'Perict.ap.Stob. 4.28.19' insert '*AP* 12.181 (Strato)'

μυθολογεύω, delete 'prob. rest. in Sapph.*Supp*. 7.4'

*μυθολόγος II 1, for 'Call.*Aet*. 3.1.55' read 'Call.fr. 75.55 Pf.'

μῦθος II 2, add 'λέγοντος δέ μου ταῦτα, ἀπεκρίνατό μοι ὅτι μύθους λέγοιμι D. 50.40' III, add 'Sch.Od. 21.71'

μυίαγρος, line 2, for 'prob.' read 'v.l. Myiacores'

μῡκάομαι 2, line 6, after 'Theoc. 22.75' insert 'of the sound made by a water-clock, Mesom. 8.28 H. (μηκ- cod.)'

μύκημα, line 2, after 'A.R. 1.1269, etc.' insert 'Longus 1.21'; add 'of the noise made by a water-clock to indicate the time, Luc.*Hipp*. 8'

μύκης, line 2, before 'Dor.' insert 'Att.nom-sg. [μ]ύκη *IG* 1³.425.38 (v BC)' II 1, after 'Hdt. 3.64' insert 'Nic.*Al* 103' 6, delete the pres. section, transferring quot. to section 1 and substitute 'πίλος καὶ δερμάτινον ὑπηρέσιον Hsch.'

+μύκλα, ἡ, in pl., black *lines* on a donkey's back or legs, Hsch.; cf. °μύκλος II.

+μύκλος, ὁ, *lustful* person, Archil. 270 W., Lyc. 771; as epith. of a donkey, ib. 816; perh. as designation of a donkey, *PTeb*. 409.7 (μοικλ-, i AD); cf. μυχλός. II (perh. different wd.) as °μύκλα, *fold* or *line* on a donkey's coat, Hsch., *EM* 594.21; cf. ἐννεάμυκλος ὄνος Call.fr. 650 Pf.

*Μύκονιάς, άδος, fem. adj., *of Myconos*, *SIG* 1024.14 (Myconos).

μυλαῖος I, for '*working in a mill*' read '*occupied at the (hand) mill*'

μυλακρίς II, add '2 *knee-cap*, Hippon. 162 W. (Poll. 2.188).'

*Μυλάντειοι θεοί· ἐπιμύλιοι (ἐπιμύλισιν cod.) Hsch.; connected with Mylas, one of the Τελχῖνες at Camirus, St.Byz. s.v. Μυλαντία (a promontory at Camirus).

*Μυλάντιος, epith. of Apollo at Camirus, *Tit. Cam*. 15 B 10, al.

μυλάσασθαι, add 'cf. mod. Chiot μουλιάζω *drench*'

μυληβόρος, for '*millstone-eating*' read '*feeding at a mill*'

*μυλινάριος, ὁ, *miller*, *IG* 4.411 (Corinth, v/vi AD).

μυλίτης II, for '*molar tooth*' read '*wisdom tooth*' and after 'Gal. 14.722' insert '*An.Ox*. 3.82.26'

μυλλός (A), add 'II καὶ παροιμία ἐπὶ τῶν ἀκουόντων καὶ ‹κωφότητα› προσποιουμένων, ἔστι δὲ καὶ κωμῳδιῶν ποιητὴς οὕτω καλούμενος Hsch.'

μυλόεις, add 'cf. Μυλόεις· ποταμὸς Ἀρκαδίας Hsch.'

μυλοκόπος, for '*millstone-worker*' read '*millstone-cutter*' and add '*PPetaus* 103.43 (ii AD), *POxy*. 3641.6 (vi AD)'

*μυλοκρῐβάνιον, τό, *milling-bakery*, *POxy*. 1890.6, 19 (vi AD).

μύλος I 2, add '*BGU* 2477.9 (i AD)' and delete 'generally, *stone*' 3, add 'Hsch.'

μυλωθρικός II, for '*IG* 2.860' read '*IG* 2².1707 (181/180 BC)'

μυλωθρός, after 'Poll. 7.180' insert 'also *mill-worker*, *Hesperia* 28.232, Ath. 14.619b'

*μυλώναρχος, ὁ, = μυλωνάρχης, *POxy*. 1890.3 (vi AD, μυλον- pap.).

μυλωνίον, add 'Aët. 2.183 (220.8 O.)'

μύμα, add 'θριδάκων τρίμμα, καὶ ὑπόχυμά τι Hsch.'

μῠνάομαι, for lemma read 'μόναμαι' and for 'Alc. 89' read 'Alc. 392 L.-P.'

μυναρός, add 'perh. for μυνδαρός, cf. μυνδός'

μυνδός, add 'Hsch.; cf. mod. Gk. μουντός *dull*'

μυξωτῆρες, add 'II sg., *vessel for pouring* oil into a lamp, Lxx *Za*. 4.12.'

*μυόγαλος, ὁ, = μύγαλος, Cyran. 42.

μυσωτίς II, for 'Dsc.-Dsc.' read 'Ps.-Dsc.'

*μυότροχον, τό, = μυοσωτίς II, Ps.-Dsc. 4.86 (μυόρτοκον codd.).

μύουρος (A) 1, after '*mouse-tailed*)' insert 'μύουρα καὶ βραχέα A.fr. 78a.29 R.' 2, line 3, before 'Ath. 14.632e' insert 'Plu. 2.611b' at end after 'μῦ-' insert 'A. l.c.'

*μύραινος· ἡ μύραινα ἀρσενικῶς Hsch. (*sea-eel*, cf. μύρος).

μυράφιον, add 'Dura⁴ 128, *POxy*. 2596 (iii AD)'

*μυρεψᾶς, ᾶ, ὁ, = μυρεψός, *MAMA* 3.712 (Corycus).

μυρεψικός, after 'ib.*Ca*. 8.2' insert 'ἀγγείον μ. Aët. 12.55' and add 'adv. -κῶς in the manner of an unguent, Asclep.Jun.ap.Gal. 13.1031.7'

μυρεψός, add 'Lxx *Si*. 38.8(7), 49.1, also written °μυροψός'

μυρηρός, add 'κύαθος *IG* 2².1424a321'

μυριᾰγωγός, add 'II *leader of ten thousand*, epith. of God, *JÖAI* 32.80 (amulet).'

μῡριᾰκις, for '*numberless times*' read 'usu. indicating a very large number' and add '*POxy*. 3063.3 (ii AD), Gal. 4.355.11'

*μῡριᾰκός, ή, όν, *of* or *belonging to the ten thousand* (in quot., full citizens), ἀρχαί *SEG* 9.1.46 (Cyrene, iv BC).

*μῡρίᾰριθμος, ον, *of countless number*, ὄχλος Ps.-Callisth. 1.19 cod.C.

μῡριάς II, add 'μου[ρι]άδεσσι λάϋς Corinn. 1(a) i 34 P. (s.v.l.)'

*μυρικόω, *make mute, strike dumb*, *Suppl.Mag*. 55 D-G 1, 8 (iii AD), perh. also ib. 95.4 (v AD).

*μυρϊοανάγωγος, ον, *that leads up ten thousand*, i.e. *a countless number*, Rott *Kleinas.Denkmäler* p. 375.

*μυρῐόκλαυστος, ον, *ten thousand times bewailed*, *GVI* 1941.1 (Thisbe, ii/iii AD).

*μυρῐονταπλάσιος, ον, *ten thousand times as great*, w. gen., Simp. *in Ph*. 479.2.

μυρίος I 2, line 6, delete 'in Ion. prose' and insert 'κραυγή μ. Hanno

Peripl. 14' **4**, add 'also sg., ὑπέρτεροι μυρίον ἄλλων *ICilicie* 49 (iii/v AD)'

μῡριοτευχής, add '**2** *having ten thousand arms*, μ. ἕδος (i.e. a *tropaeum*), *SEG* 23.451 (i AD).'

*μῡριοψήφιστος, ον, that gives ten thousand ψῆφοι (cf. *Apoc.* 2.17) Rott *Kleinas.Denkmäler* p. 375.

μυριώνυμος, add 'of Hecate, *SEG* 26.819 (iii AD)'

μύρκος, add 'cf. Lat. *murcus*'

μύρμη, after lemma insert 'in Dor. form μύρμᾱ'

μυρμηκίζω, add '**III** med., *have itching palms*, (?) interpol. in Gal. *Med.Phil.* 2.'

*μυρμηκιόεις, εσσα, εν, *warty*, Marc.Sid. 97 (*GDRK* 2 p. 22, -κώεντα codd.).

†μυρμηκολέων, οντος, ὁ, an animal not precisely identified (cf. Str. 16.4.15), but in quot. app. used simply as synonym for "lion", Lxx *Jb.* 4.11.

μύρμηξ **IV**, add 'Sch.A.R. 2.52'

μύρμος **I**, add 'Call.*fr.* 753 Pf.'

*μυροναρδοπώλης, ου, ὁ, seller of μύρον and νάρδος, *ZPE* 61.78.

μυρόπνοος, add '4.1.9 (Mel.)'

*μυροποιέω, *make perfume*, *SB* 10296.7 (i BC).

μῦρος, after 'ap.Ath. 7.312f' insert 'Ael.*NA* 14.15, μούρω *IGC* p. 98 l. 12 (Acraephia, iii/ii BC)'

*μῠρουργός, ὁ, *perfume-maker*, Heph.Astr. 2.19.16, Cypr. *mu-ro-wo-ro-ko μυροϝοργό(ς) Rantidi* 2.

*μυροψός, ὁ, *unguent-maker*, *MAMA* 3.289a, 448, 699, see also μυρεψός.

*μυρραῖος, α, ον, *of the myrrh-tree*, *AP* 4.1.29 (Mel., cj.).

*μυρσινεών, ῶνος, ὁ, *myrtle-grove*, Aq., Sm.*Za.* 1.8.

μυροΐντης **II** 2, add 'also τιθύμαλλος μ. Afric.*Cest.* p. 15 V.'

†μύρσος· κόφινος ὦτα ἔχων, ὃς καὶ ἄρριχος Hsch., cf. Call.*fr.* 756 Pf.

*Μυρτάτης, ου, ὁ, epith. of Apollo, *SEG* 23.655 (Paphos, iii BC).

μύρτον **II**, add '*AP* 7.406 (Theodorid.)'

μύρω **II** 1, line 5, for 'opt.' read 'subj.'

μῦς **I**, add 'μ. ἐλιοί *dormice*, *Edict.Diocl.* 4.38, v. ἐλειός'   **V**, for this section read '*gag, muzzle* (sts. taken as separate word), Herod. 3.85, 5.68' (deleting this from section I 1)

μύσαγμα, add '*PMag.* 4.2576, 2645'

μυσάλμαι, delete the entry.

*μυσάλμης, ου, ὁ, *one who lives very cheaply*, Eust. 1828.15, cf. 1507.2, μυσάλμαι· πολὺ πεινῶντες καὶ ‹τὰ εὐτελέστατα› (ὁτιοῦν La.) ἐσθίοντες Hsch.

μυσάρχης, delete the entry.

*Μυσάρχης, ου, ὁ, *leader of the Mysians*, Lxx 2*Ma.* 5.24.

*Μυσαχέων, gen. pl., name of a kinship group at Naupactus, *IG* 9².718.22, 28 (Locris, v BC).

μύσαχνός, after '*defiled*' insert 'Hippon. 105.10 W.' and for 'Archil. 184' read 'Archil. 209 W.'

μύσαχος, delete the entry.

*μυσαχρόν· μυσαρόν, μυσαχθές Hsch.

*Μῡσία, ἡ, *Mysia*, Hdt. 160.4, X.*An.* 7.8.8, etc.   **II** Lat. *Moesia*, *SEG* 31.1116, J.*BJ* 4.10.6, Plu.*Oth.* 4, etc.

*μυσίδιον, τό, (written μυσίδην) Iambl.*Alch.* 286.18, 288.3, dim. of μίσυ I.

Μύσιος, add 'στιχάριον M. *PWash.* 58.5 (v AD)'

*μύσκελος· στραβόπους Cyr., cf. pers. n. Μύσκελος *GDI* 345.75, Μύσκων Th. 8.85.3.

μυσπολέω, delete 'with a play on μυστιπολέω'

μυσταγωγία, add '**IV** fig., of initiation into the business of tax-farming, *PTeb.* 812.5 (ii BC).'

μύσταξ, delete 'Dor. and Lacon. .. fem.' and add 'cf. βύσταξ'

μυστάρχης, add '*SEG* 24.1050 (Moesia, ii AD)'

*μυσταρχικός (sc. βωμός), *of* or *for a μυστάρχης*, Robert *Ét.Anat.* 291 (Amastris).

*μυστηγορία, ἡ, *mystical discourse*, Procl. *in Prm.* p. 779.15.

μυστηριακός, after 'Ptol.*Tetr.* 163' insert '167 (both of persons)'

μυστηριάρχης, add '*TAM* 4(1).262'

μυστήριον **I 1**, add 'applied also to Chr. rites, Just.*Nov.* 123.31, al.'   **2**, add '**c** *hall used by* μύσται, Sardis 7(1).17.6.   **d** *tomb*, *SEG* 31.1388 (Syria, Jewish).'   **3**, add '**b** *secret sign*, *PVindob.* G30052 + 13607.11 (*Pap.Flor.* VII 337, c.AD 300).'

μύστις **1**, add 'μύστα λέων *SEG* 30.1562 (epigr. on image of lion, w. ref. to grade in Mithraism)'

†μυστίλη, ἡ, *spoon*, Aret.*CA* 1.4, 10, *CD* 3, Ath. 3.126a.   **b** a crust of bread scooped out to the form of a spoon, Ar.*Eq.* 1168, Pherecr. 113.5 K.-A., Poll. 6.87.

*μυστιονίκης, ου, ὁ, victor in some unidentified games or event, (cf. perh. °μυστικὸς ἀγών, ἱερονίκης) *POxy.Hels.* 25.31 (AD 264, sp. -νείκ-).

μυστιπόλος, add 'masc. subst. Orph.*H.* 18.18, 25.10, al.'

μύστις, add 'μύστισι σὺν βάκχαις *POxy.* 3723.13 (ii AD)'

μυστοδόκος, add '*PMag.* 20 ii 8'

*μυστρικός, written μουστρικός, ὁ, *spoon-maker*, *MAMA* 4.100 (Synnada, vi AD).

μυστρίον **II**, delete the section transferring quot. to section I; add '*PMich.*XIV 684.14 (vi AD, sp. μιστ-); cf. mod. Gk. μυστρί'

μύστρον **2**, add '*bricklayer's trowel*, *Vit.Aesop.*(G) 116'

μῡτᾰκισμός, delete 'Diom. p. 453 K.' and after '*Gloss.*, etc.' insert 'a fault of pronunciation, Diom. p. 453 K. (pl.)'

†μύτης (μύτις La.)· ἰχθὺς θήλεια ἥτις ἄνευ ἄρρενος οὐ νέμεται· καὶ ὁ ἐνεός· καὶ ὁ μὴ λαλῶν καὶ ὁ πρὸς τὰ ἀφροδίσια ἐκλελυμένος Hsch.

μύτις **II**, add 'cf. mod. Gk. μύτη'   **III**, add 'see also °μύτης'

μύττακες, after 'Ἴωνες' insert '(Λάκωνες vel Κρῆτες cj. La.)'

μυττωτός, add 'title of work by Parthenius'

μύχατος, add 'ἐνὶ μυχάτοισι δόμοιο Q.S. 13.385 codd.'

μυχή, delete the entry.

μυχθίζω **1**, for this section read 'app. *moan noisily*, A.*fr.* 461 R. (perh. referring to ἀναμυχθίζομαι A.*fr.* 473, cf. ἐκ δὲ τοῦ μύζειν καὶ ὁ μυκτὴρ λέγεται καὶ ὁ μυγμὸς καὶ τὸ μυχθίζειν παρά τε Αἰσχύλῳ καὶ ἄλλοις Eust. 440.25; see also ib. 1965.47), Call.*Dian.* 61 (cj., codd. μοχθ-)'

μυχός **1**, add 'φρένων .. ἐν μύχῳ Theoc. 29.3'

μύω **I 2**, after 'S.*fr.* 774' insert 'πρὶν μύσαι id.*Inach.* (fr. 269c.24 R.)'   add 'cf. Myc. *mu-jo-me-no* (part., prob. dat. sg.), but in sense as μνέω'

μυωξία, add 'Gr.Naz.*Ep.* 4'

μυωπάζω, add 'Heph.Astr. 3.45.2'

μυωπίζω, for pres. def. read '*spur* or *goad*'

μύωψ **I**, add 'transf., of a flower which closes when the sun is obscured, κόρκορον Nic.*Th.* 626'

*μωιδων, unexpld. wd., Pi. in Hdn.*fr.* 7 H.; cf. Hsch. μῳδεῖ· λαλεῖ, ᾄδει.

μώϊον, after 'measure of capacity' insert 'equivalent to 2 γόμοι (cf. T.Reekmans, *Sixth Century Account of Hay* (= *SB* 9920), p. 31 no. 2)'

μῶκος, delete 'cj. in Epich. 148'

*μωληθμός· μάχη Theognost.*Can.* 60 A. (-λιθ- codd.).

*μωλία, ἡ, *battle*, Hsch.; cf. μῶλος.

*μῶλον, τό, kind of garlic, Plin.*HN* 26.33; cf. μῶλυ II.

*μῶλος (B), ὁ, = Lat. *moles, mole, breakwater*, *Cod.Just.* 10.30.4 pr.

μωλυτική, delete the entry.

*μωλυτικός, ή, όν, *coming to a head*, of a tumour, prob. cj. in Praxag.ap.Orib. 44.18.2.   **2** -ή· φοβερά Hsch.

μωνιή, μωνιόν, add 'app. artificial formations after μεταμώνιος'

μώριος **3**, for this section read 'neut. μώριον· πόα τις, ἣ πρὸς φίλτρα χρῶνται Hsch.'

*μωροκυστα, epith. of a harlot, cf. κύσθος, *Dura*⁹ 1.212 (iii AD).

*μῶρον, = μόρον 1, Hsch.; = μόρον 2, Theognost.*Can.* 131; cf. Lat. *morum*.

μωρός **3**, for 'things' read 'acts, etc.'

*μωρός (B), ά, όν, μωρόν· ὀξύ, μάταιον, ἀμβλύ etc. Hsch.

*μῶρυ· ὄξος δριμύ Theognost.*Can.* 60 A., Zonar.

μώψ, add 'cf. μύωψ, νώψ'

N

νάβλα, for pres. def. read 'a Phoenician *harp*'; line 2, delete 'cj. in S.*Fr.* 849'; line 3, after 'cf. ναύλον I' insert 'ναύλα'

*ναβλᾶς, ᾶ, ὁ, player of the instrument νάβλα, *GDI* 5258 (Acrae; for reading see *La Carie* p. 283 no. 6).

*ναβλίστρια, ἡ, fem. of ναβλιστής, Heuzey-Daumet *Mission Arch.de Macédoine* no. 10.

νάερρα, add 'app. Aeol. ending -ερρα for -ειρα, (να‹έτ›ερρα cj.)'

ναεύω, for '*Leg.Gort.* 1.39' read '*Leg.Gort.* 1.40, 43' and add '*Inscr.Cret.* 4.41 iv 8, 47.31 (all Gortyn, v BC); cf. ναύω II'

ναί I 1, add 'in a dependent inf. clause, αἴ κ' αὐτὸν αἰτιῆται ναὶ ἀποδό(θ)θαι ἢ ἀποκρύψαι *Inscr.Cret.* 4.47.27 (v BC)' 2, line 3, after '*HG* 4.4.10' insert 'with petitions'; add 'E.*Ph.* 1665, Ar.*Pax* 1113, *Nu.* 784, al., *GVI* 1920 ([ν]αὶ λίτομαι), Herodes Att. ap.Philostr.*VS* 2.5.3' II 1, line 2, after '(lyr.)' insert 'Hdt. 1.159.4'

ναΐδιον, add '*IEphes.* 3327 (ii/iii AD)'

ναιετάω II, for '*to be situated, lie*' read '*to be inhabited, provide habitation*'; line 2, delete 'hence *exist*'

νάϊος, read 'νάϊος (A)'

*νάϊος (B), α, ον, non-Att.-Ion., *of a temple*, Myc. *na-wi-jo* (m. acc. sg.), cj. in Pi.*P.* 6.4.

Ναῖς (s.v. Ναϊάς), at end after 'Ναΐδες' insert 'Alcm. 63 P.'

ναΐσκος, after '*shrine*' insert '*SEG* 30.1220 (ii/i BC)'

*ναιτάω, = ναιετάω, poet. in *MAMA* 1.412 (E. Phrygia).

ναίχι, before 'S.*OT* 684' insert '*SEG* 24.73 (Attica, *c*.500 BC); add 'coupled w. εὖγε, *ARV*² 28 no. 11, 1620'

ναίω, line 1, after 'poet. verb' insert 'also in late prose, e.g. Plu. 2.606f' II 1, line 3, after '*h.Ap.* 298' insert '(dub., cj. ἔλασσαν)' III, after 'Il. 14.119' insert 'Call.*fr.* 680 Pf.' at end add '(cf. °νάω)'

Νακόρειον, read 'νᾱκορεῖον'

*νᾱκοτίλης, ου, ὁ, = νακοτίλτης, *SEG* 17.529.

νᾱκοτίλτης, add '*CEG* 626 (iv/iii BC)'

νακτός I, for 'τὰ νακτά *felt*, Hsch.' read 'νακτά· τοὺς πίλους. καὶ τὰ ἐμπίλια Hsch.' and add '*frontlet bands*, Aq.*Ex.* 13.16, *De.* 6.8' II, for this section read '*choked* (of a river), *SEG* 7.12.10 (Susa, i BC)'

νᾱκύριον, add '(νᾱκύδριον La.)'

*νάμα, *homage, reverence*, Iranian wd. used in Mithraic inscrr., νάμα θεῷ Μίθρᾳ, νάμα πατράσι Dura⁷ 87 no. 848, al.

*ναμαραν, (m. acc. sg.) perh. *candelabrum*, *Inscr.Délos* 2240, 2241, app. Semitic wd.

ναμαρᾶς, delete the entry.

*νᾱμόφορος, ον, *carried on the stream*, *CE* 47.288.

*ναννούδιον, τό, dim. of νάνος, νάννος, *lap-dog*, Sch.Luc.*Conu.* 19.

νᾶνος I, add 'as adj., of ponies, Cinna *Poet.* 9(1)' II, add 'b kind of shallow water-vessel, Varro *LL* 5.119, Paul.*Fest.* p. 176 M.'

*ναοθέσιον, τό, (?)*temple area*, PReinach 2066.36 (ii AD) in *JJP* 11/12.75.

ναολέκτης, delete the entry.

*ναοπηγός, v. °ναυπηγός.

ναός I, antepenultimate line, after 'Aeol. ναῦος' insert 'perh. in Sapph. 2.1 L.-P. (ναυον ostr., i.e. ?ναυϜον)' and for 'Alc. 9' read 'Alc. 325 L.-P.' IV, line 3, after '(Phaestus, ii BC)' insert '*Inscr.Cret.* 1.xvii 21, al.' add 'V (*Christian*) *church*, Eus.*HE* 10.2.1, Mitchell *N.Galatia* 211, *SEG* 30.1711 (AD 596).' add 'cf. Myc. adj. *na-wi-jo*, *of shrines*'

*ναοφυλακέω, *to be guardian of a temple*, *IEphes.* 3263 (ii/iii AD).

*ναοφυλακία, ἡ, *office of temple guardian*, *IEphes.* 4330.5 (*c*.iii AD).

νάοω I, for 'τὰν ἀγέλαν .. (Crete)' read '1 for a ceremony in which the ἀγέλα took the oath to the constitution, *Inscr.Cret.* 1.xix 1.24 (iii BC). 2 in order to give asylum, ib. 4.83.5 (v BC).'

νάπος, add 'II γυναικὸς αἰδοῖον Hsch.'

νάρδος, at end for '(Semitic .. *lardu*)' read 'fr. Skt. *náladam*, perh. through Semitic, cf. Hebr. *nērd*, Aram. *nārd(ēn)*, Akk. *lardu*'

*ναρδοσμῖλαξ, ακος, ἡ, uncertain whether an otherwise unattested plant, or a combination of νάρδος and σμῖλαξ, *PMed.Rez.* 8.11.

*Ναρηνός, ὁ, title of Zeus in Dacia, *SEG* 25.828, al. (Rom.imp.).

*Ναρθάκιον, τό, place-name in Thessalian Phthiotis, Xen.*HG* 4.3.8, Str. 9.5.10, etc.

*ναρθᾱκιῶντες· νάρθηξι πλήσσοντες Hsch.

ναρθηκιάω, delete the entry.

ναρθήκιον, add 'III *medicine chest*, Cic.*Fin.* 2.22, Mart. 14.78.1.'

ναρθηκοφόρος 2, add '*TAM* 5(1).822 (AD 198/9)'

νάρθηξ, line 1, after lemma insert '(originally -ᾱκ- suffix, v. °Ναρθάκιον, °ναρθᾱκιῶντες, νάθραξ)'; lines 4/6, for 'as a schoolmaster's *cane* .. Onos. 10.4' read 'used as a stick for striking, as a dummy weapon, etc., X.*Cyr.* 2.3.20, Arist.*Pr.* 948ᵃ10, Onos. 10.4; by a schoolmaster, *AP* 6.294 (Phan.); used as a splint, Hp.*Off.* 12, Gal. 10.437.18'

νάρκιον, after lemma insert '(A)'; add 'and also perh. νάρναξ'

*νάρκιον (B), τό, dim. of νάρκη II, *electric ray*, Philox.Leuc. (*b*).11 P.

ναρκισσίτης, before 'Plin.' insert 'also fem. ἡ ναρκισσῖτις'

νάρκισσος, line 2, after 'Theoc. 1.133' insert '*AP* 5.147 (Mel.), *GVI* 1409'

*Ναρνάκιος, ὁ, title of Poseidon, from Larnaca, Cyprus, *APF* 13.14 n. 2; cf. νάρναξ.

*νασαμωνῖτις, ιδος, ἡ, a precious stone, Plin.*HN* 37.175 (fr. Libyan tribal or place-name).

νασμός, add 'of a deluge of rain, Lyc. 80'

*νᾶσος, v. ‡νῆσος.

ναστήρ, add '*MAMA* 7.584 (pl.)'

+νάστης· οἰκιστὴς καὶ κύριον ὄνομα Hsch.; cf. Νάστης pers. n., Il. 2.867, °μετανάστης.

ναστός I 3, add 'neut. pl., ναστά· ψαιστά. Ῥόδιοι. καὶ Ἀττικοὶ ἄρτους καὶ ἱερὰ πέμματα Hsch.'

*ναστοφάγος, ον, *that eats* ναστοί (*cakes*), orac. in Paus. 8.42.6 (v.l. ἀναστο-).

*νατήρ [?ᾱ cf. νάτωρ], ῆρος, ὁ, *tile*, *AE* 1948/9.136 A 12, 17 (Epid., iv BC), νατῆρες· ὑπηρέται (s.v.l.). ἢ κεραμίδες Hsch.

νατῆρες, delete the entry.

ναυαρχέω II, add 'perh. also *TAM* 4(1).215.7'

Ναυαρχίς II, add '2 cult-title of Aphrodite, *IPE* 2.25 (i BC).'

*ναυδόμος, ὁ, Myc. *na-u-do-mo* (pl.), *ship-builders*.

*ναυκλάριος, α, ον, *of merchant shipping*, ναυκλαρίου Ποσειδώνος *Inscr.Délos* 2483 (i BC); cf. ναύκληρος, ναυκλήριον.

*ναυκλήρισσα, ἡ, *female boat-owner*, *BE* 1961.457 (Cos).

?ναυλεπλοῖον, τό, in formula χωρὶς ναυλεπλοίου, *free of freight charges*, PMich.VI 400, 401, etc. (iv AD), unless misreading for χωρὶς ναύλ(ων) πλοί(ων), cf. PMich.VI 399.

ναῦλον II, add '*SEG* 26.382 (Athens)'

ναυλόχιον, add 'Plu.*Them.* 9.2'

ναύλοχος II, add '**b** as a place-name, *SEG* 23.189 (Argos, iv BC), *OGI* 1.2. **2** name of a hero, *Inscr.Prien.* 196, *SEG* 33.640, al. (Rhodes, ii/i BC).'

*ναύλωσις, εως, ἡ, *hiring* or *chartering of ships*, *SB* 8754.8 (i AD), 9212.10 (iii AD).

ναυλωτικός, fem. subst., add '*Ann.Épig.* 1984.227a'

ναυμαχία, add 'II *mock naval battle* presented as a spectacle, Vell. 2.56.1, Suet.*Jul.* 39.1. **b** an artificial *lake* for this purpose, Suet. *Nero* 27.2. III prob. *game of chance*, Lucil. 14.460 K., Poll. 7.206, cf. *BCH* 79.547.'

*ναυμάχιον, τό, = °ναυμαχία II b, Acta Petr. et Paul. 79.

ναυπηγέω, line 2, after 'Pl.*Alc.* 1.107c' insert 'Arist.*EE* 1247ᵃ25'

ναυπήγιον, add '*IG* 1³.182.10 (*SEG* 26.21)'

ναυπηγός, add 'ναοπηγός *SB* 3506, ναυπᾱγός *SEG* 33.640 (Rhodes, ii/i BC)'

ναύπλιος, add 'Plin.*HN* 9.94, also adj., τῶν ν. (written ναυπλοίων) κόχλων Ps.-Democr. 357.16'

ναῦς, line 11, after 'Phryn. 147' insert 'νέας Polyaen. 4.7.6, v.l. in J.*Vit.* 33'; line 13, after 'νεός' insert '(Hdn. 2.675)' and after 'Od. 9.283' insert 'dat., νεΐ prob. in *AP* 7.637 (Antip.)'; line 17, after 'Dem.Bith. 4.6' insert 'dat. pl. νήεσιν Q.S. 3.744, 8.362' and after 'νεῦς' insert 'cf. *AP* 13.27 (Phal.)'; line 21, before ':– Dor.' insert 'cf. νηῦς, νηῦν, Hdn.Gr. 2.645, prob. in prov. in Suid. s.v. ἐγένετο (= Zen. 3.44)'; line 25, for 'f.l. .. 22.17' read '*Hymn.Curet.* 58 (prob. monosyll.)' and after 'sg.' insert 'acc. νᾶα prob. in Alc. 117(*b*).21 L.-P.' add 'II app. representing Hebr. wd. for *anus*, Lxx 1*Ki.* 5.6 (ita B, ἕδρας A).'

ναυστολόγος, add '2 *levying seamen*, Str. 8.6.15 (cf. *ZPE* 9.204).'

ναύστολος, delete the entry.

ναύτης I, add 'ὥστε μή μ' ἄγειν ναύτην *take along on a voyage*, S.*Ph.*

901, cf. Pl.*Ep.* 347a' add '**III** *owner* or *manager of a boat*, PRoss.-Georg. III 5^r.5, 10 (iii AD), PSI 948.6 (iv AD), POxy. 1947.5, 1948.7 (vi AD).'

ναυτικός I 1, add 'ναυτικὸς ὄχλος the *class of poorer citizens at Athens who rowed in the fleet*, Arist.*Pol.* 1304^a22, 1322^b7'

ναυτιώδης 1, delete 'Plu. .. 128d'   **2**, after '*disposed to nausea*' insert 'of persons, Plu. 2.128d; ὀρέξεις ib. 127a'

*****ναυτοκολυμβητής**, οῦ, ὁ, *sailor-diver*, PMich.III 174.4 (ii AD).

ναυτολόγος, delete the entry (v. °ναυστ-).

*****ναυτοτίρων**, ωνος, ὁ, *naval recruit*, PSI 781.9 (iv AD), cf. Lat. *tiro*.

*****ναυφράγιον**, τό, *shipwreck*, ν. ποιήσαντες Maecian.*Dig.* 14.2.9 (calque of Lat. *naufragium facere*).

ναυφῠλᾰκέω, add 'w. acc., θαλαμηγόν PTeb. 802.5 (ii BC)'

*****ναυφῠλάκια**, τά, *wages of* ναυφύλακες I, BCH 80.64 (Rhamnus, iii BC).

ναυφύλαξ I, add 'Ulp.*Dig.* 4.9.1.3, Suid. s.v. ναυτοδίκαι'

νάφθα, for 'Lxx *Da*. 3.64' read 'Thd.*Da*. 3.46'; add 'var. forms νεφθαρ, νεφθαι Lxx 2*Ma*. 1.36, see also ἄφθα'

*****νάω** (B), = ναίω, prob. read by Zenod. in Il. 6.34, 13.172; Lyr. ap.Clem.Al.*Strom.* 4.26.167.

*****νεάδιον**, τό, unexpld. wd., Stud.Pal. 20.233.2 (vi/vii AD).

νεάζω, line 2, after '(Lucill.)' insert 'part. AP 9.261 (Epig.)'   **III**, add '*Gp.* 2.19.1'

νεανίας I, add '**3** as an Attic cult-hero, IG 2².1358B.21, SEG 33.147.27.'

*****νεᾱνικότης**, ητος, ἡ, *youthful vigour, prowess*, Sext.*Ps.* 9.1, 109(110).3.

*****νεᾱνιότης**, ητος, ἡ, = °νεανικότης, Aq.*Ps.* 9.1, 45(46).1.

νεᾱνισκάρχης, for 'official in charge of ἔφηβοι' read 'official in charge of organization of νεανίσκοι' and add 'οἱ νεανίσκοι οἱ γυμναστικοὶ ἐτείμησαν Διόφαντον .. τὸν νεανισκάρχην SEG 29.1201 (Lydia)'

*****νεᾱνισκεία**, ἡ, *vigour*, SEG 13.261.3 (iii AD).

*****νεᾱνισκολόγος**, ὁ, app. *convener of* νεανίσκοι, SEG 17.662 (Aspendus), cf. Sch.Juv. 8.191.

νεᾱνίσκος 1, after '*young man*' insert 'freq. treated as a particular class' and add '*INikaia* 1086, SEG 29.1201'

⁺νεάοιδος, ὁ, *young singer*, prob. with unbroken voice, *treble*, AP 7.13 (Leon.Tarent. or Mel.); cf. νεαρῳδός.

Νεᾱπολίτης, add 'fem. **Νεᾱπολῖτις** AR 35 p. 112'

νεᾱροπρεπής, delete '*possessing* .. Aristid.*Rh.* 2 p. 551 S.'; add 'transf., *suited to the modern style*, Aristid.*Rh.* 2 p. 551 S.'

*****νέαρχος**, ου, ὁ, *leader of the* νέοι, IEphes. 1143, 1145 (i/ii AD).

νέᾱσις, add 'Gaius *Dig.* 50.16.30.3'

νεᾱτη, add '**2** perh. personified as a Delphic Muse, *Νήτας. Μέσσας. Ὑπάτας* SEG 30.382 (Argos, *c*.300 BC), cf. Plu. 2.744c.'

νέατος (Λ), line 1, after 'η, ον' insert '(also ος, ον Arat. 60)'; line 10, after '*to be situated*)' insert 'neut. pl. νείατα as adv., *deep down*, Nic.*Al.* 120'

νέατος (B), after 'ον' insert 'νῆτ(ος) SEG 34.940 (Camarina, iv/iii BC)'

*****νεβελ**, Hebr. wd. = *wineskin*, νεβελ οἴνου Lxx 1*Ki.* 1.24, 2*Ki.* 16.1, *Ho.* 3.2.

*****νεβεύω**, perform some function in the cult of Artemis, (?for *νεβ(ρ)εύω, cf. °νεβρίζω) Πολέμων 1.249 (Larissa), IG 9(2).1123 (Demetrias, ii BC), SEG 34.489; cf. °ἐπινεβεύω.

*****νέβρειον**, τό, name of a plant, = ἐλαφόβοσκον, Ps.-Dsc. 3.69.

νέβρειος, for 'αὐλοί .. 244' read 'Call.*Dian.* 244; ν. αὐλοί pipes *made from fawn*(*-bone*)'

⁺νεβρίζω, *dress* (initiates) *in a fawnskin* (in Sabazian revels), D. 18.259, cf. Harpocr.

*****νέβριον**, τό, dim. of νεβρός, ἴσα νεβρίοισιν prob. in Sapph. 58.16 L.-P.

νεβρίτης, before 'Orph.' insert 'rest. for νευρ- in' and after 'Orph.*L.* 748' insert '*Lapid.Gr.* 187.20'

νεβρός, add 'Theoc. 11.40, al.'

νεβροτόκος, after '*bringing forth fawns*' insert 'as subst. = *deer*'

⁺νεβροφόνος, ον, *fawn-killing*, πούς A.*fr.* 47a ii 18 R.; a kind of eagle, = πύγαργος, Arist.*HA* 218^b20; epith. of Dionysus, Nonn.*D.* 44.198.

⁺νεημελκτος, η, ον, *newly-milked*, νεημέλκτη ἐνὶ πέλλῃ, i.e. filled with new milk, Nic.*Al.* 311.

*****νεητόκος**, ον, = νεοτόκος, Nonn.*D.* 25.553, al.

νεήτομος, for '*cut* .. *young*' read '*newly castrated*'

νειάτιος, add 'rest. in Call.*fr.* 384.49 Pf.'

*****νεικεογενής**, ές, *engendered in strife*, Simp. *in Ph.* 161.12.

νεικέσσιος, add 'perh. read νείκεσσι· πολέμοις'

νεῖκλον, for 'cf. νίκλον' read 'cf. °νίκλον, °νικλεῖν'

Νειλεῖον, delete the entry.

*****Νείλεως**, v. °Νηλεύς.

*****νείλιος**, ὁ, a dull green precious stone, Plin.*HN* 37.114.

νειλοκαλάμη, add '(unless by metathesis for λινοκαλάμη)'

Νειλομέτριον, for 'rod' read 'scale'

Νειλόρυτος, add '*GVI* 766 (Tithorea)'

νειοκόρος, for '(Pancrat.)' read '(Pancrat., fem.)'

νειός 2, at end delete '*IG* 2².334.17'

νεῖος (A), for 'A.R. 1.125, Hsch.' read 'neut. as adv. *newly, lately*, Call.*fr.* 384.5 Pf., A.R. 1.125'

νεκροποιός, for '*killing*' read '*making lifeless*' and add 'Alex.Aphr.*in Top.* 376.27'

*****νεκροστολιστής**, οῦ, ὁ, prob. = νεκροστόλος, CE 26.157 (i AD).

νεκροτάφιον, τό, *tomb*, BGU 34 iv 17 (pl.).

⁺νεκροφόρος, ὁ, *one who carries out a corpse for burial*, Plu.*Cat.Ma.* 9.2, id. 2.199e, Gloss.

νεκρώδης, after '*corpse-like*' insert 'Plu.*Phoc.* 28.5'

*****νεκταίρουσιν**· κολάζουσιν Hsch. (νεκταροῦσιν La.'; cf. νεκτάρας).

νεκτάρεος, add 'also as subst. = νέκταρ, Antip.Sid., Philol. 101.104.7'

νεκτάρθη, add '(ἐξημιώθη La.)'

*****νεκύα**, ἡ, v. νέκυια IV.

⁺νεκῠδαίμων, ονος, ὁ, *spirit of a dead person* (esp. one who has died before his time), SB 4947.1 (iii AD), PMag. 4.368, al., Hesperia 54.232 no. 12; cf. νεκυοδαίμων.

νεκῠδαλος, for '*nympha*' read '*pupa*'

*****νεκύειον**, τό, *corpse*, MAMA 7.402.

νέκυια I, add 'applied to Caesar's alleged conjuring-up of outcasts of society as his entourage, Cic.*Att.* 9.10.7, 9.11.2, 9.18.2'   **III**, delete the section.

νεκυομαντεία, line 1, after 'ἡ' insert '*necromancy*, PMag. 7.285'

νεκυομαντεῖον, delete 'in pl., PMag.Lond. 121.285'

νεκύσια, add 'cj. in Plu.*Crass.* 19'

νεκυσσόος, for '*rousing the dead to life*' read 'epith. of Persephone, *speeding the dead on their way*'

Νεμέα, line 3, after 'ib. 7.82, etc.' insert 'in title of Nemean Zeus, τοῦ Διὸς τοῦ Νεμέαι SEG 30.360, 34.282'

Νεμεα, add '**Νεμέαια**, BCH 81.684'

Νεμεᾱκός, add 'Plu. 2.677b'

Νεμεάς, after 'Pi.*N.* 3.2' insert 'subst., ἡ N. *the Nemean games*, ASAA 30/32(1952/4).290 no. 66, lines 16, 17 (Rhodes)'

νεμέθω, line 1, after 'νεμέθων' insert '*devouring*'

*****νεμεσήμων**, ον, gen. ονος, *indignant, resentful*, Call.*fr.* 96.1 Pf., Nonn.*D.* 25.125.

νέμεσις, lines 8/9, delete 'πενθεῖν .. anap.)'   **B 1**, line 6, after 'S. l.c.' insert 'as a statue, AS 37.56 no. 3'; add 'Βωμὸς Νεμέσεων SEG 30.860 (Dacia, iii AD)'

νεμήϊος, for '(expld. .. νέμω)' read '= Νέμειος'

*****νέμημα**, ατος, τό, *bounty*, SB 9132.8 (iii/iv AD).

⁺νεμητής, οῦ, ὁ, *distributor, dispenser*, IEphes. 1604.4, Poll. 8.136; cf. ἀπονεμητής.

νεμήτρια, add '(Rome, iv AD)'

νέμος, add 'of a sacred grove (perh. influenced by Lat. *nemus*), Philostr.*VA* 3.1, 4.36, 7.8'   add '**II** = τὸ γυναικεῖον αἰδοῖον Hsch.   **III** = τὸ τοῦ ὀφθαλμοῦ κοῖλον (read κύλον) Hsch.'

νέμω A II 1, add '*have legal possession of*, (= Lat. *possidere*), Cod.*Just.* 10.11.8.5a, 5b, Just.*Nov.* 119.7'   **B 1**, line 1, delete '*drive to pasture*'

νέννος, lines 3/4, delete 'q.v.'; add '*Inscr.Cret.* 2.xiii 5.2'

*****νεόβακχος**, ὁ, prob. *newly initiated Bacchanal*, Sokolowski 3.90.7 (Callatis, ii BC); *worshippers of Zeus Dionysus*, AJA 66.286 (Phrygia, as adj., μύσται .. νεόβαχχοι).

νεόβλαστος, after '*sprouting afresh*' insert 'Simon. in POxy. 3965 fr. 27.16'

*****νεόβλεπτος** (cj.), ον, *newly seen*, Hsch. s.v. νεώπας.

νεογυνης, delete the entry.

νεόδμητος (A) and (B), after 'ον' insert '(also η, ον, Hsch.)'

νεόζυγης, after 'metaph.' insert 'νεοζυγὲς ἅρμα Choeril. 1.5'

νεόθηκτος, after '*newly whetted*' insert 'Plu.*Cic.* 19.2'

*****νεόθηρος**, ον, perh. *newly caught*, of Chr. converts, MAMA 6.227 (Apamea).

*****νεόθνητος**, ον, app. *who was to die young*, epigr. in SEG 36.602 (Maced.).

νεοίη, add 'pl. = ἀφροσύναι Hsch.'

Νεοκαισαρεών, add '*IEphes.* 614B.7'

*****νεοκέλαδος**, ον, *newly* or *youthfully resounding*, χορός B.*fr.* 61.2 S.-M.

*****νεοκένωτος**, ον, *newly emptied*, κεράμιον Afric.*Cest.* p. 179 V.

νεόκλωστος, for '*fresh-spun*' read '*newly woven*'

νεοκμής I, for this section read '*newly made*, Nic.*Th.* 707; *newly got, fresh*, ποίας νεοκμῆτας ib. 498 (cj.)'

νεόκμητος I, delete the section.

νεόκουφον, delete the entry (read ‡καινόκουφον).

νεόκτιτος, add 'cj. in A.*fr.* 78c.51 R.'

*****νεόλεκτρος**, ον, *newly married*, prob. in A.*fr.* 168.20 R.

*****Νεομήνιος**, v. νουμήνιος.

*****νεόπιστος**, ον, *newly believing*, Swoboda *Denkmäler* 61 (Vasada).

*****νεοποιός**, v. °νεωποιός.

*****νέοπος**, ον, perh. *uttering new things*, Trag.adesp. 654.18 K.-S.

Νεοπτόλεμος, before 'S.*Ph.* 4, 241' insert 'Pi.*N.* 7.35'

νέορτος, at end delete 'νεοργόν or -ουργόν codd. Plu.'

νέος I 1 a, add 'w. pr. n., to distinguish generations, εὐτυχῶς Ἡσυχίῳ νέῳ SEG 30.1785 (iv AD)' **II 1**, add 'w. name of a god, as title of imperial family, Μάγνητες θεὸν θεοῦ υἱὸν Τίτον Καίσαρα νέον Ἀπόλλωνα εὐεργέτην SEG 23.450; Δομετίαν νέαν Ἥραν τὴν γυναῖκα τοῦ Σεβαστοῦ IStraton. 1008; w. other celebrated names in honorific inscriptions, Ἰούλιον Νικάνορα νέον Ὅμηρον καὶ νέον Θεμιστοκλέα IG 2².3788 (Augustus), etc. **b** Νέος Σεβαστός, a month in Egypt, = Hathyr, named in honour of Tiberius, PTeb. 561 (AD 14), BGU 1.4 (iii AD).' **III**, add 'ἐ[ς] νέῳ app. for ἐκ νέου anew, SEG 9.72.127 (Cyrene, iv BC), v.l. in Theoc. 15.143' penultimate line, after 'νεϝόστατος q.v.' insert 'ΝεϜόπολις IPamph. p. 201 no. 17' at end add 'Myc. *ne-wo*'

νεοσφαγής, add 'κεφαλή Plu.Cam. 31.4'

***νεοτατεύω**, to be a member of a νεότας (v. νεότης III), Inscr.Cret. 4.164 (iii BC).

νεότης III, add 'also, body of νέοι at Tanagra, poet. in IG 7.581 (Hermes 72.233; i AD, ἐν νεότᾳ)'

νεότικτος, delete the entry.

νεοτρῑβής, for 'freshly ground' read 'newly threshed'

νεοφύτειον, after 'Gloss.' insert 'ἀμπελικὸν ν. PHamb. 68.23 (vi AD, -ιον pap.)'

*νεοφώτιστος, ον, newly enlightened, i.e. newly baptized, SEG 4.20 (Syracuse), 8.45.5 (Scythopolis, iv/v AD), BCH suppl. 8 no. 123 (Thessalonica, iv AD).

νεοχμός I, line 5, after 'Cratin. 145' insert 'A.fr. 78c.50 R. (satyr-play)'

*νεόχωστος, ον, (χώννυμι) newly heaped up, ἠρίον IG 12(2).489.9 (Lesbos).

νέπους, line 4, for 'Call.fr. 77' read 'Call.fr. 222 Pf., pl., id.fr. 66, 186'; line 8, for 'Call.fr. 260' read 'Call.fr. 533, where sense is ambiguous'

*Νερουαίδεια, τά, festival in honour of Nerva, IG 5(1).667 (Sparta, c.AD 97).

*Νερωναῖος, ὁ, a month in Cyprus beginning 2 March, Cat.Cod.Astr. 2.145.5, 146.10.

*Νερώνειος, ὁ, an Egyptian month in Caligula's calendar, POxy. 355 (AD 40/41), BGU 713.26 (AD 41/42); Νερώνειος Σεβαστός, ὁ, an Egyptian month in the time of Nero, = Choiak, BGU 1599.6 (AD 54; -ιος, PFay. 153ᵛ.8 in APF 4.98 (i AD; Νερωνι); also without Σεβαστός, PFay. 153ᵛ.10, 33 (APF 4.98f.; abbreviated).

Νερωνιανός, add '(sc. λίθος) the Neronian stone, name given to the green (πρασώδης) σμάραγδος, Socr.Dion.Lith. 26.9 H.-S.'

νεῦμα I 1, add 'λαιῆς ὑπὸ νεύματι χειρός i.e. on the left, D.P. 517; nodding in time to music, Luc.Ner. 6' **II 1**, delete the section **2**, add 'fig. v. ποδῶν Simm.Ov. 11'

νευρά 2, add 'so ψαλάξεις νευρᾶς κτύπον Lyc. 139, perh. w. play on νεῦρον v' **4** and **5**, delete these sections.

νευρίτης, for 'Orph.L. 748' read 'Lapid.Gr. 187.20'

*νευρόδετος, ον, stringed, ὄργανα Aristid.Quint. 2.19.

νεῦρον V, add 'Call.fr. 199 Pf.'

νευροσπαστέω II, delete the section.

νευρόσπαστος, add 'also fem. subst., a kind of caper berry, Plin.HN 24.121'

νευρότονον, delete the entry.

*νευρότονος, ον, gut-strung, καταπάλτης IG 2².1487.88-90.

*νευρότροπος, ὁ, sufferer from an injury to the sinews, Cyran. 105.18.

*νευρόχονδρος, ον, neuro-cartilaginous, Gal. 18(2).612.10.

νεύω I, add '**5** diverge, deviate, νεύσας .. ἄπωθεν ὁδοῦ AP 6.220.6 (Diosc.).'

*νεφάριος, α, ον, Lat. nefarius, immoral, ἐξ ἰγκέστων ἢ νεφαρίων γάμων Cod.Just. 1.3.44.3.

νεφέλη III, line 2, for 'Call.Aet. 3.1.37' read 'Call.fr. 75.37 Pf.'

νεφεληγερέτα, add 'nom. -έτης, Nonn.D. 38.203'

*νεφελίς, ίδος, ἡ, = νεφέλιον II 2, Cat.Cod.Astr. 8(3).148.11.

νεφελοειδής, add 'epith. of (Jewish) God, PMag. 4.3068'

*νέφθαρ, v. °νάφθα.

νεφόομαι, add '**2** to be formed into a cloud, of dust, Aristodem. 1.8.'

νέφος I 2, add 'ἔθανον λοιμοῦ ν[έ]φει ἐγκαταληφθείς MAMA 9.79.4 (Aezani, ii AD)' add '**III** fine hunting-net, Hsch. s.v. νέφεα; cf. νεφέλη III.'

νεφρός, add 'perh. at SEG 30.1283.4 (sg. rest.) (Didyma, vi BC). **2** nephros Adadu, the kidneys of the Syrian god Hadad, name of a gem, Plin.HN 37.186.'

νέφωσις, for 'Al.Jb. 3.5' read 'Aq.Jb. 3.5'

νέω (A), line 3, delete 'poet. νέον Alc. 143'

νέω (B), line 5, before 'S.fr. 439' insert '(app. implying the double process of spinning and weaving)'; line 6, after 'Pl.Plt. 282e' insert 'hemp, καννάβεως ἐνεσμένης εἰς ⟨σχοιν[ίο]ν Edict.Diocl. 36.9; gold (thread), χρυσοῦ ἐνησμένου ib. 30.2 (nisi leg. ἐνηγμένου)'

νέω (D), delete the entry.

νεωκορέω I 1, line 1, after 'tend' insert 'νενεωκόρηκεν BCH 83.364.45 (Thasos, i BC)'

νεωκόριον, add 'Lindos II 419.24 (i AD); νᾱκορήιον SEG 24.277.12 (Epid., iv BC)'

νεωκόρος, after 'ὁ' insert '(also ἡ, IG 11(2).287A78 (Delos, iii BC), Paus. 2.10.4)'; line 4, after 'ii AD)' insert 'Paus. 10.12.5'; line 5, after 'poet.' insert 'νειοκόρος AP 6.356 (Pancrat.)' and after 'AP 9.22 (Phil.)' insert 'νεακόρος Xanthos 15; νεοκόρος TAM 4(1).34, SEG 31.1548 (Egypt, AD 126)' **II**, add 'νεωκόρος βουλή SEG 33.1123 (Phrygia)'

νεωλκέω, for 'metaph. .. Mort. 28' read '**2** transf., pull along like a ship, haul, νεωλκῶν τὴν ὁδόν (sc. πρόβατον) Men.Dysc. 399 S., τὸ [ν]ενεω[λκημένο]ν (sc. corpse) ἐν τῇ κλίνῃ Phld.Mort. 28.1.'

νεώλκιον, add 'sg., Sch.A Il. 14.35'

νέωμα, add 'CIG 6850 (unkn. provenance, pl.)'

νεωποιεῖον, add '-ποεῖον, BCH 59.478 (Samos)'

νεωποίιον, for '= νεωποιία' read '= νεωποιεῖον'; add 'written ναοποιον in BCH 59.9 (Delph., iii BC)'

νεωποιός, line 1, after 'IG 2².1678bA14' insert 'νεοποιός IEphes. 957, (ii AD), al.' **I**, add 'νᾱποός IG 12(5).173 ii (Paros), 1016 (Naxos)'

νεώρης, add 'E.fr. 964.6'

*νεωρίδιον, τό, dim. of νεώριον, Inscr.Délos 1417Bii118, 119 (ii BC).

νεώριον, at beginning insert 'Dor. νᾱώριον IG 9(1).692.5, 11 (Corc., ii BC)' add '**2** name of a building on Delos, SEG 36.731 (c.272 BC).'

νεωρός, add 'IG 2².1.30'

νέωτα, after lemma insert '[disyll. in Theoc. 15.143 (s.v.l.)]'; add 'for Lat. designatus, οἱ ἐς ν. ἄρχοντες D.C. 56.34.2'

νεωτεροποιία, add '**2** fondness for experiment in literary style, D.H.Dem. 2.'

νεώτερος I 1, after 'Th. 3.26' insert 'w. gen., νεωτέρους ἐτῶν τριάκοντα IG 12(7).515 (ii BC)'; add 'of the New Comedy, Poll. 6.34, al. **b** the younger, junior, following a name, [Κάλλι]ππος νεώ(τερος) IG 2².2323a, Πομπήϊον τὸν νεώτερον Polyaen. 8.23.16, SEG 30.1286 (i AD), 33.589 (iii AD), etc.'

*νεώψ, νεῶπας· ἀντὶ τοῦ νεοβλέπτους, ἢ νέας Hsch.

νη-, add 'for Myc. v. °ἀνωφελής'

νή, after lemma insert 'Boeot., Arc. νεί q.v.' add '**III** without constr., in answer to a greeting, χαῖρε .. νὴ καὶ σύ Men.Sam. 129, Luc.Tim. 46, Fug. 29, al. **IV** yes (= ναί), Men.Carch. 33 S., Dysc. 510, Epitr. 1120, Sam. 385, 389, Satyr.Vit.Eur.fr. 39 xiii 23.'

*νηδεής, ές, fearless, cj. in Alcm. 26.4 P.

*νήδυιος, unexpld. wd., Call. in Suppl.Hell. 306.

νηδύς 4, add 'meton., gestation, νηδύος ἐκ τριτάτης Opp.C. 3.60'

*νηιδία, Ion. -ίη, ἡ, ignorance, SEG 36.790 (Thasos, v BC).

νήιος, line 4, for 'νήϊα alone' read 'ν. πτερά'.

νηῗς (A) I, lines 2 & 5, for '(Call.)Aet. 1.1.33' read 'fr. 178.33 Pf.'; line 4, for 'Aet. 3.1.49' read 'fr. 75.49 Pf.'; line 5, for 'Aet.Oxy. 2079.2' read 'fr. 1.2 Pf.'

νήκεστος, add 'h.Cer. 258 (cj., cod. μήκιστον)'

*νήλας, hyper-Aeol. = νηλής, Balbill. in SEG 8.716.12.

*Νηλεῖον, ὁ, temple of Νηλεύς (Νείλεως), founder of Miletus, IG 1³.84.27 (written Νελ-, 418/7 BC).

*νηλείτης, ου, ὁ, masc. adj., guiltless, Antim. 177 Wy., νηλείταις· ἀναμαρτήτοις, Iamb.adesp. 44 W.

νηλεῖτις, line 3, after 'vv. ll.' insert 'also νηλ[ει]τιε[ς] Hom. as cited in PMil.Vogl. 17 ii 9'

Νηλεύς II, for this section read '= Νείλεως (cf. Hdn. 2.450.26), founder of Miletus, IG 1³.84.4, al. (written Νελ-, 418/417 BC), Call.Dian. 226; cf. °Νηλεῖον'

*Νηληΐς, ίδος, ἡ, fem. adj., epith. of Artemis at Miletus, Call.fr. 80.18 Pf., Plu. 2.254a, Polyaen. 8.35.

νηλής I, after 'E.Cyc. 369 (lyr.)' insert 'of smoke, Hes.fr. 270 M.-W., Suppl.Hell. 1164'

νηλίπους, after 'gen. ποδός' insert 'acc. νηλίπουν A.fr. 451p.21 R.'; after 'Max.Tyr. 30.6' insert 'νηλίπουν κέλευθον A. l.c.'

νῆμα, add '**b** loosely for ὕφασμα, woven work, AP 6.286.5 (Leon. Tarent.).'

νηματικός, for 'woven' read 'spun'

νημερτής, line 1, for 'the only .. A.Pers. 246' read 'A.fr. 168.16 R. and cj. in Pers. 246'; line 3, after 'Hes.Th. 235' insert 'νύμφαι ναμερτεῖς A.fr. 168.16 R.'; line 7, before 'Sup.' insert 'ζόη Herod. 4.68 (prob.)'

νηνεμία, add 'Arist.de An. 404ᵃ20'

νηνίατον, for 'Hippon. 129' read 'Hippon. 163 W. (= Poll. 4.79)'; add 'cf. νηνίατος (cj. for νινίατος)· νόμος παιδιαρώδης καὶ Φρύγιον μέλος Hsch.'

*νηοσσόος (B), ον, driving ships, αὔρη Nonn.D. 39.177, 40.344.

νηπαθής, add 'SEG 32.1608 (Cyrene, AD 251/2)'

νηπιάζω, add '**2** to be a child, Memn. 14.1 J.'

*νηπιόεις, εσσα, εν, = νήπιος, AB 1089.

νήπιος, line 1, after 'also ος, ον Lyc. 638' insert 'SEG 30.1485 (Phrygia, AD 305/6)'

*Νηπιοτροφικός, ὁ, (sc. λόγος), on the rearing of children, title of work by Mnesith.Ath., Sch.Orib.inc. 19 Dar. (= 37 Raeder).

νηπυτία, delete the entry.

νηπῦτιος I, line 3, after '(Il.) 20.200' insert 'ἐξέτι νηπυτίης = ἐκ νέας, from childhood, A.R. 4.791' add 'cf. Myc. na-pu-ti-jo, pers. n.'

νηρείτης, add 'see also °ἀναρίτας, ἀνηρίτης'

νήριθμος, add 'epigr. in SEG 37.712 (Chios, ii AD)'

νῆρις III, delete the section.

*****νηρίς**, ίδος, ἡ, hollow rock, cavern, Hsch. (pl.).

νήρῐτος, line 3, after 'Od. 9.22' insert '(either place-name or adj. here and at 13.351) cf. pers. n. Νήριτος Od. 17.207'

νηρῐτοτρόφος, add '(or νηρῖτοτρόφος, breeding a multitude (of creatures))'

νησαῖος, after 'insular' insert 'of an island or islands'; add 'γέρων A.fr. 46a.15 R.'

νησιάς, after 'in pl.' insert 'PBaden 86.20 (i AD)'

⁺**νησίγδα**, name of a prepared food, ἐν Νυκτί (Philem. 55 K.-A.) ἀποδιδόασι μάσημά τι ποιόν Hsch.

νησίς, after 'islet' insert 'Hippon. 103.3 W.'

νησίτης, (νασῖτις), delete 'γῆ PEleph. 20.48 (iii BC)'

νησιώτης I 1, add 'fem. (-ῶτις) Ach.Tat. 1.18, Charito 5.1' **2**, delete the section **II 1**, line 3, after '(lyr.)' insert 'ἡσυχία Plu. 2.602e'

νησιωτικός, for 'of or from an island' read 'of or connected with islanders'; line 4, for 'insular situation' read 'matters, etc., concerning the islanders'; add 'adv. -κῶς, ἀρχιερασάμενον -κῶς, having been ἀρχιερεὺς νήσου SEG 23.638 (Cyprus, AD 17 or 18)'

νῆσος, line 1, after 'νᾶσος' insert '(νᾶσσος IG 12(1).70, Rhodes)' add '**3** peninsula, Hdn.Gr. 1.91.13, EM 75.1 (of Ἀλωπεκόννησος); cf. χερσονησίζω, Πελοπόννησος.'

νῆστις I 1, at end delete 'metaph., .. (lyr.)' **II 3**, delete the section. add 'see also ἄνηστις'

*****Νῆστις**, ιδος, ἡ, name of a Sicilian water-goddess, Emp. 6.3, 96.2, Alex. 323 K.-A. (= Emp. 6.3).

*****νήτος**, v. νέατος (B).

⁺**Νηφᾰλιεύς**, ὁ, sober, epith. of Apollo (opp. Dionysus), AP 9.525 (acc. -ῆα, v.l. νηφαλέον τε APL).

νηφάλιος I, penultimate line, after 'Crates Hist. 5' insert 'Hsch.'

νήφω, line 7, before 'S. OC 100' insert '(i.e. with wineless libations)'

*****νίβα**· χιόνα. καὶ κρήνην Hsch.

νίζω II, add 'αἷμα νίψαι D.fr. 20 S. (AB 1.360)'

*****νῐκᾰεις**, v. νικήεις.

νικαῖος, after 'belonging to victory' insert 'ἐφύμνιον Call.fr. 384.39 Pf., νικαίων ἔργων PLit.Lond. 62 (i BC)'

νικάριον, add 'Aët. 7.117 (394.13 O.)'

*****νῑκάς**, άδος, ἡ, figure of Victory, SEG 7.1076 (Syria).

νίκαστρον, add '(νίκατρον La.)'

νῑκάω, line 5, before 'pf.' insert 'aor. part. νικέρας CEG 321a (Eretria, c.500 BC)' **I 1**, five lines fr. end, c. dupl. acc., add 'Ἴσθμια .. ἐνίκα ἅλμα, ποδωκείην, δίσκον, ἄκοντα, πάλην APl 3 (Simon.), cf., w. triple acc., ἄνδρας .. πυγμᾶν .. Ὀλύμπια AP 6.256 (Antip.)' **4**, add 'Apoc. 5.5' **II 1 a**, line 8, after 'etc.' insert 'τοὺς νικῶντας ἐκ τοῦ θηρίου those victorious over the beast, Apoc. 15.2 (Semiticism)'

*****Νικέρως**, ὁ, a god (cf. Νίκη, Ἔρως), Inscr.Perg. 8(3).142.

νίκη II 1, add '**b** represented by a statue, IG 1³.52.B3, 323.52, etc., Νίκας δύο χαλκᾶς SEG 37.693 (Inscr.Délos 1403, ii BC). **c** as an attribute of the Roman emperors, SEG 30.1245, 31.916, etc.'

νίκημα, line 2, before 'Delph.' insert 'Satyr.Vit.Eur.fr. 39 xv 7'

νῐκητής, after 'Eust. 157.1' insert 'of the Emperor Constantine, SEG 31.1324 (Cappadocia)'

νικητικός I, add '**2** of or marking victory, τὴν νικητικὴν ψῆφον Just.Nov. 126.2.'

*****Νῑκηφόριον**, τό, a commemorative grove at Pergamum planted by Eumenes I, Liv. 32.33.5.

νῑκηφόρος II, line 5, after 'PTeb. 43.28 (ii BC)' insert 'of the Roman emperors, T&MByz 9.271 no. 3'; add 'as name of a legion (= Lat. Victrix), λεγ[εώ]νος Ϝ′ Νεικηφόρου SEG 23.317 (Delphi, AD 85)'

*****νικλεῖν**· λικμᾶν Hsch.

*****νίκλον**· τὸ λίκνον Hsch.; cf. νείκλον.

⁺**νῖκος**, εος, τό, later form for νίκη (in Lxx translating root represented by Hebr. niṣṣēaḥ "conquer", nesaḥ "eternity", etc.) victory, Lxx 1Es. 3.9, 2Ma. 10.38, BGU 1002.14 (i BC), IG 12(5).764.2 (Andros, prob. i AD; written νεῖκος), Ev.Mat. 12.20, Vett.Val. 358.5, Orph.A. 587, APl. 5.381, read by Aristarch. in Il. 12.276. **2** pre-eminence, glory, Lxx La. 3.18 (v.l. νεῖκος). **II** eternity, εἰς νῖκος, for ever, Lxx 2Ki. 2.26, Jb. 36.7, La. 5.20, etc.

*****νικύλεον**, τό, a kind of fig, Cretan, Hermonax ap.Ath. 3.76e, cf. perh. Myc. ideogram NI for figs.

*****νίκωρ**, unexpld. wd. in Sophr. 133 (Hdn.Gr. 2.938.4).

νιν 2, add 'Pae.Delph. 11'

*****Νινευδιος**, ὁ, cult-title of Zeus at Aphrodisias, BCH 9.80 no. 10, MAMA 8.410, ABSA 59.16ff. no. 19.

*****νίνισσα**, ἡ, perh. midwife, MAMA 7.554 (?or to be read as termination of a preceding pr. n.).

νιννίον, read 'νίννιον'; add 'cf. ninnium, Pl.Poen. 371 (dub. sens.)'

*****νιπτέον**, one must wash, τοὺς πόδας Aët. 16.64.

*****νίπτης**, ου, ὁ, washer, cleaner, POxy. 1917.39 (vi AD).

*****νίτρῐνος**, η, ον, of or derived from νίτρον, χοῖσκος, Inscr.Délos 1426Ai15.

*****νιτρίς**, ίδος, ἡ, some form or derivative of νίτρον, Inscr.Délos 1417Ai65.

νίτρον, penultimate line, for 'mixed with oil as a soap' read 'used in cleaning, ῥύμματι καὶ ν. Χαλαστραίῳ'

νιτροπηγικός, add 'Paul.Aeg. 7.13.17 (326.14 H.)'

*****νιτροπώλης**, ου, ὁ, seller of νίτρον, SB 3913 (Antinoopolis, Chr.), PVindob.G 14296 (Tyche 6.118, v/vi AD)

νιτρώδης I 1, after 'impregnated with ν.' insert 'alkaline' **2**, delete the section transferring quot. to section 1. **II**, add 'CIL 10.6786, 6789 (tit. Lat.)'

νῐφετός 1, add 'AP 7.8 (Antip.Sid.)'

νῐφόεις II, add 'Nic.Th. 291'

νῐφοστῐβής, for 'piled with snow' read 'walking over the snow'

*****Νοέμβριος**, ον, Lat. Novembris, IGRom. 1.176, πρὸ γ′ ἰδῶν Νοεμβρ[ίων] ib. 4.347 (AD 117), SB 10305.1 (AD 124), καλανδῶν Νοεμβρίων SEG 31.830 (Sicily, iv/v AD), etc.

νοερός, lines 1/2, delete '(ψυχαί .. (v.l., Comp.)' add '**b** intelligent, quick, ἐάνπερ γε ᾖ ὁ πυλωρὸς νοερός Aen.Tact. 28.2, Onos. 1.7.'

νοέω, line 8 and **I 3**, line 10, for 'Anacr. 10' read 'Anacr. 24 P.'

νόημα, line 2, after 'νόημα 105.3)' insert 'Aeol. **νόημμα** Sapph. 60.3 L.-P., etc.' **I 4**, for this section read 'ingenious saying, conceit, Eust. 1634.14 (referring to Epich. 87)'

νοθεύω III, after 's.v. Ἴλιον' insert 'Ath. 10.455c'

*****νοθογέννης**, ου, ὁ, app. cross-bred offspring, ψήληκες· τῶν ἀλεκτρυόνων οἱ νοθογένναι Hsch., Suid.

*****νοθολογέω**, perh. speak deceptively or disingenuously, ⟨ἐξ ἐθνῶν μὲν⟩ ὥσπερ κιλικίζειν τὸ νοθολογεῖν Suet.Blasph. 251 T.

νόθος II, line 7, for 'meretricious' read 'not genuine (opp. ἁπλοῦς)'

*****νομάδιον**, τό, guinea-fowl chick, SB 10270.33.5, 34.3 (iii AD).

⁺**νομαῖος**, α, ον, reared or growing in pastures, χίμαρος AP 6.157 (Theodorid.); ἕρπυλλον Nic.Th. 67.

νόμαιος 1, add 'ἀλάλαγμα ν. Call.fr. 719 Pf.'

*****νομάριον** (B), τό, perh. dim. of νομός or νομή, SB 7530.19 (i BC).

νομάς II, line 1, delete 'fem.'; line 2, delete 'calf of the pastures, i.e. fatted' **4**, delete the section.

νομευτικός II, for this section read 'of persons, occupied or employed in herding, νεανίσκος Plu. 2.149c; ἄνδρες Ael.NA 14.16'

νομεύω 2, delete the section.

νομή IV, add 'Lxx 2Ma. 5.14 (cod. Ven.), prob. in 3Ma. 1.5'

νομίζω I 1, transfer 'ν. θειότατον νόμον Gorg.fr. 6 D.' to section II 1.

⁺**νομικάριος**, ὁ, app. some official in a νομός, PBeatty Panop. 1.252 (AD 298), POxy. 3190.4 (iii/iv AD), PMich.inv. 439 (BASP 16.146, iv AD), etc.

νομικός, line 1, after 'ή, όν' insert '(also ός, όν SEG 26.821 (Thrace, c.100 BC)' **I 1**, add '**b** having the character of law, νομικώτεροι οἱ νόμοι Men.Rh. p. 375 S.' **2**, at end after 'Plu.Cic. 26' insert 'νομική (sc. τέχνη) jurisprudence, GVI 2021.3 (Amasia, i/ii AD)'

νόμιμος I 1, add 'ν. γάμος lawful wedlock, Mitteis Chr. II.372, VI.7 (ii AD), AP 5.267.7 (Agath.); ν. ἐπίτροπος = Lat. tutor legitimus, Modest.Dig. 27.1.10.7; ν. κληρονομία = Lat. legitima hereditas, legacy in accordance with the civil law (on intestacy), ib. 26.6.2.1; ἐμπόριον ν. officially regulated port, Peripl.M.Rubr. 4, 21, 35' **II 1**, at end, sg., add 'ἡ παροῦσα διάταξις .. κελεύει .. τὸ αὐτὸ νόμιμον κρατεῖν CodJust. 4.35.24'

*****νόμιμος** (B), ον, pasture-, perh. in IMylasa 273.2, 274.8, 275.6 (papp., νομίμου γῆς)).

νόμος (A) **2**, for this section read 'ν. τέλος pasture-dues, IG 7.2870.16 (Coronea, ii AD); so νόμιον, τό, PStrassb. 21.14 (ii AD)'

νόμισμα II, add 'spec. = χρυσοῦς I 3 (Lat. aureus), D.C. 55.12.4, Scaev.Dig. 40.4.60'

νομιστί, add 'combining sense of by law, M.Ant. 7.31'

*****νομμοκλάριος**, ὁ, perh. = Lat. nummularius, money-changer, MAMA 3.302 (Corycus).

*****νομογράφεῖον**, τό, office of the νομογράφοι, POxy. 2726.27 (ii AD).

*****νομοδεικτέω**, act as νομοδείκτης, POxy.Hels. 25.21 (AD 264).

νομοθετέω II, add 'foll. by ὥστε, CodJust. 1.11.10.5, 8.10.12.7'

νομοθέτημα I, add 'representing Lat. rogatio, D.C. 38.6.1'

νομοθέτης II, add '**2** at Rome, used for decemviri, D.S. 12.24.1, D.H. 10.57.1.'

*****νομοθετητόν**, gloss on θεμιστευτόν, Hsch.

νομομᾰθής, add 'CIJud. 113, 193, 333 (Rome, ii/iii AD)'

νομός I 3, add 'cf. °νόμος I 1 e' **II 1**, delete the section transferring quots. to section I 1

νόμος I 1 e, line 6, after 'Hdt. 9.48' insert '(in this and similar phrases νομός, distribution, shd. perh. be read); cf. νομή IV' add '**f**

νομώνης | SUPPLEMENT | νωμάω

adverbial acc., w. gen. *after the practice* of, τετράποδος νόμον Pl.*Phdr.* 250e.'

†**νομώνης**, ου, Boeot. **-ώνας**, ὁ, *official who collects dues for the use of public pasture*, *IG* 7.3171.43 (Orchomenus).

***νόνναι** v. °*νῶναι*.

†**νόννος**, v. *νέννος*.

***νοουίκιος**, ον, = Lat. *novicius*, *of recent standing*, νοουίκιον δοῦλον ἢ δούλην *SEG* 39.1180 117 (Ephesus, i AD).

νόος, line 3, after '(iamb.)' insert 'A.*Pr.* 164 (lyr.)'; line 9, after 'ibid.' insert '(gen. pl. νῶν Plot. 4.3.14)'

νορύη, add 'rest. in Lang *Ath.Agora* XXI B19'

νοσακερός, for 'Vulgar' read 'Com.'

***νοσεύομαι**, in pf. 'τὰ ἐν τῷ ὀγδόῳ μηνὶ νενοσευμένα *the illnesses suffered in the eighth month*, Hp.*Septtm.* 2'

νοσέω I 1, line 4, after 'A.*Pr.* 386' insert 'ὁ νοσέων *the patient*, Hp.*Epid.* 1.23, al.'

νοσηλεύω, before 'τινα' insert '*CEG* 37 (Athens, vi BC)'

νοσηλός, after '(q.v.)' insert 'νοσηλότερον τὸ σωμάτιον ἔχει *she is in poor health, poorly*, *POxy.* 939.26 (iv AD)'

νοσημάτιον, add '*SEG* 39.883.6 (Chios, Rom.imp.)'

νοσοκομεῖον, add '*CIG* 9256'

νοσοκομέω, before 'Iamb.' insert 'w. acc.'

νόσος, line 1, after 'Ion.' insert 'also A.*Supp.* 684 (lyr.)'   **I**, add '(εἶναι) ἐν νόσῳ *SEG* 34.657 (Maced., iii AD)'

***νοσοτροφέω**, *nurse an illness*, Jul.*Or.* 6.181d.

νοσοτυφέω, delete the entry.

***νοσσάριον**, τό, for *νεοσσάριον, dim. of νεοσσός, classed as non-Att. by Phryn. 182 (177 F.).

νοσσάς, add '*PHib.* 181.13 (iii BC)'

νόσφι, add 'in prose, *CID* I 13.7 (iv BC)'

***νόσφισμα**, ατος, τό, *stealing, peculation*, *PSI* 1120.4 (i AD).

***νοτάριος**, ὁ, Lat. *notarius*, *secretary*, Jul.*Ep.* 23.378 B, *IGRom.* 4.235.13, *BCH* 7.244 (Isauria), *Edict.Diocl.* 7.68, etc.; *PGoth.* 18.5.12, *PGrenf.* 63.16f, etc., τριβοῦνος νοταρίου *BCH* suppl. 8 no. 247 (v AD), τριβ(ούνου) νοταρίου *SEG* 28.1284 (Cilicia, vi AD).

νοτία, add '**IV** *a plant, perh. white bryony*, Plin.*HN* 24.175, Ps.-Dsc. 4.182.'

νότιος I 2, delete the section   **II**, at end after 'Comp. -ώτερος' insert 'Arat. 238, 490'

νοτίς, for 'A.*fr.* 481' read '*Trag.adesp.* 261 K.-S.'

νοτόθεν, after '*from the south*' insert '*on the south side*, *IG* 1³.426.70, *SEG* 32.161 (Athens, 402/1 BC)'

νότος II, after '*PTeb.* 164.17 (ii BC), etc.' insert 'without gen., τὰ οἰκήματα τά τε ἀπὸ νότου .. καὶ τὸ ἀπὸ βορρᾶ *SEG* 23.678 (Cyprus, ii BC)'

νουθετέω 1, before 'Pass.' insert 'w. inf. of act advised, D.Chr. 63.6'

†**νοῦθος**, adj. *soft, quiet* (of sound), Hes.*fr.* 158 M.-W.; cf. νυθός; but expld. by Hdn.Gr. 2.947 as a subst. meaning ψόφος ἐν οὔδει.

***νουμενάρια**, τά, *window-glass* (= Lat. *luminaria*), *PGot.* 7.5 (iv AD).

***νουμεράριος**, ὁ, Lat. *numerarius*, *keeper of accounts*, *SEG* 29.642 (Thessalonica), 32.1554 (Arabia, both early vi AD), *PFlor.* 295.8 (vi AD), *CodJust.* 1.42.2, 12.49.13.1.

***νούμερος**, ὁ, Lat. *numerus*, in sense of *military unit*, *IGRom.* 3.2, *BGU* 316.8 (iv AD), *BCH* 33.34 (v/vi AD), *MDAI(A)* 13.251.

νουμηνία, line 2, after 'Hdt. 6.57 (pl.)' insert '*SEG* 30.980 (Olbia, v BC)'; line 4, after '*PCair.Zen.* 167.5 (iii BC)' insert '**νομηνία** *SEG* 30.957 (Maced., AD 132); **νεμηνία** *SEG* 30.1121 (Entella, iii BC)'

νουμήνιος I, delete 'Att. contr. for νεομήνιος' and add 'as title of Apollo, *SEG* 32.337 (c.300 BC), uncontracted form Νεομήνιος, Philoch. 88b J.'

***Νουμίδαι**, οἱ, *Numidians*; Νουμίδαι Ἰουστινιανοί, body of soldiers stationed in Egypt, *BGU* 2197.7 (vi AD).

***νούμμιον**, τό, = νούμμος 3, *PKlein.Form.* 972.3 (iv AD), *POxy.* 1165.6 (vi AD); written **νούμιον**, *PKlein.Form.* 87.3, *PMasp.* 9ᵛ.24 (both vi AD).

νοῦμμος 2, for 'λίτρα .. pound' read '*twenty-fourth part of old Sicilian talent*'

†**νουνεχόντως**, v. °ἐχόντως.

νυγμή I, for '= sq.' read '*prick, puncture*'

νυκτάλωψ, add '**III** *another name for the plant* νυκτήγρετον, Plin.*HN* 21.62.'

***νύκταρχος**, ὁ, *officer of the night-watch*, *MAMA* 3.428 (Corycus).

***νυκταστράπτης**, ου, ὁ, *emitter of lightning-flashes by night*, *PMag.* 4.182.

†**νυκτερεύω**, *spend the night in the open, be out at night*, X.*Cyr.* 4.2.22, *An.* 4.4.11, 6.4.27, v. ἀθλίως Timocl. 16.1; med., of night-revellers, Timachidas ap.Ath. 699e.   **b** of things, *be left out at night*, Aen. Tact. 30.2.

νυκτηγρεσία, add 'as name of the tenth book of the Iliad, Sch.Hippon. 118 B 6 W. (cf. νυκτεγερσία)'

νυκτιβόας, add 'expl. of *bubo* in Virgil gloss., *PNess.* 1.955 (vi AD, -βόα pap.)'

νυκτιλάλος, for '*nightly-sounding*' read '*sounding by night*'

νυκτιπάταιπλάγιος, for '*nightly-roaming-to-and-fro*' read '*roaming by night*'

νυκτιπλάγκτος, for '*causing to wander .. bed*' read '*marked by nocturnal wandering*'

***νυκτιτρόμος**, ον, *trembling by night*, *Suppl.Mag.* 49.48, 57 (iii/iv AD).

νυκτιφανής, for '= foreg.' read '*appearing at night*'

***νυκτίχροος**, ον, *night-* (i.e. *dark-)skinned*, Ps.-Callisth. 83.6.

***νυκτογράφος**, ὁ, *one who writes by night*, *PMich.*II 123ᵛ ii 14, 23, al.

νυκτοπλοέω, add 'μὴ ἐξέστω αὐτῷ νυγτοπλοεῖν (sic) *POxy.* 3250.22 (AD 63)'

***νυκτοπύρετος**, ὁ, *night-fever*, *PTeb.* 275.22 (iii AD).

***νυκτοφανής**, ές, *appearing by night*, epith. of Hecate, *SEG* 26.819 (after AD 212).

νυκτοφύλαξ, add 'as a Christian office, *SEG* 29.643 (AD 532)'

νυκτοφυλάξια, for '*guard-house*' read '*name of a festival*'

νυμφαγέτης, add 'of Apollo, *SEG* 34.440 (Phocis, ii BC)'

***νυμφαγέτιος**, ον, = νυμφαγέτης, (?)Πᾶν]α Νυφαγέτιο[ν (sic) *SEG* 17.82 (Athens, iv BC).

νυμφάγωγός I 2, delete the section transferring exx. to section 1.

νυμφαῖος II, add '**2** Νυμφαῖα, τά, *festival of the Nymphs* at Apollonia, *Inscr.Délos* 1957, *Hesperia* 4.84 (Athens) (both ii BC).'

νυμφεῖος, add '**II** *of the* Νύμφαι (νύμφη II 2), *Didyma* 159 i 9 (iii AD).'

νύμφη, line 3, after 'νύμφᾱ' insert 'sp. νύφη *IG* 1³.974, 2².4650, *Kafizin* 21, al.'   **I 3**, add '*SEG* 31.1020 (Lydia, AD 82/3), 31.1037 (Lydia, AD 210/1), cf. mod. Gk. νύφη; perh. also *sister-in-law*, *SEG* 34.1221 (Lydia, AD 91/2)'   **II 1**, add 'sg., goddess at Kafizin, Cyprus, *Kafizin* 5, al.; of Isis, *SEG* 24.561 (ii AD); nymphs represented in statuary, *SEG* 24.496 (ii/iii AD)'   **2**, line 3, after '*water*' insert 'Nic.*Th.* 623, al.'

νυμφιάω, for '*mares*' read '*horses*'

νυμφικός I, after 'adv. -κῶς' insert 'Plu.*fr.* 157.6 S.'

νυμφίος I 2, after '*Jd.* 15.6' insert '(v.l. γαμβρός)'   **II**, after 'Pi.*P.* 3.16' insert '(cj.)'; after 'λέκτρα' insert 'Call.*fr.* 63.11 Pf'; for '*Epigr.Gr.* 373' read '*GVI* 1668'; add 'αἷμα Nonn.*D.* 32.34'

νυμφοκόμος, line 2, after 'Hsch.' insert 'transf., A.*fr.* 168.23 R.'

νυμφόληπτος, add '**2** *intoxicated* or *possessed by water*, ν. καὶ βάκχοι τοῦ νήφειν Philostr.*VA* 2.37.'

***νύμφος**, ὁ, *grade in Mithraic initiation*, *MDAI(R)* 49.206 (Dura), Jerome *Ep.* 107.

νῦν I 1, line 9, after 'ἀπὸ ν. *AP* 5.40 (Rufin.)' insert '*CRAI* 1982.62'   **5**, add 'νῦν ὅτε, (*it is) now that*, dub. in A.*Supp.* 630, *Th.* 705, cj. in Alex.Aet. 3.21'   add '**6** *in the present passage*, Sch.E.*Med.* 68, al., Sch.Ar.*Au* 851, al.'

νῦν δή I 1, for '*Grg.* 462b' read '*Phdr.* 250c, *Ly.* 217e' and delete '*Com.adesp.* 597, etc.'   **2**, add 'Ar.*Pax* 5, al.'   **3**, add '*in these circumstances*, Th. 6.24.2'

νύξ I 1, line 9, before 'Hdt. 7.12' insert 'Od. 15.34'; line 18, after 'Pl.*R.* 621b' insert 'νύκτα μέσην Hdt. 8.9.1'   **2**, line 7, after 'Pl.*Criti.* 117e' insert 'ὁ διὰ νυκτὸς στρατηγός *commander of the night-watch*, Laodicée p. 261 no. 3'; line 8, after 'ἐκ νυκτῶν' insert 'Od. 12.286'

Νῦσα, add '**III** pl., prob. the designation of some group of nymphs, *ABV* 39 no. 15.'

†**νύσσα**, ἡ, *marker erected at either end of a race-course*: **1** acting as the *turning-point*, Il. 23.332, 344, al., Theoc. 18.15, Nonn.*D.* 37.112, Gal.*UP* 16.4.   **b** transf., of other circular or continuous courses, Nonn.*D.* 11.165, 39.336; of celestial orbits, ib. 1.169, al.; of recurrent temporal processes, etc., ib. 3.35, 37.6, al.   **2** as the *starting-point* of a race, τοῖσι δ' ἀπὸ νύσσης τέτατο δρόμος Il. 23.758, Od. 8.121, Opp.*H.* 5.642, Lyc. 15.   **3** as the *finishing-post*, fig., ν. ἀοιδῆς ἰθύνειν Opp.*H.* 3.101, Nonn.*D.* 12.87.

νύσταγμα, for '*nap, short sleep*' read '*(period of) sleep* or *drowsiness*'

***νυσταλωπιᾶν** νυστάζειν Hsch. (but v. νυκταλωπάω).

νύχιος, line 1, delete 'Tim. .. by'; line 2, delete '*nightly*, i.e.'   add '**4** Νυχίη, ἡ, a form or title of Dione, *SEG* 34.1436 (Syria, c.iii AD).'

νύχος, add 'inscribed on gaming-board, with uncertain significance, *SEG* 23.620 (Cyprus, iii BC)'

***νώβυστρον**, τό, *a term of abuse, perh. blockhead* (νοῦς, βύω), Herod. 6.16.

νωγαλέος, add '-έον· πυρρόν Theognost.*Can.* 62 A.'

νωθής I, add '**3** *indistinct*, Arat. 228.'

***νωθραίνω**, *to be unwell*, *POxy.* 2609 (iv AD).

νώκαρ, after lemma delete 'ἄρος'   **I**, before 'Hsch.' insert 'ν.· νύσταξις, νώθεια, κακόσχολος ἔννοια'   **II**, for this section read 'νώκαρ· ὁ δυσκίνητος Suid.'

νωλεμές, after 'A.R. 2.605' insert 'cf. νωλεμέα· ἰσχυράν, νωλεμές· ἰσχυρόν Theognost.*Can.* 62 A.'

νωμάω II 2, add 'also of the voice, στονόεσσαν ἀϋτήν ν. A.R. 4.1006; med., νωμᾶται .. ἔθειραν B. 5.26 S.-M.'

219

*νωμενκλάτωρ, ορος, ὁ, Lat. *nomenclator*, *POxy*. 1244ʳ, without context (ii AD, pl.).
*Νῶναι, αἱ, Lat. *Nonae, the Nones*, Ν. Μαρτίων *IGRom*. 4.661.31 (Acmonia, i AD), *BCH* suppl. 8 no. 130 (Thessalonica, AD 469); also νόνναι Plu. 2.269d.
*νωνυμνί, adv., *without being named*, Call.*fr*. 43.55 Pf.
   νώνυμος **I 2**, add 'Arat. 370'   **II**, delete 'Call.*Aet*. .. *being named*)'
*νῶος· μωρός Theognost.*Can*. 62 A.
   νωπέομαι, for '*to be downcast*, Ion Hist. 1' read '*to be abashed*, Ion Hist. 6 J.' and add 'cf. προνωπής'
*νωπήεις, εσσα, εν, perh. = νώψ, Theognost.*Can*. 62 A.

*νώρικον or ος, Phryg. for ἀσκός, acc. to Ps.-Plu.*Fluv*. 10.2.
*νωρυμνόν· οὐχ ὑψηλόν, οὐκ ἐρυμνόν Theognost.*Can*. 62 A.
   νωτηγός, for 'ἵπποι' read 'ἡμίονοι'
   νωτιαῖος, before '*spinal*' insert 'of the back, ἐρέας λαγείας νωτιαίας *Edict.Diocl*. 1.10 (ZPE 42.283)'
   νῶτον **II 3**, for this section read '*the convex side of a shield*, *AP* 6.125, Lxx *Jb*. 15.26; also a part (?*rim*) of a wheel (translating the same Hebr. wd.), 3*Ki*. 7.33, *Ez*. 1.18, al.'
*νωφελής, ές, v. ἀνωφελής.
   νωχελής **II**, for '*abortion*' read '*that which is dilatory in moving*'
*νωχλεύω, = νωχελεύομαι, *Vit.Aesop*.(W) 76.

Ξ

ξαίνω I 1, line 6, after 'etc.' insert 'οὐ ξένουσιν (read ξαίν-) οὐδὲ νήθουσιν v.l. in *Ev.Matt.* 6.28 (cod.Sinaiticus, v. Metzger *Textual Commentary on the Greek N.T.* p. 18)'

*****ξανάα**, τά, *crippling of the fingers* caused by cold and weariness, perh. orig. from carding wool, Sch.Nic.*Th.* 383.

*****ξανθρότης**, ητος, ἡ, *golden colour*, Sch.Hes.*Th.* 350.

ξανθίζω I, add '*Trag.adesp.* 441'

ξανθοδερκής, for '*with fiery eyes*, of a dragon' read '*yellow-eyed*, δράκων'

ξανθός I 2, line 4, for 'B.*fr.* 3.4' read 'B.*fr.* 4.65 S.-M.; ξ. πεύκαις Pi.*Dith.* 2.11 S.-M.' and after 'A.*Pers.* 617' insert 'ἔλαιον E.*IT* 633, ξανθοῦ μέλιτος Lang Ath.*Agora* xxi He36 (iv AD)'; penultimate line, after 'comp. -ότερος' insert 'ξανθοτέραις ἔχῃ[..] ταῖς κόμαις δάϊδος Sapph. 98(*a*).6 L.-P.' **II 3**, add 'cf. Myc. *ka-sa-to*, pers. n.'

*****ξανθότρῐχος**, ον, *tawny-haired*, Hsch. s.v. πυρσοκόρσου λέοντος.

*****ξανθόχλωρος**, ον, *yellow-green*, ὕδωρ Zos.Alch. 142.8.

ξείρης, for 'v. ξυρίς' read 'v. ‡ξυρίς'

*****ξείτης, ξεῖτος, ξεῖθος**, = ἑξέτης, *the throw of six* at dice, *PASA* 2.88, 89.

ξενᾰπάτης, line 2, after 'ξειν-' insert 'Aeol. ξ[εν]ναπάτας Alc. 283.5 L.-P.' **2**, for 'Ibyc.*Oxy.* 1790 i 10' read 'Alc. l.c., Ibyc. 1.10 P.'

ξένη 1, add 'ὅτις ξέν[αν] γεγάμηκε *IG* 5(2).343 (Orchomenus, Arcadia, iv BC)' **2**, add 'designating a place other than one's legal residence, *PMich.*x 580.7, *POxy.* 251.11, 252.10, al.'

*****ξενηδόκος**, Ion. **ξεινη-**, ον, = ξενοδόκος, Nonn.*D.* 13.104, 18.307.

ξενηδόχος, delete the entry.

ξενηλᾰτέω, after '*banish foreigners*' insert 'Plu. 2.727e'

ξενία I, add '**4** *sojourn in an alien state*, Schwyzer 366.A4, 6 (Locris, iii BC).' **II**, before '*PSI*' insert '*PBremen* 15.4 (ii AD)' and after '(iv/v AD)' insert 'so perh. καλεῖ σε εἰς τὴν ξ. ἑαυτοῦ *POxy.* 747.1 (ii/iii AD)'

†**ξενιᾱγός**, ὁ, *bringer of* ξένια, *PZen.Pestm.* 54.32 (iii BC).

ξενίζω, line 5, after 'E.*Alc.* 1013, etc.' insert 'ξένος πεφυκὼς τοὺς ξενίζοντας σέβου Men.*Mon.* 556'

*****ξενικοκέρᾰμος**, ὁ, *foreign jar* (of wine), *SB* 10918.5 (iii AD).

ξενικός I 1, after 'D. 57.34' insert 'at Ephesus, *IEphes.* 884 i 1 (i AD)' **3**, after 'Pl.*Lg.* 702c' insert 'Διονύσια τὰ ξ. transf., of games in honour of Dionysus, *SEG* 24.1023 (Callatis, iii/ii BC)'

ξένιος II 2, neut. subst., add 'of offerings to a deity, Sophr. in *GLP* 73.18' add 'form ξένϝιος, in Myc. *ke-se-ni-wi-jo, ke-se-nu-wi-ja*'

*****ξένισμα**, τό, gloss on θαῦμα, Hsch.

*****ξενοδάϊκτος**, ον, *that murders guests* or *strangers*, *Suppl.Hell.* 991.96 (poet. word-list, iii BC).

*****ξενοδαιτῠμών**, όνος, ὁ, *one that feasts on guests* or *strangers*, E.*Cyc.* 610 (cj.).

ξενοδοκέω, add 'Pl.*R.* 419a'

ξενοδόκος I, line 4, after 'Od. 8.210' insert 'A.*fr.* 451h.3 R., Call.*fr.* 59.19 Pf.' **II**, for 'Simon. 84.7' read 'Simon.*eleg.* 87 W.²'; add '*AD* 11.61 (Larissa)'

ξενοκρῐταί, for ' = ξενοδίκαι' read '*foreign judges* in μετάπεμπτα δικαστήρια'; for '(Patara)' read '(Pinara)'; for 'title of official at Sparta' read 'description of Spartan sent as judge to Alabanda' add '**2** the title of certain Roman judges in Egypt, *POxy.* 3016 (AD 148).'

*****ξενοκυσθᾰπάτη**, ἡ, *nonce-wd.* of uncertain sexual meaning (cf. ξεναπάτης), *AP* 11.7 (Nicandr. or Nicarch., -κυστ- codd.).

ξενοκυστᾰπάτη, delete the entry.

ξενοπᾰθέω, for '*have .. feeling*' read '*feel shy* or *ill-at-ease*'

*****ξενοπάροχος**, ὁ, *official entertainer of foreigners*, Arcad.Charis.*Dig.* 50.4.18.10.

ξένος, line 5, after '(sed v. fin.)' insert 'ξῆνος, in Dor. pers. n., Schwyzer 277A (Rhodes, vi/v BC), *IG* 4.618 i 6 (Argos), *GDI* 4834 b 10, 15 (Cyrene), see also °ξηνεῖος' **A III A**, for 'στρατηγὸς ἐπὶ τῶν ξένων = *praetor peregrinus*, *IG* 9².242 (Acarnania, i BC)' **2**, delete the section transferring quot. to section 1. **B I**, add '**2** perh. as epith. of Zeus, = ξένιος, *SEG* 32.1026 (written ξεῖνο gen., Paestum, c.550 BC).' **III 2**, line 1, for '*fresh*' read '*different*'; add '(irreg. sup.)'

*****ξενοτρόφιον**, τό, prob. *payment for mercenaries*, *IMylasa* 651.6 (iii/ii BC, Cret. dialect).

ξενόω II 2, line 4, before 'ξενωθεὶς ὑπὸ' insert '*to give someone the rights of a* ξένος'

ξενών, add 'used to house the sick, *Cod.Just.* 1.3.45.1, 6.4.4.2'

ξένωσις, for '*entertainment of a guest*' read '*aberration*'

*****ξερεύω**, *become dry* (w. ξερ- for ξηρ-, cf. mod. Gk. ξερός), only found in pf. part. ἐξερευκός, -ότα, as epith. of κεράμιον (-α), referring to sun-dried produce, *SB* 9132.3, 7, 12 (iii/iv AD).

*****ξεστικῶς**, adv. app. *so as to shave the surface*, ὅρα τὸ ἐπιλίγδην ἀντὶ τοῦ ξεστικῶς καὶ ἐπιπολῆς Eust. 1119.54.

*****ξεστισμός**, ὁ, *quota of sextarii* (measures of wine due in payment), *POxy.* 2114.13 (iv AD).

ξέω II 1, add 'fig., of style, Poll. 6.140, 141'

*****ξηνεῖος**, α, ον, *relating to aliens*, κσενείαι δίκα[ι δι]κάδδε(θ)θαι *Inscr.Cret.* 4.80.8 (Gortyn, v BC).

ξηραίνω 1, pass., add 'of corn, *be ripened*, *Apoc.* 14.15'

ξήρᾱσις, for '*siccitas*' read '*desiccation*' and add 'Gal. 16.415.11'

*****ξηροκήπιον**, τό, *dry garden* (cf. ξηροὶ καρποί, for sun-drying produce, or possibly with little water supply), *PNess.* 31.20 (vi AD).

*****ξηροκόπιον**, τό, καισεκπρώπιον· δρέπανον, ξηροκόπιον Hsch.

ξηρός I 1, line 1, for 'χειμάρρους' read 'ἔκρους'; line 9, after '(iii BC)' insert 'abs., without χόρτος, *PZen.Col.* iv 95.6 (iii BC)' **2**, for this section read 'of various bodily conditions, λιμῷ γένηται ξηρός Hippon. 10 W.; ξ. κοιλίη, i.e. *costive*, Hp.*Aph.* 2.20; of limbs withered by paralysis, *Ev.Matt.* 12.10, *Marc.* 3.3, *Luc.* 6.6, al.; as a result of fear, Theocr. 24.61; externally, from lack of unguents, E.*El.* 239; adv. ξηρῶς βήττειν *have a dry* cough, Gal. 9.626.8'

ξῐφήρης, after 'Hdn. 7.5.3' insert '*POxy.* 3561 (AD 165)'

ξῐφηφορέω, add 'D.C. 53.13.7 (*ius gladii habere*)'

ξίφος I 2, for this section read 'as a symbol of judgement, Philostr.*VS* 1.25.2; of the praetorian guard, τὸ μὲν βασιλεῖον ξ. .. ἦν ἐπ' Αἰλιανῷ τότε ib. 4.42' **II**, for this section read '*pen* of the squid (τευθίς), Arist.*HA* 524ᵇ24, *PA* 654ᵃ21; *sword of the swordfish*, id.*fr.* 325, Opp.*H.* 3.558' add 'Myc. *qi-si-pe-e* (dual); see also σκίφος'

*****ξοανός**, ή, όν, app. adj. in Hsch. ξοανῶν προθύρων· ἐξεσμένων; cf. ξόανον.

ξοῖς, add '*PMich.*xv 721.57 (iii/iv AD)'

*****ξοϊτης**, ου, ὁ, prob. *one who works with an engraving-tool*, χαλκεὺς ξ. Swoboda *Denkmäler* no. 117 (Isauria).

ξουθός, add '**III** masc. as subst., *a dark yellow semi-precious stone*, Plin.*HN* 37.128.' at end add 'Myc. *ko-so-u-to*, name of an ox; also pers. n. (so later Ξοῦθος)'

ξύθος· σμάρις (a fish), Cyran. 116 (4.46.1, 2 K.).

*****ξυϊδόγλῠφος**, ον, *carved with a chisel*, *GVAK* 15.6 (Phrygia); cf. ξοῖς.

ξυλαλόη, read 'ξυλᾰλόη' and delete 'scanned .. *An.Ox.* 3.277'

ξῠλᾰμή, add '*POxy.* 1124.15 (i AD)'

ξυλᾰμητής, add '*PTeb.* 886.62, 64 (ii BC)'

*****ξυλᾰμητρον**, τό, *wages for sowing*, *PWash.Univ.* 77.25.

*****ξυλᾰμιστής**, οῦ, ὁ, *sower of green fodder plants*, *PCair.Zen.* 727.12 (iii BC).

*****ξυλέμπορος**, ὁ, *wood-merchant*, *Orientalia* 35.135 fig. 73.

*****ξύλη**, ἡ, *timber*, *SEG* 36.1087 (Sardis, 213 BC).

ξυληγός, add 'also ἡ, όν, *naves* -άς Ulp.*Dig.* 32.55.5'

*****ξυλικάριος**, ὁ, *carpenter* or *joiner*, *MAMA* 3.84, 95 (Diocaesarea), 3.731 (Corycus).

ξυλικός, penultimate line, after '*PTeb.* 8.26 (iii/ii BC) read 'ξυλικόν, τό, *a wooden construction* in a garden, perh. fencing or supports for plants, κῆπον σὺν τῷ πεπηγμένῳ ξ. *SEG* 17.545.11 (Pisidia, Rom.imp.); cf. *lignarium, pulpitum, Gloss.*'

*****ξυλῖνας**, ᾶ, ὁ, *woodcutter*, *IEphes.* 4312a (Chr.).

*****ξῠλῐνοβαστάκιον**, τό, perh. *that part of a waggon which carries the load*, *flooring*, *PMasp.* 303.15 (vi AD).

ξύλινος, after lemma insert '(sp. σύλιν- *IG* 2².1623.331, iv BC)' **I 1**, line 5, after 'Ath. 3.78d' insert 'cf. *lina xylina*, cloth *made of cotton*, Plin.*HN* 19.14' **II**, delete the section.

ξύλιον, delete the entry.

ξυλεῖομαι, add '(also expl. as form of σκύλλω or συλλέγω)'

*****ξῠλόλῠκον**, add '*ICilicie* 108.18 (v/vi AD)'

†**ξυλογλύφος**, ὁ, *wood-carver*, *BCH* 102.413 (iii AD), Hsch.

*****ξῠλοκᾰβαλ(λάριος)**, ὁ, perh. *rider armed with wooden lance* (in quot., applied to a bandit), *SEG* 35.1360.7 (Hadrianopolis, Paphlagonia, vi AD).

*****ξῠλοκάρος**, ὁ, *woodcutter*, *IPrusa ad Olympum* 149 (ii AD).

ξυλοκοπέω I, add 'Hsch. s.v. καλοκοπῆσαι' II, after 'Arr.Epict. 3.7.33' insert 'POxy. 2811 fr. 5a.4 (ii AD)'
*ξυλοκοπικός, ή, όν, for cutting wood, PTeb. 794.13 (iii BC).
*ξυλομαστίχη, ή, mastic wood, POxy. 3733.29, 3766.11 (both iv AD).
ξύλον I 2, delete the section transferring quot. to section 1. II 3 b, for 'stocks' read 'wooden frame' and add 'ξύλον ἔχοντα ἐν τοῖς ποσίν, ὥστε μὴ δυνηθῆναι ἐκ τοῦ πλοίου λαθεῖν καὶ διαφυγεῖν PKöln 281.12 (vi AD)' 5, for 'theatre' read 'court or assembly' III, line 3, for 'Call.Cer. 41' read 'Call.Cer. 40'
ξυλοπάκτων, add 'PBeatty Panop. 1.12 (AD 298)'
†ξυλοπέδη, ή, wooden frame or fetter for restraining the feet (cf. ‡ξύλον II 3 b), Aq.Jb. 13.27, 33.11, Lyd.Mag. 48.2.
*ξυλοπύριος, ον, perh. sharpened by fire, σανίς Poliorc. 271.11.
*ξυλοπωλία, Ion. -ίη, ή, sale of wood, SEG 2.579.8, 12 (Teos, iv BC).
*ξυλοσάγγαθον, τό, (or -ος, ό), an unidentified plant (recorded as commodity used in a mint), POxy. 3618.12 (iv AD); cf. ‡σάγγαθον.
*ξυλοσέλῑνον, τό, variety of celery, or sim. plant, PMil.Vogl. 302.185 (ii AD).
ξυλοσπόγγιον, add 'in Lat. inscrs., PMich.VIII 471.29, Ann.Épigr. 41.5'
*ξυλοτομέω, cut wood, PRoss.-Georg. II 19.31 (AD 141).
ξυλοτομία, after 'POxy.' insert '729.29 (ii AD)'
*ξυλοτόμιον, τό, axe for cutting wood, SB 9587 (for reading ξυλοτόμιν v. Mnemosyne 30.142, vi/vii AD).
ξυλοφθόρον, after 'τό' insert '(also -ος, ό, Hsch.)'
ξυλοχάρτια, delete the entry.
*ξυλοχάρτιος, ον, neut. pl. subst., τὰ ξυλοχάρτια paper made from papyrus, Eust. 1913.41 (BASP 9.27, 28); document written on this material, τὸ ξυλοχάρτιον κοντάκιον Steph.in Rh. 277.29.
ξυλόω II, after 'make of wood' insert 'or cover with wood' and for 'γαῦλον' read 'γαυλόν'
ξύλωμα, after 'piece of woodwork' insert 'or wooden panelling'

*ξυνέτης, ου, ό, ξυνέται· συμπολῖται Hsch.
ξυνήων I, after 'Hes.Th. 601' insert '(ὁ δεῖνα) τὴ(ν) στέγην ἐποίησεν κοὶ ξυνεῶνες Robert Hell. 9.78 (Hellespont, vi BC)' II, after 'as adj.' insert 'common to all, ξυνήονι πότμῳ Nonn.D. 12.266'
*ξύνιστρον· νόμισμα Hsch.
ξυνόω, after lemma insert 'make common, Sch.Pi.O. 7.36'
*Ξύρεος, ό, name of a god, Θεῷ Ξυρέῳ IHadr. 19; identified with Apollo, MAMA 9.60 (ii/iii AD); also Ξυρᾶς, Θεὸς Ξ. IHadr. 20.
ξύρησις, for 'baldness' read 'shaving of the head as sign of mourning'
*ξυρησίταυρος, ό, one who shaves off his pubic hair (v. ταῦρος III); pl., name of a τάγμα (set) at Sardis, Ap.Ty.Ep. 39.
*ξυρητικός, ή, όν, of shaving: -κή (sc. τέχνη), ή, An.Ox. 4.248.10.
ξυρίς, line 3, after 'ξειρίς' insert 'SEG 13.550.6 (Caesarea, Rom.imp.)'
ξυστάδες, delete the entry (v. °ξυστάς).
ξυστάλλιον, for 'ξύστρον' read 'ξύστρα'
ξυστάρχης, last line, for 'Smyrna' read 'Thyatira'
*ξυστάς, άδος, ή, closely-planted vineyard, Poll. 7.147 (v.l. ζυγάς), Hsch. (pl.); cf. συσταδόν.
ξυστήρ, line 2, delete 'graving-tool'
*ξύστης, ου, ό, = ξυστήρ, CPHerm. 127 xviii 9 (iii AD).
ξυστίς, after 'ίδος' insert '(ξύστις, ιδος, acc. to Sch.Ar.Nu. 70)'
*ξυστοπλᾰτείτης, ου, ό, resident in the *ξυστοπλατεῖα (prob. street of the ξυστός), συμβίωσις -ειτῶν ISmyrna 714 (ii AD).
ξυστός, line 2, after 'Aristias 5' insert 'Philostr.VA 4.3' I 1, for 'and statuary, Vitr. 5.11.5' read 'Vitr. 5.11.4' add 'III a carpenter's tool, Gal. 1.47.4 (s.v.l.).'
ξύστρα I, add '2 ξ. ἁλιευτική, app. instrument for scaling fish, PWürzb. 5.9, 12 (i BC).'
*ξυστρολήκῠθον· κάδη καὶ βησσία ἐλαίου λουτρικά Hsch.
ξυστρολήκυθος, delete the entry.
ξυστρωτός, after 'of pillars' insert 'SAWW 179(6).63 (Cilicia, i/ii AD)'

O

ὁ, ἡ, τό, line 10, for '*Inscr.Cypr.* 135.20 H.' read '*ICS* 217.30, al.'; line 22 (p. 1194 col. 1, line 10), after 'Sapph. 16' insert 'Thess. gen. sg. τοῖ *SEG* 31.572 (*c*.200 BC), al.'   **A VII 2**, line 2, for 'καί μοι κάλει' read 'ἀφικνοῦμαι ὥς'; line 3, for 'Pl.*Lg.* 784d ' read 'Pl.*Lg.* 784c'   **VIII 5 b**, add 'v. πρό A II 1'   at end add 'Myc. *to-jo* (gen. sg.), *to-i* (dat. pl.)'   **B II 5**, line 4, before 'freq. with advs.' insert '**6**'; line 9, after '*Or.* 1412 (lyr.)' insert 'also τὸ εἰκῇ Pl.*Grg.* 506d, al.; τὸ μόλις D.H.*Comp.* 20'   add '**7** w. numerals, indicating number of tenures of office, ὕπατος τὸ ζ' *TAM* 4(1).11 (AD 116); indicating a date, τῷ ζμσ' (sc. ἔτει) *SEG* 32.1537 (AD 184).'   **C**, line 4, for 'ὁ ἐξορύξη .. (Cyprus)' read 'Cypr. *o-e-xe o-ru-xe*, ὁ ἐξορύξε *he who expels him*, *ICS* 217.12, 25'; line 10, delete 'Com. or'

ὄα (A), add 'cf. αἶα (C)'

ὀάρισμα, for '*familiar converse*' read '*lore*'

Ὀασιτικός, add '*PCair.Zen.* 299 (iii BC), *POxy.* 2567 (iii AD), 3425 (iv AD), etc.'

*****ὄβαν**, v. °ὄουαν.

†ὄβδη, ἡ, = ὄψις, Μούσῃ γὰρ ἦλθον εἰς ὄβδην Call.*fr.* 218 Pf., unless one wd., as certainly in ποιεῖσθαι τὴν ἀπογραφὴν εἰσόβδην = *palam, in propatulo, ILampsakos* 9.42-4 (ii BC); cf. **ὄβδην** and **ἐσόβδην** cited as advs. by A.D.*Adv.* 198.7.

ὀβελεία, add 'Lang *Ath.Agora* xxi B12 (iv/iii BC, sp. ὀβελίαι)'

*****ὀβελισκοποιός**, ὁ, *maker of* ὀβελίσκοι, *IG* 1³.426.13 (Athens, v BC).

ὀβελίσκος I 3, add 'so in Dor. form ὀδελίσκος *ABSA* 61.264.13 (Epid., *c*.370 BC)'   **5**, add 'also, *bar* protecting outlet of drain' transferring here quots. fr. section IV   **IV**, delete the section.

ὀβελός I 3, for '*IG* 1².6.95, al.' read '*IG* 1³.4.A20 (Athens, v BC), 7.1739.8 (Thespiae, iii BC)'

*****ὀβιφέρι** (gen.), *wild sheep*, fr. Lat. *ovis fera*, *Edict.Diocl.* 8.25 (sp. ὀβιβέρι at *SEG* 37.335 ii 20).

ὀβολίσκος I, add '**b** perh. *water-tank*, *POxy.* 2406 (ground-plan of a house) (ii AD).'

*****ὀβολοκερε** (?for ὀβολοκέρεα = -κέρατα or -κέρεια), perh. *pin* or *skewer made of horn*, *SEG* 29.972 (label for surgeon's implements, Italy, iv BC).

ὀβολός I, line 2, for '*IG* 1².140.5, al.' read '*IG* 1³.6.C12, 237'

ὀβρίκαλα, after lemma insert 'or -οι, οἱ'; line 2, at end insert 'Phot., Ar.Byz.*fr.* 203B S.; forms ὀβρίκια, ὄβριας Poll. 5.15, see also ἰβρίκαλοι, °ὄβριχα'

*****ὀβριμάδες, αἱ**, wd. of unkn. meaning, perh. pr. n., epigr. in *BCH* 75.195 (Crete).

*****ὀβριμότοξος**, ον, *equipped with a strong bow*, Antim. 174 Wy.

*****ὄβριχα, τά**, (or -οι, οἱ), = ὀβρίκαλα, ὑστρίχων τ' ὀβρίχοισ[ι A.*fr.* 47a.809 R.

ὀβρυζιακός, add '*POxy.* 126.15, 27 (vi AD)'

ὄβρυζος, for '(cf. Lat. *obrussa*)' read '(prob. fr. Hitt. *ḫubrušḫi*, clay container, presumably used in gold refining; Lat. *obrussa* fr. Gk.), see also °βρύζα'

*****ὄγγας**· ὁ ἀπὸ πολλῶν ὑγρός Theognost.*Can.* 46 A.

ὀγδόδιον, add '(ὀγδοαῖον La.)'

ὀγδοήκοντα, add 'written ℎογδοήκοντα *Tab.Heracl.* I 43, II 70'

ὀγδοηκοντούτης, line 3, after 'Simon. 146' insert 'fem. -αέτις *AP* 7.733 (Diotim.)'

†ὄγδοος, ον, ([ὀγ]δόΓα fem., Aetol. *IG* 9²(1).152) *eighth*, Il. 7.223, etc., ὄγδον Δαμοίτας Νικομείδ[εος] *SEG* 23.273 (ii BC); as designation of a legion, ἐκλεχθεὶς ἰς ὀγδ[όαν Αὐ]γούσταν *SEG* 31.1116; ὀγδόα, ἁ, a monetary unit (cf. ἕκτη), *SEG* 26.1084 (Megara Hyblaea, vi BC); ὀγδόη (sc. ἡμέρα), ὀγδόη τῆς πρυτανείας *IG* 1³.475.284, Πανάμμοι ὀδδόα ἐφ' ἱκάδι *SEG* 31.577 (Thessaly, ii BC), Plu.*Thes.* 36; ὀγδόη (sc. μοῖρα), *eighth part*, *SEG* 33.1034 (Aeolis, iii BC), cf. ὀγδόα· ἡμιχοινικόν Hsch.; neut. (sc. ἔτος) εἰκοστὸν χὠγδοον *GVI* 1091 (Dyme, ii/iii AD).

ὀγκάομαι, delete 'Theopomp.Com. 4'; for 'Call.*Aet.Oxy.* 2079.31' read 'Call.*fr.* 1.31 Pf.'

*****ὀγκαρίζω**, = ὀγκάομαι, Aq.*Ge.* 49.14 (v.l.).

*****ὀγκάς, άδος, ὁ**, *brayer*, Theopomp.Com. 5 K.-A.

*****ὀγκιαρήσιον**, τό, perh. a coin (fr. ὀγκία, *uncia*, cf. μιλιαρήσιον), *PIand.* 103.14 (vi AD).

ὄγκος (B) **I 4**, for this section read '*the animal body with respect to its workings, system*, τῆς χολῆς ἀναχεομένης εἰς τὸν ὄγκον Ruf.*Anat.* 30, Ph. 1.391, Sor. 1.26, ταράττειν τὸν ὄγκον Plu. 2.652e, 653f, Gal. 1.272.17'   **II 3**, add 'Longin. 3.4'

ὀγκώδης (A) **I 2**, add 'of persons, D.Chr. 30.19'

ὄγμος I, add '**4** perh. *wheel-rut*, Nic.*Th.* 371.'   **II**, after '*h.Hom.* 32.11' insert 'ἠέρος ὄγμοι Call.*fr.* 335 Pf.'

ὀδαγός, add '*SEG* 23.271 (Thespiae, iii BC); as pers. n., Cypr. *o-ta-ko-se Kadmos* 29.143 (iv BC)'

ὀδαῖος II, for this section read 'ὀδαῖα, τά, *freight, merchandise*, Od. 8.163, 15.445'

*****ὀδανόν**· εὐῶδες μύρον Theognost.*Can.* 51 A.; cf. ἐδανός.

ὀδαξησμός, add 'also **ἀδαξησμός** Erot. 107.21'

ὀδάξω II, add 'abs., τῆς γὰρ ὀδαξαμένης when it has *bitten*, Nic.*Th.* 306'

ὀδάχα, read '**ὀδαχᾶς**'

ὅδε, line 5, for 'τωνδέων Alc. 126' read 'τωνδέων Alc. 130.21 L.-P.'   **I**, after '**6**' insert '*such and such*, διὰ τήνδε τὴν αἰτίαν Pl.*Phdr.* 271d, cf. 272a; πορευσόμεθα εἰς τήνδε τὴν πόλιν *Ep.Jac.* 4.13.   **b** ὅδε καὶ ὅδε, *this man and that*, A and B, D.Chr. 40.13, 33.48 (pl.).'; before 'in Arist.' insert '**c**' and at end delete 'πορευσόμεθα .. 4.13'   **III 2**, line 6, delete 'v. .. 1.2'; add 'ὅδε and οὗτος, of the same person or thing, S.*Ant.* 189, 297, Th. 1.143, etc.'   add '**4** τόδ' ἐκεῖνο E.*Med.* 98, like τοῦτ' ἐκεῖνο (v. οὗτος B III 5).'   **IV 2**, add 'also neut. pl., *SEG* 31.1525 (Egypt, iii/ii BC)'

ὀδελονόμος, after 'financial official ' insert 'at Argos, *BCH* suppl. 22 p. 235 (v BC)'

*****ὀδευτός, ή, όν**, *furnished with roads*, γῆν ὁ. ἐποίησεν D.Chr. 3.127.

ὁδηγός, add 'σκίπων *AP* 7.457 (Aristo)'; see also ‡**ὁδαγός**.

ὁδίτης, add 'ὁδῖτα (voc.) epigr. in *REG* 80.282 (Chios)'

ὁδοιπορέω, line 4, after 'S.*OT* 801' insert 'Crates Com. 16.3 K.-A.'

ὁδοιπορικός, line 2, after 'Poll. 1.181' insert 'ὁ. πήδησις, t.t. for a leap in armour on to a cantering horse, Arr.*Tact.* 43.4'

ὁδοιπόρος, add '*AP* 7.502 (Nicaenet.), *SEG* 31.379 (*c*.100 BC)'

*****ὀδοντίδας**· πολυφάγος Hsch.; cf. ὀδοντίας.

ὀδοντισμός, add 'εἶδος αὐλήσεως ὅτε ἡ γλῶττα προσβάλλεται πρὸς τὸν ὀδόντα Hsch.'

*****ὀδοντῖτις, ἡ**, a plant said to cure toothache, Plin.*HN* 27.108.

ὀδοντοτύραννος, for 'large animal, prob. *crocodile*' read 'fabulous animal, perh. based on accounts of the crocodile'

ὁδός III 3, add 'ἐκ πάσης ὁδοῦ *by every means, Cod.Just.* 1.3.45.3a; sim. διὰ πάσης ὁδοῦ ib. 1.4.34.6'

ὁδουρός, add '**III** = ὁδίτης, *traveller*, Nic.*Th.* 180.'

ὀδούς, line 1, after '*EN* 1161ᵇ23' insert '*Mech.* 854ᵃ28'   **I 1**, add 'ὀδόντες ἐλεφάντων *tusks, Didyma* 394.16 (i BC), cf. Opp.*C.* 2.493'   **II**, for '*ploughshare*' read '*tine*'

*****ὀδυρμοχαρής, ές**, *delighting in lamentation*, *IHadr.* 168 (ii/iii AD).

*****ὀδυρομένως**, adv. fr. part. pres. of ὀδύρομαι, ζῆν ὁ. *in lamentation*, Favorin.*Exil.* 10.45.

*****ὀδύρω**, act. form of ὀδύρομαι, *GVI* 969.7 (Daldis, i AD).

*****°Ὀδύσειος**, v. °Ὀδυσσεύς.

Ὀδυσσεύς, at end after 'Od. 18.353' add '°**Ὀδύσειος** Stesich. 32 i 2 P.'

ὄζαινα II, for this section read 'another name for ὀσμύλη (ὀσμύλιον, etc.), Call.*fr.* 406 Pf. (Ath. 329a)'

ὀζαινίτης, delete the entry.

†ὀζαινῖτις, ίτιδος, ἡ, a kind of nard, supposed from its name to be evil-smelling, but perh. fr. the Indian town of Ozene (Ujjain), Plin.*HN* 12.42.

ὀζαλέος, for '*branching*' read '*knobbed*'

*****ὀζηλίς**· ἡ βοτάνη Theognost.*Can.* 45 A.

ὄζος II, add '**2** rhet. term of unkn. significance, οἷον Λικύμνιος ποιεῖ ἐν τῇ τέχνῃ ἐπούρωσιν ὀνομάζων καὶ ἀποπλάνησιν καὶ ὄζους Arist.*Rh.* 1414ᵇ18.'

ὄζος, Cret., after 'Gortyn' insert 'vii/vi BC'

*****ὀζόχρωτος**, ον, app. *having knotty skin*, Heph.Astr. 3.45.2; cf. *hircosus, Gloss.*

ὄζω, line 1, after 'impf. ὤζε' insert 'Hippon. 92.10 W.'

*****ὀθεύς**· εὐνοῦχος Theognost.*Can.* 53 A. (ὄθης Zonar.).

ὅθι, line 3, after '*Phd.* 108b' insert 'Ant.Lib. 33.3, 4'

*****ὀθία**, = ὄθιζα, Zonar.

ὄθιζα, after 'cf.' insert '°ὀθία, °ὀθυσία'

223

ὄθμα, for 'Call.*Aet.* .. 37' read 'Call.*fr.* 186.29 Pf.'

ὀθνεῖος **I**, add 'ὀθνέην ὁδόν Archil. 244 W.'

*ὀθονεμπλουμάριος, ὁ, *linen-embroiderer*, *PAmst.*inv. 39 (*CE* 48.128, iv/v AD), cf. Lat. *plumarius*.

ὀθονιακός **I**, add 'written ὀθων- *BCH* suppl. 8 no. 33 (v/vi AD)'

*ὀθονιάπωλις, ιδος, ἡ, fem. of ὀθονιοπώλης, *SB* 10162.538.2 (iii/iv AD).

ὀθόνιον **1**, add '**b** ὁ. Σηρικόν *silk*, *Peripl.M.Rubr.* 64.'

*ὀθονιοπράτης, ου, ὁ, *linen-seller*, *SEG* 27.874 (Ancyra, Chr.), *IGLBulg.* 249 (Odessos, vi AD).

ὀθονιοπώλης, add '*MAMA* 3.225, 4.349'

*ὀθονιοπώλιον, τό, *linen-shop*, *PAmst.*inv. 62 (*Mnemosyne* 30.146, ii AD).

ὀθούνεκα, after lemma insert 'Ion. ὀτεύνεκα Herod. 5.20, al.' **I**, add 'E.*Hel.* 104, etc.' **II**, add 'E.*Alc.* 796'

ὀθούνεκεν, after lemma insert 'Ion. ὀτεύνεκεν, *Iamb.adesp.* 38.12 W., Herod. 7.103'; after '= foreg.' read '*because*' and after 'Timo 34' insert 'Theoc. 25.76. **II** *that*, A.R. 3.933.'

*ὀθύλλομαι, glossed by διανοέομαι, Hsch. s.v. ὠθύλλετο; cf. ὀθέω, etc.

*ὀθυσία, = ὀθίζα, Theognost.*Can.* 53 A.; cf. °ὀθία.

*ὀθύω, impf. ὤθυον, = ἄγω, Theognost.*Can.* 53 A.; cf. ὀθεύει, ὀθρεῖν.

*ὀθωνοπώλης, ου, ὁ, = ὀθονιοπώλης, *TAPhA* 90.140.18.

οἴγω, penultimate line, for 'Alc. 225 Lobel' read '*Lesb.fr.adesp.* 20 L.-P.' and continue 'ὀίγοντ' [ἴ] ἔαρος πύλ[αι Alc. 296(*b*).3 L.-P.'

οἰδημάτιον, after 'Hp.*Fract.* 5' insert 'Gal. 18(2).389.4, 390.12'; delete 'name of an eye-salve'

οἶδμα **II**, delete the section.

*οἴεον, τό, *sheepfold*, κλεῖθρα κατασκευάσαντι τοῖς οἴκοις τοῖς ἐν τῶι οἰέωι *Inscr.Délos* 290.78 (iii BC).

οἰέτεας, add 'Mosch. 2.29 G.'

ϝοιζηάζω, delete the entry.

ὀιζυρός, line 6, at end insert 'Ion Trag. 38.2 S.'; line 7, delete 'by Trag., nor'

οἴζω, add 'med. οἴζομαι dub. in S.*fr.* 269c.47 R.'

οἶις, delete 'but .. is prob.'

οἴκαδε **I**, add 'Telecl. 1.6 K.-A.; w. gen. *to the home of*, οἴ. τοῦ ξένου Eup. 99.84 K.-A.' **II**, delete the section.

οἰκεῖος **IV 2 b**, for this section read '*familiar with, at home with*, w. gen., βασιλεῖς .. καίπερ οἰκεῖοι σοφίας γεγονότες Str. 17.1.5, τοὺς οἰκείους τῆς πίστεως *Ep.Gal.* 6.10, Iamb.*VP* 30.176' **B II**, add '**5** w. gen., *in conformity with*, *ICilicie* 70 (ii BC)'.

οἰκειότης **III**, delete the section.

οἰκειόω **II 1**, line 3, delete 'abs. .. Aen.*Tact.* 24.5'

*οἴκελος, v. °οἴκυλα.

οἰκετεία **1**, after lemma insert 'ϝοικετεία *SEG* 23.566 (Crete, iv BC)'; add '*Inscr.Cret.* 1.xvi 17.16 (ii BC), Ev.Matt. 24.45, ἰατροῦ τῆς κυριακῆς οἰκετείας *IEphes.* 3233 (iii AD)'

οἰκέτης **I 2**, line 3, delete 'hence opp. δοῦλοι' and for '(Pl.*Lg.*) 853e' read '(Pl.*Lg.*) 853d' **II**, at beginning insert 'perh. *residing divinity*, ὅδε σηκὸς οἰκέταν εὐδοξίαν Ἑλλάδος εἵλετο Simon. 26.6 P.'; after '(Sparta)' insert 'Paus. 3.13.4'

*οἰκετιεύς, v. οἰκιτιεύς.

οἰκέτις, add 'οἰκέτιδα (acc.), epigr. in *SEG* 37.450 (Thessaly, ii/iii AD)'

*οἰκέτισσα, ἡ, = οἰκέτις I, *female household slave*, *TAM* 3(1).282 (Termessus; sp. ὑκαίτισσα).

*οἰκεύω, = οἰκέω, (in quot.) *colonize*, *SEG* 23.474 (Dodona, iv/iii BC).

οἰκέω, line 1, (ϝοικέω), add '*SEG* 11.244 (Sicyon, v BC), also ϝοικίω, *SEG* 23.589 (Gortyn, iii/ii BC)'; line 2, after 'Aeol. pres.' insert 'οἴκημμι Alc. 130.31 L.-P.' and for 'Alc. 69' read 'id. 328 L.-P.'; line 4, after 'Hdt. 1.1' insert 'Aeol. ἐοίκησα Alc. 130.25 L.-P.' **A 1**, line 4, after 'E. l.c.' insert 'πᾶσι τοῖς οἰκοῦσι μοναστήρια Cod.Just. 1.3.43 pr.'

οἴκημα **II 1**, for 'Isoc.' read 'Is.'

οἰκία, (ϝοικία) add '*SEG* 32.496, 34.355; also ϝυκία (all Boeotia, iv/iii BC); see also βοικία' **I 3**, delete the section.

οἰκίδιον **1**, add 'applied to a (Christian) tomb (or perh. chapel), *SEG* 30.1068 (Tenos)'

οἰκιήτης, before 'ϝοικιάτας' insert 'also at Epid.'; line 3, after '(*IG*) 262.16' insert '*SEG* 26.449.6'

οἰκίον, add 'in form οἰκίν, of a (Christian) *tomb*, *ITyr* 90'

*οἴκισμα, τό, = οἴκημα, *dwelling-place*, *SEG* 18.615.7 (Syria, iv AD).

οἰκισμός, add '*SEG* 25.486 (Boeotia, iii BC)'

*οἰκιστεία, ἡ, app. *founding*, rest. in *SEG* 12.380.9 (decree of Gela at Cos, iii BC).

*οἰκοδεσπᾶ, ἡ, = οἰκοδέσποινα, inscr. in Robert *Collection Froehner* i p. 111 (Cibyratis).

οἰκοδεσποσύνη, delete '*CIG* 2987 (Ephesus)'; for 'Keil-Premerstein .. 170.13' read '*TAM* 5(1).688 (i AD)' add '**2** *family*, *IEphes.* 622.17.'

οἰκοδεσπότης **I 1**, add '*master of an estate*, *PPhilad.* 1.48 (ii AD)'

οἰκοδεσποτικός **I**, for this section read '*of or proper to the head of a household*, Cic.*Att.* 12.44.2, τῇ συμβίῳ Δημητριάδι ζησάσῃ τὸν βίον οἰκοδεσποτικὸν ἔτη λ´ *TAM* 4(1).128'

οἰκοδομεύς, delete the entry (v. °οἰκοδόμος).

οἰκοδομέω **1 a**, add 'οἰκία ᾠκοδομημένη, i.e. not made of mud, *PAmh.* 2.51.11, 23 (i BC), *PLond.* 880.27 (ii BC)' **I 3**, line 3, delete 'cf. ἀνοικοδομέω' add '**4** = ἀνοικοδομέω, *rebuild, repair*, *IG* 11(2).161Α120 (Delos).'

οἰκοδομή **I**, line 4, after 'Plu.*Cam.* 32' insert 'γεφυρῶν οἰκοδομῆς Just.*Nov.* 131.5; fig., *building up of faith, edification*, *POxy.* 2785 (iv AD)'

οἰκοδομητός, add 'also ᾠκο- *SEG* 33.955 (Ephesus)'

*οἰκοδόμητρα, τά, *wages for building*, πάθνης *PLips.* 106.8 (i AD).

οἰκοδόμος, add 'dat. pl. οἰκοδόμεις, w. dissimilation of οι-οι to οι-ει (as in λοιπεῖς *IG* 2².1028.12, and οἴκει), *OStrassb.* 583 (iii BC)'

*οἰκοδομουργός, v. °ὁμουργός.

οἴκοθεν, after lemma insert 'ϝοίϙοθεν *SEG* 30.380 (Tiryns, vii BC)'; line 1, after 'οἴκοθε' insert 'Call.*fr.* 275 Pf.' **3**, add '*Cod.Just.* 4.21.22.4, etc.'

*οἰκονομήτρια, ἡ, *that gives effect to a dispensation*, ἡ ἡμέρα (i.e. Easter) ἡ .. τῶν πενήτων οἰκονομήτρια Ps.-Chrys.*HP* 2.7.

οἰκονομία **I 4**, add 'at Priene, *SIG* 1003.29 (ii BC)'

*οἰκονομίδιον, τό, *document of a transaction*, *POxy.* 2679 (ii AD).

οἰκονομικός, line 6, for '*the duties of domestic life*' read '*estate management*'; at end for 'also in literary sense .. Sch.Th. 1.63' read '**2** *concerning arrangement* of literary material, D.H.*Dem.* 51, *Th.* 9, Quint.*Inst.* 7.10.11; adv., *in an ordered manner*, Sch.Th. 1.63.'

†οἰκονόμισσα, ἡ, *female estate manager*, *JHS* 24.283 (v AD), *MAMA* 8.399, *INikaia* 1466.

οἰκονόμος, after 'ὁ, ἡ' insert 'written οἰκονόοι (pl.), *Hesperia* 58.118 (Crete, ii BC)' **I 1**, *steward of an estate*, add 'Mitchell *N.Galatia* 34, *INikaia* 753; transf., οὕτως ἡμᾶς λογιζέσθω ἄνθρωπος ὡς ὑπηρέτας Χριστοῦ καὶ οἰκονόμους μυστηρίων Θεοῦ 1*Ep.Cor.* 4.1, *Ep.Tit.* 1.7' **2 b**, at end for 'θεοῦ .. 1*Ep.Cor.* 4.1' read 'as the title of a Christian official, *SEG* 32.1492 (v AD), etc.'

οἰκόπεδον **1**, last line, after 'πόλεως' insert 'Demad. 26'

οἶκος **I 1**, line 3, after 'Lxx *Ge.* 31.33' insert 'or the country, town, etc., where one lives or belongs' and add 'οἱ ἐν οἴκῳ Ἀθηναῖοι X.*HG* 1.5.16, *Cyr.* 7.2.1, *An.* 2.4.8, al.' **2**, last line, after '*within*' insert 'A.*Ag.* 427' **3**, lines 2/3, delete 'Δεκελειῶν .. 33'; line 4, for '*IG* 4.1580 (Aegina)' read 'hοἶϙος *SEG* 32.356 (Aegina, *c.*550 BC)'; at end after 'funerary monument' insert 'Ἀρκι[.]άλης μηποίϝεσεν (= με ἐποίησεν) οἶϙον Δαμε[.] *SEG* 30.1058 (Naxos, vii BC), *IEphes.* 1630' add '**IV** = φρατρία (s.v. φράτρα) II 1, prob. in *IG* 2².1237.33 (Attica, iv BC), 12(5).528.15, 1061.16 (both Ceos, iii BC). **2** *guild*, τῶν ναυκλήρων *BCH* 25.36 (Amastris).' at end after 'ϝοῖκος' insert '*Kafizin* 266b, 267b (iii BC), Myc. *wo-i-ko*'

*οἰκοσιτέω, *take one's meals at home*, Luc.*Sacr.* 9 (s.v.l.).

οἰκόσιτος **II** and **III**, delete the sections.

οἰκοτραφής, after 'οἰκότριψ' insert '*Vit.Aesop.*(G) 45'

οἰκότριψ, delete 'Attic for οἰκογενής'; add '(in conjectural context)'

*οἰκοτροφής, ές, *house-bred*, κοράσιον *Delph.* 3(6).37.6.

οἰκότροφος, delete '*house-bred* .. (Priene, iv BC)'

†οἰκότως, v. εἰκότως.

οἰκουμένη **II**, add 'personified, *RA* 1987.98 (Phrygia, iii AD)'

οἰκουμενικός, after '*IG* 3.129 (iii AD)' insert '*SEG* 31.1288 (Pamphylia, iii AD)' and for '*Rev.Arch.* 1874.113' read '*SEG* 36.1051 (Miletus, ii AD)'

οἰκουρός **I**, add 'used loosely of other domestic creatures, νῆσσαι οἰκουροί Arat. 970'

οἰκτίζω **2**, line 2, after 'Din. 1.110' insert '*set forth pathetically*, Memn. 35.3 J.'

οἰκτιρμός, add '*SEG* 31.1562 (iv/v AD)'

οἰκτροπαθής, add 'transf., of feelings, πένθος epigr. in *Inscr.Cret.* 2.v 50 (i AD)'

*οἰκτροτόκεια, fem. adj. *pitiable in child-bearing*, *GVI* 467 (Amorgos, ii/iii AD).

*οἰκτρόφονος, ον, *of pitiful killing*, *ISmyrna* 522(b).10 (ii/i BC).

οἴκυλα, for '*grain*' read '*pulse*, cf. Lat. *vicia*; also οἴκυλος· ὄσπριον Theognost.*Can.* 21; οἴκελος ὁ πίσος ib. 20'

*οἰκών, ῶνος, ὁ, = οἶκος, cj. in *CIJud.* 672 (Acmonia in Phrygia).

οἶμα, add 'of a cuttle-fish, Opp.*H.* 1.312'

οἰμάω, line 1, for 'οἴμη' read 'cf. οἶμα'

οἴμη, for '= οἶμος' read '*song, poem*, etc.'; line 3, for 'οἴ. .. τέττιγι' read 'λιγυρὴν δ᾽ ἔδωκεν (sc. Φοῖβος) οἴμην to the cicada'; line 4, for 'Anacreont. 32.14' read 'Anacreont. 34.14 W.'

οἴμοι, lines 2/3, delete 'first in Thgn. and Trag. (v. infr.)' and 'and Com.' and insert '*SEG* 3.56 (Attica, *c.*540 BC), *SIG* 11 (*c.*525/500 BC), *CEG* 49 (*c.*525/500 BC), *Not.Scav.* 1899.411 (?vi BC, see Jeffery *LSAG* p. 269), Emp. 139.1 D.-K.)'

*οἰμωκτιᾶν· τὸ οἰμῶξαι Hsch., Phot.

οἰνάνθη, **II**, delete the section transferring quots. to section I. **I**,

οἰνανθίς · · · SUPPLEMENT · · · ὀκταούγκιον

add 'as fem. pers. n., SEG 25.59 (Athens, vi BC)'   **III 1**, line 3, for 'Plin.HN 21.65' read 'Plin.HN 21.167'   **2**, add 'cf. Plin.HN 10.87'   **3**, add '**b** an unguent, IStraton. 247.20, al. (ii AD).'

οἰνανθίς, for '= οἰνάνθη II' read '= οἰνάνθη I'

οἰνάρεος, for 'σποδίη' read 'σποδιῆ'

*__οἰναρχεῖον__ συμπόσιον Theognost.Can. 22.

οἰνάς I 3, add 'Myc. wo-na-si (dat. pl.), perh. vineyards'   **II**, for 'a wild pigeon .. Columba livia' read 'kind of pigeon, prob. corruption of Hebr. yônāh'   at end add 'cf. γοινάκες, γοινέες'

οἰνέμπορος, after 'wine-merchant' insert 'Ptol.Tetr. 179'; for 'Supp.Epigr. 3.537' read 'IGBulg. 1590.10'

*__οἰνεών__, ῶνος, ὁ, = οἰνών, Gloss.

†**οἰηγός**, όν, wine-carrying, PSI 568.2 (iii BC); masc. subst. wine-shipper, MAMA 3.682, 709 (Corycus), OGI 521.22 (v/vi AD).

οἰνηρός II, at end after 'IG 2².1707' for '(iii BC) read '(181/0 BC)'

οἰνικός, add 'neut. subst. contribution of wine, SEG 31.122 (i AD)'

οἰνοβρώς, for 'eaten with wine' read 'perh. consisting of juicy flesh, οἰνοβρῶτα βορήν, of the pomegranate'

*__οἰνογεύστης__, ου, ὁ, winetaster, POxy. 3517 (iii AD).

οἰνοδοτέω, add 'distribute wine (in a cult-organization), SEG 34.1107 (Ephesus)'

οἰνομετρέω, add 'SEG 25.790 (Moesia, ii BC)'

οἰνοποιέω, add 'trans., make into wine, οἰνοποιηθέντων τῶν καρπῶν PRyl. 583.7, 49 (ii BC)'

*__οἰνοποίημα__, τό, wine-making, BGU 2357 ii 9 (iii AD).

*__οἰνοποσιάρχης__, ου, ὁ, organizer of a (village) drinking-party, INikaia 1071 (i/ii AD), 726 (iii AD).

οἰνοποσίαρχος, delete the entry.

†**οἰνοπόσιον**, τό, drinking-party, TAM 4(1).16 (AD 122/3); sp. -ποσιν (unless to be taken as acc. of *οἰνόποσις), TAM 4(1).17 (AD 184/5), 4(1).68, IHistriae 57.32 (iii AD).

οἰνοποτέω, for 'Call.Aet. 1.1.12' read 'Call.fr. 178.12 Pf. (pap.; see also °ζωροποτέω)'

οἰνοπότης, add 'also -πώτης Phot. a 595 Th.'

οἰνοπώλιον, add 'Plaut.As. 200'

οἰνοπώτης, v. °οἰνοπότης.

οἶνος, line 7, after 'παρ' οἶνον' insert 'Hedyl.ap.Ath. 11.473a'; at end for 'Inscr.Cypr. 148 H.' read 'Cypr. to-wo-i-no τῶ Ϝοίνō, ICS 285' and add 'Myc. wo-no'

οἰνόσπονδος, add 'Hsch. s.v. νηφάλια ξύλα'

*__οἰνουργός__, ὁ, wine-maker, Μυρῖνος Θεοφίλῳ οἰνουργῷ PAmst. 53.1 (AD 433).

οἰνοχίτων, for 'Call.Fr.anon. 211' read 'Suppl.Hell. 1093' and for 'ib. 158' read 'Suppl.Hell. 1076'

οἰνοχοέω, lines 4/5, for 'Aeol. -όεισα .. codd. Ath.)' read 'Aeol. aor. imper. -όαισον Sapph. 2.16 L.-P.'

οἰνοχόη, line 4, for 'οἱ. θεῶν σωτήρων' read 'οἰνοχόα θεῶν σωτήρων μία'

*__οἰνοχόϊον__, τό, vessel for pouring wine, Ϝοινοχόϊα χαλκία SEG 24.361 (Thespiae, 386/0 BC).

οἰνοχόος, add '**2** -ος, ον, used for wine-pouring, δέπας AP 5.266.6 (Paul.Sil.).'

*__οἰνοχοποιός__, ὁ, maker of οἰνοχόαι, Delph. 3(4).285.3 (ii BC).

*__οἰνοχυτεῖον__, τό, wine-cellar, Dura⁷,⁸ 171 (ii AD).

†**οἶνοψ**, οπος, adj. used in Homer as epith. of the sea (conventionally translated as wine-dark), Il. 23.316, Od. 5.132, 2.421, etc.); also as epith. of oxen, βόε οἴνοπε Il. 13.703, Od. 13.32; cf. Myc. wo-no-qo-so (descr. of an ox).   **2** used to describe complexion of the skin, cf. °οἴνωπος, οἱ. Βάκχος AP 6.44 (Leon.Tarent.), Θεραπναίη .. νύμφη οἴνοπα πήχυν ἀνεῖλκε Tryph. 521.

*__Ϝοινώα__, ἁ, perh. vineyard (if not a pr. n.), πὰρ τὰς Ϝοινώας Mél.Navarre 354 (Thespiae).

*__οἴνωθρον__, v. οἴνωτρον.

οἰνωπός, line 1, for 'Semon. 180' read 'AP 7.20 ([Simon.])'; line 4, for 'but, dark-complexioned' read 'indicating a complexion midway between ὑπέρλευκος and μέλας'

οἴνωτρον, add 'also οἴνωθρον, GVI 1625.4 (Rhodes, i BC, cf. Robert Hell. 10.282 n.2)'

οἶνωψ, add '**II** Οἰνῶπες, οἱ, one of the Ionic tribes, SIG 57.1 (Miletus, v BC), etc.; sg., of a member of the tribe, ib. 798.2 (Cyzicus, i AD).'

οἰόκερως, after 'wayfarer' insert 'sec. euphem. for a robber'

οἰόκερως, add 'rest. in Call. in Suppl.Hell. 288.1'

οἴομαι, line 12, after 'ὀίω' insert '(also A.Pr. 187 (anap.))'   **VI 3**, after 'Lys. 12.26' insert '(s.v.l.)' and delete 'cf. Pl. .. Ep. 324b'   add '**4** ὡς ᾤου (after an adv.) as you thought (sc. erroneously), Pl.Ep. 319b, cf. Pers.Stoic. 100 (ap.Ath. 607c).'

οἰονεί, add 'introducing a hypothetical etymological form, κάμηλος οἰονεὶ κάμμηρος Artem. 1.4'

οἰονόμος I, delete 'ἐπ' οἰονόμοιο .. (Leon.)'   **II**, add 'perh. also neut., ἐπ' οἰονόμοιο in the sheep-pasture, APl. 230 (Leon.Tarent.)'

οἰοπέδη, delete the entry.

οἰοπόλος I, after 'Pi.P. 4.28' insert 'θεαί A.R. 4.1322, 1413' deleting these refs. fr. section II.

οἶος, line 1, for 'οἶϜος Inscr.Cypr. 135.14 H.' read 'o-i-wo-i οἶϜōι ICS 217.14'   **I 1**, add 'μαζὸν δ' ἀμφοτέροισι παρίσχεται, οἷον ἑκάστῳ, i.e. one to each for itself, Opp.H. 1.660'

οἷος III 1 a, line 2, after 'a thing' insert 'or the quality leading to, or shown in, an action'; line 5, after 'R. 415e' insert 'Thphr.Char. 1.2, al.'   **b**, line 3, after 'Is. 8.21' insert 'Sosip. 1.20'; line 4, for 'Antig.Car.ap.Ath. 7.345d' read 'Antig.Car.ap.Ath. 8.345d'; add 'cf. PEnteux. 26.3, al. (iii BC)   **3**, line 3, for 'ἐστίν' read '[ἐστίν]'   **V 2 b**, add 'οἷον ὡς Arist. 1013ᵃ4, Thphr.Od. 9'   **3**, after 'a part.,' insert 'οἷα Ἕλλησι ὁμιλήσαντα Hdt. 4.95'   **VI**, line 3, after 'Arist.EN 1114ᵇ17' insert 'οἱουδήποτε γένους SEG 33.1177 (Lycia, AD 43), PColl.Youtie 92.35 (vi AD)'; line 5, delete 'οἱοσδήπως .. AD)' (read οἱωδήποτε); before 'οἱοσδητισοῦν' insert 'οἱοσδήτις Plu. 2.1043c'; line 6, before 'οἱοσποτοῦν' insert '**οἷός περ**, see section II 1; οἱόσποτε SEG 29.250 (?iv AD)'

*__οἰριάζων__ τραχυνόμενος Theognost.Can. 23.

*__οἴριος__ ἀποστερητής Theognost.Can. 23.

οἰρών, at end delete 'cf. ἰρών'   add '**II** Cypr. i-to-i-ro-ni, perh. ἰ(ν) τ(ōι) οἰρōνι, region, ICS 217.8, 31.'

ὄϊς, line 3, after 'Call.Ap. 53' insert 'ὄϊς Opp.C. 2.377'

οἶσος, line 3, after 'Hsch.' insert 'cj. in Antim. 121 Wy.' and delete 'perh.'

οἰστεύω II, after 'shoot with an arrow' insert 'A.R. 1.759'

οἰστικός, after 'Orib.fr. 72' insert 'Procl.Inst. 63, al.'

ὀϊστός, line 2, after 'Arist.Ph. 239ᵇ30' insert 'Iamb.VP 28.140'

ὀϊστούχος, add 'SEG 37.1175 (ii AD)'

οἰστρήεις II, for 'Nonn.D. 21.188' read 'Nonn.D. 21.190'

οἴστρημα, for 'ravings of madness' read 'spur to madness'

οἰστροβολέω, for 'strike with the sting, τινα esp.' read 'sting, τινα'

οἴστρος II 2, line 2, after 'passion' insert 'Simon. 36.10 P.'

οἰσυοπλόκος, add 'also οἰσιο- Gloss.'

οἰσύπη, after 'οἴσυπος' insert '(neut. oesypum in Lat., Ov.Ars 3.213, Cels. 6.18.7.A, Plin.HN 12.74, etc.)'

*__οἰσυπλοκή__, ἡ, wickerwork, περὶ οἰσυπλοκῶν Edict.Diocl. 12.19, (perh. οἰσυ(ΐνων) πλοκῶν).

οἰφόλης, after 'lewd' insert 'τοἴφωλη (= τοῦ οἰφόλη) SEG 32.724 (vi BC)'

*__οἰφόλιος__, ὁ, prob. a title of Dionysus, in Archil. 251.5 W.

οἴφω, line 3, delete 'Mimn. 15 Diehl.'

οἴχομαι, line 3, for 'ᾤχωκε' read 'οἴχωκε' and after 'A.Pers. 13' insert 'παρ-οίχωκεν Il. 10.252'

*__οἰωνευτής__, οῦ, ὁ, = οἰωνιστής, SB 9309ᵛ.3 (iv AD).

ὄκα, line 1, after 'ὅτε' insert 'Stesich.S. 15 ii 15 (p. 160 D.)'

ὀκέλλω I 2, for this section read 'to sail a course, in quots. transf., πλόον Nic.Th. 295, στίβον ib. 321'

*__ὀκίστια__, τά, (or ?ὀκίστια) perh. harrows, IG 1³.422.135 (v BC), (cf. Lat. occare, Welsh oged, etc.).

ὄκκα, penultimate line, after 'AP 6.353.4' insert '(Noss.)'

ὄκκαβος, add 'occabus in Lat. inscr., CIL 13.1751'

Ὀκκονηνός, ὁ, cult-title of Zeus, IGBulg. 599, 718 (Ὀκον-), INikaia 1118, 1119; also perh. Οὐκονηνός SEG 32.679 (Thrace).

ὀκλάζω I 2, line 3, for 'Hld. 5.23' read 'Hld. 5.24'

ὀκνέω, line 4, after '(Il.) 20.155' insert 'w. part., D.Chr. 7.129'   **I 3**, last line, for 'ὁ. μή ..' read 'ὁ. μή w. subj.' and after 'D. 1.18' insert 'w. opt., X.An. 2.4.22; w. ind., Call.Epigr. 27.2 Pf.'

*__ὀκνηλός__, = ὀκνηρός, Theognost.Can. 62.

*__ὀκνία__, ἡ, = ὄκνος 1 2, Sch.E.Or. 708.

*__ὀκνόλακκος__, ὁ, = °ὄκνος λάκκου, POxy. 2197.130, al. (vi AD).

ὄκνος IV, add '**2** perh. kind of derrick or crane, PLond. 1164h8 (iii AD); ὄκνος λάκκου, perh. a shadoof, PMerton 41.2, PMich.XIV 682.1 (both v AD).'

ὀκρίβας, line 1, after 'Odeum' insert '(or perh. the theatre)'

Ὀκτάβαιος, ὁ, a month in Cyprus beginning 2 January, Cat.Cod.Astr. 2.148.14.

*__ὀκταβάριος__, ὁ, collector of the octavae (tax), IGChr. 10 (Hellespont, iv/v AD).

ὀκτάβλωμος, add 'cf. βλωμός'

*__ὀκταδράχμιος__, α, ον, of eight drachmas, ὀκταδραχμίας σπονδῆς Διονύσου PMich.inv. 1337 (BASP 16.194, AD 156).

ὀκτάδραχμος, add '**3** ὀκτάδραχμος, ἡ, eight-drachma tax, POxy. 1185.19 (c.AD 200).

ὀκταετής, add 'fem. Μαρκαῖνα ὀκταέτης SEG 31.564 (Delphi, ii AD)'

ὀκτάκις, after 'eight times' insert 'Plu. 2.1003f'; after 'ὀκτάκι' for 'Epigr.Gr. 356.4 (Hadriani)' read 'IHadr. 94.8 (i/ii AD)'

ὀκτακότυλος, after 'cotylae' insert 'σταμνία PSI 535.49 (iii BC)'

ὀκτάμηνος, at end after 'Arist.HA 583ᵇ33' add '(dub.)'

*__ὀκταξεστιαῖος__, ον, of eight sextarii, PMich.XV 734.16 (AD 572).

ὀκταούγκιον, add 'Cod.Just. 6.4.4.16c'

ὀκτάρουρος, add 'also ὀκτώ- PHamb. 65.15 (ii AD)'
ὀκτάς I, add 'Procl. in Ti. 2.213.31, 215.22' II, add '2 set of eight or a one-eighth measure (in quot., written ὑκτάς, in an inventory of utensils), SEG 24.361.23 (Thespiae, iv BC).'
*ὀκτασσαριαῖος, ον, τόκος ὁ. interest of eight asses a month on 100 denarii, i.e. 6% per annum, REG 19.247 (Aphrodisias).
ὀκτάχορδος, add 'octochordos at Vitr. 10.8.2'
*ὄκτις, ἡ, an unidentified bird, PAmst. 13.12 (v AD).
ὀκτώ, line 2, after '(v/iv BC)' insert 'Thess. ὄττου SEG 26.672.13, etc., (Larissa)'
ὀκτώβολοι, delete the entry.
*ὀκτώβολος, ον, at eight obols a mina, εἰσφορά IG 5(1).1432.3 (Messene, i AD).
*Ὀκτώβριος, α, ον, Lat. October, τῇ πρὸ ἐννέα καλανδῶν Ὀκτωβρίων SEG 15.815.51 (Apamea, c.9 BC); SEG 32.871 (Crete, ii AD); also Ὀκτόβριος Mitchell N.Galatia 181 (AD 145); Ὠκτόβριος SEG 28.1574 (v/vi AD).
ὀκτωκαιδεκαέτης, before 'Luc.DMort. 27.7' insert 'SEG 32.611 (Pharsalus, i BC)'
ὀκτωκαιδεκέτης, after '(Halic., iv BC)' insert 'GVI 1976.6 (Rome, ii AD), ὀκτωωκαιδεχέτης (sic) SEG 24.1239 (Egypt, iii BC)'; after 'Theoc. 15.129' insert 'fem., GVI l.c.'
*ὀκτώρουρος, v. ‡ὀκτάρουρος.
*ὀκτώχορδος, v. ‡ὀκτάχορδος.
ὀκώχιμος, delete the entry.
*ὀλαίγειος, ον, of pure goat's hair, ὀλαίγεον καρακάλλιν POxy. 3871.2 (vi/vii AD).
ὀλβιοδώτης, after 'Orph.H. 34.2' insert 'Hymn ap.Stob. 1.1.31'
*ὀλβιοτελής, ές, compd. of ὄλβιος and -τελής in indeterminable sense, Simon. 14 fr. 157.4 P.
ὀλβοθύλακος, for 'Cerc. 10' read 'Cerc. 4.2'
ὄλβος, line 6, after '(lyr.)' insert 'wealth, money, [ἐξ] ἰδίων ὄλβων TAM 4(1).223 (Rom.imp.)'
ὀλβοφόρος, for 'bringing .. θεοί' read 'wealthy, fortunate' and add 'SEG 36.694c (Berezan, late vi BC)'
ὄλεθρος, after lemma insert 'Cret. ὄλετρος Inscr.Cret. 4.51.10 (Gortyn, v BC)'
*ὀλεπίγραφος, ον, fully taxed, PThmouis col. 74.10, 81.3, 156.13.
+ὀλερός, ά, όν, alleged to be Att. for θολερός, perh. ghost-word, Hp.ap.Gal. 19.126, Hsch.; of a wind that brings dirty clouds, cj. in Str. 1.2.21, (cf. perh. ὀλός, ὁ).
*ὀλιαρχία, ἡ, v. ‡ὀλιγαρχία.
*ὀλιαρχικός, v. ‡ὀλιγαρχικός.
ὀλιγαναφορία, for 'quickness in rising' read 'low elevation of rising'
ὀλιγανάφορος, for 'quick in rising' read 'rising a little way'
ὀλιγανδρία, for 'scantiness of men' read 'scarceness of men' and add 'BGU 1835.10 (51/50 BC), ὀλιανδρ- pap.'
*ὀλιγανθής, ές, flowering briefly, ὥρη Sodalitas I 440 (Phrygia).
ὀλιγαρχία, line 1, after 'etc.' insert 'also ὀλιαρχία IG 2².448.61 (iv BC)'; line 4, after 'Pl.Ap. 32c' insert 'SEG 28.46.5'
ὀλιγαρχικός, line 1, before 'oligarchical' insert 'also ὀλιαρχικός pap. in Men.Sic. 156' 2, add 'sup., Plu.Sol. 13'
ὀλιγήριος, delete 'or perh. .. ὁρίον'
ὀλιγογράμματος, add '2 not fully literate, Just.Nov. 73.8 pr.'
ὀλιγοδρανέω, line 2, before 'also in late Prose' insert 'Ion Trag. 53 S.'; line 4, after 'pres. ind.' insert 'Anon.iamb.ap.Ath. 3.126f'
*ὀλιγοήμερος, ον, = ὀλιγήμερος, Eust. 18.8.
ὀλίγος I 1, line 4, after 'Theoc. 1.47' insert 'Cypr. ὀλίζων (v. infra VI 1), the younger, Karnak 13' and delete 'οὐκ ὀλίγης .. 2080.85' 2, add 'in adverbial sense, ὀλίγη ὡμίλει she seldom consorted with them, Arat. 115' 3, add 'cf. Call.Epig. 1.15 Pf.' IV 9, line 2, before 'to within' insert 'παρ᾽ ὀλίγον ἢ διέφευγον ἢ ἀπώλλυντο there was a narrow margin between escape and destruction, Th. 7.71.3' add '11 πρὸς ὀλίγον for short while, SEG 33.1000. 12 ὀλίγου γ᾽οὕνεκα for a little time, Ar.Lys. 74. 13 ὀλίγα πράσσειν as contradictory to πολλὰ π. (πράσσω III 4), ὁ. πρήσσε, φησίν, εἰ μέλλεις εὐθυμήσειν M.Ant. 4.24 (cf. Democr. 3).' V, add 'prob. rest. in Call.fr. 43.83 Pf., cj. ib. 80+82.21 Pf. (Add. II; -ους pap.)' VI 1, line 3, for 'the older form .. Hom.' read 'early form prob. ὀλίζων < *ὀλιγ-γων; ὀλείζων perh. in some ancient texts of Homer'; line 8, insert 'Cypr. o-li-zo-ne, ὀλίζων Karnak 13' add 'b app. in sense of ὀλίγος AP 9.521, cj. in Choeril. in Suppl.Hell. 329.1.' 2, delete 'always of Number or Quantity'
ὀλιγοστός I, line 1, at end insert 'Men.fr. 208 K.-Th.'
ὀλιγοσύλλαβος, add 'comp., Procl. in Cra. p. 34 P.'
ὀλιγοτρόφος II, delete the section.
*ὀλιγότροφος, ον, taking little nourishment, Arist.PA 682ᵃ21, Pr. 898ᵇ21.
ὀλιγοϋπνία, for 'little or short' read 'moderation in'
ὀλιγοχρόνιος I, line 2, for 'Thgn. 1020, Mimn. 5.5' read 'Mimn. 5.4'
ὀλιγωρέω, line 1, after '(Imbros, ii BC)' insert 'IG 12 Suppl. 644.34, 43 (Chalcis, iii BC)' 2, add 'also πρός τι Modest.Dig.

27.1.13.1' 3, for 'PAvrom. 1.25' read 'PAvrom. 1A25' add '4 to be worried or concerned, περὶ σοῦ PSI 1404.14 (i AD), PHerm. 14.6 (iv AD); PMerton 46.8 (vi AD).'
*ὄλιος (ὀλίς La.)· σκίουρος, ἔλειος Hsch.; cf. ὄλειρ.
ὀλίος I, line 2, for '300 BC' read 'c.300 BC'; add 'cf. °ὀλιαρχία, ὠλιώρησα s.v. ὀλιγωρέω'
*ὀλισβοδόκος, ον, receiving the ὄλισβος (perh. in sense "plectrum"), rest. in Sapph. 99 i 5 L.-P.
ὀλίσθανος, after lemma insert '(or oxyt.)'
ὀλισθάνω, line 8, after 'infr.)' insert '2 sg. ὤλισθας, GVI 1861.9 (i BC/ i AD)' I 1, add 'of a building in a state of collapse, Just.Const. Δέδωκεν 7a' 2, line 2, after 'Cra. 427b' insert 'λέξις ὁ. διὰ τῆς ἀκοῆς D.H.Comp. 22' II 2, for this section read 'fig., cause to slip, ἀναστήσονται οἱ Πέρσαι πρὸς κραταιὸν πόλεμον καὶ ὀλισθήσονται ὑπὸ Ῥωμαίων Orac.Tiburt. 112; w. abst. obj., τὰς διανοίας Lxx Si. 3.24'
ὀλισθοποιέω, add 'cf. tavefactus (transl. as if labefactus) ὀλισθοποιηθίς PNess. 1.847 (Virgil gloss., vi AD)'
ὄλισθος I 2, after 'cf.' insert '?Call.fr. 754 Pf.'
*ὀλίσθρημα, ατος, τό, intrigue, κατισχύσει βασιλείας ἐν ὀλισθρήμασιν Thd.Da. 11.21.
+ὀλκαῖον, τό, stem of a ship, A.R. 1.1314. II bucket or bowl, Antioch.ap.Poll. 6.100.
ὁλκάς 1, add 'POxy. 3342.31 (AD 204/6); transf., of the ether as carrier of the cosmos, Philol. 12'
*ὀλκεῖον II, add 'καὶ ὀλκία δὲ τὰ πηδάλια ἐν Ναυπλίῳ ὠνόμασε (sc. S.fr. 438) Poll. 10.134'
ὀλκή III 2, add 'Lang Ath.Agora xxi Hb2 (iii BC)'
*ὀλκίδιον, τό, dim. of ὀλκεῖον, PIand. 150 ii 4 (iii AD).
ὁλκός, ή, όν, I, after 'attractive' insert 'absorbent' III, for this section read 'ὁλκά· δυνατά Hsch.; θουλκότατον· βαρύτατον id.'
ὁλκός, ὁ, I 2, for 'strap, rein' read 'leather strip, thong'
+ὁλκότης· τὰ αὐτά (i.e. the glosses on ὁλκός in Hsch.), prob. weight, Hsch.
ὄλλυμι B III, line 3, for 'of the dead' read 'of the slain'
+ὅλμος, ὁ, mortar, χεῖρας ἀπὸ ξίφεϊ τμήξας ἀπό τ᾽ αὐχένα κόψας, ὅλμον δ᾽ ὣς ἔσσευε κυλίνδεσθαι δι᾽ ὁμίλου Il. 11.147, Hes.Op. 423, Hdt. 1.200, Ar.V. 201, 238, IG 1³.422.22, 25, 423.10 (written ηόλμος, v BC), καθίσας τὸν ἄνθρωπον ὀκλὰξ ἐπὶ ὅλμον δύο Hp.Haem. 4, Men.Dysc. 631, CID I 10.24, IG 12(5).872.82 (iii BC), PLille 9.9 (iii BC), etc.; prov. ὁ. ὑπὲρ κεφαλῆς Lib.Ep. 473.4. II transf., of things with a deep circular depression: 1 the part of the tripod on which the Pythia sat, Poll. 10.81; prov., ἐν ὅλμῳ κοιμᾶσθαι Plu.Prov. 2.14; ἐν ὁ. εὐνάσω Zen. 3.63. 2 drinking-vessel, Menesth. 1. 3 bulb between the main tube and mouthpiece of an aulos, Eup. 289 K.-A., cf. Poll. 4.70, and v. ὑφόλμιον II. 4 form of sundial, ὅλμου τοῦ λιθίνου ὃς ἐκαλεῖτο Ἑλληνιστὶ [γν]ώμων PHib. 27.26 (iii BC). 5 τὸ ὑπὸ ταῖς ὑπογλουτίσιν ἑκατέρωθεν κοῖλον Hsch.
*ὀλοεῖται· ὑγιαίνει Hsch.; cf. °ὀλοός (B), οὔλω, ὅλος I 2.
ὄλοισος, add '(ὄλυσος La.)'
ὁλοκάλαμος, add '2 full-grown reed, PCol. 230.5 (?iii AD).'
ὁλοκαυτέω, for 'offer whole' read 'burn an offering entire'
ὁλόκαυτος 1, after 'burnt whole' insert 'οἷς Sokolowski 2.19.84 (iv BC); for 'Call.Fr. 1.49 P.' read 'Call.fr. 228.49 Pf.'; delete 'τὸ ὁ.' and transfer 'Lxx .. (16)' after 'θυσία'
*ὁλοκίτρινος, ον, entirely citron-yellow, epith. of the λίθος ἀχάτης, Socr.Dion.Lith. 17 G.
ὁλοκληρία, line 1, after 'soundness in all parts' insert 'good health'; line 3, for '(Phrygia, i/ii AD)' read '(Lydia, ii/iii AD); Sardis 7(1).94'
ὁλόκληρος, line 11, before 'Adv.' insert 'subst. τὸ -ον Modest.Dig. 27.1.1.2; ἐξ ὁλοκλήρου in its entirety, Just.Nov. 22.44.9'
ὁλόκνημος, for 'σκελὶς' read 'σχελὶς'
*ὁλοκότινος, τό, = ὁλοκόττινος, BGU 1082.5 (iv AD).
*ὁλοκότ(τ)ινος, ὁ, Lat. solidus, a gold coin, Edict.Diocl. 30.1a, POxy. 1223.32 (iv AD), 1026.5 (v AD); also -ον, τό, PMasp. 70.1 (vi AD).
+ὁλολαμπής, ές, shining all over, coined as etym. of Ὄλυμπος, Arist. Mu. 400ᵃ8.
*ὁλόλινος, ον, made of pure linen, PMag. 36.268 (iv AD).
ὀλόλιτρος, delete the entry.
ὀλολυγή, line 3, after 'h.Ven. 19 (pl.)' insert 'Alc. 130.35 L.-P.'
ὀλόλυγμα, add '2 cause of wailing, Syria 14.385 (Mesopotamia).'
ὀλολυγών II, for this section read 'meton., frog, (sts. app. spec. a tree-frog), Eub. 102.6 K.-A., Thphr.Sign. 42, Theoc. 7.139, Arat. 948, AP 5.291.5 (Agath.); cf. qua vocem emittunt (rani) mares, cum vocantur ololygones, Plin.HN 11.173'
ὀλολύζω, add 'see also °ἐλελύζω'
ὄλολυς, add 'cf. ὄλολοι'
ὀλοός I, penultimate line, delete 'θάρσος .. Nonn.D. 13.416' and insert 'ὀλοίιος Procl.H. 5.15'

*ὁλοός (B), ή, όν, = ὅλος, καὶ ὁ ἀγαθὸς ὁλοός, ὁ φρόνιμος καὶ ὑγιής Sch.Nic.Al. 75b, Suid.
*ὁλοπίναρος, α, ον, consisting entirely of pearls, PMasp. 340ᵛ.31.
*ὁλοπόλιος, v. °ἀλοπόλιος.
ὁλοπράσινος, add 'Zos.Alch. 142.26'
ὁλόπτω, after 'tear out' insert 'τρίχα ὤλοψε Euph. in Suppl.Hell. 415c ii 16'; after 'Call.Dian. 77' insert 'fr. 573 Pf.'; after '(Antip.Sid.)' insert 'Nonn.D. 21.70'
ὁλόπυρος, after 'of unground wheat' insert 'ὁλοπύρων (sc. ἄρτων) PIand. 146 iii 8 (ii BC) and for 'esp. of wheat boiled whole' read 'ὁλόπυρος, ὁ, a dish made from this'
*ὁλορούσιος, ον, completely red, POxy. 1978.7 (vi AD).
ὅλος I 2, add 'IG 4².126 (Epid.)' 3, add 'w. gen., entirely concerned with or intent on, οὕτω τῶν .. πραγμάτων καὶ τῆς ἀληθείας ὅλοι Just.Nov. 78.4 pr.' 4, line 5, after 'Pl.Plt. 302b' insert 'abbrev. ὅ. before numbers, in total, SEG 30.1535 (ii AD), etc.; also fem., ὅλαν entirely, Inscr.Cret. 4.77B4 (Gortyn, v BC)' add 'ἐν ὅλαις altogether, in all, CPR 7.8.16, PGoodsp.Cair. 30 xxxi 12 (ii/iii AD)'
ὁλοσηρικός, after 'Cyran. 120' insert 'PVindob. G16846 (Tyche 2.6-7)'
ὁλοστήμων, add 'στιχαρομαφόριον POxy. 1978.6, al. (vi AD, -στυμ- pap.)'
ὁλόσφυρος, add 'Alcid.Od. 26'
ὁλοσχερής I 3, penultimate line, after 'Plot. 1.6.9' insert 'D.H. Din. 4'
ὁλοσώματος, add 'ἀνδριάς IG 12(7).240.29 (Amorgos, iii AD)'
ὁλοφλυκτίς, before 'ίδος' insert '(-φυκτίς Hsch., Phot.)'
ὁλόφυρσις, add 'Aret.SD 1.5'
+ὅλοφυς· οἶκτος, ἔλεος, θρῆνος Hsch.; Theognost.Can. 57 A.; Sapph. 21.3 L.-P.
ὁλόχροος, add 'Hsch. s.v. ἀμυσχρόν'
*ὁλοχωρία, ἡ, whole area or site, ἱεροῦ PSI 1145.14 (ii AD).
ὅλπη, line 1, after 'ἡ' insert 'ὄλπα SEG 26.399 (Corinth, c.580 BC)'
*Ὀλυμπεῖος, α, ον, late form of -ίειος, Olympian, μίαν ἐν τῷ Ἡρακλείῳ καὶ μίαν ἐν τῷ Ὀλυμπείῳ γυμνασίοις SEG 30.1383A (Hypaepa, AD 301).
Ὀλύμπια, for 'the Olympic games .. Zeus' read 'a religious festival with agonistic games held in honour of Olympian Zeus, at Pisa'; add 'at other places, IG 2².3162.10, 18, ibid. 16 (sp. -εια), TAM 4(1).34 (Smyrna), SEG 31.1103 (Phrygia), etc.'
Ὀλυμπιάς II 3, add '= Lat. quinquennium, lustrum, PKöln 10.4 (i BC)'
*Ὀλυμπιασταί, οἱ, name of a guild of worshippers named after a woman called Ὀλυμπιάς, MDAI(A) 25.109 (Rhodes).
Ὀλυμπίεια, after 'his festival' insert '(including agonistic games)'; delete 'later Ὀλυμπεῖα .. al.'
Ὀλυμπεῖον I, after 'Th. .. 70, al.' insert 'IG 7.1, al. (Megara, iv BC)'
Ὀλυμπικός 2, add 'Ὀλυμπικὰν στοὰν SEG 23.207 (i BC/i AD)'
Ὀλυμπιονίκη, delete 'Id. 4.17' and '(both pl.)'
*Ὀλυμπιονικία, ἡ, victory at Olympia, δύο τ' Ὀλυμπιονικίας ἀείδειν (-κας pap.) B. 4.17 S.-M.
Ὀλυμπιόνικος, for 'ib. 5.21' read 'Pi.O. 5.23' and add 'Lang Ath.Agora XXI C5 (vi BC)'
Ὀλύμπιος, line 2, after 'Men.Sam. 187' insert 'SEG 34.1093 (ii/iii AD)'; line 6, after '(Elis, vi BC)' insert 'τôι Δὶ τôι Ὀλυνπίôι SEG 31.344 (?Laconia, c.500 BC)' add '2 proper to or characteristic of the Olympian gods, Plu. 2.458c.'
*Ὀλυμπομέδων, οντος, ὁ, lord of Olympus, PHib. 172.74.
Ὄλυμπος, line 4, after 'cf. 113' insert 'Parm. 11.2'
ὄλυρα, lines 1/2, for '= ζειαί (q.v.)' read 'prob. emmer'
*ὀλυρίδιον, τό, dim. of ὄλυρα, PHib. 207.6 (iii BC).
*ὀλυροκοπία, ἡ, the milling of ὄλυραι, POxy. 3807.26.
*ὀλυροπράτης, ου, ὁ, dealer in ὄλυραι, rest. in BGU 1288.2 (ii BC).
*ὀλυρός· τραχύς Theognost.Can. 57 A.
ὅμαδος III, for 'Pi.I. 8(7).27' read 'Pi.I. 8.25a S.-M.' and insert 'Orph.L. 560'
*ὁμαίχμια, τά, = ὁμαιχμία, app. referring to taking Cretans into citizenship, Milet 2.12 (verse, c.200 BC).
ὁμαλίζω I 1, for 'cf. Damox. 2.50' read 'BGU 2354.6 (ii AD)' 2, add 'make even (in tempo), Damox. 2.50'
*ὁμαλίκιος, v. °ὁμηλίκιος.
*ὁμαλιστικός, ή, όν, used for levelling the ground, μηχανή PWarren 15.4 (ii AD).
ὁμαλός I 4, line 3, after 'Epicur.Ep. 2 p. 53 U.;' insert 'of style, D.H.Dem. 20'
ὁμάς, after 'the whole' insert 'ἐκ τῆς ἡμῶν ὁμάδος PNess. 24.4 (vi AD)' and after 'Gp. 10.2.3' insert 'καθ' ὁμάδα, of succession to an estate as a whole, Cod.Just. 6.48.1.3'
ὅμαυλαξ, after 'lands' insert '(of persons) Call.in Suppl.Hell. 238.9'; after '(Antip.)' insert '(of places) ὁ. ἀρούρας A.R. 2.787'
ὅμαυλος II, for 'γῆρας' read 'γῆρυς'
ὀμβρία, add 'II kind of meteorite, = νοτία III, Plin.HN 37.176.'

ὄμβριμος I, add 'Διὸς Ὀμβρίμου (on an altar), IHistriae 334'
ὄμβριος, add 'fem., personified as a goddess, Θεὰ ἡ Ὀμβρίος Salamine 60'
*ὀμβριστήρ, ῆρος, ὁ, = °ἐξομβριστήρ, PRyl. 583.16, 63 (ii BC).
*ὄ(μ)βρυσις, ἡ, prob. some extra charge or tax, μετὰ τῆς ὀμβρύσεως καὶ τοῦ ἀναλώματος SB 10568.7 (AD 393/4); app. an assaying charge, POxy. 3147.19 (iv/v AD).
ὁμευνέτις, add 'SEG 23.220 (Messene, i AD)'
ὁμήθης 1, for 'Call.Aet. 1.1.5' read 'Call.fr. 178.5 Pf.' 2, for 'accustomed' read 'congenial'
*ὁμηλίκιος, α, ον, Dor. ὁμᾱλίκιος, of the same age, GVI 1155.5 (Arcesine, ii/i BC).
ὁμῆλιξ, for 'ὑμᾱλιξ' read 'ὑμᾱλιξ' and add 'Sapph. 30.7, 103.11 L.-P.'
*ὅμηλυς, υδος, ὁ, companion, Nonn.D. 14.25 (pl.).
Ὁμήρειον, add 'II at Smyrna also name of a copper coin, Str. 14.1.37.'
ὁμήρης, for '= ὅμηρος' read 'mixed'
Ὁμηρικός, add 'III ὁμηρικόν, τό, some article of clothing, PRyl. 627.21 (iv AD).'
Ὁμηρομάστιξ, before 'Gal. 10.19' insert 'Vitr. 7.1.8'
Ὅμηρος, lines 1/2, after '(dub.)' insert 'Callin. 6' and before 'Hdt.' insert 'Simon. 59 P.'
ὅμηρος, add 'also neut. sg., παῖς ὢν ὅμηρον ἐδόθη βασιλεῖ, ὅ ἐστιν ἐνέχυρον Vit.Hom. 6.44'
*ὁμιλητήρ, ῆρος, ὁ, pl. as gloss on θέραπες, Hsch.
ὅμιλος, line 4, delete 'rare in Att. prose'
*ὁμιχλοφανής, ές, misty, cloudy, as a description of the sun as constituting an omen, Orac.Tiburt. 27.
ὄμμα I, add '3 the eye of gods, etc., considered as watching over human affairs, ὄ. Δίκης S.fr. 12 R., Orph.H. 62.1, SEG 11.325 (iv/v AD); ὄ. Διός Orph.H. 59.13, ὄ. δαιμόνων Trag.adesp. 499.' add 'VI ὄ. βοός = βούφθαλμον, AP 4.1.52 (Mel.).'
*ὀμμάτωρυξία, ἡ, the gouging-out of eyes, PMasp. 353.19 (vi AD).
ὄμνυμι, line 15, after '[D.] 19.318' insert 'Cypr. 1 pers. o-mo-mo-ko-ne, ὀμώμοκον, ICS 8.6' III, line 5, for 'rarely .. σιδαρέοισι' read 'w. dat., τῷ γὰρ ὄμνυτε; what do you swear by?'
ὁμοβώμιος, add 'MAMA 9.16 (i AD), Plu.fr. 157.3 S.'
*ὁμογαρικός, v. °ὠμογαρικός.
*ὁμογενέτωρ, ορος, ὁ, brother, E.Ph. 165 (lyr.).
ὁμόγραφος II, add 'SEG 35.665B.37 (Ambracia, ii BC)'
ὁμόδελφυς, for 'Call.Fr. 168 .. Fr. 1.73 P.' read 'Call.fr. 228.73, 524 Pf.'
ὁμόδουλος 3, add 'Just.Nov. 166 pr.'
*ὁμοδύναμος, ον, of the same power, Paul.Aeg. 3.46.7 (254.15 H.).
ὁμοζυγία, add '2 of married life, epigr. in IGBulg. 814 (written -ζυγέης), SEG 36.843 (Sicily, v AD).'
ὁμόζυγος I 1, add 'b ὁμόζυγος· γαμετή Hsch.' II, for 'corresponding' read 'joined in a pair'
ὁμόζυξ, add 'II κώλων, as synon. with διζύγων, v.l. in Paul.Aeg. 6.107.'
ὁμόθηλος, before 'Hsch.' insert 'Delph. 3(3).277.3 (i BC/i AD)'
ὁμόθρονος, add '2 holding the same rank, Just.Nov. 31 epilogos.'
ὅμοιος (A), line 4, after 'cf. ξυνός' insert 'or perh. levelling'; line 5, after 'νεῖκος Il. 4.444' insert 'Theoc. 22.172'
ὅμοιος (B), before 'for ὅμοιος' insert 'perh. the same as ὅμοιος (A), used by misunderstanding'
*ὁμοιόζηλος, ον, showing similar zeal, BCH suppl. 8 no. 179 (Thessalonica, iv/v AD).
ὁμοιομερής, add 'adv. -ῶς, homogeneously, Gal. 7.99.6'
ὅμοιος, line 3, Arc., add 'ὑμοῖος BCH 111.167' B 5, penultimate line, after 'E.Or. 697' insert 'folld. by ἤπερ, Aristid.Rh. 2 p. 531 S.'
ὁμοιοσχήμων, lines 4ff., form ὁμοιόσχημος, add 'ὑποδήματα ὁμοιόσχημα Leont.Byz.HP 2.3'
ὁμοιοτυπής, after 'Λητοῦς' insert 'διδύμοις' and after 'Sidyma' insert 'ii AD'
ὁμοιόω I, add '3 make equal (before the law), τὴν μητέρα .. ὠμοιώσαμεν τῷ πατρί Just.Nov. 2.3 pr.' II, after 'to be like' insert 'Aristid.Quint. 3.10'
ὁμοκλάω, add 'w. fut. inf., A.R. 4.1006'
*ὁμοκτηματικός, ή, όν, belonging to the same estate or farm, γεωργοί POxy. 1983.11 (vi AD).
*ὁμολέγω, agree, concur, PCol. 175.59 (AD 339).
+ὁμολεχής, ές, sharing the same bed, Lyr.adesp. 119.29 P.
ὁμολογία 3 c, add 'meton., amount specified in agreement, Just.Nov. 123.2.1'
Ὁμολώϊος, line 3, after '(Naupactus), etc.' insert '-ια, τά, festival at Orchomenos, IG 7.48, 3196, 3197' and after 'fem.' insert 'Ὁμολωΐς, ίδος, cult-epith. of Athena, Lyc. 520'
*ὁμομαχία, gloss on ὁμαιχμία, Hsch.
ὁμομήτωρ, after 'Orph.fr. 15' insert 'Plu. 2.482a'

ὁμονοητικός, line 2, after 'Arist.*Pol.* 1330ᵃ18' insert 'of persons, Iamb.*VP* 221'

'Ομονῶως, add 'also 'Ομόνοιος *SEG* 36.750 (Mytilene, iv BC)'

*ὁμοπαντεπόπτης, ου, ὁ, *he who sees everything at the same time*, *PMasp.* 188ᵛ.2 (vi AD).

ὁμόργνῡμι, add 'med. aor. μόρξατο, Q.S. 4.270, 374 (app. fr. false division of ἀπομόρξατο), see also ὅμαρξον'

*ὁμορρεύστης, ες, *flowing together*, Anon.Alch. 449.23.

ὁμόρροθος, after 'Theoc.*Ep.* 3.5' insert 'fig., ἀπὸ φρενὸς ὁμορρόθου Simon. 14 fr.35(b).10 P.'

*ὁμόρροπος, ον, *of the same value as, on a level with*, τοῖς ἐμοῖς παισίν *GVI* 1875.29 (Alexandria, i BC).

ὁμός, line 5, after 'Hes.*Sc.* 50' insert '(οὐκέθ' ὁμὰ codd.); οὐ καθ' ὁμὰ ζώοντες not living *together*, Ps.-Babr. Μυθικά 10 p. 217 C.' last two lines, for 'c. gen. .. 26' read 'ἑτέρων ἴχνια μὴ καθ' ὁμὰ [δίφρον ἐλ]ᾶν Call.*fr.* 1.26 Pf.; w. dat. prob. cj. in Nic.*Th.* 817'

*ὁμοσόδιον, τό, a space in a building (?assembly room, cf. ὁμόσε - ὁδός), θύραι τοῦ αὐτοῦ ὁμοσοδίου *CPR* 7.44.9 (v/vi AD).

ὁμόσπονδος, add 'Ion Trag. 53f S.'

*ὁμόστῑβοι (-βεῖς cj.)· συμπράττοντες Hsch.

ὁμόστοιχος, line 2, after 'Dam.*Pr.* 312' insert 'of girls dancing, Alcm. 33 P.'

ὁμότᾱφος, add '2 masc. subst., *member of a burial club*, Sol.*Lg.fr.* 76a R.'

*ὁμότᾱχος, ον, *having the same speed*, τὸν δρόμον οὕτως ὁμόταχον ῥυθμίζων Hld. 10.29.

*ὁμοτεμενής, ές, *sharing the same* τέμενος, *Inscr.Cret.* 2.xvii 1.16 (Lisos, iii BC).

*ὁμοτέχνης, ου, ὁ, *craft-fellow, member of the same guild*, *SB* 6266.3 (vi AD).

*ὁμοτεχνία, ἡ, *guild of workmen*, *MAMA* 9.49 (ii AD).

*ὁμοτεχνίτης, ου, ὁ, = °ὁμοτέχνης, *ZPE* 61.74 (AD 386).

ὁμότεχνος II 2, for this section read 'ὁμότεχνον, τό, *guild*, ὁ. τῶν λαναρίων *TAM* 5(1).85, ὁ. τῶν γναφέων ib. 86, ὁ. τῶν ὑφαντῶν *SEG* 33.1017 (all Saittae, ii AD)'

ὁμότης, after '*swears*' insert 'prob. of magistrates who swear to observe a law, Meiggs-Lewis 2 (Dreros, vii BC), cf.'

ὁμοτρεχής, read 'ὁμοτρεκής'

ὁμότροπος 1, line 2, after 'Pl.*Phd.* 83d' insert 'ὁ. αὐτῷ νυμφίον Men.*Dysc.* 337'

ὁμουργός, for '*mate* .. (vi AD)' read 'perh. in *SEG* 33.932 (Ephesus; or ?*[οἰκοδ]ομουργοί)'

ὁμοφῡλία, add '*SEG* 30.1723 (Egypt, i/ii AD; sp. ὁμοφιλία)'

ὁμόφωνος II 2, after 'Ptol.*Harm.* 1.7' insert '*on the same note*, of successive syllables, D.H.*Comp.* 11.63'; adv., after 'S.E.*P.* 3.239' add '*SEG* 34.1065 (Aphrodisias, vi AD)'

ὁμόχροος, add 'λαχάνοις τοὺς ἀνθρώπους ὁμόχροας Vit.*Aesop.*(G) 124'

ὁμόψῡχος I, after 'Lxx 4*Ma.* 14.20' insert 'Gorg.*Hel.* 2'

*ὄμπια, v. °ὄμπνιος.

ὀμπνηρόν, add 'also ὀμπνιηρόν, ὀπνιηρόν, v. *Suppl.Hell.* 1094'

ὄμπνιος I 1, after '*nourishing*' insert 'ὄμπνιον ὕδωρ Call.*fr.* 357 Pf.'; add 'ὄμπια (sic)· παντόδαπα τρωγάλια Hsch.' 2, for 'cf. *BCH* 11.161 (Lagina)' read 'perh. also of Hecate, *SEG* 30.1272 (Lagina, ?ii/i BC); cf. ὄμπνια Μήνη Nonn.*D.* 5.488, 38.124'

ὀμφακηρός, delete '(ὀμφακηρά ἁ' .. numeral)' and add 'ὀμφακηρά, ἡ, *rounded vessel, flagon*, *PLond.* 239.13 (iv AD), *POxy.* 1870.13 (v AD)'

ὀμφάκινος 1, for '*unripe grapes*' read '*unripe fruit*'; add '*Edict.Diocl.* 31a; κηκίδος ὀμφακίνης Gal. 14.198.17'

*ὀμφᾰλάριον, τό, some kind of vessel, *PHerm.* 23.7 (iv AD); also ὀμφᾰράριον *PVindob.Worp.* 11.6 (vi AD); cf. °ἀμφορᾰριον.

ὀμφᾰλητομία, line 1, delete '*midwifery*'

ὀμφᾰλός III, add '6 part of the νόμος κιθαρῳδικός, Poll. 4.66.'

*ὀμφᾰράριον, v. °ὀμφᾰλάριον.

ὀμφή, line 6, delete 'signified by the flight of birds'

ὀμφήεις, add '*Didyma* 497.5'

*ὁμωμότας, α, ὁ, *one who supports a person's oath by swearing with him*, .]μομότας *Inscr.Cret.* 1.xviii 5.13, 4.4.3.

ὁμωνῡμία I, after lemma insert '(Ion. -ίη *SEG* 34.1116)'; before '*AP* 6.100' insert 'ὁμωνυμίῃ παῖς πατρὸς Ἀντιφάνης'

ὁμῶς, line 2, delete '*in equal parts*, Hes.*Th.* 74'

*ὀναλόω, Thessalian for ἀναλόω, *SEG* 31.572.A7-8 (Crannon, c.200 BC).

ὄνᾱρ II, line 8, κατ' ὄναρ, add '*TAM* 4(1).67, *SEG* 31.1259 (Pisidia, ii AD)'

ὀνάριον, add 'in undetermined sense, Lang *Ath.Agora* XXI B10 (iv BC)'

ὀνᾶς II, delete the section.

*ὀνᾱς, ᾰδος, ἡ, *she-ass*, Aq., Sm.*Za.* 9.9, Al.*Ge.* 45.23, *POxy.* 3416.18 (iv AD); Gloss.

*ὀνγράφω, v. °ἀναγράφω.

ὄνε, add 'also Cypr. *o-ne* ὄνε *ICS* 306.5; *to-ne* τō(ι)νε (dat. sg.), 306.2, 7'

ὄνειαρ I 2, delete the section transferring exx. to section 1. II, line 2, for 'Call.*Epigr.* 49' read 'Call.*Epigr.* 48 Pf.'

ὀνειδίζω I, line 9, after 'S.*Ph.* 523' insert 'also w. acc. of person, τοιαῦτ' ὀνειδίζεις με id.*OC* 1002'; at end after 'Hdt. 8.143' insert 'w. pred. adj. ἐπειδή .. τυφλόν μ' ὠνείδισας S.*OT* 412' II 2, delete 'c. acc. .. *OT* 412'

ὄνειος (B), lines 5/6, for 'ὀνήϊστον .. A.R. .2.335' read 'w. acc. and inf., τἆλλα μεθέντας ὀνήϊστον πονέεσθαι θαρσαλέως A.R. 2.335'

ὀνειραιτησία, add 'sp. ὀνειρετησία *POxy.* 3298.41 (iii AD)'

*ὀνειριάζω, dream, Cyran. 1.5.13 K.

*ὀνειροκρῐτία, ἡ, = -κρισία, *PMag.* 1.330.

ὀνειροπολέω, line 4, after 'Luc.*Merc.Cond.* 20' insert 'Ach.Tat. 5.26; med. S.E.*M.* 8.57'

*ὀνειροπρόσωπος, ὁ, name of a wild poppy, Ps.-Apul. 53.13 (*oniroprosopos*).

ὄνειρος 1, add 'κατ' ὄνειρον *SEG* 36.484 (Thespiae, i/ii AD), 36.533 (Acarnania, ii AD)'

*ὀνειρόφοιτος, ον, *frequenting dreams*, i.e. *appearing frequently in the dreams* of his worshippers, epith. of Egyptian god, *APAW* 1943(14).8 (Chalcis, iii AD).

*ὀνειρωκτικός, ή, όν, *of or occurring in dreams*, θεάματα Procl. *in R.* 1.121.9 (codd. -ρακτ-); φαντασίαι Sch.Theoc. 9.16.

*ὄνη, ἡ, *she-ass*, *BGU* 228.3 (ii/iii AD), *PMich.*XI 620.218 (iii AD).

ὀνηλάτης, add 'Salamine 22 (Cyprus, i AD)'

ὀνηλᾰτικός, add 'ἐργασία *POsl.* 135.6 (iii AD)'

ὄνησις, line 9, after 'βίου ὄ.' insert 'E.*Med.* 254'

ὀνητός I, add 'cf. Myc. neut. subst. *o-na-to*, *beneficial use of land*'

ὀνήτωρ, add 'form ὀνᾱτήρ, Myc. *o-na-te-re* (pl.)'

*ὀνθομετάφορος, ὁ, *dung-transporter*, *PPrincet.* 154.3 (vi AD).

ὀνθοφόρος, add '*CPR* 10.116.5 (AD 446)'

ὀνικός, add 'σάγμα ὀ. *PSI* 527.2 (iii BC)'

*ὀνίλαμον, τό, name of an *eye-salve*, Aët. 7.106 (370.4 O.).

*ὀννα, unexpld. wd. in list of comestibles, ψωμίων τῆς οννα *BGU* 2358.10 (iv AD.).

*ὀνοθήλεια, ἡ, *female donkey*, *PNess.* 89.31, 34 (vi/vii AD, ὠνοθελ- pap.).

*ὀνοκαρδία, ἡ, precious stone of scarlet colour, Plin.*HN* 37.176.

ὀνοκένταυρα, read ὀνοκένταυρος.

ὀνοκίνδιος, add 'said by Pollux to be Doric, Poll. 7.185'

ὄνομα I, lines 4ff., form οὔνομα, add 'μὴ παρακούσῃς τὸ οὔνομα τοῦ Θεοῦ *SEG* 31.1594'; line 10, delete 'in Hom. always of a person' (cf. section II, line 1); line 17, add 'also ἐξ ὀνόματος *IHistriae* 57.28 (ii AD)' add '4 *name* considered as having magical properties, *Gp.* 20.18.' IV 1, add 'ἐν ἑνὶ ὀνόματι θανάτου Lxx *Wi.* 18.12' 2, add 'ὑπεύθυνον εἶναι Ὑπαιπηνῶν βουλῇ προστείμου ὀνόματι βφ' *IEphes.* 3829; εἰς ὄνομα *in the name of*, εἰς ὄ. Θατρῆτος *POxy.* 3903; sim. ἐπ' ὀνόματος ib. 3908. 3909 (s.vll., all AD 99)' V, add 'D. 23.36'

*ὀνομάγγων, ωνος, ὁ, *donkey-seller*, *POxy.* 3192.10, 3728.4 (iv AD), cf. Lat. *mango*.

ὀνομάζω, line 2, after 'etc.' insert 'Aeol. ὠνύμασσαν Alc. 129.8 L.-P.; unaugmented aorist ὀνόμαξαν *CEG* 894.14 (Delphi, iv BC)' IV, add '2 *put into words, express*, τὰ πραχθέντα D.H.*Th.* 26; pass., ib.'

ὄνομαι, add 'A.*Supp.* 337'

ὀνομᾰσία I 1, for '*name*' read '*nomenclature, naming*' add '3 *listing of names*, Just.*Nov.* 1.1.1.'

*ὀνόμᾰσις, εως, ἡ, *naming*, Pl.*Smp.* 199b (v.l.).

ὀνομασμός, ὁ, = ὀνομασία, Alex.Aphr. *in SE* 12.9.

ὀνομαστί II, add '*SEG* 32.1554 (Arabia, vi AD)'

ὀνομαστί, line 3, for 'Call.*Aet.Oxy.* 2080.81' read 'Call.*fr.* 43.79 Pf., Theoc. 24.78'

ὀνομαστός I, after '*named, to be named*' insert 'Arat. 381, 385'

ὀνομᾰτικός, line 3, after 'Hermog.*Id.* 1.6' insert 'ἀπαγγελία Aristid.*Rh.* 1 p. 499 S.'

ὀνομᾰτοθέτης, add 'Alcin.*Intr.* p. 160 H.'

ὀνόπορδον, before 'Hsch.' insert 'τὴν ἑλξίνην· ἔστι δὲ λάχανον ἄγριον. καὶ εἶδος κογχυλίου'

ὄνος I 1, line 5, for 'when caught .. ὀστρακίνδα' read 'when failing in a game' and add 'Pl.*Tht.* 146a' 6, add 'ὄνος (ολος pap.) ἄγειν δοκώ μοι τὴν ἑορτήν Men.*Dysc.* 550' IV, delete '= τρωξαλλίς' VII 2, after 'round' insert 'ὅ. ἀλέτας *Inscr.Cret.* 4.75*B*7 (v BC)'; line 2, after 'Alex. 13, 204' insert '*IG* 1³.422.24 (Athens, v BC)' add 'Myc. *o-no*'

ὀνοτάζω, med., add 'Ion Trag. 17 S.'

ὀνοφορβός, add 'D.Chr. 7.134'

†ὀνόφυλλον, τό, gloss on ὄνος (in sense = ὀνῖτις), Sch.Nic.*Th.* 628.

ὄντα II, add 'Just.*Const.* Δέδωκεν 7d'

*ὀντίθημι, v. °ἀνατίθημι.

ὄντως, line 6, after 'also with nouns' insert 'or adjs.' and add '*SEG* 31.286, 962'; lines 8/10, for 'not used by Th. .. earlier' read 'Arist.*EE* 1238ᵃ19; Ion. ἐόντως cj. in Hdt. 7.143.1'

ὄνυ, line 1, after 'ὅδε' insert '(also sts. in Crete, *Inscr.Cret.* 2.xii 11.3, ib. 22*B*14 (Eleutherna, vi/v BC), ib. v 20*A*5 (Oaxos, iii BC)), pron. referring back to something or somebody previously mentioned; see also °τῶνυ'; line 2, after 'nom. sg. masc.' read 'o-nu ὄνυ, *ICS* 216(b), acc. to-nu τόνυ, *ICS* 215(b)'

ὀνύδιν, for 'ὀνάριον' read 'ὀνίδιον, *little ass*'

ὄνυξ I, line 6, for 'of horses and oxen' read 'of ungulates'; line 8, delete 'metaph. .. στόνυχα)' and insert 'poet., of the point of a spear, *AP* 6.123 (Anyt.)' **1**, add 'ἀπὸ κορυφῆς ὡς ὀνύχων *SEG* 31.1492 (Alexandria, AD 608/10)' **3**, add 'σέ τε ἀκούων ἄνδρα, οἷον ἐξ ὀνύχος ἤδη ὁρῶ Philostr. *VA* 1.32' **II**, for '*anything like a claw*' read 'transf., of other pointed or claw-like objects, τρητὸς ὄνυξ πετραῖος (used for anchoring) Nonn.*D.* 3.48, οὐ πόδα .. ἀπέδιλον ὄνυξ ἐχάραξε κολώνης ib. 14.385' **III**, for '*anything like a nail*' read 'transf., of things displaying the appearance of a nail' add 'Myc. *o-nu-ka*, *o-nu-ke* (pl.), some kind of ornamentation on textiles'

*ὀνῡχανθές, οῦς, τό, kind of plant, Ps.-Apul. 10.15.

ὀνῡχίζω I, add 'in animal husbandry, *trim the hooves*, *PMil.Vogl.* 308.162'

ὀνυχισμός, add '*trimming of hooves*, *Edict.Diocl.* 7.20'

ὄνωνις, for 'Call.*fr.anon.* 366' read '*Suppl.Hell.* 1138'

*ὀξειοβᾰρής, ές, *pungent and heavy*, Ael.*NA* 7.5 (s.v.l., cj. ὀξοβ-).

ὀξίζω, add '**2** trans., *treat with vinegar*, πόδας χοίρου ὀξίσας (-ύσας cod.) *Vit.Aesop.*(G) 42.'

ὀξίς 1, after 'Ar.*V.* 1509' insert 'Iamb.*Protr.* 21θ'

ὀξυβελής II, masc. subst., add 'J.*BJ* 2.19.9.553, 3.5.2.80, al.'

ὀξυβρέχω, delete the entry.

*ὀξυγγοσάπουνον, τό, *lard-soap*, Anon.Alch. 380.18 (lemma), cf. Lat. *axungia*, *sapo*, σαπώνιον.

*ὀξυγραφία, ἡ, *tachygraphy*, Simeon Metaphrastes *Vita Luciani* 1.4.

ὀξυγράφος, add 'masc. subst. ὁ τῆς ἐξακτορίας *PLond.* 1105, *PRoss.-Georg.* v 61*B*.13 (iv AD)'

"**Ὀξυδέρκᾱ**, epith. of a goddess (cf. Ὀξυδερκώ), *IG* 4².491 (Epid.).

ὀξυδερκικός, delete the entry.

ὀξυδορκικός, for '= .. (q.v.)' read '*making the sight sharp*'; add 'Antyll.ap.Orib. 10.23.29, v.l. in Dsc. 2.163'

ὀξυδρόμος, add 'sup. Hld. 10.4'

*ὀξυζώμιον, τό, *acid liquid*, Anon.Alch. 271.2.

†ὀξύκόμινα, τά, prob. *food pickled in vinegar and cumin*, Petron. 66.7.

ὀξυόεις, add 'of a crane's beak, perh. by mistaken derivation fr. ὀξύς, ὀξυόεντι γενείῳ Nonn.*D.* 14.335'

*ὀξύορμητος, ον, *quick to start*, Heph.Astr. 3.45.3.

ὀξύπεινος, after 'Cic.*Att.* 2.12.2' insert 'adv. -ως ἔχει Men.*Dysc.* 777'

ὀξυπετής, for 'Sch.*Od.* 3.372' read '*Suppl.Hell.* 1165'

*ὀξυπίδας, ὁ, perh. non-Hellenic name of cult-object, *Inscr.Cret.* 1.v 23.16 (Arcades, ii AD).

*ὀξυπόδης, ον, ὁ, = ὀξύπους, *EA* 13.3 no. 496 (Mylasa, Rom.imp.), Hsch. s.v. καλπάζει.

ὀξυπόρος I, for '*with pointed mouth*' read '*piercing sharply*' **II**, transfer to ὀξύπορος.

ὀξύπτερος, line 3, before '*De.* 14.13' insert 'Lxx' and after it '*Cyran.* 95'

ὀξυπώγων, after '*beard*' insert '*BGU* 1080ᵛ (i AD)'

ὀξύρρινος, after 'ον' insert '*Cat.Cod.Astr.* 12.194.18'

ὀξύρροπος, for 'τὸ ὀ. τῆς πεύσεως .. Longin. 18.1' read 'τὸ ὀξύρροπον *sudden change*, τὸ ὀ. τῆς πεύσεως Longin. 18.1, τὸ ὀ. τῆς τύχης Hld. 10.2'

ὀξυρρυγχῑτικόν, for '(sc. μέτρον)' read '(sc. κεράμιον)'

ὀξύρρυγχος 2, after 'Egyptian fish' insert 'one or other species of *Mormyrus*' add '**b** app. *sturgeon*, Ael.*NA* 17.32.'

*ὀξύρ(ρ)υτος, ον, unexpld. adj. in description of siege-engine, *Poliorc.* 225.17.

ὀξύς, εῖα, ύ, line 2, delete 'and so Babr. 73.1 metri gr.' **IV**, lines 8/9, for 'ὁ. καιρός .. 6.1, al.' read 'ὁ. καιρός *fleeting* opportunity, Hp.*Aph.* 1, *PHib.* 15.42 (iii BC); *urgent crisis*, Onos. 6.1, al., Longin. 27.1'

*ὀξυτοκία, v. ὠκυτοκία.

†ὀξυτόρος, ον, *sharply piercing*, ὁ. χαλινῷ v.l. in S.*Ant.* 108, πίτυς ὀ. (i.e. sharp-leaved) *AP* 4.1.16 (Mel.).

ὀξυτυρία, delete the entry.

*ὀξύτυριος, α, ον, "*bright Tyrian*", name of a grade of purple, *Edict.Diocl.* 24.4,20, 29.19,25,31; also **-τέριος** ib. 24.20.

"**Ὀξυχία**, ἁ, epith. of Artemis, *SEG* 25.1018 (Crete, ii/iii AD).

ὄουαν (acc. sg.), represents Lat. *ovationem* (cf. εὐᾶς), Plu.*Crass.* 11; written **ὄβαν** id.*Marc.* 22.

ὀπάων I 1, add 'as the title of a hero divinity, Ὀπάονι Μελανθίῳ *SEG* 23.643 (Cyprus, i BC)'

ὀπεᾱς, line 2, after 'Hdt. 4.70' insert '[.]πέατι Hippon. 78.6 W.'

*ὀπεί, ὀπειδεί, v. °ἐπεί.

ὀπέρ, add 'so ὁοπήρ *BCH* 70.262 (Boeotia, ?iii BC); see also ὑπεραμερία s.v. ὑπερημερία'

*ὀπεράριος, ου, ὁ, Lat. *operarius*, *workman*, *ICilicie* 46 (i/ii AD).

ὀπή I, after lemma insert '(ὄψομαι, ὄπωπα)' **II**, add 'fig., *window on the mind*, Favorin.*Exil.* 16.5, 9' **III**, delete '(ὄψομαι, ὄπωπα)'

ὄπη II, line 4, after 'etc.' insert 'ὁπῆ νόμος ἀποστάτō *IG* 5(1).1155 (Laconia, v BC)'

ὀπηδέω, line 10, for 'A.*Fr.* 475' read '*Trag.adesp.* 493 K.-S.'

ὀπηδός, after 'ὁ' insert 'ἡ'; add '*Procl.Inst.* 185'

*ὀπί (A), Myc. *o-pi*, prep. w. dat. (app. = ἐπί); also in compds.

*ὀπί (B), (?) app. conditional particle, Cypr. *o-pi-si-si-ke* perh. ὀπί σίς κε (= ἐάν τις), *ICS* 217.29.

ὀπιθόμβροτος, delete 'poet. for ὀπισθόμβροτος'

*ὀπινάτωρ, ορος, ὁ, Lat. *opinator*, soldier detailed to collect money for the issue of pay and donatives, *PBeatty Panop.* 2.41, 54, al. (AD 300), ὀπιν[υ?]άτωρ *POxy.* 2114.10 (AD 316).

*ὀπινίω, ωνος, ἡ, Lat. *opinio*, *legal opinion*, *POxy.* 2130.1 (ὀπεινίω pap.), 25 (iii AD), *PSI* 1076.14 (iii AD).

ὀπιπᾶ, at end for 'πυρροπίπης' read 'πυροπίπης'

*ὀπιπάζει· εὐλαβεῖται Theognost.*Can.* 65 A.

ὀπιπεύω, at end after 'ὀπ-ωπα' add 'freq. written ὀπιπτεύω in codd.'

ὄπισθεν I 1, add 'αἱ ὄπισθε θεαί, an unidentified group of goddesses, *IEryth.* 207.21, 25; also sg., ἡ ὄπισθε θεός *Michel* 832.27'

ὀπίσθεν, line 3, after 'Arist.*HA* 500ᵇ30' insert 'without σκέλος, *IG* 2².1424a.19, 56 (iv BC); line 5, after '(Arist.)*IA* 706ᵇ1' insert 'τὰ ὀ., *back*, *buttocks*, *PLond.* 191.15 (ii AD)'

ὀπισθόδομος I, add 'ὑπισθόδομος *IG* 4.1588.9'

ὀπισθοκρηπίς, add '*IG* 2².1424a.337 (woman's shoe, Poll. 7.94)'

*ὀπισθόποινος, ον, (?)*retributive*, *Suppl.Hell.* 991.26 (poet. word-list, iii BC).

ὀπισθοπόρος, add '**2** *travelling backwards*, app. of a hand writing from right to left, ib. 4.268.'

*ὀπισθότατον· τελευταῖον, ἔσχατον Hsch.; cf. ὀπίστατος.

ὀπισθυπέρα, for '*brace of a sail*' read '*brace that maintains tension away from the helmsman in opposition to the ὑπέρα*'; add 'also ὀπισθοπ- *PZen.Col.* 100.8 (iii BC)'

*ὀπίσσωτρον, v. ‡ἐπίσωτρον.

ὀπίσω, line 2, for 'Sapph.*Supp.* 8.9' read 'Sapph. 19.10 L.-P., ὑπίσσω *Lyr.adesp.* 1A.14 P.'

*ὀπίσωθεν, adv. *from behind*, ὁ. ἀκολουθήσας *Tab.Defix.Aud.* 187.61 (ὀπίσθεν tab.), cf. Arc. 129.10.

"**Ὀπιταῖς**, ἴδος, ἡ, epith. of Artemis in Zacynthus, *IG* 9(1).600, *SEG* 14.481 (Thrace, iii BC).

ὁπλῑτοδρομέω, for 'Paus. 1.23.11' read 'Paus. 1.23.9'

ὁπλοθήκη, add '**3** perh. *compartment (on a ship) for storing tackle*, *PMich.inv.* 4001 (*ZPE* 24.83, i/ii AD).'

*ὁπλοκτυπία, ἡ, app. *fighting with weapons*, Ps.-Callisth. 14.8.

ὁπλομᾰχέω II, for '*drill-sergeant*' read '*weapon-instructor*'

ὁπλομάχης, add '*IG* 2².766.10 (iii BC)'

†ὁπλομάχος, ον, *fighting with heavy arms*, ὁπλομάχοι ἄνδρες Alciphr. 1.11, Lxx *Is.* 13.5; masc. subst., Plb. 2.65.11. **II** masc. subst., *instructor in fighting with weapons*, X.*Lac.* 11.8, Thphr.*Char.* 5.10, Teles p. 50 H., *PCair.Zen.* 298 (iii BC), *SIG* 697*E*11 (Delph., ii BC), *SEG* 26.176 (Athens, ii AD).

ὅπλον III 6, line 2, after 'Hdt. 7.218, etc.' insert 'ταλαντάρχην δι' ὅπλων *SEG* 32.660 (Thrace, ii/iii AD)'; third line from end, after '*Cyr.* 7.2.8' insert 'κατὰ τὰ ὅ. εἶναι Aen.Tact. 27.5'

ὁπλοποιία, add '**2** *place where arms are made, arms factory*, Just.*Nov.* 85.1.'

ὁπλότερος, add 'cf. Nonn.*D.* 33.343, *AP* 5.218.3 (Agath.)'

*ὁπλοφᾰνία, ἡ, *display of arms*, *AE* 1932 suppl. 20 (Phthiotid Thebes, iii BC, pl.).

ὁπλοφορέω I, delete '*BCH* .. B.C.)' and add 'in a religious rite, Διὶ Κυνθίῳ *Inscr.Délos* 1897.3 (ii BC)'

ὁπλοφόρος II, of Pallas, add '*PKöln* 245.7 (iii AD)'

*ὁπνιηρόν, v. °ὀμπνηρόν.

ὁπόθι 1, add 'also in prose, *IG* 5(2).343.*A*20 (Orchom., iv BC)'

ὅποι, after 'Ion. κοι' insert 'Aeol. ὄπποι (or ὄπποι) *IKyme* 113.14, etc.'

ὁποῖος III, delete '1076'; add '*AP* 7.295.7 (Leon.Tarent.), Lyc. 74, 182; also sg., προσώποις τισίν .. ὁποῖον δή τε παισὶ καὶ ἐγγόνοις Just.*Nov.* 1 pr. 2, 29.1'

*ὀπόκισσος, ὁ, *gum derived from the fruit of ivy*, Alex.Trall. 2.258.

*ὀποπευκέδανον, τό, *gum derived from the roots of Peucedanum (sulphur-wort)*, Alex.Trall. 1.72, 2.147.

ὀπός II, at end after 'alone' insert 'Nic.*Th.* 907'

ὁπόσος, line 3, after 'Ion. ὁκόσος' insert 'cf. *ZPE* 68.119 (Emporion, vi BC)'

ὁπότε I, line 3, for 'only ὅτε is so used' read 'ὅτε is normally used (but ὁπότε ἦσαν *Hell.Oxy.* 14.2)' add '**3** with some causal force, *in circumstances in which*, *when*, Thgn. 749, Hdt. 2.125.7, Lys. 6.23, D. 23.86, Is. 2.39, Pl.*Lg.* 895c; ὁπότε γε S.*OC* 1699, X.*Cyr.* 8.3.7.' **B**, line 2, after 'Pl.*Lg.* 895c' insert 'A.R. 1.83'

ὁπότερος, line 1, after 'Hom.' insert '(also Aeol. SEG 34.1238, c.200 BC)'

ὅπου I 1, after 'Relat.' insert 'X.An. 7.1.27'; for 'πόλεως .. cf.' read 'πόλιος ὅκου ἦν ἐπιτηδειότατον' 2, line 2, after 'S.OT924' insert 'Pl.R. 415d' B, before 'Plu. 2.427c' insert 'J.BJ prooem. 1'

*οπουπα, unexpld. designation after a name, BGU 2280a ii 14; also ουπουπα BGU 1087 iii 5 (iii AD).

ὀπτάνιον II, delete the section.

ὀπτάω 4, line 3, after 'fire of love' insert 'ὤπτηται μέγα δή τι Call.Epigr. 43.5 Pf.'

†ὀπτευτήρ, ῆρος, ὁ, one who has the oversight of, σιδήρου, of Hephaestus, cj. in Colluth. 54 (see also °κατοπτευτήρ).

ὀπτήρια 2, after 'presents' insert 'or sacrifice'; add 'sg., gloss on γενέθλιον δόσιν Sch.A.Eu. 7, also Nonn.D. 5.139; as adj., ὀπτήριον ὕδωρ ib. 6.129'

*ὀπύ, v. ὑπό.

ὅπυι, before 'SIG 56.39' insert 'SEG 30.380 (Tiryns, vii BC), SEG 26.461.5 (Sparta, v BC)'

ὀπυίω, line 1, after 'ὀπύω' insert 'Cypr. o-pu-we-ne ὀπύ(F)ēν, ICS 213a'; line 2, delete '(Hsch. .. γεγαμηκότες)' I 1, for 'marry, take to wife' read 'be married to, be the husband of' 2, for 'to be married' read 'be the wife of'

*ὀπυόλαι (ὀπυι- La.)· γεγαμηκότες Hsch.

ὀπωπή II 2, before 'Opp.C. 3.75' insert 'transf., of a leopard's spots'

ὀπώρα I, add 'personified, SEG 26.473 (Elis, iii BC)'; add 'also ὑπώρα OMich. 90.4 (iii/ii BC), POxy. 298.38 (i AD)'

ὀπωρικός 1, add 'of crops, grown for fruit, IEphes. 3217(b).36 (ii AD)'

ὅπως A I 2 b, add 'ὅπως κε οὖν .. βολλεύσαιτο ἁ βολλά Milet 3.152A.8 (ii BC)' 6, delete the section.

ὅπως ποτέ, add 'ὁπωσποτοῦν in any way whatever, Alcin.Intr. 32 p. 186 H.'

ὁραματιστής, after 'Sm.' insert '(more prob. Aq.)'

*ὁραπεία, ἡ, office of °ὁρᾶπις, SB 9346.8, 9658.12 (both ii AD).

*ὁρᾶπις, ιδος, ὁ, perh. Egyptian high priest, PRyl. 676.7 (i AD).

ὀράριον, add 'Vit.Aesop.(G) 21, Stud.Pal. 20.245.24 (vi AD); also ὠράριον, Edict.Diocl. 27.8, 23, PBodm. 29.332 (iv AD), Hsch. s.v. σιμικίνθια, Syr.Ge. 38.18'

ὅρᾱσις, add 'IV as name of bird, the Seeing-bird, of the eagle, pap. in SHAW 1923(2).17 (but perh. an unrelated Egyptian word).'

ὁρᾱτικός I, add 'of mental vision, Plot. 5.3.10' transferring here '-κὴ διάνοια id. (Ph.) 2.19' from line 3.

*ὁρατίων, ωνος, ἡ, Lat. oratio, legislative proposal of the Emperor, usu. read to the senate by his quaestor, Modest.Dig. 27.1.1.4.

ὁρᾱτός, delete 'Plu. 2.1029e'

ὁράω, lines 14/16, for 'whereas .. Gr.' read 'and always to be restored in early Att. writers; later ἑώρακα, Men.fr. 208.2 K.-Th. (proved by metre)'; lines 17/18, delete '(ἑώρακε⟨ν⟩ .. 5 D.)' II, line 1, for 'v. ὄψ' read '(v. ὄψ (B))' add 'VI pertain, refer to, w. εἰς, τὸ .. τοῦ Διὸς ἄγαλμα .. ἐς Δία πάντα ὁρῇ καὶ κεφαλὴν καὶ εἵματα καὶ ἕδρην Luc.Syr.D. 32; πάντα τὰ εἰς ἐκκλησιαστικὴν ὁρῶντα κατάστασιν Cod.Just. 1.1.7 pr.; w. πρός, 1.3.29.1; also trans., τοῦτο σὲ ὁρᾷ τὸν ληγατάριον Theophil.Antec. 2.20.20.'

*ὀρβᾶς, ᾶ, ὁ, lentil merchant, SB 9463.9 (ZPE 72.267).

ὀρβικλᾶτον, delete the entry.

*ὀρβικουλᾶτος, ον, μῆλον -ον, Lat. malum orbiculatum, a variety of apple, Androm.ap.Gal. 13.289.14; also ὀρβικλ- Diph.Siph.ap.Ath. 3.80f, Dsc. 1.115.4.

*ὄρβιον, v. °ὀρόβιον.

ὀργανάριος, add '2 waterwork engineer, CPR 14.41.5 (vi/vii AD).'

ὀργανικός, line 10, after '(Tanagra, ii BC)' insert 'ὁ. μοῦσα, opp. ᾠδικὴ D.H.Comp. 11.62'; adv. -κῶς, add '2 as affecting the organ (as a whole), Gal. 7.99.6.'

*ὀργανισμός, ὁ, apparatus, Zos.Alch. 252.15.

*ὀργανιστός, ή, όν, coming from an apparatus, ὕδατα Anon.Alch. 281.11.

ὄργανον I 2 b, at end before 'of plants' insert 'of the vocal organs, τοῖς ἡμετέροις ὁ. Aristid.Quint. 3.20, τὸ φωνητικὸν ὄ. 2.13'; for 'Id.de An.' read 'Arist.de An.' 3, for 'Simon. 31' read 'Pi.fr. 107b S.-M.' II, add '2 piece of land irrigated by one installation (cf. μηχανή I 4), PLond. 1690.9, PMasp. 87.6, 307.4 (vi AD).' III, add 'applied to techniques of literary style, τέσσαρα ὥσπερ ὄ. τῆς Θουκυδίδου λέξεως D.H.Amm. 2.2'

ὄργανος, line 1, for 'ὀργάνη χείρ' read 'ὀργάνα χείρ'; line 2, for 'BCH 52.52' read 'IG 12 suppl. 380' and for 'IG 2.1329' read 'Inscr.Délos 63 (v/iv BC), IG 2².2939'

ὀργάς 2, at end delete 'similarly .. (s.v.l.)'

*ὀργεύς, έως, ὁ, = ὀργεών, gen. pl. -έων Lys.fr. 112 S., to be read also in A.fr. 144 R. (πρῶτος ὀργεών codd.), Arist.EE 1241ᵇ25 (ὀργίων codd.), Hesperia 10.56 (Athens, c.300 BC; cf. HThR 37.82).

ὀργεών, lines 3/4, delete 'poet., .. Fr. 144'; lines 6/7, delete 'a gen. pl. .. ὀργεώνων:—'

ὀργή II 1, add 'b applied to Dionysiac frenzy, Pi.fr. 70b.20 S.-M.'

*ὀργή (B), Ionian for πίσσα acc. to Sch.Ar.Au 839.

*ὀργή (C), perh. fem. of adj. *ὀργός, opp. βέβηλος, initiated (cf. ὄργια) sc. γυνή Herod. 4.46.

ὄργια I, add '2 secret cult-objects, Theoc. 26.13, GVI 1344; sg., Clem.Al.Protr. 2.22.'

ὀργιάζω II 2, add 'pass., Euphron. 2'

ὀργιαστίς, delete the entry.

ὀργίζω II, lines 10/11, for 'ἐπί τινος D. 21.183' read '(but ἐπὶ πάντων in all cases, D. 21.183; cf. ἐπί A III 3)' add '2 feel passion against, SEG 35.26.25, 218.34, 220.19 (curse-tablets, Athens, iii AD).'

†ὀργιοφάντης, ου, ὁ, one who displays cult-objects (°ὄργια I 2) and initiates into mysteries; cf. ἱεροφάντης, AP 9.688, Orph.H. 6.11, 31.5.

ὄργυια 3, line 2, delete 'poet.'

ὀργυιαῖος, add 'τᾶν ὀργυιαιᾶν perh. of the Graces on the throne of Phidias' Zeus, Call.fr. 196.43 Pf.(-υι- scanned as short syll.)'

*ὀργώδης, ες, passionate, cj. for ὀφεώδης in Pl.R. 590a (see Jaeger Scripta Minora II 309ff.).

*ὀρδινάριος, α, ον, Lat. ordinarius, sp. ὠρδινάριος, OGI 568 (Tlos, iii AD); ὠρδενάριος, MAMA 1.168 (Laodicea Combusta, iv AD).

*ὀρδινᾶτος, ὁ, (also sp. ὠρδ-), Lat. ordinatus, ἑκατόνταρχος ὁ. = centurio ordinatus, PBeatty Panop. 2.60, 190 (AD 200); so ὠπτίων [ὠρ]δινᾶτος = optio ordinatus, SEG 31.1116 (Phrygia, Chr.).

ὀρέγω II 1, lines 2/3, delete 'ἀνδρός .. Epigr.Gr. 448.4 (Syria)' 3, after 'acc.' insert 'ἀνδρός .. ποτὶ στόμα χεῖρ' ὀρέγεσθαι Il. 24.506'; add 'GVI 270.4 (Syria)'

*ὀρεθής· ἄδικος Theognost.Can. 67 A.

ὀρειδρόμος, after 'running on the hills' insert 'Simon. 14 fr.5(b).7 P.'

ὀρεινός III, add 'also fem. subst., desert canal, SB 10541.5, 10543.6'

ὄρειος, line 1, after '(Luc.)DDeor. 20.3' insert 'Aristid.Or. 26(14).101'; after 'IG 12(7).75 (Amorgos)' insert 'epith. of Aphrodite, The Swedish Cyprus Expedition 1927-31 vol. iii p. 626 no. 12; of Dionysus, IEphes. 1267 (both ii AD)'

ὀρείχαλκος, add 'see also °ὠρόχαλκος'

*ὀρεκτύς, ύος, ἡ, ὀρεκτύν· ὀρέξεων Hsch.

*ὀρεονόμος, η, ον, = ὀρεινόμος, in quot. w. ref. to the particular hill at Kafizin, Νύμφηι ὀρεονόμηι Kafizin 8, 12, 13.

*ὀρεσιδίαιτος, ον, dwelling in the mountains, APAW 1943(14).8 (Chalcis, iii AD).

ὀρεσκῷος, after 'Od. 9.155' insert 'Alcm. 89.4 P.'

ὀρεσσιπόλος, add 'PVindob.Rainer 29801.17 (Gow Buc.Gr. p. 169)'

Ὀρέστης, add 'cf. Myc. o-re-ta, pers. n.'

ὀρεύς I, add 'ὀρέας Dialex. 2.11'

ὀρθᾱγορίσκος 1, add 'βορθαγορίσκεα'

Ϝορθασία, before 'v. sq.' insert 'also Ϝροθασία SEG 2.67 (Laconia, vi BC)'

Ὀρθεία, line 4, before 'Βωρθεία' insert 'Ϝωρθέα IG 5(1).289 (Laconia, ii AD)' and before 'Ὀρθεία' insert 'Ϝορθίη IG 5(1).1376'; line 5, after '(name of a ship)' insert 'Ϝροθαία, Ϝορϝαία IG 5(1).252a, b (Laconia, vi BC)'

ὀρθιάζω II 1, for 'set upright' read 'cause to stand, μηρῶν ῥόπαλον' add 'b abs., have a sexual erection (cf. °ὀρθιάω), Paul. Aeg. 6.70 (112.23 H.); cf. ἐξανδρόομαι III.' 2, delete the section.

†ὀρθιάω, have a sexual erection, Sch.Pi.P. 10.56, Cyran. 16 (1.5.13 K.), 26 (1.10.64 K.).'

*ὀρθιλάτης, ου, ὁ, unexpld. military office, κοιμητήριον Αὐρ(ηλίου) Γεροντίου ὀρθιλάτου BCH suppl. 8 no. 131 (AD 507).

ὄρθιος I 2, at end for 'of animals .. 10.36' read 'w. ref. to sexual erection, ὕβριν ὀρθίαν κνωδάλων Pi.P. 10.36' II 2 a, penultimate line, delete 'μελῳδία .. 1140f' b, for this section read 'ὄρθιος, ὁ, metrical foot having an arsis of four morae and a thesis of eight morae, Aristid.Quint. 1.16, 2.15, Plu. 2.1140f (but differently defined at Bacch.Harm. 101); of a melody, written in such a rhythm, τῆς ὀρθίου μελῳδίας ibid.' VI, add 'perh. different wd.; cf. Ὀρθεία'

*ὀρθόβλεψις, εως, ἡ, orthodoxy, IGLS 1801 (iii AD).

ὀρθογράφέω, add '2 to be an ὀρθογράφος, GVI 1836 (Athens, ii AD).'

*ὀρθογραφικός, ή, όν, orthographic, τὸ ὀ. = ὀρθογραφία, Sch.D.T. 302.8.

ὀρθογράφος, add '2 perh. scribe employed to make fair copies of documents, GVI 592, POxy. 3138.2 (iii AD).'

*ὀρθοεπής, ές, speaking correctly, Posidipp. in Suppl.Hell. 705.24.

*ὀρθοπλάγιος, ον, perh. having a design of straight and oblique motifs, τυλοτάπητα ὀρθοπλάκιν SB 13597 (iv AD).

ὀρθόπλουμος, for 'embroidered with feathers' read 'app. embroidered with a vertical design' and add 'Edict.Diocl. 29.12, 13'

ὀρθόπνοια, add 'Androm.ap.Gal. 12.120.8, 13.114.2; in animals, Afric.Cest. p. 31 V.'

ὀρθοπνοϊκός, add 'masc. pl., sufferers from ὀρθόπνοια, Androm.ap.Gal. 13.106.4, 113.10'

ὀρθοποδέω, for 'walk .. uprightly' read 'advance, make progress'

*ὀρθοποδία, ἡ, straight forward movement, transf., (successful) progress, PMil.Vogl. 24.8 (ii AD).

ὀρθός **II 1**, add '**b** ἡ ὀρθὴ τῆς ἐπιπέδου βάσεως *rectilinear*, i.e. enclosed by straight lines, Pl.*Ti.* 53c.' add '**3** transf., *direct*, ἐξ ὀρθοῦ *directly*, ὡς ἐξ ὀρθοῦ περὶ τὴν ἰδίαν οὐσίαν μεγάλα βλαπτόμενος *Cod.Just.* 4.21.22.2.' **III 6**, add 'of style, *tense*, D.H.*Comp.* 4.27' **V**, add 'but opp. ἐγκλινόμενα, perh. *indicative* verbs as opp. those in other moods, D.H.*Comp.* 5.37. **2** αἱ ὀρθαὶ περίοδοι (opp. αἱ ἀντεστραμμέναι), those in which the dependent clause precedes the principal clause, Sch.D.T. p. 27 H.' **VI**, at end after 'βορθαγορίσκοι = ὀ.' insert 'pers. n. Γορθαγόρας *SEG* 11.336 (Argos, vii BC)'

Ὄρθος, ὀ, better attested reading for Ὄρθρος q.v., perh. with dissimilation of ρ.

ὀρθόσημος, for '*Edict.Diocl.* 29.24, cf. 44' read '*Edict.Diocl.* 29.17, 29'

ὀρθοστάτης **I 2**, delete the section transferring exx. to section I 1.

*ὀρθόστρωτος, ον, of upright surfaces, *faced with marble*, τοίχοι, Hierocl. p. 54.16 A.; subst., πῶς ἐν ὀρθοστρώτοις οἰκῇς Arr.*Epict.* 4.7.37.

*ὀρθόσφῡρος, ον, *having straight ankles*, Hsch. s.v. τανίσφυρος.

ὀρθότης **III**, for '*the .. narrative*' read '*use of nominatives and finite verbs*, cf. °ὄρθωσις 2'

ὀρθοτομέω, for '*metaph. .. teach it aright*' read 'fig., ἐργάτην ἀνεπαίσχυντον, ὀρθοτομοῦντα τὸν λόγον τῆς ἀληθείας, i.e. not distorting it'

ὀρθόϋφος, for '*weaver .. weaving*' read '*weaver at a vertical loom*' and add '*PLaur.* 24 (v. *ZPE* 35.112), *BGU* 2471.4 (ii AD)'

ὀρθόω **III**, for 'intr. .. πλαγιάζω' read '*express by means of nominatives and finite verbs*, τὰ ἐννοήματα Aristid.*Rh.* 1.465 S. (pass.): abs., opp. πλαγιάζω ib. 2.533 S.'

ὀρθρεύω, add '**2** *rise early, make an early start*, Lxx *To.* 9.6.'

+ὀρθρίζω, *rise early, make an early start*, Lxx *Ge.* 19.2, 27, *Jo.* 3.1, 1*Ki.* 1.19, etc., *Ev.Luc.* 21.38; προῆγεν ὀρθρίζων καὶ ὀψίζων, i.e. early in the morning and late in the evening, Thd. 1*Ki.* 17.16. **2** *look diligently* for (calque of Hebr. šiḥēr), Lxx *Jb.* 7.21, πρὸς σὲ ὀρθρίζω *Ps.* 62(63).1, etc.

ὄρθρος **I**, for '*the time .. cock-crow*' read '*the period preceding daybreak while it is still dark* (see *TAPhA* 119.201ff.)' (and so throughout the section); add 'fig., ὄρθρου ἀποστέλλων, i.e. insistently (calque on Hebr.), Lxx *Je.* 25.4, 33(26).5. **2** *daybreak, first light*, J.*AJ* 1.28, Plu.*Pomp.* 36.4, Hsch. (s.v. ἔωθεν).'

ὄρθωσις **2**, for '*use of the nominative case*' read '*use of nominatives and finite verbs*'

ὁρία, add '*IG* 9²(1).177.13 (Delphi, ?iii BC)'

+ὀρῑγᾰνόεις, εσσα, εν, of ὀρίγανον, ὀριγανόεσσα .. χαίτῃ Nic.*Th.* 65.

ὀρίγᾰνον, line 1, before 'τό' insert '[usu. ῐ, but ῑ Tim.*fr.* 23 P.]'

*ὀριεντάλιος, ον, Lat. *orientalis, eastern*, λεγεώνων .. ὀριενταλίων *PBeatty Panop.* 2.187, 192 (AD 300).

ὁρίζω, line 2, after 'Hdt. 3.142' insert 'Boeot. ὁριττ[ά]ντων *IG* 7.2792, ὥριττα *SEG* 23.297' **II 2**, add '**b** abs., *establish boundary-markers, IG, SEG* ll.cc.; of line, *act as boundary*, *Tab.Heracl.* 1.13.'

*ὁρῐκοίτης, Dor. -τᾱς, ᾱ, ὁ, *lying down to sleep on the hills*, Κένταυρος *Lyr.adesp.* 6.10 P. (= B.*fr.* 66 S.-M.).

ὁρῐκός, line 3, after 'ὀρεικός' insert '(cf. βοεικός)' and delete '(interpol.)'

ὁρῐκός **2**, after '-κῶς' insert 'perh. *according to the ὅρια* (v. ὅριον I 2)'; add '(s.v.l.)'

+ὀρῑνοβάτης, ου, ὁ, *ranging the mountains*, ὀ. γαστραφέτης type of cross- or stomach-bow used in mountain-warfare, Bito 64.4.

*ὁριοδεικτία, ἡ, (in paps. usu. sp. -δικτία; ὁρωδ- *PMerton* 31.4, AD 307), *administrative district in Egypt*, *PCol.* 136.36, 51 (AD 296/8), *PCair.Isidor.* 5.5, 7.8, etc. (iii/iv AD), *PColl.Youtie* 78.7 (AD 342).

ὅριον, add 'ὅριν χω(ρίου) ἰδιοκτήτου Ἑρμοῦ *ICilicie* 33 (iii/iv AD); form ΓόρFιον, Myc. *wo-wi-ja* (pl.) ?'

ὅριος **I**, for 'ον' read 'α, ον' and add 'Διὸς Ὁρίου καὶ Ἀθηνᾶς Ὁρίας *SEG* 30.93 (Eleusis, 20/19 BC)'

*ὅρῑτις, ῐδος, ἡ, a precious stone, Plin.*HN* 37.176; cf. ὀρείτης.

ὁριχᾶται, add 'cf. ὀριγνάομαι'

*Ὁρκαμαντης, ου, ὁ, cult-title of Zeus, *SEG* 33.1118-20 (Phrygia, iii AD), *MAMA* 6.242-3.

ὁρκάνη, add 'Lycurg.*fr.* 76'

*ὅρκιολος, ὁ (or -ον, τό), Lat. *urceolus, little water-jar*, *TAM* 4(1).6, *IG* 12 suppl. 413 (Thasos, ii AD; -ίωλ), Gloss.

ὅρκιον **II 1**, line 12, after '(Il.) 4.157' insert 'cf. Alc. 129.23 L.-P.' and for 'ib. 269' read '*Il.* 4.269'

ὅρκιος **1**, comp., for 'ὁρκιωτέραν δ' ἤμην' read 'ὁρκιοτέραν δ' ἔμεν' and add '*Inscr.Cret.* 4.42*B*5 (both v BC)'

ὁρκισμός, after 'Plb. 6.33.1' insert '*SEG* 31.1594 (on an amulet)'

ὅρκος, line 8, after 'Od. 2.377' insert 'X.*An.* 2.5.7'

ὁρκωμόσια **II**, (sg. exx.), add '*SEG* 33.147.12, 52 (Attica, iv BC)'

*ὅρκωσις, εως, ἡ, *the swearing of an oath*, *SEG* 23.271.64 (Thespiae, 220/208 BC).

*ὁρκωτήριον, τό, *place of oath-taking*, *PHal.* 215 (pl., iii BC).

*Ὀρλύγιος, epith. of Zeus, *ASAA* 30/1(1952-54).262, no. 6 (Rhodes).

ὁρμᾰθός **I**, lines 5/6, delete 'perh. .. Polystr. p.9 W., cf.'

ὁρμαίνω **I 1**, add 'γάμον ὁ. A.*fr.* 47a ii 24 R. (anap.)'

*ὁρμαστρίς, ίδος, ἡ, *fiancée*, *SEG* 26.1657 (Palestine, s.v.l.).

ὁρμάω **A 1**, lines 7/8, delete 'ὁρμηθεὶς .. 8.499' **B 2 b**, after 'Th. 1.64, 2.69, al.' insert 'indicating person's original place of residence, *SEG* 26.791 (Byzantium, vi AD)'

ὁρμενος, line 2, after 'Ath. 2.62f' insert '*Edict.Diocl.* 6.11'

ὁρμή **I 2**, add 'ὁρμὴ ὕδατος *current, gush*, Lxx *Pr.* 21.1' **II 1**, add '**b** personified, Ὁρμὴ ἐπιταγὴν Φιληματίν *IG* 2².4734 (i AD), Paus. 1.17.1.'

ὅρμημα **I 1**, line 3, after 'pl.' insert 'τοῦ ποταμοῦ τὰ ὁ. *gushing streams*, Lxx *Ps.* 45(46).5' transferring 'θαλάσσης .. Ptol. 4' fr. section 2 to follow it. **2**, line 3, after 'Lxx *Ho.* 5.10' insert '*sexual impulse*, Sm.*Ez.* 23.20'

*ὅρμινθον, τό, app. the same as ὅρμινον, Sch.Nic.*Al.* 602a; also ὁρμίνθιον, ib. 601c.

ὅρμος **I 3**, for 'Luc.*Salt.* 11' read 'Luc.*Salt.* 12' **II 1**, add '*PWash. Univ.* 1.34 (i AD)' **2**, line 2, after '(lyr.)' insert 'Lib.*Ep.* 1088'

ὁρμοφύλαξ, add '*PTeb.* 370.5 (pl., ii/iii AD)'

ὀρνεάζομαι, delete '*carry the head .. birds*'; add 'ὠρνεάζετο· μετέωρον ἐπῆρε τὴν κεφαλήν Hsch.' add '**2** act., *twitter like a bird, chatter*, Aq.*Is.* 8.19 (L.-R.), prob. in Aq.*Is.* 38.14 (for ὀρνίζω).'

ὀρνεόφοιτος, for '*frequented by birds*' read 'perh. *going after birds*'

ὀρνίζω, add '(s.v.l.; v. °ὀρνεάζομαι)'

ὀρνίθειος, add 'form ὀρνίθιος, Myc. *o-ni-ti-ja-pi* (fem. instr. pl.) *decorated with birds*'

ὀρνίθιον, after 'Stratt. 58' add 'ostr. in *ZPE* 98.133 no. 22 (sp. ὠρν-, Egypt)'

*ὀρνιθοπούλλιον, τό, *young fowl*, *PKlein.Form.* 1329 (vi/vii AD).

ὀρνιθοτροφεῖον, add '*PLund* 4.11 i 15 (ii AD), Hsch. s.v. κυρτίς'

ὀρνιθοτρόφος, add '*SB* 10270.1.1 (iii AD)'

ὄρνις, line 2, after 'Hom.' insert '(in Cret. written ὄνν[ι]θα, *Inscr.Cret.* 4.41 iii 8 (Gortyn))'; line 3, delete 'in acc.'; line 4, after 'etc.)' insert 'later also nom. pl. ὄρνις Luc.*Ep.Sat.* 35'; line 8, after 'iv BC)' insert 'ὀρνίκων *PTeb.* 875.19 (ii BC)' **II 2**, at end after '(Ar.)*Av.* 719 sqq.' insert 'ὄρνιθος οὕνεκα A.*fr.* 78c.54 R.'

ὄρνῡμι **4**, line 2, after '(A.R.) 3.457' insert '*AP* 11.158 (Antip.Thess.)'

ὀρόβιον, add 'also ὄρβιον *PMich.*xi 619.1 (c.AD 182)'

ὄροβος, line 2, for 'of its seeds' read 'of the pulse'

+ὁροθέτης, ου, ὁ, *official appointed to fix boundaries*, *Inscr.Cret.* 3.iii 25, al. (i AD), *IGBulg.* 1401.9 (ii AD).

ὅρομαι, add 'Myc. part. *o-ro-me-no*'

Ὀρομπάτας, add 'cf. ὀρεμπόται'

ὅρον, delete 'ὅρος .. 130' (v. °ὅρος (B)).

ὅρος **3**, for 'in Egypt, *desert*' read 'in Egypt, spec. of the infertile hilly terrain bordering the Nile valley'

*ὅρος (B), εος, τό, wooden *implement for pressing olives*, *SEG* 11.244 (Sicyon, v BC), Poll. 7.150, 10.130; cf. ὅρον.

ὀρός **2**, add 'Nic.*Th.* 708' penultimate line, after 'οὐρός' insert 'or οὐρόν'

ὅρος, line 1, after lemma insert 'Γόρος *SEG* 25.517 (Thebes, ?v BC)'; line 4, for 'Megarian .. 885' read 'Megar. ὄρρος Schwyzer 172 (Heraclea Pontica); ὅρρους, acc. pl., *SEG* 37.576 (Maced., iv BC)' **I 1**, add 'of a race-course, *SEG* 35.218.13, 26 (Athens, iii AD)' **II b**, line 7, delete 'Thphr.*Char.* 10.9' **c**, add 'τὴν ἐκ τῶν ὅρων .. ἀσφάλειαν, i.e. sanctuary provided in sacred land, Just.*Nov.* 17.7 pr.' **III 1**, lines 1/2, delete 'ἥν .. E.*IT* 1219' add 'form ΓόρFος, Myc. *wo-wo*'

ὀροτύπος, for '*dashing down a mountain*' read '*smiting the mountain*' add '**II** of the giants, *striking with mountains* (as missiles), *Trag.adesp.* 594b K.-S., Hsch.'

ὀροφικός, after '*of or for a roof*' insert '*Inscr.Délos* 1417*A*ii17 (ii BC)'

ὄροφος **II 1**, add 'synecd., *house*, *SEG* 9.72.16 (Cyrene, iv BC)'

*ὀροφῠλᾰκέω, *to be a mountain-guard*, *La Carie* 162.9 (Apollonia Salbace, ii/iii AD).

ὀροφῠλᾰκέω, delete the entry.

*ὀροφῠλᾰκία, ἡ, *mountain guard* or *defence*, *SEG* 29.1516.8 (Lycia, ii BC).

*ὀροφῠλᾰκος, ὁ, = ὀροφύλαξ, Amyzon 2.

*ὄροχθος, ὁ, ὄρογκοι· τῶν ὀρῶν τὰ ὀγκώδη, ἃ καὶ ὀρόχθους καλοῦσιν Hsch.

*Ὀροχωρείτης, ου, ὁ, cult-title of Zeus, *SEG* 32.1271 (ii AD), 33.1158 (undated, both Phrygia).

ὄρπηξ **I 2**, for this section read '*rod or stick cut from a tree or shrub*: used for driving animals, Hes.*Op.* 468; Θεσσαλὸν ὄρπακα i.e. *lance*, E.*Hipp.* 221'

ὀρρωδέω, line 1, after 'al.' insert 'S.*fr.* 951 R.'

ὀρσοδάκνη, for '*which eats the buds of plants*' read 'born from a larva in cabbages'; at end for '(The word .. found' read '(cf. ὄρρος perh., but etym. uncertain)'

ὀρσολόπος, for 'perh. *eager for the fray, tempestuous*' read 'app. *who* ὀρτυγοκόπος *thrashes* (a fleeing enemy, cf. ὅρρος, λέπω II)' and for 'Anacr. 70' read 'Anacr. 48 P.'

ὀρτῠγοκόπος, add 'Pl.*Alc.* 1.120a'

ὀρτῡγομήτρα, add '(Hsch. explains ὀ. as ὄρτυξ ὑπερμεγέθης, and this may be the meaning in Lxx ll.cc., al.)'

ὄρτυξ, for 'ὕγος .. 245' read 'ὕγος (also ὔκος, Philem. 192 K.-A.), ὁ (also ἡ, Lyc. 401), (ῠ Att. as in δοίδυκα, κήρυκα acc. to Demetr.Ixion ap.Ath. 9.393b, but ῠ in all determinable instances)'

ὀρυγή 1, add 'b ἡ ὀρυγή ἐξέδρας the *dug-out* (i.e. underground) exedra, *Syria* 17.260 (Palmyra, iii AD).'

*ὀρυζιοπωλική (sc. τέχνη), ἡ, *rice-selling*, *PTeb.* 612 (i/ii AD).

ὀρυζοτροφέω, delete the entry; cf. ῥιζοτροφέω.

†ὀρυκτήρ, ῆρος, ὁ, *digger*, Thphr.*fr.* 30.2.

ὀρύκτης I, after 'Aesop. 99' add '*mole*, the animal who digs, (calque on Hebr.), Aq.*Is.* 2.20'

ὀρυκτός II, add 'ἁλὸς ὀρυκτοῦ *PSorb.* 35 (AD 225)'

ὄρυς, for 'Hdt. 4.192' read 'Hdt. 4.192.1'

ὀρύσσω I, for '*dig*' read '*form by digging, excavate*' III, for '*dig through* .. διορύσσειν' read '*dig up, excavate*'; add '*destroy by digging*, ὅπως μηδεὶς ὀρύσσῃ τὰς ὁδούς *Dig.* 43.10.1.2'

ὀρφάνιος, for '= foreg., *desolate*' read '*childless*'

ὀρφανιστής, add 'also at Istros, *IHistriae* 184 (iii BC)'

ὀρφανοδῐκασταί, after '*orphans*' insert 'or *guardians of orphans*'; add '*Inscr.Cret.* 4.72.12 7'

ὀρφανόομαι, add '*to be bereaved of*, Sch.Pi.*I.* 7.14'

†ὀρφανοφύλαξ, ακος, ὁ, ἡ, *publicly appointed guardian of orphans*, X.*Vect.* 2.7, *Delph.* 3(2).168.27 (ii BC); fem., *SEG* 23.353 (Naupactus, ii BC).

ὀρφνήεις, for 'poet. for ὀρφνός' read '= ὄρφνινος'

ὄρφνῑνος, add 'Orph.*A.* 965'

ὄρφνιος, delete 'but .. corrupt'

ὀρφνίτης, for 'dub. epith. of τάλαρος' read 'perh. *working at night, εἰροκόμος*'

ὀρφοβότης, after 'ὁ' insert 'also ὀρφοβώτᾱς *SEG* 39.1008.6 (Sicily, iii BC)'

ὀρφώς, add '2 an unidentified fish (sp. ὀρφός) said to be kept for omens by priests at Lycian temple of Apollo, Ael.*NA* 12.1.'

ὄρχᾰμος, for 'Ep. word' to end read 'ὀ. *στρατοῦ* A.*Pers.* 129, *ὄρχαμε γαίης* Opp.*H.* 1.70; without gen., A.*fr.* 451q R., *AP* 11.284 (Pall.)'

ὀρχέομαι, line 2 (impf.), after '(v. infr.)' insert 'ᾠρκέτο (prob. = καὶ ὠρχεῖτο)' I 2, add 'Ath. 10.454f.' II, add 'b prob. sens. obsc., *IG* 12(3).536 (Thera, viii BC).'

ὀρχηστής, line 2, after '*IG* 1².785' insert '(v BC) and after '(*IG*) 919' insert 'Dipylon vase, *c.*725 BC, perh. here in erotic sense, cf. °ὀρχέομαι II b'; add 'as adj., *πόδα AP* 7.37 (Diosc.)'

ὀρχηστοπᾰλάριος, add 'ὀρχιστοπαλαρίων πρασίνων *IEphes.* 2949'

ὄρχις I, after '*testicle*' insert 'τὸν ὄρχιν Hippon. 92.3 W.'

Ὀρχομενός, add 'Myc. *e-ko-me-no* (place-name)'

ὅς, ἥ, ὅ B IV 6 a, add 'perh. at *IG* 1³.533 (v. *SEG* 10.410)' Ab IV 2, delete 'in Att.' and after 'Ar.*Ec.* 338' insert 'Men.*Dysc.* 485, Call.*Jou.* 1.67, *Epigr.* 11.1 Pf.' 3, add 'also pl. ἅ Isoc. 12.181'

ὅς, ἥ, ὅν I, line 1, after '*his, her*' insert '*their*'; line 5, before 'sts. also in lyr.' insert 'Hes.*Th.* 71, A.R. 1.384'; last line, after 'never in Attic prose' add 'exc. τὰ ἃ δάκρυα Pl.*R.* 394a (archaizing passage)'

ὁσάκις, add 'Cret. ὁθάκις, ὁθθάκιν, ὁττάκιν, *Inscr.Cret.* 1.x 2.10 (Eltynia, v BC), vii 5 *bis* (Cnossus, iii BC), 4.73.A6 (Gortyn, v BC)'

ὅσδε, add '*IG* 7.1686 (Plataea, iv AD), *GVI* 1181, etc.'

ὁσημέραι, line 4, before 'Hyp.*Ath.* 19' insert 'Pl.*Chrm.* 176b'; line 6, after 'Ar.*Th.* 624' insert 'D.H. 1.24, al.'

ὁσία I, add 'ἐξ ὁσίης *SEG* 33.736 (epigr., Crete, ii AD)' II 2, add 'Just.*Nov.* 43 pr.' III, add 'ὁσίας χάριν Modest.*Dig.* 26.6.21'

ὅσιος I, add '3 ὅσια· ἄλφιτα δεδευμένα ἐλαίῳ καὶ οἴνῳ Hsch., cf. Suid. s.v. ἀφοσιοῦσθαι.' II 1, add 'sup. as honorific title of church dignitary, Mitchell *N.Galatia* 226 (*c.*AD 580), *SEG* 31.1446 (AD 639)' 3, add 'Διὶ ὁσίῳ ἐπηκόῳ *INikaia* 1057 (iii AD). b masc. or neut., personified as a deity, Ὁσίῳ Δικαίῳ *IHadr.* 136 (ii AD), *SEG* 31.1130, 33.1003 (both iii AD). c sup. as title of Roman emperor, τοῦ ὁσιωτάτου Αὐτοκράτορος Καίσαρος .. *IEphes.* 3217 (AD 113/120).' III, adv., add 'ᾧ τάφον οὐχ ὁσίως .. ἀνέγειρα (i.e. perh. "prematurely") *ITomi* 384 (iii/iv AD)'

ˮΟσιριασταί, οἱ, *guild of worshippers of Osiris*, σύνοδος Ὀσειριαστᾶν *Inscr.Cos* 54.1 (ii BC).

ὀσμή II, lines 3/4, after 'it occurs also in' insert 'Hippon. 92.11 W.'

ὀσμηρός 2, delete the section.

*ὀσμός, ὁ, name of a leguminous plant, Dsc. 2.147, cf. *Eranos* 53.31.

ˮΟσοραπεῖον, τό, = Σαραπεῖον, *temple of Sarapis*, *PSI* 1128.22 (iii AD).

ˮΟσορᾶπις, ιδος, ὁ, = Σάραπις, *UPZ* 19.3, al. (ii BC) (written Ὀσεράπις *SB* 5103).

ὅσος IV 1 a, after 'Arist.*Rh.*1376ᵃ34' insert 'cf. *AP* 5.216.3, 4 (Agath.), 9.581.3' V, add '3 without comp. or sup., ὅσῳ περὶ τοῦτο ἡμεῖς πονούμεθα, τοσούτῳ πᾶσαν ἐξεῦρον οἱ βουλευταὶ τέχνην .. Just.*Nov.* 38 pr. 1.' VI 2, before 'Ar.' insert 'Hdt. 1.174.3' add '3 ἐξ ὅσου *since*, Hdt. 2.98.1, 3.63.2 (v.l.).'

ὅσπερ II 1, after '(s.v.l.)' insert '*SIG* 888.10 (AD 238), etc.' 4, add 'κατὰ ταὐτά σφιν ἔστο ἅιπερ Κνοσίοις Schwyzer 83.B28 (Argos, *c.*450 BC)'

*ὅσπις, ὁ, dat. ὅσπι, Lat. *hospes, lodger*, *POxy.* 3860.10, 15, 42 (iv AD).

*ὁσπίτιον, τό, Lat. *hospitium, house*, *PLips.* 40 iii 18 (iv/v AD); ὁσπήτιον, Sch.Gen.Il. 1.396, Suid. 2 *establishment for the sick* or *destitute, hospital*, *PVindob.Worp.* 15, *PBasel* 19 (both vi/vii AD).

*ὁσπρεάχυρον, τό, *pulse-chaff*, *CRAI* (1945) p. 379 (Berytus, v AD).

ὁσπρεύω, add '*IG* 1³.252.13 (v BC)'

ὁσπρηγοί, for '*OGI* .. v/vi AD)' read '*SEG* 34.1243 (sp. ὀσπριγοί: Abydos, *c.*AD 492)'

*ὁσπρῐγίτης, ου, ὁ, *pulse-merchant*, *PKlein.Form.* 1091 (vi AD), *POxy.* 2000.14 (vi/vii AD, ωσπρ- pap.).

ὁσπριοπώλης, add 'also ὀσπρεο- *REG* 70.120.13 (Caesarea, vi/vii AD)'

*ὄσπρον, τό, perh. *decorated material*, *Tyche* 1.167 (Alexandria, vi/vii AD); cf. Hsch. ὄσπρα· ποικίλα.

ὅσσα 4, add 'Πυθίαν .. ὅσσαν Neophr. 1.2'

ὄσσομαι II 2, add 'ὀσσόμενοι φρ[εσὶ] γήρ[ας Hes.*fr.* 1.10 M.-W.'

*ὅσσος, ὁ, *pupil of the eye*, Sch.rec.E.*Ph.* 370 (interpr. of δι' ὄσσων).

ὀστᾰκός, after 'ˮΟστακος' insert '*IG* 11(2).107.8 (iii BC)'

ὅστε, line 7, after 'Eu. 25, 1024' insert 'v.l. in E.*Rh.* 972'; line 9, for 'antec.' read 'demonstr.'

*ὀστεοθήκη, ἡ, = ὀστοθήκη, *TAM* 2(3).780.1 (Lycia).

*ὀστιάριος, ὁ, Lat. *ostiarius, door-keeper, porter*, *SEG* 9.346.32 (Cyrenaica, vi AD); written ἀστ- *PFlor.* 71.518 (iv AD), *BGU* 672 (vi AD).

ὅστις, line 5, for 'Alc. 45' read 'Alc. 66 L.-P., Arc. ὄζις *SEG* 11.112 (vi/v BC), dat. ὀϜέοι *IG* 5(2).262 (v BC)'; line 6, after 'Od. 1.124' insert 'Aeol. ὀττέω *SEG* 34.1238 (*c.*200 BC)'; line 14, after '*IG* 2².1126.25' insert 'cf. Theoc. 16.68'; line 24, after 'Hes.*Op.* 31' insert 'οὕτινος Theoc. 25.35, Dor. ὥτινος id. 14.19' II 1, add 'ὅστις is less freq. in later Greek, but where used may = ὅς, τῇ ἐπαύριον ἥτις ἐστί κτλ. Ev.Matt. 27.62, cf. *POxy.* 110.3, *PFay.* 108.7 (both ii AD)' add 'Myc. neut. *jo-qi* = ὅ, τι'

ὀστογενής, v.l. for ὀστεογενής.

ὀστοθήκη, delete '*sarcophagus*' and add 'see also °ὀστεοθήκη'

ὀστολογέω, add 'Men.*Asp.* 77'

*ὀστοφάγος, ὁ, *ossuary*, graffito in *PalEQ* 1937.130 (Jerusalem, i BC/i AD).

*ὀστράκη, ἡ, = ὄστρακον, *jar*, Lang *Ath.Agora* xxi Hb12 (iv AD).

ὄστρεον, line 3, for 'ὀστρία' read 'ὄστρια' III, add 'ὀστρείου *SEG* 24.277.B60 (Epid., iv BC)'

*ὀστρῖτις, ιδος, ἡ, = ὀστρίτης, Plin.*HN* 37.177.

*ὀστροφά, v. ἀναστροφή.

*ὀσχίον, τό, dim. of (?) ὄσχος, *SEG* 7.1065 (Arabia).

ὄσχος, for 'ὦσχος' read 'ὦσχος'

ὅταν I 2, line 6, after '(*Ev.*)*Marc.* 3.11' insert '*GVI* 1113a.8 (Apamea)'; line 8, after '(s.v.l.)' insert 'cf. *PHamb.* 70.19 (ii AD)'

ὅτε, line 1, before 'Cypr.' insert 'Myc. *o-te*' and for '*Inscr.Cypr.* 135.1 H.' read '*o-te ICS* 217.1' A I, add '4 ellipt., ὅκα τὸ τέταρτον *for the fourth time*, *Inscr.Cret.* 4.184.2 (ii BC), cf. ib. 250.3 (i BC).' III 2, add 'also ὁτεδήποτε Heph.Astr. 1.24.4'

ὁτεῖος, for 'τεῖος' read 'τεῖον'

*ὁτεύνεκα, v. °ὁθούνεκα.

*ὁτεύνεκεν, v. °ὁθούνεκεν.

ὅτι A II, add '3 ὅτι may be resumed by ὡς, Hdt. 3.71.5, 9.6, etc., or vice versa; cf. ὡς B I 1.' add 'V introducing a consec. cl., οὕτως ἡμῖν τῶν βουλευτηρίων μέλει .. ὅτι δίδομεν .. Just.*Nov.* 89.2.3.'

ὁτιή, line 1, after '*because*' insert 'A.*fr.* 281a.9 R.' 2, delete '*Eq.* 360'

ὀτλεύω, add 'Procl.*H.* 1.31'

ὄτλημα, for 'Theognost.*Can.* 13' read 'Theognost.*Can.* 44 A.'

ὀτλήμων, delete '(ὁ τλήμων .. Schmidt)'

ὄτοβος, add 'ὄτοβος ἅλιμος Trag.adesp. 247 K.-S.'

*ὄφτος, v. °οὗτος.

ὄτρεα, delete the entry.

οὐ G, line 13, insert 'Cypr. *o-u-ki ICS* 306.5 may be either οὐκί or οὐχί' at end add 'Myc. *o-u-* (prefixed to vbs.)'

οὗ, οἷ, ἕ, line 2, after 'ἑοῦ' insert 'read by Zenod. in Il. 19.384, also Hes.*Th.* 401 (v.l.), A.R. 4.803'; lines 11ff. (form Ϝοι), add 'also written γοι *SEG* 8.715.7 (Egypt, ii AD), γοί· αὐτῷ Hsch.'

*οὐά (Β), ἡ, v. ὠβά.

*οὐᾶς, v. οὖς.

†οὐάτιον, τό, *pupil of the eye*, *PMag.* 5.92, 12.229.

οὐατόεις, add '3 applied to a tree with hanging branches, Hsch. s.v. οὐατόεν.'

οὐγγία, add 'κʹ ο(ὐ)γ(κίαι) (i.e. 20 oz.) Lang *Ath.Agora* xxi Hb3 (iii BC). 2 *the twelfth part of an inheritance*, Just.*Nov.* 18.1, 89.12.3.'

οὐγκιασμός (s.v. οὐγγία), add 'Zos.Alch. 164.2. **2** *assignation of shares in an inheritance*, Just.*Nov.* 107.1.'

οὐδαμόθεν, add 'ἐκείνοις οὐδαμόθεν προσήκει D. 15.26, 21.196, al.'

οὐδέ **A II 2**, line 10, after 'D. 22.4' insert 'w. ellipsis of main neg., γῆ δ' οὐδ' ἀὴρ οὐδ' οὐρανός ἦν Ar.*Av.* 694, E.*Tr.* 477, χεῖρας δὲ οὐδὲ πόδας προσθίους ἔχει Arist.*HA* 503ᵇ34; υἱὸς ἄρσην οὐδὲ θυγάτηρ Lxx *To.* 6.12 S.'

οὐδός (A), line 6, after '(Epid., iii BC) insert 'app. aspirated form ἡυπὸ τôι hοδôι τᾶς θύρας Lang *Ath.Agora* xxi B1 (vi BC)'

οὐδών, add 'cf. Lat. *udo*'

οὐδώνιον, for '*Edict.Diocl.* .. (Asine)' read '*Edict.Diocl.* 7.47'

*οὐείλλος, neut. pl. οὐείλλα, perh. = Lat. *vilia*, *Dura*⁴ 133.

*οὐεριδάριος, ὁ, Lat. *veredarius*, *imperial courier*, οὐ.· βεριδάριος Hsch.

*οὐέρνας, ὁ, Lat. *verna*, *BCH* 28.196, *MDAI(A)* 13.242.

*οὐέρτραγος, ὁ, Lat. *vertragus*, *greyhound*, Celtic wd., Arr.*Cyn.* 3.6.

οὐετερανός, line 2, after 'οὐετρανός ib. 99, 142, etc.' insert '*PBerol*.inv. 11624.16 (*JJP* 18.35, iv AD); ὀε(τρανός) *SEG* 32.1447 (Syria, AD 150)' and after 'Zonar.' insert '*IGRom.* 4.730, etc., and βετερανός *IG* 14.1470'

*οὐετρανικός, ὁ, app. *a man of veteran status*, *Princeton Exp.Inscr.* 765¹³, *LW* 3.2227, 2546.

*οὐετράριος, ὁ, Lat. *vitrarius*, *glass-worker*, *ITyr* 117.

*οὐηλάριον, τό, Lat. *velarium*, *curtain*, *SB* 7033.39 (v AD).

*οὐθένεια, v. οὐδενία.

*οὐιάτωρ, ορος, ὁ, Lat. *viator*, *agent for a Roman magistrate*, *SB* 976 (inscr., Roman period).

*οὐίγουλ, or οὐίγουλος, ὁ, = Lat. *vigil*, *watchman*, *IEphes.* 615.13 (ii AD), ἔπαρχος οὐιγούλων *SB* 9898.2, *JJP* 18.323 (*POxy.* 2231, both iii AD).

*οὐικάριος, ὁ, Lat. *vicarius*, *deputy*, *vicar*, *SEG* 34.1163 (Ephesus, Chr.); cf. °βικάριος.

*οὐικεννάλια, τά, Lat. *vicennalia*, *festival celebrated every twenty years*, *PStrassb.* 138.12 (AD 325); also sg., *POxy.* 2187.21.

*οὐικήσιμα, τά, Lat. *vicesima*, *a twentieth part*, *BGU* 388.1.7 (ii/iii AD).

*οὐι(ν)δίκτα, ἡ, Lat. *vindicta*, *form of manumission*, *PGnom.* 21.64 (ii AD).

*οὐινδικτάριος, ὁ, Lat. *vindictarius, *slave emancipated by vindicta*, *IGRom.* 3.801.20, 802.25 (Syllium).

*οὐιξιλλατίων, ωνος, ἡ, Lat. *vexillatio*, *troop*, *squadron*, *PCair.Preis.* 39 (AD 347); οὐξελλ[ατίων] *BGU* 600.13, etc.; see also °βιξ-.

*οὐιξίλλον, τό, Lat. *vexillum*, *ensign*, *SEG* 34.1306 (AD 275/6).

*οὐιόκουρος, ὁ, Lat. *viocurus*, *curator of roads*, *Ann.Épigr.* 66.376 (*IGBulg.* 884).

οὐλαί, after '**ὀλαί**' insert '(also in Epid., *BCH* 73.366 (iv BC))'

οὐλαμός **II**, after 'Plb. 6.28.3' insert '(transl. Lat. *turma*)'; add 'Sch.Lyc. 32 (cf. γόλαμος); Homeric metre indicates initial F)'

οὐλάς **II**, lines 2/3, for 'Call.*Fr.* 360 .. κεναί)' read 'Call.*fr.* 724 Pf. (οὐλαὶ ἀεὶ κεναί codd. Suid.), ib. 24.10 (prob. rest.)'; line 4, after 'Sch.Theoc. 1.53' insert 'Sch.Lyc. 183'

*οὐλένιον, v. °ὠλένιον.

οὐλή, add '**2** *a corneal opacity*, Paul.Aeg. 3.22.24 (181.1-2 H.).'

οὔλιγξ, read '**οὔλιγξ**'

*οὖλιξ· οὐρανίσκος Zonar. 1478; cf. οὖλον.

οὐλόδετον, for 'ib. 30' read 'Phot. a 1110 Th., Eust. 1162.30'

οὐλοκάρηνος **II**, delete '(cf. οὐλοκίκιννα)'

οὐλοκίκιννα, delete the entry.

*οὐλοκίκιννος, ον, *having close ringlets*, Telesill. 8 P.

οὐλόκομος, add '**II** *having thick, bushy foliage*, κίτρις Al.*Le.* 23.40.'

οὐλόμενος **I**, add 'of a fatal drink, οὐλόμενον δέπας *SEG* 33.1108 (Paphlagonia)'

οὖλος (B), line 6, after 'Hdt. 7.70' insert 'as subst., οὔλη λευκή Hsch.' **3**, line 7, after 'so perh.' insert 'οὖλον ἀείδειν *AP* 7.27 (Antip.Sid.)'; add 'cf. ἴουλος'

οὖλος (C), line 6, after 'οὖλον ἀείδειν ib. 27 (Antip.Sid.)'

οὐλοφυής, for '*rough*, *raw*, *undifferentiated* .. (τύποι χθονός)' read '"*whole-natured*" precise significance unclear)'

οὔλω, after '*h.Ap.* 466' insert 'οὖλε expld. by χαῖρε, ὑγίαινε, *PRyl.* 16(*a*) fr. 2ᵛ (iii BC)'

οὖν **II 1**, line 4, delete 'Hdt. and ' **2**, lines 2/3, for 'but only, it seems' read 'chiefly'; at end before '*AP* 12.226' insert 'Call.*Cer.* 75, *fr.* 64.5, 384.5 Pf.'

*οὐνέδων, ωνος, ὁ, Lat. *unedo*, Gal. in *CMG* 5.4.(2) 304.23.

οὕνεκα **II**, line 5, after 'ἕνεκα)' insert '*CEG* 92, οὕνεκεν χρόνου *in respect of* years, poet. in *PMich.*i 77.9 (iii BC), *IG* 2².2943.18 (iii AD), Q.S. 1.724, 4.497, al.'; delete 'It has been' to end of article.

*οὐνή, v. ὠνή.

*Οὐννικός, ή, όν, *of the Huns*, στράτευμα Procop.*Pers.* 2.4.4, *Vand.* 3.18.13, etc.

*Οὖννοι, οἱ, *the Huns*, Procop.*Pers.* 1.3.4, 2.1.14, etc.

*οὐξελλατίων, v. °οὐιξιλλ-.

*Οὐπησία, ἡ, cult-title of Artemis, *SEG* 23.208 (Messene, AD 42).

Οὖπις **I**, add 'in Thrace, Sch.Lyc. 936' **III**, before 'maiden' insert 'Hyperborean' and after 'Delos' insert 'Call.*Del.* 292'

οὐρά **I 1**, line 5, after 'Arist.*PA* 689ᵇ30, al.' insert 'Arat. 625'; for 'not used .. *HA* 504ᵃ31' read 'of birds, Arat. 600, 628, but cf. οὐρὰν μὲν οὐκ ἔχουσιν (ὄρνιθες), ὀρροπύγιον δέ Arist.*HA* 504ᵃ31'

*οὐράγιον· ἔσχατον Hsch.; cf. οὐραγός.

*Οὐράνια, τά, games celebrated at Sparta in honour of Zeus Οὐράνιος, *IG* 5(1).658.11 (i AD), *Inscr.Magn.* 180.12 (ii AD); τῶν μεγάλων Οὐ. *IG* 5(1).32B9 (ii AD); τῶν μεγίστων Οὐ. ib. 667.1 (i AD).

Οὐράνια **II**, add 'of Hecate, *SEG* 30.326 (i AD); of Μήν, *INikaia* 1515 (iii AD), *TAM* 5(1).349, Θεὰ Οὐρανία goddess with cult in Lydia, *SEG* 31.999 (AD 202/3)'

Οὐρανιάς, for '*Urania*' read '*Zeus Οὐράνιος*'

οὐράνιος **I 2**, line 4, after 'Pi.*O.* 11(10).2' insert 'Hp.*Aër.* 12'

Οὐρανίωνες, add 'in Chr. context, *BCH* suppl. 8.265'

οὐρανοειδής, add 'epith. of (Jewish) god, *PMag.* 4.3068'

οὐρανόεις **II**, for this section read '*of the roof of the mouth*, οὐρανόεσσαν ὑπήνην, perh. the inside of the upper lip, Nic.*Al.* 16'

*οὐρανολέσχης, ου, ὁ, *one whose boasts tower heaven-high*, Eust. 1687.48.

οὐρανός **I**, add '**7** *as the abode of the souls of the dead*, *IAssos* 74a.' **III**, add '*SEG* 34.1463'

οὐρανοστεγής, delete 'cf. ὑποστενάζω II'

οὐραχός **IV**, for '*stems* or *stalks*' read '*rachilla*'

*οὐρβανικιανός, ή, όν, Lat. *urbanicianus*, χειλίαρχος *IApam.* 8.8 (iii AD).

οὐρβανός, add '**2** κολλήγιον οὐρβανῶν, Lat. *collegium urbanorum*, *SEG* 37.559 (6) (Cassandreia, Maced.).'

✝οὐρεύς, ῆος, ὁ, Ion. for ὀρεύς (q.v.), *mule*, Il. 1.50, 10.84 (also understood as οὖρος *guard*, cf. Arist.*Po.* 1461ᵃ10), A.R. 3.841.

*οὐρητρίδιον, τό, dim. of οὐρητρίς, *chamber-pot*, *PMichael.* 18.11.10 (iii AD, οὐρι- pap.).

οὐριοστάτης, after '(ἵστημι)' insert 'as adj.'; add '*Trag.adesp.* 659.16 K.-S.'

οὖρος (B), add 'Cret. ὦρος, in pl., title of officials, *Inscr.Cret.* 4.184a.13 (Gortyn, ii BC); prob. in Hsch. (cod. ὤρου· ὤρια. φύλακος· [ὤρια] φύλακος cj. S.), cf. ὠρεῖον, οὐρεύω'

οὖς, line 5, for 'Simon. 37.14' read 'Simon. 38.20 P.' and add '*AP* 7.409.3'; line 10, after 'Dor.' insert 'ὦας Sophr. in *PSI* 1214a4 (= *GLP* 1.73.4)' and for 'Alcm. 41' read 'Alcm. 80 P., cf. ὦατα Balbilla in *SEG* 8.716.9'; line 24, after 'id.*Smp.* 216a' insert 'οὓς ἀνέχειν *prick up one's ears, pay heed*, Thgn. 887, A.*fr.* 126 R.' **II 1**, add 'cf. Myc. compds. ti-ri-jo-we, *three-handled*, qe-to-ro-we, *four-handled*' **4**, delete '(ὤατα Hp.)' and for 'Hp.*Cord.* 8' read 'Hp.*Morb.Sacr.* 17, *Cord.* 8 (ὤατα)' add 'see also ἆτα, ὦας'

οὐσία **I**, lines 5/7, delete 'καλῶς .. iv BC)'

οὐσιακός, after '*pertaining to an estate*' insert 'οὐσιακοῦ γεωργοῦ *PSorbonne* inv. 2364.5 (*BASP* xII 2 p. 87, AD 26)'; add 'οὐσιακά, τά, *revenues from (imperial) estates*, *PMich.*x 599.1 (AD 177)'

*οὐσιαστικός, ή, όν, *substantial*, *solid*, χαλκοῦ, σιδήρου, μολύβδου καὶ τῶν οὐσιαστικῶν μετάλλων Anon.Alch. 270.1; cf. οὐσία IV.

οὐσιότης, after '*quality of existence*' insert 'Alcin.*Intr.* p. 164.30 H.'

οὖσον, add 'cf. °σοῦσον'

*οὐσούφρουκτος, ὁ, Lat. *usufructus*, *usufruct*, Just.*Nov.* 7 pr., al.

οὔτε, add 'Myc. o-u-qe'

οὔτις, line 4, for 'only twice in E., *Fr.* 45, 325' read 'E.*fr.* 45, 325, *Alc.* 194, 293, al.' and add 'Democr. 116 D.-K.'

οὗτος **A**, line 12, after 'v BC), al.' insert 'in Crete sts. written ὅϜτος, *Inscr.Cret.* 4.3.6 (Gortyn, vii/vi BC), cf. ib. 2.xii 3.5 (Eleuthera); ἐτοῦτο (for τοῦτο, as mod. Gk.) *IKyzikos* 266' **C I**, add '**6** *such-and-such*, Pl.*Phdr.* 272a, *Prm.* 160e, Arist.*Cat.* 5ᵇ36, *Ev.Matt.* 8.9, etc.' **VII 4**, for 'at end of a formula' read 'to clinch a statement' **VIII 3**, line 3, after 'Hdt. 1.161, al.' insert 'and Antipho'; add 'τοῦτο δέ alone, or again, X.*Ath.* 3.11'

οὕτως **A I**, add '**8** καὶ οὔ. introducing a consequential action, *and so*, Acusil. 22 J. (prob.), X.*An.* 3.4.8, Arr.*Epict.* 4.8.13, al.' **IV**, delete 'in Hom. .. Hdt. 1.5'; line 6, for 'cf. 1.20' read 'ὁρῶν ὥσπερ ἂν ἄλλον τινὰ οὑτωσί D. 39.27'

ὀφειλέτης, add 'cf. Myc. *O-pe-re-ta*, pers. n.'

ὀφείλω **I 1**, lines 14/15, delete 'metaph., .. *Fr.* 126' **2**, line 3, after 'Ael. *VH* 10.5' insert 'w. gen., *to be indebted for something*, ἡ κάρτ' ὀφείλω τῶνδέ σοι A.*fr.* 78a.3 R., Ar.*Nu.* 22'; add 'ὕπνος ὀφειλόμενος *of death*, *AP* 7.78 (Dionys.), 219 (Pomp.), 419 (Mel.)' **II 3 c**, w. acc. and inf., add 'Demad. 26' add 'Myc. 3 pl. *o-pe-ro-si*, 3 pl. aor. *o-po-ro* (= ὄφλον)'

ὀφέλλω (B), line 11, before 'Pi.*P.* 4.260' insert 'μητέρα μοι ζώουσαν ὀφέλλετε (imper.) Call.*fr.* 602.3 Pf.'

ὄφελος, add 'Myc. *o-pe-ro*, *deficit*'

ὀφεώδης, add 'see also °ὀργώδης'

ὀφθαλμίζομαι, add '**III** perh. also act. -ίζω, *cast the evil eye on*, ὀφθαλμίσαι· φθονῆσαι Phot. (s.v.l., cj. -ιάσαι).'

ὀφθαλμός **VI**, add 'Gal. 18(1).837.17, 18(2).732.3, 9' add '**VIII** name of a plant, ὀφθαλμὸς Τυφῶνος (oftalmos Tifonos) Ps.-Apul.Herb. 42.17; cf. ὀφθαλμὸς Πύθωνος [Diosc.] 3.26.'

*ὀφθαουηρ, app. Gk. transcription of an Egyptian priestly title, perh. = στολιστής, BGU 2469 i 6 (ὀθφαουηρ) and 9 (ὀφθαουηρ) (ii AD).

*ὄφιδνα· δράκαινα Theognost.Can. 80 A.; cf. ὄφις, ἔχιδνα.

*ὀφικιάλιος (also ὀφφ-,) τό, Lat. officialis, BGU 657 ii 9 (ii AD), POxy. 1646.3 (iii AD), CRAI 1952.593 (Caria, iii AD), SEG 32.1554 (Arabia, vi AD).

*ὀφίκιον (also ὀφφ-,) τό, Lat. officium, official appointment, ἵνα ὀπίκια (sic) λάβῃ POxy. 3312 (ii AD), IGRom. 3.130, PSI 281.51 (ii AD).

*ὀφιόκοιλος, ὁ, precious stone, Lapid.Gr. 191.16.

ὀφιοπλόκᾰμος, after 'Orph.H. 69.12' insert '70.10'

ὀφιοφόρος, add '(lapis, -φοριος)'

ὄφις, after 'ὁ' insert '(also ἡ, Plu. 2.988a, of the serpent Python)' **I**, add 'ἐν κόλπῳ ἔχειν ὄφιν Thgn. 602' **III**, add 'but in Nonn.D. 2.290 Ὄφις Ἁμάξης is the constellation Draco'

ὀφλισκάνω, line 7, after 'And. 1.73' insert 'aor. pass. part. ὀφληθέν Just.Nov. 59.3' **I 2**, add 'pass. τὴν δίκην δικαίως ὠφλημένην D. 29.55'

ὄφρα **A I**, lines 8/9, for 'but Hom. thrice uses it' read 'in Hom.'; after 'Il.' insert '8.110' and after '16.242' insert '19.70'

ὀφρύη, add '**II** for ὀφρῦς I 1, Hp.Dieb.Judic. 2.'

ὀφρύκνηστον, add 'Archil. 58.10 W.'

ὀφρυόεις, after 'εν' insert '(neut. -όειν where long syll. required, Call.fr. 186.20 Pf.)'

ὀφρῦς **I 1**, 7 lines fr. end, after 'IA 648' insert 'cf. τὰς ὀ. ἄνες Men.Dysc. 423' **II**, add 'ornamental stone projecting above a lintel, Lib.Or. 5.51'

*ὀφρύωμα, ατος, τό, "eyebrow", name given to part of liver, Heph.Astr. 3.6.15.

*ὀφφικ-, v. °ὀφικ-.

ὀχεία **I 2**, add 'ὠχίας PSI 33.22 (iii AD)'

ὀχετεία, add 'Max.Tyr. 21.6'

*ὀχετογνώμονες, v. ‡ἐχετογνώμονες.

ὀχετόκρᾱνον, for 'Mnemos. 42.332' read 'SEG 33.336'

ὀχευτικός, after 'Thphr.fr. 183' insert 'comp. and sup., Ath. 9.391e, d '

ὀχεύω **I**, add 'fig., ἄμφω ὅπως ποταμὸς λαγόνων ῥείθροισιν ὀχεύσι TAM 3.907-8 (ii/iii AD)'; at end after 'Ph. 2.307' insert '(of bestiality)'

Ὀχεών, delete the entry (v. °Ἀντιοχεών).

ὀχή **I**, add 'Hp.Mochl. 20'

ὄχημα **I**, after 'E.Tr. 884' insert 'cf. οὗτός (sc. ὁ ἀήρ) τε γῆς ὄχημα Hp.Flat. 3; app. some form of support used in hydraulic works, τὰ φράγματα καὶ ὀχήματα SEG 32.4663 (Boeotia, i AD)'

ὀχθέω, line 1, for 'present only .. q.v.' read 'present not found '

*ὄχινος, ὁ, unexpld. wd., μυλαῖος PTeb. 793 i 26 (ii BC).

+ὀχλάζω, to be turbulent or obstreperous, Aq.Ps. 58.7, 15, Pr. 7.11, Je. 4.19.

ὀχληρία, add 'PHamb. 182.2 (249 BC)'

ὀχλίζω, line 2, delete 'by a lever' **II**, add 'cf. Nic.Al. 505'

ὀχλικός, add 'adv. -ῶς dub. in PFouad 31.2 (ii AD)'

*ὀχλοκρᾰτησία, ἡ, = ὀχλοκρατία, ὑπὸ ὀχλοκρατησίας καὶ φόνους Heph.Astr. 1.21.20.

ὄχλος **I 1**, add 'pl., crowd, D.S. 13.94; the masses, people in general, Heraclit.Ep. 7.4' **2**, add '**b** perh. referring to the non-citizen population, SEG 26.1817 (Cyrene, ii/i BC).'

ὄχμα, add 'cf. ὄχανον'

ὄχος, add 'cf. Myc. wo-ka (app. fem.) vehicle'

ὀχῠρός **2**, add 'comp., Plu.Arat. 50'

ὀχυρόω **II**, add 'μεγίστοις ἐπιτιμίοις ὀχυροῦντες τὰ παρ' ἡμῶν ὁρισθέντα Cod.Just. 1.3.38.5'

ὀχύρωσις, add 'cf. °ἐχύρωσις'

ὄψ (B), for 'gen. ὀπός' read 'nom. only'; add 'cf. Myc. compds. ka-ro-qo, po-ki-ro-qo, pers. n., o-po-qo (dual) = *ὀπώπω (cf. μέτωπον), blinkers'

*ὀψᾰρέλαιον, τό, fish-oil, Afric.Cest. p. 80 V.

ὀψάριον, after 'dim. of ὄψον' insert 'foodstuff, esp. fish' and add 'κωνοψωμίω[ν] ὀψάρει(ον) Περσικόν, ναύλου ὀψάρει(ον) κοινόν SEG 26.382 (Athens)'

ὀψαριοπωλεῖον, delete the entry.

*ὀψᾰριόπωλις, ιδος, ἡ, fish-shop, τὰς ἐν τῇ ὀψαριοπώλειδιν μαρμαρίνας τραπέζας ITrall. 77.19 (ii AD).

ὀψέ, line 1, after 'ὄψι (q.v.)' insert 'also app. Cret. ὀψῶι SEG 23.566, Pamph. *ὄψα (v. °ὀψιγενής)'

ὀψῐγενής, add 'also Ὀψαγένης IPamph. 49 (pers. n., ii BC)'

ὀψῐ́γονος **2**, line 2, after 'h.Cer. 165' insert 'Stesich 45 i 2 P.' for '**4**' read '**II** (parox.)'

ὀψίζω, add '**II** οὗ ἐὰν ὀψίσῃ, i.e. finds himself at nightfall, Lxx Si. 36.27; προῆγεν ὀρθρίζων καὶ ὀψίζων, i.e. early in the morning and late at night, Thd. 1Ki. 17.16.'

ὀψῐμᾰθής **I**, add 'adv. ὀψιμαθῶς Gal. 8.601.5'

ὄψῐμος, add 'cj. ἀψίμοθος'

ὄψιος, line 1, after 'α, ον' insert '(also ος, ον, Arat. 1027)' and before 'ὅταν' insert 'δείλης ὁ. Plb. 18.8.1'

ὄψις **I 1**, line 7, after 'id.(Th.) 7.44' insert 'δεῖ τὰ ἐκκλησιαστικὰ ἐμφυτεύειν διὰ τὴν ἀποκατάστασιν τῆς ὄψεως i.e. visible condition, Cod.Just. 1.24.5' **II 1 a**, for 'Emp. 4.10' read 'Emp. 3.10' add '**III** astrol., aspect of a planet in relation to one in a zodiacal sign on its left, Heph.Astr. 1.16; cf. ἀκτίς I 3.'

*ὀψίτευκτος, ον, late-made, Eust. 1235.17.

ὄψον **II**, for 'Ar.frr. 247, 545, cf.' read 'Ar.fr. 258 K.-A., cf. ib. 557'

+ὀψοπόνος, ὁ, cook, AP 6.306.

ὄψος, add 'PCornell 35.15, 16 (iii AD)'

ὀψοφάγος, add 'Ael.VH 1.28'

ὀψοφόρος, add 'SEG 30.1894'

ὀψωνιαστής, add 'prob. in graffito in W.Ruppel Der Tempel von Dakke 3 p. 60 no. 78'

*ὀψωνιάτωρ, ορος, ὁ, = ὀψωνάτωρ, caterer, Phot. α 239 Th.

Π

παγανός 3, add 'TAM 5(2).109.2'

παγαρχία, after 'district under a π.' insert 'POxy. 3307.1 (iv AD)'

*παγγεννήτειρα, ἡ, mother of all, PMag. 4.2556; cf. παγγενέτειρα.

πάγγεος, add 'πάγγεος .. δῆμος IHadr. 80 (ii AD)'

*παγγέραστος, ον, honoured by all, Anon.Alch. 4.5.

*παγγόνος, ὁ, epith. of Helios, procreator of all, cf. παγγενέτωρ, IG 4².529 (Epid., ?ii BC).

*πάγγωνος, ὁ, (γωνία) precious stone, Plin.HN 37.178.

πᾰγερός I, add 'transf., of death, SEG 23.137b.4 (iv BC)'

πάγη 2, add 'Iamb. VP 17.76'

πάγιος II, add 'ἐπιστῆμαι, opp. στοχαστικαί, Phld.Rh. 1.26, 59 S.'

πᾱγίς, line 1, for 'Call.Fr. 458' read 'Call.fr. 177.17 Pf.'

*Πάγκαμης, ὁ, cult-title of Heracles, SEG 26.429 (Argos, i AD).

*πάγκαρπον, τό, form of contest in the arena between men and beasts, Just.Nov. 105.1.

*παγκλυστής, οῦ, ὁ, temple-official, SB 7336.18 (iii AD, παν- pap.).

παγκοίτης, add 'Simon. in SLG 348.8'

*παγκρᾰτεύς, έως, ὁ, = παγκρατιαστής, Pi.N. 2 tit.

παγκρᾰτής 2, for 'B.fr. 10' read 'B.fr. 14.4 S.-M.' and add 'Simon. 36.5 P.'

+**παγκράτωρ**, ορος, all-powerful, Σούχῳ π. SEG 8.551.23 (i BC), θεός π. SEG 7.13, γῇ Robert Hell. 2.121 (all verse).

*παγκυκλικός, όν, adj. of uncertain application referring to celestial orbits, Anon.Astr. in PMich.III 149 x 12 (ii AD).

πάγξενος, add 'B. 11.28, 13.95 S.-M.'

*πᾱγόδετος, ον, frostbound, ὕδωρ Mesom. 10.3 H.

πάγος, after 'ὁ' insert '(also πάγος, εος, τό, v. II 1, 5)' **II 1**, after 'frost' insert 'κρ]ὐερος πάγος Alc. 286(a).3 L.-P.' **4**, after 'coagulation' insert 'στερεῷ γῆς πάγῳ Pl.Ti. 43c' **5**, for 'confused mass' read 'undivided firmament' and after 'Hp.Hebd. 6' insert 'ap.Gal. 19.73'

πάγουρος, add 'II tongs, Hsch. s.v. πυράγρη; cf. καρκίνος.'

πάγρος, for 'perh. = φάγρος' read 'an unknown bird' and add 'Ael.NA 5.48'

+**πάδος**, ἡ, a kind of tree, perh. the same as πηδός, Thphr. 4.1.3.

πᾰθεινός, for 'suffering, mournful' read 'mourning, grieving'; add 'PMich.v 234.18 (AD 43)'

πάθημα I, add '2 damage, Inscr.Cret. 4.144.13, al. (v/iv BC).'

πᾰθητικός I 2, line 3, after 'Cic.Orat. 37' insert 'of an author, Plu.Nic. 1 (sup.)'; add 'in or by feeling, τὰ δίκαια μὴ π. μόνον ἀλλ' ἐπιλογιστικῶς κατανοεῖν Phld.Rh. 2.254 S.' **II 1**, add 'Procl.Inst. 124'

πᾰθητός II 1, add 'b τὸ π. capability of emotion, Phld.D. 1.11.'

+**πᾰθῑκεύομαι**, act as a catamite, AP 11.73 (Nicarch.).

+**πᾰθῑκός**, ή, όν, in sexual sense, pathic, in Lat. form pathicus, Cat. 16.2, 57.2, Juv. 2.99.

πάθνη, after 'φάτνη' insert 'PLips. 106.8 (i AD)'

πάθος I 2 b, add 'Th. 1.106.2' **IV 2 b**, add 'AP 9.330 (Nicarch.)'

παι, for 'Inscr.Cypr. .. al.' read 'ka-sa-pa-i κάς παι ICS 217.4, i-te-pa-i iδέ παι 217.12, ta-sa-pa-i τᾶς παι 261'

Παιάν I 2, title of Apollo, add 'Ἀπόλλων Παῶνος SEG 39.427 (Boeotia, iii/ii BC)' at end add '[The first syll. is sts. short in Trag.lyr., A.Ag. 146, etc.]' and 'Myc. pa-ja-wo-ne (dat.), divine name'

+**παιᾱνίς**, ιδος, fem. adj., having the form of a paean, ἀοιδαί Pi.fr. 128c S.-M.

Παιᾱνισταί, after 'at Rome' insert 'pap. in Men.Dysc. 230 (cj. °Πανιστ-)' and add 'SEG 32.232, POxy. 3018'

παιγνία II, for ' = ἑορτή' read 'festivity, party' and add 'Herod. 3.55'

παίγνιον, before 'τό' insert '(παίχνιον in Theoc. 15.50 (PAntin.), Call.fr. 202.28, 33 Pf.; cf. mod. Gk. παιχνίδι)' **I 1**, after 'Ephipp. 24' insert 'Lang Ath.Agora XXI Hd14 (iii AD)' **III 2**, for 'comic performance' read 'diversion, amusement' and delete 'Suet.Aug. 99'

παίγνιος, for 'AP 7.12.212' read 'AP 12.212.6'

παιδάριδιον, add 'SEG 36.618 (Edessa, iii AD)'

παιδάριον, add 'πεδαρικόν ITyr 47'

παιδάριον II, after 'young slave' insert '(perh. sts. without ref. to age)'; add 'POxy. 3960.28 (AD 621)'

παιδαρίσκος, add 'Sch.Ar.Th. 291'

παιδαριώδης, after '(Sup.)' insert 'of literary work, D.H.Dem. 44; of an author, id.Is. 19'

*παιδαρίων, ωνος, ὁ, gloss on προύνικος, Hsch.; cf. °πατερίων.

παιδέρως I, add '2 pl. = παιδεραστία, IKyzikos 520 (iii/ii BC).' **II 1 a**, delete the section. **b**, add 'Paus. 2.10.5-6, cf. Plin.HN22.76; cf. pl. παιδὸς ἔρωτες Nic.fr. 75.55 G.-S.' **2**, add 'b a kind of amethyst, Plin.HN 37.123.'

παίδευμα I, for 'πόντου παιδεύματα, of fish' read 'χθονίων τ' ἀερίων τε παιδεύματα i.e. animals and birds' **II 2**, add 'πρὸς ἀρετήν D.H.Isoc. 4'

παίδευσις I, add '4 chastisement, Just.Nov. 122 pr.'

+**παιδευτής**, οῦ, ὁ, person in charge of education, educator, instructor, Pl.R. 493c, al., Lg. 811d, al.; as municipal appointment, IG 2².1011.35, TAM 5(1).700 (i AD). **2** one who imposes discipline, corrector, Ep.Hebr. 12.9.

παιδεύω II, line 15, after 'Aeschin. 3.148' insert 'τράγον ἄεθλα AP 6.312 (Anyte)'

*παιδιακόν, v. °πεδιακόν.

παιδιακός, delete the entry.

παιδικός I, line 5, after 'Lys. 21.4' insert 'ὕμνοι π. B.fr. 4.80 S.-M.' **II**, delete 'ὕμνοι .. 3.12' **III 1**, after '(i BC)' insert 'b child's garment, Edict.Diocl. 7.58, 59, Dura⁴ 97, 100 (πεδ-).'

παιδικυνηγεσία, delete the entry.

παιδιόθεν, add '(sp. πεδ-) ἐκ π. AJA 36.460 (S.Galatia)'

παιδίον III, after 'convulsions' insert 'Ruf.ap.Orib.inc. 38(20).27'

παιδισκάριον, line 2, before 'Hld. 1.11' insert 'PSI 1359.4 (ii/iii AD)' add '2 stone used in spinning, Hsch.'

παιδισκεῖος II, add '(with uncertain significance) τὸν θᾶκον σὺν τοῖς κατ' αὐτοῦ ἐπικειμένοις παιδισκήοις IEphes. 455 (ii AD)'

παιδίσκος, add 'also young slave, Ammon.Diff. 378 N.'

παιδνός, line 2, delete 'for παιδὸς χ.'

*παιδογόνιον, τό, birth of a child, παιδογονίου ἄρρενος, θηλείας PMich.v 243.5 (i AD).

*παιδογραφία, ἡ, register of birth, ἡ .. ἡλικία δείκνυται ἢ ἐξ παιδογραφιῶν ἢ ἐξ ἑτέρων ἀποδείξεων νομίμων Modest.Dig. 27.1.2.1.

παιδόθεν, delete 'Ibyc. 1.10'

*παιδοκλέπτης, ου, ὁ, boy-stealer, Call.Dieg. vii.6 (fr. 194 Pf.).

παιδομαθής, delete 'precociously quick' and add 'Quint.Inst. 1.12.9'

παιδονομέω, after 'hold office of παιδονόμος' insert 'SEG 29.527 (Thessaly, i BC), TAM 4(1).42'

+**παιδοπίπης**, ου, ὁ, one who eyes boys amorously, Ath. 13.563e; cf. παρθενοπίπης, ὀπιπεύω.

παιδοποιός 1, add 'E.Rh. 980'

παιδοτρόφος 1, for 'Simon. 12.4' read 'Simon. 3.6 P., A.fr. 47a.2.8 R. (lyr.)'

παιδουργέω, add 'Plu. in Hes. 74'

παιδουργία I, add 'Plu. in Hes. 74'

παιδοφιλέω, after 'Sol. 25' insert 'Pl.Com. 279 K.-A.' and delete 'Pass. .. 247'

παιδοφίλης, add 'AP 12.44 (Glauc.)'

παιδοφόντης, after 'Ph. 2.581' insert '(Trag.adesp. 327b K.-S.)'

*παιζόγελως, ων, playfully jesting, Cat.Cod.Astr. 12.190.9.

παίζω, line 2, for 'παιδδωᾶν' read 'παιδδωᾶν'

παίκτης, for 'dancer or player' read 'dice-player, gambler' and after '(Leon.)' insert 'Man. 4.448'

πάϊλλος, add 'Peek AV 120 (c.300 BC), Hsch.; fem. πάϊλλα, prob. in Epich. in CGFP 85.347'

*παίνουλα, v. φαινόλη.

Παιονικός, add 'as title of Dionysus, SEG 37.561 (Maced., ii AD)'

παιπάλημα, for 'piece of subtlety' read 'slang term, perh. conveying the idea of slander or innuendo w. obscene allusion (cf. Maxwell-Stuart AJP 96.11' and for 'Aeschrio 8.8' read 'AP 7.345'

παιπάλιμος, for 'Theognost.Can. 10' read 'Theognost.Can. 31 A.'

*παιπᾰλώσσω· τὸ παίζω καὶ τὸ παροινῶ Theognost.Can. 31 A.

παῖς, line 1, after 'ὁ, ἡ' insert 'acc. sg. παίδαν SEG 32.611 (Thessaly, i BC)', dat. pl. παίδεσσι, add 'SEG 23.416 (Thessaly, v BC)'; line 9, before 'I' insert 'see also πᾶς (B)' **I**, line 2, delete '(with special reference to the father, opp. τέκνον, q.v.)'; line 6, for 'Inscr.Cypr. 135.11 H.' read 'Cypr. pa-i-ta-se παῖδας ICS 217.11' **3**, after 'periphr.' insert 'παισὶν .. Αἰτναίων Pi.N. 9.30' **II**, add 'at Sparta, boy in fifth year of public education, i.e. eleven years old, Λέξεις Ἡροδότου in Stein Hdt. ii p. 465 (Berlin 1871)' **III**

παῖσκος after '(of all ages)' insert 'Hippon. 13 W.' add **IV** pl. voc., as a form of familiar address to equals, Ar.Eq. 419, Theoc. 10.52, 13.52. **b** as an exclamation, approximating to sense of παπαῖ, Men.Dysc. 500, Mis. 216, Sam. 678, 690, etc., Macho 215 G.'

*παῖσκος, ὁ, child, Kafizin 117 (221/0 BC); also πῆϊσκος Inscr.Cret. 1.x 2.5, 7.

παιφάσσω 1, line 2, for 'A.R. 4.1440' read 'A.R. 4.1442'

†παίχνιον, v. °παίγνιον.

παίω **I 1**, add 'c of a scorpion, sting, Ael.NA 5.14, 6.23, 10.23.' **4**, for 'Pax 874' read 'Pax 899'

παιωνίζω, last line, after 'παιαν-' in' insert 'B. 17.129 S.-M.'

πακτάριος, ὁ, fr. Lat. pactum, contractor, POxy. 2024.11, 2032.55 (vi AD), 138.9, 40 (vii AD).

πακτείκια, τά, fr. Lat. pacticius, perh. agreed payments in manumissions, Inscr.Perg. 8(3).44.

πάκτον, add 'Just.Nov. 120.1.1; perh. agreed sum, POxy. 3958.22, 30 (AD 614)'

*πακτονάριον, τό, dim. of πάκτων, small boat, PVindob.G 39847.463 (CPR 5.98) (iv AD), PLond. 1904.6 (v/vi AD) (written φακ-), SB 4323.9 (written πακτων-).

πακτόω 1, line 1, after 'Archil. 187' insert 'θύρην ἐπάκτωσα Hippon. 104.19 W.'

πακτωτής, add 'Bodl.Ms.Gr.Class. c. 88 (P.) (iii BC). **2** Christian church-official, ἀναγνώστης καὶ π. Inscr.Cret. 4.481 (Gortyn, v/vi AD).'

*πακτωτόν, τό, = πάκτων, BGU 1933 (ii AD).

*πάλᾱ, v. πάλη (A).

*πάλα, ἡ, Lat. pala, spade, Edict.Diocl. 15.45.

πᾰλάθη, add 'Carm.Pop. 2.6 P.'

πάλαι, line 2, after 'time' insert 'Sapph. 49 L.-P.'; line 8, delete 'cf. Eup. 11' and substitute 'Ar.Pax 414, 475'

παλαιγενής, line 3, for 'Μοίραι' read 'μοίραι'; add 'epic. in Coll.Alex. p. 82 ii 5'

παλαιμοσύνη, s.v. παλαισμοσύνη, add 'CEG 805 (iv BC)'

Πάλαιμων, after 'Hsch.' insert 'pl. Παλαίμονες, sea-gods, Call.fr. 197.19, 23 Pf.'

*παλαιοκέραμος, ὁ, old wine-jar, SB 9569.2 (pl., AD 91); cf. °καινόκεραμος.

*παλαιοπόρνη, ἡ, aged harlot, Dura⁴ 1.213 (iii AD).

παλαιός, after 'ά, όν' insert '(also ός, όν, Hp.Epid. 7.82)' **I 2**, add 'b name of kind of jasper, Epiph. in Lapid.Gr. 196.11.' **II**, add '4 applied to the original or principal sum of money involved in a transaction, Lys.fr. 6.' add 'Myc. pa-ra-jo'

*παλαιοσεβής, ές, of ancient reverence, Suppl.Hell. 974.6 (written παλησ-, iii BC (s.v.l.)).

παλαιουργός, add 'SEG 39.235 (Athens, iv BC)'

*παλαιόφυτος, ον, planted long ago, ASAE 39.292 (Panopolis).

παλαίπλουτος, for 'Quarterly .. iii AD' read 'SEG 8.269.9 (Gaza, iii/ii BC)'

παλαιοταγής, for 'that has become oily from age' read 'pressed long ago'

παλαιστή, delete 'Aeol.' to end of article.

*Παλαιστίνη, ἡ, Palestine, Hdt. 1.105.1, al.; also Π. χώρα Str. 16.4.18.

παλαιστρατιώτης, add 'Modest.Dig. 27.1.8 pr., al., SEG 33.1188 (Iconium, iii AD)'

παλαιστρίδιον, add 'Call.Dieg. viii.35'

παλαιστροφύλαξ, for 'superintendent of a wrestling-school' read 'attendant in a wrestling-school'

παλαιφάμενος, for '= sq., .. anon. 102' read 'of a tree, long spoken of, legendary, (?)Call.fr. 756 Pf.'

παλαίωμα, after 'Jb. 36.28, al.' insert '(the corresponding Hebrew wd. means "clouds, the heavens")'

παλαίωσις, last line, after 'Lxx Na. 1.15(2.1)' insert 'Sm.Ps. 71(72).7, 91(92).11'

πᾰλάμη **I 1**, add 'b applied to a bear's paws, Opp.C. 4.417.' **II**, line 2, after 'sense' insert 'παλάμαν ἔχει Alc. 249.7 L.-P.'; line 3, after 'of the gods' insert 'Κυπρογενήας παλάμαισιν Alc. 380 L.-P.'

πᾰλαμναῖος **I 2**, add 'applied to the mountain appointed for the sacrifice of Isaac, amulet in SEG 31.1594' add **III** name of month at Locri Epizephyrii, de Franciscis Locr.Epiz. 22, 24, 31.'

*παλάριος, α, ον, app. Lat. palaris, with a stake, ὅπλον παλάριν BGU 40.5 (ii/iii AD).

πᾰλάσιον, delete 'παλάθιον Suid.' and add 'παλάσια ἃ καὶ Κρατῖνος (fr. 390 K.-A.) ἰσχάδα κοπτὴν καλεῖ Poll. 6.81'

πᾰλαστή, line 4, after 'PLit.Lond. 183' insert 'Aeol. **πάλαστα** Alc. 350.6 L.-P.'

*πᾰλαστῶσαι χειροτονῆσαι Hsch.

†Παλατῖνος, η, ον, of the Palatine, D.H. 2.70. **II** palace official, TAM 4(1).255, Μημόριν Βαρδίωνος παλατίνου BCH suppl. 8.151 (v/vi AD), Cod.Just. 1.5.18.11, al.

†Παλάτιον, τό, Lat. Palatium, the Palatine hill in Rome, D.H. 1.31, etc., D.C. 53.16.5. **II** the palace or court of the Roman emperor, D.C. l.c., TAM 4(1).255, 285, PBeatty Panop. 1.260, Just.Const.Δέδωκεν pr.

*παλεός, v. παλαιός.

πᾰλευτής, for 'decoy-bird' read 'setter of decoy-traps'

πᾰλεύω **I**, after '(Ar.Av.) 1087' insert 'w. acc., decoy, Ael.NA 4.16' **II**, after 'entrap' insert 'Lyc. 405'

πᾰλέω, line 2, before 'elsewh.' insert 'to be wrecked, of a ship, πεπαληκός SB 9367.10.10'

*παλεωράφιον, τό, cobbler's shop, PAmst.inv. 62 (Mnemosyne 30.146), Gloss.

πάλη (A) **I**, add 'νικάσας Νέμεα ἀγενείους πάλαν Lindos 699a, b'

*πάληοσεβής, v. παλαιοσεβής.

παλιγγενεσία **I 1**, after 'AJ 11.3.9' insert 'cf. Memn. 40.2 J.'

παλιγκαπηλεύω, add 'prob. in Berytus 12.124.48 (Cyrene, iv BC)'

παλίγκτιστος, add 'expl. of Lat. recidivus in Virgil gloss., PNess. 1.762 (vi AD)'

πάλίμβολος, at end for 'turned or patched' read 'perh. second-hand'

πᾰλιμβουλία, for 'f.l. for -βολία' read 'perh. the giving of contrary advice'

πᾰλίμβουλος, for 'f.l. for -βολος' read 'giving contrary or untrustworthy advice' and add 'Heph.Astr. 3.16.5'

πᾰλίμπισσα, after 'Dsc. 1.72' insert '-πιττα Inscr.Délos 1441Aii 19, al. (ii BC); πᾰλίνπιττα, ἐφθὴ πίττα Hsch., παλινπίττης IG 2².1673.22 (iv BC): παλίνπισσα SIG 1171.14 (Lebena)'; delete 'cf. παλίνπιττα' (iv BC).

*πᾰλιμπλανήτης, ον, ὁ, = παλίμπλανης, Lyc. 1239.

πᾰλιμπλεκής, for 'twined or plaited back' read 'plaited back, i.e. narrowing at the top'

*πᾰλίμπλῠτος, rewashed, AP 7.708 (Diosc.).

πᾰλίμπνοος, for 'breathing' read 'breathed'; after 'Nonn.D. 37.295' insert 'al.'

†πᾰλίμποτον, τό, kind of drinking-vessel, perh. the same as ῥυτόν, Didyma 424.37, 40, PLond. 1960.16.

πᾰλίμπρᾱτος **1**, add 'cf. Call.fr. 203.55 Pf. (-πρη-)'

*πᾰλιμφροσύνη, ἡ, repentance, remorse, orac. in SEG 33.1056B.4 (Cyzicus, ii AD); παλιφ- lapis).

*πᾰλίμψηστρον, τό, = παλίμψηκτρον, SEG 33.1177 (Lycia, AD 43).

πάλιν **III**, after 'Ar.Ach. 342' insert 'Men.Dysc. 113, Macho 206 G.'

*πᾰλίνδουλος, ὁ, = ὁ πολλάκις δουλεύσας Hsch. s.v. παλιγκάπηλος.

πᾰλίνζωος, delete the entry.

*πᾰλίνζωος, ον, coming back to life, Opp.H. 1.319.

*πᾰλίνπιττα, v. °παλίμπισσα.

†πᾰλίνσοος, ον, safe again, AP 1.49.

*πᾰλιντροπή, ἡ, changing back, Zos.Alch. 196.1 (unless to be taken as two words).

παλινῳδία, add '3 repetition, Clem.Al.Paed. 3.11.60, Theol.Ar. 57 A.'

παλιουροφόρος, after 'παλίουρος I' insert '(s.v.l.)'

παλίωξις, delete '[ῑ metri gr.]'

πάλλα, add 'cf. παλίζεσθαι'

Παλλάδιον **II**, add 'also used in connection with the cults of Zeus and Athena, SEG 30.85 (Athens, i BC)'

*Παλλάδιος, α, ον, of Pallas, ἔλαιον Marc.Sid. 75.

παλλᾰκίς, line 2, after 'Od. 14.203' insert 'X.Cyr. 4.3.1'; line 4, for 'of ritual prostitution' read 'applied to a temple prostitute as being the concubine of a god'

*παλλάντιον, τό, a plant, = πολύγονον or ὀστεόκολλος, Hippiatr. 66.

Παλλάς **II**, for 'maiden-priestess' read 'temple-prostitute' and after 'Str. 17.1.46' insert '(perh. ghost-word)'

Παλλειών, add 'also Παλλεών Amyzon 36 (ii BC)'

*παλλιόλιον, τό, dim. of Lat. palliolum, small cloak, PMich.III 201.9 (i AD, παλλιώλιν), PTeb. 405.3 (ii AD, παλλιόλιν).

*παλλίολον, τό, Lat. palliolum, POxy. 3724 v 29, BGU 781.6.6 (i AD).

*πάλλιον, τό, Lat. pallium, cloak, Aesop.Prov. 120 tit. P.; in forms πάλλιν and πάλιν, Dura⁴ 97, 93, BGU 22.17 (ii AD).

πάλλω, line 8, after 'ἐφάλλομαι' insert 'though there was prob. early confusion between the vbs. through misdivision of ἐπ-ᾶλτο as ἔ-παλτο' **II**, line 6, after 'Ar.Ra. 345' insert 'throb, οἱ κρόταφοι πάλλονται Hp.Acut. 30' **III**, add '2 twitch, μηρὸς εὐώνυμος πάλλων PFlor. 391.8 (iii AD).'

*πάλμα, ατος, τό, leaping, given as etym. of name Παλλάς, POxy. 2260 ii 5 (ii AD).

παλμός **1**, add 'περὶ παλμῶν μαντική, work by Melamp.'

*πάλμυρον, τό, unexpld. wd., Theognost.Can. 131.

παλμώδης, add 'adv. -δῶς, Gal. 7.65.18 (s.v.l.)'

πάλος **2**, after 'lot' insert 'Alcm. 65 P.'

παλτός **I**, for '(lyr.)' read '(anap.)' **II**, line 2, after 'Fr. 16' insert 'παλτῷ, in spear-throwing contest, MDAI(A) 62.4 (ii BC)'

παμβᾰσιλεύς, after 'ἕως' insert 'poet. also ἧος Opp.H. 1.78'; add 'of Marcus Aurelius, Opp. l.c.; of Zeus, Orph.H. 73.3'

παμβῶτις, add 'ἐλπίς Trag.adesp. 252 K.-S. (Hsch.)'

*παμμακάριος, v. °παμμαχάριος.

παμμακάριστος, after '= foreg. (πάμμακαρ)' insert 'Antip.Sid. in Inscr.Délos 2549.11'

*παμμαχάριος, ὁ, = παγκρατιαστής, Ambros. in Ps. 36.55 (PL 14.993); prob. in Firm.Math. 8.8.1 (codd. macharios); also sp. παμμακ-, ἀθλητὰς παμμακαρίους, λωποδύτας Teuc.Bab. p. 43.27 B.

*παμμεδέων, οντος, ὁ, ruler of all, Ζεύς Nonn.D. 40.97.

*πάμμνηστος, ον, ever remembered, CIJud. 1.661 (Tortosa, vi AD), cf. πάμνηστος.

πάμμουσος, add 'as a sobriquet, Mitchell N.Galatia 370 (i AD)'

*πάμπαιδες, οἱ, app. the lowest age-group in athletic contests, IG 12(9).952.5 (Chalcis, ii BC), 7.1764.13 (Thespiae, ii/i BC), 2871.21 (ii/i BC).

*παμπάσιον, τό, entire possession, Ἠπειρωτικὰ Χρονικά 10.253 (written πανπασιον), cf. ‡παμπησία.

παμπήδην, for '(A.)Fr. 56' read 'fr. 154a.16 R.'

παμπησία, add 'non-Att.-Ion. πανπᾱσία, Ἠπειρωτικὰ Χρονικά 10.254 (Dodona, iv BC); cf. °παμπάσιον'

πάμπολυς II, after 'etc.' insert 'w. comp., πάμπολυ .. κυριώτεροι Pl.Ep. 7.345b, cf. Is. 8.33'

παμπρᾱσία, add 'also sp. παππρασια, Kafizin 266(a) (225/4 BC), 267(a) (223/2 BC)'

*παμπρᾱσίον, τό, app. = παμπρασία, unreserved sale of property, POxy. 3015.21, 27 (ii AD).

*παμφανός, prob. shining, Hdn.fr. p. 28 H.

*παμφίλητος, ον, beloved of all, Not.Scav. 1937.473 (Sicily, Chr.).

πάμφιλος, for 'prob.' read 'cj.'

*πάμφοιτος, ον, wandering everywhere, πάνφο[ιτ]ον ἄνασσαν TAM 4(1).92.

*πάμφρητρος, ον, consisting of all the phratries, παμφρήτροις .. ἀγέλαις Didyma 537.6.

*Παμφῡλαία, ἁ, of all the φῦλα, cult-title of Artemis, IG 4².503 (Epid., ii BC).

*Παμφῡλίς, ίδος, fem. adj. Pamphylian, D.P. 46, al., Nonn.D. 2.38, Π. γαίῃ GVI 815.6 (Attalea, ii AD).

πάμφῡλος I, add '2 of all the φῦλα, cult-title of Zeus, SEG 37.370 (Megara, v BC).' II, line 3, before 'gen. pl.' insert 'nom. sg. Πανφύλας SEG 13.239 (Argos, v BC)'

πάμψογος, add 'Heph.Astr. 2.15.11'

πανᾱγής I, add 'βίος SEG 37.1175.10 (verse, Pisidia, ii AD)' II, add 'accursed, Phld.Sto. 339.8'

*πᾰναγρυπνία, ἡ, complete wakefulness, PMag. 4.3274.

πᾰνᾱγῠριάρχας, etc., delete '-αγύριος'

*Πᾰνᾱγύριος, ὁ (sc. μήν), month at Amphissa, Delph. 3(3).32 (ii BC).

Πανᾰθηναϊκός, add 'τὸ π. scent (prob. sold in miniature Panathenaic amphorae), Apollon.ap.Ath. 15.688f, Plin.HN 13.6'

πανάθλιος, add 'Hld. 5.2'

πανάκεια II, add 'Πανάκιαν Σώζουσαν Inscr.Perg. 8(3).128'

πᾰνάκη, add 'II = πανάκεια II, Herod. 4.6.'

πᾰνᾰκήρᾰτος, add 'πανακήρατον ἔλλαχον εὔχος SEG 24.1243'

*πᾰναλγής, ές, full of sorrow, n. acc. pl. as adv., -έα κωκύσασα IKyzikos 518.21 (verse, i AD).

Πᾰνᾰμάρεια, add 'sp. Παναμάρια, SEG 30.1274 (Caria, c.AD 200)'

*πᾰνάμωμον, ον, gen. ονος, utterly blameless, GVI 199 (Pontus, ii/iii AD).

πάναξ, add 'Eudem. in Suppl.Hell. 412A.5'

πᾰναπηρής, for 'all-unmutilated' read 'quite unharmed, πόδες' and for 'Call.Cer. 126' read 'Call.Cer. 125'

*πᾰνάργῠρεος, α, ον, all silver, ἄμφω ταῦτα -εα Antip.Sid. in Inscr.Délos 2549.6.

*πᾰνάρεστος, ον, pleasing in every way, GVI 874.1 (Smyrna, ?ii AD).

†πᾰνάριον, τό, chest or box for keeping bread, POxy. 300.4 (i AD), 1272.8 (ii AD), S.E.M. 1.234, SB 9834b.23 (iv AD).'

*παναυτάδελφος, ον, entirely fraternal, Dioscorus in Byz.-neugr.Jahrb. 10.342.

Πᾰναχαιά, after 'Demeter' insert 'Δάματρι Παναχαιᾶι SEG 25.643 (Thessaly, Hellen.)'

*Πᾰνᾰχᾱϊκός, ή, όν, of all the Achaeans, τῶι Παναχαϊκῶι συνεδρίωι SEG 35.304 (Epid., after 68/7 BC).

πᾰνάώριος, add 'adv., παναώρια GVI 318.1 (Thessalonica, ?iii AD)'

*πανδᾱμικοί' παιδικοὶ χιτῶνες ἐν ταῖς πομπαῖς Hsch.

πανδέκτης I 1, add 'cf. Plin.HN praef. 24' add '3 medicine-chest, Sch.Ar.Pl. 711.'

*πάνδεκτος, ον, receiving all, πάνδεκτον Φερσεφόνης θάλαμον CEG 489.4 (Attica, iv BC).

*Πανδήμεια, τά, festival of Zeus Pandemos, IHadr. 128 (iii AD).

πάνδημος I, of Zeus, add 'IHadr. 125 (ii/iii AD), 126, 127; also of Dionysus, SEG 32.1243 (2 BC/AD 2)' add '2 -ον, confederation of demes, κωμῶν δύο Swoboda Denkmäler 282.' II, line 4, after 'π. ἐρασταί' insert 'ἐραστής' and after 'Pl.Smp. 181e' insert '183e' add 'b from all the town, π. ἐραστής AP 5.302.9 (Agath.).'

*πάνδοξος, ον, all-glorious, Pi.fr. 94b.8/9 S.-M.

πάνδουρος, add 'II = πανδουριστής, Rouché Aphrodisias 113 (v/vi AD), Hsch. s.v. πανδοῦρα.'

πανδύνᾰμος, add 'Plot. 5.9.9'

πάνδυρτος, for 'poet. for' read 'edd. metr. gr. for'

πανδῡσία, for 'total .. star' read 'the period during which the Pleiades and Orion are setting (at sunrise), οἱ ναυτικοί ap.Procl.in Hes.Op. 618-26'

πανδώτειρα, add 'PMag. 4.2280'

Πᾱνεῖος II, add 'τὰ ὅρια τοῦ Πανίου SEG 32.1499 (Diocletian)'

*πᾰνελευθερία, ἡ, entire freedom, IG 7.1780.7.

*Πᾰνελλάς, άδος, ἡ, the whole of Greece, Pi.Pae. 6.62 S.-M., Philod.Scarph. iii.32, Call.fr. 106 Pf.

Πᾰνελλήνιος I, add '2 title of Hadrian, SEG 32.185 (Athens).' II 2, add 'ὁ ἄρχων τοῦ Πανελληνίου SEG 28.1566 (Cyrene, AD 154)'

*πᾰνέντιμος, ον, held in all honour, ὁ π. οἶκος οὗτος BCH suppl. 8.103 (Thessalonica, v AD).

†πᾰνεορτεύω, keep a solemn festival, IPhilae 159.3 (i AD).

*πᾰνεπάρκιος, ον, all-sufficient, Suppl.Hell. 937.27, 1181.1.

*Πανεπις, ὁ, Egyptian deity, "bull of Apis", JEA 12.34 (v BC).

*πᾰνέραστος (sc. λίθος), ἡ, = πανέρως, Plin.HN 37.178.

πανεύφημος, add 'Just.Nov. 8 iusiurandum 43'

*πᾰνηγῠριαρχία, ἡ, office of πανηγυριάρχης, Didyma 157 i(c).

*πᾰνηγῠριαρχικός, ή, όν, of a πανηγυριάρχης, τιμή PASA 2.396 (Caria).

*πᾰνηγῠρίη, = πανήγυρις, SEG 8.549.24 (hymn, i BC).

πᾰνηγῠρικός I, add 'προφήτης Didyma 264.1, 238 II 2, al.; ταμίας ib. 408.3, 410.1'

πᾰνήγυρις, before 'Dor.' insert 'Aeol.' I 1, add 'b market-day, Just.Const.Δέδωκεν 8c.'

πᾰνημᾰδόν, add 'SEG 13.277 (Patrae, iv/v AD)'

πᾰνήμαρ, add 'Trag.adesp. in PKöln 241 A i 1'

*πανθάπᾱσι, = παντάπασι, PCol. 175.9,53 (AD 339).

πάνθειος I, before 'ον' insert 'α'; add 'epith. of Zeus and Athena, SEG 20.719.12, 13 (Cyrene, ii BC)'

πανθέλκτειρα, for 'Simon. 183.1' read 'AP 7.24.1 ([Simon.])'

Πανθεών, add 'at Olynthus, SEG 39.617 (iv BC)'

πανθοινία, add 'Cels.ap.Orig.Cels. 8.24'

*πανθυπακουστής, οῦ, ὁ, one who hears everything, PMag. 4.1369.

πᾰνήμερος II, for 'prob.' read 'cj.'

Πάνιος, add 'II name of month, SEG 17.829 (unkn. provenance, c.200 BC).'

*πάντρευς, εος, ὁ, title of priest at Mytilene, IG 12(2).61.3 (written -ειρ-), 102, cf. ἱρεύς.

Πανίσκος, after 'Πάν' insert 'Inscr.Délos 1416Ai51 (ii BC)'; add '(Cic.) Div. 1.14.23'

*Πανιστής, οῦ, ὁ, worshipper of Pan, Men.Dysc. 230 (παιανιστ- pap.).

πᾰνίχνιον, for 'the whole track' read 'pl., all kinds of tracks'

Πανίωνες, delete 'Πανιώνιον' to end.

*Πᾰνιώνιος, ον, of or associated with the Πανίωνες; as title of Apollo, IEphes. 814; of Hadrian, ib. 1501; ὁ κρατὴρ ὁ Π., sacred vessel at Delos, Inscr.Délos 104.129, Hyp.fr. 69. b Πανιώνιον, τό, temple and meeting-place of the Πανίωνες at Mycale, Hdt. 1.141.4, al., CIG 2909 (Mycale). c Πανιώνια, τά, festival of the Πανίωνες, Hdt. 1.148.1.

*Πανλίμνιος, ὁ, title of Apollo, SEG 29.515 (Thessaly, ii BC).

παννύχιος, before 'ον' insert 'α'

*πᾰνομφαῖος, for 'sender of ominous voices' read 'universally prophetic'

*πᾰνόπαια, ἡ, epith. of Hecate, PMag. 4.2612, 2965, v. ἀνοπαῖα 2.

*πᾰνόρφανος, ον, very much an orphan, SEG 14.563.15 (epigr., Chios, i BC).

*πᾰνόσιος, ον, pl. οἱ πανόσιοι, the number or company of the holy ones, Phld.Sto. 339.17.

πᾰνοσπρία, add 'b gloss on πάγκαρπα θύματα, Sch.S.El. 635, cf. Hsch.'

πᾰνουργέω I, add 'b διὰ σχημάτων π. play tricks with figures (of diction), Longin. 17.1, cf. 2.'

πᾰνουργία I 1, add 'b transl. Hebr. 'ormāh, elsewhere craftiness but here app. in good sense, shrewdness, Sm.Pr. 8.12.'

πᾰνοῦργος II, line 4, after 'Plu. 2.28a' insert 'Lxx Pr. 27.12, Si. 21.12'

*πανπᾱσία, v. °παμπησία.

πάνριζος, add 'also πάρριζος, π. μολεῖν Ἄιδου μέγαν κευθμῶνα poet. in Robert Collection Froehner I no. 77 (Alexandria, i BC)'

πανσᾰγία, add 'Ion. πασσᾰγίη, Call.fr. 359 Pf.'

*πανσευδί, adv., written πανσεῦδί, = πανσυδί, prob. in Inscr.Cret. 1.xxviii 7 (vi BC).

*πανσοφία, ἡ, poet. πασσοφίη, complete wisdom, Epigraphica 10.76 (Leptis Magna).

πάνσοφος, for 'most clever' read 'clever in every way'; line 3, for 'Trag.Adesp. 470.3' read 'A.fr. 181a.3 R.' and add 'SEG 30.1179 (Ostia, iii AD)'; line 4, after '(Tenos)' insert 'PHerm. 4.14 (iv AD)'

*πανσπερμεί, adv., *all seeds* (or *kinds*) *together*, φύετο στάχυς ἄμμιγα κριθαῖς πασπερμεί Lyr.adesp. 11(d).4 P.

πανσπερμία, add 'cf. πασπερμεῖον'

*πανσπέρμιον, τό, = πανσπερμία, Anon.Alch. 18.14.

πανσυδί, add 'cf. πασσυδόν, πασσυδιάζω, see also °πανσευδί'

πανσυδίᾳ II, delete the section transferring quots. to section I. add 'II *utterly*, AP 7.299 (Nicomachus).'

*πανταβροκτον, τό, app. for *παντά-βροχον, vessel for soaking dried foods, Kafizin 230(a) (225/18 BC).

παντάπᾱσι, line 1, before 'before a vowel ' insert 'usually' and add 'before a consonant, e.g. παντάπασιν καὶ .., SEG 37.1003 (Sardis, c.200 BC)'

παντάπώλης, add 'PTeb. 841 (ii BC)'

παντάχόσε, after 'Plu.Agis 14' add 'Just.Nov. 4 epilogos'

παντεβιπᾱσιν, delete the entry.

παντελής, line 1, for 'σάγην' read 'σαγὴν'

*Παντελίη, ἡ, name for Demeter, IG 4².551.1 (Epid., metr.inscr., Roman); cf. παντέλεια I, παντέλειος.

*παντευλογ(έω), uncertain reading, perh. *pray constantly*, SEG 36.970.A5 (Aphrodisias, iii AD).

†πάντεχνος, ον, *belonging to* or *dealing with all the skills*, παντέχ[νοις] Ἁφαίστου παλάμαις Pi.fr. 52i.65 S.-M.; παντέχνου πυρὸς σέλας A.Pr. 7.

παντογενής, add 'Αὗραι Orph.H. 81.1 (cj.)'

παντοδύναμος, delete 'Plot. 5.9.9' (v. °πανδύναμος)

παντοδυνάστης, add 'PMag. 12.267'

παντοθᾰλής 2, for 'BMus.Inscr. 1067.15' read 'BMus.Inscr. 1075.14'

παντοῖος II, adv., add '2 *in every way*, i.e. *absolutely*, Just.Nov. 105.2.2, 109.1.'

παντοκράτειρα, line 2, delete 'pecul.'; add 'SEG 8.548.2'

παντοκράτωρ, line 2, for 'almighty' read '*ruler over all*'; add 'as cult-title of Zeus, INikaia 1512 (ii/iii AD), 1121 (iii AD); of the Emperor Julian, BCH suppl. 8 no. 86'

πάντολμος, after 'all-daring' insert 'τὸ πάντολμον σθένος Ἡρακλέος Pi.fr. 29.4 S.-M.' and for 'shameless' read 'w. pejorative force'; add 'AP 5.218.4 (Agath.), 248 (Paul.Sil.)'

*παντοπωλικός, ή, όν, *selling all varieties of goods*, PSI 692.2, 12, 13 (AD 52/3).

†παντόπωλις, ιδος, ἡ, *of a market*, etc., *selling all kinds of goods*, τὰν παντόπωλιν στοάν SEG 23.207 (Messene, Augustus); as subst., *general dealer*, PRyl. 227.3 (iii AD).

πάντοτε, add '2 *in every instance*, Plu. 2.550b, Peripl.M.Rubr. 29.'

παντόφωνος, ον, *producing the whole range of sounds*, ὄργανα Dain Inscr.du Louvre 60.20 (Heraclea ad Latmum).

πάντρητος, for 'the part .. holes are' read 'perh. *perforated collar* enabling additional holes to be opened'

πάντροπος, add 'III *versatile*, Μουσέων π. ἦν θεράπων MDAI(A)20.228 (Rhodes, i BC).'

πάντως I, line 5, for 'A.Pr. 335' read 'A.Pr. 333'

πᾰνύπέρτατος 2, after 'supreme' insert 'Alcmaeonis fr. 3 p. 33 B.' and add 'Ζηνὶ πανυπερτάτῳ INikaia 1071 (i/ii AD)'

*πᾰνυπεύκυκλος, ον, *well-rounded all over*, π. ἀνθεμίς Poet.de herb. 134 (cj., codd. παρ-).

πᾰνῳδός, before 'ἀχώ' insert '(or Πᾱνῳδός, of the music of Pan)'

*πᾰνώλεος, ον, = πανώλης II 1, γένοισαν ἐξώλεοι καὶ πανώλεοι IMylasa 476.8 (Rom.imp.).

†πᾰνώνιος, ον, Cypr. pa-no-ni-o-se, prob. *having full rights of sale* or *enjoyment*, ICS 217 A10, B22 (also Addenda p. 415).

πάξ, after 'Diph. 96' insert 'AP 5.181 (Asclep.)'

παξαμᾶς, add 'II in μουσωτοῦ παξαμᾶ Syria 1.302 (Sidon, vi AD), π. may denote a special kind of mosaic (or the maker of it).'

*πάομαι, last line, after 'ii BC)' insert 'IG 4.752.13 (Troezen, ii BC)'

*πᾱόνιος, α, ον, *of a peacock*, Edict.Diocl. 18.9; cf. πάων, Lat. pavo.

παπειν (acc. sg. masc.) perh. local title, metr.inscr. in Robert Hell. 7.198.10 (Tarsus, ii/iii AD).

†πᾱπίας (παππ-), ου, ὁ, *dad*, childish or familiar term for *father*, Ar.V. 297, Pax 128, Men.Dysc. 856, Ephipp. 21 (all voc.); applied familiarly to an old man not one's father, Men.Dysc. 930.

*πάπος, ὁ, = πάππος I 1, MDAI(A) 27.307 (Maced., pl.), AAWW 1961.124 (Lydia, AD 233/4), SEG 30.1485 (Phrygia, AD 305/6). II something used as incense, εἰς ἀποκαυσμὸν τῶν π. Ramsay *Cities and Bishoprics* I 119 (see also SEG 6.272: perh. = *papaver*), Rev.Phil. 36.73 (Iconium), cf. perh. πάππος II, which must be a different wd. fr. sense I.

παππάζω II, add 'w. internal acc., παρακάθετο Βάκχος Ἀθήνῃ Εὔια παππάζοντα Nonn.D. 48.954'

πάπας, after 'nom. πάπας' insert 'A.fr. 47a ii 14 R. (lyr.) (prob.), Men.Dysc. 194, 204' and after 'acc. πάπαν' insert 'Men.Dysc. 648, cf. 494' add 'II (Christian) *priest*, SEG 32.1474 (Syria), PLond. 1914.25 (iv AD), al.'

παππίας, v. °πᾱπίας.

παππικός, add 'PGrenf. 55.23 (ii AD); also ός, όν, ITyr 33A.4'

*παππωνῠμικός, ή, όν, *derived from one's grandfather's name*, [ὄνομα] Sch.E.Rh. 36; adv. -ῶς Sch.Gen.Il. 9.191, Suid. s.v. Ἀλκείδης; cf. πατρω-, μητρω-, μαμμω-νυμικός, -ῶς.

παππωνῠμικῶς, delete the entry.

παππῷος, add 'CPR 7.3.7 (AD 150); perh. also cult-title of Zeus, INikaia 1513'

πᾱπῠλιών, after 'Edict.Diocl. 19.4' insert 'PMich.III 214.26 (iii AD)'

πάπυρος, at end after '‿‿‿' add 'cf. Moer. p. 311 P.'

πᾰρά, line 3, after 'Att.' insert 'once in Ion.' and after '(Paros)' insert 'SEG 30.1456 (Sinope, c.470/50 BC), ib. 366 (Argos, c.460/50 BC)' **B II 2**, line 10, for 'Pl.Phlb. 29f' read 'Pl.Phlb. 29e; παρ' ἐμαυτῷ *in my own mind*, id.Phd. 107b'; add 'παρ' Ἥραι (i.e. in or near her temple), SEG 34.282 (Nemea, iv BC)' **C I 2 b**, add 'ἀποδώσω παρὰ τὸν εὔθυνον τὸ καθῆκον IG 1³.244.B7' **III 8**, for 'of a .. possibility' read 'with expressions of possibility and permission'; transfer the reference to Arrian to follow 'An. 846' inserting 'also' before 'πεῖσαι'; add 'π. τοῦτο ἔσται καὶ ὁ κύων ἄνθρωπος S.E.P. 2.23' **F**, add 'also for πάρειμι, Cratin. 113 K.-A., Hermipp. 52 K.-A.' **G**, after 'IN COMPOS.' insert '(sts. w. apocope)' at end add 'Myc. pa-ro, prep. w. dat. = παρό'

*παραβάδην, v. °παρβάδαν.

παραβαίνω, line 3, after '*παραβάω' insert 'also w. apocope of preverb, παρβαίνοντι IG 9².609.15 (Locris, vi/v BC), part. παρβεῶντας SEG 9.3.42, 47 (Cyrene, iv BC)'

*παραβάλανεύς, έως, ὁ, *sick-nurse*, ACO 2(1).179, PIand. 154 (vi/vii AD); (cf. Lat. *parabalani*, *parabolani*).

παραβάλλω, after lemma insert '(w. apocope of preverb, e.g. παρβάλλοιτο, Schwyzer 323.C26 (Delphi, iv BC))' **B**, add '**VI** (in med.) *transgress*, εἰ δέ κα παρβάλληται Sokolowski 2.33A.9 (Dyme, iii BC), Leg.Sacr. ii.74C7, D18/9.'

παραβάσις II, line 3, after 'Plu. Comp.Ages.Pomp. 1' insert 'as t.t. in law, συνθηκῶν παραβάσεως (sc. δίκη) Poll. 8.31'

παραβάτης II, add 'ὁ π. the *Transgressor*, i.e. the Emperor Julian, Suid. s.v. Ἰουλιανός, Eust. 83.41' add '**III** *passer-by*, κεῖνος ἂν εὐδαίμων εἴη μᾶλλον παραβάτας IG 9(1).256.11 (Halae).'

παραβιάζομαι I, for this section read '*act in defiance of some constraint*, Lxx De. 1.43, Plb. 24.8.3'

*παραβίασις, εως, ἡ, *the use of violence*, Epicur.fr. 29.39.2 A.

παραβιβάζω 1, for 'put aside, remove' read '*transfer*'

*παραβιβρώσκω, *nibble at*, παραβέβρωται Vit.Aesop.(G) 45; Hsch. s.v. παρεσθίε[τα]ι.

παραβλέπω I, add '3 *see by the side*, in quot. pass., Archim.Aren. 13.'

παράβλημα I, after 'fodder' insert 'Hsch. s.v. κάπητον' **II**, add 'PRyl. 558.3 (iii BC)'

παραβολᾶνοι, delete the entry (v. °παραβαλανεύς).

παραβολεύομαι, after '(v.l. παραβουλ-)' insert 'μὴ παραβολεύεσθαι PSI 1241.27 (ii AD)'

παραβολή VI, after 'multiplication' insert 'Nicom.Ar. 2.27' and delete 'hence .. Ar. 2.27'

*παραβολινθέω, unexpld. wd., PAberd. 190.3 (i AD, pass.).

παράβολος II 2, line 7, for 'τὰ π. .. Longin. 32.4' read 'neut. subst. *audaciousness* of language, Longin. 22.4; pl., ib. 32.4'

*παραβραδύνω, *tarry*, Luc.Alex. 44 (v.l.).

παράβυστος II 1, add 'prob. in IG 2².1646.12' **2**, add 'Procop.Arc. 1.17'

παραβύω I, add 'in chariot-racing, perh. *push aside*, SEG 34.1437 (Syria, v/vi AD)'

παραβώμιος 3, add 'παραβ[ώ]μια ῥέξαι epigr. in SEG 39.855 (Patmos, iii/iv AD)'

παραγαύδιον, after 'Edict.Diocl. 19.29' insert 'Dura⁴ 93'

παραγγελία II 1, add 'Din.fr. IVa (p. 78.10 C.)' **2**, add '**b** *notice of legal proceedings*, Just.Nov. 88.1.'

παραγγέλλω III 3, add 'of the complainant, pap. in Illinois Cl.Studies 3.100' **IV 2**, for 'App.BC 1.21' read 'App.BC 1.121' and add 'w. dat., εἴ .. τις .. τῶν αἱρετικῶν .. οἱῳδήποτε δημοσίῳ φροντίσματι παραγγείλειεν Cod.Just. 1.5.18.10'

*παραγειτνιάω, *to be neighbour*, τισι Sch.E.Rh. 5.

παραγίγνομαι, line 2, after 'Plb. 3.99.2, etc.' insert 'w. apocope of preverb, **παργίνομαι**, SEG 31.575 (Larissa, 171 BC), aor. part. pass. παργενεαθέντες SEG 30.1119 (Entella, iii BC)' **I 4**, for this section read '(*have a right*) *to be present at a feast*, IG 12(7).515 (Amorgos, ii BC); παραγενόμ[εν]ος .. ἀπὸ Πλατίννας (i.e. through his descent from P.), Inscr.Cos 405' at end add 'form παρο-, prob. Myc. 3 sg. aor. med. pa-ro-ke-ne-[to]'

παράγραμμα, add 'III *play on words, pun*, Cic.Fam. 7.32.2.'

παραγραμματίζω I, for '= foreg.' read 'τινα *make a pun on the name of*'

παράγραφος, add '2 masc., *pencil for drawing lines*, CGL 3.639.3.'

†παραγράψιμος, ον, *open to objection*, S.E.M. 7.170; *disqualified*, Modest.Dig. 27.1.13.6.

παραγωγεύς, after 'introducer' insert 'or perh. official who collected a παραγώγιον from initiates'

παραγωγή I 6, after 'furnishing' insert 'Didyma 41.34'

*παραγώγιμος, ον, that is in transit, π. φορτίον SEG 14.639.B13 (Caunus, i AD).

παραδειγματίζω, delete 'Ev.Matt. 1.19' (v. δειγματίζω).

παραδείκνυμι, add '7 of a creditor, π. εἰς ἐνεχυρασίαν indicate a property as having become liable to distraint, PRyl. 176.5 (iii AD, rest. as med.), PIand. 145.3 (iii AD; act.), etc.'

παράδεισος I 3 b, add 'pl., παραδίζοισι κατοικῶ AS 5.32.24 (N. Phrygia, iv AD)'

*παραδεισοφύλαξ, ἄκος, ὁ, custodian of a παράδεισος, PCair.Zen. 690.22 (pl., iii BC).

*παραδεισών, ῶνος, ὁ, orchard, PHamb. 99.9 (i AD).

*παραδεξιόω, perh. make convenient, BGU 1844.19 (i BC).

παραδέχομαι, line 1, after 'Pl.Tht. 155c' insert 'w. apocope of preverb, παρδέξαι SEG 35.1011 (Sicily, i BC)'

*παραδηθύνω, linger upon, Orph.L. 634 (cj.).

παραδηλόω, after 'Plu.Crass. 18, etc.' insert 'w. inf., Hld. 7.9'

*παραδήλωσις, εως, ἡ, intimation, Poll. 4.33 (pl.).

παραδίδωμι, line 2, after 'AD' insert 'w. apocope of preverb, παρδῶντι Tab.Heracl. 1.106' I 4 a, add 'κατὰ τὸ παραδεδομένον according to tradition, SEG 30.622 (Thessalonica, i AD)'

παραδιώκω I, for this section read 'drive out, eject, from office, SEG 7.1.13 (Susa, i AD; pass.); reject a reading, A.D.Synt. 145.20'

παραδοξονίκης, of athletes, add 'SEG 34.1317 (Lycia, C.AD 90)'

παράδοξος II, add 'in list of kouretes, SEG 33.937 (Ephesus, ii AD)'

παράδοσις, line 2, after 'ἡ' insert 'πάρδοσις Brit.Mus.Quarterly 11.13 (on gem, ii/iii AD)'

παραδρομάδην, for 'in running or passing by' read 'in passing, briefly'

παραδρομή, add 'III perh. = παραδρομίς, CPR 7.44.2 (v/vi AD, s.v.l.). IV passage of time, CodJust. 1.3.43.13, 10.19.9.1.'

παραζηλόω I 2, for this section read 'to be overzealous, ἔν τινι Lxx Ps. 36(37).1, 7; abs. ib. 8'

+παράθερμος, ον, excessively hot-blooded or violent, D.S. 24.3, Plu. Comp.Pel.Marc. 3; transf., of actions, π. καινουργία Hierocl. p. 52 A.

παράθεσις II, add 'IG 12(7).515.77' III, delete 'IG 12(7).515.77' IV, add 'b entry on a file, PColl.Youtie 73.14 (AD 289).'

παραθέω III, add 'fig., χρυσόν δ' εὐδικίη παραθεῖ Call.fr. 384.14 Pf.'

παραθήκη, add 'b place of deposit, SEG 31.1072 (Pontus, ii/iii AD). 2 of a tablet entrusted to supernatural powers for the execution of the curse it bears, Tab.Defix.Aud. 22.39, 32.27 (both Cyprus).'

παραθρώσκω, for 'Oikonomos .. (ii BC)' read 'SEG 32.644 (Pydna, iii BC)'

παράθυμα, delete the entry.

παραιβαδόν, for 'prob. in Opp.C. 1.484' read 'cj. in Opp.C. 1.484 (cod. παραὶ βατόν)'

*παραιθου (gen.), unidentified ingredient in a medicament, POxy. 1088.15 (i AD).

παραιθύσσω II 1, delete 'λαίφεα'

παραινετήρ, add 'w. adjectival force, παραινετῆρας ἐξαντλῶ λ[όγους] Trag.adesp. in PKöln 242.2'

παραιτέομαι II 2 a, line 8, before 'delete 'Iamb.VP2.7'; add 'reject a candidate, Onos. 1.19' b, add 'abs., Modest.Dig. 26.5.21.2'

παραιτητέος 1, add 'id. VP 2.7'

*παραιτία, ἡ, responsibility, blame, PMed.inv. 68.38ʳ ii 11 (Aegyptus 69.19, i BC/i AD).

παραίτιος 1, line 3, transfer 'τῶν δ' .. A.fr. 44.7 R.' fr. section 2 to follow 'A.l.c.'; line 4, after 'αἴτιος' insert 'Plb. 4.57.10'

+παραιφάσιη, v. °παραφασία.

+παραίφασις, v. ‡παράφασις.

παρακαλέω, line 1, after 'Lxx Jb. 7.13, al.' insert 'w. apocope of preverb, παρκαλῖ SEG 32.456 (Haliartos, iii BC)'

παρακάλυμμα 1, add 'so prob. in Inscr.Délos 442A229 (ii BC)'

*παρακαπηλεύω, carry on unauthorised trade, ZPE 27.211 (Samos, iii BC).

*παρακατάγνυμι, break off in part, ῥυτὸν δίκρουνον παρακατεαγὸς τοῦ χείλ[ους] Inscr.Délos 1441Aii86 (ii BC).

παρακαταθήκη, line 1, after 'ἡ' insert '(παρκα(τ)θέκα Schwyzer 54.B1 (Tegea, but not Arc., v BC), Boeot. παρκαταθείκα IG 7.2420.34)'

παρακατάσχεσις, add 'PMasp. 295 iv 9'

παρακατατίθημι, after lemma insert 'w. apocope of preverbs, παρκαττίθεται SEG 30.1163 (Heraclea, Italy, iv/iii BC)'

παράκειμαι I, line 4, after 'Telecl. 1.7, etc.' insert 'of parallel anchors, Peripl.M.Rubr. 4' 2, delete 'cf. Plb. 5.34.7' II 2, add 'Ath. 9.409b' add 'III trans., to have put in or deposited a document (cf. ‡παρατίθημι B 2 a), ἐπὶ ῥᾳδιουργία παρακεῖσθαι αὐτὸν τὴν συγχώρησιν UPZ 162 vi 4, cf. 21, vii 3, 21, viii 2, 33 (ii BC); παρέκειτο τὴν δηλουμένην διαγραφήν SB 4512.67 (ii BC); εἴ τινα ἀπόδειξιν παράκειται UPZ 161.35 (ii BC).'

παρακελευστής, add 'app. title of an office, one who calls upon (a deity), Πολιάδος Ἀθάνας π. IG 12(2).484.17'

παρακίναιδος, delete 'f.l. in'

παρακινέω, line 2, after 'II 2)' insert 'Men.Dysc. 961; incite against the government, Plu.Pel. 6'

παρακίρναμαι, add 'act. παρεκίρνα· παρέμισγε Hsch.'

+παρακλαίω, weep beside, AP 5.103 (Rufin.); cf. παρακλαυσίθυρον; Sch.Ar.V. 977.

παρακλίνω I 3, line 4, after 'HA 540ᵃ1' insert 'Ant.Lib. 17.6' 4, delete the section.

παρακοιμάομαι 2, add 'b of sexual intercourse, Sch.Pi.P. 4.449.'

*παρακοιτάζω, lie beside, Hsch. s.v. παρευνάζων.

παρακοιτέω, abs., add 'SEG 24.154.13 (Attica, iii BC)'

παρακολλητικός, add 'Heras ap.Gal. 13.781.13'

παράκολλος I, for 'χαμεύνα, .. 10.36)' read 'χάμευνα, low couch with ornamental wood-work glued on, IG 1³.421.204 (Athens, v BC), cf. Poll. 10.36'

παρακολουθητέον, add 'Gal.Anim.Pass. 5.8 (CMG 5.4(1).1, 18.10)'

*παρακολουθητής, οῦ, ὁ, assistant, PFam.Teb. 15.103 (ii AD).

παρακολουθητικός, after 'for following or understanding' insert 'τὴν δὲ τριβὴν καὶ τὴν ἐξ αὐτῆς ἕξιν εἶναι πολλοῖς παρακολουθητικὴν Phld.Rh. 1.52 S.' 2, delete the section.

*παρακολουθήτρια, ἡ, supervisor, POxy. 3921.6, 49 (AD 219).

παρακολυμβάω, add 'part. Παρακολυμβῶσα, title of comedy by Nicostratus, Phot. α 1197 Th.'

παρακομίζω 2, after 'transport' insert 'παρακομίσεν ZPE 72.100.5 (Emporion, vi BC)'

+παρακοντίζω, throw a javelin beyond (others), Luc.Par. 61.

παρακοπτικός, add 'adv. -κῶς Gal. 17(2).454.12'

παρακόπτω II 1, add 'w. gen. χρησμῶν παρεκόπης i.e. you were diverted from the proper understanding of them, A.Ag. 1252 (cj.)'

παρακούω IV 1, add 'w. acc., μὴ παρακούσῃς τὸ οὔνομα τοῦ Θεοῦ SEG 31.1594'

παρακρέμαμαι, delete the entry.

παρακρεμάννυμι, add 'pass., Luc.Asin. 23; metaph., to be dependent, τὰ παρακρεμάμενα μέρη the dependencies of an empire, Plb. 5.35.10'

*παρακύϊσμα, ατος, τό, name of the sign ↑ (later ⸖), Sch.D.T. p. 496 H. (s.v.l.); see M (p. 1562 at end of π entries).

παρακυρόω, delete the entry.

+παράλαμψις (A), εως, ἡ, shining spot on the cornea, Hp.Prorrh. 2.20, Gal. 19.127.

*παράλαμψις (B), Dor. for ‡παράληψις.

παραλέγω III 1, delete 'Med., .. 1.101 S.'

*παραλημπτής, v. ‡παραληπτής.

παραλειπτέον, line 2, after 'Isoc.' insert '15.149'

παραληπτής, add 'steward, estate-manager, CPR 6.31.2 (cf. pp. 61 & 79) (iii/iv AD)'

παραληπτικός, add 'POxy.Hels. 41.39 (iii AD, sp. παραλημφθ-)'

παράληψις, line 1, after '-λημψις' insert 'Dor. -λαμψις, Clara Rhodos 2.175.9 (ii BC)' 2, add 'SEG 30.1274 (Caria, C.AD 200)'

παράλιος I, add '2 as title of Artemis, di Cesnola Salaminia p. 96.' III, after '= οἱ Πάραλοι' insert '(v. πάραλος II) Arist.Ath. 13.4'

παραλλαγή I 2, add 'ἐκ παραλλαγῆς γίνεσθαι τὰ σωλάρια first one side (of the road) and then the other, CodJust. 8.10.12.5b' II, add 'παρὰ τὴν προτέραν D.H.Comp. 15'

παραλληλισμός I, add 'POxy. 1916.10 (vi AD)'

παράλληλος, add '4 in music, of tetrachords, disjunct (opp. °κατάλληλος), Aristid.Quint. 1.8.'

παραλογεύομαι, add 'act., SEG 8.466.23, 41 (Egypt, i BC)'

*παραλόγια, τά, app. false arguments, PMacquarie inv. 358 (Atti Napoli III 827).

παραλογισμός I, line 2, before 'Arist.Po. 1455ᵃ13' for 'θεάτρου' read 'θατέρου' II, add 'ἐπὶ παραλογισμῷ with fraudulent intent, PColl.Youtie 12.7 (177 BC)'

παράλογος I 1, at end before 'sup.' insert 'comp. -ώτερον Plu.Tim. 1' 2, before 'adv.' insert 'comp., Plu. 2.1123a' II, add 'ἐκ παραλόγου abnormally, Pompon.Dig. 1.3.3'

πάραλος III 2, after 'cf. παραλίτης' insert '‡παράλιος III' IV, delete 'ἡ π.'

+παραλούμαι, wash beside, πάντας χρὴ παραλοῦσθαι καὶ τοὺς σπόγγους ἐᾶν Ar.fr. 59 K.-A., cf. 537; act., παραλούειν Phot., Suid.

παραλυπέω, add 'SEG 32.1149 (Ionia, AD 209)'

παράλυτος, for '= foreg.' read 'paralysed, D.H. 9.21'

παραλύω I 1 b, delete the section. I, line 10, after 'Plu.Cleom. 37' insert 'disband, ἅρματα Lxx 2Ki. 8.4'

παραμελέω, line 4, after 'neglect' delete 'a duty' and insert 'τῶν ἀγρῶν Arist.Ath. 16.5'

*παραμελορυθμοβάτας, ὁ, nonce-wd., one who ruins melody and rhythm as he goes, Pratin. 3.13 S.

παραμεμπτέον, delete the entry.

παραμένω II 2, add 'med., παραμενῶμαι ἐπεί (for ἐπί) τῆς Ἀλεξανδρείας POxy. 3396.19 (iv AD)'

*παραμίσθωμα, ατος, τό, supplementary hired hand, SEG 38.1462.23, 44 (Oenoanda, ii AD).

†παραμονάριος, ὁ, guardian of a rest-house, Aesop. 252 P.; also of a church, monastery, etc., Cod.Just. 1.3.45.3, SEG 30.1688 (sp. -μων-, Palestine, vi AD). **II** one acting as an assistant under the terms of a παραμονή (q.v., sense 1), POxy. 3960 (AD 621), SB 4490 (vii AD).

παραμονή 1, for 'obligation .. deferred' read 'obligation to remain with a person, esp. of a manumitted slave with his ex-master, for a stated period until the completion of a contract, or sim.'

παράμονος, after 'rarer form of foreg.' insert 'freq. as pers. n.' and delete '(q.v.)'

παραμυθία, add '**6** special allowance, οἷς ἡ τοιαύτη ἀφώρισται π. .. ἀποπληροῦσθαι τὴν αὐτῶν π. SEG 9.356.21, 24 (Cyrene, vi AD), BGU 1024 vii 12, PPrinceti. 96.6-7, etc., Cod.Just. 3.2.4.8. **7** interest on a mortgage, BGU 2150.12 (AD 472).'

παραμυθιακός, delete the entry (read °Φαρμουθιακός).

παραμύθιον 1, add 'ὕπνου AP 7.195 (Mel.)'

*παραναίομαι, dwell beside, ἡμῖν δὲ κακὸς παρενάσσατο γείτων Call.fr. 294.2 Pf. **2** trans., cause to dwell in, καί μιν .. σφετέρῃ παρενάσσατο χώρῃ D.P. 776.

παραναίω, delete the entry.

*παρανακαλέω, exhort, POxy. 1841.2 (vi AD).

*παράναυλον, τό, kind of transport charge, PRyl. 213.47 (ii AD).

παρανθέω, add '**III** fig., lose the bloom of beauty, cj. in X.Smp. 8.17 (based on reading of PGiss. 1 ii 4 παρανοήσῃ).'

παρανθιολογέω, for 'dub. sens. in' read 'over-harvest the flowers of'

παρανίσσομαι, after lemma insert 'or better παρανίσομαι'; add 'abs., come to aid, Arat. 426'

*παρανύμφη, woman who conducts a bride, Isid. 9.7.8; cf. mod. Gk. παρανύμφη (sponsor).

παράνυμφος I, after 'best man,' insert 'one of the dramatis personae in Ar.Ach.', and delete this reference from section II.

*παράνω, just above, prep. w. gen., PMasp. 169bis 49 (569 AD); cf. παρακάτω.

*παράνωθεν, adv. over the top, Anon.Alch. 324.11.

παραξιφίς, add 'pl., as a title of book of miscellanea, Gell.Praef. 7'

παράπαιγμα, add 'perh. inscription on a gaming counter, SEG 34.1534 (παρα..γμα, Alexandria, i AD)'

*παραπαλάριος, ὁ, mentioned among entertainers of various kinds, perh. to be connected w. Lat. palaria, the exercise of tilting against a stake, Teuc.Bab. p. 44 B.

παράπαν, for 'in correct writers .. Art.' read 'usu. joined w. art.' and after 'Th. 6.80, etc.' insert 'without art., Th. 6.18.7'

παραπατάω, after 'A.Eu. 728' insert '(codd., cj. παρηπάφησας (παραπαφίσκω))'

παραπαφίσκω, line 1, after 'παρήπαφον' insert '(but v. °παραπατάω)'

παραπείθω, line 2, for 'Il. 24.208' read 'Il. 14.208'

παραπεμπτέος II 1, add 'Gal. 14.305.15'

παραπέμπω I 3, after 'IG 12(7).53.19 (Amorgos)' insert 'escort a prisoner, PKöln 281.16 (vi AD)' **II 2**, after 'give up, omit' insert 'τὸν κατὰ μέρος λόγον Hipparch. 1.10.24'

παραπιπράσκω II, delete the section transferring quot. to section I, changing ref. to 'ISmyrna 723'.

παραπλαγιάζω, for 'Hsch.' read 'trans., divert, Hsch. s.v. παροχετεύει'

παραπλέκω II, line 2, for 'π. ἑαυτόν becurl himself' read 'w. pers. obj.'

παραπλεύριος, add 'π. ἄκανθαι, of dolphins, Sch.Pi.P. 4.29'

παραπλήξ II, after 'mad' insert 'B. 11.45 S.-M.'

παραπληξία I, delete 'IG 12(9).1179' **II**, after 'Lxx De. 28.28' insert '(quoted in IG 12(9).1179)'

παραπλομένοισι, add 'cf. περιπέλομαι)'

*παραπλωΐζω (-πλωζ- cod.), to be situated or move alongside the road, Hsch.; cf. πλόος 4 b.

*παραποδίδωμι, med., sell below market price, παρααποδόσθαι (sic) IHistriae 20 (ii BC); cf. παράπρασις.

παραποθνῄσκω, add 'Men.Dysc. 379 (s.v.l.)'

παραποίησις, after 'forgery' insert 'AS 10.71 no.124 (pl.) (Pisidia, ii AD)'

*παραπόκειμαι, to be laid up in store, PBeatty Panop. 1.208 (AD 298).

παραπομπή, add '**III** delegation of jurisdiction, εἴτε αὐτόθεν κατὰ τὴν φύσιν τῆς οἰκείας δικάσειεν ἀρχῆς εἴτε καὶ ἐκ παραπομπῆς ἡμετέρας Just.Nov. 20.7.'

παραπομπικά, add 'Cod.Just. 10.30.4.4'

παραπομπός I 1, add 'as subst., escort (= Lat. prosecutor), POxy. 3635.3 (v AD)' add '**III** app. functionary in religious processions, BCH 11.12 (Caria).'

παραπόντιος, add 'also a, ον, ἐπὶ τῆς ἀγχιθαλάσσου δὲ παραποντίας EA 14 p.22 l.34 (Ephesus, i AD)'

παραπόρφυρος, after 'edged with purple' insert 'Plu. 2.330a'

*παραπόταμος, ον, of land, riverside, POxy. 2847 i 9 (iii AD).

παράπτομαι II 1, add 'μηδὲν ὅλως παραψάμενος τῶν ἁγίων (sc. χρημάτων) BCH 56.293 (Stobi, ii/iii AD)'

παράπτωμα I 3, add 'SEG 31.1562 (iv/v AD)'

*παραράβδωσις, εως, ἡ, prob. fence or railing, AS 12.198 (Cilicia, i/ii AD).

*παράρραπτος, ον, sewn as a fringe, IG 7.2421.7 (Thebes, παρραπτ-).

παραρρέω II 2, for 'also .. etc.' read 'abs., drift from course, err'

παράρρυθμος I, add '(dancing) in irregular measure, of the Curetes, Orph.H. 31.3 [παράρυθμοι]' **II**, delete the section.

παράρρυμα, (sp. παράρ-), add 'παραρύματα λευ[κά] (?canvas), SEG 24.159(b).328; π. τρί[χινα] ib. 330 (326/5 BC)'

παράρτημα II, for 'dub. sens. in' read 'appendix, supplement'; add 'cf. mod. Gk. παράρτημα'

†παραρτίζομαι, prepare, ναῦς παραρτισάμενος Plu.Luc. 7.6; παραρτίζεσθαι· παρασκευάζεσθαι Hsch.

παραρτύω II, delete the section.

*παρασεβέω, to be impious, Sokolowski 2.33.A11 (Dyme, iii BC).

παράσειρος I, add '**2** παράσειρον, τό, app. kind of large sack, τὸ .. παράσιρον τὸ στυπέϊνον ἐν ὧι τὰς λ' ἀρτάβας τῶν ἀλφίτων κατήγαγον Pap.Lugd.Bat. xx 54.10 (iii BC).'

παρασεύω, for 'rush past' read 'speed past or alongside' and add 'Nonn.D. 37.387'

παρασημαίνομαι I, add 'also act., παρασημαίνει· παραχαράττει, παραδηλοῖ Hsch.'

παράσημον I, after 'or note' insert 'or diacritical sign' **II 1 a**, add 'distinguishing characteristic, τῆς Δημοσθένους συνθέσεως D.H.Dem. 50'

†παράσιρον, τό, v. °παράσειρος.

παράσιτος II 1, add 'OGI 195.2 (Egypt, i BC)'

*παρασκάφτης, ου, ὁ, boatman, Teuc.Bab. p. 47 B. (-σκαρ- cod.), v. Robert Hell. 1.143.

παρασκευάζω A 1, add '**b** intr., get ready, ἕκαστος ἐν τῷ καθ' ἑαυτὸν παρασκευάσει δικαστηρίῳ Just.Const.Δέδωκεν 24.' **2**, line 4, after 'D. 28.17' insert 'produce, cause, τοὺς ὄγκους καὶ τὰ καύματα Diocl.fr. 43' **3**, line 5, delete 'accustom'; line 6, delete 'accustom it not to ..' and after 'Eq. 2.3' insert 'w. acc. & inf., τὸν πόλεμον μέχρις ἀπειλῶν προκόψαι Memn. 15 J.; φρονεῖν ὑμᾶς D.Chr. 33.23' **5**, delete the section (transferred to section 2). **B I 2**, line 7, for 'Is. 8.3' read 'τοῦτον π. πράγμαθ' ἡμῖν παρέχειν Is. 8.3'

παρασκευαστικός 3, add 'τὰ π., title of treatise by Heraclid.Cum., Ath. 4.145a'

παρασκευή II, add 'in Chr. use, Friday, BCH suppl. 8 no. 270 (v/vi AD), Cod.Just. 1.4.22.1'

παρασκηνόω I, delete the section.

*παρασπαίρω, gasp beside, w. dat., θυηλαῖς Nic.fr. 62.2.

παρασπιστής, after 'companion in arms' insert 'A.fr. 303 R.'

παρασπορά 1, after 'sowing' insert 'PStrassb. 267.10 (AD 126/8)'

*παρασταθμία, ἡ, deficiency in weight, PAmst.inv. 39.21 (CE 1973.128, iv/v AD), POxy. 132 (CE l.c., vi/vii AD).

παραστάς 1, line 1, after 'ἡ' insert 'acc. sg. -στάδαν IHadr. 1 (ii AD)'; line 5, for 'also in sg.' read 'sg., part of a catapult'

παράστασις I 3, add 'delivery to a person, εἰς παράστασιν στρατηγοῦ (?for -ῷ) POxy. 2139.3, SB 10270.1, 5, 9, etc. (see BASP 11.46)' **5**, add '**b** perh. presentation of a candidate (for entry to a tribe, deme, etc.), POxy. 3463.19 (AD 58).' **II 2**, delete the section.

παραστάτις, after lemma insert '(παρστάτις, epigr. in Hesperia 23.63 (Athens))'

παραστεγάζω, delete the entry.

παράστημα, after 'Dor.' insert 'Boeot.' and after '(Epid., iii BC)' insert 'BCH 20.324 (Lebadea)'

*παραστόμιον, τό, muzzle, Hsch. s.v. φίμα (read φῖμά or φῖμός).

*παραστόμιος, ον, of or belonging to the side of the mouth, in quot. app. of an article of harness, IG 1³.422.239.

παραστρατεύομαι, add 'POxy. 2902 ii 9 (AD 272)'

*παρασυγγραφή, ἡ, breach of contract, PSI 903.22 (i AD).

*παρασυμβαίνω, occur additionally, Just.Nov. 7.2 pr.

*παρασυνάγω, hold heretical assemblies, Just.Nov. 132.

παρασύναξις, for 'clandestine religious assembly' read 'the holding of a heretical assembly' and for 'Cod.Just. 1.5.8.3, 5' read 'Cod.Just. 1.5.14, 1.5.20.2; in Lat. transliteration, ib. 1.5.8.3, 5'

παρασύνθημα, for pres. def. read 'countersign'

παρασύρω 2, before 'Plb. 16.4.14' insert 'Tim.Pers. 6'

*παρασφραγιστής, οῦ, ὁ, maker of counterfeit seals, Teuc.Bab. p. 42 B.

παρασχιστής, after 'D.S. 1.91' insert 'Ptol.Tetr. 179'

παρασωρεύω, add 'τροφὴν π. Aesop. 6 P.'

παρατατικός, adv. -ῶς, add 'in an extended sense, Julian Dig. 38.7.1, Ulp.Dig. 42.4.2.4'

παρατείνω I 2, add 'Men.Sam. 544'

*παρατέλειος, α, ον, prob. nearly full-sized, PHaw. 208.9, al. (ZPE 93.206, i AD).

*παρατενίζω, turn one's gaze aside (immodestly), π. ὀφθαλμοῖς (?)Aq.Is. 3.16 (L.-R.); cf. ἀτενίζω.

παρατίθημι **A 3**, add 'τὴν ἀσπίδα ἐπίθημα τῷ φρέατι π. Ar.fr. 306 K.-A.' **B 2 a**, lines 4/5, delete 'Plb. 3.17.10' and after 'deposit' insert 'or put into official hands'; before 'POxy.' insert 'UPZ 162 ix 7 (ii BC), Wilcken Chr. 26.35 (ii AD); pass. to be appended' **b**, after 'store up' insert 'χρήματα Plb. 3.17.10' **5**, last line, for 'v.l. in Id.Comp.' read 'cf. id.Is. 13'

†παράτιτλον, τό, marginal scholium or note, Cod.Just. 1.17.2.21, Just. Const.Δέδωκεν 21.

*παράτομος, ὁ, section, of a field, SB 9134.4, 5 (ii AD), also perh. PPrincet. 172.10 (ii AD, ZPE 70.141).

*παρατουρᾶς, ᾶ, ὁ, maker or seller of furnishing materials, ITyr 133.

παρατούριον, delete '(Lat. antepannus)'

παρατραγῳδέω, before 'Poll.' insert 'Stratt. 50 K.-A.; παρατραγῳδ- ῆσαί τί μοι ἐκ[Com. in Lex.Mess. fol. 282ᵛ4'

*παρατράπιος, η, ον, (ἀτραπός) situated by the wayside, AP 9.706.5 (Antip.).

*παρατρεπτικός, ή, όν, averting, Sch.E.Andr. 527.

παρατρέπω **7**, add 'Afric.Cest. p. 59 V.'

παρατρέχω **5**, delete 'abs., of time, Hdn. 2.12.4' add '**6** run or extend alongside, ἀλλ᾽ αὕτως λείη παραδέδρομεν Suppl.Hell. 944.8. **7** of a period of time, elapse, Hdn. 2.12.4, Cod.Just. 1.4.32.2, Just.Nov. 1.4 pr.'

παράτρητος, for 'pierced at the side .. airs' read 'of a tube, etc., having an orifice set to one side: αὐλὸς π. kind of pipe making a sound suitable for mournful music'

παρατριβή **2**, for 'Ath. 14.626e' read 'Plb. 4.21.5' and add '30.27.2'

παράτροπος **II**, delete 'where Sch. expl. παρατροπικός'

παρατροχάζω **2**, after 'APl. 4.169' insert 'SEG 34.342 (Achaea, ii/iii AD)'

†παραΰλος, ὁ, archit., collateral metal thole linking two blocks, Didyma 41.47 (ii BC).

παραυά **I**, for 'prob. in Theoc. 30.5' read 'Theoc. 30.4' add 'cf. Myc. pa-ra-wa-jo (dual), cheek-pieces of helmet'

†παραυλέω, cited by Pollux as compd. of αὐλέω without indication of sense, Poll. 4.67.

παραφαίνω **I 1**, line 3, after 'Ar.Ec. 94' insert 'γυμνὸς .. παρεφαίνετο μαζός Call.Dian. 214'

*παραφασία, ἡ, poet. παραιφασίη, consolation, comfort, Musae.fr. 22 D.-K., Epigr.Gr. 421.2, Nonn.D. 11.207, 48.133; (w. gen.) consolation for, Nonn.D. 11.365. **2** advice, persuasion, A.R. 2.324, 3.554, Euph. in Suppl.Hell. 415 i 10, poet. in Suppl.Hell. 956.3 (all pl.).

παράφασις (A) **1**, after 'Il. 11.793' insert 'Aret.SD 1.2, Them.Or. 8.106d, Nonn.D. 40.115' and after 'APl. 5.373.3' add 'πόνου AP 5.284; εὗρε σοφῶς λιμοῦ με παραίφασιν poet.ap.Orion s.v. πεσσοί, SEG 31.291 (Corinth, c.V AD)'

*παραφερνικαῖος, α, ον, = °παραφερνιμαῖος, προικιμαῖα καὶ παρα- φερνικαῖα πράγματα PNess. 33.14 (vi AD).

*παραφερνιμαῖος, α, ον, that is in addition to a dowry, extradotal, PNess. 18.25 (vi AD).

παραφέρω **III 1**, line 4, after 'Hp.Art. 12' insert 'π. τὰς κώπας Arr.An. 2.21.9' **IV**, line 3, after 'Plu. 2.432b' insert 'metaph., π. κατακέφαλα POxy. 1853.5 (vi/vii AD)'

*παραφησυχάζω, settle down in one's relations with, PCair.Isidor. 75.20 (iv AD; sp. παραπ-).

*παραφοβέω, drive aside in fright, Ion Trag. 43c S.

*παραφόρετρον, τό, charge for transport, POxy. 3169.178 (ii/iii AD).

*παράφορος **I 3**, add '-φόρως in a frenzied manner, Gal. 8.484.16, 9.188.11'

παραφρυγανισμός, for 'ib. p. 100' read 'PPetr. 2 p. 17 (iii BC)'

*παραφυλάκεια, ἡ, office of a παραφύλαξ, Robert Hell. 10.250 (Ac- monia, iii AD); also -ία, TAM 2(3).838d7 (Lycia, ii AD), IGRom. 3.649.7.

*παραφυλάκειον, τό, police- or garrison-building, TAM 3(1).14A14 (Ter- messus, ii AD).

παραφυλακή **II 1 b**, after 'garrison-duty' read 'form -ακά, SEG 26.1817 (Arsinoe Cyrenaica, ii/i BC), Notiz.Arch. 4.20.15 (Cyrene, Augustus)' add '**III** precautionary stipulation, εἴ τις δὲ παρὰ ταύτην τὴν παραφυλακὴν γυναῖκα ἐν τῷ ἰδίῳ οἴκῳ σχῇ Just.Nov. 123.29.'

*παραφυλάκια, v. °παραφυλακεία.

παραφυλάκίτης, for 'IGRom. 4.896 (Phrygia), CIG 4366x (Pisidia)' read 'Ramsay The Social Basis 106 (Cappadocia)'

παραφύλαξ, for 'watcher, guard' read 'name of an official, perh. chief of police, OGI 527 (Hierapolis); for '(Aphrodisias)' read '(Apollonia Salbace), MDAI(A) 68.23 (Samos, ii AD), IEphes. 612a, 1579, al., JHS 29.166 (ii AD)'

παραφυλάσσω **III**, after 'παραφύλαξ' insert '(q.v. with Suppl.)'; add 'OGI 485.7 (Magnesia ad Maeandrum), SEG 34.1107 (Ephesus)'

παραχειμάζω, add '**II** pass., app. to be blown off course by a storm, CPR 7.60.14 (vi/vii AD).'

παραχειμαστικός, add '**2** neut. pl., π. λεγιώνων winter-quarters or tax for maintenance in winter-quarters, MEFR 55.57 (Thyatira, ii AD).'

*παράχορδος, ον, discordant, Phot. s.v. παρακεχόρδικεν.

παραχρῆμα **1**, add 'εὐθὺς καὶ παραχρῆμα Cod.Just. 1.3.52.6'

†παραψιδάζω, perh. spatter, Hippon. 92 W.; cf. perh. ψίδες droplets.

*παρβάδαν, adv. by transgression, A.Eu. 553 (sch., codd. περβάδαν, περαιβάδαν; see παραβάτης II).

παρβαίνω, before '-βασία' insert '-βάλλω'; for 'poet. for παραβ-' read 'see παραβ-'

*παργίγνομαι, v. ‡παραγίγνομαι.

πάρδαλις **I**, add 'applied to prostitute, τὴν πόρδαλιν καλοῦσι τὴν κασαλβάδα Ar.fr. 494 K.-A.'

*παρδίδωμι, πάρδοσις, v. παραδίδωμι, ‡παράδοσις.

*παρεγγραφή, ἡ, perh. interpolation, SEG 33.1177.9, 29, 40 (Lycia, i AD).

παρέγγραφος, add 'Philoch. 119 J.'

παρεδρεία **I**, add 'οἱ ἀπὸ τῆς π., perh. = οἱ πάρεδροι PSI 1357.9 (ii AD)'

παρεδριάω, add 'παρεδρικό⟨ν⟩τι IG 10(2).1.447.10 (Thessalonica, ii/ iii AD)'

πάρεδρος **II 1**, penultimate line, after 'Hell.Oxy. 10.1' insert '(s.v.l.)'

παρείκω **I 2**, delete the section.

πάρειμι (εἰμί) **II**, add 'Ar.Eq. 330'

παρείρω, line 2, after 'Plb. 18.18.13' insert 'λόγον εἰς τοὺς Φιλιππικοὺς Did.in D. 13.17'

παρεισβαίνω, delete the entry (v. °παρεκβαίνω).

παρεισδύνω, line 2, after 'Demad. 3' insert 'Call.Epigr. 44.5 Pf.'; after 'A.D.Synt. 319.24' insert 'Hld. 1.12, 7.27'

*παρείσοδος, add 'of a stage entrance, Sch.S.Aj. 66'

παρεισπορεύομαι, for 'enter' read 'infiltrate'

παρέκ **A II 3**, after 'contrary to' insert 'Call.fr. 186.7 Pf.' **4**, after 'beside' insert 'τέρψιές εἰσι θεῆς πολλαί .. παρὲξ τὸ θεῖον χρῆμα Archil. 196a.13 W.' **B 4**, at end after 'Plb. 3.23.3' insert 'Lxx Ez 15.4'

*παρεκάτεροι, οἱ, those on each side, Vit.Aesop.(G) 25 (παρακ- cod.).

παρεκβαίνω, before 'step' insert 'Cret. παρεσβ- poet. in Inscr.Cret. 1.xxiii 3 (Phaestus, ii BC)'

παρεκδέχομαι, after 'misconstrue' insert 'Plb. 15.25.35'

*παρέκκειμαι, to be put aside, Porph.in Harm. p. 88 D.

παρεκνέομαι, after 'A.R. 2' insert '651' and after '941' add '1243'

*παρεκπέτομαι, fly out and past, παρεξέπτη Plu. 2.806e.

παρεκπίπτω, add '**II** fall to pieces, of papyri, prob. in PFam.Teb. 15.71 (i/ii AD).'

παρεκτείνω **I**, add '**b** stretch out beyond, w. gen., εἰ .. μὴ τοῦ ἀναγκαίου πουλὺ παρεξετάθης AP 9.643 (Agath.). **2** make coextensive with, τῇ ναυμαχίᾳ τὴν βύβλον π. D.H.Th. 12. **3** pass., to be strained, τὸ .. -τεταμένον intensity, Aristid.Quint. 2.10.'

παρεκτός **I**, add 'SEG 26.434'

*παρελαιν, ὁ, prob. non-Hellenic name of a cult-object, Inscr.Cret. 1.v 23.5 (Arcades, ii AD).

παρελαύνω **II 1 b**, add 'fig., Parm. 8.61'

παρέλκω **II**, abs., add 'EA 12.150 no. 3, l. 4 (Lydia, ii/iii AD)'; after 'Luc.Am. 54' insert 'παρελκύσαντος δὲ αὐτοῦ χρόνον SEG 35.1164.8 (Lydia, ii AD)'

παρεμβαίνω, for 'τεθρίππῳ .. etc.' read '**2** παρεμβεβηκὼς riding as a passenger in, τεθρίππῳ D.H. 2.34; ἐφ᾽ ἁρματείου δίφρου id. 5.47.'

παρεμβολή **III**, for this section read 'ramming at an oblique angle, Plb. 21.7.4 (cj. παραβ-)'

†παρεμβολικός, ή, όν, of or connected with a (military) camp, δεῖπνα Plu. 2.643d; of the possessions of a dead soldier, PWisc. 14 (AD 131).

*παρέμβροχος, ον, slightly tipsy, Vit.Aesop.(G) 68.

παρεμπίπτω **I 3**, add 'of a word inserted for euphony, D.H.Dem. 40'

παρέμπτωσις **2**, for 'D.H.Amm. 2.2' read 'D.H.Th. 24 (pl.)'

παρεμφερής, add 'Isid.Trag. 1'

παρεμφέρω **I**, add '**2** med., to be in association with, PBonn.inv. 2.16 (JJP 18.43).' **II 2**, for 'float in as well' read 'be suspended'

*παρεμφύομαι, add 'Alcin.Intr. p. 178.27 H.'

παρενδείκνυμαι **1**, add '**b** w. nom. and inf., assume arrogantly, Call.Dieg. vii.4 (fr. 194 Pf.).'

παρενδημέω, add 'τοῖς παρενδα[με]οντοις decr. in ZPE 101.128 (Dyme, ii BC)'

παρένθετος **II**, add 'SEG 32.1554 (Arabia, vi AD)'

παρενθήκη, add '**III** surety, guarantor (cf. Lat. intercessor), Just.Nov. 4.1.'

παρενσκάζω, app. hobble along, Aq.Is. 3.16 L.-R.

παρεντίθημι **I**, add '**3** add, μὴ ἐγγράφου γενομένου τοῦ δανείσματος μηδὲ ἐπερωτήσεως παρεντεθείσης Just.Nov. 136.4.'

παρεξαμείβω, add '**II** (tm.) παρὲκ γόνυ γουνὸς ἀμείβων perh. getting his knee past the other's knee, A.R. 2.94.'

παρεξετάζω, line 2, after 'τί τινι' insert 'D.H.Dem. 36'

*παρεξοχή, ή, wd. of uncertain meaning, Poliorc. 220.20, etc.

*παρεπικόπτω, satirize by the way, w. acc., Call.Dieg. vi.30 (fr. 192 Pf., παρεκοπτων pap.), 37.

*Παρεπιλυκάρχης, ου, ό, leader of the Παρεπίλυκοι, I.Nicolaou viiie Congrès international d'épigraphie: Communications p. 111.

*Παρεπίλυκοι, οἱ, members of an otherwise unknown society, ἔδοξεν τοῖς ἐπὶ Λυκίης καὶ Παρεπιλύκ[οις] I.Nicolaou, op. cit. s.v. °Παρεπιλυκάρχης.

†παρεπιφέρω, carry towards one's destination, Peripl.M.Rubr. 57 (cod.).

*πάρεργιον, τό, perh. waste piece, offcut or sim., rest. in PKöln 52.34, 83 (AD 263).

πάρεργος I, add '2 of persons, unimportant, Ἑλλήνων οὐχ ὁ παρεργότατος GVI 1876.4 (Termessus, ?ii AD).' II, neut. subst., add 'μᾶλλον τὸ πάρεργον ἐπεκράτησ' ἢ τοὔνομα (the subst. μανία prevailed over the name Μανία) Macho 210 G.'

παρέρχομαι III 1, add 'abs., X.Smp. 1.7'

*παρερῶ, (fut.) = παράφημι 2, Sch.Pi.O. 7.111, 117.

παρέστιος, add 'w. dat., at home in, Opp.H. 1.249'

*παρεσχαρίτης, ου, ό, one who sits by the hearth, Eust. 1564.28.

παρετικόν I, add 'Arist.EN 1172^a26'

*παρετικός, ή, όν, palsied, Asclep.Jun.ap.Gal. 13.1022.2.

παρετοιμασία, add 'apparatus or equipment, Sch.B Il. 21.490'

*παρετυμολογία, ή, allusion to etymology, An.Ox. 3.383.21.

παρευτακτέω, for 'BCH 55.439' read 'Inscr.Délos 2598'

†παρεύτακτος, ό, a former ephebe (cf. εὔτακτος), Inscr.Délos 2593.52, al., 2598.6 (ii BC), Lucil. 321, 752 M., IG 2².2998, 2999 (i BC), 2094 (ii AD); fem. πάρευτακτοι in Varro gram. 89.

†παρεφηβεία, ή, status of a °παρέφηβος, GVI 1154 (Samos, ii/i BC).

*παρέφηβος, ό, a young man of the age succeeding that of ἔφηβος, IG 12(3).339.23, 340.19 (Thera, i AD).

παρέχω, line 4, after 'Hes.Th. 639' insert 'opt. παρασχέθοι A.fr. 78a.13 R.' and after 'Ar.Eq. 321' insert '(troch.).' A II 3, and for 'δέμασδ κέντητον' read 'δέμας ἀκέντητον' III 2, line 4, after 'El. 1080' insert 'without dat., prob. in Favorin.Exil. 23.11' B, line 2, after 'Lys. 9.8' insert '1 aor. inf. παρέξασθαι IGBulg. 43.19 (Odessus, i BC)'

*παρηγόριος, ό, perh. consoler, PMich.xv 740.21 (vi AD), PMasp. 58.

παρηγορος, line 3, for 'Epigr.Gr. 344' read 'IHadr. 80 (ii AD)'

*παρηΐδιος, α, ον, of the cheeks, μᾶλα AP 9.556 (Zon.).

παρήκω I, add 'lie beside in the grave, TAM 4(1).267' III 3, add 'b ἐκ τοῦ παρήκοντος as it happens, fortuitously, Just.Nov. 1.1.4, 90.2.'

†παρῆλιξ, ικος, ό, ή, not of a suitable age: over age, too old, Plu.Alex. 32; comp. -έστερος Sor. 1.15. 2 under age, PVindob.Salomons 5.12, 25, 29 (ii AD), POxy. 1257.2 (iii AD); transf., of actions, AP 12.228 (Strat.).

πάρημαι 1, line 2, after 'only part.' insert 'in Homer' 2, add 'impf. παρείατο used as sg., κούρη δὲ π. δακρυχέουσα Call.fr. 497 Pf.'

*παρημερινός, ή, όν, = παρήμερος II, PTeb. 275.22 (iii AD).

παρήχημα, for pres. def. read 'instance of παρήχησις'

παρήχησις, for '= foreg.' read 'the use of words alike in sound but different in meaning'

παρηχητικός, for 'alliterative' read 'of or belonging to παρήχησις'; for 'Eust. 1638.17' read 'with παρήχησις, Eust. 1638.15'

παρθενεία, last line, of a man, add 'Just.Nov. 6.1.3'

παρθένειος, line 2, after 'Pi.N. 8.2' insert 'παρθενήϊα φρονεῖν id.fr. 94b.34 S.-M.'

παρθένευμα 2, for this section read '= παρθενεία, E.Ion 1473'

παρθενεύω 2, add 'Cod.Just. 1.3.52.14'

παρθένια, add 'II = παρθένεια I, τέ[κτονι πα]ρθενίων σοφῶν Ἀλκμᾶ[νι poet. in POxy. 2389 fr. 9.'

παρθενική, add '2 a plant, app. = παρθένιον 1, feverfew, Cat. 61.187.'

παρθενικός II, neut. subst., add '2 = π. χιτών, Dura⁶ 100 (pl.).'

παρθένιος I 2, add 'without ἀνήρ, bridegroom, AP 7.384.7 (Marc.Arg.)'

παρθένος I, add '6 metaph., of the number seven, Hierocl.in CA p. 465 M.' add 'see also φαρθένος'

παρθενών II, lines 2/3, for 'also, of the cella' read 'b the apartment occupied by virgin priestesses' and add 'also in other temples'

Παρθικός, add 'as title of Roman emperors, SEG 32.461, 33.1130, etc.'

πάρθυμα, delete the entry.

*παρθύμαται, nom. pl. (masc. or fem.), perh. additional sacrificial victims, Inscr.Cret. 4.65.9 (Gortyn, vi BC).

*παριαμβίζω, v. ἰαμβαυλεῖν.

*παρίαμβίς, ίδος, ή, prob. kind of music or rhythm played on the lyre, later interpreted as an instrument, Epich. 109, Apollod.ap.Hsch.

παρίαμβος III, for 'harp' read 'κιθάρα'

παρίημι, line 3, after 'HP 5.3.6' insert 'med. aor. 2 παρείμην S.OC 1666'; line 4, delete 'aor. 2 .. 1666' II 3, add 'εὖτ' ἂν τὸ νέον παρῇ let go by, S.OC 1229 (lyr.), cf. Pl.R. 460e'

Πάρινα, delete the entry.

*πάρινος, η, ον, of marble, Lxx Es. 1.6, ABSA 56.5 (Paphos, ii BC), Syria 17.260.5 (iii AD), 18.372 (both Palmyra).

Πάριος (s.v. Πάρος), after 'D.S. 2.52' add 'cf. °πάρινος. b as culttitle of Demeter, SEG 33.684 (Paros).'

παρίστημι B, after 'aor. 2' insert '(2 sg. imper. παράστα Men.fr. 110 K.-Th., Th. 28, perh. w. transference to thematic inflexion, cf. ἀπόστα)'; after 'plpf. act.' insert '(also fut. pf. παρεστήξεις Men.Dysc. 364)' add 'VIII γῆ παρεσταμένη rented or mortgaged farmland, Hsch.; cf. ἵστημι B II 2 at end.'

παριστορέω I, for 'inquire by the way' read 'learn by the way' II, add 'Did.ap.Porph. in Harm. p. 28.25 D., Sch.Theoc. 1.117'

*παρίωνικός, ή, όν, quasi-Ionic, μέτρον, of the Anacreontic verse, POxy. 220ʳ vii 7.

*παρμενίσκος, ό, ornament of a door, Inscr.Délos 1428ii67, 1429Bii13 (ii BC); also a common pers.n.

*Παρνασιάς, άδος, fem. adj., of Parnassus, E.Ion 86 (cod. Παρνησ-).

*Παρνάσιος or Παρνάσσιος, α, ον, (also ος, ον, E.IT 1244 (lyr.)), Parnassian, Pi.P. 10.8, Limen. 22, etc: also Παρνήσσιος, IG 2².1258.24 (iv BC); as cult-title of Zeus, [Δι]ὸς Παρ[ν]ησσίο SEG 34.39 (Athens, 500/480 BC), [Διὸς Πα]ρνεσίο SEG 33.244 (Attica).

*Παρνασίς (Παρνασσ-), ίδος, fem. adj., of Parnassus, Pae.Delph. 4; also Παρνησσίς A.Ch. 563.

†Παρνᾱσός or Παρνασσός, Ion. Παρνησός or Παρνησσός, ό, Parnassus, Od. 19.432, Hes.Th. 499, Th. 3.95.1, etc.

*πάρο, Aeol., = πάρεστι, Alc. 130.12 L.-P.; cf. ἐνό, ἔνο, ἐξό.

*παρό (B), Myc. pa-ro, v. ‡παρά.

παροδεύω, add '4 enter the arena, inscr. in ZPE 1.105-6 (Side, iii AD).'

παροδοιπόρος, add 'GVI 428.1 (Ephesus, ii BC)'

πάροδος (B) II b, for 'on the stage' read 'into the orchestra, IG 12(9).207.55 (prob.) (Eretria, c.290 BC)'

παροικίζω, after 'τινά τινι' insert 'AP 7.287 (Antip.), 448 (Leon. Tarent.)'

*παροίνησις, εως, ή, = παροινία, Anon. in Rh. 327.13 (pl.).

παροινία, add 'transf., rowdiness, PMich.inv. 6979 (ZPE 76.251, 215 BC)'

παροίνιος II, after 'Ph. 1.353' insert 'αὐλοί Poll. 4.80' and after 'Sch.Ar.V. 1217, 1231' insert 'τὸ π. βοήσω Anacreont. 2.8 W.'

παροίστρα I 1, line 2, after 'in front of' insert 'Arat. 306'

παροίχομαι I 2, add 'ὑπὲρ τοῦ παρωχηθέντος ὀγδόου ἔτους PMich.xi 617.17 (ii AD)'

παρολκή III, delete the section.

*πάρολκον, τό, ?towrope, OBodl. 72 (ii AD), PLond. 1164(h).10 (iii AD), POxy. 997 (iv AD), Sch.Th. 4.25.

πάρολκος I, delete the section.

*παρομφάλιος, ον, along or near the middle, v.l. for κατομφ-, Nic.Th. 290.

παρονομασία I 1, line 2, after 'Cic.de Orat. 2.63.256' insert 'D.H.Th. 48' II, for 'derivative' read 'pronoun'; delete 'by-name .. (pl.)'

παρονοματοποιέω, delete the entry (read ὀνοματοποιέω).

παροξυντικός I 2, after 'Hp.Prorrh. 1.50' insert 'Gal. 11.393.11, 14'

πάροξυς, add 'III somewhat sharp in taste, in quot. gloss on ὀμφακίας Hsch.'

*παρόρεγμα, ατος, τό, allowance, honorarium, δαμιεργοῖς SEG 9.11.19, al. (Cyrene, iv BC).

παρόρειος, for 'the form παρώρειος .. Subst.' read 'see also °παρώρειος'

παρορίζω I, delete the section.

†παρόριος (A), v. παρώρειος.

†παρόριος (B), ον, situated along a boundary, Plu. 2.366b; w. dat., τῇ Αἰθιοπίᾳ OGI 168.57 (Syene, ii BC); τὰ παρόρια space along boundaries, POxy. 1475.22 (iii AD).

παρορισμός, for 'removal of landmarks' read 'infringement of boundaries, encroachment' and add 'PGen. 99 (ii AD)'

παρορκέω, add 'SEG 26.1386 (Phrygia)'

*παρορκία, ή, perjury, MDAI(A) 29.331 (GRBS 17.265, Phrygia).

*παρορμητήριον, τό, gloss on ῥωστήριον, Phot.

*παρορμήτης, ου, ό, encourager, Hsch. s.v. τάρροθοι.

παρορμίζω, for 'Μουνυχίασιν' read '[Μουνυχίασιν]'; add 'pap. in SHAW (1923)2.23'

*παρορυγή, ή, burying, digging in alongside, σκολόπων Rh. 1.436.18 (pl.); cf. ὀρύσσω IV.

πάρος, line 1, after 'poet. Particle' insert '(also in prose, SEG 37.340.14 (Mantinea, iv BC))'

παρουλίς, line 1, for 'οὐλή' read 'οὖλον'

παρουσία I 3, delete the section adding 'S.El. 1251' to section I 1.

παροχέομαι, for 'sit beside in a chariot' read 'ride beside another in a vehicle' and add 'spec. of the groomsman at a wedding, Men.Sic. 404 S.; cf. ‡πάροχος (A)'

παροχετευτέον, add 'Gal. 10.861.9'

*παροχία, ή, = παροχή, *provision*, *SB* 9907.23 (AD 388).

πάροχος (A), line 1, for '*one who sits beside* another *in a chariot*' read '*one who rides beside* another *in a vehicle*'; line 4, for 'συμπαρέστη' read 'συμπάρεστι'

παρρησία 3, add '*freedom and fearlessness* of aspect, Sch.Pi.*N.* 10.73'

*πάρριζος, v. ‡πάνριζος.

*παρστάτις, v. ‡παραστάτις.

*παρτέλλεται· παραινεῖται (read παρανεῖται) Hsch.

παρυπάρχω, before '*attend*' insert '*partly begin*, pf. part. pass. παρυπηργμένον prob. in *IG* 2².1522.20.   **2**'   at end add 'cf. Plu. *Lib.* 5'

*παρυπερέχω, *project partly beyond*, Hero *Bel.* 88.

*παρυπεύκυκλος, ον, *somewhat round*, Poet.*de herb.* 134 (codd.).

*παρυποκρούω, perh. *offend by meddling*, Call.*Dieg.* vii.3 (*fr.* 194 Pf.).

*παρύπτιος, ον, "*concave*", used of a geometrical plane figure *with a re-entrant angle*, Papp. 652.20; cf. °ὕπτιος VII.

παρφ-, for compds. of παρά v. παραφ-.

πάρφαινε, delete the entry.

*παρφυροῦς, v. °πορφύρεος.

*παρῴδησις, εως, ή, defined as ὅταν ὁ ῥήτωρ κῶλον ἀρχαῖον τίθησι καὶ χωρίον ἑαυτοῦ i.e. when the rhetor quotes a passage of classic literature and adds his own continuation, Sch.Aristid. p. 462 D.

παρῳδός **II 2**, add '*IG* 11(2).120.48 (Delos, iii BC)'

παρωθέω **I 1**, line 2, after 'Hp.*Art.* 18' insert 'so prob. in *h.Merc.* 305 (tm.)'

παρωνύμως **II 1**, add 'in a pun, παραμύθιον ἦσθα παρωνύμιόν τε γονεῦσι *CEG* 564 (Athens, iv BC)'   **2**, for '= Lat. .. *agnomen*' read '*by-name*, Arist.*Ath.* 17.3'

*παρώρειος, ον, = παρόρειος, *PTeb.* 787.4 (ii BC), Str. 12.8.13.

παρωροφίς, after 'Poll. 1.81' add '(defined as τὸ μεταξὺ τοῦ ὀρόφου καὶ τοῦ στέγους, i.e. *cornice* or *gable*)'

πᾶς, lines 1/2, for 'Sapph. .. Alc.*Supp.* 12.6, 25.8' read 'Sapph. 31.14, 44.14, Alc. 34.6, al. L.-P., *SEG* 32.1243 (Cyme, 2 BC/AD 2)'; line 6, after 'πᾶσι' insert 'πάνσι *BCH* 109.189-194 (Crete, ii/i BC); line 10, after '*Je.* 13.11' insert 'also πᾶν τὸν χρόνον *IG* 9²(1).583.56, cf. *IG* 9(1).39.3'   **B II**, line 6, after 'v. infr. δ.' insert '**III**'   **D II 4**, add '*all the time*, ἡ γῆ νιφετῷ τὰ π. χρᾶται Hdt. 4.50.2, cf. Luc.*Asin.* 22, Ach.Tat. 5.13, 7.16'

†πᾶς (B), Cypr. pa-se, pa-sa, = παῖς, *ICS* 80.2, 92.2, 157.2.

πασίγνωστος, add 'Anon.Alch. 344.13'

πασιθέα **I**, add '= ἀρτεμισία I 2, Poet.*de herb.* 27; = παιωνία ἄρρην, ib. 144, prob. in Ps.-Dsc. 3.140 (-θέη)'

Πασικράτεια, after '(Selinus' insert 'v BC'

†Πᾱσικράτη, ή, Dor. -α, *universal queen*, θεᾶς Πασικράτας *SEG* 32.636 (Maced., AD 286); identified w. Artemis, *AE* 1910.307 (Ambracia).

*πᾱσιμέλητος, ον, *cared for by all*, *SEG* 19.794 (Pisidia).

*πασίολος, v. ‡φασίολος.

Πάσιος, add '**II** name of a month, *Hesperia* 27.75 (unkn. provenance, c.200 BC).'

πᾱσιφίλητος, add '*IG* 5(1).1494 (Messene, iii/iv AD)'

πᾱσίφιλος, for '= foreg.' read '*loving* or *friendly to all*'

*Πᾱσιχάρεα, ἁ, "*gratifying-all*", name coined by Alcm. 107 P. to suit a promiscuous woman.

*πασπερμεί, v. °πανσπερμεί.

†πασσᾱγία, v. ‡πανσαγία.

*πάσσακον· πάσσαλον Hsch. (perh. πασσάκων· πασσάλων).

πασσάλιον, add 'Poll. 7.114'

πασσαλιστής, for 'κυνδαλοπαίστης' read 'κυνδάλη'

πάσσαλος, line 1, after '(v. infr.' insert 'cf. πασσαλόφιν· τοῦ πασσάλου. ὁ δὲ σχηματισμὸς Βοιώτιος Hsch.'

πάσσον, delete the entry.

*πάσσος, ον, Lat. *passus*, πάσσος οἶνος *raisin wine* (*vinum passum*), Eust. 1178.17; also neut. (?masc.) subst., Plb. 6.11a.4, Lang *Ath.Agora* XXI Hd9 (ii AD), 12 (ii/iii AD).

*πασσοφία (-ίη), v. °πανσοφία.

*πάσσοφος, v. °πάνσοφος.

πασταί, add 'Ar.*fr.* 702 K.-A.'

παστάς **I 2**, add 'of a gymnasium, *SEG* 23.233 (Arcadia, ii BC)'   **3**, delete '*AP* 6.172'   **II**, after '*bridal chamber*' insert 'but perh. to be understood as *porch* (I 1) v. J. Roux in *REG* 74.43ff.'

πάστας, add '*BCH* 107.401 (Knossos, i AD)'

παστή, before '*case, container*' insert 'perh.' and add 'παστὰς χαλκᾶς *PFreib.* 52.4'

*παστίλη, v. σπατίλη II.

παστιλλᾶς, after '= foreg.' insert '*POxy.* 3390.3 (AD 358)'

†παστός, ὁ, *bridal canopy* or *curtain* (freq. used loosely as a symbol of marriage), LXX *Ps.* 18(19).5, *SEG* 1.567.5 (Karanis, iii BC), *AP* 5.51, 7.182, *GVI* 719, 804, etc., Luc.*DMort.* 23.3, Nonn.*D.* 5.214, Poll. 3.37.   **2** *bed-canopy* (in non-marriage context), D.Chr. 62.6; also in ritual use, Herod. 4.56, *ISmyrna* 753.

παστοφόριον, add '*PBonn*.inv. 208.3 (*JJP* 18.52; sp. παστοφοροιον, ii AD)'

παστοφόρισσα, add '*OBodl.* 1821'

παστοφόρος, for '*priests appointed for this purpose*' read '*a class of priests of unknown function*'

*παστῷος, α, ον, perh. *of the* παστός, Hdn.*fr.* p. 18 H.

*πάστωρ, ορος, ὁ, Lat. *pastor, herdsman*, *SB* 801 (ii/iii AD).

πάσχω, line 3, after '*Il.* 9.492, etc.' insert 'Lac. πάσον Alcm. 1.35 P.'   **III 1 a**, line 9, after 'with subst.' insert '*POxy.* 1121.7 (iii AD), *PGen.* 58.16 (iv AD), *Cod.Just.* 6.4.4.23'   **2 a**, add 'but also πάσχειν τι *to be affected, disturbed*, Men.*Kith.* 49, D.H. 9.3.5, Plu. 2.682b'   **III 4**, lines 3 ff. (τί παθών), add 'Ar.*Nu.* 340'

†πάσωλος, v. ‡φασίολος.

*πατάγγης, ου, ὁ, = σπατάγγης, a kind of *sea-urchin*, Poll. 6.47.

πᾰτᾰγέω **I**, add 'Pl.*Euthd.* 293d'   **II**, last line, for '*Lyr.adesp.* 121' read '(?)Call.*fr.* 761 Pf.'

*πᾰταγμός, ὁ, *striking, smack*, Rh. 3.520.30.

Παταίκεια, delete 'sg. Παταίκειον' to end.

*Παταίκειος, *belonging to the fund of the* Παταίκεια, φιάλη *Inscr.Délos* 438.2, 442*B*54 (ii BC).

*πατάκτρια, ή, *striker*, τῶν ζῴων, of a καλαῦροψ, Rh. 3.607.8.

πατάσσω **II**, line 7, after '*sting*' insert 'of a scorpion, Arist.*HA* 607ª17'

*πάταχρον, τό, *idol*, LXX *Is.* 8.21 (pl.); also masc. sg. τὸν π. ib. 37.38 (prob.). (Cf. Aram. *patakrā*).

πατελλίδιον, delete the entry.

*πᾰτελλίκιον, τό, *dish*, *CPR* 8.66.10 (v AD), *POxy.* 1901.34, 68, 2419.9 (vi AD), *Gloss.*

πατερεύω, add 'of women, app. as indication of rank, *CPR* 10.127.6 (vi AD)'

πᾰτερία, add '*POxy.* 2780.8 (vi AD)'

†πᾰτέριον, τό, dim. of πατήρ, in quots. only as voc., used in addressing an old man, Luc.*Nec.* 21; also πατερίων, *Vit.Aesop.*(G) 56 P. (*POxy.* 2083ʳ.7, iv/v AD), al.

πάτημα **I**, add '**2** perh. *trodden grapes*, used as fodder, *POxy.* 1142.3, 1156.9 (both iii AD) or *pounded spice*, *POxy.* 2570(b).5, 3733.19, 3766.102.'

πατήρ, line 2, at end insert 'dat. πατέρι *SEG* 30.1502 (Phrygia)'   **II**, add 'alone, of Zeus, *BCH* 109.189 (Cnossos, ii/i BC)'   **V**, after '*IG* 14.1272' insert '*MDAI*(*R*) 49.203 (Dura)'   **VI**, after 'similarly' insert 'πατέρα δήμου Ῥώμης, of Vespasian, *Salamine* 138'; for 'π. τῆς πόλεως .. Methone)' substitute '**2** π. τῆς πόλεως = *curator (pater) ciuitatis*, *IG* 5(1).1417.11 (Methone), *SEG* 29.1070, *PMich*.inv. 3999 (*ZPE* 75.268), *Cod.Just.* 1.5.12.7, etc. (all vi AD).'   add 'Myc. pa-te'

πατητής, before 'Hsch.' insert '*SB* 4640 (v/vi AD)'

*πᾰτίλη, v. σπατίλη.

πάτος (A) **I 1**, add '**b** *action of going, course*, Nic.*Th.* 479.'   **2**, add '*ISalamis* 45 (v AD), *Cod.Just.* 8.10.12.3a'   **3**, add 'Plu. 2.670b'

†πάτος (B), εος, τό, *robe woven for Hera*, Call.*fr.* 66.3 Pf.

πάτρα **III**, add '*SEG* 33.1016 (Lydia, AD 103/4), epigr. in 26.456.11 (Laconia, ii/iii AD)'; form πατρεία, add '*EA* 15.80 no. 31.9 (Lydia, i AD)'

πατρᾰδελφεία, for '*cousin by the father's side*' read 'collect. noun for *paternal cousins*' and transfer entry before πατραδελφεός.

πατράδελφος, before '*Is.* 4.23' insert 'ἀνεψιὸς ὢν αὐτοῖς ἐκ πατραδέλφων'

πατριά **I**, add 'πατριᾶφι (instr. pl.) *with their lineages*, *SEG* 37.340.17 (Mantinea, iv BC); cf. °πατροφιστί, ἐπιπατρόφιον'

πατριάρχης **II**, add '*SEG* 33.1298 (Palestine, iv AD)'

*πατριᾶφι, v. °πατριά.

*πατρικιᾶτος, ον, Lat. *patriciatus, patrician*, κάλτιοι π. rest. in *Edict. Diocl.* 9.7.

πατρικός, after 'ή, όν' insert 'Aeol. πάτριχος acc. to Sch.D.T. p. 532.30 H.'   **I**, add '**2** as a cult-epith. (cf. °πάτριος), Ἀρτέμι πατρικ[ῇ] *SEG* 29.1159 (Lydia).'

*πατριμώνιον, τό, Lat. *patrimonium*, *IEphes.* 3056.4 (ii AD), *PFlor.* 320.4 (written πατριμουνιον, iv AD)

πάτριος **II**, add '**2** as a cult-epith., of Μήν, *SEG* 31.1132, 1210, 1232, etc.; fem., of Hestia, *SEG* 31.728 (Delos, ii BC).'

πατρίς, line 1, after 'ίδος' insert 'Aeol. acc. πάτριν *IG* 12(2).242 (Mytilene), *SEG* 33.1037 (Cyme, ii BC)'   **II**, line 4, after 'Ar.*Pl.* 1151' insert 'for πατὴρ πατρίδος v. πατήρ VI; personified, *IStraton.* 1026 (Flavian)'; line 5, after '*UPZ* 9.5 (ii BC)' insert 'D.S. 15.11.2, etc.'

*πατρόβουλος, ὁ, *son of a member of a* βουλή, *designated by his father to succeed him*, *IEphes.* 972, 1044, *IG* 12(5).141.7 (Paros), Jul.*Ep.* 54, *PLond.* 971.7 (vi AD).

*πατρογενίδης, ου, *descended on the father's side*, τῶν ἀπὸ Ἄρδυος Ἡρακλειδῶν π. Robert *Hell.* 10.276 (Claros, ii AD).

†πατρογέρων, οντος, ὁ, *son of a member of a* γερουσία, *designated to succeed him*, *IEphes.* 26, 972, 1573.

*πατροθεῖος, ον, *hereditary*, πατροθίου καθαρουργ(οῦ) *PAlex.* (1964)32.4 (v AD).

*πατροκτασία, ή, = πατροκτονία, *PMasp.* 353.A.11 (vi AD).

πατροκτόνος **I**, add '**2** subst., *parricide*, Plu. 2.1065, *Rom.* 22.4.'

πατρομήτωρ **1**, after 'Luc.*Alex.* 58' add '(v.l., see °προμήτωρ)'

*πατρομύστης, ου, ό, *son of a* μύστης, *designated by his father to succeed him*, *ISmyrna* 731.17-8, 732.

†πατροποίητος, ό, *adoptive father*, Mitchell *N.Galatia* 358 (sp. -φοιητ-), 387 (sp. -ποητ-), *AJA* 36.460 (border of Lycaonia and Galatia), Heuzey-Daumet *Mission Arch.de Macédoine* no. 135 (Ressova).

πατροτυψία, add 'Sopat.Rh. 8.199 W.'

*πατροφιστί, adv. *with the father's name*, *SEG* 23.178.6 (Nemea, 229 BC); cf. πατριστί, πατριαστί, πατριάφι (s.v. °πατριά).

πατρωΐωχος, add 'Ἠπειρωτικὰ Χρονικά 1.255 (Dodona, v/iv BC)'

πάτρων **I**, add 'π. τῆς πόλεως *CIJud.* 619d (iii/iv AD); sim. π. τῆς μητροπόλεως *ITomis* 2.101 (iii AD)'

πατρωνεύω, line 1, after '(Delph., i BC)' insert '*IGBulg.* 314 (Mesembria)'; line 2, after 'D.S. 40.5' insert '*INikaia* 1201'

πατρωνικός, add 'also -ονικός Theophil.Antec. 3.7.2'

πατρώνισσα, add 'also -όνισσα Theophil.Antec. 3.7.3'

πατρώνυμος, add '*POxy.* 3273 (i AD)'

πατρῷος **I**, add 'also as epith. of Hermes, *SEG* 32.751 (Berezan, v BC), 30.93 (Eleusis, i BC); of Athene, *SEG* 37.295 (Epid.)'

πάτρως, line 2, for 'Stesich. 17' read 'Stesich. 51 P. (in signf. ὁ κατὰ πατέρα πρόγονος, Eust. l.c. infr.)'; line 4, after '*BCH* 11.471 (Lydia)' insert 'πάτρως *SEG* 33.1016 (Lydia, AD 103/4)'

πάτωρ, delete the entry.

*παύγλα, ή, app. = Lat. *pavicula, rammer*, παύγλα ἤτοι γλεύδια *Edict.Diocl.* 15.43.

*Παύλεια, τά, games at Alexandria Troas, Σμίνθεια Π. Buckler *Anat. Studies* 245-8.

παυνί, add 'also app. παῦνι (in broken context), Hippon. 79.16 W.'

*παῦνι, also παῦν, παοΐνι, παῦνει, month name in N.Africa, *AP* 9.383, *SEG* 33.1401, 1459, 1486 (all Cyrenaica), *POxy.* 267.39, *PHib.* 46.21, *SB* 1167, 3776, etc.

παυράκις, add 'also παυράκι, ἃ παυράκι γίνεται ἀνδρί Thgn. 859'

παυροεπής, for '*words*' read '*verses*'

παῦρος **1**, line 2, after 'Op. 538' insert 'ἰχθύες A.R. 1.573'; line 4, after 'Q.S. 7.613' insert 'sup., παύριστον τό κεν .. τις ἴδοιτο a thing of which one can see *extremely little*, Call.*fr.* 384.55 Pf.'

παυσανίας, delete '(ubi .. A.)'

*παυσικραίπαλος, ὁ or ή, *"stop-the-hangover"*, inscr. on a drinking-cup, dub. in *Hesperia* 16.240 (Corinth, ii BC).

παῦσις, add 'π. πυρετῶν Hp.*Epid.* 7.49'

*παυσιτοκεία, τά, *an offering marking the end of childbirth*, *BE* 1973.247.

Πάφιος (s.v. Πάφος), after 'abs.' insert 'Cypr. *ta-se-pa-pi-a-se* (τᾶς Παφίας), *ICS* 262 (c.500 BC)'

*Παφλᾱγόνισσα, ή, *Paphlagonian*, Εὐπορία Μανοῦ Π. *IG* 2².10052 (Piraeus, i BC)

παφλάζω, for 'Alc.*Supp.* 25.4 (p. 28 Lobel)' read 'Alc. 72.5 L.-P.'

πάχνη **1**, add 'transf., πάντα πάχνης ἦν πλέα καὶ πυρός of a sick man, Aristid.*Or.* 48(24).46'

*παχόω, *thicken*, ἔψε ἕως παχῶσαι prescription in *Hermes* 33.343 (nisi leg. παχνῶσαι).

παχυμερής **I**, add 'adv. -μερῶς, *in large portions* or *pieces*, Asclep. Jun.ap.Gal. 13.1022.9'

πᾰχύπους, add 'Hsch.'

παχύς **I 6**, add 'Λύδη καὶ παχὺ γράμμα καὶ οὐ τορόν Call.*fr.* 398 Pf.'

πᾰχύτης **II**, add '**2** *of style*, τὴν π. τῶν ποιημάτων Sch.D.P. 3 (*GGM* II p. 427ᵇ4 adn.).'

*Παχών, month name in N.Africa, *AP* 9.383, *SEG* 24.1178 (Alexandria, iii BC), 33.1395 (Cyrenaica), etc.

πε, for 'of πετ(ά) = μετά' read 'of πετ = πεδά'

†πεδά, non-Att.-Ion. prep. (corresponding in use to μετά) **A** w. gen., *with, among*, Sapph. 55.4 L.-P., Alc. 70.4, 73.10 L.-P., al., Ibyc. 40.3 P., Theoc. 28.21; following a noun, id. 29.38. **B** w. acc., *after*, Sapph. 99 i 1 L.-P., Alc. 387 L.-P., Alcm. 1.58, 17.5 P., *Leg.Gort.* 3.27, *Schwyzer* 619.20. **2** *to*, Alcm. 3 *fr.* 1.8 P., *Schwyzer* 177.5. **3** *according to*, π. θῦμον Sapph. 60.5 L.-P. **C** w. dat., *among*, τοῖς π. Ibyc. 1.46 P., unless πέδα here = μέτεστι. Myc. *pe-da.*

πεδάγρετος, add 'cj. in Alc. 358.3 L.-P.'

*πεδάλιον, τό, app. = Lat. *podium*, *JRA* suppl.2 p.28 (Aphrodisias, ii/iii AD).

*πεδάμοιρος, ον, *having a share*, *IRhod.Peraia* 351 (vi AD).

*Πεδανασσεύς, epith. of Apollo, Didyma 70 (*SEG* 30.1293).

πεδανός **I**, for this section read '*low-growing*, ῥυτῆς βλάσται Nic.*Al.* 306; *flat, squat*, ἀλκαίη, οὐρή, ἐπίγουνίς id.*Th.* 226, 289, 817'

πέδᾱρος, for 'Alc. 100' read 'Alc. 315 L.-P.' and add 'A.*Ch.* 590 (lyr., cj.)'

*πέδε, v. °πέντε.

*πέδειμι, Dor. = μέτειμι (εἰμί sum), [ἀ]ρχᾶν πεδεῖμ[εν] ..], *ASAA* 27/29(1952).112 (Acrae, v BC).

**πέδϝεις, εσσα, εν, Myc. *pe-de-we-sa*, of an ἐσχάρα, *equipped with feet.*

*πεδεπιθύω, Cret., prob. *sacrifice together with others*, *Inscr.Cret.* 4.146.3 (Gortyn, v/iv BC).

†πεδέρχομαι, Aeol. and Dor. in senses of μετέρχομαι, *pursue*, πεδελθέτω Alc. 123.13 L.-P. **2** *follow, ensue*, Pi.*N.* 7.74. **3** *beseech*, Theoc. 29.25.

πεδέχω, add '*SEG* 23.474 (Dodona, iv/iii BC)'

πέδη **I**, add '**4** *a kind of brake*, π. καὶ ἄξων *IG* 1³.422.122 (Athens, v BC); cf. τροχοπέδη. **5** *leather-covered ring* for securing the rudder or mast of a ship, Hsch.; cf. ἰστοπέδη.'

*πεδιᾱκόν, τό, *land record-book*, π. ἐπικρίσεως pap. in *Aegyptus* 15.210 (iii AD), *POxy.* 1287.2 (ii AD), *PSI* 450.69 (παιδ-; ii/iii AD).

πεδιᾱκός **I**, add 'πεδιακὴ ὁδός, *fieldpath*, *BGU* 2055.12 (ii AD)'

*πεδιαρχέω, *to be a "controller of the plain"*, (app. office connected w. the Pythian games), *BCH* 80.592 (Locris, vi/v BC).

πεδιάς **I**, add 'fem. subst., *flat land*, *BGU* 2159.8 (AD 485)'

*πεδιάσιμος, ον, *of the plain*, *POxy.* 1537.12 (ii/iii AD); cf. πεδιασιμαῖος.

πεδιεῖς, add 'Myc. *pe-di-je-we* name of a class of men; cf. deriv. *pe-di-je-wi-ja*, perh. *infantry spears*'

*πεδικόν, v. °παιδικός.

πέδιλον, add 'Myc. *pe-di-ra* (pl.)'

*πεδιόθεν, v. °παιδιόθεν.

*πέδιjος, α, ον, Cypr. *pe-ti-ja-i, situated in the plain*, *ICS* 217 B18.

πεδιοφύλαξ, add '*PVindob.Tandem* 16.23 (v/vi AD)'

†πέδοι, adv. *to the ground*, A.*Pr.* 272, Luc.*Lex.* 1.

πέδον **4**, penultimate line, for 'as also for πέδον' read '**b** πέδον *to the ground*'; add 'Call.*Del.* 227'

†πεδοσκᾰφής, *digging out the earth*, πεδοσκαφέεσσι μακέλλαις Nonn.*D.* 4.255, 12.331. **II** *hollowed out of the earth*, ib. 26.112, 30.145, al.

*πεδοτρεφής, *grown* or *nourished in the earth*, πεδοτρεφέων .. δρακόντων Nonn.*D.* 2.47, λέκτρα πεδοτρεφέων ὑμεναίων ib. 29.337, al.

πεδόρριψ, add 'Ph. 2.446'

πέζα **II**, add '**b** *stylobate* of a colonnade, *Suppl.Hell.* 978.7.'

*πεζάρχης, ου, ὁ, *leader of infantry*, *GVI* 1928.7.

πέζαρχος, add '*IG* 2².175.8 (iv BC)'

πεζίδιον, delete the entry.

πεζίς, add 'Leont.Byz.*HP* 2.2.10'

*πεζίτιον, τό, *ribbon*, Suid. s.v. ταινίαι, and so Phot. (-ζήτ-), *EM* 749.37 (-ζέτ-), *Gloss.* (-ζίδ-).

πεζός **II**, line 1, for 'πεζὸς' read 'πεζὸν'; line 2, for 'Call.*Aet.* 4.1.9' read 'Call.*fr.* 112.9 Pf.'

πεῖ, adv., after 'Sophr. 5' insert 'Theoc. 15.33 (*PAntin.*)'

*πειθηνίς, ίδος, fem. adj., = πειθήνιος II, πατρὸς π. βουλῇ *Orac.Chald.* 81.2 P.

πειθώ **II**, at end delete 'Pi. uses .. P. 3.28' and for '*I.* 4(3).72' read 'Pi.*I.* 4.90 S.-M.'

Πειθώ **I**, add '*SEG* 33.643 (Rhodes, 100 BC). **2** as epith. of Aphrodite, *IG* 9(2).236 (Pharsalus, v BC), *ABSA* 47.190 (Cnidus, iv/iii BC).'

*πεῖλα, ἡ, Lat. *pila, pier, mole*, *IEphes.* 23.17 (ii AD) (pl.).

*πειλιπής, v. πιλιπής.

πειράζω **II 1**, add 'pass., πειράζεται τὰ νήπια ποίας τινὰς ἔχει τὰς τῆς ψυχῆς διαθέσεις D.S. 2.58'

*Πειραία, τά, *the Dionysia in Piraeus*, *IG* 2².1028.16 (100 BC), etc.

πεῖραρ **I 2**, add 'π. κουροσύνας *AP* 6.281 (Leon.Tarent.)'

πειρᾰτ(ε)ῖαι, αἱ, *treacherous attacks*, Hsch. s.v. πείραι.

πειρᾱτεύω **II**, add 'πειρατευ[όν]των Θραικῶν οὐκ [ὀλίγων τὴν [.. χ]ώραν *IHistriae* 15.9 (iii/ii BC)'

πειράω **A IV 2**, for 'make an attempt on a woman's honour' read 'attempt to seduce' and add 'παῖδας ἐπείρων Ar.*Pax* 763, *V.* 1025'

*πείρημα, ατος, τό, *limit*, codd. in App.Anth. 3.186, Gr.Naz.*Carm.* 2.1.38.9.

πειρητίζω, line 1, delete 'Ep. form of πειράω' and insert 'πειρατίζω Hsch.'

πεῖρινς, line 4, after 'Hsch.' insert '(cf. dat. πυρίνθω in *PMasp.* 303.14 (vi AD))'

*πειστικόν, τό, *"room of the faithful"*, (or perh. = Lat. *posticum*), ἡ ψηφείς (mosaic work) τοῦ π., R.Stillwell *Antioch-on-the-Orontes* (1933-36) 33, 42 (v AD).

*Πειστίχη, ἡ, (πείθω) *goddess of persuasion*, epith. of Aphrodite, *Inscr.Délos* 2396, 2397; without Ἀφροδίτη, ib. 2398 (Πιστίχη).

†πειώλης, ου, ὁ, = κίναιδος, Suid., *EM* 668.36, Eust. 1684.29; cf. πεοίδης, πεώδης.

*πεκουλιάριον, τό, *personal property*, *EAM* 22 (ii/iii AD).

πεκούλιον, add '*EA* 6.63 no. 5, Just.*Nov.* 162.2.1'

πεκτήρ, add 'cf. Myc. fem. *pe-ki-ti-ra*₂, πέκτριαι'

*πεκτοραλίων, (gen. pl.), Lat. *pectorale, breastplate*, *PMich.*xv 742.5 (vi AD).

πελᾰγόστροφος, for 'roving through the sea' read 'that haunts the deep sea'

πελάζω **A II**, add 'cf., of snakes, poet. in *PKöln* 244.17 (iii AD)' **C I 3**, add 'ἐν πανηγύρει δαιμόνων πελαζόμενος Aesop.*Prov.* 39 P.'

πελᾰνός, line 4, for 'π. αἱματοσταγής .. slaughter' read 'π. αἱματοσφαγής'

πελαργός **I**, add '**b** π.· ἄγγος τι κεράμεον Hsch. cf. *PBSR* N.s. 59.177ff.'

πέλας **II**, add 'τοῖς πέλας ἀμμέων Alc. 353.1 L.-P.'

*πελᾰτεύω, to be a πελάτης, i.e. to depend on like a client, [κ]ηδεστῶν τρόπον οἷσιν [ἔ]ντροφος (-τροπος pap.) πελατεύσεις A.*fr*. 47a ii 22 R. (lyr.).

*πελεθραία, ἡ, subdivision of a πέλεθρον, *SEG* 13.395 (Larissa, iii BC); also πελεθραιαία, ib. 394; cf. πλεθριαῖος.

*πελεθραῖον, τό, = °πελεθραία, *SEG* 26.676.6 (Larissa, iii BC).

πελειάς, line 3, after 'Opp.*C*. 1.351' insert 'πελῃάσιν *GVI* 270.3 (Trachonitis, ii/iii AD)'

πελείους **2**, for this section read 'livid, Lxx *Prov.* 23.29'

*πελεκᾶς (B), ὁ, axe-maker, prob. in *OWilck*. 2.720 (i BC).

πελεκῖνος **III**, add 'a kind of fastening for a sarcophagus-lid, ἐγχωνεύσει τοὺς π. *SEG* 17.633.9 (Perge); τὸ πῶμα πελεκείνοις διεί[λ]ηται *TAM* 3.574 (Termessus). **IV** a kind of sundial, Vitr. 9.8.1.'

πέλεκυς **I 2**, add 'meton., for "lictor", id. 3.87.7' **6**, for 'prob. in *Inscr.Cypr*. 135.26 H.' read 'prob. interpr. of Cypr. *pe* standing for a sum of money, *ICS* 217 A15, B26' add '**7** an ancient weight of six or twelve minae, Hsch.'

*πελεκυφόρᾱς, α, ὁ, masc. adj., bearing (the brand of) an axe, π. ἵππον Pi.*fr*. 339a S.-M.

*πελέμαιγις, ιδος, shaking the aegis, epith. of Athena, perh. to be read rather than πολέμαιγις in B. 17.7 S.-M.

†πελιγᾶνες (-όνες Str.7 *fr*. 2), elders in Macedonian communities or colonies, *IGLS* 4.1261.22 (ii BC), cj. in Plb. 5.54.10, Hsch.

πελιδνόομαι, add 'πελιδνώθεισα Alc. 298.11 L.-P. (s.v.l.), Call.*fr*. 374 Pf.'

πέλιξ, add 'Myc. *pe-ri-ke* (pl.)'

πελιόομαι, add 'act., mark with a livid bruise, Eust. 1681.53'

*πελιωμάτιον, τό, slight contusion, *POxy*. 3195.48 (AD 331).

*πελιωπός, όν, of livid aspect, Theognost.*Can*. 69.

πέλλα, line 1, after 'Il. 16.642' insert 'Hippon. 14.1 W.' **2**, delete the section.

πελλητήρ (s.v. πελλαντήρ), before 'Clitarch.' insert 'perh. milking-pail' and after 'ibid.' insert '(Boeot.)'

*πελλοδόχος, ὁ, dub. sens., *PIand*. 17.4 (vi/vii AD).

πέλμα **II**, for 'stalk' read 'base (eye, i.e. remains of calyx)' add '**III** ground area, *ICilicie* 61 (Samos, Byz.).'

*πελμᾰτοπώλης, εω, ὁ, seller of shoe-soles, *IEphes*. 2.25, 27 (340/320 BC).

Πελοποννησιακός, add '*SEG* 26.121 (Athens, i BC)'

*πέλτα, τά, tomb or some part of a tomb (perh. platform, cf. Hitt. *palzaḫḫa*), *SEG* 6.307, 428, 434 (Lycaonia), *MAMA* 1.31, 60 (Phrygia); cf. πλάτας, πλάτος (B).

*πελτᾱφόρας, v. πελτοφόρος.

πέλτη **II 1**, for 'small light shield .. orig. Thracian' read 'small wooden or wicker shield with covering of skin'; after '*IG* 1².282.120' insert 'X.*Mem*. 3.9.2'; add 'used loosely for any shield, π. Δωρίς *AP* 7.430 (Dioscor.)'

πέλτης, delete 'salted'

†πελτίδιον, τό, v.l. for πελτάριον (in same sense), Luc.*DMort*. 12.2.

πέλτον, delete the entry (v. °πέλτα).

πέλω, add 'Myc. part. *qe-ro-me-no*'

πελωριάς, delete 'cf. *AP* 6.224 (Theodorid.)'

†πεμπᾶκι, delete the entry (v. °πεντάκις).

πεμπάμερος, for '*Inscr.Cypr*. 134 H.' read '*to-pe-pa-me-ro-ne*, τὸ πε(μ)παμέρον *ICS* 220.2'

πέμπε, add '*SEG* 26.672 (Thessaly, 200/190 BC)'

*πεμπείκοντα, = πεντήκοντα, *SEG* 26.672 (Thessaly, 200/190 BC).

*πεμπεικονταέν, = *πεντηκονταέν, fifty-one, *SEG* 26.672 (Thessaly, 200/190 BC).

πεμπταΐζω, after 'fifth day' insert '(or incorrectly, in the fifth generation)'

πέμπτος **I**, add 'in a legionary title, λε[γ](ιώνος) πέμπτης Μακεδ(ονικῆς) *ICilicie* 31 (ii AD)'

*πεμπτοστάτης, ου, ὁ, priest in the fifth grade, πεμπτοστάτου Διός *SEG* 32.1483 (Syria, i/ii AD).

*πέμφελα· δύσκολα, τραχέα, βαθέα Hsch.; cf. δυσπέμφελος.

πέμφιξ **3**, add 'φλύκταιναι πέμφιξιν ἐειδόμεναι ὑετοῖο Nic.*Th*. 273' deleting this quot. from section **4**. **4**, line 6, for 'Call.*Fr*. 483 .. 43' read 'Call.*fr*. 43.41 Pf.'

πενθᾰλέος **1**, add '-οἶσιν ὀδυρμοῖς *Bithynische St*. III 1.1'

πενθάς, for 'cf. Nonn.*D*. 14.271' read 'π. φωνῇ Nonn.*D*. 11.314'

πένθεια, for 'poet. form of πένθος' read 'perh. female mourner' and after 'lyr.' insert 's.v.l.'

πενθερά, after 'mother-in-law' insert '(at first only wife's mother, acc. to Ar.Byz.ap.Eust. 648.54)'

*πενθεράς, άδος, ἡ, = πενθερά, mother-in-law, *MAMA* 7.430.

πενθέριος, add '*SEG* 34.1224 (AD 96/7)'

πενθερίδης, for 'πενθερός' read 'πενθερά' **II**, after 'cf.' insert '*Thasos* 141.21 (iv BC)'

πενθερός, after 'father-in-law' insert '(at first only wife's father, acc. to Ar.Byz.ap.Eust. 648.54)' **II**, add '*MAMA* 8.271 (Lycaonia)'

†πενθέτηρος, ον, of five years' standing, πενθετήροις [π]ροπό[λοις] Philod.Scarph. in *SEG* 32.552.131; cf. πεντετηρικός, etc.

*πενθέτης, = πεντέτης, cj. in Alex. 125.10 (for πεμφθείς; read πένθ-ετες.)

πενθημερία, delete the entry (read πενθ(ήμερος)).

πενθημιαρτάβη, for this lemma read 'πενθημιαρτάβιον, τό'

*πενθημιδακτύλιος, α, ον, of five half-fingers' breadth, *BCH* 20.324.66 (Lebadea, ii BC).

πενθήρης, add '*Trag.adesp*. 705.1 K.-S. (iamb.trim.)'

*πενθίδιος, ον, mournful, στεναχαί, *CEG* 587 (Athens, iv BC).

πενθικός, line 2, after 'mournful' insert 'πενθικὸν μηδὲν ποιείτω μηδείς *Thasos* 141.3 (iv BC)'

πένθιμος, add 'πενθίμοις σχήμασι J.*AJ* 2.6.8, etc.; sup. διαθεὶς αὑτὸν -ώτατον ib. 19.1.18'

πένομαι, line 1, add 'pf.part. πεπεμμένοις *IG* 4.752.13 (Troezen, ii BC)'

*πένουλα, v. φαινόλη.

*πεντάβασμος, ον, having five steps or rungs, *Inscr.Délos* 1417*A*i76 (ii BC); cf. πεντέβαθμος.

*πεντάβαφος, ον, quintuple-dyed, πενταβάφου πορφύρας *PColl.Youtie* 85.3,4, *POxy*. 1978.9 (both vi AD).

*πεντᾰγωνοειδής, ές, pentagon-shaped, Poliorc. 206.15.

πεντᾰδάκτυλος **II**, for 'as Subst. = πεντάφυλλον' read 'πενταδάκτυλον, τό, = πεντάφυλλον' and add 'sp. πεντεδ- *PMag*. 2.34'

πεντᾰετηρικός, add 'masc. subst. (sc. ἀγών) *SEG* 34.1316 (Lycia, AD 90); also neut. pl., τῶν μεγάλων πενταετηρικῶν *MAMA* 9.19 (Rom.imp.)'

πεντᾰέτηρος **I**, line 3, delete 'cf. .. (vi AD)' **II**, add 'ἀρχὸν πενταέτηρον *BCH* 78.74.4 (Achaea, iv/v AD); also πεντεϜέτειρος, *Ἀρχ.Δελτ*. 14 Pl. i 27 (Thespiae, iii BC)'

πεντᾰθλέω, add 'also πενταϜεθλέω, πεντα Ϝ εθλέον νίκα *Hesperia* 28.322 (*SEG* 26.407; inscr. on ἀλτήρ, Isthmia, early vi BC)'

†πεντᾰκέλευθος, ἡ, meeting-place of five roads, orac. in Paus. 8.9.4.

πεντάκις, for 'later πεντάκι' read 'also πεντάκι' and add '*CEG* 346 (Delphi, 475/50 BC); also πενπάκι ib. 374 (Sparta, *c*.530/500 BC); erron. for πεντάκι or due to poet. influence'

*πεντᾰκορίνθιος, ὁ, app. coin of the value of five Corinthian staters, χρυσίω πεντακορινθίω (gen.) de Franciscis Locr.*Epiz*. 35 (iv/iii BC).

*πεντᾰκότυλος, ον, holding five κοτύλαι, *Inscr.Délos* 1432*Ab*ii32 (ii BC).

*πεντᾰκωμία, ἡ, union of five villages, *ABSA* 51.154 (Caralitis, AD 133), cf. ib. 156.

†πεντᾰλκία, ἡ, conjectural rest. in sense of a measure (? *πενταολκία cf. ὀλκή), *Inscr.Cret*. 4.79.6 (v BC).

πεντάμετρος, for 'Hermesian. 7.36' to end read 'Sch.D.T. 173.13; also πεντάμετρον, τό, Hermesian. 7.36, Call.*fr*. 203.31, 45 Pf.; π. ἐλεγειακόν D.H.*Comp*. 25, [τροχαϊκὸν] π. Heph. 6.2'

πεντᾰμηναῖος, for '*Epigr.Gr*. 344.17 (Bithynia)' read '*IHadr*. 80 (ii AD)'

πεντάμηνος **2**, for 'ὁ π. (sc. χρόνος)' read 'περὶ πεντάμηνον (prob. for (τὴν) πεντέμηνον περίοδον)'

πεντάμναιος, add '-μνεως, *SEG* 16.497.3 (Chios, iii/ii BC)'

*πεντᾰναυβία, ἡ, duty of performing five ναύβια, *PColl.Youtie* 21.14 (AD 80/1).

*πεντάνευρον, τό, species of plantain, An.*Boiss*. 2.395, Anon.Alch. 326.4.

πενταξεστιαῖον, delete the entry.

*πενταξεστιαῖος, ον, containing five sextarii, *PColl.Youtie* 93.8, *PSI* 881.5, *CPR* 8.63.2 (all vi AD).

*πεντᾰπάλαστος, ον, = πενταπάλαιστος, measuring five handbreadths, *Inscr.Délos* 1432*Bb*ii16 (ii BC); cf. πεντεπάλαστος.

*πεντάπλησον, aor. imper. multiply by five, perh. erron. for πενταπλωσον (i.e. °πενταπλόω) *PCair*. 10758 (vi AD), or -πλασον (-πλάζω, cf. διπλάζω).

πεντάπλοκος, add '*PSoterichos* 4.18 (AD 87), *PVindob.Boswinkel* 8.13 (AD 332)'

*πενταπλόω, multiply by five, v. °πεντάπλησον.

*Πενταπολῖται, οἱ, citizens of a πεντάπολις, *BCH* 62.37 (Maced., Rom.imp.).

*πεντάπρωτοι, οἱ, (board of) five leading men, *SEG* 32.1467 (Syria, Chr.).

πεντάρουρος, add 'also πεντεάρουρος, *PHamb*. 65.18 (ii AD)'

*πένταρχος, ὁ, member of a πενταρχία, *SEG* 31.745 (Naxos, Hellen.).

πεντάς, add 'also πεντεάς *SEG* 33.848 (Mauretania, i AD)'

*πεντασσός, ή, όν, *fivefold*, PHeid. 323 c 12 (written with the symbol ε–, cf. τετρασσός, etc., AD 310).

*πεντάστῦλος, ον, *having five posts* or *pillars*, κλίνεια πεντάστυλα PSoterichos 1.27 (AD 69), 2.23 (AD 71), PFlor. 369.2 (ii AD).

πέντε, add 'Pamph. πέδε, IPamph. 3.5 (iv BC)'

*πεντεδάκτυλον, v. °πενταδάκτυλος.

*πεντεδεκαετής, ές, = πεντεκαιδεκαετής, SEG 6.137.30 (Phrygia, iv AD).

πεντεδραχμία, add '*five-drachma tax*, Ath.Agora XIX P26.475, 479 (342 BC)'

*πεντεγέτειρος, v. °πενταέτηρος.

*πεντεκαιδεκάδρομος, ὁ, *youth who has reached fifteen years*, πεντεκαιδεκαδρομῶ (gen.) Inscr.Cret. 4.72 xi 54.

*πεντεκαιδεκάμηνος, ον, *of fifteen months*, neut. sg. as subst., π. ἔχων .. ἥρθην GVI 1244 (Athens, ii/iii AD).

*πεντεκαιδεκαμοιρία, ἡ, *space of fifteen degrees*, Heph.Astr. 3.5.62, 63, al.

*πεντεκαιδεκάπους, ποδος, *fifteen foot long*, IG 2^2.1672.156 (pl.).

*πεντεκαιδεκάστεγος, ον, *of fifteen storeys*, Poliorc. 239.2.

πεντεκαιδέκατος, add '-η, ἡ, *fifteenth day*, IMylasa 103 (ii/i BC), TAM 5(1).230 (AD 253/4)'

*πεντεκαιδεκάτροπος, ον, app. *consisting of fifteen divisions*, Procl. in Ti. 2.170.11 D.

πεντεκαιεικοσιέτης, add 'also -εικοσέτης CEG 176 (Panticapaeum, v BC), GVI 1233 (Egypt, ii/i BC)'

*πέντεκτος, η, ον, *one-fifth*, SEG 36.1116.B1 (Cyzicus, iv BC).

*πεντελίτρον, τό, *weight of five λίτραι*, Dacia 3/4.611 (Perinthus).

πεντέμηνος, delete 'τό' (cf. ‡πεντάμηνος)

πεντέπους, line 3, for 'Arr.Peripl.M.Eux. 3' read 'Arr.Peripl.M.Eux. 2'

πεντετηρικός, after '*every five years*' insert '(in alternative reckoning, *every four years*)'

πεντηκοντακάρηνος, delete the entry.

⁺πεντηκοντᾰκέφαλος, ον, (in quots. scanned -κεφᾰλος) *fifty-headed*, Hes.Th. 312, Pi.fr. 93 S.-M. (πεντηκοντο-).

πεντηκοντάπαις II, for '*Δαναός*' read 'ἀδελφός'

*πεντηκοντάπους, ποδος, *fifty foot long*, στῦλος PGiss. 69.13 (ii AD).

πεντηκόσιοι, add 'IG 12(3) suppl. 330.22 (Arcesine, ii BC)'

πεντηκοστολόγος, add 'at Byzantium, Arist.EE 1247^a19'

πεντηκοστύς, add 'at Argos, Schwyzer 90.13 (iii BC)'

*πεντημίεκτος, η, ον, *two-fifths*, SEG 36.1116.A9 (Cyzicus, iv BC).

*πεντήντα, = πεντήκοντα, CIJud. 596 (Venusia); cf. mod. Gk. πενῆντα.

*πεντόγκιον, τό, = πενταούγκιον, Epich. 9.

πεντόροβος 2, before 'IG 11(2).161B19' insert 'IG 1^3.383.176'

*πέντος, α, ον, = πέμπτος, GDI 4991 ii 39.

*πεντωβόλειος, α, ον, = πεντώβολος, τόκου πεντωβολείου PSI 1328.34 (AD 201).

*πεντωβόλιον, τό, *five obols*, δραχμᾶν δύο πεντωβόλου IG 4^2.109 ii 123 (Epid.).

πεντώβολος, line 4, delete 'δραχμᾶν .. (Epid.)'

πεντώγκιον, delete the entry.

*πεντώρυγος, add 'IG 2^2.1627.356 (iv BC)'

πεξόν, delete the entry.

*πεξός, ή, όν, Lat. *pexus*, of materials, *brushed*, Edict.Diocl. 20.12, al.; cf. πεξόν ἱμάτιον, *prosa pexa tunica*, Gloss.

πέος, after 'Ar.Ach. 158' insert 'AP 11.224 (Antip.Thess.)'

πεπαίνω, line 5, delete 'but' and substitute '*mature* meat by hanging'; line 6, for 'by being boiled with it' read 'i.e. by hanging the bird in a fig-tree'

πέπανσις, after 'fruits' insert 'etc., Hp.Hebd. 4'

πέπερι 2, add 'Peripl.M.Rubr. 49' add '3 λευκόν π. *white pepper*, SEG 37.1019 (Pergamon, ii AD).' at end, in parenth., add 'πιπέρεος SEG 37 l.c., acc. sg. πίπεριν SEG 30.326 (Athens, i AD)'

*πεπεροπαστάριον, τό, *container for pepper*, PVindob. G29709 (ZPE 64.77, v/vi AD).

πεπλᾰνημένως II, add 'Gal. 16.815.2'

πεπλογρᾰφία, for 'title .. Worthies' read 'Cicero's description of work by Varro, prob. the *Imagines*'

πεπλοθήκη, for '*wardrobe*' read '*chest for storing the (Panathenaic) peplos*'

*πεπόνιον, τό, dim. of πέπων I 2, *melon*, PRyl. 630*. 21, al. (IV AD); cf. mod. Gk. πεπόνι.

*πεπόνιος, α, ον, app. *melon-coloured*, δερμᾰτικόν π. PMichael. 18 ii 2 (iii AD).

πεπρωϊών, delete the entry.

*Πεπρώιοι or Πεπρώιοι, app. name of a χιλιαστύς, IEryth. 17.15-16 (iv BC).

*πέπτης, ου, ὁ, perh. *baker*, POxy. 3492.27 (AD 161/9).

πέρᾱ (A), add 'Myc. compd. pe-ra₃-ko-ra-i-ja, (Περαιγολαΐα) name of the Further Province'

πέρα (B), line 2, after '(lyr.)' insert 'αἴ τίς κα πέραι συναλ[λάκ]σει ἐ

ἐς πέρ[α]ν ἐπιθέντι μὲ ἀποδιδοῖ Inscr.Cret. 4.72 ix 43 (Gortyn)'; add 'πέρᾱν app. acc. and πέρᾱ (A) instr. fr. this appellat.'

περαίνω IV, add 'cf. °διαπειραίνω'

περαιόω II, lines 1/2, for 'etc., Leg.Gort. .. vii 15' read '(app. cancel the sale of a slave and return him) Inscr.Cret. 4.41 vii 15, 4.72 vii 11'

περαίτερος I, add 'sup. περαίτατον Hsch.'

περαίωσις, add 'II *end, expiry* of a period of time, PKöln 104.20 (vi AD); *completion* of a process, Cod.Just. 9.4.6.3.'

⁺πέρᾱμα, ατος, τό, act of *crossing* a river, Palladius ap.Ps.-Callisth. 3.10.   2 *place across the water*, Just.Nov. 14, 59.5 (bis).

πέρας II 2, add 'πέρατι παραδοῦναι Cod.Just. 1.3.41.29'   4, add 'rhet., = ὁρισμός II, Cratin.Jun. 7.4'   IV, at end after '(Men. Epit.) 470' insert 'Dysc. 117'

περᾱτικός, add 'SEG 24.633 (Thrace, Chr.); of kind of gum, *Bdellium peraticum*, Plin.HN 12.35'

*περάτρια· ἡ παραγγέλλουσα τὴν ὥραν ταῖς κεκτημέναις Hsch. (post Πέρδικος).

*Περγαῖος, α, ον, *of Perge* in Pamphylia, SEG 34.1305; title of Artemis, ib. 25.693, IPamph. pp. 160-1 (sp. Πρειας).

Περγηνός, after 'ή, όν' insert 'of *Pergamum*, SEG 23.362, etc.; title of Asclepius, SEG 24.978, 27.373'; line 3, before 'Suid.' insert 'Edict.Diocl. 7.38'

πέργουλος, add 'cf. σπέργουλος, σπόργιλος'

*περδικίτης, ου, ὁ, sc. λίθος, Alex.Trall. 12, Paul.Aeg. 7.17.75 (365.12-13 H.).

πέρδιξ, after '*partridge*' insert '(in quots. either *rock* or *chukar partridge*)'

πέρδομαι, after '(Ar.)Pax 335' insert 'Eup. 99.10 K.-A.'

πέρθω, line 6, after 'poet. verb' insert '(used by Arist. in pun on Πέρσαι, Rh. 1412^b2)'

περί, line 2, after 'A.v' insert 'Cypr. ta-ne-pe-re-ta-li-o-ne, τὰν πὲρ Ἐδάλιον ICS 217.27'   F, add 'cf. Myc. pe-ri-me-de, pe-ri-qo-ta, etc.'   G, for 'not in Trag. .. 634)' read 'in Trag. περέβαλον A.Ag. 1147, περεσκήνωσεν id.Eu. 634 (s.v.ll.)'

περιαγκωνίζω, after 'Lxx 4Ma. 6.3' insert 'aor. part. pass. περιαγκωνισθείς Aesop. 200 P.)'

περιαγνίζω, add 'med., περιαγνισαμένη Sokolowski 1.18.14 (Maeonia, ii BC)'

περιάγω 2 b, for 'cf. Luc. .. 187e' read '*lead in circuitous argument*, περιαγόμενος τῷ λόγῳ Pl.La. 187e, Luc.Nigr. 8'

περιαγωγεύς, before '*windlass*' insert '*capstan* or'

*περιάκτρια, ἡ, prob. *machine for changing scenery on stage* (cf. μηχαναὶ περίακτοι), SEG 9.13.13 (Cyrene, iv BC).

περίαλλος I, line 2, delete '2.217'; line 3, after '(lyr.)' insert 'w. gen., π. θεῶν A.R. 2.217'

*περιαργύρωσις, εως, ἡ, *plating with silver*, Anon.Alch. 378.15.

περίαυλον, add 'also περίαυλος, ὁ, EM 361.39'

περιβάλλω, line 1, for '(v. infr.)' read '(περέβαλον is prob. cj. for περεβάλοντο in A.Ag. 1147)'   II 1, add 'pass., ξύλον σιδήρῳ περιβεβλημένον *bound round*, App.BC 5.118'   IV 1, lines 4/6, for 'pf. Pass.' read 'plpf.' and before 'δυναστείας' insert 'pf. part.', deleting 'cf. 2.25'

περίβλεπτος, penultimate line, after 'title of honour' insert '= Lat. *spectabilis*, ranking between *illustris* and *clarissimus*, POxy. 3481.3 (AD 442), SEG 32.1554 (early vi AD), Cod.Just. 1.4.22 pr.'

*Περιβλήμαια, τά, festival at Lyttus, ἐν τοῖς Περιβλημα[ίοις] Inscr.Cret. 1.xix 1.21 (Malla, c.221 BC).

περιβολάδιον, after '*wrapper*' insert 'BGU 1848.13 (i BC)'

*περιβόλαιος, ον, *encircling*, Hsch. s.v. κόρυθα περίδρομον.

περίβολος II 1, add 'b περίβολον, τό, *surrounding wall*, Bernard Akoris 2, id. Les portes 33 (sp. -βωλον) (both i AD).'

*περιβρᾰχίων, ονος, ὁ, *armlet*, Inscr.Délos 1421Bbii10 (ii BC).

περιβρέμω, after '*round about*' insert 'περὶ δὲ βρέμει ἄχω Alc. 130.33 L.-P.'

*περιβροχή, ἡ, *moistening round about*, Gal. 18(1).571.4 (unless ἐπιβρ- is read).

περίβωτος, after 'περιβόητος' insert 'APl. 49 (Apollonid.)'

περιγηθής, after '*very joyful*' insert 'Emp. 27.4'

περιγίγνομαι II 2, add 'τὰ περιγενόμενα *surplus*, see PMich.VI 385. 42 n.'

*περίγναμπτος, η, ον, *curved*, prob. in Q.S. 1.149.

περιγραφή II 3, for '*compass of expression*' read '*sentence* or *clause embracing* a thought' and add '*structure, framework of a sentence*, D.H.Th. 26'

περιγράφω I 2, for '*being self-contained*' read 'i.e. expressed in separate sentences'

περιδαίω, add 'also act., *inflame with love*, A.R. 4.869 (in tm.)'

περιδέξιος I 1, for 'AP 12.247 (Strat.)' read 'metaph., Call.fr. 360 Pf., AP 12.247 (Strat.)'   2, after '(lyr.)' insert 'w. inf., APl. 378'

περιδέρκομαι, before 'Nonn.' insert 'intr. in' and for 'AP .. (Agath.)' read 'trans. in AP 5.289.5, 16.169.1 (both Agath.)'

*περίδερμα, gloss on ἀνθήλιον (q.v.), Hsch.
περιδῑνέω, add 'b twist out, dislocate, Plu. 2.327a.'
*περιδῑνοπλᾰνήτης, ου, ὁ, one who wanders round, PMag. 3.557.
περιδῑνος, delete 'ἡ' and for 'rover, pirate' read 'brigand'
*περιδῑνω, tread round in threshing, δινομένην πέρι βουσὶν ἐμὴν ἐφύλασσον ἄλωα Call.fr. 255 Pf.
περίδριος, ον, (δρίος) συκῇσι surrounded by a cluster of haemorrhoids, Marc.Sid. 62 (codd.).
περιδρομή II, for 'J.AJ 20.12.1' read 'J.AJ 20.11.2' and delete 'ἐκ π. Ptol.Tetr. 55' add 'IV enclosing framework, Sm.Ez 43.14.'
*περιδωμάω, build round, poet. in POxy. 2812 fr.1(a) ii 25.
περίειμι (εἰμί) III 1, add 'of surviving members of a family, Τοκης ὁ πατὴρ καὶ Οὐαδέα ἡ μήτηρ περιόντες SEG 30.612 (ii AD)'
περιέλᾱσις I, add '2 driving round in procession, dub. rest. in IG 1³.241.11.'
*περιέξ, prep. w. gen., around, πάντων τῶν π. αὐτῆς οἰκημάτων SB 6000.9 (vi AD).
*περιέξοθεν, v. °περιέξωθεν.
*περιέξωθεν or περιέξοθεν, adv., round the outside, all round, αἱ π. ἄρουραι PMich.XIII 666.11, 23, 32; PKöln 104.6, PHamb. 68.16 (all vi AD).
περιεργία 1, after 'needless questioning' insert 'Isoc. 10.2'
περίεργος II 1, after 'adv. -γως' insert 'D.H.Isoc. 3' add 'b taking particular trouble; comp. adv. -ότερον, Cod.Just. 4.21.22.2.'
περιέρπω II, add 'Call.Dieg. iv.31'
περιέρχομαι II 1, line 9, after 'Hdt. 8.106' insert 'Paus. 4.17.4, 8.53.3'
*περιεστίαρχος, ὁ, = περιστίαρχος, Poll. 8.104, Phot.
*Περιεστώ, ἡ, a goddess of healing, BE 1991.221 (Thrace, c.400 BC).
περιέσχατα, add 'D.C. 36.49 (τῶν περὶ ἔσχατα D.H. 1.79 (s.v.l.))'
περιέχω I 4, transfer 'Supp.Epigr. 3.421.33' to precede 'impers.'; add 'impers. pass., καθὰ τῇ θείᾳ ἡμῶν ταύτῃ περιέχεται νομοθεσίᾳ Cod.Just. 1.3.41.18'
*περιζᾰμενής, ές, violent, π]νείοντος Βορέαο περιζαμενὲς Διὸς αἴσῃ Hes.fr. 204.126 M.-W.
*περίζῠγος, ον, = περίζυξ, δᾳδίον τὸ π. prob. in Inscr.Délos 1442B70 (ii BC).
περίζυξ I, line 5, for 'Teos .. sens.' read 'Teos, iv BC): these exx. could be referred to °περίζυγος but nom. περίζυξ occurs'; line 6, after 'IG 2².1469.75' add '(iv BC)'
περίζωμα, line 1, for 'girdle .. loins' read 'loin-cloth'; line 4, after 'Arr.Epict. 4.8.16' insert 'by the cult-statue of a goddess, GDI 5702.23 (Samos, iv BC)'
περιζώννυμι, before 'gird' insert '(-ζωννύω Lxx Ps. 17(18).33)' add '2 metaph., τινά τι Lxx Ps. l.c., ib. 40.'
περιήγησις, after lemma insert 'Dor. περιάγησις SEG 23.178 (c.229 BC)'
*περιήκω, to lie around, surround: part. surrounding, πέτραν [τὸν ὄχθον] περιήκουσαν Philostr.VA 3.13, κύκλῳ περὶ τὸ σπήλαιον π. ἄμπελος D.Chr. 2.41. II to have come round to in the course of events, w. acc., τοῦτον τὸν ἄνδρα φαμὲν περιήκειν τὰ πρῶτα Hdt. 6.86.α´, τὰ σὲ περιήκοντα 7.16.α´.1, ἔμελλε .. δίκη περιήξειν καὶ Φιλοποίμενα Paus. 8.51.5. b w. εἰς, ἐπεὶ δ' εἰς τὸν φονέα ἡ ἀρχὴ περιήκει X.Cyr. 4.6.6, Arr.An. 4.13.4. 2 abs. of time, to have come round, καιρῷ περιήκοντι Plu.Ages. 35, ἔτει δεκάτῳ περιήκοντι Aristid.Or. 50(26).1, Parth. 30.2. III to have turned into, w. εἰς, [κεφαλαὶ] εἰς κρανία π. Philostr.Im. 2.19.
*περιηχής, ές, = περιηχητικός, Rh. 1.450.9.
περιήχησις, add '2 rumour, hearsay, PBerol.inv. 7347 (JJP 19.93).'
περιηχητικός, after 'resonant' insert 'Thphr.fr. 89.10 (comp.)'
περιθειόω, add 'med. περιθειούμενος, Plu. 2.168d'
περίθετος, for 'a mask with a wig attached' read 'app. some sort of female headdress'
*περιθεώρησις, εως, ἡ, careful consideration, Plu. 2.820a.
*περίθλασμα, ατος, τό, app. adjacent bruising, Hippiatr. 104.7 (v.l. ὑπόθ-).
περιθραύω, add '3 break at the edges, pf. part. περιτεθραυσμένον Inscr.Délos 1450A35.'
*περιθριγκίζω, fence round, σὺν τῷ περιτετριχισμένῳ τόπῳ TAM 4(1).239; cf. περιθριγκόω, θριγκός (τριγχός).
*περίθυρα, ἡ, = περίθυρον, CPR 7.44.23 (v/vi AD).
περίθυρον, for 'Ephes. 4(1) no. 28' read 'IEphes. 4128 (iv AD), 495 (vi AD), ISmyrna 849 (v/vi AD; sp. -θοιρ-)'
*περιθύτης, ου, ὁ, office-holder in the temple of Asclepius at Pergamum, π. καὶ θεραπευτής Inscr.Perg. 8(3).79, 140, 152.6 (ii AD).
*περιθῠτικός, ή, όν, of or connected with the °περιθύτης, Inscr.Perg. 8(3).140 (ii AD).
*περιθύω, perform the office of περιθύτης, BE 1971.146 (Astypalaea).
*περιθωρᾱκίζω, gird about with a breastplate (in quot. fig.), IGLS 524 (v AD).
περιιάπτω, after 'Theoc. 2.82' add '(dub., πυρὶ θυμὸς ἰάφθη PAntin.)'
περιιστάω, delete 'Chamael.ap.'

περιίστημι B I 3, delete 'come round .. Th. 1.76' and add 'cf. BCH 59.37 (Crannon, πεστάντας = περιστάντας)' II 3, add 'ἡμῖν .. ἀδοξία τὸ πλέον ἢ ἔπαινος περιέστη Th. 1.76.4'
περικαής, after 'adv. -καῶς' insert 'with a very high fever, π. πυρέττειν Gal. 7.722.17, 9.291.10'
περικάθαρμα II, add 'sg., POxy. 2331 ii 10 (iii AD)'
περικαίω I, add 'Ζώπυρος -όμενος, a comedy by Strattis, Stratt. 9, 10 K.-A.; Ἡρακλῆς π., a tragedy by Spintharus, Suid.'
*περικαλδής, ές, = περικαλλής, dub. readings in CEG 327, 335 (both Boeotian, vi BC).
*περικάλυψις, εως, ἡ, covering round, Procl.in Ti. 2.285.4 D.
*περικατάμαγμα, ατος, τό, offscouring, Hsch. s.v. περίψημα.
περίκειμαι I 1, add 'of areas, lie round, τὸν περικείμενον τόπον TAM 4(1).276'
*περικεκομμένως, with excessive concision, βαρβάρως καὶ π. εἰρῆσθαι Sch.Luc. p. 194.23.
*περικεφᾰλάδιον (written -αιδιον), τό, dim. of περικεφαλαία, Inscr.Délos 1439Aai16 (ii BC).
περικεφάλαιος II 2, for 'disorder of the oak' read 'disorder affecting the head, (of pigs)'
περικήδομαι, add 'PKöln 63.16 (?Hellen. epic)'
περικλαίω, add 'med., Call.fr. 228.68 Pf.'
περικλάω II 1, add 'b in chariot-racing, wheel round the turning-post, SEG 34.1437 (v/vi AD).'
περικλεής, after 'περικλειτός' insert 'Ibyc. 1(a).2 P.; cf. pers. n. Περικλέης, Archil. 13.1 W., etc.'
Περίκλειος, add 'also Περικλήϊος, Π. αἷμα λελογχὼς Corinth 8(1).88'
περικλειτός, after 'far-famed' insert 'B.10.19 S.-M.'
περικλήϊστος, after 'far-famed' insert 'GVI 1632.5 (Istropolis, ii/iii AD)'
*περίκλινος, ον, unexpld. adj., στάβλον LW 3.2161 (Syria, ?vi AD), perh. = περικλινής.
περικλίνω III, for 'dub. sens. in' read 'shirk, [ἐν τοῖς πράγμ]ασιν οὐ περιελκινν διὰ τὸ γῆρας MDAI(A) 31.431 (ii BC)'
περικλύμενον, after 'Dsc. 4.14' insert 'Nic.Th. 510; also -ος, ὁ, Hsch.'
*Περικλύμενος (-κλυμος cod.) ὁ Πλούτων Hsch.; cf. Κλύμενος.
περικολούω, after 'Nic.Al. 267' add '528'
περίκομος, add 'II covered with hair, hairy, Poll. 4.37.'
περικρᾱνίον (s.v. περικρᾱνιος II), add 'Gal. 8.205.14; χαλεινοῦ μουλικοῦ μετὰ τοῦ περικρανίου (Lat. frenum mulare cum capistello) Edict.Diocl. in SEG 37.335 iii 6'
περικρᾰτέω, line 3, for 'Carm.Pop. 46.24' read 'Hermocl. 24'
περικρούω 1, add 'ὃς ἂν τοῦτο τὸ μνημεῖον περικρούσῃ IKyzikos 560' 5, for 'ib. 499a' read 'Men.Dysc. 414, Plu. 2.499a'
*περικυκλόκοσμος, ὁ, prob. carved surround in the form of a garland, EA 10.101ff.
περικυλινδέω, line 5, for 'roll about' read 'revolve'
*περικῡμάτιος, ον, having a wavy border, ἱμάτιον IG 2².1514.18; cf. παρακυμάτιος.
περίλευκος, add '3 περίλευκος, ἡ, name of a kind of agate, Epiph. in Lapid.Gr. 197.5, Plin.HN 37.180.'
περίλοιπος, add 'Myc. pe-ri-ro-qo'
περιλύω I, after 'loosen round about' insert 'Plu. 2.586a'
περίμετρον, add 'II round loaf, Aq. 1Ki. 10.3.'
*περίμηρα, τά, = περιμήρια, Ps.-Callisth. 108.7.
*περιμύω, of a wound, close, μῦσεν δὲ πέρι βροτόεσσ' ὠτειλή Hom.fr. 8 D.
*περίνᾱος, ον, round the temple, Ἄμμωνι καὶ τοῖς περινάοις (sc. θεοῖς) SEG 20.719.20 (Cyrene, ii BC); cf. περινάϊος.
περινέμομαι, after 'Plu.Dio 46' add 'Cam. 34.4'
περίνησος, after 'cf.' insert 'Inscr.Délos 1442B57 (ii BC)'
περιοδεία I 1, after 'circuit' insert 'PRein. 109.14 (ii BC, -δηα pap.)'
περιοδονίκης, after 'all the great games' insert '(orig. the four panhellenic festivals, afterwards applied to different grades of multiple victor)'
περίοδος, ὁ, after lemma insert '(A)' add '2 = ‡περιοδονίκης, IG 14.1107 (Rom.imp.), IGRom. 4.1251 (Thyatira, iii AD; to be read for περὶ ὁδῶν); τρὶς π. κῆρυξ Inscr.Olymp. 243 (iii AD).'
περίοδος, ἡ, after lemma insert '(B)' I 1, add 'b detour, Modest.Dig. 27.1.10.3.' 2, for 'slow walk' read 'perambulation' II, add '2 circuitous method of investigation, Pl.Phdr. 274a, R. 504b.'
περίοικος III, for 'on the same .. of us' read 'between the same parallels of latitude but 180° to the E. or W.'
*περιοίχομαι, v. °περιοίχομαι.
περιολισθάνω, line 4, delete 'later -ολισθαίνω'
*περίοργος, ον, = περιοργής, A.Ag. 216 (cj.), codd. περίοργως, v. περιοργής.
περιορίζω I 1, add 'w. acc., περιορίσας τὴν πόλιν SEG 34.1309 (Lycia, AD 161/9)'
*περιόριον, τό, precinct surrounding a tomb, MAMA 6.83 (pl., Attouda).

*περιουλόομαι, pass., *to be cicatrized round*, Aët. 7.36 (287.10 O.).

περιουσία, add 'IV *property, estate*, Cod.Just. 1.3.52.1, 1.3.55.4; of the private property of the emperor, ib. 1.33.4, 1.34.1. **2** *possession* (opp. ownership), Theophil.Antec. 3.2.2.'

περιοχή **I 3 c**, add 'of *arguments* of plays, π. τῶν Μενάνδρου δραμάτων Suid. s.v. Ὅμηρος, Σέλλιος κτλ.' **II**, for 'section .. book' read '*passage* in a book or other writing, D.H.Th. 25' **III 2**, add '*wall of circumvallation*, Lxx Ez. 4.2'

περιπάλαξις, delete the entry.

⁺περιπᾰλάσσομαι, περιπαλαχθῆναι· περιπαλακῆναι Hsch.

περιπείρω, line 2, after '1Ep.Ti. 6.10' insert 'ἑαυτὴν π. κακοῖς Sopat.Rh. 8.30.6 W.'

*περιπέτασμα, τό, = Lat. *velamen* (in transl. of Verg.A. 1.649), PRyl. 478.144.

περιπίπτω **II 3**, add 'med., ἀώροις περιπέσοιτο συνφοραῖς EA 13.19 no. 4'

περιπλέκω **1**, med., add 'fig. Ἥλιον ὠκύμορον Κεία γαμέτην με ποθοῦσα καὶ μήτηρ τύμβῳ Ἰούλλα περιπλέκεται SEG 34.1280 (Paphlagonia, iii AD)'

περιπλήθω, add 'b of groups, *be full in number, teem*, βουκολίοισι περιπλήθουσι Theoc. 25.13.' deleting this ref. in line 3.

περιπλοκή **I 1**, for '*interlacing*' read '*embracing*'; delete 'cf. Luc. .. etc.'; add '*embrace*, Luc.Alex. 39, Stob. 3.39.32'

περίπλοος, ὁ, add '**III** perh. naval manœuvre in which a ship circled round to ram an opponent amidships, Th. 7.36.3, X.HG 1.6.31.' deleting this from section I.

περίπλῠσις, add 'π. ἐρυθρά Gal. 16.623.17'

περιπνοή, for '-πνοία' read 'περίπνοια'

*περίπνυα, ἡ, perh. = περίπνοια, app. some kind of illness (?περιπνευμονία), εὔξετο .. ὑπὲρ τῶν τέκνων διὰ τὴν περίπνυαν TAM 5(1).247 (AD 257/8).

περιποιέω **II 1**, at end delete '*make gain*' and transfer 'X.Mem. 4.2.38' to follow 'X.Mem. 2.7.3' add '**III** *make, construct round*, SEG 32.356 (Aegina, c.550 BC).'

*περιπολαρχέω, command περίπολοι, Robert Hell. 10.284 (iii BC), SEG 32.626 (Illyria, iii BC)'

περιπολέω **II 1**, add 'b w. acc. pers., *attend on*, Sch.S.OT 1322 explaining ἐπίπολος.'

*περιπολία, Ion. -ίη, ἡ, *revolution* of heavenly bodies, Hp.Hebd. 2.

περιπολλόν, add 'also περιπολλά (neut. pl.) Arat. 914'

περίπολος **2**, after 'Eup. 341' insert 'Anon.Hist. (FGrH 105) fr. 2 J.'

*περιπορπίς, ίδος, ἡ, *clasp*, cj. in A.R. 1.767.

*περιπορφῠροῦς, ᾶ, οῦν, = περιπόρφυρος 1, χιτὼν Inscr.Délos 1417Ai30 (ii BC).

*περιπρῖσμα, ατος, τό, = πρίσμα I 1, ἐλεφαντίνων π. IG 2².1408.13, 1409.6, etc.

περιπτύσσω **I 1**, at end delete 'Pass.' and transfer 'Aristaenet. 1.1' to the end of section I 3; after it add 'fig., τὴν τῶν Ἀκεράλων ἄνοιαν περιπτυσσάμενοι Just.Nov. 115.3.14'

περίπτωμα **1**, add '*sudden illness*, TAM 5(1).331, cf. Robert Hell. 10.102 n. 10'

περίρραμμα, add 'Thasos 155.8 (iv BC)'

περιρρέω **II 3**, line 4, after 'Luc.VH 2.11' insert 'τὸ πῦρ αὐτῷ περιέρρει, Call.Dieg. viii.12 (fr. 197 Pf.)' **4**, line 4, after 'Plu.Per. 16' insert '[λέξις] περιρρέουσα τοῖς νοήμασιν *overflowing* with ideas, D.H.Dem. 18'

περιρρογχάζω, add 'PCol. 242.4 (v AD)'

*περίσαος, ον, app. *remaining*, Ϝέτεα BCH 60.182.22, 24 (Thespiae, iii BC).

*περισκᾰπετεύω, *dig a ditch round*, χοῦρον περρεσκαπετευμένον πέλεθρα IIII IGC p. 11 l. 18 (Larissa, iii BC); cf. σκάπετος.

περισκέλια, for '*drawers*' read '*trousers*' and after it insert 'Hsch.'

*περισκελίδιον, τό, dim. of περισκελίς, IG 2².1534.78 (Athens, iii BC), Inscr.Délos 1409Bai98 (ii BC), CPR 5.22.4, 7 (?v AD).

περίσκεπτος **2**, delete '*admired*, Χαρίτεσσι'

περισκηπέω, add 'SEG 31.1288 (Side, AD 249/52)'

*περισκηνάω or -έω, *surround with* σκηναί, Poll. 8.20 (cod.), cf. JHS 75.117.

*περισκηνόω, aor. περεσκήνωσα, *drape round* like a tent, φᾶρος A.Eu. 634.

περισκληρύνω, add 'Gal. 18(1).391.1'

περισκοπέω **II 1**, after 'Arat. 199' insert '464, 852 (tm.)' add 'b w. gen., *look out for*, Εὔροιο περισκοπέειν ἀνέμοιο Arat. 435, 925, 987.'

*περισμῡχηρός, όν, *smouldering, covered in smoke*, Amyntas in Suppl.Hell. 44.5 (pap. -ζμυ-).

περισμύχω, after '*fire*' insert 'transf., of rust'

⁺περισπαίρω, *twitch convulsively upon* a point, barb, or blade, of the dying, γλωχῖσι Opp.H. 5.547 (prob. cj.); δουρί Q.S. 1.624; cf. λώβῃ, i.e. wound inflicted by scorpion, Nic.Th. 773: abs., dub. in Lyc. 68.

περισπασμός **I**, for '*wheeling round*' read 'in military manœuvres, *about turn*' **II**, line 5, delete 'θυμοῦ .. Ec. 2.23' add 'b *preoccupation*, πᾶσαι αἱ ἡμέραι αὐτοῦ ἀλγημάτων καὶ θυμοῦ περισπασμὸς αὐτοῦ Lxx Ec. 2.23.'

περισπάω **II**, lines 1/2, delete 'intr. .. 116.5' **III 1 a**, add 'π. [τὰ χρήματα] εἰς τὴν ἄδικον ἐπιθυμίαν προϊσταμένων IEphes. 18 b 8 (i AD)' **3**, after '*distract*' insert 'Plb. 3.116.5'

περισπογγίζω, add 'sp. περισφ- Paul.Aeg. 6.34.3 (72.20 H.)'

περισπορία, for '*suburbs*' read '*surrounding country*'

περισπουδάζω, add 'rest. in SEG 39.1056 (Neapolis, i BC)'

περισσοδάκτῠλος, for '*with .. toes*' read '*having toes of unequal length*, of a hen'

περισσός **II 1**, line 11, after '(Lucill.)' insert 'Theoc. 26.24'; add 'followed by ὑπέρ, τὸ περισσὸν ὑπὲρ τὸ τετραούγκιον Cod.Just. 6.4.4.18a' **2**, line 3, for 'A.Pr. 385' read 'A.Pr. 383'

*περισσότευκτος, ον, *elaborately worked*, βωμὸν περισσότευκτον ἀγλαόν θ' ἕδος IHadr. 132 (ii AD).

*περισσοφροσύνη, ἡ, *excessive cleverness*, Them.Or. 21.259b.

περιστᾰδόν, for 'Call.Hec. 1.1.4' read 'Call.fr. 260.14 Pf.'

περίστᾰσις **I 2 a**, add '*walled area*, PHels. 12 (ii BC)' **II 1 b**, add 'τραγῳδία ἐστὶν ἡρωϊκῆς τύχης π. Thphr.ap.Diom. p. 487.12 K. (unless π. here = περιπέτεια III 1)'

περιστᾰχῠώδης, for '*with an ear .. on it*' read '*set round spike-fashion*'

*περιστεγνοποΐα, ἡ, *building of roofed hutments*, RA 3.40 (Amphipolis, iii/ii BC).

περιστέλλω **II**, add '**2** abs., of the lips in pronouncing ω, D.H.Comp. 14.' **III 2**, after '*defend*' insert 'A.fr. 154a.18 R.'

περιστέριον, add '**III** = περιστερίς II, Hsch.'

*περιστερόπουλλος, ὁ, *small dove*, CPR 7.42 ii 1 (v AD).

περίστια, line 2, after 'Ar.Ec. 128' insert 'sg., π., τό, *purificatory offering* after return from a funeral, Hsch.'

⁺περιστιγής, ές, *spotted all over, variegated*, ἔρφος Nic.Th. 376 (v.l.); μυρμήκειον ib. 749.

περιστοιχέω, add '**2** fig., *hem in, confine*, in quot. pass., Procl.in R. 1.124 K. (nisi leg. -ισμένων).'

περιστοιχίζω, for '*surround as with toils* or *nets*' read '*surround with a fence*'

περιστολή **I 3**, for '*adornments*' read '*covering, cloak*, or sim.'

περιστροφίς, add 'Alc. 143.8 L.-P. (indeterminate sense)'

περίστρωμα, add 'SEG 28.53 (Athens, iv BC)'

*περίστρωμον, τό, or -ος, ὁ, = περίστρωμα, Dura⁴ p. 100 no. 227.

περίστῡλος **II**, add 'τὸ π. τοῦ γυμνασίου SEG 30.1535 (Lycia, after AD 152)'

περίστωον, add 'sp. περίστωον IEphes. 3239 (ii/iii AD), περίστοον ib. 3233 (iii AD)'

⁺περισυνός, prob. for περυσινός, ἐπὶ τοῖς π. ὁρίοις PFlor. 383.77, 104 (iii AD), PMasp. 128.15 (vi AD).

περισύρω **II 1**, after 'Lxx Ge. 30.37' add 'fig., Ph. 1.178' deleting this ref. fr. section 2.

περισφάλεια, delete the entry.

περισφάλλω, pass., add 'Gal. 18(1).327.9'

*περισφογγίζω, v. °περισπ-.

περισφρῖγάω, before '*gloss*' insert '*swell greatly*, A.R. 3.1258 (tm.)'

περισχοινισμός, for 'BCH 23.566 (Delphi)' read 'SEG 27.119.13 (Delphi, iii BC)'

περισῴζω, add 'in Chr. sense, *save* from damnation, Cod.Just. 1.3.41.3, 1.3.42 pr.'

*περιτειχόω, = περιτειχίζω, EA 12.109 no. 43 (Caria, Rom.imp.).

*περίτιμος, ον, *highly honoured*, Call.fr. 75.52 Pf.

περιτινάσσω, add 'med., *hop about* in pain, Aesop. 235 P.'

Περίτιος, add 'SEG 31.1003 (Lydia, AD 84/5), 32.1555 (Arabia, AD 178), 37.590 (sp. Περειτ-, Maced., AD 189)'

περιτόναιον, after 'Hp.Epid. 7.20' insert '(pl.)'

⁺περιτρέφω, pf. -τέτροφα A.R. 2.738, Nic.Th. 299, 542:— (*cause to*) *form* or *grow round*, ἀυτμὴ πηγυλὶς .. ἀργινόεσσαν ἀεὶ περιτέτροφε πάχνην A.R. 2.738; pass., περιτρέφεται κυκώωντι, i.e. it (the curds) forms as you stir, Il. 5.90, σακέεσσι περιτρέφετο κρύσταλλος Od. 14.477, τὸ περιτεθραμμένον σαρκίδιον M.Ant. 12.1, Gal. 2.504.3. **2** act., *grow round*, τῆς (ῥίζης) ἀκανθοβόλος μὲν ἀεὶ περιτέτροφε χαίτη, λείρια δ' ὡς ἴα τοῖα περιτρέφει Nic.Th. 542, 543; w. abst. subj., κραδίην .. κακὸν περιτέτροφεν ἄλγος ib. 299.

περιτρέχω **II**, add '**c** *make a rapid survey of* a subject, D.H.Din. 11.'

*περιτριγχίζω, v. °περιθριγκίζω.

περίτριμμα **I**, add 'as term of abuse, ὁ μολοβρός· αἰσχρός, ἀναιδής. περίτριμμα Hsch.'

περιτροπή, add 'κατὰ περιτροπήν Sopat.Rh. 8.132.7 W.'

περίτροπος, for 'prob. l. in Plu.Lys. 12' read 'cj.in Plu.Lys. 12 (v. παράτροπος)'

*περιτῠλόομαι, *become callous all round*, Crito ap.Gal. 13.798, Cass.Pr. 13.

περιτυλόω, delete the entry.

περιφαιδρύνω, *cleanse round*, κάρη A.R. 4.663 (v.l. ἐπι-).

περιφαίνομαι, *establish*, aberrant pf., περιπεφανούμενα ἐκφόρια BGU 2390.30 (160/59 BC).

περίφαντος II, for '*famous*' read '*conspicuous from every side*' and add 'Orph.H. 20.1'

περιφέγγω, after '*illuminate round about*' insert 'Plu.fr. 14 S. (περιφευγ- codd.)'

περιφέρεια II, for '*wandering, error*' read '*madness*'

περιφερής I 1, delete 'Hermipp. 4' **2 a**, add 'Hermipp. 73 K.-A.'

περιφθείρομαι II, for this section read '*wander about wretchedly*, Isoc.Ep. 9.10, Lycurg. 40, Men.Dysc. 101'

*****περιφιᾱλισμός**, ὁ, *digging a basin round a plant*, PCol. 179.18 (AD 300) in TAPhA 92.469, cf. ib. 93.164.

*****περιφῐμίζω**, *bind fast* by a spell, SPAW 1934.1041 (tab.defix., Boeotia; -φιμμ-); cf. °φιμόω, *muzzle*, περιφιμόω.

περιφλεγής, line 1, after '*burning*' insert 'καῦμα (*fever*) X.HG 5.3.19 (v.l. πυρι-)'

περιφλίω, read '**περιφλίω**'

+**περίφλωμα**, ατος, τό, *door-frame*, MAMA 8.498 (Aphrodisias, ii AD).

περιφορά II 5, for '*error*' read '*madness*' **IV**, after '*turntable*(?)' insert '(unless = *margin, period of grace*)' add '**V** pl., *rotating objects*, τὰς ἐπ᾽ ἄκρου σφαιροειδεῖς π., of the angler-fish's lures, Ael.NA 9.24.'

περιφόρινος, delete 'cf. περίφουρνος'

περίφουρνος, delete the entry (v. °φοῦρνος).

περίφραγμα I, add '**2** *the toils of a hunting-net*, Sch.Pi.N. 3.89 (pl.).'

περιφραδής, adv., add 'ὅππ[ως τοῦ]το περιφραδέως σὺ τελέσσῃς SEG 31.1288 (Side, AD 249/52)'

περιφροσύνη 1, for '*cunning*' read '*cleverness, skill*' and after 'Them.Or. 21.259b' insert '(dub., v. °περισσοφ-)' add '**2** *contempt*, Plu. Comp.Alc.Cor. 3.'

περίφρων I, add 'Mitchell N.Galatia 14 (Rom.imp.)'

περίχαλκος, add 'PMich.inv. 1718.18 (APF 33.57, iv BC)'

περιχαλκόω, add 'θυμιατήριον ξύλινον περικεχαλκωμένον Inscr.Délos 1442B45 (ii BC)'

περιχειρίδιον (s.v. περίχειρον), before 'Hsch.' insert 'Inscr.Délos 1417Bii46'

περίχειρον, add 'Inscr.Délos 1442B24'

*****περιχρήσιμος**, ον, *very useful*, Ps.-Callisth. 131.17.

περιχύτης, after 'Ptol.Tetr. 179' insert '(v.l.)'; add 'Stud.Pal. 22.75.39 (iii AD, περι[χ]οιτ- pap.)'

περιχωματίζω, after '*surround with a dyke*' insert 'PTeb. 775.9 (ii BC)'

περιχώριος, add 'Sch.S.OC 1059; π. [ἀγῶνες] Sch.Pi.I. 1.11'

περίχωρος, last line, after 'Ev.Matt. 14.35' insert 'SEG 36.1095 (Sardis, iii AD), Res Gestae Saporis 34 (pl.); τὸ π. IG 5(2).3.10 (Tegea, iv BC)'

περιψάω, after '*wipe clean*' insert 'Hippon. 104.18 W.'

περίψημα, add 'perh. also *something of no value*, π. σοι ποίει, Vit. Aesop.(G) 35'

περίψηφος, add '**II** subst. περίψηφον, Dor. -ψᾱφον, τό, prob. *balance, surplus*, Lindos 419.12 (i AD).'

*****περιωπής**, ές, *seeing in all directions*, διφυῆ π. .. Ἔρωτα Orph.A. 14; *seen from all directions* or *having views all round*, π. νηόν epigr. in IGBulg. 2086.

περίωπος, delete 'in Orph.A. 14' to end.

πέρνημι, line 7, after 'aor. ἔπρησα' insert '(dub., v. Meiggs-Lewis 16)'; after '**πιπράσκομαι**' for '*first* .. 224a' read 'Lys. 18.20, Pl.Sph. 224a, Phd. 69b (s.v.l.)'; after '**πιπράσκω** first found in' insert 'Thphr.fr. 98 (if the wd. is fr. Thphr.), then'

*****περοίχομαι**, *run* or *extend round*, Hes.Th. 733 (s.v.l.).

περόνη I 1, after '(IG) 2².1388.20' insert 'δίβολος περόνᾱ AP 6.282 (Theod.)'

*****Περϙοθάριοι**, οἱ, class of refugees at Locris, IG 9²(1).718.27 (v BC).

*****πέρροδων** (gen. pl.), unexpld. wd. in list mainly of foodstuffs, PWash. Univ. 52.3 (iv AD).

περσέα, for '*Mimusops Schimperi*' read '*Cordia myxa, sebesten* (a plum-like tree)'

περσέϊος (s.v. πέρσειον), add 'PMich.inv. 4001 (ZPE 24.83, i/ii AD, sp. περσοίνου)'

Περσεφόνη, line 2, for 'Φερσεφόνη' read 'Φερσεφόνᾱ'; line 3, after 'etc.' insert 'cf. Φερσοπόνη Inscr.Cret. 2.xvi 10 (ii AD)'; after 'CIG 4588' insert '(Φερσεφόνεια Hsch.)'; add 'also v.s.v. Πηριφόνα, °Πηρεφόνεια, Φερσέφασσα (where some refs. are repeated)'

*****περσή** ὀξύ(μαλα), τά, app. *peaches*, Lang Ath.Agora XXI B20.

Πέρσης I a, line 3, before 'The Greeks ..' insert 'fr. Old Iran. Pārsa' add '**2** as epith. of Zeus, TAM 5(1).267.'

Περσίδιον, v. πέρσειον.

*****περσικοποιός**, ὁ, *maker of slippers* (Περσικός 2), IG 2².11689 (iv BC).

Περσικός 5, before 'Cratin. 259' for 'ὁ Π. alone' read 'Π. ἀλέκτωρ' add '**9** as epith. of Artemis, BCH 11.447-8 no. 5, SEG 31.998, etc.'

*****Περσῖνος**, η, ον, *Persian*, σουβρικάλλιον Περσεῖνον Stud.Pal. 20.41ᵛ.3 (s.v.l., v. Tyche 2.10, ii AD).

πέρσις, for 'ἑως' read 'ιδος' and after 'ἡ' insert 'acc. -ιν Arist.Po. 1456ᵃ16'

*****περσονάλιος**, α, ον, *personal*, ταῖς περσοναλίαις (sc. ἀγωγαῖς = *actionibus in personam*) Just.Nov. 136.5 pr.

πέρσυ, after '(Cyzicus)' insert 'POxy. 1299.8 (iv AD); cf. ‡πέρυσι, mod. Gk. πέρσι'

περτέδοκε, v. °προσδίδωμι.

πέρυσι, add 'PFlor. 189.4 (iii AD), PSI 81.14 (v/vi AD); cf. ‡πέρσυ'

+**Περφερέτᾱς**, α, ὁ, epith. of Zeus in Thessaly, Liv.Ann. 3.155 (RPh. 35.128ff., Mopsium), SEG 23.444 (Larissa, both i BC); also Φερφερέτας IG 9(2).1057.1 (Mopsium, i BC).

*****πεσσά**, ἡ, part of a building, perh. the same as °πεσσός III 2, PMichael. 58A2 (vi AD).

*****πεσσάρακοντα**, v. ‡τεσσαράκοντα.

πεσσεύω, lines 3/4, for '*fortune* .. *affairs*' read 'i.e. *shifts* them *up and down like pieces on a board*'

*****πεσσοβολία**, ἡ, *casting of* πεσσοί, used as dice, PNess. 21.20, 22.10 (both sp. πεισσ-, vi AD).

+**πεσσοποιέομαι**, app. *make oneself a pessary of*, Poet.de herb. 103.

πεσσός II, line 2, after '*pessary*' insert 'Hp.Jusj.' **3**, for 'JHS 8.118 (Iasos)' read '*Iasos* 20' **III**, for this section read 'archit., *pier* (supporting an arch), Str. 16.1.5, Procop.Aed. 1.1.37. **2** perh. *terrace*, POxy. 1272.6 (ii AD).'

*****πέσωμα**, ατος, τό, *fall*, Kretschmer GV p. 122.

πεταλία I, for '*crate*(?)' read '*basket*'; after 'PCair.Zen. 99.3' insert 'SB 9091, ostr. in SB 7402.2, al.' and insert 'all' before 'iii BC' **II**, delete the section.

πέταλον I, line 4, after '(lyr.), etc.' insert '*cornstalk*, κορωνιόωντα πέτηλα βριθόμενα σταχύων Hes.Sc. 289' **II 1**, add '**b** κόλλης π. *sheet of glue*, IG 11(2).203B 97 (Delos, iii BC).'

πεταλουργός, add 'subst., Anon.Alch. 379.7'

πέτασος III, for 'OGI 510.4' read 'IEphes. 2039-41'

*****πεταστικός**, ή, όν, wd. of uncertain meaning, π. διάκρισις Physici 2.228.5.

πετασών, add 'Varro RR 2.4.10, Mart. 3.77.6, etc., Edict.Diocl. 4.8'

πετεινός, 3 lines fr. end, after 'Lycurg. 132' insert 'πετεινός, ὁ, *cock*, opp. ὄρνις (hen), Diogenian. 3.50'

πετευριστής, add 'POxy. 2860.16 (ii AD)'

πέτευρον II 2, for '*platform, stage*' read 'applied to a narrow platform at the top of a σαμβύκη II' **III**, for '*springe, trap*' read 'dub. sens. in passage where Hebr. means *depths*' **IV**, add 'Call.fr. 186.4 Pf., Lyc. 884'

πετηνίς, for 'κόρις' read 'ἀκρίς (La.)'

*****πετίτωρ**, ορος, ὁ, *applicant*, app. a Mithraic title, Dura⁷ p. 87 no. 848, cf. Lat. *petitor*.

πέτομαι, line 19, (pf. πέπτηκα), after '(v. ποτάομαι)' insert 'pf. intrans. (κατ)έπτηκα Men.Kol. 40' **II 1**, line 3, delete 'of fickle natures'; line 6, before 'of fame' insert 'τῶν μαινομένων πέτεται θυμός τε νόος τε *soar unchecked*, Thgn. 1053'

πετρακισχείλιη, add 'δραχμὰ[ς] πετρακισχιλίας BCH 60.178.24 (Thespiae, ii AD)'

*****πετρᾰκονταεττά**, = τεσσαρακονταεπτά, *forty-seven*, SEG 26.672.31 (Larissa, ii BC).

*****πετρᾰκονταπέμπε**, = τεσσαρακονταπέντε, *forty-five*, SEG 26.672.42 (Larissa, ii BC).

πετράς, add 'also Thess., SEG 31.577 (136/5 BC)'

*****πέτρες**, = τέσσαρες, *four*, SEG 26.672.43 (Larissa, ii BC).

*****Πετρησιάρχης**, εω, or -ος, ὁ, prob. ruler of a district of Ceos, IG 12(5).610.2 (Ceos, iii BC).

πέτρινος III, for this section read 'π. ἀκοντισμός, term (acc. to Arrian, Celtic) for manœuvre in which javelin is thrown from horseback while on the turn, Arr.Tact. 37.4'

*****πετροκόλαπτος**, ον, *engraved on stone*, π. ἔπος SEG 32.896 (Crete, ii/i BC).

πετρορριφής, for '*hurled from a rock*' read '*pelted with stones*'

πέτρος I 1, add '**b** *tombstone*, AP 7.465 (Heraclit.; = EG 1524 P.).' **II**, add '(app. fr. misunderstanding of Skt. *pattra* "leaf")'

πετροφυής I, add 'Τιθορεέ[α] πετροφ[υ]εῖ GVI 766'

πετρόω I 1, after '(Theodorid.)' insert 'Nonn.D. 47.591' **2**, delete the section.

πετρών, add 'Princeton Exp.Inscr. 1019'

πεττάρακοντα, add 'BCH 60.179.37 (Thespiae, iii BC)'

πέττᾰρες, add 'also Thess., SEG 31.575 (171 BC), ib. 13.394 (iii AD), etc.'

+**πεύθω**, aor. part. πεύσανς Inscr.Cret. 4.83 (Gortyn, v BC), πούσας ib. 1.xix 1.13 (Malla, iii BC), pass. πευσθένς ib. 4.83, *lay information against*, Leg.Gort. 8.55, Inscr.Cret. 4.162.7.

πεύκη II, add '**3** *oar*, Tim.Pers. 13, 76 P.'

*****πευκώδης**, ες, *covered with pine-trees*, λόφος Inscr.Olymp. 46.36 (ii BC).

πεφεισμένως, after 'cautiously' insert 'Phld.Rh. 2.140 S.'
πεφλάζει, add 'cf. παφλάζω'
*πεφοριῶσθαι· πεπαχύνθαι τὸ δέρμα τοῦ ὀφθαλμοῦ Phot., Harp.
πῇ II 3, add 'ὅππῃ μὲν .., πῇ δὲ .. SEG 32.1502 (Palestine, v AD)'
πήγανον, π. ἄγριον, add 'PColl.Youtie 4.7 (iii AD)'
*πηγάσιος, α, ον, belonging to the spring, Νάρκισ⟨σ⟩ος π. Syria 31 Pl.C (facing p. 198); cf. Πήγασος.
πηγή I 1, after 'Dor. πᾱγά' insert '(also Boeot., e.g. SEG 23.297, Lebadea, iv/iii BC)'; line 2, after 'cf.' insert 'Hes.Th. 282, Stesich. 7.2 P.'; lines 3/4, delete 'κρουνῷ .. 22.147'   II 1, after 'source' insert 'Il. 21.312, 22.147'; add 'b where π. is dist. fr. κρήνη, π. is the spring, κρήνη the artificially constructed fountain, Th. 2.15, Paus. 2.3.3, 4.31.6; cf. οὐδ' ἀπὸ κρήνης πίνω Call.Epigr. 28 Pf.'
πῆγμα I 4, add 'κατὰ π. καὶ κατὰ τὴν ἀρχαίαν συνήθιαν perh. a fixed rule or tariff (cf. πήγνυμι IV), pap. in ASNP ser. II.6.1 (v AD)'   add 'IV block (of land), ἰδιωτικῆς καὶ βασιλεικῆς πάσης οὔσης ἐν ἑνὶ πήγματι PNew York 20.11 (AD 302).'
πήγνῡμι III, add 'J.BJ 6.3.4'
πηγός II 2, add 'Antim. 145 Wy.'
πηδᾰλιοῦχος, add 'Dor. πᾱδᾰλιοῦχος, in lit. sense, Clara Rhodos 8.228 (Rhodes), SEG 33.640 (Rhodes, ii/i BC)'
πηδός, line 2, for 'πάδος (q.v.)' read '‡παδός'
πηΐσκος, delete 'Dim. of παῖς'; for 'Supp.Epigr. 2.509.5 (Crete, v BC)' read 'Inscr.Cret. 1.x 2.5, 7 (Eltynia, vi/v BC)'
πηκτίς I 1, after 'stringed instrument' insert 'a form of harp' and add 'Anacr. 28, 41 P., Diog.Ath. 1.9 S.'; after '= λύρα' insert 'AP 9.270 (Marc.Arg.), Philostr.Im. 1.10.3, Anacreont. 43.10 W.'   2, add 'IG 4.53 (Aegina, Rom.imp.)'
πηκτός III, line 4, for 'salt obtained from brine' read 'rock salt'
πηλᾰμύς, add 'also πηλᾰμίς Hsch. La. s.v. θύννον'
*πηλεύς, ὁ, wd. of unkn. meaning in damaged pap., CPR 7.44.12 (v/vi AD).
πήληξ 1, after '(Il.) 16.797' insert 'Od. 1.256'; add 'cf. Πήληξ, citizen of the Attic deme Πήληκες, Aeschin. 2.83'   add '3 ὄργανόν τι ψαλτήριον Poll. 4.61 (prob. a crested harp).'
*πηλοάρτης, ου, ὁ, (αἴρω) clay- or mortar-lifter, PRein. 2065.47, 48 (ii AD) in JJP 11/12.66, PRyl. 642.11 (iv AD).
πηλόγονος, for '= γηγενής, used of the giants' read 'used of the human race'
*πηλοκάρβων, ωνος, ὁ, kind of lute (clay), Anon.Alch. 38.1.
*πηλοπᾰτέω, add 'II tread clay, as term in pottery-making, PMich.v 241.33 (i AD).'
πηλοποιός, add 'Gal. 10.395.11'
πηλός, line 2, after 'Phryn. 38' insert 'pl., Plu. 2.993e'   I 2, after 'mire' insert 'Heraclit. 5'
πηλόω, line 3, after 'Plu. 2.980e' insert 'πηλοῦσθαι τῷ θεῷ Aristid.Or. 48(24).74'
πῆλυξ, read 'πήλυγξ' and add 'cf. σπήλυγξ'
⁺πήλωμα, ατος, τό, mud-pit (in a gymnasium), lutinae πήλωμα Char. p. 33 K.; perh. also CIG 2758 (ZPE 79.289).
πηνίκα, line 4, (πηνίκα ἄττα), after 'Ar.Av 1514' insert 'fr. 617 K.-A.'
πηνίον I, add '3 thread, Sch.E.Hec. 444, 471.'
πήνισμα, for 'woof on the spool' read 'thread (for the woof)'
πῆξις I 2, for 'fixing' read 'determination, computation'
*Πηρεφόνεια, = Περσεφόνη (Lacon.), Hsch.; cf. Πηριφόνα.
πηρίν, delete 'ἐλάφου πηρίς Hsch.'
πηρός, before 'Dor.' insert 'accented πῆρος in Att. acc. to Hdn.Gr. 1.190'; line 3, delete '(cf. .. 57)'; line 4, after 'Hp.Mul. 2.131' insert 'blind, Aesop. 37 P., AP 9.46 (Antip.Thess.)'
πῆρος, for 'dotage' read 'injury' and for 'Alc. 98' read 'Alc. 10.4 L.-P.'
*πηροφόρος, ὁ, one who carries a wallet, Hsch. s.v. θυλακοφόροι (πυρο-cod.).
πήρωσις, add 'fig., πήρωσις ἄπαις βίος AP 9.359.7 (Posidipp. or Pl.Com.; = EG 1694 P.)'
πηχυαῖος, add 'also πηχιαῖος, IG 12(2).11.14 (Mytilene, iii BC)'
πηχῡνω, for 'take in one's arms, embrace' read 'take in the crook of one's arm'; delete 'χείρεσσι .. (Rhian.)'
πῆχυς, line 1, for 'πᾰχυς Alc. 33' read 'and Dor. πᾱχυς Alc. 350.7 L.-P., Call.fr. 196.38 Pf.'; line 3, after 'πήχεως' insert 'IG 2².1013.34, al. (ii BC)'   III 1, line 1, after '(the bridge)' insert 'h.Merc. 50'
πιαίνω I, add 'Arist.Oec. 1345ᵃ3' deleting this ref. in section II 3.
*πίακλον, τό, Lat. piaculum, expiatory offering, SEG 7.351 (Dura, iii AD).
πιάτρα, after 'TAM 2 .. (Tlos)' insert '870.6; also πιέτρα ib. 847, 848'
πιβρᾱτος, after 'privatus' insert 'βαλανεῖ πιβράτου private bath attendant'
πιγκέρνης, for 'cupbearer' read 'bar-keeper' and add 'CPR 8.56.20 (v/vi AD)'
πιδῠλίς, for '= πιδακόεσσα' read 'πέτρα ἐξ ἧς ὕδωρ ῥέει' and add 'Call.fr. 67.12 Pf. (cj.)'
πιδόω, after 'gush forth' insert 'Hp.Epid. 5.16'

πιέζω I, add '2 in chariot-racing, perh. bore, μὴ πιάσωσιν, μὴ παραβύσωσιν SEG 34.1437 (Syria, v/vi AD).'   II, add '6 imprison, SB 9786.7 (iv AD).'
*πιειρῶς, adv. fr. πίειρα, richly, πιειρῶς καὶ λιπαρῶς Sch.Pi.N. 1.16.
Πιερία, add 'cf. Myc. pi-we-ri-ja-ta, pers. n.; pi-we-ri-di (dat.), fem. n.'
Πιερίδες, after 'P. 1.14, etc.' insert 'sg. Πιερίς prob. in Pratin. 3.4 S.; epigr. in SEG 37.1175.10 (Pisidia, ii AD)'
Πιερικός, after 'Hdt. 4.195' insert 'AP 7.34 (Antip.Sid.; = EG 3444 P.)'
πιεστήρ, line 2, for '(pl.)' read '(πιεστῆ[ρας], but perh. πιεστή[ρια] shd. be rest.'
πιεστήριος II, after 'press' insert 'IG 1³.425.12 (Athens, v BC; pl.)'
πῐθάκνιον, add 'also φιδάκνιον IG 1³.425.24 (Athens, v BC), 2².1627.313, 316 (Athens, iv BC)'
πῐθεών, for 'later form of πιθών' read '= πιθών' and add 'TAPhA 65.128 (Olynthus, iv BC)'
⁺πῐθήκη, ἡ, monkey-spider, also called ὀρειβάτης, ὑλοδρόμος, ψύλλα, Ael.NA 6.26.
πῐθηκίζω, after 'of flatterers' insert 'ὑπό τι μικρὸν ἐπιθήκισα Ar.V. 1290 (lyr.)'
πῐθήκιον II, for pres. def. read 'device attached to the centre of a platform between two ships supporting a siege-engine to steady it in rough weather'
πίθηκος, after 'Dor. πίθᾱκος' insert 'Naukratis ii p. 68, Ar.Ach. 907; (as pers. n. SEG 29.938, Selinous, vi BC)'
πῐθοιγίς, for 'Call.Aet. 1.1.1' read 'Call.fr. 178.1 Pf.'
πίθος I 1, at end for 'Hdn. 8.4.5' read 'Hdn. 8.4.4'
πικρία, add '4 of style, severity, D.H.Vett.Cens. 2.5.'
πικρίδιον, after 'Endivia' insert 'prob. in com. in PTeb. 693.20'
⁺πικρόλωτος, ὁ, a kind of lotus, Gal. 14.159.13.
πικρός, line 1, after 'Od. 4.406' insert 'Maced. βικρός acc. to Plu. 2.292e'
πικρότης II, add '2 of style, harshness, D.H.Pomp. 6.9.'
*πικροφᾰγία, ἡ, bitter diet, An.Boiss. 3.415.
*πικροχολία, ἡ, bitter bile, biliousness, Gal. 7.727.1.
πιλίον 1, add 'as a freedman's cap, cf. Lat. pileus, Plb. 30.18.3'
*πίλιος, α, ον, made of felt, Inscr.Délos 1441Ai13 (ii BC).
*πιλνόν· φαιόν, Κύπριοι Hsch.
πιλοποιός, add 'SEG 29.1195 (Lydia, AD 194/5)'
πιμελοσαρκοφάγος, for 'sepulchre of fat' read 'eater of fat flesh'
πιμεντάριος, after 'apothecary' insert 'MAMA 8.574 (Aphrodisias, ii/iii AD)'
Πιμπληΐδες, add 'sg., the nymph Pimpleis, mother of the Muses, Epich. 41'
πίμπλημι I, add '4 complete a period of time, τρισσὸν ἐπ' εἰκοστῶι πλήσας ἔτος GVI 842 (ii/i BC), SEG 33.1458.   5 fulfil destiny, etc., ζωᾶν πλήσατ' ἐμεῦ τέκεος SEG 37.990 (Miletus, ii BC), IEphes. 2101A.'
πίμπρημι, line 6, after 'E.Andr. 390, etc.' insert 'Πρῆσαι CEG 2.7'   I, third line from end, delete 'of wounds, .. Id.Al 438)'   II, after 'distend' insert 'inflame'; after 'Nic.Al. 477' insert 'al.' and after 'Dsc. 4.32' insert 'Nic.Al. 438, 571; med., ib. 345'
πῐνᾰκοθήκη, after 'picture-gallery' insert 'Varro RR 1.2.10, al.'
*πῐνᾰκοπλήστης, ου, ὁ, ?one who fills plates, PAmst.inv. 21 (ZPE 9.49).
πίναξ 2, after 'Lang Ath.Agora xxi B12 (iv/iii BC)' add '3, for 'Simon. 178' read 'AP 5.159.4 ([Simon.]; = EG 317 P.)'
πῐνᾱρός, after 'Eup. 251' insert 'τὰ πάλαι πιναρᾷ κεκαλυμμένα λάθᾳ ξόανα epigr. in Inscr.Délos 2548 (i BC)'
*πῐνεγχύτης, ου, ὁ, cup-bearer, Ps.-Callisth. 83.13,15.
*πιννᾱνᾶς, ᾶ, ὁ, setter of pearls, JHS 58.255 (Jewish medallion; vi AD).
πῖνον, delete 'cj. .. BC'
πίνος 2, add 'π. litterarum, Cic.Att. 14.7.2'
πῑνω, lines 9/10, imper. πίει, add 'SEG 32.31, 33.64(b) (Att., vi BC); also πῖ TAM 4(1).324 (Rom.imp.); πίον SEG 30.1129 (Alena, iii BC)'; line 20, after 'aor. ἐπόθην' insert 'Nic.Th. 622, Al. 432'; line 22, after 'EM 698.52' insert '(Alc. 401 L.-P.) πῶ also GDI 1376, 1377 (Dodona), CR 57.102 (Attic vase, vi BC); Cypr. po-ti πῶθι ICS 264.1'   III, add 'ἔρωτα πίνων Anacr. 105 P.'
*πίνωσις, εως, ἡ, ?tarnishing, Al.Pr. 25.12.
*πιπερῖτις, ιδος, ἡ, perh. pepperwort, Lepidium latifolium, (in second ref. perh. also Polygonum hydropiper), Plin.HN 19.187, 20.174.
*Πίπλειαι· αἱ Μοῦσαι ἐν τῷ Μακεδονικῷ Ὀλύμπῳ Hsch.; cf. Πίμπλεια.
πίπτω, line 12, for 'Simon. 183.7' read 'AP 7.24.7 ([Simon.]; = EG 348 P.)'   A 1, line 7, for 'Simon. l.c.' read 'AP([Simon.]) l.c.'   B I, add '3 lie down, take one's place at table, πρὸς δαῖτα E.Ion 652; cf. ἀναπίπτω 5.'   V 3, add 'πέπτωκε in receipts is freq. folld. by the payer's name in nom. and the amount in acc. (but usu. a symbol only), πέπτωκεν Θέωνι .. Τέως .. (δραχμὰς) ἑβδομήκοντα SB 1178, OWilck. 1491, 316, al. (iii BC)'   VII, last line, before 'ὅσα' insert 'fall within the range of'; add 'D.S. 3.44'
*πίπυλος, ὁ, gloss on κορυδαλλός, Sch.Theoc. 10.50.

Πῖσα, at end after 'Pi.' insert 'and B.'
Πῖσαῖος, add 'of the Olympic games, Πισαῖον ἀγῶνα SEG 23.113 (Attica, iii AD)'
πισγίς, for 'dub. sens. in' read ' = πυξίς'
***Πισιδικός**, ή, όν, Pisidian, θεῶν Πισιδικῶν SEG 32.1289 (Phrygia, Rom.imp.).
***πισκάριον**, τό, Lat. piscarium (forum), fish-market, SEG 19.115.23 (Athens, ii/iii AD).
***πισκεῖνα**, ή, Lat. piscina, basin, IHadr. 47 (i/ii AD); sp. φισκῖνα CIL 3.14894 (Salona).
***πίσσαι**, αἱ, peas (cf. πίσος), SEG 9.35 (Cyrene, ii BC).
***πισσάριον** (B), τό, gloss on ἔτνος soup, Sch.Pl.Hp.Ma. 290d; cf. πίσος, πίσινος.
***πισσία** or -εία, ή, (πιττ-), treatment with pitch, Gal. 18(2).900.5.
***πισσοκωνάω**, cover with pitch, Hsch. s.v. κωνῆσαι.
πισσόω I, add 'κούφ(ων) πεπισσωμένων PVindob. G23243 (ZPE 64.80, vi/vii AD)'
πιστάκιον, add 'PVindob.Worp 11.14 (vi AD)'
***πιστίκιον**, τό, a cereal, prob. hulled emmer (Egyptian wheat), Edict. Diocl. 1.7, PCair.Isidor. 11 iii 39 (iv AD), al., EPap 5.102-103 (nos. 24.2.4, 25.2.5, both iv AD).
†**πιστικός** (A), ή, όν, adj. describing nard, perh. = πιστικός (B), in sense genuine, Ev.Marc. 14.3, Ev.Jo. 12.3.
πιστικός (B), add '3 ὁ, trustee, custodian, CPR 8.67.19 (vi/vii AD), 85.2 (vii/viii AD; pap. has πι[].'
πίστιον, delete the entry.
***πίστιος**, a, ον, reliable, trustworthy, TAM 2.338 (Xanthus).
πίστις I 1, line 6, after 'OT 1445, etc.' insert 'εἰς πίστιν ἐλθεῖν Χρόνῳ Men.Dysc. 282' **2 c**, after 'PGnom. 180 (ii AD)' insert 'Just.Nov. 119.7; sim. κακὴ πίστις ib.' **4**, add '**b** creed, belief, CodJust. 1.1.3.2, 3.' **VI**, after 'Fides' insert 'ἐν τῷ ναῷ τῷ τῆς Π. Thasos 174E.7 (80 BC)'
†**πιστός** (A), ή, όν, of medicines, to be drunk, A.Pr. 480.
πιστός (B) **A I 1**, add 'as the title of a legion (transl. Lat. fidelis), SEG 33.1194 (Cappadocia, i/ii AD), 32.1390 (Commagene, ii AD), etc.' **B 3**, add 'οἱ πιστοί the company of believers, ITyr 15'
πιστόω III 3, at end delete 'τίς ἂν .. Opp.C. 3.355' add '**4** feel sure of, be convinced by, Onas. 13.3; τίς ἂν τάδε πιστώσαιτο, .. ὅτι Opp.C. 3.355, 417.'
***πιστρίνη**, ή, Lat. pistrina, bakery, ODouch 39, 57.
πίτνημι, lines 6/7, delete 'θαλάμων'
πιττακιάρχης, add 'PFlor. 18.1 (ii AD), Stud.Pal. 20.236.1 (v/vi AD)'
***πῐτυοφόρος**, ον, pine-bearing, expl. of pinifer in Virgil gloss., PNess. 1.605 (vi AD).
πίτυς, add 'A.fr. 78c.39 R.'
πιτύστεπτος, delete 'poet. for *πιτυόστ-'
πῐτυώδης, after 'abounding in pines' insert 'π[ι]τυώδεῖ δείρᾱι B. 12(11).39 S.-M.'
***πιτυών**, ῶνος, ὁ, pine-grove, platanonas et aerios pityonas, Mart. 12.50.1.
πίων I, add '3 of the eyes, fatty in appearance, glistening: of a snake, Nic.Th. 443; of a person in fever, τὰ λευκὰ τῶν ὀφθαλμῶν λαμυρώτατα καὶ πίονα Aret.SA 2.1, 3.10.' **III**, sup., add 'Hp.Genit. 1'
πλᾰγιάζω III 1, add '**b** use a construction other than nominative and finite verb, opp. ὀρθόω III, Aristid.Rh. 2 p. 533 S.; pass., ib. 1 p. 465 S.'
***πλᾰγιάριος**, ὁ, Lat. plagiarius, kidnapper, ?Teuc.Bab. p. 50 B.
†**πλᾰγιαυλίζω**, play the πλαγίαυλος, Eust. 1157.40.
†**πλᾰγίαυλος**, either aulos with reed mouthpiece inserted at the side or flute, Theoc. 20.29, Bion fr. 10.7 G., AP 11.34.5 (Phld.); = φώτιγξ, Juba 16 J.
πλάγιος I 1, add '**b** subst., ἡ πλαγία (sc. θύρα), side or back door, POxy. 3642.28-9 (ii AD), CPR 5.17.8 (v AD). **c** πλάγιος αὐλός, = °πλαγίαυλος, Luc.VH 2.5, Longus 1.4.3, 4.26.2, Ael.NA 6.19.' **4**, ἐκ πλαγίου, add 'transf., indicating collateral relationship, CodJust. 6.4.4.14e, f' **II**, last line, for 'Ph. 2.173' read 'Ph. 2.172'
πλαγκτύς, for 'Call.Aet. 1.2.7' read 'Call.fr. 26.7 Pf.'
†**πλᾰδάω**, to have an excess of moisture, be moist or wet, φλύκταιναι Nic.Th. 241, σκύλα 422, ῥινοί 429; of unripe corn, Ph. 1.179; of soil, A.R. 2.662; of unhealthy flesh, Hp.Aër. 10. **2** of liquids, to be thin or watery, οὐρόν Nic.Th. 708, πλαδώωντι ποτῷ id.Al. 119; cf. πῆξις πλαδῶσα Arist.HA 516ᵃ3; transf., w. ref. to looseness of the stomach, π. τὸν στόμαχον Dsc.Eup. 2.9, Gal. 13.145.10. **3** fig., of the mind, understanding, etc., to be green or unformed, Ph. 1.441, 459, 2.411. **4** in causal sense, ἐπλάδα· κατέδευεν Hsch.
***πλάδιμος**, η, ον, perh. for *πλάθιμος, moulded, decorated with moulding, ἰσώστας πλαδίμους MAMA 8.552.10.
πλᾱθά, after 'figure' insert 'AP 13.21 (Theodorid.; = EG 3148 P.)'
πλᾱθᾰνίτης, for 'baked in a mould' read 'kneaded on a board'
πλᾱθᾰνον, for 'dish or mould in which bread, cakes, etc., were baked' read 'kneading-board or tray'
***πλᾱθᾰνος**, ὁ, π. ἄγγος τι POxy. 3000 sch. 13; cf. °πλάθανον.

πλάθω, line 1, delete 'in lyr.'
πλᾰκίτης, add 'Anon.Alch. 286.26'
***πλᾰκοπρίστης**, ου, ὁ, sawyer-mason, PMerton 97 (vi AD).
***πλᾰκός**, οῦ, ὁ, = πλάξ, slab of stone, BE 1966.512 (Sicily).
***πλᾰκοτόν**, τό, (?for πλακωτόν), pavement, SEG 26.1627 (Apamea, AD 533).
πλᾰκουντάριος, add 'Vit.Aesop.(G) 63'
***πλᾰκουντοφᾰγέω**, eat cakes, Hsch. s.v. να[υ]στοφαγεῖν.
πλακοῦς II, delete the section.
πλᾰκώ, add 'Anon.Alch. 325.1'
πλᾰνάω II 4, line 4, after 'Hp.Prog. 24' insert 'πεπλ. μέτρα irregular rhythms (in prose), D.H.Comp. 25, 26'
πλᾰνης I 2, after 'planets' insert '(incl. sun and moon, Nonn.D. 5. 67-84)'
***πλᾰνησίμοιρος**, ον, causing fate to go astray, PMag. 4.1368.
πλᾰνητικός, add '**II** misleading, λόγοι Sch.E.Hipp. 486.'
***πλᾰνιτικόν**, τό, wd. occurring in list of garments, PMich.XIV 684.9 (vi AD); cf. πλανίς.
πλάσις 1, add 'D.S. 1.16.1'
πλασμᾰτώδης, add '**2** artificial, contrived, π. ἐστὶν ἡ ὑπόθεσις καὶ ἡρμοσμένη Hipparch. 1.4.6.'
πλάσσω, line 18, after 'Gal. 6.313' insert 'Poll. 6.74, 10.112, Hsch. s.v. πλάθανον' **I 2**, delete the section.
πλάστης I, at end for 'perh. = τριχοπλάστης' read 'hairdresser (cf. κεροπλάστης)'
πλάστιγξ, add '**V** μέρος τι τοῦ αὐλοῦ, καὶ σύριγγος τὸ ζύγωμα Hsch. **VI** the constellation Libra, Man. 4.242.'
πλαστός I 1, add 'μήτε πλαστὰν μήτε γραπτάν .. [εἰκόνα] Plu. 2.215a (Ἀποφθέγματα Λακωνικά)'
πλᾰτᾰγή, add 'Plu. 2.714e'
***πλᾰτᾰμος**, ὁ, object acting as boundary-marker, perh. = πλαταμών, IG 14.352 ii 12,17.
πλᾰτᾰμών 5, delete the section transferring exx. to section 1.
***πλᾰτᾰνώδης**, ες, like a plane-tree, σπίλαξ· μῶλος ὁ πλατανώδης Hsch.
πλᾰτας, after '(Aphrodisias)' insert 'also tomb, MAMA 8.538; cf. °πέλτα'
***πλᾰτειάρχης**, ου, ὁ, leader of a πλατεῖα, SEG 30.1449c (sp. πλατεάρχ-, Pontus, iii AD).
πλᾰτη I 1, for this section read 'blade of an oar, A.Ag. 695, S.Aj. 358, E.Hec. 39, Arist.PA 684ᵃ3, 13 (in describing the feet or other extremities of var. creatures). **b** use of oars, sea-voyage, S.Ph. 220, 335, E.IT 242. **c** fleet, E.IA 236, Hel. 192.' **4**, before 'AP' insert 'cj. in'
πλᾰτικός II 2, line 2, delete 'or involving breadth' and after 'id. in Cael. 579.16' insert '**b** involving extension'; line 7, adv. -κῶς, add 'POxy. 3420.12 (iv AD)'
***πλᾰτῐϝοίναρχος** [?ᾱ], ὁ, president of college of °πλατιϝοῖνοι, SEG 30.380 (Tiryns, late vii BC).
***πλᾰτῐϝοῖνοι** [?ᾱ], οἱ, members of a religious college, perh. in charge of libations of wine, SEG 30.380 (Tiryns, late vii BC).
πλάτος (A) **I 7**, delete the section. **VI**, for ' = δραχμαῖ' read 'prob. = τετράδραχμα'; add 'cf. RN 1935.1 (Delphi, i BC); cf. °πλότος'
πλάτος (B), add 'Hierapolis 322; cf. °πλάτας, °πέλτα'
***πλᾰτῠγόνατος**, ον, having broad knees, Rhetor. in Cat.Cod.Astr. 7.224.13.
***πλᾰτῠλίσγιον**, τό, rake, Ph.Bel. 100.10, Apollod.Poliorc. 220.18.
πλᾰτύς I 5, add 'πλατὺς ψυχῇ arrogant (calque on Hebr.), Aq.Pr. 28.25' **II a**, add 'Τρόφιμον .. ἡ πλατεῖα τῶν λεινουργῶν ἐτείμησαν SEG 31.1026 (ii AD)'
***πλᾰτύτοξος**, ον, of the broad bow, epith. of Apollo, CEG 331 (Boeotia, v BC).
Πλατώνειος, after 'a, ον' insert '(also -ος, -ον, GVI 1451 (Rhodes, iii BC))'
Πλατωνικός, after 'Π. φιλόσοφος' insert 'IEphes. 3901 (i AD)'
πλέγμα I 1, for 'Simon. 183.2' read 'AP 7.24.2 ([Simon.]; = EG 343 P.)'
†**πλεθρίζω**, app. extend to the length of a πλέθρον, ταῦτα πλεθρίζων, i.e. exaggerating, Thphr.Char. 23.2.
†**πλέθρισμα**· δρόμημα Hsch., Phot.
πλέθρον, add 'see also πέλεθρον'
πλειστηριάζω, add 'TAM 4(1).3.2 (sp. πληστηρ-)'
πλειστονίκης, add 'POxy.Hels. 25.21, al. (AD 264)'
πλεῖστος IV, add '**7** ἐκ πλείστου from the greatest distance, Aen.Tact. 6.5.'
πλείων A I 1, line 9, after 'id.R. 435d' insert 'ἐκ πλείονος from a fair distance, Aen.Tact. 26.1; so ἐκ π. χωρίου id. 26.9' **2**, add '**b** οἱ πλε(ί)ονες the full initiates, of the Essenes, J.BJ 2.8.9, 14.6; of the Christians, 2Ep.Cor. 2.6.' **II e**, add 'πλέονος Hdt. 3.34.2, 5.18.5' **B**, after 'FORMS' for 'Ep.' read 'poets'; line 2, for 'Call.Aet.Oxy. 2080.85' read 'Call.fr. 43.83 Pf.'; line 5, after 'iv BC)'

πλειών SUPPLEMENT πνευματώδης

insert 'neut. sg. also πλέος, ἠμ π. ἢ εἰς κατείπωσι Thasos 18.2, 10 (see also πλός)'; line 8, form πλήων, add 'SEG 32.1243 (Cyme, 2 BC/AD 2)'

πλειών, after 'Hes. Op. 617' insert '(also interpr. as seed; cf. πλειόνει)'

πλεκτή I 3, for 'fishing-basket or weel' read 'net or trap for game'

πλέκω I 1, after 'plait' insert 'or make by plaiting' II 2, after 'Pl. Hp. Mi. 369b' insert 'cf. πλέξομεν ὕμνοις [τ]ὰν .. Τρ[οῖ]αν Scol. 917(b).4 P.'

*πλέννα, πλέννα_ι μύξαι Hsch., βλένα· μύξα. οἱ δὲ διὰ τοῦ π πλένα καὶ πλέννα τὰ ἀσθενῆ καὶ δυσκίνητα id.; cf. βλέννα.

πλένναι, delete the entry.

πλεονάζω II 1, line 4, after 'id. 3.3.7' insert 'πλεονάζει δ' Ἀριστόξενος .. "ἐπηλλαγμένα" λέγων τὰ "συνημμένα" Sch.Il. 13.358, πλεονάζει δ' ἐπὶ τῆς λοιδόρου Suet. Blasph. 25, 94 T.' 2, after 'Epicur. Sent. 4' insert 'ἔν τινι Plu. 2.613c'

πλεονάκις I, add 'πληονάκι SEG 34.1198 (Lydia, ii BC)'

πλεονασμός, add '5 additional tax, BGU 2055.18 (ii AD).'

πλεύμων, line 3, for 'ὁ π.' read 'π.' II, line 2, after 'term of abuse' insert '(in sense "windbag")'

πλευρά III C, for 'side of a square or cube' read 'side of square or edge of solid figure'

*πλευρίς, ίδος, ἡ, pl., ribs of beef, Hsch. s.v. σχελίδες.

†πλευρωνία, ἡ, quoted as deriv. of πλευρά, Them. in PN 11.24, Mich. in PN 25.20.

*πλεύστης, ου, ὁ, seafarer, Rhetor. in Cat.Cod.Astr. 8(4).212.

πλέω I, add 'impers. pass., Peripl.M.Rubr. 14, 28. 2 εἰς τὸν βορέαν ἤδη πλέοντος (ἀπονεύοντος cj.) τοῦ πλοός with the course heading (northwards), etc., ib. 62.' II 2, add 'b ὀφθαλμοὶ πλέοντες swimming eyes, as a morbid symptom, Hp.Epid. 7.17.'

πλέως, line 3, for 'πλῆ .. 912' read 'πλῆ Diog.Apoll. 10 D.-K. (Hdn.Gr. 2.912)' II 2, delete 'Comp. πλειότερος .. Et.Gen.' add 'Gen' comp. πλειότερος fuller, Od. 11.359; longer, Arat. 1080; stronger, id. 644; thicker, Nic. Th. 119; π. φάρυγι Call.fr. 757 Pf., π. στόματι AP 6.350 (Crin.; = EG 5047 P.) = pleniore ore; but πλειότεροι Arat. 1005 = πλείονες.'

*πληγόω, = πλήσσω, πληγώσαι αὐτὸν πληγαῖς βιαίαις Sch.Lyc. 780.

πληθυντικός, adv. -κῶς, add 'D.H.Comp. 6'

πληθύς, line 3, delete '= δῆμος, Leg.Gort. 6.52' add 'II sum of money, Leg.Gort. 6.52. III the full of the moon, Arat. 774, 799.'

πληθύω II, line 2, after 'is trans.' insert '(fill)'

πλημμέλημα, add '(legal) wrongdoing), Just.Nov. 8.10 pr.'

πλημμύρα 1, line 3, after 'of the Nile' insert 'Ph. 2.526 (pl.)'

*πληνάριος, α, ον, comprehensive, full, ἀμεριμνίας μερικάς τε καὶ πληναρίας Just.Nov. 128.3.

*πληξίαλος, Dor. πλᾱξ-, ον, striking the sea, Simon. 14 fr.55(a).6 P.

*πληρατός, ή, όν, complete, PMilan. 48.10 (v/vi AD); cf. πλήρης (w. Lat. suffix).

*πληρέω, = πληρόω, prob. in JA 246.2/3 (Kandahar, iii BC).

πλήρης III 3, line 2, after '(Arist.Ath.) 69.1' insert 'IG 2².1641.32 (iv BC)' 6, add 'εἰς πλῆρες fully, Cod.Just. 1.3.35.3'

*πληροφορητικός, ή, όν, giving full assurance, ὅρκος Steph. in Rh. 289.18.

πληρόω III 3, add 'ἐν εἰνδ(ικτιῶνι) ε' πληρ<ο>υμ(ένῃ) SEG 34.1262 (Bithynia, v AD)' 6, add 'ἐπλήροσεν τὴν εὐχ{ι}ὴν ICilicie 118'

*πλήρω, = πληρόω, pap. in ZPE 78.146 (AD 475).

πληρωτής I, after 'one who completes' insert 'εὐδαιμονίας D.H. 1.38' II, for this section read 'a subordinate official concerned with the preparation or completion of documents, PFay. 23 intr., γραμματεύς πληρωτῶν PHamb. 59 (both ii AD); under a πριμισκρίνιος, Lyd. 3.11, 68; under a διαιτητής, Cod.Just. 2.12.27.3'

*πληρώτρια, ή, fem. of πληρωτής I, joint-lender, Hesperia suppl. 9.17 (Athens, mortgage-stone, ?iv BC).

πλησιάζω II 3, add 'abs., Plu. 2.718a, Phryn. PS 12.5, Sor. in CMG 4 p. 46.10'

πλησιαστής, add '2 adherent, follower, Simp. in Ph. 25.20.'

πλησιόμοχθος, delete the entry.

πλησίος I 3, add 'fem. subst. ἡ πλησία (sc. χώρα) the neighbourhood, ἀπὸ πλησίας Kafizin 66(a) (iii BC); sp. πλεσ- ib. 76(b)'

πλησιόχωρος, add 'sp. πλησιόχορος IG 9(2).521.34 (Larissa, iii BC)'

πλήσμα, add 'II stuffing (of cushions), TAPhA 90.140.20 (pl.) (written πλησζμ-).'

πλήσμιος, add 'adv. πλησμίως Gal. 7.751.2'

πλήσσω I 1 b, before 'Sammelb.' insert 'Arist.HA 607ª20, Ach.Tat. 2.7' 2, for 'Call.Aet. 3.1.37' read 'Call.fr. 75.37 Pf.'

πλίγμα I, after 'Hsch.' insert 'τὸ π. γνῶναι in wrestling, cj. in Gorg. 8 D.-K.'

πλινθεῖον III, add '3 perh. statue-base, τὴν Νείκην σὺν τῷ πλινθείῳ IEphes. 504 (i AD).'

πλινθευτής, after 'Poll. 7.163' insert 'PGoodsp. 30.2.9 (ii AD)'

*πλινθευτική, ή, brick-making, χειρωνάξιον πλινθευτικῆς SB 7588.3 (AD 118).

*πλινθευτικός, ή, όν, pertaining to brick-making, π. τόπος brickyard, PKöln 104.9-10 (vi AD).

*πλινθηγός, όν, used for brick-carrying, BGU 2353.6 (AD 115).

πλινθικός, add 'adv. πλινθικῶς in the (geometrical) form of a brick, Procl. in R. 2.39.28'

πλινθίς 6, add 'SEG 25.392 (Epid., iv BC)'

πλινθοειδής, after 'brick-like' insert 'PSI 1178.8 (ii AD)'

*πλινθορκία, ή, = πλινθουλκία, Stud. Pal. 22.35.20 (i AD).

πλινθουργεῖον, add 'BGU 1992a i 2 (ii BC)'

πλινθούργιον, after 'brickworks' insert 'PLond. 1166.12 (i AD)'

πλινθοφόρος 2, add 'Inscr.Délos 1415.3 (ii BC)'

πλοιαφέσια, add 'Apul.Met. 11.17'

πλοκαμίς, line 1, after 'braid of hair' insert 'Men.fr. 901 K.-Th.'

πλόκαμος I 1, line 5, for 'Call. in PSI 1092.47' read 'Call.fr. 110.47 Pf.' add '3 leaf, Anacreont. 43.6 W., AP 7.22 (Simm.; = EG 930 P.). 4 smoke ring, Poll. 2.27.'

πλοκή III 1, add 'e στίχων πλοκαί a succession of verses, Sch.E.Or. 165.'

πλόκιον II, add 'PVindob. G41866.2 (Tyche 1.89; vi AD, or headband)'

πλόκιος, delete the entry.

*πλοκοκόπιον, τό, ?garland-maker's shop, PAmst.inv. 62 (Mnemosyne 30.146).

πλόος 1, penultimate line, before 'metaph.' insert 'τριηρ[ῶν τῶ]ν ἐν ⟨τοῖς⟩ νεωρίοις καὶ τῶν ἐμ πλῶι οὐσῶ[ν IG 2².1631 (iv BC); ἀποδημῆσα[ς] [κα]τὰ πλοῦν SEG 24.1095 (Moesia, iii/ii BC), πεζοὶ καὶ κατὰ πλοῦν (?)travellers by land and sea, TAM 4(1).295'

*πλότος, εος, τό, app. Cypr. for πλάτος (A) (dat. po-lo-te-i), tablet, ostrakon, ICS 318 B vii 2.

πλουμαρικός, add 'also φλουμαρικός PSI 1082.14 (iv AD)'

πλουμάριος, add 'also φλουμάριος (written -άρης) POxy. 2421.32 (iv AD)'

*πλουμαρίσιμος, η, ον, embroidered, PAnt. 44.9 (iv/v AD), cf. Lat. plumarius, brocaded.

*πλουμάρισσα, ή, embroideress, PAberd. 59 i 7 (iv/v AD).

*πλοῦμος, ὁ, or πλοῦμον, τό, down, Edict.Diocl. 18.1, al., cf. Lat. pluma, feather, down.

Πλουτεύς, line 2, after 'Πλούτηος' insert 'SB 4313.2 (Alexandria, i/ii AD)' add 'cf. Myc. po-ro-u-te-u, pers. n.'

πλουτέω I 2, for 'πλουτίον .. IG 12(8).442.8' read 'πλουτίον τέκνων λίπετο δῶμα πό[σει] BCH 91.621 no. 81'

*πλουτηφόρος, ον, wealth-bringing, epith. of Ammon, Ps.-Callisth. 6.4.

*πλουτίνδα, adv. according to wealth, [αἱρε]τοὺς π. καὶ ἀριστίνδα IG 7.188.9 (Pagae, prob. iii BC).

*πλουτίς, ίδος, ή, (sc. ἑταιρεία) the wealthy faction at Miletus, opp. ἡ χειρομάχα, Plu. 2.298c, Eust. 1425.64.

πλουτοδότειρα, add 'of Isis, epigr. in Inscr.Cret. 4.244 (ii BC)'

πλουτοδότης, add 'of Ammon, Bernand Akôris 18 (?ii AD)'

*πλουτοδότις, ή, Giver of wealth, title of Hermuthis, SEG 8.548.1 (Egypt, i BC).

*πλουτοποιέω, enrich, Sch.Pi.O. 7.60.

πλοῦτος II, line 3, after 'Antiph. 259' insert 'also pl.; as title of comedy by Cratinus, Ath. 3.94e, etc., cf. Πλοῦτοι δ' ἐκαλούμεθ' ὅτ' [ἤρχε Κρόνος] Cratin.fr. 171 K.-A.'

πλύνω, line 8, transfer 'Hp.Acut. (Sp.) 65' to follow 'part. πεπλυμένος' II, line 2, after 'abuse' insert 'A.fr. 78a.71 R.'

*πλυσιμάριος, ὁ, laundryman, POxy. 3598.7 (iv AD), 3599.7 (v AD).

*πλυτάριος, α, ον, τὸ ἐρισκός, name given by soldiers to a clay drinking-vessel, Suid. s.v. κώθωνες.

πλωΐζω, after 'Arr.Peripl.M.Eux. 23' insert 'pass., τὰ πλοϊζόμενα parts sailed over, Peripl.M.Rubr. 45' and add 'cf. Myc. po-ro-wi-to, perh. πλώϜιστος, name of month'

*πλωϊσμόν, τό, shipment, τῷ αὐτῷ ὑπὲρ πλοεισμοῦ PVindob.Tandem 19.2 (v/vi AD), εἰς τὸ πλοεισμ<..> PMasp. 57 i 3 (vi AD).

†πλώς, πλωτός, ὁ, swimmer, i.e. fish, πλῶτες ἁλός Opp.H. 3.63; app. used spec. for the grey mullet (κεστρεύς), Epich. 44, Xenocr. ap.Orib. 2.58.29.

πλωτήρ I 1, line 2, before 'Ar.Ec. 1087' insert 'E.IT 449, Hel. 1070'

*πλώτωρ, ορος, ὁ, = πλωτήρ, AP 7.295 (Leon.Tarent.; = EG 2066 P.).

πνεῦμα III, add '2 inspiration, genius, Longin. 9.13, D.H.Th. 23, Luc.Dem.Enc. 14.' V, π. πονηρόν, add 'SEG 30.1794 (Jewish, iii AD)'

πνευμάτεμφορος, for '= πνευματόφορος' read 'expld. as ὑπὸ πνεύματος πεπληρωμένος'

*πνευματηλάτης, ου, ὁ, spirit-driver, of Cerberus, Suppl.Mag. I 42 (iii/iv AD).

πνευματοφόρος, delete 'προφῆται ib.Ze. 3.4'

*πνευμᾰτόφορος, ον, borne by the wind, light, frivolous, προφῆται Lxx Ze. 3.4.

πνευματώδης 2, for 'windy, exposed to the wind' read 'characterized by

252

wind, windy' and add 'αἱ ἅλωνες περὶ τὴν σελήνην πνευματώδεις μᾶλλον ἢ περὶ ἥλιον Thphr.*Sign.* 31'

πνευστιάω, after 'Hp.*Int.* 44' insert '*Com.adesp.* 362'

πνέω I, three lines fr. end, after 'Poll. 4.72' insert 'w. acc., ἔπνεε Δαρδανίδας piped the D. (i.e. songs celebrating them), *APl.* 7 (Alc.Mess.)'   **IV**, add 'τὴν .. πνεύσασαν .. ὕστατα *AP* 7.166 (Diosc.; = *EG* 2968 P.)'

πνιγίζω, for '= πνίγω' read 'humorous conflation of πνίγω and πυγίζω'

***πνιγμοσύνη**, ἡ, = πνῖγμα, Anon. *in Rh.* 204.20.

πνίγω I 3, add 'S.*Ichn.* 393'

***πνιχμός**, ὁ, = πνιγμός (in quot., sense 1), Nic.*Al.* 365 (v.l.).

πνοή II 2, after '*breath*' insert 'Γοργόνες .. ἃς .. οὐδεὶς εἰσιδὼν ἕξει πνοάς A.*Pr.* 800'

πόα, line 1, before 'E.*Cyc.* 333' insert 'A.*fr.* 28, 29 R.'; lines 2/3, delete 'Boeot. πύας .. iii BC'   **I 2 a**, add 'ποίη κυπάρισσος = χαμαικυπάρισσος, Nic.*Th.* 910'   **4**, line 3, delete '*meadow*, Schwyzer l.c. (pl.)' (now read γύας, v. γύης)

ποδαγρικός 1, before 'Plb.' insert '*SB* 7638.4 (iii BC)'

ποδαπός 2, add '**b** indef., *of any kind, Vit.Aesop.*(G) 32, al. (ποτ-).'

πόδαργος, add 'Myc. *po-da-ko*, description of an ox'

***ποδάριοι**, οἱ, members of an association of performers, perh. *scabillarii, TAM* 5(1).92 (ii AD), *Gloss.*

ποδάριον, add '*PHaun.* 25.10 (iv/v AD)'

ποδηγετέω, after lemma insert 'Dor. ποδᾱγ- *GVI* 1859.5 (Teos, ii/i BC)'

ποδιαῖος I 2, for '[γραμμή]' read '[δύναμις]'

ποδιστήρ II, after '*tripod*' insert 'Lxx 2*Ch.* 4.16'

ποδοκοίλιον, add 'in lists of rations, *OAshm.Shelton* 78, 80'

***ποδοκοπέω**, *strike with the feet*, Hsch. s.v. κωλαβρισθείησαν.

ποδόψηστρον, add 'Herod. 5.30'

ποδοψόφος, add 'cf. °*ποδάριοι*'

πόδωμα 1, before '*OGI* 510.5' insert '*Gerasa* 51 (AD 81/3)'

⁺**ποέχομαι**, Cypr. for προσέχομαι, part. *po-e-ko-me-no-ne adjoining, ICS* 217.19, 21.

ποηφάγος, add 'Call.*fr.* 365 Pf.'

***ποθαιρέομαι**, etc., v. °*προσαιρέομαι*.

***ποθαλόω**, v. °*προσηλόω*.

***ποθεδρεία**, ἡ, = προσεδρεία (in quot., sense 2), *IG* 12(3).247.8.

ποθεινός I 2, add 'sup. as form of address, τῷ εὐλογο[υ]μένῳ καὶ ἀληθῶς ποθεινοτάτῳ υἱ[ῷ] *PMich.*inv. 3999 (*ZPE* 75.268, vi AD)'   **III**, for 'Subst. .. paint' read 'perh. colour epith. (cf. πόθος III)'

***ποθεινότης**, ητος, ἡ, *quality of being dear, dearness*, in a complimentary address, τῇ σῇ ποθεινότητι *PKöln* 281, *PStrassb.* 279, *PHerm.Rees* 50 (all vi AD).

ποθητός, add '*SEG* 33.1475 (Cyrenaica, i/ii AD), *IG* 5(2).491 (Megalopolis, ii/iii AD)'

ποθίκω, add 'part. οἱ ποθίκοντες, *relatives*, orac.ap.D. 43.66'

πόθος II 1, after 'E.*Andr.* 824' insert '*Tr.* 116, *Hel.* 763'   add '**b** of the beloved person, Ἡράκλειτος, ἐμὸς πόθος *AP* 12.152.'

⁺**ποῖ A** *to what place?, whither?*: ποῖ πατεῖς πύλας; A.*Ch.* 732; ποῖ με χρὴ μολεῖν; S.*El.* 812; ποῖ τις φύγῃ; Ar.*Pl.* 438.   **b** (w. gen.), ποῖ φύγωμεν Ἀπίας χθονός; A.*Supp.* 777, S.*Tr.* 984.   **II** *to what point of time?, till when?, how long?*: ποῖ χρῆν ἀναμεῖναι; Ar.*Lys.* 526; ποῖ κῇχος; – ἐγγὺς ἡμερῶν τεττάρων Pherecr.*fr.* 175 K.-A.   **III** *to what end or result?, in what (final, eventual) circumstances?*: ποῖ καί ποῖ τελευτᾶν; A.*Pers.* 735; ποῖ κρανεῖ, ποῖ καταλήξει .. μένος ἄτης; A.*Ch.* 1075; ἃ δ' ὑπέσχεο ποῖ καταθήσεις; S.*OC* 227; τὸ δ' ἔνθεν ποῖ τελεύτησαί με χρή; ib. 476; (in indirect qn.), ἔνισπε ποῖ κεκύρωται τέλος A.*Supp.* 603.   **b** w. gen., ποῖ τις φροντίδος ἔλθῃ; S.*OC* 170; ποῖ φρενῶν ἔλθω ib. 310; (indirect) οὐκ ἔχω ποῖ γνώμης πέσω ib.*Tr.* 705.   **IV** in repeating indignantly another speaker's words (as ποῖος), *indeed*: .. λευκὸν ἵππον .. – ποῖ λευκὸν ἵππον; Ar.*Lys.* 193, 383.

***ποιαλίς**, v. °πυελίς.

***ποίβολος**, ὁ, *peripheral zone*, de Franciscis *Locr.Epiz.* 16, NWGk. equiv. of *πρόσβολος; cf. *περίβολος*.

ποιδέομαι, for 'προσδέομαι' read '°*προσδέω* (B)'

ποιέω A III, line 11, for 'ποιεῖσθαι .. friend' read 'κασιγνήτῳ ἶσον π. ἑταῖρον *treat* a friend *as* a brother'

ποίημα I 1, after '*work*' insert '*artefact*, τὸ Πάριόν ποίημα Κριτωνίδεω εὔχομ[αι εἶναι] *CEG* 413 (*c.*525/500 BC)'   **3**, for '*fiction*' read '*literary production in prose*' and add 'Isoc.*Ep.* 1.2'

***ποιητίκευμα**, τό, *poetical expression*, Steph.*in Rh.* 312.8.

ποιητικεύομαι II, after '*poetically*' insert 'Steph.*in Rh.* 312.5'

ποιητικός I, add '**3** gramm., of a verb, *active*, D.H.*Amm.* 2.7, 8 (i pp. 427f.).'   **II 1**, add 'comp., D.H.*Th.* 46'

⁺**ποικιλίας**, ου, ὁ, an unidentified fish of the river Aroanius, Philosteph.Hist. 20, Paus. 8.21.1.

ποικίλλω I 2, line 8, after 'Pl.*Ti.* 87a' insert '*produce a varied* utterance, ποικίλλουσα .. κρωγμὸν πολύφωνα κορώνη Arat. 1001'

***ποικιλογνώμων**, ονος, *having a versatile mind*, Heph.Astr. 3.45.5.

⁺**ποικιλόθρονος**, ον, *on a richly-worked throne* (or *adorned with variegated flowers*), Ἀφροδίτα Sapph. 1.1 L.-P. (v.l. ποικιλόφρον).

ποικιλόνωτος, add 'ποικιλόνωτον .. ἴτυν ἄστρων Nonn.*D.* 2.575; prob. in A.*fr.* 47a.26 R.'

***ποικιλόπρυμνος**, ον, *having a variegated stern, Suppl.Hell.* 991.112 (poet. wd.-list, iii BC).

ποικίλος II 1, add '**b** of mosaic work, *SEG* 26.1628 (Apamea, vi AD)'   **2**, transfer 'Od. 8.448' to section III 3.   **III 1**, add 'of deities (?Μοῖραι), θεαί .. ποικίλαι *SEG* 23.206 (Messenia, ii/iii AD)'   **3 c**, add 'as a complimentary term in an epitaph, *SEG* 25.1088(b) (Cyprus, ii/iii AD)'   add 'cf. Myc. compds. *po-ki-ro-nu-ka*, epith. of cloth; *po-ki-ro-qo*, pers. n.'

***ποικιλοτειρής**, ές, *varied with stars*, πόλον π. rest. in *IG* 2².4494.7 (i AD).

ποικιλόφρων, for 'Alc.*Supp.* 22.7' read 'Alc. 69.7 L.-P.'

⁺**ποιμανδρία**, ἡ, kind of vessel, identified w. ταναγρά, Lyc. 326 (Ποιμανδρία is said to be an old name for Τάναγρα, St.Byz.).

ποιμήν II 2, delete '*teacher*' and after 'etc.' insert 'of a sophist, Lib.*Or.* 1.25'   add 'Myc. *po-me, po-me-no* (gen.)'

ποίμνη, line 4, after 'etc.' insert 'αἰγές τε καὶ ποῖμναι καὶ βόες Philostr.*VA* 2.13'

ποίμνιον II, after '*Ev.Luc.* 12.32, al.' insert '*body* or *congregation of Christians*' and after '1*Ep.Pet.* 5.2' add '*IHadr.* 120 (iv AD)'

ποιναῖος, add '**II** perh. *expressing thanks, CEG* 356 (Isthmia, v BC).'

***ποινικάζω**, v. °*φοινικάζω*.

***ποινικαστάς**, v. °*φοινικιστάς*.

ποιολογέω, for '*put up corn in sheaves*' read '*gather grass* or *hay*'

***ποιόφυτος**, ον, *planted with grass, growing grass*, σηκοί A.*fr.* 273a.1 R.

***ποιπνύτροισι**, unkn. sens. and accent, Antim. 186 Wy.; cf. ποιπνύτροισι· σπουδαίοις Hsch. (s.v.l.).

***Ποιτρόπια**, τά, a Delphic festival, *CID* i 9.D5 (v BC); cf. Ποιτρόπιος (form dial. equiv. of προστρόπιος).

ποίφυγμα, add '**2** σχῆμα ὀρχηστικόν Hsch.'

ποιφύσσω, add '**III** trans., *frighten*, Hsch.'

***ποίωμα**, ατος, τό, *qualification, added quality*, Simp. *in Cat.* 254.12.

Πόκιος, add '*SEG* 25.606 (Doris, ii BC)'

ποκκί, for '= πρὸς τί, but' read 'of dub. origin'

πόκος I, add 'fem. acc. pl. πόκας *SEG* 23.305 (Delphi, *c.*190 BC)'   add 'Myc. *po-ka* (pl.), *fleeces*'

***πολάζω**, *to be common, abound*, Θρηικίησιν .. νήσοισι πολάζει Nic.*Th.* 482 (codd. πελ-; cf. *ἐπιπολάζω*).

πολεμιστής II, add '**2** of other things, *martial*, πολεμιστὰ (acc. masc.) σίδηρον Orph.*L.* 312.'

πόλεμος, add 'gen. alone, *in time of war*, καὶ πολέμο καὶ εἰρήνης *SEG* 31.969 (Erythrae, 351/44 BC), 30.990 (Delos, iv/iii BC).   **2** applied to rioting, *PLond.* 1912 (i AD), *POxy.* 2339 (i AD), 3065 (iii AD).'   **II 2**, delete the section.   add 'cf. Myc. *po-to-re-ma-ta*, man's name (cf. *πολεμητής*)'

πολεύω II, for '*turn up* the soil with the plough' read '*go over* (i.e. plough) land'

πολέω I, for 'intr. .. haunt' read '*range over, haunt*, w. acc.'; after 'Pers. 307' insert 'intr. *go about*'   **II**, for '*turn up* the soil with the plough' read '*go over* (i.e. plough) land'

πολιά, add 'transf., of trees, οὐδ' αὐτὰ (sc. δένδρα) γέροντος ἤδη χρόνου πολιὰ καθαύαινεν Luc.*Am.* 12'

***πολιφάναξ**, ακτος, ὁ, (cj.) *lord of the city*, Ποτιδᾶνι ΠολιϜάνα<κ>τι (sic) *ZPE* 60.90 (*IG* 4.222; Corinth, vi BC).

πολιαρχέω, add 'Thess. τολλιαρχέω *SEG* 23.437 (Crannon, iii BC) fr. °*πτολιαρχέω*'

πολιαρχία, add 'Just.*Nov.* 13 pr.'

πολίαρχος, add 'cf. *πτολίαρχος*, fr. which the Thess. form is assimilated'

***πολιάω**, *go grey, age*, οὐ πολιᾷ· οὐ γηράσκει Hsch.

***πολιδυνάστης**, ου, ὁ, *city-despot*, Plb. 5.4.3 (cj.).

Πολιεύς, line 2, after 'etc.' insert 'of Sarapis, *JHS* 21.275 (Xois, Egypt)'

πολιήτης, after 'E.*El.* 119' insert 'Ion Trag. 41 S.'

πολίητις, after 'A.R. 1.867' insert 'Posidipp. in *Suppl.Hell.* 705.1'

πολιορκέω, line 4, before '(πόλις' insert 'written **πολιουρκέω** *Marm. Par.* 113, v.l. in Str. 17.3.15, Lyd.*Mens.* p. 184 W. cod., v.l. in Procop.*Goth.* 2.24'

πολιορκία, after 'ἡ' insert 'written **πολιουρκία** *IG* 12(7).387.6 (Amorgos, iii BC), *Marm.Par.* 113'

πολιός I 1, line 5, before 'γάλα' insert 'εὔιον γέροντα of wine (w. play on 2), Men.*Dysc.* 946'   **2**, line 3, for 'Alc.*Supp.* 20.2' read 'Alc. 50.2 L.-P; σάρκες E.*Supp.* 50 (lyr.); χείρ *AP* 9.568 (Diosc.; = *EG* 2943 P.)'   add '**III** πολιά, ἡ, (sc. λίθος) a precious stone, Plin.*HN* 37.191.   Myc. *po-ri-wa* (neut. pl.)'

πολιοῦχος (A), sp. πολιόχος, add '*Inscr.Cret.* 4.171.14' πόλις, line 13, after 'An. Ox. 1.361' insert 'Thess. πόλλιος BCH 59.37 (disyll., Crannon)' **II**, line 1, after 'country' insert 'orig.'; add 'Αἴγυπτον καὶ Λιβύην τὼ πόλεε decr.ap.Crater.in Sch.T Il. 14.230'

πόλισμα, line 5, after 'Th. 1.10, 4.54' insert 'Plu.Pomp. 28.1, D.C. 40.36.4, al.'

*πόλιστος, = πλεῖστος, IG 14.645.130 (Heraclea, iv BC).

πολιταρχέω, add 'SEG 24.580 (sp. πολειττ-, Maced., ii BC), 33.520 (Maced., ii AD)'

πολιτάρχης, add 'SEG 30.568.16 (Maced., ii AD)'

*πολιταρχικός, ὁ, one who has served as πολιτάρχης, IG 10(2).162.15, 197.13 (Thessalonica, iii AD).

πολιτεία **I 3**, for pres. ref. read 'Arist.Ath. 4.3, IG 9(2).517.17 (Larissa, iii BC)'

πολίτευμα **I**, add 'ἀρετῆς ἕνεκεν .. κατὰ τὸ πολίτευμα καὶ ἐν τοῖς ἄλλοις πᾶσιν Amyzon 24'

πολιτευτής, add '2 member of the πολίτευμα (as a governing body), SEG 9.1.31, 67 (Cyrene, iv BC).'

πολιτεύω **B I**, add 'b to live doing one's civic duty, Act.Ap. 23.1, Ep.Phil. 1.27.'

πολίτης **I**, add '3 city-dweller, Just.Nov. 52.1, 89.2.2.'

πολιτικός **I 1 a**, add 'πολιτική, ἡ, perh. city tax, EA 7.74 (Sardis, 213 BC)' **b**, after 'in a town' insert 'τῇ π. ἐργασίᾳ Vit.Aesop.(G) 2'; add 'appointed or authorized by the city, ἀρτοπώλης Sardis 7(1) no. 166 (-ειτ-)' **2**, adv., add 'sup. πολιτικώτατα Arist.Ath. 40.3' **3**, add 'ἡ πολιτική, prob. the body of citizens, SEG 36.1087 (Sardis, 213 BC)' **IV**, for 'concubine, mistress' read 'prostitute, whore'

πολιτογραφία, add 'SEG 26.96 (Athens, c.220 BC)'

Πολιχνῖται, add 'IG 1³.260 viii 17 (Athens, v BC)'

πολλάκις, line 2, delete 'never in prose' **I**, add 'πολλάκι δὲ γράφων Amyzon 15 (201 BC), Phld.Ir. 9.28'

*Πολλᾱλέγων, οντος, ὁ, "Saying-much" or "Heeding-much" (cf. Οὐκαλέγων Il.3.148), name coined by Alcm. 107 P.

πολλοστός **I 1**, adv, add 'sp. πολλαστῶ[ς, PVindob.G 29.386.11 (Tyche 7.56, vi/vii AD)' **3**, transfer 'X.Mem. 4.6.7' to exx. w. οὐδέ in the following line.

*πολοειδής, ές, cylindrical, ὄργανα Anon.Alch. 275.16, 277.7.

πολυάϊκος, after 'Sch.E.Med. 10' add 's.v.l.'

*πολύανδρος **I 2**, after 'Onos. 21.5' insert 'ἔθνη -ρότατα D.Chr. 35.14'

πολυανθής, add 'πολυανθέων Ἐρώτων Anacreont. 55.7 W.'

*πολύαρνος, ον, rich in lambs, Hsch. s.v. πολύρρην.

*πολυαύχητος, ον, much-vaunted, epigr. in Lindos 177.6 (ii AD).

*πολυβιβλογενής, ές, perh. producing much papyrus, PGron.inv. 66 in ZPE 41.82 (ii AD).

*πολύβλεπτος, ον, much-observed, PMasp. 141 i^v.31 (vi AD).

*πολύβοια, fem. adj., rich in cattle, εἰρήνη π. Euph. in Suppl.Hell. 415 ii 4, cf. fr. 177.2 (Powell).

πολυβούτης, at end for 'Carm. .. K.)' read 'Carm.Naup. 2 B.'

πολύγιος, for 'Paus. 2.31.13' read 'Paus. 2.31.10 (perh. corrupted fr. ‡πολυγώνιος)'

πολύγονον, add 'also πολύγονος, ἡ, Plin.HN 27.113, Cyran. 34.19'

πολύγραμμος, add 'of kind of jasper, Plin.HN 37.118'

*πολύγραπτος, ον, fully-painted, SEG 28.541 (32.633; Maced., late Hellen.).

πολυγώνιος, add 'voc. sg., Call.fr. 114.2 Pf., prob. addressed to cult-stone representing Apollo, cf. Paus. 2.31.10 where πολυγώνιος shd. perh. be read for πολύγιος'

πολυδάκρυτος **I**, after 'Tr. 1105 (lyr.)' insert 'adv. -ως with many tears, Sch.S.OC 1646' and add 'sup. SEG 24.1075 (Moesia, iii/iv AD)'

πολυδένδρος, add 'Simon. in POxy. 3965 fr. 27.8, Theoc. 17.9'

πολυδικέω, add 'SEG 9.1.47 (Cyrene, iv BC)'

πολύδικος, after 'litigious' insert 'Heraclid.Pol. 4.5'

*πολύδινος, ον, whirling round and round, Nonn.D. 2.457.

πολύδριον, after 'πόλις' insert 'J.AJ 18.249 (s.v.l.)'

πολυδύναμος **2**, add 'JRS 40.78.24 (Cyrene, letter of Hadrian)'

πολύδωρος, after '**II**' insert 'bountiful, γῆ epigr. in MAMA 8.130 (Dinek Saray)'

⁺πολυείδεια, ἡ, = πολυειδία, Call.Dieg. ix.34 (fr. 203 Pf.).

πολυειδής, line 5, after '(Sup.)' insert 'νεώς π. having a rich variety of ornament, Aristid.Or. 50(26).28'

*πολυέλιξ, ικος, ὁ, ἡ, gloss on τετραέλιξ, Hsch.

πολύεργος **I**, delete 'perh. f.l. for ἀμπελοεργοί' **II**, add 'SEG 31.1284 (Antioch in Pisidia)'

*πολυέτηρος, ον, Ep. πουλυ-, = πολυετής II, Nonn.D. 14.103 (s.v.l.).

πολυηγερέες, before 'read' insert 'gathered from many quarters'

*πολυθαλής, ές, of many blooms, metaph., π. λοχεύματα gloss on λόχια τριθάλεια, Sch.Antim. 182 Wy.; also π. δῶρα ibid.

πολυθαύμαστος, after 'much-admired' insert 'Sch.E.Hipp. 168'

*πολυθενία, ἡ, abundance, prosperity, SEG 14.787 (Dorylaeum).

πολυθλιβής, add 'PSI 253.134 (v AD)'

*πολύθριγκος, ον, having many θριγκοί (?friezes), π. τέραμνα (houses) Dain Inscr. du Louvre 60.13 (Heraclea ad Latmum; hymn).

πολύθριδαξ, read '-θρίδαξ'

πολύθριξ **II**, add 'b a precious stone, Plin.HN 37.190.'

πολυθρόνιος, delete 'πολύθρονος' to end of article.

*πολύθρονος, ον, = πολυθρόνιος, Call.fr. 364 Pf., v.l. in Nic.Th. 875.

πολύθροος, add 'of manifold utterance, Μοῦσα Opp.C. 3.461'

*πολύκαλλος, ον, endowed with great virtue, MAMA 7.78.

*πολυκάνδηλος, ὁ, (or -ον, τό) chandelier, Corinth 8(3).618 (rest. in T&MByz 9.366 no. 74).

πολυκέφαλος, add 'sup., Max.Tyr. 38.7'

πολύκλαυστος **I**, after 'η, ον' insert 'AP 7.712 (Erinn.)' and after 'lamented' insert 'Archil. 94.3 W.' **II**, after 'causing much lamentation' insert 'ὦ πολύκλαυθ' Ἅιδη CEG 591.5 (Attica, c.350 BC)'

πολύκοινος **I**, add 'in kinship, i.e. sister to many, π. Ἀμφιτρίταν S.fr. 673'

⁺πολύκοπος, ον, accompanied by much striking or beating, ὄρχησις Ath. 1.20e (cj. πολυπρόσωπος, cf. Plu. 2.711f). **2** much-buffeted (by experience), Cat.Cod.Astr. 11(2).189.9.

πολύκοσμος, after 'much-adorned' insert 'τῆς π. χρόας Ael.NA 10.13'

πολυκρᾱνος, after 'many-headed' insert '(or huge-headed)'

πολύκροτος **I**, for 'ringing loud or clearly' read 'rowdy, noisy' **II**, add 'Call.fr. 67.3 Pf.; cf. κροτέω II 4, κρότημα'

πολυκτόνος, add 'of Hades, SEG 34.325 (c.100 BC)'

*πολυκωπίτης, ου, ὁ, one of the crew of a πολύκωπον (πλοῖον) PLond. 1712.6, 37 (AD 569); perh. also PMasp. 136.16, 287.18.

*πολύλᾱΐς, ιδος, giver of much booty, Ἀθάνα π. Alc. in SLG 262.9.

*πολύλαλητος, add '**III** much spoken-of, ἥρως epigr. in IGBulg. 796.3 (ii/iii AD; -λαλατος)'

*πολύμαστος, ἡ, many-breasted, (Dianam Ephesiam) multimammiam, quam Graeci π. vocant, Jerome in Ep.Eph. prolog. (PL 26.441).

πολυμεμφής, after 'much-blaming' insert 'Arat. 109 (v.l. περι-). **II** blameworthy, epigr. in IHadr. 36 (i BC/i AD).'

πολύμετρος **II**, after 'metres' insert 'στροφαί D.H.Comp. 26'

⁺πολύμιτος, ον, having tapestry-woven decoration in many colours, πέπλοι π. A.Supp. 432, Cratin. 481 K.-A., Plin.HN 8.196, Peripl.M.Rubr. 39, al., προσκεφάλαια SB 7033.37 (v AD).

πολυμνήστη, delete 'πολυμνάστοιο .. (Pers.)'

πολύμνηστος **I**, add 'πολυμνάστοιο .. Τισίδος AP 6.274 (Pers.)' **II**, for this section read 'much-remembered, πολυμνήστοιο Δαμαίου Nic.fr. 110 G.-S., Orph.H. 50.2; perh. also A.Ag. 1459 (lyr.)'

Πολύμνια, add 'as appellat., πολύμνια παντερπὴς κόρα Lyr.adesp. 24 P.)'

⁺πολυμνίος, α, ον, full of seaweed or waterweed, v.l. (ap.Sch.) in Nic.Th. 950; cf. πολυμνία· ἡ θάλασσα, ἡ πολὺ φυκίον ἔχουσα Hsch.

πολύμορφος **I**, after 'manifold' insert 'varied'; after 'Arist.PA 646ᵇ32' insert 'of mosaic work, SEG 26.1629 (Apamea, vi AD)'; for 'Him.Or. 34.4' read 'Him.Or. (34)35 (p. 147.33 C.), D.H.Comp. 16'

πολύμοχθος **I**, add 'Orph.H. 29.15, 37.4, al.'

πολύμυθος **I**, for 'Call.Iamb. 1.170' read 'Call.fr. 192.14 Pf.'; add 'AP 7.713 (Antip.); talkative, Call.l.c., Epigr. 16.1 Pf.' **III**, delete 'cf. Call.Epigr. 18'

πολυνεικής, after '(A.) Th. 830 (anap.)' insert 'Lyr.adesp. 96 P.'

πόλυντρα, delete '(i.e. Lat. polenta)' and add '(Aeol. for *πάλυντρα)'

πολύξυλος, add '**2** containing many sticks, or many (i.e. varying numbers of) ξύλα (v. ξύλον v) in length, δέσμαι Theb.Ostr. 144 (i AD).

πολύμματος, add 'transf., of the night sky, κυανέας [πο]λυόμματον [ποι]κίλμα νυκτ[ός] POxy. 2879 i 1 (= SLG 458)'

⁺πολυόμφαλος, ον, having many bosses or knobs, πολυόμφαλα .. πόπανα, i.e. sacrificial cakes, PDerveni ii 7 (ZPE 47, post 300 pp. 1-12, iv BC); πεδίον π., poet. expression for the formation of shields in a Roman testudo, Opp.C. 1.218.

πολυόφθαλμος **1**, add 'b also perh. large-eyed, POxy. 1380.129.'

*πολυόχλητος, ον, much-disturbed, Anon.in Rh. 34.12.

πολυπάθεια **I**, add 'SEG 35.304 (Epid., after AD 67/8)'

πολυπαίπαλος **II**, for this section read 'πολυπαί(πα)λος αἰθήρ· πεποικιλμένος. οὐχ ὁμαλός Hsch. (Suppl.Hell. 1100)'

*πολυπάλακτος, ον, sprinkled with much blood, cj. in A.Ch. 425.

πολύπαλτος, for 'Call.Sos. 4.1' read 'Call.fr. 388.1 Pf.'

πολύπαταξ, read '*πολυπάταξ*'

πολύπειρος, add '**2** = πολυπείρων 1, Orph.H. 12.13.'

πολυπείρων **1**, add 'οἴμους θηρῶν τ' οἰωνῶν τε Orph.A. 33' **2**, delete the section.

πολύπημον **II**, after 'much-suffering' insert 'Alcm. 5 fr.2 i 9 P.'

πολύπηνος, for 'thick-woven, close-woven' read 'of elaborate pattern'

πολυπλάνητος **II**, before 'A.Ch. 425' insert 'cj. in'

πολυπλήθεια, after 'Aen.Tact. 3.1' insert 'ὄχλου Men.Dysc. 166'

πολύπλοκος **2 a**, lines 4/5, delete 'neut. as .. Sign. 40'

πολυπόθητος, add 'of the dead, T&MByz 7.319 no. 12 (AD 535)'

πολυποίκιλος **1**, add 'of the sun, as having oracular significance,

πολύπους   SUPPLEMENT   πόρνη

Orac.Tiburt. 22, 23'   add '**3** of a craftsman, *versatile*, SEG 31.1284 (Antioch in Pisidia).'

πολύπους (A), after 'S.*El.* 488 (lyr.)' insert '(or *having swift feet*)'

πολύπους (B), line 9, for 'in Poets .. from' read 'poet. also' and after '**πούλυπος**' insert '*AP* 9.10 (Antip.Thess.)'; line 12, after 'Hp.*Aff.* 5 (v.l.)' insert 'acc. πώλυπον Simon. 9 P.'   add '**V** in form πῶλυψ, name of a throw in dicing, *BIFAO* 30.6 (Alexandria).'   add 'Myc. *po-ru-po-de* (dat.)'

πολυπραγμονέω 3, for '*to be curious after*' read '*busy oneself with*' and after 'etc.' insert 'σημία θεάμενος ἐπολυπράγμοσα (as if pres. -μόνω) *HThR* 27.61 (Nubia).'

πολυπραγμόνησις, delete 'dub. l. in'

***πολύπωυς**, υ, *grazed by many flocks*, πολ]υπώεος ἄγχι Πελίννη[ς Euph. in *Suppl.Hell.* 429.19.

πολύρρηνος, after 'A.*Eleg.* 3' insert '*EG* 1620 P. (Posidipp.)'

πολύρρητος, add 'as pers. n., SEG 26.577 (Orchomenos, *c.*250 BC)'

***πολύρρυπος**, ον, *full of dirt*, gloss on πουλυπινές in E.*Rh.* 716 cod. A.

πολύς, lines 4ff., acc. sg., add 'πολέα (masc.) *Suppl.Hell.* 1013'; line 8, after '(Il.) 11.708' insert 'cf. A.R. 2.898'   **I 2 b**, add '*possessing full physical strength*, ὠμογέρων ἔτι πουλὺς ἀνήρ Call.*fr.* 24.5 Pf., cf. *Epigr.* 61.1 Pf.; βούπαις οὔπω πολλός A.R. 1.760'   add 'Myc. *po-ru-* (in compds.)'

***πολῠσήσᾰμος**, ον, *containing plenty of sesame*, πλακοῦς *Vit.Aesop.*(W) 58.

***πολύσιγμος**, ον, *containing many sigmas*, Mart.Cap. 5.514.

πολύσκαλμος, for '*many-oared*' read '*with many rowlocks* (contrasted w. a small boat)' and after '*AP* 7.295' add '(Leon.Tarent.)'

πολυσκάριστος, add 'Sch.Il. 2.814 (= *POxy.* 1086.105)'

***πολυσκεδής**, ές, *widely scattered*, A.*fr.* 132c.16 R.

***πολυσκελής**, ές, *stout-limbed, stout-legged*, Clem.Al.*Strom.* 5.8 (*GCS* 52.362).

†**πολύσπορος**, ον, of men, *abounding in seed*, Ptol.*Tetr.* 72, *Cat.Cod.Astr.* 7.212; of deities controlling procreation, Πρωτόγονος Orph.*H.* 6.10, Φύσις ib. 10.19; of astrological influences, Ptol.*Tetr.* 34, Vett.Val. 4.6; adv. -ρως, *with a multiplicity of seed*, S.E.*M.* 5.58.   **2** transf., of countries, *fertile* or *teeming with inhabitants*, Ἀσιάδος πολυσπόρου E.*Tr.* 748, Πάρθων τε πολύσπορος .. αἶα Opp.*C.* 3.23.

πολυστέφᾰνος II, for this section read 'Πολυστέφανος, ἡ, a Sicilian goddess, SEG 34.959 (iii BC), *IG* 14.262, 2406.67'

πολυστεφής I, line 3, for '*wreathed with*' read '*richly crowned with*'

***πολύστρεβλος**, ον, *given to extortion*, Al.*Pr.* 28.16.

πολύστροφος 2, for '*versatile*' read '*changeable*'   add '**b** *variable*, τοῦ πάχετος μήκός τε πολύστροφον Nic.*Th.* 465.'

***πολῠσύντῠχος**, ον, perh. *gregarious*, *Cat.Cod.Astr.* 11(2).189.5.

πολυτέλεια, add '**3** lit. crit., *abundance of matter*, Phld.*Po.* 5.5.'

πολυτελής I, add '**2** lit. crit., *in rich abundance*, διανοήματα Phld. *Po.* 5.9.   Adv. πολυτελῶς *with a full treatment*, opp. εὐτελῶς, ib. 5.4.'   **II**, add 'adv. also -έως, SEG 32.1243'

***πολύτερπος**, ον, = πολυτερπής, *CEG* 452 (Corinth, *c.*580 BC), unless pers. n.

***πολύτευκτος**, ον, *much-wrought*, *GVI* 477.3 (Smyrna, i/ii AD).

***πολύτῑμιος**, ον, = πολύτιμος I, SEG 6.159.14 (Phrygia, iii AD).

πολύτιμος I, add 'γονέων πολυτείμων SEG 30.1507, μνήμην .. πολύτειμον ib. 31.1284 (both Pisidia)'

πολύτρῐχος I, add 'Heras ap.Gal. 12.430.13'

πολύτροπος III, after '*various, manifold*' insert 'Diog.Apoll. 5 D.-K.'

πολύτροφος I, delete 'Ptol.*Tetr.* 163'

***πολύτροχος**, ον, *much-rolling*, Orph.*L.* 649.

πολύυμνος, after '*famous*' insert 'δῆρις Ibyc. 1(*a*).6 P.'

***πολύφήμητος**, ον, = πολύφημος III, *famous*, *BCH* 15.455.

πολύφήτωρ, after 'Sch.' insert 'B' and add 'cf. Porph.ap.Sch.B Il. 14.200'

πολύφθογγος, add '**II** *having a full voice*, Arr.*Cyn.* 5.4.'

πολυφροσύνη, add '*ICilicie* 102 (iii AD)'

πολύφυλλος, add 'π., τό, name of a plant, Ps.-Dsc. 2.147'

πολυφωνία, add '*diversity of utterance*, Max.Tyr. 1.1 (pl.)'

***πολύχᾰρις**, ὁ, ἡ, *very grateful*, εὐχήν epigr. in SEG 16.683 (Caria, Rom.imp.).

πολύχειρ I, after 'S.*El.* 488 (lyr.)' insert '(or *having strong hands*)'

πολύχορδος, for 'Simon. 46' read '*Lyr.adesp.* 29(*b*).2 P.'

†**πολυχρήσιμος**, ον, *useful in many ways*, Zos.Alch. 215.2.

***πολυχρονέω**, *take a long time* (to rise), Ptol.*Tetr.* 132.

πολυχρονία, add '**b** *long life*, *POxy.* 465.174 (ii AD).'

πολυχρονίζω, delete 'abs. .. *Tetr.* 132'

***πολύχωρον**, τό, measure of capacity used in Egypt, *CPR* 7.20.14, 18 (iv AD).

πολύχωρος I, sup., add 'Alcin.*Intr.* p. 168.1 H.'

πολυῴδῠνος II, add 'πάτερ SEG 6.140.19 (Phrygia, iv AD)'

πόλχος, for 'coin inscription of doubtful meaning' read 'prob. pers. n.'

***πομπᾰγωγός**, ὁ, *organizer of a procession*, *SB* 9161, 9162, *BGU* 2118.4, *POxy.* 2768.5 (all iii AD).

πομπεῖον II, add '*IG* 2².1673.20 (iv BC), Plin.*HN* 35.140, Plu. 2.839c; also in Thrace, SEG 37.607 (iv BC)'

***πόμπευμα**, τό, *procession*, Ps.-Callisth. 1.10.

πομπευτής 2, for this section read '*one who takes part in a procession*, τοῖς πομπευταῖς (v.l. πομπεύουσιν) Luc.*Nec.* 16'

πομπεύω I 2, for this section read '*carry* or *escort in a ritual procession*, πομπενέτωσαν .. τὸν βοῦν ἐκ τοῦ πρυτανείου *IG* 12(7).515 (Amorgos, ii BC), αὐτήν (sc. Vestal Virgin) .. πομπεύσαντες δι' ἀγορᾶς D.H. 8.89.5'   **II 2**, after 'Procop.*Vand.* 2.9' insert 'transf., of the procession to the underworld, πομπεύων τὴν ἀδίαυλον ὁδόν epigr. in *IG* 9(2).648.11 (Larissa, Rom.imp.)' and for 'metaph.' read 'fig.'

***Πομπηϊασταί**, οἱ, guild of *worshippers of Pompeius*, *Inscr.Délos* 1641, 1797 (i BC).

πομπικός 1, add 'ὁδός Lang *Ath.Ag.* XIX L9.57 (343/2 BC)'   **2**, after 'Longin. 8.3' insert 'comp., Eun.*VS* p. 500 B.'

πόμπιμος I, add 'Plu. 2.86e'

***πομποστόλος**, ον, *of a procession*, πομποστόλον ἆμαρ ἄγει poet. in *Inscr.Délos* 2548 (i BC); **πομποστόλοι**, οἱ, *members of a procession*, *Inscr.Délos* 2607.4, 2608.4.

πομφόλυξ I, add '**2** *blister*, = πομφός, Nic.*Th.* 240.'   **IV**, before 'Paul.Aeg.' insert 'Plin.*HN* 34.128'

πονέω A, line 7, after 'A.R. 2.263' insert 'Aeol. pres. part. πονήμενοι perh. in Alc. 5.9, 119.17 L.-P.; also perh. **πονάω**, v.infra'   **II**, add 'so in pass., id. 1.752'   **B**, line 9, after 'Dor.' insert '(?Aeol.)' and after 'Pi.*O.* 6.11' insert 'pf. part. πεπονᾰμένον id.*P.* 9.93'   **I**, add '**4** act., impers., πονεῖ μοι ὅτι .. *IHadr.* 157.'

πονηρία I, after '*condition*' insert 'Hippon. 39.4 W.'

πόνος I 5, delete the section.   **III**, line 4, transfer 'πόνον .. anap.' to section I 2, line 2, after 'Hes.*Op.* 470'; add 'of poems, Call.*Epigr.* 6.1 Pf., *AP* 7.11.1 (Asclep.)'

ποντιάς, add '**2** = Ποντικός, *AP* 7.497 (Damag.).'

ποντίζω, after 'A.*Ag.* 1013 (lyr.)' insert '*fr.* 47a.16 R.'

***ποντικοφάρμᾰκον**, τό, *litharge*, Anon.Alch. 335.1.

πόντιος 1 a, add 'epith. of Aphrodite, *IHistriae* 173.2 (ii BC); of Athena, Ἀθηναίης Ποντίης SEG 28.707 (Paros, iv BC)'

***ποντόγνητος**, η, ον, = ποντογενής, *Suppl.Hell.* 991.61 (poet. wordlist, iii BC).

***ποντόγονος**, ον, = ποντογενής, *Suppl.Hell.* 991.60 (poet. word-list, iii BC) (παντ- pap.).

πόντος II 2, add 'as a Roman province, *TAM* 4(1).25'

ποντόφαρυξ, read '**ποντοφάρυξ**'

***πονωπόνηρος**, *very wicked*, comic formation in Ar.*V.* 466, *Lys.* 350 (πόνῳ πόν- cod. Rav.).

πόπανον, line 3, for 'Men. 129.4' read 'Men.*Dysc.* 450'; add 'πόπανον ῥαβδωτὸν ἐννεόμφαλον *Inscr.Perg.* 8(3).161.3'

ποππύζω II, before 'pass.' insert 'also in showing disapproval, *tut-tut*'   **IV**, for this section read '*tootle* (on a crude musical pipe), Theoc. 5.7'

***πορβιοπώλης**, v. °φορβιο-.

***πορδηκίδαι**, term of abuse, mock-patronymic formed fr. πορδή, *fr.iamb.adesp.* 33 in *Anth.Lyr.* i(3) D.³ (Hsch. s.v. πατρόθεν II).

πορεία II 3, add '**b** philos., *way*, *path*, in life, Socr.*Ep.* 29.5.'   **4**, for this section read '*means of transport*, *IG* 1³.127.34, *PRev.Laws* 50.11 (iii BC), *PGrenf.* 43.8 (ii BC)'

πορεῖον I, add '**2** pl., *transport animals*, *PCair.Zen.* 720.7 (iii BC), *PTeb.* 704.20 (iii BC), 750.6, al. (ii BC).'

***πορευτής**, οῦ, ὁ, *ferryman*, *OWilck.* 1507 (ii BC).

πορεύω I, add '**4** *convey*, *channel* information, Eur.*Melanipp.Capt.fr.* 6.17 (*GLP* 13.13).'

***πορθμάριος**, ὁ, = πορθμεύς, *ferryman*, *POxy.* 2421.8 (iv AD), *PMil.Vogl.* 188 ii 29, 33, *PSI* 808.2 (?iii AD), (all sp. προθμ-).

πορθμεῖον III, add '**2** *ferryboat tax*, *PKöln* 95.2, 15 (written προθ-, ii/iii AD), perh. also *PColl.Youtie* 31.4 (AD 199), *PRyl.* 185.6.'

πορθμεύω II, delete 'Act. intr.' and before 'τίς ἀστήρ' insert 'abs.'; add '*IG* 1³.41 (446/445 BC)'

πορθμός II, delete 'χωρεῖ .. 341c'

πορθμοφυλακία, delete the entry (only in abbrev. form, perh. read ‡πορθμεῖον'.

πόριμος II 2, delete the section transferring exx. to section I.

***Πόρισος**, epith. of Zeus, Διὸς Πορίσου Κτησίου Robert *Hell.* 10.63 (Rom.imp.), unless to be read as Πορισ⟨τ⟩οῦ.

ποριστής 1, add '**c** *purveyor, provision-merchant*, Ph. 2.525 (pl.).'

πόρκος 1, add 'perh. also π. Ἱστριεὺς τετρασκελής Lyc. 74 (kind of animal acc. to Sch. ad loc.)'

πόρνη, for 'Archil. 142' read 'Archil. 302 W.' and add 'Alc. 117(*b*).26 L.-P., al., Hippon. 104.34 W.; πόρνας (gen. sg.) SEG 37.661 (Chersonesus, iv BC)'

255

πορνοβοσκέω, after 'Ar.Pax 849' insert 'D. 59.68'

*πορνολύτᾱς, α, ὁ, perh. *one who resorts to prostitutes*, graffito in *Glotta* 40.50 (Tarentum, iv/iii BC).

Πορνόπιος, add '*SEG* 34.1238 (Aeolis, c.200 BC)'

πόρνος 1, add '*IG* 12(3).536 (vii/vi BC)'

*πορνότεκνον, τό, *child of a prostitute*, *PAlex.* 76ʳ.

πόρος I 1, add 'Th. 7.78.3, 80.6'   IV, after 'personified' insert 'as a cosmological principle, Alcm. 1.14 Sch.A, 5 fr. 2 ii 19 P.'

πόρπαξ I, add 'Critias 37 D.-K.'

πορπάω, after 'A.*Pr.* 61' insert 'med. aor. ἐπορπήσατο Hsch. s.v. περονήσατο; plpf. ἐπεπόρπηντο, id. s.v. δωριάζειν'

πόρπη, line 3, after '(E.)*Hec.* 1170' insert 'πόρπη χρυσῇ *Inscr.Délos* 439a77 (181 BC)'; line 6, for '*IGRom.* .. (Egypt)' read '*IPhilae* 159.1 (acc. πόρπαν)'

πορπίον, for pres. ref. read '*Inscr.Délos* 1417A3, 1442A53 (both ii BC)'

πορσύνω, line 3, before 'πορσαίνεσκον' insert 'πορσαίνεσκεν Hes.*fr.* 43a.69 M.-W.'   II 3, after 'pass.' insert 'στόλος πορσύνεται S.*Ph.* 781'   at end of article add 'a form πορσύνεται is found in *GVI* 1923.12 (Cyzicus, i AD); perh. πόρσυε shd. be read for πόρθυε ib. 2039.9 (Mytilene, ?i/ii AD)'

†πόρταξ, ἄκος, ὁ, *calf*, Il. 17.4 (= ἄρρην βοῦς Hsch.); cf. as pers. n., *SEG* 31.827, 32.927 (Sicily, Hellen.).

*πορταρήσις, ὁ, Lat. **portarensis, gate-watchman*, *ODouch* 31.1, 41.1, 6.

πορτᾶς, after 'dealer in calves' insert '(or *gate-keeper*, fr. Lat. *porta*)'

*πορτευθών, v. °προσέρχομαι.

*πορτί, v. πρός.

*πορτιπονέν, v. ‡προσφωνέω.

πόρτις, add 'Myc. *po-ti-pi* (instr. pl.)'

πορφύρα III, add 'granted as an honour to an agonothete, *Gerasa* 192.9, 15 (ii AD)' transferring '*IGRom.* 3.1422' fr. section IV.

*πορφυρᾶς, ᾶ, ὁ, *dealer in purple*, *MAMA* 8.562 (Aphrodisias, ii AD), *ITyr* 118b, 119, 120.

*πορφυρᾰφορία, v. ‡πορφυροφορία.

†πορφύρεος, α, ον, Att. -υροῦς, ᾶ, οῦν, Aeol. -ύριος Sapph. 54 L.-P., Alc. 45 L.-P.; dat. πορφύρωι Sapph. 98(a).4, neut. sg. πόρφυρον ?Sapph. 105(c).2, neut. pl. πορφύρα Sapph. 44.9; -ύριος also in *IG* 5(1).1390.179 (Andania, i BC); παρφυροῦς *Dura*⁴ 93,97.   I *purple-dyed*, *purple*, of stuffs, clothes, etc., π. φᾶρος Il. 8.221; χλαῖνα Od. 4.115; πέπλοι Il. 24.796; δίπλαξ 3.126, Od. 19.242; ῥήγεα Il. 24.645; τάπητες 9.200, Od. 20.151; σφαῖρα 8.273; χλάμυς Sapph. 54 L.-P.; σπάργανα, πτερά, Pi.*P.* 4.114, 183; χλανίς, χιτών Simon. 38.16 P., B. 18.52 S.-M., cf. A.*Pers.* 317, Hdt. 1.50.1, E.*Or.* 1457 codd. (lyr.), etc.; adv. -ῶς, *with purple dye*, στύφειν *PHolm.* 24.37.   **2** *purple-clad, in purple*, Luc.*Tim.* 20.   II *purple-coloured*: **1** of the sea, ἐς .. ἅλα πορφυρέην μεγάλα στενάχουσι ῥέουσαι Il. 16.391; ἀμφὶ δὲ κῦμα στείρῃ πορφύρεον μεγάλ᾿ ἴαχε νηὸς ἰούσης 1.482, Od. 2.428; π. κῦμα .. ποταμοῖο ἵσταντ᾿ ἀειρόμενον Il. 21.326, cf. Od. 11.243; θάλασσα Alc. l.c.   **2** of human complexion, *bright-red, rosy*, π. Ἀφροδίτη Anacr. 12.3 P.; στόμα Simon. 80.1 P.; παρῄδας Phryn.Trag. 13 S.; χείλη *GVI* 746 (iii/iv AD); also of birds' plumage, τὸ .. νῶτον αὐτῷ πορφυροῖς ἠγλάϊσται Ael.*NA* 17.33.   **3** of blood, αἵματι δὲ χθὼν δεύετο π. Il. 17.361; hence of death in battle, ib. 5.83, al.   **4** *dark*, of hair, *Anacreont.* 16.11 W.   **b** ὁ π. alone, of Death (perh. the *Dark One*), *AP* 11.13.2 (Ammian.).   **5** of the rainbow, Il. 17.547; to which a supernatural π. νεφέλη is compared, ib. 551, cf. Xenoph. 32.2.   **6** πορφυροῦν (sc. ἄνθος), τό, *Woodfordia floribunda* (an Indian shrub), Ctes. 45(38) J.   Form *πορφύρειος, Myc. *po-pu-re-jo*.

*πορφυροβᾰφικός or -βαπτικός, ή, όν, *connected with purple-dyeing*, ἡ π. (sc. τέχνη), Phld.*Rh.* 1.16.12.

*πορφυροβᾰφος, after '*dyer of purple*' insert 'Ion Hist. 6 J., *SEG* 25.180 (c.330/22 BC)'; delete 'Ath. 13.604b'

πορφῠροπώλης, add '*IMiletoupolis* 35 (i/ii AD), *PHerm.* 52 (AD 399), etc.'

*πορφῠρόσημος, ον, *marked with purple*, δερματίκιον σειππόεινον πορφυρόσημον *PMich.*inv. 1373 (*ZPE* 21.26).

*πορφῠρόστολος, ον, *having a purple robe*, Zos.Alch. 246.22.

πορφυροφορία, add 'also πορφῠρᾰφορία *SEG* 37.855 (Caria, i AD)'

*πορφῠρόω, *empurple*, fig., *glorify*, καὶ ἐν δόξῃ αὐτῶν πορφυρωθήσεσθε Aq.*Is.* 61.6.

πορφύρω 1, line 7, after 'id. 4.668' insert 'of storm clouds, *AP* 5.64 (Asclep.)'   **2**, line 4, before 'Q.S. 2.85' insert 'A.R. 1.461, 3.397'

*πόρω II 1, add 'τὸ πεπρωμένον μοίρης ἀπέδωκα *SEG* 32.605 (ii AD)'

Ποσειδάων, form πο-σε-δα-ο, πο-σε-δα-ονο (dat.); add 'Myc. *po-se-da-o*, *po-se-da-ono* (gen.)'; line 7, after 'B. 16.79' insert 'A.*fr.* 78c.47 R. (lyr.)'; line 14, after '(Megarian)' insert 'but ἁ *Tit.Calymn.* xii 30 (Cos, iii BC)'; add 'Ποσειδᾶν, acc. (app. conflation of Attic Ποσειδῶ and Doric Ποτ(ε)ιδᾶν, *Suppl.Hell.* 990.6)'

Ποσειδώνιος, add 'also as month-name (cf. Ποσιδάϊος, etc.), μηνὸς Ποσειδανίου *SEG* 32.871 (Crete, ii AD)'

πόσθων I, add 'as pers. n., *SEG* 35.30 (Sounion, c.500 BC)'   II, add 'Men.*fr.* 415 K.-Th.'

*ποσιαστής, οῦ, ὁ, prob. *member of* a religious association called a συμπόσιον, *BCH* 60.337 (Philippi, ii/iii AD).

Ποσιδήϊος, add 'cf. Myc. *po-si-da-e-ja*, name of a goddess'

Ποσιδηϊών, add 'sp. ποσειδι(ών) *SEG* 30.977 (Olbia, v BC)'

πόσις, ὁ, line 5, for '*Inscr.Cypr.* 93 H.' read 'Cypr. *po-si-se* πόσις, *ICS* 84.2'

πόσος II 1, ἐπὶ ποσόν, after 'Plb. 2.34.15, etc.' insert '*to a certain extent* or *degree*, *Peripl.M.Rubr.* 20'   for 'III' read 'II 3' and after 'ποσῶς' insert '*to some degree* or *extent*'

ποσσίκροτος I, add '*CEG* 785.2 (Histiaea, v BC)'

πόταγε, delete the entry.

*ποτάγω, v. προσάγω.

*ποτᾰμοδίαιτος, ον, *dwelling in rivers*, *APAW* 1943(14).8 (Chalcis, iii AD).

ποταμός I 1, add 'in a rallying cry, τοὺς ἐχθρούς σου τῷ ποταμῷ *SEG* 34.1056 (Aphrodisias, vi AD)'   II, add 'Ἱερὸς Π. *SEG* 31.933 (Caria, ii AD), Mitchell *N.Galatia* 1, 2, 3, al. (Rom.imp.)'   add 'IV ἐπὶ τοῦ ἥπατος σημεῖον Hsch.'

ποτᾰμοφόρητος, after 'river' insert '*BGU* 1216.98 (ii BC)'

ποτᾰμοφῠλᾰκίδες, delete the entry.

*ποτᾰμοφῠλᾰκίς, ίδος, ἡ, *guard- or patrol-ship on a river*, ὑπὲρ μερισμοῦ ποταμοφυλακίδος *SB* 4354.4 (AD 112), 4356 (AD 117), al.; pl., *OWilck.* 293 (i AD), *PFlor.* 91.4 (ii AD).

*ποτᾰμοφῠλαξ, ἄκος, ὁ, *river-guard*, rest. in *PAmh.* II 32.13 (ii BC), *PBremen* 11.32 (ii AD), *PLond.* 2561 (Milne *GSM* No. 715 (ποταφυλαξ)).

*ποτᾰμόω, pass., *to be formed into a river, flow together* (in quots., fig., of nations), Aq.*Je.* 28(51).44 (v.l. ποταμισθήσονται), *Is.* 2.2 (L.-R.).

ποτάομαι, line 3, for 'ποτῆται' read 'ποτῆται'

πότε II, line 3, for 'in μήποτε .. 144 H.' read '*po-te*, *ICS* 261, cf. μήποτε'; line 4, Aeol. ποτα, add '*SEG* 31.572 (c.200 BC)'   II 3, at end for 'Part.' read 'particle'; after '(dub.)' add 'ποτὲ μέν .. ἄλλοτε δέ *AP* 12.156'   III 1, add 'as first word in story, Aesop. 232 P.'

πότερος III, line 4, for '*SIG* 421' read '*IG* 9²(1).3A31'

ποτή (A), after '*h.Merc.* 544' insert 'ποτὴν ὄρνιθι ἐοικώς Arat. 278 M.'

ποτήρ, add '*PCol.* 240 (iv/v AD)'

*ποτηρία, ἡ, *drinking-cup*, ha ποτερία *IG* 9(1).303 (s.v.l., Locri, v BC).

ποτηρίδιον, add 'sp. βοτηρίδιν *SEG* 28.1586 (Rom.imp.)'

ποτήριον, for 'Alc. 52, Sapph.*Supp.* 20a.10' read '*CEG* 454.1 (Ischia, viii BC), *SEG* 32.859 (Eretria, vii BC), *Lyr.adesp.* 34 P., Sapph. 44.10 L.-P., Alc. 376 P.'

ποτής, add 'S.*fr.* 314.274 R.'

πότης, for 'usu. in fem. .. v. infr.' read 'Call.*fr.* 191.43 Pf.; fem. πότις'

ποτί, line 1, for 'Dor. for πρός' read 'dial. equiv. of πρός'; line 16, for 'also πο-' read 'also πο(τ)- w. simplified spelling'   add 'Myc. *po-si* (also in compds.), prob. *ποσί (cf. πός B) rather than *πορσί (cf. πρός, πορτί)'

*ποτι-, ποθ-, for compd. wds. w. this first element see also s.v. προσ-.

ποτίζω 1, line 2, after 'Aph. 7.46' insert 'νέκταρ ἐπότισεν Pl.*Phdr.* 247e'   **2**, for 'τοὺς ἵππους .. 247e' read 'πεπότικε μὲν γὰρ ὥσπερ ἰατρός μ᾿, ἔφη, ἃ δεῖ Macho 4 G.'   **3**, line 3, before 'also *water*' insert 'π. τοῖς ποσὶν αὐτῶν Lxx *De.* 11.10; cf. πούς I 6 k'

*ποτικαταρτίζω, Dor. for *προσ-, add *by way of equipment*, rest. in *IRhod.Peraia* 201.9 (i BC).

ποτικός 1, after 'etc.' insert 'comp., id. 2.352f'

ποτίκρᾱνον, for 'Dor. form of προσκρ- (which is not found)' read 'Dor. for *πρόσκρ- (hypothetical formation in Hsch. s.v. ποτίκρανον)'

*ποτιποιέω, *make in addition*, *SEG* 32.356 (Aegina, c.550 BC); cf. προσποιέω.

ποτισμός, for 'Call.*Fr.anon.* 121' read '*Suppl.Hell.* 1032'

ποτιστρίς, add '**2** adj., ἐν τῇ μέσῃ ποτιστρίδι διώρυγι *PCair.Zen.* 825.19 (iii BC).'

ποτιψαύω, after 'Dor.' insert 'and poet.' and after 'Pi.*fr.* 121.3' add 'S.*Tr.* 1214'

πότμος I 1, add 'Plu. 2.591c (poet. language)'   last line, after 'fr. 871.1' add 'E.*Hec.* 971, *Ion* 1605'

πότνα, after '(nom.)' insert 'π. γυναικῶν *AP* 6.287 (Antip.Thess.)'

πότνια, penultimate line, after '*Ion* 873, al.' insert 'Ar.*Ra.* 337, *Pax* 445, 975 (anap.), al.'   add 'Myc. *po-ti-ni-ja*, title of goddess'

πότος, ὁ, add '**2** *something to drink, draught*, Nic.*Al.* 59.'

*ποτουδίζω, v. °προσουδίζω.

*πουιών, ῶνος, ὁ, perh. = πουλβῖνον, *PWarren* 18.12 (iii AD).

*πουκρίς, ίδος, ἡ, kind of freshwater fish, *IGC* p. 99 B27 (Acraephia, iii/ii BC).

πουλβῖνον, add 'also φουλβῖνον (q.v.)'

*πουλβῖνος, ὁ, *bed-tick*, *Edict.Diocl.* 28.56 (sp. -βει-).

*πουλικάριος, α, ον, adj. describing kind of blanket, *Edict.Diocl.* 8.43.

*πουλλίον, τό, dim. of °πούλλος, in context perh. *cockerel*, SB 5301, 5302 (Byz.), *PColl.Youtie* 95.9 (vii AD); cf. mod. Gk. πουλί *bird*.

*πούλλος, ὁ, *chicken*, *POxy.* 1913.26 (vi AD).

*πουλυβοώτης, ὁ, ἡ, v. πολυβούτης.

*ποῦνδα, v. °φοῦνδα.

*πουρα[μ]ω (gen.), kind of freshwater fish, *IGC* p. 99 B30 (Acraephia, iii/ii BC).

*πούραυμα (for *πῦρ-), ατος, τό, perh. *brazier*, *BCH* 62.149 (Boeotia).

*πουρείνιον, τό, dim. of πουρεινίς, = πυρήνιον, *IG* 7.2421.8 (Tanagra, iii BC).

πούς **I 1**, last line, after 'Od. 4.149' insert '*AP* 5.55 (Diosc.)' **4 b**, line 6, after 'Prov.ap.Suid.' insert 'cf. Mesom. 3.9 H., Ps.-Babr. *Μυθικά* 17 p. 219 C.' **c**, after '*close at hand*' insert 'τῶν δ᾽ (ἰβίων) ἐν ποσὶ .. εἰλευμένων τοῖσι ἀνθρώποις Hdt. 2.76.1' **6 k**, lines 3/4, for 'perh. .. wheel ' read '*by the foot*, i.e. measure, (during the inundation)' and before 'τόπον' insert 'ἀπὸ π. ποτισμός *SIFC* 13.366 (cf. °ποτίζω 3)' **II 1**, add 'ἐν ποδὶ ληγούσης Πελοπηΐδος at the *extremity* of the Peloponnese, referring to the Isthmus of Corinth, Call.*fr*. 384.11 Pf.; πὰρ ποδί .. Νείλου νειατίῳ by the outermost (i.e. most easterly) *mouth* of the Nile, ib. 48; app. w. ref. to a form of interlinear glossing, τῇ κατὰ πόδα καλουμένῃ χρήσασθαι τῶν νόμων ἑρμηνείᾳ Just.*Const. Δέδωκεν* 21' **III**, for '6 fingers' read '16 fingers' at end add 'Myc. *po-de* (dat.), *po-pi* (*πόπφι, instr. pl.), *foot* of cauldron, table'

ποΰ(τ)ριν, add '(= Lat. *putris*)'

πρᾶγμα **II 8**, after '*question*' insert 'S.*fr*. 314.332 R.' **III 3**, line 5, after '= κτήματα, Hp.(*Lex* 5?)ap.Erot.' insert 'τὰ πράγματα τοῦ αὐτοῦ μοναστηρίου κινητὰ καὶ ἀκίνητα καὶ αὐτοκίνητα *Cod.Just*. 1.3.43.4; τῆς τῶν πραγμάτων διακρίσεως (transl. Lat. *separatio bonorum*) Just.*Const. Δέδωκεν* 7d '

πραγμάτεια **II 1**, add '*appurtenances of one's business*, i.e. *merchandise, wares*, Just.*Nov.*appendix 5'

πραγματεύομαι **I 2**, add 'ἐπί τι *IEphes*. 1503.10 (ii AD)'

πραγματευτής, add 'of slave (or freedman) estate-managers, *SEG* 31.1143, *INikaia* 1203 (Rom.imp.)'

πραγματευτικός **I**, after '*business*' insert 'Ptol.*Tetr.* 178'

πραγμάτικός **I 1**, line 4, delete '*men of affairs*'

+πραγμᾰτολογέω, *state one's case*, Arist.*Rh.Al.* 1438ᵇ20, Ph. 1.554, 655, D.L. 9.52.

πραιδεύω, after 'δηώσαντες' insert 'Sch.E.*Ph*. 202-15'

*πραικόκ(κ)ιον, τό, *apricot*, Orib. 1.48 (*CMG* 6.1(1).23), al., Gal. 6.811.12; also πρεκόκκιον Gal. 6.585.9, 594.6; cf. βερικόκκιον, Lat. *praecox*.

πραίκων, after '*praeco*' insert '*SEG* 36.595 (Maced., ii/iii AD)'

πραιπόσιτος, add 'in non-military context, πρεποσείτου θησαυρῶν *SEG* 32.1061 (Rome, AD 414)'

*πραῖς, ὁ, Lat. *praes*, *surety*, *SEG* 39.1180.102 (Ephesus, i AD).

*πραισεντάλιος, α, ον, Lat. *praesentalis*, *serving in the imperial palace*, Just.*Edict*. 13.2.

*πραίσεντον, τό, *troops serving in the imperial palace*, Just.*Nov.* 22 epilogos.

*πραιτέριτος, ὁ, Lat. *praeteritus*, *former comrade*, *CPR* 6.76.6.

*πραιτούρα, ἡ, Lat. *praetura*, Just.*Nov.* 13.4 pr.

*πραίτωρ, ορος, ὁ, Lat. *praetor*, *IGRom*. 3.188 (Ancyra), al., *BCH* 7.20, Porph.*Plot*. 7.35, Lyd.*Mag*. 1.27, al.

*πραιτωριανός, ή, όν, *praetorian*, Just.*Nov.* 24.1.

πραιτώριον **I**, after '*Ev.Matt.* 27.27' insert '*PPetaus* 47.44, 48.2 (sp. πλετ-) (both AD 185)' **II**, add 'ἔπαρχος τῶν ἱερῶν τῆς ἕω πραιτορίων *IEphes*. 1345'

*πραιτώριος, α, ον, Lat. *praetorius*, κοόρτης δεκάτης πραιτωρίας Φιλιππιανῆς *IEphes*. 737.6.

*πραίφεκτος, ὁ, Lat. *praefectus*, *IG* 14.680 (ii AD), *IGRom*. 1.10, Lyd.*Mag*. 2.6; cf. Plb. 6.26.5.

πρακτικός **I 1**, add '**b** *concerned with real life*, opp. μυθικός, Longin. 9.14.'

πράκτιμος, before '*SIG*' insert 'w. gen.' and add '*Delph*. 3(6).69.13 (ii/i BC)'

πρακτορεύω, add '*JRCil*. 1.19 (Side, iii AD)'

πρακτορικός, add 'πρακτορικόν, τό, *an extra charge made by* or *for the* πράκτορες, *PTeb*. 298.63 (ii AD), *BGU* 471.13, 17'

πράκτωρ **II 2**, add '**b** *court official*, = Lat. *exsecutor*, *Cod.Just*. 3.2.4.8.'

*πράμνιον, τό, *pick*, *BGU* 2359.4 (iii AD).

*πραξάγαθος, ον, *beneficent*, θεῶν π. (-αγαθαν lap.) καὶ σωτήρων *BCH* 79.340 (Arcadian Gortys, ii BC).

πράξιμος **I**, add '**b** of persons, *liable to distraint*, *Inscr.Délos* 1522.18 (i/ii AD).'

πρᾶξις, add '**IX** *certificate*, διὰ τάχους ἀποστεῖλαι τὴν πρᾶξιν μετὰ τοῦ γραμματηφόρου *PWash.Univ*. 8.6, π. ὑπομνημάτων *PFlor*. 293 (both vi AD).'

πρᾱόνως, after 'Ar.*Ra*. 856' insert 'Lys. 24.15'; for '(Formed .. -νους' read '(presupposes a comp. *πράων; cf. ἐλασσόνως)'

πρᾶος, line 4, after 'Lyr.(' insert 'first in Alc. 68.3 L.-P. (πράϋ)'; line 15, after 'cf. Phot.' insert 'ἔχει δὲ τὸ ἰῶτα' **I 3**, add 'τὸ πρᾶον *modesty*, *humility*, τὸ πρᾶον θαύμαζε τῶν κτισμάτων *SEG* 31.1472 (Arabia, AD 603)'

*πρᾱσᾶς, ᾶ, ὁ, *grower* or *seller of leeks*, *IG* 12(5).1104 (Syros, ii AD).

πράσινος **2**, add 'Epiph.Const. in *Lapid.Gr*. 194.23 (given as equiv. of σμάραγδος)' **3**, add 'τῶν νέων Πρασίνων, i.e. organized youngsters within the green faction, *SEG* 31.1493 (Alexandria, c.AD 608)'

*πράσιον **I 1**, line 3, for 'Nic.*Th*. 550' read 'Nic.*Al*. 47'

*πράσιος, ἡ, = πράσιον, Nic.*Th*. 550.

πρασοκουρίς, for '*milliped*' read '*kind of caterpillar*'

πράσσω, line 16, after 'A.*Pr*. 75' insert 'pf. part. pass. πεπράμενος *SEG* 26.121 (late i BC)' **III 3**, delete the section. **VII**, delete the section.

πρατήνιον, delete the entry.

*πρατήνιος, ον, [?ᾱ] epith. of a ram, perh. *yearling*, κριὸς π. Sokolowski 2.94.6, 11 (Camirus, iii BC, cf. also 104); cf. Phot. s.v. προτήνιον, Hsch. s.v. πρατήνιον, see also πρητήν.

πρᾱτήρ **II**, add 'so π. alone, ἔστησεν ἐπὶ τοῦ πρατῆρος *Vit.Aesop*.(G) 21'

*Πρατίνειος, ον, *of Pratinas, the reputed inventor of satyric drama*, λῆμμα Πρατίνειον epigr. in *Chiron* 19.505 (c.200 BC).

πρατοπάμπαις, for 'chief of the πάμπαιδες (v. Addenda)' read '*member of one of the age-groups in Spartan education*'

πρᾱτός, add 'οἶν πρατόν *SEG* 26.136 (Athens, iv BC), ἐν πρατῷ *for sale*, *SEG* 34.558 (Larissa, ii BC)'

πρᾶτος, line 2, delete '436'; add 'τῷ ἔτει τῷ π. *next following*, *IG* 12(3).436.7 (Thera, iv BC); as the name of a month, Πράτου μενός *AA* 4.681-7 (Phocis, v BC)'

πραϋμενής, after '*of gentle spirit*' insert 'Sch.E.*Or*. 119'

πραΰνοος, after '*of gentle mind*' insert 'πειθώ Simm. 24.10'

πρείγυς, line 4, after 'al.' insert 'title of a magistrate, *Inscr.Cret*. 4.184.13 (Gortyn, ii BC)'; line 7, for 'ib. 2562' read '*Inscr.Cret*. 4.3.7.23'; add 'see also ‡πρέσβυς'

*Πρειέτιος, v. °Πριέτιος.

*Πρείετος, v. °Πρίετος.

*πρειμ-, for transliterated compds. of Lat. *primus* v. °πριμ-.

*πρεκνόν· ποικιλόχροον ἔλαφον Hsch.; cf. περκνός, πρακνόν.

*πρέποντος, ον, (back-formation fr. πρεπόντως) *fitting*, ἔδοξε .. πρέποντον ἔμμεν *BCH* 59.37 (Crannon, ii BC).

πρεπόντως **1**, add 'δι᾽ ὅλης τῆς ἀρχῆς ἀνεστράφη π. *SEG* 33.696 (Amorgos, iii BC)'

πρεπτός, add 'S.*fr*. 314.330 R.'

πρέπω **I 2**, for this section read '*to be clearly audible*, A.*Ag*. 321, S.*fr*. 314.231 R.' **3**, delete the section. **III 3**, add 'w. inf., γείτονες οὔ τρισσαὶ μοῦνον Τύχαι ἔπρεπον εἶναι *APl*. 40 (Crin.)'

πρεσβεῖον **I 3**, add '**b** = Lat. *legatum*, *legacy*, *Cod.Just*. 1.2.5.1, 1.3.45.13, 14.'

*πρέσβειρα, after 'πρέσβυς' insert 'S.*fr*. 314.339 R.'

+πρεσβεύτειρα, ἡ, fem. of πρεσβευτής, in quot. fig., of scent, as an emissary coming to hunting-dogs, Opp.*C*. 1.464 (πρεσβύτ- codd.).

πρεσβευτής **II 2**, add 'πρεσβευτῶν Σεβαστῶν ἀντιστράτηγος *SEG* 31.908 (Aphrodisias, iii AD)'

πρεσβευτικός, add '**2** -κά, τά, *expenses of an embassy*, or perh. *court-fees*, *PMerton* 35 (iv AD).'

πρεσβεύω, after lemma insert 'πρεσγεύω *GDI* 5148' **I 1 c**, delete 'also .. 352d' **2**, line 7, for 'time' read 'value' **II 3 b**, delete the section.

πρέσβις (A), after '*ambassador*' insert 'Prisc. pp. 286, 320 D.'

πρέσβις (B) **II 1**, for 'Aesop. 107ᵇ V Chambry)' read 'Aesop. 57(II) H. (v.l. -υς) and add 'cj. for πρέσβυν in Lyc. 331'

πρέσβυς, line 1, for 'ὁ' read 'or νος (v. infra III 2 b), ὁ (also ἡ Lyc. 331 codd.)'; line 18, after '*PCair.Zen*. 447.9 (iii BC)' insert 'πρεσβυτέρα, ἡ, *old lady*, *PHaun*. 15.11 (ii BC) and after 'η, ον' insert '(πρεσβυτερώτατος *PSI* 1159.5 (ii AD), πρεσχύτατος epigr. in *ZPE* 12.175 (Egypt, v BC), πρίγιστος *Glotta* 41.65 (Crete)' **II**, add 'of a Roman senator, Plb. 21.18.7, 10; πρισγουτέρυς *SEG* 23.271 (Thespiae, iii BC)' **III 2**, line 7, after 'Ptolemaïs Hermiu, i AD)' insert 'τὸν πρεσβύτερον τῆς κώμης *JP* 18.35 (iv BC)'; line 8, after '*Ev.Matt*. 16.21, etc.' insert 'fem. πρεσβυτέρα *CIJud*. 731c (Crete, iv/v AD)' add '**b** *guardian*, πρέσβυος τῶν ὁσίων *BCH* 87.203 (Delphi, ?i AD).' 4/5 lines fr. end of article for 'Cret.' read 'Dor.' and before 'πρείγυς' insert 'πρέσγυς'

*πρεσβυτεράρχης, ου, ὁ, official at local games, π. τῶν Ὀλυμπίων καὶ ἱεροφάντης *SEG* 24.479 (Maced., iii AD).

*πρεσβυτερία, ἡ, *seniority*, εἰς πρεσβυτερίας λόγον *PMich*.v 326.11 (i AD).

*πρεσβυτέρισσα, ἡ, *elder*, (Jewish), *SEG* 27.1201 (Tripolitania, iv/v AD).

Πρέσβων, delete the entry.

*Πρεσβωνοι, οἱ, (dub. accent), *the Elders*, name of family or χιλιαστύς, *ABSA* 58.56 (Chios, v/iv BC), *SEG* 15.537.
*πρεσγεύω, v. °πρεσβεύω.
*πρέσγυς, v. °πρέσβυς.
πρηγιστεύω, add '*SEG* 32.869 (Crete, ii AD)'
πρήθω, line 2, for 'A.R. seems .. 1537' read 'pres. part. πρῆσ(σ)οντα, -οντος A.R. 4.819, 1537 (cj. πρήθ-)'
*πρήμ(ν)η, ἡ, = πρημνάς, Hsch., Phot.
*πρηρόαρχος, ὁ, ἡ, epith. of victim sacrificed at the πρηρόσια, Δαίρα ἀμνή πρηρόαρχος *IG* 1³.250*A*16.
*πρηρόσια, v. °προηρόσιος.
πρηστήρ II, delete the section transferring exx. to section I and inserting 'transf., of the hot blast produced by bellows' before 'A.R. 4.777'
πρίαμαι, add 'Myc. aor. 3 sg. *qi-ri-ja-to*'
*Πριᾱπίδιον, τό, *small image of Priapus*, *Inscr.Délos* 1442*A*4 (ii BC).
πριᾱπίζω, for '*to be lewd*' read '*behave toward like Priapus*'; delete '*to be ithyphallic*' to end of article.
Πρίᾱπος, delete '(also written Πρίεπος ..)' v. °Πρίετος
*πριβάτιος, α, ον, Lat. *privatus*, *PKlein.Form.* 1033 (vi AD).
*πριβᾶτον, τό, *private bath*, (fr. Lat.), *IGLS* 3(2).545 (v AD). II pl., *the private property* of the Emperor (Lat. *res privata*), *Cod.Just.* 1.5.18.9, Procop.*Arc.* 22.12.
*πρίγκεψ(-ιψ), ιπος, ὁ, Lat. *princeps*, (in quots., as legionary rank), Plb. 6.21.7, al., *IGRom.* 3.264, 1157, 1230, *POxy.* 1424, *BCH* 23.419.
*πριγκιπάλιος, ὁ, Lat. *principalis*, *PFlor.* 278 iii 14, iv 12 (ii AD).
*πριγκιπᾶρις, ὁ, Lat. *principalis*, π. γέγονα ἐφ᾽ ἔτους *SB* 8088.5 (ii AD).
πρίγκιπες, delete the entry.
*πριγκίπια, τά, Lat. *principia, headquarters*, *SB* 10530.4, 18 (AD 143).
Πρίεπος, delete the entry (v. °Πριέτιος).
Πρίεπος, delete the entry (v. °Πρίετος).
*Πριέτεια, τά, *festival of the god Prietos*, *SEG* 36.1155 (sp. -ηα, Nicomedia, ii AD).
*Πριέτιος, (sc. μήν), ὁ, name of a Bithynian month, *TAM* 4(1).19, 35, 59, al., (sp. Πρειέτειος, Πριέτηος. Alleged forms Πριεπ-, Περιεπ- do not exist, v. *AE* 1979.231-6).
*Πρίετος, ὁ, a Bithynian god, *TAM* 4(1).74, al. (Πρειετ-; Πρίεττος at *SEG* 36.1155); app. confused w. Πρίαπος in Luc.*Salt.* 21, Arr.*fr.* 23 J., see note at end of foreg.
*πριμάριος, ὁ, Lat. *primarius*, *BGU* 958d (iv/v AD, pl.).
*πριμικήριος, ηρος, ὁ, = °πριμικήριος, *PBodm.* 29.49, 126 (iv AD).
*πριμικήριος, ὁ, Lat. *primicerius*, *BCH* 33.34 (iv AD, written πρημηκιρις), *JHS* 11.162, *PColl.Youtie* 89.3 (AD 485), Just.*Edict.* 4.1.
*πριμιπιλάριος, ὁ, Lat. *primipilaris*, *IGRom.* 3.55 (πρειμι-), 810, etc., *INikaia* 1551 (iii AD); transf., πιθήκων π. *Vit.Aesop.*(G) 87 (-ηπη-). Also πρειμοπειλάριος, *BCH* 4.377, etc.; πριμοπιλάρ(ιν) *SEG* 34.1591 (Egypt).
*πριμιπίλον, τό, *office of primipilus*, Modest.*Dig.* 27.1.8.12, 27.1.10.5, *POxy.* 1905.10, 2001.3, *PThead.* 48, *PCol.* 141.52.
*πριμισκρίνιος, ὁ, Lat. *primiscrinius, chief secretary*, *SEG* 9.356.17, 71 (Cyrenaica, vi AD), 32.1554 (Arabia, vi AD).
πρίν, line 4, add 'see also πρίν B , Il. 1.97, Od. 19.585, al. (v. infra); by ἔστ᾽ ἄν, A.R. 2.251.' II, add '5 *sooner*, folld. by πρίν B, Il. 7.481, Od. 19.475, al. (v. infra).'
*πρῑνεύς, έως, ὁ, *ilex-grove* (possibly a place-name), *IEryth.* 151.20 (iv BC).
*πρῑνεών, ῶνος, ὁ, *ilex-grove*, Aq.*Ge.* 14.3,8.
*Πρινοφόρος, ὁ, *bearer of the holm-oak* (cult-title of Dionysus), *IG* 10.2(1).260.B2 (iii AD).
πρῑνών, delete '*IG* 1².328.1 (dub.)'
*πρῑονοποιός or πρῑονοπώλης, ου, ὁ, *maker* (or *seller*) *of saws*, rest. in *SEG* 12.84 (Athens, v BC).
*πριόρες, οἱ, Lat. *priores*, (*military*) *officers*, Just.*Nov.* 117.11.
*πρισγούτερος, v. °πρέσβυς.
πρίσμα, before 'Thphr.*HP* 5.6.3' insert 'Aen.Tact. 35'
*πρισμή, ἡ, *sawing*, ἐγλαβόντι τὴν π. τῶν ξύλων *IG* 11(2).199*A*89 (Delos, iii BC).
⁺πρισμός, ὁ, *sawing*, Theb.*Ostr.* 144.2 (πρυσμ-, i AD). 2 *gripping* as with the teeth, πρισμοῖς· ταῖς βιαίοις κατοχαῖς Hsch.
*πρισόριον, τό, app. Lat. *pressorium*, ?*clothes-press*, *PMich.*XIV 684 (*Tyche* 6.233, vi AD).
πρίων (A) 1, add 'τὸν φθόνον ἔφη πρίονα εἶναι ψυχῆς *Gnomol.Vat.* in *WS* 11.63' add '4 πρίονας χερῶν· τοὺς δεσμούς Hsch.'
⁺πρίων (B), a, nonce-wd., app. one who calls out πρίω (buy), Ar.*Ach.* 36; cf. πρίων· ἀγοράζων Hsch.; the force of a pun on πρίων (A) is not clear.
πρό A I 3, lines 5/6, delete 'E.*Alc.* 18, 645'; line 10, for 'cf.' read 'θανὼν π. κείνου E.*Alc.* 18, cf. 645' II 1, (p. 1465 b, line 4), after 'Pl.*Smp.* 173a' insert '*Trag.adesp.* 664 i 12 K.-S.'; (line 5) delete 'πρὸ τοῦ ἤ .. (Thisbe)' add 'b πρὸ τοῦ as conj. = πρίν, πρὸ τοῦ ἤ Γάιος Λοκρέτιος τὸ στρατόπεδον .. προήγαγεν *IG* 7.2225.22

(Thisbe, ii BC), π. τοῦ εἰσέλθοις *BGU* 814.14 (iii AD), π. τοῦ τις ἐνέγκῃ *PFay.* 136.6 (iv AD).' 2 b, add '*SEG* 31.830 (Sicily, iv/v AD)' D, add 'Myc. *po-ro-ko-wo* (πρόχοος), *po-ro-te-ke* (προύθηκε), *po-ro-ko-re-te* (*po-ro-* in sense of "*in place of, vice-*"), etc.'
προάγγελος II, after '*harbinger*' insert '*Trag.adesp.* 664 ii 11 K.-S.'
*προαγγελτήρ, ῆρος, ὁ, = προάγγελος II, prob. in *SEG* 9.72.137 (iv BC).
⁺προἀγοραστής, οῦ, ὁ, app. *servant who buys stores for the household*, *MAMA* 3.668a.
προαγωγή, add 'IV *rhetorical delivery*, Corn.*Rh.* p. 397 H. V *production* (of documents, etc.) *in court*, Just.*Edict.* 7.1.'
*προάγων, οντος, ὁ, title of an official in Ormele, *PASA* ii nos. 41*A*, 43, prob. in no. 89.
προαγών 1, after 'etc.' insert '(or a *rehearsal*)'
προαιρέτης, for '*steward, keeper*' read 'title of official (of uncertain function)' and add '*ODouch* 54'
προαιρέω II 1 b, add '*SEG* 23.447 (Thessaly, iii AD)'
προαισθάνομαι, add 'D. 18.63, al.'
*προακμαστικῶς, adv., *before the climax*, Aët. 5.53 (32.16 O.).
προακτικός, add 'II *advancing, promoting*, Ascl. in *Metaph.* 146.6, Phlp. in *de An.* 207.15.'
προαλίζω, add 'Aen.Tact. 17.4 (cj., cod. προσ-)'
προαλιώτας, delete the entry.
*προαναγγέλλω, aor. 1 part. προαναγγείλαντος cj. for προσαν- in Sch.Pi.*O.* 7.83.
προανάγω, add 'II *add beforehand to stock*, ὅταν τὸ ἴσον πλῆθος προαναχθῇ, Ἑλληνικά 7.179 (Chalcis, iii BC).'
προανακόπτω, add '2 *cut off in advance, preclude*, Just.*Nov.* 159.3.'
προανακρούομαι II, add 'b *begin a speech by saying*, προανακρουσάμενος ὅτι ?Duris in *POxy.* 2399.46-8; cf. ἀνακρούω II 2.'
*προανακτάομαι, *repair beforehand*, κάματον διὰ τῆς τροφῆς Porph. ap.Sch.B Il. 19.222 (προσ- cod.).
προανανεύω, *deny in advance*, *Pap.Bub.* 4 lxii 5 (iii AD).
προαναπαύω I, for pres. def. read '*prescribe rest before*' II, add '*IApam.* 133, *BE* 1980.303 (Epirus, both Chr.), *SEG* 26.790 (Byzantium, vi AD)'
προαναπλέω, add '2 *sail up* a river *first*, *PMich.Zen.* 57.12 (iii BC).'
*προανασκευάζομαι, for 'plpf.' to end of article read '*remove beforehand*, [κειμήλια] ib. 1.13.9; [ἀνδράποδα] prob. in D.H. 6.3'
*προανάσυρμα, ατος, τό, app. *indecent exposure*, Εὔβουλος δὲ ὁ κωμικὸς τὸ λαθρίδιον γέννημα προανάσυρμα παρθένου καταγελάστως ὠνόμασεν (*fr.* 138 K.-A.) Poll. 3.21.
προανατείνω, add 'med., *make apparent beforehand*, Onas. 6.11'
προανατίθημι, add '*Inscr.Perg.* 8(3).72'
προαναφώνησις I 2, add '*foreshadowing* of what is to come, Sch.T Il. 1.45, Sch.A Il. 11.604'
*πρόαντα (acc.), τόν, perh. *row of columns in front*, *IStraton.* 200.8.
προαπαγγέλλω, add '*SEG* 31.983 (Ionia, ii/i BC)'
προαπειλέω, after '*PCair.Zen.* 230.3' insert '(med.)'
προαποβάλλω, add 'εἴτε .. προαποβάλοι τὸν ἄνδρα i.e. *lose by his prior decease*, Just.*Nov.* 74.5 pr.'
προαποδίδωμι I, add '3 *act as* °προαποδότης, *GDI* 1990.9 (Delphi, 195 BC).'
προαποδότης, for '*one who .. surety*' read '*substitute for the seller, warrantor*'
*προαποθνήσκω I, add '*TAM* 4(1).217'
*προαποκλάω, *break off before*, ὁ κριὸς τῆς ἄλλης ἐμβολῆς προαπεκλάσθη Memn. 34.1 J.
*προαπόλλυμι I, after '*destroy first*' insert 'προαπολεῖ με Men.*Dysc.* 391 (προσ- pap.)'
*προαποτίλλω, *pluck out before*, Asclep.Jun.ap.Gal. 12.742.8.
*προαρραβωνίζω, *betroth beforehand*, π. αὑτὸν ἐμαυτῇ *Vit.Aesop.*(G) 30.
*πρόαρχος, ὁ, *president*, ὁ ἱερεὺς τοῦ Διὸς τοῦ Τροφωνίου καὶ πρόαρχος τοῦ βακχείου *SEG* 32.475 (Boeotia, iii AD).
προασθενέω, add 'Paul.Aeg. 3.78 (299.16 H.)'
*προαστιανός, όν, of or *belonging to the προάστιον*, θεοί *Inscr.Magn.* 309.
προάστιον, for '*suburb*' read '*land outside a town* (so also in derivatives)' add '3 *of a temple outside a town*, cf. °προαστιανός, *SEG* 8.536, 537 (Egypt, i BC; προαστιν-).'
προαστίτης, add '*POxy.* 3941.19 (AD 605)'
*πρόαστυ, εος, τό, = προάστιον, *IGBulg.* 2086.3 (epigr.).
*προάτριον, τό, *forecourt*, τὸ π. τοῦ μεγάλου γυμνασίου *TAM* 5(2).926.6.
προαυλέω, -ημα, -ία, for '*flute*' read '*aulos*'
προαφίημι, add 'Aen.Tact. 32.6'
*πρόβα, ἡ, Lat. *proba, test sample*, *SEG* 34.1243 (Abydos, c.AD 492).
προβάλλω A II 4, delete the section. B I 4, add 'And. 1.132' III 1, line 15, after 'D. 21.139' insert 'of a boxer, *to be on guard*'; line 16, after 'id. 4.40' insert 'προβεβλημένος *in the attitude of defence*, Arist.*fr.* 569'

προβᾰσῐλεύς, έως, ὁ, title of a magistrate at Argos, SEG 29.361 (c.400 BC).

προβᾰσῐλεύω, add 'Cod.Just. 1.11.9.3'

πρόβᾰσις, add 'III prob. marching in front, leading (a procession), τιμηθέντα προβάσει Delph. 3(1).555.26 (iii AD).'

προβασκᾰνία, add 'προβασκανίαν is perh. to be read for πρὸς βασκανίαν in Vit.Aesop.(W) 16'

προβασκάνιον, add 'prob. for προσβάσκανον in Vit.Aesop.(G) 16'

*προβατάγριος, ον, of a wild sheep, δέρμα Edict.Diocl. in SEG 37.335 ii 20 (where other versions have προβάτειον).

προβᾰτεύς, add 'IG 9².748.41, 43 (Locris, c.190 BC)'

*προβᾰτικόν, τό, tax on sheep, Hesperia 27.75.13 (c.200 BC).

προβάτιον, delete 'little' before 'sheep' and add 'PHerm. 27.5 (v AD)'

προβᾰτοβοσκός, after 'shepherd' insert 'IGBulg. 851 (-χός)'

προβᾰτοδόρας, for 'Procl. .. 502' read 'Sch.Hes.Op. 504'

πρόβᾰτον, line 2, after 'Hsch.' insert 'Vit.Aesop.(G) 97 (βρώμασι cod.)'   I, add 'of a pack-animal, SEG 33.1120 (Phrygia, iii AD)'

*προβᾰτονόμος, ὁ, perh. sheep-rearer, Ἰακω προβατον(όμος) SEG 36.970.B13 (Aphrodisias, iii AD).

προβᾰτώδης, for pres. def. read 'sheep-like' and for 's.v. βαίκυλος' read 's.vv. βαίκυλος, βλήχημα, βληχήματα'

*προβατωρία, ἡ, Lat. probatoria (sc. epistola), imperial letter of commendation, SEG 9.356.78 (Cyrenaica, vi AD), Lyd.Mag. 3.2; sp. προβατορ- Cod.Just. 11.8.16.1.

*προβέβαιος, ον, very firm, PMichael. 45.31 (vi AD).

*πρόβειος, α, ον, = προβάτειος, An.Ox. 2.56, An.Boiss. 3.408, PLond. 113.10.13 (vii AD), cf. CQ 33.31.

προβιβάζω 3, for 'teach' read 'inculcate in, impress on'

προβλώσκω, add 'from one's country, Ἀπαμείας πατρίδος ἐκ προμολών SEG 24.637 (Thrace, i BC)'

*προβοκάτωρ, ορος, ὁ, Lat. provocator, kind of gladiator, Robert Les gladiateurs no. 30 (Plotinopolis), 194 (Miletus, written πρω-), 291 (Cyzicus), ITomis 288 (iii AD), 30.1800 (iii AD).

πρόβολος I, add '4 face of a seal-stone, App.Anth. 3.79 (Posidipp.; = HE 20.6 G.-P.).'

προβουλή II, for 'dub. in BCH 26.168' read 'IGLS 1185'

*προβουλία, ἡ, gloss on προμηθία, Sch.E.Med. 741; προβουλίης in fragmentary context, IHadr. 159 (Rom.imp.).

προβουλικός, ή, όν, of proboulic rank, ἄνδρα ἐκ πατέρων π. AS 12.199.5 (Cilicia Trachea), SEG 35.1416 (Pamphylia, Rom.imp.).

προγᾰμιαῖος, add 'Cod.Just. 5.17.12'

προγάμιος I, add 'ἥρως π. prob. honoured by those about to wed, Swoboda Denkmäler p. 15 (Misthia (Fassiler) in Pisidia)'

*πρόγᾰμος, ον, approaching the time of marriage, about to marry, Tryph. 341, Mitchell N.Galatia 234; subst. Progamus, title of play by Caecilius.

†προγαργᾰλίζω, tickle beforehand, ὥσπερ προγαργαλίσαντες (sc. themselves) οὐ γαργαλίζονται Arist.EN 1150ᵇ22.

*προγαστρότης, ητος, ἡ, proluvies προγαστρότης, Charis. p. 46.1 B.

προγεωργός, add 'POxy. 899'ii (p. 226) (AD 200)'

προγνώμων, after 'c. gen.' insert 'or acc.'

†προγνωσία, ἡ, foreknowledge, Sch.E.Hec. 1137. 2 preliminary knowledge of a subject, Corp.Herm. 2.17b.

προγόνη, add 'rest. in TAM 3(1).338 (Pisidia); step-daughter, Sch.Lyc. 183'

*προγονία, ἡ, line of descent, prosapia ἡ π. Charis. p. 33.9 B.

προγονικός, add 'τοὺς προγονικοὺς θεούς TAM 4(1).45; adv. -ῶς, in the ancestral manner, Mitchell N.Galatia 195 (s.v.l.)'

*προγόνιον, τό, wd. of uncertain meaning in calendar of Eleusis (?sacrificial new-born lamb), IG 2².1357.30, 1363.9 (iv/iii BC).

προγόνιος, delete the entry (v. °προγόνιον).

πρόγονος I 1, delete 'SIG 1038.9 (Eleusis, iv/iii BC)'   2, add 'ἀθάνατον μνήμην παισί τε καὶ προγόνοις IEryth. 210a.4 (= CEG 858, iv/iii BC)'

†προγρᾰφή, ἡ, public notice, announcement, X.Eq.Mag. 4.9, Plb. 25.3.2, SIG 976.37 (Samos, ii BC), D.C. 47.13, 56.25.   2 notice of sale, Thphr.fr. 97.2; (cf. sense II) Plu. 2.205c.   II (= Lat. proscriptio) proscription, Str. 5.4.11, σφαγαὶ καὶ π. Plu.Brut. 27, Cic. 47, ἐπὶ θανάτῳ προγραφαί App.BC 1.2.   b warrant for arrest, BGU 372.8 (ii AD).   III advance notice, IMylasa 605 (AD 209/11); published forecast of an astronomical cycle, D.S. 12.36.   IV introduction or preamble (specifying contents, programme, etc.), Plb. 11.1a.1, BGU 780.2 (ii AD), Men.Prot. p. 16 D., Gal. 13.777.3.

προγράφω I 1, add 'INikaia 766'   III, add 'προγραφέντος τῆς βουλῆς La Carie 67.5 (Heraclea Salbace)'

*προδᾰμάζω, tame or subdue previously, Heph.Astr. 3.19.1.

†προδᾰνείζω, advance, esp. to public funds, on behalf of someone for the time being unable to pay, τὸ δὲ ἀργύριον τὸ εἰς τὴν θυσίαν προδανεῖσαι τὸν ταμίαν τοῦ δήμου IG 7.4254.38 (Oropus, iv BC), Arist.Ath. 16.2, PCair.Zen. 377 (iii BC), Plu. 2.852b; med. χρήματα εἰς τὸ θεωρικόν, χρήματα εἰς τὴν διοίκησιν π. Hyp.Dem.fr. 4.   II lend to the public funds, ὅσοι ἂν προδανείσωσιν ἄτοκα OGI 46.5 (Halic., iii BC), ἀξιωθεὶς προδανεῖσαι χρήματα ἔδω[κεν] IG 12(8).156 (Samothrace, iii BC), D.C. 51.17.   2 fig., εἰς τὴν γένεσιν τῷ πόνῳ προδανεισθεὶς χρόνος Plu.Per. 13, cf. Luc.Sacr. 3.

†προδᾰνεισμός, ὁ, advance of funds on behalf of another, IG 2².835.5-6, Milet 3 no. 138.31 (iii BC), IMylasa 601.11 (i AD).

†προδᾰνειστής, οῦ, ὁ, one who advances money on behalf of another, OGI 46.9 (pl., Halic., iii BC), IG 11(2).287A122 (Delos, iii BC), IMylasa 802.5.

προδᾰπᾰνάω 2, add 'SEG 32.453 (Boeotia, ii BC)'

προδείκνυμι, line 2, after '(v. infr.)' insert 'pf. med. (or pass.) Aeol. προδεδείχμενον Alc. 75.4 L.-P.'

προδιάζω, delete the entry.

*προδιαίρεσις, εως, ἡ, preliminary division, PFouad 35.9 (i AD).

προδιαιρέω 2, for this section read 'med., make a preliminary division, PMil.Vogl. 23.22 (ii AD)'

προδιαλαμβάνω I, add 'cj. in Th. 7.73.1'

*προδιᾰμαρτάνω, lose before, Memn. 29.1 J. (pass.).

*προδιαπορίαι, αἱ, preliminary problems, subscr. Thphr.Metaph. p. 38 R.-F. (v.l. προδιαπορήσεις).

προδιασείω, after 'stir beforehand' insert 'Men.Arg. p. 148.65 K.-Th.'

*προδιασκευάζω, plan in advance, τῶν .. γενησομένων προδιασκευάζων ἰνδάλματα Procop.Aed. 1.1.24.

*προδιαψηλάφημα, τό, preliminary fingering, testing, τὰ π. τῶν κιθαρῳδῶν Phlp.in APo. 242.14.

προδικία I, add 'SEG 25.591 (Phocis, ii BC)'

προδοματικός, add 'PMich.II 121ʳ II vii.1 (i AD)'

*προδομόνδε, adv., to the πρόδομος, rest. in Euph. in Suppl.Hell. 413.8.

προδουπέω, for 'before' read 'forward'

πρόδρομος I 2, add 'b masc. pl., name given to winds preceding the Etesian winds, Arist.Mete. 361ᵇ24, Pr. 941ᵇ7, Thphr.Vent. 11.   c name given to early figs, τάχα .. οἱ πρόδρομοι διὰ τὴν μαλακότητα τοῦ ἀέρος προτεροῦσι Thphr.CP 5.1.5, 7, al.; cf. πρόδρομα· τὰ προακμάζοντα σῦκα Hsch.'   4, for this section read 'ὅτι Μιτυληναῖοι τὸν παρ' αὐτοῖς γλυκὺν οἶνον πρόδρομον καλοῦσι, ἄλλοι δὲ πρότροπον Ath. 1.30ᵇ' add '5 πρόδρομα· τὰ ἐν τῷ ἄξονι ξύλα Hsch.'   II, delete the section.

†προεγγόνη, v. °προεκγόνη.

†προέγγονος, v. °προέκγονος.

*προεγγρᾰφεύς, έως, ὁ, some kind of petty official or clerk, Kerameikos 3.11 (i BC).

προέγκειμαι, add 'τοῖς προενκημένοις μου τέκ[νοις TAM 4(1).187'

*προεγκηδεύω, bury in before, SEG 2.602.3 (Sivrihissar).

προέδρα I, add '3 Call.fr. 43.30 Pf. (in unkn. sense).'

προεδρεύω I, add 'SEG 36.788 (Samothrace, iii BC)'

προεδρία 2, add 'b body of πρόεδροι, Hierapolis 342.'

πρόεδρος, add 'III as adj., -ος, ον, τιμαί epigr. in TAM 3(1).18 (Pisidia).'

προεῖδον II, add 'b w. acc., χρή προῖδην πλόον Alc. 249.6 L.-P.'

πρόειμι (εἰμί sum) I, add '2 t.t. in philosophy, to be prior to existence, Procl.Inst. 116, 122 D.'

προεισβάλλω II 2, add 'Gal. 9.704.10'

*προεκγόνη, ἡ, great-granddaughter, IEphes. 1066.3, 3072.14, 3274.5; TAM 2.278 (Xanthus); also προεγγόνη (cf. next), rest. in IEphes. 980.15, Cod.Just. 6.48.1.12.

*προέκγονος, ὁ, great-grandson, IGRom. 4.990 (Samos); also προέγγονος (the forms are not separable), IEphes. 3017.2, 3070.4, CIG 4380ᵇ¹.7 (Cibyra), Just.Nov. 18.4 pr.

προεκδίδωμι, add 'III give in marriage before, Sch.E.Andr. 32 (pass.).'

προεκλείπω, after 'Hp.Ep. 10' insert 'fail, be wanting, ὅτι ἐνταῦθα προεκλείπει ἡ δύναμις Sopat.Rh. 8.55.4 W.' and for 'evacuated' read 'abandoned'

προεκπίπτω II, for 'π. τὸ ἀδύνατον' read 'εἰς πᾶν προεκπίπτον ἀδύνατον'

προεκπλέω, add 'IMylasa 101.2'

προεκφέρω I 2, add 'D.H.Dem. 39'

προελευθερόομαι, add 'Cod.Just. 6.4.4.10'

προέλευσις, add '4 appearance as plaintiff, ἕκαστος τῶν ἠδικημένων τὴν π. πεποίηνται PMich.v 231.8 (i AD).'

προέλκω, after 'drag forth' insert 'Men.Dysc. 898 (προσ- pap.)'

προενδίδωμι, after 'Plu. 2.444c' add '(cj., codd. προσεν-)'

*προενθάπτω, bury in before, τῷ προεντεθαμμένῳ αὐτῆς ἀνδρί TAM 3(1).309.2 (Pisidia).

*προενθήκη, ἡ, sum previously laid by, SEG 30.1535 (Lycia, ?ii AD).

*προένοια, ἡ, previous notion, Porph.in Prm. 2.20 (cj., codd. προσ-).

*προεντῠπόω, engrave on in advance, ἀθανάτοις ἑαυτὸν προεντυπωσάμενος τοῦ βίου ὑπομνήσεσιν MAMA 8.477 (Aphrodisias).

προεξαιρέω I, after 'BC 2.64' insert 'med., προεξελόμενος ἐκ τῶν λαφύρων ὅσα .. D.H. 11.48'

προεξαριθμέομαι, add 'pass. -ηρίθμημένα enumerated above, BGU 1816.25 (i BC)'

προεξοδιάζω, add 'Salamine 22 (i AD)'
προεξορμάω II 1, add 'w. inf., *hasten prematurely to* .., J.*AJ* 19.1.16'
προεπισκέπτομαι, add 'Ptol.*Tetr.* 74'
προερευνάω, add 'pass., Aen.Tact. 27.15'
προευπορέω I, add '*SEG* 24.154 (Attica, iii BC)'
†**πρόεχμα**, ατος, τό, perh. *advantage* or *protection*, J.*AJ* 17.10.7 (s.v.l.).
***πρόζῡμα**, τό, = ζύμη, PMil.Vogl. 152 ii 36,53 (sp. προσζ-), 153 i 5, 15 (all ii AD).
προηγέομαι 5, add 'Just.*Nov.* 134.1'
προηγέτης 1, add 'applied to gods, -ῶν θεῶν Ἀρτέμιδος καὶ Ἀπόλλωνος *TAM* 2.188.7'
προηγορέω I 1, line 1, after 'Plu. 2.386b' insert '*SEG* 19.835.9 (Pisidia)'
προήγορος I, add 'cf. οἱ προήγοροι ὑπὲρ τῆς θεοῦ κατε[δι]κάσαντο, etc. *IEphes.* 2.1 (iv BC)'
*προηγός, ὁ, gloss on ἄρχων, Sch.S.*Aj.* 934.
προηγουμένως II 2, after 'Hermog.*Id.* 1.1, 7' insert 'Longin. 44.12'
*†**προηπροσιάς**, άδος, fem. adj., *of* or *connected with the* Προηρόσια, προεροσιάδον χριθόν *IG* 1³.250*A*21.
προηρόσιος, line 1, after 'ον' insert 'also **πρηρ**-'; line 5, after 'τά' insert '(not always distinguishable fr. -οσία)' and add 'προερόσια *IG* 1³.250*B*8 (c.400 BC), πρηρ- *SEG* 26.136 (Athens, 400/350 BC), Hsch.'; line 6, sg., add '*IG* 1³.250*B*18' **II**, add 'Max.Tyr. 30(24).4'
πρόθεμα I, add 'Just.*Nov.* 137'
πρόθεσις I 5, line 3, after '(*UPZ*) 31 (ii BC)' insert 'τὴν π. ἐκπληρῶσαι *PWürzb.* 4.12 (ii BC)'
*προθμάριος, ὁ, v. °πορθμάριος.
*προθμεῖον, v. ‡πορθμεῖον.
πρόθῡμα, add '2 *sacrifice on behalf of*, w. gen., E.*Hyps.fr.* 60.62 B.'
*προθύμημα, ατος, τό, *courage, spirit*, Sch.Pi.*P.* 8.61.
προθύραιος I, of Artemis, after 'Orph.*H.* 2.4, 12' insert 'cf. Sch.Antim. 182 Wy., *Inscr.Perg.* 8(3).161 (ii AD)'; line 3, before 'Procl.' insert 'rest.in'
*προθύριον, τό, dim. of πρόθυρον, ἀπὸ τῆς γωνίας τοῦ π. *Inscr.Délos* 1417*C*45 (ii BC).
πρόθῠρον, line 1, after 'τό' insert 'πρόθιουρον Schwyzer 504 (Boeotia, ii BC)'
*προθυσία, ἡ, *right of priority in sacrificing*, *SEG* 37.866 (Caria, Hellen.), Robert *Ét.épigr.* 18 (Delphi, Rom.imp.).
πρόθυσις, add '**II** *preliminary sacrifice*, *BCH* 73.366 (Epid., iv BC).'
προθύτης, for '*BCH* 24.386 (Bithynia, iii AD)' read '*INikaia* 726 (AD 288/9)'
*προίερα, τά, *preliminary rites*, *SEG* 25.445.29 (Stymphalus, ii BC).
προίημι B I 3, line 2, after 'etc.' insert 'abs. (sc. κόπρον), Macho 212 G.'
προικιμαῖος 2, before 'πράγματα' insert '*PColl.Youtie* 67.25 (iii AD)'
προίκιος, for 'Call.*Fr.* 542 = *Oxy.* 2079.34' read 'Call.*fr.* 1.34 Pf.'
*προικοφάγα, (?voc.), app. "*dowry-eater*" (applied to the coffin of a child whose dowry was spent on it), *Corinth* 8(3).630 (v/vi AD).
προικοφόρος, add '**II** πρ., ὁ, gloss on ἐεδνωτής, *Cod.Vat.Gr.* 1456.'
προικῷος, add '*PFlor.* 93.17, *PMasp.* 3.18 (both vi AD); -ῷον, τό, *dowry*, Just.*Nov.* 18.11, pl., ib. 138.4'
προίξ, line 1, after 'ἡ' insert 'also **προοίξ** *PMich.inv.* 3249 (*ZPE* 31.173.6,11, ii AD)' **II**, add 'προῖκα βουλευτής *councillor who did not have to pay an entrance-fee*, Mitchell *N.Galatia* 181.49 (AD 145)'
προΐστημι B, line 1, after 'προΰστην' insert '(Aeol. aor. inf. πρόσταν, = προστῆναι, *IEryth.* 122 (ii BC))' **II 3**, add 'προστάντα τῆς πατρίδος ἐν καιροῖς ἀναγκαίοις *SEG* 37.957 (Claros, ii BC)'
*προϊσχνόω, *thin beforehand*, Gal. 17(2).846.17, 848.8.
πρόκα, for 'in Call. in *PSI* 9.1092.52' read 'Call.*fr.* 110.52 Pf.'
*προκαθαγός, ὁ, *leader* (as epith. of gods), ἀθανάταν προκαθαγὲ μάκαρ .. Παιάν epigr. in *SEG* 24.1244 (Nubia, ii/iii AD).
προκαθαίρω, after 'Dsc.*Eup.* 1.19' insert 'Gal. 17(2).330.4' add '2 transf., *purge first* of defects, Just.*Nov.* 74.1.'
προκαθάρσιον, add 'Alcin.*Intr.* p. 182 H.'
προκαθηγεμών, add 'also ὁ, epith. of Apollo, *GDI* 3589c; of Dionysus, *SEG* 34.1289; of Heracles, *SEG* 31.1102 (both Phrygia, ii AD)'
προκαθηγέτης, add 'of Zeus, *SEG* 30.1524 (Lycia, iii AD)'
προκαθηγέτις, after '(Phaselis)' insert 'of Hecate, *TAM* 2.189*a*7'
προκαθίστημι I 2, add '**b** act., fig., *use as a screen*, τὸν νόμον π. Antipho 6.21.'
†**προκαίω**, *burn before* or *first*, Kerameikos 3.50; aor. pass., ἵνα μὴ προκαῇ (v.l. προκαυθῇ) τὸ φάρμακον Aët. 15.14 (p. 61 Z.).
προκαλίζομαι, add 'εἰς Ἀφροδίτην Nonn.*D.* 35.138'
προκαταβαίνω, add 'of boats, *descend* the river *first* or *as a preliminary* (to a sea voyage), *Peripl.M.Rubr.* 55'
προκαταβάλλω I, add '2 *pay in advance, pay on account*, *IG* 12(7).515 (ii BC).'
προκαταβολή I, add 'glossed by ἐνθήκη, Hsch.'
*προκαταγραφή, *previous contract of sale*, *PPanop.* 22.12 (iv AD).

προκατάκειμαι, add '2 *lie in the tomb before*, i.e. predecease, *TAM* 4(1).249.'
προκατάληψις I 1, add '**b** prob. *previous occupation* (of a property), *PEdfou* 5.15 (iii BC) in *CE* 14.376, (written προκαλήψ- in *JEA* 45.75.6).'
προκατάρχω II 2, add '*SB* 9786.1 (iv AD) Just.*Nov.* 112.3.2'
προκατασκευάζω II, add 'med., π. τινα πράγματα *makes* certain points *for himself in the preliminary survey*, D.H.*Is.* 15'
προκατασκευή 3, before 'Hermog.' insert 'D.H.*Lys.* 15, *Is.* 3,15'
*προκατάσκοπος, ον, *foreseeing*, gloss on ὕποπτος, Sch.E.*Hec.* 1135.
προκαταστοχάζομαι, delete the entry.
προκατατίθημι II, add '3 *bury before*, *TAM* 2.1144.3 (Lycia).'
προκαταχέω, add 'fig., τοὺς δὲ λόγων δεομένους ἢ παιδείας διὰ τὸ προκατακεχυμένον κλέος οὐκ ἐλάνθανεν Eun.*VS* p. 465 B.'
*προκατοίχομαι, *predecease*, *TAM* 2.1029, al. (Lycia).
*προκενοσπουδέω, *bustle about impatiently*, orac.in *DAW* 85.37 (Cilicia).
*προκέφαλος I, add '*Vit.Aesop.*(G) 1, Ar.*frr.* 112, 568 K.-A.'
προκήρυγμα, add '2 *public auction*, *PEnteux.* 37.3 (iii BC).'
*προκιθαριστής, οῦ, ὁ, *leading player on the cithara*, Didyma 182.17, 264.7.
πρόκληροι, delete the entry.
*πρόκληρος, ον, app. *preceding the drawing of lots*, προκλήρου οὐδεν[ὸς .. ἐξου]σίαν προπατῆσαι *SEG* 30.1382 (Lydia, AD 301), *MAMA* 8.497.
†**προκλησία**, ἡ, perh. *formal offer* or *guarantee* (?of quality), ἔνδος τᾶς προκλησίας .. ἔξος τᾶς προκλησίας *SEG* 9.11 ff. (Cyrenaica, iv BC).
*προκλυστέον, *one must purge beforehand with a clyster*, Paul.Aeg. 2.41 (113.19 H.).
*προκοιτατήριον, τό, *room* or *space at the entrance to a bedroom*, *IAskl.Epid.* 52*A*.42 (iv BC).
†**προκομιδή**, ἡ, *bringing forth, production* of a document, *Cod.Just.* 4.21.22 pr., 1, 2; cf. π.· prolatio Gloss.
προκομίζω, add '**III** προκομιόμενος, ὁ, *Spartan boy in second year of public education*, i.e. *eight years old*, 'Ἡροδότου Λέξεις in H.B. Rosén, Eine Laut- und Formenlehre der herodotischen Sprachform, (Heidelberg 1962), pp. 222 ff.'
προκόμιον I, add '*IG* 11(2).203*B*41 (iii BC)' deleting this ref. in section II.
προκοπή 2 a, lines 2/3, for 'opinion-forming' read 'self-conceit'
†**πρόκοπος**, ον, (προκόπτω) *advanced*, Sor. 1.34 (comp.), Aret.*SD* 2.4.
προκόπτω, add '**III** *pound* in a mortar *beforehand*, Gal. 14.86.5.'
*προκουράτωρ, ορος, ὁ, Lat. procurator, rest. in *BGU* 815.5 (ii AD), *PHerm.* 71.2 (v AD), *Stud.Pal.* 20.143.1 (v/vi AD).
*προκράζω, *call out before* or *first*, *BGU* 1141.48 (i BC).
προκρίνω II, add '2 of law, ruling, etc., *prejudge* another legal situation, ἀνήρ ποτε δὲ ἢ καὶ ἐλάττοσι τὴν ἡλικίαν οὐδὲ ὁ παρ᾽ ἡμῶν προκρίνει νόμος Just.*Nov.* 1.4.1.'
προκτητικός, after '*ownership*' insert '*BGU* 1148.34 (i BC)'
προκύκλιος I, for this section read '*having the nature of a procuress*, Herod. 6.90' **III**, add 'also in Aeolis, *SEG* 34.1238 (c.200 BC)'
προκυλίομαι II, add 'προεκυλινόντο τε τοῦ τεμένους X.Eph. 5.13'
Προκύων II, for this section read 'pl., transf., of the hangers-on of a critic or sim. figure, Phld.*Rh.* 1.242 S., Hippias Erythr. 1 J., πικροὶ καὶ ξηροὶ Καλλιμάχου πρόκυνες *AP* 11.322 (Antiphan.)'
πρόκωπος 1, after '(lyr.)' insert 'Luc.*Dom.* 30' **3**, delete the section.
προλάκκιον, for '*ante-chamber*' read '*tank attached to a main cistern*'
προλαμβάνω II 1, of the dead, add 'abs., *ICilicie* 36 (vi AD)'
*προλαμπάς, άδος, ἡ, wd. occurring in context w. λαμπάς in unpubl. inscr. fr. Cos relating to cult of Hermes Enagonios, v. Sokolowski 1.129.
*πρόλαμψις, εως, ἡ, *shining forth*, Procl.*in Ti.* 1.361.3 D.
προλαπαίνω, add 'Alcin.*Intr.* p. 163 H.'
πρόληψις I 1, add 'Alex.Aphr.*Fat.* 165.15, 25' add '3 *presumption of law*, Just.*Const.* Δέδωκεν 5.'
*προλιμπάνω, poet. = προλείπω, *GVI* 1752 (Demetrias, iii/ii BC).
προλογία, add '*PVindob.* G29779.43 (*WJA* N.F. 1.71)'
προλοχίζω II, before 'Plu.*Sert.*' insert 'D.H. 1.79'
*πρόμαθημα, ατος, τό, *preparatory education*, Hsch. s.v. προπαιδεύματα.
*πρόμαμα, ἡ, *great-grandmother*, *JHS* 73.34 (Caunus, i BC).
πρόμαλος, after 'ἡ' insert '(but προμάλοιο .. αἰζήεντος Theopomp.Coloph. in *Coll.Alex.* p. 28)'
προμαντεύομαι, add '2 prob. *consult the oracle first*, προθύοντα καὶ -όμενον *CID* I 9.D40 (Delphi, v/iv BC).'
*προμαρτυρία, ἡ, *evidence given beforehand*, Rh. 6.124.14.
*προμαρτυροποιέω, *testify beforehand*, *SB* 10288.1a.8 (AD 125).
πρόμαχος I 1 a, add 'X.*Hier.* 11.12'
†**προμέτρης**, ου, ὁ, *official who measures out corn allocation*, *IEphes.* 2299, 3216, Lyd.*Mag.* 1.46.
προμετωπίδιος II 2, delete 'but *chest-piece*'
*προμετώπιον, τό, perh. *façade* of a temple, Amyzon 3, 7.
†**προμετωπίς**, ίδος, ἡ, *ornament worn on an animal's forehead*, Callix. 2

J. (Ath. 5.200e, 202a). **II** part of a tomb, perh. a piece of ornamentation set above the coffin, ἡ σορὸς καὶ ἡ προμετωπὶς καὶ ἡ στήλη Inscr.Magn. 281 (i AD), 282.

προμηθέομαι, line 2, after '2 sg.' insert 'imper. aor.' and for 'Archil. (?) in PLit.Lond. 54' read 'Archil. 106 W.'

προμηνύω, for 'denounce beforehand' read 'reveal beforehand'

*__προμητάτωρ__, ὁ, Lat. *prometator, surveyor sent on ahead of a unit, Just.Nov. 130.6.

προμήτωρ II, add 'Luc.Alex. 58 (v.l. πατρομήτορος)' **III**, add '2 of Aphrodite, as mother of the imperial family, SEG 30.1254 (Aphrodisias, AD 102/116).'

προμικκ⟨ιχ⟩ιδδόμενος, delete the entry (v. °προκομίζω).

προμισθωτής, for 'Ἀρχ.Ἐφ. 1910.371' read 'SEG 33.466 (Larissa, i BC)' and add 'SEG 30.593 (Serrae, ii/iii AD)'

*__πρόμνημα__, ατος, τό, memorial, BCH 7.503.

προμολή, line 3, for 'mouth of a river' read 'approaches of a town' and before 'Opp.C. 2.134' insert 'fountain-head of a river'

*__πρόμορος__, ον, = πρόμοιρος (in quot., sense 2), GVI 1931.4 (Laconia, ii AD).

πρόμος, line 8, after '(Claudiopolis)' insert 'ξυστοῦ πρόμος SEG 23.395 (Corcyra, ii AD)'

πρόναος II, line 2, for 'BMus.Inscr. .. ii AD)' read 'IEphes. 27.272 (ii AD)' and add 'ἐν τῶι Πρόνεωι (of the Parthenon), IG 1³.292-316 (v BC)'

*__προναόω__, perh. build the façade of a temple, in quot., fig., ὅταν δέῃ τὸν ἔπαινον πολλοῖς θεοῖς προναῶσαι SEG 26.821 (Maroneia, c.100 BC).

προνήσιον, for 'veranda' read 'bench built on to the wall of a house'

*__προνικότης__, ητος, ἡ, = προυνικία, Vit.Aesop.(G) 15.

*__πρόνιννος__, ἡ, app. grandmother, SEG 37.590 (AD 189).

προνοέω A II 1, add 'w. inf., ἡ κατασκευασθῆναι .. τὸ ὑδρεκδοχ(ε)ῖον provided for the ὑ. to be built, IEphes. 695 (AD 80/1)' **3**, add 'aor. pass. προνοιηθῆναι SEG 23.447 (Thessaly, ii BC)' **B**, line 11, after 'ib. 37' insert 'imper. προνοηθήτωσαν SEG 33.1039 (Aeolis, ii BC)' and for 'Id. 26.15' read 'Lys. 26.15'

*__προνόημα__, ατος, τό, forethought, Sch.E.Hipp. 1102 (pl.).

προνοητής, add 'POxy. 2479.19 (vi AD)'

προνομή 4, after '3Ki. 10.23' insert '(the Hebrew refers to forced labour)'

προνομοθετέω, pass., add 'SEG 30.1980.80 (Athens, i BC)'

πρόνομος, ὁ (or -νομον, τό), = δικαίωμα, θεσμός, anon.ap.Suid. (pl.).

προξενέω II 1, at end delete 'π. τινί .. X.Ap. 7'

προξενητής, after 'agent' insert 'Sen.Ep. 119.1, Mart. 10.3.4'

προξενητικός, add 'fig., of the Sphinx, φόνου π. Sch.E.Ph. 1024'

*__προξενικός__, ή, όν, relating to πρόξενοι, νόμος Milet 3 no. 140.34 (iii BC).

*__προξένιον__, τό, Cypr. po-ro-xe-ni-o, gift offered to a πρόξενος, Kafizin 117b.

πρόξενος I 3, after 'witnesses' insert 'to a treaty, πρόξενοι ὁ Ζεὺς κ' Ὀπόλον κ' ὦλλοι θεοὶ πόλις Ποσειδανία Meiggs-Lewis 10 (Sybarite, vi BC)' **II**, fem., add 'IEphes. 3124'

*__προοδηγέω__, lead the way, escort, Hsch. s.v. προηγεῖται.

προοικοδόμημα, ατος, τό, frontal building, τὸ π. τῆς συνοικίας Inscr.Délos 1417 C52 (ii BC).

προοίμιον I 2, add 'ἐν προοιμίῳ τῆς ἐπιούσης πρώτης ἰνδικτιῶνος SEG 32.1554 (Arabia, vi AD)'

*__προοίξ__, v. ‡προίξ.

*__προολοφύρομαι__, lament before, Sch.Gen.Il. 1.414.

πρόοπτος, delete 'manifest' and add 'D. 3.13'

προοράω II, delete 'with pf. and plpf. pass.' and section 1 **2**, add 'w. part., προορωμένη τοῦ ζῆν καταστροφήν τινα αὐτῇ παροῦσαν Men.Pk. 132' **3**, add 'ἄγγαρος ὄντως κοὐδενὸς προορώμενος Men.fr. 349 K.-Th.'

προορύσσω, add 'τοὺς .. γυροὺς προορύττειν Thphr.HP 2.5.1'

πρόπαις, for 'boy .. year' read 'in fourth year of public education, i.e. ten years old'

*__πρόπακος__, ον, unexpl. adj. referring to bread, ἀρτίδια πρόπακα PVindob. G39847.696 (CPR 5.104, iv AD).

πρόπαππος 2, delete the section transferring quot. to section 1; add 'pl., ancestors, Call.fr. 229.5 Pf.'

πρόπαρ I 1, add 'A.Ag. 1020'

*__προπαρατυγχάνω__, happen upon previously, acquire a prior knowledge of, Sch.Pi.O. 7.98.

προπαροξύνω II, for 'have .. fever' read 'of a fever, to be provoked prematurely'

πρόπᾱς, add 'Nic.Th. 338'

προπατέω, add 'grapes, SEG 30.1382 (Lydia, AD 301)'

*__προπερίσπασις__, εως, ἡ, circumflexion of the penultimate, Eust. 341.14.

προπέτεια I, add 'Call.Epigr. 42.4 Pf.'

προπετεύομαι, add 'Just.Nov. 22.18'

προπίνα, add 'Just.Nov. 117.15 pr.'

*__προπιναρία__, ἡ, fem. of °προπινάριος, Μαρία πραπιναρέα (sic) SEG 31.1082 (Galatia, v AD).

+__προπινάριος__, ὁ, Lat. popinarius (but influenced by προπίνω), keeper of low eating-house, MAMA 3.168 (written πρω-, Corasium), CRAI 1945.379.8 (Berytus, Byz.).

προπίνω I 2, for 'take a snack before dinner' read 'take a drink before dinner' **II 1**, after 'προπινομένη ποίησις Dionys.Eleg. 1' add 'cf. qui mortalibus versus propinas Enn.Sat. 7'

*__πρόπιον__· μάντευμα Suid.; cf. θεοπρόπιον.

προπιπράσκω, add 'pass., pf. προπέπραται AA 21.20 (Milet., i AD)'

προπιστεύω, add 'II entrust beforehand, PFreib. 69.7.'

+__πρόπλασμα__, τό, sculptor's clay model, Plin.HN 35.155; transf., rough draft, Cic.Att. 12.41.4.

+__προποδέω__, lead the way in a dance, Call.fr. 228.2 Pf.

*__προποθέω__, have a previous desire for, μὴ προποθέσας ὑγιείας Gal.Thras. 29.

προπολεύω, add 'SEG 33.1056 (Cyzicus, ii AD)'

προπολέω I, add 'Critias 6.7 (cod. A Ath.)' **II**, delete the section.

προπολιόομαι, delete the entry.

*__προπολιόω__, become grey first or previously, τοὺς πώγωνας προπολιοῦντας Heph.Astr. 2.2.36, Sch.Pi.O. 4.39; med., Diod.Cron.fr. 129 D.

προπολιτεύομαι II, after 'Them.Or. 16.205c' insert 'pres. part. equiv. to Lat. principalis' and add 'POxy. 2343.17 (iii AD)'

πρόπολος, line 1, for 'ον .. before' read 'ὁ, ἡ'

πρόπομα I, add 'POxy. 2047.2 (v AD)' **II**, delete the section transferring 'Plu. 2.624c' to section I.

πρόποσις 2, for 'BMus.Inscr. 1036 (Caria)' read 'Amyzon 65.3 (Caria, ii BC)' add '4 drinking-party, συμβολική π. AP 5.134 (Posidipp.).'

προποτίζω 2, add 'Sch.Il. 11.515'

*__προπράκτωρ__, ορος, ὁ, champion, A.fr. 47a.i 5 R.

*__προπρεσβεύω__, perh. act as leader of an embassy (or in place of ambassador), IEphes. 2026 (AD 200/5), Xanthos 56.

προπτόρθιον, delete the entry.

*__προπτόρθιος__, ον, adj. applied to a sacrificial animal, prob. an indication of its age, ἔριφος προπτόρθι[ος] Sokolowski 3.18.46.

προπύλαιος I, add 'IGBulg. 1768, 1770'

*__προπύλιος__, α (Ion. η), ον, = προπύλαιος, IGBulg. 2123.

*__προπωλή__, ἡ, brokerage, Vett.Val. 4.23.

προπώλης, delete 'Vett.Val. 4.23'

προπωλητής, after 'foreg.' insert 'PStrassb. 87.16 (ii BC)'

πρόρρησις II 2, add 'of a private announcement, TAM 3(1).714 (iii AD)'

πρός, line 5, form πορτί, add 'IMylasa 660.9 in EA 19.12 (ii BC)'; lines 7/8, for 'Inscr.Cypr. 135.19 H.' read 'ICS 217.19' **C II**, for 'v. infr. III 5' read 'v. infr. III 4 and 5' **III**, add 'πρὸς ἔτος, = κατ' ἔτος, Sokolowski 2.86.5 (Lindos, c.AD 200)' **D**, line 5, after 'Hdt. 5.67' insert 'beginning sentence, without δέ or other particle, Lxx Ca. 1.16'

προσάγιος, delete the entry.

προσαγόρευσις, add '2 concr., letter, PWarren 20.9 (iii AD).'

προσαγορεύω, line 6, before 'address' insert 'Dor. ποταγορεύσαντος PP 42.114 (Cos, iii/ii BC)'

*__προσαγορία__, v. ‡προσηγ-.

προσάγω, line 2, before 'once ποσάγω' insert 'ποτάγω SEG 23.305, 570 (Delphi), SIG 1010 (Chalcedon), etc.' **I 8 a**, add 'introduce to the Mysteries, SEG 30.1980.61 (Athens, iv BC)'

προσαγωγεύς II, delete 'hence fem.' to end.

προσαγωγή II 3, for this section read 'approach to a place or thing, D.S. 13.46. **b** facility for approach, access, Plu.Aem. 13; access for ships to put in, Plb. 10.1.6, Peripl.M.Rubr. 46.'

προσαγωγικός, ή, όν, = προσαγωγός, Sch.Luc. 165.8.

*__προσαγωγίς__, ίδος, ἡ, (ὁ), Dor. ποταγωγίς Arist.Pol. 1313ᵇ13, transport boat, BGU 2400.12-13, PLille 21.8, PPetr. 3 p. 254, 257, etc. (all iii BC). **II** talebearer, informer, οἷον περὶ Συρακούσας αἱ ποταγωγίδες καλούμεναι Arist. l.c.; masc., τοὺς καλουμένους προσαγωγίδας Plu.Dio 28, 2.523a.

προσαιρέομαι, after lemma insert 'Dor. ποθαιρέομαι (καὶ αἴ τινάς κα ἄλλως τοὶ πολιανόμοι ποθέλωνται Schwyzer 62.118 (Heraclea, iv BC), aor. imper. ποθελέσθων SEG 23.305 (Delphi, c.190 BC); also ποτ- SEG 26.1817 (Cyrenaica, ii/i BC)'

προσαιτέω II, line 4, after 'c. gen.' insert '(s.v.l.)'

προσαίτησις, add '2 additional demand, POxy. 3424.1 (iv AD).'

*__προσαλεύω__, shake beforehand, PIand. 139.29 (ii AD, pass.).

προσαλίζομαι, delete the entry.

*__προσαλίζω__, assemble or collect in addition, Aen.Tact. 17.4 (cod., v. °προαλίζω).

+__προσαμαρτάνω__, commit an offence against, w. dat., SEG 32.1222 (i AD), ἴ ταύτῃ τῇ στήλῃ προσαμάρτῃ τῇ στήλῃ 33.1029 (ii AD), 34.1233 (iii AD); form ποσ- 31.1003 (i AD), 1013 (ii AD, all Lydia).

προσαμπέχω, for 'veil besides' read 'embrace'

*προσαναγγέλλω, report, make known, Cod.Just. 1.3.41.28.

προσαναγράφω, add 'also ποταναγράφω, SEG 34.282 (Nemea, late iv BC)'

προσαναδέχομαι, add 'II undertake further, Inscr.Délos 1838 (ii BC).'

προσανακλίνω, after 'lean on' insert 'Ἑρμῆν .. προσανακεκλιμένον πρὸς δενδρυφίωι Inscr.Délos 1417B.97 (ii BC)'

*προσανακομίζω, bring up in addition, PTeb. 703.190 (iii BC, pass.).

*προσαναμάσσω, in med., sully or tarnish in addition, τῆς πόλεως χαλεπῶς φερούσης ἐπὶ τῷ τὴν ἐνίων ἀγνωμοσύνην προσαναμάττεσθαι τοὔνομ᾽ αὐτῆς Ph. 2.537 (τῆς .. ἀγνωμοσύνης .. αὐτῇ cj. Mangey), Just.Nov.74.1.

προσαναμείγνυμι, add 'προσαναμέμικται γὰρ αὐτοῖς Just.Const. Δέδωκεν 8a'

προσαναπίπτω, add 'II of the arm of a torsion engine, strike against in the recoil, w. dat., Hero Bel. 91.10.'

προσαναπλέκω, add 'Just.Const. Δέδωκεν 7'

προσαναφέρω, after lemma insert 'aor. imper. προσανενενκάτωσαν SEG 31.952.8 (Ephesus, ii AD).'

*προσαναφορά, ἡ, (supplementary) report, IG 12(5).721.12 (Andros, i BC); Thess. ποτομφορά SEG 13.390.

*προσαναψύχω, chill in addition, Paul.Aeg. 2.46 (117.5 H.).

†προσανοικοδομέω, build on as an annex or support, fig. ἐλεημοσύνη .. πατρὸς .. ἀντὶ ἁμαρτιῶν προσανοικοδομηθήσεταί σοι Lxx Si. 3.14.

*προσανομολογέομαι, agree in addition, προσανωμολογήσατο δὲ καὶ Ὑβρίλαος ἔχειν .. PColl.Youtie 8.7 (224 BC).

προσαντλητέον, after 'Paul.Aeg. 3.28' add 'Aët. 16.74 (p. 119 Z.)'

προσάπαξ, add 'in a single payment, CPR 5.18.12 (vi AD)'

*προσαποβιάζομαι, fut. -βιώμαι, make an effort to produce more, PCair. Zen. 611.19 (iii BC).

προσαποκρίνομαι, add '2 give a further answer, Aristid. 2.94 J.'

*προσαπολείπω, leave out in addition, λιθείας ἧς προσαπολείπεις PCair. Zen. 771.27 (iii BC).

*προσαπολογίζομαι, state in addition, REG 101.14 (Xanthos, 206/5 BC); w. acc. and inf., UPZ 162 vi 1 (ii BC).

προσαποπέμπω, add 'SEG 34.122.70 (Eleusis, iv BC)'

*προσαποτιννύω, pay in addition as a penalty, Scaev.Dig. 50.9.6.

*προσαποτυπανίζω, attack with cudgels, PMich.inv. 6979 (ZPE 76.291, 215 BC); cf. °ἀποτυμπανίζω.

*προσαποχράομαι, embezzle, BGU 432 ii(2).2-3, 2467.21 (AD 190).

προσάσσω, delete '(unless from προειςάγω)'

*προσατιμάζω, dishonour in addition, Sch.E.Ph. 877 (pass.).

προσαυλέω, for 'perform on the flute' read 'play in accompaniment on the pipe'; line 3, after 'Plu. 2.632c' insert 'τί μοι προσαυλεῖς; Men.Dysc. 880'

προσαύλησις, for 'flute' read 'pipe'

*προσβάθρα, ἡ, perh. step, IG 2².1672.144 (pl.) (Eleusis, 327 BC).

προσβαίνω 2, add 'abs., E.Phaëth. 97 D.; of the Nile in flood, PSI 1333.18 (iii AD)'

προσβάλλω I 1, add 'c extend a structure across to, w. πρός AJPh 56.362.22 (Colophon, iv BC).'

πρόσβασις 1, add 'part of the dromos of an Egyptian temple, SEG 24.1173 (Alexandria, 29 BC)'

*Προσβατήριος, ὁ, epith. of Poseidon, SEG 30.1980 (Eleusis, i BC).

*προσβεία, προσβευτάς, προσβεύω, = πρεσβ-, SEG 25.445 (Arcadia, c.189 BC).

προσβολή, lines 1/2, delete 'e.g. .. pl.)' II 3, line 1, after 'impact' insert 'of bronze, A.Ag. 391'

*προσγαληνιάω, display calm conditions, of the sea, Hsch. s.v. ὁ Κρῆς τὴν θάλατταν.

προσγλίχομαι, add 'οὐ δεῖ θαυμάζειν ὅπως .. οὐδὲ προσγλίχεσθαι (i.e. desire to know as well) ὅπως .. Procl. in Ti. 1.80 D.'

πρόσγραφος II, add 'SEG 31.122 (Attica, ii AD)'

προσγυμναστής, for 'fellow-wrestler' read 'wrestling-partner'

προσδεκτός, add '(attested as pers. n., SEG 32.1214 (Lydia, ii AD), etc.; fem. -ή, EA 16.96 (Prusa ad Olympum)'

προσδέχομαι, line 3, after 'D.S. 15.70' insert 'ποδέξαστα (= προσδέξασθαι) SEG 36.548 (Thessaly, iii BC)'

προσδέω (A), line 1, before 'bind on' insert '(Dor. aor. inf. ποιδῆσαι IG 4².122.41 (Epid., iv BC))'

προσδέω (B) II 1, line 2, after 'Theoc. 5.63' insert 'also ποιδεῖσθαι IG 4².121.13 (Epid., iv BC)'

*προσδιασήπω, corrupt in addition, Gal. 17(1).735.8.

προσδιαστέλλω II 1, add 'pass. impers. w. acc. and inf., UPZ 118.12 (ii BC)'

*προσδιατυπόω, lay down additionally in a decree, Just.Nov. 162 epilogos.

προσδίδωμι, add 'Pamph. aor. περτέδωκε IPamph. 17 (iii BC), written -έδωκε ib. 18 (ii BC)'

προσδοκία 1, line 3, after 'in good sense' insert 'π. βιοτᾶς E.Hyps. fr. 64.108 B.' 2, add 'b personified as deity, Pap.Lugd.Bat. xxv 8 iii 14 (after AD 231).'

*προσδοτέον, one must give in addition, Gal. 10.576.6.

*προσεγκόπτω, engrave in addition, τὸ ψήφισμα ἐς τὴν στήλην GDI 5496.21 (Miletus, iv BC).

*προσεγχυμἄτιστέον, one must inject besides, Aspasia ap.Aët. 16.94 (p. 141 Z.).

προσεδρεύω 1, after 'watch the rise of the Nile' insert 'as a local function or office' and add 'Bernand Akôris 29, 30, al.'; add 'reflect upon, concentrate upon, Democr. 191 D.-K.'

*προσειλημμενῖται, οἱ, (cf. προσλαμβάνω I 2) inhabitants of the added land, Ptol.Geog. 5.4.10(8).

πρόσειμι (sum) 1, add 'τῷ προσιόντι προσεῖναι associate with him who seeks your company, Hes.Op. 353'

πρόσειμι (ibo) 2, delete 'καὶ φιλέοντα .. εἶναι)' add '6 enter on an inheritance, Just.Nov. 1.1.1.'

προσεισευπορέω, add 'IHistriae 18.24 (ii BC)'

*προσεκδέχομαι, perh. to be additional surety for, PColl.Youtie 16.27 (109 BC).

*προσεκκρέμαμαι, to be suspended in addition, Gal. 18(1).334.13.

*προσεκουτωρία, ἡ, Lat. prosecutoria, covering note, letter of authorization, CPR 7.26.17, 31 (vi AD).

προσεκπίπτω, delete 'metaph. .. Longin. 15.8'

*προσεκσκύλλω, app. ransack, PColl.Youtie 16.29a (109 BC).

προσεκτικός II, add 'Longin. 26.3 (comp.)'

προσέλευσις, add 'II appearance in court as a prosecutor, POxy. 283.19 (i AD); cf. προσέρχομαι I 5. III entrance on an inheritance, Just.Nov. 1.1.1.'

προσεμπίπρημι, before 'Lxx Ex. 22.6' insert 'Aristodem. 1.2.3 J.'

προσεμφερής, before 'X.Smp.' insert 'Ar.fr. 476 K.-A.' and after '(Sup.)' insert 'τῷ προσεμφερῆ Plat.Com. 54 K.-A.'

προσεμφέρω, add 'II pay or contribute in addition, ἕτερα νομίσματα δεκατέσσερα .. προσενενεχθῆναι PMich.XIII 659.70 (vi AD).'

*προσέναγχος, add '(dub., cj. πρὸς 'ἔμ' ἔναγχος)'

προσενίημι, add 'Gal. 11.129.1'

προσεννέπω 1, after 'll.cc.' insert 'A.Ag. 323' 2, delete the section.

προσεξαπλόω, before 'S.E.M. 1.56' delete 'dub. in'

προσεπαυξάνω, after 'further' insert 'Inscr.Prien. 107.21 (ii BC)'; add 'Robert Hell. 9.8 (Sardis); pf. part. προσεπευξηκώς (sic) IGRom. 4.293a ii 49 (Pergamum, ii BC)'

*προσεπεμβαίνω, προσεπεμβαίνει· ἐπιβαίνει, ἐπιτωθάζει, ἐπιγελᾷ Hsch.

προσεπιβάλλω II, add 'SEG 36.1087 (Sardis, 213 BC)'

προσεπιγράφω, add 'II assign responsibility for, Sopat.Rh. 8.159.6 W.'

†προσεπιδέχομαι, receive besides, PTheb.Bank 12.12 (ii BC); take upon oneself besides, Ph.fr. 63 H. (v.ll.προσενδ-, προσδ-).

προσεπιδέω, add 'Gal. 18(1).819.11, 823.16'

*προσεπιδιδάσκω, teach in addition, Them. in APo. 50.9.

προσεπιδίδωμι, after 'Pl.Sph. 222e' insert 'ἔτι προσεπέδωκε IG 2².553.9'

προσεπιπλάσσω II, for 'work into a plaster' read 'plaster on'

*προσεπισκευόω, = προσεπισκευάζω, Inscr.Cret. 3.ii 1.6 (ii BC).

προσεπισκήπτω, add '2 enjoin in addition, Just.Nov. 78.2 pr.'

προσεπιτεχνάομαι, add 'Just.Nov. 97.2'

προσεπιχέω, add 'Heras ap.Gal. 13.1045.18'

προσεπιψηφίζομαι, add 'Thasos 185.6 (i AD)'

προσέρδω, after 'sacrifice' insert 'in addition'

*προσερισμός, ὁ, rebellion, prob. in Aq. 1Ki. 15.23.

προσέρχομαι, line 3, after 'aor. .. -ηλθον' insert 'part. πορτευθών = προσελθών, Inscr.Cret. 1.xvii 11 (Lebena); pass. -ελεύσθην, part. -ελευσθείς (in sense I 5), Just.Nov. 82.7.1, Cod.Just. 2.3.18' I, add '8 enter upon an inheritance (= Lat. adire), Just.Nov. 1.1.1.'

*προσεσταλμένως, in a well girt up manner, gloss on εὐσταλέως, Gal. 18(2).692.2.

πρόσευξις, for '= προσευχή' read 'prayer, petition'; add 'BMus.Inscr. 421.2'

προσευχή I, add 'ἀναθέντα ἐκ τῶν ἰδίων ἐπὶ προσευχῇ τοῦ θεοῦ SEG 32.810 (Delos, iii/ii BC, Samaritan)'

προσεχής I 2, add 'διὰ τὸ προσεχῆ τὸν τόπον εἶναι τῷ βορέᾳ Peripl.M.Rubr. 12'

προσέχω, after lemma for 'Cypr. ποέχω (q.v.)' read 'Cypr. med. °ποέχομαι' I 3, add 'b bring to a person's attention, σοὶ γὰρ προσέχω ὅτι .. OFlorida 14.8.'

προσηγορία, add 'b concr., letter, PRoss.-Georg. III 9.7 (iv AD), PKöln 111.3 (v/vi AD); προσεκτήσαντο τὴν προσαγορίαν SEG 28.1566 (Cyrene, AD 154).'

προσήκω III 2, add 'ὥσπερ ἐστὶ προσῆκον, βάρβαρον Ἕλλησι D. 3.24, cf. Just.Const. Δέδωκεν 18, etc.' 3, οἱ προσήκοντες, add 'subordinates, PBeatty Panop. 2.98'

προσηλόω, after lemma insert 'dial. ποθᾱλόω, IRhod.Peraea 201.18 (i BC)'

προσημαίνω I, of medical symptoms, add 'Gal. 17(2).396.1'

προσήνεια, add 'w. ref. to characterization (in quot., by E.), D.Chr. 18.7'

προσήπω, for 'προσαπέντα' read 'προσαπέντων'

πρόσθεμα II, for 'πόσθη' read 'penis'

πρόσθεν, line 1, delete 'Ion.' and after 'Hdt. 1.11, al.' insert 'Isoc. 18.61, D. 20.94, Hell.Oxy. 16.5, IG 12 suppl. p. 119 n. 714.18 (Andros, iv BC)'

πρόσθεσις I, add '3 (architectural) extension, ἐψηφώθη ἡ πρόσθεσις [τοῦ] ναοῦ IGLS 1321 (vi AD).'

προσθήκη I 1 b, add 'π. κυρώσεως POxy. 3345.48 (AD 209)'

***προσθύρᾱς**, ὁ, doorkeeper, PPetaus 34.10 (c.AD 184).

***προσικτός**, ή, όν, attainable, ὁδός TAM 3(1).34D59 (Termessus); cf. ἀπρόσικτος.

προσίστημι I 2, for this section read 'weigh in addition, include in the weighing, μάγειρε, μὴ προσίστα τοῦτό μοι τοὐστοῦν Macho 20 G.' **4**, for 'Arist. l.c.' read 'Arist.Pr. 870ᵃ32' **II 1**, add 'ὅπως προσιστῆται τὸ πλῆθος πρὸς τὴν ἐπικύρωσιν τῆς χειροτονίας Arist.Ath. 41.3'

προσκαθέζομαι, line 3, for 'πολιορκία' read 'πολιορκίᾳ'

προσκαθίζω 1, add 'b transf., τὸ προσκεκαθικὸς τῆς ψυχῆς τοῖς τέκνοις its being fixed upon, Sch.S.OC1119.' **3**, delete the section.

***προσκαθιστάω**, = προσκαθίστημι, Just.Nov. 134.2.

προσκαθίστημι, for 'supply labour besides' read 'supply besides' and after '(iii BC)' insert 'κύνας προσκατέστησε SEG 24.154 (Attica, iii BC)'

πρόσκαιρος II, delete 'Ev.Matt. 13.21' add '2 of persons, concerned only with the moment, i.e. lacking staying power, Ev.Matt. 13.21, Ev.Marc. 4.17.'

***προσκαλίζω**, hoe previously, prob. in PTeb. 953.11 (ii BC).

προσκαρτερέω 4, add 'of things, ἵνα πλοιάριον προσκαρτερῇ αὐτῷ Ev.Marc. 3.9'

προσκαρτέρησις, add '2 devotion to one's profession, IG 12 suppl. 249.8 (Andros, ii BC).'

προσκαταβάλλω, add 'ἐνὶ στήθεσσιν ἐμοῖς ποτικάμβαλες αὐδ[ὴν] θέσπιν PBodm. 29 (iv AD)'

***προσκαταπλέω**, sail to a place against, w. dat., Duris 24 J.

προσκαταχωρίζω I, add 'transf., record additional grievances, SEG 29.1130 (ποτι-, Clazomenae, ii BC)'

προσκατεγγυάω, seize additionally as security, PTurku 1.46 in Tyche 6.101 (Theadelphia, ii BC).

πρόσκειμαι III 1, line 3, after 'Hdt. 1.196' insert 'προσκεῖσθαι δὲ αὐτῷ καὶ τὴν στρυπτηρίαν, Schwyzer 722.14 (Thebes-on-Mycale, iii BC)'

προσκεφαλάδιον, add 'SB 9834b.16 (iv AD)'

***προσκεφαλᾱτικόν**, τό, = προσκεφάλαιον, cushion, PKlein.Form. 1089.2 (CE 58.232, vii AD).

⁺**προσκηδής**, ές, full of sorrow, distressed, ξεινοσύνη Od. 21.35, A.R. 4.717. **II** tied by marriage or kinship, Hdt. 8.136; masc. pl. subst., AP 7.444 (Theaet.). **2** friendly, A.R. 3.588.

προσκηρύσσω I, add 'IG 1³.125.24 (405/4 BC)'

προσκινέομαι, for 'sens. obsc., of women' read 'describing a form of responsive action during sexual intercourse'

προσκληρόω, after lemma insert 'dial. ποτικλᾱρόω' add '2 choose by lot in addition, ἐκ τῶν λοιπῶν πολιτᾶν ποτικλαρώντω .. SEG 30.1119 (Entella, iii BC).'

πρόσκλησις, for 'summons to quoit-throwing' read 'summons of the discus (i.e. gong)'

προσκνυζάομαι, after 'Philostr.Her.Prooem.' insert 'of a lion, id.VA 5.42'

***προσκολλητέον**, one must stick on, Archig.ap.Gal. 12.676.16.

προσκολλητός, add '2 -όν, τό, annex, CPR 14.13.19 (form -ᾱτόν, vi/vii AD).'

πρόσκολλος, add 'adv., w. gen., adjoining, πρόσκολλα τοῦ οἴκου Stud.Pal. 10.125.3 (v/vi AD)'

***Προσκόπα**, η, (πρόσκοπος) title of Artemis, Ἀρτέμιτι Προσκόπαι SEG 39.550 (Apollonia, Illyria, iii BC).

προσκοσμέω, add '2 attach to the ranks of: med., side with, προσεκεκόσμηντο δὲ τούτοις Arist.Ath. 13.5.'

πρόσκρισις I, add 'τὴν σάρκα .. ἡμᾶς ἐν συνεχεῖ ἀποκρίσει τε καὶ προσκρίσει Alex.Aphr.Mixt. 235.22'

***προσκριτικός**, ή, όν, of or concerning assimilation, ἡ θρεπτικὴ δύναμις, ἡ μεταβλητικὴ δὴ καὶ προσκριτικὴ τῆς τροφῆς Alex.Aphr.Mixt. 234.14.

πρόσκρουσμα II, add '2 charge, imputation, Charito 4.6.2, Ach.Tat. 8.10.'

προσκύλλω, for 'molest before' read 'violate the chastity of before' and delete 'women with a past'

προσκυνέω I 2, add 'w. gen., Τύχης τῆς Δούρας SEG 7.571 (Dura)'

προσκυνητήρ, add 'Ramsay Cities and Bishoprics 1.338 (Phrygia)'

προσκυνητήριον, add 'προσκυνηητ[ήρι]α BCH 3.482 (Phrygia)'

προσκυνητός, add 'as epith. of a place of worship, προσκυνητῇ .. προσευχῇ Mitchell N.Galatia 209b (?iii AD)'

προσκυρέω 3, line 4, delete '(SB) 4208.7 (ii BC)'

⁺**προσκυρόω**, confirm the ownership or possession, τὴν προσκυρωθεῖσάν σοι παρ' ἐμοῦ δεσποτείαν (gift of part of a house), PLond. 1044.15 (ii AD), Cod.Just. 1.5.15, 1.11.9.2; cf. προσκυροῖ· βεβαιοῖ Hsch., Gloss.

***προσλάλημα**, τό, subject of talk, An.Boiss. 4.447.

προσλέγω I, after 'med. 1 aor. ἐλεξάμην' insert 'κακὰ προσελέξατο θυμῷ Hes.Op. 499'; at end delete 'metaph. .. 499'

***προσληπτήριον**, τό, unexpld. item in inventory, Inscr.Délos 104(28)b.B12.

***προσμαρτύρησις**, εως, ἡ, witness, POxy. 3807.33 (i AD).

προσμαρτύρομαι, add 'PAnt. 40.7 (iv BC)'

προσμείγνῡμι I 3, add 'Gal. 10.910.12'

προσμένω I 2, add 'SEG 26.729 (Maced., AD 195)' add '3 τὸ προσμένον the matter awaiting confirmation, Epicur. [2, 38] 6, [5] xxiv 2, 5 A., D.L. 10.34.'

πρόσμορος, add 'πρόσμορον or πρὸς μόρον, dub. sens. (perh. for πρόμορον = πρόμοιρον), Inscr.Cret. 1.xviii 177 (Lyttus)'

πρόσνευσις II, after 'of a planet' insert 'Ptol.Tetr. 4'

***προσοδαρχέω**, to be treasurer of a πρόσοδος (sense I 3), MDAI(A) 6.42 (Cyzicus).

προσοδάρχων, delete the entry.

προσοδιάζω, add '2 to be supplied with provisions, ABSA 51.154.5 (unless προσωδιάσθη is to be understood for -σθην).'

πρόσοδιος, add 'II προσόδιον, τό, income, revenue, Inscr.Cret. 1 xvii 2b.8.'

πρόσοδος II 1, add 'b fig., profit, advantage, π. ἔσεσθαί τισι epist.ap.D. 18.78.'

προσοδύρομαι, add 'complain of before a person, ὃς οὐ πρότερον αὐτῷ προσωδύρατο πάθη Just.Nov. 30.9 pr.'

***προσομιλητής**, οῦ, ὁ, disciple, Aët. 3.80 (292.13 O.).

***προσονίνημι**, benefit moreover, Numen.fr. 14 P.

προσονομάζω, add 'II name (i.e. appoint) in addition, Just.Nov. 30.4.'

προσοργίζομαι, before 'J.BJ 2.14.6 (v.l.)' insert 'Arist.Ath. 19.5'

***προσορισμός**, ὁ, inclusion within boundaries, addition to a territory, etc., MDAI(A) 72.243.22 (Samos, ii BC).

***προσορνύμι**, in pf. part. προσορωρ[ότες] rushing at him, dub. in BGU 1252.21 (ii BC).

προσουδίζω, after '(οὔδας)' insert 'dial. ποτουδίζω, aor. ἐποτούδιξε (-ιζε cod. Hsch.) Sophr. 141'

προσουσία, add '(as pers. n., Lang Ath.Agora XXI C31, c.400/390 BC, etc.)'

προσοχθίζω, after 'to be wroth with' insert 'τῷ γένει Satyr.Vit.Eur.fr. 39 xii 21; τῷ ἐπιχωρίῳ φθόνῳ fr. 39 xv 22'

πρόσοψις I, add 'b facing of a building, POxy. 2197.4, al. (vi AD).'

προσπαίζω II 2, for this section read 'play teasingly with, Luc.Dom. 24, Ael.NA 4.45; make fun of, tease, Pl.Euthd. 285a, τοὺς ῥήτορας id.Mx. 235c, τὸν Ὀδυσσέα οἱ μνηστῆρες D.Chr. 9.9'

προσπάσχω II, line 5, for 'abs., Macho 2.2' read 'w. dat. understood, Macho 468 G.'

***προσπελαστός**, ή, όν, approachable, Phot. s.v. πλατά.

προσπέμπω, add 'w. ref. to lovers' messages, Hesperia 54.225-6 no. 8'

***προσπεριέχομαι**, cling to, desire, in addition, dub. in APF 2.519 (ii BC).

***προσπεριτειχισμός**, ὁ, additional fortification of surrounding wall, εἰς τὸν προσπεριτειχισμὸν τῆς πόλεως Amyzon 35.

προσπέτομαι, line 2, after '(v. infr.)' insert 'also aor. part. προσπετάσας (sic) Aesop. 296 P.'

προσπίπτω II 5 b, for this section read 'τὰ ποτιπίπτοντα ποτὶ τὰν αἴσθησιν that strike on the senses, Archyt. 1, cf. Thphr.Sens. 5.41; σοι .. προσπιπτέτω let it (the thought) present itself to you, M.Ant. 7.19, cf. 13, 9.24, 11.7, Longin. 14.1; π. μεγαλορρημονέστερα have a more impressive effect, id. 23.2, cf. 29.1; π. δι' ἑαυτοῦ or αὐτόθεν to be self-evident, S.E.P. 2.168, M. 1.300'

***πρόσπλασις**, εως, ἡ, moulding on, adhesion, Procl.in Ti. 2.60.15 D.

προσπλάσσω, line 4, delete 'to be smeared upon'

προσπλέκω, line 4, after 'Arg. 1 Ar.Ra.' insert 'J.Ap. 1.222'

προσπλοκή 1, delete 'cf. Aq.Ex. 28.32'

⁺**πρόσπλοκος**, η, ον, plaited on, συμβολὴν προσπλόκην Aq.Ex. 28.32. **2** fig., involved in a relation with, θεῷ προσπλόκους Rhetor. in Cat.Cod.Astr. 8(4).148.

***προσπνίγω**, choke, throttle, POxy. 2331 iii 20 (verse, iii AD).

προσποιέω I, line 1, delete 'cf. Plot. 6.1.21' **2**, after 'MAMA 4.27' add '(perh. calque on Phrygian αδδακετ κακουν)' **II 4**, add 'b w. acc. and inf. pretend or claim that .., Ἑρμῆν δεδωκέναι [τοὺς νόμους] D.S. 1.94.'

προσποίησις, add '4 pretence, D.C. 42.8.'

προσποιητικός, add 'fictitious, πένθος Anon.Astr. in PMich.III 148 ii 7 (i AD)' προσπορεύομαι, line 1, ποτιπορ-, add 'SEG 26.701 (Dodona, c.205 BC)'

προσπορίζω 1, line 4, after 'POxy. 133.6 (v AD)' insert 'Cod.Just. 12.33.8.5'

προσρέω I, line 3, after 'Plu. 2.760a' insert 'abs., προσερρύη Men.Dysc. 225'

προσσημαίνω, delete 'besides'

†προσσημ(ε)ιόω, brand in addition, δεκυείροις (= a decemviris) ἐπὶ ποδὸς προσσημιωθήσεται IEphes. 215.13 (ii AD); med. προσσημειοῦμαι· adnoto, Gloss.

*προσσκυλάω, plunder in addition, dub. rest. in UPZ 6.19 (ii BC); cf. ‡προσσυλάω.

προσσπεύδω, add 'II demand urgently in addition, πλέονα μισθόν PBremen 63.16 (ii AD, written προσπ-).'

προσστερνίζομαι, add 'SEG 34.1259 (Bithynia, i AD)'

προσσυλάω, before 'UPZ 6.19' insert 'dub. in'; add 'cf. °προσσκυλάω'

προσσωρεύω II, delete 'Luc.Anach. 25' (v. °προσω-)

προστάσιμον, delete the entry.

πρόστασις IV, for this section read 'perh. balance brought forward, or an error for πρόθεσις, public notice, BGU 2467.24b (AD 190)'

προστατήριος III, add 'IG 7.2405-2406 (Thebes, 229 BC)'

προστάτης II 3, add 'οἱ πέντε π[ρο]στάται κώμης POxy. 1275.7 (iii AD)' add '4 perh. foreman, ἱστῶνος πρόστατα καλλιπέπλου INikaia 103 (epigr., ?i AD).' III, add '4 guarantor of a loan, AA 1987.681 (v BC).' IV, add '2 official in a temple, of Aphrodite, SEG 23.209 (Messene, iii BC), of Demeter, SEG 34.981 (i BC).'

†προστείχω, step forward, advance, vv.ll. in S. OC 30, 320.

*πρόστεν, = πρόσθεν, CEG 119 (Thessaly, c.450 BC).

*προστερνίδιος, ον, fitted to the front of the chest, PAnt. 67 ii 1.

*προστετυπωμένως, so as to shape by pressure, Paul.Aeg. 3.35.1 (221.8 H.).

προστηθίδιος, delete 'of horses' and add 'SEG 37.491 (Thessaly, v/iv BC)'

προστίθημι, line 7, after 'Dor.' insert 'aor. ποτθεθῆι Schwyzer 323.C39' A I 2, add 'deliver to a purchaser at an auction, Hsch. s.v. προσθεῖναι.' III, add '6 bury additionally in a grave, EA 19.37.8 (Pessinos, Rom.imp.).'

*προστικτέον, one must punctuate before, Sch.Il. 7.390.

προστιμάω, add '2 pass., of an offender, w. acc. of penalty, to be fined, Cod.Just. 10.56.1.4, 11.54.2.1.'

πρόστιμον, add '3 prob. surety, deposit, CPR 5.5.3 (iii AD).'

προστραγῳδέω I, add 'Simp.in Ph. 1015.8'

προστρέχω II 1, add 'ἀνέσεως τόπον εὑρὼν τοῖς ἀποστόλοις προσδραμών Mitchell N.Galatia 466'

προστρόπιος, add 'cf. Ποιτρόπιος, °Ποιτρόπια'

πρόστροπος, add 'III = προσάντης, Hsch.'

*πρόστρωσις, εως, ἡ, app. paving, π]ρόστρωσιν SEG 33.1277 (Jerusalem, 18/7 BC).'

προστῷον, after 'written' insert 'προστοῖον in Inscr.Délos 1417Ai 162 (ii BC); before 'IGRom.' insert 'JHS 54.142 (Delos, ii BC)'

*προσυγχωρέω, grant previously, τινι PFam.Teb. 21.22 (ii AD).

προσυντίθημι I, add 'Satyr. Vit.Eur.fr. 39 xvii 7' II, after 'Med.' insert 'prescribe besides (in teaching), D.H. Comp. 20. b'

*προσφαγιάζω, offer sacrifice at a tomb, cj. in Robert Ét.Anat. 308.

†πρόσφαγμα, ατος, τό, blood-offering made to dead before burial or before ἐκφορά, A.Ag. 1278, E.Alc. 845, Hec. 41, Plu.Comp.Thes.Rom. 1(2); pl., of a single victim, E.Hec. 265, Tr. 628. 2 sacrificial victim in general, E.IT 243.

προσφέρω A I 4, for 'address proposals, offer, etc.' read 'put forward, propose, τῶν διορθωμάτων τῶν ὕστερον ποτιφερομένων SEG 23.305 (Delphi, c.190 BC); w. noun clause, ὡς διαλέγωντι .. ποτήνεγκαν BCH 88.570 (Argos, ii/i BC); esp. w. λόγον, λόγους'

πρόσφημι, after lemma insert 'dial. ποτίφαμι, ποτέφα Stesich. S. 11.3'

προσφιλής I, add 'in epitaphs, Μάντα Ἀριστοφῶντος προσφιλὴς χαῖρε SEG 31.785 (Thasos, Rom.imp.), etc.'

*προσφοραῖος, ον, perh. concerned with προσφορά, Hdn.fr. 26 H. (= Ion Trag. 49a S.).

πρόσφορος I 2, add 'ἡδονὴν ἐμποιῆσαι προσφορώτατοι Men.Rh. 393.16 R.-W.' add '4 of judges, having jurisdiction, competent, Cod.Just. 7.51.5.1.'

προσφυή, after 'supernumerary teeth' insert 'wolf-teeth (dentes lupini)'

προσφωνέω I, add '5 call to appear in court, bring charge against, πορτιπονέν δ' ἄιπερ τὸν ἀλ(λ)ὸν Inscr.Cret. 2.v 9.8 (Axos, early v BC).' II, add '3 app. promise a thing to someone, PKöln 109.15 (iv/v AD).'

προσφωνητικός, before 'adv.' delete 'only in' and insert 'Men.Rh. 382.2, 414.31 R.-W.'

*προσχαριστήρια, v. προχαριστήρια.

προσχεθεῖν, for 'ward off from' read 'hold in front of'

πρόσχημα II 1, at end, of person, add 'Chiron 17.238 (Pamphylia)'

πρόσχολος, after 'assistant schoolmaster' insert 'IG 2².10949'

*προσχρηματίζω, to be called beside, have an additional name (cf. χρηματίζω III 1), SEG 32.1243 (2 BC/AD 2), TAM 3(1).213 (Pisidia, i/ii AD).

προσχωρέω II 1, line 3, abs., add 'ταῖς προσχωρούσαις πόλεσι Onas. 38.1'

*προσψζω, save before, Sch.E.Med. 526.

προσωνυμία I, for 'surname' read 'name (w. particular ref. to its deriv. fr. an eponym or other source)'

*προσωπικός, ή, όν, personal, προσωπικὴν ἀσφάλειαν Cod.Just. 12.60.7.2; adv. -κῶς, individually, ib. 6.4.4.19b. 2 adv., in respect of grammatical person, Choerob.in Theod. 2.29 H.

προσωπικῶς, delete the entry.

προσώπιον I, add 'Inscr.Délos 1409Bai 10 (ii BC)'

*προσωπίτης, ου, ὁ, measure, so called fr. the Prosopite Nome, POxy. 919.5 (ii AD) (s.v.l.).

?*προσωπίτιον, τό, app. = °προσωπίτης, προσωπίδιν Stud.Pal. 56.23 (ii/iii AD).

προσωπῖτις, add 'Paul.Aeg. 7.3 (196.3 H.)'

πρόσωπον, line 7, after '(Leon.)' insert Alciphr. 3.40'; line 20, after 'Ev.Matt. 22.16' insert 'ἐν προσώπῳ σου before your face, Sopat.Rh. 8.339.17 W.' III, line 5, after 'portrait' insert 'SEG 33.946 (Ephesus, i AD)' IV 2, for 'legal personality' read 'standing, social position' add '6 εἰς πρόσωπόν τινος on behalf of, to the account of, Cod.Just. 1.2.15 pr., 1.2.24.3.'

προσωρεύω, add 'Luc.Anach. 25'

πρότανις, add 'also πρώτανις, Chiron 22 p.377 (Mytilene, i BC)'

*προταρίχεια, ἡ, previous maceration, Anon.Alch. 270.2.

πρότασις I 2, delete the section transferring the quot. to section 1.

προτάσσω I 1, at end, after 'POxy. 1112.18 (ii AD)' insert 'IEphes. 3239 (ii/iii AD)'

προτατικός, add 'II of the πρότασις (I.4): προτατικὸν πρόσωπον character introduced at the beginning to explain the action of play, Donat.praef. Ter.An. i. 8, Euanthius in CGF p. 65 K.'

προτείνω I 1, add 'b stretch out for flogging, Act.Ap. 22.25.'

προτέλειος, add 'πρωτέλειος is Att. acc. to Phryn.PS p. 105 B. (πρωτο- cod.); cf. προτέλειαι in Hsch. after πρῶτα (cf. πρωπέρυσιν s.v. προπέρυσι)'

προτέλεσμα, add 'Heph.Astr. 1.20.9'

προτελευτάω, add 'ICilicie 35 (vi AD)'

προτέμνω II, for 'prune vines' read 'prepare vines before pruning, cf. Świderek La propriété foncière 67'

προτεραῖος III, delete the section.

προτερᾱσίος, add 'SEG 37.340.13 (Mantinea, iv BC)'

*προτερατεύομαι, first describe a portent, Call.Dieg. xi.21 (II p. 46 Pf.).

προτέρημα, line 3, after 'M.Ant. 1.16' insert 'τὰ ἀπὸ τύχης π. inscr. in APF 6.10 (Delos, ii BC)'

προτερικόν, delete the entry.

πρότερος and πρῶτος, line 2, add '(προτεραίτερος humorous comp. of πρότερος Ar.Eq. 1165)' A IV, add '2 προτέρως is used in Arist.Phys. 195ᵃ30 of being a cause in a prior way, cf. ὕστερος A IV.' B I 2, add 'in dates, ἔτ(ους) σλθ' μηνὸς πρώτου SEG 34.1301 (Phrygia, ii AD)' 3, add 'e in the title of a legion, SEG 31.1116 (Phrygia, Rom.imp.).' 4, add 'πρῶτος τῆς συγκλήτου (= Lat. princeps senatus), Plb. 32.6.5; π. βασιλεύς as title of Sun-god, AR 1985-6.47 (ii/i BC)' III 2, after 'Babr. 45.14' insert 'Cod.Just. 1.4.29 pr.' IV 2, add 'SEG 23.638 (Cyprus, i AD)'

προτέταρτον, add 'Inscr.Cret. 4.75.A6, 81.9 (Gortyn, v BC)'

*προτήκτωρ, ορος, ὁ, Lat. protector, PPrincet. 119.1 (iv AD), SEG 32.1553 (Arabia, undated), CIG 9448 (AD 518, sp. προτικτ-), TAM 4(1).383, al. (sp. πρω-).

προτίθημι II 3, add 'imper., introducing text of edict, πρόθες IEphes. 3217 (i AD)'

*προτιμή, ἡ, special honour, Michel 459.21 (Caria, ii BC)

*προτομάριον, τό, dim. of προτομή, PAlex. (1964) 23.8 (i/ii AD).

προτομή 1, transfer 'OGI 214.41' to section 2 3, add 'of a god, INikaia 1085 (iii AD)'

†προτονίζω, let the sail belly out against the forestay, AP 10.2 (Antip. Sid.).

†προτόνιον, τό, apron worn by sacrificing priestess, Poll. 10.191, Hsch., Phot.; cf. °προγόνιον.

πρότονοι II, delete the section transferring quots. to section I.

*προτοῦ or πρὸ τοῦ, v. ‡πρό A II 1, ὁ, ἡ, τό A VIII 5 b.

*προτρέφω, suckle, SB 10235 (i AD)

προτρέχω, line 1, after 'X.An. 1.5.2' insert 'προέδραμεν Trag.adesp. 664 i 15 K.-S.'

προτριᾱκάς, add 'Inscr.Cos 43, SEG 32.1149 (Ionia, AD 209)'

*πρότρῐτα, add 'πρότριτα παρέμμεναι SEG 34.1238 (Aeolis, c.200 BC); sg. πρότριτον on the third day before, Inscr.Cret. 4.81.5 (Gortyn, v BC); cf. προτέταρτον'

προτροπάδην, add 'Suppl.Hell. 946.5'

πρότροχος, add '2 a form of wheeled siege-engine, Poliorc. 199.13.'

*προτυπής, ές, *insistent*, φαντασία Plot. 1.2.5.20 (s.v.l.).

προτύπτω **II** and **III**, for these sections read '**II** *strike beforehand*, A.*Ag.* 132 (lyr.); *strike first*, Procop.*Vand.* 1.18.'

*προύγαμος, ον, προύγαμον βίον ζήσει, app. = πρόγαμος II, *MAMA* 7.485.

προϋωνός, add '*SEG* 29.1173 (ii AD)'

προϋπεξορμάω, delete the entry.

προϋπόκειμαι, add '**III** *to be buried first, TAM* 2.1163.3 (Lycia).'

προϋπόστασις, for ' = προΰπαρξις' read '*previous existence*'

*προϋποσυλλέγω, perh. *compile* or *collect in advance*, *SEG* 28.1566 (Cyrene, letter of Hadrian).

προϋποχρέω, add 'Gal. 18(2).524.2'

*προυρίς, v. °φρουρίς.

*προυρρά, v. ‡φρουρά.

*Προυσιακός, ή, όν, *of Prusias*, τετράχμον *Inscr.Délos* 1443 *A*i 140 (ii BC).

*Προυσιάς, άδος, ή, *upright drinking-vessel, named after Prusias, king of Bithynia*, *Didyma* 463.22 (*c*.179 BC), cf. Ath. 11.496d.

προυφήτις, delete '(metri gratia)'; add '(ii AD)'

προϋφίσταμαι, line 2, before 'but usu.' insert '(pres. -ίσταται ταύτης D.H.*Comp.* 2)'

προφανής **I**, add 'πρόφανες Alc. 132.6 L.-P.'

*προφαντάζω, *cause one to have a presentiment*, τόν ἀκροατήν Sch.B Il. 11.45.

προφασίζομαι, line 3, after 'D.C. 59.26' insert 'act. aor. part. προφασίσας, Hsch. s.v. σκήψας'

πρόφασις **I 2 c**, add '*AP* 5.53 (Diosc.), 193'

προφασιστικός, for '*reproachful*' read '*consisting of false allegations*'

προφέρω, line 1, after 'προφέρεσκον' insert 'Theoc. 25.138' **I 3 a**, add '*express*, ταὐτόν ὑποθετικῶς Demetr.*Eloc.* 296' **b**, delete 'Med.' to end of section. **4**, line 7, after '(Daphne, ii BC)' insert '*urge in objection*, w. acc. and inf., *CRAI* 1932.242.10' **5**, add 'ὁπόσσω κα προφέρηται *for whatever sum is ordained*, *SEG* 9.72.123'

προφητεύω **III**, add '*IHadr.* 6 (ii AD), al.'

προφθάνω **2**, add '**b** *to be before*, τό προφθάσαν γένος *the family in previous days*, Sch.Pi.*N.* 6.97.'

*προφθέγγομαι, *address*, *SB* 7635.6 (v/vi AD; unless προσφθ- is intended).

προφθίμενος, add 'perh. also *SEG* 34.274 (Corinth, iii AD)'

*προφιλανθρωπέω, *grant an indulgence before*, pass., ἐν τοῖς προπεφιλανθρωπημένοις *PTeb.* 124.36 (ii BC).

πρόφρων **I 2**, add 'εἰ δή μή π. γε Ποσειδάων .. ἐθέλησιν ὀλέσσαι *on purpose*, Hes.*Op.* 667, cf. Thgn. 404, *CEG* 5.5'

προφωνέω **I**, add '*bear witness, declare formally*, *Inscr.Cret.* 4.81.9 (written προπόνετō, v BC; cf. ἀποφωνέω)'

*προχειριστικός, ή, όν, *productive*, Phlp.*in de An.* 349.31.

προχειροτονία, delete 'περί .. id.*fr.* 436'

προχειροφόρος, add '*POxy.* 3197.7 (AD 111), *PPetaus* 34.24 (AD 184)'

†πρόχνυ, adv. *on one's knees*, π. καθεζομένη, i.e. *kneeling* or *crouching*, Il. 9.570; ὥς κε .. ἀπόλωνται π. κακῶς Il. 21.460, ὀλέσθαι π. Od. 14.69, perh. *perish after being brought to one's knees*, but expld. by Sch. as = παντελῶς. **2** *thoroughly*, π. γεράνδρυον A.R. 1.1118; so app. in Antim. 5 Wy. **3** *in truth*, εἰ δή π. γέρας τόδε πάρθετο δαίμων A.R. 2.249. (The aspirate is difficult but cf. γνύπετος, Skt. adj. *prajñu-* (dub. sens.); senses 2 and 3 presumably fr. misunderstanding of Homer.)

προχοή (A), lines 1/2, for '*outpouring*, i.e. *mouth* of a river' read '*flowing waters, streams* of a river'

προχοῖς **II**, for this section read '*beaker, wine-jug*, *Didyma* 426.4 (276/5 BC), *AB* 294'

πρόχοος **II**, add '*SEG* 33.1036 (Aeolis, ii BC)' add 'Myc. *po-ro-ko-wo* (pl.)'

προχορεύω, add 'also med., abs., ΠυρΓίας προχορευόμενος *CEG* 452 (Corinth, *c*.580 BC)'

πρόχρονος, add '**2** *untimely*, Thasos 334.26; adv. πρόχρονα, ib. 332.7.'

προχωρέω **I 3 a**, delete the section **b**, after '*sell*' insert '*Peripl. M.Rubr.* 6, al.'

προχώρησις, add '**III** *proceeds*, Just.*Nov.* 98.1.'

*προχωτισμός, ὁ, *pushing forward*, Simp.*in Cael.* 543.23.

προωνέομαι, add '*TAM* 4(1).366'

*προωνητικός, ή, όν, *bought previously*, προωνητικῷ δικαίῳ *on the basis of a former purchase, PColl.Youtie* 71.20 (AD 281).

πρόωρος, add 'adv. προώρως, Paul.Aeg. 3.76.1 (295.3 H.)'

πρύλις, for 'Cret.' read 'Cypr.'; after 'πυρρίχη' insert 'I'

πρύμνα **I 2**, add 'prov., πρῷρα καί πρύμνα, i.e. *all that is most important*, τῆς Ἑλλάδος D.Chr. 37.36'

πρύμνηθεν, for 'Il. 15.716 .. *IT* 1349' read 'A.*Th.* 209, Luc.*Lex.* 15; *by the stern*, π. λάβε Il. 15.716; *stern-foremost*, Arat. 348; dub. sens., E.*IT* 1349'

πρυμνόθεν **II**, after '*from the bottom*' insert 'Opp.*H.* 1.455'

πρυτάνεια (A) **I**, add '**b** *the body of* πρυτάνεις, *IG* 2².415.15 (iv BC), cf. ib. 330.8, 35 (all pl.).'

πρυτανεῖον **II 2**, add 'of other funds, τοῖς μνήμοσιν δοῦναι τούς ἄρχοντας .. ἀπό τῶν πρυτανείων δραχμάς *ΔΔΔ* *SEG* 33.679 (Paros, ii BC)'

πρυτάνειος, add '**2** as title of Hermes, *SEG* 39.870 (Ceos, v BC).'

πρυτάνευμα, for '*principate*, i.e. *prince*' read '*presidency*, i.e. *president*'; for 'Epigr. in *Rev.Phil.* 19.178' read '*Suppl.Hell.* 982.14'

πρυτανεύω **III 1**, delete 'παρά τινος' and 'by one'

*πρωβοκάτωρ, ὁ, v. °προβοκάτωρ.

πρώην, line 1, after 'πρώαν' read 'Theoc. 8.23, 14.5, Mosch. 3.69; πράν Theoc. 4.60, 5.4, 15.15' **I**, add '**3** generally *in the past, earlier*, *Cod.Just.* 1.3.41.11, 8.10.12.4.' at end delete '(The first syll.' to end of article.

*πρωηρότης or πρωϊ- (scanned ⏑⏑), εω, ὁ, (ἀρόω) *early plougher*, prob. in Hes.*Op.* 490 (cj., vv.ll. πρωτηρότης, προη-).

πρωθήβης, line 4, after 'App.*Hisp.* 65' insert 'π. φοῖνιξ *SEG* 7.195 (Syria, iv AD)'

*πρωθιερεύς, έως, ὁ, *chief priest*, *Inscr.Cret.* 1.xxii 12 (Olus, i AD); v. °πρωτοϊερεύς.

πρωθύπνιον, delete the entry.

πρωΐ, line 2, add 'πρωΐ Men.*Inc.* 58'

†πρωϊζός, όν, πρωϊζός Call.*fr.* 559 Pf. (in corrupt *fr.*); π.· προχθεσινός, ὑπόγυος *EM* 691.54. **II** neut. as adv., *the day before*, pl., χθιζά τε καί π. Il. 2.303. **2** *very early*, pl., οὕτω δή π. κατέδραθες Theoc. 18.9; sg., π. ὁδεύων *Epic.Alex.adesp.* 4.6; in indeterminate sense, τό π. Lys.*fr.* 68.

πρωΐθεν, add 'ἀπό πρωΐθεν *previously*, *SEG* 31.983 (Ionia, ii/i BC)'

πρώϊμος, delete 'cj. in Call.*fr.* 482 (*Hermes* 24.453)'; adv., comp., add 'περί τόν αὐτόν καιρόν .., προιμώτερον δέ *Peripl.M.Rubr.* 28'

πρώϊος **I 2**, line 3, πρωΐας as adv., add '*Archig.ap.Gal.* 12.444.5' **II**, line 2, after 'Hdt. 8.130' insert 'φίτυ π. S.*fr.* 889 R.' add '**III** *early in one's life*, *AP* 7.439 (Theodorid.), 716 (Dionys.).'

†πρωκίος, η, ον, *dewy*, cj. in Call.*fr.* 1.34 Pf.

πρωκτός, after '*anus*' insert 'Hippon. 104.32 W.'

*πρωρά, v. °φρουρά.

πρῷρα **I 2**, add 'of the face, = πρόσωπον, Hsch.; cf. πρωραχθής, ἀνδρόπρωρος, καλλίπρωρος'

πρωράτης, add 'Ion. -ήτης, *CID* I 7.A11 (v BC)'

*πρωταγωνιστέω, add ' of leadership in war, *JHS* 68.47.30 (Lycia, ?ii BC), *SEG* 34.1198 (Lydia, ii BC)'

*πρωταναγνώστης, ου, ὁ, *chief lector* (in a church), (?τόπος) προταναγνώστου Sardis 7(1).188 (?iv AD); also πρωτοαν- *CPR* 10.122.9 (vi AD).

*πρωταπογράφέω, *register for the first time*; med. or pass., cj. in *PFay.* 31.18 (ii AD).

πρωταπογράφομαι, delete the entry.

*πρωταπόγραφος, ὁ or -ον, τό, *list of persons registered for the first time*, ἀ[πογράφεσθαι] ἐν πρωταπογράφῳ pap. in *Aegyptus* 15.209 (iii AD).

πρωταύλης, for '*flute-player*' read '*piper*' and add '*POxy.* 2721.5 (AD 234)'

πρωτεῖος, after '*of the first quality*' insert '*Edict.Diocl.* 8.11'

*πρωτέλειος, v. °προτέλειος.

πρωτεύω **I**, lines 3ff., *hold position of* πρῶτος, add 'πρωτεύσαντα τῶν ἱερῶ[ν *SEG* 34.1107 (Ephesus); of chief official in Egyptian town or village, *CPR* 6.79.2 (v AD)'

*πρωτεπιστήμων, ονος, app. *having newly acquired the skill* or *knowledge*, ἱππηλάται π. Anon.*in Rh.* 15.18.

*πρωτέφηβος, ὁ, *leader of* ἔφηβοι, *Inscr.Délos* 1956 (i BC).

*πρωτήκτωρ, v. °προτήκτωρ.

*πρώτηλα, τά, *pair of oxen*, cf. Lat. *protelum*, dub. in *JRS* 46.46.4 (Phrygia, iii AD).

*πρωτηρότης, v. °πρωηρότης.

*πρωτοαναγνώστης, v. °πρωταναγνώστης.

πρωτοβόλος **I 1**, delete '(Rufin.)'

*πρωτογενέτωρ, ορος, ὁ, *first-begetter*, of God, *PMasp.* 188ᵛ.1.

*πρωτόγναφος, for '*fresh from the fullers*' read '*carded once only* i.e. *almost new*' and add '*PMichael.* 18 ii 3, al. (iii AD), *Peripl.M.Rubr.* 6'

*πρωτογόνατος, ον, *firstborn*, *MDAI(A)* 33.149 no.11 (Constantinople).

πρωτόγονος **I 3**, for this section read '*a name of the Orphic divinity* Φάνης, Orph.*fr.* 54, 86, *H.* 6.1. **b** fem. -η, *a name of Persephone*, Paus. 1.31.4.'

*πρωτοδρᾰκονάριος, ὁ, *chief* °δρακωνάριος, τῷ αἰδε(σι)μ(ωτάτῳ) Δομιτιανῷ (π)ρωτοδρακοναρίῳ *ITyr* 33C.

*πρωτόθετος, ον, gramm. of a word, *primitive*, opp. *derivative*, = πρωτότυπος, Ar.Byz.*fr.* 241G S.

*πρωτοϊερεύς, έως, ὁ, *chief priest* or *member of a sub-class of* ἱερεῖς, *BGU* 2469 ii 8 (ii AD), Chapouthier *Les Dioscures* 26 (Dorylaeum); v. °πρωθιερεύς.

*πρωτοκλίναρχος, ὁ, *chief president of an Isiac confraternity*, *IPhilae* 199.5 (AD 456/7).

πρωτοκλισία, after 'etc.' insert 'ἐν ταῖς ἄλλαις συνόδοις πάσαις πρωτοκλισία *Inscr.Délos* 1520.33 (ii BC)'

πρωτοκοσμέω, add 'SEG 32.869 (Crete, ii AD)'
*πρωτοκούρης, ητος, ὁ, president of a college of κουρῆτες (sense II 3), IEphes. 974, 1042a, 1061, etc. (ii/iii AD).
πρωτόκτιστος, add '2 first built, of a gate, Sch.E.Ph. 1113.'
πρωτολοχία, add 'SEG 13.403 (Maced., ii BC)'
*πρωτόμηνος, ὁ, one of the νεωποῖαι in office in the first month of the year, IEphes. 3513a (i BC/i AD); cf. σύμμηνος.
*πρωτονεωποιός, ὁ, first νεωποιός, CIG 2800, REG 19.145 (both Aphrodisias in Caria; -νεοπ-).
πρωτόπαλος, add 'SEG 32.605 (Larissa, ii AD)'
πρωτόπολις, add '2 v. °πρωτόπτολις.'
πρωτοπολίτης, after 'Gloss.' insert 'Vit.Aesop.(G) 93; sg., eminent citizen, IG 10.2(1).204 (Thessalonica; -είτης)'
πρωτοπορεία, add 'BGU 2424.7 (86 BC)'
πρωτοπρεσβύτερος, add 'SEG 19.443 (Philippi); πρωτοπρεσβοιτέρου BCH suppl. 8.238, προτο- ICilicie 119, etc.'
*πρωτοπροφήτης, ου, ὁ, chief or first prophet, Μωυσῆς π. PJen.inv. 536 (APF 27.61, v/vi AD).
*πρωτόπτολις, εως, ἡ, first, i.e. oldest, among cities, Τάρσος ἀειδομένη π. Nonn.D. 41.357.
πρωτοστάτης II, add '2 chief magistrate of a village, PCair.Isidor. 64.10 (iii AD), SB 9502.'
*πρωτοστολιστής, οῦ, ὁ, keeper of sacred vestments (of the first class), PGrenf. 44 ii 2 (ii BC), IPhilae 196, 197 (AD 452).
*πρωτοσφήν, ῆνος, ὁ, first wedge, οἱ π. Hero Stereom. 2.31, al.
*πρωτοτόκεος, ον, = πρωτοτόκος, having borne offspring for the first time, Boeot. πρᾶτο- SEG 25.556 (c.210/200 BC).
πρωτότυπος II 1, add 'Just.Nov. 4.1'
*πρωτουργικός, ή, όν, providing the initial impetus, αἰτία Procl. in R. 1.180 K.
*πρωτοφεγγής, ές, shining with first light, τῆς π. ἡμέρας Trag.adesp. 664 ii 11 K.-S..
*πρωτόχθων, gen. ονος, ὁ, ἡ, aboriginal, (of Sardis), SEG 36.1095, ἡ π. καὶ μητρόπολις τῆς Ἀσίας καὶ Λυδίας ἁπάσης ib. 1096 (iii AD), IUrb.Rom. 85.
πταῖμα, add 'TAM 4(1).1.40 (Bithynia, ii BC)'
πταῖσμα I 1, add 'b blow or bruise on the toes, Poll. 2.199; cf. °ἐπίπταισμα, πρόσπταισμα.'
πταίω II 2, line 7, transfer 'also .. S.Ph. 215 (lyr.)' to section 1.
*πτᾶσα, v. πέτομαι.
πτελέα, add 'also πελέα IG 4²(1).102.44 (Epid.); form πτελέϝα, Myc. pte-re-wa, elm-wood'
*πτεραφορία, ἡ, office of πτεροφόρος, PTeb. 298.21 (ii AD, -εία pap.).
*πτερίς I, after 'Dsc. 4.185' insert 'πτερίδα D.Chr. 7.75'
*πτεριστής, οῦ, ὁ, perh. embroiderer, IG 3.3441 (πταιρ-).
πτέρνη I 2, add 'AP 9.225.4 (Honest.), and so prob. in Call.fr. 2.4 Pf.' III, delete 'f.l.' to end of article and add 'Myc. pte-no (dual), some part of a chariot'
πτερνίζω I 1, after 'strike with the heel' insert 'of a rider urging on his horse' 2, for this section read 'trip up from behind, PVindob. Salomons 15.10 (v/vi AD); fig., as metaphor fr. wrestling, circumvent, outwit, Lxx Ge. 27.36, Ho. 12.4, Ma. 3.8, Ph. 1.125'
πτερνισμός, for 'supplanting' read 'overturning (as if by lifting the heel), in quot., fig.'
*πτέρνομαι, = πτάρνυμαι, 3 sg. πτέρνεται, gloss on χρέμπτεται, Cyr.
πτερόεις I, add 'prob. of Eros, τὸν πτερόεντα θεόν IHadr. 29 (i BC/i AD)' II, for 'winged' words read 'flighted' words'
πτερόν III 2, add 'πέμπειν χρύσεον Μουσᾶν Ἀλεξάνδρῳ π. B.fr. 20B.4 S.-M.' 3, delete the section. 4, for this section read '= σκιάδειον, Com.adesp. 1129' 8 b, for 'battlements' read 'protective structure built on to city wall, πτερὰ τὴν οἰκοδομίαν καλοῦσι ταύτην ἐπεὶ ὥσπερ ἀποκρέμασθαι τοῦ τείχους δοκεῖ' c, for 'portcullis .. gateways' read 'metal plating let down to cover city gate in the manner of portcullis' add '10 column in tables, Cat.Cod.Astr. 8(2).13. F. 47-87.'
πτεροφόρος III, for this section read 'kind of priest, = πτεροφόρας I, PGrenf. 44 ii 3 (ii BC), SEG 31.1556 (Thebes, Egypt, ii/i BC)'
πτεροφυής, add 'PMag. 12.46'
πτερύγιον II, add '11 cloudy spot in the beryl, Plin.HN 37.79.'
πτερυγοειδής, delete 'only' before 'adv.' and insert 'ἐκφύσεις Gal. 2.439.7, ὀστᾶ 441.10, 443.4'
πτέρυξ II 1, line 3, for 'tortoise' read 'turtle' 7, read 'unspecified part of a building' add '12 pl., the extremities of the world or a land, ἐπὶ πτερύγων τῆς γῆς Lxx Jb. 37.3, 38.13, Is. 24.16; imagined as four corners, Is. 11.12, Ez. 7.2.'
πτηνοπέδιλος, add 'PMag. 5.404'
πτίλον, line 1, for '(q.v.)' read '(cf. Ψίλαξ B)'; line 3, after 'Suid.)' insert 'ἁπαλόν ψ. cj. in Alcm. 3 fr.3 ii 68 P.'
πτιλωτός, add '3 decorated with feather pattern, φιάλαι IG 2².1443.135.'
*πτισανᾶς, ᾶ, ὁ, seller of barley-gruel (πτισάνη II), AP 11.351 (Pall., cj. πτιστής).

πτισάνης, delete the entry.
πτοέω, line 2, after 'Call.Dian. 191' insert 'Dor. ἐπτοάθην E.IA 586 (lyr.)' II, line 5, after '(= Thgn. 1018)' insert 'φρένας ἐπτοέαται Anacr. 1 fr.1.12 P.'; at end after 'E.IA 586 (lyr.)' insert 'Hsch. s.v. πτοιώμενον'
Πτολεμαϊκός, add '3 designation of an age group in athletics, SEG 27.1114.19 (Egypt, iii BC).'
πτολιαρχέω, in Thess. form *ττολιαρχέω, v.s.v. °πολιαρχέω.
πτόλις, line 4, for 'Inscr.Cypr. 135.1 H.' read 'po-to-li-se, ICS 217.2;' line 8, for 'πτόλιϝι Inscr.Cypr. 135 6 H.' read 'po-to-li-wi πτόλιϝι ICS 217.6'
πτόρθιος, add '(unless, reading as Sokolowski 3.11.A16 Ποσειδῶνι πτόρθι[ος], some kind of sacrificial offering, cf. °προπτόρθιος)'
*πτυάζω or πτυΐζω, winnow, λελικμημένη· κοσκινευθεῖσα, πτυασθεῖσα (πτυισ- La.) ἐπ' ἀνέμῳ Hsch.
πτύον, add 'and is restored in IG 1³.422.134 [π]τέō (dual) (v BC)'
πτύρομαι, add 'act. πτύρω as expl. of terreo in Virgil gloss., PNess. 1.778 (vi AD), Gloss.'
*πτυχίδιον, τό, tablet, PMil.Vogl. 152 ii 52.
πτύχιον I, add '2 prob. leaf of a folding door, PMerton 39.5 (iv/v AD).'
*πτύχιος, α, ον, quoted as adj. fr. πτύξ, EM 64.28; spec., folded or broken, of reeds, BGU 2210.19 (AD 617).
πτῶμα II 2, line 6, after 'Lys.fr. 203 S.' insert 'PFay. 102.20 (ii AD), BGU 2483.4 (iii AD)'
πτωμάτισμός, add 'SEG 30.1794 (iii AD)'
πτῶσις III, add '2 of the categories, mode, Arist.EE 1217ᵇ30, Metaph. 1089ᵃ27.'
πτωτικός, add '3 liable to fall (or cause falling), ὀλισθηρόν· πτωτικόν Hsch.'
πτωχικός, add '2 of a poorhouse, πτωχικοῦ πράγματος Just.Nov. 7.3 pr.'
πτωχός, line 1, delete 'S.OC 444'; lines 7/8, delete '(so πτωχός (fem.), S.OC 444)' II, after 'Ep.Gal. 4.9' insert 'νοήματα (in a comparison) D.H.Comp. 4'
πτωχοφανής, for 'like a beggar' read 'having the appearance of a beggar'
*πυάνη, ἡ, dub. sens., Sch.A Il. 12.459 (perh. = λάρναξ).
πύανος, after 'Hsch.' insert 'πύανοι, μίγμα παντοδαπῶν ὀσπρίων, Theognost.Can. 23 (cod. πτυ-)'
πύας, delete the entry.
*πυγαῖα, τά, padding to accentuate the buttocks, Hsch.
πυγίζω, add 'SEG 31.824 (Sicily, v BC), 8.574 (Egypt, iii AD), etc.'
*πυγίον, τό, dim. of πυγή, SDAW 1934.1040 (tab.defix.).
*πυγιστής, οῦ, ὁ, sodomite, SB 6872 (graffito).
πυγμάχος, add '(on accent Ath. 4.154f.'
πύελιον, after 'πύελος' insert 'Inscr.Cret. 1.xvii 12A6 (Lebena, ii/i BC)'
πύελις II, add 'also ποιαλίς TAM 2.347, 348, 706, IMylasa 468.8'
πύελος, after 'πύαλος' insert 'ποίελος, πύαιλος, ποίαλος' 4, add 'TAM 4(1).258, 361, 363, etc.'
Πυθαγόρειος, after 'Pythagorean' insert 'Pl.R. 600b, etc.'; before 'Pl.R. 530d' insert '-ειοι, οἱ'; add 'Iamb.VP 18.80, etc.'
Πυθαγορικός, before 'τὰ Π.' insert 'οἱ Π., Plu. 2.488, Num. 11.1'; after 'D.L. 7.4' insert 'a work by Aristotle, Arist.fr. 204, 205 (τὸ Π. fr. 199)'
Πυθαεύς, add 'also Πυθαιεύς SEG 11.890 (Cynuria, vi BC)'
*πυθάζω, v. °πυθαΐζω.
πυθαΐζω, add 'also πυθάζω, πυθασταὶ τοὶ πυθάξαν[τες] SEG 25.852, 853 (Telos, iii/ii BC)'
πυθαϊστής, after 'member of such a mission' insert 'SEG 21.541 ii 50, al. (Erchiae, iv BC)'; add 'also πυθασταί (pl.) SEG 28.852, 853 (Telos, iii/ii BC), cf. ‡πυθιασταί'
*πυθάρχας, α, ὁ, leader of πυθαϊσταί, SEG 25.852 (Telos, iii BC).
*πυθαρχέω, to hold the office of °πυθάρχας, Didyma 87.4.
*πυθαστάι, v. °πυθαϊστής.
Πυθία II, add 'SEG 30.1286 (Didyma)'
Πύθια, add 'also Πύτια SEG 23.566 (Crete, iv BC)'
Πυθιασταί, add 'Tit.Cam. 78.10 (i BC), cf. °πυθαϊστής'
Πυθικός I, line 4, before 'νικήσαντα' insert 'Π. αὐλός, kind of aulos, app. αὐλὸς χορικός (cf. Πυθαύλης, χοραύλης), Poll. 4.81, Aristid.Quint. 2.16; cf. τὸ .. τῶν ψιλῶν κιθαριστῶν ὄργανον, ὃ καὶ Πυθικὸν ὀνομάζεται Poll. 4.66; as an athletic category (of boys aged 12-14 years)'; line 5, for 'cf. 1064.7' read 'SIG 1065.7' add '2 Πυθικοί, οἱ, association of musicians at Saittai, SEG 29.1200.'
Πύθιον, add 'at Icarion, IG 2².4976 (iv BC), etc.'
Πυθιονίκης I, line 2, after 'Hld. 5.19' insert ' also nom. Πυθιονίκα IG 7.1888b9'
Πύθιος, add 'also Πύτιος Schwyzer 686 (Pamphylia, iv BC), 198 (Crete, ii BC); Πύττιος SEG 33.638 (Crete, ii BC)'
πυθμήν III, for 'lowest number' read 'lowest term'
Πυθόνικος, add 'GDI 1504Bb4 (Opus)'
Πυθόχρηστος II 1, add 'εἶναι Βαγαδάτην νεωκόρον τῆς Ἀρτέμιδος

πυθόχρηστον αὐτῶι γενόμενον Amyzon 2.10 (iv BC)' **2**, for 'epith. of' read 'applied to the names of gods for whom cults were approved by the Pythian oracle'

Πυθώδε, add 'Πυθόδ' ἀνέθεκ[ε] CEG 369 (Delphi, c.600/550 BC)'

Πύθωθεν, add 'IG 1³.256.4 (v BC)'

*****πυκακίνη**, ἡ, type of garment, prob. made fr. thick fabric, PVindob. G39847.881 (CPR 5.109) (iv AD).

πυκινοκίνητος, add 'Gal. 18(1).415.8, 10'

*****πυκλιή**, gloss on βαθάρα (Maced.), Hsch. (entered as if βατ-; cf. °πυρλός).

πυκνός Α ΙΙΙ, add '**3** τὸ π. density of style, D.H.Th. 24; cf. πυκνόω II b.' **V 1**, at end delete 'τὸ π. .. 24'

πυκταλεύω, after 'Sophr. 111' add '(PSI 1214a16)'

⁺πυκταλίζω, = πυκτεύω, Anacr. 1 fr.4.1 P., 51 P., perh. also πυ]κταλίζουσι Hippon. 102.8 W.

πυκτεύω, line 5, of gladiators, add 'SEG 32.605 (Larissa, ii AD)'

πυκτικός 2, after 'of or for boxers' insert 'ἱμάντας .. πυκτικούς Eup. 350 K.-A.'

⁺πυκτίς (Β), ίδος, ἡ, supposed to be some kind of animal, Ar.Ach. 879, but perh. example of πυκτίς (A) inserted for word-play w. following ἰκτίδας.

Πυλαία Ι 1, add 'IG 1³.96' **2**, for this section read 'Amphictyonic rights, D. 6.22, 8.65, cf. 5.23 (which may belong to 1)' **II**, add 'prov., app. of a trivial fiction, ἐκ πίνακος καὶ πυλαίας Plu. 2.386b'

πυλαῖος 1, add 'title of Poseidon, Ποτειδῶνι Κραναίωι Πυλαίωι SEG 15.377 (Larissa, iv BC)'

πυλαωρός, at end after 'Il. 24.681' insert 'GVI 1179.7 (Smyrna, ii BC)'

πυλεών ΙΙ, for this section read 'wreath, Alcm. 3 fr.3 ii 65 P., 60.2 P., Pamphil.ap.Ath. 15.678a; circlet, Call.fr. 80.5 Pf., Aristaenet. 1.15 (πόλεων cod.), cf. Poll. 5.96'

πύλη Ι 2, add 'of a temple, SEG 32.1260 (Paphlagonia, AD 192/3)' **4**, add 'ὁ πρὸς τῇ πύλῃ, i.e. customs officer, PRein. 95.1' **II 1 c**, after 'metaph.' insert 'ἔαρος π. Alc. 296(b).3 L.-P.'

πυλοκλειστής, for pres. ref. read 'SEG 26.1835 (Cyrene, ii AD)'

*****?πυλονόμος**, ὁ, customs officer, Κάστορος πυλωνόμο[υ] Ἑρμοπολείτου κώμης Anagennesis 3.82 (SEG 33.1363, Egypt, Chr.).

Πύλος, add 'Myc. pu-ro'

*****πυλουχίς**, ίδος, ἡ, guardian of the gate, Ἡρωΐνησι Πυλοχίσι SEG 26.136 (Athens, iv BC)'

πυλοῦχος, add '**2** of a deity, guardian of the gate, rest. πυ[λόχωι SEG 26.136 (Athens, iv BC); cf. πυλάοχος'

πυλών, add 'on a race-track, τοὺς πυλῶνας τῶν ἱππαφίων Tab. Defix.Aud. 234 (ii/iii AD)'

πυλώριον, add 'perh. also πυλωρ[ίου] SEG 25.318 (Attica, i AD)'

*****πύννος**· ὁ πρωκτός Hsch., cf. Didyma p. 100a.51 (graffito); v. πουνιάζειν.

πυξίδιον ΙΙ, after 'πυξίς' insert 'Inscr.Délos 1417B i139 (ii BC)'

πυοποιέω, add 'Heras ap.Gal. 13.766.5'

πύος, line 3, for 'supr.' read 'πύον II'

πῦρ Ι 4, add 'Πῦρ ἄφθαρτον IEphes. 1058, 1060, al.'

πύρά, τά, line 8, delete 'metaph.'

πύρά, ᾶς, **1 b**, add 'SEG 32.655 (AD 102/3)' **3**, add 'a pile of wood for burning, bonfire, Act.Ap. 28.2.'

πυράγρα 1, add 'Nic.Al. 50, Luc.DDeor. 5.4, 7.2'

*****πυραίθης** (or **πύραιθος**), ου, ὁ, fire-kindler (title of a priest), POxy. 2722.3, 6 (AD 154), 3567.3 (AD 252).

πυράκτωσις, ἡ, heating of a cautery-iron, Aët. 1.233 (98.6 O.), 7.32 (280.23, 281.24 O.).

πυραμίς ΙΙ, add '**2** perh. candle, Arist.Ph. 245ᵇ.11.'

*****πυράπτης**, ου, ὁ, fire-lighter (perh. a religious office, cf. λυχνάπτης), SEG 19.661 (Alabanda, Rom.imp.)'

πυραυγής, add '(first syll. long in Orph.H. 19.1 s.v.l.)'

πύραυνος Ι, line 1, after 'coals' insert 'Pl.Com. 142 K.-A.'

πύραυστρα, add 'Myc. pu-ra-u-to-ro (dual), fire-tongs'

*****πύργαλος**, ὁ, kind of bird (cf. πυργίτης), PAmst. 13.8 (sp. πρυγ-; v AD).

πυργηρέομαι, add 'act., πυργηροῦμεν· φυλάττομεν τὰ τείχη Hsch.'

*****Πυργία**, ἡ, goddess of the tower, title of Athena in Locris, SDAW 1935.695; cf. ἐπιπυργῖτις.

πυργίον, add '**2** part of a funerary monument, LW 1639. **3** dice-box, Sch.Aeschin. 1.59. **4** part of a trireme, Hsch.'

πυργίσκος 1, read 'perh. pillar supporting a sarcophagus, TAM 2.51, 63, 64a, b, 65, 66'

πυργίτης, after 'of a tower' insert '(cf. πύργος I 3)'

πυργοποιία, add 'de Franciscis Locr.Epiz 3, 6, 10, etc. (sp. -ποία, iv/iii BC)'

πύργος Ι 1, line 3, delete 'city walls' **3**, for 'the part of a house .. and worked' read 'any separate structure built for protection, watch-keeping, etc., esp. in the open country (estate, vineyard, etc.)' and delete remaining definitions of section. add '**b** a structure forming part of a tomb, TAM 2.245.'

*****πυργοσηκών**, ῶνος, ὁ, app. fortified enclosure, IGLS 316.

πύρδαλον, for 'Lyr.Alex.Adesp. 31' read 'Call.fr. 197.42 Pf.'

πύρεθρον, delete 'cf. πυρῖτις II'

πυρετέω, delete the entry.

πυρετός ΙΙ, add 'personified as a deity, Θεῷ Πυρετῷ Inschr.Hierap. 7'

πυρετώδης 1, add 'adv. -δῶς Gal. 18(1).551.9'

πυρήν VI, for 'Str. 4.6.10' read 'Plb. 34.10.9'

πυρία Ι 4, for '= εἰσώστη' read 'sarcophagus' and delete 'tomb-chamber'

πυρίασις, add 'Erasistr.ap.Gal. 11.200.14'

πυριάω 3, add 'Erasistr.ap.Gal. 11.206.13'

*****πυρῐδήης**, ές, flashing with fire, ὄμμα B.fr. 64.21 S.-M.

*****πυριλόχευτος**, ον, fire-born, of Dionysus, Ps.-Callisth. 55.5.

⁺πύρίνη, ἡ, stone of a fruit (= πυρήν I), v.l. in Hp.Mul. 2.138 (8.298.18 L.), Gp. 6.11, 9.18 (v.l.), Hsch.

πυρίρροθιος, delete the entry.

πυρῖτις Ι 1, for this section read 'perh. convolvulus, Nic.Th. 683; as adj., ῥιζάδα .. πυρίτιδα id.Al. 531'

πυρίφευκτος, add 'Anon.Alch. 19.19'

πυριφλεγέθων Ι, add '**2** blazing with fire, [κεραυνός] SPAW 1934.1046 (tab.defix.), SEG 34.1308 (Side, i BC/i AD).'

πυριφλεγής 1, add 'sup., Plu.Daed. 5'

*****πυρίχροος**, ον, contr. -χρους, ουν, = πυρίχρως, Xenocr.Lap. 90.

πυρκαΐα, add 'Poll. 9.156; Προμηθεὺς π. title of play by A., id. 10.64'

πυρκαϊά Ι 2, add 'Heraclit. 43 D.-K.'

*****πυρκόρος**, ὁ, prob. tender of the sacred fire (cf. νεωκόρος), AE 1934/5.140 (Atrax in Thessaly, v BC).

*****πυρλός**, gloss on βαθάρα (entered as if βατ-) (Ἀθαμᾶνες), Hsch.; cf. °πυκλιή.

*****πυροβόλησις**, εως, ἡ, sowing of wheat, BGU 1850.25 (?i BC; προβ-pap.).

πύροεις, after 'εν' insert '(poet. also ειν, Nic.Th. 748)'

*****πυροελκής**, ές, characterized by fiery ulcers, λοιμοῦ π. GVI 993.1 (Rome, ii/iii AD).

*****πυρόκοπρος**, (?)ἡ, fire of (dried) dung, Moses Alch. 301.21, 311.16.

*****πυροσπορεία**, ἡ, sowing of wheat, PMil.Vogl. 131.5, 140.34.

πυροσπορέω, add 'POxy. 1629.9 (44 BC), PVindob.Worp 2.13 (21 BC)'

*****πυροσώματος**, ον, having a fiery body, in invocation of Ἄρκτος, PMag. 7.701.

πυρπολέω ΙΙ 2, for 'Med. .. fire' read 'pass.'

*****πύρρινος**, η, ον, fiery red, BGU 2217 ii 17 (ii AD; unless to be taken as fem. subst. fire-pot, or sim.); of a garment, PVindob. G25.950 (Die Sprache 34(1).192, v/vi AD).

πυρρίχη 1, add 'χοραγήσαντα πυρρίχαι SEG 39.759.19 (Rhodes, i BC); καλεῖται δ' ἡ πυρρίχη καὶ χειρονομία Ath. 14.631c'

πυρριχίζω, add 'Ath. 14.631a'

πυρρίχιος ΙΙ, after 'πυρρίχη' insert 'D.H.Comp. 18'; add 'ῥυθμός D.H.Comp. 17'

πυρριχισμός, add '**2** the use of a pyrrhic foot at the end of a hexameter, Eust. 1577.52 (referring to Il. 12.208 (v. ὄφις)).'

πυρριχιστής, add 'glossed by λουδιώνης (Lat. ludiones), POxy. 3452.14 (ii AD)'

*****πυρροκέφαλος**, ον, red-headed, Hsch. s.v. πυρσοκόρσου λέοντος.

*****πυρροκόκκινος**, ον, scarlet, Cat.Cod.Astr. 11(2).157.7.

πυρρός Ι 2, for 'with red hair' read 'having a ruddy complexion (or sts. having red hair)' add 'cf. Myc. pu-wo (masc.), pu-wa (fem.), pers. n. and see also °φυρρός'

πυρρότης, after 'of hair' insert '(or complexion)'; add 'Hp.Aër. 20'

*****πυρρόχαλκος**, ὁ, yellow copper, Moses Alch. 310.18.

πυρρόω, dye red, Hsch. s.v. ἠρυθροδάνωται.

πυρσαυγής, before 'Orph.' insert 'dub. cj. in'

*****πυρσόπνευστος**, ον, fire-breathing, Suppl.Mag. I 42 (iii/iv AD).

πυρφλέγων, οντος, ὁ, = πυριφ., SPAW 1934.1043 (tab.defix.).

πυρφορέω ΙΙ, after 'Charito 2.4' add '(ἐπυρπό[λει POxy. 2948)'

πυρφόρος Ι b, line 4, for 'engine .. fire dart' read 'fire-basket (in an incendiary device)' and before 'Jul.Or. 2.62d' insert 'prob. fire-dart' , line 6, delete 'satyric' **2**, add 'also in the Eleusinian mysteries, SEG 30.93 (Eleusis, 20/19 BC)'

πυρώδης Ι, line 4, before 'Sup.' insert 'comp., ἀὴρ πυρωδέστερος ὤν Aen.Tact. 23.1'

πυρωπός ΙΙ, before 'Plin.' insert 'Prop. 4.10.21'

*****πύσμαδε**, adv., to interrogation, προσαγαγεῖν PCol. 175.37 (iv AD)

πυτίζω, line 1, delete 'frequently' and after 'mouth' insert 'Placit. 3.5.9'

πυτίνη, add 'Ar.fr. 880 K.-A.'

⁺πῶ (verb), v. ‡πίνω.

πώγων 1, after 'beard' insert 'Alc. 143.6 L.-P.' **4**, add 'poet. in POxy. 2817.3; also the feathered end of an arrow, Nonn.D. 16.9'

⁺πῶθι, v. ‡πίνω.

πωλάδιον, τό, = πωλάριον, PCair.Isidor. 136.6 (iii/iv AD, πολαδ‹ιο›ν pap.).

πωλάριον, add 'PMich.XI 620.218 (AD 239/40)'

πωλάς, άδος, ή, female foal, τὴν πωλάδαν (sic) ὄνον PCair.Isidor. 86.11 (iv AD).

πωλέομαι, line 4, after 'Il. 5.350' insert 'Aeol. 3 pl. πώλενται Alc. 130.33 L.-P.'

πωλέω, lines 4/5, delete 'prob. in IG 1².60.10'

πωλητήρ, add 'SEG 12.380.30 (Cos, fr. Gela-Phintias, iii BC)'

πωλητικός, add 'κατὰ τὸν πωλητικὸν νόμον IMylasa 208.13 (i BC), al.'

πωλικός 1, add 'νικάσαντα κέλητι πωλικῶι Kontorini AER 2.168 no. 74 (Rhodes, i BC)'

πωλίον, add 'III gloss on ἀκρίς, Sch.Theoc. 5.34.'

πῶλος I 2, delete 'οἱ the dog .. (Strat.)'   **3**, add 'οἱ κύνεοι π. AP 12.238 (Strat.)'   add 'Myc. po-ro (dual)'

⁺Πωλώ, οῦς, ἡ, epith. of Artemis on Thasos, IG 12 suppl. 382, 383; on Paros, suppl. 202.

πῶμα (B) **II**, after 'drinking-cup' insert 'E.Ion 1212'

πωμαρικός, ή, όν, of or concerning fruit-growing, PHamb. 222.18 (vi/vii AD).

πωμάριον, after 'orchard' insert 'POxy. 707.19 (ii AD)'; add 'π. φοινίκων PMasp. 170 (AD 564)'

πώποκα, add 'see also πήποκα'

πώποτε I, add 'followed by neg., ὅτι πώποτε οὐκ ἐρρέθη λόγος τοιοῦτος περὶ Σουσάννης Thd.Su. 27'

⁎πώπως, in any way whatever, τὰ ἀφιερωμένα πόπως (sic) SEG 32.1423.12 (Syria, AD 192/3).

⁎πώρη, ὁ δηλοῖ τὸ πένθος Sch.E.Or. 392 (expl. of ταλαίπωρος).

πῶρος 4, add '**b** stone-like tumour in other parts of the body, Orib. 45.6.1, 7.'

πῶς III 5, after 'Ach. 24' insert 'E.Hec. 1160'   **V**, delete 'π. .. Matt. 21.20'

⁎πώσποτε, indef. adv., in any way whatever, Sardis 7(1).20.16 (vi AD).

⁎πωσφόρος, v. °φωσφόρος.

πῶυ, line 5, after 'Nonn.D. 3.302' insert 'νέων ἵνα πῶυ GDRK 30.89'

Ρ

ῥαβδίον 4, for this section read 'of items in temple inventories, ῥαβδία ἀκοντίων *Inscr.Délos* 104-28*bB*17, ?λίθος φηγοειδὴς ῥαβδίωι ἠρτημένος ἀργυρῶι *IG* 2².1534.103 (iii BC)'

ῥαβδισμός, add '*PTeb.* 229 (i BC)'

ῥαβδιστής, after 'ὁ' insert '*beater* (prob. in some process of clothmaking), *PMich.*II 123ʳ vi 19, xiv 17 (i AD); *fuller*, *BGU* 2547.3 (iii AD)'

ῥαβδομαντεία, for '*a wand*' read '*rods or sticks*'; add 'Jerome *in Ezech. PL* 25.206'

ῥάβδος I 1, at end for '*divining-rod*' read '*rod or stick* used in divining'   **III**, line 4, after 'D.S. 5.37' insert 'cf. Hsch.'   **IV 1**, after '*verse*' insert '(in expl. of ῥαψῳδός) Menaechm. 9 J.'

ῥαβδουχέω, after 'badge of office' insert 'ῥαβδόχεν τὸς ἱεροποιοὶς *IG* 1³.250A9 (Attica, v BC)'

ῥαβδοῦχος, add '**3** *cattle-drover*, *POxy.* 1626.9 (iv AD); perh. also *SEG* 29.307 (Corinth, v/vi AD).'

ῥαβδωτός II, add 'of a kind of sacrificial cake, πόπανον ῥ. *Inscr.Perg.* 8(3).161.3 (ii AD)'

***ῥᾰγᾰδώδης**, ες, *cracked, fissured*, of soil, *SB* 6797.26 (iii BC).

ῥᾰγάς, line 1, after 'Ephor. 65(e) J.' insert '*SB* 6797.9 (iii BC)'

†**ῥᾰδᾰλός**, ή, όν, perh. *quivering*, read by Zenod. for ῥοδανός (q.v.) in Il. 18.576, Nicaenet. 1.4.

***ῥάδαμος**, ὁ, v. ὀρόδαμνος.

ῥαδανίζω, for 'ῥοδάνη' read 'ῥοδανίζω'

ῥᾰδινός, line 1, for 'βράδινος' read '(‡βράδινος)'

ῥᾴδιος, line 2, after '*a, ον*' insert 'also ος, ον (E.*Med.* 1375, Pl.*Plt.* 278d, D.*Ep.* 3.23)'   **B**, line 1, after 'Aeol. βραϊδίως' insert 'Alc. 129.22 L.-P.'   **2**, line 2, after '*recklessly*' insert 'βραϊδίως πόσιν ἔμβαις ἐπ' ὁρκίοισι Alc. l.c.'

ῥᾳδιουργία, add 'δίχα .. οἰασδήποτε ῥᾳδηουργίας *PColl.Youtie* 92.35 (AD 569)'

ῥᾳδιουργός, line 3, for '*knave, rogue*' read '*bandit*'

ῥᾳθῡμέω, line 3, after 'etc.' insert 'w. gen., μαθημάτων Clem.Al.*Strom.* 4.5, τοῦ νόμου *Cod.Just.* 3.2.4.7, *PMasp.* 151.194 (vi AD); w. inf., τοῦτο πρᾶξαι *Cod.Just.* 1.3.45.6'   **3**, add 'οὐδὲ γὰρ χρὴ μόνας τὰς ἐκπομπὰς κατεπείγεσθαι, ῥᾳθυμεῖσθαι δὲ τὰ στρατιωτικὰ δαπανήματα *Cod.Just.* 12.37.18'

ῥᾴθῡμος I 3, for '*slipshod*' read '*lacking in drive, perfunctory*'

***ῥαίανος**, unexpld. adj., cf. ῥαίαν, Hdn.*fr.* p. 28 H.

***ῥαιβοκερεῖς**· στρεβλοκέραται Hsch. (ῥαικακερεῖς cod.).

ῥαῖδα, for '*rhaeda*' read '*raeda*'

***ῥαιδαστίζει**· τὸ σκώπτει Theognost.*Can.* 32 A.

ῥαικακερεῖς, delete the entry.

***ῥαικάμη**· ἡ ῥᾳστώνη (*ease*), Theognost.*Can.* 32 A.; cf. ῥαθάμη.

ῥαίω II, for 'Pass., .. S.*Tr.* 268' read 'pass., A.*Pr.* 189, S.*Tr.* 268, *AP* 7.529 (Theodorid.)'

ῥακά, before 'Hebr.' insert 'Aramaic or'; add 'cf. °ῥαχᾶς'

***ῥᾰκᾴδιον**, τό, app. *rag* used as wrapping, ostr. in *ZPE* 98.138 no. 29.4 (Egypt); cf. °ῥακίδιον.

ῥάκανα, add 'cf. Lat. *rachana*, *Edict.Diocl.* 19.5, also °ῥάχνη'

***ῥᾰκίδιον**, τό, dim. of ῥάκος, *little rag*, Paul.Aeg. 2.46 (117.27 H.).

†**ῥάκιον**, τό, (ῥάκ- *GVI* 1920.9 (Athens, i AD)), *rag*, χαλκὸν ἀδόκιμον ἐν ῥ. *Inscr.Délos* 1450A103; app. of a flag, Them.*Or.* 16.210b; fig., ῥάκιόν τι τοῦ παλαιοῦ δράματος Ar.*Ach.* 415; in pl., id.*Ach.* 412, V. 128, al.; perh. for use as bandages, *IG* 1³.421.163 (Athens, v BC).

ῥάκος I 2, for 'even .. flesh' read 'of flesh torn to *rags*'

ῥᾰκώδης 1, after '*ragged*' insert 'προσκεφάλαια *IG* 11(2).147*B*13, χιτώνιον *IG* 2².1518.66'

ῥάκωσις, add 'in unexpld. context, *CPR* 7.8.58 (ii/iii AD)'

ῥάμμα (B) **3**, add '*APF* 2.2.15'

***ῥαμμάτιον**, τό, perh. *small piece of thread*, Anon.Alch. 323.3.

ῥαμματώδης, add 'ἕλιξ τὸ τῆς ἀμπέλου ῥαμματώδες gloss in Cod.Vat.Gr. 23'

Ῥαμνούσιος, add '*Ῥαμνόσιοι SEG* 35.24 (Attica, v BC)'

ῥάμφος, before 'Plu.' insert 'Call.*fr.* 647 Pf.'

ῥᾱνίζω, add '*Suppl.Hell.* 975.4 (iii BC)'

***ῥάνσις**, εως, ἡ, (ῥαίνω) *sprinkling* of temple with wine, *SB* 9199.15, 16, 19 (ii AD); written ρεανσις *Stud.Pal.* 22.183.109 (ii AD); cf. °ῥάντης.

ῥαντήρ II, after '*sprinkler*' insert '(a vessel)' and for '(Adanda)' read '(Cilicia, i/ii AD)'

ῥάντης, delete square brackets in lemma; after '*sprinkler*' insert 'of priestly official or servant, *Gnomol.Vat.* in *WS* 11.230 no. 527, cf.'; after '(ii/iii AD)' insert '*BGU* 185.10 (ii A.D; ρεαντης, cf. °ῥάνσις)'; add 'as surname or title, Paus. 5.21.12'

***ῥαντοπόλιος**, ον, *having a sprinkling of grey hairs*, gloss on μεσαιπόλιος, Sch.Gen.Il. 13.361.

ῥαντός, add 'ῥαντά, τά, *beads sprinkled* on the hair, Sext.*Ca.* 1.11'

ῥάξ 1, after '*grape*' insert 'καλὰ ῥάξ *SEG* 32.806 (Tyras, *c.*575/50 BC)'

†**ῥάπα**, v. ῥάψ.

ῥᾰπίζω I, line 5, for 'ῥεραπισμένα .. 166' read 'ῥεραπισμένῳ νώτῳ Anacr. 112 P.'

ῥάπτης, add '*TAM* 4(1).132 (Rom.imp.), *Hesperia* 41.41 no. 33 (Corinth, v/vi AD), etc.; cf. form *ῥαπτήρ, Myc. *ra-pte, ra-pte-re* (pl.)'

ῥαπτός II, after '*worked with the needle*' insert 'perh. *embroidered*, ῥ. κερβικάριον *SB* 9834 b 6 (iv AD; sp. ῥεβτόν, for pronunciation cf. mod. Gk. ῥαφτός)'

ῥάπτρια, add 'Myc. *ra-pi-ti-ra₂*'

ῥάπτω, add 'Myc. pf. part. pass. *e-ra-pe-me-na*'

***Ῥᾱρία** (s.v. Ῥάρος), add '*IG* 2².1672.119, 120, 253 (iv BC)'

***Ῥάριος**, adj. included in alphabetical list, *SB* 10769.139 (iii/iv AD), perh. pr. n., cf. Ῥάρος.

ῥάσμα, add 'ὅσσας εἴχ' Ἀχιλεὺς νῆας τόσα ῥάσματ' ὀφείλεις *PCair.Zen.* 535.2 (iii BC)'

ῥάσσω 2, add '**b** *dash down*, in pass., Euph. 9.6 vGr.'

ῥᾳστωνεύω, delete '= ῥᾳθυμέω'; for 'dub. l. in Ael.*fr.* 281' read 'Ael.*NA* 16.23'

ῥᾰφᾰνίδιον, add 'also ρεφαν- Dura 125 no. 861'

ῥᾰφάνινος, add 'also ρεφάνινος, Moses Alch. 300.17'

ῥάφανος II, add '*AP* 9.520 (Alc.Mess.)'

***ῥᾰφᾰνόσπερμον**, *radish seed*, *SB* 10532.14-15, 24 (AD 87/8).

ῥᾰφιδᾶς, add '**2** *cobbler*, *AP* 11.288 (Pall.).'

ῥᾰφιδεύς, delete the entry (v. °ῥαφιδᾶς).

ῥᾰφιδοποιός, add '*BGU* 2351.6 (ii AD)'

ῥᾰφικός, ή, όν, = ῥαπτικός, Sch.D.T. 445.22.

***ῥᾰφίον** (or ?ῥαφεῖον), τό, ῥαφίῳ τῷ κεντηρίῳ ᾧ διακεντοῦντες οἱ τεχνῖται τῶν τοιούτων ἐπιτηδείους ὀπὰς τῇ τοῦ λίνου διέρσει παρασκευάζουσι Gal. 19.134.

ῥᾰφίς II, add '*IGC* p. 98 A23 (Acraephia, iii/ii BC)'

***ῥαχᾶς**, ᾶ, ὁ, dub. sens., as nickname, *SB* 7638.7 (iii BC), cf. *Ev.Matt.* 5.22.

ῥᾰχίζω I, add 'τὰ δερόμενά τε καὶ ῥαχιζόμενα τῶν ἱερείων Philostr.*VA* 5.42'   **II**, for '*play the braggart, boast*' read '*tell outrageous untruths*'

ῥάχις I 2, add 'for ἱερὴ ῥ. see ‡ἱερός IV 10'

ῥᾰχιστής II, for '*boaster, braggart*' read '*teller of tall stories, liar*' and after 'ῥαχιστήρ .. Hsch.' add '*SB* 9806.2 (iii AD)'

***ῥάχνη**, ἡ, perh. *cloak*, *POsl.* 161.5 (late iii AD), *PGen.* 80.7 (iv AD), *CPR* 8.65.6, 7, etc. (vi AD); cf. ῥάκανα.

***ῥαχνίον**, τό, dim. of °ῥάχνη, *POxy.* 2058.22 (vi AD).

ῥάχνος, delete the entry.

ῥᾱχός, line 3, for '*thorn-hedge*' read '*growth of bushes, thicket, brushwood*' and add '*PKöln* 144.24 (ii BC), *POxy.* 3558 (after AD 150)'; line 4, delete 'in Hdt. l.c. .. *wattled fence*'

***ῥάψ**, ῥαπός, *reed*, πότ τὰς ῥάπα[ς] κυλλάς *SEG* 27.650.4 (Camarina, ii BC), ῥάπα· τὴν καλάμην Hsch.

***Ῥαψώ**, οῦς, ἡ, name of goddess or nymph, *IG* 2².4547 (Phalerum, iv BC).

***ῥαψῴδημα**· ψεῦσμα, φλυαρία Hsch.

***ῥαψῳδοτοιοῦτος**, ὁ, *sort of rhapsode*, Strato Com. 1.48 K.-A.

Ῥέα, lines 2/3, delete 'Ῥεῖα δ' .. δμηθεῖσα'

†**ῥέανσις**, v. °ῥάνσις.

†**ῥεάντης**, v. °ῥάντης.

***ῥεγεωνάριος**, ὁ, Lat. *regionarius*, *police-officer* at Antioch in Pisidia, *JRS* 2.81.

ῥέζω (A), line 9, after 'Cleonae' insert 'prob. 575/550 BC'   at end add 'form perh. *Ϝόρζω (<*wrgyō) Myc. 3 sg. *wo-ze*, inf. *wo-ze-e*, part. *wo-zo*, pass. part. *wo-zo-me-no*'

ῥεθομᾰλίδας, for 'Alc. 150' read 'Aeol. acc. to Sch.Il. 22.68'

†**ῥέθος**, εος, τό, *face, countenance*, S.*Ant.* 529 (anap.), E.*HF* 1204 (lyr.),

ῥέκος·

Call.fr. 67.13 Pf., A.R. 2.68, Lyc. 173: Aeol. in this sense acc. to Sch.Il. 22.68, cf. Sapph. 22.3 L.-P. (dub. sens.), Theoc. 29.16. **2** in pl., perh. *nose and mouth*, ψυχὴ δ᾿ ἐκ ῥεθέων πταμένη Il. 16.856. **II** in pl., fr. misunderstanding of sense I 2, *limbs, body*, ῥεθέων ἐκ θυμὸν ἕληται Il. 22.68, cf. Theoc. 23.39.

*ῥέκος· ζώνη Hsch., Phot., Suid. (prob. a dialect form of ῥάκος).

ῥέκτης, add '**2** Dor. ῥέκτας, *priest*, prob. in *IG* 14.431.'

*ῥελατωρία, ἡ, perh. *receipt brought back after delivery, POxy.* 3125.6 (AD 325), cf. Lat. *relatoria*.

ῥεμβάς, add 'Heph.Astr. 2.21.28'

ῥέμβος, add 'one who qualifies for corn dole, perh. in rotation, after completion of public service, *POxy.* 2908 iii 37; 2927.7, 21; 2928 i 1'

ῥέμω, for 'Theognost.Can. 11' read 'Theognost.Can. 32 A.'

*ῥεπαρατίων, ωνος, ἡ, Lat. *reparatio, Stud.Pal.* 20.123.33 (v AD).

*ῥεποστώριον, τό, Lat. *repos(i)torium, stand for serving courses at meal, Pap.Lugd.Bat.* XIII 6.9.

*ῥεπούδιον, τό, Lat. *repudium, Cod.Just.* 1.3.52.15, *POxy.* 129.1 (vi AD).

ῥέπω **II**, line 1, after 'trans.' for 'only' read 'ῥ. τὸν νοῦν καὶ πείθειν ἑαυτόν Epict.fr. 8 (τρέπειν Gesner); otherwise'

*ῥέσκριπτος, ὁ, *rescript, SB* 9763.33 (v AD), fr. Lat. *rescriptus, -um*.

*ῥεφᾰνίδιον, v. ‡ῥαφαν-.

*ῥεφᾰνικός, ή, όν, = ῥαφάνινος, Zos.Alch. 184.8.

*ῥεφέκλα, ἡ, plant, perh. kind of cyclamen, Anon.Alch. 13.6.

*ῥεφερενδάριος, ὁ, Lat. *referendarius*, imperial official, *Cod.Just.* 1.15.2.1, 4.59.1.1.

ῥέω **I 1**, line 24, after 'also w. gen.' insert 'Pl.*Phdr.* 230b' **2**, add 'ῥέει φάτις prob. in Nic.*Th.* 484. **b** of plague, *spread*, Hp.*Ep.* 27. **c** εὖ ῥεῖν of enterprises, *prosper*, Thgn. 639, cj. in Sol. 13.34 W. **d** of time, *pass*, πολλοῦ ῥυέντος χρόνου Memn. 14.1 J., ῥέοντι .. χρόνῳ Just.*Nov.* 1.1.4.'

*ῥέωνος· εὔωνος †ἀνδρός Hsch., cf. Theognost.Can. 32 A.

ῥηγεύς, add 'also ῥᾱγεύς, *EM* 703.28'

*ῥῆγλα, ἡ, *part of wagon* (perh. Lat. *regula*, bar), *Edict.Diocl.* 15.13; ῥῆγλαι· σιδηρᾶ ὡς ῥάβδοι Hsch.; cf. °ῥηγλίον.

*ῥηγλίον, τό, *bar of gold, Edict.Diocl.* 30.1a, cf. Lat. *regula, regularis*.

*ῥηγλοχύτης, ου, ὁ, *ingot mould*, Anon.Alch. 322.24, 325.10, (ρυ-, ρι-); cf. °ῥηγλίον.

ῥηγμίν or -μίς, line 2, for 'neither form is found' read 'these forms only in Hsch.'

ῥήγνῡμι, line 3, after 'cf. 4.22' insert '*Gnomol.Vat.* in *WS* 9.185'; line 7, after 'ῥῆξα Il. 6.6' insert 'Aeol. εὔρηξε Alc. 179.2 L.-P.' **A 4**, add 'D. 9.61' **5**, line 4, after 'Lxx *Is.* 49.13' insert 'θυμὸν ἔρρηξας ib.*Jb.* 15.13'

ῥηκτός, add 'A.R. 3.848'

*ῥημᾰτίζω, *speechify*, A.fr. 78a.30 R.

ῥημάτιον, after '*phrase*' insert 'or *word*'

ῥήν, for 'ἡ' read 'ὁ' and delete '*sheep*' add 'cf. Myc. adj. *we-re-ne-ja* (fem.)'

*Ῥηνοπότης, ου, ὁ, *one who drinks the water of the Rhine*, verse oracle in *SEG* 31.851 (Ardea, AD 78/81).

*ῥῆνος, εος, τό, = ῥήν, dub. in Hsch. s.vv. πολύρρην and ῥήνεα.

*ῥήξ, ρηγός, ὁ, Lat. *rex, king* (in the west), Procop. 5.1.26, 6.14.38. **II** pl., name of infantry unit at Rome, id. 5.23.3.

ῥηξίχθων, for '*bursting forth from the earth*' read '*breaking* or *splitting the earth*' and add '(?)of a pig, Strato Com. 1.19 K.-A. (v.l. ἐρυσίχθ- in Ath. 8.382e)' add 'also ῥηξίχθιον *Suppl.Mag.* I 12'

ῥῆον, after '*rhubarb*' insert 'Androm.ap.Gal. 14.40'

†ϝρῆτά, ά, Cypr. for ῥήτρα, *treaty, agreement*, acc. pl. *we-re-ta-se* ϝρῆτας *ICS* 217.28.

†ϝρητάομαι, *make an agreement*, Cypr. aor. *e-(u-)we-re-ta-sa-tu* ἐϝρητά-σατυ, *ICS* 217.4, 14.

*ῥητορεῖον, τό, perh. *prize for rhetoric*, Anon.*in Rh.* 98.11, 12.

ῥητορικός, add '**4** ῥ. λεξικόν, *Lex.Rhet.* ap.Eust. 200.1, al.: pl., id. 1921.57.'

ῥητός **I 1**, line 7, after 'Plb. 32.6.7' insert 'ἐκ τῶμ μὴ ῥητῆι (sc. ἡμέρᾳ), perh. *from those which do not fall on a stated* day of sacrifice, *IG* 2².1357a.25 (403/2 BC), cf. Hsch. s.v. ῥητήν' and after 'ῥητόν, τό' insert '*compact, treaty, Inscr.Cret.* 4.197.17 (τῶι .. ρη]τῶι ii BC)'; at end for 'so perh. ἀπὸ ῥητῶν .. p. ix' read 'ἀπὸ ῥετὸν perh. *in the aforesaid manner, Hesperia* 33.385 (Eretria, v BC)'

ῥήτρα, after 'Elean ϝράτρα' insert 'Cypr. ‡ϝρῆτα' **I**, add 'ἐπὶ ῥήτρῃσι λαβεῖν *receive on conditions*, Call.fr. 85.6 Pf.' **II 4**, after 'Cyr. 1.6.33' insert '*Meiggs-Lewis* 8.2 (Chios, vi BC)'

ῥηχμός, add '(= *SEG* 36.336, iv BC) perh. to be interpr. as place-name'

ῥῑγέω, line 3, delete 'Dor. 3 pl. ἐρρίγαντι Theoc. 16.77'

ῥῑγοπῠρετίον (s.v. ῥιγοπύρετος), add '*PCair.* 10263.16 (iv/v AD), *Suppl.Mag.* I 23'

*Ῥίεια, τά, *festival* of Poseidon *at Rhium* in Locris, *IG* 4.428.10 (Sicyon).

ῥίζα **I**, line 3, after '*root*' insert '(sts. extended to include *stock*)'

and add 'τὰς ἀκάνθας δὲ αὐτοῦ (sc. ῥόδου) τὰς παρὰ τὴν ῥίζαν Luc.*Hist.Conscr.* 28' add 'form *ϝρίζα, Myc. *wi-ri-za*, of wool'

*ῥιζάριον, τό, *small root*, Moses Alch. 306.5.

ῥίζις, delete 'of the elephant kind'

*ῥιζοκάλᾰμος, ὁ, *reed*, αὐξηρῶν δονάκων μεγάλων ῥιζοκαλάμων Sch. Nic.*Al.* 588b Ge.

ῥιζοκέφᾰλος, for '*of which .. root*' read '*having a bulbous root*'

*ῥιζοκρίκι(ο)ν, τό, app. *ring attached to the base*, μοχλὸν .. σιδηροῦν ῥιζοκρίκιν ἔχοντα Poliorc. 254.1.

*ῥιζοποιός, όν, *engendering roots* (or, for ῥοιζο-, *hissing*, or, for ῥυζο-, *snarling*), magical formula in *Suppl.Mag.* II 96 A 27, E ↓3 (both sp. ῥιζοπυέ), ib. F fr. A5 (sp. ῥυσοπηε).

ῥιζοφυής **II**, add 'ῥιζοφυῆς Μουσέων πτόρθος *SEG* 36.975 (Halic., late Hellen. epigr.)'

*ῥιζωτήρ, ῆρος, ὁ, *causing to strike roots*, epith. of the sun, but prob. f.l. for ῥοιήτωρ, Orph.*H.* 8.6.

ῥικνός, after 'ή, όν' insert '(cf. pers. n. ϝρικνίδας, *SEG* 36.341, Argos, vi BC)'

ῥικνώδης, add '**II** *dancing in a twisted, contorted manner* (cf. ῥικνόομαι II), of Dionysus, *AP* 9.524.18.'

ῥίκνωσις, add 'Gal. in *CMG* 5.10(2).2.172.22, 173.8'

ῥίμμα, add '**2** *outcast*, Sch.E.*Hec.* 1076.'

ῥῑνάω (B), add 'cf. καταρρινάω'

*ῥινεστήρ, ῆρος, ὁ, prob. *halter*, PTeb. 886.68 (ii BC).

ῥίνη **I**, add 'perh. also ῥίνη̄ *SEG* 30.1148 (Capua, 460/50 BC)' **II**, add 'ῥίνα *IGC* p. 98 A22 (Acraephia, iii/ii BC)'

ῥινηλάτης, for 'Poll. 2.74' read '*Trag.adesp.* 426 K.-S. (Poll. 2.74)'

ῥινοβάτος, add '*IGC* p. 98 A20 (Acraephia, iii/ii BC)'

ῥινόκερως **2**, delete the section transferring quots. to section 1.

ῥῑνός **I**, line 1, before 'Od. 5.426' insert 'pl.'; line 2, after 'etc.' insert 'Nic.*Th.* 429' add 'form *ϝρινός, Myc. *wi-ri-no*, adj. *wi-ri-ni-jo*'

ῥινοῦχος, delete '(ῥίς II)'

†ῥινωτηρία, ἡ, some part of a ship, Poll. 1.86.

ῥίον **2**, place-name, add 'Myc. *ri-jo*'

*ῥιπαρία, ἡ, *office* or *function of riparius, POxy.* 2032.50, *PMasp.* 287 iv 30, *CPR* 10.48.3 (all vi AD).

*ῥίπηεις, εσσα, εν, *stormy*, prob. in PMich.inv. 4926a (ZPE 93.157).

†ῥιπίδιον, τό, *small bellows* or *fan*, PMich.XIV 680.7 (iii/iv AD), *CPR* 10.7.45, Hdn.*Epim.* 118. **2** μέρος τῆς νεώς Hsch.

ῥιπίς **III**, add 'Ael.*NA* 13.10 S.'

*ῥῖπος, for 'ἀχύρων ῥ. .. 1.86, al.' read 'as measure of straw, ἀχύρων ῥ. *SEG* 9.11, 12, 15, 28, cf. ib. 35.1834'

ῥιπτάζω, line 6, for 'ὕφη γυναικῶν .. ἐρριπτάζετο S.fr. 210 iii 12' read 'ὕφη γυναικῶν ἀνδ[ρό]ς ἐρριπτ[ά]ζετο S.fr. 210.68 R.'

ῥίπτω, line 12, for 'ῥερίφθαι' read 'ῥερῖφθαι'

ῥίς **II**, for 'brow .. *spur of land*' read 'perh. some kind of canal, cf. *SEG* 37.758'

ῥισηγέτης (s.v. ῥισῆς), add 'written ῥισιγηγήτ- *SB* 5246.3 (iii/ii BC)'

ῥίσκος, add 'see also °ἐρίσκος'

*ῥισκοφῠλάκιον, add '*PSI* 858.30 (iii BC)'

ῥίψασπις, for '*craven*' read '(as a sign of cowardice, freq. in fig. contexts)' and add 'Lys. 11.5'

ῥιψόφθαλμος, after '*casting the eyes about*' insert 'in a covetous or lustful manner'

ῥόα, line 1, before 'ῥοιά' delete 'later' and after it insert 'Ar.*Pax* 1001' **II 2**, for '*knob .. pomegranate*' read '*ornament in the form of a pomegranate*' and delete '*tassel* .. ῥοΐσκος'

*ῥόγα, ἡ, (*money) allowance*, τὴν ῥόγαν τῶν λιμιταναίω[ν καὶ τῶν στρ]ατιωτῶν *SEG* 32.1554 (Arabia, vi AD), *CPR* 8.74.3 (vii AD); *rations, PMasp.* 76.4, 145.4 (vi AD); cf. °ῥογεύω and *Eranos* 89.121-2.

*ῥογεύω, *pay in kind*, cf. Lat. *erogare*, οἶνον *PMasp.* 76.8, 145.1, etc. (vi AD), Just.*Nov.* 130.1.

ῥόδεος **I**, after 'of roses' insert 'κάλυκες h.*Cer.* 427' **II**, after 'rosy' insert 'χείρ *AP* 9.745 (Anyt.)'

*Ῥοδιασταί, οἱ, *devotees of the goddess* °Ῥόδος, *IG* 12(1).157 (Rhodes, i AD).

ῥοδίζω **II 2**, for this section read '*decorate* the grave of a person *with roses*, at the Rosalia, *INikaia* 1422; pass., ib. 95'

ῥοδινοπόρφυροῦς, â, οῦν, read 'ῥοδινοπόρφυρος, ον (v. *Tyche* 6.234)'

ῥόδινος **I**, for 'Anacr. 83' read 'Anacr. 89.2 P., Stesich. 10.3 P. (pl.)' **II**, add '**b** subst., ῥ. ἡ, kind of hyacinth stone, Epiph.Const. in *Lapid.Gr.* 196.25.'

*?ῥο]δῐνόχρωα, perh. *rose-coloured*, PNag.Ham. 3.8 (iii/iv AD).

ῥοδισμός, for '*ceremony*' read '= Lat. *Rosalia, a festival*'

ῥοδοδάκτῠλος, add 'epith. of Helios, *SEG* 34.1294 (Phrygia)'

ῥοδόεις, add 'form *ϝροδόϝεν, Myc. *wo-do-we* (neut.)'

ῥόδον **I 1**, add '**b** flower of the ῥοδοδάφνη, Luc.*Asin.* 17. **c** of gold, etc., in a necklace, *IG* 1³.360.4 (v BC), 2².1376.6 (v/iv BC).' **II**, add '*BCH* 24.306, 60.337 (Philippi)'

Ῥόδος, add 'as a divinity, *SEG* 23.547 (Crete, 201/0 BC)'

Ῥοδοσκάρφα, delete the entry.
***ῥοδοφορία**, ἡ, *ceremony involving carrying of roses*, *POxy.* 3694.6 (AD 218/25); cf. ῥοδοφόρια.
ῥοΐδιον, after 'ῥοΐδιον' insert 'unless this is dim. of ῥοῦς, *sumach*'
ῥοιζέω, lines 5/6, after 'Arist.*HA* 535ᵇ27' insert 'of beetles, Hippon. 92.10 W.' add '**2** used Aq.*Ge.* 31.51 for *set up* (by mistake due to the ambiguity of the Hebr. wd.).'
ῥοιζήτωρ, for '(Orph.*H*.) 8.6' read 'cj. for °ῥιζωτήρ ib. 8.6'
***ῥοῖον**, τό, app. *stream*, (cf. ῥόος, ῥοία), *SEG* 23.236.3 (Phizalea, iii BC) (unless to be read as the first letters of a pr. n.).
***ῥοισπις**, wd. occurring in list of bird names, *PAmst.* 13.11 (V AD).
ῥόμβος A I, for this section read '*bull-roarer*, instrument whirled on a string to make a whirring noise, used in the worship of Rhea and Dionysus, E.*Hel.* 1362, Ar.*fr.* 315 K.-A., Archyt. 1, Diog.Ath. 1.3 S., Orph.*fr.* 34; described, Sch.Clem.Al.*Protr.* 2.17.2, cf. Hsch.; used in love-magic, Theoc. 2.30 (but expld. by sch. as = ἴυγξ 1), cf. Luc.*DMeretr.* 4.5, Propertius 2.28.35, Ov.*Am.* 1.8.7. **2** perh. *whipping-top*, *AP* 6.309 (Leon.Tarent.), M.Ant. 5.36, Sch.A.R. 1.1134, cf. 4.144.' **II**, line 6, after 'Orph.*H*. 8.7' insert 'meton., θύννων ῥ. *AP* 6.33.3 (Maec.)' **B**, add '**5** = ψωλή, *PLond.* 1821.164 (vi AD). **6** a cavalry formation, Ael.*Tact.* 19.5, Arr.*Tact.* 17.'
ῥομβωτός, for 'J.*AJ* 12.2.10' read 'J.*AJ* 12.2.9'
ῥόος, line 1, for '*Inscr.Cypr.* 135.19 H.' read 'acc. *ro-wo*, *ICS* 217.19'; line 3, after '*Peripl.M.Rubr.* 46' insert 'nom. pl. ῥόες (ρoσσ lapis) *GVI* 1684.9 (Chersonesus, i/ii AD), perh. acc. pl.: εἰς ῥοῖς ὕδατ(ος) *SB* 10549.8 (AD 204)'
ῥοπή I 2 a, add 'καὶ νόσο ῥοπᾶι *SEG* 39.1019 (Selinous, V BC)'
ῥόπτον, for '*operating table*' read 'something to which a patient is tied for an operation'
†**ῥόπτρον**, τό, (ῥέπω) part of a trap, consisting of heavy piece which falls on the victim when dislodged (cf. Eng. deadfall), Archil. 186 W., Poll. 7.115; fig., δίκης ἔπαισεν αὐτὸν ῥ. E.*Hipp.* 1172. **II** *knocker* on door, E.*Ion* 1612, Ar.*fr.* 40 K.-A., Lys. 6.1, X.*HG* 6.4.36. **2** colloq., *penis*, Hsch. **III** in pl., an instrument consisting of two small cymbal-heads struck together by shaking, used esp. by Corybantes, *rattle*, *clapper*, or sim., Corn.*ND* 30, Luc.*Trag.* 36, Orph.*frr.* 105b, 152, *AP* 6.74 (Agath.), Nonn.*D.* 9.116, al. **2** as part of a Parthian percussion device used in battle, ῥ. βυρσοπαγῆ καὶ κοῖλα περιτείναντες ἠχείοις χαλκοῖς Plu.*Crass.* 23.
***ῥοσικόν**, τό, commodity, perh. cognate of ῥούσιος, *red pigment*, *rouge*, *BGU* 2357 ii 16 (iii AD); unless for ῥωστικόν, i.e. *strengthening agent*).
***ῥόσσα**· ἡ κίχλη Theognost.*Can.* 102.
ῥοταρία, add 'cf. late Lat. *rotarius*, "of a wheel"'
ῥούδιον, add '**II** v. ‡ῥοΐδιον.'
ῥοῦς I, add '**3** shrub related to the sumach, *Coriaria myrtifolia*, Plin.*HN* 24.91.' add '**III** = ῥόον (q.v.), *dried mulberry fruit*.'
ῥούσιος I, add 'στρῶμα ῥ. *PVindob.* G25737 (*ZPE* 64.77, vi/vii AD)'
***ῥοφεῖον**, τό, item in list of kitchen vessels, Lang *Ath.Agora* XXI B12 (pl., iv/iii BC).
ῥοφέω, line 8, for 'Hippon. 132' read 'Hippon. 165 W. and is freq. in the best codd. of Hp.; also Dor., Sophr. in *PSI* 1214d 2 (κατα-)'
***ῥυγή**, ἡ, *growling*, *snarling*, Trag.adesp. 653.9 K.-S.
ῥύγχος, after 'τό' insert '(acc. ῥύ]γχον *IEryth.* 203.4 (iv BC))'
ῥύδον, after 'Od. 15.426' insert 'ῥ. ἀφνύνονται Call.*fr.* 366 Pf.'
***ῥύζα**, ἡ, (ἐρύω A), *drawing of a bow*, Hsch.
ῥυθμίζω II, after '(Arist.)*Spir.* 485ᵇ2' insert '*Cod.Just.* 6.48.1.26 (AD 528/9)'
***ῥυθμοποιός**, ὁ, = ὁ μέλη καὶ ῥυθμοὺς ποιῶν Hsch.
ῥυθμός VI, add '*Peripl.M.Rubr.* 6'
ῥυκάνη, after '*plane*' insert '(tool)'; add 'Varro *LL* 6.96'
***ῥυκάνισμα**, ατος, τό, *plane-shaving*, *Poliorc.* 223.18.
ῥῦμα (B), after '*protection*' insert 'τούτους ηὐξήσατε ῥύματα δόντες i.e. *bodyguards* Sol. 11.3 W.'
***ῥυμβάδας**· λάϊγγας τὰς διεσχισμένας Hsch.
***ῥυμεῖον**, v. °ῥυμίον.
***ῥύμιγξ**· χείμαρρος Hsch.

***ῥυμίον**, τό, dim. of ῥύμη II, *lane*, *alley*, *PMeyer* 20ʳ.5 (iii AD); also ῥυμεῖον, ἐν τυφλῷ ῥυμείῳ *SB* 9902 A i 25, ii 1 (iii AD).
ῥύμμα II, for 'Sch.Nic.*Al.* 95' read 'Nic.*Al.* 96'
ῥυμός I 2, line 2, for '*log* .. fuel' read 'perh. *spit* or other instrument used in roasting sacrificial victims' **II**, for '*trace*' read '*rein*' **IV**, for 'perh. *shelf* or *row*' read '*steelyard*, *IG* 1³.386.21, al, 387.40; meton., *amount weighed at a time*, *beam-load*'
ῥυπαρός 2, add 'λόγου μόριον D.H.*Comp.* 12' **3**, for 'of coins .. alloy' read 'of payments including the extra charge' and add '*BGU* 2554.6' **4**, for 'prob. = ἀδειγμάτιστος' read 'of grain, *unpolished*' and delete 'unwinnowed'
ῥυπάω, add 'transf., of moral turpitude, Just.*Nov.* 80.8'
***ῥυποπώλης**, ου, ὁ, *seller of rubbish*, Sch.Ar.*Pl.* 17; cf. γρυτοπώλης.
ῥύπος I 2, add 'Ar.*fr.* 931 K.-A.'
ῥυσαλέος, add 'γραῖα *GVI* 1185.1 (Palestine, ii/iii AD)'
ῥυσίπτολις, after 'ῥυσίπολις' insert 'A.*fr.* 451q.7 R. (lyr.)'
ῥύσις I, add '**b** fig., τῆς λέξεως D.H.*Dem.* 40, cf. *Comp.* 23.' **III**, after 'Math.' insert 'ῥύσει σημείου συνίσταται γραμμή Ph. 1.23'
ῥυστάζω, line 3, after 'Il. 24.755' insert 'Call.*fr.* 588 Pf.'
ῥυστήρ II 2, for this section read '= ἀρυστήρ, *ladle* of irrigating-machine (shadoof), prob. in *Aegyptus* 6.191 (pap. ρηστηρ)'
ῥύτειρα (A), ἡ, fem. of ῥυτήρ (A), Ἄρταμι ῥ. τόξων (?)Alcm. 170 P.
†**ῥύτειρα (B)**, ἡ, fem. of ῥυτήρ (B), Suid.
ῥυτήρ (B), before 'ῥυτῆρες' insert 'δήμου ῥ. *AA* 21.38 (Miletus, ii BC)'
ῥυτός II, line 4, after 'Plu.*Alex.* 67' insert 'prob. in Phld.*Rh.* 2.54 S., where the ν of τόν may be the initial of the lost wd.'
***ῥυτοφιάλιον**, τό, *small φιάλη designed to facilitate pouring*, *Ath. Asklepieion* v 71 (iii BC).
ῥύτωρ (A), add 'ῥ. τόξου the constellation *Sagittarius*, Arat. 301'
ῥωβίδας, for '*of less .. old*' read '*in the first year of public education*, i.e. seven years old'
ῥωγάς, for 'ὁ, ἡ' read 'fem. adj.' add '**II** as subst., ἡ, *cloven rock*, Nonn.*D.* 3.56, 10.175 (cod.).'
***ῥωθῶς**· σφόδρα Theognost.*Can.* 68 A.
***Ῥωμαία**, v.s.v. ‡Ῥωμαῖος.
***Ῥωμαϊκόν**, τό, *some article of clothing*, *PRyl.* 627.18 (iv AD).
Ῥωμαϊκός, line 1, after 'al.' insert 'ἀρυσᾶς .. -οῖς γράμμασιν ἐπιγεγραμμένος *Inscr.Délos* 442*B* 139 (ii BC)' add '**2** *Roman* (i.e. civil opp. ecclesiastical), 'Ρ. ἐλευθερία Just.*Nov.* 144.2.4.'
***Ῥωμαῖον**, τό, *temple of Roma*, Milet 7 p. 17 (i BC).
Ῥωμαῖος, after 'Plb. 10.36.3, etc.' insert 'applied to Roman citizens in other parts of the Empire, *SEG* 29.690 (Histria), etc.'; τὰ 'Ρ., add '*SEG* 33.1039 (Cyme, ii BC), 33.644 (Rhodes, i BC), etc.)'
Ῥωμαϊστής, add '*SEG* 31.1535 (Philae, ?i AD)'
ῥωμαλέος 2, add '**b** *strengthening*, of food, Diocl.*fr.* 133 (comp.).'
***ῥωμαλεότης**, ητος, ἡ, *physical strength*, Rh. 3.599.15.
***Ῥωμανήσιος**, ον, = Lat. *Romanensis*, archit., *of kind of construction*, *Cod.Just.* 8.10.12.5 (sp. 'Ρωμανίσιος).
Ῥώμη, add '**2** ἡ νέα 'P. i.e. Constantinople, Just.*Nov.* 70.1; Βυζαντιὰς 'P. *APl.* 56.1; ἑκατέρα 'P. Just.*Nov.* 81.1.'
ῥώννυμι II 3, lines 4/5, after 'D. 18.152, 19.248' insert 'τῇ θυσίᾳ Men.*Dysc.* 264, 520'
†**ῥῶπαες**, wd. glossed by ἀλσώδες, Arc. 143 (perh. for ῥωπᾶεν, v. ῥωπήεις).
ῥωπεύω I, for '*AP* 6.226' read 'ῥωπεύειν· ξυλεύεσθαι Suid.'
ῥωπογράφος, for '(ῥῶπος) .. *masters*' read '(ῥώψ A) *painter of forest scenery*'
ῥωποπώλης, line 3, after 'ῥοπο-' insert 'cf. Lxx 3*Ki.* 10.15 (v.l. for ἐμπόρων in A.)'
***ῥώς**, perh. = ῥῶσις, *strength*, poet. in *Pap.Lugd.Bat.* XXV 1.6.
***ῥωσιτάριον**, τό, wd. in list of dyestuffs, perh. = ῥωστήριον, *POxy.* 1922.4 (V AD).
†**ῥωσκομένως**, adv. fr. part., perh. of Ion. *ῥώσκομαι (related to ῥώομαι or ῥώννυμι), *with violent motion*, Hp.*Cord.* 1 (nisi leg. θρωσκομένως).
***ῥωστέον**, *one must strengthen*, Archig.ap.Gal. 13.167.16.
†**ῥωστήριον**, τό, glossed by παρορμητικόν, Hsch.; cf. Phot.
ῥωχμή, for '*fissures*' read '*wrinkles*'

Σ

σ' **II**, add 'τἀμὰ καὶ σ'' id.*El.* 273'
Σαβαζιασταί, add '*SEG* 33.639 (Rhodes, *c.*100 BC)'
Σαβάζιος, lines 4f., before '(Ar.)*Lys.* 388' insert 'χὠ τυμπανισμὸς χοἰ πυκνοὶ Σαβάζιοι *cries of "Sabazios!"*'; for 'θεῷ .. *CIG* 3791' read 'θεῷ Σαβαζίῳ Παγισαρανῷ *TAM* 4(1).79'; add 'Διὶ Σαουαζίῳ *INikaia* (AD 122/3)'
***Σαβαθικός**, app. = Σαβάζιος, *GVAK* 8 (Bithynia, i BC/i AD).
***σάβαθον**, v. °σάμβαθον.
+**σᾰβᾰκός**, ή, όν, *disabled, debilitated*, Hp.*Morb.* 1.31, σαβακῶν σαλμακίδων *AP* 7.222 (Phld.), Hsch.
σᾰβάκτης, read '**Σᾰβάκτης**' and for '*shatterer .. pots*' read '*Shatterer, one of a number of demons supposed to be active in a pottery*'
σάβανον, after '*towel*' insert '*Edict.Diocl.* 28.57, al.'
σαβαύτια, add '*PMich.*inv. 419.6 (*ZPE* 31.176, iii/iv AD)'
Σάβος, add '**2** place dedicated to the service of Sabazios, Sch.Ar.*Av.* 874, Harp. s.v. Σάβοι.'
***σᾰβουρᾶτος**, ον, Lat. *saburratus, filled with ballast*, σκάφος *PMerton* 46.10 (vi AD).
σάβουρος, for '*without ballast*' read 'perh. *in ballast, empty*'; add 'cf. Lat. *saburra, ballast*'
σάβυττος I, after '*hair*' insert 'ἐξυρημένος σαβύττους Eup. 313 K.-A.'
+**σᾰγάλινος**, η, ον, *of teak*, ξύλων σ. cod. *Peripl.M.Rubr.* 36.
σάγανα, add '*PPrag.* I 90.10 (vi/vii AD)'
+**σαγγαικόν** (or ?-αϊκόν), τό, app. some ritual garment, σ. βεβαμμένον κόκκινον ἐν κιβωταρίῳ *Inscr.Délos* 1416*A*58, 1417*B*i60, 1452*A*40, (perh. shoe, cf. σαγγάριος, τζάγγη).
σᾰγεσφ[όρος], for 'dub.' read 'prob. *mule*'; add 'cf. °σαγηφόρος'
***σᾰγήνιον**, τό, dim. of σαγήνη, net, *ICilicie* 108 (v/vi AD).
***σᾰγηφόρος**, ὁ, beast of burden, prob. *mule*, *AAWW* 1948.322/3 (epigr., Panopolis, Rom.imp.); cf. ‡σαγεσφ[όρος].
σᾰγίον, add '*saddle cloth* (for camel), *PMich.*xv 717.1 (iii AD)'
***σαγιττάριος**, ὁ, Lat. *sagittarius*, *SEG* 34.1598 (Egypt, AD 323), *SB* 4223.8 (pl., iv AD), *BCH* suppl. 8.63 (Maced., v/vi AD).
σάγμα II, *pack-saddle*, add '*PRyl.* 562.30 (250 BC)'
***σαγματικός**, ή, όν, *of* or *for a pack-saddle*, βελόνη *Edict.Diocl.* 16.10.
σαβαρυγά, add 'cf. °σαιθάρυγξ'
σάθων, add 'title of work by Antisthenes against Plato, whom he so nicknamed, Ath. 5.220d, D.L. 3.35'
***σαιθάρυγξ**· ἡ ταραχή Theognost.*Can.* 33 A.; cf. σαθαρυγά.
***σαῖνα**· τὸ αἰδοῖον Theognost.*Can.* 33 A.
***σαῖνος**· ὁ ἀριστερεών Theognost.*Can.* 33 A.
σαιστός, add 'ἐλαίας κλάδος Theognost.*Can.* 33 A. (σαῖστος codd.)'
σᾰκεσφόρος I, after 'Ajax' insert 'B. 13.104 S.-M.'
σᾰκηφόρος, for '*Supp.Epigr.* 4.522' read '*IEphes.* 293, 1250'
***σάκκαρος**, ὁ, perh. item of military equipment, *SB* 7181*B*36.
σακκηγία, add '*PUps.Frid* 1.12 (iii AD)'
***σακκοπάθνιον**, gloss on χλιδός, Hsch., prob. *nose-bag*, cf. φάτνη, φατνίον.
σακκοπλόκος, add '*SEG* 19.657 (Caria, Rom.imp.)'
***σακκοποιός**, ὁ, *sack-maker*, *POxy.* 3642.28 (ii AD), *PKlein.Form.* 124 (vi AD).
***σακκορ(ρ)ᾰφος**, η, ον, *used for sewing sacks*, βελόνη *Edict.Diocl.* 16.10 (σαρκο-), *SB* 7181*B*12.
σάκκος II 1, for '*Ostr. .. al.*' read '*of grain*, *OWilck.* 1091, 1096, 1101 (all ii AD); *of wine*, *OMich.* 249 (iv AD)'
***σάκκουν**, τό, (or -ους, ὁ), perh. article of women's clothing made of hair, *Pap.Lugd.Bat.* xxv 49.9, *PHamb.* 223.2 (ii AD).
σακκοφορικός, add 'subst., σ., τό, tax paid by σακκοφόροι, *PGoodsp.Cair.* 14.7 (iv AD), *POxy.* 3395.13 (AD 371)'
σάκρα, delete the entry.
***σάκρος**, α, ον, Lat. *sacer, sacred* used w. ref. to the imperial household, ἐν σάκρῳ αὐδιτορίῳ Just.*Nov.* 50 pr.; fem. pl. subst., *imperial archives*, *PSI* 481.13 (v/vi AD).
***σάκωμα**, v. ‡σήκωμα.
+**σᾰλᾰγέω**, *agitate vigorously*, colloq., of man in sexual intercourse, Luc.*Alex.* 50 (verse); in some manifestation of mourning, σαλαγεῖ (cod. σασαλ-)· θρηνεῖ Hsch.; cf. mod. Gk. σαλαγῶ *drive a flock by shouting*, σάλαγος of the noise so made.
σᾰλᾰγη, delete '(cf. σελαγή)'

*(?)**σαλακᾶς**, ᾶτος, ὁ, perh. = σαλάκων, *POxy.* 3617.11 (iii AD).
***σαλακονδεῖτον**, τό, Lat. *sal conditum, spiced salt, Edict.Diocl.* 3.9.
***σαλακωνηδα**, wd. occurring in list of bird-names, *PAmst.* 13.7 (v AD).
***σαλαμακάπηλος**, ὁ, = °σαλγαμάριος, *SB* 11077.13 (iv/v AD).
Σᾰλᾰμίνιος I, add '**2** in Cyprus, *IG* 1³.113.1, etc., Cypr. Σελαμίνιος (*se-la-mi-ni-o-se*) *ICS* 166.2, 338.'
Σᾰλᾰμίς, delete 'in Gramm.' and after 'Σᾰλᾰμίν' insert 'Eun.*VS* p. 494 B.'
σαλάριον, add 'Modest.*Dig.* 27.1.6, 11'
+**σᾰλάσσω**, *shake, make unstable*, Nic.*Al.* 457, σεσαλαγμένος οἴνῳ *APl.* 306.1 (Leon.Tarent.); *AP* 6.56 (Maced.Thess. II), 11.57 (Agath.).
***σαλάχιον**, τό, prob. = σελάχιον, *PMich.*II 127 ii 32, *Gloss.*, cf. mod. Gk. σαλάχι.
***σαλγαμαρικός**, ή, όν, *of a pickle-seller*, ἐργαστήριον σ. *CPap.Jud.* 511.11 (vi AD), *PRoss.-Georg.* III 38.11 (σαλκ-, vi AD).
***σαλγαμάριος**, ὁ, Lat. *salgamarius, seller of pickled foods*, *Corinth* 8(3).540, 551 (v/vi AD).
***σάλγαμον**, τό, *pickling material*, *PBeatty Panop.* 2.246, 286 (AD 300).
σᾰλευτός, add 'cj. in Procl. *in R.* 1.103 K.'
σᾰλεύω, line 11, after '*Si.* 28.14' insert '*disturb* a tomb, *ABSA* 51.148 (Caralitis, ii AD), etc.' **II 3**, for 'with the hip-joints far apart' read 'with a displaced hip-joint'
***Σάλιος**, ὁ, *member of the priesthood of Salii*, D.H. 2.70.1, 3.32.4, *IGRom.* 4.960 (i AD), Luc.*Salt.* 20.
***σαλκαμαρικός**, v. °σαλγαμαρικός.
***σαλοῦσα**· φροντίζουσα Hsch.
σαλπιγγωτός, add '*SB* 9319 (AD 116), 8745-7 (*c.*AD 170), etc.'
***σαλπιγκτήρ**, ῆρος, ὁ, = σαλπιγκτής, Hsch. s.v. ὀτρυντήρ.
σάλπιγξ III, for 'στρόμβος 2' read 'στρόμβος 3'
***σαλπίζηνος**, ὁ, name of a kind of opal, Socr.Dion.*Lith.* 38 H.-S.
σαλπίζω, line 7, after '*An.Ox.* 4.325' insert 'Cret. -ίνδω, inf. -ίνδε[ν] *Inscr.Cret.* 4.146.9 (Gortyn, iv BC)'
σαλτάριος, add 'also σαλτουάριος *TAM* 5(1).616 (iii AD)'
+**σαλώτιον**, τό, unidentified comestible, pl. σαλώτια δώδεκα *PRyl.* 172.15; sg., perh. *OBodl.* i P295.9. **2** item in list of vessels and implements, *POxy.* 3060.8 (ii AD), *SB* 1.25 (iii AD).
+**σάμαθον**, v. °σάμβαθον.
***σάμβαθον**, τό, kind of earthenware jar and corresponding liquid measure, σάμβαθον βωριδίων *POxy.* 2728.33, 2729.9 (iii/iv AD); also σάμφατον *PSI* 1423.23 (iv AD), σάμαθον *POxy.* 1290.1 (v AD); σάβαθον *PHaun.* 21.4 (iii/iv AD).
σαμβύκη I 1, for this section read '*small arched harp*, Aristox.*fr.* 97 W., Neanth. 5 J., Plb. 8.4.11, (?)Juba 15 J., Str. 10.3.17, Plu. 2.827a, Ath. 14.633f, Aristid.Quint. 2.16; var. form ζαμβύκη, see also °ἰαμβύκη'
σαμβυκίστρια, add '*PHib.* 270.1 (iii BC)'
***σαμκάμυκος**, ον, app. designation of kind of cloth, τῶν ἑκατὸν λίνων Σινυραιτικῶν σαμκαμύκων *CPap.Jud.* 414 (AD 21).
***Σαμναῖος**, ὁ, epith. of Apollo, *SEG* 14.688, 690.62 (Caria, ii/i BC); as an ethnic, Kontorini *AER* 2 p. 19 no. 17 (Rhodes, ii AD).
Σᾰμόθραξ, add '*SEG* 33.644 (i BC)'
***σάμφατον**, v. °σάμβαθον.
σάμψουχον, line 3, before 'Aret.' insert 'Nic. l.c., id.*fr.* 74.53, *AP* 4.1.11 (Mel.)'
***σάμψουχος**, ὁ, = σάμψουχον, *PColl.Youtie* 4.4 (iii AD).
***σάμψῡχος**, ή, = σάμψουχον, Hsch. s.v. ὑσωπίς.
***σανάπαι**, Thracian for μέθυσοι, Hecat. 34 J.
σανδάλιον III, add '(so perh. σανταλιν *SB* 7635.14, v/vi AD)'
σανδάλινος, for 'Hippon. 18' read 'Hippon. 32.5 W.'; add 'Herod. 7.125'
σανδᾰλοθήκη, add '*Inscr.Délos* 1450.139, 1451*A*38 (ii BC)'
σάνδᾰλον I, for 'Sapph. 98' read 'Sapph. 110(*a*).2 L.-P.'; add 'Call.*fr.* 631 Pf.'
***σανδάσηρος**, ὁ or ἡ, *Lapid.Gr.* 208, 1179; cf. *adfert aliquando errorem similitudo nominis sandaresi. Nicander sandaserion vocat*, Plin.*HN* 37.102.
σανδύκιον, add '*Stud.Pal.* 20.96.10'
σάνδυξ II, after '*casket*, Id.' insert 'cf. σενδούκη, -ουκι, Sch.Ar.*Pl.* 711, 809'

272

σᾰνῐδώδης, add 'ξύλον σ. κάταργον, of the ξόανον of Hera at Samos, Call.*Dieg.* iv.27 (fr. 100 Pf.)'
σᾰνῐδωτός, for 'al.' read '*Inscr.Délos* 1403*B*bii33, 1417*A*ii55 (ii BC), Hsch. s.v. φατνωτῶν, *Gloss.*'
σᾰνίς 3, for 'ship's deck' read 'in the hull of a ship'
σᾰνίσκη, for 'Herod. 4.36' read 'Herod. 4.62'
σαννάδας, delete the entry.
*σαννάς, άδος, ἡ, *wild goat*, Hsch., Cret. acc. to Polemon ap. Sch.Hippon. 118 W.
†σαννᾶς, οῦ, ὁ, *clown, buffoon*, Cratin. 489 K.-A.
*σαντᾰλι(ο)ν, v. °σανδάλιον.
†σαόπτολις, v. °σώπολις.
σάος, at end delete 'cf. σάως'
*σάπημα, ατος, τό, *decay*, οὔλων *Lapid.Gr.* 182.19 (pl.).
*σᾰπουνᾶς, ὁ, *soap-merchant*, *IGLBulg.* 105 (vi AD).
σαπρία, add '2 *filth, squalor, Vit.Aesop.*(G) 29.'
*σαπρόδης (for -ώδης), ες, of wine, *lacking a bouquet*, inscr. in *RAL* 1993 p.267.17.
*σαπρόμορφος, ον, *of squalid appearance*, *Vit.Aesop.*(G) 121.
σαπρός II 3, add 'fig., εἰρήνη σ. Ar.*Pax* 554' **4**, delete the section.
σαπρόστομος, add 'perh. in *POxy.* 3725 fr. 1 ii 9 (i/ii AD), subject-heading to *AP* 11.241 (Nicarch. II)'
*σᾰπωνίζω, *wash with soap*, Anon.Alch. 324.6.
σᾰρᾱπιακός, add 'καδίσκος σ. *Inscr.Délos* 1417*A*ii134 (ii BC); of a type of garland, *PAlex.*(1964) 22.4 (ii AD)'
*Σᾰρᾱπίδειον, τό, *temple of Sarapis*, Call.*Dieg.* vi.4 (fr. 191 Pf.).
Σᾰράπεια (s.v. Σάραπις), add 'at Naxos (written -ιηα), *IG* 12(5).38.4, 11, 18'
Σάραπις, line 1, after 'ιδος' insert '(also ιος, Robert *Hell.* 11.85, *Inscr.Délos* 1412*a*61, al.)'
σαργάνη 2, after 'basket' insert '*IG* 1³.422.150 (Athens, v BC)'
*σαρδᾰχᾰτης, ου, ὁ, kind of agate, Socr.Dion.*Lith.* 43 H.-S.
σάρδιος, add 'λίθος σάρδιος .. σαρδίῳ τῷ ἰχθύι τεταριχευμένῳ ἐοικώς Epiph.Const. in *Lapid.Gr.* 194.6'
*σάρδις, Lydian for *year*, Lyd.*Mens.* 3.20.
σαρδισμός, add 'prob. in Sch.D.T. 447.25 (σαρκασμός codd.)'
*σαρδονύχιον, τό, = σαρδόνυξ, *EA* 19.25 l. 7 (Pessinus, i/ii AD), prob. in Al.*Jb.* 28.16.
σαρήσιον, delete the entry.
*σαρκῑναρίων, ωνος, ὁ, app. *butcher*, Teuc.Bab. p. 43.3 B.
σαρκόω, add '**IV** Chr., *to endow with human nature, incarnate*, *Cod.Just.* 1.1.5.1, 1.1.6.7.'
σάρκωσις, add '**II** Chr., *incarnation*, *Cod.Just.* 1.1.5.3.'
σαρμός, add 'Hippon. 165a W.'
σάρξ, line 13, after '*Ba.* 746' insert 'so in pl., αἱ σάρκες αἱ κεναὶ φρενῶν id.*El.* 387' add '**c** ὁ ἀνὰ σάρκα *anasarca*, Aret.*SD* 2.1.7 H.'
σάρον II, for 'of sea weed' read 'in an abusive reference to Delos, Ἀστερίη, πόντοιο κακὸν σάρον' **III**, delete 'dub.' and transfer quot. to section I.
σαροννύω, add 'σαρωνύουσι *OBodl.* ii 1722 (ii AD)'
*Σαρπηδόνεια, τά, *games in honour of Sarpedon*, *SEG* 28.1248 (Xanthos).
*Σαρτιωβιάριος, ὁ, month name in Crete, *Inscr.Cret.* 1.xvi 4A.22.
σαρωνίς, for 'Poet.ap.Parth. 11.4' read 'Parth. 11.4 (= *Suppl.Hell.* 646.4)'
*σᾰρωνύω, v. °σαροννύω.
†σάσαμον, σᾰσᾰμόπαστος, etc., v. ‡σησαμ-.
σαστήρ, after '*Supp.Epigr.* 3.602' insert '*BE* 1950.151'
*σαταβαρα, ὁ, office included in list of priests fr. Cilicia, *ICilicie* 11c (ii AD).
σατραπεύω, add 'also ἐξαιθραπεύω, ἐξαιτραπεύω (qq. v.)'
σατράπης 1, add 'also ξαδράπης *SEG* 27.942 (Lycia, 337 BC)'
σᾰτῠρικός 2, add 'σατυρικαὶ κωμῳδίαι of certain Latin comedies, Nic.Dam.ap.Ath. 6.261c, Lyd.*Mag.* 1.41'
σᾰτύριον II, for 'fritillary .. graeca' read 'perh. *heart-flowered orchis*, *Serapias cordigera*'
Σᾰτῠρίσκος I, before 'Theoc.' insert '*Inscr.Délos* 104.18 (iv BC)'
Σάτυρος, for 'Τίτυρος' read 'Τίτῠρος' **II**, add 'also sg. Demetr.*Eloc.* 169'
*σαύη· ὁ κόσμος. Βαβυλώνιοι Hsch. (late Babylonian *šawê*, *heavens*; cod. σάνη).
σαυκρόπους, add 'cf. ψαυκρόπους'
σαυκρός, add 'cf. ψαυκρός'
σαυλοπρωκτιάω, for 'walk in a swaggering way' read 'walk in an affected way'
σαυνιαστής, lines 2/3, for 'Dor. -τάς' read 'dial. -τάς' and for 'or perh. .. 31.25' read 'rest. in Call.*fr.* 197.48 Pf.'
Σαυρομάτης, line 4, (Σαρματία), add 'also Σαρμαθία *SEG* 31.1116 (Phrygia, late Rom.imp.)'; line 6, (Σαρματικός), add 'as an imperial title, of Commodus, *ICilicie* 77 (AD 178)'

σᾰφής II 4, after '-εστέρως' insert 'Antipho 3.2.5'
σᾰφήτωρ, for 'a variant' read 'coined as an etymon'
*σαφρικόν, τό, *casia*, *Gloss.*
σᾰφώνιον, add '*PVindob.Worp* 11.8 (vi AD), *PAnt.* 202(a).10 (vi/vii AD)'
σάω, delete 'Med.' and for 'σαόω' read 'σάωμι (ν. σῴζω)'
σβεστήρ, add '2 as adj., ῥόος σ. Nonn.*D.* 23.291.'
σεβάζομαι 2, add 'also act. (cf. σέβω II), poet. in Favorin.*Exil.* 11.5 B.'
σέβας I, add '2 in Chr. prose, *reverence*, *Cod.Just.* 1.4.34.11.'
σέβασμα I, add 'Hld. 9.22'
σεβάσμιος I, add 'comp., σεβασμιωτέρας ἢ κατὰ ἄνθρωπον ὁμιλίας Charito 2.4; sup., Θεαῖς Νεμέσεσιν Ζμυρναίαις σεβασμιωτάταις *SEG* 32.1082 (Smyrna, iii AD)' **II**, add 'τοῦ οἴκου τοῦ σ. *SEG* 31.372 (ii AD)' add '3 σεβάσμιος, ὁ, type of coin, *PKöln* 166.10 (vi/vii AD).'
Σεβαστεῖον I, add '*SEG* 23.207 (Messene, time of Augustus)'
σεβαστικός, delete 'Iamb.*Protr.* 21 κε''
*Σεβαστολόγος, ὁ, official who pronounced panegyrics of the Emperor, *Milet* 7.65.10 (Robert *Hell.* 7.206ff.).
Σεβαστον(ε)ίκης, for 'in Imperial games' read 'in the Σεβαστά, v. σεβαστός II 4'; after '(Ancyra)' insert '*GDI* 4107 (Rhodes), *PLond.* 1178.67'
σεβαστονεικηφόρια, for 'Imperial games' read 'festival of Augustus and Athene Nicephoros'
†Σεβαστόνεως, ω, ὁ, official in the cult of Caligula, *Milet* 7.65.10 (Robert *Hell.* 7.206ff.).
σεβαστός I, for this section read 'venerable, reverend, august: of gods, Ἑστίας Σεβαστῆς *ICilicie* 44 (i/ii AD), Διὶ Σεβαστῷ *INikaia* 1129 (AD 131); of places, ἐν Σεβαστῆι Πάφωι *SEG* 23.638 (Paphos, AD 18/9). **2** θεοὶ σεβαστοί of deified emperors, *SEG* 30.1245 (Caria, i BC/i AD), 29.677 (Thrace, AD 78, etc.); without θεοί, *SEG* 30.1534 (Lycia, ii AD), *IEphes.* 691, 693 (AD 209/11).' **II 1**, fem., add '*SEG* 33.1132 (Phrygia, ii AD), etc.' **2**, add '*SEG* 34.398 (Delphi, i AD), 24.612 (Maced., ii AD)'
σεβαστοφάντης, after 'flamen Augusti' insert '*SEG* 31.901 (Caria, mid i AD)'
Σεβαστοφόροι, after '*Mens.* 4.138' insert '*BCH* 24.340.30 (Oenoanda, iii AD), Suid. s.v. Αὐγουστεῖον'
σεβένινος, add 'also σιβένινος *PLond.* 1414.13'
σεβένιον, for 'palm-fibre' read 'fibrous spathe of male date-palm'
σέβερος, for 'Theognost.*Can.* 11' read 'Theognost.*Can.* 33 A. (cf. Lat. *severus*)'
*σέβετος· ὁ κοχλίας Theognost.*Can.* 33 A.; cf. σελάτης.
*Σεβήρεια, τά, festival of Severus at Athens, *SEG* 26.184 (iii AD); also Σευήρεια *IG* 2².3169 (iii AD), *JHS* 99.161-2 no. 4 (Lycia, AD 231/5).
σέβισμα, add '*BGU* 1764.12'
σέβομαι, line 3, after '*Plot.* 12' insert 'med. aor. inf. σέψασθαι Hsch.'
σεγέστρον, add 'cf. Lat. *segestre*, ‡στέγαστρον'
σείνιοι, delete the entry (ghost word).
σειρά IV, add 'κατὰ σειράν = Lat. *in stirpes, ex stirpibus*, *Cod.Just.* 6.4.4.19b' **V**, add 'Procl.*Inst.* 97 D., etc.'
σειραίνω, after 'parch' insert 'Euph.in *Suppl.Hell.* 443.6'
σειραῖος 2, add 'σειραῖα δεσμά S.*fr.* 25'
Σειρήν I, add 'masc., ἀνέθεσαν .. Σειρῆνα ἀργύρεον *SEG* 12.391 (Samos, vi BC)' **III**, add 'but in Ael.*VH* 4.2 app. *drone*, cf. Plin.*HN* 11.48' **IV**, for this section read 'in Lxx used to translate words rendered as *ostrich* (or *desert-owl*) and *jackal*, *Jb.* 30.29, *Mi.* 1.8, *Is.* 13.21, 34.13, 43.20, *Je.* 27(50).39'
σείριος, line 4, after '*Op.* 417' insert 'Σήριον ἄστρον Alcm. 1.62 P.'; line 6, for 'of stars .. 23.62' read 'subst. neut., of all *stars*, σείρια παμφανόωντα Ibyc. 33 P.' and delete 'cf. E.*fr.* 779.8 cod. Longin.'
*σείρωμα, ατος, τό, *sediment*, defined as κάθισμα ὑδατῶδες ὑποβαλσάμου, Aët. 1.132 (67.14 O., cf. 68.1-2 O.).
*σειρώτης, ου, ὁ, liquid measure, οἴνου σιρ(ωταί) *PPrag.* I 92.1 (vi/vii AD), σ. οἴνου ἢ ἄλλου τινὸς ὑγροῦ *simussator* (i.e. *cim*-), *Gloss.*
σεισμός 2, add 'σ. πυρφόρον *whirring* of a fire-dart, Lxx *Jb.* 41.20'
*σειστημάρχης, v. °συστημάρχης.
σεκουνδαρούδης, for 'secunda rude insignis' read 'fr. Lat. *secunda rudis*'; add '*IGRom.* 4.831 (Hierapolis), *SEG* 36.595 (Maced., ii/iii AD)'
*σεκούτωρ, ορος, ὁ, Lat. *secutor*, type of gladiator, *SEG* 32.605 (Thessaly, ii AD), πρωτόπαλος σεκουτόρων D.C. 72.22.
*σέκρετον, v. °σηκ-.
*Σεκυών, v. ‡Σικυών.
σελᾰγέω II, for 'Opp.*C.* 1.210, 3.136' read 'χαλκὸν σελαγεῦντα Opp.*C.* 1.210, τοῖσι σελαγεύσιν ὀδόντες ib. 3.352'
σελᾰγίζω, after ' = σελαγέω II' insert 'Call.*fr.* 238.26 Pf.'
*σελάγινος, v. °σελγ-.
*Σελᾰμίνιος, v. °Σαλαμίνιος.
σελάσκω, for 'Theognost.*Can.* 11' read 'Theognost.*Can.* 33 A.'
σέλᾰχος, after 'τό' insert '*AP* 6.222 (Theodorid.)'

*σέλγινος, η, ον, dub. sens., ἱμάτιον σέλγινον περιπόρφυρ[ον], SEG 38.1210 (Milet., ii BC).

σεληνάριον, add 'also perh. *crescent* of land, ἐν τῷ σεληναρί(ῳ) PMich.xv 721.47 (iii/iv AD)'

σεληνῖτις, add 'PMag. 4.2360'

σεληνόβλητος, delete '*epileptic*'

*σεληνόγονος, ον, *born of the moon*, PMag. 3.331.

*σεληνοπετής, ές, *fallen from the moon*, of Musaeus, rest. in Ion ap.Phld.Piet. p. 13 (v. ZPE 57.54).

*σελία, ἡ, *seat* (= σέλλα), Swoboda *Denkmäler* 110.

σέλινον II, for 'Sch. .. 10' read '*Carm.Pop.* 6 P.' at end add 'Myc. *se-ri-no*'

σελινοφόρος, for '*Jahresh*. 18 *Beibl*. 287 (Ephesus, i BC)' read '*IEphes*. 14.20 (i BC)' add '2 adj., ἅρμα σελινοφόρον *crowned with celery*, Call.*fr*. 384.4 Pf. (οἱ γὰρ νικῶντες τὰ Ἴσθμια σελίνῳ στέφονται Sch.ad loc.).'

σελίς II, line 3, for '(Jul.)' read '(Phil.)'

*σελλάριος, ὁ, Lat. *sellarius*, gloss on κέλης, Sch.Gen.Il. 15.679, Suid. s.v. κελητίζειν.

*σελλίς, ίδος, ἡ, *chair* or *bench* (cf. σέλλα), EM 398.17.

*σελλοξύλινον, τό, perh. *wooden saddle*, CPR 14.32.19 (vii AD).

*σελλοποιός, ὁ, ?*saddle-maker*, CPR 14.32.8, 34 (vii AD).

*σελλοφόρος, ὁ, prob. the same as διφροφόρος, *litter-bearer*, Roueché *Aphrodisias* 80 (v/vi AD).

σέλμα II 1, add 'Ath.Mech. 18.12'

*σέμεστα, τά, term used w. ref. to linen, perh. either of a measure or of a manufactured item, λινᾶ σ. POxy. 3979.15 (iii AD).

σεμίδᾱλις, line 4, add 'σιμίδαλιν SEG 31.122 (Attica, c.AD 121/2)'

+σεμνεῖον, τό, room or building for private or communal worship (in quots., w. ref. to a contemplative sect), Ph. 2.475, 476, Suid., *Gloss*.

σεμνολόγημα, add 'III σχόλια· σεμνολογήματα Hsch.'

*σεμνοπολεύω, *perform sacred rites*, Dioscorus 24.16 (*GDRK* i 150).

σεμνοπρέπεια, add 'Ptol.Tetr. 206'

σεμνοπρεπής, after '*dignified*' insert 'epigr. in *TAM* 2.203.10'; after 'D.L. 8.11' insert 'Mitchell *N.Galatia* 436'

σεμνοτροπία, after 'Ptol.Tetr. 206 B.' add '(v.l.)'

*σενάτωρ, v. °σινάτωρ.

*Σεουηριανός, ή, όν, *of Severus*, TAM 4(1).25, *ICilicie* 30 (AD 222/35).

*σέρκος· ἀλεκτρυών. καὶ ἀλεκτορίδες σέλκες Hsch.

*Σευήρεια, v. °Σεβήρεια.

σεύω, line 16, after '(Dor. σώμαι Epil. 3' insert 'σώται Hsch.' II, line 6, after 'so in trag.' insert '(lyr. exc. E.*IT* 1294)'

σηκίς, add 'Sch.Ar.*Pax* 185'

σηκός III, add 'Lys. 7.2, 5, 10, al.'

*σήκρητον, τό, Lat. *secretum*, ἐν τῷ σηκρήτῳ *in court*, POxy. 1204.12 (AD 299); cf. σέκρετον· συνέδριον Hsch.

σήκωμα, after 'Dor. σάκωμα' insert 'also non-Att.-Ion., Lang *Ath.Agora* XXI Hb4 (i BC)' 1 b, after 'standard *measure*' insert '(variously 5, 6 or 8 *sextarii*)'; add 'Cypr. *sa-ko-ma* SEG 23.608 (v/iv BC)'

σηλαγγεύς, for 'gold-refiner' read 'gold-washer'

σῆμα, for 'Dor.' read 'dial.' 3, add 'b *memorial*, ξόανον .. θῆκεν σᾶμ' ἱεραπολίας epigr. in SEG 13.422 (Delos, iii BC).' 6, delete 'mostly in pl.' and add 'Arat. 72, 233, al.'

σημαίνω A I 1, at end after 'Pl.*Lg*. 682a' insert 'σ. τι w. gen. of reference, ἀστέρας οἵ κε μάλιστα τετυγμένα σημαίνοιεν .. ὡράων Arat. 12, 757' 3, line 4, after 'Th. 2.8' insert 'cf. ἐφ' ὕδατι σ. Arat. 873' B II, line 4, after '*Ath*. 8' insert 'συνθήκας *REA* 33.8 (Theangela, iv/iii BC)'

σημάντρια 2, for this section read 'σαμάντριαν .. πυρᾶς .. ἰωάν *sign*, Call.*fr*. 228.40 Pf.'

σήμαντρον, line 3, before 'metaph.' insert '2 *brand mark*, ἀνδράποδα .. σεσημασμένα τῷ δημοσίῳ σημάντρῳ X. *Vect*. 4.21.'

σημάντωρ, line 1, after 'ὁ' insert 'Dor. σᾱμ- Simon. 14.62(*a*).3 P., Pi.*Pae*. 13(*a*).24 S.-M.'

σημᾰσία IV, for '*Bull.Soc*. .. vi AD)' read 'when the water is at its highest, also the occasion of a festival Σημασία, Bernand *Akôris* 41 (iv AD)'

σημεῖα I 2, add 'κατὰ σημαίαν *by maniples*, Plb. 18.32.11'

*σημειολύτης, ου, ὁ, *interpreter of portents*, Vit.Aesop.(G) 86, Ps.-Callisth. 1.4.10 (pl.).

σημεῖον I, add '10 *signpost*, Demetr.*Eloc*. 202. b of the mark of the Nile at its highest point, Bernand *Akôris* 40, 41 (sp. σημῖον, iv AD); cf. ‡σημασία.' II 1, add 'of the image of a Syrian goddess (app. misunderstanding of wd. based on the divine name Semea or Simea), καλέεται δὲ σημήιον καὶ ὑπ' αὐτῶν Ἀσσυρίων, οὐδέ τι οὔνομα ἴδιον αὐτῷ ἔθεντο Luc.*Syr.D*. 33. b sent as proof of the genuineness of a communication, etc., PCair.Zen. 192 (iii BC), PPetaus 28.8, 17 (ii AD), POxy. 1683 (iv AD), etc.' add 'IV = Lat. *signum*, *second name*, IG 14.935; cf. °σίγνον III.'

σημειόω I 1, add 'σεσημειωμένος *patterned*, Edict.Diocl. 19.8' II 1, line 3, after 'ii 25 (i AD)' insert 'SEG 31.1560 (Egypt, Rom.imp.)'

*σημειώσιμος, ον, *entered under the seal* of, εἰς ἐφημερίδα Ἰουλίων Θέωνος καὶ Θέωνος σημιώσιμον Ματρέου POxy. 3588.3 (ii AD).

σημείωσις I, add 'SEG 33.1177 (Lycia, AD 43). 2 as a term for *pronoun*, A.D.*Pron*. 4.2.'

σημειωτός, add 'II *patterned*, φακιάλια Edict.Diocl. 29.38.'

+σημήϊον, v. ‡σημεῖον.

*σημήτωρ, ορος, ὁ, *indicator*, γραφὴν σημήτορα τύμβου GVI 233 (Cius).

+σημοθέτης, ον, poet. σᾱμ-, *making a mark*, AP 6.295 (Phan.).

*σημοφόρος, ον, *standard-bearing*, κάμαξ epigr. in SAWW 224(1).39 (Egypt, ii/i BC).

σηνοῦροι, read 'σήνουροι' and add 'cf. σαίνουρος'

σηπία, line 1, after 'ἡ' insert 'σηπίη Hippon. 166 W.'; line 2, delete 'at Athens'

σηπιάς, add 'perh. also σηπιά[δων ..] SEG 23.326 (Delphi, iii BC)'

*σηπόγαστρος, ὁ, kind of earthenware vessel, Anon.Alch. 323.18.

σήπω II 1, line 6, after 'σέσημμαι' insert 'Ps.-Luc.*Philopatr*. 20'

σήρ, delete the article.

*σηρίον· θηρίον Hsch., Lacon.; cf. σηροκτόνος.

σησαμικός, after 'pertaining to sesame' insert 'τῶι σησαμ[ι]κῶι σπόρω[ι PVat.Gr. 65 (*Tyche* 5.102)'

σησάμινος, line 3, for 'δοκοί' read 'φαλάγγων σασαμίνων' add '2 *of sissoo-wood (Dalbergia Sissoo)*, ξύλα Dsc. 1.129.'

σήσαμον I 2, add 'Men.*fr*. 709 K.-Th.' add 'form σασάμα or σάσαμον, Myc. *sa-sa-ma*'

σησαμόπαστος, for 'Dor. σασ-' read 'Dor. σᾱσ-'

*σησαμοσπορεύω, *sow with sesame*, pass. aor. subj. -ευθῆι prob. in PCair.Zen. 816.6 (iii BC).

*σησκουπλικάριος, ὁ, Lat. *sesquiplicarius*, PHamb. 39.21 (p. 169) (ii AD); also -κιάριος BGU 614.2.11/12 (iii AD).

σῆτες, add 'form prob. *τσάfετες, Myc. *za-we-te*'

σητόβρωτος, add 'PFam.Teb. 15.36 (ii AD)'

σῆψις, line 2, for 'Emp. 121' read '*Orac.Chald*. 134 P.'

σθεναρός, add 'sp. στεναρός, SEG 32.605 (Larissa, ii AD)'

σθενής, for 'κρατερός' read 'καρτερός'

*σθενόγαυρος, ον, *exulting in strength*, prob. in GVI 263.3 (Phrygia, ii AD).

*σθένος I 1, line 11, delete 'the only phrase .. (cf. infr. III)'; add 'ἀλλ' εἴτε δόλῳ ἔχουσι αὐτὴν κτησάμενοι, δόλῳ ἀπαιρεθῆναι ὑπὸ ὑμέων, εἴτε καὶ σθένεΐ τεῳ κατεργασάμενοι, σθένεΐ κατὰ τὸ καρτερὸν ἀνασώσασθαι Hdt. 3.65.6'

*σθλιβόω, [ῐ], *rub*, Anon.Alch. 323.4; cf. θλίβω.

*σιακυκα, ά, prob. non-Hellenic name of a cult-object, Inscr.Cret. 1.v 23.15 (Arcades, ii AD).

σίαλον I, add 'pl., Nic.*Th*. 86'

σίαλος, add 'form σίhαλος, Myc. *si-a₂-ro*'

Σίβυλλα, add 'II *prophetess* generally, Βαλουβουργ Σήνονι (?for Σέμνονι) σιβύλλᾳ SB 6221.8 (ostr., ii AD).'

σῑβύνη, form συβίνη, add 'SEG 31.1574 (Cyrene, ii BC)'

σιγαλόω, add 'Sch.Pi.*O*. 3.8'

*σιγάλωμα, for 'σιγαλόεντα' read 'σιγαλόεν'

σιγάω I 2, after 'E.*Alc*. 78' insert 'τὸν σιγώμενον τόπον *the place where no sound is heard*, Aen.Tact. 22.13'

σιγγλάριος I, line 3, for 'etc.' read 'ITyr 161, SEG 29.319 (Corinth, v/vi AD); in late inscrs. and papyri the wd. refers generally to a *courier*, v. REA 62.358-9' II, add 'cf. SEG 28.1624'

*σιγελλᾶτος, η, ον, Lat. *sigillatus*, *figured*, (*embroidered*) *with figures*, ἀγκονάριον ὁλοσιρικὸν σιγελλᾶτον *Tyche* 2.7 (vii AD).

σιγητής, for '*AJA* 37.262' read '*AJA* 37.269'

σίγλα, add 'II *abbreviation*, Just.*Const*. Δέδωκεν 22.'

σίγλος I 1, add 'Cypr. *si-ko-lo-ne*, ICS 309.13 (coin; at ICS 224, 368 abbrev. *si* for weight)' 2, add 'also σίγλων χρυσείων χιλιάς Alex.Aet. 4.3' II, before 'Phot.' insert '*IG* 2².1544.22 (Eleusis, iv BC)' and delete 'cf. sq.'

σῖγμα II 1, for 'Princeton .. 560' read '*IGLS* 9122 (v AD), *CIJud*. 781.4 (sp. σῖμμα; Side, Byz.); add 'of a topographical feature of Alexandria, POxy. 3756.2 (AD 325)'

σίγνον I, add '2 *of the sign of the Cross adopted by Constantine*, Salamine 238E.' II, after 'store, prison, etc.' insert 'GVI 849.3 (Phrygia, i AD), POxy. 3616.5 (iii AD), ὁστιάριον σίγνοισι βαλών poet. in PBodm. 29.131 (iv AD)' add 'III *second name*, GVI 1096.3 (Propontis, ?i AD), 446.4 (Athens, iii/iv AD); cf. °σημεῖον.'

σίδη, add 'Alcm. in *PMGF* p. 68 D.'

σίδηρεος, line 2, before 'SIG 144.14' insert 'S.*fr*. 20 R., E.*Ph*. 26, Th. 2.76.4'; line 4, after 'σιδάρεος [ᾱ]' insert '*CEG* 1 (early v BC)'; line 5, after 'Aeol. σιδάριος' insert 'SEG 24.361 (early iv BC)' I 1, add 'as the name of Roman legion (transl. *Ferrata*), λεγιώνος ἕκτης Σιδηρᾶς SEG 33.1089 (i AD), *IGRom*. 4.266 (iv AD)'

+σιδηρίσκος, ᾱ, Dor. σιδᾱρ-, some unidentified iron implement, Inscr.Cret. 4.145 (SEG 28.734, v/iv BC).

σιδηρίτης I 1, for 'Eup. 263' read 'A.*fr*. 78a.67 R., Eup. 283 K.-A.'

*σιδηροδαΐκτης, ου, ὁ, *cleaving* or *destroying with iron*, rest. in *Trag.adesp.* 720k K.-S. (σιδηρο..ίκτης *Suppl.Hell.* 991.99).

σιδηρόδετος, add 'of the smith's craft, *SEG* 31.1284 (Pisidia)'

σιδηροκόντρα, for '*Ausonia* 6.9*' read '*Inscr.Cret.* 4.305' and after 'Gortyn' insert 'iii AD'

*σιδηροπέρσης, ου, ὁ, *destroying with iron*, *Suppl.Hell.* 991.93 (poet. word-list, iii BC).

†σιδηρόπληκτος, Dor. σιδᾱρο-, ον, *struck by iron*, σιδαρόπληκτοι μὲν ὧδ' ἔχουσιν (i.e. by iron weapons), σιδαρόπληκτοι δὲ τοὺς μένουσιν .. τάφων πατρῴων λαχαί (i.e. by iron tools), A.*Th.* 911-912.

*σιδηρόροφος, ον, *iron-roofed*, Nonn.*D.* 8.137 (cj.).

*σιδηροσάνδᾰλος, ον, *wearing iron sandals*, τ[ὴν σιδη]ροσάνδαλον *Suppl.Mag.* I 49.59 (ii/iii AD).

σιδηροφάγος, add 'Anon.Alch. 344.28'

σιδηρόψῡχος, add '*PMag.* 7.356'

*σίδλι(ο)ν, v. °σιτλίον.

†σιειδής, ές, *godlike* (v. σιός), Alcm. 1.71 P.

*σικάριον, τό, fr. Lat. *sica*, *dagger*, *POxy.* 1294.8 (ii/iii AD).

Σῐκελία, add 'personified, *SEG* 31.925 (Caria, i/ii AD)'

Σῐκελός, delete 'rare in prose' and add 'Σύμμαχος Τιττάλου Σικελός *SEG* 33.455; *IG* 2².10291, 10292, al.'

*σικλάριον, τό, prob. dim. of Lat. *sicula*, *small dagger*, *CPR* 8.65.10 (vi AD).

σῐκυήρᾰτον (s.v. σικυήλατον), add 'also σικύρατον *PHamb.* 99.7 (i AD), *PPrinceton* 39.4 (iii AD)'

*σίκυια, ἡ, = σικύα, Them.*in APo.* 60.1, Suid.; prob. to be rest. in ἐπέθηκε τὰν σικ[υίαν ἐ]πὶ τὰν γαστέρα *Inscr.Cret.* 1.xvii 9 (Lebena; if an iambic line).

*σῐκυοπώλης, ου, ὁ, *gourd-seller*, *PKöln* 195.21 (ii/iii AD).

*σῐκύριον, τό, *cucumber*, *PPrinceton* 39.7 (iii AD; cf. σικύδιον).

σῐκυών, line 7, for 'Σικυώνοθε' read 'Σικυωνόθε' and add '[Σ]εκυόναθεν *IG* 9².209 (Melitea, ?v BC, v. *REG* 54.61-62'; line 8, for 'the people .. ii BC' read 'also Σεκυών, Σεκυδ[νι] *SEG* 11.257 (= 14.310, Sicyon, v BC), cf. ἡ Σικυών Σεκυών παρὰ Σικυωνίοις acc. to A.D.*Adv.* 144.20; in inscrs. at Delphi; ΣεϝυΓόνιιος *BCH* 61.57-60 (vii/vi BC), Σεκυώνιοι *GDI* 2581.273 (ii BC), but Σικυώνιοι *SIG* 31.8 (Dor., v BC), Σικυώνιος *IG* 5.1.1565 p. xxi (iv BC)'

Σῐκυώνια, add 'also sg., Macho 158 G.'

σικχάζομαι, delete the entry.

*σικχαζόμενος· σκωπτόμενος Hsch.

*σιλβος, η, ον, unexpld. description of kind of tunic, στιχαρομαφόρην (-φόριον) σιλβον *PWash.Univ.* 58.4 (v AD, ?emend to στιλβόν).

σιλεντιάριος, add 'sp. σελ-, *SEG* 26.436 (Argos, v/vi AD)'

Σῐληνός, for 'E.*Cyc.* 13, 82, 269' read 'E.*Cyc.* 539' **2**, add '**b** of fountain, orig. in the form of a statue of Silenus, *silanos* (fr. Dor. Στλᾱνός) *ad aquarum*, Lucr. 6.1265, *CIL* 8.692, etc.'

†σιληπορδέω, Dor. σιλα-, colloq. wd. popularly associated w. πέρδομαι, *behave in a vulgar manner*, Sophr. 164, Posidon. 253.46 E.-K.; cf. mod. Gk. τσιληπουρδῶ.

†σιληπορδία, ἡ, *vulgar arrogance*, Luc.*Lex.* 21, v. °σιληπορδέω.

σιλίγνιον, for '= Lat. *siligo*, *winter wheat*, ibid. (pl.)' read 'prob. = Lat. *siligineum*, *small wheat-loaf*, *CPR* 7.42 i 1, ii 3, 13 (v AD); *PKlein.Form.* 957.4 (v/vi AD), cf. *PMerton* 85.9 n.'

*σίλλῠβον, τό, *fringe, tassel*, τοὺς .. θυσάνους καὶ σίλλυβα οἱ παλαιοι καλοῦσιν Poll. 7.64; cf. σίλλυβα· κροσσοί. οἱ δὲ τὰ ἀνθέμια καὶ κορυκόσμια Hsch. **2** *kind of thistle*, Dsc. 4.155, Ruf.ap.Orib. 7.26.38.

*σῖμᾱλος, η, ον, = σιμός, Sch.Gen.Il. 15.705; as pers. n., Anacr. 41 P., al.

*σιμαις, αιτος, an unidentified food crop, *PTeb.* 419 (iii AD).

*σιμίσιον, τό, = Lat. *semissis*, *half a gold solidus*, *Inscr.Cret.* 1.xxii 65 (Olus, iv AD).

Σῖμος I, add 'Myc. *si-mo*, pers. n. (masc.)'

*σιμπλάριος, α, ον, Lat. *simplarius*, of money, *the simple amount paid*, πεντήκοντα μυριάδας σινπλαρίας ὃς ἔδωκ[εν *BCH* 76.655 (Delphi, iv BC).

σίμωρ, for 'field mouse' read 'small mouse-like animal'

σινάπιον, after 'σίναπι' insert '*Edict.Diocl.* 1.34, 35'

*σινάτωρ, ὁ, Lat. *senator*, *TAM* 4(1).367.2.

*σινγιλίων, ωνος, ὁ, Lat. *singilio*, *kind of shirt or tunic*, *Edict.Diocl.* 19.47, al.

*σινγουλάριος, -άρις, v. ‡σιγγλάριος.

*σινδόνη, ἡ, *cotton garment*, *Peripl.M.Rubr.* 6, 48, 51, al.

*σινδονίσκη, ἡ, dim. of σινδών (*garment*), Michel 832.24 (Samos, iv BC), Plu. 2.340d: v. *MDAI(A)* 68.47.24.

σινδονίσκος, delete the entry.

σινδών, line 2, after 'εἰκών' insert 'sp. σιμδώμ *ZPE* 34.131 (*SEG* 28.53 + 29.146)'

σῐνις I, delete 'ἔθρεψεν .. Ag. 718' and 'as Adj. *destroying*' **II**, after 'Corinth' insert 'B. 18.20'

σίνομαι, line 6, after 'σίνω' insert 'Plu. 2.913e'

*σινόργανον, τό, prob. *an instrument*, or part of machine, *for raising water*, *PMerton* 39.2 (iv/v AD), *POxy.* 1985.11 (vi AD).

σινπλαρία, delete the entry (v. °σιμπλάριος).

†σίντης, εω, ὁ, *predator*, of the lion, Il. 11.481, 20.165; of the wolf, 16.353; of a snake, Nic.*Th.* 625; w. fem. subst., σίνταο φάλαγγος ib. 715; acc. to Hsch. s.v. μακεσίκρανος also applied to the hoopoe. **2** *robber*, *despoiler*, Opp.*H.* 4.602, *Cat.Cod.Astr.* 7.115; humorously, of mice, Call.*fr.* 177.29 Pf.

*σῐόδμᾱτος, v. ‡θεόδμητος.

σιοειδής, for '*like σίον*' read '= θεοειδής' and add 'cf. °σιειδής'

*σιόεις, εσσα, εν, *overgrown with σίον*, *An.Ox.* 3.401.34.

*σιοκόλος, ὁ, Lacon. for θεοκόλος, Eup. 480 K.-A. (cj.).

σιοκόμος, delete the entry.

σιός, add 'σιόφιν = θεόφιν, Alcm. 12.4 P.'

σιππινόμεστος, add '(pap. σιππουν-, cf. °σίππινος)'

*σίππινος, η, ον, *made of tow*, σάκκος *PRyl.* 606.22 (iii AD); also σιππόινος (pap. σειππόεινος) *SB* 11575.11 (iii AD).

*σιπποϊνοπώλης, ου, ὁ, = στυππειοπώλης (cf. σίππιον), *PPetaus* 92.20 (ii AD).

*σιπποποιός, ὁ, *tow-maker*, *POxy.* 2799 (vi AD).

σῐπῠΐς, for 'jar' read 'jar similar to pyxis' and add '*RAL* 1965.454ff. no. 10 (Gela, iv BC)'

†σίρωμα, v. °σείρωμα.

†σιρώτης, v. °σειρώτης.

σισακικία, delete the entry (v. *Tyche* 5.180).

σῐσύμβριον I, for '*bergamot-mint*' read '*calamint*' **II**, before 'Poll. 5.101' insert 'Pherecr. 2.3 K.-A.'

*σῐσύριον, τό, dim. of σισύρα, σεισυριν *PDura* 33.13 (iii AD).

σίσυς, add 'Semon. 31b W.'

Σισυφία χθών (s.v. Σίσυφος), add '*SEG* 31.291 (Corinth, v AD)'

*Σισυφίδαι, οἱ, *descendants of Sisyphus*, χὠ τᾶς ἀσώτου Σισυφιδᾶν γενεᾶς (i.e. Odysseus) S.*Aj.* 189; poet. for *Corinthians*, Call.*fr.* 384.10 Pf.

σῐτᾰλετικός, add 'μηχανὴ σιταλητική *PMich.inv.* 257.3 (*BASP* 13.3, iii AD)'

σιτᾰποδοχεῖον, for 'Partsch' read 'Patsch'

σίταρχος (s.v. σιτάρχης), add '*SEG* 23.305 (Delphi, *c.*190 BC)'

*σιτεκλήμπτωρ, ορος, ὁ, *collector of corn*, rest. in *PAnt.* 33.24 (iv AD); cf. ἐκλή(μ)πτωρ.

σιτεύσιμος, add '*Stud.Pal.* 20.233.1 (vi/vii AD)'

σιτευτάριος, add '*Corinth* 8(3).559 (v/vi AD)'

σιτευωνέω, delete the entry.

σιτηρεσιον, add '*payment in kind*, *POxy.* 2892, al. (iii AD), *PLips.* 84 iii 20 (iv AD), *PVindob.Tandem* 19.15 (v/vi AD)'

σιτηρός, add '**IV** σιτηρά, ἡ (sc. (?)ἀποθήκη), *corn-magazine*, *SB* 8754.15, *BGU* 1742.16, 1743.13 (all i BC).'

*σιτηρουσία, ἡ, unexpld. compd. of σῖτος (unless for -ουχία), Hsch. s.v. στάχυς.

σῑτικός, line 1, after 'corn' insert 'σ. ἐμπόριον Arist.*Ath.* 51.4' add '**b** σιτική, ἡ, *corn-market*, *SEG* 8.43 (Palestine, i AD), *ICilicie* 124 (i/ii AD), *SEG* 27.947 (Tarsus, iii AD).'

σιτισμός, after '*feeding*' insert '*IGRom.* 4.144.8 (Cyzicus, i AD)'

*σῑτιστάριος, ὁ, *fattener of poultry* or *cattle*, *Corinth* 8(3).559 (v/vi AD).

*σῑτιστός, ή, όν, *fattened* (= σιτευτός), *PMil.Vogl.* 145.10 (AD 174).

σίτλα, add '*CPR* 8.65.19 (vi AD)'

σιτλίον, add 'written σίδλιν, pap. in *JEA* 20.27 (v/vi AD)'

σιτοβολεῖον (s.v. -βολών), add 'Ἑλληνικά 7.179 (Chalcis, iii BC), *IG* 9(2).243 (Thessaly)'

*σῐτοθέτης, ου, ὁ, *official concerned with supply of corn*, *SEG* 36.788 (Samothrace, iii BC).

σῐτοκάπηλος, after '*corn-factor*' insert '*SB* 10447ᵛ.5 (?iii BC), *PLond.* 44.33 (ii BC)'

σιτόκριθον, for 'mixture of wheat and barley' read 'wheat and barley together' and add '*POxy.* 2766 (AD 305), *CPR* 5.16.11 (AD 486)'

*σῑτομ(ε)ίλης, ου, ὁ, *miller*, *PLond.* 387.18 (vi/vii AD).

σῑτομετρέω 2, line 4, after '*IGRom.* 3.679 (Tlos)' insert '(ἄνδρες) σιτομετρούμενοι members of privileged class in some Lycian cities, who received special allocation of grain, *SEG* 27.938, 30.1535.8 (both ii AD), *TAM* 2.578'

σῑτομέτρης, add 'Ὀλυμπίου σιτομέτρου *ITyr* 188, 189 (Rom.imp.)'

*σῑτομετρικός, ή, όν, *of* or *concerned with measuring corn*, σειτομετρ[ι]κῶν [χρησ]τηρίων *PVindob.Salomons* 5.13 (AD 192).

σιτοπώλης, add 'σιτοπώλη (nom. pl.) *SEG* 23.271 (Thespiae, late iii BC)'

σῖτος, line 2, delete 'only' and after '*Delph.* 3(5).3 ii 19 (iv BC)' insert 'de Franciscis *Locr.Epiz.* no. 28' **I 1**, add '**b** meton., *corn-market*, *SEG* 26.72.18 (Athens, iv BC).' add 'Myc. *si-to*'

*σῑτοσπορία, ἡ, *sowing of grain*, *POxy.* 2973.25 (AD 103; sp. σειτ-).

*σῑτοτᾰμιεύω, serve as *σιτοταμίας (*corn-treasurer*), *AE* 1933 suppl. 2, *IG* 9(2).1029, 1093 (all Thessaly).

*σῑτοφάκη, = °σιτόφακον, *PRainer Cent.* 137 (vi AD).

*σῑτόφᾰκον, τό, wheat and lentils, POxy. 3406.4 (iv AD).
*σῑτοφθόρος, ον, grain-destroying, τὰ κοπροφόρα καὶ σιτοφθόρα ζῷα Sch.Nic.Al. 115a Ge.
σῖτοφΰλᾰκες I, add 'SEG 26.72 (Athens, 375/4 BC)'
**σῖτοχόγος, ό, one who dispenses grain, Myc. si-to-ko-wo.
*σῖτόχρωμος, ον, wheat-coloured, ἵππος μο[υ θ]ήλεια σειτόχρωμο[ς PMich.IX 527.8 (ii AD).
*Σιτηνή, ή, local name for a mother-goddess, IHadr. 33 (AD 160/1).
σῖτῶν, add '2 granary, SEG 9.354 (Cyrenaica, ?i AD), Gr.Naz.Carm. 2.1.11.1267.'
σῖτωνέω, for 'buy corn' read 'serve or hold office as σιτώνης' and add 'TAM 4(1).262.3'
†σῖτωνικόν, v. °σιτωνικός.
*σῖτωνικός, ή, όν, concerned with the purchase of corn, [ἐπι]δόσεις -κάς Didyma 296.4, χρήματα PASA 3.612 (Ilias); n. sg. subst., wheat-fund, Inscr.Délos 399 A 73 (ii BC), σ. καταναλισκομένου IG 4.2.8 (Aegina, i BC); n. pl., IG 2².1272.3 (iii BC), 1708 (ii BC), IGRom. 4.580 (Aezani).
σιφλός I, delete 'of fish, mad on food, greedy' and transfer 'Opp.H. 3.183' to section II, prefixing 'perh.'
*σίφνον· σιπύα Hsch. cf. id. s.v. σιπύη.
σιωπή I 1, add 'b pause between sounds, D.H.Comp. 22.'
σιωπηρός, add '2 tacit, implied, σ. ἐλευθερίαν Just.Nov. 22.11; adv. -ρῶς ib. 118.5.'
*σιωπητικόν, τό, novice's fee in mystery cult, PMich.VIII 511.3 (iii AD).
σκαιωρία, after 'mischief' insert 'Procl.CP Hom. 27.2.4, Just.Nov. 63.1'
σκαλεύω, add 'fig., Aq.Ps. 76(77).7'
*σκᾰλιδευτής, οῦ, ὁ, hoer, PCair.Zen. 816.9 (iii BC), SB 797.8, 25 (iii BC).
σκᾰλίς, delete 'or shovel' and add 'IG 1³.422.140 (v BC)'
*σκάλιστρον, τό, prob. = σκαλιστήριον, Aq.Sm.Je. 50.10.
*σκαλλόν, σκαλλίον .. οἱ δὲ σκαλλόν Hsch.
σκάλλω, add 'hoe in, Gp. 2.24.1'
*σκᾰλοβᾰτικός, ή, όν, of climbing a ladder, ἡ σ. (sc. τέχνη) Rh. 5.22.25.
σκαλώνια· τὰ ἀσκαλώνια (ἀσκωρώνια cod.) Hsch.
*σκαμβίς· θερμοποτίς Hsch.
*σκαμνοκάγκελος, ὁ, app. railing separating benches, ISmyrna 844 (iv/v AD), cf. Lat. scamnum, cancellus.
σκάμνος, add 'σκάμνον, τό, PDura 33.12 (iii AD)'
σκανδικοπώλης, for 'as Ar. called Euripides' read 'sobriquet applied in comedy to Euripides from his mother's alleged occupation'
σκᾰπᾰνεύς, add 'Cret. καπανεύς Inscr.Cret. 1 xxv 2 (Pyloros, ii BC); cf. σκάπετος, κάπετος'
σκᾰπάνη II, add 'Men.Dysc. 542'
*σκάπαρδος· ὁ ταραχώδης καὶ ἀνάγωγος Hsch. Also σκάπερδος, σ.· ὁ δυσχερής Suet.Blasph. 122 T.
σκαπέρδα, add 'acc. to Hsch., πᾶν τὸ δυσχερὲς σκαπέρδα λέγεται καὶ ὁ πάσχων σκαπέρδης'
*σκαπέρδης, v. °σκαπέρδα.
*σκαπλάριον, τό, prob. = Lat. scapulare, PRyl. 713ᵛ.2 (iv AD).
*σκάρηνα, ἡ, name of fish, IGC p. 98 B23 (Acraephia, iii/ii BC).
σκαρισμός, add 'ὄνον σκαρ(ισμῷ) ἴχνους PColl.Youtie 47.3 (AD 145)'
σκαρῖφος, after 'Hsch.' insert 'τῶν δέκα ῥητόρων σ., their score, i.e. a list of them giving the number of speeches each made, Sch.Aeschin. 2.18'
*σκαστός, v. °σχαστός.
σκατοφᾰγέω, add 'Men.Sam. 427'
σκαῦρος, for 'with deviating hoof' read 'of the feet, bent or twisted outwards'
*σκαφεία (?), ἁ, perh. a kind of σκαφεῖον, spade, σκαφ[είας] (gen.) rest. in list of temple implements, PP 42.42 (Tarentum, vi BC).
σκάφη I 4, for 'grave' read 'sarcophagus'
σκαφηφορέω, add 'Harp., Phot.
σκαφηφόρος, for 'Phot.' read 'σκαφηφόροι .. ἀντὶ τοῦ μέτοικοι· οὗτοι γὰρ ἐσκαφηφόρουν Ἀθήνησι Harp.'
σκᾰφίον (A) II 2 c, add 'Gal. 18(1).777.5'
σκᾰφιστήριον, add 'cf. Lat. caphisterium'
σκᾰφοειδής, after 'Arat. 19' insert 'of the scaphoid bone of the ankle' transferring 'Gal.UP 3.6' to follow this and adding 'Id. 2.776, cf. PLit.Lond. 167.27'
σκᾰφόπλωρος, delete the entry.
*σκαφόπρῳρος, ον, prob. round-prowed, POxy. 3031 (AD 302), PCair.Preis. 34.16 (iv AD); also rest. in BGU 812 ii 2 (ii/iii AD).
σκεδάννῡμι, line 1, after 'Nic.Al. 583' insert 'σκεδάζω, Hsch.; poet. also κεδάννυμι, κεδαίω, qq.v.'
σκεδασμός, add 'Procl.Inst. 13 D.'
*σκειρᾰφέω, v. σκιραφέω.
*Σκειρόμαντις· ὁ ἐπὶ Σκ[ε]ίρῳ μαντευόμενος Hsch.; cf. Σκίρος.
σκέλος, add 'prob. Myc. ke-re-a₂, legs of tripod-cauldron'
*σκέλος (B), εος, τό, app. some form of tax or other payment, PAvrom. 2.A9, B10 (i BC); perh. = σίγλος, σίκλος.

*σκεπάνισμός, ὁ, covering, prob. of woven material, PTeb. 1077.6 (iii BC).
*σκεπαρνᾶς, ᾶ, ὁ, maker of adzes, SEG 29.628 (Maced.).
σκεπαρνηδόν, add 'Gal. 18(2).728.12'
*σκέπαρνον I, line 2, after 'Homeric passages' insert 'A.fr. 78c.51 R.'; add 'app. set up as (temporary) memorial for the dead, τῷ ἀνδρὶ αὐτῆς ἔθηκε τὸ σκέπαρνον IGBulg. 2254 (ii/iii AD)' at end delete '[Hom. .. Σκάμανδρος.]'
σκέπη II, add '3 temporary dispensation from public duties (cf. Lat. vacatio), SEG 38.1462.102 (Oenoanda, AD 124/6).' add 'III transl. Lat. castellum, garrison, in bilingual inscr., SEG 31.952 (Ephesus, Trajan).'
σκέπτομαι II, line 6, after 'Cra. 401a' insert 'w. gen., τῶν ἁμυδις πάντων ἐσκεμμένος Arat. 1153'
*σκεπτώριον, τό, mirror, PMasp. 340ᵛ.40 (-ωριν).
σκέπω, after 'freq. in later prose' insert 'Peripl.M.Rubr. 8'; add 'imper. in invocations, θ(εοτό)κε, σκέπε SEG 30.1262 (Halic.), 1266 (Hyllarima, both Byzantine).'
*σκεπώνιον, τό, storehouse, PAberd. 191.8 (pl., iii AD).
σκευᾰγωγέω, add 'Philostr.VA 2.14'
σκευάζω I 2, line 2, after 'ib. 313' insert 'construct, put up monument, TAM 4(1).115 (Rom.imp.).'
*σκευαρίδιον, τό, small vessel, ἐν ἑτέροις σκευαριδίοις SB 11075.6, 16 (v AD).
*σκευᾶς, ᾶ, ὁ, Lat. scaeva, gladiator fighting with his left hand, Robert Les Gladiateurs no. 34 (Philippopolis), no. 178 (Iasos).
*σκευαστικός, ή, όν, artificial, prepared, τὸ ἐφήμερον σκευαστικόν ἐστι φάρμακον Sch.Nic.Al. 249b Ge.
σκευοθήκη, add 'written σχεοθ- SB 7182.46 (i BC)'
*σκεῦον, τό, = σκεῦος I, Inscr.Cret. 1.xvii 2a8 (Lebena, ii BC).
σκεῦος IV, for this section read 'sarcophagus, IEphes. 2227 (ii AD), SEG 24.568(b) (Maced., ii/iii AD), MAMA 8.580 (Aphrodisias, iii AD)'
σκέψις 3, the Sceptic philosophy, add 'personified, SEG 34.1667 (v AD)'
*σκηνᾰγωγός, ὁ, boat with an awning, BGU 1933.4 (ii BC).
σκηνίτης I 1, line 4, after 'a stall' insert 'Isoc. 17.33' 2, delete the section.
*σκηνοπάροχος, ὁ, one who furnishes stage-scenery, SEG 38.1462.44 (Oenoanda, AD 124/6).
σκηνορράφος, add '(sp. σκηνοράφος) IGBulg. 2198'
*σκηπτῐτης, ου, ὁ, precious stone, Socr.Dion.Lith. 32.1 G.
σκηπτός, line 3, for 'metaph. also of a dust-storm' read 'applied to a violent eddy of dust'; line 4, delete 'hurricane' and transfer quots. to exx. of 'thunderbolt'
σκηπτοῦχος 1, add 'as title of the emperor Theodosius III (died AD 450), BCH suppl. 8 no. 88; pl. of cosmic powers, Orph.H. 10.25'
†σκηπτοφόρος, v. °σκηπτρο-.
σκῆπτρον I, line 3, for 'used by the lame or aged' read 'used as an aid in walking' II 1, last line, delete 'S.OT 811' 2, add 'Just.Const. Δέδωκεν 23' add '3 as a symbol of divine power, invoked in imprecations, SEG 32.1222 (Lydia, i AD), 33.1029 (ii AD), etc.'
σκηπτροφόρος, for 'Delph. 3(1).510.3 (iv BC)' read 'BCH 116.589, 593 (Athenian epigr. at Delphi, iv BC)' and add 'also Dor. σκᾶπτρο- AP 7.428 (Mel., v.l. σκαπτο-). b subst., staff-bearer, of annual priests at Seleuceia Pieriae, OGI 245.22, 45 (ii BC).'
σκήπτω I 2, add 'σκήπτομαι κατὰ χειμῶνα, of a trierarch, enter a plea (that the ship was destroyed) by storm, IG 2².1631.344; also pass., of a ship about which such a plea was entered, ib. 1629.746.1.' II 1, for 'med. .. A.Eu. 801 (s.v.l.)' read 'μήτε τῇδε γῇ βαρὺν κότον σκήψητε A.Eu. 801 (cj.)' and delete 'Pass. .. IG l.c.'
σκῆψις, line 7, delete '(sc. τὰ τέκνα)'
σκιά 2, add 'mirror-image, Diogenian. 2.4'
σκῐᾱγρᾰφέω 1, add 'Bas.Sel.HP 3.29 (sp. σκιογ-)'
*σκῐᾱγρᾰφή, ή, = σκιαγραφία, Poll. 7.128 (v.l. σκιαγραμμή).
σκιάεις, add 'χθονὸς ὀμφαλὸν σκιάεντα prob. in Pi.Pae. 6.17 S.-M.'
σκιάζω, line 3, after 'pf. ἐσκίασμαι' insert 'Sapph. 2.7 L.-P.'
σκίαινα, add 'Gal. 6.724, Ath. 7.322f, Opp.H. 1.132, al.'
Σκιάποδες, add 'Alcm. 148 P.'
*Σκιέρεια, τά, festival of Dionysus at Alea, Arcadia, Paus. 8.23.1.
*σκιλλίτης, ου, ὁ, (wine) seasoned with squill, Afric.Cest. 2.6.12 V.
†σκίμπους, ποδος, ὁ, simple kind of bed or pallet, IG 1³.423.8, 425.11, Ar.Nu. 254, 709, Pl.Prt. 310c, X.An. 6.1.4; suspended as a litter, Gal. 6.150.
σκίμπτομαι, for 'press forward' read 'press, or throw, down' add 'IV act. intr., Hsch. s.vv. σκίμψαι and σκίμπτει.'
*σκινδακίζω, denominative vb. fr. *σκίνδαξ (cf. κίνδαξ, σκίναξ), σκινδακίσαι· τὸ νύκτωρ ἐπαναστῆναί τινι ἀσελγῶς Phot., Hsch. (s.v. σκίνδαρον.)
*σκινδάρ(ε)ιος· ὄρχησις οὕτω καλουμένη Hsch.
*σκίνιψ, ὁ, = σκνίψ, contemptuous term for an old miser, Suet.Blasph. 217 T. (cod. κύνιψ), cf. sciniphes Petron. 98.1.

σκιογράφεω — σκώληξ

σκιογράφεω, v. °σκιαγ-.
σκίπων, add 'as an emblem of rank, Plb. 32.1.3'
σκιράφειον, add 'Luc.*Lex.* 10'
*****σκιραφέω**, *behave deceitfully*, σκειραφεῖν· κακοπραγμονεῖν Hsch.
σκίραφος, add 'Suet.*Lud.* 1 T.'
Σκίρος, add 'see also °Σκειρόμαντις'
*****σκιρρωδῶς**, v. °σκιρώδης.
Σκίρτος, add 'also as a name for gladiators, *SEG* 30.1257 (Aphrodisias)'
σκιρώδης, add 'adv., σκιρρωδῶς *so as to form a callus*, Gal. 8.475.10, 9.163.9, 12.59.6'
Σκίρων I, add 'τοῦ Θρασκίου μὲν ἐν τῷ Πόντῳ, Σ. δὲ ἐν τῇ Ἑλλάδι καλουμένου *Peripl.M.Eux.* 4.2'
*****σκίφη**, ἡ, dub. sens., cf. perh. σκνιπός (A), καί τινος ποιητοῦ σκίφης μεστοὺς εἶναι τοὺς στίχους Crantor ap.D.L. 4.27.
*****σκιφός**· ὁ παρ' ἡμῖν λεγόμενος σκνιπός, ἢ τὸ ξίφος (v. σκιφός) Suid.
σκληρός I 2, add 'Arr.*An.* 1.17.6'
*****σκληρουργός**, ὁ, *quarryman*, *SEG* 31.1557 (Egypt), inscr. in *ZPE* 98.122 no. 11 (Egypt, i AD).
*****σκληρόχειρ**, gloss on λαϊνόχειρ, Hsch.
σκνιπός (A), add 'Sch.Ar.*Pl.* 84'
σκνίψ, for 'acc. σκνίπας [ῑ] Ezek.*Exag.* 135' read 'acc. σκνῖπας Ezek. *Exag.* 135a S.' and add 'cf. °σκίνιψ'
*****σκοδίσκος**, ὁ, wd. occurring in inventory of household furniture, *PLond.* 191.4 (ii AD).
σκοῖδος, add '*EAM* 74 (iii/ii BC)'
σκόλεφραι, after 'κατακεκαυμέναι' insert '(κατακεχυμέναι Theognost.*Can.* 71 A.)'; add 'cf. σκολοφρή'
*****σκολιοδρόμος**, ον, *running on an indirect course*, of the nymphs of underground streams, Orph.*H.* 51.4; of the moon, Man. 4.478.
σκόλιον, for '*song which went round crookedly at banquets*' read 'kind of drinking-song'
*****σκολλύφιον**, τό, app. kind of hair-style, Hsch. s.v. κόρυφος.
σκολόπενδρα 2, before 'Ael.*NA* 7.26' insert '*AP* 6.222 (Theodorid.), 223 (Antip.)'
σκολοπισμός, add '**II** *protection by palisades*, prob. in *SB* 7188.13 (ii BC); cf. σκολοπίζω.'
σκολοφρή, add 'cf. ‡σκόλεφραι'
*****σκόλυπνον**· κεκακωμένον Theognost.*Can.* 71 A.
σκολύφρα, add 'σ.· πόρνη Theognost.*Can.* 71 A.'
*****σκοπελάριος**, ὁ, *watchtower guard*, *OFlorida* 6.8 (?ii AD), *SB* 9549(4).8 (iii AD).
*****Σκοπελία**, ἡ, cult-title of Artemis, *IG* 4².505 (Epid.).
*****Σκοπελῖτις**, ιδος, ἡ, cult-title of Artemis, *SEG* 35.989.10 (Knossos, ii/i BC).
σκοπή I, add 'of observation point for divination through birds, *Inscr.Perg.* 8(3).115'
σκοπιάζω I 2, for 'Isis-worshippers' read 'fishermen' and for '(Callipolis)' read '(Parium), cf. Robert *Hell.* 9.81'
σκοραδᾶν, for '*Docum.* .. (Cyrene)' read '*SEG* 9.35 (iii BC), 41 (ii BC, both Cyrene)'
σκορδάτον, add 'perh. also *POxy.* 1923.15 (v/vi AD)'
σκορδευτής, add '*BGU* 1530.7'
σκόρδιον 1, add 'ostr. in *ZPE* 98.138 no. 29.4 (Egypt)'
*****σκορδίσκος**, ὁ, *saddle*, σκορδίσκον στρατιωτικόν *Edict.Diocl.* 10.2 (*SEG* 37.335 iii 2).
*****σκορδόω**, = σκοροδόω, Poll. 5.93.
σκόροδον I, add 'Ar.*Th.* 494, etc., perh. w. play on obscene sense "penis" (cf. *Philol.* 127.139ff.)'
σκοροδύλη· θαλάσσιος ἰχθύς. ἔνιοι κορδύλη Hsch.; cf. σκορδύλη.
σκορπέρως, τος, for '*BCH* 2.323' read '*Inscr.Délos* 1414*a*i17 (ii BC)'
σκορπίζω 1, add '*spread* manure, *PSoterichos* 1.26 (AD 69), 2.21 (AD 71), *BGU* 2354.8 (ii AD)'
σκορπιόδηκτος, add 'Heras ap.Gal. 13.786.13'
σκορπίος II, add '*IGC* p. 98 A24 (Acraephia, iii/ii BC)'
σκορπισμός, add 'ἄμμου *PMil.Vogl.* 52.106, 117 (ii AD)'
σκοτερός, after '= σκότιος' insert 'θάλαμος epigr. in *IG* 9².340.8 (Thyrrheum, ii/i BC)'
*****σκόρτια**, ἡ, Lat. *scortea*, *coat made of hide*, *Edict.Diocl.* 10.16, 16a (*SEG* 37.335 iii 16, 16a).
σκοτία III, for this section read 'Σκοτία, epith. of Aphrodite in Phaestus, Crete, *EM* 543.49; cf. perh. σκότιος 1; also in Egypt, Hsch.'
σκότιος I 1, line 8, for 'rare in Prose .. Charax 6' read 'poetic term for "illegitimate", Plu.*Thes.* 2, νόθος καὶ σκότιος id. 2.751f; παῖς σ. Charax 6'
*****σκοτοείμων**, ον, gen. ονος, *dark-clad*, χθών *Hymn.Is.* 27.
σκοτοιβόρας, add 'also σκοτοιβόρᾶς Suet.*Blasph.* 99 T.'
σκοτομήνιος, add 'νύκτ[ες .. σκοτο]μήνιοι Hes.*fr.* 66.5 M.-W.'
*****σκοτόταφρος**, ὁ, ?*hidden trench*, cod. B in Ps.-Callisth. 3.23.

σκοτουλᾶτος, v. ‡σκουτουλᾶτος.
σκοτόω, line 1, after '*Aj.* 85' insert '*fr.* 269a.30 R.'
σκουτάριος, add '*SEG* 37.1081 (Nikomedeia, iv AD)'
σκουτλάριος, delete 'or .. *flooring*'
σκουτλόω, for '*mosaics*' read '*a covering of thin plates of marble arranged in patterns (opus sectile)*'; add 'τὰς εἰσόδους καὶ ἐξόδους *AC* 2.74 (Aphrodisias; lapis κουκλόσαντα after wd. ending in -ς)'
σκούτλωσις, after '*chequered work*' insert '*decoration with opus sectile*, v. °σκουτλόω'
+**σκουτουλᾶτος**, ον, Lat. *scutulatus*, *with a checked pattern*, of material, *Peripl.M.Rubr.* 24 (written σκοτ-), *Edict.Diocl.* 20.11 (σκουτλ-), Lyd. *Mag.* 1.10.
*****σκρείβας**, α, ὁ, Lat. *scriba*, *BCH* 7.275 (ii AD), *POxy.* 59.9 (iii AD), *PColl.Youtie* 79.12 (iv AD).
*****σκριβηνδάριος**, ὁ, app. *secretary*, *SEG* 36.1335 (Beersheba, vi AD).
*****σκρίβων**, ωνος, ὁ, *officer of the imperial guard*, Agath. 3.14.5, *scribon(o)s* *SEG* 34.927 (Crete, vi AD).
σκρινιάριος, add '*MAMA* 5.309 (Nacolea, v/vi AD; sp. ἰσκρην-)'
σκύζομαι, before '*to be angry with*' insert '*to growl at*'; line 5, after '(*Il.*) 9.198' insert 'Theoc. 16.8, 25.245; also act. 'σκύζουσιν· ἡσυχῇ ὑποφθέγγονται, ὥσπερ κύνες Hsch.; cf. σκύζειν (cod. σκυζᾶν) Poll. 5.86'
σκυθάριον, for 'θάψος' read '*smoke-tree, Rhus cotinus*'
Σκύθης I 1, for 'Hes.*fr.* 55' read 'Hes.*fr.* 150.16 M.-W.'; add 'prov., ἡ ἀπὸ Σκυθῶν ῥῆσις of brutal plainness of speech, Demetr.*Eloc.* 216, D.L. 1.101' **II**, for 'τοξότης III' read 'τοξότης II'
Σκυθία, after 'Σκυθίηνδε, ib. 256' insert '**Σκυθίηθεν**, epigr. in *SEG* 39.855 (Patmos, iii/iv AD)'
Σκυθικός I, add 'as the title of a Roman legion, *SEG* 33.1194 (Cappadocia, i/ii AD), λεγεῶνος τετάρτης Σκυθικῆς *ICilicie* 87 (ii AD)' **II**, add 'also Σκυδικαί, Hsch.'
σκυλάκαινα, add 'applied to Hecate, εἰνοδία σ. θεά *AAWW* 1961.125 (Maeonia, ii AD)'
+**σκυλάκευμα**, ατος, τό, *whelp, cub*, *AP* 7.433 (Tymn.); pl., Διὸς σ. (σκυλεύματα cod.), of Amphion and Zethus, *AP* 3.7 (Inscr.Cyzic.), ib.16.91 (ii BC).
σκυλακῖτις, for '*protectress of dogs*' read '*mistress of dogs*'
σκυλακοκτόνος, add '= *Lyr.adesp.* 1029.2 P.'
σκυλακώδης, after '*dog*' insert 'φωνή *PMag.* 4.2810'
σκυλάω, delete 'dub. .. Cyzic.' and add '**2** *plunder*, τὸν ναόν *UPZ* 6.22, cf. 15 (ii BC). **3** = σκύλλω I 2, *AP* 3.6, *GVI* 1946.14 (Nicaea, Bithynia, ii/iii AD).'
σκυλεύω 1, line 7, for '*BCH* .. Cypr.' read '*SEG* 6.802 (Salamis, Cyprus, ii/iii AD)'; add 'obsc., κατευδούσης τῆς μητρὸς ἐσκύλευε τὸν βρύσσον Hippon. 70.8 W.'
σκυλήτρια, add 'Eust. 1072.64'
*****σκυλίζω**, = σκυλεύω, ἐσκούλιξε Boeot.*adesp.* 39 *fr.* 5.3 P.
σκύλλω I 1, line 4, before 'Pass.' insert '*tear* or *rend apart*'; add '*SEG* 39.340 (Corinth, Rom.imp.)' **2**, for '*AP* 3.6 (Inscr.Cyzic.)' read 'cj. in *AP* 3.6 (Inscr.Cyzic.) v. ‡σκυλάω'
σκυλτικός, add 'Heph.Astr. 2.33.15, 2.36.15,16'
*****Σκυριανός**, ή, όν, *of Scyros*; name of a kind of marble, *Edict.Diocl.* 33.14.
*****σκύρον** (B), ὦ σ., wd. of uncertain meaning, Alc. 167.3 L.-P. (cf. 58.13, 174.2), app. voc.sg. of *σκύρων.
σκύρος, for 'cf. Poll. 9.104; cf. σκίρος' read 'the centre line in the game ἐπίσκυρος, as marked out w. stone chippings, Poll. 9.104; hence perh. ἐπὶ σκύρῳ (or σκύρων; ἐπισκυρῶν cod.) πολέμοιο Call.*fr.* 567 Pf.(v. *CR* 73.101ff.)'
σκυτάλη, line 11, after 'ἀχνυμένη σκυτάλη' for '(dub. sens.)' read '(v. *CQ* N.S. 38.42ff.)' add '**VI** σκυτάλαι· αἱ ἱππικαὶ ἷλαι .. ἢ θύλακες δερμάτινοι Hsch.'
σκυτάλιον I 2, delete '*flute*' and 'perh. so in Thphr.*HP* 4.4.12' **5**, for '*BCH* 29.546 (Delos, ii BC)' read '*Inscr.Délos* 1432*Ba*i12, 1409*Aa*i111 (ii BC)'
σκύταλος, add '*BGU* 2361a i 3 (iv AD)'
*****σκυταλοφορέω**, *carry a σκυτάλη*, of heralds, Sch.Pi.*O.* 6.154a (prob. for σκυτοφορεῖν).
σκυτίς, for '*amulet*' read '*amulet-case*'
σκυτοτομεῖον, for 'Macho ap.Ath. 13.581d' read 'Macho 359 G.' and for 'v.l.' read 'cod.'
σκύφιον, after 'σκύφος' insert '*Inscr.Délos* 320*B*53 (iii BC), 1409*Aa*i105 (pl.)'
*****σκύφισμός**, ὁ, form of *operation on the scalp*, *An.Boiss.* 1.230.
σκύφος, penultimate line, before 'Anaximand.' insert 'Anacr. 88 P.'; add 'sp. σκόφος *SEG* 24.361 (Thespiae, iv BC)'
σκώη· παιδίσκη Theognost.*Can.* 71 A. cf. σκώ.
*****σκωληκοβρωσία**, ἡ, *eating by worms*, *PMasp.* 325 ιι⁷ 16 (vi AD).
*****σκωληκοέρημος**, ον, perh. *laid waste by worms*, *PTeb.* 1043.40 (ii BC).
σκώληξ, after 'ὁ' insert '(also ἡ acc. to Eust. 1504.39)'

σκῶλον II, for 'stumbling-block .. σκάνδαλον' read 'thorn in the flesh, affliction'
*σκωπελαδᾷ· τὸ σχηματίζεται Theognost.Can. 71 A.; cf. σκώψ 2.
σκωπτικός, after 'jesting' insert 'Ph. 1.215.19'
σκωπτόλης, add 'Philostr. VA 1.7'
σκωρσέλεινα, add 'written σκωρσελήνης in PMich.III 212.8 (ii/iii AD); perh. celeriac (σκώρ, σέλινον)'
σλιφομαχος, for 'Cyrenaic' read 'Arcesilas'
σμαράγδιον, add 'written ζμαράγδιον in Inscr.Délos 1409Bai102 (ii BC)'
σμάραγδος I, ζμάραγδος, add 'also in pers. n., SEG 24.1076 (Moesia, ii/iii AD)'
†Σμάραγος, ὁ, mischievous demon supposed to cause pots to break during firing, Hom.Epigr. 14.9.
*σμενία, app. Dor. for σμινύη, PP 42.42.9 (ii BC).
*σμερδαλεότης, ητος, ἡ, awesomeness, Eust. 1702.46.
σμημᾰτοθήκη, add 'IG 2².1469.97'
*σμήνιγξ, ιγγος, ἡ, = μῆνιγξ, prob. in Nic. Th. 557.
*σμηνών, ῶνος, ὁ, = σμηνιών, BCH 22.402 (ζμη-; Olymus, i BC).
*Σμικρίνης, ου, ὁ, a typical skinflint, pers. n. in com. Men.Epit., Asp., Jul.Caes. 311a, Them. Or. 34 p. 462 D.
*σμῐλεύω, carve, ἐσμιλευμένος κατακεκομμένος Hsch. (cod. ἐσμηλ-).
Σμίνθεια, add 'Illion 125.8, 14 (ii/iii AD)'
*Σμινθεῖον, τό, sanctuary of the Sminthian Apollo, IGRom. 4.246.3 (Troas).
Σμινθεύς, add 'cf. Myc. si-mi-te-u, pers. n.'
*σμῐνύδιον, τό, dim. of σμινύη, Ar.fr. 889 K.-A.
σμιρεύς, add '[σμ]ιρεύς SEG 18.743 (Cyrene, iii BC); app. abbrev. ΣΜ, SEG 9.44 (Cyrene)'
Σμισιών, after 'name of month' insert 'deriv. of Σμίνθος'
σμόρδωνες, add 'cf. pers. n. Σμόρδον SEG 28.37 (Camarina, vi BC)'
†σμῠρίζω, = μυρίζω, ἐσμυριχμένας Archil. 48.5 W.; cf. Hsch. ἐσμυριχμέναι.
σμύρις, line 3, after 'Orib. 15.1.20 codd.' insert 'σμεῖρις, εως, ἡ, IEphes. 23.18 (ii AD)'
σμύρνα, after 'ζμύρνα as in' insert 'Hyp.Ath. 6'
*σμυρνίτης, ου, ὁ, λίθος σ., precious stone, Socr.Dion.Lith. 46.1 H.-S.
σμώδιξ, line 3, after 'cf.' insert 'Lyc. 783'
σοβαρεύομαι, add 'Hsch. s.vv. σοβαρεύεσθαι and σοβαρεύεται'
σοβᾰρός II, add '3 adv. -ρῶς, colloq., tremendously, POxy. 3356.14 (i AD).'
σόγχος, delete '(where ἐξογκοῖτ' is a pun on ἐκσογκοῖτ')'
*σόκκος (B), ὁ, Lat. soccus, slipper, Edict.Diocl. 20A, 21, al.
*σόλδιον, τό, name of coin, Anon.Alch. 324.15, cf. Lat. solidus.
*σολέμνιος, α, ον, Lat. sollemnis, customary, Just.Edict. 13.13; neut. pl. subst., customary services or expenditures, ib. 13.14, 21.
σόλιον, after 'PSI 206.9 (iii AD)' insert '(sg. used to denote a pair)'
*σολίτης, ου, ὁ, sandal-maker or -seller, σο‹λ›ίτι καὶ καμισογοραστῇ Corinth 8(3).522.
σολοικιστής, add 'cf. Hsch. s.v. Βρίγες'
*σολοιτῠπ[ίη], product of forging at Soli, i.e. bronze, dub. rest. in Call.fr. 85.11 Pf. (αμφισολοιτυπ[pap.; could be divided ἀμφὶς ὀλοιτυπ[).
σολοιτῠπος II, before 'forged' insert '(proparox.)'; add '(prob. alternative expls. of wd. w. single meaning)'
*σονωπτώριον, τό, wd. in inventory of receptacles and vessels, BGU 2360.9 (iii/iv AD), perh. for συν-, cf. συνοπτάω.
*σορέλλην, νος, ὁ, app. = σορέλλη, Suet.Blasph. 213 T.
†σορίδιον, τό, little coffin or urn, Hierocl.Facet. 97; τὸ σορίδιον ἄγγος TAM 2(3).1164.2 (Lycia).
*σορικόν, τό, name of an eyesalve, Aët. 7.107 (374.9 O.) (s.v.l.).
σόριον, add 'σόριν ὑποκίμενον ITyr 31'
σοροπηγός, add 'Lib. Or. 1.225'
σορόπληκτος, add '= Suet.Blasph. 211 T.'
σορός, line 1, after 'ἡ' insert '(app. ὁ, ITyr 10)' I, transfer 'Ar. V. 1365' fr. section II, and add 'on an ossuary, SEG 31.1405 (Jericho, i AD); of a mummy, PHaun. 17.14 (ii AD)' II, for this section read 'derogatory term for old man, Macho 301 G., Luc.D.Meretr. 11'
σορώϊον, for 'cerecloth' read 'mummy-dressing'
*σουβαδιούβας, α, ὁ, Lat. subadiuva, under-assistant, POxy. 1042.13 (vi AD), BCH suppl. 8 no. 148 (Thessalonica, v/vi AD).
*σουβαλάρι(ο)ν, τό, Lat. subalare, SEG 37.335 iii 10 (Edict.Diocl. 10.10).
σοῦβος, for 'an unknown animal' read 'wild sheep or goat'
σουβρίκιον, after 'subricula' insert '(dim. of σουβρικός, cf. rica)' and add 'PMich.III 201.8 (i AD)'
†σουβρικός, ὁ, article of clothing, PCornell inv. I 11 (ZPE 22.53, AD 59), σουβρικὸς superaria, Gloss.
*σουβσκριβενδάριος, ὁ, app. under-secretary, SEG 9.356.66 (Cyrene), 32.1554A (Arabia, both vi AD); cf. °σκριβενδάριος.
*σουγγεστίων, ονος, ἡ, Lat. suggestio, the supplying of an answer to one's own question, Cod.Just. 4.59.1.1.

*σουγλάριον· ἐργαλ(ει)οθήκη Hsch.
*σουμάριον, τό, Lat. summarium, summary, Iambl.Alch. 289.9.
*σουμμάριος, ὁ, Lat. summarius, title of official on the staff of a comes, Just.Nov. 30.1.1, 30.7.1.
*σουμμαρούδης, ὁ, fr. Lat. summa rudis, chief instructor at a gladiatorial school, AS 20.39 no. 4 (BE 1971.670), SEG 36.595 (Maced., ii/iii AD).
*σοῦμμος, ὁ, Lat. summus, highest in rank, σ. κουράτωρ PBeatty Panop. 1.393, 395, 2.28; σ. εἴλης fr. Lat. summus alae, CPR 6.76.8 (ii/iii AD). 2 a position on the board in the game of τάβλη, AP 9.482.9 (Agath.).
Σουνιεύς, add 'ἐξουνιέων (i.e. ἐκ Σ.) SEG 24.226 (i BC/i AD)'
*σουπερνουμεράριος, ὁ, Lat. supernumerarius, supernumerary, (ἑκατοντάρχῳ) σουπερνουμεραρίῳ PBeatty Panop. 2.183, 264, 269, 289 (AD 300).
*σοῦφον, τό, app. kind of vessel or measure, cf. ὀκτάσουφος, τρίσ-, PMich.inv. 3725 (ZPE 61.80, iv AD).
*σουφρουμεντάριος, ὁ, Lat. *suffrumentarius, corn supply assistant, POxy. 1903.7 (vi AD).
σοῦχος, after 'ὁ' insert '(Egypt. seḥu)'
σοφία, line 1, delete 'prop.' 2, add 'Lxx Pr. 1.7' 4, for this section read 'personified in Jewish "wisdom" literature, Lxx Pr. 8, Si. 24, etc.'
σοφιστικός 1, after 'id.Sph. 224c' insert 'σ. σχήματα characteristic of sophists, Hermog.Meth. 13'
σοφίστρια, add 'Cels.ap.Orig.Cels. 5.64'
*σοφοδιδάσκαλος, ὁ, teacher of wisdom, prob. of a rabbi, inscr. in Sardis from Prehistoric to Roman Times. Results of the archaeological exploration of Sardis 1958-1975, ed. G.M.A. Hanfmann, (Cambridge Mass. 1983), pp. 183, 189 (v AD).
*σπάδη, ἡ, = σπάθη, SEG 7.376.17 (Palmyra, ii AD).
**σπαδίκινος (?), v. °σπανδίκινος.
*σπαδίκιον, τό, dim. of σπάδιξ, IG 14.956Β25 (Rome, iv AD).
σπάδων, add '2 gelding, PCair.Zen. 802.22, 28, 33 (iii BC).'
σπαθάρικόν, for 'thin upper-garment' read 'prob. closely woven veil'
*σπαθαρικός, ή, όν, of or used by a σπαθάριος, σπαθαρικὴ μάχαιρα BGU 2328.10 (v AD).
*σπαθᾶς, ᾶ, perh. cutter or swordsmith, PBerl.Bork. (Tyche 6.235).
σπαθάω II 1, add 'σπαθήσῃ (v.l. -εις) ἐπὶ τούτῳ (sc. ἔρωτι) you will waste your strength on this, Luc.Luct. 17'
σπάθη 5, add 'synecd. sword, SB 1.21 (iii AD), Just.Nov. 85.4' add '10 perh. a vase-shape, AJA 45.598 (graffito on amphora).'
σπαθίζω, add 'Aët. 15.13 Z.'
*σπάθινος, η, ον, woven or perh. close-woven: τὰ σ. articles of dress, Aq.Is. 3.19 (L.-R.); cf. σπαθαρικόν.
σπαίρω, for 'of dying fish' read 'esp. when near to death' and add 'Apollod. 3.13.6'
†σπάλαθρον, = σπάλανθρον, σκάλευθρον, etc., fire rake or poker, Poll. 10.113; fig., of an officious person, Suet.Blasph. 164 T.; prob. Myc. qa-ra-to-ro, app. a utensil.
*σπανδίκινος, η, ον, app. colour term describing a στρόφος, (perh. for *σπαδίκινος, cf. σπάδιξ 2, bay, red), SEG 38.1210.19 (Miletus, ii BC).
σπάνιος I, add 'of precious stones, λίθοις ποικιλίοις τοῖς ἀρίστοις καὶ σπανιωτάτοις πολυτελείᾳ SEG 39.1055.10 (Neapolis, AD 194); λίθος ὅ σ. the rare stone (owned only by the king of Persia), Socr.Dion.Lith. 28.1 H.-S.'
σπᾰνός, line 1, after 'Hsch.' insert 'beardless (as mod. Gk.), Anatolius in Cat.Cod.Astr. 8(3).188'; after 'Ptol.Tetr. 144' insert '(v.l.)'
Σπᾰνός 1, add '(λιθάργυρος) Σπάνη Dsc. 5.87; masc. pl. subst., ἔπαρχο]ν σπείρης β' Σπανῶν INikaia 56 (Hadrian). b ἔλαιον Σπανόν, a kind of oil, PSorb. 62.1, POxy. 2052 (vi AD); also Σπανόν, τό, Gp. 9.26, POxy. 1862.' 2, before 'grey' insert 'Spanish, referring to a certain textile colour'
*σπᾰνοτεκνία, ἡ, paucity of offspring, rest. in Cat.Cod.Astr. 2.163.9.
†σπαπιπρώτας, α, ὁ, app. a priest with some function at sacrifices, IPamph. 3.17, 24 (Sillyon, iv BC).
*σπαρακτός, ή, όν, torn, rest. in A.fr. 451s.10.2 R.
σπαρνός, for 'poet. for σπανός, σπάνιος' read 'sparse, scarce' and add 'Hes.fr. 66.6 M.-W.'
σπαρτίνη, add 'Poll. 7.114'
*σπαστήρ, ῆρος, ὁ, perh. = ἐπισπαστήρ, SEG 21.559.7 (Athens, iv BC).
σπάω III 1, lines 3/4, after 'Alex. 5, cf. 285' insert 'w. gen., ἀκράτου Ath. 613a'
*σπεγκρᾰνίς, v. °ἐπεγκρανίς.
*σπειλες, pl., perh. gut strings, BGU 2361b ii 11 (iv AD); cf. σπίλα Hsch.
σπεῖρα I 8, add 'PHib. 217.20, al. (ii AD)'
σπείραμα 1, delete 'metaph. .. App.Anth. 3.186'
σπειράομαι, line 2, delete 'πέριξ ..'
σπειράρχης, after 'σπεῖρα II 2' insert 'RA 6.31 (Amphipolis, ii BC), IGBulg. 1517.4 (Philippopolis, iii AD)'

*σπειρόω (B) (or ?σποpόω), sow, εἰ δὲ σποιρώσωμε φασούρια ἐπ' ἔτη PMich.XIII 666.33 (vi AD).

σπείρω I 3, line 5, after 'S.El. 642' insert 'cf. A.fr. 78a.65 R.'

*σπείρωμα, ατος, τό, dub. sens., Aq.Je. 50.10 (v.l.).

σπεκλάριον, add 'Cyran. 25.1, Zos.Alch. 139.2'

σπέκλον 1, add 'b prob. window-pane, PWisc. 66.1, 5 (AD 584), POxy. 1921.12, 13 (AD 621).'

σπεκουλάτωρ, line 1, before 'ὄρος' insert '(σπεκλάτωρ BE 1959.260 (Tomi))'

*σπεκταβίλιος, α, ον, Lat. spectabilis, title of the middle rank of Roman senators, cf. ‡περίβλεπτος, Just.Nov. 24.4, al.

σπένδω, line 6, after 'Plu.Rom. 19' insert 'subj. σπε͂σθέωσαι SIG 57.8 (Milet., v BC)' **I 1**, at end before 'οὗτος θεοῖσι' insert 'ὅταν οἱ κρητῆρες σπε͂σθέωσι SIG l.c., βωμὸς Διὸς .. μέλιτι σπένδεται IG 12(5).1027 (Paros, c.500 BC)'

σπέος, for 'Inscr.Cypr. 98 H.' read 'se-pe-o-se, ICS 2.2; 3.3'

σπέρμα I 1, add 'crops generally, τὰ σ. Lxx 1Ki. 8.15, SEG 33.1041 (late ii BC)' **II 1**, for 'φέρειν' read 'φέρων' and for 'to be pregnant of' read 'bearing the seed of' · add 'Myc. pe-ma, var. sp. pe-mo'

σπερμαίνω 2, after 'Hes.Op. 736' insert 'fr. 1.16 M.-W. (cf. CPh. 81.221-222)' and add 'of the female, Gal. 4.536.9, 593.10; w. internal acc., γόνιμον σπερμαίνειν id.UP 14.7 (II 302.13 H.)'

*σπευδίας, α, ό, kind of wheat having a light yield, Plin.HN 18.64.

σπεύδω, line 1, after 'σπεύσω' insert 'A.Ag. 601'

*σπευσίδωρος, ον, eagerly bringing gifts, of Prometheus, A.fr. 204b.12 R.

σπήλαιον 1, add 'in cult-title, Μητρὶ ἀπὸ σπηλαίου SEG 34.1293 (Phrygia, Rom.imp.)' **2**, delete the section.

σπιλάς (A) I, at end for 'Simon.(?) 179' read 'AP 6.217.2 ([Simon.])'

σπινθήρ, after 'Plb. 18.39.2' insert 'σπινθῆρα τῆς ψυχῆς (i.e. spark of life) εὗρεν ἐν αὐτῇ Philostr.VA 4.45'

σπινθηρίζω I, add '2 transf., sparkle, Epiph.Const. in Lapid.Gr. 195.10.'

σπῐνός, for '= ἰσχνός' read 'gaunt, lean' and add 'Ptol.Tetr. 144'

*σπλαγχνοεντεριφόρον, τό, tray for carrying σπλάγχνα and ἔντερα, subdivision σπλαγχνοεντεριφόρων Kafizin 285 (iii/ii BC).

σπλάγχνον, line 1, at end insert 'also fem. pl., σπλάνχνας SEG 31.895 (Africa)'

σπογγίτης, after 'like a sponge' insert 'ἄρτος σ. Hsch. s.v. κύστη' and delete 'only' before 'fem.'

*σπογγοκέφαλος, ό, spongehead, pers. description in CPap.Jud. 512.1 (vi/vii AD), SB 4668.6, 7 (vii AD).

+σποδησιλαύρα, term for a prostitute, σ.· ἡ πόρνη, λεγομένη οὕτω παρὰ τὸ διατρίβειν τὰ πολλὰ ἐν ὁδοῖς ἢ καὶ δημοσίᾳ συμπλέκεσθαι Suet.Blasph. 33 T., κατὰ τὴν κωμῳδουμένην σποδησιλαύραν Eust. 1088.37, Hsch.; cf. σπολάδιον.

*σπολάδιον, τό, dim. of σπολάς, rest. in IG 2².1648.20 (Delos, v. BCH 62.249f.).

σπονδαυλέω, σπονδαύλης, for 'flute' read 'pipe(s) or aulos'

*σπονδειοδάκτυλος, ό, metrical foot consisting of spondee and dactyl, Rh. 6.103.20.

*σπονδεῖον I, add '4 τὰ σπονδεῖα a festival of Heracles, SEG 39.148 (Attica, iv BC).'

*σπονδέριον, τό, perh. = σπονδάριον, PAmst. 68.25 (ii/iii AD).

σπονδή II 1, line 10, for 'less freq., .. 5.76' read 'σ. ποιεῖν arrange a truce, Th. 5.76.2, ἐκείνων (sc. τῶν θεῶν) .. σ. ποιούντων Ar.Pax 211'

*σπονδιστήριον, τό, perh. place for drink-offering, SEG 38.1321.2 (Pisidia, ii/iii AD).

σπονδοποιός, add 'Sokolowski 1.57 (Caria)'

*σπονδοφόρον, τό, libation-tray, CISem. 2.3923 (Palmyra).

*σπονδυλοκόπος, ό, app. some kind of unqualified practitioner, quack, οἱ σπονδυλοκόποι μιμοῦνται τοὺς γραμματικοὺς καὶ ὑποδεικνύσιν ἐν αὐτοῖς Sch.Hermog.Id. 2.5 in Rh. 5.536; cf. σφόνδυλος, etc.

σπορά I 1 b, add 'designating the father (opp. mother), ἡ γαστὴρ τῆς σπορᾶς προκρινέσθω Just.Nov. 162.3'

*σποράδιος, α, ον, scattered, sporadic, ἄμπ(ελοι) σπ[ο]ράδιοι καὶ ἐλάινα καὶ ἕτερα φυτά PPetaus 17.4 (c.AD 184).

σποραῖος, after '= σπόριμος' insert 'CPR 1.45.7, perh. also PMich.inv. 3761 (Tyche 1.181)'

*Σποργίλος, ό, pers. n., w. play on name of bird, prob. sparrow, Ar.Av. 300; cf. (σ)πέργουλος, mod. Gk. σπουργίτης.

σπορητός 3, add 'Gal. 17(1).18.2'

*σπορητός (B), ή, όν, = σπορευτός, ἐν ἀμπέλοις, ἐλαιῶσιν, σπορητοῖς τόπ[οις] BCTH NS 19b.109.

σπόριμος 1, fem. -η, add 'σπορίμην αὔλακα Call.fr. 22 Pf.'; line 5, for 'BCH 51.149 (Salamis Cypr.)' read 'SEG 6.802 (Salamis, Cyprus, ii/iii AD)'

σπόρος I, add '3 sown land, PLond. 1980.4 (iii BC), PEnteux. 60.4.9.' **II 4**, before 'semen genitale' insert 'Nic.Al. 582'

*σποροω, v. °σπειρόω (B).

σπουδαιότης, add 'CPR 8.20.2 (iii AD)'

σπουδάρχης, for 'σπουδαρχίας' read 'σπουδαρχίαις'

σπουδαστής, add 'also prob. student (as in mod. Gr.), CPR 8.10.10 (ii/iii AD)'

σπουδή II 1 a, add 'σπουδῇ w. gen., by the efforts of, SEG 30.1687 (Palestine), 1690 (Arabia, both vii AD)' **III**, add 'acc. as adv. (unless error for dat.) zealously, Mitchell N.Galatia 120 (AD 124)'

σπύλιον, after 'IG 2².1358' insert 'B10'

σπύλων, delete the entry.

+σπῡράμινος, η, ον, Dor. form of πυράμινος; neut. pl. as subst. wheat-flour, SEG 9.13 (Cyrene, iv BC, prose); cf. °σπυρός.

σπυράς, add 'Eup. 15 K.-A.'

σπυρίς, form σφυρίς, add 'Men.Sam. 297, PSI 543.54 (iii BC)'

+σπυρός, ό, Doric form of πυρός, IG 4².40.8, 46.38 (Epid., v BC), SIG 1026.9, 1027.11 (Cos, iv BC), SEG 9.11, 12, al. (iv BC), 18.743 (iii BC), 9.42, 43 (ii BC, all Cyrene), An.Ox. 1.362, EM 724.33.

σταβάριον, delete the entry (v. ‡ταβλάριος).

σταβλίτης, for 'official in the posting service' read 'stableman' and add 'SB 9920 ii 7.1, 9.12, 10.2 (σταυλ- pap.; vi AD)'

στάβλον, add 'for race-horses, SEG 34.1437 (Syria, v/vi AD)'

σταγματοπώλης, add 'POxy. 3748.6 (AD 319)'

*σταδιάρχης, ου, ό, one in charge of the race-course or arena: transf., of Christ (w. prothetic vowel), εἰσταδιάρχη νικηφόρε Χ(ριστ)έ SEG 32.1588 (Egypt, vi/vii AD); cf. as pr. n. of bull, Robert Les Gladiateurs 191C (Cos).

στάδιον II 3, add 'κυκλοῦντες τὸ σ. going round the walk, Lxx Su. 37'

*σταδόν, adv., (ἵστημι) in standing position, Theognost.Can. 58 A.

στάζω II 1, add 'fig., of divine anger, Lxx Je. 49(42).18, 51(44).6'

σταθμάω II 2, add 'τἀρετῇ σταθμώμενος τὰ πάντα Trag.adesp. 327 K.-S.'

*σταθμηδόν, gloss on στήδην, Sch.Nic.Al. 327a Ge.

Σταθμία, add 'of Ennodia, IG 9(2).577 (Larissa)'

σταθμός I 1, line 10, for 'A.Pr. 398' read 'A.Pr. 396'   add 'Myc. ta-to-mo, sheepfold, pillar, perh. also weight'

σταθμοῦχος I, add 'BASP 16(3).196.6 (iii AD)'   add 'III prob. subdivision of an ἄμφοδον, SB 9869 a3, b3 (AD 160).'

σταΐς I, line 1, after 'not στάς' insert 'but στάς ἄνευ τοῦ ι" in Attic acc. to Phot., dat. στατί (sic) Sch.Hippon. 118D6 W.'

*στάλαχμος or -υχμος, ό, cj. for σταλαγμός in Sapph. 37 L.-P.

στάλιξ, add 'transf., of the spines of the sargue, Opp.H. 4.606'

*σταλογράφος, v. °στηλο-.

στάλσις, add '2 gloss on στόλος, Sch.Pi.P. 8.140, Sch.Pi.N. 3.27.'

σταμίν, add 'of structures in house, epigr. in SEG 30.1272.4 (Caria, Hellen./Rom.imp.)'

σταμνίον 1, for 'Men. 129' read 'Men.Dysc. 448'

*σταμνίσκιον, τό, small jar, Phot. a 1198 Th.

*στάμνον, τό, = στάμνος, Φιλοξένης τὸ στάμνον SEG 33.1238 (Bactria, ii BC, s.v.l.).

στάμνος, line 4, after 'D. 35.32' insert 'vase, IG 1³.425.48ff.'; line 8, as a measure, add 'Lang Ath.Agora XXI Ha54'

*στάς, αδός, fem. adj., standing, στάδα λίμνην Suppl.Hell. 1055.

στάσιμος II 1 b, line 5, for 'BCH .. Cypr.)' read 'SEG 6.802 (Salamis, Cyprus, ii/iii AD)'

στάσις B I 2 a, add 'military station, garrison, Lxx Jd. 9.61' **III 2**, after 'discord' insert 'Alc. 130.26 L.-P.'

*στατίας, ό, kind of bread or cake, gloss on σποπία, Hsch.; cf. στατίας, ‡σταΐς.

*στάτιον, τό, dim. of στατός I 2, Inscr.Délos 1441Aii34 (ii BC).

+στατίων or στατιών, ωνος, ή, (ό, OGI 755.4 (Miletus)), military guard-post, TAM 2.1165 (acc. στατίωναν), SEG 2.666 (Bithynia, iii AD).   **2** association of foreigners in town or their meeting-place, OGI 595.5, al., 755.4, Robert Hell. 7.198 (Puteoli, ii AD).

στατιωνάριος, for '= stationarius, IG 14.830.32' read 'member of a °στατίων (sense 2), Robert Hell. 7.21'

*στατοῦτος, η, ον, Lat. statutus, fixed, determined, στατοῦτον ἀριθμόν Just.Nov. 16 rubric; neut. subst. fixed amount, complement, etc., ib. 3.2.1.

σταυροειδής, adv., before 'Hsch.' insert 'Anon.Alch. 321.11'

*σταυροποιία, ή, crucifixion, POxy. 2339.25 (i AD).

*σταυροφόρος, ό, bearer of the cross, Ἰησοῦ Χ(ριστ)ῷ κυρ(ίῳ) σταυροφόρῳ SEG 30.1542 (Lycaonia, iv AD).

*σταυρδευταῖος, ή, όν, ?, PColl.Youtie 87.11 (vi AD)

*σταφίδιον, τό, dried grape, raisin, PFreib. 4.67 (ii/iii AD).

στάφις I, for '(q.v.)' read '(q.v. with Suppl.); SEG 9.11, 12, al. (Cyrene, iv BC), Hp.Acut. 64, Theoc. 27.10, AP 5.304'

σταφυλή I, add '3 τοῦ ζυγοῦ τὸ μέσον Hsch.   **4** kind of sea-weed, Ael.NA 13.3 S.'

σταφυληπτόμος, after 'Nonn.D. 7.165' add '(cj. for -κόμ-)'

*σταφυλικός, ή, όν, of or concerning grapes, τῆς σταφυλικῆς προσόδου SEG 33.1041.23 (Cyme, Aeolia, ii BC).

*σταφυλινάριον, τό, dim. of σταφυλῖνος I, PRyl. 629.214 (iv AD).

σταφυλοβόλος, ον, perh. *of* or *for grape-pressing*, πλίνθοι *IG* 1³.425.39 (Athens, v BC).

σταφυλοπατητής, οῦ, ὁ, *grape-treader*, *PVindob.Salomons* 8.25 (AD 325).

σταφυλοτόμος, ον, = σταφυλητόμος, *excising the uvula*, *Suppl.Mag.* I 1 (iii AD).

στάχι, add '(Egyptian wd., Theognost.*Can.* 78)'

σταχυηκόμος, add 'epigr. in *SEG* 13.277 (Patrae, iv/v AD)'

σταχυητρόφος, add '*GVI* 720'

σταχύϊνος, η, ον, *of ears of wheat*, στέφανος *Inscr.Olymp.* 56.15.

στάχυς VII, delete the section.

στέαρ, line 1, for 'στέαρ .. sub. fin.]' read 'στέᾱρ, 'ατος, τό'; at end for '[Gen. .. στεάτιον]' read '[gen. στέατος disyll. by synizesis, Od. l.c.]'

στεατῖτις, ιδος, ἡ, a precious stone, Plin.*HN* 37.186.

στεατότριπτος, ον, *rubbed with fat*, rest. (as σταιατο-) in *PSoterichos* 4.17 (*Tyche* 3.277-8, AD 87).

στεγανόπους I, delete the section, adding 'Alcm. 148 P.' to section II.

στεγανός I, add '**3** *costive*, νηδύς Nic.*Al.* 367.' **II 1**, add 'neut. as subst., οἱ δὲ 'Ρωμαῖοι στεγανὰ ποιήσαντες παρεχείμαζον D.S.*fr.* 33 (ap. Suid. s.v.)'

στέγαξις, v. στέγασις.

⁺στεγάσιμος, ον, *used for awning* or *roofing*, πάπυροι σ. *PLond.* 1940.58 (iii BC).

στέγασις, line 3, at end for 'iv BC' read 'iii BC'

στεγάστριον, v. °στέγαστρον.

στέγαστρον 1, add 'also **στέγεστρον**, *PRyl.* 627.38 (iv AD), *Edict.Diocl.* 8.42A; cf. ‡σέγεστρον; also **στεγέστριον**, ib. 8.43A; cf. **στεγάστριον**, Hsch. s.v. λαρίεθος'

στέγη II 1, after 'room' insert 'A.*fr.* 362.3 R.'

στεγνοκοίλιος, α, ον, *suffering from constricted bowels, constipated*, Aët. 3.92 (295.9 O.).

στεγνοποΐα, ἡ, *building of shelters*, rest. at *SEG* 14.656.14-15 (Caunus, ii BC); cf. περιστεγνοποΐα.

στεγνοποιέω, after 'Med.' insert '*PHal.* 1.172 (iii BC)'

στεγονόμια, add '**2** *buildings, premises*, Just.*Nov.* 40 pr. 1.'

στεγόω, *roof over*, Sch.Pi.*P.* 4.426.

στεῖα, v. °στία.

στεῖρα (B) I, after 'Od. 10.522' insert 'Theoc. 9.3'

στελλάριον, τό, *star-shaped object*, or = στειλειάριον, *haft* (v.s.v. στελεός), *CPR* 8.66.12, 17 (vi AD).

⁺στέλμα, ατος, τό, some personal item, perh. *girdle*, listed in a magic spell, *PMag.* 7.785; cf. στέλμα· στέφος, στέμμα Hsch.

στέμμα I 1, add 'fig., ἀρετῆς .. πάσης στέμμασιν κοσμούμενος *SEG* 30.1562 (?ii AD)'

στέμφυλον, add '**III** used to translate wd. meaning "dross", Aq.*Is.* 1.22 (L.-R.), 25.'

στεμφυλουργικός, ή, όν, *of* or *for wine-pressing*, στεμφυλουργικὸν ὄργανον *POsl.* 145.3 (ii/iii AD), *POxy.* 2723.9 (iii AD).

στενακτικός, after '= sq. 2' insert 'Sch.E.*Hipp.* 415'

στεναχή, add '*SEG* 23.233 (Tegea, ii BC, verse)'

στενολεσχέω, add '**2** *use compression in discourse*, Eust. 1552.52.'

στενός II 3, line 7, after '(Hdt.) 7.175' insert 'Call.*fr.* 1.28 Pf.'

στενότης, add '**III** *financial straits*, *BCH* 59.440 (Acraephia, i AD).'

στενοχωρέω, line 1, for '*confined*' read '*short of room*' and for 'Macho ap.Ath. 13.582b' read 'Macho 396 G., σ. σταθμοῖς *PPetr.* 2 p. 28 (iii BC)'

στένω, add prose exx.: **1**, D. 18.217, 323, id.*Ep.* 3.44; Plu. 2.117a, b, 193f, Luc.*DDeor.* 6.2, *DMort.* 2.1, 27.2. **2**, w. ἐπί, D. 18.244. **3**, Luc.*Asin.* 21.

⁺στεργάνος (dub.)· κοπρῶν (κόπρων cod.) Hsch., cf. Lat. *stercus, sterculinum*.

στεργηθρον I, add 'Plin.*HN* 25.160'

στέργω III 2, add 'Lys. 33.4' **3**, add 'Isoc. 8.23, al.'

στερεός I 1 **b**, after 'σ. τάλαντα' insert '(in context app. = Ἀττικὰ τάλαντα)'; for '*SIG* 826*D*20' read '*Delph.* 3(4).279' and after this quot. insert 'τῶν στερεῶν ὀνομάτων, perh. *substantial* accounts, *POxy.* 2861 (ii AD)' add '**6** *sure, reliable*, adv. -ρεῶς, *with certainty*, φωνῆσαι Sch.Nic.*Th.* 1 C.'

στέρεσις, add '*SEG* 35.1439 (Lycia, ii AD)'

⁺στέρνα, ἡ, = στέρνον, *POxy.* 108.11, al. (ii/iii AD)'

στέρνον II 1, add 'Pi.*Pae.* 4.14 S.-M.' **2**, after 'Nic.*Th.* 924' insert 'ἐν στέρνῳ .. θυείης ib. 91'

στερνοτύπος, ον, = στερνοτυπής, γόος epigr. in *JEA* 40.119.19 (Egypt, ii BC); ἀνίαι *GVI* 1006.5 (Rhenea, i BC).

Στερόπης, add 'see also °ἀστεροπός'

στερόπτης I, after 'Ph. 1.276' add '*POxy.* 3581 (iv/v AD)'

στέρφος I, after 'cf. τέρφος, ἔρφος' add 'στρέφος'

στέρψανον, add 'cf. στρέφανον'

στεφᾶνᾶς, ᾶ, ὁ, *wreath-maker*, *ISmyrna* 504.

στεφάνη I 2 **a**, add 'perh. forming part of the decoration of a prow, *SEG* 37.692 (Delos, ii BC)'

στεφανηφορία, add '**III** *office of* στεφανηφόρος, *SEG* 30.1390 (Lydia, c.AD 150).'

στεφανηφόρος I, add 'poet., of spring, Anacreont. 55.1 W.' **III**, for this section read 'of Attic drachmas of the New Style bearing wreath on the reverse, ἀργυρίου Ἀττικοῦ -ου δραχμαί *Inscr.Délos* 1415.12; δραχμαὶ (τοῦ) στεφανηφόρου (sc. ἀργυρίου) *IG* 2².1013.33, 1028.30, al.; also δραχμὴ στεφανηφόρος, τέτραχμον -ον *Inscr.Délos* 1443*A*ii62, 1442*A*70, al. (all ii BC)'

στεφανοπωλία, ἡ, *selling of crowns* or *chaplets*, Anatolia 9.38.58.

στεφανόπωλις, after 'ίδος' insert 'Arist.*Ath.* 14.4'

στέφανος II 3, add 'Swoboda *Denkmäler* 168, 245 (Palaia Isaura)'

στεφάνωμα, line 2, after 'Thgn. 1001' insert '*Inscr.Délos* 1421*B*bii3 (ii BC)'

στεφάνωσις, add 'tab.defix. in *SEG* 34.1437 (Syria, v/vi AD)'

στεφανωτικόν (s.v. στεφανωτικός), add '*SEG* 33.1123 (Phrygia, iii AD)'

⁺στεφάριον, τό, an unidentified archit. term, *AE* 1925/6 39.6 (Oropus, ii BC).

στέφος 2, for 'of libations' read 'of offerings placed on a grave' add '**3** στέφεα· στεφῶνες Hsch.'

στέφω II 1, lines 9/10, delete 'τινος Nonn.*D.* 5.282'; line 14, after '(ii AD)' insert 'fig., στεφθήσονται γνῶσιν Thd.*Pr.* 14.18'

⁺στεφών, ῶνος, ὁ, pl., gloss on στέφεα Hsch.; *of the edge of hill* or *cliff* (cf. στεφάνη II), ὡς ὁ σ. περιφέρει κύκλῳ *IEphes.* 3.8 (c.290 BC); cf. στεφών· (τόπος) ὑψηλός, ἀπόκρημνος Hsch.

⁺στηβεύς, gloss on κουβηζός, Hsch. (prob. for στιβεύς, cf. Hebr. *kōbēs*, "fuller").

στῆθος IV, add 'X.*HG* 5.4.50'

⁺στιλλάριον, τό, *commemorative pillar*, ἐστήσαμεν τὸ στιλλάριν *IHadr.* 87 (iii AD), στηλλάριον Ramsay *Cities and Bishoprics* 1.150, *TAM* 5(1).239.

στήλη, line 1, after 'ἡ' insert 'εἰς στήλην written as one word (εἰστήλην), *IG* 12(7).515 (Aigiale, Amorgos, ii BC), *SEG* 30.546 (Maced., ii/i BC)' **II**, add '**6** *inscribed charm*, *PMag.* 4.1115, al., 5.96, al., 7.215, etc.'

στηλίδιον, add 'pl., *IG* 2².1498.11'

στηλίον, add '*IGBulg.* 679 (Nicopolis ad Istrum, iii AD)'

στηλίς, add 'στηλλείδα *SEG* 30.855 (Moesia, ii AD), στιλίδα *BCH* suppl. 8 no. 58 (Maced., iii/iv AD)'

στηλίτευμα, for '*invective*' read '*inscription on a pillar*'

⁺στηλλάριον, v. °στηλάριον.

στηλογραφέω, add 'στηλλογραφῶ *SEG* 34.656 (Maced., AD 191), ἐστηλλογράφησα *SEG* 34.1217 (Lydia, c.AD 198)'

στηλογραφία, for 'Arab.' read 'Akkadian' add '**II** *inscribing on a tablet*, *IG* 9(2).13.4, 14*a*3.'

⁺στηλογράφος, ὁ, *inscriber*, σταλογράφοι *SEG* 37.340.18 (Mantinea, iv BC).

στηλοκοπέω, before 'as a form of punishment' insert 'ἐσταλοκόπεισαν τὰ δεδο[γμένα] *SEG* 38.377 (Acraephia, c.200 BC)'

στηλόω, line 2, for 'τάφον .. (Amyntas)' read 'τίς δὲ τάφον στάλωσε; (i.e. set up with an inscription), Amyntas in *Suppl.Hell.* 43'

στημάτιον I, for 'al.' read 'Bito 47.7 M.'

στήμων II, lines 5/7, delete 'σ. .. 728' add '**III** *pole of a wagon*, *Edict.Diocl.* 15.11.'

στήριγμα 5, delete the section transferring quot. to section 1 and adding '2*Ki.* 20.21'

στηρίζω I 1, add '**b** *fix* one's eye, gaze, etc., on (in anger, etc.), στηριῶ τοὺς ὀφθαλμούς μου ἐπ' αὐτοὺς εἰς κακὰ καὶ οὐκ εἰς ἀγαθὰ Lxx *Am.* 9.4, οὐ στηριῶ τὸ πρόσωπόν μου ἐφ' ὑμᾶς *Je.* 3.12, *Ez* 6.2, etc.'

⁺στήριον· ἱεράκιον (ἱέρακι cod.); perh. read στήριον (ἢ στόριον). Σέλευκος Hsch.; cf. mod. Gk. στόρι.

στησίχορος, line 2, after 'χοροί' insert 'στεσίχορον ὕμνον ἄγοισαι, verse fr. on Attic vase, *Lyr.adesp.* 20(c) P.'

στία, add 'στεῖαι γὰρ αἱ ψῆφοι τῆς θαλάσσης Sch.Nic.*Al.* 466c Ge.'

στῑβάς 2, add '*Inscr.Perg.* 8(3).161 (ii AD)' **3**, add 'of funeral bier, *Alcmaeonis fr.* 2.2 p. 33 B.'

στιβήεις, for 'ἀγχούρως' read 'ἄγχαυρος' and for 'Call.*Hec.* .. 10' read 'Call.*fr.* 260.64 Pf.'

στῑβική, for '*PCair.Zen.* 136.247' read '*PCair.Zen.* 176.247'

⁺στιγμαῖος, α, ον, = στιγμιαῖος, Plu. 2.117e.

στιγμή I 1 and **2**, for these sections read '*tattooed mark*, D.S. 34/5.2.1; also *a natural mark* or *speck on a bird's plumage*, Alex. Mynd.ap.Ath. 9.398d. **2** *minimal mark*, *point*, οἷον ὅτι ταὐτόν .. στιγμὴ ἐν γραμμῇ καὶ μονὰς ἐν ἀριθμῷ· ἑκάτερον γὰρ ἀρχή Arist.*Top.* 108ᵇ26, *EN* 1174ᵇ12, *de An.* 427ᵃ10, Apollod.*Stoic.* 3.259. **b** the smallest division of a degree of the zodiac, *IG* 12(1).913 (Rhodes, ii/i BC).'

στιγμιαῖος, delete '117e' and 'στιγμαῖος is f.l. in Plu. 2.117e'

⁺στιλλάριον, v. °στηλάριον.

στιλπνός, add 'of clothes, SEG 31.1288 (Side, AD 249/52)'

στῖμι, add 'Peripl.M.Rubr. 49, 56'

στίξ 1, add 'ἀνιηραὶ θέρεος στίχες, i.e. swarms of flies, Opp.H. 2.448'

*στιπένδιον, τό, Lat. stipendium, year's service, ἐτῶν κε΄ ἰστοπενδίων ξ΄ JÖAI 30 Beibl. 28 (Ancyra).

στιπποκογχιστής, add 'POxy. 1980.6 (vi AD)'

στιππουργός, add 'CPR 14.5.10 (vi AD)'

στιφρός, read 'στῑφρός'

στιφρότης, read 'στῑφρότης'

στιχάριον, for 'variegated tunic' read 'kind of tunic'; delete 'also στιχαρο(sic)μαφόριον' to end.

*στῑχαρομαφόριον, τό, kind of cloak, SB 6024, 7033 (v AD), Stud.Pal. 20.275, POxy. 1978.

*στῑχαροφελώνιον, τό, kind of cloak, PMichael. 38.2 (vi AD).

στίχη, delete 'prob. in' and add 'Edict.Diocl. 22.9'

στιχολόγος, add '2 reciter of psalms, ASAA 30/32(1952/4).302.88 (Rhodes; v. BE 1956.197); MEFR 64.107 no. 23 (Caesarea Mauretaniae; v. BE 1953.261).'

στιχοπλανήτης, delete the entry.

στίχος IV, delete 'II' after 'συστοιχία'   add 'V a tax in general, CPR 8.79.2 (vii AD).'

στλεγγίς, add 'also στλιγγίς, Inscr.Délos 104.88, 113, 115 (iv BC, Attic text)'

στοιβάς, for 'v. στιβάς' read '= στιβάς 5, Ιasos 393'

στοιχάς I 1, add 'σ. νεφέλας Nonn.D. 18.282'

στοιχέω, line 4, after 'Poll. 8.105' insert '(ὅπου ἂν στειχήσω Robert Ét.épigr. 302 (iv BC))', II, add 'med. Inscr. Cret. 3 iv 9.18.   2 med., to be sufficiently available to, Modest.Dig. 34.1.4 pr.'

στοιχομῡθέω, before 'Phot.' insert 'ἐφεξῆς λέγειν' and add 'glossed by μακρηγορέω, Hsch. s.v. στοχίζῃ'

*στολάριον, τό, app. some article of clothing, Vit.Aesop.(G) 21 (s.v.l.).

στολάρχης, add 'II prob. controller of clothing, PAnt. 33.9 (iv AD), and so perh. in PCair.Zen. l.c.'

*στολᾶτα, Lat. stolata, ματρῶνα σ. POxy. 907.4 (iii AD), PFlor. 16.1 (iii AD), TAM 5(1).758 (iii AD).

στόλιον, for 'Dim. of στολή II, scanty garment' read 'garment (in dim. or pejorative sense)'

+στόλοκρος, στόλοκρον· τὸ περικεκομμένον τὰς κόμας, καὶ γεγονὸς ψιλόν, εἴτε δένδρον, εἴτε ἄνθρωπος. δηλοῖ δὲ καὶ ἀνειδὲς καὶ σκληρόν Hsch.; of a shorn boy, Anacr. 2 fr. 1.3 P.; applied to goats not having fully developed horns, Hsch. s.v. κόλον.

στόλος 3, fleet, add 'στόλου Π(οντικοῦ) SEG 33.1095'   4, λόγου σ., for 'set narrative' read 'course of argument'

στόμα I 3 b, add 'ἀπὸ στόματος προσαγορεύειν greet with a kiss, Porph.Plot. 2.17' III 2, after 'X.Ages. 11.15' insert '(s.v.l.)'

στόμαργος, add 'Myc. to-ma-ko, description of an ox'

στομαυλέω, for 'flute' read 'pipe or aulos'

στομαχικός 1, before 'πάθος' insert 'πόνος Inscr.Cret. 1.xvii 11 (Lebena, ii BC)'

*στομαχώδης, ες, irascible, Vit.Aesop.(G) 28.

στομίς I, add 'Hsch. s.v. φορβειά'

*στονύει· τὸ ὀξέως λέγει Theognost. Can. 73 A.

στόνυξ, line 5, for 'fangs' read 'claws'; line 6, for 'nail-removing prongs, i.e. nail-scissors' read 'perh. instrument for manicuring nails, clippers'

στοργή 2, add 'defixio in Hesperia 54.227 no. 9 (iii AD)'

στοργικός, after 'όν' insert 'affectionate, PMil.Vogl. 73.7 (ii AD)'

στόρνυμι, line 4, after 'A.Ag. 909' insert '(codd.)' and delete 'Com.adesp. 1211'   II, at end for 'dub. sens.' read 'floored and decked'

στουπίον, v. °στυππίον.

*στοῦπρον, τό, Lat. stuprum, illicit sexual intercourse, Cod.Just. 5.4.29.8.

στοχάζομαι II 1, add 'τί στοχασώμεθα σου; what are we to infer about you?, AP 7.422 (Leon.Tarent.)'   2, add 'Just.Const. Δέδωκεν 13'

στοχασμός I 1, after 'use of circumstantial evidence' insert 'for solving a question of fact'

στοχαστής 2, add 'πᾶς ποιητὴς ψυχαγωγίας .. σ. Agatharch. in GGM i p. 117'

στοχαστικός 1 a, add 'adv., πρὸς τὰ ἔνδοξα -κῶς ἔχειν id. Rh. 1355ᵃ17; comp., -κωτέρως Gal. 9.249.8'   2, for 'guesswork' read '‡στοχασμός'; lines 4/5, after 'Syrian.in Hermog. 2.34 R.' insert 'στάσις ib. 157' and delete 'πρὸς .. Arist.Rh. 1355ᵃ17'

*στραβόπους, ὁ, ἡ, gen. ποδος, = στραβοπόδης, gloss on μύσκιλος, Cyr.

*στράγαλος, ὁ, = ἀστράγαλος, Vit.Aesop.(G) 69.

*στραγγουρῑτία, ἡ, = στραγγουρία, POxy. 1384.30 (v AD).

*στραγγουρίωσις, εως, ἡ, = στραγγουρία, JS 1881.87.

*στρᾶτα, ἡ, Lat. strata, street, SEG 37.496 (Thessaly, iv/v AD).

στρᾰταρχικός, add 'Heph.Astr. 3.4.12'

στρατεία, add '6 service in the imperial administration (whether civil or military), Cod.Just. 4.59.1.2; spec. of military service, ἔνοπλον στρατείαν ib. 1.3.52.5.'

στράτειος, fem. -εία, add 'Ιασος 222, 223'

στρατεύω, add 'Boeot. στροτεύω, SEG 30.1980 (Orchomenos, iii BC)'

στρατηγέτης, line 2, before 'Cret.' insert 'Dor. στρατᾱγέτᾱς B.17.121, 18.7 S.-M.'

στρᾰτηγέω I 1 a, add 'so στρατηγεῖν κατὰ πόλιν, ἐπὶ τῶν ξένων to be praetor urbanus or peregrinus, IG 9²(1).242.3-5 (Thyrrheum, 94 BC)'   d, add 'στρατηγήσας ἐπ᾽ Ἴμβρον SEG 24.194(b) (Attica, i BC)'

στρᾰτηγίς, add 'as epith. of Aphrodite, IG 9²(1).256.3 (Thyrrheum, ii/i BC)'

στρᾰτηγός, add 'III ἥρως Στρατηγός at Athens, Hesperia 15.221 (c.200 BC), IG 2².1035.53 (ii AD).'

*στρατηλᾰτιανός, ὁ, officer of the staff of the στρατηλάτης, IG 10(2).1.791 (Thessalonica, iv AD or later), PSI 176.16, 183.3 (v AD), PWash.Univ. 17.5 (vi AD).

*στρατηλάτισσα, ἡ, fem. of στρατηλάτης, CPR 10.127.4 (vi AD).

στρατιά I, add '3 used as poet. equivalent of legio, στρατιῆς τε τεταγμένης ἐσθλῆς πρώτων Ἀρμενίων inscr. in Phoenix 26.183-6 (Cilicia, iv AD).'

στρατιώτης I, add '3 officer in the imperial civil service, Lib.Ep. 841.1.'

στρατιωτικός I, add 'δῶρα στρατιωτικά (= Lat. dona militaria), SEG 31.1300.15 (Lycia, in an addition to TAM 2.1201A, AD 145/6)'

στρατιῶτις 3, for 'soldier-fly' read 'colloq. term for large kind of fly'

στρατολογία, add 'SEG 31.675 (Thrace, c.AD 90/100)'

*Στρατόνικος, ὁ, month in an Asiatic calendar, Hemerolog.Flor. p. 73 (p. 10 K.).

στρατοπεδάρχης, for 'military commander' read 'commander of a force in the field' and add 'SEG 31.1590 (ii BC), Heph.Astr. 1.1.151'

στρατοπεδαρχικός, add 'IGLBulg. 240 (v/vi AD)'

στρατόπεδον I, add '3 headquarters, of the Emperor Julian at Constantinople, Jul.Ep. 46.'

στράτωρ, add 'SEG 7.951 (Arabia, iii AD)'

στρέβλωσις, add '2 vexation, harassment, Just.Edict. 7.7, CPR 10.121.5 (vi AD).'

στρεβλωτήριος, after '4Ma. 8.13' insert 'Sopat.Rh. 8.34.17 W.'

*στρειφωτήρ, ῆρος, ὁ, wd. in list of nautical building materials, cf. στροφωτήρ, PMich.inv. 4001 (ZPE 24.84, i/ii AD).

*στρέπτειρα, ἡ, spinner, βίου σ. Μοῖρα GVI 1154.9 (Samos, ii BC).

στρεπτοφόρος, add 'masc. subst., category of legionary soldier, Lyd. Mag. p. 47.18 W.'

στρέφανον, add 'cf. στέρψανον'

στρεψίμαλλος, before 'in reference' insert 'perh.'; add 'an obsc. ref. may be intended, v. Glotta 69.139'

στρίβος, for 'a weak fine voice' read 'a thin shrill cry'

*στρίγλα, ἡ, = Λάμια, Sch.Aristid. p. 42 D.

*στρικτωρία, ἡ, Lat. strictoria, shirt with long sleeves, Edict.Diocl. 22.7.

στροβιλᾶς, add 'CPR 13.29.75'

στρόβιλος 6, add 'Ps.-Hdt.Vit.Hom. 280; used as flavouring, Vit. Aesop.(G) 63'   9, for 'winch, or perh. rotating shaft' read 'perh. bell-shaped lower millstone' and add 'POxy. 3639.11 (AD 412)'

*στροβιλοφόρος (στροβηλ- cod.), ον, bearing cones, Hsch. s.v. κωνοφόρον.

*στρογγυλαῖος, α, ον, circular, of gymnasium, PKöln 52.10, 58 (AD 263).

*στρογγυλίας, v. στραβαλός.

+στρογγύλλω, make round in form or section, AP 7.726 (Leon.Tarent.); pass., of morbid growths, Aret.SA 1.8 H., Archig.ap.Gal. 8.90.

στρογγύλος I, add '6 of sputum, nummular, Hp.Prog. 14.'

*στρογγυλώδης, ες, having a round form, Alex.Aphr. in Top. 382.13.

στρογγύλωμα, delete 'or mosquito-net'

στρογγύλωσις II, for 'trench' read 'circular encampment'

στρομβεῖον, for 'Dim. of στρόμβος 5' read 'round pellet'

*στροτεύω, v. ‡στρατεύω.

*στρούκτωρ, ωρος, ὁ, Lat. structor, one who serves at table, Ath. 4.170e, Zos.Alch. 138.7.

στροφᾰλίζω, after 'στρέφω' insert 'move round rapidly, twirl'

στροφέω, add '(prob. f.l. for στρέφει)'

στρόφιγξ 4, for this section read 'applied to the conical hill at Kafizin in Cyprus, ta-i-e-pi-to-i-so-to-ro-pi-ki [νύμφαι] τᾶι ἐπὶ τῶι στροφι(γ)γι Kafizin 218b, al.; (sp. στόρφ-, ib. 50, 286, al.; στόφ-, ib. 14, 305, al.)'

στρόφις, for 'ἡ' read 'ὁ'

στροφίς, add 'acc. -ιν PP 42.42'

στροφωτήρ, add 'cf. perh. °στρειφωτήρ'

+στρυπτηρία, v. °στυπτ-.

στρυφνός III, delete 'Hp.VM 14, 15'

*στρωτή, ἡ, perh. pavement, ABSA 58.59 (Chios, i BC).

στρώτης, add 'POxy. 1951.2, 7 (v AD)'
στρωφάω I, line 10, delete 'i.e. claiming a husband's rights' and after 'A.Ag. 1224' insert 'ἐν δεμνίῳ .. [φρον]τίσιν στρωφωμένη Trag.adesp. 664.25 K.-S.'
*στύβον, τό, gloss on ἱππόφεως (spurge), Gal. 19.106.14.
*στυγνοπρόσωπος, ον, having a gloomy or glowering expression, Heph.Astr. 3.45.2.
*στυλεῖον, τό, pillar, (unless adj. στύλειος, of pillars, is understood), SEG 31.841 (Syracuse, vi BC; see also Dubois IGDS 86).
στυλίς I, at end for 'pecul. acc. .. Smyrna' read 'acc. στυλλίδαν SEG 33.1084; στυλλείδαν TAM 4(1).134 (both Nicomedia, Rom.imp.)'
στυλοβάτης, for 'base of a column' read 'continuous base supporting row of columns'
*στυλοπαραστάς, άδος, ἡ, pilaster, SEG 30.1258, Milet 1(2).102 (i BC).
στυλοπϊνάκιον, for 'pillar .. on it' read 'tablet forming part of a pillar'
*στυλοποιός, ὁ, maker of pillars, SB 4771.9.
στῦλος, line 2, after 'ib. 109 iii 92 (iii BC)' insert 'w. prothetic vowel, τοὺς ἰστύλους SEG 32.1269 (Phrygia, Rom.imp.)'
στυλόω, line 2, after '(ii BC)' insert 'ἵππων ἐστυλωμένος IG 11(2).287A166 (Delos, iii BC)'
στύμεον, for 'dub. sens.' read '= στόμιον (in quot. prob. cave)'
*στύμνιον, τό, τυπαστήριον· τὸ τῶν ἁλιέων στύμνιον Hsch.
Στυμφᾱλίς, before 'A.R. 2.1053' insert 'Pi.O. 6.84'
στυππεῖον, add 'also στουπίον Edict.Diocl. 26.2'
†στυππειοπλόκος, ὁ, tow-spinner, IG 2².1673.15, 41 (Eleusis, iv BC).
στυπτηρία, after 'Ion. -ίη' insert 'also στυπτ-' II, add '2 tax on the sale of alum, Inscr.Prien. 364.15 (iii/ii BC, form στρυπτ-).' add 'Myc. tu-ru-pte-ri-ja'
στύραξ (B), before 'X.HG 6.2.19' insert 'IG 1³.422.266'; line 2, delete 'shaft'
*στῦρόν, τό, = στύραξ (A) I, Call.fr. 43.88 Pf. (cf. στυρόν· τὸ μύρον Zonar., Theognost.Can. 130).
Στωϊκός, before 'Στοϊκός' insert 'Scanned –∪∪ by Cerc. 3 A.5 D.³'
στωμύλλω II, lines 3/4, delete 'also .. Pax 995'
σύ, line 3, after 'Boeot.' insert 'τύ CEG 326.2 (c.700/675 BC), 110 (c.500 BC)'; line 5, after 'ib. 50.27, 55.6' insert 'τύν tab.defix. in SDAW 1934.1040 (Boeotia)'; line 14, for 'Alcm. 17' read 'Alcm. 48 P., AP7.464.5 (Antip.Sid.)' p. 1659ᵃ, lines 7/8, for 'also τεί Alcm. 53' read 'also τεί (transmitted as τεί) Alcm. 70(b) P.' transferring this to the end of section I 1 III, for 'ὑμεῖς' read 'ὑμεῖς'
σύαγρειος, before 'ον' insert '(-εος, PCair.Zen. 12.52)'
σύαρτον, add 'rest. in SEG 23.326 (Delphi, iii BC)'
*σύβακχοι, οἱ, name for the φαρμακοί in Athens, Hellad.ap.Phot.Bibl. p. 534a H. (v.l. σύμβ-, q.v.).
†ὑβήνη, ἡ, case for arrows, IG 1³.350.82 (427/6 BC), Ar.Th. 1197, 1215, Sch.Ar.Th. 1197; for αὐλοί, Poll. 7.153, 10.153; cf. σ.· αὐλοθήκη ἢ τοξοθήκη. ἢ ὁ ναυτικὸς χιτών Hsch.
συβόσιον II, delete the section.
συβριασμός, for 'ἡ ἐν' read 'ὁ ἐν'
συβώτης, add 'Myc. su-qo-ta'
*σύᾱτηρ, v. °θυγάτηρ.
συγγᾱμέτης, add 'INikaia 1101 (perh. fem., iii/iv AD)'
συγγένεια I 1, add 'defined: ἡ συγγένεια ὄνομά ἐστι γενικόν, διαιρεῖται δὲ εἰς τρία, εἰς ἀνιόντας καὶ τοὺς ἐκ πλαγίου Theophil.Antec. 1.10.1; limited to cognates, Cod.Just. 6.4.4.20'
συγγένειος, add 'II app. belonging to the family, Mitchell N.Galatia 399.'
συγγενεύς, add 'TAM 4(1).284'
συγγενής II 1 b, add 'including children, Just.Nov. 115.3.12, Cod.Just. 6.4.4.23'
συγγενικός II 1, add 'of a family tomb, TAM 4(1).231, 257, al.; τὸ συγγενικόν, kindred, kinsmen, INikaia 1035 (ii AD), 1034 (iii AD), MAMA 6.24. b cult-title of Zeus, INikaia 1130 (Rom.imp.).'
†συγγέωργος, ὁ, fellow-cultivator, Ar.Pl. 223, PSI 1043.20 (ii AD). b member of association of γεωργοί, SB 7457.3 (ii BC), 8267.16, 20, al. (i BC).
συγγίγνομαι, line 2, after '-εγενόμην' insert 'Cypr. su-ne-ke-no-to (3 sg.) ICS 309 A 1'
*συγγνωστεύω, witness jointly regarding, συνγνωστεύω τὸν Θῶνιν ὡς πρόκειται PMich.XIV 676.32, 34 (AD 272).
συγγονή, add 'CPR 6.77.33 (ii/iii AD)'
*συγγόνιον, τό, unexpld. w., some kind of division within nome, νομοφύλ(αξ) β' συνγονίου Θεμίστου, PFreib. 4.62 (ii AD).
συγγράφευς I, line 4, after 'Ar.Ach. 1150' insert '(s.v.l.)'
συγγραφοδιαθήκη, add 'testamentary covenant, POxy. 1102.14 (ii AD)'
συγγράφω II 2, line 2, delete 'get speeches composed'
συγγώνιον, add 'adj. -ιος, ‡ὑπότονοι Inscr.Délos 442A229 (ii BC)'
*συγκαθάπτω, join to by grasping, σ. χεῖρά τινι Men.Dysc. 953.
συγκάθεδρος, for 'colleague' read 'adviser' and after 'συνθάκων' insert 'νομικὸς σ. ἀνθυπάτου IGRom. 1.933 (Sicca, ii AD)'

συγκαθίζω I 1, delete 'τὰ συνέδρια Hell.Oxy. 11.4' II 1, after 'sit together' insert 'ἐν τῷ συνεδρίῳ SEG 12.87.14-15 (Athens, 336 BC); τὰ συνέδρια .. ἐν τῇ Καδμείᾳ συνεκαθίζεν Hell.Oxy. 16(11).4 fin. B.'
συγκαθίστημι I 2, add 'pass. μετὰ τῶν συνκαθεσταμένων ἐνδίκων IMylasa 134.5 (ii BC)'
σύγκαιρος, for 'Anon.ap.Suid.' read 'Ael.fr. 81'
συγκαλέω 1, for 'call to council' read 'call together' and after 'Ar.Av. 201' insert 'Th. 7.60.5'
συγκαταβαίνω 4, 5 & 6, for these sections read '4 commit oneself to a battle or sim., εἰς τὸν ὑπὲρ τῶν ὅλων κίνδυνον Plb. 3.89.8, εἰς ὁλοσχερῆ κρίσιν id. 30.90.5, D.S. 12.30, 17.98, al.; w. ref. to negotiations, εἰς πᾶν Plb. 3.10.1. 5 agree or submit to, συγκαταβάντες εἰς φόρους καὶ συνθήκας Plb. 4.45.4, τοῖς ἀξιουμένοις Onas. 4.3. 6 descend to unworthy measures, dealings, etc., Plb. 26.1.3, εἰς λοιδορίαν Phld.Rh. 1.383 S.'
συγκαταβιβάζω I, add 'SEG 32.605 (Larissa, ii AD)'
συγκαταγιγνώσκω, add 'II join in thinking ill of a person, UPZ 146.17 (ii BC).'
συγκατάθεσις, add '4 perh. joint dedication, IEphes. 3239 (ii/iii AD).'
συγκατακλείω, after 'Luc.DMort. 14.4' insert 'βύκτας δ' ἐν ἀσκῷ συγκατακλείσας βοός Lyc. 738'
*συγκαταμένω, reside together, SEG 26.645 (Thessaly, c.AD 400).
*συγκαταξιόω, choose, elect, ἐκ τῶν συγκατηξιωμένων φιλτάτων IEphes. 3029.7 (ii AD).
συγκατασκευάζω, add '2 join in building, IEphes. 3712 (ii/iii AD).'
συγκατασκηνόω, for 'establish .. quarters' read 'help to settle men in their tents'
συγκατασχίζω, delete the entry.
συγκατατίθημι 1, add 'MAMA 6.264 (Phrygia, i AD)'
συγκαταφέρω, add '2 help in conveying to the grave (perh. by contributing to expenses), IGRom. 4.1453.'
συγκαταχωρίζω II, after 'BGU 578.19 (ii AD)' insert 'τῷ συνκαταχωρισθέντι μαρτυροποιήματι JJP 18.180 (ii AD)'
*συγκατηγόρημα, ατος, τό, that which in a sentence is neither subject nor predicate, Priscian.Inst. 2.15.
συγκατορύσσω, after 'τί τινι' insert 'Men.Dysc. 814'
σύγκειμαι, add 'IV to have been committed, delivered for transmission (cf. συντίθημι III), of instructions, UPZ 110.50 (ii BC).'
†συγκέλλω, bring ashore or into harbour together, transf., ἐν δ' ἄρα τῇσι (sc. κυρτίσι) στρόμβους συγκέλσαντες ὁμοῦ χήμῃσι τίθενται (sc. fishermen) Opp.H. 5.602.
συγκεραστός, add '[com]mixtum συνκεραστὸν ἢ κρᾶσις PVindob.Lat. 27 (Tyche 5.37, iv AD)'
*συγκηδεμών, όνος, ὁ, fellow ‡κηδεμών, Just.Nov. 72.2.
συγκινέω I, line 4, for 'sympathetic emotion' read 'excitement'
συγκλάω, line 4, after 'Phld.Mus. p. 23 K.' insert 'PVindob.Tandem 6.12 (v AD)'
*συγκλειδοφορέω, (of a temple official) to be a fellow key-bearer, w. dat., IStraton. 663.5, al.
σύγκλεισμα, for '3Ki. 7.29' read '3Ki. 7.16(29)' and add '4Ki. 16.17'
συγκλεισμός, add '2 fastness, refuge, Lxx 2Ki. 22.46, Mi. 7.17.' II, add 'fulfilment of a task, PMichael. 46.13, 54.4 (vi AD, -κλισμ- papp.)'
συγκληρονομέω, add 'Just.Nov. 53.6.1; perh. also μή τι[νες συνεκλη]ρ[ο]νόμησαν τῇ γυναικί; rest. in POxy. 3758.113 (AD 325)'
συγκληρονόμος, add 'Scaev.Dig. 31.88'
σύγκληρος I, add 'w. dat., Νύμφαις σ. Suppl.Hell. 978.14'
†σύγκλινος, ὁ, table-companion, Men.fr. 916 K.-Th.; pl., IG 2².2350 (iv/iii BC).
συγκλίτης, add 'SEG 31.638 (Maced.)'
συγκλύζω I 1, add 'συνεκλύσθη πόρος Ἐρυθρᾶς Θαλάσσης Ezek.Exag. 241'
†συγκολλάω, glue, cement, solder, etc., together, IG 2².1668.82 (iv BC), Luc.Alex. 14, PVindob.Salomons 2.7 (ii/iii AD); fig., Ar.V. 1041, Pl.Mx. 236ᵇ. 2 transf., join together, unite, Pl.Ti. 43a; pass., of a wound, Sor. 1.36; of a building, to be joined on to, PNess. 31.1 (vi AD).
*συγκολλήγας, ὁ, colleague, POxy. 3917.4 (rest., ii AD), PCol. 188.26 (AD 320).
συγκόλλησις, after 'Clearch. 44' insert 'PFam.Teb. 15.3, al. (ii AD)'
*συγκολλητός, όν, glued or fixed together, Ath.Asklepieion III 29 (iv BC).
συγκομίζω I 2, add 'of other foodstuffs, τὸς ταρίχος ἐς οἶκον συνκόμισον καὶ σφῆκ' ἴσα (lapis σφηκισα) SEG 37.665 (Carcinitis, c.400 BC)'
συγκοπιάω, add 'w. acc., share the work on, σ. τὴν ἐντομίδα ἐκ τῶν κοινῶν κόπων IG 10(2).478 (Thessalonica, iii AD)'
*συγκρασία, ἡ, perh. = σύγκρασις I b, Vett.Val. 248.26.
σύγκρασις II, add '2 aggregate of different classes, types, etc., in quot., in register of wheat lands, PMich.inv. 335ᵛ (ZPE 32.238, iii/iv AD).'
συγκρατύνω, add 'Gal. 17(1).821.7'
συγκρίνω II, add '2 gramm., in pass., have degrees of comparison, τῶν κυρίων οὐ συγκρινομένων A.D.Pron. 64.12.'

σύγκρῐσις **II**, add 'in the process of selection, συγκρίσεις ἀρχιτεκτόνων SEG 33.1040 (Cyme, ii BC)'

συγκροτέω **II 2 c**, add 'transf., knock together, contrive, δικαστήρια παρὰ τῇ σῇ ἐπιεικίᾳ πρός μαι συγκροτήσασε POxy. 3126 i 11 (AD 328)'

συγκρότησις **II**, add 'CRAI 1945.379.8 (Berytus, Byz.), Just.Nov. 109.1'

συγκρουσμός, add '2 perh. hammering out of a matter under dispute, συγκρουσμὸν .. ἡμῶν ἐργάζεσθαι CPR 8.30.6 (iv AD).'

†σύγκτησις, εως, ἡ, joint possession, SEG 30.1383 (Hypaepa, AD 301). **II** agglomeration of properties, Modest.Dig. 31.34.1, Scaev.Dig. 34.30.1.

συγκῠλίομαι 1, add 'b roll (down) with, v. συγκατακυλίνδομαι.'

*συγκυνηγέτις, ιδος, ἡ, fellow-huntress, τῇ Ἀρτέμιδι Sch.E.Hipp. 1130.

συγκωμαστής (after σύγκωμος), before 'Tz.H.' insert 'D.C.fr. 39.10'

*συγκωμήτης, ου, ὁ, fellow-villager, PDura 26.2, 9 (iii AD), Just.Nov. 52.1.

*συγξενόδοκος, ὁ, fellow-ξενοδόκος, SEG 29.500, 502 (Thessaly).

συγξέω, for 'by scraping .. Comp. 22' read 'stones so that they fit together, τῶν λογάδην συντιθεμένων λίθων αἱ μὴ συνεξεσμέναι βάσεις D.H.Comp. 22; metaph., of literary work, in pass., to be polished, smooth, Alcid.Soph. 20 R.'

συγχειμάζω, line 1, for 'App.BC 5.77' read 'App.BC 5.75'

συγχέω **I 1**, add 'b melt together, πολλὰ τῶν χρυσῶν ἀναθημάτων Plu. 2.401f.'

συγχίς, for 'shoe or sock' read 'sandal'

σύγχρησις, for 'common .. use' read 'commercial dealings'

συγχρισμός, before 'Paul.' insert 'SB 9199.13 (ii AD), Stud.Pal. 22.183.105 (ii AD)'

συγχρονέω **I 1**, add 'b spend time with, τινι PMich.VIII 497.15 (ii AD).'

σύγχροος **I**, add '2 perh. of blended colours, PHarris 73.44 (ZPE 37.234).'

*συγχρυλεος, α, ον, unexpld. adj. (accent unkn.), κίονες πλείονες χαμαὶ συνχρυλεαι IGC 1 p. 11, lines 22, 24 (Larissa, ?iii BC).

συγχυσμός, delete the entry (v. °συγχρισμός).

συγχωρέω **I**, add '2 move together, close up, Ar.V. 1516.' **II**, lines 1ff., delete 'get out of the way .. (anap.)'; for 'give way, yield, defer to' read 'fall in with, agree with' and delete 'Pl. l.c.' add 'b of events, conform with one's wishes or intentions, Pl.Tht. 191c.' **2**, add 'Cod.Just. 10.11.8.3' **3**, add 'τὸν δὲ τόπον συνεχώρησαν .. εἰς τὰς ὁρτασίμους ἡμέρας TAM 4(1).100'

συγχώρησις **1**, for this section read 'agreement, assent to an argument or proposition, τὴν .. σιγήν σου συγχώρησιν θήσω Pl.Cra. 435b, Lg. 770c, τὴν τῷ λόγῳ συγχώρησιν ib. 837e, Aristid.Quint. 2.10 (pl.); coupled w. συνδρομῇ Hermog.Id. 2.1. **b** agreement, consent to a course of action, IEphes. 3260.9, etc.; by the emperor, ib. 25.9, 38.8; w. κατά, κατὰ συνχώρησιν[ν τῆς] Σελευκέων προβ[ουλῆς] SEG 36.1297 (Syria, i/ii AD), IMylasa 871.' **3**, add 'Just.Nov. 139 rubric'

συζεύγνυμι **1**, add 'in building, ταύτας (sc. μεσόμνας) τρήσαντι καὶ συζεύξαντι IG 2².1673.36 (Eleusis, iv BC)'

*συζῠγικός, ή, όν, of conjugates, conjugational, Iamb. in Nic. 15.7 P.

σύζῠγος, add 'Aeol. σύνδυγος, Sapph. 213 L.-P.' **2**, after 'masc.' insert 'spouse, MAMA 7.366'

σύζυξ **II**, add '2 neut. pl. subst., consonants (cf. ἄζυγα, vowels), Nonn.D. 4.262.'

*σῠηνίς, ή, app. = sq., συηνὶς στικτή Livre d'écolier 147 (iii BC).

*σῠηνίτης, ου, ὁ, red granite quarried at Syene, Plin.HN 36.63.

σῠΐδιον, delete 'porker'

συκάζω **II**, add 'ἀποσυκάζω' add '**III** = ἐπηρεάζω, Artem. 1.73 P.'

συκαλλός (s.v. συκαλίς), add 'Edict.Diocl. 4.36'

σῡκᾰμῑνέᾱ, after 'Gal. 6.589' insert 'PFlor. 50.32, 66 (iii AD); συκαμειναί CPHerm. 7 ii 18'

συκαμινεών, delete 'PFlor. .. AD)'

σῡκάμῑνον **I**, transfer 'Lxx Am. 7.14' to section II.

σῡκάμῑνος **II**, add 'Lxx 3Ki. 10.27, Is. 9.10(9)' and after 'Ev.Luc. 17.6' insert '(perh. in sense I)'

*σῠκᾰμινών, ῶνος, ὁ, plantation of συκάμινοι, Ulp.Dig. 47.11.10.

*σῠκαμπελών, ῶνος, ὁ, vineyard with fig-trees, PNess. 32.9 (vi AD).

σῠκάριον, add 'PFlor. 176.9 (iii AD)'

σῡκέα **II**, delete the section, transferring quot. to section I.   at end add 'form prob. *σύτσα < *sūkyā, Myc. su-za'

†σῡκίδιον, τό, (young) fig-tree, τὰ νέα σ. Ar.Pax 598.

σύκινος **I 1**, line 4, after 'Ar.V. 897' insert '(perh. to be understood as sense II)' **2**, after 'metaph.' insert 'of no account, Hippon. 41 W.'

σῡκομορέα, add 'Hippiatr. ii p. 165.16 O.-H., Gp. 10.3.7'

σῡκόμορος, for 'sycamore-fig' read 'sycamore'

*σῡκοτράπεζος, ον, living on a diet of figs, i.e. cheaply, poorly, Iamb.adesp. 46 W.

σῡκοφαντέω, add '**III** guide, act as courier for, PNess. 89.22 (vi/vii AD).'

σῡλέω **II**, add 'SEG 25.606 (Doris, ii BC)'

σύλη, line 7, delete '[to fear]' **II**, line 2, after 'booty' insert 'Hedyl.ap.Ath. 11.486b'

σύλησις, add 'Ptol.Tetr. 197 B., PBerol.inv. 11624.14 (JJP 18.35, iv AD)'

*σύλινος, v. °ξύλινος.

συλλᾰβή **II 4**, add 'sg., ἄλφα σ. Herod. 3.22'

συλλαγχάνω, add 'pl., to be chosen by lot together, SEG 30.1119 (Entella, iii BC)'

συλλαμβάνω **VI 1**, add 'b w. acc., take as an associate, PBremen 9.8, PGiss. 11.12 (both ii AD).'

*συλλαμπᾰδεύω, carry a torch with, rest. in Sch.min.Il. 6.21 (cf. Alcm. 63 P.).

*συλλαυρίτης, ου, ὁ, resident in the same street, POxy. 3979.6 (sp. συνλαυρείτ-), PRyl. 606.37 (both iii AD).

συλλέγω **I 1**, add 'b contract an illness, Aristid.Or. 50(26).1 K. (pass.).' **2**, for this section read 'med., regain control of, collect oneself, one's faculties, etc., σύλλεξαι σθένος E.Ph. 850, συνειλεγμένον τὰς ἀφάς Pl.Ax. 365a, ἐκ τῆς ἀσθενείας σ. ἐμαυτόν ib. 370e; abs., αὑτήν (sc. ψυχήν) εἰς αὑτὴν ξυλλέγεσθαι καὶ ἀθροίζεσθαι Pl.Phd. 83a'

*συλλείωσις, ἡ, agglomeration, Anon.Alch. 7.5; cf. λείωσις.

*συλλέκτης, ου, ὁ, collector, συνλέκ(ται) ἀχ(ύρου) PKöln 122.2, π]ερὶ συνλέκτου οἴνου POxy. 1415.9 (both iii AD).

*συλλεκτικός, ή, όν, acquisitive, δωρεῶν Vett.Val. 46.14; abs., id. 48.19.

*συλλήμπτωρ, ορος, ὁ, = συλλήπτωρ, DAW 85.38 (Cilicia).

συλλῐθηγία, add 'OMich. 1.311, 322 (iii AD)'

συλλογεύς **1**, delete 'at Athens' to end   add '2 summoner, convener, IG 2².1257 passim, 1496.83, 114, etc., τὸ δῆμο συλλογῆς SEG 26.72 (Athens, iv BC).'

συλλογισμός **I 1**, add 'Lxx Ex. 30.12, Wi. 4.20'

σύλλῠσις, add 'BGU 1926.8, 13 (ii BC)'

σῠλόνυξ, for 'paring the nails' read '(instrument for) cleaning the nails'

συμβαίνω **III 1 b**, delete 'mostly impers.' and add 'τοιαύτη .. γὰρ ἡ ἀπορία οὖσα συνέβαινεν D. 47.44'

*σύμβακχος, add 'subst., οἱ σ., members of a Dionysiac cult association, SEG 31.983 (Ionia, ii/i BC; cf. °συβακχοί)'

συμβάλλω **II 4**, add 'πρὶν ἢ συμβάλῃ τῇ χώρᾳ Peripl.M.Rubr. 38'

*συμβαλτός, ή, όν, = συμβλητός (in quot., IV), θύρα Anon. in Rh. 184.26.

συμβᾱμᾰτικός, after 'Ptol.Tetr. 203' insert '(v.l. συμβατικός)'

συμβᾱσιλιστής, for 'fellow .. (q.v.)' read 'pl., guild or worshippers of a deified Ptolemy (in quot., Ptol. III; cf. βασιλιστής)'

*συμβετέρανος, ὁ, = Lat. conveteranus, fellow-veteran, Modest.Dig. 27.1.8.6.

συμβίβᾱσις **I**, after 'reconciliation' insert 'Ammon.Diff. 133 N.'

σύμβῐος, after 'husband' for 'Epigr.Gr. 399 (Ancyra)' read 'GVI 242 (Ancyra, i/ii AD)' and add 'SEG 30.1187, 1209 (both Italy)'

συμβιόω **1**, add 'of a wife, SEG 29.1197 (Lydia, AD 209/10), TAM 4(1).306'

συμβίωσις **I 2**, delete the section. **II**, add 'Inscr.Perg. 8(3).85, SEG 31.1010 (Lydia, ii AD), etc.'

συμβιωτής, add '**III** member of a συμβίωσις II, Ramsay Cities and Bishoprics 2.470, IG 9²(1).248.6 (Thyrrheum, ii BC), IGRom. 4.796.7 (Apamea), SEG 33.1165 (Side, iii AD), etc.'

συμβόλαιον **II 2**, line 2, after 'E.Ion 411' insert 'Men.Dysc. 469, 470'

*συμβολᾱφόρος, ὁ, bearer of sacred symbols, DAW 80.39 (Maeonia), TAM 4(1).76.

συμβολικός **2**, add 'ψῆφος AP 6.248 (Marc.Arg.); = Garl. 1422 G.-P.); n. pl. as subst., ib. 5.135' **4**, add 'sg., PGrenf. 2.41.11 (i AD)'

*συμβολογράφος, ὁ, notary, SB 5326.5 (Byz.).

σύμβολον **III 1**, line 4, delete 'clue, S.OT 221' and transfer quot. to section I 1 a (fig.); last line, transfer 'εἰράνας .. AP 6.151 (Tymn.)' to follow 'prearranged signal' in section 4.

συμβούλιον **II**, add 'sp. -ειον, POxy. 3019.8'

συμβροχέω, for '= συμβρέχω' read 'pass., of land, to be watered thoroughly' and add 'PFlor. 383.88 (iii AD)'

*συμβροχίζω, inundate, irrigate, BGU 2063.23 (ii AD), 938.8 (iv AD). **2** ret, reduplicated pf. σεσυνβρ[ο]χι[σ]μένης PColl.Youtie 80.22 (AD 315).

*συμβροχικός, ή, όν, subject to inundation, PHerm. 26.7 (v AD).

*συμβροχισμός, ὁ, retting, PLeit. 2.3. (ii AD), PColl.Youtie 68.19 (iii AD), POxy. 3256.16 (AD 317/18).

*συμμᾰλακτέον, one must soften together, Paul.Aeg. 7.3 (217.1 H.).

συμμάχομαι, at end delete '-Prose word- Poets'

σύμμᾰχος **1**, line 2, after 'ally' insert 'Archil. 108.1 W.'

συμμειρακιώδης, delete the entry.

*συμμερισμός, ὁ, distribution of shares, SEG 25.744 (Callatis, ii BC).

*συμμεταμορφόω, transform with, τὴν φαντασίαν τοῖς ὑποκειμένοις πράγμασι Procl. in R. 164 K.

συμμέτοχος, subst., add 'PKöln 144.13, 28, 33 (152 BC), also perh. spouse, EA 12.111 no. 47 (Caria)' σύμμετρος **II 1**, add 'w. gen., Aristid. 109 J. (= 1.50 L.-B.)' **4**, add 'κώμη σ. Peripl.M.Rubr. 4'

σύμμιγμα, add 'σπέρμα ἀνθρώπου ἅρπαγμα καὶ σ. τοῦ τῶν προγόνων γένους Zeno Stoic. 1.36 (ap.Gal. 19.370)'

σύμμιξις **II**, add 'Arist.Ath. 3.5'

*συμμοιχεύω, to be an adulterer with, w. dat., Heraclit.Ep. 7.3.

συμμορία **4**, for 'a company in general' read 'a group of associates' and after 'δειπνεῖν κατὰ σ. J.AJ 5.7.3' insert 'οἷον κατὰ σ. Longin. 20.1'

συμμορφίζω, add '(v. °συμφορτίζω)'

*συμπαιγνία, ἡ, app. calque of Lat. collusio, collusion, Cod.Just. 6.4.4.6, Just.Nov. 54.2 pr., Gloss.

συμπαραδίδωμι, add '**2** hand over together, Just.Nov. 6 epilogos 1.'

συμπαρακαθέζομαι, line 1, for 'aor.' read 'impf.'

συμπαρατάσσομαι, add 'also act. -παρατάσσω, set alongside, συμπαρα[τά]ξας μοι τὸν ἑαυτοῦ γραμματέα PTurku 1.12 in Tyche 6.100 (Theadelpheia, ii BC)'

συμπαρατυγχάνω, for 'Cat.Cod.Astr. 6.70' read 'Heph.Astr. 2.18.40, 43'

συμπάρειμι (εἰμί sum) **2**, add 'SEG 37.979 (Claros, ii AD)'

*συμπαρεμπλέκω, insert, include simultaneously, Anon. in Rh. 243.28, 31.

*συμπαρεμπλοκή, ἡ, simultaneous insertion, inclusion, Anon. in Rh. 243.30.

*συμπαρθενεύω, live together in celibacy, PMed.inv. 4969.11 (ZPE 93.155, v/vi AD).

σύμπας, add 'form ξύμπας, Myc. ku-su-pa'

*σύμπᾰχος, ον, thick, Poliorc. 224.6, 244.8.

*σύμπεισις, εως, ἡ, persuasion, Men.Epit. 716.

*συμπένθερος, ὁ, joint father-in-law, said of a bride's father in relation to the bridegroom's father, Sch.Od. 4.22.

συμπεραίνω **II 1**, add 'τὸ πρᾶγμα περὶ οὗ μοι ἐπεστείλατε συνεπέρανα POxy. 2862.3 (iii AD)'

συμπεραντικός, after 'tending to a conclusion' insert 'Alex.Aphr. in Top. 563.6' and delete 'only in'

*συμπεριελαύνω, round up together, τάς τε βόας καὶ τὰ πρόβατα PTeb. 729.7 (ii BC).

συμπεριποιέω, add 'Amyzon 37'

συμπερίπολος, for 'Philol. 71.92' read 'BCH suppl. 9 p. 345 no. 9' and add 'SEG 32.626 (Illyria, iii BC)'

συμπεριφέρω **II 3**, add '**b** ἀγνοίᾳ σ. go about, live one's life in ignorance, J.AJ 19.147'

συμπεριφορά **2**, line 5, before 'σ. ποιεῖσθαι χρημάτων' insert '**b** indulgence, remission' and add 'PMich.IX 568.8 (i AD), POxy. 1590 (iv AD)'

†σύμπλεγμα, ατος, τό, complex of sculptured figures, IEphes. 509, 518, 857, 858 (i/ii AD); cf. Pana et Olympum luctantes .. quod est alterum in terris symplegma nobile, Plin.HN 36.35.

συμπληγδην, delete the entry (v. °ἐκπληγδην).

*σύμπληξ, ῆγος, ὁ, ἡ, = συμπληγάς, Antim. in Suppl.Hell. 64.3 (s.v.l.).

συμπληρωτικός **1**, add 'Iamb.Protr. 21 ιγ′, Procl.Inst. 75 D.'

συμπλοκή **1**, line 4, after 'Pl.Plt. 281a, cf. 306a, al.' insert 'of scale armour, Hld. 9.15' and before 'ἡ ἁπάντων ..' insert 'fig.'; after 'Stoic. 2.284' add 'of intermarriage, REG 101.14' **2**, add 'with bandits, IHadr. 84.6 (i/ii AD). **b** quarrel, POxy. 3981.4 (AD 312).'

συμποδισμός, after 'ὁ' insert 'entanglement, Alex.Aphr. in SE 135.13'

*σύμποκος, ον, complete with fleece, i.e. unshorn, πρόβατα PThead. 22.9, 23.11 (iv AD).

συμπολιτευτής, οῦ, ὁ, fellow-citizen, Diog.Oen.fr. 39 S.

συμπολιτεύω, add '**3** have the same legal force, Just.Const. Δέδωκεν 23.'

συμπομπή, add 'σουμπομπάν SEG 32.456 (Boeotia, c.235/0 BC)'

συμπονέω, line 5, after 'S.Ant. 41' insert 'D. 2.9'

σύμπονος, add 'POxy. 1942.4 (vi AD), Just.Nov. 119.5'

σύμπορος, add 'masc. subst. ZPE 7.225 no. 19 (Miletus, verse)'

συμποσίαρχος, after '(Palmyra, iii AD)' insert 'also ostr. in BASP 20.49.11'

*συμποσιαστής, οῦ, ὁ, member of a cult-association, IGBulg. 1626 (Traiana Aug., συν-).

συμπόσιος, add 'epigr. in SEG 34.1266 (Bithynia)'

σύμπους, add 'neut. pl. τὰ ζῷα .. ἔχοντα .. οὐ διεστηκότας τοὺς πόδας , ἀλλ′ ἑστῶτα σύμποδα Sch.Pl.Men. 97d G.'

*συμπραγματευτής, οῦ, ὁ, collaborator, PMasp. 158.11 (vi AD).

σύμπρασις, εως, ἡ, joint purchase or (perh.) requisition, (ἐν συμπράσει, by way of requisition) PRyl. 215.7, PColl.Youtie 32.7 (both ii AD).

συμπράσσω **I**, line 7, after 'Plb. 28.7.2' insert 'τινί ἔς τι App.BC 5.7.31, D.C. 37.21.4'

συμπρεσβύτερος, for 'fellow-presbyter' read 'fellow-elder' and add 'PColl.Youtie 21.9 (i AD), Mitchell N.Galatia 329 (iii/iv AD)'

συμπροβαίνω, before '3 sg.' insert 'οἱ λόγοι τῷ χρόνῳ σ. advance with the advance of time, Aristid.Or. 2.103 J.'

συμπρονοέω, add 'SEG 19.790.5 (Pisidia)'

συμπροσγίγνομαι **1**, after 'Dor. συμποτιγίν-' insert 'Thess. συμπογγῖν- BCH 59.56 (Larissa, w. -γγ- for -gg-, not -ng-)'

συμπρόσειμι, for 'to be present together' read 'to be present with as a support' and add 'POxy. 1061.10 (i BC)'

*συμπροσίημι, aor. med. -ήκαντο, let approach together, πρέσβεις Did. in D. viii.10.

*συμπροστρέπω, support a supplication, συμπροστρέψω τὴν οἰκοδομὴν ποιῆσαι PMil.Vogl. 255.2 (ii/iii AD).

συμπροτρέπω, add 'med., IG 2².1039.29 (i BC)'

συμπρύτᾰνις, add 'SEG 26.694 (Epirus, c.150 BC)'

*συμπρωτοκωμήτης, ου, ὁ, associate πρωτοκωμήτης, PFlor. 296.41 (vi AD).

*σύμπρωτος, η, ον, the very first, adv. -ον Pi.fr. 128c.8 (s.v.l.).

*σύμπτωσιμος, ον, collapsing, οἰκία PGoodsp.Cair. 13.4 (iv AD).

σύμπτωσις **I**, for 'CIG 3293 (Smyrna)' read 'TAM 4(1).134 (Nicomedia)' and add 'ἐν συμπτώσει in a state of ruin or collapse, of a house, SB 9902 A, IV 2, etc. (iii AD), of a site, PFlor. 50.61'

συμφᾰνής, line 1, before 'manifest' insert 'clearly visible, Amyzon 38' and after 'evident' insert 'συμφανῶν (ξυμφώνων codd.) πράξεων Pl.Lg. 864b'

σύμφανσις, delete the entry (v. °σύμφαυσις).

*σύμφαυσις, εως, ἡ, shining together, Eust. 1060.53.

*συμφεροντής, οῦ, ὁ, perh. fellow-benefactor, καρπίων ἀνέθηκε .. εὐχὴν μετὰ συνφαιρονταῖς (sic) IG 2².4794.

συμφέρω **A I 4**, add 'w. gen., κακῶν τῶν σῶν ξυνοίσω E.fr. 909.12'

σύμφθαρσις **2**, for this section read 'breaking down into a homogeneous substance, of the digestion of food, Syrian. in Hermog. 1.66 R.; transf., of abstract things, Hermog.Id. 1.22, Iamb. in Nic. p. 80 P.'

συμφθείρω **II**, for this section read 'break down into a unified mass, φαρμάκων συντριβέντων καὶ συμφθαρέντων ἀλλήλοις Plu. 2.436b; transf., sounds, D.H.Dem. 48; συνεφθαρμένα ἀλλήλοις of π and σ forming ψ, id.Comp. 14; abstract qualities, Iamb. in Nic. p. 81 P.'

συμφῐλοκᾰλέω, add 'support in the quest of distinction, J.AJ 11.312'

*σύμφῐμος, ον, tightly closed, Iambl.Alch. 287.5.

*σύμφοιτος, ον, gloss on ὁμόφοιτος, Sch.Pi.N. 8.54.

συμφορά **II 2**, after 'misfortune' insert 'Alc. 69.2 L.-P.'

*συμφορτίζω, burden together with others, v.l. in Ep.Phil. 3.10 (v. συμμορφίζω).

σύμφραξις, add '**2** fencing round, Tit.Calymn. 52.12 (iii BC).'

σύμφρουρος **II**, add 'SEG 23.444 (Thessaly, i BC)'

συμφυγάς, add 'Isoc. 16.14, 26, etc.'

συμφυής **II**, add '**4** adv. συμφυῶς, in fusion, Simp.in de An. 59.25.' **III**, add 'Nonn.D. 14.212'

σύμφυλαξ, add 'cf. ὁ γὰρ φόβος δεινὸς δοκεῖ σ. εἶναι X.Eq.Mag. 7.7'

σύμφωνος **II 3**, line 2, before 'ἐκ συμφώνου' insert 'neut. subst., agreement, pact, Cod.Just. 1.3.55.1; also clause in an agreement, ib. 4.65.34'; line 3, after 'Cod.Just. 8.10.12' insert 'κατὰ σύμφωνον AR 1985/6.63 (W. Maced., iv BC)'

*συμψειρικός, v. ‡συψ-.

*σύμψημα, ατος, τό, pl. scrapings, Hsch. s.v. συρματὶς στρατιά.

*συμψηφισμός, ὁ, expl. as σύνθεσις ψήφων, Alex.Aphr. in Top. 9.18.

σύμψηφος **1**, add '**b** conforming to one's views, agreeable, POxy. 2711.7 (iii AD).'

*συμψιλίζω, perh. join in stripping, PVindob. G29946 i 22 (Gallo Frammenti biografici II p. 266).

σύν, lines 2 ff., w. gen., add 'MAMA 1.193, 208, 212, 217, etc.'; line 16, for 'Sapph. 75' read 'Sapph. 121 L.-P. (codd.); ξυνιείς Pi.N. 4.31 (codd.)'; line 17, after '(Locr., v BC) etc.' insert 'Cypr. ICS 217.28' **A 5**, add 'of attendant burden, σ. γήρᾳ βαρεῖς S.OT 17; σ. νόσοις ἀλγεινὸς ἐξεπέμπετο id.OC 1663' add '(form ξύν) Myc. ku-su'

συνᾰγείρω, line 2, after 'Theoc. 22.76' insert 'Aeol. 3 sg. pf. συναγάγρεται Alc. 119.10 L.-P.'

συναγέσκεο, delete the entry.

*συνᾰγορᾰνόμος, ὁ, colleague as ἀγορανόμος, MUB 26.62 (Tyre, i AD); w. dat., SEG 11.499 (Sparta, ii AD).

*συνᾰγοριστικός, v. συνηγ-.

συναγρίς, after 'sea-fish' insert 'perh. sea-bream, Dentex vulgaris; cf. mod. Gr. συναγρίδα'

συναγχικός, line 1, after 'συνάγχη' insert 'Ar.Byz.Epit. 2.178'; line 2, delete 'of .. Gal. 15.790'

σύναγχος, add 'Gal. 15.790.8'

συνάγω **II 3**, add '**b** in logic, reduce to, συνάγει (τὰ τέτταρα) εἰς τὰ δύο Arist.GC 330ᵇ20.' add '**III** perh. compel, oblige (calque of Lat. cogo), Cod.Just. 1.2.24.2.'

*συναγωνοθέτης, ου, ὁ, colleague as ἀγωνοθέτης, rest. in Inscr.Prien. 111.174, 118.4, 11 (i BC), Hesperia 28.324 (Isthmia, Rom.imp.).

συνάδελφος **I**, add 'Mitchell N.Galatia 252 (after AD 212)'

συνᾴδω **I 1**, line 3, after 'Ar.Av. 858 (lyr.)' insert '(cj. συναυλείτω)' **II**, for 'trans.' read 'w. acc. pers.'

*συναθλητής, οῦ, ὁ, fellow-athlete, MAMA 8.417.32 (dub.).

συνάθροισμα, add 'Sch.E.Hec. 100-24'
†συναΐδιος, ον, *coeternal*, *Cod.Just.* 1.1.5.1; συναΐδιος· συνυπάρχων Hsch.
συναίρεμα I, add 'sp. συνέρεμα *POxy.* 3170.23 (iii AD)'
συναιρέω II 1, line 4, after 'Sch. ad loc.)' insert 'S.*Tr.* 884' add 'III *choose together with*, αἱρεθεὶς θεωρὸς .. μετὰ τῶν συναιρεθέντων *JHS* 68.48.73, 77 (Lycia, ?ii BC).'
συναισθητικός, before 'adv.' delete 'only' and insert '*conscious* (to oneself), εἰς αὑτήν Plot. 2.2.1.10'
*συναίσθομαι, = συναισθάνομαι (in quot., III), *PLit.Lond.* 138 vii 7 (i AD).
συναιχμάλωτις, add 'Sch.S.*Tr.* 318 (ξυν-)'
συνακμάζω, for '*blossom* or *flourish at the same time*' read '*reach the peak of one's growth, development, fortune, etc.*, *at the same time*'
συνακοντίζω I, for 'Antipho 3.4.5' read 'Antipho 3.4.6 G.'
συνακροάομαι, add '2 *to be a fellow judge*, ὥστε τὸν κοιαίστωρα συνακροᾶσθαι αὐτῷ *Cod.Just.* 7.62.35.'
*συνακροατής, οῦ, ὁ, *fellow judge*, Just.*Nov.* 86 rubric.
συνακτήριον, add '*Cod.Just.* 1.5.14'
συναλίσγομαι, read συνᾱλισγέομαι (v.l. συνᾱλίσγομαι).
*συναλλάζω, v. °συναλλάσσω.
*συναλλακτήρ, ῆρος, ὁ, = συναλλακτής (in quot., II), *SEG* 34.940 (Camarina, c.300 BC).
συναλλακτής, add 'III *party in a contract*, Just.*Edict.* 7.3.'
συνάλλαξις 2, add 'ἡ τῶν καρπῶν σ. the *contract* concerning, *BGU* 1120.52 (i BC)' 3, delete the section.
συναλλάσσω, line 1, after 'Att. -ττω' insert 'Dor. -ζω, *SEG* 11.244 (Sicyon, c.500 BC)' II 2, add 'b *enter into a contract of* marriage, τῶν δευτέρων γάμων, οὓς μετὰ τόνδε ἡμῶν συναλλάξαιεν τὸν νόμον Just.*Nov.* 22.48 pr.; also abs., ib. 157.1.'
συναλοάω, add '3 *beat up, thrash*, Longus 4.29.'
συναλύω, add 'Philostr.*VA* 8.11'
*συναμφιβολεύς, έως, ὁ, *fellow-(net) fisherman*, *PCornell* 46.7 (ii AD).
συναναζεύγνῡμι, add '*BGU* 1257.20 (iii BC)'
συνανάκειμαι, add '2 euphem., of sexual intercourse, Sopat.Rh. 8.258.22 W.'
*συνανάκλῐσις, εως, ἡ, *lying down together, coitus*, tab.defix. in *SEG* 35.224.13 (Attica, iii AD).
συναναπέμπω, add '2 *admit to an inheritance equally with*, θυγατράσι δὲ συνανεπέμπετο (ἡ μήτηρ) Just.*Nov.* 22.47.2.'
συναναπίπτω, before 'concubo' insert '*lie down together*, tab.defix. in *SEG* 35.219.7, 17, 221.4 (Attica, iii AD)'
συναναρτάομαι, add 'w. dat., Socr.*Ep.* 6.10'
συνανασκευάζω, delete 'Gal.*Opt.Doctr.* 6' and transfer 'S.E.*M.* 7.214' to follow 'Phld.*Sign.* 12, al.'
συναναστρέφω II 2, delete the section.
*συνανάτᾰσις, εως, ἡ, (τείνω) *straining, effort*, Phld.*Rh.Suppl.* 45.14 S.
*συναναφύρομαι, *to frequent together*, *get mixed up with*, μὴ ἐν τῷ κο[ι]νῷ γυμνασίῳ συναναφύρωνται τοῖς .. *SEG* 28.1566 (Cyrene, AD 143); cf. ἀναφύρω.
*συνανταγωνιστής, οῦ, ὁ, *fellow-competitor*, Sch.Pi.*N.* 11 p. 185.4 D.
συναντάω II, add 'of a topographical feature, *meet* the traveller, i.e. *come next*, *Peripl.M.Rubr.* 33 C.' III, add 'τοὺν συναντακόντουν (pf. part.) φιλανθρούπουν παρ᾿ αὐτοῖο πὸτ τὰν πόλιν *SEG* 31.572 (Thessaly, c.200 BC)'
συνάντησις, after 'of heavenly bodies' insert 'Ptol.*Tetr.* 150 B.'
*συναπαιτητής, οῦ, ὁ, *fellow tax-collector*, *SB* 9840.10 (s.v.l., iv AD).
*συνάπαξ, *all at one time*, Gal. 2.381.17.
*συναπελευθέρα, ἡ, *fellow freedwoman*, *ISestos* 2 (?i BC).
συναπελεύθερος, add '*IGLS* 9032'
*συναπηχέω, *sound at the same time*, Polyaen. 8.23.2; cf. συνεπηχέω.
*συναπογέννησις, εως, ἡ, *joint procreation*, Procl. *in R.* 1.134, 2.366 K.
συναπογράφομαι, add 'III act., dub. sens., *Inscr.Cret.* 4.160B5 (Gortyn, iv/iii BC).'
*συναποκηρύττω, *banish, disown together with* another, Men.*Sam.* 509; cf. ἀποκρύπτω II.
*συναποκλίνω I, add '2 *cause to veer away with* another, w. dat., Memn. 38.5 J.'
συναποστέλλω, after lemma insert '*send out together*, ὅποι .. ἂν .. μέρος τι τῆς πόλεως συναποσταλῇ D. 4.45'
*συναποτάσσω, *assign* troops *together with a detachment*, *SEG* 34.558 (Larissa, c.150/130 BC).
*συναποχή, ἡ, *joint receipt*, *POxy.* 1891.18 (v AD).
συναποχωρέω, add 'of physical conditions, w. dat., ταυτὶ .. τὰ νοσήματα .. ξυναποχωρεῖ ποτε τῇ φύσει Philostr.*Gym.* 28'
συνάπτω A I 2, line 11, delete 'but .. Theopomp.Com. 22' II 2 b, line 2, after 'id.*Ph.* 1241' insert 'σ. τὸν λόγον Theopomp.Com. 23 K.-A.' III 3, for '*what conclusion follows*' read '*what hypothetical proposition is made*' and for 'Call.*Fr.* 70.3' read 'Call.*fr.* 393.3 Pf.'
συνάρεσκω, line 7, after 'c. dat.' insert 'Arist.*Ath.* 33.2'
*συνᾰρήγω, compd. of ἀρήγω in *Suppl.Hell.* 937.8 (fragmentary hex.).

†συναρθμέω, *cause to agree with*, σ. ἐπέεσσι A.R. 4.418.
συνᾰριθμέω I, after 'Is. 5.18' insert 'Call. *fr.* 587 Pf.'
*συναρίστιον, τό, *lunch-club*, *IG* 12(3).93, 94 (Nisyros, iii BC).
συναρμόζω I 1 a, at end, *to be joined in wedlock*, add 'also in act., Memn. 4.4 J.' add 'c *fit on*, στεφάνους .. κροτάφοισι ῥοδίνους συναρμόσαντες Anacreont. 43.2 W.'
συναρπαγή, after '= *obreptio*' insert '*Cod.Just.* 1.4.26.4' 2, for '*Cat.Cod.Astr.* 1.104' read '*Cat.Cod.Astr.* 1.106' and add '*PCarlsberg* 49ʳ i 22 (v AD; *APF* 32.17)'
συναρτύω II, for '*to be joint-*ἄρτυνος' read '*to be joint-*ἀρτύνας'
συναρχία II, in sg., add '*SEG* 23.447 (Thessaly, ii BC)'
συναρχῐεράομαι, add '*SEG* 26.784 (Thrace, AD 162/3)'
συνάρχω II 2, add 'Gal. 9.563.11'
συνασεβέω, add '*SEG* 32.1423d (Syria, AD 192/3)'
συνασπίζω, line 1, before 'Hsch.' insert 'Lxx 3*Ma.* 3.10'
†συναυλέω, *play the aulos in accompaniment* to, Ar.*Av.* 857 (cj.), συναυλεῖν τοῖς χοροῖς Ath. 14.617b. 2 of pipes, pipers, *play together*, Luc.*Dem.* 16, Longus 2.35.
συναυλία (A), for '*symphony* .. Hsch.' read '*playing together by two or more auletes*, Sch.Ar.*Eq.* 9 J. (cf. Hsch., Poll. 4.83)'
*συναυξητικός, ή, όν, *capable of increase*, φύσις Gal. 19.473.
συνᾰφή I, line 6, after 'Epicur.*Nat.* .. 9' insert 'Procl.*Inst.* 32 D., al.'; line 7, at end insert 'ἡμῖν δὲ πρὸς ἐκείνους (sc. τοὺς θεούς) γίνεται σ. Sallust. 15.3 R.'
συναφής, lines 1/2, delete 'κόλποι .. 21'; line 3, after 'Hp.*Morb.* 4.49' insert 'τὸ ἐκείνῳ (sc. τῷ πρώτῳ) σ. πλῆθος Procl.*Inst.* 149 D.'; line 4, after 'Aret.*SD* 1.7' insert 'topogr., κόλποι σ. ἀλλήλοις Arist.*Mu.* 393ᵃ21, *Peripl.M.Rubr.* 2 (w. gen.), 20, al.'
*συναφιερόω, *consecrate*, *SEG* 32.1425, 1426 (Syria, ii AD).
συναφίημι 1, add 'b *manumit together*, *SEG* 28.455 (Boeotia, iii BC).'
*συνδᾱμέτας, α, ὁ, = συνδημότης, *IRhod.Peraea* 201.15 (i BC).
συνδᾰπᾰνάω, add '*IGLS* 2707 (i AD)'
συνδᾳχονᾰφόρος, after 'Thess.' insert '(v BC)'
*συνδέξιος, ὁ, *member of a society of Mithraic* μύσται, *Dura*⁷,⁸ p. 87 no. 848, cf. Firm.*De err.prof.relig.* 5.2.
σύνδεσμος I 1, add 'σ. αὑτοῦ (sc. crocodile) ὥσπερ σμιρίτης λίθος Lxx *Jb.* 41.6(7)' V, delete the section transferring quot. to section IV. add 'VII perh. *puzzle, enigma*, Thd.*Da.* 5.12.'
συνδηλόω, line 3, for 'τῷ' read 'τοῖς'
συνδημιουργέω, after '*create*' insert 'or *fashion*' and add '*ISelge* 16'
συνδιαλέγομαι, after '*converse with* or *together*' insert '*ZPE* 77.56.7 (Cilicia, iii BC), D.S. 12.75.3'
συνδιαπνέω 2, for this section read '*exude together* with, ἱδρῶτος γενομένου συνδιέπνευσε τούτῳ τὸ ζέσαν τῆς μελαίνης Gal. 18(2).279.9'
συνδιατελέω, for '*continue with to the end*' read 'w. dat., *continue with* one *to the end*'
συνδιατίθημι, line 1, after '*help in arranging*' insert 'χσυνδιαθέσεν τὸν ἀ[γῶνα] *IG* 1³.3.8 (Marathon, v BC)'
*συνδιεξάγω, *administer together*, part. συνδιεξηχώς *SEG* 25.112.14 (Athens, early ii BC).
συνδικαστής, add '*fellow-judge*, Just.*Nov.* 53.3 pr.'
σύνδικος I 3, add '= Lat. *defensor civitatis*, *PCol.* 175 (iv AD)'
συνδιοικέω, line 3, after 'D. 24.160' insert 'Sallust. 21.1 R., Iamb.*De An.* ap.Stob. 1.49.67'
συνδοκῐμάζω, for 'J.*AJ* 20.2.2' read 'J.*AJ* 20.2.3'
*σύνδοπνος, ον, perh. *companion at table*, *Dura*⁶ p. 40.
*συνδουλίων, ωνος, ὁ, *fellow-slave*, *Vit.Aesop.*(G) 16 (pl., -ίονες cod.).
σύνδουλος 1, after 'Babr. 3.6' add '*IKios* 49 (Rom.imp.)'
συνδρομάς, for 'pecul. fem. of σύνδρομος' read '*running together, meeting*' and after 'Theoc. 13.22' insert '*of a type of door consisting of two leaves*, *IG* 1³.422.15, Poll. 10.24'
συνδρομή 2 a, add 'of rhetorical characteristics, D.H.*Dem.* 50'
συνδυάζω III, after 'Just.*Nov.* 130.7' insert '(sp. συνδοι-)'
*σύνδυασμα, ατος, τό, *combination of two elements*, Alcin.*Intr.* 10 (p. 166.3 H.).
σύνδυο, last line, for 'Plb. 8.4.2' read 'Plb. 8.6.2'
*συνεγγηράσκω, *grow old along with*, *SEG* 32.1243.21 (Aeolian Cyme, 2 BC/AD 2).
συνεγγισμός, line 3, delete 'πρὸς τὴν ἀρετήν'
*συνεγγυητής, οῦ, ὁ, *joint surety*, *MDAI(A)* 59.42 (Attica, iv BC).
*συνεγκαλέω, *charge jointly*, in pass., οἱ συνεγκαλούμενοι the *joint defendants*, *UPZ* 161.38, 56 (ii BC).
συνεδρεύω III 1, line 3, after 'Gal. 15.10' insert 'Συνεδρεύοντα, title of a book of the physician Praxagoras, Gal. 18(1).7.16'
συνέδριον 1, line 13, after 'Gal. 6.332' insert '= Lat. *consistorium principis*, Lib.*Or.* 18.154' 2, after 'X.*HG* 2.4.23' insert 'Men.*Dysc.* 177' add 'II *base for a number of statues*, *Michel* 537 (Cyzicus, i BC).'
σύνεδρος I 2, for '*sitting together, friendly*' read '*perching together*' add

συνειδός SUPPLEMENT σύννομος

'**3** *keeping company with*, ἔνθα σύνεδρος Φωσφόρῳ ἠδὲ καλῷ Ἑσπέρῳ ὄφρα πέλω *of a deceased infant*, SEG 31.846 (Italy, iii AD).' **II 2**, add 'of the Roman senate, BCH 109.597-607 (Augustus)'

*συνειδός, ότος, τό, v. σύνοιδα (sense I, p. 1721ᵃ line 14 and sense v 2).

συνειλέω, add 'εἰς αὑτὸν συνειλοῦ *withdraw into yourself*, M.Ant. 7.28'

συνείμαρται, after 'Plu. 2.569f' insert 'M.Ant. 12.3'

σύνειμι (εἰμί) **5**, after '*take part in, attend*' insert 'SEG 29.1088 (Caria, iii BC)'

σύνειμι (εἶμι) **II 3**, add 'ἐκτείνεσθαι καὶ συνιέναι Arist.*Mete.* 387ᵃ14; id.*PA* 689ᵃ30'

συνείσακτος **1**, add 'Just.*Nov.* 137.1'

*συνείσδοσις, εως, ή, app. *communal donation*, BCH 9.76 (Aphrodisias).

συνεισφορά, add '**2** transl. Lat. *collatio bonorum*, *joint contribution to estate by those emancipated by decedent in his lifetime*, Just.*Nov.* 18.6.'

*συνεκδᾰνείζω, *join in lending out*, rest. in SEG 31.122 (Attica, c.AD 121).

*συνεκκαθαίρω, *purge together*, Gal. 14.137.14.

συνεκμαρτυρέω, add '**2** med., *acknowledge jointly as one's child*, συνεκμαρτυρεῖσθαι .. τὸν γεγονότα αὐτῷ ἐκ τῆς Τνεφερσόιτος POxy.*Hels.* 35.36 (AD 151).'

συνεκπέμπω **1**, add '**b** *escort away*, in quot., the dead, Call.*fr.* 194.50 Pf. (tm.).'

συνεκπίπτω, add '**VII** *fall into* or *be included* in a schedule, ὁ δ' ἂν φανῇ ὑπολειπόμενον καὶ τούτων συνεκπεσεῖτε (i.e. πεσεῖται) τῇδε τῇ καταγραφῇ PColl.Youtie 65.75 (AD 241).'

συνεκπῠρόω, add 'Procl.*Inst.* 129 D.'

συνεκσῴζω, after 'S.*OC* 566' insert 'Men.*Dysc.* 753'

συνεκτικός **I 2**, for '*firmly gripping*' read '*holding oneself back, cautious*'

συνεκφέρω **1**, add 'SEG 33.1039 (Cyme, ii BC)'

συνελαύνω **I 2**, add 'Cod.Just. 1.3.38.2' add '**3** in chariot-racing, *drive together* with, τοὺς σὺν αὐτῷ συνελαύνοντας SEG 34.1437 (v/vi AD).'

συνελεουρέω, read συνελεορέω (cf. °ἐλεορέω)

συνελευθερόω, add '**II** *manumit several slaves together*, Cod.Just. 6.4.4.10, 11.'

συνέμφᾰσις, add 'Alcin.*Intr.* 35 (p. 189.19 H.)'

συνενεργέω, delete 'Plot. 3.4.6' (read συνεργέω) and after 'Ascl. in Metaph. 282.5' add '(cj.)'

συνεξελαύνω, add 'Just.*Nov.* 42.3 pr.'

συνεξέρχομαι **1**, add '**b** of two athletes, *step out of the ring together after a drawn fight*, Inscr.*Magn.* 180.16 (ii AD).'

*συνεπαγγέλλομαι, *join in making an ἐπαγγελία*, SEG 30.1073.14 (Chios, c.189/8 BC).

συνεπάγω **2**, lines 5/6, delete '*draw* .. cf.'

συνεπέρχομαι, add '**2** *approach together with*, in quot., of an epiphany, SEG 34.1216 (Lydia, AD 295/6).'

συνεπιβαίνω **II**, add 'abs., Men.*Dysc.* 954'

συνεπιβάλλω, add '**III** *add to an existing amount*, καὶ συνεπιβαλὶν τῷ λοιπῷ ἄνω PFreib. 53.23 (i BC). **IV** *affix a seal jointly*, συνεπιβαλλέτω τὸν δακτύλιον ὁ ἐγ [Μυ]ανίας καὶ ἐξ Ὑπνίας βούλαρχος SEG 23.305 (Delphi, c.190 BC).'

συνεπιδείκνῡμι, line 2, after 'Iamb.*Myst.* 2.7' insert 'of a ὁριοδείκτης, σ. ἀρούρας PRyl. 656.22 (AD 300)'

συνεπιδίδωμι **I 1**, add '**b** abs., *devote oneself wholeheartedly*, SEG 18.570 (Lycia, ii BC).'

συνεπικιρνάω, after '*mix with besides*' insert 'Ptol.*Tetr.* 146, Heph.Astr. 2.12.16'

συνεπινοέω **II**, add 'also act., Alcin.*Intr.* 10 (p. 164.14 H.)'

συνεπιπλέω, for '*join in a naval expedition*' read '*sail together on an expedition*' and add 'SEG 34.558 (Larissa, ii BC)'

*συνεπιτηρέω, *serve as joint-superintendent* with, τὰς ἡμέρας ἃς μόνας αὐτῷ Διδύμῳ συνεπετήρησα PMich.XI 616 (AD 182).

συνεπίτροπος, add 'PHamb. 70.2 (ii AD), IEphes. 3131'

*συνεποικιανός, ὁ, *fellow-member of a settlement*, TAM 5(1).712 (Gordus, ii AD; -επoκ-).

συνερανιστής, add 'SEG 31.122 (Attica, ii AD)'

συνεργασία, add 'ἡ σ. τῶν λινουργῶν SEG 32.1234 (Lydia, AD 192/3)'

*συνεργεπιστάτης, ου, ὁ, *joint superintendent of works*, IEphes. 674 (ii AD).

†συνέργιον, *guild of workmen*, PMich.Zen. 57.2 (iii BC), ICilicie 46.4 (i/ii AD). **2** τὸ Μέγα Συνέργιον, *a quarter in the city of Side* (see also Rev.Phil. 32.19ff.), IGRom. 3.810.9.

συνερείδω **II 2**, line 3, delete '*press on* .. Arr.*Tact.* 12.3'

συνερέτης, ου, ὁ, *fellow oarsman*, Hsch. s.v. ξυνερέται.

συνέριθος, before '*less freq.*' insert 'fem. sup. συνεριθοτάταν Βρομίῳ *most faithful handmaid to B.*, prob. in Telest. 1(c).1 P.'

συνέρχομαι **II 3**, add 'of accomplices, TAM 4(1).269' **b**, line 5, after 'PGnom. 71, al. (ii AD), etc.' insert 'σύνηλθε (for σύνελθε, on the bezel of a ring) RIB II(3).2422.35' add '**IV** w. dat., *help* (cf. Lat. *subvenio*), Just.*Edict.* 7.8.1.'

*συνέστᾱς, α, ὁ, *member of a society or college*, perh. of sim. standing to a παράσιτος, IG 9²(1).434.9 (pl.) (Acarnania, ii BC).

συνέστιος **1**, lines 2/3, delete 'ξυνέστιοι .. A.*Th.* 773 (lyr.)' **2**, for this section read 'Ζεὺς ξυνέστιος *Zeus patron of those who share the same hearth*, A.*Ag.* 704. **b** θεοὶ ξυνέστιοι *gods who share an altar*, id.*Th.* 773 (s.v.l.); perh. also συνεσ]τίοις θεοῖς PGiss. 99.26 (ii AD).'

συνευᾰρεστέω, add 'SEG 25.708, 36.590 (both Maced., Rom.imp.)'

*συνευᾰρέστησις, εως, ἡ, *consent*, Delph. 3(6).48.16 (ii BC).

*συνευφημία, ἡ, *panegyric to match another, parallel laudation*, cj. in Jul.*Or.* 3.106b.

συνεχής **I 1**, line 2, for '*of quantity*' read '*of types of measurable quantity*'; add '*consecutive in space*, ἄλλοι συνεχεῖς ὅρμοι **II 1**, line 9, delete 'τὸ σ. ἔργου (prob. for ἔργου) Anaxandr. 63'

*συνεχῖτις, ιδος, ἡ, name for γαλαξίας II, Plin.*HN* 37.162.

συνέχω **I 1**, line 10, after 'Ar.*V.* 95' insert 'συνέχει τὰς ῥίνας *holds his nose*, Sch.Ar.*Pax* 10'; line 13, for '*Is.* 52.16' read '*Is.* 52.15' **3**, add '**b** *compose, make up*, πέντε δὲ ἄλλαι (πόλεις) τὸν Πολεμωνιακὸν συνέχουσι Πόντον Just.*Nov.* 82.1 pr.'

συνηβάω, for '*παῖς ἐθέλει*' read '*παῖς ἐθέλει*' and for 'Anacr. 24' read 'Anacr. 33 P.'

συνηβολίη, for '*occurrence*' read '*meeting*'

*συνηγεμών, όνος, ὁ, *fellow commander*, SEG 31.1348 (Cyprus, iii BC).

συνηγορία **1**, add '**b** *as an official service*, Cod.Just. 1.5.18.10.'

συνήθεια **I**, add '**3** *club, guild*, τῶν πορφυροβάφων IG 10.2(1).291.1; τῶν ὄνων ABSA 18.155 (Beroea).' **III**, pl., add 'SEG 34.1243 (Abydos, c.AD 492). **2** perh. a type of dike-tax, PKöln 104.15 (vi AD), PHamb. 56 vi 15 (vi/vii AD).'

συνήθης **I**, add ''Ερμεῖ καὶ συνήθεσι, in a dedication, BCH 90.450 (Delos, ii/i BC)' **III 2**, add 'Peripl.M.Rubr. 14, 36, 38'

συνήκω **II**, for '*of walls, meet* in a point' read '*of structures, formations*, etc., *converge*'

*συνήλυσις, εως, ἡ, *meeting, assembly*, (transl. Lat. *conciliabulum*), IEphes. 4101.16 (Trajan).

συνημοσύνη **II**, delete 'sg. .. φιλημοσύνη'

συνήορος, add '**3** *brother, sister*, Hsch. s.v. ξυνάοροι.'

*συνθέαγος, ὁ, *fellow-θεαγός*, τῶν ἐξαγορείων SB 7634.13 (AD 249).

*συνθερᾰπευτής, οῦ, ὁ, *fellow-devotee* of Asclepius, Inscr.*Perg.* 8(3).28.

συνθεσία **1**, at end delete 'περὶ .. codd.'

σύνθεσις **I 2 g**, add 'of spices and incense, Lxx *Ex.* 30.32, 37; ἡ σ. θυμιάματος, τὸ θυμίαμα τῆς σ. ib. 35.28, 31.11' add '**h** *mosaic work*, τόπον τὸν ἐνθάδ' ἐκόσμησε Παῦλος τῆι πολυμόρφωι συνθέσει SEG 26.1629 (Apameia, vi AD).' add '**3** *joint tomb*, Robert *Ét. Anat.* pp. 224-5.'

σύνθετος **I 1**, add 'πλίνθος PMich.v 285-6.4 (i AD)'

συνθήκη, after lemma insert 'non-Att.-Ion. -θήκα' and add 'SEG 36.548 (Thessaly, iii BC)'

συνθηκοφύλαξ, after '*covenant*' insert 'Vit.Aesop.(G) 71'

†συνθηρᾱτής, οῦ, ὁ, *one who joins in hunting*, in quot., of hunters' horses, Philostr.*Gym.* 26; transf., *one who joins in the quest of*, τῶν φίλων X.*Mem.* 3.11.15.

συνθραύω, after 'Plu.*Arist.* 18' insert 'SEG 35.209 (c.AD 150)'

σύνθρονος, add 'of a soul in the underworld, ISmyrna 529 (verse, i AD)'

*συνθυίω, *rush together with*, τινι Sch.min.Il. 6.21.

συνθύξω, add 'perh. fut. of συντυγχάνω'

συνθῠσία **I**, add 'personified, SEG 35.1736-7 (cf. RA 1987.98, Phrygia, iii AD)'

σύνθωκος **I**, before 'Jul.*Or.* 5.166b' insert 'of the Mother of the gods' and add 'Μίνω σ. εἰμι poet. in inscr. in APF 5.164 (iii/ii BC)'

συνιαίνω, for '*cheer together*' read '*simultaneously gladden*'

*συνιερογλύφος, ὁ, *fellow hieroglyph-carver*, POxy. 1029.6 (pl., ii AD).

συνιππεύω, before 'θυέλλαις' insert 'ἀέλλαις Opp.*H.* 5.344'

*συνισθμιάζω, *join in the celebration of the Isthmian games*, A.*fr.* 78c.58 R.

συνισόομαι, add 'act., Hsch. s.v. συγκρίνει'

συνίστημι **A VI 2**, add 'Modest.*Dig.* 27.1.13.7' **B II**, add '**4** *stand by* a principle, attitude, etc., *insist on*, συστῆναι ταῖς κατηγορίαις Cod.Just. 1.4.34.18, συνίστασθαι τοῖς ἑαυτοῦ δικαίοις ib. 10.11.8.5.'

συνίστωρ **2**, line 3, after 'σώματα' insert 'αὐτοῖς'

συνισχναίνω **2**, after 'ξυνισχνᾶνεῖ' insert '(or ξυνισχάνει; codd. συνανίσχει)'

συνναύτης **II**, for '*worshippers of Isis*' read '*fishermen*' and for '(Callipolis)' read '(Parium), cf. Robert *Hell.* 9.81'

*συννεκροτάφος, ὁ, *fellow burier of the dead*, PRyl. 574.2 (pl., i BC).

συννέμω **1**, add '**b** *manage together*, αὑτη δὲ συνένεμε τὸ ἐργαστήριον PLond. 1976.12 (iii BC).'

συννομή **III**, add 'and at Lindos, Lindos 454.9'

*συννομιτεύομαι, *share quarters*, οὔτε ἐν τῇ κώμῃ συννομιτεύεται PMich.inv. 160 (ZPE 23.133, AD 162).

σύννομος (A) **I 2**, lines 5/6, for '*which lie between two seas*' read '*which consort with the sea*'

286

σύννοος 3, add 'Aen.Tact. 20.1'
σύννυμφος, add 'JRCil. 1.24 (late Rom.imp.)'
συνοδία III, delete the section.
*συνοδιακός, ή, όν, ἰνσπέσιμον σ. of assembled troops, SEG 9.356.25 (Ptolemais in Cyrenaica, vi AD).
σύνοδος (B), line 1, after 'deliberation' insert 'Alc. 130.30 L.-P.' 2, add 'personified, SEG 35.1377 (cf. RA 1987.103, Phrygia, iii AD)' 3, add 'ἡ σ. τῶν τεκτόνων SEG 29.1186 (Lydia, AD 165/6); of religious associations, SEG 32.453 (Boeotia, ii BC)'
συνόδους, at end for 'σινόδους, συνώδοντα' read 'σινόδων'
συνοικειόω 1, at end before 'Plu.Lyc. 4' insert 'SEG 38.1476.31 (Xanthos, 206/5 BC)'
συνοικέω I 2, after 'PEnteux. 91.2 (iii BC), etc.' insert 'w. acc., SEG 38.1476.29 (Xanthos, 206/5 BC)'
συνοίκησις I, add 'fig., ἡ ψυχή .. διὰ τὴν συνοίκησίν τε καὶ πρὸς τὸ σῶμα ὁμιλίαν Max.Tyr. 1.7'
συνοικιάζω, delete the entry (v. °ἐνοικιάζω).
συνοικίζω II, add 'SEG 38.1476.79 (Xanthos, 206/5 BC)'
*συνοίκιον, τό, joint lodging, PMil.Vogl. 287.5 (ii AD).
συνοικισία 1, for 'Pontus' read 'Maeander'
συνοικισμός II, after ' = foreg.' insert 'SEG 30.1120 (Entella, ?iii BC), SEG 25.445 (Stymphalus, Arcadia, c.189 BC)'
συνοικιστήρ, for 'Pi.fr. 186' read 'Epic.ap.A.D.Synt. 138.13'
συνοικιστής, after 'of a colony' insert 'SEG 12.379.9 (Cos, fr. Camarina, iii BC)'
σύνοικος 2, add 'c αἱρεῖσθαι [Δίκαν] σ. B. 15.56 S.-M.'
συνοκωχή, add 'συνοκωχά· νόσος, λοιδορία, μάχη Hsch.'
συνόμαιμος, add 'SEG 28.541 (Maced., Hellen., in quot., sister)'
*συνομβρέω, flood, cover with a deluge, pass., PMichael. 4.11 (ii AD).
συνόμευνος, for 'bedfellow' read 'consort' and add 'Κρόνου συνόμευνε Orph.H. 27.12, INikaia 1248'
συνόμιλος, before 'Hsch.' insert 'GVAK 17 (Phrygia)'
*συνομόφυλος, ὁ, ἡ, member of the same profession, POxy. 3500.12 (iii AD).
*συνονομάζω, name together with, POxy. 3498.25, 43 (AD 274).
συνοπᾱδός, add 'Iamb.Protr. 14.5 P.'
συνόριον, add 'SEG 35.665A.24, 25 (pl., Ambracia, ii BC)'
*συνορχηστής, οῦ, ὁ, fellow dancer, παιδὶ συνορχηστὴν θύρσον POxy. 3723.16 (elegy, ?ii AD).
συνουσίασμα, for 'Berl.Sitzb. 1934.1041' read 'SEG 37.389.12'
συνουσιαστής, add 'fellow member of a social club, Lys. 8 tit.'
*συνοφρυάζω, contract the brows, frown, Sch.S.Ant. 528.
συνοχή II 6, delete the section.
συνόχωκα, for 'of *συνοχόω' read 'an artificial form, perh. of συνέχω, cf. *ὄκωχα'
συντακτικός, add 'IV perh. placed in the same category, συντακτικοῖς καὶ ὑπερθέτοις PRyl. 585.29 (ii BC).' transposing entry to follow συντακτήριος
συνταμίας, before 'colleague in the quaestorship' insert 'fellow-treasurer, Didyma 393.3, 400.8'
*συνταφιαστής, οῦ, ὁ, member of a burial-society, rest. in PRyl. 590.13 (i BC).
*συντέκτων, ονος, ὁ, fellow craftsman, SEG 28.1368 (Palestine, iii AD).
συντέλεια IV, delete the section.
συντελέω I 1, line 3, delete 'σ. εἰς τὰ ἑκατόν .. X.Cyr. 6.1.50' 2, add 'b act., happen to be, CodJust. 1.3.35.3.' 4, for 'destroy' read 'exterminate (calque of Hebr. killah)' add '5 pucker up in scorn, στόμα Aq.Is. 52.16.' II 1, add 'ἐπειρᾶτο συντελεῖν αὐτῷ εἰς τὰ ἑκατὸν ἅρματα X.Cyr. 6.1.50' III 2, line 4, for 'tributaries' read 'regions united to it'
†συντέμνω, Ion. -τάμνω Hdt. 5.41.2, 7.123.1, fut. -τεμῶ, aor. -έτεμον, make or form by cutting, ξ. τὰς πρῴρας ἐς ἐλάσσον Th. 7.36.2, σ. χιτῶνας (i.e. uppers of shoes) X.Cyr. 8.2.5, σ. τὰς πλεκτάνας Alex. 187, cf. 84, Arr.An. 7.19.3; transf., συντέμνει δ' ὅρος ὑγρᾶς θαλάσσης the sea cuts short (my realm), A.Supp. 258; abs., shorten one's course, take a short route, σ. ἀπ' Ἄμπελου ἄκρης ἐπὶ Καναστραίον ἄκρην Hdt. 7.123.1. 2 cut a path, καινὴν .. καὶ ἐρήμην ἀνοδίαν Porph.Chr. 1. II curtail in quantity, degree, extent, etc., τιμᾷς σὺ μὴ σύντεμνε τὰς ἐμὰς λόγῳ A.Eu. 227, εἰς ἕν .. πάντα τὰ μέλη ξυντεμῶ Ar.Ra. 1262, τὸν ἐνιαυτὸν σ. .. εἰς μῆν' ἕνα Philippid. 25.1 K.-A. 2 cut expenses, σ. τὴν μισθοφοράν Th. 8.45.2; σ. τὰς δαπάνας εἰς τὰ καθ' ἡμέραν cut down one's expenses to one's daily wants, X.Hier. 4.9; humorously, by expressing items in the form of diminutives, Mnesim. 3.4 K.-A. 3 curtail narrative, expression, etc., ἐν βραχεῖ πολλοὺς λόγους Ar.Th. 178, cf. Aeschin. 2.31, Pl.Prt. 334d, Heph.Astr. 2.15.2; w. ellipsis of obj., ὡς δὲ συντέμω E.Tr. 441, Hec. 1180, οἶνον εἰπὲ συντεμών Antiph. 55.12 K.-A., cf. D. 39.4, Anaxil. 22.30 K.-A. 4 cut short a process, πόνους E.Rh. 450; med., πάντα τοι συντέμνεται Κύπρις .. βουλεύματα S.fr. 941.16 R.; w. ellipsis of obj., τοῦ χρόνου συντάμνοντος as time cut the matter short, Hdt. 5.41; stop a person in his course, συντέμνουσι γὰρ

θεῶν ποδώκεις τοὺς κακόφρονας Βλάβαι S.Ant. 1104. III join with a cut, ἵνα συντμηθῇ πάντα καὶ γένηται μία ἕλκωσις Heliod. ap.Orib. 44.23.69. IV join in cutting, assist in making a distinction, Pl.Sph. 227d, Plt. 261a.
συντεχνίτης, add 'PHamb. 56 v 1 (p. 207) (vi/vii AD)'
*συντηκτικῶς, of feverish patients, wastingly, Paul.Aeg. 7.19.3 (375.24 H.).
συντίμησις, before 'valuation' insert 'total'
*σύντμησις, εως, ἡ, cutting short, e.g. ἱερομηνία for ἱερονουμηνία, Sch.Pi.N. 3.4.
συντομία I, add '2 cutting of delay, dispatch, διὰ πάσης συντομίας SB 9752.3 (vii AD).'
σύντομος I 2, add 'b of speakers, writers, Aeschin. 2.51, Call.Epigr. 11 Pf., Poll. 4.20, 6.149.' 4, delete the section.
συντονάριος, after 'pedicularius' insert 'perh. referring to musician who accompanies his playing with foot-operated clapper (v. °κρούπεζα 2)'
συντρέχω I 2, before 'Hdt. 8.71' insert 'Archil. 102 W.'
Σύντριψ, for 'lubber-fiend .. kitchen' read 'demon supposed to break pots during manufacture'
σύντροφος I 1, at end before 'cf. συντρόφη' insert 'also fem., foster-sister, SEG 34.1205 (Lydia, AD 170/1), 31.991 (Lydia, AD 234/5), al.' 3, delete 'c. gen. .. S.Ph. 203, cf. σύντροπος'
συνυμνέω, after 'sing hymns together' insert 'SEG 37.961 (Claros, ii AD)'
συνυπάγω, after 'Sch.E.Or. 854' insert 'subject jointly to, τὴν τοῦ γήμαντος οὐσίαν συνυπάγει ταῖς συνθήκαις Just.Nov. 22.40' and after 'pass., to be brought under the power of' insert 'Cod.Just. 1.2.17 pr.'
συνυποδέχομαι, add '2 receive, obtain jointly, Just.Nov. 11.24.'
*συνυπομένω, endure together with, Sopat.Rh. 8.173.23 W.
συνυπουργέω, add 'BCH 7.502 (Phrygia, AD 433)'
συνυφή 1, add 'applied to the girdle of the ephod, Lxx Ex. 36.28 (39.20)'
συνωθέω, add 'III force, compel, Cod.Just. 1.3.38.3, 1.4.26.3, ὅπερ ἔλαβε .. ἀποδοῦναι συνωθείσθω Just.Nov. 123.28.'
συνωμόσιος, add 'as a title of Zeus, OGI 65 (Egypt, iii BC)'
*συνωρία, ἡ, carriage-service, οἱ μουλίωνες οἱ ἐπεστῶντες (= ἐφεστῶτες) συνωρίᾳ TAM 4(1).39.
συνώρυος, delete 'doubtful .. (Bithynia)'
*συοσκύαμος, ὁ, = ὑοσκύαμος, prob. in Nic.Al. 415.
†συππῖνᾶς, ᾶδος, ὁ, worker or dealer in tow, τῇ συμβιώσει τῶν συππινάδων ISmyrna 218.13; cf. στυππεῖον, στιππύον, σίππιον.
συρβάβυττα, delete the entry.
*σῠρήνεμος, ον, (σύρω, ἄνεμος) drawing breeze, ῥιπίδας, cj. for πυρ- in AP 6.101.2 (Phil.); cf. ‡ἀνεμόσυρις.
συρία, for 'garment' read 'stuff or material'
*Συριακός, v. °Σύρος.
σῡρίγγιον 3, add 'Gal. 13.402.15, 403.6'
σῡριγγώδης, add '2 sounding like a pan-pipe, Heph.Astr. 1.1.119, 139, al.'
σῦριγξ I 1, after 'Hes.Sc. 278' insert 'E.Ion 498, IA 1038; sg., Ion Trag. 45 S.' add 'b σ. μονοκάλαμος flute, Euph.ap.Ath. 4.184a, AB 265.21, EM 480.1.' 3, for 'mouthpiece of the αὐλός' read 'an attachment to the αὐλός allowing modification of the sound, Aristox.Harm. 21, Arist.Aud. 804ᵃ14' II 2, after '(A.)Supp. 181' insert 'but the nave itself in' 11, for 'perh. loop' read 'sheath'
σῠρίζω I, add '2 play the aulos in its top register, Aristox.Harm. 21; cf. °σῦριγξ I 3.' II, after 'hiss like a serpent' insert 'Hippon. 79.11 W.'
†Σύριος, v. ‡Σύρος.
†συρίσκος, σύρισσος, σύριχος, ὁ, v. ὑριχός.
σύρισμα, after 'the former in' insert 'Longus 2.30'
συριστής I, add 'hissing, of snakes, Opp.C. 1.521'
Σύρος, line 3, after 'v. infr.)' insert 'of Assyrians, Aristid.Or. 26(14).91'; line 16, after '(v. πύλη II 2)' insert 'Συρίᾳ Θεῷ SEG 34.885 (Chalcis); Assyrian, Σ. γράμματα, i.e., in this case, cuneiform, D.S. 2.13'
†σύρρα, phonetic sp. of σύν ῥα, v.l. in Il. 8.61; understood as simple equiv. of σύν, Orph.fr. 167.5, Hsch.
σύρροος, line 2, delete '101e' (v. °σύσσοος)
σύρροπον, delete the entry.
*σύρροπος, ον, prob. = ἰσόρροπος, of a weight which is equal to a control weight, IGLS 1272b (ii AD); also cited in Sch.D.T. p. 465.5 H.
*συρρυσμόω (Ion. for συρρυθμόω), bring into combination, συρρυσμοῦσθαι (συνρυθμοῦσθαι cod.)· συγκρίνεσθαι Hsch.
*συρτάριον, τό, instrument for making metal thread, Anon.Alch. 323.6.
σύρτης I, add '2 app. thong, παράμματα μετὰ τῶν συρτῶν Edict.Diocl. in SEG 37.335 iii 3 (μετὰ φλαγέλλου in another version).'
συρτός, ή, όν, II, after 'Cyran. 58' add 'τοῦ θηρίου συρτοῦ Orac.Tiburt. 151'

σύρω 2, line 5, after 'Plu. 2.5f' insert 'abs., ὕδωρ ἐν φάραγγι σῦρον Lxx Is. 30.28'

συσκέπτομαι, add 'Sopat.Rh. 8.44.27, 48.19 W.'

*****συσκεύασμα**, ατος, τό, *contrivance*, gloss on κάττυμα Hsch.

συσκύλλομαι, add 'POxy. 2275.19 (iv AD)'

συσπάω I, line 5, for 'fire' read 'drought'

συσσᾰρᾱπιαστής, for 'fellow-worshipper' read 'pl., *guild of worshippers*' and after 'Thasos' insert 'ii BC'

συσσίτιον I 2, add '*soldier's mess*, Aen.Tact. 27.13'

*****συσσῑτολογέω**, pf. συνεσιτολόγηκα, *to be σιτολόγος jointly*, μετά τινος PTeb. 774.8 (ii BC).

σύσσιτος, add 'D. 54.4, Aeschin. 2.22, al.; pl., of members of a religious association, SEG 32.505 (Thespiae, c.300 BC)'

*****σύσσοος**, ον, *coherent, solid*, οὐδὲ γὰρ ἔτι εἴη κα σύνσοον καὶ ἓν τὸ ζῶον Ti.Locr. 101d M.

συσταθμία, after 'Dsc. 1.54' insert 'PMerton 12.17 (i AD)'

+**συσταλτικός**, ή, όν, medic., *systolic*, Gal. 9.298.3.   **2** of music, *tending to produce lowness of spirits, depressive*, Cleonid.Harm. 13, Aristid.Quint. 1.12.

συστάς 1, add 'Ath.Agora XIX Ph9 (370/69 BC)'

σύστᾰσις B I 4, after 'conspiracy' insert 'Inscr.Perg. 160.6 (συσστ-; ii BC)'

συστάτης II, for this section read 'official, perh. only at Oxyrhynchus, who replaced the φύλαρχος at the end of the 3rd century AD, responsible for: appointments to liturgies, POxy. 1627, PFlor. 39; registration of births and deaths, PSI 164, POxy. 1551; poll-tax collection, PSI 163, PFouad I Univ. 19, etc.'

συστᾰτός 2, add 'Pl.Ti. 33a'

συστέλλω I 2 a, add 'συσταλμένος of time, *limited*, 1Ep.Cor. 7.29'   **b**, add 'Gal. 11.240.8'

συστέφομαι, add 'συνστεφθεὶς παίδων πάλην SEG 30.1524 (Lycia, iii AD)'

σύστημα 2 - 5, for these sections read '**2** *organized body* or *association*, political, ethnic, etc., Pl.Lg. 686b, Arist.EN 1168ᵇ32, Plb. 2.38.6, προσέθηκε τὴν πόλιν πρὸς τὸ τῶν Ἀχαιῶν σύστημα id. 2.41.5, cf. 6.10.14, 9.28.2, Diog.Bab.Stoic. 3.241, τὸ Ἀμφικτιονικὸν σύστημα Delph. 3(1).48.16 (i BC), ὅπου ποτὲ σύστημα τοῦ γενοῦς ἐστιν ἡμῶν J.Ap. 1.7.   **b** of a professional, administrative, etc., body, θήκη διαφέρουσα τῷ συστήματι τῶν λημενητῶν λινοπωλῶν MAMA 3.770, cf. SEG 30.1382c (Lydia, AD 301), IMylasa 144.5, al.; of priests, Plb. 21.13.11, Str. 17.1.29; τὸ δὲ σύστημα σενᾶτον προσηγόρευσεν Plu.Rom. 13, cf. Lib.Or. 11.146, σ. τῆς γερουσίας ITrall. 77.1.   **c** of a body of troops, τὸ σ. τῶν μισθοφόρων Plb. 1.81.11, cf. 5.53.3, 30.25.8, etc.; of a gang, J.AJ 20.9.4; of a ship's crew, Alciphr. 1.8.   **d** of a herd, etc., of animals, Plb. 10.27.2, 12.4.10.'

*****συστρᾱτιωτικός**, ή, όν, *of an association of συστρατιῶται*, PRyl. 585.11 (ii BC).

συστρέφω IV, add '**2** act., perh. *collect, accumulate*, ἐλαϊκὴν δὲ συστρέφειν ἕκαστ[ον] τῶν ἐλα[ι]οπωλῶν PHamb. 182.10 (iii BC).'

συστροφή I 2, add '**b** *dazzling* of the eyes, Simp. in de An. 46.1; *dizziness*, Hsch.'   **II 1**, add 'cf. Myc. ku-su-to-ro-qa, *total*'   add '**6** pl., *wrinkles*, Sch.Gen.Il. 9.503.'

*****σύστρωσις**, εως, ἡ, perh. = σύστρωμα, IGRom. 3.365.7.

*****συστυφόομαι**, = ‡συστύφω, Hsch. s.v. συνεστυφωμένοι.

συστύφω, for 'Pass.' to end read 'glossed by (συ)σκυθρωπάζω, *look gloomy*, Hsch. s.vv. συνέστυβας, -βεν, συνστύψαι'

σύστροφος, prob. error for συνίστρωρ, JÖAI 8.143 (tab.defix.).

*****συσφᾰγιάζομαι**, pass., *to be slaughtered along with*, τῇ θυγατρί Sch.E.Hec. 399.

συσφαιριστής, delete 'dat. pl. .. IG 2².4794' (v. °συμφέροντής)

συσφίγγω, line 1, after 'Herod. 5.25' insert 'so as to check the flow of blood, Ath. 2.41b'; line 2, before 'τὸ λόγιον' insert '*fasten close*'; line 3, after 'ib. 3Ki. 18.46' insert '*clench*, ib.De. 15.7'

συψειρικόν, delete the entry.

*****συψειρικός**, όν, also συμψ-, συνψ-, Lat. *subsericus, partly silk*, στίχη Edict.Diocl. 19.10, 20.1; σειρακίῳ ἐργαζομένῳ εἰς συψειρικὸν τρεφομένῳ ib. 20.9, al.

σύωδης, add 'comp. -έστερος Eun.ap.Phot.Bibl. 77'

σφᾰγίς, add '*butcher's knife*, AP 6.306 (Aristo)'

σφᾰγῖτις, add 'Hp.Acut. 9'

*****σφαγνίον**, τό, *commodity (?plant)* occurring in price lists, POxy. 37.33.24, 3766.106 (iv AD).

σφάζω 6, for 'metaph. *torment*' read 'hyperb. "*kill*"' and add 'Men.fr. 746 K.-Th.'

σφαῖρα 4, for 'a weapon of boxers .. σφαιρομαχίαι' read 'prob. *padded glove* used in boxing practice'   **6**, for 'sea-*balls*' read 'kind of zoophytes, = ἁλοσάχναι'   **8**, add 'Nic.Th. 584'   add '**10** ball used in voting, SEG 9.8 i 25 (Cyrene, 7/6 BC).'

σφαιρίον I, line 2, after 'Pl.Ep. 312d' insert 'ὑελοῦν *glass ball*, PFam.Teb. 49a ii 3 (iii AD)'

*****σφαιροληκῦθος**, ὁ, app. *globular oil-flask*, POxy. 3080.8 (ii AD).

σφαιρομᾰχέω 1, add 'fig., Men.Dysc. 605'

σφαιρομᾰχία, add '**2** *ball-contest* (description of °ἐπίσκυρος), Poll. 9.107.'

+**σφαιρωτήρ**, ῆρος, ὁ, *ornamental knob*, on a seven-branched candlestick, Lxx Ex. 25.31, 33, al.; pl., as heraldic device, Tab.Heracl. 1.184.   **2** transl. Hebr. wd. meaning 'thong' (v.l. σφυρ-), Lxx Ge. 14.23.   **b** in an inventory in unexpld. sense, PLond. 402ᵛ.22 (ii BC).

σφᾰκος I, for '*sage-apple, Salvia calycina*' read 'kind of sage, *Salvia* species which forms an edible gall-apple, cf. mod. Gk. φασκόμηλο' and add 'Men.Dysc. 517 S.'   add 'cf. Myc. adj. pa-ko-we, σφακόϜεν'

*****Σφαλεώτας**, α, ὁ, epith. of Dionysus, SEG 19.399.a3, al. (Delphi, ii BC).

σφάλλον, add 'prob. Aeol. = σφῆλον fr. σφάλλω'

σφεδᾰνός I, delete 'κάρηαρ Nic.Th. 642' and add '**2** perh. *having a rough surface*, κάρηαρ Nic.Th. 642.'

σφειδρόν, for 'Theognost.Can. 12' read 'Theognost.Can. 37 A.'

σφεῖς A I 3, add 'Arc. σφεῖσιν SEG 37.340.15, (Mantinea, iv BC)'   **II**, line 7, for 'Riv.Ist.Arch. 2.19' read 'Inscr.Cret. 4.83.2 (v BC)'   **B I 3**, line 6, after 'al.' insert 'And. 3.11; πλεῖν ἐπὶ σφᾶς αὐτούς *against themselves* or *their own country*, Th. 8.86.4.'; add 'as 2 pers. pl., σφᾶς αὐτοὺς ἐπεφόβησθε And. 2.8'   add 'Myc. pe-i (dat. pl. σφέhι)'

σφεκλάριος, delete the entry.

*****σφεκλαρᾶς**, ᾶ, ὁ, = Lat. *specularius, glazier*, SEG 7.197 (Berytus, v/vi AD).

σφέκλον, add 'Olymp. in Grg. p. 252.7 W.'

σφέλας III, for this section read '*block for pounding or chopping*, ἢ σφέλᾳ ἢ ὅλμῳ κεάσας Nic.Th. 644'

σφενδάμνινος, for 'metaph. .. *hearts of oak*' read 'fig., in uncertain significance'

σφενδονάω I 1, add 'generally, *pelt*, σ. βώλοις λίθοις Men.Dysc. 120'   **II 2**, for '*move like a swing*' read '*move like a sling*'

σφενδονήτης, for 'Ἀρχ.Δελτ. 14 Pl. iv 26' read 'SEG 23.271'

σφηκιά, before 'Lxx Ex. 23.28' insert '*swarm of wasps*'

*****σφήκιον**, τό, perh. *rafter*, SEG 37.665.3 (N. coast of Black Sea, c.400 BC; see also BE 1989.478); cf. σφηκίσκος, ἡμισφήκιον.

+**σφηκισμός**· εἶδος αὐλήσεως εἰρημένον ἀπὸ τῆς ἐμφερείας τῶν βομβῶν Hsch.

σφήνωσις 1, add 'Gal. 18(2).594.2'

σφίγγω II 1, add 'ἡ δὲ ἀμέρεια σφίγγουσα καὶ συσπειρῶσα (τὴν ἑκάστου δύναμιν) Procl.Inst. 86 D.'

Σφίγξ, line 2, for 'where .. Sch.' read 'v.l.'   at end, form Σφίξ, add 'CEG 120 (?c.450 BC)'

σφόδρα I 4, line 5, after 'Pl.Phd. 100a' insert 'οὐδέπω γεγονὼς σ. εἴκοσιν ἔτη *not quite twenty*, Pl.Alc. 1.123d'

σφοδρός II, line 5, after '-ότερον' insert 'Thphr.Sign. 32'

σφόνδυλος II 1, add 'perh. also Aen.Tact. 36.2'   **4**, delete the section.

σφρᾱγίς I 1, add 'fig., of the final moment, ἐν δ' ὥραις ὀλίγαις Μοιρῶν γὰρ σφραγεῖδες ἐπῆλθον GVI 1166 (Rome, iii AD)'   add '**VI** name of a section of the νόμος κιθαρῳδικός, Poll. 4.66.'

*****σφράγιστρα**, τά, app. *fee for attaching a seal* or *warrant* to a sacrificial victim, PMoen inv. 5 (CE 54.275).

*****Σφρᾱγίτιδες** [?ᾱ], αἱ, nymphs associated w. prophecy on Mt. Cithaeron, Plu.Arist. 19.6, id. 2.628f, cf. the cave Σφραγίδιον, Paus. 9.3.9; cf. Σφραγιτίδων.

σφρᾱγῖτις, ιδος, ἡ, kind of medicament, cf. σφραγίς v, τῇ Πολυείδου σφραγιτίδι Aët. 9.38.18.

*****σφῠρίζω**, *pound, crush*, Anon.Alch. 329.23, 27.

+**σφῠρίς**, v. σπυρίς.

σφῠρόν I, add 'prov., ἀνώτερον τοῦ σ. λέγειν *talk of matters outside one's knowledge*, Ath. 8.351a, cf. *ne supra crepidam sutor*'

σφῠρωτήρ, before 'Lxx' insert 'v.l. in'; for 'cod. .. cett.)' read '(cf. °σφαιρωτήρ)'

σφωΐτερος II, delete ' = σφέτερος' and section 1.   **4**, after '1.1286' insert '4.454'

σχάδιον, after ' = ἰσχάδιον' insert 'PRyl. 629.32 (iv AD), al.'

+**σχᾰδοκάρυα**, v. °ἰσχαδοκάρυον.

*****σχαστός**, ή, όν, perh. *out of line, askew*, SEG 39.442.36 (Oropus, iv BC; sp. σκαστ-; cf. ἄσχαστος).

σχεδιάζω, add '**3** *approach*, Hsch.; cf. σχέδιος I 1.'

*****σχεδιαστής**, οῦ, ὁ, *improviser*, Teuc.Bab. p. 44 B.

σχεδικός, for '*riddling*' read '*of* or *belonging to σχέδη* (°σχέδος 2)' and for '*riddle-composers*' read '*makers of σχέδη*'

*****σχεδιοναύτης**, ου, ὁ, *raft-navigator*, SEG 28.1040 (Nicomedia).

σχεδογρᾰφία, for '*art of parsing*' read '*writing of σχέδη* (°σχέδος 2)'

σχεδόθεν, add 'σχεδόθεν δὲ κακῷ κακόν .. ἀκειομένη, *shortsightedly trying to cure one wrong with another*, A.R. 4.1081'

σχέδος, for '*riddle*' read 'in later Gk., *account of a word* w. ref. to

288

origin, sp., etc., *An.Boiss.* 2.349; hence **2** *riddle* based on alternative sps. and division of wds.'

σχεδουργός, for '*riddle-maker*' read '*maker of σχέδη (°σχέδος 2)*'

*****σχέθω I 2**, line 2, after '(Od.) 14.490' insert 'εὔνοον θῡμὸν σκέθοντες Alc. 129.10 L.-P.'

σχελίς, add 'cf. σχέλος = σκέλος (q.v.)'

σχέμα II, delete the section.

σχερός II, for 'Theognost.*Can.* 12' read 'Theognost.*Can.* 38 A.'

σχετλιασμός, add '**2** gramm., *interjection* (of indignation), Charis. p. 470.19 B. (given as equiv. of Lat. *interiectio*).'

σχετλιαστικός, line 2, delete 'A.D. .. al.'; at end for 'ἐπίρρημα' read '°ἐπίρρημα D.T. 642.2, A.D.*Pron.* 34.30, *Adv.* 127.19'

σχῆμα 5, add '**b** *calling*, τὸ μοναχικὸν σχῆμα *Cod.Just.* 1.3.52.9, τοῦδε τοῦ σχήματος (sc. office of καγκελλάριος) Lyd.*Mag.* 3.38.' **8 d**, add 'also of dice used in prophecy, Paus. 7.25.10'

*****σχηματίζω**, *shape with protuberances and hollows*, i.e. opp. smooth, Alex.Aphr. *in Sens.* 58.12.

*****σχηματιστικός**, ή, όν, *fitted for giving shape*, ἐνέργεια Procl. *in Ti.* 2.216.22 D.

σχίζω I 1, line 6, after 'Pi. l.c.' insert '*SB* 9123.7 (ii BC)'

σχῖνος I, for 'trodden on by goats' read 'browsed upon by goats, Eup. 13.4 K.-A.' **II**, delete 'Thphr.*CP* 5.6.10, *Sign.* 55'

σχίσμα I, add 'of the *vulva*, Ruf.*Onom.* 110' **III**, delete the section.

σχισμή, add '*split*, Heraclid.Pont.ap.Sch.Pi.*O.* 6.119'

*****σχοινάριον**, τό, perh. *ball of twine*, PMich.inv. 4001 (*ZPE* 24.84, i/ii AD), *PRyl.* 627.170 (iv AD).

*****σχοινεύομαι**, app. *measure by σχοῖνοι (σχοῖνος III)*, Hsch. s.v. καίνυσθαι.

σχοινιά II, for '*CIG* (add.) 2056g' read '*IGBulg.* 57.7 (Tiberius)'

σχοινίον I 2 c, add '*PGen.* 99 (ii AD).

*****σχοίνιος**, α, ον, prob. = σχοίνινος, ῥῦμα PZen.Col. 43.3 (iii BC).

σχοινίς (A) **I 2**, for this section read '*raised band of decoration*, Didyma 39.17; on silver cup, ib. 424.55. **b** perh. *enclosure wall*, *TAM* 2.850.'

σχοινίων II, delete '*effeminate*' and for '*flute*' read '*aulos*'

σχοινοειδής, for 'J.*AJ* 12.2.9' read 'J.*AJ* 12.2.8'

*****σχοινοκοπέω**, *cut rushes*, Hsch. s.v. καλοκοπῆσαι.

σχοῖνος, line 6, after 'σ. εὐώδης' insert '*AP* 4.1.26 [Mel.]' and delete 'Hp.*Mul.* .. *Nat.Mul.* 33'; transfer 'Aret.*CA* 2.8' to follow 'Dsc. 1.17' **I 3**, add 'Lxx *Ps.* 138(139).3' **III**, add '**b** as a measure of area, *Tab.Heracl.* 1.19, al.'   add 'Myc. *ko-no*, var. *ko-i-no*'

*****σχοινουλκία**, ἡ, *survey*, PMerton 5.33 (ii BC).

*****σχοινωτός**, ή, όν, perh. *rope- or rush-patterned*, ψέλια PMichael. 18Ai5 (iii AD).

σχολάζω II, add '**2** *to be annulled*, θεσπίζομεν ὥστε πᾶσαν ἐκποίησιν πραγμάτων .. τοῖς σεβασμίοις οἴκοις διαφερόντων .. σχολάζειν *Cod.Just.* 1.2.17.1, Just.*Nov.* 120.11; *to be idle, be in abeyance*, *Cod.Just.* 10.16.13.1. **3** *of period of time, expire*, Just.*Nov.* 111.1.'

σχολάριος, add '*SEG* 37.1076 (Bithynia)'

σχολεῖον II, for this section read '*place for sitting or resting, exedra*, *BCH* 10.414 (Thyatira), IEphes. 2407'

σχολή I, add '**4** *absence* of, σχολὴν .. τῆς τῶν πραγμάτων νομῆς Just.*Nov.* 167.1.' **III**, add 'κόμ(ης) σχολῆ(ς) γεντιλίων ἰωνιόρων *Chiron* 6.305 no. 12 (Phrygia, vi AD)'

σχολικός I 1, at end after 'D.H.*Comp.* 22' insert '*academic, artificial*, πάθη Longin. 3.5, cf. 10.7' **2**, delete the section.

*****σωβῆρις**· ναῦς, πορθμίς. ἔνιοι τὴν παλαιάν Hsch.

*****σώειλον**, τό, unexpld. wd. in *CPR* 7.44.22 *bis* (v/vi AD).

σώζω, line 17, after 'ἐσώσθη only in Hsch.' insert 'σωσθείς (sp. σοσθίς) *SEG* 32.1084 (Spain, v AD)'   forms, **2**, line 3, before 'fut.' insert '2 sg. impf. ἐσάους Inscr.*Délos* 1658' **3**, for 'Call. l.c.' read 'Call.*Lav.Pall.* 142, Cer. 134, *fr.* 112.8 Pf.' **I 3**, line 7, after 'Procl.*Hyp.* 5.10' insert 'cf. Hipparch. 2.3.23; σ. τὴν ὑπόθεσιν Arist.*Cael.* 306ᵃ30'   add '**b** part. σῳζόμενος, *being currently in force*, τοῦ σῳζομένου κανόνος *Cod.Just.* 1.2.24.5.' **II 2**, add '*Peripl.M.Rubr.* 56 (cod.)' **7 b**, add '*SEG* 31.1310 (Lycia); fem. Μοίραις σῳζούσαις *SEG* 24.902 (Thrace)'

σωκέω 2, add 'cf. σωκῶ· ἀντὶ τοῦ ἰσχύω. οἱ δὲ νεώτεροι ἀντὶ τοῦ σῴζω Sch.S.*El.* 119'

⁺σωλάριον, τό, Lat. *solarium, sun-terrace*, archit., as a feature of tombs in Asia Minor, IEphes. 1645, 2200b, al., ISmyrna 191.7, 8, 212.3, al.; attached to a private dwelling, *Cod.Just.* 8.10.12.5.

σωλήν 4, delete the section transferring exx. to section 2 **6**, add 'Sophr. 24'

*****σωληνοποιός**, ὁ, *pipe-maker*, IEphes. 4315.

σῶμα I 2, lines 3/4, for 'σῴζειν or -εσθαι' read 'σῴζεσθαι' and delete 'D. 22.55'; add 'fig., of an institution, εἰς τὸ τῆς πόλεως σῶμ' ἀποβλέψαντες Din. 1.110, Hyp.*Dem.* col. 25' **3**, line 2, before 'Pl.*Grg.* 493a' insert '*CEG* 10 (v BC)'; line 3, after 'Pl.*R.* 328d ' insert '**b** *person*, opp. *property*, [τὸ σ.] σῶσαι *keep it untouched*, D. 22.55'; add 'opp. ὄνομα, Hdt. 1.139, E.*Or.* 390'

σωμάλοιφος, add '(cj. **σωμάλοιφος**· κατηλιμμένος. **σωμάτια**· τὰ σκύτινα αἰδοῖα)'

σωμάτειον, add '**2** *a body*, i.e. component individual, ὡς ἐλλειπόντων δῆθεν τοῖς ἀριθμοῖς σωματείων *Cod.Just.* 1.2.20.'

*****σωμάτεμπόριον**, τό, *stock of slaves for sale*, Vit.*Aesop.*(G) 15, al.

σωματίζω II 2, add '**b** *register for taxation*, *BGU* 2488.3 (ii AD), PWarren 3.13 (vi AD).'

*****σωματομιξία**, ἡ, *sexual intercourse*, Ps.-Callisth. 3.12 (pl.).

σωματοφύλαξ, after '*bodyguard*' insert '*SEG* 31.80 (Athens, 307/1 BC)'

*****σωμόβουβλον**, τό, *beef*, *Stud.Pal.* 20.250.3 (vi/vii AD).

*****σωμφιακός**, ή, όν, adj. describing some aspect of embalming, τὴν σωμφιακὴν τέχνην PHeid. 327.8 (AD 99).

*****Σώπολις**, εως, ὁ, = σωσίπολις, name of a deity, IEphes. 1060, 1233. **2** as an honorific title (artificial epic form σαόπτ-), Δῆμος ὁ Μιλήτοιο σαόπτολιν ὧδε γεραίρει Βιτιανὸν στήληι *SEG* 29.1139 (Miletus, iv/v AD).

*****σωράκιον**, τό, dim. of σώρακος, PMil.Vogl. 61.26 (ii AD); v. σωρακίς II.

σωροβόλιον, for '*LW* .. (Mylasa)' read '*IMylasa* 814.10, 253.5, 256.2, 7'

σῶς (A), last line, after 'ΣαFοκλέϜης' insert '*ICS* 383 (Abytos)'

*****σωσικόλωνος**, ἡ, *saviour of the coloni*, epith. of Artemis, Inscr.*Délos* 2377.

*****σωσικόσμιος**, ὁ, member of an Alexandrian phyle founded by Nero, fr. an epith. of the Emperor, *SB* 10780.4, 10894.6, 15, 11165.1 (all ii AD); v. °σωσίκοσμος.

*****σωσίκοσμος**, ὁ, *saviour of the world*, Suppl.*Mag.* 49.61 (ii/iii AD).

*****σωσίνεως**, ω, ὁ, *saviour of ships*, epith. of Poseidon, *CIRB* 30 (Panticapaeum, i BC).

σωσίπολις, add 'of a tutelary hero, Paus. 6.20.2.4, etc.; of a human benefactor, *REG* 72.213 (Aenus, Rom.imp.)'

*****σωστεύματα**· τὰ τοῦ τροχοῦ ξύλα Hsch. (σωστρ- Musurus; cf. s.v. σωτεύματα).

σωτήρ I 2 b, add 'of Heracles, *SEG* 33.555 (Epirus, ii BC)' **3**, line 4, after 'al.' insert 'pl. of Ptolemy I and Berenice, Call.*Del.* 166, cf. *APF* 5.156 (Egypt, iii BC; θεοὺς σ.)' **II**, for 'γονῆς .. *Th.* 225' read 'A.*Ag.* 664, S.*OT* 80 1, *Ph.* 1471, E.*Med.* 360 (anap.)' add '**IV** *grain of salt*, Hsch.'

σωτηρία I 1, add 'in opening formula, Ἀγαθᾶι τύχαι καὶ ἐπὶ σωτηρίαι *SEG* 32.623 (Illyria, ii/i BC)'

σωτήριος II 2, add 'οἱ ἀγωνοθέται τῶν μεγάλων Σωτηρίων καὶ Ῥωμαίων *SEG* 33.1039 (Cyme, ii BC)'

*****σωφρόνη**, ἡ, = σωφροσύνη, dub. cj. at E.*Ba.* 1002 (as pr. n., Men.*Epit.*, *GVI* 683, etc.).

*****σωφρονητικός**, ή, όν, = σωφρονικός 2, Rh. 1.231.6.

σωφρονισμός, add '**2** euphem. *punishment, chastisement*, *Cod.Just.* 1.1.5.4, 4.20.15 pr.'

σωφροσύνη, add '**4** -ης ἄρχων, poet. for σωφρονιστής II, *IG* 2².3768.13 (iii AD).'

289

T

ταβελλάριος, add 'BGU 2355.3 (ii/iii AD)'
*ταβέρνα, ἡ, Lat. taberna, shop, stall, Scaev.Dig. 20.1.34.1, SEG 33.1123 (iii AD).
*ταβερνάριος, ὁ, Lat. tabernarius, shopkeeper, MAMA 3.311.
*ταβέρνιον, τό, small shop or stall, PRyl. 627.293, al. (iv AD).
*τάβης, v. °τάπης.
*ταβλάριον, v. °ταβουλάριον.
ταβλάριος, add 'also sp. ταβουλάριος MAMA 7.524 (pl.), PHamb. 31.17 (pl., ii AD), POxy. 2268.14 (v AD), 3867.8 (vi AD), Mitchell N.Galatia 474, Just.Edict. 9.4; cf. °ταβουλάριον'
*ταβλοειδής, ές, adv. -ῶς, app. in the form of a tablet, Anon.Alch. 325.7.
*ταβουλάριον, τό, Lat. tabularium, archives, POxy. 2116.10 (iii AD); perh. also ταβλάριον, ἐν [τῷ τ]αβλαρίῳ POxy. 676.38 (iii AD; v. ZPE 91.81).
†τᾱγά, ἁ, office of ταγός, hence time when there is a ταγός (i.e. wartime), κὲν ταγᾶ κὲν ἀταγίαι SIG 55 (Thessaly, v BC); (also expld. as special sense of †ταγή, mobilisation); perh. in dub. reading officers in charge, 'Ελλάδος ἥβας ξύμφρονα ταγάν A.Ag. 110.
ταγεύω I 3, add 'of a female ταγός, SEG 34.481 (Thessaly, c.150 BC)'
ταγή 1, for 'κἂν ἐκ τᾶς ταγᾶς' read 'καί κ' ἐκ τᾶς ταγᾶς'   3, add 'INikaia 1071, Ps.-Callisth. Hist.Alex.Magn. pp. 35.19, 45.7, 57.13 K.' after '6' insert 'tribute paid to the Great King by the satrapies, Arist.Oec. 1345ᵇ25, cf. Hsch. τ. βασιλική δωρεά'
τάγμα III, add 'd staff of an official, Just.Nov. 13.4 pr.   e club (at Sardis), Ap.Ty.Ep. 39, 40, 41.'
ταγός, line 2, delete 'ξύμφρονε .. A.Ag. 110 (lyr.)'   II 2, at end for 'Inscr.Cypr. 116, 170 H.' read 'alleged Cypr. exs. are dub., v. ICS 258'
ταθρίσιον, delete the entry (v. °θρίσσιον).
ταινιόω, line 2, after 'X.HG 5.1.3' insert 'SEG 24.156(a) (Eleusis, c.238/7 BC)'
τακτικός I 1, delete 'adv. comp. .. Sch.E.Ph. 1141' (v. °τατικός) add 'III -ός, ὁ, Ephebic official at Thebes, IG 7.2440 (?i BC).'
τακτόμισθος, for 'PLond.ined. 2243' read 'PLond. 1986'
τακτός, adv. τακτῶς, for 'v.l. in Plot. 3.1.2' read 'cj. in Plot. 3.1.3'
*τακτώριος, ὁ, name of a variety of σμάραγδος, treatise περὶ λίθων in WS 20.319.
τᾰλᾱεργός, add 'of a spinning-woman, A.R. 4.1062'
ταλαμών, v. °τελαμών.
*τᾰλαντάρχης, ου, ὁ, title given to one who has power over the life and death of gladiators (w. ref. to Il. 22.209 ff.), τὸν ἱερέα καὶ ταλαντάρχην δι' ὅπλων SEG 32.660 (ii/iii AD).
τᾰλαντιαῖος I 1, add 'τοῦ ἐράνου τοῦ ταλαντιαίō SEG 23.96 (iv BC)'
τᾰλαντόομαι, add 'act. fut. ταλαντώσω, Hsch.'
τᾰλαπείριος 1, line 3, after '(Od.) 14.511' insert 'Πέργαμον Ibyc. 1(a).8 P.'; add 'Trag.adesp. 599 K.-S.'
τᾰλᾰρος 1, add 'fig., Μουσέων τ. of the Museum, Timo 12'   2, delete the section.
τάλας, before 'τάλαινα' insert '(hyper-Ion. τάλης Herod. 3.35, 7.88 (but τάλας 5.55))'; line 2, after 'also' insert 'gen. τάλαντος, Hsch.'
†τᾰλᾰσήϊος, ον, of or related to wool-working, τ. ἔργα A.R. 3.292, Orph.fr. 178, τ. ἱδρώς Nonn.D. 6.142.
τᾰλᾰσία, for 'wool-spinning' read 'wool-working'   add 'Myc. ta-ra-si-ja, portion of raw material for manufacture'
τᾰλᾰσιουργέω, for 'spin wool' read 'work with wool'
τᾰλᾰσιουργικός, for 'of .. wool-spinning' read 'of .. wool-working'
τᾰλᾰσιουργός, for 'wool-spinner' read 'wool-worker' and add 'IG 2².1554.2, SEG 25.180.8, 28, al. (Athens, iv BC)'
†τᾰλις, ἡ, gen. τάλιδος S.Ant. 629, acc. τάλιν Call.fr. 75.3 Pf., dat. τάλῑ (τάληι codd.) Sch.Il. 14.296, betrothed woman; cf. τάλις· ἡ μελλόγαμος παρθένος καὶ κατωνομασμένη τινί, οἱ δὲ γυναῖκα γαμετήν, οἱ δὲ νύμφην Hsch. (Aeol. acc. to Sch.S. l.c.).
ταμεία, add 'also sp. ταμία Wolters Kabirenheiligtum 79, TAM 2(3).838 f 7 (Lycia)'
τᾰμία, after 'Il. 24.302' insert 'Ar.fr. 305.2 K.-A.'
τᾰμιακός, line 2, after 'fiscalis' insert 'τ. λόγος POxy. 1414.8 (iii AD)'
τᾰμίας II 2, add 'Plb. 1.52.7, al., IG 14.951 (AD 78)'
τᾰμιεῖον, add 'w. ἱερώτατον, of the imperial treasury, IEphes. 3712 (ii/iii AD), INikaia 766 (iii AD)'
*τᾰμιεύς, έως, ὁ, store-keeper, St.Byz. s.v. ταμιεῖον, cf. SB 5223.8.

τᾰμιευτικός I, add '2 of a steward, βυβλία PCornell 1.16 (iii BC).'   add 'III belonging to the Elatean ταμίαι, χρήματα IG 9(1).144.8 (Elatea, ii AD).   IV -ός, ὁ, = ταμίας, TAM 2(3).845.7 (Lycia).'
ταν, delete 'only Att.' and add 'Pi.fr. 215b.4, Epich. 87'; line 3, for 'not in Ar.' read 'Ar.Pl. 66, al.'
Ταναγραῖος, add 'fem. Ταναγραία SEG 31.473 (Boeotia)'
*Ταναγρεύς, έως, ὁ, = Ταναγραῖος, IG 2².10405 (iii BC).
*Ταναγρίδαι· οἱ Ταναγρεῖς Hsch.
τᾰνᾱήκης, line 1, after 'edge' insert 'tapering to a point'; lines 2/3, delete 'II tall' transferring exx. to section I.
τᾰνᾱός, after 'h.Cer. 454' insert 'διέβα ταναοῖς πο[σί] Alcm. 3.70 P.' add 'Myc. ta-na-wa, of wheels, prob. thin'
τᾰνῠήκης I, add 'ἰοί B.fr. 20D.7 S.-M.'
τᾰνυκνήμις, add '2 of mountains, having extensive spurs, Nonn.D. 13.67; cf. °βαθυκνήμις.'
*τᾰνυπρήων, ωνος, having a towering reach, τανυπρήωνος .. ['Ε]λικ[ῶνος Suppl.Hell. 938.5 (hex.).
τᾰνυπτέρυξ, line 3, gen. pl., add 'νώτων AP 9.59 (Antip. Thess.)'
τᾰνύσφυρος, add '(v. τανίσφυρος)'
τᾰνύφυλλος I, delete 'prob. .. 426.7'   at end add '(v. τανίφυλλος)'
τᾰνύω I, add 'w. ref. to crucifixion, τὸ ξύλον .. ἐν ᾧ Χριστὸς μέλλει τανύεσθαι Orac.Tiburt. 72'   II, add 'τὰ σκέλεα τανύσας τῆς γυναικός Hp.Steril. 244'
ταξείδιον, add 'also ταξίδιον, journey, PMich.VIII 501.24 (pl., ii AD); cf. mod. Gk. ταξίδι'
ταξίλοχος, for 'Arist.Pepl. 9' read 'Arist.Pepl. 34'
ταξιόομαι, delete the entry.
*ταξιόω, post, station as for battle: med., μή ποτε σφετέρας ἄτερθε ταξιοῦσθαι δαμασιμβρότου αἰχμᾶς Pi.O. 9.78.   2 transf., place in a rank, ἵνα .. ταξιώσις με μετὰ τῶν ἐγκλε[κτ]ῶν σου Mitchell N.Galatia 87 (Byz.).
*ταξιπᾰτέω, perh. trample on (w. ἐπί), Procl.CP Hom. 29.33.
τάξις I 4 c, for this section read 'the body of subordinates of a high military or civil official (= Lat. officium), Lib.Or. 27.7, Cod.Just. 1.5.12.15, 1.5.18.10, 6.48.1.10, BCH suppl. 8 no. 148 (Thessalonica, v/vi AD)'   IV 1, add 'b category, class of relatives, in order of entitlement to inheritance, Cod.Just. 6.48.1.5.'
τᾰπεινός 4, add 'T&MByz 7.312 no. 6 (Thessalonica, c.AD 491/7)'
τάπης, add 'τάβης Edict.Diocl. 7.62'
*τάπητᾶς, ᾶ, ὁ, rug-merchant, POxy. 3044.5, 3045.3 (iv AD).
Ταραντίναρχος, for 'Ἀρχ. Δελτ. 14 Pl. iv 19' read 'SEG 23.271.19'
τάραξις, add 'Choerob. in Theod. 1.141.32'
Τάρας, pr. n. of the river-hero, add 'SEG 34.1020, 1021 (ii AD), Inscr.Perg. 8(3).132'
τᾰραχή 4, add 'D.C. 37.31.1, 41.3.3'
*τᾰρᾰχίζεσθαι, gloss on θολερεῖν, Hsch.
τάραχος, add 'of digestive upsets, Hp.VM 14, Flat. 14'
ταρβέω, line 1, after 'Hdn.Gr. 2.930' insert 'Aeol. τάρβημι Alc. 302.12 L.-P.'
*ταρθωτής, οῦ, ὁ, app. for ταρσωτής, revetter of a dike with brushwood, POxy. 1053.25 (vi/vii AD).
*τᾰριχηρά, ἡ, tax on pickling business, OMRL 67.26 (i AD; sp. -χιρ-).
τᾰρίχιον, add 'Pherecr. 26 K.-A., Axionic. 4.16 K.-A.'
τᾰριχοπώλης, add 'SEG 25.180 (Athens, iv BC)'
*τᾰριχόπωλις, ιδος, ἡ (sc. ἀγορά; cf. ἰχθυόπωλις), = ταριχοπωλεῖον Gnomol.Vat. 340 (WS 11.44).
τάριχον (s.v. τάριχος), delete 'Philippid. 9' and 'Axionic. 3.15'
τάριχος II, line 2, after 'fish' insert 'Epich. 162' and after 'Ach. 967' insert 'SEG 37.665 (N. shore of Black Sea, c.400 BC)'
ταρσικάριος, before 'PLips. 26.9' insert 'SB 8268 (ii/iii AD)'
*ταρσίκιον, τό, perh. crate for drying figs, POxy. 2273.8 (iii AD).
ταρσός II 3, line 2, before 'Babr.' insert 'Mosch. 2.61'; line 4, delete 'of a peacock's tail, Mosch. 2.60'   add '7 λίθος ὁ κάτω τιθέμενος ἐν τῷ ἱπνῷ Hsch.'
*ταρταροειδής, ές, = ταρταρώδης, of the sun viewed as an omen, ὁ τρίτος ἥλιος αἱματοειδής, παμμεγέθης, πῦρ φλέγον Orac.Tiburt. 24, 25.

290

Τάρταρος I, add 'transf., of a river-bottom, τάρταρον ἰλυόεσσαν Nic. l.c. (Th. 203)'

*ταρτημοριαῖος, α, ον, shortened form of τεταρτημοριαῖος, τ. τινα καλοῦσιν οἷον διχάλκου ἄξιον Phot. s.v. ταρτημόριον (cf. EM 747.18).

*Τάρφη· πόλις Λοκρίδος, οἱ δὲ σποδός, τέφρα. ἢ βλαστός Hsch.

*ταρφήεντα ἐντάφια· τεφρώδη Hsch. (ταρφίεντα cod.; cf. °Τάρφη).

*τάρχεα· τὰ νενομισμένα τοῖς νεκροῖς Sch.A Il. 7.85, cf. Sch.Gen.ibid.; Sch.B has ταρχύματα, Sch.T ταρχῶα.

τάσσω, line 1, after 'Pl.Prt. 262e, etc.' insert 'Boeot. τάδδεσθη (pass. inf.) SEG 32.496 (Thespiae, c.250 BC)'; line 4, after 'pass. sense)' insert 'E.Supp. 521 (ἐπι-)' I 2, line 9, add 'in fig. phr., τῆς ὑστέρας τεταγμένος Pl.Criti. 108c; also act., Isoc. 12.180' add '3 organize a state, ὁ δὲ (sc. Νουμᾶς) αὐτὴν (sc. τὴν πόλιν) νόμοις τάξας τε καὶ κατακοσμήσας Just.Nov. 47 pr.'

*τάταῖ, an exclamation of pain, Herod. 3.79; cf. ἀτταταῖ; tatae, Plaut.St. 771.

τᾱτί, add 'perh. also θαθί Lang Ath.Agora xxi F11'

τᾱτικός 2, add 'adv. comp., with greater force, Sch.E.Ph. 1141'

Τᾱΰγετον, add 'also τὰ Ταΰγετα ὄρη Iamb.VP 19.92'

*ταυράς, άδος, ἡ, cow, XVIII Int.Cong.Pap. II 83 (Tyche 4.237).

Ταυρεασταί, add 'at Istria, IHistriae 60, 61 (iii/ii BC); also Ταυριασταί ib. 57 (ii AD).'

ταυρεία, add '3 helmet of ox-hide, Hsch.'

Ταυρεών, after 'Herod. 7.86, etc.' insert 'name of first month at Olbia, SEG 30.977, 980 (v BC)'

*Ταύρια· ἑορτή τις ἀγομένη Ποσειδῶνος Hsch.; cf. ταύρεος II.

ταυρινάδαι, delete the entry.

*ταυρινᾱς, άδος, ὁ, shoemaker, MAMA 6.234 (Apamea); συνεργασία πλήθους -άδων IEphes. 2080, 2081 (ταυρειν- lapides).

ταυροβολικός, ή, όν, connected with the ταυροβόλιον, ara taurobolica, CIL 9.1538 (AD 228).

ταυροβόλιον, add '2 bullfight, as an entertainment, Illion 12.14 (i BC), TAM 2.508 (Lycia, i BC).'

ταυροθύσια, for 'Supp.Epigr. 4.180.6' read 'IKeramos 7.6, 9.10'

ταυροκαθάπτης, add 'IG 2².3156 (Athens)'

ταυροκαθάψια, add 'SEG 32.660 (Thrace, ii/iii AD)'

†ταυρόκερως, ω (also dat. -κέρωτι Euph. 14.1 vGr.), ὁ, ἡ, bull-horned, of Dionysus, E.Ba. 100 (lyr.), Euph. l.c., Orph.H. 52.2; of Io, Agatharch.ap.Phot.Bibl. 443a.25; of Μήνη, Orph.H. 9.2; of Attalus of Pergamum, orac.ap.D.S. 34/35.13, Paus. 10.15.3.

ταυρομαχία, delete 'IGRom. 3.631.14 (Xanthus)'

*ταυρομάχια, τά, fighting with bulls, TAM 2.287.14.

Ταυροπόλια, for 'Ἀρχ.Ἐφ. .. iv BC)' read 'SEG 34.103 (Attica, iv BC)' and add 'SEG 31.615 (Maced.)'

ταυροπόλος, add 'SEG 31.614 (Maced., 179/1 BC); Ταυροπόλαι· ἡ Ἄρτεμις καὶ ἡ Ἀθηνᾶ Hsch.'

ταῦρος I 2, add 'cf. ταύροι· οἱ παρὰ Ἐφεσίοις οἰνοχόοι Hsch.'

ταυροσφαγέω, for 'cut a bull's throat' read 'slaughter a bull or bulls'

*ταυροτάφος, ὁ, one who prepares (sacred) bulls for burial, PMich.inv. 855 (ZPE 27.147, i AD).

*ταυροτρόφος, ὁ, bull-breeder, Roueché Performers and partisans no.44 (Aphrodisias, Chr.).

*Ταυροφόνια, τά, festival at Mylasa, IMylasa 201.

ταυροχόλια, for 'Κυζικῷ' read 'Κυζίκῳ'

*ταυρόω, v. ταυρόομαι 2.

ταύτᾳ, after lemma insert '(or ταυτᾷ)' and add 'cj. in Theoc. 15.18 (perh. in origin ταυτᾶ instr.)'

ταὐτότης III, delete 'maintenance of identity, Plot. 1.2.7' (v. αὐλότης)

ταφή 2, for this section read 'burial place, tomb, pl., Hdt. 4.71.3, 5.63.4, S.Aj. 1090, 1109; sg., SB 6028, al. (i BC), INikaia 767 (iv AD); μιᾷ καὶ μόνῃ ταφῇ, i.e. urn or tomb which would require no renewal, GVI 1975.24. b mummy, POxy. 736.13 (i AD), Wilcken Chr. 499 (ii/iii AD); of the covering of a mummy, whether linen wrappings or mummy-case, δευτέρα τ. PGiss. 68.7 (ii AD), APF 4.133 (ii AD).'

*τᾰφία, ἡ, tombstone, SEG 3.674.A4 (Rhodes, ii BC); also ταφίη MAMA 9.89.1 (Phrygia, Rom.imp.)

τάφιος II, delete 'but also .. (Rhodes, ii BC)'

τάφος, add 'Dor. τράφα, IRhod.Peraea 353A.8 (c.200 BC)'

τάχα II, line 7, after 'with aor. ind.' insert 'S.Ph. 305'

*τᾰχύγουνος, ον, having swift legs, Nonn.D. 1.91, 9.155, al.

*τᾰχυδιάνοιος, gloss on λειόμερος (q.v.), Hsch.

*τᾰχύλογος, ον, speaking rapidly, Physiogn. 1.332.11, 17.

τᾰχύμαχης, for 'ὠκυβόας' read 'ὠκυβόαι'

τᾰχύπλοος, for 'ὠκύπλους' read 'ὠκύαλος'

τᾰχύπους, add 'ταχύπουν κέλευθον Trag.adesp. 127.6 K.-S. (codd.)'

τᾰχύς B adv. I, line 2, after 'etc.' insert 'soon, Sapph. 1.21, 23 L.-P.' add 'b early, Plu. 2.178e.' C I 3, line 4, for 'Men. 402.16' read 'Men.fr. 333.16 K.-Th., AP 11.23 (Antip.Thess.)' II 3, as adv., add 'ἵνα .. τὰν ταχίσταν συντελεσθοίσιν IKyme 13.15 (ii BC)'

*τᾰχύω, make haste, hurry, Suppl.Mag. 34 (vi AD).

*ταωνίτης, ου, ὁ, precious stone, = ταώς II, Lapid.Gr. 168.25.

τε A II, lines 16/17, delete 'διάνδιχα .. Il. 8.168' add 'Myc. -qe'

*τέαφη, ἡ, sulphur, Anon.Alch. 323.9 (written τι- ib. 327.11).

*τέγεα, v. °τένεα.

Τεγεάτης, after 'Ion. -ήτης' insert 'Arc. -ᾱτᾱς SEG 33.319 (iii BC)'

*τεγκοσχ(..), prob. type of reed, SB 9699.491 (i AD).

*τεγοποιέω, cover with a roof, w. acc., Inscr.Délos 444B95 (ii BC).

τέθηπα 1, line 5, for 'amazed, astonied' read 'dazed'

†τεινεσμώδης, ες, affected by τεινεσμός, Hp.Epid. 1.26.γ', προθυμίαι -ώδεες Aret.SA 2.5, cf. Sor. 2.20; suffering from τεινεσμός, Gal. 12.325.12. Adv. -δῶς Ruf.ap.Orib. 8.24.24.

*τειρωνολογέω, recruit for the army, BE 1960.230 (IGBulg. 517).

τειχεσιπλήτης, add 'Orph.H. 65.2'

τειχήρης, add 'Aen.Tact. 1.3, 23.4'

*τειχητός, ή, όν, walled or fortified, εἰς τὸν τειχητὸν [.. Ath.Agora xix P26.534 (iv BC).

τειχίζω II, add 'of non-physical protection, τὰ κοινὰ πράγματα τῇ τοῦ φιλανθρώπου θεοῦ χάριτι τειχίζεσθαι πεπιστεύκαμεν CodJust. 1.3.42 pr.'

τειχισμός, add 'SEG 25.486 (Oropus, iii BC), REG 101.14 (Xanthos, 206/5 BC)'

τειχοδομέω, add '2 fortify with a wall, Θήβη ἐτειχοδομήθη Sch.E.Ph. 287.'

τειχομαχέω, line 5, for 'App.Hann. 92' read 'App.Hisp. 92'

*τειχοποίης, v. °τειχοποιός.

τειχοποιία, add 'SEG 36.1277 (Edessa, iii AD)'

τειχοποιός, add '-ποίης SEG 32.795 (Olbia, c.325/300 BC)'

*τειχοσεισμοποιός, όν, causing earthquakes destructive of walls, PMag. 4.183.

τεκμαίρομαι A I, for 'assign, ordain' read 'mark out, indicate, designate'; line 10, for 'with a notion of foretelling' read 'of a prophet' III, for this section read 'indicate (to an observer), show, ὁλκόν, οὖρον (= ὅρον) D.P. 101, 135, 178; πόμα Nic.Al 105, τοῦ μὲν ὑπὲρ κυνόδοντε δύω χροΐ τεκμαίρονται ἰὸν ἐρευγόμενοι id.Th. 231.'

τέκμαρ, line 1, delete 'never elsewhere' and insert 'A.R. 3.493, etc., Opp.H. 4.26, Q.S. 3.503, etc., Synes.Hymn. 1.491; sp. τέκμορ SEG 31.1203 (Rom.imp.)' I 2, add 'a principle in Alcman's cosmology, τέκμωρ Alcm. 5 fr. 2 ii 14 sqq. P., v. CQ NS 13.155-6'

*τεκνοθεσία, ἡ, placing in adoption, PZen.Col. 58.9 (iii BC), παρὰ Ἰσιδώρας τῆς Ἀπολλωνίο(υ) .. κατὰ δὲ τεκνοθεσίαν Διονυσίο(υ) POxy. 3271.4 (i AD).

τεκνοποιΐα I, add 'PPanop. 28.5 (AD 329)'

*τεκνοτρόφος, ον, bringing up children, ἀνὴρ ἀξιόλογος καὶ τ. IG 12(7).394.5 (Amorgos); prob. subst. in GVI 977.8 (Mauretania, ii/ iii AD).

τεκνοφάγος, add 'SEG 31.1285 (Pisidia, Rom.imp.), Thphl.Ant.Autol. 1.9.3, 3.3.16'

τέκνωμα, add 'Dosiad.Ara 4'

τέκτων, add '5 kind of spider, Hsch.' add 'Myc. te-ko-to-ne (pl.)'

τελᾰμών I 2, after 'E.Ph. 1669' insert 'κατεδήσατο τελαμῶνι τοὺς ὀφθαλμοὺς αὐτοῦ Lxx 3Ki. 21(20).38' 3, add 'Aq.Is. 3.18' II 2, add 'sp. ταλαμών Robert Hell. 7.32ff. (Bithynia).'

τέλειος I 2 a, add 'τέλειοι αὐλοί full-sized, the second largest of five sizes of aulos, Aristox.fr. 101 W., Arist.Aud. 804ᵃ11, Ath. 4.176f, Poll. 4.81' 3 b, line 9, after '(ii AD)' insert 'τ. ἀριθμητικόν complete land-tax, SB 4415.4 (ii AD); εἰς τέλειον absolutely, CodJust. 11.1.1 pr.' VII 3, add 'IHadr. 77 (i/ii AD)'

τελειόω II 4, add 'TAM 4(1).356'

*τελέσεργος, ὁ, dub. l. for τελεσιουργός 3, Διὸς τ. IG 12 suppl. 380.2 (ΤΕΛΣΕΕΡ- lapis; Thasos, v BC).

τελεσιουργός 3, add 'of Athena, IG 2².4338a'

*Τέλεσσαι, αἱ, name of goddesses at Cyrene, PP 15.294 (ii BC).

τελεστήριον II, for 'thank-offering for success' read 'offerings to a sanctuary' add '2 = Att. τέλη (v. τέλος I 6), Wolters Kabirenheiligtum 27 no. 4 (iii/ii BC) (τελεστειρ- lapis).'

τελεστής, add 'Myc. te-re-ta, title of an official'

τέλεστρα, for 'admission to priesthood' read 'initiation'

τελεσφορία, for 'initiation in the mysteries' read 'mystery rites' and delete 'Cer. 130' add 'III ripening of fruit, etc., περὶ καρπῶν τελεσφορίας ABSA 49.15 (Dorylaeum; τηλ-); χαλεπὴν .. τελεσφορίην rest. in Call.fr. 85.13 Pf.'

Τελεσφόρος III, add 'Inscr.Perg. 8(3).125, 126'

τελεταρχέω, delete the entry.

*τελετουργός, όν, working by means of a rite, θεοί Procl. in R. 2.153.23 K.

τελευταῖος I 1, add 'sup. Peripl.M.Rubr. 16, 18, 21' 2, add 'τει τελευτ[αίαι ἐμέραι] IG 1³.364.22 (433/2 BC)'

τελευτάω II 1, delete 'ἐλπίδες E.Ba. 908 (lyr.)' 2 b, line 3, after 'A.Supp. 211' insert 'ἐλπίδες τ. ἐν ὄλβῳ E.Ba. 908 (lyr.) (s.v.l.)'

τελέω I 7, line 1, transfer 'ὅτε δὲ .. Od. 5.390' to section 5. 8, after 'S.El. 147 (lyr.)' insert 'epigr. in Phoenix 26.183 (iv AD)'; after τελίαμβος **SUPPLEMENT** τετραετής

'to be' insert 'ἐπὶ τῇ ἐξουσίᾳ τελοῦντα Cod.Just. 1.2.24.1' **III 3**, add 'cf. τελεῖν τὰν Ἀφροδίταν Anacreont. 36.16 W.'

*τελίαμβος, ὁ, dactylic hexameter having an iambus in place of the final spondee, *novam potius hanc speciem quam miuron existimant versum et teliambon appellant* Mar.Vict. (Aphth.) in *Gramm.Lat.* 6.68 K.

τέλμα **I**, add '*Ath.Agora* xix P26.161, L6.140, 142, L10.40 (iv BC)'

*τελμάτιον, τό, dim. of τέλμα ι, Simp. *in Cael*. 66.9.

τέλος **I 2**, line 4, after 'Semon. 1.1' insert 'cf. Archil. 298 W., Alc. 200.10 L.-P.' **III 2**, add 'epigr. in *SEG* 23.278 (Thebes, iii BC)'

τέλσον, add 'transf., of the edge of a threshing-floor, Nic.*Th.* 546'

Τελχίνιος **I**, after 'Paus. l.c.' insert "Ἥραι Θελχινίαι Sokolowski 3.18.7 (Erchiae, iv BC)'

τέλωρ, for 'τελώριον' read 'πελώριον'

τεμαχοπώλης, add 'perh. at *SB* 10258 i 14 (iv AD)'

τέμαχος, line 6, after 'Paul.Aeg. 7.11' insert 'τ. γάλακτος i.e. of cheese, Al. 1*Ki.* 17.18'

τεμενίζω, add '**b** οἱ τεμενίζοντες members of a religious association at Miletus, *SEG* 30.1339 (Miletus, ii/i BC).'

τεμένιος **I**, add 'subst., τεμένια, τά, *REA* 44.35.45 (Olus, ii BC)' **II**, add 'also of Artemis, *SEG* 25.938 (Naxos, iv BC)'

τεμενίτης, add '**2** pl., members of a religious association at Miletus, Τεμενεῖται οἵδε *SEG* 30.1340 (Miletus, ii/i BC).'

τέμενος, after lemma insert 'Cypr. nom. *te-me-no-se ICS* 265.2 (in quot., sense II)' **III**, after '*temple*' insert 'Pl.*Ax*. 367c' at end add 'Myc. *te-me-no* (sense I)'

τέμνω (A), line 10, after 'Il. 19.197' insert 'Aeol. part. τόμοντες Alc. 129.15 L.-P.' **I 3**, pass., add '*TAM* 4(1).367' **III 2**, after 'so in Med., 9.580' insert 'A.R. 1.868' **IV 3**, line 4, after 'Th. 2.19, 55' insert '*POxy*. 3575.11 (iv AD)'

Τέμπεα **II**, add 'sg. τέμπος, Sch.E.*Ph.* 600, *EM* 527.46'

*τέμπλον, τό, Lat. *templum*, (in quot., unexpld. archit. term), τὸ τένπλον τοῦ τίχους τοῦτο *ICilicie* 24 (v/vi AD).

τεναγῖτις, after 'ιδος' insert 'acc. τεναγῖτιν'

τένεα, add 'cf. τέγε(α)· κόρυζα, Κῷοι id.'

*τένται, v. τέλομαι.

τένων **I**, line 7, for 'for the foot' read 'in ref. to the sole of the foot'; last line, delete 'τένοντα .. 62.3' add '**b** neck, Lyc. 1112, Luc.*Cat.* 19, Babr. 62.3, Ael.*NA* 2.39, al.' **II**, after 'mountain-*ridge*' insert 'δύσβατον ἀμφὶ τ. Nonn.*D.* 11.193'

τέξις, add '*birth, nativity*, Ptol.*Tetr.* 105'

τεός, line 3, after 'E.*Heracl.* 911' insert '(cj.)'

*τερατοειδῶς, adv., *in monstrous form, as a prodigy*, κατελθεῖν Sch.E.*Ph.* 806.

Τερβινθεύς, add 'also Τερμ- *SEG* 36.1047 (iii BC), 1048 (ii BC, both Miletus)'

τερεβινθίζω, add '*terebinthizusa*, a variety of the stone *iaspis*, Plin.*HN* 37.116'

τερετίζω **I**, for '*hum*' read '*warble, trill,* or sim.' and add '**4** of αὐλοί, Philostr.*VA* 6.36.'

*τερετίστρια, ἡ, *chirruper*, of the cicada, *Vit.Aesop.*(W) 99.

τέρην, line 1, after 'εν' insert '(ειν Hsch.)'

τέρθρον, add 'τίς .. ἦλθεν ἐπὶ τέρθρον θυράων; Apollod.*Lyr.* 1 P.'

*τερμαστής, οῦ, ὁ, *boundary-commissioner*, *SEG* 35.665.B35, 43 (Epirus, ii BC).

τερματίζω, after '*limit, bound*' insert 'Hippon. 103.3 W.'

Τερμέρειον **2**, delete the section.

*Τερμινθεύς, v. °Τερβινθεύς.

τέρμινθος **I 1**, add 'reported to grow parasitically on the olive, Thphr.*CP* 2.17.14. **b** used w. ref. to the pistachio-tree, *Pistacia vera*, φασὶ δ' εἶναι καὶ τέρμινθον, οἳ δ' ὅμοιον τερμίνθῳ, .. id.*HP* 4.4.7; also its fruit, θυμίαμα καὶ στακτὴν καὶ τερέμινθον καὶ κάρυα Lxx *Ge.* 43.11.' **II** & **III**, for these sections read '**II** τέρμινθος· φυτὸν ἐμφερὲς τῷ λίνῳ ἐξ οὗ πλέκουσιν παρ' Ἀθηναίοις [παρ]ορμιάς Hsch., Phot., *EM* 753.10.'

τερμιόεις, line 1, for '*fringed*' read '*having a decorated border*'; add 'form τερμιδϝεις, Myc. *te-mi-dwe, te-mi-dwe-ta* (neut. pl.), of wheels'

τέρος, for 'dub. .. 65 H.' read 'dub. in Cypr. *i-te-re-i* perh. ἴ(ν) τέρει *ICS* 244'

τερπνός **II**, for 'Call.*Fr.* 256 .. 1218*c*6' read 'Call.*fr.* 369, 536 Pf.; hyper-characterized adv. -ίστατα ib. 93.3'

τέρτα, delete the entry.

*τέρτος, ᾱ, ον, Aeol. for τρίτος, τέρτον τόνδε Alc. 129.7 L.-P., cf. Choerob.in *An.Ox*. 2.275, Hsch. s.v. τέρτον; perh. also Boeot. τορτ- (in pers. n. Τορτέας *SEG* 23.271, iii BC).

Τερφεῖος, add '*IKyme* 13 i, ii, *SEG* 33.1039 (Aeolian Cyme, all ii BC)'

†τέρφος, εος, τό, = στέρφος, *outer covering, integument, skin*, of snake, Nic.*Th.* 323, of chestnut, id.*Al*. 268.

τέρχνος for '*twig, young shoot*' read '*shoot, young plant* or *sapling*, Call.*VB* in *Suppl.Hell*. 257.25'; line 3, for '*plants* .. 9 H.' read '(*te-re-ki-ni-ja*) *plants*, *ICS* 217.9, al.'

τερψίχορος, add 'σοφία *CEG* 797 (Larissa, iv BC)'

τεσσᾰρᾰκαιδεκέτης, before '-δεκαετής' insert '-δεχετής *GVI* 1709 (Thessalonica, ii/i BC), cf. °ἔτος; also τεττ- *SEG* 34.1271 (τετρασκαιδεκέτης lapis)'; delete 'fem. .. Arist.*Ath*. 56.7'

*τεσσᾰρᾰκαιδεκέτις, ιδος, fem. of τεσσαρακαιδεκέτης, Arist.*Ath.* 56.7, *GVI* 1461.4.

τεσσᾰρᾰκαιεικοσίπους, delete the entry.

τεσσᾰρᾰκοντα, after 'τεσσεράκοντα (q.v.)' insert 'τεΤαράfοντα *IEphes.* 1 A1, B8 (vi BC)'; line 3, form τετρώ-, add '*SEG* 34.940 (Sicily, iv/iii BC)'; line 6, after 'πετταράκοντα (q.v.)' insert 'Aeol. πεσ-(σ)αράκο[ντα] *EA* 11 p. 4 (Assos, late vi BC)' and after 'indecl.' insert 'exc. Aeol. gen. τεσσ[ερ]ακόντων Schwyzer 688*C*14 (Chios, v BC)'

τεσσᾰρᾰκονταχοίνικος, add '*PMich*.III 145 iii 7 (ii AD)'

τεσσᾰρᾰκόσιοι, add 'also τεττᾰρᾰκ- *Inscr.Délos* 1401*C*8 (ii BC)'

τεσσᾰρᾰκοστόγδοος, add 'τεσσαρακοστόγδοον, τό, *forty-eighth part*, *SB* 10497.23-4 (τεσσαρακοσθωγδον pap., AD 213).'

τεσσᾰρᾰκοστός **I**, add 'τεττ- *Inscr.Magn.* 16 (207/203 BC)' **II**, add 'μισθωτὴν τεσσερακοστῆς ἐρεῶν *POxy.* 3104 (AD 228)'

τεσσᾰράριος, add 'Ἐπίγονος τεσσεράρις *SEG* 26.1853 (Cyrenaica, i AD)'

τέσσᾰρες, line 19, after 'Theoc. 14.16' insert 'τέζαρα *SEG* 19.618 (Metapontum, v BC)'; add 'τέσσαρες ἄνδρες = Lat. *quattuorviri*, Xanthos 50.2 (AD 96/9)'

τεσσᾰρεσκαίδεκα, add 'τεττ- *IG* 2².1673 (Eleusis, iv BC)'

τεσσᾰρεσκαιδεκᾰσύλλᾰβος, add 'Sch.Theoc. 29 *prooem.*'

τεσσᾰρεσκαιδέκᾰτος, add 'neut. as adv., *for the fourteenth time, AS* 36.139 no. 1 (5/4 BC)'

†τεσσερακαιεβδομηκοντούτης, ες, *seventy-four years old*, Ἅ(σ)σων τε-(σ)σερακαιεβδοκοντότης ἐὼν τὰς οἰκίας ἐχσεποίησεν *IG* 12(5).219 (Paros, vi BC).

τεσσεράκοντα, add 'τεσεράκαντα *SEG* 31.883 (Tridentum, v AD)'; delete 'but also' to end of article.

*τεσσεράριος, v. ‡τεσσαρ-.

*τεστάτωρ, ορος, ὁ, Lat. *testator*, Just.*Nov.* 19.9, 159 pr.

τετᾰνός **I**, line 2, for '*smooth*' read '*taut, stretched*' and for 'πῆχυς' read 'πρίων'; line 4, after 'Crito ap.Gal. 12.825' insert 'μελίχρως, τετανός , στρογγυλοπρόσωπος *SB* 7169.17 (ii BC)'

τέτᾰνος **I**, add 'humorously, in obsc. sense, κᾆτ' ἐντέξῃ τέτανον τερπνὸν τοῖς ἀνδράσι καὶ ῥοπαλισμούς Ar.*Lys.* 553' **II**, delete the section.

*τετᾰνοτρῐχῖνος, ον, = τετανόθριξ, *PMich*.v 298 introd. 6 (i AD).

τεταρταῖος **2**, after 'Id.*Ti.* 86a' insert 'πυρετοὶ τ. πολυχρόνιοι Hp.*Aër.* 7'; add 'personified as a god in Samos, *ZPE* 3.149ff.'

*τεταρτολογία, ἡ, *payment of duty of one-fourth*, *PVindob.* G40822 (*BASP* 23.74).

τέταρτος **I**, add 'in the title of a legion, λεγεῶνος τετάρτης Σκυθικῆς *ICilicie* 87 (ii AD)' **II**, add '**3** coin, Πτολεμαϊκὰ τ. *Inscr.Délos* 444*B*32, al. (ii BC); *quarter of a mina*, *SEG* 31.1626; *of a stater*, *BE* 1989.324 (Didyma, ii BC), cf. Hsch. τέταρτον ἥμισυ· τὸ τέταρτον ἡμιστάτηρον.'

*τεταρτώνης, ου, ὁ, *collector of the quarter-tax*, *Syria* 22.263, 264 (Palmyra, ii AD).

τέτμω **1**, penultimate line, after 'opt.' insert 'τέτμοι Call.*Del.* 159'

†τετρᾰβόλος, ον, *shedding the teeth a fourth time*, ὄνοι θήλειαι τετραβόλοι *PSI* 79.10 (iii AD).

*τετρᾰγκαθος, ὁ, *gum tragacanth*, Anon.*Alch.* 331.9.

τετρᾰγραμμος, add '**2** = τετραγραμμιαῖος, Hsch. s.v. στατήρ.'

τετρᾰγωνικός, before 'Iamb. *in Nic.*' insert 'Hero *Dioptr.* 28 (p. 280 line 2)'

*τετρᾰγώνιον, τό, *square object*, *CPR* 8.66.1 (vi AD); perh. *a square container* (or ?*square coin*), Just.*Nov.* 105.2.1, 3.

*τετρᾰγώνιος, ον, = τετράγωνος ι 1, *CPR* 8.66.8 (vi AD) and ι 4, Ptol.*Tetr.* 115.

*τετρᾰγωνίτης, ου, epith. of Hermes, cf. °τετράγωνος, *PASA* 3.342*D*66 (Pisidia).

τετράγωνος **I 2**, add '**c** cubic measure of wood, prob. a cubic ξύλον (v. ξύλον v) *PBremen* 15.12 (ii AD), *POxy*. 669.21 (iii AD); so prob. τὴν ὄργυαν τὴν τετράγωνον *Inscr.Délos* 290.167 (iii BC).' add '**V** epith. of Hermes, Babr. 48.1, *PMag.* 5.401, 7.669, *TAM* 3(1).34*D*71 (Pisidia).'

*τετρᾰδία, ἡ, dub. sens., κοιμητήριον Σεργίου μικροῦ τετραδίας Ἰάννου νέου *IG* 3.3486.

τετρᾰδισταί **I**, for this section read 'members of a θίασος who met for celebration on the fourth day of the month, Alex. 260.1 K.-A., Ath. 14.659d (Men.*Kol.fr.* 1 S.)'

τετράδραχμον, line 1, delete '*silver*'

τετράδωρος, add 'Call.*fr*. 196.27 Pf.'

τετραετηρικός, add '*TAM* 2(1).307.6'

τετραετηρίς, for '*CIG* 2741.22, 2812' read '*MAMA* 8.506, 519' and add '*SEG* 35.1415 (Pamphylia, ii/iii AD)'

τετραετής **I**, add 'Men.*Sic*. 355 S., Call.*Cer.* 58, τ. δαμάλην *Suppl.Hell.*

396.2 (Dorieus), *AP* 6.155 (Theodorid.), 6.356 (Pancrat.)' **II**, add '[τ]ε[τ]ραετῆ .. ἀλικία[ν *CEG* 691.2 (Rhodes, iv BC)'

τετρακαιεικοστός, before '*PFay*. 82.12' insert '-όν, τό, *twenty-fourth part*' and add '*SB* 10359.4 (ii AD)'

τετρακαιεξηκοστόν, after '*part*' insert '*BGU* 234.12, 17 (ii AD)'

***τετρακεράτιον**, τό, name of a tax, cf. °δικεράτιον, *REG* 70.120 (Palestine, vi/vii AD).

τετράκι, before '*MAMA* 4.157' insert 'w. ref. to the repetition of a name through four generations' and add '*SEG* 34.1302 (Pamphylia), al.'

***τετρακοντόγδοος**, ον, *forty-eighth*, *SB* 9760.4 (vii AD).

τετρακόσιοι, form -κάτιοι, add 'τζετρακάτιαι *IG* 5(2).159.10, (Tegea, v BC, engraved by Arcadian)'

τετρακτύς II, for '6:8:9:10' read '6:8:9:12'

τετρακωμία, add '*INikaia* 726 (AD 288/9)'

***τετράλιτρος**, ον, of a ξέστης, *containing four λίτραι*, *SB* 9751.2 (vii AD).

τετραμαίνω, after '= τρέμω' insert 'Archil. 23.9 W.'

τετραμηνιαῖος, add '*SEG* 32.1554 (Arabia, vi AD); τ. βρέβια lists of supplies *for four months*, *Cod.Just.* 1.42.2'

***τετραμηνιακός**, ή, όν, neut. pl. subst., *supplies for four months*, *SEG* 9.356.82 (Ptolemais in Cyrenaica, vi AD).

τετράμηνος, line 4, ἡ πρώτη τ., add '*OStrassb.* 178.2, *OHeid.* 254 (both ii BC)'

τετράμορος, for '*four parts*' read '*a quarter*' and after 'Nic.*Th.* 106' insert 'ταμίσοιο ib. 712'

τετραούγκιον, for '= *triens* (i.e. *coin of four unciae*)' read '*four-twelfths*, i.e. *one-third*, of an estate, = Lat. *triens*' and add 'Just.*Nov.* 134.10.2'

τετραπάλαστος 1, add '(sp. -παλαιστ-) *PSoterichos* 1.21 (AD 69), 2.17 (AD 71), *IG* 12(2).11.15' **2**, add 'Hsch. s.v. τετράφυον'

τετράπεδος I, transfer '*IG* 4²(1).119.14, al. (Epid.)' to section II after '*four feet*' and for 'Plb. 8.4.4' read 'Plb. 8.6.4'

τετραπλάσιος, after 'cf. 756e' insert 'τούτῳ (ἀποδώσεις) τὸ τετρᾰπλάσειον ἢ ὅσου πέπραται τὰ ὑπάρχοντα *PWisc.* 81.7 (ii AD)'

τετραπλεθρία, for '(Corcyra)' read '(Crete, iii BC)'

***τετραπλεῖ**, adv., *Inscr.Cret.* 4.41.4.2 (Gortyn, v BC), cf. τετραπλῆ.

***τετράποδί**, adv., *on all fours*, ἅτε βρέφος ἑρπύζουσι Nic.*Al.* 543.

τετράπολος, add '*PCornell* 39.6 (iii/iv AD)'

τετράπορς, delete 'Cret. nom. .. (Gortyn)'

τετράπους I 1, line 2, after 'Pl.*Ti.* 92a' insert 'τετράπουν μῖμον ἔχων θηρός E.*Rh.* 255 (lyr.)' add 'Cret. τετράπος, Schwyzer 181 iii 7 (Gortyn), Myc. *qe-to-ro-po-pi* (instr. pl.) (both in sense I 2)'

τετραπρόσωπος, add 'Μήτηρ *MAMA* 5.101, θεά *ABSA* 49.13 (both Dorylaeum), *PMag.* 4.2818'

τετραπυργία, after 'Str. 17.3.22' insert 'also in Lydia, *TAM* 5(1).230 (iii AD)'

τετράπωλον, after 'τέθριππον' insert '*PLond.* 1912.45 (i AD)'

τετράς I, after '**2**' read '*the fourth day* of the first, middle, or last decad of the month' and add 'μέχρι τετράς ἐπιλάμψει (?-ψῃ) δεκάδι (i.e. the fourteenth day) Ezek.*Exag.* 178 S.' add '**5** in general, *fourth*, Guarducci *EG* 4.363 (c.AD 300). **6** *group of four*, τετράδας ἐς πίσυρας κρηνῶν προχέεις σέο κάλλος *SEG* 32.1502 (v AD).'

τετράσπαστος, delete the entry.

***τετραστάσιος**, ον, *of four times the value, weight*, etc., *CID* II 62 iiA 5.

τετράστοον, add '*Ann.Épigr.* 1990.957 (Aphrodisias, iv AD; sp. -στω-)'

***τετράστροφος**, ον, *consisting of four strophes*, Serv. in *Gramm.Lat.* 4.468.21 K.

τετράτομος 2, for this section read '*of a papyrus roll*, *consisting of four τόμοι*, χάρτας τετρατόμους *PLond.ined.* 2134 (*JHS* 55.95, ii AD); -ον, τό, such a *roll*, *PFreib.* 53.25'

τέτρατος, add 'Thess. πετριτ-, παρ πέτριτεν ἔτες, i.e. *every fourth year*, *SEG* 37.494.10 (Metropolis, late iii BC)'

***τετραΰφαντος**, ον, *having a quadruple web*, ἱστοῦ τ. *PMasp.* 6ᵛ.49.

***τετράφυλλος**, ον, *of four leaves*, στεφάνιον χρυσοῦν τετράφυλλον ἄστατον *Inscr.Délos* 1416*A*i57 (ii BC).

τέτραχα, add '῾Ινδικήτας μεμερισμένους τέτραχα Str. 3.4.1; Nonn.*D.* 2.248, etc.; *in four ways*, Opp.*H.* 3.72. **b** *four times*, *APl.* 336 (vi AD).'

***τετραχάλκιον**, τό, = °τετράχαλκον, *IEphes.* 13 ii 27.

***τετράχαλκον**, τό, *quadruple chalcus*, a coin = half of an obol, *BMC Ionia* 340; also -ος, ὁ, Hsch. s.v. ἵπποπορ.

τετραχοίνικος, after '*holding four χοίνικες*' insert '*PTeb.* 796.11 (ii BC)'

τετράχους, after '*holding four χόες*' insert 'τετρ]άχουν Lang *Ath.Agora* XXI Ha13 (iii BC)'

***τετρωβολεῖος**, ον, = τετρώβολος, *PStrassb.* 52.13 (ii AD).

τετρώβολος II, after '**τετρώβολον**' insert '(often written τετροβ- in papyri and ostraca)'

***τετρῶος**, ὁ, *throw of four* at dice, *TAM* 3(1).34*D*67 (Termessus); pl., *BCH* 8.502, 503; also sg. τετρώ ib. 502 (Phrygia).

τεῦγμα, for '*work*' read '*structure* or *artefact*' and add 'χρυσέοις τεύγμασιν Didyma 118.9 (ii BC)'

τευμάομαι, for 'Antim. 3' read 'Antim. 3.2 W.' and add '*Suppl.Hell.* 1044'

τεῦχος IV, add 'τ. δημόσιον *BCH* 60.131 (Delphi, i AD), *Delph.* 3(6).35.19, al.' at end add 'Myc. *te-u-ke-pi* (instr. pl.) *equipment*'

τεύχω I 1, add 'ἐλεγεῖον, -α τ. *CEG* 819.13, *SEG* 28.1245.B19' **2**, τετυγμένος, add '*fixed, definite*, Arat. 12, 757' at end add 'Myc. pf. part. *te-tu-ko-wo-a₂* = τετυχϝόha'

τέφρα, at end delete 'Gr. θέπτανος'

τεφράς, for 'τέττιξ' read 'cicada'

***τεφρίας**, ου, ὁ, kind of serpentine (ὀφίτης), also called Μεμφίτης, Plin.*HN* 36.56.

***τεφρῖτις**, ιδος, ἡ, precious stone, Plin.*HN* 37.184.

τεχνάζω II 3, line 3, after '*Philopatr.* 26' insert 'in aor. part. *AP* 6.4 (Leon.Tarent.)'

τεχνάρχης, add 'w. ref. to an association of sculptors, *Salamine* 43 (ii/iii AD)'

τέχνασμα II, add 'Aen.Tact. 37.8'

τεχνήεις II, after '*Q.S.* 8.296' insert 'τ. καὶ δεινὸν καὶ πόριμον Gorg. 11a.25 D.-K., τεχνήεντι νόῳ *Suppl.Hell.* 938.11, δόλῳ Nonn.*D.* 37.202'

τεχνικός II, before 'Thphr.*Lap.* 55' insert 'Pl.*Lg.* 889a'

τεχνίτης III, add 'γόης καὶ τ. ἄνθρωπος, opp. ἰδιώταις ἀνθρώποις, Luc.*Peregr.* 13'

τῇ, after lemma insert 'Cypr. τᾶ (*ta*) *ICS* 346, 347'

τηθία, after '*old woman*' insert 'Men.*Mis.* 13, Ar.Byz. *fr.* 243C S.'

Τηθύς I, as type of very old woman, for 'Call. .. 174' read 'Call.*fr.* 194.52 Pf.' and add 'Suet.*Blasph.* 216 T., *Com.adesp.* 57 D.'

***τηλέγνωτος**, ον, *recognized from afar*, A.*fr.* 204c R. (dub.).

***τήλεμος**, ον, perh. *distant*, Theognost.*Can.* 64.

τηλέπορος 1, add 'Nonn.*D.* 17.339, 37.692'

***Τηλέφεια**, ἡ, name of a tetralogy of plays relating to the Telephus legend, Σοφοκλῆς ἐδίδασκε Τηλέφειαν *IG* 2².3091.8 (early iv BC).

τηλεφόρος, for 'Nonn.*D.* 19.149' read 'Nonn.*D.* 18.262 cod., 19.149 cod.'

***τῆλις**, delete the entry.

τηλόθεν 1, add 'σήμηνε .. *Il.* 23.359' (fr. section 2) **2**, for this section read 'w. gen., *far from*, Πελειάδων μὴ τηλόθεν Pi.*N.* 2.12, τοῦ Τελαμῶνος τηλόθεν οἴκου S.*Aj.* 204, E.*HF* 1112'

τηλοῦ 1 a, add '*Trag.adesp.* 77 K.-S.' **2**, add 'E.*fr.* 884'

τηνεσμός, for 'f.l. for τεινεσμός' read '= τεινεσμός'

***τηνεσμώδης**, ες, = τεινεσμώδης, Sch.Nic.*Al.* 382 Ge.

τηρέω II 1, after '*observe*' insert 'ἀνάλωσιν Thgn. 903' **III 1**, add '*keep* a regulation, rule, etc., *IEphes.* 3217 (AD 113/20)'

τητινός, add 'ἐ[φ]ήβων τητινῶν *SEG* 29.806 (Euboea, 120/100 BC)'

***Τιβέριος**, ὁ, name of an Asiatic month beginning Oct. 24, *Hemerolog.Flor.* 79 (22 K.).

***Τιβηνός**, ὁ, wine, app. named after the village of Τίβας in Lydia, Gal. 6.806.7, 14.16.9, al.

***τιβιάζομαι**, app. *play the tibia*, *Mim.adesp.* 15.6, 42.

τιγρήιος, delete the entry.

***τιθασμός**, ὁ, *nursing at the breast*, Procl.*CP Hom.* 26.11.

τίθημι, lines 25/26, after 'ἔθηκαν' insert '(E.*HF* 590)' and after 'Attic insert 'inscrr.' **A III 2**, after 'E.*Ph.* 576' insert 'βόστρυχον Call.*fr.* 110.8 Pf.' **B II 1**, at beginning delete 'when Med. is more freq. than Act.' at end add 'Myc. 3 aor. *te-ke* (in sense B I 1)'

τιθηνήτειρα, add 'ὦ Βάκχοιο τιθηνήτειρα *SEG* 26.683 (Thessaly, iii BC)'

τιθύμαλλίς 3, delete the section.

τιθύμαλλος, for 'used for poisoning water in warfare' read 'used as poison' and add 'Ael.*NH* 1.58'

τίκτω, line 6, aor. 1 ἔτεξα, add 'subj. τέξῃ Hes.*fr.* 343.8 M.-W., *Epigr.Gr.* 706.2'; lines 13f., aor. pass. ἐτέχθην, add '*IHadr.* 80 (ii AD)'

τίλλω I 2, line 6, delete 'as a description of an idle fellow'

τίλτρον, delete '*PRyl.* .. (iv AD)'

***τιμάδα**, adv., app. *in proportion to one's rank*, *SEG* 23.566 (Crete, iv BC).

τιμαῖος, add 'as cult-epith. of Zeus, Διὶ Τειμαίῳ *TAM* 5(1).267'

τιμάξιος, after '*worthy of honour*' insert '*PAmh.* 153ᵛ.1 (vi/vii AD)'

†**τιμαχεῖον**, τό, place where the τίμαχος (τιμοῦχος) exercises his office, *SEG* 9.5 (Cyrene, 109/8 BC); perh. also *SEG* 9.18 (iv BC), 9.33 (iii BC).

τιμάω I, line 20, τιμᾶν τινα τάφῳ, γόοις, etc., add '*SEG* 31.1003 (Lydia, i AD), Mitchell *N.Galatia* 28 (Rom.imp.), etc.'; add '*honour* with a bequest, λεγάτῳ .. τιμηθέντες *Cod.Just.* 1.5.15, 1.5.18.4, al.' **III**, line 7, for 'sentence of death' read 'condemnation'

***τιμευτικός**, ή, όν, *of or concerned with valuation*, τειμευτητικῶν νόμων *SEG* 39.1180.98 (Ephesus, i AD).

τιμή, line 1, add 'dial. τιμά *GVAK* 13' **I 4**, add 'of a monument,

IEphes. 3233 (iii AD)' **II 2**, before 'valuation' insert '*the process of evaluating*, πάντως οὐ σὴ αὕτη ἡ τιμή Pl.Grg. 497b; τὰς τιμὰς τὰς τῶν πολλῶν ἀνθρώπων ib. 526d' **III**, line 5, delete 'οὐ σὴ .. 497b'

τῐμήεις, line 1, delete 'acc. τιμήϜεντα .. (Sicily)' **1**, add 'ἥβη τιμήεσσα Mimn. 5.5 W.'

τίμημα **1**, add 'pl., rest. in SEG 19.835.3 (Pisidia)' **4**, add 'τὸ τείμημά σου τὸ πολιτικόν app. your *fine* due to the city, POxy. 3105 (iii AD)'

τιμητεία, add 'SEG 30.1442 (iii AD)'

τιμητεύω, after 'D.C. 41.14' insert 'in Bithynia, MDAI(A) 12.178 (Prusias, iii AD), cf. °βουλογραφέω'; add 'ISalamis 11.5 (c.AD 60)'

τιμητός, add '**IV** as honorific epith. of emperors, Xanthos 49 (ii AD).'

τιμίζω, *declare the value of*, οὗ δὲ ἂν ἀναγράψηται καὶ τιμίζηται εἰσαγωγήν SEG 39.1180.50 (Ephesus, i AD).

τίμιος **I**, sup., add 'SEG 31.1281 (ii AD)' **II 3**, add 'adv. τιμίως *honourably*, IHadr. 71'

τιμουλκέω, after 'Hsch.' insert 'where also τῑμουλκέω'

τιμογράφέω, add 'w. gen., *assess, estimate*, PCol. 98.4 (iii BC), PCornell inv. 1.34 (BASP 22.87, i AD)'

*Τιμοθεαστής, οῦ, ὁ, *follower of Timotheus*, perh. the citharode of Miletus, rest. in Didyma 181.5 (iii AD).

*Τῑμουλκέω, v. °τῑμιουλκέω.

τιμουχέω, add 'SEG 31.985D (also fr. Teos, written τιμοχ-, 480/450 BC)'

τιμοῦχος **I**, add 'epith. of Aphrodite, IG 12(5).222 (Paros, ii BC)' **II**, line 3, after 'Teos' insert 'SEG 31.985D (480/450 BC, written τιμοχ-)'

τῑμωρέω **I 1**, line 6, transfer 'Democr. 261' to section II 1.

τίναγμα, add 'SEG 15.853.11 (Alexandria, ii BC)'

τινάσσω, penultimate line, for 'poet. Verb' read 'rare in early prose but'

τινθᾰλέος, after '= sq.' insert 'λοετρά Call. in Suppl.Hell. 287(b).5' and for 'Epic. .. 7.7' read 'Dionys.Bassar.fr. 81.4 L.'

τίνω, lines 4/5, for 'Inscr.Cypr. 135.12 H.' read 'ICS 217.12'; line 6, after 'etc.' insert 'Cypr. subj. *pe-i-se* πείσε, ICS 306'

τίπτε, line 1, after 'τί ποτε;' insert '*why ever?*'

τίρων, add 'SEG 24.1218 (Egypt, iv/v AD); also sp. τήρ- SEG 31.1116 (Phrygia, Chr.)'

τίς, line 4, for 'Cypr. .. 10 H.' read 'Cypr. *si-se* σις ICS 217.10, al.; perh. acc. pl. *si-na-se* σινας ICS 10a.4' **A II 11**, add '**d** after neg. τί ποτε *anything at all*, οὐ δύναμαι (sic) τί ποτε πρᾶξαι SB 9616ᵉ 6 (vi AD).' **B**, line 7, after 'Pi.N. 7.57' insert '(but prob. in h.Cer. 404)' **I 8 c**, line 2, after 'etc.' insert 'w. inf., AP 5.178.2 (Mel.)' **II 1 a**, add 'SEG 31.1003 (Lydia, AD 84/5)'

†ἰσάνη, ἡ, = πτισάνη, POxy. 736.51 (i AD).

τιταίνω **I 3**, add 'of a charioteer, Nonn.D. 37.290'

Τιτάν **I**, add '**2** as adj., Τιτῆνι σιδήρῳ, of the knife with which the Titans killed Dionysus, Nonn.D. 6.174.'

Τιτανίς, add 'pl., Acus. 7 J.'

τίτλος **I 1**, add 'also τίτλον, τό, CIG 8621.10 (v AD), MAMA 7.484; τίτουλος, ὁ, REA 64.59 (Vienne (Isère)); τίτυλος, ὁ, Theognost.Can. 61' add '**III** *tax*, CPR 8.54.2 (v AD), Cod.Just. 10.19.9 pr.'

*τίτουλος, v. °τίτλος.

τιτρώσκω, line 4, for '*Historia* .. (Gortyn)' read 'Inscr.Cret. 1.x 2.1, 9 (Eltynia, vi/v BC)'

*τίτυλος, v. °τίτλος.

τίω **I**, at end after 'id.Ag. 706 (lyr.)' insert 'τίεσκε μύθους Trag.adesp. 268 K.-S.'

*Τλᾱπολέμεια, τά, *games held in honour of Tlepolemos at Rhodes*, SIG 1067.8 (ii AD).

τλήμων **III 2**, after '*miserably*' insert 'prob. in Anacr. 2 fr.1.7 P.'

*τληπενθής, Dor. τλᾱ-, ές, *sorrowful*, Νιόβα B.fr. 20D.4 S.-M.

τμήγω, line 2, after 'aor. 1 ἔτμηξα' insert 'Il.Pers. 4.4 B.'; line 3, after '(prob.)' insert 'also Aeol., τμάξα Balbill. in SEG 8.716.9'

τμῆμα, add '**5** geom., *division (degree)*, Ptol.Alm. 1 p. 31 H., al.'

τμητικός **3**, add '**b** *solemnly sworn*, ὅρκος Steph.in Rh. 289.18.'

*τόγα, ἡ, Lat. *toga*, IHadr. 1 (ii AD).

*τόγε (i.e. τό γε), *at least*, δοκοῦντος τοῦ ἀφισταμένου τῷ μετὰ τῶν ἄλλων διαίτης ἀναχωρεῖν τόγε ἐπὶ τῷ συνοικεσίῳ Cod.Just. 1.3.52.15, 1.3.55.4.

τοῖος **I 1**, add '**b** *such as this* (that follows), *the following*, ἐφθέγξατο τοῖα Call.Del. 108, Cer. 97; τοίην ἀρχήν Batr. 8; τ. μῦθον ib. 77.' **4** *introducing a reason for something said*, πάντα δ᾽ ἐνίκα ῥηιδίως· τοίη οἱ ἐπίρροθος ἦεν Ἀθήνη Il. 4.390, cf. S.Aj. 562, Ant. 124.' **V**, add 'also neut. pl. τοῖα Nic.Th. 429'

τοιοῦτος, line 4, delete 'Ag. 315' **6**, for this section read 'introducing a reason for something said, S.Tr. 46; βασιλεὺς οὐδεὶς ἄλλος .. ἔσται· τοιαῦτα ἔχω φάρμακα Hdt. 3.85.2' and renumber the pres. section 6 as 7; after 'S.OT 1327' insert 'App.Anth. 5.17 (Hedyl.)'

τοιχίον, add 'pl., prob. in Inscr.Cret. 4.85 (Gortyn, v BC)'

*τοιχιοποιός, ὁ, *official responsible for wall-construction*, de Franciscis Locr.Epiz. p. 17 no. 3.

*τοίχισμα, ατος, τό, *wall, side* (of a ship), Poll. 1.120 (?interp.).

*τοιχογράφος, ὁ, *wall-painter*, Edict.Diocl. 7.8.

**τοιχοδόμος, ὁ, (cf. τοιχοδομέω), Myc. *to-ko-do-mo*, *builder, mason*.

τοῖχος **3**, add 'of a bed, Artem. 1.74'

τοιχωρύχος, after 'ὁ' insert 'also ἡ, Men.Dysc. 588'

*τοκάριον, τό, *interest*, PHaun. 41.6 (iv AD).

τοκεών, for 'elsewh. .. 1.137' read 'Call.fr. 191.72 Pf.'

τόκιον, add 'SEG 37.422 (Phocis, v BC), SB 5344.9'

τόκος **II 4**, for this section read '*homoeophonic substitute for Hebr. tôk* (*oppression*), Lxx Ps. 71(72).14, Je. 9.6'

τόλμα, line 9, after '(O.) 13.11' insert 'and prob. Aeol., cf. Sapph. 24(b).6 L.-P.'

*τολμηρία, ἡ, *audacity*, UPZ 196.66 (ii BC).

τομή **II**, add '**8** transf., *cutting out, excision*, τελεία τομῇ πάσης τῆς κατ᾽ αὐτοῦ ζητήσεως PMich.XIII 659.212, 238 (vi AD).' **III 4**, add 'δεῖν ᾠήθημεν γενικῷ νόμῳ τῇ .. παρούσῃ ζητήσει δοῦναι τομήν Just.Nov. 2.3.pr.'

τονθορύζω, after '(Opp.C.) 3.169' insert '*squeal*, of pigs, Vit.Aesop.(G) 48'

*τονομυρικόω, *make dumb the voice of*, τονομυρίκοσον αὐτόν Suppl.Mag. 55.17 (iii AD).

τονωτικός, after 'Gal. 6.577' insert 'adv. -κῶς, τ. θεραπεύειν *treat bracingly*, Aët. 12.46 O.'

τοξαλκέτης, after '= sq.' insert 'epith. of Apollo' and add 'applied to Artemis by Antisthenes of Paphos, Philol. 101.105.15'

τοξαρχέω, add 'SEG 23.481 (Illyricum, ii/i BC)'

τοξάρχης, (s.v. τόξαρχος), add 'SEG 24.1095 (Moesia, iii/ii BC)'

τοξία, for '= τοξίτις' read '*associated with the bow*, epith. of Artemis' and add '(sp. τοκοσίαν)'

τοξικός **1**, add 'BGU 2085.13 (AD 119)'

τοξοδάμας, add 'Μίνως B. 26.12 S.-M.'

τοξοεργός, v. °τοξουργός.

τοξοποιός, add 'SB 10558.3 (vi/vii AD), PLond. 1028 (p. 276).20 (vii AD)'

τοξότης, add 'Myc. *to-ko-so-ta*, subst. or pers. n.'

*τοξουργός, ὁ, Cypr. *to-ko-zo-wo-ro-ko* τοξοϜοργό(ς), *bow-maker*, ICS 352d (Addenda p. 421); Myc. *to-ko-so-wo-ko*.

*τοπειώδης, ες, of a *landscape*, -ες, τό, *a landscape scene*, Vitr. 5.6.9.

*τοπιάρι(ο)ς, ὁ, Lat. *topiarius*, *ornamental gardener*, TAM 5(1).53, 524 (ii AD).

τοπικός **I 2**, after 'PFlor. 58.8 (iii AD)' insert 'Peripl.M.Rubr. 12, δικαστήρια JÖAI 4 Beibl. 37 (Phrygia)'

τόπιον **I**, add 'τόπεν ICilicie 92 (v/vi AD)'

*τοπογραμμᾰτικός, *of or for a* τοπογραμματεύς; τ. κλῆρος, a κλῆρος whose profits supported a τοπογραμματεύς, BGU 2437.2, 52 (i BC).

τοπογράφος, for '*topographer*' read '*painter of landscapes*' and add 'Inscr.Délos 2618.17 (?i BC)'

τόπος **I 1**, add 'applied to localities other than towns, Cod.Just. 1.2.25, ἐν οἰῳδήποτε τόπῳ ἢ πόλει ib. 4.59.1 pr.; *place reserved for particular purpose*, τόπος Κυριακοῦ βρακαρίου IGChr. 262' **5**, add 'ITyr 8; also of a religious sanctuary, SEG 32.453 (Boeotia, ii BC), IEphes. 3418a' **9**, add 'SEG 34.1437 (Syria, v/vi AD)' add '**IV** *rank, position*, TAM 4(1).32, 40, 42, Cod.Just. 12.33.8.3.'

τοποτηρητής, add '**2** *delegate, representative*, Just.Nov. 8.4, 128.19.'

*τοποφύλαξ, ακος, *guardian of a place* (in quot., a cemetery), CIG 9546 (Italy, Chr.).

*τόρανος, ὁ, *instrument used for drilling wells*, AE 1948/9.133ff.

τορευτός, add 'τορευτὰ ἀργυρώματα Peripl.M.Rubr. 24'

τορεύω **II 1**, add 'ἀργυρώματα τετορευμένα Peripl.M.Rubr. 28'

*τόρμα (B), Lat. *turma*, *troop, squadron*, Syria 22.219 (Arabia, ii AD).'

τορνευτός **I**, for 'ποτήρια Men. 977' read 'ποτήριον Men.fr. 921 K.-Th. (v.l.)'

*τορνοσύνθετος, ον, *built in a circle*, Anon.Alch. 39.15.

τόρος, after 'etc.' insert 'IG 1³.422.141, 426.22 (Athens, v BC)'

*τορύνιον, τό, gloss on ‡ροταρία, Hsch.

τόσος, add 'Myc. *to-so*'

τοσόσδε, add 'Myc. *to-so-de*'

τότε **I 1**, line 8, before 'Pl.R. 557b' insert 'E.El. 42, Ar.Ra. 169'

*τοτη, τοτοτε, *words written on black-figure epinetron, perh. representing sound of trumpet*, BCH 108.99ff.

*τοῦμα· στόμα Hsch.

*τοῦρτα, ἡ, Lat. *torta*, *twisted loaf, twist*, Erot. s.v. ἄρτον ἐγκρυφίαν.

*τουρτίον, τό, dim. of °τοῦρτα, PRyl. 629.26 (iv AD).

τουτᾰκῖς **I 2**, delete the section.

*τουταυτοῦν (= τοῦτ(ο) αὐτὸ οὖν), *exactly the same thing*, PMich.XI 624.29 (vi AD).

✝**τοφιών**, ῶνος, ὁ, *burial-ground* (= ταφεών, cf. ἐντοφήϊα, etc.), *Tab. Heracl.* 1.137.

τραγαλέον, add 'cf. τρηγαλέον, ὑρειγαλέον, ῥωγαλέος, τρ- and ὑρ- being prob. written for Ϝρ; cf. Ϝρῆξις = ῥῆξις'

***τραγεῖον**· πόας εἶδος Hsch.; cf. *τράγιον*.

***Τράγιος** (B), title of Apollo at Tragia in Naxos, St.Byz. s.v. *Τραγία*.

τραγοπρόσωπος, add '*PMag.* 13.31'

***τρᾱκάς**, = *τριακάς* II 1, μεινὸς [Ἀγαγυ]λλίοι τρακάδι Schwyzer 614.6 (Thessaly, ii BC).

***τρᾱκισχίλιοι**, αι, α, *three thousand*, πέλεθρα [τρ]ακισχίλια *SEG* 26.672 (Larissa, 200/190 BC).

***τρᾱκοντα**, (= *τριάκοντα*), *thirty*, πέλεθρα ἑκατὸν τράκοντα δύα *SEG* 26.672.10 (Larissa, 200/190 BC).

τρακταΐζω II, add '2 *treat, discuss*, Just.*Nov.* 111.1, 118.4.'

***τρακτατίων**, ωνος, ἡ, Lat. *tractatio*, *tax-list*, *PMasp.* 329 ii 5, al. (vi AD).

***τρακτᾶτον**, τό, *administration*, τοῦ τρακτάτου τῆς ἐπαρχίας *Cod.Just.* 12.49.13.1.

***τράκτυλος**, ὁ, given as alternative name for *μαλάβαθρον*, Gal. 19.735.

τράπεζα I, add '3 (*Christian*) *altar*, *Cod.Just.* 1.4.35.4.' II 1, line 2, after 'Plu. 2.70e' insert 'μεταξὺ τῶν τραπεζῶν *SEG* 26.72.5, 46 (Athens, 375/4 BC)' III 3, add '*TAM* 4(1).87 (ii AD)' at end add 'Myc. *to-pe-za* (form **torpedza*)'

τρᾰπέζιον I, after '*table*' insert 'rest. in *IG* 1³.426.28 (Athens, v BC)'

τραπεζῑτεία, add 'λαβεῖν τὴν τραπεζιτίαν *PHarris* 54.4 (*ZPE* 18.256, vi AD)'

τρᾰπεζῑτης I 1, add '*PMich.*XIV 681.2 (v AD)' add '3 *assayer*, Clem.Al.*Strom.* 2.15.4, 6.81.2, al., Arr.*Epict.* 3.3.3.'

τρᾰπεζοποιός, after '*table*' insert '*IG* 1³.422.73 (Athens, v BC)'

τρᾰπητός, add 'cf. Lat. *trapetus*'

***τράππαγον**, τό, type of Indian *galley*, μακρῶν πλοίων ἃ λέγεται τράππαγα καὶ κότυμβα *Peripl.M.Rubr.* 44.

***τραύλισμα**, ατος, τό, *imperfect speech* as of children, νηπιάχοις τραυλίσμασι *GVI* 977.1 (Mauretania, ii/iii AD).

τραύξανα, after 'Suid.)' insert '= °*τρώξανα*, *Edict.Diocl.* 14.12' and delete 'Cf. *τρώξανον*'

τρᾰφερός II, add 'coupled w. διερή Dionys.Bassar.*fr.* 34 (b/c).3 L.'

τράφω, after 'Hes.*Th.* 480' insert '(Byz. cj., *τρεφ-* codd. plerique)'; add 'aor. ἔθραψα epigr. in *Inscr.Cret.* 1.xxii 58 (Olus, ii/iii AD)'

τρᾱχηλίζω II 2, add 'fig., *get a neck-lock* (on the wind), i.e. twist to the right quarter, *Peripl.M.Rubr.* 57'

τρᾱχήλιον, add '2 *collar*, *PMich.*xv 752.42 (ii AD).'

τρᾱχηλοειδής, add 'Gal. 18(2).350.3'

τρᾱχηλος I 1, line 3, after '*throat*' insert 'Hippon. 103.1 W.'

τρᾱχηλώδης, add 'Gal. 18(2).349.13, 351.2'

τρᾱχύς, line 2, after 'Theoc. 25.74' insert 'τρηχύν fem. acc. sg., id. 25.256' I, add '5 of coins, *in mint condition*, τὴν .. τιμὴν ἀπόστειλόν μοι ἐν μαρσιππίῳ ἐσφραγισμένον ἐν τραχαίοις (perh. for τραχέοις) *POxy.* 2728.28 (iii/iv AD)'

τρᾱχών, as pr. n., add 'Str. 16.2.16, 20, *IEphes.* 3157 (ii AD)'

τρεῖος, add 'cf. *BCH* 8.501 ff.'

τρεῖς, line 4, add 'neut. pl. τρά *SEG* 26.675 (Larissa, ii BC)'; line 5, after 'τρίς' insert '*IG* 4.1588.23 (Aegina, v BC)'; line 9, after 'Pi.*N.* 7.48' insert 'τρεῖς ἄνδρες (= Lat. *tres viri*) *triumvirate*, *SEG* 31.952 (Ephesus, ii AD); indicating the day of the month, τῇ τρισὶ τοῦ Δίου μηνός *SEG* 32.638 (Maced., iii AD)'; line 14, after '(cf. τριάζω)' insert 'so τὰ τρί᾽ ἡ δάφνη κεῖται Call.*fr.* 194.80 Pf.' at end add 'Myc. *ti-ri-si* (dat.)'

τρεισκαιδέκατος, add 'fem. subst. *thirteenth day*, Od. 19.202, Hes. l.c.; ἐπὰν ἐπιτελῶσιν τὰς θυσίας .. τῇ τρισκαιδεκάτηι *SEG* 33.1039.38 (Aeolis, ii BC), *Amyzon* 36 (ii BC)'

***τρεισκαιδεκάπολις**, v. °*τρισ-*.

***τρεκινάριος**, ὁ, Lat. *trecenarius*, *centurion in command of the 300 speculatores in the praetorian guard*, *IGRom.* 3.1432.

τρελλός, ή, όν, *crazy*, as pers. n., *IG* 2².12552 (*Τρέλλος*, iv BC), cf. mod. Gk. adj. for meaning.

τρέπεδδα, line 3, after 'BC)' insert 'also **τράπεδδα**, *IG* 7.3172.170, 173 (Orchomenus, iii BC)' add '2 *table*, *SEG* 24.361 (Boeotia, iv BC).' and delete '(Not from .. for τράπεζα)'

τρέπω, line 11, delete 'once' and after 'Pl.*Cra.* 395d' insert '(προ-) id.*Prt.* 348c'

τρέφω, line 5, after 'Plb. 12.25ʰ.5' insert 'also τετράφηκα (ἀνα-) *TAM* 2(3).1104.16 (Lycia)' II 1, add 'aor. part. as subst., *foster-parent*, οἱ θρέψαντες *SEG* 33.1060, Ἀνδρομέδαν τὴν ἑαυτοῦ θρέψαν 34.1223 (Lydia, AD 218/9); pass., *foster-child*, Μουσονίῳ τῷ υἱῷ μου [καὶ ..] Σακέρδωτι νέῳ τραφέντι μου *TAM* 4(1).262' III 1, add 'perh. οἱ θρέψαντες class of people in a religious association, *SEG* 30.622 (Thessalonica, i AD)'

τρέχω, line 7, after '(v. infr.)' insert 'Aeol. subj. δρό[μωμεν] Alc. 6.8 L.-P.' I 2, add 'b transf., of time, *Cod.Just.* 1.4.32.2.'

✝**τρήρων**, ωνος, ὁ, ἡ, epith. of, or name for, *dove* (cf. τρέω), τρήρωσι πελειάσιν Il. 5.778, 22.140, 23.853, Od. 12.63, h.*Ap.* 114, Ar.*Au.* 575, A.R. 3.541; τ. alone, Moero 1.3; in fig. contexts (oracular) referring to women, λεύσσω θέοντα γυμνὸν ἐπτερωμένον τρήρωνος εἰς ἅρπαγμα Lyc. 87, 423. **b** applied to other birds, κέπφοι τρήρωνες Ar.*Pax* 1067.

***τριᾱκᾰδίζω**, ἐτριακάδιξεν· εἰς τριακάδας ἐνέγραψε, Σικελοί Hsch. (cj., ἐτριέκοψεν cod.).

τριᾱκάς III, for this section read '*group of thirty persons*, a division of the population in some cities: at Sparta, Hdt. 1.65.5; at Athens, *IG* 2².1214.18; at Phlius or Corinth, *SEG* 30.990 (c.325/275 BC)' at end add 'see also °*τρακάς*'

τριᾱκάσιοι, add 'also at Mantinea, *SEG* 37.340.24 (iv BC)'

***τριᾱκᾰτιαρχέω**, *to be a τριακατιάρχας*, athematic part. -αρχέντες, *SB* 9937.1 (Cyrene, iii BC).

τριᾱκις, after '*IG* 5(1).222' insert '(c.530/500 BC)'

***τριᾱκονθετηρίς**, ίδος, ἡ, = τριακονταετηρίς, *SEG* 18.633.3 (217 BC).

τριᾱκονθήμερος 2, for 'τριακονθήμερον, τό' read 'τριακονθήμερος, ἡ'

τριᾱκοντα, delete '[In late Epigr. .. etc.]' and add 'see also °*τράκοντα*'

τριᾱκοντᾰδραχμοι, add '2 τρ., οἱ, a census-class, Schwyzer 366A22 (iii BC).'

***τριᾱκοντᾰμεροι**, οἱ, prob. a *board of officials* performing duties *for thirty days*, *IG* 14.256.27 (Phintias Geloorum); cf. πεντάμεροι.

τριᾱκοντάμερος, delete the entry.

***τριᾱκονταμναῖος**, add '*worth thirty minae*, ἔρανος *GDI* 1772.17 (Delphi, ii BC)'

τριᾱκοντόριον, add '*PCol.* 115j'

***τριᾱκόσθεκτος**, η, ον, *thirty-sixth*, *PMich.*III 186.39, 187.42, 43 (both i AD); sp. τριακοστ- *PMich.*III 186.8.

τριᾱκοσιαστός, for '*JHS* 33.338' read '*SEG* 30.568'

***τριᾱκοσιόδραχμα**, τά, *loans of 300 drachmas*, *IG* 1³.248.28, 35 (Rhamnus, v BC), cf. °*διακοσιόδραχμα*.

τριᾱκόσιοι, line 2, after 'Il. 11.697' insert '(τρῑη-)' 1, add 'sg., τριηκοσίης δρόμον Ἠοῦς *passage of three hundred days*, Nonn.*D.* 25.308 (-κοστῆς cod.), cf. διακόσιοι' II 2, after 'Id. 19.295' insert 'at Thasos, *Thasos* 7.8 (v BC)'

***τριᾱκοστέκτος**, v. °*τριακόσθεκτος*.

τριᾱκοστός II, add '2 *the thirtieth day*, κατὰ τριακοστήν Mitchell *N.Galatia* 257 (AD 140).'

τριᾱντα, add 'ἔ[θ]ανε πέντε κὲ τρειάντα ἐτῶν *TAM* 4(1).132 (Rom. imp.)'

***τριᾱνταήμερος**, ἡ, *period of thirty days*, *Vit.Aesop.*(G) 120.

***τριᾱντόφυλλος**, ον, *having thirty leaves*, Anon.Alch. 331.7.

τριάριοι, add 'Plb. 2.33.4, 14.8.5, al.; sg. ὀπτίων τριάρες *SEG* 31.1116 (Phrygia, late Rom.imp.)'

τριάς, add '5 in wrestling, *triple throw*, *SEG* 30.1616 (Cyprus, iii AD).'

***τριαστής**, οῦ, ὁ, *victor in the triple throw* (type of wrestling), *POsl.* 85.6 (iii AD), cf. τριάζω, τριακτήρ.

***τρῐβαῖον**, τό, prob. = τριβαία, *Dacia* 1988.146ff. (c.AD 200).

***τρῐβάκηλος**, ὁ, *the Thrice-Effeminate*, title of comedy by Naevius, Donat. in Ter.*Adelph.* 521 (tribaselo codd.).

τρῐβακός I 2, delete the section transferring quot. to section 1.

Τριβαλλοί, add 'sg. as pers. n., Lang *Ath.Agora* xxi F62 (v BC)'

τρῐβάς I, for pres. def. read '"*masculine" lesbian*' add 'III fem. adj., *worn*, τύλη *PFam.Teb.* 49aii4 (iii AD).'

***τρίβιβλος**, ον, *consisting of three books*, Gal. 1.408.5.

***τρίβλιον**, v. °*τρύβλιον*.

τρῐβολεκτράπελος, for 'deal in *coarse rude jests*' read '*rattle on about thorny, abstruse matters*'

τρίβολος II, for this section read '*threshing-board* (perh. confused w. Lat. *tribulum*), Ph.*Bel.* 85.37, τριβόλους ἀχυρότριβας *AP* 6.104 (Phil.), Longus 3.30, τ. ξύλινος *Edict.Diocl.* 14.41. 2 transl. Hebr. wd. denoting some cutting instrument, Lxx *2Ki.* 12.31.' VI, add 'cf. τρίβολον ἄκοντα· τρίαιναν Hsch.'

τρῐβος II 1, add 'fig. τρίβῳ κατέτριψαν ἄνθος Ἀργείων A.*Ag.* 197' **2**, add 'b *hinge*, *BGU* 2359.1 (iii AD).' **3**, delete the section.

***τριβοῦνος**, ὁ, Lat. *tribunus*, *tribune*, *IGRom.* 3.279, *PFlor.* 89.6, *SEG* 34.1515 (Arabia, vi AD); τριβ(ούνου) νοταρίου *ICilicie* 91 (vi AD).

τρίβων (A), for '*worn garment, threadbare cloak*' read '*cloak* (sts. w. implication of being worn or threadbare, cf. τρίβω II), *IG* 1³.422.120 (Athens, v BC)'

τρίβων (B) 2, delete the section.

τριβωνάριον, add '*POxy.* 3617.12 (iii AD)'

τρῑγέρων, after 'A.*Ch.* 314 (anap.)' insert 'of wine, *AP* 9.409 (Antiphan.); of persons, *AP* 7.295 (Leon.Tarent.), 421.6 (Mel.)'

τρίγληνος I, for this section read '*having three eyeballs*: in quots. as epith. of ear-rings consisting of a cluster of three round stones, ἕρματα τρίγληνα Il. 14.183, Od. 18.298; cf. ‡*τρίκοκκος*'

τριγλῖτις, add '2 *precious stone*, *triglitis mulli .. colore cognominatur*, Plin.*HN* 37.187.'

τρίγλῠφος II, at end before 'pl.' insert 'also neut., ἡ δὲ τῆς κρηπίδος καὶ τοῦ τριγλύφου (ποίησις) ἀτελής Arist.*EN* 1174ᵃ26'

τρίγωνος I 1, add 'of type of harp (cf. II 2), τ. πηκτίδες Diog.Ath. 1.9 S., τ. ψαλτήρια Arist.Pr. 919ᵇ12'   **II 2**, for this section read '*triangular harp*, masc., S.fr. 239, 412 R., Eup. 88.2, 148.4 K.-A., Ar.fr. 255 K.-A., Pherecr. 47 K.-A., Pl.R. 399c; neut., Pl.Com. 71.13 K.-A. (s.v.l.), Arist.Pol. 1341ᵃ41 (s.v.l.), Aristox.fr. 97 W.'   add '**6** masc., name for the plant μηδική (v. Μηδικός II 1), Ps.-Dsc. 2.147, Hsch.'

⁺τρίδουλος, ον, *triply a slave*, i.e. *a slave through and through*, A.fr. 78c.5 R., οὐδ᾽ ἐὰν τρίτης ἐγώ μητρὸς φανῶ τρίδουλος S.OT 1063, Ach.Tat. 8.1.2; literally, of one descended through three generations of slaves, Theopomp.Hist. 244.   **II** *consisting of three slaves*, ζεῦγος τ. Ar.fr. 580 K.-A.

τριέλικτος, line 2, after 'Orac.ap.Hdt. 6.77' insert 'τ. ἀλωή *triple* halo, Arat. 816'

τριετήρης, delete the entry.

*****τριετηρία**, ἡ, = τριετηρίς 2, Inscr.Cret. 4.146.7 (iv BC; sp. τριϜετηριαν).

τριέτης II, add 'Aristonous 1.37'

*****τριετίρης**, ενος, ὁ, prob. a *third-year* (ε)ίρήν, i.e. fifteen years old, IG 5(1).1120 (Geronthrae, v BC; v. AC 27.105f.). (τριτ- shd. perh. be read, cf. τριτίρενες, πρωτείρης.)

τρίζω, line 3, after 'Il. 2.314' insert 'fut. τρίσω Sm.Is. 38.14, τριζήσω Aq.Am. 2.13'   **2**, line 8, for 'Call.Hec. 1.4.14' read 'Call.fr. 260.68 Pf., Aq. l.c.'

*****τριήδαρχος**, ὁ, perh. for τριη⟨κά⟩δαρχ-, POxy. 43ᵛ ii 27 (iii AD); cf. τριακάδαρχος.

*****τριημιπηχιαῖος**, α, ον, = τριημίπηχυς, *a cubit and a half long*, IG 12(2).11.14.

*****τριημιστατήρ**, ῆρος, ὁ, *one and a half staters*, SEG 34.122 (Eleusis, 333/2 BC).

τριημιστατήρα, delete the entry.

τριπραύλης, for '*flute-player*' read '*piper*'

τριηριτικός, line 2, after 'IG 2².1629.70, 100, 134' insert 'κρατήρ ib. 1424a.153, 1425.361, 1649.3'

⁺τριηροποιός, ὁ, *builder of triremes*, IG 1³.153, 182 (both v BC), Arist.Ath. 46.1.

⁺τριθάλεια, ἡ, epith. of Artemis, Antim. 182 Wy. (v.l.), cf. τριθαλλίαι (read τριθαλείαι)· μεγάλως τοῦ θάλλειν αἴτιαι Hsch.

*****τρικάμαρος**, ον, *having three vaults*, PNess. 22.19 (vi AD).

*****τρικάρανοστρεφής**, οῦς, of Cerberus, *turning three heads*, Suppl.Mag. 42 (iii/iv AD).

*****τρικέλευθος**, ον, *having three ways*, SEG 25.449 (Arcadia, iii/ii BC).

*****τρικέλλαρον**, τό, *name of some implement*, POxy. 1290.5 (v AD).

*****τρικενναλικός**, ή, όν, *of a tricennial festival*, SB 10988.2 (vi AD), cf. late Lat. tricennalis, °ούικεννάλια.

*****τρικίναιδος**, ὁ, app. *thoroughgoing catamite*, perh. in POxy. 3724 fr. 1 ii 29 (list of epigrams, i AD).

⁺τρίκογχος, ον, archit., *having three niches*, τὸ τρίκονχον σίγμα (i.e. Σ-shaped) IGLS 9122 (AD 488).

τρίκοκκος, after '*with three grains or berries*' insert '(in quot., as gloss on °τρίγληνος)'

τρικόλωνος, add 'οἱ Τρικόλωνοι, app. as name of an estate, Lang Ath.Agora XXI 14 (iv AD)'

τρικόνητος, for 'cf. ἐπικονέω, κονή' read 'cf. °δια-, °ἐπι-'

⁺τρίκορυς, υθος, *having triple helmets* (the exact significance is unclear), Κορύβαντες E.Ba. 123.

τρικότυλος I, add 'TAPhA 79.184 (Attic vase, v BC)'

τρίκροτος, for '*rowed with triple stroke*' read '*rowed by three banks of oars*'

*****τρίκτοινοι**, οἱ, *name of association in Rhodes*, ASAA 17/18 (1939-40).149.18.

⁺τρικτύαρχος, v. °τριττύαρχος.

⁺τρικύλιστος, ον, perh. *carried on three wheels* (i.e. *a three-wheeled chair*), Epicur.fr. 125 (CPh 35.183).

*****τρίλλη**, ἡ, *part of the body* (in quot., w. ref. to horses), Hippiatr. 2.239, cf. perh. δρῖλος, Simon Ath.ap.Suid.

⁺τρίλλιστος, ον, (for τρισ-λιστος), *thrice*, i.e. *earnestly, addressed in prayer or entreated*, νύξ Il. 8.488; as epith. of Demeter, Call.Cer. 138; adv. -τως AP 5.271 (Maced.Thess. II).

τριμερής, after 'adv. τριμερῶς' insert '*in three instalments*, Cod.Just. 10.16.13.5, 7'   add '**II** τριμερές, τό, *a third*, PLond. 1674.56, PMasp. 45.2 (both vi AD).'

τρίμηνος, line 3, delete 'τὸ τ.'

τριμιτάριος, before 'PLond.' insert 'PAnt. 33.10 (iv AD)'; at end add 'also -ία, ἡ, Ζωσίμη τριμιταρία Not.Scav. 1895.482 no.159 (Syracuse).'

τριόδιον, after ' = foreg.' insert 'BGU 958e (iii/iv AD)'

τριοδίτης II 1 a, after 'Chariclid. 1' insert 'Hesperia 6.391.18 (tab. defix., Athens)'

τρίοδος I 2, add '**b** as adj., epith. of Hecate, SEG 26.819 (Thrace, after AD 212).'

τρίοδους II 1, line 4, for 'cf. AP 11.126' read 'Opp.H. 4.639'

τριούγκιον, after ' = *quadrans*' insert 'as a proportion of an inheritance, Cod.Just. 6.4.4.16'

*****τρίοψ**, ὁ ὑπὸ τῶν Πυθαγορικῶν ἐν Δελφοῖς τρίπους ⟨οὕτως καλούμενος⟩ Hsch., cf. Τρίοπος.

τρίπεδος, add 'Inscr.Cret. 4.30 (?vi BC; indeterminable context)'

τριπετής, delete the entry.

*****τρίπλευρος II**, add 'sg., ASAA NS 33/4 (1955-6).165 no. 14 (Ialysus)'   add '**III** τρίπλευρον, τό, *spherical triangle*, Menelaus ap.Papp. 6.476.16.'

*****τρίπλησον**, aor. imper. *multiply by three*, perh. error for τρίπλωσον (τριπλόω), PCair. 10758 (vi AD), cf. °πεντάπλησον.

τριπλόος, after 'PRev.Laws 19.14 (iii BC)' insert 'ἀποτεινέτω προσ[τ]είμιον τὸ τριπλοῦν SEG 31.122 (Attica, ii AD)'

τριποδίσκος, add 'Myc. ti-ri-po-di-ko (pers. name)'

τρίπους, add '(form τρίπος) Myc. ti-ri-po, ti-ri-po-de (dual) (in sense IV 1)'

*****τρίππος**, ὁ, app. *team of three horses*, trigae τρίππος Charis. p. 35 B., cf. τρίιππον.

τρισέληνος 1, before 'AP 9.441' insert 'POxy. 2331 ii 4a (CR 71.189, iii AD)'

*****τρίσελλος**, ον, *having three seats*, PFay. 117.17 (ii AD).

*****τρισεύμοιρος**, ον, *thrice fortunate*, Anon.Alch. 28.14.

*****τρισήρως**, ὁ, Myc. ti-ri-se-ro-e (dat.), *divine name*.

*****τρισκαιδεκάπολις**, ιδος, ἡ, *confederacy of thirteen* Ionian *cities*, Didyma 356.7 (ii AD).

τρισκακοδαίμων, for 'Men. 404.1' read 'Men.Pk. 400, Epit. 19, al.: adv., τρισκακοδαιμόνως ἔχω dub. l. in id.Dysc. 523'

τρισπερίοδος, delete the entry.

*****τρισσάδιος**, α, ον, = τρισσός, πεντάδα τρισσαδίην ἐτέων, IG 10.2(1).447.5 (Thessalonica, ii/iii AD).

τρισσός, add 'cf. Ps. 79(80).6.   **VI** = τριστάτης, Aq.Ex. 14.7, al.'

⁺τριστάτης, ου, ὁ, app. calque of Hebr. šālīš, *military officer of high rank*, Lxx Ex. 14.7, 15.4, 4Ki. 10.25; esp. *officer attending on the king*, ὁ τ. αὐτοῦ 4Ki. 15.25, cf. Hsch. (Expld. in Sch.Od. 3.324 as, in pl., *the three who stood* on an Egyptian chariot).

τρίστιχος, add 'IGRom. 1.1162 (i AD, -ικ-)'

τρίστοος, add 'fem. subst., *atrium of church*, τὴν πᾶσαν ἐκαλέργησεν (for ἐκαλλιέργησεν) τρίστωον DOP 6.87 (Nicopolis, vi AD)'

τρισχίλιοι II, add 'X.HG 2.3.51, al., Lys. 25.22'

τριταῖος, add '**IV** app. in non-temporal sense (v. SEG 36.1538), SEG 36.903 (Italy).'

τρίτατος, add 'τριτάτην ἤματος (app. *on the third day of her illness*) ὀλλυμένην GVI 662'

*****τριτεῖος**, α, ον, *of third-class quality*, ἐρέα Edict.Diocl. 21.4; cf. τριτεῖα.

τριτημόριος II 2, add '**b** *as a weight*, τριτε(μόριον) SEG 31.154(a) (Athens, v BC); Ath.Agora x LW17, al.'

τριτήμορον, add 'Inscr.Délos 104.73 (iv BC); perh. used as adj., ληκύθια τριτάμορα ZPE 12.267 (Sicily, v BC)'

τριτίρενες, add '(Cf. °τριετίρης)'

*****τριτόδιος**, ὁ, epith. of Zeus, SEG 33.453, 454 (Atrax, Thessaly, v, iii BC).

τρίτομος, add 'τρίτομον, τό, *papyrus roll consisting of three sections*, PFreib. 53.25'

τρίτος II 2, τρίτη (a weight), add 'SEG 31.154(g) (Athens)'   **IV**, add 'τὸν ἀπὸ τρίτου ἑκατοστῆς ἀποδιδότω τόκον (i.e. ½ percent) Just.Nov. 22.44.4; sim. ἐκ τρίτου ib. 7'

*****τριτοστολιστής**, οῦ, ὁ, app. στολιστής *of the third rank*, PSelect. 21.5.

τρίτρα, after '*times*' insert 'or *a third of*' and for 'GDI .. (Gort.)' read 'Inscr.Cret. 4.43Aᵇ9 (Gortyn, v BC)'

τριττύς III, for '*a third of the* φυλή' read 'app. *a tripartite civil division*, understood as a third of a φυλή'

*****τρίχας**, ᾶ, ὁ, *hairdresser*, PAmst.inv. 62.54 (Mnemos. ser. 4 30.146, ii AD).

τριχιάω, add '**III** τριχιῶν, ὁ, calque of Hebr. śa ʿīr "*hairy one*" (in context, *a goat-formed demon*), Aq.Is. 13.21, Aq., Sm., Thd.Is. 34.14 (cf. τριχίας 1).'

τρίχιον, add 'item in inventory of travelling equipment, PRyl. 627.171 (iv AD; ?*wig*, ed.)'

τριχοίνικος, neut. as subst., add 'IEphes. 3437'

*****τριχόπλαστος**, *having one's hair set in curls*, Hsch. s.v. κίκιννα.

⁺τρίχορδος, ον, *having three strings*, λύρα D.S. 1.16.1, κιθάρα St.Byz. 130.20; *based on three notes*, of the music of Olympus and Terpander, Plu. 2.1137b.   **II** subst., τ., ὁ, *a three-stringed instrument*, prob. *lute*, Anaxil. 15 K.-A.; later neut., τρίχορδον, τό, Poll. 4.60.

*****τρίχορροια**, ἡ, *shedding of the hair*, Cyran. 63.16, 74.4 K.

*****τρίχους**, ουν, *holding three* χόες, cj. in PErasm. 19.4 (ii BC).

τρίχωμα, add 'of the sacred hair of Isis, παρὰ τοῖς τριχώμασι ἐν Κοπτῷ PMich.VIII 502.5 (ii AD); in title, Ἴσιδι τριχώματος θεᾷ μεγίστηι SEG 18.704 (32.1583, Egypt, AD 105); of a coral reef, Agatharch. 108'

τρίχωρος I, add 'κλεῖθρον τρίχωρον PMag. 4.2337'

τρίχωσις **I** 1, add '**b** *plumage*, Sch.Pi.*P.* 4.380, 381.'

τρῐωβολεῖος, add 'applied to a person as an indication of worthlessness, *Act.Alexandr.* iv 13.18'

*τρῐωβόλιον, τό, = τριώβολον, dub. in Steph. *in Rh.* 286.22 (pl.).

*τρῐώβολος, ὁ, = τριώβολον, *Inscr.Délos* 1429*B*ii25 (ii BC).

τρῐωδέομαι, add 'Procl.*in R.* 2.21.16, 2.22.2 K., al.'

τρίωτον, delete the entry (v. °τρίωτος).

*τρίωτος, ον, Cret. δρίωτος, *Inscr.Cret.* 4.145.6 (Gortyn, v/iv BC), *three-handled*, καναστ(ραῖ)ον *Inscr.Cret.* l.c.; -ον, τό, *three-handled jar, BGU* 544.17 (ii AD).

τροπαϊκόν, add '*SEG* 32.599 (Thessaly, i BC/i AD); also -κός, ὁ, *SB* 10288.1a.7 (τροπαιεικος pap., Palestine, AD 125)'

τροπαῖον, add 'of a monument for a successful athlete, *SEG* 34.1316 (Lycia, c.AD 90).   **2** *memorial monument, SEG* 36.1260 (Paphos, end ii AD).'

τροπαιοῦχος, Ζεὺς τ., add '*ICilicie* 16; also Ἑρμῆς τ. ib. 17 (both early iii AD)'; at end add 'transf., of deeds, τροπαιούχοισιν ἐπ᾽ ἔργοις Orph.*H.* 33.4'

τροπαιοφόρος **I** 2, before 'a coin' insert 'neut. subst.'

τρόπις, add 'form *τόρπις Myc. *to-qi-de* (dat.), prob. *spiral*; deriv. adj. *to-qi-de-we-sa* (fem.), *torqᵘidwessa, cf. Lat. *torqueo*'

τρόπος **II**, line 6, after 'etc.' insert 'ὁ γενεαλογικός, ὁ πραγματικός τ. [τῆς ἱστορίας] *kind*, Plb. 9 *fr.* 1.1.4, 2.4'   **4 b**, line 2, after 'Id.*Lg.* 638c' insert 'κατὰ τ. ἔχειν, εἶναι, *to be all right, be in order*, Men.*Dysc.* 134, 215' add '**c** w. gen., *by way of*, κατ᾽ ἰσχύος τρόπον A.*fr.* 281a.20 R.'

τροπόω (A), add 'pass., ἐτροπώθη ὁ πόλεμος *reached a turning-point*, Lxx 3*Ki.* 22.35'

τρούλλα, for '*BMus.Inscr.* 980' read '*Salamine* 30' and add 'also τρυλλα *BGU* 2360.3'

τροφεία, before 'ἡ' insert '(also τροφέα *GDI* 2254.6 (Delphi, i BC))'

τροφεύς 1, add 'transf., πάντα νικᾷ ὁ τροφεὺς τῆς ὅλης οἰκουμένης *Suppl.Mag.* 7'   **4**, add '*BMC Phrygia* p. 399 (Synnada, i AD), cf. D.Chr. 48.10'

τροφεύω, add 'w. acc., τοῦτον, γύναι, τρόφευε Ezek.*Exag.* 29'

τρόφιμος **II** 1, add 'cf. *IG* 2².6731'

*Τροφωνιάς, άδος, fem. adj., *of Trophonius*, Τροφ]ωνιάδος γᾶς *SEG* 23.297 (Lebadea, iv/iii BC).

Τροφώνιος, Ζεὺς τ., add '*SEG* 32.475 (Boeotia, after AD 213)'

τροχάδια, for '*walking-shoes*' read '*running-shoes*' and add '*PMon.* 142.17 (vi AD)'

τροχιά **II**, for '*the round of a wheel*' read '*wheel*' and transfer 'Nic.*Th.* 816' to section I after '*rut*'

*τροχινός, ή, όν, *round, liable to roll*, εὐτροχάλοιο· τοῦ τροχινοῦ Sch. Nic.*Al.* 134d Ge. (s.v.l.).

τροχίσκος 1, add '**b** *item of temple equipment*, μέτρα χαλ(κᾶ) ἱερατικὰ β᾽ σὺν τροχίσκῳ σιδηρῷ α᾽ *POxy.* 3473.25 (ii AD).'   **2**, for '*troche* or *trochisk*' read '*ellipsoid*'   **3**, for '*ear-ring*' read '*pendant*'

τροχός **B I** 1, after 'E.*Med.* 46' insert 'or *hoop*, cf. section A II)'

*τροχωτός, όν, *round*, P*Harris* 88.20 (v AD).

τρύβλιον **I**, add 'Lang *Ath.Agora* xxi B12 (iv/iii AD), *Kafizin* 49; sp. τρίβλ- *Inscr.Délos* 104.16, 23 (364 BC)'

τρυγάω **1** 1, lines 3/4, for 'metaph. .. Ar.*Pax* 1338 (lyr.)' read 'w. sexual innuendo, τρυγήσομεν αὐτήν (sc. Ὀπώραν) Ar.*Pax* 1338'

τρύγη **I** 1, add 'cf. τρύγη· ὁ πυρός, καὶ ἡ κριθή, καὶ πᾶς ἄλλος καρπός, καὶ ποιὰ βοτάνη Hsch.'

τρυγών, line 1, after 'ἡ,' insert 'also ὁ, Lxx *Ca.* 2.12'

*τρῠγώνι(ο)ν, τό, a bird, dim. of τρυγών, P*Amst.* 13.4 (v AD).

*τρύεινος, η, ον, perh. = θρύινος, *of reed rope*, *BGU* 2361a ii 6 (iv AD).

τρύξ **I**, after '*PTeb.* 555 (ii AD)' insert 'fig. ἔοικας ὦ πρεσβῦτα νεοπλούτῳ τρυγί Ar.*V.* 1309'; cf., in pun w. sense II 1, συνεκποτέ᾽ ἐστί σοι καὶ τὴν τρύγα.— ἀλλ᾽ ἔστι κομιδῇ τρὺξ παλαιὰ καὶ σαπρά id.*Pl.* 1086'   **II** 4, delete 'metaph. also .. 1086'

τρύπανον **IV** 1, delete the section.   **2**, append this section to section I, adding '**3** term of opprobrium in list of words app. indicating noisy chatterers, Suet.*Blasph.* 21, 167 T.'

τρῠσίππιον, add 'wheel-shaped, acc. to Hsch. s.v. ἵππου τροχός'

τρῠφεροδίαιτος, after 'foreg.' insert 'Ptol.*Tetr.* 166'

τρύφος, add 'Call.*fr.* 261.1 Pf. (= *Suppl.Hell.* 289.1), A.R. 1.1168'

Τρῳάδεύς, add '*Inscr.Perg.* 8(3).74'

τρωγλοδύνων, before 'Batr.' insert 'interpol. in'

τρώγω, line 4, before 'ἐν-' insert 'ἀπο-, δια-' and after 'παρα-' insert 'ὑπεκ-'   **III**, add 'ἀθήραν τρώγειν *eat porridge* or *gruel*, perh. prov., *be spoon-fed* or *pampered, SB* 10567.36 (iii AD)'

τρωκτός **I**, transfer '*eatable*, Hdt. 2.92' to section II.

τρωξαλλίς, line 1, after 'Alex. 15.12' insert 'Ael.*NA* 6.19'

*τρώξανα, τά, *dry twigs*, Thphr.*CP* 3.2.2, cf. ‡τραύξανα.

τρώξανον, delete the entry.

τυγχάνω **A I** 3 **a**, line 13, after 'ἄν' insert 'or ἦν' and after '*it may be*' insert 'E.*Ph.* 765, *Or.* 780'   **B II** 2 **d**, add 'X.*Cyn.* 1.8'

†τυί (or τυῖ)· ὧδε, Κρῆτες Hsch., i.e. *hither* (cf. υἱ).

†τυῖδε, (written as τυίδε, rather than τυῖδε, in papyri of Sapph. and Alc., but always scanned disyll.), Aeol. adv., *to this place, hither*, Sapph. 1.5 L.-P., al., perh. at Alc. 142.3 L.-P., cj. in Hes.*Op.* 635, Theoc. 28.5, *Epigr.Gr.* 988.3 (Balbilla), Sch.T *Il.* 14.289.

τύκισμα, after 'E.*Tr.* 814' insert '(codd. τεκ-, τυκτ-)'; after '*fr.* 125.3, cf. *HF* 1096' insert '(codd. τειχ-)'

τυλάριον 1, after 'τύλη' insert '*cushion*'; add 'also τύλαρον, *PSI* 825.17 (iv/v AD, pl.).'

*τυλοεργός, ὁ, *cushion-* or *mattress-maker*, P*Iand.* 150iii3 (iii AD).

τύλος 3, add '*Tab.Defix.* 74.17'   add '**5** = τύλη 2 and 3, *pad* or *cushion* for horses, dual [τ]υλο prob. to be read at *IG* 1³.421.187, v. *SEG* 33.20 (414/13 BC).'

*τυλοτάπης, ητος, ὁ, perh. kind of cushion, P*Ryl.* 627.36 (iv AD), *SB* 13597 (v AD).

τύλωσις **II**, add 'but see *BCH* 80.516-18'

τυμβαύλης, for '*flute*' read '*pipe*'

τύμβιος, after '(Macedonia)' insert 'λώβην .. τύμβιον *TAM* 3(1) p. 361, *SEG* 33.111 (Paphlagonia)'

τύμβος 2, add 'Pl.*Lg.* 872b, Plu.*Alex.* 72.5, Luc.*Charid.* 22'   add '**III** *altar*, Lyc. 313 (of Apollo), 613 (of Hera).'

τυμβοχοέω, add '*Suppl.Hell.* 1002'

*τυμβωρυχέω, τό, *grave-robbing*, τυμβωρυχίου ἐγκλήματι ὑπεύθυνος ἔσται I*Kalch.* 73.4.

τυμβωρύχος, for '*CIG* 2826, al.' read '*MAMA* 8.544, 547, 550, al.' and add '*SEG* 32.1423D (Syria, ii AD)'

*τῡμολῑτική, ἡ, kind of jar or sim. vessel, τὴν ἑτέραν τυμολειτικήν P*Oxy.Hels.* 46.3 (i/ii AD), *POxy.* 1759.8 (ii AD), 1760.13-14 (ii AD).

τυμπᾰνικός, after 'ὕδρωψ' insert 'Plin.*HN* 25.60'

τύμπᾰνον **I**, for '*kettle-drum*' read '*frame drum, tambour*'   **II** 1, add 'meton., ὦ τύμπανα· ὦ ἐπιτήδειοι τυμπανισθῆναι Hsch.'   **3**, delete the section transferring quot. to section 1

*τύν, v. °σύ.

Τυνδᾰρίδης, add 'Τινδαρίδαι, perh. dat. dual without final ν, to be read at *IG* 5(1).937, v. *SEG* 36.354 (Kythera, iv BC)'

τυννός, for '*so small, so little*' read '*small, little*' and for 'Call.*Fr.* 420' read 'Call.*fr.* 471 Pf.'

τῠπικός, after '2' insert '*carved in relief*, [σύμπαντι τῷ τ]υπικῷ [κόσμῳ] *TAM* 3.21 (early iii AD)'

†τύπιον, τό, dim. of τύπος, in quot., *small moulded figure, IG* 2².1534.205, al., 11(2).161.B119 (Delos, iii BC).

τῠπίς, for 'Call. in *PSI* 9.1092.50' read 'Call.*fr.* 110.50 Pf.'

τύπος **IV**, add '**2** *drawing, painting*, γραπτοὶ τ. E.*fr.* 764, *AP* 7.730 (Pers.); τ. alone, *APl.* 136, 143.'   **V**, delete 'γραπτοὶ τ. .. *AP* 7.730 (Pers.)'

τυπόω **II** 1, add '**b** *paint, portray, APl.* 138.'   add '**3** *assign to a type, classify*, Aristox.*Harm.* 4.'

τύπτω **I** 1, add 'inscribed on an astragalus as an injunction, τύπτ[ε] *SEG* 30.949 (Olbia)'

τύπωμα 1, add 'ἀνέθηκε δὲ καὶ τύπωμα χρύσεον I*Hadr.* 1 (ii AD)'

τύπωσις, before 'Lyc.' insert 'τόρμα' and add 'φιάλη Didyma 426.8, 435.5, 436.7'

*τῠρᾶς, ᾶ, ὁ, *cheesemonger, SEG* 26.1673 (Palestine, iv AD).

*Τυρβηνός, ὁ, cult-name of Apollo, Hsch.

Τύριος, add 'as epith. of Heracles, *RIB* 1129'

*τύριος, ὁ, = *tetradrachm, SB* 10305.5; perh. also ib. 10304.10 (both ii AD).

τῡροκνηστις, add 'τυροκνάσσιδες τρῖς *SEG* 24.361 (Thespiae, iv BC)'

τῡροποιός, add 'I*Tyr* 43'

τῡροπώλης, add '*SEG* 25.180 (Athens, iv BC)'

*τῠροπῶλις, ιδος, ἡ, *cheesemonger, SB* 10447ᵛ i 12 (iii BC).

τῡρός, add 'Myc. *tu-ro₂*, perh. τυρροί *cheeses*'

τύρσις, delete 'also .. *fortified house*'

τῠφλάγκιστρον, before '*blunt hook*' insert 'a surgical instrument'

τυφλός **I** 2, for 'τοξεύματα' read 'βέλη' and for '*HF* 199' read 'B. 5.132 S.-M.'   **4**, after 'metaph.' insert 'ἐλπίδες A.*Pr.* 250'   **II** 1, lines 1/2, for 'ἐλπίδες A.*Pr.* 252' read '*unseen*' and after 'Id.*Fr.* 593.6 (lyr.)' insert 'of weapons, E.*HF* 199'

*τυφομένως, *smoulderingly*, Gal. *de Crisibus* 122.8 A.

*τῦφος, τό, term of abuse for an old man, cf. τυφογέρων, τῦφος, etc., Suet.*Blasph.* 209 T.

Τῡφῶν **I**, lines 2-6, for 'represented .. Sch.Pl.*Phdr.* 230a' read 'another form of Typhoēus, Typhos, Hes.l.c., *h.Ap.* ll.cc., A.*Pr.* 354, *Th.* 493, etc.'; line 8, identified w. Set, add '*Hesperia* 54.214 no. 1 (iii AD)'

*Τῡφωνιακός, ή, όν, = Τυφωνικός, P*Mag.* 7.468.

Τῡφῶς, as appellat., add 'E.*Ph.* 1154'

τῠχαῖος **II**, add '**2** neut. pl. subst., *Genitalia* τύχαια, Charis. p. 37 B.; cf. τύχεια.   **b** *small statuettes of Τύχη*, ἔμπορος τυχαίων *IG* 14.419 (Messene, iii AD).'   **III**, delete the section.

τύχη, line 2, after 'iii BC)' insert 'Cypr. also dat. *to-ka-i Kafizin* 133a, 177'   **III** 1, add 'Ἀπόλλωνι .. ὑπὲρ Κλέονος τοῦ υἱοῦ ἀνέθεκεν ἐν

τώμεντον   SUPPLEMENT   **τώς**

τύχηι *SEG* 23.621 (Cyprus, iii BC), *Kafizin* 113a (iii BC), 177, al.' **4**, add 'τυχἀγαθᾶι *Delph.* 3(2).137' **IV 1**, add 'νικᾷ ἡ τύχη τοῦ Δόρου (a charioteer) *SEG* 31.1486, 1492 (both Alexandria, AD 608/10)'; personified, add '*SEG* 31.537 (Delphi, c.280 BC), etc.'

*****τώμεντον**, τό, Lat. *tomentum, wool, flock, Edict.Diocl.* 18.7.

*****τώνᾱ**, ἁ, Cret. for ζώνη, Hsch.

*****τῶνυ**, adv., *thence*, τῶνυ ἁι ἁ ὀδὸς ἐπὶ τὸ .. *CRAI* 1985.255 (Crete, c.500 BC) cf. ὄνυ.

*****τώρα**, adv., *now*, *Syria* 23.179.37 (iii AD), as mod. Gk., < τῇ ὥρᾳ (ταύτῃ).

τώς I, add 'Alcm. 1.46 P.'   **II**, line 2, after 'A. *Th.* 637' insert '(cj. in *Ag.* 242)'

Y

*ὑαινίτης, ου, ὁ, kind of precious stone, cf. ὑαίνιος, Socr.Dion.Lith. 53 H.-S.

ὑακίνθινος, line 1, after 'Od. 6.231' insert 'ἄρουραι Anacr. 1. fr. 1.7 P.' and at end add 'ἐν ποτηρίοις σμαραγδίνοις καὶ ὑακινθίνοις Ps.-Callisth. 63.21, cf. 63.33'

*ὑακίνθιος, ον, perh. sp. for *-θειος, consisting of or like the stone ὑάκινθος, λιθάριον ὑ. SEG 37.1001.3-4 (Lydia, ii/iii AD).

Ὑάκινθος, add 'SEG 25.1110(f) (Cyprus, Hellen.)'

ὑάκινθος I 1, for 'wild hyacinth .. Scilla bifolia' read 'plant, perh. Hyacinthus orientalis'   II, for 'aquamarine' read 'sapphire'

ὑαλᾶς I, add 'also ὑελ- POxy. 3428.14 (written οἰελᾶ, iv AD)'

ὑάλεος, add 'cf. Myc. adj. we-a₂-re-jo, prob. decorated with rock crystal'

ὑαλοειδής 1, for 'crystalline lens of the eye' read 'retina'   2, for 'topaz' read 'peridot'

ὕαλος, line 6, after 'Apoc. 21.18' insert 'penultimate syll. long in Mesom. 13.1, 5 H. (ὑέλο- cod.; ὑελλ- is v.l. in Hdt. 3.24.1, Luc. VH 2.11)'

ὑαλουργός, form ὑελ-, add 'POxy. 3265.5, 3742.3 (both early iv AD)'

ὑαλοψός, add 'PBaden 97.35 (vii AD)'

ὑβρίζω II 4, add 'οὐδὲν τῶν ὑβρισμένων no ostentatious gift, Ael. VH 1.31 D.'

ὑβριστής I 1, add 'E.Andr. 977, Supp. 575, 728'

+ὑβριστοδίκαι, οἱ, nonce-wd., "outrageous jurymen", title of comedy by Eupolis, Eup. p. 466 K.-A.

ὑγιαίνω I 4, line 4, after 'BMus.Inscr. 1123a (inc.loc.)' insert 'cf. καθ᾽ ἣν ὑγιαίνομεν ὥρην at the time of day when we say farewells, AP 12.177 (Strat.)'   II, after 'pass.' insert 'Hp.Morb. 1.20, 21'

ὑγίεια A, line 4, after '(Herod.) 4.20' insert '(perh. ῡ here and in Call.fr. 203.21 Pf.)'   II, add 'b generally any gift received by the sacrificer, πᾶν τὸ ἐκ θεοῦ φερόμενον εἴτε μύρον εἴτε θαλλός Hsch., cf. θαλλός III.'   III, for this section read 'cure, medicine, ὕπνος δὲ πάσης ἐστὶν ὑγιεία νόσου Men.Mon. 783.   2 name of a medicine, Alex.Trall. 5.4 (II p.159.3 P.); also, of plaster, Androm.ap.Gal. 13.932.3, Heras ap.Gal. 13.766.6.'   B, after 'Call.Com. 6 (hex.)' insert 'SEG 30.1330 (Rom.imp.); identified with Athena, IG 1³.506 (v BC)'

ὑγιής II, for 'sound in mind' read 'sound (as applied to the mind, character, etc.)' and delete 'virtuous'   III 2, after 'POxy. 1031.18 (iii AD), etc.' insert 'Corinth 8(3).486 (iii AD)'

*ὕγιος, v. ὕγειος.

ὑγιόω, add 'PMasp. 283 I 15, ὑγιοῦν· τὸ σα[ρρ]οῦν· τὸ θεραπεύειν Hsch.'

ὑγρασία, add 'euphem. urine, Lxx Ez 7.17, 21.12'

ὑγροβαφής, add 'see also °ὑδροβαφής'

*ὑγρομαντεία, ἡ, water-divination, Cat.Cod.Astr. 8(2).143.

ὑγρόπορος, for ' = ὑγροκέλευθος' read 'going through the water'

ὑγρός I 1, add 'b as quality of the air, αἰθήρ Pi.N. 8.41, E.Ion 796; ἀήρ Emp. B 38.3 D.-K., Arist. GC 330ᵇ4, Mete. 348ᵇ28.'

ὑδάτινος II, delete the section transferring quots. to section III.

ὑδατόεις I, add 'νέφη ὑδατόεντα Theoc. 25.89'   II, for 'transparent .. fine' read 'of garments, flowing' and for 'cf. ὑδάτινος II' read 'cf. °ὑδάτινος'

*ὑδατομαντεία, ἡ, water-divination, Tz.Alleg.Il. 18.195.

ὑδατοτρεφής, add '(-τροφής Hsch. and codd. at Ath. 2.41a in a citation of Od. l.c.)'

ὑδατώδης I, add 'adv. -δῶς Gal. 16.761.15'

ὑδατώλενος, for 'dub. sens.' read 'having watery arms' and add '(ii/i BC)'

ὑδερώδης, add 'b suffering from dropsy, Gal. 12.177.7, 13.224.2.'

*ὑδραγώγημα, ατος, τό, irrigation-channel, Sch.Gen.Il. 21.257.

+ὑδραγωγία, ἡ, system of irrigation, Arist.PA 668ᵃ14, Duris 89 J.; transf. of veins, etc., Pl.Ti. 77e, cf. ὑδρεία I 2.

ὑδραγωγός I, add 'τόπος where water gathers, Horap. 1.49 S.'   II 2 a, add 'gloss on ὀχετηγός, Sch.bT Il. 21.257'

*ὑδραλετάριος, ὁ, app. operator of a water-mill, Teuc.Bab. p. 46.7 B.

*ὑδραλετᾶς, ᾶ, ὁ, water-mill engineer, Sardis 7(1).169 (iv/v AD); delete ὑεραλέτης II.

ὑδράλμη, add 'Anon.Alch. 348.9'

ὑδραντικός, delete the entry.

*ὑδραντλητικός, ή, όν, for water-pumping, ὑδραντλητικὴ παροχία SB 9907.23 (AD 388); prob. in PFlor. 58.10-11.

ὑδραύλης, add 'IEphes. 1601a.8'

*ὑδραυλικός, ή, όν, hydrostatic, machinas hydraulicas, Vitr. 1.1.9, 9.8.4; hydraulicis organis, Plin.HN 7.125, Suet.Nero 41.2; also neut. subst. τὸ -όν, hydrostatic organ, Aristocles ap.Ath. 4.174c, Hero Spir. 1.42.

ὕδραυλις, for 'hydraulic' read 'hydrostatic' and delete 'so τὸ -αυλικόν .. 1.42'

ὑδρεῖον II, add '2 some part of public baths, SEG 26.784 (Thrace, AD 162/3), IEphes. 435.'

+ὑδρεκδοχεῖον, τό, water-tank, IEphes. 695.9 (i AD), ib. 424 (ii AD), MAMA 8.449.4 (the last two sp. -εγδ-).

+Ὕδρεος, ὁ, Syrian deity, Inscr.Délos 2155, 2160; also Ὕδρειος, ib. 2087 (all after mid-ii BC)

ὕδρευμα, after 'tank' insert 'Inscr.Cret. 3.iv 18 (iii BC)'

ὑδρία II 2, delete 'esp. in law courts, etc.'

ὑδρίσκη, add 'written ἑδρύσκη, PMich.Teb. 121ʳ II ii 8 (i AD), etc.'

+ὑδροβαφής, ές, ὑδροβαφές (?sc. ἱμάτιον) ὃ νῦν ψυχροβαφὲς καλοῦσιν a garment dyed with only (cold) water added to the dyestuff, Poll. 7.56 (v.l. ὑγρο-).

*ὑδροβάφος, ὁ, perh. one who carries out ceremonial immersions, cf. καταλουστικοί, IEphes. 3414, 3415 (ii/iii AD, cf. BE 1982.293).

ὑδροβόλος, for 'Epigr.Gr.' 1036' read 'TAM 5(2).92 (revised text)'

*ὑδρογέρων, οντος, ὁ, another name for ἠριγέρων, Apul.Herb. 76.22 (cj.).

*ὑδροδότης, ου, ὁ, provider of water, SB 9653 i 16 (ii AD).

ὑδροκόμος, for 'prob. well-bucket' read 'prob. drive-cable of water-wheel'

*ὑδρολογία, ἡ, work on the water-supply, CPR 8.22.49 (AD 314).

ὑδρομαντεία, add 'Plin.HN 37.192'

ὑδρομέλι, add 'SEG 32.1601 (Nubia, iv/v AD)'

*ὑδρομίκτης, ου, ὁ, seller of wine mixed with water, SEG 26.817 (sp. -μήκτ-; Thrace).

+ὑδροπαροχία (-εια) ἡ, supply of water for irrigation, IEphes. 4337.10 (i AD), POxy. 137.22, PVindob.Salomons 9.8 (both vi AD).

*ὑδροπαροχικός, όν, of the supply of water for irrigation, POxy. 3582.6 (v AD).

ὑδροπάροχος, add 'PVindob.Salomons 9.9 (vi AD)'

ὑδροπίσσιον, τό, (?) liquid pitch, CRAI 1945.378 (Berytus, Byz.).

ὑδροπότης, delete 'used .. fellow' and add 'Ath. 2.44b, SEG 25.774 (Moesia, Rom.imp.), 31.238 (GVI 1841, Athens, iii AD); also -πώτης Macho 46 G., Phot. a 595 Th.'

ὕδρος I, add 'ὕδρον ἐν Λέρνῃ Hippon. 102.10 W. (= Ὕδρα)'   add 'Myc. u-do-ro, water-pot (cf. ὑδρία)'

ὑδροσκόπιον 2, add 'Cat.Cod.Astr. 8(2).113.17'

ὑδροφόβος II, add 'Heras ap.Gal. 13.431.16'

*ὑδροφορείη, ἡ, poet. = ὑδροφορία, κεκασμένη ὑδροφορείη Didyma 344.3.

ὑδροφορία II, add 'ostr. in CE 61.275 (i/ii AD)'

ὑδροφόρια, lines 1/2, for 'a festival .. Sch.Pi.N. 5.81' read 'race in honour of Apollo at Aegina run by carriers of amphorae, Call.Dieg. viii.32 (fr. 198 Pf.), v. °ἀμφορίτης'; lines 3/4, delete 'name of .. Διηγήσεις viii.32'

ὑδροφόρος II, of priestess, add 'SEG 30.1286 (Didyma, i AD), 36.1060 (Miletus)'   add '2 ὑδροφόρον, τό, vessel for carrying water, Kafizin 267 (223/222 BC).'

ὕδρω, delete 'apparently' and add 'IG 14.1890.11'

ὕδρωψ, line 1, after 'ὁ' insert '(also ἡ Nic.Th. 467)'

ὕδωρ, line 2, for 'Call.fr. 475' read 'Call.fr. 268 Pf.'; acc. pl. ὑδάτη Nautarum Cantiuncula 4'   I 1, add 'b water-supply, IEphes. 3217 (ii AD), SEG 35.189 (v/vi AD).'   5, add 'perh. also Lang Ath.Agora xxi Hd 16 (iii AD)'   add '6 pl., ὕδατα, τά, urine, Sm.Ez 7.17, 21.12.'

ὕειος, add 'form perh. *ὑήϝιος, cf. Myc. we-e-wi-ja'

*ὑελᾶς, v. °ὑαλᾶς.

ὑέτιος I 1, add 'Θεὸς Ὑέτιος SEG 39.958 (Eleutherna)'

ὑεύχομαι, delete the entry (v. ICS 181).

+ὕϝαις, dub. in Cypr. u-wa-i-se za-ne, app. in perpetuity, for ever, ICS 217.10, 23, 28.

*ὑϊδεύς, έως, ὁ, = υἱδοῦς, TAM 5(1).786, Hsch.; also υἱιδεύς Isoc.Ep. 8.1.

υἰδιον (B), add 'τῷ φιλ]τάτῳ υἱιδίῳ SEG 30.606 (Maced., ii/iii AD)'

*ὑϊδός, ὁ, = υἱδοῦς, Hsch.

*υἱή, ἡ, daughter, SB 101 (i AD): also Aeol. ὐά Schwyzer 625 (Mytilene, ?i AD).

υἱήν, add 'cf. Myc. we-je-we υἱῆϜες app. vines'

†υἱιδεύς, v. °ὑϊδεύς.

υἰκός, after 'ἱερεῖον υἱκόν .. Milet. 7.18' insert 'neut. subst. υἱκόν, τό, SEG 31.122 (Attica, c.AD 121/2)'

υἱοθεσία, sp. ὐο-, add 'SEG 30.1007, AR 1985/6.99 (both Rhodes)'

υἱοποιία, add 'rest. in SEG 23.317 (Delphi, i AD), Just.Nov. 89.7'

υἱός, line 6, for 'ύύς .. 686' read 'hυύς IG 1³.783, 791, 865, hῦς SEG 23.38 (Attica, vi BC)' 3, add 'υἱοὶ τῶν συμμίξεων children of mixed race, Lxx 4Ki. 14.14 (mistranslation of Hebr. phr. meaning "hostages")', deleting this quot. fr. section 4 add '10 heir, SB 9902A ii 13 (iii AD).'

υἱωνός, add 'also ὑωνός SEG 33.1016 (Lydia, AD 103/4), PFlor. 71.235 (iv AD).'

†ὑκερός, v. °ἑκυρός.

*ὑκτάς, v. °ὀκτάς.

ὑλαῖος I, add 'τὴν ὄρνειθαν ὑλήαν (unidentified bird), OFlorida 15.4 (ii AD)'

ὑλακτέω I, after 'Eup. 207' insert 'Herod. 6.14' II, transfer 'hence Vespasian .. D.C. 66.13' to section I 1.

*Ὑλάτης, ὁ, Cypr. u-la-ta-i, ICS 2.4, 85.1, al.; epith. of Apollo, Bernand Les Portes no. 47 (iii BC), Lyc. 448, Nonn.D. 13.444.

*ὑλιᾶσθαι· κινεῖσθαι Theognost.Can. 21 A.

*ὑλιελινᾶτες, unexpld. wd. in list of tradesmen, PAmst.inv. 21 (ZPE 9.49).

*ὑλινόμος, v. ‡ὑλονόμος.

*ὕλιος, α, ον, muddy, app. in place-name, Πλακὸς Ὑλίας (gen.) IG 9²(1).609 (Locr., v BC).

ὑλιστήριον, add 'also ὑλιστάριον SB 9483.13 (ii AD)'

ὑλίτης, before 'v. ὑλήτις' insert 'BGU 2430.23 (ὑλείτης pap., i BC)'

*Ὑλλεῖς, οἱ, one of the three Dorian tribes, Tyrt. 19.8 W., Hdt. 5.68.2, Ath.Agora XVII 4.3 (458 BC); fem. Ὑλλίδες, αἱ, Tit.Calymn. 88.18, al. (ii BC).

ὑλομανέω 2, add 'Philostr.VS 2.32.2'

ὑλονόμος, for 'Simon.(?) 179.7' read 'AP 6.217.7 ([Simon.])' and add 'also ὑλινόμος' IAskl.Epid. 167

ὑλοξιδής, after 'woodcutter' insert 'or glass-maker (for ὑελο-), but the second element is unexpld., v. ICilicie p. 221 n. 2'

ὑλοτομικός, add 'Alex.Aphr.in Top. 237.25'

*ὑλώνης, ου, ὁ, buyer of wood, Ath.Agora XIX L8.103, 141 (Oropus, c.330 BC).

ὑλωρέω, add 'SEG 34.564 (Thessaly, c.200/190 BC)'

ὑμέναιος I, add 'also app. ὑμέναια, τά, Mitchell N.Galatia 118' II, line 1, for 'Ὑμήν' read 'Ὡμήν'

ὑμενώδης I, add '2 app. thin as a membrane, filmy, χιών SEG 29.1477.'

ὑμέτερος I, line 7, for 'sts.' read 'normally'; add 'in addresses to a single person with whom others are associated, Sol. 19.2 W., Call.Del. 204' deleting these refs. fr. section II.

ὑμήν, line 2, after 'Arist. .. 519ᵇ4, al.' insert 'POxy. 3195.42 (AD 331)'

†ὑμνἄγωγός, ὁ, leader of hymns in Eleusinian worship, SEG 30.90 (Eleusis, 20/19 BC).

ὑμνέω I 1, line 9, after 'c. dupl. acc.' insert 'praise in song, τὰ νομισθέντα γὰρ αἰεὶ Διόνυσον ὑμνήσω E.Ba. 72'; at end delete 'impers. .. 1203.5'

ὕμνησις, after 'praising' insert 'Pi.Pae. 12.5 S.-M. (rest.)'

*ὑμνηστός, ή, όν, = ὑμνητός (cf. ὑμνήστρια), celebrated, praised, Εὐλάλις ἀνη[ρ] ὑμν(ι)οστος (sic ed.) Mitchell N.Galatia 323 (iii/iv AD)

*ὑμνίδιος, ον, app. produced by (insects') wings (fr. syncopated dim. of ὑμήν), ὑμνιδίῳ .. πατάγῳ AP 7.198.6 (Leon.Tarent., s.v.l.).

*ὑμνοδιδασκάλέω, act as ὑμνοδιδάυκαλος, La Curie II p. 216.

ὑμνολογέω, add 'epigr. in SEG 30.1367 (Smyrna, late Hellen.)'

ὑμνοποιός, after 'E.Rh. 651' insert 'μάθησις IG 12(7).449.7 (Amorgos, ii BC)'

ὑμνοπόλος II, for 'Simon. 184' read 'AP 7.25.2 ([Simon.])'

ὑμνῳδέω I, add 'οἱ ὑμνῳδήσαντες κοῦροι SEG 37.962 (Ionia, ii AD)'

ὑμνῳδός, add 'also sg. CEG 578'

*ὕμως, Aeol. for ὅμως, Sapph. 58.21 L.-P.

*ὑναφορέω, Cypr. for ἀναφορέω, Kafizin 266b (= ICS 231).

*Ὕνγιος, v. °Ἴνγυος.

ὑνεύχομαι, delete the entry (v. ICS 181, Kourion 25).

*ὑνιερόω, v. °ἀνιερόω.

ὕννος, add 'ὕνοι pl. perh. in AAWW 1948.322-3 (Panopolis, Rom.imp., see also BE 1958.139)'

†ὑντίθημι, Cypr. for ἀνατίθημι, aor. u-ne-te-ke ὑνέθεκε ICS 181; also Arc., 3rd pl. imper. aor. ὑνθεάντω SEG 25.447 (Aliphera, iii BC).

ὑοβοσκός, after 'Arist.HA 603ᵇ5' insert 'POsl. 160.3 (iii AD)'

ὑπαγκαλίζω, for 'Pass.' read 'med.'

ὑπαγκάλισμα, after 'S.Tr. 540' insert 'Διὸς ὑ. σεμνόν (i.e. Hera) E.Hel. 242'

*ὑπἄγορᾱνόμος, ὁ, deputy agoranomos, AR 1989/90.33 (Messene, late Rom.).

*ὑπἄγωνοθετέω, act as sub-ἀγωνοθέτης, rest. in IGRom. 4.850 (Laodicea ad Lycum), v. REA 62.296.

†ὑπάετος or -αίετος, ὁ, kind of eagle, = ὀρειπέλαργος, Arist.HA 618ᵇ34, Ant.Lib.fab. 20.6 (cj.), cod. oxyt., cf. °γυπαιετούς.

*ὑπαίθρειος, ον, = ὑπαίθριος, cj. in S.Ant. 357.

ὑπακούω II 5, delete fr. 'ὑπάκουουσι' to end of section IV, delete the section.

*ὑπακωνίδιον, τό, dim. of °ὑπακώνιον, PSI 1355.4 (ii BC).

*ὑπακώνιον, τό, perh. article of clothing, PSI 1355.6 (ii BC).

*ὑπἄλαζών, όνος, ὁ, something of a charlatan, Men.Asp. 375 S.

ὑπἄλεύομαι, after 'ib. 760' insert 'AP 7.472b (Leon.Tarent.), SEG 31.1288 (Side, AD 249/252)'

ὑπαλλακτέον, add 'Gal. 17(1).98.18'

ὑπαλλακτικός, before 'adv.' for 'only in' read 'gramm., involving an interchange of relation, ὑπαλλακτικήν (στάσιν) Quint.Inst. 3.6.47'

ὑπἄμάω, add 'prob. also A.fr. 273a.3 R. (in tm.)'

*ὑπαμπετίν, written ὑπανπετίν, adv., app. going round under, SEG 35.991.B8,10 (Lyttos, v BC), cf. περιαμπετίξ.

†ὑπανατέλλω, rise somewhat, begin to rise, of a star, Ael.NA 14.24, of a fountain, ib. 15.4'

ὑπαναφύομαι, add 'also act., τραχύτητας ὑπαναφύει Ael.NA 10.13'

ὕπανδρος I, add 'θυγατέρας ὑπάνδρους SEG 24.911 (iv AD)'

*ὑπανελεύθερος, ον, somewhat deficient in liberality, niggardly, Gallo Framm.Biogr. II 274.

ὑπανίημι II, line 2, delete 'so .. al.'

ὑπανίσχω, delete 'slowly'

ὑπαντάω I 1, add 'also in pass., ὑπηντήθη τῷ στρατηγῷ Vit.Aesop.(G) 65'

ὕπαρ, for 'Schwyzer 686.2' read 'in sense A II 4, Ϝίλσιιος ὕπαρ for the sake of preventing, IPamph. 3.2 (iv BC)'

*ὑπαρχιτεκτονικόν, τό, budgetary money of the ὑπαρχιτέκτων, CID II.1 ii 46 (iv BC).

ὕπαρχος I 2 b, add 'ὁ τῶν ἀννόνων ὑ. = Lat. praefectus annonae, Just.Nov. 82.2 pr.'

ὑπάρχω B III 2, for 'to be devoted to' read 'give support to'

ὑπασπιστής 2, add 'SEG 31.1574 (Cyrene, late ii BC); sg., member of this unit, SEG 13.403 (= EAM 87, ii BC)'

ὑπασχολέομαι, add 'BGU 1159.23 (i BC/i AD)'

ὑπάτη, add '2 name of a Delphic Muse, SEG 30.382 (Argos, c.300 BC), cf. Plu. 2.744c, see also °μέση I 2.'

*ὑπἄτικιᾱνός, ή, όν, associated with an officer of consular status, ταξεωτῶν Mir.Demetr. p. 157.19 L.

*ὑπάτισσα, ἡ, wife or widow of a consul, POxy. 2243a.86 (vi AD), CIG 9008.

ὕπατος II, add 'νόος ὕπατος λόγῳ καὶ διανοίας Archyt.ap.Iamb.Protr. 4' III 1, after 'Mon.Anc.Gr. 5.1' insert 'Jul.Caes. 332b'

ὑπαυλέω, for 'flute' read 'aulos'

†ὕπαυλις, εως, ἡ, part of house, perh. covered yard, οἰκίδιον .. ἐν ᾧ ὑ. καταπεπ[τωκυῖα PMich.XII 627.7 (AD 298).'

ὑπαυλισμός, for 'flute' read 'aulos'

*ὕπαυτα, v. ὑπό C III 2, line 20.

ὑπειδόμην, line 1, after 'ὑπιδόμενος' insert '(ὑφ- BCH 10.301.20 (Alabanda, ii BC))'

ὑπειλέομαι, of bandage, add 'Gal. 18(1).789.5'

ὑπεισέρχομαι I 4, add 'succeed to in place of another, ἵνα τοῦ πρώτου παραιτουμένου ὁ μετὰ ταῦτα βαθμὸς ὑπεισέρχηται CodJust. 6.4.4.20a; εἰς τὰ τῶν ἰδίων γονέων δίκαια Just.Nov. 118.3'

†ὑπεκδέχομαι, have under oneself, μαστῷ πόρτιν ὑ., v.l. in AP 9.722 (Antip.Sid.).

ὑπέκθεσις, add 'cf. ὑπέκθεσις· ὑπόθεσις Hsch.'

ὑπεκκαλύπτω, for 'uncover from below or a little' read 'expose by removing the top covering'

ὑπεκπροθέω, add '2 run out before another's advance, Emp. B 35.12 D.-K.'

ὑπεκπροφεύγω, add 'INikaia 1045 (iii AD)'

ὑπεκτρώγω, after 'gnaw secretly away' insert 'in quot., fig., i.e. filch'

ὑπεκφέρω II 2, add 'Lesb.Rh. 2.7' IV, delete the section.

ὑπελάσσων, add 'Just.Nov. 115.5 pr.'

*Ὑπελλαῖος, -αία, epith. of Zeus and Athena, SEG 20.719.A16 (Cyrene, ii BC).

*ὑπέμφασις, εως, ἡ, perh. indication, κατ᾽ ἀντίθεσιν καὶ ἀπόθεσιν καὶ ὑ. Zos.Alch. 134.13.

ὑπεναντίος II, add 'also neut. sg., w. gen., ὑπεναντίον τῶν νόμων Just.Nov. 134.4' ὑπεξαίρεσις, line 6, delete ': hence' and insert 'exception, Cod.Just. 4.35.24'

*ὑπεξᾰκούω, listen fully to (w. gen.), of arbitrators, prob. in Inscr.Perg. 245.6 (rest., ὑπεξ[lapis, ii BC).

ὑπεξέρχομαι II, add 'Cod.Just. 1.5.18.11'

ὑπεξουσιότης, add 'Just.Nov. 81 pr., al.'

ὑπεπιστάτης, add 'SB 4638, PBatav. 4 ii 14'

ὑπέρ, line 2, after 'Arc. ὁπέρ (q.v.)' insert 'Pamph. ‡ὑπάρ; ὑπερί SEG 33.1152 (ii AD), TAM 4(1).370 (v. SEG 30.1438, Nicomedeia, Chr.)' **A I**, add '4 above, in excess of, MAMA 8.252B (iii AD).' **B III**, add 'ὑπὲρ ἥμισυ Κᾶρες ἐφάνησαν Th. 1.8.1' **IV**, add '2 later than, Cod.Just. 9.47.26.3, 10.11.8.7a.' **V**, add 'SEG 25.447 (Arcadia, iii BC), 23.207 (Messene, Augustus)' **C**, add 'Cod.Just. 10.11.8.4b (s.v.l.)'

ὑπέρα II, delete the section.

*ὑπεράγιος, α, ον, supremely holy, MAMA 7.190 (Hadrianoupolis).

*ὑπεραιρέω, take in addition, τὸ ἐπιτιμηθέν IG 11(2).199A74, cf. 84, 85 (Delos, iii BC).

ὑπεραισχύνομαι, add 'w. aposiopesis, ὑπεραισχύνομαι γυναιξὶν ἐν ταὐτῷ - Men.Dysc. 871'

ὑπεραιώρησις, add 'Gal. 18(2).488.11'

ὑπεράλλομαι I, add '2 abs., jump further (than others), Luc.Anach. 8.'

ὑπερανατείνομαι, add 'act. intr., Hsch. s.v. ὑπερτενῇ'

*ὑπερανατέλλω, = ὑπερτέλλω, Hsch.

ὑπεράνω 4, add 'BCH 60.119 (Delphi, i BC), SEG 31.825 (Sicily, ii BC)'

*ὑπεράξιος, ον, well worthy, Ael.NA 4.29.

ὑπεραπαιτέω, add 'Cod.Just. 3.10.1.2'

*ὑπεραρθρισμός, ὁ, hyperbatic use of the article, Sch.D.T. 460.13.

ὑπεράριθμος, after 'supernumerary' insert 'PTeb. 703.155 (iii BC)'

ὑπέραρσις, for 'exaltation' read 'high-water mark'

ὑπεράστειος, add '2 exceedingly refined, over-elegant, ὄψιν ὑπεράστειος Men.Mis. A93 p. 353 S.'

ὑπεραυγέω, add '2 irradiate from above, τὴν γῆν Eudox. 20.9.'

ὑπερβαίνω III, after 'protect' insert 'Aq., Sm., Thd.Is. 31.5, perh. also IT yr 75'

ὑπέρβᾰσις I, add 'Aq.Ex. 12.11'

*ὑπερβάτης, ου, ὁ, housebreaker, PMich.inv. 3736 (Anagennesis 4.141, iv/v AD).

Ὑπερβερεταῖος II, add 'Schwyzer 590.9 (Larissa, 214 BC), ICilicie 109 (AD 99), SEG 32.1537 (Arabia, AD 184), etc.; sp. -βερταῖος SEG 34.1208 (Lydia, AD 111/2); μη(νὸς) Ὑπερβερτέου SEG 31.991 (Lydia, AD 234/5)'

ὑπερβιβασμός, add 'An.Par. 4.31.28'

ὑπέρβιος II, line 3, ὑ. ἦτορ ἔχειν, add 'Hes.Th. 139, 898'

*Ὑπερβόϊος, v. °Ὑπερβώϊος.

ὑπερβολή I 4, add 'qualifying adv. εὖ Macho 168 G.'

ὑπερβολία, add 'II dat. (-βολίῃ) = ὑπερβολῇ, exceedingly, MAMA 8.208.'

ὑπερβώϊα, add 'also -βόϊα'

*Ὑπερβώϊος (or Ὑπερβῷος), ὁ, name of month in Crete, τὸ Ὑπερβοΐο μηνός SEG 23.530 (vii BC).

ὑπεργεμίζω, add 'PHib. 182 xiii 183 (iii BC)'

*ὑπέργω, ὑπηργμένη in unkn. sense in fragmentary inscr., SEG 24.139 (Athens, ii BC).

+ὑπερδᾰπᾰνάω, overspend, BGU 1838 (ii BC), PLond. 1171.21 (i BC).

ὑπερδάπανον, delete the entry.

ὑπερδέξιος, line 1, after 'ον' insert 'sts. fem. -α, Ion. -η, of Athena' **II**, add '4 possessing superiority, epith. of Apollo, Plu.Arat. 7.2; of Zeus, IGC p. 11, line 12 (Larissa, ii BC); of Athena, IG 12(1).22 (Rhodes, Hellen.); of Zeus and Athena, Thasos 124 (ii/i BC), SEG 15.517 A II 5 (Paros, iii BC), cf. Robert Hell. 10.63 ff., 295 and v. °καθυπερδέξιος.'

ὑπερδῐκέω, add 'SEG 34.1238.48 (Aeolis, c.200 BC)'

ὑπερείδω I 2, delete the section transferring quot. to section II.

ὑπερέπαρσις, for 'excessive exaltation' read 'raising up (to safety)'

ὑπερεπιθυμέω, for 'Porph.Plot. 19' read 'Longin.ap.Porph.Plot. 19.25'

*ὑπερεπιτηδείως, very suitably, Com.adesp. 22.38 D.

*ὑπερεχής, ές, that exceeds the proper, normal, etc., amount, excess, neut. as subst., τὸ ὑ. τοῦ τόκου καὶ τοῦ ἡμιολίου IG 12(7).515.35 (Amorgos, ii BC).

ὑπερήκω, add '2 jut out beyond, w. gen., τούτου .. ὑπερήκει τοῦ κόλπου ἀκρωτήριον Peripl.M.Rubr. 40.'

ὑπερημερία, line 2, after 'Lebad.' insert 'iv BC'

+ὑπερηνορέος, ον, app. arrogant, Aeol. ὑπερᾶν-, dub. in Theoc. 29.19.

ὕπερθεν, (ὕπερθε), line 1, at end insert 'E.Ion 1153 (trimeters)'

ὑπέρθετος, add 'title of official in Egypt (Lat. superpositus), σ]υν-τακτικοὺς καὶ ὑπερθέτοις PRyl. 585.29 (ii BC)'

ὑπέρθῠρον, add 'Carm.Pop. 2.14 P.'

*ὑπερί, v. °ὑπέρ.

ὑπεριδρύω, before 'in pass.' insert 'set or establish above, Procl. in R. 1.174 K.'

ὑπερίστᾰμαι 1, after 'and pf. act.' insert 'cf. also ὑπερεστήξει· ὑπερ-σταθήσεται Hsch.'

Ὑπερίων, add 'MAMA 1.390'

+ὑπερκᾰκέω, ὑπερκακεῖν· ὃ νῦν ἐκκακεῖν Hsch.

ὑπερκεράω, line 2, for 'stretch beyond' read 'curve beyond'

*ὑπερλαμβάνω, unexpld. wd. in broken text, POxy. 2344.15 (iv AD).

ὑπέρλαμπρος I 1, add 'of an appearance of the sun considered as omen, Orac.Tiburt. 23'

ὑπερμᾰζάω, add 'Poll. 7.24'

ὑπερμεγέθης, line 5, after 'Cyr. 1.6.8' insert 'of youths, perh. oversized, ephebic list in JEA 37.89.50 (Memphis, iii AD)'

ὑπερμενής, add 'epith. of Θεοὶ Σωτῆρες, RIB 461'

*ὑπερμεριμνάω, to be exceedingly concerned about, ὑγίαν PBerl.Zill. 14.3 (vi AD), cf. Gnomon 22.143ff.

ὑπερμετρέω, add 'ὑπερμετρῆσαι rest. in PAnt. 55 fr.(b)ʳ 10'

*ὑπερμηρίδια, τά, parts of a sacrificial victim above the thigh, BCH 113.449 (Crete, iv BC); perh. also in SEG 15.564 (cf. ib. 39.954; Dreros, c.600 BC).

*ὑπερνήχομαι, swim past or over, ὑπερνήχεται· ὑπερέχει, ὑπερβαίνει Hsch.

*ὑπέρνομος, ον, supralegal, προαίρεσις An.Boiss. 2.45.

ὑπεροικοδομέω, add 'Berytus 33.52'

*Ὑπεροῖος, v. ὑπερῷος 2.

+ὕπερον, τό, (perh. also ὕπερος, ὁ, Hes.Op. 423 v.l.), pestle, λεήναντες ὑπέροισι Hdt. 1.200, Plb. 1.22.7, Luc.Philops. 35, Poll. 1.245, ὕπερα σιδηρᾶ id. 7.107 (cf. .. ἔροις σιδηροῖς, the mutilated title of a comedy in IG 14.1097), 10.114, EM 779.48, Lang Ath.Agora XXI B19, IG 1³.422.265, 425.105 (Athens, v BC), PRyl. 167.14; prov., of never-ending and ineffectual labour, ὑπέρῳ μοι περιτροπή γενήσεται Pl.Com. 1 K.-A., Pl.Tht. 209d (cf. Philem. 30 K.-A.), Plu. 2.1072b; used as lever for stretching dislocated joints, Hp.Fract. 13, Art. 78, al.; as a club, Plu.Alex. 63.9; as a staff, Luc.Demon. 48. **II** like πηνίον, pupa of geometrid moth, Arist.HA 551ᵇ6.

ὑπέροπλος I, add 'comp. -ότεροι A.fr. 168.4 R. (s.v.l.)'

*ὑπεροπτεία, ἡ, haughtiness, cj. in S.Ant. 130.

*ὑπερόπτις, ιδος, ἡ, fem. of ὑπερόπτης, acc. -ιν Rh. 1.559.6.

ὑπερόριος I 1, add 'adv. -ίως, ἐξορίζομαι Cod.Just. 11.41.7'

+ὕπερος, ὁ, v. °ὕπερον, τό.

ὑπέροχος, after 'Ion. ὑπείρ-' insert 'Ὑπάροχος pers. n. BCH 114.460 (Doris, c.500 BC)'

ὑπερπᾰθής, for 'grievous' read 'affected by great emotion' and for 'Phleg.fr. 36.1 J.' read 'Phleg.fr. 36.1.5 J.' and insert 'Alcin.Intr. p. 184.22 H.'; after 'adv. -θῶς' insert 'excessively, Ptol.Tetr. 188 B., cf.'

ὑπερπαίω, last line, before 'Supp.' insert 'surplus'; add 'so ἐκ τῶν ὑπερπαιόντων IGBulg. 1565'

*ὑπερπαρέχω, provide on behalf of, SEG 30.380 (Tiryns, vii BC).

*ὑπέρπεδον· ὅρος, βουνός, ἔπαρμα γῆς Phot.

ὑπέρπικρος, add 'Men.Dysc. 129'

ὑπερπράξιον, add 'ZPE 56.89 (wax tablet, Egypt, v/vi AD)'

ὑπερπράσσω, add 'Just.Nov. 17.4'

ὑπέρπῠρος 2, add 'σίτῳ οὐπερπούρῳ SEG 25.556 (Boeotia, c.210/200 BC); n. subst., τὰ οὐπέρπουρα SEG 32.456 (Boeotia, iv BC)'

ὑπερσαρκέω, add 'Androm.min.ap.Gal. 13.729.10'

ὑπερτέλειος 1, for 'beyond completeness or perfection' read 'larger than °τέλειος (I 2 a), designation of a type of aulos'

ὑπερτερέω, add 'Certamen 149'

ὑπέρτερος I 1, add 'b w. gen., on the upper part, τᾶς τῆνω φλιᾶς καθ' ὑπέρτερον Theoc. 2.60; in the parts above, καθ' ὑπέρτερα γαίης Arat. 498.' **II**, add 'older, Mitchell N.Galatia 392'

ὑπερτίθημι I 3 a, line 3, after 'D.S. 13.3' insert 'οὐδὸν ὑπερθεμένη APl. 58'

ὑπερφιλοσοφέω, add 'Philostr.VA 7.37'

*ὑπέρφορβος, ὁ, exceedingly bountiful, τὸ νῶϋν ἄζομαι θεῶν πατ[έρα] βροτῶν πάσας γενεᾶς ὑπ[έρ]φορβον SEG 36.350 (Epid., Rom.imp., new text of IG 4².134, which has [πολύ]φορβον).

*Ὑπερφορεύς, έως, ὁ, epith. of Zeus, SEG 20.719A (Cyrene, ii BC).

*ὑπερφρονητής, οῦ, ὁ, despiser, An.Boiss. 5.340.

ὑπέρφρων 1, add 'of persons, τοὺς ἄγαν ὑπέρφρονας Trag.adesp. 521 K.-S.'

ὑπερχᾰρής, add 'SEG 23.206 (Messene, AD 2/3)'

ὑπερχειλής, for 'over the brim, running over' read 'full to the brim'

*ὑπερχράομαι, use to excess, Sch.B Il. 1.193.

ὑπέρχρεως, add '2 of estates, encumbered, v.l. for ὑπόχ- (q.v.) in Is. 10.16, 17.'

*ὑπερχύννω, overflow, Socr.Dion.Lith. 37 G.

*ὑπεστιοῦχος, ὁ, functionary in the cult of Hestia, IEphes. 1078.16, cf. °ἑστιοῦχος.

ὑπεύθῠνος II, add '4 defendant, BGU 2173.3 (AD 498), Cod.Just. 2.2.4 pr.'

ὑπευλᾰβέομαι, for 'to be somewhat afraid' read 'to be cautious or wary of' and add 'συμβολὴν μάχης Memn. 29.1 J.'
ὑπήνη **2**, add 'οὐρανόεσσαν ὑπήνην perh. the inside of the upper lip, Nic.Al. 16'
*ὑπηνῆτις, ιδος, of the upper lip, θρίξ An.Boiss. 4.431.
ὑπηρεσία **I**, for 'body of rowers, ship's crew' read 'collective term for the assistants to the trierarch over and above the rowers'; line 7, delete 'crews' **II 2**, add 'εἰς ἀπρεπεῖς αὐτῷ (sc. ὕδατι) ὑπηρε[σίας καταχρωμένους] IEphes. 3217 (ii AD) and then in naval contexts' and transfer fr. section I quots. Plb. 5.109.1, 1.25.3, Gp. 18.9.3.
ὑπηρετέω **I**, for 'as a rower' read 'serve as a member of a °ὑπηρεσία' **II 3 d**, add 'in other cults, Mitchell N.Galatia 204 (AD 227)'
ὑπηρέτης **II 1**, add 'temple servant, SEG 23.209 (Messene, iii BC)'
*ὑπήχησις, εως, ἡ, subterranean noise, Paus. 7.24.8 (cj., v. ὑφήγησις II).
*ὑπισθόδομος, v. °ὀπισθόδομος.
*ὑπιωγή, ὑπιωγαί· ὑπαγωγαί, ὑποδρομαὶ τῆς πέτρας διὰ σκέπην, σκεπηνὰ μέρη Hsch.
*ὑπνέω, = ὑπνόω, Anon.Fig. p. 172.11 S.
ὑπνοδώτις, delete the entry.
ὕπνος **I 3**, after '(Plu.) Alex. 50' insert 'καθ' ὕπνου Sardis 7(1).94'
ὑπνόω **II**, add 'AP 5.23 (Call.), 5.184 (Mel.)'
*ὑπνωτέον, one must sleep, An.Boiss. 3.327.
ὑπό, at the end of first paragraph add 'Myc. u-po' **C II**, add '**2** subject to conditions, etc., ὑπὸ ἐκκλησιαστικὸν ἐπιτίμιον Cod.Just. 1.4.29.10; ὑπὸ τὸν ἔσχατον κίνδυνον ib. 9.47.26.2.' **III**, add 'in the time of, ἡυπὸ δὲ Ἐχεμένε ἔφορο[ν] Schwyzer 12.66 (Sparta, v BC' add '**V** ὑπό τι up to a point, somewhat, ὑπό τι ἄτοπα Pl.Grg. 493c; ὑ. τι ἀσεβῆ id.Phdr. 242d; ὑ. τι μικρὸν ἐπιθήκισα Ar.V. 1290 (lyr.), ὑ. τι ⟨δὴ⟩ σκυθρωπάσας Macho 247 G.'
ὑποβάλλω **I 3**, add 'Cod.Just. 1.1.5.4, 1.3.29.1' add '**4** ὑ. ψήφους cast voting-pebbles surreptitiously, Arist.Ath. 68.3, cf. Call.fr. 85.8 Pf.' **III**, add '**2** propose (as an official measure), οἱ κωμάρχαι τῆς Θεαδελφίας ὑπέβαλον, ὡς σου ἀπενεγκόντος .. PPrag. I 108.3 (Theadelphia, iii AD).'
ὑπόβασις, add '**IV** ὑπόβασις· ὁ ἐνδότατος χιτών, ἢ περίζωμα Hsch.'
ὑποβιβλιοθηκοφύλαξ, delete the entry.
*ὑποβιβλιοφύλαξ, ἄκος, ὁ, sub-librarian, BGU 660.9, 14 (ii AD).
ὑπόβλητος, line 3, after 'OC 794' insert 'PMich.III 174.9, 11 (ii AD)'
ὑπογάστριον **II**, add 'IGC p. 99 B5 (Acraephia, iii/ii BC)'
ὑπογραμματεύς, add 'of an ephebic officer, SEG 26.184, 188, 194, 198 (iii AD)'
ὑπογραμμᾰτεύω, add 'τοῦ Λυκίων ἔθνους SEG 17.711.8-9 (Balbura, ii AD)'
ὑπογραφεύς **2**, for 'person .. another' read 'writer acting on behalf of an illiterate'
ὑπογραφή **I 1**, add '**d** signature, Cod.Just. 1.2.24.9.'
ὑπογράφω, add '**VI** register, enter as, (cf. ἐπιγράφω III 4) ἐὰν ἰδιωτικὴν μὲν ὑπογραφόμενοι τύχην, δύνανται βασανισθῆναι Cod.Just. 4.20.15.1.'
ὑπόγυιος **II**, neut. as adv., add 'Gal. 7.949.8, 8.858.2, 9.520.9'
ὑποδακρύω **2**, add 'Gal. 11.318.11'
ὑπόδειγμα **I 2**, add 'συγκρίσεις ἀρχιτεκτόνων γέγοναν μεθ' ὑποδειγμάτων SEG 33.1040 (Cyme)'
*ὑποδειλικός, όν, unexpld. adj., in quot. of a shirt, καμίσιν ὑποδειλικόν PMich.XI 607.31 (AD 569), perh. cf. ὑποδειρίς.
*ὑποδείριον, gloss on ὑποδέραιον, Hsch.
ὑποδεσμεύω, add 'med., Ister 36 J.'
ὑποδεσμός **I**, add 'Edict.Diocl. 8.7 (pl.)'
ὑποδέχομαι, add 'also act. ὑποδέχω POxy. 3400.5, 18, 23 (iv AD)'
*ὑποδηλωτέον, one must indicate, Aristox.Harm. p. 4 M.
ὑπόδημα, add 'sg. used w. force of pl., PMich.VIII 477.27 (ii AD)' add '**b** applied to °κρούπεζα 2, Poll. 7.87, 10.153.'
ὑποδηματοποιός, add 'IG 2².1559.48'
*ὑποδηματοπώλης, εω, ὁ, sandal-seller, IEphes. 2.29 (see also SEG 36.1011, iv BC).
+ὑποδηματουργός, ὁ, sandal-maker, JHS 22.124 (v. Robert Castabala 34, Iconium).
*ὑποδημιουργός, ὁ, official or magistrate ranking below δημιουργός, DAW 44(6).27 no. 59 (v. Robert Castabala 34, Cilicia).
*ὑποδημόσιον, τό, the sub-office of the public archives, Hierapolis 341.
ὑποδιᾱκονέω, for 'serve under another .. (loc.inc.)' read 'serve under a διάκονος, SEG 37.527 (Epirus, Rom.imp.)'
ὑποδιάκονος, before 'MAMA 3.462' insert 'subdeacon in the Chr. Church' and add 'ITyr 36, SEG 30.1701 (v AD), Cod.Just. 1.4.34.5, etc.'
ὑποδιάκων, for '= ὑποδιάκονος' read 'subdeacon (cf. °ὑποδιάκονος)' and add 'SEG 24.899, etc.'
ὑπόδικος, line 6, after 'iii BC' insert 'w. gen. of the penalty, κεφαλῆς SEG 9.8.66 (Cyrene, i AD)'; line 10, before 'ὑ. ἀσεβείας' insert 'liable to an action before'; line 11, after 'Pl.Lg. 868d' insert 'Ath.Agora XIX L4A.96 (363 BC)'

ὑποδίπλωσις, add 'applied to the scaly formation of a crocodile's skin, double layer, Sm.Jb. 41.5'
+ὑπόδουπος, understood as adj. "reverberating" by ed. in Hdn.Gr. 2.947, quoting Hes.fr. 158 M.-W. (ὕπο δοῦπος cod.).
ὑποδοχεῖον **II**, after 'socket of door-hinge' insert 'or perh. groove for metal rollers'
ὑποδοχή **IV 2**, add 'transf., ἀρθρῖτις καὶ ποδάγρα πολλῶν ἄλλων κακῶν ὑποδοχαί εἰσιν Ruf.ap.Orib. 45.30.62' **3**, for this section read 'philos., the receptacle or space in which things are created, πάσης .. γενέσεως ὑποδοχήν Pl.Ti. 49a, 51a, Plot. 2.4.1, al., ὑποδοχαὶ τῶν μαθηματικῶν εἰδῶν Iamb.Comm.Math. 3 (p. 14.9 F.), cf. ὕλην τε καὶ ὑ. ib. 4 (p. 16.20 F.; cod. ἀπο-)'
ὑποδράξ, for 'Call.Fr.anon. 63' read 'Call.frr. 194.101, 374.1 Pf.'
+ὑποδρομέω, Aeol. ὑπα-, run under, w. dat., λέπτον .. χρῶι πῦρ ὑπαδεδρόμηκεν Sapph. 31.10 L.-P.
ὑπόδρομος (A) **2**, delete the section.
ὑπόδροσος, for 'somewhat dewy' read 'wet with dew'
ὑποζεύγνῡμι **I b**, add 'to be inferior to, ὑπέζευκται .. Ὁμήρῳ AP 7.409.9 (Antip.Sid.)'
*ὑπόζω, begin to smell, PRyl.Zen. 11.24 (v. Aegyptus 14.119; cj. -ζ(ε)οντα).
ὑπόθεμα, after '**II**' insert '= ὑποθήκη II, PMich.III 173.11, al. (iii BC), PTeb. 891.5 (ii BC)'
ὑποθετικός **III**, for 'by way of suggestion' read 'by way of precept'
ὑποθήκη **I**, line 6, delete 'instructions, Cic.Att. 2.17.3' and at end, after '(Gal.) 6.405' add 'cf. ad me ab eo quasi ὑποθήκας adferes, Cic.Att. 2.17.3' **II**, add 'of the object given as security, Cod.Just. 1.3.45.12'
ὑποθηκιμαῖος, add 'pl. -αῖα, τά, mortgaged property, POxy. 2411.50 (ii AD)'
*ὑποϊεράρχης, ου, ὁ, under-hierarch, dub. in Wolters Kabirenheiligtum p. 30 no. 5a (cf. p. 79).
*ὑποϊερεύς, v. °ὑφιερεύς.
*ὑποικίδιος, ον, belonging, functioning, etc., within the house, ὑ. κιθάραι paraphrase of φόρμιγγες ὑπωρόφιαι Sch.Pi.P. 1.188.
ὑποικοδομέω, after 'IG 2².463.114' insert '(iv BC), 11(2).287A60, 61 (Delos, iii BC)'
ὑποκάθαρσις, add 'Gal. 11.85.3'
ὑποκάθημαι **I**, add '**3** to be situated below, τὰ ὑποκαθήμενα πεδία POxy. 3167.5 (AD 195/198).'
ὑποκαθίημι **I**, add '**2** secretly cause to act, suborn to act, PTeb. 820.30 (iii BC).'
*ὑποκᾰλᾰθηφόρος, ἡ, assistant woman basket-bearer, IEphes. 1072.9, cf. °καλαθηφόρος.
ὑποκατάστασις, add 'appointing of a substitute heir or legatee in the event of the original ones not accepting, Cod.Just. 1.3.52.13, Just.Nov. 1.1.3'
ὑποκαταστάτης, for 'substitute' read 'assistant °καταστάτης'
ὑποκατάστατος **I**, add '**2** substitute heir (cf. °ὑποκατάστασις), Just.Nov. 22.44.9.'
ὑπόκειμαι **I 2**, add 'to be before the audience in the theatre, Sch.Ar.Nu. 889c H.' **II 7**, after 'to be pledged or mortgaged' insert '(whether the creditor or debtor is in possession)'
ὑποκεντέω, for 'τινὰ δόρασιν' read 'γεφύρας δόρασιν'
*ὑποκῆρυξ, ῡκος, ὁ, assistant κῆρυξ, rest. in SEG 18.83.6, v. BE 1961.269 (Athens, ii AD), but dub., v. S. Follet Athènes au II e et au III e siècle, Paris 1976, pp. 280-281.
*ὑποκῐθαριστής, οῦ, ὁ, supporting lyre-player, PMich.inv. 4682.15 (Illinois Class.Stud. 3.135, ii/iii AD).
*ὑποκλείδιον, τό, perh. clasp of a belt, SEG 38.1210.21 (Miletus, ii BC).
ὑποκλέπτω **I 1**, pass., add 'w. retained acc., τοὺς εὐνὰς ὑποκλεπτομένους S.El. 114' **2**, delete the section. **II 1**, after 'keep secret' insert 'ζῆλον Nonn.D. 1.71, al.' **2**, for this section read 'evade, ζῆλον AP 5.269.5 (Agath.), ὄμμα ib. 290 (Paul.Sil.); cheat, beguile, μόχθον Nonn.D. 18.58, μερίμνας ib. 42.215'
ὑποκλύω, after 'A.R. 3.477' add '(v.l. ὑπ- for ἐπέκλυες)'
*ὑποκόκκινος, η, ον, scarlet-tinged, PGrenf. II 28.5 (ii BC, ὑποκκιν- pap.).
ὑποκόλπιος **I 2**, add 'transf., ὑποκόλπιον τοῦ χοροῦ· ὑτάσεως χῶραι αἱ ἄτιμοι Hsch.'
*ὑποκολπόω, spread round stealthily, αὐτίκα οἵ γε ἠρέμα τὰ κέρα ὑποκολπώσαντες περιβάλλουσιν ἅπαντας Agath. 3.22.6.
ὑπόκοπρος, for 'slightly faecal' read 'containing faecal matter'
*ὑποκορυφαῖος, ὁ, assistant κορυφαῖος, rest. in BGU 347 i 13, ii 11 (ii AD).
ὑπόκρημνος, delete the entry.
ὑποκρίνομαι **II 4**, add 'ὑπεκρίνατο μανίαν Ael.VH 13.12'
ὑπόκρισις **II 1**, add 'transf., imitation, ἐρεθίζομαι πρὸς αὐτά[ν] ἁλίου δελφῖνος ὑπόκρισιν Pi.fr. 140b.15 S.-M.' **4**, delete the section.
*ὑποκρύσταλλος, ον, crystalline, PWash.Univ. inv. 181, 221 (ZPE 74.85, ii/iii AD).

ὑποκυανίζω SUPPLEMENT ὑπουργία

*ὑποκυανίζω, to have a darkish blue colour, λίθος ἕτερος λάγγουρος· οὗτος ὑποκυανίζων Socr.Dion.Lith. 41.1 G.

ὑπολανθάνω, add 'II perh. go in ignorance of, πάνθ' ὑπολανθάνετε τὰ βίου συνεχῶς μυστήρια σεμνά IUrb.Rom. 1169 (iii/iv AD).'

ὑπολείβω, after 'A.Ag. 69 (anap.)' insert '(codd., edd. ἀπο-)' and for 'pass.' read 'med.'

*ὑπόλειος, ον, rather soft or smooth, μειράκιον .. ὑπόλειον Men.Sic. 201.

ὑπόληψις II 3 b, delete the section.

ὑπολογή II, add 'Hesperia 9.68.105 (= revision of IG 2².463, 307/6 BC)'

*Ὑπολυμπιδία, fem. adj. (worshipped) at the foot of Olympus, Ἀφροδείτηι Ὑπολυμπιδίαι SEG 34.630 (Maced., ii/iii AD).

†ὑπολύριος, ον, placed under the lyre, δόναξ Ar.Ra. 232, δόνακα δέ τινα ὑπολύριον οἱ κωμικοὶ ὠνόμαζον, ὡς πάλαι ἀντὶ κεράτων ὑποτιθέμενον ταῖς λύραις Poll. 4.62 (cf. h.Merc. 47ff.).

ὑπομάλακος, add 'Ptol.Tetr. 162'

ὑπομαρτῠρέω, delete the entry (v. Glotta 50.95).

ὕπομβρος, after 'Ph.Bel. 82.28, 97.27' insert 'rain-affected, damp, τὸν σῖτον PMich.inv. 3207 (ZPE 100.76, ii BC)'

ὑπομένω II 2, add 'b await in hope, ὁ γὰρ βλέπει, τίς ὑπομένει; v.l. in Ep.Rom. 8.24; abs., Lxx La. 3.21.' add 'III undertake, shoulder a task or office, SEG 18.27.3 (Attica, ii BC), SEG 31.122 (Attica, c.AD 121/2).'

ὑπομήκης, for '= ὑπόμακρος' read 'elongated'

†ὑπομηλίς, ίδος, ἡ, perh. service-berry, Sorbum torminale, Pallad.Agric. 13.4; of gold objects in imitation of the fruit, used to adorn Artemis, Didyma 432.17 (iii BC), al.

ὑπομιμνήσκω I, add '5 summon to court, Cod.Just. 3.2.4 pr., 10.11.8.5b.'

ὑπόμνημα, line 1, insert 'Dor., Aeol. ὑπόμναμα SEG 30.1122 (Entella, iii BC), 32.1243 (Cyme), etc.' I 2, for this section read 'funerary monument, SEG 33.1059, 1060 (Cyzicus, early Rom.imp.), etc., 24.1075 (Tomi, iii/iv AD).' II 4, add 'b registry of public documents, Cod.Just. 1.3.45.14, 4.21.22.7.' IV, add 'addressed to a king, SEG 13.403 (Maced., iv BC).'

ὑπομνηματίζομαι, add 'perh. also act., SEG 30.82 (Athens, c.AD 230)'

ὑπομνηματικός, add 'II subst., ὑπομνηματικός, ὁ, commentator, St.Byz. s.v. Ἄβιοι.'

ὑπομνηματισμός, lines 1/2, delete 'of a shopping list' and for 'PFreib. .. UPZ 62.12' read 'PFreib. 53'

ὑπομνηματογρᾰφέω II, add 'so act., enter in minutes, PMerton 26.16 (iii AD)'

ὑπόμνησις, add '5 notification of a summons, Cod.Just. 2.2.4 pr., 10.11.8.4.'

*ὑπομόλυβδος, ον, containing a mixture of lead, of adulterated coinage, SEG 26.72 (Athens, iv BC).

ὑπομονή II 1, add 'Lxx 4Ma. 1.11, al.; enduring to do, αἰσχρῶν ἔργων Thphr.Char. 6.1' III, for this section read 'hope, Lxx Ps. 9.19, 61.5, etc. 2 that in which one's hope is placed, Lxx Jb. 14.19, ὑπομονὴ Ἰσραὴλ κύριε Je. 14.8, 17.13.'

ὑπόμυξος, for 'somewhat charged with mucus' read 'characterized by mucus'

†ὑπομυξώδης, ες, having a mucous quality, Gal. 18(1).363 14.

ὑποναίω, add '(unless to be taken as two wds.)'

*ὑπονᾱκόρεω, serve as ὑπονακόρος, Lindos 301 (i BC).

*ὑπονᾱκόρος, ὁ, sub-warden of a temple, Lindos 295 (i BC).

ὑπόνομος II 3, add 'cf. OGI 483.163 (Pergamum, ii AD)'

ὑπονόστησις, of the Nile, add 'PBeatty Panop. 2.8, 46 (AD 300)'

ὑπόξυλος, transfer 'A.Fr. 286' to follow 'counterfeit' in section 2

ὑποοπλομάχος, for 'Hesperia 2.507' read 'in an ephebic catalogue, SEG 33.158.50'

*ὑποπαραδείκνῡμι, set down in a report or (?)sell by the procedures of παράδειξις, PTeb. 1101.11 (AD 114).

†ὑποπετρίδιος, dwelling under rocks, ὄνειροι Alcm. 1.49 P. (wrongly expld. by Hdn. 2.237 as = ὑπόπτερος).

ὑποπιθηκίζω, delete the entry (v. °πιθηκίζω).

ὑπόπιμελος, for 'somewhat fat' read 'containing fat, fatty'

ὑποπίπτω I 2, line 4, after 'Isoc. 7.12' insert 'Peripl.M.Rubr. 16'

ὑποπόδιον, add 'ὑ. διπλοῦν, a device used for beating time with the foot, cf. °βάταλον, Sch.Aeschin. 1.126'

ὑποπορφῠρίζω, add 'Epiph.Const. in Lapid.Gr. 196.23'

*ὑποπρακτικός, ή, όν, subordinate, Ptol.Tetr. 182 (s.v.l.; cf. ‡ὑποτακτικός in parallel passage in Heph.Astr. 2.19.21).

ὑποπρό II, for 'before' read 'just previously'

ὑποπτάζομαι, add 'act., suspect, POxy. 2274.8 (iii AD) as corrected in TAPhA 87.68'

ὑποπτεύω, add 'IV in mystery rites, hold the rank of °ὑπόπτης, SEG 29.799 (Samothrace, late Hellen.).'

ὑπόπτης, add 'II a rank of initiate (app. below μύστης), SEG 29.799 (Samothrace, late Hellen.); cf. ἐπόπτης II.'

ὕποπτος, add 'also ὕποπτος BE 1967.582 (Phrygia)'

ὑπόπῠος I, after 'Hp.VC 15' insert 'Nat.Hom. 12'

*ὑποπύργιον, τό, rest., name of a tax, ἀπὸ ἐπικεφ(αλαίου) κ(αὶ) ὑπο- πυρ(γίου) REG 70.120.9 (Caesarea, vi/vii AD).

ὑπόπῠρος 1, for 'with .. secret fire' read 'fiery' 2, after 'metaph.' insert 'fevered'

ὑπορθόω, add 'vines, PMich.XIII 666.17 (vi AD)'

ὑπορρᾰφή, add 'written ὑπορᾰφή Edict.Diocl. 7.48, 50, 51 (see also SEG 37.335)'

*ὑπόρραψις, εως, ἡ, = ὑπορραφή, Edict.Diocl. 7.49.

ὑπόρρυσις I 2, add 'Heras ap.Gal. 13.775.2'

ὑπορύσσω, line 2, after '(Hdt.) 6.18' insert 'Aen.Tact. 32.8, 37.7, al.'

*ὑπορώξ or ὑπορρώξ, ωγος, ἡ, perh. underground passage, prob. in Aq.Is. 2.19 (L.-R.).

ὑποσείω I, for this section read 'set in violent motion, twist back and forth, οἱ δέ τ' ἔνερθεν ὑποσσείουσιν ἱμάντι Od. 9.385. 2 cause to have a fit of trembling, Hp.Coac. 159; PMerton 59.6 (ii BC). 3 agitate, shake, v.l. in Gal. 6.481.15 (v. ὑποσήθω); ἄρτους ὑποσείων (under the nose of dogs) Ael.NA 7.13 (codd.).' II, delete the section

ὑποσήπω, after 'ib. (Ael.NA) 1.51' add 'intr., ib. 15.18'

ὑποσιώπησις, after 'silence' insert 'Ptol.Tetr. 192'

ὑπόσκληρος, add 'Gal. 13.421.17'

ὑπόσομφος 1, before 'Them.Or. 18.222d' insert 'fig. τὴν πόλιν ἀντὶ λαγαρᾶς καὶ ὑποσόμφου μεστὴν ἐποίησεν ἀγλαΐας'

ὑποσόριον, add 'SEG 30.1395 (Philadelphia)'

ὑποσπασμός, for 'drawing secretly away' read 'withdrawing (of claims)'

*ὑποσπειρίτης, ου, ὁ, plinth, base, IGChr. 10 (Panderma, iv/v AD).

*ὑποσσόομαι, contemplate (mentally), ὑποσσόμενοι πραπίδεσσιν hex. in POxy. 3535 i 16 (ii AD).

*ὑπόσταθμον, τό, base, stand, καρδόπου IG 1³.422.35 (Athens, v BC).

*ὑπόσταξις, εως, ἡ, discharge, παρακολουθεῖ δὲ ἀεὶ τοῖς τὸ ἀκόνιτον πεπωκόσιν .. ὑγρὰ ὑπόσταξις Sch.Nic.Al. 24d Ge. (cf. ἀπόσταξις).

ὑπόστᾰσις A 2, for this section read 'upward pressure, τοῦ κύματος (v.l. πνεύματος) Arist.Mete. 368ᵇ12, Hp.Off. 3' add '4 κοιλίας ὑ. constipation, Hp.Coac. 108 (v.l. ἐπι-), 295.' B II 4, add 'Φοινίκων τοὺς γείτονας προσέλαβεν εἰς τὴν αὐτὴν ὁρμήν τε καὶ ὑ. Ael.fr. 59' add 'VII sum or amount on deposit, BGU 432 ii(2).6 (AD 190; see BGU 2467).'

ὑποστάτης II, delete 'Theol.Plat. 3.7'

†ὑποστᾰτίς, ιδος, ἡ, that creates, αἰτία Procl.Theol.Plat. 3.7.

ὑποστέλλω I 2, add 'b reduce one's diet, Erasistr.ap.Gal. 11.201.7, PPrincet. 114.43 (BASP 12.77), cf. usage in section II 1.'

†ὑπόστεμα, v. ‡ὑπόστημα.

ὑποστενᾰχίζω, delete 'Δι''

ὑπόστη, after 'part of a tomb' insert 'perh. = loculus' and for 'BCH 12.280, 281' read 'SEG 16.696'

ὑπόστημα, add 'VI (written -στεμα), property in land (cf. ὑπόστασις B VI), PWürzb. 18.12 (iv AD), PMichael. 33.7 (v AD).'

ὑπόστολοι, for 'officials .. Thessaly' read 'an organized group of worshippers, esp. of the Egyptian gods' and add 'SEG 36.583 (Maced., 67/66 BC)'

ὑπόστροφος, delete 'cf. ὑπόφορος'

*ὑποστύλωσις, εως, ἡ, under- or supporting colonnade, Poliorc. 224.5.

*ὑποσφρᾱγίς, ῖδος, ἡ, app. seal at foot of document, περίστροφος· ὁ τῆς ὑποσφραγίδος τόπος Hsch.

ὑποτάκτης, add 'an ephebic official, IG 2².2051.101 (ὑβο- lapis)'

ὑποτακτικός 4, insert 'subordinate (in quot., opp. αὐθεντικός)' Heph.Astr. 2.19.21 (cf. °ὑποπρακτικός); before '-τακτικόν' prefix 'b'

ὑποτάσσω line 1, fut. ὑποταγήσομαι, for 'Cyran. 15' read '1Ep.Cor. 15.28, Cyran. 1.4.49 K.' I, add 'IG 12(7).515 (Amorgos, ii BC)'

ὑποτελής I, add '2 subst., subject (of a ruler), Cod.Just. 1.4.26.11.'

ὑποτίθημι I 1, add 'pass., to be presented to sight of audience, Sch.Ar.Nu. 1 H., cf. °ὑπόκειμαι I 2' VII, add '3 enter in one's accounts, IG 2².1228.5 (ii BC).'

ὑποτίμησις, add '3 underestimating, toning-down, ἡ γὰρ ὑ., φασίν, ἰᾶται τὰ τολμηρά Longin. 32.3.'

ὑποτίμητος, delete the entry.

*ὑπότονος, ὁ, perh. = ὑποτόναιον, threshold, Inscr.Délos 442A229 (ii BC).

ὑποτρίζω, line 1, for 'cats' read 'γαλέαι' 2, after 'of things' insert 'emit a sharp sound'

ὑπότροφος, add 'II subst., ἡ, under-nurse, in list of temple officials, Inscr.Magn. 117.10 (ii AD).'

ὑποτρύγος, add 'Gal. 19.149.17'

ὑποτρύζω, for '[ὑποτρύ]ζουσιν ἀοιδῇ .. 1219.1' read 'θεσπεσίη λάλον ὕμνον ὑποτρύζοντος ἀοιδῇ Nonn.D. 39.359, cf. 17.374'

ὑπουργέω 1, add 'b w. gen., serve, be employed in, Ἀκακίου ὑπουργοῦντος σιτικῆς ITyr 16.'

*ὑπούργησις, εως, ἡ, assistance, ICilicie 113 (AD 590).

ὑπουργία 1, for '(Theangela)' read 'JÖAI 11.71 (both Theangela,

303

iii/ii BC)' **2**, before 'pl.' insert '*work performed as a task*, Lib.*Or.* 11.89' add '**3** *job, office, Cod.Just.* 1.2.24.15.'

ὑποφέρω, line 2, after 'Il. 5.885' insert 'aor. subj. ὑποίσω *AP* 7.26.7 (Antip.Sid.)' **I 1**, add '*be borne* or *go underneath*, τῷ ἡλίῳ περὶ αὐτὴν (sc. γῆν) ὑποφερομένῳ Plu. 2.1006e' **V 2**, line 1, for '*bring .. BC* 5.6' read '*cause to fall, bring down*, App.*BC* 5.6 (in numbers); εἰς θρύψιν Lib.*Or.* 59.149'; line 3, after '(Hp.) 17.ιγ'' insert 'Arist.*Ath.* 25.1, 36.1'

*ὑποφεύξιμος, ον, *providing a way of escape*, Sch.A.R. 1.246.

ὑποφήτης, add '*BCH* 116.280.31 (Colophon, ii BC)'

ὑποφθέγγομαι 2, for '*reply*' read '*speak as if from the abdomen*' and transfer here fr. section 1 'ἐντὸς ὑ. .. Pl.*Sph.* 252c'

ὑποφοινίσσομαι, for 'pass.' read 'med.' and add 'also act., Hsch. s.v. ψαιθόν'

ὑπόφορος I, transfer 'τισι' to precede 'Plu. 2.774c'

*ὑποφυάς, άδος, ἡ, ?*undergrowth*, Hsch. s.v. μόλσον.

ὑποφυλακία, add '*TAM* 2(1).189, 284'

*ὑπόφυλλα, τά, app. *part of an olive crown*, *IG* 2².1476.11, *SEG* 38.143.18, *Ath.Asklepieion* v 29 (iii BC, s.v.ll.).

*ὑποφυλλόω, *strip off leaves at the base* of a vine, *SB* 10768.22 (iii BC).

*ὑποχειριστής, οῦ, ὁ, *administrative assistant*, *PVindob.Tandem* 22.6 (ὑποχιρ-, AD 64).

ὑποχεύς, add 'app. *some device for catching fish*, cf. ὑποχή, *POxy.* 3268.10, 3269.2 (iii AD)'

*ὑποχλωρίζω, perh. *to be yellowish*, Epiph.Const. in *Lapid.Gr.* 194.11.

*ὑπόχνοος, ον, *downy*, of a peach, Rh. 1.523.19.

*ὑποχορήγημα, ατος, τό, app. *something furnished* or *supplied*, *SEG* 30.1073.16 (Chios, 189/8 BC), unless to be divided as two wds., ὑπὸ χορηγημάτων.

†ὑποχραίνω, *make dirty underneath*, μὴ πόδες ἱμερόεντες ὑποχραίνοιντο κονίης Colluth. 232.

*ὑποχρηστεύω, *serve as* ὑποχρήστης (*assistant oracle-giver*), *Didyma* 353.18, 381.14.

*ὑπόχρῑσις, εως, ἡ, *smearing underneath*, Anon.Alch. 379.13 (ὑπόχρησιν codd.).

ὑπόχρῡσος III, delete '*gleaming* .. 1.31' add '**IV** *gleaming with gold*, μῆλα Philostr.*Im.* 1.31.'

*ὑποχρύσωμα, τό, perh. either *thin adhesive layer used as base for applying gold-leaf* or *gold paint*, *PKöln* 52.12, 60 (AD 263).

*ὑπόχρως, ῥυπόεις· ὑπόχρως Sch.Nic.*Al.* 470c Ge.

ὑποχωρέω I 1, line 6, after 'νέμεσθαι ὑποχωροῦντας' insert 'i.e. *walking backwards*' add '**4** of a coastline, *recede*, *Peripl.M.Rubr.* 12, 15, 29.'

ὑποχώρησις I 2, before '*CIG* 3705' insert 'perh. *latrine*'

*ὑποψάλτης, ου, ὁ, in Chr. worship, *subchanter, succentor*, *ITyr* 222B.

*ὑποψῡχρόομαι, *begin to grow cold*, Sch.Pi.*N.* 10.137.

ὕπτιος II, add 'ὑπτίου ποδός *of the upturned sole of the foot*, S.fr. 501.2 R.', add '**VII** ὕπτιον, τό, math., a *quadrilateral with no parallel sides*, Papp. 652.20; cf. °παρύπτιος.'

*ὑπύ, v. ὑπό.

*ὑπώρα, v. ‡ὀπώρα.

ὑπωρόφιος 2, line 3, (ὑπωρυφία), after '(Epid., iv BC)' insert 'also Aeol., *SEG* 30.1040 (Cyme, ii BC)'

ὑρῐχός, line 3, after 'συρίσκος' insert 'Poll. 7.174'

†ὕρχη, ἡ, *earthenware jar*, Ar.*Vesp.* 676, ὕρχας οἴνου Poll. 10.74 (Ar.fr. 435 K.-A.), ὕ. ταρίχου *PSI* 428.8 (iii BC), ὑπογαστρίων ὕ. ib. 84; ὕρχη· ἐφ᾽ ἧς τὰ φορτία φέρουσιν οἱ ναῦται Hsch.; perh. also τὸν (sc. λέβητα) ἐπ᾽ ἡυρϲεϰᾱι *SEG* 30.500(b) (Delphi, vi BC); Aeolic acc. to Poll. 6.14, cf. αὐτίκα τὸ ὕρχας Αἰολικὸν ὂν ψιλοῦται.

ὗς (A), add '**IV** *pudenda muliebria*, Macho 332 G.'

ὑσγῑνον, line 5, after 'ἰσγένης ib.(*Edict.Diocl.*) 24.9-12' insert 'also γισγίνης ib. 19.41A'

ὑσγῑνόσημος, after '*Edict.Diocl.* 29.36, al.' insert 'γισγινοσήμων ib. 29.10'

Ὕσπορος, add '(app. a Hellenized form of a foreign name, also Ὑπόβαρος, Plin.*HN* 37.39 (= Ctes. 450 p.500 J.))'

ὑσσωπῑτης, add 'perh. also *SEG* 30.956 (abbrev., Olbia, iii BC)'

†ὕσταριν, adv., accent uncertain, Elean equiv. of ὕστερον, *afterwards*, Schwyzer 424 (iv BC).

ὑστέρημα, add '*SB* 9257ʳ.6 (232 BC)'

*ὑστερόμητις, -ιος or -ιδος, masc. adj., *late in counsel*, Nonn.*D.* 13.540.

ὑστερόποτμος, add 'also applied to a second marriage, app. after the death of a first spouse, Hsch.'

ὕστερος A II 1, line 5, δεκάτῃ ὑ., add 'δεκάτει ὕστεραι *SEG* 32.110 (Athens, 273/2 BC)' **IV 1**, for this section read 'adv. ὕστερον, of place, *behind*, ὁπαδεῖν .. ὕ. *Trag.adesp.* 493 K.-S.; w. gen., ὕ. τῶν ἱππέων γίγνεσθαι X.*Cyr.* 5.3.42' **2 a**, add 'τῶι ὕστερον *SEG* 37.340.12-13 (Mantinea, iv BC)' **3**, ἐξ ὑστέρου, add 'Hp.*Coac.* 418, *PHels.* 10.15 (ii BC)' add '**4** of logical relationship, ὑστέρως *in a secondary* or *posterior way*, Arist.*Ph.* 195ᵃ30, cf. °πρότερος A IV 2.'

ὕφαιμος, line 5, for 'Men.*Epit.* 479' read 'Men.*Epit.* 900 S.'

ὑφαίνω II, line 8, after 'Hymn.*Is.* 14' add 'πολλοὶ γὰρ φθόνῳ δίκας οὐκ ἀδικήματός τινος .. ὑφαίνοντες *Cod.Just.* 8.10.12.7'

ὕφαλμος, after '*somewhat salt*' insert '*brackish*' and add 'Paul.Aeg. 7.3 (231.8 H.)'

ὕφᾰλος I 1, line 1, after '*under the sea*' insert 'ὕ. τείρετο, of Danae, A.fr. 47a.2.31 R. (anap.)' **II**, after '*somewhat salt*' insert '*brackish*'

ὑφάπτω I 2, add 'E.*Ba.* 778'

*ὑφέννῡμι, app. *wear as an undergarment*, pass. plpf. στάδιον ὑφέεστο χιτῶνα Call.fr. 293 Pf., cf. ὑποέστης.

ὕφεσις, add '**IV** *act of passing up to*, ἡ ἐκ τῆς ἀριστερᾶς χειρὸς ἐς τὴν δεξιὰν ὕφεσις τῶν ἀκοντίων ὀξεῖα Arr.*Tact.* 38.3.' deleting this quot. fr. section 1 1.

ὑφέσπερος, for '*AP* 5.304 (better divisim)' read '*AP* 5.305'

*ὑφηγηλάζω, *go before, guide*, Arat. 893 (tm.). .

ὑφήγησις II, add '(v. °ὑπήχησις)'

ὑφηγητικός, after '*fitted for guiding*' insert 'Alcin.*Intr.* 6 p. 158 H.'

*ὑφῆλιξ, ικος, ὁ, ἡ, *below in age*, *BE* 1974.368 (Callatis).

*ὑφιερεύς, έως, ὁ, *assistant priest*, *SEG* 16.452 (Delos, 109/8 BC), *IG* 5(2).49 (Tegea, AD 78); also ὑποϊερεύς, *IMylasa* 544.5.

ὑφίημι II, line 2, after 'c. gen.' insert 'δρόμῳ Alc. 117(b).6 L.-P.'

*ὑφιππάρχης, ου, ὁ, *deputy to a* ἱππάρχης, *SEG* 16.864 (Abu Simbel).

ὑφιστάνω, add 'Procl.*Inst.* 25 D.'

ὑφίστημι A I 4, line 2, delete '*treat as* .. 5.1.4' **B I**, add '**3** w. dat., *stand in the way* of, i.e. *block from sight*, νύκτα δὲ γαῖα τίθησιν ὑφισταμένη φάεεσσι Emp. B 48 D.-K.' **V**, add 'cf. Hp.*Coac.* 281'

ὑφόλμιον II, for '*part* .. II 5' read '*cup-shaped mouthpiece of an aulos, in which the reed was inserted*'

†ὕχηρος, ἁ, Cypr. prob. = ἐπίχειρον II 1, *wage*, gen. sg. u-ke-ro-ne ὐχέρον *ICS* 217.5, 15 (Idalion, v BC).

ὑψαύχην 2, add 'ἡ διέπεις ὄχθους ὑψαύχενας ἀκρωρείους Orph.*H.* 32.4' **3**, add 'also in prose, Him.*Or.* 18.5, 47.1, 63.6 C., Sopat.Rh. 8.188.13 W.'

*ὑψηλοποιέω, *raise, pile up*, Sch.E.*Or.* 402.

*ὑψηλόπορος, ον, *moving on high*, Hsch. s.v. ὑψιφοίτης.

ὑψηλός 1, add '**b** *on the high sea*, *Peripl.M.Rubr.* 57, cf. °ὕψος I 2. **c** ὑψηλόν, τό, perh. *long robe*, *PHarris* 109.5 (iii/iv AD), *POxy.* 2054.4 (vii AD).'

*ὑψηλοταπεινότης, ητος, ἡ, *difference between the top and the bottom*, Anon.Alch. 435.17.

*ὑψηλόφρονος, ον, *lofty-minded*, *PMag.* 5.482.

ὑψῐβίας, add 'rest. in *IG* 4²(1).129.5'

ὑψίβρομος, delete the entry (v. °ὑψίδρομος).

ὑψίγονος, add 'as pers. n., e.g. *SEG* 25.771 (Moesia, Rom.imp.)'

*ὑψίδρομος, ον, *travelling on a lofty course*, Orph.*H.* 19.1, Nonn.*D.* 38.310. **b** of water, *running at a high* or *flood level*, ib. 13.523, 23.253.

*ὑψιεπής, ές, *high-flown*, ὑψιεπής rest. in *PMich.inv.* 6 (*ZPE* 93.166, iii AD).

*ὑψικέλευθος, ου, masc. adj., *treading a lofty path*, *PMag.* 2.89.

*ὑψικέρᾱς, ᾱτος, *having lofty horns*, ὑψικέρατα πέτρας *a high-peaked rock*, Pi.fr. 325 S.-M.: acc. fem., ὑψικέραν βοῦν B. 16.22 S.-M.; also -κέρης, ητος, Choerob.in *Theod.* 1.166 H.

*ὑψικέραυνος, ον, *making lightning on high*, Hsch. (gloss defective).

†ὑψικέρης, v. °ὑψικέρας.

ὑψίκερως, delete 'metaplast. acc.' to end; v. °ὑψικέρας.

ὑψίκρημνος I, add '*Trag.adesp.* 445a K.-S.'

ὑψιπᾰγής, for '*high-built, towering*' read 'perh. *ice-capped*'

†ὑψιπόδης, ου, masc. adj., *standing to a great height*, Nonn.*D.* 20.81, 37.686.

ὑψίπολις, for 'citizen .. ἄπολις' read '*having a lofty city*'

ὑψίπους, for '*high-footed* .. *lofty*' read '*standing in the heights*, i.e. of heavenly nature'

ὑψίπυλος, after 'Il. 16.698' insert 'Ibyc. 1(a).14 P.'

*ὑψίτυπος or -τῠπος, ον, perh. *striking at a great height*, *Lyr.adesp.* 7(d).7 P.

ὑψιφᾱής, add '*BKT* 5.2 p. 143.4'

*ὑψιφρονέω, *to have proud thoughts*, Sch.Pi.*P.* 2.91.

†ὑψιχαίτᾱς, ᾱ, ὁ, perh. *having proud* or *splendid locks*, Pi.*P.* 4.172 S.-M.

ὕψος I, add '**2** *high (open) sea*, *Peripl.M.Rubr.* 33. **3** astrol. = ὕψωμα I 2, opp. βάθος, Vett.Val. 241.25.'

ὑψόω, before '*lift high*' insert '(ὑψέω in Hp.*Praec.* 7, v. infra)'

ὔω I 2, lines 6/7, for 'prayer .. earth' read '*cry at the Eleusinia*'

†ὑωνός, v. ‡υἰωνός.

Φ

φαάντερος, after 'more brilliant' insert 'Call.fr. 238.16 Pf.'

φάβα, Lat. faba, add 'BGU 2359.10 (iii AD)'

φᾱβάτον, τό, bean flour or cake, PRyl. 630. 406 (iv AD), [φ]οβάθου (sic) PPrag. i 90.9 (vi/vii AD).

*φαβρικήσιος, ὁ, armourer, φαβρικησίου καὶ δουκηναρίου SEG 26.1314, 1320 (both Sardis, iv/vi AD).

*φάβριξ, ικος, ἡ, workshop, PBeatty Panop. 1.214 (AD 298), SEG 24.911.6 (Thrace, iv AD), cf. Lat. fabrica.

*φᾰγεδαινίζω, afflict with ulcerous sores, Aq. 1Ki. 5.6, 7.10.

*φᾰγίον, τό, food, Mim.adesp. 15.34 (ed. proparox., but cf. mod. Gk. τὸ φαγί), also prob. at ODouch 34.

*φᾱγός, v. φηγός.

φᾰεινός, line 2, after 'φαεννός' insert '(pers. n. Φαβεννός SIG 422.7, Delphi, w. -β- for -F-)'

φᾰεσίμβροτος, after '(Eleusis, ii/iii AD)' insert 'of Christ, PBodm. 29.168 (iv AD)'

φαιδῐμόεις, add 'GVAK 11.2'

φαίνω, line 11, after 'Sophr. 83' insert 'Aeol. πέφαννε Alc. 206.5 L.-P.'; line 16, (fut. φανήσομαι), add 'Pl.Hp.Ma. 300d' **B II 1**, four lines fr. end delete 'ὡς' before 'ἀγαθοί'

φαιοχίτων, for '(where .. causa)' read 'app. φᾰϊο-'

*φαιώδης, ες, dark in complexion, Vit.Sapph. in POxy. 1800 fr. 1.22.

φάκη, for 'pearls before swine' read 'of an incongruous juxtaposition of the precious and the common'

φάκιον, add 'II (?)small flask, Pap.Lugd.Bat. xxv 13.12 (Byz.).'

*φακτιωνάριος, ὁ, Lat. factionarius, leader of a faction; φ. Ἀλεξανδρίας καλλιείνων leader of the Blue faction at Alexandria, PCair.Isidor. 58.13 (iv AD); Ἀμμω(νίου) το(ῦ) καλοῦ φακ(τιωναρίου) SEG 34.1562 (Egypt).

*φάκτον, τό, μέτρον παρὰ Ἀρκάσι, κοτύλαι ἀττικαὶ τρεῖς Cyr.; cf. Myc. pa-ko-to (dual), name of a vessel.

φακτονάριον, delete the entry (v. °πακτ-).

*φαλαγγικός, ὁ, soldier in a phalanx, πεζοὶ μὲν ἐν μέσοισι καὶ φαλαγγικοί Ezek.Exag. 198 S.

*φᾰλαίνιος, ὁ, resembling a whale (as nickname), Διόγνητος ὁ φ. ἐπικαλούμενος Hist.adesp. in POxy. 2399.35 (i BC).

φᾰλακρός I 1, line 4, after 'E.Cyc. 227' insert 'cf. A.fr. 47a.24 R.; τὸ -όν, S.fr. 171.3, 314.368 R.'

*Φαλερνός, ὁ, Falernian wine, Lang Ath.Agora xxi He27 (iii AD).

*φᾰλητάριον, τό, dim. of φάλης, Mim.adesp. 15.29.

φαλίπτει, add 'cf. φαλός II'

φάλλαινα II, after 'Sch.)' insert 'III ἡ ἐν τῇ κεφαλῇ θρίξ Hsch.'

*φαλλίων, ωνος, = φαλλοφόρος, Suid. s.v. Φαλῆς.

*φαλωδός, όν, singing phallic songs, Atil.Fort. p. 293.23.

*φᾱμιστός, for 'Bull. .. i AD)' read 'GVI 1861 (lapis -σθα; Leontopolis, i AD)'

φᾰνερός I 5 b, line 6, for 'rarely .. Ages. 5.7' read 'more rarely ἐν φ., Th. 4.73.2, X.Ages. 5.7, AP 12.66'

*φᾰνίζω, = φανερόω, Stud.Pal. 20.75 (iv AD).

φᾰνός, ή, όν I 1, add 'neut. pl. well-lit parts, X.Oec. 9.3'

*φαντήρ, ῆρος, ὁ, epith. of Zeus, that displays or brings to light, SEG 17.406 (Chios, iv BC).

φάος I 3, pl. eyes, add 'Call.Dian. 53, 71, 211, Nic.Al. 24, al.'

*φαοσφόρος, v. φωσφόρος.

φᾰραγγίτης, add 'epith. of (prob.) Heracles, [Ἡρακ]λῆ Φαραγγείτῃ SEG 24.1037 (Moesia, ii/iii AD)'

φαρκίς, after 'wrinkle' insert 'Simon. in POxy. 3965 fr. 27.15'

φάρμᾰκον I 1, add 'φ. δηλητήρια SIG 37A1 (Teos, v BC), 985.18 (Philadelphia), 1180.2 (Cnidus)' **4**, delete 'φ. δηλητήρια .. v BC)' at end add 'perh. Myc. pa-ma-ko'

φαρμᾰκοποσία 1, after 'Hp.' insert 'Nat.Hom. 7'

*Φαρμουθιακός, ή, όν, of the Egyptian month Φαρμοῦθι (March-April), Φ. ἐργασία POxy. 1631.13, 3354.12 (both iii AD).

φάρξις, add '= φράξις II'

φᾶρος, at end add '(form φάρϜος) Myc. pa-we-a (pl.)'

φάρος, ὁ (or ἡ), add 'II medic., lozenge, Asclep.Jun.ap.Gal. 13.97.4.'

φάρος, τό, for 'Alcm. 23.61' read 'Alcm. 1.61 P.' and for 'Antim.Eleg. p. 293 B.' read 'Antim. 119 Wy.'

*φαρσάγγιον, τό, = παρασάγγης, parasang, Cat.Cod.Astr. 7.102.30.

+φάρω, v. ‡φέρω.

*φᾱσᾱνάριος, v. °φασιανάριος.

φάσγᾰνον, after 'Pi.N. 1.52ᵇ' insert 'A.Ag. 1262'; at end add 'Myc. pa-ka-na (pl.), swords or daggers'

φασήλιον 1, after 'dim. of φάσηλος' insert '(bean)' add 'II small boat, Βερενίκης φ. ἀγωγῆς διακοσίων PRyl. 576.7 (ii BC).'

φάσηλος, line 1, for parenthesis read '(fem. in Colum. 10.377)' **II**, delete 'hence Lat. phaselus' and add 'App.BC 5.95'

*φᾱσῐᾱνάριος, ὁ, pheasant-farmer, Dig. 32.1.66; in form φασαν- IG 10(2).1.857 (iii AD), Corinth 8(3).561 (v/vi AD).

φᾱσιᾱνός, add 'also φασιανὴ θήλεια, the female bird, Edict.Diocl. 4.19'

φᾱσίολος, add 'πάσωλος Gloss.'

φάσις (B) II 4, after 'sentence' insert 'Thd.Su. 55'

φασκία, after 'Dura⁴ 93 (iii AD)' insert 'pl., PRyl. 627 ii 41, as some kind of garment, ζεῦγος φασκιῶν PKöln Ketouba 17'

*φασκιάρια, τά, fr. Lat. fascis, bundle, SEG 37.1186.43 (Takina, AD 212/213).

*φασκίδιον, τό, perh. dim. of φασκία, bandage, PWarren 18.16 (iii AD).

*φασκίς, gloss on διάφυσον, Hsch. (διάφυσος La.), also pl. φασκίδες, gloss on βασκευταί, id.

*φᾰσούλιον, = φασήλιον (in quot., sense 1); sp. φασούρ-, PMich.XIII 666.33 (vi AD), cf. mod. Gk. φασόλι, φασούλι.

φάσσα, at end for 'Luc.Sol. 7 coined a masc. form φάττος' read 'masc. φάττος suggested in a reductio ad absurdum, Luc.Sol. 7'

*φᾱταρχ-, v. °φρᾱτρῐαρχ-.

*φᾰτίζω III, before 'Pass.' insert 'give a name to, Rhian. 13'

*φᾰτις II 2, add 'AP 7.352 (?Mel.)'

φατνωματικός, for 'in form .. (cf.' read 'τὸ φ. (in form παθνω-, cf.'

*φατριασμός, v. φρατριασμός.

*φατρῖται, οἱ, members of a φ(ρ)άτρα, IG 5(2). 446.8 (Megalopolis, i BC).

φάττος, for 'φάσσα' read '‡φάσσα'

φαῦλος, line 3, add 'Hdt. 2.173.2' **II 5**, after 'Hp.Aph. 2.32' insert 'Macho 71 G.'

φαυροφόρος, for 'Call.Fr.anon. 132 .. cod. B)' read 'Suppl.Hell. 1042'

φαυστήρ, for pres. def. read 'window' and add 'cf. φωστήρ II'

φαύω, add 'Cypr. Kafizin 47'

*Φεβρουάριος, Lat. Februarius, February, also Φεβράριος, etc., Plu.Rom. 21.3, Num. 18.2, al., Lyd.Mens. 1.17, SEG 30.1212 (Rome), Mitchell N.Galatia 466, IG 14.142.

*φέγγασπις, ιδος, ἡ, having a shield of light, of Athena, poet. in PKöln 245.9 (iii AD).

φέγγος I 1 d, for 'Sosiph. 3.1' read 'Sosiph. 3.3 S.'

Φειδιακός, for 'made by Phidias' read 'made in the style of Phidias' and add 'neut. subst. JRCil. 1.271'

φείδομαι IV, line 9, after 'from doing' insert 'Call.Epigr. 1.13 Pf.'

φειδωλία II, for 'τόξου' read 'τόξων'

φελλεύς II, delete the section transferring quots. to section I.

*φελλοχᾰλαστέω, release the cork floats from fishing nets, IParion 5.10.

*φεμινάλια, τά, Lat. feminalia, trousers or leggings, Hsch. (s.v. ἀναξυρίδες; sp. φημ-); also φιμινάλια, Phot. a 1578 Th. (s.v. ἀναξυρίδας), PVindob.G 41673.2 (Tyche 1.88, vi/vii AD).

*φενέστρα, ἡ, Lat. fenestra, window, PLond. 481ᵛ.28 (iv AD).

*φενίκουλα, ἡ, Lat. *faenicula, perh. some kind of hay-implement or equipment, Edict.Diocl. 15.21.

φεννίς, add 'Eust. 1554.35'

φερέζυγος, add '2 benched, νᾶα φ[ερ]έσδυγον Alc. 249.3 L.-P. (cf. ζυγόν III 1).'

*φερεντάριοι, Lat. ferentarii, τάγμα στρατιωτικόν Hsch.

φερέοικος I, add 'of Byzas, Cadmus, Nonn.D. 3.365, 4.33'

φερεπτόλεμος, for 'Jahresh. 18 Beibl. 35' read 'GVAK 47.2'

*φερέπῡρος, ον, wheat-bearing, Didyma 496B5 (ii AD).

φερέσβιος, add 'A.fr. 204b.12 R., Ion Trag. 7 S.'

φερνά (= φερνή II), for 'portion of victim' read 'portion of offerings'

*φερνίκουλον, τό, dowry, PMich.inv. 1373 (ZPE 21.26, iii AD).

*φερνιμαία, ἡ, app. some legal process involving a dowry, SEG 33.1177 (Lycia, AD 43).

*φερνῳός, ον, of a dowry, BE 1970.512 (Lydia).

*Φερρόφαττα, v. °Περσεφόνη.

+Φερσέφασσα, v. Περσεφόνη.

+Φερσεφόνα, -όνεια, -όνειος, v. Περσεφόνη.

***Φερσοπόνη**, v. °Περσεφόνη.

***Φερσσοφάσα**, v. °Περσεφόνη.

φέρτατος I 2, (form φέριστος), delete 'mostly voc.' and add 'SEG 31.291 (Corinth, late Rom.imp.)'

†**Φερφερέτᾱς**, v. °Περφερέτας.

φέρω, line 1, φάρω, add 'EM 114.20'   **I**, line 7, after 'E.Hec. 762' insert 'med. [τὴν] ὠδεῖνα μίην γα[στρὸς] ἐνεγκαμένη Mitchell N.Galatia 149'

Φετιάλιοι, line 2, after 'sg.' insert '**Φητιᾶλις** Mon.Anc.Gr. 4.7'

Φετταλός, v. Θεσσαλός.

φηγῐνέος, read '**φηγῐνεος**' and after 'Maec.' insert 'or Maccius'

***φηγοειδής**, ές, acorn-shaped, Ath.Asklepieion IV 126.103 (iii BC).

***φηλάρρην**, ενος, ὁ, sham-male, φ. μοιχός Sen.Contr. 1.2.23.

φῆληξ, add 'Ar.fr. 541 K.-A.'

φήμη I 1, line 10, delete 'φήμη .. Hel. 820' add '**b** vehicle of a prophetic voice, oracle, E.Hel. 820, Pl.Lg. 738c.'

φημί, line 21, after 'Hdt. 3.153' insert 'Antipho 5.51, Isoc. 5.119'; line 32, after 'Pi.N. 9.43' insert 'Arc. subj. 3 sg. φάτοι SEG 37.340.21 (Mantinea, iv BC)'; line 33, for 'A.R. 2.500' read 'A.R. 1.988, 4.555   **II**, add 'w. acc., φησὶ τούσδε τοὺς στίχους Certamen 95'   **III**, line 4, after 'οὐ φημί' insert '(so οὐκέτι, οὔπω, etc. φ.)'; line 10, before 'id.(Pl.)Phdr.' insert 'Ar.Nu. 1325'   **IV**, add 'w. acc. of person, Ἑρμῆς γάρ νιν ἔφησε θεαῖς Ταθνηι προπολεύειν CEG 860 (Cnidos, vi BC)'   at end add '(form φαμί) Myc. 3 sg. pa-si'

φημίζω I 3, for 'Call.Aet. 3.1.14, 58' read 'Call.fr. 75.14, 58 Pf.'   **4**, delete 'ὁ ἐφήμισεν .. xi 3'

φήρ, line 2, after 'sg.' insert 'Alc. 286(b).3 L.-P.'

φήρεα, delete the entry.

***φήρε(ι)ος**, α, ον, belonging to satyrs, satyr-like, neut. pl. subst. swelling of glands near or beneath the ears (such as are attributed to satyrs), Hp.Epid. 6.3.6; perh. also in Gal. 19.151.

φθείρ I 1, after 'louse' insert 'or tick'   **2**, after 'lice' insert 'or other parasites'   **III**, add 'Sch.Lyc. 1383'

φθείρω II 1, after '1' insert '(cf. ἀνα-, εἰσ-, προσ-, συμπερι-)'

***φθῑνάριον**, = φθίνα II, kind of olive, SB 10727.2 (ii/iii AD).

φθῐνοπωρίς I, add 'φ. ὧραι rest. in Call.fr. 43.40 Pf. (addenda)'

φθίω II, add 'τό τε φέρον καὶ τὸ ἀλλοιοῦν καὶ τὸ αὖξον ἢ φθίνον Arist.Ph. 243ᵃ39'

Φθιώτης (s.v. Φθία), line 2, delete 'cf. E.Tr. 575' and after 'as adj.' insert 'Φθιώτας .. ναούς E.Tr. 575'

φθογγή, line 1, after 'poet. form of φθόγγος' insert '(also in late prose, e.g. Plu. 2.613e, Cat.Ma. 13.4, Crass. 23.8)'

φθοϊκός, ή, όν, consumptive, Asclep.Jun.ap.Gal. 13.101.13.

φθοΐς, before 'ὁ' insert '(also φθοῖς Eust. 1753.2)'; line 3, before 'acc.' insert 'nom. and ' and after 'φθοῖς' insert 'Clem.Al.Protr. 2.19'

φθόνος I 1 b, add 'AP 7.117 (Zenodotus)'

φθορία, delete the entry.

-φῐ, -φῐν, delete the article.

φιάλη, after lemma insert 'non-Att.-Ion. φιάλᾱ, Cypr. pi-a-la, ICS 177 (vi BC)'   **I 2**, line 2, after 'libations' insert 'Sapph. 44.29 L.-P.' at end for 'The form φιέλη .. p. 389 P.' read 'forms: φιέλη IMylasa 301.16, ILabraunda 92.1-2, cf. Moer. p. 389 P.; φιhάλα, φιhέλα Myc. pi-a₂-ra, pi-je-ra₃ (pl.)'

φιᾰληφόρος, after 'Locrian priestess' insert 'an office originally held by a boy, ὁ φιαληφόρος' and add 'fem., pl., at Athens in the cult of the Great Mother, IG 2².1328.10 (ii BC)'

φιᾱλιον, add 'also φιέλιον IMylasa 897 (in EA 13.9)'

φιάλλω, add 'cf. ἐφίαλεν· ἐπεχείρησεν and ἠφίαλεν· ἐπεχείρησεν Hsch.'

φίβλα, delete 'Suppl.Epigr. 2.776 (Dura)'

***φιβλᾱτώριον**, τό, cloak fastened by a fibula, Edict.Diocl. 19.65, al., φιβουλατώριον ib. 19.24, al.; φιβλατώριον περιβόλαιον Περσικόν Suid., cf. °ἰμφειβλατώριον.

***φίβλον**, τό, equiv. of Lat. fibula, SEG 7.371 (Dura, ii AD), 2.776 (iii AD).

***φιδάκνιον**, v. ‡πιθάκνιον.

***φιδεικομ(μ)ισσάριος**, α, ον, Lat. fideicommissarius, ἐπιστολή PMasp. 151.54, 312.25 (both vi AD).

***φιδεικόμμισσον**, τό, Lat. fideicommissum, Just.Const. Δέδωκεν 6; sp. φιδικ- Cod.Just. 6.48.1.1.

***φίδνα**· ῥίζα ἡ Ἀχίλλειος καλουμένη Hsch.

***φίκατι**, v. °εἴκοσι.

φικιδίζω, delete the entry (v. Aristopho 3 K.-A.).

***φίκις**, ιδος, ἡ, app. buttocks, POxy. 3070.5, perh. PHeid. 190 fr. 1.75, v. ZPE 30.36 (for accent: Hdn.Gr. 1.88.35 codd. [Κίκις L.]).

†**φικιῶ**, desiderat. fr. same stem as φίκις, Suid. (v. ZPE 52.56).

***φικοπήδᾱλος**, for 'dub. sens.' read 'having a rudder in the shape of a fig-leaf' and add 'PCair.inv. 10580/10488 (ZPE 20.158)'

***φῐλᾰβάρσακος**, ὁ, friend of (the town) Abarsacus, IApam. 103.1 (Pylai).

***φῐλάγγελοι**, οἱ, members of a religious (app. Chr.) society, Φιλανγέλων συνβίωσις SEG 31.1130 (Phrygia, iii AD).

***φίλαγρίππας**, ου, ὁ, member of a cult of Agrippa, Robert Hell. 11/12.226.

φῐλᾰδέλφεια, add 'neut. sg. also of a building in commemoration of Φ., Inscr.Délos 400.38 (ii BC)'

φῐλᾰδελφία, add 'ICilicie 16 (AD 209/11)'

φῐλᾰδελφία, add 'epith. of the Nymph at Kafizin, Kafizin 300 (225/218 BC); fem. -α IPamph. 154'

***φῐλᾰδελφοσύνη**, ἡ, love of brother or sister, An.Boiss. 4.408.

φῐλᾰθήναιος, add 'Isoc.Ep. 5.2'

†**φῐλάκανθις** (cod. -θίς), ιδος, fem. adj. bony, of fish, AP 6.304 (Phan.).

***φιλάκουον**, τό, Egyptian name for κληματῖτις, Ps.-Dsc. 4.180.

φῐλάκρᾱτος, line 2, for 'Simon. 183.5' read 'AP 7.24.5 ([Simon.])'

***φῐλαλληλία**, after 'mutual love' insert 'Diog.Oen.fr. 21.8 S.'

φῐλαναγνώστης, after 'of reading' insert 'D.S. 1.77.1'

φῐλανδρία I, add 'TAM 4(1).124'   **II**, for this section read 'excessive desire (of women) for men, E.Andr. 229, cf. Hermog.Id. 2.5'

φῐλανθρωπεύομαι I 1, add 'abs., Men.Dysc. 573'

φῐλάνθρωπος I 1, add 'as an honorific description, SEG 31.901 (Caria, i AD)'   **2**, add 'of Asclepius, SEG 37.1019 (Pergamum, ii AD)'

φῐλάοιδος, line 1, delete 'or singers'; line 2, after 'musical' insert 'φιλάοιδον λιγύραν χελύνναν Sapph. 58.12 L.-P.' and after '(Antip.Sid.)' insert 'Nonn.D. 1.415'

***φῐλᾰπᾰμεύς**, έως, ὁ, friend of Apamea, τὸ συνέδριον τῶν φιλαπαμέων IApam. T 8 (Perinthos).

φῐλᾰπεχθημοσύνη, after 'D. 54.37' insert 'Isoc. 15.315'

φῐλᾰπεχθήμων, after 'D. 24.6' insert 'of words' and transfer 'Isoc. 8.65' to follow, adding 'id. 12.249, al.'

φίλαυλος, for 'flute' read 'aulos'

φῐλᾰχαιός, add 'cf. pers. n. Φιλάχαιος IG 5(2).159 (Tegea, Lacon. dial., v BC)'

φῐλεκᾱγᾰθία, delete the entry.

***φῐλέννομος**, ον, loving those who keep the law, φιλόθεος φ. Χριστοῦ MAMA 1.237 (Laodicea Combusta).

***φῐλεξᾰπάτης**, ου, ὁ, ἡ, fond of deceit, AP 5.164 (Asclep.).

φῐλέρημος, line 3, for 'AP 5.8' read 'AP 5.9'

φῐλέρως, after 'amorous' insert 'AP 5.206 (Leon.Tarent.), Nic.fr. 16 G.-S.'

†**φῐλεύνος**, ον, libidinous, Anacreont. 1.7 W.   **2** devoted to one's marriage(-bed), SB 10162(546).2 (iii/iv AD).

φῐλεύτακτος, for 'devoted to discipline' read 'well disciplined'

***φῐλεφέσιος**, ὁ, loving the Ephesians, IEphes. 1381a.11, 1545.6.

***φῐληγορία**, ἡ, poet. -ίη, friendly speech, GVI 1864 (Athens, ii/iii AD).

φῐληδέω, at end after 'c. part.' insert 'φ. διέλκων Ar.l.c.'

Φιλήσιος I, add 'Arr.Peripl.M.Eux. 2.2'

φῑλήτης, line 8, for 'Call.Hec. 1.4.11' read 'Call.fr. 260.65 Pf.'; line 11, after 'the papyri of' insert 'Hippon. 79.10, 102.12 W.'

φῐλητός I, add 'γυνὴ φ. IG 12(3).910; sup. -τότατος GVI 1899.2 (Pisidia, ii/iii AD)'

φίλητρον, delete 'f.l. .. (Crates)'

φῐλήτωρ I 1, for '(Cretan)' read '(Cretan acc. to Str. l.c.), Call.fr. 23.4 Pf.'   **2**, delete '(τῷδε .. darling)'

φῐλιᾱκός 1, add 'PRyl. 28.99 (iv AD)'

φῐλικός, add '**2** φ. μέλος a love-song, Theoc. 10.22.'

φίλιος I 2, add 'of the Muses, Mitchell N.Galatia 146'

Φιλίππειος I, add 'epith. of Zeus, IG 12(2).526 (Eresus, end iv BC)'

φίλιππος, add '?Alcm. 168 P., B. 3.69 S.-M.'

***φῐλοβάναυσος**, ον, fond of vulgarity, φιλοβαναύσικους, φιλοβαναύσους, ἀνευφράντους Heph.Astr. 2.15.12.

***φῐλοβάσανος**, ον, fond of torment, Ptol.Tetr. 161.

φῐλοβάσκανος, delete the entry (v. °φιλοβάσανος).

***φῐλόβροτος**, ον, fond of mortals, MAMA 7.582; cf. φιλόμβροτος.

†**φῐλογέρων**, ό, loyal to the Elders (γερουσία), Hesperia suppl. 6.163 no. 52.5 (Apamea, ii — IGRom 4.783).

***φῐλόγῠνος**, ον, = φιλογύνης, Lys.fr. 122 S.

φῐλοδέμνιος, for 'loving the bed' read 'libidinous'

***φῐλοδεσποτεύομαι**, love acting as a despot, Anaxil. 42 K.-A.

***Φῐλοδιόνῡσοι**, οἱ, friends of Dionysus, name of an association, Didyma 502.1 (ii BC).

φῐλόδοξος, line 2, delete 'Pl.R. 480a'   add '**2** fond of (mere) belief, (opp. true knowledge), Pl.R. 480a.'

φιλοζέφυρος, add 'as epith. of Arsinoe, Hedylus ap.Ath. 11.497d'

φῐλόζωος I 2 b, delete the section transferring quots. to section 2 a.

***φῐλοθάλᾰμος**, ον, loving the bridal-chamber, epith. of Aphrodite, Ps.-Callisth. 12.16.

***φῐλοθάλασσος**, loving the sea, OA 6.46 (Cyprus, iii/iv AD); also -θάλαττος, Cat.Cod.Astr. 7.201.1.

φῐλόθεος, add 'as honorific epith., SEG 32.501 (Thespiae, i BC/i AD), TAM 5(1).457'

φῐλοθύτης, add 'Philostr.VA 4.19, 5.21'

*φῐλοΐδιος, α, ον, *loving one's own*, in quot., of a good wife, *DAW* 1896(6) no. 178.

*φῐλοικος, ον, *loving one's home*, τρόπον φίλανδρον [καὶ φίλ]οικον *SEG* 29.1199 (Lydia, AD 212).

φῐλοκάθᾰρος, delete 'Ptol.*Tetr.* 63'

φῐλοκᾰλέω **5**, pass., add '*SEG* 31.1476 (Arabia, vi AD)'

φῐλοκέρτομος, add '*Trag.adesp.* 365a K.-S.'

φῐλοκτίστης, add '*SEG* 34.1053 (Caria, vi AD), ib. 36.1341 (Caesarea Maritima, vi AD)'

φῐλοκύνηγος, add 'as member of a guild of huntsmen' and transfer '*Supp.Epigr.* 3.499' to follow, adding '*SEG* 32.1218 (Lydia, iii AD)'

φῐλοκωθωνιστής, delete 'f.l. for κωθωνιστής' and add '(cf. °κωθωνιστής)'

φῐλόκωμος, for 'Simon. 183.5' read '*AP* 7.24.5 ([Simon.])'

φῐλόλᾱος, add '**2** *loving his people* (in epitaph of a Jew), *CIJud.* 1.203.'

*φῐλολβος, *loving prosperity*, Εἰρήνα *Hymn.Curet.* 40.

*φῐλόλεκτρος, ον, *devoted to one's marriage* (*bed*), *IUrb.Rom.* 1284 (ii/iii AD).

φῐλόλογος **II 2**, add 'as honorific description in an epitaph, *TAM* 4(1).232 (Rom.imp.)'

φῐλολοίδορος, add 'φιλολοιδόροιο γλώττης *Anacreont.* 42.11 W.'

φῐλολουτρέω, add '*Gal.* 11.34.7'

φῐλόλουτρος, add '*Gal.* 11.137.8'

*φῐλολύκιος, ον, *friend of the Lycians*, *BCH* 83.498.24 (i AD).

φῐλομᾰλακος, after 'Ptol.*Tetr.* 162' add '(v.l. °ὑπομάλακος)'

φῐλόμβροτος, add 'also Φιλόμροτος pers. n., *SEG* 24.405 (Thessaly, v BC)'

φῐλομέτριος, for '*loving moderation*' read '*friend of the poor*'

φῐλόμολπος, after 'song' insert '*Stesich.* 16.10 P.'; add '*Call.Del.* 197'

+φῐλομόνᾰχος, v. °φιλομονόμαχος.

*φῐλομονόμᾰχος, ον, *fond of individual combats*, Ptol.*Tetr.* 180, *Cat.Cod.Astr.* 8(2).86 (cod. φιλομόναχος).

φῐλόμωμος, after 'Ptol.*Tetr.* 162' insert '(v.l. °ἐπίμωμος)'

φῐλονῑκέω **1**, add '**b** pass., of words, *to be uttered contentiously*, Pl.*Lg.* 907c.'

φῐλονῑκία **1**, line 2, after 'in bad sense' insert 'Simon. 36.11 P. (pl.)'

φῐλοξενία, after 'hospitality' insert 'Thgn. 1358, B. 3.16 S.-M.'

φῐλόπαις **I**, for 'Simon. 183.6' read '*AP* 7.24.6 ([Simon.])'

φῐλόπαππος, delete '-παπποι θεοί .. 176' (v. *ASAE* 19.49)

φῐλοπᾰτωρ, add 'of Cleopatra, *SEG* 24.1217 (Egypt, 39 BC)'; at end add 'cf. Myc. *pi-ro-pa-ta-ra*, fem. pers. n. (*Φιλοπάτρα)'

*φῐλοπένης, ητος, ὁ, ἡ, *loving the poor*, *CIJud.* 1.203 (Rome).

φῐλοπευστέω, add '*PRyl.* 624.10'

*φῐλόπιλος, v. °φιλόφιλος.

*φῐλόπιστος, ον, app. *lover of the faithful*, Mitchell *N.Galatia* 271.

*φῐλοπλία, ἡ, perh. *association of young men supporting gladiators*, *IEphes.* 2226.3.

φῐλόπλος, delete '*Ephes.* .. 70' add '**b** masc. subst., member of a °φιλοπλία, *IEphes.* 3055, 3070.'

*φῐλοποίμνιος, ον, *loving the flock*, κύων Theoc. 5.106.

φῐλόπολις **II**, add 'as a complimentary term in an epitaph, *SEG* 33.1087 (Bithynia, AD 215/217)'

φῐλοπολίτης, add 'as an honorific term in an inscr., *SEG* 31.901, 913 (Aphrodisias, i AD), *IEphes.* 1390.3'

*φῐλοπραγμᾰτία, ἡ, *meddlesomeness*, *POxy.* 2267.6 (iv AD).

+φῐλορήτωρ, ορος, ὁ, ἡ, *fond of orators and oratory*, Cic.*Att.* 1.13.5, Phld.*Rh.* 2.218.15 S.

*φῐλόρκειος, ον, *ready to make vows*, *Kafizin* 102.

φῐλορώμαιος, after '*friend to the Romans*' insert 'Mitchell *N.Galatia* 188 (43/40 BC)'

φῐλος, delete the entry.

φῐλος, line 1, delete 'also ος, ον, Pi.*O.* 2.93' (but cf. *ZPE* 35.264-5)   **I**, line 8, after 'Ar.*Nu.* 1168 (lyr.)' insert 'also Theoc., Bion; in prose, Longin. 6'; lines 19/20, for '(in bad sense, *Lac.* 2.13)' read 'φίλος *Ἡρᾶς SEG* 32.847.B55 (graffito, Thasos, iv BC)'   add '**e** as a member of an association, *comrade, fellow*, *SEG* 29.1188, 1195 (Lydia, ii AD), *TAM* 5(1).93 (iii AD).'   **IV 1**, add 'sup. as masc. or fem. pers. n., cf. Myc. *pi-ri-ta*'   **4**, for '*Call.Fr.* 146' read 'as pers. n. Φιλοτέρα *Call.fr.* 228.43 Pf.' and add 'Φιλοτέρα *SEG* 31.526 (Boeotia, iv BC)'

φῐλοσάρᾱπις, add 'also **φιλοσέραπις** (dat. -πι) *IG* 12(5).712.25 (Syros)'

φῐλοσοφέω **II 3**, add 'Macho 374 G.'

φῐλοσοφία **3**, add 'Hp.*VM* 20'

φῐλόσοφος **I 1**, add 'perh. as official title, *IG* 2².791.30 (Attica, iii BC), Λ. Φλ. Ἀρριανὸ[ν] ὑπατικὸν φιλό[σο]φο[ν] *SEG* 30.159 (Attica, ii AD)'

φῐλοσυγγενής, after 'p.56 A' insert '*TAM* 5(1).12.12 (AD 80/81).

φῐλοτεκνία, add '*ISelge* 15.9-10 (Rom.imp.)'

φῐλότεκνος, add '*SEG* 28.1493 (Egypt, i/ii AD), *ISelge* 15.21 (Rom.imp.)'

φῐλότεχνος, line 2, after 'Ath. 15.700c, etc.' insert 'of a smith, *SEG* 31.1284 (Antioch in Pisidia)'

φῐλότης **2**, after 'Il. 3.73, cf. 94, 323' insert 'ἐπὶ φιλότατι *Olympia Bericht* 7.207 (Sybaris at Olympia, vi BC)'

φῐλοτήσιος, line 1, for 'Dor.' read 'dial.'   **II**, add 'also (?)neut. φιλοτάσιον *CEG* 445 (Boeotia, vi BC); φιλ]οτέσιον Lang *Ath.Agora* XXI C6 (*c*.500 BC)'

φῐλοτῑμέομαι **III**, add 'Θεῷ Πρειέτῳ Δημόφιλος .. τὸν βωμὸν ἐφιλοτειμησάμην *TAM* 4(1).77'

φῐλοτῑμητέον, add '**b** *one must contend*, Gal. 8.553.16.'

φῐλοτῑμία **I 4**, add '**b** *fund of staple commodities* or *money* out of which disbursements were made to professionals and craftsmen as honoraria, *PLond.* 1305 (*Tyche* 5.63ff.; vi AD).   **c** *donation, bounty*, *Cod.Just.* 1.3.44.3, Just.*Nov.* 22.23.'

*φῐλότιμος, ον, = φιλότιμος, only as adv. -ίως *Inscr.Cret.* 4.168.10 (iii BC, unless fr. φιλότιμος, w. analogical adv. suffix).

φῐλότῑμος **I 2**, add 'φ. τῆς δοχῆς Macho 106 G.'   **II**, comp., -ότερον, add '*SEG* 32.794.11 (Olbia, 325 BC)'

φῐλότρῠφος, delete the entry.

*φῐλοφαρές, τό, name for the plant πράσιον, Dsc. 3.105.1.

φῐλόφῐλος, add 'written φιλόπιλε (voc.) *SEG* 30.1769 (inscr., Egypt, ii/iv AD)'

φῐλόψῑλος, for 'Alcm. 152' read 'Alcm. 32 P.' and after 'ψιλεύς' add '(so Phot., Suid., but perh. error for φιλόψιλος = *φιλόπτιλος, *loving down*)'

φῐλοψῡχέω, line 3, for 'φ. ὑπὲρ τῆς ἀρετῆς' read 'ὑπὲρ δὲ τῆς ἀρετῆς οὐ φιλοψυχήσαντες'

φίλτᾰτος, add '*SEG* 30.1470 (Phrygia, iii AD), *TAM* 4(1).352 (Chr.)'

φιλτροκατάδεσμος, for '*PMag.Par.* 1.296' read '*PMag.* 4.296, *Suppl.Mag.* 38.8 (ii AD)'

*φιλτροπόσιμος, ον, *having the qualities of a love-potion*, Cyran. 3.37.23 K., al.; neut. as subst., *love potion*, ib. 1.22.29, al.

φῐλυπόστροφος, add 'in non-medical context, *Corp.Herm.fr.* 23.67 N.-F.'

φῐλῳδός, add 'cf. φιλάοιδος'

*φῑμά, ά, = φιμός, Hsch.

*φιμινάλια, τά, v. °φεμινάλια.

φῑμός **III**, add '*qui .. mitteret in phimum talos* Hor.*Sat.* 2.7.17'

φῑμόω, line 4, after 'Ev.Matt. 22.34' insert 'by a spell, *Tab.Defix.Aud.* 22.42 (Cyprus), al.; τὰ στόματα πάντων ib. 15.24 (Syria)'

φίμωσις **II**, for '*stopping up an orifice*' read '*contraction of an orifice* (so that it cannot be opened)'

φῑμωτικός, after 'silencing' insert 'παραθήκην φ. τινος *Tab.Defix.Aud.* 22.39, 32.27 (both Cyprus)'

-φῐν, delete the entry.

*φίντων, ωνος, ὁ, *beloved*, (cf. φίντατος, pers. n. Φιντεία, Φιντίας), A.*fr.* 47a.802 R.

φίσκος **II**, add '*Just.Nov.* 117.8.2; perh. of other treasuries, *SEG* 29.980 (Italy, late Rom.imp.)'

φῐτρός **I**, for '*Call.Fr.* 246 (= *PSI* 11.1218*a*2)' read '*Call.fr.* 177.2, 785 Pf.'

φῑτυ, before 'Ar.' insert 'S.*fr.* 889 R.'

*φλᾰγέλλα, ἡ, Lat. *flagella*, *whip*, φλαγέλλας μαστιγωθῆναι *POxy.* 2339.10 (i AD), see also φραγέλλα.

φλαγέλλιον, add 'see also ‡φραγέλλιον'

*φλάγελλον, τό, Lat. *flagellum*, *whip, lash*, *Edict.Diocl.* 10.3, 18.

φλᾱμέντας, lines 1/2, for '*IG* .. φλάμινα' read 'gen. φλαμένος *IGRom.* 3.1332 (Bostra); acc. φλαμίνα'; after '(Pisidia)' insert 'φλαμ. (abbrev.) *IG* 2².5206'; add 'φλαμένιος *AE* (1934/5) παρ. 15 (Thebes)'

φλαῦρος, line 1, after 'first in Sol. 13.15' insert '(dub. in Alc. 59(*a*) L.-P.)'

φλάω, line 2, for '(Ahrens, φλασῶ codd.)' read '(v.l. φλασῶ)'; line 4, for '(Ahrens, φλάσαιμι codd.)' read '(v.l. φλάσαιμι)'

φλεβοτομέω, pass., add '*Inscr.Perg.* 8(3).139 (Rom.imp.)'

φλεγμαίνω **II 3**, add 'of literary style, παχὺ καὶ φλεγμαῖνον Eust. 285.27'

φλεγμᾰτόεις, add '(= *Suppl.Hell.* 1116)'

Φλεύς, after '(Hdn.Gr.) 2.911' insert 'dat. Φλεῖ *IEryth.* 207.61 (ii BC)'

φλέψ **3**, add '**b** *vein in a stone*, Epiph.Const. in *Lapid.Gr.* 196.2.'   add '**4** of the caverns of the underworld, βυθίων φλέβα πᾶσαν ἐναύλων Nonn.*D.* 36.103.'

φληνᾰφία, add 'perh. φληναφεῖ[α *POxy.* 2802.3'

φλήνᾰφος **I**, transfer 'Amelius ap.Porph.*Plot.* 17' to section II.

φλήνᾰφώδης, add 'Sch.Hes.*Op.* 160'

φλῐά, line 1, delete '(later .. Oropus, i BC)' (τὰ στεφάρια is now read, v. °στεφάριον)   **1**, line 3, (in sg.), after 'Theoc. 23.18' insert '*Call.Epigr.* 42.6 Pf.'; line 4, (παρὰ φλιῇ), for '*Call.Iamb.* 1.220' read '*Call.fr.* 194.24 Pf.' and add 'ἐπὶ φλιῆς ib. 91'   **2**, transfer pres. quots. to section 1 and for this section read 'used to translate Hebr.

wd. usu. understood as "lintel", Lxx *Ex.* 12.7, 22, 23' at end add 'Myc. *pi-ri-ja-o* (gen. pl.)'

Φλιάσιος, for 'cf. Φλυήσιος' read 'see also °Φλυήσιος'

φλόγινος, add '**III** φλόγινος (sc. λίθος), ή, a precious stone, Plin.*HN* 37.179.'

φλογίον, add 'ostr. in *BASP* 23.28'

*φλογοῦχος, ὁ, *flame-holder, SEG* 16.741 (Pergamum, Rom.imp.).

*φλογοφόρος, ον, *flashing*, of the moonstone, *Physici* 2.204.16.

φλόγωσις 1, add 'pl., Hld. 8.11'

*Φλοιά, ά, Lacon. name for Persephone, Hsch.

*Φλοιάσιος, v. Φλιάσιος.

φλοιός, line 2, for 'Call.*Fr.* 101' read 'Call.*fr.* 73 Pf. (cj.)'

φλόξ I 3, for 'of the *heat* of the sun' read 'of the *fire* of the sun' and after 'S.*Tr.* 696' insert 'Simon. 76.3 P., Orph.*H.* 69.10'; for 'the *blade* of a sword' read 'of a flashing sword' **II**, add '*AP* 4.1.51 (Mel.)'

φλόος, line 4, for 'cf.' read '*bast*'; line 5, delete 'the *slough* of' and for 'φλους' read 'φλοῦς'

*φλουμάρης, v. °πλουμάριος.

†φλουμαρικός, v. °πλουμαρικός.

φλυᾱρία, add 'applied to a libellous statement, *Cod.Just.* 1.5.12.17'

φλυζάκιον, delete 'cf. φυσάκια'

*Φλυήσιος, ὁ, title of Hermes, Hsch., cf. τὸν Φλυησίων Ἑρμῆν Hippon. 47 W. and ‡Φλιάσιος.

φλύω, line 8, for 'φλύζειν' read 'φλύζων'

*φοβερόμματος, ον, *having terrible eyes*, voc., *PMag.* 5.437, Bonner *Magical Amulets* 168.

*φοβερόφθαλμος, ον, *having terrible eyes*, gloss on γοργῶπιν (acc.), Hsch.

*φοβεσάνωρ, νορος, ὁ, ἡ, *man- or warrior-scaring*, θυμός *GVI* 1918 (Crete, ii/i BC).

φοβέω A I, add 'Th. 7.30.1'

φόβος II 1, line 9, after '*Act.Ap.* 9.31' insert 'of a woman for her husband, *SEG* 35.1427.5 (pl., Side, iii AD)'

φοιβάω I 1, after '*purify*' insert 'E.*Phaëth.* 57 D.'

Φοίβη, of Artemis, add '*SEG* 32.1068 (Italy, c.AD 150)'

*Φοιβίη, ή, (Ion.) title of three-headed female deity, perh. Dione as Hecate, *SEG* 34.1436 (Syria, c.iii AD).

φοῖβος I, add 'comp. φοιβότερος, orac. in *SEG* 33.1056.C6 (Cyzicus, ii AD)'

*φοιδερατικός, ή, όν, *connected with a treaty* or *alliance*, Just.*Nov.* 148.2, al.

*φοιδερᾶτος, ὁ, Lat. *foederatus, bound* (to Rome) *by a treaty, Cod.Just.* 1.5.12.17, 12.37.19.2, Just.*Nov.* 117.11.

*φοῖδες, v. φωῖς.

*φοινῑκάζω, *write, act as scribe*, ποινικάζεν (inf.) *SEG* 27.631.A5 (Lyttos, Crete, c.500 BC), cf. ἐκφοινίσσω, °φοινικαστάς, °φοινικογραφέω.

Φοινῑκαῖος, add '*SEG* 26.704.5, 37.510 (both Dodona, ii BC), *IG* 2².951 add. (Ambracia, ii BC), etc., cf. °Φοινίκη.'

*φοινῑκαστάς, ᾶ, ὁ, *scribe*, ποινικαστάν (acc.) *SEG* 27.631.A11, B1, al. (Lyttos, Crete, c.500 BC), cf. °φοινικιστής.

Φοινίκη, line 4, for 'metaph. .. Diehl²' read 'perh. w. ref. to blushing, ..]νικεω αἰδώς Erinn. in *Suppl.Hell.* 401.34'

Φοίνκη, add **IV** 'title of Athena at Corinth, Lyc. 658 and Sch.'

*φοινικηγός, όν, wd. of uncertain meaning in list of building supplies, *BGU* 2361a ii 4 (iv AD) (φυνικηγός pap.).

φοινικήϊος II, delete 'γράμματα' and 'Φ. alone'

Φοινικίζω, for 'of unnatural vice' read 'of cunnilingus'

Φοινικικός, after 'ή, όν' insert '(freq. var. **Φοινῑκός**)'

φοινίκινος I a, line 3, after 'Id. 8.2' insert 'of honey, *Edict.Diocl.* 3.12' add '**III** = φοινίκεος, red, Hsch. s.v. φοίνικι φαεινόν.'

φοινίκιον III, for this section read 'item in a list of medical supplies, *Inscr.Cret.* 4.145 (Gortyn, ?v/iv BC)' add 'Myc. *po-ni-ki-jo*, name of a spice'

φοινίκιος I, delete 'φ. οἶνος' to end of section add 'Myc. *po-ni-ki-ja* (fem.), of chariots, *crimson* or *of palm-wood*'

φοινικιστής II, for '*wearer of purple .. rank*' read 'perh. *secretary, scribe*, cf. °φοινικαστάς'

*Φοινῑκῖτις, ιδος, ἡ, a precious stone, Plin.*HN* 37.180, 188.

*φοινικογραφέω, *act as secretary* or *scribe*, *SEG* 31.985D (Teos, v BC), cf. °φοινικάζω, φοινικογράφος.

*φοινικοεάνος, ον, *purple-robed*, φοινικοεάνων .. Ὠρᾶν Pi.*fr.* 75.14 S.-M. (rest.).

φοινικοπάρῃος, add 'Dor. φοινικοπάραος *Suppl.Hell.* 991.34 (poet. word-list, iii BC)'

*φοινικόπαστιλλος, ὁ, (?)*red dye from palm-tree*, Anon.Alch. 346.10.

*φοινικόπρωρος, ον, *having a red prow, Suppl.Hell.* 991.35 (poet. word-list, iii BC).

φοινικόπτερυξ, read **φοινῑκοπτέρυξ** and for 'Lyr. in *Mitteil.* .. 139' read '*Lyr.adesp.* 11(*f*).2 P.'

*Φοινικός, v. °Φοινικικός.

*φοινικόχλοος· ξανθόχλοος Hsch.

Φοῖνιξ A I 2, after 'Od. 15.417' insert 'of Europa, as daughter of Phoenix, B. 17.54 S.-M.'; last line, for 'ἄμπεχος' read 'ἄμπελος' and add 'cf. Hsch.' **B I 2**, line 2, after 'Pi.' insert 'B.'; line 4, after 'P. 1.24' insert 'B. 18.56 S.-M.' **IV**, for 'like a *guitar*' read 'type of lyre' and add 'Alc.ap.Cyr. in cod. Matrit. (*GRBS* 9.272)' **X**, for 'εὐρύνοτος' read 'εὐρόνοτος a south-easterly wind' **XII**, add 'Asclep.Jun.ap.Gal. 12.776.4, 10' at end add 'Myc. *po-ni-ke* (dat.), *po-ni-ki-pi* (instr. pl.), prob. *palm-tree*'

φοίνιος, at end for 'Rare in Com.' read 'com. only paratrag.'

φοινίσσω I 1, lines 4/5, delete '*empurple*, μόρον S.*Fr.* 395' and transfer ref. to section II.

φοινός 1, add 'app. masc. subst., *red poppy*, Lang *Ath.Agora* xxi B19'

φοῖς, read '**φοΐς**' and for 'φώς' read 'φωΐς'

φοιτᾱλέος, after 'E.*Or.* 327 (lyr.)' insert 'φοιτᾱλ- A.*Pr.* 598, E. l.c.'; after '*roaming wildly about*' insert 'Euph.*fr.* 97 vGr.'

φοιτάω, line 1, after '(Hdt.) 7.126' insert 'Call.*fr.* 194.32, 202.67 Pf. (v. Add. in vol. II)' **II**, add '**2** of decrees, etc., *to be issued, Cod.Just.* 1.3.52.11, 1.3.55.2.'

*φοιτητικός, ή, όν, *wandering, moving*, Sch.E.*Ph.* 1024.

φοῖτος, add 'Sch.A.R. 4.55'

φόλετρον, add '*PMich.*xv 741.5, 11 (vi AD)'

*φολιᾶτον, τό, Lat. *foliatum*, a perfumed oil made from aromatic leaves, *Edict.Diocl.* 36.88; see also φουλιᾶτα.

*φολιᾶτος, ον, Lat. *foliatus, bearing foliage*, ἀρσενίκιον Anon.Alch. 318.7.

φολίς, line 3, after 'Epic.ap.Sch.Nic.*Th.* 257' insert '(= *Suppl.Hell.* 1166)'

*φολλατώριον, τό, Lat. *fullatorium, fuller's shop*, *PLond.* 191.5 (ii AD).

φόλλις II, add '**2** sum of money equal to 12,500 denarii, *PBeatty Panop.* 2.302 (AD 300).'

φονεύς, add 'see also φονής'

φονεύω 1, line 4, after 'cf. *El.* 34' insert 'οὐ φονεύσεις Lxx *Ex.* 20.15' **2**, after 'of an animal' insert 'considered legally as a murderer' add '**II** *slaughter* animals, *Peripl.M.Rubr.* 4.'

φόνιος II 3, add 'adv. φονίως, Sch.E. *Tr.* 539'

φόνος, line 15, after 'D.S. 19.8' insert 'ἐν φόνῳ .. Lxx *Nu.* 21.24' fr. section II.

φορά B 3, line 6, after 'D.S. 16.54' insert 'Men.*Pk.* 533 S.'

*φοράριος, ὁ, perh. *shopkeeper*, fr. Lat. *forarius, *IGRom.* 3.93 (written -ις, Sinope), cf. °φόρος (B).

φορβειά I, for '*halter .. manger*' read '*halter* for horses, camels, etc.' and add '*PCair.Zen.* 781.12, *PHib.* 211.17 (both iii BC), *PMich.*xv 717 (iii AD)' **II**, for this section read 'kind of halter used by players of wind instruments to support the mouthpiece against the lips, Ar.*V.* 582, Plu. 2.456b; in fig. phr., app. w. ref. to lack of restraint, ἀγρίαις φύσαισι φορβειάς ἄτερ S.*fr.* 768 R.' at end add 'form *φορβηϝια, Myc. *po-qe-wi-ja*'

φορβή, add 'Myc. *po-qa*'

*φορβιοπώλης, ου, ὁ, *seller of φόρβιον*, or perh. *of fodder, POxy.* 1037.4 (v AD; πορβιο-).

φόρετρον, add '**2** *transport, carriage, BGU* 2269.9 (AD 138/9). See also ‡φόλετρον.'

φορεύς, add '**IV** *shield-strap*, Hsch.'

φορέω I 1, add '**b** prob. of road or door, *lead to* (as φέρω A VII 1), Call.*fr.* 504 Pf.'

φόριμος II, add 'also τὸ φ., Plin.*HN* 35.184, Gal. 12.917.7'

*φορίνιον, τό, see quot.; φορίνιον γὰρ λέγεται τὸ παχυνθέν δέρμα τοῦ ὀφθαλμοῦ Phot. s.v. πεφορινῶσθαι.

φορμαλεία, line 2, for 'perh.' read 'prob.' and after 'φρουμαρία' insert '*list of supplies to be delivered*' add '**2** *official receipt for supplies*, *PLond.* 1663.25 (vi AD), cf. *PMasp.* l.c.'

*φόρμη (φώρμη, φούρμη), ή, Lat. *forma, quality* or *class*, ταπήτια διπρόσωπα δύο τῆς πρώτης φούρμας *PMich.*xiv 680.11 (iii/iv AD), πρώτης φόρμης *Edict.Diocl.* 8.2, al. **II** *cobbler's last*, περὶ φορμῶν καλικαρικῶν ib. 9.1.

φόρμιγξ 1, add 'Φ. ἀστερόεσσα the constellation *Lyra*, Nonn.*D.* 1.257; ἡδυμελῆ .. οὐρανίῃ Φόρμιγγι τεὴν σύριγγα συνάψω ib. 467'

*φορμίδιον, gloss on φορμίον, Hsch.

φορμός 1, transfer 'prov., .. Arist.*Rh.* 1385ᵃ 28' to end of section 2. **3**, delete the section adding quots. to section 2.

φορμοφόρος, add 'rest. in Schwyzer 230 (ed. pr. ἑρμο-, v. *ZPE* 33.88-9)'

φορολόγος, add '*POxy.* 3273.3 (i AD), *SEG* 32.676 (Thrace, iii AD)'

φόρον, for 'cf.' read 'as a court of law' adding '*PMasp.* 312.7 (vi AD), *Cod.Just.* 4.20.15.3, 8.10.12.8'

φόρος 1, add '**b** *rent paid in kind, PMich.*xv 727.3 (iv/v AD).'

*φόρος (B), ὁ, *forum, market-place*, φ. Θεοδοσιανός *IEphes.* 1534 (Chr.), φόρος· ὁ τόπος, ἐν ᾧ πωλητήριον Suid., cf. ‡φόρον.

*φορτηγέσιον, τό, app. *the business of carrying merchandise* or the *equipment* for it, *SEG* 26.845.3 (Berezan, vi/v BC).

308

†φορτηγός, όν, *carrying loads* or *cargoes*, of ships, ἄκατοι Critias 2.11 W.; νῆες Plb. 1.52.6, 5.68.4, etc.; πλοῖα D.S. 14.55, 20.85; of men, *engaged in transporting cargoes*, ναυβάτην A.fr. 263 R., Metag. 4.4 K.-A. (hex.); subst., *porter*, Thgn. 679, Cratin. 171.73 K.-A., cf. Poll. 7.131.

φορτίον 1, line 2, for 'Sapph.*Supp.* 9.13' read 'Sapph. 20.13 L.-P.' and delete '*Pl.* 352'   add '**b** pl., *contents of a building, fixtures*, *POxy.* 242.16, 243.27, *PMich.*x 584.25 (all i AD).'   **2 a**, add 'so perh. sg. in Ar.*Pl.* 352 is *piece of goods* offered, *bargain* offered'   **3**, delete 'ἔρωτος .. Anacr. 170'

φόρτος I 1, line 4, after 'φ. ἔρωτος' insert 'dub. sens., Anacr. 115 P.'

φορω, delete 'prob. abbreviation of'; add (cf. demotic and Coptic πορο).

*Φορωνιάς, άδος, ἡ, epith. of Hera, epigr. in *Lindos* 698 (*c.*200 BC).

φόσσᾱτον, for '*CIG* .. Anastasii)' read '*SEG* 9.356.36 (Ptolemais in Cyrenaica, vi AD)'

φουλβῑνον, add 'cf. πουλβῖνον'

φουλιᾱτα, add 'see also °φολιᾶτος, -ον'

*φούλλων, ωνος, ὁ, Lat. *fullo, fuller, launderer, Edict.Diocl.* 22.1.

*φοῦνδα, ἡ, Lat. *funda, belt for carrying money, Ann.Épigr.* 1907.22 (Aphrodisias), *PHamb.* 10.34, 38; φ.· *ventralis* Gloss. (cf. mod. Gk. = *tassel*); also **ποῦνδα** *PMich.*inv. 3163.42 (*TAPhA* 92.258, iii AD).

*φοῦρκα, ἡ, Lat. *furca*, in quot., *forked frame for a cart, Edict.Diocl.* 15.9.

φούρμη, v. °φόρμη.

*φουρνάκιον, τό, *small oven*, Anon.Alch. 367.15, 17, 19.

*φουρνέλλον, τό, *oven, furnace*, Anon.Alch. 321.9, al.

φοῦρνος, add '**II** περὶ φούρνων οἰκεί(ων) app. kind of slipper, *Edict.Diocl.* 9.20 (περιφορίνων 20A., q.v.).'

*φουσκάριος, ὁ, *seller of φούσκα, PLond.* 1028.5 (vii AD).

*φράγδην, adv., *as a defence, defensively*, Batr. 266.

φραγέλλη, add 'see also °φλάγελλα'

φραγέλλιον I, add '-ιν μουλιωνικόν *Edict.Diocl.* in *SEG* 37.335 iii 18, cf. φλαγέλλιον'   **II**, delete the section transferring quot. to section I (cf. *ZPE* 94.285)

*φραγηλίτης, ὁ, app. for φραγελλίτης, kind of policeman employed to guard church property, *ITyr* 10.

φρᾱδάτηρ, add '*SEG* 34.940 (Camarina, iv/iii BC)'

φρᾱδή, after lemma insert 'dial. φραδά *IG* 5(2).261.15 (v. infra under sense II)'

*φρᾱδητός, ή, όν, *known*, Sch.Pi.*N.* 3.45 (explaining φράδασε in Pi. *N.*3.26 S.-M.).

φράδμων, add 'also φράσμων, φ.· προσέχων Hsch.'

φράζω, after lemma insert 'Boeot. φράττω Corinn. 34 P.'

*φρασμοσύνη, ἡ, *injunction*, ἀνέθεκε .. μάντειον (i.e. μάντεων) φρασμοσύνᾱι *CEG* 243 (Athens, v BC); cf. φραδμοσύνη.

*φράσμων, v. °φράδμων.

φράσσω, line 6, after 'Il. 15.566' insert 'subj. φαρξώμεθα Alc. 6.7 L.-P.'

*φρᾱτερικός, ή, όν, *of a phratry*, φ. γραμματεῖον D. 44.41.

φρᾱτορικός, delete the entry.

φρᾱτριαρχέω, add 'also **φᾱταρχέω** de Franciscis *Locr.Epiz.* 20; see also φρητ αρχέω'

*φρᾱτριάρχιον, τό, *temple of a phratry*; in form φᾱτάρχιον, de Franciscis *Locr.Epiz.* 14, 16.

φρᾱτρίαρχος, add 'also **φᾱταρχος**, de Franciscis *Locr.Epiz.* 8, 22, 23, 34; see also φρήταρχος'

φρᾱτρικός, after '= φρατριακός' insert 'app. *belonging to a phratry*, *SEG* 26.676 (Larissa, 200/190 BC)'

φράτριος III, add 'at Scepsis, *JÖAI* 3.55, where see other refs.'

φράττω, add 'also Boeot. for °φράζω'

φρέαρ, line 1, before 'Ep.' insert 'φρῆρ *Tit.Cam.* 64a.1, *PLond.* 1948.7 (iii BC); φρῆν *SEG* 39.1002.4 (Camarina, ii/i BC)'   **1**, add 'Lang *Ath.Agora* XXI K1 (vi BC)'   **2**, delete '**b** .. 810' transferring the ref. to line 2 of section 2, after 'Th. 2.48, 49'   at end read '(cf. ‡στέᾱρ)'

φρεᾱτία, φρεᾱτιον, add 'see also φρητία, °φρήτιον'

*φρένησις, εως, ἡ, = φρενῖτις, Cels. 3.18.1.

φρενοβλᾰβής, add '*Trag.adesp.* 625.45 K.-S.'

φρενοβλᾰβία, add 'Heph.Astr. 2.31.16 (pl.)'

φρενοδᾰλής, for '-δᾱλ-' read '-δᾰλ- [prob. -δᾰλ-]'

φρενώλης, after '*frenzied*' insert 'Hippon. 77.5 W.'

*φρεωρῠχικός, ή, όν, *for digging wells*, ἐργαλεῖον φ. Hsch. s.v. τόρος.

*φρήν, v. °φρέαρ.

*φρῆρ, v. °φρέαρ.

*φρήτιον, τό, = φρεᾱτιον, *IG* 14.217 (Sicily, pl.).

*φριγιδάριον, τό, Lat. *frigidarium, cooling-room* in a bath, *SEG* 26.784 (φριγδ-; Thrace, AD 162/3).

†φρῑμαγμός, ὁ, *sound uttered by animals in a state of excitement, snorting*: by horses, Lyc. 244; by goats, Poll. 5.888, D.H.*Comp.* 16.

φρῑμάσσομαι, for '*snort and leap: wanton*' read '*snort in excitement*' and after 'of goats' insert 'Pi.*fr.* 332 S.-M.'

φρίξ II 2, after 'Hp.*Morb.* 2.68' insert 'Nic.*Th.* 778'

*φριξοχαίτης, ου, ὁ, *having bristling hair*, S.*fr.* 10d.5 R.

φριξωποβρόντᾱξ, delete the entry.

*φριξωποβρονταξαστράπτης, ου, ὁ, *hurler of frightful thunder and lightning*, *PMag.* 5.19.

φρονέω IV, line 13, ζῶν καὶ φρονῶν, add 'fem. ζῶσα προνοῦσα (sic) Mitchell *N.Galatia* 242 (Rom.imp.)'

*φροντάριον, τό, (cf. Lat. *frontalia*), *ornament for (?horse's) forehead*, in pl., pap. in *JÖByz* 33.13.5 (vi AD).

φροντίζω II 2, line 16, for '*to be concerned* or *anxious* about' read '*give consideration* or *attention* to'

φροντίς III 2, add '*SEG* 32.1554 (Arabia, vi AD)'

*φροντιστεία, ἡ, prob. *office of φροντιστής* II 1, *SEG* 19.882 (Syria, ii AD).

φροντιστής II 1, add 'as title of Zeus, *INikaia* 1141 (ii AD)'; at end, as transl. of Lat. *procurator*, add 'Modest.*Dig.* 26.6.2.5'

φροῦδος 2, add '*POxy.* 3069.19 (iii/iv AD)'

*φρουμαλεία, v. ‡φρουμαλεία.

†φρουμεντάριος, α, ον, Lat. *frumentarius, of military personnel, concerned with victualling, SIG* 830.5 (Delphi, ii AD), *IG* 10(2).207.6 (Thessalonica, iii AD), *SEG* 31.905 (Caria, iii AD).

φρουρά II, add 'sp. πρωρά *SEG* 25.447 (Arcadia, iii BC), προύρρα *SEG* 37.494 (Matropolis, Thessaly, iii BC)'

†φρουρίς, ίδος, fem. adj., *watch-, guard*-: as subst. sc. ναῦς, *guard-ship*, *IG* 1³.21.85, Th. 4.13.2, X.*HG* 1.3.17; also προυρίς, αἱ πρ. καλούμεναι πύλαι (at Abdera), Call.*Dieg.* ii.34 (*fr.* 90 Pf.) (perh. pr. n.).

φρουρός, add 'see also προυρός'

φρύαγμα II, after 'insolence' insert 'Men.*fr.* 333.13 K.-Th.'

φρύγανον I, add 'applied to driftwood, Ael.*NA* 5.23'

*Φρυγιᾰκός, ή, όν, *Phrygian*, Macho 191 G., Str. 10.3.15, *Edict.Diocl.* 19.53, 62.

*Φρυγικός, ή, όν, *Phrygian*, D.H. 1.29, St.Byz. s.v. Φρυγία.

Φρύγιος, line 1, after 'ος, ον' insert 'Arist.*Pol.* 1276ᵇ9' and after 'Luc.*Harm.* 1' insert 'etc.'   **I 1**, add 'Μητρὶ Φρυγε[ίηι] *the Phrygian goddess, Ilasos* 229.2'

φρυκτός 2, add 'cf. ἐπὶ φρυκτῷ παρίηι *CID* I 13.15-16 (precise interpr. uncertain, Delphi, iv BC); as sacrifice, perh. rest. at ib. 7.7-8 (Andrian law, v BC)'

*φρῡνεός, ὁ, = φρύνη, Pratin. 3.10 S. (φρυναιου cod. A of Ath. 14.617e), *EM* 801.29.

φρύνη III, for 'nickname .. complexion' read 'as fem. pers. n.' and add 'Φρυνᾶν *SEG* 31.824 (Gela, v BC)'

*φρύνινον, τό, = φρύνιον 1, kind of plant, Paul.Aeg. 7.3 (254.7 H.).

*φρῡνίτης, ου, ὁ, kind of precious stone, Socr.Dion.*Lith.* 52 H.-S.

*φρῡνοποπείον, τό, perh. kind of skillet, *SEG* 24.361 (Boeotia, iv BC).

*φρούριον, v. φρούριον.

*φυγαδεία, v. ‡φυγαδεύω.

*φῠγᾰδευτής, οῦ, ὁ, *that which drives away* (snakes), of the stone σιδηρίτης, Orph.*Lith.Kerygm.* 15.25 G.

φῠγᾰδεύω, after 'Elean φυγαδείω .. (iv BC)' insert 'also aor.subj. φυγαδεύαντι ib. line 6'   **II**, transfer 'Plb. 10.22.1' to precede 'fut.'

φύγεθρον, add 'Hp.*Aff.* 35'

*φῠγελίτης, ου, ὁ, name of a Lydian wine, from Phygela or Pygela, Dsc. 5.10.

φυγή II, line 10, after 'pl.' insert 'Alc. 129.12 L.-P.'

φυγοπτόλεμος, for 'poet. for *φυγοπόλεμος' read 'also φυγοπόλεμον (acc.) Hsch. s.v. ξύξηλιν'

φύη, delete the entry.

†φῡκάριον, τό, *rouge obtained from seaweed*, Hsch. s.v. ἄφυκα, Zonar.; sp. φουκάριον *PMich.*VIII 508.6 (ii/iii AD), cf. φῦκος II.

φῡκίον I 1, for '= φῦκος I' read '*seaweed*' and add 'Hippon. 75.2, 115.10 W.'

φῦκος II, add 'Lxx *Wi.* 13.14'

φύλαγμα I 2, add 'of the tomb, φύλαγμα σωμάτων epigr. in *TAM* 4(1).303 (Rom.imp.)'

φυλᾰκή II 1, line 9, after 'Id. 8.39' insert 'Θησεὺς .. διὰ πάσης ἦν φυλακῆς τῷ πατρὶ Αἰγεῖ Call.*Dieg.* x.21 (*fr.* 230 Pf.)'

φυλᾰκίς, add 'φ. τριημολίαι *Hesperia* 11.292 no. 57 (Athens, iii AD, v. Robert *Hell.* 2.124)'

φυλᾰκῑτικός, line 2, after '*pertaining to police*' insert 'φ. κλῆρος *PTeb.* 808.3 (?ii BC)'

φῠλᾰκός I, last line, for 'Call.*Hec.* 1.2.12' read 'Call.*fr.* 260.28 Pf.'   **II**, after '(Il. 2.695, etc.)' insert 'Φυλάκα, title of Demeter, *IG* 9(2).573, *SEG* 17.288 (Larissa, i BC or later)'

*φῠλακσία, ἡ, perh. *registration of members of a tribe*, *PHarris* 64.7, al. (iii/iv AD).

φῠλακτέος II 1, add 'Gal. 10.838.2'

φῠλακτήριον 1, for '*guarded post, fort, castle*' read '*fortified guard post*' and add '*SEG* 24.154 (Attica, iii BC); fig., Pl.*R.* 424d'   **2**, for 'amulet' (line 7) to end read 'magical inscription worn as an amulet, PMag. 4.1626, 7.298, 580'

φῠλακτικός II, after 'Adv. -κῶς' insert 'Men.Dysc. 95'

φῠλάκτωρ, for 'Bull. .. 244' read 'GVI 1861'

*φυλαουργέω, v. °φυλλα-.

φῠλαρχέω, line 3, c. gen., add 'φυλαρχήσας φυλῆς Ποσειδωνιάδος TAM 4(1).223, 299 (Rom.imp.)'

φῠλαρχία, add 'II = πομπή τις Hsch.'

φύλαρχος I a, for 'CIG 3773' read 'TAM 4(1).42' **II**, after 'Hdt. 5.69' add 'Ar.Au 799, Pl.Lg. 755c, al., Lys. 15.5, Arist.Ath. 30.2, al., etc.'

φῠλάσσω, line 12, imper., add 'φεφύλαχσο Schwyzer 538 (Acraephia, vi BC)'

*φυλατός· ἡ λέξις παρὰ Βλαίσῳ (Blaes.fr. 5). σημαίνει δὲ ᾠδήν Hsch. (app. Italic wd.).

φῠλετικός I 2, delete 'ἡ φ. .. BC 3.30'; v. °φυλέτις.

*φῠλέτις, ιδος, fem. adj., φ. ἐκκλησία, = Lat. comitia tributa, App.BC 3.30.

φυλλάζω, add 'frondentesq(ue)· καὶ φυλλαζούσας Virgil gloss. in PNess. 1.851'

*φυλλαουργέω, strip the foliage (from vines), PMich.XIII 666.17 (written φυλα-, vi AD).

*φυλλάω, = φυλάσσω, (cf. mod. Gk. φυλάγω), φύλλατε ἀὴ τὸ σὸ δοῦλδ SEG 36.1269 (Armenia, late Rom.imp.).

*φυλλίδιον, τό, dim. of φυλλίς, salad, σάκκος φυλιδίων BGU 2359.5 (iii AD).

Φυλλικός (s.v. φυλλικός II), add 'SEG 23.412 (189/8 BC), 32.599 (Augustus)'

φυλλίνης, add '2 εἶδος τι κυκεῶν‹ος› Hsch.'

φύλλον I 2, for 'of flowers, petal' read 'flower'; add 'AP 6.154 (Leon.Tarent. or Gaet.)' **II**, after 'PTeb. 38.3, 78.4 (ii BC)' insert 'κατὰ φύλλον according to crops, w. ref. to taxes, BGU 1120.20 (ii BC; v. ZPE 19.284)' add '**III** something shaped like a leaf, applied to segment of the covering of a ball, AP 14.62.'

φύξηλις, for 'cowardly' read 'apt to run away, fugitive' and add 'poet., of thunderbolt, Nonn.D. 1.320'

φύξιμος I, line 4, after 'Id. 9.29.4' insert 'τὸν να[ὸν] φ. Chiron 19.134.48 (Sardes, i BC)'

φύξιος 2, add '**b** repellent, φ. ὀδμήν Nic.Th. 54 (cj.).'

φύος, add 'PSI 892.82 (?iv AD)'

+**φῠρᾱματικά**, τά, perh. some form of interior decoration, cf. κονιατικά, MAMA 8.498.26 (Aphrodisias).

φύρασις, line 2, for 'mixture' read 'kneading'

φύρκος, delete 'Dor. φοῦρκος'

+**φυρός**, v. °φυρρός.

+**φυρόχρωμος**, ον, perh. tawny-coloured, of cow, PBaden 19.5 (ii AD), v. °φυρρός.

*φυρρός, ά, όν, app. = πυρρός, tawny, of a camel, PVindob.Worp 9.5 (AD 158); also **φυρός**, PLond. 1132b.5 (AD 142), of an ox, PGen. 48.8 (AD 346).

φῡσάω I, line 11, after 'Antiph. 117 (troch.)' insert 'w. ὡς, boast that, φυσήσας ὡς κλαύσοιτο ὁ Φρυνίων εἰ .. D. 59.38'

*φύσγων, v. °φύσκων.

*φύσελος, ὁ, or -ον, τό, wind in the stomach, Sch.Nic.Al. 287 Ge.; cf. φύσαλος.

φύσησις, add '2 swelling, of waves, Sch.A.R. 1.1167.'

φῦσιγξ I, add '(unless erron. for φαῦσιγξ)'

*φῠσιδρόμος, ον, (sp. φυσιτρ-), nature-roaming (cf. φύσις VII), epith. of a demon, Suppl.Mag. 49.57 (iii/iv AD).

φύσις III, line 2, before 'κατὰ φύσιν' insert 'ἐκ τῆς φύσεως in the course of nature, ὡς ἐκ τῆς φύσεως τὸν θάνατον ἴσον ἔσχεν SEG 26.821 (Thrace, c.100 BC)' add '**b** ἐν φύσει in living form, opp. the state of those not yet born, μὴ μόνον οἱ ἐν φύσει .. ἀλλὰ καὶ οἱ κυοφορούμενοι Cod.Just. 6.4.4.21, 6.48.1.2.' **VII 2**, for 'testes' read 'vagina and anus'

*φῠσίσοφος, ον, wise by nature, μέλισσα Rh. 3.530.

*φῠσιτρόμος, v. °φυσιδρ-.

φύσκων, for 'Alc. 37B' read 'Alc. 429 L.-P.; written φύσγων id. 129.21 L.-P.'

φυστή, line 1, for 'kind of light pastry or puff' read 'cake made of coarse meal and wine'; line 2, for '(Leon.)' read '(Leon.Tarent.), Teles p. 4.12 H.'

φῠτᾰλιά I, line 3, after '(Il.) 20.185' insert 'Inscr.Cret. 4.43Ba 2 (v BC)' and add 'AP 6.44 (Leon.Tarent.)' **II**, delete 'also of the vine .. (?))' and for 'ib.' read 'AP 6.44' at end add 'Myc. pu-ta-ri-ja'

φύτευσις, for '= φυτεία' read 'planting' and add 'φιτεύσιος δαφνέων SEG 24.277.A29 (Epid., iv BC)'

φυτεύω I 4, for this section read 'implant, ταύτην (sc. ψυχὴν) μὴ φυτεῦσαι εἰς μηδεμίαν θήρειον φύσιν Pl.Phdr. 248d'

φυτηκόμος, add 'also as adj., causing plants to grow, μαρμαρυγὴν πέμπουσα φυτηκόμον Nonn.D. 7.303'

*φυτήκομος, ον, covered with vegetation, φ.· σύνδενδρος τόπος Hsch.

φύτλον, for pres. ref. read 'orac. in TAM 4(1).92'

*φυτοεργείη, ἡ, cultivation of plants, Gr.Naz.Carm. 1.2.1.257 (PG 37.542a), cf. ‡φυτουργία.

φυτοεργός, add 'Nonn.D. 47.58, al.'

φυτόν I 1, add '**b** stem of wood, μονόδροπον φυτόν Pi.P. 5.42.' at end add 'Myc. pu-ta (pl.)'

φυτός I, for this section read 'naturally-formed, πυάλους δύω, μίαν μὲν φοιτήν (sic) TAM 4(1).276.8 (iii/iv AD)'

φυτοσπορία, add '2 generation of offspring, IHadr. 166 (Rom.imp.).'

φυτοσπόρος, after 'S.Tr. 359' insert 'SEG 23.220 (Messene, i AD)'

φυτουργία 1, add 'PKöln 144.29 (152 BC)' at end add 'see also °φυτοεργείη'

φύω B II 2, lines 3/4, delete 'πολλῷ γ' ἀμείνων .. A.Pr. 337'

φωῖς, add 'Cratin. 226 K.-A.'

Φωκαΐς, as a coin, add 'Inscr.Perg. 8(3).161.32 (ii AD)'

*φωκαρία, ἡ, Lat. focaria, housekeeper, concubine, BGU 614.13 (iii AD).

*φωκάριον, τό, concubine, BGU 600.21 (ii/iii AD); φωκάριν PPrincet. 57 (ii AD).

*φώλαρχος, ὁ, app. some kind of cult official, PP 18.385-6 (Velia, Italy; see also SEG 30.1225, cf. perh. Φωλευτήριος).

φωλεός I, add 'Hippon. 86.4 W.' **II**, add 'Call.fr. 68.2 Pf.'

*Φωλευτήριος, ὁ, epith. of Apollo, IHistriae 105 (iii BC, see also SEG 30.798; cf. perh. °φώλαρχος).

φωνέω I 3, for '(written πωνίω)' read '3rd sg. pres. part. πωνίοντες, etc., cf. Cret. forms of ἀποφωνέω, °προσφωνέω' **III**, add 'pass., to be proclaimed, AP 7.430 (Diosc.)'

φωνή II 2, after 'language' insert 'or dialect'; at end after 'cf. 409e' insert '(or perh. w. ref. to orthography, v. REG 80.234 ff.)'

φωνήεις 1, add 'poet., of a lyre, Sapph. 118 L.-P.; of a song, Pi.O. 9.2; transf., of speech, B.fr. 26.2 S.-M.' **2, 3 & 4**, delete these sections.

φωράω I, add 'search a person, Plu. 2.248f'

φώριος 1, add '**b** of a robber, ἐπήλυσις Call.fr. 331 Pf.'

φωσφόριον II, add 'IGBulg. 1731.29 (Thrace, iv/iii BC)' **III**, add 'PVindob.Tandem 26.14 (AD 143)'

φωσφόρος, line 2, after 'Call.Dian. 204, etc.' insert 'w. dissimilation of aspirates, πωσφ- Hesperia 4.138 (sense II, pl., Athens, ii/iii AD), SEG 26.413 (lamp signature, Isthmia, ii AD), 31.1285 (Pisidia, Rom.imp.)'

*φωταθυρίς, v. ‡φωτοθυρίς.

φωτεινός I, add 'sup., Orac.Tiburt. 22'

φῶτιγξ, for 'flute' read 'transverse pipe' and add 'app. distd. fr. πλαγίαυλος by Nicom.Harm. 4'

+**φωτιστήριον**, διαφανῆ φ. λυχνικά gloss on Lat. luminaria, Gloss. **2** baptistery, SEG 8.318 (Mt. Nebo, vi AD), 30.1697, 33.1270 (Palestine, vi AD), 31.1476 (Arabia, vi AD).

φωτοθυρίς, add 'also φωταθ-, PMil.Vogl. 99.12, PLond. 1179.62 (both ii AD)'

φωτοφόρος, add '2 fem. subst., app. lampstand, SEG 16.741 (written φωιτοφόρος, Pergamum, Rom.imp.).'

Χ

Χ, line 3, before 'stands' insert 'more often'

χαβότια, for 'dub. sens., perh. *honey-pots*' read 'perh. *small pots*, cf. °καβίδιον'

χαβῶνες, for 'ἀπό' read 'ἀπὸ'; add 'see also χανών, °χαμῶνας'

*χαβώνιον, v. °καβόνιον.

χάζω B 2, line 5, for 'nor in truth .. hit him' read 'of a stone, i.e. it nearly hit the man' and after '(Il.) 16.736' insert '(s.v.l.)'

*χαιμαφάριον, τό, app. some kind of oil or lubricant in medical use, perh. *camphor*, PRyl. 529.13 (iii AD).

χαιρετίζω, add 'SEG 26.730 (Maced., ii/i BC)'

χαίρω III 1 a, lines 6/7, transfer 'S.Ph. 462' to section 2 a; add 'cf. w. χαίρω as response, κῆρυξ Ἀχαιῶν χαῖρε .. – χαίρω A.Ag. 538; also opt., χάροις (sic) παροδεῖτα INikaia 767' **b**, for this section read 'inf. after vb. of speaking, χαίρειν δὲ τὸν κήρυκα προὐννέπω S.Tr. 227; ὅτι προσειπών τινα χαίρειν οὐκ ἀντιπροσερρήθη X.Mem. 3.13.1; inf. alone, πόλλα μοι τὰν Πωλυανάκτιδα παῖδα χαίρην Sapph. 155 L.-P., Pl.Ion 530a; Κῦρος Κυαξάρῃ χαίρειν X.Cyr. 4.5.27, Theoc. 14.1' **c**, delete the section. **2 a**, add 'inf. after vbs. of speaking, MDAI(A)56.131 (= GVI 1344.1) (Miletus, Hellen.), Luc.Dem.Enc. 50' deleting these refs. from section 2 c, together w. 'Sapph. 86, Ar.Pl. 322, Eup. 308, X.HG 4.1.31'

χαίτη 1, add 'χαίτην (-ας) σείειν *shake one's locks*, Anacr. 77 P., E.Med. 1191; poet., of trees, Call.Del. 81, Anacreont. 18.12 W.' **3 b**, for 'hedgehog' read 'porcupine' **5**, for this section read '*tuft* of the papyrus plant, βύβλος .. ἐπ' ἄκρῳ χαίτην ἔχουσα Str. 17.1.15; *thistledown*, Theoc. 6.16'

*χάκαξ (?), gen. χάκακος, name of a freshwater fish, IGC p. 99 B 30 (Acraephia, iii/ii BC) (cf. χάραξ IV).

χαλάδριον, after '**χαλάτριον**' insert 'POxy. 3354.16 (iii AD)'; add 'also **χελάδριον**, POxy. 1142.13 (iii AD); **χαράδριον** PTeb. 815 fr. 2 iii 74 (iii AD); see also χαλατριόομαι'

χάλαζα II 1, add '**b** *eruption of pimples*, Nic.Th. 252, 778.'

χαλαζήεις II, for '*whose sting causes an icy chill*' read '*causing* (by its sting) *skin eruptions*'

*χαλάζησις, ἡ, *pimple* or *tubercle* in flesh of swine, Alex.Aphr.Pr. 3.139 bis.

*χαλαζίτης, ου, ὁ, *stone resembling hailstone* (= χαλαζίας, χαλάζιος), Orph.Lith.Kerygm. 25 H.-S.

χαλάζωσις, add 'CIL 13.10021 (181)'

*χαλαίβασις· ἀπὸ τοῦ χαλαρῶς βαδίζειν Suet.Blasph. 66 T.

χάλασις 3, for 'University .. 3(2).58' read 'SEG 8.647.3 (Egypt, iv AD)'

χάλασμα 2, add '*gap* between part of the surface of pavement blocks and the bed on which they rest, IG 7.3073.114 (Lebadea, ii BC); delete from section 3).' add '**3** *over-measure* (leaving an allowance for error), of land, PGiss. 36.17 (ii BC).'

*χαλαστάριον, τό, prob. *necklace*, dim. of χαλαστόν, PKöln 166.4 (-ταριν pap.) (vi/vii AD).

+**χαλβανίς**, ίδος, *made from the plant all-heal*, Nic.Th. 938, subst. Androm.ap.Gal. 14.41 (= GDRK 2.62.164).

+**χαλβανόρυτον μέλι**, τό, kind of honey, made from the resinous juice of all-heal (χαλβάνη), Alex.Aphr.Pr. 3.2.

χαλεπός B II 1, lines 6/7, for 'χ. λαμβάνειν περί τινος Th. 6.61' read 'also χαλεπῶς λαμβάνειν, w. ellipse of obj., Th. 6.61.1'

χαλέπτω I 1, add 'Κύπριδα χ. AP 5.263.5 (Agath.); τὴν παῖδα ib. 300.3 (Paul.Sil.)' **II**, at beginning, delete '*provoke* .. (Agath.)' and add 'Call.Cer. 48' **III**, for this section read 'act. intr., *to be angry*, Nonn.D. 16.34; w. dat., Bion fr. 14.2 G.'

*χαλικός, ή, όν, *of cement*, χαλικὴ ἡμμιτία on a tile, app. indicating a measure, SEG 16.848 (Caesarea).

χαλινός I 2, add 'of a string of words full of difficult sounds, used for practice in articulation, Quint.Inst. 1.1.37' **III**, after 'part of the tackle of a ship' insert 'perh. *parrel*' **IV 1**, add 'of a snake's mouth, Nic.Th. 234' **2**, delete the section.

χάλις I, for '*Docum*. .. 101' read 'SEG 26.1835 (dub. l.)' and add 'Nonn.D. 15.25'

+**χαλκάνθη**, ἡ, = χάλκανθον, Heras ap.Gal. 13.558.8, v.l. in Dsc. 3.80.

χάλκειος, add 'Myc. *ka-ke-ja-pi* (fem. instr. pl.)'

χαλκέλατος, for 'poet. for' read 'var. of'

+**χαλκεόγομφος**, ον, *fastened with bronze nails*, ἐν ἀτερπέϊ δούρατι χαλκεογόμφῳ Simon. 38.10 P.

*χαλκεόζωνος, ον, *girt with bronze*, of Heracles, Suppl.Hell. 1033, cf. χαλκόζωνος.

χάλκεος II, add 'form χάλκιος, Myc. *ka-ki-jo*'

*χαλκεότευκτος, ον, *fashioned of bronze*, εἰκόνι .. χαλκεοτεύκτῳ SEG 18.137.B4 (Isthmus, iii AD), see also χαλκότευκτος.

χαλκεόφωνος, add 'χαλκεόφωνον ἐπισπέρχουσαν ἀοιδὴν Μελπομένην AP 9.505.15'

χαλκεύς, add 'Myc. *ka-ke-u*'

χαλκεών 1, add 'serving as repository for bronze implements, SEG 11.244 (sp. χαλκιών, Sicyon, vi/v BC)'

χαλκηδόνιον, add '**II** prob. *chalcedony*, PRyl. 627.162 (iv AD), χαλκεδ-).'

*χαλκηδόνιος (sc. λίθος), ὁ, *chalcedony*, Sm.Is. 54.12.

Χαλκιδικός IV, for this section read 'neut. subst., app. *room* or *recess* built into the end of a basilica, Vitr. 5.1.4'

χαλκίον I 1-3 for these sections read '**1** *bronze* (object), IG 1³.510, ἐν τῷ χαλκίῳ (sc. of a shield) ἐνορῶ .. Ar.Ach. 1128. **2** *bronze cauldron*, Ar.fr. 345 K.-A., Eup. 99.41, 272 K.-A., X.Oec. 8.19, IEphes. 3757; χ. θερμαντήριον IG 1³.421.96, Gal. 13.663.15; χ. ἐγλουτήριον IG 4.39.18 (Aegina); of the cauldron at the oracle of Zeus at Dodona, used prov. for a chatterbox, Men.fr. 60.3 K.-Th. **b** *bronze basin* used in the game of κότταβος, Poll. 6.110.'

*χαλκιτάριον, τό, *chalcite*, Anon.Alch. 5.8.

χαλκόδετος, add 'Myc. *ka-ko-de-ta* (neut. pl.)'

+**χαλκοκέραυνος**, ον, epith. of the sea, app. *bronze-thundering*, A.fr. 192.3 R. (s.v.l.).

*χαλκοκορώνη, ἡ, dub. sens., Rh. 6.90.30.

*χαλκοπυρίτης, ου, ὁ, *copper pyrites*, Anon.Alch. 16.6.

χαλκοπώλης, add 'SEG 32.239 (Athens, c.400 BC)'

χαλκός, line 1, for 'GDI 5011.4 (iii BC)' read 'Inscr.Cret. 4.162.3 (Gortyn, iii BC)'; line 11, before 'alloyed' insert 'χ. Μαριεύς fr. Marion in Cyprus, IG 2².1675.17' **II 4**, add 'humorously, of a moneyed man, AP 9.241 (Antip.Thess.)'

*χαλκοστεγίς (or -στεγής -ές), *bronze-roofed*, Gramm.Lat. 4.197.22 K. (Appendix Probi).

χαλκότευκτος, add 'see also °χαλκεότευκτος'

χαλκοτυπική, add 'w. τέχνη added, PSI 871.12 (i AD)'

χαλκοῦς II, add 'sobriquet of a moneyed man, Ἀριστομήδης ὁ χαλκοῦς λεγόμενος Did.in D. 9.52; cf. °χαλκός II 4'

χαλκόχρους, add 'ὄνον ἄρρενα χαλκόχρωον POxy. 3143.10 (AD 305)'

χαλκωμάτιον, add 'SB 9834a4, 9 (iii AD)'

χαλκωρύχος, after '*miner*' insert 'SB 7200.19 (ii AD, -ορ- pap.)'

*χαλκωτός, ή, όν, app. *covered with bronze*, Inscr.Délos 104-28.bB.15.

*χαλτουλάριος, v. ‡χαρτουλάριος.

*χάλχη, v. κάλχη.

χἄμᾰθεν I, add '**2** legal t.t. = Lat. *de plano*, *informally*, Modest.Dig. 27.1.13.10.'

χἄμαί I 1, line 4, after 'Hdt. 4.67' insert 'IG 1³.474.103'; line 5, after 'IG 2².1672.305' insert 'χ. ἐγένοντο Call.Epigr. 43.4 Pf.'; add 'τὸ χαμαί *the ground*, τοῦτον τὸν τόπον ἐξέκλεισεν τὸ χαμὲ ὀρύξας SEG 36.935 (Rome)'

χαμαιδικαστής, add 'PLips. 64, PLond. 980 (both iv AD), Cod.Just. 7.15.5.4; also χαμο- Hsch. s.v. σήλεκτος'

χἄμαιευνάς, (s.v. -εύνης), add 'θύμβρη Nic.Th. 532; see also χαμευνάς'

*χἄμαιλίχων, οντος, ὁ, kind of fishing-net, Sch.Th. 7.25.

χἄμαιριφής I, for this section read '*thrown* or *knocked to the ground*, Eust. 1279.45; of a building, epigr. in SEG 9.189 (Cyrene, ii AD); of a person, PCair.Isidor. 63.25 (AD 296; sp. χαμερ-). **2** of infants, *exposed*, *abandoned*, Hsch. s.v. ὑποβολιμαῖον, EM 781.36. **b** = *collecticius*, Gloss.'

*χαμέτρυος, v. χαμαίδρυς.

*χἄμευνάδιος, α, ον, *sleeping on the ground*, Eleg.adesp. in Suppl.Hell. 958.17 (s.v.l.).

χἄμευνάς II, for '*lair*' read '*bed on the ground*'

χἄμευνίον, after 'χαμεύνη' insert 'Hippon. 62 W.'

*χἄμεύρετος, ον, *found exposed*, *foundling*, of babies, Just.Nov. 153.

*χἄμοκέντησις, εως, ἡ, *floor-mosaic*, BE 1966 p. 386 no. 229 (two inscrs.; Eurytania, Aetolia; v/vi AD).

χαμοσόριον, τό, *tomb hollowed out of the ground* or *rock*, *BCH* suppl. 8.231, 232 (Maced., iii/iv AD), *MAMA* 3.27, 30, al.

χάμψα, for 'ὁ' read 'ἡ (or -ης, ὁ)' and delete 'cj. in A. *Supp.* 878 (lyr.)'

χαμῶνας· στέαρ ἢ τὰ ἐκ στέατος τικτόμενα Hsch., cf. χαβῶνες, χαυών. (Prob. Semitic, cf. Akk. *kamānu*, Hebr. *kawwān* "sacrificial cake").

χανάκτιον· τὸ μωρόν, Δωριεῖς Hsch.

χανδάνω II, line 6, after '*h. Ven.* 252' insert 'Q.S. 12.328'

χανδόν, for 'Call.*Aet.* 1.1.11' read 'Call.*fr.* 178.11 Pf.'

χάόω, after 'Simp.*in Epict.* p. 47 D.' add 'tab.defix. in *SPAW* 1934.1043'

*χαράβδη, ἡ, a disease of corn, Hsch.

*χαράγιον, τό, = χάραγμα 2, *Cat.Cod.Astr.* 8(2).165.14.

χάραγμα, add '**5** *loaf* (cf. χαραγμή), *PWash.Univ.* 56.16 (v/vi AD).'

χαράδριον (B), v. ‡χαλάδριον.

χάραδρος, add 'cf. Myc. *ka-ra-do-ro*, place-name'

χαράκίας III, add '*IGC* p. 98 A33 (Acraephia, iii/ii BC)'

χαράκισμός, for '*JHS* 33.338 (Maced., iii AD)' read '*SEG* 30.568 (Maced., ii AD)'

χαράκίτης 1, for this section read 'perh. a bird, *the Egyptian great reed warbler*, typically noisy and aggressive, βυβλιακοί (v.l. βιβλ-) .. χ. Timo in *Suppl.Hell.* 786.2'

χαράκοβολία, for '*forming a palisade*' read '*erection of a palisade*'

*χαράκοκόπος, ον, *used for cutting stakes*, δρέπανα *PCair.Zen.* 851a26 (iii BC).

χαράκτήρ I 1, line 2, for '*IPE* 1².16*A* 14' read '*IPE* 1².32*A*18' **II 1**, line 4, after '(Arist.)*Ath.* 10.2' insert 'ἔχον τὸν αὐτὸν χαρακτῆρα τῶι Ἀττικῶι *SEG* 26.72.9' **2**, line 6, after '*brand*' insert 'on a slave, *PHib.* 198.87 (iii BC)'

χαρακτηρίζω 2, after 'Iamb.*Comm.Math.* 4' add 'Procl.*Inst.* 102, 121, 158 D., al.'

χαράκών, for 'perh. *vineyard containing staked vines*' read 'perh. *paling*, *palisade*' and add '*CPR* 7.38.10n. (iv AD)'

χαράσσω II 1, line 3, after '(Apollonid.)' insert '*PMil.Vogl.* 69.B38, 83, 84 (ii AD), v. °ἐκτάσσω'

χαρίεις, line 1, for '*Mon.Piot* 2.138' read '*CEG* 326'; line 3, after 'dat. -εντι' insert 'fem. pl. χαρίεσσιν Orph.*H.* 46.5'

χαρίζω II 3, add 'of the gift of a slave to a deity, *SEG* 34.656 (Maced., ii AD)'

χάρις V 2, add 'θῦμα ἐκ τριῶν ποπάνων συγκείμενον, τινὲς δὲ πλακούντων εἴδη, καὶ ἀρτοχάριτας καλεῖσθαι Hsch.' **VI 1**, line 4, after 'A.*Ch.* 266' insert 'ἔμαν χάριν Alc. 304 i 7 L.-P.' **2 a**, after 'κατὰ χάριν Pl.*Lg.* 740c' insert '*Cod.Just.* 1.1.7.5'

χαρισμός, delete '*gratifying*' and add 'Phld.*Mort.* 21'

*χαριστεία, ἡ, *thank-offering*, ἀρὰν καὶ χαριστήιαν *SEG* 23.593 (Gortyn, i BC), cf. χαριστεῖον, -ήιον.

χαριστέον I, add 'Isoc. 19.22'

χαριστήριος II, neut. subst., add 'χαριστέριο(ν) τόδε *Kafizin* 6; pl. χαριτήρια (sic), ib. 278'

χαριστίων, add 'in a list of kitchen equipment, *PAlex.* (1964)31.6 (iii/iv AD)'

χαρῑτήσιον III, for '(Orchom. .. lapis)' read '(-είσια lapis), *SEG* 34.356 (both Orchom., in Boeotia)' **IV**, add '*SEG* 28.953.57.62 (Kyzikos, i AD)'

χαρίτινος, after 'dub. sens.' insert 'perh. the same as °χάρτινος'

χαρίτώπης, delete the entry.

*χαρίτωπις, ιδος, fem. adj., *of charming aspect*, *IG* 2².12828.7.

*χαρουχάριος or *χαρουχᾶς, v. ‡καρουχάριος.

*Χάροψ, οπος, Boeotian deity, *SEG* 23.102.c.i, ii, 28.457 (vi BC, see also 36.428); identified w. Heracles, *AD* 1916.218ff. (iii/ii BC), *SEG* 28.455 (iii BC), Paus. 9.34.5.

χαρταλάμι(ο)ν, τό, perh. *belt*, *SB* 9754.3 (AD 647).

χαρταρίδιον, τό, dim. of χαρτάριον, *PMich.*VIII 510.23 (ii/iii AD).

χαρτάριον, after '*small piece of papyrus*' insert '*PMich.*II 123ᵛix 30 (i AD)'

*χαρτάριος, ὁ, app. = χαρτουλάριος, *BGU* 466.12 (ii/iii AD), *GVI* 477.2 (χαρτάρις, Smyrna, i/ii AD).

*χαρτᾰτικόν, τό, *payment for an official document, clerk's fee*, *SEG* 9.356.18, 79 (edict of Anastasius, Cyrenaica, vi AD; pl.); also **χαρτιατικόν**, *CIG* 5187c21 (Ptolemais), Ulp.*Dig.* 48.20.6 (Lat. *chartiaticum*).'

χάρτης 1, add '**b** *official document*, Just.*Nov.* 8.1, *Cod.Just.* 4.21.16 pr., 4.21.22.11.'

χαρτιατικά, read χαρτιατικόν, v. °χαρτατικόν.

*χάρτινος, η, ον, *of* or *for papyrus*, ἐκ τεύχους χαρτίνου *SEG* 32.1149 (Magnesia on the Maeander, AD 209).

χαρτίον, add '**2** perh. *tax on papyrus*, *BGU* 2370.66 (i BC).'

χάρτισμα, delete the entry (v. ‡χόρτασμα).

χάρτος 2, after 'of persons' insert 'μὴ χαρταὶ γενώμεθ᾽ ἐχθροῖς Call.*fr.* 194.98 Pf.'

*χαρτουλάριον, τό, perh. *secretariat*, ἐν τῷ χαρτουλαρίῳ *POxy.* 3960.21, 26 (AD 621).

χαρτουλάριος, add 'sp. χαλτουλάριος Teuc.Bab. p. 47.28 B.'

*χᾰρυβδεύω, *to fish in a* °χάρυβδις 2 b, *POxy.* 3269.4, (iii AD), 3270.10 (iv AD).

χάρυβδις 2, add '**b** *pool in a river*, *POxy.* 3267.5.'

*χαρχᾰρόπεπλος, v. °καρχαρό-.

χάρων I, after 'Lyc. 455' insert 'Call.*fr.* 339 Pf.'

χάσιος, add 'cf. χάϊος'

*χάσκαξ, ᾱκος, ὁ, = *pathicus*, Eust. 1909.54 (cf. °χαυνόπρωκτος).

χάσκω I 2, add 'w. περί, *develop an appetite* for, κεχηνέναι περὶ τὰς ἐπιθυμίας Clem.Al.*Paed.* 2.10.102'

χάσμα I, add (fr. section II) 'χάρυβδις .. ἅρμα περιβαλοῦσα χάσματι E.*Supp.* 501' **II**, for this section read '*gaping mouth* of an animal, Plu. 2.670c, Σκύλλης χάσμασιν *AP* 11.379 (Agath.); w. defining gen., χ. φάρυγος *AP* 6.218 (Alc.), χ. ὀδόντων Anacreont. 24.4 W.; synecd. *gaping-mouthed head*, E.*HF* 363 (lyr.), Rh. 209, Plu. 2.366a'

*χασμωδιώδης, ες, *with hiatus*, Rh. 3.544.11.

*χαυλιαστής, οῦ, ὁ, dub., perh. an occupational designation, ἀπὸ δομεστίκων χαυλιαστίς *MAMA* 5.5 (Dorylaeum, iv/v AD).

*χαυνόπρωκτος, ον, nonce-wd., humorous term of abuse, *loose-arsed*, Ar.*Ach.* 104, 106.

χαυνότης, line 2, after 'X.*Oec.* 19.11' insert 'Plu.*Sert.* 17.3'

*χαυνοτρίβωνες, οἱ, app. members of some guild (?clay-moulders), *SEG* 32.931 (Sicily, iv/iii BC).

χαυνών, v. ‡χαυών.

χαύνωσις II 1, for '*making confused, mystification*' read '*making a case out of nothing*'

χαυών, for '*kavvân*' read '*kawwān*'; add 'cf. χαβῶνες'

χεῖ, after '*IG* 2².1491.33' insert '(τὸ χεῖ perh. *crosswise*)'

*χειλογρᾰφία, v. °χειρο-.

χεῖλος I 2, add 'Hippon. 118.3 W., Call.*fr.* 194.82 Pf.' **II**, line 4, after 'Ar.*Ach.* 459' insert 'of a ship, Eup. 353 K.-A., *AP* 7.215 (Anyt.)'; penultimate line, for 'τείχους' read 'τεύχους'

*χείλωμα (B), ατος, τό, = ‡χήλωμα.

χειλωμάτιον, v. °χηλωμάτιον.

*χειλῶνες· τῶν ἀλεκτρυόνων τινές Hsch. (perh. *χίλωνες, i.e. *fatted cockerels*).

χειμάρροος II 1, add '*AP* 7.411 (Diosc.)' **2**, delete the section transferring quot. to section 4.

χειμάω, for 'ῥιγέω' read 'ῥιγόω'

*χειμώδης, ες, *stormy*, Sch.E.*Rh.* 247.

χειμωνικός I, add 'πωμάριον *PKlein.Form.* 951 (v/vi AD)'

χειμωνόθεν, for 'in a storm' read '*from stormy weather*'

χείρ, line 15, for 'Aeol. .. Theoc. 28.9' read 'Aeol. acc. χέρρ', Alc. 58.21 L.-P.; pl., χέρρας, Theoc. 28.9' **I 3**, add 'of a horse, *SEG* 34.1437 (Apamea, v/vi AD)' **II 4**, add 'in making an agreement, ἔδοσαν χεῖρας *SEG* 30.568 (Maced., ii AD)' **6 g**, add '*SEG* 26.701 (Dodona, *c*.205 BC)' **i**, add 'ἔχων μετὰ χεῖρα τὴν Ἀνθίαν holding A. *by the hand*, X.*Eph.* 1.12.1' **IV**, add '**2** *dominion, rule*, Lxx 2*Ki.* 8.3, 1*Ch.* 18.3. **3** in Roman law = *manus*, as the power of a paterfamilias, Modest.*Dig.* 27.1.8.8, 11.' **V**, line 4, delete 'δεδωμάτωμαι .. A.*Supp.* 958' **VII 4**, add '(transl. Hebr. *yad*) *signpost*, Lxx *Ez.* 21.20 (25)'

χείρα, ἡ, = χειράς I, Hsch.

*χειράγρα, ἡ, *pain in the hand* (incl. gout), Gal. 13.1026.13, Ptol.*Tetr.* 153 (pl.); cf. Lat. *cheragra* (*cheir-, chir-*), Cels. 1.9.1, Plin.*HN* 24.188, etc.

χειραγρικός, add 'cf. *podagrici pedibus suis male dicunt, chiragrici manibus*, Petr. 132.14'

χειραπτέω, add '*misuse*, *SB* 9066 i 14 (ii AD)'

*χείραργος, ον, *having a useless hand*, gloss on κολόχειρ, Hsch.

*χειρέμβολον, τό, perh. *official receipt of goods for transport*, Ulp.*Dig.* 4.9.1.3.

*χειρετέροπλος, ον, (χείρ, ἕτερος, ὅπλον) *having one hand armed*, of a gladiator, epigr. in *ITomis* 288 (ii/iii AD).

χειρίδιον, add '**2** kind of *loose sleeve* (cf. χειρίς 2), *PMich.*xv 752.42 (χιρίδ- pap., ii AD).'

χειρίζω II 3, line 2, after '(iii AD)' insert 'also act., χειρίσαι πρεσβευτήν Thasos 170.27 (ii/i BC)'

*χειρικός, ή, όν, (in quots., χερ-), *manual*, ἔργα *POxy.* 1692.5 (ii AD), *PGiss.* 56.11 (vi AD), ἐργασία *PHamb.* 23.22 (vi AD).

χειρισμογράφος, after '*Stud.Pal.* 20.81.4 (iv AD)' insert '(cf. *ZPE* 22.103, χιρισ- pap.)'

χειριστής, after '*manager, administrator*' insert 'minor official who acted as agent for the πράκτωρ in the collection of dues' and add '*PMich.*XII 640.1 (i AD), *IEphes.* 3239 (ii/iii AD)'

*χειρίτεχνος, ον, *hand-made*, ἔρια κερίθεκνα *Inscr.Cret.* 4.75*B*4 (Gortyn, v BC), cf. χειροτεχνία, -τεχνος.

χειροβοσκός, add 'S.*fr.* 164a R.'

χειρογραφή, ἡ, *contract of loan*, *PMich.*II 123ᵛiii 12, al. (i AD).

χειρογραφία 2, add 'sp. χειλο- by dissim., *PVindob.Worp* 17.1 (AD 58)'

χειροδάκτυλος, ὁ, *fingerstall*, χειροδάκτυλοι ἀργυροῖ Anon.Alch. 366.2.

χειροκμής, ῆτος, ὁ, *manual labourer*, Steph. *in Rh.* 270.16 (pl.), cf. χειρόκμητος.

χειρονίπτριον, τό, dim. of χειρόνιπτρον, *hand-basin*, POxy. 3860.35 (iv AD; sp. -νίπτιν).

χειρόνιπτρον I, after 'Eup. 118.1' insert '*IG* 1³.405.5 (*c.*412 BC)' and add 'sp. χερο- Eust. 1353.53'

χειρονιψάτης, perh. *basin for washing hands*, BGU 2360.1 (iii/iv AD).

χειρονομέω I, add 'of a pantomimic actor, *SEG* 31.1072 (Pontus, ii/iii AD)'

χειρονόμος, for 'one who .. posture-master' read '· ὀρχηστής' and after 'Hsch.' add 'but *shadow-boxer*, Didyma 179.3'

χειροπόνιον, τό, *expense of manufacture*, σὺν χειροπονίοις καὶ πάσαις δαπάναις *SEG* 33.1179 (Lycia, i/ii AD), cf. χειροπόνια.

χειρότεχνος, ὁ, = χειροτέχνης, POxy. 38.17 (i AD).

χειροτονέω II 1 b, add 'spec., *ordain by laying on hands, CPR* 5.11.4 (iv AD)'

†χειροτονητής, οῦ, ὁ, *one who appoints* or *elects to a position*, POxy. 2894 ii 37, 2936 ii 9, 12, al. (iii AD); = *creator*, Gloss.

χειροτονία II, add '**4** *power of appointing, disposition, Cod.Just.* 1.3.38.6.'

χειρουργικός I 2, add 'χειρουργικὸν μέρος τῆς τέχνης Gal. 18(2).667.7'

χειροχρήστης, ου, ὁ, *scribe*, pap. in *ZPE* 99.118 (Arabia, ii AD).

χειροψέλ(λ)ιον, τό, *bracelet*, POxy. 3491.5 (ii AD), PKöln 166.20, 22 (vi/vii AD).

χελάδριον, v. ‡χαλάδριον.

χέλειον I, delete '*crab's shell*' and transfer 'Nic.*Al.* 561 .. χέλιον cod.)' to follow '*testudinum*' **II**, for '*Philol.* 90.137' read '*Suppl.Hell.* 415.1.24'

χελιδονιακός, ή, όν, *shaped like a swallow's tail, chelidoniacus gladius,* Isid.*Etym.* 18.6.7.

χελιδόνιος, after 'also ος, ον' insert 'Macho 427 G., Dsc. 5.32' **II 1**, at end delete 'χελιδόνια (sc. σῦκα) .. Epigen. 1.2)'

χελιδονισμός, add 'Eust. 1914.16'

χελιδονοειδής, ές, *coloured like the swallow*, PMasp. 6ᵛ.83 (vi AD).

χελιδών III 5, delete '(with play on Ar.*Lys.* 770 (hex.))' and 'cf. Juv. 6.365(6)'   add '**IV** *kind of (russet-coloured) fig*, Ar.*fr.* 581.4 K.-A.'

χελιχελώνη, add 'cf. °κορικορώνη'

χελύνη II, add 'Dor. χελῡνᾶ Call.*fr.* 196.22 Pf.'

χελύνιον, for '*Mitteil.* .. 160' read '*PRain.* (*NS*) 1.28 (p. 160)'

χελώνη I, χελώνη χερσαία, add '*Peripl.M.Rubr.* 3, 30'   **III 6**, for '*JHS* 10.82' read '*TAM* 2.448'   add '**10** *ship's keel*, Hsch.'

χέραδος, line 5, after 'Pi. l.c.' insert '(cf. Sch. ad loc.)'

χερικός, v. °χειρικός.

Χέρνασος, for 'prob.' read 'Ep. Χέρνησος'; after 'Χερσόνησος' insert 'A.R. 1.925 (v.l.), 4.1175'; for '*Docum.* .. 94' read '*SEG* 9.76 (Cyrene, iv BC)'

χερνιβόξεστον, add '*PWash.Univ.* 59.14 (v AD)'

χέρνιψ, add 'Myc. ke-ni-qa (acc.), also deriv. ke-ni-qe-te-we (pl.)'

χερόνησος, delete 'A.R. 1.925' (v. °Χέρνασος)

†χερόνιπτρον, v. ‡χειρόνιπτρον.

χερσαῖος I, add '**2** *consisting of dry land*, PPetaus 25.22 (ii AD).'

χερσιμῑμάς, άδος, ἡ, *hand-mime actress*, SB 10769.204.

χερσοκαλαμία, ἡ, *land overgrown with reeds*, PMich.v 310.7, al. (i AD).

χερσονομή, add '*PPetaus* 43.23 (ii AD)'

χερσοπαράδεισος, ὁ, *garden land*, POxy. 3205.11, 52, 58 (iii/iv AD).

χέρσος line 8, after '(Pi.)*N.* 1.62' insert '9.43'   **II**, line 3, delete 'ἐν κονίᾳ .. ib. 9.43'; cf. κόνιος I.

Χέρχνος, ὁ, v. °κέρχνος A.

χέω I 3 b, add '*APl.* 119 (Posidipp.)'

χηβάδις, ἡ, *ewer* or sim. *vessel*, Anon.Alch. 322.16.

χηλή II 1, line 3, after 'sg.' insert '*App.Anth.* 3.80 (Posidipp.)'

χήλωμα, add '**II** *box, chest*, POxy. 1294.5 (χειλ-, ii/iii AD), cf. χηλός.'

χηλωμάτιον, τό, dim. of χήλωμα, *small box, chest*, POxy. 1294.3 (χειλ-, ii/iii AD).

χηλωτός, ή, όν, adj. app. fr. χηλή or χηλός (*chest*), PRyl. 627.63 (iv AD).

χηνάγριον 1, delete '*young*' and add '*PMil.Vogl.* 305.22 (ii AD)'

χηνᾶς, ᾶ, ὁ, *goose-keeper*, SB 5377 (*ZPE* 20.231).

†χηνιάζω, *make a noise like a goose*, of an incompetent piper, Diph. 78 K.-A.'

χηνίς, ίδος, ἡ, perh. *gosling*, *IG* 11(2).224*A*11 (Delos, iii BC).

χηνοβοσκία, gloss on χηνοβοσκός, Hsch.

χηνόπους, ποδος, ὁ, ἡ, app. *flat-footed*, of a woman, *IG* 12(3).388.

χηραμός, line 6, after 'Hom.' insert 'and Lyc.'

†χήρειος, α, ον, *bereaved of a relative:* of a husband, λέκτρα *AP* 9.192 (Antiphil.); Ion. χηρήϊος, of children, οἶκος Antim. 81 Wy.

χηρεύω II, after 'E.*Cyc.* 440' insert '(dub. l.)' and for 'χηρεύσει' read 'χηρεύει'

χῆτος, add 'Hes.*Th.* 605, Philostr.*VA* 2.39'

χθεσῑνός, add 'of bread, Aët. 3.177 (349.28 O.)'

χθιζός, last line, after 'Il. 19.195' insert 'A.R. 4.1397, *AP* 9.305 (Antip.Thess.)'

χθονικός, ή, όν, = χθόνιος, Ἑρμῆν χθονικόν *Tab.Defix.* 107.3.

χθόνιος I, add '**2** *characteristic of* or *suitable to the underworld*, i.e. *gloomy, severe*, Anacr. 60 P., ῥυσμοί id. 71.2 P., cf. Suet.*Blasph.* 98 T.'

χίασμα 1, add '*PFlor.* 233 (*ZPE* 78.98)'

χιδροβρόχον, τό, app. *vessel for soaking χῖδρα*, Kafizin 219 (223/222 BC).

χίδρον, delete '*unripe*'

χιθών, v. °χιτών.

χιλᾶς, ᾶ, ὁ, perh. a trade designation (?*dealer in fodder*), *SEG* 36.970.B29 (Aphrodisias, iii AD).

χιλεύω I, add 'χειλεύει στρατόν Hsch.'

χιλιαγωγός, ὁ, *leader of a thousand*, epith. of God, *JÖAI* 32.80 (amulet).

χιλιαρχία II 3, for '*AJA* .. BC)' read '*Sardis* 7 no. 1i6 (iii/ii BC)'

χιλιαστήρ, add '*BCH* 59.478 (Samos)'

χίλιοι 3, add 'also masc. (sc. στατῆρες), *SEG* 32.794 (Olbia, *c.*325 BC)'

χῑλιοπλᾱσίων, for '= foreg.' read '*a thousand times as much (as many, as great)*'

χῑλός 1, line 1, before '*green*' insert '(also χιλόν, τό, Hsch.)'   **2**, add 'pl., Nic.*Th.* 569'

χιλωκτός, ή, όν, or χιλωκτόν, τό, [?ῑ] perh. *fodder-bag*, PMich.VI 421.24 (i AD).

χίλωμα I, add '*PLond.* 190.45 (?iii AD)'

χίμαρος, after 'Ar.*Eq.* 661' insert '*SEG* 33.167.20 (Attica, 440/30 BC)'

χιονοβροχοπᾱγής, ές, *snow- and rain-congealing*, PMag. 4.1358.

χιονοδροσοφερής, ές, *bringing snow and dew*, PMag. 4.1362.

χιονοειδής, ές, *snowy in appearance*, of the sun appearing as an omen, Orac.Tiburt. 27 (cf. χιονώδης).

†χιονόομαι, *to be snowed on*, Lxx Ps. 67(68).15 (see also χιονίζω).

Χῖος I 1 b, add '(see also χῖον, ἡμίχιον)'   **II**, line 3, after '*ace-dot*' insert '*Epigr.Gr.* 1038 (χεῖ-)'; line 5, for '(Poll.) 205' read '(Poll.) 9.100'

χιρίδιον, v. °χειρ-.

χιρισμογράφος, v. ‡χειρ-.

χιτών I 1, add 'sp. χιθών, BGU 816.18'   **2**, add 'as ceremonial garment for goddess, *IEphes.* 2 (*c.*340/320 BC)'   at end add 'Myc. ki-to, ki-to-ne (pl.)'

Χιτώνη, add '**Κιθωνέα**, Hsch.'

χιτωνίσκιον, add '*IG* 2².1515.20; also χιθωνίσκιον, ib. 1516.7'

χλαῖνα, line 10, delete '*of husband and wife*'

χλᾰμῠδαρχικός, ὁ, app. *one in charge of military cloaks*, PPanop. 17.2, 18.2 (both AD 329).

χλᾰμῠδηφορέω, *wear a chlamys* (as an ἔφηβος), *SEG* 28.1458 (Antinoopolis, ii AD), *JEA* 37.87.2 (Memphis, iii AD); also **χλαμυδοφ-** PMich.VI 426.18 (ii/iii AD); glossed by Θετταλίζειν, Poll. 7.46; also written χλαμυρο-, POxy. 2895.6 (iii AD).

χλᾰμῠδηφόρος, add 'as epith. of Hermes, PMag. 5.403'

χλᾰμῠδοφορέω, v. °χλαμυδηφορέω.

χλανίαι· περιβολαί Hsch. (cf. χλανίτιδες, κλανίον, etc.).

χλανίδιον (form χλάνδιον), add '*SEG* 38.1210.22 (Miletus, ii BC)'

χλανιδοφόρος, ον, *wearing a* χλανίς, Archipp. 50 K.-A.

χλαρ(ον), perh. a kind of gem, PRyl. 627.159 (iv AD).

χλεύη, for 'Aeschrio 8' read '*AP* 7.345 (Aeschrio)'

†χλῆδος, ὁ, app. *rubble, debris*, A.*fr.* 16 R., D. 55.22, 27; transf., ἀργυρίου χλῆδον λαβών Crates Com. 31 K.-A.

χλιαίνω, line 10, delete 'al.' and after '(Mel.)' insert '*AP* 12.136'

χλιαροπᾱγής, ές, *easily fusible*, Anon.Alch. 31.3.

χλιαρός, add '**b** neut. pl. subst., *warm baths*, *SEG* 32.1502, 1503 (Palestine, v AD).'

χλιάω, delete '(sed leg. χλιαρόν)' and for 'ἀτὰρ .. Hesperia 5.95' read 'μὴ κούρας ἅτ᾽ ἀηγὸς (= ἀρωγὸς) ἀφάσσων στέρνα πόθῳ χλιάοι Hesperia 5.95'

χλιβίον, v. ‡κλουβίον.

χλιδώδιον, τό, app. = sq., *SEG* 39.163 (Attica, iv BC).

χλιδών, for 'Plu. 2.145a (prob. l.)' read 'Plu. 2.317f'

χλιδώνιον, add '*IG* 2².1457.8'

χλόϊα, before '*IG* 2².949.7' insert '*IG* 1³.250.A26, B31 (v BC)'

χλόος, add '**II** pl., *green foodstuffs*, PMich.VIII 496.17 (ii AD).'

χλούβιον, v. ‡κλουβίον.

χλωραίνω, add '-κόν, τό, *green fodder*, τῷ θ᾽ (ἔτει) χλωρικῷ ξυλ(αμῆσαι) POxy. 3911.25 (AD 199)'

χλωρόπαστος, ὁ, *shot with green*, name of a stone, Socr.Dion.*Lith.* 39.1 G.

χλωρός I 1, lines 8/9, for 'ἡ .. scenery' read 'ἡ or τά, *something required for depicting a river* (perh. personified) *in a mime*, perh. *green stain* or *green draperies*' and add '**b** χλωρά, τά, *green crops*, opp.

313

πυρός, POxy. 501.16 (ii AD).'   **III**, lines 6/7, delete 'of fish .. 7.309b'

χλωρότης I, for 'greenness' read 'green or yellowish colour'; line 2, after 'Lxx Ps. 67(68).14' insert 'Plu. 2.952c; of the pallor of gold mixed w. silver, ib. 395d'   **II**, delete the section.

*χλωροφαγία, ἡ, green fodder, PLond. 1165.3 (ii AD), PMasp. 87.13, BGU 2139.14 (AD 432).

*χλωροφόρος, ον, bearing (green) fodder, χ. (sc. γῆ) PMich.II 123ʳiii.6 (i AD), PTeb. 553 (i/ii AD).

χνοάω, line 1, for 'cheeks' read 'breasts'

χνόος, line 4, after 'Od. 6.226' insert 'Hippon. 115.9 W.'   **II 1**, add 'applied to the first coat of a foal, AP 6.156 (Theodorid.)'

χνόος (B), v. °κνόος.

χοανεύω **II**, add '2 melt down metal objects, Just.Nov. 120.10.'

*χοάνιον, τό, dim. of χοάνη, rest. in IG 1³.427.57 (Athens, v BC).

*χοαχυτίς, ίδος, ἡ, fem. of χοαχύτης, keeper of mummies, UPZ 189.4 (ii BC).

χοδέαντες, delete the entry (v. °χοδέω).

*χοδέω, defecate, = χοδιτεύω (v. χοδιτεύειν), Sophr. in PSI 1214d5.

†χοΐδιον, τό, small pouring vessel, SEG 21.557.9 (Athens, iv BC).

χοιρίσκος, after 'dim. of χοῖρος' insert 'IG 5(1).1390.68 (Andania, i BC)'

*χοιρόγυνος, ὁ, a Nile fish, PYale 56 (100 BC), cf. χοῖρος II.

*χοιροθυσία, ἡ, sacrifice of young pig, POxy. 3866.3 (χυρ-, vi AD).

*χοιροκαταγωγεύς, έως, ὁ, pig-drover, cj. in BGU 92 (ii AD; v. ‡καταγωγεύς).

χοῖρος, line 1, after 'Ach. 764' insert 'SEG 33.147 (Athens, v or iv BC), Herod. 8.2'

χοιροσφάγειον, add 'PCair.Cat. 10703 (Aegyptus 68.38; vi/vii AD)'

*χοιρότριψ, ιβος, ὁ, obsc. compd. of χοῖρος I 2 and -τριψ (τρίβω), Hdn.Gr. 1.246.26.

*χοιροτρόφος, ὁ, swineherd, Hsch. s.v. συβώτης.

*χοΐσκιον, τό, dim. of χοΐσκος, IG 2².1533.102.

†χοΐσκος, ὁ, pouring vessel, (dim. of χοῦς A), IG 2².1533.115 (Athens, iv BC), Inscr.Délos 1426Aii15 (ii BC).

χολέδρα **2**, add 'written χολέτρα, POxy. 3285.32, 35, 37 (ii AD)'

χολή **I 2**, sg., add 'Nic.Th. 561' deleting this ref. fr. section II.   **3**, line 1, for 'Poets' read 'Com.'

*χολιάζω, make angry, SEG 38.1233 (ii AD).

χολοβαφής, add 'Alex.Aphr. in SE 48.22; also κολοβαφής, Hsch. s.v. κολοβάφινα'

χολοβάφινος, add 'also κολοβ-, ἀγαλμάτιον κολοβάφινον ἐν ναϊδίῳ Inscr.Délos 1416Ai12 (ii BC), βοιδάριον ib. 17, 1423Baii22, δακτυλίδιον 1439Bbái94, cf. Hsch.'

*χολοποιέω, produce bile, Vit.Aesop.(G) 3.

χόλος **I**, add '2 snake's venom, AP 7.172.6 (Antip.Sid.).'

*χονδήν, adv., in capacity, φιδάκνας ἀμφορέων χ. SEG 21.644.19 (Attica, iv BC) (acc. of *χονδή, cf. χανδάνω).

*χονδρίασις, ἡ, the condition of swelling with clots of milk, of women's breasts, Aët. 16.36 (p. 52 Z.).

*χονδρόγαλα, τό, mash of milk and crushed grain, SEG 32.1243 (Cyme, Aeolia, 2 BC/AD 2).

*χονδροκόπος, ὁ, grinder of groats, CE 42.356 (iii BC).

χονδροσύνδεσμος, delete the entry (read χόνδρῳ σύνδεσμος).

χοοφορία, add 'PSoterichos 1.24 (AD 69), 2.20 (AD 71)'

χοραυλέω, -αύλης, for 'flute' read 'aulos'

χορεῖος **I**, after 'of .. dance' insert 'ῥυθμός (cf. II) Men.Dysc. 951'; delete 'cf. Ael.NA 2.11'; epith. of Dionysus, add 'SEG 30.86 (ii AD)'

χορεύω, line 3, delete 'E.Ion 1084 (lyr.)'   **I 3**, read 'fig., take part (in a particular group) Pl. Tht.173c; w. dat., to be a votary of, παιδείᾳ Phld.Rh. 1.141, φιλοσοφίᾳ ib. 2.271.'   **2**, add 'in erotic contexts, Nonn.D. 8.228, 16.67, σὲ γὰρ τέκε χάλκεος Ἄρης Κύπριδος ἐν λεχέεσσιν Ἐρωτοτόκοιο χορεύων ib. 34.117'   **III**, line 2, after '(lyr.)' insert 'so prob. in A.fr. 204b. 1 R. (lyr.), unless in sense II 2'

*χορηγεσία, ἡ, Dor. χοραγ-, = χορηγία I 1, epigr. in Lindos 197f6 (ii BC).

*χορηγητῶς, adv., Dor. χοραγ-, on a grand scale, AS 17.81.9 (Bithynia, ii AD).

χορηγία **I 1**, add 'ποιησάμ[ε]νος τὰν χοραγίαν SEG 32.1243.45 (Cyme, 2BC/AD 2); personified, ἄνασσα Χοραγία SEG 30.133 (Athens, after AD 130)'   **II 2 b**, add 'furniture of the Temple, Lxx 2Es. 5.3; pl., supplies, Diog.Oen. 64 ii 1 S.'

χορῑτεία, after ' = χορεία' insert '(unless a mistake for that word)'

*χοροδιδασκαλέω, train a chorus: fig., ἐντέχνως χ. ἐν φιλοσοφίᾳ Phld. Lib. col. 3 line 9.

χοροιτυπία, sg., add 'Panyas. 16.15 B.'

*χοροστατέω, before 'Hsch.' insert 'IGRom. 1.562 (Nicopolis ad Istrum)'

χορτάριον, add '2 hay, BGU 625.33 (ii/iii AD), POxy. 1862.37 (vii AD).'

χόρτασμα **1**, after 'fodder, forage for cattle' insert 'PHamb. 27.17 (iii BC)'

χορτάχυρον, for 'chopped hay' read 'grass straw' and add 'PMich.XII 650.27 (iii AD)'

*χόρτη, v. °χώρτη.

χορτηγός, add 'ὑποζύγια PCair.Zen. 292.480 (iii BC)'

*χορτόβρωμα, ατος, τό, grazing, PSI 1327⁷7 (ii AD).

χορτοκοπή, add 'PSI 1327⁶6 (ii AD)'

*χορτοκόπιον, add 'IGBulg. 1401.7 (ii AD), IMylasa 257.9'

*χορτοπάτημα, ατος, τό, threshed straw, POxy. 2985.2, 6, 2986.7, 8 (both ii/iii AD).

χόρτος **II 1**, add 'χ[όρτ]ου ἀρακίνου ἀρουρῶν δέκα τριῶν PVindob.Worp 3.19 (AD 321)'

*χορτωνέω, purchase food, PCair. 10311.9 (APF 33.5; iii AD).

*χούζιον, τό, perh. a kind of gourd, ICilicie 108.6 (v/vi AD).

χοῦς (A), line 1, delete 'Nic.Th. 103'; line 2, before 'measure of capacity' insert 'vessel or'

χραισμέω **2**, add 'b w. dat. of thing, have an antidote for, χραισμήσεις ὀφίεσσι Nic.Th. 551.'

χραίσμημα, delete the entry.

χραίσμησις, delete 'Nic.Th. 926'

*χραύω **II**, for this section read 'med. w. gen., Cypr. to-ka-ra-u-o-me-no-ne τὸ(ν) χραυόμενον Ὀ(γ)καντος ἀλϝō be adjacent to, adjoin, ICS 217.9; app. variant to-ka-ra-u-zo-me-no-ne ICS 217.18'

χράω (B), line 5, after 'Hermesian. 7.89' insert 'A.R. 1.302'   **C III 4 a**, ἔς τι, add 'E.Med. 821, D. 19.30'; πρός τι, add 'Hdt. 4.87.2, Lys. 24.24'; ἐπί τι, add 'Pl.Grg. 508b'   **VII**, at end delete 'Hsch.' and 'has' and after 'χρησιμεύσει' insert 'Hsch.; also perf., Phryn. 206'

χρεία, line 2, for 'Call. in PSI 11.1216.43' read 'Call.fr. 195.33 Pf., Cretan χρῆια BCH 109.189ff.'   **V**, line 3, for 'Theon Prog. 5, etc.' read 'Men.Rh. 392.31 R.-W.'; line 4, delete 'Aristipp. etc.'

†χρειᾱκός, ὁ, app. employee, agent, or sim., Peripl.M.Rubr. 16, BGU 14 ii 9 (iii AD); in the service of a temple, pap. in Mélanges Desrousseaux p. 199 (ii AD).

†χρεονόμος, ὁ, title of an official, SEG 25.447 (Arcadia, iii BC), ABSA 26.166 (Sparta, ii AD).

†χρεοφυλάκιον, -φυλάκέω, etc., v. ‡χρεω-.

χρεῦμα· ῥεῦμα, ὕδωρ Hsch. (prob. conflated fr. χεῦμα and ῥεῦμα).

χρεών **II**, add 'μετέστη εἰς τὸ χρεών i.e. died, Mitchell N.Galatia 223 (AD 165)'

*χρεωστικός, ή, όν, of a χρεώστης, ἀσφάλεια BGU 472.2.11 (ii AD).

χρεωφυλακικός, add 'and YClS 3.26-47'

χρεωφυλάκιον, for 'office .. is kept' read 'office of the °χρεωφύλαξ' and add 'τῶν ἐν τῶι χρεοφυλακίωι βυβλίων Salamine 90 (ii BC)'

χρεωφύλαξ, for 'keeper of the register of public debtors' read 'keeper of documents relating to debts' and add 'Salamine 90 (ii BC)'

χρή **I 1**, add 'b εἰ χρή if it is proper, if one should, ὦ μῶρος, εἰ χρή δεσπότας εἰπεῖν τόδε E.Med. 61, IT 1288, resumed by χρὴ δέ id.El. 300, HF 141.'

χρῄζω, line 5, after 'SIG 56.23 (Argos, v BC)' insert 'SEG 34.282 (Nemea, iv BC)'

χρῆμα **II 1**, line 6, for 'τεκμαίρει .. O. 6.74' read 'χ. ἕκαστον everything, Pi.O. 6.74, 9.104'   **III**, for 'oracle' read 'oracular pronouncement or decree' and add 'Swoboda Denkmäler 107'

χρηματῐσις, after '-ῑσις' insert '(-ιξις Inscr.Cret. 4.232.3 (Gortyn, ii BC))'

χρηματισμός **I 5**, add 'INikaia 1071 (i/ii AD)'

χρηματιστής, add '**III** giver of oracles, Θεοὶ χρηματισταί Pap.Lugd.Bat. xxv 8 i 1 (after AD 231).'

*χρηματοθήκη, ἡ, treasury, τὰς ἱερὰς χρηματοθήκας JEA 56.179 (63 BC).

†χρηματοφύλαξ, ακος, ὁ, treasurer, PErasm. 10 (ii BC), PRyl. 586.9, al., (i BC); used to translate Lat. praefectus aerarii, Vett.Val. 38.34 (38.6 P.).

χρήσιμος **I 1**, add 'w. inf., ἀπολεμεῖν χ. Pl.Phdr. 260b'

χρησιμότης, add 'Epigraphica 10.76.23 (Leptis Magna, ii/iii AD)'

χρῆσις **II**, add 'oracular saying, Inscr.Perg. 8(3).34.16'

χρησμοδοτέω, line 1, after 'Poll. 1.17' insert 'prophesy, ἀληθείας Ramsay Cities and Bishoprics 2.566 (iv AD)'

χρησμολόγος **II**, add 'BCH 116.279.4 (Colophon, ii BC)'

χρηστεύομαι, add 'Lxx Ps.Sal. 9.6'

χρήστης **II 2**, line 2, after 'D. 32.12' insert 'AP 7.732 (Theodorid.)'

χρηστομαθής, for 'an adept in polite' read 'desirous of'

*χρηστομουσία, ἡ, an excellent abode of learning, Γάδαρα χ. RA 35 (1899/2).49 (Palestine).

χρηστός **II 6**, delete the section.   **III**, line 1, after 'Hp.Art. 32' insert 'Arist.EE 1214ᵃ21'   add '**IV** in treaty between Sparta and Tegea χρηστὸν ποιεῖν was interpr. by Arist. (fr. 592) as ἀποκτιννύναι, and χ. there may be good as euphemism for dead.'

*χρηστοσύνη, ἡ, goodness, IG 9².662 (Locris, ?iii BC).

*χρίθη, ἡ, v. ‡κριθή.

χρῖσμα **III**, add 'La Carie p. 363'

χρίω, line 9, delete 'only in late poets, as' **I 1**, line 4, after 'Il. 23.186' insert 'w. gen., ἐλαίου στέρνα χρίουσιν Achae. 4 S.'

χροιά II, line 7, for 'μεμεγμένας' read 'μεμιγμένας'

*χροῖα, τά, app. used as pl. to χρώς II, Emp. 71.3 D.-K.

χροΐζω I, line 3, after 'of a woman' insert 'Call.fr. 21.4 Pf.'

*χρονέω, = χρονίζω (in quot., sense I 4), χρονέεσκε AP 5.77 (Rufin.).

χρόνιος I 3, add 'ὕλας τὰς χρονίας Salamine 204 (v AD)' add '**6** pertaining to the lapse of time, χ. παραγραφή Cod.Just. 1.3.45.12.'

*χρόνισις, εως, ἡ, expenditure of time, ἐπί τινι πράγματι χ. Anon. in Rh. 176.29.

χρονογραφία I, delete 'αἵ χ. καὶ ἡ Ἀτθίς'; add 'cf. Hesperia 26.164.23'

χρονοκράτορία, add 'Heph.Astr. 2.30 tit., 2.31 tit., al.'

χρονοκράτωρ, for 'Ptol.Tetr. 209' read 'Ptol.Tetr. 290' and add 'Heph.Astr. 2.26.18, 19, 2.29.1, 11'

χρόνος I 3 a, add 'τίνα χρόνον; at what time? (if the following words are sound), Call.Del. 1' **b**, add 'χρόνου for a while, Ael. NA 4.45, ἤδη χρόνου ib. 5.33' **IV**, add 'χρόνον ἐμποεῖν τῷ πράγματι Men.Dysc. 186'

χρυσαλλίς II, delete 'old' before 'name'

*Χρυσαμπελῖται, οἱ, name of a guild at Attouda, MAMA 6.84.

χρύσαμπυξ, lines 2/3, delete 'epith. .. Od.'; line 4, after 'B. 5.13' insert '**2** with golden headband, of the gods' horses, Il. 5.358, 363, al.'

χρυσάνθεμον 4, delete the section.

χρυσανθής I, add 'κιθών PMich.II 121ʳ iv i 3 (i AD)'

χρυσάνθινα, add 'perh. also SEG 31.1044 (Sardis)'

*χρυσάνθιον, τό, sulphate, Maria Alch.ap.Zos.Alch. 146.13, Anon.Alch. 15.13.

Χρυσαορεῖς, (s.v. χρυσάορος), add 'Amyzon 16 (iii/ii BC)'

*Χρυσαορικός, ή, όν, of or belonging to the Χρυσαορεῖς, Amyzon 28 (ii BC).

†χρυσαστράγαλος, ον, having gold bosses, φίαλαι Sapph. 192 L.-P.

χρυσάφιον, add 'PMich.inv. 1363 (BASP 16(3).196; iii AD), χρυσάφιν PSI 836.13 (vi AD), cf. mod. Gk. χρυσάφι'

*χρυσειδής, ές, = χρυσοειδής, Hymn.Is. 109.

*χρύσειον, read '**χρύσειον**' and add 'Pl.Lg. 742d'

*χρυσελάτης, ου, ὁ, goldbeater, Edict.Diocl. 30.5; also -ηλάτης, Anon. Alch. 379.8.

*χρυσέλατος, ον, = χρυσήλατος, epigr. in Mansel Ausgrabungen in Side (1951).54.

χρυσέμπαικτος, delete the entry (v. °χρυσέμπαιστος).

*χρυσέμπαιστος, ον, embossed with gold, BGU 781 iv 1 (i AD).

*χρυσεόκαρπος, v. ‡χρυσόκαρπος.

χρυσεόκυκλος, add 'epith. of Horus, PMag. 4.460'

χρυσεομίτρης and -μίτρα, add 'also -μιτρος, ον, AS 9.104 (Pisidia, ii AD)'

χρύσεος I 3, add '= Lat. aureus (cf. °χρύσινος), Just.Nov. 121.1' **III 1 a**, add 'epith. of Zeus, Robert Hell. 10.105; χ. Παρθένος, partner of Hosion Dikaion, ib. 107; of Artemis, INikaia 1501 (Rom.imp.)' **b**, before 'Luc.Laps. 1' insert 'Men.Dysc. 675'

*χρυσεόστροφος, ον, wearing a golden girdle, Ibyc. 1(a).40 P.

*χρυσετήσιος (sc. λίθος), a stone also known as αἱματίτης, Anon.Alch. 8.1.

*χρυσηλάτης, v. °χρυσελάτης.

χρυσήλατος, add 'cf. °χρυσέλατος'

χρυσήνιος, add 'of Apollo, TAM 4(1).48 (Rom.imp.), PMag. 2.91'

χρυσίζω, **I** add 'τῷ χρώματι χλωροί καὶ χρυσίζοντες (ὄφεις) Peripl.M.Rubr. 40'

χρύσινος I, add '= Lat. aureus (cf. °νόμισμα II), Cod.Just. 6.4.4.10'

χρυσῖτις I, add 'Plin.HN 33.106 (chrysitis, -im)' **II 1**, add 'cf. Plin.HN 37.179'

*χρυσογραμμία, ἡ, writing in letters of gold, Anon.Alch. 327.1.

*χρυσόδεσμος, ον, having gold fastenings, Hsch. s.v. χρυσάμπυκες.

*χρυσοδότης, ου, ὁ, distributor of gold, epigr. in TAM 3(1).127.5 (Termessus); of Ammon, SEG 26.1721 (Egypt).

χρυσόθειρ, line 3, fem. -έθειρα, add 'Ibyc. 1(a).9 P.'

χρυσόζωνος, for 'Hes.fr. 278.4' read 'Suppl.Hell. 1168.4'

χρυσόκαρπος, add 'also χρυσεό-, κλάδα χρυσεόκαρπον Suppl.Hell. 1056'

*χρυσοκέλευθης, ου, ὁ, travelling a golden path, χρυσοκέλευθα, voc., of Apollo, PMag. 2.91.

χρυσοκέλευθος, delete the entry.

χρυσοκόλλα I, for 'basic copper carbonate' read 'or other green copper mineral' add '**2** see quot.: τὸ .. δι᾽ οὔρου παιδὸς σκευαζόμενον φάρμακον ὃ καλοῦσιν ἔνιοι χρυσοκόλλαν, ἐπειδὴ πρὸς τὴν τοῦ χρυσοῦ κόλλησιν αὐτῷ χρῶνται Gal. 12.286.15.'

χρυσόκομος, after 'golden-haired' insert 'χρυσόκομ᾽ Ἄπολλον CEG 308' and after '(Mnasalc.)' insert '(s.v.l.)'

*χρυσοκοράλλιον, τό, dim. of χρυσοκόραλλος, Moses Alch. 307.5.

*χρυσοκόσμητος, ον, adorned with gold, Sch.E.Rh. 382.

*χρυσολάμπετος, ον, sparkling with gold, ῥάβδῳ Hippon. 79.7 W.

χρυσολαμπίς II, add 'Socr.Dion.Lith. 45 G.'

*χρυσολευκόλιθος, ον, (made) of gold and white marble, Sch.D.T. 378.11; also χρυσεο- Hdn. 2.849.1.

χρυσόλιθος, for 'topaz' read 'yellowish precious stone, perh. peridot' and add 'Peripl.M.Rubr. 39, 49, 56'

χρυσολύρης, delete 'Pi.Pae. 5.41'

χρυσόμαλλος, add 'τὸ χρυσόμαλλον Αἰήτου δέρος Trag.adesp. 37a K.-S.'

χρυσομανής, add 'μελέτη AP 10.76.4 (Paul.Sil.)'

χρυσομίτρης, add 'also χρυσόμιτρος, Hsch. s.v. χρυσάμπυκας'

*χρυσόμιτρος, v. ‡χρυσομίτρης.

*χρυσονεστριεύς, έως, ὁ, gold-spinner, Edict.Diocl. 30.6.

*χρυσονομέω, serve as ‡χρυσονόμος II or treasurer, SEG 30.1340 (Miletus, ii/i BC), al.

*χρυσονομία, ἡ, office of ‡χρυσονόμος II, Didyma 486 (188/7 BC).

*χρυσονόμος II, add 'title of a treasurer at Miletus, SEG 30.1343 (49/8 BC)'

χρυσόπαστος, add '**2** subst., χρ., ὁ, precious stone, Epiph.Const. in Lapid.Gr. 197.19.

*χρυσόπεζα, ἡ, gold-footed or -sandalled, POxy. 2444.3.9 (perh. Pi.).

*χρυσοπέταλον, τό, gold-leaf, Anon.Alch. 377.7 (lemma).

χρυσοποιός, add 'SB 10769.206 (iii/iv AD)'

χρυσόπτερος I, add 'χρυσόπτερε παρθένε (the Muse), Stesich. 16.11 P.'

χρυσοπτερύγος, add 'IKourion 104.16 (ii AD)'

χρυσορόης, add 'also χρυσορόας NC 144(1984).52ff. no. 43 (Hierapolis, coin of Annia Faustina)'

χρυσόροφος, add 'SEG 36.1099 (Sardis, v/vi AD)'

χρυσός 2, add '**b** a coin, = Lat. aureus (cf. °χρύσινος), Cod.Just. 3.2.2 pr. 1.' **3**, after 'Pi.O. 7.50' insert 'applied to a lover, BCH 106.3 ff.' at end for '(Borrowed .. yellow)' read 'Myc. ku-ru-so, also app. as adj. (Semitic borrowing, cf. Akk. ḫurāṣu, Hebr. ḥārûṣ, etc.)'

*χρυσοσημέω, embroider with gold, CPR 8.65.6, 8, 14, 15 (vi AD).

*χρυσόσκαλμος, ον, with golden oar-pins, of a ship, BIFAO 70.13.

*χρυσοτέλεια, ἡ, = Lat. aurum coronarium, REG 70.120 (Caesarea, vi/vii AD).

χρυσότυπος, before 'E.El. 470' for 'κράνος' read 'χρυσοτύπῳ κράνει'

χρυσουργός, add 'form χρυσοϜοργός, Myc. ku-ru-so-wo-ko'

χρυσοφαής, add 'Nonn.D. 42.495'

*χρυσοφόρμιγξ, ιγγος, (prosody uncertain) possessing or associated with a golden lyre, Ἀπόλλων Simon. 6.fr.1(a).5 P.

χρυσοφόρος I 2, add 'IEphes. 3263 (ii/iii AD)'

*χρυσόφυλλον, τό, kind of chrysolite, Epiph.Const. in Lapid.Gr. 197.16.

χρυσοχαλίνωτος, ον, gloss on χρυσάμπυξ, Sch.Gen.Il. 5.358.

†χρυσόχαλκος, ὁ, app. = aurichalcum, a copper alloy, SEG 39.1180.67 (Ephesus, i AD).

*χρυσοχεύς, έως, ὁ, goldsmith, PCol. 214.6 (AD 85/86).

*χρυσοχοοποίησις, εως, ἡ, fusion of gold, Comarius Alch. 291.11.

χρυσοχόος 2, add 'PPetaus 69.34'

χρυσόχροος, add 'χρυσόχ⟨ροε⟩ Παιάν rest. in epigr. in SEG 24.1244.2 (Nubia, ii/iii AD)'

χρυσυποδέκτης, add 'ITyr 90 (χρυσουπο-, Chr.)'

χρυσῶπις, add '**III** χ. (sc. λίθος), ἡ, precious stone, Plin.HN 37.156.'

†χρυσώρυφος, v. ‡χρυσωρύχος.

χρυσωρύχος, for 'Supp. .. iv AD' read 'GVI 1170 (Phrygia, iv AD, lapis χρυσώρυφα' and at end delete 'cf. χρυσώρυφος'

χρύσωσις, after 'gilding' insert 'Inscr.Délos 290.231, 234 (iii BC)'

χρώζω 1, line 1, after 'E.Ph. 1625' insert 'pass., ὡς μάτην κεχρώσμεθα κακοῦ πρὸς ἀνδρός id.Med. 497' deleting this passage in section 3.

χρῶμα II 2, add '**b** τὰ χ. the paints, i.e. painting, Chor. p. 280 F.-R.'

χρωμάτινος, after 'coloured' insert 'or for colouring' and add 'SEG 36.267 (61/0 BC)'

*χρωματογράφος, ὁ, painter, Anon.in Rh. 26.17.

*χρωμάτωτός, όν, coloured, στιχάριν ἔμπλουμον χρωματωτόν Stud.Pal. 20.275.4 (vi AD).

χρώς I 2, lines 3/4, after 'Pherecr. 30' insert 'εἰς χρόα κειράμενοι AP 7.446 (Hegesipp.)'

χρωτίζω, add '**2** med., = χροΐζομαι, Mim.adesp. 1.36 C. (without dat.); fut. pass. ib. 26.'

χυδαῖος II, add '**3** geom., square (as opp. to ἀγελαῖος cubic), POxy. 3455.8, al. (iii/iv AD).'

χύδην, line 1, for 'Call.fr. 1.11 P.' read 'Call.fr. 228.11 Pf.' **I**, after 'Pl.Phdr. 264b' insert 'IG 2².1491.7 (Athens, iv BC)'

χῦμα 3, add 'quantity (of powdered, granular, or sim. substance), κασίας χύμα πλεῖστον Peripl.M.Rubr. 10'

*χυματίζω, gloss on κλύζω, Sch.Nic.Al. 140a Ge.

χυμός I 2, line 1, after 'Hp.VM 18' insert 'Gorg.Hel. 14'

*χυτάργυρος, ὁ, liquid silver, Anon.Alch. 16.14.

χύτλον 1, for '*Berl.Sitzb.* 1927.161' read '*SEG* 9.72.49'
χύτρα, line 4, add 'Ϟύτρα Lang *Ath.Agora* xxi K2 (vi BC)'   **I 1**, at end delete 'children .. χυτρίζω'
χυτρόγαυλος, add 'Men.*Dysc.* 505, 506'
χύτρος I 1, add '*SEG* 35.113 (Attica, *c*.300 BC)'   **II 2**, add 'τοὺς Κύθρους *IG* 2².2130.69 (ii AD)'
χωλός II 2, after 'Demetr.*Eloc.* 301' insert 'Call.*fr.* 203.14, 66 Pf.'
χωλόω, for 'Did. ad D. 11.22' read 'Did. *in D.* 13.6 P.-S.'
χώλωμα, delete '(Hp.*Art.*) 64' and add 'applied also to impairment or deformity of the arms, Hp.*Art.* 64'
χώλωσις, after '*lameness*' delete 'Hp.*Art.* 66' and add 'applied also to impairment or deformity of the arms, Hp.*Art.* 66'
χῶμα I 4, for '*mole .. jetty*' read '*mole* or *causeway* projecting into the sea   **5**, delete the section.   **II**, after 'Hdt. 1.93, 9.85' insert 'A.*Supp.* 870 (acc. to sch. = ἄκρα)'   **III**, for '*mass of soil* in which roots are found' read '*deposit of soil*'   add '**b** applied to a dunghill, ἐκ χώματος κοπρηγείας *PHaun.* 24.5 (i/ii AD).'
*χωμᾰτεία, ἡ, *work on the dikes*, *PSI* 901 (i AD; -ηα).
χωμᾰτεκβολεύς, add 'Πιᾶς χωματοεγβολ(εύς) *PPetaus* 88.8 (*c*.AD 185)'
*χωμᾰτεπιμελητεία, ἡ, *superintendence of dikes*, *POxy.* 3508.23, 25 (AD 70).
χωμᾰτεπιμελητής, before '*BGU*' insert '*PColl.Youtie* 21.10 (AD 80/81)'
*χωμᾰτοεγβολεύς, -έως, ὁ, v. ‡χωματεκβ-.
χωμᾰτοφῠλαξ, add '*PPetaus* 44.28 (AD 184)'
χωνεία, add '*PMich.*xv 706.8 (ii/iii AD).   **2** perh. *cementing* (unless to be taken as misspelling of κονία), χάλικα μεταφέροντες εἰς χωνίαν ληνῶν *PMich.*xi 620.132, 136 (iii AD).'
χωνευτήρ, add 'Moses Alch. 311.17'
χωνευτός, for 'al.' read 'subst., χωνευτόν, τό, ib.*Jd.* 18.20, *Is.* 42.17 (pl.)'
χώρα I 2, line 5, after 'Paus. 5.17.6' insert '*enclosed space* between two main beams of a building, *room*, *IG* 2².1668.77 (iv BC), *ITrall.* 147'   add '**6** χ. λαμβάνειν, *have place, be possible*, *Cod.Just.* 1.3.45.4; sim. w. εἶναι, οὐκ εἶναι χώραν ἐκκλήτῳ ib. 1.4.29.3, 6.4.4.11a.'   **II 1**, add 'transl. Lat. *territorium*, *IEphes.* 4101.17 (Trajan)'   **2 a**, add 'so perh. ἀγεώργητοι μενοῦσιν αἱ χῶραι *IGLS* 1998.30 (Hama, i AD)'
χωράζω, add '*SEG* 23.201 (Messenia, ii BC)'
χωράφιον, add '*Inscr.Cret.* 4.338.6 (i/ii AD), cf. mod. Gk. χωράφι'
*χωρεῖον, τό, = χωρίον, in quot. transl. Lat. *forum*, *SEG* 31.952 (Ephesus, ii AD).
χωρεπίσκοπος, for '*coadjutor* or *suffragan-bishop*' read '*superintendent of country districts* at a distance from the bishop's seat' and add '*SEG* 30.1675 (sp. χωροεπ-, Syria, vi AD), *Cod.Just.* 1.3.38.2'
χωρέω II 1, add '**c** χωρεῖν ἐπί τινα *take legal action* against, *Cod.Just.* 1.3.55.4.'   **5**, add '*to be transferred* or *paid out*, χωροῦντος το[ῦ ἡμι]σέου μέρους εἰς τὸ δημόσιον *SEG* 24.614 (Maced., ii AD).'   **III 1**, line 10, before '*to be capable of*' insert '*to be big enough to grasp, take in* (mentally), τὴν τοῦ θεοῦ δύναμιν Longin. 9.9'
χωρητικός 1, add 'neut. sg. subst., *capacity*, ἅ καὶ εὔστερνα ὡς πλατέα διὰ τὸ χ. καλεῖ Simp.*in de An.* 68.10'
χωρίδιον, after 'Lys. 19.28' insert 'Men.*Dysc.* 23' and after '[ῐ in' insert 'Men. l.c.'
χωρίον 2, after '*town*' insert 'or *village*' and add 'Mitchell *N.Galatia* 84; fortified, *Amyzon* 19'   **3**, after '*IG* 1².325.10' insert 'Mitchell *N.Galatia* 34; pl., *rural areas*, Just.*Nov.* 73.9, 85.3 pr.'
⁺**χωρισμός**, ὁ, *separation*, λύσις καὶ χ. ψυχῆς ἀπὸ σώματος Pl.*Phd.* 67d, Thphr.*CP* 6.7.3; χ. τῆς ἀφέδρου, ἀκαθαρσίας, *menstrual discharge*, Lxx *Le.* 12.2, 18.19; of abst. things, Arist.*EN* 1175ª20, Plot. 4.7.8; *separation* of a person from an association, ὁ ἀπὸ θεοῦ χ. Hierocl. *in CA* 24 p. 103.24 K.   **II** *going away, departure*, Plb. 5.16.6, D.S. 2.60, 17.10.
χωριστός I, delete the section transferring quots. to section II.
χωρίτης 1, after 'Muson.*Fr.* 11 p. 60 H.' insert 'Alciphr. 3.70.1'
χῶρος II, add '**5** subdivision of τριττύς, *BCH* 78.317 (Ceos, iv BC).'
*χωροφῠλᾰκέω, *act as a* χωροφύλαξ, *SEG* 23.305 (Delphi, *c*.190 BC).
*χώρτη, ἡ, = Lat. *cohors*, χ. ἑνδεκάτης ὀρβανῆς *JÖAI* 4.207 (near Selymbria, i/ii AD), cf. *IGRom.* 3.359 (Sagalassus, Pisidia), *SEG* 30.818 (Moesia, ii/iv AD), etc.; στρατιώτης χόρ(της) η' πρ(αιτωρίας) *IG* 14.1661 (Rome); also κώρτη *IUrb.Rom.* 134 (iii AD).
⁺**χῶς**, v. χοῦς (A).
*χώστρα, ἡ, *melting-pot*, Anon.Alch. 271.22, 287.25.
χωφόριον, add '*POxy.* 3511.24 (iv AD)'

ψ

ψάγδᾶν, for 'or **σάγδας**' read 'also **ψάγδας** (Hsch.), **σάγδας**'; line 6, for 'nom. .. Hsch.' read 'σάγδης, app. fem.gen. Ath. 15.690e, σάγδας 16.691c.'

ψαιστός, add 'AP 6.300, 334 (Leon.Tarent.)'

*ψάκαστρον, v. βουτόρος.

*ψάλαγμα, ατος, τό, *light touch*, ψαλάγματα· ψηλαφήματα Hsch.

ψαλάσσω, add 'Ion Trag. 13a K.-S.'

*ψάλιος, ὁ, see quot.: σημαίνει γὰρ ἡ μὲν (λέξις, sc. ψάλιος) τὸ τιθασευόμενος ἵππος Sch.E.Ph. 793.

+**ψᾰλίς**, ίδος, ἡ, *U-shaped cutting tool* or *shears*, Ar.fr. 332.1 K.-A., AP 6.307 (Phan.), 11.368 (Jul.Antec.), Poll. 2.32, 10.140, perh. PTeb. 331.13 (ii AD).   **II** *U-shaped bracelet*, S.fr. 407a R.   **2** *U-shaped band* for affixing hangings to columns or the like, Lxx Ex. 27.10, 11, 30.4, 37.6.   **III** *vaulted chamber* or *passage*, στενὴν δ' ἔδυμεν ψαλίδα S.fr. 367 R., ψαλίδα προμήκη λίθων ποτίμων Pl.Lg. 947d, SEG 27.119.30, 34 (Delphi, iii BC), Ph.Bel. 80.46, D.S. 2.9, Str. 17.1.42, MAMA 8.435 (Aphrodisias, sp. ψελίς), IGLS 438.4, POxy. 2804.221 (vi AD).   **IV** glossed by ταχεία κίνησις (alternative explanation), Sch.Pl.Lg. 947d G.

ψᾰλίττεται, add '(perh. *run a U-shaped course*, i.e. round a turning-point and back)'

ψάλλω II 2, add 'Chr., *sing psalms*, Cod.Just. 1.3.41.24'

*ψαλμοποιός, prob. = ψαλμῳδός, SB 10769.211 (iii/iv AD).

*ψαλταναγνώστης, ου, ὁ, *intoner*, MAMA 6.237 (Apamea); also ψαρτ- SEG 30.1323 (Ephesus).

*ψαλτήρ, ῆρος, ὁ, = ψάλτης, Hsch.

ψαλτήριον, for '*stringed instrument, psaltery, harp*' read '*stringed musical instrument*, prob. usu. some form of *harp*, but perh. also extended to include a *psaltery*'; for 'Hippias' read 'Alcid.'

ψάλτης, add '**2** *psalm-singer, cantor*, Cod.Just. 1.3.44 pr., POxy. 3958.11 (AD 614).'

ψάμμη, add 'cf. ψαμμήν· ἄλφιτα Hsch.'

ψάμμος, line 1, before '*sand*' insert 'Aeol. °ψόμμος'

*ψαμμόχωστος, ον, = ἀμμόχωστος, *piled up with sand*, POxy. 1911.89 (vi AD).

ψάρος or **ψᾶρος**, add '**2** *a sea fish*, Cyran. 4.75.1 K., cf. ψόρος.'

*ψαρταναγνώστης, v. °ψαλτ-.

ψαυκροπόδης, add '(v. °Suppl.Hell. 1122)'

ψαυκρός, add '(v. °Suppl.Hell. 1123)'

ψαῦσις, after 'Democr. 11' insert 'Hp.Vict. 1.23'

ψαύω I 5, add '**b** ψ. οὐρανοῦ *grasp heaven*, i.e. reach the supreme heights, Sapph. 52 L.-P., Plu.Demetr. 22, Ael.VH 12.41; cf. ἡμιθέων ψ. Synes.Ep. 142.55.'   add '**7** *put one's hand to, attempt*, Herod. 4.75, Plb. 3.32.5, 18.53.1.'

*ψάφεα· ψωμία Hsch., cf. ψάθεα.

*ψᾰφοτρῐβέων· περὶ τοὺς λόγους (cj. λογισμούς) τριβομένων Hsch.

ψέλιον I 1, after '*armlet* or *anklet*' insert '(U-shaped, i.e. in the form of an interrupted circle)'   **2**, for pres. def. read 'a kind of shears'   penultimate line, within parentheses, insert 'see also °ψελιοῦχος'   at end add 'see also σπέλλιον'

*ψελιοῦχος, ον, *bridle-chain holding*, ἅρμα ψ. Suppl.Hell. 996.10 (techno-paignion, i BC).

ψελιοφόρος, add 'in pl., gloss on Lat. wds. βραχιᾶτοι/ἀρμιλλίγεροι Lyd.Mag. 47.19 W.'

ψελλός, add 'cf. Myc. pe-se-ro, pers. n.'

*ψενδύλοι (cod. ψελύνοι)· σπόνδυλοι Hsch.

ψευδής I 1, line 5, delete 'ψ. λόγοι Hes.Th. 229'

ψευδογρᾰφία, for '= foreg.' read '*fallacy in geometry or arithmetic*'

*ψευδογρᾰφικός, ή, όν, *of or belonging to fallacious proof*, ἡ ψ. (sc. τέχνη) Alex.Aphr. in SE 195.18.

*ψευδοδιάτοιχος, ὁ, archit., *false parpen*, MDAI(I) 19-20.238.25.

*ψευδοδοξάζω, *suppose erroneously*, Plb. 10.2.3 (v.l.).

ψευδοδοξέω, after 'Plb. 10.2.3' insert '(v.l., v. °ψευδοδοξάζω)'

*ψευδοκᾰμῑνάριος, ὁ, *false potter*, Σωτήριχε, κίναιδε ψευδοκαμινάρι nonce-wd. in SEG 39.1062 (Rhegium, i BC/i AD)'

*ψευδοπαρηχητικός, ή, όν, *connected with* ψευδοπαρήχησις, Eust. 1586.21.

*ψευδορήτωρ, ορος, ὁ, *false orator* or *rhetorician*, Rh. 6.577.6.

ψεῦδος I 1, add 'pl., personified, Hes.Th. 229'   **II 2**, add 'App.Anth. 3.79 (Posidipp.)'

ψευδοσέλινον, add 'Cat.Cod.Astr. 7.234.4'

ψευδοσοφιστής, add 'Cyran. 1.12.29 K. (pl.)'

ψεύδω B I 3, add 'w. cogn. acc., ψεῦδος λυσιτελέστερον .. ἐψεύσατο Pl.Lg. 663d'

ψεῦσμα, after 'τό' insert 'also ψεῦμα Numen. l.c.' and after 'Plu.Art. 13' add 'Numen.fr. 27.70, 71 P.'

ψήκτρα, add 'implement used by a barber, made of reed, perh. *comb*, AP 6.307 (Phan.).'

*ψηφηδᾰκέω, *bite* (i.e. *injure*) *with one's vote*, Ar.Ach. 376 (cf. POxy. 856 i 23).

ψηφίζω, line 2, after 'Th. 2.24' insert 'med., ἐψαφίξατο SEG 31.572 (Thessaly, c.200 BC)'   **III**, line 2, after 'also found in' insert 'Inscr.Cret. 4.78.1 (Gortyn, v BC), al.'   add '**IV** app. (as ψηφόω) *adorn with mosaic*, SEG 32.1468 (Syria, Chr.).'

ψηφίον, add '**2** *mosaic*, CIJud. 803 (Syria, iv AD), SEG 32.1440 (Syria, vi AD); pl., SEG 30.1715 (Arabia, vi AD).'

ψηφίς I 1, add '**b** as collect. sg., *shingle*, Hippon. 128.3 W.'   **2**, add '**b** *pebble for voting*, πά]ντες ὑπὸ ψηφίδα κακὴν βάλον Call.fr. 85.8 Pf., cf. ψῆφος II 5 and °ὑποβάλλω.'   **3**, after '*tessellated work*' insert '*mosaic*' and add 'SEG 24.1197(e) (Egypt, iv AD), 31.1396 (Syria, v AD)'

ψήφισμα, form ψάφισμα, add 'also Aeol., IKyme 13, etc.'   add '**III** *mosaic work*, SEG 32.1517 (Palestine, vi AD).'

*ψηφιωτής, οῦ, ὁ, *maker of mosaics*, IGChr. 226(5) (sp. ψιφιοτῶν).

*ψηφοθεσία, ἡ, *laying of a mosaic pavement*, ISmyrna 733.5 (ii/iii AD), SEG 20.462.5 (Palestine).

ψηφοθεσμία, add 'unless error for °ψηφοθεσία'

ψηφοθέτης, after '*pavements*' insert 'Edict.Diocl. 7.7'

ψηφολογέω, add '**2** *calculate by counters*, Anon. in SE 2.39.'

ψηφολογικός, for '*juggling*' read '*legerdemain*'

*ψηφοπεριβομβήτρια, ἡ, app. *cup constructed with cavity containing clay pellets, causing it to make rattling sound*, Eub. 56 K.-A. (cf. JHS 90.200).

ψῆφος I 2, add 'of a pearl, Ael.NA 15.8'   add '**3** applied to the pupil of the eye (or the eyeball), w. play on sense II 1, Artem. 1.26 (p. 32 P.).'   **II 5**, line 11, after 'Antipho 5.47' insert 'περὶ τούτου ψήφου διενεχθείσης *a vote having been taken*, BCH 86.58.25 (Maced., ii AD)'   add '**8** *judgement* as a faculty, ψ. ὀρθήν Lib.Ep. 19.10.'

*ψηφοφορικός, ή, όν, *accustomed to manipulation of counters*, Anon. in SE 2.38.

ψηφόω II, add 'SEG 23.653 (Cyprus, v/vi AD), 31.1473 (Arabia, vi AD)'

ψήφωσις, add '**II** *adorning with mosaics*, IGLS 1320 (Apamea, AD 391), 770 (Antioch, ?vi AD), SEG 8.21 (near Ptolemais, vi AD).'

ψηχρός, after '*fine*' insert 'Nic.Th. 559 (v.l. ψῆγμα)'

ψήχω I 2, for 'δέρην μέτωπά τ'' read 'δέρην (sc. ταύρου)'

ψιά, add 'perh. also Archil. 48.20 W.'

ψιάθιον, delete 'perh.' and add 'BGU 812 i 5, 8 (iii/ii BC)'

ψιαθοπλόκος, add 'CPR 13.11.39 (iii BC), SB 9375.21 (ψιαθω-, ii AD)'

ψίαθος III, for 'perh. *sack*' read '*rush-basket*' and add 'BGU 2334.5 (ψειάθοισιν pap., iv AD)'

*ψιθωμία〈ν〉· Λάκωνες τὸν ἀσθενῆ Hsch.

ψιλῆται, after 'Eust. 1222.53' insert '(ψιλῖται ib. 907.38)'

ψίλιον, after '*armlet*' insert 'SEG 37.994 (Priene, vii BC)'

ψῑλότης II 2, for '(pl.)' read '(pl.; opp. δασύτητες); Phld.Po. 2.18 (opp. πρόσπνευσις)'

ψιμείον, v. °ψιμμίον.

*ψιμμίον, τό, = ψιμύθιον, *white lead*, Zos.Alch. 248.11, perh. also ψιμείον Stud.Pal. 20.96.9 (iv AD).

ψιμύθιον, add 'ψημίθι(ο)ν PVindob. G39847.936 (CPR 5.110, iv AD)'

ψιμυθοειδής, add 'Zos.Alch. 111.9'

*ψιττάκινος, (η), ον, *green-coloured*, of cloth, SB 9122.10 (i AD); collyrium psittacinum *a colore ita dictum*, Scrib.Larg. 27.

*ψιχόμαλλον, τό, perh. f.l. for *ψιλο-, *smooth (?)woollen garment*, Stud. Pal. 20.245.15 (vi AD), cf. ZPE 76.114.

ψοαλγικός, ή, όν, *suffering from pain in the loins*, Aët. 12.70.

*ψοθέω, = ψοφέω, Call.fr. 194.106 Pf.

+**ψόμμος**, ὁ, Aeol. for ψάμμος, *sand* or *dust*, Alc. 306 fr. 14 ii 2, 6 L.-P.; cf. ψόμμος· ἀκαθαρσία, καπνός Hsch.

ψόρος, add 'SEG 23.326 (Delphi, iii BC)'

ψοφοειδής, for 'φωνήεντα' read 'τὰ ψ. = those consonants which are not mutes, *continuants*'

***ψυγμογνᾰφεύς**, έως, ὁ, *dry-cleaner*, PFlor. 388.80 (i/ii AD).

ψυγμός II, add 'ἔψυξαν ἑαυτοῖς ψυγμούς Lxx *Nu.* 11.32' **III**, delete the section.

ψυδρεύς, add 'also in Epirus, SEG 32.623 (Buthrotus, ii/i BC)'

ψυδρός, before 'Lyc.' insert 'Simon. in POxy. 3965 fr. 26.16'

ψύλλα II 1, add 'cf. Ael. *NA* 6.26'   **3**, delete the section.

ψύλλιον, add 'also **ψύλλιος**, ὁ or ἡ, Cyran. 1.23.1, 2.45.4 K.'

ψῡχᾰγωγέω II 1, add '**b** *divert, amuse*, Alciphr. 3.18, Jul. *Or.* 1.40a.   **c** *console*, αὑτούς ib. 8.244b, cf. 248c.'

ψῡχάριον, add 'SEG 33.947 (Ephesus, i BC)'

ψῡχή I 2, line 4, after 'endearing name' insert 'Theoc. 24.8, Macho 223 G.'   **IV 3**, line 5, after 'Arist. *EN* 1168ᵇ7' insert 'perh. also E. *Or.* 1046'

ψυχογονικός, after '*of* or *for ψυχογονία*' insert 'ψ. διαιρέσεις Procl. in *R.* 2.192.26'

ψῡχολέτης, add 'of the devil, PYale inv. 1336 (*BASP* 22.333ff., iv/v AD)'

ψῡχολῐπής, after '*lifeless*' insert '*GVI* 1154 (ii BC)'

***ψῡχόλυτρος**, ὁ, *one whose soul is ransomed*, PBerl. 17612.8 (*APF* 21.78, vi AD).

ψῡχομαντεῖον, add 'pl., *rites of necromancy*, Cic. *Div.* 1.132'

ψῡχορρᾰγέω, for '*break loose*' read '*break up, collapse*'

ψῡχοστᾰσία, add 'as alternative name for νέκυια as part of Odyssey (bk. 11 or 24), Philostr. *VA* 8.7.7, *Her.* 51.7'

ψῡχότροφον, delete the entry (v. °ψυχρότροφον).

***ψύχρανσις**, ἡ, *chilling*, Alex. Aphr. *Febr.* 12.8 (opp. θέρμανσις).

ψῡχροβᾰφής II, for this section read 'of dyes, *used in cold immersion*, ὥσπερ τῶν ἀνθῶν τὰ μὲν ψυχροβαφῆ τὰ δὲ θερμοβαφῆ Thphr. *Od.* 22'

***ψῡχ⟨ρ⟩οθερμοφύσησος**, epith. of Typhos, *breathing cold and hot*, PMag. 4.183.

ψῡχροποτέω, add 'Gal. 17(2).199.12'

ψῡχρός I, line 9, transfer 'of a snake, Theoc. 15.58' to precede 'esp. of dead things' and add 'Thgn. 602'

***ψῡχρότροφον**, name for the plant κέστρον 1, cf. Plin. *HN* 25.84.

ψῡχροφόρον, after '*Gloss.*' add 'τῶν δύο ψ. POxy. 896.11 (iv AD)'

ψύχωσις, after 'M. Ant. 12.24' insert 'pl., Procl. *Inst.* 63 D.'

***ψωλοκοπέομαι**, *to be affected with priapism*, Lucilius 304 M.; act. in causative sense, ψωλοκοπῶ τὸν ἀναγιγνώσκοντα in margin of PLond. 604 *B* col. 7 (i AD).

***ψωμή**, ἡ, = ψωμός, Anon. Alch. 16.7 (pl.).

⁺**ψωμίον**, τό, dim. of ψωμός, *piece of bread*, PTeb. 33.14 (ii BC), *Ev.Jo.* 13.26, M. Ant. 7.3, D.L. 6.37, *BGU* 2357 iii 12 (iii AD).

ψωμός, after 'Od. 9.374' insert 'Hippon. 75.4 W.'

ψώρα I, for '*scurvy*' read '*scabies*' and add 'Hp. *Aff.* 35'

ψωρανθεμίς, for ' = λιβανωτίς' read 'kind of λιβανωτίς (A)'

ψωράω, add 'Friedländer *Epigrammata* p. 164'

ψωρός II, add 'perh. also SEG 32.771 (Berezan, vi BC, unless for ψωλός)'

ψώχω, for '*rub small*' read '*rub so as to break up, separate*, etc.' and before '(pass.)' insert 'Diog. Oen. fr. 9 ii 13 S.'

Ω

ὤ and ὦ I, add '2 w. acc., ὦ τὸν Ἄδωνιν Sapph. 168 L.-P., *GVI* 1386, ὦ ἐμὲ δειλάν Call.*Lav.Pall.* 89.'

ὦ add III, 'perh. also in Myc. *o-*, *how* (usu. prefixed to verbs)'

ᾤα (A) I, for this section read '*covering for the body made from sheepskin, a loincloth* or sim., περιζωσάμενος ᾤαν λουτρίδα κατάδεσμον ἥβης Theopomp.Com. 38 K.-A., Hermipp. 56, 76 K.-A., στέγασμα, εἴ τι βλέεστε, ἀποπέμψαι ἢ ᾤας ἢ διφθέρας ὡς εὐτελεστά(τα)ς καὶ μὴ σισυρωτάς *SIG* 1259 (Athens, iv BC); cf. τὸ δέρμα ᾧ ὑποζώννυνται αἱ λουόμεναι γυναῖκες ἢ οἱ λούοντες αὐτάς, ᾤαν λουτρίδα ἔξεστι καλεῖν .. οὕτω δὲ τὴν μηλωτὴν ἐκαλοῦν ἴσως ἀπὸ τῆς ὄιος Poll. 10.181, Hsch.' II, at the beginning insert 'perh. orig. different word'

Ὠγύγιος 1, *primeval*, add 'νόμου ὠγυγίου Orph.*H.* 59.10, 64.10' 2, add 'πελειάων ὠγύγιον τέμενος epigr. in *SEG* 39.1673.10 (Arabia, ii AD)'

ὧδε II 1, penultimate line, ὧδε καὶ ὧδε, *this way* and *that*, add 'Call.*Epigr.* 28.2 Pf.' 2, for 'τηνεῖ δρύες .. Theoc. 1.106' read 'Theoc. 5.45'; add 'Call.*Epigr.* 47.4 Pf., Ὑγινὴ ὧδε κῖτε *SEG* 26.376 (Athens), 33.1315 (Egypt, i BC/i AD)'

*ᾤδησις, εως, ἡ, *singing*, *POxy.* 3555.13 (i/ii AD).

+ὠδινολύτης, ου, ὁ, name given to a fish (ἐχενηΐς II) from its supposed power of facilitating childbirth, Plin.*HN* 32.6.

ὠδίνω I 2, after 'Lxx *Ca.* 8.5' insert 'cf. τέκνα μου, οὓς πάλιν ὠδίνω *Ep.Gal.* 4.19'; add 'transf., of a fountain, ὠδείνουσι τεὸν μένος ὄβριμον ἠνεκ[ὲς αἰέν *SEG* 32.1502 (Palestine, c.AD 455)' add '3 causal, *cause to go into labour*, [φωνὴ Κυρίου] ὠδίνοντος ἐλάφους Aq.*Ps.* 28(29).9'

ὠδίς I 2, line 6, delete 'τοῦ ᾠοῦ .. 560ᵇ22'

*ᾠδοδιδάσκαλος, ὁ, *singing-master*, *OGI* 56.70 (iii BC).

ὠθέω, line 10, for '(plpf.) ἑώκει' read 'ἐώκει' II 1, add 'fig., πενίην *AP* 6.117 (Pancrat.); ἀμαθίαν Plu. 2.47f'

ὠθίζω, add 'III *refuse, deny* an applicant, ἀλόγως παραιτησάμενος ὠθισθήσεται *Cod.Just.* 3.1.12.1.'

*Ὠκεάνη (v.l. -μη), ἡ, alleged to be the oldest name of *the Nile*, D.S. 1.19.4 (cf. 1.12.6 and *Ὠκεανός* IV).

Ὠκεάνης, delete the entry.

Ὠκεανός IV, add 'οἱ Αἰγύπτιοι Ὠκεανὸν νομίζουσι τὸν Νεῖλον D.S. 1.12.6, *BMus.Inscr.* 1077 (Sudan)'

*ᾠκοδομητός, v. ‡οἰκο-.

*Ὠκτώβριος, v. Ὀκτώβριος.

ὠκύπλοος, for '(Hsch. s.v. ὁ) ὠκύκλοος' read 'ὠκύπλοος'

ὠκύς I, add '3 of (the passage of) time, βίου ὠκυτάτου *TAM* 5(1).201; comp. adv. ὠκυτέρως Gal. 9.454.17.'

ὠκυτόκιος II, add 'b title of a collection of synonyms by Telephus of Pergamum, designed for speedy selection, Suid. s.v. Τήλεφος Περγαμηνός.'

*Ϝωλά, v. °βουλή.

*ὠλεία, ἡ, *destruction*, *JRCil.* 2.234.

*Ὠλέναδε, *to Olenos*, Ὠ· ὡς ἄγραδε Hsch. (v. *Suppl.Hell.* 1126).

ὠλένη 1, line 8, delete 'is dub. l.'

*ὠλένιον, τό, dim. of ὠλήν, *mat*, used in brick-making, *BGU* 2361a i 11, ii 9 (iv AD).

ὠλιτόφρονας, v. °ἀλιτόφρων.

ὠμαλία, line 1, after 'average' insert 'only in phrase ἐφ' ὡμαλίαν *IG* 2².1673.8 (iv BC)'

ὠμηστής 1, line 6, after ' = ὠμάδιος I' insert 'Alc. 129.9 L.-P.'

+ὠμία, ἡ, *shoulder*, Lxx 1*Ki.* 9.2. 2 transf., *the "shoulder" of a building* formed by the convergence of the roof and upright wall, Lxx 3*Ki.* 6.13(8), 2*Ch.* 23.10, al. b *shoulder-piece* or *flange* to support a sacred vessel, Lxx 3*Ki.* 7.17. II *bend* in a river, *PTeb.* 828.9 (ii BC).

*ὠμίς, ίδος, ἡ, kind of sea fish, Cyran. 1.24.1, al. K.

ὠμοβόειος, line 6, after 'etc.' insert '*IG* 2².1471.57'

ὠμοβόρος, before 'A.R. 1.636' insert 'S.*fr.* 10g.13a/b.3 R.'

*ὠμογάρικός, ή, όν, neut. pl. subst., perh. vessels *containing* °ὠμόγαρον, *PGot.* 17ᵛ.2 (sp. ὁμο-).

*ὠμόγαρον, τό, *uncooked garum*, *PVindob.Worp* 11.7 (vi AD).

ὠμογέρων I, after 'Il. 23.791' insert 'Call.*fr.* 24.5 Pf.'

*ὠμόδαιτος, ον, *feasting on raw meat*, prob. cj. in Ps.-Callisth. 58.11 K.

ὠμόλινον, add 'III *linen sack*, *Vit.Aesop.*(G) 34 P.'

*ὠμόπλινθος, ὁ, *unbaked brick*, *PLond.* 1708.91, *PKlein.Form.* 1092 (both vi AD).

ὦμος I 2, add 'b of meat, *SEG* 31.416 (Boeotia, iv BC).' add '4 object used in wine or oil production, perh. part of a *press* or *carrying-pole*, *PFlor.* 233.2, *PRyl.* 236.23 (both iii AD), cf. τοῖς ἀπ' ὤμων ἐμβαλλομένοις ἐργάταις *PHels.* 4D ii 10 (ii AD, unless ὦμος here is to be taken literally).'

ὠμός I 4, add 'comp., Alc. 119.16 L.-P.'

ὠμοτάριχος, for 'prob. *pickled flesh of the tunny's shoulder*' read '*raw pickled fish*'

*ὠμοτόκετος, ὁ, *premature child-birth*, *PFouad* 75.5 (i AD); -τοκητ-).

*ὠνανθεσία, ἡ, *manumission through sale*, *SEG* 36.518.15, 23 (Delphi, c.AD 100); cf. ‡ἱερανθεσία.

*ὠνάρχης, ου, ὁ, title of a Milesian officer, perh. = ἀρχώνης, *Didyma* 315.7.

ὠνέομαι, line 10, after 'D. 37.5)' insert 'Ion. pf. inf. ὠνῆσθαι *ZPE* 68.121.4 (lead letter, Emporion, v BC)' II, line 2, for 'dub. in pres. since' read 'pres. *Cod.Just.* 1.3.52.11'; line 3, after 'part.' insert 'ὀνονημένος *SEG* 12.391 (see also 34.868) (Samos, vi BC)'

ὠνή, line 1, after 'ii BC' insert 'οὐνή *TAPhA* 65.125, 128, 130 (Olynthus, iv BC)' II 2, add 'also *TAPhA* 65 l.c., *TAM* 4(1).276 (Rom.imp.)' III, add 'Cypr. *o-na* ὠνά *ICS* 299.7'

*ὠνηνικός, ή, όν, prob. f.l. for °ὠνητικός, πράσεις *PMichael.* 45.7 (vi AD).

+ὠνητικός, ή, όν, *of* or *concerning sale*, *PPanop.* 21.22 (AD 315), *PHerm.* 18.16 (AD 323); adv. -κῶς, ὠ. ἔχειν, *to be disposed to buy*, Ph. 2.465, 468.

*ὠοΐ (or ὠοί), cry of distress, *POxy.* 3722 *fr.* 17 ii 7 (lemma in Anacr. commentary, ii AD).

+ὠοιοί, var. of ὠαιαί, A.D.*Adv.* 127.28, 31.

ᾠόν 4, line 2, after 'Dinon 14' insert 'cf. *Inscr.Délos* 1417*A*ii140'

*ᾠοπώλης, ου, ὁ, *egg-seller*, *POxy.* 83.4 (iv AD; written ὀω-).

ᾠοσκοπία, delete '*divination from them*'

ᾠοσκοπικά, for 'a treatise thereon' read 'a treatise, app. *on divination from eggs*'

*Ὠπ, ὁ, a Syrian divinity, *IGLS* 1.230, 6.2916.

Ὦπις, for 'cf. Hdt. 4.35' read '2 name of a Hyperborean maiden at Delos, Hdt. 4.35, Apollod. 1.4.5.'

*ὠποθηκάριος, v. °ἀποθηκάριος.

ὥρα (C) A I 3, line 7, after 'Theoc. 15.74' insert 'ἐρεῖτε τάδε Εἰς ὥρας· καὶ σὺ ὑγιαίνων καὶ ὁ οἶκός σου Lxx 1*Ki.* 25.6' II 2 b, add 'πρὸ μιῆς ὥρης perh. = πρὸ πρώτης ὥρας Call.*fr.* 550 Pf.' B I 1, at end delete 'freq. .. etc.' 4, line 6, (αὐτῆς ὥρας), add 'Plu. 2.239b'; line 14, after 'Plu. 2.784b' insert 'καθ' ὥραν also, *at this very moment*, καθ' ὥραν γράφω ἐπιστολὴν *POxy.* 3150.17 (vi AD)' C, add '*SEG* 33.115 (Athens, iii BC)'

ὡραΐζω II 2, delete '(leg. ὡράζεθ')' and for 'Men. 855' read 'Men.*fr.* 788 K.-Th.'

ὡραῖος I 2, add 'b ὡραῖα· νεκύσια, οἱ δὲ δαιμόνια Hsch.' V, add 'Gal. 13.162.4'

*ὡράϊσμα, ατος, τό, *youth*, Sch.Pi.*N.* 8.1.

ὡρακιάω, after 'swoon away' insert 'S.*fr.* 120 R.'

*ὡράριον, v. °ὀράριον.

*ὡρδινάριος (ὠρδεν-), α, ον, v. °ὀρδινάριος.

*Ὠρειόνια, v. °Ὠριόνια.

*Ὠρείτης, ου, ὁ, title of Apollo, perh. *of Oreus* in Euboea, *Anecd.Stud.* 267 (v. *ZPE* 54.132).

*Ὠριόνια, τά, festival in honour of Orion, [ἀγωνοθ]έτης Ὠρειονίων *IG* 12 suppl. 646.16 (Chalcis, perh. fr. Tanagra, iii AD).

ὥριος (A) I, line 1, after 'ος, ον' insert '*GDI* 1775.22 (Delphi, ii BC)'; line 3, delete 'Theoc. 7.62' II 1, line 2, after 'Hes.*Op.* 392, 422' insert 'κρύος ib. 543' at end add 'ταῖς -οις (sc. ἡμέραις) *GDI* l.c.'

ὡρίτης, delete 'Ἀπόλλωνος .. *Anecd.Stud.* 267'

ὠρολόγιον, line 6, for '*CIG* .. (loc. incert.)' read '*SEG* 37.527 (Nicopolis, Epirus, Rom.imp.)'

ὦρος (A), for 'Call.*Fr.* 150 .. 28)' read 'Call.*fr.* 177.28 Pf.'

ὦρος (B), add 'also Cret. for °οὖρος (B)'

ὦρος (C), delete '(C)' and transpose to follow (new) ὦρος (C)

ὦρος (D), read 'ὦρος (C)' and delete '(better ὧρος)'

ὡροσκοπεῖον I, after '-σκόπιον' insert 'SEG 18.739 (Cyrene, ii BC), IEphes. 3223A'

***ὡρόχαλκος**, ὁ, = ὀρείχαλκος, brass, Peripl.M.Rubr. 6, PGiss. 47.6 (ii AD).

ὤρυγμα, add 'roaring of a lion, Sm., Thd.Is. 5.29'

ὠρύομαι, for '[ῡ]' read '[ῡ exc. Pl.Com. 138.2 K.-A., D.P. 83]'

ὡς Ab 1, add 'θεὸς ὥς X.Cyn. 1.6 (perh. in paraphrase of poetry)' **B III 3**, after 'omitted' insert 'in Hdt.'

ὡσάν I 3, add 'Gal. 6.360.15, cf. mod. Gk. σάν'

***ὡσκοφόροι**, v. °ὠσχοφόροι.

***ὡσομοίως**, or better **ὡς ὁμοίως**, perh. = ὡσαύτως, Delph. 3(3).421.6 (i AD).

ὥστε A II, add 'τὸν ὄγδοον, ὥστε Κόροιβον, οὐ συναριθμέομεν Call.fr. 587 Pf.'

ὠσχός, read ὤσχος.

ὠσχοφόροι, add 'Ath.Agora XIX L4a.21, 49 (363/2 BC)'

ὦτε, after 'A.D.Pron. 48.28' insert 'also Arc., SEG 37.340.12 (ἀφωτε, Mantinea, iv BC)'

ὠτειλή II, line 2, at end insert 'E.Supp. 945 (pl.)'; penultimate line, after 'ii.488)' insert '**ὠτέλλα** (-η cod.) Theognost.Can. 111'

ὠφέλεια, add '**III** Ὠφελία, personified as a divinity, Pap.Lugd.Bat. XXV 8 iii 15 (iii AD).'

ὠφέλημα I 1, add 'τὸν μηδὲν ὠφέλημα applied iron. to a useless person, Macho 385 G.'

ὠφέλησις, add 'Diog.Apoll. B 2 D.-K., Phld.Po. 13.23 J.'

ὠφέλιμος, after 'Pl.R. 607d' insert 'Men. 98c'

***Ὠφέλιος**, ὁ, helper, cult-title of Zeus, IHadr. 10 (ii/iii AD).